AF715136

The Vegetable Kingdom

Or, The Structure, Classification, and Uses of Plants Illustrated upon the Natural System

JOHN LINDLEY

CAMBRIDGE
UNIVERSITY PRESS

CAMBRIDGE
UNIVERSITY PRESS

University Printing House, Cambridge, CB2 8BS, United Kingdom

Cambridge University Press is part of the University of Cambridge.
It furthers the University's mission by disseminating knowledge in the pursuit of education, learning and research at the highest international levels of excellence.

www.cambridge.org
Information on this title: www.cambridge.org/9781108077224

© in this compilation Cambridge University Press 2015

This edition first published 1847
This digitally printed version 2015

ISBN 978-1-108-07722-4 Paperback

This book reproduces the text of the original edition. The content and language reflect the beliefs, practices and terminology of their time, and have not been updated.

Cambridge University Press wishes to make clear that the book, unless originally published by Cambridge, is not being republished by, in association or collaboration with, or with the endorsement or approval of, the original publisher or its successors in title.

Selected botanical reference works available in the CAMBRIDGE LIBRARY COLLECTION

al-Shirazi, Noureddeen Mohammed Abdullah (compiler), translated by Francis Gladwin: *Ulfáz Udwiyeh, or the Materia Medica* (1793) [ISBN 9781108056090]

Arber, Agnes: *Herbals: Their Origin and Evolution* (1938) [ISBN 9781108016711]

Arber, Agnes: *Monocotyledons* (1925) [ISBN 9781108013208]

Arber, Agnes: *The Gramineae* (1934) [ISBN 9781108017312]

Arber, Agnes: *Water Plants* (1920) [ISBN 9781108017329]

Bower, F.O.: *The Ferns (Filicales)* (3 vols., 1923–8) [ISBN 9781108013192]

Candolle, Augustin Pyramus de, and Sprengel, Kurt: *Elements of the Philosophy of Plants* (1821) [ISBN 9781108037464]

Cheeseman, Thomas Frederick: *Manual of the New Zealand Flora* (2 vols., 1906) [ISBN 9781108037525]

Cockayne, Leonard: *The Vegetation of New Zealand* (1928) [ISBN 9781108032384]

Cunningham, Robert O.: *Notes on the Natural History of the Strait of Magellan and West Coast of Patagonia* (1871) [ISBN 9781108041850]

Gwynne-Vaughan, Helen: *Fungi* (1922) [ISBN 9781108013215]

Henslow, John Stevens: *A Catalogue of British Plants Arranged According to the Natural System* (1829) [ISBN 9781108061728]

Henslow, John Stevens: *A Dictionary of Botanical Terms* (1856) [ISBN 9781108001311]

Henslow, John Stevens: *Flora of Suffolk* (1860) [ISBN 9781108055673]

Henslow, John Stevens: *The Principles of Descriptive and Physiological Botany* (1835) [ISBN 9781108001861]

Hogg, Robert: *The British Pomology* (1851) [ISBN 9781108039444]

Hooker, Joseph Dalton, and Thomson, Thomas: *Flora Indica* (1855) [ISBN 9781108037495]

Hooker, Joseph Dalton: *Handbook of the New Zealand Flora* (2 vols., 1864–7) [ISBN 9781108030410]

Hooker, William Jackson: *Icones Plantarum* (10 vols., 1837–54) [ISBN 9781108039314]

Hooker, William Jackson: *Kew Gardens* (1858) [ISBN 9781108065450]

Jussieu, Adrien de, edited by J.H. Wilson: *The Elements of Botany* (1849) [ISBN 9781108037310]

Lindley, John: *Flora Medica* (1838) [ISBN 9781108038454]

Müller, Ferdinand von, edited by William Woolls: *Plants of New South Wales* (1885) [ISBN 9781108021050]

Oliver, Daniel: *First Book of Indian Botany* (1869) [ISBN 9781108055628]

Pearson, H.H.W., edited by A.C. Seward: *Gnetales* (1929) [ISBN 9781108013987]

Perring, Franklyn Hugh et al.: *A Flora of Cambridgeshire* (1964) [ISBN 9781108002400]

Sachs, Julius, edited and translated by Alfred Bennett, assisted by W.T. Thiselton Dyer: *A Text-Book of Botany* (1875) [ISBN 9781108038324]

Seward, A.C.: *Fossil Plants* (4 vols., 1898–1919) [ISBN 9781108015998]

Tansley, A.G.: *Types of British Vegetation* (1911) [ISBN 9781108045063]

Traill, Catherine Parr Strickland, illustrated by Agnes FitzGibbon Chamberlin: *Studies of Plant Life in Canada* (1885) [ISBN 9781108033756]

Tristram, Henry Baker: *The Fauna and Flora of Palestine* (1884) [ISBN 9781108042048]

Vogel, Theodore, edited by William Jackson Hooker: *Niger Flora* (1849) [ISBN 9781108030380]

West, G.S.: *Algae* (1916) [ISBN 9781108013222]

Woods, Joseph: *The Tourist's Flora* (1850) [ISBN 9781108062466]

For a complete list of titles in the Cambridge Library Collection please visit:
www.cambridge.org/features/CambridgeLibraryCollection/books.htm

INTERIOR OF A BRAZILIAN FOREST ON THE ORGAN MOUNTAINS.

THE VEGETABLE KINGDOM;

OR,

The Structure, Classification, and Uses of Plants,

ILLUSTRATED UPON THE NATURAL SYSTEM.

BY

JOHN LINDLEY, PH. D., F.R.S., & L.S.,

PROFESSOR OF BOTANY IN THE UNIVERSITY OF LONDON, AND IN THE ROYAL INSTITUTION OF GREAT BRITAIN.

"Methodum intelligo naturæ convenientem quæ nec alienas species conjungit, nec cognatas separat."—*Raii Sylloge, præf.*, p. 15.

WITH UPWARDS OF FIVE HUNDRED ILLUSTRATIONS.

Second Edition,

WITH CORRECTIONS AND ADDITIONAL GENERA.

LONDON:
PUBLISHED BY BRADBURY & EVANS, WHITEFRIARS.

MDCCCXLVII.

PREFACE.

THIS work originated in a desire, on the part of the Author, to make his countrymen acquainted with the progress of Systematical Botany abroad, during the previous quarter of a century. When it first appeared, the science was so little studied that the very names of some of the best writers on the subject were unfamiliar to English ears. In our own language there was nothing whatever; and the Natural System of arranging plants, although occasionally mentioned as a something extremely interesting, was currently regarded as the fond speculation of a few men with more enthusiasm than sound judgment; and this, too, was the opinion expressed by persons who stood at the head of English Botany, in the estimation of many British Naturalists. The Author had himself severely experienced the want of some guide to this branch of Natural History, and he felt anxious to relieve others from the inconvenience which he had encountered; the more especially after he had undertaken the responsibility of filling the Botanical Chair in the then London University. At that time, too, there was nothing of foreign origin which could be advantageously consulted; for Bartling's *Ordines* had not reached England, Perleb's *Lehrbuch* was unknown, and both it and Agardh's *Classes* were of too slight a texture to be generally useful to any except Botanists themselves.

The importance of the Natural System in a practical country like Great Britain was too manifest to leave any doubt in the mind of the Author that the good sense of his countrymen would lead to its universal reception when once placed within their reach. Nor has he been disappointed. Fifteen years have sufficed to render the once popular, but superficial and useless, system of Linnæus a mere matter of history. *Fuit Ilium.*

The Natural System of Botany being founded on these principles, that all points of resemblance between the various parts, properties, and qualities of plants shall be taken into consideration; that thence an arrangement shall be deduced in which plants must be placed next each other which have the greatest degree of similarity in those respects; and that consequently the quality of an imperfectly known plant may be judged of by that of another which is well known, it must be obvious that such a method possesses great superiority over artificial systems, like that of Linnæus, in which there is no combination of ideas, but which are mere collections of isolated facts, having no distinct relation to each other. The advantages of the Natural System, in applying Botany to useful purposes, are immense, especially to medical men, who depend so much upon the vegetable kingdom for their remedial agents. A knowledge of the properties of one plant enables the practitioner to judge scientifically of the qualities of other plants naturally allied to it; and therefore, the physician acquainted with the Natural System of Botany, may direct his inquiries, when on foreign stations, not empirically, but upon fixed principles, into the qualities of the medicinal plants which have been provided in every region for the alleviation of the maladies peculiar to it. He is thus enabled to read the hidden characters with which Nature has labelled all the hosts of species that spring from her teeming bosom. Every one of these bears inscribed upon it the uses to which it may be applied, the dangers to be apprehended from it, or the virtues with which it has been endowed. The language in which they are written is not indeed human; it is in the living hieroglyphics of the Almighty, which the skill of man is permitted to interpret. The key to their meaning lies enveloped in the folds of the Natural System, and is to be found in no other place.

The great obstacle to the adoption of the Natural System of Botany in this country was the supposed difficulty of mastering its details; but of that difficulty it may be observed, in the first place, that it is only such as it is always necessary to encounter in all branches of human knowledge; and secondly, that it has been much exaggerated by persons who have written upon the subject without understanding it.

It has been pretended that the characters of the Natural classes of plants are not to be ascertained without much laborious research; and that not a step can be taken until this preliminary difficulty is overcome. But it is hardly necessary to say, that in natural history many facts which have been originally discovered by minute and laborious research, are subsequently ascertained to be connected with other facts of a more obvious nature; and of this Botany offers perhaps the most striking proof that can be adduced. One of the first questions to be determined by a student of Botany, who wishes to inform himself of the name, affinities, and uses of a plant, seems to be, whether it contains spiral vessels or not, because some of the great divisions of the vegetable kingdom are characterised by the presence or absence of those minute organs. It is true that careful observation, and multiplied microscopical analyses, have taught Botanists that certain plants have spiral vessels, and others have none; but it is not true, that in practice so minute and difficult an inquiry needs to be instituted, because it has also been ascertained that plants which bear flowers have spiral vessels, and that such as have no flowers are usually destitute of spiral vessels, properly so called; so that the inquiry of the student, instead of being directed in the first instance to an obscure but highly curious microscopical fact, is at once arrested by the two most obvious peculiarities of the vegetable kingdom.

Then, again, among flowering plants two great divisions have been formed, the names of which, Monocotyledons and Dicotyledons, are derived from the former having usually but one lobe to the seed, and the latter two,—a structure much more difficult to ascertain than the presence or absence of spiral vessels. But no Botanist would proceed to dissect the seeds of a plant for the purpose of determining to which of those divisions it belongs, except in some very special case. He knows from experience that the minute organisation of the seed corresponds with a peculiar structure of the stem, leaves, and flowers, the most highly developed, and most easily examined parts of vegetation; a Botanist, therefore, prefers to examine the stem, the flower, or the leaf of a plant, in order to determine whether it is a Monocotyledon or a Dicotyledon, and rarely finds it necessary to anatomise the seed.

The presence or absence of albumen, the structure of the embryo, the position of the seeds or ovules, the nature of the fruit, the modifications of the flower, are not to be brought forward as other difficult points peculiar to the study of the Natural System, because, whatever system is followed, the student must make himself acquainted with such facts, for the purpose of determining genera. The common Toad-flax cannot be discovered by its characters in any book of Botany, without the greater part of this kind of inquiry being gone through.

In the determination of genera, however, facility is entirely on the side of the Natural System. Jussieu has well remarked "that whatever trouble is experienced in remembering, or applying the characters of Natural Orders, is more than compensated for by the facility of determining genera, the characters of which are simple in proportion as those of Orders are complicated. The reverse takes place in arbitrary arrangements, where the distinctions of classes and sections are extremely simple and easy to remember, while those of genera are in proportion numerous and complicated."

But really all considerations of difficulty ought to be put aside when it is remembered how much more satisfactory are the results to which we are brought by the study of Nature philosophically, than those which can possibly be derived from the most ingenious empirical mode of investigation.

Such were the motives which led to the publication, in 1830, of the first edition of the present work, under the name of an *Introduction to the Natural System of Botany*. No one would have more readily than the Author transferred the labour to another hand, if any other had been found. Indeed, he confesses that it was because the most capable of those whom he knew belonged to the class of men described by Lord Bacon, who "object too much, consult too long, adventure too little, repent too soon, *and seldom drive business home,*" that he undertook a task for which no man's abilities are in reality high enough. He could not but feel that: "To think nothing done while anything remains to be done is a good rule for perseverance, but to think that nothing should be done while a main thing remains undone, would be a most idle and thriftless maxim. If there be a good presently practicable, it may be done without any desertion of another good not so immediately attainable. And in effecting all secondary amendments, we have the satisfaction of feeling assured that there is a link between all real improvements, and that every sound reform is a step to others, though the connexion may not be broadly distinguishable."

The *Introduction to the Natural System* was originally written in illustration of the popular system of De Candolle; but daily experience showed the insufficiency of that system, and the necessity of forming sub-divisions of the primary groups of plants higher than their so-called Natural Orders became so apparent, as to lead to serious attempts to carry out a plan of Alliances, in imitation of a few continental writers. These attempts were embodied in the second edition of the present work, which appeared in 1836, under the name of *A Natural System of Botany.* Notwithstanding some glaring defects in the method then proposed, and a host of errors of a less manifest description, the views of the Author were favourably received by those best able to judge of their value. On the other hand, they have been severely criticised by writers who show a singular want of knowledge of the true bearing of such works. Those persons have imagined that a natural classification of plants is something which is suddenly to start into existence, perfect in all its parts, and their criticisms betray a total ignorance of the difficulties by which such a subject is surrounded. The Natural System of Botany may be likened to the plan of a vast edifice, at the construction of which many are labouring. Certain courts and quadrangles are easily set out; a particular style of architecture is agreed upon, and it may be even settled irrevocably in what places the state apartments and cellars are to be stationed. But when further details are to be discussed, many unsatisfactory attempts must be made by the architects, and many an awkward arrangement of the rooms proposed, before a final plan can be produced. If perfection in such small matters is impracticable, if it is impossible so to arrange all the details of even an edifice as to satisfy all critics, how much more hopeless must be the task of classifying the infinite works of the creation! To demand perfection in a work of that nature is little less than impious; for perfection is the attribute, not of man, but of his Maker.

The Author may now be equally charged with inconsistency in not adhering to his former plan of classification after having promulgated it. But he is not conscious of having ever pretended that it even approached permanency.—See *Natural System,* p. xiii. In fact, there is no such thing as stability in these matters. Consistency is but another name for obstinacy. All things are undergoing incessant change. Every science is in a state of progression, and of all others the sciences of observation most so. Since 1836 the views of the Author have, of course, been altered in some respects, although they have experienced but little modification in others. This is inevitable in such a science as that of Systematic Botany, where the discovery of a few new facts or half a dozen fresh genera may instantly change the point of view from which a given object is observed. The Author cannot regard perseverance in error commendable, for the sake of what is idly called consistency; he would rather see false views corrected as the proof of their error arises. His object, and, he thinks he may say that of every one else who has turned his attention to this question of late, has not been to establish a system of his own, which shall be immutable, but to contribute to the extent of his ability towards that end. He indeed must be a very presumptuous person, having a microscopically small acquaintance with his subject, who should even dream of being able to accomplish such a purpose All that we can do is to throw our pebbles upon the heap, which shall hereafter, when they have sufficiently accumulated, become the landmark of Systematical Botany.

Having stated thus much by way of preface, it only now remains to explain the plan of the work in its new form. Its object is to give a concise view of the state of Systematical Botany at the present day, to show the relation or supposed relation of one group of plants to another, to explain their geographical distribution, and to point out the various uses to which the species are applied in different countries. The names of all known genera, with their synonyms, are given under each Natural Order, the numbers of the genera and species are in every case computed from what seems to be the best authority, and complete Indices of the multitudes of names embodied in the work are added, so as to enable a Botanist to know immediately under what Natural Order a given genus is stationed, or what the uses are to which any species has been applied. Finally, the work is copiously illustrated by wood and glyphographic cuts, and for the convenience of Students, an artificial analysis of the system is placed at the end. Some of these points demand a few words of comment.

In offering to the public a view of the present state of Systematical Botany, the Author has pursued the plan developed in the succeeding pages, of first taking certain characters common to very extensive assemblages of plants, by means of which Classes have been constituted; and, secondly, of breaking up those Classes into minor groups called Alliances, whose common characters are also more extensive than those of Natural Orders, and under which the Natural Orders are themselves assembled. Very short characters have been proposed, under the name of Diagnoses, for both Alliances and Orders; these are intended to express the prevailing tendency observable in each group, but do not include casual exceptions, for which the reader is referred to the descriptions immediately following the Diagnosis. The Alliances are the most important feature in the arrangement; and it is to be hoped will be found much better limited than they formerly were. The serious fault committed in the Author's former work, of founding Alliances upon single Natural Orders, has been avoided in every case except that of Palms, which in reality seem to form an Alliance by themselves. The name Alliance has been preserved in preference to that of class, family, circle, cohort, &c., because it is not susceptible of two interpretations, as is the case with all the others; it is employed as an English equivalent for the Latin term *nixus*, which some have imagined was a misprint for *nexus*, but which was used in the sense of Cicero, and intended to express a tendency to assume some particular form of structure. If any one should inquire why no synonyms have been quoted to these Alliances, concerning which so many Botanists have lately occupied themselves, the Author's answer is, that they have hitherto been much too little agreed upon, except in a few very special cases, and that an examination of their history would involve an inquiry which must extend back to the Anthemides of Cæsalpinus, and which belongs to the history of Systematical Botany rather than to its actual condition. The whole practice, indeed, of quoting synonyms is carried by Botanists beyond useful limits. It is in many cases a matter of courtesy rather than of utility; and for this reason, as no one is bound to be courteous to himself, the Author has very generally refrained from making references to his own writings, except when some real necessity for doing so appeared to exist. He may also state in this place, that throughout the present work he has struck out many of the citations given in the last edition, conceiving it useless again to occupy space with the names of authorities which can be always found by those who are desirous to search for them.

In pointing out the affinities of plants the opinions of the most judicious systematists have been consulted; among these the names of Arnott, Auguste de St. Hilaire, Bennett, Bentham, Ad. Brongniart, Brown, Cambessédes, Decaisne, the De Candolles, Endlicher, the Hookers, the Jussieus, Martius, Miers, and Richard, stand in the first rank. In addition to the short discussion upon this subject which always follows the paragraph descriptive of a Natural Order there is appended to the list of genera a plan of indicating affinity now adopted for the first time. It consists of printing the name of the Order under discussion in capital letters; placing right and left of it in small Roman letters the names of those Orders which are supposed to be in nearest alliance to it; and above and below it in italic type the names of such as are only analogous, or at least have a more distant affinity. The idea of this is borrowed from Mr. Strickland's excellent paper on the true method of discovering the Natural System in Zoology and Botany, printed in the *Annals of Natural History*, vol. vi. p. 184.

The uses to which plants are applied has been re-examined with great care, and principally re-written. This part was originally intended as a mere sketch of so vast and important a subject, and in truth it is little more even now. It is, however, materially enlarged, and the Author hopes better arranged. In preparing it great numbers of works have been consulted, and most especially the special treatises of Dierbach, Fée, Geiger, Guibourt, Martius, Nees v. Esenbeck, Pereira, Richard, and Royle, together with the capital condensation published by Endlicher in his *Enchiridion.* The Author was also strongly advised by one whose opinion has great weight with him, to introduce among the properties of plants an account of their proximate principles and ultimate constituents. But after a full consideration of the subject, he has come to the conclusion that it is not expedient to do so. In the first place, such matters belong to Chemistry, and not to Botany; secondly, it does not appear possible to connect them with any known principle of botanical classification; and, moreover, the extremely unsteady condition of the opinions of chemists themselves upon the result of their own researches, and the uncertainty at present connected with the details of organic chemistry, would render the introduction of the supposed results of chemists embarrassing rather than advantageous. If it is true, as appears to be admitted, that such principles as Caffeine and Theine are identical, and that oils of Anise and Tarragon are chemically undistinguishable, it is clear that these substances can have no connexion with structure, or Botanical classification, if indeed they are not altogether artificial products produced by chemical processes, like Dr. Fownes's furfurol—a vegeto-alkali resulting from the distillation of bran, sulphuric acid, and water.

In forming the lists of genera, the Author is called upon to acknowledge the great assistance that he has derived from those of Professor Endlicher, which indeed he has ventured to take as the foundation of his own, making however considerable additions and material changes in some, and entirely re-writing others;

in which troublesome but necessary task he has been most essentially assisted by the Rev. M. J. Berkeley, who furnished the list of Fungals, and by Mr. Bentham, to whom he is indebted for those of Leguminous and Labiate plants and of Figworts. The reader will perceive that according to the custom of Botanists the names of genera which the Author adopts, are printed in Roman letters, and succeeded by others indented and printed in italics. The latter are either synonyms, or subgenera which do not at present appear to be of importance enough to be regarded as true genera.

In computing the number of species, attention has been paid not only to published statements, but also to such appearances of undescribed species as the Author's own herbarium indicates, assisted occasionally by a little guess-work, where Natural Orders have not been recently examined with care, or where species have been notoriously founded upon trifling and unimportant characters. He does not however doubt that the numbers are in all cases too low. All they pretend to is as near an approach to truth as, under existing circumstances, is possible.

The illustrations are partly original, partly derived from other authorities. It would have been more useful if a larger number could have been introduced; but costly embellishments are not possible beyond a certain limit. Should the present work be favourably received, others may be inserted hereafter in the numerous blanks that have been left among the pages.

Finally, the artificial analysis of Orders given in former editions has again been improved, and is now adapted to the volume in its new dress. It is, however, no longer placed at the beginning of the work, but will be found immediately before the indices. It has been gratifying to the Author to know that this table is habitually consulted by some of the most experienced Botanists.

There is still another point in which the Author has endeavoured to effect some improvement, and that is the nomenclature. Since the days of Linnæus, who was the great reformer of this part of Natural History, a host of strange names, inharmonious, sesquipedalian, or barbarous, have found their way into Botany, and by the stern but almost indispensable laws of priority are retained there. It is full time, indeed, that some stop should be put to this torrent of savage sounds, when we find such words as Calucechinus, Oresigenesa, Finaustrina, Kraschenninikovia, Gravenhorstia, Andrzejofskya, Mielichoferia, Monactineirma, Pleuroschismatypus, and hundreds of others like them, thrust into the records of Botany without even an apology. If such intolerable words are to be used, they should surely be reserved for plants as repulsive as themselves, and instead of libelling races so fair as flowers, or noble as trees, they ought to be confined to Slimes, Mildews, Blights, and Toadstools. The Author has been anxious to do something towards alleviating this grievous evil, which at least need not be permitted to eat into the healthy form of Botany clothed in the English language.

No one who has had experience in the progress of Botany, as a science, can doubt that it has been more impeded in this country by the repulsive appearance of the names which it employs than by any other cause whatever; and that, in fact, this circumstance has proved an invincible obstacle to its becoming the serious occupation of those who are unacquainted with the learned languages, or who, being acquainted with them, are fastidious about euphony, and Greek or Latin purity. So strongly has the Author become impressed with the truth of this view, that on several occasions he has endeavoured to substitute English names for the Latin or Greek compounds by which the genera of plants are distinguished. Upon turning over the late volumes of the *Botanical Register* many such instances will be found, in imitation of the well-known and usual English words, Houndstongue, Loosestrife, Bugloss, Soapwort, Harebell, &c. He cannot, however, boast of any success in these feeble attempts at reforming a great evil; nor, perhaps, ought he to have expected it. If such English names are not universally adopted, it is to be suspected that the circumstance is traceable to the indifference of the public to partial and inconsiderable changes, which are unseen in the ocean of Botanical nomenclature. That they are important must be admitted; that the person most careless as to the difficulties of articulation would prefer to speak of a Fringe-Myrtle rather than of a Chamælaucium, or of a Gritberry than of a Comarostaphylis, will probably be allowed on all hands; and therefore the Author does not confess discouragement at failure; but would rather invite suggestions as to more probable means of success. Mere translation is neither necessary nor desirable in all cases. Many Latin names have, from custom, been adopted into the English language, and no wisdom would be shown in attempting to alter such words as Dahlia, Crocus, Ixia, or even Orchis. Others again are so easily sounded, and so much in harmony with the English tongue, that nothing could be gained by interfering with them; such as Penæa, Hugonia, Parkia, Mimosa, Arbutus, &c. And, finally, there is a large class of scientific words which are best Englished by an alteration of their foreign terminations; for example, Melanthium may be changed to Melanth; Desmanthus to Desmanth; Lecythis to Lecyth; Myrospermum to Myrosperm; and such an alteration would at once possess the great advantage of rendering English plural terminations possible. Melanthiums, Desmanthuses, Lecythises, &c., sound offensively to classical ears; Melanthia, Desmanthi, Lecythides, are, if not pedantic, at least beyond the skill of uneducated readers; but Desmanths, Melanths, and Lecyths, are formed by the ordinary English plural termination without difficulty.

It is, however, to be feared that a long time will elapse before these views are carried out in such a manner as to insure their adoption. But in the meanwhile a commencement of the plan is practicable, and the Author hopes it will meet with support. The names by which the great groups of plants are known are few in number, and very often in use. There is certainly no reason why we should not at once English them; the practice, indeed, is already adopted to some extent by the substitution of the words Monocotyledons, Dicotyledons, Exogens, Endogens, Cryptogams, Phænogams, &c., for Monocotyledones, Dicotyledones, Exogenæ, Endogenæ, Cryptogamæ, Phænogamæ, &c. It is even carried further by speaking of Rosaceous plants instead of Rosaceæ, Orchidaceous or Orchideous plants instead of Orchidaceæ, or Orchideæ, &c. But these amended names are still too long, and too un-English in sound to be in favour with the world which lies without the narrow circle of mere systematists; and no valid reason seems to exist for not immediately reforming that part of the nomenclature of Botany. The attempt has been already made in the Author's *School Botany*, where it will be found that by availing himself of well-known English names, or of the English word "wort," or by merely remodelling the terminations, a uniform English nomenclature has been secured for all the common European Natural Orders of plants. Thus for Nymphæaceæ, Ranunculaceæ, Tamaricaceæ, Zygophyllaceæ, Elatinaceæ, are substituted Water-Lilies, Crowfoots, Tamarisks, Bean-Capers, and Water-Peppers; for Malvaceæ, Aurantiaceæ, Gentianaceæ, Primulaceæ, Urticaceæ, Euphorbiaceæ, are employed Mallowworts, Citronworts, Gentianworts, Primworts, Nettleworts, Spurgeworts; and the terms Orchids, Hippurids, Amaryllids, Irids, Typhads, Arads, Cucurbits, are taken as English equivalents for Orchidaceæ, Haloragaceæ, Amaryllidaceæ, Iridaceæ, Typhaceæ, Araceæ, and Cucurbitaceæ. The principles kept in view in effecting those changes have been also observed throughout the present work, so that standard English names for Classes and Orders are now no longer wanting. The Author confidently believes that every intelligent reader will admit that such names as Urn-mosses, Taccads, False Hemps, Pepperworts, Bristleworts, Chenopods, Hydrocharads, Scale-mosses, Birthworts, and Fringe-Myrtles are preferable to Bry-a-ce-æ, Tac-ca-ce-æ, Da-tis-ca-ce-æ, El-a-ti-na-ce-æ, Che-no-po-di-a-ce-æ, Des-vaux-i-a-ce-æ, Hy-dro-cha-ri-da-ce-æ, Jun-ger-man-ni-a-ce-æ, A-ris-to-lo-chi-a-ce-æ, Cha-mæ-lau-ci-a-ce-æ, and other sesquipedalian expressions.

University College, London.
October, 1845.

CONTENTS.

INTRODUCTION.

THAT part of the material world which bears the name of the Vegetable Kingdom, consists, like the Animal, of a vast multitude of species, whose outer and inner forms alike offer a prodigious diversity of modifications of one common simple plan of structure. Organic vesicles, usually extending into tubes of various kinds, exclusively constitute what we call Vegetation; but this simplicity of nature is attended by very complex details of arrangement, as is shown in trees, whose framework is knit together by countless myriads of such vesicles and tubes, entangled with an astonishing intricacy of simple arrangement.

Any living combination whatsoever of such vesicles constitutes a plant; but as the combinations themselves are countless, so are the resulting external forms; for, although two or three words may suffice to express all combinations whatsoever in their most general sense, as when the name of thallus is given to the simplest expansion of vegetable matter, while all the more complex forms are included under the name of axis and its appendages, yet ingenuity is exhausted in the attempt to distinguish by appropriate terms the manifold external forms assumed by that axis and the parts which it bears.

Hence it is that wherever the eye is directed it encounters an infinite multitude of the most dissimilar forms of vegetation. Some are cast ashore by the ocean in the form of leathery straps or thongs, or are collected into pelagic meadows of vast extent; others crawl over mines and illuminate them with phosphorescent gleams. Rivers and tranquil waters teem with green filaments, mud throws up its gelatinous scum, the human lungs, ulcers, and sordes of all sorts bring forth a living brood, timber crumbles to dust beneath insidious spawn, corn crops change to fetid soot, all matter in decay is seen to teem with mouldy life; and those filaments, that scum-bred spawn and mould, alike acknowledge a vegetable origin. The bark of ancient trees is carpeted with velvet, their branches are hung with a grey-beard tapestry, and microscopical scales overspread their leaves; the face of rocks is stained with ancient colours, coeval with their own exposure to air; and those too are citizens of the great world of plants. Heaths and moors wave with a tough and wiry herbage, meadows are clothed with an emerald mantle, amidst which spring flowers of all hues and forms, bushes throw abroad their many-fashioned foliage, twiners scramble over and choke them, above all wave the arms of the ancient forest, and these too acknowledge the sovereignty of Flora. Their individual forms too change at every step. With every altered condition and circumstance new plants start up. The mountain side has its own races of vegetable inhabitants, and the valleys have theirs; the tribes of the sand, the granite, and the limestone are all different; and the sun does not shine upon two degrees on the surface of this globe the vegetation of which is identical: for every latitude has a Flora of its own. In short, the forms of seas, lakes, and rivers, islands and peninsulas, hills, valleys, plains, and mountains, are not so infinitely diversified as that of the vegetation which adorns them.*

Botanists have gathered together these endless forms, have studied and arranged them, and calculated their numbers, which amount to more than 92,000 species: a mighty host whose ranks are daily swelled by new recruits.

This vast assemblage has not been gathered together in a few years; it is coeval with man, and we cannot but feel that the study of the distinctions between one plant and another commenced with the first day of the creation of the human race. The name indeed of Botany is modern; but its antiquity dates from the appearance of our first parents. We may assume it as a certain fact that the Vegetable Kingdom was the first to engage the attention of man, for it was more accessible, more easily turned to useful purposes, and more directly in contact with him than the Animal. Plants must have yielded man his earliest food, his first built habitation; his utensils and his weapons must alike have been derived from the same source. This could not fail to produce experience, and especially the art of distinguishing one kind of plant from another, if it were only as a means of recognising the useful and the worthless species, or of remembering those in which such qualities were most predominant. This would involve from the very beginning the contrivance of names for plants, together with the collection of individuals into species; and the mental process by which this was unconsciously effected gradually ripened into the first rude classifications that we know of. By placing together individuals identical in form and the uses they could be applied to, species were distinguished; and by applying a similar process to the species themselves, groups analogous to what we now call genera were obtained. The last step was to constitute classes, which were recognised under the well-known names of "grass, and herbs yielding seed, and fruit trees yielding fruit."

---

* It is in the tropics that the prodigious diversity of appearance among plants is most strikingly exemplified. The beautiful forest scene, given as a frontispiece to this work, is copied from a plate in the *Flora Brasiliensis* of Dr. Von Martius, who describes it thus: "The landscape is divided into two unequal parts by a tree (*) rising to the height of 70 or 80 feet; it is Eschweilera angustifolia. It is overrun with ropes which cling around it, or hang down in various festoons; these ropes yield a milky white or yellowish juice when wounded, and probably belong to the Dogbanes or Asclepiads; other twiners, decorated with fine, large, beautifully green leaves, consist of species of Banisteria, Smilax, Serjania and Bignonia, voluptuously intertwined and entangled. A little above there is a tuft of the large leaves of Anthericum glaucum, and from the summit of all hangs down some unknown kind of Bromelwort. On the left stands a slender Acacia, whose bark is embraced by some parasitical climber; then comes the Couratari legalis, a high tree, whose timber is used in house-building; it forms a stem 60 or 70 feet high without a branch, and then spreads into a hemispherical head; owing to the slowness of its growth it is overrun with epiphytes. In front of the Acacia is a low tree with a close head and a shining bark; that is a Ficus americana, and Banisterias are shooting downwards from among its branches. Before this lie the bones of some fallen giant of the forest, overspread with great tufts of Anthericum and Epiphyllum phyllanthus. Close by, some Psychotria expands its large leaves and wide branches. A Heliconia and a Phrynium start from the mud and marshy foreground; a great patch of Anthericum umbellatum flourishes on the rotten trunk, and just in front is a group of Agarics, such as we see in the woods of Europe. The tall tree on the right of Eschweilera, with a smooth bark and pinnated leaves, is an Inga; next it is a small bush of Leandra scabra, behind which is a thicket of Palicuria and Renealmia nutans, backed by the Eriodendron leiantherum. The beautiful Palm to the right of them is Geonoma Pohliana. The foreground on the right is occupied by Ficus longifolia, conspicuous with its ample foliage, and loaded with epiphytes of various kinds, especially with Anthericum glaucum, umbellatum, and longifolium, and Caladium auritum. These and different kinds of Bilbergia have also taken possession of the rotten trunks in the neighbourhood. Near these is the white-barked Cecropia peltata, with large green leaves hoary with down on the under side." The cable-like climbers on the extreme right are not named by Dr. Von Martius.

But as human intelligence advanced, and a knowledge of things increased, such rude distinctions were improved, and when no means existed of appreciating the value of minute or hidden organs, the functions and existence of which were unknown, objects were at first collected into groups, characterised by common, external, and obvious signs. Theophrastus had his water-plants and parasites, pot-herbs and forest trees, and corn-plants; Dioscorides had aromatics, and gum-bearing plants, eatable vegetables and corn-herbs; and the successors, imitators, and copiers of those writers, retained the same kind of arrangement for ages. It was not till 1570 that Lobel, a Fleming, improved the ancient modes of distinction, by taking into account characters of a more definite nature than those which had been employed by his predecessors; but he was soon succeeded by others, among the most distinguished of whom were Cæsalpinus, an Italian who wrote in 1583, the celebrated Tournefort, and especially our countryman, John Ray, who flourished in the end of the seventeenth century. The latter added much to the knowledge of his predecessors, and had so clear and philosophical a conception of the true principles of classification, as to have left behind him in his *Historia Plantarum* the real foundation of all those modern views which, having been again brought forward at a more favourable time by Jussieu, are generally ascribed exclusively to that most learned Botanist and his successors. Ray, however, laboured under the great disadvantage of being too far in advance of his contemporaries, who were unable to appreciate the importance of his views or the justness of his opinions; and who therefore, instead of occupying themselves with the improvement of his system, set themselves to work to discover some artificial method of arrangement, that should be to Botany what the alphabet is to language, a key by which the details of the science may be readily ascertained. With this in view, Rivinus invented, in 1690, a system depending upon the formation of the corolla; Kamel, in 1693, upon the fruit alone; Magnol, in 1720, on the calyx and corolla; and finally, Linnæus, in 1731, on variations in the stamens and pistil. The method of the last author has enjoyed a degree of celebrity which has rarely fallen to the lot of human contrivances, chiefly on account of its clearness and simplicity; and in its day it effected a large amount of good.

It was soon, however, perceived by those who studied the Vegetable Kingdom profoundly, that no improvement could be made in the knowledge of its true nature, of the best manner of arranging it, or even of the purposes to which it might be applied, unless the philosophy of the subject was investigated; and this became daily more apparent as the materials collected by botanical travellers accumulated. It was found that the few thousand ill-examined plants which inhabit Europe gave a most imperfect idea of the vegetation of the globe; that methods of classification which were tolerable so long as species were few, became useless, or an incumbrance as the number increased, and that no real progress in Botany, as a branch of science, could be hoped for so long as a few arbitrary signs were taken as the basis of all arrangement. The older Botanists knew little of vegetable physiology; and of the laws of vegetable structure they had at the most but a glimmering perception. Yet those subjects are the foundation of all sound principles of classification. The recognition of that fact immediately led to the investigation of new branches of knowledge, in which discoveries were daily made, and it has terminated in a universal adoption of the principles of Ray, improved and extended by the admirable views of Jussieu, as developed in his *Genera Plantarum secundum Ordines*

Naturales disposita,—a book of wonderful sagacity and most profound research.

Since the appearance of that work Botany has assumed a new position in the ranks of science, and the evidence from which conclusions are to be drawn has multiplied beyond all that could have been anticipated. Twenty thousand species at the utmost could have been known to Jussieu in 1789; we have seen that the number actually on record at the present day amounts to more than 92,000. Vegetable Anatomy, the foundation of Vegetable Physiology, was at the former period in the state in which it had been left by Grew and Malpighi; it has since engaged the attention of the most acute and indefatigable observers, now armed with optical instruments of surprising excellence. The resources of Chemistry and Natural Philosophy have been enlisted in its cause; and the result is the accumulation of a prodigious mass of facts, the best mode of arranging which is the great problem that modern science has to solve.

That no artificial mode of classifying the vast materials of Botany could satisfy the human mind was clearly perceived and fully admitted by Linnæus himself, when he declared a Natural System to be the *primum et ultimum in botanicis desideratum* (Phil. Bot. § 77). That no insuperable obstacle to its attainment could exist in the nature of things became evident the moment that the work of Jussieu was before the world. That Botanist for the first time proposed distinctive characters for the groups of genera, which he called Natural Orders, and those characters were framed with such skill that a large proportion of his distinctions is still unaffected by the progress of modern discovery. The manner in which he obtained the distinctions of his Natural Orders was thus described by himself :—" *C'est ainsi que sont formées les familles très naturelles et généralement avouées. On extrait de tous les genres qui composent chacune d'elles les caractères communs à tous, sans excepter ceux qui n'appartiennent pas à la fructification, et la réunion de ces caractères communs constitue celui de la famille. Plus les ressemblances sont nombreuses, plus les familles sont naturelles, et par suite le caractère général est plus chargé. En procédant ainsi, on parvient plus sûrement au but principal de la Science, qui est, non de nommer une plante, mais de connoître sa nature et son organisation entière.*"

The Natural Orders thus obtained were bound together into a system by adopting the important distinctions of Acotyledons, Monocotyledons, and Dicotyledons, and then by subdividing the two latter into Classes mainly characterised by the insertion of the stamens or the condition of the corolla; as will be more particularly explained hereafter.

It was not, however, to be expected that the views of Jussieu should be just in all respects, or that his scanty materials would enable him to form a plan of classification sound and perfect in all its parts. On the contrary, his system abounded in errors and imperfections, and, in fact, the latter years of his life were occupied in striving to improve and consolidate it. The same object has been sought by great numbers of those who have succeeded him, and every few years of late have witnessed the production of some scheme of classification which, although founded essentially upon the groundwork of Jussieu, differed nevertheless in numerous details. In another place, the principal of these schemes will be mentioned. It will be for the present sufficient to say that, beginning with Brown in 1810, and ending with Adolphe Brongniart in 1843, the mass of suggestions and improvements which has been collected renders comparatively easy the task of applying Jussieu's principles of classification to the vast multitudes of species now forming the Vegetable Kingdom.

The true principles of classification, however much they may have been amplified and refined upon, were in reality expressed by Ray, when he defined a Natural System to be that *which neither brings together dissimilar species, nor separates those which are nearly allied.* However much the words of this definition may have been varied, it still retains the very meaning given to it by its author. *A species,* said Jussieu, *consists of individuals very much alike in all their parts, and retaining their resemblances from generation to generation. Those species are to be associated which correspond in the greater number of their characters ; but one constant is of more importance than several inconstant characters. On these two axioms hangs the whole principle of Natural classification.*—(Genera Plantarum Præf.) And then he proceeded to show how a group of species combined upon this principle forms a Genus, of Genera an Order, and of Orders a Class; the same rules of combination being observed throughout, with this difference only, that the larger the group the fewer the characters by which it is limited (*Quò generalior enim extat plantarum ordinatio quælibet, eò paucioribus utitur signis definientibus*).

But it is far more easy to lay down principles than to put them in execution. The definition of Ray is perfect, but its application is surrounded with difficulty. The very first point to settle in attempting to carry out his views is by what rule the dissimilarity or alliance of species is to be determined. In fact, very different ideas of likeness or unlikeness are entertained by different observers. The common people can see no difference of moment between a Daphne, and a Cherry, and a Rhododendron, but call them all Laurels, although a Botanist fails to perceive their resemblance. On the other hand, there seems to the vulgar eye no connection between the Hemp plant and the Mulberry tree, and yet the Botanist brings them into close alliance. Nor are these conflicting views confined to the ignorant and the uneducated ; such differences of opinion may be found among Botanists themselves. For instance, Linnæus joined Arum with Phytolacca under his Piperitæ, and Convolvulus with Viola under his Campanacei, combinations which modern Botanists entirely repudiate ; and in like manner the association of Hugonia with Chlenads by Endlicher, of Nepenthes with Birthworts by Brown, of Planes with Witch Hazels by Adolphe Brongniart, of Vines with Berberries by the Author of this work, of Spurgeworts with Heathworts and Chenopods by Fries, are so many modern instances of peculiar views from which other Botanists withhold their assent.

It is therefore of the first importance to settle with something like precision what it is that constitutes likeness among plants, or, as it is technically called, their affinity.

The reason why the vulgar commit mistakes in judging of natural affinity is, because they draw their conclusions from unimportant circumstances, the chief of which are size, form, and colour. The similitude of size gave rise to the old notion that all trees made a class by themselves ; which is as if in a classification of animals the horse, the lion, and elephant were placed in a different part of the animal kingdom from the rat, the cat, and the goat. Form is another of the false guides which lead to error ; if all round-leaved or square-stemmed plants are to be associated, so ought glass to be classed with the diamond when it is cut to the same shape. Colour is less a source of mistake, and yet it is sometimes unconsciously employed by the superficial observer, as when he calls all yellow-flowered Composites Marigolds, and all white-flowered vernal bushes Thorns. It must be evident to the most careless thinker that such resemblances are trifling.

That which really determines affinity is correspondence in structure. It may be said that those plants are most nearly related which correspond in the greatest number of points, and those the most distantly in which we find the fewest points of correspondence ; and this must be true when we remember that if every point in the structure of any two plants is found to be alike, then those two must be identical. But it will be obvious that an examination of all plants through every detail of their organisation is impracticable ; it has never in fact been accomplished in any one case. Experience must have shown that the organs of vegetation are of very different degrees of value in determining resemblance in structure, that some are of paramount importance, others of less consequence, and others of comparative insignificance. Hence the relative value of characters forms a most important part of the study of the Botanist ; it is in fact the pivot upon which all the operations of a systematist must turn.

The only intelligible principle by which to estimate their respective value is according to their known physiological importance ; regarding those organs of the highest rank which are most essential to the life of the plant itself ; placing next in order those with which the plant cannot dispense if its race is to be preserved ; assigning a still lower station to such organs as may be absent without considerable disturbance of the ordinary functions of life ; and fixing at the bottom of the scale those parts, or modifications of parts, which may be regarded as accessory, or quite unconnected with obviously important functions.

The first office which all organised beings have to perform is that of feeding ; for it is thus only that their existence is maintained. The second is that of propagating, by means of which their species is perpetuated. These being functions of the highest importance, it is reasonable to conclude that the organs provided for their proper execution must be of the highest importance also, and hence that they are beyond all others valuable for the purposes of classification. And, again, because the power of feeding must come before that of propagating, it might be conjectured beforehand that the organs destined for the former operation would afford the first elements of a Natural method. But since the action of feeding is very simple in the Vegetable Kingdom, because of the similar modes of life observable among plants, while, on the contrary, the act of propagation is highly diversified, on account of the very varied nature or structure of the parts by which it is accomplished ; so might we conjecture that the organs of nutrition would afford but few distinctions available for purposes of classification, while those of fructification would furnish many. And such is the fact. Hence it is that the great classes of plants are principally distinguished by their organs of growth, and that in the numerous minor groups such peculiarities are comparatively disregarded, their chief distinctions being derived from their parts of reproduction. These principles are more fully expressed in the following axioms :—

1. Peculiarities of structure which are connected with the manner in which a plant is developed are *physiological ;* those which are connected with the manner in which parts are arranged are *structural.* Physiological characters are of two kinds, viz., those which are connected with the *mode of growth* (*the organs of vegetation*), and those which regulate *reproduction* (*the organs of fructification*). Physiological characters are of greater importance in regulating the natural classification of plants than structural

2. All modifications of either are respectively important, in proportion to their connection with the phenomena of life.

3. If we allow ourselves to be steadily guided by these considerations, we shall find that the internal or anatomical structure of the axis, and of the foliage, is of more importance than any other character; because these are the circumstances which essentially regulate the functions of growth, and the very existence of an individual.

4. That next in order is the internal structure of the seed, by which the species must be multiplied. Thus the presence of an embryo, or its absence, the first indicating a true seed, the latter a spore, are most essential circumstances to consider. And so also the existence of albumen in abundance round the embryo, or its absence, must be regarded as a physiological character of the highest value: because, in the former case, the embryo demands a special external provision for its early nutriment, as in oviparous animals; while, in the latter case, the embryo is capable of developing by means of the powers resident in itself, and unassisted, as in viviparous animals.

5. Next to this must be taken the structure of the organs of fructification, by whose united action the seed is engendered; for without some certain, uniform, and invariable action on their part, the race of a plant must become extinct. Thus we find that the structure of the anthers, placentæ, and ovules, are more uniform than that of the parts surrounding them, while their numbers are variable; and the condition of the filament, which appears of so little importance in a physiological point of view, is also inconstant. So also the texture and surface and form of the pericarp, which acts as a mere covering to the seeds, is not to be regarded in these inquiries, and, in fact, differs from genus to genus; as, for instance, between Pyrus and Stranvæsia, or Rubus and Spiræa, in the truly natural Rosaceous Order.

6. On the other hand, the floral envelopes seem to be unconnected with functions of a high order, and to be designed rather for the decoration of plants, or for the purpose of giving variety to the aspect of the vegetable world; and, consequently, their number, form, and condition, presence or absence, regularity or irregularity, are of low and doubtful value, except for specific distinction. There seems, indeed, reason to expect that every Natural Order will, sooner or later, be found to contain within itself all the variations above alluded to. Even in the cases of regularity and irregularity we already know this to be so; witness Veronica and Scoparia in Figworts, and Hyoscyamus in Nightshades, Delphinium in Crowfoots, and Pelargonium in Cranesbills.

7. The consolidation of the parts of fructification is a circumstance but little attended to in a general point of view, except in respect to the corolla; but as it seems to indicate either the greatest change that the parts can undergo, or, where it occurs between important and usually unimportant organs, that in such cases the latter become essential to the former, it probably deserves to be regarded with great attention. For instance, the presence or absence of the corolla is often a point of little moment, and is, we know, a very fluctuating circumstance. This is especially true of those Natural Orders in which the stamens and petals are separated; as in Roseworts, Rhamnads, Onagrads, &c. On the other hand, when the stamens, which are indispensable organs, adhere to the petals, the latter are more constantly present, as in Figworts, Acanthads, Nightshades, &c.

There are also certain other principles which experience tells us the systematist must keep in view; and most especially that of regarding of importance whatever appears to be constant in its nature among nearly allied species. Nothing which is thus constant can be considered unimportant, for everything constant is dependent upon or connected with some essential function. Therefore all constant characters, of whatever nature, require to be taken into account in classifying plants according to their natural affinities. Of this nature are the internal structure of stems and leaves, the anatomical condition of tissue, the organisation of the anther, pollen, and female apparatus, and the interior of the seed.

On the other hand, whatever points of structure are variable in the same species, or in species nearly allied to each other, or in neighbouring genera, are unessential to the vital functions, and should be set aside, or be regarded as of comparative unimportance. Hence the badness of the Monopetalous, Polypetalous, and Apetalous divisions of Jussieu, depending upon the mere presence or absence, and union or disunion, of petals. The genus Fuchsia, for example, has petals highly developed; but in F. excorticata they are absent, and yet the plant differs no otherwise from the rest of the genus: the same is true of species of Rhamnus. Again, the Rue has the petals separate; and Correa, very nearly allied to it, has them combined.

All classifications in which the foregoing principles are observed are natural; and that will be the most stable in which they are employed with the greatest skill. Some writers, indeed, maintain that there cannot be more than one really natural system, any more than one planetary system; and in a certain sense this may be true, inasmuch as we must suppose that one plan only has been observed in the creation of living things, and that a natural system is the expression of that plan. But, on the other hand, it must not be forgotten that such a plan may be represented in many ways; and that although the order of nature is in itself settled and invariable, yet that human descriptions of it will vary with the mind of the describer. A universal history is a collection of events; but it is not necessary that all universal histories should follow the same order of narration. The events themselves are unalterable, but the way of combining them and causing them to illustrate each other is manifold.

In natural science, indeed, the mode of arranging the matter is susceptible of infinitely more variation than history: because in the latter subject time is an inflexible leader who cannot be lost sight of. But in natural science there is no beginning and no end. It is impossible, from the nature of things, that any arrangement should exist which shall represent the natural relations of plants in a consecutive series. It is generally admitted by those who have turned their attention to a consideration of the manner in which organised beings are related to each other, that each species is allied to others in different degrees, and that such relationship is best expressed by rays (called affinities) proceeding from a common centre (the species). In like manner, in studying the mutual relationship of the several parts of the Vegetable Kingdom, the same form of distribution constantly forces itself upon the mind; Genera and Orders being found to be apparently the centre of spheres, whose surface is only determined by the points where the last traces of affinity disappear. But although the mind may conceive such a distribution of organised beings, it is impossible that it should be so presented to the eye, and all attempts at effecting that object must of necessity fail. If in describing the surface of a sphere we are compelled to travel in various directions, continually returning back to the point from which we started; and if in presenting it to the eye at one glance we are compelled to project it upon a plane, the effect of which is to separate to the greatest distance some objects which naturally touch each other; how much more impossible must it be to follow the juxtaposition of matter in treating of the solid contents of a sphere!

An arrangement, then, which shall be so absolutely correct an expression of the plan of nature as to justify its being called *the* Natural System is a chimera.* All that the Naturalist can do is to carry into effect the principles above explained, with a greater or less amount of skill; the result of which will be *a* Natural System.

When Linnæus attempted to form a Natural System, he merely threw together such genera as he knew into 67 groups, which he called Fragments, and which were equivalent to the Natural Orders of Modern Botany. Jussieu advanced a step further, by forming 15 Classes, under which he placed 100 Natural Orders. At a later period the name Class was reserved for the three great divisions of Acotyledons, Monocotyledons, and Dicotyledons; and the Orders were collected into smaller groups called Sub-classes; and thus, by degrees, the necessity of forming three grades of distinctive characters superior to genera was recognised. But our countryman, Dr. Robert Brown, whose sagacity is not the least remarkable part of his scientific character, long ago pointed out the insufficiency of even this amount of subdivision, and proposed the combination of Natural Orders into groups intermediate between Orders and Sub-classes. The necessity of this measure is now universally acknowledged; attempts have been made for some years, by various Botanists, to work out the problem; and I think it must be conceded that a real advance has thus been made, by the efforts of various independent observers, to the accomplishment of so very desirable an object. To such attempts the present work is an addition.

The leading idea which has been kept in view in the compilation of it has been this maxim of Fries: *Singula sphæra (sectio) ideam quandam exponit, indeque ejus character notione simplici optimè exprimitur.* I cannot but think that the true characters of all natural assemblages are extremely simple; nothing can be more certain than that their value diminishes in proportion to their complexity. If two objects are not to be distinguished by a few simple circumstances, they can hardly be called distinguishable at all. In the highest groups or classes it is always so, (see p. 4;) and there is no apparent reason why the same rule should not obtain in groups of a minor rank. Nevertheless, we find that this is too often lost sight of, and that long details of structure are substituted for precise words of distinction.

It may be, and certainly is in some measure, true, that insuperable difficulties are, in the present state of our knowledge, opposed to strict definitions of Natural Orders, and *à fortiori* of their Alliances, &c. But that is no reason why we should not endeavour to render their distinctive characters as precise as the nature of the subject will permit. Vague distinctions, which are at once the bane and opprobrium of Natural History, are so repulsive to the understanding as to deter the mass of mankind from giving it their attentive study. And it is not too much to assert that this vagueness arises more frequently out of the prejudices or mistiness of the Naturalist's own mind than out of things themselves. It will constantly happen that two groups may stand, by common consent, in the nearest conceivable relation to each other; it is quite possible, by one way of arranging

* Systema illud naturæ ipsius absolutum (quod mera empiria captant!) mens humana capere non potest; est quoddam supra naturale cujus clavem, manibus v. ingenio humano non prensandam, summus tantum tenet Naturæ auctor.—*Fries Corpus Florarum*, p. xvii.

them, to render their distinctions nugatory, and by another, clear and precise. Now, if the supposed groups are really as closely allied, as for this argument we may assume them to be, it can be of no possible importance theoretically, whether a given Genus or Order is placed in the one or the other. The near consanguinity of the two does away with all importance in such a case. In Physical Geography it is of no consequence whether London is stationed in Middlesex or Surrey; and in like manner, in Theoretical Botany, the place of a given Order may be equally indifferent. But it may be of great consequence practically, because a definition of limits may be possible or not, according to the arrangement. For example, let us take the Solanal and Bignonial Alliances. These touch at the Orders of Nightshades and Figworts respectively. If Nightshades are placed in the Bignonial Alliance because of their intimate relation to Figworts, no apparent means remain of clearly defining what is meant by the Bignonial Alliance. If, on the other hand, Figworts are stationed in the Solanal Alliance, then the distinctive characters of that Alliance are also rendered obscure and difficult, or impossible of application. But place Nightshades in the Solanal, and Figworts in the Bignonial Alliance, and the language of Botanists affords as clear a discrimination as can be wished for. And so of other cases. Indeed, I am so persuaded of this, that in my opinion all instances of confused and vague characters are only so many proofs of Botanists not having clearly understood the plants that they have endeavoured to classify.*

It will, perhaps, be alleged that the doctrine just inculcated is directly opposed to the first principles of a Natural System: but such is not the case. No absolute limits, in fact, exist, by which groups of plants can be circumscribed. They pass into each other by insensible gradations, and every group has apparently some species which assumes in part the structure of some other group. Two countries are separated by a river whose waters are common to both banks: in a geographical division of territory the river may be assigned to either the left bank or the right bank, but such an arrangement is arbitrary; and yet the interior of the countries is unaffected by it. So with the groups of plants; it cannot be of any possible consequence whether an intermediate or frontier plant be assigned to one group or another, and convenience alone should be considered in such a matter. This long since led me to offer the following observations, the justice of which, much more experience entirely confirms:—"All the groups into which plants are thrown are in one sense artificial, inasmuch as Nature recognises no such groups. Nevertheless, consisting in all cases of species very closely allied in nature, they are in another sense natural. But as the Classes, Sub-classes, Alliances, Natural Orders, and Genera of Botanists, have no real existence in nature, it follows that they have no fixed limits, and consequently that it is impossible to define them. They are to be considered as nothing more than the expression of particular *tendencies* (nixus), on the part of the plants they comprehend, to assume a particular mode of development. Their characters are only a declaration of their prevailing tendencies."

We must not, however, deceive ourselves with the expectation that by this or any other expedient definitions in Botany will become possible. Mathematical precision is unknown in such subjects, and exceptions occur

* No Botanist will regard this as an offensive remark. It is the misfortune, not the fault, of men of science, that they cannot investigate everything with their own eyes, and that they are compelled, from the vastness of their subject, to take much of all they study upon trust. In Botany this is most especially the case; for who has ever been able to examine one-tenth of all the plants he speaks of, with minute accuracy?

to all known rules. "When Zoology," says Mr. Milne Edwards, "is only studied in systematic works, it is often supposed that each class, each family, and each genus, present to us boundaries precisely defined, and that there can be no uncertainty as to the place to be assigned, in a natural classification, to every animal the organisation of which is sufficiently known. But when we study this science from Nature herself, we are soon convinced of the contrary, and we sometimes see the transition from one plan of structure to an entirely different scheme of organisation take place by degrees so completely shaded one into the other that it becomes very difficult to trace the line of demarcation between the groups thus connected."—*Ann. Sc. Nat.* 1840, *Sept.* Ray long ago pointed this out in a very remarkable passage, which cannot be too often quoted.

"Verum quod alias dixi illud hic repeto et inculco, non sperandam à me Methodum undequaque perfectam et omnibus suis numeris absolutam, quæ et plantas in genera ità distribuat ut universæ species comprehendantur, nullâ adhuc anomalâ et sui generis reliquâ, et unumquodque genus notis suis propriis et characteristicis ità circumscribat, ut nullæ inveniantur species incerti, ut ita dicam, laris, et ad plura genera revocabiles. Nec enim id patitur natura rei. Nam, cùm Natura (ut dici solet) non faciat saltus, neque ab extremo ad extremum transeat nisi per medium, inter superiores et inferiores, rerum ordines nonnullas mediæ et ambiguæ conditionis producere solet, quæ de utroque participent, et utrosque velut connectant, ut ad utrum pertineant omninò incertum sit. Præterea eadem alma parens in methodi cujuscunque angustias coerceri repugnat, sed ad libertatem et αὐτονομίαν suam nullis legibus obnoxiam ostentandam, in unoquoque rerum ordine nonnullas species creare solet, tanquam exceptiones à regulis generalibus, singulares et anomalas."—(RAII, *Hist. Plant.* vol. i. Præf.) Linnæus did but copy this when he asserted that Nature makes no leaps (*Natura non facit saltus.*—Phil. Bot. 77.)

This doctrine has, however, been lately called in question by no less eminent a writer than M. Alphonse De Candolle, who requires that absolute limits should be assigned to all groups of whatever degree. "If," he says, "we cannot state in what respect two families differ permanently and *universally*, those two families are but one. Two pieces of land which touch each other form one island, and not two; but two pieces of land which are separated by an arm of the sea, form two islands, and not one." —*Annales des Sciences*, series 3, vol. 1. p. 254. But this is a kind of reasoning wholly inapplicable to Natural History, for the reasons so admirably given by Ray, and is contrary to all experience. If the groups limited by M. Alphonse De Candolle himself are examined by this standard they alone suffice to demonstrate how visionary are such expectations. Mr. Bentham has satisfactorily answered the learned Botanist of Geneva. "We Botanists," he says, "cannot be so mathematically exact as geographers, and where an isthmus is very narrow, we must class the peninsula with the island. How often does it happen that two large Orders, say of five hundred to two thousand or three thousand species, totally distinct from each other in all those species by a series of constant characters, are yet connected by some small isolated genus of a dozen, half a dozen, nay a single species, in which these very characters are so inconstant, uncertain, or variously combined as to leave no room for the strait through which we ought to navigate between the two islands."—*London Journal of Botany*, 4. 232. It would be very convenient to find that the views of M. Alphonse De Candolle were practicable, but in truth they are quite Utopian.

While, however, the impracticability of absolute definitions is thus insisted upon, there can be no doubt that much more precision may be introduced than is too frequently found among them. Exceptions, although to some extent inevitable, are not uncommonly apparent, not real. It will frequently be found that a particular species is at variance with the definition of its Genus, or of a Genus with that of its Order, or of an Order with that of its Alliance; but, upon a full examination of all the structure of such supposed exceptions, it will turn out that they are misplaced, and do not in fact belong to the station which they occupy. Exceptions of this kind were formerly very common, but they are disappearing under the diligent criticism of modern observers. The genus Rhynchotheca may be taken as an example. The great feature of the Cranesbills is their beaked torus and folded-up embryo, and it is by that circumstance that they are essentially distinguished from their neighbours. But Rhynchotheca was described as having a beaked fruit and straight embryo; it therefore formed an apparent exception to the definition of Cranesbills. Investigation of the plant has however shown that its beak belongs to the carpels and not to the torus; and, therefore, it is merely an Oxalid, with a tendency towards the structure of a Cranesbill.

The manner in which the foregoing principles have been applied to practice has differed greatly, and the result has been schemes of various degrees of merit, some of which have dropped still-born from the press, while others continue to enjoy a well-deserved reputation. It would be alike unjust to their authors and the public to omit all mention of even the most obscure of these, each of which has been the result of much thought and patient study, and has doubtless contributed something to the progress of systematic science. But it would be beyond the object of the present sketch to treat them all at length, nor would the student derive any advantage from doing so. While, therefore, the following pages will be occupied by some account of every plan for a Natural classification of which I have any knowledge,* since the year 1789 inclusive, and of those of Ray and Linnæus of an earlier date, such as are comparatively unimportant will be dismissed in a few words, and those only which have been really employed in practice will be stated at length. In order to render the latter more useful, references are given to the pages in the present work where an account of each Order may be found; so that those who are accustomed to the use of other systems may not experience inconvenience from the arrangement proposed in the work now submitted to their consideration.

* I do not, however, include the arrangements of the German Naturphilosophists; not, indeed, from any disrespect to those learned men, but because I must confess my inability to master their ideas, or to comprehend how their views are made applicable to any intelligible classification. The student will, I believe, find full information upon the subject in Oken's *Lehrbuch der Naturphilosophie*, edition of 1843. See also Reichenbach's *Conspectus Regni Vegetabilis*, 1828, the same author's *Flora Germanica Excursoria*, 1830-2, and Schultz *Naturliches System des Pflanzenreichs*, 1832.

NATURAL SYSTEMS.

[Where references are given after the names of Orders, in this part of the present work, they refer to the page where such Orders are to be found in the succeeding sheets].

1703. Ray, John.—(*Methodus Plantarum emendata et aucta*).

Here we have the germ of the present methods of natural arrangement. In fact the first divisions of the Vegetable Kingdom, proposed by Ray, are identical with those of Jussieu. Like him, he proceeded from the more imperfect to the most highly organised forms; the only difference being that he placed Dicotyledons before Monocotyledons. The author's words are "Floriferas dividemus in *dicotyledones*, quarum semina sata binis foliis anomalis, seminalibus dictis, quæ cotyledonum usum præstant, è terrâ exeunt, vel in binos saltem lobos dividuntur, quamvis eos supra terram foliorum specie non efferant; et *monocotyledones* quæ nec folia seminalia bina efferunt nec lobos binos condunt. Hæc divisio ad arbores etiam extendi potest: siquidem palmæ et congeneres hoc respectu eodem modo a reliquis arboribus differunt quo monocotyledones à reliquis herbis."

His plan was this :—

Plants are either
Flowerless, or
Flowering; and these are
Dicotyledones, or
Monocotyledones.

Among the genera of Ray, which were what we now call Natural Orders, were Fungi, Mosses, Ferns, Composites, Cichoraceæ, Umbellifers, Papilionaceous plants, Conifers, Labiates, &c., under other names, but with limits not very different from those now assigned to them.

1751. Linnæus, Charles.—(*Philosophia Botanica*).

"Plantæ omnes utrinque affinitatem monstrant, uti Territorium in mappa geographica."

The following is the Natural distribution first proposed by Linnæus, under the name of Fragments. Many of his groups were taken from his predecessors; others were contrived by himself. At a later period they underwent some alteration; but the list now given will serve to show the learned author's plan. He never assigned any characters to these Fragments.

1. Piperitæ. Arum, &c. Piper, Phytolacca.
2. Palmæ. Corypha, &c., Cycas.
3. Scitamina. Musa, Canna, Amomum, &c.
4. Orchideæ. As now.
5. Ensatæ. Iris, &c., Xyris, Eriocaulon, Aphyllanthes.
6. Tripetaloideæ. Butomus, Alisma, Sagittaria.
7. Denudatæ. Crocus, &c.
8. Spathaceæ. Leucoium, Amaryllis, &c.
9. Coronariæ. Ornithogalum, Scilla, &c.
10. Liliaceæ. Lilium, Tulipa, &c.
11. Muricatæ. Bromelia, &c.
12. Coadunatæ. Anona, Magnolia, &c., Thea.
13. Calamariæ. Scirpus, &c., Juncus?
14. Gramina. As now.
15. Coniferæ. Abies, Pinus, &c.
16. Amentaceæ. Pistacia, Alnus, Populus, Juglans, Quercus, &c.
17. Nucamentaceæ. Xanthium, Iva, &c.
18. Aggregatæ. Statice, Protea, Hebenstreitia, Brunia, Valeriana, Boerhaavia, Circæa? &c.
19. Dumosæ. Viburnum, Rondeletia, Cassine, Rhus, Ilex, Callicarpa, Lawsonia, &c.
20. Scabridæ. Ficus, &c.
21. Compositæ. As now, nearly.
22. Umbellatæ. As now.
23. Multisiliquæ. Modern Crowfoots.
24. Bicornes. Azalea, Myrsine, Memecylon, Santalum, &c.
25. Sepiariæ. Jasminum, Ligustrum, Brunfelsia, &c.
26. Culminiæ. Tilia, Bixa, Dillenia, Clusia, &c.
27. Vaginales. Polygonum, Laurus, &c.
28. Corydales. Melianthus, Epimedium, Fumaria, Monotropa? &c.
29. Contorti. Rauwolfia, Vinca, Asclepias, &c.
30. Rhœades. Papaver, Podophyllum, &c.
31. Putamineæ. Capparis, &c.
32. Campanacei. Convolvulus, Lobelia, Viola, &c.
33. Luridæ. Solanum, &c., Celsia, Digitalis.
34. Columniferæ. Camellia, Gossypium, Mentzelia, &c., but chiefly Mallowworts.
35. Senticosæ. Roseworts exclusively.
36. Comosæ. Spiræa, Filipendula, Aruncus.
37. Pomaceæ. Punica, Pyrus, &c., Ribes.
38. Drupaceæ. As now.
39. Arbustiva. Philadelphus, and Myrtleblooms.
40. Calycanthemæ. Œnothera, &c., Lythrum, Glaux, Rhexia.
41. Hesperideæ. Citrus, Styrax, Garcinia.
42. Caryophyllei. Cloveworts, with Frankenia and Scleranthus.
43. Asperifoliæ. The modern Borageworts.

44. STELLATÆ. Galium, &c., Hedyotis, Spigelia, Cornus? Coffea, &c.
45. CUCURBITACEÆ. Passiflora and Cucurbits.
46. SUCCULENTÆ. Cactus, Mesembryanthemum, Sedum, Oxalis, Fagonia, &c. &c.
47. TRICOCCÆ. Cambogia, Euphorbia, &c., Cliffortia, Sterculia, &c.
48. INUNDATÆ. Hippuris, Elatine, Ruppia, Typha, &c.
49. SARMENTACEÆ. Vitis, Hedera, Houstonia, Ruscus, Smilax, Menispermum, Aristolochia, &c.
50. TRIHILATÆ. Sapindus, Malpighia, Begonia, Berberis? &c.
51. PRECIÆ. Part of modern Primworts.
52. ROTACEÆ. Gentiana, Lycimachia, Anagallis, &c.
53. HOLERACEÆ.—Spinacia, &c., Herniaria, Callitriche, Petiveria, &c.
54. VEPRECULÆ. Rhamnus, &c., Lycium, Daphne, &c.
55. PAPILIONACEÆ. As now.
56. LOMENTACEÆ. Leguminous plants, with jointed pods, Cæsalpinieæ and Mimoseæ.
57. SILIQUOSÆ. Crucifers.
58. VERTICILLATÆ. Labiates.
59. PERSONATÆ. Figworts, Sesamum, Justicia, Bignonia, Verbena, &c.
60. PERFORATÆ. Hypericum, Cistus, Telephium.
61. STATUMINATÆ. Ulmus, Celtis, Bosea.
62. CANDELARES. Rhizophora, Mimusops, Nyssa.
63. CYMOSÆ. Lonicera, Loranthus, Ixora, Cinchona? &c.
64. FILICES. As now.
65. MUSCI. As now.
66. ALGÆ. Nearly as now.
67. FUNGI. As now.
68. VAGÆ. All his doubtful genera.

At a later period Nos. 7, 10, 11, 17, 26, 27, 36, 38, 39, 60, 61, 62 and 63, were cancelled; and four added, viz.

GRUINALES. Cranesbills.
CALYCIFLORÆ. Osyris, Trophis, Hippophäe, Elæagnus.
HEDERACEÆ. Hedera and Vitis, &c.
MISCELLANEÆ. A curious mixture.

1789. JUSSIEU, Antoine Laurent de.—(*Genera Plantarum secundum ordines naturales disposita, juxta methodum in horto regio Parisiensi exaratum, anno* MDCCLXXIV).

Adopting the views of Ray as to primary divisions, Jussieu applied them to the system of Tournefort, which had been in common use in France from the year 1694, and which was by far the best suited for the state of knowledge of the age in which it was promulgated. To this he added the position of the stamens with respect to the ovary, and thus constructed his 15 classes in the following manner :—

				CLASS.
Acotyledones.				I.
Monocotyledones.		Stamina hypogyna.		II.
		——— perigyna.		III.
		——— epigyna.		IV.
Dicotyledones.	Apetalæ.	Stamina epigyna.		V.
		——— perigyna.		VI.
		——— hypogyna		VII.
	Monopetalæ.	Corolla hypogyna.		VIII.
		——— perigyna.		IX.
		——— epigyna.	Antheris connatis.	X.
			Antheris distinctis.	XI.
	Polypetalæ.	Stamina epigyna.		XII.
		——— hypogyna.		XIII.
		——— perigyna		XIV.
	Diclines irregulares.			XV.

Under each of these classes he arranged his Natural Orders as follows, usually deriving their name from some genus, which he regarded as a good illustration of their general structure.

Class I.

1. Fungi, 29
2. Algæ, 8
3. Hepaticæ, 58
4. Musci, 64
5. Filices, 74
6. Naiades, 143

Class II.

7. Aroïdeæ, 127
8. Typhæ, 126
9. Cyperoïdeæ, 117
10. Gramineæ, 106

Class III.

11. Palmæ, 134
12. Asparagi, 200
13. Junci, 191
14. Lilia, 200
15. Bromeliæ, 147
16. Asphodeli, 200
17. Narcissi, 155
18. Irides, 159

Class IV.

19. Musæ, 163
20. Cannæ, 165
21. Orchides, 173
22. Hydrocharides, 141

Class V.

23. Aristolochiæ, 792

Class VI.

24. Elæagni, 257
25. Thymeleæ, 530
26. Proteæ, 532
27. Lauri, 535
28. Polygoneæ, 502
29. Atriplices, 512

Class VII.

30. Amaranthi, 510
31. Plantagines, 642
32. Nyctagines, 506
33. Plumbagines, 640

Class VIII.

34. Lysimachiæ, 644
35. Pediculares, 681
36. Acanthi, 678
37. Jasmineæ, 650
38. Vitices, 663
39. Labiatæ, 659
40. Scrophulariæ, 681
41. Solaneæ, 618
42. Boragineæ, 655
43. Convolvuli, 630
44. Polemoniа, 635
45. Bignoniæ, 675
46. Gentianeæ, 612
47. Apocineæ, 599
48. Sapotæ, 590

Class IX.

49. Guaïacanæ, 595
50. Rhododendra, 453
51. Ericæ, 453
52. Campanulaceæ, 689

Class X.

53. Cichoraceæ, 702
54. Cinarocephalæ, 702
55. Corymbiferæ, 702

Class XI.

56. Dipsaceæ, 699
57. Rubiaceæ, 761
58. Caprifolia, 766

Class XII.

59. Araliæ, 780
60. Umbelliferæ, 773

Class XIII.

61. Ranunculaceæ, 425
62. Papaveraceæ, 430
63. Cruciferæ, 351
64. Capparides, 357
65. Sapindi, 382
66. Acera, 387
67. Malpighiæ, 388
68. Hyperica, 405

69. Guttiferæ, 400
70. Aurantia, 457
71. Meliæ, 463
72. Vites, 439
73. Gerania, 493
74. Malvaceæ, 368
75. Magnoliæ, 417
76. Anonæ, 420
77. Menisperma, 307
78. Berberides, 437
79. Tiliaceæ, 371
80. Cisti, 349
81. Rutaceæ, 469
82. Caryophylleæ, 496

Class XIV.

83. Sempervivæ, 344
84. Saxifragæ, 567
85. Cacti, 746
86. Portulaceæ, 500
87. Ficoideæ, 525
88. Onagræ, 724
89. Myrti, 734
90. Melastomæ, 731
91. Salicariæ, 574
92. Rosaceæ, 563
93. Leguminosæ, 544
94. Terebintaceæ, 465
95. Rhamni, 581

Class XV.

96. Euphorbiæ, 274
97. Cucurbitaceæ, 311
98. Urticæ, 258
99. Amentaceæ, 248
100. Coniferæ, 226

1810. BROWN, Robert.—(*Prodromus Floræ Novæ Hollandiæ, &c.*)

In this work the system of Jussieu is principally followed, but the Classes are omitted, and the sequence of the Orders is changed. The author states that he regards most of the Orders of Jussieu as being truly natural, but his classes, as the latter candidly admits, often artificial, and apparently founded upon doubtful principles. It was the intention of Dr. Brown to publish a second volume of his work, and then to explain his views upon this and other subjects; but that intention has not yet been carried into execution. It is here that we find the importance of the æstivation of the flower pointed out, and applied to the characters of Natural Orders. Those characters have been a model for succeeding writers.

1813. DE CANDOLLE, A. P.—(*Théorie Élémentaire de la Botanique, ou Exposition des Principes de la Classification Naturelle et de l'Art de décrire et d'étudier les Végétaux*).

In this work is to be found the explanation of the principles which guided its clear-minded author to the construction of a method of arrangement which has now almost superseded all others, partly because of its easiness and simplicity, and most especially because it is that which has been followed in the author's *Prodromus*, or celebrated description of species. He himself explains the course he has taken, to the following effect:—"I place Dicotyledons first, because they have the greatest numbers of distinct and separate organs. Then, as I find families where some of these organs become consolidated, and consequently seem to disappear, I refer them to a lower rank. This principle gives me the following series:—

1. Dicotyledons; polypetalous and hypogynous.
2. ———; ———and perigynous.
3. ———; monopetalous and perigynous.
4. ———; ——— and hypogynous.
5. ———; apetalous, or with a single perianth.
6. Monocotyledons; phænogamous.
7. ——— ; cryptogamous.
8. Acotyledons; leafy and sexual.
9. ——— ; leafless and without any known sexes.

I have adopted this series partly because I think it that which is least removed from a natural sequence, and partly because it is convenient and easy for study. But let no one imagine that I attach the least importance to it. The true science of general Natural History consists in the study of the symmetry peculiar to each family, and of the relation which these families bear to each other. All the rest is merely a scaffolding, better or worse suited to accomplish that end."—p. 206, *first edition.*

At this time De Candolle made no attempt to combine the Natural Orders in Alliances; but at a later period (1819), in a second edition of the *Théorie*, he proposed a few such groups, under the name of Cohorts, as will be seen by the following list of his Orders, taken from the edition of 1819. In that of 1844, published by his son after his death, these Cohorts are all broken up, and considerable alterations are made in the sequence of the Natural Orders. I, however, prefer publishing his plan of forming Alliances, rather than his last list, even although that does give his latest views of affinity.

I. VASCULAR OR COTYLEDONOUS PLANTS; that is to say, furnished with cellular tissue and vessels, and whose embryo is provided with one or more cotyledons.

1. Exogens or Dicotyledons; that is to say, where the vessels are arranged in concentric layers, of which the youngest are the outermost, and where the embryo has opposite or verticillate cotyledons.

A. Perianth double; that is, where the calyx and corolla are distinct.

THALAMIFLORÆ

Petals distinct, inserted on the receptacle.

Cohort I. Carpels numerous, or stamens opposite the petals.

1. Ranunculaceæ, 425
2. Dilleniaceæ, 423
3. Magnoliaceæ, 417
4. Anonaceæ, 420
5. Menispermeæ, 307
6. Berberideæ, 437
7. Podophylleæ, 430
8. Nymphæaceæ, 409

Cohort II. Carpels solitary or consolidated, placentæ parietal.

9. Papaveraceæ, 430
10. Fumariaceæ, 435
11. Cruciferæ, 351
12. Capparideæ, 357
13. Flacourtianeæ, 327
14. Passifloreæ, 332
15. Violaceæ, 338
16. Polygaleæ, 375
17. Resedaceæ, 356
18. Droseraceæ, 433
19. Frankeniaceæ, 340
20. Cistineæ, 349

Cohort III. Ovary solitary, placenta central.

21. Caryophylleæ, 496
22. Lineæ, 485
23. Malvaceæ, 368
24. Chlenaceæ, 486
25. Byttneriaceæ, 363
26. Sterculiaceæ, 360
27. Tiliaceæ, 371
28. Elæocarpeæ, 371
29. Sapindaceæ, 382
30. Hippocastaneæ, 382

31. Aceraceæ, 387
32. Malpighiaceæ, 388
33. Hippocraticeæ, 584
34. Hypericineæ, 405
35. Guttiferæ, 400
36. Marcgraviaceæ, 403
37. Sarmentaceæ, 439
38. Geranieæ, 493
39. Cedreleæ, 461
40. Meliaceæ, 463
41. Hesperideæ, 457
42. Camellieæ, 396
43. Olacineæ, 443
44. Rutaceæ, 469

Cohort IV. Fruit gynobasic.

45. Simaroubeæ, 476
46. Ochnaceæ, 474

Calycifloræ.

Petals free or more or less united, always perigynous or inserted on the calyx.

47. Frangulaceæ, 581
48. Samydeæ, 330
49. Zanthoxyleæ, 472
50. Juglandeæ, 292
51. Terebinthaceæ, 465
52. Leguminosæ, 544
53. Rosaceæ, 563
54. Salicariæ, 574
55. Tamariscineæ, 341
56. Melastomeæ, 731
57. Myrtineæ, 734
58. Combretaceæ, 717
59. Cucurbitaceæ, 311
60. Loaseæ, 744
61. Onagrarieæ, 724
62. Ficoïdeæ, 525
63. Paronychieæ, 510
64. Portulaceæ, 500
65. Nopaleæ, 746
66. Grossulaceæ, 750
67. Crassulaceæ, 344
68. Saxifrageæ, 567
69. Cunoniaceæ, 571
70. Umbelliferæ, 773
71. Araliaceæ, 780
72. Caprifolieæ, 766
73. Lorantheæ, 789
74. Rubiaceæ, 761
75. Opercularieæ, 761
76. Valerianeæ, 697
77. Dipsaceæ, 699
78. Calycereæ, 701
79. Compositæ, 702
80. Campanulaceæ, 689
81. Lobeliaceæ, 692
82. Gesnerieæ, 671
83. Vaccinieæ, 757
84. Ericineæ, 453

Corollifloræ.

Petals united into an hypogynous corolla, or not attached to the calyx.

85. Myrsineæ, 647
86. Sapoteæ, 590
87. Ternstromieæ, 396
88. Ebenaceæ, 595
89. Oleineæ, 616
90. Jasmineæ, 650
91. Strychneæ, 602
92. Apocyneæ, 599
93. Gentianeæ, 612
94. Bignoniaceæ, 675
95. Sesameæ, 669
96. Polemonideæ, 635
97. Convolvulaceæ, 630
98. Boragineæ, 655
99. Solaneæ, 618
100. Antirrhineæ, 681
101. Rhinanthaceæ, 681
102. Labiatæ, 659
103. Myoporineæ, 665
104. Pyrenaceæ, 663
105. Acanthaceæ, 678
106. Lentibularieæ, 686
107. Primulaceæ, 644
108. Globularieæ, 666

B. Monochlamydeæ.

Perianth simple, or whose calyx and corolla form only one envelope.

109. Plumbagineæ, 640
110. Plantagineæ, 642
111. Nyctagineæ, 506
112. Amaranthaceæ, 510
113. Chenopodeæ, 512
114. Begoniaceæ, 318
115. Polygoneæ, 502
116. Laurineæ, 535
117. Myristiceæ, 301
118. Proteaceæ, 532
119. Thymeleæ, 530
120. Santalaceæ, 787
121. Elæagneæ, 257
122. Aristolochieæ, 792
123. ? Euphorbiaceæ, 274
124. Monimieæ, 298
125. Urticeæ, 260
126. Piperitæ, 515
127. Amentaceæ, 254
128. Coniferæ, 226.

2. Endogens or Monocotyledons; that is to say, plants whose vessels are arranged in bundles, the youngest being in the middle of the trunk, and whose embryo is furnished with solitary or alternate cotyledons.

A. Phanerogams.

Fructification visible, regular.

129. Cycadeæ, 223
130. Hydrocharideæ, 141
131. Alismaceæ, 209
132. Orchideæ, 173
133. Drymyrhizeæ, 165
134. Musaceæ, 163
135. Irideæ, 159
136. Hæmodoraceæ, 151
137. Amaryllideæ, 155
138. Hemerocallideæ, 200
139. ? Dioscoreæ, 214
140. Smilaceæ, 215
141. Liliaceæ, 200
142. Colchicaceæ, 198
143. Junceæ, 191
144. Commelineæ, 188
145. Palmæ, 133
146. Pandaneæ, 130
147. Typhaceæ, 126
148. Aroïdeæ, 127
149. Cyperaceæ, 117
150. Gramineæ, 106

B. Cryptogams. Fructification hidden, unknown or irregular.

151. Naïades, 143
152. Equisetaceæ, 61
153. Marsileaceæ, 71
154. Lycopodineæ, 69
155. Filices, 78

II. Cellular or Acotyledonous Plants; that is to say, composed of cellular tissue only, not furnished with vessels, and whose embryo is without cotyledons.

A. Foliaceæ, having leaf-like expansions, and known sexes.

156. Musci, 64
157. Hepaticæ, 58

B. Aphyllæ, not having leaf-like expansions, and no known sexes.

158. Lichenes, 45
159. Hypoxyla, 29
160. Fungi, 29
161. Algæ, 8

1825. Agardh, Carl von.—(*Classes Plantarum*).

This is a duodecimo pamphlet of 22 pages, with a coloured map, and is a recapitulation of the views of classification promulgated by its author between 1821 and 1826, in his *Aphorismi Botanici*. The object is to group Natural Orders in Classes, that is to say, in divisions subordinate to the primary ramifications of a system, and equivalent to my Alliances. "Classes," says Bishop Agardh, "should be formed by the same rules and on the same principles as Genera and Orders; and therefore not by the breaking up of higher groups, but by the gathering together of lower groups. Yet, up to this time, all the so-called natural classes of plants have been formed upon an opposite principle, with the exception of the arrangement of Batsch.——We must distinguish, with Linnæus, between the character of a plant and its affinity. The former is derived from the latter, and not *vice versâ*. Plants will sometimes agree in very few characters, which nevertheless are bound together by the strongest possible affinity. For instance, Ceratonia is very different from Leguminous plants, and Fraxinus from Jasmines; yet they are nearly allied."

Agardh's primary divisions are nine; namely,

1. Acotyledons.
2. Pseudocotyledons.
3. Cryptocotyledons.
4. Phanerocotyledons; incomplete.
5. ————————; complete, hypogynous, monopetalous.
6. ————————; ————, ———— polypetalous.
7. ————————; ————, discigynous, monopetalous.
8. ————————; ————, ————, polypetalous.
9. ————————; ————, perigynous.

But he adds, that the perigynous and discigynous structures run together, and that no fixed difference can be found between the monopetalous and polypetalous conditions.

The Classes or Alliances which are formed within these primary groups are contrived without sufficient regard to the definitions which precede them, and by which alone they are to be recognised. In fact, the principle of disregarding characters and trusting merely to (presumed) affinity, is carried to such a length as to diminish the value of the groups; and hence, no doubt, Agardh's method has never been adopted, notwithstanding its merits in some respects.

He describes, in the following words, what he conceives to be the fundamental principles of natural classification :—

"Forma normalis in omnibus plantis non æque perspicua, sed sæpissime in quacumque sectione sensim magis magisque prominet et explicatur, ita ut in quibusdam plantis perfectissima appareat, et in aliis vix perspicienda.

"Forma normalis constantior cernitur in fructificatione, h. e. in flore et fructu, quam in habitu, tam quia in unum tantum finem illa explicatur, cum organa vegetationis indirecte etiam florem et fructum præparare debent, quam etiam quia partes vegetationis individuum tantum servant, fructus vero formam normalem perennem tueri debet.

"Sequitur tamen sæpissime habitus fructificationem, ita ut plantæ quæ flore et fructu non differant, habitu etiam quodam generali conveniant. Non autem semper nec necessario.

"Hinc systema in fructificatione nititur.

"Ceterum observandum, quod fructus jamdudum plantam quamvis non explicitam continet, et quod planta antequam flos et fructus eam coronet, non perfecta est.

"In sectione vero illa, quam speciem vocamus, non fructus solus characteres præbet, quia in omnibus notis, præter quod e causis accidentalibus pendeat, convenire debent individua ejusdem speciei.

"Affinitas plantarum componitur secundum nostram sententiam tam e multitudine characterum quorumcumque in quibus conveniunt, quam ex eorum præstantia et prominentia.

"Sic sufficit vel levis nota in flore et fructu, si multis notis habitualibus conveniunt plantæ; et quo pauciores notæ præstantiorum partium communes sunt, eo pluribus convenire debent in partibus minoris momenti. Sic etiam quo magis prominet character quidam, eo minus dilaceranda sectio, etiam si pluribus aliis notis differunt plantæ sub ea inclusæ."

1826. PERLEB, C. J.—(*Lehrbuch der Naturgeschichte der Pflanzenreichs.*)

See this author's Clavis, 1838. p. xlix.

1827. DUMORTIER, B. C.—(*Florula Belgica.*)

The following is the system of this author, who does not appear to have given any account of its principles. His Orders are equivalent to Alliances. His Staminacia begins with Conifers and ends with Lemnads, and is the only part concerning which I find any details :—

CLASSES.	SUB-CLASSES.	DIVISIONS.	ORDERS.
Staminacia	Corticalia	Simplitegmia	1. Julitegmia 2. Fructitegmia 3. Thalamitegmia
		Tubifloria	4. Thalamitubia 5. Fructitubia
		Ungulifloria	6. Fructungulia 7. Calicungulia 8. Thalamungulia
	Decorticalia	Bitegmia	9. Thalamifloria 10. Fructifloria 11. Caliciflora
		Solitegmia	12. Fructaulia 13. Thalamaulia
Pollinacia	Capsellia		14. Ecalyptria 15. Calyptria
	Ecapsellia		16. Scutellinea 17. Funginia
Fluidacia	Soligrania		18. Granulinia.
	Plurigrania		19. Cocculinia 20. Fartinia.

CHARACTERS OF THE ORDERS.

1. Julitegmia.—Flowering scales, placed on a catkin.
2. Fructitegmia.—Floral envelope one, epigynous.
3. Thalamitegmia.—Floral envelope one, hypogynous.
4. Thalamitubia.—Tube of a monopetalous corolla hypogynous.
5. Fructitubia.—Tube of a monopetalous corolla epigynous.
6. Fructungulia.—Claws of a polypetalous corolla epigynous.
7. Calicungulia.—Claws of a polypetalous corolla perigynous.
8. Thalamungulia.—Claws of a polypetalous corolla hypogynous.
9. Thalamifloria.—Corolla hypogynous.
10. Fructifloria.—Corolla epigynous.
11. Calicifloria.—Corolla perigynous.
12. Fructaulia.—Floral envelope one, epigynous.
13. Thalamaulia.—Floral envelope one, hypogynous.

1830. BARTLING, Fr. Th.—(*Ordines Naturales Plantarum, eorumque Characteres et Affinitates, adjectâ generum enumeratione*).

In this work the Vegetable Kingdom is divided into 8 principal divisions, and 60 subdivisions or Alliances, called by the author Classes. The latter are furnished with detailed characters drawn up in the same manner as those of the Orders, and to the whole is prefixed an abridgment of the plan of classification. The synonyms of the Alliances are slightly given ; but it is remarkable that they do not contain any allusion to the anterior works of Perleb and Agardh. As this work is the first in which considerable details are introduced into the characters of Alliances, it seems worth stating, at length, its nature, which is as follows :—

Plants are

CELLULAR . . . . . . or . . . . . . VASCULAR.

Homonemeous or *Heteronemeous.* *Cryptogamic* . . . . or . . . *Phanerogamic.*

Monocotyledonous . . . or . . . Dicotyledonous.

With the embryo inclosed in a vitellus, or With the embryo naked.

Apetalous. *Monopetalous.* *Polypetalous.*

Homonemeous	*Heteronemeous*	*Cryptogamic*	Monocotyledonous	With the embryo inclosed in a vitellus	*Apetalous.*	*Monopetalous.*	*Polypetalous.*
Fungi.	,,	Rhizocarpæ.	Glumaceæ.	,,	Coniferæ.	Aggregatæ.	Lorantheæ.
Lichenes.	,,	Filices.	Juncineæ.	,,	Amentaceæ.	Compositæ.	Umbellifloræ.
Algæ.	Musci.	Lycopodineæ.	Ensatæ.	,,	Urticineæ.	Campanulinæ.	Cocculinæ.
		Gonyopterides.	Liliaceæ.	,,	Fagopyrineæ.	Ericineæ.	Trisepalæ.
			Orchideæ.	,,	Proteinæ.	Styracinæ.	Polycarpicæ.
			Scitamineæ.	,,	Salicinæ.	Myrsineæ.	Rhœadeæ.
			Palmæ.	,,		Labiatifloræ.	Peponiferæ.
			Aroideæ.	Aristolochieæ.		Tubifloræ.	Cistifloræ.
			Helobiæ.	Piperinæ.		Contortæ.	Guttiferæ.
			Hydrocharideæ.	Hydropeltideæ.		Rubiacinæ.	Caryophyllinæ.
						Ligustrinæ.	Succulentæ.
							Calycifloræ.
							Calycanthinæ.
							Myrtinæ.
							Lamprophyllæ.
							Columniferæ.
							Gruinales.
							Ampelideæ.
							Malpighinæ.
							Tricoccæ.
							Terebinthinæ.
							Calophytæ.

Class I. FUNGI.

Coniomycetes, 29
Gasteromycetes, 29
Pyrenomycetes, 29
Hymenomycetes, 29

Class II. LICHENES.

Coniothalami, 45
Hymenothalami, 45
Pyrenothalami, 45

Class III. ALGÆ.

Nostochinæ, 18
Confervaceæ, 14
Florideæ, 23
Fucaceæ, 20

Class IV. MUSCI.

Hepaticæ, 58
Bryaceæ, 64

Class V. RHIZOCARPÆ.

Salviniaceæ, 71
Marsileaceæ, 71
Isoëteæ, 71

Class VI. FILICES.

Polypodiaceæ, 78
Osmundaceæ, 78
Ophioglosseæ, 77

Class VII. LYCOPODINEÆ.

Lycopodiaceæ, 69

Class VIII. GONIOPTERIDES.

Characeæ, 26
Equisetaceæ, 61

Class IX. GLUMACEÆ.

Gramineæ, 106
Cyperaceæ, 117

Class X. JUNCINÆ.

Restiaceæ, 121
Juncaceæ, 191
Xyrideæ, 187
Commelinaceæ, 188

Class XI. ENSATÆ.

Burmanniaceæ, 171
Hypoxideæ, 150
Hæmodoraceæ, 151
Irideæ, 159
Amaryllideæ, 155
Bromeliaceæ, 147

Class XII. LILIACEÆ.

Asphodeleæ, 200
Colchicaceæ, 198
Smilaceæ, 215
Dioscoreæ, 214

Class XIII. ORCHIDEÆ.

Orchideæ, 173

Class XIV. SCITAMINEÆ.

Amomeæ, 165
Cannaceæ, 168
Musaceæ, 163

Class XV. PALMÆ.

Palmæ, 133

Class XVI. AROIDEÆ.

Callaceæ, 193
Orontiaceæ, 193
Typhaceæ, 126

Class XVII. HELOBIÆ.

Najadeæ, 143
Podostemeæ, 482
Alismaceæ, 209
Butomeæ, 208

Class XVIII. HYDROCHARIDEÆ.

Hydrocharideæ, 141

Class XIX. ARISTOLOCHIEÆ.

Balanophoreæ, 89
Cytineæ, 91
Asarineæ, 792
Tacceæ, 149

Class XX. PIPERINÆ.

Saurureæ, 521
Piperaceæ, 515
Chlorantheæ, 519

Class XXI. HYDROPELTIDEÆ.

Cabombeæ, 412
Nymphæaceæ, 409
Nelumboneæ, 414

Class XXII. CONIFERÆ.

Cycadeæ, 223
Abietinæ, 226
Cupressinæ, 226
Taxinæ, 230

Class XXIII. AMENTACEÆ.

Casuarineæ, 249
Myriceæ, 256
Betulaceæ, 251
Cupuliferæ, 290
Ulmaceæ, 580

Class XXIV. URTICINÆ.

Monimieæ, 298
Artocarpeæ, 269
Urticeæ, 260

Class XXV. FAGOPYRINÆ.

Polygoneæ, 502
Nyctagineæ, 506

Class XXVI. PROTEINÆ.

Laurineæ, 535
Santalaceæ, 787
Elæagneæ, 257
Thymelææ, 530
Proteaceæ, 532

Class XXVII. SALICINÆ.

Salicinæ, 254

Class XXVIII. AGGREGATÆ.

Plantagineæ, 642
Plumbagineæ, 640
Globularieæ, 666
Dipsaceæ, 699
Valerianeæ, 697

Class XXIX. COMPOSITÆ.

Calycereæ, 701
Synanthereæ, 702

Class XXX. CAMPANULINÆ.

Goodenovieæ, 694
Stylideæ, 696
Lobeliaceæ, 692
Campanulaceæ, 689

Class XXXI. ERICINEÆ.

Vaccinieæ, 757
Ericeæ, 453
Epacrideæ, 448

Class XXXII. STYRACINÆ.

Styraceæ, 592
Ebenaceæ, 595
Sapoteæ, 590

Class XXXIII. MYRSINEÆ.

Ardisiaceæ, 647
Primulaceæ, 644

Class XXXIV. LABIATIFLORÆ.

Lentibulariæ, 686
Scrophularinæ, 681
Orobancheæ, 609
Gesnerieæ, 671
Sesameæ, 669
Myoporinæ, 665
Selagineæ, 666
Verbenaceæ, 663
Labiatæ, 659
Acanthaceæ, 678
Bignoniaceæ, 675

Class XXXV. TUBIFLORÆ.

Polemoniaceæ, 635
Hydroleaceæ, 638
Convolvulaceæ, 630
Cuscuteæ, 633
Solanaceæ, 618
Hydrophylleæ, 638
Borragineæ, 655

Class XXXVI. CONTORTÆ.

Gentianeæ, 612
Asclepiadeæ, 623
Apocyneæ, 599
Loganieæ, 602

Class XXXVII. RUBIACINÆ.

Lygodysodeaceæ, 761
Rubiaceæ, 761
Caprifoliaceæ, 766
Viburneæ, 766

Class XXXVIII. LIGUSTRINÆ.

Jasmineæ, 650
Oleineæ, 616

Class XXXIX. LORANTHEÆ.

Lorantheæ, 789

Class XL. UMBELLIFLORÆ.

Umbelliferæ, 773
Araliaceæ, 780
Hederaceæ, 780
Hamamelideæ, 784

Class XLI. COCCULINÆ.

Berberideæ, 437
Menispermeæ, 307

Class XLII. TRISEPALÆ.

Myristiceæ, 301
Anonaceæ, 420

Class XLIII. POLYCARPICÆ.

Magnoliaceæ, 417
Dilleniaceæ, 423
Pæoniaceæ, 425
Ranunculaceæ, 425

Class XLIV. RHŒDEÆ.

Tremandreæ, 374
Polygaleæ, 375
Resedaceæ, 356
Fumariaceæ, 435
Papaveraceæ, 430
Cruciferæ, 351
Cappărideæ, 357

Class XLV. PEPONIFERÆ.

Samydeæ, 330
Homalineæ, 742
Passifloreæ, 332
Turneraceæ, 347
Loaseæ, 744
Cucurbitaceæ, 311
Grossularieæ, 750
Nopaleæ, 746

Class XLVI. CISTIFLORÆ.

Flacourtianeæ, 327
Marcgravieæ, 403
Bixineæ, 327
Cistineæ, 349
Violarieæ, 338
Droseraceæ, 433
Tamariscineæ, 341

Class XLVII. GUTTIFERÆ.

Sauvagesieæ, 343
Frankeniaceæ, 340
Hypericineæ, 405
Garcinieæ, 400

Class XLVIII. CARYOPHYLLINÆ.

Chenopodieæ, 512
Amaranthaceæ, 510
Phytolacceæ, 509
Scleranthеæ, 528
Paronychieæ, 499
Portulaceæ, 500
Alsineæ, 496
Sileneæ, 496

Class XLIX. SUCCULENTÆ.

Ficoideæ, 525
Crassulaceæ, 344
Saxifragaceæ, 567
Cunoniaceæ, 571

Class L. CALYCIFLORÆ.

Halorageæ, 722
Lytharieæ, 574
Onagrariæ, 724
Rhizophoreæ, 726
Vochysieæ, 379
Combretaceæ, 717

Class LI. CALYCANTHINÆ.

Granateæ, 734
Calycantheæ, 540

Class LII. MYRTINÆ.

Memecyleæ, 731
Melastomaceæ, 731
Myrtaceæ, 734

Class LIII. LAMPROPHYLLÆ.

Camelliaceæ, 396
Ternstrœmiaceæ, 396
Chlenaceæ, 486

Class LIV. COLUMNIFERÆ.

Tiliaceæ, 371
Sterculiaceæ, 360
Büttneriaceæ, 363
Hermanniaceæ, 363
Dombeyaceæ, 363
Malvaceæ, 368

Class LV. GRUINALES.

Geraniaceæ, 493
Lineæ, 485
Oxalideæ, 488

Class LVI. AMPELIDEÆ.

Sarmentaceæ, 439
Leeaceæ, 439
Meliaceæ, 463
Cedreleæ, 461

Class LVII. MALPIGHINÆ.

Malpighiaceæ, 388
Acerineæ, 387
Coriarieæ, 475
Erythroxyleæ, 391
Sapindaceæ, 382
Hippocastaneæ, 382
? Rhizoboleæ, 398
Tropæoleæ, 366

Class LVIII. TRICOCCÆ.

Stackhouseæ, 589
Euphorbiaceæ, 274
Empetreæ, 285
Bruniaceæ, 785
Rhamneæ, 581
Aquifoliaceæ, 597
Pittosporeæ, 441
Celastrineæ, 586
? Hippocrateaceæ, 584
? Staphyleaceæ, 381

Class LIX. TEREBINTHINÆ.

Ochnaceæ, 474
Simarubeæ, 476
Zanthoxyleæ, 472
Diosmeæ, 469
Rutaceæ, 469
Zygophylleæ, 478
Aurantiaceæ, 457
Amyrideæ, 459
Connaraceæ, 468
Cassuvieæ, 465
? Juglandeæ, 292

Class LX. CALOPHYTÆ.

Pomaceæ, 559
Rosaceæ, 563
Dryadeæ, 563
Spiræaceæ, 563
Amygdaleæ, 557
Chrysobalaneæ, 542
Papilionaceæ, 544
Swartzieæ, 544
Cæsalpineæ, 544
Mimoseæ, 544

1830. LINDLEY, JOHN.—(*An Introduction to the Natural System of Botany, &c.*)

This was a slight modification of De Candolle's plan, with the apetalous and polypetalous plants thrown together, and consequently with a different sequence of the Natural Orders. No attempt was made at forming the minor groups, now called Alliances.

Class I. VASCULARES, or Flowering Plants.
 Sub-class 1. Exogens or Dicotyledons.
  Tribe 1. Angiospermæ.
   § 1. Polypetalous, apetalous, and achlamydeous plants.
   § 2. Monopetalous plants.
  Tribe 2. Gymnospermæ.
 Sub-class 2. Endogens or Monocotyledons.
  Tribe 1. Petaloideæ.
  Tribe 2. Glumaceæ.

Class II. CELLULARES, or flowerless plants.
  Tribe 1. Filicoideæ; or Fern-like plants.
  Tribe 2. Muscoideæ; or Moss-like plants.
  Tribe 3. Aphyllæ; or Leafless plants.

1832. HESS. J.—(*Uebersicht der Phanerogamischen naturlichen pflanzenfamilien mit einer kurzen charakteristik derselben*).

This is essentially an imitation of the method of De Candolle, with some changes in the sequence of Orders. No attempt is made at forming groups higher than Natural Orders, and it cannot be said that the work has contributed to the progress of Natural classification. The great object of the author seems to have been to form a good series.

1832. SCHULTZ, Carl Heinrich.—(*Natürliches System des Pflanzenreichs nach seiner inneren organization*).

In some respects this is like the system of De Candolle. The author first breaks up the Vegetable Kingdom into *Homorgana*, which have an exclusively cellular construction, and *Heterorgana*, which are formed with spiral vessels, and laticiferous vessels in addition. These are evidently the Cellular and Vascular plants of De Candolle. His Heterorgana he divides into *Synorgana* and *Dichorgana*, the first having all the forms of tissue dispersed through a common cellular mass, the latter having them separated in the form of bark and wood; Synorgana are therefore Endogens, and Dichorgana Exogens. The principal peculiarity consists in laticiferous vessels or cinenchyma being made a mark of classification, a certain number of flowering plants being thus combined with flowerless, under the name of Homorgana florifera; viz., Charads, Naiads, Hornworts, Podostemads, Seawracks, Hydrocharids, Lemnads, &c. Another peculiar feature is the formation among Synorgana, or Endogens, of a Class called *Synorgana dichorganoidea*, which is regarded as intermediate in nature between Synorgana and Dichorgana. This Class is divided into 2 groups, of which the first consists of Peppers, Saururads, and Chloranths, the second of Nyctagos, Waterstars, Hippurids, Amaranths, Cycads, Waterlilies, &c. The plan of this classification is as follows:—

				Class
Vegetabilia.	Homorgana	1. sporifera	Class	I. Homorgana rhizospora.
			—	II. Homorgana phyllospora.
			—	III. Homorgana caulospora.
		2. florifera	—	IV. Homorgana florifera.
	Heterogana. Synorgana	1. sporifera	—	V. Synorgana sporifera.
		2. florifera	—	VI. Synorgana gymnantha.
			—	VII. Synorgana coronantha.
			—	VIII. Synorgana palmacea.
			—	IX. Synorgana dichorganoidea.
	Heterogana. Dichorgana (omnia florifera)		—	X. Dichorgana lepidantha.
			—	XI. Dichorgana perianthina.
			—	XII. Dichorgana anthodiata.
			—	XIII. Dichorgana siphonantha.
			—	XIV. Dichorgana petalantha monocarpa.
			—	XV. Dichorgana petalantha polycarpa.

1833. LINDLEY, John.—(*Nixus Plantarum*).

This was an attempt, in imitation of Agardh and Bartling, to reduce the Natural Orders into groups subordinate to the higher divisions. Such groups were called *Nixus* (tendencies). The author threw aside the distinctions between perigynous and hypogynous insertion as uncertain and leading to bad grouping; insisted upon the value of albumen as a primary character, and objected to the general principle that the sections of plants are to furnish their character, and not a character the section. Finally, he maintained that no sections are capable of being positively defined, except such as depend upon physiological peculiarities; and that all other collections of species, by whatever name they are known, whose distinguishing marks are dependent upon structure alone, merely exhibit tendencies to resemblance in certain points, for which tendencies definitions are impracticable.

Keeping these principles in view, the following was the arrangement:—

		Classes.	
Sexuales	Vasculares	I. EXOGENÆ.	ANGIOSPERMÆ.
		II. EXOGENÆ.	GYMNOSPERMÆ.
		III. ENDOGENÆ.	
	Evasculares	IV. RHIZANTHEÆ.	
V. ESEXUALES.			

Class I. EXOGENÆ.

Sub-class I. POLYPETALÆ.

Cohort 1. ALBUMINOSÆ; embryo much smaller than the albumen.

N. 1. *Ranales.*	Nelumboneæ	§ Schizandreæ	Escallonieæ
Ranunculaceæ	Cephaloteæ	Dilleniaceæ	Bruniaceæ
§ Sarracennieæ	N. 2. *Anonales.*	N. 3. *Umbellales.*	N. 5. *Pittosporales.*
Papaveraceæ	Myristiceæ	Umbelliferæ	Vites
§ Fumariaceæ	Magnoliaceæ	Araliaceæ	Pittosporeæ
Nymphæaceæ	Wintereæ	N. 4. *Grossales.*	Olacineæ
§ Podophylleæ	Anonaceæ	Grossulaceæ	? Dionæa
§ Hydropeltideæ			

Cohort 2. GYNOBASICÆ; carpels arranged round an elevated axis.

N. 1. *Rutales.*	Zygophylleæ	Tropæoleæ	N. 3. *Coriales.*
Ochnaceæ	Xanthoxyleæ	Geraniaceæ	Coriarieæ
Simarubaceæ	N. 2. *Geraniales.*	Oxalideæ	N. 4. *Florkeales.*
Rutaceæ	Hydrocereæ	Balsamineæ	Limnantheæ
§ Diosmeæ			

Cohort 3. EPIGYNÆ; ovary inferior, generally with an epigynous disk.

N. 1. *Onagrales.*	Salicariæ	Philadelpheæ	N. 4. *Cucurbitales.*
Onagraceæ	N. 2. *Myrtales.*	N. 3. *Cornales.*	Cucurbitaceæ
§ Circæaceæ	Memecyleæ	Hamamelideæ	Loaseæ
§ Halorageæ	Myrtaceæ	Corneæ	Cacteæ
Combretaceæ	Melastomaceæ	Loranthеæ	Homalineæ
Alangieæ	Lecythideæ		N. 5. *Begoniales.*
Rhizophoreæ			Begoniaceæ

Cohort 4. PARIETALES; placentæ parietal.

N. 1. *Cruciales.*	N. 2. *Violales.*	N. 3. *Passionales.*	N. 4. *Bixales.*
Cruciferæ	Violaceæ	Passifloreæ	Bixaceæ
Capparideæ	Samydeæ	Papayaceæ	
Resedaceæ	Moringeæ	Flacourtiaceæ	
	Droseraceæ	Malesherbiaceæ	
	Frankeniaceæ	Turneraceæ	

Cohort 5. CALYCOSÆ; calyx incompletely whorled; two of the sepals being exterior.

N. 1. *Guttales.*	N. 2. *Theales.*	Hippocastaneæ	Chlænaceæ
Guttiferæ	Ternströmiaceæ	Polygaleæ	Cistineæ
Rhizoboleæ	N. 3. *Acerales.*	Vochyaceæ	Reaumurieæ
Marcgraaviaceæ	Acerineæ	N. 4. *Cistales.*	N. 5. *Berberales.*
Hypericineæ	Sapindaceæ	Lineæ	Berberideæ

Cohort 6. SYNCARPÆ; carpels consolidated, and none of the characters of the other Cohorts.

N. 1. *Malvales.*
- Sterculiaceæ
- Malvaceæ
- Elæocarpeæ
- Tiliaceæ
- Dipterocarpeæ

N. 2. *Meliales.*
- Meliaceæ
- Cedreleæ
- Humiriaceæ
- Aurantiaceæ
- Spondiaceæ

N. 3. *Rhamnales.*
- Rhamneæ
- Chailletiaceæ
- Tremandreæ
- Nitrariaceæ
- Burseraceæ

N. 4. *Euphorbiales.*
- Euphorbiaceæ
- Stackhouseæ
- Fouquieraceæ
- Celastrineæ
- § Hippocrateaceæ
- § Staphyleaceæ
- Malpighiaceæ
- § Erythroxyleæ

N. 5. *Silenales.*
- Portulaceæ
- Sileneæ
- Alsineæ
- Tamariscineæ
- Illecebreæ

Cohort 7. APOCARPÆ; carpels distinct, or separable, or solitary, and none of the preceding characters.

N. 1. *Rosales.*
- Rosaceæ
- § Pomaceæ
- § Sanguisorbeæ
- § Amygdaleæ
- Leguminosæ
- § Swartzieæ
- § Cæsalpinieæ
- § Mimoseæ
- Connaraceæ
- Chrysobalaneæ
- Calycantheæ

N. 2. *Saxales.*
- Baueraceæ
- Cunoniaceæ
- Saxifrageæ

N. 3. *Ficoidales.*
- Ficoideæ

N. 4. *Crassales.*
- Crassulaceæ
- Galacineæ

N. 5. *Balsamales.*
- Amyrideæ
- Anacardiaceæ

Sub-class II. INCOMPLETÆ.

Cohort 1. TUBIFERÆ; calyx tubular, often like a corolla, without the characters of the other Cohorts.

N. 1. *Santalales.*
- Santalaceæ

N. 2. *Daphnales.*
- Elæagneæ
- Thymeleæ
- Hernandieæ
- Aquilariaceæ

N. 3. *Proteales.*
- Proteaceæ

N. 4. *Laureales.*
- Lauraceæ
- Cassytheæ

N. 5. *Penæales.*
- Penæaceæ

Cohort 2. CURVEMBRYÆ; embryo curved round albumen, or horseshoe-formed, or spiral; calyx rarely tubular.

N. 1. *Chenopodales.*
- Amarantaceæ
- Chenopodiaceæ
- Phytolacceæ

N. 2. *Polygonales.*
- Polygoneæ

N. 3. *Petivales.*
- Petiveriaceæ

N. 4. *Sclerales.*
- Scleruntheæ
- Nyctagineæ

N. 5. *Cocculales.*
- Menispermeæ

Cohort 3. RECTEMBRYÆ: calyx very imperfect; embryo straight.

N. 1. *Amentales.*
- Cupuliferæ
- Betulineæ

N. 2. *Urticales.*
- Urticeæ
- § Ceratophylleæ
- § Artocarpeæ
- Stilagineæ
- Empetreæ
- Myriceæ
- Juglandeæ

N. 3. *Casuarales.*
- Casuarineæ

N. 4. *Ulmales.*
- Ulmaceæ

N. 5. *Datiscales.*
- Datisceæ
- Lacistemeæ

Cohort 4. ACHLAMYDEÆ; both calyx and corolla deficient.

N. 1. *Piperales.*
- Chlorantheæ
- Saurureæ
- Piperaceæ

N. 2. *Salicinales.*
- Salicineæ
- Plataneæ
- Balsamiferæ

N. 3. *Monimiales.*
- Monimieæ
- Atherospermeæ

N. 4. *Podostemales.*
- Podostemeæ

N. 5. *Callitrales.*
- Callitrichineæ

Cohort 5. COLUMNIFERÆ; stamens monadelphous.

N. 1. *Nepenthales.*
- Nepentheæ

N. 2. *Aristolochiales.*
- Aristolochiæ

Sub-class III. MONOPETALÆ.

Cohort 1. POLYCARPÆ; hypogynous (rarely epigynous) with a polycarpous ovary.

N. 1. *Brexiales.*
- Brexiaceæ

N. 2. *Ericales.*
- Pyrolaceæ
- Ericeæ
- Vaccinieæ
- Epacrideæ

N. 3. *Primulales.*
- Primulaceæ
- Myrsineæ
- Sapoteæ
- Ebenaceæ
- § Styraceæ
- Ilicineæ

N. 4. *Nolanales.*
- Nolanaceæ

N. 5. *Volvales.*
- Cuscuteæ
- Convolvulaceæ
- Polemoniaceæ
- Hydroleaceæ

Cohort 2. EPIGYNÆ; epigynous, with a 2- or many-celled ovary.

N. 1. *Campanales.*
- Lobeliaceæ
- Campanulaceæ
- ? Belvisieæ
- Columelliaceæ

N. 2. *Goodeniales.*
- Stylideæ
- Goodenoviæ
- Scævoleæ

N. 3. *Cinchonales.*
- Cinchonaceæ
- Lygodysodiaceæ

N. 4. *Capriales.*
- Caprifoliaceæ

N. 5. *Stellales.*
- Stellatæ

Cohort 3. DICARPÆ; hypogynous, regular-flowered, with a dicarpous ovary.

N. 1. *Gentianales.*
- Gentianeæ
- Spigeliaceæ
- Apocyneæ
- Asclepiadeæ

N. 2. *Oleales.*
- Oleaceæ
- Jasmineæ

N. 3. *Loganiales.*
- Loganiaceæ
- Potaliaceæ

N. 4. *Echiales.*
- Boragineæ
- Ehretiaceæ
- § Heliotropiceæ
- Cordiaceæ
- Hydrophylleæ

N. 5. *Solanales.*
- Solaneæ
- Cestrineæ

Cohort 4. PERSONATÆ; hypogynous, irregular-flowered, with a dicarpous ovary.

N. 1. *Labiales.*	Stilbineæ	N. 3. *Scrophulales.*	N. 4. *Acanthales.*
Labiatæ	N. 2. *Bignoniales.*	Scrophulariaceæ	Acanthaceæ
Verbenaceæ	Bignoniaceæ	Orobancheæ	N. 5. *Lentibales.*
Myoporineæ	Pedalineæ	Gesnereæ	Lentibulariæ
Selagineæ	Cyrtandraceæ		

Cohort 5. AGGREGATÆ; ovary 1-celled.

N. 1. *Asterales.*	N. 2. *Dipsales.*	N. 3. *Brunonialcs.*	Globularineæ
Calycereæ	Dipsaceæ	Brunoniaceæ	N. 5. *Plumbales.*
Compositæ	Valerianeæ	N. 4. *Plantales.*	Plumbagineæ
		Plantagineæ	

Class II. GYMNOSPERMÆ.

Cycadeæ
Coniferæ
Taxineæ
Equisetaceæ

Class III. ENDOGENÆ.

Cohort 1. EPIGYNÆ; stamens distinct, ovary inferior.

N. 1. *Amomales.*	N. 2. *Narcissales.*	Tacceæ	N. 4. *Bromeliales.*
Scitamineæ	Hypoxideæ	N. 3. *Ixiales.*	Bromeliaceæ
Marantaceæ	Amaryllideæ	Irideæ	N. 5. *Hydrales.*
Musaceæ	Hæmodoraceæ		Hydrocharideæ
	Burmanniæ		

Cohort 2. GYNANDRÆ; anthers united, ovary inferior.

Orchideæ
Cypripedieæ
Apostasieæ

Cohort 3. HYPOGYNÆ; flowers on a plan of 3, coloured, ovary superior.

N. 1. *Palmales.*	Melanthaceæ	N. 3. *Commelales.*	Alismaceæ
Palmæ	Gilliesieæ	Commelinaceæ	N. 5. *Juncales.*
N. 2. *Liliales.*	Asphodeleæ	N. 4. *Alismales.*	Junceæ
Pontedereæ	Liliaceæ	Butomeæ	Philydreæ

Cohort 4. IMPERFECTÆ; flowers herbaceous, or imperfect, or none; or finally of two parts and coloured, with a superior ovary.

N. 1. *Pandales.*	N. 2. *Arales.*	N. 4. *Smilales.*	N. 5. *Fluviales.*
Cyclantheæ	Aroideæ	Dioscoreæ	Fluviales
Pandaneæ	Acoroideæ	Smilaceæ	Juncagineæ
	N. 3. *Typhales.*	Roxburghiaceæ	Pistiaceæ
	Typhace æ		

Cohort 5. GLUMACEÆ; scale-like bracts in place of a perianth.

Gramineæ
Cyperaceæ
Desvauxieæ
Restiaceæ
§ Eriocauloneæ
Xyrideæ

Class IV. RHIZANTHEÆ.

Rafflesiaceæ
Cytineæ
Balanophoreæ
Cynomorieæ

Class V. ESEXUALES.

N. 1. *Filicales.*	Ophioglosseæ,	N. 3. *Muscales.*	N. 4. *Charales.*
Polypodiaceæ	N. 2. *Lycopodales.*	Musci	Characeæ
Gleicheniaceæ	Lycopodiaceæ	Andræaceæ	N. 5. *Fungales.*
§ Parkeriaceæ	Marsileaceæ	Jungermanniaceæ	Fungi
Osmundaceæ	Salvinieæ	Hepaticæ	Lichenes
Danæaceæ			Algæ

1834. HORANINOW, Paul.—(*Primæ lineæ Systematis Naturæ, nexui naturali omnium evolutionique progressivæ per nixus reascendentes superstructi.*)

Here the Vegetable Kingdom is divided into 4 Circles, viz.—

Circle 1. Sporophoræ (or Acotyledons).
2. Pseudospermæ (containing Gymnosperms and Rhizanths).
3. Coccophoræ (or Monocotyledons).
4. Spermophoræ (or Dicotyledons).

Each of these is broken up into classes. Water-lilies, Sarraceniads, Peppers and their allies, with Nepenthes, are placed in the third circle; while Cistusrapes and Taccads stand in the fourth. The classes are in some instances extremely large, as, for example, the Thalamopetaleæ, which contain 58 Orders, and are the equivalent of the Thalamifloral section of De Candolle. By this author, as by some of the German Naturalists, Fungals and Algals are expelled from the Vegetable Kingdom, and form a part of a kingdom of Phytozoa; for Mr. Horaninow divides the organic world into Vegetables, Phytozoa, Animals, and Man.

1835. FRIES, Elias.—(*Corpus Florarum provincialium Sueciæ.*)

In this work the author has given a general scheme of arrangement according to his own peculiar views, and has applied it to the Flora of Scania. He prefaces his plan with an exposition of his ideas as to the manner of constructing a Natural System, and, among other things, maintains that it is more likely to be perfected by a small number of good observations clearly expressed than by a multitude of them. He regards germination as the first in rank of all the phases of vegetable life, manner of growth second, of flowering third, and of fruiting lowest of all, observing that the latter is the last stage of metamorphosis, beyond which there is nothing but the seed, whose constitution has nothing to do with that of the fruit. The seed is the beginning of germination. He regards the fruit as of importance in distinguishing Orders, and employs three forms of it, to which paramount importance is assignable. These are 1, simple, with a central placenta; 2, apocarpous, with the carpels disjoined; and 3, syncarpous, in which the carpels are all consolidated. The first he divides into *a*, with one stigma, and *b*, with two or more stigmas. The following is the general plan of his system, in which those numbers and letters have the value just assigned to them.

Class I. DICOTYLEDONS.

† Perianth genuine, complete, with a thickened disk for the insertion of the petals and stamens. Stamens inserted,

A. on the Corolla. I. COROLLIFLORÆ.

a. *epigynous.* I. SEMINIFLORÆ.	b. *amphigynous.* II. ANNULIFLORÆ.	c. *hypogynous.* III. TUBIFLORÆ.
1. *a.* Synanthereæ	1. *a.* Campanulaceæ	1. *a.* Solanaceæ
Dipsaceæ	*b.* Gesnerieæ	Personatæ
b. Valerianeæ	Polemoniaceæ	*b.* Gentianeæ
2. Rubiaceæ	2. Boragineæ	2. Asclepiadeæ
3. Caprifoliaceæ	Labiatæ	3. Primulaceæ
	3. Hydroleaceæ	

B. on the Receptacle. II. THALAMIFLORÆ.

a. *epigynous.* IV. DISCIFLORÆ	b. *amphigynous.* V. BASIFLORÆ.	c. *hypogynous.* VI. COLUMNIFLORÆ.
1. *a* Corneæ	1. *a.* Berberideæ	1. *a.* Cistineæ
b. Celastrineæ	*b.* Cruciferæ	*b.* Tiliaceæ
Malpighiaceæ	Papaveraceæ	Hypericineæ
2. *a.* Araliaceæ	2. *a.* Nymphæaceæ	2. *a.* Gruinales
b. Umbelliferæ	*b.* Ranunculaceæ	*b.* Malvaceæ
3. Lorantheæ	3. Balsamineæ	3. Caryophylleæ

C. on the Calyx. III. CALYCIFLORÆ.

a. *epiperigynous.* VII. FAUCIFLORÆ.	b. *amphigynous.* VIII. TORIFLORÆ.	c. *hypogynous.* IX. CENTRIFLORÆ.
1. *a.* Calycanthema	1. *a.* Leguminosæ	1. *a.* Ericinæ
b, Rhamneæ	Drupaceæ	*b.* Empetreæ
Ribesieæ	*b.* Pomaceæ	Aquifoliaceæ
2. Succulentæ	2. Senticosæ	2. Euphorbiaceæ
3. Portulacaceæ	3. Paronychieæ	3. Polygoneæ
		Chenopodeæ

†† Apetalous. IV. INCOMPLETÆ, with the disk not thickened or staminiferous.

Perianth

a. *gamosepalous, concentrated.*	b. *squamaceous, imbricated.*	c. *none or doubtful.*
X. BRACTEIFLORÆ.	XI. JULIFLORÆ.	XII. NUDIFLORÆ.
1. *a.* Vepreculæ	1. *a.* Fraxineæ	1. *a.* Chloranthеæ
b. Aristolochiæ	*b.* Juglandineæ	*b.* Piperaceæ
Cucurbitaceæ	Amentaceæ	2. Saurureæ
2. Artocarpeæ	2. Salicineæ	Callitrichineæ
3. Urticeæ	3. Myriceæ	3. Naiadeæ
Balanophoreæ	Coniferæ	Ceratophylleæ
? Lycopodiaceæ	Equisetum	Chara

Class II. MONOCOTYLEDONS.

Perianth

† *complete in 2 rows.*		†† *incomplete or 0.*
a stamens epigynous.	*b.* stamens amphigynous.	*c.* stamens hypogynous.
XIII. FRUCTIFLORÆ.	XIV. LILIIFLORÆ.	XV. SPADICIFLORÆ.
1. *a.* Orchideæ	1. *a.* Liliaceæ	1. *a.* Callaceæ
b. Irideæ	*b.* Melanthaceæ	*b.* Orontiaceæ
Narcisseæ	2. Alismaceæ	2. Potamogetoneæ
2. Hydrocharideæ	3. Juncaceæ	3. Cyperaceæ
3. Valisneria		††† *bracteate, valvate.*
		XVI. GLUMIFLORÆ.
		Gramineæ
This series is conspicuous for its fruit, epigynous, retrogressive.	This series is conspicuous for its flowers, central, amphigynous.	This series is conspicuous for its vegetation, progressive, hypogynous.

Class III. CRYPTOGAMS, or NEMEÆ.

A. HETERONEMEÆ.

Germinating threads

a. *solitary, simple.*	b. *several, ramifying.*
XVII. FILICES.	XVIII. MUSCI.

B. HOMONEMEÆ.

Gonidia

a. *present.* Colour herbaceous.	b. *absent.* Colour metallic.
XIX. ALGÆ.	XX. FUNGI.
This series is conspicuous for its vegetation, and progressive.	This series is conspicuous for its fruit, and retrogressive.

1835. MARTIUS, C. Fr. Ph. v.—(*Conspectus Regni Vegetabilis secundum characteres morphologicos præsertim carpicos in classes ordines et familias digesti, &c.*)

The motto prefixed to this treatise, "Ye shall know them by their fruit," explains the principles upon which Dr. Von Martius has constructed his system. He assumes that "because the fruit and its seed, or the parts analogous to them, constitute the crown and end of the whole nature and vitality of plants, on that very account it must be superior to the other parts in dignity." Accordingly its variations are scrutinised with much care, and many new terms are proposed for the sake of expressing those variations with great precision.

Two primary divisions of the Vegetable Kingdom are admitted, viz.—1. *Primitive Vegetation*, consisting of all known plants except Fungals, which form of themselves the other division called 2, *Secondary Vegetation.*

Primitive vegetation is separated into the following classes, viz.: I. Ananths, or flowerless plants; II. Loxines, or Monocotyledons; III. Tympanochetes, or Gymnogens; IV. Orthoines, or Dicotyledons. Each of the more extensive classes is broken up into certain sub-classes and series, under which are stationed Cohorts (or Alliances), in which the Natural Orders are finally marshalled. As the plan, which is very artificial, has never been adopted, it will be sufficient to give the Cohorts of one of the subdivisions, for which purpose a portion of the second Sub-class of Orthoines may be selected.

Cohort 1. *Monocarpæ scabrifoliæ.*—Urticeæ, Moreæ, Artocarpeæ, Ulmaceæ, Stilagineæ, Hensloviaceæ.
Cohort 2. *Haplocarpæ columniferæ.*—Myristiceæ.
Cohort 3. *Haplocarpæ chromanthæ.*—Thymelæeæ, Elæagneæ, Anthoboleæ, Osyrideæ, Illigereæ, Hernandieæ, Aquilarineæ, Proteaceæ, Santalaceæ, Nyssaceæ.
Cohort 4. *Polyplocarpæ chromanthæ.*—Penæaceæ.
Cohort 5. *Haplocarpæ auxanthæ.*—Chenopodiaceæ, Petiveriaceæ, Nyctaginere, &c.

1836. BROMHEAD, Sir E. French, Bart.

This author's system first appeared in the *Edinburgh Journ. Apr.* 1836, and has since been more than once revised to embrace the later discoveries of the science. The last published revision was in the *Mag. Nat. Hist. July,* 1840. The writer proposes to proceed wholly by induction. The families are collected into Alliances, designated by a termination in *ales,* from some characteristic or well-known family contained in the assemblage. Each family is placed in that Alliance in which it may meet the greatest number of families of admitted affinity to it, the character being subsequently deduced from the assemblage so constituted, and used as a test of admissibility in the more doubtful cases. —See *Mag. Nat. Hist. April,* 1838. A sketch of characters for the whole series of Alliances as they stood in 1838 appeared in the *Edinb. Phil. Journ. April and July* of that year. He considers it an advantage that above 60 of his Alliances are to be found indicated or adopted with more or less accuracy by other Botanists. He has given some of these synonyms in the *Phil. Mag. July,* 1837, and in the *Mag. Nat Hist. July,* 1840. The author arranges with great care the contents of each Alliance in the order of the immediate affinities and transitions, and then places each Alliance between the two Alliances into which it passes. He considers himself to have thus established by induction a continuous series of Alliances, commencing with Algals and ending with Fungals, in which each family in a continuous succession stands between the two families of nearest affinity. The system thus resulting presents the aspect of two parallel races meeting in the Rhizanths, and presenting in their progress, at equal distances from the commencement, analogous Alliances, such, for instance, as Rosales and Fabales, Boraginales and Lamiales, Geraniales and Rutales, &c. In the Alliances, and in the grouping of the Alliances, the system accords with the quinary method; but to this the author does not bind himself, remarking that quinary combinations very frequently occur, and that he has extended them for the sake of convenience, by leaning towards that method in cases where the limits of families are ambiguous.

He considers the theory of the circulation of organic forms to be confirmed by his method, but does not look on them as closed or re-entering circles. He would rather compare them to the approach of the returning parts of a spiral or to the similarity of the opposite ends of a fusiform figure.

The subjoined table of his Alliances shows their succession, but the transitions and contents of the Alliances could not be exhibited without giving his tables at length.

RACE OF THE ALGÆ.

A.—Nostocales.
B.—Fucales, rhodomelales, ulvales, charales, osmundales.
C.—Ephedrales, myricales, ulmales, piperales, haloragales, œnotherales, myrtales, rosales, saxifragales, cucurbitales, portulacales, chenopodiales, polemoniales, boraginales, solanales, gentianales, apocynales, cinchonales, sambucales, cornales, geraniales, cistales, brassicales, nymphæales, aristolochiales.
C. C.—Alismales, restiales, agrostidales, cocoales, typhales.
C. C. C.—Cytinales.

RACE OF THE FUNGI.

A.—Mucorales.
B.—Auriculariales, lycoperdales, usneales, jungermanniales, lycopodiales.
C.—Cupressales, betulales, rhamnales, euphorbiales, æsculales, hypericales, limoniales, fabales, violales, passiflorales, homaliales, elæagnales, acanthales, lamiales, rhinanthales, ericales, campanulales, asterales, dipsacales, myrsinales, rutales, malvales, laurales, magnoliales, menispermales.
C. C.—Asparagales, juncales, orchidales, zingiberales, narcissales.
C. C. C.—Cytinales.

1836. LINDLEY, John.—(*A Natural System of Botany, &c., second edition.*)

The arrangement here adopted was nearly the same as that proposed in the *Nixus Plantarum* (see p. xli.) An attempt was also made to reform the nomenclature of the Natural System, by making all the names of divisions of the same value end in the same way. The Orders were distinguished by ending in *aceæ*, the Sub-orders in *eæ*, the Alliances in *ales,* and certain combinations, called groups, in *osæ*. It was conceived that certain advantages and conveniences would attend the establishment of uniformity in these matters. Botanists do not, however, appear to be as yet disposed to entertain this opinion, and the terminations have not been generally adopted, in part, no doubt, because of the difficulty of adapting them to Greek and Latin compounds.

1836-1840. ENDLICHER, Stephen.—(*Genera Plantarum secundum ordines naturale disposita.*)

Upon this system has been published the most important systematical work that has appeared since the *Genera Plantarum* of Jussieu, in 1789. It commences with plants of the simplest kind, and closes with what the author regards as most complicated, viz., leguminous plants. It has been executed with great skill, but is too much dependent upon mere theoretical considerations, and is difficult to use in consequence of the looseness of the characters assigned to what the author names Classes, which are equivalent to my Alliances. The following are the details of his system :—

No opposition of stem and root. No spiral vessels. No sexes. Spores lengthening in all directions . . . . THALLOPHYTA.

Born without soil: feeding by the surface: fructification vague . . PROTOPHYTA.

Born on languid or decaying organisms: feeding from within: developing all the organs at once. HYSTEROPHYTA.

Opposition of stem and root. Spiral vessels. Sexes in the more perfect . . . CORMOPHYTA.

Stem growing at the point only, using the lower part only for conveying fluids. ACROBRYA.

Sp.-vessels 0. Both sexes present. Spores loose in spore-cases . . . *Anophyta.*

Sp.-vessels . Male sex missing. Spores loose in one or many-celled spore-cases. *Protophyta.*

Both sexes present. Seeds embryoless, of many spores . . *Hysterophyta.*

Stem growing at the circumference . . . . AMPHIBRYA.

Stem growing at both point and circumference . . . . ACRAMPHIBRYA.

Ovules naked, receiving impregnation immediately by the foramen . . . *Gymnosperma.*

Perianth 0, rudim. or simple, calycine or coloured, free or adherent . . *Apetala.*

Perianth double, outer calycine inner corolline, monopet. occasionally abortive. *Gamopetala.*

Perianth double, outer calycine inner corolline, parts distinct or united by the base of the stamens, occasionally abortive. *Dialypetala.*

Region I. THALLOPHYTA.

Section 1. PROTOPHYTA.

Class 1. *Algæ.*

Diatomaceæ, 12
Nostochinæ, 18
Confervaceæ, 14
Characeæ, 26
Ulvaceæ, 18
Florideæ, 23
Fucaceæ, 20

Class 2. *Lichenes.*

Coniothalami, 45
Idiothalami, 45
Gasterothalami, 45
Hymenothalami, 45

Section 2. HYSTEROPHYTA.

Class 3. *Fungi.*

Gymnomycetes, 29
Hyphomycetes, 29
Gasteromycetes, 29
Pyrenomycetes, 29
Hymenomycetes, 29

Region II. CORMOPHYTA.

Section 3. ACROBRYA.

Cohort I. ANOPHYTA.

Class 4. *Hepaticæ.*

Ricciaceæ, 57
Anthoceroteæ, 60
Targioniaceæ, 58
Marchantiaceæ, 58
Jungermanniaceæ, 59

Class 5. *Musci.*

Andræaceæ, 63
Sphagnaceæ, 64
Bryaceæ, 64

Cohort II. PROTOPHYTA.

Class 6. *Equiseta.*

Equisetaceæ, 61

Class 7. *Filices.*

Polypodiaceæ, 78
Hymenophylleæ, 80
Gleicheniaceæ, 80
Schizæaceæ, 80
Osmundaceæ, 81
Marattiaceæ, 82
Ophioglosseæ, 77

Class 8. *Hydropterides.*

Salviniaceæ, 71
Marsileaceæ, 71

Class 9. *Selagines.*

Isoeteæ, 71
Lycopodiaceæ, 69

Class 10. *Zamiæ.*

Cycadeaceæ, 223

Cohort III. HYSTEROPHYTA.

Class 11. *Rhizantheæ.*

Balanophoreæ, 89
Cytineæ, 91
Rafflesiaceæ, 93

Section 4. AMPHIBRYA.

Class 12. *Glumaceæ.*

Gramineæ, 106
Cyperaceæ, 117

Class 13. *Enantioblastæ.*

Centrolepideæ, 120
Restiaceæ, 121
Eriocauloneæ, 122
Xyrideæ, 187
Commelynaceæ, 188

Class 14. *Helobiæ.*

Alismaceæ, 209
Butomaceæ, 208

Class 15. *Coronariæ.*

Juncaceæ, 191
Philydreæ, 186
Melanthaceæ, 198
Pontederaceæ, 206
Liliaceæ, 200
Smilaceæ, 215

Class 16. *Artorhizæ.*

Dioscoreæ, 214
Taccaceæ, 149

Class 17. *Ensatæ.*

Hydrocharideæ, 141
Burmanniaceæ, 171
Irideæ, 187
Hæmodoraceæ, 151
Hypoxideæ, 154
Amaryllideæ, 155
Bromeliaceæ, 147

Class 18. *Gynandræ.*

Orchideæ, 173
Apostasiaceæ, 184

Class 19. *Scitamineæ.*

Zingiberaceæ, 165
Cannaceæ, 168
Musaceæ, 163

Class 20. *Fluviales.*

Naiadeæ, 143

Class 21. *Spadicifloræ.*

Aroideæ, 127
Typhaceæ, 126
Pandaneæ, 130

Class 22. *Principes.*

Palmæ, 133

Cohort I. GYMNOSPERMÆ.

Class 23. *Coniferæ.*

Cupressinæ, 226
Abietinæ, 226
Taxineæ, 230
Gnetaceæ, 232

Cohort II. APETALA.

Class 24. *Piperitæ.*

Chloranthaceæ, 519
Piperaceæ, 515
Saurureæ, 521

Class 25. *Aquaticæ.*

Ceratophylleæ, 263
Callitrichineæ, 284
Podostemeæ, 482

Class 26. *Julifloræ.*

Casuarineæ, 249
Myriceæ, 256
Betulaceæ, 251
Cupuliferæ, 290

Section 5. ACRAMPHIBRYA.

Ulmaceæ, 580
Celtideæ, 580
Moreæ, 266
Artocarpeæ, 269
Urticaceæ, 260
Cannabineæ, 265
Antidesmeæ,
Plataneæ, 272
Balsamifluæ, 253
Salicineæ, 254
Hensloviaceæ, 570
Lacistemeæ, 329

Class 27. *Oleraceæ.*

Chenopodeæ, 512
Amarantaceæ, 510
Polygoneæ, 502
Nyctagineæ, 506

Class 28. *Thymeleæ.*

Monimiaceæ, 298
Atherospermeæ, 300
Laurineæ, 535
Gyrocarpeæ, 535
Santalaceæ, 787
Daphnoideæ, 530
Aquilariaceæ, 579
Elæagneæ, 257
Penæaceæ, 577
Proteaceæ, 532

Class 29. *Serpentariæ.*

Aristolochiaceæ, 792
Nepenthaceæ, 287

Cohort III. GAMOPETALÆ

Class 30. *Plumbagines.*

Plantagineæ, 642
Plumbagineæ, 640

Class 31. *Aggregatæ.*

Valerianeæ, 697
Dipsaceæ, 699
Compositæ, 702
Calycereæ, 701

Class 32. *Campanulinæ.*

Brunoniaceæ, 657
Goodeniaceæ, 694
Lobeliaceæ, 692
Campanulaceæ, 689
Stylidieæ, 696

Class 33. *Caprifolia.*

Rubiaceæ, 761
Lonicereæ, 766

Class 34. *Contortæ.*

Jasmineæ, 650
Oleaceæ, 616
Loganiaceæ, 602
Strychneæ, 602
Apocyneæ, 599
Asclepiadeæ, 623
Gentianeæ, 612
Spigeliaceæ, 602

Class 35. *Nuculiferæ.*

Labiatæ, 659
Verbenaceæ, 663
Stilbineæ, 607
Globulariaceæ, 666
Selagineæ, 666
Myoporaceæ, 665
Cordiaceæ, 628
Asperifoliæ, 655

Class 36. *Tubifloræ.*

Convolvulaceæ, 630
Polemoniaceæ, 635
Hydrophylleæ, 638
Hydroleaceæ, 638
Solanaceæ, 618

Class 37. *Personatæ.*

Scrophualarineæ, 681
Acanthaceæ, 678
Bignoniaceæ, 675
Gesneraceæ, 671
Cyrtandreæ, 671
Pedalineæ, 669
Orobancheæ, 609
Utricularinæ, 686

Class 38. *Petalanthæ.*

Primulaceæ, 644
Myrsineæ, 647
Sapotaceæ, 590
Ebenaceæ, 595

Class 39. *Bicornes.*

Epacrideæ, 448
Ericaceæ, 453
Vaccinieæ, 757

Cohort IV. DIALYPETALA.

Class 40. *Discanthæ.*

Umbelliferæ, 773
Araliaceæ, 780
Ampelideæ, 439
Cornaceæ, 782
Loranthaceæ, 789
Hamamelideæ, 784
Bruniaceæ, 785

Class 41. *Corniculatæ.*

Crassulaceæ, 344
Saxifragaceæ, 567
Ribesiaceæ, 750

Class 42. *Polycarpicæ.*

Menispermaceæ, 307
Myristicaceæ, 301
Anonaceæ, 420
Schizandraceæ, 305
Magnoliaceæ, 417
Dilleniaceæ, 423
Ranunculaceæ, 425
Berberideæ, 316

Class 43. *Rhœades.*

Papaveraceæ, 430
Cruciferæ, 351
Capparideæ, 357
Resedaceæ, 356
Datisceæ, 316

Class 44. *Nelumbia.*

Nymphæaceæ, 409
Sarracenieæ, 429
Cabombeæ, 412
Nelumboneæ, 414

Class 45. *Parietales.*

Cistaceæ, 349
Droseraceæ, 433
Violaceæ, 338
Sauvagesiaceæ, 343
Frankeniaceæ, 340
Turneraceæ, 347
Samydaceæ, 330
Bixaceæ, 327
Homaliaceæ, 742
Passifloraceæ, 332
Malesherbiaceæ, 335
Loasaceæ, 744
Papayaceæ, 301

Class 46. *Peponiferæ.*

Nandhirobeæ, 311
Cucurbitaceæ, 311
Begoniaceæ, 318

Class 47. *Opuntia.*

Cactaceæ, 746

Class 48. *Caryophyllineæ.*

Mesembryaceæ, 525
Portulacaceæ, 500
Caryophylleæ, 496
Phytolaccaceæ, 509

Class 49. *Columniferæ.*

Malvaceæ, 368
Sterculiaceæ, 360
Buttneriaceæ, 363
Tiliaceæ, 371

Class 50. *Guttiferæ.*

Dipterocarpeæ, 393
Chlenaceæ, 486
Ternstromiaceæ, 396
Clusiaceæ, 400
Marcgraaviaceæ, 403
Hypericaceæ, 405
Elatinaceæ, 480
Reaumuriaceæ, 407
Tamariscineæ, 341

Class 51. *Hesperides.*

Humiriaceæ, 447
Olacaceæ, 443
Aurantiaceæ, 457
Meliaceæ, 463
Cedrelaceæ, 461

Class 52. *Acera.*

Acerinæ, 387
Malpighiaceæ, 388
Erythroxyleæ, 391
Sapindaceæ, 382
Rhizoboleæ, 398

Class 53. *Polygalinæ.*

Tremandreæ, 374
Polygaleæ, 375

Class 54. *Frangulaceæ.*

Pittosporeæ, 441
Staphyleaceæ, 381
Celastrineæ, 586
Hippocrateaceæ, 584
Ilicineæ, 579
Rhamneæ, 581
Chailletiaceæ, 583

Class 55. *Tricoccæ.*

Empetreæ, 285
Stackhousiaceæ, 589
Euphorbiaceæ, 274

Class 56. *Terebinthinæ.*

Juglandeæ, 292
Anacardiaceæ, 465
Burseraceæ, 459
Connaraceæ, 466
Ochnaceæ, 474
Simarubaceæ, 476
Xanthoxyleæ, 472
Diosmeæ, 469
Rutaceæ, 469
Zygophylleæ, 478

Class 57. *Gruinales.*

Geraniaceæ, 493
Lineæ, 485
Oxalideæ, 488
Balsamineæ, 490
Tropæoleæ, 366
Limnantheæ, 366

Class 58. *Calycifloræ.*

Vochysiaceæ, 379
Combretaceæ, 717
Alangieæ, 719
Rhizophoreæ, 726
Philadelpheæ, 753
Œnothereæ, 724
Halorageæ, 722
Lythrarieæ, 754

Class 59. *Myrtifloræ.*

Melastomaceæ, 731
Myrtaceæ, 734

Class 60. *Rosifloræ.*

Pomaceæ, 559
Calycantheæ, 540
Rosaceæ, 563
Amygdaleæ, 557
Chrysobalaneæ, 542

Class 61. *Leguminosæ.*

Papilionace, 544
Swartzieæ, 544
Mimoseæ, 544

1838. LINDLEY, John.—(*Article "Exogens" in the Penny Cyclopedia.*)

In this place the author's views, as explained in previous works, were considerably modified so far as regards Exogens. He proposed in the first place to abandon altogether the old divisions of Polypetalous, Monopetalous, and Apetalous plants, and to reconstruct the whole fabric of Exogenous classification, upon the following principles:—

In the first place, the Orders whose embryo is furnished with an excessive quantity of albumen (a great physiological distinction), were formed into an Albuminous group. ——The remainder of Exogens then consists of Orders in which some have the sexes in distinct flowers, and others hermaphrodite flowers. As we know of no character intimately connected with the reproduction of the species which is upon the whole so important as this, a Diclinous group was established, as had formerly been done by Jussieu.——The hermaphrodite Orders were then separated into those with the calyx, corolla, and stamens confluent at the base with each other and with the ovary, that is, having an inferior ovary, and those in which those parts are distinct, either altogether or at least from each other, the former constituting an Epigynous group.——Finally, the remainder of the Orders were divided into those with a monopetalous corolla combined with an ovary upon a binary plan (Dicarpous), and those which, if monopetalous, have the ovary simple or complex (Polycarpous).

The following table will put this in a clearer point of view:—

Albumen extremely abundant; embryo minute	1. ALBUMINOSÆ.
Albumen absent, or in small quantity.	
Sexes in the same flower.	
Ovary inferior	2. EPIGYNOSÆ.
Ovary superior.	
Flowers, if monopetalous, not with a dicarpous ovary . . .	3. POLYCARPOSÆ.
Flowers monopetalous, with a dicarpous ovary	4. DICARPOSÆ.
Sexes in different flowers	5. DICLINOSÆ.

Each of these groups would form a series by itself, the sequence of which ought to be natural, and to exhibit various lateral analogies with other groups. And thus the three Monopetalous, Apetalous, and Polypetalous divisions were exchanged for five others founded upon totally different principles. It will be seen that this scheme has been partly adopted in the present volume.

1838. PERLEB, C. J.—(*Clavis Classium ordinum et familiarum, atque Index generum regni vegetabilis.*

This author admits nine Classes, each of which is subdivided into 48 Orders, which are themselves the equivalents of Alliances, and under these are arranged 330 Natural Orders, which he calls Families. Professor Perleb states that most of the Alliances employed in this book were proposed by him in his work entitled *Lehrbuch der Naturgeschichte des Pflanzenreichs*, published in 1826, which I have not seen.

The *Clavis* deserves to be studied. The Alliances are often well constructed, but not having the genera arranged under them, they are extremely troublesome to use; and this is no doubt the reason why the work has attracted so little notice among Botanists. Sir Edward Bromhead has analysed it (*Mag. of Nat. Hist., new series,* 1840, p. 329), and speaks of it as "a work of very great value." Professor Perleb's Classes are the following:—

Cellulares or Acotyledons			leafless (usually with a thallus; fruit imperfect . .	I. PROTOPHYTA.
			leafy; fruit perfect, capsular.	II. MUSCOSÆ.
Vasculares or Cotyledonous.	Endogens or Monocotyledons		cryptogamous	III. FILICINÆ.
			phanerogamous	IV. TERNARIÆ.
	Exogens or Dicotyledons.	perianth simple, often incomplete, sometimes 0		V. MONOCHLAMYDEÆ.
		Perianth double (both calyx and corolla). Corolla monopetalous	Corolla hypogynous	VI. THALAMANTHÆ.
			Corolla perigynous	VII. CALYCANTHÆ.
		Corolla pleiopetalous	petals perigynous	VIII. CALYCOPETALÆ.
			petals hypogynous	IX. THALAMOPETALÆ.

1839. LINDLEY, John.—(*Botanical Register, p.* 77, *Miscellaneous Matter.*)

On this occasion the author directed his attention to an extension of the primary Classes of plants, which he proposed to raise to 8, in the following manner:—

STATE I. SEXUAL OR FLOWERING PLANTS.

Division 1. Exogens.	*Cyclogens.*	Class I. Exogens.
		Class II. Gymnogens.
		Class III. Homogens.
Division 2. Endogens.	*Spermogens.*	Class IV. Dictyogens.
		Class V. Endogens.
		Class VI. Sporogens. (*Rhizanths.*)

STATE II. ESEXUAL OR FLOWERLESS PLANTS.

Division 3. Acrogens.	—	Class VII. Cormogens.
		Class VIII. Thallogens.

To what extent these views can be sustained will be discovered in the present volume.

1839. BASKERVILLE, Thomas.—(*Affinities of Plants, with some Observations upon Progressive Development.*)

The author of this tract was a very young man, with little experience; but he possessed strong perceptive powers, and would doubtless have distinguished himself had life been spared to him. But he died almost as soon as his little book saw the light. In the main he adopted the scheme of Orders in the *Nixus Plantarum*, p. xli.; but he criticised that arrangement with some skill, and avoided many of its worst errors. Baskerville's main purpose was to establish a theory of progressive development in the Vegetable Kingdom, and to show by maps and other schemes all existing affinities. The following observations deserve to be quoted:—

"Before we endeavour to establish any plan of affinity, it will be necessary to make a few observations upon a subject bearing closely upon that, namely, the respective rank or dignity of plants, and the means we possess of ascertaining the same. That this is no easy matter will appear when we reflect that imperfection is impossible in any work of supreme intelligence: our ideas of one plant having a station above that of another will not be drawn from any positive defect observable in the lowest, but from excellency we fancy to discover in the higher being. A Moss or Lichen is as perfectly fitted to the conditions it is intended to fulfil, and its organs as completely adapted to that purpose as the stately Palm, or magnificent forest tree. To imagine one plant, therefore, more noble than another, we merely imply that we consider its organisation, either by its complexity or some other character, to raise the plant possessing such qualifications above the surrounding species. When our investigations are confined to plants upon, or nearly upon, the same level, the problem is so intricate that it scarcely admits of solution; but when we take species separated by a long interval, the sum of additional properties enables us to decide with more certainty; yet the amount of difference is so trifling, and probably so exquisitely compensated for, that the balance is by no means so great as might be expected. In consequence of this it does not appear that any one has as yet been able to suggest what ought properly to be considered as the highest kind of plant; and the same difficulty would occur with regard to the lowest, were it not decided by the degree of proximity to the animal kingdom.

"It will be seen, therefore, that this kind of study is essentially comparative, and our proper attainment of it dependent upon the extent of our acquaintance with the vegetable species and their organisation, and on a proper interpretation of the importance of the characters which we construct from these, which, as character scarcely ever maintains an equal value in all its relations, lays open another source of difficulty."—p. 39.

1841. TRAUTVETTER, Ernst Christian.—(*De Novo Systemate Botanico.*)

This is a speculative disquisition upon the philosophical way of classing plants. The author begs that he may be understood to have executed his task not like a Botanist, but like a philosopher (non botanico sed philosophico munere perfungi). He divides the Vegetable Kingdom into *semi-plants* and *true plants;* the former into *Favi* or Acotyledons, and *Trunculi* or Monocotyledons; and the latter into *Herbs* and *Trees.* The views of the author cannot be given better than in his own words:—"Flagrant naturæ venatores nova semper et incognita visendi cupiditate. Nos vero antiquitatis alumni aliter sumus affecti." The treatise will be found in the *Bulletin de la Société Impériale des Naturalistes de Moscou,* 1841, p. 509.

1843. BRONGNIART, Adolphe.—(*Énumération des Genres de Plantes cultivés au Muséum d'Histoire Naturelle de Paris, suivant l'Ordre établi dans l'école de Botanique en* 1843).

The apetalous division of Jussieu is abandoned on the ground that the Orders belonging to it are an imperfect state of polypetalous Orders, (called after Endlicher dialypetalous). The impracticability of a lineal natural arrangement is insisted upon. Rules are to be formed upon *à posteriori* not *à priori* considerations. Albumen is regarded of high value, especially the difference between farinaceous albumen, and that which is fleshy, oily, and horny, which last are taken to be slight modifications of each other. Finally, the direction of the embryo is regarded of more importance in its relation to the pericarp than to the hilum. The following are the details of the system:—

Division 1. CRYPTOGAMÆ. No sexual organs, &c.
Branch 1. AMPHIGENÆ. No distinct axis or appendages, &c.
Branch 2. ACROGENÆ. Distinct axis and appendages, &c.

Division 2. PHANEROGAMÆ. Sexual organs evident, &c.
Branch 3. MONOCOTYLEDONS. Embryo with one cotyledon, &c.
Ser. 1. *Albuminosæ.* Albumen.
Ser. 2. *Exalbuminosæ.* No albumen.

Branch 4. DICOTYLEDONS. Embryo with two cotyledons, &c.
Sub-branch 1. *Angiospermæ.* Ovules in an ovary.
Ser. 1. *Gamopetalæ.* Monopetalous.
§ 1. *Perigynæ.* Stamens and corolla inserted on a calyx adhering to the ovary.
§ 2. *Hypogynæ.* Stamens and corolla inserted under the ovary.
Ser. 2. *Dialypetalæ.* Petals distinct.
§ 1. *Hypogynæ.*
§ 2. *Perigynæ.*
Sub-branch 2. *Gymnospermæ.* Ovules naked.

Division 1. CRYPTOGAMÆ.

Branch 1. AMPHIGENÆ.

Class 1. *Algæ.*

Zoosporeæ, 8
Aplosporeæ, 8
Choristosporeæ, 8

Class 2. *Fungi.*

Hyphomycetes, 29
Gasteromycetes, 29
Hymenomycetes, 29
Scleromycetes, 29

Class 3. *Lichenoideæ.*

Lichenes, 45

Branch 2. ACROGENÆ.

Class 4. *Muscineæ.*

Hepaticæ, 58
Musci, 64

Class 5. *Filicinæ.*

Filices, 78
Marsileaceæ, 71
Lycopodiaceæ, 69
Equisetaceæ, 61
? Characeæ, 26

Division 2. PHANEROGAMÆ.

Branch 3. MONOCOTYLEDONS.

Ser. 1. ALBUMINOSÆ.

⁂ Perianth 0, or sepals glumaceous. Albumen farinaceous.

Class 6. *Glumaceæ.*

Gramineæ, 106
Cyperaceæ, 117

Class 7. *Juncineæ.*

Restiaceæ, 121
Eriocauloneæ, 122
Xyrideæ, 187
Commelynaceæ, 188
Juncaceæ, 191

Class 8. *Aroideæ.*

Araceæ, 127
Typhaceæ, 126

⁂ Perianth 0, or double, sepaloid or petaloid. Albumen not farinaceous.

Class 9. *Pandanoideæ.*

Cyclantheæ, 130
Freycinetieæ, 130
Pandaneæ, 130

Class 10. *Phœnicoideæ.*

Nipaceæ, 133
Phytelephasieæ, 133
Palmæ, 133

Class 11. *Lirioideæ.*

Melanthaceæ, 198
Liliaceæ, 200
Gilliesiaceæ, 196
Amaryllidaceæ, 155
Hypoxidaceæ, 154
Astelieæ, 191
Taccaceæ, 149
Dioscoreæ, 214
Iridaceæ, 159
Burmanniaceæ, 171

⁂ Perianth double, the innermost or both petaloid. Albumen farinaceous.

Class 12. *Bromelioideæ.*

Hæmodoraceæ, 151
Vellosieæ, 151
Bromeliaceæ, 147
Pontederiaceæ, 206

Class 13. *Scitamineæ.*

Musaceæ, 163
Cannaceæ, 168
Zingiberaceæ, 165

Ser. 2. EXALBUMINOSÆ.

Class 14. *Orchioideæ.*

Orchidaceæ, 173
Apostasieæ, 184

Class 15. *Fluviales.*

Hydrocharideæ, 141
Butomeæ, 208
Alismaceæ, 209
Naiadeæ, 143
Lemnaceæ, 124

Branch 4. DICOTYLEDONS.

Sub-branch 1. *Angiospermæ.*

Ser. 1. GAMOPETALÆ.

⁂ Perigynous.

Class 16. *Campanulinæ.*

Campanulaceæ, 689
Lobeliaceæ, 692
Goodeniaceæ, 694
? Stylidieæ, 696
? Calycereæ, 701
Brunoniaceæ, 657

Class 17. *Asteroideæ.*

Compositæ, 702

Class 18. *Lonicerinæ.*

Dipsaceæ, 699
Valerianeæ, 697
Caprifoliaceæ, 766

Class 19. *Coffeinæ.*

Rubiaceæ, 761

⁂ Hypogynæ.

† Anisogynæ.
* Isostemoneæ.

Class 20. *Asclepiadaceæ.*

Spigeliaceæ, 602
Loganiaceæ, 602
Apocynaceæ, 599
Asclepiadaceæ, 623
Gentianaceæ, 612

Class 21. *Convolvulineæ.*

Polemoniaceæ, 635
Nolaneæ, 654
Convolvulaceæ, 630

Class 22. *Asperifoliæ.*

Cordiaceæ, 628
Boragineæ, 655
Hydrophyllaceæ, 638
? Hydroleaceæ, 638

Class 23. *Solanineæ.*

Cestrineæ, 618
Solaneæ, 618

** Anisostemoneæ.

Class 24. *Personatæ.*

Scrophulariæ, 681
Utriculariæ, 686
Orobancheæ, 609
Gesnerieæ, 671
} albuminous.

Cyrtandreæ, 671
Bignoniaceæ, 675
Pedalineæ, 669
Acanthaceæ, 678
} exalbuminous.

Class 25. *Selaginoideæ.*

? Jasmineæ, 650
Globulariæ, 666
Selagineæ, 666
Myoporineæ, 665

Class 26. *Verbenineæ.*

Verbenaceæ, 663
Labiatæ, 659
Stilbineæ, 607
? Plantagineæ, 642

†† Isogynæ.

Class 27. *Primulineæ.*

Primulaceæ, 644
Myrsinaceæ, 647
Theophrasteæ, 647
Ægicereæ, 647
Plumbagineæ, 640

Class 28. *Ericoideæ.*

Epacrideæ, 448
Ericaceæ, 453
Pyrolaceæ, 450
? Monotropeæ, 452
? Brexiaceæ, 573

Class 29. *Diospyroideæ.*

Ebenaceæ, 595
? Oleaceæ, 616
Ilicineæ, 597
Empetreæ, 285
Sapoteæ, 590
? Styraceæ, 592

Ser. 2. DIALYPETALÆ.

§ 1. *Hypogynæ.*

† Flowers complete.

A. Calyx permanent.

* Polystemoneæ.

Class 30. *Guttiferæ.*

exalbuminous:
Clusiaceæ, 400
Marcgraaviaceæ, 403
Hypericineæ, 405
Reaumuriaceæ, 407
? Tamariscineæ, 341

albuminous:
Cistineæ, 349
Bixaceæ, 327
Ternstromiaceæ, 396
Chlenaceæ, 486
Dipterocarpeæ, 393

Class 31. *Malvoideæ.*

Tiliaceæ, 371
Malvaceæ, 368
Sterculiaceæ, 360
Buttneriaceæ, 363

* * Oligostemoneæ.

Class 32. *Crotonineæ.*

Antidesmeæ, 259
Forestiereæ, 283
Euphorbiaceæ, 274

Class 33. *Polygalineæ.*

? Tremandreæ, 374
Polygaleæ, 375

Class 34. *Geranioideæ.*

Balsamineæ, 490
Tropæoleæ, 366
Geraniaceæ, 493
? Limnantheæ, 366
? Coriariaceæ, 474
Lineæ, 485
Oxalideæ, 488
Zygophylleæ, 478

Class 35. *Terebinthineæ.*

Rutaceæ, 469
Diosmeæ, 469
Ochnaceæ, 474
Simarubeæ, 476
Xanthoxyleæ, 472
Anacardieæ, 465
? Connaraceæ, 468

Class 36. *Hesperideæ.*

Burseraceæ, 459
Aurantiaceæ, 457
Cedreleæ, 461
Meliaceæ, 463
Ximeneæ, 443
Nitrariaceæ, 388
? Humiriaceæ, (*bis*) 447
Erythroxyleæ, 391

Class 37. *Æsculineæ.*

Malpighiaceæ, 388
Acerineæ, 387
Hippocastaneæ, 382
? Rhizoboleæ, 398
Sapindaceæ, 382
Vochysieæ, 379

Class 38. *Celastroideæ.*

Viniferæ, 439
Hippocrateaceæ, 584
Celastraceæ, 586
Staphyleaceæ, 381
Pittosporeæ, 441

Class 39. *Violineæ.*

? Sauvagesieæ, 343
Violaceæ, 338
Droseraceæ, 433
Frankeniaceæ, 340

B. Calyx deciduous.

* Albumen none or thin.

Class 40. *Cruciferineæ.*

Resedaceæ, 356
Capparidaceæ, 357
Cruciferæ, 351

** Albumen thick, fleshy, or horny.

Class 41. *Papaverineæ.*

Fumariaceæ, 435
Papaveraceæ, 430

Class 42. *Berberineæ.*

Berberideæ, 437
Lardizabaleæ, 303
Menispermaceæ, 307

Class 43. *Magnolineæ.*

Schizandreæ, 305
Myristicaceæ, 301
Anonaceæ, 420
Magnoliaceæ, 417

Class 44. *Ranunculineæ.*

Dilleniaceæ, 423
Ranunculaceæ, 425
? Sarracennieæ, 429

*** Albumen double, the outer farinaceous.

Class 45. *Nymphæineæ.*

Nelumboneæ, 414
Nymphæaceæ, 409
Cabombeæ, 412

† † Flowers incomplete. Never a corolla.

Class 46. *Piperineæ.*

Saurureæ, 521
Piperaceæ, 515

Class 47. *Urticineæ.*

Urticeæ, 260
Artocarpeæ, 269
Moreæ, 266
Celtideæ, 580
Cannabineæ, 265

Class 48. *Polygonoideæ.*

Polygoneæ, 502

§ 2. *Perigynæ.*

† Embryo curved round farinaceous albumen.

Class 49. *Caryophyllineæ.*

Nyctagineæ, 506
Phytolacceæ, 509
Chenopodeæ, 512
Baselleæ, 524
Amaranthaceæ, 510
Sileneæ, 496
Alsineæ, 496
Paronychiæ, 499
Portulaceæ, 500

Class 50. *Cactoideæ.*

Mesembryanthemeæ, 525
Cacteæ, 746

† † Albumen fleshy or horny.

Class 51. *Crassulineæ.*

Crassulaceæ, 344
Elatineæ, 480
Datisceæ, 316

Class 52. *Saxifragineæ.*

Francoaceæ, 451
Philadelpheæ, 753
Saxifragaceæ, 567
Ribesiæ, 750

Class 53. *Passiflorineæ.*

Loaseæ, 744
Papayaceæ, 321
Turneraceæ, 347
Malesherbiæ, 335
Passifloreæ, 332
Samydeæ, 330
Homalineæ, 742

Class 54. *Hamamelineæ.*

Plataneæ, 272
Balsamifluæ, 253
Hamamelideæ, 784
Alangieæ, 719
Bruniaceæ, 785

Class 55. *Umbellinæ.*

Umbelliferæ, 773
Araliaceæ, 780
Corneæ, 782
? Garryaceæ, 295

Class 56. *Santalinæ.*

? Ceratophylleæ, 263
? Chloranthaceæ, 519
Loranthеæ, 789
Santalaceæ, 787
Olacineæ, 443

Class 57. *Asarineæ.*

? Balanophoreæ, 89
Rafflesiaceæ, 93
Cytineæ, 91
Nepentheæ, 287
Aristolochiaceæ, 792

††† Albumen 0, or little.

Class 58. *Cucurbitineæ.*

Begoniaceæ, 318
Nhandirhobeæ, 311
Cucurbitaceæ, 311
Gronovieæ, 744

Class 59. *Œnotherineæ.*

Halorageæ, 722
Œnothereæ, 724
Melastomaceæ, 731
Lythraceæ, 574
? Rhizophoreæ, 726
Memecyleæ, 731
Combretaceæ, 717
? Myrtaceæ, 734

Class 60. *Daphnoideæ.*

Gyrocarpeæ, 535
Lauraceæ, 535
Hernandiaceæ, 535
Thymelaceæ, 530

Class 61. *Proteineæ.*

Proteaceæ, 532
Elæagnaceæ, 257

Class 62. *Rhamnoideæ.*

Penæaceæ, 577
Rhamneæ, 581
Stackhousieæ, 589

Class 63. *Myrtoideæ.*

Myrtaceæ, 734
Lecythideæ, 739
Granateæ, 734
Calycantheæ, 540
? Monimieæ, 298

Class 64. *Rosinæ.*

Pomaceæ, 559
Neuradeæ, 563
Spiræaceæ, 563
Rosaceæ, 563
Amygdaleæ, 557
Chrysobalanaceæ, 542

Class 65. *Leguminosæ.*

Papilionaceæ, } 544
Cæsalpinieæ, } 544
Mimoseæ, } 544
? Moringeæ, 536

Class 66. *Amentaceæ.*

Juglandeæ, 292
? Salicineæ, 254
Quercineæ, 290
Betulineæ, 251
Myriceæ, 256
Casuarineæ, 249

Sub-branch 2. *Gymnospermæ.*

Class 67. *Coniferæ.*

Gnetaceæ, 232
Taxineæ, 230
Cupressineæ, 226
Abietineæ, 226

Class 68. *Cycadoideæ.*

Cycadeæ, 223

The great faults of this arrangement, in bringing Amentaceous into contact with Leguminous plants, in separating Chloranths from Pepperworts, Myrtleblooms from Hippurids, and many such instances, need not be insisted on. Such a system cannot be founded on sound principles. It has, however, merits, and is decidedly the most forward step that the Botanists of the Modern French School have yet taken. The abandonment of the Apetalæ of Jussieu is more especially important.

1843. MEISNER, Carl Friedrich.—(*Plantarum vascularium genera secundum Ordines naturales digesta, eorumque differentiæ et affinitates tabulis diagnosticis expositæ.*)

In the beginning of this large and useful work Professor Meisner intended to follow nearly the order observed by De Candolle in his *Prodromus;* and accordingly he commenced without any plan for throwing the Natural Orders into higher groups. But as he advanced in his labour he found the inconvenience of neglecting the latter, and, as early as p. 13, he commenced with his Class Malpighinæ. His final views are given in a Conspectus diagnosticus, the skeleton of which is the following :—

A. VASCULAR PLANTS.

I. DICOTYLEDONS.

† Diplochlamyds.

* Dialypetalous or Polypetalous.

I. THALAMIFLORALS.

Class 1. *Polycarpicæ.*
Ranunculaceæ, 425
Dilleniaceæ, 423
Magnoliaceæ, 417
Anonaceæ, 420
Menispermaceæ, 307
Berberidaceæ, 437

Class 2. *Nymphæoideæ.*
Nelumboneæ, 414
Hydropeltideæ, 412
Nymphæaceæ, 409
Sarraceniaceæ, 429

Class 3. *Rhœadeæ.*
Papaveraceæ, 430
Fumariaceæ, 435
Cruciferæ, 351
Capparideæ, 357
Resedaceæ, 356

Class 4. *Polygalinæ.*
Tremandreæ, 374
Polygaleæ, 375

Class 5. *Parietales.*
Pittosporeæ, 441
Frankeniaceæ, 340
Tamariscineæ, 341
Podostemeæ, 482
Droseraceæ, 433
Violarieæ, 338
Cistineæ, 349
Bixaceæ, 327
Samydeæ, 330
Homalineæ, 742

Class 6. *Caryophyllinæ.*
Caryophylleæ, 496
Sclerantheæ, 528
Paronychieæ, 499
Portulaceæ, 500
Elatineæ, 480

Class 7. *Columniferæ.*
Malvaceæ, 368
Buttneriaceæ, 363
Sterculiaceæ, 360
Tiliaceæ, 371

Class 8. *Lamprophyllæ.*
Dipterocarpeæ, 393
Chlænaceæ, 486
Ternstrœmiaceæ, 396
Guttiferæ, 400
Marcgraviaceæ, 403
Hypericineæ, 405
Rhizoboleæ, 398

Class 9. *Malpighinæ.*
Hippocastaneæ, 382
Sapindaceæ, 382
Malpighiaceæ, 388
Acerineæ, 387
Erythroxyleæ, 391
Hippocrateaceæ, 584
? Coriarieæ, 475

Class 10. *Hesperides.*
Humiriaceæ, 447
Olacineæ, 443
Melioideæ, 463
Aurantiaceæ, 457
Ampelideæ, 439

Class 11. *Gruinales.*
Geraniaceæ, 493
Lineæ, 485
Oxalideæ, 488
Ledocarpeæ, 488
Vivianaceæ, 365
Balsamineæ, 490
Tropæoleæ, 366

Class 12. *Rutaceæ.*
Zygophyllaceæ, 478
Ruteæ, 469
Diosmeæ, 469
Zanthoxylaceæ, 472
Simarubeæ, 476
Ochnaceæ, 474
? Pittosporeæ, 441

Class 13. *Terebinthaceæ.*
Juglandeæ, 292
Amyrideæ, 459
Cassuvieæ, 465
Spondiaceæ, 459
Burseraceæ, 459
Connaraceæ, 468

II. CALYCIFLORALS.

Class 14. *Leguminosæ.*
Leguminosæ veræ, 544
Moringeæ, 366

Class 15. *Rosifloræ.*
Rosaceæ, 563
Calycantheæ, 540
Myrtineæ, 734

Class 16 *Calycanthemæ.*
Melastomoideæ, 731
Lythrarieæ, 574
Onagraceæ, 724
Combretaceæ, 717
Rhizophoraceæ, 726
Vochysieæ, 379

Class 17. *Corniculatæ.*
Saxifragaceæ, 567
Crassulaceæ, 344
Surianeæ
Francoaceæ, 451
Ficoideæ, 525

Class 18. *Peponiferæ.*
Papayaceæ, 321
Turneraceæ, 347
Malesherbiaceæ, 335
Passifloraceæ, 332
Belvisieæ, 728
Loaseæ, 744
Grossularieæ, 750
Cacteæ, 746
Cucurbitaceæ, 311

Class 19. *Frangulaceæ.*
Celastrineæ, 586
Rhamneæ, 581
Bruniaceæ, 785
Aquilarineæ, 579
Chailletiaceæ, 583

Class 20. *Umbellifloræ.*
Hamamelideæ, 784
Umbelliferæ, 773
Araliaceæ, 780
Corneæ, 782
Alangieæ, 719
Loranthaceæ, 789

* * Monopetalous.

a. Fruit inferior.

Class 21. *Rubiacineæ.*
Rubiaceæ, 761
Lygodysodeaceæ, 761
Caprifoliaceæ, 766

Class 22. *Aggregatæ.*
Valerianeæ, 697
Dipsaceæ, 699
Compositæ, 702
Calycereæ, 701

Class 23. *Campanulineæ.*
Stylideæ, 696
Brunoniaceæ, 657
Goodeniaceæ, 694
Lobeliaceæ, 692
Campanulaceæ, 689
Pongatieæ, 689

b. Fruit superior.

Class 24. *Ericineæ.*
Vaccinieæ, 757
Ericaceæ, 453
Monotropeæ, 452
Epacrideæ, 448

Class 25. *Ligustrinæ.*
Columelliaceæ, 759
Bolivariaceæ, 612
Jasmineæ, 650
Oleaceæ, 616

Class 26. *Plantagoideæ.*
Plantagineæ, 642
Plumbagineæ, 640
Salvadoraceæ, 652

Class 27. *Petalanthæ.*
Primulaceæ, 604
Myrsineæ, 647

Class 28. *Styracineæ.*

Styraceæ, 592
Ebenaceæ, 595
Sapoteæ, 590
Aquifoliaceæ, 597

Class 29. *Contortæ.*

? Roussæaceæ
Loganiaceæ, 602
? Gentianaceæ, 612
Apocynaceæ, 599
Asclepiadeæ, 623

Class 30. *Tubifloræ.*

Cuscuteæ, 633
Diapensiaceæ, 606
? Retziaceæ, 618
Polemoniaceæ, 635
Hydroleaceæ, 638
Hydrophylleæ, 638
Convolvulaceæ, 630
Solanaceæ, 618
Nolanaceæ, 654
Erycibeæ, 595
Cordiaceæ, 628
Ehretiaceæ, 653
Borragineæ, 655

Class 31. *Labiatifloræ.*

Labiatæ, 659
Verbenaceæ, 633
Acanthaceæ, 678
Pedaliaceæ, 669
Bignoniaceæ, 675
Cyrtandraceæ, 671
Gesneriaceæ, 671
Scrophularineæ, 681
Stilbineæ, 607
Myoporineæ, 665
Selagineæ, 666
Orobancheæ, 609
Utricularieæ, 686
Globularieæ, 666

†† Monochlamyds.

Class 32. *Oleraceæ.*

Petiveriaceæ, 509
Polygonaceæ, 502
Eriogoneæ, 502
Nyctagineæ, 506
Chenopodiaceæ, 512
Amarantaceæ, 510
Phytolacceæ, 509

Class 33. *Daphnoideæ.*

Monimieæ, 298
Atherospermeæ, 300
Laurineæ, 535
Gyrocarpeæ, 535
Grubbiaceæ
Nyssaceæ, 592
Helvingiaceæ, 296
Santalaceæ, 787
Anthoboleæ
Phalerieæ
Aquilarineæ, 579
Thymeleæ, 530
Hernandieæ, 530
Proteaceæ, 532
Penæaceæ, 577
Elæagneæ, 257
Myristiceæ, 301

Class 34. *Serpentariæ.*

Aristolochiaceæ, 792
Nepentheæ, 287
? Sarracennieæ, 429

Class 35. *Tricoccæ.*

Begoniaceæ, 318
Euphorbiaceæ, 274
Stackhousiaceæ, 589
Empetreæ, 285

Class 36. *Julifloræ.*

Cupuliferæ, 290
Gunneraceæ, 780
Cynocrambeæ
Garryaceæ, 295
Datisceæ, 31
Putranjiveæ
Forestiereæ
Scepaceæ, 283
? Henslowiaceæ, 569
Lacistemeæ, 329
Balsamifluæ, 253
Plataneæ, 272
Antidesmeæ, 259
Salicineæ, 254
Batideæ, 286
Celtideæ, 580
Urticaceæ, 260
Moreæ, 266
Artocarpeæ, 269
Trewiaceæ, 274
Cannabineæ, 265
Betulaceæ, 251
Ulmaceæ, 580
Myriceæ, 256
Casuarineæ, 249

Class 37. *Piperinæ.*

Chlorantheæ, 519
Piperaceæ, 515
Saurureæ, 521

Class 38. *Coniferæ.*

Gnetaceæ, 232
Cupressineæ, 226
Abietineæ, 226
Taxineæ, 230
Cycadeæ, 223

II. MONOCOTYLEDONS.

Class 39. *Rhizantheæ.*

Balanophoreæ, 89
Cytineæ, 91
Rafflesiaceæ, 93

Class 40. *Spadicifloræ.*

Palmæ, 133
Pandanaceæ, 130
Typhaceæ, 126
Aroideæ, 127

Class 41. *Helobiæ.*

Najadeæ, 143
Alismaceæ, 209
Butomeæ, 208
Hydrocharideæ, 141

Class 42. *Gynandræ.*

Orchideæ, 173
Apostasieæ, 184

Class 43. *Scitamineæ.*

Zingiberaceæ, 165
Canneceæ
Musaceæ, 163

Class 44. *Ensatæ.*

Burmanniaceæ, 171
Irideæ, 159
Hæmodoraceæ, 151
Hypoxideæ, 154
Amaryllideæ, 155
Bromeliaceæ, 147

Class 45. *Conorariæ.*

Pontederaceæ, 206
Liliaceæ, 200
Dioscoreaceæ, 214
Ophiopogoneæ, 200
Taccaceæ, 149
Melanthaceæ, 198
Juncaceæ, 191
Philydreæ, 186

Class 46. *Enantioblastæ.*

Commelynaceæ, 188
Mayaceæ, 189
Xyrideæ, 187
Eriocauleæ, 122
Restiaceæ, 121
Centrolepideæ, 120

Class 47. *Glumaceæ.*

Cyperaceæ, 117
Gramineæ, 106

B. CELLULAR PLANTS.

III. ACOTYLEDONS.

1843. HORANINOW, Paul.—(*Tetractys Naturæ, seu systema quadrimembre omnium naturalium.*)

In this work the views of the author, as expressed nine years before in his *Primæ lineæ* (p. xliv.), are repeated with some modifications of detail. His 4th Circle, or Spermophoræ, are called Euspermæ, and the number of the Alliances, called Orders, much increased. They are, moreover, distinguished by the termination *astra*, as Rutastra, Araliastra, &c. No distinctive characters are proposed for any of the groups, so that means are not afforded by the learned author of judging of the principles which have guided him in the details of his classification.

1844. JUSSIEU, Adrien de.—(*Cours Élémentaire d'Histoire Naturelle: Botanique.*)

This little work contains all the Natural Orders of plants now admitted, arranged on the plan of Jussieu, by his son. It is therefore the most recent exposition of the views of the learned authors. In addition to the names, an analysis of their distinctive characters is introduced in the original, to which a student may be usefully referred. The arrangement is not however extracted, because it is merely artificial, and contrived for the purpose of finding a plant easily; in which respect it may be compared to the Artificial Analysis affixed to the present work.

1845. LINDLEY, John.—(*The Vegetable Kingdom, &c.*)
The following is the system employed in the present Work :—

CLASSES.

Asexual, or Flowerless Plants.

Stems and leaves undistinguishable . . . . I. THALLOGENS.
Stems and leaves distinguishable . . . . II. ACROGENS.

Sexual, or Flowering Plants.

Fructification springing from a thallus . . . . III. RHIZOGENS.
Fructification springing from a stem.
Wood of stem youngest in the centre; cotyledon single.
Leaves parallel-veined, permanent; wood of the stem always confused . IV. ENDOGENS.
Leaves net-veined, deciduous; wood of the stem, when perennial, arranged in a circle with a central pith . . . . V. DICTYOGENS.
Wood of stem youngest at the circumference, always concentric; cotyledons 2 or more.
Seeds quite naked . . . . VI. GYMNOGENS.
Seeds inclosed in seed-vessels . . . . VII. EXOGENS.

Class I. THALLOGENS.

ALLIANCES OF THALLOGENS.

1. ALGALES.—*Cellular flowerless plants, nourished through their whole surface by the medium in which they vegetate; living in water or very damp places; propagated by zoospores, coloured spores, or tetraspores.*
2. FUNGALES.—*Cellular flowerless plants, nourished through their thallus (spawn or mycelium); living in air; propagated by spores, colourless or brown, and sometimes inclosed in asci; destitute of green gonidia.*
3. LICHENALES.—*Cellular flowerless plants, nourished through their whole surface by the medium in which they vegetate; living in air; propagated by spores usually inclosed in asci; and always having green gonidia in their thallus.*

NATURAL ORDERS OF THALLOGENS.

ALLIANCE 1. ALGALES, p. 8.

Crystalline, angular, fragmentary bodies, brittle, and multiplying by spontaneous separation	1. *Diatomaceæ* or *Brittleworts*, p. 12
Vesicular, filamentary or membranous bodies, multiplied by zoospores generated in the interior at the expense of their green matter .	2. *Confervaceæ* or *Confervas*, p. 14
Cellular or tubular unsymmetrical bodies, multiplied by simple spores formed externally	3. *Fucaceæ* or *Seaweeds*, p. 20
Cellular or tubular unsymmetrical bodies, multiplied by tetraspores	4. *Ceramiaceæ* or *Rosetangles*, p. 23
Tubular symmetrically branched bodies, multiplied by spiral coated nucules, filled with starch	5. *Characeæ* or *Charads*, p. 26

ALLIANCE 2. FUNGALES, p. 29.

Spores generally quaternate on distinct sporophores. Hymenium naked	6. *Hymenomycetes*, *Agaricaceæ*, or *Toadstools*.	p. 29
Spores generally quaternate on distinct sporophores. Hymenium inclosed in a peridium.	7. *Gasteromycetes*, *Lycoperdaceæ*, or *Puffballs*.	
Spores single, often septate, on more or less distinct sporophores. Flocci of the fruit obsolete or mere peduncles	8. *Coniomycetes*, *Uredinaceæ*, or *Blights*.	
Spores naked, often septate. Thallus floccose	9. *Hyphomycetes*, *Botrytaceæ*, or *Mildews*.	
Sporidia contained (generally eight together) in asci	10. *Ascomycetes*, *Helvellaceæ*, or *Morels*.	
Spores surrounded by a vesicular veil, or sporangium. Thallus floccose	11. *Physomycetes*, *Mucoraceæ*, or *Moulds*.	

ALLIANCE 3. LICHENALES, p. 45.

Nucleus breaking up into naked spores . . .	12. *Graphidaceæ*, or *Letter-Lichens*.	p. 45
Nucleus bearing asci; thallus homogeneous, gelatinous or cartilaginous	13. *Collemaceæ* or *Jelly-Lichens*.	
Nucleus bearing asci; thallus heterogeneous, pulverulent or cellular	14. *Parmeliaceæ* or *Leaf-Lichens*.	

Class II. ACROGENS.

ALLIANCES OF ACROGENS.

4. MUSCALES.—*Cellular (or vascular). Spore-cases immersed or calyptrate (i. e. either plunged in the substance of the frond, or inclosed within a hood having the same relation to the spores as an involucre to a seed-vessel).*
5. LYCOPODALES.—*Vascular. Spore-cases axillary or radicle, one or many-celled. Spores of two sorts.*
6. FILICALES.—*Vascular. Spore-cases marginal or dorsal, one-celled, usually surrounded by an elastic ring. Spores of but one sort.*

NATURAL ORDERS OF ACROGENS.

ALLIANCE 4. MUSCALES, p. 54.

1. HEPATICÆ.

Spore-cases valveless, without operculum or elaters	15. *Ricciaceæ*, or *Crystalworts*, p. 57
Spore-cases valveless or bursting irregularly, without operculum, but with elaters . . .	16. *Marchantiaceæ*, or *Liverworts*, p. 58
Spore-cases opening by a definite number of equal valves, without operculum, but with elaters	17. *Jungermanniaceæ*, or *Scalemosses*, p. 59
Spore-cases peltate, splitting on one side, without operculum, and with an elater to every spore	18. *Equisetaceæ*, or *Horsetails*, p. 61

2. MUSCI.

Spore-cases opening by valves, with an operculum, without elaters	19. *Andræaceæ*, or *Splitmosses*, p. 63
Spore-cases valveless, with an operculum, without elaters	20. *Bryaceæ*, or *Urnmosses*, p. 64

ALLIANCE 5. LYCOPODALES, p. 68.

Spore-cases 1-3-celled, axillary; reproductive bodies similar	21. *Lycopodiaceæ*, or *Clubmosses*, p. 69
Spore-cases many-celled, radicle (or axillary); reproductive bodies dissimilar	22. *Marsileaceæ*, or *Pepperworts*, p. 71

ALLIANCE 6. FILICALES, p. 74.

Spore-cases ringless, distinct, 2-valved, formed on the margin of a contracted leaf	23. *Ophioglossaceæ*, or *Adders' Tongues*, p. 77
Spore-cases ringed, dorsal or marginal, distinct, splitting irregularly	24. *Polypodiaceæ*, or *Ferns*, p. 78
Spore-cases ringless, dorsal, connate, splitting irregularly by a ventral cleft	25. *Danæaceæ*, or *Danæads*, p. 82

Class III. RHIZOGENS.

ALLIANCE THE SAME AS THE CLASS, p. 83.

Ovules solitary, pendulous; fruit one-seeded.	26. *Balanophoraceæ*, or *Cynomoriums*, p. 89
Ovules 00, parietal; fruit many-seeded; calyx 3-4-6-parted; anthers opening by slits . .	27. *Cytinaceæ*, or *Cistusrapes*, p. 91
Ovules 00, parietal; fruit many-seeded, calyx 5-parted, anthers opening by pores. . . .	28. *Rafflesiaceæ*, or *Patmaworts*, p. 93

Class IV. ENDOGENS.

ALLIANCES OF ENDOGENS.

* Flowers glumaceous; (that is to say, composed of bracts not collected in true whorls, but consisting of imbricated colourless or herbaceous scales).

7. GLUMALES.

** Flowers petaloid, or furnished with a true calyx or corolla, or with both, or absolutely naked; ♂ ♀ (that is having sexes altogether in different flowers, without half-formed rudiments of the absent sexes being present).

8. ARALES.—*Flowers naked or consisting of scales, 2 or 3 together, or numerous, and then sessile on a simple naked spadix; embryo axile; albumen mealy or fleshy. (Some have no albumen).*
9. PALMALES.—*Flowers perfect (with both calyx and corolla), sessile on a branched scaly spadix; embryo vague, solid; albumen horny or fleshy. Some Palms are* ⚥.
10. HYDRALES.—*Flowers perfect or imperfect, usually scattered; embryo axile, without albumen.—aquatics. (Some are* ⚥.)

***—Flowers furnished with a true calyx and corolla, adherent to the ovary; ⚥.

11. NARCISSALES.—*Flowers symmetrical; stamens 3 or 6, or more, all perfect; seeds with albumen; (Some Bromeliaceæ have a free calyx and corolla).*
12. AMOMALES.—*Flowers unsymmetrical; stamens 1 to 5, some at least of which are petaloid; seeds with albumen.*
13. ORCHIDALES.—*Flowers unsymmetrical; stamens 1 to 3; seeds without albumen.*

*** Flowers furnished with a true calyx and corolla, free from the ovary; ☿.

14. XYRIDALES.—*Flowers half herbaceous, 2-3-petaloideous; albumen copious.*
15. JUNCALES.—*Flowers herbaceous, dry, and permanent, scarious if coloured; albumen copious. (Some Callas have no albumen).*
16. LILIALES.—*Flowers hexapetaloideous, succulent, and withering; albumen copious.*
17. ALISMALES.—*Flowers 3-6-petaloideous, apocarpal; albumen none. (Some Alismads are absolutely ♀ ♂).*

NATURAL ORDERS OF ENDOGENS.

ALLIANCE 7. GLUMALES, p. 105.

Ovary 1-celled, with 2 or more distinct (or united) styles; ovule ascending; embryo lateral, naked	29. *Graminaceæ*, or *Grasses*, p. 106
Ovary 1-celled, with 2 or more (distinct or) united styles; ovule erect; embryo basal. .	30. *Cyperaceæ*, or *Sedges*, p. 117
Ovaries several (sometimes united) with 1 style to each; ovule pendulous; glumes only; styles 1-2; anthers 1-celled; embryo terminal.	31. *Desvauxiaceæ*, or *Bristleworts*, p. 120
Ovary 1-2-3-celled, with 2 or 3 styles always; ovule pendulous; glumes only; styles 2-3; anthers 1-celled; embryo terminal	32. *Restiaceæ*, or *Cordleafs*, p. 121
Ovary 2-3-celled, with 1 style to each cell; ovule pendulous; a membranous 3-lobed cup within the glumes; anthers 2-celled; embryo terminal	33. *Eriocaulaceæ*, or *Pipeworts*, p. 122

ALLIANCE 8. ARALES, p. 123.

Flowers 2 or 3, of which one only is ♀. Spadix 0. Ovary one-celled. Ovules erect. Embryo slit	34. *Pistiaceæ*, or *Lemnads*, or *Duckweeds*, p. 124
Flowers 00, on a naked spadix. Calyx scaly or hairy. Anthers with long filaments. Ovule solitary, pendulous. Seed adherent to the pericarp. Embryo slit.	35. *Typhaceæ*, or *Typhads*, or *Bulrushes*, p. 126
Flowers 00, naked, on a solitary spadix covered by a single hooded spathe. Anthers sessile. Seed loose. Embryo slit, axile. .	36. *Araceæ*, or *Arads*, p. 127
Flowers 00, naked or scaly, on a spadix covered by many spathes. Anthers stalked. Seeds loose. Embryo solid, minute . . .	37. *Pandanaceæ*, or *Screwpines*, p. 130

ALLIANCE 9. PALMALES, p. 133. — 38. *Palmaceæ*, or *Palms*, p. 133

ALLIANCE 10. HYDRALES, p. 140.

Stamens epigynous; ovary adherent	39. *Hydrocharidaceæ*, or *Hydrocharads*, p. 141
Stamens hypogynous; ovary free; pollen globose	40. *Naiadaceæ*, or *Naiads*, p. 143
Stamens hypogynous; ovary free; pollen confervoid	41. *Zosteraceæ*, or *Sea-wracks*, p. 145

ALLIANCE 11. NARCISSALES, p. 146.

Flowers tripetaloideous, 6-leaved, imbricated. Albumen mealy	42. *Bromeliaceæ*, or *Bromelworts*, p. 147
Flowers half tripetaloideous, tubular. Albumen fleshy	43. *Taccaceæ*, or *Taccads*, p. 149
Flowers hexapetaloideous, tubular, scarcely imbricated. Stamens 3, opposite the petals, or 6; anthers turned inwards. Radicle remote from the hilum, which is naked . .	44. *Hæmodoraceæ*, or *Bloodroots*, p. 151
Flowers hexapetaloideous, much imbricated. Stamens 6; anthers turned inwards. Radicle remote from the hilum, which is often strophiolate	45. *Hypoxidaceæ*, or *Hypoxids*, p. 154
Flowers hexapetaloideous, much imbricated. Stamens 6, or more; anthers turned inwards. Radicle next the hilum.	46. *Amaryllidaceæ*, or *Amaryllids*, p. 155
Flowers hexapetaloideous. Stamens 3, opposite the sepals; anthers turned outwards. .	47. *Iridaceæ*, or *Irids*, p. 159

ALLIANCE 12. AMOMALES, p. 162.

Stamens more than 1; (anthers 2-celled, no vitellus)	48. *Musaceæ*, or *Musads*, p. 163
Stamen but 1; anther 2-celled; embryo in a vitellus	49. *Zingiberaceæ*, or *Gingerworts*, p. 165
Stamen but 1; anther 1-celled (halved), no vitellus	50. *Marantaceæ*, or *Marants*, p. 168

ALLIANCE 13. ORCHIDALES, p. 170.

Flowers regular. Stamens free, perigynous. .	51. *Burmanniaceæ*, or *Burmanniads*, p. 171
Flowers irregular, gynandrous. Placentæ parietal	52. *Orchidaceæ*, or *Orchids*, p. 173
Flowers regular, half-gynandrous. Placentæ axile	53. *Apostasiaceæ*, or *Apostasiads*, p. 184

ALLIANCE 14. XYRIDALES, p. 185.

Sepals 0. Petals 2. Stamens 3, of which 2 are abortive. Embryo axile, in fleshy albumen.	54. *Philydraceæ*, or *Waterworts*, p. 186
Sepals 3. Petals 3. Stamens 3, fertile. Carpels opposite sepals. Placentæ parietal. Embryo minute, on the outside of fleshy albumen	55. *Xyridaceæ*, or *Xyrids*, p. 187
Sepals 3. Petals 3. Stamens 6 (or 3). Carpels opposite sepals. Placentæ axile. Embryo trochlear, half immersed in fleshy albumen	56. *Commelynaceæ*, or *Spiderworts*, p. 188
Sepals 3. Petals 3. Stamens 3; (anthers 1-celled). Carpels opposite petals. Placentæ parietal. Embryo minute, on the outside of fleshy albumen	57. *Mayaceæ*, or *Mayacs*, p. 189

ALLIANCE 15. JUNCALES, p. 190.

Flowers scattered. Embryo minute, undivided.	58. *Juncaceæ*, or *Rushes*, p. 191
Flowers spadiceous. Embryo axile, with a conspicuous cleft on one side	59. *Orontiaceæ*, or *Orontiads*, p. 193

ALLIANCE 16. LILIALES, p. 195.

Perianth surrounded by a calycine involucre, the inner bracts of which are coloured and petaloid	60. *Gilliesiaceæ*, or *Gilliesiads*, p. 196
Perianth naked, flat when withering. Anthers turned outwards; styles distinct; albumen fleshy	61. *Melanthaceæ*, or *Melanths*, p. 198
Perianth naked, flat when withering. Anthers turned inwards. Styles consolidated. Albumen fleshy	62. *Liliaceæ*, or *Lilyworts*, p. 200
Perianth naked, circinate when withering. Anthers turned inwards. Albumen mealy .	63. *Pontederaceæ*, or *Pontederads*, p. 206

ALLIANCE 17. ALISMALES, p. 207.

Flowers 3-petaloideous. Placentæ many-seeded, netted and parietal	64. *Butomaceæ*, or *Butomads*, p. 208
Flowers 3-petaloideous. Placentæ few-seeded, simple, and axile, or basal. Embryo solid .	65. *Alismaceæ*, or *Alismads*, p. 209
Flowers scaly. Placentæ few-seeded, simple and axile, or basil. Embryo slit on one side, with a very large plumule	66. *Juncaginaceæ*, or *Arrow-grasses*, p. 210

Class V. DICTYOGENS.

ALLIANCE THE SAME AS THE CLASS, p. 211.

Flowers ♂ ♀. Perianth free. Carpels 00, one-seeded	67. *Triuridaceæ*, or *Tailworts*, p. 213
Flowers ♂ ♀. Perianth adherent. Carpels consolidated, several-seeded	68. *Dioscoreaceæ*, or *Yams*, p. 214
Flowers ♂ ♀. Carpels several, quite consolidated. Placentæ axile. Flowers hexapetaloideous	69. *Smilaceæ*, or *Sarsaparillas*, p. 215
Flowers ⚥. Carpels several, quite consolidated. Placentæ parietal. Flowers 3-6-petaloideous	70. *Philesiaceæ*, or *Philesiads*, p. 217
Flowers ⚥. Carpels several, half-consolidated. Placentæ axile. Flowers 3-petaloideous. .	71. *Trilliaceæ*, or *Parids*, p. 218
Flowers ⚥. Carpels solitary, simple, many-seeded, with long-stalked anatropal seeds and a basal placenta	72. *Roxburghiaceæ*, or *Roxburghworts*, p. 219

Class VI. GYMNOGENS.

ALLIANCE THE SAME AS THE CLASS, p. 221.

Stem simple, continuous. Leaves parallel-veined, pinnate. Scales of the cone antheriferous	73. *Cycadeaceæ*, or *Cycads*, p. 223
Stem repeatedly branched, continuous. Leaves simple, acerose. Females in cones . . .	74. *Pinaceæ*, or *Conifers*, p. 226
Stem repeatedly branched, continuous. Leaves simple, often fork-veined. Females solitary. Membrane next the nucleus inclosed. Anthers 2-celled, opening longitudinally . . .	75. *Taxaceæ*, or *Taxads*, p. 236

Stem repeatedly branched, jointed. Leaves simple, net-veined. Membrane next the nucleus tubular, protruded. Anthers 1-celled, opening by pores . . . . } 76. *Gnetaceæ*, or *Joint Firs*, p. 232

Class VII. EXOGENS.

ALLIANCES OF EXOGENS.

SUB-CLASS I. DICLINOUS EXOGENS.

Flowers ♂ ♀, without any customary tendency to ⚥.

18. AMENTALES.—*Flowers in catkins, achlamydeous or monochlamydeous; carpels superior; embryo small, with little or no albumen.*
19. URTICALES.—*Flowers scattered, monochlamydeous; carpel single, superior; embryo large, lying in a small quantity of albumen.*
20. EUPHORBIALES.—*Flowers scattered, monodichlamydeous; carpels consolidated, superior; placentæ axile; embryo surrounded by abundant albumen. (Albumen occasionally absent).*
21. QUERNALES.—*Flowers in catkins, monochlamydeous; carpels inferior; embryo amygdaloid, without albumen.*
22. GARRYALES.—*Flowers monochlamydeous, sometimes amentaceous; carpels inferior; embryo minute, in a large quantity of albumen.*
23. MENISPERMALES.—*Flowers monodichlamydeous; carpels superior, disunited; embryo surrounded by abundant albumen.*
24. CUCURBITALES.—*Flowers monodichlamydeous; carpels inferior; placentæ parietal; embryo without albumen.*
25. PAPAYALES.—*Flowers dichlamydeous; carpels superior, consolidated; placentæ parietal; embryo surrounded by abundant albumen.*

SUB-CLASS II. HYPOGYNOUS EXOGENS.

Flowers ⚥, or ♂ ⚥ ♀; stamens entirely free from the calyx and corolla.

26. VIOLALES.—*Flowers monodichlamydeous; placentæ parietal or sutural; embryo straight, with little or no albumen.*
27. CISTALES.—*Flowers monodichlamydeous; placentæ parietal or sutural; embryo curved or spiral, with little or no albumen.*
28. MALVALES.—*Flowers monodichlamydeous; placentæ axile; calyx valvate in æstivation; corolla imbricated or twisted; stamens definite or* 00; *embryo with little or no albumen.*
29. SAPINDALES.—*Flowers monodichlamydeous, unsymmetrical; placentæ axile; calyx and corolla imbricated; stamens definite; embryo with little or no albumen. (Stamens rarely* 00).
30. GUTTIFERALES.—*Flowers monodichlamydeous; placentæ axile; calyx imbricated; corolla imbricated or twisted; stamens* 00; *embryo with little or no albumen. (Stamens sometimes definite in number).*
31. NYMPHALES.—*Flowers dichlamydeous; placentæ axile or sutural; stamens* 00; *embryo on the outside of a very large quantity of mealy albumen. (A part have no albumen).*
32. RANALES.—*Flowers monodichlamydeous; placentæ sutural or axile; stamens* 00; *embryo minute, inclosed in a large quantity of fleshy or horny albumen.*
33. BERBERALES.—*Flowers monodichlamydeous, unsymmetrical in the ovary; placentæ sutural, parietal, or axile; stamens definite; embryo inclosed in a large quantity of fleshy albumen.*
34. ERICALES.—*Flowers dichlamydeous, symmetrical in the ovary; placentæ axile; stamens definite; embryo inclosed in a large quantity of fleshy albumen. (Stamens occasionally adherent to the corolla).*
35. RUTALES.—*Flowers monodichlamydeous, symmetrical; placentæ axile; calyx and corolla imbricated, if present; stamens definite; embryo with little or no albumen. (Occasionally* ♂ ♀).
36. GERANIALES.—*Flowers monodichlamydeous, symmetrical; placentæ axile; calyx imbricated; corolla twisted; stamens definite; embryo with little or no albumen.*
37. SILENALES.—*Flowers monodichlamydeous; placenta free, central; embryo external, curved round a little mealy albumen; carpels more than one, completely combined into a compound fruit. (Some slightly perigynous, others* ♂ ♀).
38. CHENOPODALES.—*Flowers monochlamydeous; placentæ free, central; embryo external, either curved round or applied to the surface of a little mealy or horny albumen; carpels solitary, or, if more than one, distinct. (Some slightly perigynous, others* ♂ ♀).
39. PIPERALES.—*Flowers achlamydeous; embryo minute, on the outside of a large quantity of mealy albumen. (Occasionally* ♂ ♀).

SUB-CLASS III. PERIGYNOUS EXOGENS.

Flowers ⚥, or ♂ ⚥ ♀; stamens growing to the side of either the calyx or the corolla; ovary superior, or nearly so.

40. FICOIDALES.—*Flowers monodichlamydeous; placentæ central or axile; corolla, if present, polypetalous; embryo external, and curved round a small quantity of mealy albumen.*
41. DAPHNALES.—*Flowers monochlamydeous; carpel solitary; embryo amygdaloid, without albumen.*
42. ROSALES.—*Flowers monodichlamydeous; carpels more or less distinct: placentæ sutural; seeds definite; corolla, if present, polypetalous; embryo amygdaloid, with little or no albumen.*
43. SAXIFRAGALES.—*Flowers monodichlamydeous; carpels consolidated; placentæ sutural or axile; seeds* 00; *corolla, if present, polypetalous; embryo taper, with a long radicle, and a little or no albumen.*

44. RHAMNALES.—*Flowers monodichlamydeous; carpels consolidated; placentæ axile; fruit capsular, berried, or drupaceous; seeds definite; embryo amygdaloid, with little or no albumen.*
45. GENTIANALES.—*Flowers dichlamydeous, monopetalous; placentæ axile or parietal; embryo minute, or with the cotyledons much smaller than the radicle, lying in a large quantity of albumen.*
46. SOLANALES.—*Flowers dichlamydeous, monopetalous, symmetrical; placentæ axile; fruit 2-3-celled; embryo large, lying in a small quantity of albumen. (Occasionally achlamydeous or polypetalous).*
47. CORTUSALES.—*Flowers dichlamydeous, monopetalous, symmetrical; placenta free, central; embryo lying among a large quantity of albumen. (Occasionally monochlamydeous, or polypetalous).*
48. ECHIALES.—*Flowers dichlamydeous, monopetalous, symmetrical, or unsymmetrical; fruit nucamentaceous, consisting of several one-seeded nuts, or of clusters of them separate or séparable; embryo large, with little or no albumen. (Very rarely hypogynous!)*
49. BIGNONIALES.—*Flowers dichlamydeous, monopetalous, unsymmetrical; fruit capsular or berried, with its carpels quite consolidated; placentæ axile, or parietal, or free central; embryo with little or no albumen.*

SUB-CLASS IV. EPIGYNOUS EXOGENS.

Flowers ⚥ or ♂ ⚥ ♀; stamens growing to the side of either the calyx or corolla; ovary inferior or nearly so.

50. CAMPANALES.—*Flowers dichlamydeous, monopetalous; embryo with little or no albumen.*
51. MYRTALES.—*Flowers dichlamydeous, polypetalous; placentæ axile; embryo with little or no albumen. (Occasionally monochlamydeous).*
52. CACTALES.—*Flowers dichlamydeous, polypetalous; placentæ parietal; embryo with little or no albumen.*
53. GROSSALES.—*Flowers dichlamydeous, polypetalous; seeds numerous, minute; embryo small, lying in a large quantity of albumen.*
54. CINCHONALES.—*Flowers dichlamydeous, monopetalous; embryo minute, lying in a large quantity of albumen.*
55. UMBELLALES.—*Flowers dichlamydeous, polypetalous; seeds solitary, large; embryo small, lying in a large quantity of albumen.*
56. ASARALES.—*Flowers monochlamydeous; embryo small, lying in a large quantity of albumen.*

NATURAL ORDERS OF EXOGENS.

ALLIANCE 18. AMENTALES, p. 248.

Ovary 1-celled. Ovule 1 or 2, ascending. Radicle superior	77. *Casuarinaceæ*, or *Beefwoods*, p. 249
Ovary 2-celled. Ovule 1, pendulous. Radicle superior	78. *Betulaceæ*, or *Birchworts*, p. 251
Ovary 2-celled. Ovules 00. Seeds winged.	79. *Altingiaceæ*, or *Liquidambars*, p. 253
Ovary 1-celled. Ovules 00. Seeds cottony.	80. *Salicaceæ*, or *Willowworts*, p. 254
Ovary 1-celled. Ovule 1, erect. Radicle superior	81. *Myricaceæ*, or *Galeworts*, p. 256
Ovary 1-celled. Ovule 1, ascending. Radicle inferior	82. *Elæagnaceæ*, or *Oleasters*, p. 257

ALLIANCE 19. URTICALES, p. 258.

Radicle superior. Ovules twin, suspended. Embryo straight, albuminous. Anthers 2-lobed, with vertical fissures	83. *Stilaginaceæ*, or *Antidesmads*, p. 259
Radicle superior. Ovule solitary, erect. Embryo straight, albuminous. Juice limpid. Stipules small, flat	84. *Urticaceæ*, or *Nettleworts*, p. 260
Radicle inferior. Embryo exalbuminous. Plumule many-leaved, large	85. *Ceratophyllaceæ*, or *Hornworts*, p. 263
Radicle superior. Ovule solitary, suspended. Embryo hooked, exalbuminous	86. *Cannabinaceæ*, or *Hempworts*, p. 265
Radicle superior. Ovules solitary, suspended. Embryo hooked, albuminous	87. *Moraceæ*, or *Morads*, p. 266
Radicle superior. Ovule solitary, erect or suspended. Embryo straight, exalbuminous. Juice milky. Stipules large, convolute	88. *Artocarpaceæ*, or *Artocarpads*, p. 269
Radicle inferior. Embryo albuminous. Plumule minute. Juice limpid. Stipules large, deciduous	89. *Platanaceæ*, or *Planes*, p. 272

ALLIANCE 20. EUPHORBIALES, p. 273.

Ovules definite, suspended, anatropal. Radicle superior	90. *Euphorbiaceæ*, or *Spurgeworts*, p. 274
Ovules definite, suspended, campylotropal. Radicle inferior, albumen mealy	**Gyrostemoneæ*, p. 282
Ovules definite, suspended, anatropal. Radicle superior. ♂ amentaceous	91. *Scepaceæ*, or *Scepads*, p. 283
Ovules definite, suspended, amphitropal. Radicle superior	92. *Callitrichaceæ*, or *Starworts*, p. 284
Ovules definite, ascending, anatropal. Radicle inferior	93. *Empetraceæ*, or *Crowberries*, p. 285
Ovules solitary, ascending. ♀ naked, combined into a succulent cone	**Batideæ*, p. 286
Ovules 00, ascending. Radicle inferior. Seeds scobiform	94. ? *Nepenthaceæ*, or *Nepenths*, p. 287

ALLIANCE 21. QUERNALES, p. 289.

Ovary 2- or more celled. Ovules pendulous or peltate.	95. *Corylaceæ*, or *Mastworts*, p. 290
Ovary 1-celled. Ovule solitary, erect . . .	96. *Juglandaceæ*, or *Juglands*, p. 292

ALLIANCE 22. GARRYALES, p. 294.

Flowers amentaceous. Leaves opposite, exstipulate	97. *Garryaceæ*, or *Garryads*, p. 295
Flowers fascicled. Leaves alternate, stipulate.	98. *Helwingiaceæ*, or *Helwingiads*, p. 296

ALLIANCE 23. MENISPERMALES, p. 297.

Albumen copious, solid. Seeds pendulous; embryo small, external. Stamens perigynous.	99. *Monimiaceæ*, or *Monimiads*, p. 298
Albumen copious, solid. Seeds erect. Anthers opening by recurved valves	100. *Atherospermaceæ*, or *Plume-Nutmegs*, p. 300
Albumen copious, ruminated. Sepals united into a valvate cup	101. *Myristicaceæ*, or *Nutmegs*, p. 301
Albumen copious, solid. Seeds parietal; embryo minute	102. *Lardizabalaceæ*, or *Lardizabalads*, p. 303
Albumen copious, solid. Seeds pendulous; embryo minute, internal. Stamens hypogynous	103. *Schizandraceæ*, or *Kadsurads*, p. 305
Albumen sparing, solid. Seeds amphitropal; embryo large	104. *Menispermaceæ*, or *Menispermads*, p. 307

ALLIANCE 24. CUCURBITALES, p. 310.

Fruit pulpy. Placentæ strictly parietal. Monopetalous	105. *Cucurbitaceæ*, or *Cucurbits*, p. 311
Fruit dry. Placentæ strictly parietal. Apetalous	106. *Datiscaceæ*, or *Datiscads*, p. 316
Fruit dry. Placentæ projecting and meeting in the axis. Monodichlamydeous	107. *Begoniaceæ*, or *Begoniads*, p. 318

ALLIANCE 25. PAPAYALES, p. 320.

Corolla monopetalous; ♀ without scales . .	108. *Papayaceæ*, or *Papayads*, p. 321
Corolla polypetalous; ♀ with scales in the throat	109. *Pangiaceæ*, or *Pangiads*, p. 323

ALLIANCE 26. VIOLALES, p. 326.

Flowers scattered, apetalous or polypetalous. Petals and stamens both hypogynous. Leaves dotless, or with round dots only	110. *Flacourtiaceæ*, or *Bixads*, p. 327
Flowers in catkins, apetalous, scaly, polygamous. Stamens unilateral	111. *Lacistemaceæ*, or *Lacistemads*, p. 329
Flowers scattered, apetalous, tubular, hermaphrodite. Leaves marked with both round and linear transparent dots. (Stamens perigynous).	112. *Samydaceæ*, or *Samyds*, p. 330
Flowers polypetalous or apetalous, coronetted. Petals perigynous, imbricated. Stamens on the stalk of the ovary. Styles simple, terminal. Seeds arillate. Leaves stipulate . .	113. *Passifloraceæ*, or *Passionworts*, 332
Flowers polypetalous, coronetted. Petals perigynous, imbricated. Stamens on the stalk of the ovary. Styles simple, dorsal. Seeds without aril. Leaves without stipules . .	114. *Malesherbiaceæ*, or *Crownworts*, p. 335
Flowers polypetalous. Calyx many-leaved. Petals perigynous. Anthers 1-celled. Fruit stipitate, consolidated, siliquose. Seeds without albumen. Stamens perigynous. .	115. *Moringaceæ*, or *Moringads*, p. 336
Flowers polypetalous. Calyx many-leaved. Petals hypogynous. Stamens all perfect; anthers crested, and turned inwards. Fruit consolidated. Seeds albuminous	116. *Violaceæ*, or *Violetworts*, p. 338
Flowers polypetalous. Calyx tubular, furrowed. Petals hypogynous, unguiculate. .	117. *Frankeniaceæ*, or *Frankeniads*, p. 340
Flowers polypetalous. Calyx many-leaved. Petals hypogynous. Styles distinct. Fruit consolidated. Seeds 00, basal, comose, without albumen	118. *Tamaricaceæ*, or *Tamarisks*, p. 341
Flowers polypetalous. Calyx many-leaved. Petals hypogynous. Stamens partly sterile and petaloid; anthers opposite the petals, naked, turned outwards. Fruit consolidated. Seeds albuminous	119. *Sauvagesiaceæ*, or *Sauvageads*, p. 343
Flowers polypetalous or monopetalous. Calyx many-leaved. Petals hypogynous. Fruit follicular, apocarpous	120. *Crassulaceæ*, or *Houseleeks*, p. 344
Flowers polypetalous. Petals perigynous, contorted. Styles forked. Leaves exstipulate.	121. *Turneraceæ*, or *Turnerads*, p. 347

ALLIANCE 27. CISTALES, p. 348.

Stamens not tetradynamous, generally indefinite. Flowers $\sqrt[3]{}$ or $\sqrt[5]{}$. Seeds with albumen. Fruit closed up	122. *Cistaceæ*, or *Rock Roses*, p. 340
Stamens tetradynamous. Flowers $\sqrt[4]{}$ . . .	123. *Brassicaceæ*, or *Crucifers*, p. 351
Stamens not tetradynamous, definite. Flowers not tetramerous. Seeds without albumen. Fruit usually open at the point	124. *Resedaceæ*, or *Weldworts*, p. 356
Stamens not tetradynamous. Flowers $\sqrt[4]{}$. Seeds without albumen. Fruit closed up .	125. *Capparidaceæ*, or *Capparids*, p. 357

ALLIANCE 28. MALVALES, p. 359.

Stamens columnar, all perfect. Anthers 2-celled, turned outwards	126. *Sterculiaceæ*, or *Sterculiads*, p. 360
Stamens monadelphous, in most cases partly sterile. Anthers 2-celled, turned inwards .	127. *Byttneriaceæ*, or *Byttneriads*, p. 363
Stamens free. Disk none. Seeds with albumen. Embryo curved. Petals permanent. Calyx ribbed	128. *Vivianiaceæ*, or *Vivianads*, p. 365
Stamens free. Disk none. Seeds without albumen. Embryo amygdaloid	129. *Tropæolaceæ*, or *Indian Cresses*, p. 366
Stamens columnar, all perfect. Anthers 1-celled, turned inwards	130. *Malvaceæ*, or *Mallowworts*, p. 368
Stamens free, on the outside a disk. Seeds with albumen. Embryo straight	131. *Tiliaceæ*, or *Lindenblooms*, p. 371

ALLIANCE 29. SAPINDALES, p. 373.

Flowers complete, partially symmetrical. Calyx valvate. Anthers 2- 4-celled, opening by pores	132. *Tremandraceæ*, or *Poreworts*, p. 374
Flowers complete (irregular), unsymmetrical. Petals naked. Anthers 1-celled, opening by pores. Seeds carunculate	133. *Polygalaceæ*, or *Milkworts*, p. 375
Flowers complete, unsymmetrical, very irregular. Petals naked. Anthers opening longitudinally. Carpels 3. Seeds winged. (In one case the ovary is adherent) . . .	134. *Vochyaceæ*, or *Vochyads*, p. 379
Flowers complete, partially symmetrical. Calyx imbricated. Ovules ascending. Stigmas simple. Leaves opposite, with stipules.	135. *Staphyleaceæ*, or *Bladder Nuts*, p. 381
Flowers complete, unsymmetrical. Petals usually with an appendage or 0. Anthers opening longitudinally. Carpels 3. Seeds usually arillate, wingless	136. *Sapindaceæ*, or *Soapworts*, p. 382
Flowers apetalous. Carpel solitary	137. *Petiveriaceæ*, or *Petiveriads*, p. 386
Flowers complete, unsymmetrical. Petals naked or 0. Anthers opening longitudinally. Carpels 2. Seeds without an aril	138. *Aceraceæ*, or *Maples*, p. 387
Flowers complete, partially symmetrical. Calyx imbricated. Petals naked, stalked. Ovules hanging by cords. Stigmas simple. Embryo usually convolute	139. *Malpighiaceæ*, or *Malpighiads*, p. 388
Flowers complete, partially symmetrical. Calyx imbricated. Petals with an appendage. Ovules sessile, pendulous. Stigmas capitate. Embryo straight	140. *Erythroxylaceæ*, or *Erythroxyls*, p. 391

ALLIANCE 30. GUTTIFERALES, p. 392.

Leaves simple, alternate, with large convolute stipules. Flowers symmetrical. Petals equilateral. Calyx unequal, permanent, winged. Anthers beaked. Fruit one-celled, one-seeded	141. *Dipteraceæ*, or *Dipterads*, p. 393
Leaves simple, alternate, without stipules or with very small ones. Flowers symmetrical. Petals equilateral. Anthers versatile. Seeds few or single. Stigmas on a long style . .	142. *Ternströmiaceæ*, or *Theads*, p. 396
Leaves digitate, opposite. Flowers symmetrical. Petals equilateral. Stigmas sessile. Seeds solitary. Embryo with an enormous radicle	143. *Rhizobolaceæ*, or *Rhizobols*, p. 398
Leaves simple, opposite, without stipules. Flowers symmetrical. Petals equilateral. Anthers adnate, beakless. Seeds solitary or few. Stigmas sessile, radiating	144. *Clusiaceæ*, or *Guttifers*, p. 400
Leaves simple, alternate, without stipules. Flowers unsymmetrical. Petals equilateral. Anthers versatile. Seeds innumerable, minute. Stigmas sessile	145. *Marcgraviaceæ*, or *Marcgraviads*, p. 403
Petals oblique, glandular. Seeds numerous, naked. Styles long, distinct	146. *Hypericaceæ*, or *Tutsans*, p. 405
Petals oblique, glandless. Seeds few, shaggy. Styles long, distinct	147. *Reaumuriaceæ*, or *Reaumuriads*, p. 407

ALLIANCE 31. NYMPHALES, p. 408.

Carpels united into a many-celled fruit, with dissepimental placentæ	148. *Nymphæaceæ*, or *Waterlilies*, p. 409
Carpels distinct. Albumen copious. Torus absent	149. *Cabombaceæ*, or *Watershields*, p. 412
Carpels distinct. Albumen 0. Torus honey-combed, very large	150. *Nelumbiaceæ*, or *Waterbeans*, p. 414

ALLIANCE 32. RANALES, p. 416.

Carpels distinct. Stipules large, convolute. Corolla imbricated. Albumen homogeneous.	151. *Magnoliaceæ*, or *Magnoliads*, p. 417
Carpels distinct. Stipules 0. Corolla valvate. Albumen ruminate	152. *Anonaceæ*, or *Anonads*, p. 420
Carpels distinct. Stipules 0. Corolla imbricated. Albumen homogeneous. Seeds arillate	153. *Dilleniaceæ*, or *Dilleniads*, p. 423
Carpels distinct. Stipules 0. Corolla imbricated. Albumen homogeneous. Seeds without an aril	154. *Ranunculaceæ*, or *Crowfoots*, p. 425
Carpels consolidated. Calyx permanent. Placentæ axile	155. *Sarraceniaceæ*, or *Sarraceniads*, p. 429
Carpels consolidated. Calyx deciduous. Placentæ usually parietal	156. *Papaveraceæ*, or *Poppyworts*, p. 430

ALLIANCE 33. BERBERALES, p. 432.

Flowers regular and symmetrical. Placentæ parietal. Stamens alternate with the petals, or twice as many	157. *Droseraceæ*, or *Sundews*, p. 433
Flowers irregular and unsymmetrical. Placentæ parietal. Stamens opposite the petals	158. *Fumariaceæ*, or *Fumeworts*, p. 435
Flowers regular, symmetrical. Placentæ sutural. Stamens opposite the petals. Anthers with recurved valves	159. *Berberidaceæ*, or *Berberids*, p. 437
Flowers regular, symmetrical. Placentæ axile. Stamens opposite the petals. Anthers opening longitudinally	160. *Vitaceæ*, or *Vineworts*, p. 439
Flowers regular, symmetrical. Placentæ axile and parietal. Stamens alternate with the petals. Ovules ascending or horizontal. Corolla imbricated	161. *Pittosporaceæ*, or *Pittosporads*, p. 441
Flowers regular, symmetrical. Placentæ axile. Stamens alternate with the petals. Ovules pendulous. Corolla valvate	162. *Olacaceæ*, or *Olacads*, p. 443
Flowers regular, symmetrical. Placentæ axile. Stamens alternate with the petals if equal to them in number. Ovules pendulous. Corolla imbricated	163. *Cyrillaceæ*, or *Cyrillads*, p. 445

ALLIANCE 34. ERICALES, p. 446.

Flowers polypetalous. Stamens all perfect, monadelphous. Anthers 2-celled, with a long membranous connective	164. *Humiriaceæ*, or *Humiriads*, p. 447
Flowers monopetalous. Stamens all perfect, free. Seeds with a firm skin. Anthers 1-celled, opening longitudinally	165. *Epacridaceæ*, or *Epacrids*, p. 448
Flowers half-monopetalous. Stamens all perfect, free. Seeds with a loose skin. Embryo at the base of the albumen	166. *Pyrolaceæ*, or *Winter-greens*, p. 450
Flowers polypetalous. Stamens half-sterile and scale-like, free. Seeds with a firm skin	167. *Francoaceæ*, or *Francoads*, p. 451
Flowers half-monopetalous. Stamens all perfect, free. Seeds with a loose skin or wing. Embryo at the apex of the albumen . . .	168. *Monotropaceæ*, or *Fir-rapes*, p. 452
Flowers monopetalous. Stamens all perfect, free. Seeds with a firm or loose skin. Anthers 2-celled, opening by pores	169. *Ericaceæ*, or *Heathworts*, p. 453

ALLIANCE 35. RUTALES, p. 456.

Fruit consolidated, succulent, indehiscent. Petals imbricated. Stamens free, or nearly so. Leaves dotted	170. *Aurantiaceæ*, or *Citronworts*, p. 457
Fruit consolidated, hard, dry, somewhat valvular. Petals valvate. Stamens free. Leaves generally dotted	171. *Amyridaceæ*, or *Amyrids*, p. 459
Fruit consolidated, capsular. Stamens deeply monadelphous or free. Seeds numerous, winged	172. *Cedrelaceæ*, or *Cedrelads*, p. 461
Fruit consolidated, berried, or capsular. Stamens deeply monadelphous. Seeds few, wingless. Leaves dotless	173. *Meliaceæ*, or *Meliads*, p. 463
Fruit apocarpous. Ovule single, suspended by a cord rising from the base of the carpel .	174. *Anacardiaceæ*, or *Anacards*, or *Terebinths*, p. 465

Fruit apocarpous. Ovules collateral, ascending, orthotropal, sessile	175. *Connaraceæ*, or *Connarads*, p. 468
Fruit finally apocarpous, few-seeded, with the pericarp separating in two layers. Ovules sessile, pendulous. Flowers ☿	176. *Rutaceæ*, or *Rueworts*, p. 469
Fruit finally apocarpous, few-seeded, with the pericarp separating in two layers. Ovules sessile, pendulous. Flowers ♂-☿-♀ .	177. *Xanthoxylaceæ*, or *Xanthoxyls*, p. 472
Fruit finally apocarpous, one-seeded, with the pericarp not laminating, and a succulent conical torus	178. *Ochnaceæ*, or *Ochnads*, p. 474
Fruit finally apocarpous, one-seeded, with the pericarp not laminating, and a dry inconspicuous torus. Albumen wanting. Leaves alternate, without stipules	179. *Simarubaceæ*, or *Quassiads*, p. 476
Fruit finally apocarpous, few-seeded, with the pericarp not laminating, and a dry inconspicuous torus. Albumen present. Leaves opposite, with stipules	180. *Zygophyllaceæ*, or *Beancapers*, p. 478
Fruit finally apocarpous, many-seeded. Flowers polypetalous	181. *Elatinaceæ*, or *Water-peppers*, p. 480
Fruit finally apocarpous, many-seeded. Flowers apetalous, very imperfect	182. *Podostemaceæ*, or *Podostemads*, p. 482

ALLIANCE 36. GERANIALES, p. 484.

Flowers symmetrical. Styles distinct. Carpels longer than the torus. Seeds with little or no albumen	183. *Linaceæ*, or *Flaxworts*, p. 485
Flowers regular, unsymmetrical, with a permanent cup-like involucre. Stamens monadelphous. Albumen abundant	184. *Chlænaceæ*, or *Chlenads*, p. 486
Flowers symmetrical. Styles distinct. Carpels longer than the torus. Seeds with abundant albumen	185. *Oxalidaceæ*, or *Oxalids*, p. 488
Flowers very irregular and unsymmetrical, without an involucre. Stamens distinct. Albumen none	186. *Balsaminaceæ*, or *Balsams*, p. 490
Flowers usually symmetrical. Styles and carpels combined round a long beaked torus . .	187. *Geraniaceæ*, or *Cranesbills*, p. 493

ALLIANCE 37. SILENALES, p. 495.

Calyx and corolla usually both present and symmetrical (4 and 4, or 5 and 5), the latter conspicuous. Ovules amphitropal. Leaves opposite, without stipules	188. *Caryophyllaceæ*, or *Cloveworts*, p. 496
Calyx and corolla usually both present and symmetrical (4 and 4, or 5 and 5), the latter rudimentary. Ovules amphitropal. Leaves with scarious stipules	189. *Illecebraceæ*, or *Knotworts*, p. 499
Calyx and corolla both present and unsymmetrical (2 and 5), the latter usually conspicuous. Ovules amphitropal. Leaves alternate, succulent, without stipules . . .	190. *Portulaceæ*, or *Purslanes*, p. 500
Calyx only present, but often coloured. Ovules orthotropal. Nut usually triangular . . .	191. *Polygonaceæ*, or *Buckwheats*, p. 502

ALLIANCE 38. CHENOPODALES, p. 505.

Sepals united into a long (often coloured) plaited tube, which separates from its base, the latter becoming hard, and forming a spurious pericarp	192. *Nyctaginaceæ*, or *Nyctagos*, p. 506
Sepals separate, flat. Stamens alternate with the sepals or 00. Carpels several (or 1) . .	193. *Phytolaccaceæ*, or *Phytolaccads*, p. 509
Sepals separate or nearly so, flat. Stamens opposite the sepals. Anthers often 1-celled. Ovary 1, often several-seeded. (Flowers scarious, surrounded by imbricated bracts) .	194. *Amarantaceæ*, or *Amaranths*, p. 510
Sepals separate, or nearly so, flat. Stamens opposite the sepals. Anthers 2-celled. Ovary 1, always one-seeded. (Flowers herbaceous, naked)	195. *Chenopodiaceæ*, or *Chenopods*, p. 512

ALLIANCE 39. PIPERALES, p. 514.

Carpel solitary. Ovule erect. Embryo lying in vitellus. Leaves opposite or alternate, with or without stipules	196. *Piperaceæ*, or *Pepperworts*, p. 515
Carpel solitary. Ovule suspended. Embryo naked. Leaves opposite, with intermediate stipules	197. *Chloranthaceæ*, or *Chloranths*, p. 519
Carpels several, distinct. Ovule erect. Embryo lying in vitellus. Leaves alternate, with stipules	198. *Saururaceæ*, or *Saururads*, p. 521

ALLIANCE 40. FICOIDALES, p. 523.

Petals absent. Sepals distinct. Fruit inclosed in a membranous or succulent calyx. Carpel single, solitary. Seed erect	199. *Basellaceæ*, or *Basellads*, p. 524
Petals numerous, conspicuous. Carpels several, consolidated	200. *Mesembryaceæ*, or *Ficoids*, p. 525
Petals absent. Carpels several, consolidated	201. *Tetragoniaceæ*, or *Aizoons*, p. 527
Petals absent. Sepals united into a tube. Carpel single, solitary. Fruit inclosed in the hardened calyx tube	202. *Scleranthaceæ*, or *Scleranths*, p. 528

ALLIANCE 41. DAPHNALES, p. 529.

Anthers bursting lengthwise. Apetalous or polypetalous. Ovule solitary, suspended. Calyx imbricated	203. *Thymelaceæ*, or *Daphnads*, p. 530
Anthers bursting lengthwise. Apetalous. Ovules erect. Calyx valvate.	204. *Proteaceæ*, or *Proteads*, p. 532
Anthers bursting by recurved valves. Leaves perfect. Fruit naked	205. *Lauraceæ*, or *Laurels*, p. 535
Anthers bursting by recurved valves. Leaves mere colourless scales. Fruit buried in a succulent permanent calyx	206. *Cassythaceæ*, or *Dodder-laurels*, p. 538

ALLIANCE 42. ROSALES, p. 539.

Flowers consisting of numerous imbricated scales. Cotyledons convolute	207. *Calycanthaceæ*, or *Calycanths*, p. 540
Flowers polypetalous (or apetalous), nearly or quite regular. Carpel solitary. Style proceeding from the base of the ovary . .	208. *Chrysobalanaceæ*, or *Chrysobalans*, p. 542
Flowers polypetalous (or apetalous), papilionaceous or leguminous. Carpel solitary, with the style proceeding from the apex of the ovary	209. *Fabaceæ*, or *Leguminous plants*, p. 544
Flowers polypetalous, regular, drupaceous. Carpel solitary, with the style proceeding from the apex of the ovary	210. *Drupaceæ*, or *Almondworts*, p. 557
Flowers polypetalous, regular. Carpels adhering to the calyx by their back . . .	211. *Pomaceæ*, or *Appleworts*, p. 559
Flowers apetalous. Carpel solitary, inclosed in a hardened calyx-tube forming a false pericarp	212. *Sanguisorbaceæ*, or *Sanguisorbs*, p. 561
Flowers polypetalous. Carpels free from the calyx, and quite or nearly so from each other	213. *Rosaceæ*, or *Roseworts*, p. 563

ALLIANCE 43. SAXIFRAGALES, p. 566.

Styles distinct. Leaves alternate	214. *Saxifragaceæ*, or *Saxifrages*, p. 567
Styles distinct. Leaves opposite, without stipules	215. *Hydrangeaceæ*, or *Hydrangeads*, p. 569
Styles distinct. Leaves opposite, with large interpetiolar stipules	216. *Cunoniaceæ*, or *Cunoniads*, p. 571
Styles consolidated. Calyx many-leaved. Albumen 0. Leaves alternate	217. *Brexiaceæ*, or *Brexiads*, p. 573
Styles consolidated. Calyx tubular, permanent, with the petals in the margin. Albumen 0. Leaves opposite	218. *Lythraceæ*, or *Loosestrifes*, p. 574

ALLIANCE 44. RHAMNALES, p. 577.

Flowers apetalous. Ovary composed of 4 carpels. Calyx tubular, with definite divisions. Cotyledons consolidated . . .	219. *Penæaceæ*, or *Sarcocollads*, p. 577
Flowers apetalous. Ovary composed of 2 carpels. Calyx tubular, with a definite number of divisions. Cotyledons amygdaloid	220. *Aquilariaceæ*, or *Aquilariads*, p. 579
Flowers apetalous. Ovary composed of 2 carpels. Calyx imperfect, and irregularly divided at the edge. Cotyledons thin and leafy	221. *Ulmaceæ*, or *Elmworts*, p. 580
Flowers polypetalous. Calyx valvate. Stamens opposite petals. Seeds erect . . .	222. *Rhamnaceæ*, or *Rhamnads*, p. 581
Flowers polypetalous. Calyx valvate. Stamens alternate with petals. Seeds pendulous	223. *Chailletiaceæ*, or *Chailletiads*, p. 583
Flowers polypetalous. Calyx imbricated. Stamens (3) monadelphous	224. *Hippocrateaceæ*, or *Hippocrateads*, p. 584
Flowers polypetalous. Calyx imbricated. Stamens (5) distinct	225. *Celastraceæ*, or *Spindle-trees*, p. 586
Flowers monopetalous. Stamens episepalous	226. *Stackhousiaceæ*, or *Stackhousiads*, p, 589
Flowers monopetalous. Stamens epipetalous. Ovules ascending. Radicle short. Cotyledons amygdaloid	227. *Sapotaceæ*, or *Sapotads*, p. 590

Flowers monopetalous. Stamens epipetalous. Ovules, in part at least, suspended. Radicle long. Cotyledons leafy . . .	228. *Styracaceæ*, or *Storaxworts*, p. 592

ALLIANCE 45. GENTIANALES, p. 594.

Stipules 0. Stigmas simple, sessile, radiating	229. *Ebenaceæ*, or *Ebenads*, p. 595
Stipules 0. Stigmas simple, at the end of a manifest style. Placentæ axile. Seeds definite, pendulous. Corolla imbricated .	230. *Aquifoliaceæ*, or *Hollyworts*, p. 597
Stipules 0. Stigmas collected into a massive head, expanded at the base in the form of a ring or membrane, and contracted in the middle. (Albumen sometimes 0) . .	231. *Apocynaceæ*, or *Dogbanes*, p. 599
Leaves opposite, with intervening stipules .	232. *Loganiaceæ*, or *Loganiads*, p. 602
Stipules 0. Stigmas simple, at the end of a manifest style. Placentæ axile. Seeds indefinite, peltate. Stamens interpetalous	233. *Diapensiaceæ*, or *Diapensiads*, p. 606
Stipules 0. Stigmas simple, at the end of a manifest style. Placentæ axile. Seeds definite, erect. Corolla valvate. Flowers unsymmetrical	234 *Stilbaceæ*, or *Stilbids*, p. 607
Stipules 0. Stigmas simple, at the end of a manifest style. Placentæ parietal. Flowers didynamous	235. *Orobanchaceæ*, or *Broomrapes*, p. 609
Stipules 0. Stigmas simple, at the end of a manifest style. Placentæ parietal. Flowers regular	236. *Gentianaceæ*, or *Gentianworts*, p. 612

ALLIANCE 46. SOLANALES, p. 615.

Stamens free, 2 or 4	237. *Oleaceæ*, or *Oliveworts*, p. 616
Stamens free, 5. Placentæ axile. Embryo terete	238. *Solanaceæ*, or *Nightshades*, p. 618
Anthers and stigma consolidated into a column	239. *Asclepiadaceæ*, or *Asclepiads*, p. 623
Stamens free, 5. Placentæ axile. Cotyledons leafy, folded longitudinally	240. *Cordiaceæ*, or *Sebestens*, p. 628
Stamens free, 5. Placentæ basal. Cotyledons leafy, doubled up	241. *Convolvulaceæ*, or *Bindweeds*, p. 630
Stamens free, 5. Placentæ basal. Embryo filiform, spiral	242. *Cuscutaceæ*, or *Dodders*, p. 633
Stamens free, 5. Placentæ axile. Cotyledons straight, plano-convex	243. *Polemoniaceæ*, or *Phloxworts*, p. 635

ALLIANCE 47. CORTUSALES, p. 637.

Stamens alternate with the petals. Styles 2. Inflorescence circinate	244. *Hydrophyllaceæ*, or *Hydrophyls*, p. 638
Stamens opposite the petals. Fruit membranous, one-seeded. Styles 5. Stem herbaceous	245. *Plumbaginaceæ*, or *Leadworts*, p. 640
Stamens alternate with the petals. Style 1. Inflorescence straight	246. *Plantaginaceæ*, or *Ribworts*, p. 642
Stamens opposite the petals. Fruit capsular, many-seeded. Style 1. Stem herbaceous	247. *Primulaceæ*, or *Primworts*, p. 644
Stamens opposite the petals. Fruit indehiscent, drupaceous. Style 1. Stem woody	248. *Myrsinaceæ*, or *Ardisiads*, p. 647

ALLIANCE 48. ECHIALES, p. 649.

* Regular-flowered Orders, passing from Solanales.

Flowers regular, $\sqrt[2]{}$, unsymmetrical. Stamens 2. Fruit 2-lobed. Stigma naked	249. *Jasminaceæ*, or *Jasminworts*, p. 650
Flowers regular, symmetrical. Stamens 4. Fruit simple. Stigma naked	250. *Salvadoraceæ*, or *Salvadorads*, p. 652
Flowers regular, symmetrical. Stamens 5. Stigma naked. Nuts 4, confluent. Inflorescence circinate	251. *Ehretiaceæ*, or *Ehretiads*, p. 653
Flowers regular, symmetrical. Stamens 5. Nuts 5 or $\sqrt[5]{}$. Stigma naked. Inflorescence straight	252. *Nolanaceæ*, or *Nolanads*, p. 654
Flowers regular, symmetrical. Stamens 5. Nuts 4 or $\sqrt[2]{}$. Stigma naked. Inflorescence circinate	253. *Boraginaceæ*, or *Borageworts*, p. 655
Flowers regular, symmetrical. Nut solitary. Stigma indusiate. (Stamens hypogynous !)	254. *Brunoniaceæ*, or *Brunoniads*, p. 657

** Irregular-flowered Orders, passing into Bignonials.

Flowers irregular, unsymmetrical. Nuts 4. Ovule erect	255. *Lamiaceæ*, or *Labiates*, p. 659
Flowers irregular, unsymmetrical. Nuts confluent. Ovules erect	256. *Verbenaceæ*, or *Verbenes*, p. 663
Flowers irregular, unsymmetrical. Nuts confluent. Ovules pendulous. Anthers 2-celled	257. *Myoporaceæ*, or *Myoporads*, p 665

Flowers irregular, unsymmetrical. Nuts confluent. Ovules pendulous. Anthers 1-celled	258. *Selaginaceæ*, or *Selagids*, p. 666

ALLIANCE 49. BIGNONALES, p. 668.

Placentæ parietal. Fruit bony or capsular. Embryo amygdaloid. Radicle short . .	259. *Pedaliaceæ*, or *Pedaliads*, p. 669
Placentæ parietal. Fruit capsular or baccate. Embryo with minute cotyledons. Radicle long	260. *Gesneraceæ*, or *Gesnerworts*, p. 671
Placentæ parietal. Fruit succulent, hard-shelled. Embryo amygdaloid. Radicle short.	261. *Crescentiaceæ*, or *Crescentiads*, p. 673
Placentæ axile. Seeds winged, sessile, without albumen. Cotyledons large, leafy	262. *Bignoniaceæ*, or *Bignoniads*, p. 675
Placentæ axile. Seeds wingless, attached to hard placental processes, without albumen. Cotyledons large, fleshy	263. *Acanthaceæ*, or *Acanthads*, p. 678
Placentæ axile. Seeds albuminous. Cotyledons scarcely larger than, or not so large as, the radicle	264. *Scrophulariaceæ*, or *Figworts*, p. 681
Placenta free, central. Seeds minute, without albumen. Cotyledons much smaller than the radicle	265. *Lentibulariaceæ*, or *Butterworts*, p. 686

ALLIANCE 50. CAMPANALES, p. 688.

Ovary 2- or more-celled. Anthers free, or half united. Stigma naked. Corolla valvate. regular	266. *Campanulaceæ*, or *Bellworts*, p. 689
Ovary 2- or more-celled. Anthers syngenesious. Stigma surrounded by hairs. Corolla valvate, irregular.	267. *Lobeliaceæ*, or *Lobeliads*, p. 692
Ovary 2- or more-celled. Anthers syngenesious or free. Stigma indusiate. Corolla induplicate	268. *Goodeniaceæ*, or *Goodeniads*, p. 694
Ovary 2- or more-celled. Stamens and style united into a column. Corolla imbricated .	269. *Stylidiaceæ*, or *Styleworts*, p. 696
Ovary 1-celled. Corolla imbricated. Anthers free. Ovule pendulous. Albumen none .	270. *Valerianaceæ*, or *Valerianworts*, p. 697
Ovary 1-celled. Corolla imbricated. Anthers free. Ovule pendulous. Seeds albuminous.	271. *Dipsacaceæ*, or *Teazelworts*, p. 699
Ovary 1-celled. Corolla valvate. Anthers syngenesious. Ovule pendulous. Seeds albuminous	272. *Calyceraceæ*, or *Calycers*, p. 701
Ovary 1-celled. Corolla valvate. Anthers syngenesious. Ovule erect. Albumen none .	273. *Asteraceæ*, or *Composites*, p. 702

ALLIANCE 51. MYRTALES, p. 716.

Ovary 1-celled. Ovules pendulous. Leaves dotless. Seeds without albumen. Cotyledons convolute	274. *Combretaceæ*, or *Myrobalans*, p. 717
Ovary 1-celled. Ovules pendulous. Leaves dotless. Seeds albuminous. Cotyledons flat.	275. *Alangiaceæ*, or *Alangiads*, p. 719
Ovary 1-celled. Ovules ascending. Leaves dotted. Embryo fused into a solid mass .	276. *Chamælauciaceæ*, or *Fringe Myrtles*, p. 721
Ovary with more than one cell. Flowers polypetalous or apetalous. Calyx open, minute. Stamens definite. Ovules pendulous. Cotyledons minute. (Occasionally one-celled) .	277. *Haloragaceæ*, or *Hippurids*, p. 722
Ovary with more than one cell. Flowers polypetalous or apetalous. Calyx valvate. Stamens definite. Ovules horizontal or ascending. Cotyledons flat, much larger than the radicle	278. *Onagraceæ*, or *Onagrads*, p. 724
Ovary with more than one cell. Flowers polypetalous. Calyx valvate. Stamens indefinite. Cotyledons flat, much shorter than the radicle, which germinates before the fruit falls .	279. *Rhizophoraceæ*, or *Mangroves*, p. 726
Ovary with more than one cell. Flowers monopetalous, coronetted. Calyx valvate. Stamens indefinite, monadelphous. Cotyledons amygdaloid	280. *Belvisiaceæ*, or *Napoleonworts*, p. 728
Ovary with more than one cell. Flowers polypetalous. Calyx imbricated. Stamens definite. Anthers rostrate. Leaves usually dotless	281. *Melastomaceæ*, or *Melastomads*, p. 731
Ovary with more than one cell. Flowers polypetalous or apetalous (or valvate). Calyx imbricated. Stamens 00. Anthers oblong. Leaves usually dotted	282. *Myrtaceæ*, or *Myrtleblooms*, p. 734
Ovary with more than one cell Flowers polypetalous. Calyx valvate or imbricated. Stamens 00, in part collected into a fleshy hood. Anthers oblong. Leaves dotless . .	283. *Lecythidaceæ*, or *Lecyths*, p. 739

ALLIANCE 52. CACTALES, p. 741.

Sepals and petals distinct. Stamens opposite the petals. Styles separate. Ovules pendulous	284. *Homaliaceæ*, or *Homaliads*, p. 742
Sepals and petals distinct. Stamens scattered. Styles confluent. Ovules pendulous. Seeds albuminous	285. *Loasaceæ*, or *Loasads*, p. 744
Sepals and petals numerous, undistinguishable. Stamens scattered. Styles confluent. Ovules horizontal. Seeds without albumen.	286. *Cactaceæ*, or *Indian Figs*, p. 746

ALLIANCE 53. GROSSALES, p. 749.

Fruit pulpy. Placentæ parietal	287. *Grossulariaceæ*, or *Currantworts*, p. 750
Fruit capsular. Placentæ axile. Style and stamens definite. Calyx imbricated . . .	288. *Escalloniaceæ*, or *Escalloniads*, p. 752
Fruit capsular. Placentæ axile. Styles disunited. Stamens 00. Calyx valvate . . .	289. *Philadelphaceæ*, or *Syringas*, p. 743
Fruit pulpy or fibrous. Placentæ axile. Style 1. Stamens 00. Calyx imbricated . . .	290. *Barringtoniaceæ*, or *Barringtoniads*, p. 754

ALLIANCE 54. CINCHONALES, p. 756.

Stamens epigynous; anthers opening by pores.	291. *Vacciniaceæ*, or *Cranberries*, p. 757
Stamens epipetalous, bursting longitudinally; anthers sinuous. Flowers unsymmetrical .	292. *Columelliaceæ*, or *Columelliads*, p. 759
Stamens epipetalous, bursting longitudinally; anthers straight. Leaves with interpetiolar stipules	293. *Cinchonaceæ*, or *Cinchonads*, p. 761
Stamens epipetalous, bursting longitudinally; anthers straight. Fruit consolidated. Leaves without stipules	294. *Caprifoliaceæ*, or *Caprifoils*, p. 766
Stamens epipetalous, bursting longitudinally; anthers straight. Fruit didymous. Leaves verticillate, without stipules	295. *Galiaceæ*, or *Stellates*, p. 768

ALLIANCE 55. UMBELLALES, p. 772.

Fruit didymous, with a double epigynous disk.	296. *Apiaceæ*, or *Umbellifers*, p. 773
Fruit not didymous, without a double epigynous disk, 3- or more-celled. Pentamerous flowers. Corolla valvate. Leaves alternate, without stipules. Anthers turned inwards, opening lengthwise	297. *Araliaceæ*, or *Ivyworts*, p. 780
Fruit not didymous, without a double epigynous disk, 2- or more celled. Tetramerous flowers. Corolla valvate. Leaves opposite, without stipules	298. *Cornaceæ*, or *Cornels*, p. 782
Fruit not didymous, without a double epigynous disk, 2-celled. Corolla imbricated. Leaves alternate, with stipules. Anthers with deciduous valves	299. *Hamamelidaceæ*, or *Witch-Hazels*, p. 784
Fruit not didymous, without a double epigynous disk, 3- (or 1-) celled. Corolla imbricated. Leaves alternate, without stipules. Anthers turned outwards, opening lengthwise	300. *Bruniaceæ*, or *Bruniads*, p. 785

ALLIANCE 56. ASARALES, p. 786.

Ovary 1-celled. Ovules definite, with a coated nucleus	301. *Santalaceæ*, or *Sandalworts*, p. 787
Ovary 1-celled. Ovules definite, with a naked nucleus	302. *Loranthaceæ*, or *Loranths*, p. 789
Ovary 3-6-celled. Ovules 00	303. *Aristolochiaceæ*, or *Birthworts*, p. 792

THE

VEGETABLE KINGDOM.

WHEN the Animal Kingdom is studied as a vast whole, and not merely in the highly-developed classes of Mammals, Birds, and Reptiles, the naturalist perceives forms with which he is most familiar gradually changing, organs which are indispensable to the highest orders of Animals disappearing, the limbs ceasing to be formed, all the internal structure of the body simplified, and, at last, nothing left but pulpy and seemingly shapeless masses, such as inhabit shells. Let his power of vision be enlarged, and the microscope discovers to his amazement, that the Animal Kingdom has not ceased with the soft-bodied creatures at which his inquiry had stopped, but that a new and vast field of observation opens before him, teeming with myriads of forms, which are, as it were, the beginning of another kingdom of nature. Nevertheless, he soon finds that the smallness of the size of these creatures is no hindrance to their possessing the peculiar attributes of animal life. Though bones, and muscles, and external limbs, with veins, arteries, and nerves, may have disappeared, or become too fine for human vision, yet there is still left the animal motion, and the power of hunting for prey, of feeding by a mouth and by the destruction of other species, which is one of the great marks of animal structure. He sees that cells, although so small that the acutest vision and the most powerful instruments are alone sufficient to detect them, are the recipients of a stomach, of eyes, of a mouth. He perceives in such bodies all those elements of activity, by which the Animal Kingdom is in general so well distinguished from the passive Region of plants.

And hence it is that those who deal in generals only, without descending to particulars, pronounce with a voice of authority that the Animal and Vegetable Kingdoms are sundered by decisive characteristics. The zoologist declares that the power of spontaneous motion, and the feeding by a stomach, are qualities confined to the Animal Kingdom. But numerous plants move with all the appearance of spontaneity; the spores of those Confervæ which are sometimes called Zoosporous, swim in water with great activity; the filaments of Zygnemata combine with the energy of animal life; and as for a stomach, it is impossible to say, that the whole interior of a living independent cell is not a stomach. Chemists once referred to the presence of nitrogen as a certain characteristic of animals; but plants abound in nitrogen. With more reason they now appeal to the existence of starch in plants, an organic compound unknown among the animal creation. And this is perhaps the best mark of distinction that has hitherto been found; for it is universally present in plants, and has enabled Mr. Payen to confirm by chemical evidence the vegetable nature of certain productions till lately regarded as Zoophytes, and therefore as belonging to the Animal Kingdom. (*Ann. Sc. Nat.* 2. *ser.* xx. 65.)

But it has been long ago asserted by Bory de St. Vincent, and others, that there exist in nature organised bodies which are animal at one period of their lives, and vegetable at another! This, if true, would for ever put an end to the possibility of distinguishing the two kingdoms when they shall each have arrived at their lowest forms. Its truth has, however, been denied. On the contrary, Kützing, in his recent magnificent work on Algæ, insists that it happens in his Ulothrix zonata, (Fig. I.) He asserts that in the cells of that plant there are found minute animalcules, with a red eye-point, and a transparent mouth-place; that they are not in fact, distinguishable from Ehrenberg's Microglena monadina; these bodies, however, are animals only for a time. At last they grow into vegetable threads, the lowest joint of which still exhibits the red eye-point. This phenomenon, which Kützing assures us he has ascertained beyond all possibility of doubt, puts an end to the question of, whether animals and plants can be distinguished at the limits of their two kingdoms, and sufficiently accounts for the conflicting opinions that naturalists entertain as to the nature of many of the simpler forms of organisation.

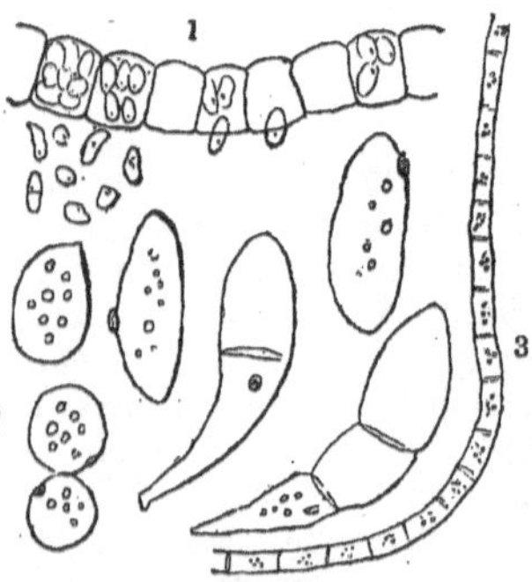

Fig. I.

Such being the case, it is not worth attempting to decide, whether the lowest forms of structure, to be presently mentioned, belong to the one Kingdom or the other. It will be sufficient that they have been regarded as plants by many eminent naturalists.

It is in this microscopical cellular state of existence that the Animal Kingdom ends, and the Vegetable commences. It is from this point that the naturalist who would learn how to classify the Kingdom of Plants must take his departure. He perceives that those species which consist of cells, either independent of each other (*Protococcus*, *Uredo*), or united into simple threads (*Conferva*, *Monilia*), are succeeded by others in which the threads collect into nets (*Hydrodictyon*), or plates (*Ulva*), or the cells into masses (*Laminaria*, *Agaricus*); peculiar organs make their appearance, and at last, as the complication of structure increases, a leaf and stem unfold as distinctly limited organic parts.

Those simpler plants which exist without the distinction of leaf and stem, are also destitute of flowers; they are equally without the breathing-pores so abundantly formed in the skin of more complex species, and they multiply by the spontaneous formation in their interior, or upon their surface, of reproductive spheroids called spores. Among the many names that Botanists have given such plants, that of THALLOGENS is here preferred. A thallus is a fusion of root, stem and leaves, into one general mass; and that is much the nature of these elements of Vegetable structure.

Fig. I. ULOTHRIX ZONATA, after Kützing.—1. A portion of the plant discharging its vegeto-animalcules; 2. the latter much enlarged, and in various states of progress into a thread; 3. a young thread, or plant, three or four days old, much less magnified.

Beyond Thallogens are found multitudes of species, which like the former are not furnished by nature with flowers, but which otherwise approach closely to the higher forms of structure, occasionally acquiring the stature of lofty trees. They have breathing-pores in their skin; their leaves and stem are distinctly separated; in some of them, those spiral threads which form so striking a portion of the internal anatomy of a more perfect species, exist in considerable abundance; and finally, they multiply by reproductive spheroids, or spores, either formed without the agency of sexes, or, if the contrary shall be proved, at all events not possessing bodies constructed like stamens on the one hand and embryos on the other. Their stem, however, does not increase in diameter; it only grows at the end, and hence it has given to such plants the name of ACROGENS.*

The changes which thus occur in the races of Thallogens and Acrogens, represent the progress of development in the remainder of the Vegetable Kingdom. A sphere, called a pollen grain, protrudes a tube into a soft pulpy receptacle in the interior of an ovule; there the new plant takes its birth, at first in the form of a cell, which by degrees forms a thread (the suspensor), then generates a cellular mass (the young embryo), and eventually becomes a mass of cells arranged in the form of stem and leaves (the perfect embryo, with its cotyledons, radicle, and plumula). But this is not the end of growth; it is rather the beginning. A loftier destiny awaits such plants; flowers are to be formed, seeds to be fertilised, and this is to be effected by a complex apparatus unknown in Acrogens or Thallogens.

Foremost among the more perfect races comes a most anomalous collection of species, called RHIZOGENS, or Rhizanths. These plants, leafless and parasitical, have the loose cellular organisation of Fungi; a spiral structure is usually to be found among their tissue only in traces. Some of them spring visibly from a shapeless cellular mass which stands in place of stem and root, and seems to be altogether analogous to the thallus of Fungi; and it is probable, that they all partake in this singular mode of growth. Their flowers are like those of more perfect plants; their sexual apparatus is complete; but their embryo, which is not furnished with any visible radicle or cotyledons, appears to be a spherical or oblong homogeneous mass. Rhizogens seem, in fact, of an intermediate nature between Fungal Thallogens and Endogens.

The remainder of the Vegetable Kingdom consists of plants having flowers, and propagated by seeds; that is to say, by bodies procreated by the mutual action of two manifest and undoubted sexes. Such plants are therefore called Phænogamous or SEXUAL.

Sexual plants are themselves divisible into two unequal masses. Of these masses one consists of species whose germination is endorhizal, whose embryo has but one cotyledon, whose leaves have parallel veins, and whose trunk is formed of bundles of spiral and dotted vessels guarded by woody tubes; which bundles are arranged in a confused manner, and are reproduced in the centre of the trunk. These are ENDOGENS.

The other mass is composed of innumerable races having an exorhizal germination, an embryo with two or more cotyledons, leaves having a network of veins, and a trunk consisting of woody bundles composed of dotted and woody tubes, or of woody tubes alone, arranged around a central pith, and either in concentric rings, or in a homogeneous mass, but always having medullary plates, forming rays from the centre to the circumference, and

* Thallogens and Acrogens together constitute the ACOTYLEDONES of *Jussieu*, the EXEMBRYONATÆ or ARHIZÆ of *Richard*, the AGAMÆ, CRYPTOGAMÆ, or ÆTHEOGAMÆ of *others*, the NEMEA of *Fries*.

reproduced in the circumference of the trunk, whence their name of EXOGENS.

Among Exogens there are, however, two totally different modes in which the influence of the pollen is communicated to the seed. The larger part of this great class consists of plants provided with the apparatus called style and stigma, through which pollen-tubes are introduced into the ovary during the act of fertilisation. But others are so constructed that the pollen falls immediately upon the ovules, without the introduction of any intermediate apparatus; a peculiarity analogous to what occurs among reptiles in the Animal Kingdom: and, as was to have been anticipated, the plants in which this singular habit occurs prove, upon being collected together, to form a group having no direct affinity with those among which they had been previously associated. Hence Exogens have been broken up into 1. *Exogens* proper, or those having an ovary, style, and stigma; and 2. GYMNOGENS, which have neither.

Among Endogens no difference has been remarked in the mode of propagation, but a material peculiarity has been noticed in the manner of growth. In the great mass of the class the stem and root are formed in a similar way, or there is no considerable difference between them, and the leaves have no articulation with the stem; but in a part of them the root is exactly like that of an Exogen without concentric circles, and the leaves fall off the stem by a clean fracture, just as in that class. Such fundamental distinctions have given rise to the separation by me of Endogens into 1. Endogens proper, and 2. DICTYOGENS.

This gives us for the whole Vegetable Kingdom the following

CLASSES.

Asexual, or Flowerless Plants.

Stems and leaves undistinguishable . . . . **I. THALLOGENS.**
Stems and leaves distinguishable . . . **II. ACROGENS.**

Sexual, or Flowering Plants.

Fructification springing from a thallus . . **III. RHIZOGENS.**
Fructification springing from a stem.

Wood of stem arranged in a confused manner, youngest in the centre: cotyledon single.

Leaves parallel-veined, permanent; Root much like the stem internally . . . **IV. ENDOGENS.**

Leaves net-veined, deciduous; Root with the wood in a solid concentric circle . . **V. DICTYOGENS.**

Wood of stem arranged in a concentric or uniform manner, youngest at the circumference; cotyledons 2 or more.

Seeds quite naked . . . . . **VI. GYMNOGENS.**
Seeds enclosed in seed-vessels . . **VII. EXOGENS.**

CLASS I. THALLOGENS.

Anandræ, *Link. in Berl. Mag. III.* Cellulares, *DC. Fl. Fr.* 1. 68. (1815). Acotyledoneæ, *Agardh. Aph.* 72. Homonemeæ, *Fries Syst.* 1825. Aphyllæ, *Ed. prim.* Cryptophyta, *Link. Handb.* 163. Thallophyta, *Endl. Gen.* p. 1. Amphigenæ, *Ad. Brong. Enumeration, p.* xi. (1843).

The whole of the plants stationed in this class are remarkable for the extreme simplicity of their structure. They have no wood, properly so called, although in the case of some sea-weeds and Fungi they must acquire considerable age. Those spirally-coated tubes which the old anatomists called tracheæ, because of their respiratory office, are unknown among them, unless occasionally in the form of local cells connected with the reproductive organs only; and consequently upon the surface of even the most perfect of them there is no sign of the organic apertures in the skin called stomates or breathing-pores. They are mere masses of cells. On their surface nothing is discoverable which can be regarded as analogous to leaves; for even in such sea-weeds as Hypnea, which resemble mosses in appearance, and in some of the Lichens which seem leafy, the exact symmetry which, without exception, characterises true foliage is wanting. In Chara alone, which is wholly leafless, do we find a symmetrical arrangement even of the divisions of the axis. Their mode of reproduction is not by pollen and ovules, or by sexual apparatus, as it is usual to call those parts, of which there is no sign, but by a special disintegration and solidification of some part of their tissue, spontaneously effected in various ways according to their kinds. It is true that such names as Antheridia and Pistillidia are met with in the writings of Cryptogamic Botanists, from which it might be inferred that something analogous at least to sexes was observable among such plants; but these are theoretical expressions, and unconnected with any proof of the parts to which they are applied performing the office of anthers and pistils. If it should be assumed, as it has been by some, that they do represent sexual organs, it is to be remembered that it is a mere assumption unsupported by sufficient evidence. Even in Charas, in whose globule some writers have seen a true anther, so little reason is there to suppose that it deserves such a name, that, on the contrary, an observer, worthy of credit, assures us that he has seen it grow. So entirely, in the simplest forms of Thallogens, is all trace of sexes missing, that in some of them their reproductive matter has been regarded by certain writers as altogether of an ambiguous nature. In their opinion, it is even uncertain whether this matter will reproduce its like, and whether it is not a mere representation of the vital principle of vegetation, capable of being called into action either as a Fungus, an Alga, or a Lichen, according to the particular conditions of heat, light, moisture, and medium, in which it is placed; producing Fungi upon dead or putrid organic beings; Lichens upon living vegetables, earth, or stones; and Algæ where water is the medium in which it is developed. Kützing, (*Ann. des Sc. n. s. vol.* ii. *p.* 225), endeavours to maintain the following propositions connected with this subject: "1st, the formation of organic matter can only take place by means of the previously dissolved elements of other organic principles; 2nd, simple globules, such as Cryptococcus, Palmella, and Protococcus, can give birth to different formations according to the influence of light, air, and temperature; 3rd, we must regard all the forms of lower Algæ as vegetations of a very simple structure, and distinguish them from each other, notwithstanding that in certain circumstances they may raise themselves to vegetations of a higher form; for in other circumstances they can exist and multiply independently; 4th, the same superior formation may be produced by primitive formations of altogether different kinds."

It is not easy to settle the limits of the alliances of Thallogens. Linnæus and Jussieu had but two divisions, viz., Algæ (including Lichens) and Fungi; and they have been followed by some modern botanists, particularly Fries and Wahlenberg. Others have been satisfied with separating the Lichens from Algæ, which, indeed, was virtually done by most of those who acknowledged but two divisions; and with admitting three equally distinct groups. Some, on the contrary, have sought to multiply the orders, as De Candolle and others, by introducing a tribe called Hypoxyla; Greville by adopting the latter, Gastromyci, Byssoideæ, and Epiphytæ, and proposing a new group under the name of Chætophoroideæ; and finally, Adolphe Brongniart, who carries the number of groups in this division of Acotyledones as far as 12, viz. Lichens, Hypoxyla, Fungi, Lycoperdaceæ, Mucedineæ, Uredineæ, Fucaceæ, Ulvaceæ, Ceramiaceæ, Confervæ, Chaodineæ, and Arthrodieæ; part of which have originated with himself, and others with Bory de St. Vincent. It is clear, however, that these groups are of very unequal degrees of importance, and that after all they must be reduced under the three great forms whose existence is universally recognised.

In what way those forms can be best defined is a very difficult question. It has been said that Algæ are aquatics, while Lichens and Fungi are terrestrial; but Fungi will develop in water, when they assume the form of Algæ. Lichens have been characterised by their shields, or reproductive disks containing spores lying in the fusiform spore-cases called asci; but a whole division of Lichens consist of genera without such asci. Then as to Fungi, they have been characterised by the want of a thallus, which is essential to Lichens; but the mycelium or spawn of Fungi is really a thallus; and it is impossible to distinguish by that character the genus Verrucaria of Lichens from Sphæria of Fungi. According to two of the most skilful of our modern systematists, the following are the distinctions of the three great groups:—

AGARDH (1821).

1. ALGÆ. Aquatic plants, filamentous, lamelliform, or leafy, intensely and brightly coloured, including spores, which are either contained in pericarps or scattered over the surface.
2. FUNGI. Fugacious, pulverulent, flocculent, crustaceous or fleshy plants, arising out of the destruction of organic matter (or capable of doing so), whitish, or coloured, not green, with their spores immersed.
3. LICHENS. Perennial plants, crustaceous, laminated or filiform, not of a leaf-green, including spores plunged in a thallus as well as in shields.

ADOLPHE BRONGNIART (1843).

1. ALGÆ. Frond cellular, living in fresh or salt water (rarely in very moist air), fixed by suckers or little roots.
2. FUNGI. Thallus filamentous (or Mycelium), developed on land or in dead or living organi-bodies, producing reproductive organs externally.
3. LICHENS. Frond of various forms, living in air, fixed by cellular fibrils, without a thallus developed in subjacent bodies. Fructification, occupying limited spaces on the surface of the frond, formed of thecæ mixed with paraphyses.

Neither of these definitions is however satisfactory; they hold indeed in many cases; but many Fungi have not a filamentous thallus; again some Lichens (especially if Collema be included) have a filamentous thallus, and some species are all but aquatic, *e.g.* Verrucaria submersa. In Algals again, in the terrestrial Vaucheriæ, the terrestrial Sphærozyga, &c., the fruit is developed in free air; so also in Botrydium, Trentepohlia, and some others.

Mr. Berkeley finds that "the main distinction between Fungi and Algals (including Lichens) consists in the fact that Fungi are universally nourished by the matrix by means of their mycelium, while Lichens and Algals are nourished at the expense of the medium in which they vegetate. In a few cortical species of Lichens, indeed, there is a very intimate connection between the bark and stroma, but then in these cases there are the green gonidia which do not exist in Fungi. It is true that moulds will vegetate in fluids; but as soon as they assume their normal form, there is a distinction between the immerged and free portion."

Following these views, I venture to propose the following as the characteristic mark of the

Alliances of Thallogens.

Algales.—*Cellular flowerless plants, nourished through their whole surface by the medium in which they vegetate; living in water or very damp places; propagated by zoospores, coloured spores, or tetraspores.*

Fungales.—*Cellular flowerless plants, nourished through their thallus (spawn or mycelium); living in air; propagated by spores colourless or brown, and sometimes inclosed in asci; destitute of green gonidia.*

Lichenales.—*Cellular flowerless plants, nourished through their whole surface by the medium in which they vegetate; living in air; propagated by spores usually inclosed in asci, and always having green gonidia in their thallus.*

ALLIANCE I. *ALGALES.*—THE ALGAL ALLIANCE.

Algæ, *Juss. Gen.* 5. (1788) ; *DC. Fl. Fr.* 2. 2. (1815) ; *Agardh Synops. Alg.* (1817) ; *Species Alg.* (1821-1828) ; *Syst. Alg.* (1824) ; *Greville Alg. Brit.* (1830) ; *Hooker, Brit. Fl. vol.* 2. *pt.* 1. (1833) ; *Agardh JG. Algæ Maris Mediterranei ; Decaisne in Ann. Sc. Nat.* 2 *ser. vols.* 17 *&* 18, *passim ; Kützing, Phycologia Generalis. Endlicher, Gen. Suppl.* 3.—Phycei, *Acharius* (1807 ?).—Thalassiophyta, *Lamouroux Ann. Mus.* 20. (1812) ; *Gaillon in Dict. des Sc.* 53. 350. (1828).—Hydrophyta, *Lyngb. Tentam.* (1819).—Arthrodieæ, *Bory in Dict. Class.* 1. 591. (1822).—Hydronemateæ, *Nees in Nov. Act. Nat. Cur.* 11. 509. (1823) ; *Ann. des Sc.* 13. 439. (1828).—Chaodineæ, Confervæ *and* Ceramiarieæ, *Bory in Dict. Class.* 3. *and* 4. (1823).—Chætophoroideæ, *Greville Fl. Edin.* 321. (1824).—Hydrophycæ, *Fries Syst. Orb. Veg.* 320. (1825).—Nemazoaires, *Gaillon in Ann. Sc. Ser.* 2. 1. 44. (1834) ; Phycées, *Mont. Dict. Univ. d'Hist. N. sub. Algis* (1843).

DIAGNOSIS.—*Cellular flowerless plants, nourished through their whole surface by the medium in which they vegetate ; living in water or very damp places ; propagated by zoospores, coloured spores, or tetraspores.*

It is here that the transition from animals to plants, whatever its true nature may be, occurs ; for it is incontestable, as the varying statements of original observers testify, that no man can certainly say whether many of the organic bodies placed here belong to the one kingdom of nature or the other. Whatever errors of observation may have occurred, those very errors, to say nothing of the true ones, show the extreme difficulty, not to say impossibility, of pointing out the exact frontier of either kingdom. If those ambiguous marine productions, which Pallas considered to be plants, but which Lamarck and much later writers have mostly placed among Zoophytes, have been shown by Kützing and Decaisne to be merely sea-vegetables coated with calcareous matter, we have in that fact another testimony to the near approach of the two realms being through the Algal alliance. Indeed, if any faith is to be placed in the observations of Kützing and Hornschuch, the one is capable of giving birth to the other. The former of these writers mentions (*Ann. Sc. Nat.* 2. *ser.* 5. 376) a very extraordinary fact, if it be one. He cut to pieces the marine animal called Medusa aurita, washed the pieces carefully in distilled water, put them into a bottle of distilled water, corked it close, and placed it in a window facing the east. The bits of Medusa soon decomposed, and emitted a very offensive odour, during which time no trace of Infusoria was discoverable. After a few days the putrid smell disappeared, and myriads of Monads came forth. Shortly after the surface of the liquid swarmed with extremely small green points, which eventually covered the whole surface ; similar points attached themselves to the sides of the bottle ; seen under a microscope they appeared to be formed of numberless Monads, united by a slimy mass ; and at last, after some weeks, the Conferva fugacissima of Lyngbye developed itself in perfection.

Reissek, of Vienna, goes still further. He professes to have observed the green colouring matter of ordinary flowering plants metamorphosed into confervæ ; such forms were even witnessed by him proceeding from the pollen cells of plants (*Bot. Zeit.* 1844. *July* 19). Kützing also believes that the lower forms of Algals are capable of being changed into more highly organised species, or even into species belonging to different families of the higher cellular plants. With regard to these astounding statements I cannot do better than avail myself of the excellent remarks of the Rev. M. J. Berkeley, than whom no one has a more intimate knowledge of the subject in question. In Taylor's Annals of Natural History, vol. xiv. p. 434, he observes, "that such observations cannot be considered conclusive, apart from all prejudice either way, till a certain number of bodies ascertained to be precisely of the same nature be isolated, and the changes of these observed with every possible precaution to avoid error. At present it seems that there is not by any means sufficient proof that the objects in question really arise from germs of the same nature. The second remark we would make is, that there appears too often in treatises of this description to be great indistinctness as to the notion of what a species really is. We know that in the course of development higher bodies go through a vast variety of phases which resemble very closely true substantial species which have arrived at their full development ; but we are not therefore to suppose, that in passing through their phases the production has really consisted of such a number of real species. In the sense of Agardh this may be true enough ; for when he pronounces the vessels and cells of phænogamous plants to be Algæ, his meaning appears to be, however strongly he expresses himself, merely that they are representatives of Algæ, and resemble them in structure.

"We would remark, also, that the real difficulty of the case does not depend on the question as to the difference of animal and vegetable life. These evidently in certain parts of the creation are so intimately combined, that it is quite impossible to say where the one ceases and the other begins ; and there is really no reason why we should be incredulous as to the possibility of the same object being at one time endowed more especially with animal, and at another with vegetable life. Late observations on the reproductive bodies of some Algæ show that their motion is produced by vibratile cilia, exactly in the same way as in certain animals. But it is exceedingly difficult to imagine the transformation of one real species into another. The same species may assume a vast variety of forms according to varying circumstances, and it is highly instructive to observe these changes; but that the same spore should under different circumstances be capable of producing beings of an almost entirely different nature, each capable of reproducing its species, is a matter which ought not to be admitted generally without the strictest proof."

For what wise purpose the Creator has filled the sea and the rivers with countless myriads of such plants, so that the Flora of the deep waters is as extensive as that of dry land, we can only conjecture; the uses to which they are applied by man are, doubtless, of but secondary consideration; and yet they are of no little importance in the manufactures and domestic economy of the human race. One of the most curious facts connected with them is their property of growing occasionally upon living animals, which they destroy; this is the case with Achlya prolifera, to be hereafter noticed.

Their history and classification have occupied the attention of some of the most acute botanists of the present day. Bishop Agardh and his son, Greville, Harvey, Decaisne, and Kützing, deserve to be especially named as most excellent and skilful investigators of a very obscure and difficult subject. It is those only who have made the subject their peculiar study who can determine which of the classifications proposed by these authors has the strongest claim on attention. I, at least, am unable to decide; and therefore I have preferred to employ the arrangement made use of by Endlicher in his last Supplement, as that which is most likely to be permanently employed for some years to come. Those who wish to acquaint themselves with the views of the great Algologists of the day should consult the younger Agardh's *Algæ Maris Mediterranei, &c.* (1842); Greville's *Algæ Britannicæ* (1830); Harvey's *Manual of British Algæ* (1841); Decaisne's papers in the *Annales des Sciences Naturelles*, 2 Series, vol. xvii. (1842); Kützing's *Phycologia generalis, oder Anatomie, Physiologie und Systemkunde der Tange* (1843), a most elaborate work, illustrated with eighty exquisite plates; the *Kieselchaligen Bacillarien oder Diatomeen* by the same author, with three plates, 1844, which we regret to say we know only by name; the younger Agardh's *Adversaria in Systemata Algarum hodierna*, 1844, and various papers of *Dr. Montagne.*

Natural Orders of Algals.

Crystalline, angular, fragmentary bodies, brittle, and multiplying by spontaneous separation . . . . 1 Diatomaceæ.

Vesicular, filamentary or membranous bodies, multiplied by zoospores generated in the interior at the expense of their green matter . . . 2 Confervaceæ.

Cellular or tubular unsymmetrical bodies, multiplied by simple spores formed externally . . . . 3 Fucaceæ.

Cellular or tubular unsymmetrical bodies, multiplied by tetraspores . 4 Ceramiaceæ.

Tubular symmetrically branched bodies, multiplied by spiral coated nucules, filled with starch . . . . 5 Characeæ.

*** For the information of those who may wish to know something of the system of Kützing, which I do not adopt, the following list is extracted from his great work, to which the reader is referred for an explanation of the peculiar views of its author.

* I. CLASS.—ISOCARPEÆ.

Tribus I.—*Gymnospermeæ.*

Order I.—EREMOSPERMEÆ.

Subordo I.—*Mycophyceæ.*

I. Cryptococceæ.—Cryptococcus, Ulvina, Sphærotilus.

II. Leptomiteæ.—Hygrocrocis, Sirocrocis, Leptomitus, Mycothamnion, Chamænema, Nematococcus, Chionyphe.

III. Saprolegnieæ.—Saprolegnia, Mycocœlium.

IV. Phæonemeæ.—Stereonema, Phæonema.

Subordo II.—*Chamæphyceæ.*

V. Desmidieæ.—Closterium, Microtheca, Pentasterias, Euastrum, Xanthidium, Staurastrum, Crucigenia, Merismopœdia, Scenodesmus, Tessarthra, Micrasterias, Sphærastrum, Gomphosphæria, Desmidium, Didymoprium.

VI. Palmelleæ.—Protococcus, Microhaloa, Botryocystis, Microcystis, Botrydina, Polycoccus, Palmella, Inoderma, Coccochloris, Glœocapsa, Tetraspora, Palmoglœa.

VII. Hydrococceæ.—Actinococcus, Entophysalis, Hydrococcus, Hydrurus, Helminthonema.

Subordo III.—*Tiloblasteæ.*

A. Glœosipheæ.

a.) Asemospermeæ.

VIII. Oscillarieæ.—Spirulina, Oscillaria, Actinocephalus, Phormidium, Hydrocoleum, Chthonoblastus.

IX. Leptotricheæ. — Leptothrix, Asterothrix, Symphyothrix, Synploca, Dictyothrix, Entothrix, Inactis.

b.) Mesospermeæ.

X. Limnochlideæ.—Limnochlide.

XI. Nostoceæ.—Nostoc, Hormosiphon, Anabæna, Sphærozyga, Cylindrospermum, Spermosira, Nodularia.

XII. Scytonemeæ. — Drilosiphon, Scytonema, Synchæta, Symphyosiphon, Sirosiphon.

c.) Paraspermeæ.

XIII. Lyngbyeæ. — Siphoderma, Amphithrix, Leibleinia, Lyngbya, Blennothrix.

XIV. Calotricheæ. — Tolypothrix, Calothrix, Hypheothrix, Schizothrix, Schizodictyon, Dictyonema.

d.) Hypospermeæ.

XV. Mastichotricheæ. — Merizomyria, Mastichothrix, Mastichonema, Schizosiphon, Geocyclus.

XVI. Rivularieæ.—Physactis, Heteractis, Chalaractis, Ainactis, Limnactis, Rivularia, Dasyactis, Euactis.

B. Dermatosipheæ.

a.) Endospermeæ.

XVII. Hormidieæ.—Hormidium, Goniotrichium, Allogonium, Glœotila, Schizogonium, Schizomeris, Bangia.

XVIII. Ulotricheæ. — Ulothrix, Stygeoclonium.

XIX. Conferveæ.—Dedogonium, Psichohormium, Conferva, Spongopsis, Rhizoclonium, Sphæroplea, Cladophora, Crenacantha, Ægagropila, Spongomorpha, Periplegmatium, Pilinia, Fischeria.

XX. Zygnemeæ. Mougeotia, Sirogonium, Staurospermum, Spirogyra, Zygnema, Zygonium.

XXI. Hydrodictyeæ. Hydrodictyon.

b.) Ectospermeæ.

XXII. Protonemeæ. Gongrosira, Protonema.

XXIII. Chantransieæ. Chroolepus, Chantransia, Chlorotylium.

XXIV. Draparnaldieæ. Draparnaldia.

XXV. Ectocarpeæ. Ectocarpus.

XXVI. Sphacelarieæ. Sphacelaria, Halopteris, Stypocaulon, Ballia, Chaetopteris, Cladostephus.

Subordo IV.—*Dermatoblasteæ.*

XXVII. Ulvaceæ.—Phyllactidium, Protoderma, Prasiola, Ulva.

XXVIII. Phycoserideæ. — Phycoseris, Diplostromium, Phycolapathum.

XXIX. Enteromorpheæ.—Enteromorpha, Chlorosiphon, Stictyosiphon, Dictyosiphon.

Subordo V.—*Cœloblasteæ.*

XXX. Vaucherieæ. — Botrydium, Vaucheria, Bryopsis, Valonia.

XXXI. Caulerpeæ.—Caulerpa.

XXXII. Codieæ.—Codium, Rhipozonium, Halimeda, Corallocephalus, Rhipocephalus.

XXXIII. Anadyomeneæ.—Anadyomene.

XXXIV. Polyphyseæ. — Acetabularia, Polyphysa.

XXXV. Dasycladeæ.—Cymopolia, Dasycladus, Ascothamnion.

XXXVI. Chareæ.—Nitella, Charopsis, Chara.

Ordo II.—CRYPTOSPERMEÆ.

XXXVII. Lemanieæ. — Thermocœlium, Lemania, Halysium.

XXXVIII. Chætophoreæ.—Chætophora, Chætoderma, Thorea.

XXXIX. Batrachospermeæ. Batrachospermum.

XL. Liagoreæ.—Liagora.

XLI. Mesoglœaceæ. — Cladosiphon, Myriactis, Phycophila, Corynophlæa, Corynephora, Mesoglœa, Chordaria.

Ordo III.—PYCNOSPERMEÆ.

XLII. Chordeæ.—Chorda, Spermatochnus, Halorhiza.

XLIII. Encœlieæ. — Encœlium, Halodictyon, Striaria.

XLIV. Dictyoteæ. — Dichophyllium, Cutleria, Stœchospermum, Spatoglossum, Haloglossum, Halyseris, Stypopodium, Phycopteris, Zonaria, Phyllitis.

XLV. Sporochneæ. — Sporochnus, Carpomitra, Desmarestia, Arthrocladia.

XLVI. Laminarieæ. — Phlœorhiza, Laminaria, Hafgygia, Phycocastanum, Alaria, Costeria, Agarum, Thalassiophyllum, Lessonia, Macrocystis, Nereocystis.

Tribus II.—*Angiospermeæ.*

XLVII. Fuceæ.—Splachnidium, Durvillæa, Hormosira, Ecklonia, Himanthalia, Fucus, Carpoglossum, Physocaulon, Scytothalia, Phylospora, Sirococcus.

XLVIII. Cystosireæ.—Treptacantha, Halerica, Phyllacantha, Cystosira, Hormophysa, Halidrys, Pycnophycus.

XLIX. Sargasseæ. — Pterocaulon, Sargassum, Turbinaria, Carpophyllum, Phycobotrys.

L. Halochloæ.—Blossevillea, Spongocarpus, Halochloa, Myagropsis, Carpacanthus, Sirophysalis, Coccophora, Scaberia, Carpodesmia.

II. CLASS.—HETEROCARPEÆ.

Tribus III.—*Paracarpeæ.*

Ordo I.—TRICHOBLASTEÆ.

LI. Callithamnieæ.—Callithamnion, Griffithsia, Halurus, Phlebothamnion, Wrangelia, Spyridia, Ptilota.

LII. Ceramieæ. — Hormoceras, Gongroceras, Echinoceras, Acanthoceras, Ceramium, Centroceras, Microcladia.

Ordo II.—EPIBLASTEÆ.

LIII. Porphyreæ. — Porphyra, Hildenbrandtia, Peyssonelia.

LIV. Spongiteæ. — Hapalidium, Pneophyllum, Melobesia, Spongites.

LV. Corallineæ.—Amphiroa, Corallina, Jania.

Ordo III.—PERIBLASTEÆ.

LVI. Gymnophlæaceæ.—Gymnophlæa, Helminthora, Naccaria.

LVII. Chætangieæ.—Chætangium, Thamnoclonium, Sarcophycus.

LVIII. Halymenieæ.—Myelomium, Halymenia, Dumontia, Halarachnion, Catenella.

LIX. Caulacantheæ.—Caulacanthus, Acanthobolus.

LX. Gigartineæ. — Iridæa, Chondrodictyon, Grateloupia, Mastocarpus, Chondros, Chondracanthus, Euhymenia, Constantinea, Callophyllis, Sarcophyllis, Solieria, Furcellaria, Gigartina.

LXI. Rhynchococceæ.—Rhynchococcus, Calliblepharis.

LXII. Cystoclonieæ. — Cystoclonium, Hypnophycus.

LXIII. Gelidieæ. — Acrocarpus, Echinocaulon, Gelidium, Ctenodus.

LXIV. Sphærococceæ.—Bowiesia, Sphærococcus, Trematocarpus.

LXV. Tylocarpeæ.—Tylocarpus, Oncotylus, Pachycarpus, Phyllotylus, Coccotylus, Phyllophora, Acanthotylus.

Tribus IV.—*Choristocarpeæ.*

Ordo IV.—AXONOBLASTEÆ.

LXVI. Dasyeæ.—Dasya, Eupogonium, Trichothamnion.

LXVII. Polysiphonieæ. — Polysiphonia, Helicothamnion, Halopithys, Digenea, Bryothamnion, Physcophora, Alsidium.

LXVIII. Chondrieæ. — Lophura, Carpocaulon, Chondria, Acanthophora.

Ordo V.—CŒLOBLASTEÆ.

LXIX. Chondrosipheæ.—Bonnemaisonia, Chondrothamnion, Chondrosiphon, Halosaccion.

LXX. Champieæ.—Champia, Lomentaria, Gastroclonium.

Ordo VI.—PLATYNOBLASTEÆ.

LXXI. Delesserieæ. — Æglophyllum, Schizoglossum, Inochorion, Cryptopleura, Phycodrys, Hypoglossum, Delesseria.

LXXII. Botryocarpeæ.—Neuroglossum, Botryocarpa.

LXXIII. Amansieæ.—Polyzonia, Amansia.

LXXIV. Rytiphlæaceæ.—Rytiphlæa, Dictyomenia.

LXXV. Carpoblepharideæ. — Carpoblepharis, Odonthalia.

LXXVI. Plocameæ.—Plocamium, Thamnocarpus, Thamnophora.

LXXVII. Claudieæ.—Claudea.

ORDER I. DIATOMACEÆ. BRITTLEWORTS.

Diatomaceæ, *Agardh, Syst.* xii. (1824); *Kützing, in Linnæa,* 8. 529; *Part of* Chaodineæ and Fragillariæ, *Bory in Dict. Class.* 3. *and* 4. (1823); *Endlich. Gen. I.; Ralfs. in Ann. Nat. Hist.* 11 *et seq.*

DIAGNOSIS.—*Crystalline, angular, fragmentary bodies, brittle, and multiplying by spontaneous separation.*

Crystalline fragmentary bodies, generally bounded by right lines, rarely included in curved lines, flat, stiff, brittle, usually nestling in slime, uniting into various forms, and then separating again.

Those who have ever examined the surface of stones constantly moistened by water, the glass of hothouses, the face of rocks in the sea, or of walls where the sun never shines, or the hard paths in damp parts of gardens after rain, cannot fail to have remarked a green mucous slime with which such places are covered. This slime consists of Algals in their simplest state of organisation; they have been called Chaodineæ by Bory de St. Vincent, whose account of them is to the following effect: "The slime resembles a layer of albumen spread with a brush; it exfoliates in drying, and finally becomes visible by the manner in which it colours green or deep brown. One might call it a provisional creation waiting to be organised, and then assuming different forms, according to the nature of the corpuscles which penetrate it or develop among it. It may further be said to be the origin of two very distinct existences, the one certainly animal, the other purely vegetable. This matter lying among amorphous mucus consists, in its simplest state, of solitary, spherical corpuscles; these corpuscles are afterwards grouped, agglomerated, or chained together, so producing more complex states of organisation. Sometimes the mucus, which acts as the basis or matrix of the corpuscles, when it is found in water, which is the most favourable medium for its development, lengthens, thickens, and finally forms masses of some inches extent, which float and fix themselves to aquatic plants. These masses are at first like the spawn of fish, but they soon change colour, and become green, in consequence of the formation of interior vegetable corpuscles. Often, however, they assume a milky or ferruginous appearance; and if in this state they are examined under a microscope, they will be found completely filled with the animalcules called Naviculariæ, Lunulinæ, and Stylariæ, assembled in such dense crowds as to be incapable of swimming. In this state the animalcules are inert. Are they developed here, or have they found their way to such a nidus, and have they hindered the development of the green corpuscles? Is the mucus in which they lie the same to them as the albuminous substance in which the eggs of many aquatic animals are deposited? At present we have no means of answering these questions."

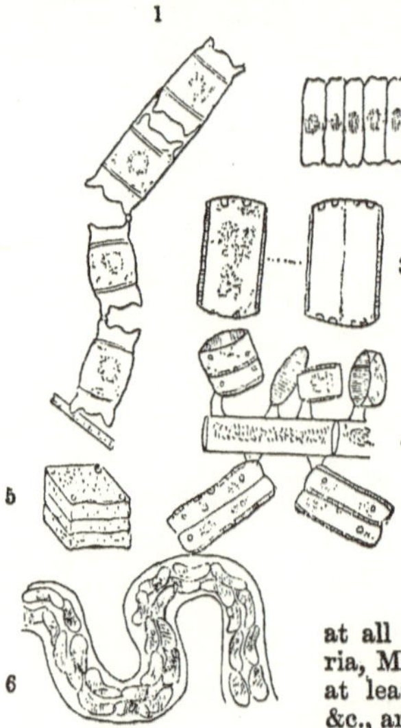

Fig. II.

These form, no doubt, the extreme limits of the Vegetable and Animal Kingdom. Their regular form, and the power of separating into distinct particles, which the most of them have, are almost as much the attributes of the mineral, as of the vegetable, or even animal kingdom. Agardh includes them among plants. Kützing asserts that their life is as much animal as vegetable; and that, at all events, Achnanthes, Gomphonema, Exilaria, Fragilaria, Meloseira, Schizonema, Micromega, and Berkleya, are at least plants, if Frustulia, Cymbella, Navicula, Surirella, &c., are animalcules. He has also recently ascertained, that the frustules of Micromega are metamorphosed into green globular spores. Dr. Dickie of Aberdeen has observed something of the same kind. Mr. Ralfs, who has paid great attention to the history of these

Fig. II.—1. Biddulphia; 2. Grammonema; 3. Eunotia; 4. Achnanthes; 5. Amphitetras; 6. Gloionema, a production once referred to this order, but determined by Mr. Berkeley to be the eggs of an insect.

doubtful creatures, observes, that one division of them, the Cymbelleæ, rapidly become putrid, have a siliceous covering, and consequently their form is not altered in drying, and they are not destroyed by fire. When in perfection they are generally brownish, and not unfrequently become greenish when dry; they are usually of either a quadrilateral or prismatic form, and often marked with streaks and dots. The Desmidieæ, on the contrary, putrify very slowly, have not a siliceous coat, and therefore alter their shape in drying. When in perfection they are generally of an herbaceous green colour, and most frequently have the fragments divided into two portions resembling each other in form, but sometimes differing much as to size. "This division is marked in Desmidium mucosum merely by a shallow groove passing round the joint, and in Desm. Swartzii by notches in the angles, by which it is rendered still more apparent; whilst in Euastrum the two portions are connected only by a central chord." (*Ann. N. H.* 11. 448.) In another place (*Ib.* 13. 377) this accurate observer recognizes the universal presence of starch among the Desmidieæ, which, not being an animal product, seems to settle the question of the vegetable nature of at least that portion of Brittleworts.

Natives of still waters, and oozy places in the northern parts of the world.

The uses of these plants to man are unknown.

GENERA.

Suborder I. CYMBELLEÆ. — Individuals quite free, angular, siliceous.

Frustulia, *Ag.*
Cyclotella, Ktz.
Haplotella, Ktz.
Cymbella, Ktz.
Navicula, Bory.
Styllaria, *Ag.*
Rhabdium, Wallr.
Meridion, *Ag.*
Licmophora, *Ag.*
Exilaria, Grev.
Psygmatella, Kütz.
Biddulphia, *Gray.*
Gomphonema, *Ag.*
Cymbophora, Ktz.
Paltonophora, Ktz.
Sphenophora, Ktz.
Eunotia, *Ehr.*
Cocconema, *Ehr.*
Achnanthes, *Bory.*
Striatella, *Ag.*
Amphitetras, *Ehr.*
Isthmia, *Ag.*
Diatoma, *DC.*
Bacillaria, Ehr.
Tabellaria, *Ralfs.*
Tessella, *Ehr.*
Fragilaria, *Lyngb.*
Nematoplata, Bory.
Temachium, Wallr.
Grammonema, *Ag.*
Tetracyclus, *Ralfs.*
Lysigonium, *Lk.*
Melosira, Ag.
Gaillonella, Bory.
Vesiculifera, Hass.
Oncobyrsa, *Ag.*

Suborder II. HYDROLINEÆ.—Individuals enclosed in tubes, angular.

Encyonema, *Ktz.*
Dickiea, *Berk.*
Hydrolinum, *Link.*
Schizonema, Ag.
Monema, Grev.
Girodella, Gaill.
Spermogonia, Bonnem.
Homœocladia, *Ag.*
Gloiodictyon, *Ag.*
Hydrurus, *Ag.*
Cluzella, Bory.
Corradorus, Gray.
Micromega, *Ag.*
Caleothrix, Desv.

Suborder III. DESMIDIEÆ.—Individuals cylindrical.

Pentasterias, *Ehr.*
Xanthidium.
Euastrum, *Ehr.*
Cosmarium, Menegh.
Eutomia, Harv.
Heterocarpella, Turp.
Odontella, *Ehr.*
Closterium, *Nitzsch.*
Tetmemorus, *Ralfs.*
Micrasterias, *Ag.*
Staurastrum, *Meyen.*
Pediastrum, *Meyen.*
Sphærastrum, *Meyen.*
Helierella, *Bory.*
Potareus, Rafin.
Crucigenia, *Morren.*
Selenaca, *Nitsch.*
Heliactis, *Ktz.*
Scenedesmus, *Meyen.*
Tessarthra, *Turp.*
Echinella, *Ach.*
Desmidium, *Ag.*

NUMBERS. GEN. 45. Sp. 457.

Acrita.

POSITION. ——— DIATOMACEÆ.—Confervaceæ-Palmelleæ.

ORDER II. CONFERVACEÆ.—CONFERVAS.

Confervaceæ, *Endl. gen. Suppl. III., p.* 10. Zoospermeæ, *J. Agh. Alg. Med.* 1. Synsporeæ and Zoospermeæ, *Decaisne in Ann. Sc. N.* 2 *ser.* 18. 305.

DIAGNOSIS.—*Vesicular, filamentary or membranous bodies, multiplied by zoospores generated in the interior, at the expense of their green matter.*

Water plants, not commonly of a green colour, but occasionally olive, violet, and red; inhabiting the ocean in some instances, but more commonly found in fresh water; some of them even belonging to both kinds of fluids; some found in mud, others floating freely, most attached, in some way, on rocks or as parasites. Cells solitary or many, globose, elliptical, cylindrical, or tubular; sometimes variously branched; sometimes formed in slimy matter in which they are scattered, or irregularly heaped, or placed one above the other in a regular series forming an articulated frond; some disposed in several rows and forming a thin layer, or some combined in the form of a net. Their mode of growth by a subdivision of the cells, of ramification by a lateral extension of such cells, a dividing partition being eventually formed. The *propagation* by *sporidia* (internal cells, or a gelatinous substance which organizes itself into cells,) found in each cell, singly, or in a definite, or indefinite number, formed from the colouring matter of one or more cells, or sometimes by the copulation of distinct individuals, and discharged by the opening or absorption of the mother cell.—*Endl.*

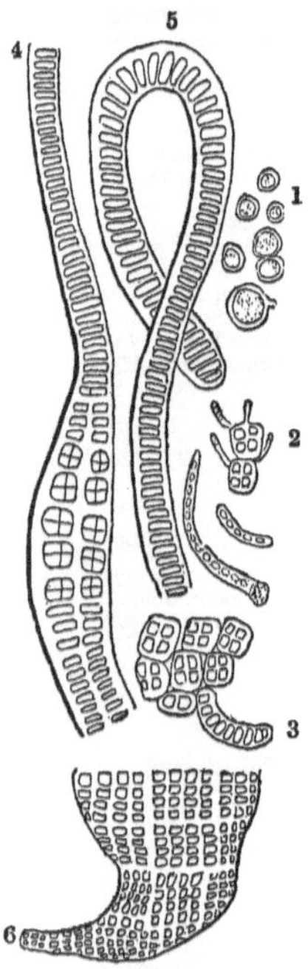

Fig. III.

If doubts exist as to the Vegetable nature of the last order, or of some part of it, no question arises as to what that of Confervas is. Its genera are now admitted on all hands to be plants, since M. Decaisne's important discovery of the vegetable nature of several things which had been previously regarded as Zoophytes. Nevertheless, it is curious to see how much, at one period at least of their existence, they have of an animal nature, if the power of moving from place to place is to be taken as an indication of such a quality. It seems incontestable, notwithstanding the denial of Mohl and others, that many of the Conferva tribe, especially of the genera Conferva, Ulva, and their near allies, produce in their tubular threads reproductive bodies, or spores, which after a time acquire a power of rapid, and quasi-voluntary motion while in the inside of their mother; that by degrees, and in consequence of their constantly tapping against the soft side of the cell that holds them, they escape into the water; that when there they swim about actively, just like animalcules; and at last retreating to a shady place, attach themselves to a stone or some other body, lose their locomotive quality, and thenceforward germinate and grow like plants.—(*J. Ag. Ann. Sc. Nat.* 2 *ser. vol.* 6.)* It is

* "The filaments of Conferva ærea," says the younger Agardh, "are, as is well known, articulated or divided at equal distances into little compartments (joints), which have no communication among themselves other than what results from the permeability of the dissepiments. The green matter contained in these joints appears at first altogether homogeneous, as if it were fluid; but in a more advanced state it becomes more and more granular. The granules are, at their formation, found adhering to the inner surface of the membrane, but they soon detach themselves, and the irregular figure which they present at first passes to that of a sphere. These granules congregate by degrees in the middle of the joint, into a mass, at first elliptical, but which at length becomes perfectly spherical. All these changes are conformable to phenomena known in vegetable life; those which are to follow have more analogy with the phenomena of animal life. At this stage an important metamorphosis exhibits itself, by a motion of swarming (un mouvement de fourmillement) in the green matter. The granules of which it is composed detach themselves from the mass, one after another, and having thus become free, they move about in the vacant space of the joint with an extreme rapidity. At the same time, the exterior membrane of the joint is observed to swell in one point, till it there forms a little mammilla, which is to become the point from which the moving granules finally issue. By the extension of the membrane for the formation of the mammilla, the tender fibres of which it is composed separating, cause an opening at the end of the mammilla, and it

Fig. III.—1. Protococcus viridis; 2. the same beginning to develop; 3. the same more advanced; 4 & 5. Schizogonium murale; 6. A fragment of Ulva (Prasiola) furfuracea (Kützing).

even asserted by M. Thuret, that in Conferva glomerata and rivularis, the spores have special organs of motion, of the nature of ciliæ or tentacula, and that it is by their rapid action that the spores swim so freely in fluid.—(*Ibid.* xix. 267.) Motions of another kind have been noticed in the Oscillatorias; and in the species called Zygnemas, they are so extraordinary as to approach nearly to the act of copulation in animals. In the language of M. Decaisne, " the spores of these plants result from the coupling of two

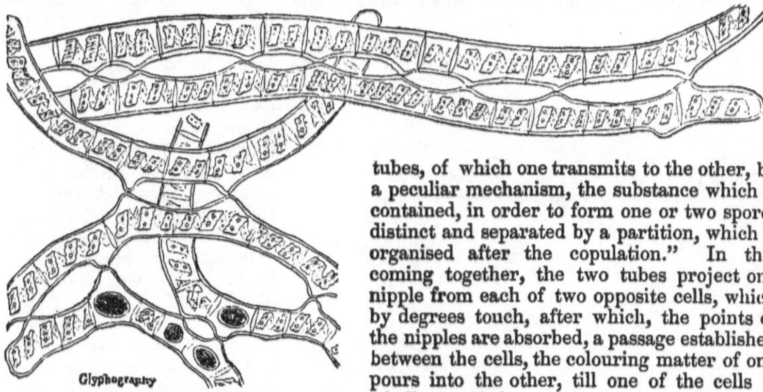

Fig. IV.

tubes, of which one transmits to the other, by a peculiar mechanism, the substance which it contained, in order to form one or two spores distinct and separated by a partition, which is organised after the copulation." In this coming together, the two tubes project one nipple from each of two opposite cells, which by degrees touch, after which, the points of the nipples are absorbed, a passage established between the cells, the colouring matter of one pours into the other, till one of the cells is wholly emptied.

Meyen states, that the red and green Snowplants, which have been described as Confervæ, and assigned to the genus Protococcus, are nothing more than the animalcules called Enchelis sanguinea, and Pulvisculus. But this does not affect the genus Protococcus, which contains productions respecting whose vegetable nature no doubt is entertained.

Hydrodictyon utriculatum has the appearance of a green net. According to M. Areschoug, the cells of this plant, when nearly ripe, contain a number of active spherical granules, which in the process of reproduction become elliptical, and are attached by their extremities, when an articulation is soon produced, so as to form pentagons or hexagons. Each granule becomes a cell of the new Hydrodictyon. (*De Hydr. utric. dissertatio.*)

---

is by this passage that the granules escape. At first they issue in a body, but soon those which remain, swimming in a much larger space, have much more difficulty in escaping, and it is only after innumerable knockings (titubations) against the walls of their prison, that they succeed in finding an exit. From the first instant of the motion one observes that the granules or sporules are furnished with a little beak, a kind of anterior process, always distinguishable from the body of the seed by its paler colour. It is on the vibrations of this beak that the motion, as I conceive, depends; at least, I have never been able to discover any ciliæ. However, I will not venture to deny the existence of these, for with a very high power of a compound microscope one sees the granules surrounded with a hyaline border, as we find among the ciliated Infusoria on applying a glass of insufficient power. The sporules, during their motion, always present this beak in front of their body, as if it served to show them the way; but when they cease to move, by bending it back along the side of their body, they resume the spherical form, so that before and after the motion one sees no trace of this beak. The motion of the sporules before their exit from the joint consists principally in quick dartings along the walls of the articulation, knocking themselves against them by innumerable shocks; and in some cases we are almost forced to believe that it is by this motion of the sporules that the mammilla is formed. Escaped from their prison they continue their motion for one or two hours, and retiring always towards the darker edge of the vessel sometimes they prolong their wandering courses, sometimes they remain in the same place, causing their beak to vibrate in rapid circles. Finally, they collect in dense masses, containing innumerable grains, and attach themselves to some extraneous body at the bottom or on the surface of the water, where they hasten to develop filaments like those of the mother plant. The spherical sporules elongate at first into egg-shaped bags, attached to the strange body by the narrowest end. Their development only consists in a continual expansion, without emitting any root. The green internal matter divides in the middle by a partition, which appears at first sight as a hyaline mucilage, but which gradually changes into a complete diaphragm. It is thus, by successive divisions of the joint first formed, that the young plant increases. The position of the mammilla in each joint is uncertain, at least I have seen it very different in neighbouring joints. The exit of the sporules does not take place at the same time in the different joints. One often sees those of one of the articulations already escaped, while in the neighbouring one they are not yet completely formed. Commonly the uppermost joints empty themselves first, so that it is not rare to see all the upper part of a filament entirely transparent, whilst the lower part continues still to develop. In this manner the formation and dissemination of the seeds continue during the whole summer, and thus a single filament suffices for the formation of an infinite quantity of sporules. If one remembers that each joint contains perhaps many hundred of spores, it is not astonishing that the water becomes perfectly covered with them; so that we might readily take for a Protococcus, or other simple Alga, what are only the spores of a Conferva. I suspect that from such a mistake have arisen the theories of metamorphosis proposed by many modern algologists."

Fig. IV.—Spirogyra quinina (Kützing).

Confervas are more frequently found in the temperate parts of the world than within the tropics, occupying both salt and fresh water, but more especially the latter, and several species are common to both. One of them, the Tiresias ericetorum, grows on the ground, but in places that are very damp, and often inundated; others among the oscillating species cover the humid surface of rocks or earth, and the interstices in the pavement of cities; some even grow in hot springs of a very high temperature. Ulva thermalis lives in the hot springs of Gastein, in a temperature of about 117° Fahr. Dr. Lankester speaks of Oscillatorias found in the sulphuretted hydrogen water of Harrowgate (*Ann. N. H.* vii. 107); and Calothrix nivea is said to have occurred there also. They often give a peculiar colour to large bodies of water. The Red Sea has derived its name from the abundance of Trichodesmium erythræum which floats in it, and concerning which MM. Evernor Dupont and Montagne have given a curious account.* Dunal states that the crimson colour of the salt-water tanks on the coast of the Mediterranean is owing to the presence of Protococcus salinus and Hæmatococcus salinus, two of the most simple of this order. Hæmatococcus Noltii stains crimson the marshes of Sleswick.

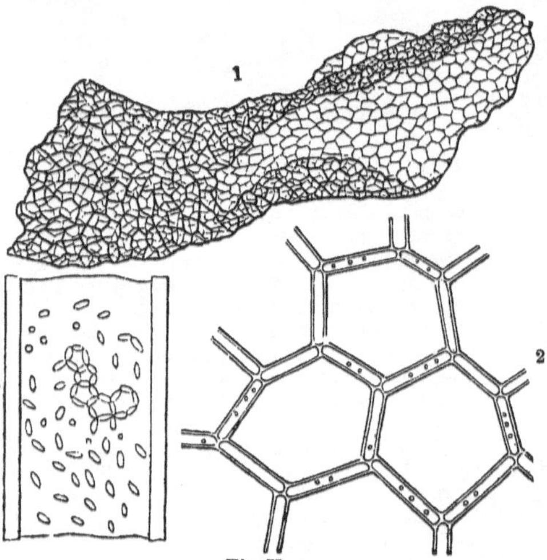

Fig. V.

Dr. Drummond ascertained that the Irish lake of Glaslough, which is remarkable for its peculiar greenness, owes its colour to the presence of his Oscillatoria æruginescens. (*Ann. N. H.* i. 1.) The green of the Grand-canal docks near Dublin has been found to arise from the presence of a Sphærozyga (Trichormus *Allm.*) and in like manner Mr. Thompson found that the water of Ballydrain lake is coloured green by Sphærozyga (Anabaina) spiralis, and that in the same place broad verdigris patches proceed from collections of Aphanizomenon incurvum. (*Ann. N. Hist.* v. 83.) It has also occurred that acres of inundated meadow land have been clothed to the depth of an inch with a thick entangled layer of Conferva crispa, which then forms a texture not unlike that of some woollen fabric, whence it has gained the name of water-flannel. Confervæ sometimes attack diseased animal tissue. Mr. Goodsir has described such an instance in the case of a gold-fish. (*Ann. Nat. Hist.* ix. 336.) It has been ascertained that this is of very common occurrence, and that the plant which makes the attack is the Achlya prolifera. This production has been carefully

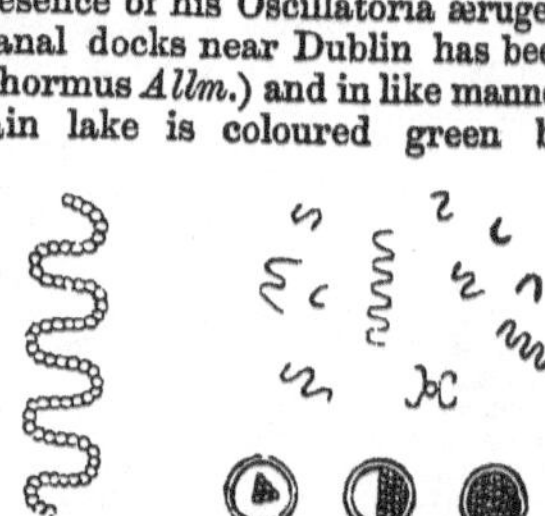
Fig. VI.

* "On the 8th July 1843, I entered the Red Sea by the straits of Babelmandel, on board the Atalanta steamer. On the 15th the burning sun of Arabia suddenly awoke me with its brilliancy unannounced by the dawn. I was leaning mechanically out of the poop windows, to catch a little of the fresh air of night before the sun had devoured it, when, imagine my surprise to find the sea stained red as far as the eye could reach behind the vessel.——If I was to attempt to describe this phenomenon, I would say that the surface of the ocean was entirely covered with a close thin layer of fine matter, the colour of brickdust, but slightly orange. Mahogany sawdust would produce such an appearance.—When put into a white glass bottle, it became in the course of a day deep violet, while the water itself had become a beautiful rose colour. This appearance extended from Cosseir, off which we were at daybreak on the 15th May, to Tor, a little Arabian village, which we made about noon the next day, when it disappeared, and the sea became blue as before. During this time we must have passed through about 256 miles of the red plant." *Comptes rendus*, xix. 171.—Similar appearances have been mentioned by Mr. Darwin; and Mr. Hinds, when at anchor off Libertad in the Pacific, and at the Abrolhos, perceived large quantities of another species of Trichodesmium, which exhaled a most disagreeable smell. To this cause, or one of the same kind, is probably referable the phenomenon mentioned in the *Colombo Herald* of May 14, 1844: "The sea to the southward of Colombo, and, more lately, opposite the fort itself, has presented a very uncommon appear-

Fig. V.—1. Hydrodictyon utriculatum; 2. portion of full-grown plant; 3. portion of a joint in which the granules have commenced to dispose themselves in pentagons, the rudiments of the new plant.
Fig. VI.—Sphærozyga spiralis.

examined by Dr. Unger. When arrived at its full growth, it consists of transparent threads of extreme fineness, packed together as closely as the pile of velvet; they greatly resemble, in general appearance, certain kinds of mouldiness. These threads are terminated by an extremity about $\frac{1}{1200}$ of an inch in diameter, consisting of a long single cell, within which is collected some green mucilage intermixed with granules. Dr. Unger assures us that at this time no starch is present, but the whole of the green matter is of the nature of gum, as is proved by the action of iodine upon it. The contents of the cell are seen to be in constant motion, directing themselves in lines such as are represented at Fig. 5. While this is going on, the end of the cell continues to grow, and at the same time the contents collect at the extremity, and distend it into a small head in form resembling a club, immediately after which a chamber is formed, and then the first stage of fructification is accomplished. The next change is observed to take place in the granular matter of the clubhead, which itself enlarges, while the contents gain opaqueness, and by degrees arrange themselves in five or six-sided meshes, which are in reality the sides of angular bodies, that are rapidly forming at the expense of the mucilage above mentioned,

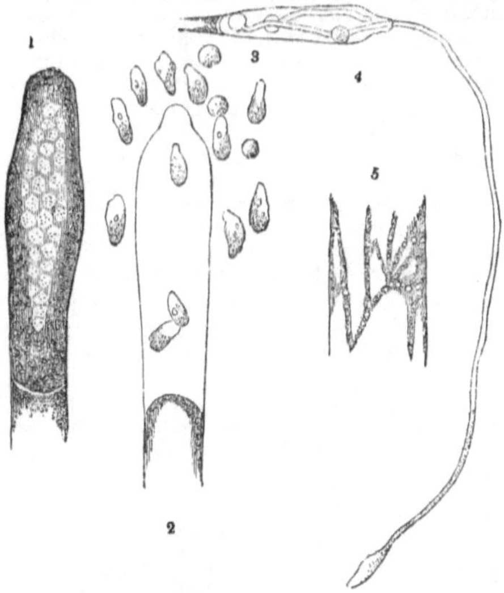

Fig. VII.

which has disappeared. It is not the least surprising part of this history, that all the changes above mentioned take place in the course of an hour or an hour and a half, so that a patient observer may actually witness the creation of this singular plant. At this time all the vital energy seems directed towards changing the angular bodies in the inside of the clubhead into propagating germs or spores. Meanwhile the clubhead grows, and gives them a little room, and they in their turn alter their form and become oval. Then it is that is witnessed the surprising phenomenon of spontaneous motion in the spores, which, notwithstanding the narrow space in which they are born, act with such vigour that at last they force a way through the end of the clubhead. At first one spore gets out into the water, then another, and another, till at last the clubhead is emptied. All this takes place with such rapidity that a minute or two suffice for the complete evacuation of the clubhead or spore-chamber. The spores, when they find their way into the water, are generally egg-shaped, and swim *with their small end* foremost; but they are often deformed, in consequence of the narrowness of the hole through which they have had to pass. It even happens that they stick fast in the hole, and perish there. They are extremely small, their breadth not exceeding the 1896th

ance for some days past. Instead of its usual brightness, the surface has been to a considerable extent covered with what appears to the naked eye a sort of nasty froth or scum, emitting a fœtid smell. In the mornings, when it has been usually calm, this scum has presented itself in broad belts and fields, and by the afternoon, after being exposed to the sea-breeze, it is broken down into streaks, lying in the direction of the wind, which, if it blows pretty fresh, disperses it altogether. We have examined some of this unusual substance in a tumbler of salt water, and were not a little surprised to find, that while it floated on the surface, in the form of a scum, some parts of a yellowish-green, and some of a purplish-brown colour, it tinged the whole water of a beautiful violet. We afterwards found that the whole water in the bucket, in which it was brought from the sea, had acquired the same colour; and, indeed, it appeared to us the other day, when it was very abundant, as if the sea itself had been stained of this beautiful tint. We found, on minute inspection, that it consisted of an infinite multitude of small spindle-shaped bodies, each of which, in its turn, was a bundle of small threads jointed but unbranched, and seemingly very brittle. We have no doubt but it is a vegetable production in the sea, something similar to the green substance which covers stagnant pools of fresh water. The most remarkable and unpleasant feature is its fœtid odour. When we read in books of voyages, of ships sailing for so many hours through seas of a blood-colour, and similar wonders, we are apt to suppose the author is taking the liberty of a traveller; but witnessing such a phenomenon as this, is calculated to prepare us for giving them more credit."

Fig. VII. Achlya prolifera.—1. The club-shaped spore-chamber; 2. the same emptied of its spores; 3, 4. a spore-chamber much less magnified, containing two germinating spores, and a dead one; 5. a piece of the thread at an early period, with the lines of motion.

part of an inch. Their small end is the most transparent, and it is curious to see how constantly this is pushed forwards in the rapid evolutions made in the water by these living particles. This sort of *quasi* animal life does not last long—a few seconds, some minutes, or at the most half-an-hour. They often die: Unger assures us that he has seen them in the agonies of death, and *struggling convulsively* (!), with all the appearance of animal life.

Porphyra laciniata and vulgaris are stewed, and brought to our tables as a luxury, under the name of Laver; and even the Ulva latissima, or green Laver, is not slighted in the absence of the Porphyræ. Ulva compressa, a common species on our shores, is regarded, according to Gaudichaud, as an esculent by the Sandwich Islanders. Common Nostoc, commonly called star-jelly, a trembling gelatinous plant, that springs up suddenly after rain, is by superstitious persons supposed to possess virtue as a vulnerary, and in pains of the joints; oyster green or Ulva lactuca (the βρύον θαλάσσιον of Dioscorides) is sometimes employed in scrofula; the ancients used it in inflammations and gouty affections; its taste is so bitter and salt that it is usually given with lemon juice.

The Confervals found in many thermal springs, mostly species of Sphærozyga, are used empirically as external applications to goître, enlarged glands, &c. Henry has examined the Confervals in the springs of Vichy, Neris, and Vaux, and found small quantities of an alkaline iodide in each. (*Chem. Gaz.* 1844, p. 447.)

GENERA.

Suborder I. — *Palmelleæ.* Cells somewhat globose or elliptical, free, and more or less distinct, or collected by means of a slimy layer into a frond.

Tribe 1. Protococcidæ. —*The slimy substratum obsolete.*

Protococcus, *Ag.*
Sphærella, Somm.
Coccophysium, Link.
Globulina, Turp.
Protosphæria, Turp.
Hæmatococcus, *Ag.*
Gloiococcus, Shutt.
Chlorococcum, *Grev.*
Globulina, Turp.
Protosphæria, Turp.
Pleurococcus, *Menegh.*
Hormospora, *Breb.*
Stereococcus, *Kütz.*

Tribe 2. Coccochloridæ. —*The slimy substratum evident.*

Palmella, *Lyngb.*
Priestleya, Meyen.
Chaos, Bory.
Phytoconis, Bory.
Coccodea, Pal.
Merrettia, Gray.
Sarcoderma, Ehr.
Coccochloris, *Spr.*
Microcystis, *Kütz.*
Bichatia, Turp.
Anacystis, *Menegh.*
Oncobyrsa, *Ag.*
Hydrococcus, Kütz.
Micraloa, *Biass.*
Hydrothrombium, Ktz.
Botrydina, *Brebiss.*

Suborder II.—*Nostocheæ.* Cells somewhat globose or elliptical, coalescing into a simple or branched thread; united into several rows by means of a slimy substratum of various forms.

Nostoc, *Vauch.*
Linkia, Mich.
Undina, Fries.
Hydrococcus, Link.
Thrombium, Wallr.
Monormia, Berkel.
Sphærozyga, *Ag.*
Anabaina, Bory.
Trichormus, Allm.
Anhaltia, *Schwabe.*

Suborder III. — *Oscillatoreæ.* Cells tubular, naked or furnished with a slimy or gelatinous layer, continuous, but seeming to be jointed in consequence of interruptions of the colouring matter.

Tribe 1. Rivularidæ.—*Tubes proceeding singly, or in pairs, from a transparent globule; collected into a frond by means of a gelatinous layer.*

Gloiotrichia, *J. Ag.*
Rivularia, *Roth.*
Lynckia, Lyngb.
Gaillardotella, Bory.
Stylobasis, Schw.
Stypnion, Raf.
Zonotrichia, *J. Ag.*
Diplotrichia, *J. Ag.*

Tribe 2. Oscillatoridæ. —*Tubes cylindrical, free, or woven into a frond, falsely jointed in consequence of the ringed or streaked appearance of the colouring matter.*

Oscillatoria, *Bosc.*
Oscillaria, Bosc.
Trichophora, Bonnem.
Spirogyra, Nees.
Spirulina, Turp.
Loten, Adams.
Trichodesmium, *Ehrenb.*
Microcoleus, *Desmaz.*
Vaginaria, Bory.
Merizomyria, Poll.
Calothrix, *Ag.*
Hempelia, Meyen.
Ulothrix, Kütz.
Dillwynella, Bory.
Lyngbya, *Ag.*
Cyclosperma, Bonnem.
Humida, Gray.
Scytonema, *Ag.*
Percursaria, Bonnem.
Sphæroplea, *Ag.*
Cadmus, Bory.
Sphærogona, Link.
Sphæroplethia, Duby.
Beggiatoa, *Trev.*

Suborder IV. — *Conferveæ.*—Cellules resembling joints, arranged in a net, or more frequently in simple or branched threads, separate, or combined by common slime.

Tribe 1. Hydrodictidæ.—*Cells tubular, combined by their pointed extremities into a net-like frond.*

Hydrodictyon, *Roth.*
Microdictyon, *Decaisne.*
Dictylema, Raf.
Talerodictyon, *Endl.*

Tribe 2. Zygnemidæ.—*Cells tubular, united by their truncated extremities into jointed threads, which are at first distinct, and then, by the aid of transverse tubelets which discharge the colouring matter, brought into copulation.*

Mougeotia, *Ag.*
Serpentinaria, Gray.
Conjugata, Lk.
Zygnema, *Ag.*
Agardhia, Gray.
Globulina, Lk.
Stellulina, Lk.
Lucernaria, Roussel.
Diadema, Pal.
Tyndaridea, Bory.
Leda, Bory.
Spirogyra, *Lk.*
Choaspis, Gray.
Salmacis, Bory.

Tribe 3. Confervidæ.—*Cells tubular, united by their truncated extremities into free, simple, or branched threads.*

Myxonema, *Fries.*
Myxotrix, Fries.
Nematrix, Fries.
Conferva, *Fries.*
Polysperma, Vauch.
Chloroniton, Gaill.
Hormiscia, *Fries.*
Nodularia, *Mertens.*
Aphanizomenon, *Morren.*
Tiresias, *Bory.*
Œdogonium, Lk.
Draparnaldia, *Bory.*
Charospermum, Lk.
Leptomitus, *Ag.*
Saprolegmia, Nees.
Pythium, Nees.
Sphærotilus, Kg.
Achlya, *Nees.*
Hydronema, Carus.
Hygrocrocis, *Ag.*

Tribe 4. Chætophoridæ.—*Cells tubular, adhering by truncated extremities in jointed branched threads coalescing into a gelatinous frond.*

Chætophora, *Schrank.*
Myriodactylon, Desv.
Hydrocoryne, *Schwab.*
Coleochæte, *Breb.*

Suborder V.—*Siphoneæ.* Frond either monosiphonous, that is, consisting of a single cell, usually branched in various ways, with the branches continuous or jointed, distinct or variously united; or pleiosiphonous, consisting of many tubular cells, placed in contact, branched, and variously united or held together by means of intercellular matter.—Marine plants usually covered with calcareous incrustations.

Tribe 1. Caulerpidæ.—*Frond monosiphonous, continuous, variously branched, and filled with the reticulated fibres of the continuous branch.*

Caulerpa, *Lamx.*
Chauvinia, Bory.
Tricladia, *Dec.*

Tribe 2. Acetabularidæ.—*Frond monosiphonous, jointed, with radiating or flabelliform branches at the end; the branches continuous, separate, or combined.*

Polyphisa, *Lamx.*
Acetabularia, *Lamx.*
  Acetabulum, Tourn.
  Callopilophorum, Don.
  Olivia, Bert.
Rhipidosiphon, *Mont.*

Tribe 3. Halymedidæ.—*Frond polysiphonous, made up of tubes which are continuous or jointed, and branched more or less densely.*

Udotea, *Lamx.*
  Flabellaria, Link.
  Rhipozonium, Kütz.
Avrainvillæa, *Dec.*
Halymeda, *Lamx.*
Penicillus, *Lamx.*
  Nesea, Lamx.
  Coralliodendron, Ktz.
Espera, *Dec.*
Anadyomene, *Lamx.*
? Dictyosphæria, *Dec.*
Tetraspora, *Dec.*
  Pexisperma, Raf.
Bangia, *Lyngb.*
  Spermagonia, Bonnem.
  Prasiola, Menegh.
Stigonema, *Ag.*
  Girardia, Gray.
Zignoa, *Trevis.*
  Percursaria, Bory.
  Enteromorpha, J. Ag.
Ulva, *Ag.*
  Enteromorpha, Lk.
  Ilea, Fries.
  Hydrosolen, Mart.
  Tubularia, Rouss.
  Fistularia, Grev.
  Ulvastrum, D. C.
  Halithridax, Targ.
  Ramularia, Rouss.
  Phylloma, Wigg.
  Tepposa, Lk.
Porphyra, *Ag.*

NUMBERS. GEN. 66. Sp. 368. (Endl.)

POSITION.—Diatomaceæ. CONFERVACEÆ.—Fucaceæ.

ORDER III. FUCACEÆ.—SEAWEEDS.

Phyceæ, *Endl. Gen. Supp.* iii. p. 19. (1843).—Aplosporeæ, *Decaisne in Ann. Sc. Nat.* 2 *ser.* 17, 305.

DIAGNOSIS.—*Cellular or tubular unsymmetrical bodies, multiplied by simple spores formed externally.*

Plants sometimes inhabiting fresh water, but more frequently salt water; the former approaching closely to Confervas. Frond either monosiphonous, consisting of a single cell, which is sometimes uninterruptedly branched, or more commonly polysiphonous, composed of several cells, various in form, placed one above the other, or interwoven, barked or barkless, jointed or continuous, thread-shaped, or of various figures, and not uncommonly divided into a sort of trunk and leaflike blade. Mode of growth by division of the cells; of branching by lateral increase or a vague proliferousness. Mode of propagation by spores, contained in superficial cells, which are often bladdery (and called VESICLES), growing singly out of thin colouring matter, consisting of a single nucleus clothed by its proper cellular membrane (or EPISPORE), and discharged by the opening of a transparent mother cell (or PERISPORE). VESICLES (or original mother cells) scattered through the whole frond, or seated in particular parts of it, (often the points of the branches), sometimes on a peculiar receptacle, naked, or supported by small branches.—(*Endlicher.*)

1

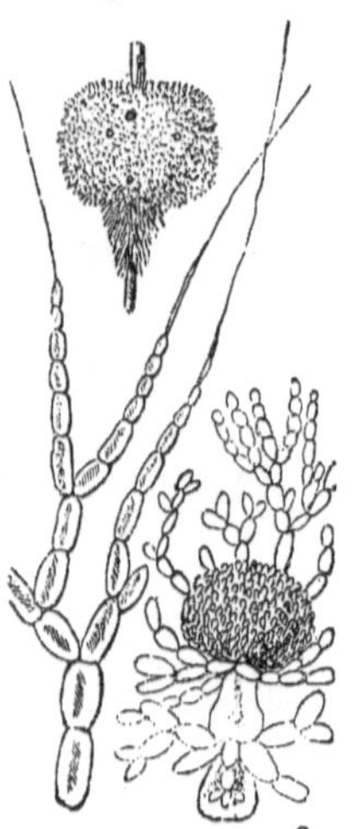

2

3

Fig. VIII.

The reproductive bodies of these plants distinguish them from others of the alliance. In the words of Decaisne "they are simple, and result neither from a modification of green matter, nor from its concentration in a pre-existing cell; their structure is quite peculiar. In the beginning they are little warts, invested by a very thin membrane, placed close over an inner sac filled with green granules." (The black or brown colour assigned to them by Mr. Harvey is a mistake arising out of imperfect observation.) "All the spores are external, that is to say, inserted on the surface of a vesicle upon which they are generated. They are never found in the interior of the frond as in Confervas; and if in Seaweeds they can be compared, in consequence of their being contained in a common chamber or conceptacle, to the spores of certain Rosetangles, it can only be to the corpuscles enclosed in the organs named Ceramidia by the younger Agardh, which however never have the double integument of Seaweeds. In most of the latter the spores appear at the base of certain flocks or filaments, which are simple or jointed, thread-shaped or dilated, or more or less filled with green matter; these flocks are wanting however in the greater part of the Dictyotidæ, and their use is wholly unknown. There is no reason to suppose them male organs." Decaisne, indeed, in one place, treats as an absurdity Donati's calculation that a single individual of a Cystoseira (Acinaria) bears 545,000 male flowers and 1,728,000 females.

The younger Agardh, however, has within a few months expressed his deliberate opinion that in the Rosetangles (his Florideæ) organs analogous to sexes are present. "I am very much inclined," he says, "to adopt the opinion that the two sorts of fructification observable among them are the first attempts at the agency which in higher plants perform the office of sexes, without however having their qualities established, and each capable of producing a new plant without the aid of the other." See his pamphlet called *In systemata Algarum hodierna Adversaria* (p. 8,) in which the reader will find abundant criticism of the views of Kützing and others concerning the Algal alliance.

M. Decaisne seems also to have altered his opinion upon this subject, for (*Comptes Rendus*, Nov. 11, 1844,) he and M. Thuret now describe what they suppose to be sexual organs in Fucus serratus, and other species, to which they even apply the Linnean names Monœcious and Diœcious. They describe the conceptacles of the males as being filled with articulated filaments bearing numerous antheridia in the form of vesicles containing red granules. "These antheridia are expelled by the orifice of the conceptacles; if we examine them with a microscope, we see issue from one of their extremities transparent somewhat pear-shaped bodies, each enclosing a red granule. Every one of such bodies is furnished with very thin ciliæ, by means of which it moves with very great

Fig. VIII.—1. Batrachospermum moniliforme; 2. portion of a branch; 3. summit of a branch, bearing a cluster of spores. (*Decaisne.*)

activity." Such bodies are regarded as analogous to the spiral threads of mosses and other cryptogamic plants. Indeed, according to M. Thuret, such threads are also furnished with ciliary locomotive organs. But what proof is there that these curious bodies are pollen?

One of the most remarkable plants of the order is the Hydrogastrum, which Endlicher describes as a perfect plant, with root, stem, bud, and fruit, in imitation of the most highly developed races, *but all produced by the branching of one single cell.*

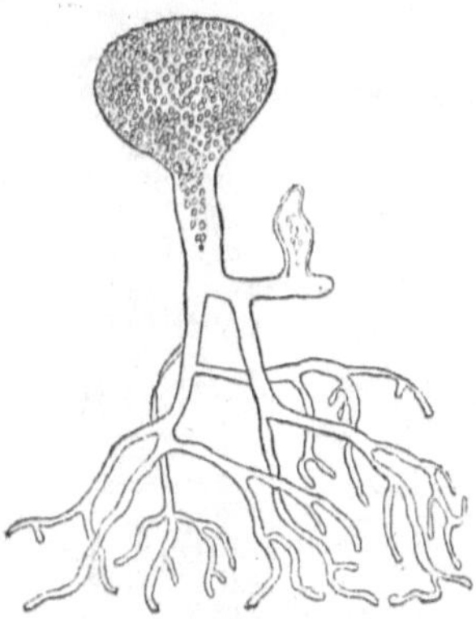

Fig. IX.

Professor Morren thinks that he has ascertained that the animalcule called Rotifer vulgaris, is actually generated in the cells of Vaucheria clavata. He lives in certain protuberances formed on the stem of that plant, travels quite at his ease within them, traverses the partitions, displaces the colouring matter. (*Ann. Nat. Hist.* vi. 346.)

Like all this alliance the Seawracks have no particular geographical limits, but occur wherever the ocean or rivers spread themselves over the land. They are, however, remarkable for the enormous space which single species of them occasionally occupy; some of them forming subaqueous forests in the ocean, emulating in their gigantic dimensions the boundless element that enfolds them. Scytosiphon filum, a species common in the North Sea, is frequently found of the length of 30 or 40 feet; in Scalpa Bay, in Orkney, according to Mr. Neill, this species forms meadows, through which a pinnace with difficulty forces its way. Lessonia fuscescens is described by Bory de St. Vincent as 25 or 30 feet in length, with a trunk often as thick as a man's thigh. But all these, and indeed every other vegetable production, is exceeded in size by the prodigious fronds of Macrocystis pyrifera. "This appears to be the sea-weed reported by navigators to be from 500 to 1500 feet in length: the leaves are long and narrow, and at the base of each is placed a vesicle filled with air, without which it would be impossible for the plant to support its enormous length in the water; the stem not being thicker than the finger, and the upper branches as slender as common packthread." This plant, and Durvillæa utilis, was seen by Dr. Joseph Hooker in lat. 61° S. in large vegetating patches, wherever the water was free of icebergs; and Scytothalia Jacquinotii as low as 63° S.

Some of the species are eatable, owing doubtless to the large quantity of gelatinous matter that they secrete. The young stalks of Laminaria digitata and saccharina are eaten under the name of "tangle." In Asia, Sargassum acanthocarpum and pyriforme, with Laminaria bracteata, and in the Sandwich Islands, Sargassum cuneifolium, are also used for food. When stripped of the thin part, the beautiful Alaria esculenta forms a part of the simple fare of the poorer classes of Ireland, Scotland, Iceland, Denmark, and the Faroe Islands. The large Laminaria potatorum of Australia furnishes the aborigines with a proportion of their 'instruments, vessels, and food.' On the authority of Bory de St. Vincent, the Durvillæa utilis and other Laminaridæ constitute an equally important resource to the poor on the west coast of South America. In some of the Scottish islands, horses, cattle, and sheep, feed chiefly upon Fucus vesiculosus during the winter months; and in Gothland it is commonly given to pigs. Fucus serratus also, and Scytosiphon filum, constitute a part of the fodder upon which cattle are supported in Norway. In the manufacture of kelp, for the use of the glass-maker and soap-boiler, Seawracks take their place among the more useful vegetables. The species most valued for this purpose are, Fucus vesiculosus, nodosus, and serratus, Laminaria digitata and bulbosa, Himanthalia lorea, and Scytosiphon filum. It is principally, indeed, because of the quantity of soda which they contain that they are found so useful as manures. In medicine they have been occasionally employed, as, for instance, Fucus vesiculosus in Europe against scrofula, Sargassum vulgare in Portuguese India against calculus, and Sarg. bacciferum with some Laminarias in South America against tumours and strangury. But whatever medical value they possess seems to be owing to the presence of Iodine, which may be obtained either from the plants themselves, or from kelp. French kelp, according to Sir Humphry Davy, yields more Iodine than British; and, from some experiments made at the Cape of Good Hope, Ecklonia buccinalis is found to contain more than any European sea-weed. Iodine is known to be a

---

Fig. IX. Hydrogastrum, (*Endlich. grundz*, p. 54, f. 47.)

powerful remedy in cases of goître. The burnt sponge formerly administered in similar cases, probably owed its efficacy to the Iodine it contained; and it is also a very curious fact, that the stems of a sea-weed are sold in the shops, and chewed by the inhabitants in South America, wherever goître is prevalent, for the same purpose. This remedy is termed by them Palo Coto (literally, goître-stick), and consists of fragments of the Sargassum bacciferum and Laminarias above alluded to. Iodine is principally obtained in Europe from the ashes of the Fuci vesiculosus, nodosus, ceranoides, and serratus.

GENERA.

Suborder I.—*Vaucheriæ.* Frond mono-or pleiosiphonous, without bark. The utricles forming a lateral branchlet, proceeding either from the upper joint of the branch, or occasionally from the lowest.

Tribe 1. Hydrogastridæ—*Frond produced from a single vesicle or tube, rarely from several that are continuous and loosely interwoven.*

Hydrogastrum, *Desv.*
Botrydium, Wallr.
Rhizococcum, Desmaz.
Vaucheria, *D. C.*
Ectosperma, Vauch.
Bryopsis, *Lamx.*
Valonia, *Ginnan.*
Physydrum, Raf.?
Codium, *Stack.*
Lamarkia, Olivi.
Agardhia, Cabrera.
Spongodium, Lamx.

Tribe 2. Dasycladidæ.—*Frond monosiphonous, continuous, or jointed, with verticillate branches, which are fastigiate, jointed, and have the last joint transformed into a vesicle.*

Chamædoris, *Mont.*
Dasycladus, *Agh.*
Myrsidium, Raf.
Neomeris, *Lamx.*
Cymopolia, *Lamx.*

Tribe 3. Ectocarpidæ.—*Threads jointed, consisting of a single row of cells, variously branched. Vesicles derived from one joint, either at the end of the branches, or of the laterals.*

Leiblinia, *Endl.*
Desmarestella, Bory.
Chantransia, *Fries.*
Audrienella, Bory.
Genicularia, Rons.
Ectocarpus, *Lyngb.*
Lyngbya, Gaillon.
Macrocarpus, Bonnem
Opospermum, Raf.?
Calospermum, Raf.?
Pylaiella, Bory.
Lyngbyella, Bory.
Bulbochæte, *Agh.*

Tribe 4. Batrachospermidæ.—*Frond polysiphonous, composed of a primary thread surrounded by parallel accessory ones. Vesicles terminal or lateral, clustered.*

Batrachospermum, *Roth.*
Charospermum, Lk.
Draparnaldia, Bory.
Monilina, Bory.
Thorinia, Bory.
Lemanina, Bory.
Gelatinaria, Roussel.
Torularia, Bonnem.
Liagora, *Lamx.*
Actinotrichia, *Decaisne.*
Galaxaura, *Lamx.*
Dichotomaria, Lamk.
Alysium, Agh.
Microthoe, Dec.
Thorea, *Bory.*
Polycoma, Palis.
Myriocladia, *J. Agh.*
Ægira, Fries?

Tribe 5. Chordaridæ.—*Frond polysiphonous, with flocks proceeding in all directions from the medullary substance, free in the circumference.*

Cruoria, *Fries.*
Myrionema, *Grev.*
Elachista, Aresch.
Mesogloia, *Agh.*
Chordaria, *Agh.*
Leathina, *Gray.*
Corynephora, Agh.
Clavaletta, Bory.
Liebmannia, *J. Agh.*

Suborder II.—*Halysereæ.* Frond polysiphonous, barked, jointed, or continuous. Vesicles scattered over the surface of the frond, or collected into heaps.

Tribe 1. Sphacelaridæ.—*Frond jointed; vesicles lateral, solitary.*

Sphacelaria, *Lyngb.*
Delisella, Bory.
Lyngbyella, Bory.
Myriotrichia, *Harvey.*
Cladostephus, *Agh.*

Tribe 2. Dictyotidæ.—*Frond continuous, membranous. Vesicles supported by flocks, collected in heaps, or scattered over the upper surface of the frond.*

Halyseris, *Targ.*
Neurocarpon, Web.
Dictyopteris, Lamx.
Polypodioides, Stack.
Dictyosiphon, *Grev.*
Dictyota, *Lamx.*
Zonaria, *J. Agh.*
Stifftia, Nardo.
Zanardinia, Nardo.
Padina, *Adans.*
Trattinickia, Web.
Padinella, *Aresch.*
Cutleria, *Grev.*
Arthrocladia, *Duby.*
Elaionema, Berk.
Scytosiphon, *Agh.*
Chorda, Stack.
Filum, Stack.
Chordaria, Lk.
Soranthera, *Postels.*
Punctaria, *Grev.*
Asperococcus, *Lamx.*
Encœlium, Agh.
Hydroclathrus, Bory.
Striaria, *Grev.*
Carmichaelia, Grev.
Stilophora, *J. Agh.*
? Hildenbrandia, *Nardo.*
Ralfsia, Berk.

Tribe 3. Laminaridæ.—*Frond continuous, coriaceous, sometimes bearing bladders. Vesicles scattered, or collected in heaps, supported by flocks, growing on both sides of the frond.*

Lessonia, *Bory.*
Macrocystis, *Agh.*
Nereocystis, *Postels.*
Ecklonia, *Hornem.*
Laminaria, *Lamx.*
Gigantea, Stack.
Saccharina, Stack.
Musæfolia, Stack.
Polyschidia, Stack.
Palmaria, Lk.
Laminastrum, Duby.
Fasciata, Gray.
Capea, *Montagn.*
Haligeria, *Dec.*
Alaria, *Grev.*
Orgya, Stackh.
Thalassiophyllum, *Post.*
Agarum, *Grev.*
Myriotrema, Lapyl.
Costaria, *Grev.*

Tribe 4. Sporochnidæ.—*Frond continuous, between cartilaginous and membranous, flocks formed astride a capitate receptacle, bearing the vesicles.*

Sporochnus, *Agh.*
Desmarestia, *Lamx.*
Desmia, Lyngb.
Dichlora, Grev.
Trinitaria, Bory.
Hippurina, Stack.
Hyalina, Stack.
Flagellaria, Stack.

Suborder III.—*Fuceæ.* Frond polysiphonous, often bladdery. Vesicles seated in hollow conceptacles formed of a folding in of the frond, pierced by a pore, and surrounded by flocks; conceptacles scattered or collected upon a receptacle.

Tribe 1. Lemanidæ.—*Frond hollow, wholly converted into a receptacle.*

Lemanea, *Bory.*
Nodularia, Link.
Gongycladon, Link.
Trichogonea, Palis.
Vertebraria, Rouss.

Tribe 2. Fucidæ.—*Conceptacles not collected upon a receptacle.*

Fucus, *L.*
Cervina, Gray.
Halidrys, Stack.
Bifurcaria, Stack.
Ozothalia, *Dec. and Th.*
Pelvetia, *Dec. and Th.*
Carpodesmia, *Grev.*
Myriodesma, *Dec.*
Himanthalia, *Lyngb.*
Lorea, Stack.
Xiphophora, *Montagn.*
Splachnidium, *Grev.*
Durvillæa, *Bory.*
Hormosira, *Endl.*
Moniliformia, Lamx.
Monilia, A. Rich.
Castraltia, *A. Rich.*
Scaberia, Grev.

Tribe 3. Cystoseiridæ.—*Conceptacles or receptacles distinct from the frond.*

Coccophora, *Grev.*
Halidrys, *Lyngb.*
Siliquaria, Gray.
Blossevillea, *Dec.*
Cystophora, J. Agh.
Cystoseira, *Agh.*
Acinaria, Targ.
Machaia, Gray.
Catenaria, Raf.
Ascophylla, Stack.
Ericaria, Stack.
Monilifera, Stack.
Sargassum, *Rumph.*
Baccularia, Gray.
Halochloa, Kütz.
Myagropsis, Kütz.
Spongocarpus, Kütz.
Turbinaria, *Bory.*
Carpacanthus, *Kütz.*
Phyllospora, *Agh.*
Carpophyllum, *Grev.*
Marginaria, *A. Rich.*
Scytothalia, *Grev.*
Stackhousia, Lamx.
Seirococcus, *Grev.*
Polyphacum, *Agh.*
Osmundaria, Lamx.

NUMBERS. GEN. 81. Sp. 452. (Endl.)

POSITION.—Confervaceæ. FUCACEÆ.—Ceramiaceæ.

ORDER IV. CERAMIACEÆ.—ROSETANGLES.

FLORIDEÆ. *J. Agardh, Alg. Med.* 54. (1842); *Endl. Gen. Supp.* iii. 33.—CHORISTOSPOREÆ, *Decaisne in Ann. Nat. Hist.* 2 *ser.* 17, 306. (1842).

DIAGNOSIS.—*Cellular or tubular unsymmetrical bodies, multiplied by tetraspores.*

Seaweeds of a rose or purplish colour, seldom olive or violet. Their cells long and tubular, or round and short, or polygonal; sometimes arranged in a single row; sometimes disposed in several parallel rows, and of equal length, forming an articulated frond; sometimes in several rows, and of unequal length, when they constitute a cellular frond. The propagation by means of spores (called also Sphærospores and Tetraspores), formed in fours (or threes), within a transparent perispore, or mother cell, and collected in bodies of many different forms and structure.*

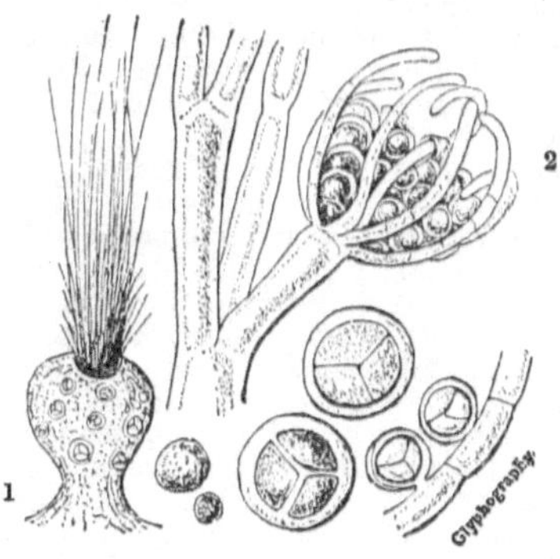

Fig. X.

The subdivision of the reproductive bodies or tetraspores into four, or occasionally three particles, is the great feature of this natural order, and at once distinguishes it from the rest of the alliance. M. Decaisne lays great stress upon this point, first used by himself for systematical purposes, and he attaches quite a secondary value to the various modes in which such spores are grouped. To rank those modes more highly *c'était sacrifier évidemment une foule des considérations de la plus haute valeur à un caractère qui n'a d'autre importance que d'être plus visible, et par suite plus facile à saisir que le premier.* It is, however, a very striking peculiarity of the Rosetangles, that they should have so much greater a variety of fructification than their allies, and this, in connection with the quaternary structure of their spores, seems to indicate their being the highest form of the Algal alliance.

Although the subdivision of the spores by four is of uniform occurrence among these plants, yet it takes place in different ways, and is subject to certain modifications, concerning which the language of M. Decaisne is instructive. "I have shewn," he says, "in another place, that the sphærospores, or quaternary reproductive bodies, which M. Kützing has perhaps better called Tetraspores, offer three modifications. They are either little spheres, which divide into four wedge-shaped particles with a round base (Delesseria, Ceramium, &c.); or oblong bodies, which are cut across into four distinct spores (Hypnea, Catenella, &c.); or, finally, oblong bodies, which divide vertically and transversely, so as to form segments of cylinders, rounded at one extremity, and truncate at the other, as in Peysonnelia. The mode of formation, and the essential organisation of these spores, is the same in each type, whether the tetraspores project beyond the tissue, or are organised in the interior of the frond. When young, the tetraspores show no exterior membrane, but appear as a reddish spherule, the development of which may be followed

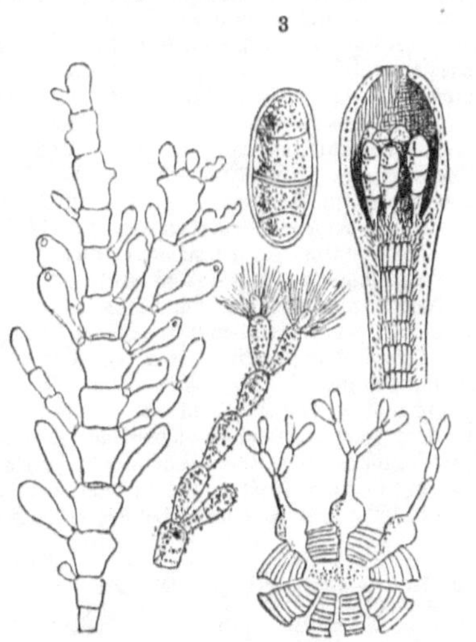

Fig. XI.

* For the explanation of the terms invented to express these forms, see *Decaisne in Ann. des Sc. Nat.* 2 *ser.* 17, 348.

Fig. X.—1. Chondria obtusa; 2 Griffithsia sphærica; 3. Gr. corallina.
Fig. XI.—Magnified branch of Corallina officinalis; 2. a section of its spore case (ceramidium) with the tetraspores in situ; 3. a tetraspore; 4. Cymopolia barbata; 5. a cross section of the stem of Dasycladus clavæformis, showing its rings of growth.

in the different species of Griffithsia. We see them enlarge for a certain space of time, and present the appearance of a rose-coloured globule ; but at a more advanced period the external envelope dilates, becomes transparent, and the central body, considerably increased in size, tends to separate into four parts or distinct spores, each invested with a special envelope, and of the most brilliant carmine colour. This structure brings to mind, with some slight differences, that of pollen grains." And then M. Decaisne goes on to explain how, by a stoppage of growth, or by interior multiplication, the quaternary character of these bodies may be affected.

According to Endlicher, the maximum of this order is found in the ocean between 35° and 48° N. lat. They are entirely marine. Towards the pole and the equator they diminish in numbers, and are comparatively rare in the southern hemisphere. Rhododermis Drummondi covers the rocks of caves with patches of a dark blood or brick-red colour.

It is among the genera of this order that occur the seaweeds whose gelatinous qualities render them valuable as food. Many species are so used among Indian nations. Of them Plocaria tenax, and candida, are the principal ; and the material out of which the swallows construct the esculent nests which are so highly valued by the Chinese, is supposed to be a sort of Gelidium. The British Plocaria compressa, and Chondrus crispus (or Carrageen moss), have been found to possess similar qualities ; and another species of the order, on the south-west coast of New Holland, furnishes a jelly of great excellence. Rhodomenia palmata, the dulse of the Scots, dillesk of the Irish, and saccharine Fucus of the Icelanders, is consumed in considerable quantities throughout the maritime countries of the north of Europe, and in the Grecian Archipelago ; Iridæa edulis is still occasionally used, both in Scotland and the south-west of England. Laurentia pinnatifida, distinguished for its pungency, and hence called Pepperdulse, is eaten in Scotland ; and even now, though rarely, the old cry, "Buy dulse and tangle," may be heard in the streets of Edinburgh.

But it is not to mankind alone that such marine Algals have furnished luxuries, or resources in times of scarcity. Several species are greedily sought after by cattle, especially in the north of Europe. Rhodomenia palmata is so great a favourite with sheep and goats, that Bishop Gunner named it Fucus ovinus. One species is invaluable as a glue and varnish to the Chinese. This is the Plocaria tenax, the Fucus tenax of Turner's *Historia Fucorum.* Though a small plant, the quantity annually imported at Canton from the provinces of Fokien and Tchekiang is stated by Mr. Turner to be about 27,000 lbs. It is sold at Canton for 6*d.* or 8*d.* per pound, and is used for the purposes to which we apply glue and gum-arabic. The Chinese employ it chiefly in the manufacture of lanterns, to strengthen or varnish the paper, and sometimes to thicken or give a gloss to silks or gauze. It seems probable that this is the principal ingredient in the celebrated gummy matter called Chin-chon, or Hai-tsai, in China and Japan Windows made merely of slips of Bamboo, crossed diagonally, have frequently their lozenge-shaped interstices wholly filled with the transparent gluten of the Hai-tsai On the southern and western coasts of Ireland, our own Chondrus crispus is converted into size, for the use of house-painters.

In medicine we are not altogether unindebted to Rosetangles. The Plocaria Helminthochorton, or Corsican Moss, as it is frequently called, is a native of the Mediterranean, and had once a considerable reputation as a vermifuge. To Hypnea musciformis similar qualities are ascribed in the Greek Archipelago. Several species furnish Iodine, which gives them an odour of violets. Rytiphlœa tinctoria yields a red dyeing matter, the *Fucus* of the ancients. The Plocaria candida, or Fucus amylaceus, has been found to consist of pectine, gum, and starch, with a pretty considerable quantity of inorganic matter, especially sulphate of lime. (*Ch. Gaz.* 1843, 638.) The Tsantjan or Kanten (called Fucus cartilaginosus), used in China as a substitute for the edible birds'-nests, seems to have a similar composition.

GENERA.

Suborder I. — *Cerameæ.* Frond tubular, jointed. Favellæ containing a loose mass of semi-transparent granules in a gelatinous envelope. Tetraspores external.

Callithamnion, *Lyngb.*
Ballia, *Harvey.*
Griffithsia, *Agh.*
  Plumaria, Lk.
  Polychroma, Bonnem.
Wrangelia, *Agh.*
Spyridia, *Harv.*
Bindera, *J. Agh.*
Ceramium, *Adams.*
  Boryna, Gratel.
  Dictyderma, Bonnem.
Ptilota, *Agh.*
  Plumaria, Stackh.
Microcladia, *Grev.*
? Haplolegma, *Mont.*

Suborder II.—*Cryptonemeæ.* Frond cellular. Favellidia containing a firm mass of compact granules within a gelatinous envelope. Tetraspores globose or oblong, formed out of cells of the circumference.

a) Gloiocladidæ.
Crouania, *J. Agh.*
Dudresnaya, *Bonnem.*
Naccaria, *Endl.*
  Chætospora, Agh.
  Capillaria, Stackh.
Gloiocladia, *J. Agh.*
Gloiopeltis, *J. Agh.*
Nemalion, *Targ.*
  Helminthora, Fries.

b) Nemastomidæ.
Catenella, *Grev.*
Endocladia, *J. Agh.*
Iridæa, *Bory.*
  Nemastoma, J. Agh.
  Dilsea, Stackh.

c) Spongiocarpidæ.
Furcellaria, *Lamx.*
  Fastigiaria, Stackh.
Polyides, *Agh.*
  Spongiocarpus, Grev.
Rhododermis, *Harv.*
Thuretia, *Dec.*

Peysonnellia, *Dec.*
Squamaria, Zanard.
Pterigospermum, Targ.
Phyllophora, *Grev.*
Prolifera, Stackh.
Membranifolia, Stack.
Stenogramma, *Harv.*
Chondrus, *Grev.*
Polymorpha, Stackh.
Gymnogongrus, *Mart.*
Ahnfeldia, Fries.
Dasyphlæa, *Mont.*

d) Gasterocarpidæ.
Dumontia, *Lamx.*
Halymenia, *Agh.*
Kallymenia, *Agh.*
Constantinea, Postels.
Ginannia, *Mont.*

e) Coccocarpidæ.
Cryptonemia, *J. Agh.*
Gelidium, *Lamx.*
Suhria, *J. Agh.*
Grateloupia, *Agh.*
Phoracis, Raf.
Gigartina, *Lamx.*
Mammillaria, Stack.
Chrysymenia, *J. Agh.*

f) Ctenodontidæ, *Mont.*
Ctenodus, *Kütz.*
Nothogenia, *Mont.*

Suborder III.—*Lomentareæ.* Frond cellular. Ceramidia having pear-shaped granules at the base of a cup-shaped envelope, which finally bursts by a pore. Tetraspores scattered within the branches.

Lomentaria, *Lyngb.*
Chylocladia, Grev.
Gastridium, Grev.
Kaliformia, Stackh.
Sedoidea, Stackh.
Champia, *Agh.*
Mertensia, Roth.
Laurencia, *Lamx.*
Cornea, Stackh.
Osmundia, Stackh.
Asparagopsis, *Mont.*
Lictoria, J. Agh.
Bonnemaisonia, *Agh.*
Capillaria, Stackh.
Calocladia, Grev.
Bowiesia, Grev.
Thysanocladia, *Endl.*
Delisea, Lamx.
Mammea, J. Agh.
Lenormandia, Mont.

Suborder IV.—*Rhodomeleæ.* Frond jointed. Ceramidia as before. Tetraspores enclosed in transformed branches or Stichidia.

Dasya, *Agh.*
Stichocarpus, Agh.
Rhodonema, Martius.
Asperocaulon, Grev.
Grateloupia, Bonnem.
Ellisius, Gray.
Gaillona, Bonnem.
Baillouviana, Gris.
Polysiphonia, *Grev.*
Hutchinsia, Agh.
Grammita, Bonnem.
Corradoria, Mart.
Vertebrata, Gray.
Dicarpella, Bory.
Brongniartella, Bory.
Gratiloupella, Bory.
Heterosiphonia, *Mont.*
Alsidium, *Agh.*
Amphibia, Stackh.
Bostrychia, Mont.
Helicothamnium, Kütz.
Digenea, *Agh.*
Rhodomela, *Agh.*
Fuscaria, Stackh.
Acanthophora, *Lamx.*
Pollexfexia, *Harv.*
Dictyomenia, *Grev.*
Volubilaria, Lamx.
Spirhymenia, Dec.
Carpophyllum, Suhr.
Botryocarpa, *Grev.*
Odonthalia, *Lyngb.*
Fimbriaria, Stackh.
Rytiphlœa, *Agh.*
Polyzonia, *Suhr.*
Leveillea, *Dec.*
Amansia, *Lamx.*
Heterocladia, *Dec.*
*Corallineæ.
Corallina, *Tourn.*
Titanephyllum, Nardo.
Jania, *Lamx.*
Haliptilon, *Dec.*
Amphiroa, *Lamx.*
Arthrocardia, *Dec.*
Eurytion, *Dec.*
Cheilosporum, *Dec.*
Melobesia, *Lamx.*
Agardhia, Mengh.
Lithophyllum, Philip.
Spongites, Kütz.
Nullipora, Lam.
**Anomalophylleæ.
Dictyurus, *Bory.*
Calidictyon, Grev.
Hemitrema, *R. Br.*
Martensia, Her.
Claudea, *Lamx.*
Lamourouxia, Agh.
Oneillia, Agh.
? Thaumasia, *Agh.*

Suborder V. — *Sphærococceæ.* Frond cellular. Coccidia enclosing closely-packed oblong granules arising from the base, within a spherical cellular envelope which finally bursts: Tetraspores in indefinite heaps, scattered over the frond.

Hypnea, *Lamx.*
Plocaria, *Nees.*
Gracilaria, Grev.
Helmintochortos, Lk.
Rhodomenia, *Grev.*
Palmaria, Stackh.
Bifidia, Stackh.
Ciliaria, Stackh.
Heringia, *J. Agh.*
Sphærococcus, *Grev.*
Coronopifolia, Stackh.

Suborder VI. — *Delesseriæ.* Frond cellular. Coccidiæ as before. Tetraspores in definite heaps, or collected in Sporophylls.

Plocamium, *Grev.*
Nereidea, Stackh.
Thamnophora, *Agh.*
Aglaophyllum, *Mont.*
Nitophyllum, Grev.
Papyracea, Stackh.
Dawsonia, Bory.
Wormskioldia, Spreng.
Hymenena, *Grev.*
Delesseria, *Lamx.*
Hydrolapatha, Stackh.
Membranoptera,
Solieria, *J. Agh.*
Acropeltis, *Mont.*
? Hydropuntia, *Mont.*

NUMBERS. GEN. 88. Sp. 682. (Endl.)

POSITION.—Fucaceæ. CERAMIACEÆ.—Characeæ.

ORDER V. CHARACEÆ.—CHARAS.

CHARACEÆ, *Rich in Humb. et Bonpl. N. G. Pl.* 1, 45. (1815) ; *A. Brong. in Dict. Class.* 3. 474. (1823) ; *Grev. Fl. Edin.* xvii. (1824) ; *Endlich. Gen.* iv. ; *Schnitzl. ic.*—CHAREÆ, *Kützing, Phycologia*, p. 313.

DIAGNOSIS.—*Tubular symmetrically branched bodies, multiplied by spiral-coated nucules, filled with starch.*

Water plants composed of an axis, consisting of parallel tubes, which are either transparent or encrusted with carbonate of lime, and of regular whorls of symmetrical tubular branches. Organs of reproduction, lateral, round, succulent, brick-red globules, and axillary nucules. The globules, consisting of triangular valves, enclosing centripetal tubes and slender annular threads ; the nucules having two coats, of which the external is transparent and usually surmounted by five teeth ; the internal firm, spirally-ribbed, filled with starch granules of various sizes.

The genera of which this little order is composed are among the most obscure of the vegetable kingdom, in regard to the nature of their reproductive organs ; and accordingly we find them, under the common name of Chara, placed by Linnæus among Cryptogamous plants near Lichens ; then referred by the same author to Phænogamous plants, in Monœcia Monandria ; retained by Jussieu and De Candolle among Naiads, by Brown at the end of Hydrocharaceæ, and by Leman in Halorageæ ; referred to Confervas by Von Martius, Agardh, and Wallroth ; and finally admitted as a distinct order, upon the proposition of Richard, by Kunth, De Candolle, Adolphe Brongniart, Greville, Hooker, and others. Such being the uncertainty about the place of these plants, it will be useful to give a rather detailed account of their structure, in which I avail myself chiefly of Ad. Brongniart's remarks in the place above referred to, and of Agardh's observations in the *Ann. des Sciences*, 4. 61.

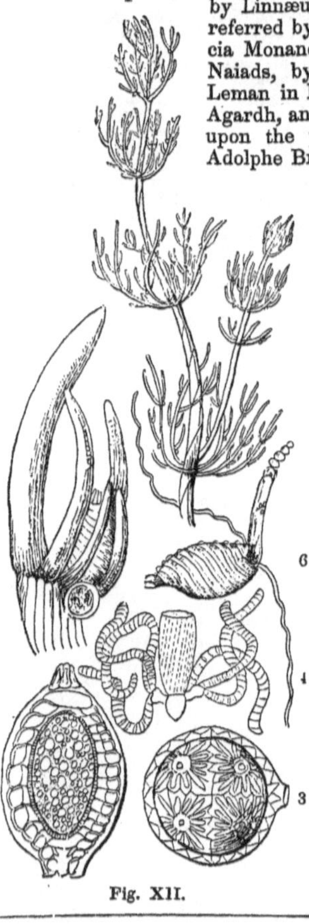

Fig. XII.

Charas are aquatic plants, found in stagnant fresh or salt water ; always submersed, giving out a fetid odour, and having a dull greenish colour. Their stems are regularly branched, brittle, and surrounded here and there by whorls of smaller branches. In Nitella the stem consists of a single transparent tube with transverse partitions ; Agardh remarks that it is so like the tubes of some Algals, as to offer a strong proof of the affinity of the orders. In Chara, properly so called, there is, in addition to this tube, many other external ones, much smaller, which only cease to cover the central tube towards the extremities. In the axils of the uppermost whorls of these branchlets the organs of reproduction take their origin ; they are of two kinds, one called the nucule, the other the globule ; the former has been supposed to be the pistil, the latter the anther.

The nucule is described by Greville as being " sessile, oval, solitary, spirally striated, having a membranous covering, and the summit indistinctly cleft into five segments ; the interior is filled with minute sporules. *Fl. Edin.* xvii. This is the general opinion entertained of its structure. But Brongniart describes it thus :—Capsule unilocular, monospermous ; pericarp composed of two envelopes : the outer membranous, transparent, very thin, terminated at the upper end by five spreading

Fig. XII.—1. Chara vulgaris ; 2. a portion of a branch with a nucule and globule ; 3. the globule more magnified ; 4. the spiral tubes of the latter ; 5. a nucule cut open ; 6. a nucule in germination.

teeth; the inner hard, dry, opaque, formed of five narrow valves, twisted spirally." *Dict. Class.* l. c. He founds his opinion of the nucule containing but one germinating body upon the experiments of Vaucher, of Geneva, who ascertained that if ripe nucules of Chara, which have fallen naturally in the autumn, are kept through the winter in water, they will germinate about the end of April; at that time a little body protrudes from the upper end between the five valves, and gradually gives birth to one whorl of branches, which produce a second. Below these whorls the stem swells, and little tufts of roots are emitted. The nucule adheres for a long time to the base of the stem, even when the latter has itself begun to fructify. Hence it is reasonable to conclude that the nucule is really one-seeded. Brongniart remarks, that it is true, when a fresh nucule of Chara is cut across, an infinite number of little white grains are squeezed out; but if these were really all reproductive particles, how would they ever find their way out of the nucule, which is indehiscent? he considers them rather of the nature of albumen. And he is the more confirmed in his opinion, because in Pilularia, the thecæ of which also contain many similar grains, but one plant is produced by each theca. These grains have been ascertained by the observations of Kützing to be really starch, iodine colouring them violet; yet Endlicher describes them as spirally-striated spores. Finally, Amici has described (*Ann. des Sc.* 2.) the nucule in another way. He admits it to be one-seeded, but he considers the points of the five valves to be stigmata, and the valves themselves to be at once pericarp and style. These observations seem to show that the five valves of the nucule, as they are called, are a whorl of leaves, straight at first, and twisted afterwards; and that the nucule itself is analogous to the bud of flowering plants.

The globule is described by Greville as "a minute round body, of a reddish colour, composed externally of a number of triangular (always ?) scales, which separate and produce its dehiscence. The interior is filled with a mass of elastic transversely undulated filaments. The scales are composed of radiating hollow tubes, partly filled with minute coloured spherical granules, which freely escape from the tubes when injured." Vaucher describes them as "tubercles formed externally of a reticulated transparent membrane, containing, in the midst of a mucilaginous fluid, certain white articulated transparent filaments, and some other cylindrical bodies, closed at one end, and appearing to open at the other. These latter are filled with the red matter to which the tubercles owe their colour, and which disappears readily and long before the maturity of the nucule." The account of the globule by Agardh is at variance with both these. "Their surface," he remarks, "is hyaline, or colourless; under this membrane is observed a red and reticulated or cellular globe, which has not, however, always such an appearance; often, instead of this reticulated aspect, the globe is colourless, but marked by rosettes or stars, the rays of which are red or lanceolate. In the figures given by authors, one finds sometimes one of these forms, sometimes the other. I have myself found them both on the same species; and I am disposed to believe that the last state is the true kernel of the globule, concealed under the reticulated scale. (When the globule is very ripe, one may often succeed, by means of a slight degree of pressure, in separating it into several valves, as is very well shown in Wallroth's figures, tab. 2. f. 3. and tab. 5. These valves are rayed, and no doubt answer to the stars, of which mention has been made.) The kernel contains some very singular filaments; they are simple (I once thought I saw them forked), curved and interlaced, transparent and colourless, with transverse striæ, parallel and closely packed, as in an Oscillatoria or Nostoc; but what is very remarkable, they are attached, several together, to a particular organ formed like a bell, which is itself also colourless, but filled with a red pigment. This bell, to the base of which on the outside they are fixed, differs a little in form in different species. It is slender and long in Chara vulgaris, thicker in C. firma, shorter in C. delicatula, and shorter still in C. collabens. I have not succeeded in determining the exact position of these bells in the kernel. I have often thought they were the same thing as the rays of the rosettes or stars upon the globule above mentioned; whence it would follow that they are placed near the surface, while the filaments have a direction towards the centre. The bells are not numerous; they often separate from the filaments, and readily part with their pigment, which renders it difficult to observe them, and has caused them to be overlooked." That these globules, whatever their nature may be, have no resemblance in structure to anthers, is clear from these descriptions, whichever may be eventually admitted. Nevertheless Fritsche, the patient investigator of pollen, regards them as anthers! Wallroth says he has sown them, and that they have germinated; but this observation requires to be verified.

In the annular or chambered threads of Chara are found in abundance little spiral bodies having an active motion when discharged into water, and resembling entirely the so-called animalcules in mosses, &c. M. Thuret, who finds tentacula in the spores of Confervas, ascribes a similar moving apparatus to these bodies, adding that they are turned brown by iodine and not dissolved by ammonia as animalcules are. (*Ann. Sc. N.* 2 *ser.* 14, 65.) They are probably analogous to the elastic spires of Equisetum.

There are two other points deserving of attention in Charas; 1st, the calcareous incrustation of some species; and 2dly, the visible and rapid motion of the sap in the articulation of the stem.

Of the genera, Nitella is transparent and free from all foreign matter; but Chara contains, on the outside of its central tube, a thick layer of calcareous matter, which renders it opaque. This incrustation appears, from the observations of Greville (*Fl. Edin.* 281), not to be a deposit upon the outside, and of an adventitious nature, but a result of some peculiar economy in the plant itself; and according to Brewster, it is analogous to the siliceous deposit in Equisetum, exhibiting similar phenomena.

Whatever is known of the motions of the fluids of vegetables has been necessarily a matter of inference, rather than the result of direct observation; for who could ever actually see the sap of plants move in the vessels destined to its conveyance? It is true that it was known to botanists that a certain Abbé Corti, of Lucca, had, in 1774, published some remarkable observations upon the circulation of fluid in some aquatic plants, and that the accuracy of this statement had been confirmed by Treviranus so long ago as 1817; but the fact does not seem to have attracted general attention until the publication, by Amici, the celebrated professor at Modena, of a memoir in the 18th volume of the *Transactions of the Italian Society*, which was succeeded by another in the 19th. From all these observers it appears, that if the stems of any transparent species of Chara, or of any opaque one, the incrustation of which is removed, are examined with a good microscope, a distinct current will be seen to take place in every tube of which the plant is composed, setting from the base to the apex of the tubes, and returning at the rate, in Chara vulgaris, of about two lines per minute (*v. Ann. des Sc.* 2. 51. line 9); and according to Treviranus this play is at any time destroyed by the application of a few drops of spirit, by pressure, or by any laceration of the tube. Such is the nature of the singular phenomena that are to be seen in Charas. Those who are anxious to become acquainted with the details of Amici's observations will find his first paper translated in the *Annales de Chimie*, 13. 384, and his second in the *Ann. des Sc.* 2. 41; that of Treviranus is to be found in the latter work, 10. 22. The observations made upon Chara circulation by the foregoing authors have been much extended by the careful inquiries of Solly, Slack, and Varley, whose remarks are to be found in the Transactions of the Society of Arts, vol. 49, p. 177, and vol. 50, p. 171; and by Donné, Dutrochet, and others, in the *Ann. Sc. Nat.* 2 *ser. vol.* 9, *pp.* 5, 65, 80, *and* 10, *p.* 346. As however they relate to physiological and not to systematical questions I forbear to dwell upon them in this place.

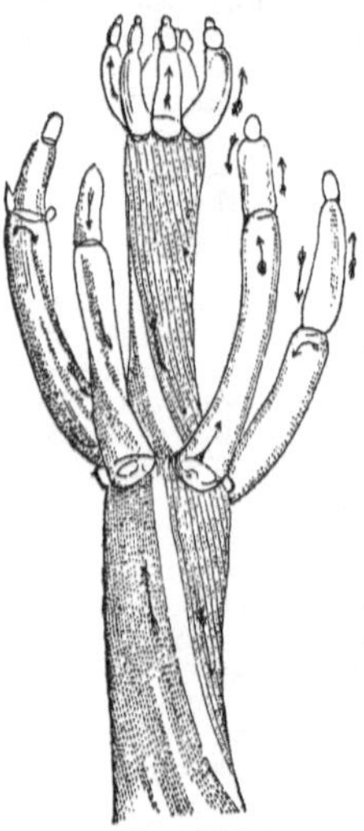

Fig. XII.*

The creation of plants of this order would appear to have been of a very recent date, compared with that of Ferns and Palms, or even Algals, if we are to judge by their fossil remains, called Gyrogonites, which are found for the first time in the lower freshwater formation, along with numerous Dicotyledonous plants resembling those of our own æra. In the recent Flora of the world they make their appearance everywhere in stagnant waters, in Europe, Asia, and Africa, in North and South America, in New Holland, and in either India. They are most common in temperate countries.

We can scarcely claim any knowledge of their uses. Their stems, often encrusted with lime in the state of carbonate according to some, and of the phosphate according to others, are probably useful as a manure. The fetid effluvium arising from them is regarded as very unhealthy, and one of the sources of the malaria of the Campagna of Rome.

GENERA.

Chara, *L.* Nitella, *Ag.* Charopsis, *Kütz.*

NUMBERS. GEN. 3. Sp. 35.

POSITION.—Ceramiaceæ. CHARACEÆ. *Fluviales.* / *Equisetaceæ.*

Fig. XII.*— A magnified view of Nitella, with the motion of its sap shown by arrows.

ALLIANCE II. *FUNGALES.*—THE FUNGAL ALLIANCE.*

Fungi, *Juss. Gen.* 3. (1789); *DC. Fl. Fr.* 2. 65. (1815); *Nees das System der Pilze und Schwämme,* (1817); *Fries Syst. Mycolog.* (1821); *Syst. Orb. Veg.* (1825); *Elench. Fung.* (1828); *Adolphe Brongn. in Dict. Class.* 5. 155. (1824); *Grev. Scott. Crypt. Fl.* 6. (1828); *Hooker British Flora,* 457. (1830); *Berk. in Id. vol.* 2. *pt.* 2. (1835); *Montagne in Hist. de Cuba Bot. p.* 239. (1838-1842), *translated, with Notes, in Ann. of Nat. Hist. vol.* 9. *p.* 1. *by Berk.* (1842); *Corda Anleitung,* (1842). —Epiphytæ, *Link; Grev. Fl. Edin.* xxv. (1824).—Gasteromyci, *Grev. Fl. Edin.* xxiv. (1824).—Mycetes, *Spreng. Syst.* 4, 376. (1827).—Uredineæ, Mucedineæ, *and* Lycoperdaceæ, *Ad. Brongn. in Dict. Class. l. c.* (1824).—Byssaceæ, (in part) *Fr. Syst. Orb. Veg.* (1825).

DIAGNOSIS.—*Cellular flowerless plants, nourished through their thallus (spawn or mycelium); living in air; propagated by spores colourless or brown, and sometimes inclosed in asci; destitute of green gonidia.*

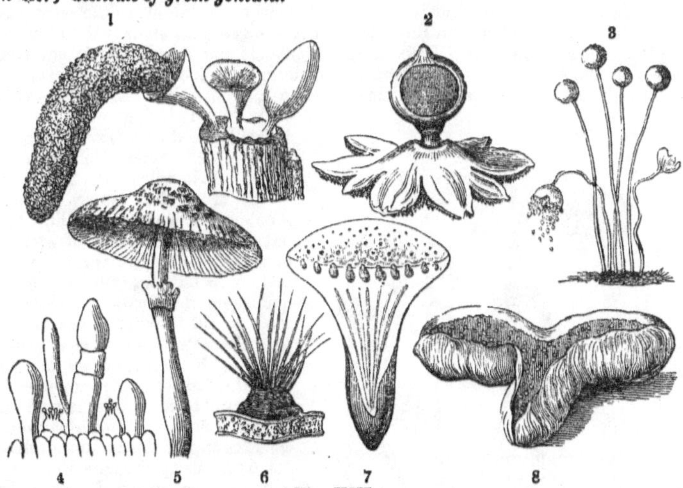

Fig. XIII.

Plants consisting of a congeries of cellules or filaments, or both variously combined, increasing in size in the more perfect species by addition to their inside, their outside undergoing no change after its first formation, chiefly growing upon decayed organic substances, or soil arising from their decomposition, frequently ephemeral, and variously coloured, never accompanied as in Lichens by reproductive germs of a vegetable green called gonidia; nourished by juices derived from the matrix. Fructification either spores attached externally, and often in definite numbers, to the cellular tissue, and frequently on peculiar cells called sporophores or basidia, which are in many cases surmounted by fine processes which immediately support the spores, and called spicules or sterigmata; or inclosed in membranous sacs or asci, and then termed sporidia. Vessels of the latex have been observed in Agaricus fœtens, by Corda. Spiral filaments, like the elaters of Jungermannia, exist in Trichia and Batarrea. They were first detected by the younger Hedwig, and described afterwards by Kunze and Corda. Mr. Berkeley detected them in the latter genus, and has very recently observed them, but very sparingly in Podaxon. The spores of fungi germinate either by a simple elonga-

* It is impossible to look at the huge mass of genera collected by Botanists under the name of Fungi, without perceiving that they in truth consist of groups equivalent to those called Natural Orders in the Algal Alliance, as well as in other parts of this arrangement. And if I had such an acquaintance with the subject as would justify my doing so, I should have presumed to break up the members of this Alliance into similar orders. It would, however, be presumptuous in me, with whom Fungi have never been a special study, to disturb the arrangements of those learned men who have made this investigation the business of their lives.

The following admirable account of the Alliance has been most kindly prepared by the Rev. M. J. Berkeley, whose knowledge of the species is unequalled in this or any other country. This gentleman permits me to state, that in his opinion the divisions here called orders may be regarded as Natural Orders, in the sense in which that term is applied to Algals.

Fig. XIII.—1. Arcyria flava; 2. Geastrum multifidum; 3. Mucor caninus; 4. Hymenium of an Agaric; 5. Agaricus cepæstipes; 6. Vermicularia trichella; 7. Vertical section of Hypoxylon punctatum; 8. Angioridium sinuosum. From Greville's Cryptogamic Flora, with the exception of No. 4.

tion of the episporium, or by the protusion of the inner membrane which exists in most cases, and is easily separated from the outer in the asci of many species of Sphæria. Fungals absorb oxygen and exhale carbonic acid. They abound in nitrogen.

Fungals are distinguished from Lichens by their more fugitive nature, their more succulent texture, their want of a thallus or expansion independent of the part that bears the reproductive matter, but more especially, as Fries has pointed out in his *Lichenographia Europœa*, in their never containing germs distinct from the fructifying bodies of a vegetable green so constant in Lichens. Many species indeed of Sphæria accord very closely in their mode of fructification, producing like the Lichens distinct nuclei in the centre of their substance, which at length burst through the cortical layer, though the fructifying disc is not exposed. In the Phacidiacei, however, the cups sometimes approach very nearly to the shield of Lichens ; so nearly, indeed, that they are occasionally mistaken for one another.

From Algals there is, as regards structure, scarcely any palpable difference ; but the most obvious distinction between Fungals and the two great divisions just mentioned consists in their mode of growth. Lichens and Algals do not derive nutriment from the substance on which they grow, but from the medium in which they are generated. Both are produced occasionally on the hardest subtances, from which it is impossible that they should derive much nutriment.* Fungals, on the contrary, live by imbibing juices impregnated with the peculiar principles of their matrix. It is true that many species of moulds will vegetate in liquids without any peculiar point of attachment, but these in general are in a very anomalous condition, and are in consequence often referred to Algals ; but as soon as they begin to revert to their true characters, there is a distinction between the free and submerged portion, the former being supported by the juices imbibed by the latter. A few species indeed of Fungals may almost be called aquatic, such as Cantharellus lobatus, Agaricus epichysium, Peziza clavus, Vibrissea truncorum, Leotia uliginosa ; but in most of such cases it will be observed, that it is not the habit of the whole genus but merely exceptional ; and in all there is an attachment to a matrix, from which it is highly probable that a portion at least of their nutriment is derived, especially in an early stage of growth. In fact, these cases having been stated by way of anticipating objections, it is rather the medium in which Fungals and Algals are developed that distinguishes them, than any peculiarity in their own organisation. While there is so near an approximation of these families to each other, particularly in the simplest forms, it is important to remark that, "with a single exception," perhaps, no spontaneous motion has been observed in Fungals, which, therefore, cannot be considered so closely allied to the Animal Kingdom as Algals, notwithstanding the presence of nitrogen in them, and the near resemblance of the substance by chemists called Fungine, to animal matter. Molecular motion, indeed, takes place in the particles which give consistence to the milk of the lactescent Agarics, but this is very different from that which has been so repeatedly observed in Algals, and which is produced in many instances by minute cilia which invest the reproductive bodies exactly as in the Animal Kingdom. Spontaneous motion has, however, been observed in Achlya prolifera, which is possibly a species of Mucor developed in water ; *Linn.* 1843, p. 129.

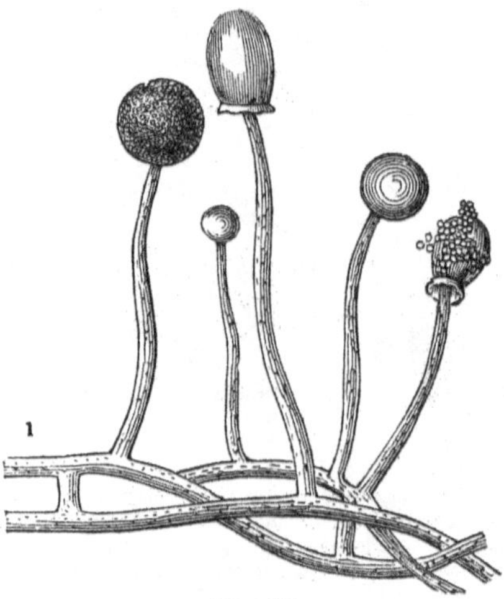

Fig. XII.

Fungals are almost universally found growing upon decayed animal or vegetable substances, and scarcely ever, except in the lower groups, upon living bodies of either

* It is, however, to be remembered, that observation has shown that Lichens corrode the hard bodies on which they grow, from which it is, perhaps, to be inferred, that they do to a certain extent really feed upon them.

Fig. XII.—Mucor mucedo, very highly magnified, exhibiting 1, the spawn or mycelium.

Kingdom ; in which respect they differ from Lichens, which very commonly grow upon

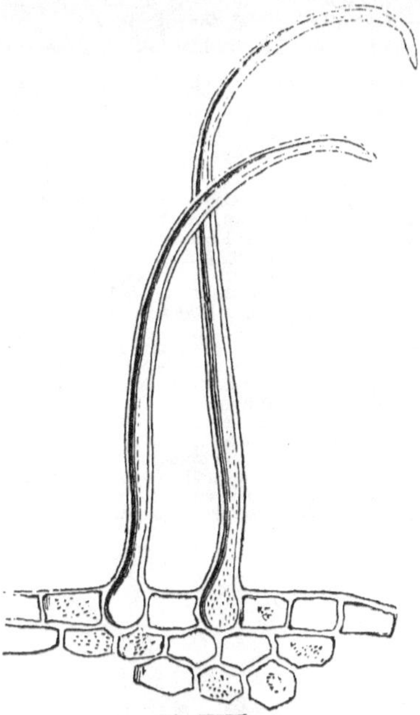

Fig. XIII.

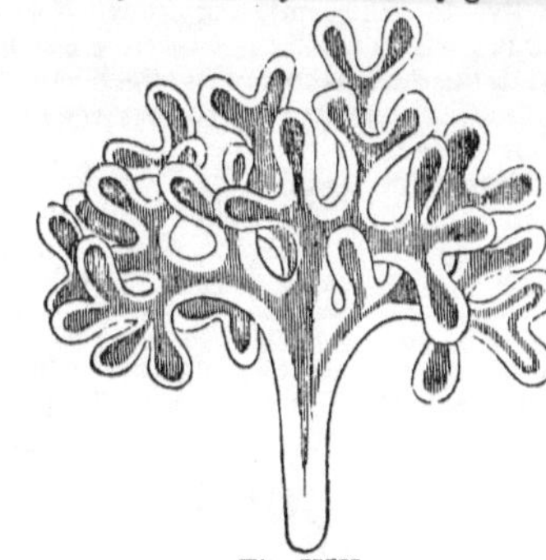

Fig. XIV.

the living bark of trees. The more simply organised species are not confined to dead or putrid substances, as is shown by their attacking various plants when in a state of perfect life and vigour ; for it has been incontestably proved by the discoveries of Léveillé and Corda, that the extensive tribe of Epiphyllous Fungi really belong to this division, and are not mere anamorphoses of the cellular tissue, as is the case with some productions usually referred to Fungi, as Erineum, Taphrina, &c.* Many observations, also, have been made of late years on the development of Fungi on living animal tissues. Of this

* It is not merely alterations of the epidermis of plants which assume the appearance of Fungi ; galls also, or tubercles caused by the attacks of insects, bear occasionally a wonderful resemblance to such bodies · so much so indeed, that they have been referred to them even by good botanists, on a hasty and superficial inspection. For here, as in other branches of the creation, we observe somewhat of that wonderful analogy by which, in each distinct class or even division of natural productions, the same, or extremely similar forms are repeated, though accompanied by an organisation totally different ; and it is this amongst other circumstances which makes it so absolutely necessary to examine into the intimate structure of the works of the creation, before venturing to pronounce upon their proper place in the system. Several of these galls have been figured by Mr. Curtis in his interesting entomological articles in the "Gardeners' Chronicle ;" such, for example, as Oak-spangles, produced by Diplolepis lenticularis; Oak-currants, by Cynips Quercus pedunculi, Woolly-oak galls, which owe their origin to the puncture of Cynips Quercus ramuli ; Elm-galls, brought on by the attacks of the Aphis ; in the case of galls, however, it is but a superficial examination which can possibly deceive, for

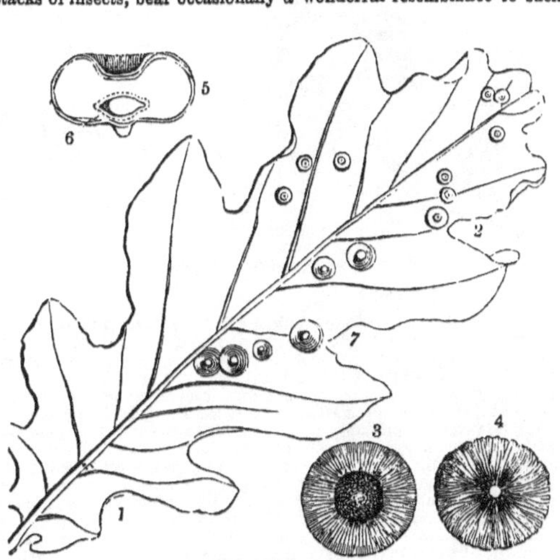

Fig. XV.

Fig. XIII.—Erineum Juglandis. Fig. XIV.—Erineum botryocephalum (*Corda*).
Fig. XV.—Oak Spangles.—3. Upper side ; 4. under side ; 2. silk button galls ; 6. a section of one with a larva in the interior. See Curtis in *Gardeners' Chronicle*, 1843, p. 52.

nature are the Guêpes végétantes of the West Indies; the Muscardine, which is so destructive to silkworms, and on which so many excellent Memoirs have been written; the mould, which so often causes the death of the common house-fly in autumn; and above all, the curious instances which have been recorded of the development of moulds in the mucous membrane of the viscera of vertebrate animals, and in certain cutaneous disorders in man.

Mouldiness, for instance, has been found by M. Deslongchamps on the internal surface of the air-cells of an Eider-duck while alive; and Mr. Owen observed a similar growth in the lungs of a Flamingo.—*Ann. Nat. Hist.* viii. 230. Col. Montagu had previously remarked it in the same situation in the Scarp-duck.—*Ib.* ix. 131. Gruby observed the

even where the little grub which produced them has vanished, the total absence of all parts of fructification will at once decide the point. If, for instance, the cup-shaped gall, which is so common on Oak leaves, be the object in question, any one who has once examined the hymenium of a Peziza, and observed the fructifying cells arranged vertically like the pile of velvet, with their row of eight mostly elliptic sporidia, cannot for a moment be deceived. It does, however, sometimes happen that galls are extremely like Fungi; a remarkable instance of which has been figured in the Transactions of the Linnean Society. It was sent by Mr. Macleay, from Cuba, on the leaf of some plant of the Natural Order Ochnaceæ. In this case there is not merely an extraordinary development of the external cellular tissue, but the gall is formed within the substance of the leaf, and after a time bursts through the skin, and presents a little ovate body with a crenate border, and within this an operculum which is perforated, or at least apparently perforated in the centre, so as to present a very close resemblance to some strange para-

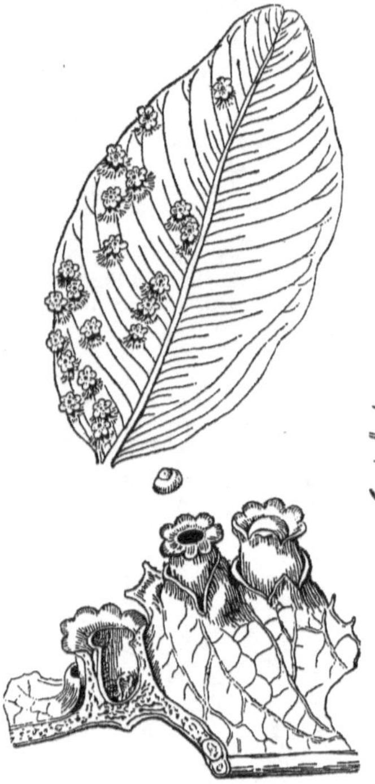

Fig. XVI.

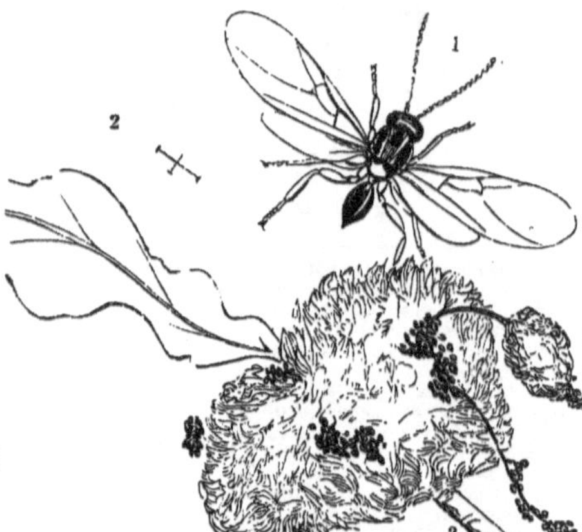

Fig. XVII.

site. And, as if to make the resemblance to some Fungus more close, the gall appears to make an abortive attempt to penetrate the opposite surface of the leaf, almost exactly in the way which is observable in the curious production which is sometimes so injurious to Pear-trees. But even in this case, where there is no trace of the inclosed grub or pupa, the texture of the walls of the gall is so different from that of Fungals that it can scarcely deceive, on any moderately accurate examination.

There is yet another production, referred to Fungals by Bernhardi, and after him by Fries and others, which, however, is probably to be regarded neither as a disease nor parasite. These are the tuberous bodies so common on the roots of leguminous plants. Their exact nature and use at present is not known; but a Memoir on them has been prepared some time by M. Desmazières. They appear a very few days after the germination of the seeds, and are accompanied by a little bed of vessels, in which they are nestled. At an early stage of growth, the contents of their cells become blue, when treated by iodine, which is not the case when their pulpy contents have acquired a salmon-coloured hue, when in some cases the granules are simple and oblong, in others forked. There can be little doubt that they are of some importance to the plant, though they are not, like common tubers, destined for the reproduction of the species, as they pass through the phases of vegetation in a short time, and soon become ruptured and discharge their contents. No insect has ever been observed in them, nor indeed does it at all appear that they are of the nature of galls. It is possible that in very dry situations, and in time of drought, the nutriment collected in them is serviceable to the plant; but this is very doubtful.

Fig. XVI.—Galls on the leaf of an Ochnaceous plant.
Fig. XVII.—Woolly Oak-gall, produced by Cynips Quercus ramuli.—(*Curtis.*)

crusts of Tinea favosa and Porrigo lupinosa to be accompanied by moulds, *Comptes Rend. Aug.* 1841; and these observations have been extended by Dr. Bennett, *Trans. Roy. Soc. Ed.*, vol. xv., Part 2, p. 277, who has also observed a mould growing on the lining membrane or cheesy matter of tubercular cavities in the lungs of man; as also the development of a mould on the skin of living gold-fish. Much information will be found on the subject in the place above quoted.

In their simplest form Fungi are little articulated filaments, composed of simple cellules placed end to end; such is the mouldiness that is found upon various substances, the mildew of the Rose-bush, and, in short, all the tribes of Mucor and Mucedo; in some of these the joints disarticulate, and appear to be capable of reproduction; in others, spores collect in the terminal joints, and are finally dispersed by the rupture of the cellule that contained them. In a higher state of composition, Fungi are masses of cellular tissue of a determinate figure, the whole centre of which consists of spores attached, often four together, to the cellular tissue, which at length dries up, leaving a dust-like mass intermixed more or less with flocci, as in the puff-balls, or sporidia contained in membranous tubes or asci, like the thecæ of Lichens, as in the Sphærias. In their most complete state they consist of two surfaces, one of which is even and imperforate, like the cortical layer in Lichens; the other separated into plates or cells, and called the hymenium, to whose component cells, which form a stratum resembling the pile of velvet, the spores are attached by means of little processes, and generally in fours, though occasionally the number is either less or greater. Many of these cells remain barren; but after a time there is a succession of fertile cells constantly making its appearance above the surface of the hymenium; and, what is more remarkable, the spicules or sterigmata, which support and give rise to the spores, have been observed by Corda to produce a succession of fruit, a new spore being produced where the old one had fallen. This, he informs us, is very easy of observation in Agaricus pluteus. Besides the barren and fertile cells, other bodies are observed which have been supposed by authors to perform the office of anthers. These have long been known in the dunghill Agarics, but they appear to be pretty generally distributed. The true structure of the more perfect Fungi has only been recognised within a few years, though Müller, half a century since, gave a correct figure of it in Agaricus comatus, and there are indications of it scattered through many works. Léveillé's Memoir in Annales des Sciences Naturelles, that of Berkeley in the Annals of Nat. History, of Phœbus in Nova Acta Cæs. Leop., and those of Berkeley and Tulasne in the Ann. of Nat. Hist. and Ann. des Sc. Nat. on the fructification of Lycoperdons, as also that of the Messrs. Tulasne on Hypogæous Fungi, may be consulted on this subject.

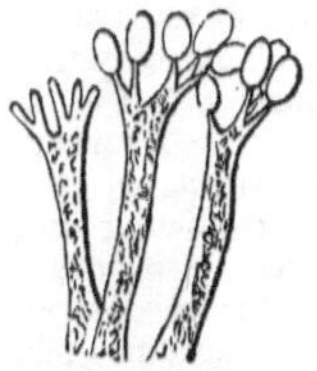

Fig. XXI.

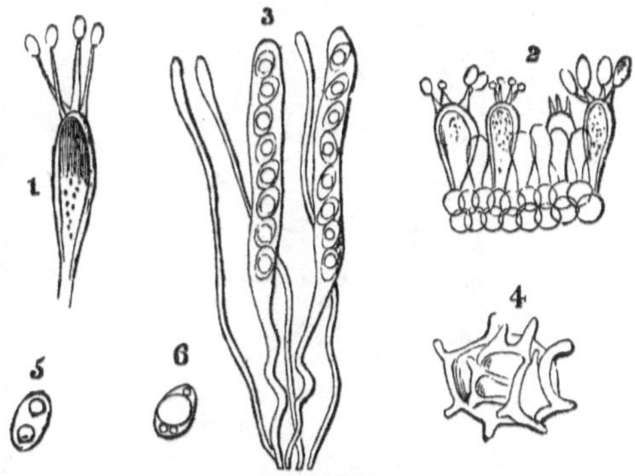

Fig. XXII.

Upon this kind of difference of structure, Fungi have not only been divided into distinctly marked tribes, but it has been proposed to separate certain Orders from them under the name of Byssaceæ, Gasteromyci, and Hypoxyla: the first comprehending the filamentous Fungi found in cellars, and similar plants; the second Lycoperdons and the like; and the third species which approach Lichens in the formation of a distinct nucleus for the sporules, such as Sphæria. But Fries considers the first as a distinct group, and the two last as Fungi.

Some writers have questioned the propriety of considering Fungi as plants, and

Fig. XXI.—Botrytis curta.

Fig. XXII.—1. Spore-stalk of Agaricus elixus, with its four long sterigmata and small spores; 2. spore-stalks of Ag. semiovatus, with spores in various states of development; 3. asci and sporidia of Helvella elastica; 4. sporidium of Tuber magnatum (Piedmontese Truffle), *from a sketch by Dr. Montagne;* 5. sporidium of Peziza aurantia, with its two nuclei; 6. single sporidium of Helvella elastica, with a large globose nucleus.

have proposed to establish them as an independent Kingdom, equally distinct from animals and vegetables; others have entertained doubts of their being more than mere fortuitous developments of vegetable matter, called into action by special conditions of light, heat, earth, and air — doubts which have been caused by some remarkable circumstances connected with their development, the most material of which are the following: they grow with a degree of rapidity unknown in other plants, acquiring the volume of many inches in the space of a night, and are frequently meteoric, that is, spring up after storms, or only in particular states of the atmosphere. It is possible to increase particular species with certainty, by an ascertained mixture of organic and inorganic matter exposed to well-known atmospheric conditions, as is proved by the process adopted by gardeners for obtaining Agaricus campestris, a process so certain, that no one ever saw any other kind of Agaricus produced in Mushroom-beds, except a few of the dunghill tribe, where raw dung has been placed near the surface of the bed; this could not happen if the Mushroom sprang from seeds or sporules floating in the air, as in that case many species would necessarily be mixed together; Fungi are often produced constantly upon the same kind of matter, and upon nothing else, such as the species that are parasitic upon leaves: all which is considered strong evidence of the production of Fungi being accidental, and not analogous to that of perfect plants. Fries, however, whose opinions must have great weight in all questions relating to Fungi, argues against these notions in the following manner: "The sporules are so infinite (in a single individual of Reticularia maxima I have reckoned above 10,000,000), so subtile (they are scarcely visible to the naked eye, and often resemble thin smoke), so light (raised, perhaps, by evaporation into the atmosphere), and are dispersed in so many ways (by the attraction of the sun, by insects, wind, elasticity, adhesion, &c.), that it is difficult to conceive a place from which they can be excluded." I give his words as nearly as possible, because they may be considered the sum of all that has to be urged against the doctrine of equivocal generation in Fungi; but without admitting, by any means, so much force in his statement as is required to set the question at rest. In short, it is no answer to such arguments as those just adverted to. It seems to me that a preliminary examination is necessary into the existence of an exact analogy between all the plants called Fungi; a question which must be settled before any further inquiry can be properly entered upon. That a number of the fungus-like bodies found upon leaves are mere diseases of the cuticle, or of the subjacent tissue, is by no means an uncommon opinion; that many more are irregular and accidental expansions of vegetable tissue in the absence of light, is not improbable; and it is already certain that no inconsiderable number of the Fungi of botanists are actually either, as various Rhizomorphas, the deformed roots of flowering plants growing in cellars, clefts of rocks, and walls; or mere stains upon the surface of leaves, as Venularia grammica; or the rudiments of other Fungi, as many of Persoon's Fibrillarias. Those who are anxious to inquire into these and other points, are referred to Fries'

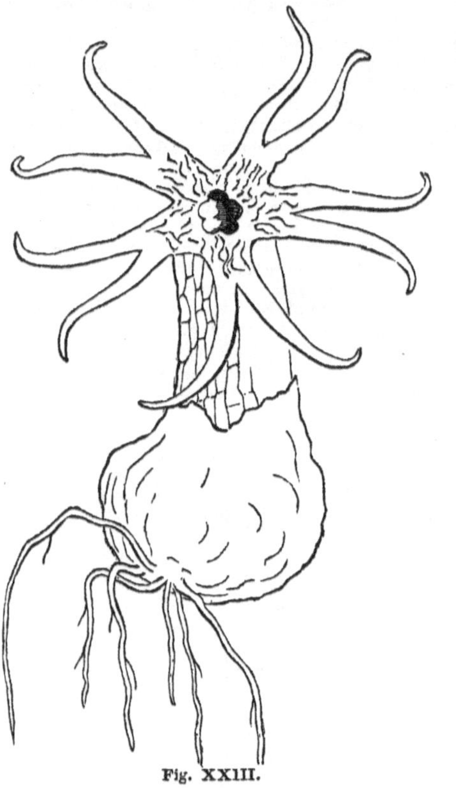

Fig. XXIII.

Fig. XXIII.—Aseroe pentactina.

works generally, to the various writings of Nees von Esenbeck, and to the Scottish Cryptogamic Flora of Greville. In the ensuing list of genera, I have chiefly availed myself of the writings of Fries. The disposition, however, of the genera has been modified in conformity with recent discoveries as to the real structure of the more highly organised species, and the numerous discoveries of Corda, where their affinities were at all clear, have been recorded. That it must be a matter of extreme difficulty to form any precise opinion concerning Fungi, without long experience, will be apparent from the observations of Fries upon the genus Thelephora. (*Elenchus*, p. 158.) He asserts that out of mere degenerations or imperfect states of Th. sulphurea, the following genera, all of which he has identified by means of unquestionable evidence, have been constructed; viz., Athelia of Persoon, Ozonium of Persoon, Himantia of Persoon, Sporotrichum of Kunze, Alytosporium of Link, Xylostroma, Racodium of Persoon, Ceratonema of Persoon, and some others. Th. Fr. Nees von Esenbeck also assures us that the same fungoid matter which produces Sclerotium mycetospora in the winter, develops Agaricus volvaceus in the summer. It would thus seem that the opinions of those who have asserted that the species or genus of a Fungus depends not upon the seed from which it springs, but upon the matrix by which it is nourished, are at least specious; especially if we take the above fact in connection with the experiments of Dutrochet, who obtained different genera of Mouldiness at will, by employing different infusions. He says that certain acid fluids constantly yield Monilias, and that certain alkaline mixtures equally produce Botrytis. *Ann. des Sc.* 2 *ser.* 1. 30. For a description of the gradual development of an Agaric, see this ingenious observer's Memoir in the *Nouv. Ann. du Mus.* vol. iii. p. 76. For the views of Unger upon spurious Fungi, which he considers nothing but morbid conditions (eruptions) of vegetable matter, see the *Ann. des Sc.* vol. ii. *n. s.* 209; and Berkeley's remarks thereupon, in *Hook, Brit. Fl.* vol. ii. pt. 2, p. 361.

Since, however, the remarks of Unger were published, Léveillé and Corda, almost at the same time, and quite independently of each other, made their discovery of the Mycelium of Uredines and Pucciniæ, and Corda has succeeded in making many germinate. Unger's speculations, therefore, must be considered as much invalidated, at least so far as their being mere transformations of the cellular tissue, as is the case in Erineum. Whether animal and vegetable bodies are ever produced without pre-existent germs, belongs to quite another question. And, as regards the genera Ozonium, Himantia, &c., they are now regarded by all good mycologists as mere barren states, or anamorphoses of other species; and the same is probably true of many of the more anomalous Fungi; and the observations of Léveillé, in the Annales des Sciences Naturelles, go very far to prove that the whole genus Sclerotium belongs to the same category. Some of them, as Acrospermum cornutum, and Sclerotium mycetospora, are undoubtedly mere forms, and have no right whatever to be considered as species; others arise from the condensation of the filamentous tufts of moulds; others, as S. lotorum, are little excrescences upon the roots, and the celebrated Ergot is produced by the action of a minute parasite. There is indeed a difficulty about such species as Sclerotium scutellatum; but there is little doubt that, in the main, Léveillé's observations, even though from the nature of the subject the proof is not rigorous, are founded in fact. Some supposed species of Uredo are merely the young of Puccinia, Aregma, &c.; but there are also true species of the genus. See *Henslow, Journ. of Roy. Soc. Ag.* 1841, vol. ii. p. 2.

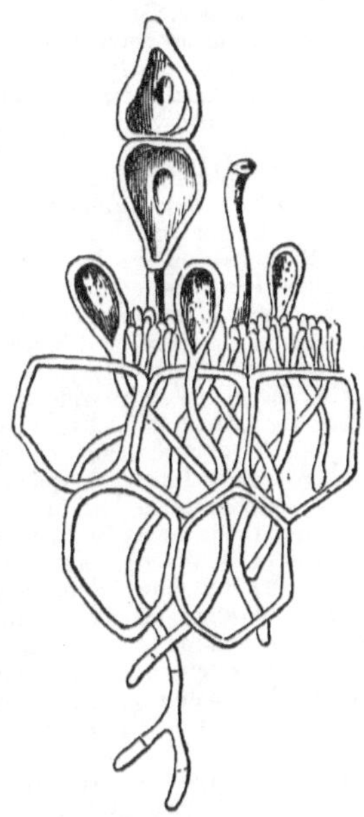

Fig. XXIV.

Kützing, in his Prize Essay on the Transformation of Plants, asserts that from one and the same organic material, even when it has acquired form and colour, different vegetables may be developed, which, according to the circumstance of the surrounding medium, are Algals, Fungi, Lichens, or Mosses; and that even the spores of these, when pro-

Fig. XXIV.—Puccinia graminis (common Mildew), with its spawn or mycelium penetrating the cell of the plant on which it grows.

duced, are capable of generating plants belonging to different Orders. This has been long a favourite theory in Germany, but it has not been so fully developed before. *Natuurkundige Verhandelingen van de Holl. Maatsch. der Wetensch. te Haarlem. Tweede Verz.* 1. *Deel.*

The subject, as regards the possible development of Algals, &c., from Infusoria, has been rehandled by the same author in a Memoir just published at Nordhausen. Those who are not convinced by his reasonings, will at least be ready to acknowledge the great research and patience with which they have been followed out. His observations are entitled to the greater attention, because he is well acquainted with the various forms assumed by cellular plants, though his great work on Algals scarcely shows him to have accurate notions as to the limits of genera and species.

The Fungi by which most extra-tropical countries are inhabited are so numerous, that no one can safely form even a conjecture as to the number that actually exists. If they are ever fortuitous productions, the number must be indeterminable; if many are mere diseases, and the remainder fixed species, then the knowledge of their nature must be reduced to a more settled state before any judgment upon their number can be formed. Fries discovered no fewer than 2000 species within the compass of a square furlong in Sweden; of Agaricus alone above 1000 species are described; and of the lower tribes the number must be infinite. Sprengel, however, does not enumerate in his *Systema Vegetabilium* more than between 2700 and 2800; but when we consider that his genus Agaricus does not go beyond number 646, although 1000 at least are described, it is not improbable that the rest of his enumeration is equally defective, and that the number of described Fungi perhaps amounts to between 4000 and 5000. Of tropical species we know but little; their fugitive nature, the difficulty of preserving them, and perhaps the incuriousness of travellers, as well as their scarcity in the damp parts of equinoctial countries, have been the causes of the proportion in such climates between Fungi and other plants being unknown. Mr. Berkeley has taken occasion, from the publication of a list of Java Fungi by Junghuhn, to institute a comparison between those of Java and the Philippine collection made by Mr. Cuming. Neither list can, indeed, be considered as complete, but in both cases the proportion of Fungi remaining to be described is probably much the same. Parts of the Philippines are situated in a degree of latitude in the northern hemisphere exactly corresponding with that of Java in the southern. The number of species described by Junghuhn is 113, that collected by Cuming about 40. Of these only $\frac{1}{8}$ of the species are common to the two localities, and out of these four are species of Polyporus common to all tropical countries. Of Junghuhn's Fungi $3=\frac{1}{37}$ are Coniomycetes, $9=\frac{1}{12}$ are Hyphomycetes, $7=\frac{1}{16}$ Gasteromycetes, $18=\frac{1}{6}$ Pyrenomycetes, $10=\frac{1}{11}$* Discomycetes, and 66, or above $\frac{1}{2}$, Hymenomycetes. In Mr. Cuming's collection there are no species of the first, second, and fifth Families; of the remaining Families $1=\frac{1}{40}$ belongs to Gasteromycetes, $5=\frac{1}{6}$ are Pyrenomycetes, and 33, or more than $\frac{3}{4}$, are Hymenomycetes. It will be observed that the proportion of Pyrenomycetes is the same, and there is even a greater proportion of Hymenomycetes in the Philippines. Of the Hymenomycetes in Java, 40 are Polypori; in the Philippines, 19, taking the genus in its widest sense. There is now an opportunity of contrasting with these the Fungi of Cuba, which have been so well worked out by Dr. Montagne. The species of that island, as far as at present recorded, are 115, of which $4=\frac{1}{28}$ are Coniomycetes, $10=\frac{1}{11}$ Hyphomycetes, $9=\frac{1}{13}$ Gasteromycetes, $25=\frac{1}{5}$ Pyrenomycetes, $8=\frac{1}{14}$ Discomycetes, and $59=\frac{1}{2}$ Hymenomycetes. The proportion of Pyrenomycetes is nearly the same as in Java and the Philippines, and the predominance of Hymenomycetes is equally striking. Of this number 28, or $\frac{1}{4}$, are European species; whereas among the Philippine Fungi there are but 2, while in Java there are 42. Of these the greater part are very common species. With the exception of European species, 5 only are common to Cuba and Java, and 4 to Cuba and the Philippines; and these, with one exception, species universally distributed. The species which forms an exception is Micropeltis applanata, which, as it is a minute Epiphyllous plant, may possibly have been overlooked in other countries. The number of Fungi peculiar to Cuba is very large. Cuba, then, has but little in common with Java and the Philippines, when the cosmopolites and European species are excepted. Several species, however, are identical with those of North and South America, extending in one instance even as far as Juan Fernandez; and there are one or two isolated species which call to mind Mauritius, Ceylon, and Australia. The genus Polyporus, as usual, predominates, counting 31 species, of which 8 are European; or, if Favolus and Hexagonia be included, the number

* It will be observed that in the list of genera given below, the Discomycetes and Pyrenomycetes are comprised in one group under the name of Ascomycetes. The Discomycetes correspond with the three first Suborders.

amounts to 35. When the climates are at all analogous, and the range of the thermometer at certain seasons similar, it is astonishing how great a resemblance, and even identity, there is between the Fungi of very distant portions of the globe. North America produces far the greater part of the European species, with a certain portion peculiar to itself. Hundreds of the same species of Sphæria and Agaricus occur in that country which are found with us. The curious genus Mitromyces, which seemed peculiar to that country, has been found in Java, Van Diemen's Land, and New Holland. And it would, perhaps, be difficult to point out any specific group peculiarly characteristic of the country. But the same resemblance exists, to a great extent, also in the southern hemisphere. In the island of Juan Fernandez, which was so carefully investigated by Bertero, scarce a third of the species differ from European Fungi. The same is the case in the Flora of New Zealand and Australia, from whence I possess a large quantity of species; and though there are many new forms, and some belonging to genera not hitherto found in Europe, a large proportion of the species are identical. In the genus Agaricus the species in countries of every variety of climate are often identical. The African Mycology is remarkable for the varied forms it produces amongst the puff-balls and allied genera, especially in that tribe which is called Podaxineæ. They commence at the south of Europe, in the environs of Marseilles; abound at the Cape of Good Hope, and form a very remarkable feature still in the Fungi of Swan River. Two species of the African genus Secotium occur at the Swan River; and possibly a third, and a very beautiful species, occurs in New Zealand. A species of Podaxon was found by Dr. Hooker at Porto Praya, identical with the East Indian species. A single imperfectly known species occurs in the warmer parts of North America. The genus Clathrus, which is perhaps the most beautiful amongst Fungi, though unknown in the more northern latitudes, has a most extensive geographical range. A line, running obliquely from the Isle of Wight through Germany, defines its northern limits: two species, one of the allied genus Ileodictyon, occur at the Swan River; and a magnificent species of that genus occurs in New Zealand, and is eaten by the natives. On the whole, then, it will be seen that the geographical limits of Fungi are by no means so definite as those of Phænogamous plants. Some species are found in every part of the globe; and several tropical forms are either universally dispersed, or occur in spots separated from each other by many thousands of miles. In the genus Polyporus every country seems to have species peculiar to itself; and from the number of new forms which daily occur, the genus seems to be almost co-extensive with Agaricus. It is in this genus, probably, if in any, that the species will be found to follow the most nearly a geographical arrangement.

A large volume might be written upon the qualities and uses of Fungi. They may be said to be important, either as food or as poison, or as parasites destructive to the plants upon which they grow. As food, the most valuable are the Agaricus campestris, or common Mushroom, the various species of Helvella or Morel, and Tuber or Truffle; but a considerable number of other kinds are used for food in various parts of the world, of which a useful account will be found in De Candolle's excellent *Essai sur les Propriétés Médicales des Plantes*, in Persoon's work, *Sur les Champignons comestibles*, in a paper by Greville in the 4th volume of the *Transactions of the Wernerian Society*, and in Roque's *Hist. des Champ. comestibles et veneneux*, ed. 2, 1841. A long list might be given of works on the subject, some of them like those of Vittadini, Phœbus, and Krombholz, very admirably got up.

About half a dozen species only are eaten in London, and in Paris none are permitted to appear in the markets except the common Truffle, Morel, and Mushroom, the latter being cultivated to a very considerable extent in the ancient quarries which run under parts of the city.

It is necessary to exercise the utmost care in employing Fungi the nature of which is not perfectly well ascertained, in consequence of the resemblance of poisonous and wholesome species, and the dreadful effects that have followed their incautious use. But the greatest caution and knowledge will not always avail, for it appears that some species which are in general perfectly wholesome, sometimes produce very disastrous consequences. A family at Cambridge a few years since suffered from eating mushrooms; a part of what were gathered were submitted to the writer of the present remarks, and proved to be Ag. personatus, a species sold sometimes in the London markets, and ascertained by Mrs. Hussey, who has paid great attention to the subject, to be most excellent for food. The case perhaps is similar to that of the prejudicial effects sometimes experienced by persons after eating mussels, and may be considered as a mere exception.

It is true that many kinds are named by Pallas as being commonly used by the Russians, which are plentiful in countries where they are not employed for food; but, in the first place, it is not perhaps quite certain that poisonous and wholesome species are not confounded under the same name; in the next place, climate may make a difference; and lastly, much depends upon the mode in which they are cooked. Upon this subject Delile observes, that it was ascertained by Paulet, in 1776, that salt and vinegar removed every deleterious principle from that most poisonous plant the Agaricus bulbosus; that it is the universal practice in Russia to salt the Fungi, and that this may be the cause of their harmlessness, just as the pickling and subsequent washing of the poisonous Agaric of the Olive renders it eatable in the Cevennes; but that, nevertheless, it is much wiser to run no risk with unknown Fungi, even taking such precautions—a remark to which he was led by the lamentable death of a French officer and his wife, in consequence of breakfasting off some poisonous Agarics, which were nevertheless eaten by other persons in the same house with impunity. It was probable that in that case a difference in the cooking was the cause of the difference in the effect of the Fungi; but it was a sufficient ground for distrusting all Fungi except the cultivated ones. So strongly did the late Professor L. C. Richard feel the prudence of this, that although no one was better acquainted with the distinctions of Fungi, he would never eat any except such as had been raised in gardens in mushroom beds. One of the most poisonous of our Fungi is the Amanita muscaria, so called from its power of killing flies when steeped in milk. Even this is eaten in Kamchatka, with no other than intoxicating effects, according to the following account by Langsdorf, as translated by Greville, from whom I borrow it:—

"This variety of Amanita muscaria is used by the inhabitants of the north-eastern parts of Asia in the same manner as wine, brandy, arrack, opium, &c. is by other nations. Such Fungi are found most plentifully about Wischna, Kamchatka, and Wilkowa Derecona, and are very abundant in some seasons, and scarce in others. They are collected in the hottest months, and hung up by a string in the air to dry; some dry of themselves on the ground, and are said to be far more narcotic than those artificially preserved. Small deep-coloured specimens, thickly covered with warts, are also said to be more powerful than those of a larger size and paler colour. The usual mode of taking the Fungus is, to roll it up like a bolus, and swallow it without chewing, which, the Kamchatkadales say, would disorder the stomach. It is sometimes eaten fresh in soups and sauces, and then loses much of its intoxicating property; when steeped in the juice of the berries of Vaccinium uliginosum, its effects are those of strong wine. One large, or two small Fungi, are a common doze to produce a pleasant intoxication for a whole day, particularly if water be drank after it, which augments the narcotic principle. The desired effect comes on from one to two hours after taking the Fungus. Giddiness and drunkenness result in the same manner as from wine or spirits; cheerful emotions of the mind are first produced; the countenance becomes flushed; involuntary words and actions follow, and sometimes at last an entire loss of consciousness. It renders some remarkably active, and proves highly stimulant to muscular exertion: by too large a dose, violent spasmodic effects are produced. So very exciting to the nervous system in many individuals is this Fungus, that the effects are often very ludicrous. If a person under its influence wishes to step over a straw or small stick, he takes a stride or a jump sufficient to clear the trunk of a tree; a talkative person cannot keep silence or secrets; and one fond of music is perpetually singing. The most singular effect of the Amanita is the influence it possesses over the urine. It is said that, from time immemorial, the inhabitants have known that the Fungus imparts an intoxicating quality to that secretion, which continues for a considerable time after taking it. For instance, a man moderately intoxicated to-day will, by the next morning, have slept himself sober, but (as is the custom) by taking a teacup of his urine he will be more powerfully intoxicated than he was the preceding day. It is, therefore, not uncommon for confirmed drunkards to preserve their urine as a precious liquor against a scarcity of the Fungus. The intoxicating property of the urine is capable of being propagated; for every one who partakes of it has his urine similarly affected. Thus, with a very few Amanitæ, a party of drunkards may keep up their debauch for a week. Dr. Langsdorf mentions, that by means of the second person taking the urine of the first, the third of the second, and so on, the intoxication may be propagated through five individuals."

It is universally known that the common Agaric is cultivated with as much certainty by good gardeners as any other vegetable. The excellent Boletus edulis has been partially cultivated in the south of France by inclosing a portion of a wood, and watering the ground with water in which the tubes had been steeped. Borch raised Tuber Borchii from the sporidia about the year 1780, and the growth of the common Truffle has been attempted with more or less success. Mr. Drummond has sent over the spawn of a large variety of Agaricus campestris from the Swan River, which he says is as far superior to the common mushroom as the improved peas to the old varieties, and it has been submitted to Mr. J. Henderson, but it is feared that it is too old to run.

Polyporus fomentarius has been artificially produced in Germany, but merely by placing wood in a favourable situation, and keeping it well moistened. Five or six crops were obtained in the year. (*Rom. and Uster. Mag.* iv. p. 182.)

A curious species which grows on the living branches of the South American beeches, and which has been described by Mr. Berkeley in the Transactions of the Linnean Society, under the name of Cyttaria Darwinii, forms a principal part of the food of the natives of Tierra del Fuego during many months of the year.

Fungi are much used in Australia by the natives, especially of the genus Boletus. The large truffle Mylitta australis, Berk., which attains a weight of more than two pounds, is known under the name of native bread. The marsupial animals are particularly fond of Fungi, and some species they hunt for so greedily, devouring them before they burst through the earth, that it is very difficult to obtain a well-grown specimen.

Mr. Backhouse also informs us that Fungi are much used by the natives. Two to which he particularly alludes are probably Polyporus portentosus, Berk., a species which could only be eaten in the absence of all other food, and a species of Cyttaria hitherto unrecorded by botanists.

One or two species are used in medicine. Sphæria sinensis, Berk., described in Hook, Lond. Journ. of Bot., is a celebrated remedy amongst the Chinese, and is much praised in Du Halde's book, but probably without reason.

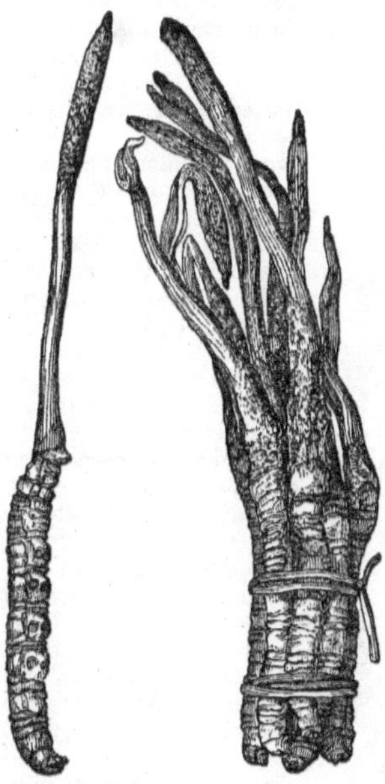

Fig. XXIV.*

Many Fungi were admitted into the old Pharmacopœias, as Exidia auricula Judæ, Polyporus officinalis, Tremella mesenterica, but at present they are little if at all used.

Lysurus mokusin is considered by the Chinese as an excellent remedy in gangrenous ulcers. It is also eaten, but is often poisonous. The jelly-like volva of the nearly allied genus Ileodictyon is eaten in New Zealand.

Ergot of rye is well known for its specific action on the uterus, and is in consequence one of the most valuable remedies of the modern Pharmacopœia. It is, however, said to be uncertain. It is unhappily no less notorious for the dreadful effects it produces on the human frame when it exists in considerable quantities in bread-corn, causing the most terrible ulcers and gangrenes, which at length destroy the limbs. Similar effects have been experienced from the use of mouldy provisions. Interesting details on the subject will be found in Burnett's Outlines, and in Professor Henslow's Report on the Diseases of Wheat, in the Journal of the Royal Society of Agriculture, 1841, vol. ii. part 1. Copious details will also be found in Phœbus's Deutschlands Cryptogamische Gewächse. On the real nature of Ergot Smith and Quekett's Memoir, Linn. Tr. xviii. p. 452, 3, and xix. p. 137, should be consulted. Corda has lately confirmed the observations of Messrs. Smith and Quekett; and more recently a Memoir on the subject has been published by Fée.

Of parasitical Fungi, the most important are those which are called dry rot, such as Polyporus destructor, Merulius lacrymans and vastator, &c., which are the pest of wooden constructions: next to these come the blight in corn, occasioned by Puccinia graminis; the smut and ergot, if they are really anything more than the diseased and disorganised tissue of the plants affected; the rust, which is owing to the ravages of Uredos and Pucciniæ; and finally, in this class is to be included what we call mildew, minute simple articulated Mucors, and Mucedos. The effects of different moulds on bread, preserves, &c., are but too well known. In some cases, however, as in cheese, provisions are thought to be improved by them. The decay of fruit, according to the observations of Mr. Hassall, appears to be in great measure produced by them. The genus Rhizomorpha (which it may be observed is a spurious genus, consisting of imperfectly developed Sphæriæ, Polypori, &c.) vegetates in dark mines far from the light of day, and is remarkable for its phosphorescent proper-

Fig. XXIV.*—Sphæria sinensis. The right hand figure represents the manner in which it is made up for sale.

ties. In the coal mines near Dresden the species are described as giving those places the air of an enchanted castle; the roof, walls, and pillars, are entirely covered with them, their beautiful light almost dazzling the eye. The light is found to increase with the temperature of the mines. *Ed. P. J.* xiv.

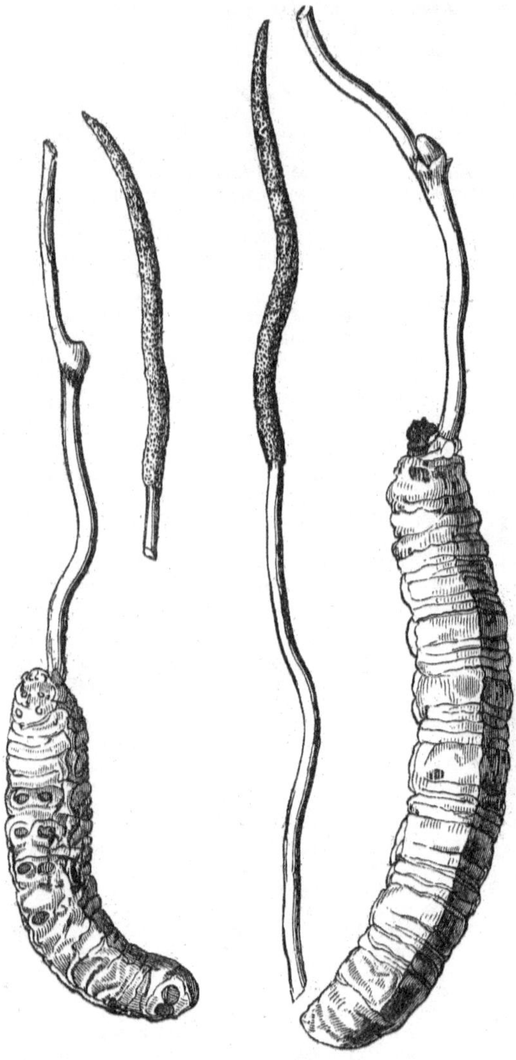

Fig. XXV.

Several species of genuine Fungi have been observed to be phosphorescent in various parts of the world. Agaricus Gardneri, Berk., which grows on a sort of Palm called Pintada in Brazil is highly luminous. Such also is the case with Agaricus olearius in the South of Europe, as observed by Delile. (*Arch. de la Bot.* vol. ii. p. 519.) Mr. Drummond has found two or more luminous species at the Swan River, (*Hook. Lond. Journ. of Bot.* ii. p. 263;) and Rumphius observed the same phenomenon in Amboyna.

It is a most remarkable circumstance, and one which deserves particular inquiry, that the growth of the minute Fungi, which constitute what is called mouldiness, is effectually prevented by any kind of perfume. It is known that books will not become mouldy in the neighbourhood of Russia leather, nor any substance, if placed within the influence of some essential oil. Polyporus fomentarius, or an allied species, is used in India as a styptic, as well as for Amadou. It is also employed by the Laplanders and others as Moxa. (*Ainslie*, i. 5.) The Boleti, when wounded, heal much in the same manner as the flesh of animals. (*Edin. Philosoph. Journ.* xiv. 369.)

A very curious phenomenon takes place in several species of the genus Boletus, and analogous appearances present themselves in other genera. The flesh, when broken, changes very rapidly from yellow or white to deep blue, and if the juice be squeezed out, though at first colourless, it quickly becomes blue. Professor Robinson of Armagh has ascertained that this is not a chemical action, but believes it to arise from some change in the molecular arrangement. Tannin, though prejudicial to most vegetables, is not so always to Fungi. A species of Rhizomorpha is often developed in tan-pits. The greatest proper heat met with by Dutrochet in the Vegetable Kingdom, with the

Fig. XXV.—Sphæria Robertsii, growing from the caterpillar of a New Zealand moth called Hepialus virescens.

exception of that of the spadix of Arum, was in Boletus æneus. (*Ann. des Sc. Nat.* Feb. 1840.) Fungine, which was considered as a simple body, has been shown by Payen to consist of *cellulose* and a fatty matter. Payen communicated to Dr. Montagne, as the result of his analysis, the following list of substances which enter generally into the composition of Fungi: 1. Water. 2. Cellulose, constituting all the solid part of the membranes of the tissue. 3. Three nitrogenized substances; one insoluble in water; a second soluble, coagulable by heat; a third soluble in alcohol. 4. Fatty matter analogous to wax. 5. Fatty substances, one fluid at an ordinary temperature, the other solid, crystallisable at the same temperature. 6. Sugar. 7. Matter capable of being turned brown by the action of free air. 8. An aromatic substance. 9. Traces of sulphur. 10. Traces of salts of silex and potash. (*Ann. of Nat. Hist.* vol. ix. p. 294.) Some species, as Agaricus cantharellus, Clavaria coralloides, and Agaricus piperatus, contain a sweet sugary matter, which, according to Liebig, is *Mannite.*—(*Annalen,* Feb. 1844.) M. Bonjean, is of opinion that the poisonous qualities of Ergot are owing to an oily acrid principle. (*Journ. de Ch. Med.*) Unlike other plants, Fungi, instead of purifying the air by robbing it of its carbonic acid and restoring the oxygen, vitiate it by exhaling carbonic acid and absorbing oxygen. This has been proved experimentally by Dr. Marcet of Geneva, and will probably explain the cause of Fungi being so universally destitute of green colouring matter, which we know results from the decomposition of carbonic acid. Certain Fungi in an imperfect state are said by Caignard-Latour, Schwann, and others, to be connected with the process of fermentation. The curious circumstance that in certain bakehouses all the bread becomes ropy, and though sometimes prevented from assuming this condition by repeated washing of the walls and floor with chloride of lime, the evil is occasionally so obstinate as to prove the ruin of the establishment, is probably dependent on the same cause. Dutrochet believes that he has witnessed the growth of a Penicillium from the globules of milk. (*Caignard-Latour, L'Instit.* Feb. 1837; *Meyen Jahresb.* 1838; *Dutrochet Ann. des Sc. Nat. N. S. Zool.* vol. viii.)

*** Although the Fungal Alliance is not here formally broken up into NATURAL ORDERS, yet the following may be regarded as their names and peculiar characters:—

Spores generally quaternate on distinct Sporophores, Hymenium naked .	6. HYMENOMYCETES, or AGARICACEÆ.
Spores generally quaternate on distinct Sporophores, Hymenium inclosed in a Peridium	7. GASTEROMYCETES, or LYCOPERDACEÆ.
Spores single, often septate, on more or less distinct Sporophores, Flocci of the fruit obsolete or mere peduncles	8. CONIOMYCETES, or UREDINACEÆ.
Spores naked, often septate. Thallus floccose	9. HYPHOMYCETES, or BOTRYTACEÆ.
Sporidia contained (generally eight together) in asci	10. ASCOMYCETES, or HELVELLACEÆ.
Spores surrounded by a vesicular veil, or Sporangium. Thallus floccose .	11. PHYSOMYCETES, or MUCORACEÆ.

GENERA.

COHORS I.—*Sporiferi.*

Ordo I.—HYMENOMYCETES.

Suborder I. *Agaricini,* Fr.

Agaricus, *L.*
*Amanita, *Fr.*
*Lepiota, *Fr.*
*Armillaria, *Fr.*
*Tricholoma, *Fr.*
*Clytocybe, *Fr.*
*Omphalia, *Fr.*
*Collybia, *Fr.*
*Mycena, *Fr.*
*Pleurotus, *Fr.*
*Volvaria, *Fr.*
*Pluteus, *Fr.*
*Entoloma, *Fr.*
*Clitopilus, *Fr.*
*Eccilia, *Fr.*
*Leptonia, *Fr.*
*Nolanea, *Fr.*
*Pholiota, *Fr.*
*Hebeloma, *Fr.*
*Flammula, *Fr.*
*Naucoria, *Fr.*
*Galera, *Fr.*
*Crepidotus, *Fr.*
*Psalliota, *Fr.*
*Hypholoma, *Fr.*
*Psilocybe, *Fr.*
*Psathyra, *Fr.*
*Panæolus, *Fr.*
*Psathyrella, *Fr.*
Coprinus, *P.*
Bolbitius, *Fr.*
Cortinarius, *Fr.*
*Phlegmacium, *Fr.*
*Myxacium, *Fr.*
*Inoloma, *Fr.*
*Dermocybe, *Fr.*
*Telamonia, *Fr.*
*Hygrocybe, *Fr.*
Paxillus, *Fr.*
Gomphidius, *Fr.*
Stylobates, *Fr.*
Hygrophorus, *Fr.*
Lactarius, *P.*
Russula, *P.*
Cantharellus, *Adans.*
Phlebophora, *Lev.*
Heliomyces, *Léveill.*
Pterophyllus, *Léveill.*
Nyctalis, *Fr.*
Marasmius, *Fr.*
Lentinus, *Fr.*
Panus, *Fr.*
Xerotus, *Fr.*
Trogia, *Fr.*
Schizophyllum, *Fr.*
Lenzites, *Fr.*
Hymenogramme, *Mont. & Berk.*
Junghuhnia, *Corda.*

Suborder II. *Polyporei,* Fr.

Boletus, *Dill.*
Ceriomyces, Batt.
Hypodrys, Pers.
Suillus, Mich.
Polyporus, *Mich.*
Trametes, *Fr.*
Dædalea, *Pers.*
Cyclomyces, *Klotzch.*
Hexagonia, *Fr.*
Favolus, P. B.
Favolus, *Fr.*
Glæoporus, *Mont.*
Laschia, *Fr.*
Merulius, *Hall.*
Epichysium, *Tode.*
Porothelium, *Fr.*

Suborder III. *Hydnei,* Fr.

Fistulina, *Bull.*
Hydnum, *L.*
Manina, Scop.
Hericium, *Fr.*
Sistotrema, *Fr.*
Irpex, *Fr.*
Radulum, *Fr.*
Phlebia, *Fr.*
Grandinia, *Fr.*
Odontia, *Fr.*
Kneiffia, *Fr.*

Suborder IV. *Auricularini,* Fr.

Craterellus, *Fr.*
Thelephora, *Ehr.*
Cladoderris, *P.*
Stereum, *Lk.*
Hypolyssus, *Berk.*
Auricularia, *Fr.*
Cora, *Fr.*
Dictyonema, *P.*
Midotis, *Fr.*
Corticium, *Fr.*
Guepinia, *Fr.*

Cyphella, *Fr.*
Hypochnus, *Ehb.*

Suborder V. *Clavati.*

Sparassis, *Fr.*
Clavaria, *L.*
Calocera, *Fr.*
Crinula, *Fr.*
Typhula, *Fr.*
Pistillaria, *Fr.*

Suborder VI. *Tremellini.*

Tremella, *Dill.*
Coryne, *Nees.*
Exidia, *Fr.*
Næmatelia, *Fr.*
Dacrymyces, *Fr.*
Lemalis, *Fr.*
Hymenula, *Fr.*

Ordo II.—GASTEROMYCETES.

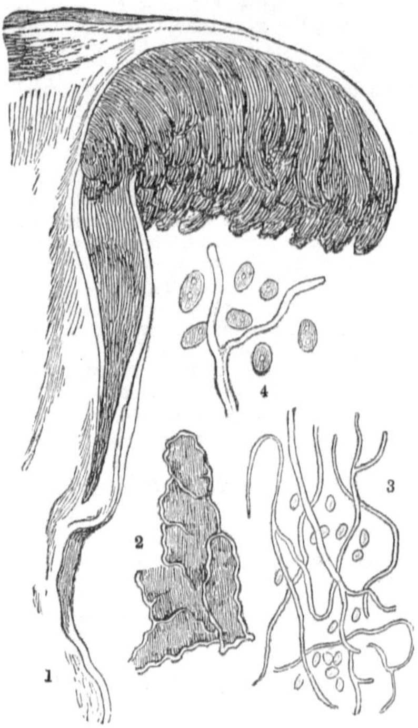

Fig. XXVI.

Suborder I. *Podaxinei,* Mont.

Montagnea, *Fr.*
Gyrophragmium, *Mont.*
Polyplocium, *Berk.*
Secotium, *Kze.*
Podaxon, *Desv.*
Cauloglossum, *Fr.*
Cycloderma, *Kl.*

Suborder II. *Hypogæi,* Berk.

Gautieria, *Vitt.*
Splanchnomyces, *Corda.*
Hymenangium, *Kl.*
Octaviana, *Tul.*
Melanogaster, *Cord.*
Hyperhiza, *Bosc.*
Hydnangium, *Wallr.*
Hysterangium, *Vitt.*

Suborder III. *Phalloidei,* Fr.

Phallus, *L.*
Aseroe, *Labill.*
Calathiscus, *Mont.*
Lysurus, *Fr.*
Simblum, *Kl.*
Clathrus, *Mich.*
*Laternea, *Turp.*
*Coleus, *Cav. & Sech.*
*Clethria, *Brown.*
Ileodictyon, *Tul.*

Suborder IV. *Trichogastres.*

Batarrea, *P.*
Tulostoma, *P.*
Lycoperdon, *Tourn.*
Scolecotrichum, *Berk.*
Phellorinia, *Berk.*
Broomeia, *Berk.*
Geaster, *P.*
Plecostoma, *Desv.*
Myriostoma, *Desv.*
Diploderma, *Lk.*
Bovista, *Dill.*
Hippoperdon, *Mont.*
Mycenastrum, *Desv.*
Scleroderma, *P.*
Polysaccum, *Desp.*
Polyangium, *Lk.*
Ciliciocarpus, *Corda.*
Arachnion, *Schwein.*
Polygaster, *Fr.*
Mitremyces, *Nees.*
Cenococcum, *Fr.*
Pilacre, *Fr.*

Subord., V. *Myxogastres.*

Lycogala, *Mich.*
Reticularia, *Bull.*
Æthalium, *Lk.*
Ptychogaster, *Corda.*
Spumaria, *P.*
Diderma, *P.*
Polyschismium, *Corda.*
Didymium, *P.*
Tripotrichia, *Corda.*
Trichamphora, *Jungh.*
Physarum, *P.*
Angioridium, *Grev.*
Trichoscytale, *Corda.*
Craterium, *Trent.*
Stegobolus, *Mont.*
Stegasma, *Corda.*
Diachea, *Fr.*
Stemonitis, *Gled.*
Dictydium, *Schrad.*
Cribraria, *Schrad.*
Arcyria, *Hill.*
Trichia, *Hall.*
Perichæna, *Fr.*
Licea, *Schrad.*
Cirrholus, *Mart.*

Suborder VI. *Nidulariacei.*

Nidularia, *P.*
Cyathus, *Hall.*
Crucibulum, *Tul.*
Sphærobolus, *Tode.*
Thelebolus, *Tode.*
Atractobolus, *Tode.*

Ordo III.—CONIOMYCETES, *Fr.*

Suborder I. *Sphæronemei,* Corda.

Microthyrium, *Desm.*
Coniothyrium, *Corda.*
Sacidium, *Nees.*
Leptostroma, *Fr.*
Phoma, *Fr.*
Leptothyrium, *Kze.*
Actinothyrium, *Kze.*
Apiosporium, *Kze.*
Microthecium, *Corda.*
Cryptosporium, *Kze.*
Sphæronema, *Fr.*
Aposphæria, *Berk.*
S. acuta, Hoffm.
Acrospermum, *Tode.*
Diplodia, *Fr.*
Hendersonia, *Berk.*
Lichenopsis, *Schwein.*
Pyrenotrichum, *Mont.*
Vermicularia, *Tode.*
Phlyctidium, *Not.*
Septoria, *Fr.*
Dilophosporium, *Desm.*
Neottiospora, *Desm.*
Pestalozzia, *Not.*
Angiopoma, *Lév.*
Prosthemium, *Kze.*
Asteroma, *D. C.*
Couturia, *Castg.*
Bryocladium, *Kze.*

Suborder II. *Melanconiei,* Corda.

Melanconium, *Lk.*
Stegonosporium, *Corda.*
Stilbospora, *P.*
Seimatosporium, *Corda.*
Asterosporium, *Kze.*
Cytispora, *Fr.*
Ceuthospora, *Grev.*
Nemaspora, *P.*
Coryneum, *Kze.*
Selenosporium, *Corda.*
Bactridium, *Kze.*
Botryospora, *Schwein.*
Myriocephalum, *Not.*
Hyperomyxa, *Corda.*

Suborder III. *Phragmotrichacei.*

Endotrichum, *Corda.*
Schizoxylon, *Fr.*
Schizothecium, *Corda.*
Pilidium, *Kze.*
Excipula, *Fr.*
Seiridium, *Nees.*
Phragmotrichum, *Kze.*
Endotrichum, *Corda.*
Schizoxylon, *P.*

Suborder IV. *Torulacei.* Cord.

Torula, *P.*
Conoplea, *P.*
Ceratospora, *Schwein.*
Clasterispora, *Schwein.*
Speira, *Corda.*
Dictyosporium, *Corda.*
Gyrocerus, *Corda.*
Helicomyces, *Corda.*
Bispora, *Corda.*
Septonema, *Corda.*
Trimmatostroma, *Corda.*
Alternaria, *Corda.*
Dicoccum, *Corda.*
Sporidesmium, *Lk.*
Coniothecium, *Corda.*
Hymenopodium, *Corda.*
Echinobotrys, *Corda.*
Spilocæa, *Fr.*

Suborder V. *Pucciniæi.*

Xenodochus, *Schlecht.*
Aregma, *Fr.*
Triphragmium, *Lk.*
Puccinia, *P.*
Gymnosporangium, *Lk.*
Podisoma, *Lk.*

Suborder VI. *Cæomacei,* Corda.

Uredo, *P.*
Pileolaria, *Castg.*
Ustilago, *Lk.*
Sporisorium, *Ehb.*
Testicularia, *Kl.*
Tuburcinia, *Fr.*
Cronartium, *Fr.*
Rœstelia, *Reb.*
Graphiola, *Poit.*
Æcidium, *Gmel.*

Fig. XXVI.—1. Polyplocium inquinans, divided vertically, natural size; 2. flocci and spores; 3 and 4, the same more highly magnified.—*Berkeley.*

Ordo IV.—HYPHOMYCETES.

Suborder I. *Isariacei*, Corda.
Isaria, *Hill.*
Podosporium, *Schwein.*
Ceratocladium, *Corda.*
Anthina, *Fr.*
Pterula, *Fr.*
Scorias, *Fr.*
Dacrina, *Fr.*
Ceratium, *A. & S.*
Byssocaulon, *Mont.*

Suborder II. *Stilbacei.*
Graphium, *Corda.*
Stilbum, *Tode.*
Corallodendron, *Jungh.*
Ceratopodium, *Corda.*
Hyalopus, *Corda.*
Doratomyces, *Corda.*
Periconia, *Tode.*
Phycomyces, *Kze.*
Tubercularia, *Tode.*
Periola, *Fr.*
Ciliciopodium, *Corda.*
Chœtostroma, *Corda.*
Volutella, *Tode.*
Blennoria, *Moug.*
Fusarium, *Lk.*
Illosporium, *Mart.*
Epicoccum, *Lk.*
Sphærospora, *Schwein.*

Suborder III. *Dematiei*, Fr.
Stachybotrys, *Corda.*
Cephalotrichum, *Fr.*
Rhopalomyces, *Corda.*
Sporocybe, *Fr.*
Œdemium, *Fr.*
Myxotrichum, *Kze.*
Gonatotrichum, *Nees.*
Helminthosporium, *Lk.*
Exosporium, *Lk.*
Blastotrichum, *Corda.*
Leptotrichum, *Corda.*
Mystrosporium, *Corda.*
Stemphylium, *Wallr*
Septosporium, *Corda.*
Trichægum, *Corda.*
Amphitrichum, *Corda.*
Triposporium, *Corda.*
Helicoma, *Corda.*
Helicosporium, *Corda.*
Cladotrichum, *Corda.*
Dematium, *P.*
Polythrincium, *Kze.*
Cladosporium, *Lk.*
Helicotrichum, *Nees.*
Macrosporium, *Fr.*
Arthrinium, *Kze.*
Goniosporium, *Lk.*
Sporophleum, *Nees.*
Camptoum, *Lk.*

Suborder IV. *Mucedines*, Fr.
Aspergillus, *Mich.*
Botrytis, *Mich.*
Chætopsis, *Grev.*
Streptothrix, *Corda.*
Campsotrichum, *Ehb.*
Menispora, *P.*
Sporophleum, *Nees.*
Polyactis, *Lk.*
Cladobotryum, *Nees.*
Gonatobotrys, *Corda.*
Botryosporium, *Corda.*
Clonostachys, *Corda.*
Sceptromyces, *Corda.*
Verticillium, *Nees.*
Pteronospora, *Corda.*
Actinocladium, *Ehb.*
Gliocladium, *Corda.*
Acmosporium, *Corda.*
Corethropsis, *Corda.*
Cephalothecium, *Corda.*
Haplotrichum, *Lk.*
Cephalosporum, *Corda.*
Brachycladium, *Corda.*
Arthrobotrys, *Corda.*
Penicillium, *Lk.*
Coremium, *Corda.*
Rhodocephalus, *Corda.*
Briarea, *Corda.*
Stysanus, *Corda.*
Monilia, *Hill.*
Dactylium, *Nees.*
Dendryphium, *Corda.*
Rhinotrichum, *Corda*
Sporodum, *Corda.*
Gonatorhodius, *Corda*
Sporotrichum, *Lk.*
Acremonium, *Lk.*
Oidium, *Lk.*
Lanosa, *Fr.*
Fusidium, *Lk.*

Suborder V. *Sepedoniei*, Fr.
Asterophora, *Ditm.*
Sepedonium, *Lk.*
Zygodesmus, *Corda.*
Monotospora, *Corda.*
Amphiblistrum, *Corda.*
Collarium, *Lk.*
Fusisporium, *Lk.*
Epochnium, *Lk.*
Scolicotrichum, *Lk.*
Mysothecium, *Ditm.*
Psilonia, *Fr.*
Gyrothrix, *Corda.*
Tricholeconium, *Corda,*
Aleurisma, *Lk.*
Dendrina, *Fr.*

COHORS II.—*Sporidiiferi.*

Ordo V.—ASCOMYCETES, *Berk.*

Suborder I. *Elvellacei*, Fr.
Morchella, *Dill.*
Helvella, *L.*
Verpa, *Sw.*
Geoglossum, *P.*
Mitrula, *Fr.*
Spathulea, *Fr.*
Leotia, *Hill.*
Rhizina, *Fr.*
Peziza, *Dill.*
Desmazierella, *Lib.*
Solenia, *P.*
Ascobolus, *Pers.*
Agyrium, *Fr.*
Stictis, *P.*
Cryptodiscus, *Corda.*
Melittiosporium, *Corda.*
Cryptomyces, *Grev.*
Propolis, *Corda.*
Bulgaria, *Fr.*
Cyttaria, *Berk.*
Vibrissea, *Fr.*
Sarea, *Fr.*
Ditiola, *Fr.*

Suborder II. *Tuberacei*, Fr.
Tuber, *Mich.*
Choiromyces, *Vitt.*
Pachyphlæus, *Tul.*
Choir. Melanoxanthus, Tul.
Hydnobolites, *Tul.*
Balsamia, *Vitt.*
Picoa, *Vitt.*
Genea, *Vitt.*
Sphærosoma, *Klotzsch.*
Endogone, *Lk.*
Elaphomyces, *Nees.*
Mylitta, *Fr.*

Suborder III. *Phacidiacei*, Fr.
Stegia, *Fr.*
Patellaria, *Fr.*
Tympanis, *Tod.*
Dermea, *Fr.*
Cenangium, *Fr.*
Cordierites, *Mont.*
*Sclerodermis, *Fr.*
*Triblidium, *Fr.*
*Clithris, *Fr.*
*Heterosphæria, *Grev.*
Glonium, *Muhl.*
Lophium, *Fr.*
Actidium, *Fr.*
Cliostomum, *Fr.*
Rhytisma, *Fr.*
Phacidium, *Fr.*
Hysterium, *Fr.*
Sporomega, *Fr.*
Ailographum, *Lib.*
Hysterographium, *Corda.*
Labrella, *Fr.*

Suborder IV. *Sphæriacei*, Fr.
Hypocrea, *Fr.*
Acrosphæria, *Corda.*
Thamnomyces, *Ehb.*
Hypoxylon, *Bull.*
Sphæria, *L.*
Stigmea, *Fr.*
Saccothecium, *Mont.*
Splanchnonema, *Corda.*
Melanospora, *Corda.*
Haplosporium, *Mont.*
Pemphidium, *Mont.*
Micropeltis, *Mont.*
Discosia, *Lib.*
Cheilaria, *Lib.*
Dothidea, *Fr.*
Corynelia, *Fr.*

Suborder V. *Perisporiacei*, Fr.
Lasiobotrys, *Kze.*
Erysiphe, *Hedw. Fil.*
Perisporium, *Fr.*
Chætomium, *Kze.*
Meliola, *Fr.*

Suborder VI. *Onygenei*, Berk.
Spadonia, *Fr.*
Onygena, *P.*

Ordo VI.—PHYSOMYCETES, *Berk.*

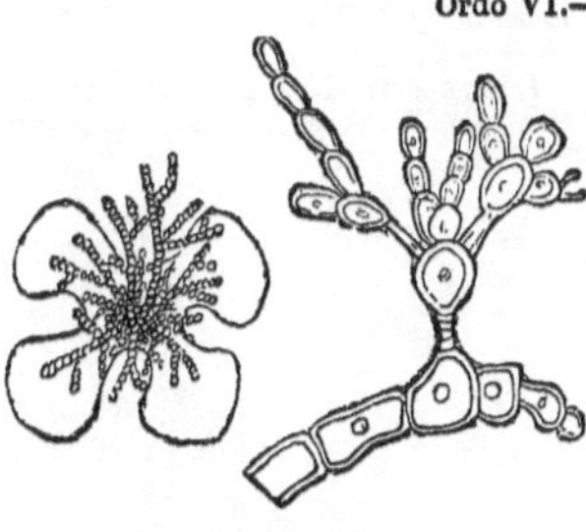

2

Fig. XXVII.

Suborder I. *Antennariei*, Corda.
Antennaria, *Lk.*
Pleuropyxis, *Corda.*
Pisomyxa, *Corda.*

Suborder II. *Mucorini*, Fr.*
Phycomyces, *Ag.*
Ascophora, *Tode.*
Pilobolus, *Tode.*
Pycnopodium, *Corda.*
Chordostylum, *Tode.*
Hydrophora, *Tode.*
Mucor, *Mich.*
Chionyphe, *Thien.*
Sporodinia, *Lk.*
Calyssosporium, *Corda.*
Hemiscyphe, *Corda.*
Crateromyces, *Corda.*
Rhizopus, *Ehb.*
Acrostalagmus, *Corda.*
Thelactis, *Mart.*
Helicostylum, *Corda.*
Endodromia, *Berk.*
Diamphora, *Mart.*
Didymocrater, *Mart.*
Syzigites, *Ehb.*
Melidium, *Eschw.*
Eurotium, *Lk.*
Myriococcum, *Fr.*
Caulogaster, *Corda.*
Ægerita, *P.*
Dichosporium, *Nees.*

* The mode of fructification is exactly intermediate between that of Agaricus and Sphæria. The reproductive bodies appear to spring from some definite point, and the same point produces a succes-

Fig. XXVII. Antennaria Robinsonii.—1. The evolution of a spore; 2. a portion of the threads from a germinating spore.

Genera not sufficiently known.
Papularia, *Fr.*
Phyllædium, *Fr.*
Hypodermium, *Lk.*
Schizoderma, *Kze.*
Protomyces, *Unger.*
Gymnosporium, *Corda.*
Leucosporium, *Corda.*
Chromosporium, *Corda.*
Conisporium, *Lk.*
Coccularia, *Corda.*
Entomyclium, *Wallr.*
Myxosporium, *Corda.*
Fusoma, *Corda.*
Apotemnoum, *Corda.*
Ramularia, *Unger.*
Athelia, *P.*
Acrothamnium, *Nees.*
Alytosporium, *Lk.*
Capillaria, *P.*
Circinotrichum, *Nees.*
Plecotricum, *Corda.*
Miainomyces, *Corda.*
Chrysosporium, *Corda.*
Chromelosporium, *Corda*
Myxonema, *Corda.*
Melanotrichum, *Corda.*
Memnonium, *Corda.*
Merosporium, *Corda.*
Coccotrichum, *Corda.*
Didymaria, *Corda.*
Scolicotrichum, *Kze.*
Myxocladium, *Corda.*
Soredosporium, *Corda.*
Azozma, *Corda.*
Mydonotrichum, *Corda.*
Macroon, *Corda.*
Coccosporium, *Corda.*
Diplosporium, *Lk.*
Mydonosporium, *Corda.*
Gliotrichum, *Eschw.*
Balanium, *Wallr.*
Gongylocladium, *Wallr.*
Ospriosporium, *Corda.*
Trichostroma, *Corda.*
Medusula, *Corda.*
Spondylocladium, *Mart.*
Acrophyton, *Eschw.*
Clisosporium, *Fr.*
Tipularia, *Chev.*
Asterothecium, *Wallr.*
Amphisporium, *Lk.*
Hyphelia, *Fr.*
Trichoderma, *P.*
Ostracoderma, *Fr.*
Ostracococcum, *Wallr.*
Myrosporium, *Corda.*
Cylichnium, *Wallr.*
Goupilia, *Merat.*
Diploderma, *Lk.*
Anixia, *Fr.*
Ceratophora, *Humb.*
Hydnocaryon, *Wallr.*
Ascospora, *Fr.*
Hercospora, *Fr.*
Coccobolus, *Fr.*
Ostropa, *Fr.*
Hypospila, *Fr.*
Gibbera, *Fr.*
Valsa, *Fr.*
Podostromium, *Kze.*
Collacystis, *Kze.*
Pyrenium, *Tode.*
Acinula, *Fr.*
Sclerococcum, *Fr.*
Sarea, *Fr.*
Phymatostroma, *Corda.*
Melanostroma, *Corda.*
Gliostroma, *Corda.*
Dermosporium, *Lk.*
Chroostroma, *Corda.*
Crocysporium, *Corda.*
Myxacium, *Wallr.*
Myxomphalon, *Wallr.*
Hirneola, *Fr.*
Amphicorda, *Fr.*
Epichysium, *Tode.*
Gyrolophium, *Kze.*
Sporendonema, *Desm.*
Cælosporium, *Lk.*
Dryophilum, *Schwein.*
Malacharia, *Fée.*

Spurious Genera.
Rhizomorpha, *Ach.*
Byssus, *L.*
Mycoderma, *P.*
Mycomater, *Fr.*
Tophora, *Fr.*
Herpotrichum, *Fr.*
Fibrillaria, *P.*
Himantia, *P.*
Capillaria.
Ozonium, *Lk.*
Chætosporium, *Corda;*
Erineum, *P.*
Septotrichum, *Corda.*
Physoderma, *Wallr.*
Cephaleuros, *Kze.*
Sphinctrina, *Fr.*
Sclerotium, *P.*
Rhizoctonia, *Fr.*
Spermœdia, *Fr.*
Pachyma, *Fr.*
Nosophlæa, *Fr.*
Peribotryon, *Fr.*
Ectostroma, *Fr.*
Institale, *Fr.*
&c. &c. &c.

NUMBERS.—GEN. 598. Sp. 4000? *M. J. B.*

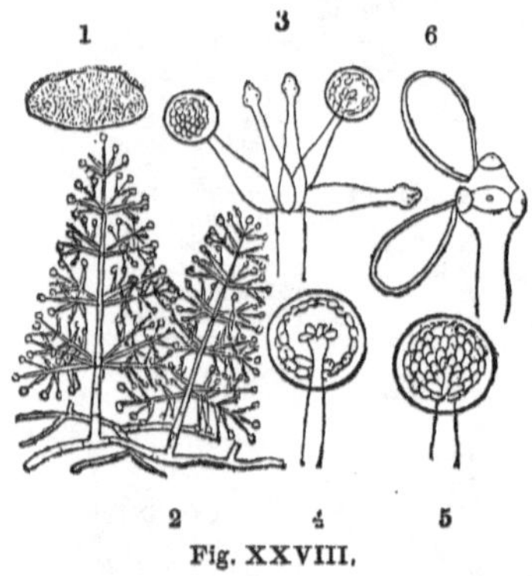

Fig. XXVIII.

sion of spores. At least this is the case, according to Corda, in some genera. Should this structure be found to prevail generally, it would become a question whether they should not be associated with the sporiferous fungi, the vesicle being regarded simply as a veil. I am myself inclined to this view, but at present think it savours too much of theory to venture to propose it.

Fig. XXVIII. Acrostalagmus cinnabarinus.—1. A patch, the natural size; 2. plants very highly magnified; 3 a portion of the fructification still more magnified; 4, 5. spores contained in gelatinous heads; 6. a point of a branch with two spores remaining upon it.

ALLIANCE III. *LICHENALES*.—THE LICHENAL ALLIANCE.

Algæ, § 3. Lichenes, *Juss. Gen.* 6. (1789).—Lichenes, *Hoffm, Enumerat. Lichenum,* (1784) ; *Acharius Prodr. Lichen.* (1798) ; *Id. Methodus,* (1803) ; *Id. Lichenogr. Univers.* (1810) ; *DC. Fl. Fr.* 2. 321. (1815) ; *Fries in Act. Holm.* (1821) ; *Agardh Aph.* 89. (1821) ; *Eschweiler Syst. Lich.* (1824) ; *Wallroth Naturgesch. der Flechten.* (1824) ; *Grev. Flora Edin.* xix. (1824) ; *Meyer über die Entwickelung, &c. der Flecht.* (1825) ; *Fée Méth. Lich.* (1825) ; *Fries Syst. Orb. Veg.* 224. (1825) ; *Martius in Bot. Zeitung,* 193. (1826) ; *Fée in Dict. Class.* 9. 360. (1826) ; *Fries Lichenogr. Europæa.* (1831) ; *Eschw. in Mart. Fl. Bras.* 1. 51. (1833) ; *Hooker Brit. Fl.* vol. ii. pt. 1. 129. (1833) ; *Endlich. Gen.* p. 11 ; *Link Ausgew. Anatom. Botan. abbild. fasc.* 3. — Graphideæ, *Chevalier Hist. des Graphidées.* (1824, &c.)

DIAGNOSIS.—*Cellular flowerless plants, nourished through their whole surface by the medium in which they vegetate; living in air; propagated by spores usually inclosed in asci, and always having green gonidia in their thallus.*

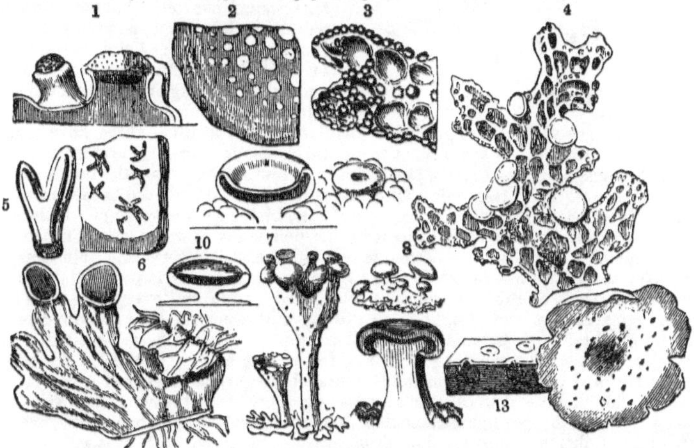

Fig. XXIX.

Perennial plants, often spreading over the surface of the earth, or rocks, or trees, in dry places, in the form of a lobed and foliaceous, or hard and crustaceous, or leprous substance, called a thallus. This thallus is formed of a cortical and medullary layer, of which the former is simply cellular, the latter both cellular and filamentous ; in the crustaceous species the cortical and medullary layer differ chiefly in texture, and in the former being coloured, the latter colourless ; but in the fruticulose or foliaceous species, the medulla is distinctly floccose, in the latter occupying the lower half of the thallus, in the former enclosed all round by the cortical layer. Reproductive matter of two kinds ; 1, spores naked, or lying in membranous amylaceous tubes (thecæ) immersed in nuclei of the medullary substance, which burst through the cortical layer, and colour and harden by exposure to the air in the form of little discs called shields ; 2, the separated cellules of the medullary layer of the thallus. These, called gonidia, or gongyli, are

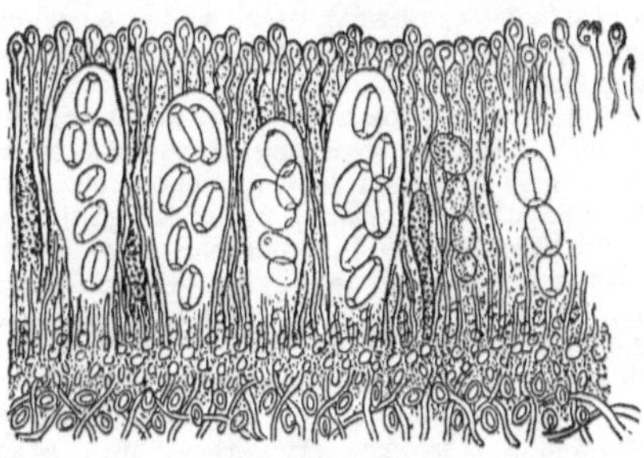

Fig. XXX.

Fig. XXIX.—1. Shields of Variolaria amara ; 2, a portion of the thallus of the same plant ; 3. a piece of the thallus of Sticta pulmonacea, with lacunæ and soredia ; 4. thallus of the same, bearing shields ; 5. shield of Opegrapha scripta ; 6. thallus of the same ; 7. shields, young and old, of Lecanora perella ; 8. shields of Bæomyces rufus ; 9. part of thallus of Peltidea canina ; 10. section of a shield of Sticta pulmonacea ; 11. Podetia of Cenomyce coccinea ; 12. section of a shield of Bæomyces rufus ; 13. shields of Endocarpon miniatum ; 14. thallus of the same. Chiefly from Greville's Flora Edinensis.

Fig. XXX.—Section of a shield of Parmelia parietina. *Link.*

universally of a green colour, and either lie singly or in clusters beneath the cortical layer of the thallus, or break out in clusters called soredia, or in cups called cyphelia.

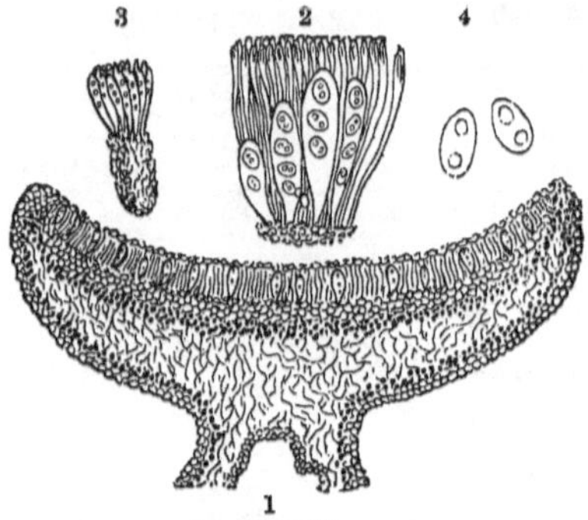

Fig. XXXI.

Nothing can be more varied than the appearance of Lichens. If the grey, and yellow, and brown stains upon old walls, ancient churches, and other buildings are carefully examined, those appearances will always be found to arise from minute Lichens having taken possession of the surface of the stones, to which they adhere, drawing their food from the atmosphere; small shields are scattered over their surface, sometimes round, but not unfrequently like dark clefts or lines, giving the Lichen the appearance of being covered with broken letters. Others are found on trees and pales, forming broad patches of various colours, some being of the richest golden yellow; others spread upon the ground in plantations and heaths—these have usually a much larger growth; some again hang from the branches of venerable trees, which they clothe with a shaggy beard of grey; and, finally, a few start up upon the heath, grey and deformed, but eventually fashioning themselves into tiny goblets, the border of which is studded with crimson shields. According to Fries, Lichens "are types of Algals born in the air, interrupted in their development by the deficiency of water, and stimulated into forming a nucleus by light. No Lichen is ever submersed (Verrucaria submersa is an exception); there is none of which the vegetation is not interrupted by the variable hygrometrical state of the atmosphere; and, finally, none that ever develop in mines, caverns, or places deprived of light. On this account, their shields are more rare in the fissures of mountains, or in shady groves, than in places fully exposed to light. In wet places, also, their shields are not produced; for so long as they are under the influence of water they are hardly distinguishable from Hydrophyceæ (forms of Algals); as, for instance, Collema, &c. But these plants, when exposed to the sun, do perfect their shields, as is found by Nostoc lichenoides, foliaceum, &c., which, when dry, are ascertained to be Collema limosum, flaccidum, &c., surcharged with water." By being acquainted with this rule, the same author says, he has succeeded in discovering many Swedish Lichens with shields, which have for many years been constantly found sterile; as Parmelia conoplea, lanuginosa, gelida, &c.; and he even asserts that he has succeeded artificially in inducing sterile Lichens to become fruitful, as Usnea jubata, and others.—*Plant. Hom.* 224. Lichens consist, according to Eschweiler, of a medullary and a cortical layer of tissue, of which the former is imperfectly cellular or filamentous, and bursts through the latter in the form of shields (apothecia), which contain a nucleus, consisting of a flocculent gelatinous substance, among which lie the cases of sporules. These cases (thecæ) are transparent membranous tubes, either simple or composed of several placed end to end, which either lie free in the nucleus, or are themselves contained in other membranous cases (asci). In the beginning Lichens are stated to be in all cases developed in humidity, and to be, in fact, at that time, mere Phyceæ or Confervæ; but as soon as the humidity diminishes, the under part dies, and an inert leprous crust is formed, which ultimately becomes the basis of the plant. Hence Lichens consist of two distinct sorts of tissue,—living cellules forming the vegetating part, and dead cellules the cohesion of which is lost; when separate, the former is Palmella botryoides, and the latter Lepraria. Of these two sorts of matter, the leprous is incapable of perpetuating the Lichen, while every part of the living stratum has been ascertained to become reproductive matter. See Fries, as above quoted, and Meyer *Ueber die Entwickelung, &c., der Flechten.* The investigations of the latter are exceedingly interesting. By sowing Lichens, he arrived at some curious conclusions, the chief of which are, that, like other imperfect plants, they may owe their origin either to an elementary, or a reproductive, generating power—the latter capable of development like the plant by which they are borne: that decomposed vegetable, and some inorganic, matter, are equally capable of assuming organisation under the influence of water and light; and that the pulverulent matter of Lichens is

Fig. XXXI.—Section of the shield of Parmelia tiliacea; the green gonidia are the black dots beneath the skin; 2. a portion of the same more magnified, showing the spore-cases and paraphyses; 3. a morsel of the shield of Cladonia coccifera; 4. spores of Parmelia parietina.

that which is subject to this kind of indefinite propagation, while the spores lying in the shields are the only part that will really multiply the species. He further says, that he has ascertained, by means of experiments from seed, that supposed species and even some genera of Acharius, are all forms of the same; as, for instance, Lecanora cerina, Lecidea luteo-alba, and others, of the common Parmelia parietina.

The distinction between Lichens and Fungals has already been fully explained by Mr. Berkeley (p. 30). It is, therefore, only necessary, in this place, to give a few details concerning the geographical distribution and uses of the order, or cluster of orders, which Botanists combine under the name of Lichens.

Pulverulent Lichens are the first plants that clothe the bare rocks of newly-formed islands in the midst of the ocean, foliaceous Lichens follow these, and then Mosses and Liverworts. (*D'Urville, Ann. Sc.* 6. 54.) They are found upon trees, rocks, stones, bricks, pales, and similar places; and the same species seem to be found in many different parts of the world: thus, the Lichens of North America differ little from those of Europe. They are not met with on decaying matter, where they give way to fungi; but they often occupy the surface of living plants, especially their bark. In the tropics they lay hold of evergreen leaves. Their chosen climate is one that is temperate and moist; aspects to the north or west are also their favourite resort, for they shun the rays of the noon-day sun. No place seems to be a more constant haunt than the surface of sand-stone rocks, and buildings, in cool and moist countries. They are met with, in one place or other, from the equator to the pole, and from the sea-shore to the limits of eternal snow. The finest species are found near the equator; the most imperfect, such as the crustaceous genera, which can hardly be distinguished from the rocks they grow upon, are chiefly observed on mountain-tops, and near the pole. The Idiothalami are most abundant in tropical America.

Lichens have been remarked by De Candolle to possess two distinct classes of characters, the one rendering them fit for being employed as dyes after maceration in urine, the other making them nutritive and medicinally useful to man. Braconnot has ascertained that oxalate of lime exists in great abundance in Lichens, particularly in those which are granular and crustaceous. The common Variolaria, which is found upon almost every old beech-tree, contains rather more than twenty-nine per cent. (*Ed. P. J.* 13. 194.) Lichens that grow on the summit of fir-trees have been found by John, of Berlin, to contain an uncommon proportion of oxide of iron, a curious illustration of the peculiar powers which various plants possess of separating the inorganic matters presented to them in their food. (*Ibid.* 2. 394.) Of those used in dyeing, the principal crustaceous kinds are, Lecanora perella, the Orseille de terre, or Perelle d'Auvergne of the French, Lecanora tartarea (or Cudbear), hæmatomma and atra, Variolaria lactea, Urceolaria scruposa and cinerea, Isidium Westringii, Lepraria chlorina; of the foliaceous species, Parmelia saxatilis, omphalodes, encausta, conspersa, and parietina, Sticta pulmonacea, Solorina crocea, and Gyrophora deusta and pustulata; but the most important are Roccella tinctoria and fusiformis, the dye of which makes litmus, and is largely used by manufacturers under the name of Orchall, or Archill, or Orseille des Canaries; there are other species capable of being employed in a similar manner, as Usnea plicata, Evernia prunastri, Alectoria jubata, Ramalina scopulorum, and several Cenomyces. Dr. Robert Thomson finds the common yellow pale Lichen (Parmelia parietina) to contain a peculiar colouring matter, called Parietin, of a bright yellow. This is heightened by a drop of nitric, muriatic, or sulphuric acid; while minute quantities of ammonia, or other alkalies, change it to a rich red inclining to purple.

Agardh considers Lichens more nearly allied to Fungals than to Algals: he remarks, that if Sphærias or Pezizas had a thallus, they would be Lichens, and that the same part is all that determines such genera as Calycium, Verrucaria, or Opegrapha to be Lichens, and not Fungi. He adds, that all the transitions from Algals to the state of Lichens, which have been detected by modern inquirers, are mere degenerations into the form of the Lichen tribe, and by no means into Lichens themselves.

According to Fries, Lichens have the vegetation of Algals, and the fruit of Fungals. (*Systema*, 52.)

Fries refers Byssaceæ to Lichens with the following short character:—"Aerial, perennial, constantly growing, with a filamentous texture; consisting of solid fibres (either few or several glued together with a common bark), unchanged and permanent. Fructification homogeneous, growing externally, and naked." *Syst. Orb. Veg.* 291. Some of these plants appear to be meteoric productions; on one occasion they are said to have suddenly overrun all the leaves of pines on the side next the wind in the neighbourhood of Dresden; on another, on the 29th of Aug. 1830, to have in an instant spread over the sails and masts of a ship at Stockholm; and Fries is disposed to consider the cobweblike matter, that overruns the grass in the mornings of spring and autumn, of this nature, and not of an animal origin. See *S. O. Veg.* 318.*

The nutritive properties of Lichens probably depend upon the presence of an amylaceous substance analogous to gelatine, which, according to Berzelius, occurs in the form of pure starch or amylaceous fibre, to the amount of 80.8 per cent. in Cetraria islandica. This plant, which is the Iceland Moss of the shops, is slightly bitter as well as mucilaginous, and is frequently used as a tonic, demulcent, and nutrient; Cetraria nivalis, Sticta pulmonaria, and Alectoria usneoides, will all answer the same purpose. Tripe de Roche, on which the Canadian hunters are often forced to subsist, is the name of various species of Gyrophora; several kinds of Lecanora inhabit even the deserts of Asia in large quantities, and are eaten by the nomade tribes of those regions. The Rein-Deer Moss, which forms the winter food of that animal, is Cenomyce rangiferina. Parmelia parietina, Borrera furfuracea, Evernia prunastri, Cenomyce pyxidata and coccifera, are reputed astringents and febrifuges, and Peltidea aphthosa an anthelmintic. Alectoria Arabum (Oschnah) is said to be sedative and soporific. Peltigera canina was once regarded as a specific in hydrophobia. Sticta pulmonacea is used in Siberia for giving a bitter to beer, and in this country is employed, under the name of Lungs of the Oak, as a nourishing diet for weak persons. Evernia vulpina, called Ulfmossa by the Swedes, is believed by that people to be poisonous to wolves; but this requires confirmation. See *De Cand. Essai Méd.* 318, and *Agardh, Aph.* 94.

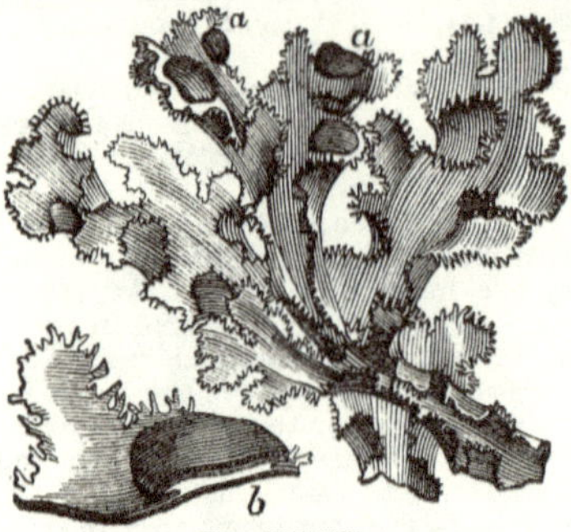

Fig. XXXII.

According to the chemists, Lichens contain several peculiar principles; such as Cetrarine, Picrolichenine, Stictine, and Varioline, which are bitter; and the colouring matters called Orcine, Erythrine, Parmelochromine (also called Vulpuline and Vulpinic acid), Strychnochromine, Strychnerythrine, Lecanorine, &c.: and finally, from Usnea florida,

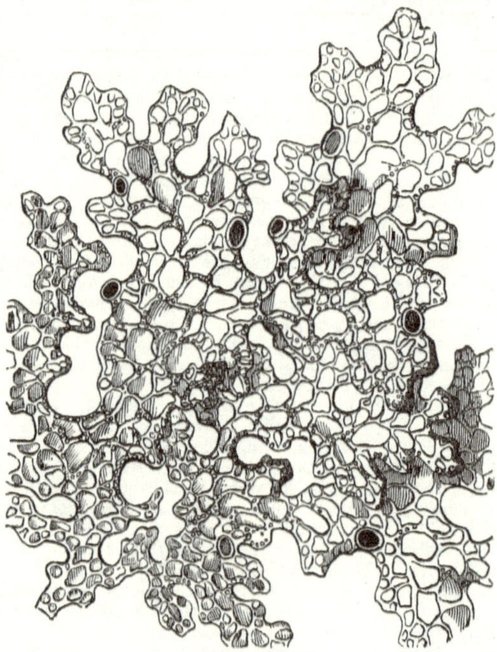

Fig. XXXIII.

* Nothing, however, can be more heterogeneous than the mass of genera collected by Fries under the unfortunate name of BYSSACEÆ. Many of them are spurious genera, others true Fungals, a few anomalous Lichens, and a small portion not easily arranged under Fungals, Algals, or Lichens. An excellent notice will be found by Dr. Montagne in the History of Cuba, and the latest information on the subject in the article Byssacées in the new Dict. d'Hist. Nat. By excluding such genera as Cilicia and Cœnogium, which the example of Parmelia gossypina will justify us in uniting with Lichens, we have remaining a very natural, though small group, which may be distinguished under the name of Collemaceæ; and Dr. Montagne, who has lately had some correspondence on the subject with

Fig. XXXII.—Cetraria islandica: *a a.* its shields; *b.* a shield magnified and divided vertically.
Fig. XXXIII.—Sticta pulmonaria, or Lungs of the Oak.

hirta and plicata, Parmelia fraxinea and farinacea, and various others, M. Knop has obtained a substance called Usnine or Usnic acid. This author finds the sulphur-coloured and yellowish-green lichens are especially rich in usnine, for instance, Lecidea geographica and Parmelia sarmentosa. Usnine acts a conspicuous part by its various metamorphoses and combinations in the alterations of colour of many lichens. In all lichens however it is accompanied by yellow or green resins, which in common with it partake of the property of becoming red by ammonia and exposure to the air; this red colouring however is destroyed by sulphuretted hydrogen. Usnine occurs in the thallus as well as in the fruit-discs. The shields of the Cladoniæ contain near the fruit-bearing vesicles quill-shaped cylindrical cells, which are coloured pale red at the base, but darker towards the apex by a colouring substance, which dissolves in ammonia and potash with a wine-red, in sulphuric acid with a carmine-red colour; the sulphatic solution is precipitated by water; the alkaline solution is not decolorized by sulphuretted hydrogen. The nearly scarlet-red fruit-discs of the Cladoniæ become brown and blackish-brown with age. In fact, the fruit-discs of the lichens containing usnine are precisely similar in colour to the thallus, or brown, reddish-brown and carmine-red. The sulphur-yellow lichens contain most usnic acid, and indeed in a free state; the other colours are probably produced by the action of the alkalies and earths of the vegetable salts in the lichens, the ammonia of the rain-water assisting the chemical action of the usnic acid, which is otherwise insoluble in water. In this manner the green, red and brown colours may originate. The silver-white Cladonia rangiferina probably contains the usnic acid in the state of an earthy salt. Lecidea geographica is sometimes sulphur-yellow, sometimes yellowish-green. If some pure yellow specimens be suspended in a glass over a solution of carbonate of ammonia, they become covered with carmine-red globules, after frequent washing entirely lose the usnic acid, and finally become grayish-white like dead lichens. The Parmeliæ and Usneæ continue of a brilliant green colour in shady and moist places, but when exposed to the heat of the sun they become brownish-black; if treated as above with ammonia and dried, they likewise present similar colours. The fruit-discs of the Cladoniæ also turn brown under similar treatment. The cause of all these changes is the usnic acid, which itself is of a yellow colour, but becomes oxidized in combination with bases by exposure to the air, forming various coloured compounds. *Chem. Gaz.* 1844. 182.

Fig. XXXIII*.

In this, as in the Fungal alliance, I have forborne formally to break up Lichens into several Natural orders, and have preferred to leave the task to others more skilled than myself in this branch of Botany; but it is not to be doubted, that hereafter the pro-

Mr. Berkeley, is not averse to this reduction. The Collemaceæ have strictly the thallus of an Alga, and the fruit of a Lichen. The following genera are comprised in the group:—

Collema, *Ach.*	Synalyssis, *Fr.*	Omphalidium, *Mey. et Flotw.*	Lichina, *Ag.*
Leptogium, *Fr.*	Myxopuntia, *Mont. et Dur.*	Paulia, *Fée.*	Myriangium, *Mont. and Berk.*

It is better for the present, in a matter confessedly so difficult, to throw out the above in the form of a hint, rather than to propose a distinct natural order. But every thing seems to indicate the necessity of placing them apart under some kind of denomination.

Fig. XXXIII*.—Parmelia tiliacea.

priety of carrying out the principles of ordinal division recognised elsewhere, will be introduced among Lichens. In the meanwhile the evidence that has been collected seems to point to such a mode of grouping as is indicated in the following proposed

NATURAL ORDERS OF LICHENALS.

Nucleus breaking up into naked spores . . . . 12. GRAPHIDACEÆ.
Nucleus bearing asci; thallus homogeneous, gelatinous or cartilaginous . 13. COLLEMACEÆ.
Nucleus bearing asci; thallus heterogeneous, pulverulent or cellular . . 14. PARMELIACEÆ.

GENERA.*

I.—*Coniothalameæ.* Shields open; the nucleus breaking up into naked spores.

* Pulveraridæ.

Arthronia, *Ach.*
Incillaria, *Fr.*
Arthronaria, *Fr.*
Pulveraria, *Ach.*
Lepraria, Achar.
Pulina, Adans.
Phytoconis, Bory.
Leptuberia, Rafin.

** Calycidæ.

Coniocarpon, *DC.*
Conioloma, Flörk.
Trachilia, *Fr.*
Calycium, *Pers.*
Cyphelium, Ach.
Acolium, Fée.
Coniocybe, *Ach.*
Sclerophora, Chev.
Fulgia, Chev.

II.—*Idiothalameæ.* Shields closed at first, afterwards open; the nucleus gelatinous, made up of naked spores.

* Graphidæ.

Coniangium, *Fr.*
Ustalia, *Fr.*
Pyrochroa, Eschw.
Platygramma, Meyer.
Sclerophyton, *Eschw.*
Lecanactis, *Eschw.*
Lecanotis, Rehb.
Opegrapha, *Pers.*
Hysterina, Ach.
Oxystoma, Eschw.
Scaphis, Eschw.
Leucogramma, Meyer.
Graphis, Eschw.
Fissurina, Fée.
Graphis, *Fr.*
Leiorreuma, Eschw.
Platygramma, Meyer.

** Glyphidæ.

Medusula, *Eschw.*
Sarcographa, Fée.
Asterisca, Meyer.
Chiodecton, *Ach.*
Glyphis, *Ach.*

*** Limboridæ.

Urceolaria, *Ach.*
Polystroma, Clement.
Thelotrema, *Ach.*
Hymenoria, Ach.
Anthrocarpum, Meyer.
? *Pyrenula*, Fée.
? *Ascidium*, Fée.
Limboria, *Ach.*
Gyrostomum, *Fr.*
Cliostomum, *Fr.*

**** Pyxinidæ.

Umbilicaria, *Hoffm.*
Lasallia, Merat.
Gyrophora, Achar.
Gyromium, Wahlenb.
Capnia, Vent.
Pyxine, *Fr.*

III.—*Gasterothalameæ.* Shields always closed, or opened by the irregular separation of the thallodial covering. Nucleus enclosed, containing asci, deliquescing or shrivelling up.

* Verrucaridæ.

Diorygma, *Eschw.*
Pyrenothea, *Fr.*
Leprantha, Dufour.
Thrombium, Wallr.
Gelatinaria, Flörk.
Pyrenastrum, *Eschw.*
Parmentaria, Fée.
Verrucaria, *Pers.*

** Trypethelidæ.

Sphaeromphale, *Reichb.*
Segestria, Fr.
Segestrella, Fr.
Mycoporum, *Meyer.*
Porothelium, Eschw.
Porodothium, Fr.
Astrothelium, *Eschw.*
Tripethelium, *Spr.*
Bathelium, Achar.
Ophthalmidium, Eschw.
Ocellularia, Meyer.

*** Endocarpidæ.

Pertusaria, *DC.*
Porina, Ach.
Porophora, Meyer.
Sagedia, *Ach.*
Stigmatidium, Meyer.
Enterographa, Fée.
Endocarpon, *Hedw.*
Dermatocarpon, Eschw.

**** Sphaerophoridæ.

Siphula, *Fr.*
Dufourea, Ach.
Sphaerophoron, *Pers.*
Coralloides, Hoffm.

IV.—*Hymenothalameæ.* Shields open; nucleus forming a disk, permanent and bearing asci.

* Ephebidæ.

Micarea, *Fr.*
Ephebe, *Fr.*
Coenogonium, *Ehrenb.*

** Lecideidæ.

Lecidea, *Ach.*
Catillaria, Achar.
Echinoplaea, Fée.
Myriotrema, Fée.
Rhizocarpon, Ramond.
Patellaria, *Pers.*
Biatora, Fr.
Lepidoma, Ach.
Psora, Hoffm.
Circinaria, Fée.
Pulveraria, Willd.
Verrucaria, Hoffm.
Sphaerothallia, Nees.
Baeomyces, *Pers.*
Sphyridium, *Flot.*
Cladonia, *Hoffm.*
Cenomyce, Achar.
Capitularia, Flörk.
Scyphophorus, DC.
Helopodium, DC.
Cladonia, Ach.
Schasmaria, Ach.
Ceraunia, Ach.
Pyxidium, Schreb.
Pyxidaria, Bory.
Pycnothelia, Achar.
Stereocaulon, *Schreb.*
Thamnium, Vent.

*** Parmeliadæ.

Gyalecta, *Ach.*
Dirina, *Fr.*
Cilicia.
Cænogium.
Parmelia, *Fr.*
Lecanora, Achar.
Squamaria, DC.
Urceolaria, Fr.
Phlyctis, Wallr.
Patellaria, Fr.
Psora, Fr.
Placodium, Fr.
Zeora, Fr.
Amphiloma, Fr.
Panaria, Delis.
Lobaria, Hoffm.
Physcia, Fr.
Hagenia, Eschw.
Imbricaria, Fr.
Platisma, Hoffm.
Sticta, *Schreb.*
Pulmonaria, Hoffm.
Reticularia, Baumg.
Crocodia, Link.
? *Plectocarpon*, Fée.
Peltigera, *Willd.*
Peltidea, Ach.
Antilyssus, Hall.
Erioderma, Fée.
Solorina, Ach.
Sommerfeltia, Flörk.
Nephroma, Achar.

**** Usneidæ.

Cetraria, *Ach.*
Physcia, DC.
Cornicularia, Hoffm.
Coelocaulon, Link.
Roccella, *DC.*
Ramalina, *Achar.*
Platyphyllum, Vent.
Evernia, *Ach.*
Borrera, Ach.
Bryopogon, Lk.
? *Neuropogon*, Nees.
Usnea, *Hoffm.*
Reichenbachia, Spr.

NUMBERS. GEN. 58. Sp. 2400. (*Fee.*)

POSITION.—Marchantiaceæ.—LICHENALES.—Fungales.

* Arranged principally according to Endlicher.

Class II. ACROGENS.

PSEUDOCOTYLEDONEÆ, *Agardh, Aph.* 72. (1821).—HETERONEMEA, *Fries, Syst. Orb. Veg.* 1. 30. (1825). ACROBRYA, *Mohl. in Mart. Pl. Crypt. p.* 56; *Endlich. Gen. p.* 42. (*in Part.*)—ACROGENÆ, *Ad. Brongn. Enum. p.* xiii. (1843).

WITH this class a great advance in structure is accomplished. The simplicity which is so remarkable in Thallogens is exchanged for a complicated apparatus of many kinds. All the species have stomates or breathing-pores on their surface : in the great majority there is a distinct stem and leaves, the latter of which are always arranged with perfect symmetry; and in those species which approach Thallogens, (as the Crystalworts, which stand close upon Lichens) the thallus has all the texture of leaves, although a separate stem is refused to them. There is, however, no trace of flowers, properly so called; and yet in the involucre of many Liverworts, and in the spore-cases of Mosses, an arrangement of leaves occurs, which appears to be the forerunner of the flowers of more perfect plants. Sexes, however, are wholly missing; that is to say, nothing can be found which resembles the anthers and pistil of flowering plants, except in some vague external circumstance : there is no evidence to show that any one order of Acrogens possesses organs which require to be fertilised the one by the other in order to effect the generation of seeds. Hence those reproductive bodies of Acrogens which are analogous to seeds are called spores. Mr. Griffith takes, however, a very different view of this question, and assigns true sexes to Acrogens.

He thinks it probable that we have at least three modifications of the phenomenon of fecundation "among the higher acotyledonous plants. In one the male influence is applied to the apex of a pistillum, in the second to a nucleus without the intervention of a pistillary apparatus. In the third the male influence is exerted on a frond itself, and is followed by the development of the young capsule from a point in the substance of the frond corresponding to and sometimes distant from the place to which the male influence has been applied. This is founded on observations made on Anthoceros in 1836, from which it would appear that the place of exsertion of the future capsules is pointed out by a slight protuberance, over the apex of which a flake of matter like the so called male matter of Musci and Salvinia is spread, sending down to some distance within the frond a tube-like process, which causes the dislocation of the cells of the tissue with which it comes into contact. The future capsule is stated in his notes not to be appreciably pre-existent, and its situation is only pointed out by a bulbiform condensation of the tissue of the frond. The young capsule during its development ascends along the same line, and pushes before it a corresponding cylindrical body of the tissue of the frond, the calyptra of authors." But, it seems to me, that this very complexity of action is more like variations in self-propagation, than phenomena of fecundation, which, among the plants in which that action certainly takes place, is subject to no such modifications.

A large number of Acrogens have no true spiral vessels, which are confined to the more highly developed forms, such as Ferns, Clubmosses, and Horsetails; but there is a very general tendency to the production of spiral threads in their cells. This has been long known to exist in the bodies called elaters among Liverworts, and traces of it have been recognised in the leaves of certain mosses, such as Sphagnum.

"So far as I am aware," says Schleiden, "the occurrence of a spiral formation has been observed in the reproductive organs of Hepaticæ only in the elaters or fruit-valves. But it is not less strikingly developed in the organs of vegetation in Marchantiaceæ. The parenchyma of the leaf of Marchantia polymorpha and Fegatella conica consists almost entirely of cells whose partitions appear distinctly porous, or (especially in M. polymorpha) beautifully thickened with net-work. This thickening of the partitions of the cell takes place to so great a degree in the older parts and in the proximity of the midrib, that by transverse sections the pore-channels may be plainly recognised. Amongst mosses, the true Dicrana, for example D. Schraderi, spurium, &c., are distinguished by the cells of the leaf having very thick sides, and their partitions evidently pierced by very wide, or funnel-shaped pore-channels, just as happens in the epidermis of many phanerogamous plants; and still more conspicuously do these spiral and porous formations display themselves in Sphagneæ, and in the nearly related group of Leucophaneæ established by Hampe."—(*Ann. Nat. Hist.* v. 73.) The same tendency is still more remarkably apparent in a curious formation of loose short spiral threads generated in the cells of the bodies called Antheridia, and elsewhere; which, because of an apparently spontaneous motion when they are floating in water, have been thought to be animalcules of the genera Spirillum or Vibrio.

In general, Acrogens are plants of very small stature. But in Ferns they occasionally acquire the size of trees; always however growing with a simple stem in such cases, unless when their growth is interrupted by accident. If they branch naturally, they do so in a forking manner. Their stem, instead of increasing by the deposition of matter originating in the leaves, appears to be a mere extension of one common vegetating point, which becomes cylindrical and long, when it is capable of being acted upon by the influence of light. It may be regarded indeed as a mere combination of the bases of leaves, gradually evolved one from the bosom of the other.

The orders of Acrogens seem to resolve themselves into three Alliances, of which the lowest in organization in some respects is the highest in others. This which is named the Muscal, inasmuch as it includes the true Mosses, has no spiral vessels, no veins to its leaves, and its species are of diminutive size; but it has reproductive organs of two very distinct kinds, and its spore-cases are usually elaborately provided with elaters at least, and often with a complicated arrangement of rudimentary leaves. The two others have a far larger stature, are abundantly furnished with scalariform or true spiral vessels in their stem, but their reproductive organs are of the most simple kind, and never assume different forms in the same individual. The one called the Lycopodal Alliance has scaly leaves and pulverulent spores, always of two sorts, contained in cases which usually open by definite valves; the other, called the Filical Alliance, has thin expanded veiny leaves and granular spores of only one kind enclosed in cases which burst irregularly.

The affinities of Acrogens are well ascertained. Riccia and its neighbours are closely allied to Lichens. Horsetails may be looked upon as an approach towards the structure of Ephedra among Gnetaceæ, or of Casuarina in Galeworts. The Clubmosses evidently approach Coniferous Gymnogens in their small scale-like imbricated leaves and coniferous fructification.

Ferns themselves have in their foliage the peculiar veining of certain genera belonging to the order of Yews in Gymnogens; they also approach Cycadaceous Gymnogens in their simple cylindrical stems and gyrate foliage, which bears the fructification on the margin. Nor are the Urn Mosses (Bryaceæ) without their resemblance to the order of Yews when we compare some of the larger species with the little Dacrydia of New Zealand, which are only a few inches high.

Alliances of Acrogens.

Muscales.—*Cellular* (*or vascular*). *Spore-cases immersed or calyptrate* (*i. e. either plunged in the substance of the frond, or enclosed within a hood having the same relation to the spores as an involucre to a seed-vessel.*)

Lycopodales.—*Vascular. Spore-cases axillary or radical, one or many celled. Spores of two sorts.*

Filicales.—*Vascular. Spore-cases marginal or dorsal, one-celled, usually surrounded by an elastic ring. Spores of but one sort.*

ALLIANCE IV.—*MUSCALES.*—THE MUSCAL ALLIANCE.

CELLULARES FOLIACEÆ, *DC. Theor. Elem.* 249. (1819).—PSEUDOCOTYLEDONEÆ, Class I. *Agardh, Aph.* 103. (1822).—HETERONEMEA, *Fries Syst. Orb. Veg.* 33. (1825) *in part.*—ACOTYLEDONES, Class 2. *Ad. Brongniart in Dict. Class.* 5. 159. (1824).—CRYPTOGAMICÆ, 2d Circle, *T. F. L. Nees v. Esenbeck and Ebermaier Handb. der Med. Bot.* 1. 18. (1830).—HEPATICÆ *and* MUSCI, *Endlicher Gen.* 42. *and* 46.

DIAGNOSIS.—*Cellular or vascular Acrogens, with the spore-cases either plunged in the substance of the frond, or enclosed in a cap-like hood.*

NEXT after the Algal series follows that which derives its name from Mosses, presenting at one point a structure nearly as simple as that of Lichens, and at another a complexity of organization unknown elsewhere among Acrogens. The Crystalworts (Ricciaceæ), by which the series begins, are mere lobes of green or purple parenchyma floating in water or spreading over mud, and multiplied by reproductive particles (spores) generated in hollow flask-like cases. Then follow masses of species gathered together under the names of Liverworts (Marchantiaceæ) and Scalemosses (Jungermanniaceæ), whose stems and leaves are, in the majority of instances distinctly separate, and among whose spores are formed elastic threads with a powerful hygrometric quality and of unknown use. Finally the ranks are closed by Splitmosses (Andræaceæ), and Urn-Mosses (Bryaceæ), which have in all cases a distinct axis of growth, symmetrical leaves, and a complicated reproductive apparatus formed by the adhesion of leaves in rings or whorls: in emulation, as it were, of flowers, in the more completely organized classes of Endogens and Exogens.

In the opinion of a large number of modern observers there are two sexes in all these plants, the one bearing the name of Antheridia (or false anthers), and the other of Pistillidia (or false pistils). That such organs exist is certain; the question is whether or not they are to be looked upon as connected with sexual qualities. Those who regard them in that light have naturally taken the imbedded oblong antheridia of Marchantia, and the stalked reticulated ones of Jungermannia, for anthers; but Hooker, in his beautiful *Monograph* of the latter genus, and also in his *British Flora*, (p. 459,) is unsatisfied as to their nature. Greville, in the *Flora Edinensis*, is in a similar state of uncertainty; and Agardh admits nothing more in them than a resemblance to male organs, adopting the opinion that they are a particular form of gemmules. Mirbel considers the cups or baskets of Marchantia to be filled with little buds, and the peltate receptacles to be male flowers, while the stalked receptacles are masses of pistils. (See his admirable Memoir, tt. vi. et vii.) On the other hand Greville and Arnott, in the fourth volume of the *Transactions of the Wernerian Society*, speak thus positively against the sexuality of the organs in question:—

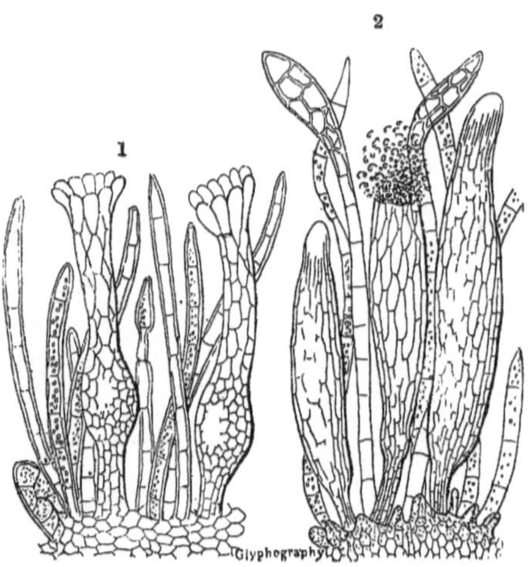

Fig. XXXIV.

"What the organs really are, in the plants under review, which the accurate Hedwig so well figured and described under the name of stamens, we leave to others to decide; but we cannot help entering our protest against those bodies called Stamina and Pistilla (the young thecæ) being regarded in a similar light with the same organs in more perfect plants. 'Though,' says Sprengel, 'I have formerly been a zealous advocate for Hedwig's *Theory of the Fructification of Mosses*, it has nevertheless appeared to me an insurmountable objection, that the supposed anther can again produce buds and strike roots; which is certainly the case with regard to the disks of Polytrichum commune,

Fig XXXIV.—1. Young spore-cases and paraphyses of Mnium cuspidatum. 2. Antheridia and paraphyses of Polytrichum commune.—*Link.*

Bartramia fontana, Bryum palustre, undulatum, cuspidatum, punctatum, and with those of Tortula ruralis. In Bryum argenteum we see the buds containing the supposed anthers constantly drop off, strike root, and produce new plants; this I have observed myself times out of number. Still more in point is the experiment first made by David Mees, of sowing the stellulæ of Polytrichum commune, containing merely club-shaped bodies, when he found that plants came up, which in their turn produced fruit. Another excellent naturalist, Dr. Roth, has made similar observations with regard to Hypnum squarrosum and Bryum argenteum. It is more probable, therefore, that these supposed anthers are mere gemmæ, produced by the superabundance of the juices, and hence surrounded by succulent filaments.'" Fries also, in his *Plantæ homonemeæ*, xxxi., expresses himself thus, "Musci sunt esexuales et in dicta organa masculina meras esse gemmas vix dubium videatur."

Nevertheless, in the face of this evidence, Adolphe Brongniart retains a belief in the sexuality of Mosses, and in the male functions of the axillary bodies; and he says, with justice, that it appears from Brown's mode of describing Mosses, that he entertains a similar opinion. Dr. Taylor also thinks that the Liverworts show the presence of two sexes in the most evident manner. (*Linn. Trans.* xvii. 375.) That the flask-like bodies called pistillidia are female organs he considers proved by the germination of the dark brown particles (spores) that are contained within them. He admits that no direct evidence exists to show that the antheridia are male organs; but he says that they discharge a viscid whitish liquor, which is rapidly dissolved in the air, uniformly precede the pistillidia, and have fulfilled their office before the seeds (spores) are ripe. Dr. Montagne follows on the same side (*Ann. Sc. Nat.* 2 *Ser.* ix. 100), with the sweeping assertion that "no body now-a-days (1838) doubts that Mosses and Liverworts have two sexes." Mr. Valentine, in two elaborate papers (*Linn. Trans.* xvii. 465, and xviii. 499), denies the sexuality of some plants at least of the Muscal Alliance; justly observing, however, that the experiments mentioned by Sprengel and Mees are unsatisfactory, there being no proof in them that it was the antheridia which grew; it might have been the whole mass of the stellate disks in which the antheridia occur. Mr. Valentine relies upon the very important fact, first remarked by himself, that the pistillidum, in which the spores are produced, is not in existence at the time when the antheridia are in action. Like Mohl and Agardh, he maintains that the spores, although equivalent to seeds, are almost identical with pollen grains. "The only difference," he adds, "that I can find between pollen and sporules is, that the coat of the latter is of a more rigid and opaque texture. From this difference it is that the sporules rarely burst in a sudden manner upon the application of water; but when they do, the moving particles are discharged loose in the water, precisely in the same manner as are those of pollen."

Upon this point however Mr. Griffith observes, that "it is to be borne in mind, that whereas pollen is the result of a simple separation constituting a primary and independent process; in Musci, Hepaticæ, Salvinidæ, the spores, otherwise so similar to pollen, are the result of a secondary process, dependent on a primary one which appears to be remarkably analogous to phanerogamic fecundation."

Finally, Unger in his account of the anatomy of Riccia (*Linnæa*, xiii. 13), states that antheridia and pistillidia are alike at first, that the contents of the first are lost, of the second retained, and that the first perishes while the second is permanent,—whence it is reasonable to presume that the emission from the antheridia is a necessary condition for the formation of spores. He therefore regards them as male and female.

It seems clear from all these statements, that the question of sexes in the Muscal Alliance is undecided. There is no doubt that two very different sorts of organs exist among its species; but it does not appear to me that we have sufficient evidence at present to show that the antheridia are male organs. So far as they are concerned we have conjecture and nothing more. All that is proved is: 1. That the spores are bodies which reproduce the plant, and are, therefore, analogous to seeds; and 2. That the structure of the antheridia and pistillidia is wholly at variance with that of anthers and pistils properly so called.

Mr. Griffith, nevertheless, in an elaborate Memoir on Azolla and Salvinia, published in the Calcutta Journal of Natural History, adopts in the fullest extent the opinion that Acrogens have sexes, as will appear hereafter. It is, however, to be remarked that the question is not, whether there may not be in such plants as these some trace of a male and female principle, or certain organs in which it is probable that such a principle resides; but whether there is any such structure as that which we know to be sexual in all the classes of plants higher than Acrogens. And I must confess, after reading Mr. Griffith's very learned and ingenious observations, that my opinion remains unshaken as to the existence of most essential differences between Acrogens and other plants in all that regards the organs of reproduction.

A remarkable point of structure in Liverworts is the spiral filament, or elater, as it is called, lying among the sporules within the spore-case. This consists of a single fibre, or of two, twisted spirally in different directions, so as to cross each other, and contained within a very delicate, transparent, perishable tube. They have a strong elastic force, and have been supposed to be destined to aid in the dispersion of the sporules,—a most inadequate end for so curious and unusual an apparatus. It is more probable that they are destined to fulfil, in the economy of these plants, some function of which we have no knowledge.

One of the most extraordinary points in the history of the Muscal Alliance, is the fact that in the cells of the antheridia are generated bodies having what seems to be spontaneous motion, and apparently of the same nature as the spermatic animalcules of animals. This unexpected fact has been fully and correctly described by Meyen, (*Ann. Sc. Nat. N. S.* x. 319), who has found the same creatures (?) in the corresponding organs of Chara and Marchantia. Unger has also published an elaborate Memoir upon this singular subject. (*Ann. Sc. N. S.* xi. 257 and 274.) He describes the spiral threads of Sphagnum thus :—" These animalcules consist of a thick and swollen body having a slender threadlike appendage. The length varies between the 0.0025 and 0.0020 or $\frac{1}{400}$ to $\frac{1}{500}$ of a line Vienna measure. The length of the appendage is about $4\frac{1}{4}$ longer than the body, so that the total length of the animal may be stated to be the 0.01 of a line." It is to be observed by those who may search for such bodies that they can only be found just when the antheridium is completely formed, and that a magnifying power of at least 600 diameters is required for their detection. Unger regards them as analogous to the genus of animalcules called Spirillum. It is so improbable that animals should be generated in the cells of plants, unless accidentally, that we cannot but entertain grave doubts whether, notwithstanding their locomotive powers, these bodies are really any thing more than a form of vegetable matter ; and it is worth considering if they may not after all be a diminutive representation of the clavate processes surrounding the spore of Equisetum, and perhaps of the elaters found in the spore-cases of Liverworts. This is certain, that the spores and elaters of Equisetum, when at rest, have very much the appearance of the Spirilla in the antheridium of an Urn Moss or a Chara ; and since it has been proved that the spiral filaments of Equisetum arise from the splitting of a cell in which a spore is generated, there seems no reason why a similar action should not take place in cells that are destitute of spores. As to the motion, how are we to tell that it is not a hygrometrical action ? There is as active a motion in the elaters of Equisetum as in the spirilla of Mosses, only it arises in the former from drying and in the latter from floating in water. Nägeli has lately found the spiral threads of Liverworts in the leaves of Ferns.

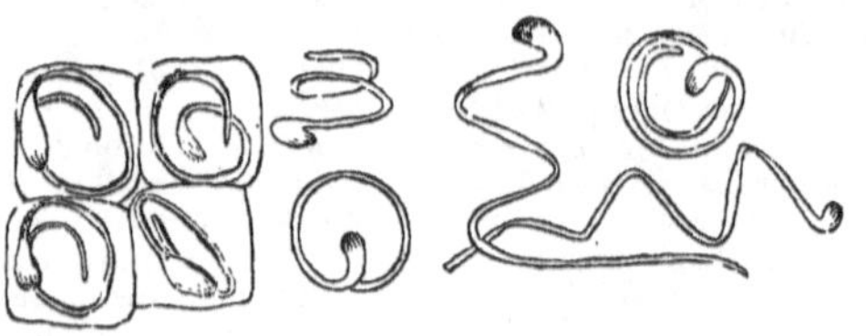

Fig. XXXV.

Equisetum may be regarded as a link between this alliance and Chara on the one hand, while its high degree of composition brings it into the neighbourhood of Ferns and Clubmosses.

By some Botanists the orders of the Muscal Alliance are separated into two great groups, Hepaticæ and Musci ; of which the former are without an operculum and have for the most part elaters, while the latter have an operculum and always want elaters. But such distinctions seem to be of hardly sufficient importance to be employed for higher purposes than the distinction of Natural orders.

NATURAL ORDERS OF MUSCALS.

1. HEPATICÆ.

Spore-cases valveless, without operculum or elaters . . 15. RICCIACEÆ.

Spore-cases valveless or bursting irregularly, without operculum, but with elaters . . . . } 16. MARCHANTIACEÆ.

Spore-cases opening by a definite number of equal valves, without operculum, but with elaters . . . . . . } 17. JUNGERMANNIACEÆ.

Spore-cases peltate, splitting on one side, without operculum, and with an elater to every spore . . . . . } 18. EQUISETACEÆ.

2. MUSCI.

Spore-cases opening by valves, with an operculum, without elaters . . . . } 19. ANDRÆACEÆ.

Spore-cases valveless, with an operculum, without elaters . 20. BRYACEÆ.

ORDER XV. RICCIACEÆ.—CRYSTALWORTS.

RICCIEÆ, *Nees Leberm.* 86; *Bischoff in Nov. Act.* xvii. 2. 964; *Lindenb. ibid.* xviii. 412.—RICCIACEÆ, *Endl. Gen.* xvii.

DIAGNOSIS.—*Spore-cases valveless, without operculum or elaters.*

Terrestrial herbs, of diminutive size, inhabiting mud or water, swimming or floating, usually annual, their leaves and stems blended into a frond of a cellular structure, creeping, green or purple underneath, with a distinct epidermis, and a cavity of air-passages beneath it in some species. Antheridia immersed in the frond, with their mouth projecting in the form of a papilla, or a slender cone. Pistillidia in the frond of the same or a different individual, immersed or superficial, sessile or stalked; the common involucre either missing or scaly; the involucels none, or in the form of little bladders perforated at the point. Spore-cases membranous, united to a calyptra, or distinct from it, globose, bursting irregularly when ripe. Spores triangular, pyramidal, and half round, without elaters.—*Endl.*

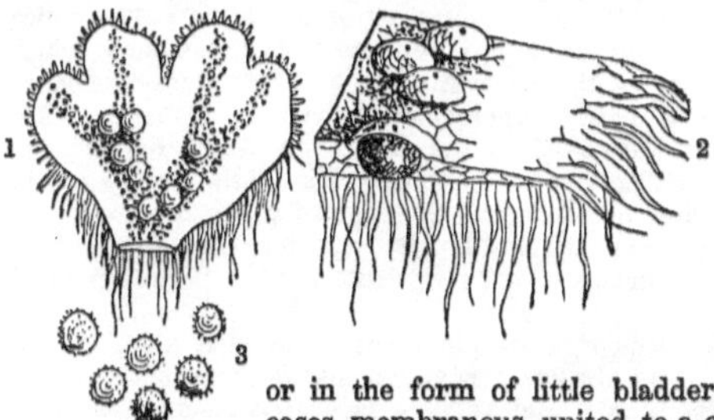

Fig. XXXVI.

These little plants form a plain transition from Thallogens to Acrogens. They have that combination of leaves and stem into what is called a frond, which is characteristic of Lichens, and their spores may be not unaptly compared to the tetraspores of the Rose-tangle order. But, on the other hand, their spores are collected in large numbers within organs resembling the pistils of phænogamous plants; they have a distinct axis of growth, and an epidermis is distinctly formed with stomates for breathing with. (See *Lindenberg*, l. c.) While, however, for the latter reasons, they are to be regarded as more elevated in the scale of organization than Lichens, or similar plants, they are inferior to Liverworts and Scale-mosses, because of the absence of those spiral springs called elaters, by which, in the latter orders, the spores are dispersed; and to Split-mosses and Urn-mosses, because they want the complicated apparatus which is added to the spore-cases of those orders, under the form of either an operculum, or peristome. According to Endlicher, the Crystalworts pass through Corsinia into the tribe of Liverworts, and by Sphærocarpus into that of Scalemosses. There is a detailed account, by Unger, of the anatomy of Riccia glauca, in the *Linnæa*, vol. xiii. p. 1. The genus Duriæa is regarded by Messrs. Bory and Montagne as forming the nearest transition to Liverworts; they describe it as fructifying under water, which is very seldom the case with the other Crystalworts. *Ann. Sc. N.* 3 *ser.* i. 225.

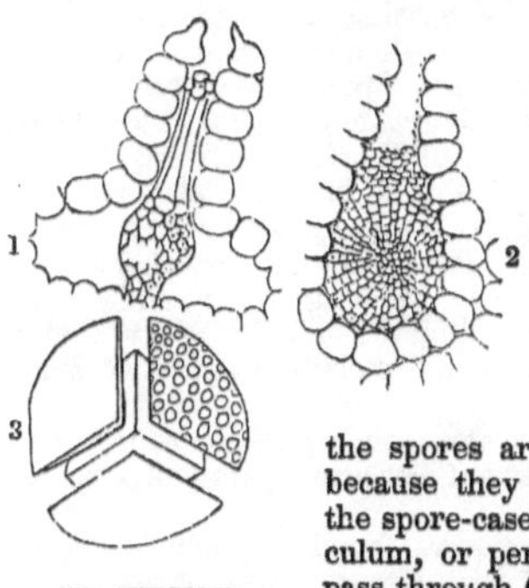

Fig. XXXVII.

Of the species hitherto known, two-thirds have been observed in Europe, and the remainder in various parts of the world. Several species in North America, the Cape of Good Hope, and Brazil, appear to be very similar to those of Europe.—*Endl.*

The uses of Crystalworts are unknown.

GENERA.

Duriæa, *B. & Mont.*	Hemiseumata, *Bisch.*	Sphærocarpus, *Mich.*	*Güntheria*, Trevir.
Riccia, *Mich.*	*Ricciocarpus*, Cord.	Oxymitra, *Bisch.*	*Brissocarpus*, Bisch.
Lichenoides, *Bisch.*	*Salviniella*, Hübn.	*Ruppinia*, Corda.	*Tesselinia*, Dum. part.
Ricciella, *A. Braun.*	*Hemna*, Raf.	Corsinia, *Raddi.*	

NUMBERS. GEN. 8. Sp. 29.

Lichenaceæ.
POSITION. Bryaceæ.—RICCIACEÆ.—Marchantiaceæ.
Ceramiaceæ.

Fig. XXXVI.—1. Riccia natans, a lobe magnified; 2. a portion of it, showing the spore-cases cut open; 3. spores.

Fig. XXXVII.—Riccia glauca. 1. A young spore-case; 2. an antheridium; 3. spores as they lie in the mother cell. (*Unger.*)

ORDER XVI. MARCHANTIACEÆ.—LIVERWORTS.

Hepaticæ, *Juss. Gen.* 7. (1789); *DC. Fl. Fr.* 2. 415. (1815); *Agardh Aph.* 104. (1822); *Nees ab Esenb. in Martius, Fl. Bras.* 1. 295. (1833); *Hooker's British Flora*, vol. ii. p. 97. (1833); *Bischoff de Hepaticis in Act. Acad. Nat. Cur.* xvii. pars 2. (1836); *Ann. des Sc.* 2. *ser.* 4. 309. (1836).—Marchantiaceæ *and* Targioniaceæ, *Ed. pr. Endl. Gen.* xx.—Marchantieæ *and* Targionieæ, *Nees Lebermoose*, 84.—Marchantieæ, *Taylor in Linn. Trans.* 17. 377.

DIAGNOSIS.—*Spore-cases valveless, or bursting irregularly, without operculum, but with elaters.*

Plants growing on the earth or trees in damp places, composed entirely of cellular tissue, emitting roots from their under-side, and consisting of an axis or stem which is leafless, but bordered by membranous expansions, which sometimes unite at their margins, so as to form a broad lobed frond, having a distinct epidermis pierced by stomates. Antheridia either immersed in the frond, or placed on disk-like sessile or stalked peltate receptacles. Pistillidia lurking within involucres, either placed below the edge of the frond, or on the edge or under-side of stalked heads. Spore-cases stalked, opening by irregular fissures, or by separate teeth. Spores globose, with elaters.

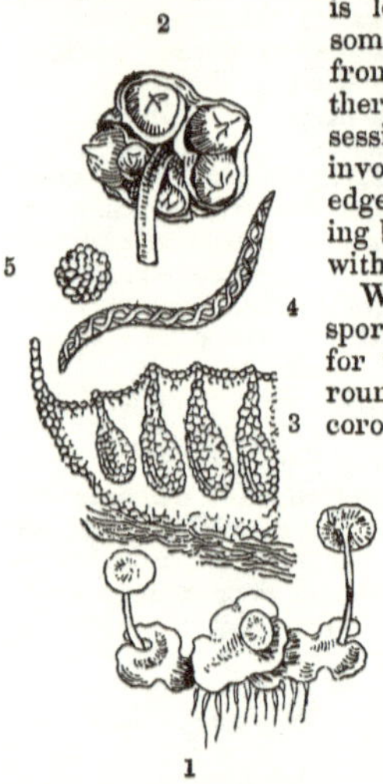

Fig. XXXVIII.

With these plants organization advances another step. To the spores of the Crystalworts are added spiral threads or elaters for their dispersion; and various lacerated membranes surrounding the spore-cases seem to be imitating the calyx and corolla of perfect plants. There is still, however, a want of true leaves, which are fused, with the stem, into a frond. The principal part of the order has the spore-cases raised on a long stalk, and clustered into a head; but this character is missing in Targioneæ, which Endlicher regards as a distinct order. In these plants, as in Mosses and Charas, each cell of what are called the antheridia contains a body resembling an animalcule of the genus Vibrio, which moves about rapidly in water, as soon as it is liberated from its birth-place. Germination takes place by an universal increase and enlargement of the spore, which becomes lobed, as it were, by the swelling of the cellules, and is afterwards nourished by the emission of a radicular fibre. The original development of Ferns and Liverworts is much the same. *Fl. Bras.* i. 299.

The Liverworts differ from Crystalworts in having elaters and involucrate spore-cases, and from Scalemosses or Jungermanniaceæ, in the want of power to separate their spore-cases into distinct valves.

Natives of damp shady places in all climates; two were found in Melville Island. The only atmospheric condition to which they cannot submit is excessive dryness.

Little is known of their uses. De Candolle thinks it probable that the larger kinds will be found to resemble foliaceous Lichens in their qualities. A few are slightly fragrant, with a subacrid taste. They have been employed in liver complaints, but their use seems a mere superstition. It is, however, alleged that Marchantia hemisphærica has really proved advantageous in dropsical affections.

GENERA.

Suborder I. MARCHANTEÆ.—Spore-cases capitate. Involucels membranous, regularly slit.

Grimaldia, *Radd.*
Pleurochiton, Radd.
Syndonisce, Radd.
Mannia, Cord.
Duvalia, *Nees.*
Petalophyllum, *Nees.*
Fimbriaria, *Nees.*
Hypenantron, Cord.
Dictyochiton, Cord.
Fegatella, *Radd.*
Conocephalus, Vaill.
Cynocephalum, Wigg.
Lunularia, *Michel.*
Sedgwickia, Bowd.
Plagiochasma, *Lehm.*
Otiona, Cord.
Sedgwickia, Bisch.
Aitonia, Forst.
Ruppinia, L. f.
Antrocephalus, *Lehm.*
Rebouillia, *Radd.*
Astercla, Palis.
Rhakiocarpon, Cord.
Achiton, Cord.
? *Mesoregma*, Cord.
Sauteria, *Nees.*
Hampea, Nees.
Dumortiera, *Nees.*
Hyrophila, Mack.
Hygrophyla, Tayl.
? *Spathysia*, Nees.
Marchantia, *March.*
Astromarchantia, Nees
Chlamidium, Cord.
Preissia, *Nees.*
Chomiocarpon, Cord.

Suborder II. TARGIONEÆ.—Spore-cases submarginal, solitary. Involucels wanting.

Targionia, *Michel.*
Cyathodium, *Lehm.*
? *Carpobolus*, Schwein.

NUMBERS. GEN. 15. Sp. 20?

Equisetaceæ.
POSITION. Ricciaceæ.—MARCHANTIACEÆ.—Jungermanniaceæ.
Lichenaceæ.

Fig. XXXVIII.—1. Marchantia commutata, natural size; 2. a head of spore-cases; 3. a section of the disk which bears the spore-cases; 4. elater; 5. granular spore.

ORDER XVII. JUNGERMANNIACEÆ.—SCALEMOSSES.

Hepaticarum, § § Jungermanniaceæ *et* Lejeuniaceæ, *Dumort. Comment. Botan.* 112. (1822).—Jungermanniaceæ *et* Anthocerineæ, *Id. Syllog. Jungerm.* 6. (1831).—Hepaticarum § *of most other Authors.*—Jungermanniaceæ, *Nixus Pl.* 24. (1833).—*Nees v. Esenb. Naturgeschichte der Europæischen Lebermoose, vol.* i. (1833).—*Endl. Gen.* xxi.

DIAGNOSIS.—*Spore-cases opening by a definite number of equal valves, without operculum, but with elaters.*

Creeping moss-like plants, either with imbricated very cellular leaves surrounding a central axis, or with the leaves and axis all fused into one common leafy expansion.

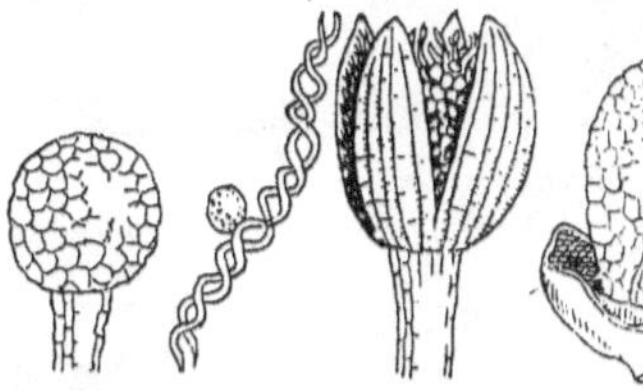

4 3 1 2

Fig. XXIX.

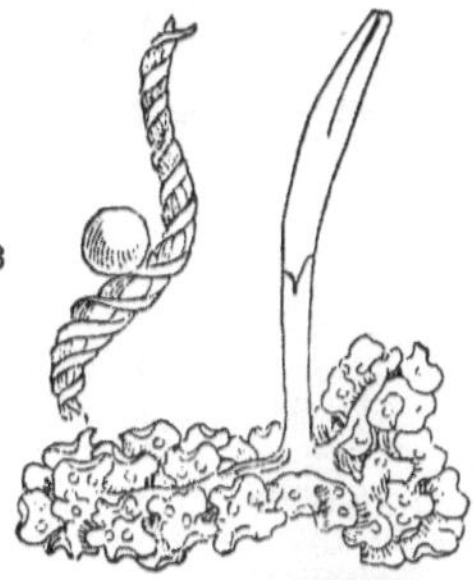

3

1

Fig. XLI.

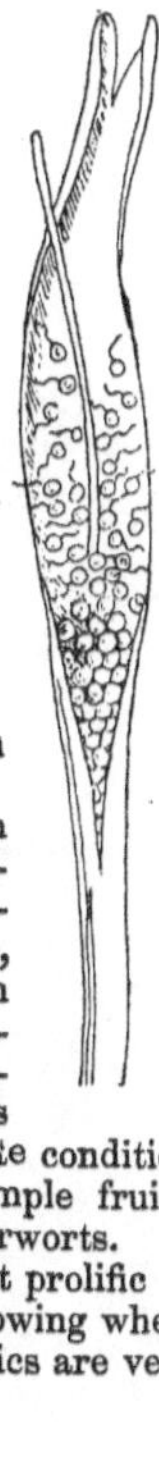

2

Antheridia scattered, free, or immersed. Pistillidia solitary, with both involucre and involucel. Spore-cases without an operculum, 4-parted, or 4-valved, with or without a central columella. Spores mixed with elaters.

Fig. XL.

Here the approach to a higher organization becomes more manifest. Instead of a frond consisting of a stem and the leaves not distinguishable from it, we have, in the majority of cases, well-defined separate symmetrical leaves; and in the section Anthoceroteæ there is a central columella, which is evidently a transition to the structure of the Urnmosses. These Scalemosses differ from the Liverworts in the regularly valvate condition of the spore-cases, and in their long-stalked simple fruits. In Blasia and others, the habit is that of the Liverworts.

Shady woods in hot climates appear to be most prolific in these plants, which, however, seem capable of growing wherever the climate will produce Lichens. The tropics are very rich in them.

Their uses are unknown.

GENERA.

Suborder I. JUNGERMANNEÆ.—Spore-cases 1 or 4 valved without a columella.

Metzgeridæ.

Metzgeria, *Radd.*
Echinomitrium, Cord.
Echinogyna, Dumort.
Fasciola, Dumort.

Aneuridæ.

Trichostylium, *Cord.*
Aneura, *Dumort.*
Römeria, Radd.
Metzgeria, Cord.
Sarcomitrium, Cord.

Haplolaenidæ.

Blasia, *Michel.*
Symphyogyna, *Nees et Mont.*
Pellia, *Radd.*
Scopulina, Dumort.

Diplomitridæ.

Hollia, *Endl.*
Blytia, Endl.
Diplolaena, Dumort.
Dilaena, Dumort.
Cordaca, Nees.
Diplomitrium, Cord.

Codonidæ.

Fossombronia, *Radd.*
Codonia, Dumort.

Jubulidæ.

Lejeunia, *Lib.*
Phragmicoma, *Dumort.*
Frullania, *Nees.*
Jubula, Dumort.
Bryopteris, Nees.
Schulthesia, Radd.
Physananthus, *Lœbg.*
Ptychanthus, *Nees.*
? *Frullanioides*, Radd.
Madotheca, *Dumort.*
Bellincinia, Radd.
Antoiria, Radd.
Porella, Dicks.
Radula, *Dumort.*
Radulotypus, Dumort.
Candollea, Radd. part.

Fig. XXXIX.—1. Spore-case of Jungermannia hyalina ripe and bursting; 2. the same, very young and covered with its calyptra; 3. Elater and spore; 4. Antheridium.
Fig. XL.—Jungermannia bidentata.
Fig. XLI.—1. Monoclea crispata, a little magnified; 2. Spore-case and columella; 3. Elater and spore.

Ptilidæ.
Ptilidium, *Nees.*
Blepharozia, Dumort.
Trichocolea, *Nees.*
Thricholea, Dumort.
Thricolea, Dumort.

Mastigophoridæ.
Sendtnera, *Endl.*
Mastigophora, Nees.
Schisma, *Dumort.*

Trichomanidæ.
Physiotium, *Nees.*
Herpetium, *Nees.*
Mastigophora, Nees.
Pleuroschisma, Dum.
Lepidozia, *Dumort.*
Mastigobryum, *Nees.*
Pleuroschismatypus, Dum.
Calypogeia, *Radd. part.*
Cincinnulus, Dumort.

Geocalycidæ.
Gongylanthus, *Nees.*
Geocalyx, *Nees.*
Saccogyna, *Dumort.*
Syckorea, Cord.

Jungermannidæ.
Gymnoscyphus, *Cord.*
Cheiloscyphus, *Cord.*
Marsupella, Dum.part.
Harpanthus, *Nees.*
Gymnanthe, *Taylor.*
Lophocolea, *Nees.*
Jungermannia, *Dill.*
Aplozia, Dumort.
Lophozia, Dumort.
Cephalozia, Dumort.
Anthelia, Dumort.
Blepharostoma, Dum.
Odontoschisma, Dum.
Plagiochila, *Nees* et *Mont.*
Radulæ sect., Dumort.
Scapania, Dumort.
Candollea, Radd.
? *Notarisia*, Coll.

Gymnomitridæ.
Alicularia, *Cord.*
Mesophylla, Dumort.
Acrobolbos, *Nees.*
Sarcoscyphus, *Cord.*
Marsupia, Dumort.
Marsupella, Dumort.
Gymnomitrium, *Nees.*
Acolea, Dumort.
Haplomitrium, *Nees.*
Mniopsis, Dumort.

Suborder II. ANTHOCEROTEÆ.—Nees. Spore-cases pod-shaped, split on one side, or 2-valved, with a columella.

Anthoceros, *Mich.*
Anthocerites, Corda.
Monoclea, *Hook.*
Cladobryum, *Nees.*

NUMBERS. GEN. 42. Sp. 650 ?

POSITION. Marchantiaceæ.—JUNGERMANNIACEÆ.—Andræaceæ.

ORDER XVIII. EQUISETACEÆ.—HORSETAILS.

Equisetaceæ, *DC. Fl. Fr.* 2. 580. (1805); *Agardh Aph.* 119. (1822); *Kaulfuss Enum. Filicum*, 1. (1824); *Adolphe Brongniart Hist. Veg. Foss.* 99. (1828.)—*Endl. Gen.* xxv. *Link filic. sp.* p. 9.

DIAGNOSIS.—*Spore-cases peltate, splitting on one side, without operculum, and with an elater to every spore.*

Leafless branched plants with a striated fistular stem, in the cuticle of which silex is secreted; the articulations separable and surrounded by a membranous toothed sheath. Stem fistular, with many longitudinal cavities in its circumference; chiefly consisting of cellular substance, but coated externally with a layer of hard woody tubes, from which plates of a similar nature project towards the centre, partially dividing the longitudinal cavities from each other. Stomates arranged longitudinally on the cuticle. Spiral vessels very small but abundant. Spore-cases opening inwards by a longitudinal slit, attached to the lower face of peltate scales, which are collected into terminal cones. Spores, oval grains, wrapped round with a pair of highly elastic clavate elaters.

Fig. XLII.

The remarkable plants known by the vulgar name of Horsetails, seem to have no very decided affinity to any existing order. With Ferns their relation is not obvious. In the arrangement of their reproductive organs they have a striking resemblance to Zamia, and in their general aspect to Ephedra or Casuarina. Their germination is that of Cellular plants, and approaches nearly to Urnmosses. The structure of their stem is well described by Ad. Brongniart in his *History of Fossil Vegetables*, as are, indeed, other parts of their organisation: see Tables 11 and 12 of that work. This ingenious writer entertains the opinion that the green body, which is known to be the spore, is a naked ovule, and the four swollen filaments that surround it four grains of pollen united in pairs to the base of the ovule. In the last edition of this work I adopted M. Brongniart's view, and accordingly placed Equisetum with Coniferæ, an error so very obvious, as to have called forth rebukes, which were richly deserved. The development of the swollen filaments has been carefully observed by Mohl, Henderson, and others, who have demonstrated that they are really produced by the spiral splitting of the cell in which the spore is formed; in fact, they appear quite analogous, as Mr. Griffith has stated, to the elaters of Marchantia and its allies, to which the order bears, perhaps, a nearer relation than to any other plant. To regard Horsetails as a high form of the Muscal alliance seems to me more expedient than to station them with Ferns and Clubmosses, to which they seem to have no immediate affinity. The resemblance between the peltate scales of Equisetum and the heads of spore-cases in Marchantia, is too obvious not to strike the most unpractised observer. Link calls these scales Sporidochia.

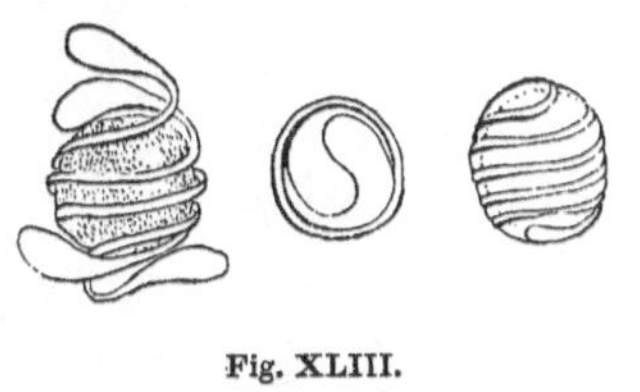

Fig. XLIII.

The germination of the spores has been explained, both by Agardh and Bischoff. The former (*Aphor.* 120) describes it thus: from three to fourteen days after they are sown, they send down a filiform, hyaline, somewhat clavate, simple root, and protrude a confervoid, cylindrical, obtuse, articulated, torulose thread, either two-lobed (in E. pratense) at the apex, or simple (in E. palustre). Some days after, several branches grow out and are agglutinated together, forming a body resembling a bundle of confervoid threads, each of which pushes out its own root. The account of Bischoff (*Nov. Act. Acad. N. Cur.* 14. t. 44.) is not materially different: he finds the confervoid threads, or numerous processes of cellular development, go on

Fig. XLII.—Equisetum arvense. 1. A peltate disk seen from the side.
Fig. XLIII.—Equisetum; its spores wrapped round by elaters.

growing and combining, until a considerable cellular mass is formed; then this mode of development ceases, and a young bud is created, which springs up in the form of the stem of the Equisetum, at once completely organised, with its air-cells, its central cavity, and its sheaths, the first of which is formed before the elongation of the stem, out of the original cellular matter.

Horsetails are found in ditches and rivers in most parts of the world, within and without the tropics.

None are of importance in a medical point of view; they are said to be slightly astringent and stimulating, and have even been recommended as diuretics and emmenagogues; they are, however, not now employed. In economical purposes they are found to be useful for polishing furniture and household utensils—a property which is due to the presence of a great quantity of silex in their cuticle. According to the observations of John of Berlin, they contain full thirteen per cent. of siliceous earth. The ashes have been found by chemists to contain half their weight of silica. The quantity of silex contained in the cuticle of Equisetum hyemale is so great, that Sivright succeeded in removing the vegetable matter and retaining the form. On subjecting a portion of the cuticle of Equisetum hyemale to the analysis of polarised light under a high magnifying power, Brewster detected a beautiful arrangement of the siliceous particles, which are distributed in two lines parallel to the axis of the stem, and extending over the whole surface. The greater number of the particles form simple straight lines, but the rest are grouped into oval forms, connected together like the jewels of a necklace, by a chain of particles forming a sort of curvilinear quadrangle, these rows of oval combinations being arranged in pairs. Many of those particles which form the straight lines do not exceed the 500th of an inch in diameter. Brewster also observed the remarkable fact, that each particle has a regular axis of double refraction. In the straw and chaff of Wheat, Barley, Oats, and Rye, he noticed analogous phenomena; but the particles were arranged in a different manner, and displayed figures of singular beauty. From these data it is concluded that the crystalline portions of silex and other earths, which are found in vegetable tissues, are not foreign substances of accidental occurrence, but are integral parts of the plant itself, and probably perform some important function in the process of vegetable life. A very large quantity of starch is found during winter in the rhizomes; in whose cells, during the month of October, the particles may be seen in active motion, passing up one side, and retreating by the other, much in the same way as in Chara. This I have often noticed in Equisetum fluviatile.

GENUS.

Equisetum, L.

NUMBERS. GEN. 1. Sp. 10

Characeæ.

POSITION. Marchantiaceæ.—EQUISETACEÆ. ————

Gnetaceæ.

ORDER XIX. ANDRÆACEÆ.—SPLITMOSSES.

Andræaceæ, *Nixus Pl.* 24. (1833); *Endl. Gen.* xxii.

DIAGNOSIS.—*Spore-cases opening by valves, with an operculum, without elaters.*

Branching moss-like reddish or brown plants, with imbricated ribbed or ribless leaves. Spore-case with a calyptra, seated on a fleshy apophysis, splitting longitudinally into four equal valves whose summits are always bound together by the persistent operculum. Peristome 0. Spores surrounding a central columella.

Linnæus considered the only genus of which this order consists, the same as Jungermannia; more recent observers have withdrawn it to associate with Urnmosses. It hardly, however, belongs more to the one than the other; if it agrees with Urnmosses in having an operculum, it disagrees in having a valvular spore-case; and if it accords with the Scalemosses in the latter circumstance, it differs from them in the former, and in the want of elaters.

Natives of cold and temperate regions, especially on rocks in bleak places, as high as the limits of eternal snow, where they form a close mat.

Their uses are unknown.

GENERA.

Andræa, *Ehr.* Acroschisma, *Hook. fil.* *Petrophila*, Brid.

NUMBERS. GEN. 2. Sp. 13.

POSITION.—Jungermanniaceæ.—ANDRÆACEÆ.—Bryaceæ.

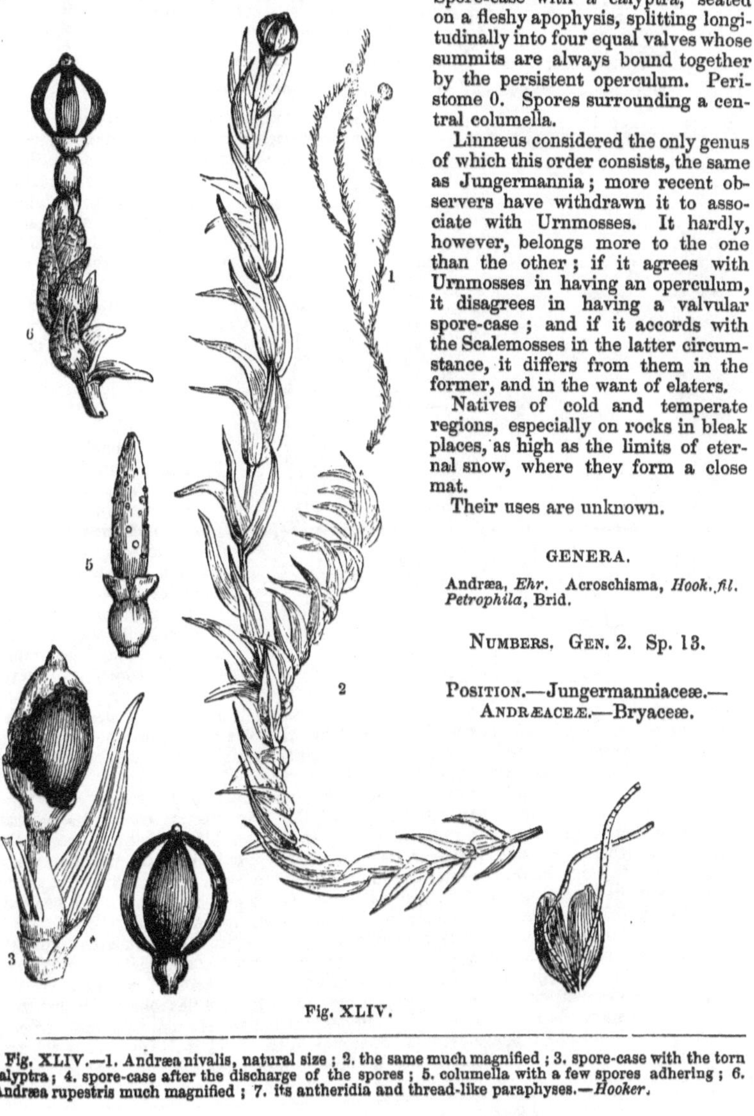

Fig. XLIV.

Fig. XLIV.—1. Andræa nivalis, natural size; 2. the same much magnified; 3. spore-case with the torn calyptra; 4. spore-case after the discharge of the spores; 5. columella with a few spores adhering; 6. Andræa rupestris much magnified; 7. its antheridia and thread-like paraphyses.—*Hooker.*

ORDER XX. BRYACEÆ.—URNMOSSES.

Musci, *Juss. Gen.* 10. (1789); *Hedwig Descr. et Adumb.* (1787-1797); *Bridel Muscolog. recentiorum* (1797-1803); *Hedw. Species Muscor. Frondos.* (1801); *Palisot Prodrome des* 5 *et* 6 *Fam. de l'Ætheogam.* (1805); *Bridel Suppl.* (1806-1819); *Weber Tabul. Musc. Frondos.* (1813); *DC. Fl. Fr.* 2. 438. (1815); *T. F. L. Nees de Muscor. Propag.* (1818); *Hooker and Taylor, Musc. Brit.* (1818); *Hooker, Musci Exotici* (1818-1820); *Agardh Aphor.* 105. (1822); *Greville and Arnott in Wern. Trans.* 4. 109. *&c.* (1822); *Nees v. Esenbeck, Hornschuch, and Sturm, Bryolog. Germ.* (1823); *Grev. Fl. Edin.* xiii. (1824); *Ad. Brongn. in Dict. Class.* 11. 248. (1827); *Hooker. Brit. Fl.* 1. 459. (1830).—Bryaceæ, *Ed. pr.* (1836); *Endl. Gen.* xxiv.—Sphagnaceæ, *Endl. Gen.* xxiii.

DIAGNOSIS.—*Spore-cases valveless, with an operculum, without elaters.*

Erect or creeping, terrestrial or aquatic, cellular plants, having a distinct axis of growth, destitute of a vascular system, and covered with minute imbricated, entire, or serrated leaves. Reproductive organs of two kinds, viz. 1. Antheridia, which are axillary, cylindrical or fusiform stalked sacs, containing a multitude of spherical or oval particles emitted upon the application of water, and coiled up bodies which move in water with activity; 2 Pistillidia, or flask-like bodies inclosed within a convolute bract, which is eventually carried up upon the point of the spore-case. Spore-cases, or ripened pistillidia, hollow urnlike vessels, seated upon a seta or stalk, covered by a membranous calyptra, closed by a lid or operculum, beneath which are one or more rows of cellular rigid processes, called collectively the peristome, and separately teeth, which are always some multiple of four, and combined in various degrees; the centre of the theca is occupied by an axis or columella, and the space between it and the sides of the theca is filled with sporules. Sporules in germination protruding confervoid filaments, which afterwards ramify, and form an axis of growth at the point of the ramifications.

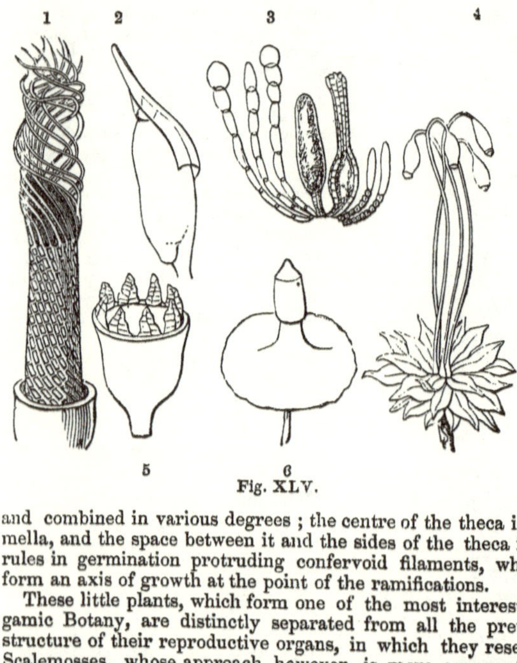

Fig. XLV.

These little plants, which form one of the most interesting departments of Cryptogamic Botany, are distinctly separated from all the previous tribes by the peculiar structure of their reproductive organs, in which they resemble no others, except the Scalemosses, whose approach, however, is more apparent than real. In their organs of vegetation they are strikingly similar to many Clubmosses, to which, perhaps, an approach is made by Sphagnum, whose spore-case has no peristome, on which account, indeed, that genus is regarded as a distinct Natural Order by Endlicher.

For a long time Urnmosses were considered to be destitute of stomates; but first Treviranus, and afterwards Valentine, distinctly proved those organs to be present; (*Linn. Trans.* 18, 239). In addition to such apertures, some of the cells of certain species of Sphagnum are pierced with large round openings; and Rœper has observed, that such perforated cells are the habitation of the animalcule called Rotifer vulgaris. (*Flora*, 1838, p. 17.) Mohl has observed similar openings in the cells of Leucobryum vulgare, (Dicranum glaucum,) and Octoblepharum albidum; he thinks they are formed subsequently to the construction of the cells. *Ann. Sc. N. s.* xiii. 108. Schleiden confirms

Fig. XLV.—1. Peristome of Tortula ruralis; 2. Theca of Ceratodon purpureus; 3. Supposed representatives of sexual organs in Meesia longiseta; 4. Bryum roseum; 5. Peristome of Octoblepharum albidum; 6. Apophysis and theca of Splachnum luteum.

this, and adds to the list of porous Urnmosses, Octoblepharum cylindricum, Didymodon sphagnoides, and Leucobryum minus, albidum, and longifolium.

Mr. Griffith (*Calc. Journ.* v.) strenuously advocates the sexuality of the Antheridia and Pistillidia, regarding the former as a true male apparatus, and the latter as a pistil containing an ovule. I do not know that he has anywhere adduced proof of the validity of this opinion; and it is difficult to comprehend upon what evidence that theory depends; it may, however, be presumed, that he considers the spores to be analogous to embryos, formed in vast numbers. This admirable observer thinks, that evidence in favour of fecundation in some way in Mosses and Liverworts, is afforded by the breaking up of the tissue, terminating and closing what he calls the style, that is to say, the point of the pistillidium, subsequently to the application of a particular matter, whereby the style becomes a canal, opening externally by a browning observable in the orifice of this canal, extending downwards until it reaches the cavity of the (his) ovary, and by a corresponding enlargement of a cell (his ovule) existing in that cavity. Mr. Valentine, however, does not regard these appearances as connected with fecundation.

An uninitiated person, reading the definition of a genus of Urnmosses, might suppose that to be the tribe in which an approach to the animal creation most nearly takes place. Unacquainted with the exact meaning of the Latin words employed by Bryologists, he might understand by the peristomium a jaw, by the calyptra a nightcap, and by the struma a kind of goître; and when he saw that teeth belonged to this jaw, he would naturally conclude that it was really a vegeto-animal of which he was reading. Struck with the evident absurdity of giving such names to parts of plants, without at the same time explaining their real nature, I formerly ventured to call the attention of naturalists to the subject by the following paragraph in the *Outlines of the First Principles of Botany*.

"The calyptra may be understood to be a convolute leaf; the operculum another; the peristomium one or more whorls of minute flat leaves; and the theca itself to be the excavated distended apex of the stalk, the cellular substance of which separates in the form of sporules."

The reasoning upon which I conceived this hypothesis to be sustained, was the following:—Every one agrees in describing the calyptra as a membrane arising from between the leaves and the base of the young spore-case, and as enveloping the latter, but having no organic connexion with it: when the stalk of the spore-case lengthens, no corresponding extension of the parts of the calyptra takes place; so that it must be either ruptured at its apex (as in Jungermannia), or at the base; and in the latter case it would necessarily be carried up upon the tip of the spore-case, which it originally enveloped. Now, what can be more reasonable than that such an organ, situated as thus described, should be one of the last convolute leaves of the axis which the spore-case terminates, bearing the same relation to the latter as the convolute bractea to the flower of Magnolia, or, to speak more precisely still, as the calyptriform bracteæ to the flower of Pileanthus? If the calyptra be anatomically examined, especially in such genera as Tortula and Dicranum, no difference in its tissue and that of the leaves will be observable; and that very common tendency to dehisce on one side only as the diameter of the theca increases, which characterises the dimidiate calyptra, may be understood to be a separation at the line where the margins of the supposed leaf united; in the mitriform calyptra this separation at a given line does not take place, and the consequence is an irregular laceration of its base. The analogy of the calyptra being of this nature, the next inference would naturally be, that the part it contains corresponds with a flower-bud. Upon this supposition, the external series of parts belonging to this supposed bud would be the operculum; the adhesion of this organ to the spore-case, which would answer to the apex of the axis, or to the tube of the calyx of flowering plants, would be analogous to what occurs in Eucalyptus, or perhaps more exactly to that of Eschscholtzia. As to the number of the parts, in a state of cohesion, of which it is made up, it will be observed that in the paragraph above quoted, it is stated to be one only. My reason for adopting this conclusion was the absence of any trace of division upon its surface or in the substance of its tissue, and also the apparent identity of nature between it and the calyptra when both are young, in the Tortula and Dicranum genera already cited. With regard to the peristomium:—The teeth, as they are called, occupy one or more whorls; they are evidently not mere lacerations of a membrane, because they are in a constant and regular number in each genus, and that number is universally some multiple of 4, as the floral leaves of flowering plants are ordinarily of 3, 4, or 5; they have the power of contracting an adhesion with each other by their contiguous margins, as the floral leaves of flowering plants; they alter their position from being inflexed with their points to the axis, to being recurved with their points turned outwards,—exactly as happens in flowering plants; the teeth of the inner peristomium often alternate with those of the outer, thus conforming to the law of alternation prevalent in the floral leaves of flowering plants; and, finally, if we compare the various states of the leaves of Buxbaumia aphylla with the tooth of other Urnmosses, it is impossible not to be struck with the great similarity in the anatomical structure of the two. These considerations led me to the conclusion, that the calyptra, operculum, and teeth of Urnmosses, are all modified leaves; and hence that the spore-case is to be considered more analogous to a flower than to a seed-vessel. With regard to the membrane, or epiphragma, which occasionally closes up the orifice of the spore-case, it may be considered as formed by the absolute cohesion of the leaves of the peristome, just as the operculum of the genus Eudesmia is formed by the cohesion of petals; and this is confirmed, first, by Calymperes, in which the membrane ultimately separates into teeth, and by the fact that the horizontal membrane exists most perfectly in such genera as Polytrichum and Lyellia, in which there is no distinct peristome. As to the internal structure of this curious apparatus we may regard the spore-case as the hollow apex of the axis, the sporules as a partial dissolution of its cellular tissue, and the columella as the unconverted centre. That the end of the axis or growing point of plants frequently becomes much more thickened than the spore-case of Urnmosses, requires no illustration for those who are acquainted with Eschscholtzia, Rosa, or Calycanthus. That tissue is frequently disintegrated for particular purposes, is proved by the production of pollen out of the cellular tissue of an anther, and by the general law of propagation that seems to prevail in all the lower alliances of plants; the same phenomenon may be therefore expected in Urnmosses. That the columella should be left in this dissolution of the tissue might be expected, from its being a continuation of the seta or axis of development, the tissue of which is more compact, and of course less liable to separation, than the looser tissue that surrounds it; this is analogous to the separation of the pollen from the connective of most plants, or from parts only of the anther of all those genera which, like Viscum, Ægiceras, or Rafflesia, have what are called cellular anthers.

Mr. E. Quekett has lately proved the general accuracy of these views by the discovery of a monstrous moss, in which common leaves take the place of the spore-case, its peristome, and other apparatus. As this is a very curious subject, I extract at length his observations, with a few unimportant omissions:—"Soon after Mr. Ward made known his plan of growing plants in closely-glazed cases I had constructed a small case, in which were placed various Mosses, both in fruit, and having the tendency to form fruit. Among the number was a mass of Tortula fallax, showing, at the time, the early condition of the seta, capped with a calyptra. After watching the progress of the plants, it was discovered that the Tortula, which, when placed in it, showed every tendency to produce fruit, now presented, instead of fruit advancing to maturity, a miniature forest of elevated stems, leafy above and below, but in the intermediate portion, destitute of leaves; in fact, all appearance of capsules approaching maturity was dissipated. On placing some of the plants under the microscope, it was evident that the specimens were furnished with the usual leaves at the base of the plant,—the setá existed, and presented the usual brown colour, quite destitute of leaves, but in the place of the capsule, there was a continued elongation of the seta, of a green colour, bearing several green leaves, varying in number in different specimens, being generally from about twelve to twenty. It appears that the capsule had scarcely commenced to be formed, when the elements of the modified leaves, (which I conceive would have otherwise formed the capsule and peristome), having received an increased degree of heat, combined with more moisture than is natural to these plants, occasioned by the structure of the case, and by its position, instead of being converted into the ordinary capsule and peristome, the matters which entered the plants were not appropriated to the development of organs of reproduction, but underwent a change into a state fitting them apparently for the purposes of nutrition."

Mr. Quekett objects, however, to that part of the theory which assumes the spore-case to be the hollowed apex of the axis; he considers the theca and operculum to be the representatives of a consolidated calyx; the corolla to be the lining membrane, whose fringed edge constitutes a peristome, which is either single or double, and appears to be the representative of the reproductive apparatus; and the columella to be the receptacle, torus or axis on which these several organs are arranged.

Fine illustrations of the Anatomy of Urnmosses will be found in Link's *Ausgew. Anat. Bot. Abbild. Fasc.* 4.

Urnmosses are found in all parts of the world where the atmosphere is humid: but they are far more common in temperate climates than in the tropics. They are among the first vegetables that clothe the soil with verdure in newly-formed countries, and they are the last that disappear when the atmosphere ceases to be capable of nourishing vegetation. The first green crust upon the cinders of Ascension consisted of minute

Mosses; they form more than a quarter of the whole Flora of Melville Island; and the black and lifeless soil of New South Shetland is covered with specks of Mosses struggling for existence. How they find their way to such places, and under what laws they are created, are mysteries that human ingenuity has not yet succeeded in unveiling. The Sphagna occupy vast tracts of morass with their spongy stems and leaves.

The slight astringency and diuretic qualities of Polytrichum and others caused them to be formerly employed in medicine, but they are now disused. In the economy of man they perform but an insignificant part; but in the economy of nature, how vast an end! Sphagnum forms part of the food of the reindeer; and in the polar regions the inhabitants dry it and make it into a sort of bread "miseræ vitæ delicias." *Endl.*

GENERA.

- Archidium, *Brid.*
- Phascum, *L.*
  - *Pyxidium*, Ehrh.
- Pleuridium, *Brid.*
- Bruchia, *Schwaegr.*
  - *Voitia*, Moug. et Nestl.
  - *Saproma*, Brid.
- Physedium, *Brid.*
- Voitia, *Hornsch.*
- Gymnostomum, *Hedw.*
  - *Pottia*, Ehrh.
  - *Anodontium*, Brid.
- Physcomitrium, *Brid.*
- Hymenostomum, *R. Br.*
- Hymenostylium, *Brid.*
- Pyramidium, *Brid.*
  - *Pyramidula*, Brid.
- Hyophila, *Brid.*
  - *Rottleria*, Brid.
- Entosthymenium, *Brid.*
- Schistidium, *Brid.*
  - *Harrisonia*, Adans.
- Grimmia, *Ehrh.*
- Hydropogon, *Brid.*
- Dryptodon, *Brid.*
- Racomitrium, *Brid.*
- Holomitrium, *Brid.*
- Orthotheca, *Brid.*
- Cinclidotus, *Palis.*
- Tetraphis, *Hedw.*
  - *Tetrodontium*, Schw.
  - *Tetracmis*, Brid.
  - *Georgia*, Ehrh.
  - *Tetrapilis*, Hedw.
- Syrrhopodon, *Schwaegr.*
  - *Cleisostoma*, Brid.
- Campylopus, *Brid.*
  - *Thysanomitrion*, Schw.
- Oedipodium, *Schwaegr.*
- Orthodon, *Bory.*
- Eremodon, *Brid.*
  - *Cyrtodon*, R. Br.
  - *Dissodon*, Grev. et Arn.
- Aplodon, *R. Br.*
- Splachnum, *L.*
  - *Pycnapophysium*, Rchb.
  - *Apophysis*, Hedw.
  - *Discapophysium*, Rchb.
  - *Cystapophysium*, Rchb.
  - *Apodanthus*, La-Pyl.
  - *Sciadophysium*, Endl.
- Systylium, *Hornsch.*
- Scouleria, *Hook.*
- Wardia, *Hook. et Harv.*
- Tridontium, *Hook.*
- Raineria, *Notar.*
- Tayloria, *Hook.*
  - *Phrissotrichia*, Brid.
- Brachysteleum, *Reichenb.*
  - *Brachypodium*, Brid.
  - *Glyphomitrion*, Brid.
- Glyphomitrium, *Schw.*
  - *Griffithia*, R. Br.
- Orthotrichum, *Hedw.*
  - *Navia*, Borkh.
  - *Trachytrichum*, Röhl.
- Ulota, *Mohr.*
  - *Leiotheca*, Brid.
- Cryptocarpon, *Dozy.*
- Macromitrium, *Brid.*
- Schlotheimia, *Brid.*
  - *Schizodon*, Swartz.
- Orthodontium, *Schw.*
- Zygodon, *Hook.*
  - *Amphidium*, Nees.
  - *Gagea*, Radd.
- Codonoblepharum, *Schw.*
- Weissia, *Hedw.*
  - *Afzelia*, Ehrh.
  - *Swartzia*, Ehrh.
  - *Cavanillea*, Barkh.
  - *Brachyodus*, Fürnr.
  - *Brachyodontium*, Fürn.
- Discelium, *Brid.*
- Catascopium, *Brid.*
  - *Melania*, Brid.
- Coscinodon, *Spr.*
  - *Anacalypta*, Röhl.
  - *Trimatium*, Fröhl.
- Mielichoferia, *Hornsch.*
  - *Eurybasis*, Brid.
  - *Oreas*, Brid.
  - *Auchenangium*, Brid.
- Calymperes, *Sw.*
  - *Cryphium*, Palis.
- Octoblepharum, *Hedw.*
- Campylodontium, *Schw.*
- Leucophanes, *Brid.*
- Oncophorus, *Brid.*
- Trematodon, *Rich.*
- Dicranum, *Hedw.*
- Ceratodon, *Brid.*
  - *Aegiceras*, Green.
- Trichostomum, *Hedw.*
- Didymodon, *Hedw.*
  - *Ditrichium*, Timm.
- Pilipogon, *Brid.*
- Plaubelia, *Brid.*
- Desmatodon, *Brid.*
- Leucoloma, *Brid.*
- Barbula, *Hedw.*
  - *Mollia*, Schrank.
  - *Streblotrichum*, Palis.
  - *Tortula*, Hedw.
- Syntrichia, *Web. et Mohr.*
- Encalypta, *Hedw.*
  - *Leersia*, Hedw.
- Cynodontium, *Brid.*
  - *Cynontodium*, Hedw.
- Ptychostomum, *Hornsch.*
  - *Brachymenium*, Hook.
- Hemisinapsium, *Brid.*
- Cladodium, *Brid.*
- Bryum, *L.*
  - *Webera*, Hedw.
  - *Trentepohlia*, Hoffm.
  - *Polla*, Adans.
- Cynclidium, *Swartz.*
  - *Amblyodon*, Palis.
- Leptostomum, *R. Br.*
- Leptotheca, *Schwaegr.*
- Megalangium, *Brid.*
  - *Macrothecium*, Brid.
  - *Acidodontium*, Schw.
- Pohlia, *Hedw.*
  - *Amphirhinum*, Green.
  - *Lagenium*, Brid.
- Paludella, *Ehrh.*
- Mnium, *Dill.*
  - *Orthopyxis*, Palis.
- Aulacomnion, *Schw.*
  - *Gymnocephalus*, Schw.
  - *Fusiconia*, Palis.
  - *Bryum*, Hedw.
  - *Peromnion*, Schwaegr.
- Arrhenopterum, *Hedw.*
  - *Maschalarrhen*, Spr.
- Bartramia, *Hedw.*
  - *Cephaloxis*, Palis.
- Philonotis, *Brid.*
  - *Cyrtopodium*, Brid.
  - *Cryptosete*, Hook.
- Glyphocarpus, *R. Br.*
- Plagiopus, *Brid.*
- Conostomum, *Sw.*
- Entosthodon, *Schwægr.*
- Funaria, *Hedw.*
  - *Koelreutera*, Hedw.
  - *Strephedium*, Palis.
- Meesia, *Hedw.*
  - *Ambliodum*, Palis.
- Diplocomium, *Web.*
  - *Tristichis*, Ehrh.
- Timmia, *Hedw.*
  - *Omphalophora*, Brid.
- Polytrichum, *L.*
  - *Pogonatum*, Palis.
  - *Catharinea*, Ehrh.
  - *Oligotrichum*, DC.
  - *Atrichum*, Palis.
  - *Callibryum*, Web.
- Psilopilum, *Brid.*
- Lyellia, *R. Br.*
  - *Gomphophorus*, Brid.
- Buxbaumia, *Hall.*
  - *Saccophorum*, Palis.
  - *Hippopodium*, Röhl.
- Diphyscium, *Web. et M.*
  - *Hymenopogon*, Palis.
- Dawsonia, *R. Br.*
  - *Triplocoma*, La-Pyl.
  - *Stylocomium*, Brid.
- Hypnum, *Linn.*
- Stereodon, *Brid.*
- Fabronia, *Raddi.*
  - *Stereophyllum*, Brid.
- Maschalanthus, *Schulz.*
  - *Pterigynandrum*, Hed.
  - *Pterogonium*, Sw.
  - *Maschalocarpus*, Spr.
- Leptohymenium, *Schw.*
  - *Haplohymenium*, Schw.
- Anacamptodon, *Brid.*
- Pylaisaea, *Desv.*
- Leskea, *Hedw.*
- Omalia, *Brid.*
- Hemiragis, *Brid.*
  - *Helicodontium*, Schw.
- Isothecium, *Brid.*
- Anoectangium, *Hedw.*
  - *Hedwigia*, Hook.
  - *Erpodium*, Brid.
- Endotrichum, *Dozy.*
- Symphysodon, *Dozy.*
- Neckera, *Hedw.*
  - *Eleutheria*, Palis.
  - *Cyrtopus*, Brid.
  - *Distichia*, Brid.
  - *Cryptopodia*, Röhl.
- Actinodontium, *Schw.*
  - *Aetinodon*, Brid.
- Daltonia, *Hook.*
  - *Macrodon*, Arn.
- Dendropogon, *Schimp.*
- Rhegmatodon, *Brid.*
- Sclerodontium, *Schwaegr.*
- Prionodon, *K. Mull.*
- Leucodon, *Schwaegr.*
  - *Fuscina*, Schrank.
  - *Cecalyphum*, Palis.
- Pterobryon, *Hornsch.*
- Leptodon, *Web.*
  - *Lasia*, Brid.
- Dicnemon, *Schwaegr.*
  - *Eucnemis*, Brid.
  - *Hollia*, Sieb.
- Astrodontium, *Schwaegr.*
  - *Plaubelia*, Brid.
- Symphyodon, *Mont.*
- Antitrichia, *Brid.*
  - *Anomodon*, Hook.
- Climacium, *Web.*
  - *Porotrichum*, Brid.
  - *Zygotrichia*, Brid.
- Trachyloma, *Brid.*
- Hookeria, *Smith.*
  - *Pterygophyllum*, Brid.
  - *Hypopterygium*, Brid.
  - *Helicophyllum*, Brid.
  - *Cyathophorum*, Palis.
  - *Chaetephora*, Brid.
- Racopilum, *Palis.*
  - *Aubertia*, Palis.
- Cryphaea, *Brid.*
  - *Pilotrichum*, Palis.
  - *Meteorium*, Brid.
- Erpodium, *Brid.*
- Carovaglia, *Endl.*
  - *Esenbeckia*, Brid.
- Lepidopilum, *Brid.*
  - *Trachypodium*, Brid.
- Fontinalis, *L.*
- Spiridens, *Nees.*
- Rhizonium, *Brid.*
- Schistostega, *Web.*
  - *Dicksonia*, Ehrh.
- Drepanophyllum, *Rich.*
- Phyllogonium, *Brid.*
- Phyllogium, *Brid.*
- Eustichia, *Brid.*
- Octodiceras, *Brid.*
- Fissidens, *Hedw.*
  - *Fuscinia*, Schrank.
  - *Schistophyllum*, Palis.
- Sphagnum, *Dill.*

NUMBERS. GEN. 44. Sp. 1100?

POSITION. Jungermanniaceæ.—BRYACEÆ.—Andræaceæ.

ALLIANCE V.—*LYCOPODALES.*—THE LYCOPODAL ALLIANCE.

DIAGNOSIS.—*Vascular Acrogens, with axillary or radical one—or many-celled spore-cases, and spores of two sorts.*

The formation of leaves, which in the Muscal Alliance had become complete, is in this group carried still further; for the leaves are now capable of generating spore-cases in their axils. That tendency to form spiral vessels which in Muscales is confined to the cellular tissue, with the single exception of the Horsetails, is now a characteristic of this Alliance, the axis containing in all cases spiral tubes in abundance. The larger of the Clubmosses seem to imitate Coniferous Gymnogens in their manner of growth, and in their tendency to collect their spore-cases in cones. The Pepperworts evidently exhibit an approach to that system of converting leaves into seed-vessels which is so generally characteristic of flowering plants. Here too it would seem that we have a great approach to the manner in which sexual organs are formed in the more perfect classes.

NATURAL ORDERS OF LYCOPODALS.

Spore-cases 1-3-*celled, axillary; reproductive bodies similar* . . 21. LYCOPODIACEÆ.

Spore-cases many-celled, radical (or axillary); reproductive bodies dissimilar . . . . 22. MARSILEACEÆ.

ORDER XXI.—LYCOPODIACEÆ.—CLUBMOSSES.

Lycopodineæ, *Swartz Synopsis Filicum* (1806); *R. Brown Prodr.* 164. (1810); *Agardh Aph.* 112. (1822); *Greville Flor. Edin.* xii. (1824); *Martius Ic. pl. crypt.* 37. (1834).—Lycopodiaceæ, *DC. Fl Fr.* 2. 257. (1815); *Ad. Brongn. in Dict. Class.* 9. 561. (1826); *Link. Filic. Sp.* 155; *Endl. Gen.* xxxvi.

DIAGNOSIS.—*Lycopodal Acrogens, with 1-3-celled axillary spore-cases, and the reproductive bodies all of the same nature.*

Usually moss-like plants, with creeping stems and imbricated leaves, the axis consisting of one solid cord of annular vessels, or of a reticulated column of such vessels intersected by cellular tissue; or stemless plants, with erect subulate leaves, and a solid corm. Spore-cases 1-3-celled, axillary, sessile, either bursting by distinct valves, or indehiscent, and containing either minute powdery matter, or sporules, marked at the apex with three minute radiating elevated ridges upon their proper integument, or irregularly tuberculated.

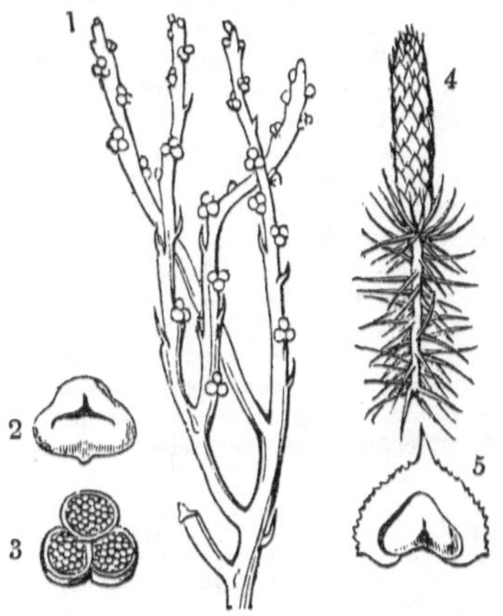

Fig. XLVI.

Intermediate as it were between Ferns and Coniferæ on the one hand, and Ferns and Mosses on the other; related to the first of those tribes in the want of sexual apparatus, and in the abundance of annular ducts contained in their axis; to the second in the aspect of the stems of some of the larger kinds; and to the last in their whole appearance, Lycopodiaceæ are distinctly characterised by their organs of reproduction. These are generally considered to be of two kinds, both of which are axillary and sessile, and have from 1 to 3 regularly dehiscing valves, the one containing a powdery substance, the other bodies much larger in size, which have been seen to germinate. In conformity with the theory that all plants have sexes, the advocates of that doctrine have found anthers in the former, and pistils in the latter; but, as in other similar cases, this opinion is entirely conjectural, and founded upon no direct evidence: all that we really know is, that the larger bodies do germinate, and, if we are to credit Wildenow, the powdery particles grow also. He says he has seen them. I think it is hardly to be doubted that the latter are the abortive state of the former. Link, however, takes quite a different view of the matter, and regards the larger bodies as Antheridia, while the smaller he calls spores. (*Ausgew. Anat. Bot. Abbild. fasc.* 4. *t.* 4.) According to Salisbury, in the *Linnean Transactions*, vol. 12. tab. 19, Lycopodium denticulatum emits two cotyledons upon germinating; but, supposing this observation, which requires confirmation, to be exact, it is much more probable that the two little scales so emitted are primordial leaves than analogous to cotyledons. The genus Phylloglossum is remarkable for having the foliage, and mode of growth of Isoetes combined with the fructification of a Lycopodium, and offers a strong argument

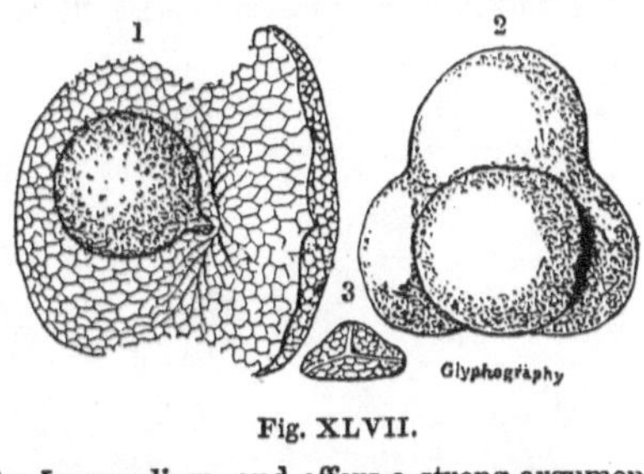

Fig. XLVII.

Fig. XLVI.—1. Bernhardia dichotoma; 2. its spore-case; 3. the same, cut across; 4. Lycopodium annotinum; 5. its spore-case, with the scale to which it is axillary.
Fig. XLVII.—1. Spore-case of Lycopodium denticulatum opened; 2. antheridium; 3. spore.—*Link.*

to those who would place the former genus in this natural order. It is said to have quite the appearance of Plantago pusilla.

According to Ad. Brongniart, the stem of a Lycopodium is almost identical, anatomically, with the root of Ferns.

In geographical distribution these follow the same laws as Ferns, being most abundant in hot humid situations in the tropics, and especially in small islands. As they approach the north they become scarcer; but even in the climate of northern Europe, in Lapland itself, whole tracts are covered with Lycopodium alpinum and Selaginoides.

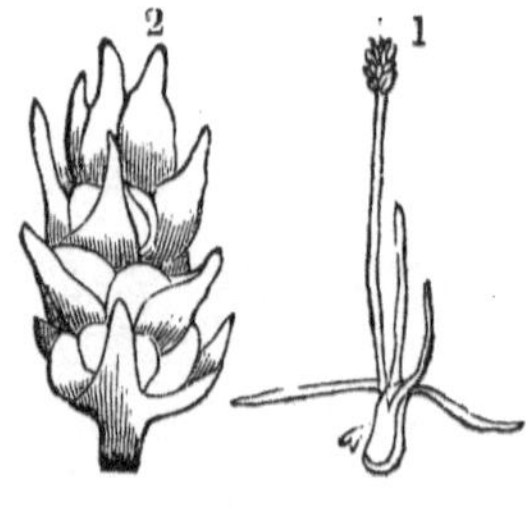

Fig. XLVIII.

The powder contained in the spore-cases of Lycopodium clavatum and Selago is highly inflammable; shaken out and collected it is employed under the name of Lycopode, or vegetable brimstone, on the Continent, in the manufacture of fireworks, and in pharmacy to roll up pills, which when coated with it may be put into water without being moistened. The plant of Lycopodium clavatum has long been used as an emetic, and that of L. Selago as a cathartic; but it is said that if the dose is not small it is followed by faintness and convulsions; it is regarded as a powerful irritant, and has been externally employed for keeping blisters open, and as a counter-irritant in cases of inflamed eyes. The most remarkable plant of the order, however, is the Yatum condenado (Yatum Great Devil, and condenado accursed,) which appears to be the Lycopodium rubrum of Chamisso. Sir W. Hooker, who calls it L. catharticum, states that it acts most vehemently as a purgative, and has been administered successfully in Spanish America in cases of elephantiasis. According to Vastring, Clubmosses are likely to become of importance in dyeing; he asserts, that woollen cloths boiled with Lycopodiums, especially with L. clavatum, acquire the property of becoming blue when passed through a bath of Brazil wood. Lycopodium Phlegmaria is reputed an aphrodisiac. So also the rock-lily, a name sometimes given to Selaginella convoluta, Spring, also called Lycopodium squamatum, a plant remarkable for its hygrometrical properties, rolling up into a ball when dry and unrolling again when damped, is asserted by Martius, who found it abundantly in the provinces of Bahia and Pernambuco, to act upon the mucous membrane, especially of the uropoetic system. "Potentiam virilem amissam ejus decocto reduci posse perhibent, quo jure nescio." He, however, advises a full trial to be made of these and the East Indian species.

GENERA.

Tmesipteris, *Bernh.*	*Tristeca*, Palis.	*Lepidotis*, Palis.	*Diplostachyum*, Palis.
Psilotum, *Swartz.*	Lycopodium, *Linn.*	*Chamaeclinis*, Mart.	*Gymnogynum*, Palis.
Bernhardia, Willd.	*Selago*, Hook. et Gren.	Selaginella, *Spring.*	Phylloglossum, *Kunze.*
Hoffmannia, Willd.	*Huperzia*, Bernh.	*Stachygynandrum*, Ps.	

NUMBERS. GEN. 1. Sp. 200. (*Hooker.*)

Coniferæ.

POSITION.—Ophioglossaceæ.—LYCOPODIACEÆ.—Marsileaceæ.

Fig. XLVIII.—Phylloglossum Drummondii: 1. whole plant, natural size; 2. spike magnified.

ORDER XXII. MARSILEACEÆ.—PEPPERWORTS.

Rhizocarpæ, *Batsch. Tab. Aff.* (1802); *Agardh Aph.* 111. (1822).—Rhizospermæ, *Roth. DC. Fl. Fr.* 3. 577. (1815).—Hydropterides, *Willd. Sp. Pl.* 5. 534. (1810).—Marsileaceæ, *R. Brown Prodr.* 166, (1810); *Grev. Fl. Edinens.* xii. (1824); *Ad. Brongn. in Dict. Class.* 10. 196. (1826); *DC. and Duby,* 542. (1828); *Martius, Ic. Pl. Crypt.* 121. (1834); *Endl. gen.* xxxiv.—Salvinieæ, *Juss. in Mirb. Elém.* 853. (1815).—Salviniaceæ, *Bartl. Ord. Nat.* 15. (1830); *Martius, Ic. Plant. Crypt.* 123. (1834); *Ed. Pr. Endlich. gen.* xxxiii.—Isoeteæ, *Rich. Bartl. Ord.* 16. *Endlich. gen.* xxxv.—Salvininæ and Azollinæ, *Griffith in Calcutta Journ.*, vol. v.

DIAGNOSIS.—*Lycopodal Acrogens, with many-celled radical spore-cases, and the reproductive bodies of two different kinds.*

Stemless plants, creeping, or floating; leaves usually stalked, sometimes sessile and scaly, occasionally destitute of lamina, and rolled up in vernation. Reproductive organs enclosed in involucres, and of two kinds; *the one,* clustered and stalked, or crowded confusedly without stalks, and distinct from the second, or mixed with it, or in contact with it; *the other,* simple oval bodies, sometimes having a terminal nipple, from which germination uniformly proceeds. [Stem and leafstalks filled with longitudinal cells. A central simple fascicle of vessels composed of scalariform ducts and prosenchyma, enclosing in the middle a quantity of elongated cells containing starch. Leaves with nerves, veins and stomates.—*Martius.*]

The Order to which Pilularia and Marsilea belong consists of floating or creeping plants, often having the circinate vernation of Ferns, with their reproductive organs in close cases, called involucres, springing either from the root, or from the petioles of the leaves. These involucres contain oval bodies of two kinds, one of which has been called anther, and the other capsule. Figures of Marsilea vestita and polycarpa have been published by Hooker and Greville, at t. 159 and 160 of their noble *Icones Filicum.* From these, and the more detailed observations of Esprit Fabre, it is clear that the involucre of that genus consists of an involute leaf analogous to the carpellary leaf of flowering plants.

Esprit Fabre has also shown, (*Ann. Sc. Nat.* 2 *ser.* 7. 221, 9. 115 *and* 381, *and* 12. 255,) that on the side of a mucilaginous cord, which I regard with Braun as a midrib, proceeding from the involucre when it opens, there arise oblong plates bearing two sorts of bodies packed closely, sometimes intermixed, but sometimes separated, so that each occupies a different side of the plates (which are leaflets). He regards these two sorts of bodies as anthers and ovules, and says, that their mutual position is such, that the side which bears the ovules is above that which bears the anthers. The "ovules" are from 10 to 15 on each side, whitish, semitransparent, ovoid, obtuse at one end, and terminated at the other by a nipple. The "anthers" are little flattened parallelopipedons, rounded at each end. "They consist of a membranous sac, very thin and transparent, in which you see numerous pollen grains. The latter are spherical or elliptical, often pointed on one side. When you crush them beneath the microscope, spermatic granules of extreme smallness are seen to come out." Germination of this species takes place, according to the same observer, from the nipple at the point. He thinks, that the two sorts of bodies are certainly anthers and ovules, because, if they are left apart in water they putrefy, while, on the other hand, if mixed together in water, he has seen the sides of the "anthers" burst, and the "grains of

1

2

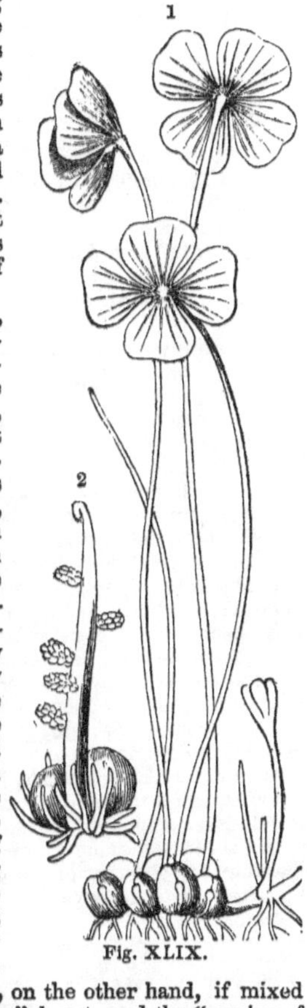

Fig. XLIX.

3.

A

Fig. XLIX.—1. Growing plant of Marsilea pubescens; 2. an involucre opened by 2 valves, from which rises a leaf whose lateral leaflets are loaded with spores; 3. an involucre which has opened, and from which the sporiferous leaf is disengaging itself; at A is seen the side which Fabre regards as anthers.

pollen" collect about the nipple at the surface of the water, after which the "ovules" fall to the bottom, where, at the end of seven or eight days germination commences. These observations, however, require to be repeated; for Braun (*Flora*, 1839, p. 297,) and Griffith each regards both sorts of bodies as sporules. Fabre's experiment calls to mind those of Professor Savi of Pisa, upon Salvinia, another plant of this Order. He put into different vessels, 1st, the seeds alone; 2d, the male globules alone; and 3d, both mixed. In the first two vessels nothing appeared; in the 3d, the seeds rose to the surface of the water and fully developed. But Duverney has since published a dissertation upon this plant, in which he states that, having repeated the experiments of Savi, he has not obtained the same results, and that the seeds, when separated from the supposed male organs, developed perfectly.

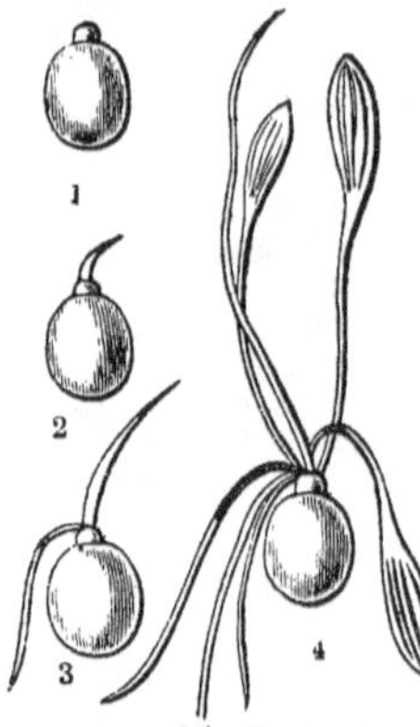

Fig L.

The structure of Pilularia is analogous. From the very correct and careful observations of Valentine, (*Linn. Trans.* 18.483,) it has apparently been proved, that the so-called anthers of that plant are, as I formerly suggested, nothing but abortive spores.

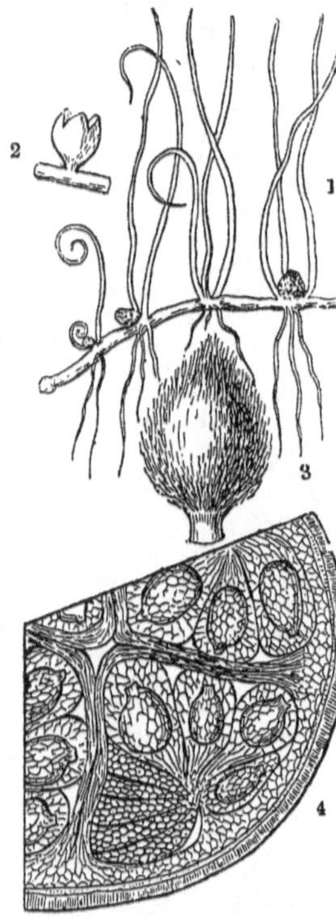

LI.

Following Jussieu, Salvinia and Azolla were separated in the last edition of this work as a distinct Natural Order, a view that Endlicher has since taken. But upon a full consideration of the structure of these plants, or of what is known of it, it does not appear to justify the separation. Like Pilularia and Marsilea, they have two distinct kinds of reproductive bodies enclosed in involucres, and that seems to be the main feature by which Pepperworts are known as an Order from Lycopodiaceæ. For the same reason it appears better to combine with them Isoetes, instead of regarding that too as the type of still another Order. Mr. Griffith does not include Isoetes among these plants; but I cannot assent to the propriety of erecting every genus in this curious Order into a Suborder.

The genera Salvinia and Azolla have been the subject of some elaborate observations by Mr. Griffith, (*Calcutta Journal*, vol. v.), who elevates each into a Suborder, and throws an entirely new light upon their structure. He regards them as having true sexes, the male being certain necklace-shaped threads found at an early stage, in contact with what he denominates an orthotropous ovulum. But strange to say, this so called ovulum, instead of giving birth to an embryo, becomes the parent of reproductive bodies of two totally different kinds, having not even the smallest resemblance the one to the other, although the matrix out of which they are evolved is identical at an early period of the organisation. I regret that Mr. Griffith's most curious memoir only reached me as this sheet was going to press, so that it was impossible to have cuts prepared to illustrate his observations, for which the reader is referred to the work above quoted. All I can do is to give in a note the substance of his descriptions of Salvinia and Azolla.*

* *Salvinia verticillata*.—Male organs? articulated hairs on the stalks of the ovula; each joint containing a nucleus and a brownish fluid; Ovula nearly sessile, concealed by the roots, and partly covered

Fig. L.—Marsilea pubescens in different states of germination; advancing from 1. the spore, up to 4. the perfect young plant.

Fig. LI.—1. Pilularia globulifera; 2. spore-case, natural size, bursting; 3. the same younger and magnified; 4. a section of the spore-case, showing the large and small spores, (after Valentine).

Delile has published an account of the germination of Isoetes setacea, from which it appears that its sporules sprout upwards and downwards, forming an intermediate solid body, which ultimately becomes the stem, or corm ; but it is not stated whether the points from which the ascending and descending axes take their rise are uniform. In Pilularia Mr. Valentine finds, that germination takes place invariably from a fixed point. Delile points out the great affinity that exists between Isoetes and Lycopodium, particularly in the relative position of the two kinds of reproductive matter. In Lycopodium, he says the pulverulent spore-cases occupy the upper ends of the shoots, and the granular spore-cases the lower parts : while, in Isoetes, the former are found in the centre, and the latter at the circumference. If this comparison is good, it will afford some evidence of the identity of nature of these bodies, and that the pulverulent ones are at least not anthers, as has been supposed ; for in Isoetes the pulverulent inner bodies have the same organization, even to the presence of what has been called their stigma, as the outer granular ones ; so that, if Isoetes has sexes, it will offer the singular fact of its anther having a stigma. The anatomy of Isoetes is described by Mohl in the *Linnæa*, xiv. 181.

The Pepperworts evidently approach the Clubmosses through Isoetes, which is sometimes referred to the one Order, sometimes to the other. Their genus Azolla appears to bring them into contact with Jungermanniaceæ. According to Mr. Griffith, Marsilea evidently appears to connect Salvinia with Ferns ; " its important differences from Salvinia consist in the capsules, which correspond to the secondary capsules of that family, being developed within the substance of a modified leaf, in their occurring mixed with each other, and in the spores of the pedicellate capsules not becoming imbedded in apparently cellular masses."

All are inhabitants of ditches or inundated places. They do not appear to be affected by climate so much as by situation, wherefore they have been detected in various parts of Europe, Asia, Africa, and America ; chiefly however in temperate latitudes.

Uses unknown.

GENERA.

Pilularia, *Linn.*	Azolla, *Lam.*	Isoetes, *Linn.*
Marsilea, *Linn.*	*Carpanthus*, Raf.	*Calamaria*, Dill.
Lemna, Juss.	*Rhizosperma*, Meyen.	
Zaluzianskia, Neck.	Salvinia, *Michel.*	

NUMBERS. GEN. 5. Sp. 24.

Filices.

POSITION.—Lycopodiaceæ.—MARSILEACEÆ.—Jungermanniaceæ.

---

with hairs ; tegument open at the top ; mature reproductive organs solitary, or in racemes of 3-5, about the size of a pea, covered with brown rigid hairs. The upper ones of each raceme, (or lowest as regards general situation,) contain innumerable sphærical bodies, of a brownish colour and reticulated cellular surface, terminating capillary simple filaments. These again contain a solid whitish opaque body. The other, which occupies the lowest part of the raceme, and which is the first and often the only one developed, is more oblong, containing 6-18 larger, oblong-ovate bodies, on short stout compound stalks : colour brown, surface also reticulated. Each contains a large, embossed, opaque, ovate, free body, of a chalky aspect : it is three-lobed at the apex, and contains below this a cavity lined by a yellowish membrane, filled with granular and viscid matter and oily globules.

Azolla pinnata.—The growing points present a number of minute confervoid filaments, the assumed male organs, which at certain periods may be seen passing into the foramen, the ovula becoming resolved into their component cells within the cavity of that body ; organs of reproduction in pairs, attached to the stem and branches, one above the other, concealed in a membranous involucrum ; ovula atropous, oblong-ovate, with a conspicuous foramen and nucleus, around the base of which are cellular protuberances ; capsules of each pair either difform—in which case the lowest one is oblong-ovate, the upper globose—or both of either kind, generally perhaps the globose, presenting at the apex the brown remains of the foramen, and still enclosed in the involucrum ; upper half generally tinged with red ; the oblong-ovate capsule opens by circumcision ; with the apex separate the contents, which consist of a large yellow sac contained in a fine membrane, the remains of the nucleus (or the secondary capsule.) The sac is filled with oleaginous granular fluid, and surmounted by a mass of fibrous tissue, by which it adheres slightly to the calyptra ; on the surface of the fibrous tissue are 9 cellular lobes (the three upper the largest), which when pulled away, separate with some of the fibrous tissue, and so appear provided with radicles. The globose capsule has a rugose surface from the pressure of the secondary capsules within ; these are many in number, spherical, attached by long capilliform pedicels to a central much branched receptacle ; each contains two or three cellular masses, presenting on their contiguous faces two or three radiciform prolongations. In their substance may be seen imbedded numerous yellow grains, the spores.

ALLIANCE VI.—*FILICALES.*—THE FILICAL ALLIANCE.

Filices, *Juss. Gen.* 14. (1789); *Swartz Synops. Filicum* (1806); *Willd. Sp. Pl.* vol. v.; *R. Brown Prod.* 145; *Agardh Aph.* 115. (1822); *Kaulfuss Enum.*; *Hooker and Greville Icones Filicum*; *Blume, Fl. Javæ*; *Schott's Genera Filicum*; *Mohl et Martius Plantæ Cryptogamicæ Brasilienses*, p. 40. (1834); *Hooker Species Filicum*; *Brongniart, Veg. Fossiles*, p. 141; *Presl. Tentamen pteridographiæ*; *J. Smith in Hooker Journ. Bot.*; *Endl. gen.* p. 58; *Hooker and Bauer, Genera Filicum*; *Link, Filicum Species.*

DIAGNOSIS.—*Vascular Acrogens, with marginal or dorsal one-celled spore-cases, usually surrounded by an elastic ring; and spores of only one kind.*

These are leafy plants, producing a rhizome, which creeps below or upon the surface of the earth, or rises into the air like the trunk of a tree; this trunk consists of a woody cylinder, of equal diameter at both ends, growing at the point only, containing a loose cellular substance which often disappears; it is coated by a hard, cellular, fibrous rind, which is much thicker next the root than at the apex, and it is itself composed of the united bases of leaves. Wood, when present, consists almost exclusively of large scalariform or dotted ducts, imbedded in hard plates of thick-sided elongated tissue, which usually assumes an interrupted sinuous appearance, but occasionally, according to Brown, forms a complete tube in Dipteris, Platyzoma, and Anemia. Leaves coiled up in vernation, with annular ducts in the vascular tissue of their petiole, either simple or divided in various degrees, traversed by simple, dichotomous, or netted veins of equal thickness, which are composed of elongated cellular tissue, with occasional ducts; cuticle frequently with stomates. Reproductive organs consisting of spore-cases arising from the veins upon the under surface of the leaves or from their margin, either pedicellate, with the stalk passing round them in the form of an elastic ring, or sessile and destitute of such a ring; either springing from beneath the cuticle, which they then force up in the form of a membrane (or indusium), or from the actual surface of the leaves. Spores arranged without order within the spore-cases. Sometimes the leaves are contracted about the cases, so as to assume the appearance of forming a part of the reproductive organs, and sometimes the place of spore-case is supplied by the depauperated lobes of the leaves.

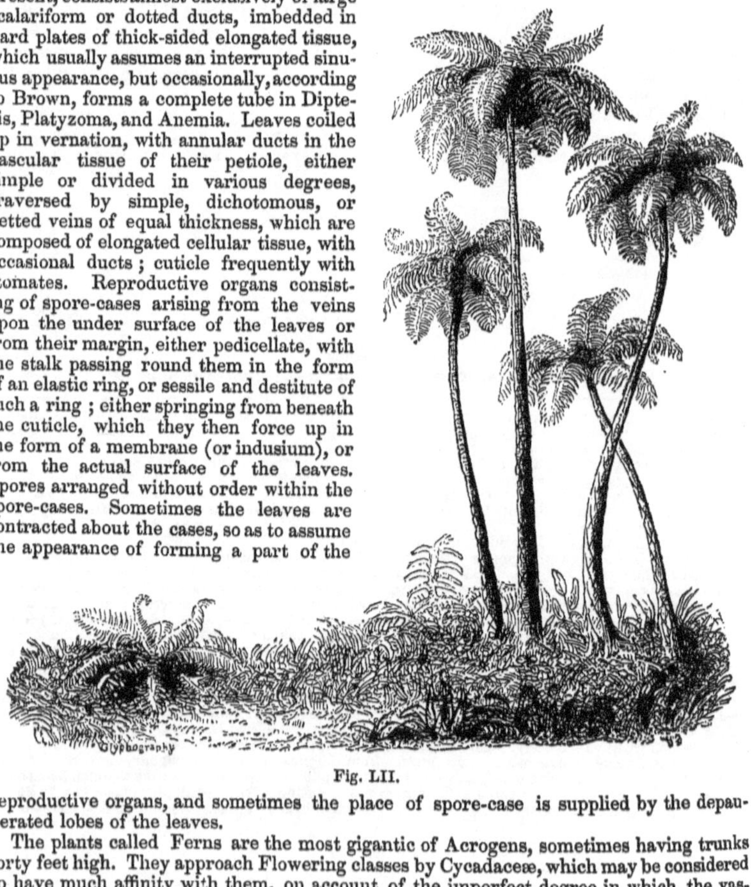

Fig. LII.

The plants called Ferns are the most gigantic of Acrogens, sometimes having trunks forty feet high. They approach Flowering classes by Cycadaceæ, which may be considered to have much affinity with them, on account of the imperfect degree in which the vascular system of that Order is developed, of their pinnate leaves with a gyrate vernation,

Fig. LII.—Tree-Ferns, from *Blume*.

and their naked ovules borne upon the margin of contracted leaves, as the spore-cases of Ferns are upon the leaves of Osmunda. To Coniferous Gymnosperms they also advance very closely through Salisburia, whose leaves might be mistaken for those of a Fern. The affinity of Ferns with Equisetum, consists more in a want of flowers, and the presence of annular vessels, than in any similarity of habit. Clubmosses are readily known by their axillary spore-cases dehiscing by regular valves. Pepperworts are so very different, that it is difficult to find points of comparison between them, except the gyrate leaves of some of the genera.

The organ in Ferns which deserves the most particular attention is the theca, or case that contains the reproductive matter. By many it is named capsule; but as that kind of pericarp is essentially connected with the power of conveying fertilisation from the male apparatus to the ovules, and implies the existence of a certain definite relation between the various parts that it contains, nothing of which kind is found in the spore-case of Ferns, it is not necessary to insist upon the impropriety of applying such a name. Easy as it is to show that the spore-case is not analogous to a capsule, it is far less so to demonstrate with what organs or modifications of organs it really has an analogy. I am not, indeed, aware that this had been attempted, all botanists seeming to consider it a special organ, until, in the *Outlines of the First Principles of Botany*, I ventured to hazard the following theory: "The thecæ may be considered minute leaves, having the same gyrate mode of development as the ordinary leaves of the tribe; their stalk the petiole, the annulus the midrib, and the theca itself the lamina, the edges of which are united." I was led to this opinion, first, by the persuasion that there was no special organ in Ferns to perform a function which in flowering plants is executed by modifications of leaves; and, secondly, by the examination of viviparous species. Observation has shown us that the leaves of flowering plants have the power of producing leaf-buds from their margin or any point of their surface; and in certain kinds of Grasses it has been found that they can produce flower-buds also. In Ferns, which are exceedingly subject to become viviparous, the young plants often grow from the same places as the spore-cases, or from the margin; and in a viviparous Fern, of which a morsel was given me by Dr. Wallich, the young plants form little clusters of leaves in the place of sori. Upon examining these young plants, it appears that the more perfect, though minute, leaves are preceded by still more minute primordial leaves or scales, the cellular tissue of which has nearly the same arrangement as the cellules of the spore-case; and the resemblance between the midrib of one of these scales and the ring of a Polypodium is striking. It is, however, necessary to add, what is only implied in the little work from which the foregoing extract is taken, that this explanation applies only to the gyrate Ferns. With regard to those with striated spore-cases, or with what is called a broad transverse ring, they may either be considered not to have the midrib of the young scale, out of which the case is supposed to be formed, so much developed; or the case may be still considered a nucleus of cellular tissue, separating both from that which surrounds it and also from its internal substance, which latter assumes the form of sporules, in the same way as the internal tissue of an anther separates from the valves under the form of pollen. This conjecture seems confirmed by the anatomical structure of those striated cases which consist of a cluster of spore-like areolæ of cellular tissue at the base and apex, connected by extended cellules of the same description, as in Gleichenia; and is far from being weakened by such cases as those of Parkeria. In Ophioglossum another kind of provision is made for the production of spores, which in that genus seem to have no spore-case beyond the involute contracted segments of the leaf which bears them. What are called the thecæ in Ophioglossum seem more analogous to the involucre of Marsilea.

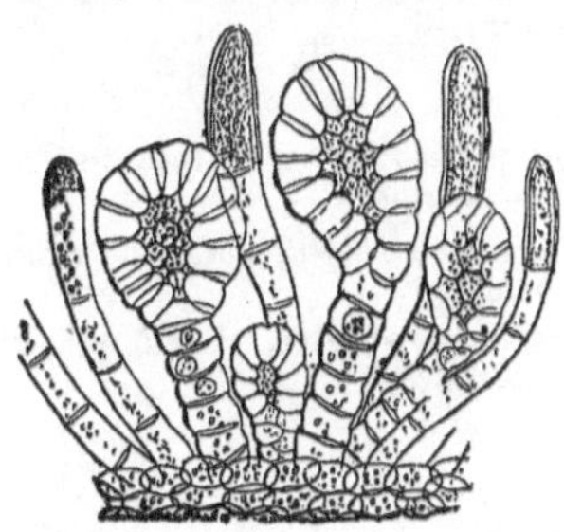

Fig LIII.

It has been thought that sexes occur in these plants, and different parts have been pointed out as the anthers; more especially little threads which contain a grumous matter, sometimes exuded in the form of a crust, and spring up among the spore-cases. Some probability seems to have been given to the presence of anthers by what has been considered an occurrence of Mule ferns, principally belonging to the genus Gymnogramma, some account of which will be found in the *Gardeners' Chronicle* 1844, p. 500; but it does not appear to me that there is good evidence to show that such instances are

Fig. LIII.—Young spore-cases and antheridia of Polypodium effusum. *Link.*

connected with hybrid action; and I agree with Link, in his first view of the question that the function of the Antheridia *nondum sit perspecta et declarata*, an opinion which he has, however, since abandoned in favour of those bodies being anthers. They may be bodies analogous to anthers; but if so they have none of their structure.

Nägeli has lately mentioned that the spiral threads, with an active motion, already mentioned under Mosses, also exist in some Ferns. He found them abundantly in the germinating leaf of Aspidium augescens, and elsewhere, traced their development, and determined that they are produced among the earliest cells that go to the composition of a fern-leaf. (*See Schleiden and Nägeli, Zeitschrift für Wissensch. Bot.* s. 1. 168. *t.* 4.)

The stems of Ferns, when arborescent, are objects of great interest to the botanist, partly on account of their rarity, secondly, because of their singular structure, and especially because they offer the highest form of development in Flowerless Plants. It has not been till lately that they have been well understood; they have now, however, received full illustration from Mohl, in Martius's beautiful *Icones Plantarum Cryptogamicarum*. One of the most interesting of them is that of the Baranetz or Barometz, called also the Scythian Lamb, in which, by cutting off the leaves, except a small portion of the stalk, of a woolly-stemmed species, and turning it upside-down, simple people have been persuaded that there existed in the deserts of Scythia creatures half animal half plant.

The veins of the leaves of Ferns have been sometimes described as dichotomous; it is only, however, in a certain number that this peculiarity occurs. In some they are simple, in others they are collected in lozenge-shaped meshes, and in some they are still differently arranged. Langsdorf and Fischer seem to have been the first to pay attention to these peculiarities, which have been admirably applied to the characters of genera by Adolphe Brongniart and Presl, who have shown them to be of the first importance in distinguishing genera.

Bory de St. Vincent elevates Ferns to the rank of a class, intermediate between Monocotyledons and Acotyledons; but at the same time he attaches no importance to the descriptions of those writers who, having seen the germination of the sporules, have attempted to prove an identity between them and Monocotyledons in that respect. He justly observes, that the irregular unilateral scale which has been seen to sprout forth upon the first commencement of their growth is extremely different from the cotyledon of Monocotyledons, which pre-exists in the seed and never quits it, but swells during germination, and acts as a reservoir of nutriment for the young plantlet. He most properly regards it as an imperfectly developed primordial leaf.

In some modern books of Botany Ferns are broken up into several distinct natural orders, which in my opinion are not to be maintained. But it does appear that three essentially distinct groups exist among them. Of these the largest portion consists of what were once named "dorsiferous ferns," in all which the spore-case is furnished with an elastic ring or band; in two other groups, of inconsiderable extent, the spore-cases have no such band. In one of them the cases are often immersed in the tissue of the back of the leaf, and partially, or entirely, united by their touching edges into many-celled bodies; in the other, the spore-cases appear to be nothing more than an alteration of the edge of a contracted leaf. Hence arise the three following orders:—

NATURAL ORDERS OF FILICALS.

Spore-cases ringless, distinct, 2-valved, formed on the margin of a contracted leaf . . . . } 23. OPHIOGLOSSACEÆ.

Spore-cases ringed, dorsal or marginal, distinct, splitting irregularly . . . . } 24. POLYPODIACEÆ.

Spore-cases ringless, dorsal, connate, splitting irregularly by a ventral cleft . . . . } 25. DANÆACEÆ.

ORDER XXIII. OPHIOGLOSSACEÆ.—ADDERS' TONGUES.

Ophioglosseæ, *R. Br. l. c.* 163. (1810); *Agardh Aph.* 113. (1822); *Mart. ic. Pl. Crypt.* 39. (1834); *Link. Filicum Species*, p. 15; *Endl. Gen.* xxxii.

DIAGNOSIS.—*Filical Acrogens, with ringless, distinct, 2-valved spore-cases, formed on the margin of a contracted leaf.*

Stem erect, or pendulous, with a cavity in the middle, instead of pith, and two or three woody bundles placed round it in a ring. Below, the stalks of the leaves and the spike become blended together. Leaves with netted veins sometimes forked. Spore-cases collected into a spike formed out of the sides of a contracted leaf, 2-valved, without any trace of an elastic ring. Spores resembling fine powder.

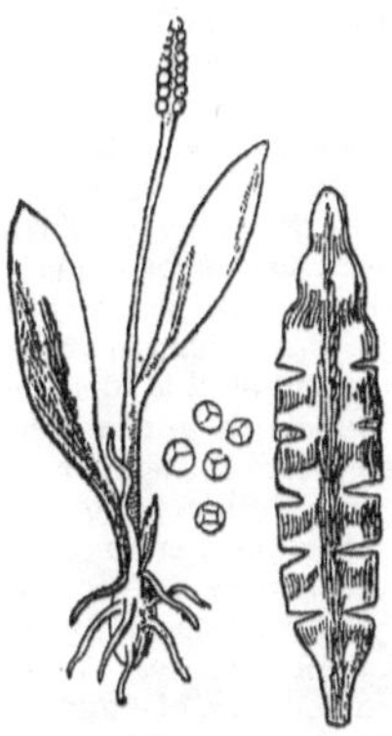

Fig. LIV.

These little plants exhibit a manifest transition to Clubmosses, with which they closely agree in the valvate nature of their spore-cases; but in the latter they are axillary, while in the former they are planted on the margin of a contracted leaf. The curious little genus Phylloglossum seems to be an imitation among Clubmosses of the habit of Adders' tongues. Link finds, in the hollow stem, whose cavity is surrounded by woody bundles, a structure intermediate between that of Clubmosses and Horsetails.

Adders' tongues are most abundant in the islands of tropical Asia, occurring however in the West Indies, and by no means uncommon in temperate latitudes of both worlds. In the tropical parts of Africa, and in Barbary, they seem unknown; at the Cape of Good Hope and in Tasmannia they are uncommon.

The herbage of the order is mucilaginous, whence the species have been employed in broths. Ophioglossum vulgatum and Lunaria botryoides have been used in medicine as vulneraries, but they seem to possess that quality as little as the magical virtues once ascribed to them. Helminthostachys dulcis is regarded in the Moluccas as a slight aperient, is used as a pot-herb, and its young shoots as asparagus. The Haytians fancy Botrychium cicutarium to be an alexipharmic.

GENERA.

Ophioglossum, *Linn.*	Helminthostachys, *Klf.*	Botrychium, *Swartz.*
Ophioderma, *Blum.*	*Botryopteris*, Presl.	*Botrypus*, Rich.
	Ophiala, Desv.	

NUMBERS. GEN. 4. Sp. 25?

	Lycopodiaceæ.
POSITION.—Polypodeæ (Osmundidæ).	OPHIOGLOSSACEÆ.—Danæaceæ.
	Equisetaceæ.

Fig. LIV.—Ophioglossum lusitanicum.

ORDER XXIV. POLYPODIACEÆ.—FERNS.

Gyratæ, *Swartz Synopsis Filicum*, (1806).—Filices veræ, *Willd. Sp. Pl.* 5. 99. (1810.)—Polypodiaceæ, *R. Brown Prodr.* 145. (1810); *Agardh Aph.* 116. (1822); *Kaulfuss Enumeratio*, 55. (1824); *Bory in Dict. Class.* 6. 586. (1824); *Martius Ic. Pl. Crypt.* 83. (1834).

DIAGNOSIS.—*Filical Acrogens, with ringed spore-cases, growing on the back or edge of the leaves, distinct, and splitting irregularly.*

The vast number of plants of the Filical Alliance, collected under this head, are so much alike in many respects, that to separate them into distinct natural orders seems to me contrary to all the rules that govern Botanists in their limitation of such groups. The great mark by which they are known is the presence on the spore-cases of a ring or band of coarse meshes, distinctly different from the tissue of their sides, and too strong to be broken through when the case opens to discharge its contents. Whether the band is vertical or horizontal, complete, incomplete, or otherwise, seems unconnected with any physiological peculiarities that can be pointed out, and to be of no greater importance than for the subordinate purposes of classification. The order consists for the most part of species bearing their spore-cases on the back of leaves, usually named fronds; with the exception of the suborder called Hymenophylleæ, a group of thin, delicate, membranous species, whose leaves open their edges for the protrusion of a vein, over whose surface the spore-cases are arranged. But, independently of all other reasons for regarding the Hymenophylleæ as a mere form of the great order of Ferns, the existence of such genera as Cibotium, Deparia, &c., among Ferns not Hymenophylleous, forbids our attaching much importance to that peculiarity. A very remarkable deviation from the common plan of structure seems at first sight to occur in Osmundeæ and Schizeæ, in which the spore-cases are collected together upon contracted leaves, after the manner of the Adders' tongues; but such plants have no combining character, occurring among Hymenophylleæ as well. The passage of the true Ferns into neighbouring orders is not very gradual. If we regard them as resting on the one hand upon Danæa-worts, they can scarcely be said to touch Adders' tongues on the other, unless the great character of the ringed spore-cases is left out of consideration, and then Osmundeæ may be taken as the connecting link.

1

2

Fig. LV.

The following proportions borne by Ferns to other plants in different latitudes will serve to give some idea of the manner in which they are geographically distributed. There is an enormous disproportion between Ferns and the rest of the Flora in certain tropical islands, such as Jamaica, where they are 1-9th of the Phænogamous plants; New Guinea, where D'Urville found them as 28 to 122; New Ireland, where they were as 13 to 60; and in the Sandwich Islands, where they were as 40 to 160; and it is clear, from the collections of Wallich, that Ferns must form a most important feature in the Indian Archipelago. Upon continents, however, they are far less numerous: thus, in equinoctial America Humboldt does not estimate them higher than 1-36th; and in New Holland Brown finds them 1-37th. They decrease in proportion towards either pole: so that in France they are only 1-63d; in Portugal, 1-116th; in the Greek Archipelago, 1-227th; and in Egypt, 1-971st. Northwards of these countries their proportion again augments, so that they form 1-31st of the Phænogamous vegetation of Scotland; 1-35th in Sweden; 1-18th in Iceland; 1-10th in Greenland; and 1-7th at North Cape. (See a very good paper upon this subject by D'Urville, in the *Ann. des Sc. Nat.* 6. 51.; also *Brown's Appendix to the Congo Voyage*, 461.) Brown has observed (*Flinders*, 584), that it is remarkable, that although arborescent Ferns are found at the southern extremity of Van Dieman's Island, and even at Dusky Bay in New Zealand,

Fig. LV.—1. Part of the leaf of Aspidium Lonchitis; 2. a magnified view of a morsel of Asp. exaltatum.

in nearly 46° south latitude, yet they have in no case been found beyond the northern tropic. For an excellent account of the geographical distribution of Tree Ferns, see *Martius Icones Plantarum Cryptogamicarum*, p. 81.

The leaves generally contain a thick astringent mucilage, with a little aroma, on which account many are considered pectoral and lenitive, especially Adiantum pedatum and Capillus Veneris; but almost any others may be substituted for them. Capillaire is so called from being prepared from the Adiantum Capillus Veneris, a plant which is considered to be undoubtedly pectoral and slightly astringent; though its decoction, if strong, is, according to Ainslie, a certain emetic. The Peruvian Polypodium Calaguala, Acrostichum Huacsaro, and Polypodium crassifolium, are said to be possessed of important medicinal properties, especially the former; their effects are reported to be solvent, deobstruent, sudorific, and antirheumatic; antivenereal and febrifugal virtues are also ascribed to them. The leaves of Adiantum melanocaulon are believed to be tonic in India. (*Ainslie*, 2. 215.) The tubes of the pipes of the Brazilian negroes are manufactured from the stalk of Mertensia dichotoma, which they call Samanbaya. The stem of many species is both bitter and astringent; whence that of several, especially Aspidium Filix Mas, and Pteris aquilina, has been employed as an anthelmintic; and Nothochlæna piloselloides has been used in India to subdue sponginess in the gums. They have also been given as emmenagogues and purgatives. Osmunda regalis has been employed successfully, in doses of 3 drachms, in the rickets. The rhizomes of Nephrodium esculentum are eaten in Nipal, according to Buchanan. Diplazium esculentum, Cyathea medullaris, Pteris esculenta, and Gleichenia Hermanni, are also occasionally employed for food in different countries. Speaking of Pteris esculenta, the Tasmannian fern-root, Mr. Backhouse says, "Pigs feed upon this root where it has been turned up by the plough; and in sandy soils, they will themselves turn up the earth in search of it. The Aborigines roast it in the ashes, peel off its black skin with their teeth, and eat it with their roasted kangaroos, &c. in the same manner as Europeans eat bread. The root of the Tara-fern possesses much nutritive matter; yet it is to be observed, that persons who have been reduced to the use of it, in long excursions through the bush, have become very weak, though it has prolonged life." Pteris aquilina and Aspidium Filix Mas have been used in the manufacture of beer, and Aspidium fragrans as a substitute for tea. *Agdh.* The fragrance which gives its name to the latter species occurs occasionally elsewhere. Polypodium phymatodes is employed, along with Angiopteris evecta, in preparing the cocoa-nut oil of the South Sea islands; Aneimia tomentosa smells of myrrh, and Mohria thurifera of benzoin.

GENERA.

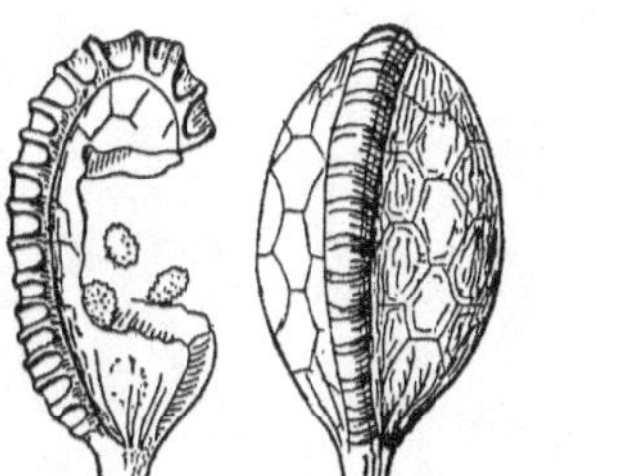

Fig. LVI.

I.—*Polypodeæ.* Endl. Gen. xxvi. Spore-cases stalked, with a vertical ring; spores roundish or oblong.

Acrostichum, *L.*
Polybotrya, *H. B.*
Egenolphia, Schott.
Olfersia, *Radd.*
Elaphoglossum, Schott.
Rhipidopteris, Schott.
Stenochlæna, *J. Sm.*
Lomagramma, *J. Sm.*
Aconiopteris, *Presl.*
Stenosemia, *Presl.*
Campium, *Presl.*
Platycerium, *Desv.*
Alcicornium, Gaudich.
Cyrtogonium, *J. Sm.*
Photinopteris, *J. Sm.*
Poecilopteris, *Eschw.*
Bolbitis, Schott.
Gymnopteris, *Presl.*
Hymenolepis, Kaulf.
Leptochilus, Kaulf.
Anapausia, Presl.
Hemionitis, *Linn.*
Antrophyum, *Kaulf.*
Loxogramma, *Blum.*
Polytaenium, *Desv.*
Leptogramma, *J. Sm.*
Gymnogramma, *Desv.*
Neurogramma, *Presl.*
Calomelanos, *Presl.*
Ceterach, *Adans.*
Grammitis, *Swartz.*
Xiphopteris, *Kaulf.*
Micropteris, Desv.
Chilopteris, *Presl.*
Synammia, Presl.
Cryptogramma, *R. Br.*
Diblemma, *J. Sm.*
Selliguea, *Bory.*
Diagramma, Blum.
Microgramma, *Presl.*
Stegnogramma, *Blum.*
Sphærostephanos, *J. Sm.*
Meniscium, *Schreb.*
Tænitis, *Swartz.*
Pleurogramma, *Presl.*
Tæniopsis, *J. Sm.*
Pteropsis, *Presl.*
Chilogramme, Blum.
Monogramma, *Commers.*
Cochlidium, Kaulf.
Adenophorus, *Gaudich.*
Amphoradenium, *Desv.*
Nothochlaena, *R. Br.*
Cincinalis, Desv.
Drymoglossum, *Presl.*
Polypodium, *Linn.*
Ctenopteris, Blum.
Dicranopteris, Blum.
Phegopteris, Presl.
Lastræa, Bory.
Goniopteris, *Presl.*
Pleocnemia, *Presl.*
Amblia, *Presl.*
Goniophlebium, *Blum.*
Marginaria, *Presl.*
Pleurogonium, *Presl.*
Cyrtophlebium, *R. Br.*
Campyloneurum, Presl
Phlebodium, *R. Br.*
Dictyopteris, *Presl.*
Phymatodes, *Presl.*
Anaxetum, *Schott.*
Pleuridium, Presl.
Dryostachyum, *J. Sm.*
Drynaria, *Presl.*
Dipteris, Reinw.
Microsorus, Link.
Aglaomorpha, *Schott.*
Psygmium, Presl.
Pleopeltis, *H.* et *B.*
? *Paragramme*, Bl.
Niphobolus, *Kaulf.*
Candollea, Mirb.
Pyrrhosia, Mirb.
Cyclophorus, Desv.
Cyclophorus, *Presl.*
Scyopteris, *Presl.*
Lecanopteris, *Reinw.*
Onychium, Reinw.
Calymodon, *Presl.*
Cheilanthes, *Swartz.*
Hypolepis, *Presl.*
Ochropteris, *J. Sm.*
Lonchitis, *Linn.*
Adiantum, *Linn.*

Fig. LVI.—Spore-cases of Hymenocystis caucasica.

Hewardia, *J. Sm.*
Cassebeera, *Kaulf.*
Platyloma, *J. Sm.*
Doryopteris, *J. Sm.*
Pteris, *Linn.*
Allosorus, *Bernh.*
Ceratodactylis, J. Sm.
Phorolobus, Desv.
Amphiblestra, *Presl.*
Litobrochia, *Presl.*
Campteria, *Presl.*
Monogonia, *Presl.*
Jamesonia, *Hook.*
Salpichlæna, *J. Sm.*
Blechnum, *Linn.*
Sadleria, Kaulf.
Acropteris, Link part.
Haplopteris, Presl.
Lomaria, *Willd.*
Stegania, R. Br.
Vittaria, *Smith.*
Struthiopteris, *Willd.*
Onoclea, *Linn.*
Angiopteris, Mitch.
Calypterium, *Bernh.*
Ragiopteris, *Presl.*
Neottopteris, *J. Sm.*
Asplenium, *Linn.*
Onopteris, Bernh.
Belvisia, Mirb. part.
Acropteris, Link.
Thamnopteris, *Presl.*
Darea, Juss.
Caenopteris, Berg.
Hemidictyum, *Presl.*
Allantodia, *R. Br.*
Doodia, *R. Br.*
Woodwardia, *Smith.*
Scolopendrium, *Smith.*
Antigramma, *Presl.*
Camptosorus, *Link.*
Onychium, *Kaulf.*
Leptostegia, Don.
Diplazium, *Swartz.*
Callipteris, Bory.
Anisogonium, *Presl.*
Digrammaria, *Presl.*
Oxygonium, *Presl.*
Didymochlaena, *Desv.*
Monochlaena, Gaud.
Hippodium, Gaudich.
Tegularia, Reinw.
Ceramium, Reinw.
Hysterocarpus, Langs.
Nephrolepis, *Schott.*
Nephrodium, *Rich.*
Oleandra, *Cav.*
Neuronia, Don.

Ophiopteris, Reinw.
Dryopteris, *Adans.*
Lastræa, Presl.
Thelypteris, *Schott.*
Arthrobotrys, *Presl.*
Aspidium, *Swartz.*
Psidopodium, Neck.
Polystichum, *Roth.*
Tectaria, Cav.
Rumohria, Radd.
Phanerophlebia, *Presl.*
Fadgenia, *Hooker.*
Cyclodium, *Presl.*
Cyrtomium, *Presl.*
Sagenia, *Presl.*
Bathmium, *Presl.*
Cystopteris, *Bernh.*
Acrophorus, *Presl.*
? *Leucostegia*, Presl.
Lindsæa, *Dryand.*
Schizolomia, *Gaud.*
Hymenotomia, Gaud.
Isoloma, *J. Sm.*
Dictyoxiphium, *Hooker.*
Synaphlebium, *J. Sm.*
Odontosoria, *J. Sm.*
Davallia, *Smith.*
Microlepia, *Presl.*
Saccoloma, *Kaulf.*
Humata, *Cav.*
Pachypleuria, *Presl.*
Colposoria, *Presl.*
Wibelia, Bernh.
Odontosoria, *Presl.*
Stenolobus, *Presl.*
Prosaptia, *Presl.*
Cystidium, *J. Sm.*
Dicksonia, *Herit.*
Balantium, *Kaulf.*
Culcita, *Presl.*
Sitolobium, *Desv.*
Denstaedtia, Bernh.
Leptopleuria, *Presl.*
Patania, *Presl.*
Pæsia, *St. Hil.*
Cibotium, *Kaulf.*
Pinonia, Gaudich.
Deparia, *Hook* et *Grev.*
Woodsia, *R. Br.*
Physematium, Kaulf.
Diacalpe, *Blum.*
Hymenocystis, *C. A. Mey.*
Hypoderris, *R. Br.*
Sphæropteris, *R. Br.*
Peranema, Don.
Podeilema, R. Br.
Prionopteris, Wall.

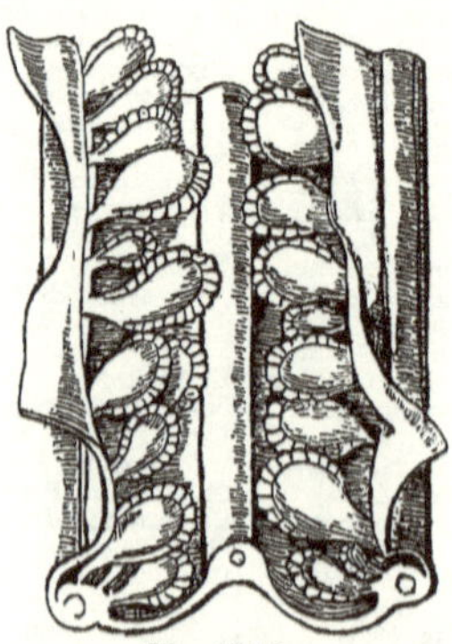

Fig. LVII.

II. — *Cyatheæ.* *Kaulf. Enum.* (1824); spore-cases with a vertical ring, usually sessile, on a more or less elevated receptacle; spores 3-cornered or 3-lobed.

Thyrsopteris, *Kunz.*
Panicularia, Coll.
? *Chonta*, Molin.
Schizochlæna, *J. Sm.*
Hemitelia, *R. Br.*
Cnemidaria, *Presl.*
Alsophila, *R. Br.*
Haplophlebia, *Mart.*
Dicranophlebia, *Mart.*
Metaxya, *Presl.*
? *Amphidesmium*, Scht.
Trichopteris, Park.
Trichipteris, *Presl.*
Chnoophora, Kaulf.
Arachniodes, *Blum.*
Gymnosphæra, *Blum.*
Cyathea, *Smith.*
Sphæropteris, Bernh.
Schizocæna, *J. Smith.*
Notocarpia, *Presl.*
Disphemia, *Presl.*
Cnemidaria, *Presl.*
Matonia, *R. Br.*

III. *Parkereæ.* *Hooker, exot. fl.* p. 147. (1825); spore-cases very thin, surrounded by a broad imperfect, sometimes obsolete ring.

Ceratopteris, *Brongn.*
Ellobocarpus, Kaulf.
Teleozoma, R. Br.
Cryptogenis, Rich.
Furcaria, Desv.
Cryptogramma, Grev.
Parkeria, *Hooker.*

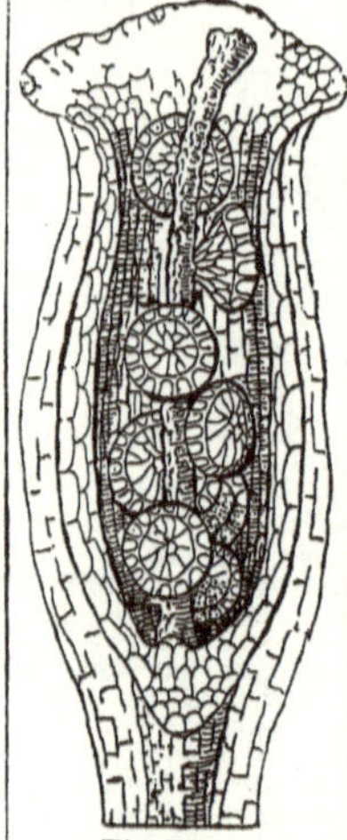

Fig. LVIII.

IV. *Hymenophylleæ.* *Endl. prod. Norf.* 16. (1833); *Martii ic. pl. crypt.* 102. (1834); *Endl. Gen.* xxxvii.— Spore-cases marginal, placed upon the surface of a vein extended beyond the edge of the leaf, with a complete horizontal ring; spores convexo-tetraedral.

Hymenophyllum, *Smith.*
Trichomanes, *Linn.*
Didymoglossum, Desv.
Hymenostachys, Bory.
Feea, Bory.
Lecanium, Presl.
Cardiomanes, Presl.
Ragatellus, Presl.
Cephalomenes, Presl.
Neurophyllum, Presl.
Microgonium, Presl.
Abrodictyum, Presl.
Meringium, Presl.
Hemiphlebium, Presl.
Leptocyonium, Presl.
Myrmecostylum, Presl.
Ptychophyllum, Presl.
Sphærocyonium, Presl.
Hymenoglossum, Presl.
Loxsoma, *R. Br.*

V.—*Gleicheneæ.* Schismatopterides, *Willd. l. c.* 69. (1810).—Gleicheneæ, *R. Br. l. c.* 160. (1810); *Kaulfuss l. c.* 36. (1824). — *Bory, l. c.* (1824). — Pleurogyratæ, *Bernh.*—Gleicheniaceæ, *Mart. ic. pl.* 105. (1834); *Endl. gen.* xxviii.; spore-cases dorsal, with a transverse occasionally oblique ring, nearly sessile, and bursting lengthwise internally; spores oblong, or kidney-shaped.

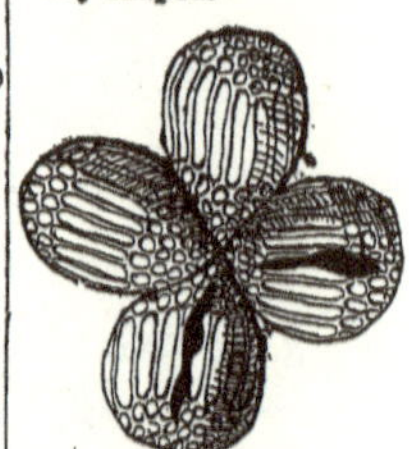

Fig. LIX.

Gleichenia, *Smith.*
Mertensia, *Willd.*
Dicranopteris, Bernh.
Sticherus, *Presl.*
Platyzoma, *R. Br.*
Calymella, *Presl.*

VI.—*Schizæeæ.* *Mart. ic. pl. crypt.* 113. (1834); *Endl. gen.* xxix.; spore-cases dorsal, with a complete terminal contracted ring; spores pyramidal or conical.

Aneimia, *Swartz.*
Ornithopteris, Bernh.
Anemidictyon, *J. Sm.*

Fig. LVII.—Onychium lucidum. Fig. LVIII.—Spore-cases and cup of Trichomanes radicans. Fig. LIX.—Spore-cases of Mertensia flexuosa.

Schizæa, *Smith.*
Rhipidium, Bernh.
Lophidium, Rich.
Actinostachys, *Wall.*
Lygodium, *Swartz.*
Hydroglossum, Willd.
Ugena, Cav.
Cteisium, Rich.
Ramondia, Mirb.
Odontopteris, Bernh.
Gisopteris, Bernh.
Vallifilix, Thouars.
Lygodictyon, *J. Sm.*
Mohria, *Swartz.*

VII. — *Osmundeæ.* Osmundaceæ, *R. Br. l. c.* 161. (1810); *Agardh, l. c.* 115. (1822); *Kaulfuss, l. c.* 42. (1824); *Endl. gen.* xxx.; Acrogyratæ, *Bernh.*; spore-cases dorsal, or panicled, stalked, with a broad dorsal incomplete ring, opening vertically; spores oblong or roundish.
Osmunda, *Linn.*
Aphyllocarpa, Cav.
Todea, *Willd.*

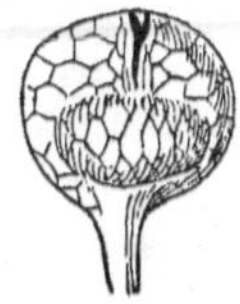

Fig. LXI.

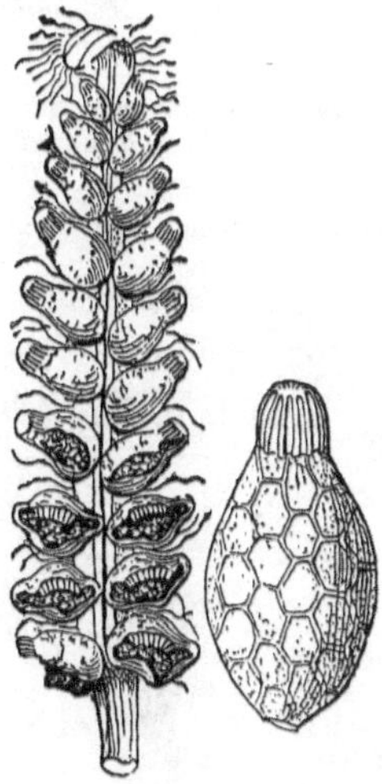

Fig. LX.

NUMBERS. GEN. 183. Sp. 2000.

POSITION.—Danæaceæ.—POLYPODIACEÆ.—Ophioglossaceæ.

Fig. LX.—Schizæa dichotoma; 1. its spore-case. Fig. LXI.—Spore-case of Todea Frazeri.

ORDER XXV. DANÆACEÆ.—DANÆAWORTS.

Agyratæ, *Swartz. Synops.* (1806).—Poropterides, *Willd. l. c.* 66. (1810).—Danæaceæ, *Agardh, l. c.* 117. (1822).—Marattiaceæ, *Kaulf. l. c.* 31. (1824); *Bory, l. c.* (1824); *Mart. ic. pl. crypt.* 119. (1834); *Endl. gen.* xxxi.; *Link. filic. sp.* p. 31.

DIAGNOSIS.—*Filical Acrogens, with ringless dorsal spore-cases, combined in masses, and splitting irregularly by a central cleft.*

With all the habit of Dorsiferous Ferns, these plants are widely distinguished by the peculiar nature of their spore-cases, which are neither like those of Ferns nor Adder's-tongues. To the latter they approach the nearest, but instead of being connected with, and perhaps fashioned out of, the margin of a contracted leaf, they appear sunk within, or more rarely seated upon, the back of the leaflets. The entire want of that clas tic ring, which, in some state or other so strikingly characterises true ferns, gives them a far stronger title to be regarded as a distinct order, than the trifling differences which have in the eyes of some botanists elevated little groups of the latter to that dignity.

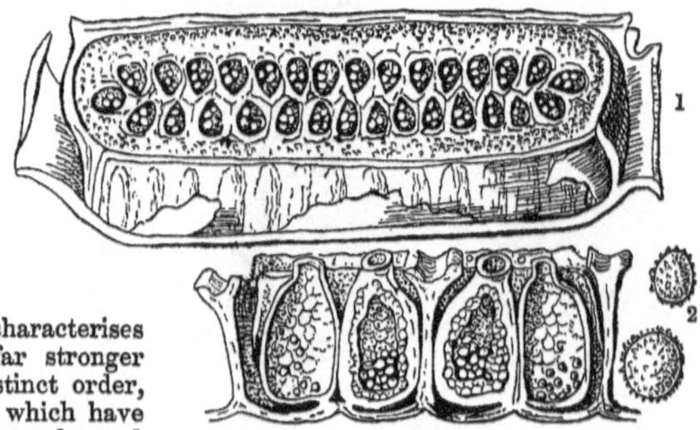

Fig. LXII.

In addition to this, their spore-cases are always united more or less by their inner faces, as if in anticipation of the prevailing tendency among the carpels of flowering plants. For this reason they may be regarded as the highest form of the highest Alliance among Acrogens.

The few known species of the Order are all tropical in both hemispheres. Some form trees.

The bruised leaves of the fragrant Angiopteris evecta, an arborescent species, are said to be employed in the Sandwich Islands to perfume the cocoa-nut oil. The rhizome of Marattia alata is eaten by the Sandwich Islanders in time of scarcity, according to Mr. Hinds: this would appear to be the Nehai, and not the former plant.

GENERA.

Kaulfussia, *Blum.*	*Clementea*, Cav.	Eupodium, *J. Sm.*	*Myriotheca*, Comm.
Angiopteris, *Hoffman.*	Danæa, *Smith.*	Marattia, *Swartz.*	*Celanthera*, Thouin.

NUMBERS. GEN. 5. Sp. 15. (J. Smith.)

POSITION.—Polypodiaceæ.—DANÆACEÆ.—Ophioglossaceæ.

Fig. LXII.—Danæa alata. 1. Collection of spore-cases; 2. sections of the same and spores, 2 of which are highly magnified.

CLASS III.—RHIZOGENS.

Rhizantheæ, *Blum. Fl. Javæ,* (1828); *Endlicher Meletemata,* p. 10. (1832); *Ed. prior,* p. 389; *Endl Gen.* p. 72.

These are parasitical plants destitute of true leaves, in room of which they have cellular scales. Their stem is either an amorphous fungous mass, or a ramified mycelium, sometimes, perhaps always, appearing to be lost in the tissue of the plants on which it grows; and is very imperfectly supplied with spiral vessels, which in some instances seem to be wholly deficient. No instance of green colour is known among them; but they are brown, yellow, or purple. They are furnished with true flowers, having genuine stamens and carpels, and surrounded by a trimerous or pentamerous calyx, or absolutely naked. Their ovules appear to be constructed upon the same plan as in other flowering plants. The true nature of their seeds is in most species quite unknown; by some they are described as breaking up into a mass of spores, by others as consisting of a cellular nucleus abounding in grumous corpuscles (*Endl.*), and in general they may be regarded as too small for exact observation; but it is certain that in some instances they have a minute undivided embryo enclosed in mucilagino-granular albumen.

At this point of the Vegetable Kingdom we find a most curious assemblage, which, with many of the peculiarities of Endogens, seems to be an intermediate form of organisation between them and Thallogens. They have no relation to Acrogens, although they follow at this place, but they agree with Endogens in the presence of sexes, and sometimes in the ternary structure of their flower; they have, however, scarcely any spiral vessels, and their seeds appear, as far as they have been examined, either, as some say, to want the cotyledons and axis of other flowering plants, or to lose themselves in a mass of pulp, from which they are almost undistinguishable. In their amorphous succulent texture, in their colour, often in their putrid odour when decaying, in the formation of a mycelium or spawn, which is evident in Helosis, and is with good reason suspected to exist in others, and in their parasitical habits, these plants resemble Fungals, while in their flowers and sexes they accord with Arumworts, or similar Endogens.

Rhizogens all agree in being of a fungus-like consistence, and in their habits of living parasitically on the roots of other plants. They very generally stain water, or spirit, of a deep blood-red colour. Their forms are exceedingly diversified; some have the aspect of a Mushroom, or develop a head like that of a Bullrush (Typha): others push forth a thyrse of flowers, or an elegant panicle; while some have their bloom in a head like that of some Cynaraceous plant. In Helosis and Langsdorffia the rhizome, which is horizontal and branched, and which at intervals throws up perpendicular flowering stalks, is quite analogous to the spawn * of

* The existence of a mycelium has also been adverted to by Dr. Brown. (*Linn. Trans.* xix. 232.) He suggests that in Rafflesia the earliest effort of the seed, after being deposited in its proper nidus, may consist in the formation of a cellular tissue extending laterally under the bark of the stock. He remarks that in Pilostyles and Cytinus, where the plants are closely approximated, their possible origin from a common base or thallus, is rendered the more probable by the parasites in the former genus, which is diœcious, being produced generally, perhaps always, in groups of the same sex, and by those groups, which are often very dense, not unfrequently surrounding the branch of the stock. He adds, however, that this view is not sustained by sufficient observation, but that there are circumstances in both genera favourable to the hypothesis, especially in Pilostyles.

Fungals. In Cynomorium, Scybalium and Balanophora, this part is wanting, and in its room the roots of those genera emit roundish deformed tubers collected in a circle upon the roots of other plants, and growing into them by some unknown process. Blume says, "that at the period of germination of Balanophoreæ there is produced from the roots of the Fig on which they grow an intermediate body, of a fleshy nature and intimately combined with its superficial woody layers, and that this intermediate body is penetrated by their spiral vessels, which render it woody." He moreover adds, that "several seeds of Balanophoreæ germinate on nearly the same points of the Fig-root; hence this woody body, or luxuriant product of the juices that are sucked out, has generally an irregular form, and the plants proceeding from such tubers grow out in different directions, much in the same manner as the tubers of a Potato generate their offsets: with this difference, however, that in a Potato the eyes of the plant are in the circumference, while in Balanophora they are placed in the centre, and on that account the intermediate body where the offsets break out, has necessarily a conical extension." Something of the same kind occurs in Scybalium, whose tubers are expanded in an irregular form about the root of some unknown tree, are fleshy, and composed even in the substance of the stalk of somewhat irregular cells and no spiral vessels. In the room of leaves these plants have scales, which differ from true leaves in the want of colour, a character common to all other plants parasitical on roots. A vertical stalk (stipes), sometimes terminated by a solitary head of flowers, sometimes bearing several heads variously arranged upon the stalk, is found in all the genera of Balanophoraceæ; which moreover agree in this that the flower-heads, which at first are sessile on the rhizome and concealed by many rows of imbricated scales, resemble the leafy rosette of a Sempervivum without colour, or rather the very small bud of a Rafflesia. The genuine species of Helosis show on their rhizome roundish conical buds seated on a very short stalk, or altogether sessile, enclosing the rudiments of the future head within a very thin involucre, as a fungus within the volva; this latter after a time splits into three or more segments, and emits the flower-head enlarged and furnished with a stalk, which is altogether naked except at the base, where it is surrounded by the scale-like segments of the withering involucre. This is the most simple form of involucre, which in the other genera becomes more and more complicated, and finally runs into numerous series of imbricated scales which clothe the stipes more or less completely. In those genera which grow upon the bark of the stems of trees, there are some diversities of structure in the organs of vegetation that are very remarkable. Blume tells us that Rafflesia Patma appears upon the creeping roots or stems of Cissus scariosa in the form of solitary or clustered hemispherical dilatations, which look like excrescences or expansions of the root. These excrescences are something of the nature of leaf-buds, consisting of layers of scales and a more solid centre. As the latter increase in size they burst through the wrapper by tearing it irregularly from the apex towards the base, and develop themselves in the form of numerous scales, at first flesh-coloured, then brownish, and finally deep purple, which surround the flowers. As soon as these parts are exposed, richly nourished as they are by the humid air that surrounds them, they grow with such rapidity that it is reported that Rafflesia, which, when full-blown, is a yard across, and when unexpanded, is as large as a middle-sized cabbage, only takes about three months for its complete formation. Brugmansia has a similar mode of development.

At one time it was believed that Rhizogens agreed with Fungals in the total want of spiral vessels. That, however, was a mistake. Spiral vessels do exist among them. Brown says that he has discovered them in Rafflesia, in which he originally failed to perceive them, and in several other cases. Martius also found them in Langsdorffia, in the form of bundles lying in the rhizome, stem and branches, and Mohl in similar parts of Helosis, but in small quantity compared to the mass of the plants. Brown adds that "the vascular system of all these parasites is uniform and more simple than that of the far greater part of Phænogamous plants; that the spiral, or slight modifications of it, is the only form of vessel hitherto observed in any of them; and that the large tubes or vessels, with frequent contractions, corresponding imperfect diaphragms, and variously marked surface, which have received several names, as vasa porosa, punctata, vasiform cellular tissue, dotted ducts, &c., and which are so conspicuous in the majority of arborescent Phænogamous plants, have never been observed in any part strictly belonging to these parasites. (*Linn. Trans.* vol. xix. 231.) He, however, does not attach systematical importance to this curious fact.

The flowers are in general formed upon some symmetrical plan, the proportions varying from genus to genus. But in a singular deformed genus called Sarcophyte the flowers are not reducible to symmetry, as far as has yet been observed. It has not, however, been examined in a philosophical manner.

The seeds of most Rhizogens appear to have baffled the inquiries of those who have had the best opportunities of examining them. Even the seeds of the common Cytinus Hypocistis of the South of Europe are wholly unknown; to the disgrace of Mediterranean Botanists. Blume describes the seeds of Brugmansia Zippelii as containing, 1° a grumous substance which under a powerful microscope exhibits a lax cellular tissue, formed of roundish cells, which become angular by mutual pressure, and are filled with grumous matter; among these are dispersed, 2° threads or tubes, very numerous, very tender, long, entangled without order, usually forked, sometimes irregularly branched. Upon these threads Blume makes the following remarks:—"The tender tubes, principally visible in the ripe spores of Rhizogens, may be considered analogous to those lowest forms of vegetation which belong to the genus Mycoderma, *Pers.*, of the family Hydronemateæ. The parietes of the fruit of Brugmansia are seldom covered with spores when the pericarp is closed up, but they constantly are when the plant is decaying: a circumstance which is attributable to the facility with which the spores separate from their stalks, and to the cellulo-gelatinous matter in which the fruit abounds. It is worthy of remark, however, that the spores are attached in the same manner as the seeds of more perfect plants, although they are altogether different from them in structure. They are, indeed, to be compared only to the unfecundated ovules of Phænogamous plants, which in the latter are more completely evolved after impregnation, but in Rhizogens, as in other Cryptogamous plants, only after germination. That the ovules of Rhizogens, while inclosed in the pericarp of their mother, ever arrive at the development of an embryo, seems to be altogether untrue. For I have over and over again examined numerous specimens at different stages of formation; the observations have been repeated under the eyes of Reinwardt, and the brothers Nees von Esenbeck; as also by Meyen, so celebrated as an anatomist, upon specimens preserved in spirits of wine, so that I can deny the possibility of any error. How, indeed, can we suppose Rhizogens to be plants furnished with an embryo, when they exhibit only the simplest form of cellular organization in all their parts." (*Fl. Javæ, Rhizantheæ,*

p. 23.) Endlicher describes the seeds of Scybalium fungiforme as being a "nucleus compositus e telâ cellulosâ, corpusculis sporidiiformibus e cellulis angulatis conflatis et massulis grumosis immixtis, filis tenuissimis earundem particulas connectentibus farctâ." But these are perhaps mistakes arising from inaccurate observation. At least I can positively confirm the statement of the elder Richard (*Mem. Mus.* viii. t. xxi.), who gives to Cynomorium coccineum an embryo. I find in that plant that the seed consists of a mucilaginous mass filled with angular particles, which are doubtless loosely cohering cells. They contain starch in a very minutely globular state, but are chiefly composed of gum. On one side of this seed is a globular embryo, looking like a speck, but found, when properly examined, to be a globose mass of cells, destitute of starch, enclosed within the albumen, and apparently undivided on any part of its surface. It is, however, difficult to speak positively upon this point, on account of its smallness, and I am not sure that it is not very slightly 2-lobed. Francis Bauer too ascertained the ovules of Rafflesia Arnoldi to have the ordinary structure, a strong indication that the seeds would not be so anomalous as has been represented, and he found an undivided embryo in the seed of the same plant, (*Linn. Trans.* xix. t. xxv.) a circumstance confirmed by the observations of Brown. Ferdinand Bauer, too, found in Hydnora Africana what seems to be a central embryo (*Ibid.* t. xxx.) of the same nature.

Such being the principal facts that have been ascertained with regard to these singular parasites, it only remains to notice some of the views entertained regarding them by systematic botanists. Dr. Robert Brown, who, aided by the microscopical drawings of the two Bauers, has had peculiar advantages for considering the question, appears to be opposed to the idea of regarding Rhizogens as a peculiar class. He considers the Patma worts (Rafflesiaceæ) as being unquestionably allied to Birthworts, and therefore as a form of Exogens. The reasons, however, which have led this botanist to form such an opinion, require to be stated with much more detail before they can claim serious attention. His objections to regarding Rhizogens as a peculiar class are more definite. He denies the absence of spiral vessels, which he himself and others once supposed to be a characteristic of some at least among them, and asserts that the vascular texture of Rhizogens is not essentially different from that of any perfectly developed Phænogamous plants. But, as was stated in the last edition of this work, the true question to be considered is, not as to the presence or absence of spiral vessels, but as to their abundance. In Exogens or Endogens *equally developed* they would be most copious, and would exist in all the foliaceous organs; and it is no argument against the importance of this circumstance, to say, that spiral vessels have no existence in certain Endogens, as Lemna, for instance; for in that and similar cases the small degree in which such plants are developed, may be considered to account for the absence of spiral vessels; just as in a common Exogen, the spiral system does not make its appearance until the general development of the individual has made some progress.

So, indeed, in Ferns and other Acrogens of high degree, we have no right to say that the vascular system is absent; on the contrary, in the centre of the stem of Clubmosses, and in the soft parts of that of Ferns, either spiral or scalariform vessels exist in abundance; but they do not make their appearance in the foliaceous organs as in more perfect plants.

Brown also attaches no importance to the supposed homogeneity of the embryo of Rhizogens, because the same structure, he says, exists in Orobanche and Orchisworts. But with regard to Orobanche, that plant has a slightly two-lobed embryo lying in a mass of albumen, so that I do not see how it can be brought into comparison with that of Rhizogens ; and as to Orchisworts, we have no right to say that their embryo is essentially different from that of common Endogens, except in its smallness.

More recently, Mr. Griffith has adopted the views of Brown, and endeavoured, by new arguments, to show that Rhizogens cannot be regarded as a peculiar class in the Vegetable Kingdom. He is of opinion that " in the construction of the group, a remarkable diversity of characters has been sacrificed to an appearance resulting from parasitism on roots, and to an assumed absence of an ordinary form of vegetable embryo." He asserts, that these plants are not similar in their parasitism, and that in those which he has examined there would appear to be two remarkably different types of development of the embryo. He thinks, moreover, that such a class is opposed to the system of Nature, a chief point of the plan of which consists in an extensive interchange of characters, either positively by structure, or negatively by imitation of structure. The want of uniformity in opinion of the founders of the group regarding its rank or value, is incompatible with any group of the system of Nature. And he is persuaded that Rhizogens are an entirely artificial class, not even sanctioned by practical facility, which is the only merit of an artificial association, and " a retrograde step in the course of philosophical botany." This being the case, it was of course necessary to show where the genera of Rhizogens can be stationed, if they are not collected into one common class, as is here proposed. Accordingly, Mr. Griffith suggests, that the genus Mystropetalon may be " the homogeneous-embryo-form of that order, which he takes to include Proteaceæ, Santalaceæ, &c., and which nearly agrees with Professor Lindley's alliance Tubiferæ." The tendency of Sarcophyte is, he thinks, towards Urticaceæ, and he also considers Balanophora as the homogeneous-embryo-form of Urticaceæ, forming a direct passage in one, and usually the more perfect, structure to Musci and Hepaticæ." Finally, he stations his genus Thismia between Taccaceæ and Burmanniaceæ.—(*Proceedings of the Linnæan Society, No. XXII.*, p. 220.)

I think that this kind of argument affords strong evidence in favour of the propriety of constructing the class here again proposed. The forced resemblances which are sought after by both Dr. Brown and Mr. Griffith, in themselves indicate the weakness of the arguments by which it is proposed to do away with the class of Rhizogens. The fact is, that the species which constitute it have no real relation to any other parts of the system. It is true that the genera differ very much from each other in the details of their fructification ; though not at all more than the genera of other classes ; but the character of the order does not depend upon the fructification. It depends wholly upon the great peculiarity in the manner of growth, already pointed out ; and the fructification is connected with questions of quite a subordinate degree. All the classes of plants depend equally upon such considerations ; and, therefore, Rhizogens are a class. I am indeed surprised, that so acute a botanist as Mr. Griffith—one of the very few men who combine with minute accuracy of observation great general views—should not himself perceive how much his position is weakened by comparisons like the following, the justice of which, however, I am far from disputing. He particularly directs attention to the resemblance between the pistil of Cynomoriums and that of Mosses, or more especially to that of some evaginulate Liverworts, and to the effects produced by the action of the pollen on their styles. "Indeed," he observes, "in the development of the female organ, in the continuous surface of the style before fecundation, and in its obvious perforation after, Balanophora presents a direct affinity to a group of plants with which otherwise it has not a single analogy." In another genus, called Phæocordylis, he finds that the hairs in which the fruits are imbedded present a remarkable analogy with the paraphyses of Drepanophyllum and certain Neckeræ, and also with the bodies which he suspects to be the male organs of Ferns. Surely this is a class of peculiarities which indicate a group of a much lower rank than that of Exogens or Endogens. It is easy, indeed, to get rid of many of these characteristics by a particular mode of reasoning; but, since the same mode of reasoning would equally destroy every class now recognised among plants, I do not think it necessary to examine it particularly.

There is an account of Rhizogens by Endlicher in his *Meletemata*, which contains a summary of all that was in 1832 known concerning them. For further information the reader is referred, to Blume's *Flora Javæ;* Martius' *Nova Genera*, &c., vol. 3; Brown's *Observations on Rafflesia*, in the 13th and 19th volumes of the *Linnean Society's Transactions;* Griffith, in the *Proceedings* of the same learned body, and the various works quoted at the head of the following natural orders:

NATURAL ORDERS OF RHIZOGENS.

Ovules solitary, pendulous; fruit one-seeded	26. BALANOPHORACEÆ.
Ovules 00, *parietal; fruit many-seeded; calyx*, 3-4-6-*parted; anthers opening by slits*	27. CYTINACEÆ.
Ovules 00, *parietal; fruit many-seeded, calyx* 5-*parted, anthers opening by pores*	28. RAFFLESIACEÆ.

ORDER XXVI. BALANOPHORACEÆ.—CYNOMORIUMS.

Balanaphoreæ, *Rich. in Mém. Mus.* 8. 429. (1822); *Endlicher Meletemata*, p. 10. (1832); *gen.* xxxix. *Meisner*, p. 366; *Junghuns in nov. act.* xviii. *suppl.*; *Griffith, Proceedings Linn. Soc. No.* xxii.

DIAGNOSIS.—*Stems amorphous, fungoid; peduncles scaly: flowers in spikes; ovules solitary, pendulous; fruit one-seeded.*

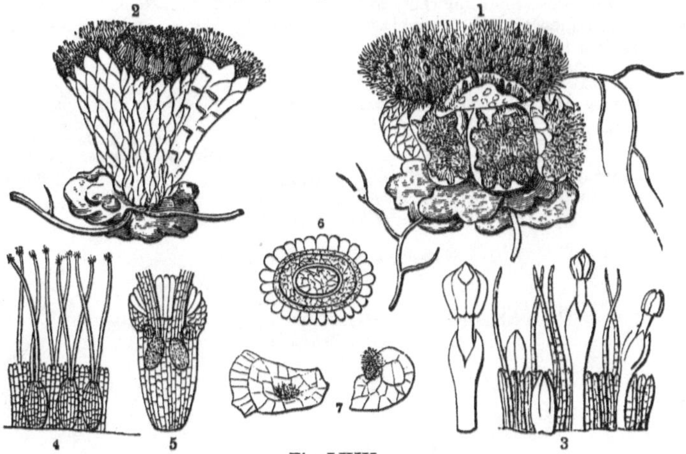

Fig. LXIII.

Fungus-like plants, parasitical upon roots, with fleshy, horizontal, branched stems, and peduncles covered by imbricated scales. Flowers monœcious, collected in dense heads, which are roundish or oblong, usually bearing both male and female flowers, but occasionally having the sexes distinct; the receptacle covered with scales or setæ variable in form, here and there bearing also peltate thick scales; rarely naked. ♂ flowers pedicellate; calyx deeply 3-parted, equal, spreading, with somewhat concave segments; stamens 1-3 (seldom more), epigynous, with both united filaments and anthers; the latter 3. ♀ ovary inferior, 1-2-celled, 1-2-seeded, crowned by the limb of the calyx, which is either marginal and nearly inverted, or consists of from 2 to 4 unequal leaflets; ovule pendulous; style 1, seldom 2, filiform, tapering; stigma simple, terminal, rather convex. Fruit 1-celled. Seed, in Cynomorium coccineum, solitary, consisting of mucilaginous albumen, in which angular cells of gum and starch-granules are loosely arranged; embryo very minute in proportion to the albumen, roundish, whitish, enclosed beneath the skin, undivided.

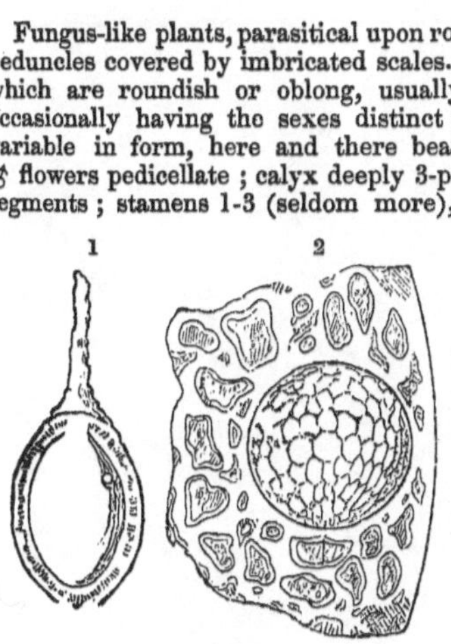

Fig. LXIV.

The solitary ovules of these plants, suspended from the apex of the cells, distinguish them positively from Patmaworts and the Cistusrapes. It is, however, to the latter that they are most similar in habit. In all cases they seem to have an amorphous fungous development in the first instance, out of

Fig. LXIII.—Scybalium fungiforme. 1. A male plant; 2. a female; 3. male flowers with hairs between them; 4. females; 5. a vertical section of a female, with the two pendulous ovules; 6. a section across a ripe fruit; 7. seeds?

Fig. LXIV.—Cynomorium coccineum. 1. A section of the ripe fruit, showing the embryo on the right of the albumen; 2. a portion of the nucleus very highly magnified, showing the embryo and the angular cells among which it lies. N.B. These cells are separated by the pressure of a compressorium.

which the scaly peduncles start up. Junghuns assures us, that his Rhopalocnemis has altogether the appearance of Phallus impudicus, and that it would certainly be taken for a toadstool if it were not for its flowers. The species put on very extraordinary forms, one of which is represented here, and several of which have been given by Junghuns.

All are tropical in both worlds, with the exception of one found in Malta. About as many occur in Asia as in America, and several are known from the Cape of Good Hope.

They seem, as far as anything is known of them, to be styptics. The Cynomorium coccineum, or Fungus melitensis of the apothecaries, long had a great reputation in that way; and various kinds of Helosis have had a similar character. Sarcophyte, a Cape plant, is said to have an atrocious odour. Pöppig says, that Ombrophytum, which in Peru springs up suddenly after rain, in the manner of the toadstool, is called Mays del Monte, in consequence of its resemblance to a kind of Maize, and is quite insipid, on which account it is cooked and eaten like Fungi.

GENERA.

Tribe I.—Sarcophytidæ.	*Archimedea*, Leandr.	Balanophora, *Forst.*	Helosis, *Rich.*
Sarcophyte, *Sparrm.*	Ombrophytum, *Pöpp.*	Phæocordylis, *Griff.*	*Caldasia*, Mut.
Ichthyosma, Schlecht.			*Lathræophila*, Leandr.
		Tribe IV.—Helosidæ.	Langsdorfia, *Mart.*
Tribe II.-Lophophytidæ.	Tribe III.—Cynomoridæ.		? *Thonningia*, Schmch.
		Cynopsole, *Endl.*	? *Hæmatostrobus*. Scht.
Lophophytum, *Schott* et *Endl.*	Cynomorium, *Michel.*	Scybalium, *Schott* et *Endl.*	Rhopalocnemis, *Jungh.*
	Mystropetalum, *Harv.*		

NUMBERS. GEN. 12. Sp. 30.

Fungales.
POSITION. ———. BALANOPHORACEÆ.—Cytinaceæ.
Muscales?

ORDER XXVII. CYTINACEÆ.—CISTUSRAPES.

Cytineæ, *Adolph. Brongn. in Ann. des. Sc. Nat.* 1. 29. (1824); *Endlicher Meletemata*, p. 13. *Gen.* xl. *Meisner*, p. 367. *R. Brown in Linn. Trans.* xix.—Pistiaceæ, *Agardh. Aphor. Bot.* p. 240. (1826).—Aristolochiæ, § Cytineæ, *Link Handb.* 1. 368. (1829).—Hydnoreæ, *R. Br. Linn. Trans.* 19. (1844).

DIAGNOSIS.—*Flowers in spikes at the end of a scaly stem, with a 3-6-parted calyx, anthers opening by slits, and innumerable ovules growing over parietal placentæ.*

Flowers ⚥, or ♂ ♀, solitary and stemless, or clustered at the top of a stalk covered with imbricated scales, the males uppermost, the females lowermost, in the axil of a bract, and supported on each side by a bractlet. Perianth tubular-campanulate, with a spreading 3-6-lobed limb, the segments imbricated, the exterior alternating with the bractlets or induplicate and valvate. Anthers sessile, 2-celled; their cells distinct, opening longitudinally; four dissepiment-like membranes in Cytinus alternate with the segments of the perianth, and join its tube with the column. ♀ Perianth as in the males, but epigynous. Ovary inferior, 1-celled, with vertical or parietal placentæ, covered by innumerable ovules; style cylindrical, joined to the tube of the perianth by septiform processes, with a thick stigma, or free, and consisting of several styles, each having a free stigmatic apex. Fruit berried, leathery, one-celled, with innumerable seeds buried in pulp, and having a hard leathery skin firmly attached to the nucleus. Seed in Hydnora, with a small undivided embryo in the centre of cartilaginous albumen, and in Cytinus exalbuminous according to Brown.

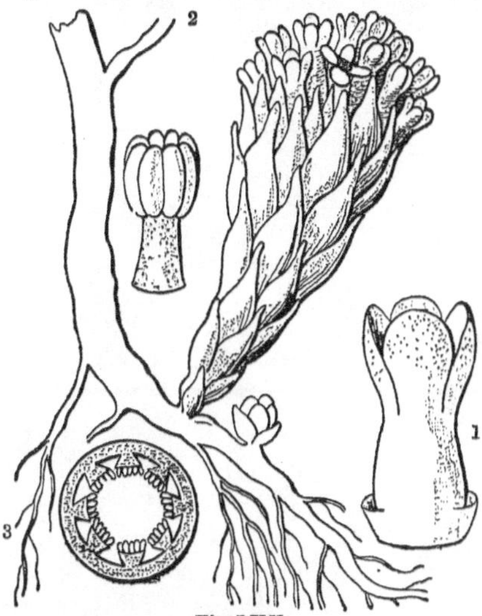

Fig. LXV.

In these we have a near approach to the common condition of Endogens, both in structure and habit, if we compare Cytinus with some Bromelworts.

But the appearance of Hydnora is so peculiar that we know nothing to contrast it with, except some such Fungus as a Geaster, like which it grows half-buried in the soil. Its innumerable seeds distinguish it from Patmaworts, as well as its caulescent habit and slit anthers.

The history of this extraordinary plant has been fully given by Ferdinand Bauer and Dr. Brown, in the 19th vol. of the *Linnean Transactions*, from which place the accompanying cuts are taken. The genus is regarded by Brown as the type of a peculiar Order: and perhaps with justice. But for reasons elsewhere given, I demur to the formation of all Orders that depend upon a single genus.

Cytinus is parasitical on the roots of Cistus in the South of Europe; the rest are from the Cape of Good Hope, where Hydnora is parasitical on the roots of succulent Euphorbias, and of Cotyledon orbiculatum.

Hydnora Africana (*Jackals Kost* or *Kawimp*), smells like decaying roast-beef, or some fungus (*Harvey*); when roasted it is eaten by the African savages. Cytinus Hypocistis (ὑποκίστις Diosc.) contains gallic acid, and according to Pelletier, has the property of precipitating gelatine without containing tannin; its extract is still officinal in the South of Europe, under the name of Succus Hypocistidis; it is blackish, sub-acid, astringent, and is employed in hæmorrhages and dysentery.

Fig. LXV.—Cytinus Hypocistis. 1. A flower; 2. a head of anthers; 3. a transverse section of the ovary.

GENERA.

Cytinus, *L.*	*Aphyteia*, L.	*Hyobanche*, Sparrm.
Hypocistis, Tourn.	Hypolepis, *Pers.*	*Thyrsine*, Gled.
Hydnora, *Thunb.*	*Phelypæa*, Thunb.	? Thismia, *Griff.*

NUMBERS. GEN. 4. Sp. 7.

Fungales.
POSITION.—Rafflesiaceæ.—CYTINACEÆ.—Balanophoraceæ.
Bromeliaceæ ?

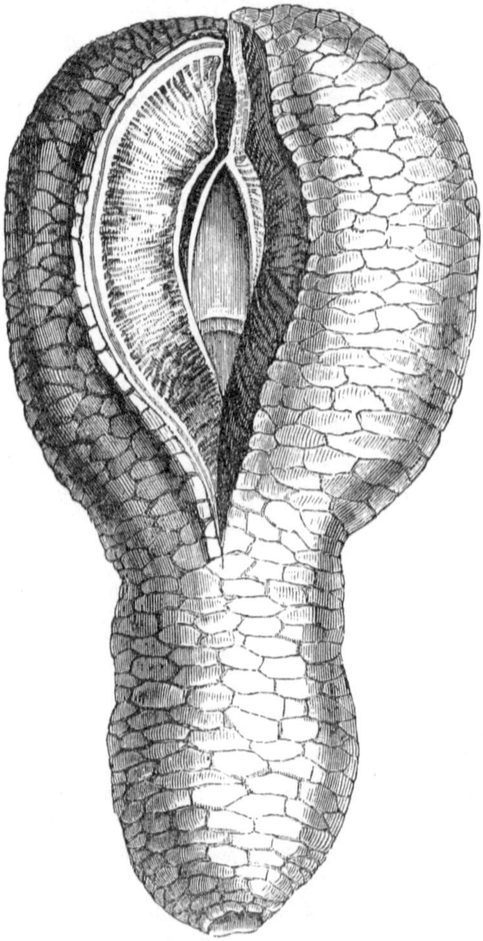

Fig. LXVI.

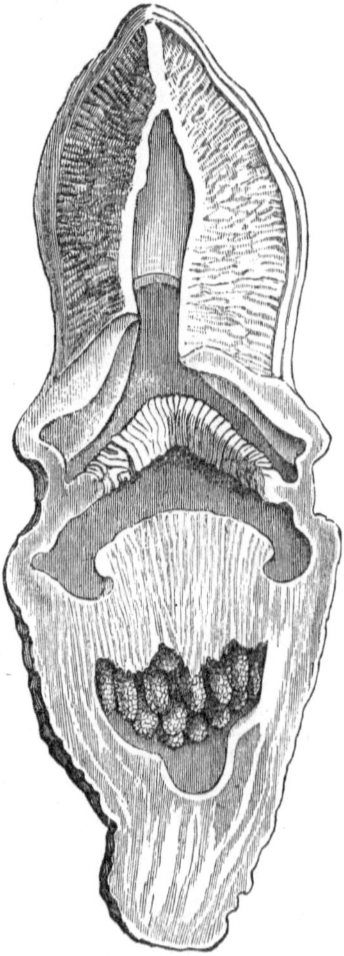

Fig. LXVII.

Fig. LXVI. - A plant of Hydnora Africana. Fig. LXVII.—A longitudinal section of it.—*Ferd. Bauer.*

ORDER XXVIII. RAFFLESIACEÆ.—PATMAWORTS.

Rafflesiaceæ, *Endlicher Meletemata*, p. 14. (1832); *Gen.* xli. *Meisner*, p. 367; *R. Brown in Linn. Trans.* 19. 241.

DIAGNOSIS.—*Stemless and stalkless; flowers 5-parted, sessile on the branches of trees, solitary, with anthers opening by pores, and innumerable ovules growing over parietal placentæ.*

Stemless plants, consisting merely of flowers growing immediately from the surface of branches, and immersed among scales; flowers hermaphrodite, or diœcious. Perianth superior, globose or campanulate; the limb 5-parted, with the segments imbricated or doubled inwards in æstivation; the throat surrounded by calli, which are either distinct or run together into an entire ring. Column (synema) hypocrateriform or sub-globose, adhering to the tube of the perianth; anthers numerous, distinct, or somewhat connate, adhering by the base, in one row; 2-celled, with the cells opposite, and each opening by a vertical aperture, or concentrically many-celled with a common pore. Ovary inferior, 1-celled, with many-seeded parietal placentæ; styles conical, equal in number to the placentæ, run together within the column, but projecting beyond it, and then distinct. Fruit, an indehiscent pericarp, with an infinite multitude of seeds. [Embryo undivided, with or without albumen.—*R. Brown.*]

1

2

3

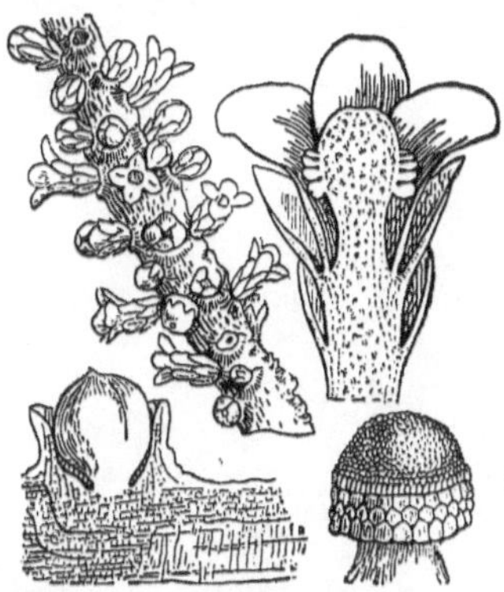

Fig. LXVIII.

These extraordinary plants have no stems whatever, but consist of flowers only, supported by scales in room of leaves. Among them is the very remarkable species described by Brown in the 13th vol. of the *Linnean Society's Transactions*, under the name of Rafflesia, to which those may be referred who are desirous either of knowing what is the structure of one of the most anomalous of vegetables, or of finding a model of botanical investigation and sagacity, or of consulting one of the most beautiful specimens of botanical analysis which Francis Bauer ever made. They differ from the Cistusrapes in having no proper stem, in their anthers being porous, and in their flower, which constitutes the whole plant, being divided by 5, like Exogens, instead of 2 or 3, like Endogens. An affinity has been suggested with Birthworts, to which this Order seems to have no immediate relationship.

Natives of the East Indies, on the stems of Cissi; or of South America, on the branches of leguminous plants.

Rafflesia Patma is employed in Java as a powerful styptic, in relaxation or debility of the urino-genital apparatus, and Brugmansia seems to possess similar qualities.

GENERA.

*RAFFLESEÆ, *R. Br.*	Brugmansia, *Blume.*	**APODANTHEÆ, *R. Br.*	Pilostyles, *Guillem.*
Rafflesia, *R. Br.*	*Zippelia*, Rchb.	Apodanthes, *Poit.*	*Frostia*, Bert.
Sapria, *Griffith.*	*Mycetanthe*, Rchb.		

NUMBERS. GEN. 5. Sp. 16.

Aristolochiaceæ ?
POSITION.—Balanophoraceæ—RAFFLESIACEÆ.—Cytinaceæ.

Fig. LXVIII.—Pilostyles Berterii ♂. 1. A vertical section of a flower; 2. a young flower bursting through the bark; 3. a head of stamens.

CLASS IV.—ENDOGENS.

Monocotyledones, *Juss. Gen.* 21. (1789); *Desf. Mém. Inst.* 1. 478. (1796). Endorhizeæ, *Rich. Anal.* (1808). Monocotyledoneæ *or* Endogenæ, *DC. Théorie*, 207. (1813). *Meisner*, p. 353. Cryptocotyledoneæ *or* Graniferæ, *Agardh.* 73. (1821).—Amphibrya, *Endl. Gen.* p. 76.—Teleophyta, *Schleiden.*

Having now passed in review the absolutely sexless plants, called Thallogens, and all that class which, under the title of Acrogens, comprehends a numerous race among whom the existence of a double sex is conjectured to exist, and having, moreover, disposed of the curious Rhizogens, which, to a fungal mode of growth join a complete sexual apparatus, we pass to Endogens, or Monocotyledons.

Here we find a vast multitude of species, with extremely diversified habits, among whom occurs every attribute supposed to be connected with the most perfect structure. Leaves and stems are distinctly separated; spiral vessels, breathing-pores, and sexes, are in a condition that admits of no further complication; and we find in the great majority everything which constitutes as elaborate an arrangement of parts as we have any knowledge of in the vegetable kingdom.

This great class bears the name of Endogens, in consequence of its new woody matter being constantly developed in the first instance towards the interior of the trunk, only curving outwards in its course downwards. That palm-trees grow in this way was known so long since as the time of Theophrastus, who distinctly speaks of the differences between endogenous and exogenous wood.* But that this peculiarity is also extended to a considerable part of the vegetable kingdom is a modern fact, the discovery of which we owe to the French naturalists Daubenton and Desfontaines. The path being thus opened, the inquiry has subsequently, and more particularly of late years, been much extended, and the result is the conviction that all those numerous races to which Jussieu applied the name of Monocotyledoneæ, agree essentially in this manner of growth. We may take the palm-tree as typical of the endogenous structure. In the beginning the embryo of a palm consists of a cellular mass of a cylindrical form, very small and not at all divided. As soon as germination commences a certain number of cords of ligneous fibre begin to appear in the radicle, deriving their origin from the plumule. Shortly afterwards, as soon as the rudimentary leaves of the plumule begin to lengthen, spiral and dotted vessels appear in the tissue in connection with the ligneous cords; the latter increase in quantity as the plant advances in growth, shooting through the cellular tissue, and keeping parallel with the outside of the root. At the same time the cellular tissue increases in diameter to make room for the ligneous cords (or woody bundles, as they are also called). At last a young leaf is developed with a considerable number of such cords in connection with its base, and, as its base passes all round the plumule, these cords are consequently connected equally with the centre which that base surrounds. Within this a second leaf gradually unfolds, the cellular tissue increasing horizontally at the same time; the ligneous cords, however, soon cease to maintain anything like a parallel

---

* Ἔχει δὲ τὴν μήτραν, τὰ μὲν μεγάλην καὶ φανερὰν, ὡς Πρίνος, Δρῦς, καὶ τὰ ἄλλα τὰ προειρημένα· τὰ δὲ, ἀφανεστέραν, οἷον, Ἐλαία, Πύξος· οὐ γάρ ἐστιν ἀφορισμένην οὕτω λαβεῖν· ἀλλὰ καὶ φασί τινες οὐ κατὰ τὸ μέσον, ἀλλὰ κατὰ τὸ πᾶν ἔχειν, ὥστε μὴ εἶναι τόπον ἀφωρισμένον· διὸ καὶ ἔνια οὐδ' ἂν δόξειεν ὅλως ἴχειν· ἐπεὶ καὶ τοῦ Φοίνικος οὐδεμία φαίνεται διαφορὰ κατ' οὐδέν. *Theophr. Hist.* I. 8.

direction, but form arcs whose extremities pass upwards and downwards, losing their extremities in the leaf on the one hand, and on the other in the roots, or in the cellular integument on the outside of the first circle of cords ; at the same time the second leaf pushes the first leaf a little from the centre towards the circumference of the cone of growth. In this manner leaf after leaf is developed, the horizontal cellular system enlarging all the time, and every successive leaf, as it forms at the growing point, emitting more woody bundles curving downwards and outwards, and consequently intersecting the older arcs at some place or other ; the result of which is that the first formed leaf will have the upper end of the arcs which belong to it longest and much stretched outwardly, while the youngest will have the arcs the straightest ; and the appearance produced in the stem will be that of a confused entanglement of woody bundles in the midst of a quantity of cellular tissue. As the stem extends its cellular tissue longitudinally while this is going on, the woody arcs are consequently in proportion long, and in fact usually appear to the eye as if almost parallel, excepting here and there, where two arcs intersect each other. As in all cases the greater number of arcs curve outwards as they descend, and eventually break up their ends into a multitude of fine divisions next the circumference where they assist in forming a cortical integument, it will follow that the greater part of the woody matter of the stem will be collected near the circumference, while the centre, which is comparatively open, will consist chiefly of cellular tissue ; and when, as in many palms, the stem has a limited circumference, beyond which it is its specific nature not to distend, the density of the circumference must, it is obvious, be proportionably augmented. It is however a mistake to suppose that the great hardness of the circumference of old palm wood is owing merely to the presence of augmenting matter upon a fixed circumference ; this will account but little for the phenomenon. We find that the woody bundles next the circumference are larger and harder than they originally were, and consequently we must suppose that they have the power of increasing their own diameter subsequent to their first formation, and that they also act as reservoirs of secretions of a hard and solid nature, after the manner of the heartwood of exogens.

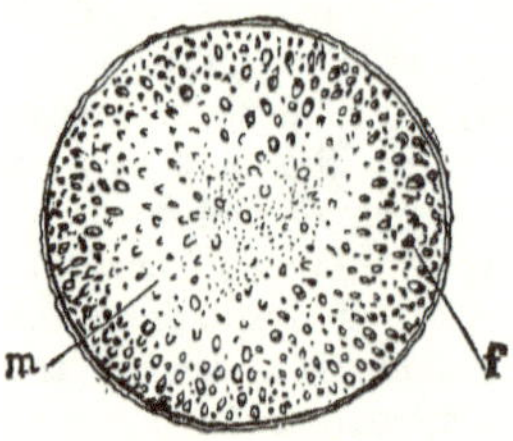

Fig. LXIX.

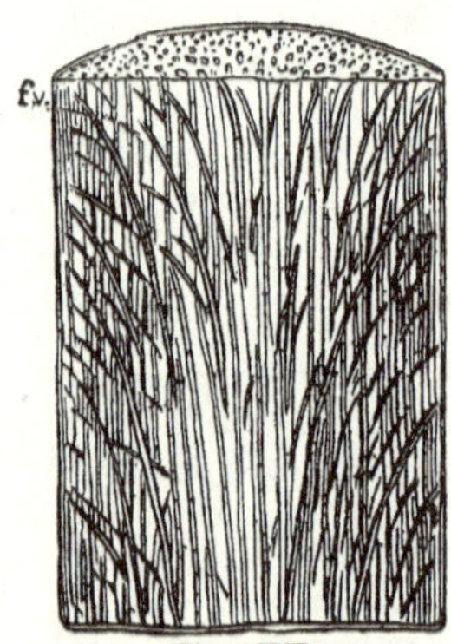

Fig. LXX.

When the growth of the stem of an endogen goes on in this regular manner, with no power of extending horizontally beyond a specifically limited diameter, a trunk is formed, the sections of which present the appearances shown in the accompanying cut. There is a number of curved spots crowded together in a confused way, most thick and numerous at the circumference, comparatively small and thinly placed at the centre ; and the only regular structure that is observable with the naked eye is that the curves always present their convexity to the

---

Fig. LXIX.—Horizontal section of a Palm stem.
Fig. LXX.—Longitudinal section of a Palm tree.

circumference. When there is no limited circumference assigned by nature to an Endogen, then the curved spots, which are sections of the woody arcs, are much more equally arranged, and are less crowded at the circumference. Never is there any distinct column of pith, or medullary rays, or concentric arrangement of the woody arcs ; nor does the cortical integument of the surface of endogenous stems assume the character of bark, separating from the wood below it ; on the contrary, as the cortical integument consists very much of the finely divided extremities of the woody arcs, they necessarily hold it fast to the wood, of which they are themselves prolongations, and the cortical integument can only be stripped off by tearing it away from the whole surface of the wood, from which it does not separate without leaving myriads of little broken threads behind.

This is the apparent and general structure of the most perfect among Endogens. It is of course modified exceedingly according to the nature of particular individuals, and may even be reduced to nullity, as is the case in Lemna, Tillandsia usneoides, Naiads, and similar plants.

Schleiden, who treats this subject in a merely anatomical manner, thus describes the peculiarities of Endogens or Monocotyledons, and the manner in which they differ from Exogens or Dicotyledons.

In all plants, he says, the woody bundles, whose development always proceeds from the interior to the exterior, are either limited or unlimited in their growth. Commonly every woody bundle consists of three different physiological parts ; firstly, of a tissue of extreme delicacy, capable of rapid development, in which new cells are continually generated and deposited in various ways, in two different directions, viz. next the circumference, in the shape of a peculiar kind of lengthened cellular tissue with very thick walls, the liber ; and next the centre, in the form of annular, spiral, reticulate, and porous vessels : secondly, of woody cells, which are either uniform in appearance, or different, and form the wood, properly so called. Up to a certain period the development of the vascular system in Monocotyledons and Dicotyledons proceeds upon the same plan ; but in Monocotyledons (Endogens) the active, thin, solid, delicate cellular tissue, suddenly changes ; the partitions of its cells become thicker ; their generating power ceases ; and when all the surrounding cells are fully developed, they assume a peculiar form, ceasing to convey gum, mucilage, and other kinds of thick formative sap.

From this cause all further development of vascular bundles is rendered impossible, and therefore Schleiden calls the woody bundles of such plants "limited." In Dicotyledons (Exogens), on the contrary, this tissue retains, during the whole lifetime of the plant, its vital power of formation ; continues to develop new cells ; and so increases the mass, ceaselessly augmenting both the exterior (liber), and the interior faces (wood), for which reason Schleiden calls such woody bundles "unlimited." This, he continues, happens according to the climate and nature of the plant : either pretty continuously, as in Cactaceæ ; or by abrupt periodical advances and cessations, as occurs in forest trees of Europe. In the latter, the stem forms an uninterrupted tissue, from the pith to the bark, during every period of life, and the bark is never organically separate from the stem ; what is considered their natural separation in the spring, is only a rent produced by tearing the delicate tissue already spoken of, which is present, even during winter, and constitutes the foundation of new annual zones, although compressed, and filled with gum, starch, and other secretions. In the spring, being expanded and swollen by the new current of sap, it is deprived of its contents by their solution.

This difference between limited and unlimited woody bundles affords, in Schleiden's opinion, the only universal distinction between Endogens and Exogens. In the annual Exogens the unlimited woody bundle, checked in its further development by the death of the plant, has, it is true, in so far some similarity to the limited one of Endogens; yet, sufficient research shows the difference distinctly, for the formative layer in the former constantly retains to the last moment its generating power. (*See Annals of Natural History*, iv. 236.)

The distinction between Endogens and Exogens, whether it be as we have first described it, or such as Schleiden states, is so obvious and universally recognised, that one would have thought them beyond the reach of controversy. Nevertheless, M. de Mirbel has very recently (*Comptes Rendus, Oct.* 1844, *p.* 699) asserted, that, according to his theoretical views of their structure, a great number of Monocotyledons are Exogens, more especially Dracæna, Phœnix, Chamærops, and Bromelia. Meneghini, moreover, long since pointed out the fact that Yucca gloriosa arranges its woody bundles in concentrical circles, (*Ricerche sulla Struttura del Caule nelle Piante Monocotiledoni, Padova* 1836) and the same tendency is discoverable in some other Endogens allied to Yucca. But the mere gathering together the woody bundles into imperfect rings, does not in any degree invalidate the distinction between Endogens and Exogens, because their whole manner of growth is different. The fibrovascular tissue which forms the wood of Yucca gloriosa itself, is in fact present in the form of arcs, just as much as in a Palm-tree.

In many of the larger kinds of Endogens the stem increases principally by the development of a single terminal bud, a circumstance unknown in Exogens, properly so called. In many however, as all grasses, the ordinary growth takes place by the full development of axillary buds in abundance.

In general there is so great a uniformity in the structure of an endogenous stem, that the common cane or asparagus illustrates its peculiarities sufficiently. There are, however, anomalous states that require explanation.

Grasses are endogens with hollow stems strengthened by transverse plates at the nodes. This is seen in the bamboo, whose joints are used as cases to hold rolls, or in any of our indigenous species. In this case the deviation from habitual structure is owing to the circumference growing faster than the centre, the consequence of which is the tearing the latter into a fistular passage, except at the nodes, where the arcs of ligneous tissue, connected with the leaves, cross over from one side of the stem to the other, and by their entanglement and extensibility form a solid and impenetrable diaphragm. That this is so is proved by the fact, that the stems of all grasses are solid, or nearly so, as long as they grow slowly; and that it is when the rapidity of their development is much accelerated that they assume their habitual fistular character. In the sugar-cane grass the hollowness of the stem is indeed unknown. Independently of that circumstance, their organisation is sufficiently normal.

Xanthorhæa hastilis has been shown by De Candolle to have an anomalous aspect. When cut through transversely, the section exhibits an appearance of medullary rays proceeding with considerable regularity from near the centre to the very circumference. (*Organographie Végétale*, t. vii.) But such horizontal rays are not constructed of muriform cellular tissue like real medullary processes, but are composed of ligneous cords lying across the other woody tissue; they are in fact the upper ends of the woody arcs pulled from a vertical into a horizontal direction by the growth of the stem and the thrusting of the leaves to which they belong from the centre to the circumference. Such a case throws great light upon the real nature of the more regular forms of endogenous wood.

Other appearances are owing to imperfect development, as in some of the aquatic species of this class. Lemna, for example, has its stem and leaves fused together into a small lenticular cavernous body; and in Zannichellia and others, a few tubes of lengthened cellular tissue constitute almost all the axis.

By far the most striking kind of anomaly in the stem of Endogens is that which occurs in Barbacenia, and which was originally noticed in the first edition of this work, p. 334. In an unpublished species of Barbacenia from Rio Janeiro, allied to B. purpurea, the stems appear externally like those of any other rough-barked plant, only that their surface is unusually fibrous and ragged when old, and closely coated by the remains of sheathing leaves when young. Upon examining a transverse section of this stem it is found to consist of a small firm pale central circle having the ordinary endogenous organisation, and of a large number of smaller and very irregular oval spaces pressed closely together but having no organic connection; between these are traces of a chaffy ragged tissue which seems as if principally absorbed and destroyed. A vertical section of the thickest part of this stem exhibits, in addition to a pale central endogenous column, woody bundles crossing each other or lying parallel, after the manner of the ordinary ligneous tissue of a palm stem, only the bundles do not adhere to each other, and are not embodied as usual in a cellular substance. These bundles may be readily traced to the central column, particularly in the younger branches, and are plainly the roots of the stem, of exactly the same nature as those aerial roots which serve to stay the stem of a screw pine (Pandanus). When they reach the earth the woody bundles become more apparently roots, dividing at their points into fine segments, and entirely resembling on a small scale the roots of a palm-tree. The central column is much smaller at the base of the stem than near the upper extremity. A figure of this structure will be found under the order Hæmodoraceæ.

The age of endogenous trees has been little studied. When the circumference of their stem is limited specifically, it is obvious that their lives will be limited also; and hence we find the longevity of palms inconsiderable when compared with that of exogenous trees. Two or three hundred years are estimated to form the extreme extent of life in a date-palm and in many others. But where, as in the Dragon Trees, the degree to which the stem will grow in diameter is indefinite, the age seems, as in Exogens, to be indefinite also: thus a famous specimen of the Dracæna Draco, of Oratava in Teneriffe, was an object of great antiquity so long ago as A. D. 1402, and is still alive.

Important as the character furnished by the internal manner of growth of an Endogen obviously is, it is much enhanced in value by its being found very generally accompanied by peculiarities of organisation in other parts. The leaves have in almost all cases the veins placed in parallel lines, merely connected by transverse single or nearly single bars. Straight-veined foliage is therefore an external symptom of an endogenous mode of growth. When such an appearance is found in Exogens it is always fallacious, and is found to be owing to the excessive size and peculiar direction of a few of the larger veins, and not to be a general character of all the venous system; as is sufficiently obvious in Rib-grass, Gentian, and many more.

The flowers too of Endogens have in most cases their sepals, petals, and stamens corresponding with the number three, or clearly referrible to that type; and the pistil usually participates in the same peculiarity. Where such a proportion exists in Exogens, it is usually confined to the sepals and petals by themselves, or to the pistil by itself, not extending to the other organs. In Endogens it is almost universal in all the whorls of the flower, although sometimes obscured by the abortion, dislocation, or cohesion of particular parts, as happens in the whole of the extensive natural order of grasses.

Fig. LXXI.—Yucca aloifolia.

The effect of the manner of growth in Endogens is to give them a very peculiar appearance. Their trunks frequently resemble columns rising majestically with a plume of leaves upon their summit; and the leaves, often very large—the fan-shaped leaves of some palms are from 20 to 30 feet wide—have most commonly a lengthened form, resembling a sword blade if stiff, or a strap if weak and broad. These peculiarities are connected with others belonging to endogenous vegetation in its most rudimentary condition. The embryo of an Endogen is, in its commonest state, a small undivided cylinder, which protrudes from within its substance a radicle from one end and a plumule from a little above the radicle; in other cases its embryo has a slit on one side, in the cavity of which the plumule reposes; or, finally, the embryo is a flat plate as in grasses, with the plumule and radicle attached to its face near the base. In the latter case the flat plate is a solitary cotyledon, which, in the second instance, is folded together so as to give the embryo the appearance of being slit, and which in the first, or most habitual, condition is not only folded up, but united at its edges into a case entirely burying the plumule and cotyledon. Hence the embryo of an endogen is called monocotyledonous; a name that is really unexceptionable, notwithstanding the occasional appearance of a second rudimentary cotyledon, as occurs in common wheat. M. Adrien de Jussieu has endeavoured to show that the slit, which is generally supposed to be peculiar to the Arums and their allies, is of general occurrence in the endogenous embryo. (*Ann. Sc. N. Ser.* xi. p. 341.)

It has already been stated that the radicle is protruded in germination from within the substance of the embryo; the base of the radicle is consequently surrounded by a minute collar formed of the edges of the aperture produced by the radicle upon its egress. For this reason Endogens are called endorhizal.

Hence the great natural class of plants forming the subject of these remarks has five most important physiological peculiarities, by all which combined, or usually by each of which separately, the class may be characterised.

1. The wood is endogenous.
2. The leaves are straight-veined.
3. The organs of fructification are ternary.
4. The embryo is monocotyledonous.
5. The germination is endorhizal.

It may however be readily supposed that, viewed as a large class of plants, Endogens are essentially characterised only by the combination of these five peculiarities, and that occasional deviations may occur from every one of them. Thus in Naias, Caulinia, Zannichellia, and others, which constitute a part of what Professor Schultz names Homorganous floriferous plants, the whole organisation of the stem is so imperfect that the endogenous character is lost; but their true nature is nevertheless sufficiently indicated by their straight veins, monocotyledonous embryo, &c. The examples of a concentrical arrangement of the woody bundles, above alluded to, may be regarded as instances of endogenous development tending towards the exogenous, and are usually looked upon as cases of transition from one form to the other—perhaps not very correctly. Of a similar nature are the resemblances between the columnar Cycadaceous Gymnosperms and Palms, between the livid, fœtid, one-sided calyx of Aristolochia and the equally livid, fœtid, one-sided spathe of Araceous Endogens, or, in another point of view, between such lenticular plants as Lemna in Endogens, with the leaves and stems fused, as it were, together, and similar forms of stem and leaf among Marchantiaceous Acrogens.

Really intermediate forms of vegetation connecting Endogens with other classes, are extremely uncommon. One of the most striking is that which occurs between Ranunculaceæ and Nymphæaceæ on the part of Exogens, and Alismaceæ and Hydrocharaceæ on that of Endogens; if Ranunculus lingua, or better R. parnassifolius, is contrasted with Alisma plantago, or Damasonium, leaving out of consideration subordinate differences, it will be found that there is little of a positive nature to distinguish them except the albuminous dicotyledonous seeds of the former as compared with the exalbuminous monocotyledonous seeds of the latter; and the resemblances between Hydropeltis and Hydrocharis in the other case are so very great, that Schultz and others actually refer them to the same class.

Endogens probably contain more plants contributing to the food of man, and fewer poisonous species in proportion to their whole number, than Exogens. Grasses, with their floury albumen, form a large portion of this class, to which have to be added Palms yielding fruit, wine, sugar, sago; Arums, Arrow-roots, Amaryllids, &c., producing arrow-root; the nutritious fruit of Plantains; the aromatic secretions of Gingers; and Orchisworts, forming salep. Among the deleterious species we have no inconsiderable number among Amaryllids, Arums, Melanths, and even Lilies.

In this, as in all other large groups, the extremes of development are so far apart, that one would be tempted to doubt the possibility of their being mere forms of each other, were it not certain that numerous traces exist in the vegetable kingdom of a frequent tendency to produce the typical structure of a natural association of whatever kind in both an exaggerated and degraded state, if such figurative terms may be employed in science. For instance, the genus Ficus contains some species creeping on the ground like diminutive herbaceous plants, and others rising into the air to the height of 150 feet, overspreading with the arms of their colossal trunks a sufficient space of ground to protect a multitude of men; the type of organisation in the willow is in like manner represented on the one hand by the tiny Salix herbacea, which can hardly raise its head above the dwarf moss and saxifrages that surround it; and on the other by Salix alba, a tree sixty feet high. Then among natural orders we have the Rosal structure exaggerated, on the one hand, into the arborescent Pomaceæ, and degraded, on the other, into the apetalous imperfect Sanguisorbeæ; the Myrtal type, highly developed in Myrtus, and almost obliterated in Streamworts (Halorageæ); the Urtical, in excess in Artocarpus, and quite imperfect in Ceratophyllum; Grasses, presenting the most striking differences of perfection between the moss-like Knappia, and Bamboos a hundred feet high; and the Lilial in equally different states of development, when Asparagus is compared with the Dragon-tree, or an autumnal squill with an arborescent Yucca. So, in like manner, we find at one extreme of the organisation of the class of Endogens, Palms, Plantains, and arborescent Liliaceous species, and at the other, such submersed plants as Potamogeton, Zannichellia, and Duckweed, the latter of which has not even the distinction of leaf and stem, and bears its flowers, reduced to one carpel and two stamens, without either calyx or corolla—and therefore at the minimum of reduction, if to remain flowers at all—in little chinks in its edges.

The classification of Endogens is not a subject upon which there is any very great diversity of opinion among botanists. If the natural orders are sometimes not distinctly limited, they are, upon the whole, grouped much better than those of Exogens; and although it may be expected that some changes have still to be introduced into this part of systematic botany, yet there seems no probability of the limits of the natural orders themselves being disturbed to any considerable extent.

The principles of classification here adopted are the following:—

In the first place, all those numerous species whose flowers are like grasses are stationed by themselves, and constitute the Glumal alliance. They are not perhaps so close upon flowerless plants as some hereafter to be mentioned, but they form, as a whole, the lowest condition of structure to which a great mass of Endogens is reduced. Their flowers may be regarded as made up of scales, analogous to bracts, without any thing that can be strictly called either calyx or corolla being ever present. These have in many instances the sexes separated; but their glumaceous structure overrules this peculiarity.

Next to them seem to be stationed Bulrushes; plants with scales too for their floral envelopes, but arranged in rings, and so falling within the definition of at least a calyx. Their sexes are disunited, and that important circumstance associates them with Palms, Arums, and other arborescent tropical plants, together with a small group of water plants, or Hydrals. This separation of the sexes appears to be a mark of very great importance, when it is complete; and must not be confounded with another kind of separation, in which flowers of one sex have the other sex present in an imperfect condition, and often become actually hermaphrodite. All such cases, although set down in books as monœcious or diœcious, are by no means diclinous, and are excluded from the division containing the Aral Alliance, with the exception of Palms, in which flowers are occasionally altogether hermaphrodite, and which, therefore, form a real exception to the prevailing character of this part of the classification.

The remainder of Endogens are typically hermaphrodite, the number of exceptions to that character being very few. One division of them has the ovary adherent to the calyx and corolla, the other has that organ free, a portion of the Narcissal Alliance having both characteristics. The line of orders thus associated is closed by the Alismal Alliance, some of whose species are almost exogenous as has been already mentioned, while others, being truly diclinous, carry the circle of affinity back to the Hydral Alliance.

Alliances of Endogens.

I. Glumales.—*Flowers glumaceous; (that is to say, composed of bracts not collected in true whorls, but consisting of imbricated colourless or herbaceous scales).*

II. *Flowers petaloid, or furnished with a true calyx or corolla, or with both, or absolutely naked;* ♂ ♀ *(that is, having sexes altogether in different flowers, without half-formed rudiments of the absent sexes being present).*

Arales.—*Flowers naked or consisting of scales, 2 or 3 together, or numerous, and then sessile on a simple naked spadix; embryo axile; albumen mealy or fleshy. (Some have no albumen.)*

Palmales.—*Flowers perfect (with both calyx and corolla), sessile on a branched scaly spadix; embryo vague, solid; albumen horny or fleshy. Some Palms are* ☿.

Hydrales.—*Flowers perfect or imperfect, usually scattered; embryo axile, without albumen—aquatics. (Some are* ☿.)

III.—*Flowers furnished with a true calyx, and corolla, adherent to the ovary;* ☿.

Narcissales.—*Flowers symmetrical; stamens 3 or 6, or more, all perfect; seeds with albumen; flowers unsymmetrical. (Some Bromeliaceæ have a free calyx and corolla.)*

AMOMALES.—*Flowers unsymmetrical; stamens* 1 *to* 5, *some at least of which are petaloid; seeds with albumen.*

ORCHIDALES.—*Stamens* 1 *to* 3; *seeds without albumen.*

IV. *Flowers furnished with a true calyx and corolla, free from the ovary;* ⚥.

XYRIDALES.—*Flowers half herbaceous,* 2-3-*petaloideous; albumen copious.*

JUNCALES.—*Flowers herbaceous, dry, and permanent, scarious if coloured; albumen copious.* (*Some Callas have no albumen.*)

LILIALES.—*Flowers hexapetaloideous, succulent, and withering; albumen copious.*

ALISMALES.—*Flowers* 3-6-*petaloideous, apocarpal; albumen none.* (*Some Alismaceæ are absolutely* ♀ ♂.)

ALLIANCE VII. *GLUMALES.*—THE GLUMAL ALLIANCE.

DIAGNOSIS.—*Glumaceous Endogens.*

The great mass of herbage known by the name of Sedges and Grasses, constitutes perhaps a twelfth part of the described species of flowering plants, and at least nine-tenths of the number of individuals composing the vegetation of the world; for it is the chief source of that verdure which covers the earth of northern countries with a gay carpet during the months of winter. Such forms of vegetation are provided by nature with true flowers, that is to say, with stamens and pistils, the action of the former of which upon the latter is indispensable for the creation of a seed; but there is little trace of the calyx and corolla, which are commonly characteristic of the more perfect races of plants; not that floral envelopes are wanting, but they do not assume the whorled or ringed position of the parts which form a calyx and corolla; they merely consist of minute green or brown bracts placed one over the other, and sometimes appearing to be united by their edges. There is also great simplicity in their pistil, but one ovule being formed in each cavity, whatever number of carpels (indicated by the stigmas) may be employed in the construction of it. Their foliage is as simple as it can be to have any considerable degree of development, consisting of fine thread-shaped veins running side by side from one end of the leaf to the other.

It is usual to restrict the term glumaceous to Grasses and Sedges; but there seems no intelligible reason why the Cordleafs (Restiaceæ,) Pipeworts (Eriocaulaceæ,) and Bristleworts (Desvauxiaceæ,) should be omitted, for they have precisely the same habit and the same substitution of imbricated scales for calyx and corolla. It is only among the Pipeworts that we have the beginning of a calyx, in the form of a membranous tube surrounding the ovary. They do not, however, indicate a more complex condition; rather less so indeed than in Grasses and Sedges; for their pistils are perfectly simple, while those of the latter are invariably formed by the coalition of at least 2 carpellary leaves for each cavity of the ovary.

Two divisions may be formed among the orders, viz.:—

1. Ovule erect or ascending; pistil compound.—*Graminaceæ* and *Cyperaceæ.*
2. Ovules pendulous; pistil simple.—*Desvauxiaceæ, Restiaceæ, Eriocaulaceæ.*

The first set touch Palms, the latter Rushes; the whole, in consequence of their spiked-inflorescence, scaly floral envelopes, and great tendency to a separation of the sexes, pass naturally into Bulrush worts (Typhaceæ).

NATURAL ORDERS OF GLUMALS.

Ovar. 1-*celled, with* 2 *or more distinct* (*or united*) *styles; ovule ascending; embryo lateral, naked*	29. GRAMINACEÆ.
Ovar. 1-*celled, with* 2 *or more* (*distinct or*) *united styles; ovule erect, embryo basal*	30. CYPERACEÆ.
Ovar. several (*sometimes united*) *with* 1 *style to each; ovule pendulous; glumes only; st.* 1-2; *anth.* 1-*celled; embryo terminal*	31. DESVAUXIACEÆ.
Ovar. 1-2-3-*celled, with* 2 *or* 3 *styles always; ovule pendulous; glumes only; st.* 2-3; *anth.* 1-*celled; embryo terminal* . . .	32. RESTIACEÆ.
Ovar. 2-3-*celled, with* 1 *style to each cell; ovule pendulous; a membranous* 3-*lobed cup within the glumes; anthers* 2-*celled; embryo terminal*	33. ERIOCAULACEÆ.

ORDER XXIX. GRAMINACEÆ.—GRASSES.

Gramina, *Juss. Gen.* 28. (1789).—Gramineæ, *R. Brown Prodr.* 168. (1810); *Palisot de Beauv. Agrostol.*; *Kunth in Mem. Mus.* 2. 62; *Id. in N. G. et Sp. Humb. et Bonpl.* 1. 84; *Turpin in Mém. Mus.* 5. 426; *Trinius Fundam. Agrostol.*; *Dumortier Agrost. Belg.*; *Trinius Diss. de Gram. Unifl. et Sesquif.*; *De la Harpe in Ann. Sc.* 5. 335. 6. 21; *Raspail in Ann. des Sc.* 4. 271. 422. 5. 287. 433. 6. 224. 384. 7. 335; *Nees v. Esenbeck Agrostol. Brasil.*; *Kunth Enum. pl. vol.* 1 *et* 2; *Endl. Gen.* xlii. *Meisner*, p. 414.

DIAGNOSIS.—*Glumal Endogens, with split-sheathed leaves, a one-celled ovary, and a lateral naked embryo.*

Evergreen herbs, occasionally having stems of considerable size and living for many years. Rhizoma, fibrous or bulbous. Stem cylindrical, usually fistular and closed at

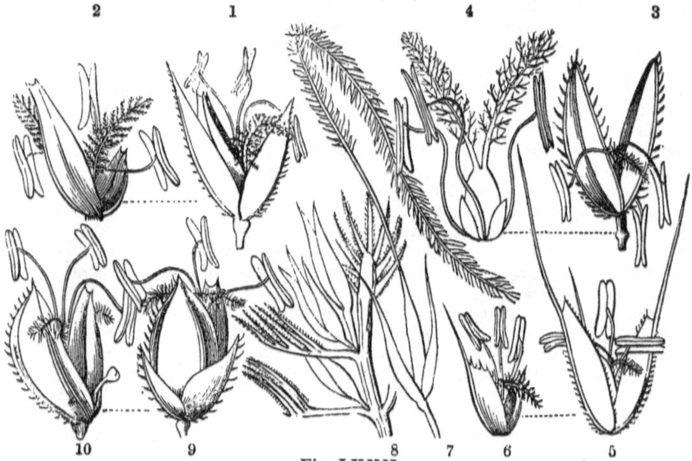

Fig. LXXII.

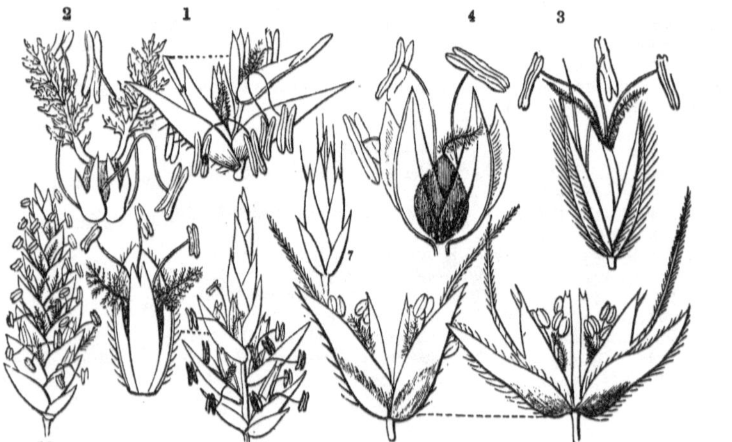

Fig. LXXIII.

Fig. LXXII.—1. Locusta of Agrostis alba; 2. paleæ and stamens, &c. of the same; 3. paleæ of Leersia oryzoides; 4. pistil, stamens, and hypogynous scales of the same; 5. locusta of Polypogon monspeliensis; 6. paleæ, &c. of the same; 7. locusta of Stipa pennata; 8. rachis, bracteæ, and florets of Cynosurus cristatus; 9. locusta of Cynodon dactylon; 10. paleæ, and abortive floret of the same.

Fig. LXXIII.—1. Locusta of Corynephorus canescens; 2. paleæ, &c. of the same; 3. locusta of Phalaris aquatica; 4. locusta of Alopecurus pratensis; 5. locusta of Aira caryophyllea; 6. floret of the same; 7. locusta of Festuca duriuscula; 8. locusta of Glyceria fluitans; 9. floret of the same; 10. locusta of Eragrostis poæformis.

the joints, covered with a coat of silex, sometimes solid. Leaves narrow and undivided, alternate, with a split sheath, and a membranous expansion (ligula) at the junction of stalk and blade. Flowers green in little spikes called locustæ, arranged in a spiked, racemed or panicled manner. Flowers usually ☿, sometimes monœcious or polygamous; consisting of imbricated bracts, of which the most exterior are called glumes, the interior immediately enclosing the stamens paleæ, and the innermost at the base of the ovary scales. Glumes usually 2, alternate; sometimes single, most commonly unequal. Paleæ 2, alternate; the lower or exterior simple, the upper or interior composed of 2 united by their contiguous margins, and usually with 2 keels, together forming a kind of dislocated calyx. Scales 2 or 3, sometimes wanting; if 2, collateral, alternate with the paleæ, and next the lower of them; either distinct or united. Stamens hypogynous, 1, 2, 3, 4, 6, or more, 1 of which alternates with the 2 hypogynous scales, and is therefore next the lower paleæ; anthers versatile. Ovary simple; styles 2 or 3, very rarely combined into one; stigmas feathery or hairy; ovule ascending by a broad base, anatropal. Pericarp usually undistinguishable from the seed, membranous. Albumen farinaceous; embryo lying on one side of the albumen at the base, lenticular, with a broad cotyledon and a developed plumula; and occasionally, but very rarely, with a second cotyledon on the outside of the plumula, and alternate with the usual cotyledon.

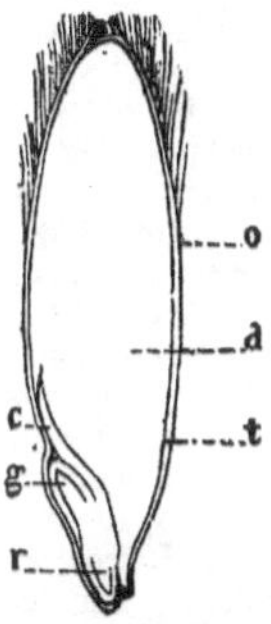

Fig. LXXV.

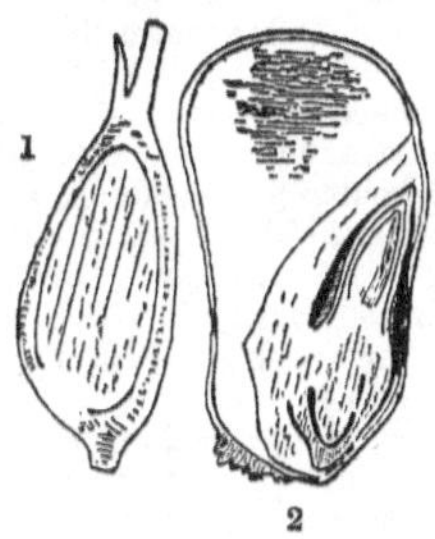

Fig. LXXIV.

This most important Order offers great singularities in its organisation, although it is one in which, formerly, botanists the least suspected anomalies to exist. They found calyx and corolla and nectaries here with the same facility as they found them in a Ranunculus; and yet such organs exist in no one genus of Grasses. Their so-called flowers consist of green scales, not placed in whorls, but arranged one above the other, and are undoubtedly constructed of bracts alone. Not a trace is discoverable among them of calyx or corolla, properly so called, unless certain scales usually present, next the ovary, are to be so considered. Brown's account of their construction is still the best that has been published. He says,—

"The natural or most common structure of Gramineæ is to have their sexual organs surrounded by the floral envelopes, each of which usually consist of two distinct valves; but both of these envelopes are, in many genera of the order, subject to various degrees of imperfection or even suppression of their parts. The outer envelope, or gluma of Jussieu, in most cases containing several flowers with distinct and often distant insertions on a common receptacle, can only be considered as analogous to the bracteæ or involucrum of other plants. The tendency to suppression in this envelope appears to be greater in the exterior or lower valve; so that a gluma consisting of one valve may, in all cases, be considered as deprived of its outer or inferior valve. In certain genera with a simple spike, as Lolium and Lepturus, this is clearly proved by the structure of the terminal flower or spicula, which retains the natural number of parts; and in other genera not admitting of this direct proof, the fact is established by a series of species showing its gradual obliteration, as in those species of Panicum which connect that genus with Paspalum. On the other hand, in the inner envelope, or calyx of Jussieu, obliteration first takes place in the inner or upper valve; but this valve having, instead of one central nerve, two nerves equidistant from its axis, I consider it as composed of two confluent valves, analogous to what takes place in the calyx and corolla of many irregular flowers of other classes; and this confluence may be regarded as the first step towards its obliteration, which is complete in many species of Panicum, in Andropogon, Pappophorum, Alopecurus, Trichodium, and several other genera. With respect to the nature of this inner or proper envelope of Grasses, it may be observed, that the view of its structure now given, in reducing its parts to the usual ternary division of Monocotyledons, affords an additional argument for considering it as the real perianthium. This argument, however, is not conclusive, for a similar confluence takes place between the two inner lateral bracteæ of the greater part of Irideæ; and with these, in the relative insertion of its valves, the proper envelope of Grasses may be supposed much better to accord than with a genuine perianthium. If, therefore, this inner envelope of Grasses

Fig. LXXIV.—1. Section of ovary of Penicillaria spicata; 2. section of grain of rice, showing the lateral embryo.
Fig. LXXV.—Section of an Oat; *o* ovary, *t* testa, *a* albumen, *r* radicle, *g* plumule, *c* cotyledon.

be regarded as consisting merely of bracteæ, the real perianthium of the order must be looked for in those minute scales, which, in the greater part of its genera, are found immediately surrounding the sexual organs. The scales are, in most cases, only two in number, and placed collaterally within the inferior valve of the proper envelope. In their real insertion, however, they alternate with the valves of this envelope, as is obviously the case in Ehrharta and certain other genera ; and their collateral approximation may be considered as a tendency to that confluence which uniformly exists in the parts composing the upper valve of the proper envelope, and which takes place also between these two squamæ themselves, in some genera, as Glyceria and Melica. In certain other genera, as Bambusa and Stipa, a third squamula exists, which is placed opposite to the axis of the upper valve of the proper envelope, or, to speak in conformity with the view already taken of the structure of this valve, opposite to the conjunction of its two component parts. With these squamæ the stamina in triandrous Grasses alternate, and they are consequently opposite to the parts of the proper envelope ; that is, one stamen is opposed to the axis of its lower or outer valve, and the two others are placed opposite to the two nerves of the upper valve. Hence, if the inner envelope be considered as consisting of bracteæ, and the hypogynous squamæ as forming the perianthium, it seems to follow, from the relation these parts have to the axis of inflorescence, that the outer series of this perianthium is wanting, while its corresponding stamina exist, and that the whole or part of the inner series is produced while its corresponding stamina are generally wanting. This may, no doubt, actually be the case ; but as it would be, at least, contrary to every analogy in Monocotyledonous plants, it becomes in a certain degree probable that the inner or proper envelope of Grasses, the calyx of Jussieu, notwithstanding the obliquity in the insertion of its valves, forms in reality the outer series of the true perianthium, whose inner series consists of the minute scales, never more than three in number, and in which an irregularity in some degree analogous to that of the outer series generally exists. It is necessary to be aware of the tendency to suppression existing, as it were, in opposite directions in the two floral envelopes of Grasses, to comprehend the real structure of many irregular genera of the order, and also to understand the limits of the two great tribes into which I have proposed to subdivide it. One of these tribes, which may be called Paniceæ, comprehends Ischæmum, Holcus, Andropogon, Anthistiria, Saccharum, Cenchrus, Isachne, Panicum, Paspalum, Reimaria, Anthenantia, Monachne, Lappago, and several other nearly related genera ; and its essential character consists in having always a locusta of two flowers, of which the lower or outer is uniformly imperfect, being either male or neuter, and then not unfrequently reduced to a single valve. Ischæmum and Isachne are examples of this tribe in its most perfect form, from which Anthenantia, Paspalum, and Reimaria, most remarkably deviate, in consequence of the suppression of certain parts : thus Anthenantia (which is not correctly described by Palisot de Beauvois) differs from those species of Panicum that have the lower flower neuter and bivalvular, in being deprived of the outer valve of its gluma ; Paspalum differs from Anthenantia in the want of the inner valve of its neuter flower, and from those species of Panicum whose outer flower is univalvular, in the want of the outer valve of its gluma ; and Reimaria differs from Paspalum in being entirely deprived of its gluma. That this is the real structure of these genera may be proved by a series of species connecting them with each other, and Panicum with Paspalum. The second tribe, which may be called Poaceæ, is more numerous than Paniceæ, and comprehends the greater part of the European genera, as well as certain less extensive genera peculiar to the equinoctial countries ; it extends also to the highest latitudes in which Phænogamous plants have been found ; but its maximum appears to be in the temperate climates, considerably beyond the tropics. The locusta in this tribe may consist of 1, 2, or of many flowers ; and the 2-flowered genera are distinguished from Paniceæ by the outer or lower flower being always perfect, the tendency to imperfection in the locusta existing in opposite directions in the two tribes. In conformity with this tendency in Poaceæ, the outer valve of the perianthium in the single-flowered genera is placed within that of the gluma, and in the many-flowered locusta the upper flowers are frequently imperfect. There are, however, some exceptions to this order of suppression, especially in Arundo Phragmites, Campulosus, and some other genera, in which the outer flower is also imperfect : but as all of these have more than two flowers in their locusta, they are still readily distinguished from Paniceæ." *Brown in Flinders*, 580.

According to this view, in a locusta of several florets, the scales at its base, or glumes, are bracts, and each floret consists of a calyx formed of one sepal remote from the rachis, and two cohering by their margins and next the rachis ; the little hypogynous scales are the rudiments of two petals, and the stamens alternate with these in the normal manner. This may be rendered more clear by the following diagram, in which the triangle A B B represent the outer series, or paleæ, or calyx, A being the inferior valve, and B B the superior, formed of two sepals united by their contiguous margin at *x*. If the triangle C D D be understood to represent the next series, the position of the parts will be at the three angles; and in reality the two scales that are usually developed do occupy the places D D; while the third, whenever it is superadded, is stationed at C. The triangle E E F indicates by its angles the normal position of the first series of stamens, which are actually so situated, the stamen F which is opposite the sepal A alternating with the rudimentary petals D D. The objection to this is, that the parts of the supposed calyx or paleæ are not inserted upon the same plane, or truly verticillate, and consequently do not answer exactly to what is required in a floral envelope; and it is on this account that Turpin rejects Brown's opinion, giving the paleæ the name of spathelle, and considering them bracts of a second order. Kunth entertains a somewhat different view of the nature of the floral envelopes, considering the hypogynous scales to be analogous to the ligula, and the normal state of Grasses to be hexandrous. See *Enumeratio*, vol. i. p. 3, 4.

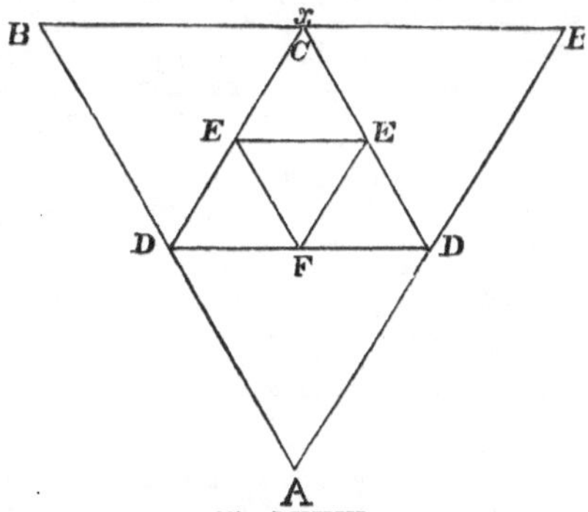

Fig. LXXVI.

Raspail, in a memoir upon the structure of Grasses, hazards a theory, that the midrib of the bracts of Grasses is an axis of development in cohesion with the bracts, and that when it separates, as in Phleum, Bromus, or Corynephorus, it is attempting to revert to the functions of ulterior development, for which it is more especially destined. Among other things, he states (*Ann. des Sc.* 4. 276. E) that he should not be surprised one day to find some Grass in which the midrib of the lower palea actually became a new axis bearing other florets. I mention this for the sake of remarking that such a case is known, without however admitting that it is any confirmation of Raspail's views, which are at variance with the laws of vegetable development, for reasons which are so obvious, as to render it altogether unnecessary to give them here. I have a monstrous Barley, the Hordeum Ægiceras of Royle, cultivated as Wheat in the Himalayeh mountains, specimens of which I communicated in 1830 to M. Kunth and others, in which the midrib of the lower palea actually becomes saccate towards the apex, bearing an imperfect floret, with stamens, ovary, and hypogynous scales in its cavity. The well-known tendency to a special development of buds in the margins of certain leaves, in Ferns, and according to the observations of Turpin, in the whole substance of certain monocotyledonous leaves, leaves nothing in this fact to excite surprise or to give rise to new theories; but it is worth mentioning as the only instance upon record of a flower-bud with sexual apparatus being developed under such circumstances.

The embryo is here described in conformity with the views that are most commonly taken of its nature; that is to say, it is considered to consist of a dilated lenticular cotyledon applied to the albumen on one side, and bearing a naked plumule on the other side, next the testa. It is proper, however, to remark, that the opinion of the late L. C. Richard, that the part commonly called cotyledon is a peculiar process, and that the plumule is a body contained within the apparent plumule, has been adopted by Nees v. Esenbeck, in his Agrostologia Brasiliensis, but with some difference. Richard considered the cotyledon to be a part of the radicle, to which he gave the name of macropodal, in consequence of its great supposed enlargement in Grasses and some other families; Nees v. Esenbeck, on the contrary, seems to entertain the opinion that this cotyledon is a special organ, for which he retains Richard's name of hypoblastus, although he does not adopt the view that botanist took of its nature. But I think if we consider the improbability of any special organ being provided for Grasses, which is not found elsewhere, and if we consider how nearly alike are the embryos of Grasses and certain Arumworts, in which the plumule lies within a cleft of the cotyledon, it is impossible to doubt the identity of the hypoblastus of Richard and Nees v. Esenbeck, and the cotyledon of other Monocotyledons. Indeed, the latter himself appears, in one place, to hesitate about the accuracy of distinguishing them, when he says (p. 9), "Tum vero hypoblastus pars quædam habenda est cotyledoni analoga, magisque ad interiora seminis quam ad externam corculi evolutionem spectans."

In some Grasses a portion of the inflorescence assumes a nearly bony texture. "This change takes place in Coix, in the involucre; in Chionachne and Sclerachne, in the outer valve of the glume of the female locusta; and in Tripsacum, in the rachis of the spike." *Bennett in Horsfield's Plantæ Javanicæ*, p. 19; where the systematic reader will find some curious and important details relating to the structure and affinities of the genera of Grasses.

The stem of Grasses seems to be so much at variance in structure with that of other Endogens, as to have led Agardh to remark, that it is the least monocotyledonous of all Monocotyledonous plants. It is probable, however, that its peculiarity does not depend so much upon any specific deviation from the ordinary laws of growth, as upon a separation of the parts at an early period of their growth. The stem of a Grass, it must be remembered, exists in two different states,—that of the rhizome, and of the straw: the rhizome, which is the true trunk; and the straw, which may be considered a ramification of it. The rhizome grows slowly, and differs in no respect from the stem of other Monocotyledons, as is evident in that of the Bamboo. The straw, on the contrary, which grows with great rapidity, is fistular, with a compact impervious diaphragm at each articulation; a fact which must be familiar to every one who has examined corn, or the joint of a Bamboo. In the beginning, when this straw was first developed, it was a solid body like the rhizome, only infinitely smaller; but in consequence of the great rapidity of its development, the cellular tissue formed more slowly than the woody vascular bundles which it connects, and in consequence a separation takes place between the latter and the former, except at the articulations, where, by the action of the leaves, and their axillary buds, is formed a plexus of vessels, which, growing as rapidly as the straw, distends, and therefore never separates in the centre. Something analogous to this occurs in the flowering stem of the common Onion among Monocotyledons, and in Umbelliferæ among Dicotyledons. The stem of Grasses is not, however, always hollow; in the Sugar Cane it is solid, as in common Endogens.

The relation that exists between Palms and Grasses will be adverted to in speaking of the former order: Nees v. Esenbeck considers Grasses to be a sort of Palms of a lower grade. In reality, the habit of the genera Calamus and Bambusa is nearly alike; the inflorescence of Grasses may be considered to be the same as that of Palms, the floral envelopes of the latter taken away, and only their bracts remaining; and, finally, the leaves are formed upon exactly the same plan, with this difference only, that those of Grasses are undivided. With Sedges, however, it is that Grasses are most properly to be compared. While a manifest tendency, at least to the degree of verticillation requisite to constitute a calyx, evidently takes place in the paleæ of Grasses, Sedges are destitute of all trace of such a tendency, unless the opposite connate glumes of the female flowers of Carex, or the hypogynous scales of certain Schœni and others, be considered an approach to the production of a perianth. For this reason, Grasses may be considered plants in a higher state of evolution than Sedges. Independently of this difference, the orders are usually known by the stems of Grasses being hollow, those of Sedges solid; the leaves of Grasses having a ligula at the apex of their sheath, which is split, while the sheath of Sedges is not split, and is destitute of this ligula; and, finally, the embryo of Grasses is external, lateral, and with a naked plumule, while that of Sedges is undivided and enclosed within the base of the albumen.

As nothing can be uninteresting which is connected with the habits of a tribe of such vast importance to man, I extract the following account of the geographical distribution of Grasses by Schouw, from Jameson's *Philosophical Journal for April*, 1825:—" The family is very numerous: Persoon's *Synopsis* contains 812 species, 1-26th part of all the plants therein enumerated. In the system of Rœmer and Schultes there are 1800; and, since this work, were it brought to a conclusion, would probably contain 40,000 in all, it may be assumed that the Grasses form a 22nd part. It is more than probable, however, that in future the Grasses will increase in a larger ratio than the other phanerogamic plants, and that perhaps the just proportion will be as 1 to 20, or as 1 to 16. Greater still will be their proportion to vegetation in general, when the number of individuals is taken into account; for, in this respect, the greater number, nay perhaps the whole of the other classes, are inferior. With regard to locality in such a large family, very little can be advanced. Among the Grasses there are both land and water, but no marine, plants. They occur in every soil, in society with others, and alone; the last to such a degree as entirely to occupy considerable districts. Sand appears to be less favourable to this class; but even this has species nearly peculiar to itself. The diffusion of this family has almost no other limits than those of the whole vegetable kingdom. Grasses occur under the equator; and Agrostis algida was one of the few plants which Phipps met with on Spitzbergen. On the mountains of the south of Europe, Poa disticha and other Grasses ascend almost to the snow-line; and, on the Andes, this is also the case with Poa malulensis and dactyloides, Deyeuxia rigida and Festuca dasyantha.

" The greatest differences between tropical and extra-tropical Grasses appear to be the following:—1. The tropical Grasses acquire a much greater height, and occasionally assume the appearance of trees. Some species of Bambusa are from 50 to 60 feet high. 2. The leaves of the tropical Grasses are broader, and approach more in form to those of other families of plants. Of this the genus Paspalus affords many examples. 3. Separate sexes are more frequent in the tropical Grasses. Zea, Sorghum, Andropogon, Olyra, Anthistiria, Ischæmum, Ægilops, and many other genera, which only occur in the torrid zone, and are there found in perfection, are monœcious, or polygamous. Holcus is perhaps the only extra-tropical genus with separate sexes. 4. The flowers are softer, more downy, and elegant. 5. The extra-tropical Grasses, on the contrary, far surpass the tropical in respect of the number of individuals. That compact grassy turf, which, especially in the colder parts of the temperate zones, in spring and summer, composes the green meadows and pastures, is almost entirely wanting in the torrid zone. The Grasses there do not grow crowded together, but, like other plants, more dispersed. Even in the southern parts of Europe, the assimilation to the warmer regions, in this respect, is by no means inconsiderable. Arundo donax, by its height, reminds us of the Bamboo ; Saccharum Ravennæ, S. Teneriffæ, Imperata arundinacea, Lagurus ovatus, Lygeum spartum, and the species of Andropogon, Ægilops, &c. by separate sexes, exhibit tropical qualities. The Grasses are also less gregarious, and meadows seldomer occur, in the south than in the north of Europe. The generality are social plants.

"The distribution of cultivated Grasses is one of the most interesting of all subjects. It is determined, not merely by climate, but depends on the civilisation, industry, and traffic of the people, and often on historical events. Within the northern polar circle, agriculture is found only in a few places. In Siberia grain reaches at the utmost only to 60°, in the eastern parts scarcely above 55°, and in Kamtschatka there is no agriculture even in the most southern parts (51°). The polar limit of agriculture on the North-west coast of America appears to be somewhat higher ; for, in the more southern Russian possessions (57° to 52°), barley and rye come to maturity. On the east coast of America it is scarcely above 50° to 52°. Only in Europe, namely, in Lapland, does the polar limit reach an unusually high latitude (70°). Beyond this, dried fish, and here and there potatoes, supply the place of grain. The grains which extend farthest to the north in Europe are barley and oats. These, which in the milder climates are not used for bread, afford to the inhabitants of the northern parts of Norway and Sweden, of a part of Siberia and Scotland, their chief vegetable nourishment. Rye is the next which becomes associated with these. This is the prevailing grain in a great part of the northern temperate zone, namely, in the south of Sweden and Norway, Denmark, and in all the lands bordering on the Baltic ; the north of Germany, and part of Siberia. In the latter another very nutritious grain, buck-wheat, is very frequently cultivated. In the zone where rye prevails, wheat is generally to be found ; barley being here chiefly cultivated for the manufacture of beer, and oats supplying food for the horses. To these there follows a zone in Europe and western Asia, where rye disappears, and wheat almost exclusively furnishes bread. The middle, or the south of France, England, part of Scotland, a part of Germany, Hungary, the Crimea and Caucasus, as also the lands of middle Asia, where agriculture is followed, belong to this zone. Here the vine is also found ; wine supplants the use of beer ; and barley is consequently less raised. Next comes a district where wheat still abounds, but no longer exclusively furnishes bread, rice and maize becoming frequent. To this zone belong Portugal, Spain, part of France on the Mediterranean, Italy, and Greece ; further, the countries of the East, Persia, northern India, Arabia, Egypt, Nubia, Barbary, and the Canary Islands ; in these latter countries, however, the culture of maize or rice towards the south, is always more considerable, and in some of them several kinds of sorghum (doura) and Poa Abyssinica come to be added. In both these regions of wheat, rye only occurs at a considerable elevation ; oats, however, more seldom, and at last entirely disappear ; barley affording food for horses and mules. In the eastern parts of the temperate zone of the Old Continent, in China and Japan, our northern kinds of grain are very unfrequent, and rice is found to predominate. The cause of this difference between the east and the west of the Old Continent appears to be in the manners and peculiarities of the people. In North America, wheat and rye grow as in Europe, but more sparingly. Maize is more reared in the Western than in the Old Continent, and rice predominates in the southern provinces of the United States. In the torrid zone, maize predominates in America, rice in Asia, and both these grains in nearly equal quantity in Africa. The cause of this distribution is, without doubt, historical ; for Asia is the native country of rice, and America of maize. In some situations, especially in the neighbourhood of the tropics, wheat is also met with, but always subordinate to these other kinds of grain. Besides rice and maize, there are, in the torrid zone, several kinds of grain, as well as other plants, which supply the inhabitants with food, either used along with them, or entirely occupying their place. Such are, in the New Continent, yams (Dioscorea alata), the manihot (Jatropha manihot), and the batatas

(Convolvulus batatas), the root of which, and the fruit of the pisang (Banana Musa), furnish universal articles of food. In the same zone, in Africa, doura (sorghum), pisang, manihot, yams, and Arachis hypogæa. In the East Indies, and on the Indian Islands, Eleusine coracana, E. stricta, Panicum frumentaceum; several palms and Cycadeæ, which produce the sago; pisang, yams, batatas, and the breadfruit (Artocarpus incisa). In the islands of the South Sea, grain of every kind disappears, its place being supplied by the bread-fruit tree, the pisang, and tacca pinnatifida. In the tropical parts of New Holland there is no agriculture, the inhabitants living on the produce of the sago, of various palms, and some species of Arum. In the high lands of South America there is a distribution similar to that of the degrees of latitude. Maize, indeed, grows to the height of 7200 feet above the level of the sea, but only predominates between 3000 and 6000 of elevation. Below 3000 feet it is associated with the pisang, and the above-mentioned vegetables; while, from 6000 to 9260 feet, the European grains abound: wheat in the lower regions, and rye and barley in the higher; along with which Chenopodium Quinoa, as a nutritious plant, must also be enumerated. Potatoes alone are cultivated from 9260 to 12,300 feet. To the south of the tropic of Capricorn, wherever agriculture is practised, considerable resemblance with the northern temperate zone may be observed. In the southern parts of Brazil, in Buenos Ayres, in Chile, at the Cape of Good Hope, and in the temperate zone of New Holland, wheat predominates; barley, however, and rye, make their appearance in the southernmost parts of these countries, and in Van Diemen's Land. In New Zealand the culture of wheat is said to have been tried with success; but the inhabitants avail themselves of the Acrostichum furcatum as the main article of sustenance. Hence it appears, that, in respect of the predominating kinds of grain, the earth may be divided into five grand divisions, or kingdoms. The kingdom of rice, of maize, of wheat, of rye, and lastly of barley and oats. The first three are the most extensive; the maize has the greatest range of temperature; but rice may be said to support the greatest number of the human race."

It is a very remarkable circumstance, that the native country of wheat, oats, barley, and rye, should be entirely unknown; for although oats and barley were found by Col. Chesney apparently wild on the banks of the Euphrates, it is doubtful whether they were not the remains of cultivation. This has led to an opinion, on the part of some persons, that all our cereal plants are artificial productions, obtained accidentally, but retaining their habits, which have become fixed in the course of ages. This curious subject has been discussed in the *Gardeners' Chronicle* for 1844, p. 555, 779, &c., whither the reader is referred for further information.

Fig. LXXVII.—Setaria Germanica.

The uses of this most important tribe of plants, for fodder, food, and clothing, require little illustration. The abundance of wholesome fæcula contained in their seeds renders them peculiarly well adapted for the sustenance of man; and if the Cereal Grasses only, such as Wheat, Barley, Rye, Oats, Maize, Rice, and Guinea Corn, are the kinds commonly employed, it is because of the large size of their grain compared with that of other Grasses; for none are unwholesome in their natural state, with the exception of Lolium temulentum, a common weed in many parts of England, the effects of which are undoubtedly deleterious, although perhaps exaggerated; of Bromus purgans and catharticus, said to be emetic and purgative; of Bromus mollis, reported to be unwholesome, and of Festuca quadridentata, which is said to be poisonous in Quito, where it is called Pigonil. To these must be added Molinia varia, injurious to cattle, according to Endlicher; and a variety of Paspalum scrobiculatum, called Hureek in India, (*Graham's Bombay Plants*, p. 234), which is perhaps the Ghohona Grass, a reputed Indian poisonous species, said to render the milk of cows that graze upon it narcotic and drastic. (*Madras Journal*, 1837, p. 107). It is however uncertain how far the injurious action of some of these may be owing to mechanical causes, which, in the case of the species of Calamagrostis and Stipa seem to be the cause of mischief in consequence of their roughness and bristles. In their qualities the poisonous species seem to approach the properties of putrid Wheat, which is known to be dangerous.

Among corn plants less generally known may be mentioned Eleusine coracana, called Natchnee, on the Coromandel coast, and Nagla Ragee, or Mand, elsewhere in India; Phalaris canariensis, which yields the canary seed; Zizania aquatica or Canada Rice; Paspalum scrobiculatum, the Menya or Kodro of India, a cheap grain, regarded as unwholesome; Setaria germanica, yielding German millet; Panicum frumentaceum, called Shamoola, in the Deccan; Setaria italica, cultivated in India under the name of Kala kangnee or Kora kang; Panicum miliaceum, a grain called Warree in India; and P. pilosum, called Bhadlee. Penicillaria spicata or Bajree; Andropogon Sorghum or Durra, Doora, Jowaree or Jondla; and Andropogon saccharatus or Shaloo, are also grown in India for their grain. A kind of fine-grained corn, called, on the west of Africa, Fundi or Fundungi, is produced by Paspalum exile; and finally, both the Teff and Tocusso, Abyssinian corn plants, are species of this order; the former Poa abyssinica, the latter Eleusine Tocusso, (*Linnæa*, 1839). Even Stipa pennata is said to produce a flour much like that of Rice.

The value of Grasses as fodder for cattle is hardly second to that of their corn for human food. The best fodder Grasses of Europe are usually dwarf species, or at least such as do not rise more than 3 or 4 feet above the ground, and of these the larger kinds are apt to become hard and wiry; the most esteemed are Lolium perenne, Phleum and Festuca pratensis, Cynosurus cristatus, and various species of Poa and dwarf Festuca, to which should be added Anthoxanthum odoratum for its fragrance. But the fodder Grasses of Brazil are of a far more gigantic stature, and perfectly tender and delicate. We learn from Nees von Esenbeck, that the Caapim de Angola of Brazil, Panicum spectabile, grows 6 or 7 feet high: while other equally gigantic species constitute the field crops on the banks of the Amazons. In New Holland the favourite is the Anthistiria australis or Kangaroo Grass; in India the A. ciliata is also in request. But the most common Indian fodder Grass appears to be Doorba, Doorwa, or Hurryalee, Cynodon Dactylon. Gama Grass, Tripsacum dactyloides, has a great reputation as fodder in Mexico; and attention has lately been directed to the Tussac-grass of the Falklands, Festuca flabellata, a species forming tufts 5 or 6 feet high, and said to be unrivalled for its excellence as food for cattle and horses. (See *Gardener's Chronicle*, 1843, p. 131).

The fragrance of our sweet Vernal Grass (Anthoxanthum), is by no means confined to it. Other species are Hierochloe borealis, Ataxia Horsfieldii, and some Andropogons; their odour is said to be owing to the presence of benzoic acid. The most famous species are Andropogon Iwarancusa and Schœnanthus, the latter the Lemon Grass of English gardens; A. Calamus aromaticus, which Dr. Royle considers the plant of that name described by Dioscorides, and the "sweet cane" and "rich aromatic reed from a far country" of Scripture; and the Anatherum muricatum, called Vetiver by the French, and Khus in India, where its fragrant roots are employed in making tatties, covers for palanquins, &c.

This fragrance is connected with aromatic secretions which have in part recommended Grasses to the notice of medical practitioners. The last mentioned plant (Anatherum muricatum), is said to be acrid, aromatic, stimulating, and diaphoretic; another species, A. Nardus, is called, because of its quality, Ginger Grass, or Koshel. The roasted leaves of Andropogon Schœnanthus are used in India, in infusion, as an excellent stomachic. An essential oil of a pleasant taste is extracted from the leaves in the Moluccas; and the Javanese esteem the plant much as a mild aromatic and stimulant. (*Ainslie*, ii. p. 58.) The former is one of the Grass oils of Nemaur, called in India Ivarancusa, and described in *Brewster's Journal*, ix.p. 333. Many others partake of the same qualities. But it is not merely for their aroma that Grasses are used medicinally. A cooling drink is employed in India from the roots of Cynodon Dactylon. The hard stony fruits of Coix Lachryma (Job's-tears), have been supposed to be strengthening and diuretic ; and the latter quality has been recognised in many others, especially the common Reeds, Phragmites arundinacea and Calamagrostis in Europe, Perotis latifolia in the West Indies, and the Brazilian species of Gynerium. A decoction of Eleusine indica is employed in Demerara, in the convulsions of infants, according to Schomburgk. Donax arundinaceus is astringent and subacrid. The creeping roots of the Quitch or Quick Grass, Triticum repens, of Tr. glaucum and junceum and Cynodon Dactylon and lineare, have some reputation as a substitute for Sarsaparilla. A decoction of the root of Gynerium parviflorum is used in Brazil to strengthen the hair.

Sugar is a general product of Grasses. Gynerium saccharoides, a Brazilian Grass, derives its name from that circumstance. It exists in great quantity in the Sugar-cane (Saccharum officinarum) ; Maize so abounds in it that its cultivation has been proposed in lieu of the Sugar-cane ; and it is probable that the value of other species for fodder depends upon the abundance of this secretion.

For economical purposes Grasses are often of much importance. The strong stems of the Bamboo are employed instead of timber and cordage. The Arundo arenaria and Elymus arenarius (Marrum Grasses) are invaluable species for keeping together the blowing sands of the sea-coast, by their creeping suckers and tough entangled roots. The first is employed in the Hebrides for many economical purposes, being made into ropes for various uses, mats for pack-saddles, bags, hats, &c. Some of the Reeds of Brazil, called Taquarussa, are living fountains: they grow from 30 to 40 feet high, with a diameter of six inches, form thorny impenetrable thickets, and are exceedingly grateful to hunters; for, on cutting off such a Reed below a joint, the stem of the younger shoots is found to be full of a cool liquid, which quenches the most burning thirst. Reeds and other coarse species furnish in Europe the materials for thatching. The reeds (sometimes 16 feet long), from which the Indians of Esmeralda form the tubes whence they blow the arrows poisoned with the deadly Urari or Woorali, are single internodes of the Arundinaria Schomburgkii. (*Linn. Trans.* xviii. p. 562.) A coarse but good sort of soft paper is manufactured in India from the tissue of the Bamboo, and the very young shoots of that plant are eaten like Asparagus.

Besides these things the inorganic products are remarkable. That the cuticle contains a large proportion of silex, is proved by its hardness, and by masses of vitrified matter being found whenever a hay-stack or heap of corn is accidentally consumed by fire. In the joints of some Grasses a perfect siliceous deposit is found, particularly in a kind of Jungle Grass mentioned in a letter from Dr. Moore to Dr. Kennedy of Edinburgh. It is also said that Wheat-straw may be melted into a colourless glass with the blow-pipe, without any addition. Barley-straw melts into a glass of a topaz yellow colour. The siliceous matter of the Bamboo is often secreted at the joints, where it forms a singular substance called tabasheer, of which see a very interesting account in *Brewster's Journal*, viii. p. 268. It was found by Turner that the tabasheer of India consisted of silica containing a minute quantity of lime and vegetable matter. Sulphur exists, in combination with different bases, in Wheat, Barley, Rye, Oats, Maize, Millet, and Rice.

For an account of the disease called Ergot, see p. 39, in the Fungal Alliance. It seems to be found in all Grasses, but most abundantly in Rye and Maize. When mixed with flour, in any quantity, it causes a mortification of the limbs, and the most horrible poisoning. Medical men have however found it to exercise a decidedly powerful stimulant effect upon the uterus, on which account it is now frequently and successfully employed by European practitioners in cases of difficult parturition.* The ergot

* Ergot is a disease which causes the grain of Rye to lengthen, harden, turn black, and form horns or spurs upon the ears. Where Rye is the food of man or of cattle, most dreadful consequences have followed the use of the spurred grains. Some curious observations have lately been made upon it by M. Bonjean. He says that the action on animals is extremely similar to that of morphine, although it in fact contains no trace of that substance. The first effect is to produce a loss of appetite and stupefaction; when it begins to act, dogs howl frightfully until they are completely under its influence, and then lie down and groan. In fowls the comb and crop become black. It appears that the Ergot which breaks with a white fracture is quite as dangerous as that which is violet; but until it is quite ripe it has no dangerous action; six or eight days are sufficient for its maturity, and even its being very old, hard, and dry seems in no way to impair its venomous qualities. M. Bonjean adds that Ergot contains two principles entirely different: one, of an oily nature, is venomous; the other, of a watery character, is harmless, but produces the extraordinary medical effects for which Ergot is employed—in particular in stopping the most frightful cases of hemorrhage. He asserts that the watery part, which he calls hæmostatic extract, may be prepared without difficulty, and that he has administered as much as 2 drachms of it, which is equal to 9 or 10 drachms of the Ergot, without any dangerous consequences.

The best Ergot is obtained from Rye which is grown on dry, airy, elevated regions, and where the

of Maize is, according to Roulin, very common in Colombia, and the use of it is attended with a shedding of the hair, and even the teeth, of both man and beast. Mules fed on it lose their hoofs, and fowls lay eggs without shell. Its action upon the uterus is as powerful as that of Rye ergot, or perhaps more so. The country name of the Maize thus affected is Maïs peladero. This statement however requires confirmation.

GENERA.

I.—*Oryzeae.*

- Leersia, *Sol.*
  - *Asprella*, Schreb.
  - *Homalocenchrus*, Mieg.
- Blepharochloa, *Endl.*
- Potamochloa, *Griff.*
- Oryza, *Linn.*
  - *Maltebrunia*, Kunth.
- Potamophila, *R. Br.*
- Hydrochloa, *P. Br.*
  - *Hydropyrum*, Lk.
  - *Melinum*, Lk.
- Zizania, *L.*
- Hygroryza, *Nees.*
- Caryochloa, *Trin.*
  - *Arrozia*, Schrad.
- Luziola, *Juss.*
- Ehrharta, *Thunb.*
  - *Trochera*, Rich.
- Tetrarrhena, *R. Br.*
- Microlæna, *R. Br.*
- Diplax, *Sol.*
- Pharus, *P. Br.*
- Leptaspis, *P. Br.*

II.—*Phalareæ.*

- Lygeum, *L.*
- Zea, *L.*
- Coix, *L.*
  - *Lithagrostis*, Gærtn.
- Chionanche, *R. Br.*
- Sclerachne, *R. Br.*
- Polytoca, *R. Br.*
- Cornucopiæ, *Linn.*
- Crypsis, *Ait.*
  - *Antitragus*, Gærtn.
  - *Heleochloa*, Host.
- Mibora, *Adans.*
  - *Sturmia*, Hopp.
  - *Chamagrostis*, Borkh.
- Alopecurus, *L.*
  - *Colobachne*, Palis.
  - *Tozzettia*, Savi.
- Limnas, *Trin.*
- Beckmannia, *Host.*
  - *Joachimia*, Ten.
  - *Bruchmannia*, Nutt.
- Phleum, *L.*
  - *Stelephurus*, Adans.
  - *Chilochloa*, Palis.
  - *Achnodonton*, Palis.
  - *Achnodon*, Lk.
- Fingerhuthia, *Nees.*
- Chondrolæna, *Nees.*
  - *Prionachne*, Nees.
- Hilaria, *H. B. K.*
  - *Hexarrhena*, Presl.
- Phalaris, *Linn.*
- Digraphis, *Trin.*
  - *Baldingera*, Gærtn.
  - *Typhoides*, Mönch,
- Holcus, *L.*
- Reynaudia, *Kunth.*
- Despretzia *Kunth.*

III.—*Paniceæ.*

- Reimaria, *Flügg.*
- Paspalum, *L.*
  - *Axonopus*, Röm. et Sch.
  - *Ceresia*, Pers.
  - *Garnotia*, Brongn.
- Milium, *L.*
  - *Miliarium*, Mnch.
  - *Leptocoryphium*, Nees.
- Amphicarpum, *Rafin.*
- Olyra, *L.*
  - *Lithachne*, Palis.
  - *Raddia*, Bertol.
- Strephium, *Schrad.*
- Thrasya, *Kunth.*
- Eriochloa, *Kunth.*
  - *Œdipachne*, Lk.
  - *Helopus*, Trin.
- Urochloa, *Palis.*
  - *Axonopus*, Palis.
  - *Coridochloa*, Nees.
- Rhynchelytrum, *Nees.*
- Panicum, *Linn.*
  - *Digitaria*, Scop.
  - *Dactylon*, Vill.
  - *Syntherisma*, Schrad.
  - *Hymenachne*, Palis.
  - *Streptostachys*, Palis.
  - *Monachne*, Palis.
  - *Aulaxanthus*, Ell.
  - *Aulaxia*, Nutt.
  - *Thalasium*, Spr.
  - *Trichachne*, Nees.
  - *Otachyrium*, Nees.
- Ichnanthus, *Palis.*
- Bluffia, Nees.
- Isachne, *R. Br.*
  - *Meneritaria*, Herm.
- Stenotaphrum, *Trin.*
  - *Rottboella*, Sw.
- Acratherum, *Lk.*
- Berghausia, *Endl.*
  - *Miquelia*, Nees.
- Melinis, *Palis.*
  - *Suardia*, Schrank.
  - *Tristegis*, Nees.
- Thysanolæna, *Nees.*
- Chætium, *Nees.*
- Oplismenus, *Palis.*
  - *Orthopogon*, R. Br.
  - *Hippagrostis*, Rumph.
  - *Echinocloa*, Palis.
- Berchtoldia, *Presl.*
- Chamærhaphis, *R. Br.*
- Pennisetum, *Rich.*
- Setaria, *Palis.*
- Gymnothrix, *Palis.*
  - *Cataterophora*, Steudel.
  - *Beckera*, Fres.
- Penicillaria, *Sw.*
- Cenchrus, *Linn.*
  - *Panicastrella*, Michel.
- Trachyozus, *Reichenb.*
  - *Trachys*, Pers.
  - *Trachystachys*, Dietr.
- Anthephora, *Schreb.*
  - *Colladoa*, Pers.
- Lappago, *Schreb.*
  - *Tragus*, Hall.
- Lopholepis, *Decaisn.*
  - *Holboellia*, Wall.
- Latipes, *Kunth.*
- Echinolæna, *Desv.*
- Navicularia, *Bertol.*
- Thouarea, *Pers.*
  - *Microthouarea*, Thouars.
- Spinifex, *Linn.*
- Neurachne, *R. Br.*

IV.—*Stipeæ.*

- Oryzopsis, *Rich.*
  - *Dilepyrum*, Raf.
- Greenia, *Nutt.*
- Piptatherum, *Palis.*
  - *Urachne*, Trin.
- Lasiagrostis, *Lk.*
- Dichelachne, *Endl.*
- Orthoraphium, *Nees.*
- Macrochloa, *Kunth.*
- Stipa, *Linn.*
  - *Nasella*, Trin.
  - *Piptochætium*, Presl.
  - *Aristella*, Trin.
  - *Jarava*, Ruiz et Pav.
- Eriocoma, *Nutt.*
- Streptachne, *R. Br.*
- Aristida, *Linn.*
  - *Chætaria*, Palis.
  - *Curtopogon*, Palis.
  - *Pseudachne*, Endl.
  - *Streptachne*, Kunth.
  - *Arthratherum*, Palis.
- Stipagrostis, *Nees.*

V.—*Agrosteæ.*

- Mühlenbergia, *Schreb.*
  - *Podosæmum*, Kunth.
  - *Trichochloa*, Trin.
  - *Dilepyrum*, Michx.
  - *Brachyelytrum*, Palis.
- Clomena, *Palis.*
- Lycurus, *H. B. K.*
- Coleanthus, *Seid.*
  - *Schmidtia*, Tratt.
  - *Willibalda*, Sternb.
- Phippsia, *R. Br.*
- Colpodium, *Trin.*
- Cinna, *L.*
- Epicampes, *Presl.*
- Echinopogon, *Palis.*
- Sporobolus, *R. Br.*
  - *Heleochloa*, Palis.
  - *Agrosticula*, Raddi.
  - *Calotheca*, Steud.
- Agrostis, *Linn.*
  - *Trichodium*, Auct.
  - *Vilfa*, Auct.
  - *Anemagrostis.*
  - *Apera*, Palis.
- Gastridium, *Palis.*
- Nowodworskya, *Presl.*
  - *Raspailia*, Presl.
  - *Chætotropis*, Kunth.
- Polypogon, *Desf.*
- Chæturus, *Lk.*
- Ægopogon, *Willd.*
- Pereilema, *Presl.*

VI.—*Arundineæ.*

- Sericura, *Hassk.*
- Calamagrostis, *Adans.*
- Deyeuxia, *Clar.*
- Lachnagrostis, *Trin.*
- Pentapogon, *R. Br.*
- Ammophila, *Host.*
  - *Psamma*, Palis.
  - *Amagris*, Rafin.
- Arundo, *Linn.*
  - *Donax*, Palis.
  - *Scolochloa*, Koch.
  - *Trichoon*, Roth.
- Ampelodesmos, *Lk.*
- Graphephorum, *Desv.*
- Phragmites, *Trin.*
  - *Czernya*, Presl.
- Amphidonax, *Nees.*
- Gynerium, *H. B. K.*

VII.—*Pappophoreæ.*

- Amphipogon, *R. Br.*
- Diplopogon, *R. Br.*
  - *Dipogonia*, Palis.
- Trirhaphis, *R. Br.*
- Pappophorum, *Schreb.*
- Enneapogon, *Desv.*
- Polyrhaphis, *Trin.*
- Euraphis, *Trin.*
  - *Corethrum*, Vahl.
- Cottæa, *Kunth.*
- Echinaria, *Desf.*
  - *Panicastrella*, Mönch.
- Cathestecum, *Presl.*

VIII.—*Chloreæ.*

- Microchloa, *R. Br.*
- Schœnefeldia, *Kunth.*
- Cynodon, *Rich.*
  - *Digitaria*, Juss.
  - *Fibigia*, Kölr.
  - *Capriola*, Adans.
  - *Cabrera*, Lagasc.
- Dactyloctenium, *Willd.*
- Eustachys, *Desv.*
  - *Schultesia*, Spr.
- Chloris, *Sw.*
  - *Apogon*, Endl.
  - *Euchloris*, Kunth.
  - *Actinochloris*, Panz.
  - *Geopogon*, Endl.
  - *Tetrapogon*, Desf.
- Leptochloa, *Palis.*
  - *Leptostachys*, Meyer.
  - *Oxydenia*, Nutt.
  - *Diplachne*, Palis.
- Eleusine, *Gærtn.*
- Harpochloa, *Kunth.*
- Ctenium, *Panz.*

soil is sandy or chalky in character. When its form is somewhat long, and it is of a very dark colour, or if it has been gathered in plains or damp valleys, it is of inferior quality. On chemical analysis, according to the experiments of Vauquelin, Wiggers, and others, it yields nearly half its weight in oil, resin, wax, fatty matter and gum, all hydrogenous principles, and a little albumen, and nitrogenous extract. If the season has been a wet one, or if the Ergot has been gathered in moist places, these principles lose their relative proportions; and the spurred Rye, approaching nearer in quality to good grain, contains but few oleo-resinous principles. It is worthy of remark, that this parasitic grain is only met with on the finest plants of Rye in shady places, or towards the ends of fields recently cleared of wood, and where the carbonic principles and a rich soil abound.—*Chemical Gazette.*

Campuloa, Desv.
Campulosus, Palis.
Monocera, Elliot.
Monothera, Raf.
Melanocenchris, *Nees.*
Chondrosium, *Desv.*
Actinochloa, Willd.
Bouteloa, Lagasc.
Opizia, *Presl.*
Spartina, *Schreb.*
Limnetis, Rich.
Trachynotia, Michx.
Ponceletia, Thouars.
Eutriana, *Trin.*
Atheropogon, *Mühlenb.*
Aristidium, *Endl.*
Heterostega, *Desv.*
? *Enteropogon*, Nees.
Triplathera, *Endl.*
Triathera, *Desv.*
Gymnopogon, *Palis.*
Polyodon, *H. B. K.*
Pentarhaphis, *H. B. K.*
Polyschistis, *Presl.*
Triæna, *H. B. K.*
Triplasis, *Palis.*
Pleuraphis, *Torrey.*
Bromidium, *Nees.*

IX.—*Aveneæ.*

Hierochloe, *Gmel.*
Disarrhenum, Lab.
Dimeria, Raf.
Anthoxanthum, *L.*
Ataxia, *R. Br.*
Podopogon, *Ehrenb.*
Corynephorus, *Palis.*
Weingärtneria, Bernh.
Deschampsia, *Palis.*
Campella, Lk.
Dupontia, *R. Br.*
Aira, *L.*
? *Periballia*, Trin.
? *Poidium*, Nees.
Airopsis, *Desv.*
Trisetaria, *Forsk.*
Lagurus, *L.*
Trisetum, *Kunth.*
Colobanthus, *Trin.*
Rostraria, *Trin.*
Kœleria, Lk.
Trichæta, *Palis.*
Acrospelion, *Bess.*
Ventenata, Köl.
Avena, *Linn.*
? *Leptopyrum*, Rafin.
Gaudinia, Palis.
Arthrostachya, Lk.
Arrhenatherum, *Palis.*
Tristachya, *Nees.*
Monopogon, Presl.
Anisopogon, *R. Br.*
Trichopterya, *Nees.*
Eriachne, *R. Br.*
Achneria, Palis.
Brandtia, *Kunth.*
Danthonia, *DC.*
Sieglingia, Bernh.
Triodia, Palis.
Tripogon, Röm. et Sch.
Triathera, Roth.
Pentameris, Palis.
Chætobromus, *Nees.*
Uralepis, *Nutt.*
Diplocea, Rafin.
Windsoria, Nutt.
Tricuspis, Palis.
Tridens, Rom. et Sch.
Triodia, *R. Br.*
Pommereulla, *Lin. fil.*

X.—*Festuceæ.*

*Bromidæ.

Sesleria, *Ard.*
Oreochloa, Lk.
Psilathera, Lk.
Poa, *L.*
Æluropus, *Trin.*
Brizopyrum, Lk.
Distichis, Raf.
Eragrostis, *Palis.*
Megastachya, Palis.
Dissanthelium, *Trin.*
Tetrachne, *Nees.*
Phalaridium, *Nees.*
Centotheca, *Desv.*
Glyceria, *R. Br.*
Devauxia, Palis.
Hydrochloa, *Lk.*
Exydra, Endl.
Lophochlæna, *Nees.*
Pleuropogon, *R. Br.*
Eatonia, *Raf.*
Reboulea, Kunth.
? *Chondrachyrum*, Nees
Catabrosa, *Palis.*
Cœlachne, *R. Br.*
Briza, *L.*
? *Neuroloma*, Raf.
Chascolytrum, *Desv.*
Calotheca, *Kunth.*
Anthochloa, *Nees.*
Melica, *L.*
Bulbilis, Rafin.
Molinia, *Mönch.*
Airochloa, *Lk.*
Kœleria, *Lk.*
Collinaria, Ehrh.
Ægialitis, Trin.
Ægialina, Schult.
Lophochloa, Rchb.
Schismus, *Palis.*
Hemisacris, Steud.
Wangenheimia, *Monch.*
Dactylis, *L.*
Lasiochloa, *Kunth.*
Urochlæna, *Nees.*
Cynosurus, *L.*
Chrysurus, *Palis.*
Lamarckia, *Monch.*
Pterium, Desv.
Ectrosia, *R. Br.*
Lophatherum, *Brongn.*
Elytrophorus, *Palis.*
Echinalysium, Trin.
Plagiolytrum, *Nees.*
Festuca, *Linn.*
Sclerochloa, Palis.
Sphenopus, Trin.
Catapodium, Lk.
Brachypodium, Palis.
Vulpia, Gmel.
Mygalurus, Lk.
Schedonorus, Palis.
Amphibromus, *Nees.*
Bromus, *Linn.*
Ceratochloa, Palis.
Libertia, Lej.
Michelaria, Dumort.
Orthoclada, *Palis.*
Uniola, *Linn.*
Chasmanthium, Lk.
Trisiola, Raf.
Diarrhena, *Palis.*
Diarina, Raf.
Rœmeria, Zea.
Corycarpus, Zea.

** Bambusidæ.

Arundinaria, *Rich.*
Miegia, Pers.
Ludolfia, Willd.
Triglossum, Fisch.
Macronax, Rafin.
Arthrostylidium, *Ruppr.*
Phyllostachys, *Sieb.*
Streptogyna, *Palis.*
Chusquea, *Kunth.*
Rettbergia, Raddi.
Platonia, Kunth.
Dendragrostis, Nees.
Merostachys, *Spreng.*
Guadua, *Kunth.*
Nastus, *Juss.*
Stemmatospermum, Pal.
Schizostachyum, *Nees.*
Bambusa, *Schreb.*
Arundarbor, Bauh.
Dendrocalamus, Nees.
Beesha, *Rheed.*
Melocanna, Rop.
Streptochæta, *Nees.*
Lepideilema, Trin.

XI.—*Hordeæ.*

Lolium, *Linn.*
Cræpalia, Schrank.
Triticum, *Linn.*
Spelta, Endl.
Agropyrum, Palis.
Trachynia, Lk.
Secale, *Linn.*
Elymus, *Linn.*
Psammochloa, Endl.
Cuviera, Kœl.
? *Sitanion*, Raf.
Gymnostichum, *Schreb.*
Asprella, Humb.
Hystrix, Mönch.
Hordeum, *Linn.*
Zeocriton, Palis.
Critesium, Rafin.
Ægilops, *L.*
Polyantherix, *Nees.*
Pariana, *Aubl.*

XII.—*Rottboelleæ.*

Nardus, *Linn.*
Psilurus, *Trin.*
Asprella, *Host.*
Monerma, Palis.
Lepturus, *R. Br.*
Myurus, Endl.
Micrurus, Endl.
Monerma, Palis.
Syurus, Endl.
Pholiurus, Trin.
Oropetium, *Trin.*
Ophiurus, *Gærtn.*
Hemarthria, *R. Br.*
Lodicularia, Palis.
Vossia, *Wall. et Griff.*
Mnesithea, *Kunth.*
Thyridostachyum, Nees.
Rottböella, *R. Br.*
Hemipus, Endl.
Stegosia, Lour.
? *Cymbachne*, Retz.
Cœlorhachis, Brongn.
Ratzeburgia, *Kunth.*
Aikinia, Wall.
Xerochloa, *R. Br.*
Tripsacum, *Linn.*
Manisuris, *Linn.*
Peltophorus, Desv.

XIII. *Andropogoneæ.*

Perotis, *Ait.*
Xystidium, *Trin.*
Leptothrium, *Kunth.*
Zoysia, *Willd.*
Epiphytis, Trin.
Matrella, Pers.
Osterdamia, Neck.
Dimeria, *R. Br.*
Haplachne, Presl.
Arthraxon, *Palis.*
Pleuroplitis, Trin.
Lucæa, Kunth.
Eriochrysis, *Palis.*
Plazerium, Willd.
Saccharum, *Linn.*
Phragmites, Adans.
Saccharophorum, Neck.
Tricholæna, Schrad.
Eriopogon, Endl.
Imperata, *Cyrill.*
Pogonatherum, *Palis.*
Homeoplitis, Trin.
Erianthus, *Rich.*
Ripidium, Trin.
? *Microstegium*, Nees.
Eulalia, *Kunth.*
Leptatherum, *Nees.*
Apocopis, *Nees.*
Elionurus, *Kunth.*
Anthistiria, *Linn.*
Themeda, Forsk.
Perobachne, *Presl.*
Androscepia, Brongn.
Diectomis, *Kunth.*
Apluda, *Linn.*
Diectomis, *Palis.*
Batratherum, *Nees.*
Hologamium, *Nees.*
Lepeocercis, *Trin.*
Anatherum, *Palis.*
Cymbopogon, Spr.
Hypogynium, Nees.
Agenium, Nees.
Trachypogon, *Nees.*
Schizachyrium, Nees
Pithecurus, Willd.
Sorghum, Pers.
Blumenbachia, Kol.
Andropogon, *Linn.*
Pollinia, Spr.
Chrysopogon, Trin.
? *Rhaphis*, Loureir.
Centrophorum, Trin.
Heteropogon, *Pers.*
Ischæmum, *Linn.*
Sehima, Forsk.
Meoschium, Palis.
Colladoa, Cav.
Spodiopogon, Trin.
Arundinella, Raddi.
Goldbachia, Trin.
Riedelia, Trin.
Thysanachne, Presl.
Pogonopsis, *Presl.*
Thelepogon, *Roth.*
Arthropogon, *Nees.*
Zeugites, *P. Br.*
Alloteropsis, *Presl.*
Blyttia, *Fries.*

Doubtful Genera.

Pterium, *Desv.*
Rytachne, *Desv.*
Xenochloa, *Lichtenst.*
Caryochloa, *Spr.*
Heterelytron, *Jungh.*
Aristaria, *Jungh.*

NUMBERS. GEN. 291. Sp. 3800 ?

Bryaceæ ?
POSITION.—Restiaceæ.—GRAMINACEÆ.—Cyperaceæ.
Palmaceæ.

ORDER XXX. CYPERACEÆ. SEDGES.

Cyperoideæ, *Juss. Gen.* 26. (1789).—Cyperaceæ, *R. Brown Prodr.* 212. (1810) ; *Lestiboudois, Essai ; Nees von Esenbeck in Linnæa*, 9. 273 ; *Endl. Gen.* xliii. ; *Meisner*, p. 110 ; *Kunth. Enum. vol.* 2 ; *Nees ab Esenb. in Fl. Bras. fasc.* 4.

DIAGNOSIS.—*Glumal Endogens with whole leaf-sheaths, a one-celled ovary, and an embryo enclosed within the base of the albumen.*

Grass-like herbs, growing in tufts and never acquiring a shrubby condition. The stems are never hollow, and seldom have any partitions at their nodes ; they are frequently angular, and are sometimes enlarged at the base into corms or tubers. The leaves are narrow or taper, and, when they wrap round the stem in the form of a sheath, never have that sheath slit. Flowers ☿ or ♂ ♀, consisting of imbricated solitary bracts, of which the lowermost are often empty, very rarely enclosing other opposite bracts at right angles with the first, and called glumes. Calyx none. Stamens hypogynous, definite, 1, 2, 3, 4, 5, 6, 7, 10, 12 ; anthers fixed by their base, entire, 2-celled. Ovary 1-seeded, often surrounded by bristles called hypogynous setæ ; ovule erect, anatropal ; style single, trifid, or bifid ; stigmas undivided, occasionally bifid. Nut crustaceous or bony. Albumen fleshy or mealy, of the same figure as the seed ; embryo lenticular, undivided, enclosed within the base of the albumen ; plumule inconspicuous.

1 3 2

Fig. LXXVIII.

Sedges so nearly resemble Grasses in appearance, that the one may be readily mistaken for the other by incurious persons ; they are, however, essentially distinguished by many important points of structure. In the first place, their stems are usually angular, not round and fistular ; there is no diaphragm at the articulations ; their flowers are destitute of any other covering than that afforded them by a single bract, in the axil of which they grow, with the exception of Carex, Uncinia, and Diplacrum, in which 2 opposite glumes are added ; and, finally, the seed has its embryo lying in the base of the albumen, within which its cotyledonar extremity is enclosed, and not on the outside, as in Grasses ; a very important fact, which it is the more necessary to point out, since Brown describes it (*Prodr.* 212) as lenticular and placed on the outside of the albumen. The additional glumes above adverted to form what Linnæan botanists call the nectary or aril ! Brown mentions a case where these glumes, which he calls a capsular perianth, included stamens instead of a pistil. According to Turpin, rudiments of the

Fig. LXXIX.

Fig. LXXVIII.—Scirpus lacustris. 1. A flower surrounded with hypogynous bristles ; 2. a seed ; 3, a section of it, showing the lenticular embryo.
Fig. LXXIX.—Utricle or additional glumes of Carex rivularis.

latter sometimes appear in different species of Mariscus. Sedges approach Cordleafs (Restiaceæ) in the peculiar state of the flowers and in general habit. They are, however, clearly distinguished from that order by their seeds being erect not pendulous, and by their more complicated ovary, which is always formed by 2 or 3 carpellary leaves, although enclosing only one ovule, while Cordleafs have but one carpellary leaf to each ovule. The sheaths of the leaves of Cordleafs are slit, like those of Grasses. Sedges stand then in the same relation to Cordleafs as Buckwheats to Chenopods. The species are extremely difficult to determine, and the distinctive characters of the genera were unsatisfactory, until Professor Nees v. Esenbeck rearranged the Order in the place above quoted.

Found in marshes, ditches, and running streams, in meadows and on heaths, in groves and forests, on the blowing sands of the sea shore, on the tops of mountains, from the arctic to the antarctic circle, wherever Phænogamous vegetation can exist. Humboldt remarks, that in Lapland Sedges are equal to Grasses; but that thence, from the temperate zone to the equator, in the northern hemisphere, the proportion of Sedges to Grasses very much diminishes. As we approach the Line, the character of the order also changes: Carex, Scirpus, Schœnus, and their allies, cease to form the principal mass, the room of which is usurped by multitudes of species of Cyperus, by Kyllinga, Mariscus, and the like, genera comparatively unknown in northern regions, or at least not forming any marked feature in the vegetation. A few species are common to very different parts of the world, as Scirpus triqueter, Eleocharis capitata, and Fuirena umbellata, to New Holland and South America, and several Scirpi to Europe and the southern hemisphere.

While Grasses are celebrated for their nutritive qualities, and for the abundance of fæcula and sugar they contain, Sedges are little less remarkable for the frequent absence of those principles: hence they are scarcely sought for by cattle. The roots of Carex arenaria, disticha, and hirta, have diaphoretic and demulcent properties, on which account they are called German Sarsaparilla. Those of Cyperuses are succulent, and filled with a nutritive and agreeable mucilage. In Cyperus longus (the κυπειρος of Hippocrates) a bitter principle is superadded, which gives its roots a tonic and stomachic quality. The tubers of Cyperus hexastachyus or rotundus are said by General Hardwicke to be administered successfully in cases of cholera by Hindoo practitioners, who call the plant Mootha. Those of C. pertenuis, or Nagur-Mootha, are, when dried and pulverised, used by Indian ladies for scouring and perfuming their hair. The root of Cyperus odoratus has a warm aromatic taste, and is given in India, in infusion, as a stomachic. The root of Scirpus lacustris is astringent and diuretic, and was once officinal. Remirea maritima, a common plant in tropical America, is said to be powerfully diaphoretic and diuretic; and the same qualities are ascribed to Kyllinga odorata and Hypoporum nutans. The leaves of Cotton-grasses, Eriophorum, were once used in diarrhœa, and the spongy pith of the stem to destroy tape-worms. Cyperus Iria has a reputation in India as a useful medicine in suppression of the menses, and in colic. The root of Kyllinga triceps is employed in the East Indies in diabetes, and as a stomachic, for which its acridity combined with some aroma has recommended it. The root of Scleria lithosperma is supposed upon the Malabar coast to have antinephritic virtues. The tubers or corms of Cyperus esculentus, (the μαλινοθαλλη of Theophrastus), called by the French Souchet comestible or Amande de terre, are used as food in the south of Europe, and are employed in the preparation of orgeat; Dr. Royle adds, that when roasted they have been proposed as a substitute for coffee and cocoa. The Chinese cultivate several species for food, especially the Pi-tsi or Scirpus tuberosus, which Nees v. Esenbeck regards as a bulbous form of Limnochloa plantaginea. And Dr. Royle informs us (*Illustr.* p. 413), that the Cyperus bulbosus of Vahl (C. jemenicus L.), called Sheelandie-aresee in Madras, and Puri-drempa by the Telingas, has tubers which when roasted or boiled taste like potatoes, and would be valuable for food if they were not so small. Scirpus dubius of Roxburgh, (the Allikee of the Telingas) is given on the same authority as having tubers, which the natives say are as good as yams.

The Papyrus of the banks of the Nile, Papyrus antiquorum, of which boats, paper, and ropes are made, is a plant of this family; it is said to be called Babeer in Syria, and is described by the Arabians (*Avic.* c. 543), by the name Fafeer and Burdee: the former evidently of the same origin as the Greek and Syrian names. A species of the genus Papyrus (P. corymbosus, N. ab E., P. Pangorei Arnott) is hardly of less use in India, being extensively employed for making the mats so much used there for covering the floors of rooms, and which are also so much esteemed in Europe. Dr. Ainslie says that a species, called Rora and Toonghi, which he refers to C. textilis of Thunberg, is employed in the peninsula for the same purpose. Some of the species of Scirpus, especially S. lacustris, are sometimes substituted for rushes in making baskets and chair bottoms, &c.; Cyperus textilis is employed in making ropes, and as the Papyrus of Egypt was by the ancients. The species of Eriophorum, called Cotton-grass in England, from having their fruit clothed at the base with a silky or cotton-like substance, of which paper and wicks of candles have been made, and pillows stuffed, has a species (E. comosum, Wall., cannabinum, nob.), Bhabhur of the natives, of which the leaves, previous to the plant flowering, are in the Himalayas extensively employed for rope-making. Cyperus inundatus probably, with other species, helps much to bind and protect the banks of the Ganges from the rapidity of the stream and the force of the tides; as in Holland Carex arenaria is carefully planted on the dikes, where its far-extending roots, by mutually interlacing with each other, fix the sand and give strength to the embankment." (*Royle, Illustr.* p. 415.) Cyperus Hydra, called Nut-grass in the West Indies, is said to be a pest there, overrunning the Sugar-cane plantations, and rendering them barren.

GENERA.

I.—Cariceæ.

Carex, *Mich.*
Vignea, Palis.
Schelhammeria, Mönch.
Scuria, Rafin.
Triodia, Rafin.
Trasus, Gray.
Uncinia, *Pers.*
Hoppia, *Nees.*
Schoenoxyphium, *Nees.*

II.—Elyneæ.

Trilepis, *Nees.*
Dilepis, Endl.
Fintelmannia, Kunth.
Elyna, *Schrad.*
Fröhlichia, Wulff.
Kobresia, *Willd.*

III.—Sclereæ.

Diplacrum, *R. Br.*
Ptychocarya, *R. Br.*
Scleria, *Berg.*
Cylindropus, Nees.
Pteroscleria, *Nees.*
Schizolepis, *Schr.*
Ophryoscleria, *Nees.*
Macrolomia, *Nees.*
Osmoscleria, *Nees.*
Mastigoscleria, *Nees.*
Acrocarpus, *Nees.*
Cephalocarpus, *Nees.*
Cryptanguina, *Schr.*
Lagenocarpus, *Nees.*
Chondrolomia, *Nees*
Trachyloma, *Nees.*
Hymenolytrum, *Schr.*
Becquerela, *Brongn.*
Calyptrocarya, *Nees.*
Hypoporum, *Nees.*
Anogyna, *Nees.*
Aulacorhynchus, *Nees.*

IV.—Rhynchosporeæ.

* *Rhynchosporidæ.*

Morisia, *Nees.*
Mitrospora, *Nees.*
Haplostylis, *Nees.*
Pterotheca, *Presl.*
Calyptrostylis, *Nees.*
Ephippiorhynchium, *Nees*
Cephaloschœnus, Nees.
Diplochæte, *Nees.*
Ceratóschœnus, *Nees.*
Rhynchospora, *Vahl.*
Chætospora, *R. Br.*
Carpha, Banks & Sol.
Streblidia, Lk.
Asterochæte, Nees.
Cyathocoma, *Nees.*
Eucyathocoma, Fenzl.
Ideleria, Kunth.
Trianoptiles, *Fenzl.*
Ecklonia, Steud.
Nemochloa, *Nees.*
Nomochloa, Palis.
Pleurostachys, Brongn.
Machærina, *Vahl.*
Buekia, *Nees.*
Lepidosperma, *Labill.*
Lepidotosperma, Röm. et Sch.
Sclerochætium, Nees.
Oreobolus, *R. Br.*

** *Schœnidæ.*

Spermodon, *Palis.*
Triodon, Rich.
Psilocarya, *Torr.*
Astroschœnus, *Nees.*
Ptilochæta, *Nees.*
Dichromena, *Rich.*
Zosterospermum, *Palis.*
Echinoschœnus, *Nees.*
Haloschœnus, *Nees.*
Elynanthus, *Palis.*
Vincentia, Gaud.
Chapelliera, *Nees.*
Baumea, Gaudich.
Schœnus, *Linn.*
Torulinium, Desv.
? *Schœnopsis*, Lestib.
Gussonea, Presl.
Gymnoschœnus, *Nees.*
Isochœnus, *Nees.*
Remirea, *Aubl.*
Miegia, Schreb.

V.—Cladeæ.

Cladium, *P. Br.*
Lamprocarya, *R. Br.*
Morelotia, Gaud.
Melachne, Schrad.
Didymonema, Presl.
Epiandria, Presl.
Gahnia, *Forst.*
Caustis, *R. Br.*
Evandra, *R. Br.*

VI.—Chrysitricheæ.

Chrysithrix, *Linn. fil.*
Pandanophyllum, *Hassk.*
Lepironia, *Rich.*
Chondrachne, R. Br.
Chorizandra, *R. Br.*

VII.—Hypolytreæ.

Hemicarpha, *Nees.*
Lipocarpha, *Nees.*
Hypælyptum, R. Br.
Hypelytrum, Lk.
Platylepis, *Kunth.*
Hypolytrum, *Rich.*
Beesa, Palis.
Albikia, Presl.
Diplasia, *Rich.*

VIII.—Fuireneæ.

* *Melanocranidæ.*

Melanocranis, *Vahl.*
Hypolepis, Palis.
Sickmannia, *Nees.*
Anosporum, *Nees.*

** *Hemichlænidæ.*

Hemichlæna, *Schrad.*
Acrolepis, *Schrad.*
Hypophialium, Nees.
Pleurachne, *Schrad.*

*** *Ficinidæ.*

Fuirena, *Rottb.*
Vaginaria, L. C. Rich.
Vauthiera, *A. Rich.*
Ficinia, *Schrad.*
Schœnidium, Nees.
Oxycaryum, *Nees.*
Blepharolepis, *Nees.*
Oncostylis, *Mart.*
Fimbristylis, *Vahl.*
Trichelostylis, Lestib.
Dichelostylis, Palis.
Echinolytrum, Desv.

IX.—Scirpeæ.

Isolepis, *R. Br.*
Holoschœnus, Lk.
Eleogiton, Lk.
Trichelostylis, *Lestib.*
Dichostylis, Palis.
Nemum, *Palis.*
Helothrix, *Nees.*
Scirpus, *L.*
Pterolepis, *Schrad.*
Malacochæte, Nees.
Hymenochæte, Palis.
Elytrospermum, C. A. Meyer.
Blysmus, *Panz.*
Bæothryon, *Nees.*
Eleocharis, *R. Br.*
Eleogenus, Nees.
Chætocyperus, Nees.
Scirpidium, Nees.
Androtrichum, *Brongn.*
Androcoma, *Nees.*
Eriophorum, *L.*
Linagrostis, Lam.
Trichophorum, Pers.

X.—Cypereæ.

Dulichium, *Rich.*
Pleuranthus, Rich.
Comostemum, *Nees.*
Diclidium, *Schr.*
Cyperus, *Linn.*
Torreya, Rafin.
Papyrus, *Willd.*
Kyllingia, *Linn.*
Mariscus, *Vahl.*
Adulpa, Bosc.
Courtoisia, Nees.
Opetiola, Gærtn.
Tryocephalon, Forst.
Abilgaardia, *Vahl.*
Iria, Rich.
Leptoschœnus, *Nees.*

Uncertain Genera.

Mapania, *Aubl.*
Diaphora, *Lour.*
Haplostemum, *Rafin.*
Diplarrhinus, *Rafin.*
Distichmus, *Rafin.*
Tetraria, *Palis.*
Catagyna, *Palis.*
Tricostularia, *Nees.*

NUMBERS.—GEN. 112. Sp. 2000.

POSITION.—Graminaceæ.—CYPERACEÆ.—Restiaceæ.
Acoraceæ.
Typhaceæ.

ORDER XXXI. DESVAUXIACEÆ.—BRISTLEWORTS.

Desvauxieæ, *Nixus Plantarum, p.* 23. (1833), *a § of* Restiaceæ ; *Bartl. Ord. Nat. p.* 36 ; *Martius Conspectus, No.* 38.—Centrolepideæ, *Desvaux in Ann. des Sc.* 13. 36. (1828) ; *Endl. Gen.* xliv. ; *Meisner, Gen. p.* 409 ; *Kunth. Enum.* 3. 487.

DIAGNOSIS.—*Glumal Endogens, with several ovaries (sometimes consolidated), a pendulous ovule,* 1-2-*stamens,* 1-*celled anthers, and terminal embryo.*

Little tufted herbs, resembling small Scirpi. Leaves setaceous, sheathing at the base. Scapes filiform, undivided, naked. Flowers enclosed in a terminal spathe. Glumes one, in front, or two somewhat opposite each other. Paleæ 0, or one or two tender scales parallel with the glumes. Stamen 1, very rarely 2 ; anther simple. Ovaries from 1 to 18 attached to a common axis, distinct or partially united, 1-celled, with a single stigma to each ; ovules solitary, orthotropal. Fruit as many 1-seeded utricles, opening longitudinally ; seed pendulous ; embryo lenticular, placed within the extremity most remote from the hilum.

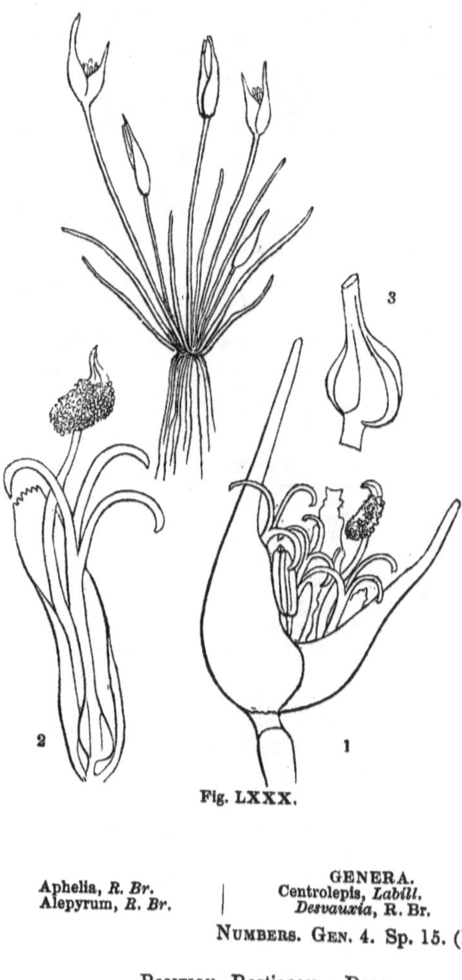

Fig. LXXX.

The main distinction of this Order consists in the ovaries, which are variable in number, and usually distinct from each other round a common axis, in the manner of a Ranunculus. Occasionally they are partially united ; in all cases they change to little one-seeded utricles. The stamen, which is usually solitary, has a second added in the genus Gaimardia, which does not seem to be otherwise different. Aphelia has only one carpel, and this is regarded by Endlicher as a near approach to Sedges ; but it is really very different, for the single ovary of that order is evidently made up of from 2 to 3 carpels enclosing a single ovule ; while in Aphelia, as in all the Order of Bristleworts, the ovary consists of but a single carpel.

All inhabit the South Sea Islands ; and nearly all New Holland.

They are of no known use.

GENERA.

Aphelia, *R. Br.*	Centrolepis, *Labill.*	Gaimardia, *Gaudich.*
Alepyrum, *R. Br.*	*Desvauxia*, R. Br.	

NUMBERS. GEN. 4. Sp. 15. (Kunth.)

POSITION. Restiaceæ.—DESVAUXIACEÆ.—Eriocaulaceæ.

Fig. LXXX.—Centrolepis fascicularis. 1. A head of flowers ; 2. a single flower separated ; 3. an ovary with the style cut off.—*Endlicher.*

ORDER XXXII. RESTIACEÆ.—CORDLEAFS.

Restiaceæ, *R. Brown, Prodr.* 243. (1810); *Kunth in Humb. N. G. et Sp.* 1. 251. (1815); *Agardh Aph.* 156. (1823) *a § of* Junceæ; *Nees v. Esenbeck, in Linnæa,* 5. 627. (1830) *et* 7. 614. (1832); *Endl. Gen.* xlv.; *Meisner, Gen. p.* 408; *Kunth Enum.* 3. 381.—Elegieæ, *Beauv. in eod. loc.* (1828).

DIAGNOSIS.—*Glumal Endogens, with a* 1-3-*celled ovary, a pendulous ovule,* 2-3 *stamens,* 1-*celled anthers, and terminal embryo.*

Herbaceous plants or under-shrubs. Leaves simple, narrow, or none. Culms naked, or more usually protected by sheaths, which are slit, and have equitant margins. Flowers generally aggregate, in spikes or heads, separated by bracts, and most frequently unisexual. Glumes 2-6, seldom wanting. Stamens 2 to 3, attached to 4 or 6 glumes and opposite the innermost; anthers usually unilocular and peltate. Ovary 1- or more-celled, cells monospermous; styles or stigmata never fewer than 2, although the ovary be 1-celled; and otherwise equal in number to the cells of the ovary; ovules pendulous. Fruit capsular, or nucamentaceous. Seeds inverted; albumen of the same figure as the seed; embryo lenticular, on the outside of the albumen, at that end of the seed which is most remote from the hilum.

Fig. LXXXI.

According to Brown, the principal character distinguishing this order from Rushes and Sedges consists in its pendulous seed and lenticular embryo placed at the extremity of the seed opposite to the umbilicus. From Rushes it also differs in the order of suppression of its stamina, which, when reduced to 3, are opposite to the inner glumes; and most of its genera are distinguishable from both these Orders, as well as from Commelinaceæ, by their simple or unilocular anthers.—(*Flinders,* 579.) But in truth it is essentially distinguished from the order of Rushes by its glumaceous flowers, as well as by the characters already named. If the glumes are absent, it is then only to be known from Sedges by the pendulous ovules, terminal embryo, and by the sheaths of its leaves being slit. The tripetaloid flower and polyspermous fruit of Xyris, a genus formerly referred here, are characters indicating a far superior degree of evolution, and sufficient to separate it as the representative of a peculiar order; a measure which Brown anticipated when he remarked (*Prodr.* 244.), that the genus Xyris, although placed by him at the end of Restiaceæ, is certainly very differentf rom the other genera, in the inner segments of the perianth being petaloid, with the stamens proceeding from the top of their ungues, and in its numerous seeds. Pipeworts are known by their having a membranous sheath between the glumes and ovary, and thus indicating an approach to the petaloid Orders, especially to Xyrids.

All are extra-European, and chiefly found in the woods and marshes of South America, New Holland, and southern Africa. They have not been found in America.

The tough wiry stems of some species are manufactured into baskets and brooms. Willdenowia teres is employed for the latter purpose, and Restio tectorum for thatching.

GENERA.

Rhodocoma, *Nees.*
Leptocarpus, *R. Br.*
Loxocarya, *R. Br.*
Chætanthus, *R. Br.*
Hypolæna, *R. Br.*
Cucullifera, Nees.
Dovea, *Kth.*
Willdenowia, *Thunb.*
Nematanthus, Nees.
Hypodiscus, *Nees.*
Leucoplocus, *Nees.*
Mesanthus, *Nees.*
Anthochortus, *Nees.*
Ceratocaryum, *Nees.*
Lepidanthus, *Nees.*
Anarthria, *R. Br.*
Lyginia, *R. Br.*
Lepyrodia, *R. Br.*
Thamnochortus, *Berg.*
Staberoha, *Kunth.*
Elegia, *Thunb.*
Chondropetalum, Rottb.
Restio, *Linn.*
Calorophus, Labill.
Calopsis, *Palis.*
Cannomois, *Palis.*
Boeckhia, *Kunth.*

NUMBERS. GEN. 23. Sp. 171 (*Kunth.*)

POSITION. Cyperaceæ.—RESTIACEÆ.—Eriocaulaceæ.

Fig. LXXXI.—Lepyrodia hermaphrodita. 1. A flower: 2. the same when the fruit is ripe; 3. the ovaries; 4. the fruit in a state of dehiscence; 5. the seed, cut vertically, of Restio dichotomus.

ORDER XXXIII. ERIOCAULACEÆ.—PIPEWORTS.

Eriocauloneæ, *L. C. Richard in H. B. K. Nov. Gen. et Sp. Pl.* 1. 251. (1815); *Desvaux in Ann. Sc.* 13. 36.; *Martius in Act. Acad. Cæs. Nat. cur.* 17.; *Endl. Gen.* xlvi.; *Meisner. gen. p.* 407; Eriocauleæ, *Kunth. enum.* 3. 493.; *Act. Acad. Wissench. Berlin, Febr.* 1841.

DIAGNOSIS.—*Glumal Endogens, with a 2-3-celled ovary, a pendulous ovule, 2-celled anthers, a terminal embryo, and a 3-lobed cup within the glumes.*

Perennial marsh-plants, with linear cellular spongy leaves sheathing at the base. Flowers capitate, bracteate, very minute, ♂ ♀. Glumes two, unilateral, or 3. A membranous tube, with 2 or 3 teeth or lobes, surrounds the ovary. Ovary superior 3- or 2-celled; ovules solitary, orthotropal; style very short; stigmas as numerous as the cells of the ovary. Dehiscence of the capsule loculicidal. Seeds solitary, pendulous, coated with wings or rows of hairs. Embryo more or less lenticular, lying upon the albumen at the end of the seed most remote from the hilum.

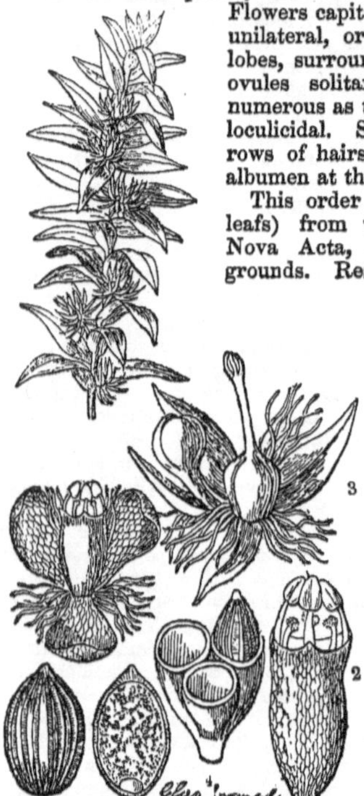

Fig. LXXXII.

This order is usually combined with Restiaceæ (or Cordleafs) from which, in a memoir in the 17th vol. of the Nova Acta, Von Martius separates it on the following grounds. Restiaceæ: Flowers in spikes. Calyx glumaceous ▽. Stamens in a single row, 1-3, opposite the petals; anthers generally 1-celled. Seeds without rows of hairs. Eriocaulaceæ: Flowers in heads, unisexual. Calyx sepaloideous △. Stamens 3, 6, 2, 4; if in two rows, with the inner row most developed; anthers 2-celled. Seeds solitary, with rows of hairs. The most important distinctions seem to consist in the presence among the Pipeworts of a membranous tube, which may be regarded as the most distinct approach, in the Glumal Alliance, to the corolla of the petaloid series, and in the anthers being 2-celled, not 1-celled; a further indication of a higher order of development. Xyrids, with a perfect corolla, may be regarded as the link which connects these plants with some of the more perfect orders of Endogens.

Many remarkable species are figured by Bongard in *Memoirs of the Imperial Academy of St. Petersburgh, 6th series*, 1. *p.* 601., &c.

A large number of species is collected under this head; all of which are amphibious or aquatic. According to Endlicher, two-thirds are found in the tropics of America, and half the remainder in the north of New Holland. A few occur in North America, and one is found in Great Britain, in the isle of Skye.

Eriocaulon setaceum, boiled in oil, is said to be a popular remedy for the itch in the East Indies.

GENERA.

Lachnocaulon, *Kth.*	*Randalia*, Petiv.	Tonina, *Aubl.*	Cladocaulon, *Gardn.*
Eriocaulon, *L.*	*Sphærochloa*, Palis.	*Hyphydra*, Schreb.	Stephanophyllum, *Guill.*
Dupatya, Fl. flum.	Leucocephala, *Roxb.*	Philodice, *Mart.*	? Symphachne, *Palisot.*
Nasmythia, Huds.	Paepalanthus, *Mart.*		

NUMBERS. GEN. 9. Sp. 200 (Kunth).

Juncaceæ.
POSITION.—Restiaceæ.—ERIOCAULACEÆ. ———
Xyridaceæ.

Fig. LXXXII.—Tonina fluviatilis. 1. ♂ flower; 2. centre of do.; 3 ♀ an ♂ flowers; 4. section of ripe fruit; 5. seed; 6. section of do.—*Martius.*

Alliance VIII. *ARALES.*—The Aral Alliance.

Diagnosis.—*Unisexual petaloid or naked flowered endogens, with a simple naked spadix, and an embryo in the axis of mealy or fleshy albumen.*

It is here that we find the lowest structure known among flowering plants. Lemna, in the Lemnad order, has a lenticular frond, in a cleft of whose edge lurk a couple of flowers, one ♂ and the other ♀, enclosed in a membranous bag. In Pistia, of the same Lemnad order, the leaves are separated from the stem, the flowers are more separated, and the ♂ has the beginning of a calyx. In Ambrosinia, also associated with Lemna, a complete bearded spathe is formed, and the ♂ is of a more complicated structure. From these plants we pass into the Arads, with naked flowers growing in dense spikes or spadixes, and they lead, on the one hand to the palm-like Screw Pines, and on the other to the sedgy Typhads, by means of which, especially the former, a communication is effected with the princely Palms. By another transition, into the Orontiaceæ of hermaphrodite hypogynous Endogens, a passage is formed into Lilyworts on the one hand and Peppers on the other. In fact, as I stated long since, the Aral alliance, and more especially the Araceous order, is the centre of a large system whose rays pierce very remote parts of the vegetable kingdom. Through Lemna this alliance passes into the Hydral by way of the Naiads. The *Spadicifloræ* of modern botanists, or *Spadicicarpæ* of Blume (Rumphia 2. 74) are nearly the same plants, except that Meisner includes Palms among them, to which there seems some objection.

Natural Orders of Arals.

Flowers 2 or 3, of which one only is ♀. Spadix 0. Ovary one-celled. Ovules erect. Embryo slit	Pistiaceæ.
Flowers 00, on a naked spadix. Calyx scaly or hairy. Anthers with long filaments. Ovule solitary, pendulous. Seed adherent to the pericarp. Embryo slit	Typhaceæ.
Flowers 00, naked, on a solitary spadix covered by a single hooded spathe. Anthers sessile. Seed loose. Embryo slit, axile . .	Araceæ.
Flowers 00, naked or scaly, on a spadix covered by many spathes. Anthers stalked. Seeds loose. Embryo solid, minute	Pandanaceæ.

ORDER XXXIV. PISTIACEÆ.—LEMNADS, OR DUCKWEEDS.

PISTIACEÆ, *Rich. in Humb. et Bonpl. N. G. et Sp.* 1. 81. (1815); *Lindl. in Hooker's Fl. Scot.* 2. 191. (1821); *Synops.* 251. (1829); *Endl. Gen. p.* 233; *Meisner, p.* 363; *Kunth. enum.* 3. 7; *Blume, Rumphia* 2. 76.—LEMNACEÆ, *DC. and Duby*, 532. (1828); *Endl. Gen. p.* 232; *Meisner, p.* 363; *Kunth. enum.* 3. 2. *Schleiden in Linnæa*, xiii. 384; *Hoffman in Tydschr. v. nat. Gesch. Leyden* (1838).

DIAGNOSIS.—*Aral Endogens with* 2 *or* 3 *flowers, of which one only is* ♀, *no spadix, a one-celled ovary, erect ovules, and a slit embryo.*

Floating or land plants, with very cellular, lenticular, or lobed fronds or leaves, some of them wholly destitute of spiral vessels, except perhaps in the pistil. Flowers appearing from the margin of the fronds, 2 or 3, naked, enclosed in a spathe, but without a spadix. ♂ Stamens definite, often monadelphous (pollen globose, muricated, with a single aperture in Lemna *Schleiden*): ♀ Ovary 1-celled, with one or more erect ovules; style short; stigma simple; ovules anatropal, hemianatropal, or atropal. Fruit membranous or capsular, not opening, 1- or more-seeded. Seeds with a fungous testa, and a thickened indurated foramen; embryo either in the axis of fleshy albumen, and having a lateral cleft for the emission of the plumule, or at the apex of the nucleus, covered in by a hardened endostome.

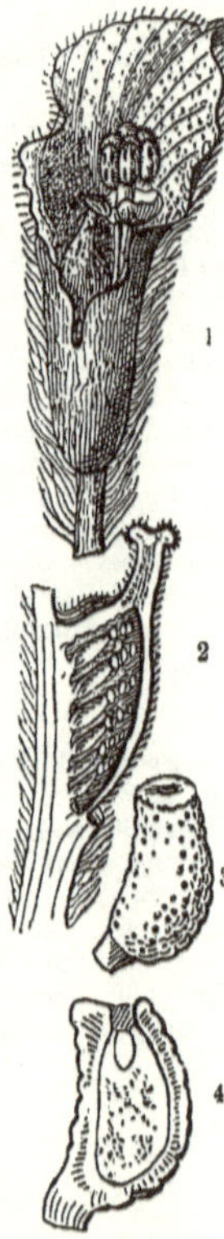

Fig. LXXXIII.

The common Duckweed (Lemna) may be regarded as being the most simple of all Phænogamous plants. Its stem and leaves are fused into a minute lenticular frond, which pullulates by openings in its sides; its roots are simple fibres, tipped by a calyptra, which Schleiden regards as a peculiar organ, and its flowers are two in number, one male and the other female, lying concealed in a slit of the frond; they have neither calyx nor corolla, but are enclosed in a delicate membranous bag. Lemna is indeed but one remove from a Crystalwort (Riccia, p. 57); species of which have even been mistaken for Lemnas by some authors, according to Schleiden.

All the true Lemnas are almost entirely destitute of spiral vessels, which the same author found abundantly in the old L. polyrhiza, now called Spirodela. A Lemna indeed may be said to consist of a small plate of cellular tissue, and a couple or three flowers. There is however in the fresh water of tropical countries a very common floating plant, called Pistia, which may be regarded as a Lemna with the leaves and stem separated, and the flowers more highly developed; there being a distinct spathe for the inflorescence, and a kind of cup-like calyx to the male flower. And then again the Mediterranean gives birth to Ambrosinia, a little land plant, with leaves of an ordinary kind, and a small spathe inclosing a couple of flowers, of which the uppermost has many monadelphous stamens, perfectly destitute of a calyx, and an ovary which is like that of Pistia. If we disregard the simplicity of this structure, and consider the organisation as if it belonged to plants of a more highly developed character, it will be found that these are really nothing but Arads, the spadix of which is reduced to two or three flowers of different sexes. But while the accuracy of this view of the nature of the Duckweed order is generally acknowledged, it must be borne in mind that this very reduction of parts is inconsistent with the notion of Arads, properly so called; and hence the necessity of constituting a particular order. I find from an examination of seeds of Pistia, most kindly procured from India for me by Dr. Wallich, that the embryo is a minute body lying within the apex of the albumen; in Lemna it occupies the axis; in both there is a fungous testa, with a remarkable induration of the foramen of the secundine. The embryo of Pistia is very minute, and, as far as I can see, solid; but Horkel says it is slit, and in Lemna there is certainly a cleft on one side for the emission of the plumule, just as in Arads. Most

Fig. LXXXIII.—Pistia Stratiotes. 1. A spathe with its 2 flowers; 2. a section of a seed-vessel; 3. a seed; 4. the same cut perpendicularly.

modern systematists regard Pistiaceæ and Lemnaceæ as distinct sub-orders of Arads, from which I separate them on account of their want of spadix, Ambrosinia connecting them with the curious genus Cryptocoryne. By some oversight, both Adrien de Jussieu and Endlicher regard Lemnaceæ as exalbuminous.

Lemna inhabits the ditches of the cooler parts of the world; Pistia the tropics; Ambrosinia the basin of the Mediterranean.

Pistia Stratiotes grows in water-tanks in Jamaica, where, according to Browne, it is acrid, and in hot dry weather impregnates the water with its particles to such a degree as to give rise to the bloody flux. A decoction of the same plant is considered by the Hindoostanees as cooling and demulcent, and they prescribe it in cases of dysuria. The leaves are also made into a poultice for hæmorrhoids. *See also Martius Mat. Med. Bras.* 97.

GENERA.

Lemna, *L.*
Wolffia, *Hork.*
Horkelia, Rchb.
Telmatophace, *Schleid.*
Spirodela, *Schleid.*
Pistia, *L.*
Zala, Lour.
Ambrosinia, *L.*
Ucria, Targ.

NUMBERS. GEN. 6. Sp. 20.

Ricciaceæ.
POSITION.—PISTIACEÆ.—Araceæ.
Naiadaceæ.

ORDER XXXV. TYPHACEÆ.—TYPHADS OR BULRUSHES.

Typhæ, *Juss. Gen.* 25. (1789).—Aroideæ, § 3. *R. Brown Prodr.* 338. (1810).—Typhineæ, *Agardh Aph.* 139. (1823). *Kunth. enum.* 3. 88. (1841).—Typhaceæ, *DC. and Duby*, 482. (1828) ; *Richard in Arch. de Bot. vol.* 1. *p.* 193 ; *Endlich. gen.* lxxiii. ; *Meisn. p.* 360.—Typhoideæ *and* Sparganioideæ, *Link. Handb.* 1. 132. 133. (1829), *as sections of* Cyperaceæ.

DIAGNOSIS.—*Aral Endogens, with numerous flowers on a naked spadix, a scaly or hairy calyx, long filaments, a solitary pendulous ovule, a seed adherent to its pericarp, and slit embryo.*

Herbaceous plants, growing in marshes or ditches. Stems without nodes. Leaves rigid, ensiform, with parallel veins. Flowers ♂ ♀, very closely arranged upon a spatheless spadix. Sepals = mere scales, 3 in number or more ; sometimes a mere bundle of hairs. Petals wanting. ♂ : Stamens 3 or 6; anthers wedge-shaped, attached by their base to long filaments, which are sometimes monadelphous. ♀ : Ovary single, superior, 1-celled ; ovule solitary, pendulous, anatropal ; style short ; stigmas simple, linear. Fruit dry, not opening, 1-celled, 1-seeded, made angular by mutual pressure. Seed pendulous, with a membranous skin adhering to the pericarp. Embryo in the centre of mealy albumen, straight, taper, with a cleft in one side, in which the plumule lies ; radicle next the hilum.

1 3

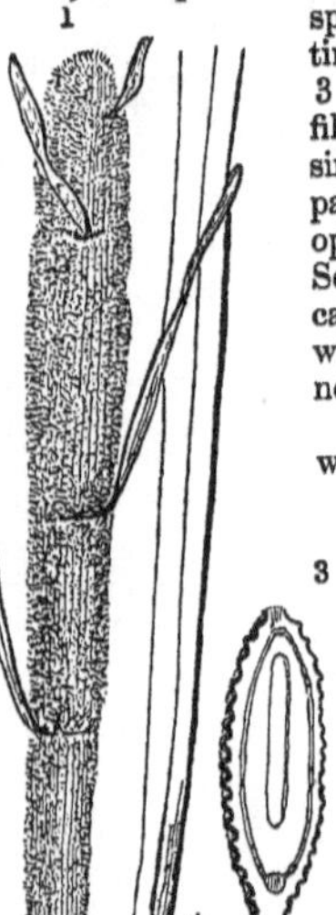

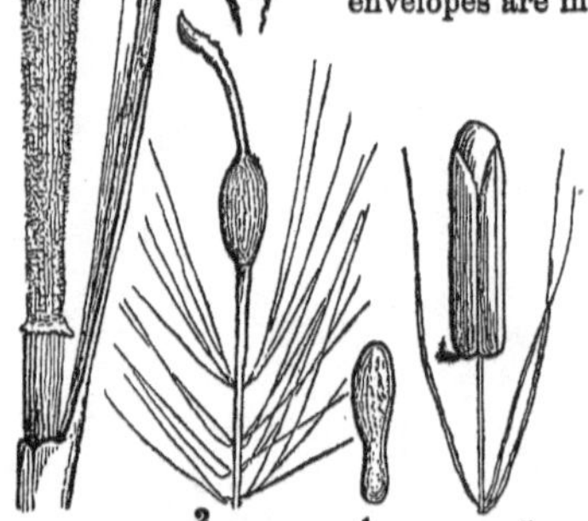

2 4 5

Fig. LXXXIV.

Jussieu, following Adanson, distinguishes these from Arads, with which Brown re-unites them, retaining them, however, in a separate section. They are generally regarded as a distinct tribe by most writers, and seem sufficiently characterised by their calyx being 3-sepaled and half-glumaceous, or a mere bundle of long hairs, by their lax filaments, wedged anthers, solitary pendulous ovules, and peculiar habit. Agardh refers Bulrushes to glumaceous Monocotyledons, on account of the analogy between the calyx of Typha and the hypogynous hairs of Eriophorum, a genus of Sedges ; and a similar view of their affinity has been taken by Link ; and in fact they do appear to constitute a direct transition from the glumaceous to petaloid Endogens, for although their floral envelopes are mere scales, yet they are arranged in regular whorls. In habit they are hardly distinguishable from Sedges. In another point of view they may be looked upon as diminutive species of Screw-pines (Pandanaceæ), and Kunth so considered them formerly : but their simple fruit, solitary ovules, and the slit in the side of their embryo, offer sufficient marks of distinction.

Found commonly in the ditches and marshes of the northern parts of the world, but uncommon in tropical countries : one species occurs in St. Domingo, and another in New Holland. Two are described from equinoctial America.

They are of little known use. The powdered flowers have been used as an application to ulcers. The pollen of Typha is inflammable, like that of Lycopodium, and is used as a substitute for it. De Candolle remarks that it is probable the facility of collecting this pollen which is the real cause of its use, and that any other kind would do as well. The rhizomes of Typha abound in starch, are somewhat astringent and diuretic, and are employed in the east of Asia in dysentery, gonorrhœa, and the measles.

GENERA.

Typha, *L.*	Sparganium, *L.*
	Platanaria, Gray.

NUMBERS. GEN. 2. Sp. 13. (Kunth.)

POSITION. ——— *Acoraceæ.* / TYPHACEÆ.—Pandanaceæ. / *Cyperaceæ.*

Fig. LXXXIV.—1. Typha latifolia ; 2. its fruit ; 3. a section of the seed ; 4. the embryo ; 5. a stamen. —*Nees v. Esenbeck.*

ORDER XXXVI. ARACEÆ.—ARADS.

AROIDEÆ, *Juss. Gen.* 23. (1789); *R. Brown Prodr.* 333; *Blume, Rumphia* 1. 74; *Endl. Gen.* lxxii.; *Meisner*, p. 360; *Kunth enum.* 3. 1; *Martius in Bot. Zeitung*, 1831, p. 449. *Richard in Arch. de Bot.* i. 11.—ARACEÆ, *Schott Meletemata*, 16. (1832).

DIAGNOSIS.—*Aral Endogens, with numerous naked flowers on a solitary spadix covered by a simple hooded spatha, sessile anthers, loose seeds, and a slit axile embryo.*

Herbaceous plants, frequently with a fleshy corm; or shrubs; stemless or arborescent, or climbing by means of aerial roots. Leaves sheathing at the base, convolute in the bud, usually with branching veins; sometimes compound! often cordate. Spadix generally enclosed in a spathe. Flowers ♂ ♀, naked, arranged upon the surface of a spadix, within a spathe. ♂: Stamens definite or indefinite, hypogynous, very short; anthers 1- 2- or many-celled, ovate, turned outwards. ♀, at the base of the spadix, Ovary free, 1-celled, very seldom 3- or more-celled, and many-seeded; ovules erect or parietal, sessile, or attached to long cords, orthotropal, campylotropal, or occasionally anatropal; stigma sessile. Fruit succulent. Seeds pulpy; embryo in the axis of fleshy or mealy albumen, straight, taper, with a cleft in one side, in which the plumule lies; (radicle obtuse, usually next the hilum, occasionally at the opposite extremity. *R. Br.*) Albumen sometimes wanting.

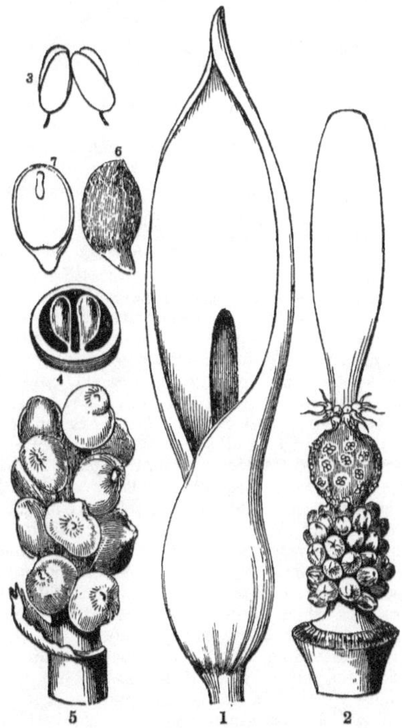

Fig. LXXXV.

The hooded spathe of the order of Arads affords a character not to be mistaken, and, connected with their diclinous naked flowers, gives them their most essential diagnosis; Bulrushes are distinguished by their long anthers and want of spathe; Screw-Pines by their solid embryo and compound fruit; and Duckweeds by their reduction to the simplest state in which flowering plants can exist. The whole of these Orders, taken together, are known by their general tendency to develop their flowers upon a spadix, by their want of floral envelopes, or by those parts not assuming the distinct forms of calyx and corolla, but existing only in the state of herbaceous scales. With the exception of Screw-Pines, they are all also known by their plumule lying within a cleft of the embryo; a structure found in few other monocotyledonous plants, except Naiads, in which the embryo is otherwise widely different, and the hermaphrodite Orontiaceæ, which are so much like Arads in all but the combination of their sexes.

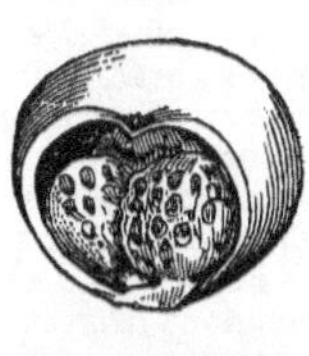

Fig. LXXXVI.

Fig. LXXXVII.

Natives of all tropical countries abundantly, but of temperate climates rarely. In

Fig. LXXXV.—1. Spathe of Arum maculatum; 2. its spadix loaded with flowers; 3. an anther; 4. a transverse section of an ovary; 5. a cluster of ripe fruits; 6. a seed; 7. a section of the same, showing the embryo.

Fig. LXXXVI.—A single fruit divided vertically, so as to show the seeds *in situ*.

Fig. LXXXVII.—A perpendicular section of one of the seeds.

cold or temperate climates they are usually herbaceous, while in tropical countries they are often arborescent and of considerable size, clinging to trees by means of their aerial roots, which they protrude in abundance. In America, according to Humboldt (*Distr. Géogr.* 196), their principal station is on the submontane region, between 1200 and 3600 feet of elevation, where the climate is temperate and rains abundant.

An acrid principle generally pervades this Order, and exists in so high a degree in some of them as to render them dangerous poisons. The most remarkable is the Dumb Cane, or Dieffenbachia Seguina, a native of the West Indies and South America, growing to the height of a man: this plant has the property, when chewed, of swelling the tongue, and destroying the power of speech. Hooker relates an account of a gardener, who "incautiously bit a piece of the Dumb Cane, when his tongue swelled to such a degree that he could not move it; he became utterly incapable of speaking, and was confined to the house for some days in the most excruciating torments." The same excellent botanist adds, that it is said to impart an indelible stain to linen. P. Browne states, that its stalk is employed to bring sugar to a good grain when it is too viscid, and cannot be made to granulate properly by the application of lime alone; Cryptocoryne ovata is used for the same purpose. The leaves of Colocasia esculenta excite violent salivation and a burning sensation in the mouth, as I have myself experienced. Milk in which the acrid root of Arum triphyllum has been boiled has been known to cure consumption. *DC.* Notwithstanding this acridity, the flat under-ground corms, called roots, and the leaves of many Arads, are harmless, and even nutritive when roasted or boiled; as for instance, those of Caladium bicolor, pœcile and violaceum, Colocasia esculenta, himalensis, antiquorum, mucronata, and others, which, under the names of Cocoa root, Eddoes, and Yams, are common articles of food in hot countries. Nevertheless the juice of Caladium bicolor is cathartic and anthelmintic. Whole fields of Colocasia macrorhiza are cultivated in the South Sea Islands, under the name of Tara or Kopeh roots. The corms of the Arum maculatum are commonly eaten by the country people in the Isle of Portland; they are macerated, steeped, and the powder obtained from them is sent to London for sale under the name of Portland Sago. They are universally cultivated in India, and known there under the names of Kuchoo and Gaglee. Arum nymphæifolium, which Dr. Roxburgh considers only a variety of C. antiquorum, is but rarely cultivated in Bengal. Arum indicum, Mankuchoo and Man-guri of the Bengalese, is a species much cultivated about the huts of the natives for its esculent stems and small pendulous tubers. Arum campanulatum, now Amorphophallus, Ol of the Bengalese, and which deserves to be called the Telinga Potato, is also much cultivated, especially in the Northern Circars, according to Dr. Roxburgh, where it is highly esteemed for the wholesomeness and nourishing quality of its roots. In the Himalayas, the species which is called Colocasia himalensis forms the principal portion of the food of the hill-people. *Royle.* (Medicinally, the root in its recent state is stimulant, diaphoretic, and expectorant.) A similar starchy substance is yielded by Xanthosoma sagittifolia (Chou caraib), Peltandra virginica, and the huge and hideous Amorphophalli of the Indian Archipelago. The spadixes of some species have a fetid putrid smell; others, such as Arum cordifolium, Italicum, and maculatum, are found to disengage a sensible quantity of heat at the time when they are about to expand. The emanations from Arum Dracunculus are extremely inconvenient; when in flower they produce dizziness, head-ache, and vomiting. A writer in the Annals of Chemistry says that he was attacked with violent head-ache and sickness after gathering about 40 of the spadixes. Amorphophallus orixensis having exceedingly acrid roots, is, when fresh, applied in India by the natives in cataplasm to excite, or bring forward tumours. Dr.

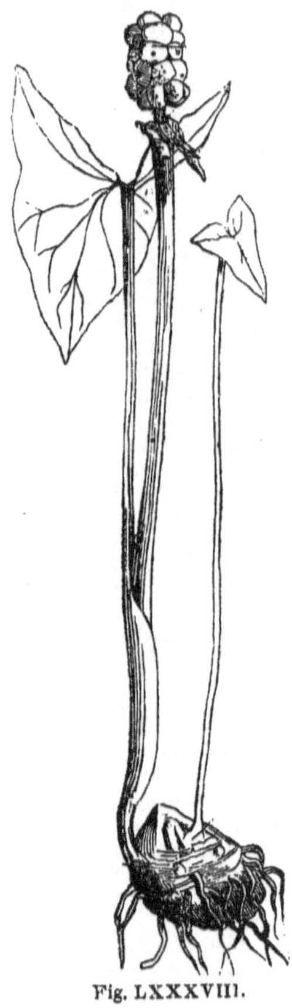

Fig. LXXXVIII.

Fig. LXXXVIII.—Arum maculatum.

Roxburgh pronounces it to be certainly a most powerful stimulant ; other species are likewise employed, as A. montanum, Roxb., (macrorhizon, Ainslie). The plant called by the latter Dracontium polyphyllum is exhibited internally when its acrimony has been subdued ; it is considered antispasmodic, and is also said to be useful in asthmatic cases. An emmenagogue is said to be prepared from it in the Society Islands. Agardh considers that the acrid principle, which, notwithstanding its great fugacity, has been obtained pure, is no doubt of great power as a stimulant. *Aph.* 133. The Colocasias are remarkable for being milky. Various species of Philodendron have a turbid acrid juice, and are found useful in cleansing foul ulcers ; they are also employed for many other purposes in Brazil. See *Martius Mat. Med. Bras.* 96, who mentions Dracontium polyphyllum, Arisæma Pythonium, and Monstera Adansonii, as caustics.

GENERA.

I.—*Cryptocoryneæ.* Stamens distinct from the pistils, which are several, whorled round the base of the spadix, and there combined into a many-celled ovary.

Cryptocoryne, *Fisch.*
Stylochaeton, *Lepr.*

II.—*Dracunculeæ.* Stamens and pistils numerous, with rudimentary organs interposed. Spadix naked at the end. Cells of the anthers larger than the connective.

Arisarum, *Tournef.*
Arisaema, *Mart.*
Biarum, *Schott.*
Homaid, Adans.
Ischarum, Blume.
Arum, *Linn.*
Gigarum, Caesalp.
Eminium, Blume.
Typhonium, *Schott.*
Stauromatum, Schott.
Theriophonum, Blum.
Dracunculus, *Tournef.*
Pythonium, *Schott.*
Thomsonia, Wall.
Amorphophallus *Blume.*
Candarum, Reichenb.
Pythion, Mart.

III.—*Caladieæ.* Stamens and pistils numerous, contiguous or separated by the rudimentary bodies. Spadix usually naked at point. Cells of anthers with a very thick connective.

Remusatia, *Schott.*
Gonatanthus, *Kl*
Colocasia, *Ray.*
Caladium, *Vent.*
Peltandra, *Rafin.*
Rensselaeria, Beck.
Lecontia, Torr.
Xanthosoma, *Schott.*
Acontias, *Schott.*
Syngonium, *Schott.*
Culcasia, *Palis.*
Denhamia, Schott.
Philodendron, *Schott.*
Calostigma, Schott.
Meconostigma, Schott.
Sphincterostigma, Scht.

IV.—*Anaporeæ.* Stamens and pistils numerous, contiguous, usually having the rudimentary bodies intermixed with the pistils. Point of spadix rarely naked. Cells of the anthers immersed in a very thick fleshy connective.

Spathicarpa, *Hook.*
Dieffenbachia, *Schott.*
Pinellia, *Tenor.*
Atherurus, Blum.
Hemicarpurus, Nees.
Aglaonema, *Schott.*
Homalonema, *Schott.*
Richardia, *Kunth.*
Zantedeschia, Spr.

NUMBERS. GEN. 26. Sp. 170.

Orontiaceæ.
POSITION.—Lemnaceæ.—ARACEÆ.—Typhaceæ.
Palmaceæ.

ORDER XXXVII. PANDANACEÆ—SCREWPINES.

Pandaneæ, *R. Brown, Prodr.* 340. (1810); *De Cand. Propr. Méd.* 278. (1816); *Agardh Aph.* 133. (1822); *Gaudichaud in Ann. des Sc.* 3. 509. (1824); *Schott et Endlicher Meletemata, p.* 15. (1832). *Endl. gen.* lxxiv.; *Meisner, p.* 359; *Kunth Enum.* 3. 93; *Bennett in Horsfield, Pl. Jav.* 32; *Blume Rumphia*, 1. 155.—Cyclantheæ, *Poiteau in Mem. Mus.* 9. 34. (1822); *Schott et Endlicher, Meletemata, p.* 15. (1832); *Martius Conspectus, No.* 22. (1835).—Cyclanthaceæ, *ed. pr.*—Freycinetieæ, *Ad. Brongn. tableau* xv. (1843).

DIAGNOSIS.—*Aral Endogens, with numerous naked or scaly flowers, arranged on a spadix covered by many spathes, stalked anthers, loose seeds, and a solid minute embryo.*

Trees or bushes, sometimes sending down aerial roots, sometimes weak and decumbent. Leaves imbricated, in three rows, long, linear-lanceolate, amplexicaul, with their margins almost always spiny; or pinnated, or fan-shaped; the latter being true leaves, the former, perhaps, mere leaf-stalks. Floral leaves smaller, often coloured, and spathaceous.

Flowers ♂ ♀ or polygamous, naked, or furnished with a few scales, arranged on a wholly covered spadix. ♂: Stamens numerous. Filaments with single anthers; anthers 2-4-celled. ♀: ovaries usually collected in parcels, 1-celled; stigmas as many as the ovaries, sessile; ovules solitary, attached to the suture, or very numerous, and springing from as many parietal placentæ as there are styles, anatropal. Fruit either fibrous drupes, usually collected in parcels, each 1-seeded; or many-celled berries, with polyspermous cells. Albumen fleshy, with a minute embryo at the base next the hilum, not slit on one side.

Although this Order is certainly very distinct from Arads, it is by no means easy to define its limits. Blume says it is principally known by its numerous spathes to each spadix, and its narrow, sessile, 3-rowed leaves, spiny at the back and edge, (*Rumphia* 2. 155); but this applies only to Pandaneæ proper, for the Cyclantheous division has the flabellate or pinnate foliage of Palms, and to all appearance establishes the connection between the Aral and Palmal Alliances.

The species of Pandanus and Freycinetia have the aspect of gigantic Bromelias, bearing the flowers of a Sparganium. While there is no analogy with the former in structure beyond the general appearance of the foliage; the organisation of the fructification bears so near a resemblance to the latter as to have led to the combination of Screwpines and Typhads by botanists of the first authority. But when we contrast the naked flowers, the compound highly-developed fruit, the spathaceous bracts, the entire embryo, and the arborescent habit of the for-

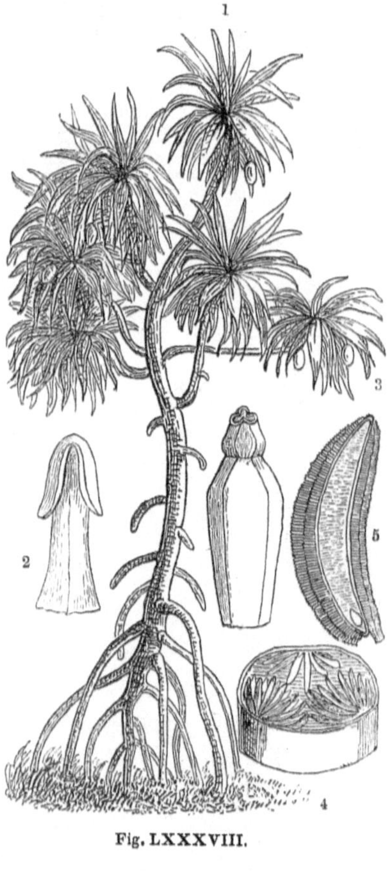

Fig. LXXXVIII.

Fig. LXXXVIII.—1. A Pandanus; 2. a stamen of Freycinetia imbricata; 3. an ovary of ditto; 4. the transverse section of the same; 5. a perpendicular section of its seed.—*Blume*

mer, with the half-glumaceous flowers, the simple fruit, the want of spathaceous bracts, the slit embryo, and the herbaceous sedgy habit of the latter, it is difficult to withhold our assent from the proposition to separate them. Brown remarks (*Prodr.* 341), that these have no affinity with Palms beyond their arborescent stems. But, on the contrary, Cyclantheæ, which, following Poiteau and others, I formerly adopted, have, with the structure of Pandaneæ proper, the foliage of Palms, and are in reality a connecting link between the two Orders. At least, Carludovica evidently is so, as is shown by Hooker's figure in the Botanical Magazine, t. 2951, and Cyclanthus seems to have no peculiarity beyond a curious spiral arrangement of its ♂ and ♀ flowers in alternate rows.

Mr. Bennett has pointed out an error made by Gaudichaud, who places the embryo at the apex of semitransparent albumen. He states, that it is certainly at the base, as indeed Blume has shown in a beautiful figure of Freycinetia imbricata. Screw-pines are remarkable among arborescent monocotyledons for their constant tendency to branch, which is always effected in a dichotomous manner. Their leaves have also a uniform spiral arrangement round the axis, so as to give the stems a sort of corkscrew appearance before the traces of the leaves are worn away. The Chandelier Tree of Guinea and St. Thomas's derives its name (Pandanus Candelabrum) from this peculiar tendency to branching. According to *Fée* (1. 223), Nipa ought to be referred here, and not to Palms, an opinion adopted by Kunth, but not by Endlicher. A figure of it will be found at p. 133, in a sketch of the vegetation of Palms. The Tagua plant, or Vegetable ivory, referred hither by Endlicher and others, seems to be a true Palm. According to Mr. Bennett, the seeds of Freycinetia and Pandanus have such an abundance of raphides in their testa, that those crystals are conspicuous to the naked eye.

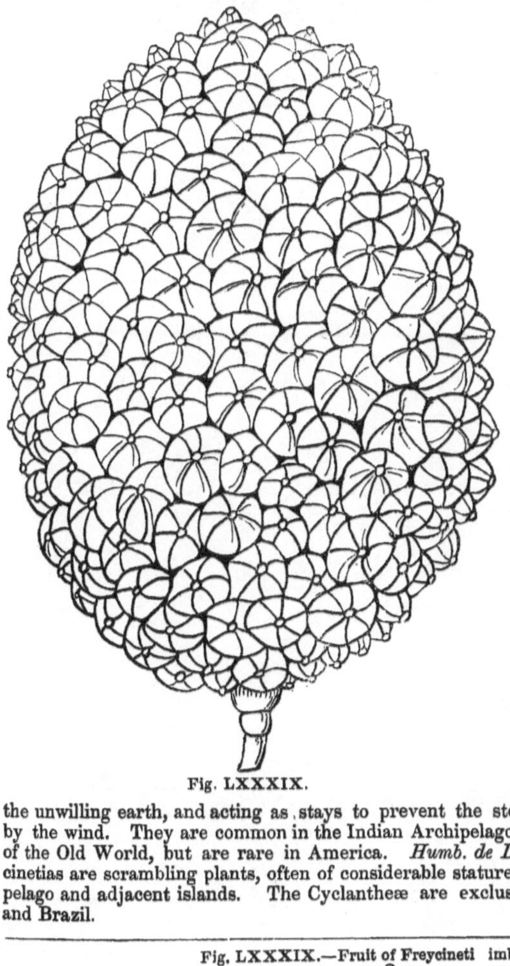

Fig. LXXXIX.

The Screw-pines are abundant in the Mascaren Islands, especially the Isle of France, where, under the name of Vaquois, they are found covering the sandy plains. There they have peculiar means given them by nature to subsist in such situations in the shape of strong aerial roots, which are protruded from the stem, and descend towards the earth, bearing on their tips a loose cup-like coating of cellular integument, which preserves their tender newly-formed absorbents from injury until they reach the soil, in which they quickly bury themselves, thus adding at the same time to the number of mouths by which food can be extracted from the unwilling earth, and acting as stays to prevent the stems from being blown about by the wind. They are common in the Indian Archipelago, and in most tropical islands of the Old World, but are rare in America. *Humb. de Distr. Geogr.* 198. The Freycinetias are scrambling plants, often of considerable stature, found in the Indian Archipelago and adjacent islands. The Cyclantheæ are exclusively American, from Peru and Brazil.

Fig. LXXXIX.—Fruit of Freycineti imbricata

The seeds of Pandanus are eatable. The flowers of Pandanus odoratissimus are fragrant and eatable, and are reckoned in India aphrodisiac. The juice of Nipa, as it flows from the pounded spadices, furnishes one of the inferior kinds of Palm wine. Some plant of this Order is probably the "Palm" mentioned by Mr. Drummond as having a fruit which the natives of the Swan River find wholesome when fermented for some time, but which without preparation, produces violent vomiting and other dangerous symptoms. *Hook. Journ.* 356. The fruit of several is also an article of food. The leaves are used for thatching and cordage, and their juice is employed in diarrhœa and dysentery. The immature fruit is reputed emmenagogue.—*Humb.* l. c.

GENERA.

I.—*Pandaneæ.* Leaves simple. Flowers naked.

Pandanus, *Linn. fil.*
Arthrodactylis, Forst.
Keurva, Forsk.
Marquartia, *Hassk.*
Freycinetia, *Gaudich.*

II.—*Cyclantheæ.* Leaves flabellate or pinnate. Flowers usually furnished with a calyx.

Carludovica, *Ruiz et Pav.*
Ludovia, Pers.
Salmia, Willd.
Nipa, *Rumph.*
Cyclanthus, *Poit.*
Cyclosanthes, Pöpp.
Wettinia, *Pöpp.*

NUMBERS. GEN. 7. Sp. 75.

POSITION.—Araceæ.—PANDANACEÆ.—Typhaceæ.
Palmaceæ.

ALLIANCE IX. *PALMALES.*—THE PALMAL ALLIANCE.

DIAGNOSIS.—*Unisexual (or bisexual) Endogens, with perfect flowers, seated on a branched scaly spadix, and a minute embryo lodged below the surface of horny or fleshy albumen.*

At this point the vegetative force of Endogens acquires its maximum power, resulting for the most part in trees of gigantic stature, always forming wood, and occasionally arriving at dimensions wholly unknown among other plants, as in the instance of some of the Calami or Rotangs, which Rumphius assures us are sometimes as much as 1200 or even 1800 feet long, (*Rumphia*, 2. 158.) A writer in the Linnæa, (14. 263) asserts that Palms are nearer Arads than to any other order of Endogens, and I think him so right that I should have followed Meisner in including Palms in the Aral alliance, if it had not been for their perfect floral envelopes, the uniformly indeterminate station of their embryo, and the tendency that exists among them to form ☿ flowers; which circumstances bring them close upon Lilyworts, and seem to indicate a higher organ of organization than we find among the incomplete-flowered, constantly ♂ ♀ Aral plants.

The general opinion of Botanists seems to be in favour of regarding Palms as one natural order, an opinion to which it does not seem at present desirable to object. It may however be observed, that the scaly-fruited genera, called Lepidocaryinæ by Martius, Calameæ by Kunth, and Calaminæ by Griffith, offer in that

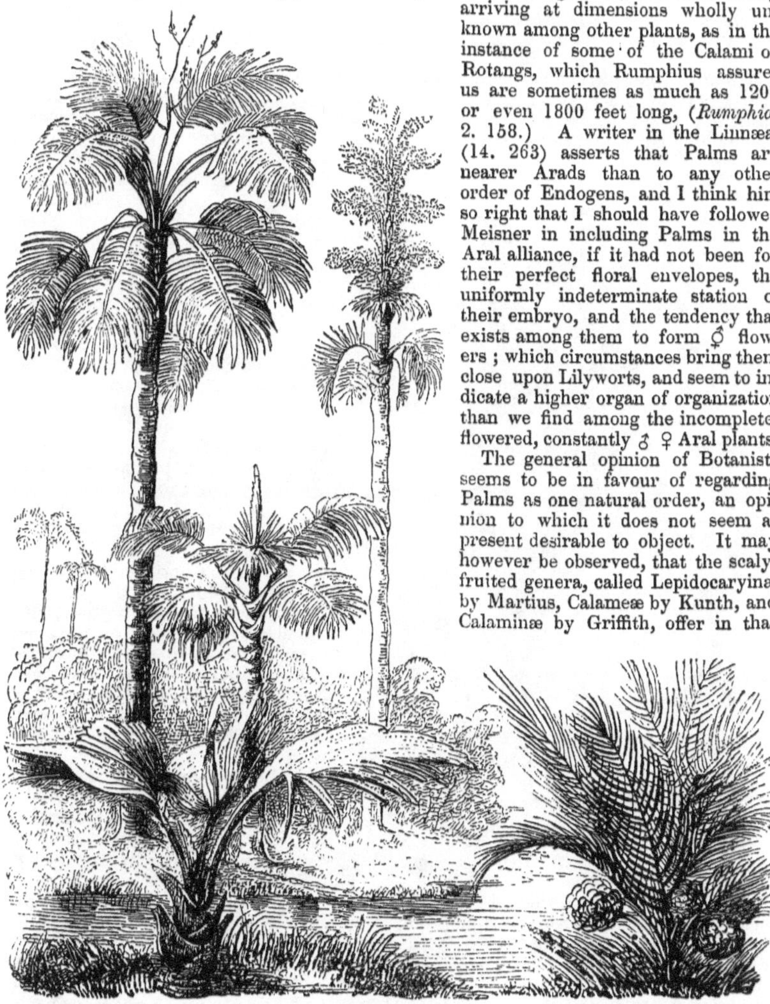

1 Fig. XC. 2

circumstance, and also in most instances in their habit, a very considerable deviation from the condition of the other genera, and seem to indicate the existence of at least one natural order to be struck off the true Palms.

Fig. XC.—Palm Trees (Blume). 1. Corypha Gebanga; 2. Nipa fruticans (see **Pandanaceæ**).

ORDER XXXVIII. PALMACEÆ.—PALMS.

Palmæ, *Juss. Gen.* (1789) ; *R. Brown, Prodr.* 266. (1810) ; *Von Martius Palm. Bras.* (1824 to 1836) ; *Id. Programma* (1824) ; *Bartl. Ord. Nat.* 63. (1830) ; *Endl. gen.* lxxv. ; *Royle Illustrations, p.* 399 ; *Blume Rumphia, vol.* 2. *passim ; Kunth. enum.* 3. 168 ; *Meisner, p.* 354. *Griffith in Calcutta Journal of Natural History, vol.* ?.—Phytelephanteæ, *Martius Conspectus, No.* 21. (1835).

DIAGNOSIS.—*Unisexual (or bisexual) Endogens, with perfect flowers, seated on a branched scaly spadix, and a minute embryo lodged below the surface of horny or fleshy albumen.*

Trunk arborescent, simple, occasionally shrubby, sometimes branched, rough with the dilated half-sheathing bases of the leaves or their scars ; in the Rotangs flagelliform, and extremely long ; occasionally armed with stiff spines. Leaves clustered, terminal, usually very large, pinnate or flabelliform, plaited, with parallel simple veins ; in some cases eroded and wedge-shaped. Spadix scaly, terminal, often branched, enclosed in a 1- or many-valved spathe, which is often woody. Flowers small, supported by scaly bracts, ♂ ♀, or occasionally ☿. Sepals 3, colourless, fleshy or leathery, persistent. Petals 3, often larger, and sometimes deeply connate. Stamens inserted into the base of the perianth, usually definite in number, opposite the segments, to which they are equal in number, seldom 3 ; sometimes indefinite in number. Ovary free, usually composed of 3 carpels, completely united, or partially so ; occasionally of 2 or 1 only. Ovules solitary, very rarely 2, erect, orthotropal, or anatropal in various degrees. Styles continuous with the carpels. Fruit drupaceous, or nut-like, or berried, often with a fibrous rind. Seed filling the cavity in which it grows, often reticulated. Albumen cartilaginous, often ruminate, frequently furnished with a central or ventral cavity ; embryo lodged in a particular cavity of the albumen, usually at a distance from the hilum, dorsal and indicated by a little nipple, taper or pulley-shaped ; plumule concealed, scarcely visible ; the cotyledonar extremity becoming thickened in germination, and either filling up a pre-existing cavity, or one formed by the liquefaction of the albumen in the centre.

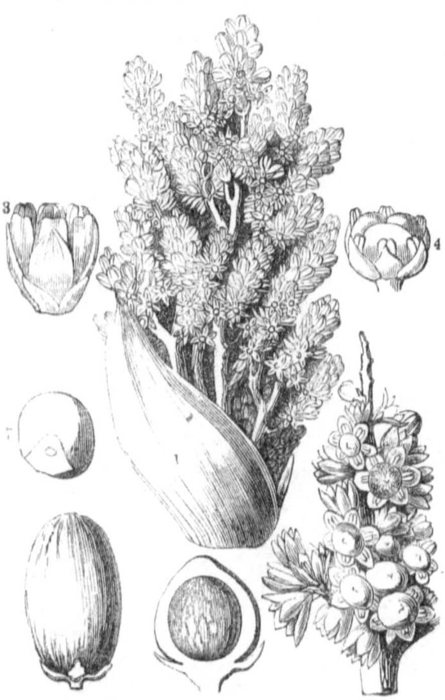

5 6 Fig. XCI. 2

The race of plants to which the name of Palms has been assigned is, no doubt, the most interesting in the vegetable kingdom, if we consider the majestic aspect of their towering stems, crowned by a still more gigantic foliage ; the character of grandeur which they impress upon the landscape of the countries they inhabit ; their immense value to mankind, as affording food, and raiment, and numerous objects of economical importance ; or, finally, the prodigious development of those organs by which their race is to be propagated. A single spathe of the Date contains about 12,000 male flowers ; Alfonsia amygdalina has been computed to have 207,000 in a spathe, or 600,000 upon a

Fig. XCI.—1. Inflorescence of Chamærops humilis, in its spathe ; 2. a portion of the same, with the fruit ripening ; 3. a male flower ; 4. a female flower ; 5. a ripe fruit ; 6. a section of another variety, showing the seed ; 7. a seed with a portion of the surface cut away, to display the embryo.

single individual; while every bunch of the Sejo Palm of the Oronoco bears 8000 fruits They are very uniform in the botanical characters by which they are distinguished, especially in their fleshy colourless 6-parted flowers, enclosed in spathes, their minute embryo lying in the midst of albumen and remote from the hilum, and their arborescent stems with rigid, plaited or pinnated, inarticulated leaves, called fronds; but their aspect and habits are extremely various. To use the words of the most accomplished traveller of our own, or any age;—" While some (Kunthia montana, Aiphanes Praga, Oreodoxa frigida) have trunks as slender as the graceful reed, or longer than the longest cable, (Calamus Rudentum, 500 feet), others (Jubæa spectabilis and Cocos butyracea) are 3 and even 5 feet thick; while some grow collected in groups (Mauritia flexuosa, Chamærops humilis), others (Oreodoxa regia, Martinezia caryotæfolia) singly dart their slender trunks into the air; while some have a low caudex (Attalea amygdalina), others exhibit a towering stem 160-180 feet high (Ceroxylon andicola); and while one part flourishes in the low valleys of the tropics, or on the declivities of the lower mountains, to the elevation of 900 feet, another part consists of mountaineers bordering upon the limits of perpetual snow." To which may be added, that while many have a cylindrical undivided stem, the Doom Palm of Upper Egypt, and an allied species, the Hyphæne coriacea, are remarkable for their dichotomous repeatedly-divided trunk. The Calami, or Rotangs, and the siliceous secretions of their leaves, indicate an affinity with Grasses, which would hardly be anticipated, if the grasses of our European meadows were compared with the Cocoa Nuts of the Indies, but which becomes more apparent when the Bamboo is placed by the side of the Cane. The Rattan Palms, called by Rumphius Palmijunci, are described as inhabitants of dense forests, where the rays of the sun can hardly penetrate, in which situations they form spiny bushes which obstruct all passage into those jungles, rising to the tops of the highest trees and falling again, so as to resemble a prodigious length of cable, adorned however with the most beautiful leaves, pinnated or terminating in graceful tendrils.

Fig. XCII.

Von Martius, the great illustrator of this noble family, speaks thus of their habits and geographical arrangement:—" Palms, the splendid offspring of Tellus and Phœbus, chiefly acknowledge as their native land those happy regions seated within the tropics, where the beams of the latter forever shine. Inhabitants of either world, they hardly range beyond 35° in the southern, or 40° in the northern hemisphere. Particular species scarcely extend beyond their own contracted limits, on which account there are few countries favourable for their production in which some local and peculiar species are not found; the few that are dispersed through many lands are chiefly Cocos nucifera, Acrocomia sclerocarpa, and Borassus flabelliformis. It is probable that the number of species thus scattered over the face of nature will be found to amount to 1000 or more. Of these not a few love the humid banks of rivulets and streams, others occupy the shores of the ocean, and some ascend into alpine regions; some collect into dense forests, others spring up singly or in clusters over the plains." *Progr.* 6. The testimony of Von Martius is confirmed by Humboldt, who also asserts that there must be an incredible number still to discover in equinoctial regions; especially if we consider

Fig. XCII.—Sagus Rumphii. 1. a flower; 2. the same opened; 3. a section of an ovary; 4. a section of a seed of Sagus filaris; 5. fruit and remains of spadix.—*Blume.*

how little is yet known of Africa, Asia, New Holland, and America. He and Bonpland discovered a new species in almost every 50 miles of travelling, so narrow are the limits within which their range is confined. A different opinion appears to be entertained by Schouw, a respectable Danish writer upon botanical geography, whose views deserve to be quoted, although he is far from having had such personal means of judging as Humboldt and Von Martius. He seems to consider that we are acquainted already with the greater part of the Palms; for he says, "it appears from the reports of travellers that such Palm woods as those of South America are less frequent in other parts of the world. Africa and New Holland seem to be less favourable to this tribe, for on the Congo, Smith found only from 3 to 4 Palms; in Guinea we know merely of the same number; and of the other African Palms, 6 belong to the Isles of Bourbon and France; New Holland has, in the torrid zone, three species, while Forster's *Prodromus* of the Flora of the South Sea Islands contains four." It is, however, not to be forgotten that Blume and Griffith have alone added 65 new species to the list of Indian Palms. Blume is of opinion that great numbers still remain to be discovered "in immensis illis et fertilissimis regionibus quarum pleræque primitivâ atque intactâ vegetatione conteguntur, neque unquam ab Europæis lustratæ sunt." The most northern limit of Palms is that of Chamærops Palmetto in N. America, in lat. 34°-36°, and of Chamærops humilis in Europe, near Nice, in 43°-44° N. lat. They are found in the southern hemisphere as low as 38° in New Zealand. "It is remarkable that no species of Palm has been found in South Africa, nor was any observed by M. Leschenault on the west coast of New Holland, even within the tropic." *Brown in Flinders*, 577.

Wine, oil, wax, flour, sugar, salt, says Humboldt, are the produce of this tribe; to which Von Martius adds, thread, utensils, weapons, food, and habitations. The most remarkable is the Cocoa Nut, of which an excellent account will be found in the *Trans. of the Wernerian Society*, vol. v. The root is sometimes masticated instead of the Areca Nut; of the small fibres baskets are made in Brazil. The hard case of the stem is converted into drums, and used in the construction of huts; the lower part is so hard as to take a beautiful polish, when it resembles agate; the reticulated substance at the base of the leaf is formed into cradles, and, as some say, into a coarse kind of cloth. The unexpanded terminal bud is a delicate article of food; the leaves furnish thatch for dwellings, and materials for fences, buckets, and baskets; they are used for writing on, and make excellent torches; potash in abundance is yielded by their ashes; the midrib of the leaf serves for oars; the juice of the flower and stems is replete with sugar, and is fermented into excellent wine, or distilled into a sort of spirit, called Arrack; or the sugar itself is separated, under the name of Jagery. The value of the fruit for food, and the delicious beverage which it contains, are well known to all Europeans. The fibrous and uneatable rind is not less useful: it is not only used to polish furniture and to scour the floors of rooms, but is manufactured into a kind of cordage, called Coir rope, which is nearly equal in strength to hemp; and which Roxburgh designates as the very best of all materials for cables, on account of its great elasticity and strength. Finally, an excellent oil is obtained from the kernel by expression. The juice which flows from the wounded spathes of Borassus flabelliformis, Raphia vinifera, Mauritia vinifera, the Cocoa Nut, and other Palms, is known in India by the name of Toddy. Independently of the grateful qualities of this fluid as a beverage, it is found to be the simplest and easiest remedy that can be employed for removing constipation in persons of delicate habit, especially European females. According to Roxburgh, Caryota urens is highly valuable to the natives of the countries where it grows in plenty. It yields them, during the hot season, an immense quantity of this toddy, or palm wine. The best trees will yield at the rate of 100 pints in the twenty-four hours. The pith, or farinaceous part of the trunk of old trees, is said to be equal to the best Sago; the natives make it into bread, and boil it into thick gruel; these form a great part of the diet of the people whose country it inhabits, and during famines they suffer little while those trees last. Roxburgh found it highly nutritious. He ate the gruel, and thought it fully as palatable as that made of the Sago we get from the Malay countries (Sagus lævis). *Fl. Ind.* 3. 625.

The finest Sago is prepared from Sagus lævis and genuina, trees forming immense forests on nearly all the Moluccas, and so rich in starch that each individual is reckoned to furnish from 600 to 800 lb. of Sago (*Rumphia*, 2. 148); a similar substance is however yielded by Caryota urens, Phœnix farinifera, and many others.

The Saguerus saccharifer (or Arenga saccharifera) is one of the most important of the Order. Blume describes it (*Rumphia*, vol. 2. p. 126) as being from 20 to 25 feet high, and very common in the islands of the Indian Archipelago, the Moluccas and Philippines, where it is of the greatest value on account of its saccharine secretions. This juice is obtained continually from the spadixes in large quantities, by wounding and pounding them while on the trees; it yields by fermentation an intoxicating beverage, and, when boiled, a kind of sugar, consumed for various purposes. When the trees are exhausted by the incessant draining of their fluids, Sago of good quality is obtained from the trunk,—as much as 150 to 200 lbs. weight from a single tree. The timber is extremely hard, and fit for building purposes; and the leaf-stalks yield annually from 4 to 7 lbs. of the strong black fibres, resembling horsehair, called Gomutie, which are extensively used in the manufacture of cables and various kinds of rope; they are also employed for stitching together thatch, for making brooms and for similar purposes. (Are these the vegetable bristles now so largely imported for making brooms?) The midribs of the side leaves are converted into pens called Pansuri, and the fine arrows which the Indians blow from their long tubes. Finally, there is at the base of the leaves a fine woolly material (Baru) much employed in caulking ships, as stuffing for cushions, and as tinder. Their "Cabbage" is moreover eatable, like that of the West Indian Cabbage Palm, Areca oleracea, whose huge terminal bud is known by this name. Egyptian Bdellium, a gum-resinous substance, formerly employed as a diuretic and diaphoretic, is obtained from Hyphæne thebaica. Besides the Saguerus already mentioned, very considerable quantities of sugar are procured from Phœnix sylvestris, a kind of wild date, which Dr. Roxburgh computed to furnish annually in Bengal 100,000 cwt. of date sugar.

The well known Betel Nut is the fruit of Areca Catechu, and remarkable for its narcotic or intoxicating power; from the same popular fruit is prepared a kind of spurious Catechu. It seems to me however doubtful whether the intoxicating effect of the Betel nut is not owing to the Piper leaf in which it is wrapped when eaten, rather than to any special property of its own.

Blume tells us that the Asiatic nations would rather forego meat and drink than their favourite Areca nuts; whole ship-loads of which are annually exported from Sumatra, Malacca, Siam, and Cochinchina. They contain a large quantity of tannin, which has caused them to be employed in some part of India for dyeing cotton cloths. The leafstalks, spathes, and timber are employed for many domestic purposes, and in Malabar an inebriating lozenge is prepared from the sap. (*Rumphia*, 2. 67.) In the opinion of this author, the practice of chewing the nuts, although offensive to Europeans, is really very conducive to health in the damp and pestilent regions of India, where the natives live upon a spare and miserable diet. As to the Brazilian Palms, Martius states that the kernel of various species of Attalea, when rubbed in water, form an emulsion used in medicine, both externally and internally. The juice of the unripe fruit of Cocos schizophyllus is employed in slight ophthalmic attacks.

The fruit of a few of them is eatable; as, for example, the Date Palm, Phœnix dactylifera, which furnishes the most important part of their food to the tribes of the desert; some other species of Phœnix eaten in India; the Cocoa Nut, too well known to require description; and the Doom Palm, Hyphæne thebaica, which is called in Egypt the Gingerbread Tree, because of the extreme resemblance of its brown mealy rind to that sort of cake; Zalacca edulis, a kind of Cane, with a juicy, pulpy covering to its seeds, much esteemed by the Burmese; and a few others of less importance. In some, however, the fruit is extremely acrid.

The fruit of Saguerus saccharifer is of that nature, exciting severe inflammation in the mouth of those who chew it; it was the basis of the "infernal water" which the Moluccans used in their wars, to pour over their enemies; nevertheless, the unripe albumen forms a beautiful kind of sweetmeat, which the Chinese and Indian nobles drink with their tea; it is prepared by soaking in lime-water and boiling in refined sugar. The same acridity occurs in the fruit of Caryota urens and some others.

Oil and wax are only of less common occurrence than farinaceous secretions. Palm oil, of which enormous quantities are employed in Europe as a sort of grease, and in soap and candle making, is chiefly obtained from Elais guineensis and melanococca, and these trees are also said to yield the best kind of Palm wine. Œnocarpus Bacaba and many Cocoinæ are other species whose fruit contains oil. The Ceroxylon andicola, or Wax Palm of Humboldt, has its trunk covered by a coating of wax, which exudes from the spaces between the insertion of the leaves. It is, according to Vauquelin, a concrete inflammable substance, consisting of 1-3d wax and 2-3ds resin. It is a very remarkable fact, first noticed by Brown (*Congo*, 456), that the plants of this order whose fruit affords oil belong to a tribe called by him Cocoinæ, which are particularly characterised by the originally trilocular putamen having its cells when fertile perforated opposite the seat of the embryo, and, when abortive, indicated by foramina cæca. A species called Carnauba, in Brazil, throws off waxy scales from its leaves.

Cocoa-nut oil is imported into England in considerable quantities, and it is surprising that it is not more generally used in England; for, instead of the detestable smell of fish-oil, it has rather an agreeable odour; and it is readily consumed in open glass vessels, with floating or standing wicks, whatever the temperature of the air may be.

The natural secretion of the fruit of Calamus Draco constitutes the best D'jurnang or Dragon's Blood, a dark coloured inodorous insipid resin; a second and rather inferior kind is produced from the fruit from which the natural secretion has been removed by heat and bruising; the third and most inferior kind appears to be the refuse of the last process. It is doubtful whether this article is procured from the plant by incisions, as has been supposed.—*Griffith.*

The roots of the American Palmetto have been found to contain a large quantity of tannin.

There seems no end to the economical purposes to which the products of Palms are applied. Their huge and hard-skinned leaves are universally employed as thatch All the hard-wooded sorts furnish excellent timber. The Brazilian Indians, especially the Puris, Patachos, and Botocudos, manufacture their best bows from the wood of a species of Cocoa-nut, called the Airi, or Brejeuba. Palmyra wood is produced by Borassus flabelliformis. Among those best known in Europe are the Rattans, belonging to various kinds of Cane, and so much valued for their flexibility on the one hand, and flinty hardness on the other. Palm walking-sticks (under the name of Penang lawyers), are also very extensively used in England. Mention has already been made of the valuable horse-hair-like bristles obtained from Saguerus saccharifer. Fibrous matter is also procured from Sagus filaris, a Malay plant, whose bristles are dried and used for sewing linen garments. Ropes and strings are prepared in Affghanistan from the Maizurrye Palm, a species of Chamærops, according to Mr. Griffith.

Thousands of boys and girls are employed in Java in weaving into baskets and bags the young leaves of the Gebang Palm (Corypha Gebanga, *Bl.*), one of the most useful of all the species of India; its pith furnishes a sort of Sago; its leaves are used for thatch and broad-brimmed hats; fishing-nets and linen shirts are woven from its fibres; ropes from its twisted leaf-stalks; the root is both emollient and slightly astringent: sliced, it is used in slight diarrhœas, and Waitz even says that it is a most valuable remedy for the periodical diarrhœas which, in the East Indies, attack Europeans out of health.—*Rumphia*, 2. 60.

Finally, the hard albumen of some species is turned to use in manufactures. The Hyphæne furnishes materials for rosaries; and Date kernels have been used by the turner; but the most celebrated is the Vegetable Ivory. This is the produce of a tree found on the banks of the river Magdalena, resembling Palms in its leaves, which equal those of the Cocoa-nut in dimensions, in its torulose scaly stem, and, finally, in the remarkable structure and weight of its fruit.—*Humb.* The Spanish Botanists Ruiz and Pavon also met with it in the groves of Peru in the hotter parts of the Andes, and named it Phytelephas macrocarpa. The natives of Columbia call it Tagua, or Cabeza de Negro (Negro's head), in allusion, we presume, to the figure of the nut. Almost all we know about it is contained in the following memorandum, published by the Spanish writers above mentioned. "The Indians cover their cottages with the leaves of this most beautiful Palm. The fruit at first contains a clear insipid fluid, by which travellers allay their thirst; afterwards this same liquor becomes milky and sweet, and it changes its taste by degrees as it acquires solidity, till at last it is almost as hard asi vory. The liquor contained in the young fruits becomes acid if they are cut from the tree and kept some time. From the kernels the Indians fashion the knobs of walking-sticks, the reels of spindles, and little toys, which are whiter than ivory, and as hard, if they are not put under water—and if they are, they become white and hard again when dried. Bears devour the young fruit with avidity." The toys prepared from it by the turner are well known in the London shops, and are much admired for their beautiful texture.

For further details concerning the useful qualities of this interesting race, see *Dr. Royle's Work* in the place above quoted.

GENERA.

I. Areceæ.

Chamædorea, *Willd.*
Nunnezharia, Ruiz et Pav.
Nunnezia, Willd.
Hyospathe, *Mart.*
Morenia, *Ruiz et Pav.*
Kunthia, *H. et B.*
Hyophorbe, *Gærtn.*
Sublimia, Commers.
Leopoldinia, *Mart.*
Euterpe, *Mart.*
Œnocarpus, *Mart.*
Oreodoxa, *Willd.*
Pinanga, *Rumph.*
Kentia, *Blume.*
Oncosperma, *Blume.*
Areca, *Linn.*
Euterpe, Gærtn.
Dypsis, *Noronh.*
Noronha, Thours.
Seaforthia, *R. Br.*
Orania, *Zippel.*
Arausiaca, Bl.
Ptychosperma, *Lab.*
Drymopalæus, Zipp.
Harina, *Hamilt.*
Orania, Bl.
Wallichia, Roxb.
Wrightia, Roxb.
Iriartea, *Ruiz et Pav.*
Ceroxylon, H. et B.
Cyrtostachys, *Bl.*
Calyptrocalyx, *Bl.*
Iguanura, *Bl.*
Saguerus, *Rumph.*
Arenga, Lab.
Gomutus, Rumph.
Caryota, *Linn.*

II. Calameæ. — (Lepidocaryeæ, *Martius;* Calaminæ, *Griffith.*)

* *Pinnated.*

Calamus, *Linn.*

Palmijuncus, Rumph.
Zalacca, *Reinw.*
Plectocomia, *Mart.*
Ceratolobus, *Blume.*
Dæmonorops, *Blume.*
Calamosagus, *Griff.*
Eugeissona, *Griff.*
Raphia, *Palis.*
Sagus, *Gærtn.*
Metroxylon, Rottb.

* * *Fan-leaved.*

Mauritia, *Linn. fil.*
Lepidocaryum, *Mart.*

III. Borasseæ.

* *Fan-leaved.*

Borassus, *Linn.*
Lontarus, Rumph.
? *Pholidorpus*, Blume.
Lodoicea, *Labill.*
Latania, *Commers.*
Cleophora, Gærtn.
Hyphæne, *Gærtn.*
Cucifera, Delil.
Douma, Lam.

* * *Pinnated.*

Bentinckia, *Berry.*
Keppleria, Mart.
Geonoma, *Willd.*
Gyncstum, Poiteau.
Vouay, Aubl.
Manicaria, *Gærtn.*
Pilophora, Jacq.

IV. Corypheæ.

§ 1. *Sabalidæ.*

Corypha, *Linn.*
Taliera, Mart.
Gembanga, Blume.
Livistona, *R. Br.*
Licuala, *Rumph.*
Saribus, *Rumph.*
Bissula, Rumph.
Pericycla, *Blume.*
Brahea, *Mart.*
Copernicia, *Mart.*
Caranaiba, Marc. Piso.
Cryosophila, *Blume.*
Sabal, *Adans.*
Chamærops, *Linn.*
Chamæriphes, Pont.
Phœnix, Cav.
Trithrinax, *Mart.*
Rhapis, *Linn fil.*
Thrinax, *Linn, fil.*

§ 2. *Phœnicidæ.*

Phœnix, *Linn.*
Elate, Ait.

V. Cocoeæ.

* *Spiny.*

Desmoncus, *Mart.*
Aititara, Marcgr.
Bactris, *Jacq.*
Guilielma, *Mart.*
Martinezia, *Ruiz et Pav.*
Acrocomia, *Mart.*
Astrocaryum, *C. W. G Meyer.*
Toxophœnix, Schott.

* * *Unarmed.*

Attalea, *H. B. K.*
Elaïs, *Jacq.*
Alfonsia, Kunth.
Cocos, *Linn.*
Langsdorfia, Raddi.
Syagrus, *Mart.*
Diplothemium, *Mart.*
Maximiliana, *Mart.*
Jubæa, *H. B. K.*
Molinæa, Berter.
Orbignya, *Mart.*

? Alagoptera, *Nees.*
Phytelephas, *Ruiz et Pav.*
Elephantusia, Willd.
Nipa, *Thunb.*
Nypa, Rumph.

NUMBERS. GEN. 73. Sp. 400.—(1000 Martius.)

Position —— *Pandanaceæ.* / PALMACEÆ. / *Juncaceæ.* ——

ALLIANCE X. *HYDRALES.*—THE HYDRAL ALLIANCE.

DIAGNOSIS—*Unisexual aquatic Endogens, with perfect or imperfect flowers, not arranged on a spadix, and without albumen.*

The essential character of the Hydral Alliance consists in its ♂ ♀ flowers and exalbuminous seeds ; it is therefore necessary to expel all those genera, which, like Potamogeton, have been placed among the Naiads although they are ⚥ ; for in truth there is nothing except the diclinous character which can distinctly divide the Arrow-grasses from the Naiads. Among the Frogbits, however, a couple of genera occur which are described as being truly ⚥ and yet cannot be referred to any other Order, and they therefore constitute real exceptions to the otherwise positive distinction. The Hydrals are all, as their name indicates, strictly aquatic, no instance of a land-plant occurring among them. They divide into three well-marked Orders, namely :

Stamens epigynous ; ovary adherent . . . . 39. HYDROCHARIDACEÆ.
Stamens hypogynous ; ovary free ; pollen globose . . . 40. NAIADACEÆ.
Stamens hypogynous ; ovary free ; pollen confervoid . . 41. ZOSTERACEÆ.

The genera of these Orders demand, however, a much more careful examination than they have yet had, and considerable changes may be expected among them ; for it is uncommon to find in the same Order so much diversity of condition as occurs among the Naiads and Frogbits as at present constituted.

ORDER XXXIX. HYDROCHARIDACEÆ.—HYDROCHARADS.

Hydrocharides, *Juss. Gen.* 67. (1789.)—Hydrocharideæ, *DC. Fl. Fr.* 3. 265. (1815); *R. Brown Prodr.* 344. (1810); *Richard in Mem. Mus.* vol. 1. 365. (1815); *Agardh Aph.* 127. (1822), *Endl. gen.* lix. *Meisner, p.* 365.—Vallisneriaceæ *and* Stratioteæ, *Link Handb.* 1. 281. (1829).—Anacharideæ, *Endl. gen. p.* 161.

DIAGNOSIS.—*Hydral Endogens with epigynous stamens and an adherent ovary.*

Floating or water-plants. Leaves with parallel veins, sometimes spiny. Flowers enclosed in a spathe, ♂ ♀ (or occasionally ⚥). Sepals 3, herbaceous. Petals 3, petaloid, occasionally absent. Stamens definite or indefinite. Ovary adherent, composed of several carpels, and 1- 6- 8- 9-celled; stigmas 3-6; ovules indefinite, anatropal, often parietal. Fruit dry or succulent, indehiscent, with 1 or more cells. Seeds without albumen; embryo undivided, orthotropal, with a plumule more or less lateral and generally manifest.

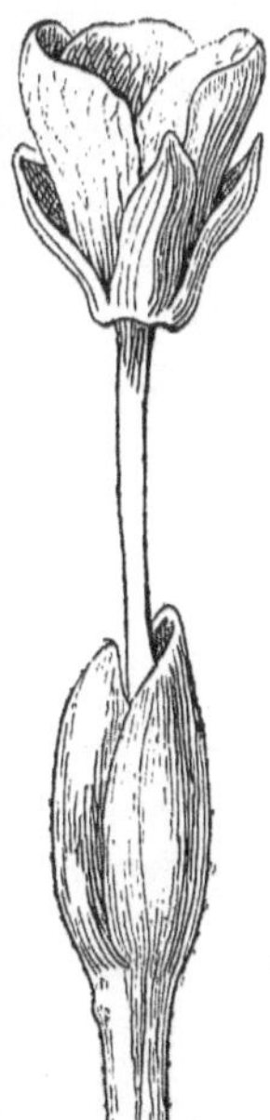

Such appear to be the essential characteristics of this singular group of plants, whose inflorescence lives and passes through all the stages of its existence under water, except just at the time when fertilization is necessary, when the flowers rise above the surface for a few hours. Darwin has celebrated the so-called phenomena connected with this function in Valisneria spiralis, (see his *Loves of the Plants*); but they are greatly in need of more accurate investigation. Mr. Quekett, in an elaborate memoir on this plant, (*London Phys. Journ.* 1. 65,) considers that a part at least of the statements are fabulous.

It is not easy to determine what is the immediate affinity of Hydrocharads. Their exalbuminous seeds and diclinous flowers distinguish them from Bromeliaceæ, to which their adherent ovary, and the habit of the Water-soldier (Stratiotes) seems to approach them; from Naiads, their indefinite seeds and adherent ovary equally divide them. By their tripetaloideous flowers, with an inferior ovary, they are separated from Alismads, with which some agree in habit and want of albumen, but from which they differ in their carpellary leaves being definite, not indefinite. Commelynaceæ are at once recognised by their superior trilocular ovary. Agardh refers here Trapa! Linnæus placed Hydrocharads along with Palms! in his natural arrangement. Hydrocharis Morsus Ranæ has been compared, and not unaptly, to a pigmy Nymphæa. Perhaps, taking into account their diclinous flowers, the universal presence of a spathe and their aquatic nature, they may be regarded as approaching to Arads through Lemnads.

Natives of fresh water in Europe, North America, and the East Indies. One species is found in Egypt (Damasonium indicum), and two Vallisnerias in New Holland. A few occur in estuaries of the sea.

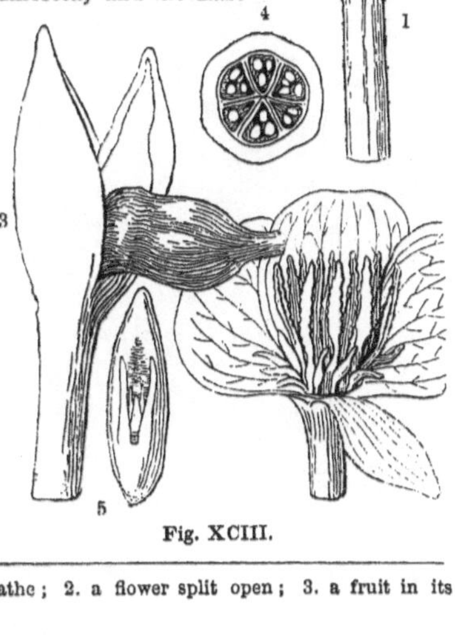

Nothing is known of their uses, unless that the fruit of Enhalus is eatable, and its fibres capable of being woven, according to Agardh (*Aph.* 128). The Jangi of Hindostan, called Vallisneria alternifolia by Roxburgh, Hydrilla by Hamilton, is one of the plants used in India for supplying water mechanically to sugar in the process of refining it, "as clay is used in the West Indies to permit the slow percolation of water."—*Royle.* The herbage of Hydrocharis Morsus Ranæ is mucilaginous and slightly astringent. Ottilia and Boottia are eaten in India as potherbs.

Fig. XCIII.

Fig. XCIII.—Stratiotes aloides. 1. a flower and spathe; 2. a flower split open; 3. a fruit in its spathe; 4. a section of the fruit; 5. an embryo.

GENERA.

Tribe I.— Valisnereæ
Ovary 1-*celled*.

Udora, *Nutt.*
Elodea, L. C. Rich.
Philotria, Rafin.
Anacharis, *Rich.*
Hydrilla, *Rich.*
?*Hydrospondylus*, Hskl.

Lagarosiphon, *Harvey.*
Vallisneria, *Michel*
Physkium, Loureir.
Blyxa, *Thouars.*
Saivala, Wall

Tribe II. — Stratioteæ.
Ovary 6- 8- 9-*celled.*

Stratiotes, *Linn.*
Aloides, Boerh.
Enhalus, *L. C. Rich.*
Ottelia, *Pers.*
Damasonium, Schreb.
Hymenotheca, Salisb.
Bootia, *Wall.*

Limnobium, *L. C. Rich.*
Hydromystria, F.G.W. Meyer.
Jalambicea, Llav. et Lex.
Hydrocharis, *Linn.*
Stratiotes, Dillen.

NUMBERS. GEN. 12. Sp. 20 ?

Pistiaceæ.
POSITION. ——— HYDROCHARACEÆ.—Naiadaceæ.
Bromeliaceæ.

ORDER XL. NAIADACEÆ.—NAIADS.

Naiades, *Juss. Gen.* 18. (1789) *in part.*—Fluviales, *Vent. Tabl.* 2. 80. (1799); *Kunth Enum.* 3. 111.—Potamophilæ, *Rich. Anal. Fr.* (1808).—Potameæ, *Juss. Dict. Sc. Nat.* 43. 93. (1826).—Naiadeæ, *Agardh Aph.* 125. (1822); *Endl. gen.* lxxi. *Meisner*, p. 363.—Fluviales, *Rich. Mém. Mus.* 1. 364. (1815).—Hydrogetones, *Link. Handb.* 1. 282. (1829).

DIAGNOSIS.—*Hydral Endogens, with hypogynous stamens, a free embryo, and globose pollen.*

Water-plants, inhabiting both the ocean and fresh water. Leaves very cellular, with parallel veins, and membranous interpetiolar stipules. Flowers inconspicuous, often arranged in terminal spikes, ♂ ♀. Perianth of 2 or 4 pieces, often deciduous, rarely wanting. Stamens definite, hypogynous. Ovaries 1 or more, superior; stigma simple; ovule solitary, pendulous and orthotropal or campylotropal, or erect and anatropal. Fruit dry, very rarely opening by regular valves, 1-celled, 1-seeded. Seed erect or pendulous; albumen none; embryo with a greatly enlarged radicle, and a latent cleft for the emission of the plumule.

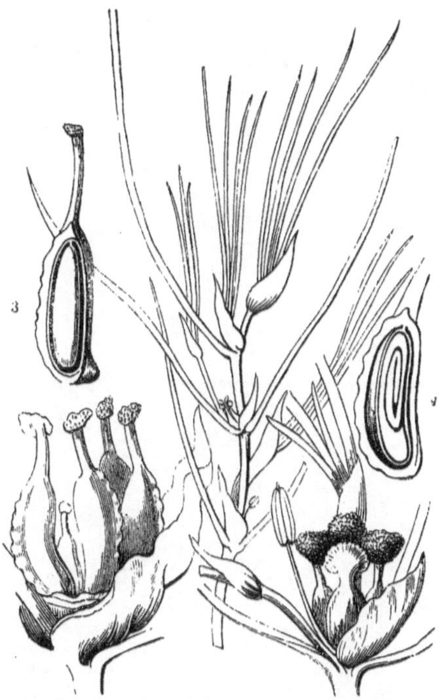
3 4 2 Fig. XCIV. 1

In this Order we have the nearest approach to the great class of Thallogens. Many of them live under water. The perianth is reduced to a few imperfect scales, and there is in some of the genera either a total absence of spiral vessels, or that form of tissue exists in a very rudimentary state. Pollini asserts, according to De Candolle, that spiral vessels do exist in them; but Amici, on the other hand, maintains that there is no trace of them, at least in Caulinia. The manifest affinity of Naiads to Arrow-grasses determines a relation on the part of the former to Arads, which is confirmed by the tendency to produce a rudimentary spathe in some of them, and by their undoubted resemblance to the Duckweeds. It is remarkable that Adanson was aware of this relationship between Arads and Naiads, to which, however, Jussieu, whose Naiades are a very heterogeneous assemblage, did not assent. The species of the Order, as now circumscribed, are generally translucent cellular plants, destitute of stomates, having no epidermoidal layer, and perishing rapidly upon exposure to air. Amici has seen the sap circulate in the transparent joints of Caulinia fragilis, which he states is the unknown plant upon which Corti made observations relating to the same subject. See Amici in *Ann. des Sc.* 4. 42. Mr. Griffith has remarked that, although the difference between the development of the vegetable carpel leaf and vegetable ovulum is in general sufficiently apparent, an exception has appeared to him to be presented by Naias, in which the future pistil seems to be derived from an annular growth round a central body, which subsequently becomes the ovule!

Fig. XCIV.—Zannichellia palustris. 1. A flower; 2. a cluster of ripe ovaries; 3. an ovary opened to exhibit the ovule; 4. a vertical section of a seed, showing the folded up embryo.

Common in extra-tropical countries, either inhabiting fresh water, or the shores of the ocean, but also found near the equator.

Their uses are unknown.

GENERA.

Caulinia, *Willd.*	Phyllospadix, *Hook.*	Tetroncium, *W.*	Althenia, *Petit.*
Ittnera, Gmel.	Zannichellia, *Michel.*	*Cathanthes*, Rich.	*Bellevalia*, Delil.
Najas, *Willd.*	Lilæa, *H.B.K.*	Halodule, *Endl.*	Epigynanthus, *Blume.*
Fluvialis, Michel.	*Heterostylus*, Hook.	*Diplanthera*, Thouars.	

NUMBERS. GEN. 9. Sp. 16.

Juncaginaceæ.
POSITION.—Hydrocharidaceæ.—NAIADACEÆ.—Zosteraceæ.
Thallogens.

ORDER XLI. ZOSTERACEÆ.—SEA WRACKS.

Zosterinæ.—*Nees ab Esenb.* ex Kunth.—Zostereæ.—*Kunth. enum.* 3. 115. (1841).—Posidonieæ.—*Id.*

DIAGNOSIS.—*Hydral Endogens with hypogynous stamens, a free ovary, and confervoid pollen.*

Marine plants resembling sea weeds and living among them. Leaves grassy, thin, sheathing at the base. Flowers very minute, absolutely naked, or surrounded by 3 scales, ♂ ♀, arranged within herbaceous spathes. ♂ Anthers definite in number, one or two-celled, sessile; pollen filamentous, resembling fine confervæ. ♀ Ovary free, one-celled; ovule solitary, pendulous, campylotropal; or parietal with the foramen downwards; stigmas 1 or 2, capillary. Fruit drupaceous, one-seeded. Seed pendulous; albumen 0; embryo antitropal or homotropal, with a very large radicle, and a highly developed plumule lying in its cavity.

If we are to find anywhere a positive intercalation of flowering with flowerless plants it is here, where with naked flowers, but distinct sexes, we have the pollen in a condition that may be well compared to the elaters of Marchantia and its allies, and totally different from all that is known in other flowering plants. The habit too is quite that of sea weeds. It therefore seems expedient to separate these genera from the Naiads, which are an Order higher in organization, and in fact differ in nothing from the common types of flowering structure, except in their simplicity. The manner in which fertilization takes place among these plants is unknown. Zostera marina, whose flowers of both sexes are inclosed in a spathe filled with air, offers indeed no insuperable difficulty to the supposition that in such a situation, although the plants are under water, yet the flowers may be in a dry medium; but, as Vaucher has observed, this does not assist us to understand how fertilization is effected in Zostera maritima which is diœcious. Does the confervoid pollen float to the place where it is wanted?

The bottom of the ocean is the locality of these plants, which occur from the North Sea to the Mediterranean, the Indian Ocean and the coasts of Arabia. One species indeed, Amphibolis zosteræfolia, is seen on the shores of New Holland, and another in the West Indies.

They can scarcely be said to form any part of the vegetation subdued by man, except in the case of the Sea wrack, Zostera marina, which is a common material for packing, and for stuffing cottagers' cushions, and has also been used for tumours, owing apparently to the iodine of the sea weeds that are gathered with it.

GENERA.

Cymodocea, *König.*	Thalassia, *Sol.*	Posidonia, *Kön.*
Amphibolis, *Agh.*	Zostera, *L.*	*Kernera*, W.
Graumullera, Rchb.		*Caulinia*, DC.

NUMBERS. GEN. 5. Sp. 12 (Kunth).

POSITION. *Ceramiaceæ.* ———.—ZOSTERACEÆ.—Naiadaceæ. *Marchantiaceæ.*

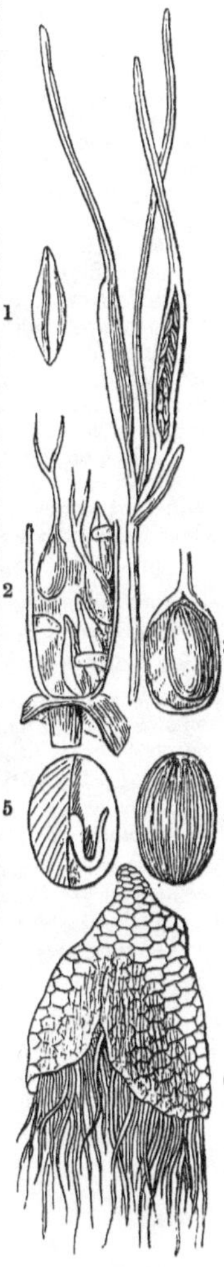

Fig. XCV.

Fig. XCV.—Zostera Noltii. 1. An anther; 2. a portion of a spathe opened, to show the ♀ and ♂ flowers; 3. a section of the ovary; 4. a seed; 5. the same cut in half, to show the plumule; 6. an anther opened and discharging its confervoid pollen.—*Nees* v. *Esenbeck.*

ALLIANCE XI. *NARCISSALES.*—THE NARCISSAL ALLIANCE.

DIAGNOSIS.—*Epigynous petaloid Endogens, with symmetrical flowers, 3 or 6 stamens, and albuminous seeds.*

From the Hydral Alliance and its higher forms, such as the Water Soldiers (Stratiotes), we pass, by an easy transition, to the Narcissals, which may be regarded as hermaphrodite Hydrals growing on dry land, and having albumen in their seeds. This transition is effected by the Bromelworts (Bromeliaceæ), which have quite the same habit, and in addition a tripetaloid flower. This point being settled, the remainder of the Alliance consists of plants which might be regarded as Lilials, if their ovary were not adherent; for it is difficult to separate the Irids from Melanths or the Amaryllids from Lilyworts, by any other precise character.

The principal difficulty in limiting this Alliance arises out of the Bromelworts, some of whose genera have the ovary absolutely free: but such plants are not at all like any other part of the system, and if their calyx is free, it is so fleshy or permanent as to have all the external appearance of being adherent to the ovary.

While however there is, as has been stated, a gentle passage from Hydrals into Narcissals, we find, on the other hand, the Aral Alliance provided here with its representative in the form of the Taccads, which have much the habit of some Arads, and nevertheless an adherent ovary and almost tripetaloideous flower. These plants have also a very evident resemblance to Orontiaceæ.

NATURAL ORDERS OF NARCISSALS.

Flowers tripetaloideous, 6-leaved, imbricated. Albumen mealy	42. BROMELIACEÆ.
Flowers half tripetaloideous, tubular. Albumen fleshy	43. TACCACEÆ.
Flowers hexapetaloideous, tubular, scarcely imbricated. Stamens 3, opposite the petals, or 6; anthers turned inwards. Radicle remote from the hilum, which is naked	44. HÆMODORACEÆ.
Flowers hexapetaloideous, much imbricated. Stamens 6; anthers turned inwards. Radicle remote from the hilum, which is often strophiolate	45. HYPOXIDACEÆ.
Flowers hexapetaloideous, much imbricated. Stamens 6, or more; anthers turned inwards. Radicle next the hilum	46. AMARYLLIDACEÆ.
Flowers hexapetaloideous. Stamens 3, opposite the sepals; anthers turned outwards	47. IRIDACEÆ.

ORDER XLII. BROMELIACEÆ.—BROMELWORTS.

Bromeliæ, *Juss. Gen.* 49. (1789); *Dict. Sc. Nat.* 5. 347. (1817).—Bromeliaceæ, *Lindl. in Bot. Reg.* fol. 1068. (1827); *Bartl. Ord. Nat.* 46. (1830); *Schult. f. in Röm. and Sch. Syst. Veg. vol.* 7. (1830); *Endl. Gen.* lxv.; *Meisn. p.* 395. Tillandsieæ, *Adr. Juss. Cours Elém. tab.* 3.

DIAGNOSIS.—*Narcissal Endogens with tri-petaloideous six-leaved flowers having imbricated divisions, and mealy albumen.*

Stemless or short-stemmed plants, with rigid channelled leaves often covered with cuticular scales, and spiny at the edge or point. Flowers with gay colours, in racemes or panicles. Calyx 3-parted or tubular, persistent, never withering, more or less cohering with the ovary, usually herbaceous, sometimes coloured. Petals 3, coloured, withering or deciduous, equal or unequal, rigidly imbricated. Stamens 6, inserted into the tube of the calyx and corolla; anthers opening inwards. Ovary 3-celled, many-seeded; ovules anatropal; style single; stigma 3-lobed, or entire, often twisted. Fruit capsular or succulent, 3-celled, many-seeded. Seeds innumerable in most cases, always numerous, with a leathery skin, or tapering into a slender thread; embryo taper, curved or straight, minute, lying in the base of mealy albumen, with the radicle next the hilum.

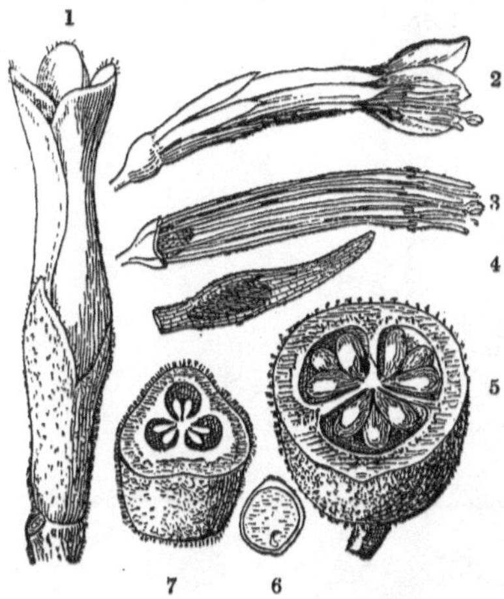

Fig. XCVI.

Stratiotes among the Hydrocharads has so much the foliage of this Order as to render it probable, taking the fructification also into account, that the nearest affinity of the Bromelwort Order is with the former. It is, however, essentially distinguished by its seeds having mealy albumen. This circumstance also cuts it off from the Amaryllids and Hypoxids. The habit of Bromelworts is peculiar; they are hard dry-leaved plants, often with a scurfy surface; the species are frequently capable of sustaining long drought without inconvenience. There can be no doubt about the Order belonging to an epigynous series, and yet the whole race of Tillandsias has the ovary free; but it is never, I believe, wholly so, but has always so much union to the calyx at the base as will show its adherent tendency. Besides, the sepals are always as fleshy, to the last, as if they were absolutely incorporated with the ovary. Nevertheless, Adrien de Jussieu regards the free genera as a peculiar Order, which he calls Tillandsieæ.

All, without exception, are natives of the continent or islands of America, whence they have migrated eastwards in such numbers as to have established themselves as part of the present Flora of the west coast of Africa, and some parts of the East Indies. They are all capable of existing in a dry hot air without contact with the earth; on which account they are favourites in South American gardens, where they are suspended in the buildings, or hung to the balustrades of the balconies; situations in which they flower abundantly, filling the air with fragrance.

The most remarkable species is the Pine Apple, or Ananas, which is well known for the sweetness and fine aromatic flavour of its fruit; in its wild state, however, and unripe, its fruit is excessively acid, burning the gums. In the West Indies it is employed, along with Bromelia Pinguin and others, to destroy intestinal worms, and to promote the secretion of urine. Tillandsia usneoides hangs down from the trees in the woods of tropical America like long dry beards, and is used for stuffing birds, and in the

Fig. XCVI.—1. Flower of Bromelia fastuosa; 2. a flower of Pitcairnia ringens; 3. the stamens of the same; 4. its seed; 5. a cross section of the seed of Bromelia Pinguin; 6. a section of its seed; 7. a cross section of the ovary of Bromelia fastuosa.

preparation of an ointment used against hæmorrhoids. Puya chilensis yields an extract used in healing broken bones ; a transparent gum flows from the spike of Puya lanuginosa. A yellow colour is extracted in Brazil from the root of Billbergia tinctoria. Ropes are made in Brazil from a species of Bromelia, called Grawatha ; and very fine muslin has been manufactured from the fibres of the common Pine Apple.

GENERA.

Ananassa, *Lindl.*	Aræococcus, *Brongn.*	*Amalia*, Hort. hispan.	Dyckia, *Schult. fil.*
Ananas, Tournef.	Cryptanthus, *Klotsch.*	*Strepsia*, Nutt.	Encholirium, *Mart.*
Bromelia, *Linn.*	Brocchinia, *Schult. fil.*	Caraguata, *Plum.*	Pouretia, *Ruiz* et *Pav.*
Karatas, Plum.	Pitcairnia, *Herit.*	*Devillea*, Bert.	*Puya*, Molina.
Ananas, Gaertn.	*Hepetis*, Swartz.	Guzmannia, *Ruiz* et *Pav.*	*Renealmia*, Feuill.
Aechmea, *Ruiz* et *Pav.*	*Spirastigma*, Herit.	Bonapartea, *Ruiz* et *Pav.*	*Achupalla*, Humb.
Oechmea, Juss.	Vriesia, *Lindl.*	*Acanthospora*, Spr.	Hechtia, *Klotsch.*
Billbergia, *Thunb.*	Neumannia, *Brongn.*	*Misandra*, Dietr.	Dasylirion, *Zucc.*
Hohenbergia, *Schult. fil.*	Tillandsia, *Linn.*	Navia, *Mart.*	? *Roulinia*, Brongn.
Acanthostachys, *Klotsch.*	*Renealmia*, Plum.	Cottendorfia, *Schult. fil.*	

NUMBERS. GEN. 23. SP. 170.

Hydrocharidaceæ.
POSITION.—Hæmodoraceæ.—BROMELIACEÆ.—Hypoxidaceæ.

Fig XCVI.*

Fig. XCVI.*—Æchmea fulgens.—*Paxton.*

ORDER XLIII. TACCACEÆ.—TACCADS.

Taccеæ, *Presl. Reliq. Hænk.* 1. 149. (1830); *Bartl. Ord. Nat.* 82. (1830).—Taccaceæ, *Key, &c.* 70. (1835); *Ed. prior.* ccxxxix.; *Endl. Gen.* lviii.—*Meisner, p.* 403.

DIAGNOSIS.—*Narcissal Endogens with tubular half-tripetaloideous flowers and fleshy albumen.*

Large perennial herbs, with a tuberous root. Leaves all radical, stalked, undivided or pedatifid, the segments pinnatifid and entire, with curved parallel veins. Stipules 0.

2 3 4

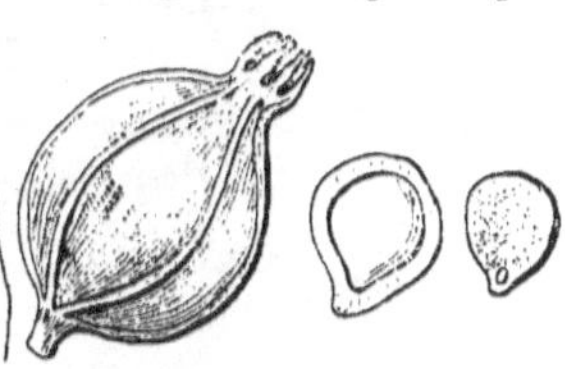

1 Fig. XCVII.

Flowers placed on the top of a simple taper or angular furrowed scape, in umbels, ☿, regular, surrounded by undivided bracts forming an involucre. Perianth adherent, with a cylindrical ribbed tube; limb petaloid, the petals rather the longest, persistent. Stamens 6, inserted into the base of the segments of the perianth, distinct; filaments dilated, petaloid, hooded at the apex; anthers inserted below the points of their filaments in their concavity, 2-celled, the cells distinct. Ovary composed of 3 connate carpels, 1-celled, or half 3-celled, with 3 parietal polyspermous placentæ; ovules ascending and anatropal, or horizontal and amphitropal; styles 3, connate; stigmas connate at the base, radiating, 2-lobed. Pericarp berried, indehiscent, 1-celled, or half 3-celled, many-seeded. Seeds lunate or somewhat ovate, striated. Albumen fleshy. Embryo placed inside the albumen in the region of the hilum, or remote from it.

Personally I have had no opportunity of examining critically the plants which compose this small Order. They are in some respects like Arads, in others like Gingerworts (Tacca lævis); but certainly have nothing to do with Dicotyledons. Blume has the following remarks upon Tacca. *Enum.* 1. 82. "The genus Tacca offers the type of a new family between Araceæ and Aristolochiaceæ. To the former it approaches closest in habit, especially in the leaves, but it is very different from them in the structure of the parts of fructification. For in no species of true Araceæ is a corolline perianth, properly so called, to be found; what we have the custom of calling so in Dracontium and others, is nothing but scales, and not even a calycine integument; the perianth is, moreover, superior in Tacca. By this superior perianth the affinity with Aristolochiaceæ is evident; but from those too Tacca differs in the situation of the stamens, which are not as in that Order adherent to the pistil with the anthers opening outwards, but are placed on the perianth itself with the anthers turned inwardly." In Tacca it is probable that there are several germinating points upon the embryo, analogous to the double or triple plumule of Dracontium; hence embryos of such a kind may be said to be tubers formed in the fruit itself. Brown long since stated (*Prodromus*, 1810) that a relation is established between Arads and Birthworts by means of Tacca. See also *Agardh's Aphorisms*, 245. For my own part, however, this resemblance to Birthworts seems so very slight as to be unworthy of notice. The true relation is with the Arads, or at least with those ☿ plants which are now separated under the name of Orontiaceæ, of which these seem to be an epigynous form. Endlicher compares them with Yams, to which they appear to have even less resemblance than to the Birthworts.

Fig. XCVII.—1. Tacca integrifolia; 2. fruit of T. pinnatifida; 3. seed of do. with half the testa removed; 4. section of its albumen and embryo.—*Gærtner.*

Found in damp maritime places and woods in the hotter parts of India, the South Sea Islands, and the tropical parts of Africa.

"The plants of this family are possessed of some degree of acridity, both in their tubers and in their herbaceous parts, as Rumphius informs us that the tubers of T. pinnatifida, dubia, and montana are rasped and macerated for four or five days in water, and a fecula is separated in the same manner that sago is, and like it employed as an article of diet by the inhabitants of the Malayan and Molucca Islands. In Otaheite and other Society Islands, they make cakes of the meal of the tubers of T. pinnatifida, which are the Tacca youy of some navigators; they form an article of diet in China and Cochin China, as also in Travancore, where Dr. Ainslie informs me they attain a large size, and that the natives eat them with some acid to subdue the acrimony."—*Royle.*

GENERA.

Tacca, *Forst.* | Ataccia, *Presl.*

NUMBERS. GEN. 2. Sp. 8.

Orontiaceæ.
POSITION ——— TACCACEÆ.—Bromeliaceæ.
Araceæ.

ORDER XLIV. HÆMODORACEÆ.—BLOOD ROOTS.

Hæmodoraceæ, *R. Brown, Prodr.* 299. (1810); *Agardh. Aphor.* 170. (1823); *Endl. Gen.* lxii.; *Meisner, p.* 396.—Vellozieæ, *D. Don in Edinb. Ph. Journal.* (1830).

DIAGNOSIS.—*Narcissal Endogens with hexapetaloideous tubular flowers,* 3 *stamens opposite the petals or* 6, *anthers turned inwards, and radicle remote from the hilum which is naked.*

Herbaceous plants with fibrous perennial roots and permanent sword-shaped equitant leaves, which are mostly in two ranks. Flowers ☿. Perianth usually more or less woolly, adherent; the sepals and petals in many cases undistinguishable and united into a (cylindrical) tube. Stamens arising from the sepals and petals, either 3 and opposite the petals, or 6; anthers bursting inwardly. Ovary with the cells 1- 2- or many-seeded, adherent, usually 3-celled, occasionally 1-celled, with a placenta occupying only a point of the axis; style simple; stigma undivided; ovules amphitropal. Fruit covered by the withered perianth, capsular, valvular, seldom indehiscent, somewhat nucamentaceous, with the placenta easily separable from the dissepiments, if any. Seeds either definite or indefinite, fixed by the base or peltate, winged or wrinkled and angular. [Embryo lying in cartilaginous albumen, short, straight, with the radicle usually remote from the hilum. *Endl.*]

Fig. XCVIII.

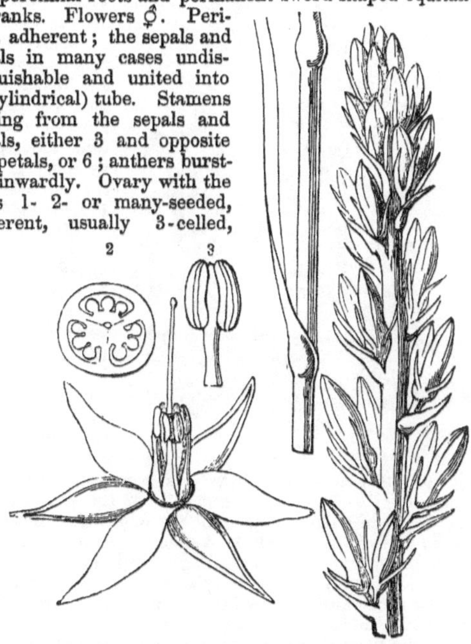

Fig. XCIX.

The distinction between these and Amaryllids consists in their perianth not having the regular equitant position of sepals and petals which is found in the latter, in their constantly equitant leaves, and in their flowers, which have frequently a woolly surface, and a small limb compared with the tube. From Irids they are divided

---

Fig. XCVIII.—Blancoa canescens. 1. a flower and ovary of Conostylis æmula opened.

Fig. XCIX.—Hæmodorum spicatum. 1. A flower spread open; 2. a cross section of the ovary; 3. an anther.

by the number of their stamens, and by their anthers turning inwards, or, if their stamens are reduced to three, then, by those organs being opposite the petals; and by their simple stigma. Dr. Herbert includes all the hexandrous genera in Amaryllids; and limits the Order to those having 3 stamens and an adherent ovary; but, although it may be very difficult to express in satisfactory language the exact differences between the Blood-roots and Amaryllids, yet I think there can be no doubt of their real distinctness, and that the diagnosis now assigned to them does sufficiently characterize them.

In Brazil, Southern Guiana, and also in the Mascaren islands, there occurs a race of these plants which may be compared to the Conestyles of New Holland on a gigantic scale. Martius, who calls them Vellozias, describes them as perennial Lilies, with their trunks closely covered by the withered remains of leaves, branching by forks, and bearing at their points tufts of leaves in the manner of a Yucca or Dracæna; some of them are from 2 to 10 feet high, with a trunk sometimes as thick as a man's body. I find the structure of that trunk most curious. It consists of a central slender subcylindrical column, which never increases in diameter after its first formation, and which has the ordinary monocotyledonous structure. Outside of the column are arranged great quantities of slender fibrous roots, which cohere firmly by their own cellular surface, and form a spurious kind of wood, which is extremely like that of some kinds of Palm wood, only it is developed by constant additions to the very outside of the stem. Something analogous occurs in Pandanus, but it is in some tree ferns only that this mode of growth is exactly repeated. Don proposed to make an Order of the Vellozias; but till their structure and that of the Bloodroots shall have been thoroughly investigated this step is premature.

Fig. C.

As to Wachendorfia and its allies, with triandrous flowers, and free ovary, Mr. Herbert looks upon it as the type of an Order (Wachendorfiaceæ) quite unconnected with Hæmodorum and Conostylis, and he is possibly right; but in the meanwhile, as we know very little of these genera, it seems most expedient to dismiss them from the Blood-roots and station them in reserve among the Lilies. Endlicher states that the genera of this Order have the cells of the ovary opposite the petals, and this, if so, would certainly be an important characteristic; but I cannot confirm the statement: it is in truth very difficult to determine such a point in the majority of the genera, whose sepals and petals are all apparently on the same plane. The true Hæmodoraceæ are smooth and dissimilar in habit to Conostylis and its allies; wherefore a couple of additional sub-Orders may be conveniently admitted here, for which better characters may be hereafter found.

The species occur in North America sparingly, and the Cape of Good Hope; several are described from the more temperate parts of New Holland, and a good many Vellozias and Barbacenias occur in Brazil and the Mascaren islands. A Barbacenia (Alexandrinæ) growing from 10 to 12 feet high has also been noticed by Sir R. Schomburgk in the Southern parts of British Guiana.

De Candolle remarks, that the red colour found in the roots of Lachnanthes tinctoria in North America, where it is used for dyeing, prevails in Hæmodorum, and deserves to be studied in the rest of the Order. The natives of the Swan River live on the roots of such plants, especially of Hæmodorum paniculatum and spicatum, and Anigozanthus floridus, which are mild and nutritious when roasted, but acrid when raw. *Hook. Journ.* 2. 355. One of the most intense bitters known is Aletris farinosa. It is used in infusion as a tonic and stomachic, but large doses produce nausea and tendency to vomit. It has also been employed in chronic rheumatism.

Fig. C.—Landscape with Vellozias; *Martius.*

GENERA.

I.—*Hæmodoreæ.* Perianth smooth, short.
Hæmodorum, *Sm.*
Phlebocarya, *R. Br.*

II.—*Conostyleæ.* Perianth woolly, long.
Dilatris, *Berg.*
Lachnanthes, *Elliot.*
Heritiera, Gmel.
Gyrotheca, Salisb.
Lanaria, *Thunb.*
Argolasia, Juss.
Augea, Retz.
Anigosanthus, *Labill.*
Anigozia, Salisb.
Anægosanthus, Reich.
Schwægrichenia, Spr.
Androstemma, *Lindl.*
Conostylis, *R. Br.*
Blancoa, *Lindl.*
Aletris, *Linn.*
Tribonanthes. *Endl.*

III.?—*Vellozieæ.*
Vellozia, *Mart.*
Xerophyta, Comm.
Campderia, A. Rich.
Radia, A. Rich.
Barbacenia, *Vandelli.*
Visnca, Steud.

NUMBERS. GEN. 13. SP. 50.

Liliaceæ.
POSITION.—Iridaceæ.—HÆMODORACEÆ.—Amaryllidaceæ.

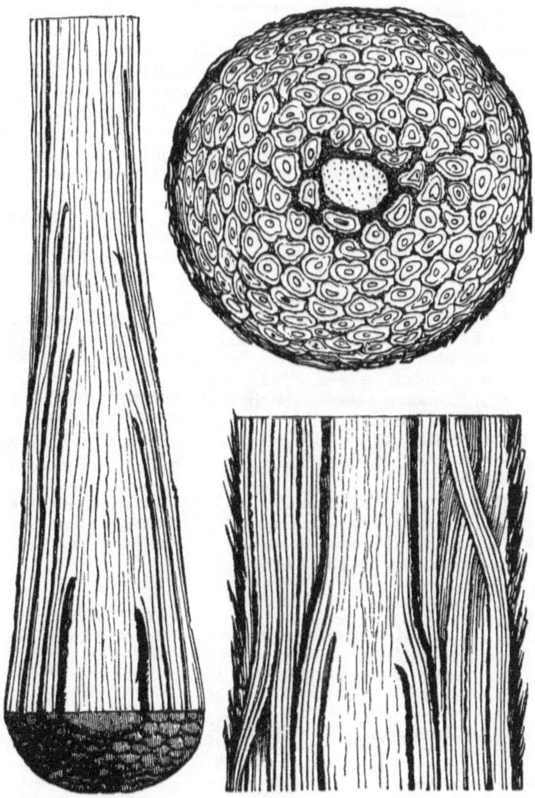

1

3

2

F g. CI.

Fig. CI.—Sections of the stem of a Brazilian Vellozia; 1. transversely; 2, 3. longitudinally.

ORDER XLV. HYPOXIDACEÆ.—HYPOXIDS.

Hypoxideæ, *R. Br. in Flinders* (1814); *Agardh Aph.* 164 (1823); *Ed. prima. No.* 235 (1830); *Endl. Gen.* lxiii. *Meisner, p.* 397.

DIAGNOSIS.—*Narcissal Endogens with hexapetaloideous flowers which are much imbricated, 6 stamens with anthers turned inwards, and a radicle remote from the hilum, which is often strophiolate.*

Herbaceous plants with a tuberous or fibrous perennial root. Leaves always growing from the root and crown, nowhere else, linear, entire, plaited, of a dry texture. Scapes simple or branched, occasionally very short. Flowers complete, ☿. Perianth petaloid, adherent to the ovary, 6-parted, with the sepals coarser than the petals. Stamens 6, inserted into the base of the segments of the perianth; filaments distinct; anthers turned inwards, 2-celled, erect, opening lengthwise. Ovary adherent, 3-celled, with the cells opposite the sepals; style terminal, simple; stigmas distinct or combined, [crowned by an operculum formed by the base of the style.—*Herbert*]; ovules 00, axile, amphitropal. Fruit indehiscent, dry or berried, 1- 2- 3-celled; seeds 00, roundish, with a lateral hilum, and a beaked strophiole. Embryo in the axis of fleshy albumen, straight, with the radicle remote from the hilum, and directed upwards.

1

2

Fig. CII.

As far as habit goes, these are very different from the Amaryllids, for their leaves are harsh and hairy, and although dwarf, they have no bulbs. But when we look to the fructification there is but little to connect with the difference in the vegetation. It is true that the sepals are much coarser in texture than the petals, but that is of small importance; and in truth it is the position of the embryo, remote from the hilum, and that alone, by which the Order is to be certainly known; for the beaked strophiole, which is often found near the hilum, is of small importance. As to the texture of the seed-skin, formerly relied upon in distinguishing some of the Orders of Endogens, experience and reason equally reject it as an ordinal character.

The whole number of Hypoxids is inconsiderable. What are known inhabit the Cape of Good Hope, New Holland, the East Indies, the tropics of America, and the warmer parts of the United States.

The roots of Curculigo orchioides are somewhat bitter and aromatic, and are employed in the East Indies in gonorrhœa. The tubers of Curculigo stans are eaten in the Marianne islands; those of Hypoxis erecta are employed by the aborigines of North America in healing ulcers, and against intermittents.

GENERA.

Curculigo, *Gærtn.* *Molineria*, Colla. Forbesia, *Eckl.*	Hypoxis, *L.* *Schnitzleinia*, Steud.	*Niobæa*, W. Pauridia, *Harv.*

NUMBERS. GEN. 4. Sp. 60.

Orchidaceæ.
POSITION.—Hæmodoraceæ.—HYPOXIDACEÆ.—Amaryllidaceæ.
Apostasiaceæ.

Fig. CII.—1. Seed of Curculigo orchioides; 2. a perpendicular section of it.—*Gærtner.*

ORDER XLVI. AMARYLLIDACEÆ.—AMARYLLIDS.

Narcissi, the second section, *Juss. Gen.* 54. (1789).—Amaryllideæ, *R. Brown Prodr.* 296. (1810); *Herbert, Appendix to the Bot. Mag.* (1821); *Id. Amaryllid,* (1837); *Endl. Gen.* lxiv.; *Meisner, p.* 393. —Narcisseæ *Agardh. Aph.* 173. (1823).

DIAGNOSIS.—*Narcissal Endogens with hexapetaloideous much imbricated flowers,* 6 *or more stamens with the anthers turned inwards, and the radicle next the hilum.*

Generally bulbous plants, sometimes fibrous-rooted, occasionally with a tall, cylindrical, woody stem. Leaves ensiform, with parallel veins, rarely expanded at the sides into an oval lamina with a narrow stalk. Flowers usually with spathaceous bracts. Scape not spadiceous. Calyx and corolla confounded, adherent, regular, coloured, the former overlapping the latter. Stamens 6, arising from the sepals and petals, sometimes cohering by their dilated bases into a kind of cup; sometimes an additional series of barren stamens is present, often forming a cup which surmounts the tube of the perianth; anthers bursting inwardly. Ovary 3-celled, the cells opposite the sepals, many-seeded, or sometimes 1- or 2-seeded; ovules anatropal; style 1; stigma 3-lobed. Fruit either a 3-celled, 3-valved capsule, with loculicidal dehiscence, or a 1-3-seeded berry. Seeds with either a thin and membranous, or a brittle and black or a thick and fleshy testa; albumen fleshy or corneous; embryo nearly straight, with its radicle turned towards the hilum.

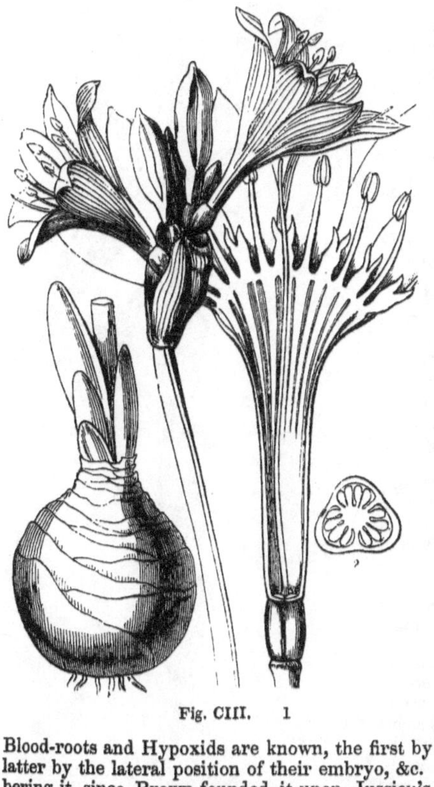

Fig. CIII. 1

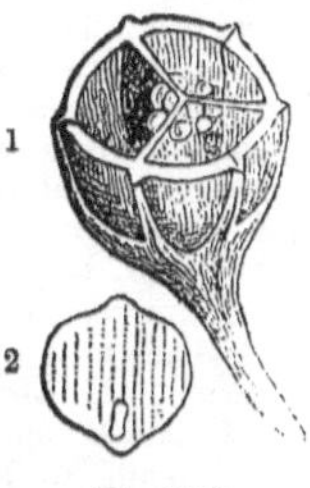

Fig. CIV.

The only Orders with which this need be compared are the Lilies, from which it is known by its inferior ovary: the Irids, which are distinguished by being triandrous, with the anthers turned outwards; and the Blood-roots and Hypoxids are known, the first by the nature of their albumen, and the latter by the lateral position of their embryo, &c. No one has ever thought of dismembering it, since Brown founded it upon Jussieu's 2d section of Narcissi; and it can scarcely be said to comprehend an anomalous genus, unless Clivia and Doryanthes be so considered, on account of their fascicled roots, Agave and Fourcroya, the stems of which are woody, and Gethyllis, because of its being polyandrous. The latter deviation from the ordinary character of the Order will probably be considered of less importance, if we bear in mind the polyandrous structure of some Blood-roots, and especially if, in the first place, the genuine Amaryllidaceous genus Phycella be attended to, which has a tendency to produce additional stamens; and if, secondly, the coronet of Narcissus itself be borne in mind, which is in fact an organ representing an extra number of stamens. I have elsewhere remarked (*Bot. Reg.* 1341.) that this is connected with a strong tendency in the whole Order to form another set of male organs between the perianth

Fig. CIII.—Pancratium maritimum. 1. a flower cut open, and showing that there is a bifid tooth, forming a coronet or cup, between each stamen; 2. a transverse section of the ovary.

Fig. CIV.—Alstrœmeria Pelegrina. 1. A section of its capsule; 2. a perpendicular section of its seed.

and those stamens that actually develope. Hence a curious instance is exhibited, to which several parallels may, however, be found in other families, of the force of development being generally confined to a series of organs originating within those which should be formed according to the ordinary laws of structure. Of course, in all such Orders a multiplication of the usual number of stamens is more to be expected than where this peculiar circumstance does not exist.

The learned investigator of the Order, the Honourable and very Rev. W. Herbert, Dean of Manchester, includes in it the whole Narcissal Alliance, to which he adds the Yams; for his reasons for which the reader is referred to the elaborate monograph above quoted. The remarkable difference in habit between the bulbous species, like Narcissus, and the arborescent kinds, such as Agave and Littæa, is precisely analogous to what occurs among the Lilies, and does not appear to be connected with differences in the fructification. Dr. Joseph Hooker is of opinion that Brown is right in regarding Campynema as belonging to Melanths; but its inferior ovary is against this view, notwithstanding its separate styles. It is probably an osculant genus.

A very few only are found in the North of Europe and the same parallel; these are plants of the genera Narcissus and Galanthus. As we proceed south they increase. Pancratium appears on the shores of the Mediterranean; Crinums and Pancratiums abound in the West and East Indies; Hæmanthus is found for the first time with some of the latter on the Gold Coast; Hippeastra show themselves in countless numbers in Brazil, and across the whole continent of South America; and, finally, at the Cape of Good Hope the maximum of the Order is beheld in all the beauty of Hæmanthus, Crinum, Clivia, Cyrtanthus, and Brunsvigia. A few are found in New Holland, the most remarkable of which is Doryanthes.

This is one of the few monocotyledonous Orders in which poisonous properties occur. They are principally apparent in the viscid juice of the bulbs of Hæmanthus toxicarius and some neighbouring species, in which the Hottentots are said to dip their arrow-heads, and Amaryllis Belladonna, which is said to be employed for poisoning in the West Indies, (*Endl.*); but this is no doubt a mistake, and the statement applies to some other bulbs of the Order—for the Belladonna is a Cape plant; probably to Hippeastra, which Martius tells us have poisonous bulbs. The bulbs of Leucoium vernum, of the Snowdrop and Daffodil, have for ages been known as emetic; and it has recently been shown by Loiseleur Deslongchamps that a similar power exists in Narcissus Tazetta, odorus and Poeticus, and in Pancratium maritimum.

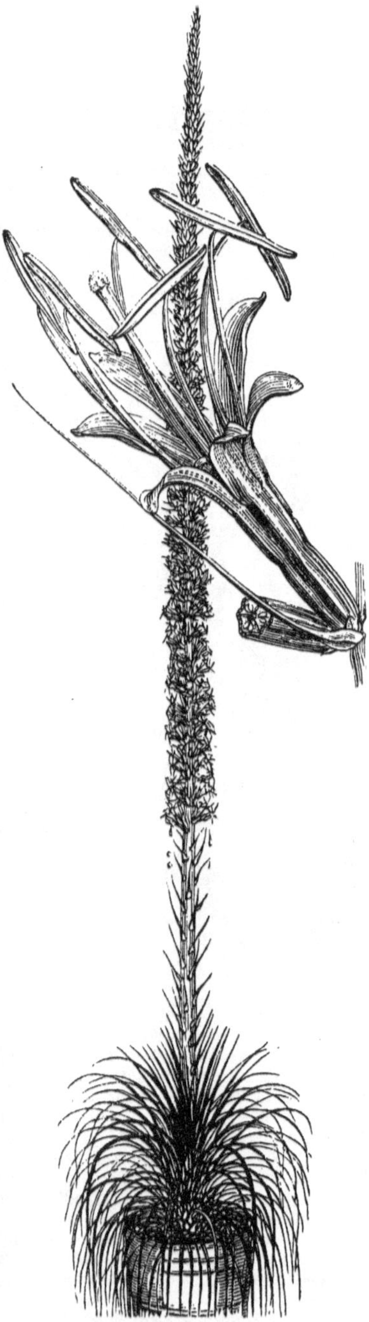

Fig. CV.

Fig. CV.—Littæa geminiflora.

The flowers of Narcissus Pseudo-Narcissus are not only emetic, but a dangerous poison, occasionally producing serious consequences in infants which are allowed to swallow them. De Candolle considers the principle found in Amaryllids analogous to that of the Squill (*Essai*, p. 290). Oporanthus luteus is purgative, Alströmeria salsilla diaphoretic and diuretic, Amaryllis ornata astringent. *Agardh Aph.* 178. A kind of arrow-root is prepared from the succulent roots of Alströmeria pallida and others, in Chile. Bomarea Salsilla is employed as a substitute for Sarsaparilla. Agave Americana, the American Aloe, which is said to flower once only in a hundred years, a gardener's fable, forms impenetrable hedges with its hard and spiny leaves; its fibre and that of some neighbouring species, especially the Pita plant, is extremely tough, and forms excellent cordage; its root is diuretic and antisyphilitic, and is even brought to Europe mixed with Sarsaparilla. "The species of Agave are not alone ornamental as plants and useful as hedges, but are important for their products. The roots, as well as the leaves, contain ligneous fibre (pita thread), useful for various purposes: this is separated by bruising and steeping in water, and afterwards beating.—The Mexicans also made their paper of the fibres of Agave leaves laid in layers. The expressed juice of the leaves evaporated, is stated by Long, in his Hist. of Jamaica, to be also useful as a substitute for soap.

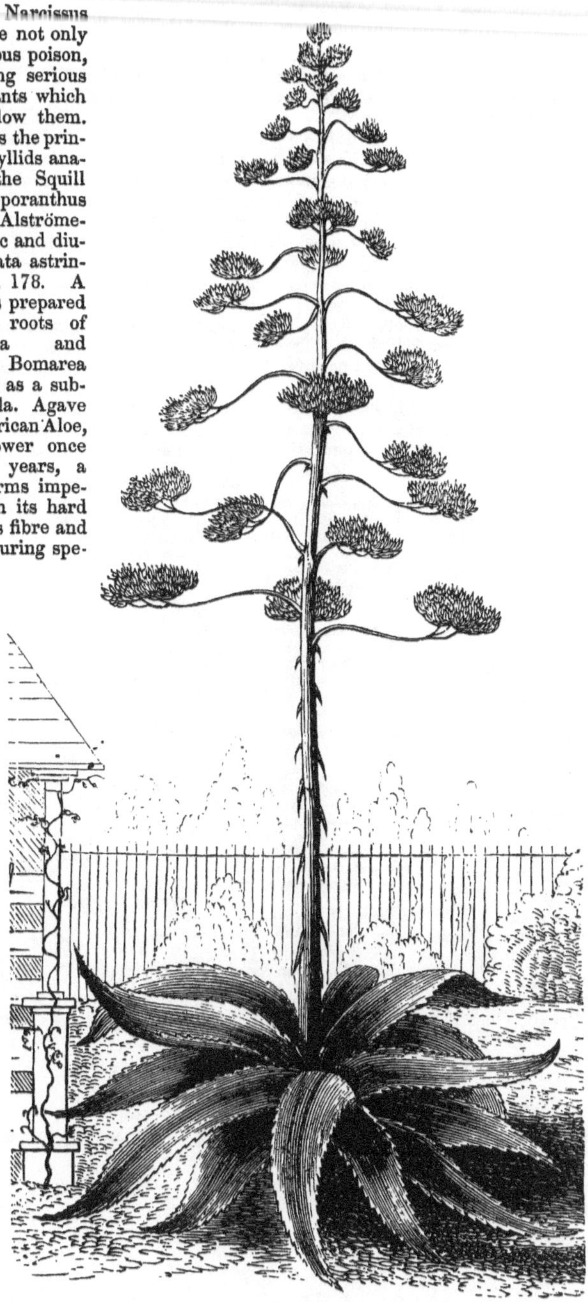

Fig. CVI.

Fig. CVI.—Agave Americana.

But the most important product of Agave, and especially of A. Americana, the species now most common in the South of Europe, is the sap, which exudes upon the cutting out of the inner leaves, just before the flower-scape is ready to burst forth. Of this a very full account is given by Humboldt, in his Political History of New Spain, book iv. c. 9. The species is A. Americana, called metl by the Mexicans, and Maguay de Cociuza in Caraccas. Pittes and maguey-metl are varieties of A. Americana, which is stated to be common everywhere in Æquinoctial America, from the plains even to elevations of between 9000 and 10,000 feet. A. Mexicana is also, by some authors, called maguei-metl, and also manguai; and A. Vivipara is theometl or manguei divinum. In Cumana and Caraccas, A. Cubensis is called maguey de Cocay. Humboldt informs us, that the first (A. Mexicana) is extensively cultivated in the interior table-land of Mexico, and, indeed, extends as far as the Aztec language. The juice of the Agave is of a very agreeable sour taste. It easily ferments on account of the mucilage and sugar it contains, when it is called pulque by the Spaniards. This vinous beverage, which resembles cider, has an odour of putrid meat, extremely disagreeable; but the Europeans, who have been able to get over the aversion which this fetid odour inspires, prefer the pulque to every other liquor. A very intoxicating brandy is formed from the pulque, which is called mexical or aguardiente de maguey. The government drew from the Agave juice a net revenue of £166,497 in three cities."—*Royle.* Agave saponaria is a powerful detergent; its roots are employed in Mexico as a substitute for soap. A cold infusion of the leaves of Chæradodia Chilensis is purgative and diuretic; it is called Thekel, in Chile.—*Molina.*

GENERA.

Tribe I. — Amarylleæ. *Bulbs, without a coronet in the flower.*

Galanthus, *Linn.*
? Erangelia, Renealm.
Leucojum, *Linn.*
Nivaria, Mönch.
Acis, *Salisb.*
Erinosma, *Herb.*
Lapiedra, *Lagasc.*
Carpolyza, *Salisb.*
Hessea, Berg.
Gethyllis, *L.*
Papiria, Thunb.
Ixiolirion, *Fisch.*
Bravoa, *Llav.*
Cœtocapnia, Lk. et Otto.
Sternbergia, *Waldst. et Kit.*
Oporanthus, *Herb.*
Haylockia, *Herb.*
Cooperia, *Herb.*
Sceptranthus, Grah.
Amaryllis, *Linn.*
Lilio-Narcissus, Tour.
Belladonna, Sweet.
Callirhoe, Link.
Zephyranthes, *Herb.*
Argyropsis, *Herb.*
Pyrolirion, *Herb.*
Habranthus, *Herb.*
Sprekelia, *Heist.*
Hippeastrum, *Herb.*
Amaryllis, Sweet.
Coburgia, Herb.
Leopoldia, Herb.
Vallota, *Herb.*
Lycoris, *Herb.*
Strumaria, *Jacq.*
Hessea, *Herb.*
Nerine, *Herb.*
Galathea, Herb.
Brunsvigia, *Heister.*
Imhofia, *Herb.*
Buphane, *Herb.*
Boophane, Herb.
Ammocharis, *Herb.*
Griffinia, *Ker.*
Crinum, *Linn.*
Hæmanthus, *Linn.*
Tristegia, Rchb.
Polystegia, Rchb.
Cyrtanthus, *Ait.*
Timmia, Gmel.
Cyrtanthus, Herb.
Monella, Herb.
Gastronema, *Herb.*
Coleophyllum, Klotsch.

Tribe II. — Narcisseæ. *Bulbs, with a coronet in the flower.*

Phycella, *Lindl.*
Placea, *Miers.*
Eucrosia, *Ker.*
Carpodetes, *Herb.*
Liperiza, *Herb.*
Calliphruria, *Herb.*
Eurycles, *Salisb.*
Proiphys, Herb.
Calostemma, *R. Br.*
Vagaria, *Herb.*
Tapeinanthus, *Herb.*
Chlidanthus, *Herb.*
Clinanthus, *Herb.*
Urceolina, *Rchb.*
Urceolaria, Herb.
Collania, Schultz.
Coburgia, *Sweet.*
Phædranassa, *Herb.*
Stenomesson, *Herb.*
Chrysiphiale, Ker.
Sphærotele, Presl.
Elisena, *Herb.*
Liriope, Herb.
Liriopsis, Rchb.
Pancratium, *Linn.*
Hymenocallis, *Salisb.*
Schizostephanium, Rchb.
Halmyra, Salisb.
Tiaranthus, Herb.
Choretis, *Herb.*
Ismene, *Herb.*
Callithaume, Herb.
Narcissus, *Linn.*
Ajax, Haw.
Diomedes, Haw.
Queltia, Haw.
Schizanthes, Haw.
Ganymedes, Haw.
Philogyne, Haw.
Hermione, Haw.
α. Jonquillia, DC.
β. Tazetta, DC.
Chloraster, Haw.
Corbularia, *Haw.*

Tribe III.—Alströmerieæ. *Fibrous rooted. Sepals different in form from the petals.*

Chæradodia, *Herb.*
Alströmeria, *L.*
Collania, *Herb.*
Sphærine, *Herb.*
Bomarea, *Mirb.*

Tribe IV. — Agaveæ. *Fibrous rooted. Sepals and petals alike.*

Clivia, *Lindl.*
Imatophyllum, Hook.
Himantophyllum, Spr.
Campynema, *Labill.*
Campylonema, Poir.
Doryanthes, *Correa.*
Agave, *L.*
Littæa, *Tagl.*
Bonapartea, W.
Fourcroya, *Vent.*

NUMBERS. GEN. 68. Sp. 400.

Melanthaceæ.
POSITION.—Iridaceæ.—AMARYLLIDACEÆ.—Hypoxidaceæ.
Liliaceæ.

ORDER XLVII. IRIDACEÆ.—IRIDS.

Irides, *Juss. Gen.* 57. (1789).—Ensatæ, *Ker in Ann. of Botany*, 1. 219. (1805).—Irideæ, *R. Brown Prodr.* 302. (1810); *Ker. Gen. Irid.* (1827); *Bartl. Ord. Nat.* 44. (1830); *Meisner*, p. 391.—Iridaceæ, *Ed. pr.* ccxl.; *Endl. Gen.* lxi.

DIAGNOSIS.—*Narcissal Endogens with 3 stamens opposite the sepals, and anthers turned outwards.*

Herbaceous plants, or very seldom under-shrubs, usually smooth; the hairs, if there are any, simple. Roots tuberous or fibrous. Leaves equitant and distichous in most genera. Inflorescence terminal, in spikes, corymbs, or panicles, or crowded, sometimes radical. Bracts spathaceous, the partial ones often scarious; the sepals occasionally rather herbaceous. Calyx and corolla adherent or coloured, their divisions either partially cohering, or entirely separate; sometimes irregular, the 3 petals being occasionally very short. Stamens 3, arising from the base of the sepals; filaments distinct or connate; anthers bursting externally lengthwise, fixed by their base, 2-celled. Ovary 3-celled, cells many-seeded; ovules anatropal; style 1; stigmas 3, often petaloid, some-

Fig. CVII.

Fig. CIX.

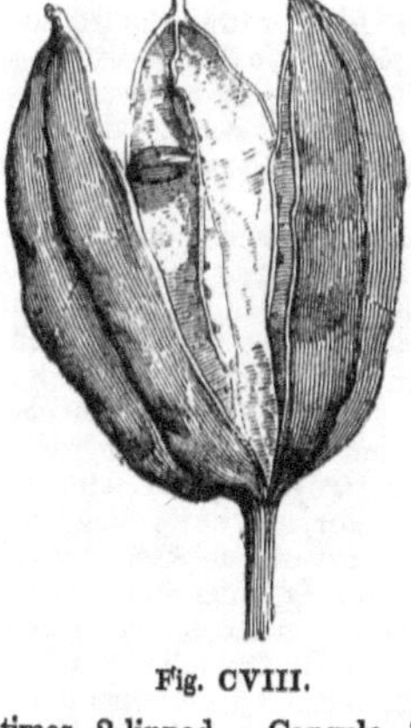

Fig. CVIII.

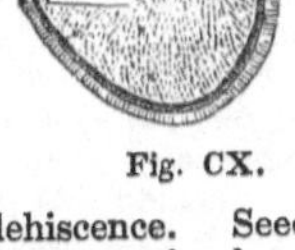

Fig. CX.

times 2-lipped. Capsule 3-celled, 3-valved, with a loculicidal dehiscence. Seeds attached to the inner angle of the cells, sometimes to a central column becoming loose,

Fig. CVII.—Diagram of an Iris.
Fig. CVIII.—Ripe capsule of an Iris.
Fig. CIX.—Iris germanica.
Fig. CX.—Vertical section of its seed.

spheroidal, angular, oblong, or winged; albumen horny, or densely fleshy; embryo inclosed within it, the radicle being uniformly next the hilum.

This Order differs from that of Amaryllids essentially, in being triandrous, with the anthers turned outwards; from Orchids, to which it approaches nearly in some respects, in not being gynandrous; in the nature of the seeds and placentæ, in all the anthers being distinct; from Gingers and Arrowroots the three perfect stamens divide it, independently of the structure of the leaves, which are extremely different. Blood-roots, which are often triandrous with equitant leaves, have the anthers bursting inwardly, and when triandrous their stamens are opposite the petals. The Iris represents the general structure of the Order; but a departure from the form of perianth found in that genus takes place in Crocus, the flower of which is extremely like that of Gethyllis and Oporanthus among Amaryllids on the one hand, and of Colchicum among Melanths on the other; the latter is known by their superior triple ovary. The dilated stigma found in Iris is characteristic of only a part of the Order; in Crocus the stigma is rolled up instead of being spread open, and in many genera it is absolutely thread-shaped. Brown observes, that Burmannia appears at first sight to agree with Irids, especially in its equitant leaves, coloured superior triandrous perianth, and 3 dilated stigmas; it cannot, however, be united with them, on account of its fertile stamens being opposite the inner segments of the perianth, and alternating with an equal number of sterile ones, because of the transverse dehiscence of the anthers, and also the structure of the seeds. In Xyris some resemblance with this Order is discoverable, especially in the disposition of the leaves, the triandrous flowers, and anthers turned outwards; but that genus is very distinct in its free perianth, the outer segments of which are glumaceous, and the inner distinctly petaloid, in the ungues bearing the stamens at their apex, in the sterile alternate stamens, and especially in the structure of the seed.—*Prodr.* 302. The whole Order is greatly in want of a good critical examination; but much caution is required in forming the genera, especially in deriving characters from the seeds, for they are both round, and fleshy, and thin, in the genus Iris.

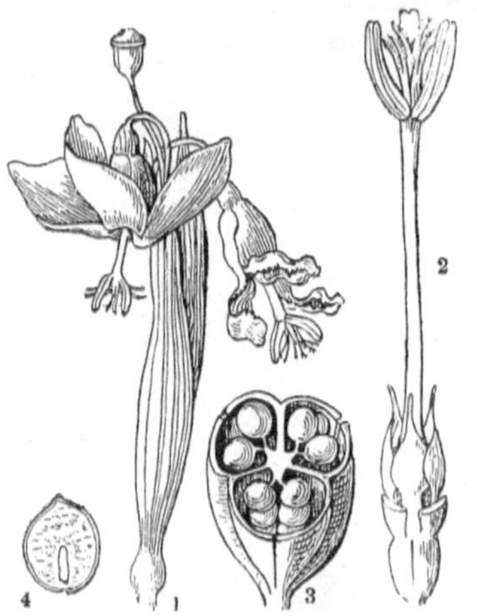

Fig. CXI.

The Irids are principally natives either of the Cape of Good Hope, or of the middle parts of North America and Europe. A few only are found within the tropics, and the Order is generally far from abundant in South America, if compared with the numbers that exist at the Cape. The genera Marica and Moræa appear to occupy the same station in hot climates that Iris, a closely related genus, does in cooler latitudes. Crocus, among the most conspicuous of the Order, occurs only in Europe and Asia. None of the Cape or New Holland forms appear in America.

More remarkable for their beautiful fugitive flowers than for their utility. The rhizome of some of them is slightly stimulating, as the violet-scented Orris root, the produce of Iris Florentina. Various species of Sisyrinchium, Ferraria, Libertia, and the Irises pseud-acorus, tuberosa, versicolor, and verna, are used as diuretics, purgatives, and emetics, but some of them are apt to produce distressing nausea like sea-sickness, with a prostration of strength. The substance called Saffron is the dried stigmas of Crocus sativus; its colouring ingredient is a peculiar principle, to which the name Polychroite has been given; it possesses the properties of being totally destroyed by the action of the solar rays, of colouring in small quantity a large body of water, and of forming blue and green tints when treated with sulphuric and nitric acid, or with sulphate of iron. In moderate doses this substance stimulates the stomach, and in large quantities excites the vascular system. Moreover it seems to have a specific influence on the cerebro-spinal system, as it affects, it is said, the mental faculties, a result which De Candolle considers analogous to that produced by the petals of certain odorous flowers. "In modern practice it is little used, except as a colouring ingredient; on the Continent it is employed

Fig. CXI.—1. Spathe and flowers of Rigidella immaculata; 2. the petals, stamens, &c. of it; 3. a cross section of the capsule of Pardanthus Chinensis; 4. a perpendicular section of its seeds.

as an agreeable stimulant in many culinary preparations and liqueurs. In a medicinal point of view it is frequently used to assist the eruption of exanthematous diseases; on the same principle that bird-fanciers give it to birds in the moult. It has been used as a carminative, antispasmodic and emmenagogue."—*Pereira.* Sicilian saffron is obtained from Crocus odorus, according to Gussone. According to Gray, the roasted seeds of Iris pseud-acorus very nearly approach Coffee in quality.—*Suppl. Pharmac.* 237. Iris sibirica is regarded as an antisyphilitic; Iris fœtidissima, the ξυρις of Dioscorides, has some reputation as a cure for scrofula. Gladiolus segetum has been fancied an aphrodisiac, a reputation doubtless obtained from its acrid qualities, which seem to occur in the whole Order, as far as they have been examined. Nevertheless, we are told that the Hottentots eat the tubers or corms of various species, whose starch renders them nutritious. Those of Trichonema edule are eaten by the natives of Socotra, as we learn from Welstead. According to Endlicher, the purple flowers of Iris germanica and sibirica, treated with lime, furnish a green colour (Liliengrün), "much used by artists." The stem of Witsenia maura is said to abound in rich saccharine juice.—*Bot. Reg.* l. 5. Some Brazilian Irids are purgative, among which Martius particularly enumerates Ferraria purgans and cathartica, and Sisyrinchium galaxioides.

GENERA.

Sisyrinchium, *L.*
Bermudiana, Tourn.
Syorhynchium, Hffmsg.
Orthrosanthus, *Sweet.*
Solenomelus, *Miers.*
Crukshanksia, Miers.
Symphyostemon, *Miers.*
Eleutherine, *Herb.*
Psythirisma, *Herb.*
Echthronema, *Herb.*
Eriphilema, *Herb.*
Calydorea, *Herb.*
Glumosia, *Herb.*
Tecophilæa, *Bert.*
Phyganthus, Popp.
Pöppigia, Kunze.
Libertia, *Spr.*
Renealmia, R. Br.
Nematostigma, Dietr.
Cipura, *Aubl.*
Marica, Schreb.
? *Trimeriza*, Salisb.
? *Hydastylis*, Salisb.
? *Galathea*, Salisb.
Hymenostigma, *Hochst.*
Vieusseuxia, *Roche.*
? *Freuchenia*, Eckl.
Plantia, *Herbert.*
Trimezia, *Herbert.*

Moræa, *Linn.*
Homeria, Vent.
? *Dietes*, Salisb.
Diplarrhena, *Labill.*
Iris, *Linn.*
Xiphion, Tournef.
Hermodactylus, Tourn.
Sisyrinchium, Tournef.
Isis, Tratt.
Herbertia, *Sweet.*
Cypella, *Herb.*
Phalocallis, *Herb.*
Alophia, *Herb.*
Trifurcaria, *Herb.*
Hydrotænia, *Lindl.*
Beatonia, *Herb.*
Tigridia, *Juss.*
Rigidella, *Lindl.*
Ferraria, *Linn.*
Pardanthus, *Ker.*
Belemcanda, Rheede.
Aristea, *Soland.*
Cleanthe, Salisb.
? *Bobartia*, Linn.
Wredowia, Eckl.
Witsenia, *Thunb.*
Nivenia, Vent.
Genlisia, Rchb.
Sophronia, Lichtenst.

Tapeinia, Commers.
Patersonia, *R. Br.*
Genosiris, Labill.
Galaxia, *Thunb.*
Ovieda, *Spreng.*
Lapeyrousia, Pourr.
Peyrousia, Sweet.
Meristostigma, Dietr.
Anomatheca, *Ker.*
Anomaza, Lawson.
Babiana, *Ker.*
Acaste, Salisb.
Acidanthera, *Hochst.*
Gladiolus, *Tournef.*
Hebea, Pers.
Lemonia, Pers.
Homoglossum, Salisb.
Synotia, Sweet.
Streptanthera, Sweet.
Bertera, Sweet.
Antholyza, *Linn.*
Cunonia, Buttn.
Anisanthus, Sweet.
Petamenes, Salisb.
Watsonia, *Mill.*
Micranthus, Pers.
Phalangium, Houtt.
Meriana, Trev.

? *Neuberia*, Eckl.
Sparaxis, *Ker.*
Montbretia, *DC.*
Hexaglottis, Vent.
Tritonia, Ker.
Waizia, Rchb.
Houttuynia, Houtt.
Freesa, Eckl.
Bellendenia, Rafin.
Morphixia, *Ker.*
Ixia, *Linn.*
Hyalis, Salisb.
Eurydice, Pers.
Agretta, Eckl.
Diasia, *DC.*
Aglæa, Pers.
Melasphærula, Ker.
Phalangium, Burm.
Hesperantha, *Ker.*
Hesperanthus, Salisb.
Geissorhiza, *Ker.*
? *Weihea*, Eckl.
? *Spatalanthus*, Sweet.
Trichonema, Ker.
Romulea, Maratti.
Nemastylis, *Nutt.*
Gelasine, *Herb.*
Crocus, *Tournef.*

NUMBERS. GEN. 53. Sp. 550.

Orchidaceæ.
POSITION.—Hæmodoraceæ.—IRIDACEÆ.—Amaryllidaceæ.

ALLIANCE XII. *AMOMALES.*—THE AMOMAL ALLIANCE.

DIAGNOSIS.—*Epigynous petaloid Endogens, with unsymmetrical flowers, from* 1 *to* 5 *stamens, some of which are abortive, and albuminous seeds.*

In the Narcissal Alliance, the series was terminated by the Irids, many of whose genera have a singularly irregular corolla : as, for example, Babiana ; there was, however, even in these last, an exact symmetry in the number of parts of which the flowers consist. In this Alliance that symmetry is wholly lost, the number of perfect stamens, as represented by anthers, being reduced to one, or even half a one, and not exceeding five in any instance. At the same time the development of the foliage takes a new direction. In the majority of Narcissals the leaves are absolutely sword-shaped, and their veins consequently run in parallel lines ; and even when, as sometimes happens, their leaves become widened, the veins still converge at the point. But in the Amomal Alliance the veins always diverge ; the result of which is a foliage of quite another character, to which, among Endogens, some Lilyworts offer the only resemblance. When such leaves acquire a large size, they are frequently split into lateral ribands.

NATURAL ORDERS OF AMOMALS.

Stamens more than 1 ; (*anthers* 2-*celled, no vitellus*) . . . . . 48. MUSACEÆ.
Stamen but 1 ; *anther* 2-*celled, embryo in a vitellus* . . . . . 49. ZINGIBERACEÆ.
Stamen but 1 ; *anther* 1-*celled* (*halved*), *no vitellus* . . . . . 50. MARANTACEÆ.

ORDER XLVIII. MUSACEÆ.—MUSADS.

Musæ, *Juss. Gen.* (1789).—Musaceæ, *Agardh Aph.* 180. (1823); *Ach. Rich. Nouv. Elém. ed.* 4. 436. (1828); *Endlicher Prodr. Fl. Norf.* 34. (1833); *Endl. Gen.* lxx.; *Lestiboudois in Ann Sc. Nat.* 2. *ser.* 17. 257.; *Meisner, p.* 389.

DIAGNOSIS.—*Amomal Endogens with more stamens than one.*

Stemless or nearly stemless plants, with leaves sheathing at the base, and forming a kind of spurious stem, often very large, their limb separated from the taper petiole by a round tumour, and having fine parallel veins diverging regularly from the midrib towards the margin. Flowers spathaceous. Perianth 6-parted, adherent, petaloid, in 2 distinct rows, more or less irregular. Stamens 6, inserted upon the middle of the divisions, some always becoming abortive; anthers linear, turned inwards, 2-celled, often having a membranous petaloid crest. Ovary inferior, 3-celled, many-seeded, rarely 3-seeded; ovules anatropal; style simple; stigma usually 3-lobed. Fruit either a 3-celled capsule, with a loculicidal dehiscence, or succulent and indehiscent. Seeds sometimes surrounded by hairs, with an integument which is usually crustaceous; embryo orthotropal, oblong-linear, or mushroom-shaped, with the radicular end touching the hilum, having pierced through the mealy albumen.

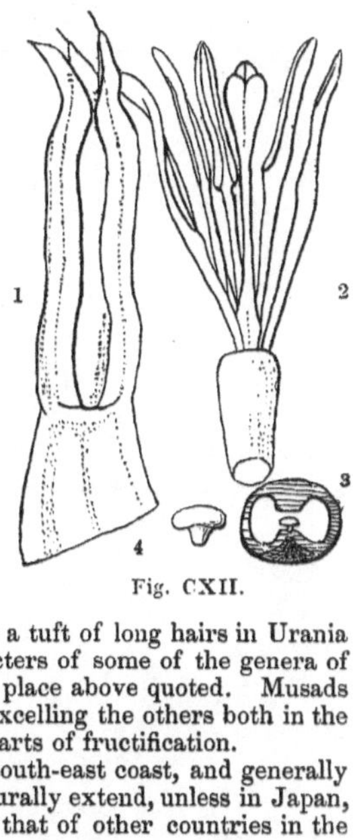

Fig. CXII.

The relationship of this Order will be pointed out under Gingerworts and Marants, with which the Musads are strictly related. The flower of Musa is well described in the Appendix to the *Congo Expedition*, 471., in a note; that of Strelitzia is pentandrous and exceedingly irregular, and is admirably illustrated in Bauer's drawings, published some years since by Ker, under the title of *Strelitzia Depicta*. The hilum of the seed gives rise to a tuft of long hairs in Urania and Strelitzia. For remarks upon the distinctive characters of some of the genera of Musads, see *Endl. Prodr.* p. 34, and *Lestiboudois* in the place above quoted. Musads are doubtless the most perfect of the Amomal Alliance, excelling the others both in the size at which they arrive, and the completeness of their parts of fructification.

Natives of the Cape of Good Hope, the islands of its south-east coast, and generally of the plains of the tropics, beyond which they do not naturally extend, unless in Japan, the climate of which seems to be much at variance with that of other countries in the same latitude.

They are most valuable plants, both for the abundance of nutritive food afforded by their fruit, called in the tropics Plantains and Bananas, and for the many domestic purposes to which the gigantic leaves of some species are applied. The latter are used for thatching Indian cottages, for a natural cloth from which the traveller may eat his food, as a material for basket making, and finally they yield a most valuable flax (Musa textilis), from which some of the finest muslins of India are prepared. The stems are formed of the united petioles of the leaves, which are remarkable for the vast quantity of spiral vessels they contain: these exist in such numbers as to be capable of being pulled out by handfuls, and are said to be collected in the West Indies and sold as a kind of tinder.—*Dec. Org.* 38. The number of threads in each convolution of these spiral vessels varies from 7 to 22.—*Ibid* 37. The young shoots of the Banana are eaten as a delicate vegetable. The root of Heliconia Psittacorum, the fruit of the Bihai, and the seed of Urania speciosa or Ravenala, a magnificent Palm-like plant, called by the French *Arbre du Voyageur*, are said to be eatable; its pulpy aril, of the

Fig. CXII.—Musa paradisiaca. 1. A flower; 2. the stamens, style, and stigma; 3. a section of a seed; 4. the embryo.

most brilliant blue colour, yields an essential oil. The juice of the fruit and the lymph of the stem of Musa are slightly astringent and diaphoretic. The juice of the fruit of Urania is used for dyeing.—*Agdh.*

GENERA.

I.—*Heliconeæ.* Seeds solitary. Fruit a capsule bursting through the partitions. Heliconia, *Linn.* *Bihai*, Plum.	II.—*Uraneæ.* Seeds numerous in each cell. Fruit berried, or, if capsular, bursting through the cells. Musa. *Tournef.*	Strelitzia, *Banks.* ? *Heliconia*, Gærtn. Ravenala, *Adans.* *Urania*, Schreb.

NUMBERS. GEN. 4. Sp. 20.

Liliaceæ.
POSITION.—Zingiberaceæ.—MUSACEÆ.—Marantaceæ.
Palmaceæ?

ORDER XLIX. ZINGIBERACEÆ.—GINGERWORTS.

Cannæ, *Juss. Gen.* 62. (1798), *in part.*—Drymyrhizeæ, *Vent. Tabl.* (1799); *DC. Ess. Méd.* 281. (1816). —Scitamineæ, *R. Brown, Prodr.* 305. (1810); *Agardh Aph.* 182. (1823); *Rosc. Monogr.; Blume Enumeratio, p.* 39. (1827); *Lestiboudois in Ann Sc.* 2. *ser.* 15. 305.—Zingiberaceæ, *Rich. Anal. Fr.* (1808); *Ed. pr.* ccxxxiii.—*Endl. Gen.* lxviii.; *Meisner, p.* 388.—Amomeæ, *Juss. in Mirbel's Elém.* 854. (1815); *Ach. Rich. Nouv. Elém. ed.* 4. 438. (1828).—Alpiniaceæ, *Link Handb.* 1. 228. (1829), *a sect. of* Scitamineæ.

DIAGNOSIS.—*Amomal Endogens with one stamen, a two-celled anther, and a vitellus round the embryo.*

Aromatic tropical herbaceous plants. Rhizome creeping, often jointed. Stem formed of the cohering bases of the leaves, never branching. Leaves simple, sheathing, their

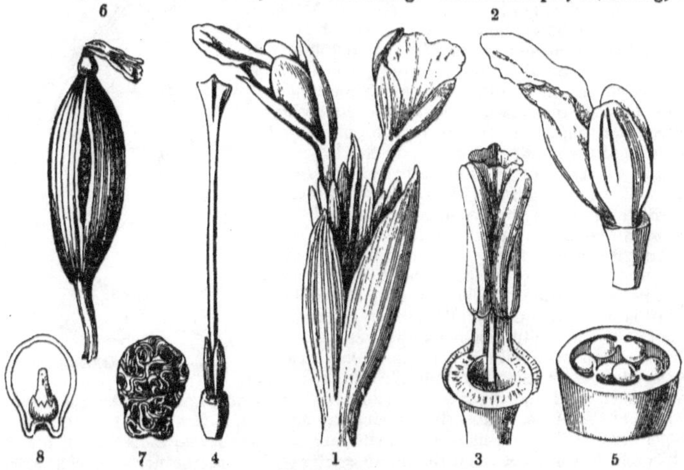

Fig. CXIII.

lamina often separated from the sheath by a taper neck, and having a single midrib, from which very numerous, simple, crowded veins diverge at an acute angle. Inflorescence either a dense spike, or a raceme, or a sort of panicle, terminal or radical. Flowers arising from among spathaceous membranous bracts, in which they usually lie in pairs. Calyx superior, tubular, 3-lobed, short. Corolla tubular, irregular, with 6 segments in 2 whorls; the outer 3-parted, nearly equal, or with the odd segment sometimes differently shaped; the inner (sterile stamens) 3-parted, with the intermediate segment (labellum) larger than the rest, and often 3-lobed, the lateral segments sometimes nearly abortive. Stamens 3, distinct, of which the 2 lateral are abortive, and the intermediate one fertile; this placed opposite the labellum, and arising from the base of the intermediate segment of the outer series of the corolla. Filament not petaloid, often extended beyond the anther in the shape of a lobed or entire appendage. Anther 2-celled, opening longitudinally, its lobes often embracing the upper part of the style. Pollen globose, smooth. Ovary 3-celled, sometimes imperfectly so; ovules several, anatropal, attached to a placenta in the axis; style filiform, stigma dilated, hollow. Fruit usually capsular, 3-celled, many-seeded, [sometimes by abortion 1-celled]; occasionally berried (the dissepiments generally central, proceeding from the axis of the valves, at last usually separate from the latter, and of a different texture.—*R. Br.*) Seeds roundish, or angular, with or without an aril (albumen floury, its substance radiating, and deficient near the hilum, *R. Br.*); embryo inclosed within a peculiar membrane (vitellus,

Fig. CXIII.—1. Flowers of Kæmpferia pandurata; 2. the inner row of the corolla seen in profile; 3. the anther, inclosing the apex of the style between its lobes; 4. the style and stigma, with two abortive stamens at the base; 5. a transverse section of the ovary; 6. ripe fruit of Ceylon Cardamoms, Elettaria Cardamomum Zeylanicum of Pereira; 7. a seed; 8. the same cut through to show the embryo seated in vitellus.

R. Br. Prodr.; membrane of the amnios, *Ibid. in King's Voyage*, 21), with which it does not cohere.

Formerly the Gingerworts and Marants were united in one tribe called Canneæ : hence it is certain that they are at least more nearly related to each other than to anything else, and that whatever is the affinity of the one will be that of the other. Taking the vegetation into account, these two tribes are exceedingly nearly allied to Musads, in which is found the same kind of leaf, the veins of which are closely set, and diverge from the midrib to the margin, being connected by very weak and imperfect intermediate veins ; the leaves have also the same distinct petiole, often with a thickened rounded space at the apex ; Musads are, however, pent- or hexandrous, with a calyx and corolla of the same texture. Irids are the next Order with which Gingerworts may be compared, agreeing in their superior flowers, which have sometimes an approach to the irregularity of Alpinia, and also in the triple number of their stamens ; but while these organs are all developed in Irids, two are abortive or deformed in Gingerworts and Marants. Bromelworts have been identified with them of old, but their resemblance consists chiefly in the distinction of calyx and corolla, and their inferior ovary. To Orchids, to which the flowers of Mantisia bear much resemblance, they are related in consequence of the reduction of their three stamens to one by the abortion of two : but the cohesion of the stamens and style in the latter, and the want of any distinction between calyx and corolla, sufficiently separate them, besides which the series which produces the stamens in Orchids answers to the sterile stamens or inner limb of the corolla in the Gingerworts. There is a volume consecrated to plants of this kind by Roscoe, who first remodelled the genera and reduced them within fixed limits. Between the embryo and the albumen is interposed a fleshy body enveloping the former : this has been called a process of the rostellum by Correa, a cotyledon by Smith, a vitellus by Gærtner and Brown, a central indurated portion of the albumen by Richard. It is now known to be the innermost integument of the ovule, unabsorbed during the advance of this body to maturity.

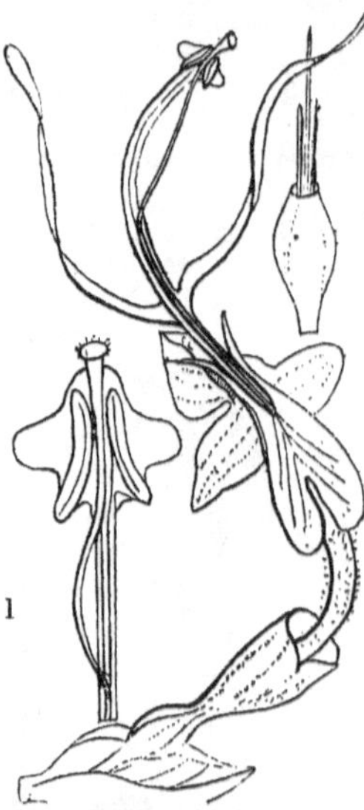

Fig. CXIV.

Independently of the presence of this vitellus, the most remarkable part of the structure of Gingerworts depends on the number of divisions of the floral envelopes, which consist of a tubular calyx, and of two more series instead of one. Brown, struck with this unusual deviation from the ordinary organization of Monocotyledons, was disposed to consider the calyx an accessory part (*Prodr.* 305) ; but Lestiboudois' explanation appears more satisfactory. According to this botanist Gingerworts are really hexandrous, like the nearly-related Musads ; but of their stamens the outer series is petaloid, and forms the inner limb of the corolla, and of the inner series of stamens the central one only developes, the lateral ones appearing in the form of rudimentary scales. This notion of Lestiboudois is confirmed by Marants, in which the inner stamens (even that which is antheriferous) become petaloid like the outer : thus showing that in these plants there is a strong and general tendency in the filaments to assume the state of petals.

All are tropical, or nearly so. By far the greater number inhabit various parts of the East Indies ; some are found in Africa, and a few in America. They form a part of the singular Flora of Japan.

They are generally objects of great beauty, either on account of the high development of the floral envelopes, as in Hedychium coronarium and Alpinia nutans ; or because of the rich and glowing colours of the bracts, as in Curcuma Roscoeana. They are, however, principally valued for the sake of the aromatic stimulating properties of the root or rhizome, such as are found in Ginger (Zingiber officinale), Galangale (Alpinia racemosa and Galanga), Zedoary (Curcuma Zedoaria and Zerumbet), and some other species of the latter genus. Many more species are used in a similar manner. The warm and pungent roots of the greater and lesser Galangale are not only used by the Indian doctors in cases of dyspepsia, but are also considered useful in

Fig. CXIV.—A flower of Mantisia saltatoria ; 1. style, stigma, and anther ; 2. ovary, style, and abortive stamens.

coughs, given in infusion. A bad sort of Galangale is obtained from Alpinia pyramidata, *Bl.*, and Allughas, with which are often mixed Alpinia nutans and Kæmpferia Galanga. The seeds of many partake of the properties of the root. Cardamoms are the seeds of several plants of this Order. On the eastern frontiers of Bengal the fruit of Amomum aromaticum is used. Malabar Cardamoms are produced by Elettaria Cardamomum; Ceylon Cardamoms, an inferior sort, by Elettaria major. Grains of Paradise, a sort of hot acrid seed, used to give a pungent flavour to spirituous liquors, belong principally to Amomum Grana Paradisi, but Amomum angustifolium, macrospermum, maximum, and Clusii are, according to Dr. Pereira, also the parents of an inferior description of this seed. Others are known for their dyeing properties, such as Turmeric. This substance, obtained from Curcuma longa, is cordial and stomachic; it is also considered by the native practitioners of India an excellent application in powder for cleaning foul ulcers. The fruit of Globba uviformis is said to be eatable. Generally, in consequence of the presence of the aromatic oil that is so prevalent in the Order, the roots or rhizomes, although abounding in fæcula, are not fit for the preparation of arrow-root; but an excellent kind is prepared in Travancore, in the East Indies, from Curcuma angustifolia.

A species of Curcuma is supposed by Von Martius to furnish the astringent Mexican drug called Cascara de Pingue, which abounds in tannin. What is called Cascara de Lingue is the bark of some tree.—*Chem. Gaz.* 1844. 263. The American Renealmias are stated by Pöppig to have aromatic leaves which, when bruised, are employed in pains of the limbs. The roots of Costi are very bitter, and have had a great reputation as tonics, but they are out of use. The roots of Alpinia aromatica and Paco seroca are sweetly aromatic, and are employed in Brazil as carminatives and stomachics.—*Martius.* All the Brazilian Costi have a sub-acid mucilaginous juice, which is used in nephritic diorders and gonorrhœa.—*Id.* According to Roxburgh the pendulous tubers of Curcuma rubescens and several other species yield a very beautiful pure starch, like Arrow-root, which the natives of the countries where the plants grow prepare and eat. In Travancore this flour or starch forms a large part of the diet of the inhabitants. Such Arrow-root, obtained from C. angustifolia, is commonly sold in the markets of Benares. See *Flora Medica* for further information concerning these plants.

GENERA.—*Much in need of re-examination.*)

Globba, *Linn.*
Catimbium, Juss.
Colebrookia, *Don.*
Ceranthera, Horn.
Hura, König.
Sphærocarpus, Gawl.
Manitia, Gieseke.
Ceratanthera, *Hornem.*
Mantisia, *Curt.*
Zingiber, *Gærtn.*
Jägera, Gieseke.
Dietrichia, Gieseke.
Casumunar, Colla.
Lampujang, Rumph.
Curcuma, *Linn.*
Zerumbet, Rumph.
Stissera, Gieseke.
Erndlia, Gieseke.
Kæmpferia, *Linn.*
Soncorus, Rumph.
Trilophus, *Lestib.*
Roscöea, *Smith.*
Amomum, *Linn.*
Cardamomum, Rumph.
Marenga, Salisb.
Alexis, Salisb.
Hornstedtia, Retz
Meistera, Gieseke
Wurfbainia, Gieseke.
Greenwaya, Gieseke.
Paludana, Gieseke.
Etlingera, Gieseke.
Elettaria, *Rheed.*
Matonia, Sm.
Cardamomum, Salisb.
Geanthus, Reinw.
Donacodes, *Blume.*
Diracodes, *Blume.*
Hedychium, *König*
Gandsulium, Rumph.
Gamochilus, *Lestib.*
Renealmia *Linn.*
Alpinia, Plum.
Gethyra, Salisb.
Peperidium, *Lindl.*
Alpinia, *Linn.*
Zerumbet, Jacq.
Costus, Pers.
Ethanium, Salisb.
Allughas, Linn.
Buekia, Gieseke.
Catimbium, Lestib.
Leptosolena, *Presl.*
Gastrochilus, *Wall.*
Hellenia, *Willd.*
Albina, Gieseke.
Martensia, Gieseke.
Heritiera, Retz.
Languas, König.
Monolophus, *Wall.*
Cenolophon, *Blume.*
Costus, *Linn.*
Tsjana, Gmel.
Planera, Gieseke.
Banksia, Konig.
Hellenia, Retz.
Glissanthe, Salisb.
Jacuanga, *Lestib.*
Monocystis, *Lindl.*
Kolowratia, *Presl.*
Nyctophylax, *Zippel.*
Hitchenia, *Wall.*

NUMBERS. GEN. 29. Sp. 247.

POSITION.—Musaceæ.—ZINGIBERACEÆ.—Marantaceæ.
Orchidaceæ.

ORDER L. MARANTACEÆ.—MARANTS.

Cannæ, *Juss. Gen.* 62. (1789) *in part.*—Cannæ, *R. Brown, Prodr.* 1. 307. (1810).—Canneæ *or* Maranteæ, *Brown in Flinders*, (1814).—Cannaceæ, *Agardh Aph.* 181. (1823); *Link Handb.* 1. 223. (1829), *a sect. of* Scitamineæ; *Endl. Gen.* lxix.; *Lestiboudois in Ann. Sc.* 2 *ser.* 17. 205.; *Meisner, p.* 389.

DIAGNOSIS.—*Amomal Endogens, with one stamen, half an anther, and no vitellus.*

Herbaceous tropical plants, destitute of aroma. Rhizome often tuberous, and abounding in starch. Stem often branching. Leaves, inflorescence, and flowers, as in Gingerworts. Calyx superior, of 3 sepals, short. Corolla tubular, irregular, with the segments in 2 whorls the outer 3-parted, nearly equal: the inner very irregular; one of the lateral segments usually coloured, and formed differently from the rest; sometimes by abortion fewer than 3. Stamens 3, petaloid, distinct, of which one of the laterals and the intermediate one are either barren or abortive, and the other lateral one fertile. Filament petaloid, either entire or 2-lobed, one of the lobes bearing the anther on its edge. Anther 1-celled, opening longitudinally. Pollen round (papillose in Canna coccinea, smooth in Calathea zebrina.) Ovary 1-3-celled; ovules solitary, erect, and campylotropal, or numerous, anatropal, and attached to the axis of each cell; style petaloid or swollen; stigma either the mere denuded apex of the style, or hollow, cucullate, and incurved. Fruit capsular, as in Gingerworts. Seeds round, without aril; albumen hard, somewhat floury; embryo straight, naked, its radicle, lying against the hilum.

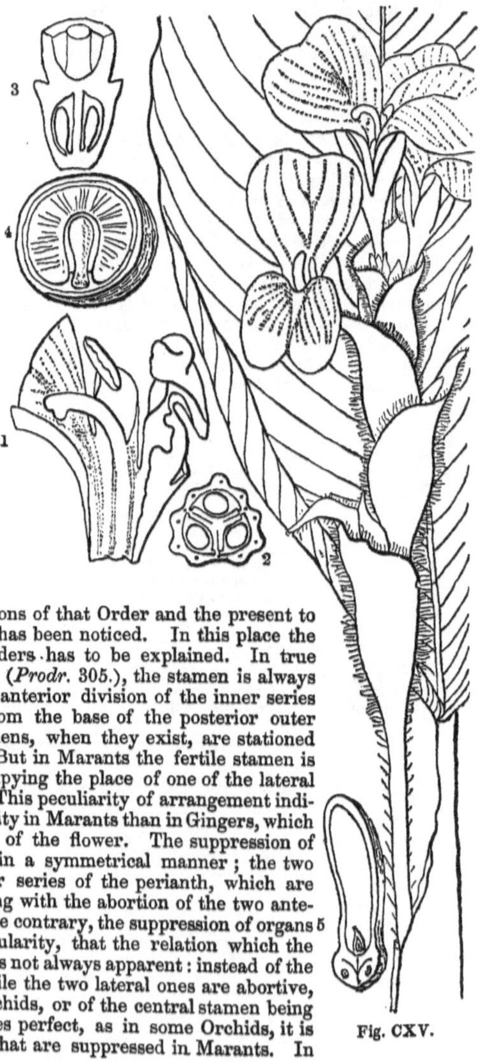

Under Gingerworts, the relations of that Order and the present to other monocotyledonous groups has been noticed. In this place the distinction between the two Orders has to be explained. In true Gingers, as Brown has observed (*Prodr.* 305.), the stamen is always placed opposite the labellum or anterior division of the inner series of the corolla, and proceeds from the base of the posterior outer division; while the sterile stamens, when they exist, are stationed right and left of the labellum. But in Marants the fertile stamen is on one side of the labellum, occupying the place of one of the lateral sterile stamens of Gingerworts. This peculiarity of arrangement indicates a higher degree of irregularity in Marants than in Gingers, which also extends to the other parts of the flower. The suppression of organs takes place in the latter in a symmetrical manner; the two posterior divisions of the inner series of the perianth, which are occasionally absent, corresponding with the abortion of the two anterior stamens. In Marants, on the contrary, the suppression of organs takes place with so much irregularity, that the relation which the various parts bear to each other is not always apparent: instead of the central stamen being perfect while the two lateral ones are abortive, as in Gingerworts and most Orchids, or of the central stamen being abortive and the two lateral ones perfect, as in some Orchids, it is the central and one lateral one that are suppressed in Marants. In

Fig. CXV.

Fig. CXV.—Calathea villosa; 1. a flower cut open; 2. a transverse section of the ovary; 3. a perpendicular section of it; 4. a section of the seed of Canna; 5. a section of its embryo.

the perianth of Canna only the most external part within the calyx can properly be called corolla; the remainder of the segments being attempts to produce barren petaloid stamens analogous to what is called the inner limb of the corolla in Gingerworts; and the characters upon which botanists found their specific distinctions depend upon the degree to which this development of petaloid abortive stamens extends. When, for instance, they describe some as having an inner limb of 2 or of 3, or of 4 or of 5 segments, they should rather say 2, 3, 4, or 5 stamens are partially developed.

Perhaps it will be possible to put the relative structure of Gingerworts and Marants in a clearer light by the following diagrams, in which the triangle C, C, C represents the calyx, the angles corresponding with the position of the sepals; the triangle P, P, P the corolla; R, *r*, *r* an outer series of petaloid stamens, of which *r*, *r* are rudimentary only; and S, *s*, *s* the inner series of stamens, of which S is the fertile and fully developed one.

GINGERWORTS.

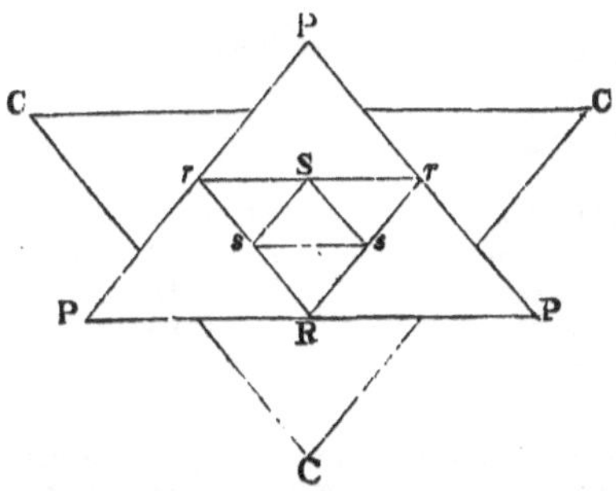

MARANTS.

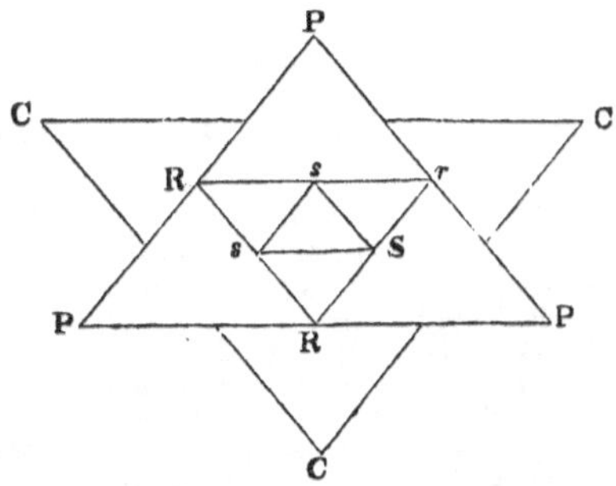

The greater part are found in tropical America and Africa; several are natives of India; none are known in a wild state beyond the tropics.

While Gingerworts are valued for their aromatic heating principle, the Marants are esteemed on account of the fæcula, which abounds in the rhizome and root of both tribes, the Gingerworts being destitute of that principle: on this account it is collected as a delicate article of food, both from Maranta arundinacea, Allouyia, and nobilis, in the West Indies, and also from Maranta ramosissima in the East. The fleshy corms of some Cannas are reported to be eaten in Peru, and a sort of Arrow-root called tous les mois is extracted in the West Indies from some species supposed to be C. Achiras. The seeds of others, called Indian shot, have been used as a substitute for Coffee, and yield a purple dye. A tough fibre is obtained from Phrynium dichotomum; and the leaves of the South American Calatheas are worked into baskets, whence their name. The juice of Maranta arundinacea is said to be efficacious in poisoned wounds; it is acrid when fresh, reddening the skin, and exciting saliva when chewed. The tubers of Maranta Allouyia, cooked with pepper and salt, are eaten in the West Indies. Martius says that the tubers of Canna aurantiaca, glauca, and others, are diuretic and diaphoretic, and are not unlike Orris-root in action.

GENERA.

Thalia, *Linn.*	*Phyllodes*, Loureir.	Canna, *Linn.*
Peronia, DC.	Calathea, *G. F. W. Meyer*	*Cannacorus*, Tournef.
Maranta, *Plum.*	*Göppertia*, Nees.	
Phrynium, *Willd.*	Myrosma, *Linn. fil.*	

NUMBERS. GEN. 6. Sp. 160.

Liliaceæ.
POSITION.—Zingiberaceæ.—MARANTACEÆ.——
Orchidaceæ.

ALLIANCE XIII. *ORCHIDALES.*—THE ORCHIDAL ALLIANCE.

DIAGNOSIS.—*Epigynous Endogens, with 1 to 3 stamens, and seeds without albumen.*

At this point there is an abrupt break in the series of direct affinity. No gradual change can be traced from other natural Orders to that of the Orchidal Alliance, which is distinguished by the embryo not only having no albumen, but being a solid homogeneous body, equally destitute of any visible radicle or cotyledon. In the majority the structure is what Linnæus called Gynandrous ; that is to say, the stamens, and style, and stigma, are blended together into one solid body, named a column ; in two, however, of the natural Orders of which it consists, the stamens are perfectly free. If we neglect the condition of the seeds, we then may find a variety of approaches to other Orders, as, for example, to the Irids, in which Gladiolus seems to be an imitation of the structure of an Orchis ; or to Sisyrinchium, to which Thelymitra or Paxtonia offer some analogy ; or to the Hypoxids, of which Apostasias and Tropidia have much the aspect ; or to Gingerworts, whose close heads of imbricated bracts are imitated in Evelyna. The Burmanniads are remarkable for their perfect symmetry, among hundreds of species whose prevailing character is want of symmetry.

NATURAL ORDERS OF ORCHIDALS.

Flowers regular. Stamens free, perigynous . . . . 51. BURMANNIACEÆ.
Flowers irregular, gynandrous. Placentæ parietal . . . . . . 52. ORCHIDACEÆ.
Flowers regular, half-gynandrous. Placentæ axile . . . . . . 53. APOSTASIACEÆ.

ORDER LI. BURMANNIACEÆ.—BURMANNIADS.

Burmanniæ, *Spreng. Syst.* 1. 123. (1825); *Reichenb. Conspect.* 60. (1828), *a sect. of* Amaryllideæ.—Burmanniaceæ, *Blume Enum. Pl. Jav.* 27. (1827); *Bartl. Ord. Nat.* 41. (1830); *Schult. f. in Röm. et Sch. Syst. Veg.* 7. lxxiii. (1830); *Endl. Gen.* lx.; *Meisner, p.* 390.; *Miers in Linn. Trans.* 18. 552. —Tripterelleæ, *Nuttall in Act. Philadelph.* 7. 23.

DIAGNOSIS.—*Orchidal Endogens, with regular flowers and free perigynous stamens.*

Herbaceous plants, with tufted radical acute leaves, or none; a slender naked stem; and terminal flowers, sessile upon a 2- or 3-branched rachis, or solitary. Flowers ♂. Perianth coloured, tubular, adherent, membranous, with 6 teeth, the 3 inner of which (petals) are minute, the 3 outer larger, sometimes with a wing or keel at the back. Stamens 3, inserted in the tube opposite the petals; anthers sessile, 2-celled, opening transversely, with a fleshy, simple, or 2-lobed connective. Ovary adherent, 1- 3-celled, with 3 placentæ, which are either simple and parietal, or double and axile; the cells of the 3-celled genera opposite the sepals; ovules innumerable; capsule surmounted by the persistent perianth, 1- 3-celled, bursting vertically, or horizontally, or not at all, or by one fissure, into a boat-shaped pericarp. Seeds innumerable, very minute, with the testa loose or fitting tight; apparently with a solid nucleus, and no albumen; style single; stigma 3-lobed. Capsule covered by the withered perianth, or 3-celled, bursting irregularly.

This is a most singular race, which has been well illustrated by Mr. Miers, who has been the first to point out its relationship to the Orchids. This he has shown to consist in the minute seeds, parietal placentæ, in many cases peculiar condition of the capsule, and the nucleus loose in the middle of a net-like testa. To this I think may be added the organization of the kernel of the seed, which is, to all appearance, in Burmannia, Apteria, and Dictyostega, exactly like that of Orchids. Mr. Miers, however, describes the nucleus of the latter genus as being suspended by a thread in the middle of the testa; I find it, on the contrary, ascending. Two very different forms are pointed out by Mr. Miers, the Burmannieæ, with 3 cells in the ovary and an axile placentation, and the Apterieæ, with 1 cell and 3 parietal placentæ; this peculiarity is not however accompanied by any other, and may, for the present, be regarded as of secondary importance. The single genus upon which the Order was founded, was placed by Jussieu in Bromelworts; Brown stationed it as a doubtful genus at the end of Rushes, with the remark, that it is extremely distinct both in flower, fruit, and inflorescence, and not really allied to any other known plant, but more nearly related to Xyris and Philydrum than to either Bromelia or Hypoxis. Von Martius, who has beautifully illustrated the Brazilian species, refers them to Hydrocharads. Blume, who has added two new genera, merely remarks that "the Order is known from Juncaceæ by its tubular perianth, which is petaloid instead of glumaceous, and by the structure of the fruit; it is well distinguished from Irids by the station of the stamina, and the transverse dehiscence of the anthers." —*Enum.* p. 27.

In reality the Order must be considered to connect Orchids and Irids.

Natives of marshy grassy places in the tropics of Asia, Africa, and America. Burmannia is found as far to the north as Virginia in North America.

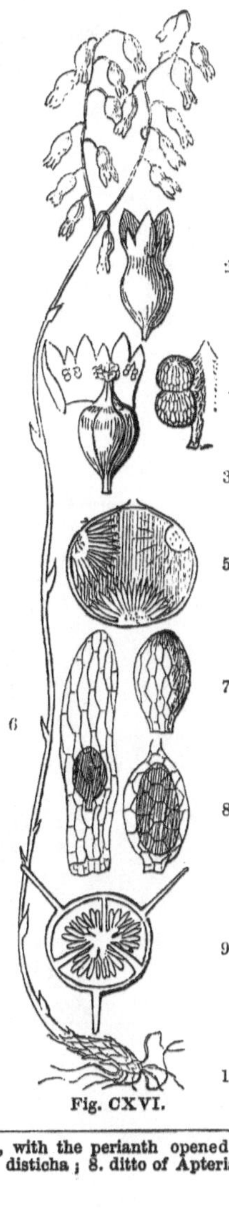

Fig. CXVI.

Fig. CXVI.—1. Dictyostegia orobanchoides; 2. a flower; 3. the same, with the perianth opened; 4. half an anther; 5. section of ovary; 6. seed; 7. seed of Burmannia disticha; 8. ditto of Apteria setacea; 9. transverse section of the ovary of a Burmannia from Ceylon.

Apteria setacea is slightly bitter and very astringent. A similar flavour, something like that of Green Tea, is discernible in Burmannia cærulea.—*Nuttall.*

GENERA.

I.—*Apterieæ.* Miers. Ovary 1-celled. Placentæ parietal.

Gymnosiphon, *Blume.*
Apteria, *Nutt.*
Dictyostega, *Miers.*
Cymbocarpa, *Miers.*
Stenoptera, *Miers.*

II.—*Burmannieæ.* Miers. Ovary 3-celled. Placentæ parietal.

Burmannia, *L.*
Tripterella, Rich.
Vogelia, Gmel.
Maburnia, Thouars.
Gonyanthes, *Blume.*

NUMBERS. GEN. 7. Sp. 30.

Iridaceæ.
POSITION.—Apostasiaceæ.—BURMANNIACEÆ.—Orchidaceæ.

ORDER LII. ORCHIDACEÆ.—ORCHIDS.

Orchides, *Juss. Gen.* 64. (1789).—Orchideæ, *R. Brown Prodr.* 309. (1810); *Rich. in Mém. Mus.* 4. 23. (1818); *Bauer, Francis, and Lindley, Illustrations of Orchidaceous Plants; Id. Genera and Species of Orch.* (1830): *R. Brown Observations on the Sexual Organs, &c. of Orchideæ and Asclepiadeæ* (1831); *Endl. Gen.* lxvi.; *Meisner, Gen. p.* 367.—Vanillaceæ, *Ed. pr.* ccxliv.

DIAGNOSIS.—*Orchidal Endogens, with irregular gynandrous flowers and parietal placentæ.*

Herbaceous plants or shrubs, always perennial, occurring all over the world, except in the very coldest regions, or those where everlasting dryness reigns; in temperate countries terrestrial, in warmer latitudes growing on trees (epiphytes), or fixing themselves to stones. Their roots are fibrous and fasciculated, fleshy or resembling tubers, and these filled with starch, or horny nodules of bassorin. Stem none, or long and annual, or perennial and woody, forming a rhizome, or jointed branches. Leaves flat, terete or equitant, generally sheathing, membranous, coriaceous, or hard, never lobed, occasionally bordered by cartilaginous teeth, their veins parallel, almost never slightly reticulated. Flowers ☿, irregular, extremely variable in form, solitary, clustered, spiked, racemose, or panicled, always supported by a solitary bract; very often most gratefully fragrant, sometimes fetid, and not unfrequently scentless. Perianth adherent, variable, herbaceous or coloured, membranous or fleshy, permanent and withering, or deciduous; its parts arranged in two rows, rarely in 3, free or adhering in various ways; very often resupinate in consequence of a twist in the ovary. Sepals (which, morphologically speaking, are petals) 3, equal at the base, or variously extended or expanded there; the two lateral standing in front when the ovary is twisted, and the third then dorsal, or

Fig. CXVII.

Fig. CXVIII.

Fig. CXVII.—Herminium monorchis.
Fig. CXVIII.—Accidental manner of producing a jointed stem in Aspasia epidendroides.

next the axis; occasionally surrounded by a calyculus (or true calyx). Petals (which are to be regarded as sterile stamens) usually 3; very rarely one only, placed between the sepals: the lateral usually similar to the dorsal sepal; the third, called the lip (*labellum*), usually larger than the petals, and quite unlike them in form; horned or furnished with various appendages, free or adherent to some other body, occasionally moveable as if spontaneously; now and then contracted so as to form two separate parts, of which the lowest is called the hypochil, the highest the epichil, and the middle one the mesochil; sometimes furnished with a single or double appendage, derived from the stigma. Column consisting of the stamens and style consolidated into a central body, so that the latter stands next the lip and the former next the dorsal sepal, sometimes petaloid, and occasionally extended far beyond the perianth (corymbis). Stamens 3, opposite the sepals, the central only being perfect, except in Cypripedium, when the central is abortive and the two lateral perfect; anthers occasionally one-celled; usually two-celled, with the cells separated by 2 or 4 partitions; standing erect at the end of the column, or turned down flat upon it, or altogether dorsal; pollen powdery, or collected into grains, or adhering in wedges tied together by an elastic material, or consolidated into masses of a waxy texture and fixed number, the masses either free or adhering by a caudicle to a gland belonging to the apex (or rostellum) of the stigma. Ovary adherent, 1-celled, composed of 6 carpels, of which 3, opposite the petals, have didymous polyspermous parietal placentæ without stigmas, and 3 opposite the sepals have as many stigmas but no placentæ; style never distinct, except in Cypripedium and some Neottieæ; stigmas usually confluent in a hollow (or prominent) mucous disk; the dorsal stigma having on the upper edge one or two glands, which are separate in Vandeæ and Neotteæ; often extended into a beak (rostellum), or hollowed out into pouches, or sometimes drawn out into 2 parallel or diverging arms; the lateral stigmas usually obsolete, but sometimes united to the base of the lip in the form of an appendage or pair of plates. Capsule very rarely fleshy, indehiscent and pod-shaped, usually breaking up into 6 dry woody rigid valves with horizontal cells, of which 3 only bear seeds. Seeds innumerable, very minute, with a loose netted skin, very rarely with a hard crustaceous one, sometimes expanded into a circular wing; embryo solid, fleshy, without albumen; chalaza at the apex of the seed, and therefore the radicle next the hilum.

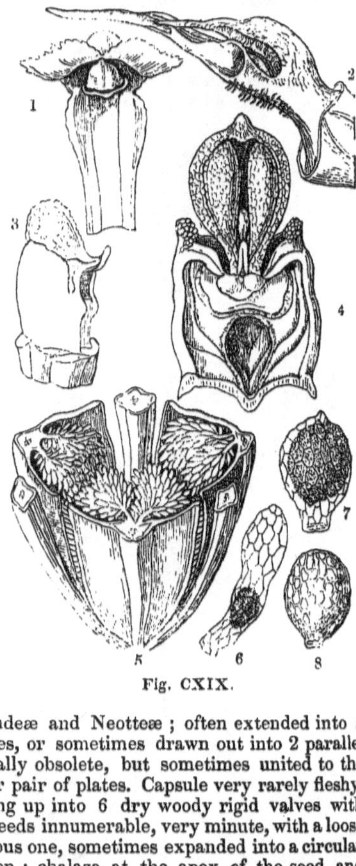

Fig. CXIX.

The general structure of Orchids, briefly embodied in the foregoing description, has been treated of at such length in the prefatory matter of the *Illustrations of Orchidaceous Plants*, that it is unnecessary to do more than refer the reader to that work. I must, however, take the opportunity of correcting one part of the theoretical view which was there taken of the structure of the column. While, in common with Dr. Brown, I regarded the stigma as really consisting of three parts, usually in a state of confluence, I also supposed the position of the stigmata to be opposite the petals; being led to that conclusion by the constant position of the stigmatic arms of Ophrydæ. That opinion I afterwards retracted, in consequence of the position of the stigmas in Cypripedium, which C. spectabile shows most clearly to be opposite the sepals; and therefore the stigmatic arms of Ophrydæ are to be understood as side lobes of that stigma which is opposite the dorsal sepal. This circumstance, however, only confirms the accuracy

Fig. CXIX.—1. Column of Arethusa; 2. of Stenorhynchus; 3. of Brassia maculata; 4. of Orchis mascula; 5. section of capsule of Ophrys apifera; 6. seed of Ophrys; 7. of Pterygodium atratum; 8. of Vanilla aromatica.

of my view of the true nature of the stamens, which are certainly all opposite the lobes of the stigma in Cypripedium. While, however, the untenableness of the first opinion concerning the relation borne by the stigmas to the other parts of the flower, is thus admitted, there remains a difficulty that opposes itself to the view I now take in common with Brown, and which must not be overlooked. It is that the placentiferous pieces of the ovary are not opposite the stigmas, but alternate with them, while the seedless pieces of the ovary are in a line with the stigmata! This seems to show that the ovary is composed of 6 carpellary leaves, of which three bear stigmas without ovules, and three bear ovules without stigmas. However paradoxical this may appear, it is by no means incompatible with the due performance of the functions of fertilisation; for the carpellary leaves do not adhere into a solid mass, either in the ovary or in the style. On the contrary they form a cavity open from the stigmatic apex down to the ovules, and the whole of that cavity is lined with a lax conducting tissue, which may nevertheless be exclusively furnished by 3 stigmas only, and may become so confluent with the placentæ as to form a perfect channel of communication for the pollen tubes in their descent into the ovules.

The Order owes its chief peculiarities to the following circumstances: firstly, to the consolidation of stamens and pistil into one common mass, called the column; secondly, to the suppression of all the anthers, except one in the mass of the Order, or two in Cypripedeæ; thirdly, to the peculiar condition of its pollen, and the anther which contains it; and fourthly, to the very general development of one of the inner leaves of the perianth or petals in an excessive degree, or in an unusual form. These peculiarities are in most cases so striking, and are all so strongly manifested in the same flower, that the inexperienced botanist may be unable to discover their real character. We find, however, that the true nature of each part is indicated by special cases of structure occurring in different parts of the Order. Thus in Cypripedium not only are two lateral stamens furnished with anthers, while the central stamen is antherless, but the stigma and style separate from the filaments nearly to the base, and the triple nature of the former is distinctly shown, together with the relation of its lobes to the other parts of the flower. The pollen, which has so anomalous an appearance in its waxy or sectile state, presents the usual appearance of that substance in Goodyera, and many Neotteæ. And the irregularity of the labellum disappears in such genera as Thelymitra, Paxtonia, Macdonaldia, Hexisea, and some others, whose flowers are almost as regular as those of a Sisyrinchium. It is indeed to the latter genus, more nearly than to any other, that Orchids seem to approach in structure, unless to Gingerworts; so that they may be supposed to pass into Irids through Thelymitra and Sisyrinchium on the one hand, and into Gingerworts through Phrynium and such a genus as Evelyna on the other. With regard to Apostasiads, their relation to that Order does not appear to be greater than to either of the two now mentioned; and in the absence of all evidence as to the connecting links which join Orchids and Apostasiads it seems unnecessary to advert further to the subject. It may, however, be observed that Apostasia has apparently as much claim to be regarded as a diandrous monadelphous Hypoxid, standing, perhaps, in the same relation to that Order as Gilliesia to Lilyworts, as it has to be regarded as a trilocular Orchid with the gynandrous organization lost.

It is not necessary to enter, in this place, into a history of the gradual alteration that has taken place in the views of botanists with regard to the structure of the sexual apparatus of these most curious plants, or to explain what degree of ignorance was shown by those who mistook masses of pollen for anthers, or a column of stamens for a style; such errors could only have occurred at a period when the laws of organization were unknown. They have been corrected, in a more or less perfect manner, by various writers; most completely by Brown in his *Prodromus*, published in 1810, and subsequently by the late most accurate and indefatigable Richard. But long before the publication of any rational explanation of the structure of Orchids, while botanists were in utter darkness upon the subject, it had been investigated by a man unrivalled in his day for the perfection of his microscopical analyses, the beauty of his drawings, and the admirable skill with which he followed Nature in her most secret workings; and let me add, which is a still rarer quality, the generous disinterestedness with which he communicated to his friends the result of his patient and silent labours. Sketches were executed by the late Francis Bauer, between 1794 and 1807, in which the most material part of what has been published since that period is distinctly shown; and it has been my good fortune to be the humble means of giving some of these remarkable productions of the pencil to the world, in the *Illustrations of the Genera and Species of Orchidaceous Plants.*

If the column of an Orchidaceous plant is examined, it will be found to consist of a fleshy body stationed opposite the lip, bearing a solitary anther at its apex, and having in front a viscid cavity, upon the upper edge of which there is often a slight callosity, called the rostellum. This cavity is the stigma, and the rostellum is the point by which the pollen masses are secured when any adhesion between them and the stigma takes place. Hence such a plant would appear to be monandrous ; it will be seen, however, in Gingerworts and Marants, the only other monandrous Orders of Endogens, that, while only one perfect stamen is developed, two others exist in a rudimentary state ; so that the ternary number prevalent in Monocotyledons is not departed from. So it is in Orchids : the column does not consist of a single filament cohering with a style, but of three filaments firmly grown together, the central of which is antheriferous, and the lateral sterile. This is proved by the frequent presence of callosities, or processes in the place of the sterile stamens ; by imperfectly-formed anthers occasionally appearing at the side of the perfect one ; and, if any further evidence were wanted, by monsters, in which a regular structure is exchanged for the ordinary irregularity. Such an instance in Orchis latifolia is described by Achille Richard, in the *Mémoires de la Soc. d'Hist. Nat. of Paris*, in which the flowers were perfectly triandrous, with no trace of irregularity in any part of the floral envelopes ; and other cases of a similar nature are by no means uncommon, and have been occasionally mentioned.

Orchids are remarkable for the unusual figure of their irregular flowers, which sometimes represent an insect, sometimes a helmet with the visor up, and are so various in form that there is scarcely a common reptile or insect to which some

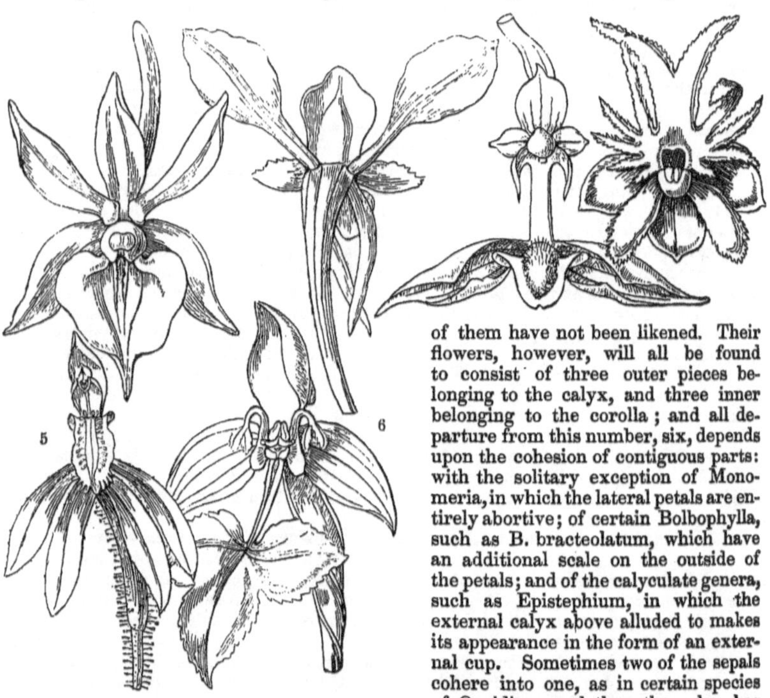

Fig. CXX.

of them have not been likened. Their flowers, however, will all be found to consist of three outer pieces belonging to the calyx, and three inner belonging to the corolla ; and all departure from this number, six, depends upon the cohesion of contiguous parts: with the solitary exception of Monomeria, in which the lateral petals are entirely abortive; of certain Bolbophylla, such as B. bracteolatum, which have an additional scale on the outside of the petals; and of the calyculate genera, such as Epistephium, in which the external calyx above alluded to makes its appearance in the form of an external cup. Sometimes two of the sepals cohere into one, as in certain species of Oncidium, and then the calyx has the appearance of consisting of but two sepals ; sometimes the lateral petals are connate with the column, as in Gongora and and Lepanthes, and then the column appears furnished with two wings. In nearly the whole Order the odd petal, called the lip, arises from the base of the column, and is opposite it ; but in the Cape genus Pterygodium, the lip sometimes grows from the apex of the column, and sometimes is stalked and turned completely over between the fork of the inverted anther, and thus seems to belong to the back of the column. Nor is the anther less subject to modification, although constant to its place : sometimes it stands erect, the line of dehiscence of its lobes being turned towards the lip ; sometimes it is turned upside down, so that its back regards the lip ; often it is prone upon the apex of the column, where a niche is excavated for its reception. The pollen is not

Fig. CXX. 1. Angræcum eburneum; 2. Diuris; 3. Drymoda picta; 4. Oberonia Griffithiana; 5. Caladenia; 6. Disa spathulata.

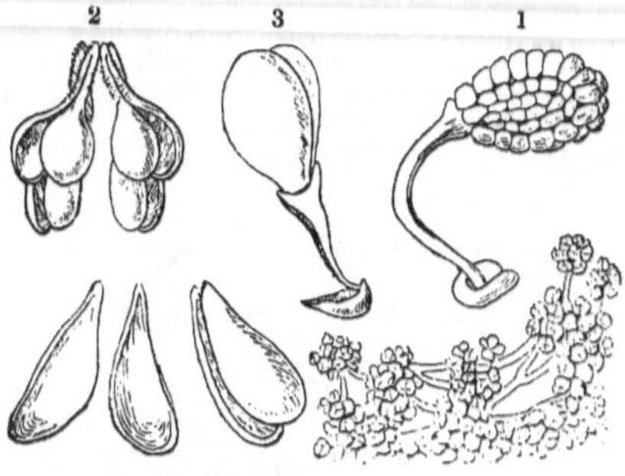

Fig. CXXI.

less curious: now we have it in separate grains, as in other plants, but cohering to a meshwork of cellular tissue, which is collected into a sort of central elastic strap; now the granules cohere in small angular indefinite masses, and the central elastic strap, becoming more apparent, is found attached to a glandular process of the stigma, which is often inclosed in a peculiar pouch especially destined for its protection; again, the pollen combines into larger masses, which are definite in number, and attached to another modification of the elastic strap; and finally a complete union of the pollen takes place, in solid waxy masses, without any distinct trace of this central elastic tissue. Such is a part of the singularities of Orchidaceous plants, and upon these the distinctions of their tribes and genera are naturally founded. Whoever studies them must bear in mind that their fructification is always reducible to 3 sepals, 3 petals, a column consisting of 3 stamens grown firmly to one another, and to a single style and stigma; and, with this view, he will have no difficulty in understanding the organization of even the most anomalous Cape species. In the last edition of this work an Order called Vanillaceæ was proposed, about which I shall only say that its introduction would have been much better omitted.

Professor Link has shown that beyond all doubt the nucleus of the seed in this Order is a naked embryo, with an excessively enlarged radicula. See his beautiful figures in the *Ausgewälte Anatomisch-botanische Abbildungen fasc.* 2. *t.* vii. Here we again have a structure analogous to that of Nymphæa and Nelumbium.

Among the most singular circumstances connected with this Order is the manner in which, upon the same spike, flowers of extremely different structure are produced. This was first noticed in Demerara by Sir R. Schomburgk, who published in the *Linn. Transactions* (17. 551.) an

Fig. CXXII.

Fig. CXXI.—1. Pollen masses of Ophrys apifera; 2. of Phaius Tankervillia; 3. of Brassia maculata; 4. of Malaxis paludosa; 5. Pollen of Stenorhynchus speciosus.

Fig. CXXII.—2. Cycnoches ventricosum; 4 and 5. C. Egertonianum; the others intermediate forms.

account of the production of Monachanthus viridis, Myanthus barbatus, and a Catasetum, 3 supposed genera, upon the same spike; and he expressed his opinion that the Catasetum was the female of these, because he found it producing seeds abundantly, while Monachanthus was uniformly sterile. Afterwards a similar specimen made its appearance in the garden of his Grace the Duke of Devonshire at Chatsworth, and has been figured in the *Botanical Register, fol.* 1951. And still more lately two species of Cycnoches, ventricosum and Egertonianum have appeared in company, as represented in the accompanying figure (CXXII.)

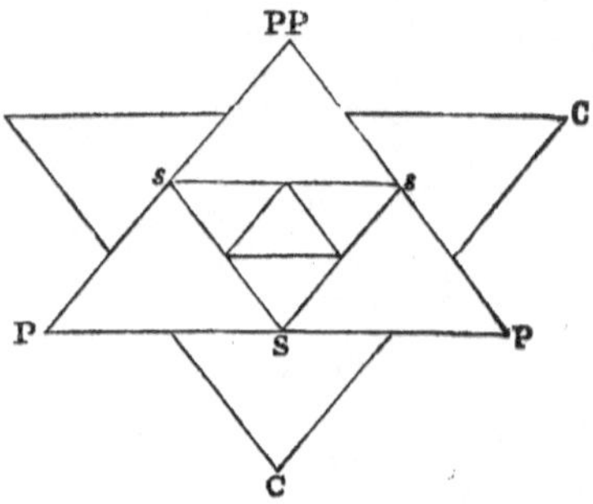

Such cases shake to the foundation all our ideas of the stability of genera and species, and prepare the mind for more startling discoveries than could have been otherwise anticipated.

If the accompanying diagram be compared with those employed to illustrate the distinctions of Marants and Gingerworts, p. 169, the relation borne to those Orders by Orchids will be distinctly seen. In the diagram the parts are arranged as they are in nature before the ovary twists; that is, with the lip next the axis, or uppermost, and the stamen undermost. Let C, C, C represent the outer series of floral envelopes or calyx, and PP, P, P the inner, or corolla, of which PP is the labellum: then the position of the single fertile stamen will be at S, and the sterile ones at *s*, *s*; that is to say, in the situation of the supernumerary petaloid stamens of Gingerworts and Marants, while the second series of stamens, to which the fertile stamen of these Orders belongs, is not developed in Orchids.

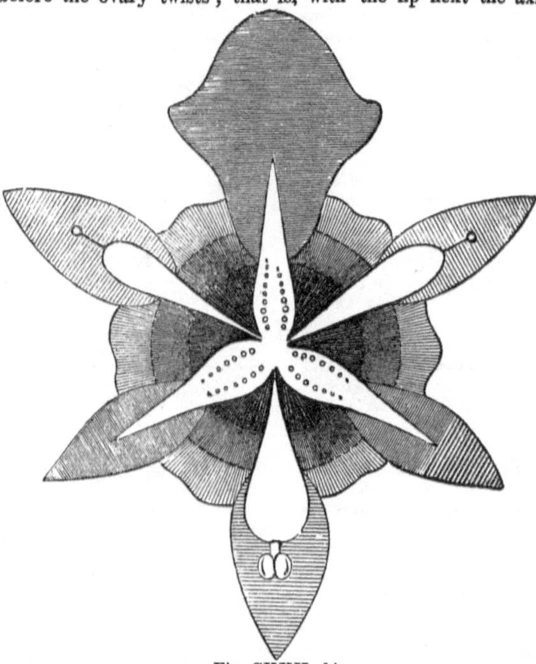

Fig. CXXII. *bis.*

But although this is the apparent structure of the Order, it is more probable that the parts called sepals are the true petals, because Epistephium and others have a calyculus exterior to the apparent calyx. In that point of view the apparent petals will be sterile stamens, as among the Marants, and the nature of the parts will be shown by the above projection.

In classifying this Order the most important characters appear to reside in the pollen, which in many is consolidated into firm waxy masses of a definite number in each species, and in others is either in its usual loose powdery condition, or is collected in granules or small wedges, the number of which is far too great to be counted. Of those with waxy pollen masses some (Malaxeæ) are destitute of any visible processes by which the masses are brought into contact with the stigma; others (Epidendreæ) have strap-shaped cau-

Fig. CXXII. *bis.*—A theoretical projection of all the parts of an Orchid flower, showing what is supposed to be their true relative position.

dicles, which are either bent down upon the masses themselves, or serve to hold them together, without, however, forming any organised union with the stigma; while the remainder (Vandeæ) have a caudicle, which adheres firmly to a gland found on the upper margin of the stigma, and separating freely from that organ. The genera with powdery, granular, or sectile pollen cannot be classified so conveniently by modifications of that part, but are readily divided into 3 natural tribes by peculiarities in the anther. In some (Ophreæ) the anther is erect, not hinged to the column but continuous with it, and stands above the stigma, the pollen masses having their points directed to the base of the lobes of the anther; in others (Arethuseæ) the anther is hinged to the column, upon the end of which it is placed transversely like a lid; and in others (Neotteæ) it is also hinged to the column, but is stationed at its back so as to be nearly parallel with the stigmatic surface. If to this we add that Cypripedeæ have two anthers, while all the others have one only, we find the Order divided into seven tribes, of which the following is a tabular view.

I. Anther one only.
- A. *Pollen masses waxy.*
  - *a.* No caudicle or separable stigmatic gland — I. *Malaxeæ.*
  - *b.* A distinct caudicle, but no separable stigmatic gland — II. *Epidendreæ.*
  - *c.* A distinct caudicle, united to a stigmatic gland — III. *Vandeæ.*
- B. *Pollen powdery, granular, or sectile.*
  - *a.* Anther terminal, erect — IV. *Ophreæ.*
  - *b.* Anther terminal, opercular — V. *Arethuseæ.*
  - *c.* Anther dorsal — VI. *Neotteæ.*

II. Anthers two — VII. *Cypripedeæ.*

Among many other remarkable peculiarities the irritability of the labellum must not be passed over in silence. This is extremely striking in various species of Pterostylis, in the genus Megaclinium, and in many Bolbophylls, especially barbigerum and Careyanum. But some of the Swan River species are still more singular. In Caleana nigrita, Mr. Drummond describes the structure to the following effect. The column is a boat-shaped box, resembling a lower lip; the labellum forms a lid that exactly fits it, and is hinged on a claw which reaches the middle of the column; when the flower opens, it (the labellum) turns round within the column, and falls back, so that, the flower being inverted, it stands fairly over the latter. The moment a small insect touches its point, the labellum makes a sudden revolution, brings the point to the bottom of the column, passing the anther in its way, and thus makes prisoner any insect which the box will hold. When it catches an insect it remains shut while its prey continues to move about; but if no capture is made the lid soon recovers its position. Another plant, Drakæa elastica, has a single flower placed at the end of a slender smooth erect scape, from twelve to eighteen inches high, and its labellum, which is hammer-headed, and placed on a long arm with a moveable elbow-joint in the middle, is stated by Mr. Drummond to resemble an insect suspended in the air, and moving with every breeze. Another plant of this description is Spiculæa ciliata, whose rusty flowers when spread open may be compared to long-legged spiders, the lip with a long solid lamina looking like their body, while an appendage at its apex, which is apparently moveable, is not unlike the head of such a creature.

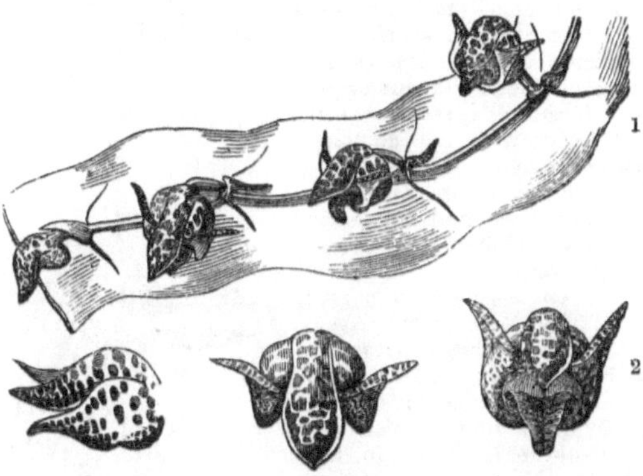

Fig. CXXIII.

Orchids are found in almost all parts of the world, except upon the verge of the frozen

Fig. CXXIII. *bis.*

Fig. CXXIII.—Megaclinium Bufo; 1. a portion of a spike magnified; 2. flowers in various positions, more magnified.

Fig. CXXIII. *bis.*—Spiculæa ciliata, its flower.

zone, and in climates remarkable for dryness. In Europe, Asia, and North America, they are seen growing everywhere, in groves, in marshes, and in meadows; in the drier parts of Africa they are either rare or unknown; at the Cape of Good Hope they abound in similar situations as in Europe; but in the hot damp parts of the West and East Indies, in Madagascar, and the neighbouring islands, in the damp and humid forests of Brazil, in the warm mild parts of Central America, and Western Mexico, in the damp tropical parts of India, and on the lower mountains of Nipal, the Orchidaceous plants flourish in the greatest variety and profusion, no longer seeking their nutriment from the soil, but clinging to the trunks and limbs of trees, to stones and bare rocks, where they vegetate among ferns and other shade-loving plants, in countless thousands. Of the epiphytal class, one only is found so far north as South Carolina, growing upon the branches of the Magnolia, if we except the species from Japan, a country which has a climate peculiar to itself, among regions in the same parallel of latitude. The most southern stations are those of Earina mucronata in New Zealand, in lat. 35° S., and of Gunnia australis in Emu Bay, Van Diemen's Land, lat. 41° S. Ample details respecting their distribution in Australia are given by A. Cunningham in the *Botanical Register* for 1843 *t.* 37.

It often happens that those productions of nature which charm the eye with their beauty, and delight the senses with their perfume, have the least relation to the wants of mankind, while the most powerful virtues or most deadly poisons are hidden beneath a mean and insignificant exterior: thus Orchids, beyond their beauty, can scarcely be said to be of known utility, with a few exceptions. The nutritive substance called Salep has been prepared from the subterraneous succulent roots of Orchis mascula and many others of the Ophreous division; and in India from the tubers of a species of Eulophia; it consists almost entirely of a chemical principle called Bassorin. The root of Bletia verecunda is said to be stomachic. Some of the South American species, such as the Catasetums, Cyrtopodiums, &c., contain a viscid juice, which being inspissated by boiling, becomes a kind of vegetable glue used for economical purposes in Brazil. The viscidity of the tuber of Aplectrum hyemale is such that it is called Putty-root in the United States, and is used for cementing broken earthenware.

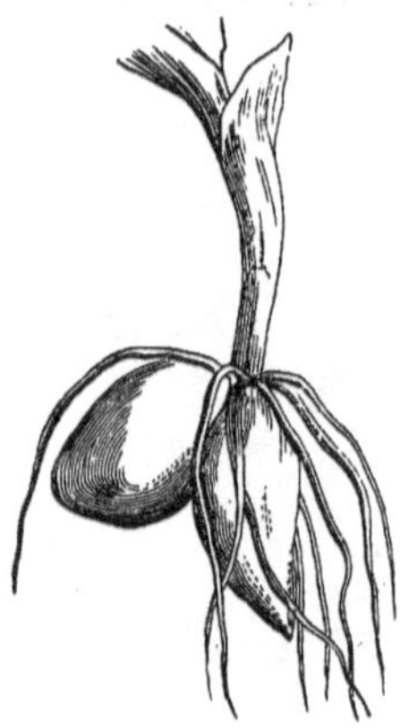

Fig. CXXIV.

Other medical qualities have been assigned to other species, but they seem to be of no importance; thus, Arethusa bulbosa is employed in the United States in toothache and bringing tumours to a head, Spiranthes diuretica as a diuretic in Chile, where also Chloræa disoides is fancied to promote the flow of milk. Cypripedium pubescens is used in North America as a substitute for Valerian, C. guttatum in Siberia against epilepsy. Vanilla is one of the most delightful aromatics known; it is used in the manufacture of chocolate, of liqueurs, and of various articles of confectionery. The substance called by this name in the shops is the dried fruit of Vanilla planifolia, and other species; it contains a great quantity of essential oil, and a good deal of benzoic acid. Dr. Bird says that the effluvium of Vanilla intoxicates the labourer who gathers it.—*Peter Pilgrim*, 1. 234. See *Linnæa*. 4. 573, for some account of the cultivation of the plant in Mexico. Vanilla claviculata is bitter as well as fragrant, and its leaves are regarded in the West Indies, where it is called Liane à blessures, as a vulnerary, and antisyphilitic. In New Holland many species are eaten by the natives, who find their starchy roots a good article of diet. Mr. Backhouse describes the Gastrodia sesamoides as having a root like a series of kidney potatoes, terminating in a branched, thick mass of coral-like fibres. It is eaten by the aborigines of Tasmannia, and is sometimes called native potato; but its tubers are watery and insipid.

P. Browne states that the corm of Bletia verecunda is "bitterish and attended by a clamminess that leaves a light prickly warmth behind it; but this wears off soon, leaving the palate free from every sensation but that of the bitter. When dried it may be used with great propriety as a stomachic." According to Sir R. Schomburgk the expressed juice of Epidendrum bifidum is a purgative, taken in doses of a table spoonful at a time; it is also reckoned in Tortola an anthelmintic, and diuretic, &c. —*Linnæa*, ix. 512.

Fig. CXXIV.—Orchis mascula roots in the state in which they are dried as salep.

GENERA.

[In the following list, I have cast all the genera into natural subdivisions, preparatory to a further revision of them; but I have not attempted to settle very exactly their order of sequence. In fact, a great number demand a very careful revision, and others, to which an * is prefixed, I have never had the opportunity of examining.]

I.—MALAXEÆ.

PLEUROTHALLIDÆ. *Lindl. in Bot. Reg.* 1842. *misc.* p. 67.

Pleurothallis, *R. Br.*
Rhynchopera, Klotzch.
Myoxanthus, Popp. et Endl.
Specklinia, Lindl.
Centranthera, Scheidw.
Acianthera, Scheidw.
Dialissa, *Lindl.*
Stelis, *Swz.*
Humboldtia, Fl. Peruv.
Lepanthes, *Swz.*
Restrepia, *Kth.*
* ? Cadetia, *Gaudich.*
Physosiphon, *Lindl.*
Masdevallia, *Fl. Per.*
Stenoglossum, *H. B. K.*
Octomeria, *R. Br.*

LIPARIDÆ.

Liparis, *Rich.*
Sturmia, Rchb.
Alipsa, Hffmg.
Cestichis, Thouars.
Distichis, *Thouars.*
Dendrochilum, *Bl.*
* Osyricera, *Bl.*
* Chrysoglossum, *Bl.*
Oberonia, *Lindl.*
Ensifera, Bl.
* Titania, *Endl.*
Empusa, *Lindl.*
Empusaria, Rchb.
* Platystylis, *Bl.*
* Gastroglottis, *Bl.*
Microstylis, *Nutt.*
Crepidium, Bl.
Monorchis, Mentz.
Achroanthes, Raf.
Pterochilus, Hook.
Dienia, *Lindl.*
Pedilea, Lindl.
Malaxis, *Swz.*
* Nephelaphyllum, *Bl.*
Calypso, *Salisb.*
Cytherea, Salisb.
Norna, Wall.
Orchidium, Swz.

DENDROBIDÆ.

Dendrobium, *Swz.*
Grastidium, Bl.
Ceraia, Lour.
Keranthus, Lour.
Bontia, Petiv.
? *Sarcostoma*, Bl.
§ *Stachyobium*, Lindl.
§ *Ceratobium*, Lindl.
§ *Pedilonum*, Bl.
§ *Onychium*, Bl.
§ *Desmotrichum*, Bl.
§ *Dendrocoryne*, Lindl.
* Macrostomium, *Bl.*
Aporum, *Bl.*
Schismoceras, Presl.
* § Diploconchium, *Schauer.*
Oxystophyllum, *Bl.*
* ? Diglyphis, *Bl.*
Diglyphosa, Bl.
Monomeria, *Lindl.*
* Epicrianthes, *Bl.*
? Drymoda, *Lindl.*
Bolbophyllum, *Thouars.*
Diphyes, Bl.
Tribrachia, Lindl.
Odontostylis, Bl. (f. Endl.)
Gersinia, Neraud. (f. Endl.)
Macrolepis, A. Rich.
§ *Anisopetalum*, Hooker.
? Sunipia, *Lindl.*
Trias, *Lindl.*
* Thelychiton, *Endl.*
* Cochlia, *Bl.*
* Lyræa, *Lindl.*
Megaclinium, *Lindl.*
Cirrhopetalum, *Lindl.*
Zygoglossum, Reinw.
Ephippium, Bl.
? *Sestochilus*, Kuhl et Hass.
Bryobium, *Lindl.*
Conchidium, *Griff.*
Mycaranthes, *Bl.*
Phreatia, *Lindl.*
Eria, *Lindl.*
Dendrolirium, Bl.
Pinalia, Hamilt.

CORALLORHIZIDÆ.

Corallorhiza, *Haller.*
Aplectrum, *Nuttall.*
* Aphyllorchis, *Blume.*

II.—EPIDENDREÆ.

CŒLOGYNIDÆ.

* ? Acanthoglossum, *Bl.*
Cœlogyne, *Lindl.*
Chelonanthera, Bl.
Panisea, *Lindl.*
Pleione, *Don.*
Gomphostylis, Wall.
Trichosma, *Lindl.*
Dilochia, *Lindl.*
Pholidota, *Lindl.*
Ptilocnema, Don.
Crinonia, Bl.
Otochilus, *Lindl.*
Earina, *Lindl.*

ISOCHILIDÆ.

Isochilus, *R. Br.*
Hexisea, *Lindl.*
? *Elleanthus*, Presl.
Diothonea, *Lindl.*
Gastropodium, *Lindl.*

LÆLIADÆ.

Epidendrum, *L.*
§ *Hormidium*, Lindl.
§ *Epicladium*, Lindl.
§ *Encyclium*, Hooker.
§ *Diacrium*, Lindl.
§ *Aulizeum*, Lindl.
§ *Osmophytum*, Lindl.
§ *Lanium*, Lindl.
§ *Spathium*, Lindl.
§ *Amphiglottium*, Salis.
§ *Euepidendrum*, Lindl.
Seraphyta, Fisch.
Physinga, *Lindl.*
Ponera, *Lindl.*
Nemaconia, Knowles.
* ? Aspegrenia, *Pöpp. et Endl.*
Hexadesmia, *Brongn.*
Hexopia, Batem.
Dinema, *Lindl.*
Sophronitis, *Lindl.*
Alamania, *Llave.*
Hartwegia, *Lindl.*
Arpophyllum, *Llave.*
Barkeria, *Knowles.*
Broughtonia, *R. Br.*
? Chysis, *Lindl.*
Lælia, *Lindl.*
Amalia, Rchb.
Cattleya, *Lindl.*
Schomburgkia, *Lindl.*
Tetramicra, *Lindl.*
Leptotes, *Lindl.*
Brasavola, *Lindl.*

BLETIDÆ.

Phaius, *Lour.*
Pachyna, Salisb.
Tankervillia, Link.
Arundina, *Bl.*
Evelyna, *Pöpp. et Endl.*
Bletia, *R. et Pav.*
Gyas, Salisb.
Thiebaudia, Colla.
* Mitopetalum, *Bl.*
Tainia, Bl.
Spathoglottis, *Bl.*
Paxtonia, *Lindl.*
* Collabium, *Bl.*
Cytheris, *Lindl.*
Pesomeria, *Lindl.*
Ipsea, *Lindl.*
* ? Pachystoma, *Bl.*
Apaturia, *Lindl.*
? Cremastra, *Lindl.*
Ania, *Lindl.*
* ? Callostylis, *Bl.*
Tylostylis, Bl.
* ? Ceratium, *Bl.*
Cylindrolobus, Bl.
* ? Trichotosia, *Bl.*
* ? Plocoglottis, *Bl.*
* ? Pachychilus, *Bl.* (End.)

III.—VANDEÆ.

SARCANTHIDÆ. — *Lindl. in Bot. Reg.* 1843. *misc.* p. 12.

Eulophia, *R. Br.*
Galeandra, *Lindl.*
Corydandra, Rchb.
Cyrtopera, *Lindl.*
Lissochilus, *R. Br.*
Doritis, *Lindl.*
Luisia, *Gaud.*
Pseudovanda, Lindl.
Mesoclastes, Lindl.
Birchea, A. Rich.
Vanda, *R. Br.*
Fieldia, Gaud.
Renanthera, *Lour.*
Arachnis, Blume.
Nephranthera, Hassk.
Arachnanthe, Blume.
Phalænopsis, *Bl.*
Diplocentrum, *Lindl.*
* Microsaccus, *Blume.*
Camarotis, *Lindl.*
Chiloschista, *Lindl.*
Gunnia, *Lindl.*
Micropera, *Lindl.*
Saccolabium, *Lindl.*
Saccochilus, Blume.
Gastrochilus, Don.
Robiquetia, Gaudich
Gussonea, A. Rich.
Rhynchostylis, Blume.
Carteretia, A. Rich.
Sarcochilus, *R. Br.*
* Tæniophyllum, *Blume.*
Cleisostoma, *Blume.*
Polychilos, Kuhl et Hass.
* Ceratostylis, *Blume.*
* Ephippium, *Blume.*
* Ceratochilus, *Blume.*
Omæa, Blume.
* Echioglossum, *Blume.*
Sarcanthus, *Lindl.*
* Pteroceras, *Hass.*
Agrostophyllum, *Blume.*
* Adenoncos, *Blume.*
Œceoclades, *Lindl.*
Trichoglottis, *Bl.*
Aërides, *Loureir.*
Dendrocolla, Blume.
Cucula, Blume.
Tubera, Blume.
Fornicaria, Blume.
Pilearia, Lindl.
Ornithochilus, Wall.
* Schœnorchis, *Blume.*
Aëranthus, *Lindl.*
Cryptopus, *Lindl.*
Beclardia, A. Rich.
Œonia, *Lindl.*
Angræcum, *Thouars.*
Aërobion, Spr.
Mystacidium, *Lindl.*
Microcœlia, *Lindl.*
Appendicula, *Bl.*
Metachilum, Lindl.
Podochilus, *Bl.*
Platysma, Bl.
Placostigma, Bl.
Apista, Bl.
Hexadesmia, R. Br.
? *Blumea*, Meyer.
* Cryptoglottis, *Bl.*
* Glomera, *Bl.*
* Thelasis, *Bl.*
Tetrapeltis, *Wall.*
* ? Conchochilus, *Hsskl.*
* ? Todaroa, *A. Rich.*

CRYPTOCHILIDÆ.

Cryptochilus, *Wall.*
Acanthophippium, *Bl.*
* ? Anthogonium, *Wall.*

BRASSIDÆ.

Cymbidium, *Swz.*
Bolbidium, *Lindl.*
Grammatophyllum, *Bl.*
Gabertia, Gaud.
* Stauroglottis, *Schauer.*
Bromheadia, *Lindl.*
Ansellia, *Lindl.*
Aganisia, *Lindl.*
Epiphora, *Lindl.*
Aspasia, *Lindl.*
? Acriopsis, *Bl.*
Trichopilia, *Lindl.*
Helcia, *Lindl.*
Nanodes, *Lindl.*
Pilumna, *Lindl.*
Dipodium, *R. Br.*
? *Armodorum*, Kuhl et [H.
Dichæa, *Lindl.*
Fernandezia *R. et Pav.*
Lockhartia, Hooker.
Phymatidium, *Lindl.*
Leochilus, *Knowles.*
Oncidium, *Swz.*
Cyrtochilum, H. B. K.
Odontoglossum, *H. B. K.*
§ *Trymenium*, Lindl.
Brassia, *R. Br.*

Miltonia, *Lindl.*
Macrochilus, Knowles.

PACHYPHYLLIDÆ.
Nasonia, *Lindl.*
Centropetalum, *Lindl.*
Pachyphyllum, *H. B. K.*

MAXILLARIDÆ. *Lindl. in Bot. Reg.* 1843, *misc. p.* 12.
Stanhopea, *Frost.*
Ceratochilus, Lodd.
Houlletia, *A. Brongn.*
Peristeria, *Hooker.*
Eckardia, Rchb.
Acineta, *Lindl.*
Lacæna, *Lindl.*
* ? Cuitlauzina, *Llav.*
Govenia, *Lindl.*
Eucnemis, Lindl.
Angidium, Lindl.
Batemannia, *Lindl.*
Gongora, *Fl. Peruv.*
Acropera, *Lindl.*
Coryanthes, *Hook.*
Chæanthe, *Lindl.*
Malachadenia, *Lindl.*
Cœlia, *Lindl.*
Ornithidium, *Salisb.*
Trigonidium, *Lindl.*
* Psittacoglossum, *Llav.*
Stenia, *Lindl.*
Promenæa, *Lindl.*
Grobya, *Lindl.*
Warrea, *Lindl.*
Huntleya, *Lindl.*
Zygopetalum, *Hooker.*
Bifrenaria, *Lindl.*
Stenocoryne, *Lindl.*
Maxillaria, *Fl. Per.*
§ * ? *Nothium*, Lindl.
§ * *Xylobium*, Lindl.
§ * *Dicrypta*, Lindl.
Heterotaxis, Lindl.
Lycaste, *Lindl.*
Anguloa, *Fl. Per.*
Camaridium, *Lindl.*
* Siagonanthus, *Pöpp et Endl.*
Scuticaria, *Lindl.*
Scaphyglottis, *Pöpp et E.*
Cladobium, Lindl.
Colax, *Lindl.*
Paphinia, *Lindl.*
Polystachya, *Hooker.*
*?Orchidofunkia, *A.Rich.*
Clynhymenia, A. Rich.
* ? Galeottia, *A. Rich.*

CATASETIDÆ. *Lindl. in Bot. Reg.* 1842. *p.* 22.
Catasetum, *Rich.*
Monachanthus, Lindl.
§ *Myanthus*, Lindl.
Mormodes, *Lindl.*
Cyclosia, Klotzsch.
Clowesia, *Lindl.*
Cycnoches, *Lindl.*
Cyrtopodium, *R. Br.*
Tylochilus, Nees.

NOTYLIDÆ.
Notylia, *Lindl.*
Cirrhæa, *Lindl.*
Zygostates, *Lindl.*
Dactylostyles, Scheidw.
Ornithocephalus, *Hook.*
? Trophianthus, *Scheidw.*
Cryptarrhena, *R. Br.*
Macradenia, *R. Br.*
Sutrina, *Lindl.*
Telipogon, *H. B. K.*
Trichoceros, *H. B. K.*
* * *
Trizeuxis, *Lindl.*
Quekettia, *Lindl.*

IONOPSIDÆ.
Rodriguezia, *R. et Pav.*
Gomeza, R. Br.
* Scelochilus, *Klotzsch.*
Burlingtonia, *Lindl.*
Ionopsis, *H. B. K.*
Iantha, Hook.
Cybelion, Spreng.
* Diadenium, *Pöpp et En.*
Comparettia, *Pöpp et En.*
Trichocentrum, *Pöpp et E.*
Acoidium, Lindl.

CALANTHIDÆ.
Calanthe, *R. Br.*
Centrosia, A. Rich.
Alismorchis, Thouars.
Amblyglottis, Blume.
Styloglossum, Kuhl et Hass.
* Limatodes, *Bl.*
* Ghiesbrechtia, *A. Rich.*
* Tipularia, *Nutt.*
Anthericlis, Ref.
Geodorum, *Jacks.*
Otandra, Salisb.
Cistella, Bl.

IV.—OPHREÆ.

SERAPIADÆ.
Orchis, *L.*
§ *Herorchis*, Lindl.
§ *Androrchis*, Endl.
Anacamptis, *Rich.*
Nigritella, *Rich.*
Aceras, *R. Br.*
Loroglossum, Rich.
Himantoglossum, Spr.
Serapias, *L.*
Helleborine, Pers.
Ophrys, *Swartz.*
Hemipilia, *Lindl.*
Glossaspis, *Spreng.*
Glossula, Lindl.
Perularia, *Lindl.*
Bartholina, *R. Br.*
Lathrisia, Swz.

SATYRIADÆ.
Pachites, *Lindl.*
Satyrium, *Swz.*
Diplectrum, Rich.
Satyridium, *Lindl.*
Aviceps, *Lindl.*

GYMNADENIDÆ.
Aopla, *Lindl.*
Herminium, *R. Br.*
Arachnites, Hoffm.
§ *Chamorchis*, Rich.
Chamærepes, Spr.
Gymnadenia, *R. Br.*
Sieberia, Spr.
Platanthera, *Rich.*
Mecosa, Bl.
Peristylus, *Blume.*
Benthamia, A. Rich.
Habenaria, *W.*
Dissorhynchium, Schauer.
?*Centrochilus*, Schauer.
Ate, *Lindl.*
Bonatea, *W.*
Bilabrella, Lindl.
Stenoglottis, *Lindl.*
Diplomeris, *Don.*
Diplochilus, Lindl.
Paragnathis, Spreng.
Bicornella, *Lindl.*
Cynorchis, *Thouars.*
?*Amphorchis*, Thouars.
Cœloglossum, *Lindl.*
Ommatodium, *Lindl.*

HOLOTRICHIDÆ.
Holothrix, *Rich.*
Saccidium, *Lindl.*
Monotris, *Lindl.*
Scopularia, *Lindl.*
Tryphia, *Lindl.*
Bucculina, *Lindl.*

DISIDÆ.
Disa, *Berg.*
§ *Repandra*, Lindl.
§ *Phlebidia*, Lindl.
§ *Vaginaria*, Lindl.
§ *Pardoglossa*, Lindl.
§ *Coryphæa*, Lindl.
§ *Stenocarpa*, Lindl.
§ *Oregura*, Lindl.
§ *Trichochila*, Lindl.
§ *Disella*, Lindl.
Monadenia, *Lindl.*
Schizodium, *Lindl.*
Penthea, *Lindl.*
Forficaria, *Lindl.*
Herschelia, *Lindl.*
Brachycorythis, *Lindl.*
Brownleea, *Harv.*

CORYCIDÆ.
Pterygodium *Swz.*
Corycium, *Swz.*
Disperis, *Swz.*
Dipera, Spreng.
Dryopeia, Thouars.
Ceratandra, *Lindl.*
§ *Hippopodium*, Harv.
§ *Evota*, Lindl.
Calota, Harv.
Arnottia, *A. Rich.*

V.—ARETHUSEÆ.

LIMODORIDÆ.
Chloræa, *Lindl.*
Epipactis, Feuill.
Asarca, *Lindl.*
Gavilea, Pöpp.
Asarca, Pöpp.
Bipinnula, *Commers.*
Limodorum, *Tournef.*
Cephalanthera, *L.C.Rich.*
Macdonaldia, *R. Gunn.*
Eriochilus, *R. Br.*
Diplodium, Swartz.
Caladenia, *R. Br.*
Calonema, Lindl.
Leptoceras, *R. Br.*
Glossodia, *R. Br.*
Elythranthe, Endl.
Lyperanthus, *R. Br.*
Microtis, *R. Br.*

ACIANTHIDÆ.
Acianthus, *R. Br.*
Chiloglottis, *R. Br.*
Cyrtostylis, *R. Br.*
Corysanthes, *R. Br.*
Calcearia, Bl.
Corybas, Salisb.
Steleocorys, Endl.
Pterostylis, *R. Br.*

CALEYIDÆ.
Caleya, *R. Br.*
Caleana, R. Br.
Drakæa, *Lindl.*
Spiculæa, *Lindl.*

POGONIDÆ.
Pogonia, *Juss.*
Triphora, Nutt.
Nervilia, Commers.
Odonectis, Rafin.
Isotria, Rafin
* Didymoplexis, *Griff.*
Codonorchis, *Lindl.*
Arethusa, *Gronov.*
* Haplostellis, *A. Rich.*
Cleistes, *Rich.*
Calopogon, *R. Br.*
Cathea, Salisb.
Crybe, *Lindl.*

GASTRODIDÆ.
Gastrodia, *R. Br.*
Epiphanes, Blume.
Ceratopsis, *Lindl.*
* Gamoplexis, *Falc.*
Epipogium, *Gmel.*

VANILLIDÆ.
* Cyathoglottis, *Pöpp et Endl.*
Sobralia, *Ruiz et Pav.*
Epistephium, *H. B. K.*
Erythrorchis, *Blume.*
Cyrtosia, *Blume.*
Vanilla, *Swartz.*
* Pogochilus, *Falcon.*

VI. NEOTTEÆ.

CRANICHIDÆ, *Lindl.*
Ponthieva, *R. Br.*
Schœnleinia, Klot.
Pterichis, *Lindl.*
Acræa, *Lindl.*
Cryptostylis, *R. Br.*
*Zosterostylis, *Blum.*
Gomphichis, *Lindl.*
Stenoptera, *Lindl.*
Altensteinia, *H. B. K.*
Cranichis, *Swartz.*
Tripleura, *Lindl.*
*Chlorosa, *Blum.*
*Rophostemon, *Blum.*
Cordyla, Blume.
*Galeoglossum, *A. Rich.*
*Ocampoa, *A. Rich.*
Prescottia, *Lindl.*
Decaisnea, Brongn.

LISTERIDÆ, *Lindl.*
Listera, *R. Br.*
Diphyllum, Raf.
Neottia, *R. Br.*
Neottidium, Lk.
Calochilus, *R. Br.*
Epipactis, *Hall.*
Serapias, Pers.

SPIRANTHIDÆ, *Lindl.*
Cnemidia, *Lindl.*
Decaisnea, Lindl.
Spiranthes, *L. C. Rich.*
Ibidium, Salisb.
Cyclopogon, Presl.
Gyrostachys, Pers.
Stenoptera, *Presl.*
Sarcoglottis, *Presl.*
Cordylestylis, *Falc.*
Stenorhynchus, *Rich.*
Sauroglossum, *Lindl.*
Pelexia, *Poit.*
Synassa, Lindl.

PHYSURIDÆ, *Lindl.*
*Plexaure, *Endl.*
Chloidia, *Lindl.*
Zeuxine, *Lindl.*
Adenostyles, Blume.
Cionisaccus, Kuhl.
*Chæradoplectron, *Schr.*
Monochilus, *Blume.*
Haplochilus, Endl.
Cheirostylis, *Blume.*
Myoda, *Lindl.*
Hæmaria, *Lindl.*
Hylophila, *Lindl.*
Ætheria, *Blum.*
Platylepis, A. Rich.
Goodyera, *R. Br.*
Leucostachys, Hffg.
Gonogona, Lk.
Tussaca, Rafin.
*Eucosia, *Blume.*

Georchis, *Lindl.*
*Macodes, *Blum.*
Tropidia, *Lindl.*
  Ptychochilus, Schauer.
Ulantha, *Hook.*
Anæctochilus, *Blume.*
  Anœcochilus, Blume.
  Chrysobaphus, Wall.
  Orchipedum, Kuhl.
*Galera, *Blume.*
Physurus, *L. C. Rich.*
  Microchilus, Presl.
  Erythrodes, Blume.
  Psychechilos, Kuhl.
Baskervilla, *Lindl.*
Herpysma, *Lindl.*

DIURIDÆ, *Lindl.*

Diuris, *Smith.*
Orthoceras, *R. Br.*
Prasophyllum, *R. Br.*
Burnettia, *Lindl.*
Genoplesium, *R. Br.*

THELYMITRIDÆ, *Lindl.*

Thelymitra, *Forst.*
Epiblema, *R. Br.*

VII. CYPRIPEDEÆ.

Cypripedium, *Linn.*
  Criosanthes, Rafin.
  Arietinium, Beck.

GENERA about which nothing certain is known

*Hysteria, *Reinw.*
Corymbis, *Thouars.*
*Thrixspermum, *Lour.*
*Scaredederis, *Thouars.*
*Oxyanthera. *A. Brongn.*
*Galeola, *Lour.*
*Callista, *Lour.*
*Acronia, *Presl.*
*Scleropteris, *Scheidw.*
*Macrostylis, *Kuhl et Hass.*
*Amblostoma, *Scheidw.*

NUMBERS. GEN. 394. Sp. 3000 ?

Iridaceæ.
POSITION.—Apostasiaceæ.—ORCHIDACEÆ.—Burmanniaceæ.
Zingiberaceæ.

Fig. CXXIV. *bis.*

Fig. CXXIV. *bis.*—Manner of growth of Odontoglossum grande. *Paxton.*

ORDER LIII. APOSTASIACEÆ.—APOSTASIADS.

Apostasieæ, *Lindl. Nixus Plantarum*, p. 22. (1833); *Blume in Ann. Sc. Nat. Ser.* 2. 2. 91. (1834); *Endl. Gen.* lxvii.; *Meisn.* p. 387.

DIAGNOSIS.—*Orchidal Endogens with regular half-gynandrous flowers, and axile placentæ.*

Perennial herbaceous plants. Stem simple or branched. Leaves firm, thin, sheathing at the base. Flowers in simple or compound terminal racemes. Calyx and corolla each consisting of 3 similar pieces. Anthers 2 or 3, sessile upon a short column, erect, 2-celled, opening longitudinally; pollen cohering in 3s or 4s according to Mr. Bauer (*Illust. Fruct.* t. 15),—in single oval grains with a longitudinal furrow according to Mr. Griffith (*Letter dated Merqui Dec.* 28, 1834) and Blume. Ovary 3-celled, with 3 polyspermous placentæ in the axis; ovules with their integuments very distinct and much shorter than the protruded nucleus (*Griffith*); style filiform, with a slightly 3-lobed stigma as long as the anthers, and adhering with their filaments into a short column. [Capsule 3-celled, 3-valved; the valves bearing the dissepiment in the middle, but cohering at the apex and base. Seeds very numerous, minute, ovate, and with a skin fitting the nucleus, or scobiform with a membranous testa loose at each end.—*Blume.*]

Fig. CXXV.

Very closely allied to Orchids, from which they differ essentially in having a 3-celled fruit, with loculicidal dehiscence, and in the style being altogether free from the stamina for the principal part of its length. At the same time the structure is gynandrous enough to afford a clear distinction from the Burmanniads. There are many admirable observations upon Apostasia itself in Brown's *Observations on the organs and mode of fecundation in Orchideæ and Asclepiadeæ*, and some further information is given by Blume in the place above quoted. The Order seems as if connecting Orchids and Hypoxids. If Rhyncanthera is correctly represented by Blume, its 3-locular ovary will refer it here, while the structure of its column would keep it in Orchids. The essential character is, however, framed without reference to it.

Found in damp woods in the hotter parts of India.

No uses have been assigned to any of them.

GENERA

Apostasia, *Bl.* *Mesodactylus*, Wall.	Neuwiedia, *Bl*	?Rhyncanthera, *Bl.*

NUMBERS. GEN. 3. Sp. 5.

POSITION.—Orchidaceæ.—APOSTASIACEÆ.—*Hypoxidaceæ.*

Fig. CXXV.—Apostasia odorata; 1. a flower; 2. the stamens and style; 3. a cross section of the ovary; 4. a seed.

ALLIANCE XIV. *XYRIDALES.*—THE XYRIDAL ALLIANCE.

DIAGNOSIS.—*Hypogynous bisexual tripetaloid Endogens, with copious albumen.*

It is in this Alliance that, among Endogens with a free ovary, the first distinct separation of a corolla from the calyx takes place, in the form of (2 or) 3 petals. Hence the essential character of the Alliance is its tripetaloideous condition. In the absence of that circumstance it is not to be distinguished from Juncals on the one hand, or Lilials on the other. The Waterworts (Philydraceæ) seem to have anticipated the tripetaloideous organization by forming petals before sepals, and hence they present the anomaly of a flower with a very conspicuous corolla having no calyx, the office of which appears to be performed by spathaceous bracts. Xyrids resemble Sedges with a corolla, and are no doubt akin to Pipeworts (Eriocaulaceæ). Spiderworts are analogous to Parids among Dictyogens, and as for the Mayacs they may be compared to Leptanthus among Pontederas, or to Potamogeton among the Arrow-grasses.

NATURAL ORDERS OF XYRIDALS.

Sepals 0. *Petals* 2. *Stamens* 3, *of which* 2 *are abortive. Embryo axile, in fleshy albumen*	54. PHILYDRACEÆ.
Sepals 3. *Petals* 3. *Stamens* 3 *fertile. Carpels opposite sepals. Placentæ parietal. Embryo minute, on the outside of fleshy albumen*	55. XYRIDACEÆ.
Sepals 3. *Petals* 3. *Stamens* 6 (*or* 3). *Carpels opposite sepals. Placentæ axile. Embryo trochlear, half immersed in fleshy albumen*	56. COMMELYNACEÆ.
Sepals 3. *Petals* 3. *Stamens* 3 ; (*anthers one-celled*). *Carpels opposite petals. Placentæ parietal. Embryo minute, on the outside of fleshy albumen*	57. MAYACEÆ.

ORDER LIV. PHILYDRACEÆ.—WATERWORTS.

Philydreæ, *R. Br.* (1832?); *Lindl. Nixus*, 22. (1833); *Endl. gen.* lii.; *Meisner, p.* 406; *Kunth Enum.* 3. 379.

DIAGNOSIS.—*Xyridal dipetalous Endogens without a calyx, with 3 stamens of which 2 are abortive, and an embryo in the axis of fleshy albumen.*

Root fascicled-fibrous. Stems erect, simple, leafy, often woolly. Leaves ensiform, somewhat cellular, equitant with their half-sheathing bases. Spikes terminal, simple or divided. Flowers alternate solitary, sessile, subtended by a spathaceous persistent bract, yellow, scentless. Calyx abortive. Corolla 2-leaved, coloured, withering. Filaments 3, united at the base, inserted into the base of the lower leaf of the perianth; the lateral ones petaloid and sterile; anther with distinct cells. Ovary superior; style simple; stigma capitate; ovules numerous, on narrow, parietal or axile placentæ, horizontal, anatropal. Capsule 3-celled, 3-valved; the valves having the partition in their middle. Seeds numerous, minute, horizontal; their skin thick; with the embryo in the axis of fleshy albumen.

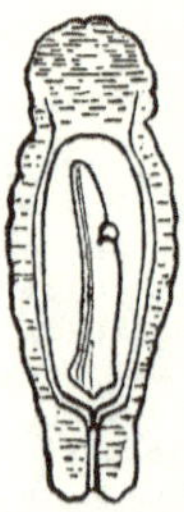

Fig. CXXVI.

These are herbaceous plants, having the great spathaceous bracts of a plant of the Musads, combined with the habit of Sedges; and at the same time having a flower like that of a Spiderwort, minus its calyx and one petal. It is uncertain what the exact analogy of its petaloid divisions may be; but they appear to belong to the corolla. Brown regards the Waterworts as having some relation to Burmannia, and even to Orchids, on account we presume of the constant abortion of 2 out of the 3 stamens. Their nearest relationship, however, is plainly with Xyrids and Spiderworts, from the former of which they differ in the want of a glumaceous calyx, and from both in the large embryo lying in the axis of the albumen.

Fig. CXXVII.

The only plants of this Order yet discovered are found in New Holland, Cochinchina, and China.

Nothing is known of any use to which they may be applied.

GENERA.

Philydrum, *Banks.*	*Garciana*, Lour.	Hetæria, *Endl.*

NUMBERS. GEN. 2. Sp. 2.

Orchidaceæ.
POSITION.—Commelynaceæ.—PHILYDRACEÆ.—Xyridaceæ.

Fig. CXXVI.—A seed of Philydrum lanuginosum, divided perpendicularly so as to show the embryo.
Fig. CXXVII.—1. Hetæria pygmæa; 2. a flower; 3. the fertile stamen and two lateral sterile ones; 4. a cross section of the ovary.

ORDER LV. XYRIDACEÆ.—XYRIDS.

Xyrideæ, *Kunth in Humb. N. G. et Sp.* 1. 255. (1815) *a sect. of* Restiaceæ; *Agardh Aphorism*, 158. (1823); *Desvaux in Ann. des Sc.* 13. 49. (1828); *Endl. Gen.* xlvii.; *Meisner, p.* 407; *Kunth Enum.* 4. *p.* 1.—Rapateæ, *Endl. l. c.*

DIAGNOSIS.—*Xyridal Endogens, with* 3 *sepals opposite the carpels,* 3 *petals,* 3 *fertile stamens, parietal placentæ, and a minute embryo on the outside of fleshy albumen.*

Herbaceous sedgy plants with fibrous roots. Leaves radical, ensiform, or filiform, with enlarged scarious sheathing bases. Flowers in terminal, imbricated, scaly heads. Sepals 3, glumaceous. Petals 3, thin, long, and coloured, united into a monopetalous corolla. Fertile stamens 3, inserted upon the claws of the petals; anthers turned outwards, 2-celled; sterile stamens alternate with the petals. Ovary single, 1-celled, with parietal placentæ; ovules numerous, orthotropal; style trifid; stigmas obtuse, multifid, or undivided. Capsule 1-celled, 3-valved, many-seeded, with parietal placentæ. Seed with the embryo on the outside of the fleshy albumen, and at the end most remote from the hilum.

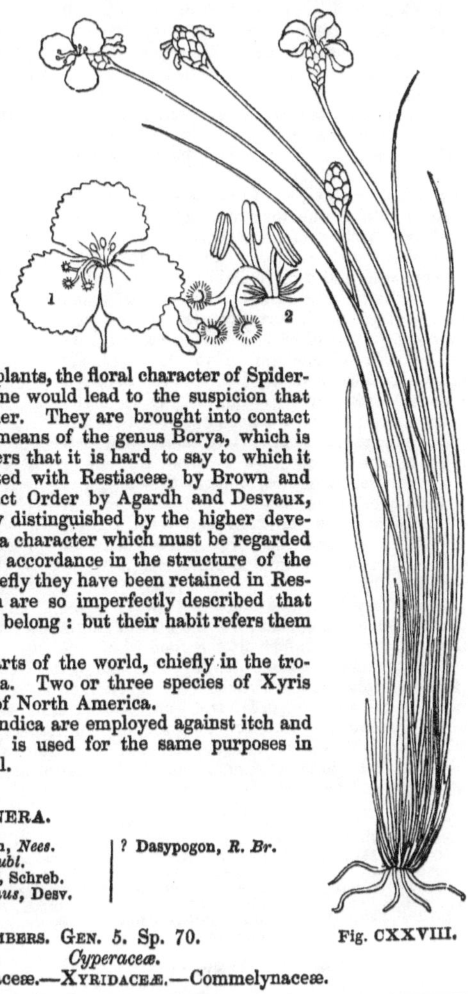

These plants join to the habit of Sedges and other glumaceous plants, the floral character of Spiderworts; and this circumstance alone would lead to the suspicion that they form a peculiar natural Order. They are brought into contact with the Aphyllanth Lilies by means of the genus Borya, which is so intermediate between the Orders that it is hard to say to which it belongs. The Xyrids were united with Restiaceæ, by Brown and others, but separated as a distinct Order by Agardh and Desvaux, and they appear to be essentially distinguished by the higher development of their floral envelopes, a character which must be regarded as more important than the mere accordance in the structure of the seed, in consequence of which chiefly they have been retained in Restiaceæ. Rapatea and Dasypogon are so imperfectly described that it is impossible to say where they belong: but their habit refers them either here or to the Rushes.

All are natives of the hotter parts of the world, chiefly in the tropics of America, Asia, and Africa. Two or three species of Xyris are found in the southern states of North America.

The leaves and root of Xyris indica are employed against itch and leprosy in India; X. americana is used for the same purposes in Guiana, and X. vaginata in Brazil.

GENERA.

Xyris, *Linn.*	? Acoridium, *Nees.*	? Dasypogon, *R. Br.*
Abolboda, *H. et B.*	Rapatea, *Aubl.*	
Chloerum, Willd.	*Mnasium*, Schreb.	
	Spathanthus, Desv.	

NUMBERS. GEN. 5. Sp. 70.

Fig. CXXVIII.

Cyperaceæ.
POSITION.—Mayaceæ.—XYRIDACEÆ.—Commelynaceæ.

Fig. CXXVIII.—Xyris operculata; 1. a flower seen in front; 2. a style, stigmas and stamens.

ORDER LVI. COMMELYNACEÆ.—SPIDERWORTS.

Ephemereæ, *Batsch. Tab. Affin.* 125. (1802) *in part.*—Commelyneæ, *R. Brown Prodr.* 268. (1810); *Richard in Humb. Bonpl. N. Gen.* 1. 258. (1815); *Agardh Aph.* 168. (1823); *Kunth. Enum.* 4. 34. —Commelynaceæ, *Ed. prior. Endl. Gen.* xlviii.; *Meisner, Generaæp.*406.—Flagellarieæ, *Endl. Gen. p.* 131.

DIAGNOSIS.—*Xyridal Endogens, with 3 sepals opposite the carpels, 3 petals, 6 (or 3) stamens, axile placentæ, and a trochlear embryo half immersed in fleshy albumen.*

Herbaceous plants. Leaves flat, narrow, usually sheathing at the base. Sepals 3, distinct from the petals, herbaceous. Petals coloured, sometimes cohering at the base. Stamens 6, or a smaller number, hypogynous, some of them either deformed or abortive; anthers 2-celled, turned inwards. Ovary 3-celled, with few-seeded cells; style 1; stigma 1. Capsule 2- or 3-celled, 2- or 3-valved, the valves bearing the dissepiments in the middle. Seeds often twin, inserted by their whole side on the inner angle of the cell, whence the hilum is linear, with a papilla covering over the embryo; embryo pulley-shaped, antitropal, lying half-buried in a cavity of the albumen remote from the hilum; albumen densely fleshy.

The Spiderworts are plants which exhibit a transition from the first remove out of the regions of sedge-like plants to the true Lilies. In other words, while Xyrids are glumaceous herbs with their perfectly-formed petals, there are Xyrids with the glumaceous structure gone, and the Liliaceous peculiarities gained: all but the long axil embryo and the petaline condition of the calyx. Brown compares them with Rushes, observing that they are very different both in habit and structure; agreeing better with Restiaceæ in the situation of the embryo and the sheathing leaves, although otherwise quite distinct; they have scarcely any affinity with Palms, except in the trochlear embryo, remote from the hilum, and indicated in both Orders by an external papilla. The Spiderworts may also be compared with Alismads, which are equally tripetaloideous, and with Mayacs, which have 1-celled anthers, a wholly cellular structure, and, as they say, the carpels opposite the petals.

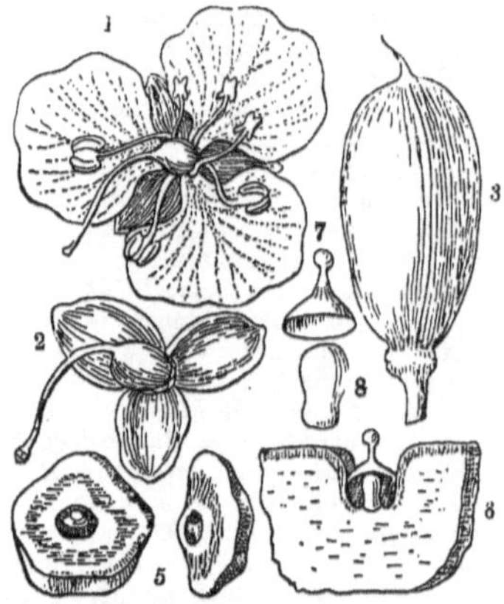

Fig. CXXIX.

Chiefly found in the East and West Indies, New Holland, and Africa. A few occur in North America, but none in Northern Asia or Europe.

Concerning their uses there is little to relate. The fleshy rhizomes of Commelyna cœlestis, tuberosa, angustifolia and striata, contain a good deal of starch mixed with mucilage, and are therefore fit for food when cooked. The Chinese employ those of C. medica in cough, asthma, pleurisy, strangury, and dysury. Tradescantia diuretica has a similar application in Brazil. A decoction of Cyanotis axillaris is drunk in the East Indies in cases of tympanis, and Tradescantia Malabarica is administered in the same country, boiled in oil, as a remedy for itch and leprosy. Murdannia scapiflora is said by Dr. Royle "to have some repute in Hindoo Materia Medica." Commelyna Rumphii is held in India to be emmenagogue. The leaves of Flagellaria indica are said to be astringent and vulnerary.

GENERA.

Commelyna, *Dillen.*
Hedwigia, Medik.
Lechea, Lour.
Ananthopus, Raf.
Aneilema, *R. Br.*
Aphilax, Salisb.

Palisota, *Reichenb.*
Pollia, *Thunb.*
Aclisia, E. Mey.
Lamprocarpus, Blum.
Callisia, *Löffl.*
Hapalanthus, Jacq.
Murdannia, *Royle.*

Tinnantia, *Scheidw.*
Tradescantia, *Linn.*
Ephemerum, Tournef.
Spironema, *Lindl.*
Cyanotis, *Don.*
Zygomenes, Salisb.
Lampra, *Benth.*

Campelia, *Rich.*
Zanonia, Plum.
Dichorisandra, *Mik.*
Cartonema, *R. Br.*
Forrestia, *A. Rich.*
Flagellaria, *L.*

NUMBERS. GEN. 16.—SP. 260.

POSITION.—*Liliaceæ.*—COMMELYNACEÆ.—Xyridaceæ.
Bromeliaceæ.

Fig. CXXIX.—Aneilema crispatum; 1. a flower; 2. the calyx and pistil; 3. the capsule; 4, 5. seeds; 6 a section of ditto showing the embryo; 7. the papilla; 8. the embryo. *Fred. Bauer.*

ORDER LVII. MAYACEÆ.—MAYACS.

Mayaceæ, *Kunth. Enum.* iv. 30 (1843).

DIAGNOSIS.—*Xyridal Endogens, with 3 sepals alternate with the carpels, 3 petals, 3 stamens, 1-celled anthers, parietal placentæ, and a minute embryo on the outside of fleshy albumen.*

Moss-like plants, creeping over damp places, almost entirely destitute of spiral vessels. Leaves very narrow, pellucid, undivided. Flowers ☿, small, white, pink, or violet. Sepals valvate ? herbaceous. Petals much longer, imbricated. Stamens 3, inserted into the base of the sepals; anthers 1-celled, adhering by the base to a thread-like filament, opening at the point only. Carpels 3, alternate with the sepals, combined into a 1-celled ovary; placentæ 3, parietal; ovules sessile, horizontal, orthotropal; style thread-like; stigma simple. Capsule membranous, covered by the permanent sepals and petals, 1-celled, 3-valved; seeds attached to the middle of the valves, roundish, ribbed, terminated by a conical tubercle. Albumen shaped like the seed, composed of angular crystal-like cells, arranged in a radiant manner. Embryo very minute, antitropal, half plunged in the vertex of the albumen.

Fig. CXXX.

Such appears to be the structure, according to Kunth, and Schott, and Endlicher, of a few plants which are separated from the Spiderworts by the former of these botanists. They are very little known, and demand a fresh examination, but in the meanwhile appear to be distinguished from the Spiderworts by their peculiar habit, their 1-celled anthers, and their carpels being opposite the petals (according to Schott and Endlicher), while, on the other hand, the Xyrids are separated by their monopetalous glumaceous capitate flowers and 2-celled anthers. There is, however, but little other difference, unless the valvate calyx of the Mayacs and the position of their carpels should afford additional characteristics. This, however, is to be noted, that the figures given by the last mentioned botanists are at variance with their account of the position of the carpels. No spiral vessels were detected by Schleiden in the leaves and stems of Mayaca fluviatilis, except in the flower-stalks.— *Wiegm. Arch.* v. 231.

The few species that are described inhabit the marshes of America, from Brazil up to Virginia.

They are of no known use.

GENERA.

Mayaca, *Aubl.*
Biaslia, Vand.
Syena, Schreb.
Colletia, Flor. Flum.

NUMBERS. GEN. 1. Sp. 4.

POSITION.—Commelynaceæ.—MAYACEÆ.—Xyridaceæ.

Fig. CXXX.—Mayaca Vandellii: 1. a flower; 2. a cross section of its ovary; 3. a seed vessel; 4. two seeds, one of which is cut perpendicularly in order to show the embryo.

ALLIANCE XV. *JUNCALES.*—THE JUNCAL ALLIANCE.

DIAGNOSIS.—*Hypogynous, bisexual, scaly or scarious flowered Endogens, with abundant albumen.*

This and the Xyridal Alliance stand on the same line in the scale of organization. They both consist of Endogens, which are equally related to Orders of a very low and very high structure. The Juncals approach Grasses and their allies in the glumaceous character of their calyx and corolla, the Xyrids in that of their calyx and bracts. Some of them are absolutely without floral envelopes, the majority have those organs in the form of inconspicuous scales, and when colour or a petaline condition appears among them, the parts in which it occurs are dry and sapless, as if they were mere membranes or attempts at the organs they represent. The Rushes have a very minute embryo, wholly destitute of all appearance of a plumule; Orontiads have the cleft of an Arum, through which a plumule is easily found. The great exception to their character consists in the absence of albumen from the seeds of a few genera among the Orontiads; but such plants are readily known by their spadiceous inflorescence from the exalbuminous Alismal Alliance.

NATURAL ORDERS OF JUNCALS.

Flowers scattered. Embryo minute, undivided . . . . 58. JUNCACEÆ.
Flowers spadiceous. Embryo axile, with a conspicuous cleft on one side. 59. ORONTIACEÆ.

ORDER LVIII. JUNCACEÆ.—RUSHES.

Junci, *Juss. Gen.* (1789), *in part.*—Junceæ, *DC. Fl. Fr.* 3. 155. (1815); *R. Brown Prodr.* 257. (1810).—Juncaceæ, *Agardh Aphor.* 156. (1823), *in part; Endl. Gen.* li.; *Meisner, Gen. p.* 405. *Kunth. Enum.* 3. *p.* 295. Kingiaceæ, Calectasieæ, Xerotideæ, *Endl. l. c.*

DIAGNOSIS.—*Juncal Endogens, with scattered flowers and a minute undivided embryo.*

Herbaceous plants, with fascicled or fibrous roots. Leaves fistular, or flat and channelled with parallel veins. Inflorescence often more or less capitate. Flowers generally brown or green, in umbels, racemes, or long compact spikes, or even panicles. Calyx and corolla forming an inferior, 6-parted, more or less glumaceous or cartilaginous, perianth. Stamens 6, inserted into the base of the segments; sometimes 3, and then opposite the calyx. Anthers 2-celled, turned inwards, opening longitudinally, or by pores at the points. Ovary 1- or 3-celled, 1- or many-seeded, or 1-celled and 3-seeded; style 1; stigmas generally 3, sometimes only 1; ovules anatropal. Fruit capsular, with 3 valves, which have the dissepiment in their middle, sometimes destitute of valves, and 1-seeded by abortion. Seeds with a thin skin; albumen firm, fleshy, or cartilaginous; embryo very minute, included, near the hilum.

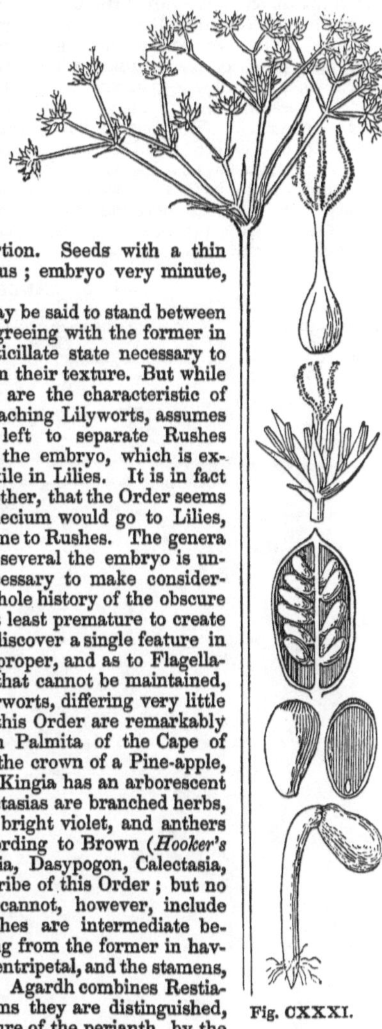
2
1
3
4
5
Fig. CXXXI.

This Order, in its most genuine state, may be said to stand between petaloideous and glumaceous Endogens, agreeing with the former in the floral leaves having assumed the verticillate state necessary to constitute a perianth, and with the latter in their texture. But while a glumaceous confused calyx and corolla are the characteristic of one part of the Order, another part, approaching Lilyworts, assumes a petaloid state; so that little is finally left to separate Rushes from the latter, except the difference in the embryo, which is extremely small in Rushes, and large and axile in Lilies. It is in fact by this last character, more than by any other, that the Order seems to be distinguished; for otherwise, Narthecium would go to Lilies, and all the Aphyllanthous Lilies would come to Rushes. The genera are in great need of careful revision; of several the embryo is unknown, and it may be found hereafter necessary to make considerable alteration among them: but till the whole history of the obscure genera shall have been cleared up, it is at least premature to create more Orders for their reception. I do not discover a single feature in Xerotes which can divide it from Rushes proper, and as to Flagellaria, equally made the usurper of a throne that cannot be maintained, it seems a mere runaway from the Spiderworts, differing very little from Aneilema. Some of the species of this Order are remarkably unlike European Rushes. The Prionium Palmita of the Cape of Good Hope, has the look of an Aloe, or of the crown of a Pine-apple, mounted upon a thick black spongy stem. Kingia has an arborescent stem terminated by a tuft of leaves. Calectasias are branched herbs, with dry, permanent, starry flowers, of a bright violet, and anthers opening by pores, like a Solanum. According to Brown (*Hooker's London Journal*, 2. 494.), the genera Kingia, Dasypogon, Calectasia, Xerotes, and Baxteria, form a peculiar tribe of this Order; but no character is assigned to such tribe. I cannot, however, include Dasypogon. Brown remarks, that Rushes are intermediate between Restiaceæ and Asphodeleæ, differing from the former in having an included embryo, a radicle usually centripetal, and the stamens, when there are only 3, opposite the sepals. Agardh combines Restiaceæ and Rushes.—*Aph.* 157. From Palms they are distinguished, independently of their habit, by the texture of the perianth, by the constant tendency to produce more than 1 ovule in each cell, and by the embryo never

Fig. CXXXI.—Juncus acutiflorus; 1. a flower; 2. the pistil; 3. a perpendicular section of the ovary; 4. seeds; 5. a seed germinating.

being remote from the hilum. Juncus is an instance of a monocotyledonous plant having distinct pith. "Xerotes, in the structure and appearance of its flowers, and in the texture of albumen, has a considerable resemblance to Palms, but it wants the peculiar characters of the seed, and also the habit of that remarkable Order."—*Brown in Flinders*, 578.

Chiefly found in the colder parts of the world, some even in the coldest, two existing in the ungenial climate of Melville Island. Several, however, are known in the tropics. Eight are mentioned as inhabiting the tropical parts of New Holland alone. According to Humboldt they constitute $\frac{1}{400}$ of the flowering plants in the equinoctial zone; in the temperate zone, $\frac{1}{90}$; in the frozen zone $\frac{1}{25}$; in North America, $\frac{1}{150}$; in France, $\frac{1}{88}$. In Sicily, according to Presl, they do not form more than $\frac{1}{308}$.

Only employed for mechanical purposes, as the Rush and others for making the bottoms of chairs, &c.; the pith of the same for the wick of common candles. One species is cultivated in Japan like Rice, entirely for making floor-mats.—*Thunb.* The blanched portion of the base of the inner leaves of some Rushes, and of Astelia alpina, a sedgy plant, which grows on the sand-hills of the coast of Tasmannia, and has the mature leaves an inch wide, and of a deep green, are eatable, and of a nutty flavour. The flowers resemble those of Rushes. They grow in clusters, on a stem as flat and broad as the leaves.—*Backhouse.* The roots of Luzula campestris, and several Rushes, have a popular reputation as diuretics, and are used as such in the north of Europe and China. The herbage of Narthecium ossifragum was once regarded as a vulnerary. Susum, a Java plant, supposed to be near Xerotes, has anthelmintic roots employed in veterinary practice. Dr. J. Hooker observes, that in some species of this Order the outer membrane of the seeds forms with water a transparent jelly similar to what is seen on the moistened grains of some Composite plants.

GENERA.

Luzula, *DC.*
Luciola, Smith.
Prionium, *E. Mey.*
Juncus, *DC.*
Distichia, *Nees.*
Rostkovia, *Desv.*
Marsippospermum, Desv.
Narthecium, *Moehr.*
Abama, Adans.
Cephaloxys, *Desv.*
Susum, *Bl.*
Xerotes, *R. Br.*
Lomandra, *Lab.*
Astelia, *Sol.*
Hamelinia, A. Rich.
? *Funkia*, W.
Hanguana, Bl.
Kingia, *R. Br.*
Baxteria, *R. Br.*
Calectasia *R. Br.*

NUMBERS. GEN. 13. Sp. 200.

Liliaceæ.
POSITION.—————.—JUNCACEÆ.—Orontiaceæ.
Cyperaceæ.

ORDER LIX. ORONTIACEÆ.—ORONTIADS.

Orontiaceæ, *R. Brown*, *Prodr.* 337. (1810); *Endl. Gen. p.* 239. *Adr. Juss. Cours. Elém. p.* 506.—Callaceæ, *Endl. Gen. p.* 239. (1836); *Meisner, p.* 360.—Acoroideæ, *Agh. Aph.* 133. (1822); *Schott. Meletem*, 22, (1832).—Acorinæ, *Link Handb.* i. 144.—Acoraceæ, *ed. pr.* cclxii.

DIAGNOSIS.—*Juncal Endogens, with spadiceous flowers, and an axile embryo with a lateral cleft.*

Herbaceous plants, with broad entire or deeply divided leaves, which however are occasionally sword-shaped and equitant. Some of them are stemless, others scramble over trees, to which they adhere by creeping roots; a few are aquatics. Flowers ☿, on a simple spadix, furnished with a spathe, white, green, or purple. Calyx and corolla absent, or consisting of 4, 5, 6, 8 scales. Stamens of the same number, either hypogynous or perigynous; anthers 2-celled, opening longitudinally or transversely. Ovary free, with 1 or more cells; ovules erect, anatropal or campylotropal, or pendulous and orthotropal; stigma capitate, sessile, or furnished with a subulate style. Fruit a berry. Embryo slit on one side, in the axis of fleshy, or horny, or mealy albumen. (Albumen absent in Scindapsus, Dracontium, Symplocarpus, Orontium.—*Endl.*)

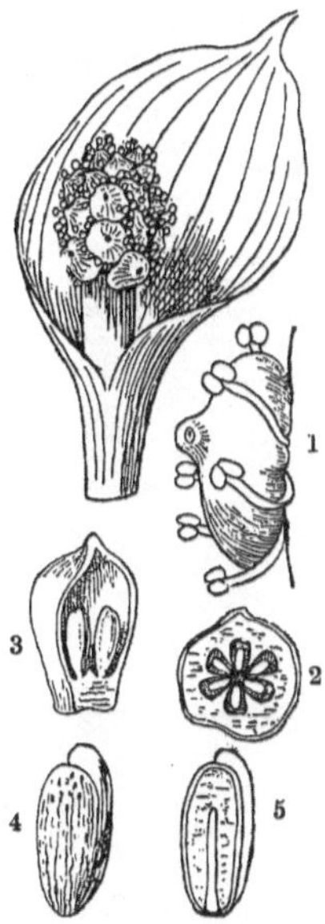

Fig. CXXXII.

The greater part of these plants have the habit of Arads, with which they are usually associated, and from which in fact they differ only in having hermaphrodite flowers, which have usually a scaly perianth. For this reason other Botanists separate them, and it seems more especially desirable to do so, because there is no tendency among them to a separation of the sexes. Acoreæ are indeed usually regarded as the type of a peculiar Order; and if this opinion is correct, the Orontiads must certainly accompany them, for they differ in nothing except the form of their leaves, which, in Acoreæ, are sword-shaped and straight-veined. In fact, Acorus seems to bear the same relation to Orontiads as Pandanus and Freycenetia to Cyclanths. Blume considers these plants to be allied, on the one hand, by Pothos to Peppers and Saururaceæ, and, on the other, to Lilyworts.—*Rumphia* 2. 74. in which he is probably right; for Aspidistreæ form a connecting link between Orontiads and Lilies. Brown has remarked that in Dracontium polyphyllum and fœtidum, in which there is no albumen, the plumule consists of imbricated scales, and that it is sometimes double or even triple. In the former of these plants the external scales, in germination, quickly wither away, when other internal and larger ones appear, and remain for some time round the base of the primordial leaf, before the development of which no rootlets are emitted.—*Prodr.* 334. A similar economy has been noticed by Du Petit Thouars, in his genus Ouvirandra in Alismads.

The plants of this Order chiefly occupy woodland stations within the tropics of both hemispheres, but many are found in colder latitudes; for Symplocarpus is common in the swamps of the United States; Calla palustris inhabits the deep muddy frozen marshes of S. Lapland, in 64° N., and on the Andes, Pothos pedatus and quinquenervius rise to the height of 8400 feet above the sea.

The fresh leaves of Monstera pertusa are employed by the Indians of Demerara as vesicatories or rubefiants in cases of dropsy. The root and seeds of Skunk Cabbage, Symplocarpus fœtidus, a most fœtid species, are powerful antispasmodics and expectorants; they have considerable reputation in N. America, as palliatives in paroxysms of asthma. Dracontium polyphyllum, said to be the Labaria plant of Demerara, is reputed to possess similar properties. Orontium aquaticum is acrid when fresh, but its dried root can be eaten without inconvenience. The corm of the beautiful Richardia africana, with its snowy spathe and golden spadix, was formerly officinal under the name of Radix Ari Æthiopii. The rhizomes of Calla palustris, although acrid and caustic in the highest degree, are, according to Linnæus, made into a kind of bread in

Fig. CXXXII.—Calla palustris; 1. a flower; 2. a section of the ovary; 3. a perpendicular section of the ripe fruit; 4. a seed; 5. its longitudinal section.—*Nees.*

high estimation, called Missebrœd, in Lapland. This is performed by drying and grinding the roots, afterwards boiling and macerating them till they are deprived of their acrimony, when they are baked like other farinaceous substances. The plant has the credit of being a very active diaphoretic. The fruit of Scindapsus officinalis, cut into transverse pieces and dried, is an article of some importance in Hindoo Materia Medica, is called Guj-pippul, and sold by the druggists under that name.—*Roxb.* Pothos scandens is employed in India as a remedy for putrid fevers. The rhizome of Acorus Calamus contains an aromatic bitter principle, which has caused the plant to be regarded as medicinal. In cases of chronic catarrh and humid asthma benefit has been received from its exhibition. In Constantinople it is made into a confection, is considered a good stomachic, and eaten freely during the prevalence of epidemic diseases. It is in this country chiefly employed by perfumers, in the manufacture of hair powder, on account of the fragrance of the essential oil which is mixed with its farinaceous substance. Dr. Pereira says, that although it is rarely employed in medicine, it might frequently be substituted for other more costly aromatics ; it is adapted to cases of dyspepsia, or as an adjunct to tonics or purgatives.

GENERA.

Tribe I.—Calleæ. *Flowers naked. Ovules erect.*

Calla, *Linn.*
Monstera, *Adans.*
Heteropsis, Kth.
Scindapsus, *Schott.*
* Rhaphidophora, *Hassk.*

Tribe II. — Orontieæ. *Flowers with a regular perianth. Leaves plane, entire, palmate or pinnated. Ovules pendulous.*

Pothos, *Linn.*
Lasia, *Loureir.*
Anthurium, *Schott.*
Spathiphyllum, *Schott.*
Dracontium, *Linn.*
Symplocarpus, *Salisb.*
Ictodes, Bigelow.
Spathyema, Rafin.
Orontium, *Linn.*

Tribe III. — Acoreæ. *Flowers with a regular perianth. Leaves ensiform, equitant. Ovules pendulous.*

Gymnostachys, *R. Br.*
Acorus, *Linn.*

NUMBERS. GEN. 13. Sp. 7C.

Piperaceæ.
POSITION.—Juncaceæ.—ORONTIACEÆ.—Liliaceæ.
Araceæ.

ALLIANCE XVI. *LILIALES.*—THE LILIAL ALLIANCE.

DIAGNOSIS.—*Hypogynous, bisexual, hexapetaloid Endogens, with copious albumen.*

These are the centre of the division of Endogens with complete flowers free from the ovary. They are known from the Xyrids by their sepals and petals being all equally coloured; from the Juncals by their tender highly developed flowers; and from the Alismals by their abundant albumen. To Palms they often approach in habit, and even in the separation of their sexes; but the genera described by botanists as monœcious or diœcious seem to be never truly diclinous, the distinct rudiments of one sex always accompanying the perfect state of another. By the Gilliesiads they seem to show a tendency to assume the glumaceous condition; Pontederads are evidently on the limits of Alismals, by their genus Leptanthus; Juncals are brought into the closest proximity by the Aphyllanths among Lilies, and so are Amaryllids by means of the Conanthereæ of the same great Order. Their undoubted accordance with Dictyogens, in many essential particulars, enables them to extend their frontier to that of the vast mass of Exogens; and their wood, which does certainly, in Yucca and Dracæna, arrange itself in circles, confirms the tendency of the Lilials towards a junction with the same class.

NATURAL ORDERS OF LILIALS.

Perianth surrounded by a calycine involucre, the inner bracts of which are coloured and petaloid	60. GILLIESIACEÆ.
Perianth naked, flat when withering. Anthers turned outwards; styles distinct; albumen fleshy	61. MELANTHACEÆ.
Perianth naked, flat when withering. Anthers turned inwards. Styles consolidated. Albumen fleshy	62. LILIACEÆ.
Perianth naked, circinate when withering. Anthers turned inwards. Albumen mealy	63. PONTEDERACEÆ.

ORDER LX. GILLIESIACEÆ.—GILLIESIADS.

Gilliesieæ, *Lindl. in Bot. Reg.* 992. (1826); *Hooker in Bot. Mag.* 2716. (1827).—Gilliesiaceæ, *Ed. pr.* ccxlix. *Endl. Gen. p.* 152; *Meisner, Gen. p.* 398.

DIAGNOSIS.—*Lilial Endogens with a calyx-like involucre, the inner bracts of which are coloured and petal-like.*

Small herbaceous plants, with tunicated bulbs. Leaves grasslike. Flowers umbellate, somewhat spathaceous, inconspicuous, hermaphrodite, surrounded by bracts the outer of which are petaloid and herbaceous, the inner starved and coloured. Perianth minute, either a single lip-like lobe, or an urceolate 6-toothed body. Stamens 6, either all fertile, or 3 sterile and nearly obliterated. Ovary superior, 3-celled; style 1; stigma simple. Capsule 3-celled, 3-valved, with a loculicidal dehiscence, many-seeded. Seeds attached to the axis, by means of a broad hollow neck; testa black and brittle; embryo curved in the midst of fleshy albumen.

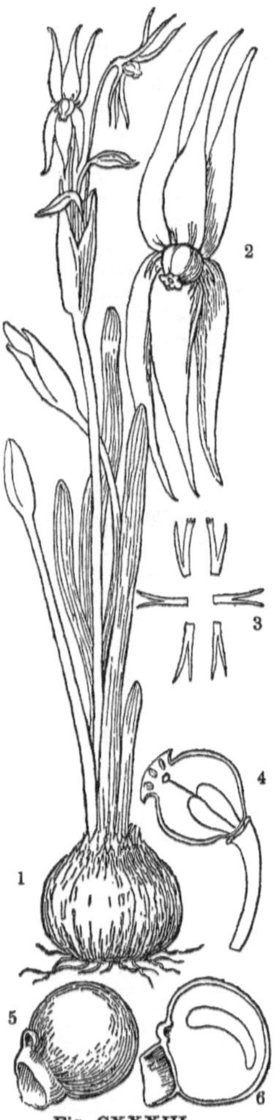

Fig. CXXXIII.

To the following account of these plants, originally given in the *Botanical Register*, when speaking of Gilliesia, little has to be added.

"The whole structure of this plant is so peculiar, that I scarcely know whether the description of the parts of fructification above given will not be considered more paradoxical than just; and yet, if the analogies the various organs bear to those of other plants be carefully considered, their structure will scarcely admit of any other interpretation. With respect to the five petaloid leaves, which are here described as bracts, and which bear a considerable degree of resemblance to a perianth, it may be observed, that this appearance is more apparent than real; they neither correspond in insertion nor in number with the segments of a monocotyledonous perianth, nor do they bear the same relation to the parts contained as a perianth should bear. The three outer are not inserted on the same line, but are distinctly imbricated at the base; and the two inner do not complete the second series, as would be required in a regular monocotyledonous perianth. But if we were to admit, for a moment, the possibility of these bracts being segments of a perianth, what explanation could be given of the setiform processes proceeding from their base, or of the central fleshy slipperlike body from within which the stamens proceed? The former bear no determinate relation to the other parts of the flower in their insertion; they are subject to much diversity of form and number, being sometimes eight, consisting of two unequal subulate bodies proceeding from the edges of each lateral segment, the outermost of the two being wider than the innermost, and being, moreover, not unfrequently a manifest process of the margin of the segment itself; sometimes having their number reduced to four by the suppression of the exterior processes of each lateral segment; and occasionally having the outer processes suppressed on one segment, and not suppressed on the other. In the many flowers which have been under examination, the processes, moreover, were always constituted of cellular tissue alone, without either spiral or tubular vessels. These circumstances being considered, it will scarcely be proposed, we presume, to identify them with abortive stamina. If they are, notwithstanding what has been advanced, determined to be the perianth itself, what

Fig. CXXXIII.—1. Miersia chilensis; 2. its flower; 3. the interior coloured petaloid bracts; 4. a perpendicular section of the perianth *(from a sketch by Mr. Miers)*; 5. a seed of Gilliesia graminea; 6. a section of the same.

becomes of the outer segments, which had previously been referred to the perianth? for it would be difficult to trace any analogy between the structure of Gilliesia and of those genera in which a third series is added to the usual ternary division of Monocotyledons. But none of the peculiarities adverted to are opposed to those bodies being referred to depauperated or reduced bracts. With respect to the central body from which the stamens proceed, this body, which might be conveniently disposed of by referring it to what Linnæan botanists call a nectarium, consists of a fleshy slipperlike lobe, with or without two auricles at the base, and within which the cup of stamens is inserted. The relation it bears, as regards insertion, to the parts which have been already noticed, is very obscure; it is always opposite the solitary external bracts; but whether it is anterior with respect to the common axis of inflorescence, or posterior, has not at present been ascertained. The reasons which have been offered for the view here taken of the parts surrounding this body, make it obvious that it must be considered the perianth. It manifestly bears an intimate relation to the stamens, being obliterated in the same direction and degree as they are. In this view, then, the petaloid segments are considered perfect bracts, the subulate interior processes abortive bracts, and the fleshy central labelloid body the perianth. However paradoxical this description of Gilliesia may appear, it will probably be found more deserving of attention if compared with Miersia. In Miersia the bracts are six in number, of which two are interior and four exterior, a still more valid reason against their being segments of a perianth. The subulate processes assume a more regular form, and a more constant mode of insertion, but still bear no very apparent relation to the bracts, and the fleshy labelloid central body is represented by an urceolate six-toothed cup, within the orifice of which six fertile stamens are included. In Miersia, therefore, the perianth, which was in Gilliesia subject to a certain degree of imperfection, in which the stamens also participated, is in the usual regular form of many Monocotyledons, no irregularity occurring in the stamens. As there can be no doubt of the affinity between Gilliesia and Miersia, and as there can also be little doubt that the central body of the latter genus is a perianth, it will follow, that as the supernumerary appendages of that genus are external with respect to the perianth, and are therefore neither perianth nor stamens, so also will the analogous appendages of Gilliesia not be perianth. And the central body having been ascertained to be perianth, all the parts which surround it will necessarily be bracts, or modifications of them.

"The natural affinity of these two genera is obscure. Their black, brittle seeds, large axile embryo, tunicated bulbs, spathaceous inflorescence, and general appearance, place them near Lilyworts, with some genera of which, especially Muscari and Puschkinia, Miersia at least agrees in the structure of perianth; but there is no genus among the Lilies to which the fructification of Gilliesiads can be otherwise compared. If the one-flowered species of Schoenus, in which a single naked flower is surrounded by several imbricated scales, be admitted as a form of inflorescence analogous to that under consideration, it may perhaps be allowable to carry this comparison yet further, and to suggest an identity of origin and function between the depauperated bracts of Gilliesia and the hypogynous setæ of Scirpus and other Sedges."

But although such plants may be analogous in structure to the Gilliesiads, as well as to Cordleafs, to which they were also compared in the work above quoted, yet no doubt can exist, that they form a most curious part of the Lilial Alliance.

Chilian bulbs, of no known size.

GENERA.

Gilliesia, *Lindl.* Miersia, *Lindl.*

NUMBERS. GEN. 2. Sp. 5.

POSITION. ——— GILLIESIACEÆ.—Liliaceæ.
Cyperaceæ.

ORDER LXI. MELANTHACEÆ.—MELANTHS.

Melantheæ, *Batsch. Tab. Aff.* (1802).—Colchicaceæ, *Dec. Fl. Fr.* 3. 192. (1815); *Ess. Méd.* 298. (1816); *Bartl. Ord. Nat.* 51. (1830).—Melanthaceæ, *R. Brown, Prodr.* 272. (1810); *Endl. Gen.* liii. *Meisner, Gen. p.* 404. *Kunth. Enum.* 4. 136. *A. Gray, Lyceum, N. York, vol.* 4. (1837).—Veratreæ, *Salisb. in Hort. Trans.* 1. 328. (1812); *Agardh Aphor.* 166. (1823).—Merenderæ, *Mirb. according to De Candolle.*—Anguillareæ, *Don. in Linn. Trans.* 18. 513.

DIAGNOSIS.—*Lilial Endogens with a naked perianth, flat when withering, anthers turned outwards, distinct styles, and fleshy albumen.*

Bulbous, tuberous, or fibrous-rooted plants, extremely variable in appearance; in the Colchiceæ stemless, with the flowers half subterranean like a Crocus; in the Veratreæ, with spiked, racemose, panicled, branching, or simple herbaceous stems. Flowers not unfrequently ♂ ⚥ ♀, white, green, or purple. Calyx and corolla both alike, free, petaloid, in 6 pieces, or, in consequence of the cohesion of the claws, tubular; the pieces generally involute in æstivation. Stamens 6; anthers turned outwards. Ovary 3-celled, many-seeded; style 3-parted; stigmas undivided; [ovules orthotropal, semi-campylotropal, semi-anatropal or anatropal, *Endl.*] Capsule generally divisible into 3 pieces; sometimes with a loculicidal dehiscence. Seeds with a membranous testa; albumen dense, fleshy or cartilaginous; embryo very minute, inclosed, extremely uncertain in its position.

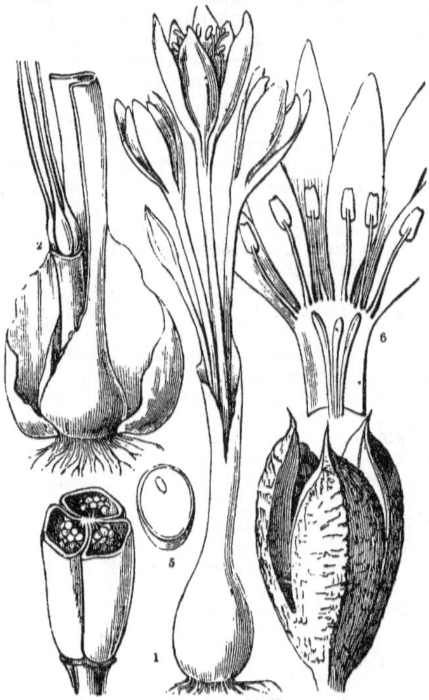

Fig. CXXXIV.

The plants of this Order have in some cases the appearance of Crocuses, in others that of small Lilies. Brown considers its station to be between Lilyworts and Rushes, from both which it is known by its tripartible fruit, and anthers turned outwards. The latter character gives the Melanths their distinctive character, more than anything else, and, combined with their separable carpels, generally renders their identification free from difficulty. Don has well observed that "the genus Colchicum establishes an evident relationship, through Sternbergia and Crocus, between Melanths, Amaryllids, and Irids; Disporum joins them to Smilaceæ, and Tofieldia to Rushes, while a comparison of the structure of Uvularia and Erythronium fully makes out their affinity with Liliaceæ."

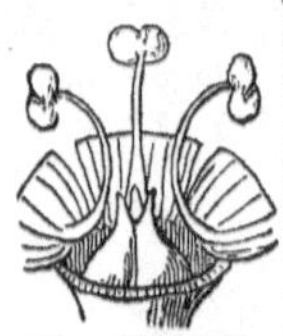

Fig. CXXXV.

Frequent at the Cape of Good Hope, not uncommon in Europe, Asia, and North America, and existing within the tropics of India and New Holland, this Order appears to be confined within no geographical limits; it is, however, far more abundant in northern countries than elsewhere.

Few Orders of plants are more universally poisonous than this, whose qualities are conspicuously indicated by Colchicum and Veratrum. The corm and seeds of the

Fig. CXXXIV.—Colchicum autumnale. 1. A corm in flower; 2. The same stripped of its outer coats, and showing the ovaries after the floral envelopes are cut away; 3. a transverse section of the ovaries; 4. a ripe capsule; 5. a section of a seed; 6. the flower cut open, to show the stamens and the 3-parted style.

Fig. CXXXV.—Section of the centre of the flower of Veratrum nigrum.

former are well-known to be acrid, cathartic, narcotic, and diuretic; the latter is a nauseous, dangerous emetic. These properties are owing to a peculiar alkaline principle, called Veratria, which acts with singular energy on the membrane of the nose, exciting violent sneezing, though taken in very minute quantity. When received internally in very small doses, it produces excessive irritation of the mucous coat of the stomach and intestines; and a few grains are found fatal to the lower animals. Veratrum viride of North America is an acrid, emetic, and powerful stimulant, followed by sedative effects. Serious accidents have followed the incautious use of Meadow Saffron, Colchicum autumnale; it is only a few months since a woman was poisoned by the sprouts of Colchicum roots, which had been thrown away in Covent Garden market, and which she mistook for onions. White Hellebore, Veratrum album, the *ἱππομανη* of the ancients, is used by gardeners to destroy the Gooseberry Caterpillar, and similar noxious insects. Asagræa officinalis, an Alpine Mexican plant, yields most of the Cevadilla, Cebadilla or Sabadilla seeds of commerce, which were formerly used to destroy pediculi, and as anthelmintics, and have also been employed in chronic rheumatism and paralysis, and in neuralgic cases, but are now chiefly consumed in the manufacture of Veratria. Helonias frigida, called in Mexico Savoeja, is allied to this, and is a well-known poison, stupefying the horses that feed upon it. The root of Helonias dioica in infusion is anthelmintic, but its tincture is bitter and tonic; when chewed it excites the saliva and produces vomiting; the N. Americans call it Blazing Star and Devil's Bit. A decoction of Helonias bullata is given in obstructions of the bowels. Amianthium muscætoxicum is said to poison cattle which feed upon its foliage in the autumn, whence the United States Americans call it Fall Poison; they employ it to destroy flies. Uvularias are said to be simply astringent; the bruised leaves of Uvularia grandiflora are a popular remedy in the United States for the bite of the rattlesnake. The Hermodactyls of the Arabians, formerly so celebrated for soothing pains in the joints, were corms of the Colchicum variegatum, a species found in the Mediterranean. Dr. Royle found them in the bazaars of India, where they bear names traceable to the χολχικον and εφεμερον of the Greeks.

GENERA.

VERATREÆ.
Tofieldia, *Huds.*
Narthecium, Ger.
Helonias, Willd.
Heritiera, Schrank.
Isidrogalvia, R. et P.
Hebelia, Gmel.
Conradia, Raf.
Leptilix, Raf.
Triantha, Nutt.
Pleea, *Rich.*
Xerophyllum, *Rich.*
Helonias, *Linn.*
Abalon, Adans.
Chamaelirium, Willd.
Ophiostachys, Del.
Diclinothrys, Raf.
Asagræa, *Lindl.*
Sabadilla, Brandt.
Schœnocaulon, *A. Gr.*
Amianthium, *A. Gr.*
Amiantanthus, Kth.
Cyanotris, Raf.
Chrosperma, Raf.
Veratrum, *Tournef.*
Stenanthium, A. Gr.
Anticlea, *Kth.*
Zygadenus, *Rich.*
Leimanthium, Willd.
Burchardia, *R. Br.*
Erythrostictus, *Schlecht.*
Ornithoglossum, *Salisb.*
Lichtensteinia, Willd.
Cymation, Spr.
Anguillaria, *R. Br.*
Melanthium, *Linn.*
Criocephalus, Schlecht.
Meliglossum, Schlecht.
Dipidax, Laws.
Androcymbium, *Willd.*
Cymbanthes, Salisb.
Wurmbea, *Thunb.*
Bæometra, *Salisb.*
Kolbea, Schlecht.
Jania, Schult. f.

UVULAREÆ, *A. Gray.*
Schelhammera, *R. Br.*
Kreysigia, *Reichenb.*
Tripladenia, Don.
Uvularia, *Linn.*
Tricyrtis, *Wall.*
Compsanthus, Spreng.
Disporum, *Salisb.*
Drapiezia, Blum.
Lethea, Noronh.
Prosartes, *Don.*
Hekorima, *Raf.*
Streptopus, *Rich.*

COLCHICEÆ, *Nees.*
Monocaryum, *R. Br.*
Bulbocodium, *Linn.*
Merendera, Ram.
Geophila, Berg.
Colchicum, *Tournef.*
Hermodactylus, R. Br.
? *Leucocrinum*, Nutt.
? *Geanthia*, Raf.
? Weldenia, *Schult. f.*
Rugendasia, Schiede.

NUMBERS. GEN. 30. Sp. 130.

Iridaceæ.
POSITION.——————MELANTHACEÆ.—Liliaceæ.
Juncaceæ.

ORDER LXII. LILIACEÆ.—LILYWORTS.

Lilia, *Juss. Gen.* 48. (1789).—Narcissi, *the first sect. Ibid.* 54. (1789).—Hemerocallideæ, *R. Brown Prodr.* 295. (1810).—Liliaceæ, *DC. Théor. Elém.* 1. 249. (1813); *Endl. Gen.* lv.; *Meisner*, p. 398; *Kunth, Enum.* 4. 215; *Ann. sc.* 2. *ser.* 18. 290.—Tulipaceæ, *DC. Ess. Méd.* 297. (1816); *Bernh. in Botan. Zeit. Oct.* 1835.—Coronariæ, *Agardh Aphor.* 165. (1823).—Asparagi *and* Asphodeli, *Juss.* (1789).—Asphodeleæ, *R. Brown Prodr.* 275. (1810); *Kunth, Enum.* 4. 280.—Alliaceæ, Aloinæ, Hyacinthinæ, Dracænaceæ, *Link Handb.* vol. 1. (1829).—Asparaginæ, *Ib.*—Asparageæ, *DC. and Duby*, 458. (1828).—Asparagineæ, *Ach. Rich. Dict. Class* 2. 20. (1822); *Nouv. Elém. ed.* 4. 430. (1828).—Convallariaceæ, *Link Handb.* 184. (1829.)

DIAGNOSIS.—*Lilial Endogens with a naked perianth, flat when withering, anthers turned inwards, consolidated styles, and fleshy albumen.*

Herbaceous plants, shrubs or trees, with bulbs, or tubers, or rhizomes, or fibrous roots. Leaves narrow, with parallel veins, only in a very small number expanded into a broad blade with diverging veins; never articulated with the stem. Flowers large and showy, or small and green, with all kinds of intermediate gradations; in nearly all cases ☿; never, perhaps, truly ♂ ♀. Calyx and corolla confounded, coloured alike, regular or nearly so, occasionally cohering in a tube. Stamens 6, inserted into the sepals and petals. Anthers opening inwards. Ovary free, 3-celled, many-seeded; style 1; stigma simple, or 3-lobed; ovules anatropal or amphitropal. Fruit succulent, or dry and capsular, 3-celled. Seeds packed one upon another in 1 or 2 rows; embryo with the same direction as the seed, in the axis of fleshy albumen, or uncertain in direction and position, occasionally very minute.

Fig. CXXXVI.

Fig. CXXXVII.

The beautiful creations which constitute the Order of Lilies would seem to be well known to all the world; for what have been so long admired and universally cultivated as they? Nevertheless, there are few great groups of plants which have been more neglected by the exact botanist, or which stand more in need of his patient attention. The best proof of the justice of this assertion is to be found in the unsteady and conflicting views of botanists as to its limits, or the subordinate groups which it contains. While one writer breaks Lilyworts up into a number of distinct Orders, another refuses to recognise the limits assigned to them by his predecessor, and prefers a new arrangement, just as unsatisfactory as that which it succeeds. We have seen the classification of Jussieu and Brown break down beneath a rigorous scrutiny; it has been succeeded by schemes of Bartling, Endlicher, Kunth, Meisner, Bernhardi and others, all alike unsatisfactory; and I doubt whether we can be truly said to know more about the true characteristics and exact structure of a very large proportion of this Order, than we did twenty years ago. Genera in plenty have been added, but a good combination of

Fig. CXXXVI.—1. Section of seed of Asphodelus ramosus; 2. of Tulipa hortensis.

Fig. CXXXVII.—Arthropodium paniculatum. 1. A flower magnified; 2. a ripe capsule; 3. a transverse section of it; 4. a vertical section of a seed.

them is still wanted. Under these circumstances it seems to me better not to meddle with the supposed Orders or Suborders that have of late years been proposed, but to gather together, in tolerably natural groups, under the Order of Lilies, everything that does not belong to the other parts of the Lilial Alliance. It will be a task hereafter for some botanist, with ample materials and good general views, to study the details of the structure of these interesting plants, and out of those details to form an intelligible and solid classification. In the meanwhile, a few general remarks upon such groups as are here adopted are all that it will be useful to bring forward.

The favourite distinctions among the majority of systematic botanists are those by which the Liliaceæ, Asphodeleæ, and Asparageæ or Smilaceæ of authors are said to be known. Brown thought to distinguish them by their seeds and fruit; ascribing to the first a spongy and dilated or winged seedcoat and a capsule, rarely a berry; to the second, a black brittle seedcoat; and, to the third, a membranous seedcoat and berry. With regard to the colour of the seedcoat or its texture, I must remark firstly, that one would be slow to recognise such a peculiarity as a valid distinction even of genera, and that as an ordinal characteristic, it is still less admissible; that exceptions to such a character appear, as might be expected, in all directions, and prove it to be wholly illusory. By the great botanist just mentioned, the distinction of Smilaceæ was strengthened by adding to its character an embryo remote from the hilum, and it is probable that this circumstance deserves more attention than it has hitherto received; nevertheless, Streptopus, which is expressly named by Brown as one of his Smilaceæ, has the embryo next the hilum; so that this character also is untenable. Bartling, who retains Smilaceæ, adds to the distinction of the Order a minute embryo, but then he admits such genera as Asparagus and Drymophila, in which the embryo is the same as that of Asphodeleæ. Bernhardi assigns to his Tulipaceæ anthers attached to the filament by a fine point lodged in a narrow canal, and an inflorescence without membranous spathes; or, as Jussieu expressed it, *Flores nudi*, while he gives Asphodeleæ, anthers attached to the filament by a broad base, and membranous bracts, combining moreover under the name of Alliaceæ, the Asphodeleæ, Hypoxids, Rushes, Amaryllids and others, a proposition in which I think no judicious botanist would concur. But the character derived from the anther of Tulipaceæ, if valid, which Kunth denies, is trifling; and as to the peculiarities asserted to exist in the inflorescence of these plants, such membranous bracts do not exist in Eucomis among Asphodeleæ more than in Fritillaria persica among Tulipaceæ, while the Gageas have all the habit of the former group, and if it were otherwise it would be idle to propose such a character for the mark of a natural Order. M. Adrien de Jussieu has lately reduced these Orders to two, viz. Liliaceæ and Smilaceæ, giving the former an undivided style and parallel veined leaves, while the latter have a triple style and reticulated leaves. In this respect he appears to adopt the views which are taken in the present work. That good and high grounds of distinction will one day be found for some at least of the groups here admitted is probable; but they have not yet been discovered, nor is it likely that they will be until the true nature of the ovules, the position of their foramen, the direction of the embryo, and similar circumstances shall have been inquired into with scrupulous accuracy. In the meanwhile the following may be taken as the chief peculiarities of the sections now admitted.

TULIPEÆ are the Lilia of Jussieu, a couple of his genera being excluded, and they may be justly regarded as the type of the Order of Lilies. Bulbs, annual stems little or not at all branched, flowers usually large and gaily coloured, without membranous spathes, but axillary to leaves but little changed, the calyx and corolla and their parts scarcely united, although often arranged in a tube, anthers swinging lightly by the fine drawn point of a stiff filament, and finally a dry seed vessel, separate the group from all that follow. They are among the gayest of our garden flowers, as Tulips, Fritillaries, and Dogs' Tooth Violets testify; one of them indeed, the Lilium chalcedonicum, a plant that covers the plains of Syria with its scarlet flowers, is most memorable from having been selected by our Saviour as the subject of allusion in his sermon on the Mount. The Gloriosa, a tuberous plant from India, hardly belongs to them.

The HEMEROCALLEÆ or DAY LILIES, differ from the last in nothing except their calyx and corolla being so joined to each other as to form a tube of conspicuous length, and in their want of a bulb in many instances. The Agapanthus, so commonly cultivated in vases for decorating architectural gardens, and the fragrant Tuberose, are the more remarkable among them; but Funkia, Hemerocallis, Blandfordia, and the Velthiemias and Tritomas, are also species of familiar occurrence. Phormium, which yields the celebrated flax of N. Zealand, with its hard perennial leaves and panicles of yellow flowers, must be considered to connect the present division with that of Aloes.

There is so little to separate ALOINEÆ, or ALOES, from the Day Lilies, that scarcely anything can be named except their succulent foliage; and even that disappears in

Yucca, which has the hard leaves of Phormium, with which however its distinct sepals and petals forbid its being associated.

With the SCILLEÆ or SQUILLS, we reach a division of the Order, abounding in beautiful species, all of which are bulbous, with annual stems. Their peculiarity resides in the anthers not being so lightly attached to the filaments as in Tulipeæ, and in the leaves from whose axils the flowers proceed, acquiring a membranous condition.

CONANTHEREÆ are Squills with the ovary partially adhering to the calyx and corolla, and springing from tubers, not bulbs. They offer a direct transition to Amaryllids.

ANTHERICEÆ or ASPHODELS, agree with the last in having tubers or fleshy fascicled roots and not bulbs, but their ovary is free; they are therefore tuberous or fibrous rooted Squills. Chrysobactron, a genus gathered by Dr. Joseph Hooker, in Auckand and Campbell's islands, is described as diœcious, but apparently is polygamous. The fruit in these three last Orders is a capsule.

APHYLLANTHEÆ are plants with the habit of Rushes, and the bracts so membranous and closely imbricated, as to give the appearance of Xyrids when the flowers are past. They seem to form a connection between Lilies and some plants of the Juncal or Xyridal Alliances. The genera have been very insufficiently examined. Xanthorrhæas, called Grass Trees in New Holland, are very different in habit from the remainder; their shrubby stems, which emulate small Palm trees in appearance, bear tufts of long wiry foliage at their extremities, from the midst of which rise very long cylindrical spikes of densely compacted flowers, resembling Bullrushes (Typha). By this genus the Aphyllanths completely join Rushes, for the genus Kingia, included in Rushes, because of its minute embryo, has entirely the aspect of a Xanthorrhæa.

The reason for referring WACHENDORFEÆ hither, have been given in speaking of the Bloodroots (p. 152). They are plants with ensiform or plaited leaves of a hard texture, fibrous roots, with flowers usually in panicles and by no means remarkable for size or bright colouring. If it is really true that their carpels are opposite the petals, as is said, they will undoubtedly have to be removed from their present station.

ASPARAGEÆ are Lilies with a succulent fruit. They consist of plants extremely dissimilar in appearance, the common Asparagus and the Lily of the Valley being associated under this title. In general their leaves are broad; in the genus Cordyline they even acquire the expanded form and diverging veins of the Amomal Alliance. Their stems, although among the dwarfest that the Lilies comprehend, are in the common Asparagus branched and of considerable size, and in the Dragon-trees they acquire the dimensions and age of large trees. A tendency to the separation of sexes occurs here on the part of the genus Ruscus; but it is not carried so far as to constitute a diclinous structure. According to Von Martius (*Choix.* p. 21.), the position of the sepals in Lilyworts (in which he includes Asphodeleæ) is ▽ with respect to the axis; while in Asparageæ it is △. He also finds throughout the Liliaceous Order that the petaline stamens are larger and more perfect than the sepaline, it being the latter moreover which disappear when there is any deficiency in the usual number of stamens.

With respect to ASPIDISTREÆ, concerning whose structure we have very insufficient information, they are principally known by a large mushroom-shaped stigma. Their foliage is that of Gingerworts; their flowers are dingy purple or green, with a campanulate perianth, on whose sides the stamens are inserted. In many respects they are very like Orontiads, to which, perhaps, they ought to be referred.

In like manner the OPHIOPOGONEÆ, or TEATWORTS, have a foliage hardly belonging to Lilies, Peliosanthes Teta resembling a Ginger more than a plant of this Order. They are remarkable for their seeds bursting through the sides of the ovary at a very early period, growing freely though exposed to air, and finally acquiring the succulent appearance of a tuber. It is very uncertain whether they have any real claim to the rank of Lilies.

If we suppose that the doubtful members of this great Order are removed, we shall find that its most immediate relations are as follows. From the Melanths it chiefly differs in its anthers being turned inwards, and its carpels quite consolidated. To the Amaryllids it approaches so nearly that there is perhaps nothing to separate them except its free ovary; and the group Conanthereæ exhibits a structure intermediate in this respect. With Rushes Lilyworts are brought into close contact by means of the Aphyllanths as has been already stated. Towards Arads they extend in the direction of Orontiads, through the intermediate group of Aspidistreæ. Finally, it is here that Dictyogens are reached by means of the Asparageæ, which, by most botanists, are actually made to comprehend the genus Smilax and the Parids. For the affinity of Lilies and Palms, the reader is referred to the observations under the latter Order.

The geographical limits of the Order are as wide as its differences of structure. Upon the whole, however, the species are much more abundant in temperate climates than in the tropics, where they chiefly exist in an arborescent state. Aloes are mostly found in the southern parts of Africa; one species is a native of the West Indies, and two or three more of Arabia and the East. Dracænas, the most gigantic of the Order, attain their largest size in the Canaries; a D. Draco there is described as being between 70 and 75 feet high, 46½ feet in circumference at the base, and was known to have been a very ancient tree in the year 1406. The northern Flora comprehends for the most part plants of the genera Scilla, Hyacinthus, Allium, and Ornithogalum. In the East Indies Lilyworts are rare; in New Holland they form a distinctly marked feature of the vegetation, and in New Zealand they are represented by the Phormium or Flax-bush.

A very considerable number are employed for the purposes of mankind. Among the most extensively useful are those whose fibre is strong enough to furnish cordage. Such are Phormium tenax, the New Zealand Flax, whose toughness rivals that of Hemp, and the Sansevieras, a race of hard-leaved perennial plants, found all over the tropics of Africa and India, from which a yet stronger substance is obtained under the name of African Hemp, or Bowstring Hemp. The Yuccas too yield a tenacious fibre, but it is of comparative unimportance. Several species have been used as food from the most remote antiquity; those chiefly belong to Allium. The Onion, Garlic, and Leek, says Dr. Royle, called in Arabic Busl, Som, and Korras, seem to be alluded to in the earliest parts of the Bible (Numbers, ch. xi. v. 12), as the names there used are very similar to these. All are cultivated in gardens in India, as well as Allium ascalonicum and A. tuberosum. The bulbs of Allium leptophyllum are eaten by the hill-people, and the leaves are dried and preserved as a condiment. Chives, Shallots, and Rocambole are other species of the Alliaceous race. The bulbs of Camassia esculenta are eaten by the North American Indians under the name of Quamash, and those of Lilium pomponium are roasted and eaten in Kamtchatka, where it is as commonly cultivated as the Potato with us. Erythronium Dens canis is said to furnish a part of the diet of the Tartars. The Cordyline Ti (Dracæna terminalis), or Ti plant, affords an important part of the food of a Sandwich islander. Its great woody roots are baked, when they become sweet and nutritious. Bruised, mixed with water, and fermented, it forms an intoxicating beverage; distilled, an ardent spirit is readily obtained; boiled before fermentation, a rich syrup, capable of being a substitute for sugar, is the result. Cattle, sheep, and goats are fond of the leaves, which furnish thatch for houses, and are woven into a kind of cloth. Its truncheons take root when stuck in the ground, and form a valuable permanent hedge.—*Bot. Reg.* 1. 1749. Mr. Drummond says that the tops of different species of Xanthorrhæa furnish all kinds of cattle with valuable fodder, in the Swan River colony, *Hooker Journ.* 2. 328; and we learn from Mr. Backhouse that the base of the inner leaves of the Grass-tree of Tasmannia is not to be despised by the hungry. The aborigines beat off the heads of these singular plants by striking them about the top of the trunk with a large stick; they then strip off the outer leaves and cut away the inner ones, leaving about an inch and a half of the white tender portion, joining the trunk: this portion they eat raw or roasted; and it is far from disagreeable in flavour, having a milky taste, slightly balsamic. There are some other species of Grass-tree in the colony, the base of the leaves of which may also be used as food: those of the dwarf Grass-tree, Xanthorrhæa humilis, which is abundant about York Town, may be obtained by twisting the inner leaves firmly together, and pulling them forcibly upwards; but care is required not to cut the fingers, by slipping the hand. Even in Europe the young shoots of Polygonatum (Solomon's Seal), and others, have been substituted for Asparagus, and the annual cultivation of the latter for kitchen purposes is known to every one.

Fig. CXXXVIII.

Aloes and Squills indicate the value of some Lilies in medicine. The acrid matter which thus renders them valuable as purgatives or emetics, is found in a considerable number of species. The bulb of the Urginea or Scilla maritima, and Pancratium, (the Σκιλλη and Παγκρατιον of Dioscorides) is nauseous and acrid, acting either as an emetic, purgative, expectorant, or diuretic, in proportion to the dose in which it is given; its properties are said to be due to a peculiar principle, called by Vogel, Scillitin. It is curious

Fig. CXXXVIII.—Xanthorrhæa hastilis.

that in India a species very closely allied to the Mediterranean plant, and called Scilla indica by Dr. Roxburgh, is substituted for the Urginea maritima, and Iskeel given as its Greek name; the bulb is also used by weavers in preparing their thread.—*Royle.* According to Theodore Martius the bulbs of Ledebouria hyacinthoides are also used as a substitute for Squills in the East Indies; Ainslie states that they are employed in cases of strangury and fever in horses. Both leaves and roots of Erythronium americanum are emetic; so are the bulbs of Muscari moschatum, various Gageas, Hyacinths, and Ornithogalums. As purgatives, the Aloes are in most extensive use; it is, however, exclusively from the arborescent species, especially A. vulgaris, soccotrina, purpurascens, and spicata, that the drug is collected. Similar qualities reside in Bulbine planifolia, the roots of common Asparagus, Lily of the Valley, the capsules of Yuccas, &c. As may be supposed, the peculiar secretions which produce actions like these will, when a little modified, become diuretic; and thus we have a long list of species to which this quality is attributed. Foremost are Alliums, whose bulbs abound in free phosphoric acid; then follows Asparagus, notorious for its singular effect upon the urine, many of the emetic species, and the roots of Asphodelus ramosus, Asphodeline lutea, Anthericum ramosum, the berries of Smilacina racemosa, &c. According to Dr. Dieffenbach, the root of Phormium tenax is an excellent substitute for Sarsaparilla, acting as a purgative, diuretic, sudorific, and expectorant.—*Chem. Gaz.* 1842. 150. Then, when these acrid principles become concentrated, we have virulent poisons. Such are Gloriosa superba, and the fetid bulb of the Crown Imperial, whose very honey is said to be emetic as it distils from the flowers.

Resinous matters are yielded in abundance by some species, whence they have been found useful in dysenteries. Of these the most celebrated is Dragon's-blood, a tonic astringent resin, sometimes employed in diarrhœa and passive hæmorrhages; it is yielded in part by Dracæna Draco, from the surface of the leaves, and from the cracks in its trunk; this is, however, scarcely known to modern druggists, who sell the astringent resin of Pterocarpus. A fragrant brownish yellow resin, called Botany Bay gum, when burnt smelling like Benzoin, flows in abundance from Xanthorrhæa arborea. It is probable that some such secretion occurs in Dianella odorata, whose powdered roots are said by Blume to be made into fragrant pastiles. The roots of Dracæna terminalis and ferrea are said to be useful astringents, to which may be added Streptopus amplexifolius and Ruscus hypoglossum, both of which have been employed in gargles.

A few miscellaneous instances of useful Liliaceous plants still remain to be added. The roots of Asparagus racemosus and adscendens are both employed medicinally in North India; those of the latter, conical in form and semi-transparent, are considered a good substitute for salep.—*Royle.* Polianthes tuberosa, or the Tuberose, is well known for its delicious fragrance. This plant emits its scent most strongly after sunset, and has been observed in a sultry evening, after thunder, when the atmosphere was highly charged with electric fluid, to dart small sparks, or scintillations of lucid flame, in great abundance from such of its flowers as were fading. The roots of Sanseviera have been employed as remedies for gonorrhœa, pains of the joints, and coughs. The bulbs of Erythronium Dens canis have been regarded as aphrodisiac and anthelmintic. Oil of Lilies was prepared by infusing the flowers of Lilium candidum in oil. Tulbaghia, a Cape genus, smelling like Garlic, is boiled in milk and prescribed in phthisical complaints. Asparagus owes its remarkable qualities to the presence of a peculiar principle called Asparagin, which is said to be more abundant in Asparagus acutifolius than in the species commonly cultivated. The flowers of Cordyline reflexa are said to be emmenagogue. A decoction of the root of Dianella odorata is administered in Java in gonorrhœa, dysury, and fluor albus, according to Blume. The Butchers' Brooms (Rusci) of Europe, were once celebrated as aperients and diuretics, on account of their bitter, subacrid, mucilaginous roots, especially Ruscus aculeatus, the ὀξυμυρσίνη of Dioscorides. The Arabian writers called the fruit Rhabâbath, out of which, according to Endlicher, the Latinobarbarous word Cubeba has been corrupted. Ruscus hypophyllum had considerable reputation as a stimulant of the uterus. The seeds of these Rusci are very horny, and when roasted are said to furnish a pleasant substitute for coffee. The bulbs of the common Tulip are sometimes substituted fraudulently for Colchicums; large quantities have been thus imported from Naples; they are readily known by being true bulbs, while the Colchicum has a corm. It is not a little remarkable that the Yuccas, like some species of Fourcroya, have the property of producing tubers although they have arborescent stems.

GENERA.

I.—TULIPEÆ, *DC.*

Erythronium, *Linn.*
Dens Canis, Tournef.
Tulipa, *Tournef.*
Orithya, *Don.*
Gagea, *Salisb.*
Ornithoxanthum, Lk.
Bulbillaria, *Zucc.*
Iphigenia, *Kth.*
Plecostigma, *Traut.*
Hornungia, *Bernh.*
Lloydia, *Salisb.*
Rhabdocrinum, Rchb.
Nectarobothrium, Led.
Calochortus, *Pursh.*
Cyclobothra, *Don.*
Eucrinum, *Nutt.*
Fritillaria, *Linn.*
Petilium, Linn.
Imperialis, Juss.

Rhinopetalum, *Fisch.*
Lilium, *Linn.*
Amblirion, Rafin.
Martagon, Tourn.
Cardiocrinum, *Endl.*
Clinostylis, *Hochst.*
? Gloriosa, *L.*
Methonica, Herm.

II. HEMEROCALLEÆ, *R. Br.*

Hemerocallis, *L.*
Funkia, *Spr.*
Hosta, Tratt.
Bryocles, Salisb.
Niobe, Salisb.
Saussurea, Salisb.
Libertia, Dumort.
Agapanthus, *Herit.*
Mauhlia, Thunb.
Abumon, Adans.
Polianthes, *Linn.*
Blandfordia, *Sm.*
Veltheimia, *Gled.*
Tritoma, *Ker.*
Tritomanthe, Hffsg.
Tritomium, Lk.
Rudolpho-Romeria, Steud.
Kniphofia, Mnch.
Phormium, *Forst.*
Chlamidia, Banks.
Sanseviera, *Thunb.*
Acyntha, Commel.
Salmia, Cav.

III. ALOINEÆ, *Link.*

Aloë, *Tournef.*
Apicra, Haw.
Catevala, Medik.
Haworthia, Duval.
Gasteria, Duval.
Ripidodendron, Willd.
Kumara, Medik.
Bowiea, Haw.
Pachydendron, Haw.
Agriodendron, Haw.
Lomatophyllum, *Willd.*
Phylloma, Ker.
Yucca, *Linn.*

IV. SCILLEÆ, *Bartl.*
Allieæ, *Link.*

Allium, *Linn.*
Moly, Mönch.
Mönchia, Medik.
Saturnia, Maratti.
Ophioscorodon, Wallr.
Codonoprasum, Rchb.
Gethioides, Column.
Schœnoprasum, Kunth.
Porrum, Tournef.
Cepa, Tournef.
Scorodoprasum, Michel.
Nectaroscordum, *Lindl.*
Caloscordum, *Herb.*
Milla, *Cav.*
Hesperoscordum, *Lindl.*
Pseudoscordum, *Herbert.*
Nothoscordum, Kth.
Ornithogalodeum, G. Don.
Calliprora, *Lindl.*
Camassia, *Lindl.*
Cyanotris, Raf. ms.
Scilla, *Linn.*
Urginea, *Steinh.*
Stellaris, Mönch.
Squilla, Nees.
Ornithogalum, *Lk.*
Chlorogalum, *Lindl.*
Albuca, *Linn.*
Myogalum, *Lk.*
Albucea, Rchb.
Honorius, Gray.
Nolina, *Rich.*
Nolinea, Pers.
Uropetalum, *Ker.*
Pollemannia, Berg.
Zuccagnia, Thunb.
Dipcadi, Mönch.
Litanthes, *Harv.*
Muscari, *Tournef.*
Botryanthus, Kth.
Bellevalia, *Lap.*
Hyacinthus, *Linn.*
Eratobotrys, *Fenzl.*
Puschkinia, *Adams.*
Adamsia, Willd.
Strangweia, Bertoloni.
Barnardia, *Lindl.*
Ledebouria, *Roth.*
Bessera, *Schult.*
Pharium, Herbert.
Leucocoryne, *Lindl.*
Brodiæa, *Sm.*
Hookeria, Salisb.
Dichelostemma, Kth.
Triteleja, *Hook.*
Seubertia, Kth.
Tristagma, *Pöpp.*
Agraphis, *Link.*
Lachenalia, *Jacq.*
Cœlanthus, *W.*
Peribæa, *Kth.*
Polyxena, *Kth.*
Drimia, *Jacq.*
Idothea, *Kth.*
Massonia, *Linn.*
Daubenya, *Lindl.*
Eucomis, *Herit.*
Basilœa, Juss.

V. CONANTHEREÆ, *Don.*

Zephyra, *Don.*
Cummingia, *Don.*
Conanthera, *Ruiz et Pav.*
Pasithea, *Don.*

VI. ANTHERICEÆ, *Bartl.*

Sowerbæa, *Smith.*
Anemarrhena, *Bung.*
Eremurus, *Bieberst.*
Henningia, *Kar.*
Ammolirion, *Kar.*
Asphodelus, *Linn.*
Asphodeloides, Mönch.
Bidwillia, *Herb.*
Asphodeline, *Rchb.*
Chrysobactron, *Hook. f.*
Cyanella, *Linn.*
Anthericum, *Linn.*
Phalangium, Juss.
Czackia, Andr.
Allobrogia, Tratt.
Liliastrum, Lk.
Bulbine, Linn.
Anthericum, Juss.
Bulbinella, *Kth.*
Trachyandra, *Kth.*
Arthropodium, *R. Br.*
Dichopogon, *Kth.*
Chlorophytum, *Ker.*
Hartwegia, Nees.
Trichopetalum, *Lindl.*
Bottionœa, Colla.
Stypandra, *R. Br.*
Simethis, *Kth.*
Thysanotus, *R. Br.*
Chlamysporum, Salisb.
Cæsia, *R. Br.*
Chloopsis, *Blume.*
Tricoryne, *R. Br.*
Echeandia, *Orteg.*
Tulbaghia, *Linn.*
Herreria, *R. et Pav.*
Eriospermum, *Jacq.*

VII. APHYLLANTHEÆ, *Endl.*

Alania, *Endl.*
Laxmannia, *R. Br.*
Borya, *Labill.*
Daviesia, Lam.
Baumgartenia, Spr.
Aphyllanthes, *Tournef.*
Johnsonia, *R. Br.*
Xanthorrhæa, *Sm.*

VIII. WACHENDORFEÆ, *Herbert.*

Hagenbachia, *Nees.*
Xiphidium, *Aubl.*
Wachendorfia, *Burm.*
Pedilonia, Presl.
Schiekia, *Meisn.*
Lophiola, *Ker.*

IX. ASPARAGEÆ, *Lindl.*

Dianella, *Lam.*
Diana, Commers.
Excremis, Willd.
? Duchekia, *Kostel.*
Rhuacophila, *Blume.*
Eustrephus, *R. Br.*
Geitonoplesium, *A. Cunningh.*
Luzuriaga, R. Br.
Asparagus, *Linn.*
Oncus, *Lour.*
Myrsiphyllum, *Willd.*
Cordyline, *Commers.*
Charlwoodia, Sweet.
Dracæna, *Vandell.*
Störkia, Crantz.
Œdera, Crantz.
Tœtsia, Medik.
Drymophila, *R. Br.*
Polygonatum, *Tournef.*
Axillaria, Rafin.
Convallaria, *Desf.*
Brachypetalum, *Nutt.*
Smilacina, *Desf.*
Majanthemum, Mönch.
Unifolium, Hall.
Evallaria, Neck.
Bifolium, Fl. wett.
Clintonia, Rafin.
Sigillaria, Rafin.
Tovaria, Neck.
Luzuriaga, *Ruiz et Pav.*
Callixene, *Commers.*
Enargea, Sol.
Ruscus, *Tournef.*
Danaida, Link.
Danaë, Medik.

X. ASPIDISTREÆ, *Endl.*

Rhodea, *Roth.*
Tupistra, *Ker.*
Aspidistra, *Ker.*
Macrogyne, Lk. et Otto.

XI. OPHIOPOGONEÆ, *Endl.*

Ophiopogon, *Ait.*
Flüggea, Rich.
Slateria, Desv.
Polygonastrum, Mönch.
Liriope, Lour.
Sanseviella, Rchb.
Bulbospermum, *Blume.*
Peliosanthes, *Andr.*
Teta, Roxb.

NUMBERS. GEN. 133. Sp. 1200.

Dictyogens.
POSITION.—Orontiaceæ.—LILIACEÆ.—Melanthaceæ.
Juncaceæ.

ORDER LXIII. PONTEDERACEÆ.—PONTEDERADS.

Pontedereæ, *Kunth in Humb. et Bonpl. N. G.* 1. 211. (1815); *Agardh Aph.* 169. (1823).—Pontederaceæ, *Ach. Rich. Nouv. Elém. ed.* 4. 427. (1828); *Endl. Gen.* liv. *Meisn. Gen. p.* 398. *Kunth Enum.* 4. 119.

DIAGNOSIS.—*Lilial Endogens with a naked perianth, circinate when withering, anthers turned inwards, and mealy albumen.*

Aquatic or marsh-plants. Leaves sheathing at the base, with parallel veins, in the larger species arrow-headed, cordate or dilated. Flowers either solitary or in spikes or umbels, spathaceous, frequently blue, sometimes yellow. Perianth tubular, coloured, 6-parted, more or less irregular, with a circinate æstivation. Stamens arising from the calyx, 6, or 3 opposite the petals; anthers turned inwards, opening lengthwise. Ovary free, more or less completely 3-celled, many-seeded; style 1; stigma simple; ovules anatropal. Capsule 3-celled, occasionally acquiring an adhesion to the perianth, 3-valved, with loculicidal dehiscence. Seeds indefinite, attached to a central axis, ascending; hilum small; embryo with its radicle rather enlarged, orthotropal, in the axis of somewhat mealy albumen.

Fig. CXXXIX.

The aquatic plants comprehended under this name are essentially distinguished by the divisions of their flowers being rolled inwards after flowering, to which may be added mealy albumen, and an indefinite number of seeds. For this reason a plant called Reussia, which seems to want the first and last characters, appears to have no business among them. They were referred to Spiderworts by Salisbury, and are considered nearly related to that Order by Achille Richard, who, however, separates them, suggesting their being referable to Lilyworts. There can be no doubt of their close relation to the latter Order, from which they are principally known by their irregular flowers, mealy albumen, and perianth rolling inwards after expansion. Leptanthus, however, if it is really one of the Order, has all the habit of a Potamogeton, and establishes a connection with the Arrow-grasses. Hooker, who has given an excellent figure of Eichhornia speciosa (*B. M. t.* 2932), states that each fibre of the roots has a calyptrate covering at the extremity, similar to that found on the roots of the Duck-weed.

Water-plants found exclusively in North and South America, the East Indies, and tropical Africa.

Very little is known of their uses. Monochoria vaginalis is employed in Indian pharmacy in liver-complaints and disorders of the stomach. Rubbed down in butter and drank, it is thought to remove redness of the eyes; powdered and mixed with sugar it is administered in asthma; and when chewed, is said to relieve toothache; brayed with milk it is given in fever; and, finally, when young, is eaten as a pot-herb.—*Endlicher.*

GENERA.

Heteranthera, *Ruiz et Pav.*
Buchozia, Flor. Flum.
Heterandra, Palis.

Leptanthus, *L. C. Rich.*
Schollera, Willd.
Eichhornia, *Kth.*

Pontederia, *Linn.*
Unisema, *Rafin.*
Monochoria, *Presl.*

NUMBERS. GEN. 6. Sp. 30.

POSITION.—*Juncaginaceæ.*—PONTEDERACEÆ.—Liliaceæ. *Mayaceæ?*

Fig. CXXXIX.—Eichhornia speciosa; 1. pistil; 2. transverse section of the ovary.—*Hooker.*

ALLIANCE XVII. *ALISMALES.*—THE ALISMAL ALLIANCE.

DIAGNOSIS.—*Hypogynous, (bisexual), tri-hexapetaloideous Endogens, with separate carpels and no albumen.*

These stand in the same relation to hermaphrodite hypogynous Endogens as Orchidals and Hydrals to the Alliances with which they are respectively associated. The want of albumen is their great feature. They are however known, in addition, by their carpels not having any tendency to combine; so that they are to Endogens almost what the Crowfoots are to Exogens. And it is to be observed that if it were not for their monocotyledonous embryo there would be no distinguishing such plants as Alisma from certain Ranunculi, represented by Ranunculus parnassifolius. A very few are ♂ ♀; such however occur only among the Alismads, and are not liable to be mistaken for any other plants than Hydrals, with none of the Orders in which can they be properly associated. Arrow-grasses offer the lowest organization in the Order, and may be regarded as an Alismal form of Naiads.

This Alliance seems to close the class of Endogens, and to stand on the limits of Exogens, in consequence of the intimate and unquestionable relation between Alismads and Crowfoots.

NATURAL ORDERS OF ALISMALS.

Flowers 3-petaloideous. Placentæ many-seeded, netted and parietal	64. BUTOMACEÆ.
Flowers 3-petaloideous. Placentæ few-seeded, simple, and axile, or basal. Embryo solid	65. ALISMACEÆ.
Flowers scaly. Placentæ few-seeded, simple and axile, or basal, slit on one side, with a very large plumula	66. JUNCAGINACEÆ.

ORDER LXIV. BUTOMACEÆ.—BUTOMADS.

Butomeæ, *Richard in Mem. Mus.* 1. 364. (1815); *Endl. Gen.* 1. *Meisner, Gen. p.* 365. *Kunth. Enum.* iii. 162.

DIAGNOSIS.—*Alismal Endogens with 3-petaloideous flowers, and many-seeded netted and parietal placentæ.*

Aquatic plants. Leaves very cellular, with parallel veins, often yielding a milky juice. Flowers in umbels or solitary, conspicuous, purple, or yellow, or white. Sepals 3, usually herbaceous. Petals 3, petaloid. Stamens definite or indefinite, hypogynous, some of them occasionally abortive. Ovaries free, 3, 6, or more, either distinct or united into a single mass; stigmas the same number as the ovaries, simple; ovules 00, anatropal or campylotropal, attached to a parietal network. Achenia or follicles many-seeded, either distinct and rostrate, or united in a single mass. Seeds minute, very numerous, attached to the whole of the inner surface of the fruit; albumen none; embryo with the same direction as the seed.

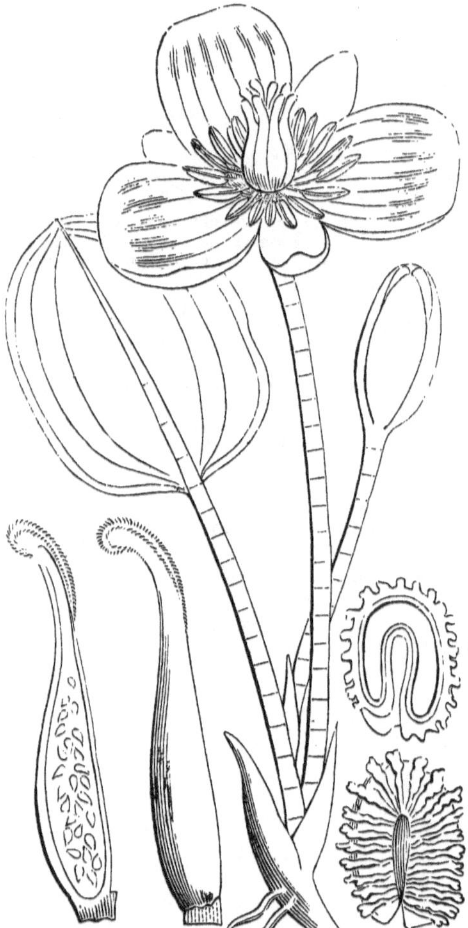

5

4

3 2 1

Fig. CXL.

These water plants are readily known by their placenta extending over the whole lining of the fruit, which is formed either of separate or concrete carpels. In this respect there is an evident analogy with Water lilies, which Limnocharis resembles in the structure of its fruit. De Candolle says that no Endogens are lactescent; but some of these yield milk in abundance. Limnocharis offers a singular example of a large conspicuous open hole in the apex of its leaf, apparently destined by nature as an outlet for superfluous moisture, which is constantly draining from it.

The species are natives of the marshes of Europe and Siberia, the North Western provinces of India, and equinoctial America.

Butomus umbellatus is acrid and bitter; its rhizome and seeds have been regarded emollient, refrigerant and solvent, and were once officinal under the name of Radix et Semina Junci floridi. The roasted rhizome is eaten in the North of Asia.

GENERA.

Butomopsis, *Kth.* *Tenagocharis*, Hochst.	Butomus, *Tourn.* Hydrocleis, *Rich.*	Limnocharis, *H. et B.*

NUMBERS. GEN. 4. Sp. 7.

POSITION. *Nymphæaceæ.* ———BUTOMACEÆ.—Alismaceæ.

Fig. CXL.—1. Hydrocleis Commersoni; 2. one of the carpels; 3 the same opened to show the placentæ; 4. seed of Limnocharis Plumieri; 5. a section of the same.

ORDER LXV. ALISMACEÆ.—ALISMADS.

Alismaceæ, *R. Brown Prodr.* 342. *in part* (1810); *Rich in Mém. Mus.* 1. 365. (1815); *Juss. Dict. Sc. Nat.* 1. 217. (1822); *Endl. Gen.* xlix.; *Meisner, Gen.* p. 364. *Kunth Enum.* 3. 147.—Alismoideæ, *DC. Fl. Fr.* 3. 188. (1805).

DIAGNOSIS.—*Alismal Endogens with 3-petaloideous flowers, few-seeded simple and axile or basal placentæ, and a solid embryo.*

Floating or swamp plants, very rarely annual, usually having a creeping fleshy perennial rhizome. Flowers in umbels, racemes or panicles, ☿, very rarely truly ♂ ♀. Leaves either narrow and strap-shaped, or expanded into a broad blade, always however with the veins parallel. Sepals 3, herbaceous. Petals 3, petaloid. Stamens definite or indefinite; anthers turned inwards. Ovaries superior, several, 1-celled; ovules erect or ascending, solitary, or 2 attached to the suture at a distance from each other, anatropal or campylotropal. Styles and stigmas the same number as the ovaries. Fruit dry, 1- or 2-seeded. Seeds without albumen, hooked; embryo shaped like a horse-shoe, undivided, with the same direction as the seed.

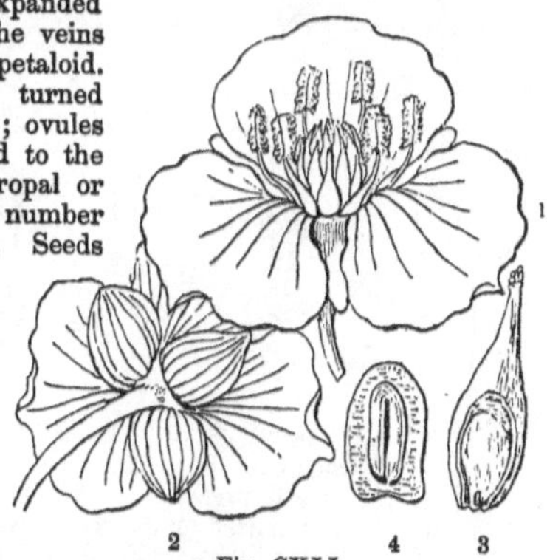

Fig. CXLI.

This Order is to Endogens what Crowfoots are to Polypetalous Exogens, and is in like manner recognised by its disunited carpels and hypogynous stamens. Such plants as Ranunculus parnassifolius are hardly distinguishable from Alismads by external characters. Arrow-grasses are known by their imperfect floral envelopes, and straight embryo having a lateral slit for the emission of the plumule. The plants belonging to Alismads, Hydrocharads, Naiads, Arrow-grasses, and Butomads, have all a disproportionately large radicle, whence their embryos were called by the late L. C. Richard, macropodal. The truly diclinous flowers of Sagittaria constitute a great and unusual exception to the otherwise hermaphrodite structure of this Order.

Chiefly natives of the northern parts of the world. Several Sagittarias and Damasoniums inhabit the tropics, the former those of both hemispheres.

Many have a fleshy rhizome, which is eatable; such are Alisma and Sagittaria: a species of the latter, S. sinensis, is cultivated for food in China; its herbage is acrid. Alisma Plantago and Sagittaria sagittifolia are among the plants foolishly recommended in hydrophobia; the rhizome of the former, deprived of acridity by drying, is eaten by the Kalmucks. Various Brazilian Sagittarias are very astringent; and their expressed juice is even employed in the preparation of ink.—*Martius, mat. m. br.* 47.

GENERA.

Alisma, *Juss.*
Echinodorus, Rich.
Sagittaria, *Linn.*
§ *Lophiocarpus*, Kth.
Damasonium, *Juss.*
Actinocarpus, R. Br.

NUMBERS. GEN. 3. Sp. 50.

POSITION.—*Commelynaceæ*. / *Ranunculaceæ*.—ALISMACEÆ.—Butomaceæ.

Fig. CXLI.—1. Flower of Alisma ranunculoides seen in front; 2. the same from the rear; 3. a section of the ovary; 4. a section of a seed.

Order LXVI. JUNCAGINACEÆ.—Arrow-grasses.

Juncagineæ, *Rich. Anal. Fr.* (1808); *Mém. Mus.* 1. 364. (1815); *Endl. Gen. p.* 127; *Meisner, p.* 364; *Kunth Enum.* 3. 141.—Potamogetoneæ, *Rchb. Fl. Excurs.* 1. 6. (1830).

Diagnosis.—*Alismal Endogens with scaly flowers, few-seeded simple axile or basal placentæ, and an embryo slit on one side, with a very large plumule.*

Herbaceous aquatic or marsh plants, whose leaves have in all cases parallel veins, whether they are narrow and grassy, or broad and quite different from the leaf-stalk.

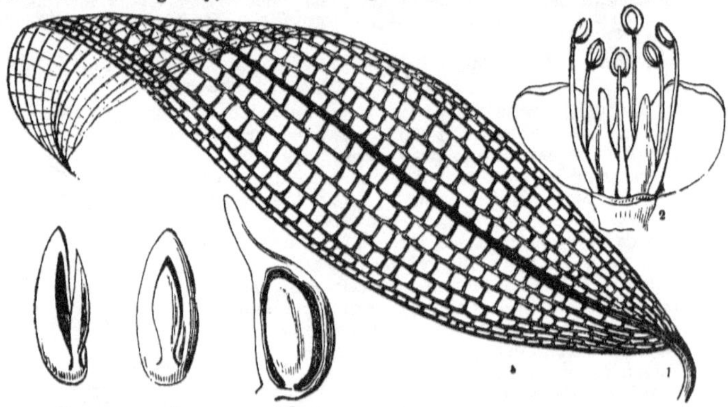

Fig. CXLII.

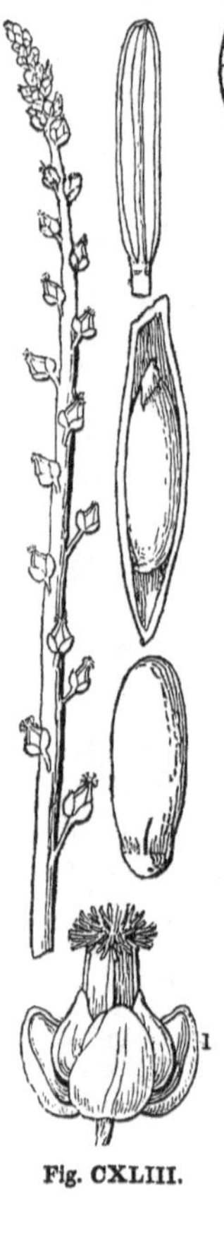

Fig. CXLIII.

Flowers ☿, of no conspicuous appearance, white or green, in spikes or racemes. Sepals and petals small and much alike. Stamens 6; anthers usually turned outwards and opening longitudinally. Carpels 3, 4, or 6, free, united or distinct; ovules 1 or 2, approximated at their base, erect or pendulous. Fruit dry, 1- or 2-seeded. Albumen wanting; embryo having the same direction as the seed, with a lateral cleft for the emission of the plumule.

With the exception of their flowers being ☿, there is little to separate these plants from the Naiads, in whose Order some of them have been included by all botanists before this time; for the old distinction of pendulous ovules in the Naiads, and erect ones in Arrow-grasses, fails in consequence of Caulinia, and Naias itself, having them erect. The plumule lying within a cleft on one side of the embryo indicates a decided tendency on the part of these plants to Arads, and the incomplete condition of their floral envelopes confirms the relationship. The genus Scheuchzeria is a transition from Arrow-grasses to Rushes.

Marshy places in most parts of the world may be expected to indicate traces of this Order, which is found in Europe, Asia, and North America, the Cape of Good Hope, and equinoctial America. Potamogetons occur in ditches and swamps as far north as Iceland.

Triglochin has a salt taste. The root of Potamogeton natans is said to be eaten in Siberia; the foliage of others is regarded as styptic.

GENERA.

Triglochin, *Linn.* *Juncago*, Tournef. *Tristemon*, Raf.	Scheuchzeria, *Linn.* Ruppia, *L.* Potamogeton, *L.*	*Peltopsis*, Raf. Aponogeton, *L.* *Spathium*, Lour.	Cycnogeton, *Endl.* Ouvirandra, *Thouars.* *Hydrogeton*, Pers.

Numbers. Gen. 7. Sp. 44.

Naiadaceæ.
Position.——— .—Juncaginaceæ.—Alismaceæ.
Juncaceæ.

Fig. CXLII.—1. Leaf of Ouvirandra fenestralis; 2. a flower cut open; 3. section of a ripe carpel of O. Bernieriana; 4, 5. embryo in different positions: the thicker part is the cotyledon, the smaller the plumule.

Fig. CXLIII.—Triglochin palustre. 1. A flower; 3. a ripe fruit; 4. one ripe carpel opened, and exhibiting a seed; 5. embryo.

CLASS V.—DICTYOGENS.

Retosæ, *Ed. pr. p.* 358. (1836).—Dictyogens, *Bot. Reg.* 1839. *Misc. p.* 76.

There is among the plants referred by Jussieu to his Monocotyledons, and consequently by later Botanists to Endogens, a small number of species whose foliage and habit of growth are so very peculiar, that the reference of them to Endogens is wholly dependent upon their conformity in the structure of the embryo. They have a broad net-veined foliage, which usually disarticulates with the stem, and in some cases the small green flowers are very nearly the same as those of such plants as Menispermum, among Exogens. For these reasons I have endeavoured to show that they ought to be regarded as a transition class partaking somewhat of the nature of Endogens and also of that of Exogens. And if we regard merely the foliage, the distinction seems admissible, for no Endogens possess such a character except a few Arads, otherwise widely different. The nearest approach to this structure, with which I am acquainted, occurs in Lilium giganteum, but the leaves of that plant have a flat foliaceous petiole and do not disarticulate. The broad-leaved Amaryllids like Griffinia, Eurycles, &c., are totally different; their leaves not only having no articulation with the stem, but having no reticulations between the ribs, further than what arises from the anastomosing of the fine parallel secondary veins which connect them.

It is not, however, in the leaves alone that a distinction is found between Endogens and Dictyogens. If the annual branches of a Smilax are examined, there is nothing indeed in their internal structure at variance with that of a stem of Asparagus; they are exactly Endogenous; but in the rhizome of the whole genus (take the Sarsaparilla of the shops for instance) the wood is disposed in a compact circle, below a cortical integument, and surrounding a true pith; in that of Smilax aspera the woody matter is disposed in the form of a cylinder, inclosing a centre of soft cellular matter; and the vessels of the cylinder have an evident tendency to arrange themselves in lines forming rays from the centre. In Dioscorea alata the stem itself is formed of eight fibrovascular wedges placed in pairs, with their backs touching the bark, surrounding a central pith and having wide medullary plates between them; in fact, when the stems of this plant are in a state of decay, the eight fibrovascular wedges may be pulled asunder, like those of a Birthwort or a Menisperm. In the curious Testudinaria elephantipes the structure of the stem is of nearly the same kind; several bundles of fibrovascular tissue form a circle surrounding a pith, and pierced with broad medullary processes. Lapageria and Philesia have each a zone of wood below their bark, and a central pith in which the common fibrovascular bundles of Endogens are disposed; a tendency to which is also observable in Smilax. It therefore seems that the peculiarities in the foliage of these plants are accompanied by others equally remarkable in the structure of the stem; indeed I do not see why the stem of a common Yam has not as good a title to be regarded Exógenous as Endogenous. Schleiden indeed has remarked that he believes it to be the regular structure of the roots of Endogens to have a simple circle of fibrovascular closed bundles; and this seems to be sometimes the case. But I do not find it in Strelitzia, or even in the arborescent Aloes, and when it does become evident it is unaccompanied by any peculiarity of the foliage. But, in the perennial stem of Dictyogens the bundles are what this Anatomist calls unlimited, that is to say, they go on growing for years together as in Exogens.

The principal difficulty about admitting the class of Dictyogens seems to me to consist in the small number of genera and species which it comprehends, and in the absence of any evidence as to the stem of Triurids or even Parids having the anatomical structure here assigned to it. These objections are undoubtedly deserving of serious consideration; but on the other hand it must be borne in mind that the plants collected under Dictyogens agree well with each other, and ill with any alliances of Endogens.

The Natural Orders of Dictyogens are poor in species, and can hardly be considered as established on recognised characters. The following are the distinctions, as far as they can be at present pointed out.

NATURAL ORDERS OF DICTYOGENS.

Flowers ♂ ♀. *Perianth free. Carpels* 00, *one-seeded* . . .	67. TRIURIDACEÆ.
Flowers ♂ ♀. *Perianth adherent. Carpels consolidated, several-seeded*	68. DIOSCOREACEÆ.
Flowers ♂ ♀. *Carpels several, quite consolidated. Placentæ axile. Flowers hexapetaloideous*	69. SMILACEÆ.
Flowers ⚥. *Carpels several, quite consolidated. Placentæ parietal. Flowers* 3-6-*petaloideous*	70. PHILESIACEÆ.
Flowers ⚥. *Carpels several, half consolidated. Placentæ axile. Flowers* 3-*petaloideous*	71. TRILLIACEÆ.
Flowers ⚥. *Carpels solitary, simple, many-seeded, with long-stalked anatropal seeds and a basal placenta*	72. ROXBURGHIACEÆ.

ORDER LXVII. TRIURIDACEÆ.—TAILWORTS.

Triuraceæ, *Gardner in Linn. Trans.* 19. 160. (1843).

DIAGNOSIS.—*Dictyogens with unisexual flowers, a free perianth, and numerous* 1-*seeded carpels.*

Little perennial herbs with a creeping rhizome. Leaves solitary, distant from the flowering scape, on long stalks, reticulated, entire, with sheathing scales at their base. Scape a little branched, scaly at the base. Flowers regular ♂ ♀, with 1-flowered bracteate stalks. Perianth corolline, 3- 6-parted, spreading, permanent, with the segments extended into long tails, at the base valvate in æstivation, the tails curled inwards before expansion. ♂ Stamens 3- 6 ?—; anthers turned outwards, with disunited cells inserted on a large central fleshy receptacle. ♀ Ovaries 00, sessile on the receptacle, packed close together, disunited; ovules solitary ? ; styles rather on one side, awl-shaped, or thickened at the apex and obliquely truncate. Fruit unknown.

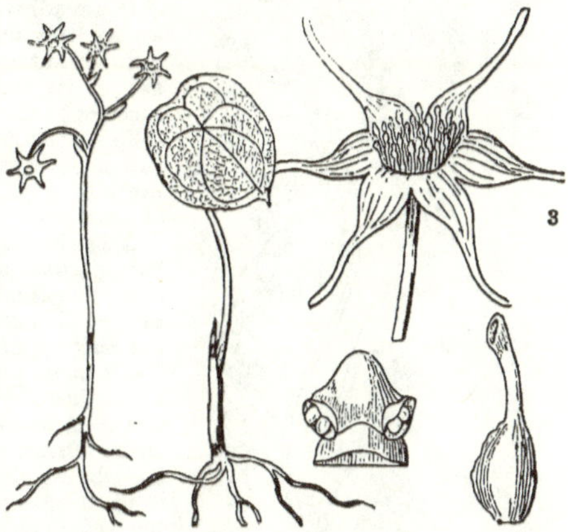

1 2 3 4 5

Fig. CXLIV.

It is in these words that Mr. Gardner describes the present Order, which appears to have escaped the attention of everybody except himself and Mr. Miers. He considers it to be allied to Menisperms in some degree, but in reality to belong to the present class, and he distinguishes the Order thus: "From Smilaceæ by the ovaries being free and numerous, not 3 and cohering; from Dioscoreaceæ by the same characters, and by their being superior, not inferior; from Roxburghiaceæ by their habit, diœcious flowers, and very numerous ovaries; while from all of them they are still further distinguished by their extrorse anthers." Till the structure is more exactly determined, it would be unprofitable to speculate further on the affinities of these curious little things.

The species inhabit the woods of Brazil, in moist shady places. Their uses, if any, are unknown.

GENERA.

Triuris, *Miers.* | Peltophyllum, *Gardner.*

NUMBERS. GEN. 2. Sp. 2.

POSITION. ——— TRIURIDACEÆ.—Trilliaceæ.
Menispermaceæ ?

Fig. CXLIV.—1. Peltophyllum luteum; 2. its leaf; 3. its flower; 5. its carpel; 4. anther of Triuris hyalina.—*Miers.*

ORDER LXVIII. DIOSCOREACEÆ.—YAMS.

Dioscoreæ, *R. Brown Prodr.* 294. (1810); *Agardh Aphor.* 169. (1823); *Ach. Rich. Nouv. Elem.* 434. (1828); *Endl. Gen.* lvii.—Dioscoreaceæ, *Ed. Pr.* cclvii.; *Meisner, p.* 404.

DIAGNOSIS.—*Dictyogens with unisexual flowers, an adherent perianth, and consolidated several-seeded carpels.*

Twining shrubs, with large tubers either above or below ground. Leaves alternate, occasionally opposite, with reticulated veins. Flowers small, spiked, with from 1 to 3 bracts each, ♂ ♀. Calyx and corolla confounded, herbaceous, adherent. ♂ Stamens 6, inserted into the base of the sepals and petals; anthers turned inwards, bursting longitudinally. ♀ Ovary adherent, 3-celled, with 1- or 2-seeded cells; style deeply trifid; stigmas undivided; ovules suspended, anatropal. Fruit leafy, compressed, with two of its cells sometimes abortive; occasionally succulent. Seeds two in each cell, or by abortion solitary, compressed, winged or wingless, or in the succulent species roundish; embryo small, near the hilum, lying in a large cavity of cartilaginous albumen.

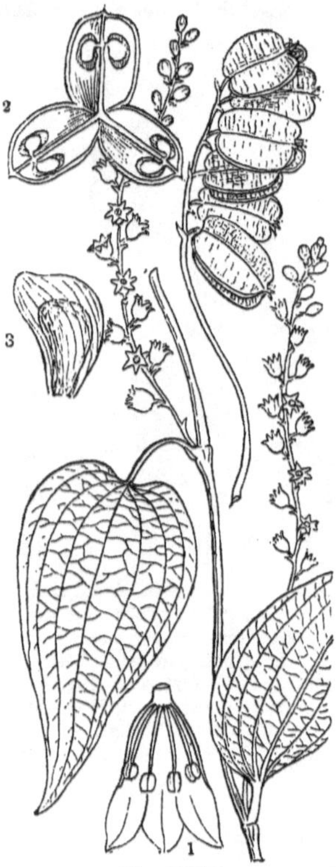

Fig. CXLV.

According to Brown this Order is separable from Sarsaparillas by the threefold character of inferior ovary, capsular fruit, and albumen having a large cavity. Tamus is, however, between the two Orders, agreeing with Sarsaparillas in its baccate, with Yams in its inferior fruit.—*Prodr.* 294. Endlicher says it has no obscure resemblance to Birthworts, and it is probably in this place that that singular Order finds one of its nearest relationships; in fact, the woody tissue of the common Yam arranges itself in the stem very much in the same manner as the wedges of Aristolochia.

Although the genera are few in number the species are numerous, and are found exclusively in tropical countries of either hemisphere, if Tamus be excluded, which is a native of Europe and the temperate parts of Asia.

An acrid principle exists in the plants of this Order, and when concentrated renders them dangerous. Tamus communis, for example, has a large fleshy root, so acrid as to have been formerly employed for stimulating plaisters, while the tubers of Dioscorea triphylla and dæmona have dreadfully nauseous qualities even after being carefully cooked. Nevertheless, this principle is more generally so much diffused as to be of no importance: hence the principal part of the species belonging to the genus Dioscorea produce what are called Yams, large, fleshy, farinaceous tubers, which form as important an article of food in tropical countries as the Potato in Europe. The young suckers of Tamus communis (the αμπελος μελαινα of Dioscorides, and οβρυὰ of the modern Greeks,) and also of T. cretica, are eaten in Greece like Asparagus, as we learn from Sibthorp; but Endlicher says, that unless they are well boiled (diligentius coquantur) they are powerfully purgative, and even emetic.

GENERA.

Tamus, *L.*	Rajania, *L.*	Dioscorea, L.	Oncus, *Lour.*
Tamnus, Juss.	*Janraja*, Plum.	*Ubium*, Rumph.	Podianthus, *Schnitzl.*
Testudinaria, *Salisb.*			

NUMBERS. GEN. 6. Sp. 110.

POSITION. ———— DIOSCOREACEÆ.—Smilaceæ.
Aristolochiaceæ.

Fig. CXLV.—Dioscorea adenocarpa; 1. a section of its male flower; 2. a seed vesse open; 3. a seed.

ORDER LXIX. SMILACEÆ.—SARSAPARILLAS.

Smilaceæ, *Ed. prior.* cclvi. (1836).

DIAGNOSIS.—*Dictyogens with bisexual or polygamous hexapetaloideous flowers, several consolidated carpels, and axile placentæ.*

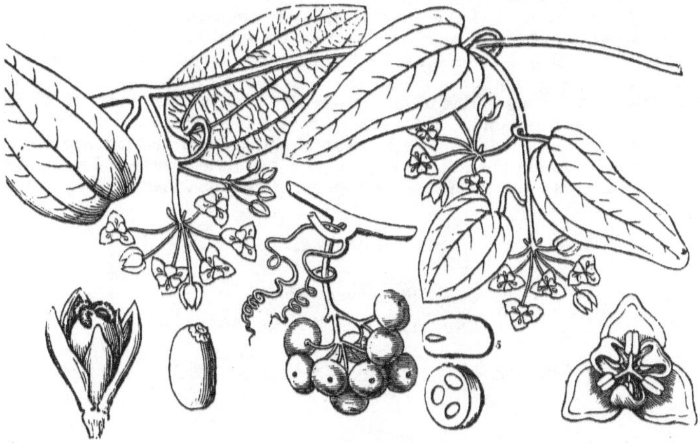

2 4 3 1

Fig. CXLVI.

2 1

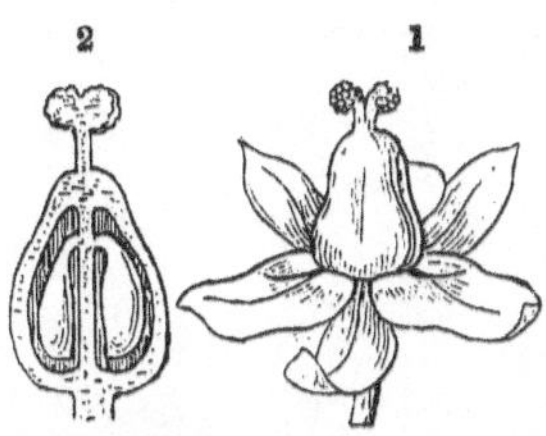

Fig. CXLVII.

Herbaceous plants or under-shrubs, with a tendency to climb, and sometimes having fleshy tubers. Stems scarcely woody. Leaves reticulated. Flowers ☿ or ♀ ☿ ♂. Calyx and corolla both alike, free, 6-parted. Stamens 6, inserted into the perianth near the base; seldom hypogynous. Ovary 3-celled, the cells 1- or many-seeded; style usually trifid; stigmas 3; ovules orthotropal. Fruit a roundish berry. Albumen between fleshy and cartilaginous; embryo very small, usually distant from the hilum.

From what has been already said in a previous page, it is obvious that the Order of Sarsaparillas, as I understand it, is very different from that of other botanists. Its nearest affinities are with Lilies on the one hand, from which its reticulated leaves and quasi-exogenous rhizomes distinguish it, and on the other with Parids, whose tripetaloideous flowers afford a clear mark of distinction.

The species are found in small quantities in most parts of the world, especially in the temperate and tropical parts of Asia and America.

The diuretic demulcent powers of Sarsaparilla are well known. This drug is the produce of many species of Smilax; as S. Purhampuy, a Peruvian species highly extolled by Ruiz; Smilax medica, which furnishes the Sarsaparilla of Vera Cruz; S. siphilitica, which, according to Dr. Pereira, yields the Lisbon or Brazilian sort; and S. officinalis, which the same acute pharmacologist suspects to be the Jamaica Sarsaparilla, the best in the English market. Dr. Hancock maintains that the only Sarsaparilla to be relied upon for medical use, is that of the Rio Negro. Martius says (*Flora Bras*) that S. papyracea (officinalis, Pöpp.) yields Brazilian or Lisbon Sarsaparilla, and that many others are of medical value. It also appears that a considerable quantity of Smilax glycyphylla, of excellent quality, has been imported from New Holland; the leaves of

Fig. CXLVI.—Smilax glycyphylla; 1. a male flower seen from above; 2. a female flower; 3. a transverse section of an ovary; 4. a seed; 5. a section of a seed, showing the embryo.
Fig. CXLVII.—1. Flower of Smilax brasiliensis; 2. perpendicular section of its ovary.

this plant are known under the name of Sweet Tea. Various Asiatic species, such as S. zeylanica, glabra, perfoliata, and leucophylla, are reputed to be little different in their qualities from the American species. Smilax excelsa and aspera are common substitutes in the south of Europe; according to Dr. Walsh (*Hort. Trans.* vi. 41), the root of both S. aspera and S. excelsa, which abound on the hills and in the woods on both sides of the Bosphorus, is used in decoction, like Sarsaparilla, for which it is sometimes substituted. Nees and Ebermaier say that it sometimes comes into the market under the name of Italian Sarsaparilla, but that it has little resemblance to the genuine drug. Smilax China has a large fleshy root, the decoction of which is supposed to have virtues equal to that of Sarsaparilla, in improving the health after the use of mercury. According to the Abbé Rochon, the Chinese often eat it instead of Rice, and it contributes to make them lusty.—*Ainslie*, 1. 70.

Roxburgh informs us that the large tuberous rhizomes of S. lanceæfolia are much used by the natives of India, and are not to be distinguished from China root. The juice of the fresh tuber is taken inwardly for the cure of rheumatic affections, and the refuse, after extracting the juice, is laid over the parts most painful. American China root is reported to belong to this plant; but several species seem to be mixed together by botanists under this name. Elliot says that he believes Smilax Pseudo-China to be the one generally preferred in medicine as an alterative, and that it forms the basis of many diet drinks among the unlicensed faculty of the United States. From the tubers, with maize, sassafras and molasses, the negroes of Carolina manufacture a very pleasant beer.

GENERA.

Smilax, *L.*
Ripogonum, *Forst.*

NUMBERS. GEN. 2. Sp. 120.

Menispermaceæ.
POSITION.—Dioscoreaceæ.—SMILACEÆ.—Trilliaceæ.
Liliaceæ.

ORDER LXX. PHILESIACEÆ.—PHILESIADS.

Philesieæ, *Ed. pr. under* cclviii. (1836); *Endl. Gen. p.* 157.—Philesiaceæ, *Endl. Ench. p.* 91. (1841).

DIAGNOSIS.—*Dictyogens with bisexual trihexapetaloideous flowers, consolidated carpels and parietal placentæ.*

Twining or upright shrubs, with ribbed or 1-nerved coriaceous, deciduous, reticulated, leaves. Flowers large, showy, ☿, solitary, scaly at the base; either tripetaloideous with the calyx coloured, membranous and short, or hexapetaloideous, with the sepals and petals equal and similar. Stamens 6, inserted into the base of the perianth; anthers linear, opening longitudinally. Ovary 1-celled, free, with 3 parietal placentæ; style long, club-shaped; stigmas 3; ovules 00, orthotropal, enveloped in mucilage. Fruit succulent. Nothing more known with certainty, except that the seeds of Lapageria are obovate, horny, and buried in pulp, according to the Flora Peruviana.

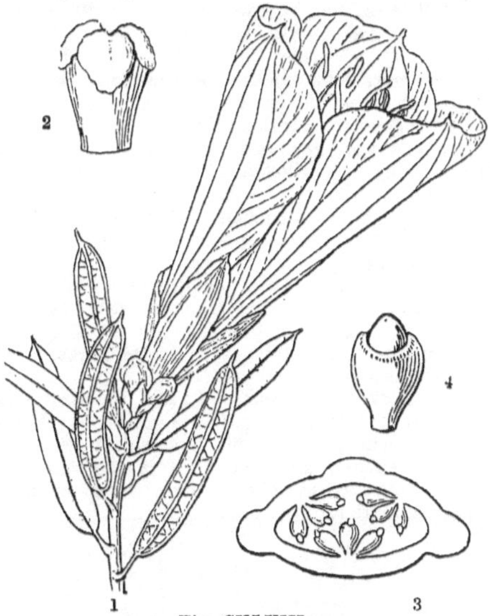

Fig. CXLVIII.

In the last edition of this work I regarded these plants as forming a part of the Roxburgh-worts; but the discovery by Mr. Griffith, that the carpel of these plants is quite simple, and a further consideration of the parietal placentæ, orthotropal ovules, and hexamerous flowers of the Philesiads, has decided me to separate them, in the belief that recruits may be hereafter found for them. Very little is known about them at present; no one has analysed their seeds, and it is even doubtful whether the two genera here brought together are so closely allied as is supposed. For my part, I only know the ovules of Philesia. Lapageria looks like a Smilax bearing the flowers of a Bomarea.

Chili, especially its southern provinces, produces all we as yet know of the species of these plants.

Lapageria rosea, a most beautiful twiner, is said to have sweet eatable berries, and a root like Sarsaparilla in quality.

GENERA.

Philesia, *Commers.*
Campia, Domb.
Lapageria, *Ruiz et Pav.*

NUMBERS. GEN. 2. Sp. 2.

POSITION.—Smilaceæ.—PHILESIACEÆ.—Roxburghiaceæ.
Amaryllidaceæ.

Fig. CXLVIII.—1. Philesia buxifolia; 2. stigma of Lapageria rosea; 3. its ovary cut across; 4. its ovule.

Order LXXI. TRILLIACEÆ.—Parids.

Trilliaceæ, *DC. Propr. Med.* 294. (1816); *A. Gray, Ann. Lyc. N. York*, 4. 106.—Parideæ, *Link Handb.* 1. 277. (1829); *Endl. Gen. p.* 153; *Meisn. Gen. p.* 403.

Diagnosis.—*Dictyogens with bisexual tripetaloideous flowers, half consolidated carpels and axile placentæ.*

Simple-stemmed herbaceous plants with tubers or rhizomes, and verticillate membranous netted leaves. Flowers large, terminal, solitary, ☿. Sepals 3, herbaceous. Petals 3, much larger, coloured, or herbaceous. Sometimes one-fourth is added to their parts. Stamens 6-10; filaments subulate; anthers linear, with cells on their edges, and the connective extended beyond them. Ovary free, 3- 5-celled; styles as many, distinct; stigmas inconspicuous; ovules 00, in 2 rows, anatropal, ascending. Fruit succulent, 3- 5-celled. Seeds 00, with a leathery brownish skin; embryo minute, in fleshy albumen.

These plants have been generally included in Sarsaparillas, from which they differ somewhat as Spiderworts from Lilies.

They are found in thickets in the temperate parts of Europe, Asia, and North America.

Paris quadrifolia is reckoned a narcotic acrid poison. The root of Medeola virginica is emetic and diuretic. Trillium cernuum and sessile have rhizomes that are violently emetic, and their fruit is suspicious; the juice of the berries mixed with alum gives a blue colouring matter.

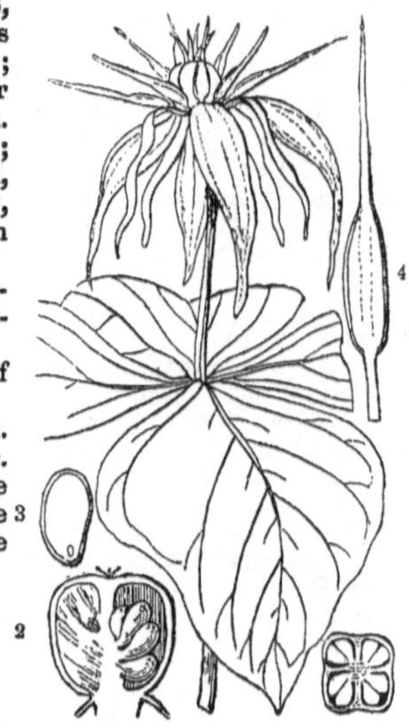

Fig. CXLIX.

GENERA.

Paris, *Linn.*
Demidovia, *Hoffm.*
Trillium, *Mill.*
  Phyllantherum, Rafin.
  Delostiyls, Rafin.
? Medeola, *Gronov.*
  Gyromia, Nutt.

Numbers. Gen. 4. Sp. 30.

Melanthaceæ.
Position.—Smilaceæ.—Trilliaceæ.—Roxburghiaceæ.
Commelynaceæ.

Fig. CXLIX.—Paris quadrifolia. 1. A transverse section of an ovary; 2. perpendicular section of the ripe fruit; 3. longitudinal section of a seed; 4. an anther.

ORDER LXXII. ROXBURGHIACEÆ.—ROXBURGHWORTS.

Roxburghiaceæ, *Wall. Plant. As. Rar.* 3. 50. (1832); *Lindl. Nixus*, 23. (1833); *Endl. Gen. p.* 157. *Meisner, Gen. p.* 402.; *Griffith in Calc. Journ. Nat. Hist. p.* 143.

DIAGNOSIS.—*Dictyogens with bisexual flowers, solitary simple many-seeded carpels, with long stalked anatropal seeds, and a basal placenta.*

Twining shrubs with tuberous roots? Leaves reticulated and coriaceous, with parallel secondary veins connecting several primary ribs. Flowers large and showy, solitary, fœtid. Perianth of 4 large petaloid divisions. Stamens 4, hypogynous; anthers adnate, opening inwards, pointed, with connectives projecting far beyond the cells, which separate from the latter as far as their bases. Ovary superior, 1-celled, with 2 polyspermous placentæ arising from the very base of the pericarp; style none; stigma somewhat pencil-shaped; ovules 00, anatropal. Pericarp 1-celled, 2-valved, with 2 clusters of seeds at the base. Seeds attached to long cords covered with loose hairs just below the seeds; embryo taper, in the axis of fleshy albumen, with the plumule lying within a slit.

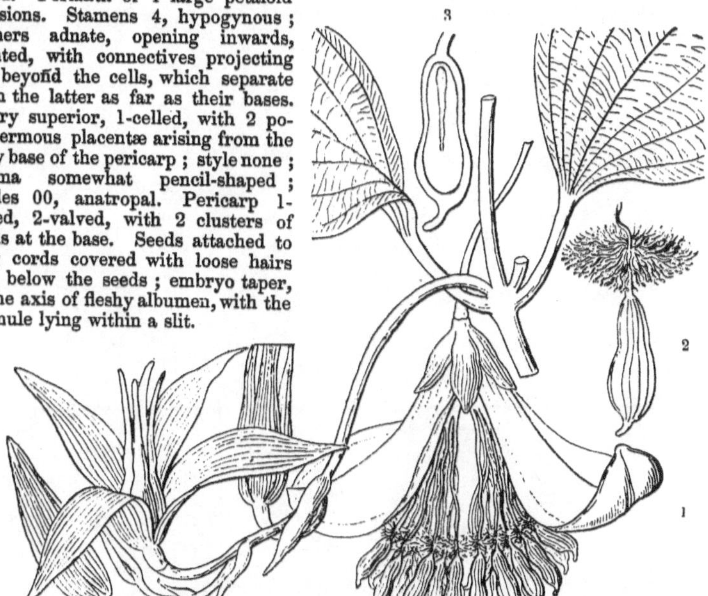

Fig. CL.

The affinity of this singular genus is not sufficiently marked to enable botanists to refer it to any known Natural Order: by Endlicher it is placed at the end of Sarsaparillas, and there can in fact be no doubt about its relation to the Parids, which that botanist includes in the Smilaceous Order. I, however, formerly regarded it as more nearly allied to Arads than to anything else, and Mr. Griffith has so far agreed with that opinion as to consider it certainly one of the class of which Arads are the type: in which he has apparently been influenced by the discovery of a slit on one side of the embryo. But this character has lost its value ever since the discovery by Adr. de Jussieu that a slit embryo is found very generally in Endogens; and a diclinous spadiceous inflorescence is indispensable to Arads; so that this view of the affinity of Roxburghia can hardly be maintained. It would rather appear to be the type of an Order for recruits to which we have still to look. In the meanwhile it may be looked upon as a tendency towards Arals on the part of Dictyogens. Roxburghia is said to have stems 100 fathoms long. Mr. Griffith regards the pistil as consisting, beyond all doubt, of one carpel only, as "is indicated by the obliquity of the ovary." Its double nature, as above

Fig. CL.—Roxburghia gloriosoides; 1. the fruit; 2. a seed; 3. the same divided longitudinally.

described, is the result of its maturation, and has no existence in the young state of the organ.

The plants of this small Order are natives of the hotter parts of India.

The roots of Roxburghia, previously prepared with limewater, are candied with sugar and taken with tea. Their flavour is insipid.—*Roxb.*

GENERA.

Roxburghia, *Dryand.*
Stemona, Lour.
Ubium, Rumph.

NUMBERS. GEN. 1. Sp. 4.

POSITION.—Smilaceæ.—ROXBURGHIACEÆ.—Trilliaceæ.
Araceæ.

CLASS VI.—GYMNOGENS.

Synorhizæ, *Rich. Anal. du Fr. Eng. ed.* (1819).—Phanerogames Gymnospermes, *Ad. Brongniart Veget. Foss.* 88. (1828).—Gymnospermæ, *Nixus Plantarum.* 21. (1833); *Ed. pr. p.* 310.

The plants comprehended in this class have nearly an equal relation to flowering and flowerless plants. With the former they agree in habit, in the presence of sexes, and in their vascular tissue being complete; with Ferns and Clubmosses, among the latter, some also accord in habit, in the peculiar gyrate vernation of the leaves of some Cycads, in their spiral vessels being imperfectly formed, and in the sexes being less complete than in other flowering plants; the females wanting a pericarpial covering, and receiving fertilisation directly through the foramen of the ovule, without the intervention of style or stigma, and the males sometimes consisting of leaves imperfectly contracted into an anther bearing a number of pollen-cases upon their surface. So great is the resemblance between Club-mosses and certain Conifers, that I know of no obvious external character, except size, by which they can be distinguished. Gymnogens are known from most other Vasculares by the vessels of their wood having large apparent perforations or disks. It is not, however, on this account to be understood that they differ in growth from other Exogens; on the contrary, they are essentially the same, deviating in no respect from the plan upon which Exogenous plants increase, but having a kind of tissue peculiar to themselves.

At this point of the vegetable kingdom there is a plain transition from the highest form of organization to the lowest. Gymnogens are essentially Exogens in all that appertains to their organs of vegetation; they have concentric zones in their wood, a vascular system in which spiral vessels are found, and a central pith; but they are analogous to reptiles in the animal kingdom, inasmuch as their ova are fertilized by direct contact with the male principle. The two most remarkable of the Orders are Conifers and Cycads. Of these, the former is connected with Clubmosses among Acrogens by means of the extinct genus Lepidodendron (see *Fossil Flora*, vol. 2. t. 98), and their branches are sometimes so similar to those of certain Lycopods themselves, as to leave no doubt of their relation. Compare, for instance, Lycopodium Phlegmaria, and Cunninghamia sinensis. Some Cycads have the gyrate vernation of the leaves of true Ferns, along with the inflorescence of Conifers; and their mode of forming their trunk, although essentially the same as that of Exogens, yet resembles the growth of Acrogens in lengthening by a terminal bud only. While, however, the class of Gymnogens is thus distinctly marked by the most important physiological peculiarities, it approaches the highest forms of vegetation by that portion of it which bears the name of Joint-firs, plants, with all the structure of their class, but with the manner of growth of Chloranths and Beefwoods, which will be found in a future part of this classification.

THE NATURAL ORDERS OF GYMNOGENS.

Stem simple, continuous. Leaves parallel-veined, pinnate. Scales of the cone antheriferous	73. CYCADEACEÆ.
Stem repeatedly branched, continuous. Leaves simple, acerose. Females in cones	74. PINACEÆ.
Stem repeatedly branched, continuous. Leaves simple, often fork-veined. Females solitary. Membrane next the nucleus inclosed. Anthers 2-*celled, opening longitudinally*	75. TAXACEÆ.
Stem repeatedly branched, jointed. Leaves simple, net-veined. Membrane next the nucleus tubular, protruded. Anthers 1-*celled, opening by pores*	76. GNETACEÆ.

ORDER LXXIII. CYCADEACEÆ.—CYCADS.

Cycadeæ, *Rich. in Pers. Synops.* 2. 630. (1807); *Brown Prodr.* 346. (1810); *Kunth in Humb. et Bonpl. Nov. Gen. et Sp.* 2. 1. (1817); *R. Brown in King's Voyage*, (1825); *Rich Mémoire*, 195. (1826); *Ad. Brongniart in Ann. des Sc.* 16. 589. (1829); *Meisner, Gen. p.* 353; *Miquel in Linnæa*, 17. 675. —Cycadeaceæ, *Ed. prior*, (1836); *Endl. Gen.* xxxviii.

DIAGNOSIS.—*Gymnogens with a simple continuous stem, parallel-veined pinnate leaves, and antheriferous cone-scales.*

Small trees or shrubs, sometimes resembling Palm trees in their aspect. The stems are either simple and cylindrical, or spheroidal, or dichotomously branched, and in all cases strongly marked with the lozenge-shaped scars of broad woody leaf-stalks. Internally the stem consists of a mass of pith, traversed by woody bundles, and rings of woody matter, sometimes irregularly disposed, sometimes collected into regular and numerous concentrical circles, always pierced by medullary plates. The wood consists of glandular woody tissue and spiral vessels. The leaves are pinnated, hard and woody, perennial, generally circinate when young, but in some instances flat; the leaflets have fine simple veins, and are placed somewhat obliquely on their petiole, from which they finally disarticulate. (Miquel regards these leaves as a sort of branch, "rami scissi."—*Linnæa*, *l.c.*) Flowers ♂ ♀, perfectly destitute of all trace of calyx and corolla. ♂ collected in terminal cones, consisting of scales covered over their lower side with anthers which are one-celled, often collected in twos and threes, and split longitudinally. Pollen hyaline,

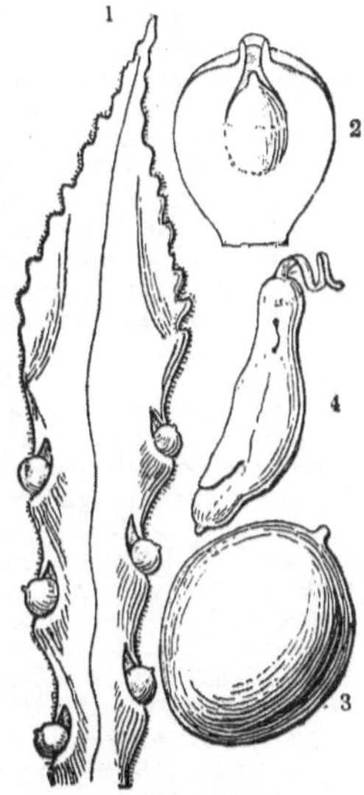

Fig. CLI.

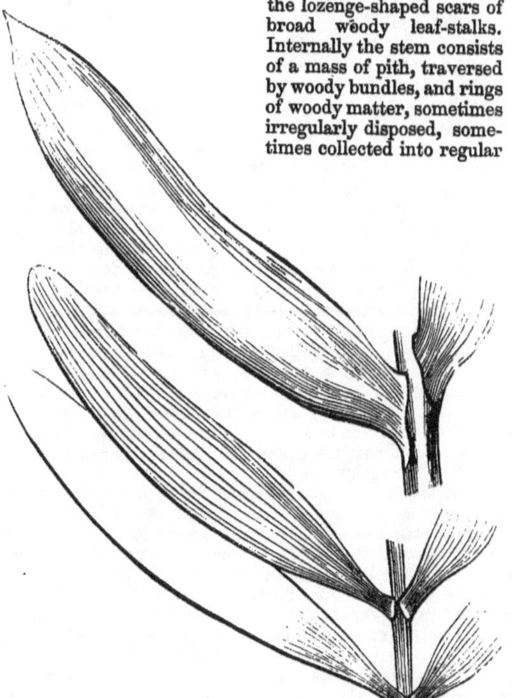
Fig. CLII.

Fig. CLI.—Cycas circinalis; 1. a portion of a female frond; 2. section of the naked ovule; 3. ripe fruit; 4. embryo.
Fig. CLII.—Leaves of Zamias.

angular, collected in masses. ♀ consisting of naked ovules, placed beneath peltate scales, or at the base of flat ones, or on the margins of contracted leaves. Seeds hard or spongy-coated nuts, with one or more embryos suspended by a long funiculus in a central cavity of large white fleshy or mealy albumen ; the cotyledons unequal, more or less connate ; radicle superior.

One of the botanists who originally noticed the plants that constitute this Order referred them to Ferns ; an opinion to which Linnæus, having first adopted the idea of Adanson that they were related to Palms, finally acceded. He was followed by others, until, after some suggestions by Ventenat that the genera Cycas and Zamia ought to form a particular tribe, the present Order was finally characterised by the late L. C. Richard in Persoon's *Synopsis*, in 1807, with the observation that it was intermediate between Ferns and Palms. The opinion of the affinity to Ferns seems to have been thus generally adopted in consequence of the striking resemblance on the part of most species in the mode of developing their leaves ; but the supposed relation to Palms was suggested rather by a vague notion of some general similarity, as, for instance, in their cylindrical trunks, than by any precise knowledge of the structure of Cycads. It is only within a few years that more accurate inquiries have determined the real nature of their affinities. In 1825, the publication of Brown's remarks upon the ovule, in which he demonstrated the similarity of conformation between the flowers of Cycads and Conifers, suggested new ideas of the affinities of both Orders ; and the determination, in 1829, by Adolphe Brongniart, of the resemblance between them in the structure of the vessels of their wood, while it decided the near relation of Conifers and Cycads, confirmed the proximity of the latter to Ferns, and showed the inaccuracy of the ideas formerly held of a close resemblance between the latter and Palms. With regard to the nature of the evidence by which their strict relation to Conifers is established, it may be observed, that they both are dicotyledonous in seed, both have naked ovules constructed in a similar remarkable manner, and borne in both cases not upon an ordinary axis of growth, but upon the margin or face of metamorphosed leaves ; that they have the same peculiar form of inflorescence, the same kind of male flowers, the same constant separation of sexes ; that there is a like imperfect formation of spiral vessels ; and that they both agree in having the vessels of their wood marked with circular disks ; a character which, if not confined to them, is uncommon elsewhere. The difference between the cylindrical simple stem of Cycads and the branched conical one of Conifers arises from the terminal bud only of the former developing, its axillary ones all being uniformly latent, unless called into life by some accidental circumstance, as in the case recorded in the *Horticultural Transactions*, 6. 501 ; while in Conifers a constant tendency to a rapid evolution of leaf-buds takes place in every axil. With regard to their foliage, on which the difference of aspect chiefly depends, the leaves of Firs are minute and undivided, while those of Cycads are very large and pinnated ; in Conifers there is a tendency to a higher development in the scales of the cones, while in Cycads there is a corresponding contraction, firstly in Cycas itself, and especially in Zamia, in which it takes place to exactly the same point as the evolution of Conifers. To this it may be added that the cones of Araucaria, among Firs, and of Dion among Cycads, are almost undistinguishable.

Natives of the tropics and temperate parts of America and Asia ; but not found in equinoctial Africa, although they exist at the Cape of Good Hope and in Madagascar. —*Brown Congo*, 464. Dion edule occurs in Mexico, where it seems to be common in some places. According to Mr. Bunbury, Zamias are among the forms of vegetation that characterise the eastern part of the colony of the Cape of Good Hope, especially the great tract of thicket extending along the Caffer frontier.—*Lond. Journ. Bot.* 2. 40. Upon the west coast of New Holland a Zamia, supposed to be Macrozamia spiralis, grows to the height of 30 feet. The undoubted remains of Cycads attest their having once formed a considerable portion of the vegetation of Great Britain !

All the species abound in a mucilaginous nauseous juice. With this, however, is mixed, in many instances, a very considerable quantity of starch, whence they are common articles of food in the countries where they grow. At the Cape of Good Hope various species of Encephalartos are called Cafferbread. The great seeds of Dion edule furnish a kind of Arrowroot in Mexico. A similar material of excellent quality is extracted in the Bahamas and other West India islands from Zamia pumila and other dwarf species. In Japan a kind of sago is procured from the cellular substance occupying the interior of the stem of Cycas revoluta. This is said by Thunberg to be held in the highest esteem ; soldiers are able to exist for a long time upon a very small quantity of it, and it is contrary to the laws of Japan to take the trees out of the country. The nuts are also eatable. So also is a sort of sago extracted from Cycas circinalis, whose fruit is eaten in the Moluccas, and a kind of flour of bad quality is procured from the kernels pounded in a mortar. It is supposed that the account given by Rheede of true sago being the produce of the plant is a mistake. This species also yields a clear transparent gum something like tragacanth, which when dried in the air, coagulates into a gummy mass which is applied to malignant ulcers, in which it excites suppuration in an incredibly short space of time.—*Blume.*

GENERA.

Cycas, *L.*
Dion, *Lindl.*
Zamia, *L.*
? *Arthrozamia*, Rchb.
Encephalartos, *Lehm.*
Macrozamia, *Miq.*
Dipsacozamia, *Lehm.*

NUMBERS. GEN. 6. Sp. 45.

POSITION.—Pinaceæ.—CYCADEACEÆ.—————— *Filicales.* / *Palmaceæ.*

Fig. CLIII.

Fig. CLIII.—Male plant of Cycas revoluta; *a* one of the scales viewed from above; the same presenting the lower face, where the anthers grow.

ORDER LXXIV. PINACEÆ.—CONIFERS.

Coniferæ, *Juss. Gen.* 411. (1789); *Brown in King's Voyage, Appendix*, (1825); *Rich. Monogr.* (1826). —Abietinæ et Cupressinæ, *Rich. l. c.* (1826); *Bartl. Ord. Nat.* 94 *et* 95. (1830); *Endl. Gen.* lxxvi. *and* lxxvii.; *Meisner, p.* 352.—Cunninghamiaceæ, *Siebold, Fl. Jap. tt.* 101, 102.—Conaceæ, *Lindl. Key, No.* 232. (1835).

DIAGNOSIS.—*Gymnogens with a repeatedly branched continuous stem, simple acerose leaves, and females in cones.*

These are noble trees or evergreen shrubs, with a branched trunk abounding in resin. Wood with the ligneous tissue marked with circular disks. Leaves linear, acerose or lanceolate, entire at the margins; sometimes fascicled in consequence of the non-development of the branch to which they belong; when fascicled, the primordial leaf to which they are then axillary is membranous, and enwraps them like a sheath. Flowers ♂ ♀, naked. ♂ monandrous or monadelphous; each floret consisting of a single stamen, or of a few united, collected in a deciduous amentum, about a common rachis; anthers 2-lobed or many-lobed, bursting longitudinally; often terminated by a crest, which is an unconverted portion of the scale out of which each stamen is formed; ♀. in cones. Ovary spread open, and having the appearance of a flat scale destitute of style or stigma, and arising from the axil of a membranous bract. Ovule naked; in pairs or several, on the face of the ovary, inverted, and consisting of 1 or 2 membranes open at the apex, together with a nucleus. Fruit consisting of a cone

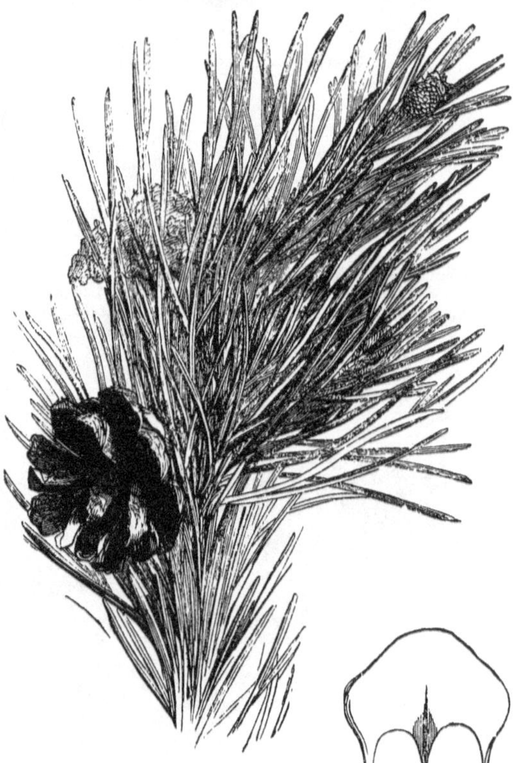

Fig. CLIV.

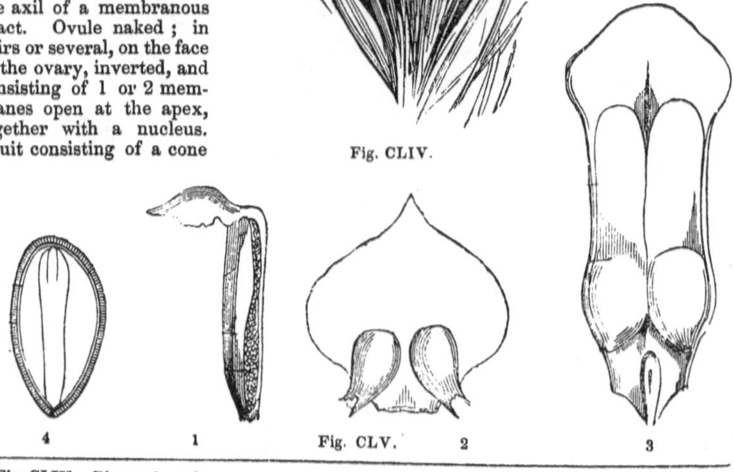

4 1 Fig. CLV. 2 3

Fig. CLIV.—Pinus sylvestris.
Fig. CLV.—1. side view of an anther; 2. carpellary scale and pair of inverted ovules; 3. inside of ripe scale and seeds; 4. section of the seed, minus the wing at its base.

formed of the scale-shaped ovaries, become enlarged and hardened, and occasionally of the bracts also, which are sometimes obliterated, and sometimes extend beyond the scales in the form of a lobed appendage. Seed with a hard crustaceous integument. Embryo in the midst of fleshy oily albumen, with 2 or many opposite cotyledons; the radicle next the apex of the seed, and having an organic connection with the albumen.

With the exception of Orchids, there is perhaps no Natural Order the structure of which remained so long and universally misunderstood as that of Conifers. This has arisen from the anomalous nature of their organisation, and from the investigations of botanists not having been conducted with that attention to logical precision which is now found to be indispensable. It is not expedient to enter upon an inquiry into the ideas that botanists have successively entertained upon the subject. Those who are desirous of informing themselves upon that point will find all they can desire in the Appendix to Captain King's *Voyage to New Holland*, and in Richard's *Mémoires sur les Conifères et les Cycadées.* It may, however, be useful to advert briefly to the principal theories which have met with advocates. These are, firstly, that the female flowers consist of a bilocular ovary having a style in the form of an external scale, an opinion held by Jussieu, Smith, and Lambert; secondly, that they have a minute cohering perianth, and an external additional envelope called the cupule: this view was taken by Schubert, Mirbel, and others; thirdly, that they have a monosepalous calyx cohering more or less with the ovary, contracted and often tubular at the apex, with a lobed, or glandular, or minute entire limb, an erect ovary, a single pendulous ovule, no style, and a minute sessile stigma: this explanation is that of Richard, published in his Memoir upon the subject in 1826. It appears, however, from the observations of Brown, that the female organ of Conifers is a naked ovule, the integuments of which have been mistaken for floral envelopes, and the apex of whose nucleus has been considered a stigma. About the accuracy of this view there is at this time no difference of opinion. These female organs, or naked ovules, originate from the larger scales of the cone towards their base, and occupy the same relative place in Conifers and in Zamia, a genus of Cycads. Now, as there cannot be any doubt of the perfect analogy that exists between the scales of the cone of Zamia and the fruit-bearing leaves of Cycas, the former differing from the latter only in each being reduced to 2 ovules, and to an undivided state; so there can be no doubt of the equally exact analogy between the scales of Conifers and Zamia, and therefore, the former would be called reduced leaves if the general character of the tribe was to produce a highly developed foliage; but as the foliage of Conifers is in a much more contracted state than the scales of their cones, the latter must be understood to be the leaves of Conifers in a more developed state than usual. That the scales of the cone really are metamorphosed leaves, is apparent not only from this reasoning, but from the following facts. They occupy the same position with respect to the bracts as the leaves do to their membranous sheaths; they surround the axis of growth as leaves do, and usually terminate it; but in some cases, as in the Larch, the axis sometimes elongates beyond them, and leaves them collected round it in the middle. In Araucaria they have absolutely the same structure as the ordinary leaves; and finally, they sometimes assume the common appearance of leaves, as is represented in Richard's Memoir, tab. 12., in the case of a monstrous Abies. The scales of the cones of Conifers and cone-bearing Cycads are therefore to these Orders, what carpellary leaves are to other plants. Schleiden does not, however, admit the scales of the cone of Abieteæ to be expanded carpellary leaves. He regards them as no other than the axillary buds of carpellary leaves; they, he says, cannot be the latter, because *folium in axilla folii* is without example in the whole vegetable world.—*Ann. Sc. N. S.* xii. 374. We would ask this ingenious anatomist what the fruit of Salix is but folium in axilla folii?

With regard to the male flowers, it is obvious that in the Larch, the Cedar of Lebanon, the Spruce, and the like, each anther is formed of a partially converted scale, analogous to the indurated carpellary scale of the females; and therefore, each amentum consists of a number of monandrous naked male flowers, collected about a common axis. Some botanists, however, consider each male catkin as a single monadelphous male flower, which is impossible. But in Araucaria, these cavities occupy one side only of an ordinary flat scale. In this genus, and such others as agree with it in structure, the anthers may be considered to consist of an uncertain number of lobes, and in this respect to recede from the usual structure of the male organs of plants: in Conifers, the anthers of which are normal, we have 2; in Juniperus, the like number; in Cunninghamia, but 3; in Agathis, 14; and in Araucaria, from 12 to 20. Brown remarks, what is certainly very remarkable, that in Cunninghamia the lobes of the anther agree in number, as well as insertion and direction, with the ovules.—*King's Appendix*, 32. The same author has noticed a very general tendency in some species of Pinus and Abies to produce several embryos in a seed, (*4th Report of Brit. Assoc.* 1835, *p.* 596:) where also are some curious remarks upon the origin of the embryo in such plants.

Conifers are broken up by many modern botanists into 2 Orders, Abieteæ and Cupresseæ, the distinctive characters of which are given below. But I regard the cones as the true mark of Conifers, and consequently, such groups as mere divisions of the same Natural Order. Recently, Mr. Bennett has given the weight of his authority in favour of the separation of the two groups, relying upon the pollen of Abieteæ having a curved oval form, dark granular extremities, and an intermediate band; while Cupresseæ have spheroidal grains whose outer coats are ruptured and thrown off, in consequence of the great capacity for absorbing moisture possessed by the mucous matter surrounding the inner coat. But however beautiful this distinction may be in theory, it is by no means clear that it is of value in practice. Indeed, Mr. Bennett admits, that "it is not always a safe criterion in systematic arrangement;" and a comparison of his own statements with those of Mohl and others does not increase confidence in its importance. I, however, admit two well-defined groups, one of which has the ovules inverted and the others erect.

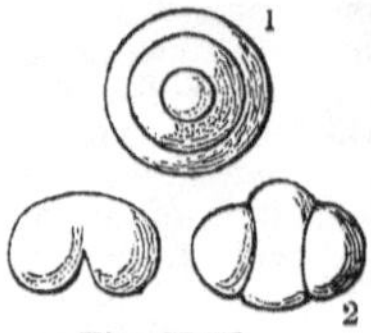

Fig. CLVI.

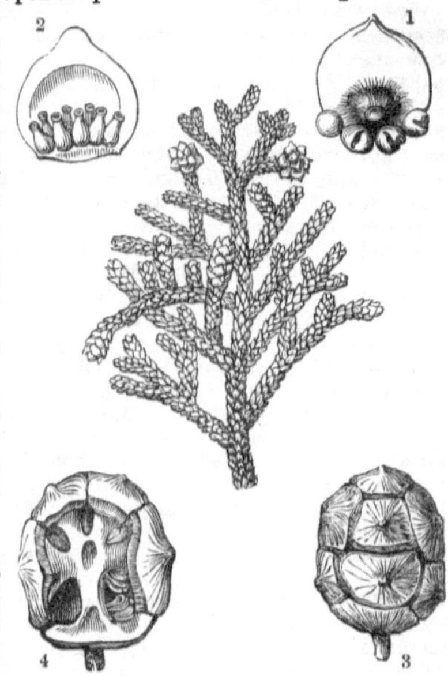

Fig. CLVII.

Natives of various parts of the world, from the perpetual snows and inclement climate of arctic America, to the hottest regions of the Indian Archipelago. The principal part of the Order is found in temperate countries; in Europe, Siberia, China, and the temperate parts of North America, the species are exceedingly abundant, and have an aspect very different from that of the southern hemisphere. In the former we have various species of Pines, the Larch, the Cedar, Spruce, and Juniper; the place of which is supplied in the latter by Araucarias, Podocarps, Dammars, Eutassas and Dacryds. A Callitris (quadrivalvis) is found on Atlas, and a true Araucaria (Bidwillii) in New Holland. In New Zealand the Dacryds are sometimes no bigger than Mosses.

No Order can be named of more universal importance to mankind than this, whether we view it with reference to its timber or its secretions. Gigantic in size, rapid in growth, noble in aspect, robust in constitution, these trees form a considerable proportion of woods or plantations in cultivated countries, and of forests where nature remains in temperate countries in a savage state. Their timber, in commerce, is known under the names of Deal, Fir, Pine, and Cedar, and is principally the wood of the Spruce, the Larch, the Scotch Fir, the Weymouth Pine, and the Virginian Cedar: but others are of at least equal, if not greater value. Pinus palustris is the Virginian Pine, so largely employed in the navy for masts. The Stone Pine, and Pinus halepensis (*πευκη*, *Diosc.*) are extensively used by the Greeks in ship-building. The gates of Constantinople, famous for having stood from the time of Constantine to that of Pope Eugene IV., a period of 1100 years, were of Cypress. The wood of Juniperus oxycedrus is supposed to have been that from which the images of their gods were carved by the Greeks; and finally, the Deodar wood of India is all but imperishable. The Norfolk Island Pine is an immense tree, known to botanists as Eutassa (Araucaria) excelsa; the Huon Pine of Tasmannia is Microcachrys tetragona; the Kawrie Tree of New Zealand, or Dammara australis, attains the height of 200 feet, and yields an invaluable light compact wood, free from knots, from which the finest masts in the navy are now prepared. But they are both surpassed by the stupendous Pines of north-west America, one of which, P. Lambertiana, is reported to attain the height of 230 feet, and the other, Abies Douglasii, to equal or even to exceed it. The latter is probably the most valuable of the whole for its timber. Their secretions consist of various kinds of resinous matter. Oil of turpentine, common and Burgundy pitch, are obtained from Pinus sylvestris; Hungarian balsam from Pinus

---

Fig. CLVI.—Pollen of, 1. Juniperus virginiana; 2. Pinus sylvestris.

Fig. CLVII.—Cupressus sempervirens; 1. a scale of a male cone with pollen; 2. a scale of a female cone with naked ovules; 3. a ripe cone; 4. the same with one of the scales removed.

Pumilio; a most fragrant resin from Araucaria brasiliensis; a hard brittle resin like copal from Dammara australis; Bourdeaux turpentine from P. Pinaster; Carpathian balsam from P. Pinea; Strasburg turpentine from Abies pectinata (P. Picea *L.*), our Silver Fir; Canadian balsam from Abies balsamea, or the Balm of Gilead Fir. The common Larch yields Venetian turpentine; a saccharine matter called Manna of Briançon exudes from the branches, and when the Larch forests in Russia take fire a gum issues from the trees during their combustion, which is termed Gummi Orenbergense; and which is wholly soluble in water like gum-arabic. Liquid storax is thought to be yielded by the Dammar Pine. Sandarach, a whitish yellow, brittle, inflammable, resinous substance, with an acrid aromatic taste, is said by Thomson to exude from Juniperus communis; but upon the authority of Brongniart and Schousboe, it is the tears of Callitris quadrivalvis. I have seen a plank two feet wide of this Sandarach tree, which is called the Arar Tree in Barbary. The wood is considered by the Turks indestructible, and they use it for the ceilings and floors of their mosques. The substance from which spruce beer is made is an extract of the branches of the Abies canadensis, or Hemlock Spruce, and of Abies nigra. Great tanning powers exist in the bark of the Larch; as great, it is said, as in the Oak. The stimulating diuretic powers of the Savin, Juniperus Sabina, are well known, and are partaken of in some degree by the common Juniper, the diuretic berries of which are an ingredient in flavouring gin; and by the Thuja occidentalis, and Taxodium distichum. Cypress was even once regarded febrifugal, and its oil as anthelmintic. The fetid oil of Juniperus oxycedrus is employed in veterinary practice. The large seeds of many are eatable. Those of the Stone Pine of Europe, Pinus Pinea (the πιτυς, *Diosc.*), Pinus Cembra, Pinus Lambertiana and Gerardiana, and Araucaria Dombeyi, are all eatable when fresh; and Mr. Bidwill found the natives of Moreton Bay feeding on the seeds of the Araucaria Bidwillii.

GENERA.

Suborder I. ABIETEÆ. —*Ovules inverted; pollen oval, curved.*

Pinus, *Linn.*
Abies, *Tournef.*
Picea, Link.
Larix, Tournef.
Cedrus, Mill.
Cunninghamia, *R. Br.*
Belis, Salisb.
Arthrotaxis, *Don.*
Microcachrys, *Hook, fil.*
Sciadopitys, *Zucc.*
Araucaria, *Juss.*
Dombeya, Lam.
Colymbea, Salisb.
Eutassa, *Salisb.*
Altingia, Loud.
Dammara, *Rumph.*
Agathis, Salisb.

Suborder II. CUPRESSEÆ.—*Ovules erect; pollen spheroidal.*

Juniperus, *Linn.*
Thuiæcarpus, *Trautv.*
Thuja, *Tournef.*
Biota, Don.
Platycladus, Spach.
Cyparissa, Don.
Cryptomeria, *Don.*
Thujopsis, *Zucc.*
Cupressus, *Tournef.*
Chamæcyparis, Spach.
Retinispora, *Zucc.*
Callitris, *Vent.*
Parolinia, *Endl.*
Pachylepis, Brongn.
Taxodium, *L. C. Rich.*
Schubertia, Mirb.
Condylocarpus, Salisb.
Chamæpeuce, *Zucc.*

NUMBERS. GEN. 20. Sp. 100.

POSITION.—Cycadeaceæ.—PINACEÆ.—Taxaceæ.

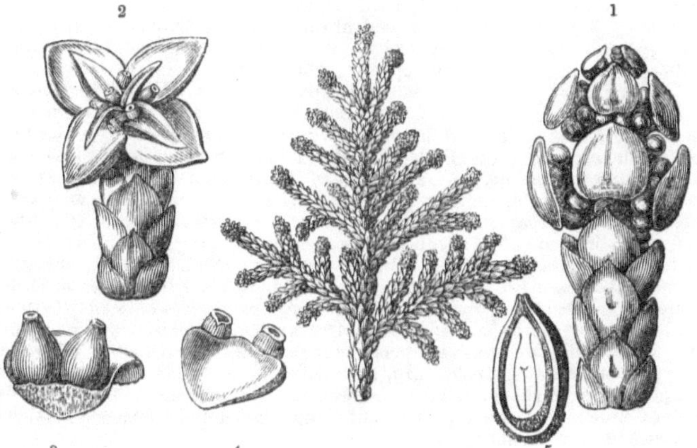

Fig. CLVIII.

Fig. CLVIII.—Thuja orientalis; 1. a magnified fragment of a branch bearing a cone of male flowers; 2. a portion of a female branch; 3, 4. scales with naked ovules; 5. a vertical section of a ripe seed.

ORDER LXXV. TAXACEÆ.—TAXADS.

Taxineæ, *Rich. Conif.* 124. (1826); *Bartl. Ord. Nat.* 95. (1830); *Martius Conspectus, No.* 58. (1835); *Endl. Gen.* lxxviii.; *Meisner*, p. 353.—Taxaceæ, *Ed. pr.* (1836).

DIAGNOSIS.—*Gymnogens with repeatedly branched continuous stems, simple leaves often fork-veined, solitary females, 2-celled anthers opening longitudinally, and the membrane next the nucleus inclosed.*

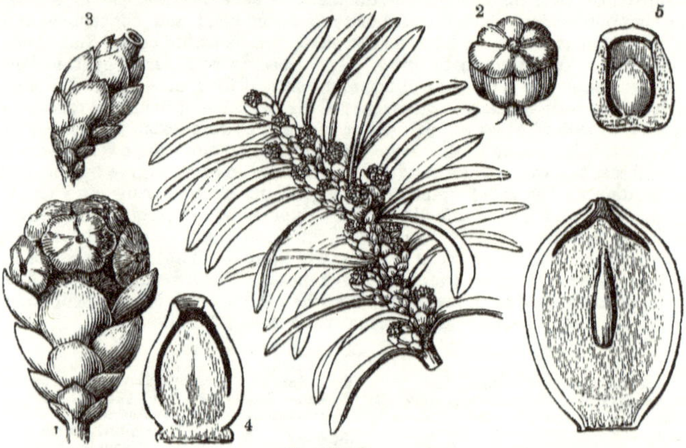

Fig. CLIX.

Trees or shrubs with continuous, unarticulated branches. Wood having the ligneous tissue marked with circular disks. Leaves usually narrow, rigid, entire and veinless, evergreen, alternate or distichous; sometimes dilated and lobed, and in those cases having forked veins of equal thickness. Flowers ♂ ♀, naked, but surrounded by imbricated bracts. ♂ Stamens several; filaments usually monadelphous; anthers combined or distinct, opening longitudinally. ♀ solitary. Ovules naked, the foramen at their apex, their outer skin becoming finally hard. Seed usually supported or surrounded by a succulent imperfect cup-shaped pericarp. Albumen fleshy. Embryo straight, dicotyledonous, either antitropal or orthotropal.

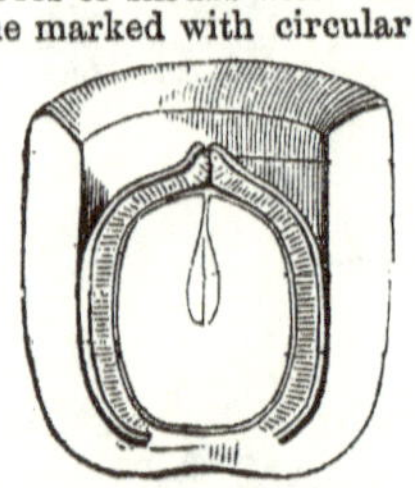

Fig. CLX.

Yews are separated from Conifers by their fruits not being collected in cones, each ovule growing singly, unprotected by hardened scales; so that this is a degree of organization yet lower than that of Conifers themselves. It is also to be observed, that in this Order the leaves do not always preserve the veinless needle-shaped state of Conifers, but expand and form veins, which are then forked and of uniform thickness, just as in Ferns. To me it appears that this deviation on the part of many genera from the Coniferous form of fruit, is a good practical distinction. Mr. Bennett, however, is of opinion that Taxads should not form a distinct Natural Order, but ought to be associated with Conifers; at least such I presume to be the bearing of his observations in Horsfield's *Plantæ Javanicæ*, p. 37. In the opinion of this excellent botanist, Taxus belongs to Cupresseæ, while Podocarpus and Dacrydium should be associated with Abieteæ, an opinion to which he seems to be led, in part at least, by considerations connected with the pollen of those plants. What these peculiarities of the pollen are, is explained at p. 228. But I see no reason why two kinds of pollen should not be comprehended under the Order of Taxads as well as under Conifers; and the importance of distinctions in the pollen of plants appears to me to be at least very doubtful.

Fig. CLIX.—Taxus baccata loaded with male flowers; 1. a male flower; 2. an anther; 3. a female flower; 4. a vertical section of an ovule; 5. of a ripe fruit; 6. of a ripe seed, showing the embryo.—N.B. 4. and 6. are the same part in youth and age; 5. is the ripe ovule, with an accessory cup.

Fig. CLX.—Perpendicular section of the ripe fruit of Taxus, together with the cup-shaped pericarp, which rises round it after the pollen has taken effect upon the ovule.

These plants occur in the milder climates of a great part of the world, and hence they are found in elevated situations within the tropics. The common Yew is the only species known in Europe; and it is common in the North of Asia. The majority belong to Asia or its dependencies. Dacrydium and Phyllocladus are abundant in New Zealand. Of Podocarp, the richest of any in species, three are found at the Cape of Good Hope.

Fig. CLXI.

Yews and their allies are resinous like Conifers, and often valuable for their timber, as evinced by the common Yew, which is unsurpassed for durability and elasticity. Podocarpus cupressina (Chomoro) is one of the best timber trees of Java. The Dacrydium taxifolium, or Kakaterro of New Zealand, acquires a height of 200 feet.—*Ed. Ph. Journ.* 13. 378; its branches may be manufactured into a beverage resembling in antiscorbutic qualities the well-known spruce beer. Podocarpus Totarra furnishes the most valuable timber in New Zealand; and it is said that the possession of the trees has been the cause of wars among the savage natives. The leaves of the common Yew are fetid, very poisonous, especially to horses and cows. (Rex Cativolus Taxo, cujus magna in Gallia Germaniaque copia est, se exanimavit. *Cæsar.*) The berries are not dangerous. The seeds are said to be unwholesome. On the authority of an Italian physician it is stated that Yew-leaves, when administered in small doses to man, have a power similar to that of Digitalis over the action of the heart and arteries, reducing the circulation, and if persisted in too long, or given in too large doses, as certainly fatal. Yew is, however, reported to have one decided advantage over Digitalis, by its effects not accumulating in the system; so that it is a much more manageable and more efficient remedy.—*Burnett.* The bark of Phyllocladus trichomanoides yields a red dye. The fruits of Salisburia, a tree of great beauty, now common in Europe, are about as large as Damsons, and both resinous and astringent; their kernels are thought by the Japanese to promote digestion. The nuts of Caryotaxus are very astringent, and are employed by the Japanese interpreters, "ad coercendam urinam," when they are likely to be detained for a long time in the Imperial Council Chamber.

GENERA.

Taxus, *L.*	Caryotaxus, *Zucc.*	*Thalamia*, Spreng.	Cephalotaxus, *Zucc.*
Podocarpus, *L'Her.*	Nageia, *Gärtn.*	*Robertia*, L. C. Rich.	Salisburia, *Smith.*
Dacrydium, *Sol.*	Phyllocladus, *L. C. Rich.*	*Brownetera*, L. C. Rich	*Ginko*, Kämpf
Torreya, *Arnott.*			

NUMBERS. GEN. 9. Sp. 50.

Polypodiaceæ.
POSITION.—Gnetaceæ.—TAXACEÆ.—Pinaceæ.

Fig. CLXI.—Phyllocladus rhomboidalis; 1. a spike of ♂; 2. an anther; 3. the inflorescence of the ♀, with a pair of flowers.

ORDER LXXVI. GNETACEÆ.—JOINT FIRS.

Gneteæ, *Blume, in Ann. Sc.* 2. *Ser.* 2. 105. (1834).—Gnetaceæ, *Lindl. in Bot. Reg.* 1686. (*July,* 1834); *Endl. Gen.* lxxix.; *Meisner, p.* 352.

DIAGNOSIS.—*Gymnogens with repeatedly branched jointed stems, simple net-veined leaves, 1-celled anthers opening by pores, and the membrane next the nucleus protruded.*

Small trees very much branched, or sarmentose shrubs, secreting watery, not resinous

1

Fig. CLXII.

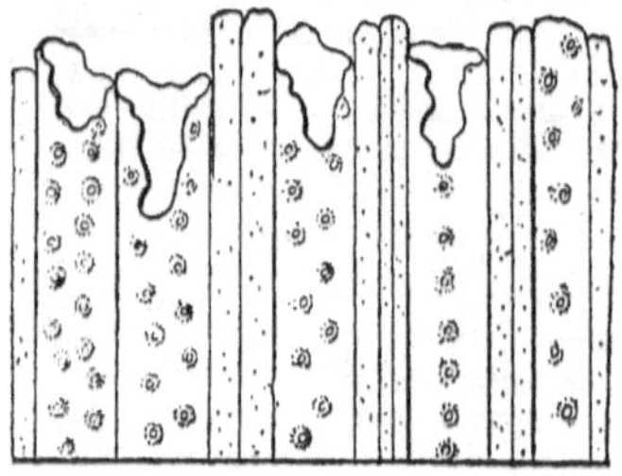

Fig. CLXIII.

matter, with opposite or clustered branches, and thickened separable articulations. Leaves opposite, entire, with anastomosing, reticulated veins; sometimes very minute and scale-shaped. Wood with the ligneous tissue marked with circular disks. Flowers ♂ ♀, arranged in catkins or heads, surrounded by opposite decussating scales which are connate at the base, or altogether consolidated into a horizontal ring. ♂ Calyx 1-leaved, transversely slit at the end, projecting from its bottom a monadelphous filament bearing 1-celled anthers, bursting longitudinally and centrally, so as to form a pore. Pollen (in Gnetum, simple, smooth, oblong, *Griffith*), in Ephedra ellipsoid, with 6 longitudinal furrows. ♀ altogether naked, or sheltered by a false calyx consisting of two scales, more or less combined, each of which surrounds one or two flowers. Ovary 0. Ovule pointed by a style-like process formed from a third membrane surrounding the nucleus. Seed drupaceous, before maturity pierced at the point and terminated by a style-shaped protruded process; finally pointless. Seed-coat thickish, either altogether leathery, or shelly, or fibrous internally, and succulent externally; in Gnetum lined by acicular woody tissue. Embryo dicotyledonous, in the middle of fleshy albumen; radicle superior.

Conifers and Cycads present features so peculiar that their separation from all other Orders is a point concerning which there can be no difference of opinion. It is indeed difficult to trace a plain transition from them to the other parts of the Vegetable Kingdom in which perfect sexes are present. There exist, however, a few plants, not very similar to each other in appearance, bearing the names Gnetum and Ephedra, in which we find precisely the structure and habit that would be wished for by a theorist searching for evidence to bring Gymnogens into communication with true Exogens; for one of them has all the appearance of a Chloranth, and the other of a Casuarina; and yet both retain the true peculiarities of Gymnogens. These are called Gnetaceæ, and may in English be termed Joint Firs, for they are closely allied to Conifers, but are distinctly known by their stems being jointed at every node. In these plants there is little tendency to form cones, and in the genus Gnetum the development of the ovule is so peculiar that botanists at one time, myself included, supposed that the real ovule was in truth an ovary pierced at the summit, for it consists of an exterior shell of considerable thickness and of a green colour; within which is a thinner envelope through which passes a tubular projection fringed at the point, and within these lies a nucleus, as is represented in the accompanying figure of the young ovule of Gnetum Brunonianum, copied from an unpublished drawing by Mr. Griffith. So that this sort of ovule has 3 distinct integuments, clear of the nucleus. It is to Mr. Griffith that I owe the knowledge of the true nature of these plants. In a most elaborate unpublished Memoir

Fig. CLXII.—Gnetum Gnemon: 1. a section of an ovule showing the three membranes, of which the innermost protrudes in the form of a stigma.

Fig. CLXIII.—A thin section of the wood of Gnetum Gnemon, highly magnified, after A. Brongniart.

on the structure of Gnetum, he shows that in reality the whole of the apparatus belongs to the ovulum. In that Memoir (dated Aug. 4, 1835), which unfortunately did not reach me till after the publication of the last edition of this work, there is the following description of the development of the parts of this extraordinary structure.

"At a period long before the exsertion of the anthers, the ovules, which lie upon the male flowers, are generally of an oblong form, and consist of a central cellular solid body, inclosed in two envelopes. The outermost of these is fibro-cellular and divided longitudinally on the upper face, or that nearest the axis; the fissure extending nearly to the base of the ovule.* The inner or second envelope is cellular, and is divided irregularly towards its apex.

"This envelope does not at this period entirely inclose the nucleus: the points of some of the laciniæ or divisions project occasionally beyond the apex of the outer envelope. The nucleus is an oval or oblong cellular body, rounded off at its apex, which is composed of lax cellular tissue.

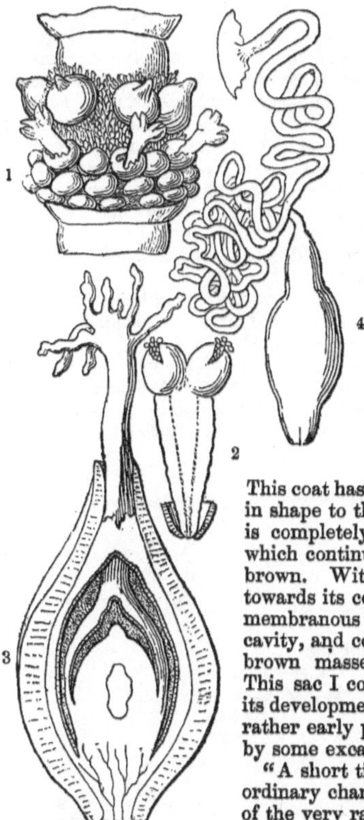

Fig. CLXIV.

"The next change consists in the commencement of the obliteration of the longitudinal fissure, existing along the posterior face of each outer envelope, and of an extension of the inner coat over the nucleus, the apex of which becomes more or less depressed: the centre of the depression, however, projecting in the form of a cone of a very slight elevation. At the time of flowering, or of the exsertion and dehiscence of the anthers, the fissure originally existing along the upper face of the outer coat has disappeared; with the exception of a small portion at the apex of the ovule, which remains unclosed throughout. The ovules are at this period in some species oblique. The inner envelope is generally entirely inclosed within the outer; the points of its laciniæ reach, however, to the opening existing in the apex of this latter, and occasionally, but by no means universally, project beyond it to a short distance. This coat has undergone scarcely any change, and corresponds in shape to the cavity of the outer envelope. The nucleus is completely covered by both integuments, and its apex, which continues of the same form, is occasionally tinged with brown. Within its substance, which is entirely cellular, and towards its centre, there exists a small cavity, lined with a membranous sac, attached apparently to the apex of the cavity, and containing a number of minute grumous-looking brown masses arranged without any obvious regularity. This sac I consider to be the amnios, with which it agrees in its development and subsequent disappearance; it exists at a rather early period, and is developed within a cavity formed by some excavating process.

"A short time after the fall of the male flowers an extraordinary change will be found to have occurred, consisting of the very rapid and apparently sudden development of a new membrano-cellular envelope between the second coat and the nucleus. This new formation, which I may term the additional coat, envelopes the nucleus pretty closely, and is continued upwards beyond the apex of the nucleus into a cylindrical tubular process; the mouth of the tube being laciniate or fimbriated. At the period now referred to, its apex barely projects beyond the outer envelope. During its development no particular change has taken place either in the original integuments or nucleus. At a somewhat later period, the ovules, except in the instance quoted in the note, hitherto concealed by the involucre, will be found exposed, and the outer coat to have become of a green colour.†

"* This division is perhaps similar to that which Brown states to take place in Dacrydium."

"† In one species, G. Brunonianum, the ovules are at an early period exposed, owing to the obsoleteness of the annulate involucre."

Fig. CLXIV.—Analysis of Gnetum, from sketches by Mr. Griffith; 1. a ring of ♂ and ♀ flowers; 2. a ♂; 3. a perpendicular section of a ♀, showing the 3 membranes overlying the nucleus; 4. an embryo extracted, with its long funicle.

The opening through its apex is distinct, and its direction vertical. The second envelope continues unchanged. The tubular prolongation of the additional or third envelope now projects through the openings in the original coats to a considerable distance. The mouth of the tube is also rather dilated, and the fringes of its margin spread out irregularly and to various extents. The whole of the tubular prolongation has become tinged with brown, in some cases approaching to black. It is to this stage or period that the descriptions of those authors who attribute a style and stigma to this genus apparently refer. Both Dr. Brown and Professor Lindley must likewise advert to this period when they state the nucleus to be surrounded with three envelopes."

There can be no doubt, then, that in reality Gnetum is as truly naked-seeded as Conifers themselves.

Independently of the singular organisation of its ovule, the genus Gnetum is remarkable for some other peculiarities. Its seed, which resembles a drupe, has within the outer fleshy integument, a layer of needle-like woody tissue of a very remarkable nature, freely separating when disturbed, and looking much like the hairs of Cowhage. The embryo, according to Mr. Griffith, is attached to an "enormously long tortuous and spirally but irregularly twisted cellular funiculus, the cells of which are much elongated and twisted. Its length varies, when moderately pulled out, from 3½ to 5 inches, the whole length of the seed being about an inch. This funicle, as well as the extremely similar one of Cycas, has the property of contracting when immersed in water."—*MSS.* p. 15. Although belonging to the same category as Conifers and Taxads, the Joint Firs are very distinctly separated from them ; for they have a calyx for the male flowers, and their anthers burst by pores, not longitudinally, to say nothing of the peculiarities of the ovule.

Natives of the temperate parts of Europe, Asia, and South America, and in the case of Gnetum, of the hottest parts of India and Guiana.

The interior of the pericarp of Gnetum urens is lined with stinging hairs ; the seeds are eaten ; the stem exudes a transparent gum, and when cut across yields an abundance of clear transparent tasteless water, which may be drank.—*Aubl.* In Amboyna the seeds of Gnetum Gnemon are eaten roasted, boiled, or fried, and the green leaves form a favourite vegetable in lieu of Spinach ; they are, however, very tasteless.—*Rumph.* The branches and flowers (Amenta Uvæ maritimæ, *Off.*) of the Asiatic Ephedras were formerly kept in the shops as styptics. The fruit is said to be mucilaginous, eatable, sub-acid, and slightly pungent.

GENERA.

Ephedra, *Linn.*
Gnetum, *Linn.*
Thoa, Aubl.
Gnemon, Rumph.
Ula, Rheede.

NUMBERS. GEN. 2. Sp. 15.

Chloranthac
POSITION.—Pinaceæ.—GNETACEÆ.—Taxaceæ
Casuarinaceæ.

CLASS VII. EXOGENS.

Dicotyledones, *Juss. Gen.* 70. (1789); *Desf. Mem. Inst.* 1. 478. (1796).—Exorhizeæ *and* Synorhizeæ, *Rich. Anal.* (1808).—Dicotyledoneæ *or* Exogenæ, *DC. Theor. p.* 209. (1813).—Phanerocotyledoneæ *or* Seminiferæ, *Agardh. Aph.* 74. (1821).—Anthophytæ *and* Carpophytæ, *Oken.*—Dichorgana, *Schultz.*—Phylloblastæ, *Reichenbach.*—Homogens, *Lindl. in Bot. Reg.* 1839.—Acramphibrya, *Endl. Gen. p.* 258. (1837).—Synechophyta, *Schleiden.*

BY common consent the plants to which botanists formerly gave the name of Dicotyledons, and which now bear that of Exogens, are recognised as the most completely formed of all the Vegetable Kingdom. In the more highly organised species they possess a degree of vitality unknown except among Gymnogens. A century or two terminates the life of an Endogenous tree, unless in a few rare cases; while many Exogens may have been the monarchs of their forests even at the commencement of the Christian era. This arises from their peculiar manner of growth, which insures a renovation of their vigour with each succeeding year; and it is in allusion to this circumstance that their name has been contrived.

Exogens, or outward growers, are so called because, as long as they continue to grow they add new wood to the outside of that formed in the previous year; in which respect they differ essentially from Endogens, whose wood is constructed by successive augmentations from the inside. All the trees of cold climates, and the principal part of those in hot latitudes, are exogenous. In an Exogen of ordinary structure the embryo consists of a cellular mass, in which there is usually no trace of woody or vascular tissue; but as soon as germination commences fine ligneous cords are seen proceeding from the cotyledons towards the radicle meeting in the centre of the embryo, and forming a thread-like axis for the root. As the parts grow the ligneous cords are increased in thickness and number, and having been introduced among the cellular mass of the embryo, are separated from each other by a portion of the cellular substance, which continues to augment both in length and breadth as the woody cords extend. By degrees the plumule or rudimentary stem becomes organised, and having lengthened a little, forms upon its surface one, two, or more true leaves, which gradually expand into thin plates of cellular substance traversed by ligneous cords or veins converging at the point of origin of the leaves. If at that time the interior of the young plant is again examined, it will be found that more ligneous cords have been added from the base of the new leaves down to the cotyledons, where they have formed a junction with the first wood, and have served to thicken the woody matter developed upon the first growth. Those ligneous cords which proceed from the base of the leaves do not unite in the centre of the new stem, there forming a solid axis, but pass down parallel with the outside, and leave a small space of cellular tissue in the middle; they themselves being collected into a hollow cylinder, and not uniting in the middle until they reach that point where the woody cords of the cotyledons meet in order to form the solid centre of the root. Subsequently the stem goes on lengthening and forming new leaves: from each leaf may again be traced a formation of woody matter disposed concentrically as before, and uniting with that previously formed: a cylinder of cellular substance being always left in the middle. The solid woody centre of the root proceeds in its growth in a corresponding ratio, lengthening as the stem lengthens, and increasing in diameter as the leaves unfold and new woody matter is produced. The result of this is, that when the young Exogen has arrived at the end of its first year's growth it has a root with a solid woody axis, and a stem with a hollow woody axis surrounding cellular tissue, the whole being covered in by a cellular integument. But as the woody cords are merely plunged into a cellular basis, the latter passes between them in a radiating manner, connecting the centre with the circumference by straight passages, often imperceptible to the naked eye, but always present. The annexed diagram illustrates this.

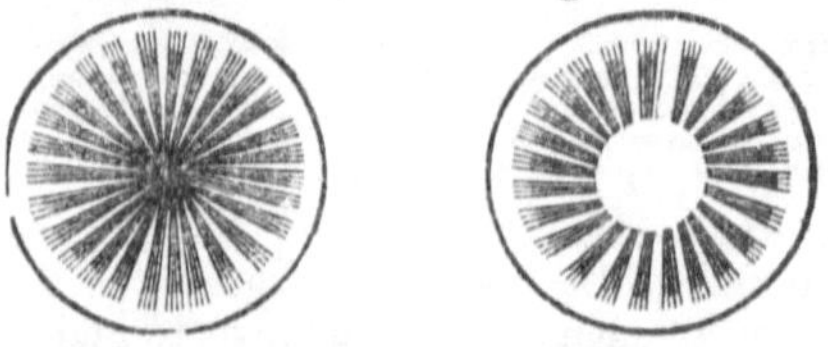

Root. Fig. CLXV. Stem.

Here we have the origin of pith in the central cellular tissue of the stem, of wood in the woody axis, of bark in the cellular integument, and of medullary processes in the radiating passages of cellular tissue connecting the centre with the circumference.

The woody axis is not, however, quite homogeneous at this time. That part which is near the centre contains vessels of different kinds, particularly dotted vessels (bothrenchyma); the part next the circumference is usually destitute of vessels, and consists of woody tissue exclusively: of these two parts that with the vessels belongs to the wood, properly so called, and serves as a mould on which future wood is added; the other belongs to the bark, separates under the form of liber, and in like manner serves as a mould within which future liber is disposed.

At the commencement of a second year's growth the liber separates spontaneously from the true wood; a viscid substance called cambium is secreted between them; and the stem again lengthens, forming new leaves over its surface. The ligneous cords in the leaves are prolonged into the stem, passing down among the cambium, and adhering in part to the wood and in part to the liber of the previous year, the former again having vessels intermingled with them, the latter having none. The cellular tissue that connected the wood and liber is softened by the cambium, and grows between them horizontally while they grow perpendicularly, extending to make room for them, and consequently interposed between the woody cords of which they each consist, forming in fact a new set of medullary processes terminating on the one hand in those of the first year's wood, and on the other in those of the first year's liber. This addition of new matter takes place equally in the stem and in the root, the latter extending and dividing at its points, and receiving the ends of the woody cords as they diverge from the main body. The following figure illustrates this, and shows, when compared with the last, what difference there is in the appearance of the stem of an Exogen one and several years old.

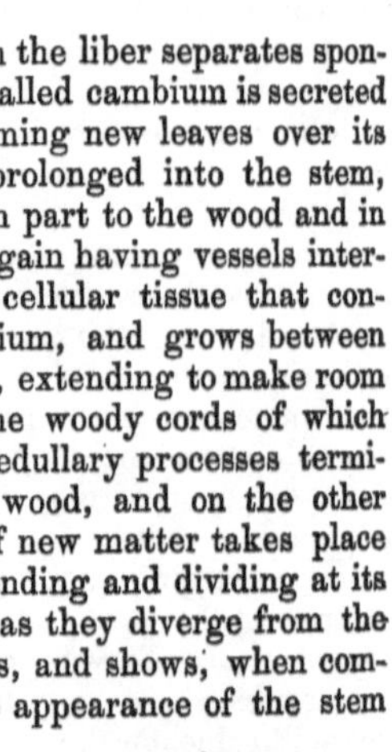

Fig. CLXVI.

And thus, year after year, the Exogen goes on, forming zone upon zone of wood, which is permanent, and zone within zone of bark which perishes at the outside, but is renovated at the inside, as the stem increases in diameter.

If this account is compared with what has already been stated concerning Endogens, it must be obvious that the stem of these two great classes is formed from the very beginning in an essentially different manner. Endogens have no cylindrical column of pith; their woody arcs are never collected into a cylinder, through the sides of which the cellular tissue passes

Fig. CLXV.—Ideal section of the root and stem of an Exogen one year old.
Fig. CLXVI.—Real section of the stem of an Exogen six years old.

in the form of medullary processes; and the woody matter of their bark, so to call their cortical integument, is not parallel with that of the wood and spontaneously separable from it. The only way in which the growth of the stem of Exogens corresponds with that of Endogens is that in both classes the woody matter is connected with the leaves; and in both a cellular substance is the foundation of the whole structure. Nevertheless, attempts have been made by some modern physiologists to identify the two, and to show that the one is very little different from the other.

It is not, however, to be supposed that the manner of growth in Exogens, is in all cases exactly what has been thus described as its normal condition. On the contrary, a great variety of modifications has been found to exist, dependent in part upon an excessive development of cellular matter, and in part upon the formation of angles, lobes, or sinuosities, upon the loss of concentric rings of wood for which a great homogeneity of structure is substituted, and upon the production of irregular zones of cellular matter resembling bark, between the zones of wood. Cases of this kind have attracted the attention of most modern botanists. Several have been noticed in my Introduction to Botany, in the *Penny Cyclopædia*, art. Exogens, and by Decaisne, Adrien de Jussieu, Schultz, Gaudichaud and Schleiden; but they have not been applied successfully to systematical purposes. In a sketch of a possible plan of extending the classes of plants at the expense of Exogens (*Bot. Reg.* 1839. *Misc.* p. 76), I have suggested the formation of a group to be called Homogens, to which it has been proposed to unite Birthworts, Nepenths, Lardizabalads, Menisperms, Peppers, and several other Orders. The character upon which reliance was placed was the remarkable nature of the wood of these plants, which never have more than one zone of woody matter, to whatever age they may have arrived. M. Decaisne has however shown (*Mémoire sur les Lardizabalées*), that although this peculiarity is extremely striking in some cases, as for example, in Aristolochia labiosa, yet that it is not constant in even the same Order, A. Clematitis having annual zones; and that in Menispermads, while there is a great departure from the ordinary structure of Exogens, except Aristolochia so far as regards the liber, the wood is regularly zoned in many instances, although the dotted vessels are wanting.

Nevertheless, although from the very imperfect state of information concerning the true structure of the stems of plants, I am unable to offer, for retaining this division, such reasons as would be satisfactory, yet I think it will be recognised hereafter, either wholly or in part; at least I am persuaded that the time will come when the internal structure of the stem will be far more extensively consulted than it now is, and be made the basis of good and important systematic divisions. Schultz preceded me in this attempt, in preparing his *Synorgana dichorganoidea* (*Naturliches System des Pflanzenreichs*, *p.* 319, 1832), to which he referred Piperaceæ, Saururaceæ, Chloranthaceæ, Nyctaginaceæ, Callitrichaceæ, Hippuridaceæ, Myriophyllaceæ, Amarantha-ceæ, Cycadeaceæ, Nymphæaceæ, Nelumbiaceæ, and Diphylleiaceæ, and his proposition, like mine, has fallen to the ground. But although the genera he collects under Diphylleiaceæ, namely, Diphylleia, Podophyllum, and others, are in no wise different from the ordinary state of herbaceous Exogens, yet it must be admitted that Hippurids and several of the others offer less resemblance to that plan of organisation. It is difficult to say whether Schleiden contemplates the possibility of any similar division; but it is worthy of notice that he, in his paper *On the Anatomico-physiological Differences in the Structure of Stems*, translated in the *Annals of Botany* (iv. 240) collects Peppers, Nyctaginaceæ, Amaranths, and Chenopods, by the common character of their stems having several sets of fibrovascular bundles, as in Endogens. It is however evident, as is stated in the proper places of this work, that the character proposed by Schleiden is no more universal than that which has been mentioned by myself; and therefore I think it more prudent to defer for the present an attempt at maintaining the Class of Homogens, and to leave it to be determined by future and very extended inquiries, whether such a group really exists in nature, what are its limits, and how they are to be defined. It seems probable that some such a group does exist, or at least that in the stems of Dicotyledonous plants there are modifications of structure of the very highest importance, to which attention has been hitherto insufficiently directed.

If Exogens are distinctly known from Endogens by their peculiar manner of growth and by the arrangement of their woody matter, they are not less clearly defined by external marks.

Their leaves have the veins ramifying from the midrib, or ribs if there are several, in so intricate a manner as to give the appearance of irregular network. Their veins never run parallel with each other without ramifications; for if, as sometimes happens, they appear to do so, it will be found that the appearance is confined to the principal veins or ribs, and that the secondary veins between them ramify in the usual way. The leaves are moreover in most cases articulated with the stem, leaving behind them a clean scar when they die, not rotting away and hanging upon the stem in the form of a ragged sheath, as is common in Endogens. Moreover, they are frequently furnished with stipules, an unusual circumstance in Endogens.

The flowers of Exogens are usually constructed upon a quinary type, that is to say, have five sepals, five petals, and five stamens, or some power of that number; now and then they vary to a type of four, or they exceed the number five; but we rarely find the ternary structure of Endogens present in them. If, as in Crowfoots, Berberids, Anonads, and other Orders, the sepals and petals follow a ternary type, the number three is lost in the stamens or the ovary. The Natural Order of Menispermads is the only one among Exogens in which the ternary type regularly pervades all the parts of the flower.

In their manner of growth they rarely resemble Endogens. The consequence of the ramification of the veins is to give their leaves a broad and rounded figure, the effect of which upon their general appearance is to produce the round-headed aspect that we recognise in all the trees naturally inhabiting this country. In no known instance does the stem grow by the development of a single terminal bud; so that we never find in this class the columnar aspect of Palm-trees, unless the genus Theophrasta be considered an exception.

The differences between Exogens and Endogens, thus strongly marked in the stem, leaves, and flowers, are connected with others in the embryo. In Exogens of the common kind this organ has two lobes, held together by a minute central body, the upper end of which, between the lobes, is the plumule or rudimentary stem, the lower the radicle or rudimentary root; the lobes themselves, or cotyledons, are rudimentary leaves. This structure is readily seen in a hazel-nut or a garden-bean; the deviations from it are few and unimportant as compared with those of Endogens. Three or a greater number of cotyledons may be present in a whorl, instead of two opposite to each other. Or one of the two cotyledons may be much smaller than the other, as in Trapa; or they may be deeply lobed, as in the garden-cress. But in all these cases the deviations are obviously reconcileable with the typical character of being dicotyledonous.

When the embryo of an Exogen germinates, the radicle simply lengthens at its point, without having to break through the coat of the embryo; on this account Exogens have been named exorhizal.

Hence the class of Exogens has five important, and, in some measure, independent characters, by which its limits are settled.

1. The wood is exogenous.
2. The veins of the leaves are netted.
3. The fructification is formed upon a quinary or quaternary type.
4. The embryo is dicotyledonous.
5. The germination is exorhizal.

Exogens have received other appellations in allusion to such characters; they are commonly called Dicotyledones, and Exorhizæ is another but less common appellation; moreover, they are the Phanerocotyledoneæ of Agardh, the Anthophytæ and Carpophytæ of Oken's school, the Dichorgana of Schultz, the Phylloblastæ of Reichenbach; not to mention names still more obscure.

In consequence of imperfect development, and the abortion or multiplication of parts, many deviations occur from the above characters. But, as in Endogens, so in these, such anomalies do not cause any real difficulty in distinguishing Exogens from other plants. Suppose the stem to be so slightly formed, as in Mossweeds (Podostemaceæ), or the aquatic Hippurids, as not to arrive at a state in which the exogenous arrangement is perceptible, we have the dicotyledonous embryo, and the typical number of the floral organs to guide us. Let the leaves appear as scales, as in Lathræa, Orobanche, and the like; still there is the embryo or again the floral proportions. If the fructification is absolutely ternary as in Menispermads, the organisation of the stem, leaves, and embryo reveals the true nature of such plants. Or if the embryo is undivided, as in Cuscuta, and at the same time the veins of the leaves deficient, and all this with an incomplete formation of woody matter, then the number of parts in the flower remains to prevent our falling into error. It is therefore always to be remembered, that the limits of this great class are not exclusively determined by one single character, but by a combination of five; a part of which may be occasionally exceptional or undiscoverable.

But while the class of Exogens is thus distinctly circumscribed, it is found to approach the limits of other classes at various points. It evidently touches Gymnogens by means of Beefwoods (Casuarinaceæ); Endogens are represented by Crowfoots, some of the species of Ranunculus having a striking resemblance to Alismads, and perhaps by Peppers, which seem to have a tendency to Arads. Menispermads may almost be mistaken for Sarsaparillas (Smilaceæ), and thus a connection is established with Dictyogens; Mossweeds (Podostemaceæ) may be regarded as analogous to Liverworts among Thallogens; it is not unreasonable to regard Hippurids as an exogenous form of Arrowgrasses, or Callitriche as the analogue of Lemna, and the whole Nymphal Alliance certainly comes very near to Hydrocharads.

The different methods of classifying Exogens have been considered in the introductory part of this work. That which is here adopted is founded on the following considerations.

The office of reproduction is, after that of sustaining life, the most essential in the economy of both plants and animals, and therefore the modifications which are found in the organs of reproduction may be expected to furnish the best characters for classification, after those of nutrition. The latter have been already employed as the foundations of the classes, as far as they appear susceptible of being so applied; the former, consisting of the stamens and pistil, have been little used for the classes, and appear to present as many modifications as are required for secondary divisions. That was the opinion of Linnæus, who adopted them in the construction of the Classes and Orders of his sexual system; but he mainly relied upon their number, which is a circumstance of little or no importance, and where that was done his classification proved useless; but in those parts of the system in which he made use of other circumstances, as in his Monadelphia, Diadelphia, Tetradynamia, Didynamia, Syngenesia, &c., his divisions ceased wholly or in part to be artificial, and although in some instances modified, still correspond essentially with the Natural Orders of modern botanists.

Nor did the importance of the stamens and pistil escape the keen eye of Jussieu, who relied upon them very much in the construction of his ingenious system. In the first place, he separated from all other Exogens those which have the stamens in one flower and the pistil in another, and he called them Diclinous, and by this process he brought together a collection of Natural Orders, corresponding with the Monœcious and Diœcious plants of Linnæus. No one can doubt that this was a judicious step, and upon the whole the plants collected in the diclinous division resemble each other more than they resemble anything else; but he excluded a large number of truly diclinous plants, which are scattered over other parts of his classification, and this has led to the idea that the distinction itself was a bad one, an opinion in which I formerly concurred; but a more careful examination of it since, and an extended acquaintance with the Vegetable Kingdom, has entirely convinced me that we have no available characters for breaking up Exogens into primary groups or sub-classes superior to those of separated and united sexes, that is, to diclinism and hermaphroditism. Not that they are without exceptions; to employ the forcible language of Jussieu himself: "*Ut in præcedenti serie nonnullas diclinis hermaphroditis commixtas plantis admittit exceptio, sic in diclinium ordines quædam irrepunt hermaphroditæ, consentiente aut jubente naturâ quæ stabiliores interdum eludit regulas, nonnunquam instabilis ipsa aut abstrusis legibus obtemperans.*"—*Gen. Pl.* 384. But if what are called polygamous plants, that is to say, such as have a rudimentary pistil in the male flowers, and rudimentary stamens in the female flowers, are regarded as being hermaphrodite, as they surely are, and the idea of a diclinous structure is limited to cases of a total separation of the stamens and pistil, these exceptions are reduced to a small and unimportant number, of no moment in a classification. For this reason, then, the diclinous sub-class of Jussieu is still preserved; increased by modern discoveries and improved by the expulsion of such plants as Piper, Gnetum, Ulmus, and others which belong to hermaphrodite Orders, or have other affinities than those suggested by Jussieu.

In this way Exogens are broken up into 2 groups, the one Diclinous and the other Hermaphrodite. The latter is divided by almost everybody into Polypetalous, Monopetalous, and Apetalous sub-classes; following the old systematists, who knew of little beyond external characters, and had small acquaintance with any plants except those of Europe. But all experience shows, what reason seems to indicate, that no great natural combinations can be effected by such distinctions. Exceptions to the constancy of such characters are endless; there is probably not one polypetalous Order that is not also apetalous, and many of them are even monopetalous, of which Rueworts, Houseleeks, Anonads, Leguminous plants, Milkworts, and many more, afford familiar examples. The apetalous Orders are occasionally polypetalous, as in many genera of Buckwheats and Daphnads. The monopetalous structure becomes polypetalous in all but a very few cases, even indeed in such natural Orders as the Primworts; and it even disappears altogether, as in Oliveworts and Primworts. Nor is it probable that characters derived from the calyx and corolla should be of the very highest value; for, in the first place, those organs are physiologically identical, their distinction having no real existence except in certain special instances; and, in the next place, the importance of them to the act of reproduction can hardly be considerable, when we find that plants are multiplied quite as well in their absence as in their presence, and even that, as in the Violet, some Leguminous plants, the common Apple, &c., which habitually produce them, seeds are matured as freely when they are partially away as when in a state of high development. For this reason, the calyx and corolla are here rejected as organs suited for distinguishing the primary groups, or the Sub-classes, of Exogens.

We are not, however, justified in assuming that the calyx and corolla are never of any high importance in plants; and, therefore, while they are objectionable as forming the basis of a classification *per se*, they are recognised as having a real value in connection with the stamens. If the stamens have no adhesion to either calyx or corolla, then it may be assumed that the latter organs may be dispensed with; and for this reason the first Sub-class of hermaphrodite Exogens is characterised by the stamens standing entirely clear of the floral envelopes, or being, in the language of Jussieu, Hypogynous. But if there is any adhesion between the stamens and either the calyx or corolla, it may equally be assumed that the one organ is in some way necessary to the other; for this reason the Perigynous character is admitted as a valid mark of a Sub-class; not, however, a slight and inappreciable adhesion, but a real and manifest union of the parts; and it is considered immaterial whether the stamens grow on the petals or the calyx, provided they grow on one of them.

Beyond this we have that further degree of adhesion, to which Jussieu gave the name of Epigynous: consisting of a union not only of the calyx or corolla to the stamens, but of all those organs to the sides of the ovary. This, in which it may be supposed that a higher degree of necessity for the incorporation of the floral organs exists than in the former case, is taken as the distinctive mark of a third Sub-class of hermaphrodite Exogens. So that the Sub-classes are established on the following grounds:—

Flowers absolutely unisexual	I. Diclinous.
Flowers hermaphrodite.	
Stamens not adhering to either calyx or corolla	II. Hypogynous.
Stamens adhering to either calyx or corolla	III. Perigynous.
Stamens, calyx, and corolla all adhering to the side of the ovary	IV. Epigynous.

This, it may be said, is essentially the old plan of Jussieu; but there is this material difference between the method now proposed and that of the great chief of the French school: that what he treated as a secondary character is made primary; while his primary distinction, of polypetalous, monopetalous, and apetalous structure, is treated quite as a subordinate consideration, as it surely deserves to be.

If the classification thus obtained be attentively studied, it will be found to offer many entirely new combinations, while others of universally recognised truth are not disturbed by it. Of these new combinations there are few to which any serious objection seems to apply, and it is believed that the larger part of them are more opposed to our prejudices than to truth. Not that I have the presumption to suppose that they will meet the universal approval of Botanists. What method of classification ever has or ever can? So long as there are many points of view from which a survey may be taken of the Vegetable Kingdom, so long will there be conflicting opinions as to the way in which the objects that meet the eye can best be grouped.

In former attempts at redistributing the natural Orders of Exogens, I had proposed to throw into one Sub-class all those in which the embryo is very small as compared with the albumen in which it is imbedded; and I still think that this peculiarity is of as much importance among plants as the being oviparous or viviparous among animals. But, although I do not at present see a reason for retracting my former opinion upon that subject, yet I do see that the time is hardly come for carrying out such a principle satisfactorily. And, therefore, instead of employing it for the character of a Sub-class, it has only been used as a means of limiting Alliances.

Although, from the complicated nature of the affinities of plants, no hope can be reasonably entertained of securing an unbroken line of transition from one end to the other of the series in which the various groups must necessarily be treated of, yet it will be found that the method here proposed offers very few considerable gaps in the chain of relationship. Commencing with the Amental Alliance, which seems to stand in near relation to the Joint-firs (Gnetaceæ) among Gymnosperms, the passage to the Urtical and Euphorbial is too plain to require explanation: of the latter the Quernal and Garryal may be regarded as epigynous forms,—the first without albumen, the second with an abundance of it. Nutmegs, in the Menispermal Alliance, then fit in; and the twining Menispermads may be taken as an anticipation of Cucurbitals, of which the Papayal Alliance is an offset, a little out of the direct line of succession.——Even to the latter, however, an analogue is found among Violals, in the form of Bixads and Samyds; thence Turnerads conduct us directly into the Cistal Alliance. At this point we quit the debateable ground of affinities, and, passing successively through Malvals, Sapindals, Guttiferals, we reach the Nymphal Alliance through Tutsans. Here, however, the chain is evidently broken, and probably the sequence is wrong. The Water-shields (Cabombaceæ), among Nymphals, pass directly into the Ranal Alliance by way of the Crowfoots, whence Poppyworts join Fumeworts in the Berberal Alliance. At this place Cyrillads appear to form a connecting link with Humiriads among Ericals, and the latter pass directly into the Rutal Alliance by the intervention of such plants as Correa. From Rutals the passage is easy to the Geranial, Silenal, and Chenopodal Alliances, which suddenly stop with the Peppers; this is, however, a doubtful case of affinity, although such a plant as Batis may seem to justify the approximation. At the point now reached the Perigynous Sub-class is penetrated by way of the Ficoidal Alliance, which might be almost united with the Chenopodal. Scleranths, among Ficoidals, seem to present a transition to Salvadorads in the Daphnal Alliance, of which again a part of the Rosal Alliance is almost a polypetalous form. From

Rosals to Saxifragals, and then by way of Brexia to Rhamnads, is but a step. At this point the Gentianal Alliance is entered by way of Hollyworts, and we quit it by moving from Gentianworts into the Solanal Alliance. The Cortusal, Echial, and Bignonial Alliances may be passed without an obstacle ; and thus we reach the end of the Perigynous Sub-class.——Gesnerworts, in the Bignonial Alliance, fit on to Goodeniads among the Campanals of the Epigynous Sub-class ; these join Myrtals, through Myrobalans on the one hand, and Napoleonworts on the other. From Myrtals we pass to the Cactal Alliance, which may be theoretically considered a parietal condition of the former, so near do the Onagrads of the former approach the Loasads of the latter group. This brings us to Barringtoniads and other Orders collected in the Grossal Alliance. The Cinchonals are entered by way of Bilberryworts, and quitted through the Stellate plants, which evidently touch Umbellifers in the Umbellal Alliance. At this point a passage is effected into the last Alliance, that of Asarals, by way of Witch Hazels and Sandalworts, till the whole line is finally closed by the Birthworts. These singular plants, with their ternary flowers, appear to have an incontestable relationship to Yams among Dictyogens, and thus the circle of affinities eventually returns into itself.

Each of the Sub-classes consists of Alliances which have also in many instances a strong lateral relation ; so that in order to obtain a clear idea of their mutual correspondence it is necessary to place them side by side as well as in succession. This is very obvious in the following instances :—

Diclinous.	*Hypogynous.*	*Perigynous.*	*Epigynous.*
Urticales, . . .	Chenopodales, . .	Ficoidales, . . .	———
Euphorbiales, . .	Malvales, ...	Daphnales, . . .	———
Menispermales, .	Ranales, ...	Saxifragales, . .	Myrtales.
Cucurbitales, . .	Violales, ...	Bignoniales, . . (*Crescentia.*)	Campanales.

This abundantly shows how hopeless it is to express the real affinities of plants by any other means than a map, or some such contrivance ; and that all sequences will of necessity be inadequate to explain in any considerable degree the position in which natural Orders really stand with relation to each other.

Alliances of Exogens.

Sub-Class 1. DICLINOUS EXOGENS.

Flowers ♂ ♀, without any customary tendency to ⚥.

Amentales. — *Flowers in catkins, achlamydeous or monochlamydeous ; carpels superior ; embryo small, with little or no albumen.*

Urticales. — *Flowers scattered, monochlamydeous ; carpel single, superior; embryo large, lying in a small quantity of albumen.*

Euphorbiales. — *Flowers scattered, monodichlamydeous ; carpels consolidated, superior; placentæ axile; embryo surrounded by abundant albumen. (Albumen occasionally absent.)*

Quernales. — *Flowers in catkins, monochlamydeous; carpels inferior ; embryo amygdaloid, without albumen.*

Garryales. — *Flowers monochlamydeous, sometimes amentaceous ; carpels inferior ; embryo minute, in a large quantity of albumen.*

Menispermales.—*Flowers monodichlamydeous; carpels superior, disunited; embryo surrounded by abundant albumen.*

Cucurbitales. — *Flowers monodichlamydeous; carpels inferior; placentæ parietal; embryo without albumen.*

Papayales. — *Flowers dichlamydeous; carpels superior, consolidated; placentæ parietal; embryo surrounded by abundant albumen.*

Sub-Class II. HYPOGYNOUS EXOGENS.

Flowers ⚥, or ♂ ⚥ ♀; stamens entirely free from the calyx and corolla.

Violales. — *Flowers monodichlamydeous; placentæ parietal or sutural; embryo straight, with little or no albumen.*

Cistales. — *Flowers monodichlamydeous; placentæ parietal or sutural; embryo curved or spiral, with little or no albumen.*

Malvales. — *Flowers monodichlamydeous; placentæ axile; calyx valvate in æstivation; corolla imbricated or twisted; stamens definite or* 00; *embryo with little or no albumen.*

Sapindales. — *Flowers monodichlamydeous, unsymmetrical; placentæ axile; calyx and corolla imbricated; stamens definite; embryo with little or no albumen.* (*Stamens rarely* 00.)

Guttiferales. — *Flowers monodichlamydeous; placentæ axile; calyx imbricated; corolla imbricated or twisted; stamens* 00; *embryo with little or no albumen.* (*Stamens sometimes definite in number.*)

Nymphales. — *Flowers dichlamydeous; placentæ axile or sutural; stamens* 00; *embryo on the outside of a very large quantity of mealy albumen.* (*A part have no albumen.*)

Ranales. — *Flowers monodichlamydeous; placentæ sutural or axile; stamens* 00; *embryo minute, inclosed in a large quantity of fleshy or horny albumen.*

Berberales. — *Flowers monodichlamydeous, unsymmetrical in the ovary; placentæ sutural, parietal, or axile; stamens definite; embryo inclosed in a large quantity of fleshy albumen.*

Ericales. — *Flowers dichlamydeous, symmetrical in the ovary; placentæ axile; stamens definite; embryo inclosed in a large quantity of fleshy albumen.* (*Stamens occasionally adherent to the corolla.*)

Rutales. — *Flowers monodichlamydeous, symmetrical; placentæ axile; calyx and corolla imbricated, if present; stamens definite; embryo with little or no albumen.* (*Occasionally* ♂ ♀.)

Geraniales. — *Flowers monodichlamydeous, symmetrical; placentæ axile; calyx imbricated; corolla twisted; stamens definite; embryo with little or no albumen.*

Silenales. — *Flowers monodichlamydeous; placenta free, central; embryo external, curved round a little mealy albumen; carpels more than one, completely combined into a compound fruit.* (*Some slightly perigynous, others* ♂ ♀.)

CHENOPODALES. — *Flowers monochlamydeous; placentæ free, central; embryo external, either curved round or applied to the surface of a little mealy or horny albumen; carpels solitary, or, if more than one, distinct. (Some slightly perigynous, others ♂ ♀.)*

PIPERALES. — *Flowers achlamydeous; embryo minute, on the outside of a large quantity of mealy albumen. (Occasionally ♂ ♀.)*

SUB-CLASS III. PERIGYNOUS EXOGENS.

Flowers ⚥, or ♂ ⚥ ♀; stamens growing to the side of either the calyx or corolla; ovary superior, or nearly so.

FICOIDALES. — *Flowers monodichlamydeous; placentæ central or axile; corolla, if present, polypetalous; embryo external, and curved round a small quantity of mealy albumen.*

DAPHNALES. — *Flowers monochlamydeous; carpel solitary; embryo amygdaloid, without albumen.*

ROSALES. — *Flowers monodichlamydeous; carpels more or less distinct; placentæ sutural; seeds definite; corolla, if present, polypetalous; embryo amygdaloid, with little or no albumen.*

SAXIFRAGALES. — *Flowers monodichlamydeous; carpels consolidated; placentæ sutural or axile; seeds* 00; *corolla, if present, polypetalous; embryo taper, with a long radicle, and a little or no albumen.*

RHAMNALES. — *Flowers monodichlamydeous; carpels consolidated; placentæ axile; fruit capsular, berried, or drupaceous; seeds definite; embryo amygdaloid, with little or no albumen.*

GENTIANALES. — *Flowers dichlamydeous, monopetalous; placentæ axile or parietal; embryo minute, or with the cotyledons much smaller than the radicle, lying in a large quantity of albumen.*

SOLANALES. — *Flowers dichlamydeous, monopetalous, symmetrical; placentæ axile; fruit* 2-3-*celled; embryo large, lying in a small quantity of albumen. (Occasionally achlamydeous or polypetalous.)*

CORTUSALES. — *Flowers dichlamydeous, monopetalous, symmetrical; placenta free, central; embryo lying among a large quantity of albumen. (Occasionally monochlamydeous, or polypetalous.)*

ECHIALES. — *Flowers dichlamydeous, monopetalous, symmetrical, or unsymmetrical; fruit nucamentaceous, consisting of several one-seeded nuts, or of clusters of them separate or separable; embryo large, with little or no albumen. (Very rarely hypogynous!)*

BIGNONIALES. — *Flowers dichlamydeous, monopetalous, unsymmetrical; fruit capsular or berried, with its carpels quite consolidated; placentæ axile, or parietal, or free central; embryo with little or no albumen.*

Sub-Class IV. **EPIGYNOUS EXOGENS.**

Flowers ☿ or ♂ ☿ ♀; stamens growing to the side of either the calyx or corolla; ovary inferior or nearly so.

Campanales. — *Flowers dichlamydeous, monopetalous; embryo with little or no albumen.*

Myrtales. — *Flowers dichlamydeous, polypetalous; placentæ axile; embryo with little or no albumen. (Occasionally monochlamydeous.)*

Cactales. — *Flowers dichlamydeous, polypetalous; placentæ parietal; embryo with little or no albumen.*

Grossales. — *Flowers dichlamydeous, polypetalous; seeds numerous, minute; embryo small, lying in a large quantity of albumen.*

Cinchonales. — *Flowers dichlamydeous, monopetalous; embryo minute, lying in a large quantity of albumen.*

Umbellales. — *Flowers dichlamydeous, polypetalous; seeds solitary, large; embryo small, lying in a large quantity of albumen.*

Asarales. — *Flowers monochlamydeous; embryo small, lying in a large quantity of albumen.*

⁂ The following artificial arrangement of the Alliances of Exogens will render it more easy to compare their characters.

Sub-class I.—DICLINOUS EXOGENS.

- *a.* Albumen abundant.
  - Ovary inferior . . . Garryales.
  - Ovary superior.
    - Carpels several, disunited. Menispermales.
    - Carpels consolidated.
      - Placentæ axile . . Euphorbiales.
      - Placentæ parietal . Papayales.
- *b.* Albumen wanting, or in moderate quantity.
  - Ovary inferior.
    - Placentæ axile . . Quernales.
    - Placentæ parietal . . Cucurbitales.
  - Ovary superior.
    - Flowers amentaceous . Amentales.
    - Flowers not amentaceous Urticales.

Sub-class II.—HYPOGYNOUS EXOGENS.

- *a.* Albumen abundant.
  - Flowers achlamydeous . Piperales.
  - Flowers monodichlamydeous.
    - Embryo external . . Nymphales.
    - Embryo internal.
      - Stamens 00 . . . Ranales.
      - Stamens definite.
        - Flowers unsymmetrical in the ovary } Berberales.
        - Flowers symmetrical in the ovary . } Ericales.
- *b.* Albumen wanting, or in small quantity.
  - Embryo external.
    - Carpels solitary, or distinct . . . . } Chenopodales.
    - Carpels consolidated . Silenales.
  - Embryo internal.
    - Placentæ axile or central.
      - Flowers unsymmetrical. Sapindales.
      - Flowers symmetrical.
        - Calyx valvate . . Malvales.
        - Calyx imbricated.
          - Stamens 00 . . Guttiferales.
          - Stamens definite.
            - Corolla twisted . Geraniales.
            - Corolla imbricated . . } Rutales.
    - Placentæ parietal or sutural.
      - Embryo curved or spiral . . . . } Cistales.
      - Embryo straight . . Violales.

Sub-class III.—PERIGYNOUS EXOGENS.

- *a.* Albumen abundant.
  - Placenta free, central . Cortusales.
  - Placentæ axile or parietal . Gentianales.
- *b.* Albumen wanting, or in small quantity.
  - Embryo external . . Ficoidales.
  - Embryo internal.
    - Flowers unsymmetrical.
      - Fruit capsular or baccate . . . } Bignoniales.
      - Fruit nucamentaceous. Echiales.
    - Flowers symmetrical.
      - Monochlamydeous.
        - Carpels solitary . Daphnales.
        - Carpels consolidated. Rhamnales.
      - Dichlamydeous.
        - Polypetalous.
          - Carpels consolidated. Seeds definite . . } Rhamnales.
          - Carpels apocarpous. Rosales.
          - Carpels consolidated. Seeds 00 } Saxifragales.
        - Monopetalous.
          - Capsular or baccate. . . } Solanales.
          - Nucamentaceous . Echiales.
          - Tricoccous . . Rhamnales.

Sub-class IV.—EPIGYNOUS EXOGENS.

- *a.* Albumen abundant.
  - Monochlamydeous . . Asarales.
  - Dichlamydeous, monopetalous . . } Cornales.
  - Dichlamydeous, polypetalous.
    - Seeds solitary . . . Umbellales.
    - Seeds 00 . . . Grossales.
- *b.* Albumen wanting, or in small quantity.
  - Placentæ axile.
    - Monopetalous . . . Campanales.
    - Polypetalous . . . Myrtales.
  - Placentæ parietal . . Cactales.

SUB-CLASS I. DICLINOUS EXOGENS.

The plants thus named never, or at least very rarely, have bisexual flowers, but consist of species in which the stamens constantly appear in one kind of flower, and the pistil in another. They appear to constitute the nearest approach that can be found to Gymnosperms, to which the whole Amental Alliance might in fact be referred, if the carpels would open and present the ovules naked to the action of the pollen. It is indeed impossible to overlook, on the one hand, the close resemblance which exists between the cones of an Abies among Conifers, and the female catkins of a Betula in the Order of Birchworts ; or, on the other, the vegetation of a Coniferous Ephedra, and an Amental Casuarina.

These and similar Orders must be regarded as the simplest forms of structure which Diclinous Exogens present, their condition reaching its lowest state in Hornworts (Ceratophyllaceæ). At this part of the Sub-class we have so entire a transition to the Chenopodal Alliance by means of certain Diclinous Chenopods, which form an exception to the general condition of their Order, as to make it clear that the Hypogynous Sub-class stands parallel with the unisexual Orders.

If we advance along the line of Diclinous Alliances, towards those forms whose organisation is the highest, such as Menispermads, Cucurbits, and Papayals, we shall find that all the others have also lateral affinities of a not less manifest description. Thus Spurges, Juglands, and Papayals pass into the Perigynous Rhamnads, Anacards, and Passion-flowers ; Spurges and Nutmegs stand on the limits of the Hypogynous Byttneriads and Anonads ; while the Epigynous Sandalworts and Loasads are closely approached by the Diclinous Helwingiads and Cucurbits.

These facts show, that, although in one direction such a series of affinities may be perceived, as that of which use has here been made, yet that it must be considered to be a very imperfect expression of the relationship borne by the Diclinous to the bisexual Alliances and Orders.

It would be possible to break up the Diclinous Alliances into Hypogynous, Perigynous, and Epigynous clusters ; and to some it may appear that such a distribution would have been more logical than what is now proposed; and perhaps that view is correct. But, upon the whole, it is doubtful whether the advantages of that plan would have been equal to some of its disadvantages.

ALLIANCE XVIII. *AMENTALES.*—THE AMENTAL ALLIANCE.

DIAGNOSIS.—*Flowers diclinous, in catkins, achlamydeous or monochlamydeous; carpels superior; embryo small, with little or no albumen.*

About the near alliance of the mass of genera here collected, no reasonable doubt can be entertained; and, in fact, they are associated in almost all systems of classification. Their strictly unisexual flowers, amentaceous inflorescence, and incomplete calyx, afford the most obvious marks of identification. To this, however, the Order of Oleasters offers an exception; those plants are almost universally referred to the vicinity of Daphnads (Thymelaceæ), among Perigynous Exogens, because the Elæagnus is taken as the type of their structure. I confess, however, that the latter genus seems to be far from offering a correct idea of this peculiar Order, which is much better represented by Hippophaë and Shepherdia. Indeed, it is open to question whether the genus Elæagnus itself would not fall more properly into the ranks of Daphnads. Upon that supposition, no doubt could be entertained of Oleasters finding their most natural station here.

Independently of the relations borne by Beefwoods (Casuarinaceæ) to the Joint-firs among Gymnogens, it is evident that other strong lateral affinities present themselves. These are more especially manifest between Liquidambars and the Planes of the Urtical Alliance, and between Galeworts and the Crowberries of the Euphorbial Alliance.

If we attempt to trace a passage from Order to Order in the Alliance itself, it will be observed that the winged fruit of Beefwoods is of the same nature as that of their successors the Birchworts; that the latter are imitated by the Liquidambars, which may be almost regarded as polyspermous Alders. If we suppose the two carpels of Liquidambars to lose their partition, and the seeds to be covered with hairs, Willowworts would be the result of the change. At this point the series is interrupted, for there is nothing at present known to connect either Galeworts or Oleasters with Willows; these Orders are rather to be regarded as a modification of Beefwoods and Birchworts, by the substitution of a fleshy for a membranous pericarp. Their true relation will be best expressed thus:

Joint-firs.
Beefwoods—Birchworts—Liquidambars—Willowworts.
Galeworts.
Oleasters.

NATURAL ORDERS OF AMENTALS.

Ovary 1-*celled.* *Ovule* 1 *or* 2, *ascending.* *Radicle superior* . . . 77. CASUARINACEÆ.
Ovary 2-*celled.* *Ovule* 1, *pendulous.* *Radicle superior* . . . . . 78. BETULACEÆ.
Ovary 2-*celled.* *Ovules* 00. *Seeds winged* . . . . 79. ALTINGIACEÆ.
Ovary 1-*celled.* *Ovules* 00. *Seeds cottony* . . . . 80. SALICACEÆ.
Ovary 1-*celled.* *Ovule* 1, *erect.* *Radicle superior.* . . . . 81. MYRICACEÆ.
Ovary 1-*celled.* *Ovule* 1, *ascending.* *Radicle inferior* . . . . . 82. ELÆAGNACEÆ.

ORDER LXXVII. CASUARINACEÆ.—BEEFWOODS.

Casuarineæ, *Mirb. in Ann. Mus.* 16. 451. (1810) ; *R. Brown in Flinders*, 2. 571. *Endlich. Gen.* lxxxvi. ; *Meisner*, p. 351.

DIAGNOSIS.—*Amental Exogens, with a* 1-*celled ovary*, 1 *or* 2 *ascending ovules, and a superior radicle.*

Branching weeping trees, with jointed shoots, the internodes of which are striated. Leaves 0 ; in their room short, toothed, ribbed sheaths. Flowers in spikes, ♂ ♀, each with a single bract. ♂ in spikes. Flowers whorled about the articulations of the jointed rachis. Bracts 2, membranous, right and left of a two-leaved calyx, the sepals of which stand fore and aft, and adhere at their points, and at the time of flowering are separated from their bases and carried up by the stamen in the form of a calyptra to the anther. Stamen 1 ; filament subulate ; anther erect, two-celled, with parallel contiguous cells opening by a longitudinal fissure. ♀: In dense heads. Rachis not jointed. Calyx 0. Ovary one-celled, with one obliquely ascending orthotropal ovule, or two standing side by side. Styles 2, united at base. Caryopsides winged, collected in a cone, hidden in thickened bracts, sessile. Seed erect, coated densely with spiral vessels, without albumen ; radicle superior.

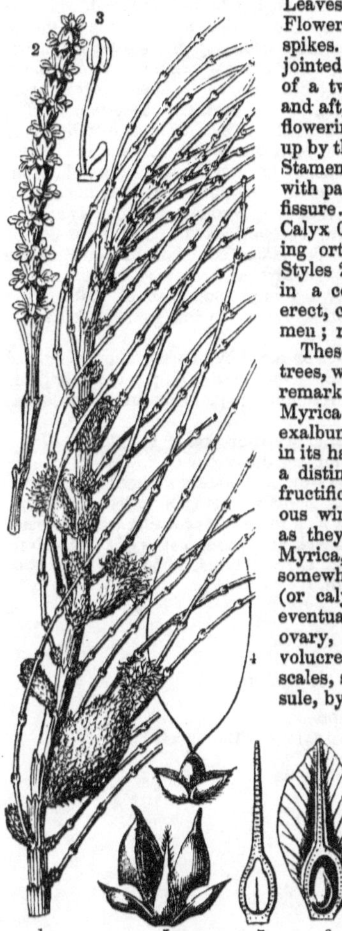

Fig. CLXVII.

These are jointed, leafless, tropical, or sub-tropical trees, with all the appearance of an Equisetum. Blume remarks, that "Casuarina is undoubtedly related to Myrica in its ovaries, its single erect ovule, and its exalbuminous inverted embryo ; but it differs so much in its habit, that it is better, with Mirbel, to consider it a distinct family, which differs from Galeworts in its fructification, especially in its achenia with membranous wings included between two lateral scales, which, as they grow up, are collected into a compact cone. Myrica, on the contrary, has distinct drupes, each somewhat immersed in a somewhat fleshy involucre (or calyx), which, although at first hypogynous, is eventually, after fecundation, extended beyond the ovary, with which it is conglutinated. Of such an involucre there is no trace in Casuarina, since the lateral scales, surrounding each achenium like a 2-valved capsule, by no means answer to the calycine involucre of Galeworts, but rather to those inferior bractlets which we observe at the base of the drupes in that Order."

The peculiar jointed leafless stems of these plants necessarily suggest a relationship to Ephedra among Gymnosperms ; and it is indeed probable, that Casuarina offers more distinctly than any other Exogen the passage from Angiospermous to Gymnospermous Orders. Endlicher describes the ovule as pendulous ; Blume, as erect. Neither are right. At least, the half-grown ovule is obliquely ascending from a little way up the side of the ovarian cavity, with a large foramen at the apex.

Fig. CLXVII.—1. Casuarina ♀ ; 2. ♂ ; 3. ♂ flower ; 4. ♀ flower ; 5. the ripe valves of the calyx, from which the fruit has been taken ; 6. a section of the half ripe ovary ; 7. a section of the fruit showing the seed and embryo.

Gœppert has examined the timber anatomically, (*Ann. Sc. N.*, 2 *ser.*, 18. 1.) He finds it to consist of woody bundles separated by medullary rays in the usual way, and divided by interrupted concentrical bands of cellular substance. There is no trace of any tendency to form annual growth; for the appearance of it, caused by the concentrical bands above mentioned, is illusory.

Brown, in the Appendix to *Flinders's Voyage*, has the following observations on the structure of this remarkable genus:—"In the male flowers of all the species of Casuarina, I find an envelope of four valves, as Labillardière has already observed in one species, which he has therefore named C. quadrivalvis. But as the two lateral valves of this envelope cover the others in the unexpanded state, and appear to belong to a distinct series, I am inclined to consider them as bracteæ. On this supposition, which, however, I do not advance with much confidence, the perianthum would consist merely of the anterior and posterior valves; and these, firmly cohering at their apices, are carried up by the anthera, as soon as the filament begins to be produced, while the lateral valves or bracteæ are persistent; it follows from it, also, that there is no visible perianthum in the female flower; and the remarkable economy of its lateral bracteæ may, perhaps, be considered as not only affording an additional argument in support of the view now taken of the nature of the parts, but also as in some degree again approximating Casuarina to Coniferæ, with which it was formerly associated. The outer coat of the seed or caryopsis of Casuarina consists of a very fine membrane, of which the terminal wing is entirely composed; between this membrane and the crustaceous integument of the seed, there exists a stratum of spiral vessels, which Labillardière, not having distinctly seen, has described as an 'integumentum arachnoideum;' and within the crustaceous integument there is a thin proper membrane, closely applied to the embryo, which the same author has entirely overlooked. The existence of spiral vessels, particularly in such quantity, and, as far as can be determined in the dried specimens, unaccompanied by other vessels, is a structure at least very unusual in the integuments of a seed or caryopsis, in which they are very seldom at all visible; and have never, I believe, been observed in such abundance as in this genus, in all whose species they are equally obvious."

These are for the most part Australasian trees or scrubby bushes, chiefly confined to the more temperate latitudes of that vast island. One species only, C. equisetifolia, is recorded as inhabiting the tropics of the Indian Archipelago; and another, C. nodiflora, is met with in New Caledonia.

Notwithstanding their want of leaves, these plants are remarkable for the excellence of their timber, which is hard, heavy, and resembling the colour of raw beef, whence their Colonial name. The heavy war clubs of the New Hollanders are said to have been fashioned out of it. The bark of C. equisetifolia is slightly astringent; that of C. muricata is said to be employed in India, in infusion, as a tonic. According to Backhouse, (*Visit to Australia, App.* xxxvii.), the young branches and young cones of C. quadrivalvis, or she-oak, when chewed, yield a pleasant acid, extremely useful to persons in want of water. Cattle are also exceedingly fond of them.

GENUS.

Casuarina, *L.*

NUMBERS. GEN. 1. Sp. 20. (*Endl.*)

Gnetaceæ.

POSITION.—Myricaceæ.—CASUARINACEÆ.—Betulaceæ.

ORDER LXXVIII. BETULACEÆ.—BIRCHWORTS.

Amentaceæ, *Juss. Gen.* 407. (1789) *in part.*—Betulineæ, *L. C. Richard in A. Richard, Elém. Bot., ed.* 4., 562. (1828.)—Betulaceæ, *Bartl. Ord. Nat.* 99. (1830); *Endl. Gen.* lxxxviii.; *Meisner*, p. 351.

DIAGNOSIS.—*Amental Exogens, with a 2-celled ovary, a solitary pendulous ovule, and a superior radicle.*

Trees or shrubs. Leaves alternate, simple, with the primary veins often running straight from the midrib to the margin; stipules deciduous. Flowers ♂ ♀, amentaceous, with small scales for their calyx, which are sometimes arranged in a whorl so as to simulate a real calyx (Alnus). ♂ Stamens distinct, opposite the calycine scales; anthers 2-celled. ♀ Ovary free, 2-celled; ovules solitary, pendulous, anatropal; style single, or none; stigmas 2. Fruit thin, indehiscent, by abortion 1-celled, combined with the scales into a sort of cone. Seeds pendulous; albumen none; embryo straight; cotyledons flat; radicle superior.

Fig. CLXVIII.

1

2

3

4

Fig. CLXIX.

The various kinds of Birch and Alder alone make up this Order, which is distinctly defined, among the Amental Alliance, by its fruit consisting of two carpels, in each of which there is but one pendulous ovule. If they had albumen, they might be regarded as Urticaceous plants with pendulous seeds and double carpels. Their nearest approach to other Orders is to Liquidambars, which have a little albumen, and numerous amphitropal ovules. In the male flowers of several species there is a distinct approach to the formation of a four-leaved membranous calyx. The leaves have the same venation as Mastworts, which, however, have an adherent calyx, and thus are distinguished by a well-marked character, independently of their cupule or involucre.

Inhabitants of the woods of Europe, Northern Asia, the Himalayas, and North America, and even making their appearance on the mountains of Peru and Columbia, and in the antarctic regions. They appear capable of existing up to the last limits between land and eternal snow.

The species are usually timber trees, with deciduous leaves; their bark is astringent, and sometimes employed as a febrifuge; but they are chiefly valued for their importance as ornaments of a landscape. Their wood is often light, and of inferior quality, but that of the Black Birch of North America is one of the hardest and most valuable we know. The bark of this species has a singular acrid taste; it contains a balsamic oil, and a peculiar resinous substance called Betuline, or Birch Camphor. The oil extracted from the Common Birch is employed in dressing Russia leather, and gives it its well known smell. From Betula papyracea the North American Indians strip off the thick tough bark, and manufacture it into boats, shoe-soles, and various domestic utensils. The sap of the Common Birch (B. alba) is obtained in the spring by tapping the trees, and, on account of

Fig. CLXVIII.—♂ and ♀ catkins of Betula alba.

Fig. CLXIX.—Betula lenta. 1. ♂ flowers; 2. ♀ flowers; 3. perpendicular section of a ripe fruit; 4. transverse section of it.

the quantity of sugar it contains, ferments, and is converted into an agreeable sparkling wine, much valued in the North of Europe; it has been found to contain free acetic acid and some saline matters; Birch wine has a popular reputation as a remedy for stone and gravel. Betula nigra and lenta furnish the North Americans with sugar of as good quality as that extracted from the Sugar Maple. The bark of the Common Alder is bitter and astringent, and has been employed for gargles, and with success in cases of ague. The leaves and female catkins are employed by dyers and tanners in some countries.

GENERA.

Betula, *L.*
Pterocaryon, Spach.
Apterocaryon, Spach.
Betulaster, Spach.
Alnus, *L.*
Alnaster, Spach.
Clethropsis, Spach.

NUMBERS. Gen. 2. Sp. 65.

POSITION.—Myricaceæ.—BETULACEÆ.—Altingiaceæ.
Pinaceæ.

ORDER LXXIX. ALTINGIACEÆ.—LIQUIDAMBARS.

Balsamaceæ, *ed. pr.*—Balsamifluæ, *Blume Fl. Javæ; Endl. Gen.* xcviii.; *Meisner, p.* 347.

DIAGNOSIS.—*Amental Exogens, with a 2-celled ovary and numerous winged seeds.*

Tall trees, yielding balsam. Leaves alternate, simple or lobed, with glandular serratures at the edges. Stipules deciduous. Female catkins on longer stalks than the males, and below them, the males surrounded by a deciduous 4-leaved involucre. Flowers ♂ ♀. Catkins unisexual, roundish. ♂: Anthers numerous, oblong, nearly sessile; with no calyx, but mixed with a few minute scales, and covering the common receptacle. ♀: Ovaries 2-celled, collected into a globe, each surrounded by a few scales: styles 2, long: ovules indefinite, attached to the dissepiment, amphitropal. Fruit a cone composed of hard connected scales, in the cavities of which lie obconical, 2-lobed, 2-celled capsules. Seeds numerous, or solitary by abortion, compressed, membranous, winged, peltate, attached to the middle of the dissepiments; embryo inverted, in the midst of fleshy albumen; cotyledons leafy; radicle short, superior.

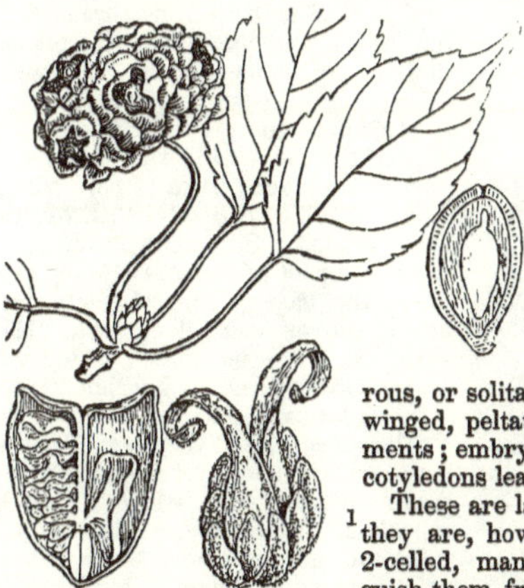

2 1 3

Fig. CLXX.

These are large trees with the appearance of Planes; they are, however, known from that Order by their 2-celled, many-seeded capsules, which equally distinguish them from all the Amental Alliance, in which it seems necessary to retain them, notwithstanding the presence of a small quantity of albumen in their seeds. They may be regarded as a connecting group, touching Planes on the one hand and Willowworts on the other, and standing intermediate between the latter Order and Birchworts. Their balsamic products have no parallel among similar plants, except in a slight degree in Willowworts.

The tropics of India, and the warmer parts of North America and the Levant, are the countries of this order.

A fragrant resin called Storax is yielded by several species of Liquidambar. That from North America, the produce of Liquidambar styraciflua, abounds in Benzoic acid. —(*Endl.*) The principal part, however, of what liquid Storax is used in this country is obtained from Trieste, and is probably collected from L. orientale, the Xylon Effendi, or Lord Wood, of the Cypriots. The bark of these plants is hot, bitter, and stomachic. What liquid Storax comes from the Malayan Archipelago is no doubt derived from Liquidambar Altingia, a lofty tree, 150 to 200 feet high, with a reddish brown, compact heavy wood, of very close grain, and extremely fragrant.

GENERA.

Liquidambar, *L.*
Altingia, Nor.

NUMBERS. GEN. 1. Sp. 3.

POSITION.—Salicaceæ.—ALTINGIACEÆ.—Betulaceæ.
Platanaceæ.

Fig. CLXX.—Liquidambar Altingia. 1. ♀ flower; 2. a section of the ripe fruit; 3. a section of a seed.—*Blume.*

ORDER LXXX. SALICACEÆ.—WILLOWWORTS.

Amentaceæ, *Juss. Gen.* 407. (1789) *in part.*—Salicineæ, *L. C. Richard in Ach. Richard. Elém. de la Bot., ed.* 4., 560 ; *Endl. Gen.* xcix. ; *Meisner, p.* 348.

DIAGNOSIS.—*Amental Exogens, with a 1-celled ovary, and numerous cottony seeds.*

Trees or shrubs. Leaves alternate, simple, with deliquescent primary veins, and frequently with glands on the edge or stalks ; stipules deciduous or persistent. Flowers ♂ ♀, naked or with a membranous cup-like calyx, amentaceous. ♂ : Stamens distinct or monadelphous ; anthers 2-celled. ♀ : Ovary superior, 1-celled ; ovules numerous, erect, anatropal, at the base of the cell, or adhering to the lower part of the sides ; style 1 or 0 ; stigmas 2 or 4. Fruit leathery, 1-celled, 2-valved, many-seeded. Seeds either adhering to the lower part of the axis of each valve or to the base of the cell, very small, covered over with long silky hairs springing from their base ; albumen 0 ; embryo erect ; radicle inferior.

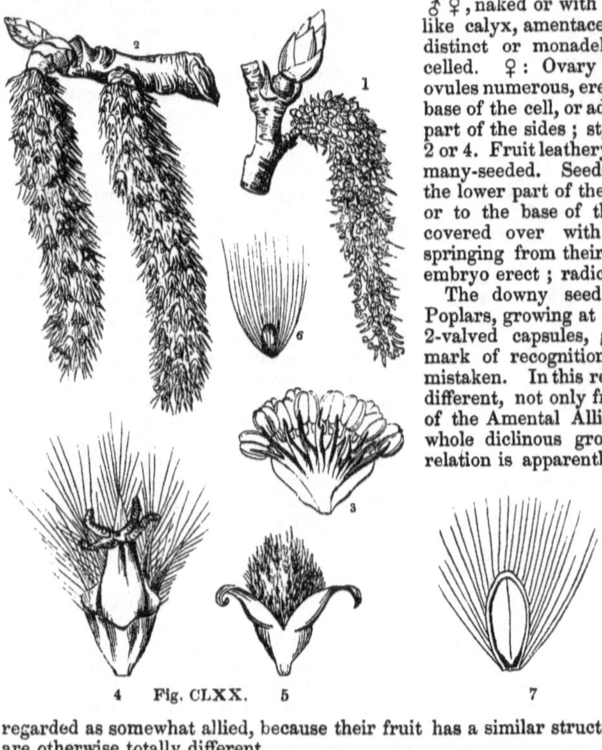

Fig. CLXX.

The downy seeds of Willows and Poplars, growing at the base of leathery 2-valved capsules, give such plants a mark of recognition which cannot be mistaken. In this respect they are quite different, not only from the remainder of the Amental Alliance, but from the whole diclinous group. Their nearest relation is apparently, on the one hand, with Liquidambars, which, like them, have a many-seeded double fruit, and on the other with the Birchworts, with which they agree in their naked male flowers and double fruit. Tamarisks have been regarded as somewhat allied, because their fruit has a similar structure ; but the plants are otherwise totally different.

Natives, generally, of the same localities as Birchworts, but extending even further to the north than those plants. The most northern woody plants that are known are the Willows, Salix arctica and polaris. The Order is found sparingly in Barbary, and there is a species of Willow even in Senegal.

They are valuable trees, either for their timber or for economical purposes, the Willow, the Sallow, and the Poplar being the representatives of the Order. The bark is usually astringent, tonic, and stomachic ; that of Populus tremuloides is known as a febrifuge in the United States ; of P. tremula and alba, in Europe. A crystallisable principle, called Salicine, has been obtained from Salix helix, which, according to Majendie, arrests the progress of a fever with the same power as sulphate of quinia. The best species to prepare it from are said to be S. fragilis, pentandra, Russelliana, vitellina, and purpurea (the bitterest of all) in Europe, and eriocephala, nigra, and conifera, in the United States. Populine, a substance resembling Mannite, has been obtained from the leaves and bark of some poplars. Poplar buds, especially those of P. nigra, balsamifera, and

---

Fig. CLXX. — Populus. 1. nigra ♂ ; 2. tremula ♀ ; 3. a male flower ; 4. a female flower ; 5. a ripe capsule ; 6. a seed ; 7. the same more magnified, and split to show the embryo.

candicans, are besmeared in winter with a resinous, balsamic, bitter, aromatic exudation, which, under the name of Tacamahac, is said to be diuretic and antiscorbutic ; they are also formed into an ointment for tumours, wounds, and burns, and are the basis of a balsam and tincture used for colic, &c. The sweet-scented male catkins of Salix ægyptiaca are employed in the preparation of a medicated water called Kalaf, which has a celebrity in the East for its cardiac and sudorific qualities. The same reputation once attached to our Salix alba and rosmarinifolia.

The use of Osiers for wicker-work, of Sallows for charcoal making, is well known. Excellent cricket-bats are made from the light wood of Salix alba ; arrows from the Aspen (Populus tremula); and various turneryware, and other even grained, soft implements from the Poplars, which are white-wooded. They have also been used for coarse flooring, but have no strength or durability.

GENERA.

Salix, *L.*
Populus, *L.*

NUMBERS. GEN. 2. Sp. 220.

POSITION.—Betulaceæ.—SALICACEÆ.—Altingiaceæ.

ORDER LXXXI. MYRICACEÆ.—GALEWORTS.

Myriceæ, *Rich. Anal. du Fr.* (1808); *Blume Fl. Javæ; Bartl. Ord. Nat.* 98. (1830); *Endlich. Gen.* lxxxvii; *Meisner, p.* 351.

DIAGNOSIS.—*Amental Exogens, with a* 1-*celled ovary and a single erect seed, with a superior radicle.*

Leafy shrubs, or small trees, covered with resinous glands and dots; leaves alternate, simple, with or without stipules. Flowers ♂ ♀, amentaceous, naked. ♂ Stamens 2 to 8, generally in the axil of a scale-like bract; anthers 2- or 4-celled, opening lengthwise. ♀ Ovary 1-celled, surrounded by several hypogynous scales; ovule solitary, erect, orthotropal; stigmas 2, subulate, or dilated and petaloid. Fruit drupaceous, often covered with waxy secretions, and, with the hypogynous scales of the ovary, become fleshy and adherent. Seed solitary, erect; embryo without albumen; cotyledons 2, plano-convex; radicle short, superior.

Fig. CLXXII.

The fragrant Gales are just half way between the Urtical and Amental Alliances. With Nettleworts they exactly agree, except in wanting albumen and having catkins; with the Amental Alliances they correspond in all essential particulars, but stand distinctly marked by the perfect simplicity of their fruit, in which they agree with Beefwoods only. In their solitary erect ovule, superior radicle, often dilated stigmas, and aromatic secretions, so uncommon among plants in this neighbourhood, they nearly approach Juglands, but are distinguished by their free ovary. Looking at plants not belonging to the Diclinous group, they may be compared with Peppers, because of their erect ovules, 1-celled ovary, and naked flowers; but the resemblance is distant.

Found in the temperate parts of North America, the tropics of South America, the Cape of Good Hope, and India. One species only inhabits the swamps of Europe.

Aromatic shrubs, or trees of considerable size. Comptonia asplenifolia possesses astringent and tonic properties, and is much used in the domestic medicine of the United States, in cases of diarrhœa. Benzoic and tannic acids, with a resinous matter, occur in its aromatic bark. Wax is obtained in great abundance from the berries of Myrica cerifera, and other species. The fruit of Myrica sapida is about as large as a cherry, and, according to Buchanan, is pleasantly acid and eatable in Nepal. Myrica Gale yields an ethereal oil of a yellow colour, feeble odour, and mild taste, which after a while becomes slightly warm. Its leaves were formerly used against the itch, and in Sweden as a substitute for hops in brewing. The root of Myrica cerifera is said to be emetic, or drastic in large doses.

GENERA.

Myrica, *L.* | *Gale*, Tourn. | Comptonia, *Banks.* | Clarisia, *R. et P.*

NUMBERS. GEN. 3. Sp. 20?

Empetraceæ.
Juglandaceæ.
POSITION.—Betulaceæ.—MYRICACEÆ.—Casuarinaceæ.
Urticaceæ.

Fig. CLXXII.—Comptonia asplenifolia; 1. ♀ of Myrica cerifera guarded by its scale; 2 and 3. the same divested of the scale and cut perpendicularly; 4. ♂ of ditto; 5. Fruit of Myrica Gale; 6. a perpendicular section; 7. a section of the seed.

ORDER LXXXII. ELÆAGNACEÆ.—OLEASTERS.

Elæagni, *Juss. Gen.* 75. (1789).—Elæagneæ, *Ach. Rich. Monogr.* (1823); *Bartl. Ord. Nat.* 113; *Endl. Gen.* cxi.; *Meisner, p.* 329.

DIAGNOSIS—*Amental Exogens, with a* 1-*celled ovary, a single ascending ovule, an inferior radicle, and flowers occasionally* ☿, *or scattered.*

Trees or shrubs, usually covered with leprous scurf. Leaves alternate, or opposite, entire, without stipules. Flowers axillary, often fragrant, in catkins or even panicles. Flowers ♂ ♀, rarely ☿. ♂ Flowers amentaceous, each in the axil of a scale-like bract. Sepals 2-4, sometimes united in a cup; stamens 3, 4, or 8, sessile; anthers 2-celled.—♀ and ☿ Calyx free, tubular, with a fleshy disk, which often closes it up, persistent; the limb entire, or 2-5-toothed. Ovary free, simple, 1-celled; ovule solitary, ascending, stalked, anatropal; stigma simple, subulate, glandular. Fruit crustaceous, inclosed within the calyx become succulent. Seed erect; embryo straight, surrounded by very thin fleshy albumen; radicle short, inferior; cotyledons fleshy.

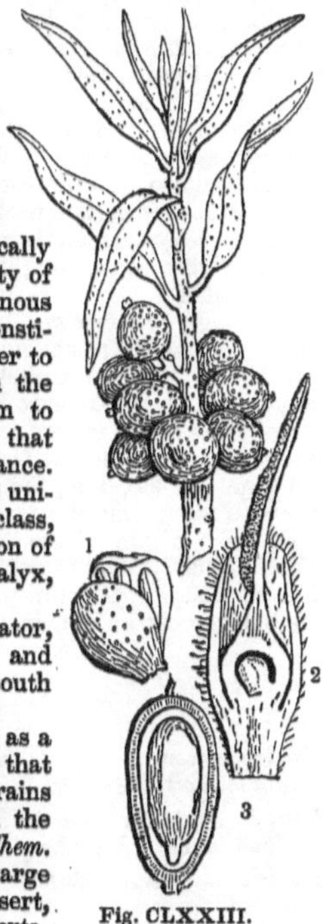

Fig. CLXXIII.

These plants are regarded by most botanists as being typically hermaphrodite, and hence they are referred to the vicinity of Daphnads; Jussieu himself excluded them from his Diclinous division. But when we consider that out of the genera constituting them, all except Elæagnus are ♂ ♀ it seems better to station them here, as one of the connecting links between the ♂ ♀ and ☿ races. Indeed, the Diclinous genera seem to approach closely to Galeworts, for the quantity of albumen that surrounds their embryo is too inconsiderable to be of importance. Supposing that the Order of Oleasters were not regarded as unisexual, it would then, no doubt, stand in the Perigynous Sub-class, where it would be known from Daphnads by the position of its ovule; and from Proteads, by the valvate irregular calyx, and dehiscent fruit of that Order.

The whole of the northern hemisphere, down to the equator, is occupied more or less by this family, from Canada and Japan to Guiana and Java; they are comparatively rare south of the line.

The fruit of Hippophaë rhamnoides is occasionally eaten as a sauce with fish. Professor Santagala has, however, found that it contains a fatty matter of a narcotic nature. Twelve grains given to a moderate-sized dog, in a few hours prostrated the strength of the animal in a most remarkable manner.—*Chem. Gaz.* 1844, 121. That of Elæagnus orientalis is almost as large as a Jujube, and is known in Persia as an article of the dessert, under the name of Zinzeyd; the drupes of E. arborea, conferta, and others, are eaten in Nepal. The flowers of Elæagnus orientalis and angustifolia are highly fragrant, and abound in honey which is esteemed as a remedy for malignant fevers in some parts of Europe.

GENERA.

Shepherdia, *Nutt.*	Hippophaë, *Linn.*	Elæagnus, *Linn.*
Lepargyreia, Nutt.	Conuleum, *L. C. Rich.*	

NUMBERS. GEN. 4. SP. 30.

POSITION.—Myricaceæ.—ELÆAGNACEÆ.—*Thymelæaceæ.*

Fig. CLXXIII.—Hippophaë rhamnoides. 1. a ♂ flower; 2. a perpendicular section of a ♀; 3. a section of a ripe fruit.—*Richard.*

Alliance XIX. *URTICALES.*—The Urtical Alliance.

Diagnosis.—*Diclinous Exogens, with scattered monochlamydeous flowers, single superior carpels, and a large embryo lying in a small quantity of albumen.*

The main distinction between this and the Amental Alliance consists in the presence of albumen, and the flowers not being arranged in catkins. The former character, however, fails in several instances, especially in some Artocarpads and Hempworts; so that in reality the amentaceous inflorescence is the only difference that can be at present pointed out to separate two Alliances, which nevertheless appear to be really distinct if regarded as wholes. They touch most closely among the Planes and Artocarpads, which may be referred indifferently to the one Alliance or the other, for both have a quasi-amentaceous inflorescence, and Liquidambars agree with Planes in having albumen, while Artocarpus itself is said to differ from the mass of its order in the want of it. Artocarpus cannot however be separated from Ficus, nor Platanus from Artocarpus, and this seems to justify the place assigned to Artocarpads and Planes in this arrangement.

The Orders themselves do not always rest upon such distinctions as a botanist would wish to discover; this is most especially the case with Hempworts and Morads, which might very well be united. But they may be allowed to remain for the present, because we really know so little about the plants of the Urtical Alliance, that any final distribution of the genera must be premature. It is much to be wished that some one would seriously examine the heaps of undescribed obscure plants related to this part of the vegetable kingdom, to be found in all large herbaria; it would be hardly possible to render a more welcome service to systematical botany.

So many plants of the Chenopodal and Silenal Alliances are ♂ ♀, especially of the former, where Atriplex alone forms a large mass of exceptions to the usually ⚥ structure, that we must suppose this to be one of the most remarkable instances of contact between the hypogynous and diclinous sub-classes.

Natural Orders of Urticals.

Radicle superior. Ovules twin, suspended. Embryo straight, albuminous. Anthers 2-lobed, with vertical fissures . . .	83. Stilaginaceæ.
Radicle superior. Ovule solitary, erect. Embryo straight, albuminous. Juice limpid. Stipules small, flat	84. Urticaceæ.
Radicle inferior. Embryo exalbuminous. Plumule many-leaved, large	85. Ceratophyllaceæ.
Radicle superior. Ovule solitary, suspended. Embryo hooked, exalbuminous	86. Cannabinaceæ.
Radicle superior. Ovules solitary, suspended. Embryo hooked, albuminous	87. Moraceæ.
Radicle superior. Ovule solitary, erect or suspended. Embryo straight, exalbuminous. Juice milky. Stipules large, convolute	88. Artocarpaceæ.
Radicle inferior. Embryo albuminous. Plumule minute. Juice limpid. Stipules large, deciduous	89. Platanaceæ.

ORDER LXXXIII. STILAGINACEÆ.—ANTIDESMADS.

Stilagineæ, *Agardh's Classes*, 199. (1824); *Von Martius Hort. Reg. Monac.* (1829).—Antidesmeæ, *Sweet Hort. Brit. ed.* 2. 460. (1830); *Endl.* xcvi.; *Meisner*, p. 347.

DIAGNOSIS.—*Urtical Exogens, with 2-lobed anthers splitting vertically, twin suspended ovules, a straight albuminous embryo, and superior radicle.*

Trees or shrubs. Leaves alternate, simple, coriaceous, undivided or toothed. Stipules twin, deciduous. Flowers minute, in axillary scaly spikes. Flowers ♂ ♀. Calyx 2- 3- or 5-parted. Corolla 0. ♂ Stamens 2, or more, arising from a tumid receptacle; filaments capillary; anthers innate, usually 2-lobed, with a fleshy connective and vertical cells opening transversely. ♀ Ovary free, 1- 2-celled, often with a conspicuous disk; ovules anatropal, suspended in pairs; stigma sessile, 3- 4-toothed. Fruit drupaceous. Seed suspended, sometimes perforated by processes of the putamen; embryo green, with foliaceous cotyledons, lying in the midst of copious fleshy albumen; radicle short, superior.

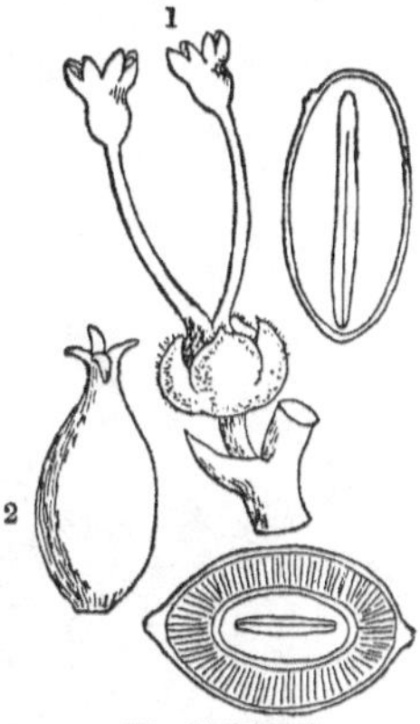

Fig. CLXXIV.

An obscure Order, whose limits are not ascertained. Judging from the genera Stilago and Antidesma, it is very near Nettleworts, from which it is chiefly distinguished by a great cushion-shaped disk, unelastic filaments, and anthers split into 2 lobes, which burst transversely at the apex. But Falconeria is said to have a 2-celled fruit, and therefore would approach very nearly to the Euphorbial Alliance. On the other hand, Pyrenacantha, referred hither by Endlicher, because of its two collateral pendulous ovules, is a milking plant, and wants the peculiar anthers of Antidesmads; but its pierced albumen and embryo are so similar to those of Phytocrene (or Gynocephalium), that it had better perhaps be referred to the Artocarpads. The male Antidesmas have much the inflorescence of East Indian Mastworts.

Natives of the East Indies and Madagascar.

These plants appear to be destitute of noxious qualities. Their succulent currant-like drupes are eaten by the natives of the countries where they grow; those of Antidesma pubescens are mentioned by Roxburgh, who also states that the shining deep red fruit of Stilago Bunius is sub-acid and palatable. The leaves of that plant are acid and diaphoretic; and, when young, are boiled with potherbs and given in India in cases of syphilis. The leaves of Antidesma alexiteria are among the imaginary remedies for serpent bites.

GENERA.

Antidesma, *L.*
Stilago, *L.*
Falconeria, *Royle.*

NUMBERS. GEN. 3. Sp. 20?

POSITION.—Urticaceæ.—STILAGINACEÆ.—Artocarpaceæ.
Corylaceæ.

Fig. CLXXIV.—Stilago lanceolata. 1. ♂ flower; 2. half ripe fruit; 3. a transverse section of the fruit and seed; 4. a perpendicular section of a seed.

ORDER LXXXIV. URTICACEÆ.—NETTLEWORTS.

Urticeæ, *Juss. Gen.* 400. (1789); *Gaudichaud in Freyc. Voyage, p.* 503. (1826); *Bartl. Ord. Nat.* 105. Urticaceæ, *Endlich. Prodr. Norf.* 37; *Gen.* xciv.; *Meisner, p.* 348.

DIAGNOSIS.—*Urtical Exogens, with small flat stipules, limpid juice, a solid erect ovule, a straight albuminous embryo, and superior radicle.*

Trees, shrubs, or herbs; never milky. Leaves alternate, usually covered either with asperities or stinging hairs, with membranous stipules, which are deciduous or convolute in vernation. Flowers herbaceous, inconspicuous, ♂ ♀, (occasionally ⚥

2 3 1 4

Fig. CLXXV.

intermixed) scattered, or clustered, or in catkins, or close heads. Calyx membranous, lobed, persistent. ♂ Stamens definite, distinct, inserted into the base of the calyx, and opposite its lobes; anthers often curved inwards in æstivation, and turned backwards with elasticity when bursting. ♀ Ovary superior, simple; ovule solitary, erect; stigma simple, fringed. Fruit a simple indehiscent nut, surrounded either by the membranous or fleshy calyx. Embryo straight, with fleshy albumen; cotyledons flat; radicle superior.

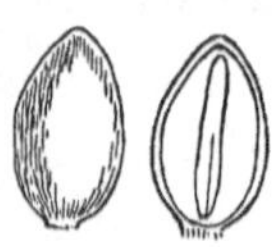

Fig. CLXXVI.

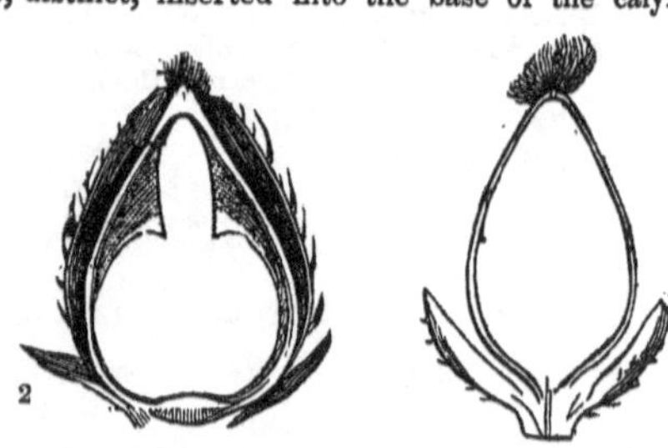

2 1

Fig. CLXXVII.

Nettleworts, as now circumscribed, consist almost entirely of rough-leaved plants, which, although they occasionally acquire the stature of trees, have nevertheless little more than a herbaceous texture, their wood being remarkable for its lightness, sponginess, and profusion of cellular tissue. Their great distinction consists in their having a

Fig. CLXXV.—1. Branch of Procris splendens; 2. cluster of male and female flowers; 3. a male flower about to expand; 4. the same expanded.

Fig. CLXXVI.—Parietaria officinalis; one of the lenticular fruits both whole and divided perpendicularly to show the embryo.

Fig. CLXXVII.—1. A section of the ovary of Urtica dioica; 2. the same when ripe, after the embryo is developed.

single erect ovule in a simple carpel, the foramen of which is at the apex, so that when the seed is ripe the embryo is necessarily inverted, its radicle pointing upwards. (In the second edition of this work the position of the radicle was misstated, owing to some accident.) Nettleworts will then be easily known from Morads and Hempworts, which have a hooked embryo, and from Antidesmads, which have pendulous ovules. What differences exist between them and the Artocarpads are mentioned under that Order. The flowers are occasionally, in part, hermaphrodite, although the greater number are absolutely unisexual, and, on this account, they must be regarded as entirely conterminous with Chenopods in the hypogynous sub-class. They will, however, be found to differ not only in their habitually diclinous flowers, but also in their embryo being enclosed in albumen and not external to it as in Chenopods.

Independently of the resemblances borne by Nettleworts to Chenopods as well as to other Orders in the Urtical Alliance, we must not lose sight of their very close affinity to the hypogynous Buckwheats, some of which are ♂ ♀. This has been already alluded to at p. 258, and will be further noticed hereafter.

The species are widely dispersed over every part of the world; appearing in the most northern regions, and in the hottest climates of the tropics; growing now upon dry walls, where there is scarcely nutriment for a Moss or a Lichen, and inhabiting the dampest recesses of the forest. Many follow the steps of man, flourishing on rubbish heaps and waste places around his dwellings.

All the more important of the old Urticaceous Order having been removed from this place, the qualities of the few that remain are of little interest. Excessive causticity in the limpid juice is their chief characteristic, as is exemplified in the common stinging Nettles, Urtica dioica, urens, and pilulifera, which are, however, not to be compared for an instant with some of the E. Indian species. Leschenault de la Tour thus describes the effect of gathering Urtica crenulata in the Botanic Garden at Calcutta:—"One of the leaves slightly touched the first three fingers of my left hand: at the time I only perceived a slight pricking, to which I paid no attention. This was at seven in the morning. The pain continued to increase; in an hour it had become intolerable; it seemed as if some one was rubbing my fingers with a hot iron. Nevertheless, there was no remarkable appearance; neither swelling, nor pustule, nor inflammation. The pain rapidly spread along the arm, as far as the armpit. I was then seized with frequent sneezing, and with a copious running at the nose, as if I had caught a violent cold in the head. About noon I experienced a painful contraction of the back of the jaws, which made me fear an attack of tetanus. I then went to bed, hoping that repose would alleviate my suffering; but it did not abate; on the contrary, it continued nearly the whole of the following night; but I lost the contraction of the jaws about seven in the evening. The next morning the pain began to leave me, and I fell asleep. I continued to suffer for two days; and the pain returned in full force when I put my hand into water. I did not finally lose it for nine days." A similar circumstance occurred, with precisely the same symptoms, to a workman in the Calcutta Garden. This man described the sensation, when water was applied to the stung part, to be as if boiling oil was poured over him. Another dangerous species was found by the same botanist in Java (U. stimulans), but its effects were less violent. Both these seem to be surpassed in virulence by a Nettle called Daoun Setan, U. urentissima, or devil's leaf, in Timor; the effects of which are said, by the natives, to last for a year, or even to cause death. In some species the acrid fluid is so abundant that it is spontaneously discharged from the whole surface of the leaf. According to Endlicher the causticity of Nettle juice is owing to the presence of bicarbonate of ammonia. The foliage of Bœhmeria caudata is used advantageously in Brazil in baths, as a relief for hæmorrhoidal complaints, and in the same country an extract of Pilea muscosa is regarded as a remedy for dysuria. The tenacity of the fibres of some species is such that cordage has been successfully manufactured from them; the stalks of Urtica cannabina were even expected, at one time, to be equal in strength to Hemp itself. Urtica tenacissima, called Caloose in Sumatra, yields an extremely tough cordage in that island.—*Roxb.* Flogging with nettles has been employed in cases of arthritis, paralysis, &c. Nettles when very young and tender are commonly used as an ingredient in broth by the English peasantry, who consider that they purify the blood. The tubers of Urtica tuberosa are esculent and nutritious; the natives eat them raw, boiled, or roasted.—*Roxb.* The herbage and seeds of Urtica membranacea, an Egyptian plant, are regarded as emmenagogue and aphrodisiac. Several Parietarias, especially P. erecta and diffusa, have had some reputation as refrigerants and diuretics. The leaves when dried have been used in polishing mirrors. A decoction of Urtica dioica strongly salted, will coagulate milk without giving it any unpleasant flavour; the whole plant is esteemed astringent and diuretic.—*Burnett.*

GENERA.

Urtica, *Tournef.*	*Vaniera*, Loureir.	Böhmeria, *Jacq.*	Pouzolzia, *Gaudich.*
Urera, *Gaudich.*	Malaisia, *Blanco.*	*Duretia*, Gaudich.	*Memorialis*, Hamilt.
Laportea, *Gaudich.*	Schychowskya, *Endl.*	Procris, *Commers.*	Rousselia, *Gaudich.*
Fleurya, *Gaudich.*	Pilea, *Lindl.*	Neraudia, *Gaudich.*	Soleirolia, *Gaudich.*
Girardinia, *Gaudich.*	*Dubrueilia*, Gaudich.	Parietaria, *Tournef.*	*Helxine*, Requ.
Elatostema, *Forst.*	*Haynea*, Schum.	Freirea, *Gaudich.*	Forskolea, *Linn.*
Langeveldia, Gaudich.	Pellionia, *Gaudich.*	Thaumuria, *Gaudich.*	*Cuidbeja*, Forsk.
Sciophila, Gaud.	Splitgerbera, *Miq.*	Gesnouinia, *Gaudich.*	? Australina, *Gaudich.*
Sciobia, Rchb.			

NUMBERS. GEN. 23. Sp. 300. There having been no recent enumeration of the species this is merely a rough estimate. As I find 216 actually described, the number now estimated is probably much too low both for genera and species.

Chenopodiaceæ.
POSITION.—Moraceæ.—URTICACEÆ.—Cannabinaceæ.
Polygonaceæ.

ORDER LXXXV. CERATOPHYLLACEÆ.—HORNWORTS.

Ceratophylleæ, *Gray's Arrangement of Brit. Fl.* 2. 554.; *DC. Prodr.* 3. 73. (1828); *Asa Gray, Ann. Lyc. N. York*, 4. 48. (1837); *Schleiden in Linnæa*, 11. 540. (1837); *Endl. Gen.* lxxxiii.

DIAGNOSIS.—*Urtical submersed Exogens, with an inferior radicle, an exalbuminous embryo, and a large many-leaved plumule.*

Submersed herbs, with dichotomous, cellular, verticillate leaves. Flowers monœcious. Calyx inferior, many-parted. ♂ Stamens from 12 to 20; filaments wanting; anthers 2-celled. ♀ Ovary superior, 1-celled; ovule solitary, pendulous, orthotropal; style pervious; stigma filiform, oblique. Nut 1-seeded, indehiscent, terminated by the hardened stigma. Seed pendulous, solitary; albumen 0; embryo with 2 cotyledons; plumule many-leaved; radicle inferior.

It would be difficult to name a plant concerning whose affinity more different opinions have been entertained. Possessed of the most simple organisation, it scarcely presents a single salient point for consideration, with the exception of its plumule being very highly developed. The number of cotyledons, although only two, as Schleiden has shown, appears to be four; and, in consequence of this, Richard placed it near Conifers, with which it seems to me to have no kind of affinity. Nevertheless, Schleiden leans towards Richard's view; in order to support which, he calls the male flower an amentum. De Candolle urges its relation to Hippuris and Myriophyllum, among Hippurids, from which it differs in its superior ovary; and he inquires whether Naias, which, according to some, is dicotyledonous, does not belong to the same Order. Agardh actually places it among Naiads.

Fig. CLXXVIII.

I formerly suggested the possibility of its being a degeneration of Nettleworts, and as bearing the same relation to them as Mossweeds and Starworts to some unknown Order, or as Lemnads to Arads. This opinion Schleiden has criticised with incivility, forgetting that it was formed upon the assumption that the radicle of Ceratophyllum is superior, an error of De Candolle's, of which I was not then aware. And even now, I do not see where it can be better placed, unless near Chenopods. It agrees with the Urtical Alliance in having incomplete, unisexual flowers, a free ovary consisting of a single carpel and a solitary ovule; but it differs in having the radicle inferior, and no albumen. The high development of the plumule may however, in this instance, as in Waterbeans, (Nelumbiaceæ), be a compensation for the want of albumen, enabling the embryo to germinate without assistance, as soon as it is exposed to the fitting conditions.

Hornworts may also be compared with Chenopods, on account of their incomplete flowers, inferior radicle, simple free carpel, and single ovule; and the unisexual flowers and want of albumen would not be opposed to such a comparison. But the unisexuality of Chenopods is the exception, not the rule, and the straightness of the embryo of Ceratophyllum is as much at variance with the characteristic mark of that Order, as its

Fig. CLXXVIII.—Ceratophyllum submersum.—1. ♂ flower; 2. ♀ ditto; 3. perpendicular section of ovary, the style being cut off; 4. fruit; 5. embryo.

inferior radicle is with most of the Urtical Alliance. Such being the case, it seems to me that until some better suggestion shall have been offered than that of stationing Hornworts near the Gymnospermous Conifers, they may be regarded as anomalous plants of the Urtical Alliance.

A singular view is that of Dr. Asa Gray, who would place Ceratophyllum near Nelumbium, because of its highly developed plumule, for it would be difficult to point out any other resemblance ; to this opinion he has been led by the account of the development of the embryo given by M. Adolphe Brongniart *Ann. Sc.* 12. 253, which is only in part correct.

Found in ditches in Europe, North America, Northern Asia, Senegal, Barbary, and India.

These plants have no known use : they have sometimes the heavy smell of Chara.

GENUS.

Ceratophyllum, *L.*

NUMBERS. GEN. 1. Sp. 1. according to Schleiden ; 6 according to Chamisso.

Chenopodiaceæ.

POSITION.—Urticaceæ.—CERATOPHYLLACEÆ.——————

Halorageæ.

ORDER LXXXVI. CANNABINACEÆ.—HEMPWORTS.

Cannabineæ, *Endl. Gen.* xcv. (Oct. 1837); *Meisner, p.* 348.

DIAGNOSIS.—*Urtical Exogens, with a solitary suspended ovule, and a hooked exalbuminous embryo, with a superior radicle.*

Herbaceous, rough-stemmed, watery plants, with alternate lobed stipulate leaves, and small inconspicuous flowers. Flowers ♀ ♂. ♂ in racemes or panicles. Calyx herbaceous, scaly, imbricated.' Stamens few, opposite the sepals; filaments filiform; anthers terminal, 2-celled, opening longitudinally. ♀ in spikes or cones. Sepal single, enwrapping the ovary. Ovary free, 1-celled; ovule solitary, pendulous, campylotropal; stigmas 2, subulate, sessile. Fruit indehiscent, with a single suspended seed. Embryo without albumen, hooked or spirally coiled; radicle superior, lying against the back of the cotyledons.

These plants, formerly regarded as a division of Nettleworts, differ from that Order in having their seeds suspended, their embryo coiled up, and in wanting albumen. To the Artocarpads they approach in technical characters, differing chiefly in their embryo; but they have no milky juice, and are widely different in appearance. From Morads they are hardly distinguishable except by the absence of albumen.

Hempworts are found wild in the temperate parts of the Old World, in the northern hemisphere. The Hemp inhabits the cooler parts of India, whence it has been transported to Europe; the Hop occurs wild in the South Eastern provinces of Europe.

Fig. CLXXIX.

The valuable fibre called Hemp, is produced by Cannabis sativa, which is hardly less celebrated for its narcotic qualities. In the elegant language of Endlicher, "Emollitum exhilarat animum, impotentibus desideriis tristem stultam lætitiam provocat et jucundissima somniorum conciliat phantasmata." The Turks employ it under the names of Hadschy and Malach. Linnæus speaks of its vis narcotica, phantastica, dementens, anodyna, et repellens. Even the Hottentots use it to get drunk with, and call it Dacha. The Arabians name it Hashish. The Brazilian savages delight in its use. It appears to owe its narcotic properties to the presence of a resin which is not formed in Europe. This resin exudes in India from the leaves, slender branches, and flowers; when collected into masses it is the churras or cherris of Nepal. Its odour is fragrant and narcotic, its taste slightly warm, bitterish, and acrid.—*Pharm. Journ.* 1. 489. The imbricated heads of the common Hop, Humulus Lupulus, participate in this quality, and in like manner are used for the purpose of producing intoxication, in the preparation of beer. Their scales are scattered over with resinous spherical glands, which are easily rubbed off, and have a powerful agreeable odour, and bitter taste; they appear to consist of an acrid, ethereal oil, an aromatic resin, wax, extractive, and a bitter principle called Lupuline. By pressure Hopheads yield a green, light, acrid oil, called Oil of Hops. Its young shoots are eaten as Asparagus, and the roots have been employed as a substitute for Sarsaparilla.

GENERA.

Cannabis, *Tournef.* | Humulus, *Linn.* | *Lupulus*, Tournef.

NUMBERS. GEN. 2. SP. 2.

POSITION.—Urticaceæ.—CANNABINACEÆ.—Moraceæ.

Fig. CLXXIX.—Humulus Lupulus. 1. ♂ flower; 2. ♀; 3. section of ovary; 4. ripe fruit; 5. section of it, showing the embryo; 6. a grain of lupuline.

ORDER LXXXVII. MORACEÆ.—MORADS.

Moreæ, *Endl. Prodr.* 40, (1833); *Gaudich, in Freycinet*, 509; *Meisner, p.* 350; *Endl.* xcii.—Pholeosantheæ, *Blume Bijdr.* 436, (1825).—Sycoideæ, *Link Handb.* 1. 292. (1829).

DIAGNOSIS.—*Urtical Exogens, with solitary suspended ovules, and a hooked albuminous embryo with a superior radicle.*

Trees or shrubs, with a milky juice, sometimes climbing. Leaves of various forms and texture, very commonly lobed and rough, with large stipules often rolled up, inclosing the younger leaves, and leaving a ringed scar when they drop off. Flowers

Fig. CLXXX.

very inconspicuous, ♂ ♀, collected in heads, or spikes, or catkins. ♂ calyx 0, or 3-4-parted, imbricated. Stamens 3-4, inserted into the base of the calyx and opposite its segments; filaments generally shrivelled on the inner face; anthers 2-celled, opening lengthwise. ♀ sepals 3-4-5, sometimes in two rows. Ovary 1-celled, occasionally (by accident ?) 2-celled. Ovules solitary, pendulous, or amphitropal, with the foramen uppermost; style terminal, bifid, with the lobes often unequal. Fruit, small nuts or utricles, 1-seeded, inclosed within a succulent receptacle, or collected in a fleshy head formed by the consolidated succulent calyx. Seed solitary, with a thin brittle integument. Embryo lying in fleshy albumen, hooked, with the radicle long, superior, folded down towards the cotyledons.

The whole of the genera of this Order have either a remarkably enlarged receptacle, upon or within which the flowers are arranged, as is seen in Ficus, and even more strikingly in Dorstenia, or a tendency towards its formation is indicated, when the flowers are gathered into heads of a spheroidal form, as in the Mulberry and Osage Orange (Maclura). In this manner the Order of Morads passes into that of the Artocarpads, from which indeed it hardly differs except in having an abundance of albumen, and a hooked slender embryo. Strictly speaking, however, albumen occurs in the Artocarpads in Phytocrene, which certainly must belong to them, and in Pyrenacantha, which must, I think, be also referred thither, notwithstanding its somewhat different habit. In the last edition of this work, Batis was referred to the present Order; but I now see, that while the species so named by Roxburgh certainly stand next to Morus, the West Indian plant to which the designation properly applies must be stationed elsewhere.

The tenacity of life in some plants of this family is remarkable. A specimen of Ficus australis lived and grew suspended in the air, without earth, in one of the hothouses in the Botanic Garden, Edinburgh, for eight months, without experiencing any apparent inconvenience.

None of the Morads are European, for the Mulberries and common Fig have been brought from the East. The species inhabit the temperate and tropical latitudes of both

Fig. CLXXX.—Morus alba. 1. A male flower; 2. clusters of females; 3. a female flower separate; 4. the same with a part of the calyx cut away; 5. a vertical section of a ripe achænium; 6. a cluster of fruit consisting of succulent calyxes enclosing achænia.

hemispheres, often constituting vast forests, in the case of the various species of Fig, which in all hot countries have generally very thick trunks, with extremely strong boughs, and a prodigious crown. Travellers say, that the colossal wild Fig-trees are among the most grateful presents of Nature to hot countries: the shade of their magnificent head refreshing the traveller when he reposes under their incredibly wide-spreading branches and dark green shining foliage. In India, two of the species have historical celebrity. Of these the Banyan-tree, so remarkable for its vast rooting branches, is Ficus indica; the Pippul or sacred Fig of the same country, readily known by its rootless branches, and its heart-shaped leaves with long attenuated points, is Ficus religiosa. Blume also relates, that a Ficus microcarpa, which he planted before the door of his house in Java, had in seven years covered a space of above 60 (square ?) feet with its dense shade. And he describes a sacred specimen of enormous stature, growing in the same island, at a place called Batu-Tulies, from whose huge branches he gathered as many as 34 species of parasites and epiphytes, which were not, however, half what might have been collected. The genus Ficus, indeed, is one of those which travellers describe as most conducing to the peculiarities of a tropical scene. Mr. Hinds (*Ann. N. H.* xv. 100) points out the complex appearance of the main stem of many species; their immense horizontal branches, their proportionate lowness, and the vast number of smaller stems in every stage of development, some just protruding from the horizontal limbs, others hanging midway between the leafy canopy and earth, displaying on each thick rounded extremity an enormous spongiole, while many reach the soil, and having attained strength and size act as columns to sustain the whole structure. "The tropical forest abounds with these in every variety of growth and apparent distortion."

Caoutchouc is furnished by many of this Order in great abundance; all the India Rubber of continental India is obtained from Ficus elastica; in Java, other species yield this substance of excellent quality, as do F. Radula, elliptica, and prinoides in America. Their milky fluid is in some instances bland, and actually employed as a beverage; for of the different plants which have been occasionally brought to Europe under the name of Cow-trees, most are Figs. One of these has been figured by M. De Candolle, under the name of Ficus Saussureana, (*Mém. de la Soc. Phys. de Genève*); and others have been described by M. Desvaux, *Ann. Sc.* 18. 309. The juice is, however, in many cases excessively acrid; that of Ficus septica is emetic, and of F. toxicaria and Dæmona, a virulent poison; indeed the milky juice of the cultivated Fig itself possesses considerable acridity, causing a burning sensation in the throat when chewed. In some species the juice assumes a resinous character, when discharged from parts attacked by Cocci, as is the case with F. indica, benghalensis, and Tsjela, which form a sort of gum-lac in the East Indies. Notwithstanding the prevalence of an acrid secretion, the fruit of many species appears to decompose it and convert it into sugar, or some other substance; hence we have the eatable Fig of the shops from the acrid Ficus Carica, and a fruit of inferior quality, but still eatable, from F. religiosa, Benjamina, pumila, auriculata, Rumphii, benghalensis, aspera, Granatum, and the Egyptian Sycomorus, whose imperishable wood is said to have been used in the construction of the cases in which the mummies are inclosed. On the other hand the common Mulberry, Morus nigra, has an agreeable sub-acid succulent fruit, for the sake of which it was long since introduced from Persia; and that of the White Mulberry, and other species, both Asiatic and American, is eatable though not esteemed; but these fruits are not entirely harmless, causing diarrhœa if indulged in too freely, and their roots are both cathartic and anthelmintic, thus indicating the presence in their system of the acrid secretions of the Order. There is also a Brazilian Ficus anthelmintica. The Mulberries contain mannite and succinic acid, according to the chemists. Among other uses of less extensive application are the following:—Dorstenia contrayerva, brasiliensis, opifera, and others, have bitterish roots, and a remarkable overpowering odour, with a little pungency. They are supposed to be antidotes to the bites of venomous animals, and certainly possess stimulant, sudorific, and tonic qualities; but they lose them by keeping, and soon become inert; they are also emetic, and are

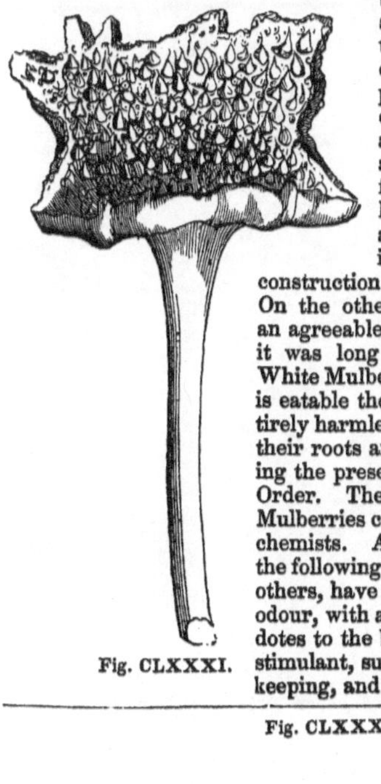

Fig. CLXXXI.

Fig. CLXXXI.—Dorstenia contrayerva.

employed for the same purposes as Aristolochia Serpentaria. A kind of paper is manufactured from Broussonetia papyrifera, whose fruit is succulent and insipid. The fruit of Maclura aurantiaca, (the Osage Orange), is as large as the fist, orange-coloured, and filled with a yellow fœtid slime, with which the native tribes smear their faces when going to war. The wood of Maclura tinctoria is the dyewood called Fustick; it contains morine, a peculiar colouring matter; its fruit is pleasant, and used in North American medicine, for the same purposes as the black Mulberry in Europe. According to Martius, both it and other species of the same genus yield fustick in Brazil. It is to be observed, that the latter name is also given to the wood of Rhus Cotinus. The seeds of Ficus religiosa are supposed by the doctors of India to be cooling and alterative. The bark of Ficus racemosa is slightly astringent, and has particular virtues in hæmaturia and menorrhagia; the juice of its root is considered a powerful tonic. The white glutinous juice of Ficus indica is applied to the teeth and gums, to ease the toothache; it is also considered a valuable application to the soles of the feet when cracked and inflamed; the bark is supposed to be a powerful tonic, and is administered by the Hindoos in diabetes. Is it not possible that the Indian poison with which the Nagas tip their arrows, of the tree that produces which nothing is known, may belong to this tribe? See, for an account of its effects, *Brewster's Journal*, 9. 219.

GENERA.

Epicarpurus, *Blume.*
  Albrandia, Gaudich.
Morus, *Tournef.*
  Ampalus, Boj.

Batis, *Roxb.*
? Fatoua, *Gaudich.*
Broussonetia, *Vent.*
  Papyrius, Lam.

Maclura, *Nutt.*
Sycomorphe, *Miq.*
Ficus, *Tournef.*
  Erosma, Both.

Dorstenia, *Plum.*
  Sychinium, Desv.
? Kosaria, *Forsk.*

NUMBERS. GEN. 8. Sp. 184.

POSITION.—Cannabinaceæ.—MORACEÆ.—Artocarpaceæ.

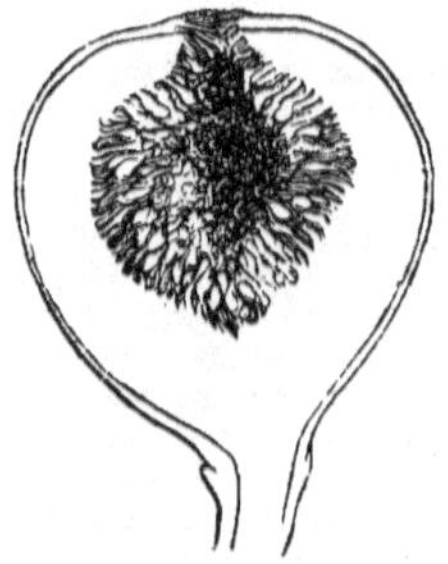

Fig. CLXXXII.

Fig. CLXXXII.—Perpendicular section of the succulent hollow receptacle of Ficus Carica.

ORDER LXXXVIII. ARTOCARPACEÆ.—ARTOCARPADS.

Artocarpeæ, *R. Brown in Congo*, (1818); *Blume By. Dr.* 479; *Ed. prim. No.* 80, (1830); *Bartl. Ord. Nat.* 104; *Endl.* xciii.; *Meisner, p.* 349; *Bennet in Horsfield, p.* 48.

DIAGNOSIS.—*Urtical Exogens, with milky juice, large convolute stipules, solitary erect or suspended ovules, a straight exalbuminous embryo, and superior radicle.*

Trees or shrubs, abounding in milky juice. Leaves alternate, simple, often lobed, with large deciduous stipules. Flowers ♂ ♀, always collected into dense heads of some kind.

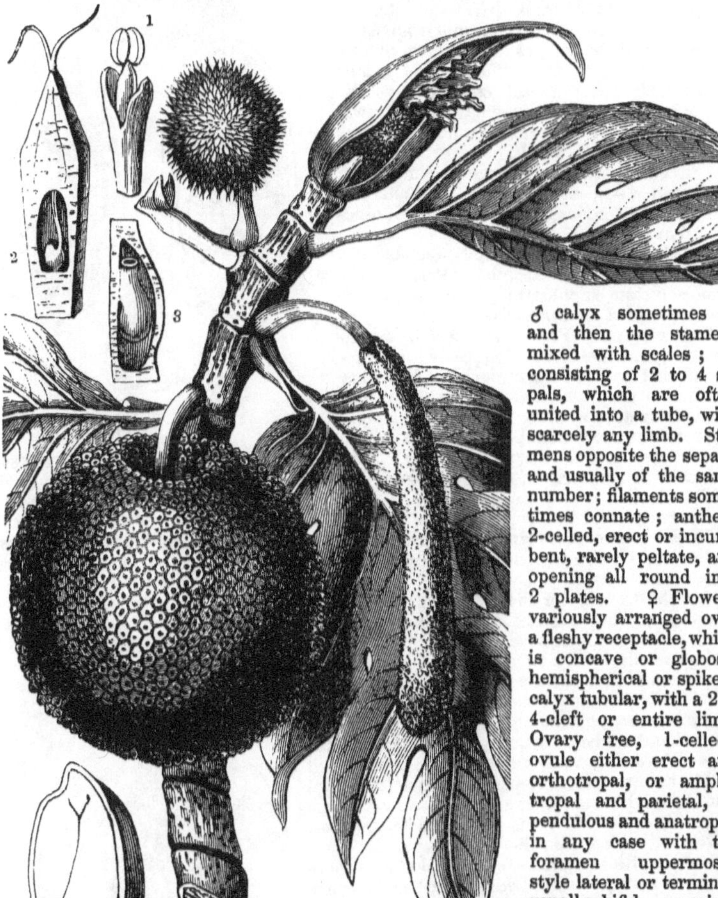

Fig. CLXXXIII.

♂ calyx sometimes 0, and then the stamens mixed with scales; or consisting of 2 to 4 sepals, which are often united into a tube, with scarcely any limb. Stamens opposite the sepals, and usually of the same number; filaments sometimes connate; anthers 2-celled, erect or incumbent, rarely peltate, and opening all round into 2 plates. ♀ Flowers variously arranged over a fleshy receptacle, which is concave or globose, hemispherical or spiked; calyx tubular, with a 2 to 4-cleft or entire limb. Ovary free, 1-celled; ovule either erect and orthotropal, or amphitropal and parietal, or pendulous and anatropal, in any case with the foramen uppermost; style lateral or terminal, usually bifid, occasionally undivided with a simple lateral or radiating stigma. Fruit variable, surrounded by a fleshy involucre, or composed of consolidated fleshy calyxes, within which lies a multitude of nuts. Seed erect, parietal, or pendulous. Embryo with much or very little albumen, straight, with the radicle directed towards the vertex of the ovary; cotyledons thick and fleshy, when the albumen is deficient, thin when it is abundant, often very unequal.

Fig. CLXXXIII.—Artocarpus incisa, with a ripe fruit, a head of ♀, and a pendulous club-shaped spike of ♂ flowers. 1. A ♂ flower; 2. a ♀ cut out of the globular head; 3. a section of the ovary, showing the position of the ovule; 4. a section of a seed; partly after *Hooker*.

The massive heads into which the fruits of the Breadfruit tree are collected represent the typical condition of the genera of this Order, whose milky juice has long since suggested its separation from Nettleworts; an opinion, however, in which it was difficult to agree, so long as the Fig and its allies were associated with it by that character. Now, however, that such plants have been more carefully studied, it appears that the old Urticaceous Order should rather be regarded as an Alliance, of which the Artocarpads form one of the Orders. In that point of view the Artocarpads will be distinguished from Hempworts and Morads by their straight embryo with large cotyledons, and from Antidesmads by their anthers and solitary ovules. From Nettleworts the difference is rather one of habit than of real structure, as far as our information at present goes. Brown, indeed, who first proposed the Order, stated that the ovule was erect, which, however, is not the case in either Artocarpus or Maclura, both which have a suspended ovule. Endlicher, on the other hand, relies upon the absence of albumen; but a trace of it occurs in Artocarpus, and in Phytocrene it is extremely abundant, to say nothing of Pyrenacantha. Perhaps the large convolute stipules may form a further characteristic of Artocarpads.

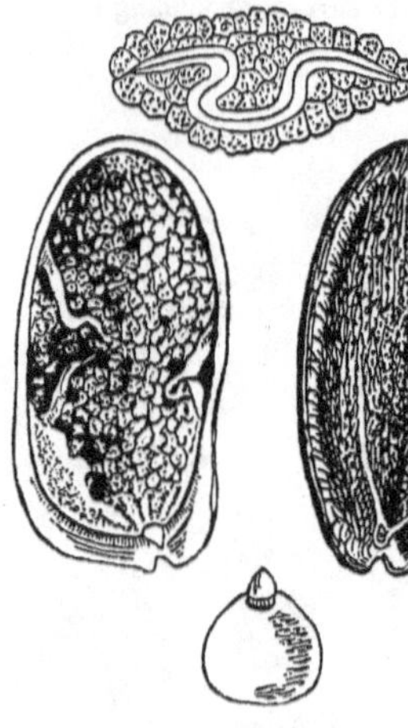

Fig. CLXXXIV.

With respect to Phytocrene, which is considered by M. Decaisne identical with Gynocephalium, I find that it is remarkable for a very large quantity of granular albumen, which Blume says is altogether wanting in Gynocephalium; I therefore retain it as a distinct genus.

The Order is not without anomalies. Phytocrene and Pyrenacantha have copious albumen. In Antiaris the ovary adheres to the involucre. It is doubtful whether all yield milk.

The tropics, and the tropics only, of both worlds, are the stations of these plants.

The most important plant of the Order is the Breadfruit, Artocarpus incisa, the most virulent the Upas tree, Antiaris toxicaria. Like Morads the species afford caoutchouc and an eatable fruit.

The edible quality of the Breadfruit appears to be owing to the presence of a large quantity of starch in its succulent heads. The Jack, Artocarpus integrifolia, has a similar quality, but is inferior. The venom of the Antjar poison, Antiaris toxicaria, is due to the presence of that most deadly substance strychnia; notwithstanding the exaggerated statements that have been made regarding this tree, the Upas of the Javanese, there remains no doubt that it is a plant of extreme virulence, even linen fabricated from its tough fibre being so acrid as to verify the story of the shirt of Nessus; for it excites the most distressing itching if insufficiently prepared.

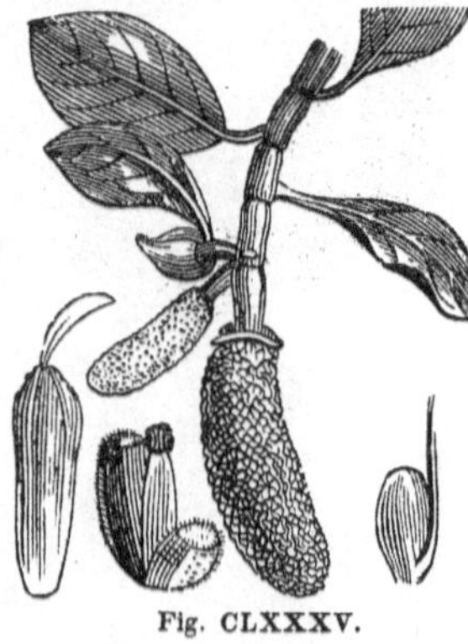

Fig. CLXXXV.

However, the seeds are always wholesome; those of a plant nearly allied to Cecropia, called Musanga by the Africans of the Gold Coast, as well as of Artocarpus, are eatable as nuts. The famous Cow Tree, or Palo de Vaca, of South America, which yields a copious supply of a rich and wholesome milk, as good as that of the cow, is a species of Brosimum. It has been analysed by various chemists, especially Mr. Edward Solly, who found in it as much as 30·57 per cent. of galactin. —See *Phil. Mag., Nov.* 1837. Brosimum alicastrum abounds in a tenacious gummy milk; its leaves and young shoots are much eaten by cattle, but when they become old they cease to be innocuous. The roasted nuts are used instead of bread, and have much

Fig. CLXXXIV.—1. Nut of Phytocrene; 2. the same, showing the seed in its interior; 3. a cross section of the seed, showing the cotyledons and granular albumen; 4. the club-shaped radicle.
Fig. CLXXXV.—Artocarpus integrifolia.

the taste of Hazel nuts. The milkiness of the sap is in itself an evidence of the presence of caoutchouc, and accordingly the tree Ule of Papantla, from which caoutchouc is obtained in that country, is supposed to be Castilloa elastica, a plant of this Order. A similar substance is obtained from Cecropia peltata, a very common tropical tree. The bark of this plant, remarkable for its stems being hollow between the nodes, is astringent, and used in diarrhœa and gonorrhœa. The light porous wood is employed by the American savages to give them light by friction.

From a species of Antiaris (called by Mr. Nimmo Lepurandra saccidora), sacks are made in Western India by the following singular process. "A branch is cut corresponding to the length and diameter of the sack wanted. It is soaked a little, and then beaten with clubs till the fibre separates from the wood. This done, the sack formed of the bark is turned inside out, and pulled down till the wood is sawed off, with the exception of a small piece left to form the bottom of the sack." These sacks are in general use. A specimen of them was exhibited to the Linnæan Society some years ago. Here there is no trace of the virulence of the Upas tree, and notwithstanding the fatal character of that species, others appear to be also inert. In the province of Martaban, Dr. Wallich found his Water Vine (Phytocrene), whose singular soft and porous wood discharges when wounded a very large quantity of pure and tasteless fluid, which is quite wholesome, and is drunk by the natives. This is an extraordinary exception to the usual character of the Order, and if the plant be really destitute of milk, it will break down very much the limit between Artocarpads and Nettleworts, unless, indeed, Phytocrene is out of its place, which its copious albumen (?) leads one to suspect. Martius says that the fruit of Pourouma bicolor is sub-acid, and worth cultivation, although mucilaginous. Snake-wood, or Bois de Lettres, so called because of the markings which it presents, is obtained from the Brosimum, called by Aublet Piratinera guianensis, a tree 60 or 70 feet high, whose beautiful timber is so hard that it can only be felled by the American axe. —*Schomb.*

GENERA.

Brosimum, *Swartz.*
Piratinera, Aubl.
Galactodendrum, Hum.
Antiaris, *Leschen.*
Lepurandra, Nimmo.
Olmedia, *Ruiz et Pav.*
Macquira, Aubl.
Trymatococcus, *Pöpp.*
Sorocea, *St. Hil.*
Pourouma, *Aubl.*
Cecropia, *Linn.*
Musanga, *Chr. Sm.*
Coussapoa, *Aubl.*
Myrianthus, *Palis.*
Artocarpus, *Linn.*
Sitodium, Banks.
Rademachia, Thunb.
Soccus, Rumph.
Polyphema, *Loureir.*
Iridaps, Commers.
Conocephalus, *Blume.*
Gynocephalium, *Blume.*
Phytocrene, *Wall.*
? Natsiatum, *Hamilt.*
Trophis, *P. Br.*
Streblus, Loureir.
Achymus, Soland.
Pyrenacantha, *Hooker.*
* * *
Bruea, *Gaudich.*
Perebea, *Aubl.*
Bagassa, *Aubl.*
Castilloa, *Cerv.*
Aporosa, *Blume.*
Sciaphila, *Blume.*

NUMBERS. GEN. 23. SP. 54 ?

POSITION.—Moraceæ.—ARTOCARPACEÆ.—Platanaceæ.

ORDER LXXXIX. PLATANACEÆ.—PLANES.

Plataneæ, *Lestiboudois*, according to *Von Martius, Hort. Reg. Monacensis, p.* 46. (1829.); *Endl. Gen.* xcvii.; *Meisner, p.* 347.

DIAGNOSIS.—*Urtical Exogens, with deciduous sheathing stipules, capitate flowers, limpid juice, an inferior radicle, albuminous embryo, and minute plumule.*

Deciduous trees or shrubs. Leaves alternate, palmate, or toothed, with scarious sheathing stipules. Catkins round, pendulous. Flowers ♂ ♀, amentaceous, naked; the sexes in distinct catkins. ♂ Stamens single, without any floral envelope, but with several small scales and appendages mixed among them; anthers linear, 2-celled. ♀ Ovary 1-celled, terminated by a thick awl-shaped style, having the stigmatic surface on one side; ovules solitary, or two, one above the other, suspended, orthotropal. Nuts, in consequence of mutual compression, clavate, with a persistent recurved style. Seeds solitary, or rarely in pairs, pendulous, elongated; testa thin; embryo long, antitropal, taper, lying in the axis of very thin albumen; radicle inferior.

This group of trees or large shrubs, formerly comprehended in the Order once called Amentaceæ, is particularly known by its round heads of flowers, its 1-celled ovary, containing 1 or 2 pendulous ovules, and its embryo lying with the radicle downward, by which it is distinguishable from both Birchworts, Galeworts, and Artocarpads, with all which, especially the latter, it has a close affinity. From the latter, indeed, it is chiefly known by the want of calyx, the inferior radicle, the presence of albumen, and the absence of milk; the habit of the two Orders is much the same. Bartling even combines Platanus with that Order, and it must be confessed that the grounds of separating the two are not strong. The simple carpel of the Planes refers it rather to the Urtical than the Amental Alliance: they may be regarded as the connecting link between Artocarpads and Liquidambars, agreeing most with the former on account of the simplicity of their fruit.

Noble timber-trees, natives of Barbary, the Levant, and North America, and extending even into Cashmere.

They are chiefly cultivated for the sake of their noble appearance; their broad, shady, palmated leaves being equalled in this country by those of no hardy trees except the Sycamore and its fellow species. The timber is firm and close grained, but brittle, perishable, and only fit for indoor work. That of P. orientalis is said, however, to be in request in the East for cabinet work, and even to have been used in ship-building. The timber of P. occidentalis is redder, but warps, and will not bear exposure to weather. No use is made of any other part of these plants.

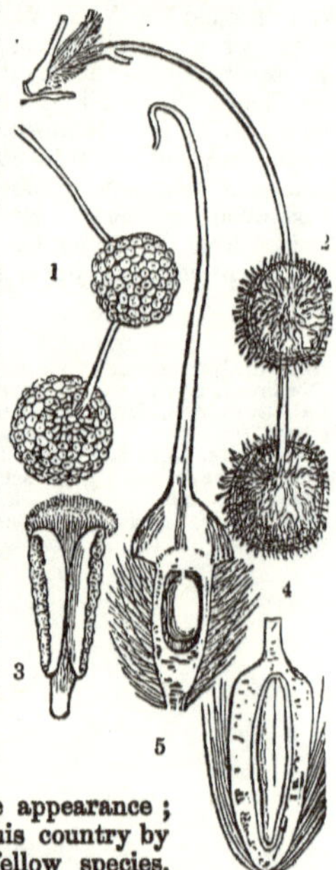

Fig. CLXXXVI.

GENUS
Platanus, *L.*

NUMBERS. GEN. 1. Sp. 6?

POSITION.—Artocarpaceæ.—PLATANACEÆ.——
Altingiaceæ.

Fig. CLXXXVI.—Platanus orientalis. 1. The ♂ inflorescence; 2. the ♀; 3. an anther; 4. a perpendicular section of an ovary; 5. a perpendicular section of a ripe fruit.

ALLIANCE XX. *EUPHORBIALES.*—THE EUPHORBIAL ALLIANCE.

DIAGNOSIS.—*Diclinous Exogens, with scattered monodichlamydeous flowers, superior consolidated carpels, axile placentæ, and a large embryo surrounded by abundant albumen.*

The main difference between the Euphorbial and Urtical Alliances consists in the compound consolidated pistil of the former and the simple one of the latter. In other respects they are much the same. Euphorbials may be regarded then as a higher form of Urticals, and accordingly we find their lateral affinities also pointing to groups with a more complicated structure; as for example to Rhamnads in the perigynous, and Malvads in the hypogynous Sub-class. They touch Urticals by such a genus as Eremocarpus among Spurgeworts; and Scepads also bring them to the borders of the Amental Alliance. Starworts are to Euphorbials what Hornworts are to Urticals. Into Garryals they pass by way of the Helwingiads, which, if their embryo was not so small, would be almost an inferior fruited form of Spurgeworts. The only doubtful part of the Alliance is the Batids, whose structure is ill ascertained, and the Nepenths, whose indefinite scobiform seeds are very unlike anything else in the Alliance. But it seems difficult to find any better place for the last Order.

NATURAL ORDERS OF EUPHORBIALS.

Ovules definite, suspended, anatropal. Radicle superior.	90. EUPHORBIACEÆ.
Ovules definite, suspended, campylotropal. Radicle inferior. Albumen mealy	*GYROSTEMONEÆ.
Ovules definite, suspended, anatropal. Radicle superior. ♂ *amentaceous*	91. SCEPACEÆ.
Ovules definite, suspended, amphitropal. Radicle superior . . .	92. CALLITRICHACEÆ.
Ovules definite, ascending, anatropal. Radicle inferior	93. EMPETRACEÆ.
Ovules solitary, ascending. ♀ *naked, combined into a succulent cone* . .	*BATIDEÆ.
Ovules 00, *ascending. Radicle inferior. Seeds scobiform* . . .	94. ? NEPENTHACEÆ.

ORDER XC. EUPHORBIACEÆ.—SPURGEWORTS.

Euphorbiæ, *Juss. Gen.* 385. (1789).—Euphorbiaceæ, *Ad. de Juss. Monogr.* (1824); *Endl.* ccxliii.; *Meisner, p.* 336; *Klotzsch in Erichs. Archiv.* 7. 175. (1841).—Trewiaceæ, *Ed. prior. p.* 174.—Pseudantheæ, *Endl. p.* 328.—Anthoboleæ, *Endl. p.* 328?—Putranjiveæ, *Endl. p.* 287.

DIAGNOSIS.—*Euphorbial Exogens, with definite suspended anatropal ovules, scattered flowers, and tricoccous fruit.*

Trees, shrubs, or herbaceous plants, often abounding in acrid milk. Leaves opposite or alternate, simple, rarely compound, often with stipules. Flowers axillary or terminal,

Fig. CLXXXVII.

arranged in various ways, sometimes inclosed within an involucre resembling a calyx. Flowers ♂ ♀. Calyx inferior, with various glandular or scaly internal appendages; (sometimes wanting.) Corolla either consisting of petals or scales equal in number to the sepals, or absent, or sometimes more numerous than the sepals; sometimes monopetalous. ♂ Stamens definite or indefinite, distinct or monadelphous; anthers 2-celled, sometimes opening by pores. ♀ Ovary free, sessile, or stalked, 1- 2- 3- or more celled; ovules solitary or twin, suspended from the inner angle of the cell; styles equal in number to the cells, sometimes distinct, sometimes combined, sometimes none; stigma compound, or single with several lobes. Fruit generally tricoccous, consisting of 3 carpels splitting and separating with elasticity from their common axis, occasionally fleshy and indehiscent. Seeds solitary or twin, suspended, often with an aril; embryo inclosed in fleshy albumen; cotyledons flat; radicle superior.

No group of plants can illustrate better than this the entangled nature of botanical affinities; for it claims kindred in an almost equal degree with Nettleworts, because of its unisexuality, and with Rhamnads and Mallowworts when that circumstance is left out of consideration. By the school of Jussieu it is considered an apetalous Order, with a tendency to form a corolla; by myself and others it is regarded as a polypetalous Order, losing its petals in a part of the species.

The reason for considering Spurgeworts as an apetalous Order is because of the want of a corolla in the genera with which European Botanists are most familiar. But if, instead of considering the imperfectly developed genera of Europe as typical of the true structure of the Order, we look to those of tropical countries, we find that the apetalous character by no means holds good with them. In Aleurites, for example, the petals are as much developed as in a Malvaceous plant; the same thing occurs in Jatropha, Elæococca, and others; and, in fact, upon looking through the genera described by Adrien de Jussieu in his Monograph, it appears that out of 61 genera no fewer than 32 have petals. The tendency of the Order is, therefore, at least as great to form petals as to want them. Now if this be so, and the separation of sexes be disregarded, it will be found that it is with Mallowworts, on the one hand, and Rhamnads, on the other, that they

---

Fig. CLXXXVII.—Andrachne telephioides. 1. A male flower; 2. a female flower; 3. a pistil with the scales at its base; 4. a transverse section of an ovary; 5. a ripe seed; 6. a vertical section of it.

most agree, and especially with the former. Their habit, indeed, and general appearance, are in certain instances so much alike that one might easily mistake some Crotons, Aleurites, &c., for Mallowworts ; the starry structure of their hairs, their monadelphous stamens, the definite number of ovules in a definite number of united carpels, are all further points of resemblance. As to the relationship of Spurgeworts to Rhamnads it was long ago perceived by Jussieu, and has been since adverted to by Adolphe Brongniart (*Monogr. des Rhamn. p.* 35.). Brown, too, in omitting Spurgeworts from the apetalous Orders, in his Prodromus, may be conjectured to have entertained a similar opinion; and Auguste de St. Hilaire inquires whether they are not intermediate between Mallowworts and Menispermads. A writer in the *Linnæa* (14. 250.), would place this Order next Byttneriads, and it is no doubt to that form of the Malval Alliance that it approaches nearest.

But if, with Jussieu, we consider the separation of sexes a great physiological character, the Order of Spurges will join that of Nettles, through Eremocarpus, a curious Californian plant lately discovered by Mr. Bentham, which indeed might be referred indifferently to Spurgeworts or Artocarpads. And so again with Antidesmads, their character is very little different from that of such drupaceous Spurgeworts as Sarcococca. Nor can their close connection with the Garryal Alliance be overlooked ; for Helwingiads are scarcely more than Spurgeworts with an adherent ovary and minute embryo.

Misled by imperfect information, I formerly proposed a group called Trewiaceæ ; but it has been shown by Klotzsch, who has had the opportunity of examining authentic materials concerning Trewia, that it is really a tetracoccous genus of the present Order, nearly allied to Rottlera, (*Wiegm.* vii. 257). Although there does not appear to be any considerable affinity between this tribe and Cucurbits, yet it is to be noted that several genera have a scrambling habit, that the number 3 prevails in the ovary of both Orders, and that the genus Peripterygium has, according to Hasskarl, the habit of a Momordica. There seems to be nothing in Putranjiva to distinguish it from a drupaceous Spurgewort.

In general the structure of Spurgeworts is very uniform, and upon the whole there are few extensive Orders in which it is less liable to exceptions. Some, however, of a very striking kind do occur. Foremost among the instances of anomalous structure is the common Spurge, (Euphorbia), in which the involucre is so much like a calyx, and sometimes seems to be even augmented with petals, that a student finds it difficult to believe that it is

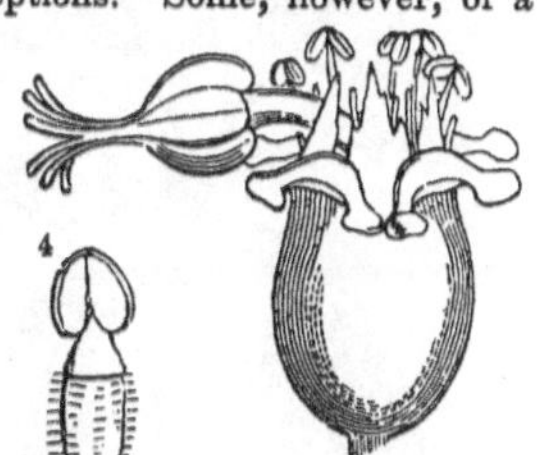

Fig. CLXXXIX.

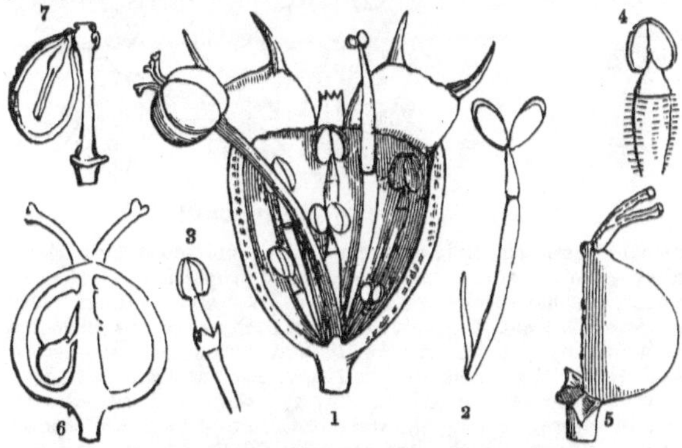

Fig. CLXXXVIII.

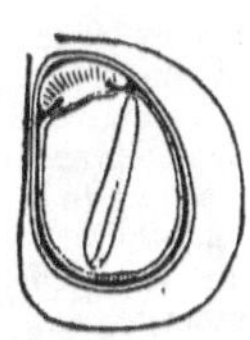

Fig. CXC.

really composed of numerous naked ♂ flowers surrounding a ♀ equally destitute of calyx and corolla. The real history of the structure is however proved in many ways, and especially by such plants as Monotaxis, in which there is also a number of ♂ flowers surrounding one ♀, but each is furnished with a calyx, and the cup-like involucre which disguises Euphorbia is reduced to a few scales. Besides these instances of anomalous, structure, we find the carpels reduced to 2 in Mercurialis, &c., or even 1 in Eremocarpus and Peripterygium, or increased to as many as 9 in Anisonema, or 15 in Hura. The

Fig. CLXXXVIII.—1. The involucre of a Euphorbia, containing monandrous male florets, surrounding a long-stalked female ; 2, 3, 4. male florets of different species, with the articulation that separates the filament from the pedicel ; 5. a carpel separate ; 6. a vertical section of an ovary ; 7. a vertical section of a ripe seed, showing the central column and an embryo in the midst of albumen.

Fig. CLXXXIX.—The involucre of Euphorbia Lathyris.

Fig. CXC.—Perpendicular section of the seed of Euphorbia Lathyris.

fruit, moreover, which generally splits with elasticity, becomes a drupe in Sarcococca and others. Finally, in a few rare instances the albumen is said to be missing.

This extensive Order, which probably does not contain fewer than 2500 species, either described or undescribed, exists in the greatest abundance in equinoctial America, where about 3-8ths of the whole number have been found; sometimes in the form of large trees, frequently of bushes, still more usually of diminutive weeds, and occasionally of deformed, leafless succulent plants, resembling Indian Figs in aspect, but differing from them in every other particular. In the Western world they gradually diminish as they recede from the equator, so that not above 50 species are known in North America, of which a very small number reaches as far as Canada. In the Old World the known tropical proportion is much smaller, arising probably from the species of India and equinoctial Africa not having been described with the same care as those of America; not above an eighth having been found in tropical Africa, including the islands; a sixth is perhaps about the proportion in India. A good many species inhabit the Cape, where, and in the North of Africa, they often assume a succulent habit; and there are almost 120 species from Europe, including the basin of the Mediterranean: of which 16 only are found in Great Britain. and 7 in Sweden.

Fig. CXCI. Fig. CXCII.

A very large proportion of these plants is venomous. The poisonous principle resides chiefly in their milky secretion, and is most powerful in proportion as that secretion is abundant. The hairs of some species are stinging. The bark of various species of Croton is aromatic, as Cascarilla; and the flowers of some, such as Caturus spiciflorus, give a tone to the stomach. Many of them act upon the kidneys, as several species of Phyllanthus, the leaves of Mercurialis annua, and the root of Ricinus communis. Several are asserted by authors to be useful in cases of dropsy; some Phyllanthuses are emmenagogue. The bark of several Crotons, the wood of Croton Tiglium and common Box, the leaves of the latter, of Cicca disticha, and of several Euphorbias, are sudorific, and used against syphilis: the root of various Euphorbias, the juice of Commia, Anda, Mercurialis perennis, and others, are emetic; the leaves of Box and Mercurialis, the juice of Euphorbia, Commia, and Hura, the seeds of Ricinus, Croton Tiglium, &c. &c., are purgative. Many are dangerous, even in small doses, and so fatal in some cases, that no practitioner would dare to prescribe them; as, for example, Manchineel. In fact, there is a gradual and insensible transition, in this Order, from mere stimulants to the most dangerous poisons. The latter have usually an acrid character, but some of them are also narcotic, as those Phyllanths the leaves of which are thrown into water to intoxicate fish. Whatever the stimulating principle of Spurgeworts may be, it seems to be volatile, because application of heat is sufficient to dissipate it. Thus the

Fig. CXCI.—Eremocarpus setigerus.—*Bentham.* 1. a young pistil; 2. a ripe fruit after dehiscing.

Fig. CXCII.—Monotaxis tridentata. 1. a ♀ flower surrounded by several ♂s; 2. a ♂ apart; 3. a stamen; 4. a sepal; 5. a ♀ apart; 6. a transverse section of the ovary.—*Endlicher.*

starchy root of the Manihot or Cassava, which when raw is a violent poison, becomes wholesome nutritious food when roasted. In the seeds of some the albumen is harmless and eatable, but the embryo itself is acrid and dangerous. Many of the species furnish Caoutchouc, that most innocuous of all substances, produced by the most poisonous of all families, which may be almost said to have given a new arm to surgery, and which has become an indispensable necessary of life ; it exists in Artocarpads and elsewhere, but is also the produce of species of Spurgeworts.

The properties of this Order are so important, that the object of this work would be unfulfilled if I did not, in addition to the foregoing general view, add a detailed list of the qualities of the most remarkable species named by writers.

Among milking species, the first to be noticed are the Cactus-shaped kinds, inhabiting Africa chiefly, but also found in India. It is said that King Juba discovered Euphorbia in Barbary, and named it after his physician, who was brother to Musa. The plant of King Juba is referred to Euphorbia officinarum, *Linn.;* a many-angled succulent species growing in tufts armed with double spines, and now found in the North of Africa ; others, however, believe it to have been E. antiquorum, a triangular branching species whose angles are sinuous and spiny, and which appears to be widely dispersed through Africa. The gum resin Euphorbium, now found in our shops, an acrid poison, is partly gathered in Africa from those two species, and partly in the Canaries, from E. canariensis ; it flows from the wounded stems, and is collected in leather bags. It is an extremely acrid inflammable substance, producing severe inflammation of the nostrils, if those who powder it do not guard themselves from its dust ; according to chemists, it consists of wax, myricine, phyteumacolla, and various salts. In India it is mixed with the oil expressed from the seeds of Sesamum orientale, and used externally in rheumatic affections, internally in cases of obstinate constipation. Orfila regards it as a poison. It is little used in Europe. The Arabs make up violent diuretic pills, by rubbing over the juice of E. antiquorum with flour ; yet their camels will eat the branches of the plant when cooked. The juice of E. heptagona, virosa, and cereiformis, African species, furnishes the Ethiopians, and E. cotinifolia, the wild Brazilians, with a mortal poison for their arrows. That of the leaves of Euphorbia nereifolia is prescribed by the native practitioners of India, internally as a purge and deobstruent, and externally, mixed with Margosa oil, in such cases of contracted limb as are induced by ill-treated rheumatic affections. The leaves have, no doubt, a diuretic quality. E. tribuloides, one of the least of the Cactus-shaped species, is regarded as a diaphoretic in the Canaries, where it grows wild. Of the leafy Euphorbias great numbers have been found to possess a milk with purgative or emetic qualities. Endlicher mentions E. Esula (Wolfsmilch of the Germans), Cyparissias, amygdaloides, whose roots have been the basis of some celebrated quack fever mixtures, Helioscopia, our commonest weed, (τιθύμαλος and ἡλιοσκόπιος), Peplus, Peploides (the πέπλος of Hippocrates), palustris, pilosa, Chamæsyce, Peplis, (the πέπλιον of Hippocrates, and πεπλὶς of Dioscorides), spinosa (ἱπποφαές,) dendroides (τιθύμαλος μὲγας, Hippoc.), Aleppica, and Apios ; all plants having more or less reputation as purgatives. In America there are also employed for the same purpose, E. buxifolia in the West Indies ; papillosa in Brazil, a species apt to produce dangerous superpurgations, and called Leiteira or Lechetres ; laurifolia in Peru ; portulacoides in Chili ; and Tirucalli in India ; the fresh acrid juice of the latter is used as a vesicatory ; it is common in the Madras Presidency, and makes an excellent hedge, which may be formed with very little trouble. A trench must be dug where it is intended to be, at the beginning of the rainy season ; in this, cuttings being placed, and the earth pressed about them, they establish themselves without further trouble. No cattle will touch the leaves, and in one year it becomes a tolerable fence.

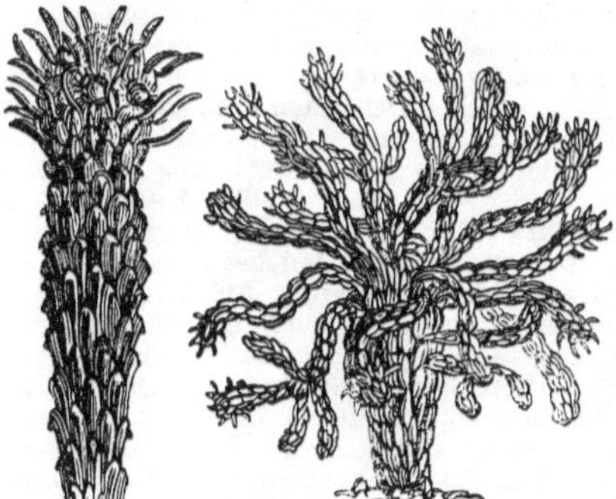

Fig. CXCIII.

Among syphilitic remedies are Euphorbia parviflora and hirta in India, and E. linearis in America. E. hiberna also, before the introduction of mercury, was frequently administered in England against venereal affections ; the Spaniards use E. canescens for such purposes to this day.

Fig. CXCIII.—Euphorbia Caput Medusæ.

The roots of some are emetic. According to Deslongchamps, the powdered root of E. Gerardiana vomits easily in doses of 18 or 20 grains. The root of Euphorbia Ipecacuanha is said, by Barton, to be equal to the true Ipecacuanha, in some respects superior; and not unpleasant either in taste or smell. E. Pithyusa in the Mediterranean is also esteemed. Euphorbia thymifolia is somewhat aromatic and astringent, and is prescribed in India in the diarrhœa of children, and as a vermifuge. In the same way is employed E. hypericifolia, a plant of tropical America, which is astringent and somewhat narcotic. Nevertheless E. balsamifera has no such qualities, and is eaten when cooked. E. mauritanica is also employed as a condiment, but its acridity is by no means inconsiderable; they say it is used to adulterate Scammony. The sap of E. phosphorea shines with a phosphorescent light in a warm night in the ancient forests of Brazil.

The genus Pedilanthus stands nearest to Euphorbia, and is not less potent in its quality; P. tithymaloides has an acrid bitter milk; a decoction of the dried shrub of it and P. padifolius (called Jewbush) is employed in syphilitic cases, and in amenorrhœa; the root is emetic. Some of the trees again are among the most poisonous of all that tropical countries produce. The juice of Excæcaria Agallocha, and even its smoke when burnt, affects the eyes with intolerable pain, as has been experienced occasionally by sailors sent ashore to cut fuel, who, according to Rumphius, having accidentally rubbed their eyes with the juice, became blinded, and ran about like distracted men, and some of them finally lost their sight. This juice is described as being thick, nauseous, and a violent purgative. The smoke of the burning branches is said to injure the eyesight. Agallochum or Aloes wood, an inflammable, fragrant, resinous substance, has been supposed to belong to this plant, but is really produced by quite a different race. *See* AQUILARIACEÆ. The famous Manchineel tree, Hippomane Mancinella, is said to be so poisonous that persons have died from merely sleeping beneath its shade. This is doubted, indeed, by Jacquin, who, however, admits its extremely venomous qualities; but it is by no means improbable that the story has some foundation in truth, particularly if, as Ad. de Jussieu truly remarks, the volatile nature of the poisonous principle of these plants is considered, and the various degrees of susceptibility of such influences in the human constitution. The juice of Manchineel is pure white, and a single drop of it falling on the skin burns like fire, forming an ulcer often difficult to heal. The fruit, which is beautiful, and looks like an apple, is turgid with a similar fluid, but in a milder form; the burning it causes in the lips of those who bite it guards the careless from the danger of eating it. The juice of Hura crepitans is stated to be of the same fatal nature as that of Excæcaria; its seeds are said to have been administered to negro slaves as purgatives, in number not exceeding 1 or 2, with fatal consequences. The juice of Sapium aucuparium is reputed poisonous. A case is mentioned by Tussac of a gardener whose nostrils became swollen and seized with erysipelatous phlegmasis, in conseqence of the fumes only of this plant. The sap of Commia cochinchinensis is white, tenacious, emetic, purgative, and deobstruent. Cautiously administered, it is said to be a good medicine in obstinate dropsy and obstructions.

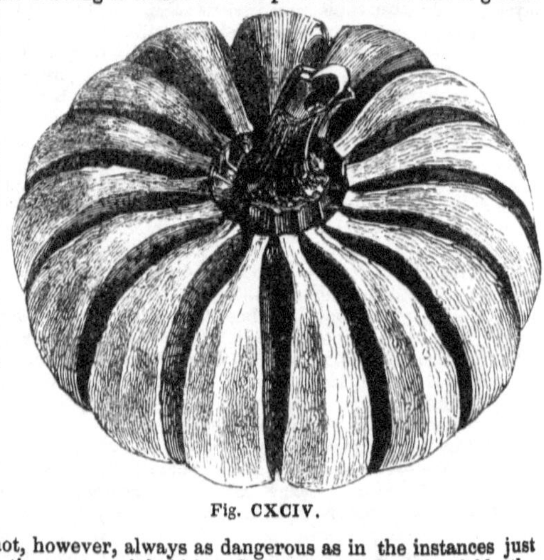

Fig. CXCIV.

The juice of this Order is not, however, always as dangerous as in the instances just given. That of Siphonia elastica, a tree inhabiting Guayana and Brazil, yields the bottle India Rubber, which is known in Europe; in preparing it the natives smear clay moulds with repeated layers of the juice, at the same time drying it in smoke. Aleurites triloba, whose seeds will be mentioned presently, exudes a gummy substance which the natives of Tahiti chew; A. laccifera furnishes gum lac in Ceylon; and the secretions of certain Crotons, viz. Draco and sanguiferum, become a similar red substance in the tropical parts of America.

Among the crowd of emetic and purgative plants having more or less reputation

Fig. CXCIV.—Fruit of Hura crepitans.

in the countries where they grow, many are found in the Euphorbiaceous Order. The root of Stillingia sylvatica is regarded in Carolina and Florida as a specific in syphilitic maladies; the same reputation attaches to Cnemidostachys Chamælea in India, and Jatropha officinalis (Raiz de Tihu) in Brazil. The Tragias volubilis of America, and involucrata, cannabina, and Mercurialis of Asia, are noted for their solvent, diaphoretic and diuretic qualities. More especially the root of Tragia involucrata is reckoned, by the Hindoo doctors, among those medicines which they conceive to possess virtues in altering and correcting the habit in cases of cachexia, and in old venereal complaints attended with anomalous symptoms. The Mercurialis or Dog Mercury of Europe is another active genus. M. annua has a nauseous taste and is slightly purgative; M. perennis is much more active, sometimes producing violent vomiting, diarrhœa, a burning heat in the head, convulsions, and death; M. tomentosa, a Mediterranean shrub, is used in hydrophobia; it is vulgarly believed to this day that if women eat the male individuals of this plant, which is diœcious, they will conceive boys, and if the female girls; when boiled with other vegetables it acts as a mild purgative (English Mercury must not be confounded with these plants. *See* CHENOPODS). Omphalea triandra, a Guayana plant, has a white juice which turns black in drying, and is used in place of ink. In Cicca disticha, an Indian bush, the root is violently purgative, a decoction of the leaves diaphoretic. A decoction of Croton perdicipes, called Pe de Perdis, Alcamphora, and Cocallera, in different provinces of Brazil, is much esteemed as a cure for syphilis, and as a useful diuretic. The root of another species, called Velame do Campo, C. campestris, has a purgative root, also employed against similar disorders. It is, however, more common to find balsamic juices in the American Crotons, among which balsamifer is employed in Martinique in the preparation of the liqueur called Eau de Mantes; frankincense is extracted from C. thurifer and adipatus on the banks of the Amazons; C. humilis is used for its aromatic qualities in medicating baths in the West Indies; at the Cape of Good Hope the fragrant C. gratissimus is used by the Koras as a perfume; the balsam of C. origanifolius is mentioned among the substitutes for copaiva; its leaves and bark are considered diaphoretic and antispastic; finally, C. niveus is a vulnerary.

The most important, however, among the aromatic Spurgeworts are the plants that yield Cascarilla, a valuable bitter, tonic, aromatic, stimulant bark, imported from the West Indies. This drug has been at one time referred to C. Eleuteria, a Bahama shrub, at another to C. Cascarilla, a Jamaica bush, called, from its appearance, Wild Rosemary. As a good deal of controversy has been raised respecting this matter, it is as well to state that the question is now set at rest in consequence of the Hon. J. C. Lees, Chief Judge in the Bahamas, having sent home specimens of the Cascarilla tree, with the bark itself and the leaves adhering to it. It proves to be this species, concerning which Mr. Lees has favoured me with the following note: "The plant is scarcely known here by the name of Cascarilla, but is commonly called Sweet Wood Bark, and often Eleuthera Bark, because it is chiefly gathered on the island of Eleuthera. It is the only bark receiving the name of Cascarilla exported from the Bahamas, where the tree grows in great abundance." It is, however, certain that the C. pseudo-China furnishes Cascarilla in Mexico, where it is called Quina blanca, and Copalche Bark; and C. nitens, cascarilloides, micans and suberosus, seem to be little inferior to the C. Eleuteria itself.

The bark of the Asiatic Bridelias is astringent; so is that of Stylodiscus trifoliatus, whose wood is of a red colour, as is the bark; the former is employed for masts and spars of small vessels in Java.—*Horsfield.* The common Box tree, Buxus sempervirens, has a bark with qualities similar to those of Guaiacums, for which it has been substituted; the leaves are bitter, and very purgative; nevertheless, it is alleged that they have been used as a substitute for Hops.—*Endl.* They say that in some parts of Persia where Box trees abound, camels cannot be employed because it is found impracticable to prevent their browsing on the leaves, which kill them. The root and bark of Codiæum variegatum are acrid, and excite a burning sensation in the mouth if chewed: but the leaves are sweet and cooling. The root, leaves, and young shoots of Phyllanthus Niruri are considered, in India, deobstruent, diuretic, and healing; the leaves are very bitter, and a good stomachic. Some other species, particularly P. urinaria, are powerful diuretics. The bruised leaves of P. Conami are used for inebriating fishes. The boiled leaves of Plukenetia corniculata are said to be an excellent potherb, for which purpose the plant is cultivated in Amboyna. Acalypha Cupameni, an Indian herb, has a root which, bruised in hot water, is cathartic; a decoction of its leaves is also reported to be laxative. The flowers of Caturus spiciflorus are spoken of as a specific in diarrhœa, either taken in decoction or in conserve.

The oil of the seeds is perhaps the most important part of the useful products of this Order. It is often among the most valuable of known diuretics and purgatives. Croton

Tiglium, and Pavana, two East Indian trees, whose seeds were formerly called Grana molucca, stand at the head of this class of medicines; their oil is so acrid as to blister the skin, and it will even act when externally applied to the abdomen. Next to these comes Ricinus communis, the castor oil plant, an annual in Europe, a tree in Africa, conspicuous with its broad palmate leaves, which have given it the name of Palma Christi, and spiny capsules, whose use is traceable into remote antiquity, under the name of Semina cataputiæ majoris; it is found that the albumen of this plant has little activity, but that the virulence resides mainly in the embryo and seed-coats; so that the activity of the oil will depend upon the amount of pressure, &c. to which the seeds may have been subjected; when long boiled their oil is found poisonous. In like manner the seeds of Omphalea are said to be eatable if the embryo is extracted, but if this is not done, to be too cathartic for food. Mr. W. Macleay calls this nut "most delicious and wholesome," and speaks of it as the Cobnut or Hognut of Jamaica. Similar qualities reside in the seeds of Hura crepitans, the Sand Box tree, and Curcas purgans (Jatropha purgans, *L.*); the latter plant is remarkable for the fierce acridity of its seeds, which are commonly called Purging-nuts. An expressed oil is obtained from them, which is reckoned a valuable external application in itch and herpes; it is also used, a little diluted, in chronic rheumatism. The varnish used by the Chinese for covering boxes is made by boiling this oil with oxide of iron. The leaves are considered as rubefacient and discutient; the milky juice is supposed to have a detergent and healing quality, and dyes linen black. In like manner Curcas multifidus produces a purgative oil called Pinhoen, under which name it reaches Europe from South America. From the seeds of Jatropha glauca the Hindoos prepare, by careful expression, an oil which, from its stimulating quality, they recommend as an external application in cases of chronic rheumatism and paralytic affections. Euphorbia Lathyris, called in English gardens the Caper-bush, to which it has no resemblance, was one of the plants which Charlemagne in his Capitularies commanded to be cultivated in all monastic gardens, for the sake of its purgative seeds, which were called Semina Cataputiæ minoris; they are acrid like Tiglium, and not mild like those of Ricinus. The capsules of this plant are reported to intoxicate fish. Euphorbia hibernica is extensively used by the peasantry of Kerry for poisoning, or rather stupefying fish, in the same manner as the exotic E. piscatoria. So powerful are its qualities that a small creel or basket, filled with the bruised plant, suffices to poison the fish for several miles down a river.—*Hooker, Brit. Fl. ed.* 4. p. 326. The Anda of Brazil is famous for the purgative qualities of its seeds, which are called Purga da Paulistas, and are fully as powerful as those of the Palma Christi. The Brazilians make use of them in cases of indigestion, in liver complaints, the jaundice, and dropsy. The bark, roasted on the fire, passes as a certain remedy for diarrhœa brought on by cold. According to Marcgraaf, the fresh bark steeped in water communicates to it a narcotic property which is sufficient to stupefy fish. The seeds are either eaten raw, or are prepared as an electuary; they yield an oil, which is said by M. Auguste de St. Hilaire to be drying and excellent for painting; in short, much better than nut oil. The Cape colonists collect the fruit of Hyænanche globosa, an anomalous plant of this Order, and kill hyænas with mutton rubbed with the powder. The seeds of Stillingia sebifera, a Chinese tree, common in most tropical countries, are enveloped in a fatty matter, from which candles are prepared; a mild oil is also furnished by them. Two species of Elæococca, the one E. verrucosa from Japan, the other E. vernicia from China, furnish oil by pressure of their seeds; the former for burning, the latter for painters' work; both too acrid to be used as food.

Nevertheless, some have an eatable fruit; that of Anda and Omphalea has been already mentioned. Aleurites triloba, a Molucca tree, has much reputation for its nuts, which are reported to be aphrodisiac; and the seeds of Conceveiba guianensis are said to be delicious. The succulent fruit of Cicca disticha and racemosa is sub-acid, cooling, and wholesome; its leaves are sudorific, and its seeds cathartic. The capsules of Cluytia collina are poisonous, according to Roxburgh. Emblica officinalis also, has an acrid fruit, which in India is made into a pickle; when ripe and dry it is astringent, and has been employed, under the name of Myrobalani Emblici, against diarrhœa, dysentery, and cholera.

It is not a little remarkable, that here, as in so many other cases, we should find in a very dangerous Natural Order such an abundant secretion of starch as renders certain species useful for food when the acrid matter is removed. This is most especially the case with the Mandioc plant, Manihot utilissima, *Pohl*, (Jatropha manihot, *L.*,) a shrub about 8 feet high, extensively cultivated for food all over the tropical parts of the world. Of this plant the large root, weighing as much as 30lbs., is full of venomous juice, which if taken internally produces death. The roots are rasped, the pulp well bruised, and then thoroughly washed, after which the mark is placed on iron plates to be heated. In this way the venom is washed out or driven off, and the residue becomes Cassava.

The powder which floats off in the water is a very pure starch, which, when it settles down, becomes Tapioca. Manihot Aipi, Pohl assures us, has a harmless root.

Cnidoscolus quinquelobus (Jatropha urens, *L.*) is covered with hairs which sting severely, as indeed occurs elsewhere in this Order. The juice of its branches and seeds is diuretic. The root of another species, C. herbaceus, is used in the same way as Mandioc in Mexico and Carolina.

A few yield dyes. Turnsole, a well known purple drug, which becomes blue upon the application of ammonia, is the inspissated juice of Crozophora tinctoria (ἡλιοτρόπιον μικρόν) found in the southern parts of Europe. Its juice is acrid, and its seeds cathartic, as in others of this Order. Similar colours are found in other species of Crozophora, in some Crotons, Argythamnia, Ditassa and Claoxylon. The seed-vessels of Rottlera tinctoria are covered with a mealy powder which gives a scarlet colour, as also does its root. Maprounea brasiliensis, or the Marmeleiro do Campo of Brazil, yields a black dye, which is, however, fugitive; a decoction of its root is also administered in derangement of the stomach; according to Auguste de St. Hilaire, it is destitute of milky juice.

The timber of Buxus sempervirens is remarkable for its hardness and compactness, whence its value to wood engravers. There is reason to believe that the timber imported from the coast of Africa, under the name of African Teak, belongs to some tree of this Order. For further information as to the uses of Spurgeworts, see *Martius Materia Medica Braziliensium.*

GENERA.

I. — PROSOPIDOCLINEÆ, *Klotzsch.* Ovule solitary. Seeds with an arillus, and no albumen (?). Involucre globose, bladder-like, opening on one side, finally dropping off, containing from 3 to 6 flowers. Flowers diœcious, apetalous.

Schismatopera, *Kl.*
Spixia, *Leandr.*
Pera, *Mutis.*
Peridium, *Schott.*

II.—EUPHORBIEÆ. Ovule solitary. Seeds albuminous. Flowers monœcious, apetalous; ♂ and ♀, mixed, in a cup-shaped involucre.

Pedilanthus, *Neck.*
Crepidaria, Haw.
Tithymaloides, Tournf.
Euphorbia, *Linn.*
Tithymalus, Tournef.
Euphorbium, Isn.
Keraselma, Neck.
Athymalus, Neck.
Treisia, Haw.
Dactylanthus, Haw.
Medusea, Haw.
Galorhœus, Haw.
Esula, Haw.
Anisophyllum, Haw.
Lopadocalyx, *Kl.*
Poinsettia, *Graham.*
Anthostema, *Adr. Juss.*
Dalechampia, *Plum.*

III.—HIPPOMANEÆ. Ovule solitary. Flowers apetalous, in spikes; bracts 1-many-flowered.

Maprounea, *Aubl.*
Ægopricon, Linn. f.
Adenopeltis, *Bert.*
Colliguaja, *Molin.*
Dactylostemon, *Kl.*
Gymnarrhen, Leandr.
Excœcaria, *Linn.*
Gymnanthes, *Swartz.*
Gussonia, *Spreng.*
Sebastiania, *Spreng.*
Adenogyne, *Kl.*
Sennefeldera, *Kl.*
Actinostema, *Mart.*
Sarothrostachys, *Kl.*
Styloceras, *Adr. Juss.*
Commia, *Lour.*
Synaspisma, *Endl.*
Hura, *Linn.*
Psilostachys, *Turcz.*
Hippomane, *Linn.*
Mançanilla, Plum.
Pachystemon, *Blum.*
Omalanthus, *Adr. Juss.*
Carumbium, Reinw.
Stillingia, *Gard.*
Stillingfleetia, Boj.
Sapium, *Jacq.*
Triadica, Lour.
Cœlebogyne, *J. Sm.*
Microstachys, *Adr. Juss.*
Cnemidostachys, Adr. Juss.

IV.—ACALYPHEÆ. Ovule solitary. Flowers apetalous, in clustered spikes or racemes.

Tragia, *Plum.*
Schorigeram, Adans.
Traganthus, *Kl.*
Leucandra, *Kl.*
Cnesmone, *Blum.*
Leptorhachis, *Kl.*
Bia, *Kl.*
Plukenetia, *Plum.*
Sajor, Rumph.
Botryanthe, *Kl.*
Hedraiostylus, *Hassk.*
Pterococcus, Hassk.
Ceratococcus, Meisn.
Anabæna, *Adr. Juss.*
Mercurialis, *Linn.*
Linozostis, Endl.
Trismegista, Endl.
Acalypha, *Linn.*
Cupameni, Adans.
Usteria, Dennst.
? Caturus, Linn.
Galurus, Spreng.
Adenocline, *Turcz.*
Mappa, *Adr. Juss.*
Macaranga, *Thouars.*
Panopia, Noronh.
Monospora, *Kl.*
Ctenomeria, *Harv.*
Claoxylon, *Adr. Juss.*
Erythrochilus, Reinw.
Conceveiba, *Aubl.*
Cladogynos, *Zippel.*
Platygyna, *Mercier.*
Simmondsia, *Nutt.*
Aparisthmium, *Endl.*
Conceveibum, L.C.Rich.
Omphalea, *Linn.*
Omphalandria, P. Br.
Duchola, Adans.
Hecatea, *Thouars.*
Cleidion, *Blum.*
Alchornea, *Soland.*

V.—CROTONEÆ. Ovule solitary. Flowers usually having petals, in clusters, spikes, racemes or panicles.

Cephalocroton, *Kl.*
Trachycaryon, *Kl.*
Calyptrostigma, *Kl.*
Garcia, *Rohr.*
Mabea, *Aubl.*
Siphonia, *Rich.*
Hevea, Aubl.
Elateriospermum, *Blum.*
Anda, *Marcgr.*
Johannesia, Velloz.
Andiscus, Fl. Fl.
Aleurites, *Forst.*
Ambinux, Commers.
Telopea, Soland.
Camirium, Rumph.
Givotia, *Griff.*
Ostodes, *Blum.*
Elæococca, *Commers.*
Dryandra, Thunb
Vernicia, Lour.
Abasin, *Kämpf.*
Jatropha, *Kunth.*
Adenorhopium, Pohl.
Curcas, *Adans.*
Bromfeldia, Neck.
Castiglionia, R. et Pav.
Cnidoscolus, *Pohl.*
Bivonea, Raf.
Jussievia, Houst.
Manihot, *Plum.*
Janipha, Kunth.
Mandiocca, Link.
Aypi, Bauh.
Camagnoc, Aubl.
Ricinus, *Tournef.*
Spathiostemon, *Blum.*
Baloghia, *Endl.*
Ricinocarpus, *Desf.*
Echinosphæra, Sieb.
Röperia, Spreng.
Amperea, *Adr. Juss.*
Mozinna, *Orteg.*
Loureira, Cav.
Hemicyclia, *W. et Arn.*
Gelonium, *Roxb.*
Erythrocarpus, Blum
Codiæum, *Rumph.*
Phyllaurea, Lour.
Tetrorchidium, *Pöpp.*
Rottlera, *Roxb.*
Mallotus, Lour.
Adisca, Blum.
Trewia, *L.*
Tetragastris, Gærtn.
Adriania, *Gaudich.*
Cheilosa, *Blum.*
Acidoton, *Swartz.*
Baliospermum, *Blum.*
Beyeria, *Miq.*
Hæmatospermum, *Wall.*
Hendecandra, *Eschsch.*
Asterogyne, Benth.
Adelia, *Linn.*
Bernardia, Houst.
Crotonopsis, *L. C. Rich.*
Leptemon, Raf.
Friesia, Spreng.
Peripterygium, *Hassk.*
Heterochlamys, *Turcz.*
Serophyton, *G. B.*
Eremocarpus, *G. B.*
Engelmannia, *Kl.*
Geisseleria, *Kl.*
Pilinophytum, *Kl.*
Croton, *Linn.*
Ricinoides, Tournef.
Cascarilla, Adans.
Tridesmus, Lour
Aroton, Neck.
Luntia, Neck.
Cinogasum, Neck.
Brunsvia, Neck.
Ricinocarpus, Boerh.
Julocroton, *Mart.*

Podostachys, *Kl.*
Astræa, *Kl.*
Ocalia, *Kl.*
Eutropia, *Kl.*
Cleodora, *Kl.*
Timandra, *Kl.*
Medea, *Kl.*
Crozophora, *Neck.*
Tournesolia, Scop.
Chiropetalum, *Adr. Juss.*
Caperonia, *St. Hil.*
Cavanilla, Fl. Flum.
? *Schinza*, Dennst.
Ditaxis, *Vahl.*
Monotaxis, *Brongn.*
Argythamnia, *P. Br.*
Ateramnus, P. Br.
Philyra, *Kl.*
Trigonostemon, *Blum.*
Trigostemon, Blum.
Ryparia, *Blum.*
Ryparosa, Blum.

VI. — PHYLLANTHEÆ. Ovules in pairs. Stamens in the centre of the flower.

Cyclostemon, *Blum.*
Enchidium, *Jack.*
Briedelia, *Willd.*
Heydia, Dennst.
Cluytia, *Ait.*
Clutia, Boerh.
Altora, Adans.
Cratochwila, Neck.
Andrachne, *Linn.*
Eraclissa, Forsk.
Limeum, Forsk.
Arachne, Neck.
Sauropus, *Blum.*
Agyneia, *Linn.*
Leiocarpus, *Blum.*
Micranthea, *Desf.*
Pseudanthus, *Sieb.*
Menarda, *Commers.*
Xylophylla, *Linn.*
Genesiphylla, Herit.
Phyllanthus, *Linn.*
Niruri, Adans.
Nymphanthus, Lour.
? *Cathetus*, Lour.
? *Breynia*, Forst.
Melanthesa, *Blum.*
Asterandra, *Kl.*
Kirganelia, *Juss.*
Ardenghelia, Commers.
Emblica, *Gärtn.*
Cicca, *Linn.*
Cheramela, Rumph.
? *Tricaryum*, Lour.
Leptonema, *Adr. Juss.*
Anisonema, *Adr. Juss.*
Glochidonopsis, *Adr. Jus.*
Glochidion, *Forst.*
Bradleia, Banks.
Gynoon, *Adr. Juss.*
Scepasma, *Blum.*
Epistylium, *Swartz.*
Eriococcus, *Hassk.*
Stylodiscus, *Benn.*
Poranthera, *Rudg*

VII.—BUXEÆ, Ovules in pairs. Stamens inserted beneath the sessile rudiment of an ovary.

Flüggea, *Willd.*
Richeria, *Vahl.*
Amanoa, *Aubl.*
Lithoxylon, *Endl*
Securinega, *Commers.*
Discocarpus, *Kl.*
Geblera, *Fisch. et Mey.*
Colmeiroa, *Reuter.*
Savia, *Willd.*
Actephila, *Blum.*
Tricera, *Swartz.*
Crantzia, Swartz.
Buxus, *Tournef.*
Pachysandra, *Michx.*
Thecacoris, *Adr. Juss.*
Microelus, *Wight et Arn.*
Adenocrepis, *Blum.*
Drypetes, *Vahl.*
Sarcococca, *Lindl.*
Putranjiva, *Wall.*
Nageia, *Gærtn.*
Hyænanche, *Lamb.*
Toxicodendron, Thunb.

Doubtful Genera.

Podocalyx, *Kl.*
? Anthobolus, *R. Br.*
Meborea, *Aubl.*
Tephranthus, Neck.
Ægotoxicum, *Ruiz et Pav.*
Æxtoxicum, Id.
Margaritaria, *Linn. f.*
Suregada, *Roxb.*
Hexadica, *Lour.*
Homonoia, *Lour.*
Cladodes, *Lour.*
Echinus, *Lour.*
? *Ulassium*, Rumph.
Lascadium, *Raf.*
Rhytis, *Lour.*
Baccaurea, *Lour.*
Lumanaja, *Blanc.*
Lunasia, *Blanc.*
Dovyalis, *E. Mey.*
Desfontenea, *Fl. Fl.*
Mainea, *Fl. Fl.*
Plagiopteron.

NUMBERS. GEN. 191. Sp. 2500?

Rhamnaceæ.
Byttneriaceæ.
POSITION.—Empetraceæ.—EUPHORBIACEÆ.—Scepaceæ.
Helwingiaceæ.

GYROSTEMONEÆ. (*Endl. Gen. p.* 978; *Meisner, Gen. p.* 322). Trees or shrubs inhabiting New Holland. Leaves alternate, entire, feather-veined, without stipules. ♂ Calyx 6-7-lobed. Corolla 0. Stamens indefinite, distinct, with anthers bursting longitudinally. ♀ Calyx cup-shaped, 6-7-lobed. Corolla 0. Carpels 00, collected round a flat central torus, 2-seeded, with suspended campylotropal ovules. Fruit composed of several membranous cases, arranged in a ring. Seed with a strophiolate hilum. Embryo hooked round mealy albumen, with linear incumbent cotyledons and a slender inferior radicle.—Two genera and three species at present constitute this group, about whose relationship we have at present no certain evidence. Because of their unisexual imperfect flowers, numerous consolidated carpels, and suspended ovules, they are related to Spurgeworts, and especially to the genus Hura; but they cannot be referred to them if Endlicher's statement is correct, that the albumen is mealy and the radicle inferior. Hooker and others station it in the Urtical Alliance, but its composite fruit, mealy albumen, and inferior radicle, do not justify that opinion. Desfontaines and Endlicher regard it as a form of the Mallowworts, but the separated sexes, free stamens, peculiar albumen, and apetalous flowers are unfavourable to that supposition. Finally, Meisner after referring it to Lindenblooms, has come to the conclusion that it ought to be associated with Phytolaccads; and if the flowers were not unisexual this opinion would have great weight: for it must be admitted that the plants under consideration have much the structure of that Natural Order. Nevertheless it seems for the present most advisable to associate Gyrostemonads with unisexual Orders, among which they may be looked upon as a passage to the Phytolaccads in the Chenopodal Alliance.

GENERA.

Gyrostemon, *Desf.*
Codonocarpus, *Cunningh.*

NUMBERS. GEN. 2. Sp. 3.

ORDER XCI. SCEPACEÆ.—SCEPADS.

Scepaceæ, *Ed. pr. p.* 171. (1836); *Endl. Gen. p.* 288; *Meisner, p.* 347.—Forestiereæ, *Endl. p.* 288; *Meisner, p.* 257.

DIAGNOSIS.—*Euphorbial Exogens, with ♂ amentaceous flowers, definite suspended anatropal ovules, and a superior radicle.*

Trees. Leaves coriaceous, alternate, with membranous stipules which form the scales of the buds. Flowers ♂ ♀—♂ amentaceous. Calyx 4-5-leaved, imbricated, very minute and membranous. Stamens 2-5; filaments short, straight, not elastic; anthers 2-celled, opening by longitudinal parallel sutures; connective inconspicuous. ♀ in short axillary racemes. Calyx of six sepals in two whorls, free; the inner ones in one species at least 3-lobed. Ovary 2-celled; style 0; stigma with two short emarginate lobes, or with 4 equal fringed ones; ovules in pairs, collateral, pendulous, anatropal, with a broad scale projecting from the placenta and covering over the foramen; their ends often buried in hairs projecting from the base of the cell. [In Lepidostachys Roxburghii the capsule is round, two-celled, 4-valved; the endocarp thin, tough, and separable from the friable sarcocarp. Seeds single or two, enveloped in a succulent aril; embryo green in the axis of albumen, with obovate cotyledons and a radicle next the hilum.—*Roxb.*]

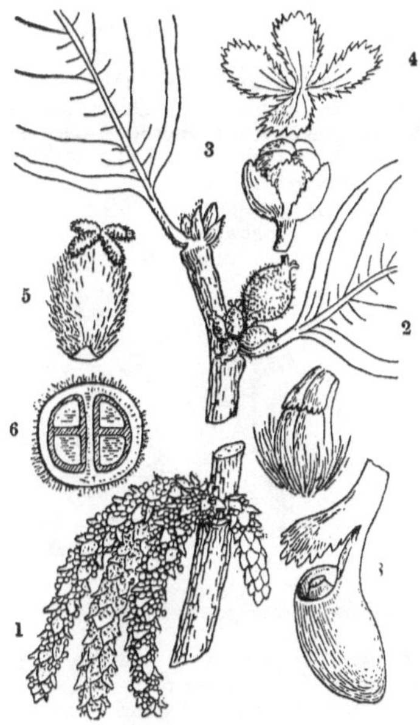

Fig. CXCV.

Here it is that the Euphorbial Alliance touches the Amental; for the plants of the Scepad order may be regarded as Amentaceous Spurgeworts. In their male state they have much the aspect of Mastworts or Birchworts, and one of them has actually been considered an Alnus by Roxburgh; but the females have more the appearance of Antiaris, or of some such Urtical genus. The fruit, which is very remarkable, I only know from Roxburgh's account, the substance of which is quoted from the *Flora Indica.* The manner in which the plates of the placenta overlap the foramen (I believe not till after impregnation) is exceedingly curious; these are no doubt what ultimately become the aril. In the genus Scepa the ends of the ovules are buried in a thick mass of hairs proceeding from the placentary suture near the base of the cell. Forestiera does not appear to differ from Scepads more than the genera of Spurgeworts from each other. It has no aril, and its fruit is indehiscent; but it is amentaceous. Piptolepis of Bentham, placed next it by Endlicher, seems to me very different, on account of its hermaphrodite flowers.

Natives of the tropical forests of India.

The wood of the Kokra, Lepidostachys Roxburghii, is very hard, and is used for various economical purposes.

GENERA.

Scepa, *Lindl.*
Lepidostachys, *Wall.*
Hymenocardia, *Wall.*
Forestiera, *Poir.*
Bigelowia, Smith.
Borya, Willd.
Adelia, L. C. Rich.

NUMBERS. GEN. 3. Sp. 6.

Betulaceæ.
POSITION.—Euphorbiaceæ.—SCEPACEÆ.——————

Fig. CXCV.—Scepa villosa. 1. a cluster of ♂ catkins; 2. a ♀ branch; 3. a ♂ flower; 4. its calyx; 5. an ovary; 6. a cross section of it; 7. the ovules in situ; 8. one of them forced downwards.

ORDER XCII. CALLITRICHACEÆ.—STARWORTS.

Callitrichineæ, *Link. Enum.* 1. 7. (1821); *Lavielle in Ann. Soc. Linn. Par. p.* 229; *DC. Prodr.* 3. 71; *Ed. pr.* (1836); *Endl.* lxxxiv.; *Meisn. p.* 336.

DIAGNOSIS.—*Euphorbial aquatic Exogens, with definite suspended anatropal ovules, and a superior radicle.*

Small aquatic herbaceous plants, with opposite, simple, entire leaves. Flowers axillary, solitary, very minute. Flowers unisexual, monœcious, naked, with 2 fistular coloured bracts. ♂ Stamen hypogynous, single, rarely 2; filament filiform, furrowed along the middle; anther reniform, 1-celled, 2-valved; the valves opening fore and aft. ♀ Ovary solitary, 4-cornered, 4-celled; ovules solitary, attached to the axis, suspended, amphitropal; styles 2, right and left, subulate; stigmas simple points. Fruit 4-celled, 4-seeded, indehiscent. Seeds peltate; embryo inverted in the axis of fleshy albumen; radicle very long, curved, superior; cotyledons very short.

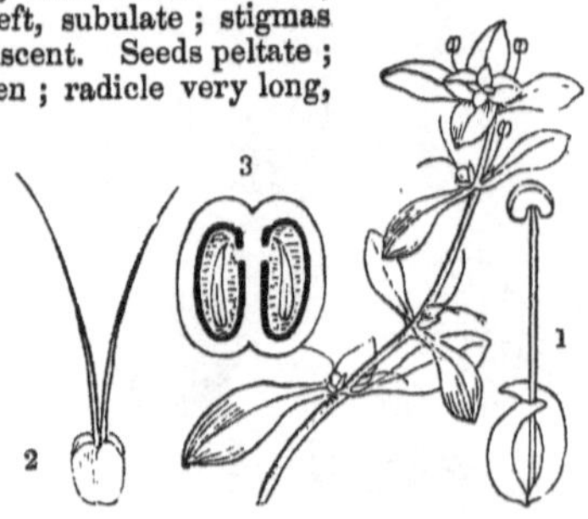

Fig. CXCVI.

I have formerly remarked, that "the affinity of this Order to other dicotyledons appears to be of the same nature as that borne by Lemna to Monocotyledons: they each exhibit the lowest degree of organisation known in their respective classes." Brown considers the Order allied to Hippurids: an opinion in which Botanists seem disposed to concur. The great objection to it is this; Hippurids are a reduced form of the exalbuminous Onagrads, with the petals often absent, and the calyx sometimes diminished to what seems a mere rim; but in reality, in consequence of the ovary being adherent, the whole of the tube of the calyx as well as its rim remains adhering to the ovary, so that the calyx is not in fact materially diminished; but Starworts are absolutely destitute of a calyx and are albuminous. These circumstances, and the unisexual flowers of the Order, seem to point to a widely different station, and accordingly, in the last edition of this work, it was arranged among the Incomplete Orders—in the neighbourhood of Mossweeds. It must, however, be confessed that its relation to these plants is one of analogy rather than of affinity. Nevertheless, Endlicher places it in the same situation, remarking, however, that it is perhaps an aquatic form of Spurgeworts. And in this he seems to be right; at all events it differs so little from that Order, except in its indehiscent fruit and amphitropal ovules, that unless we should hereafter be able to employ internal structure for high systematical divisions, it is in the Euphorbial Alliance that this plant will remain. It is doubtful indeed whether it ought, in the present state of our knowledge, to be regarded as an independent Order.

Natives of still waters in Europe and North America.

The uses are unknown.

GENUS.

Callitriche, *L.*

NUMBERS. GEN. 1. Sp. 6.

Ceratophyllaceæ?

POSITION.—Euphorbiaceæ.—CALLITRICHACEÆ.————

Halorageæ.

Fig. CXCVI.—Callitriche verna. 1. a ♂ flower; 2. a ♀; 3. a perpendicular section of the ripe fruit.

Order XCIII. EMPETRACEÆ.—Crowberries.

Empetreæ.—*Nutt. Gen.* 2. 233; *Don. in Edinb. New Phil. Journ.* (1826); *Hooker in Bot. Mag.* t. 2758. (1827); *Endl.* ccxli.; *Meisner*, p. 336.

Diagnosis.—*Euphorbial Exogens, with definite ascending anatropal ovules, and an inferior radicle.*

Small arid shrubs with heathlike evergreen aves without stipules, and minute flowers in their axils. Flowers ♀ ♂. Sepals, hypogynous persistent imbricated scales, the innermost of which are sometimes petaloid, or even combined into a monopetalous corolla (as in Oakesia). ♂ Stamens equal in number to the inner sepals, and alternate with them; anthers roundish, 2-celled, the cells distinct, bursting longitudinally. ♀ Ovary free, seated in a fleshy disk, 3- 6- or 9-celled; ovules solitary, anatropal, ascending; style 1; stigma radiating, the number of its rays corresponding with the cells of the ovary. Fruit fleshy, seated in the persistent calyx, 3- 6- or 9-celled; the coating of the cells bony. Seeds solitary, ascending; embryo taper, in the axis of fleshy watery albumen; radicle inferior.

This little group can in nowise be separated from Spurgeworts, from which indeed it is scarcely distinguishable by any positive character except the ascending seeds and inferior radicle. In habit too it quite corresponds with such heath-like genera of Spurgeworts as Micranthea and Pseudanthus, which do not seem to differ from that Order.

A very small group, comprising a few species from the North of Europe, North America, the South of Europe, and the Straits of Magellan.

The leaves and fruit are slightly acid. The black berries of the Crowberry, Empetrum nigrum, sub-acid and unpleasant to the taste, are eaten in the arctic parts of Europe, and are regarded there as scorbutic and diuretic; the Greenlanders prepare a fermented liquor from them. The white berries of the Camarinheira (Corema) are employed by the Portuguese in preparing an acidulous beverage, which the domestic physicians esteem in fevers.—*Endl.*

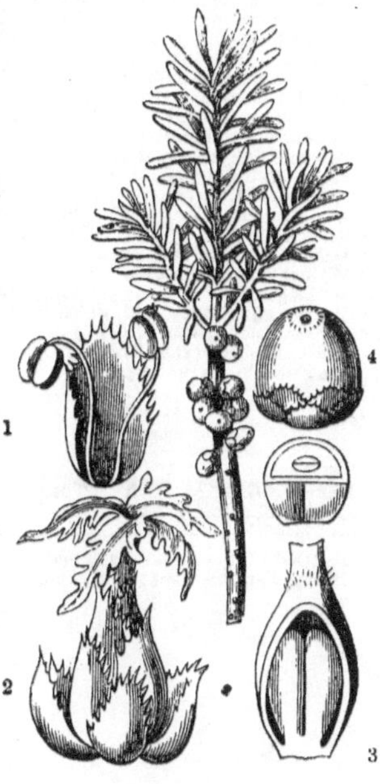

Fig. CXCVII.

GENERA.

Empetrum, *L.*
Corema, *Don.*
Ceratiola, *Michx.*
Oakesia, *Tuckerm.*
Tuckermannia, Klotzsch.

Numbers. Gen. 4. Sp. 4.

Myricaceæ.
Position.—Euphorbiaceæ.—Empetraceæ.————

Fig. CXCVII.—Ceratiola ericoides. 1. a ♂ flower; 2. a ♀; 3. a view of the ovary, with its side removed to show the ovules; 4. ripe fruit; 5. section across a seed.—*Hooker*.

BATIDEÆ.—(*Martius Conspectus, No. 70, p. 18; Meisner, Gen. p. 349.*) Shrubs inhabiting salt marshes, with opposite succulent leaves, having no stipules. Flowers ♂ ♀ in spikes. ♂ Scales of cone 1-flowered, closely imbricated. Calyx a scale rolled up with its back next the axis, and the edges united, so as to form an oblique membranous cup. Stamens 4, longer than the calyx, 2-celled, opening lengthwise; with flattened filaments. ♀ Flowers absolutely naked? or composed of succulent scales arranged in a short 4-rowed cone, and completely united to the back of the ovaries. Calyx 0. Ovary 5-6-celled, buried in the substance of the axis; stigma sessile, roundish; ovules solitary, erect. Fruit succulent; otherwise unknown.

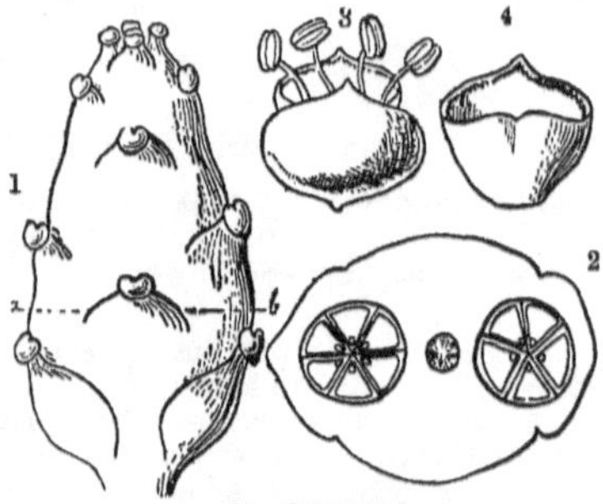

Fig. CXCVIII.

Such appears to be the structure of a small West Indian plant, which has but little attracted the notice of systematical writers, and whose fruit is still unknown. The only point which I am doubtful about is the presence or absence of scales to the female flowers; there is an external appearance of them: but I am not able to separate them from the ovary within their axil. It would be premature to offer any observations upon the affinity of the plant until the structure of the ovules and seeds shall have been ascertained. For the present it must be sufficient to remark, that what writers call the seeds of Batis, are certainly the separable cells of a succulent ovary, the ovules of which are abortive; and it seems probable that this abortion is so common as to render the real seeds difficult to discover. Martius places the plant between Podostemaceæ and Salicaceæ; Meisner following Urticaceæ; Endlicher among his unsettled genera, without a remark. In the last edition of the present work this genus was absolutely placed among the Order of Nettleworts, with the remark that "Batis has a common urticaceous fruit, and it agrees with many genera of the Order in its embryo having the radicle turned down upon the cotyledons." This remark applied to the Batis aurantiaca of Wallich, which I had inadvertently assumed to belong to the genus in which that learned Botanist had placed it. I now find, however, that the shrubs called Batis by Roxburgh and Wallich belong to a totally different genus, allied to Morus, and therefore the remark now quoted falls to the ground. For the present, until we know something more of this plant it may be stationed in the Euphorbial Alliance, with which its diclinous flowers and compound free ovary seem to associate it. It may even belong to the Order of Crowberries.

The salt marshes of the West Indies abound in this plant, which is sometimes gathered for the purpose of mixing with West Indian pickles. Its ashes yield barilla in abundance.

GENUS.

Batis, *P. Br.*

NUMBERS. GEN. 1. Sp. 2? (There is in Sir W. Hooker's Herbarium, a Texan plant in too young a state for examination, but which may be a second species of Batis).

Fig. CXCVIII.—Batis maritima. 1. a ♀ cone; 2. an ideal section of it through *a b*; 3. a ♂ with its bract and stamens; 4. the same without either.

ORDER XCIV. NEPENTHACEÆ.—NEPENTHS.

Aristolochiæ, § Nepenthinæ, *Link. Handb.* 1. 369. (1829).—Nepenthaceæ, *Ed. Pr.* (1836).—Nepentheæ, *Meisn. p.* 334; *Endl.* cxv.

DIAGNOSIS.—*Euphorbial Exogens? with an infinite multitude of scobiform seeds, having an inferior radicle.*

Herbaceous or half-shrubby caulescent plants. Leaves alternate, slightly sheathing at the base, with a dilated foliaceous petiole, pitcher-shaped at the end, which is articulated with a lid-like lamina. Stem without concentric zones, with an abundance of spiral vessels in the wood, pith, and bark, and also with a dense layer of the same between the wood and the bark. Racemes terminal, dense, many-flowered. Flowers diœcious. Calyx 4-leaved, inferior, oppositely imbricated in æstivation. ♂ Stamens cohering in a solid column, bearing at the apex about 16 anthers, collected in various directions in one head; anthers 2-celled, opening longitudinally and externally. ♀ Ovary free, four-cornered, 4-celled, with an indefinite number of ascending ovules attached to the sides of the dissepiments; stigma sessile, simple. Fruit capsular, 4-celled, 4-valved, with the seeds sticking to the sides of the dissepiments, which proceed from the middle of the valves. Seeds indefinite, ascending, very minute, fusiform, with a lax outer integument; nucleus oblong, much less than the seed, lying about the middle of the outer integument, suspended by the chalaza; embryo in the midst of fleshy albumen, with 2 cotyledons placed face to face; radicle turned towards the hilum.

The relation that is borne by the highly curious plants which this Order contains was not even guessed at until Adolphe Brongniart pointed out a resemblance between them and Cistusrapes, which had not before been suspected, but which he considered so important as to justify him in placing the two Orders together. But it is impossible to agree in this conclusion. To say nothing of the extreme dissimilarity in habit between these plants, the structure of their fruit appears to be essentially different; and the seeds of Cytinus being unknown, the resemblance between it and Nepenthes is reduced to a similarity in the arrangement of the anthers, which cannot in the present case be considered of much importance, as it in some degree depends upon the unisexuality of the flowers of both genera. A better approximation of the Order has been made by Brown, who points out a relation to Birthworts; as to which the structure of the wood in some respects confirms his views. Like many in that Order, it is zoneless, although plainly exogenous; but it has this in particular to characterise it, that the system of spiral vessels is developed in a degree unknown in any other plants. Endlicher adopts the same view as does A. Brongniart; and I

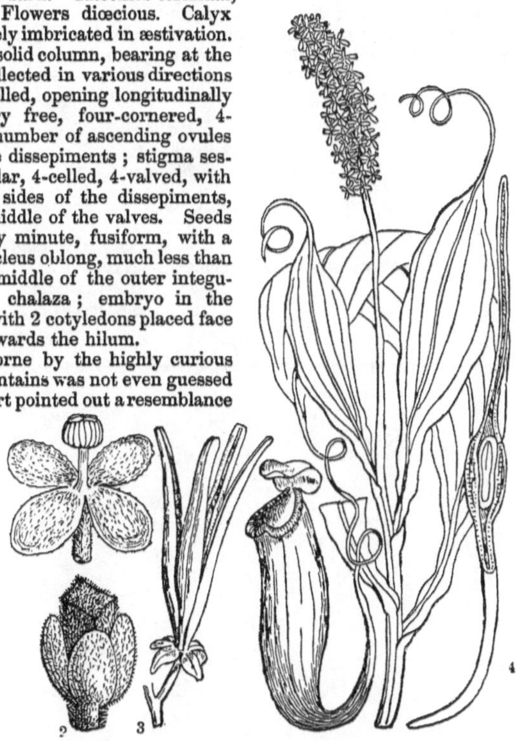
2 3 4

Fig. CXCIX.

Fig. CXCIX.—Nepenthes distillatoria. 1. ♂; 2. ♀; 3. ripe fruit; 4. a section of a seed very highly magnified.

have formerly coincided with those Botanists. But the adherent ovary of Birthworts, their highly developed calyx, axile placentation, and hermaphrodite flowers, are serious difficulties in the way of a close contact between them and Nepenths, unless the peculiar structure of the wood, the consideration of which I for the present abandon, should lead to the final establishment of the Class of Homogens, in which case Nepenths and Birthworts will be brought into contact, or at least a near neighbourhood. For the present, the true position of this Order must be regarded as an undetermined point. In the meanwhile it may be observed, that to station it in the Euphorbial Alliance will be to violate as few affinities as by taking any other course. Its points of agreements are its unisexual flowers, albuminous seeds, incomplete floral envelopes, and climbing habit. Its great disagreement consists in its indefinite seeds, and peculiar woody structure, which is, however, in some respects without example.

There is a good account of the germination of Nepenthes, in Jameson's Journal for April 1830, from which it may be concluded that the long loose tunic of the seed is intended to act at first as a buoy, to float the seed upon the surface of the water, and afterwards as an anchor, to keep it fast upon the mud until it can have struck root.

Natives of swamps in the East Indies and China.

Properties unknown. The water contained in the unopened pitcher of a plant which flowered in the Botanic Garden of Edinburgh, was found by Dr. Turner " to emit, while boiling, an odour like baked apples, from containing a trace of vegetable matter, and to yield minute crystals of superoxalate of potash on being slowly evaporated to dryness."

GENUS.

Nepenthes, L.
Phyllamphora, Lour.

NUMBERS. GEN. 1. Sp. 6 ?.

Sarraceniaceæ ?
POSITION.—Euphorbiaceæ?.—NEPENTHACEÆ.————
Aristolochiaceæ ?
Menispermaceæ ?

ALLIANCE XXI. *QUERNALES.*—THE QUERNAL ALLIANCE.

DIAGNOSIS.—*Diclinous Exogens, with amentaceous monochlamydeous ♂ flowers, an inferior fruit, and an amygdaloid embryo without albumen.*

The Alliance, which comprehends the common Oak and the Beech tree of Europe, is one whose limits are in no degree invaded. The truly diclinous epigynous flowers, the ♂ of which are uniformly arranged in catkins, and the exalbuminous seeds with a large amygdaloid embryo, offer marks of recognition not to be mistaken. That the Walnut is nearly allied to the Oak seems incontestable, although it is often placed in a very different part of the system. Its diclinous epigynous amentaceous flowers, and superior radicle, are entirely those of the Mastworts, and the crumpled cotyledons of Quercus Skinneri are an imitation of those of the Walnut itself. Indeed if the Walnut had a many-celled fruit and a cupule, there would be no very good reason for separating it from Mastworts, except its resinous juices.

At this point the Diclinous Sub-class touches the Perigynous, where the Terebints, having in some instances diclinous flowers, as is the case with Pistacia, come up to the very limits of Juglands. On the other hand, the Myrobalans in the Myrtal Alliance are not very different from hermaphrodite Mastworts, and establish a less close, but well-marked, approach on the part of the Epigynous Sub-class.

The transition from the Quernal Alliance seems to be formed by Garrya itself, whose flowers are so much like those of Juglans, although the habit is different, that if it were not for the minute embryo and large mass of albumen in Garrya it might take its place in the Quernal Alliance.

NATURAL ORDERS OF QUERNALS.

Ovary 2- *or more celled. Ovules pendulous or peltate* . . 95. CORYLACEÆ.
Ovary 1-*celled. Ovule solitary, erect* . . . . 96. JUGLANDACEÆ.

ORDER XCV. CORYLACEÆ.—MASTWORTS.

Castaneæ, *Adans. fam.* 366. (1763).—Cupuliferæ, *Rich. Anal. du Fr.* (1808); *Blume Flora Javæ, Endlich.* lxxxix.; *Meisner, p.* 346.—Corylaceæ, *Mirb. Elem.* 906. (1815).—Quercineæ, *Juss. in Dict. Sc. Nat. vol.* 2. *Suppl.* (1816).

DIAGNOSIS.—*Quernal Exogens, with* 2 *or more cells in the ovary, and pendulous or peltate ovules.*

Trees or shrubs. Leaves with stipules, alternate, simple, often with veins proceeding straight from the midrib to the margin. Flowers ♂ ♀; ♂ amentaceous, ♀ aggregate or amentaceous. ♂ Stamens 5 to 20, inserted into the base of scales or of a membranous valvate calyx, generally distinct. ♀ Ovary crowned by the rudiments of an adherent calyx, seated within a coriaceous involucre (cupule) of various figure, with several cells and several ovules, the greater part of which are abortive; ovules twin or solitary, pendulous or peltate; stigmas several, sub-sessile, distinct. Fruit a bony or coriaceous 1-celled nut, more or less inclosed in the involucre. Seeds solitary, 1, 2, or 3; embryo large, with plano-convex fleshy cotyledons and a minute superior radicle.

The trees or bushes which constitute this Order are among the most important that are known in the Flora of Europe. They are readily recognised by their amentaceous flowers and peculiarly veined leaves; from all other plants they are distinguished by their apetalous superior rudimentary calyx, fruit inclosed in a peculiar husk or cup, and nuts containing but 1 cell and 1 or 2 seeds, in consequence of the abortion of the remainder. They are akin to Willowworts and Birchworts, from which the superior calyx and, in the former case, very often the veining of their leaves, distinguish them. To Nettleworts they are nearly allied, but differ in their many-celled ovary, pendulous ovules, and superior calyx. At first sight, in consequence of their leaves never being pinnate, their relationship to Juglands escapes notice; but the discovery that some at least have the same kind of wrinkled and 4-lobed cotyledons, as for instance certain Oaks and Synædrys, has called attention to the fact. Quercus Skinneri, a kind of Oak from Guatemala, shows this in a striking manner; and upon considering all their points of structure, no doubt seems to remain about the Oak and Walnut really belonging to the same Natural Alliance.

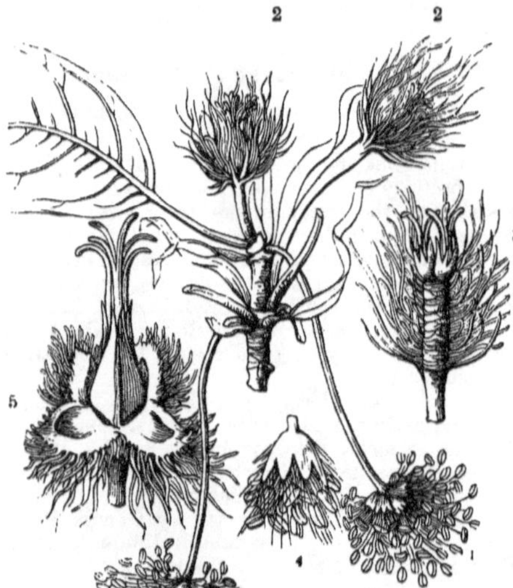

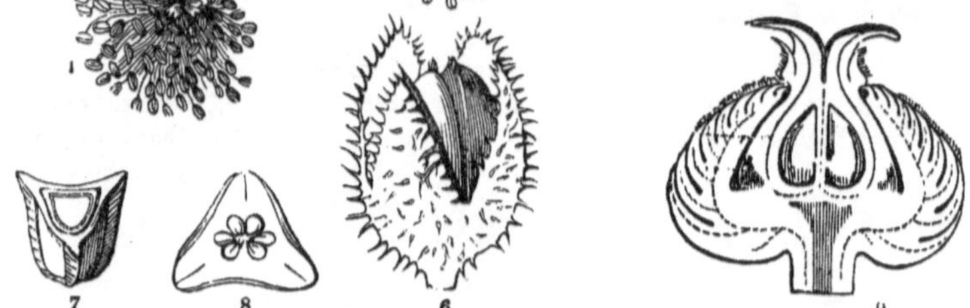

Fig. CC.

Fig. CC.—Fagus sylvatica; 1. ♂ catkins; 2. ♀ do.; 3. the latter, with the scales of the involucre stripped off to show the ovaries at the apex; 4. a ♂ flower; 5. a half-grown ♀ with the involucre, now consisting of consolidated scales, forced back; 6. a ripe involucre opening and exposing the nuts; 7. a transverse section of a ripe nut; 8. the same of a young ovary (from *Nees*); 9. a vertical section of the ♀ flower of Quercus pedunculata.

Inhabitants of the forests of all the temperate parts of the continent both of the Old and New World; extremely common in Europe, Asia, and North America; more rare in Barbary and Chile, and the southern parts of South America; and wanting at the Cape. The species which are found within the tropics of either hemisphere are chiefly Oaks and Chestnuts, which abound in the high lands, but are unknown in the valleys of equatorial regions. The most southern genus is the Beech, of which many species occur in the lower parts of South America, and in Van Diemens Land, and New Zealand. Of the former, Fagus procera is said to be a larger tree than the Araucaria itself, in whose country it grows wild.

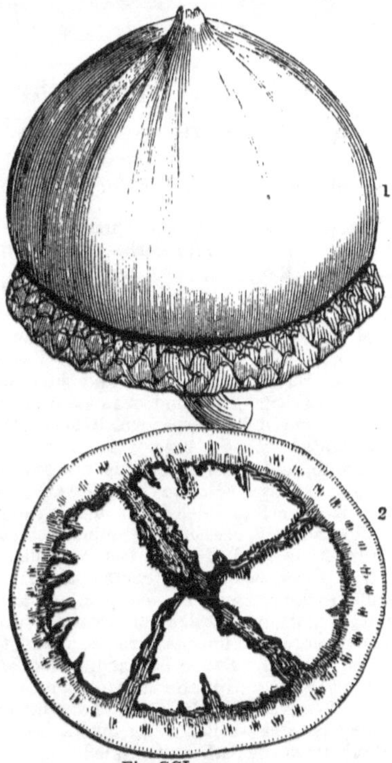

Fig. CCI.

An Order which comprehends the Oak, the Hazel Nut, the Beech, and the Spanish Chestnut, can scarcely require much to be said to a European reader of its properties, which are of too common a use to be unknown even to the most ignorant. Whatever excellence may be found in the timber of the European species is not at all inferior in that of hotter countries. Blume tells us that his Lithocarpus javensis is called Passan-Batu, or Stone-oak, because of its hardness. The leaves of Quercus falcata are employed, on account of their astringency, externally in cases of gangrene; and the same astringent principle, which pervades all the Order, has caused them to be employed even as febrifuges, tonics, and stomachics. Cork is the bark of Quercus Suber; it contains a peculiar principle called Suberin, and an acid called the Suberic. The galls that writing ink is prepared from are the produce of the Quercus infectoria, from which they derive their astringency. The acorns of a species known in the Levant under the name of Velonia (Quercus Ægilops) are imported for the use of dyers. The fixed acids, called Quercitannic and Gallic, which have the power of guarding animal and vegetable fibre from decay, are abundant in many of the Oaks, whose bark is therefore invaluable for tanning. The yellow dyeing bark, called Quercitron, belongs to Q. tinctoria. The husks of the common Beech-tree yield a narcotic extractive, called Fagine. The sweetness of Spanish Chestnuts and Filberts is not confined to the nuts of those trees; the other species of Castanea and Corylus resemble them in that respect, as do the Beech and many sorts of Oak, especially Q. gramuntia, whose acorns are the Belotes of Spain, and a variety of Q. sessiliflora, which is believed to be the Æsculus of Virgil. The bark of the Oak has been employed as a coarse kind of febrifuge. In hot weather a large quantity of saccharine matter is secreted by the leaves of Q. mannifera, in Koordistan, where it is made into sweetmeats. Oil is obtained from the seeds of some species, such as the Beech and Hazel-nut.

GENERA.

Carpinus, *L.*	Fagus, *L.*	Castanea, *Gærtn.*	*Suber*, Tourn.
Ostrya, *Scop.*	*Calucechinus*, H. & J.	Quercus, *L.*	Lithocarpus, *Bl.*
Corylus, *L.*	*Calusparassus*, H. & J.	*Ilex*, Tourn.	Synædrys, *Lindl.*

NUMBERS. GEN. 8. Sp. 265.

POSITION.—Juglandaceæ.—CORYLACEÆ. — *Betulaceæ.* / *Urticaceæ.*

Fig. CCI.—1. Acorn and cupule of Quercus Skinneri, natural size; 2. cross section of the acorn, showing the lobed embryo.

ORDER XCVI. JUGLANDACEÆ.—JUGLANDS.

Juglandeæ, *D.C. Theorie*, 215. (1813); *Kunth. in Ann. Sc. Nat.* 2. 343; *Blume, Fl. Jav.*; *Bartl. Ord. Nat.* 397.; *Endlich.* ccxliv.

DIAGNOSIS.—*Quernal Exogens, with* 1 *cell in the ovary, and* 1 *solitary erect ovule.*

Trees, with a watery or resinous juice. Leaves alternate, pinnated, usually undotted. Stipules none. Flowers herbaceous, inconspicuous. Flowers imperfect, ♂ ♀; the ♂ in catkins, the ♀ in terminal clusters; occasionally both mingled in 1 panicle. ♂ Calyx adherent to a scale-like bract, 2- 3- 6-parted, with membranous unequal segments. Stamens 3, or a considerable number, with short free filaments and erect 2-celled anthers. ♀ either terminal, clustered, and surrounded with a few small bracts of the bud, or in loose racemes inclosed in a 1-flowered involucre, which is cup-shaped at the base, united with the base of the calyx, growing with its growth, and finally acquiring unequal wing-like expansions. Calyx adherent to the ovary, with a minute limb, in from 3 to 5 deciduous or shrivelling divisions. Corolla usually 0, occasionally minute petals. Ovary adherent, 2- or 4-celled at the base, 1-celled at the apex, with a short column on which the ovule is seated. Ovule solitary, erect, on the point of the central column, orthotropal. Styles 1 or 2, very short; stigmas 2-4, seldom more, and unequal, fringed; sometimes sessile, discoid, 4-lobed. Drupe 1-stoned, naked, or in an adherent involucre; with the sarcocarp usually separating from a 2-valved or valveless stone, which is 2-4-celled at the base and 1-celled at the apex. Seed erect, without albumen, smooth or wrinkled, 2- or 4-lobed at the base, and partly divided by partial dissepiments which cut into it. Cotyledons fleshy, oily, sinuous. Radicle very short, superior.

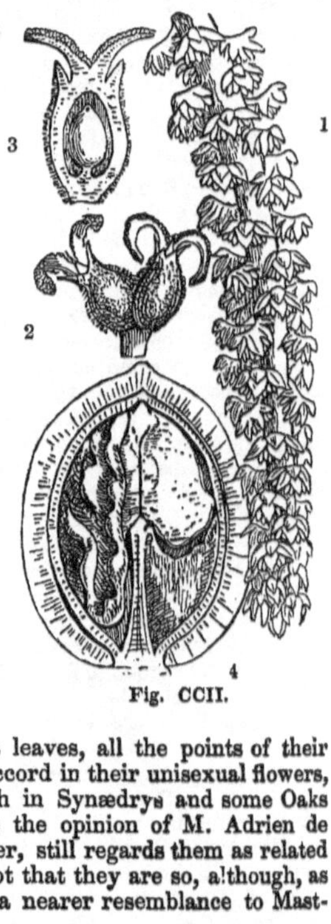

Fig. CCII.

Almost everybody refers these fine trees to the neighbourhood of the plants called Terebints by Jussieu; to which, however, their affinity is obscure. On the contrary, with the single exception of their terebinthinous leaves, all the points of their structure seem to point to Mastworts, with which they accord in their unisexual flowers, adherent calyx, and large exalbuminous embryo, which in Synædrys and some Oaks is also 4-lobed and wrinkled. This too seems to be the opinion of M. Adrien de Jussieu (*Cours élémentaire*, p. 510). Endlicher, however, still regards them as related to the Terebints through Pistacia, and there is no doubt that they are so, although, as has been already stated (p. 289), they seem to have a nearer resemblance to Mastworts.

Chiefly found in North America; a few are East Indian; one species, the common Walnut, is a native of Persia and Cashmere; another, of Caucasus; and a third, of the West India Islands.

The bark is acrid and purgative: so is the rind of the fruit of the common Walnut, notwithstanding its astringency. This quality is not confined to J. regia, but gives its name to the J. cathartica of the United States. The seed of the Walnut is esteemed for its sweetness and wholesome qualities. It abounds in oil, of a very drying nature, and valuable for domestic purposes. Mr. Vigne says that above 12,000 ass loads of Walnut kernels are annually appropriated to the oil press in Cashmere, where Walnut oil is preferred to Linseed oil, and is chiefly employed in cookery and for burning in lamps. This oil possesses such qualities as fairly entitle it to introduction into Europe, and if divested of its mucilage, it might, perhaps, compete with oil of Olives, at least for medi-

Fig. CCII.—Juglans regia. 1. a ♂ catkin; 2. a pair of ♀ flowers; 3. perpendicular section of a ♀ flower; 4. perpendicular section of a ripe Walnut.

cinal purposes. The fruit of several kinds of Hickory is eaten in America. The timber of all is valuable; that of J. regia and nigra for its rich deep brown colour when polished, and that of Carya alba, the common Hickory, for its elasticity and toughness. The seeds of Carya amara are too bitter to be eaten, but, combined with oil of Chamomile, are found useful in colic. The Engelhardtias are very resinous; E. spicata, a large Java tree, as much as 200 feet high, has a pale brown wood, hard and heavy, and used in Java for cart wheels, which are cut out of a single horizontal slab.

GENERA.

Juglans, *L.*
Carya, *Nutt.*
Hicorius, Rafin.
Pterocarya, *Nutt.*

Engelhardtia, *Lesch.*
Pterilema, Rndwt.
Dammara, Rumph.

NUMBERS. GEN. 4. Sp. 27.

Garryaceæ.
POSITION.——————.—JUGLANDACEÆ.—Corylaceæ.
Anacardiaceæ.

ALLIANCE XXII. *GARRYALES.*—THE GARRYAL ALLIANCE.

DIAGNOSIS.—*Diclinous Exogens, with monochlamydeous, sometimes amentaceous, flowers, an inferior fruit, and a minute embryo lying in a large quantity of albumen.*

If we consider this Alliance conterminous with the Quernal on the one hand, because of the approach by Garrya to Juglans, so on the other must it stand in near relation to the Euphorbial, in consequence of including Helwingia, which may be considered as being almost a Spurgewort with an inferior ovary. It appears however to be sufficiently limited by its minute embryo, copious albumen, inferior ovary, and diclinous flowers. The former of these circumstances brings it near the Menispermal Alliance, in which alone among Diclinous Orders does this peculiar embryo occur.

Helwingia has a lateral relation to Sandalworts (Santalaceæ), in the Epigynous Sub-class.

NATURAL ORDERS OF GARRYALS.

Flowers amentaceous. Leaves opposite, exstipulate. . . . . . . 97. GARRYACEÆ.
Flowers fascicled. Leaves alternate, stipulate. . . . . 98. HELWINGIACEÆ.

Order XCVII. GARRYACEÆ.—Garryads.

Garryaceæ, *Lindl. in Bot. Regist.* 20. *t.* 1686. (July 1834); *Endl. p.* 288; *Meisn. p.* 346.

Diagnosis.—*Garryal Exogens, with amentaceous flowers, and opposite leaves, without stipules.*

Shrubs. Leaves opposite, without stipules. Flowers arranged in pendulous amentaceous racemes, within connate bracts. Wood without distinct concentric zones, or dotted ducts. Flowers unisexual, amentaceous. ♂ Sepals 4. Stamens 4, alternate with the sepals, not elastic. ♀ Calyx superior, two-toothed. Ovary one-celled; styles 2, setaceous; ovules 2, pendulous, with funiculi as long as themselves. Pericarp berried, indehiscent, two-seeded. Embryo very minute, in the base of fleshy albumen.

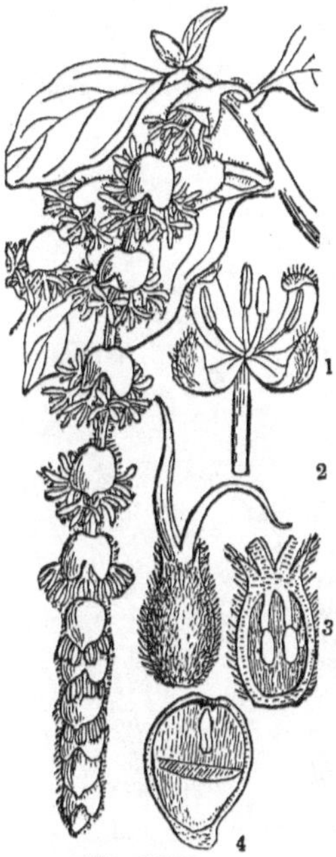

Fig. CCIII.

In its amentaceous inflorescence, imperfect flowers, superior calyx, and mode of germination, this Order is similar to Mastworts, from which it differs most essentially in its wood without concentric circles or vasiform tissue (dotted vessels), its opposite exstipulate leaves, simple fruit, and minute embryo lying in a great mass of albumen. The latter characters bring it near Peppers and their allies, especially Chloranths, with which its zoneless wood (for Chloranthus has no annual zones), simple fruit, and opposite leaves, also agree; but the stipules of Chloranths, together with their naked bisexual flowers, and articulated stems, distinctly separate that Order. Nettleworts and Antidesmads may also be compared with Garrya on account of their imperfect unisexual flowers, somewhat amentaceous inflorescence, and simple fruit; but their superior fruit, alternate leaves, and more perfectly formed wood, are important points of difference. Gnetum again may be taken into comparison on account of its opposite exstipulate leaves, amentaceous unisexual flowers appearing from the axils of connate bracts, minute embryo lying in a great mass of albumen, and imperfect zoneless wood, which in both cases is chiefly constituted of woody fibre (the sides of which are marked with numerous brownish granules), and of annular and reticulated vessels lying scattered sparingly among the tubes of woody fibre. On the other hand, these plants have entirely the appearance of Viburnums or Dogwoods. I formerly referred Garrya to the Urtical Alliance, in which Endlicher follows, placing it among his Julifloræ. Its adherent calyx, however, and more complicated fruit, afford an opportunity of associating it with Helwingiads, which agree in having a minute embryo in the base of solid albumen, and bring the whole into immediate contact with the Euphorbial Alliance.

These shrubs are all found in North America, in temperate latitudes, or in the West Indies.

Their uses are unknown.

GENERA.

Garrya, *Douglas*,
Fadgenia, *Endl.*

Numbers. Gen. 2. Sp. 6.

Juglandaceæ.
Position. ——— Garryaceæ.—Helwingiaceæ.
Caprifoliaceæ.

Fig. CCIII.—Garrya elliptica. 1. a ♂ flower; 2. a ♀; 3. a perpendicular section of the ovary; 4. a section of the seed.

ORDER XCVIII. HELWINGIACEÆ.—HELWINGIADS.

Helwingiaceæ, *Decaisne, Ann. Sc. Nat.* 2. *ser.* 6. 69. (1836); *Endl. p.* 328.

DIAGNOSIS.—*Garryal Exogens, with fascicled flowers, and alternate leaves with stipules.*

A shrub. Leaves alternate, serrate, without deciduous stipules. Flowers fascicled on the midrib of the leaves. Flowers unisexual. Calyx simple, 3-4-parted, with ovate spreading segments, which are deciduous in the females; æstivation valvate. ♂ Stamens 3-4, alternate with the sepals. Anthers continuous, roundish, turned inwards, 2-celled. Pollen smooth. ♀ Ovary adherent to the calyx, crowned by an epigynous disk, 3-4-celled, with one ovule in each cell. Ovules pendulous from the inner angles, anatropal. Style very short. Stigmas 3-4, short, awl-shaped, diverging. Drupe surmounted by the remains of the styles and disk, 3- or 4-celled, scarcely dehiscent (at last loculicidal, *Sieb.*); the cocci one-seeded. Seeds suspended by a short cord. Embryo minute, in the end of solid fleshy albumen; radicle superior.

Although this Order appears to be composed at present of only a single genus, yet it is one of those obscure apetalous unisexual plants, of which few have yet engaged the attention of Botanists, and it is almost sure to find companions hereafter; and even in the absence of this probability, its characters are so well marked as to justify its establishment. M. Decaisne seems inclined to refer it to the neighbourhood of Witch-hazels, rather than to that of Spurgeworts, with which he, however, compares it. But on the one hand, the minute embryo and unisexual flowers remove it far from the former Order; and again, its inferior fruit, unisexuality, and seeds, bring it near to Garryads, with which it seems more fit to be associated. With the Santalaceous Order, to which it has been referred, it has an indirect affinity, as is shown by its inferior fruit, small embryo, valvate calyx, and definite stamens.

The only known species inhabits Japan.

The mountaineers of Japan employ the young leaves of Helwingia rusciflora as an esculent vegetable.—*Siebold.*

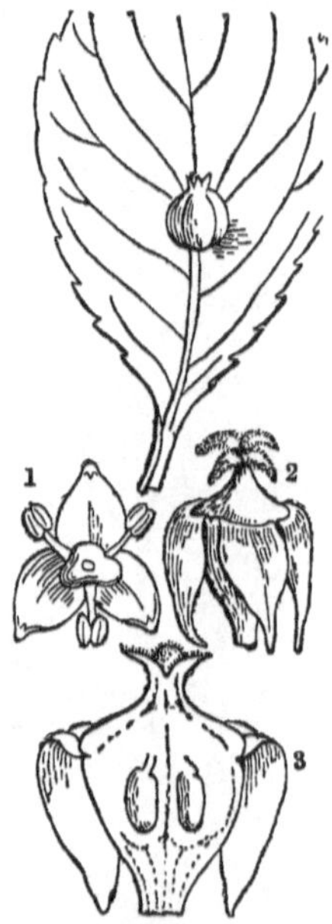

Fig. CCIV.

GENUS.

Helwingia, *Willd.*

NUMBERS. GEN. 1. SP. 1.

	Euphorbiaceæ.
POSITION. ———	HELWINGIACEÆ.—*Garryaceæ.*
	Santalaceæ.

Fig. CCIV.—Helwingia ruscifolia.—*Siebold.* 1. a ♂ flower; 2. a ♀; 3. a perpendicular section of the latter.

ALLIANCE XXIII. *MENISPERMALES.*—THE MENISPERMAL ALLIANCE.

DIAGNOSIS.—*Diclinous Exogens, with monodichlamydeous flowers, superior disunited carpels, and an embryo surrounded by abundant albumen.*

This Alliance stands in the same relation to others in the Diclinous series as Ranals to the Hypogynous, or Rosals to the Perigynous Sub-classes. Its combining character resides in the disunited carpels, abundant albumen, and diclinous flowers. In the Alliances just mentioned, the carpels are sometimes reduced to unity, and the same circumstance occurs in the present instance, where most of the Nutmegs are absolutely simple in their carpellary structure; but as such instances are regarded as exceptional in Ranals and Rosals, being unaccompanied by other points of difference, so among Nutmegs the same conclusion must be adopted, the more especially since Hyalostemma, which I regard as a genuine genus of the Order, possesses more carpels than one, as its habitual character.

The relation of the Orders now collected in the same Alliance will hardly be disputed. Their combining characters are apparently solid, and their passage from one to the other sufficiently well marked. Plume Nutmegs and Monimiads are by some Botanists regarded as the same Order; Monimiads pass into true Nutmegs by means of Tetratome, and the remarkable peculiarity observable in the thin divergent cotyledons of the embryo is common to both Orders. Nutmegs are brought into contact with Menispermads by their trimerous flowers, and by the ruminated albumen of Anomospermum; finally, the strict relation of Menispermads, Kadsurads, and Lardizabalads is unquestionable.

In its external relations this Alliance is very remarkable. Its twining or scrambling habit and unisexual flowers so nearly approach those of Cucurbits in some instances, that even so acute a Botanist as Dr. Blume has referred the genus Gynostemma, a true Cucurbit, to the Order of Menispermads. To the Ranal Alliance it passes directly by means of the genus Hyalostemma, which will be regarded as a Nutmeg or an Anonad, according to the different points of view in which the question of affinities is regarded. And even to Dictyogens it cannot but be regarded as a near approach, if we compare the trimerous Menispermads with Smilax.

NATURAL ORDERS OF MENISPERMALS.

Albumen copious, solid. Seeds pendulous; embryo small, external. Stamens perigynous	99. MONIMIACEÆ.
Albumen copious, solid. Seeds erect. Anthers opening by recurved valves	100. ATHEROSPERMACEÆ.
Albumen copious, ruminated. Sepals united into a valvate cup.	101. MYRISTICACEÆ.
Albumen copious, solid. Seeds parietal; embryo minute . . .	102. LARDIZABALACEÆ.
Albumen copious, solid. Seeds pendulous; embryo minute, internal. Stamens hypogynous.	103. SCHIZANDRACEÆ.
Albumen sparing, solid. Seeds amphitropal; embryo large. .	104. MENISPERMACEÆ.

ORDER XCIX. MONIMIACEÆ.—MONIMIADS.

Monimieæ, *Juss. in Ann. Mus.* 14. 130. (1809); *Bartl. Ord. Nat.* 103.; *Endl. Gen.* cv.

DIAGNOSIS.—*Menispermal Exogens, with perigynous stamens, pendulous seeds, and a minute embryo on the outside of copious fleshy albumen.*

Aromatic trees or shrubs. Leaves opposite, without stipules. Flowers axillary, ♂ ♀. Calyx somewhat globose, divided at the border, sometimes into more rows than one, in which case the segments of the latter are petaloid and imbricated. ♂ Stamens indefinite, covering all the inside of the tube of the calyx; filaments often with a pair of scales at the base; anthers 2-celled, bursting longitudinally. ♀ Ovaries several, superior, 1-celled, distinct, inclosed within the tube of the calyx, each with its own style and stigma; ovule solitary, anatropal, pendulous. Fruit consisting of several 1-seeded nuts, inclosed within the enlarged calyx. Seed pendulous; embryo [small, at the end of an abundant fleshy albumen, to which it is wholly external, its thin diverging cotyledons being applied to the surface of the albumen; testa very fleshy; radicle superior in Ruizia fragrans or Boldoa].

Fig. CCV.

The plants which constitute this Natural Order have been stationed by different Botanists in various parts of their Natural arrangements. Being shrubs with apetalous flowers and an aromatic quality, they have been placed near Laurels (Lauraceæ), with which they also correspond in their ovaries containing but one ovule. Their flowers being apetalous and the sexes disunited, others have referred them to the vicinity of Nettleworts (Urticaceæ), with which, moreover, some species of Citrosma correspond in habit. The true station, however, is evidently among unisexual Orders, with a very large quantity of albumen, where they may be very naturally associated with Nutmegs and their allies. In fact, Mr. Gardner's Tetratome elliptica has so much the appearance of a Nutmeg, that it has been laid into herbaria as such. The extremely aromatic quality of these Monimiads is a strong confirmation of the propriety of this view. Their numerous carpels bring them also into contact with Kadsurads, another aromatic Order. The structure of the calyx of Boldoa, the gradual transition of its segments into petaloid leaves, and the disunited carpels, indicate some analogy to Calycanths, but the minute embryo and disunited sexes forbid us to regard the connection between these plants and Monimiads as being of an intimate kind.

Brown says (*Flinders*, 553.) that what is here called, with Jussieu, a calyx, is more properly an involucre; a view that I formerly adopted, not having had the opportunity of examining specimens for myself. Now, however, that good materials have been acquired by me, I no longer concur with him in that opinion.

In most books the embryo is said to be in the axis of fleshy albumen. How far this may be true in other genera I am unable to ascertain, but it is certainly not so in Boldoa fragrans; which, as was partly stated long ago by Correa de Serra, has the very curious structure above described. Is it possible that the thick fleshy radicle has been taken for an embryo, and that the thin diverging cotyledons have been overlooked?

Fig. CCV.—Boldoa fragrans. 1. a section of the ripe fruit; 2. the embryo shown separately.

Most of these Monimiads are found in the forests of South America; a few only occur in the Mauritius, Madagascar, Java, New Zealand, and New Holland.

All the parts of the bark and leaves exhale an aromatic odour, which is compared by travellers to that of Laurels or Myrtles. Boldoa, the Boldu of Chili, produces an aromatic succulent fruit, which is eaten by the natives; both the wood and leaves are very fragrant; the former makes a kind of charcoal, which is preferred beyond all other kinds by the smiths of Chili; the bark is used by tanners.

GENERA.

Ambora, *Juss.*	Kibara, *Endl.*	Tetratome, *Pöpp.*	*Ruizia*, Pav.
Tambourissa, Sonner.	*Brongniartia*, Blume.	Hedycarya, *Forst.*	*Peumus*, Pers.
Mithridatea, Commers.	Citrosma, *Ruiz et Pav.*	Boldoa, *Juss.*	Mollinedia, *Ruiz et Pav.*
Monimia, *Thouars.*			

NUMBERS. GEN. 8. Sp. 40?

Urticales.
POSITION.—Myristicaceæ.—MONIMIACEÆ.—Atherospermaceæ.
Piperaceæ.

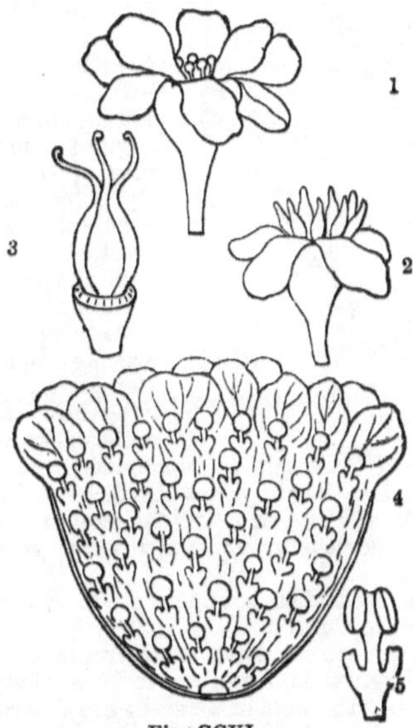

Fig. CCVI.

Fig. CCVI.—The details of the flowers of Boldoa fragrans after Endlicher. 1. a ♂; 2. a ♀; 3. the carpels; 4. a section of a ♂; 5. a stamen.

Order C. ATHEROSPERMACEÆ.—Plume Nutmegs.

Atherospermeæ, *R. Brown in Flinders*, 553. (1814); *Arnott in Edinb. Encycl.* 130.

Diagnosis.—*Menispermal Exogens, with anthers opening by recurved valves.*

Trees. Leaves opposite, without stipules. Flowers axillary in short racemes, with large deciduous bracts, ♂ ♀ (or ⚥). Calyx tubular, divided at the top into several segments, usually placed in two rows, the inner of which is partly petaloid; to these are superadded in the ♀ flowers some abortive stamens in the form of scales. Stamens in the ♂ very numerous in the bottom of the calyx; in the ♀ fewer, and arising from the orifice of the calyx; anthers adnate, 2-celled, bursting with a valve which separates from the base to the apex; filaments with a pair of scales at their base. Ovaries several, usually indefinite, each with a single erect ovule; styles simple, arising either from the side or the base; stigmas simple. Nuts inclosed in the tube of the calyx, with the adherent styles converted into feathery awns. Seed solitary, erect; embryo minute, erect, at the base of soft fleshy albumen, with divaricating cotyledons; radicle inferior.

Fig. CCVII.

Although the anthers of this Order are the same as those of Laurels and Berberries, and notwithstanding that it agrees with the former in its aromatic odour, yet it seems to stand in the nearest relationship to Monimiads, with which it is even combined by Jussieu, Bartling, and Endlicher. It differs, however, in the position of the ovule, and the structure of the anthers, and is probably a nucleus around which other genera will be hereafter collected.

The Australian continent produces two of the genera; Laurelia belongs to Chile.

All the species seem to be fragrant. The wood of Doryphora Sassafras, called Sassafras in New Holland, is said to smell like Fennel. The nuts of Laurelia are described as possessing the fragrance of the Nutmeg. Mr. Backhouse gives the following account of Atherosperma moschata. "This forms a very beautiful tree in many parts of the colony, attaining to a height of 150 feet, and is from 6 to 7 feet in circumference. Its mode of growth resembles many Coniferæ, in being conical, and in having all its branches of the same year's growth, radiating from one point on the trunk. A decoction of the bark, either when in its green state or after having been dried, is used in many remote parts of the colony as a substitute for tea, and, when taken with plenty of milk, has a pleasant taste. Its effects are, however, slightly aperient."

Fig. CCVIII.

GENERA.

Atherosperma, *Labill.*	*Thiga*, Molina.
Laurelia, *Juss.*	Doryphora, *Endl.*
Pavonia, Ruiz.	

Numbers. Gen. 3. Sp. 4.

Lauraceæ.

Position.—Monimiaceæ.—Atherospermaceæ.—Myristicaceæ.

Calycanthaceæ.

Fig. CCVII.—Atherosperma moschata.

Fig. CCVIII.—1. carpel; 2. stamen of Doryphora Sassafras.—*Endlicher.*

ORDER CI. MYRISTICACEÆ.—NUTMEGS.

Myristiceæ, *R. Brown, Prodr.* 399. (1810); *Bartling, Ord. Nat.* 244; *Martius Conspectus*, No. 78; *Endl.* clxxiii; *Meisner, p* 329.

DIAGNOSIS.—*Menispermal Exogens, with ruminated albumen, and a valvate cup-shaped calyx.*

Tropical trees, often yielding a red juice. Leaves alternate, without stipules, not dotted, quite entire, stalked, coriaceous. Inflorescence axillary or terminal, in racemes, glomerules, or panicles; the flowers very small, often each with one short cucullate bract. Calyx coriaceous, mostly downy outside. Flowers completely unisexual. Calyx trifid, rarely quadrifid, with valvular æstivation. ♂. Filaments either separate or completely united in a cylinder. Anthers 3-12 or more, 2-celled, turned outwards, and bursting longitudinally: either connate or distinct. ♀ Calyx deciduous. Carpels solitary, or many, with a single erect anatropal ovule; style very short; stigma somewhat lobed. Fruit baccate. Albumen ruminate, between fatty and fleshy; embryo small, orthotropal; cotyledons diverging; radicle inferior.

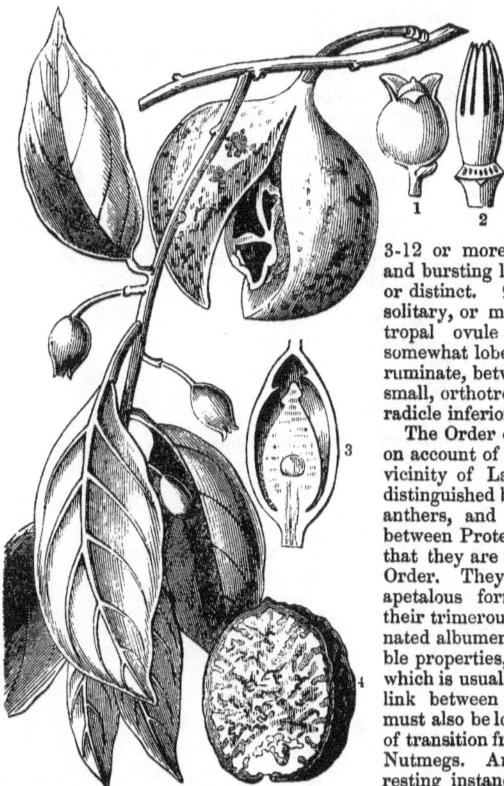

Fig. CCIX.

The Order of Nutmegs is usually placed, on account of their apetalous flowers, in the vicinity of Laurels, from which they are distinguished by the structure of their calyx, anthers, and fruit. Brown places them between Proteads and Laurels, remarking, that they are not closely akin to any other Order. They may also be regarded as an apetalous form of Anonads, with which their trimerous flowers, arillate seed, ruminated albumen, minute embryo, and sensible properties, closely ally them. Bocagea, which is usually considered as a connecting link between the latter and Berberids, must also be looked upon as one of the cases of transition from Anonads to Virola among Nutmegs. Another and much more interesting instance is afforded by Wallich's genus Hyalostemma, which would be almost an involucrated Myristica if it had an aril. That plant, which has unisexual apetalous flowers, and a trifid calyx surrounded by an involucre of six subulate bracts, was regarded as a Uvaria by Roxburgh, and may be indifferently regarded as Anonaceous if its numerous carpels are considered, or Myristicaceous if its unisexual flowers and simple trifid calyx are allowed to have weight.

While, however, all these relationships may be allowed their due importance, it seems impossible to disjoin Nutmegs from the Menispermal Alliance, because of their strictly unisexual flowers. The diverging cotyledons of their embryo bring them up to Monimiads, while the ruminated albumen finds its parallel in the genus Anomospermum in Menispermads.—*See* p. 308.

Fig. CCIX.—Myristica fragrans.—*Blume.* 1. a flower; 2. a column of stamens; 3. a section of a ♀ showing the ovary and ovule; 4. a section of a nutmeg with the embryo at the base of the albumen.

Natives exclusively of the tropics of India and America, and most common in the former.

Their bark abounds in an acrid juice, which is viscid and stains red; the rind of the fruit is caustic. The aril and albumen of Myristica moschata, the former known under the name of Mace, and the latter of Nutmeg, are important aromatics. An aromatic fruit is also borne by other species. The coarse, strong-smelling Nutmegs of Santa Fé are from the Myristica Otoba. Another species is the M. tomentosa, and a third the M. officinalis, which is reckoned in Brazil an energetic tonic. In the Indian Archipelago Myristica spuria is employed as a substitute, and also a species in the Philippines called Dooghan, Dungan, or Gonogono; in Madagascar, M. acuminata and madagascariensis, and in Brazil M. Bicuiba (Bicuiba or Vicuiba) or officinalis. The seeds also abound in oil. Virola sebifera yields a fatty oil upon simple immersion in hot water; the common Nutmeg furnishes a similar secretion, and also a fluid oil. From the white Mace of M. Otoba is prepared an ointment used against the itch in Colombia. The red Mace of Pyrrhosa tingens, an Amboyna plant, when rubbed between the fingers is mucilaginous, and stains them fiery red; by the addition of lime it yields a red pigment, with which the natives stain their teeth.—*Blume.* The aromatic quality, although so common in this Order, sometimes deserts their fruit. In Myristica fatua the fragrance is very slight and soon disappears, and in others it is scarcely perceptible. It must, however, not be supposed that the insipid Nutmegs are inert. Mr. Hinds states that in New Guinea, where the latter are common, persons who ate as many as two were soon after surprised by a violent evacuation of the bowels, and disturbance of the stomach. A single one produced nausea, sensation of fullness, and flatus.—*Lond. Journ. Bot.* 1. 675. This corresponds with the qualities of the common Nutmeg, which can only be used safely in very small quantities; in excess it produces oppression of the chest, intense thirst, headache, and even delirium and fatal apoplexy.—*Endl.* The Dungan of the Philippines, already mentioned, yields a crimson juice which is collected from incisions in the trunk, and used as a substitute for Dragon's Blood.—*Endl.* See *Blume's Rumphia,* 1. p. 179.

GENERA.

Myristica, *Linn.*
Virola, *Aubl.*
  Sebophora, Neck.
Knema, *Lour.*
Pyrrhosa, *Blum.*
  Horsfieldia, Willd.
Hyalostemma, *Wall.*

NUMBERS. GEN. 5. Sp. 35?

Anonaceæ.
POSITION.—Schizandraceæ.—MYRISTICACEÆ.—Menispermaceæ.
Euphorbiaceæ.

ORDER CII. LARDIZABALACEÆ.—LARDIZABALADS.

§ of Menispermaceæ, *DC. Prodr.* 1. 95. (1824); *Bartl Ord. Nat.* 343 (1830); *Endl. p* 828. (1830).—Lardizabaleæ, *Decaisne, Mémoire* (1837); *Endl. Ench.* clxxiv.

DIAGNOSIS.—*Menispermal Exogens, with parietal seeds and a minute embryo in abundant solid albumen.*

Twining smooth shrubs. Leaves alternate, compound, without stipules. Racemes solitary or clustered. Flowers coloured white, lilac, deep purple, or pale yellow, sometimes fragrant. ♂ Sepals 3 or 6, in 2 rows, deciduous. Petals 6, in 2 rows, opposite the sepals, the inner smaller, or gland-like, or 0. Stamens 6, opposite the petals; filaments united into a tube, or even distinct; anthers turned outwards, rarely inwards, 2-celled, opening by a longitudinal slit. Pollen when wet spherical, 3-furrowed. Rudiments of ovaries 2 or 3 in the centre; ♀ as before but larger, with 6 very imperfect stamens. Carpels distinct, 3, rarely 6-9, one-celled, with a short style and a simple stigma. Ovules many, rarely single, sunk in sockets on the inner surface of the ovary. Fruit short-stalked, berried, many-seeded, rarely one-seeded; sometimes follicular and opening lengthwise. Seeds with a cartilaginous integument; albumen between fleshy and horny, large; embryo minute, with its inferior radicle turned towards the hilum; very rarely almost as large as the seed, with flat cotyledons.

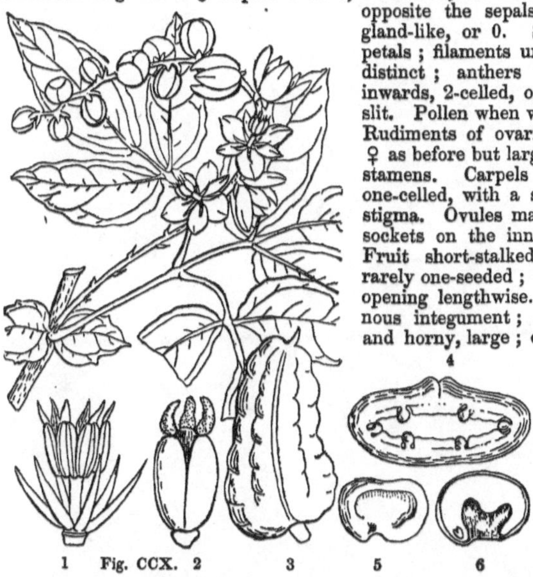

Fig. CCX.

This small Order has been admirably illustrated by M. Decaisne. It consists of twining plants, with much the appearance of Menispermads, from which they are readily known by their leaves being compound, and their ovules sunk in niches in the inside of the ovary, with the single exception of a Madagascar plant called Burasaia, which probably, as M. Decaisne suggests, hardly belongs to the Order. This Botanist regards them as otherwise allied on the one hand to Kadsurads, which have simple leaves and a great many carpels, whose ovules are not parietal, on the other to Berberids, whose foliage and flowers issuing from scaly buds are very similar; and finally to Anonads through Bocagea. No doubt they entirely participate in the relationship of Menispermads, whatever that may be. Stauntonia has the placentation of Flacourtia, according to Griffith.

Two of the genera inhabit the cooler parts of South America; the remainder are from the temperate parts of China. Burasaia is the only tropical form.

These plants appear to be harmless. The fruits of Holböllia are eaten in the hills of India, according to Royle. Those of Stauntonia hexaphylla have a sweetish watery taste, and are eaten by the country people of Japan, who also employ their juice as a domestic remedy for ophthalmia.—*Siebold* In like manner Akebia quinata produces

Fig. CCX.—Lardizabala triternata.—*Decaisne.* 1. petals and stamens of ♂ flower; 2. carpels; 3. fruit of a Lardizabala; 4. a cross section of it; 5. a seed; 6. a section of it, showing the embryo.

fruits used by the same people as an emollient medicine.—*Ib.* The branches of Lardizabala are extremely tough, and are employed as cordage in Chile, by merely passing them through fire and then leaving them for some hours in water.—*Decaisne.* According to Thouars, the fruit of the doubtful genus Burasaia abounds in mucilage.

GENERA.

§ 1. MADAGASCAR.
Burasaia, *Thouars.*

§ 2. ASIATIC.
Akebia, *DC.*
Holboellia, *Wall.*
Stauntonia, *DC.*
Parvatia, *DC.*

§ 3. AMERICAN.
Lardizabala, *Ruiz et Pav.*
Boissiera, Domb.
Thouinia, Domb.
Boquila, *DC.*

NUMBERS. GEN. 7. Sp. 15.

Anonaceæ.
POSITION.—Schizandraceæ.—LARDIZABALACEÆ.—Menispermaceæ.
Berberidaceæ.

ORDER CIII. SCHIZANDRACEÆ.—KADSURADS.

Schizandraceæ.—*Blume Bijdr.* 21. (1825); *Endl.* clxxv.; *Meisner, p.* 5.

DIAGNOSIS.—*Menispermal Exogens, with hypogynous stamens, pendulous seeds, and a minute embryo, inclosed in copious solid albumen.*

Scrambling shrubs. Leaves alternate, simple, entire or toothed, without stipules, often with pellucid dots. Wood (in Sphærostema propinquum) without annual rings, composed of glandular-sided woody tubes, arranged in rays, and separated by fine

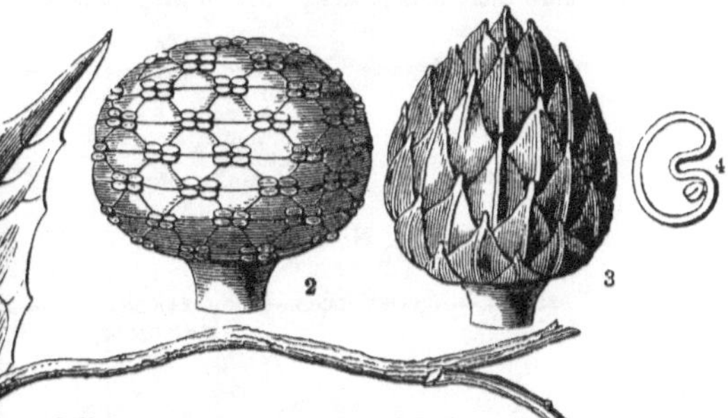

medullary processes, among which lie disorderly numerous annular and reticulated vessels. Flowers small, solitary or axillary, with imbricated bracts. Flowers unisexual, solitary or clustered. Sepals 3- 6, the outer smaller. Petals 3- 9, hypogynous. ♂ Stamens indefinite, collected into a close cluster, either monadelphous or free. Anthers turned outwards, 2-celled, with a thickened connective, sometimes united lengthwise, sometimes at the point. ♀ Carpels 00, free or adherent, 1-celled; stigma sessile; ovules 2, pendulous, one above the other. Fruit numerous, berried, collected in a head, usually curved, 1 or 2-seeded, pulpy inside, and thus divided by spurious dissepiments. Seeds nestling in pulp, somewhat crescent-shaped, compressed, with a vertical hilum next the apex of the fruit. Embryo in the base of abundant solid fleshy albumen, very small; radicle next the hilum.

The ternary number, which is found in the parts of the flower, have led authors to refer these plants to Anonads, from which, nevertheless, they are removed by their imbricated sepals and petals, unisexual flowers, and solid albumen. Their habit too is at variance with that of the last mentioned Order. They have been again compared to Magnoliads, from which they differ in their unisexual flowers, and in the form of their stamens, which in Magnoliads are never monadelphous; moreover, the different nature of the vegetation and their different properties form a great distinction between the two Orders; for Magnoliads are erect trees or shrubs with gemmaceous stipules and entire leaves, and bitter aromatic properties, while on the other hand Kadsurads are trailing shrubs, destitute

Fig. CCXI.

Fig. CCXI.—Kadsura japonica. 1. a calyx; 2. a head of stamens; 3. a pistil; 4. a section of a seed.—*Siebold*

of stipules, constantly having toothed leaves, but having no aromatic or bitter properties; on the contrary, they abound in vegetable mucus. It is near Menispermads that they seem most to demand a place, notwithstanding the very different views that have formerly been held upon the subject. The unisexual flowers, with the parts on a ternary plan, scrambling habit, disunited carpels, and copious albumen, now appear of more importance than the hypogynous insertion of their stamens and the polypetalous flowers.

The few species hitherto discovered belong to the continents and islands of India, Japan, and the hotter provinces of North America.

The species abound in mucus, and appear to be quite insipid. The fruit of some are eaten. Siebold describes that of Kadsura japonica as being viscid, tasteless, and uneatable; he adds, that by boiling a sort of mucilage is obtained from its branches and applied to the fabrication of Broussonetia paper; it is also employed by the Japanese women to cleanse their hair of the pomatum they so largely employ.

GENERA.

Kadsura, *Juss.*
Sarcocarpum, Blum.
Sphærostema, *Blum.*
Schizandra, *L. C. Rich.*
Hortonia, *Wight.*
? Mayna, *Aubl.*

NUMBERS. GEN. 5. Sp. 12.

POSITION.—Myristicaceæ.—SCHIZANDRACEÆ.—Lardizabalaceæ.
Anonaceæ.

ORDER CIV. MENISPERMACEÆ.—MENISPERMADS.

Menispermeæ, *Juss. Gen.* 284. (1789) ; *DC. Syst.* 1. 508.—Menispermaceæ, *DC. Prodr* 1. 95. (1824) ; *Wight and Arnott. Prodr.* 1. 11 ; *Endl.* clxxii. ; *Meisner*, *p.* 5 ; *Wight Illustr.* 1. 19.

DIAGNOSIS.—*Menispermal Exogens, with amphitropal seeds, and a large embryo in a moderate quantity of solid albumen.*

Shrubs, with a flexible tough tissue, and sarmentaceous habit ; their wood often without zones. Leaves alternate, entire. Flowers small, usually racemose, and diœcious.

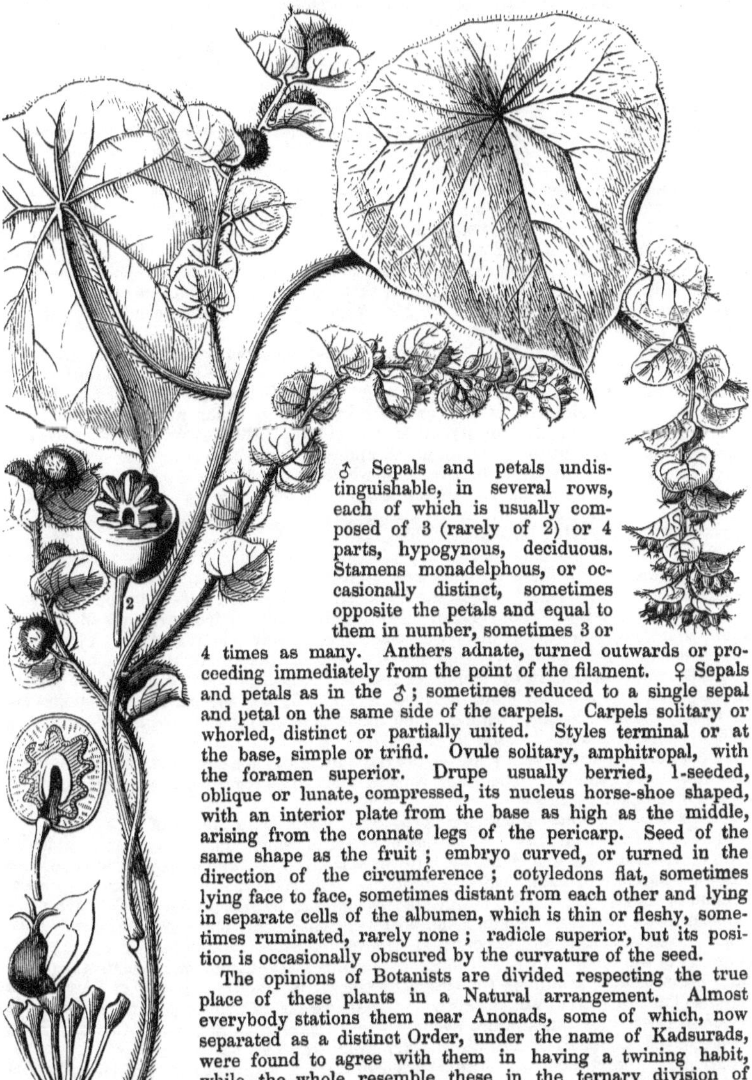

Fig. CCXII.

♂ Sepals and petals undistinguishable, in several rows, each of which is usually composed of 3 (rarely of 2) or 4 parts, hypogynous, deciduous. Stamens monadelphous, or occasionally distinct, sometimes opposite the petals and equal to them in number, sometimes 3 or 4 times as many. Anthers adnate, turned outwards or proceeding immediately from the point of the filament. ♀ Sepals and petals as in the ♂ ; sometimes reduced to a single sepal and petal on the same side of the carpels. Carpels solitary or whorled, distinct or partially united. Styles terminal or at the base, simple or trifid. Ovule solitary, amphitropal, with the foramen superior. Drupe usually berried, 1-seeded, oblique or lunate, compressed, its nucleus horse-shoe shaped, with an interior plate from the base as high as the middle, arising from the connate legs of the pericarp. Seed of the same shape as the fruit ; embryo curved, or turned in the direction of the circumference ; cotyledons flat, sometimes lying face to face, sometimes distant from each other and lying in separate cells of the albumen, which is thin or fleshy, sometimes ruminated, rarely none ; radicle superior, but its position is occasionally obscured by the curvature of the seed.

The opinions of Botanists are divided respecting the true place of these plants in a Natural arrangement. Almost everybody stations them near Anonads, some of which, now separated as a distinct Order, under the name of Kadsurads, were found to agree with them in having a twining habit, while the whole resemble these in the ternary division of their flowers. De Candolle also points out a resemblance with Sterculiads, consisting in the monadelphous stamens, and

Fig. CCXII.—Cissampelos tropæolifolia. 1. a ♀ flower ; 2. a portion of a fruit, with the seed laid bare ; 3. a perpendicular section of a fruit.

peltate leaves. According to Aug. de St. Hilaire, the Order is related to Spurgeworts through Phyllanthus, the male flowers of which are in certain species absolutely the same as those of Cissampelos. It is also thought to approach Mallowworts by those genera which, like Caperonia, have stipulate leaves, and distinct caducous petals separated from the calyx by the torus. But, on the other hand, the small flowers having no distinct separation of calyx and corolla, the disunited sexes, the large embryo, and the peculiar habit, are quite at variance with a series so high in the scale as the Anonal Alliance; and the resemblance of the species to Smilax is such as to persuade a Botanist that the two Orders cannot but have some more intimate affinity than would be suspected from their widely different embryo. To me it seems clear that whatever lateral affinity there may be between Menispermads and the great hermaphrodite Orders with hypogynous stamens and disunited carpels, yet that their direct relationship is with Spurges and similar Orders. At the same time, the separation of Kadsurads from the Anonal, and its insertion among the unisexual Alliances, tends to show the justice of the views of Botanists so long as that change was not effected. I once relied upon the zoneless wood of several species to remove the position of Menispermads altogether from among Exogens; but M. Decaisne has shown (*Mém. Lardizab.*) that this peculiarity is not constant, such plants as Cocculus lancifolius and Cissampelos Pareira having wood with numerous concentric zones. As I do not propose to insist, for the present, upon the value of the woody structure for systematical purposes, the reader is referred to M. Decaisne's Treatise and Gaudichaud's *Organogr.* xxviii. f. 12 and 13, for information upon this point. It may be as well, however, to add that, according to Endlicher, the root of Cissampelos Caapeba can hardly be distinguished from the Pepper of that name. A very curious and unexpected approach to Nutmegs is furnished by Mr. Miers' genus Anomospermum, which has ruminated albumen.

One of the most singular facts connected with the Menispermaceous Order, consists in the position of the seed, which is altered materially from that of the ovule in the progress of the growth of the fruit. According to Aug. de St. Hilaire, the ovule of Cissampelos is attached to the middle of the side of a straight ovary, which after fecundation gradually incurves its apex until the style touches the base of the pericarp, when the two surfaces being thus brought into contact unite, and a drupe is formed, the seed of which is curved like a horse-shoe, and the cavity of which is divided by a spurious incomplete dissepiment, consisting of two plates: the attachment of the seed is at the top of the false dissepiment, on each side of which it extends equally. Several of the species of Cocculus are so remarkably tenacious of life, that if even a large branch be broken at a considerable distance from the ground, the upper portion immediately throws out a slender filiform root, which speedily re-establishes the connection with the soil, and preserves the plant. Such a root has been seen 8 feet long, and not thicker than a common packthread.—*Wight.* The hermaphrodite genera Spirospermum and Agdestis cannot be regarded as belonging here; their true position is unsettled. Phytocrene or Gynocephalium too, must have another station; see the URTICAL ALLIANCE.

This Order is common in the tropics of Asia and America, but uncommon out of those latitudes: all Africa contains but 5, North America 6, and Siberia 1. The species are universally found in woods, twining round other plants. Cocculi are most common in the Old World, Cissampeli in the New.

Active narcotic and bitter qualities prevail among the species; the former in excess, rendering them poisonous; the latter, causing them to be regarded as valuable tonics. A few are mucilaginous.

The roots of Cocculus palmatus, a herbaceous plant, with the air of a Bryony, found on the coast of Eastern Africa, is largely in use as a tonic, under the name of Kalumba or Calumba root. In this plant, the bitterness is mixed with mucilage, and contains a somewhat narcotic principle called Calumbine. The roots of Cocculus peltatus in Malabar, and of C. flavescens in the Moluccas, are found to be substitutes for Calumba; it is also said that a root called Radix Lopeziana belongs to some plant of this Order.—*Endl.* In fact, medical men in all the countries where the Order is found believe in the powerful action of the roots. In Brazil those of Cocculus platyphyllus, and cinerascens (called Butua), Cissampelos ovalifolia (or Orelha de Onça), in the West Indies of Cissampelos Pareira (Pareira brava) and Caapeba, in Madagascar of Cissampelos mauritianus, in Ceylon of Coscinium fenestratum (Woniwol or Venivel), in Senegal of Cocculus Bakis are employed as tonics or diuretics. The Brazilians also use the roots of Cissampelos glaberrimus, and ebracteatus, against serpent bites. An intoxicating spirit is distilled from the root of Cissampelos obtecta.—*Royle.* That of Cissampelos glabra is said to be extremely acrid.—*Roxb.*

The bark, wood, and leaves, are also employed for the same purposes. The stalks and leaves of Cocculus cordifolius are employed in an infusion called in Bengal Páchana, much in use as a tonic; while an extract of the stem, called Pálo, is regarded as a diuretic. The bark of Chondrodendron convolvulaceum (Uva del Monte) is employed as a febrifuge in Peru. Endlicher states, that the bark of some species is used for dyeing yellow. Wight says, that extract of Guluncha, so much recommended in India as a febrifuge, may be prepared from the bruised stems of both Cocculus verrucosus and cordifolius. The young shoots of the latter are a powerful emetic. The wood and bark of Coscinium indicum are regarded as furnishing, in infusion, an excellent stomachic.

It is, however, the fruit of this Order to which most importance attaches. That of Anamirta Cocculus furnishes the dangerous Cocculus indicus seed, in which resides picrotoxine, a most venomous principle; and in the pericarp is found the not less formidable alcaloid menispermine. From those seeds a fatty oil is expressed. Forskåhl states, that from the acrid berries of Cocculus Cebatha, a spirit is distilled in Arabia called Khumr-ool-majnoon.

GENERA.

Menispermum, *Tournef.*
Trilophus, Fisch.
Pselium, *Lour.*
Cocculus, *DC.*
Abuta, Aubl.
Baumgartia, Mönch.
Androphylax, Wendl.
Wendlandia, Willd.
Braunea, Willd.
Tiliacora, Colebr.
Epibaterium, Forst.
Limacia, Lour.
Fibraurea, Lour.
Nephroia, Lour.
Cebatha, Forsk.
Leaeba, Forsk.
Columba, Commers.
Bagalatta, Roxb.
Odontocarya, *Miers.*
Anomospermum, *Miers.*
Chondrodendrum, *Ruiz et Pav.*
Meniscosta, *Blum.*
Jödes, *Blum.*
Trichoa, *Pers.*
Batschia, Thunb.
Abuta, Pöpp.
Coscinium, *Colebr.*
Pereiria, Lindl.
Anamirta, *Colebr.*
Stephania, *Lour.*
Clypea, Blum.
Cissampelos, *Linn.*
Caapeba, Plum.
Chasmanthera, *Hochst.*
Cyclea, *Arnott et Wight.*

NUMBERS. GEN. 11. Sp. 175.

Smilaceæ.
POSITION.—Schizandraceæ.—MENISPERMACEÆ.—Lardizabalaceæ.
Cucurbitaceæ.

Fig. CCXIII.

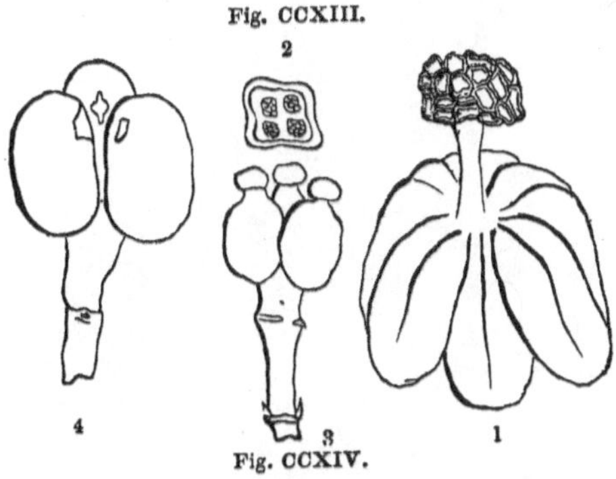

Fig. CCXIV.

Fig. CCXIII.—Cissampelos Pareira; 1. a ♂ flower; 2. a ♀ flower; 3. horseshoe formed drupe; 4. a vertical section of a drupe, showing the embryo and albumen; *a* is the true apex of the fruit, brought to the base as above described.

Fig. CCXIV.—Anamirta Bauerana; 1. ♂; 2. anther; 3. ♀ without the calyx; 4. fruit.

ALLIANCE XXIV. *CUCURBITALES.*—THE CUCURBITAL ALLIANCE.

DIAGNOSIS.—*Diclinous Exogens, with monodichlamydeous flowers, inferior fruit, parietal placentæ and embryo without a trace of albumen.*

The plants of this Alliance differ from all others in their diclinous flowers, combined with an inferior ovary, whose placentæ are more or less manifestly parietal. They approach Menispermads in their scrambling habit, and Passionworts in their placentation. In the greater part the stigmas are horseshoe-shaped. They differ from Papayals in their inferior ovary and exalbuminous seeds, and from the scrambling genera of Euphorbials in the same circumstances.

NATURAL ORDERS OF CUCURBITALS.

Fruit pulpy. Placentæ strictly parietal. Monopetalous . . . . 105. CUCURBITACEÆ.
Fruit dry. Placentæ strictly parietal. Apetalous . . . . . 106. DATISCACEÆ.
Fruit dry. Placentæ projecting and meeting in the axis. Monodichlamydeous . . . . 107. BEGONIACEÆ.

ORDER CV. CUCURBITACEÆ.—CUCURBITS.

Cucurbitaceæ, *Juss. Gen.* 393. (1789) ; *Aug. St. Hil. in Mem. Mus.* 9. 190-221. ; *DC. Prodr.* 3. 297.—*Schrad. in Linnæa*, 12. 401. ; *Endl. Gen.* ccii. ; *Meisner*, *p.* 126. ; *Wight's Illustr.* 2. *p.* 24 ; *Arnott in Hooker's Journal*, 3. 271.—Nàndhirobeæ, *Aug. de St. Hil.* l. c. (1823) ; *Turpin Dict. des Sc. Atlas.* ; *Endl. Gen.* cci.

DIAGNOSIS.—*Cucurbital Exogens, with monopetalous flowers, strictly parietal placentæ, and pulpy fruit.*

Roots annual or perennial, fibrous or tuberous. Stem brittle, climbing by means of tendrils formed by abortive stipules ? Leaves usually palmated, or with palmate ribs, very succulent, covered with numerous asperities, sometimes ternate. Flowers white, red or yellow ; occasionally small and herbaceous. Flowers ♂ ♀. Calyx 5-toothed, sometimes obsolete. Corolla 5-parted, scarcely distinguishable from the calyx, very cellular, with strongly marked reticulated veins, sometimes fringed. ♂ Stamens 5, inserted on the corolla, and alternate with its segments, rarely 3 or 2, either distinct, or monadelphous, or so combined that 4 join in pairs and the fifth remains free ; anthers 2-celled, very long and sinuous. ♀ Ovary adherent, 1-celled, with 3 parietal placentæ, which often project into the cavity, and unite there into a solid central column, while the ovules remain attached to the free edges ; ovules occasionally only one and pendulous, usually horizontal, anatropal ; style short ; stigmas very thick, velvety, lobed or fringed. Fruit more or less succulent, crowned by the scar of the calyx. Seeds flat, ovate, enveloped in a skin, which is either juicy, or dry and membranous ; testa coriaceous, often thick at the margin, sometimes winged ; embryo flat, with no albumen ; cotyledons foliaceous, veined ; radicle next the hilum.

Fig. CCXV.

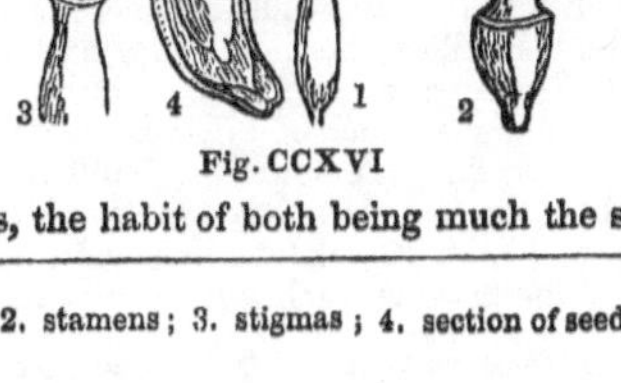

Fig. CCXVI

Cucurbits are placed by Auguste de St. Hilaire and De Candolle between Myrtles, to which they appear to have little affinity, and Passionworts, to which they are so closely allied, that they scarcely differ, except in their sinuous stamens, adherent ovary, unisexual flowers, and exalbuminous seeds, the habit of both being much the same. By

Fig. CCXV.—Bryonia dioica.
Fig. CCXVI.—Coccinia indica. 1. a flower ; 2. stamens ; 3. stigmas ; 4. section of seed. — *Wight.*

the former of these two writers a very particular account of the structure of the Order has been given in the *Mémoires du Muséum*. He adopts the opinion of Jussieu, that the apparent corolla of these plants is really a calyx, considering the apparent calyx to be merely certain external appendages. In discussing the affinities of the Order, which he does much at length, he remarks, that Carica (now the type of the Order Papayads) should be excluded; that the tendrils of Cucurbits are transformed stipules, but scarcely analogous to the stipules of Passionflowers; that there is an affinity between the Order and Bellworts, manifested in the perigynous insertion of the stamens, the inferior ovary, the single style with several stigmas, the quinary division of the flower connected with the ternary division of the fruit, and, finally, some analogy in the nature of the floral envelopes. He, however, chiefly insists upon their affinity with Onagrads, with which, including Myrobalans, they agree in their definite perigynous stamens, single style, exalbuminous seeds, fleshy fruit, and occasionally in the unisexual flowers and climbing stem, being connected in the latter point of view with Onagrads through Gronovia, a climbing genus then referred to that Order. He also points out the further connection that exists between Cucurbits and Onagrads through Loasads, which, with an undoubted affinity to the latter, have the habit of the former, especially in the genus Gronovia which has just been named. With regard to the supposed affinity of Cucurbits to Myrtles, this is founded upon the characters of a small group, called Nandhirobeæ, consisting of plants having the habit of Cucurbits, but some resemblance in the form of the fruit to that of Lecythids, which border closely upon Myrtles.

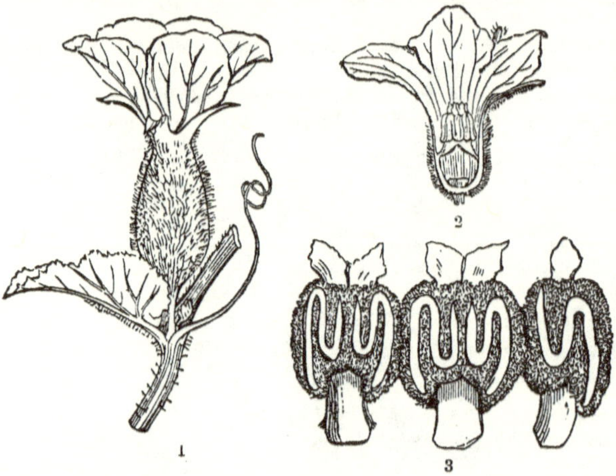

Fig. CCXVII.

The true affinity of Cucurbits seems, however, to be with diclinous Orders. Into Lardizabalads they run through Zanonia; the relation to Spurgeworts is indicated by the climbing habit and the ternary plan of structure observable in the ovary of both Orders, coupled with their disunited sexes; and then to Papayads they belong in everything except their adherent ovary and exalbuminous seeds.

The plants called Nhandirobeæ by Auguste de St. Hilaire, whom Endlicher follows, do not appear to differ essentially from other Cucurbits. Zanonia indeed, with its panicles of small flowers and capsules opening at the point with 3 valves, has a peculiar habit approaching Tetrameles, and so connecting this Order with Datiscads; but it is associated with the genus Feuillea, which seems to be a Gourd and nothing else, notwithstanding its axillary tendrils. The characters relied upon to distinguish Nhandirobeæ as an Order are, 1. a 3-celled ovary; 2. the position of the ovules; 3. the distinct styles; 4. the oblong anthers and axillary not lateral tendrils. But there does not appear to be any difference between the placentation of Zanonia and common Cucurbitaceous plants; its 3 cells are formed by the adhesion of 3 projecting parietal placentæ; 2. if the ovules of Feuillea are ascending, those of Zanonia are horizontal; 3. the styles of Luffa are hardly united, and if it were otherwise, such a character would not be entitled to much attention; 4. the anthers of Zanonia and Feuillea are not alike, and those of the latter genus do not seem to be essentially different from those of Telfairia, Zehneria, and Mukia. As to the supposed axillary stipules of Nhandirobeæ I can only say that they are represented to be lateral in M. Turpin's figure of Feuillea hederacea, drawn expressly to illustrate this supposed Order, and that they are certainly so in Z. cissoides and clavigera.

According to M. Payer, the tendrils of Cucurbits are the two lateral fibrovascular bundles out of the 3 which each leaf forms in its axil.—*Ann. Sc.* 3 *ser.* III. 164.

Fig CCXVII.—Cucumis Melo. 1. ♀ flower and leaf; 2. a section of a ♂ flower; 3. the stamens.

The anther lobes of the Order are occasionally not sinuous, and in a few cases the ripe fruit opens by valves at the point.

The placentation of this Order has needlessly perplexed many Botanists, who have supposed that it is essentially different from that which prevails in other plants. The notion of De Candolle and others has been that in Cucurbits the carpellary leaves are not curved inwards, but outwards, their midrib being in the axis, not circumference, of the fruit. This view has lately been advocated by Dr. Wight in his *Illustrations of Indian Botany* and elsewhere ; and seems to have been taken from the peculiar appearance of such fruits as the Cucumber when cut transversely ; in which case the placentæ do certainly appear as if they were out of their ordinary position ; but if the fruit of these plants is examined early enough it is evident that the illusion arises from 3 parietal placentæ, with revolute seed-bearing edges projecting forward into the cavity, where they adhere. In the garden Cucumber, for example, when half an inch long, the placentæ are exactly as in this cut, (CCXVIII. fig. 1) and have no adhesion. There is, therefore, no ground for regarding the Cucurbitaceous structure at variance with general rules. There is, however, a great peculiarity in the fruit of some of them, such as Luffa fœtida, which, when ripe, appears to consist of horizontal fibres forming a singular entangled mass ; these are visible in the young ovary in the form of semi-transparent concentrical lines which take a somewhat perpendicular direction in the placentæ ; thus apparently proving that part to be a portion of the carpellary leaves and not an independent part of the axis, as Schleiden's theory would suggest.

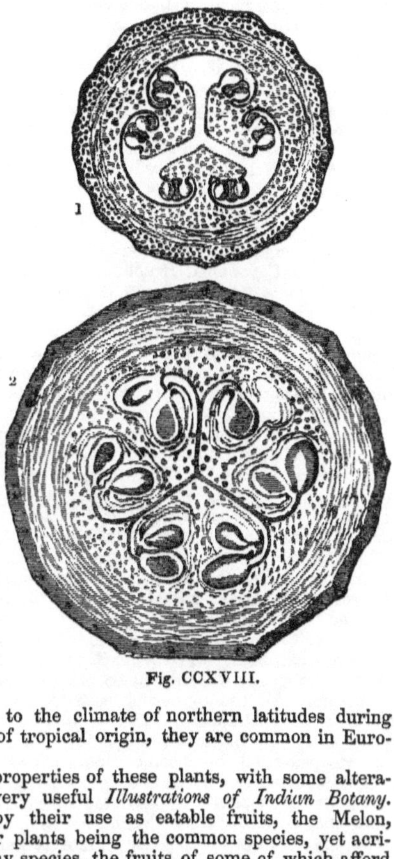

Fig. CCXVIII.

Natives of hot countries in both hemispheres, chiefly within the tropics ; a few are found to the north in Europe and North America, and several are natives of the Cape of Good Hope. India appears to be their favourite station ; a good many occur in Peru and Brazil, but are little known ; one is found in Norfolk Island, and they are met with in Australia. Those which are annuals readily submit to the climate of northern latitudes during the summer, and hence, although mostly of tropical origin, they are common in European gardens.

I borrow the following account of the properties of these plants, with some alterations and additions, from Dr. Wight's very useful *Illustrations of Indian Botany.* Although we best know the Cucurbits by their use as eatable fruits, the Melon, Cucumber, Vegetable Marrow, and similar plants being the common species, yet acrimony and a drastic tendency pervade many species, the fruits of some of which afford cathartics of remarkable power, acting, in even small doses, with great energy on nearly the whole line of the alimentary canal. Generally speaking, however, this intensity of power is of rare occurrence, though the property is found more or less active in every part of the plant ; mildly in the roots of some and the leaves and young shoots of others, but in greatest intensity in the pulp surrounding the seed. The seeds themselves do not partake of the property, being, in nearly all, mild and oily. There is reason to believe that some at least, if not all the edible sorts, owe their freedom from poisonous properties to cultivation, for some in the wild state are found to possess them in much activity. The Lagenaria vulgaris, or Bottle Gourd, may be cited as an example of this, it being recorded that some sailors were poisoned by drinking beer that had been standing in a flask made of one of those Gourds ; and Dr. Royle mentions a somewhat

Fig.—CCXVIII.—1. a section of a very young Cucumber, long before the expansion of the flower ; 2. the same at a period subsequent to the setting of the fruit.

similar case, where symptoms of cholera were induced by eating the bitter pulp. The fruit of many of the species of Cucumis, the genus to which the Melon and Cucumber belong, are powerfully cathartic; among these C. Hardwickii and C. pseudocolocynthis may be enumerated as the chief; but even the Cucumbers, especially the less highly cultivated varieties of India, are sometimes known to prove strongly aperient in susceptible constitutions. C. Colocynthis (now Citrullus), the source of the drug Colocynth, affords one of the most valuable medicinal agents derived from the Order. The Melon, C. Melo, and C. utilissimus, so far as I have been able to learn, is free from it. The fruit of some species of Luffa is violently cathartic, such as L. amara and L. Bindaal of Roxburgh, and the Brazilian Luffas purgans and drastica mentioned by Martius; yet, those of L. acutangula (Cucumis acutangulus, *Ainslie*) are a favourite potherb of the natives of India, and are esteemed very wholesome. Some of the species of Bryonia, especially B. alba and B. dioica, partake of the cathartic properties of the family in great intensity. Curiously enough, the juice of their root is strongly cathartic, and is often employed as such, while the young shoots are so free from the property, that they are used as potherbs, and are reported to resemble Asparagus in flavour. The purgative properties of Bryony root have been long known, and in the opinion of some modern writers have fallen into unmerited neglect, they being fully equal in power, even when dried and powdered, to Jalap, and when recent much more so. Bryonia americana and africana are said to have similar properties. Yet the root of B. abyssinica, when cooked, is said to be eaten without danger. The root of Bryonia epigæa was once supposed to be the famous Calumba; (see MENISPERMACEÆ). Nearly allied to these plants appear to be various species of little known genera, Trianosperma, Wilbrandia, and Cayaponia, Brazilian drastics of great energy. Trianosperma ficifolia, indeed (Bryonia ficifolia, *Lam.*) is a species of great reputation for its activity as a purgative and purifier of the blood. But of all those yet mentioned, none approach the spirting Cucumber, Ecbalium agreste (Momordica Elaterium, *L.*, σίκυς ἄγριος), in the concentrated virulence of this quality. It is a native of the hotter parts of Europe, and remarkable for the force with which its poisonous pulp is suddenly expelled from the interior of the fruit, when it is quite ripe and the stalk is loosened. An ingenious explanation of this curious phenomenon has been given by Dutrochet in his *Nouvelles Recherches sur l'Endosmose*. A few grains of Elaterium, a drug prepared from the pulp of this plant, have been known occasionally to bring on symptoms of poisoning; a case is recorded by Dr. Christison, where a person, after carrying a specimen in his hat, was attacked with headache, succeeded by colic pains and frequent bilious vomiting and purging. Such being the predominant quality of the family, it is well to be cautious in the use of even the best known species.

Many, however, are in use as potherbs, among which may be mentioned with just encomiums the red Gourd, Cucurbita maxima, the flesh of which, when boiled, somewhat resembles in taste a tender Carrot; the Water Melon, Cucurbita citrullus, so highly esteemed for the cool refreshing juice of its large fruit; the white Gourd (Benincasa cerifera, or Cucurbita pepo), which Ainslie informs us is presented at every native marriage feast, being supposed to insure prosperity to the wedded pair; the Vegetable Marrow (Cucurbita ovifera), justly esteemed one of our finest culinary vegetables, and a few others. All the numerous cultivated varieties of the Melon and Cucumber are known to be wholesome. Some, if not all the Indian species of Momordica, seem equally safe. M. Balsamina, a species with a singular warted fruit, and M. Charantia, when steeped in oil, have some reputation as vulneraries. In a green state they form an agreeable pickle. It should be observed, however, that the fruit of a plant called Neurosperma cuspidata by Rafinesque, which is generally supposed to be Momordica Balsamina, is, according to that author, a dangerous poison, or in small doses a hydragogue. Momordica operculata, a plant common in the southern provinces of Brazil, quite answers to the character given of this Neurosperma, so far as its drastic qualities go.

A waxy substance is secreted by the surface of the fruit of Benincasa cerifera. The fruit of several species of Trichosanthes, especially that of T. anguina, are in daily use in India, even among Europeans, dressed in curries; but those of T. palmata are not employed, and are considered poisonous by the natives. Those of Coccinia indica (Momordica monadelpha, *Roxb.*), so common in every Indian hedge, are eaten by the natives in their curries, and when fully ripe (quite red and pulpy) seem to afford a favourite repast to many birds.

The seeds of all the species are oily, and capable of forming very readily an emulsion; those of Telfairia pedata (Joliffia africana, *D.C.*), an African plant, are as large as Chestnuts, and said to be as excellent as Almonds, having a very agreeable flavour; when pressed they yield an abundance of oil, equal to that of the finest Olives. The pulp is excessively bitter, and produced a violent headache when only applied to the tongue. De Candolle remarks, that the seeds of this family never participate in the property of the pulp that surrounds them. But this is a mistake, for the oily seeds of Feuillæa cordifolia, a West Indian shrub, are intensely bitter, and act violently both as emetics and purgatives. The seeds of Feuillæa trilobata yield a fatty oil, used instead of ointment, in pains of the joints. The Americans employ the oil of both species for lamps. The Bandolier fruit (so called, no doubt, from the form of its seed-vessels), or Zanonia indica, has the taste and smell of Cucumber. The bruised leaves are infused in baths, and, mixed with butter, serve for the preparation of an antispastic liniment. The seed of some Cucurbitaceous plants, called Giraumont seeds, are said to destroy tape-worm.—*Endl.*

The seeds of Anisosperma Passiflora (Fava de S. Ignacio, Castanha do Jobotà) contain a bitter oil, mixed with a bland sebaceous matter and resin, and are regarded in Brazil as valuable stomachics; in large doses they purge. Those of Hypanthera Guapeva, another Brazilian climber, have similar qualities.

GENERA.

I.—NHANDIROBEÆ. Anthers not sinuous. Placentæ adhering in the axis of the fruit. Seeds numerous.

Telfairia, *Hook.*
Joliffia, Bojer.
Ampelosicyos, Thouars.
Feuillæa, *L.*
Nhandiroba, Plum.
Zanonia, *L.*
Alsomitra, *Bl.*
Actinostemma, *Griff.*
Anisosperma, *S. Mans.*
Hypanthera, *S. Mans.*

II.—CUCURBITEÆ. Anthers sinuous. Placentæ adherent in the axis of the fruit. Seeds numerous.

Coniandra, *Schrad.*
Cyrtonema, *Schrad.*
Melothria, *Linn.*
Sicydium, *Schlecht.*
Bryonopsis, *Arn.*
Æchmandra, *Arn.*
Zehneria, *Endl.*
Pilogyne, *Schrad.*
Anguria, *Linn.*
Psiguria, *Neck.*
Rhynchocarpa, *Schrad.*
Bryonia, *Linn.*
? *Solena*, Lour.
? *Cucumeroides*, Gärtn.
Karivia, *Arn.*
Mukia, *Arn.*
Diclidostigma, *Kze.*
Schizostigma, *Arn.*
Trianosperma, *Torr et Gr.*
Alternasemina, Manso.
Wilbrandia, *Manso.*
Cayaponia, *Manso.*
Citrullus, *Neck.*
Colocynthis, Tournef.
Rigocarpus, Neck.
Ecbalium, *L. C. Rich.*
Momordica, *Linn.*
Elaterium, Tournef.
Amordica, Neck.
Poppya, Neck.
Muricia, Lour.
Neurosperma, Raf.
Luffa, *Tournef.*
Turia, Forsk.
Trevouxia, Scop.
Benincasa, *Savi.*
Lagenaria, *Ser.*
Cucumis, *Linn.*
Melo, Tournef.
Cucurbita, *Linn.*
Pepo, Tournef.
Melopepo, Tournef.
Coccinia, *Wight et Arn.*
Trichosanthes, *Linn.*
Anguina, Mich.
Ceratosanthes, Juss.
Involucraria, *Ser.*
Gymnopetalum, *Arn.*
Apodanthera, *Arn.*
Elaterium, *Jacq.*
Echinocystis, *Torr. et A. Gr.*
Cephalandra, *Schrad.*
Cyclanthera, *Schrad.*
Discanthera, *Torr. et A. Gr.*

Little known.

Schizocarpum, *Schrad.*
Rytidostylis, *Hook et Arn.*

III.—SICEÆ. Placentæ not projecting into the cavity. Seed solitary from the top of the cell.

Sicyos, *Linn.*
Sicyoides, Tournef.
Badaroa, Berter.
Sechium, *P. Br.*
Chayota, Jacq.
Gynostemma, *Blum.*

Genera quite doubtful.

? Erythropalum, *Blum.*
Zucca, *Commers.*
Thladiantha, *Bung.*
Pentaclathra, *Bertol.*
Perianthopodus, *Silv. M.*
Dromophylla, *Silv. M.*
Druparia, *Silv. M.*

NUMBERS. GEN. 56. Sp. 270. But this number is too low, in consequence of the South American species having been little investigated.

Passifloraceæ.
POSITION.—Datiscaceæ.—CUCURBITACEÆ.—Begoniaceæ.
Loasaceæ.

ORDER CVI. DATISCACEÆ.—DATISCADS.

Datisceæ, *R. Brown in Denham*, 25. (1826); *Bartl. Ord. Nat.* 419. (1830); *Endl. Gen.* clxxxiv.; *Meisner*, p. 346.

DIAGNOSIS.—*Cucurbital Exogens with apetalous flowers, strictly parietal placentæ, and dry fruit.*

Herbaceous branched plants; or trees of considerable size. Leaves alternate, cut, simple, or compound, without stipules. Flowers in axillary racemes or terminal panicles, ♂ ♀. Calyx of the ♂ divided into 3-4 pieces; of the ♀ adherent 3- 4-toothed. ♂ Stamens 3-7; anthers 2-celled, membranous, linear, bursting longitudinally. ♀ Ovary 1-celled, with 3-4 polyspermous parietal placentæ; ovules anatropal; stigmas equal in number to the placentæ, and opposite the lobes of the calyx. Fruit capsular, opening at the vertex, 1-celled, with polyspermous parietal placentæ. Seeds enveloped in a membranous finely reticulated integument, with a cupulate membranaceous strophiole; embryo straight, without albumen, its radicle very long, turned towards the hilum. Cotyledons very short.

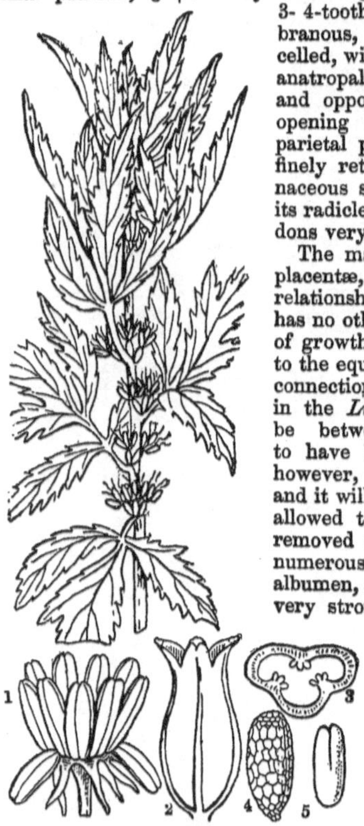

Fig. CCXIX.

The many-seeded capsule of this genus, with parietal placentæ, and open at the apex, naturally suggested its relationship to Reseda, with which, however, it really has no other point in common. The foliage and manner of growth of Datisca cannabina has in like manner led to the equally wrong conclusion that it might have some connection with Hempworts. An anonymous writer in the *Linnæa* (xiv. 262) has suggested its station to be between Cucurbits and Loasads. This seems to have been a close approach to the truth. It is, however, with Begonia that it corresponds most nearly, and it will have to follow the fate of that Order, whether allowed to retain the station now assigned to it or removed to some other place. The unisexual flowers, numerous minute seeds, orthotropal embryo without albumen, and adherent calyx of these two Orders, afford very strong marks of relationship; to which may be added the triple placentation of two out of three of the known Datisceous genera. It is true, indeed, that Datisca and Tricerastes are said to have albumen; but I can find none in Datisca nepalensis when fully ripe, and therefore it may be doubted whether it exists at that time in Tricerastes, or Datisca cannabina. To this it may be added that the naked mode of flowering in loose terminal panicles, and the oblique leaves of Tetrameles, are equally characteristic of Begonia.

Fresenius asserts (*Linnæa*, 1839) that female plants of Datisca cannabina are capable of bearing seed, although entirely cut off from the males. He regards this property to depend upon a mere act of vegetable increment, which, upon the supposition that an embryo is a bud, is not inconceivable. Tetrameles, the Weenong of Java, and Jungle Bendy of Bombay, is remarkable as being a large tree in this very small Order, consisting otherwise of annual stemmed herbaceous plants.

The very few species of which the Order consists are scattered over North America, Siberia, Northern India, the Indian Archipelago, and the south-eastern corner of Europe.

Fig. CCXIX.—Tricerastes glomerata.—*Presl.* 1. ♂ of Datisca cannabina; 2. its fruit; 3. a cross section of it; 4. a seed; 5. its embryo.

Datisca is bitter and purgative; it is occasionally used in Italy in fevers, as well as gastric and scrofulous complaints. Chemists have found in its roots a kind of starch analogous to Inuline. They call it Datiscine.

GENERA.

Datisca, *L.*
Tetrameles, *R. Br.*
Anictoclea, Nimmo.
Tricerastes, *Presl.*

NUMBERS. GEN. 3. Sp. 4. excluding Datisca hirta, which Mr. Bennett has ascertained to be Rhus Typhinum.

POSITION.—Cucurbitaceæ.—DATISCACEÆ.—Begoniaceæ.
Loasaceæ?

ORDER CVII. BEGONIACEÆ.—BEGONIADS.

Begoniaceæ, *R. Brown in Congo*, 464. (1818); *Endl. Gen.* cciii.; *Meisner*, p. 336.

DIAGNOSIS.—*Cucurbital Exogens with dry fruit and placentæ projecting and meeting in the axis.*

Herbaceous plants or succulent under-shrubs, with an acid juice. Leaves alternate, toothed, rarely entire, oblique at the base. Stipules large, scarious. Flowers pink, in cymes. Flowers ♂ ♀. Calyx adherent. Sepals coloured; in the ♂ 4, 2 within the others and smaller; in the ♀ 5, imbricated, two smaller than the rest, or 8, of which 4 are petaloid. ♂ Stamens indefinite, distinct, or combined into a solid column; anthers collected in a head, 2-celled, continuous with the filaments, clavate, the connective very thick, the cells minute, bursting longitudinally. ♀ Ovary adherent, winged, 3-celled, with 3 large placentæ meeting in the axis; ovules anatropal; stigmas 3, 2-lobed, sessile, somewhat spiral. Fruit membranous, capsular, winged, 3-celled, with an indefinite number of minute seeds; bursting by slits at the base on each side of the wings. Seeds with a transparent thin testa marked by reticulations, which are oblong at the sides and contracted at either extremity; embryo very cellular, without albumen, with a blunt round radicle next the hilum.

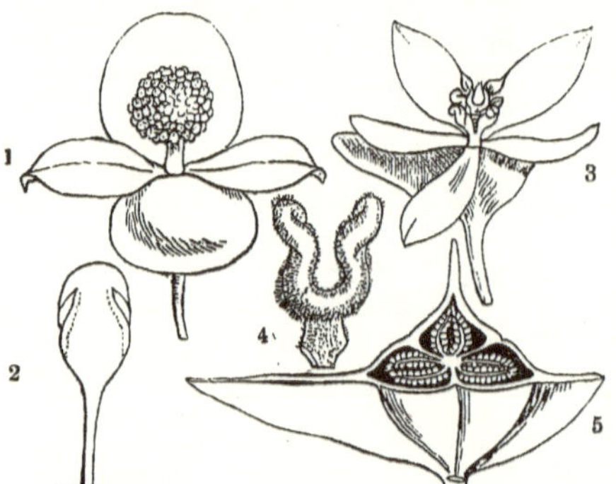

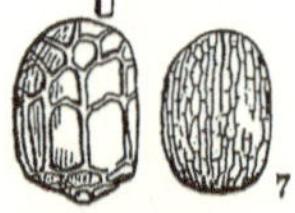

Fig. CCXX.

It is not a little curious that the opinions of Botanists concerning the affinity of these well-known plants should remain so undecided up to the present day. I formerly supposed the Order related to Hydrangea from some resemblances in its seeds, &c.; others have approximated it to Buckwheats on account of the stipules, 3-cornered fruit, and coloured calyx. Link places it near Umbellifers; Von Martius next Scævolaceæ; Meisner with Spurgeworts; and the tendency to the production of 4 in the sepals and petals, as evinced by Eupetalum, has led to the opinion that it may be related to the epigynous Myrtal Alliance, its seeds being indefinite and destitute of albumen. And that some near relation does exist between such plants and Begonia, is rendered more probable by Bertolonia maculata, which has the winged fruit and much the habit of that genus. Nevertheless, these are but distant points of approach; and the real affinities seem to be with Cucurbits, with which Begoniads accord in the unisexual flowers, peculiar stigmas, and even ternary number of the carpels. The discovery by Mr. Hartweg of Begoniads scrambling up trees and shrubs to the height of 25 feet, renders the resemblance almost complete. To Datisca the relationship seems to be well made out.

The main objection to the association of Begoniads and Cucurbits in the same Alliance arises from the great apparent difference in their placentation; that of Begoniads being axile, and of Cucurbits parietal. But a careful examination of the ovary of Diploclinium Evansianum, acuminatum, incarnatum, &c., shows that distinction to be one of words rather than of essential structure. The ovary of such Begoniads consists of 3 carpels, whose dorsal suture is winged, and whose margins turn inwards for a considerable distance, each margin forming a plate or placenta, over which the ovules are arranged. This, with the exception of the wing proceeding from the dorsal suture, is the structure of Cucumis, as figured at p. 313, fig. 1, with this difference, however, that the inflexed edges of the carpellary leaves adhere in Diploclinium at a very much more early period than in Cucumis. In Diploclinium acuminatum, when the flower-buds hardly project beyond their fringed bracts, these inflexed edges are easily separable from each other and from

Fig. CCXX.—Diploclinium Evansianum. 1. a ♂ flower; 2. an anther; 3. a ♀ flower; 4. a stigma; 5. a cross section of the ovary; 6. a seed; 7. an embryo

the soft torus which rises up between them and holds them together. I cannot however add that I have ever succeeded in finding the placentæ absolutely separate, as is the case in a young Cucumber. If the true Begonias, which like B. Meyenii, coccinea, &c., have a solid, not 2-lobed placenta, are boiled for an instant in caustic potash, it becomes evident that the real structure is the same, and that there also the placentation is exactly the same as in Cucumis, except that the inflexed edges adhere into a solid wedge, and that the matter of the torus which rises between them and holds them together is more copious. The ovary of Begonia Meyenii may even be taken as an excellent illustration of the true nature of that of a Cucurbit.

Common in the West Indies, South America, and the East Indies. Brown remarks, that no species has been found on the continent of Africa, though several have occurred in Madagascar and the Isles of France and Bourbon, and one in the island of Johanna.

The roots are astringent and slightly bitter. Those of two species are employed in Peru with success in cases of a flux of blood, or in other visceral diseases in which astringents are employed. They are also said to be useful in cases of scurvy, and in certain fevers. B. malabarica, and tuberosa, and several more, are used as pot-herbs. The root of B. grandiflora and tomentosa is bitter and very astringent. Some are said to be drastic purgatives in Mexico (*Endl.*) ; and if so, this is an additional point of resemblance between them and Cucurbits.

GENERA.

Begonia, *L.* Eupetalum, *Lindl.* Diploclinium, *Lindl.**

NUMBERS. GEN. 2. Sp. 159. (*Walpers*).

Melastomaceæ?
POSITION.—Cucurbitaceæ.—BEGONIACEÆ.—Datiscaceæ.

* This genus includes those Begonias which have a double placenta.

ALLIANCE XXV. *PAPAYALES.*—THE PAPAYAL ALLIANCE.

DIAGNOSIS.—*Diclinous Exogens with dichlamydeous flowers, superior consolidated carpels, parietal placentæ, and embryo surrounded by abundant albumen.*

If the plants referred to this Alliance had no albumen and an inferior ovary, they would be Cucurbitals; if their flowers were bisexual and coronetted they would do for Passionworts; if their fruit were simple and their ovules orthotropal, or at least with the foramen uppermost, they would fall into the Order of Nettleworts. They seem evidently to join the Violal Alliance, the whole of which, if the flowers were diclinous, might have been brought into the closest contact with Papayals, as will be sufficiently evident if Pangiads are compared with Bixads.

NATURAL ORDERS OF PAPAYALS.

Corolla monopetalous; ♀ without scales . . . . 108. PAPAYACEÆ.
Corolla polypetalous; ♀ with scales in the throat . . . . 109. PANGIACEÆ.

ORDER CVIII. PAPAYACEÆ.—PAPAYADS.

Papayæ, *Agardh Classes*. (1824).—Cariceæ, *Turpin in Atl. du Dict. des Sc. Nat.* (?)—Papayaceæ, *Von Martius Conspectus, No.* 169. (1835); *Endl.* cc.; *Meisner, p.* 123; *Wight. Illustr.* 2. 33.—Modecceæ, *Endl. p.* 927.

DIAGNOSIS.—*Papayal Exogens, with monopetalous flowers, having no scales in the throat of the females.*

Trees or shrubs, sometimes yielding an acrid milky juice. Leaves alternate, lobed, on long taper petioles. Flowers in axillary racemes or solitary, unisexual. Calyx inferior, minute, 5-toothed. Corolla monopetalous, with 5 lobes. ♂ Stamens definite, epipetalous; anthers erect, splitting longitudinally, occasionally partly imperfect. ♀ Ovary free, 1-celled, with 3 to 5 parietal polyspermous placentæ; stigma 3- 5-lobed, lacerated. Fruit succulent, or dehiscent, 1-celled, with parietal placentæ. Seeds enveloped in a loose mucous coat, with a brittle pitted testa; embryo in the axis of fleshy albumen, with flat cotyledons and a taper radicle turned towards the hilum.

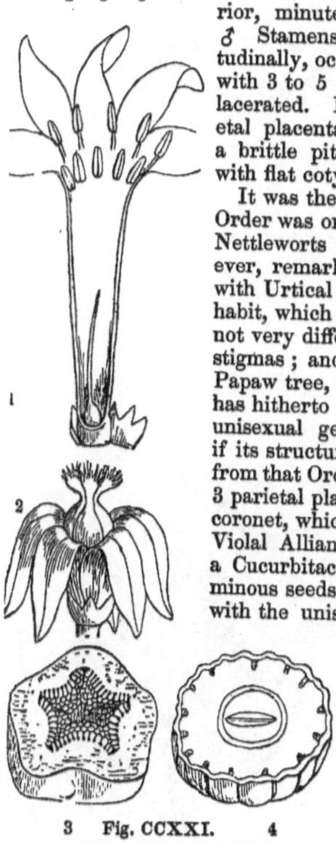

Fig. CCXXI.

It was the opinion of Jussieu that the genus upon which this Order was originally founded held a sort of middle station between Nettleworts and Cucurbits. Auguste de St. Hilaire has, however, remarked upon this subject, that the only relation it has with Urtical plants consists in the separation of sexes, milky juice, habit, which is like that of some species of Ficus, foliage, which is not very different from that of Cecropia, and the position of its stigmas; and to these he attached little importance. But the Papaw tree, instead of standing in the system almost alone, as it has hitherto done, appears to be in reality the associate of all the unisexual genera hitherto referred to the Passionworts; for if its structure be scrutinised carefully it will be found to differ from that Order in nothing except having a fruit with 5 instead of 3 parietal placentæ, in its separate sexes, and the absence of the coronet, which in some form or other is so characteristic of the Violal Alliance. On the other hand, it may be regarded as a Cucurbitaceous plant with a free ovary, 5 placentæ, and albuminous seeds; and in that point of view it equally claims kindred with the unisexual Passionworts. The opinion of Jussieu then seems to have been right, as it has so often proved to be in difficult cases.

The species of Carica are natives of South America, and unknown, except as objects of cultivation, beyond that continent; the other genera belong to the temperate parts and tropics of the Old World.

The fruit of the Papaw (Carica Papaya) is eaten, when cooked, and is esteemed by some persons; but it appears to have little to recommend it. Its great peculiarities are, that the juice of the unripe fruit is a most powerful and efficient vermifuge (the powder of the seed answers the same purpose), and that a constituent of this juice is fibrine, a principle formerly supposed peculiar to the animal kingdom and to Fungals. The tree has, moreover, the singular property of rendering the toughest animal substances tender, by causing a separation of the muscular fibre; its very vapour even does this; newly-killed meat suspended among the leaves, and even old hogs and old poultry, when fed on the leaves and fruit, become tender in a few hours. See an excellent account of the Papaw by Hooker in the *Bot. Mag.* 2898. Dr. Wight observes that the seeds, when chewed, yield, in a very marked degree, the pungency and flavour of Tropæolum majus. The excessive acridity which renders the Papaw an active vermifuge, is indicated by the disgusting and overpowering odour of its roots, which smell like decaying Radishes. The leaves are used by negroes

Fig. CCXXI.—Carica Papaya. 1. ♂ flower; 2. ♀ flower; 3. section of ovary; 4. section of seed.

to wash linen, instead of soap. The Carica digitata, (Chamburu), a Brazilian plant, is regarded by the natives of Mayna as a deadly poison, and with as much awe as the Upas tree by the Javanese. Pöppig says that the juice which spirted over his skin when he cut the tree, caused itching on the face, and drew a few blisters on his hands; the male flowers of this plant have the disgusting smell of human excrement. It is worthy of remark that the fruits of the plant, although handsome, scentless, and insipid, are untouched by birds or other animals except an ant belonging to the genus Atta. The root of Modecca palmata, a native of tropical Asia, rubbed down with oil, is regarded as a corroborant; mixed with Cocoa-nut milk it is used for pain in the chest. The leaves of M. integrifolia, boiled with butter, are used for piles; its juice is thought to assist labour.—*Endl.*

Vauquelin, who analysed the juice of the Papaw, says that no doubt can be entertained of its being a highly animalised substance; although it is not exactly like any animal matter known to him. It most resembles animal albumen, dissolving, like it, in water. Its solution is coagulated by heat, by acids, alkalies, the metallic salts, and infusion of nut-galls; and by distillation it yields the same products as animal substances.

GENERA.

* *Fruit succulent, indehiscent.*

Carica, *Linn.*
Papaya, Tournef.
Vasconcella, *St. Hil.*
Tetrapathea, *Raoul.*

** *Fruit capsular.*

Modecca, *Linn.*
Microblepharis, Wight et Arn.
Blepharanthus, Smith.
Paschanthus, *Burch.*
Kolbia, *Palis.*
Ceratiosicyos, *Nees.*
Acharia, *Thunb.*

NUMBERS. GEN. 8. SP. 25.

Cucurbitaceæ.
POSITION.——PAPAYACEÆ.—Pangiaceæ.
Passifloraceæ.

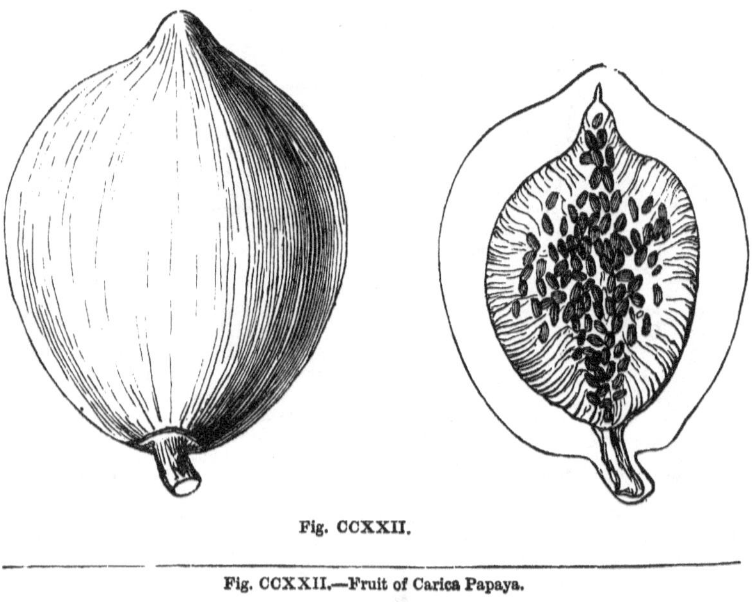

Fig. CCXXII.

Fig. CCXXII.—Fruit of Carica Papaya.

ORDER CIX. PANGIACEÆ.—PANGIADS.

Pangiaceæ, *Blume in Ann. Sc. N. ser.* 2. 88. (1834); *Bennett in Horsfield, pl. Javan. p.* 208; *Endl. Gen. p.* 922.

DIAGNOSIS.—*Papayal Exogens, with polypetalous flowers, which have scales in the throat, of the females.*

Trees. Leaves alternate, stalked, entire, or somewhat lobed. Flowers axillary, solitary or fascicled, or in few-flowered racemes, ♂ ♀. Sepals 5, rarely 2, 3, or 4. Petals 5, rarely 6. Scales as many, opposite the petals. ♂ Stamens 5 or 00; not a rudiment of ♀. ♀ Sterile stamens equal in number to the petals, rarely more. Ovary free, 1-celled; ovules 00, attached to 2-6 parietal placentæ. Capsules succulent, indehiscent, 1-celled. Seeds 00, large; albumen abundant, oily; embryo nearly as large as the albumen; radicle protruded; cotyledons generally leafy and veined. —*Bennett.*

What the distinction is between these plants and Papayads, except that the last are monopetalous, and have no faucial scales in the ♀ flowers, it is hard to say. Mr. Bennett throws no light upon the matter, and I am unable to supply any.

The species are found in the hotter parts of India. All are poisonous plants. The seeds of Gynocardia odorata are employed extensively by the natives of India in the cure of cutaneous disorders. When freed from the integuments they are beaten up with clarified butter, into a soft mass, and in this state applied thrice a day to the parts affected. Chaulmoogra and Petarkura are the native names. Hydnocarpus venenatus, a Cingalese plant related to this Order, bears a poisonous fruit, which when eaten occasions giddiness and dangerous intoxication. It is used in Ceylon for

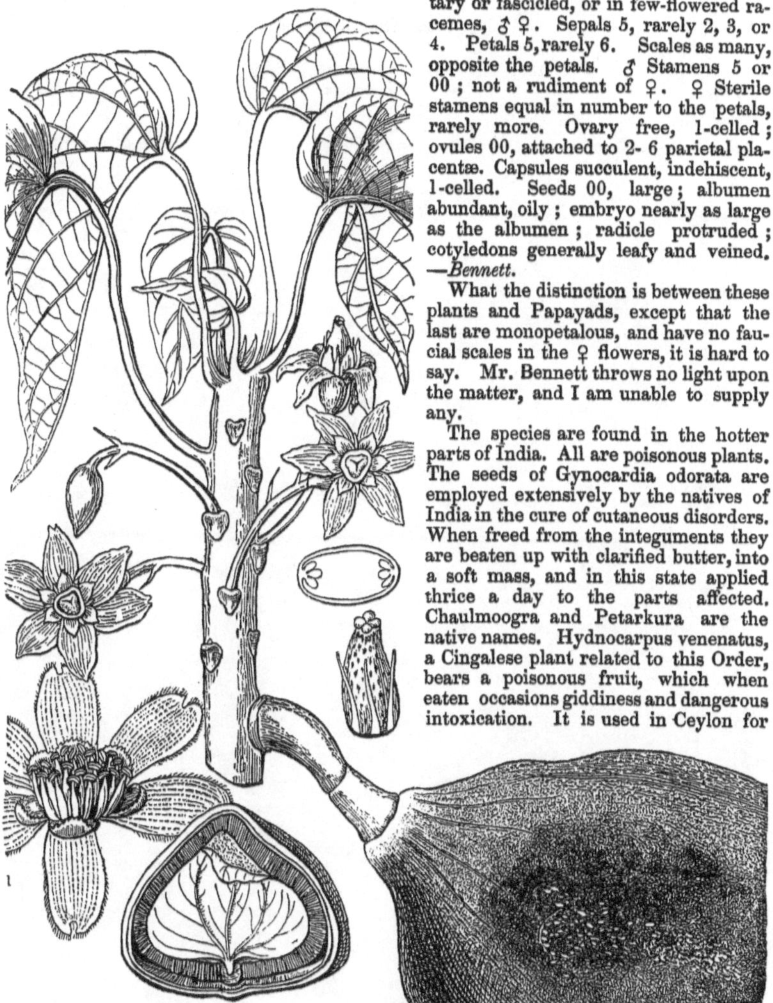

Fig. CCXXIII.

poisoning fish, which afterwards become so unwholesome as to be unfit for food. According to Rumphius, the plant Pangi has a hard, solid wood, and its kernels are boiled, then cut to pieces and macerated in cold water to remove the noxious narcotic qualities;

Fig. CCXXIII.—Pangium edule.—*Horsfield.* 1. a ♂ flower; 2. an ovary with its sterile stamens; 3. a cross section of an ovary; 4. a perpendicular section of a seed.

after which they are dried to be used as a condiment. The bark thrown into water poisons fish; the juice of the leaves is used for destroying vermin, and in cutaneous diseases, and cows die from feeding on them. The oil of the seeds is however employed for frying. Dr. Horsfield adds that these seeds are rarely used, and curry containing them operates as a cathartic on persons unaccustomed to them.

GENERA.

Pangium, *Reinw.*	*Chaulmoogra*, Roxb.	Hydnocarpus, *Gærtn.*
Gynocardia, *R. Br.*	*Chilmoria*, Hamilt.	*Munnicksia*, Dennst.

NUMBERS. GEN. 3. Sp. 4.

POSITION. ——— PANGIACEÆ.—Papayaceæ.
Bixaceæ.

SUB-CLASS II. HYPOGYNOUS EXOGENS.

The hypogynous insertion of the stamens has been regarded by the French school of Botanists as one of very great systematical importance; and it does seem to collect together a large mass of plants the genera of which have a great resemblance to each other. If we assume that the entire separation of the calyx and corolla from the stamens is an indication of those organs being in hypogynous plants of less importance than usual, then the character acquires a physiological value not previously assigned to it. And such appears to be the case; for it is only among hypogynous Exogens that we find a total absence of floral envelopes, as in the Piperal and Chenopodal Alliances; it is among them that the presence of petals seems to be of least moment, as the character of a Natural Order; for in 12 Alliances out of 14, petals are either constantly or frequently absent, and in one only are they often combined into a tube; in all other cases such a circumstance is exceptional.

It is, however, found that in some cases plants with a perigynous insertion of the stamens will nevertheless combine with hypogynous Alliances; as happens in the case of Samyds among Violals, and here and there in the Erical, Silenal, and Chenopodal Alliances; but these again seem to be mere exceptional instances not affecting the general value of the hypogynous character, even where it is certain that the Orders in which such exceptions occur are rightly placed. Experience shows too that all natural groups of plants come in contact here and there; and in such instances exceptions to habitual structure make their appearance. It will be found, moreover, that the perigynous Orders or genera here and there introduced among the hypogynous series refuse to associate with any part of the perigynous Sub-class. Thus Samyds, a perigynous form of Violals, have no *locus standi* in any perigynous Alliance, while their affinity to the hypogynous Violals is of an obvious nature.

The sequence observed in the arrangement of the Alliances is chiefly objectionable on account of the presence of Berberals in contact with Ericals; but if we regard Cyrillads and Pittosporads as Berberals, then the Erical Humiriads join them perfectly; but these approximations have not yet received the sanction of Botanists, and depend for their justification upon giving a higher value than customary to the presence of a small embryo in copious albumen.

There can be no doubt about the closeness of the relationship borne by the diclinous Papayals to the hypogynous Violals, and therefore it is with the latter that the hypogynous series is made to begin. The transition from Violals to Cistals, thence to Malvals, to Sapindals, and to Guttiferals is so much in conformity to the views generally entertained by Botanists, that no objection to it is anticipated. The next step to Nymphals is more open to criticism; but if Tutsans are taken as an extreme form of Guttiferals, there is no difficulty in admitting the justice of bringing Nymphals into contact with them. The next relationship, that of Ranals, is obvious; their Poppyworts possess a genus, Hypecoum, which stands on the limits of Fumeworts among Berberals; thence either Cyrillads or Olacads will join the chain to Humiriads among Ericals. In the latter Alliance Heathworts themselves come distinctly in contact with such monopètalous Rueworts as Correa. Geranials join Rutals by means of Oxalids, and Silenals through Flaxworts; and, finally, the Chenopodal and Piperal Alliances are apparently degraded forms of the Silenal Alliance. Piperals ought, however, to be regarded as a lateral offset from Chenopodals rather than as an Order in the direct line of succession to the perigynous Sub-class.

Alliance XXVI. *VIOLALES.*—The Violal Alliance.

Diagnosis.—*Hypogynous Exogens, with monodichlamydeous flowers, parietal or sutural placentæ, and straight embryo with little or no albumen.*

If we except Moringads, Tamarisks and Houseleeks, which are doubtful members of this Alliance, the present group seems quite natural; and those Orders themselves appear to find no better station, as will be shown when speaking of their respective affinities. The parietal placentation is without example among Hypogynous Alliances, except in Cistals, whose curved or spiral embryo seems to distinguish them perfectly.

Natural Orders of Violals.

Flowers scattered, apetalous or polypetalous. Petals and stamens both hypogynous. Leaves dotless, or with round dots only. .	110. Flacourtiaceæ.
Flowers in catkins, apetalous, scaly, polygamous. Stamens unilateral	111. Lacistemaceæ.
Flowers scattered, apetalous, tubular, hermaphrodite. Leaves marked with both round and linear transparent dots. (Stamens perigynous)	112. Samydaceæ.
Flowers polypetalous or apetalous, coronetted. Petals perigynous, imbricated. Stamens on the stalk of the ovary. Styles simple, terminal. Seeds arillate. Leaves stipulate . . .	113. Passifloraceæ.
Flowers polypetalous, coronetted. Petals perigynous, imbricated. Stamens on the stalk of the ovary. Styles simple, dorsal. Seeds without aril. Leaves without stipules	114. Malesherbiaceæ.
Flowers polypetalous. Calyx many-leaved. Petals perigynous. Anthers 1-celled. Fruit stipitate, consolidated, siliquose. Seeds without albumen (Stamens perigynous)	115. Moringaceæ.
Flowers polypetalous. Calyx many-leaved. Petals hypogynous. Stamens all perfect; anthers crested, and turned inwards. Fruit consolidated. Seeds albuminous	116. Violaceæ.
Flowers polypetalous. Calyx tubular, furrowed. Petals hypogynous, unguiculate	117. Frankeniaceæ.
Flowers polypetalous. Calyx many-leaved. Petals hypogynous. Styles distinct. Fruit consolidated. Seeds 00, basal, comose, without albumen.	118. Tamaricaceæ.
Flowers polypetalous. Calyx many-leaved. Petals hypogynous. Stamens partly sterile and petaloid; anthers opposite the petals, naked, turned outwards. Fruit consolidated. Seeds albuminous.	119. Sauvagesiaceæ.
Flowers polypetalous or monopetalous. Calyx many-leaved. Petals hypogynous. Fruit follicular, apocarpous	120. Crassulaceæ.
Flowers polypetalous. Petals perigynous, contorted. Styles forked. Leaves exstipulate	121. Turneraceæ.

ORDER CX. FLACOURTIACEÆ.—BIXADS.

Flacourtianeæ, *Richard in Mém. Mus.* 1. 366. (1815); *DC. Prodr.* 1. 255; *Wight Illustr.* 1. 36; *Bennett in Horsfield's pl. Jav. p.* 187.—Flacourtiaceæ, *Ed. pr.* 1.—Bixineæ, *Kunth. Diss. Malv.* p. 17. (1822); *DC. Prodr.* 1. 259; *Wight Illustr.* 1. 38.—Bixaceæ, *Ed. pr.* liv. (1836); *Endl. Gen.* cxcv.

DIAGNOSIS.—*Violal Exogens, with scattered apetalous or polypetalous flowers, hypogynous petals and stamens, and dotless or round-dotted leaves.*

Shrubs or small trees. Leaves alternate, simple, on short stalks, without stipules, usually entire, and leathery, very often marked with transparent dots. Peduncles axillary, many-flowered. Sepals from 4-7, cohering slightly at the base. Petals equal to the latter in number and alternate with them, or wanting. Stamens hypogynous, of the same number as the petals, or twice as many, or some multiple of them. Ovary roundish, sessile, or slightly stalked, free, 1 or more celled, with 2 or more parietal placentæ, which are either simple or branched; style either none or filiform; stigmas several, more or less distinct; ovules attached to the surface or sides of the placentæ, and never to the axis in those genera whose ovary has several cells. Fruit 1-celled, either fleshy and indehiscent, or capsular, with 4 or 5 valves, the centre filled with a thin pulp. Seeds 00, usually enveloped in a pellicle formed by the withered pulp; albumen fleshy, somewhat oily; embryo straight, in the axis, with the radicle turned to the hilum, and therefore usually superior; cotyledons flat, foliaceous.

1 2 3 4

Fig. CCXXIV.

The two supposed Natural Orders now brought together, as suggested by several writers, and especially by Mr. Bennett and Professor Endlicher, have never possessed any valid claim to be distinguished. The differences between them were derived from the mode of placentation, which in Bixa and its allies is parietal in lines, while in Flacourtia it spreads like a net all over the inner surface of the fruit. But intermediate structures annul this characteristic. It was also supposed that the presence among the allies of Flacourtia of certain barren stamens or scales, would assist in dividing the latter from Bixa,—in fact, establishing a direct affinity between the first and Passionworts; but those scales belonged to genera now referred to Pangiads. Taken as a Natural Order, Bixads form a group readily known from Samyds by their hypogynous stamens and dotless leaves, or at least by all their dots being round if they are present; from Passionworts by the petals if present being hypogynous, and the total absence of all sign of a coronet. Because of their indefinite stamens, and the valvate calyx of some genera, they have been compared to Lindenblooms: but there

Fig. CCXXIV.—Bixa Orellana.—*Wight.* 1. a pistil and two stamens; 2. a cross section of the ovary; 3. a ripe fruit; 4. a cross section of a seed.

seems to be only a remote analogy with that Order. The frequent tendency to a polygamous structure shows their affinity to Lacistemads.

Almost all these plants are natives of the hottest parts of the East or West Indies, and Africa. Two or three species are found at the Cape of Good Hope, and one or perhaps two in New Zealand.

The pulp of Oncoba is sweet, and eaten in Nubia. The fruits of some of the Flacourtias are eatable and wholesome. Those of F. Ramontchi, a Madagascar species, are much like black plums; of F. sapida and sepiaria have a pleasant refreshing subacid taste; and the berries of a species of Roumea, found in the jungles of Ceylon, are much prized at Colombo. The young shoots and leaves of Flacourtia cataphracta, which have the taste, but not the bitterness, of Rhubarb, are considered astringent and stomachic, and are prescribed, in the Circars, in cases of diarrhœa and general debility; in Bahar, a cold infusion is used in hoarseness. The infusion of F. sepiaria is considered useful in bites of snakes; the bark rubbed with oil, and made into a liniment, is employed against gout on the Malabar coast.—*Wight.* Aphloia theiformis, a shrub inhabiting the Isle of France, has an emetic bark. Lætia apetala secretes in tropical America a balsamic resin, becoming white in contact with air, like Sandarach.

The seeds of Bixa Orellana are angular, and covered with an orange-red waxen pulp or pellicle. The latter substance is the Arnotto of the shops; it is separated from the seeds by washing. It is chiefly used in the preparation of chocolate; but was reckoned an antidote to the poison of the manioc or Janipha Manihot. Farmers employ it to stain their cheeses, and dyers for a reddish colour. Martius says that the seeds are cordial, astringent, and febrifugal.

GENERA.

I.—Bixeæ. *Style simple. Fruit splitting.*

Bixa, *Linn.*
Echinocarpus, *Blum.*
Trichospermum, *Blum.*
Lindackeria, *Blum.*
Xylotheca, *Hochst.*
Denhamia, *Meisn.*
Leucocarpon, A. Rich.

II.—Prockeæ. *Style simple. Fruit not splitting.*

Carpotroche, *Endl.*
Mayna, Radd.
Oncoba, *Forsk.*
Lundia, Thonn.
Phoberos, *Lour.*
Rhinanthera, Blum.
Limonia, Gärtn.
Scolopia, Schreb.
Eriudaphus, Nees.
Dasyanthera, Presl.
Ludia, *Lam.*
Laetia, *Löffl.*
Thamnia, P. Br.
Hellwingia, Adans.
Prockia, *P. Br.*
Thiodia, *Benn.*
Lightfootia, Swartz.
Aphloia, *Benn.*
Neumannia, A. Rich.
Xylotheca, *Hochst.*
Ascra, *Schott.*
Trilix, *L.*
Zuelania, *A R.*
Banara, *Aubl.*
Bosca, Fl. Flum.
? *Xyladenus*, Desv.
Azara, *Ruiz et Pav.*
Kuhlia, H. B. K.
Lilenia, Bert.
Almeja, Endl.
Pineda, *Ruiz et Pav.*
Christannia, Presl.
? Leonia, *Ruiz et Pav.*
Steudelia, Mart.

III.—Flacourteæ. *Styles or stigmas several, Fruit succulent.*

Flacourtia, *Commers.*
Stigmarota, Lour.
Rhamnopsis, Reichenb.
Cræpaloprumnon, Endl.
Roumea, *Poit.*
Koelera, Willd.
Bessera, Spreng.
Limacia, Dietr.
Hisingera, *Hellen.*
Xylosma, *G. Forst.*
Myroxylon, J.R. Forst.
Lunania, *Hook.*
Melicytus, *Forst.*

IV. — Erythrospermeæ. *Styles several. Fruit splitting.*

Kiggellaria, *Linn.*
Erythrospermum, *Lam.*
? Tachibota, *Aubl.*
Salmasia, Schreb.

NUMBERS. GEN. 31. Sp. 85.

Pangiaceæ.
POSITION.—Samydaceæ.—FLACOURTIACEÆ.—Lacistemaceæ.
Tiliaceæ ? ?

ORDER CXI. LACISTEMACEÆ.—LACISTEMADS.

Lacistemeæ, *Martius*, *N. G. et Sp. Pl.* 1. 154. (1824); *Endl. Gen.* c.; *Meisn. p.* 347.

DIAGNOSIS.—*Violal Exogens, with amentaceous scaly apetalous polygamous flowers, and unilateral stamens.*

Small trees or shrubs. Leaves simple, alternate, with stipules. Flowers disposed in clustered axillary catkins, ☿, or ♂ ♀ by abortion. Calyx in several narrow divisions, free, covered over by a dilated bract. Corolla wanting. Disk somewhat fleshy, surrounding the stamens, or in front of them, sometimes hardly visible. Stamen 1, hypogynous, standing on one side of the ovary, with a thick 2-lobed connective, at the apex of each of whose lobes is placed a single cell of an anther, bursting transversely. Ovary superior, seated in a fleshy disk, 1-celled, with several anatropal ovules attached to 2-3-parietal placentæ; stigmas 2 or 3, sessile or on a style. Fruit capsular, 1-celled, splitting into 2 or 3 valves, each of which bears a placenta in its middle. Seed usually, by abortion, one to each valve, suspended, with a fleshy aril; integument crustaceous; albumen fleshy; embryo inverted, with plane cotyledons and a superior straight cylindrical radicle.

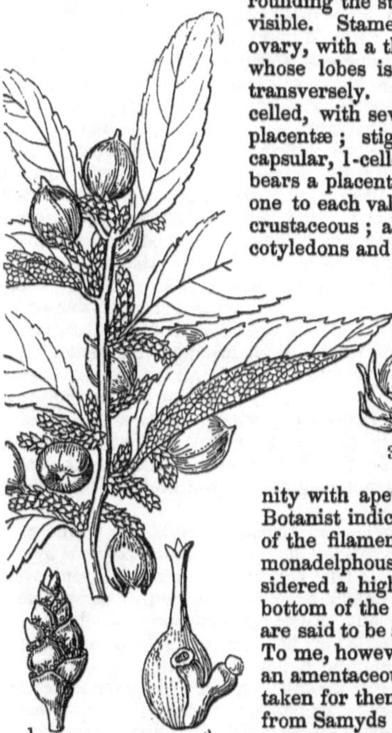

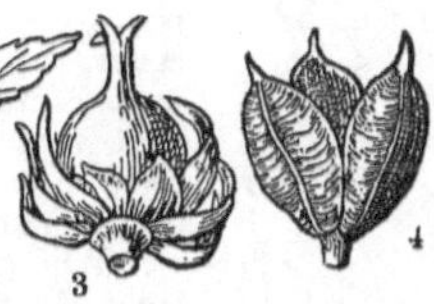

Fig. CCXXV.

Von Martius, the founder of this Order, which he divides from Nettleworts, speaks of it thus: "The peculiar character consists in the presence of a distinct perianth, while the amentaceous inflorescence is an indication of an affinity with apetalous Orders of a lower grade." The same Botanist indicates its relation to Chloranths in the structure of the filament, and to Samyds in that of the fruit, "the monadelphous stamens of both which may be perhaps considered a higher kind of evolution of the fleshy disk in the bottom of the flower of Lacistema." In habit the species are said to be something like Peppers, but more arborescent. To me, however, they look much more like Casearias with an amentaceous inflorescence, and they might easily be mistaken for them, when not in flower. They differ, however, from Samyds in their leaves not being dotted, in their scaly, not perfect tubular and half-coloured, calyx, and their curious unilateral stamens. No doubt they are a transition form from the more perfect to the diclinous Orders, as is sufficiently indicated by their polygamous flowers.

Natives of low places in woods in equinoctial America.

Their properties are unknown.

GENERA.

Synzyganthera, *Ruiz et Pav.*	Lacistema, *Swartz.*
Didymandra, Willd.	*Nematospermum*, L. C. Rich.

NUMBERS. GEN. 2. Sp. 6.

Piperaceæ.
POSITION. ———. LACISTEMACEÆ.—Samydaceæ.
Betulaceæ.

Fig. CCXXV.—Lacistema serrulatum.—*Martius.* 1. amentum in flower; 2. pistil and stamen; 3. pistil and calyx; 4. fruit in its state of dehiscence.

ORDER CXII. SAMYDACEÆ.—SAMYDS.

Samydeæ, *Vent. Mem. Inst.* 2. 142. (1807); *Gærtn. fil. Carp.* 3. 238. 242. (1805); *Kunth. Nov. Gen.* 5, 360. (1821); *DC. Prodr.* 2. 47. (1825); *Endl. Gen.* cxciv.; *Meisn. p.* 72.

DIAGNOSIS.—*Violal Exogens, with scattered apetalous tubular hermaphrodite flowers, perigynous stamens, and both round and linear transparent dots in the leaves.*

Trees or shrubs. Leaves alternate, often somewhat distichous, simple, entire or toothed, evergreen, with stipules, usually with pellucid markings, which are both linear and oblong. Peduncles axillary, solitary, or numerous. Sepals 4-5, more or less cohering at the base, usually coloured inside; æstivation somewhat imbricated, very seldom completely valvate. Petals 0. Stamens arising from the tube of the calyx, 2, 3, or 4 times as many as the sepals; filaments monadelphous, either all bearing anthers, or alternately shorter, villous or ciliated, and alternately bearing ovate 2-celled erect anthers. Ovary superior, 1-celled; style 1, filiform; stigma capitate, or slightly lobed; ovules 00, attached to parietal placentæ, ascending, half anatropal. Capsule coriaceous, with 1 cell and from 3 to 5 valves, many-seeded, the valves dehiscing imperfectly, often somewhat pulpy inside, and coloured. Seeds fixed to the valves, without order, on the papillose or pulpy part, with a fleshy aril and excavated hilum; albumen oily or fleshy; embryo large, in the middle of it; cotyledons ovate; radicle pointing to the extremity remote from the hilum.

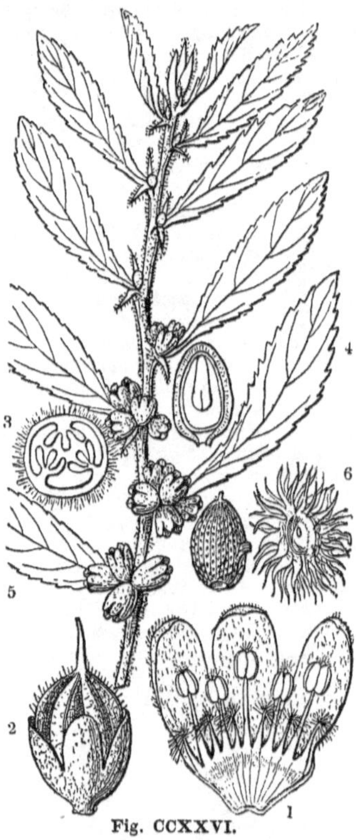

Fig. CCXXVI.

This Order, although petals are unknown in it, was placed in Polypetalous Exogens by De Candolle, who regarded a petaloid layer covering the inner surface of the sepals as analogous to a corolla. Although this cannot be admitted as true, yet it may be taken as evidence of a tendency to assume a corolline state. According to authors its apetalous flowers and parietal placentation approximate it to Bixads, its dotted leaves to Amyrids, near which De Candolle stations it, and its perigynous stamens to Roseworts, with which its alternate stipulate leaves also ally it. Its fruit, as in Casearia parviflora, is sometimes remarkably like that of Violetworts. In habit the Order approaches Smeathmannia among Passionworts. The difficulty of coming to any satisfactory conclusion in this matter, arises from the stamens having a manifestly perigynous insertion; and if this circumstance is to be regarded as of the usual importance, it is certain that Samyds have no title to a place among the Violal Alliance. If, however, we regard it as exceptional in the present instance, we then find the Order very naturally associated, by the force of all its other characters, with those among which it is now placed. Its composite fruit, with distinct parietal placentation, is much the same as that of many Bixads on the one hand, and of Lacistemads on the other; and its sterile stamens appear to offer a plain indication of a tendency to acquire the coronetted structure of Passionworts. Brown observes, that Samyds are especially distinguished by their leaves having a mixture of round and

Fig. CCXXVI.—Casearia grandiflora.—*A. St. Hilaire.* 1. part of a calyx split open; 2. the pistil half grown; 3. section across the ovary; 4. section of seed; 5. seed; 6. aril, removed from the seed.

linear pellucid dots, which distinguish them from all the other families with which they are likely to be confounded.

Samyds are all tropical and principally American. Little is known of the African or Asiatic species.

The bark and leaves are said to be slightly astringent. In Brazil the leaves of Casearia ulmifolia are applied to wounds, and their juice is drnnk by the sick; it is said to be a most certain remedy against the bite of the most noxious serpents, and is called Marmaleiro do Mato. A decoction of the leaves of Casearia lingua, called by the Brazilians Cha de Frade and Lingua de Fin, is also used internally in inflammatory disorders and malignant fevers. Casearia astringens bark is mucilaginous and somewhat acrid; it is used in Brazil as a poultice or lotion for badly healed ulcers, and is said by Martius to be wonderfully efficacious as a cleanser and stimulant of the raw flesh. Casearia Anavinga, an Indian species, is bitter in all its parts; the leaves are used in medicated baths; the pulp of the fruit is very diuretic. The root of Casearia esculenta is bitter and purgative; but its foliage is eatable.

GENERA.

Samyda, *Linn.*	*Anavinga*, Rheed.	*Pitumba*, Aubl.	*Crateria*, Pers.
Guidonia, Plum.	*Iroucana*, Aubl.	*Melistaurum*, Forst.	*Lindleya*, Kunth.
Mongezia, Fl. Flum.	*Langleia*, Scop.	*Piparea*, Aubl.	Euceræa, *Mart.*
Casearia, *Jacq.*	*Athenæa*, Schreb.	*Chætocrater*, Ruiz et Pav.	Candelabria, *Hochst.*
Antigona, Fl. Flum.	*Bedousia*, Dennst.		? Periclistia, *Benth.*
Hexanthera, Endl.			

NUMBERS. GEN. 5. Sp. 80.

Homaliaceæ.
POSITION.—Passifloraceæ.—SAMYDACEÆ.—Flacourtiaceæ.

ORDER CXIII. PASSIFLORACEÆ.—PASSIONWORTS.

Passifloreæ, *Juss. Ann. Mus.* 6. 102. (1805); *Id. Dict. des Sciences Nat.* 38. 48.; *DC. Prodr.* 3. 321; *Achille Richard Dict. Class* 13. 95. (1828); *Endl. Gen.* cxcvii.; *Meisner, Gen. p.* 124.

DIAGNOSIS.—*Violal Exogens, with polypetalous or apetalous coronetted flowers, perigynous imbricated petals, stamens on the stalk of the ovary, simple terminal styles, arillated seeds, and stipulate leaves.*

Herbaceous plants or shrubs, usually climbing, very seldom erect. Leaves alternate, with foliaceous stipules, often glandular. Flowers axillary or terminal, often with a 3-leaved involucre. Sepals 5, sometimes irregular, combined in a tube of variable length, the sides and throat of which are lined by filamentous or annular processes, apparently metamorphosed petals. Petals 5, arising from the throat of the calyx, on the outside of the filamentous processes, occasionally wanting, sometimes irregular, imbricated in æstivation. Stamens 5, monadelphous, rarely indefinite, surrounding the stalk of the

1

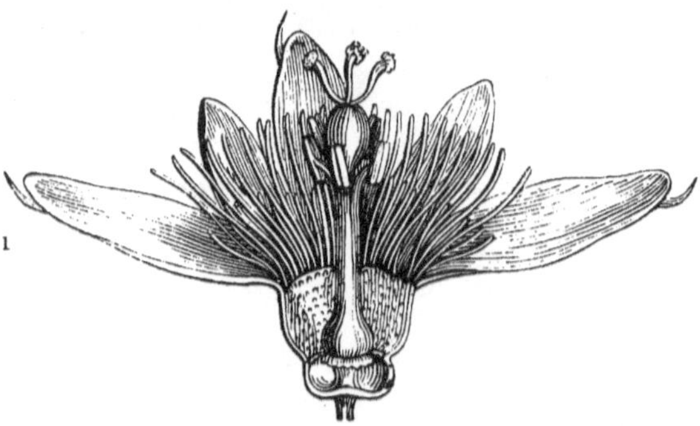

2

Fig. CCXXVII.

ovary; anthers turned outwards, linear, 2-celled, bursting longitudinally. Ovary seated on a long stalk, superior, 1-celled; styles 3, arising from the same point, clavate; stigmas dilated; ovules 00, anatropal, parietal, often inserted on long stalks. Fruit stalked, 1-celled, with 3 parietal polyspermous placentæ, sometimes 3-valved. Seeds attached in several rows to the placentæ, with a brittle sculptured testa surrounded by a pulpy aril; embryo straight, in the midst of fleshy thin albumen; radicle turned towards the hilum; cotyledons flat, leafy.

The real nature of the floral envelopes of this remarkable Order is a question upon which Botanists entertain different opinions, and their ideas of its affinities are consequently at variance. According to Jussieu (*Dict. des Sciences*, 38. 49.), the "parts

Fig. CCXXVII.—Passiflora; 1. section of a flower; 2. of its ripe fruit.

taken for petals are nothing but inner divisions of the calyx, usually in a coloured state, and wanting in several species;" and, therefore, in the judgment of that venerable Botanist, the Order is apetalous. De Candolle adopts the same view of the nature of the floral envelopes as Jussieu; but he nevertheless considers the Order polypetalous; a conclusion which I confess myself unable to understand, upon the supposition of the inner series of floral envelopes being calyx. Other Botanists, and I think with justice, consider the outer series of the floral envelopes as the calyx, and the inner as the corolla, for two principal reasons. In the first place, they have the ordinary position and appearance of calyx and corolla, the outer being green, and the inner coloured; and, in the second place, there is no essential difference between the calyx and corolla, except the one being the outer, and the other the inner of the floral envelopes. And if the real nature of these parts is to be determined by analogy, an opinion in which I do not, however, concur, the great affinity, as I think, of the Order with Violetworts would confirm the idea of its being polypetalous rather than apetalous. The nature of the filamentous appendages, or coronet, or rays as they are called, which proceed from the orifice of the tube, and of the membranous or fleshy, entire or lobed, flat or plaited, annular processes which lie between the petals and the stamens, is ambiguous. I am disposed to refer them to a peculiar form of petals, rather than to the stamens, for the reasons which I have assigned in the *Hort. Trans.* vol. 6, p. 309, for understanding the normal metamorphosis of the parts of fructification to be centripetal. There can, at least, be no doubt of their being of an intermediate nature between petals and stamens. With regard to the affinity of Passionworts, Jussieu, swayed by the opinion he entertained of their being apetalous, and De Candolle, who partly agreed and partly disagreed with Jussieu in his view of their structure, both assigned the Order a place near Cucurbits, and there can be doubt that Cucurbits are really little more than Passionworts with separate sexes and inferior fruit; but when we consider the stipitate fruit, occasionally valvular, the parietal placentæ, the sometimes irregular flowers, the stipulate leaves, and the climbing habit of these plants, it is difficult not to admit their greater affinity with Capparids or Violetworts, the dilated disk of the former of which is probably analogous to the innermost of the annular processes of Passiflora. That the fleshy covering of the seeds in this Order is a real aril, is clear from the seeds of a capsular species nearly related to P. capsularis, a drawing of which, by Ferdinand Bauer, exists in the Library of the Horticultural Society. In this plant the apex of the sculptured testa is uncovered by the aril. Smeathmannia forms a connecting link between Passionworts and Samyds.

Crownworts (Malesherbiaceæ) are perhaps not very distinct; their differences, such as they are, are noticed in the proper place. Passionflowers are the pride of South America and the West Indies, where the woods are filled with their species, which climb about from tree to tree, bearing at one time flowers of the most striking beauty, and of so singular an appearance, that the zealous Catholics who discovered them, adapted Christian traditions to those inhabitants of the South American wilderness; and at other times fruit, tempting to the eye and refreshing to the palate. One or two extend northwards into North America. Several are found in Africa and the neighbouring islands; and a few in the East Indies.

As far as we have any knowledge of the uses of these plants they appear, notwithstanding their eatable fruit, to possess active and rather dangerous qualities. Passiflora quadrangularis, whose fruit is the great Granadilla sometimes seen in our hot-houses, has an emetic root (*Martius*), and is powerfully narcotic, on which account it is said by Mr. Burnett, on the authority of a French writer, to be cultivated in several French settlements for the sake of its root. It is said to owe its activity to a peculiar principle called Passiflorine. P. Contrayerva is said to be alexipharmic and carminative. According to Browne, a tincture of the flowers of P. rubra, formed by infusion in wine or spirits, is used in the leeward parts of Jamaica, under the name of Dutchman's Laudanum, as a safe narcotic. P. fœtida, and some allied species, are esteemed as emmenagogues, and are thought to be serviceable in hysteria; the infusion of the flowers is also taken as a pectoral medicine in the West Indies. The foliage is used in Brazil in poultices, against erysipelas and inflammatory affections of the skin. The bitter and astringent leaves of P. laurifolia have some reputation as anthelmintics. P. pallida, maliformis, and incarnata are employed in cases of intermittent fevers. Murucuja ocellata, a West Indian climber, is said to be anthelmintic, diaphoretic, and antihysteric. Among the species whose fruit is eaten, the most important are Passiflora filamentosa, pallida, lutea, coccinea, maliformis, laurifolia, edulis, incarnata, and serrata, Tacsonia mollisima, tripartita and speciosa, and the Madagascar shrub called Paropsia edulis.

GENERA.

Ryania, *Vahl.*	Passiflora, *Juss.*	*Decaloba*, DC.	Disemma, *Labill.*
Patrisia, L. C. Rich.	*Granadilla*, Tournef.	*Granadilla*, DC.	Tacsonia, *Juss.*
Smeathmannia, *Sol.*	*Tetrapathea*, DC.	*Anthactinia*, Bory.	*Distephana*, Juss.
Bülowia, Schum.	*Cieca*, Medik.	Dysosmia, *DC.*	*Distephia*, Salisb.
Paropsia, *Noronh.*	*Astephananthes*, Bory.	Astrophea, *DC.*	? Vareca, *Gärtn.*
Thompsonia. *R. Br.*	*Monactineirma*, Bory.	Murucuja, *Tournef.*	
Deidamia, *Thouars.*	*Balduina*, Raf.		

NUMBERS. GEN. 12. Sp. 210.

Papayaceæ.
POSITION.—Samydaceæ.—PASSIFLORACEÆ.—Malesherbiaceæ.
Capparidaceæ.

ORDER CXIV. MALESHERBIACEÆ.—CROWNWORTS.

Malesherbiaceæ, *Don in Jameson's Journal*, 321. (1826); *Ed. pr.* lii.; *Endl. Gen.* cxcviii.; *Meisner Gen. p.* 193.—Passifloreæ, § Malesherbieæ, *DC. Prodr.* 3. 337. (1828.)

DIAGNOSIS.—*Violal Exogens, with polypetalous coronetted flowers, perigynous imbricated petals, stamens on the stalk of the ovary, simple dorsal styles, seeds without aril, and leaves without stipules.*

Herbaceous or half-shrubby plants. Leaves alternate, lobed, without stipules. Flowers axillary or terminal, solitary, yellow or blue. Calyx tubular, membranous, inflated, 5-lobed, the lobes with an imbricated æstivation. Petals 5, alternate with the segments of the calyx, persistent, with a convolute æstivation, arising from without a short membranous rim or coronet. Stamens 5 or 10, perigynous; filaments filiform, distinct, or connected with the stalk of the ovary; anthers versatile. Ovary superior, stipitate, 1-celled, with parietal placentæ; ovules 00, pendulous, anatropal; styles 3, filiform, very long, arising from distant points of the apex of the ovary; stigmas clavate. Fruit capsular, 1-celled, 3-valved, membranous, more or less many-seeded. Seeds attached to placentæ arising either from the axis of the valves, or from their base; testa brittle, with a fleshy crest, and no aril; embryo taper, in the midst of abundant fleshy albumen, with the radicle next the hilum.

According to Don, by whom these plants were first considered the rudiments of an Order, "they agree on the one hand with Passionworts, and on the other with Turnerads;" and I am persuaded that this is their true position. From the former they differ in the insertion of their styles at the back, not on the apex of the ovary, in their taper embryo, want of aril and of stipules, and altogether in their habit: from Turnerads, to which their habit quite allies them, they differ in the presence of a membranous coronet within the petals, in the remarkable insertion of the styles, and in the want of all trace of an aril. In their thin-sided fruit they approach Smeathmannia in Passionworts. Their tubular, somewhat furrowed calyx is not altogether different from that of Frankeniads.

All are natives of Chili and Peru.

Their uses are unknown.

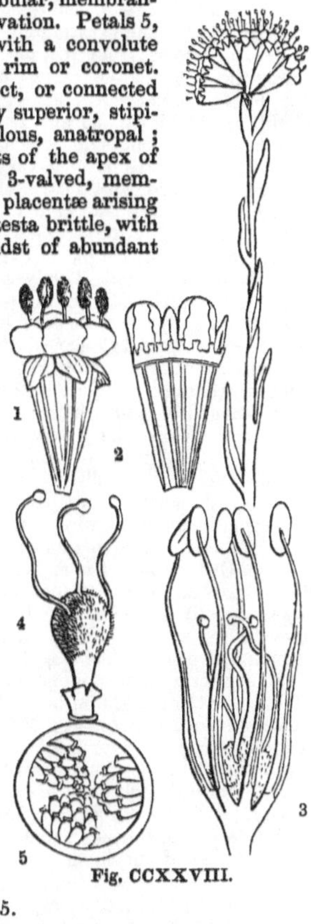

Fig. CCXXVIII.

GENERA.

Malesherbia, *Ruiz. et Pav.* | Gynopleura, *Cav.*

NUMBERS. GEN. 2. Sp. 5.

POSITION.—Turneraceæ.—MALESHERBIACEÆ.—Passifloraceæ.

Fig. CCXXVIII.—Malesherbia fasciculata; 1. a flower; 2. a part of the calyx seen from within, showing 2 petals and a portion of the coronet; 3. the stamens and pistil; 4. the pistil apart; 5. a section of the ovary.—*Endlicher.*

ORDER CXV. MORINGACEÆ.—MORINGADS.

Moringeæ, *R. Brown in Denham, p.* 33. (1826); *Bartl. Ord. Nat.* 425. (1830); *Decaisne in Ann. Sc. N. S.*, 4. 203. (1835); *Endl. Gen. p.* 1321.; *Meisn. Gen. p.* 78.; *Wight and Illustr.* 1. *f.* 75.

DIAGNOSIS.—*Violal Exogens, with a many-leaved calyx, perigynous petals and stamens, 1-celled anthers, stipitate consolidated siliquose fruit, and exalbuminous seeds.*

Trees, with 2- 3- pinnated leaves, whose leaflets very readily drop off, and thin, deciduous, coloured stipules. Flowers irregular, white, in loose panicles. Sepals 5, petaloid, nearly equal, deciduous; the tube lined with a fleshy disk; æstivation slightly imbricated. Petals 5, visibly unequal, the uppermost of which is ascending. Stamens 8 or 10, arising from the top of a disk lining the tube of the calyx; 5 opposite the sepals, sometimes sterile; filaments slightly petaloid, callous and hairy at the base; anthers simple, 1-celled, with a thick convex connective. Ovary stipitate, superior, 1-celled, with 3 parietal placentæ bearing numerous suspended anatropal ovules; style filiform, terminal, obliquely recurved; stigma simple. Fruit a long pod-like capsule, with 3 valves, and only 1 cell; the valves bearing the seeds along their middle. Seeds numerous, half buried in the fungous substance of the valves, sometimes winged; embryo amygdaloid, without albumen; radicle straight, superior (turned to the hilum), very small; cotyledons fleshy, plano-convex.

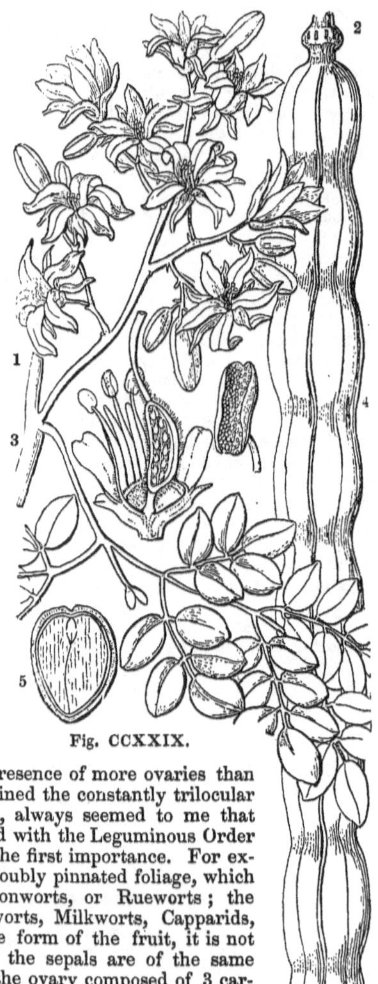

Fig. CCXXIX.

This is a little group of small trees, with an appearance so peculiar that one hardly knows with what to compare them. It however seems generally admitted that they resemble some plants of the Leguminous Order; and it is to the vicinity of those that all Botanists, except myself, seem agreed in referring them, moved thereto by their pinnated leaves, with glands between the leaflets, declinate decandrous perigynous stamens, and pod-like fruit. De Candolle, who did not overlook the anomalous structure of Moringa as a Leguminous plant, accounted for the compound nature of its fruit upon the supposition, that although unity of carpels is the normal structure of Leguminosæ, yet the presence of more ovaries than one, in a few instances in that Order, explained the constantly trilocular state of that of Moringa. It has, however, always seemed to me that the resemblance which Botanists have found with the Leguminous Order are trifling, while the discrepancies are of the first importance. For example, the habit of the plant consists in a doubly pinnated foliage, which would do as well for Roseworts, or Citronworts, or Rueworts; the declinate stamens may be found in Rueworts, Milkworts, Capparids, and many others; and as to the pod-like form of the fruit, it is not worth a thought. The objections are, that the sepals are of the same texture as the petals, the anthers 1-celled, the ovary composed of 3 carpels which have not the power of turning inward their sides so as to form dissepiments, and that the attachment of the carpels is strictly parietal. It is true that the latter circumstance will not be so much at variance

Fig. CCXXIX.—1. Moringa pterygosperma; 2. its fruit; 3, the section of a flower of M. aptera; 4. its anther; 5. a section of its seed.—*Wight and Decaisne.*

with the Leguminous structure as it appears to be if it should be proved that sutural and parietal placentation are of the same nature, which seems to be the fact; but, connected as it is with the other points of difference, and considering that it is parietal placentation in excess, it appears to be of considerable moment. This has always led me to regard the Moringads as a member of some great parietal Alliance, and as claimants of a nearer affinity with Violetworts than with any other Order; and to this opinion I adhere, for the following reasons; the stamens are definite in number, the corolla is manifestly irregular, the placentation is parietal, and the flowers are not isomeric, the parts of the fruit being 3, while those of the calyx, corolla, and stamens are 5. The main objection to this view is derived from the stamens being perigynous; and it will be seen from the altered arrangements introduced into the present volume, that I now attach much more importance to that circumstance than formerly. But it must be remembered that Moringa is not at all more perigynous than Verrucularia and others among Malpighiads, or than Reseda among the Crucifers, or than Escholtzia among Poppyworts; and that, in fact, it may be very well regarded as standing in the same relation to Violetworts as Escholtzia to Poppyworts. While, however, the parietal placentation seems to turn the scale in favour of the near affinity of Moringads to Violetworts, there can be little doubt that they also approach the anisomerous Sapindal Alliance, especially Milkworts, in their declinate stamens, 1-celled anthers, and petaloid calyx.

The species are natives of the East Indies and Arabia.

The root of the Moringa pterygosperma has a pungent odour, with a warm, biting, and somewhat aromatic taste, very like Horseradish; it is used as a stimulant in paralytic affections and intermittent fever; it is also employed as a rubefacient. Dr. Wight suggests that it would greatly increase the activity of sinapisms. He adds that a large quantity of gum, resembling Tragacanth, exudes from wounds in the bark. The seeds of this plant, called by the French Pois Quéniques and Chicot, have been used in venereal affections. They are the Ben-nuts of old writers, from which the oil of Ben was extracted, formerly more famed than at present. It is chiefly used by perfumers as the basis of various scents, and by watchmakers, because it does not readily freeze. The flowers, leaves, and tender seed-vessels, are eaten by the natives of India in their curries.

GENERA.

Moringa, *Burm.*	*Anoma*, Lour.	*Alandina*, Neck.
Hyperanthera, Forsk.	*Hypelate*, Smith.	*Balanus*, Endl.

NUMBERS. GEN. 1. Sp. 4.

Fabaceæ.

POSITION.——————MORINGACEÆ.—Violaceæ.

Polygalaceæ.

ORDER CXVI. VIOLACEÆ.—VIOLETWORTS.

Violarieæ, *DC. Fl. Fr.* 4. 801. (1805); *Juss. Ann. Mus.* 18. 476. (1811); *DC. Prodr.* 1. 287. (1824); *Bartl. Ord. Nat.* 283. (1830); *Endl. Gen.* cxc.; *Meisner Gen.* 20; *Wight Illustr.* 1. 142.—Violaceæ, *Lindl. Synops.* 35. (1829.)

DIAGNOSIS.—*Violal Exogens, with polypetalous flowers, a many-leaved calyx, hypogynous petals, stamens all perfect, anthers crested and turned inwards, consolidated fruit, and albuminous seeds.*

Herbaceous plants or shrubs. Leaves simple, usually alternate, sometimes opposite, stipulate, with an involute vernation. Inflorescence various. Sepals 5, persistent, with

Fig. CCXXX.

Fig. CCXXXI.

Fig. CCXXXII.

an imbricate æstivation, usually elongated at the base. Petals 5, hypogynous, equal or unequal, usually withering, and with an obliquely convolute æstivation. Stamens 5, alternate with the petals, occasionally opposite them, inserted on a hypogynous disk, often unequal; anthers 2-celled, bursting inwards, either separate or cohering, and lying close upon the ovary; filaments dilated, lengthened beyond the anthers; two, in the irregular flowers, generally furnished with an appendage or gland at their base. Ovary 1-celled, many-seeded, or rarely 1-seeded, with 3 parietal placentæ opposite the 3 outer sepals; style single, usually declinate, with an oblique hooded stigma; ovules anatropal. Capsule of 3 valves, bearing the placentæ in their axis. Seeds 00, or definite, roundish or winged, often with a tumour at their base; embryo straight, erect, in the axis of fleshy albumen.

The Violetworts are distinctly defined by their definite stamens, whose anthers turn inwards, and extend their connective into a crest; but the irregularity of their flowers,

---

Fig. CCXXX.—Corynostylis Hybanthus. 1. a set of stamens, each having the connective lengthened beyond the anther, in the form of a scale; 2. a spurred petal; 3. a transverse section of an ovary, showing the three parietal placentæ; 4. a ripe fruit; 5. an embryo.
Fig. CCXXXI.—Side view of the flower of Viola tricolor. Fig. CCXXXII.—Its fruit.

although a very common circumstance, is a mere peculiarity of certain genera. Rock Roses (Cistaceæ), by some associated with them, are very different in their indefinite stamens, curled embryo, and orthotropal ovules. So also the Sundews (Droseraceæ), another race to which they approach, are far separated by their minute embryo in the midst of profuse albumen, by their numerous styles, circinate leaves, and want of stipules. Passionworts, to which the baccate genera of Violetworts, and especially Corynostylis (Calyptrion, *DC.*), which has a twining stem, undoubtedly approach, are distinguished by a multitude of characters, among the more striking of which are their petals adhering to the tube of the calyx and the long stalk of the ovary.

Of the two Sub-orders recognised among these plants, Violeæ chiefly consist of European, Siberian, and American plants; a few only being found within the tropics of Asia. They are abundant in South America, where the forms are, however, materially different from those of the more temperate parts of the world, most of the species being shrubs while the northern Violets are uniformly herbaceous, or nearly so. Alsodineæ are exclusively South American and African, with the exception of Pentaloba, which belongs to the Malayan Flora.

The roots appear to be more or less emetic, a property which is strongly possessed by the South American species, and in a less degree only by those of Europe. Hence they form part of the herbs known under the name of Ipecacuanha. Ionidium parviflorum, and others, called Cuchunchully in Peru, are violent purgatives and emetics, and have a great reputation as a cure for the disease called Cocobay, in Jamaica, or Mal de S. Lazaro in Spanish America, the Elephantiasis tuberculata; they are used by the Spanish Americans, and I. Poaya by the Brazilians, as a substitute for Ipecacuanha. The root of another species called Poaya, Poaya de praia, and Poaya branca, the Ionidium Itubu of Kunth, is commonly sold as true Ipecacuanha, to which it approaches very nearly in its properties; at Pernambuco it is esteemed the very best remedy that can be employed in dysentery; and the inhabitants of Rio-Grande-do-Norte consider it a specific against gout. The foliage of the Conohoria Lobolobo is used in Brazil for the same purposes as Spinach with us. Boiled, it becomes mucilaginous. Viola canina is reputed a powerful agent for the removal of cutaneous affections; and Anchietea salutaris, a creeping bush, with the smell of Cabbage, and a nauseous taste, is accounted by the Brazilians not only a purgative, but also a remedy against similar maladies. A. de St. Hilaire remarks, that this notion deserves attention, as connected with the depurative properties ascribed in Europe to Viola canina, of which, although Anchietea is botanically related to it, there is nothing in the appearance which would have led the Portuguese settlers to attribute the virtues of the one to the other. The petals of Viola odorata are used as a laxative for children, one drachm operating pretty freely; the seeds possess similar properties; the root is emetic and purgative. The aqueous tincture of the flowers is a useful chemical test: uncombined acids changing the blue to red and alkalies to green. The Romans had a wine made of violet flowers, and it is said they are still used in the preparations of the Grand Signor's sherbet. By some the flowers are considered anodyne; they certainly induce faintness and giddiness in particular constitutions, as I have witnessed. Triller mentions a case in which they produced apoplexy. When bruised, the leaves of Viola tricolor smell like Peach kernels, hence they have been supposed to contain prussic acid. They were once esteemed efficacious in the cure of cutaneous disorders, and are still employed in Italy in tinea capitis. Viola ovata is said to be a remedy for the bite of the rattlesnake.

GENERA.

I.—VIOLEÆ.

Viola, *Linn.*
Erpetion, *DC.*
Mnemion, Spach.
Cittaronium, Rchb.
Hybanthus, Jacq.
Solea, Spreng.
Pigea, *DC.*
Jonidium, *Vent.*
Pombalia, Vand.
Noisettia, *Kunth.*
Bigelovia, DC.
Violæoides, Michx.
Anchietea, *St. Hil.*
Noisettia, Mart. et Zucc.
Schweiggeria, *Spreng.*
Glossarrhen, Mart. et Zucc.
Corynostylis, *Mart. et Zucc.*
Calyptrion, Ging.
Amphirrhox, Spreng.
Amphirrhoge, Rchb.
Spathularia, St. Hil.
Bradleia, Fl. Flum.

II.—ALSODEÆ.

Alsodeia, *Thouars.*
Alsodea, Mart. et Zucc.
Conohoria, Kunth.
Dripax, Noronh.
Physiphora, Soland.
Conohoria, Aubl.
Riana, Aubl.
Passoura, Aubl.
Rinorea, Aubl.
Ceranthera, Palis.
Passalia, Soland.
? *Prosthesia*, Blum.
Tetrathylacium, *Pöpp.*
Pentaloba, *Lour.*
? *Vareca*, Roxb.
Hymenanthera, *R. Br.*

NUMBERS. GEN. 11. Sp. 300.

Droseraceæ.
POSITION.—Passifloraceæ.—VIOLACEÆ.—Frankeniaceæ.
Cistaceæ.

ORDER CXVII. FRANKENIACEÆ.—FRANKENIADS.

Frankeniaceæ, *Aug. de St. Hilaire, Mém. Plac. Centr.* 39. (1815); *DC. Prodr.* 1. 349; *Endl. Gen.* cxcii.; *Meisner, Gen.* 22.

DIAGNOSIS.—*Violal Exogens, with polypetalous flowers, a tubular furrowed calyx, and hypogynous unguiculate petals.*

Herbaceous plants or under-shrubs. Stems very much branched. Leaves opposite, exstipulate, with a membranous sheathing base; often revolute at the edge. Flowers sessile in the divisions of the branches, and terminal, embosomed in leaves, usually pink. Sepals 4-5, united in a furrowed tube, persistent, equal. Petals alternate with the sepals, hypogynous, unguiculate, often with appendages at the base of the limb. Stamens hypogynous, either equal in number to the petals, and alternate with them, or having a tendency to double the number; anthers roundish, versatile, opening longitudinally. Ovary superior; style filiform, 2- 3- or 4-fid; ovules 00, anatropal, attached to parietal placentæ, and usually arising from long stalks. Capsule 1-celled, inclosed in the calyx, 2- 3- or 4-valved, many-seeded. Seeds very minute; embryo straight, erect, in the midst of albumen (divided into two plates, *Gærtn. fil.*) with a very short inferior radicle.

Fig. CCXXXIII.

Allied on the one hand to Cloveworts, from which they are distinguished by their different placentation, and by the form of their embryo; and on the other to Violetworts, which differ in having a loculicidal, not septicidal, dehiscence. Their great feature is the presence of a long furrowed calyx, within which the petals are inserted below the ovary, by means of long stalks. The petals, moreover, have generally a scaly appendage. Wormskioldia is a very anomalous plant. It seems more nearly allied to this than any other Order, and cannot possibly belong to Droseraceæ, in which it is placed by Achille Richard provisionally. It seems to indicate a relation between Frankeniads, on the one hand with Moringads, and on the other with Capparids. The nearest approach to the tubular calyx of Frankeniads is to be found in Crownworts (Malesherbiaceæ).

This Order is chiefly found in the north of Africa and south of Europe. Two species are natives of the Cape of Good Hope, one of South America, four of New Holland, and three of temperate Asia. None have been found in tropical India or North America.

Endlicher says that Frankeniads are mucilaginous and slightly aromatic. The leaves of Beatsonia portulacifolia are used in St. Helena as tea.

GENERA.

Frankenia, *Linn.*
Nothria, Berg.
Franca, Michel.

Beatsonia, *Roxb.*
Anisadenia, *Wall.*

Wormskioldia, *Thonn.*
Tricliceras, DC.

Schumacheria, Spr.
Streptopetalum, Hocht.

NUMBERS. GEN. 4. Sp. 24.

Caryophyllaceæ.
POSITION.—Violaceæ.—FRANKENIACEÆ.—Sauvagesiaceæ.

Fig. CCXXXIII.—Frankenia ericifolia.—*Webb.* 1. a flower; 2. its stamens, &c.; 3. a perpendicular section of the ovary; 4. a section of a seed.

ORDER CXVIII. TAMARICACEÆ.—TAMARISKS.

Tamariscineæ, *Desvaux, in a Dissertation read before the French Institute* (*in* 1815,) *according to the Ann. Sc. Nat.* 4. 344. (1825); *A. St. Hil. Mém. Mus.* 2. 205. (1816); *Ehrenb. in Annales des Sciences*, 12. 68. (1827); *DC. Prodr.* 3. 95. (1828); *Endl. Gen.* ccxxi.; *Meisner, p.* 129; *Wight, Illustr.* 1. *t.* 24.

DIAGNOSIS.—*Violal Exogens, with polypetalous flowers, a many-leaved calyx, hypogynous petals, distinct styles, consolidated fruit, and* 00 *basal comose seeds without albumen.*

Shrubs or herbs, with rod-like branches. Leaves alternate, resembling scales, entire, usually with pits on the surface. Flowers in close spikes or racemes. Calyx 4- or 5-parted, persistent, with an imbricated æstivation. Petals inserted into the base of the calyx, withering; with an imbricated æstivation. Stamens hypogynous, either equal to the petals in number, or twice as many, distinct or monadelphous. Anthers turned inwards, 2-celled, opening longitudinally. Ovary superior; styles 3; ovules numerous, ascending, anatropal. Capsule 3-valved, 1-celled, many-seeded; placentæ 3, either at the base of the cavity, or along the middle of the valves. Seeds erect or ascending, comose; albumen none; embryo straight, with an inferior radicle.

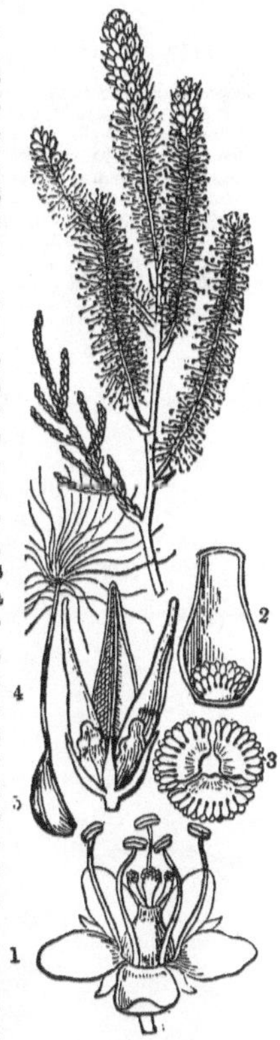

Fig. CCXXXIV.

Botanists are divided in opinion as to the proper place, in the Natural system, of the Tamarisk, that common but beautiful bush, and its allies. De Candolle stations it near Purslanes, from which its straight embryo and want of albumen remove it; others have suggested an affinity to Lythrads, or even to Onagrads; Meisner adopts the view of De Candolle, which I too have formerly followed. Endlicher is inclined to station Tamarisks next to Reaumuriads, with which they not only agree in habit, but in very many respects of structure. The main differences consist in Reaumuriads having a many-celled fruit, axile placentæ, mealy albumen round the embryo, and petals with unequal sides, while Tamarisks have a 1-celled fruit, with a basal and partially parietal placentation, no albumen, and their petals are equal-sided. Endlicher is also of opinion that a tendency towards Lythrads is observable among these plants. I think, however, that, notwithstanding the resemblances with Reaumuriads, the true place of the Order must be in this Violal Alliance, where it may perhaps be regarded as a near ally of Sauvageads and Houseleeks. The habit of some of the latter is not very different from that of Tamarisks. The most important distinctions are the total absence of albumen in Tamarisks, and the axile or sutural placentation of Houseleeks. The presence of albumen is of less consequence than usual in an Alliance whose embryo is so highly developed. The placentation is however of greater importance, and more than anything else throws doubt upon the affinity now suggested.

The species are exclusively confined to the northern hemisphere, and even to its eastern half, that is, to the Old World, on which they extend as far as the Cape de Verds. They usually grow by the sea-side, but occasionally by the edges of rivers and torrents. The maximum of species and of individuals also is found in the basin of the Mediterranean. The Order appears bounded on the south by the 8th or 9th parallel of N. lat., and on the north by that of 50° and 55° in Siberia, Germany, and England.

Their bark is slightly bitter, astringent, and probably tonic. Tamarix gallica and africana are remarkable for the quantity of sulphate of soda which their ashes contain. Ehrenberg found that the Manna of Mount Sinai is produced by Tamarix mannifera. This substance, being analysed by Mitscherlich, was ascertained to contain no crystallisable Mannite, but to consist wholly of pure mucilaginous sugar.* The galls of Tamarix

* Ehrenberg considers it as an exudation produced by a species of Coccus (manniparus) which inhabits the tree, and this is confirmed by Mr. Malcolmson, who in a note I received from him some time since,

Fig. CCXXXIV.—Tamarix. 1. a flower; 2. a view of the interior of the ovary; 3. placentæ seen from above; 4. a ripe seed-vessel split open; 5. a seed.

indica, dioica, Furas, and orientalis are highly astringent, and are used both in medicine and dyeing. Myricaria germanica, a common shrub in our gardens, has a balsamic astringent bitter bark, which was formerly officinal. Myricaria herbacea is used as tea among the Monghols, and its woody tissue is considered to be tonic.

GENERA.

Tamarix, *L.*	Myricaria, *Desv.*	Trichaurus, *Arn.*

NUMBERS. GEN. 3. Sp. 43.

Lythraceæ?
POSITION.—Crassulaceæ ?.—TAMARICACEÆ.—Frankeniaceæ.
Reaumuriaceæ.

---

observes that the Persian manna known by the name of Gen, is formed by an insect in that way, and is not found on the upper branches or leaves, but only on the larger branches covered by those minute insects, and none is formed near wounds or cracks in the bark. This was particularly observed by Colonel Frederick in Persia, in a latitude not much south of Mount Sinai, and his account corresponds with that of a traveller who saw it in the same country both on a Tamarisk and on the small Oak of Kermanshaw. It is remarkable that the secretion should be unknown in Egypt and Arabia, where the T. gallica would seem to be common. Forskahl, who says it is the Tarfa of the Arabs, takes no notice of any manna being produced by it, and Mr. Malcolmson informs me that he could gain no intelligence of manna being produced by the Tamarisk in any of the south and west coasts of Arabia and Upper Egypt. He observed the trees frequently secreting salt, but not sugar. I must however add, that the plant which this gentleman found the Arabs calling Tarfa, was T. orientalis, not T. gallica, as appeared from the specimens he brought home. The bark of T. gallica is slightly bitter and astringent.—*Flora Medica.*

ORDER CXIX. SAUVAGESIACEÆ.—SAUVAGEADS.

Violaceæ, § Sauvageæ, *DC. Prod.* 1. 315. (1824).—Sauvagesieæ, *Bartl. Ord. Nat.* 289. (1830); *Endl. Gen.* cxci.; *Meisn. Gen. p.* 21.—Sauvagesiaceæ, *Von Martius Conspectus, No.* 238. (1835).

DIAGNOSIS.—*Violal Exogens, with polypetalous flowers, a many-leaved calyx, hypogynous petals, stamens partly sterile and petaloid, anthers opposite the petals, naked, and turned outwards, consolidated fruit, and albuminous seeds.*

Smooth shrubs or annual herbs, with a terete, simple, or branched stem. Leaves alternate, simple, shining, feather-veined, nearly sessile, with fringed permanent stipules. Flowers perfect, regular, white, pink, violet or yellow, generally in terminal panicles or racemes, and on slender thread-shaped stalks. Sepals 5, equal or unequal, imbricated. Petals 5, twisted in æstivation, deciduous. Stamens hypogynous, definite and opposite the petals, or 00, all fertile, in more rows than one, of which the innermost alone is fertile, the exterior assuming the appearance of petaloid scales. Anthers turned outwards, 2-celled, opening lengthwise. Ovary free, 1-celled, with 3 parietal placentæ, sometimes 3-celled at the base and 1-celled at the apex; style terminal and stigma simple or nearly so; ovules parietal, anatropal. Capsule 3-valved, 1-celled or 3-celled at the base, with the seeds attached to the edges of the valves. Seeds small, oblong, pitted, with a straight embryo in the axis of fleshy albumen, and the radicle next the hilum.

Fig. CCXXXV.

Among the other differences between these plants and Violetworts may be mentioned their stamens, when definite in number, being opposite the petals, the anthers not having a membranous termination, the presence of 5 hypogynous scales representing sterile stamens, the fruit having a septicidal dehiscence, so that the seeds adhere to the edges and not the centre of the valves, and the strongly ribbed and imbricated calyx. The last character brings them near Tutsans, with which they accord in habit, but they differ in their stipules and decidedly parietal placentation. They are also said to approach Sundews; but this is by no means clear Endlicher points out their affinity with Frankeniads, from which, however, they are easily distinguished by their polysepalous calyx, stipules, and anthers turned outwards.

Almost nothing is known of their uses. Sauvagesia erecta, the herb of St. Martin, is very mucilaginous. It has been used in Brazil for complaints in the eyes, in Peru for disorders of the bowels, and in the West Indies as a diuretic, or rather in cases of a slight inflammation of the bladder.

GENERA.

Sauvagesia, *Linn.*
  Sauvagea, Neck.
  Iron, P. Br.

Lavradia, *Velloz.*
Luxemburgia, *St. Hil.*
  Plectanthera, Mart. et Zucc.

NUMBERS. GEN. 3. Sp. 15.

POSITION.—Violaceæ.—SAUVAGESIACEÆ.—Frankeniaceæ.
Hypericaceæ.

Fig. CCXXXV.—Lavradia Vellozii.—*A. St. Hilaire.* 1. an expanded flower; 2. the stamens with exterior petaloid scales; 3. a seed; 4. a section of it.

Order CXX. CRASSULACEÆ.—House-leeks.

Sempervivæ, *Juss. Gen.* 207. (1789).—Succulentæ, *Vent. Tabl.* 3. 271. (1799).—Crassulæ, *Juss. Dict. des Sc. Nat.* 11. 369. (1818).—Crassulaceæ, *DC. Bull. Philom. n.* 49. *p.* 1. (1801); *Fl. Fr. ed.* 3. *v.* 4. *p.* 271. (1805); *Mémoire* (1828); *Prodr.* 3. 381. (1828); *Endl. Gen.* clix.; *Meisn. p.* 134.—Sedeæ, *Spreng.*

Diagnosis.—*Violal Exogens, with polypetalous or monopetalous flowers, a many-leaved calyx, hypogynous petals, and follicular apocarpous fruit.*

Succulent herbs or shrubs. Leaves entire or pinnatifid; stipules none. Flowers usually in cymes, sessile, often arranged on one side only, along the divisions of the cymes. Sepals from 3 to 20, more or less united at the base. Petals inserted in the bottom of the calyx, either distinct or cohering in a monopetalous corolla. Stamens inserted with the petals, either equal to them in number and alternate with them, or twice as many, those opposite the petals being shortest, and arriving at perfection after the others; filaments distinct, subulate; anthers of 2 cells, bursting lengthwise. Hypogynous scales several, 1 at the base of each carpel, sometimes obsolete. Carpels of the same number as the petals, opposite to which they are placed around an imaginary axis, 1-celled, tapering into stigmas, sometimes consolidated; styles continuous with the ovaries; ovules sutural, 00, or definite in number, horizontal or pendulous, anatropal. Fruit consisting of several follicles, opening by the suture, or collected into a capsule of several cells opening at the back. Seeds attached to the margins of the suture, variable in number; embryo straight, in the axis of fleshy albumen, with the radicle pointing to the hilum.

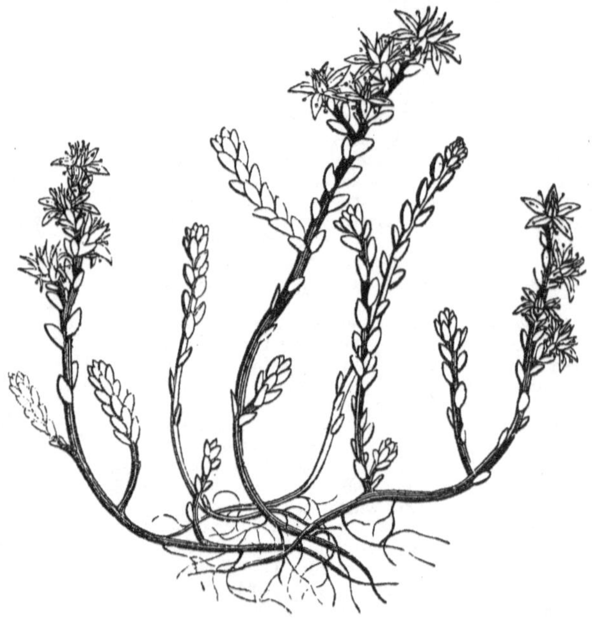

Fig. CCXXXVI.

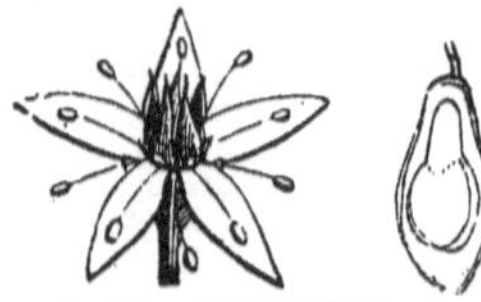

1 Fig. CCXXXVII. 2

All these plants are remarkable for the succulent nature of their stems and leaves, in which they resemble many other and very different Orders. De Candolle suggests that their real affinity is with Saxifrages through Penthorum, and with Knotworts (Illecebraceæ) through Tillæa. In both those Orders the hypogynous scales of Houseleeks are wanting.

Fig. CCXXXVI.—Sedum acre. Fig. CCXXXVII.—1. its flower; 2. its seed cut open.

Are not these bodies analogous to the scales out of which the stamens of Beancapers spring? If so, an unsuspected affinity exists between these Orders. To me it appears that if we were to resolve the fruit of a Sauvagesia, or any other of this Violal Alliance, into its component parts, the result would be what we find in Sedum and Crassula. Endlicher entertains a similar opinion, considering the Houssleeks certainly allied to Turnerads. De Candolle observes (*Mémoire*, p. 5.) that there is no instance of a double flower in the Order, although it might have been expected from their analogy in structure with Cloveworts. Sempervivum tectorum exhibits almost constantly the singular phenomenon of anthers bearing ovules instead of pollen. Adolphe Brongniart has remarked that in certain Houseleeks no medullary rays are to be found. He describes the woody cylinder of Sempervivum as consisting of little parcels of annular and spiral vessels immediately around the pith, on the outside of which are placed fusiform woody fibres with very fine 4-sided dots, arranged in radiating rows, and intermingled with some parcels of annular and reticulated vessels. These fibres are all in contact, are entirely destitute of medullary processes, and are only interrupted in order to leave a passage for the vascular bundles belonging to the leaves, and for the cellular tissue that accompanies them. M. Brongniart states, however, that this structure is not of constant occurrence in the Order of Houseleeks. On the contrary, he describes the Crassula portulacacea in the following words: " In this plant it may be said, notwithstanding the large size at which it arrives in a few years, that there is no woody zone at all; in it, that very hard tissue, which is found in regular concentric circles in other Houseleeks, and which consists of dotted woody fibre and vessels, is entirely wanting; the stem in fact contains nothing more than bundles of the medullary sheath, composed entirely of spiral vessels, false tracheæ, with annular and reticulated vessels; but these bundles increase and multiply, so that they may be from 40 to 50 in an old stem, while there is not more than 20 or 24 in a young branch. They then are 2 or 3 millimetres thick, in the direction of the rays, instead of half a millimetre. Finally, the cellular space which they surround, or the pith, itself augments from 4 or 5 millimetres to 3 or 4 centimetres. So that every part continues to grow, whether cellular or vascular; but the bundles of the medullary sheath, thus increased in number and size, still remain entirely composed of annular vessels or false spirals, without intermixture of woody fibre, and are separated by hard medullary processes. Thus we have in this Order an example of essential differences in the anatomical structure of the trunk."—*Obs. on Sigillaria, Arch. Mus.* 1. 437. Schleiden found in an old stem of an Echeveria an entire uniform mass of wood, formed of parenchyma without vessels, and scattered therein were vertical cords of very thin-sided parenchyma, in the midst of which ran spiral vessels, most of which might still be unrolled (*Wiegman*, 1839); and he suspects that it may belong to the whole of this Natural Order. I do not, however, find it in Echeveria lurida, whose succulent stem has a very large pith, and a ring of extremely imperfect wood, among which spiral vessels are distributed with great irregularity.

It appears, from De Candolle's researches, that of the 272 species of which he supposed the Order to consist, 133 are found at the Cape of Good Hope, 2 in South America beyond the tropics, 2 in the same country within the tropics, none in the West Indies or the Mauritian Islands, 8 in Mexico, 7 in the United States, 12 in Siberia, 18 in the Levant, 52 in Europe, 18 in the Canaries, 1 in Southern Africa beyond the limits of the Cape, 9 in Barbary, 3 in the East Indies, 4 in China and Japan, and 2 in New Holland. To these are to be added several species from the Himalayas. They are found in the driest situations, where not a blade of grass nor a particle of moss can grow, on naked rocks, old walls, sandy hot plains, alternately exposed to the heaviest dews of night and the fiercest rays of the noon-day sun. Soil is to them a something to keep them stationary, rather than a source of nutriment, which in these plants is conveyed by myriads of mouths, invisible to the naked eye, but covering all their surface, to the juicy beds of cellular tissue which lie beneath them.

Refrigerant and abstergent properties, mixed sometimes with a good deal of acridity, distinguish them. The fishermen of Madeira rub their nets with the fresh leaves of the Ensião or Sempervivum glutinosum, by which the nets are rendered as durable as if tanned, provided they are steeped in some alkaline liquor. Malic acid exists in Sempervivum tectorum combined with lime. Kalanchoe brasiliensis appears to form an exception to the general acrid and stimulating properties of the Order. The Brazilians use it as a refrigerant; and this is the common quality of the Order. Sedum ochroleucum, the αειζωον το μικρον of Dioscorides, and Sempervivum tectorum are notable instances; Sedum Telephium is another, and also astringent: its leaves boiled in milk are used by country people in diarrhœa. Its acridity on the other hand gave its name to Sedum acre, a rubefacient emetic and purgative. Bryophyllum calycinum is considered a vulnerary. The herbage of Crassula tetragona, boiled in milk, is used at the Cape of Good Hope against dysentery; that of Rhodiola rosea is an esculent among the Greenlanders.

GENERA.

I. CRASSULEÆ.

Tillæa, *Mich.*
Bulliarda, DC.
Helophytum, Eck. et Z.
Dasystemon, *DC.*
Telmissa, *Fenzl.*
Septas, *Linn.*
Crassula, *Haw.*
Gomara, Adans.
Sarcolipes, Eck. et Zh.
Petrogeton, Eck. et Zh.
Tetraphyle, Eck. et Zh.
Disporocarpa, C.A.M.
Pyrgosea, Sweet.
Turgosea, Haw.
Globulea, *Haw.*
Thisantha, *Eckl. et Zeyh.*
Grammanthes, *DC.*
Vauanthes, Haw.
Cyrtogyne, *Haw.*
Rochea, *DC.*
Danielia, DC.
Larochea, Haw.
Franciscaria, DC.
Franciscæa, DC.
Kalosanthes, Haw.
Dietrichia, Tratt.
Kalanchoë, *Adans.*
Verea, Willd.
Bryophyllum, *Salisb.*
Crassouvia, Comm.
Physocalycium, Vest.
Cotyledon, *DC.*
Pistorinia, *DC.*
Umbilicus, *DC.*
Orostachys, Fisch.
Cotyle, DC.
Cotylephyllum, Link.
Mucizonia, DC.
Rosularia, DC.
Echeveria, *DC.*
Pachyphytum, Kl.
Sedum, *Linn.*
Rhodiola, Linn.
Anacampseros, Tourn.
Procrassula, *Gris.*
Aithales, *Webb et Berth.*
Sempervivum, *Linn.*
Jovibarba, DC.
Monanthes, DC.
Chronobium, DC.
Aichryson, Webb et B.
Æonium, Webb et Bth.
Greenovia, Webb et B.
Petrophye, Webb et B.

II. DIAMORPHEÆ.

Diamorpha, *Nutt.*
Penthorum, *Linn.*

NUMBERS. GEN. 22. Sp. 450.

Caryophyllaceæ.
POSITION.—Sauvagesiaceæ.—CRASSULACEÆ.—Turneraceæ.
Saxifragaceæ.

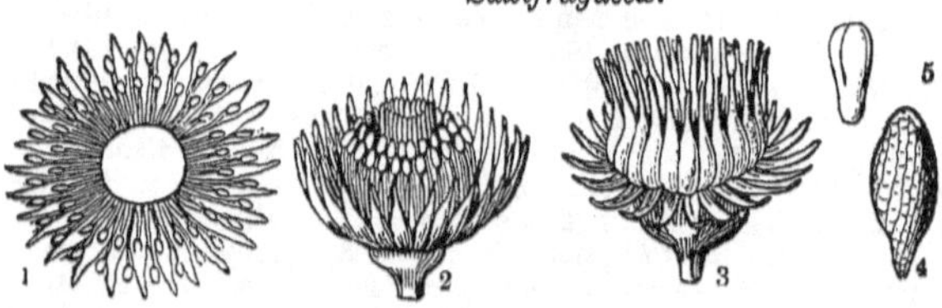

Fig. CCXXXVIII.

Fig. CCXXXVIII.—Greenovia (Sempervivum aureum.)—*Webb.* 1. petals and stamens; 2. flower seen from one side; 3. ripe fruit; 4. seed; 5. its embryo.

ORDER CXXI. TURNERACEÆ.—TURNERADS.

Loaseæ, § of Turneraceæ, *Kunth. N. G. et. Sp.* 6. 123. (1823).—Turneraceæ, *DC. Prodr.* 3. 345.; *Endl. Gen.* cxciii.; *Meisner, p.* 123.

DIAGNOSIS.—*Violal Exogens, with polypetalous flowers, perigynous contorted petals, forked styles and exstipulate leaves.*

Herbaceous plants, having sometimes a tendency to become shrubby, with a simple or occasionally stellate pubescence. Leaves alternate, without stipules, most commonly with 2 glands on the petiole. Flowers axillary, their pedicel either distinct or cohering with the petiole; with 2 bractlets. Petals yellowish, rarely blue. Calyx inferior, often coloured, with 5 equal lobes, imbricated in æstivation, lined by a very thin plate. Petals 5, inserted into the tube of the calyx, equal, with a twisted æstivation. Stamens 5, inserted into the tube of the calyx below the petals, with which they are alternate; filaments distinct; anthers oblong, erect, 2-celled. Ovary superior, 1-celled, with 3 parietal placentæ; ovules 00, anatropal; styles 3, cohering more or less, and simple, split, branched or multifid at the apex. Capsule 3-valved, 1-celled, opening from the point about as far as the middle, the valves bearing the placentæ in their median line. Seeds with a thin membranous strophiole on one side, crustaceous, reticulated; embryo slightly curved, in the middle of fleshy albumen; radicle turned towards the hilum; cotyledons somewhat plano-convex, acquiring the texture of leaves when germinating.

1 2 3 4

Fig. CCXXXIX.

The little herbaceous plants forming this Order were oddly placed by De Candolle between Loasads and his Fouquieraceæ, the former a group of plants with an inferior ovary, and the latter an imaginary Order, one of whose two genera seems to be a Cantua. Others station it in the vicinity of Rock Roses, from which it differs in the calyx, in the insertion of the stamens, and in the approximation of the radicle to the hilum, agreeing with them in habit. With Mallowworts the Order corresponds in the twisted æstivation of the corolla, and in habit. But with Passion-worts and Loasads there is most in common: the presence of glands upon the ends of the petioles of Turnerads is a confirmation of their affinity to the former. They are distinguished from Loasads by their fruit being superior, and by their definite stamens; the former character is, however, weakened by the nearly superior fruit of some Loasads. The hypogynous petals of Frankeniads sufficiently distinguish that Order, to say nothing of their unguiculate petals. The forked styles of Turnerads are very peculiar.

Natives exclusively of the West Indies and South America. There seems no good reason for supposing Turnera trioniflora to be a native of Japan, as has been asserted.

The herbage of some of them is rather aromatic. Turnera opifera is astringent, and is employed in Brazil against dyspepsia.—*Martius.* Turnera ulmifolia is considered a tonic and expectorant.

GENERA.

Turnera, *Plum.* *Pumilea*, P. Br. *Bohadschia*, Presl.	Piriqueta, *Aubl.* *Burghartia*, Neck. *Burkardia*, Scop.

NUMBERS. GEN. 2. Sp. 60.

POSITION.—Frankeniaceæ.—TURNERACEÆ.—Malesherbiaceæ.
Loasaceæ. (above) / *Cistaceæ.* (below)

Fig. CCXXXIX.—Turnera genistoides.—*St. Hil.* 1. a flower cut open; 2. a section of the ovary; 3. a seed; 4. a section of it.

ALLIANCE XXVII. *CISTALES.*—THE CISTAL ALLIANCE.

DIAGNOSIS.—*Hypogynous Exogens, with monodichlamydeous flowers, parietal or sutural placentæ, and a curved or spiral embryo, with little or no albumen.*

If we consider the Violal Alliance to be closed by Turnerads, that of Cistals will necessarily commence with Rock Roses, which have much the habit of the former, but which are distinctly separated by their convolute embryo and orthotropal seeds, to say nothing of divers other characters. If the Rock Roses are regarded as an Order with indefinite stamens, they will join Capparids, but if the genera with definite stamens are assumed to be the point of departure onwards, then it is into Crucifers that the line will pass. The parietal placentation of Rock Roses is universal; and though the number of placentæ is never reduced to two, yet if Fig. CCXLI., 2, in the following page has one of its placentæ removed and the other two brought into contact, we shall have the silicle of a Crucifer. There is no distinct passage from Crucifers into Weldworts, which may be regarded as an anomalous form of Capparids rather than in direct succession from Crucifers; but to Capparids themselves Crucifers pass by the whole division of Cleomeæ among the former, some of which are actually hexandrous. The stipitate fruit of Capparids brings us easily to Sterculiads in the next Alliance.

Supposing these views to be just, then the mutual relation of the Orders included in the Cistal Alliance may be thus expressed:

Turneraceæ.—Cistaceæ : Brassicaceæ : Capparidaceæ.—*Sterculiaceæ.*
Resedaceæ.

NATURAL ORDERS OF CISTALS.

Stamens not tetradynamous, generally indefinite. Flowers $\sqrt[3]{}$ *or* $\sqrt[5]{}$. *Seeds with albumen. Fruit closed up* . . .	122. CISTACEÆ.
Stamens tetradynamous. Flowers $\sqrt[4]{}$	123. BRASSICACEÆ.
Stamens not tetradynamous, definite. Flowers not tetramerous. Seeds without albumen. Fruit usually open at the point	124. RESEDACEÆ.
Stamens not tetradynamous. Flowers $\sqrt[4]{}$. *Seeds without albumen. Fruit closed up*	125. CAPPARIDACEÆ.

ORDER CXXII. CISTACEÆ.—ROCK-ROSES.

Cisti, *Juss. Gen.* 294. (1789).—Cistoideæ, *Vent. Tabl.* 3. 219. (1799).—Cistineæ, *DC. Prodr.* 1. 263. (1824).—Cistaceæ, *Ed. Pr.* lxix. (1836); *Endl. Gen.* clxxxviii.; *Meisner, p.* 8; *Spach in Ann. Sc. n. s.* 6. 365.

DIAGNOSIS.—*Cistal Exogens, with trimerous or pentamerous flowers, stamens usually* 00 *and never tetradynamous, closed up fruit and albuminous seeds.*

Shrubs or herbaceous plants. Branches often viscid. Leaves entire, opposite or alternate, stipulate or exstipulate, generally feather-veined, but sometimes fan-veined. Racemes usually unilateral. Flowers white, yellow, or red, very fugacious. Sepals 3-5, continuous with the pedicel, persistent, unequal, the three inner with a twisted æstivation. Petals 5, very rarely 3, hypogynous, fugitive, often crumpled in æstivation, and twisted in a direction contrary to that of the sepals. Stamens definite or indefinite, hypogynous, distinct; anthers 2-celled, opening longitudinally. Ovary free, 1- or many-celled; ovules orthotropal, (very rarely anatropal, *Spach*); style single; stigma simple. Fruit capsular, usually 3- or 5-valved, occasionally 10-valved, either 1-celled with parietal placentæ in the axis of the valves, or imperfectly 5- or 10-celled with dissepiments proceeding from the middle of the valves, and touching each other in the centre. Seeds definite or 00. Embryo inverted, either spiral or curved, in the midst of mealy, or somewhat horny albumen. Radicle remote from the hilum.

Fig. CCXL.

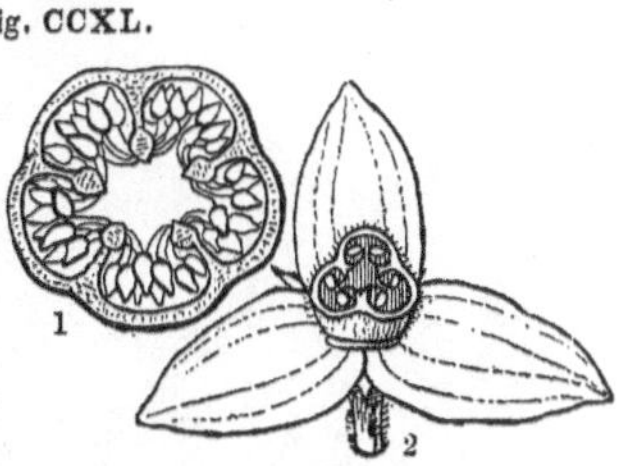

Fig. CCXLI.

These plants are perfectly distinguished from Violetworts, with which they were formerly confounded, by their annular and inverted embryo; from Bixads by this last character, by their mealy albumen, habit, and not having the leaves ever dotted; from Tutsans, by the latter character, and the structure of the fruit; they are also akin to Poppyworts by the genus Dendromecon. None of their affinities, or of others that may have been mentioned by other Botanists, appear, however, so strong as with Crucifers and Capparids, to which their curved embryo and parietal placentation bring the Rock Roses very near. From all the Cistal Alliance they are, moreover, known by the presence of albumen in some abundance.

A remarkable plant, found in Asia, Africa, and South America, and named Cochlospermum, seems to offer the most highly developed form of this Order, from which it differs in very little except its habit. Botanists usually place it in the Theads, but its parietal placentæ, anisomerous flowers, and curved embryo lying in the midst of albumen, seem fatal objections to that association. In fact it has no resemblance to the

Fig. CCXL.—Cistus Berthelotianus. 1. a vertical section of ovary and calyx; 2. a seed cut through; the pointed end being the true apex.

Fig. CCXLI.—1. a section of the ovary of Cistus Berthelotianus; 2. calyx and divided ovary of Helianthemum canariense.—*Webb*

Theads except in its indefinite stamens. An anonymous writer in the *Linnæa*, whose views are often judicious, would place the Rock Roses in the neighbourhood of Mesembryaceæ, Nyctaginaceæ, and Polygonaceæ, and next Portulacaceæ: an opinion evidently formed upon the supposed importance of a curved embryo and mealy albumen.

South Europe and the north of Africa are the countries that Rock Roses chiefly inhabit. They are rare in North America, extremely uncommon in South America, and scarcely known in Asia.

The species have no marked properties, except that the resinous balsamic substance, called Ladanum, is obtained from Cistus creticus, and others; it has been much esteemed as a stimulant and emmenagogue; it has also been recommended in chronic catarrh. Helianthemum vulgare had once some reputation as a vulnerary, but it is now forgotten. The trunk of Cochlospermum Gossypium yields the gum Kuteera, which in the north-western provinces of India is substituted for Tragacanth.—*Royle.* A decoction of the roots of Cochlospermum insigne, called in Brazil Butua do curvo, is employed in internal pains, especially such as are produced by falls or accidents; it is also asserted to heal abscesses already commenced. C. tinctorium is used in cases of amenorrhœa, and also as a yellow dye.

GENERA.

Fumana, *Spach.*
Cistus, *Tournef.*
Halimium, Dunal.
Ladanium, Spach.
Rhodocistus, Spach.
Erythrocistus, Dunal.
Ledonia, Spach.
Stephanocarpus, Spch.
Helianthemum, *Tournef.*
Brachypetalum, Dun.
Aphananthemum, Sph.
Eriocarpum, Dun.
Pseudocistus, Dun.
Rhodax, Spach.
Argyrolepis, Spach.
Tuberaria, Dun.
Lecheoides, Dun.
Crocanthemum, Spach.
Heteromeris, Spach.
Trichasterophyllum, Willd.
Lechea, *Linn.*
Lechidium, Spach.
Hudsonia, *Linn.*
Tæniostoma, *Spach.*
Cochlospermum, *Kunth.*
Wittelsbachia, Mart.
Maximiliania, Schk.

NUMBERS. GEN. 7. Sp. 185.

Sterculiaceæ.
POSITION.—Brassicaceæ.—CISTACEÆ.—Capparidaceæ.
Hypericaceæ.

ORDER CXXIII. BRASSICACEÆ.—CRUCIFERS.

Cruciferæ, *Juss. Gen.* 237. (1789); *DC. Mémoire sur les Crucifères; Syst.* 2. 139; *Prodr.* 1. 131; *Bartl. Ord. Nat.* 261; *Endl. Gen.* clxxxi.; *Meisner, Gen. p.* 9.

DIAGNOSIS.—*Cistal Exogens, with tetramerous flowers and tetradynamous stamens.*

Herbaceous plants, annual, biennial, or perennial, very seldom suffruticose. Leaves alternate. Flowers usually yellow or white, seldom purple, without bracts, generally in racemes. Sepals 4, deciduous, imbricate or valvate. Petals 4, cruciate, alternate with the sepals. Stamens 6, of which two are shorter, solitary, and opposite the lateral sepals; occasionally toothed; and four longer, in pairs, opposite the anterior and posterior sepals, generally distinct, sometimes connate, or furnished with a tooth on the inside. Disk with various green glands between the petals and the stamens and ovary. Ovary superior, unilocular, with parietal placentæ usually meeting in the middle, and forming a spurious dissepiment. Stigmas 2, opposite the placentæ. Fruit a silique or silicule, 1-celled, or spuriously 2-celled; 1- or many-seeded; dehiscing by two valves separating from the replum; or indehiscent. Seeds attached in a single row by a funiculus to each side of the placentæ, generally pendulous. Albumen none. Embryo with the radicle folded upon the cotyledons, which are occasionally slit or lobed.

Fig. CCXLII.

This Order is among the most natural that are known, and its character of having what Linnæan Botanists call tetradynamous stamens is scarcely subject to exception. It has a near relation to Capparids, with which it agrees in the number of the stamens of some species of that Order, in the fruit having two placentæ and a similar mode of dehiscence, and in the quaternary number of the divisions of the flower. To Poppyworts it is thought to approach in the unusual number of the petals and in the structure of the fruit of some genera of that Order, such as Glaucium and Chelidonium; with the siliquose-fruited Fumeworts it has also some analogy, and even with the whole of that Order in the number of its petals, supposing the common opinion of the nature of the floral envelopes of Fumeworts to be correct, or in the binary division of its flower, from which the quaternary is only a slight deviation, upon the hypothesis I have suggested in speaking of that Order. But the totally different structure of the seed forbids Crucifers to be associated in the same group with the latter.

Crucifers may be said to be characterised by their deviation from the ordinary symmetry observable in the relative arrangement of the parts of fructification of other plants,—deviations which are of a very interesting nature. Their stamens are arranged thus: two stand opposite each of the anterior and posterior sepals, and one opposite each of the lateral sepals; there being 6 stamens to 4 sepals, instead of either 4 or 8, as would be normal. Now in what way does this arise? Is the whorl of stamens to be considered double, one of the series belonging to the sepals, and one to the petals, and, of these, a part imperfect? I am not aware of any such explanation having been offered, nor do I know of a better one. It appears to me that the outer series is incomplete, by the constant abortion of the stamens usually belonging to the anterior and

Fig. CCXLII.—Erucastrum Canariense. 1. a flower; 2. the stamens; 3. the siliqua, with the valve separating from the replum; 4. a transverse section of a seed; 5. a perfect seed.

posterior sepals, the two pairs that remain belonging in fact to the four petals. But it is in their fruit that the great peculiarity consists.

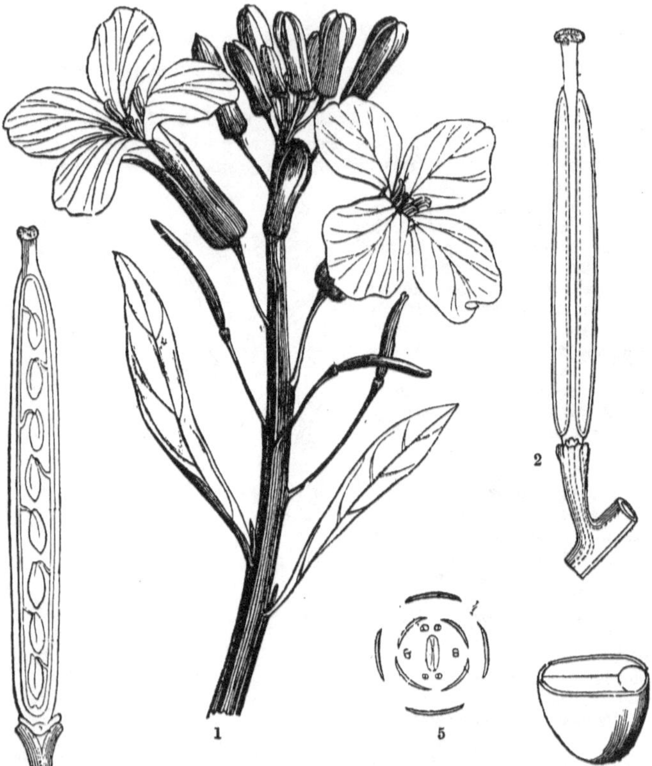

Fig. CCXLIII.

Fig. CCXLIV.

Since the placentæ are opposite the lobes of the stigma in this Order, it is difficult to reconcile the fruit with any general theory of structure. Either it is in reality composed of four carpels, two of which are abortive, as was first suggested by me in the *Botanical Register*, fol. 1168, or each of the two lobes of the stigma is composed of two half lobes belonging to different carpels, as in Poppyworts. In any view, the dissepiment which cuts off the interior of the fruit into two cells must be considered spurious, and a mere expansion of the placentæ.

The opinions of Botanists are much divided as to this matter; M. Kunth agrees with me in considering the fruit composed of four carpels. And a variety of evidence has gradually collected in favour of this theory. M. Alph. De Candolle has shown that the common Wall-flower is occasionally 4-celled (*Monstruosités Végétaux*, 15. t. 5.) There is a genus called Tetracellion, which derives its name from the same circumstance. Mr. Barker Webb has published an account of a Canary shrub, named Parolinia, in which the valves are constantly extended into stigmas. But Mr. Howell (*Ann. N. Hist.* x. 254.) adopts Brown's view of the subject, and, because of the supposed affinity of Poppyworts, concludes that the fruit of Crucifers is only composed of two carpels. He does not, however, offer any direct proof of the correctness of this opinion.

Almost all Crucifers are destitute of bracts, and have the

Fig. CCXLIII.—1. Cheiranthus cheiri; 2. its stigma; 3. the same with one valve off; 4. a cross section of a seed; 5. a diagram to illustrate the position of the parts of the flower.
Fig. CCXLIV.—Fruits of Parolinia ornata, after *Webb*.

calyx imbricated in æstivation; but Brown has noticed (*Denham*, p. 7.) that in Savignya and Ricotia it is valvate.

Linnæus divided the Order, which is the same as his Tetradynamia, by the form of the fruit, under two heads, bearing the names of Siliquosa and Siliculosa. More recently divisions have been founded upon the nature of the plicature of the cotyledons and the position of the radicle with respect to them. It is difficult to say what degree of importance really deserves to be attached to these characters, which are however at present in general use.—See *Torrey in Ann. Lyc. N. York*, iv. 90.

This is an Order eminently European; 166 species are found in northern and middle Europe, and 178 on the northern shore or islands of the Mediterranean; 45 are peculiar to the coast of Africa, between Mogador and Alexandria; 184 to Syria, Asia Minor, Tauria, and Persia; 99 to Siberia; 35 to China, Japan, or India; 16 to New Holland and the South Sea Islands; 6 to the Isle of France and the neighbouring Islands; 70 to the Cape of Good Hope; 9 to the Canaries or Madeira; 2 to St. Helena; 2 to the West Indies; 41 to South America; 48 to North America; 5 to the islands between North America and Kamtchatka; and 35 are common to various parts of the world. This being their general geographical distribution, it appears that, exclusive of species that are uncertain, or common to several different countries, about 100 are found in the southern hemisphere, and about 800 in the northern, or 91 in the New, and the rest in the Old World. Finally, if we consider them with regard to temperature, we shall find that there are,—

In the frigid zone of the northern hemisphere			205
In all the tropics (and chiefly in mountainous regions)			30
In the temperate zone	of the northern hemisphere	548	634
	of the southern ditto	86	

Such were the calculations of De Candolle in 1821. Although requiring considerable modification, especially in the Asiatic and North American numbers, which are much too low, they serve to give a general idea of the manner in which this Order is dispersed over the globe.

The universal character of Crucifers is to possess antiscorbutic and stimulant qualities, combined with an acrid flavour. The officinal species are among the commonest of all plants, and only require to be named. They are found to contain a great deal of nitrogen, to which it is supposed is due their animal odour when rotting. Mustard, Cress, Horseradish, and many others, are extremely stimulating and acrid. The seeds of Sinapis chinensis are considered by Hindoo and Mahometan practitioners as stimulant, stomachic, and laxative. The seeds of one species of Arabis (chinensis, *Rottler*) are prescribed by the Indian doctors as stomachic and gently stimulant; but they apprehend its bringing on abortion if imprudently given. When the acrid flavour is dispersed among an abundance of mucilage, various parts of these plants become a wholesome food; such as the root of the Radish and the Turnip, the herbage of the Water-cress, the Cabbage, and the Sea-kale. According to Muller the Water-cress contains iodine. Sulphur exists in the oils of Mustard and Horseradish to the extent of about 30 per cent.—*Ch Gaz.* 1843, p. 674. The oil of the seeds is one of their more important products. That from Rape is in very general use, and the residue, rich in nitrogen, is largely employed by the farmer as manure, or cattle feed, under the name of Oil-cake. Another of the oil plants is Camelina

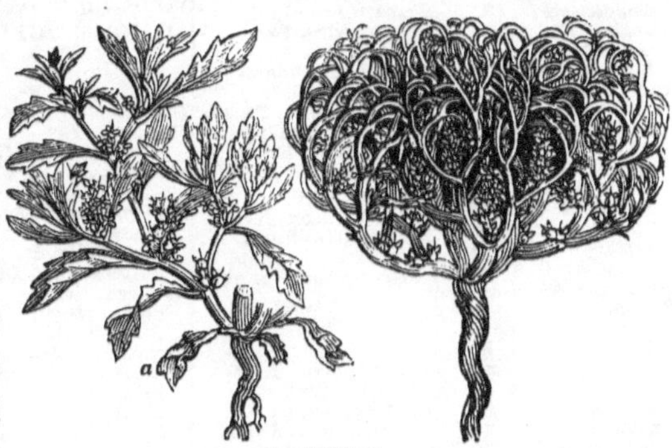

Fig. CCXLV.

sativa, or Gold of Pleasure; but its cake is said to be too acrid for cattle, (*Gardeners' Chronicle*, 1843, p. 678;) brooms are made from the dry haulm. Cochlearia officinalis, or Scurvy-grass was once in great repute as an antiscorbutic. It is stimulant and diuretic if eaten fresh, but becomes inert when dried. Cardamine pratensis is said to

Fig. CCXLV.—Anastatica hierochuntina.

be stimulant, diaphoretic and diuretic. The dried flowers have been a popular remedy for epilepsy in children. The great fleshy root of Crambe tatarica, sometimes called Tartar bread, is eaten in Hungary, cooled and sliced with oil, vinegar, and salt, or even when boiled. Isatis tinctoria, or Woad, was formerly a favourite blue dye in this country. Numerous species are celebrated for their beauty, of which the Wall-flower, Stock, Honesty, and Rocket, are every-day examples. Finally, one of the Order possesses strongly-marked hygrometrical qualities. This plant, the Anastatica hierochuntina or Rose of Jericho, is an annual, found wild in the Egyptian deserts, and when full grown contracting its rigid branches into a ball, which is soon caught up by the wind and hurried from place to place. But as soon as it is exposed to water the branches relax and spread flat, as if its life was renewed. Some superstitious tales are told of it, among which it is said to have first bloomed on Christmas eve, to salute the birth of the Redeemer, and paid homage to His resurrection by remaining expanded till Easter.—See *Gardeners' Chronicle*, 1842, p. 363.

GENERA.

I. PLEURORHIZEÆ.
(0 =).

ARABIDÆ.

Matthiola, *R. Br.*
Leucoium, Mönch.
Pachynotum, DC.
Luperia, DC.
Pinaria, DC.
Acinotum, DC.
Triceras, Andrzeiows.
Notoceras, *R. Br.*
Diceratium, Ait.
Parolinia, *Webb.*
Andrzeiowskya, *Reichnb.*
Macroceratium, DC.
Cheiranthus, *R. Br.*
Schelhammeria, Herit.
Cheiri, Adans.
Psilostylis, Andrz.
Dichroanthus, Webb et Ber.
Jodanthus, *Tor. et A. Gr.*
Clausia, *Trotzk.*
Oudneya, *R. Br.*
Nasturtium, *R. Br.*
Cardaminum, Mönch.
Sisymbrium, Magnol.
Bæumerta, Fl. Wetter.
Brachylobos, Allion.
Radicula, Dill.
Roripa, Scop.
Caroli-Gmelina, Fl. Wetter.
Clandestinaria, DC.
Barbarea, *R. Br.*
Streptanthus, *Nutt.*
Euclisia, Nutt.
Turritis, *Dill.*
Pachyneurum, *Bung.*
Arabis, *Linn.*
Abazicarpus, Andrz.
Campylocarpus, C. A. M.
Turritella, C. A. M.
Cardaminopsis, C. A. M.
Leptostylis, C. A. Mey.
Catalobus, C. A. Mey.
Stevenia, *Fisch. et Adam.*
Parrya, *R. Br.*
Neuroloma, Andrz.
Leisopora, C. A. Mey.
? *Ermannia*, Cham.
Phœnicaulis, *Nutt.*
Macropodium, *R. Br.*
Cardamine, *Linn.*
Pteroneuron, *DC.*
Dentaria, *Tournef.*
Leavenworthia, *Torr.*

ALYSSIDÆ.

Lunaria, *Linn.*
Brachypus, *Led.*
Ricotia, *Linn.*
Scopolia, Adans.
Farsetia, *Torr.*
Alyssum, Adans.
Cyclocarpæa, DC.
Fibigia, Medik.
Meniocus, *Desv.*
Berteroa, *DC.*
Alysson, Medik.
Mönchia, Roth.
Stevena, Andrz.
Aubrietia, *Adans.*
Vesicaria, *Lam.*
Alyssoides, Medik.
? *Physaria*, Nutt.
Coluteocarpus, *Boiss.*
Glyce, *Lindl.*
Koniga, Adans.
Clypeola, Neck.
Octadenia, R. Br.
Lobularia, DC.
Schiwereckia, *Andrz.*
Aurinia, *Desv.*
Psilonema, *C. A. Mey.*
Alyssum, *Linn.*
Adyseton, Scop.
Odontarrhena, *C. A. Mey.*
Ptilotrichum, *C. A. Mey.*
Clypeola, *Linn.*
Fosselinia, Scop.
Jonthlaspi, Tournef.
Orium, Desv.
Bergeretia, Desv.
Peltaria, *Linn.*
Bohatschia, Crantz.
Petrocallis, *R. Br.*
Zizia, Roth.
Draba, *Linn.*
Odontocyclus, *Turcz.*
Erophila, *DC.*
Gansblum, Adans.
Cochlearia, *Linn.*
? *Rhizobotrya*, Tausch.
Kernera, Medik.
Armoracia, Rupp.
Raphanis, Mönch.
Grællsia, *Boiss.*

TETRAPOMIDÆ, *Turcz.*

Holargidium, *Turcz.*
Tetrapoma, *Turczan.*
Tetracellion, Turczan.

SELENIDÆ.

Selenia, *Nutt.*

THLASPIDÆ.

Didymophysa, *Boiss.*
Thlaspi, *Dillen.*
Pachyphragma, DC.
Pterolobium, Andrz.
Carpoceras, Link.
Nomisma, DC.
Neurotropis, DC.
Pterotropis, DC.
Lyrocarpa, *Harv.*
Brossardia, *Boiss.*
Teesdalia, *R. Br.*
Guepinia, Bart.
Iberis, *Linn.*
Arabis, Adans.
Pseudo-Thlaspi, Magn.
Thlaspidium, Andrz.
Cynocardamum, *Webb et Berth.*
Heldreichia, *Boiss.*
Zygopeltis, Fenzl.
Biscutella, *Linn.*
Jondraba, Medik.
Thlaspidium, Medik.
Dithyrea, *Harv.*
Diastrophis, *Fisch. et Mey.*
Megacarpæa, DC.
Crenularia, *Boiss.*
Moriera, *Boiss.*

CREMOLOBIDÆ.

Cremolobus, *DC.*
Menonvillea, *DC.*

ANASTATICIDÆ.

Morettia, *DC.*
Nectouxia, DC.
Anastatica, *Gärtn.*
Hierocontis, Adans.

EUCLIDIDÆ.

Euclidium, *R. Br.*
Soria, Adans.
Ochthodium, *DC.*
Bunias, Desv.

CAKILIDÆ.

Cakile, *Tournef.*
Chorispora, *DC.*
Chorispermum, R. Br.
Cordylocarpus, *Desf.*

II. NOTORHIZEÆ.
(0‖).

SISYMBRIDÆ.

Malcolmia, *R. Br.*
Citharelma, *Bung.*
Hesperis, *Linn.*
Hesperidium, DC.
Deilosma, Andrz.
Arabidium, C. A. Mey.
Plagioloba, C. A. Mey.
Dontostemon, *Andrz.*
Andreoskia, DC.
Hesperidopsis, DC.
Tonguea, *Endl.*
Pachypodium, Webb. et Berth.
Sisymbrium, *Linn.*
Erysimum, Tournef.
Velarum, DC.
Kluckia, Andrz.
Chamæplium, Wallr.
Norta, Adans.
? *Psilostylum*, DC.
Leptocarpæa, DC.
Descurainia, Webb. et Berth.
Descurea, Guett.
Sophia, Hall.
Hugueninia, Reichenb.
Kibera, Adans.
Alliaria, Adans.
Arabidopsis, DC.
? *Halimolobus*, Tausch.
Drabopsis, *Koch.*
Tropidocarpum, *Hook.*
Erysimum, *Linn.*
Agonolobus, C. A. Mey.
Cuspidaria, Link.
Cheiropsis, C. A. Mey.
Cheirinia, Link.
Erysimastrum, C. A. M.
Conringia, Heist.
Gorinkia, Presl.
Crantzia, Lagasc.
Tetracme, *Bung.*
Tetraceratum, DC.
Smelowskia, *C. A. Mey.*
Taphrospermum, *C. A. M.*
Braya, *Sternb. et Hopp.*
Syrenopsis, *Jaub.*
Leptaleum, *DC.*
Christolea, *Camb.*
Thelypodium, *Endl.*
Pachypodium, Nutt.
Stanleya, *Nutt.*
Podolobus, Rafin.
Warea, *Nutt.*

CAMELINIDÆ.

Syrenia, *Andrz.*
Stylonema, DC.
Menkea, *Lehm.*
Camelina, *Crantz.*
Myagrum, DC.
Leiolobium, DC.
Stenopetalum, *R. Br.*
Eudema, *H. B. K.*
Mathewsia, *Hook.*
Platypetalum, *R. Br.*
Eutrema, *R. Br.*
Aphragmus, *Andrz.*
Orobium, Reichenb.
Oreas, Cham.
Platyspermum, *Hook.*

LEPIDIDÆ.

Capsella, *Vent.*
Marsypocarpus, Neck.
Rodschiedia, Gärtn.
Bursa, Guett.
Hymenolobus, *Nutt.*
Ionopsidium, *Reichenb.*
Bivonæa, *DC.*
Eunomia, *DC.*
Hutchinsia, *R. Br.*
Noccæa, Reichenb.
Nasturtiolum, Gray.
Iberidella, *Boiss.*
Lepidium, *R. Br.*
Kandis, Adans.
Cardaria, Desv.
Cardiolepis, Wallr.
Jundzillia, Andrz.
Ellipsaria, DC.
Bradypiptum, DC.
Cardamon, DC.
Nasturtium, Bœrh

Lepia, Desv.
Lasioptera, Andrz.
Dileptium, Raf.
Nasturtioides, Medik.
Senckenbergia, Fl.Wet.
Lepidiastrum, DC.
Physolepidium, *Schrk.*
Hymenophysa, *C.A. Mey.*
Æthionema, *R. Br.*
Campyloptera, *Boiss.*
Hexaptera, *Hook.*
?*Dispeltophorus*, Lehm.

ISATIDÆ.

Tetrapterygium, *Fisch. et Mey.*
Isatis, *Linn.*
Glastum, DC.
Sameraria, Desv.
Pachypterygium, *Bung.*
Pachypteris, Kar.
Tauscheria, *Fisch.*
Chastoloma, *Bunge.*
Texieria, *Jaub.*
Glastaria, Boiss.
Boreava, *Jaub.*
Neslia, *Desv.*
Vogelia, Med.
Rapistrum, Hall.
Myagrum, *Tournef.*
Deltocarpus, Herit.
Sinistrophorum, Schrk.
Traillia, *Lindl.*
Lachnoloma, *Bunge.*

ANCHONIDÆ.

Goldbachia, *DC.*
Anchonium, *DC.*
Sterigma, *DC.*
Sterigmostemon, M. B.
Anthrolobus, Stev.
Morisia, *Gay.*
Cryptospora, *Kar.*

III. ORTHOPLOCEÆ.
(0 > >).

BRASSICIDÆ.

Sinapidendron, *Lowe.*
Disaccium, DC.
Brassica, *Linn.*
Brassica, Tournef.
Rapa, Tournef.
Napus, Tournef.
Sinapis, *Tournef.*
Sinapistrum, Reichnb.
Rhamphospermum, Andrz.
Bonnania, Presl.
? *Hirschfeldia*, Mönch.
Douepea, *Cambess.*
Erucastrum, *Presl.*
Micropodium, DC.
Güntheria, Andrz.
Orychophragmus, *Bung.*
Moricandia, *DC.*
Diplotaxis, *DC.*
Eruca, *Tournef.*
Euzomum, Link.

VELLIDÆ.

Vella, *DC.*
Boleum, *Desv.*
Stroganovia, *Kar.*
Stubendorfia, *Schr.*
Carrichtera, *DC.*
Succowia, *Medik.*
Savignya, *DC.*

PSYCHIDÆ.

Schouwia, *DC.*
Psychine, *Desf.*

ZILLIDÆ.

Zilla, *Forsk.*
Muricaria, *Desv.*
Calepina, *Adans.*

RAPHANIDÆ.

Crambe, *Tournef.*
Rapistrum, *Boerh.*
Schrankia, Medik.
Condylocarya, Bess.
Arthrolobus, Andrz.
Didesmus, *Desv.*
Euarthrocarpus, *Labill.*
Raphanistrum, *Tournef.*
Dondisia, Neck.
Ormycarpus, Neck.
Durandea, Delarbr.
Raphanus, *Tournef.*

FORTUYNIDÆ, *Boiss.*

Fortuynia, *Shutt.*

IV. SPIROLOBEÆ.
(0 ‖ ‖).

BUNIADÆ.

Bunias, *R. Br.*
Erucago, DC.
Lælia, Adans.

ERUCARIDÆ.

Erucaria, *Gärtn*
Cycloptychis, *E. M.*

V. DIPLECOLOBEÆ
(0 ‖ ‖ ‖).

SENEBIERIDÆ.

Senebiera, *Poir.*
Nasturtiolum, Medik.
Carara, Medik.
Coronopus, Hall, DC.
Cotyliscus, Desv.
Monoploca, *Bunge.*
Cycloptychis, *E. M.*
Brachycarpæa, *DC.*

SUBULARIDÆ.

Subularia, *DC.*
Consana, Adans.

HELIOPHILIDÆ.

Heliophila, *N. Burm.*
Trentepohlia, Roth.
Carponema, DC.
Leptormus, DC.
Orniscus, DC.
Selenocarpæa, DC.
Orthoselis, DC.
Pachystylum, DC.
Lanceolaria, DC.
Carpopodium, DC.
Chamira, *Thunb.*

SCHIZOPETALIDÆ.

Schizopetalon, *Hook.*
Peyreymondia, *Barnéoud.*

?Redowskia, *Cham. et Sch.*
Schimpera, *Stud. et Hoch.*
Discovium, *Raf.*

NUMBERS. GEN. 173. Sp. 1600.

Papaveraceæ.
POSITION.—Cistaceæ.—BRASSICACEÆ.—Capparidaceæ.
Fumariaceæ.

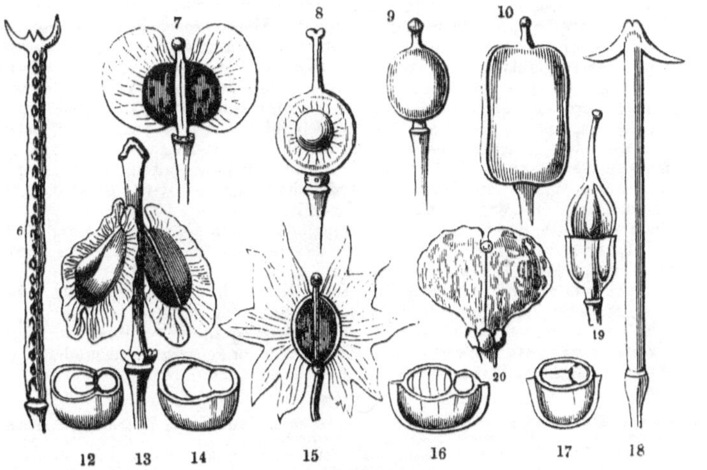

Fig. CCXLVI.

Fig. CCXLVI.—Fruits of various genera. 6. silique of Mathiola livida; 7. silicule of Thlaspi latifolium; 8. silicule of Alyssum spathulatum; 9. silicule of Schivereckia podolica; 10. silicule of Farsetia; 12. seed of Didesmus Ægyptius cut across; 13. silicule of Menonvillea linearis; 14. seed of Lepidium africanum; 15. silicule of Æthionema cristatum; 16. seed of Heliophila crithmifolia; 17. seed of Mathiola oxyceras; 18. silique of Mathiola oxyceras; 19. silicule of Didesmus ægyptius; 20. silicule of Senebiera serrata.

ORDER CXXIV. RESEDACEÆ.—WELDWORTS.

Resedaceæ, *DC. Théor. ed.* 1. 214. (1813); *Aug. de St. Hil. Ann. Soc. Roy. Orl. vol.* 13.; *Endl. Gen.* clxxxiii.; *Meisner, Gen. p.* 18; *Wight Illustr.* 1. 36.

DIAGNOSIS.—*Cistal Exogens, with definite not tetradynamous stamens, not tetramerous flowers, exalbuminous seeds, and fruit usually open at the point.*

Soft herbaceous plants, or in a few instances small shrubs, with alternate entire or pinnately divided leaves, and minute gland-like stipules. Flowers in racemes or spikes. Calyx many-parted. Petals broad fleshy plates, having lacerated appendages at the back, unequal. Disk hypogynous, 1-sided, glandular. Stamens definite, inserted into the disk; filaments erect; anthers 2-celled, opening longitudinally. Ovary sessile, 3-lobed, 1-celled, many-seeded, scarcely closed, usually with 3-6-parietal placentæ, sometimes surrounding a free central ovule-bearing body. Stigmas 3, glandular, sessile. Ovules amphitropal or campulitropal. Fruit dry and membranous, or succulent, opening at the apex; or apocarpous, with empty carpels surrounding a central placenta; or even hooded and 1-seeded. Seeds several, reniform; embryo taper, arcuate, without albumen; radicle next the hilum.

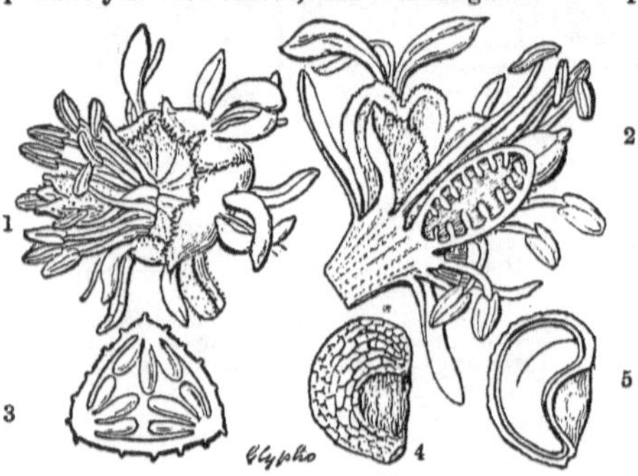

Fig. CCXLVII.

The flowers of these plants, of which the common Mignonette may be taken as the type, differ in many respects from those of other Orders, especially in the presence of a very large glandular 1-sided plate, out of which the stamens grow, and in the petals bearing a great resemblance to that disk. This led me, in the *Collectanea Botanica*, and in the first edition of this work, to describe the structure of Weldworts, as consisting of an apparent calyx which was really an involucre, while the petals are abortive male flowers, and the disk a calyx of one central bisexual flower. I am, however, now convinced, by the arguments of Henslow, that this theory was erroneous, and I accordingly revert to the old view of the organisation and affinities of the Order. These latter are chiefly with Capparids, with which the seeds, the great disk out of which the stamens arise, and the parietal placentæ, agree.

All these plants are weeds inhabiting Europe, the adjoining parts of Asia, the basin of the Mediterranean, and the adjacent islands. A very few occur in the North of India, the Cape of Good Hope, and California.

Little more is known of their uses than that Reseda luteola, called Weld, yields a yellow dye, and that the Mignonette (R. odorata) is among the most fragrant of plants. They were once regarded as sedative, as is indicated by the word Resēda. They are generally sub-acrid; nevertheless Reseda Phyteuma, the ὄχίστρα of the modern Greeks, is eaten as a kitchen esculent in the Greek Archipelago.

GENERA.

Ochradenus, *Delil.*	*Eresda*, Spach.	*Elimia*, Nutt.	*Sesamoides*, Tournef.
Reseda, *Linn.*	Oligomeris, *Cambess.*	Holbpetalum, *Turcz.*	*Sesamella*, Reichenb.
Luteola, Tournef.	*Resedella*, Webb et B.	Astrocarpus, *Neck.*	Caylusea, *St. Hil.*

NUMBERS. GEN. 6. Sp. 41.

POSITION.————RESEDACEÆ.—Capparidaceæ.

Fig. CCXLVII.—Reseda mediterranea. 1. a flower seen from above, much magnified; 2. a section of the same, showing the great disk on one side of the ovary, and within which the stamens arise; 3. a cross section of the ovary; 4. a seed; 5. a section of it.

ORDER CXXV. CAPPARIDACEÆ.—CAPPARIDS.

Capparideæ, *Juss. Gen.* 242. (1789); *Ann. Mus.* 18. 474; *DC. Prodr.* 1. 237; *Endl. Gen.* clxxxii.; *Meisner, Gen. p.* 17; *Wight. Illustr.* 1. 33.

DIAGNOSIS.—*Cistal Exogens, with stamens not tetradynamous, tetramerous flowers, exalbuminous seeds, and a closed up fruit.*

Herbaceous plants, shrubs, or even trees, without true stipules, but sometimes with spines in their place. Leaves alternate, stalked, undivided, or palmate. Flowers solitary or racemose. Sepals 4, either nearly distinct, imbricated, or valvate, equal, or unequal, or cohering in a tube, the limb of which is variable in form. Petals 4, or even 8, imbricated, or 0, cruciate, usually unguiculate and unequal. Stamens almost perigynous, very seldom tetradynamous, most frequently arranged in some high multiple of a quaternary number, definite or 00, placed upon a large hemispherical disk, or at the apex of a stalk-like torus; anthers turned inwards, opening longitudinally. Disk greatly developed, sometimes as a fleshy, hemispherical, roundish, or stalk-like body, sometimes as a nectariferous glandular plate of various forms, sterile on one side and anther-bearing on the other. Ovary stalked, or sessile, 1-celled, with 2 or more parietal placentæ; ovules amphitropal or campylotropal; style 0, or filiform; stigma generally round. Fruit either podshaped and dehiscent, or baccate, 1-celled, very rarely 1-seeded, most frequently with polyspermous placentæ. Seeds generally reniform, without albumen, but with the lining of the testa tumid, attached to the margin of the valves; embryo curved; cotyledons foliaceous, flattish; radicle taper, short or long, turned to the hilum.

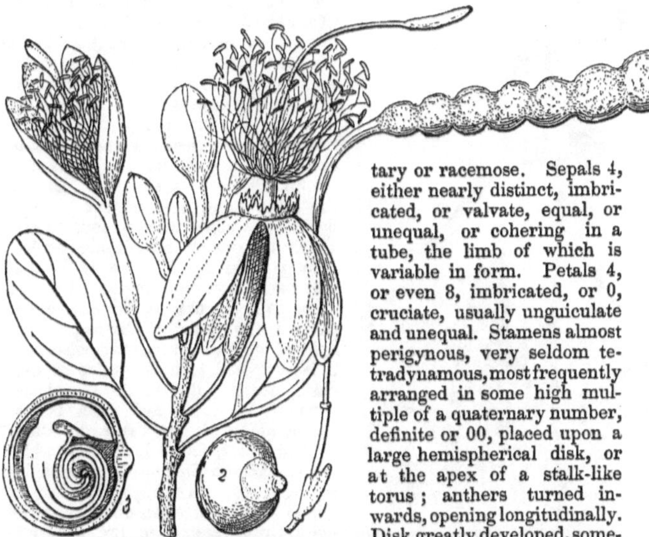

Fig. CCXLVIII.

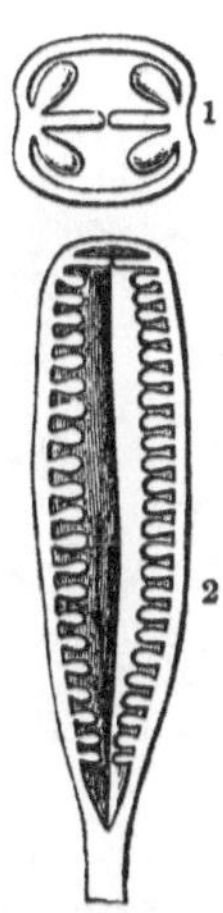

Fig. CCXLIX.

Distinguished from Crucifers by their stamens being often indefinite, if definite never tetradynamous, or scarcely ever, and by their reniform seeds. They are related to Passionworts in their stipitate ovary, and fleshy indehiscent fruit with parietal polyspermous placentæ; and to Bixads in the structure of their fruit, parietal placentæ, and indefinite stamens; from these last they are known by their narrow placentæ, exalbuminous seeds, and peculiar habit; and from the former by a number of obvious characters. Brown remarks (*Denham*, 15,) that some species of Capparis, of which C. spinosa is an example, have as many as 8 placentæ. Aug. de St. Hilaire and Moquin Tandon state that Capparids are referable to a tetrandrous type, which is very possible. But the explanation they give, or the proofs they offer of this, are less clear than could be desired. (See *Ann. des Sc.* 20. 321).

Capparids are chiefly found in the tropics and in the countries bordering upon them, where they abound in almost every direction. Of the capsular species, a single one, Cleome violacea, is found in Portugal; another, Polanisia graveolens, occurs as far to the north as Canada; and one or two others are met with in the southern provinces of the United States. Of the fleshy-fruited kinds, the common Caper, Capparis spinosa, a native of the most southern parts of Europe, is that which approaches the nearest to the north. Africa abounds in them.

Fig. CCXLVIII.—Mœrua angolensis.—*Delessert.* 1. a ripe fruit; 2. a seed; 3. a section of it.
Fig. CCXLIX.—Capparis Sinclairii. 1. a transverse; 2. a longitudinal section of its ovary.

De Candolle compares Capparids with Crucifers in regard to their sensible qualities; and they no doubt resemble each other in many respects; for instance, the flower-buds of the Caper (Capparis spinosa, rupestris in Greece, Fontanesii in Barbary, and ægyptiaca in Egypt) are stimulant, antiscorbutic, and aperient, and form a well-known pickle; the bark of the root passes for a diuretic; and some species of Cleome and Polanisia have a pungent taste, like that of mustard; the root of Cleome dodecandra is used as a vermifuge in the United States. The bark of the root of Cratæva gynandra (the Garlick Pear) blisters like Cantharides; so does that of Capparis cynophallophora, amygdalina, and ferruginea. Polanisia icosandra acts as a vesicatory, and is used in Cochin China as a sinapism. But on the other hand the pungent principle becomes in some cases so concentrated as to be dangerous. Colicodendron Yeo is said by Martius to be dangerous to mules and horses. There is a plant called Fruta de Burro, found in the neighbourhood of Carthagena, the fruit of which is extremely poisonous; it is supposed to be a species of Capparis, nearly allied to the C. pulcherrima of Jacquin; and must not be confounded with the Fruta del Burro of Humboldt, found in Guiana, which is a valuable medicinal plant, belonging to Anonads. Although they are in general plants of small dimensions, yet from Cratæva excelsa the people of Madagascar, who call it Vouēn pouēn, cut planks as much as four feet broad, according to Bojer.—*Ann. Sc. N. N. S.* xx. 58. The bruised leaves of Cratæva Tapia are used in Brazil against inflammation; its bark is bitter and tonic. Capparis Sodada has a narcotic odour, and its acrid stimulating fruits are employed by women to produce fecundity. The root of Cadaba indica is said to be aperient and anthelmintic. The juicy berries of Cratæva Nurvala are said to be agreeable.

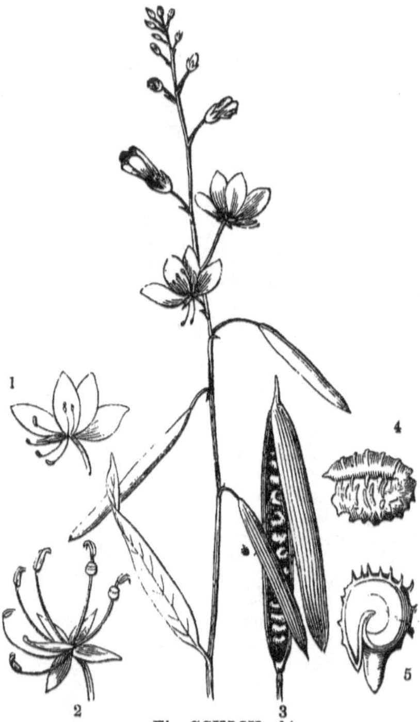

Fig. CCXLIX. *bis.*

GENERA.

I. CLEOMEÆ.—Fruit a capsule.

Cleomella, *DC.*
Gynandropsis, *DC.*
Gymnogonia, R. Br.
Cleome, *DC.*
Sinapistrum, Mönch.
Pedicellaria, DC.
Atalanta, Nutt.
Peritoma, DC.
Siliquaria, Forsk.
Roridula, Forsk.
Rorida, Röm. et Sch.
Dactylæna, *Schrad.*
Physostemon, *Mart. et Zucc.*
Polanisia, *Raf.*
Corynandra, Schrad.
Ranmanissa, Endl.
Cyrbasium, *Endl.*
Cristatella, Nutt.
Isomeris, *Nutt.*
Dipterygium, *Decaisne.*
Pteroloma, St.

II. CAPPAREÆ.—Fruit a berry.

Schepperia, *Neck.*
Macromerum, Burch.
Atamisquea, *Miers.*
Cadaba, *Forsk.*
Stromia, Vahl.
Desmocarpus, Wall.
Thylacium, *Lour.*
Niebuhria, *DC.*
Boscia, *Lam.*
Podoria, Pers.
Streblocarpus, *Arnott.*
Maerua, *Forsk.*
Colicodendron, *Mart.*
Calanthea, DC.
? Quadrella, DC.
Capparis, *Linn.*
Sodada, Forsk.
Hombach, Adans.
Lindackera, Sieb.
Capparidastrum, DC.
Cynophalla, DC.
Uterveria, Bert.
Breyniastrum, DC.
Breynia, Plum.
Busbeckea, *Endl.*
Morisonia, *Plum.*
Cratæva, *Linn.*
Othrys, Noronh.
Ritchiea, *R. Br.*
Steriphoma, *Spreng.*
Römeria, Tratt.
Stephania, Willd.
Tovaria, *Ruiz et Pav.*
? Singana, *Aubl.*
Sterebeckia, Schreb.
? Hermupoa, *Löffl.*
? Roydsia, Roxb.

NUMBERS. GEN. 28. Sp. 340.

Passifloraceæ.
POSITION.—*Brassicaceæ.*—CAPPARIDACEÆ.—*Resedaceæ.*
Flacourtiaceæ.

Fig. CCXLIX. *bis.*—Physostemon lanceolatum. 1. a flower of the natural size; 2. the calyx, stamens, and ovary; 3. the ripe fruit, with one valve separating; 4. a seed; 5. the same cut vertically, to show the incurved embryo.

ALLIANCE XXVIII. *MALVALES.*—THE MALVAL ALLIANCE.

DIAGNOSIS.—*Hypogynous Exogens, with monodichlamydeous flowers, axile placentæ, valvate calyx, an imbricated or twisted corolla, definite or* 00 *stamens, and embryo with little or no albumen.*

This Alliance is little else than the old Order of Malvæ or Columniferæ, with the addition of Indian Cresses and Vivianiads. The limit of it is determined by the valvate calyx in combination with a twisted or much imbricated corolla, if any corolla is present. The Sterculiads, with their long-stalked ovary, indicate an approach to the very edge of Capparids, whose calyx is sometimes valvate; and at the other end of the series Lindenblooms are not unaptly brought into contact with the Poreworts (Tremandraceæ), which are also valvate.

It is not, however, with Sapindals that Malvals are most allied; they claim a much nearer affinity to Geranials, which differ in their imbricated calyx more than in any thing else, and to which the Indian Cresses form a direct transition. But I have not succeeded in bringing together the Alliances of Hypogynous Exogens, so as to allow Geranials to follow or precede Malvals. There is also another kind of relationship on the part of Malvals, equally strong, and equally impossible to describe lineally; namely, with Spurgeworts, which Sterculiads and Byttneriads (especially the latter) quite touch at several points.

The real position of the Malval Alliance may therefore be expressed somewhat in the following manner:—

Sterculiaceæ.—Byttneriaceæ.—Vivianiaceæ.—Tropæolaceæ.—Malvaceæ.—Tiliaceæ.
Euphorbiaceæ. *Geraniaceæ.*

NATURAL ORDERS OF MALVALS.

Stamens columnar, all perfect. Anthers 2-celled, turned outwards. 126. STERCULIACEÆ.

Stamens monadelphous, in most cases partly sterile. Anthers 2-celled, turned inwards . . . . 127. BYTTNERIACEÆ.

Stamens free. Disk none. Seeds with albumen. Embryo curved. Petals permanent. Calyx ribbed. . . . . 128. VIVIANIACEÆ.

Stamens free. Disk none. Seeds without albumen. Embryo amygdaloid . . . . 129. TROPŒOLACEÆ.

Stamens columnar, all perfect. Anthers 1-celled, turned inwards. . 130. MALVACEÆ.

Stamens free, on the outside a disk. Seeds with albumen. Embryo straight. . . . . 131. TILIACEÆ.

ORDER CXXVI. STERCULIACEÆ.—STERCULIADS.

Sterculiaceæ, *Vent. Malm.* 2. 91. (1799); *Endl. Meletem. p.* 30.; *Gen.* ccx.; *Meisner, Gen. p.* 28.—Bombaceæ, *Kunth. Diss. Malv. p.* 5. (1822); *DC. Prodr.* 1. 475.; *A. St. Hilaire Fl. Br. Merid.* 1. 275; *Ed. pr. No.* 26. (1830); *Wight Illustr.* 1. *p.* 66.

DIAGNOSIS.—*Malval Exogens with columnar stamens all perfect, and 2-celled anthers turned outwards.*

Large trees or shrubs. Hairs, if present, stellate. Leaves alternate, simple or compound, sometimes digitate, often toothed, with free deciduous stipules. Inflorescence variable. Flowers regular or irregular, frequently ♂ ♀ by abortion. Calyx either naked or surrounded with an involucre, consisting of 5 sepals, more or less united at the base, with a valvate or nearly valvate æstivation, except where the calyx is irregularly ruptured. Petals 5, (or none), hypogynous, convolute in æstivation. Stamens indefinite, monadelphous in various ways; anthers 2-celled, turned outwards, sometimes anfractuose. Pistil consisting of 5, or rarely 3, carpels, either distinct or cohering into a single ovary, often seated upon a column-like axis. Styles equal in number to the carpels, distinct or united; ovules orthotropal or anatropal, erect if definite; sometimes indefinite. Fruit capsular, with 3 or 5 cells, or even drupaceous or berried, or composed of distinct follicles, opening by the ventral suture long before the ripening of the seeds. Seeds ovate or angular, sometimes winged or woolly; albumen oily or fleshy, rarely wanting; embryo straight or curved; cotyledons either foliaceous, flat, and plaited, or rolled round the plumule, or else very thick, but this only in the seeds without albumen; radicle next the hilum, or at the opposite end of the seed, or even transverse.

Fig. CCL.

These have the columnar stamens of Mallowworts, and therein exhibit a near approach to that Order; but their anthers are 2-celled, and turned outwards. Sterculiads also lie on the borders of Byttneriads, from which they are readily distinguished by their columnar stamens not being partially sterile, and by the anthers being turned outwards. The Sub-Order Bombaceæ is remarkable for having a tough leathery calyx, which sometimes splits irregularly so as to hide the true manner in which the sepals are arranged. The fruit of Sterculia often exhibits beautiful illustrations of the real nature of that form of fruit which Botanists call the follicle, and helps to demonstrate that it, and hence all simple carpels, are formed of leaves, the sides of which are inflexed, and the margins dilated into placentæ bearing ovules. In Firmiana platanifolia, in particular, the follicles burst and acquire the form of coriaceous leaves, bearing the seeds upon their margin.

According to Dr. R. Brown, the Sub-Order Sterculeæ is remarkable for the different positions taken by the radicle within the seed; although in the majority it is at the extremity most remote from the hilum, yet in others it is next the hilum, and in some transverse with respect to that part, an unusual circumstance in the same Natural Order.—*Pl. Javan.* p. 224.

Nearly all the known species are tropical, or at least natives of very warm climates. They are extensively scattered over the world, the Sterculeæ preferring India and Africa, the Bombaceæ America; Helictereæ seem to be unknown in Africa. The Baobab trees are from Senegal, where they are remarkable for their enormous size and prodigious longevity, estimated, but no doubt incorrectly, to amount in certain instances to some thousand years. The various species of Bombax and Ceiba are prodigious American

Fig. CCL.—Helicteres brevispira.—*A. St. Hilaire.* 1. a column of stamens; 2. an anther; 3. a pistil; 4. a ripe fruit.

forest trees, with huge buttresses projecting from their colossal trunks. A few of the Helictereæ are remote from the latitudes usually assigned by nature for the habitation of Sterculiads, extending as far to the southward as Tasmannia and New Zealand.

Sterculiads, like the Orders most nearly related to them, are chiefly remarkable for the abundance of mucilage they contain. The seeds of Sterculia tomentosa, acuminata, the true Kola spoken of by African travellers, when chewed or sucked, render the flavour of water, even if half putrid, agreeable. Those of the Chichà, Sterculia Chicha, and lasiantha, are eaten as nuts by the Brazilians. So are those of Sterculia nobilis in Asia. The Gum Tragacanth of Sierra Leone is produced by S. Tragacantha. S. urens in Coromandel yields a gum which is exceedingly like Tragacanth, and has been imported as such into England. The pod of S. fœtida is, according to Horsfield, employed in gonorrhœa in Java; the leaves are considered repellent and aperient; a decoction of the fruit is mucilaginous and astringent. The bark of a species of Sterculia is employed in the Moluccas as an emmenagogue; and the seeds of all that genus are filled with an oil, which may be expressed and used for lamps. There is a slight acridity in the seeds of Sterculia. It is said that the seeds of Pterygota alata are narcotic.

Bombax and its allied genera are more remarkable for their noble aspect than for their utility. They are, however, not without interest. The seeds of many are enveloped in long hairs, like those of the true Cotton: it is found, however, that this wool cannot be manufactured, in consequence of no adhesion existing between the hairs. The woolly coat of the seeds of the Arvore de Paina (Chorisia speciosa), and several species of Eriodendron and Bombax, is employed in different countries for stuffing cushions, and for similar domestic purposes. Bombax pentandrum, the Cotton Tree of India, yields a gum, which is given in conjunction with spices in certain stages of bowel complaints. The bark of such trees is, however, reported to be emetic; this is more especially the case with Salmalias and the American species of Bombax. The honey of the flowers of Salmalia malabarica is said to be purgative and diuretic. One of the largest trees in the world is the Adansonia, or Baobab Tree, the trunk of which has been found with a diameter of 30 feet; but its height is not in proportion. It is emollient and mucilaginous in all its parts. The leaves dried and reduced to powder constitute Lalo, a favourite article with the Africans, which they mix daily with their food, for the purpose of diminishing the excessive perspiration to which they are subject in those climates; and even Europeans find it serviceable in cases of diarrhœa, fevers, and other maladies. The fruit is, perhaps, the most useful part of the tree. Its pulp is slightly acid and agreeable, and frequently eaten; while the juice is expressed from it, mixed with sugar, and constitutes a drink which is valued as a specific in putrid and pestilential fevers. The dried pulp is mixed with water, and administered, in Egypt, in dysentery. It is chiefly composed of gum, like Gum Senegal, a sugary matter, starch, and an acid which appears to be the malic. The fruit of the Durian (Durio zibethinus), is considered one of the most delicious productions of nature; it is indeed fœtid, and therefore disagreeable to those who are unaccustomed to it, but it universally becomes in the end a favourite article of the dessert; it is found in the islands of the Indian Archipelago, where it is cultivated extensively. Ochroma Lagopus, a West Indian tree, is used for many purposes. Its light wood is used instead of cork, its bark is antisyphilitic, the woolly lining of its fruit is applied to various purposes, and its wounded trunk discharges abundance of gum. The Handplant of Mexico, or Manita, (Cheirostemon platanoides), has no petals, but a large angular calyx, resembling a leather cup, from the centre of which rises up a column, bearing 5 narrow curved anthers with a curved style in the middle; these have considerable resemblance to a hand furnished with long claws. Helicteres Sacarolha, called by the latter name only in Brazil, is used against venereal disorders: a decoction of the root is administered. It is supposed that its effects depend upon its mucilaginous properties. Myrodia angustifolia is said by Martius to have similar qualities.

GENERA.

I. BOMBACEÆ.—Leaves palmate or digitate. Flowers perfect.

Adansonia, *Linn.*
Baobab, P. Alpin.
Ophelus, Lour.
Pachira, *Aubl.*
Carolinea, Linn. f.
Chorisia, *H. B. K.*
Eriotheca, *Schott. et Endl.*
Eriodendron, *DC.*
Campylanthera, Schott.
Gossampinus, Rumph.
Erione, Schott et Endl.
Bombax, *Linn.*
Ceiba, Mart. et Zucc.
Salmalia, *Schott. et Endl.*
Cavanillesia, *Ruiz et Pav.*
Pourretia, Willd.
Durio, *Rumph.*
Ochroma, *Swartz.*
Cheirostemon, *Humb.*
Cheiranthodendron, Lavrad.
Neesia, *Blum.*
Esenbeckia, Blum.
Cotylephora, Meisn.
? Montezuma, *Moç. et Sess.*
? Hampea, *Schlecht.*

II. HELICTEREÆ.—Leaves simple. Flowers perfect.

Plagianthus, *Forst.*
Asterotrichion, *Kl.*
Blepharanthemum, Kl.
Hoheria, *A. Cunningh.*
Myrodia, *Schreb.*
? *Lexarza*, Llav.
Quararibea, Aubl.
Gerberia, Scop.
Matisia, *Humb. et Bonpl.*
Methorium, *Schott.*
Helicteres, *Linn.*
Orthocarpæa, DC.
Spirocarpæa, DC.
Isora, Schott. et Endl.
Alicteres, Neck.
Orthothecium, Schott.
Ungeria, *Schott. et Endl.*
Reevesia, *Lindl.*

III. STERCULEÆ. — Leaves simple or palmate. Flowers unisexual by abortion.

Heritiera, *Ait.*
Balanopteris, Gärtn.
Sutherlandia, Gmel.
Samandura, Linn.
Atunus, Rumph.
Sterculia, *Linn.*
Clompanus, Rumph.
Ivira, Aubl.
Theodoria, Neck.
Chichæa, Presl.
Mateatia, Fl. Flum.
Southwellia, Salisb.
Balanghas, Burm.
Cavalam, Rumph.
Cavallium, Sch. et Endl.
Triphaca, Lour.
Astrodendron, Dennst.
Brachychiton, *Schott.*
Pæcilodermis, Id.
Trichosiphon, Id.
Hildegardia, *Sch. et Endl.*
Cola, Bauh.
Lunana, DC.
Edwardia, Raf.
Bichy, Lunan.
? *Culhamia*, Forsk.
Scaphium, *Sch. et Endl.*
Firmiana, *Marsigl.*
Erythropsis, Lindl.
Pterygota, *Scht. et Endl.*
Tetradia, *R. Br.*
Pterocymbium, *R. Br.*
Courtenia, *R. Br.*
Micrandra, *R. Br.*

NUMBERS. GEN. 34. Sp. 125.

Ternstromiaceæ.
POSITION.—Malvaceæ.—STERCULIACEÆ.—Byttneriaceæ.
Capparidaceæ.

ORDER CXXVII. BYTTNERIACEÆ.—BYTTNERIADS.

Byttneriaceæ, *R. Brown in Flinders*, 2. 540. (1814); *Kunth Diss. p.* 6.; *DC. Prodr.* 1. 481.; *Aug. St. Hil. Fl. Bras. Mer.* 1. 139. (a § of Malvaceæ); *Endl. Gen.* ccxi.; *Meisn. Gen. p.* 32; *Wight Illustr.* 1. 72.—Hermanniaceæ and Dombeyaceæ, *Bartl. Ord. Nat.* 341.—Philippodendreæ, *Endl. Gen.* 1004.

DIAGNOSIS.—*Malval Exogens, with monadelphous stamens, in most cases partly sterile, and 2-celled anthers turned inwards.*

Trees, shrubs, or undershrubs, occasionally with a climbing habit; their surface usually covered with stellate or forked hairs, occasionally with scurfs. Leaves alternate, simple, feather-veined or hand-veined, commonly notched at the edge; stipules deciduous, in a few instances 0. Flowers often in clusters, but also in spikes or panicles. Calyx herbaceous, membranous, or leathery, 4- 5-lobed, valvate in æstivation. Corolla 0, or consisting of as many petals as there are lobes to the calyx, either flat, but twisted in æstivation, or arched and drawn out into a strap; folded inwards at the edges and valvate in æstivation, either permanent or deciduous, often adhering to the tube of stamens. Stamens hypogynous, definite and opposite the petals, or twice as many and half only fertile and opposite the petals, or 00, as many being barren as there are sepals, and opposite them; almost always united into a cup or tube; anthers turned inwards, 2-celled, opening lengthwise, very rarely by a pore or cleft near the point. Ovary free, sessile, or on a short stalk, composed of from 4 to 10 carpels arranged round a central column, or reduced to one only; ovules 2 in each cell, anatropal, ascending, or nearly horizontal, or even pendulous; styles terminal, consolidated; stigmas equal in number to the cells. Fruit generally a capsule, splitting through the cells or resolving itself into its original elements by dividing at the partitions; seeds sometimes winged, but generally round. Embryo generally lying in a small quantity of fleshy or mucilaginous albumen, straight or bowed; cotyledons fleshy or generally leafy, entire or split, plaited or folded up, occasionally spiral: radicle straight or curved, next the hilum.—*Chiefly from Endlicher.*

Fig. CCLI.

Fig. CCLII.

Byttneriads are often united with Sterculiads, from which their slightly monadelphous stamens, with the anthers turned inwards, and, excepting the Sub-orders Hermanneæ and Eriolæneæ, the stamens partially imperfect, sufficiently divide them. Their two-celled anthers and not columnar stamens distinguish them from Mallowworts. The tendency to a loss of petals, an abortion of the stamens, and even a separation of the

Fig. CCLI.—Byttneria celtoides.—*A. St. H.* 1. an expanded flower; 2. cup of stamens; 3. abortive stamen; 4. pistil.

Fig. CCLII.—Melochia graminifolia.—*A. St. H.* 1. flower; 2. stamens and pistil; 3. ripe fruit; 4. a coccus; 5. section of a seed.

sexes, which is so frequently observable in Sterculiads and Byttneriads, is an indication of a lower degree of organisation than occurs among Mallowworts, and clearly brings the Malval into contact with the Euphorbial Alliance.

These are wholly tropical, or from temperate climates. The Lasiopetaleæ are Australasian, Hermanneæ are South African, Dombeyeæ African and Asiatic, Eriolæneæ exclusively Asiatic, and Philippodendreæ from New Zealand, (not Nepal, as has been stated by Mr. Poiteau); Byttnereæ are both Asiatic and American.

Beyond all other products Cacao or Cocoa, the chief ingredient in Chocolate, is remarkable in this Order. It is the seed of Theobroma Cacao, a small tree of which whole forests occur in Demerara. An ardent spirit is distilled from the pulp of the fruit. The Waltheria Douradinha is used in Brazil as a remedy for venereal disorders, for which its very mucilaginous nature renders it proper. The fruit of Guazuma ulmifolia is filled with a sweet and agreeable mucilage, which the Brazilians suck with much pleasure. In Martinique the young bark is used to clarify sugar, for which the copious mucilage it yields when macerated qualifies it. In the same island the infusion of the old bark is esteemed as a sudorific, and useful in cutaneous diseases. The bark of Kydia calycina is applied in India to the same purpose. The fibrous tissue of the bark of many species is so tough as to be well adapted for manufacturing into cordage; this is more especially the case with Microlæna spectabilis, and Abroma augustum. The bark of Dombeya spectabilis is made into ropes in Madagascar. The Pterospermums are all mucilaginous.

GENERA.

I. LASIOPETALEÆ.

Seringia, *Gay.*
Gaya, Spreng.
Guichenotia, *Gay.*
Thomasia, *Gay.*
Leucothamnus, *Lindl.*
Lasiopetalum, *Smith.*
Corethrostylis, *Endl.*
Keraudrenia, *Gay.*
Sarotes, *Lindl.*
Fleischeria, *Steud.*

II. BYTTNEREÆ.

Rulingia, *R. Br.*
Commersonia, *Forst.*
? *Medusa*, Lour.
Jürgensia, Spreng.
Abroma, *Jacq.*
Ambroma, Linn. f.
Hastingia, König.
Byttneria, *Löffl.*
Chætæa, Jacq.
Heterophyllum, Boj.
Telfairia, Newm.
Ayenia, *Linn.*
Dayenia, Mill.
Herrania, *Goudot.*
Lightia, Schomb.
Theobroma, *Linn.*
Cacao, Tournef.
Guazuma, *Plum.*
Bubroma, Schreb.
Kleinhovia, *Linn.*
Actinophora, *Wall.*
Pentaglottis, *Wall.*

III. HERMANNEÆ.

Waltheria, *Linn.*
Lophanthus, Forst.
Astropus, Spreng.
Melochia, *Linn.*
Riedlea, *Vent.*
Riedleia, DC.
Altheria, Thouars.
Lochennia, Arn.
Physodium, *Presl.*
Hermannia, *Linn.*
Mahernia, *Linn.*

IV. DOMBEYEÆ.

Ruizia, *Cav.*
Astyria, *Lindl.*
Pentapetes, *Linn.*
Moranda, Scop.
Brotera, *Cav.*
Sprengelia, Schult.
Vialia, Vis.
Assonia, *Cav.*
Königia, Commers.
Vahlia, Dahl.
Dombeya, *Cav.*
Paulowilhelmia, *Hochst.*
Xeropetalum, *Delil.*
Leeuwenhoeckia, E. Mey.
Melhania, *Forsk.*
Astrapæa, *Lindl.*
Hilsenbergia, Boj.
Glossostemon, *Desfont.*
Trochetia, *DC.*
Pterospermum, *Schreb.*
Velaga, Adans.
Velago, Gärtn.
Pterolæna, DC.
Kydia, *Roxb.*

V. ERIOLÆNEÆ.

Eriolæna, *DC.*
Schillera, *Reichenb.*
Microlæna, Wall.
Wallichia, DC.
Jackia, Spreng.
? Visenia, *Houtt.*
Wisenia, Gmel.
Aleurodendron, Reinw.
Glossospermum, Wall.
? Exitelia, *Blum.*
Maranthes, Blum.

VI. PHILIPPODENDREÆ.

Philippodendron, *Poit.*

NUMBERS. GEN. 45. Sp. 400.

POSITION.—Sterculiaceæ.—BYTTNERIACEÆ.—Tiliaceæ.
Euphorbiaceæ.

ORDER CXXVIII. VIVIANIACEÆ.—VIVIANIADS.

Vivianiaceæ, *Klotzsch in Linnæa*, 10. 433. (1836); *Meisner, Gen.* 58; *Endl. Gen. p.* 1169.

DIAGNOSIS.—*Malval Exogens, with free stamens, no disk, seeds with albumen, a curved embryo, permanent petals, and ribbed calyx.*

Herbaceous or half-shrubby plants. Leaves opposite or whorled, entire or notched, without stipules, often covered beneath with a hoary down. Flowers in panicles or corymbs, white, red, or pink. Calyx ten-ribbed, with 5 valvate divisions. Petals 5, hypogynous, furnished with claws, often drying up and remaining permanently round the seed-vessel, with a twisted æstivation. Stamens 10, hypogynous; those opposite the sepals inserted into a fleshy gland; filaments distinct; anthers 2-celled, opening lengthwise. Ovary free, 3-celled; stigmas 3, sessile; ovules 2 in each cell, attached to the central axis, one ascending, the other suspended. Capsule 3-lobed, 3-celled, splitting through the cells; the valves bearing the partitions in the middle. Seeds roughish, containing a curved embryo lying among a large quantity of fleshy albumen; cotyledons linear; radicle next the hilum.

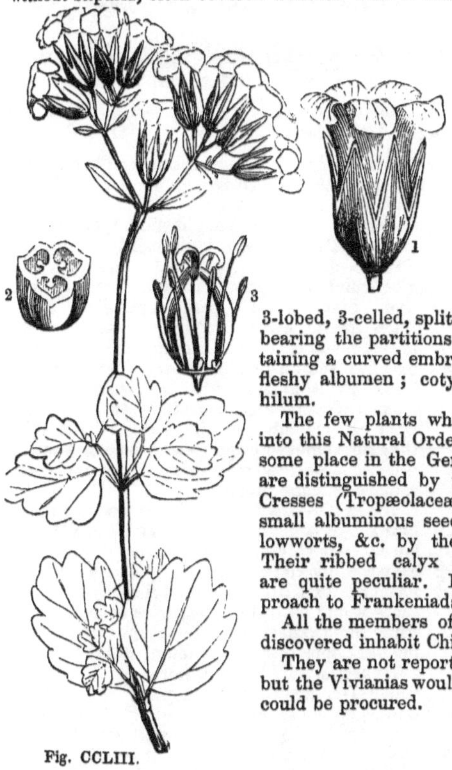

Fig. CCLIII.

The few plants which recent writers have combined into this Natural Order, have been generally referred to some place in the Geranial Alliance, from all which they are distinguished by their valvate calyx; from Indian Cresses (Tropæolaceæ) they are distinguished by their small albuminous seeds, and regular flowers; from Mallowworts, &c. by their not having columnar stamens. Their ribbed calyx and permanent withering corolla are quite peculiar. It may be that we have here an approach to Frankeniads.

All the members of this Order which have yet been discovered inhabit Chili and South Brazil.

They are not reported to possess any useful properties; but the Vivianias would be pretty greenhouse plants if they could be procured.

GENERA.

Cæsarea, *Cambess.*
Viviania, *Cav.*
 Macræa, Lindl.
 Xeropetalon, Hook.
Cissarobryon, *Pöpp.*
Linostigma, *Klotsch.*

NUMBERS. GEN. 4. SP. 15.

POSITION.—Tiliaceæ.—VIVIANIACEÆ.—Tropæolaceæ.

Fig. CCLIII.—Viviania crenata; 1. a flower; 2. a section of the ovary; 3. stamens and ovary.

ORDER CXXIX. TROPÆOLACEÆ.—INDIAN CRESSES.

Tropæoleæ, *Juss. Mem. Mus.* 3. 447. (1817); *DC. Prodr.* 1. 683. (1824); *Endl. Gen.* cclviii.—Limnantheæ, *R. Br. in Lond. and Edinb. Philosoph. Mag. July* 1833; *Lindley Bot. Reg. t.* 1673. (1834); *Nixus Plantarum, p.* 11. (1833); *Martius Conspectus, No.* 272. (1835); *Endl. Gen.* cclix.

DIAGNOSIS.—*Malval Exogens, with free stamens, no disk, seeds without albumen, and an amygdaloid embryo.*

Smooth herbaceous plants, of tender texture and with an acrid taste, trailing or twining. Leaves alternate, without stipules. Peduncles axillary, 1-flowered. Flowers yellow, scarlet, orange, or even blue! Sepals 3-5, the upper one with a long distinct spur; æstivation usually valvate, or very slightly overlapping. Petals 1-5, hypogynous, equal or unequal, with a convolute æstivation, sometimes partially abortive. Stamens 6-10, perigynous, distinct; anthers innate, erect, 2-celled. Ovary 1, 3-cornered, made up of 3 or 5 carpels; style 1; stigmas 3-5, acute; ovules solitary, erect, or pendulous. Fruit indehiscent, the pieces separable from a common axis, sometimes winged. Seeds large, without albumen, filling the cavity in which they lie; embryo large; cotyledons 2, straight, thick, consolidated into a single body, or distinct; radicle next the hilum.

Fig. CCLIV.

Indian Cresses form an Order standing on the limits between the Malval and Geranial Alliances. Its valvate calyx is almost the only character which determines its preference for the former; for if that were imbricated and ribbed there would be little to separate Indian Cresses from the Cranesbills. Tropæolum majus has the very spur of a Pelargonium, only in the latter the spur is adnate to the flower-stalk. Limnanths, which Dr. Brown first proposed as a distinct Order, do not seem to be naturally distinguished, and, considering the very small extent of the Order of Indian Cresses, are far better combined with them. If the leaves of Limnanthes Douglasii and Tropæolum majus are chewed, their flavour is so similar that one is hardly able to distinguish them. The principal difficulty in the way of stationing Limnanths with Indian Cresses, consists in the perigynous insertion of the stamens of the former; but in this instance other considerations must, I think, outweigh that circumstance. Perhaps Limnanths should

Fig. CCLIV.—1. Chymocarpus pentaphyllus; 2. a longitudinal section of its flower; 3. ovary of Tropæolum majus; 4. a vertical section of a carpel, showing the position of the ovule; 5. a perpendicular section of a seed.

be regarded as an approach to Rueworts on the one hand, and Nolanads on the other, because of its deeply lobed pistil ; but this is probably a similarity of but little importance.

All are natives of the temperate parts of North and South America.

The fleshy fruit of Tropæolum majus is acrid, and possesses the properties of Cress ; and De Candolle remarks, that the caterpillar of the Cabbage butterfly feeds exclusively upon Crucifers and Tropæolum. The root of T. tuberosum is eaten in Peru. Chymocarpus is used in Brazil as an antiscorbutic, under the Portuguese name of Chagas da Miuda. Limnanthes has all the peculiar pungency of a Tropæolum.

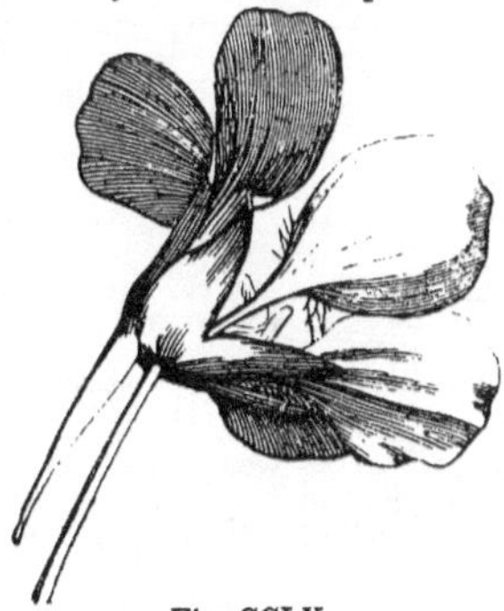

Fig. CCLV.

GENERA.

I. TROPÆOLEÆ.—Flowers irregular. Ovules pendulous.

Tropæolum, *L.*
Magallana, *Cav.*
Chymocarpus, *Don.*

II. LIMNANTHEÆ.—Flowers regular. Ovules erect.

Limnanthes, *R. Br.*
Flörkea, *W.*

NUMBERS. GEN. 5. SP. 43

POSITION.—Vivianiaceæ.—TROPÆOLACEÆ.— *Geraniaceæ.* / *Nolanaceæ.*

Fig. CCLV.—Flower of Tropæolum majus, showing the spur.

ORDER CXXX. MALVACEÆ.—MALLOWWORTS.

Malvaceæ, *Juss. Gen.* 271. (1789) *in part.; Brown in Voy. to Congo, p.* 8; *Kunth. Diss. p.* 1; *DC. Prodr.* 1. 429. (1829); *Endl. Gen.* ccix.; *Meisner, Gen. p.* 26.; *Wight. Illustr.* 1. *p.* 55.

DIAGNOSIS.—*Malval Exogens, with columnar stamens all perfect, and 1-celled anthers turned inwards.*

Herbaceous plants, trees, or shrubs. Leaves alternate, more or less divided, stipulate. Hairs stellate if present. Peduncles usually axillary. Flowers showy, often inclosed in an involucre of various forms. Sepals 5, very seldom 3 or 4, more or less united at the

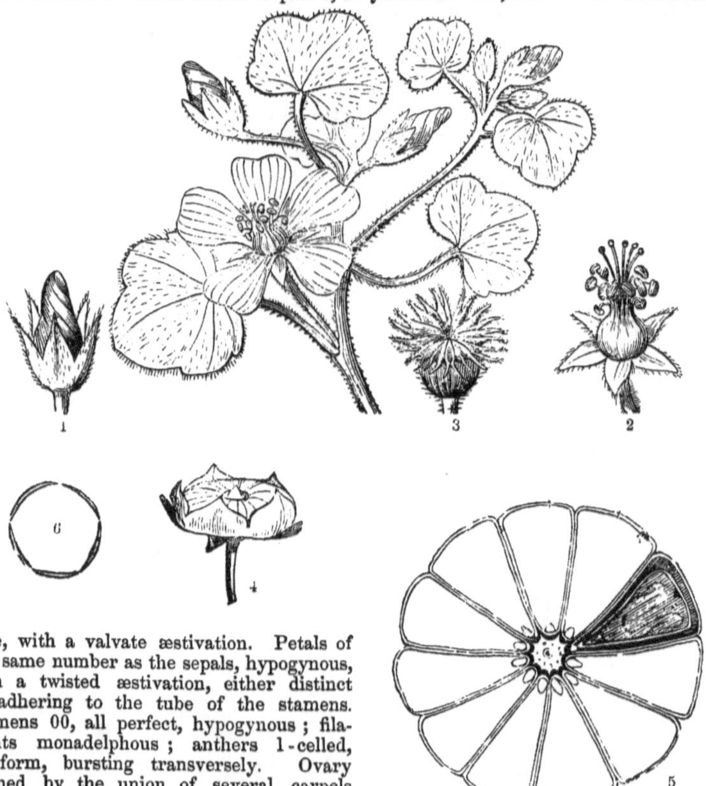

Fig. CCLVI.

base, with a valvate æstivation. Petals of the same number as the sepals, hypogynous, with a twisted æstivation, either distinct or adhering to the tube of the stamens. Stamens 00, all perfect, hypogynous; filaments monadelphous; anthers 1-celled, reniform, bursting transversely. Ovary formed by the union of several carpels round a common axis, either distinct or united; ovules definite or indefinite, attached to the inner angle of the cells, amphitropal or semianatropal; styles the same number as the carpels, either united or distinct; stigmas variable. Fruit either capsular or baccate, its carpels being either monosper-

Fig. CCLVI.—Abutilon macropodum. 1. an unexpanded flower; 2. the stamens and styles; 3. a ripe fruit, consisting of many carpels, whose upper extremities are free and radiant; 4. a ripe fruit of the Malva sylvestris, natural size; 5. a transverse section of the same fruit, from which all the seeds have been taken except one, which is seen at C; 6. a section of a calyx, showing its valvate structure.

mous or polyspermous, sometimes united in one, sometimes separate or separable into distinct cocci or follicles; dehiscence either loculicidal or septicidal. Seeds sometimes hairy; albumen none, or fleshy and in small quantity; embryo curved, with twisted and doubled cotyledons.

The relation of Mallowworts with Sterculiads, Lindenblooms, and Byttneriads, is clearly indicated by their general accordance in structure, combined with the valvate æstivation of their calyx. A less immediate affinity is indicated with Crowfoots by means of the curious genus Malope, whose carpels are separate and collected in considerable numbers over a central torus, although its organisation is in all other respects that of Mallowworts. There also seems to be a considerable degree of relationship between them and Cranesbills, Chlenads, and Flaxworts (Linaceæ). To the first they approach by their monadelphous stamens and crumpled embryo, to the second by their involucres and columnar stamens, to the third by their twisted corollas, and mucilaginous properties. The whole of these Orders are, however, sufficiently distinguished by the characters severally assigned to them in their proper places. Theads (Ternströmiaceæ) are another Order to which Mallowworts have been occasionally compared; but the slightly monadelphous condition of the stamens in that Order is very different from their columnar structure in Mallowworts, and there is little else in common between them.

These plants are found in great abundance in the tropics, plentifully in the hotter parts of temperate regions, but gradually diminishing to the north. Thus in Sicily they form $\frac{1}{88}$ of the flowering plants, in France $\frac{1}{143}$, in Sweden $\frac{1}{233}$, in Lapland they are unknown, in the temperate parts of North America $\frac{1}{125}$, in the equinoctial parts of the same continent $\frac{1}{47}$; or taking into account only the vegetation of the valleys, they, according to Humboldt, form $\frac{1}{50}$ of the flowering plants in the tropics, $\frac{1}{200}$ in the temperate zone, and are not found in the frigid zone. But these calculations no doubt include Sterculiads.

The uniform character of the Order is to abound in mucilage, and to be totally destitute of all unwholesome qualities. The use to which Mallows and Marshmallows are applied in Europe is well known. The whole plant of the latter, especially the root, yields in decoction a plentiful, tasteless, colourless mucilage, salutary in cases of irritation. It is used as a demulcent for children, and is a favourite medicine with the French, who employ it constantly in poultices, lozenges, &c., under the name of Guimauve. The flowers of the gaudy Hollyhock (Althæa rosea, μαλαχη, *Diosc.*) are officinal in Greece for the same purposes. Similar properties are possessed by extra-European species. Sida cordifolia mixed with rice is used to alleviate the bloody flux. Emollient fomentations are prepared from Sida mauritiana by the Hindoo doctors. The flowers of Bençao de Deos, Abutilon esculentum, are used in Brazil as a boiled vegetable. A decoction of Sphæralcea cisplatina is administered in the same country in inflammation of the bowels, and is generally employed for the same purposes as the Marshmallow in Europe. Pavonia diuretica is prescribed in Brazil as a diuretic; but is supposed to act rather as an emollient. The chewed leaves of another species, S. carpinifolia, are applied in Brazil to the punctures of wasps. And finally, to omit a great quantity of plants having similar qualities, it is sufficient to name the Abelmoschus esculentus, whose fruit, called Ochro, Gombo, Gobbo, Bandikai, &c., is a favourite ingredient in soup, to which it imparts its mucilaginous quality. The wood is always very light, and of little value. Rocket-sticks are obtained from the light straight stems of Sida micrantha. The bark is often so tenacious as to be manufactured into cordage. Malva crispa was found by Cavanilles to be fit for this purpose; and several species of Hibiscus are employed in like manner in tropical countries. From the fibres of the bark of Hibiscus arboreus the whips were manufactured with which the negro slaves were lashed in the West India Islands; the plant is called Mohoe or Mohaut. Sida abutila is said to be cultivated in China, as we know Hibiscus cannabinus, or Sun, is in India, as a substitute for hemp. The bark of this plant is called Waak; a coarse kind of oil is expressed from the seeds. Various other species are named as furnishing serviceable fibres. The petals of some are astringent; this property exists in Malva Alcea and in Hibiscus Rosa sinensis, of which the Chinese make use to blacken their eyebrows and the leather of their shoes. The leaves of Althæa rosea are said to yield a blue colouring matter not inferior to indigo. A decoction of the root and stem of Urena lobata is employed in Brazil as a remedy in windy colic; the flowers are used as an expectorant in dry and inveterate coughs. The bark furnishes good cordage. A few species, such as Hibiscus Sabdariffa and suratensis, &c., are slightly acid. The musky seeds of Abelmoschus moschatus are considered cordial and stomachic, and by the Arabians are mixed with coffee. In the West Indies these seeds, called Gumbo musqué, reduced to powder, and steeped in rum, are regarded as a potent remedy against serpent bites. The root of Sida lanceolata is intensely bitter, and is considered a valuable stomachic. The Cotton of commerce is the hairy covering of the seeds of several species of Gossypium. For an excellent account of this plant, and the various species used in commerce, see *Royle's Illustr.* p. 84., and *Wight's Illustr.* The young leaves and seeds of Gossypium vitifolium are employed in Brazil in dysentery, and steeped in vinegar are applied to the head in hemicrania.—*Martius.*

GENERA.

I. MALOPEÆ. — Carpels distinct, crowded into a heap.

Palava, *Cavan.*
Palavia, Mönch.
Malope, *Linn.*
Kitaibelia, *Willd.*

II. MALVEÆ. — Fruit consisting of several cocci. An involucre.

Lavatera, *Linn.*
Axolopha, DC.
Olbia, Medik.
Savinionia, Webb et B.
Navæa, Webb et Bth.
Stegia, Mönch.
Althæa, *Cav.*
Althæastrum, DC.
Ferberia, Scop.
Alcea, Linn.
Malva, *Linn.*
Nuttallia, Dieks.
Callirhoë, Nutt.
Anthema, Medik.
Sphæralcea, *St. Hil.*
Phymosia, Desv.
Sphæroma, DC.
? *Meliphlea*, Zucc.
Modiola, *Mönch.*
Haynea, Reichenb.
Urena, *Linn.*
Pavonia, *Cav.*
Typhalea, DC.
Malache, Trew.
Cancellaria, DC.
Pentaspermum, DC.
Thorntonia, Reichnb.
Columella, Commers.
Malvaviscoides, Endl.
Anotea, DC.
Lopimia, Nees et Mart.
Pentameris, E. M.
Lebretonia, Schrank.
Goethea, *Nees et Mart.*

III. HIBISCEÆ.—Fruit a capsule. An involucre.

Kosteletzkya, *Presl.*
Hibiscus, *Linn.*
Ketmia, Tournef.
Furcaria, DC.
Cremontia, *DC.*
Sabdariffa, *DC.*
Polychlæna, *Don.*
Trionum, *Medik.*
Bombycella, *DC.*
Malvaviscus, *Dill.*
Achania, Sw.
Fugosia, *Juss.*
Cienfugosia, Cav.
Cienfuegia, Willd.
Redoutea, Willd.
Serræa, *Cav.*
Senræa, Willd.
Senra, DC.
Dumreichera, Steud.
Abelmoschus, *Medik.*
Bamia, R. Br.
Hymenocalyx, Zenk.
Manihot, DC.
Lagunaria, *Don.*
Paritium, *Adr. Juss.*
Pariti, Rheed.
Azanza, Moç. et Sess.
Thespesia, *Cav.*
Malvaviscus, Gärtn.
Decaschistia, *Wight et A.*
Gossypium, *Linn.*
Xylon, Tournef.

IV. SIDEÆ.—Fruit consisting of cocci or a capsule. No involucre.

Anoda, *Cav.*
Cristaria, *Cav.*
Sida, Linn.
Napæa, Linn.
Stevartia, Forsk.
Malvinda, Medik.
Periptera, DC.
Dictyocarpus, Wight.
Gaya, *H. B. K.*
Malachra, *Linn.*
Abutilon, *Gärtn.*
Lawrencia, *Hook.*
Bastardia, *Kunth.*
Lagunea, *Cav.*
Solandra, Murr.
Triguera, Cav.
Wissadula, *Medik.*
? Ingenhouzia, *Moç. et Ses.*

NUMBERS. GEN. 37. Sp. 1000.

Geraniaceæ.
POSITION.—Sterculiaceæ.—MALVACEÆ.—Byttneriaceæ.
Chlænaceæ.

ORDER CXXXI. TILIACEÆ.—LINDENBLOOMS.

Tiliaceæ, *Juss. Gen.* 290. (1789) *in part.; Kunth. Malv. Diss. p.* 14. (1822); *DC. Prodr.* 1. 503. (1824); *Lindl. Coll. p.* 54. (1829); *Endl. Gen.* ccxii.; *Meisner, Gen. p.* 36.—Elæocarpeæ, *Juss. Ann. Mus.* 11. 223. (1808); *DC. Prodr.* 1. 519. (1824); *Arnott. Prodr. Penins. Ind.* 1. 81. (1834).—Maquinæ, *Mart.*—Aristoteliaceæ, *Endl.*

DIAGNOSIS.—*Malval Exogens, with free stamens on the outside of a disk, albuminous seeds, and straight embryo.*

Trees or shrubs, very seldom herbaceous plants. Leaves simple, stipulate, toothed, alternate. Flowers axillary, usually perfect. Sepals 4 or 5, distinct or united, with a

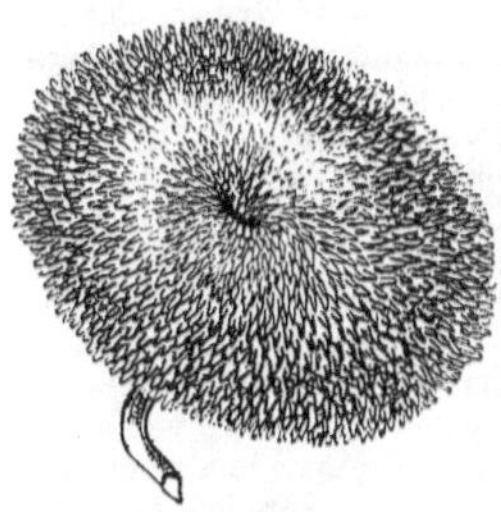

Fig. CCLVII.

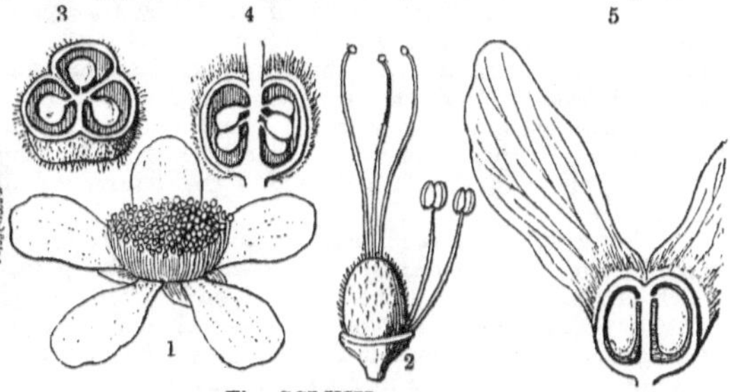

Fig. CCLVIII.

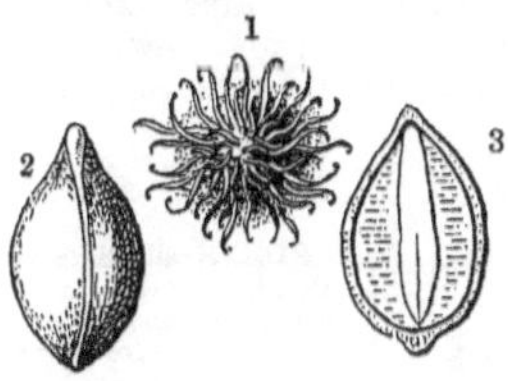

Fig. CCLIX.

valvate æstivation. Petals 4 or 5, entire, usually with a little pit at their base, or wanting; most commonly the size of the sepals. Stamens generally 00, hypogynous, distinct, sometimes surrounded at the base by the lobed and enlarged border of the stalk of the pistil; anthers 2-celled, dehiscing longitudinally or by pores, the outer stamens sometimes abortive and petaloid. Ovary single, composed of from 2 to 10 carpels, which are sometimes disunited; style 1; stigmas as many as the carpels; ovules attached to the inner angle, either free or 00, in two rows, pendulous, horizontal, or ascending, anatropal. Fruit dry or pulpy, often prickly, sometimes winged, with several cells, or one only by abortion. Seeds solitary or numerous; embryo erect in the axis of fleshy albumen, with flat foliaceous cotyledons and a radicle next the hilum.

Although this Order is apparently limited by the character assigned to it in the Diagnosis, yet it includes so many instances of anomalous structure, that some reasonable doubt must be entertained as to its being really so natural an assemblage as it seems to be. The petals are sometimes absent. A genus called Diplophractum is remarkable for having a fruit with several spurious cells, and the placentæ in the circumference instead of the axis. Apeiba has sometimes as many as 24 cells in the fruit. Brown notices the existence of an African genus of this Order (Christiania, *DC.*), remarkable in having a calyx of 3 lobes, while its corolla consists of 5 petals; the fruit composed of 5 single-seeded capsules connected only at the base. The genus Aristotelia, sometimes placed near Homaliads or Philadelphads, seems to have most affinity with this Order, notwithstanding that its calyx is not valvate; it has formed the type of an Order called Maquinæ by Martius, and Aristoteliaceæ by Endlicher. In most respects Lindenblooms resemble Sterculiads, Mallowworts, and the Orders allied to them, more especially in the valvate æstivation of their calyx. They are sufficiently known by their glandular disk and distinct stamens, with 2-celled anthers.

The principal part of the Order is found within the tropics all over the world, forming mean weed-like plants, or shrubs, or trees, with handsome, usually white or pink, flowers. A small number are peculiar to the northern parts of either hemisphere, where they form timber-trees.

Fig. CCLVII.—Fruit of Apeiba aspera.—*Gærtner.*

Fig. CCLVIII.—Berrya ammonilla.—*Wight.* 1. a flower; 2. the ovary and two stamens; 3. a cross section of the ovary; 4. a perpendicular section of it; 5. a portion of its fruit.

Fig. CCLIX.—Triumfetta cordifolia. 1. a fruit; 2. a seed; 3. a section of it.

They have all a mucilaginous, wholesome juice. The leaves of Corchorus olitorius are used in Egypt as a pot-herb. The berries of some of them are succulent and eatable. The species are most remarkable for the toughness of the fibres of their inner bark, which are used for various economical purposes. Fishing lines and nets, rice bags or gunny, and a coarse kind of linen called tat, are made in India of Corchorus capsularis; and the Russian mats of commerce are manufactured from the Tilia. The bark of Luhea grandiflora is used in Brazil for tanning leather. The wood of Luhea divaricata, which is white and light, but very close grained, makes good musket-stocks, and wooden soles for shoes; the Brazilians call all such Açoita cavallos, because the sticks they use for driving their cattle are obtained from them. The flowers of Tilia, separated from the bracts, are used in infusion, according to Host, with much success in vertigo and spasms; they promote perspiration and alleviate coughs; but if the bracts and fruits are mixed with the flowers, the infusion then becomes astringent, and confines the bowels. Some species of Grewia, as G. sapida, asiatica, &c., bear pleasant acid berries, much used in the manufacture of sherbet. The wood of Grewia elastica, called Dhamnoo, affords timber highly valued for its strength and elasticity, and therefore much used for bows, the shafts of carriages, &c. The excellent light timber called Trincomalee wood, employed in the construction of the Massoola boats of Madras, is furnished by Berrya Ammonilla. The berries of Aristotelia Maqui are eatable, and made into wine; the tough bark makes the strings, and the wood the sides of musical instruments.

The leaves of Vallea cordifolia are used for dyeing yellow. The furrowed, sculptured, bony fruit of the Elæocarps, being freed from its pulp, forms handsome necklaces, which are not uncommonly set in gold, and sold in the shops. The name Julpai, or Olive, is applied to the fruit of some species of Elæocarp, which is eaten; while that of others is dried and used in the curries of the natives of India, and is also pickled. Roxburgh did not succeed in extracting any oil from the fruit. Dr. Horsfield says that the bark of one of the Java Elæocarps is bitter and used as an anthelmintic.

The mucilaginous, and at the same time astringent, properties of the leaves and fruit of certain Triumfettas, called Carapixo da Calcada in Brazil, which grow everywhere in that country, especially on the road-side and in the vicinity of dwellings, render them serviceable in injections for inveterate gonorrhœas.

GENERA.

I. TILEÆ.—Corolla 0, or the petals entire. Anthers opening longitudinally.

SLOANIDÆ.

Hasseltia, *H. B. K.*
Ablania, *Aubl.*
Trichocarpus, Schreb.
Dasynema, *Schott.*
Adenobasium, Presl.
Myriochæta, DC.
Foveolaria, DC.
Sloanea, *Linn.*
Sloana, Plum.
Gynostoma, DC.
Oxyandra, DC.

GREWIDÆ.

Vantanea, *Aubl.*
Lemniscia, Schreb.
Apeiba, *Aubl.*
Aubletia, Schreb.
Lühea, *Willd.*
Brotera, Flor. Flum.
Allegria, Moç. et Sess.
Mollia, *Mart et Zucc*
Schlechtendalia, Sprng.
Heliocarpus, *Linn.*
Montia, Houst.
Entelea, *R. Br.*
Sparmannia, *Thunb.*
Clappertonia, *Meisn.*
Honkenya, Willd.
Corchoropsis, *Sieb. et Zuc.*
Corchorus, *Linn.*
Antichorus, Linn. f.
Caricteria, Scop.
Coreta, P. Br.
Mærlensia, DC.
Ganja, Rumph.
Triumfetta, *Plum.*
Lappula, DC.
Bartramia, Gärtn.
? *Porpa*, Blum.
Tilia, *Linn.*
Lindnera, Reichenb.
Brownlowia, *Roxb.*
Humea, Roxb.
Christiana, *DC.*
Grewia, *Juss.*
Nehemia, Endl.
Mallococca, Forst.
Chadara, Forsk.
Siphomeris, Boj.
Microcos, Linn.
Arsis, Lour.
Damine, Endl.
Vincentia, Boj.
Belotia, *A. Rich.*
Diplophractum, *Desf.*
Columbia, *Pers.*
Colona, Cav.
Berrya, *Roxb.*
Espera, Willd.
Muntingia, *Linn.*
Calabura, Plukn.
Trilix, *Linn.*
Jacquinia, Mut.
Bancroftia, *Macfad.*
Aristotelia, *Herit.*

II. ELÆOCARPEÆ.—Petals lacerated. Anthers opening by a transverse valve at the apex.

Elæocarpus, *Linn.*
? *Adenodus*, Lour.
Lochneria, Scop.
Ganitrus, Gärtn.
? *Craspedum*, Lour.
Monocera, *Jack.*
Diceras, Endl.
Dicera, Forst.
Beythea, *Endl.*
Friesia, *DC.*
Acronodia, *Blum.*
Acrozus, Spreng.
Vallea, *Mut.*
Tricuspidaria, *Rz. et Pav.*
Tricuspis, Pers.
Crinodendron, *Molin.*

NUMBERS. GEN. 35. Sp. 350.

POSITION.—Malvaceæ.—TILIACEÆ.————
Tremandraceæ.

ALLIANCE XXIX. *SAPINDALES.*—THE SAPINDAL ALLIANCE.

DIAGNOSIS.—*Hypogynous Exogens, with monodichlamydeous unsymmetrical flowers, axile placentæ, an imbricated calyx and corolla, definite stamens, and little or no albumen.*

In every Order, comprehended under this Alliance, the flowers are more or less unsymmetrical, and in several of them such irregularity occurs in more than one series of the floral organs. Thus, the Poreworts, which are among the most symmetrical, are pentamerous except the ovary, which consists of 2 carpels, Bladder-nuts have similar proportions, while Malpighiads and Erythroxyls combine a 3-merous ovary with a 5-merous flower.

The Orders seem to be all bound up in close relationship, with the exception of Vochyads, which are but little known and whose true station is therefore doubtful. Petiveriads, though generally disunited from Sapindals by a long interval, can hardly be regarded as anything more than an apetalous, very simple form of Soapworts.

The passage into Guttiferals is not through Erythroxyls, which stand last in the following series, but through Soapworts, which are extremely near the Guttiferal Rhizobols.

NATURAL ORDERS OF SAPINDALS.

Flowers complete, partially symmetrical. Calyx valvate. Anthers 2-4-celled, opening by pores	132. TREMANDRACEÆ.
Flowers complete (irregular), unsymmetrical. Petals naked. Anthers 1-celled, opening by pores. Seeds carunculate. .	133. POLYGALACEÆ.
Flowers complete, unsymmetrical, very irregular. Petals naked. Anthers opening longitudinally. Carpels 3. Seeds winged. (In one case the ovary is adherent)	134. VOCHYACEÆ.
Flowers complete, partially symmetrical. Calyx imbricated. Ovules ascending. Stigmas simple. Leaves opposite, with stipules	135. STAPHYLEACEÆ.
Flowers complete, unsymmetrical. Petals usually with an appendage or 0. Anthers opening longitudinally. Carpels 3. Seeds usually arillate, wingless.	136. SAPINDACEÆ.
Flowers apetalous. Carpel solitary	137. PETIVERIACEÆ.
Flowers complete, unsymmetrical. Petals naked or 0. Anthers opening longitudinally. Carpels 2. Seeds without an aril	138. ACERACEÆ.
Flowers complete, partially symmetrical. Calyx imbricated. Petals naked, stalked. Ovules hanging by cords. Stigmas simple. Embryo usually convolute	139. MALPIGHIACEÆ.
Flowers complete, partially symmetrical. Calyx imbricated. Petals with an appendage. Ovules sessile, pendulous. Stigmas capitate. Embryo straight	140. ERYTHROXYLACEÆ.

ORDER CXXXII. TREMANDRACEÆ.—POREWORTS.

Tremandraceæ, *R. Brown in Flinders*, p. 12. (1814); *DC. Prodr.* 1. 343. (1824); *Endl. Gen.* ccxxxii.

DIAGNOSIS.—*Sapindal Exogens, with partially complete symmetrical flowers, a valvate calyx, and 2-4-celled anthers opening by pores.*

Slender heath-like shrubs, with their hairs usually glandular. Leaves alternate or whorled, without stipules, entire or toothed. Pedicels solitary, axillary, 1-flowered. Flowers often large and showy. Sepals 4 or 5, equal, with a valvate æstivation, slightly cohering at the base, and deciduous. Petals equal in number to the sepals, with an involute æstivation, wrapping up the stamens in pairs, much larger than the calyx, and deciduous. Stamens hypogynous, distinct, 2 before each petal, and therefore either 8 or 10; anthers 2- or 4-celled, opening by a pore at the apex. Ovary 2-celled; ovules from 1 to 3 in each cell, anatropal, with a hooked apex, pendulous; styles 1 or 2. Fruit capsular, 2-celled, 2-valved; dehiscence loculicidal. Seeds pendulous, ovate, with a hooked appendage at the apex, but with none about the hilum; embryo cylindrical, straight, in the axis of fleshy albumen, and about half as long, the radicle next the hilum.

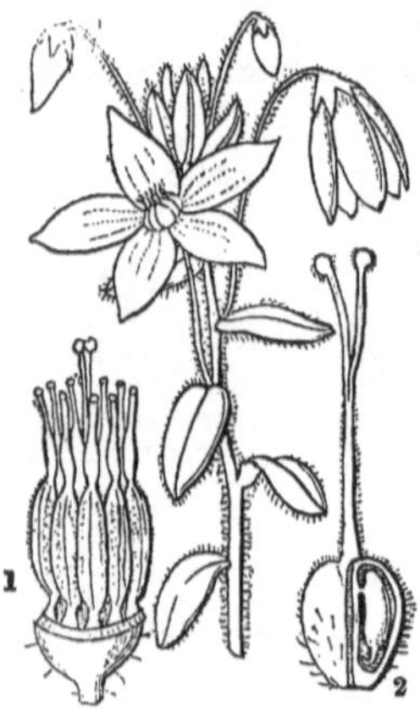

Fig CCLX.

There is little to divide these plants from Milkworts, except their regular symmetrical flowers, and valvate calyx. They want the caruncula of that Order, in room of which they have the chalazal end of the seed extended into a hooked process. Their stamens being opposite the petals in pairs may, taken with the valvate calyx, be regarded as an indication of some tendency towards Rhamnads, and the general condition of their flower is much like that of Pittosporads, except in the great development of the embryo.

All are natives of New Holland.

Their properties are unknown.

De Candolle placed them between Milkworts and Pittosporads, Meisner between Frankeniads and Milkworts, Endlicher in his Polygalinæ consisting of Poreworts and Milkworts only, and Adolphe Brongniart takes the same view of their affinity.

GENERA.

Tetratheca, *Sm.*
Tremandra, *R. Br.*
Platytheca, *Steetz.*

NUMBERS. GEN. 3. SP. 16.

	Tiliaceæ.
POSITION.—————	TREMANDRACEÆ.—Polygalaceæ.
	Pittosporaceæ.

Fig. CCLX.—Tetratheca hirsuta. 1. the stamens; 2. the pistil, with one of the cells laid open.

ORDER CXXXIII. POLYGALACEÆ.—MILKWORTS.

Polygaleæ, *Juss. Ann. Mus.* 14. 386. (1809); *R. Brown in Flinders; Juss. Mem. Mus.* 1. 385.; *DC. Prodr.* 1. 321.; *Aug. de St. Hilaire and Moquin-Tandon Mém. Mus.* 17. 313.; *Wight's Illustr.* 1. 46.; *Endl. Gen.* ccxxxiii.—Krameriaceæ, *Martius, Ed. pr.* 87.—Trigoniaceæ, *Martius Consp.* 247.; *Endl. Gen. p.* 1080.—Moutabeæ, *Endl. Ench. p.* 365.—Soulameæ, *Id. p.* 570.

DIAGNOSIS.—*Sapindal Exogens, with complete (irregular), unsymmetrical flowers, naked petals, 1-celled anthers opening by pores, and caruncùlate seeds.*

Shrubs or herbaceous plants, sometimes twiners. Leaves generally alternate, sometimes opposite, mostly simple, and always destitute of stipules. Flowers usually racemose, often small and inconspicuous, but showy in many species of Polygala.

2 3 4 1

Fig. CCLXI.

Pedicels with 3 bracts. Sepals 5, very irregular, distinct, often glumaceous; 3 exterior, of which 1 is superior and 2 anterior; 2 interior (the wings) usually petaloid, and alternate with the upper and lower ones. Petals hypogynous, usually 3, of which 1 is anterior and larger than the rest (the keel), and 2 alternate with the upper outer, and lateral inner sepals, and often connate with the keel; sometimes 5, and then the 2 additional ones minute and between the wings and the lower sepals. Keel sometimes entire, and then either naked or crested; sometimes 3-lobed, and then destitute of a crest. Stamens hypogynous, 8 usually combined in a tube, unequal, and ascending; sometimes 4, and distinct; the tube split opposite the upper sepal; anthers clavate, innate, mostly 1-celled and opening at their apex, sometimes 2-celled. Disk either absent or present, regular or irregular. Ovary superior, compressed, with 2 or 3 cells, which are anterior and posterior, the upper one occasionally suppressed; ovules solitary, very rarely twin, pendulous, anatropal; style simple, curved, sometimes very oblique and cucullate at the apex, which is also entire or lobed; stigma simple. Fruit usually opening through the valves; occasionally indehiscent, membranous, fleshy, coriaceous, or drupaceous, winged, or apterous. Seeds pendulous, with a caruncula next the hilum, naked or enveloped with hairs; the outer integument crustaceous, the inner membranous; albumen abundant, fleshy, rarely reduced to a thin gelatinous plate; embryo straight, or slightly curved, with the radicle next the hilum.

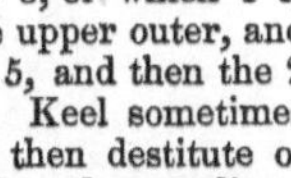

Fig. CCLXII.

The structure of this Order has been explained by Aug. de St. Hilaire and Moquin-Tandon, from whose Memoir, above quoted, the foregoing character is extracted.

Fig. CCLXI.—Polygala erioptera. 1. an entire flower seen from the side; 2. the same cut open to exhibit the stamens; 3. the pistil; 4. a section of a ripe seed; in the middle is the embryo; at the apex, which represents the real base, is seen a caruncula.

Fig. CCLXII.—Anthers of Polygala vulgaris, expanded.

Milkworts are remarkable, among other things, for the irregularity of their flowers, which is such as to obscure, in a great measure, the relative position of the sepals and petals. The calyx apparently consists of but three pieces, which are usually green, and like sepals in their common state ; but their real number is 5, the two coloured lateral petal-like bodies sometimes lying within the apparent sepals, being in reality part of the series of the calyx. The corolla is mostly monopetalous, and, if carefully examined, formed of 3 pieces ; namely, the keel and two petals, all blended together. We have, therefore, an abortion of two petals, according to the laws of alternation. But this is not all ; there is not only an abortion of two petals, but of those two which would, if present, be found right and left of the keel. The monopetalous corolla is, therefore, formed by the cohesion of the two posterior and the anterior petal of a pentapetalous corolla, of which the two lateral petals are suppressed. The keel has an appendage of an anomalous character, called technically a crest, and often consisting of one or even two rows of fringes or divisions, originating not from the margin, but from within it, and sometimes cohering in a common membrane at their base. Aug. de St. Hilaire has shown that this crest is nothing more than the deeply-lobed middle segment of a keel, with these lobes in such a state of cohesion that the central lobe is pushed outwards, while the lateral ones cohere by their own margins and with its back. The stamens are only 8 : two therefore are suppressed. This relative position of the fifth sepal and petal respectively, was first indicated by Brown.

Milkworts are stationed by De Candolle between Sundews and Poreworts (Tremandraceæ), and in the immediate vicinity of Violetworts. With the latter they were thought to be related on account of their hypogynous stamens, irregular flowers, and cucullate stigma ; and with Poreworts on account of the caruncula of their seed. To Fumeworts they approach in the general aspect of their flowers, and in little more. Leguminous plants are, notwithstanding their perigynous stamens, an Order with which Milkworts seem at first sight to have some affinity ; the irregularity of corolla is of a similar nature in both ; there is in Leguminous plants a tendency to suppress the upper lateral petals in Erythrina, as in Polygala, and the ascending direction of the style with a cohesion of stamens are characters common to both Orders. Many additional observations are made to the same effect by St. Hilaire and Moquin-Tandon, who, moreover, compare the Order with Rueworts ; but those authors appear to have finally decided upon the true position of Milkworts being in the vicinity of Soapworts ; remarking that "the calyx of the latter is unequal, the corolla very irregular, and the ovary of Schmidelia usually 2-celled and 2-seeded, like that of Polygala. Moreover, a great part of the genera of that Order have, with a calyx of five divisions, a corolla with four petals, and the place of the fifth is manifestly vacant. This suppression is not exactly the same as what is observed in the corolla of Polygala, where there are only 3 petals with 5 sepals ; but the suppression has more analogy with what concerns the stamens, since with a quinary number in the calyx each Order has eight antheriferous filaments." In this view I fully agree. The unsymmetrical flowers, more especially manifested in the reduction of the number of carpels to 2 or 3 in a structure otherwise quinary, the definite ovules, the twining habit of Comesperma, the samaroid fruit of Securidaca, and it may even be added, the deleterious qualities of some Polygalas, together with the saponaceous secretions of the Monninas, are all arguments of the strongest kind in favour of Milkworts and Soapworts belonging to the same Alliance.

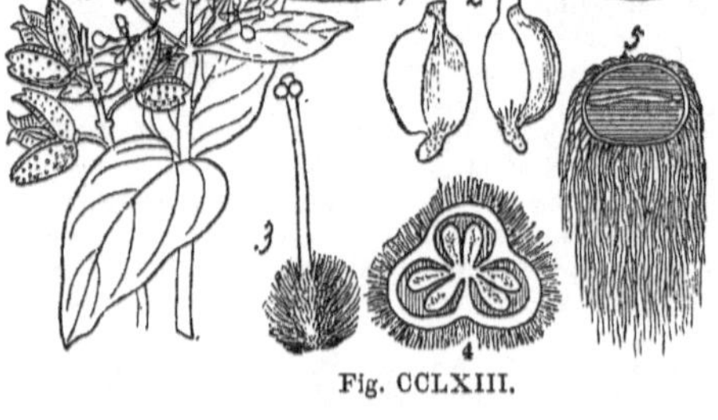

Fig. CCLXIII.

Certain anomalous genera, belonging as I think to this Order, have been elevated to the rank of Natural Orders. Of these Trigonia, a genus of tropical American trees, has been divided from Milkworts because of its leaves being opposite and having sti-

Fig. CCLXIII.—Trigonia crotonoides.—*A. de St. Hilaire.* 1. a flower seen from the side; 2. the corolla, with the petals displayed in their natural position; 3. a pistil; 4. cross section of the ovary; 5. cross section of a seed.

pules, of some supposed difference in the relative position of the largest petal, the anthers opening longitudinally, and the presence of some fleshy glands between the ovary and stamens; I cannot, however, concede anything like ordinal importance to these circumstances. Trigonia may be regarded either as an approach on the part of Milkworts to the Sapindaceous structure, as is indicated by the longitudinal dehiscence of its anthers, the greater symmetry of its flowers, and its 3-valved fruit; or as actually a member of the Soapworts, approaching Milkworts. The supposed relation between it and Spindle trees or Leguminous plants, which M. Cambessédes suggests, appears to be a very slight indication of analogy.—*See Fl. Bras. Mer. v.* 2. *p.* 112.

Krameria has much higher claims to separation. The whorls of its flowers are so completely dislocated that it is difficult to determine the relative position of the parts; there is no trace of the quasipapilionaceous structure generally characteristic of Milkworts, its ovary is imperfectly 2-celled, and it is said that no albumen exists in its seeds. Certainly these are points of moment. Nevertheless, its definite hypogynous stamens, porous anthers, unsymmetrical flowers, definite pendulous ovules, bur-like fruit, which resembles that of Salomonia, and in some degree its habit, are conformable to the Milkwort structure; and in the absence of all trace of the existence of other genera approaching this kind of organisation, it seems expedient to regard it as a mere exception to the usual structure of an Order whose general condition is in many respects very anomalous. It, too, may be regarded as assisting to bring into contact the Milkworts and Soapworts, for Krameria cytisoides has ternate leaves.

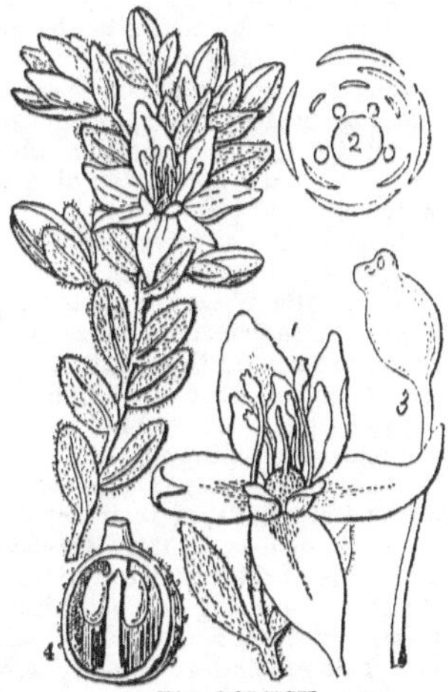

Fig. CCLXIV.

Soulamea is another instance of the elevation of a solitary genus into a Natural Order. This is a Molucca plant, also without albumen in its seeds, and having a regular trimerous flower with 2-celled anthers. It may perhaps be considered as an instance of the usual irregular flower of Milkworts assuming a regular type.

Lastly, of the genus Moutabea, promoted by Endlicher, who stations it near Storaxworts, regarding it as monopetalous, it may be said with tolerable confidence that it has not a single feature that can justify its separation from Milkworts. Like Xanthophyllum its petals are equal-sized, and as for their adhesion into a tube, that is no more than what occurs in all the Polygalas, whose stamens hold together parts which under ordinary circumstances are distinct. So entirely, indeed, does Moutabea agree with Polygala, that it even has its eight 1-celled anthers opening at the point, in combination with a 5-petalled corolla. The berries of Moutabea longifolia are said to be eatable; so are those of Mundia spinosa. A tubular calyx exists in Moutabea, but that will hardly be insisted upon as a ground for forming it into a Natural Order.

Most of the genera are limited to one or two of the five parts of the globe; thus Salomonia is only found in Asia, Soulamea in the Moluccas, Muraltia at the Cape of Good Hope, and Monnina and Badiera in South America. Comesperma is found both in Brazil and New Holland, and, what is very remarkable, there is in the former country a species of the Cape genus Mundia. Polygala itself occurs in four of the five parts; under the torrid zone and in temperate climates, at Cayenne, and on the mountains of Switzerland; it is, however, very unequally distributed. This genus inhabits almost every description of station—dry plains, deep morasses, woods, mountains, cultivated and barren soils. Comesperma is only known in Brazil and Australia. Monnina and Krameria inhabit open places in the temperate parts of South America.

Milkworts offer, as has been stated, considerable diversities of structure, and therefore, as might have been anticipated, the purposes to which they are applicable are by no means uniform. The greater part are bitter, and in their roots milky. To this category may be referred the following cases. Polygala amara is a European perennial, all the parts of which are extremely bitter; it is much extolled in pulmonary complaints and spitting of blood. P. vulgaris and major have similar uses, but are inferior in energy. A strong bitter taste pervades all the parts of Polygala rubella, a North American

---

Fig. CCLXIV.—Krameria cistoidea.—*Hooker*. 1. an expanded flower; 2. a diagram, showing the relative position of the parts; 3. a stamen; 4. a perpendicular section of an ovary.

species; in small doses its infusion is found useful as a tonic and stimulant to the digestive organs; in large doses it opens the body and excites diaphoresis. In a Molucca plant, the Soulamea amara, called by Rumphius Rex amaroris, all the parts, especially the roots and fruit, have an intense bitterness (horrenda amarities *Rumf.*). They are employed in the Malayan Archipelago with extraordinary success in cholera and pleurisy, and are regarded as being most valuable as a febrifuge.

Others are distinguished for their emetic, purgative, and diuretic action. Of these the most celebrated is a North American herb called Snake-root, Polygala senega; of this plant the root is somewhat acid and acrid. It acts as a sudorific and expectorant in small doses, and as an emetic and cathartic in large ones. It is employed in pneumonia, asthma, croup, dropsy, chronic rheumatism, and especially in such uterine complaints as amenorrhœa. Dr. Archer has extravagantly praised it in cynanche trachealis. Chemists refer the action to the presence of a peculiar principle called Polygaline or Senegine. P. sanguinea and purpurea in North America, Chamæbuxus in Europe, P. paniculata, a very common West Indian annual, P. serpentaria of the Cape, P. crotalarioides in the Himalayas, all appear to participate in these qualities, and it is not a little remarkable that the whole of these plants have the reputation of being antidotes to snake bites; the oppression of breathing observable in such cases appears to be certainly relieved by them. Some are mere emetics; such as P. poaya, used successfully in Brazil in bilious fevers, and P. glandulosa and scoparia, Mexican species. P. thesioides, called Chinchin in Chili, is said to have a powerfully diuretic root. Badiera diversifolia, a little West Indian bush, is said to rival Guaiacum in its peculiar qualities. Finally, these principles become so concentrated in P. venenosa, called Katu-tutun in Java, as to render that plant a poison; it is dreaded by the natives, who say that its heavy noxious odour, or even the touch, produces violent sneezing and severe headache.

Among plants whose uses are not reducible to either of the foregoing heads may be mentioned the following. The drupes of Mundia spinosa, a Cape shrub, are eatable. The bark of the root of Monnina polystachya and salicifolia, when fresh, pounded and moulded into balls, or their dry bark, is detergent; it readily froths when agitated in water, and is used by the Peruvians as a substitute for soap; the ladies of Peru ascribe the beauty of their hair to the use of its infusion, and the silversmiths of Huanuco employ it for cleansing and polishing wrought silver. It is also used with great success in the cure of dysenteries and irritating diarrhœas in Peru, where it is preferred to Quassia. This saponaceous quality is, among other things, an indication of the relation borne by Milkworts to Soapworts (Sapindaceæ). P. tinctoria is used by the dyers in Arabia. The wood of Xanthophyllum, a genus of trees of considerable size, is said to be valuable. The Kramerias, anomalous plants inhabiting the temperate parts of South America, and called Rhatany-roots, are intensely astringent. The infusion of their roots is blood-red, and is employed to adulterate Port wine; in Peru, an extract is formed from K. triandra, which is a mild, easily assimilated, astringent medicine, possessed of great power in passive bloody or mucous discharges; and also in weakness of the digestive organs, muscular debility, and even in intermittent and putrid fevers. The powder forms, along with charcoal, an excellent tooth powder; and an infusion is used as a gargle and wash. Such other species as have been examined seem to be identical in their nature.

GENERA.

Salomonia, *Lour.*
Polygala, *Linn.*
  Psychanthus, Raf.
  Blepharidium, DC.
  Clinclinia, Feuill.
  Timutua, DC.
  Senega, DC.
  Chamæbuxus, Dill.
  Triclisperma, Raf.
Brachytropis, *DC.*
Badiera, *DC.*
  Penæa, Plum.
Comesperma, *Labill.*
Catocoma, *Bth.*
Muraltia, *Neck.*
  Heisteria, Berg.
Mundia, *Kunth.*
  Nylandtia, Dumort.
  Vascoa, DC.
Monnina, *Ruiz et Pav.*
  Hebeandra, Bonpl.
Carpolobia, *G. Don.*
Lophostylis, *Hochst.*
Securidaca, *Linn.*
Krameria, *Löffl.*
Xanthophyllum, *Roxb.*
  Jackia, Blum.
Soulamea, *Lam.*
  Cardiocarpus, Reinw.
Trigonia, *Aubl.*
  Mainea, Fl. Flum.
Moutabea, *Aubl.*
  Cryptostomum, Schreb.
  Acosta, Ruiz et Pav.
? Bredemeyera, *Willd.*
? Hymenanthera, *R. Br.*

NUMBERS. GEN. 19. Sp. 495.

POSITION.—Tremandraceæ.—POLYGALACEÆ.—Sapindaceæ.
Violaceæ?

ORDER CXXXIV. VOCHYACEÆ.—VOCHYADS.

Vochysiaceæ, *Mart. Nov. Gen.* 1. 123. (1824) ; *Endl. Gen.* cclx. ; *Meisner*, 119.—Vochysieæ, *A. St. Hil. Mém. Mus.* 6. 265. (1820) ; *DC. Prodr.* 3. 25. (1828).

DIAGNOSIS.—*Sapindal Exogens, with complete, unsymmetrical, very irregular flowers, naked petals, anthers opening longitudinally, 3 carpels, and winged seeds.*

Trees or shrubs. Branches opposite, when young 4-cornered. Leaves opposite, sometimes towards the extremities of the branches alternate, entire, with glands or 2 stipules at the base. Flowers usually in terminal panicles or racemes. Sepals 4-5, combined at the base, very unequal, the two outer the smallest, the two in front the largest, imbricated in æstivation, the upper one much the largest and spurred. Petals 1, 2, 3 or 5, alternate with the segments of the calyx, and inserted into their base, unequal. Stamens 1-5, usually opposite the petals, rarely alternate with them, arising from the bottom of the calyx, for the most part sterile, 1 of them having an ovate fertile 4-celled anther. Ovary free, or partially adherent, 3-celled ; ovules in each cell, solitary or twin, or 00, attached to the axis, amphitropal, with the foramen uppermost ; occasionally 1-celled, with 2 anatropal ovules rising from the base ; style and stigma 1. Capsule 3-cornered, 3-celled, 3-valved, the valves bursting along their middle, with a central columella ; occasionally indehiscent, 1-celled, 1-seeded, and crowned by the sepals grown out into wings. Seed without albumen, erect, usually winged ; embryo straight in the capsular genera, with large leafy cotyledons and a short superior radicle; in the monospermous fruit, orthotropal, cylindrical, with semi-cylindrical cotyledons, and a short inferior radicle.

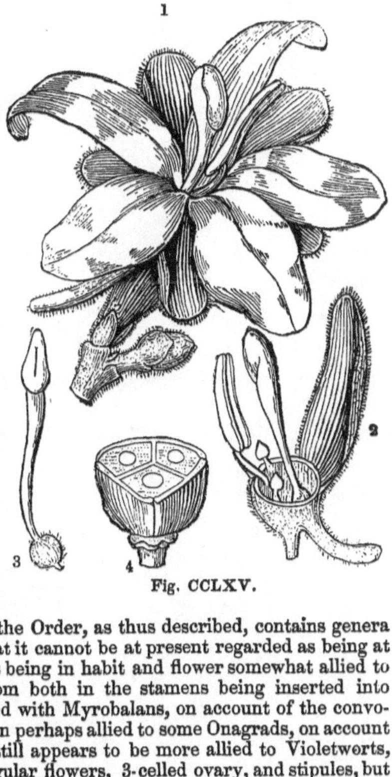

Fig. CCLXV.

Such is the character that Botanists give to a most curious race of trees and shrubs, which few have had the opportunity of studying, but which are remarkable for the beauty of their large and gaily coloured flowers. It seems, however, certain that the Order, as thus described, contains genera which must hereafter be separated, and that it cannot be at present regarded as being at all well limited. De Candolle speaks of it as being in habit and flower somewhat allied to Guttifers or Margraviads, but distinct from both in the stamens being inserted into the calyx ; perhaps more directly connected with Myrobalans, on account of the convolute cotyledons and inverted seeds ; and even perhaps allied to some Onagrads, on account of the abortive solitary stamen. To me it still appears to be more allied to Violetworts, an affinity strongly pointed out by the irregular flowers, 3-celled ovary, and stipules, but also to be yet nearer Milkworts, from which the calcarate flowers and ascending ovules principally distinguish it. The main difficulty in associating it with any Alliance to which these Orders belong, consists in the stamens being truly perigynous. But there is no perigynous Alliance to which it seems referable, and the peculiar proportion of the 3-celled ovary to the 5-parted calyx and corolla, strongly indicates the true affinity to be with the Sapindal Alliance.

Natives of equinoctial America, where they inhabit ancient forests, by the banks of streams, sometimes rising up mountains to a considerable elevation. They are often trees with large spreading heads.

Fig. CCLXV.—Salvertia convallariodora.—*St. Hilaire.* 1. an expanded flower ; 2. a portion of the calyx, with the stamens ; 3. a pistil ; 4. a transverse section of the ovary.

Little is known of any use to which they can be applied. Their flowers are reputed to be very sweet, and some are said to have a resinous juice. The Itaballi, or Copai yé wood of Guiana, a hard but not very durable timber, is obtained from Vochya guianensis, according to Schomburgk.

GENERA.

Callisthene, *Mart. et Zuc.*
Callisthenia, Spreng.
Amphilochia, *Mart.*
Agardhia, Spreng.
Qualea, *Aubl.*

Schüchia, *Endl.*
Vochysia, *Juss.*
Vochy, Aubl.
Vochya, Vandell.

Salmonia, Neck.
Cucullaria, Schreb.
Struckeria, Fl. Fl.
Salvertia, *St. Hil.*

? Erisma, *Rudge.*
Debræa, Röm. et Schlt.
Dittmaria, Spreng.
? Lozania, *Seb. Mut.*

NUMBERS. GEN. 8. Sp. 51.

Violaceæ.
POSITION.—Polygalaceæ.—VOCHYACEÆ.—Sapindaceæ.

ORDER CXXXV. STAPHYLEACEÆ.—BLADDER-NUTS

Celastrineæ, § Staphyleaceæ, *DC. Prodr.* 2. 2. (1825).—Staphyleaceæ, *Lindl. Synopsis*, 75. (1829); *Endl. Gen.* ccxxxv.

DIAGNOSIS.—***Sapindal Exogens, with partially complete, symmetrical flowers, an imbricated calyx, ascending ovules, simple stigmas, opposite leaves, with stipules.***

Shrubs. Leaves opposite (rarely alternate), pinnate, with both common and partial deciduous stipules. Flowers in terminal, stalked racemes, sometimes ♀-⚥-♂. Sepals 5, connected at the base, coloured, with an imbricated æstivation. Petals 5, alternate, with an imbricated æstivation, inserted in or around a free crenated saucer-shaped disk. Stamens 5, alternate with the petals, perigynous. Ovary 2- or 3-celled, free, with the carpels more or less distinct; ovules several, horizontal or ascending, anatropal; styles 2 or 3, cohering at the base. Fruit membranous or fleshy, indehiscent or opening internally, often deformed by the abortion of some of the parts. Seeds ascending, roundish, with a bony testa; hilum large, truncate; albumen little or none; cotyledons thick; radicle short, next the hilum.

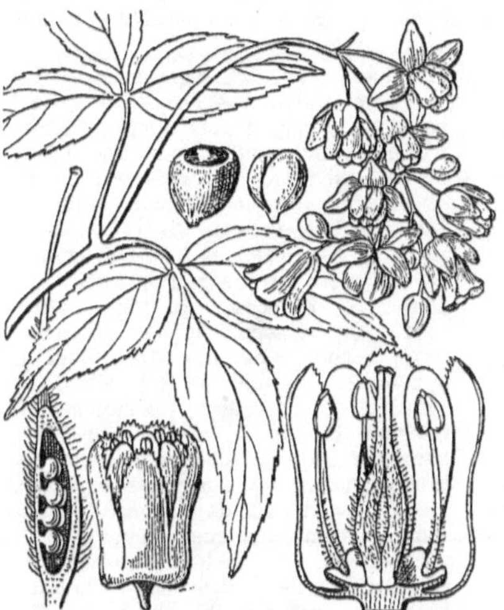

3 1 Fig. CCLXVI. 2

Combined with Spindle-trees by De Candolle, but distinguished by Ad. Brongniart (*Mém. sur les Rhamnées*, p. 16), this Order appears to be essentially characterised by its opposite pinnated stipulate leaves, and to be far more closely allied to Soapworts, from which it is distinguished by the number of its sepals, petals, and stamens being alike.

The very few species which belong here are irregularly scattered over the face of the globe. Of the genus Staphylea, 1 is found in Europe, 1 in North America, 1 in Japan, 2 in Jamaica, 1 in Peru; and of Turpinia, 1 is Mexican, and 1 East Indian.

Very little is known of their uses. The Bladder-nuts are handsome trees of small size; their seeds are oily, rather austere, and slightly purgative. The inner bark of the root of Euscaphis staphyleoides, a Japan plant, is bitter and astringent, and is used in dysentery and chronic diarrhœa, according to Siebold.

GENERA.

Turpinia, *Vent.*	Staphylea, *Linn.*
Dalrympelea, Roxb.	*Stayhylodendron*, Tournef.
Euscaphis, *Sieb. et Zucc.*	*Bumalda*, Thunb.

NUMBERS. GEN. 3. Sp. 14.

POSITION. ——— STAPHYLEACEÆ.—Sapindaceæ.
Celastraceæ.

Fig. CCLXVI.—Staphylea Bumalda.—*Delessert.* 1. a flower; 2. a perpendicular section of it; 3. a section of its ovary.

ORDER CXXXVI. SAPINDACEÆ.—SOAPWORTS.

Sapindi, *Juss. Gen.* 246. (1789).—Sapindaceæ, *Juss. Ann. Mus.* 18. 476. (1811); *DC. Prodr.* 1. 601. (1824); *Cambessèdes in Mém. Mus.* 18. 1. (1829); *Endl. Gen.* ccxxx.; *Wight Illustr.* 1. *p.* 141.—Æsculaceæ, *Ed. pr.* lxii.—Hippocastaneæ, *DC. Théorie, Ed.* 2. 244. (1819); *Prodr.* 1. 597. (1824); *Endl. Gen. p.* 1075.—Castaneaceæ, *Link Enum.* 1. 354. (1821).—Millingtonieæ, *Jack in Malay. Misc.* 2. 32; *Hooker Journal*, 377.—Millingtoniaceæ, *Wight and Arnott in Ed. Ph. Journ.* 15. 177. (1833); *Prodr. Penins.* 115. (1834); *Royle Illustr. p.* 139. (1835); *Wight Illustr.* 1. *t.* 53.—Meliosmeæ, *Endl. Gen. p.* 1074.

DIAGNOSIS.—*Sapindal Exogens, with complete, unsymmetrical flowers, petals usually with an appendage, anthers opening longitudinally, 3 carpels, and usually arillate, wingless seeds.*

These are for the most part trees of considerable size, or twining shrubs bearing tendrils, or, though seldom, climbing herbs. Their timber has frequently several distinct axes of growth. Leaves alternate, compound, very rarely simple, with or without stipules, often marked with lines or pellucid dots. Flowers in racemes, or racemose panicles, small, white or pink, seldom yellow, ♂-⚥-♀. Calyx more or less deeply 4-5-parted, or 4-5-leaved, with an imbricated æstivation. Petals 4-5, or occasionally absent, alternate with the sepals, hypogynous, sometimes naked, sometimes with a doubled appendage in the inside; æstivation imbricated. Disk fleshy; sometimes occupying the base of the calyx, regular, nearly entire, expanded between the petals and stamens; sometimes glandular, incomplete, the glands stationed between the petals and stamens. Stamens 8-10, rarely 5-6-7, very seldom 20, sometimes inserted into the disk, sometimes into the receptacle between the glands and the pistil; filaments free or combined just at the base; anthers turned inwards, bursting longitudinally. In the ♂ there is a very small rudiment of a pistil, or none. Ovary 3-celled, rarely 2-4-celled, the cells containing 1, 2, 3, very seldom more, ovules. Style undivided, or more or less deeply 2- or 3-cleft. Ovules anatropal, sessile when solitary, erect, or ascending, rarely suspended; when double, the upper ascending, the lower suspended. Fruit sometimes capsular, 2-3-valved, sometimes extended at the back into a wing and becoming a key (samara), sometimes fleshy and indehiscent. Seeds usually with an aril; the outer integument crustaceous or membranous, the interior pellucid. Albumen 0. Embryo seldom straight, usually curved, or spirally twisted. Radicle next the hilum. Cotyledons incumbent, sometimes combined into a thick mass.

2

1 3 4

Fig. CCLXVII.

This Order is composed of a great diversity of species, which assume appearances widely different from each other; so that Botanists have not unnaturally supposed that it really contains the elements of several distinct Natural Orders. Thus the Horse-chesnuts have been separated because of their opposite leaves, and a singular peculiarity of the ovules, which are both erect and suspended in the same cell; and Meliosmeæ

Fig. CCLXVII.—Sapindus senegalensis. 1. an expanded flower; 2. a petal; 3. the ovaries before fertilisation; 4. a vertical section of a ripe drupe, showing the embryo.

have been set apart because of their fruit being a drupe, their ovules all suspended, and their stamens reduced to two only in a fertile condition. There does not, however, appear to be in these cases such differences from the true Soapworts as can stamp the supposed Orders with authority; and, as might have been expected, the progress of discovery does not sanction the separation by adding new members to such groups. The true character of Soapworts resides in their unsymmetrical flowers, (the stamens never agreeing in number or power with the sepals,) in their anthers bursting longitudinally, and in the petals having an appendage, while the seeds have an aril and the embryo is curved or spiral. But none of the latter characters are constant, and consequently the definition of the Order becomes very difficult. From Maples they hardly differ. At least, the characters usually pointed out as distinguishing them are fallacious in practice. The opposite leaves of Maples are found in Æsculus and others, and that genus has not appendages on its petals more than Acer itself, and a whole race of the Soapworts has samaroid fruit, which is the more obvious mark of the Order of Maples. To Milkworts they are no doubt akin in the singular combination of 8 stamens with 5 unequal sepals, and an uncertain number of petals; and also in their aril, which may be compared to the caruncula of Milkworts, although somewhat different in its origin. The dried leaves resemble, as De Candolle remarks, those of Connarads. Their climbing habit and tendency to produce tendrils indicate a relation to Vines, which, however, is not very near. Malpighiads are known with certainty by their symmetrical flowers, although they too have the "keys" or samaræ that are so common among Soapworts. Petiveriads are certainly very near this Order; but, in addition to their constant want of petals, their carpel is always solitary, and absolutely simple.

A very general character of the Soapworts is to have their embryo either curved, or twisted spirally. This occurs in a remarkable manner in the nut of a Demerara tree, called the Snake-nut, in consequence of the large embryo resembling a snake coiled up. Sir R. Schomburgk, who first described this production in the *Annals of Natural History*, vol. 5. p. 204, has called the tree Ophiocaryon paradoxum. The accompanying figure represents it in a germinating condition. Another peculiarity resides in the trunk of such as have a climbing habit. These remarkable plants possess several distinct woody axes, held together by masses of cortical matter, so that they resemble several thick-barked stems, forced together with violence. Instances of this structure have been figured by Gaudichaud, at Plate XIII. of his *Recherches sur l'Organographie.*

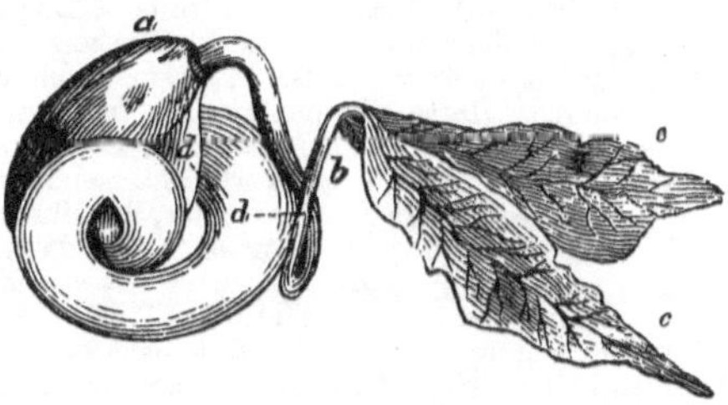

Fig. CCLXVIII.

Natives of most parts of the tropics, but especially of South America and India. Africa knows many of them, but they are wanting in the cold regions of the north. None are found wild in Europe. Dodonæa represents the Order in New Holland; Horse Chesnuts in the north of India, Persia, and the United States.

It is singular that while the leaves and branches of many of these plants are unquestionably poisonous, the fruit of others is valuable as an article of the dessert. Thus the Longan, the Litchi, and the Rambutan, fruits among the more delicious of the Indian archipelago, are the produce of different species of Nephelium. Pierardia sativa and dulcis, to which belong the Rambeh and Choopa of Malacca, and Hedycarpus malayanus producing the Tampui, are other fruit trees of the Order. The fruit of Schmidelia edulis is known at desserts in Brazil, under the name of Fruta de paraô; it is said to have a sweet and pleasant taste. Various species of Sapindus are mentioned as fruit trees. The blacks of Senegal highly value the berries of S. senegalensis; the fruit of S. esculentus is very fleshy, and much esteemed by the inhabitants of Certaô, by whom it is called Pittomba. Melicocca bijuga, a West Indian tree, is now cultivated in Brazil for its agreeable subacid vinous berries. The fruit of Pappea capensis is called Wild Prunes at the Cape of Good Hope; its seeds abound in oil. The succulent aril of the Akee tree (Blighia or Cupania sapida), of Paullinia subrotunda, and Schleichera trijuga, are also articles of food in their respective countries.

Nevertheless, these fruits belong to a race eminently dangerous; and, as in other

Fig. CCLXVIII.—Germinating seed of Ophiocaryon paradoxum. *a.* radicle; *b.* cauliculus; *cc.* cotyledons, which, when at rest and unexpanded, are folded down upon *d.*

cases, appear to be parts in which the deadly juices of the branches and leaves are too much diffused among watery matter to be dangerous. For example, although the fruit of Sapindus senegalensis is eaten, its seeds are known to be poisonous; those of the eatable Nephelia are so bitter as to excite suspicion as to their nature; and it is asserted that both the fruit and leaves of the Buck-eye, or American Horse Chesnut, Æsculus ohiotensis, are a mortal poison, both to man and animals. In no part of this Order is the narcotic quality more developed than in the genus Paullinia. Of all the species, P. pinnata is supposed to be the worst; bark, leaves, and fruit abound in an acrid principle, and the Brazilian blacks prepare from them an insidious poison, which slowly but certainly destroys life. Martius suggests that the nature of this poison should be inquired into, and experiments made as to whether it may not be advantageously administered in hydrophobia aud insanity. A venom for their arrows is prepared by the savages of Guiana from Paullinia Cururu; P. australis and Serjania lethalis are together supposed to furnish the Lecheguana honey, which has been found a most dangerous food. (See *Edinb. Ph. Journ.* 14. 269, and *Plantes Rémarquables*, p. 192.) From P. Cupana an inebriating drink is prepared on the banks of the Oronoco. The leaves of Magonia pubescens and glabrata, called Tinguy in Brazil, are used for stupefying fishes; their bark is employed for healing sores in horses, caused by the stings of insects. Serjania triternata is also employed as a fish poison. The roots of the American Horse Chesnut are held to be poisonous.

Some are used in medicine as astringents. The root of Schmidelia serrata is employed in India to stop diarrhœa. The bark of Schleichera trijuga is rubbed up with oil in the same country to cure the itch. The bark of the Horse Chesnut, Æsculus Hippocastanum, has been recommended as a valuable febrifuge in intermittent and other fevers; a decoction has been recommended in gangrene, and its powder as an errhine. Its young leaves are aromatic, and have been used instead of Hops in brewing beer, according to Endlicher. The fruits of Blighia (or Cupania) sapida, boiled down with sugar and cinnamon, are used in diarrhœa.

A saponaceous principle exists in a remarkable degree in certain species. The seeds of the common Horse Chesnut are not free from it. The acrid fruits of Sapindus saponaria, inæqualis, and others, lather freely in water, and are used in the West Indies instead of soap; "a few of them will cleanse more linen than 60 times their weight of soap." Pounded and thrown into water, they intoxicate fish. A tincture of the berries has been recommended in chlorosis. The distilled water of the flowers of Blighia sapida is regarded by negro women as a cosmetic, probably owing to the presence of the saponaceous matter just alluded to.

Notwithstanding these qualities, a food called Guarana bread is prepared by the Brazilian savages from the seeds of Paullinia sorbilis. Martius, who has investigated the nature of this substance, says that oblong or round cakes of it are sold all over Brazil as an indispensable requisite for travellers, and a cure for many disorders. His brother Theodore found them to be composed chemically of an astringent matter, forming a green precipitate with iron, resin, fat oil of a green colour, gum, starch, vegetable fibre, and a white crystalline bitter substance, which he called Guaranene, and which appears to be identical with Theine and Caffeine. The Brazilians pound this bread in water, sweeten it, and esteem it as a stomachic, febrifuge, and aphrodisiac. Martius regards it as a substance of considerable activity ("nobile remedium"), and adds, "Appetitum venereum movet, spermatis vero fœcunditatem diminuere dicitur."

In addition to the uses already indicated, Soapworts present occasionally other qualities. The root of Cardiospermum Halicacabum is diaphoretic, diuretic, and aperient. Its leaves are cooked as a vegetable in the Moluccas. The seeds of the Horse Chesnut are an excellent sheep-food,* and have been recommended as a good substitute for Coffee. The Dodonæas are somewhat aromatic; the leaves of D. viscosa are used in baths and fomentations; the wood of D. dioica is carminative; D. Thunbergiana is said to be slightly purgative and febrifugal. The branches of Plösslea floribunda, a Cape plant, are covered with a gummy exudation. The timber of some of the South African trees of the Order appears to be valuable. That of Pteroxylon utile is said to be as hard and handsome as Mahogany; its sawdust makes the workmen sneeze, wherefore they call

* Whilst I was at Geneva in the autumn of 1837, I observed every one collecting carefully the fruit of the Horse Chesnut, and on inquiry I learnt that the butchers and holders of grazing-stock bought it readily at a certain price per bushel. I inquired of my butcher, who himself kept a very extensive grazing farm, and he told me it was given to those sheep in particular that were fattening. The Horse Chesnuts were well crushed; something in the way, so I understood, that Apples are, previous to cider being made. They are crushed or cut up in a machine kept solely, in Switzerland, for that purpose; then about two pounds' weight is given to each sheep morning and evening. Sheep eat the food greedily; it must be portioned out to them, as too much would disagree with them, it being of a very heating nature. The butcher told me that it gave an excel'ent rich flavour to the meat. The Geneva mutton is noted for being as highly flavoured as any in England or Wales.—*Gardeners' Chronicle*, 1843, p. 737.

it Nieshout; it is found to burn rapidly, though green, and is used by the Hottentots for lighting their fires. Hippobroma alatum, commonly called Pardepis, is extensively employed for timber at the Cape of Good Hope.

GENERA.

I. SAPINDEÆ. — Leaves alternate. Ovules generally solitary. Embryo curved, or occasionally straight.

Cardiospermum, *Linn.*
Corindum, Tournef.
Erythrophila, *E. M.*
Urvillea, *H. B. K.*
Serjania, *Plum.*
Seriana, Schumach.
Toulicia, *Aubl.*
Ponæa, Schreb.
Bridgesia, *Bert.*
Tripterocarpus, Meisn.
Paullinia, *Linn.*
Semarillaria, R. et Pav.
Cururu, Plum.
Enourea, *Aubl.*
Natalia, *Hochst.*
Schmidelia, *Linn.*
Allophyllus, Linn.
Ornitrophe, Juss.
Toxicodendron, Gärtn.
Azamaza, Hochst.
Aporetica, Forst.
Gemella, Lour.
Usubis, Burm.
Nassavia, Fl. Flum.
Valenzuelia, *Bert.*
Irina, *Blum.*
Prostea, *Camb.*
Lepisanthes, *Blum.*
Sapindus, *Linn.*
Pappea, Eckl.
Erioglossum, *Blum.*
Matayba, *Aubl.*
Ephielis, Schreb.
Ernstingia, Neck.
Moulinsia, *Camb.*
Cupania, *Plum.*
Trigonis, Jacq.
Vouarana, Aubl.
Molinæa, Juss.
Gelonium, Gärtn.
Tina, Röm. et Schult.
Stadmannia, Lam.
Mischocarpus, Blum.
Guioa, Cav.
Blighia, König.
Akeesia, Tuss.
Harpulia, Roxb.
Bonnania, Raf.
Dimereza, Labill.
Diplopetalon, Spreng.
Ratonia, DC.
Erioglossum, Guill. et Perr.
Digonocarpus, Fl. Fl.
Trigonocarpus, Fl. Fl.
Aphania, *Blum.*
Talisia, *Aubl.*
? *Acladodea*, R et Pav
Nephelium, *Linn.*
Euphoria, Commers.
Scytalia, Gärtn.
Dimocarpus, Lour.
Pometia, Forst.
Li-tchi, Sonner.
Thouinia, *Poit.*
Thyana, Hamilt.
? *Varg asia*, Bert
Hypelate, *P. Br.*
Sphærococca, DC.
Exothea, Macf.
Melicocca, *Linn.*
Oococca, DC.
Casimira, Scop.
Schleichera, *Willd.*
Cussambium, Rumph.
Koon, Gärtn.
? Pierardia, *Jack.*
Pierandia, Blum.
? Hedycarpus, *Jack.*

II. HIPPOCASTANEÆ.— Leaves opposite. Ovules 2 in each cell, one ascending, the other suspended. Embryo curved with great fleshy consolidated cotyledons.

Ungnadia, *Endl.*
Æsculus, *Linn.*
Hippocastanum, Tourf.
Pavia, *Boerh.*
Macrothyrsus, Spach.
Calothyrsus, Spach.

III. DODONEÆ.—Leaves alternate. Ovules 2 or 3 in each cell. Embryo rolled spirally.

Kœlreuteria, *Lam.*
Cossignia, *Cambess.*
Llagunoa, *Ruiz. et Pav.*
Amirola, Pers.
Diplopeltis, *Endl.*
Dodonæa, *Linn.*
Alectryon, *Gärtn.*
Evonymoides, Soland.
Ophiocaryon, *Schomb.*

IV. MELIOSMEÆ. Leaves alternate. Flowers extremely irregular. Stamens 5, of which 2 only are fertile. Ovules 2 in each cell, both suspended. Embryo folded up. Fruit a drupe.

Meliosma, *Blum.*
Millingtonia, Roxb.
Wellingtonia, Meisn.

Anomalous Genera.

Plösslea, *Endl.*
Xanthoceras, *Bung.*
Magonia, *St. Hil.*
Phæocarpus, M. et Zuc.

Doubtful Genera.

Valentinia, *Swartz.*
Racaria, *Aubl.*
Eustathes, *Lour.*
Pedicellia, *Lour.*
Ptæroxylon, *Eckl. et Zey.*
Hippobroma, *Eck. et Zey.*
Tarrietia, *Blum.*
Deinbœllia, *Schumach.*
Hornschuchia, *Nees.*

NUMBERS. GEN. 50. Sp. 380.

Vitaceæ.
POSITION.—Polygalaceæ.—SAPINDACEÆ.—Aceraceæ.

ORDER CXXXVII. PETIVERIACEÆ.—PETIVERIADS.

Petiverieæ, *Agardh Classes*, (1825); *Endl. Gen. p.* 975.—Petiveriaceæ, *Link Handb.* 1. 392. (1829); *Ed. pr.* clix.; *Meisn. Gen. p.* 316.

DIAGNOSIS.—*Sapindal Exogens, with apetalous flowers and a solitary carpel.*

Under-shrubs or herbaceous plants, with an alliaceous odour. Leaves alternate, entire, with distinct stipules, often with minute pellucid dots. Flowers racemose or panicled. Calyx of several distinct leaves. Stamens between perigynous and hypogynous, either indefinite, or, if equal to the segments of the calyx, alternate with them. Ovary superior, 1-celled; style one; stigma lateral; ovule erect. Fruit 1-celled, indehiscent, dry, either wingless, wedge-shaped and spiny at the point, or extended at the back into a narrow flat wing (samara). Seed erect without albumen; embryo straight or curved; cotyledons convolute; radicle inferior.

According to Brown and Endlicher these plants are only a section of Phytolaccads. They are, however, distinguished by the presence of stipules, and by their straight exalbuminous embryo with spiral cotyledons. Their habit too is adverse to this approximation, while the key-like fruit of Seguiera and its inflorescence suggests a relationship to Soapworts, which does not seem removed by a comparison of the exact structure of the two. It is true that the latter Order in general has petals, and that Petiveriads have none; but then we have many apetalous genera among Soapworts. In both the seeds are erect, the exalbuminous embryo rolled up, the radicle inferior; and even in the number of their stamens they correspond, if we compare Seguiera with Prostea. In fact, instead of separating these Petiveriads from Soapworts by a long interval, they might almost be regarded as an apetalous form of that Order, with carpels reduced to one. It is to be observed that Petiveria and Seguiera are not entirely like one another, and that these remarks apply to Seguiera only.

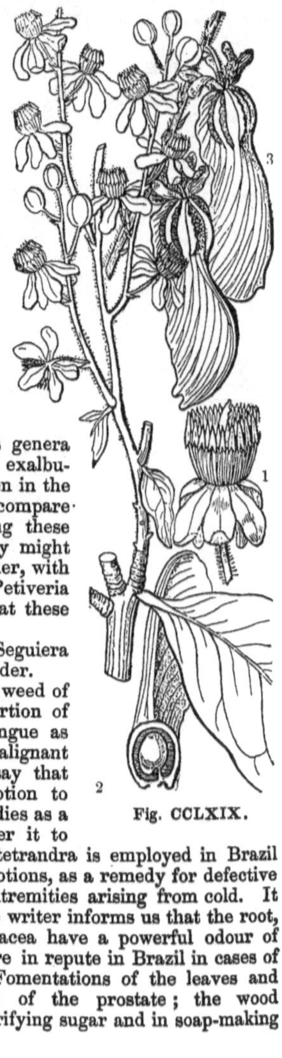

Fig. CCLXIX.

West Indian or tropical American plants; for the Seguiera asiatica of Loureiro probably does not belong to the Order.

All the parts of Petiveria alliacea, the Guinea-hen weed of the West Indies, are excessively acrid; a small portion of the leaves chewed is said by Burnett to render the tongue as dry and black and rough as it appears in cases of malignant fever. The negroes consider it a sudorific, and say that vapour baths or fumigations of it will restore motion to paralysed limbs. The roots are used in the West Indies as a remedy for toothache; the negresses also administer it to procure abortion.—*Schomb. in Linnæa*, ix. 511. P. tetrandra is employed in Brazil under the name of Raiz de Pipi in warm baths and lotions, as a remedy for defective contractibility of the muscles, or in paralysis of the extremities arising from cold. It has an intense alliaceous odour.—*Martius.* The same writer informs us that the root, wood, and all the herbaceous parts of Seguiera alliacea have a powerful odour of garlic or asafœtida; baths impregnated with them are in repute in Brazil in cases of rheumatism, dropsy, and hæmorrhoidal affections. Fomentations of the leaves and young branches are employed to alleviate tumours of the prostate; the wood abounds in potash, and the ashes are employed in clarifying sugar and in soap-making in Brazil.

GENERA.

Petiveria, *L.* | Seguiera, *L.* | Gallesia, *Casar.*

NUMBERS. GEN. 3. Sp. 10.

Phytolaccaceæ.
POSITION.————PETIVERIACEÆ.—Sapindaceæ.

Fig. CCLXIX.—Seguiera floribunda. 1. a flower; 2. a perpendicular section of an ovary; 3. samaras.

ORDER CXXXVIII. ACERACEÆ.—MAPLES.

Acera, *Juss. Gen.* 50. (1789) ; *Ann. Mus.* 18. 477. (1811).—**Acerineæ**, *DC. Théorie, ed.* 2. 244. (1819) ; *Prodr.* 1. 593. (1824) ; *Endl. Gen.* ccxxvii. ; *Meisner, Gen. p.* 56.

DIAGNOSIS.—*Sapindal Exogens, with complete unsymmetrical flowers, petals naked or* 0, *anthers opening longitudinally,* 2 *carpels, and seeds without an aril.*

Trees. Leaves opposite, simple, usually with palmate veins, rarely pinnate, without stipules. Flowers often polygamous, in axillary corymbs or racemes. Calyx divided into 5, or occasionally from 4 to 9 parts, with an imbricated æstivation. Petals equal in number to the lobes of the calyx, imbricated, inserted round an hypogynous disk, or 0. Stamens inserted upon the disk, generally 8, not often any other number, always definite. Ovary free, 2-lobed ; style 1 ; stigmas 2 ; ovules in pairs, amphitropal, pendulous. Fruit formed of two nuts, which are indehiscent, with a narrow wing at the back (samaroid) ; each 1-celled, with 1 or 2 seeds. Seeds ascending, with a thickened lining to the testa ; albumen none ; embryo curved, with foliaceous wrinkled cotyledons, and an inferior radicle.

Fig. CCLXX.

These plants differ from Soapworts in their fruit having but 2 carpels, the petals never being furnished with scales, and their opposite leaves. The distinction is however scarcely satisfactory, even when the want of an aril is added. From Malpighiads their unsymmetrical flowers, inferior radicle, glandless calyx and palmate-veined leaves, decidedly divide them.

Europe, the temperate parts of Asia, the north of India, and North America, are the stations of this Order, which is unknown in Africa and the southern hemisphere.

The species are only known for the sugary sap of Acer saccharinum and others, from which sugar is extracted in abundance, and for their light useful timber. It is said, however, that their juices become acrid as the season advances. The bark is astringent, and yields the dyer reddish brown and yellow colours.

GENERA.

Acer, *Linn.*
Negundo, *Mönch.*
Negundium, Raf.
Dobinea, *Hamilt.*

NUMBERS. GEN. 3. Sp. 60.

POSITION.—Petiveriaceæ.—ACERACEÆ.—Sapindaceæ.

Fig. CCLXX.—1. Acer circinatum.—*Hooker*. 2. flower of A. campestre, *Gærtner* ; 3. its samara ; 4. the same, with the seed laid bare ; 5. the embryo unfolded.

ORDER CXXXIX. MALPIGHIACEÆ.—MALPIGHIADS.

Malpighiaceæ, *Juss. Gen.* 252. (1789); *Ann. Mus.* 18. 479; *DC. Prodr.* 1. 577; *Endl. Gen.* ccxxviii.; *Adrien de Jussieu, Monogr.* (1843); *Wight Illust.* 1. 136.—Nitrariaceæ, *Ed. pr.* No. cxlix. (1830).

DIAGNOSIS.—*Sapindal Exogens, with complete, partially symmetrical flowers, an imbricated calyx, naked stalked petals, ovules hanging by cords, simple stigmas, and usually a convolute embryo.*

Trees or shrubs, often having a climbing habit. The leaves usually opposite or whorled, rarely alternate, simple, usually entire, generally stalked, and having glands on

Fig. CCLXXI

the stalk or under side; stipules generally short and deciduous, occasionally larger, and intrapetiolar. If there are any hairs they are fixed by their middle, and sometimes are stiff and brittle. The inflorescence is variable. The flowers ☿ or ♂-☿-♀, red, or more commonly yellow, rarely white, and very rarely blue; in a few instances abortive green flowers are intermixed with the perfect ones. Calyx 5-parted, with conspicuous glands at the base of one or all the segments, very rarely without glands; in æstivation quincuncial, seldom valvate. Petals 5, unguiculate, with a convolute æstivation. Stamens mostly double the number of the petals, often monadelphous, usually with a fleshy connective that projects beyond the lobes of the anthers. Carpels generally 3, rarely 2, very rarely 4, altogether or partially consolidated, often crested at the back; ovules solitary, orthotropal, rising up from a long pendulous cord, with which they form a sort of hook; styles distinct or united; stigmas the same number, simple, capitate, truncate, or variously expanded. Fruit very various; a drupe, or a woody nut, or samaroid, the wings of different forms and in different positions. Seed suspended obliquely by a short cord below the apex; albumen 0; embryo with a short superior radicle and

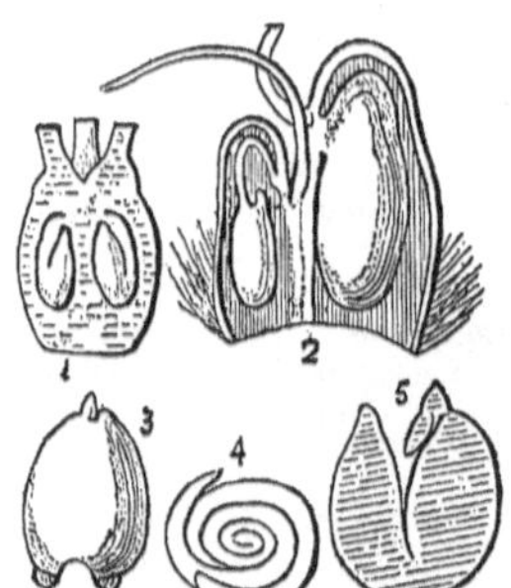

Fig. CCLXXIII.

Fig. CCLXXII.

Fig. CCLXXI.—Diplopteris paralias; 1. a flower-bud, showing the double glands of the calyx; 2. an expanded flower; 3. the carpels; 4. ripe fruit of Ryssopteris timorensis.
Fig. CCLXXII.—Jubelina riparia, after *A. de Jussieu.*
Fig. CCLXXIII.—1. Section of ovary of Malpighia; 2. of Coleostachys; 3. embryo of Burdachia; 4. of Byrsonima; 5. of Brachypterys.—*A. de Jussieu.*

longer cotyledons, which are straight, and equal, or unequal, curved, or plaited, or even rolled up, very thick or leafy.

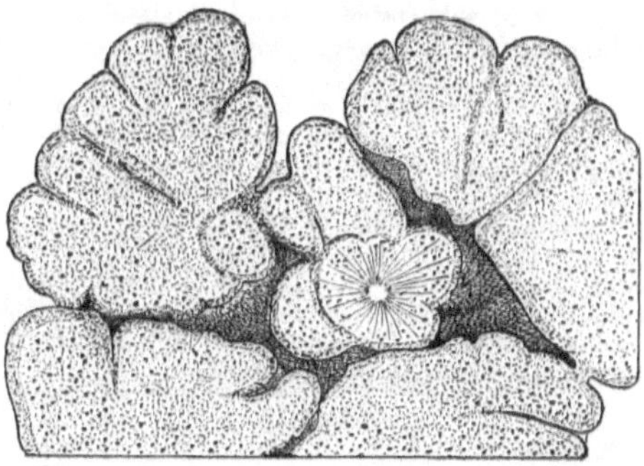

Fig. CCLXXIV.

This remarkable Order has been treated at great length, and with infinite skill, by M. Adrien de Jussieu, from whom the principal part of the following and previous remarks is borrowed. Among the most striking peculiarities of the race is the presence upon the calyx of certain glands of such large size, sometimes, as to constitute a considerable part of the whole calycine apparatus; it is very remarkable that when these glands are missing, it is those which are next the outside of the inflorescence that disappear. They are secreting organs, and according to Payen, their exudations are (in certain Malpighias) of the nature of a fatty oil containing a fluid substance, besides one that is concrete. Another very remarkable circumstance is, the general tendency that has been observed among the stems of the climbing kinds, to assume appearances quite anomalous among Exogens. In these instances there is in the beginning the usual formation of a woody circular zone round pith, but immediately afterwards the woody matter is deposited in the most irregular lobed and zoneless ribs. Many details relating to this matter have been given by M. Adrien de Jussieu. The distinctions between this Order and others in its alliance, will be evident upon comparing the Diagnosis already given; and will be further explained under the other Orders.

The genus Nitraria, consisting of a few salt plants from the west of Asia, the north of Africa, and N. Holland, appears to be not essentially distinct from the present Order, of which it has the unsymmetrical ovary, peculiar ovules, and drupaceous fruit. Its principal distinctions consist in the entire consolidation of its styles, and in the stamens being collected in parcels lodged in the cavity of concave petals. It has, however, given rise to a supposed Order,* originally suggested in the first edition of the present work, No. 149, (1830), as possessing some affinity on the one hand with Goosefoots, and the other with Buckthorns (Rhamnaceæ).

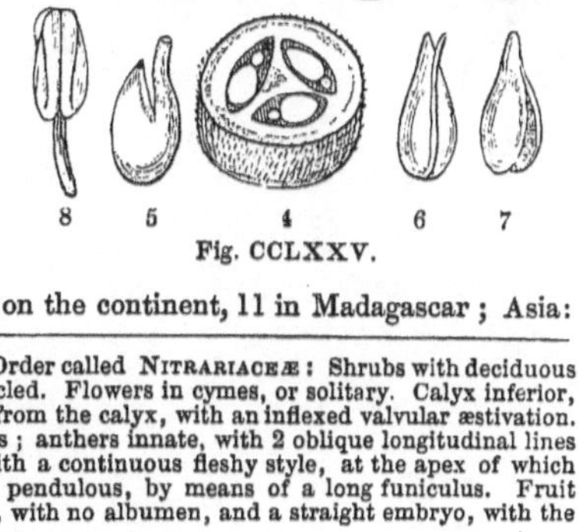

Fig. CCLXXV.

The following is the distribution of the Order, according to M. A. de J.; Africa: 14 on the continent, 11 in Madagascar; Asia:

* The following is the character of the supposed Order called NITRARIACEÆ: Shrubs with deciduous succulent alternate leaves, which are sometimes fascicled. Flowers in cymes, or solitary. Calyx inferior, 5-toothed, fleshy. Corolla of five petals, which arise from the calyx, with an inflexed valvular æstivation. Stamens 3 times the number of the petals, perigynous; anthers innate, with 2 oblique longitudinal lines of dehiscence. Ovary superior, 3- or more celled, with a continuous fleshy style, at the apex of which are as many stigmatic lines as there are cells; ovules pendulous, by means of a long funiculus. Fruit drupaceous, opening by 3 or 6 valves. Seeds solitary, with no albumen, and a straight embryo, with the radicle next the hilum.

Fig. CCLXXIV.—Wood of Heteropterys anomala.—*A. de Jussieu.*

Fig. CCLXXV.—Nitraria Schoberi. 1. an expanded flower; 2. the calyx and pistil; 3. a perpendicular section of the ovary; 4. a cross section of it; 5. an ovule; 6. a seed; 7. an embryo; 8. an anther.

Arabia, India, and Ceylon 16, Indian Archipelago, China and Polynesia 14; West Indies 56; Mexico 61; South America 408, of which 290 come from Brazil. They are nearly all tropical.

Of the uses of the species of this Order little can be said. A large number are beautiful trees or climbers with gaudy flowers; and they seem to be generally astringent. Byrsonima bark is of common employment by tanners in Brazil, under the name of Murici. The wood of some kinds, especially Byrsonima verbascifolia, is bright red. The fruit of the Malpighias and Byrsonimas is eaten in the West Indies; the hairs of a few are painfully pungent. The bark of Byrsonima crassifolia, or Malpighia Moureila, according to Aublet, is employed in Cayenne as a febrifuge. That of the Chapara Manteca, Byrsonima crassifolia, is astringent, and is used in infusion or decoction taken inwardly, as an antidote to the rattlesnake bite; it is also employed successfully as a remedy for abscesses in the lungs. It is said that Alcornoco bark is the produce of Byrsonima laurifolia, rhopalæfolia, and coccolobæfolia. The acid astringent berries of Byrsonima spicata (Bois-tan) are prescribed in dysentery. It is said that the seed of Bunchosia armeniaca, a Peruvian tree, is poisonous.

GENERA.

MALPIGHEÆ, *A. de J.*

Malpighia, *Plum.*
Byrsonima, *Rich.*
Burdachia, *A. de J.*
Carusia, Mart.
Coleostachys, *A. de J.*
Lophanthera, *A. de J.*
Pterandra, *A. de J.*
Verrucularia, *A. de J.*
Galphimia, *Cav.*
Thryallis, L.
Spachea, *A. de J.*
Meckelia, Mart.
Bunchosia, *Rich.*
Malacmæa, Gris.
Echinopteris, *A. de J.*
Dialla, *Griseb.*
Heladraia, *A. de J.*
Thryallis, *Mart.*
Nitraria, *L.*

II. BANISTEREÆ, *A. de J.*

Lophopterys, *A. de J.*
Brachypterys, *A. de J.*
Stigmaphyllon, *A. de J.*
Ryssopterys, *Blume.*
Banisteria, *L.*
Peixotoa, *A. de J.*
Heteropterys, *Kth.*
Tricomaria, *Hk. et Arn.*
Acridocarpus, *Guillem.*
Anomalopteris, G. Dn.

III. HIREÆ, *A. de J.*

Tristellateia, *Pet. Th.*
Zymum, Noronh.
Hiptage, *Gærtn.*
Gærtnera, Schreb.
Molina, Cav.
Succowia, Dennst.
Triaspis, *Burch.*
Flabellaria, Cav.
Aspidopterys, *A. de J.*
Triopterys, *L.*
Tetrapterys, *Cav.*
Hiræa, *Jacq.*
Mascagnia, Bert.
Diplopterys, *A. de J.*
Jubelina, *A. de J.*
Dinemandra, *A. de J.*
Dinemagonum, *A. de J.*

IV. GAUDICHAUDEÆ, *A. de J.*

Gaudichaudia, *Kth.*
Aspicarpa, *Lga.*
Acosmus, Desv.
Camarea, *St. Hil.*
Janusia, *A. de J.*
Schwannia, *Endl.*
Fimbriaria, St. Hil.

GENERA INSUFFICIENTLY KNOWN.

Caucanthus, *Forsk.*
Platynema, *Wight et A.*
Bembix, *Lour.*

NUMBERS. GEN. 42. Sp. 555.

POSITION.—Aceraceæ.—MALPIGHIACEÆ.—Sapindaceæ.

ORDER CXL. ERYTHROXYLACEÆ.—ERYTHROXYLS.

Erythroxyleæ, *Kunth in Humb. N. G. Am.* 5. 175. (1821); *DC. Prodr.* 1. 573. (1824); *Endl. Gen.* ccxxix.; *Meisn. Gen. p.* 56.; *Martius Beiträge zur Kenntniss der g. Erythroxylon* (1840); *Wight Illustr.* 1. 135.

DIAGNOSIS.—*Sapindal Exogens, with complete, partially symmetrical flowers, an imbricated calyx, petals with an appendage, sessile pendulous ovules, capitate stigmas, and a straight embryo.*

Shrubs or trees; young shoots often compressed and covered with acute imbricated scales. Leaves alternate, usually smooth; stipules within the petioles. Flowers small, whitish or greenish. Peduncles axillary, solitary or clustered, emerging from numerous imbricated scale-like bracts. Sepals 5, combined at the base, persistent. Petals 5, hypogynous, broad at the base, with a plaited scale there, equal, the margins lying upon each other in æstivation. Stamens 10, monadelphous; anthers innate, erect, 2-celled, dehiscing lengthwise. Ovary 3-celled, with 2 cells spurious; styles 3, distinct, or united almost to the point; stigmas 3, capitate; ovule solitary, pendulous, anatropal, not suspended by a cord. Fruit drupaceous, 1-seeded. Seed angular; albumen between fleshy and mealy, or 0; embryo straight, central; cotyledons plano-convex; radicle superior, taper, straight.

The Erythroxyls are distinguished from Malpighiads by their flowers growing from amongst small imbricated scales, having no glands on the calyx, a pair of parallel membranous plates on the petals, capitate stigmas, and ovules which are truly anatropal, without any cord to connect them with the placentæ. These marks are, however, hardly sufficient for the characteristics of a Natural Order, and it would perhaps be better to merge the Order in the Malpighiads, as has been done with Nitraria. An elaborate account of the genus will be found in Martius's Memoir, above quoted.

Chiefly West Indian and South American. A few are found in the East Indies, several in the Mauritius and Madagascar, and one in New Holland. Brazil within the tropics is their favourite haunt.

The wood of some is bright red; that of E. hypericifolium, is the Bois d'huile of the Isle of France. A permanent reddish brown dye is obtained from the bark of Erythroxylum suberosum, called in Brazil Gallinha choca and Mercurio do campo. E. areolatum, a shrub found near Carthagena, is said to have some medical value; its young branches are refrigerant, its bark tonic, from the juice of the leaves is prepared an ointment employed against scald head, and the sub-acid juice of its fleshy fruit is purgative and diuretic. The bark of the root of E. anguifugum is regarded as an alexipharmic in Brazil; that of E. campestre is employed in the same country as a purgative.—*Martius mat. m. Bras.*

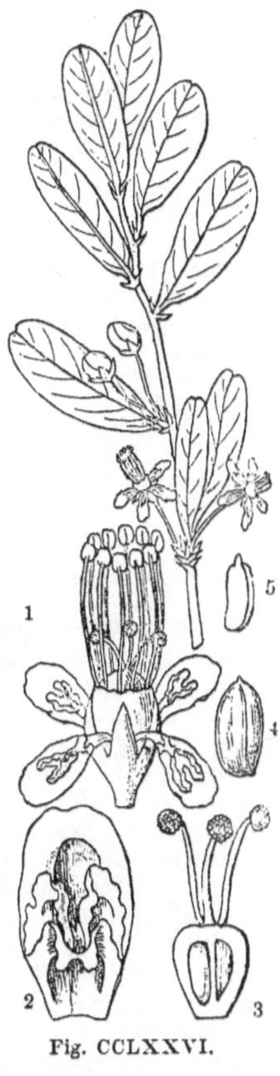

Fig. CCLXXVI.

Erythroxylon Coca is a plant much used by the miners of Peru for its remarkable power in stimulating the nervous system, in which respect it quite resembles opium. Its leaves are chewed with a small mixture of finely powdered chalk. No effects that have been ascribed to the immoderate use of opium are exceeded by what seems the consequence of chewing the Coca leaf. See a curious account of this plant in *Pöppig's Reise in Chile.*

GENUS.

Erythroxylon, *Linn.*
Venelia, Commers.
Roelana, Commers.
Steudelia, Spreng.
Sethia, Kunth.

NUMBERS. GEN. 1. Sp. 75.

POSITION.————.—ERYTHROXYLACEÆ.—Malpighiaceæ.

Fig. CCLXXVI.—Erythroxylon.—*Martius.* 1. a flower; 2. one of the petals; 3. the pistil with 3 capitate stigmas; 4. a drupe; 5. the embryo.

ALLIANCE XXX. *GUTTIFERALES.*—THE GUTTIFERAL ALLIANCE.

DIAGNOSIS.—*Hypogynous Exogens, with monodichlamydeous flowers, axile placentæ, an imbricated calyx, an imbricated or twisted corolla,* 00 *stamens, and an embryo with little or no albumen.*

The true passage from Sapindals into this Alliance is from Soapworts to Rhizobols; for the habit of Æsculus in the former is the same as that of most of the latter, and they nearly correspond in their structure. But Rhizobols have an indefinite number of stamens. It is in that respect indeed that Guttiferals principally differ from Sapindals, and the former may be almost regarded as a polyandrous form of the latter. It is however customary to find no want of symmetry in the calyx, corolla, and stamens of Guttiferals, while the reverse is generally characteristic of Sapindals. These too have seldom, if ever, resinous secretions; while, on the other hand, those are often remarkable for their abundance of resin.

The near relationship of all the Orders here collected is undisputed. They lean towards the diclinous structure in some Guttifers, and approach the diclinous series in the Dipterads, which have a strong analogy to Mastworts.

NATURAL ORDERS OF GUTTIFERALS.

Leaves simple, alternate, with large convolute stipules. Flowers symmetrical. Petals equilateral. Calyx unequal, permanent, winged. Anthers beaked. Fruit one-celled, one-seeded . .	141. DIPTERACEÆ.
Leaves simple, alternate, without stipules or with very small ones. Flowers symmetrical. Petals equilateral. Anthers versatile. Seeds few or single. Stigmas on a long style	142. TERNSTRÖMIACEÆ.
Leaves digitate, opposite. Flowers symmetrical. Petals equilateral. Stigmas sessile. Seeds solitary. Embryo with an enormous radicle	143. RHIZOBOLACEÆ.
Leaves simple, opposite, without stipules. Flowers symmetrical. Petals equilateral. Anthers adnate, beakless. Seeds solitary or few. Stigmas sessile, radiating	144. CLUSIACEÆ.
Leaves simple, alternate, without stipules. Flowers unsymmetrical. Petals equilateral. Anthers versatile. Seeds innumerable, minute. Stigmas sessile	145. MARCGRAVIACEÆ.
Petals oblique, glandular. Seeds numerous, naked. Styles long, distinct	146. HYPERICACEÆ.
Petals oblique, glandless. Seeds few, shaggy. Styles long, distinct	147. REAUMURIACEÆ.

ORDER CXLI. DIPTERACEÆ.—DIPTERADS.

Dipterocarpeæ, *Blume Bijdr. p.* 222. (1825); *Fl. Javæ* (1829); *Wight and Arnott, Prodr. Fl. Ind. Penins.* 1. 83. (1834); *Endl. Gen.* ccxiii.; *Meisner Gen.* 35.; *Wight Illustr.* 1. *t.* 36, 37.

DIAGNOSIS.—*Guttiferal Exogens, with simple alternate leaves, large convolute stipules, symmetrical flowers, equilateral petals, an unequal, permanent, winged calyx, beaked anthers, and a* 1-*celled* 1-*seeded fruit.*

Gigantic trees, abounding in resinous juice. Leaves alternate, involute in vernation, with veins running out from the midrib to the margin; stipules deciduous, oblong, convolute, terminating the branches with a taper point. Flowers usually large; the racemes terminal and panicled, or axillary and solitary, or several from the same leaves, or from the axils, often one-sided. Calyx tubular, 5-lobed, unequal, persistent, and afterwards enlarged, naked at the base; æstivation imbricated. Petals hypogynous, sessile, often combined at the base; æstivation contorted. Stamens indefinite, hypogynous, distinct, or slightly and irregularly polyadelphous; anthers innate, subulate, opening longitudinally towards the apex; filaments dilated at the base. Ovary superior, without a disk, 3-celled; ovules in pairs, pendulous; style single; stigma simple. Fruit coriaceous, 1-celled by abortion, 3-valved or indehiscent, surrounded by a calyx having tough leafy enlarged permanent divisions which crown the fruit. Seed single, without albumen; cotyledons plano-convex, or more commonly twisted and crumpled; radicle superior.

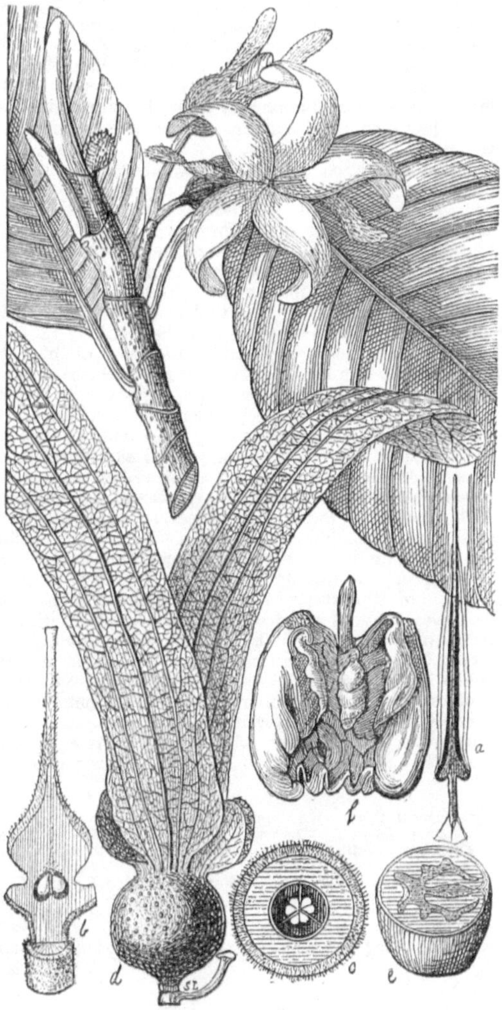

Fig. CCLXXVII.

These trees, which are apparently unknown in Europe in a living state, are described by Dr. Wight as deserving cultivation for ornamental purposes, for the sake of their majestic size, handsome forms, the beauty of their clustered flowers, and the richly coloured wings of their curious fruit. They form a remarkable Order, which is one of those whose limits are best defined, and yet it appears to participate in the affinities of plants which cannot be brought into its vicinity by any

Fig. CCLXXVII.—Dipterocarpus trinervis.—*Blume.* *a* an anther; *b* a perpendicular, *c* a transverse, section of an ovary; *d* a fruit; *e* section of seed of Dryobalanops camphora; *f* its embryo unfolded.—*Gärtner.*

of the schemes for classification which Botanists have hitherto employed. It has, for example, the peculiar rolled-up stipules which occur in Magnoliads; while the Oak is strikingly like Dipterads in foliage, in the germination of the seeds, which takes place underground without the cotyledons rising into the air, and in a constant tendency to lose the major part of the ovules in the process of maturing one; it is also to be remarked that the hard cupule or involucre of Mastworts (Corylaceæ) is much like the hardened calyx of these Dipterads. It is herein, indeed, that the great feature of the latter resides; we have nothing elsewhere exactly like the long wing-like lobes of their calyx. Botanists generally contrast Dipterads with the Elæocarpeous division of Lindenblooms, but their valvate calyx, diskless flowers, and peculiar fruit indicate a distant relationship only. The resinous juice, compound superior ovary, drupaceous fruit, numerous long anthers, irregular coloured calyx, and single exalbuminous seed, ally Dipterads, as Blume remarks, to Guttifers, from which their stipules and the æstivation of the corolla abundantly distinguish them.

Only found in India, and especially in the eastern islands of the Indian Archipelago, where, according to Blume, they form the largest trees of the forest. Shorea robusta limits the northern distribution of the Order, being found all along the foot of the Himalayas.

All the species seem filled with balsamic resin, which assumes various forms. Dryobalanops camphora yields the hard Camphor of Sumatra; this substance is found in a concrete state in cavities and fissures in the heart of the tree; it is less volatile than the common camphor of commerce; the same tree, which is fully described in Blume's *Flora Javæ*, also yields the camphor-oil of Borneo and Sumatra; the latter is supposed to be camphor in a partially formed state. Shorea robusta produces a balsamic resin used in the temples of India under the name of Ral or Dhoona; Saul, the best and most extensively used timber in India, is produced by the same tree. Vateria indica furnishes the resin called in India Copal (in England known by the name of Gum animi), and very nearly approaching the true resin of that name; in its recent and fluid state it is used as a varnish (called Piney varnish) in the south of India, and, dissolved by heat, in closed vessels, is employed for the same purpose in other parts of India; it is extremely tenacious and solid, but melts at a temperature of 97½ Fahr. Dr. Wight tells us that the natives obtain it by the simple process of cutting a notch in the tree, sloping inwards and downwards; the resin collects there and soon hardens. Under the name of Piney Dammar this most useful substance is applied in India to many purposes; it forms an excellent varnish, and on the Malabar coast is made into candles which "diffuse in burning an agreeable fragrance, give a clear bright light, with little smoke, and consume the wick so as not to require snuffing. Some of these candles, that were sent home, were highly prized and sold for very high prices" (*Wight*), but their importation was stopped by the excessive duties that were levied upon them. The resin of Dipterocarpus trinervis is found an excellent material for plaisters; and made into tincture, or formed into an emulsion with yolk of egg, it acts upon the mucous membranes like Balsam of Copaiva.—*Blume*. The natives of Java smear the leaves of the Plantain with this resin, and so form torches, which are said to yield a white light and to produce a not unpleasant smell. Other kinds of resin are furnished by other species; as, by Shorea robusta and Tumbugaia, the dhoona or dammer pitch, generally used in India for marine purposes, and as incense; by various species of Dipterocarpus, the balsam called by the natives of India Gurjun, by the Cinghalese Dhoonatil, and by the English Wood-oil. This also is used like Balsam of Copaiva.

GENERA.

Dipterocarpus, *Gärtn.*	Anisoptera, *Korth.*	Isauxis, *Arn.*	Vatica, *Linn.*
Pterygium, Corr.	Dryobalanops, *Gärtn. f.*	*Seidlia*, Kostel.	*Shorea*, Roxb.
? *Caryolobis*, Gärtn.	Vateria, *Linn.*	*Retinodendron*, Korth.	Hopea, *Roxb.*

NUMBERS. Gen. 7. Sp. 47.

Tiliaceæ.
POSITION.—Ternströmiaceæ.—DIPTERACEÆ.—Clusiaceæ.
Corylaceæ.

LOPHIRACEÆ.—*Endl.* Under this name Mr. Endlicher proposes to establish an Order, of which the following is the description. "Trees from tropical Africa, having a pyramidal form, many branches, and a dry bark. Leaves alternate, stalked, quite entire, with raised veins, and a jointed stalk; stipules very small and deciduous, planted on each side of the leaf-stalk at the base. Flowers perfect, regular, axillary and terminal, panicled, yellow, with straggling flower-stalks which are jointed above the base, and furnished with 2 very small bracts at the articulation. Sepals 5, the 3 inner smaller and concave, the two outer opposite, larger, and finally expanded into a pair of wings. Petals 5, hypogynous, without claws, their points twisted together in æstivation, eventually spreading flat. Stamens hypogynous, indefinite, nearly in two rows; filaments filiform, short; anthers 2-celled, their cells linear, opposite, parallel, adnate, opening at the point by a lateral cleft. Disk 0. Ovary conical, one-celled; ovules 00, long, curved backwards, hooked, placed upon a thick free basal placenta; stigmas 2, very small, twisted, reflexed. Nut leathery, spindle-shaped, contracted at the base, and consolidated with the enlarged calyx, one-celled, and by abortion one-seeded. Seeds erect, with a thin membranous skin. Embryo without albumen; cotyledons amygdaloid, plano-convex; radicle very short, immersed, inferior.——The solitary genus which constitutes this Order is allied to nothing yet known. It is very different from Dipterocarpeæ (Dipteraceæ), with which it is associated because of its two enlarged calyx-leaves, and yet it can scarcely be excluded from the Guttiferous class."—*Enchiridion*, p. 526.——In his Guttiferous class Mr. Endlicher includes Dipteraceæ, Chlænaceæ, Ternstromiaceæ, Clusiaceæ, Marcgraaviaceæ, Hypericaceæ, Elatinaceæ, Reaumuriaceæ, Tamaricaceæ. It must be confessed that none of those present any marked resemblance to Lophira, which is the Scrubby Oak of Sierra Leone, except Dipterads and Guttifers. To the irregular fruit of the former that of Lophira is quite similar, but its ovary is one-celled, with a crowd of ovules upon a free central placenta, its seed is solitary with the radicle downwards, and the cotyledons are plano-convex, all points of difference from Dipterads, which have an ovary with 3 cells, a pair of pendulous ovules in each, a seed with the radicle upwards, and crumpled cotyledons. Moreover Lophira wants the large stipules of Dipterads. On the other hand, its foliage is so like that of Calophyllum, a genus of Guttifers, that the one might be mistaken for the other, except that the leaves of Lophira are alternate; but in all the structure of the fruit the genus differs from the Guttiferous Order. Nevertheless, although Lophira is so different from Dipterads it is to be observed that it agrees with that Order not alone in its peculiar calyx; for in both cases the ovules are anatropal, and consequently the radicle is directed to the hilum, and in Lophira there is an evident tendency to produce the long anthers which are so characteristic of Dipteraceæ. The late M. Guillemin regarded it as being absolutely a Dipterad, because "of the convolute æstivation of the petals, the length of the 2 sepals extended into membranous wings, one of them being moreover out of all proportion to the others, the alternate leaves furnished with little deciduous stipules, and the dry corky bark not filled with milky secretions."—See *Floræ Senegambiæ Tentamen*, p. 110.

GENUS.
Lophira, *Banks.*

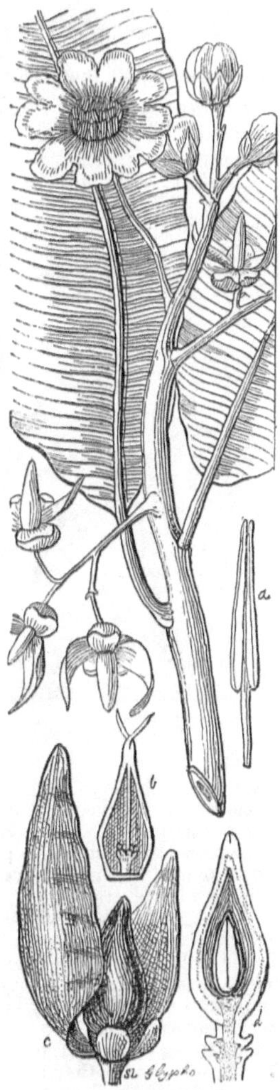

Fig. CCLXXVIII.

Fig. CCLXXVIII.—Lophira alata.—*Decaisne.* *a* an anther; *b* a perpendicular section of an ovary; *c* a fruit; *d* a perpendicular section of a fruit.

ORDER. CXLII. TERNSTRÖMIACEÆ.—THEADS.

Ternströmieæ, *Mirbel. Bull. Philom.* 381. (1813).—Ternströmiaceæ, *DC. Mém. Soc. H. N. Genev. vol.* 1 (1823); *Prodr.* 1. 523. (1824); *Cambessédes Mémoire*, (1828); *Endl. Gen.* ccxv.; *Meisn. Gen. p.* 40.; *Wight Illustr.* 1. *p.* 94.—Theaceæ, *Mirb. Bull. Phil.* (1813).—Camellieæ, *DC. Théor. Elém. ed.* 1. (1813); *Prodr.* 1. 529. (1824).

DIAGNOSIS.—*Guttiferal Exogens, with simple alternate leaves, without stipules or with very small ones, symmetrical flowers, equilateral petals, versatile anthers, few or single seeds, and stigmas on a long style.*

Trees or shrubs. Leaves alternate, coriaceous, generally without stipules, usually undivided, now and then with pellucid dots. Peduncles axillary or terminal, articulated

3 4 2 1

Fig. CCLXXIX.

at the base. Flowers generally white, seldom pink or red, occasionally polygamous. Sepals 5 or 7, imbricated in æstivation, concave, coriaceous, deciduous, the innermost often the largest. Petals 5, 6, or 9, not equal in number to the sepals, often combined at the base. Stamens 00, hypogynous; filaments filiform, monadelphous or polyadelphous, or distinct; anthers versatile, or adnate, 2-celled, opening longitudinally; ovary superior, with several cells; styles from 3 to 7, filiform, more or less combined; ovules pendulous, or erect, or peltate. Capsule 2-7-celled and capsular, with the dehiscence taking place in various ways; sometimes coriaceous and indehiscent; usually with a central column. Seeds attached to the axis, large, very few; albumen none, or in very small quantity; embryo straight, bowed, or folded back, the radicle turned to the hilum; cotyledons very large, often filled with oil, occasionally plaited lengthwise; an aril sometimes present.

This Order originated in 1813, with Mirbel, who separated some of its genera from Citronworts, where they had been placed by Jussieu, and at the same time founded another closely allied Order, under the name of Theads. These opinions were substantially adopted by Kunth and De Candolle, the latter of whom, moreover, formed several sections among the genera. Since that time the Theads have attracted the attention of several Botanists, especially of M. Cambessédes, whose views are generally adopted. He, however, combines under this Order genera with axile and parietal placentation, with truly albuminous and exalbuminous seeds, with large amygdaloid embryos, and those whose embryo is too small to be easily found among its copious albumen, to say nothing of other differences of considerable moment. It is therefore difficult to suppose that such an arrangement can be maintained; and at least we must, I think, remove a genus called Saurauja, consisting of about 30 Asiatic trees or shrubs, in which there is a tendency to form a monopetalous corolla, an infinite number of minute seeds, a very small embryo lying in the base of abundant albumen, and anthers opening by pores; it

Fig. CCLXXIX.—Kielmeyera rosea. 1. the pistil; 2. a transverse section of it; 3. a ripe fruit; 4. embryo.

has, in fact, the habit of a Clethra and seems to bring into contact the Ranal and Erical Alliances. Abstracting this genus and Cochlospermum, which is transferred to the Cistal Alliance, a better limited group remains, of which the Camellia may be taken as the type, and which differs from Guttifers in having alternate leaves, versatile anthers, and a long style, without any tendency to form the flowers on a quaternary plan.

Although the plants of this Order which are known in European gardens are chiefly from China or North America, they form but an inconsiderable part of the whole : 7 or 8 are all that are contained in the first of these countries, and 4 in the latter; while between 60 and 70, all beautiful trees or shrubs, are natives of the woods of South America: about a score are known in the East Indies, and one in Africa.

Their properties are ill understood, but little being known of the greater part of the species. The tea which is so extensively consumed by Europeans is produced by two or three species of Thea: its slightly stimulating properties become narcotic in very hot latitudes, as at Penang. For a most valuable account of this plant, see *Royle's Illustr.*, p. 107. An excellent table oil is expressed from the seeds of Camellia oleifera. The different species and varieties of Camellia japonica are the glory of gardeners. The leaves of Kielmeyera speciosa are employed in Brazil for fomentations, for which they are well adapted, on account of the mucilage in which they abound. The bark of Gordonia is used by tanners in the United States.

GENERA.

Anneslea, *Wall.*
? Dicalyx, *Lour.*
Sariava, Reinw.
? Visnea, *Linn. f.*
Mocanera, Juss.
Reinwardtia, *Korth.*
Ternströmia, *Mut.*
Toanabo, Aubl.
Tonabea, Juss.
Dupinia, Neck.
Amphania, Banks.
Sarosanthera, *Korth.*
Adinandra, *Jack.*
Eurya, *Thunb.*
Geeria, Blum.
Cleyera, *Thunb.*
Hoferia, Scop.
Mukopf, Kämpf.
Sukaki, Kämpf.
Freziera, *Swartz.*
Erotium, Soland.
Lettsomia, *Ruiz et Pav.*
Ventenatia, *Pal.*
? Microsemma, *Lab.*
Ploiarium, *Korth.*
Laplacea, *H. B. K.*
Hæmocharis, Salisb.
Wikströmia, Schrad.
Lindleya, Nees.
Bonnetia, *Mart. et Zucc.*
Kieseria, Nees.
Archytæa, *Mart. et Zucc.*
Ixionanthes, *Jack.*
Kielmeyera, *Mart. et Zucc.*
Martineria, Fl. Flum.
Catostemma, *Benth.*
Ochthocosmus, *Benth.*
Caraipa, *Aubl.*
Marila, *Swartz.*
Monoporina, J. S. Presl.
Scyphæa, C. B. Presl.
Anisosticte, Bartl.
Mahurea, *Aubl.*
Bonnetia, Schreb.
Stuartia, *Catesb.*
Malachodendron, Cav.
Gordonia, *Ellis.*
Lasianthus, Catesb.
Polyspora, Sweet.
Franklinia, Marsh.
Lacathea, Salisb.
Closaschima, Korth.
Antheeischima, Korth.
Schima, *Reinw.*
Pyrenaria, *Blum.*
Camellia, *Linn.*
Sasanqua, Nees.
Kissi, Endl.
Thea, *Linn.*
? Godoya, *Ruiz et Pav.*
Godovia, Pers.
? Leucoxylon, *Blum.*
? Euryanthe, *Schlechtend.*

NUMBERS. GEN. 33. Sp. 130.

Sterculiaceæ.
POSITION.—Clusiaceæ.—TERNSTROMIACEÆ.—Hypericaceæ.
Dilleniaceæ.

ORDER CXLIII. RHIZOBOLACEÆ.—RHIZOBOLS.

Rhizoboleæ, *DC. Prodr.* 1. 599. (1824); *Cambessédes in Aug. St. Hil. Fl. Bras. Merid.* 1. 322. (1827); *Endl. Gen.* ccxxxi.

DIAGNOSIS.—*Guttiferal Exogens, with digitate opposite leaves, symmetrical flowers, equilateral petals, sessile stigmas, solitary seeds, and an embryo with an enormous radicle.*

Trees of very large size. Leaves opposite, digitate, coriaceous, with a jointed stalk and no stipules. Flowers large, regular, arranged in racemes, with their stalks jointed at the base and below the apex. Sepals 5 or 6, more or less combined, imbricated in æstivation. Petals 5 to 8, equal-sided but unequal, thickish, arising along with the stamens from a hypogynous disk. Stamens extremely numerous, slightly monadelphous, arising in a double row from a disk, the innermost being shorter and often abortive; anthers roundish, 2-celled, opening lengthwise. Ovary superior, 4 or 5, or even many-celled; styles as many as the cells; stigmas minute; ovules solitary, attached to the axis by their middle, semianatropal, with the foramen uppermost. Fruit formed of several combined nuts, part of which are sometimes abortive; each nut indehiscent, 1-seeded, 1-celled, with a thick double putamen. Seed reniform, without albumen, with a funicle which is dilated into a spongy excrescence; radicle very large, constituting nearly the whole of the almond-like substance of the nut, with a long 2-edged caulicle, having two small cotyledons at the top, and lying in a furrow of the radicle.

Fig. CCLXXX.

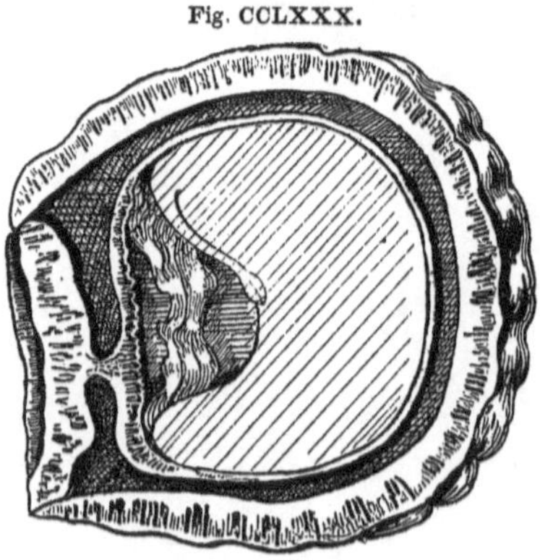

Fig. CCLXXXI.

This very distinct Order De Candolle thought allied to Soapworts in its hypogynous flowers and its fruit; and especially to Æsculus on account of its opposite compound palmate leaves; but in that genus the radicle is small, and the cotyledons very large, while in Rhizobols the radicle is enlarged, and the cotyledons small. It however appears to be with Guttifers that Rhizobols best agree. "In these two Orders we find the leaves opposite and articulated at their base, hypogynous petals, a similar æstivation, numerous hypogynous stamens, and exalbuminous seeds. The large flowers of Caryocar call to mind those of most Guttifers; its inflorescence is nearly that of Moronobea; its fruit has a relation to that of Mammea, and presents, in that genus, as in several others of the same Order, a single seed in each cell."—*Camb. in Aug. St. H. Fl. Bras.* 1. 323. Endlicher traces a resemblance between them

Fig. CCLXXX.—Anthodiscus trifoliatus. 1. a flower bud; 2. a flower; 3. a perpendicular section of the pistil.
Fig. CCLXXXI.—Caryocar butyrosum; a section of one of the lobes of its fruit.

and Terebinths, through the intervention of Mangifera among the former. Their great peculiarity is the seeds having a radicle of enormous size, compared with the cotyledons. If it were not for that, the Order could not be satisfactorily distinguished from Guttifers.

A few large trees, found in the forests of the hottest parts of South America, constitute the whole of this Order.

It is from trees belonging to it that are produced the Souari (or Suwarrow) Nuts of the shops, the kernel of which is one of the most delicious fruits of the nut kind that is known. An oil is extracted from them not inferior to that of the Olive. They chiefly come from Caryocar butyrosum, the wood of which is said to be of much value for ship-building. These nuts must not be confounded with what are called Brazil Nuts, which are the seed of Bertholletia excelsa, a genus of the Myrtal Alliance. The timber of Caryocar butyrosum (Pekea tuberculosa) is excellent for ship-timber, mill-work, planks, &c., according to Schomburgk, who also speaks of another timber tree of this Order, known under the name of Cakaralli or Kukaralli, whose bark consists of numerous layers, which the Indians, by beating, separate till they are as thin as satin paper, when they use them as wrappers for cigars. Is not this the very different Lecythis ollaria ?—*See* LECYTHS.

GENERA.

Caryocar, *Linn.*	*Pekea*, Aubl.
Rhizobolus, Gärtn.	*Souari*, Aubl.
Acanthocarya, Arruda.	Anthodiscus, *G. W. F. Mey.*

NUMBERS. GEN. 2. Sp. 8.

POSITION.—Clusiaceæ.—RHIZOBOLACEÆ.——————
Sapindaceæ.

ORDER CXLIV. CLUSIACEÆ.—GUTTIFERS.

Guttiferæ, *Juss. Gen.* 243. (1789); *DC. Prodr.* 1. 557. (1824); *Meisner, p.* 42; *Wight Illustr.* 1. 114; *Cambessédes, Mémoire* (1828).—Clusiaceæ, *Ed. pr.* lv. (1836).

DIAGNOSIS.—*Guttiferal Exogens, with simple opposite leaves, without stipules, symmetrical flowers, equilateral petals, adnate beakless anthers, solitary or few seeds, and sessile radiating stigmas.*

Trees or shrubs, occasionally parasitical, yielding resinous juice. Leaves without stipules, opposite, coriaceous, entire, with a strong midrib, and often with the lateral veins running through to the margin. Flowers usually numerous, axillary, or terminal,

Fig. CCLXXXII.

white, pink, or red, articulated with their peduncle, ⚥ or ♂ ♀ by abortion. Sepals 2, 4, 5, 6, or 8, imbricated by alternate pairs, usually persistent, round, membranous, frequently unequal and coloured like petals. Petals hypogynous, equal in number to the sepals, or the same power, and sometimes passing insensibly into them. Stamens numerous, either distinct, or combined in one or more parcels, hypogynous, rarely definite; filaments of various lengths; anthers adnate, bursting inwards, sometimes very small, occasionally bursting outwards, sometimes 1-celled, and sometimes opening by a pore or transversely; even immersed in a fleshy receptacle. Disk fleshy, occasionally 5-lobed. Ovary solitary, superior, 1- or many-celled; ovules solitary, orthotropal or anatropal, (*Endl.*), erect, or ascending, or numerous and attached to central placentæ; style none, or very short; stigma peltate, or radiate. Fruit either dry or succulent, 1- or many-celled, 1- or many-seeded, dehiscent or indehiscent. Seeds frequently nestling in pulp; their coat thin and membranous; always wingless; very frequently with an aril; albumen none; embryo straight; cotyledons thick, inseparable; radicle either turned to or from the hilum.

Their opposite coriaceous leaves, broken-whorled calyxes, equilateral petals, indefinite stamens, and sessile radiant stigmas, must be regarded as the main features of the Guttifers, to which may usually, though not always, be added the binary arrangement of their calyx and corolla. If these are neglected the Order merges in that of Tutsans. Dr. Wight has indeed proposed to send into that Order Clusia and all the other genera having the calyx and corolla arranged in fives; but to this proposition there are great objections; not the least of which must be the destruction of the precise character of both the Orders. The reader is, however, referred to that excellent Botanist's work above quoted, for an explanation of the reasons which have led him to this conclusion. It is not a little remarkable, that a strong tendency to the separation of sexes should be found among plants so high in the scale of organisation as these are.

Fig. CCLXXXII.—Cambogia gutta. 1. a ♀ flower, with the sterile stamens surrounding the pistil; 2. a ♂ flower; 3. an anther, which opens by throwing off a cap, in consequence of transverse dehiscence; 4. a transverse section of the ovary.

Cambessédes remarks, that "Guttifers differ from Tutsans in their branches, their leaves, and their articulated peduncles; in the normal number of the parts of their flowers, which appears to be two and its multiples, instead of five which obtains in Tutsans; in their anthers united the whole length with the filament, and not articulated at its summit; in their seeds, which often have an aril, and are solitary in each cell of the ovary, a character found in no Tutsans (the monospermous cells of the fruit of some Vismias is due to abortion); finally, in the structure of the embryo, which is different in the two Orders. Tutsans, moreover, have the carpels often nearly distinct. Marcgraaviads are distinguished by their alternate leaves, the singular form of their lower bracts, their petals frequently united, their unsymmetrical flowers, and by their seeds being very small, and exceedingly numerous." Royle remarks that Guttifers are in some respects allied to Ebonyworts, as may be seen by comparing species of Garcinia with some kinds of Diospyrus.

Fig. CCLXXXIII.

All natives of the tropics, the greater part of South America; a few are from Madagascar and the continent of Africa. They generally require situations combining excessive heat and humidity.

An acrid, purgative, yellow gum resin appears to be a very general secretion of the various species of this Order. In one of its forms it becomes the gamboge of commerce, a substance well known because of its use as a pigment, and as an active medicine dangerous in over-doses. The best gamboge comes in the form of pipes from Siam, and this is conjectured to be the produce of Garcinia cochinchinensis; another kind, in lumps, has been said to be derived from Cambogia gutta, called also Hebradendron cambogioides; but Dr. Wight's last experiments are not favourable to this supposition, and he expressly states that the tears of Cambogia gutta "are a substance altogether distinct from true gamboge." Roxburgh says he received frequent samples of the gamboge of his Garcinia pictoria from a correspondent at Tellicherry, and uniformly found it, even in its crude unrefined state, superior in colour, while recent, to any other kind he had tried, but not so permanent as that from China. Dr. Royle confirms this statement. The yellow juice, however, of Xanthochymus pictorius is said to be of very inferior quality.

The seeds of Calophyllum inophyllum yield an oil, and a resin exudes from the roots, which is supposed by some authors to be the same as the Tacamahaca of the Isle of Bourbon. The true East India Tacamahaca is produced by C. Calaba; and Maynas resin is referred to the same species. Martius states that C. brasiliense also yields an acrid aromatic lemon-scented resin. The Hog Gum tree of Jamaica is stated by Dr. Bancroft to be a plant of this Order allied to Ochrocarpus and Garcinia. The gum is a resinous substance, burning with a smoke and yielding an aromatic agreeable odour.—*Hook. Journ.* 4. 144. Dr. Macfadgen asserts that this Hog gum is the same as the mani or oanani of Brazil, and therefore belongs to Moronobea coccinea, to which he refers it. It is largely used in the West Indies for the same purposes as pitch, and also in the form of pills, as a substitute for balsam of copaiva. Endlicher, on the contrary, refers the Hog gum to Clusia flava. Balsam of Maria comes from Verticillaria acuminata; and a great many more furnish similar balsamic substances. In the West Indies the juice of Mammea is employed to destroy the chiggers (Culex penetrans), little insects which attack the naked feet, introducing themselves into the flesh below the toe-nails. The Butter and Tallow-tree of Sierra Leone, which owes its name (Pentadesma butyracea) to the yellow greasy juice its fruit yields when cut, belongs to this Order. The flowers of Clusia insignis weep a considerable quantity of resin from the disk and

Fig. CCLXXXIII.—Arrudea clusioides.—*A. St. Hilaire.*

stamens ; so much indeed, that Von Martius says he obtained an ounce from two flowers ; this resin, rubbed down with the butter of the Chocolate-nut, the Brazilian women employ to alleviate the pain of a sore breast. A few are cultivated for their timber. Calophyllum angustifolium, the Piney-tree, furnishes the straight spars called Peon at Penang, and in the islands to the eastward of the Bay of Bengal, and the Mesuas are said to have excessively hard timber. Of these last plants the root and bark are bitter and aromatic, and powerfully sudorific, their leaves mucilaginous, their unripe fruit aromatic, acrid, and purgative ; the blossoms of Mesua ferrea occur in the bazaars of India under the name of Nagkesur, being used in medicine and esteemed for their fragrance. Lastly, the fruit of many species acquires great excellence and is highly esteemed in tropical desserts. The Mammee Apple, or Wild Apricot of South America, is said to rival the Mangosteen : its seeds are anthelmintic ; its flowers yield, by distillation, a stomachic spirit called Eau de Créole : and a wine is obtained by fermenting its sap. The large berries of Platonia insignis (called Pacoury-uva in Brazil), are very sweet and delicious, while their seeds have the taste of Almonds. The Mangosteen itself, produced in the Straits of Malacca by Garcinia Mangostana, has the reputation of being the finest of all fruits ; it resembles a middle-sized Orange, and is filled with a sweet and most delightful pulp. It is generally thought that this tree will not thrive beyond the hot and damp atmosphere of Malacca : but Dr. Wight states that it has been introduced into the gardens of Courtallum, where it had already begun to bear in the year 1840. *Illust.* 1. 115. Garcinia cornea, Kydiana and pedunculata are mentioned as other species whose fruit is brought to table, but they are represented to be very inferior ; that of G. pedunculata is said to be the nearest approach to the Mangosteen.

GENERA.

I.—CLUSIEÆ.

Tovomita, *Aubl.*
Marialva, Vand.
Marialvea, Mart.
Beauharnoisia, Ruiz. et Pav.
Micranthera, Chois.
Bertolonia, Spreng.
Ochrocarpus, Thouars.
Chrysochlamys, *Pöpp.*
Verticillaria, *Ruiz et Pav.*
Chloromyron, Pers.
Havetia, *H. B. K.*
Renggeria, *Meisn.*
Schweiggera, Mart.
Rengifa, *Pöpp.*
Quapoya, *Aubl.*
Xanthe, Schreb.
Clusia, *Linn.*
Triplandron, *Benth.*
Arrudea, *St. Hil.*

II.—MORONOBEÆ.

Chrysopia, *Noronh.*
Moronobea, *Aubl.*
Symphonia, Linn. f.
Blackstonia, Scop.
Aneuriscus, Presl.

III.—GARCINIEÆ.

Mammea, *Linn.*
Garcinia, *Linn.*
Mangostana, Rumph.
Oxycarpus, Lour.
Brindonia, Thouars.
Xanthochymus, *Roxb.*
Stalagmitis, Mun.
? Discostigma, *Hassk.*
Pentadesma, *R. Br.*
Cambogia, *L.*
Hebradendron, Grah.
Gynotroches, *Bl.*
Platonia, *Mart.*

IV.—CALOPHYLLEÆ.

Mesua, *Linn.*
Rhyma, Scop.
Nagassarium, Rumph.
Calophyllum, *Linn.*
Bintagor, Rumph.
Calysaccion, *Wight.*
Kayea, *Wall.*
Apoterium, *Blum.*

?Rheedia, *Linn.*
Van-Rheedia, Plum.
?Stelechospermum, *Blum.*
?Macanea, *Juss.*
Macahanea, Aubl.
?Macoubea, *Aubl.*
?Souala, *Blanc.*

NUMBERS. GEN. 30. Sp. 150.

Ebenaceæ.
POSITION.—Hypericaceæ.—CLUSIACEÆ.—Ternströmiaceæ.

ORDER CXLV. MARCGRAVIACEÆ.—MARGRAVIADS.

Marcgraviaceæ, *Juss. Ann. Mus.* 14. 397. (1809); *DC. Prod.* 1. 565. (1824); *Endl. Gen.* ccxvii.; *Meisn. Gen.* 44.

DIAGNOSIS.—*Guttiferal Exogens, with simple alternate leaves without stipules, unsymmetrical flowers, equilateral petals, versatile anthers, sessile stigmas, and innumerable minute seeds.*

Trees or shrubs, sometimes climbing and rooting. Leaves alternate, simple, coriaceous, entire, without stipules. Flowers regular, in umbels, racemes, or terminal spikes, usually furnished with bracts which are sometimes bag-shaped or hooded. Sepals from 2 to 7, usually coriaceous and imbricated. Corolla hypogynous; sometimes monopetalous, calyptriform, entire, or torn at the point; sometimes consisting of five imbricated petals. Stamens usually indefinite, inserted either on the receptacle or on a hypogynous membrane; filaments dilated at the base; anthers long, innate, 2-celled, bursting inwards. Ovary single, superior, usually furrowed, 3 or many-celled; style single; stigma simple or capitate; ovules numerous, attached to the projecting lobes of a central placenta, ascending, with the foramen downwards. Fruit supposed to be usually succulent; but also capsular, coriaceous, and consisting of several valves which separate slightly; dissepiments proceeding from the middle of the valves, but not meeting in the centre, so that the fruit becomes 1-celled. Seeds very minute and numerous, nestling in pulp, [oblong, blunt at each end, straight or incurved, with the outer skin hardish and netted, with a lateral hilum. Embryo without albumen, incurved, between club-shaped and cylindrical, with very short obtuse cotyledons, and a long conical acute radicle, which is inferior, contiguous to the hilum, and parallel with it.—*Endlicher.*]

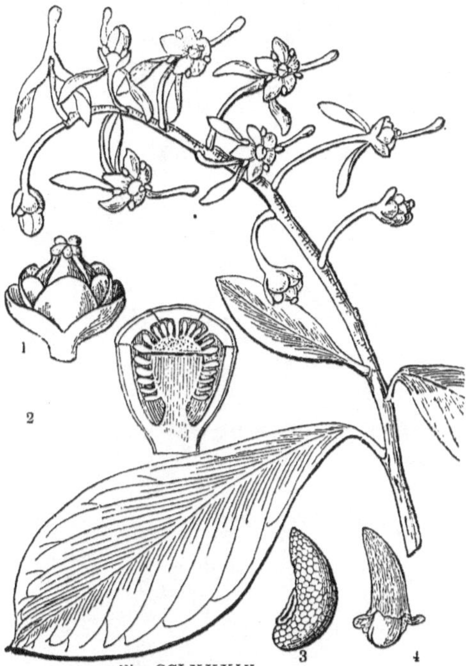

Fig. CCLXXXIV.

The true station of this Order is not clearly made out. It approaches Ebonyworts in its monopetalous corolla cut round at the base, in the anthers attached by their base, and the alternate leaves; Heathworts in the anthers and disk of the genus Antholoma; Tutsans and Guttifers in the hypogynous stamens, the polypetalous corolla of some genera, placentation, and numerous seeds; wherefore Jussieu stationed the Order near Clusia. And this view of the relationship of Margraviads is generally accepted. Indeed, Endlicher says, that the species hardly differ from Guttifers except in their alternate leaves and versatile anthers. But we really know very little about them. Some of the genera are remarkable for the singular condition of their bracts, which assume the appearance of hoods, pouches, or spurs. Turpin has somewhere remarked, that such bracts offer a clear explanation of the conversion of a degenerated leaf into an ovule.

Fig. CCLXXXIV.—Ruyschia amazonica.—*Martius.*—1. a calyx and pistil; 2. a section of the ovary; 3. a seed; 4. the same, with a portion of the testa torn open to show the cotyledons. [*N. B.* Figs. 3 and 4 are reversed in the cut.]

All the species occur in equinoctial America, except the doubtful genus Antholoma, which is a native of New Caledonia.

They are handsome and curious plants, remarkable for their singular cucullate bracts. The stem, root, and leaves of Marcgravia umbellata are regarded in the West Indies as diuretic and antisiphylitic.

GENERA.

Ruyschia, *Jacq.* *Souroubea*, Aubl. *Surubea*, Mey.	*Loghania*, Scop. Norantea, *Aubl.*	*Ascium*, Schreb. *Schwarzia*, Fl. Flum.	Marcgravia, *Plum.* ? Antholoma, *Labill.*

Numbers. Gen. 4. Sp. 26.

Ebenaceæ.
Position.—Clusiaceæ.—Marcgraviaceæ.—Hypericaceæ ?

ORDER CXLVI. HYPERICACEÆ.—TUTSANS.

Hyperica, *Juss. Gen.* 254. (1789).—Hypericineæ, *Chois. Prodr. Hyp.* 32. (1821); *DC. Prodr.* I. 541. (1824); *Endl. Gen.* ccxviii.; *Meisner, Gen., p.* 44; *Wight Illustr.* 1. *t.* 43; *Spach. in Ann. Sc. Nat. ser.* 2. v. 157. 349.—Hypericaceæ, *Ed. Pr.* lviii.—Eucryphieæ, *Endl. Ench. p.* 528.

DIAGNOSIS.—*Guttiferal Exogens, with oblique glandular petals, numerous naked seeds, and long distinct styles.*

Herbaceous, or even occasionally annual, plants, shrubs, or trees, having a resinous juice, and often with angular branches. Leaves opposite, entire, without stipules, occasionally alternate, sometimes crenelled, usually impressed with transparent dots, and bordered with black glands. Flowers in most instances yellow, sometimes red or white, regular, with various forms of inflorescence. Sepals 4 or 5, free from the ovary, persistent, so arranged as to have two exterior to the others, separate or partially united. Petals of the same number as the sepals, unequal-sided, twisted spirally in æstivation, bordered with black dots, sometimes having a fleshy scale or a hollow at their base.

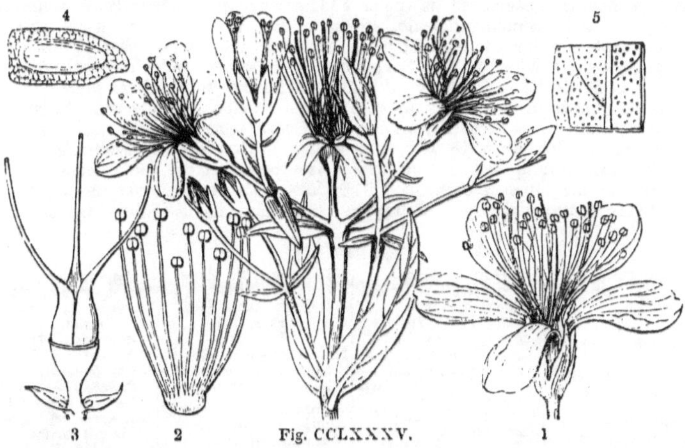

Fig. CCLXXXV.

Stamens hypogynous, almost always 00, sometimes distinct, occasionally monadelphous but almost always polyadelphous; sometimes having fleshy glands intervening between the bundles of stamens; filaments filiform; anthers 2-celled, opening lengthwise, frequently surmounted by a gland. Carpels 3 to 5, partially united round a placenta, which forms the axis, and introduces its arms into their cavity; styles as many as the carpels, usually distinct, but occasionally cohering at the base; stigmas capitate or truncate, rarely 2-lobed; ovules 00, (rarely definite,) generally horizontal, rarely ascending, occasionally pendulous, anatropal, or, in some instances, amphitropal. Fruit sometimes 1-celled, but in most instances either a dry or fleshy capsule, of many valves and many cells; the edges of the former being curved inwards. Seeds minute, usually tapering, attached to a placenta in the axis, or adhering to the inner edge of the dissepiments; embryo straight or curved, with an inferior radicle and no albumen.

The unequal-sided petals, and dark glands upon their edge, offer in most cases a ready means of recognising this Order, which moreover commonly possesses polyadelphous stamens. Its long styles, and distinctly apocarpous fruit, afford a further means of recognition. Keeping these characters in view, no doubt can be entertained of the two genera Eucryphia and Carpodontos, separated by Mr. Endlicher, being genuine members of the Order of Tutsans; for the inequality of the petals is distinctly visible in the latter genus. Nor does it appear desirable to separate from the Tutsans the curious genus

Fig. CCLXXXV.—Hypericum floribundum; 1. an entire flower; 2. a bundle of stamens; 3. a pistil with 3 carpels; 4. a seed laid horizontally and cut through, to show the embryo and netted testa; 5 a piece of a leaf with transparent dots.

Parnassia, whose fringed glands can scarcely be doubted to represent phalanges of sterile stamens, and consequently, indicate a tendency to the production of an indefinite number of polyadelphous stamens, which is one of the characteristics of Tutsans. If indeed the seeds of Parnassia were really parietal, as they are described to be, that would be a reason for removing it to some other place : but its exalbuminous seeds forbid its being stationed among Sundews, and it has nothing in common with Saxifrages except its habit. I believe, however, that in Parnassia, as in Hypericum, the placentæ are truly axile and projected into the cavities of the ovary, which closes over them and adheres to them ; and it is certain that the petals are in some species very unequal-sided, while the anthers of others are tipped by the glands of Tutsans, and the petals themselves, if they have not projecting glands possess immersed glands, in no inconsiderable quantity.

Tutsans are very generally spread over the surface of the earth, inhabiting mountains and valleys, marshes and dry plains, meadows and heaths. The following is the distribution of the species, according to Choisy :—Europe, 19 ; North America, 41 ; South America, 21 ; West Indies, 1 ; Asia, 24 ; New Holland, 5 ; Africa and the neighbouring islands, 7 ; Azores and Canaries, 5 ; common to Europe and Asia, 4 ; common to Europe, Asia, and Africa, 1.—*Choisy Prodr.* 1821. Many have, however, to be added for Asia and South America.

The juice of many is slightly purgative and febrifugal. In the European species this yellow juice is in small proportion to the essential oil, and the rest of the vegetable matter, and they have been used as tonics and astringents ; especially H. perforatum, (ασκυρον,) and Androsæmum officinale. Some of the American species are possessed of a more copious yellow juice, and more energetic properties ; that obtained from Vismia guianensis, a Mexican and Surinam tree, is known in commerce and called American Gummi Gutta. So also the Vismias micrantha and laccifera yield red sticks of a drastic gum-resin analogous to gamboge.—*Martius.* Hypericum hircinum is fœtid. A gargle for sore throats is prepared in Brazil from the Hypericum connatum, commonly called Orelha de Gato. A decoction of the leaves of another species, Hypericum laxiusculum, or Allecrim brabo, is reputed in the same country to be a specific against the bites of serpents. The United States people prepare a stomachic tincture from Elodea virginica. Cratoxylon Hornschuchia is slightly astringent and diuretic.

GENERA.

I. HYPERICEÆ. — No glands between the stamens.

Ascyrum, *Linn.*
Isophyllum, Spach.
Hypericum, *Linn.*
Eremosporus, Spach.
Drosanthe, Spach.
Webbia, Spach.
Holosepalum, Spach.
Milleporum, Spach.
Adenosepalum, Spach.
Drosocarpium, Spach.
Coridium, Spach.
Crossophyllum, Spach.
Olympia, Spach.
Campylopus, Spach.
Psorophytum, Spach.
Androsæmum, Allion.
Eremanthe, Spach.
Campylosporus, Spach.
Norysca, Spach.
Roscyna, Spach.
Myriandra, Spach.
Brathydium, Spach.
Brathys, Mut.
Receveura, Fl. Flum.
Sarothra, Linn.
Eucryphia, *Cav.*
Carpodontos, *Lab.*

II. ELODEÆ. — Glands alternating with the bundles of stamens.

Parnassia, *L.*
Elodea, *Adans.*
Martia, Spreng.
Triadenia, Spach.
Vismia, *Velloz.*
Coapia, Piso.
Psorospermum, *Spach.*
Haronga, *Thouars.*
Harongana, Lam.
Arongana, Pers.
Hæmocarpus, Noronh.
Eliæa, *Cambess.*
Cussonia, Commers.
Lanigerostemma, Chpl.
Ancistrolobus, *Spach.*
Tridesmis, *Spach.*
Cratoxylon, *Blum.*
Hornschuchia, Blum.

NUMBERS. GEN. 13. Sp. 276.

Saxifragaceæ.
POSITION.—Clusiaceæ.—HYPERICACEÆ.—Reaumuriaceæ.

ORDER CXLVII. REAUMURIACEÆ.—REAUMURIADS.

Reaumurieæ, *Ehrenberg in Ann. des. Sc.* 12. 78. (1827).—Reaumuriaceæ, *Ed. pr.* lxx.; *Endl. Gen.* ccxx.; *Meisner, p.* 129.

DIAGNOSIS.—*Guttiferal Exogens, with oblique glandular petals, a few shaggy seeds, and long distinct styles.*

Small shrubs with fleshy scale-like leaves, which are alternate, have no stipules, and are overspread by resinous sunk glands. Calyx 5-parted, surrounded externally by imbricated bracts. Petals 5, hypogynous, unequal-sided, sometimes having a pair of membranous plates planted upon their middle. Stamens definite or indefinite, hypogynous, monadelphous or polyadelphous, with or without a hypogynous disk; anthers ovate, turned inwards, and bursting longitudinally. Carpels free, 2- 4- 5, partially separate from each other, surrounding a central placenta which passes into the base of each; ovules 2 or 4, ascending, anatropal; styles filiform, or subulate. Fruit capsular, with 2 to 5 valves and as many cells, unless the number is diminished by abortion. Seeds shaggy, definite, erect; embryo straight, surrounded by a small quantity of mealy albumen; radicle next the hilum.

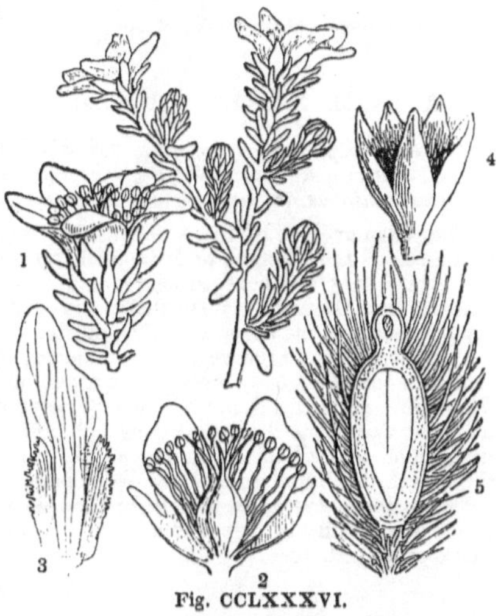

Fig. CCLXXXVI.

Ehrenberg suggested (*Ann. des Sc.* 12. 78.) that Reaumuria and Hololachna might constitute a little group, to be called Reaumuriaceæ. At that time the true relations of plants were ill understood, and if he had referred the genera he knew to Tutsans, he would never have had his opinion called in question. In fact there is nothing to distinguish these Orders except that Reaumuriads have shaggy seeds, and appendages at the base of the petals, which appear to be destitute of glands. They have no affinity with Ficoids or Tamarisks.

Natives of the coast of the Mediterranean and the salt plains in the milder parts of northern Asia.

It seems that these plants abound in saline matter, a circumstance that is doubtless owing to the situations in which they grow. Reaumuria vermiculata is used at Alexandria as a cure for itch. Its bruised leaves are applied externally, and a decoction is administered internally.

GENERA.

Hololachna, *Ehrenb.*	Reaumuria, *Hasselq.*	Eichwaldia, *Ledeb.*

NUMBERS. GEN. 3. SP. 4.

POSITION.—Hypericaceæ.—REAUMURIACEÆ.————

Fig. CCLXXXVI.—Reaumuria hypericoides. 1. a flower and its bracts; 2. the same divided perpendicularly; 3. a petal; 4. capsule; 5. seed divided perpendicularly and much magnified.—*Schnitzlein.*

ALLIANCE XXXI. *NYMPHALES.*—THE NYMPHAL ALLIANCE.

DIAGNOSIS.—*Hypogynous Exogens, with dichlamydeous flowers, axile or sutural placentæ, 00 stamens, and an embryo on the outside of a large quantity of albumen; or, if exalbuminous, the seeds have a very large plumule.*

The singular fact of the embryo lying on the outside of a large mass of albumen would enable the Botanist to distinguish Nymphals with certainty from all those Orders with which they are here associated, if it were not for the Waterbeans, which appear to have no albumen at all. With them, however, it would seem as if an enormous plumule compensated for the absence of this substance. The species are among the most highly developed of any in the vegetable kingdom, if we only regard their flowers; but the total absence of a woody stem places them, on the other hand, among less noble allies. They differ from the Ranal Alliance principally in their embryo, and seem to run close upon the Crowfoots themselves, through both the Waterbeans and Watershields. They have no obvious alliance with any part of Guttiferals, except Guttifers themselves; but they touch the Ranal Alliance at every point.

The stamens are often attached to the sides of the ovary, and are even not liberated in some cases till the very summit of it; but this seems a mere modification of the hypogynous structure.

NATURAL ORDERS OF NYMPHALS.

Carpels united into a many-celled fruit, with dissepimental placentæ. 148. NYMPHÆACEÆ.
Carpels distinct. Albumen copious. Torus absent. . . . . . . 149. CABOMBACEÆ.
Carpels distinct. Albumen 0. Torus honey-combed, very large. . 150. NELUMBIACEÆ.

ORDER CXLVIII. NYMPHÆACEÆ.—WATERLILIES.

Nymphæaceæ, *Salisbury, Ann. Bot.* 2. p. 69. (1805) ; *DC. Propr. Med. ed.* 2. p. 119. (1816) ; *Syst.* 2. 39. (1821) ; *Prodr.* 1. 113. (1824) ; *Wight's Illustrations*, p. 24 ; *Endl. Gen.* clxxxvii. ; *Meisner Gen.* 6.

DIAGNOSIS.—*Nymphal Exogens, with a many-celled fruit and dissepimental placentæ.*

Herbs, with peltate or cordate fleshy leaves, arising from a prostrate trunk, growing in quiet waters. Flowers large, showy, often sweet-scented. Sepals usually 4, free, rarely adherent ; petals numerous, imbricated, often passing gradually out of the last into stamens ; the former persistent, the latter deciduous, and inserted upon the disk, sometimes forming a monopetalous corolla. Stamens numerous, inserted above the petals into the disk, filaments petaloid ; anthers adnate, bursting inwards by a double longitudinal cleft. Disk large, fleshy, surrounding the ovary more or less. Ovary polyspermous, many-celled, with radiating stigmas, alternate with the dissepiments ; ovules numerous, anatropal, attached to the sides of the dissepiments. Fruit many-celled, indehiscent. Seeds very numerous, attached to spongy dissepiments. Albumen farinaceous. Embryo small, on the outside of the base of the farinaceous albumen, inclosed in a fleshy vitellus ; cotyledons fleshy, concave ; plumule oblique.

Fig. CCLXXXVII.

The opinions of Botanists are divided concerning the true nature of the structure of these beautiful plants, and consequently as to their proper station in a Natural System. This has been caused by some peculiarities in the embryo on the one hand, and by the want of any resemblance in the internal condition of the stem and that of Exogens. Richard supposed the vitellus, or amniotic sac, in which the embryo is inclosed, to be a cotyledon, enveloping a two-lobed plumule ; and hence the Order was referred to Endogens, or Monocotyledons, and placed in the vicinity of Hydrocharads. But it is now well known that Richard's cotyledon is a vitellus, analogous to that of Peppers, Gingerworts, and others ; and that what Richard and his followers denominated plumule, is a 2-lobed embryo, whence the Order is more generally placed in Exogens, or Dicotyledons. Even Von Martius, who once adhered to the opinion that Waterlilies are monocotyledonous, and nearly related to Hydrocharads, (see *Hortus Regius Monacensis*, p. 25.) now places the Order near Crowfoots (see *Conspectus, No.* 188). Those who are curious to examine the different opinions on this subject are referred to *De Candolle's Memoir*, in the first volume of the Transactions of the Physical and Natural History Society of Geneva.

Fig. CCLXXXVII.—Nymphæa alba. 1. the pistil ; 2. a perpendicular section of a seed ; 3. half an embryo, showing the great plumule lying in the cavity of one cotyledon.

It seems, however, desirable to state, in this place, what the reasons are which have led so many modern Botanists to place the Order in the class of Exogens. If the rhizome of Nymphæa is examined it will be found to consist principally of cellular tissue, with a very confused distribution of fibrovascular bundles among it, not at all like that of Exogens, but more resembling what occurs in succulent Endogens. But, according to Mirbel's examination of the anatomy of the roots of Nuphar luteum, in the *Annales du Museum*, vol. 16, pl. 20, the bundles of fibres are there placed in a concentric circle, the youngest being outermost. Secondly, the leaves are those of Dicotyledons, and so is their convolute vernation, which is not known in Monocotyledons, together with their insertion and distinct articulation with the stem. Thirdly, the flowers of Waterlilies have so great an analogy generally with Dicotyledons, and particularly with those of Magnoliads, and their fruit with Poppyworts, that it is difficult to doubt their belonging to the same group.

It is not possible to refuse assent to the importance of some, at least, of these considerations; but I do not think that they quite dispose of the question as to where, in a Natural System, Nymphæa and its allies are to be placed. To the foliage little value can be assigned, for it is sufficiently like that of Hydrocharis. Nor does the structure of the root of Nuphar prove the stem to be an anomalous form of Exogens; for the circle of fibrovascular bundles found there is the common character of the roots of Endogens, as Schleiden first pointed out, and has no resemblance to that of Exogens. The argument derived from internal structure is therefore more in favour of Waterlilies being Endogens than Exogens. The true ground for considering them Exogens is certainly confined to the two-lobed embryo. It seems to have been forgotten that when Brown and Brongniart proved Richard's cotyledon to be nothing more than the amniotic sac, they did not also prove, as a necessary consequence, that the so-called plumule of Richard was a dicotyledonous embryo. It may be monocotyledonous, notwithstanding its vitellus. Certainly its two lobes are very like those of Exogens; but I find that in Nymphæa alba the lobes are not suddenly contracted at their base like true cotyledons, (nor are they in Nelumbium,) and, moreover, that the plumula is, in that plant, placed in an oblique direction with respect to the lobes; so that, in fact, the embryo of Nymphæa is much like a modification of such monocotyledonous embryos as those of Aponogeton, Cymodocea, and Posidonia.—See *Ann. Sc. n. s.* xi. *t.* 17. Indeed, I perceive no reason why it should not be regarded as having one split cotyledon, rather than two distinct ones. The principal mass of the nucleus in the seeds of Orchids appears from the researches of Professor Link to be an analogous case. In these plants the nucleus is a spheroidal cotyledon, from whose surface the radicle and plumule respectively protrude. We have only to imagine it elevated on each side, and we should have the two-lobed body of this Order. For the present, however, I am not prepared to disturb existing arrangements; though I much suspect that it will be done by some other Botanist. Indeed M. Ad. Brongniart has lately declared that the position of Waterlilies appears to him very doubtful.—*Enumeration* xxv. De Candolle assigns as a further reason for considering Waterlilies to be Dicotyledons, that they are lactescent, a property not known in Monocotyledons. But in this he is mistaken; Limnocharis, a genus belonging to Butomads is lactescent. Finally, Mr. Hassal appeals to the condition of the pollen of Waterlilies, which he thinks proves them to be undoubted Monocotyledons. The pollen grain of Nymphæa is described by this observer as being oval, hispid, with a furrow down one side, and emitting a single pollen tube, which marks he regards as characteristic of Endogens.

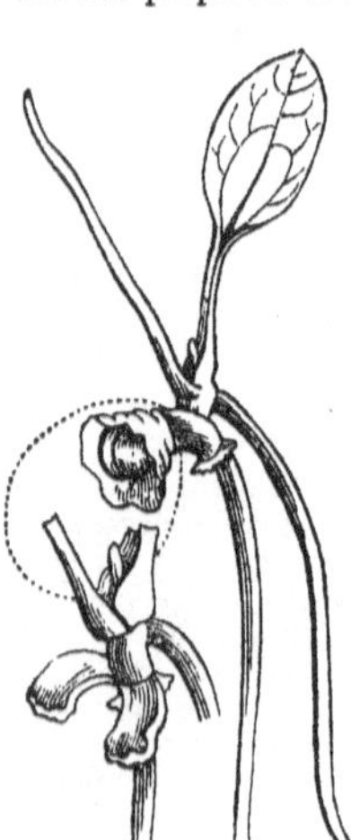

Fig. CCLXXXVII. *bis.*

The germination of Nymphæa alba is not exactly either exogenous or endogenous. The radicle is clearly endorhizal, as in the latter; but the cotyledons lengthen their bases to allow the plumule to escape, just as in an acorn; and this is perhaps one of the strongest arguments in favour of the lobes of the embryo being really cotyledons.

Supposing this Order to be exogenous and dicotyledonous, its immediate affinity will be with Poppyworts, with some genera of which it agrees in the very compound nature of the fruit, from the apex of which the sessile stigmas radiate, in the presence of narcotic principles and a milky secretion, and in the great breadth of the placentæ. Waterlilies are also considered akin to Magnoliads, with which they agree

Fig. CCLXXXVII. *bis.*—Germination of Nymphæa alba.

in the imbricated nature of the petals, sepals, and stamens; to Waterbeans their resemblance is evident; with Crowfoots they are connected through the tribe of Pæonies, with which they agree in the dilated state of the disk, which, in Pæonia papaveracea and Moutan, frequently rises as high as the top of the ovaries, and in the indefinite number of their hypogynous stamens; but in Crowfoots the placentæ only occupy the suture of each of the carpels of which the fruit is made up; so that in Nigella, in which the carpels cohere in the centre, the seeds are attached to the axis, while in Waterlilies the placentæ occupy the whole surface of each side of the individual carpels, of which the fruit is composed.

On the other hand, if we consider Waterlilies as a part of the Endogenous class, we shall be at no loss to find strong affinities for them in that series; as for example with Hydrocharads, and more particularly with the Alismal Alliance, whose indefinite carpels, habit, and peculiar placentation, are very important points of resemblance.

Independently of the circumstances to which allusion has just been made, this Order is remarkable in some other respects. It offers one of the best examples which can be adduced of the gradual passage of petals into stamens, and of sepals into petals: if attentively examined, the transition will be found so insensible that many intermediate bodies will be seen to be neither precisely petals nor stamens, but both in part. The development of the torus, which is so remarkable in Waterbeans, is here represented by a similar enlargement of the disk, which in some, as in Nuphar, is merely a hypogynous expansion, out of which grow the stamens and petals; in others, as Nymphæa, elevates itself as high as the top of the ovary, to the surface of which it is adnate, and as the stamens are carried up along with it, we have these organs apparently proceeding from the surface of the ovary; in the genus Barclaya, the petals are also carried up with the stamens, on the outside of which they even cohere into a tube, so that in this genus we have a singular instance of an inferior calyx and a superior corolla in the same plant. In Victoria the sepals are also adnate to this disk, and thus a half-adherent ovary is produced. In Nymphæa alba, the seeds are inclosed in a true arillus; but M. Planchon (*Mém. sur les Arilles*, p. 18) has shown that no such integument exists in Nuphar luteum.

Floating plants, inhabiting the whole of the northern hemisphere, occasionally met with at the southern point of Africa, but generally rare in the southern hemisphere; on the continent of South America they are represented by Victoria.

This Order has the reputation of being antiaphrodisiac, sedative, and narcotic—properties not very clearly made out, but generally credited. Dr. Wight has, however, well observed that these are quite imaginary qualities, assumed to exist in consequence of the habitation of Waterlilies "in the midst of cool and placid waters, combined with the chaste whiteness of their flowers." The Turks prepare a cooling drink from the flowers of Nuphar luteum, which they call Pufer ciceghi. Their stems are certainly bitter and astringent, for which reason they have been prescribed in dysentery. They contain a considerable quantity of starch, and after repeated washings, are capable of being used for food without danger. The seeds are eagerly sought after in times of scarcity, by the wild people in whose countries they grow. They taste like Poppy-seeds, and are used either boiled or raw like Millet. Victoria, the most gigantic and beautiful of water plants, is said to be on that account called Water Maize in South America. Euryale seeds are in like manner a favourite food among the Indians and Chinese. The large quantity of starch contained in them accounts for this. The rhizomes of various species of Nymphæa are esteemed by the negroes of Senegal, who are said to roast and eat them like Potatoes. In India the farinaceous seeds are eaten either in a raw state, or after having been roasted in heated sand. It is said by Fée that the rhizomes of Nymphæa alba are better than Oak-galls for dyeing gray; they have also been long employed advantageously for tanning leather; and a tolerable sort of beer has been prepared from them. The leaves of Nuphar luteum are reported to be styptic.—*Endl.*

GENERA.

Tribe 1. Euryalidæ.—*Tube of the calyx adherent to the disk. Petals distinct.*

Euryale, *Salisb.*
Anneslea, Andr.
Victoria, *Lindl.*

Tribe 2. Nupharidæ.—*Calyx and petals both distinct.*

Nymphæa, *Neck.*
Castalia, Salisb.
Leuconymphæa, Boerh.
Cyanea, DC.
Lotos, DC.
Castalia, DC.
Nuphar, *Smith.*
Nymphosanthus, Rich.
Nenuphar, Hayn.

Tribe 3. Barclayidæ.—*Calyx free. Corolla adhering to the disk, monopetalous.*

Barclaya, *Wall.*

NUMBERS. GEN. 5. Sp. 50.

Papaveraceæ.
POSITION.—Cabombaceæ.—NYMPHÆACEÆ.—Nelumbiaceæ.
Alismales.

ORDER CXLIX. CABOMBACEÆ.—WATERSHIELDS.

Cabombaceæ, *Torrey and Gray*, 1. 54. (1838). — Cabombeæ, *Rich. Anal. Fr.* (1808); *Endl. Gen.* clxxxvi —Podophyllaceæ, § Hydropeltideæ, *DC. Syst.* 2. 36. (1821); *Prodr.* 1. 112. (1824).—Hydropeltideæ, *Schleid. in Wiegm. Arch.* 5. 230.

DIAGNOSIS.—*Nymphal Exogens, with distinct carpels, abundant albumen, and no visible torus.*

Aquatic plants, with floating peltate leaves. Flowers axillary, solitary, yellow, or purple. Sepals 3 or 4, coloured inside. Petals 3 or 4, alternate with the sepals. Stamens definite or indefinite, hypogynous, arising from an obscure torus; anthers linear, turned inwards, continuous with the filament. Carpels 2 or more, terminated by a short style. Ovules orthotropal, pendulous. Fruit indehiscent, tipped by the hardened style. Seeds definite, pendulous; embryo minute, two-lobed, inclosed in the fleshy sac of the amnios, at the apex of the nucleus, and external to an abundant fleshy albumen.

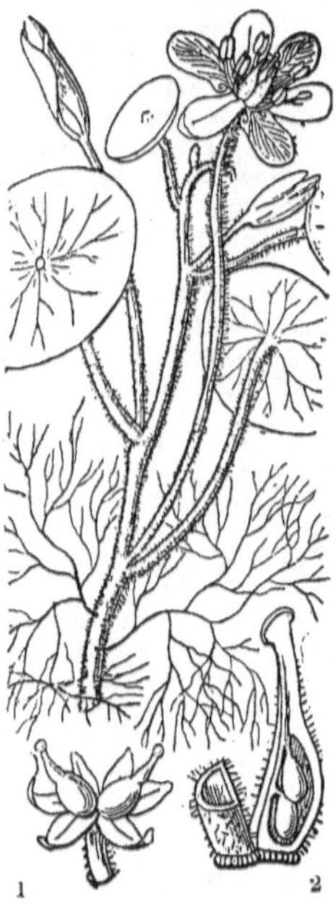

Fig. CCLXXXVIII.

There can be no doubt about the near relationship of these plants to Waterlilies, with which they correspond in having a minute embryo inclosed in a vitellus, and from which they only differ in having disunited carpels, and a very small number of sutural ovules. From Waterbeans, with which they correspond in their disunited carpels, they are distinguished by their abundant albumen, minute embryo, nearly total want of torus, and having more seeds than one in each carpel. Their relationship to Podophyls is much more remote, nor can they belong to the same Alliance, although they have been combined with that Order by De Candolle. Richard, who regarded Waterlilies as Monocotyledons, referred these plants also to that great class; but he misunderstood the structure of their embryo, which has been well illustrated by Schleiden. Nuttall describes the young leaves and flowers, together with the other parts exposed to the air, to be covered "with an inconspicuous flocculent pubescence, immersed in a gelatinous substance." This Schleiden states to be a remarkable state of the epidermis, which consists of a very thick layer of well-defined insoluble gelatine, in which the cells of the epidermis are introduced.—*Wiegm.* 5. 230. The same author says, that not a trace of spiral vessels can be found in any of the submersed parts. I find the stem to consist of a mass of small cellular tissue surrounding 15 or 16 large air-tubes, and some smaller ones, in the centre of which is a pair of woody bundles, crescent-shaped in their transverse section, with the convexity directed inwards. These bundles consist of thin-walled elongated tissue, in the middle of which is a solitary tube of larger size, apparently also an air-tube, for I can find no trace of spiral structure in it.

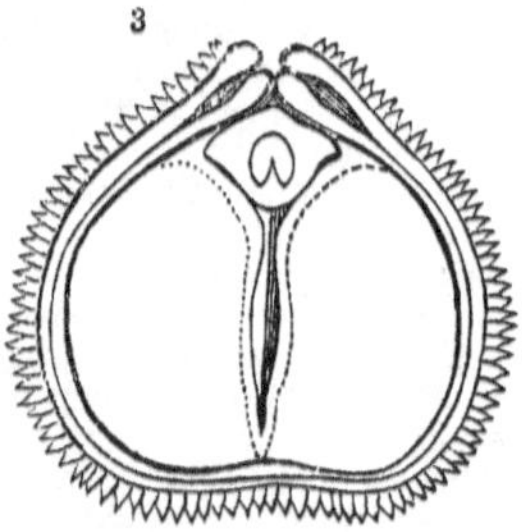

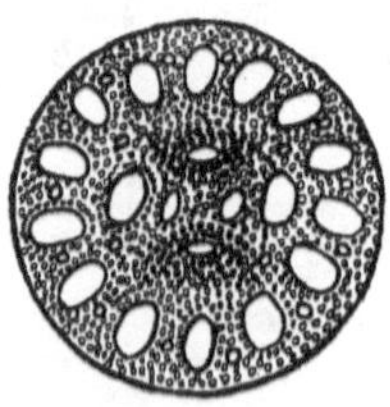
Fig. CCLXXXIX.

Fig. CCLXXXVIII.—Cabomba aquatica. 1. the pistil and calyx; 2. sections of the carpels *(Turpin)*; 3. a section of its seed.—*Schleiden.*
Fig. CCLXXXIX.—Section of the stem of Hydropeltis purpurea.

American water-plants, found from Cayenne to New Jersey, and also on the coast of New Holland, beyond the tropics, according to Endlicher.

Hydropeltis purpurea is said to be nutritious, but slightly astringent. The leaves are employed as a remedy for phthisis and dysentery.

GENERA.

Cabomba, *Aubl.*
Nectris, Schreb.
Hydropeltis, *Mich.*
Brasenia, Pursh.
Rondachime, Bosc.

NUMBERS. GEN. 2. Sp. 3.

Ranunculaceæ.
POSITION.—Nymphæaceæ.—CABOMBACEÆ.—Nelumbiaceæ.

ORDER CL. NELUMBIACEÆ.—WATER BEANS.

Nymphæaceæ, § Nelumboneæ, *DC. Syst.* 2. 43. (1821); *Prodr.* 1. 113. (1824)—Nelumboneæ, *Martius Conspectus, No.* 187. (1835); *Endl. Gen.* clxxxix.—Nelumbiaceæ, *Ed. Pr. Wight Illustr.* i. *t.* 9.

DIAGNOSIS.—*Nymphal Exogens, with distinct carpels immersed in a large honeycombed torus, and without albumen.*

Herbs, with peltate, fleshy, floating leaves arising from a prostrate trunk, growing in quiet waters. [The rhizome growing at the point, with bundles of vessels forming a net-like cylinder, from whose outer and inner part bundles pass to the leaves and lateral flowers.—*Unger.*] Sepals 4 or 5. Petals numerous, oblong, in many rows, arising from without the base of the torus. Stamens numerous, arising from within the petals, in several rows; filaments petaloid; anthers adnate, bursting inwards by a double longitudinal cleft. Torus fleshy, elevated, excessively enlarged, inclosing in hollows of its substance the carpels, which are numerous, one-seeded, with a very short style and simple stigma. Ovule single, suspended from the point of a cord rising from the base of the cavity, anatropal. Nuts numerous, half buried in the hollows of the torus, in which they are, finally, loose. Seeds solitary, rarely 2; albumen none; embryo large, with two fleshy cotyledons and a highly developed plumule, inclosed in its proper membrane.

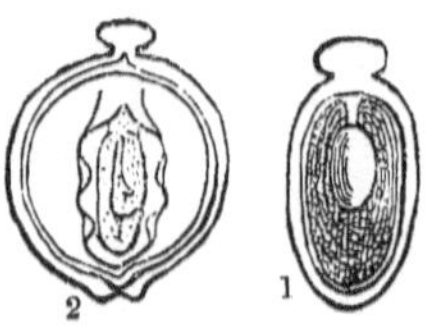

Fig. CCXC.

This beautiful race of water plants offers one of the most striking exceptions to the usual importance of albumen as a general mark of affinity; for, although undoubtedly a member of the Nymphal Alliance, it has not a trace of albumen. Its cotyledons, however, are crammed with starch, and it has a plumule so completely organised, that it is ready to perform all the functions of growth the instant that germination is excited, and thus that necessity for a separate magazine of food, which is so great with the feeble Nymphæaceous embryo, does not here exist. The nature of what is here called the proper membrane of the plumule is not explained by Botanists. Richard regarded it as a cotyledon, the apparent cotyledons being in his view a two-lobed radicle. Ad. Brongniart refers it to the sac of the amnios, which seems inadmissible. De Candolle regarded it as a stipule; but it is found in connection only with the first leaf of the plumule, while, if De Candolle is right, it ought to be present at the base of the second leaf also. The singular enlargement of the torus, which constitutes so striking a feature in these plants, is probably a less important circumstance than their large exalbuminous embryo.

Natives of stagnant or quiet waters in the temperate and tropical regions of the northern hemisphere, both in the Old and the New World; most abundant in the East Indies. They were formerly common in Egypt, but are now extinct in that country, according to Delile.

Chiefly remarkable for the beauty of the flowers. The fruit of Nelumbium speciosum is believed to have been the Egyptian Bean of Pythagoras, and the flower that Mythic Lotus, which so often occurs on the monuments of Egypt and India. The nuts of all the species are eatable and wholesome. The root, or more properly the creeping stem,

Fig. CCXC.—Nelumbium speciosum. 1. a section of its young carpel, 2. a section of the same when ripened into a bean, and showing the structure of the seed.

is used as food in China. Dr. Roxburgh relates that the tender shoots of the roots (rootstock), between the joints, are eaten by the natives of India, either simply boiled or in their curries. The seeds are eaten raw, roasted, or boiled. Nuttall states that the tubers of Nelumbium luteum resemble those of the Sweet Potato, are as farinaceous and agreeable when boiled, and are used for food by the American Indians. Endlicher says that the milky viscid juice of the leaf-stalks and flower-stalks is employed as a remedy against sickness and diarrhœa, and that the petals, which smell of Anise flowers, are slightly astringent and used like Rose flowers. Dr. Wight informs us that the leaf and flower-stalks abound in spiral vessels, which are carefully extracted in India and formed into those wicks "which on great and solemn occasions are burnt in the lamps of the Hindoos placed before the shrines of their gods." Similar wicks are prepared from some Nymphæas, but are not considered so sacred.

GENUS.

Nelumbium, *Juss.*
Nelumbo, Gærtn.
Cyamus, Salisb.

NUMBERS. GEN. 1. Sp. 3, at least.

POSITION.—Nymphæaceæ.—NELUMBIACEÆ.—Cabombaceæ.

ALLIANCE XXXII. *RANALES.*—THE RANAL ALLIANCE.

DIAGNOSIS.—*Hypogynous Exogens, with monodichlamydeous flowers, sutural or axile placentæ,* 00 *stamens, and a minute embryo inclosed in a large quantity of fleshy or horny albumen.*

Under this name are collected some of the most common, and at the same time the most highly developed species of the Vegetable Kingdom. In general they are characterised by the presence of a distinct calyx and corolla; but it is by no means uncommon to find these organs so blended together as to be undistinguishable, while in other instances the corolla is wholly wanting, and it even occurs occasionally that neither one nor the other is present. In appearance Ranals are singularly different even in the same Order; as, for example, in the Crowfoots, under which arrange themselves the common Crowfoot, the Aconite, Thalictrum, and Xanthorrhæa. But although there is so much diversity of appearance among them, nevertheless they certainly form a well compacted group, no one member of which can be spared, as will be seen by examining the remarks made under each Order. In general they have an indefinite number of stamens, but the genus Bocagea presents a very remarkable exception to that rule. They pass into the Berberal Alliance by the Poppyworts, some of which resemble Sarraceniads, and others the common forms of the Crowfoot Order. A clear case of transition to the Erical Alliance also seems to be established by the genus Saurauja, which to the disunited styles of Ranals and their indefinite stamens, adds the minute indefinite seeds, porous anthers, and monopetalous corolla of Heathworts themselves; that genus may be regarded as a Clethra, with the indefinite stamens of Tetracera, or as a Tetracera with the monopetalous corolla, minute seeds, and porous anthers of a Clethra. To Umbellifers in the Epigynous series they pass by way of their genus Thalictrum, whose whole habit is that of the former Order, and whose fruit would, if it adhered to the calyx, be nearly that of an Umbellifer.

NATURAL ORDERS OF RANALS.

Carpels distinct. Stipules large, convolute. Corolla imbricated. Albumen homogeneous	151. MAGNOLIACEÆ.
Carpels distinct. Stipules 0. *Corolla valvate. Albumen ruminate.*	152. ANONACEÆ.
Carpels distinct. Stipules 0. *Corolla imbricated. Albumen homogeneous. Seeds arillate*	153. DILLENIACEÆ.
Carpels distinct. Stipules 0. *Corolla imbricated. Albumen homogeneous. Seeds without an aril*	154. RANUNCULACEÆ.
Carpels consolidated. Calyx permanent. (*Placentæ axile*) . .	155. SARRACENNIACEÆ.
Carpels consolidated. Calyx deciduous. (*Placentæ usually parietal*)	156. PAPAVERACEÆ.

ORDER CLI. MAGNOLIACEÆ.—MAGNOLIADS.

Magnoliæ, *Juss. Gen.* 280. (1789.)—Magnoliaceæ, *DC. Syst.* 1. 439. (1818); *Prodr* 1. 77. (1824); *Blume Fl. Jav.*; *Endl. Gen.* clxxvi.; *Meisner Gen. p.* 3.; *Wight Illustr.* 1. 9.—Wintereæ, *R. Brown in De Cand. Syst.* 1. 548. (1818.)—Illicieæ, *DC. Prodr.* 1. 77. (1824), *a section of* Magnoliaceæ.

DIAGNOSIS.—*Ranal Exogens, with distinct carpels, (usually) large convolute stipules, an imbricated corolla, and homogeneous albumen.*

Fine trees or shrubs. Scales of the leaf-bud formed of stipules either placed face to face or rolled up. Leaves alternate, sometimes with pellucid dots, coriaceous, articulated distinctly with the stem; usually with deciduous stipules which, when young, are rolled together like those of Ficus, and mark the branches with ringed scars where they fall off. Flowers solitary, often strongly odoriferous, usually ⚥, but in Tasmannia ♂-⚥-♀. Trochodendron has neither calyx nor corolla. Sepals 3-6 (rarely 2-4), deciduous. Petals 3 or more, imbricated, hypogynous, in several rows. Stamens 00, distinct, hypogynous; anthers adnate, long. Carpels several, arranged upon a torus above the stamens, 1-celled; ovules anatropal, one or more, ascending or suspended; styles short, stigmas simple. Fruit either dry or succulent, consisting of numerous carpels, which are either dehiscent or indehiscent, distinct or partially connate, always numerous, often collected in a cone upon a lengthened axis, sometimes terminated by a membranous wing. Seeds solitary, or several, attached to the inner edge of the carpels, from which, when ripe and open, they often hang suspended by a delicate umbilical cord, often enveloped in an aril. Embryo minute, at the base of fleshy albumen.

In this Order are included some of the finest trees and shrubs in the world, strikingly beautiful in their flowers and foliage, often of very considerable size, and in the majority of cases emitting the most fragrant odours. They are closely allied to Dilleniads, which have a quinary arrangement of the calyx and corolla and no stipules, and to Anonads, which also have a ternary arrangement of the floral envelopes and a very great similarity in general structure; but in that Order the petals have a valvate æstivation, and no stipules are present, besides which the albumen is ruminated like that of a Nutmeg. In Winter's Barks, or Wintereæ, which do not seem to possess any solid distinction from Magnoliads, the wood has been observed to present the singular circular disks which are so abundant and remarkable in Coniferous plants.* The genus Tasmannia, called in books diœcious, seems

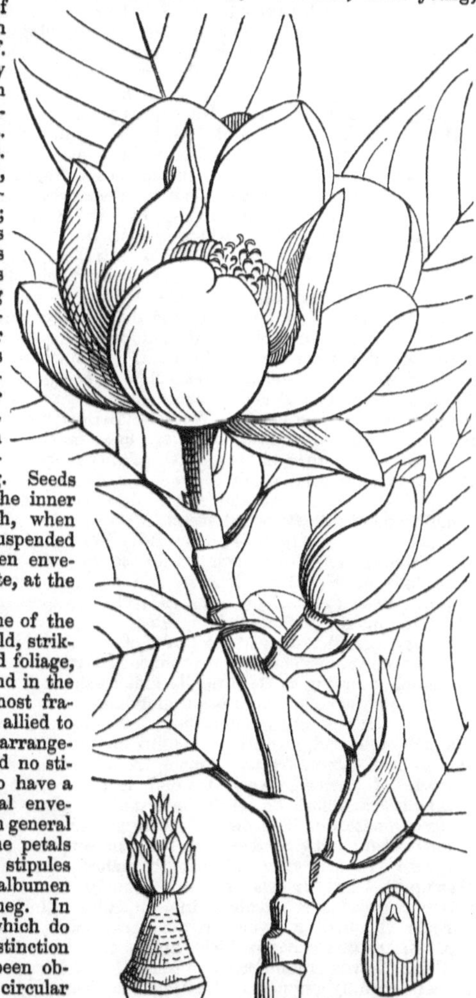

1 Fig. CCXC. *bis.* 2

* This has been denied or confirmed, I hardly know which, by Goeppert, who, in a Memoir on the subject says in one place that the woody tubes of Drimys Winteri are constructed "comme nous les voyons chez les Araucaria," and in another he calls this a resemblance "remarquable sans doute, mais qu'on ne saurait confondre avec celle des Conifères."—*Ann. Sc. Nat.* 2. *ser.* 18. 320.

Fig. CCXC. *bis.*—Magnolia glauca. 1. the pistil; 2. a section of the seed.

rather to be polygamous, and therefore has no claim to be regarded as an exception to the hermaphrodite character of this Order; the tendency, however, in that genus to unisexuality corroborates the opinion of some Botanists, that Magnoliads approach certain diclinous Orders included in the Urtical Alliance, as is indicated by their large convolute stipules, which are very like those of Figs and other genera of Morads. Tasmannia is, indeed, quite an anomalous plant. It is so nearly related to the aromatic Winter's Bark, Drimys Winteri, from which its unisexual flowers and solitary carpels chiefly distinguish it, that it must follow the affinity of that plant. For this reason it seems necessary to associate it with the Order of Magnoliads rather than with that of Kadsurads or Anonads. The three Orders are generally distinguished by the following characters:—Magnoliads are bisexual, have stipules of large size, and their flowers have an imbricated æstivation. Kadsurads resemble them in all things, except the want of stipules, and their flowers being absolutely unisexual. Anonads are bisexual like Magnoliads, but they have no stipules, their corolla is valvate, and their albumen ruminate. Moreover Magnoliads are astringent sub-aromatic trees or bushes; Anonads are similar in quality, but they are more aromatic; Kadsurads are scrambling plants with no aroma. If we regard the aromatic quality of Tasmannia, it will belong to either Magnoliads or Anonads; but from the former it differs in the want of stipules, from the latter in its imbricated corolla, and from both in its unisexual flowers. On the other hand it has the unisexual flowers of Kadsurads, but not their habit nor their mucilaginous qualities. Its unisexual flowers, however, point strongly in the direction of Kadsurads; but then it is not separable from Drimys, which is bisexual, and, moreover, its own flowers are in reality in many cases furnished with a central carpel. Tasmannia must then be regarded as having a manifest tendency towards hermaphroditism, while no such attribute is known among Kadsurads. For these reasons it will be stationed along with Drimys among bisexual Natural Orders, and then will necessarily fall into the ranks of Magnoliads; for its imbricated corolla and homogeneous albumen are at variance with the most essential peculiarity of Anonads. It, however, like Drimys itself, wants the stipules of Magnoliads, in which respect it is exceptional to the usual character of that Natural Order, and must be regarded as a genus stationed on the frontier between Kadsurads and Magnoliads. The small perigynous Order of Calycanths is moreover so like Illicium in appearance, and there is so much resemblance between them in their separate carpels, that, although their affinity is by no means direct, yet we must suppose that some cross relationship exists between them. According to Blume, the umbilical cord, which is so remarkably extensible in some of these plants, is wholly composed of a multitude of delicate spiral vessels.

The focus of the Order is undoubtedly North America, where the woods, the swamps, and the sides of the hills abound with the species. Thence they straggle, on the one hand, into the West India Islands, and on the other, into India, through China and Japan. Brown remarks (*Congo*, 465), that no species have been found on the continent of Africa, or any of the adjoining islands.

The general character of the plants of this Order is to have a bitter tonic taste, and fragrant flowers. The latter produce a decided action upon the nerves; Magnolia tripetala, according to De Candolle, induces sickness and headache; and on the authority of Barton, Magnolia glauca is so stimulating as to produce paroxysms of fever, and even an attack of inflammatory gout. The bark has been found to be destitute of tannin and gallic acid, notwithstanding its intense bitterness. None of the species can be said to have eatable fruits. Among the most fragrant are the Tsjampac or Champaca, a species of Michelia so called, which is the delight of the people of Hindostan; the Magnolia grandiflora, one of the noblest of evergreen trees; Magnolia pumila, well known in green-houses for its brownish-green flowers; while the Yulan, Magnolia conspicua, is unrivalled among northern trees for the surpassing brilliancy of its large and snow-white flowers upon gray and naked branches. As tonics many have great value. The Swamp Sassafras, or Beaver tree (Magnolia glauca), has a bitter and aromatic bark, resembling and even rivalling in its qualities Cinchona. It is particularly useful in chronic rheumatism, whether the bark, seeds, or cones are employed. The same qualities are recognised in Liriodendron tulipifera, the seeds of Magnolia Yulan, called in China Tsin-y, grandiflora, and others. All the parts of Michelia Tsjampaca appear to be powerfully stimulant. Of Magnolia Frazeri (auriculata, *Bartr.*), and M. acuminata, both called Cucumber-trees in the United States, the bitter and somewhat aromatic infusion of the green cones in whisky or brandy is extensively used against intermittent fevers, and also in rheumatic affections. The tonic qualities of these plants are partly owing to their aromatic secretions, which sometimes become very intense. The Aromadendron elegans of Java is one of the most remarkable, and has a great local reputation as a stomachic, antihysteric, and carminative. Michelia montana bark is compared for efficacy to Cascarilla, but it is less bitter. Michelia gracilis bark smells strongly of Camphor. The whole plant of Illicium anisatum, especially the fruit, has a pleasant aromatic flavour of Anise, sweetish and rather pungent. It is reckoned a stomachic and carminative among the Chinese, and is used as a spice in their cookery. The fruit is aromatic and carminative, and by distillation yields an oil which has most of the properties of oil of Anise, for which it is often substituted. It is chiefly used in the fabrication of liqueurs. Illicium floridanum and other species have similar spicy qualities. The seeds of Illicium religiosum are so fragrant that the Chinese burn them in their temples. Drimys Winteri yields the Winter's Bark, which is known for its resemblance to that of Cinnamon. A bark called Melambo Bark, possessing similar properties, is described by Cadet in the *Journal de Pharmacie*, 1815, p. 20; but it is very uncertain whether it belongs to any plant of this Order. The bark of Drimys granatensis, called Casca d'Anta in Brazil, is much used against colic. It is tonic, aromatic, and stimulant, and resembles, in nearly all respects, the Drimys Winteri, or Winter's Bark. Similar in their nature are Drimys axillaris and Tasmannia aromatica, one a New Zealand and the other a New Holland tree, whose fruit is occasionally used as pepper by the settlers in Tasmannia. Many are valuable for their timber. Michelia Doltsopa is one of the finest trees in Nipal, yielding an excellent fragrant wood, much used in that country for house-building.—*Don, Prodr.* 226. Magnolia excelsa has a valuable timber called Champ, at first greenish, but soon changing into a pale yellow; the texture is fine. Manglietia glauca has a white solid wood which is largely employed in Java, and supposed to prevent the decay of corpses put into coffins made of it. Another valuable timber of the same country is that of Aromadendron elegans. Blume remarks that Magnoliads are absolutely known from Dilleniads by their bitter aromatic properties; the latter never being anything beyond styptics.

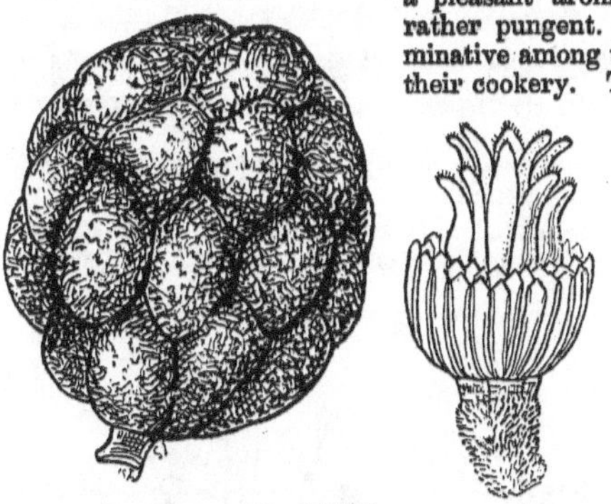

2 Fig. CCXCI. 1

GENERA.

I. MAGNOLIEÆ. — Carpels arranged in a cone. Leaves not dotted, or scarcely.	*Liriopsis*, Spach.	II. WINTEREÆ.—Carpels whorled, in a single row. Leaves with pellucid dots, and often with no stipules.	*Magallana*, Commers.
Talauma, *Juss.*	*Yulania*, Spach.	Tasmannia, *R. Br.*	*Canella*, Domb.
Blumia, Nees.	*Tulipastrum*, Spach.	Drimys, *Forst.*	*Boique*, Molin.
Magnolia, Plum.	*Lirianthe*, Spach.	*Wintera*, Murr.	Illicium, *Linn.*
Aromadendrum, *Blum.*	Manglietia, *Blum.*	*Winterana*, Sol.	*Skimmi*, Kämpf.
Magnolia, *Linn.*	Michelia, *Linn.*		*Badianifera*, Linn.
Gwillimia, Rottl.	*Champaca*, Rheed.		*Cymbostemon*, Spach.
	Sampaca, Rumph.		Trochodendron, *Sieb.etZ.*
	Liriodendron, *Linn.*		*Gymnanthus*, Jungh.
	Tulipifera, Herm.		? Temus, *Molin.*

NUMBERS. GEN. 11. Sp. 65.

Moraceæ.
POSITION.—Anonaceæ.—MAGNOLIACEÆ.—Dilleniaceæ.
Schizandraceæ.
Monimiaceæ?

Fig. CCXCI.—1. stamens and pistil; 2. fruit, of Aromadendron elegans.

ORDER CLII. ANONACEÆ.—ANONADS.

Anonæ, *Juss. Gen.* 283. (1789.)—Anonaceæ, *Rich. Anal. Fr.* 17. (1808) ; *Dunal. Monogr.* (1817) ; *DC. Syst.* 1. 462. (1818) ; *Prodr.* 1. 83. (1824) ; *Bl. Fl. Jav.* ; *Alph. De Cand. in Mem. Phys. Genev.* (1832) ; *Wight Illustr.* 1. 17 ; *Endl. Gen.* clxxiv. ; *Meisner, Gen. p.* 4.—Glyptospermæ, *Vent. Tab.* 3. 75. (1799.)

DIAGNOSIS.—*Ranal Exogens, with distinct carpels, no stipules, a valvate corolla, and ruminate albumen.*

Trees or shrubs. Leaves alternate, simple, almost always entire, without stipules. Flowers usually green or brown, axillary, solitary, or 2 or 3 together, shorter than the

Fig. CCXCII.

leaves ; the peduncles of abortive flowers sometimes indurated, enlarged, and hooked. Sepals 3, persistent, usually partially cohering. Petals 6, hypogynous, in two rows, coriaceous, with a valvate æstivation, sometimes united into a monopetalous corolla, very rarely absent. Stamens indefinite, covering a large hypogynous torus, packed closely together, very rarely definite ; filaments short, more or less angular ; anthers adnate, turned outwards, with an enlarged 4-cornered connective, which is sometimes nectariferous. Carpels usually numerous, closely packed, separate or cohering, occasionally definite ; styles short ; stigmas simple ; ovules solitary, or a small number, erect or ascending, anatropal. Fruit consisting of a number of carpels, which are either succulent or dry, sessile or stalked, 1- or many-seeded, distinct or concrete into a fleshy mass. Seeds attached to the suture in one or two rows, sometimes furnished with an aril ; testa brittle ; embryo minute, in the base of hard, fleshy, ruminate albumen.

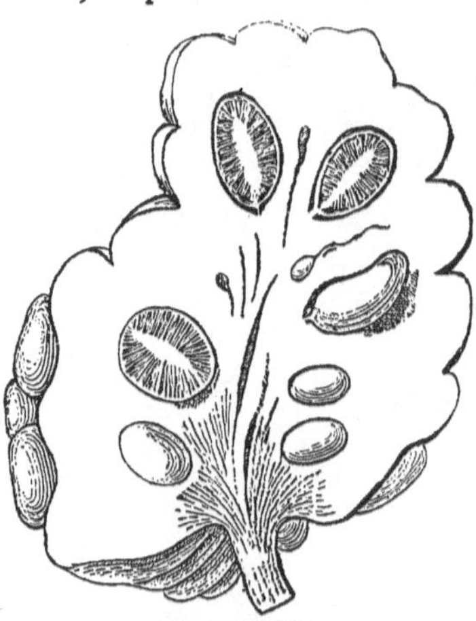

Fig. CCXCIII.

Monodora has a solitary carpel. In Anona palustris the ovaries are not distinct. The stamens and carpels are definite in Bocagea. The flowers are pentamerous in Hentschelia.

Fig. CCXCII.—Anona furfuracea. 1. an expanded flower ; 2. a vertical section of ♂ and ♀ apparatus, which latter occupies the centre ; 3. a vertical section of a carpel ; 4. ditto of a ripe seed, showing the ruminated albumen and embryo.
Fig. CCXCIII.—Section of ripe fruit of Anona squamosa.—*Martius.*

The corolla of these plants is so frequently monopetalous that it affords one of the most striking instances that can be found of the worthlessness of the monopetalous structure as a fundamental mark of distinction. And none of the affinities point in the direction of monopetalous Orders. No doubt can be entertained of the close resemblance of this Order to Magnoliads, from which, however, it differs in the want of stipules, in its valvate corolla, and in the form of the anthers: agreeing in the ternary division of the parts of fructification, and the indefinite stamens and ovaries. An affinity has been pointed out with Menispermads; but it appears to be weak. The great feature of the Order is its ruminated albumen, to which there is no exception, and very few parallels. The parietal insertion of ovules, ascribed to the Order by De Candolle, is not universal. The ovules are erect in Anona, Guatteria, and Anaxagorea. A remarkable plant is described by Brown, in the Appendix to *Flinders's Voyage*, under the name of Eupomatia laurina, in which the stamens are manifestly perigynous, and the tube of the calyx (?) coherent with the ovaries. This plant affords one of the most remarkable exceptions we know of to habitual structure. It is no doubt analogous to Eschscholtzia among Poppyworts and Rosa in Roseworts. I have remarked in Anona laurifolia that the pollen is arranged in two distinct rows in each cell of the anther, and that when that organ bursts, the grains of pollen fall out, cohering in a single row, so as to have the appearance of a necklace. Anonads are connected with Berberids through Bocagea. I also think there can be no doubt of the alliance of the Order to Nutmegs; as has been indicated by Blume and fully admitted by Endlicher.

The tropics of the Old and New World are the natural land of these plants: thence they spread, in a few instances, to the northward and the southward. Some of them, useful to man, such as the Custard-apple, the Cherimoyer, and others, have been carried by colonists far from their native stations.

Their general character is, to have a powerful aromatic taste and smell in all the parts. The bark of Uvaria tripetaloidea yields, being tapped, a viscid matter, which hardens in the form of a fragrant gum. The flowers of many species, especially of Artabotrys odoratissima and Guatteria virgata, are exceedingly sweet. The dry fruits of others are very aromatic; those of Xylopia aromatica are the Piper æthiopicum of the shops, and are commonly used as pepper by the African negroes. The leaves of Artabotrys are regarded as invaluable in Java against cholera. The Polyalthias of Java are employed in Java with advantage as aromatics of great energy, especially their roots. The leaves of Anona squamosa have a heavy disagreeable odour, and the seeds contain a highly acrid principle fatal to insects, on which account the natives of India use them powdered and mixed with the flour of Gram, or Cicer arietinum, for occasionally washing their hair. Xylopia sericea, a large tree found in forests near Rio Janeiro, where it is called Pindaïba, bears a highly aromatic fruit, with the flavour of pepper, for which it may be advantageously substituted. Its bark is tough, and readily separated into fibres, from which excellent cordage is manufactured. Blume remarks that the Javanese species require, because of their powerful properties, to be employed with caution; for if they are administered for too great a length of time, or in too large doses, they produce vertigo, hæmorrhage, or even abortion, in pregnant women. The carpels are chewed after dinner in Java for dispelling flatulence. Xylopia glabra, we are told, is called Bitter-wood in the West Indies, because of the presence of well-marked bitterness in every part. The wood, bark, and berries are said to taste like Orange seeds. The wild pigeons that feed on the berries are said to acquire their flavour, and sugar hogsheads made of the wood are reported to render their contents uneatable, even by cockroaches. Of some species the fruit is succulent and agreeable, containing a sugary mucilage, which predominates over the slight aromatic flavour that it possesses. Of this kind are the delicious Custard-apples of the East and West Indies, the Cherimoyer of Peru, and others. In Uvaria triloba an acid is present of a very active nature, according to Duhamel; but this is not certain. Its leaves are used to bring languid abscesses to a head; its seeds are said to be emetic. The Anona sylvatica, called Araticu do mato, in Brazil, has a light white wood, very fit for the use of turners, and for the same purposes as the Lime-tree of Europe. Its fruit is described as good for the dessert. The wood of the root of A. palustris is employed in Brazil for corks. Martius has remarked that many species of Xylopia strike root with great facility, even though the smallest pieces are committed to the earth. The strong elastic wood called Lancewood by the coachmakers, the Yari yari of Guiana, is stated by Schomburgk to be obtained from Duguetia quitarensis. Martius found the specific gravity of the wood of a species of Guatteria, called Pindaiba preta to be 0·839 after being kept for 20 years in a dry room. See that author's *Flora Brasiliensis* for many interesting particulars concerning the plants of this Order. The Indians on the Orinoco, particularly in Atures and Maypura, have an excellent febrifuge, called Frutta de Burro, which is the fruit of Uvaria febrifuga, or Xylopia grandiflora, according to Martius. The Calabash Nutmeg, Monodora Myristica, is a rival of the true Nutmeg for aromatic qualities; it is not, however, quite certain that it belongs to this place.

GENERA.

I. BOCAGEÆ.—*Endl.*

Bocagea, *St. Hil.*
Poppowia, *Endl.*
Orophea, *Blum.*
Miliusia, *Alph. DC.*
Saccopetalum, *Benn.*

II. XYLOPEÆ.—*Endl.*

Polyalthia, *Blum.*
Oxymitra, Blum.
Kentia, Blum.
Goniothalamus, Blum.
Xylopia, *Linn.*
Bulliarda, Neck.
Xylopicron, P. Br.
Embira, Marcgr.
Pindaiba, Piso.
Ibira, Marcgr.
Habzelia, Alph. DC.
Waria, Aubl.
Cœlocline, *Alph. DC.*
Patonia, *Wight.*
Uvaria, *Linn.*
Unona, Linn. f.
Krokeria, Neck.
Mitrephora, Blum.
Asimina, Adans.
Orchidocarpum, L. C. Rich.
Porcelia, Ruiz et Pav.
Melodorum, Lour.
Trigyneia, Schlecht.
Desmos, Lour.
Marenteria, Noronh.
Hexalobus, *Alph.* DC.

III. ANONEÆ.—*Endl.*

Anaxagorea, *St. Hil.*
Artabotrys, *R. Br.*
Guatteria, *Ruiz et Pav.*
Cananga, Aubl.
Aberemoa, Aubl.
Oxandra, *A. Rich.*
Duguetia, *St. Hil.*
Cardiopetalum, Schl.
Anona, *Linn.*
Guanabanus, Plum.
§ *Atta*, Martius.
Rollinia, *St. Hil.*
Monodora, *Dun.*
Lobocarpus, *Wight et Arn.*

NUMBERS. GEN. 20. Sp. 300.

POSITION.—Magnoliaceæ.—ANONACEÆ.—*Myristicaceæ.* / *Berberidaceæ.*

ORDER CLIII. DILLENIACEÆ.—DILLENIADS.

Dilleniaceæ, *DC. Syst.* 1. 395. (1818); *Prodr.* 1 67; *A. St. H. Fl. Bras.* 1 23; *Endl. Gen.* clxxvii.; *Meisn. Gen.* 2; *Wight Illustr.* 1. 6.

DIAGNOSIS.—*Ranal Exogens, with distinct carpels, no stipules, an imbricated corolla, homogeneous albumen, and arillate seeds.*

Trees, shrubs, or under-shrubs, rarely herbaceous plants. Leaves usually alternate, almost always without stipules, very seldom opposite, most commonly coriaceous, and with strong veins running straight from the midrib to the margin, entire or toothed, often separating from the base of the petiole, which remains adhering to the stem. Flowers solitary, in terminal racemes, or in panicles, often yellow. Sepals 5, persistent, 2 exterior, 3 interior. Petals 5, imbricated, deciduous, hypogynous in a single row. Stamens 00, hypogynous, arising from a torus, either distinct or polyadelphous, and either placed regularly around the pistil or on one side of it; filaments dilated either at the base or apex; anthers adnate, 2-celled, usually bursting longitudinally, always turned inwards. Ovaries definite, more or less distinct, with a terminal style and simple stigma; ovules ascending, anatropal, solitary, or several. Fruit consisting either of from 2 to 5 distinct carpels, or of a similar number cohering together, (now and then one carpel only is present;) the carpels either baccate or 2-valved, pointed by the style. Seeds fixed in a double row to the inner edge of the carpels, either several or only 2, occasionally solitary by abortion; surrounded by a pulpy aril. Testa hard. Embryo minute, lying in the base of solid fleshy albumen.

Fig. CCXCIV.

These are nearly akin to Magnoliads, from which they are distinguished by their want of stipules and the quinary arrangement of the parts of fructification; also to Crowfoots, from which their persistent calyx, stamens, and whole habit, in general divide them. They are universally characterised by the presence of an aril round their seeds. The most genuine form of the Order is known by the veins of the leaves running straight from the midrib to the margin. Some of the genera are remarkable for having the stamens developed only half way round the pistil, so that the central part of the flower has a one-sided appearance. In this respect they tend towards Pittosporads, where Cheiranthera has also declinate stamens. To Anonads they also approach in a variety of ways, especially in the genus Acrotrema, whose albumen is irregularly indented upon the surface, as if it were approaching to a ruminated state.

The genus Saurauja is usually stationed among Theads (Ternstromiaceæ), from which its minute embryo, indefinite seeds, and very copious albumen remove it. From Dilleniads it differs in the want of an aril, and in little else that can be regarded as essential; for its styles, which are divided to the very base, afford conclusive evidence as to its having a tendency to disunite its carpels. If it were not for that circumstance, and its indefinite stamens, it might be placed among Heathworts, of which it has the embryo, the minute indefinite seeds, a tendency to form a monopetalous corolla, and anthers opening by pores. I can scarcely doubt that it forms a complete transition from the Ranal to the Erical Alliance.

The larger part of this Order is found in Australasia, India, and equinoctial America; a comparatively small number is known from equinoctial Africa.

The plants of the Order are generally astringent. The Brazilians make use of a decoction of Davilla rugosa and Tetracera Breyniana and oblongata, in swellings of the

Fig. CCXCIV.—Candollea tetrandra. 1. stamens and pistil; 2. pistil; 3. section of a half-ripe carpel, showing the arillate seeds.

legs and other parts, very common maladies in hot and humid parts of South America. Davilla elliptica is also astringent, and furnishes the vulnerary called Sambaïbinha in Brazil. In Curatella Sambaïba the same astringent principle recommends its decoction as an excellent wash for wounds; this plant is also used by tanners in Brazil. The young calyxes of Dillenia scabrella and speciosa have a pleasantly acid taste, and are used in curries by the inhabitants of Chittagong and Bengal. Tetracera Tigarea (Liane rouge) is diaphoretic and diuretic, and has the reputation of being an antisyphilitic. The acid juice of Dillenia speciosa fruit is, according to Rheede, when added to syrup, considered useful as a cough mixture. The ripe fruits are said to be laxative, and even to produce diarrhœa. Almost all Delimeæ have the foliage covered with asperities, which are sometimes so hard that the leaves are even used for polishing.

The Indian species are in almost all cases plants of great beauty. Dr. Wight speaks of them as remarkable, not less for the grandeur of their foliage than the magnificence of their flowers. He adds, that several species of Dillenia are large trees, and afford hard, durable, valuable timber.

GENERA.

I. DILLENEÆ. — Connective of anthers equal, or narrow at the point. Asiatic and Australian.

Capellia, *Blum.*
Colbertia, *Salisb.*
Reifferscheidia, *Presl.*
Dillenia, *Linn.*
Syalita, Adans.
Actinidia, *Lindl.*
Wormia, *Rottb.*
Clugnia, Commers.
Lenidia, Thouars.
Schumacheria, *Vahl.*
Pleurodesmia, Arn.
Adrastea, *DC.*
Hibbertia, *Andr.*
Burtonia, Salisb.
Cistomorpha, Caley.
Saurauja, *W.*
Palava, R. & P.
Apatelia, DC.
Scapha, Nor.
Vanalphimia, Lesch.
Marumia, Reinw.
Reinwardtia, Nees.
Blumia, Spreng.
Trochostigma, *Sieb.*
Pleurandra, *Labill.*
Candollea, *Labill.*
Pachynema, *R. Br.*
Hemistemma, *Commers.*
Aglaja, Noronh.
Acrotrema, *Jack.*
Tetracarpæa, *Hook.*

II. DELIMEÆ. — Connective of anthers dilated at the point. Chiefly American.

Curatella, *Linn.*
Pinzona, *Mart. et Zucc.*
Doliocarpus, *Roland.*
Calinea, Aubl.
Soramia, Aubl.
Mappia, Schreb.
Othlis, Schott.
Empedoclea, *St. Hil.*
Davilla, *Velloz.*
Hieronia, Flor. Flum.
Delima, *Linn.*
Tetracera, *Linn.*
Tigarea, Aubl.
Rhinium, Schreb.
Euryandra, Forst.
Assa, Houtt.
Wahlbomia, Thunb.
Röhlingia, Dennst.
Trachytella, *DC.*
Actæa, Lour.
Calligonum, Lour.
? Recchia, *Moç. et Sess.*

NUMBERS. GEN. 26. Sp. 200.

Pittosporaceæ.
POSITION.—Ranunculaceæ.—DILLENIACEÆ.—Magnoliaceæ.
Ericaceæ.

ORDER CLIV. RANUNCULACEÆ.—CROWFOOTS.

Ranunculi, *Juss. Gen.* (1789).—Ranunculaceæ, *DC. Syst.* 1. 127. (1818); *Prodr.* 1. 2; *Bartling Ord.* 253. *Endl. Gen.* clxxviii.; *Meisner Gen. p.* 1; *Wight Illustr.* 1. *p.* 1.—Podophyllaceæ, § Podophylleæ, *DC. Syst.* 2. 32. (1821); *Prodr.* 1. 111.—Podophylleæ, *Mart. Conspect. No.* 171. (1835).

DIAGNOSIS.—*Ranal Exogens, with distinct carpels, no separate stipules, an imbricated corolla, homogeneous albumen, and seeds without an aril.*

Herbs, or rarely shrubs. Leaves alternate or opposite, generally much divided, with the petiole dilated and forming a sheath half clasping the stem. Stipule-like processes occasionally present. Hairs, if any, simple. Inflorescence variable. Flowers usually conspicuous; if apetalous, then with the sepals large and gaily coloured. Sepals 3-6, hypogynous, deciduous, generally imbricate in æstivation, occasionally valvate or duplicate. Petals 3-15, hypogynous, in one or more rows, distinct, sometimes deformed, in some cases missing. Stamens 00, (very rarely definite,) hypogynous; anthers adnate. Carpels numerous, 1-celled or united into a single many-celled pistil; ovary one or more-seeded, the ovules sutural; styles simple; ovules anatropal. Fruit either consisting of dry akenia; or baccate with one or more seeds; or follicular with one or two valves. Seeds albuminous; when solitary, either erect or pendulous. Embryo minute. Albumen horny.

1

4 3 2

Fig. CCXCV.

Under the name of Crowfoots is collected a very considerable number of plants, differing from each other materially in the nature of their calyx and corolla, but very similar otherwise. Some of them have perfectly distinct sepals and petals, in others these parts seem completely blended together, as in Caltha and Anemone; in others it is manifest that the former only are present, as in Clematis. These too, which have their parts quite distinct, vary greatly from the real Crowfoots in their nature, the calyx or corolla being extended into spurs, and assuming a very irregular condition in various ways, as in the Clematis and Larkspurs. It is, however, very interesting to find the spurred irregular-flowered plants of this Order assimilated with the regular spurless species by means of Ranunculus acaulis, an Antarctic species, the petals of which have a socket in their middle, evidently anticipating the spurs of Aquilegia, &c.

The Order has a strong affinity with some which are widely apart from each other. Its most immediate resemblance is with Dilleniads, Magnoliads, and their

---

Fig. CCXCV.—1. Ranunculus bulbosus; 2. pistil of Ranunculus reptans; 3. a perpendicular section of one of its carpels; 4. a similar section of its ripe achænium.

allies, to which it approaches in the position, number, and structure of its parts of fructification generally, differing however in abundance of particulars; as from Dilleniads in the want of an aril, a deciduous calyx, and whole habit; from Magnoliads in the want of true stipules; from Poppies and Water-lilies in the distinct, not concrete, carpels, watery, not milky fluids, and acrid, not narcotic properties. To Berberids it approaches so very closely that Podophyllum is by many authors placed in that Order; from which, however, it differs essentially in its stamens not bursting by recurved valves; it, however, evidently forms a connecting link between the two Orders. More distant analogy may be traced with Roseworts, with which Crowfoots strikingly agree in their numerous carpels, floral divisions, and indefinite stamens; but differ in their stamens being hypogynous instead of perigynous, in the presence of large albumen surrounding a minute embryo, want of true stipules, and acrid properties. With Umbellifers they accord in the last particular, and also in their sheathing leaves, habit, and abundant albumen, with a minute embryo; but those plants differ in their calyx being concrete with the ovary, and in their stamens being invariably definite.

Fig. CCXCVI.

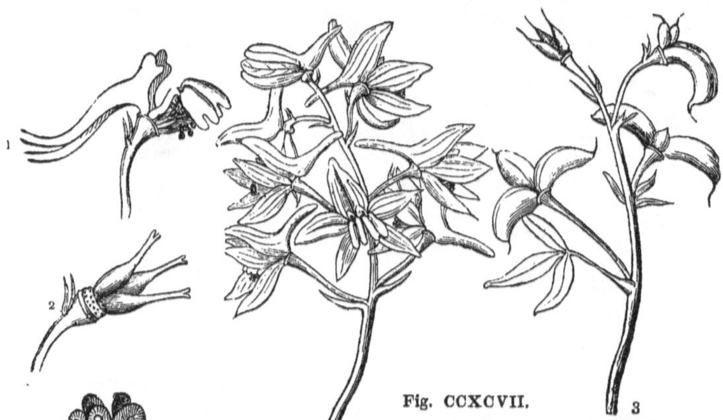

Fig. CCXCVII.

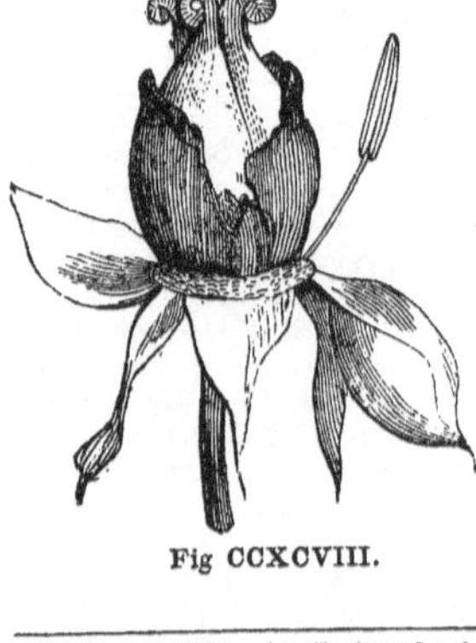

Fig CCXCVIII.

Another analogy has been indicated by Botanists between this Order and Water-plantains, which agree in their numerous carpels, habit, and sometimes in a ternary structure of their flowers; but Water-plantains are monocotyledonous. An instance is described of the polypetalous regular corolla of Clematis viticella being changed into a monopetalous irregular one, like that of Labiatæ.—*Nov. Act. Acad. N. C.* 14, p. 642, t. 37. The genus Pæonia is remarkable for producing in one of its species, the Moutan, the largest form of disk known in the vegetable kingdom.

The largest proportion of this Order is found in Europe, which contains more than 1-5th of the whole; North America possesses about 1-7th, India 1-25th, South America 1-17th; very few are found in Africa, except upon the shores of the Mediterranean: eighteeen species have, according to De

Fig. CCXCVI.—Aquilegia vulgaris.
Fig. CCXCVII.—Delphinium tricorne. 1. petals and stamens; 2. carpels; 3. a branch of ripe fruit.
Fig. CCXCVIII.—The ovary of Pæonia Moutan, surrounded by its broken disk.

Candolle, been discovered in New Holland. They characterise a cold damp climate, and are, when met with in the tropics, found inhabiting the sides and summits of mountains: in the lowlands of hot countries they are almost unknown.

Acridity, causticity, and poison, are the general characters of this suspicious Order, which, however, contains species in which those qualities are so little developed as to be innoxious. The caustic principle is, according to Krapfen, as cited by De Candolle, of a very singular nature; it is so volatile that, in most cases, simple drying, infusion in water, or boiling, are sufficient to dissipate it: it is neither acid nor alkaline: it is increased by acids, sugar, honey, wine, spirit, &c. and is only effectually destroyed by water and vegetable acids. The leaves of Knowltonia vesicatoria are used as vesicatories in Southern Africa. Ranunculus glacialis is a powerful sudorific: Aconitum Napellus and Cammarum are diuretic. The Hepatica, Actæa racemosa, and Delphinium consolida, are regarded as simple astringents. The roots and leaves of several Hellebores are drastic purgatives; of the perennial Adonises, according to Pallas, emmenagogues; and of several Aconites, especially Napellus and ferox, acrid in a high degree. The black Hellebore of the ancients was H. officinalis rather than H. niger, (see *Bot. Reg.* 1842, *t.* 34 & 58). The root of an Aconite of India, one of the substances called Bikh, or Bish, is a most virulent poison. According to Hamilton, the Bishma, or Bikhma, is a strong bitter, very powerful in the cure of fevers: the Bish, Bikh, or Kodoya Bikh, has a root possessing poisonous properties of the most dreadful kind, whether taken into the stomach, or applied to wounds: the Nir Bishi, or Nirbikhi, has no deleterious properties, but is used in medicine. For some important information on this Bikh, Vish, Visha, or Ativisha, which Wallich considers his Aconitum ferox, see *Plant. As. Rar.* vol. i. p. 33, tab. 41, and especially *Royle's Illustrations*, 40. Ranunculus flammula and sceleratus are powerful epispastics, and are used as such in the Hebrides, producing a blister in about an hour and a half. Their action, is, however, too violent, and the blisters are difficult to heal, being apt to pass into irritable ulcers. Beggars use them for the purpose of forming artificial ulcers, and also the leaves of Clematis recta and flammula. The root of Ranunculus Thora is reported to be extremely acrid and poisonous, its juice having been formerly used by the Swiss hunters of wild beasts to envenom their javelins, whose wound by that means became speedily fatal and incurable. The root of Hydrastis canadensis has a strong and somewhat narcotic smell, and is exceedingly bitter; it is used in North America as a tonic, under the name of Yellow-root. The root of Coptis trifolia, or Gold-thread, is a pure and powerful bitter, devoid of anything like astringency; it is a popular remedy in the United States for aphthous affections of the mouth in children. The wood and bark of Xanthorhiza apiifolia are a very pure tonic bitter. The shrub contains both a gum and resin, each of which is intensely bitter. The seeds of Nigella sativa were formerly employed instead of pepper; those of Delphinium Staphisagria are vermifugal, caustic, drastic, and emetic; those of Aquilegia simply tonic. It is supposed that a pungent seed used by the Affghans under the name of Siah dana, for flavouring curries, is the Black Cumin of Scripture, and a species of Nigella.—*Royle*. Pæony seeds are emetic and cathartic; the root has the credit of being antispasmodic. The black berries of the Baneberry, Actæa spicata, are poisonous, the roots antispasmodic, expectorant, astringent; they are reported to have afforded very marked relief in cases of catarrh. Similar qualities are assigned to Botrophis actæoides (Actæa racemosa, *L.*), whose nauseous, astringent bitter roots are regarded, in the United States, as a remedy for the bite of the rattlesnake. For further details, see *Endlicher's Encheiridion*.

The fruit of the May-apple (Podophyllum peltatum) is very acid, whence one of its names is Wild Lemon, and it may be eaten; but if other parts are used it is a sure and active cathartic. The rhizome is administered in fine powder. The leaves are poisonous, and the whole plant narcotic.

Dr. Wight remarks that, notwithstanding their ancient reputation, the whole Order, with a few exceptions, has fallen into disuse; Hellebore being almost the only evacuant retained, and that, from the uncertainty of its operation, is seldom used.—*Illustr.* 1. 3. Dr. Fleming has indeed shown that of all the European Aconites, one only, A. Napellus, is of any value; the remainder, including A. Cammarum, being feeble and unimportant in their action. This A. Cammarum is what the London College of Physicians directs to be employed exclusively, under the name of A. paniculatum!

GENERA.

I. CLEMATEÆ. — Calyx valvate, or induplicate,

Clematis, *Linn.*
Viticella, Dillen.
Stylurus, Raf.
Trigula, Noronh.
Clematopsis, Boj.
Viorna, Pers.
Atragene, *DC.*
Naravelia, *DC.*

II. ANEMONEÆ.—Calyx usually coloured, in æstivation imbricated. Achenia one-seeded, tailed. Seed inverted.

Cyrtorhyncha, *Nutt.*
Thalictrum, *Tournef.*
Tripterium, DC.
Physocarpum, DC.
Physocarpidium, Rchb
Syndesmon, Hoffmans.
Anemonanthe, Spach.
Anemone, *Hall.*
Pulsatilla, Tournef.

Campanaria, Endl.
Preonanthus, Ehrh.
Anemanthus, Endl.
Pulsatilloides, DC.
Asteranemia, Reichb.
Anemonanthea, DC.
Oriba, Adans.
Anemonospermos, DC.
Homalocarpus, DC.
Hepatica, *Dill.*
Knowltonia, *Salisb.*
Anamenia, Vent.
? *Thebesia*, Neck.
Hamadryas, *Commers.*
Hydrastis, *Linn.*
Warneria, Mill.
Adonis, *Dill.*
Sarpedonia, Adans.
Consiligo, DC.
Adonanthe, Spach.
Callianthemum, *C.A.M.*
Myosurus, *Dill.*
Aphanostemma, *St. Hil.*

III. RANUNCULEÆ. — Calyx in æstivation, imbricated. Achenia without tails; seed erect.

Casalea, *St. Hil.*
Ranunculus, *L.*
Batrachium, DC.
Ranunculastrum, DC.
Krapfia, DC.
Cyprianthe, Spach.
Thora, DC.
Hecatonia, Lour.
Philonotis, Reichenb.
Echinella, DC.
Ceratocephalus, *Mönch.*
Ficaria, *Dillen.*
Scotanum, Adans.
Oxygraphis, *Bung.*

IV. HELLEBOREÆ.—Calyx, in æstivation, imbricated. Fruit many-seeded follicles.

Caltha, *Linn.*
Nirbisia, G. Don.
Psychrophila, DC.
Populago, DC.
Thacla, Spach.
Trollius, *Linn.*
Geisenia, Raf.
Hegemone, *Bunge.*
Eranthis, *Salisb.*
Koellea, Biria.
Robertia, Merat.
Helleborus, Mönch.
Helleboroides, Adans.
Helleborus, *Adans.*
Helleboraster, Mönch.
Isopyrum, *Linn.*
Olfa, Adans.
Thalictrella, A. Rich.
Leptopyrum, Reichenb.
Enemion, *Raf.*
Coptis, *Salisb.*
Chrysa, Raf.
Chrysocoptis, Nutt.
Pterophyllum, Nutt.
Garidella, *Tournef.*
Nigella, *Tournef.*
Erobatos, DC.
Aquilegia, *Tournef.*
Delphinium, *Tournef.*
Consolida, DC.
Aconitella, Spach.
Delphinellum, DC.
Phledinium, Spach.
Delphinastrum, DC.
Staphisagria, DC.
Aconitum, *Tournef.*
Anthora, DC.
Lycoctonum, DC.
Cammarum, DC.
Napellus, DC.
Pæonia, *L.*
§ *Onæpia*, Lindl.

V. ACTÆEÆ. — Calyx coloured, imbricated. Fruit succulent, indehiscent, one or many-seeded.

Trautvetteria, *Fisch. et M.*
Actæa, *Linn.*
Christophoriana, Tourn.
Botrophis, *Raf.*
Macrotys, Raf.
Pityrosperma, *Sieb.*
Actinospora, *Turcz.*
Cimicifuga, *Linn.*
Xanthorrhiza, *Marsh.*
Zanthorhiza, Herit.
Podophyllum, *L.*

NUMBERS. GEN. 41. Sp. 1000.

Apiaceæ.
Berberidaceæ.
POSITION.—Papaveraceæ.—RANUNCULACEÆ.—Dilleniaceæ.
Alismaceæ.

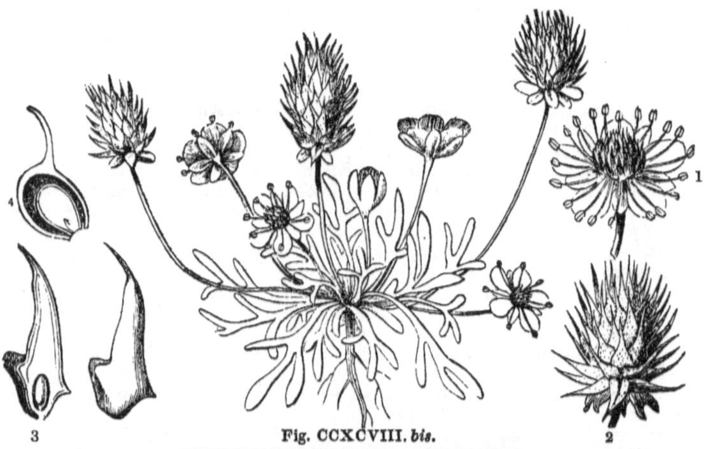

Fig. CCXCVIII. *bis.*

CEPHALOTEÆ, (*R. Brown, Phil. Mag.* (1832).—Cephalotaceæ, *Lindl. Key*, No. 5. (1835); *Ed. pr.* No 5). A stemless herb with exstipulate leaves, among which are mingled operculate pitchers. Scape simple, bearing a compound terminal spike. Flowers small. Calyx coloured, six-parted, with a valvate æstivation. Corolla 0. Stamens 12, those opposite the sepals shortest, inserted into the edge of a deep glandular perigynous disk; anthers with a thick granular connective. Carpels 6, distinct, one-seeded; ovule erect. Akenia membranous, opening by the ventral suture, surrounded by the persistent calyx and stamens. Seed solitary (very seldom two) erect. Embryo minute, in the base of the axis of a fleshy friable somewhat oily albumen.—The single species on which this imaginary Order has been founded is a native of the marshes of King George's Sound in New Holland. It is allied, according to Labillardiere, to Roseworts, and according to Jussieu, to Houseleeks; according to Brown, the Order should be placed between Houseleeks and Francoads. Its very copious albumen and apocarpous fruit seem, however, to fix it far from the former of those Orders, and to place it unquestionably in the Ranal Alliance, from which it forms a transition to Francoads in the Berberal Alliance, and through those plants to Sarraceniads, in which the leaves are in like manner transformed into pitchers. The difficulty that Botanists have found in deciding where to place it, has arisen out of the apparently perigynous station of its stamens, which are represented as growing from the outer edge of a deep glandular perigynous disk. But if, as seems probable, that disk is a mere expansion of the footstalk, analogous to what occurs in Eschscholtzia, then all difficulty about the station is removed, and the genus will fall into the ranks of the Crowfoots; a probability somewhat increased by its valvate æstivation, which is like that of Clematis. GENUS. Cephalotus, *R. Br.* Gen. 1. Sp. 1.

Fig. CCXCVIII. *bis.* - Ceratocephalus orthoceras. 1. flower; 2. ripe fruit; 3. ovaries of Ranunculus Krapfia; 4. section of carpel and seed of the same.

ORDER CLV. SARRACENIACEÆ.—SARRACENIADS.

Sarraceniеæ, *Turpin in Dict. des Sc. c. ic.* (?); *De la Pylaie in Ann. Linn. Par.* 6. 388. t. 13. (1827); *Hooker Fl. Boreal. Am. p.* 33. (1829); *Endl. Gen. p.* 901; *Meisner, p.* 334.

DIAGNOSIS.—*Ranal Exogens, with consolidated carpels, a permanent calyx, and axile placentæ.*

Herbaceous perennial plants, living in bogs. Roots fibrous. Leaves radical, with a hollow urn-shaped petiole, at whose apex is articulated the lamina, which fits on like a lid. Scapes each having 1 or more large flowers, of a more or less herbaceous colour, or white. Calyx 4- 6-leaved, broken-whorled, much imbricated, without a corolla; or consisting of 5 persistent sepals, often having a 3-leaved involucre on the outside, and 5 hypogynous, unguiculate, concave petals. Stamens 00, hypogynous; anthers oblong, adnate, 2-celled, bursting internally and longitudinally. Ovary free, 3- 5-celled, with polyspermous placentæ in the axis; style simple, truncate, or expanded into a large peltate plate with 5 stigmatic angles; ovules anatropal. Capsule with 2- 5 cells. Seeds very numerous, minute, slightly warted or winged, covering large placentæ, which project from the axis into the cavity of the cells; albumen abundant; embryo cylindrical, lying near the base of the seed, with the radicle turned to the hilum.

Fig. CCXCIX.

The genus Sarracenia, inhabiting the bogs of North America, bears the strange name of Side-saddle Flower, in allusion to the singular tubular leaves of itself and its ally. So long as the former alone was known, no clear idea could be formed of its affinity, and a large peltate plate which terminates the style and leaves a stigma beneath each of its 5 angles, was thought to be essential to the Order which it represents; but the discovery in Guiana, by Sir R. Schomburgk, of a very curious genus in which the stigma is reduced to a truncated point, shows that opinion to be unfounded. The same fact also proves that the floral envelopes are subject to great diversity of condition, consisting, in Sarracenia itself, of 5 sepals and 5 distinct petals, but reduced in Heliamphora to 4, 5, or perhaps 6, imbricated segments, standing in the place of both calyx and corolla. This deviation from what may be termed the typical structure of the Order is quite analogous to what occurs among Crowfoots, where Ranunculus may be compared to Sarracenia and Caltha to Heliamphora. This leads to the supposition that it is in the neighbourhood of the Ranal Alliance that Sarraceniads are to be placed; and in fact Poppyworts, which are Ranals with completely consolidated carpels, must be taken as the nearest connection of these singular plants.

The pitchers appear to be secreting organs, for they are lined by hairs of a very singular nature, as is mentioned in Mr. Bentham's Memoir on Heliamphora, in *Linn. Trans.* xviii. p. 429; but their physiological action remains to be ascertained.

The species are confined to the bogs of North America, with the exception of Heliamphora nutans, found in Guayana.

Their uses are unknown.

GENERA.

Sarracenia, *Linn.* | *Coilophyllum*, Moris. | *Bucanaphyllum*, Pluk. | Heliamphora, *Benth.*

NUMBERS. GEN. 2. Sp. 7.

Droseraceæ.
POSITION.—Papaveraceæ.—SARRACENIACEÆ.—Ranunculaceæ.

Fig. CCXCIX.—Heliamphora nutans. 1. the stamens and pistil; 2. the latter separate; 3. a cross section of the ovary; 4. a perpendicular section of a seed.

ORDER CLVI. PAPAVERACEÆ.—POPPYWORTS.

Papaveraceæ, *Juss. Gen.* 236. (1789) *in part; DC. Syst.* 2. 67. (1818); *Prodr.* 1. 117. (1824); *Bernhardi in Linnæa.* 8. 401. (1833); *Endl. Gen.* clxxx.; *Meisner, p.* 7; *Wight Illustr.* 1. 27.

DIAGNOSIS.—*Ranal Exogens, with dimerous or trimerous flowers, consolidated carpels, deciduous calyx, and usually parietal placentæ.*

Herbaceous plants or shrubs, often with a milky juice. Leaves alternate, simple or divided, without stipules. Peduncles long, 1-flowered; flowers never blue. Sepals 2 (or 3), deciduous. Petals hypogynous, either 4 (or 6), or some multiple of that number, usually crumpled before expansion, occasionally 0. Stamens hypogynous, 00; anthers 2-celled, innate. Ovary 1-celled, with parietal placentæ; which in Romneya adhere in the axis; style short, or none; ovules 00, anatropal. Fruit 1-celled, either pod-shaped, with parietal or sutural placentæ, or capsular, with several placentæ. Seeds numerous; albumen between fleshy and oily; embryo minute, straight, at the base of the albumen, with plano-convex cotyledons.

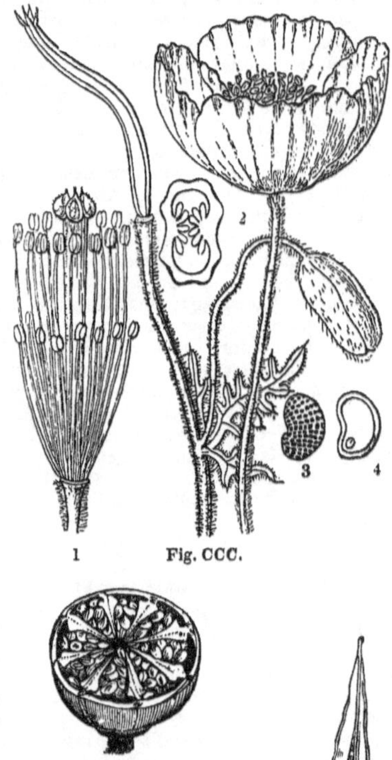

1 Fig. CCC.

The common Redweed of the corn fields offers a good representation of the general character of the plants of this Order, whose appearance is varied principally by the flowers being white or yellow, and occasionally by their being collected into dense panicles, when they are greatly reduced in size, and even in the number of their parts, Bocconia having no petals. In this state they approach the Crowfoots through Thalictrum. In general also their carpels are completely consolidated, but in the curious genus Platystemon, they are as distinct as in a Crowfoot, and in fact that genus would be referable to Ranunculaceæ if it were not for its 2 sepals, no such number being known in that Order.

The siliquose-fruited genera, such as Glaucium and Eschscholtzia, have been supposed to indicate the near affinity of this Order to Crucifers; but the totally different structure of their seeds is such as to neutralise what little affinity may be indicated by the form of the fruit. Through Papaver the Order approaches Water Lilies. To Rock-roses an unexpected relationship has been established by the discovery of Dendromecon. The greatest affinity is, however, with Crowfoots, from which it is sometimes extremely difficult to know this Order, without ascertaining that the juice is milky and narcotic. Platystemon is the connecting link between the two Orders. Bernhardi indeed denies that true Poppyworts are universally lactescent plants, and he quotes Hunnemannia, Eschscholtzia, and Glaucium, as instances to the contrary; but in reality

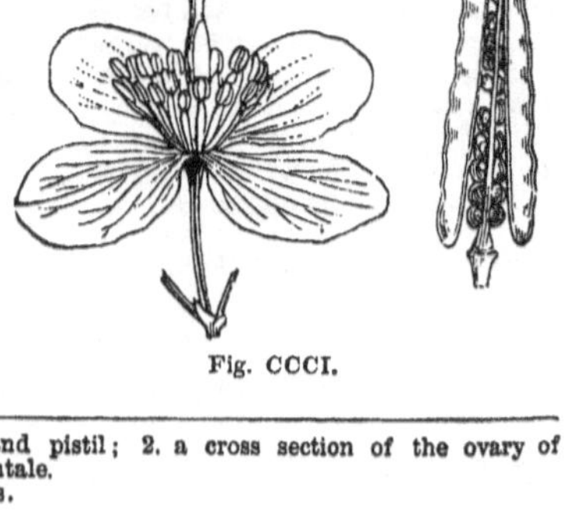
Fig. CCCI.

Fig. CCC.—Romeria refracta. 1. its stamens and pistil; 2. a cross section of the ovary of Eschscholtzia californica; 3, 4, seeds of Papaver orientale.
Fig. CCCI.—Flower and fruit of Chelidonium majus.

they are all furnished with milk, as every gardener well knows. The anomalies in the Order are of little importance, with the exception of Eschscholtzia, which has its stamens arising from the throat of a bell-shaped excavation of the flower-stalk, analogous to what occurs in the Rose, and which gives the stamens the appearance of being perigynous instead of hypogynous. A comparison of the structure of Poppyworts and Crucifers, by Mirbel, is to be found in the *Ann. des Sc.* 6. 266. A plant called Romneya, found in California by Coulter, offers a very remarkable structure. It approaches in many respects very near Argemone ; but its placentæ meet in the axis and divide the cavity of the ovary into many distinct cells, in which respects it agrees with Sarraceniads ; moreover, the ovules are distributed over the whole surface of the dissepiments, a character proper to Water-lilies. Thus the genus Romneya, whose seeds indeed are unknown, forms a link between all the three Orders just mentioned.

Europe, in all directions, is the principal seat of Poppies, almost two-thirds of the whole Order being found in it. Two species only are, according to De Candolle, peculiar to Siberia, three to China and Japan, one to the Cape of Good Hope, one to New Holland, and six to Tropical America. Several are found in North America, beyond the tropics ; and it is probable that the Order will yet receive many additions from that region. Most of them are annuals. The perennials are chiefly natives of mountainous tracts. They are unknown in a wild state within the tropics.

Every one knows what narcotic properties are possessed by the Poppy, and this character prevails generally in the Order. The seed is universally oily, and generally in no degree narcotic. The oil obtained from the seeds of Papaver somniferum is found to be perfectly wholesome, and is, in fact, consumed on the Continent in considerable quantity. It is also employed extensively for adulterating olive oil. Its use was at one time prohibited in France by decrees issued in compliance with popular clamour ; but it is now openly sold, the government and people having both grown wiser. Meconopsis napalensis, a Nipal plant, is described as being extremely poisonous, especially its roots. The Sanguinaria canadensis, or Puccoon, is emetic and purgative in large doses, and in smaller quantities stimulant, diaphoretic, and expectorant. The seeds of Argemone mexicana, called Fico del inferno by the Spaniards, are said to be narcotic, especially if smoked with tobacco, and purgative. They are used in the West Indies as a substitute for ipecacuanha ; and the juice is considered by the native doctors of India as a valuable remedy in ophthalmia, dropt into the eye and over the tarsus ; also as a good application to chancres. It is purgative and deobstruent. The Brazilians administer the juice of this plant, their Cardo santo, to persons or animals bitten by serpents, but, it would appear, without much success. The juice of Chelidonium majus is a violent acrid poison. It has been regarded officinally as stimulating, aperient, diuretic, sudorific, and a powerful deobstruent. It is a popular remedy for warts, and has been employed successfully in opacities of the cornea. The narcotic principle of opium is an alkaline substance, called Morphia. The same drug contains a peculiar acid, called the Meconic ; and a vegetable alkali, named Narcotine, to which the unpleasant stimulating properties are attributed by Magendie. To these principles chemists add Codeine, Thebaine or Caramorphine, Narceine, Meconine, and Comenic and Pyromeconic acids. It is curious that the native country of the Opium Poppy should be unknown. Homer records its virtues when Gorgythion was wounded in the breast, and oppressed by the weight of his helmet :—

> " *μήκων δ' ὡς ἑτερώσε κάρη βάλεν, ἥτ' ἐνὶ κήπῳ,*
> *καρπῷ βριθομένη νοτίῃσί τε εἰαρινῇσιν·*
> *ὡς ἑτερωσ' ἤμυσε κάρη πήληκι βαρυνθέν.*"

Ancient Latin songs record its cultivation in the gardens of Tarquinius superbus, and Charlemagne thought it worthy of a place in his Capitularies. See the *Encheiridion* of Endlicher, who derives Mag or Magei from *μηκων* or *μακων*, the name of the Opium Poppy.

GENERA.

Bocconia, *Plum.*	Meconopsis, *Viguier.*	*Rhœadium*, Spach.	Hunnemannia, *Sweet.*
Macleaya, *R. Br.*	*Cerastites*, Gray.	*Argemonidium*, Spach.	Dendromecon, *Benth.*
Sanguinaria, *Linn.*	Papaver, *Tournef.*	Closterandra, *Belang.*	Platystigma, *Benth.*
Chelidonium, *Tournef.*	*Calomecon*, Spach.	Rœmeria, *Medik.*	Meconella, *Nutt.*
Stylophorum, *Nutt.*	*Meconium*, Spach.	Glaucium, *Tournef.*	Platystemon, *Benth.*
Argemone, *Tournef.*	*Meconidium*, Spach.	Eschscholtzia, *Cham.*	Romneya, *Harv.*
Ecthrus, Lour.	*Meconella*, Spach.	*Chryseis*, Talbot.	

NUMBERS. GEN. 18. Sp. 130.

Nymphæaceæ.
POSITION.—Sarraceniaceæ.—PAPAVERACEÆ.—Ranunculaceæ.
Fumariaceæ.

ALLIANCE XXXIII. *BERBERALES.*—THE BERBERAL ALLIANCE.

DIAGNOSIS.—*Hypogynous Exogens, with monodichlamydeous flowers, unsymmetrical in the ovary, sutural, parietal, or axile placentæ, definite stamens, and embryo inclosed in a large quantity of fleshy albumen.*

The combination in the same Alliance of Epimedium and Vines, or of Fumitories and Berberries, may at first appear paradoxical. But the sequence of affinities shows that this association is truly natural. The Berberal Alliance is connected with the Ranal by means of Fumeworts, which are so nearly related to Poppyworts, that some Botanists refuse to separate them as independent Orders. The affinity of Fumitories and Epimedium with the plants generally associated with Fumeworts under the name of Nandineæ is obvious ; to the latter all Botanists ally the true Berberids. The passage from Berberids proper to Vines is by no means difficult to perceive, and thence Vines may be regarded as passing into Pittosporads by means of the climbing fleshy-fruited Billardieras in the latter Order.

The characteristic marks of the Berberal Alliance are its unequal-parted flowers, definite number of stamens, and minute embryo, lying inclosed in hard horny albumen. The only exception to this distinction is found in Berberis itself, whose embryo is much larger than in the remainder of the Alliance, but in that genus the long radicle and small cotyledons proclaim its relationship to be with the Orders characterised by the large quantity of their albumen. From the Erical Alliance they differ in little except the number of parts in the flower being unequal ; that is to say, although the stamens, corolla, and calyx, may correspond in the number of their parts, yet the ovary is at variance with them in that respect. For this reason the Sundews are stationed here, although their habit is rather that of the Erical Alliance, to which they may be regarded as a transition. The parietal placentæ of Sundews are also in conformity with that portion of the Berberals which constitute the Fumeworts.

The true passage from Ranals is at once into Fumeworts ; but Sundews being as much a modification of the structure of Poppyworts as Fumeworts themselves, the two Orders stand on the same level, and in a lineal arrangement must necessarily interfere, by the one taking a precedence to which it is not entitled.

NATURAL ORDERS OF BERBERALS.

Flowers regular and symmetrical. Placentæ parietal. Stamens alternate with the petals, or twice as many.	157. DROSERACEÆ.
Flowers irregular and unsymmetrical. Placentæ parietal. Stamens opposite the petals	158. FUMARIACEÆ.
Flowers regular, symmetrical. Placentæ sutural. Stamens opposite the petals. Anthers with recurved valves.	159. BERBERIDACEÆ.
Flowers regular, symmetrical. Placentæ axile. Stamens opposite the petals. Anthers opening longitudinally.	160. VITACEÆ.
Flowers regular, symmetrical. Placentæ axile and parietal. Stamens alternate with the petals. Ovules ascending or horizontal. Corolla imbricated	161. PITTOSPORACEÆ.
Flowers regular, symmetrical. Placentæ axile. Stamens alternate with the petals. Ovules pendulous. Corolla valvate	162. OLACACEÆ.
Flowers regular, symmetrical. Placentæ axile. Stamens alternate with the petals if equal to them in number. Ovules pendulous. Corolla imbricated	163. CYRILLACEÆ.

ORDER CLVII. DROSERACEÆ. SUNDEWS.

Droseraceæ, *DC. Théorie*, 214. (1819); *Prodr.* 1. 317. (1824); *Endl. Gen.* clxxxix.; *Meisner*, *p.* 22.

DIAGNOSIS.—*Berberal Exogens, with regular symmetrical flowers, parietal placentæ, and stamens alternate with the petals, or twice as many.*

Delicate herbaceous plants, often covered with glands. Leaves alternate, with stipulary fringes and a circinate vernation. Peduncles, when young, circinate. Sepals 5, persistent, equal, with an imbricated æstivation. Petals 5, hypogynous, imbricated. Stamens distinct, withering, either equal in number to the petals and alternate with them, or 2, 3, or 4 times as many. Ovary single; styles 3-5, either wholly distinct, or slightly connected at the base, bifid or branched. Ovules 00, parietal, or attached to a placenta at the base, anatropal. Capsule of 3 or 5 valves, which bear the placentæ either in the middle or at their base, and sometimes turn in their edges so as to form almost perfect dissepiments. Seeds either naked or furnished with an aril. Embryo minute, in the base of fleshy albumen.

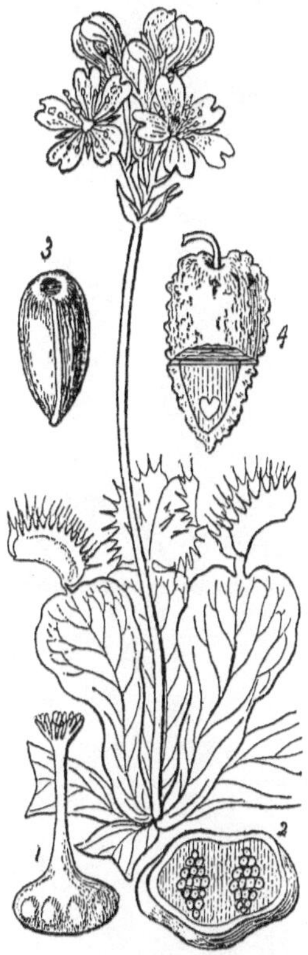

Fig. CCCII.

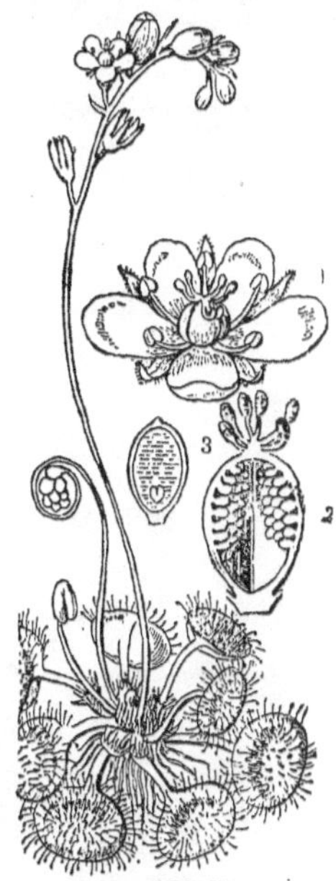

Fig. CCCIII.

These plants are generally supposed to be nearly allied to Violetworts, from which their circinate vernation, several styles, and exstipulate leaves, distinguish them. They are also no doubt related to Tutsans, among which Parnassia accords with Sundews. Rock-roses (Cistaceæ) are also named as approaching Sundews, and so are Turnerads, the parietal placentation of these Orders having led to the comparison. But if we regard the minute embryo and copious albumen of Sundews as the first point of importance in their structure, then they must be removed from immediate relation to all the Orders already mentioned, and will fall into either the Berberal or Erical Alliance. They will correspond with the former in the number of parts in their ovary not agreeing with that of the surrounding parts, and with Fumeworts in their parietal placentation; on the other hand they will claim affinity with Ericals in their general appearance. Aldrovanda, a water plant, inhabiting the ditches in the South of Europe, is remarkable for its whorled, cellular, shell-like leaves.

At the Cape of Good Hope, in South America, North America, New Holland, China, Europe, Madagascar, the East Indies, wherever there are marshes or morasses, these plants are found. Drosophyllum lusitanicum grows on the barren sands of Portugal.

The common Droseras are rather acid, slightly acrid, and according to some, poisonous to cattle. The Drosera communis of Brazil is said by A. de St. Hilaire to be poisonous to sheep. Drosera lunata has viscid leaves with glandular fringes, which close upon

Fig. CCCII.—Dionæa muscipula. 1. its pistil; 2. a sectional view of it showing the placentæ; 3. a seed; 4. the same without its crustaceous skin, and opened so as to show the embryo.

Fig. CCCIII.—Drosera rotundifolia. 1. a flower; 2. a perpendicular section of the ovary; 3. a perpendicular section of a seed.

flies and other insects that happen to alight upon them. It is probable it would yield a valuable dye. It is also believed that some of the Swan River species of Drosera might be turned to account in that way; for every part of D. gigantea stains paper of a brilliant deep purple, and when fragments are treated with ammonia they yield a clear yellow. The bulbs of D. erythorhiza and stolonifera have similar dyeing qualities; they have been stated by Dr. Milligan to be eatable, but that is a mistake, according to Drummond. The irritability of the glandular hairs which clothe the leaves is one of the peculiar features of the Order, and reaches its maximum in the curious genus Dionæa, whose leaves, bordered by stiff teeth, and divided into two halves, are furnished on each half with 3 minute bristles arranged in a triangle, which bristles are extremely irritable, and when touched cause the two sides of the leaf to collapse with such considerable force, that they cannot be separated again without employing violence: they, however, spontaneously open again in a short time.

GENERA.

Drosera, *Linn.*
Rorella, Rupp.
Ros-Solis, Tourn.
Esera, Neck.

Aldrovanda, *Monti.*
Byblis, *Salisb.*
Drosophyllum, *Link.*
Dionæa, *Ellis.*

Roridula, *Linn.*
Iridion, Burm.
Sondera, *Lehm.*

NUMBERS. GEN. 7. Sp. 90.

Pyrolaceæ.
POSITION.—Fumariaceæ.—DROSERACEÆ.—Berberidaceæ.
Violaceæ.

ORDER CLVIII. FUMARIACEÆ.—FUMEWORTS.

Fumariaceæ, *DC. Syst.* 2. 105. (1821); *Prodr.* 1. 125. (1824); *Endl. Gen. p.* 858; *Meisner, p.* 8.

DIAGNOSIS.—*Berberal Exogens, with irregular unsymmetrical flowers, parietal placentæ, and stamens opposite the petals.*

Herbaceous plants, with brittle stems and a watery juice. Leaves usually alternate, multifid, often with tendrils. Flowers purple, white, or yellow. Sepals 2, deciduous.

1 3 2 5

Fig. CCCIV.

Petals 4, cruciate, very irregular. Stamens 4, distinct, hypogynous, or 6, in 2 parcels, opposite the outer petals, very seldom all separate; anthers membranous, the outer of each parcel 1-celled, the middle one 2-celled. Ovary free, 1-celled; ovules horizontal, amphitropal; style filiform; stigma with two or more points. Fruit various; either an indehiscent 1- or 2-seeded nut, or a 2-valved or succulent indehiscent polyspermous pod. Seeds horizontal, shining, crested. Albumen fleshy. Embryo minute, out of the axis; in the indehiscent fruit straight; in those which dehisce somewhat curved.

Any one who compares Fumaria with Epimedium, or Aceranthus with Hypecoum, will see their very near resemblance, and thus will be led to admit, what at first sight seems inadmissible, the affinity of Fumeworts and Berberids. De Candolle remarks that "Fumeworts are very near Poppies, on account of their two-leaved deciduous calyx, of the structure of the fruit of those species in which it splits, and of their fleshy albumen; but they differ, firstly, in their juice being watery, instead of milky; secondly, in their petals being usually irregular, and in adhering to each other; thirdly, in their

Fig. CCCIV.—Fumaria officinalis. 1. a flower seen from below; 2. the same from the side; 3. the pistil, stamens, and a portion of the bagged upper petal; 4. a parcel of anthers, inaccurately drawn, for the two at the sides should be half anthers; 5. the fruit.

diadelphous stamens, which bear indifferently 1- and 2-celled anthers." I am, however, inclined to suspect that the floral envelopes of Fumeworts are not rightly described. I am by no means sure that it would not be more consonant to analogy to consider the parts of their flower divided upon a binary plan; thus understanding the outer series of the supposed petals as calyx, and the inner only as petals; while the parts now called sepals are perhaps more analogous to bracts; an idea which their arrangement, and the constant tendency of the outer series to become saccate at the base, which is not uncommon in the calyx of Crucifers, but never happens, as far as I know, in their petals, would seem to confirm. Of this, some further evidence may be found in the stamens. Those organs are combined in two parcels, one of which is opposite each of the divisions of the outer series, and consists of one perfect 2-celled anther in the middle and two lateral 1-celled ones: now, supposing the lateral 1-celled anthers of each parcel to belong to a common stamen, the filament of which is split by the separation of the two parcels, we shall find the number of stamens of Fumeworts to be 4, one of which is before each of the divisions of the flower; an arrangement that is precisely what we should expect in a normal flower consisting of 2 sepals and 2 petals, and the reverse of what ought to occur if the divisions of the flower were really all petals, as has been hitherto believed. M. Gay, however, objects to this view, and considers the stamens of a Fumitory to be essentially of the same nature as those of a Crucifer, and therefore truly 6 (*Ann. Sc. Nat., ser.* 2. 18. 216.), an opinion in which I am quite unable to concur, for reasons that need not be here explained. It is sufficient to say that Hypecoum negatives M. Gay's theory.

The economy of the sexual organs of Fumitories is remarkable. The stamens are in two parcels, the anthers of which are a little higher than the stigma; the two middle ones of these anthers are turned outwards, and do not appear to be capable of communicating their pollen to the stigma; the four lateral ones are also naturally turned outwards, but by a twist of their filament their face is presented to the stigma. They are all held firmly together by the cohesion of the tips of the flower, which, never unclosing, offer no apparent means of the pollen being disturbed, so as to be shed upon the stigmatic surface. To remedy this inconvenience, the stigma is furnished with two blunt horns, one of which is inserted between and under the cells of the anthers of each parcel, so that without any alteration of position on the part of either organ, the mere contraction of the valves of the anthers is sufficient to shed the pollen upon that spot where it is required to perform the office of fecundation. At first sight Fumeworts are entirely unlike Poppies, and common observers would scarcely suspect their close relationship. But the seeds, and very often the fruit, of these plants are so much the same, and the genus Hypecoum is so exactly intermediate between the two, that there is not much to object to those who look upon Fumeworts as an irregular form of Poppyworts with definite stamens. The latter circumstance, by itself, perhaps, would not be very important, but taken with the former it sanctions the propriety of regarding them as independent Natural Orders.

Fumeworts offer every gradation, from monospermous to polyspermous fruit, and between indehiscence, as in Fumaria itself, and dehiscence, as in Corydalis.

Their principal range is in the temperate latitudes of the northern hemisphere, where they inhabit thickets and waste places. Two are found at the Cape of Good Hope.

The usual character of Fumeworts is, to be scentless, a little bitter, in no degree milky, and to act as diaphoretics and aperients. The tuber of Corydalis tuberosa has been found to contain a peculiar alkali called Corydalin. C. bulbosa has a tuber which is somewhat aromatic, extremely bitter, slightly astringent and acrid, and was formerly used as a substitute for Birthworts in expelling intestinal worms, and as an emmenagogue. Dicentra Cucullaria has been employed in North America in the same way; and Corydalis Capnoides seems to possess similar properties.

GENERA.

I. HYPECOEÆ.—Stamens distinct.
Hypecoum, *Tournef.*
Mnemosilla, Forsk.
Chiazospermum, *Bernh.*
Pteridophyllum, *Sieb.*

II FUMARIEÆ. — Stamens diadelphous.
Dactylicapnos, *Wall.*
Dicentra, *Borkh.*
Diclytra, DC.
Eucapnos, Bernh.
Capnorchis, Borkh.
Cucullaria, Raf.
Bicucullata, March.
Macrocapnos, Royle
Adlumia, *Raf.*
Bicuculla, Borkh.
Phacocapnos, *Bernh.*
Corydalis, *DC.*
Capnogorium, Bernh.
Capnoides, Boerh.
Neckeria, Scop.
Borkhausenia, Fl.Wet.
Capnites, *Endl.*
Bulbocapnos, *Bernh.*
Leonticoides, *DC.*
Discocapnos, *Cham.et Sch.*
Sarcocapnos, *DC.*
Cysticapnos, *Boerh.*
Capnocystis, Juss.
Fumaria, *Tournef.*
Sphærocapnos, DC.
Platycapnos, *DC.*

NUMBERS. GEN. 15. Sp. 110.

Papaveraceæ
POSITION.—Droseraceæ.—FUMARIACEÆ.—Berberidaceæ.
Brassicaceæ.

ORDER CLIX. BERBERIDACEÆ.—BERBERIDS.

Berberideæ, *Vent. Tabl.* 3. 83. (1799); *DC. Syst.* 2. 1. (1821); *Prodr.* 1. 105. (1824); *Endl. Gen.* clxxix. *Meisner Gen.* 6.

DIAGNOSIS.—*Berberal Exogens, with regular symmetrical flowers, sutural placentæ, stamens opposite the petals, and recurved anther-valves.*

Shrubs or herbaceous perennial plants, for the most part hairless, but very often spiny. Leaves alternate, compound, usually without stipules. Flowers solitary, race-

Fig. CCCV.

mose or panicled. Sepals 3-4-6, deciduous, in a double row, surrounded externally by petaloid scales. Petals hypogynous, either equal to the sepals in number, and opposite to them, or twice as many, sometimes with an appendage at the base in the inside. Stamens equal in number to the petals, and opposite to them; anthers with two cells, opening with a valve from the bottom to the top. Carpel solitary, free, 1-celled; style rather lateral; stigma orbicular; ovules anatropal, attached to the suture, numerous, or in pairs, ascending or suspended. Fruit berried or capsular. Seeds crustaceous or membranous; albumen between fleshy and horny; embryo minute, occasionally as long as the axis of the albumen.

Fig. CCCV.—Berberis vulgaris. 1. stamen; 2. perpendicular section of a pistil with one stamen and one petal adhering; 3. cross section of the fruit; 4. 5. perpendicular section of the seed of B. vulgaris and B. Aquifolium.

Among the conflicting opinions of Botanists who have referred these plants to many different places, it appears clear that they are in fact allied, as Auguste de St. Hilaire affirmed, to Vines, with which they so nearly agree in fructification that if a Berberry had two consolidated carpels and anthers opening longitudinally it would be almost a Vine. While, however, the Berberry itself touches the Vine, some plants of its family show a very different tendency, and are so organised as to resemble very nearly the Fumeworts; these are the Sub-order Nandineæ, in which Epimedium has all the habit and much of the structure of a Fumaria. Some Botanists fancy that Podophyllum should stand here: but the main distinction between Berberids and Crowfoots consists in the recurved anther-valves of the former, and as Podophyllum has not such valves, it must go to Crowfoots. In the singular structure of their anthers there is a striking analogy with Laurels, Plume Nutmegs (Atherospermaceæ), and Witch Hazels, Orders not otherwise akin to Berberids. Caulophyllum thalictroides offers one of the few instances of seeds being absolutely naked, that is to say, not covered by any integument originating in the pericarp. In this plant the ovary is ruptured at an early stage by the expansion of the ovule, which, having been impregnated, continues to grow, and ultimately arrives at maturity, although deprived of its pericarpial covering. The spines of the common Berberry are a curious state of leaf, in which the parenchyma is absorbed, and the ribs are indurated. They, as well as all the simple leaves of the other species, are articulated with the petiole, and are therefore compound leaves reduced to a single leaflet; wherefore the supposed genus Mahonia does not differ essentially from Berberis in foliage any more than in fructification. Berberids are related to Anonads through the genus Bocagea; and their ovary is described as being sometimes strikingly like that of Davilla in Dilleniads. Some of the pinnated species of Berberis have stipules.

Natives of mountainous places in the temperate parts of the northern hemisphere, and of South America as far as the Straits of Magellan; none in Africa, Australasia, or the South Sea Islands. They are very common in the northern provinces of India.

The berries of Berberis vulgaris and other species are acid and astringent, and form with sugar an agreeable refreshing preserve. Their acid is the oxalic (malic, *Royle.*) The stem and bark of the Berberry are excessively astringent, and are employed for that reason by dyers, who also obtain from them a bright yellow colour. Dr. Royle has ascertained that the λυκιον ινδικον of Dioscorides was a Berberry; to this day an extract of the root, stem, and branches of Indian Berberries is employed in cases of ophthalmia, and it is said with great advantage. The fruits of B. asiatica are dried in the sun like raisins. The somewhat bitter leaves of Epimedium alpinum were formerly regarded as sudorific and alexipharmic; the same properties are ascribed to the roots of Caulophyllum thalictroides, whose seeds have been employed as a substitute for Coffee. The leaves of Bongardia Chrysogonum are eaten in the East like Sorrel. The root of Leontice Leontopetalum is used at Aleppo as a substitute for soap; and is regarded by the Turks as a corrective of overdoses of Opium. The tubers of Bongardia Rauwolfii are eaten, both boiled and roasted, in Persia.

GENERA.

§ 1. BERBERIDEÆ.

Berberis, *L.*
Mahonia, Nutt.
Odestema, Raf.

§ 2. NANDINEÆ.

Epimedium, *L.*
Nandina, *Thunb.*
Leontice, *L.*
Leontopetalum, Tourn.
Achlys, *DC.*
Caulophyllum, *Michx.*
Diphylleja, *Michx.*
Jeffersonia, *Bart.*
Croomia, *Torrey.*
Aceranthus, *Morren.*
Vancouveria, *Morren.*
Bongardia, *Meyer.*

NUMBERS. GEN. 12. Sp. 100.

Araliaceæ.
POSITION.—Fumariaceæ.—BERBERIDACEÆ.—Vitaceæ.
Ranunculaceæ.

ORDER CLX. VITACEÆ.—VINEWORTS.

Vites, *Juss. Gen.* 267. (1789).—Sarmentaceæ, *Vent. Tabl.* 3. 167. (1799).—Viniferæ, *Juss. Mem. Mus.* 3. 444. (1817).—Ampelideæ, *Kunth in Humboldt, N. G. et Sp.* 5. 223. (1821); *DC. Prodr.* 1. 627. (1824); *Endl. Gen.* clxiv.; *Meisner Gen.* 51.; *Wight Illustr.* 1. 149.; *Royle Illustr.* 144.—Leeaceæ, *Bartling Ord. Nat. p.* 354. 1830.

DIAGNOSIS.—*Berberal Exogens, with regular symmetrical flowers, axile placentæ, stamens opposite the petals, and anthers opening longitudinally.*

Scrambling, climbing shrubs, with tumid separable joints, or erect bushes; the woody tissue abounds with dotted ducts of large size, which, at certain seasons, pour forth sap in unusual quantity. Leaves with or without stipules at the base, the lower opposite, the upper alternate, simple or compound. Peduncles racemose, sometimes by abortion changing to tendrils, often opposite the leaves. Flowers small, green, arranged in thyrses, umbels, or panicles. Calyx small, nearly entire at the edge. Petals 4 or 5, inserted on the outside of a disk surrounding the ovary; in æstivation turned inwards at the edge, in a valvate manner, and often inflected at the point. Stamens equal in number to the petals, and opposite them, inserted upon the disk, sometimes sterile by abortion; filaments distinct, or slightly cohering at the base; anthers ovate, versatile. Ovary superior, 2-6-celled; style 1, very short; stigma simple; ovules erect, definite, anatropal. Berry round, often by abortion 1-celled, pulpy. Seeds 4 or 5, or fewer by abortion, bony, erect; albumen hard; embryo erect, about one-third the length of the albumen; radicle inferior.

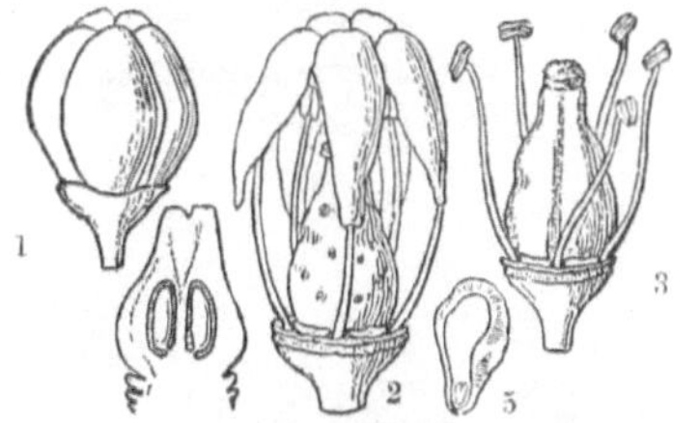

Fig. CCCVI.

The main point of distinction in this Order is, independently of general facts, the stamens being opposite the petals; and by this circumstance it is known among its allies in the same way as Rhamnads, Primworts, &c., among theirs; and, perhaps, Vines ought to be regarded as having a certain amount of relation to Rhamnads, though they have none to Primworts. They have, however, other very strong, though not direct affinities. If the Vine is compared with Aralia racemosa, the relationship of the present Order to it will be too obvious to be mistaken. Suppose that Aralia racemosa had an adherent calyx, erect ovules, with stamens opposite the petals, and it would be a Vitis. A remarkable character in Umbellifers is their petals turned inwards at the points; this occurs also in Ampelopsis quinquefolia; in foliage there is no material difference between them, and even a trace of similarity between the sensible properties of Vineworts and Umbellifers may be perceived in the acrid berries of some species of Cissus. The propriety of placing Leea along with Vineworts has been questioned, and that plant has either been referred absolutely to Meliads, or erected into a distinct Order, as by Von Martius.

Fig. CCCVI.—Vitis vinifera. 1. a flower; 2. the same casting its petals; 3. the pistil and stamens; 4. a section of the ovary; 5. of the seed.

Adrien de Jussieu has, however, in his Dissertation upon Meliads, satisfactorily shown (p. 33) that the genus ought not to be divided from Vineworts. The tumid joints, which separate from each other by an articulation, along with the many other points of agreement in their fructification, approximate the Order to Cranesbills ; the habit and inflorescence to Caprifoils, through Hedera. The tendrils of the Order are the branches of inflorescence, the flowers of which are abortive. A singular variety of Vitis vinifera, with capsular fruit and loculicidal dehiscence, is described in the *Linnæa*, 5. 493. One of the most curious of all plants is Pterisanthes, which bears innumerable flowers on a thin flattened wing-like receptacle. It is well figured and described in the *Linnæa*, vol. 1844. t. viii.

The species are inhabitants of woods in the milder and hotter parts of both hemispheres, especially in the East Indies. None are wild in Europe. As to the Grape Vine, which follows the steps of civilised man everywhere, it is considered certain that its native country is the shores of the Caspian, in lat. 37°, where it is called Dewaz. But it is worth inquiry whether the Vitis indica is not also a wild form of the same plant. For much information regarding these matters, see *Royle*, in the place above quoted.

Acid leaves, and a fruit like that of the common Grape, are the usual characters of this Order. The sap or tears of the Vine are a popular remedy in France for chronic ophthalmia, but they are of little value. The leaves, on account of their astringency, are sometimes used in diarrhœa. But the dried fruit, called Raisins and Currants (Corinths), and wine, are the really important products of the Grape ; products which are, however, yielded by no other of the Order, if we except the Fox-grapes of North America, which scarcely deserve to be excepted. The acid of Grapes is chiefly the tartaric ; malic acid, however, exists in them. The sugar contained in Grapes differs slightly from common sugar in composition, containing a smaller quantity of carbon. The leaves of Cissus cordata and C. setosa are described as being acrid, and useful in bringing indolent tumours to suppuration. The berries of the latter are also acrid, as indeed are those of some other species. Both leaves and fruit of Cissus tinctoria abound in a green colouring matter, which soon becomes blue, and is highly esteemed by the Coroados and other Brazilian Indians as a dye for cotton fabrics.—*Martius.*

GENERA.

I. VITEÆ.—Petals distinct. Stamens distinct. Ovules in pairs. Tendrils.	II. LEEÆ.—Petals united at base. Stamens monadelphous. Ovules solitary. Tendrils 0.
Cissus, *Linn*, *Sælanthus*, Forsk *Columellia*, Lour. *Botria*, Lour. *Cayratia*, Juss. *Ingenhousia*, Dennst. *Irsiola*, P. Br. *Ampelopsis*, L. C. Rch. Vitis, *Linn.* Pterisanthes, *Blum.*	Leea, *Linn.* *Aquilicia*, Linn. *Otilis*, Gärtn. ? Geruma, *Forsk.* ? Lasianthera, *Palis.* ? Bersama, *Fresen.*

NUMBERS. GEN. 7. Sp. 260.

Araliaceæ.
POSITION.—Berberidaceæ.—VITACEÆ.—Pittosporaceæ.
Rhamnaceæ.

ORDER CLXI. PITTOSPORACEÆ.—PITTOSPORADS.

Pittosporeæ, *R. Brown in Flinders' Voyage*, 2. 542. (1814); *DC. Prodr.* 1. 345. (1824); *Ach. Rich. in Dict. Class* 13. 643. (1828); *Endl. Gen.* ccxxxiv.; *Meisner, Gen.* 66. *Putterlick, Synopsis Pittosporearum*, 1839.

DIAGNOSIS.—*Berberal Exogens, with regular symmetrical flowers, axile and parietal placentæ, stamens alternate with the petals, ascending or horizontal ovules, and imbricated petals.*

Trees or shrubs. Leaves simple, alternate, without stipules, usually entire, sometimes serrated. Flowers terminal or axillary, ☿, with imbricated æstivation. Sepals 4-5, deciduous, either distinct or partially cohering. Petals 4-5, hypogynous, sometimes slightly cohering. Stamens 5, hypogynous, distinct, alternate with the petals. Anthers two-celled, opening longitudinally or by a pore. Ovary single, distinct, with the cells or the placentæ 2 or more in number, and many-seeded; style 1; stigmas equal in number to the placentæ; ovules horizontal or ascending, anatropal. Fruit capsular or berried, with many-seeded cells, which are sometimes incomplete. Seeds often covered with a glutinous or resinous pulp; embryo minute, near the hilum, lying in fleshy albumen; radicle rather long; cotyledons very short.

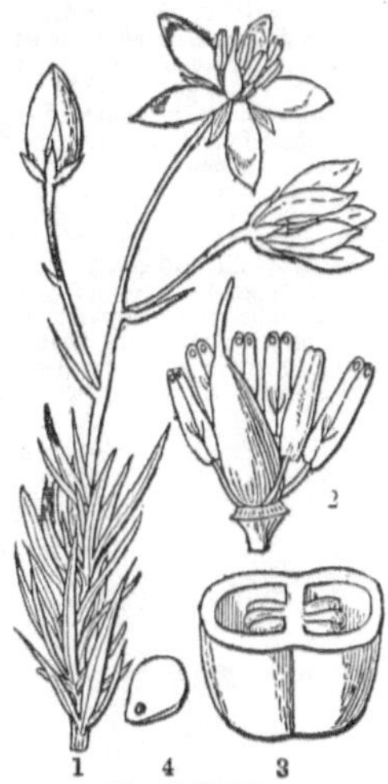

Fig. CCCVII.

Brown, in establishing this as an Order, remarks that it is widely different from Rhamnads and Spindletrees, but he seems to have been unable to point out its real affinity; De Candolle places it between Milkworts and Frankeniads; according to Achille Richard, it is very near Rueworts, to which he thinks it allied by a crowd of characters. Endlicher puts it into his Frangulaceous group. To me, however, it appears that the great mass of albumen in the seeds, the minute embryo, and the general accordance of the flowers with the structure of Vineworts, which is further established by the succulent fruit and climbing habit of Billardiera, seem to place Pittosporads in the same Alliance as the Vine and Berberry. The little genus Cheiranthera forms a transition from Pittosporads to Dilleniads, at once curious and unexpected.

Chiefly New Holland plants. A few occur in Africa and the adjacent islands, and one in Nipal. Brown remarks that Pittosporum itself has been found not only in New Holland, but also in New Zealand, Norfolk Island, the Society and Sandwich Islands, the Moluccas, China, Japan, and even Madeira. They seem to be unknown in America.

The berries of Billardiera are eatable; but they have a resinous odour, and a bitter subacrid taste. The bark of Pittosporum Tobira has a resinous smell, and this resinous quality seems very general in the Order. Mr. Backhouse states that Billardiera mutabilis has a green cylindrical fruit, becoming of a lighter green, or amber colour, when ripe, possessing a pleasant subacid taste; but the seeds are numerous and hard.

GENERA.

Citriobatus, *A. Cunningh.*	Marianthus, *Hügel.*	*Campylanthera*, Hook.	?Stachyurus, *Sieb. et Zucc.*
Pittosporum, *Soland.*	Cheiranthera, *Cunning.*	Billardiera, *Smith.*	?Koeberlinia, *Zucc.*
Schoutensia, *Endl.*	Sollya, *Lindl.*	*Labillardiera*, Röm. et	
Bursaria, *Cav.*	Pronaya, *Hügel.*	—— Schult.	
Oncosporum, *Putterl.*	*Spiranthera*, Hook.		

NUMBERS. GEN. 12. Sp. 78.

Dilleniaceæ.
POSITION.—Olacaceæ.—PITTOSPORACEÆ.—Vitaceæ.
Tremandraceæ.

Fig. CCCVII.—1. Cheiranthera linearis; 2. its pistil and stamens; 3. a cross section of its ovary; 4. a seed of Pittosporum undulatum, cut across to show the minute embryo.

CANELLACEÆ, (*Von Martius, Nov. Gen. et Sp.* 3. 163. (1829); *Conspectus, No.* 300. (1835); *Ed. pr. under Guttiferæ, p.* 75, *Endl. Gen. p.* 1029). This name has been given to a supposed Order of plants represented by Canella alba, a common West Indian aromatic shrub, with evergreen, coriaceous, obovate, alternate, stalked leaves, no stipules, and corymbs of purple flowers. The calyx is leathery, and consists of 3 blunt, tough, permanent, concave sepals, which imbricate each other. The petals are 5, twisted in æstivation. Within these stands a tough truncated hypogynous cone, whose upper half, on the outside, bears about 20 linear parallel 2-celled anthers, which open longitudinally and touch each other. Its ovary is ovate, and tapers into a stiff style, whose end is emarginate. According to Botanical writers, the stigma is permanent and 2-lobed, while the ovary is 3-celled, with more ovules than are attached to the central angle. But I can find no such structure; on the contrary, although the stigma is very slightly emarginate, yet the ovary does not offer a trace of even two cells, but is absolutely one-celled, with 2 or 3 half anatropal ovules hanging by long cords from a little below the dome of the cavity. Gærtner has figured what purports to be the fruit of this plant, representing it to have 3 cells, of which 2 are abortive, and 2 or 3 seeds in the perfect cell, somewhat rostrate, consisting of hard homogeneous albumen, and containing a very small curved cylindrical embryo, lying obliquely with the radicle turned towards the rostrum. But this fruit can hardly belong to Canella alba, if it is correctly drawn; and yet, from the appearance of the calyx in Gærtner's figure, and from his having obtained his fruits out of the Banksian collection, one can scarcely doubt that they really do belong to Canella; in which case we may assume that the seed-vessel has been incorrectly observed. Upon this supposition Canella can have nothing to do with Guttifers, from which in fact its alternate leaves and general appearance remove it. Nor can it be possibly associated, as Martius proposes, with Platonia, a Brazilian tree with the look of a Clusia; neither is it by any means clear that the Canella axillaris of Nees, now called Cinnamodendron, is allied to it. However, as I have had no opportunity of examining the fruit of Canella, or of seeing Cinnamodendron, it must be enough to suggest that the plants referred to Canellaceæ are much more like Pittosporum than Clusia, and to refer the question of their real place in the natural system to Botanists who possess the means of studying both in all their details. Canella alba is often called Wild Cinnamon in the West Indies, on account of its warm aromatic fragrant qualities. Its bark yields by distillation a warm aromatic oil, reckoned carminative and stomachic. It is often mixed with oil of cloves in the West Indies. In this country it is principally employed as an aromatic addition either to tonics or purgatives in debilitated conditions of the digestive organs. Canella bark has also been employed in scurvy. It is exported in considerable quantities from the Bahamas, under the name of White Wood Bark, and is so called in consequence of the whiteness of the trees after they have been stripped. It is only the inner smooth white bark that is exported, the rough outside rind being easily separated by beating. Cinnamodendron axillare, a Brazilian tree, is also aromatic; its bark is used with success as a tonic and antiscorbutic; it is prescribed in low fevers, and is made into gargles in debility of the tonsils.

GENERA.

Canella, *P. Br.*
Winterania, Linn.
?Cinnamodendron, *Endl.*

NUMBERS. GEN. 2. Sp. 3.

ORDER CLXII. OLACACEÆ.—OLACADS.

Olacineæ, *Mirb. Bull. Philom. n.* 75. 377. (1813); *DC. Prodr.* 1. 531. (1824); *Bartl. Ord. Nat. p.* 423. (1830); *Endl. Gen.* ccxxiii.; *Bentham in Linn. Trans.* 18. 676; *Wight. Illustr.* 1. *t.* 40.

DIAGNOSIS.—*Berberal Exogens, with regular symmetrical flowers, axile placentæ, stamens alternate with the petals, pendulous ovules, and valvate corolla.*

Trees or shrubs, often spiny. Leaves simple, alternate, entire, without stipules; occasionally altogether wanting (rarely compound). Flowers small, axillary, often fragrant. Calyx small, entire, or slightly toothed, finally becoming, in many cases, enlarged. Petals definite, hypogynous, valvate in æstivation, either altogether separate, or cohering in pairs by the intervention of stamens, often having thick matted hairs along the middle vein or on some other part. Stamens usually part fertile, part sterile; the former varying in number from 3 to 10, hypogynous, usually cohering with the petals, and alternate with them; the latter opposite the petals, to which they in part adhere, their upper end resembling an appendage; filaments compressed; anthers innate, oblong, 2-celled, bursting longitudinally. Ovary free, or partially adherent, seated in a disk, which is sometimes small and sometimes thickened and united with the calyx; 1-celled, or occasionally imperfectly 3- 4-celled, or 3-celled out of the centre; ovules 2, 3 only, or even 1 only, either pendulous from the apex of free placentæ or adherent to the ovary or the spurious dissepiments, pendulous, anatropal.—*Benth.* Style filiform; stigma simple. Fruit somewhat drupaceous, indehiscent, frequently surrounded by the enlarged calyx, 1-celled, 1-seeded. Seed pendulous; albumen large, fleshy; embryo small, in the base of albumen or in the axis, with very short cotyledons, its radicle near the hilum.

Fig. CCCVIII.

If we neglect the internal structure of the seed the present Order will stand, as De Candolle supposes, near Citronworts; if we suppose the tendency of the corolla to be towards a monopetalous condition with epipetalous stamens, then it must be stationed, with Jussieu, near Sapotads; those who undervalue the perfect adhesion of the calyx and the ovary will pronounce the nearest affinity to be with Sandalworts; but if the condition of the embryo and albumen are considered, a very different view will be entertained of its affinity, and Humiriads, if they really have a small embryo and copious albumen, will be fixed upon as the true point of nearest resemblance. To the latter opinion I adhere; and I am glad to find that Mr. Bentham joins in it. In fact, if it were not for the great dilated connective of the Humiriads, their somewhat imbricated corolla, more numerous stamens and balsamic secretions, I hardly know how they could be distinguished. They obviously agree with the Berberal Alliance, in the anisomerous structure of their flowers, and must be regarded as near allies of Canellaceæ, if indeed that supposed Order does not in reality belong to them. See p. 442.

A small Order, consisting of tropical or nearly tropical shrubs, chiefly found in the East Indies, New Holland, and Africa. One only is known in the West Indies. A few are from the Cape of Good Hope.

It is often said that the wood of Heisteria coccinea is the Partridge-wood of the

Fig. CCCVIII.—Apodytes dimidiata. 1. a flower; 2. a corolla spread open; 3. a pistil; 4. section of an ovary; 5. section of ripe fruit.

cabinet makers, but this is certainly a mistake, as is shown in the *Penny Cyclopædia*, article Partridge-wood. The drupes of Ximenia americana have a sweet aromatic taste, but are a little rough to the palate. They are eaten in Senegal. The flowers are very sweet scented. Olax zeylanica has a fœtid wood with a saline taste, and is employed in putrid fevers ; its leaves are used in salad.

GENERA.

* OLACEÆ, *Benth.*
Heisteria, *L.*
Ximenia, *Plum.*
Heymassoli, Aubl.
Rottböllia, Scop.
Tetanosia, Rich.
Hypocarpus, *A. DC.*
Olax, *L.*
Fissilia, Commers.
Spermaxyrum, Lab.
Roxburghia, König.
Schœpfia, *Wall.*

** OPILEÆ, *Benth.*
Opilia, *Roxb.*
Groutia, Guillem.
Cansjera, *Gmel.*

*** ICACINEÆ, *Benth.*
Icacina, *A. Juss.*
Gomphandra, *Wall.*
Apodytes, *E. M.*
Leretia, *Vell.*
Emmoteca, Desv.
Pogopetalum, *Benth.*
? Pseudaleia, *Thouars.*
?Pseudaleioides, *Thouars.*
? Platea, *Blum.*
? Stemonurus, *Blum.*
Anacalosa, Blum.
? Quilesia, *Blanc.*
? Lepionurus, *Bl.*
Parastemon, *A. DC.*
Tripetaleia, *Sieb.*
Liriosma, *Popp.*

NUMBERS. GEN. 21. Sp. 48.

Santalaceæ.
POSITION.—Cyrillaceæ.—OLACACEÆ.—Pittosporaceæ.
Humiriaceæ.

ORDER CLXIII. CYRILLACEÆ.—CYRILLADS.

Cyrilleæ, *Torrey and Gray, Fl. Bor. Am.* 1. 256. (1838); *Endl. Ench. p.* 578.

DIAGNOSIS.—*Berberal Exogens, with regular symmetrical flowers, axile placentæ, stamens alternate with the petals if equal to them in number, pendulous ovules, and an imbricated corolla.*

Shrubs, with evergreen simple leaves without stipules. Flowers usually in racemes. Calyx 4-5-parted. Petals 5, distinct, hypogynous, with an imbricated æstivation. Stamens 5 or 10, hypogynous. Anthers bursting lengthwise. Ovary 2-3-4-celled, always composed of some number of carpels different from that of the calyx, corolla and stamens; ovules solitary, pendulous; style short; stigma with as many lobes as there are cells of the ovary. Fruit a succulent capsule, or a drupe. Seeds inverted. Embryo in the axis of a very large quantity of albumen, with a very long superior radicle.

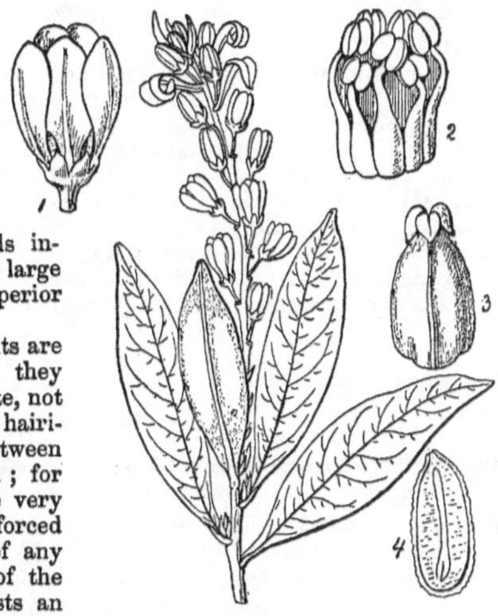

Fig. CCCIX.

There can be no doubt that these plants are nearly related to Olacads, from which they are principally known by their imbricate, not valvate petals, destitute of all traces of hairiness. That being so, the connection between Olacads and Heathworts is established; for Ledum and Clethra in the latter come very near Cyrillads; these are, however, forced into a different Alliance by the want of any definite proportion between the whole of the parts of the flower. Endlicher suggests an affinity between Cyrillads and Hollyworts. The genus Pickeringia, now regarded as a sub-genus of Ardisia, seems to connect this Order with Ardisiads.

They are all inhabitants of North America.

Nothing has been recorded of any uses to which they could be applied.

GENERA.

Cyrilla, *Linn.*
Mylocaryum, *W.*
Cliftonia, Sol.
Walteriana, Fraz.
Elliottia, *Muhl.*

NUMBERS. GEN. 3. Sp. 5.

POSITION.—Pittosporaceæ.—CYRILLACEÆ.—Olacaceæ.
? *Myrsinaceæ.*
Ericaceæ.

Fig. CCCIX.—Mylocaryum ligustrinum. 1. a flower; 2. stamens; 3. ovary; 4. section of seed.

Alliance XXXIV. *ERICALES.*—The Erical Alliance.

Diagnosis.—*Hypogynous Exogens, with dichlamydeous flowers, symmetrical in the ovary, axile placentæ, definite stamens, and embryo inclosed in a large quantity of fleshy albumen.*

The striking resemblance in interior structure between the seeds of Wintergreens, Fir-rapes, Francoads, and Sundews is such as to render it improbable that they should not be placed by nature in very close affinity; and it is only the want of correspondence in the number of the floral organs of the latter which has led to its being detained on the borders of the neighbouring Berberal Alliance.

Ericals join Berberals by way of Humiriads, which are very like Olacads, and they evidently pass into Rutals through the assistance of Correa among Rueworts; nor is this a feeble indication of consanguinity.

Among more distant affinities, one with Dilleniads is very remarkably established by means of the curious genus Saurauja in that Order.

Natural Orders of Ericals.

Flowers polypetalous. Stamens all perfect, monadelphous. Anthers 2-celled, with a long membranous connective	164. Humiriaceæ.
Flowers monopetalous. Stamens all perfect, free. Seeds with a firm skin. Anthers 1-*celled, opening longitudinally*	165. Epacridaceæ.
Flowers half-monopetalous. Stamens all perfect, free. Seeds with a loose skin. Embryo at the base of the albumen	166. Pyrolaceæ.
Flowers polypetalous. Stamens half-sterile and scale-like, free. Seeds with a firm skin	167. Francoaceæ.
Flowers half-monopetalous. Stamens all perfect, free. Seeds with a loose skin or wing. Embryo at the apex of the albumen	168. Monotropaceæ.
Flowers monopetalous. Stamens all perfect, free. Seeds with a firm or loose skin. Anthers 2-*celled, opening by pores* . . .	169. Ericaceæ.

ORDER CLXIV. HUMIRIACEÆ.—HUMIRIADS.

Humiriaceæ, *Adrien de Jussieu in Aug. de St. Hil. Flora Bras. Merid.* 2. 87. (1829); *Martius Nov. Gen.* 2. 147. (1826); *Endl. Gen.* ccxxii.; *Meisner, p.* 47.

DIAGNOSIS.—*Erical Exogens, with polypetalous flowers, perfect monadelphous stamens, and 2-celled anthers with a long membranous connective.*

Trees or shrubs with balsamic juice. Leaves alternate, simple, coriaceous feather-veined, without stipules. Flowers in terminal or axillary cymes, or corymbs. Calyx imbricated, in 5 divisions. Petals imbricated, regular, alternate with the lobes of the calyx. Stamens hypogynous, 4 times or many times as numerous as the petals, monadelphous; anthers 2-celled, with a fleshy connective extended beyond the 2 lobes. Ovary superior, usually surrounded by an annular or toothed disk, 5-celled, often furnished with a transverse partition, with from 1 to 2 suspended anatropal ovules in each cell; style simple; stigma 5-lobed. Fruit drupaceous, with 5 cells on the same plane, or with secondary cells near the apex, sometimes with fewer, on account of the abortion of a part. Seed with a membranous integument; embryo narrow, orthotropal, sometimes lying in fleshy albumen; radicle long, superior.

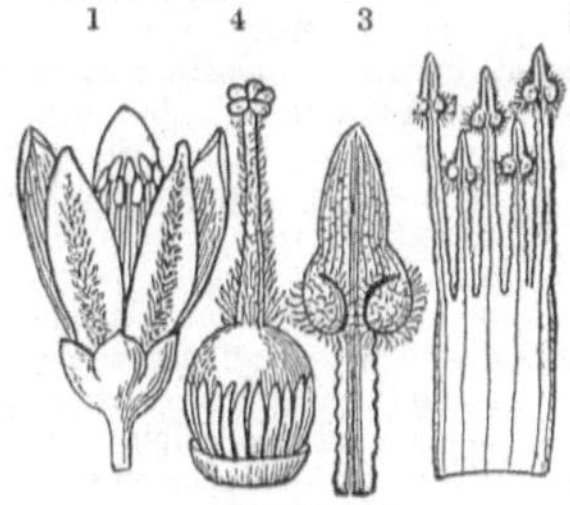

Fig. CCCX.

The affinities of Humiriads cannot be satisfactorily discussed until their seeds shall have been more exactly examined. As the evidence at present stands, there is nothing to show that all the genera now collected really belong to the same group. Helleria, for instance, is said to have no albumen. They differ from Meliads much in habit, and in many respects in fructification, especially in having the æstivation of the corolla quincuncial, not valvate, and the stamens sometimes indefinite; the anthers of Humiriads, as Von Martius observes, are very different from those of Meliads in the great dilatation of their connective; their albuminous seeds and slender embryo are also at variance with Meliads. In the latter respect, and in their balsamic wood, they agree better with Storaxworts, as also in the variable direction of the embryo. Besides these points of affinity, Von Martius compares Humiriads with Chlænads, on account of both Orders containing definite and indefinite monadelphous stamens, several stigmas, partially abortive cells, inverted albuminous seeds, and a singular complicated vernation, by which two longitudinal lines are impressed upon each leaf. To me it formerly appeared that the real affinity is with Citronworts; as is indicated by their inflorescence, the texture of their stamens, their disk, their winged petioles, and their balsamic juices. But this cannot be, if their seeds are really albuminous, as is stated. Assuming the latter to be correct, they will form a connecting link between the Erical and Berberal Alliances, because of their resemblance to Olacads. If really connected with Heathworts, it must be through some such genus as Clethra.

All are natives of the tropical parts of America.

Humirium floribundum, when the trunk is wounded, yields a fragrant liquid yellow balsam, called Balsam of Umiri, resembling the properties of Copaiva and Balsam of Peru. The juice of Humirium balsamiferum and floribundum has a reddish colour, and smells of Storax; an ointment prepared from it is used for pain in the joints, and internally as a remedy for blennorhœa and attacks of tænia.

GENERA.

Saccoglottis, *Mart.*	*Myrodendron*, Schr.	*Wernisekia*, Scop.	Vantanea, *Aubl.*
Humirium, *Mart.*	*Humiri*, Aubl.	Helleria, *Nees et Mart.*	*Lemniscia*, Schreb.
Humiria, Juss.			

NUMBERS. GEN. 4. Sp. 10.

POSITION.—Ericaceæ.—HUMIRIACEÆ.—— *Olacaceæ.* / *Aurantiaceæ.*

Fig. CCCX.—Humirium crassifolium. 1. its flower; 2. a part of its stamens; 3. an anther; 4. pistil.

ORDER CLXV. EPACRIDACEÆ.—EPACRIDS.

Epacrideæ, *R. Brown Prodr.* 535. (1810); *Link Handb.* 1. 601. (1829), *a* § *of* Ericeæ; *DC. Prod.* 7. 734; *Endl. Gen.* clx.

DIAGNOSIS.—*Erical Exogens, with monopetalous flowers, perfect free stamens, seeds with a firm skin, and* 1-*celled anthers opening longitudinally.*

Shrubs or small trees, their hair, when present, being simple. Leaves alternate, very rarely opposite, entire or occasionally serrated, usually stalked; their bases sometimes dilated, cucullate, overlapping each other and half sheathing the stem, without a midrib, but with the veins simple and parallel, or radiating from the base. Flowers white or purple, seldom blue, either in spikes or terminal racemes, or solitary and axillary; the calyx or pedicels with 2 or several bracts, which are usually of the same texture as the calyx. Calyx 5-parted (very seldom 4-parted), often coloured, persistent. Corolla hypogynous, monopetalous, either deciduous or withering, sometimes capable of being separated into 5 pieces, its limb with 5 (rarely 4) equal divisions, sometimes, in consequence of the cohesion of the segments, bursting transversely; the æstivation valvate or imbricated. Stamens equal in number to the segments of the corolla, and alternate with them; very seldom fewer in number. Filaments arising from the corolla, or hypogynous. Anthers one-celled, with a single receptacle of pollen, which forms a complete partition sometimes having a border; undivided, opening longitudinally. Pollen either nearly round or formed of 3 connate grains. Ovary sessile, usually surrounded at the base with 5 distinct or connate scales; with several, rarely a single, cell; ovules solitary and pendulous, or 00; style 1; stigma simple, or occasionally toothed. Fruit drupaceous, baccate, or capsular. Seeds with albumen. Embryo taper, straight in the axis, more than half as long as the albumen; the cotyledons very short, the radicle superior in the drupaceous species, variable in the capsular.

This Order differs from Heathworts chiefly in the structure of the anther; but that organ being one of the principal features of Heathworts, any material deviation from it acquires an unusual degree of consequence. In the latter Order the anther consists of 2 cells, usually furnished with peculiar appendages; in Epacrids it is 1-celled, with no appendages whatever. In some other respects Epacrids are different from Heathworts: their stamens very commonly adhere to the sides of the corolla, a circumstance which is at variance with the Erical and all the neighbouring Alliances, and their leaves have veins with quite the structure of Endogens, so that although the two Orders have but slender verbal distinctions, they are in reality extremely dissimilar.

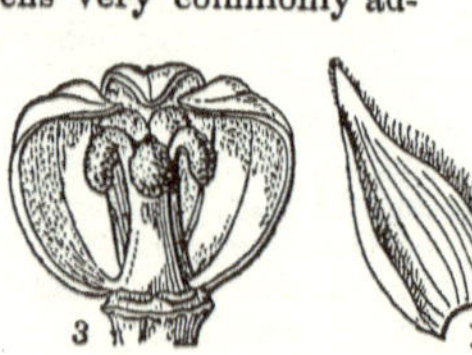

3 1 2

Fig CCCXI.

All natives of the Indian archipelago, or Australasia, or Polynesia, where they abound as Heaths do at the Cape of Good Hope. It is remarkable that only 1 or 2 Heathworts are found in the countries occupied by Epacrids.

The species are chiefly remarkable for the great beauty of their flowers and the singular structure of their leaves, as above described. All the fruits of the berry-bearing section, especially those of Lissanthe sapida, are esculent; but the seeds are too large, and the pulpy covering too thin, to render them very available for food; Astroloma humifusum, the Tasmanian Cranberry, is found all over that colony. It has a fruit of a

Fig. CCCXI.—Dracophyllum scoparium.—*Hooker.* 1. a sepal; 2. a flower with its bract; 3. section of a seed-vessel.

green or whitish colour, sometimes slightly red, about the size of a black currant, consisting of a viscid, apple-flavoured pulp inclosing a large seed. This fruit grows singly on the trailing stems of the plant, which resembles Juniper, bearing beautiful scarlet blossoms in winter. The fruit of Styphelia adscendens, a small prostrate shrub, resembles in appearance and character that of Astroloma humifusa. Leucopogon Richei, called Native Currant, is a large dense shrub, growing only on the sea-coast, and attaining to a height of from four to seven feet. The berries are small, white, and of a herby flavour. In D'Entrecasteaux's voyage in search of La Perouse, a French naturalist named Riche was lost for three days on the south coast of New Holland, and supported himself principally upon the berries of this plant ; in commemoration of which circumstance it has received its specific name.—*Backhouse.*

GENERA.

I. STYPHELIEÆ.— One-seeded.

Conostephium, *Benth.*
Styphelia, *Smith.*
Soleniscia, DC.
Astroloma, *R. Br.*
Ventenatia, Cav. part.
Stomarrhena, DC.
Stenanthera, *R. Br.*
Brachyloma, *Sonder.*
Melichrus, *R. Br.*
Ventenatia, Cav. part.
Melidepas, Endl.
Cyathodes, *R. Br.*
Lissanthe, *R. Br.*
Androstoma, *Hook fil.*
Leucopogon, *R. Br.*
Perojoa, Cav.
Peroa, Pers.
Monotoca, *R. Br.*
Acrotriche, *R Br.*
Trochocarpa, *R. Br.*
Decaspora, *R. Br.*
Pentachondra, *R. Br.*
Needhamia, *R. Br.*
Oligarrhena, *R. Br.*

II. EPACREÆ.— Many-seeded.

Epacris, *Smith.*
Lysinema, *R. Br.*
Julieta, Leschen.
Lebetanthus, *Endl.*
Allodape, Endl.
Prionotes, *R. Br.*
Cosmelia, *R. Br.*
Andersonia, *R. Br.*
Atherocephala, DC.
Ponceletia, *R. Br.*
Sprengelia, *Smith.*
Poiretia, Cav.
Cystanthe, *R. Br.*
Pilitis, *Lindl.*
Richea, *R. Br.*
Dracophyllum, *Lab.*
Dacryanthus, Endl.
Sphenotoma, *R. Br.*

NUMBERS. GEN. 30. Sp. 320.

POSITION.—Ericaceæ.—EPACRIDACEÆ.—

ORDER CLXVI. PYROLACEÆ.—WINTERGREENS.

Pyroleæ, *Lindl. Coll. Bot. t.* 5. (1821).—Pyrolaceæ, *Ed. pr.* clxiv. (1836); *Endl. Gen. p.* 760. *DC. Prodr.* 7.580.

DIAGNOSIS.—*Erical Exogens, with half monopetalous flowers, the stamens free and all perfect, loose-skinned seeds, and an embryo at the base of the albumen.*

Herbaceous plants, rarely under-shrubs. Stems round, naked; in the frutescent species leafy. Leaves simple, entire or toothed. Flowers in terminal racemes, or solitary. Sepals 5, persistent, inferior. Corolla slightly monopetalous, hypogynous, regular, deciduous, 4- or 5-parted, with an imbricated æstivation. Stamens hypogynous, twice as numerous as the divisions of the corolla, those opposite the petals sometimes without anthers; anthers 2-celled, opening by pores. Ovary superior, 4- or 5-celled, many-seeded, with an hypogynous disk; style 1, declinate; stigma slightly indusiate. Fruit capsular, 4- or 5-celled, dehiscent, with central placentæ. Seeds indefinite, minute, with a loose winged skin; embryo minute, at the base of a fleshy albumen, placed across the principal axis of the exterior skin.

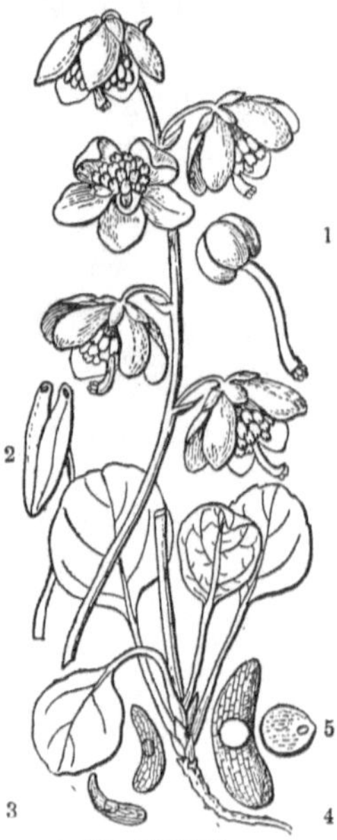

Fig. CCCXII.

Wintergreens are usually considered a portion of the Order of Heathworts, but their habit is so different, that I cannot hesitate to separate them, especially as their minute embryo and declinate styles are real marks of difference. Cladothamnus fruticosus forms a passage to Heathworts, and Pyrola aphylla to Fir-rapes. An approach to the indusiate stigma of Goodeniads occurs in that of P. aphylla and some others.

Natives of Europe, North America, and the northern parts of Asia, in Fir woods, or similar situations.

Chimaphila umbellata is a most active diuretic; it is also found to possess valuable tonic properties. The leaves, which are bitter-sweet, applied to the skin, act as slight vesicatories. C. maculata, a very closely allied species, is asserted by some American practitioners to be wholly inert. It is said to be a palliative in strangury and nephritis, and to alleviate the ardor urinæ. It appears to possess a narcotic action. But this is contrary to the statement of Pursh, who says it has active properties; and therefore Wood and Bache are of opinion that it probably possesses the same qualities as Chimaphila corymbosa. Pyrola rotundifolia had once a great reputation as a vulnerary.

GENERA.

Cladothamnus, *Bung.* *Tolmiæa*, Hook. Chimaphila, *Pursh.* *Chimaza*, R. Br. *Pseva*, Raf	Shortia, *Torr et Gr.* Pyrola, *Tournef.* *Moneses*, Salisb. *Bryophthalmum*, E. Mey.	Galax, *Linn.* *Erythrorhiza*, L. C. Rich. *Solenandria*, Palis.	*Blandfordia*, Andr. *Viticella*, Mich. *Belvedera*, Gronov.

NUMBERS. GEN. 5. Sp. 20.

POSITION.—Monotropaceæ.—PYROLACEÆ.—Francoaceæ.
Gentianaceæ.
Orobanchaceæ.

Fig. CCCXII.—Pyrola chlorantha.—*Hooker*. 1. a pistil; 2. an anther; 3. seeds; 4. a seed much more magnified, with the nucleus cut through; 5. a section of the nucleus, showing the embryo.

ORDER CLXVII. FRANCOACEÆ.—FRANCOADS.

Galacineæ, *Don in Edinb. New Phil. Journ. Oct.* 1828. *Ed. Pr. No.* 146. (1830).—Francoaceæ, *Ad. de Juss. Ann. Sc. Nat.* 25. 9. (1832); *Lindl. in Bot. Reg. fol.* 1645 (1834); *Key to Bot.* 47. (1835) *DC. Prodr.* 7. 777.; *Endl. Gen. p.* 812.

DIAGNOSIS.—*Erical Exogens, with polypetalous flowers, the stamens free, half sterile and scale-like, and tight-skinned seeds.*

Stemless herbaceous plants, with lobed or pinnated leaves, without stipules. Stems scape-like, with a racemose inflorescence. Petals persistent for a long time. Calyx deeply four-cleft. Petals 4, inserted near the base of the calyx. Stamens sub-hypogynous, four times as numerous as the petals, alternately rudimentary. Ovary superior, with 4 cells opposite the petals; ovules numerous; stigma 4-lobed, sessile. Capsule membranous, 4-valved, with a loculicidal or septicidal dehiscence. Seeds numerous, minute, with a minute embryo in the base of fleshy albumen.

The importance of the character derived from the presence of a very minute embryo in the base of a large quantity of albumen not having been taken into account, Botanists do not seem to have judged correctly of the true position of Francoads in a natural system. They stand near Saxifrages according to Don, Roseworts in the opinion of De Candolle, Houseleeks according to Adrien de Jussieu, Hooker, and Endlicher. It is true, that looking to the separation of the carpels of Francoa when ripe, and its abortive stamens, a case in favour of the approximation of the Order to Houseleeks (Crassulaceæ) may seem to be made out; but then Tetilla does not separate its carpels, but divides them through the back; and moreover, there is no resemblance either in habit, or in the proportions of the flowers, or in the structure of the fruit, or in the organisation of the seeds between that Order and Francoads. There can be no doubt that the real affinity of these plants is with Dionæa, which chiefly differs in its unilocular fruit, anisomerous flowers, and the want of sterile stamens. Its seeds are absolutely the same in all essential respects.

All the species hitherto discovered are Chilian.

The juice of the Francoas is said to be regarded in Chili as cooling and sedative; the root dyes black. Tetilla is called in the same country Teta de capra and Culantrillo; according to Pöppig, the leaf-stalks are eaten as a remedy for dysentery, and are remarkable for their astringency.

Fig. CCCXIII.

GENERA.

Francoa, *Cav.*	Tetilla, *DC.*	*Anarmosa*, Miers.
Llaupanke, Feuill.	*Dimorphopetalum*, Bert.	*Tetraplasium*, Kunze.

NUMBERS. GEN. 2. SP. 5.

POSITION.—Pyrolaceæ.—FRANCOACEÆ.—— *Crassulaceæ.* / *Droseraceæ.*

Fig. CCCXIII.—Francoa appendiculata. 1. stamens and pistil; 2. cross section of the ovary; 3. perpendicular section of the seed.

ORDER CLXVIII. MONOTROPACEÆ.—FIR-RAPES.

Monotropeæ, *Nutt. Gen.* 1. 272. (1818); *Endl. Gen.* p. 760.; *DC. Prodr.* 7. 779.

DIAGNOSIS.—*Erical Exogens, with half monopetalous flowers, free stamens all perfect, loose-skinned or winged seeds, and an embryo at the apex of the albumen.*

Parasites growing on the roots of Pines and other trees. The stems brown or almost colourless, with no true leaves, but covered with scales. Flowers in terminal spikes or racemes. Sepals 4, 5, membranous, tapering, distantly arranged in a broken whorl. Petals the same number, either imbricated and saccate at the base, or combined into a monopetalous corolla. Stamens 8- 10, hypogynous, sometimes alternating with 10 hypogynous recurved glands; anthers 2-celled, sometimes opening longitudinally, the cells becoming confluent by the rolling back of the short anterior valves, and producing the appearance of a bilabiate anther; sometimes parallel-celled with bristles at the base. Ovary round, 4- 5-furrowed, articulated, with a short cylindrical style, terminating in a succulent funnel-shaped stigma; 4- 5-celled at the base, 1-celled, with 5 parietal placentæ at the apex. Fruit a dry capsule, splitting through the cells and bearing the placentæ on the middle of the valves. Seeds 00, with a loose skin, or winged at the end; embryo minute, undivided, inclosed within the apex of fleshy albumen.

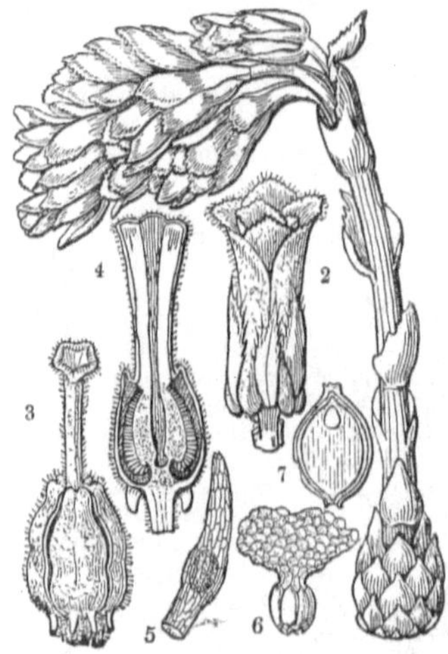

Fig CCCXIV.

The dehiscence of the anthers separates these from Wintergreens, as well as their leafless, scaly, and parasitical habit; besides which, there is a difference in the position of the embryo, that organ being at the apex of the albumen in Fir-rapes, and at its base in Wintergreens. The curious leafless Pyrola called P. aphylla exhibits, among Wintergreens, the peculiar scaly brown aspect of Fir-rapes, and thus connects the two Orders.

Natives of Europe, Asia, and North America, in cool places, especially in Fir woods.

Several species smell of Violets or Pinks. In Germany the powder of Monotropa Hypopithys is given to sheep when attacked by coughs. The North American Indians are said to employ Pterospora andromedea as an anthelmintic and diaphoretic.

GENERA.

Monotropa, *Nutt.*	*Orobanchoides*, Tournf.	Schweinitzia, *Ell.*	? Corallophyllum, *Kunth.*
Hypopithys, *Dill.*	Pterospora, *Nutt.*	*Monotropsis*, Schwein.	? Pholisma, *Nuttall.*

NUMBERS. GEN. 6. Sp. 10.

Orobanchaceæ?
POSITION.—Pyrolaceæ.—MONOTROPACEÆ.——

Fig. CCCXIV.—Monotropa Hypopithys. 2. a flower; 3. a pistil; 4. the same divided perpendicularly; 5. a seed.—*Nees.* 6. seed of Pterospora andromedea; 7. a section of it.

ORDER CLXIX. ERICACEÆ.—HEATHWORTS.

Ericæ, *Juss. Gen.* 159 (1789).—Ericeæ, *R. Brown Prodr.* 557. (1810).—Rhododendra, *Juss. Gen.* 158, (1789).—Ericineæ, *Desv. Journ. Bot.* 28. (1813); *Don in Edinb. Phil. Journal, p.* 150. (1834); *Klotzsch in Linnæa*, vol. 9. 67. *Litt.* (1835).—Rhodoraceæ *and* Ericaceæ, *DC. Fl. Fr.* 3. 671. *and* 675. (1815).—Ericaceæ, *Ed. Pr.* clxvi. (1836); *Endl. Gen.* clxi.; *DC. Prodr.* 7. 580; *Meisner, p.* 244.

DIAGNOSIS.—*Erical Exogens, with monopetalous flowers, free stamens all perfect, loose-skinned or tight-skinned seeds, and 2-celled anthers opening by pores.*

Shrubs or under-shrubs. Leaves evergreen, rigid, entire, whorled, or opposite, without stipules. Inflorescence variable, the pedicels generally bracteate. Calyx 4- or

3 Fig. CCCXV.

5-cleft, nearly equal, inferior, persistent. Corolla hypogynous, monopetalous, 4- or 5-cleft, occasionally separable into 4 or 5 pieces, regular or irregular, often withering, with an imbricated æstivation. Stamens definite, equal in number to the segments of the corolla, or twice as many, hypogynous, or scarcely inserted into the base of the corolla; anthers 2-celled, the cells hard and dry, separate either at the apex or base, where they are furnished with some kind of appendage, and dehiscing by a pore. Ovary surrounded at the base by a disk, or secreting scales; many-celled, many-seeded; style 1, straight; stigma 1, undivided or toothed, or 3-cleft, with an indication of an indusium. Fruit capsular, many-celled, with central placentæ; dehiscence various. Seeds indefinite, minute; testa firmly adhering to the kernel; embryo cylindrical, in the axis of fleshy albumen; radicle much longer than the cotyledons and next the hilum.

Fig. CCCXVI.

This Natural Order contains some of the most beautiful plants of which we have any knowledge. They were formerly separated into two Orders by Jussieu, who distinguished Ericæ and Rhododendra by the dehiscence of their capsule; a character not now esteemed of ordinal importance, and consequently abandoned. Heathworts differ from Bilberries and Bellworts in their superior ovary, from Epacrids in the 2-celled anthers, from Wintergreens and Fir-rapes in the structure of their seeds and habit, and from all the Orders of which Figworts and Gentianworts

Fig. CCCXV.—Rhododendron albiflorum; 1. a calyx and pistil, with all the stamens removed save one; 2. an anther; 3. a ripe capsule burst; 4. a vertical section of a seed.
Fig CCCXVI.—Arctostaphylos pungens. 1. a stamen; 2. a cross section of an ovary.

may be considered the representatives, in the stamens not growing upon the petals, and in the cells of the ovary agreeing in number with the lobes of the calyx and corolla. The genus Saurauja among Dilleniads, has very much the structure of a Clethra. In Horsfield's *Plantæ Javanicæ*, p. 86, mention is made of the peculiar nature of the stigma in these plants, which Mr. Bennett justly compares to the indusium of Goodeniads. I have endeavoured to show that this rim is nothing more than the points of carpellary leaves separated from the stigma, which is itself a prolongation of the placenta. See *Botanical Register*, 1840, t. 9, and some observations on Babingtonia in the same work.

Heathworts are most abundant at the Cape of Good Hope, where immense tracts are covered with them ; they are common in Europe and North and South America, both within and without the tropics ; less common in Northern Asia and India, and almost unknown in Australasia, where their place is supplied by Epacrids. Although found in tropical countries, as for example, Java, it is only in their highlands.

It is worthy of note that although Botanists do not now admit the two sections of this Order to be of the same value as was assigned to them by Jussieu, yet that there is a considerable difference in the nature of their secretions. Ericeæ are to a large extent inert, there not being, in the whole of the vast genus Erica, a single instance of a medicinal species, for Erica arborea, once held to be an alexipharmic, seems to have been a merely superstitious medicine. Calluna vulgaris, the common Heather, is however astringent, and is employed both by fullers and dyers ; its tough branches are the common material out of which brooms are made in this country, and the flowers are peculiarly grateful to bees. We do, however, find among the Ericeæ species to which useful qualities cannot be denied. Some are astringents ; as Arctostaphylos Uva ursi, believed to be a decided palliative in nephritic paroxysms ; it is also employed in dysuria, catarrhus vesicæ, leucorrhœa and gonorrhœa. Its action is slow, and it therefore requires to be given for a considerable period ; although the effects are uncertain they sometimes give astonishing relief.—*Pereira*. The fruit of Gaultheria procumbens, a little dwarf North American evergreen, contains an aromatic, sweet, highly pungent volatile oil, which is antispasmodic and diuretic. A tincture has been useful in diarrhœa. Coxe states that the infusion is serviceable in asthma. It is used in North America as tea ; and brandy in which the fruit has been steeped is taken in small quantities in the same way as common bitters. The oil is known under the name of Oil of Wintergreen, and is used by druggists to flavour syrups, and also by perfumers.

The berries of the succulent-fruited kinds are usually grateful, and sometimes used as food. G. procumbens and Shallon, Arctostaphylos alpina, and Brossæa coccinea, are examples of this. In Van Diemens Land the G. hispida, or Waxcluster, bears snow-white berries, with a flavour by no means unpleasant ; in taste it is said to resemble the Gooseberry, but it is somewhat bitter ; but according to some, the G. antipoda is said to have more merit as a fruit. The Arbutus Unedo (κομαρος of Dioscorides) bears a red fruit something like a Strawberry, whence the plant has been familiarly named the Strawberry-tree ; its bark and leaves are astringent. A wine is made from the fruit in Corsica, but it is reported to be narcotic, if taken in quantity. A. Andrachne is stated to have similar qualities. In some instances this narcotic quality is so concentrated that the plants become poisonous. The shoots of Andromeda ovalifolia poison goats in Nipal. It is stated by Dr. Horsfield that a very volatile heating oil, with a peculiar odour, used by the Javanese in rheumatic affections, is obtained from another species of Andromeda. A. polifolia, a small shrub, found wild in the bogs of the North of Europe, is an acrid narcotic, and proves fatal to sheep. Similar properties have been observed in the United States in A. mariana, and others. It is however in the Rhododendreæ that dangerous narcotic qualities are most prevalent. The leaves of Ledum latifolium and palustre infused in beer render it unusually heady, producing headache, nausea, and even delirium. They have nevertheless been used, it is said, with advantage in tertian agues, dysentery, and diarrhœa. The leaves of Kalmia latifolia are poisonous to many animals, and are reputed to be narcotic, but their action is feeble. Bigelow states that the flesh of pheasants which have fed upon the young shoots is poisonous to man, and some cases of severe illness are on record which have been ascribed to this cause alone. The flowers exude a sweet honey-like juice, which is said when swallowed to bring on intoxication of a phrenitic kind, which is not only formidable in its symptoms but very lengthened in its duration.—*Burnett*. Rhododendron chrysanthum, a Siberian bush, is one of the most active of narcotics. Pallas and Koelpin assert that a strong decoction of the leaves is of the greatest service in chronic, but dangerous in acute, rheumatism. Its value as a means of removing arthritic complaints has also been highly spoken of. Finally, Pallas mentions an inveterate case of nervous sciatica, which had brought the patient to a state of lameness and deplorable emaciation, which was completely cured by perseverance in the use of the leaves for two years. No subsequent inconvenience was experienced, nor any signs of habitual drunkenness, although the dose was as much as 4 fluid ounces of the concentrated infusion daily. It has been said that the common evergreen shrub, Rhododendron ponticum, was the plant from whose flowers the bees of Pontus collected the honey that produced the extraordinary symptoms of poisoning described as having attacked the Greek soldiers in the famous retreat of the 10,000. Xenophon says that after eating it the men fell stupefied in all directions, so that the camp looked like a battle-field covered with corpses. But the Russian traveller Pallas is of opinion that Azalea pontica was the real cause of the mischief. He says that the effects of the Euxine honey are like those of Lolium temulentum, and occur in a country where no Rhododendron grows. The natives are well aware of the deleterious qualities of the plant, and it is related that goats which browse on the leaves, before the pastures are green, suffer in consequence, and moreover that cattle and sheep perish. R. maximum is said by some writers to be a mere astringent, and by others to be certainly a poison. The Swiss R. ferrugineum is another narcotic; an oil is obtained from its buds, which in Piedmontese medicine is called Olio di Marmotta, and is used in pains of the joints. The flowers of R. arboreum are eaten by the hill people of India, and are formed into a jelly by European visitors. The ferruginous leaves of R. campanulatum are used as snuff by the natives of India, as, we are informed by De Candolle, is in the United States the brown dust that adheres to the petioles of Kalmias and Rhododendrons. Loiseleuria procumbens has some reputation as an astringent medicine.

GENERA.

I. ERICEÆ.—Fruit loculicidal, rarely septicidal or berried. Buds naked.

* ERICIDÆ.

- Salaxis, *Salisb.*
  - *Coccosperma*, Klotsch.
- Lagenocarpus, *Klotsch.*
- Scyphogyne, *Brongn.*
  - *Tristemon*, Klotsch.
  - *Omphalocaryon*, Klots.
- Codonostigma, *Klotsch.*
- Coilostigma, *Benth.*
  - *Thamnium*, Klotsch.
- Codonanthemum, *Klot.*
- Syndesmanthus, *Klotsch.*
  - *Macrolinum*, Klotsch.
- Sympieza, Lichtenst.
  - *Microgomphus*, Benth.
- Simocheilus, *Benth.*
  - *Plagiostemon*, Klotsch.
  - *Thamnus*, Klotsch.
  - *Thoracosperma*, Klot.
  - *Octogonia*, Klotsch.
  - *Pachycalyx*, Klotsch.
- Acrostemon, *Klotsch.*
  - *Comacephalus*, Klotsch.
- Grisebachia, *Klotsch.*
- Finckea, *Klotsch.*
- Eremia, *Don.*
  - *Poderemia*, Benth.
  - *Micreremia*, Benth.
  - *Hexastemon*, Klotsch.
- Microtrema, *Klotsch.*
- Blæria, *Linn.*
- Ericinella, *Klotsch.*
- Philippia, *Klotsch*
  - *Eleutherostemon*, Klot.
- Bruckenthalia, *Reichenb.*
- Erica, *Linn.*
  - *Ectasis*, Benth.
  - *Callicodon*, Benth.
  - *Desmia*, Don.
  - *Polydesmia*, Benth.
  - *Chromostegia*, Benth.
  - *Eriodesmia*, Don.
  - *Amphodea*, Salisb.
  - *Geissostegia*, Benth.
  - *Gigandra*, Salisb.
  - *Pelostoma*, Salisb.
  - *Didymanthera*, Benth.
  - *Syringodea*, Benth.
  - *Eurylepis*, Benth.
  - *Callibotrys*, Salisb.
  - *Pleurocallis*, Salisb.
  - *Evanthe*, Salisb.
  - *Chona*, Don.
  - *Octopera*, Benth.
  - *Dasyanthes*, Benth.
  - *Batridium*, Salisb.
  - *Stellanthe*, Benth.
  - *Myra*, Salisb.
  - *Ceramus*, Salisb.
  - *Euryloma*, Don.
  - *Platyloma*, Benth.
  - *Callista*, Don.
  - *Cyatholoma*, Benth.
  - *Platyspora*, Salisb.
  - *Lamprotis*, Don.
  - *Eurystegia*, Benth.
  - *Trigemma*, Salisb.
  - *Oxyloma*, Benth.
  - *Pseuderemia*, Benth.
  - *Pachysa*, Don.
  - *Anaclasis*, Benth.
  - *Hermes*, Benth.
  - *Diphilus*, Salisb.
  - *Loxomeria*, Salisb.
  - *Eremocallis*, Salisb.
  - *Pyronium*, Salisb.
  - *Gypsocallis*, Salisb.
  - *Ceramia*, Don.
  - *Ephebus*, Salisb.
  - *Orophanes*, Salisb.
  - *Leptodendron*, Benth.
  - *Heliophanes*, Salisb.
  - *Lophandra*, Don.
  - *Melastemon*, Salisb.
  - *Eurystoma*, Benth.
  - *Polycodon*, Benth.
  - *Elytrostegia*, Benth.
  - *Arsace*, Salisb.
  - *Chlorocodon*, Benth.
- Pentapera, *Klotsch.*
- Macnabia, *Benth.*
  - *Nabea*, Lehm.
- Calluna, *Salisb.*

ANDROMEDIDÆ.

- Menziesia, *Smith.*
  - *Bryanthus*, Gmel.
  - *Phyllodoce*, Salisb.
  - *Daböecia*, Don.
  - *Boretta*, Neck.
  - *Arcimbalda*, Endl.
  - *Candollea*, Baumg.
- Andromeda, *Linn.*
  - *Cassiope*, Don.
  - *Polifolia*, Buxb.
  - *Cassandra*, Don.
  - *Lyonia*, Reichenb.
  - *Dipluria*, Raf.
  - *Baumannia*, DC.
  - *Chamædaphne*, Buxb.
  - *Zenobia*, Don.
  - *Leucothöe*, Don.
  - *Cassiphone*, Rchb.
  - *Maria*, DC.
  - *Agauria*, DC.
  - *Pieris*, Don.
  - *Agarista*, Don.
- Oxydendron, *DC.*
- Lyonia, *Nutt.*
  - *Xolisma*, Raf.
- Clethra, *Linn.*
  - *Cuellaria*, Ruiz et Pav.
  - *Tinus*, Linn.
  - *Volkameria*, R. Br.
  - *Junia*, Adans.
- Epigæa, *Linn.*
  - *Memecylon*, Mitch.
- Gaultheria, *Linn.*
  - *Gautiera*, Kalm.
  - *Chiogenes*, Salisb.
  - *Glycyphylla*, Raf.
  - *Phalerocarpus*, Don.
  - *Amphicalyx*, Blum.
  - *Diplicosia*, Blum.
  - *Shallonium*, Raf.
- Pernettia, *Gaudich.*
- Arbutus, *Tournef.*
  - *Unedo*, Link.
- Enkyanthus, *Lour.*
  - *Melidora*, Salisb.
- Arctostaphylos, *Adans.*
  - *Uva ursi*, Tournef.
  - *Mairania*, Neck.
- Comarostaphylis, *Zucc.*

II. RHODODENDREÆ.—Fruit capsular, septicidal. Buds scaly, and resembling cones.

- Azalea, *Linn.*
  - *Loiseleuria*, Desv.
  - *Chamæcistus*, Gray
  - *Chamæledon*, Link.
- Osmothamnus, *DC.*
- Kalmia, *Linn.*
- Rhodothamnus, *Reichb.*
  - *Adodendron*, Neck.
  - *Chamæcistus*, Don.
  - *Ledum*, Michel.
- Rhododendron, *Linn.*
  - *Anthodendron*, Reich.
  - *Theis*, Salisb.
  - *Pentanthera*, Don.
  - *Rhodora*, Linn.
  - *Beverinckia*, Salisb.
  - *Vireya*, Blum.
  - *Booram*, Endl.
  - *Buramia*, DC.
  - *Hymenanthes*, Blum.
- Befaria, *Mut.*
  - *Bejaria*, Adr. Juss.
  - *Acuna*, Ruiz et Pav.
- Leiophyllum, *Pers.*
  - *Ammyrsine*, Pursh.
  - *Fischera*, Swartz.
  - *Dendrium*, Desv.
- Ledum, *Linn.*
  - *Dalia*, Adans.

NUMBERS. GEN. 42. Sp. 850.

Rutaceæ.
POSITION.—Pyrolaceæ.—ERICACEÆ.—Epacridaceæ.
Dilleniaceæ.

Alliance XXXV. *RUTALES.*—The Rutal Alliance.

Diagnosis.—*Hypogynous Exogens, with monodichlamydeous symmetrical flowers, axile placentæ, an imbricated calyx and corolla, definite stamens, and an embryo with little or no albumen.*

The larger part of this Alliance might and even has been regarded as one Natural Order, and by all Botanists the members of it are placed in very close relationship, with the exception of Waterpeppers and Podostemads. The two latter are, however, so very like degraded forms of Rueworts, that I cannot but regard them as standing in the same relation to Rutals as Hippurids to the Myrtal Alliance. They are, however, in tracing affinities, to be looked upon as mere lateral offshoots from some of the higher Orders, and not as either terminating a line or completing a circle of affinities.

Strictly speaking, the Rutal Alliance touches the Erical by means of Rueworts themselves, among which Correa assumes the appearance of an Andromeda; in like manner it does not pass into the Geranial Alliance by way of Podostemads, but through Beancapers, which stand close to Oxalids, or even Cranesbills, of which they have the stipules.

Terebinths approach the genus Juglans in the Diclinous series, through such of their unisexual genera as Pistacia.

Natural Orders of Rutals.

Fruit consolidated, succulent, indehiscent. Petals imbricated. Stamens free, or nearly so. Leaves dotted	170. Aurantiaceæ.
Fruit consolidated, hard, dry, somewhat valvular. Petals valvate. Stamens free. Leaves generally dotted	171. Amyridaceæ.
Fruit consolidated, capsular. Stamens deeply monadelphous or free. Seeds numerous, winged	172. Cedrelaceæ.
Fruit consolidated, berried or capsular. Stamens deeply monadelphous. Seeds few, wingless. Leaves dotless	173. Meliaceæ.
Fruit apocarpous. Ovule single, suspended by a cord rising from the base of the carpel	174. Anacardiaceæ.
Fruit apocarpous. Ovules collateral, ascending, orthotropal, sessile	175. Connaraceæ.
Fruit finally apocarpous, few-seeded, with the pericarp separating in two layers. Ovules sessile, pendulous. Flowers ☿.	176. Rutaceæ.
Fruit finally apocarpous, few-seeded, with the pericarp separating in two layers. Ovules sessile, pendulous. Flowers ♂ - ☿ - ♀.	177. Xanthoxylaceæ.
Fruit finally apocarpous, one-seeded, with the pericarp not laminating, and a succulent conical torus.	178. Ochnaceæ.
Fruit finally apocarpous, one-seeded, with the pericarp not laminating, and a dry inconspicuous torus. Albumen wanting. Leaves alternate, without stipules	179. Simarubaceæ.
Fruit finally apocarpous, few-seeded, with the pericarp not laminating, and a dry inconspicuous torus. Albumen present. Leaves opposite, with stipules	180. Zygophyllaceæ.
Fruit finally apocarpous, many-seeded. Flowers polypetalous . .	181. Elatinaceæ.
Fruit finally apocarpous, many-seeded. Flowers apetalous, very imperfect	182. Podostemaceæ.

ORDER CLXX. AURANTIACEÆ.—CITRONWORTS.

Aurantiaceæ, *Corr. Ann. Mus.* 6. 376. (1805); *Mirb. Bull. Philom.* 379. (1813); *DC. Prodr.* 1. 536. (1824); *Endl. Gen.* ccxxiv.; *Wight Illustr.* 1. *t.* 42.

DIAGNOSIS.—*Rutal Exogens, with consolidated succulent indehiscent fruit, imbricated petals, free or nearly free stamens, and dotted leaves.*

Trees or shrubs, almost always smooth, and filled everywhere with little transparent receptacles of volatile oil. Leaves alternate, often compound, always articulated with the petiole, which is frequently winged. Spines, if present, axillary. Calyx urceolate or campanulate, somewhat adhering to the disk, short, 3- or 5-toothed, withering. Petals 3 to 5, broad at the base, sometimes distinct, sometimes slightly combined, inserted upon the outside of an hypogynous disk, slightly imbricated at the edges. Stamens equal in number to the petals, or twice as many, or some multiple of their number, inserted upon the same hypogynous disk; filaments flattened at the base, sometimes distinct, sometimes slightly combined in one or several parcels; anthers terminal, innate. Ovary free, many-celled; style 1, taper; stigma slightly divided, thickish; ovules solitary, twin, or 00, pendulous or occasionally horizontal, anatropal. Fruit pulpy, one or more-celled, sometimes with a leathery rind replete with receptacles of volatile oil, and even separable from the cells; cells often filled with pulp. Seeds attached to the axis, sometimes numerous, sometimes solitary, usually pendulous, occasionally containing more embryos than one; raphe and chalaza usually very distinctly marked; albumen 0; embryo straight; cotyledons thick, fleshy; radicle very short.

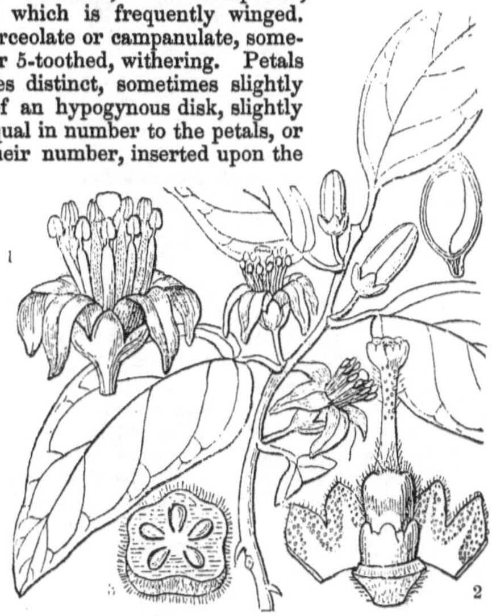

Fig. CCCXVII.

These are readily known by the abundance of oily receptacles which are dispersed over all parts of them, by their deciduous petals, compound leaves, often with a winged petiole, imbricated petals, and succulent or pulpy fruit. They are nearly related to Amyrids on the one hand, and to various genera of Rueworts on the other, but differ from the first in their pulpy fruit and imbricated petals, and from the latter in their consolidated juicy fruit. It is more difficult to distinguish them from Xanthoxyls, unless attention is paid to the fruit, the apocarpous structure of the ovary, and the polygamous flowers. Luvunga is remarkable for having the climbing habit of Xanthoxyls, and the fruit of Citronworts. The raphe and chalaza are usually distinctly marked upon the testa, and sometimes beautifully. The genus Citrus is very subject to a monstrous separation of the carpels, which produces what are called horned Oranges, and fingered Citrons, the last of which is the genus Sarcodactylis of the younger Gærtner (t. clxxxv.), or to a multiplication of the normal number of carpels, in which case Orange is formed within Orange.

Fig. CCCXIX.

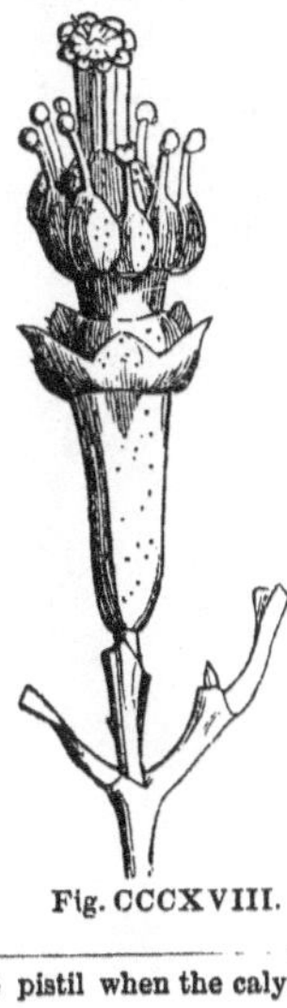

Fig. CCCXVIII.

Fig. CCCXVII.—Micromelum monophyllum.—*Wight.* 1. a flower; 2. the pistil when the calyx is rolled back; 3. a cross section of an ovary; 4. longitudinal section of a seed.
Fig. CCCXVIII.—A young Orange, with a row of supernumerary carpels.
Fig. CCCXIX.—The fruit produced by this.

Citronworts are almost exclusively found in the East Indies, whence they have in some cases spread over the rest of the tropics. Two or three species are natives of Madagascar; one is described as found wild in the woods of Essequebo; and Prince Maximilian of Wied Neuwied speaks of a wild Orange of Brazil, called Laranja da terra, which has by no means the delicious refreshing qualities of the cultivated kind, but a mawkish sweet taste. This is called by Martius Citrus Aurantium efferata; but must have been introduced. Limonia laureola is remarkable as the only plant of this family found on the tops of cold and lofty mountains, where it is for some months of the year buried under the snow. The Hill people of India call it Kidar-patri and Kuthar-chara, and fancy that it is by feeding on its leaves that the Musk acquires its peculiar flavour.—*Royle.*

The wood is universally hard and compact; the leaves abound in a volatile, fragrant, bitter, exciting oil; the pulp of the fruit is always more or less acid. The Orange, Lemon, Lime, Shaddock, Pompelmoose, Forbidden Fruit, and Citron, Indian fruits, some of which have now become so common in other countries as to give a tropical character to a European dessert, are the most remarkable products of the Order. To this must be added the excellence of their wood, and the fragrance and beauty of their flowers. The fruits just mentioned are not, however, its only produce. The Wampee, a fruit highly esteemed in China and the Indian archipelago, is the produce of Cookia punctata. The berries of Glycosmis citrifolia are delicious; those of Triphasia are extremely agreeable. The productiveness of the common Orange is enormous. A single tree at St. Michael's has been known to produce 20,000 Oranges fit for packing, exclusively of the damaged fruit and the waste, which may be calculated at one-third more. The juice of the Lime and the Lemon contains a large quantity of citric acid. Oranges contain malic acid. A decoction of the root and bark of Ægle Marmelos is supposed, on the Malabar coast, to be a sovereign remedy in hypochondriasis, melancholia, and palpitation of the heart; the leaves in decoction are used in asthmatic complaints, and the fruit a little unripe is given in diarrhœa and dysentery. Roxburgh adds, that the Dutch in Ceylon prepare a perfume from the rind; the fruit is most delicious to the taste, exquisitely fragrant and nutritious, but laxative; the mucus of the seed is a good cement for some purposes. The leaves of Bergera Königii are considered by the Hindoos as stomachic and tonic; an infusion of them toasted stops vomiting; the green leaves are used raw in dysentery; the bark and root internally as stimuli. The young leaves of Feronia elephantum have, when bruised, a most delightful smell, much resembling Anise; the native practitioners of India consider them stomachic and carminative; its gum is very like Gum Arabic. Orange-leaves are sometimes prescribed to hysterical females instead of Tea. Oil of Neroli and Napha water, two delicious perfumes, are distilled from Orange-flowers; Cedrati, a variety of the Lime, is another perfume in much esteem. See further, *Royle's Illustrations*, p. 129.

GENERA.

Atalantia, *Corr.*
Triphasia, *Lour.*
Limonia, *Linn.*
  Winterlia, Dennst.
Glycosmis, *Corr.*
Sclerostylis, *Blum.*
Rissoa, *Arn.*

Bergera, *Kön.*
Murraya, *Kön.*
  Chalcas, Lour.
  Marsana, Sonn.
Cookia, *Sonner.*
  Quinaria, Lour.
  ? *Aulacia*, Lour.

Acronychia, *Forst.*
Clausena, *Burm.*
Micromelum, *Blum.*
Paramignya, *Wight.*
Luvunga, *Hamilt.*
  Lavanga, Meisn.
Feronia, *Corr.*

Ægle, *Corr.*
  Belou, Adans.
Citrus, *Linn.*
  Sarcodactylis, Gærtn.
Papeda, *Hassk.*
? Chionotria, *Jack.*
? Severinia, *Tenor.*

NUMBERS. GEN. 20. Sp. 95.

POSITION.—Amyridaceæ.—AURANTIACEÆ.—Xanthoxylaceæ.

ORDER CLXXI. AMYRIDACEÆ.—AMYRIDS.

Terebintaceæ, *Juss. Gen.* 368. (1789) *in part; tribes* 4 & 5. *DC. Prodr.* 2. 81. (1825).— Amyrideæ, *R. Brown in Congo*, 431. (1818); *Kunth in Ann. Sc. Nat.* 2. 353. (1824).—Amyridaceæ, *Ed. pr.* cxviii.; *Prodr.* 2. 81. (1825).—Burseraceæ, *Kunth in Ann. Sc. Nat.* 2. 333. (1824); *Endl. Gen.* ccxlvi.; *Meisner, Gen. p.* 77.—? Balaniteæ, *Endl. Ench. p.* 547. (1841).

DIAGNOSIS.—*Rutal Exogens, with consolidated, hard, dry, and somewhat valvular fruit, valvate petals, free stamens, and generally dotted leaves.*

Trees or shrubs, abounding in balsam or resin. Leaves alternate or opposite, ternate or unequally pinnate, occasionally with stipules, and pellucid dots. Flowers axillary or terminal, in racemes or panicles, sometimes unisexual by abortion. Calyx persistent, somewhat regular, with from 2 to 5 divisions. Petals 3-5, inserted below a disk arising from the calyx; æstivation usually valvate, sometimes imbricated. Stamens twice as many as the petals, all fertile. Disk orbicular or annular. Ovary 1- 5-celled, superior, sessile in or upon the disk; style solitary and compound; stigmas as many as the cells of the ovary, and where there is but one cell capitate; ovules in pairs, attached to the apex of the cell, anatropal, collateral. Fruit hard and dry, 1- 5-celled, with its outer part often splitting into valves. Seeds without albumen; cotyledons either wrinkled and plaited, or amygdaloid; radicle superior, straight, turned towards the hilum.

These are plants with the appearance of Oranges, and in the instance of Amyris itself, with the dotted leaves of that Order; nor have they any positive mark of distinction, except their fruit forming a shell whose husk eventually splits into valve-like segments. In general, however, the petals have a valvate æstivation; and Amyris, which wants that character, has only a one-celled ovary. The genera collected under this name are by no means perfectly known, and demand a scrupulous revision. Copaifera and Myrospermum, placed here in the last edition of this work, belong to the Leguminous Order. In referring the genus Balanites hither, I do so without having had the opportunity of examining its fruit, the seeds of which are said to be albuminous. Its calyx is certainly not valvate, as it has been described to be, but is truly imbricated.

What species have as yet been ascertained are exclusively natives of tropical India, Africa, and America.

It is here that we find the trees yielding myrrh and frankincense, besides which the species have all an abundance of fragrant resinous juice. The resin of Boswellia is used in India as frankincense, and also as pitch. It is hard and brittle, and, according to Roxburgh, is boiled with some low-priced oil to render it soft and fit for use. The native doctors prescribe it, mixed with ghee (clarified butter), in cases of gonorrhœa, and also in what they call Ritta Kaddapoo, which signifies flux accompanied with blood. The wood is heavy, hard, and durable. Boswellia serrata, called Libanus thurifera by Colebrooke, produces the gum-resin Olibanum, a substance chiefly used as a grateful incense, but which also possesses stimulant, astringent, and diaphoretic properties. Arabian frankincense has also been said to be the produce of the same tree, but this is very uncertain. Myrrh, or Hobali,

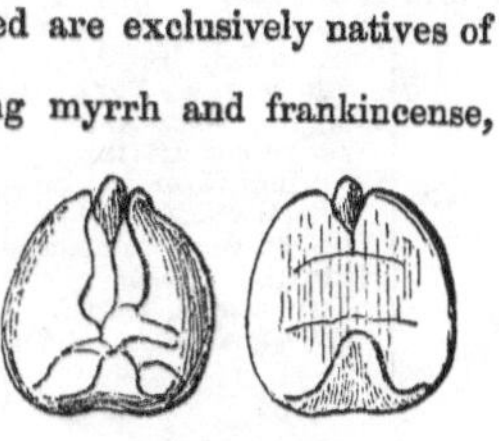

Fig. CCCXXI. Fig. CCCXX.

Fig. CCCXX.—Marignia obtusifolia.—*Delessert.* 1. a flower; 2. the same divided vertically; 3. a fruit; 4. a section of the same.

Fig. CCCXXI.—Embryo of Elaphrium excelsum.—*Turpin.*

is obtained on the Abyssinian coast from Balsamodendron Myrrha, a dwarf shrub called Kerobeta by the natives. Balm of Mecca, Beshan (perhaps the origin of the word Balsam), the Balessan of Bruce, is yielded by B. Opobalsamum.—Harris in *Chem. Gaz.* 1844. 148. B. Gileadense is also said to furnish it. A species of Balsamodendron is also mentioned by Mr. Griffith as being one of the most cultivated plants in Afghanistan for its aromatic and stimulant properties; it is called Schnee.—*Ann. Nat. Hist.* x. 194.

A kind of coarse resin is obtained from Boswellia glabra, and is used, boiled with oil, for pitching the bottom of ships. Bursera paniculata, called Bois de Colophane in the Isle of France, gives out, from the slightest wound in the bark, a copious flow of limpid oil of a pungent turpentine odour, which soon congeals to the consistence of butter, assuming the appearance of camphor. The gum of Canarium commune has the same properties as those of the Balsam of Copaiva; the three-cornered nuts are eaten in Java both raw and dressed, and an oil is expressed from them, which is used at table when fresh, and for burning when stale. The raw nuts are, however, apt to bring on diarrhœa.

Among fragrant products of less moment may be named Bdellium, the resin, in Africa, of the Niouttont or Balsamodendron africanum, and in India of B. Roxburghii, supposed to be the same plant as Commiphora madagascariensis; Tacamahac from Elaphrium tomentosum; Incense-wood from Icica guianensis; American Elemi in part from Icica Icicariba; American Balm of Gilead from Icica Carana; Balsam of Acouchi from Icica Aracouchini; Chibou or Cachibou resin from Bursera gummifera; Resin of Carana from Bursera acuminata; Beaume a cochon or Beaume à sucrier, a substitute for Copaiva, from Hedwigia balsamifera; Resin of Coumia from Icica ambrosiaca. Finally, it is said that Amyris toxifera is poisonous; that Amyris Plumieri, and another species called by Dr. Hamilton A. hexandra, yield a part of the Gum Elemi of commerce; and that the wood of Amyris balsamifera, a Jamaica tree, furnishes one of the sorts of Lignum Rhodium. Picramnia ciliata, a Brazilian tree, has a bitter subacrid bark, which is administered successfully as a substitute for Cascarilla, according to Martius. The layers of the liber of a species of Amyris were found by Cailliaud to be used by the Nubian Mahometans as paper, on which they write their legends. Icica altissima furnishes the Curana, Samaria, Acuyari, Mara, or Cedarwood of Guiana, one variety of which is red, the other white, according to Sir R. Schomburgk. It is light, easily worked, and very aromatic. This distinguished traveller states that one of his canoes, 42 feet long and 5½ feet wide, had been made from a single tree of this species. The leaves of Balanites ægyptiaca, a tree cultivated in Egypt under the Negro name of Soum, and the Arabic Hilelgie or Haledsch, are slightly acid, and have the reputation of being anthelmintic. The unripe flesh of its drupes is acrid, extremely bitter, and violently purgative; but when ripe it is eaten without inconvenience. A fat oil, called Zachun, is pressed from its seeds. The fruits are said to be mixed in commerce with Myrobalans.

GENERA.

I. BURSERIDÆ. — Ovary with more cells than one.

Boswellia, *Roxb.*
Libanus, Colebr.
Ploesslea, Endl.
Protium, *Burm.*
Balsamodendron, *Kunth.*
Heudelotia, A. Rich.
Nioutout, Adans.
Balsamea, Gled.
Balessam, Bruce.
Commiphora, Jacq.
Elaphrium, *Jacq.*
Icica, *Aubl.*
Bursera, *Jacq.*
Trattinickia, *Willd.*
? *Dacryodes*, Vahl.
Marignia, *Commers.*
Dammara, Gärtn.
Canarium, *Linn.*
Pimelea, Lour.
Colophonia, Commers.
Hedwigia, *Swartz.*
Tetragastris, Gärtn.
Schwägrichenia, Reic.
? *Coproxylon*, Tuss.
? *Knorria*, Moç. et Sess.
Garuga, *Roxb.*
Kunthia, Dennst.
Hemprichia, *Ehrenb.*
? Balanites, *Del.*
? Fagarastrum, *Don.*
? Picramnia, *Swartz.*
? Methyscophyllum, *Eckl. et Zeyh.*
? Tapiria, *Juss.*
? *Joncquetia*, Schreb.
? Loureira, *Meisn.*
? *Toluifera*, Lour.
? Triceros, *Lour.*
? Barbylus, *P. Br.*
? Pachylobus, *G. Don.*

II. AMYRIDÆ. — Ovary one-celled.

Amyris, *Linn.*
Elemifera, Plum.
Lucinium, Plukn.

NUMBERS. GEN. 22. Sp. 45.

POSITION.—Anacardiaceæ.—AMYRIDACEÆ.—Aurantiaceæ.

ORDER CLXXII. CEDRELACEÆ.—CEDRELADS.

Cedreleæ, *Brown in Flinders*, 64. (1814).—Meliaceæ, § Cedreleæ, *DC. Prodr.* 1. 624. (1824).—Cedrelaceæ, *A. de Jussieu Mémoire* (1830); *Ed. pr.* lxxviii.; *Endl. Gen.* ccxxvi.; *Meisner p.* 47.

DIAGNOSIS.—*Rutal Exogens, with consolidated capsular fruit, deeply monadelphous or free stamens, and numerous winged seeds.*

Trees, with timber which is usually compact, scented, and beautifully veined. Leaves alternate, pinnated, without stipules. Flowers in terminal panicles. Calyx 4-5-cleft. Petals 4-5, longer. Stamens 8-10; the filaments either united into a tube (Swietenieæ), or distinct (Cedreleæ), and inserted into an hypogynous disk. Style and stigmas simple. Cells of the ovary equal in number to the petals, or fewer (3), with the ovules ascending or pendulous, anatropal, 4, or often more, imbricated, in two rows. Fruit capsular, with the valves separable from the thick axis, with whose angles they alternate. Seeds flat, winged; albumen thin or none; embryo orthotropal, straight; cotyledons flattish or fleshy; radicle very short, next the hilum.

Nearly related to Meliads, in whose affinities they participate, and chiefly distinguished by their winged and indefinite seeds. Flindersia, a genus established by Brown in the Appendix to Captain Flinders' *Vogage*, differs from Cedrelads both in the insertion of its seeds, which are erect, in the dehiscence of its capsules, and also in having moveable dissepiments: these last, however, Brown considers as segments of a common placenta, having a peculiar form. Flindersia, and Chloroxylon are distinct from the rest of the Order, in having the leaves dotted with pellucid glands, in which respect they serve to connect Cedrelads with Citronworts, and, notwithstanding

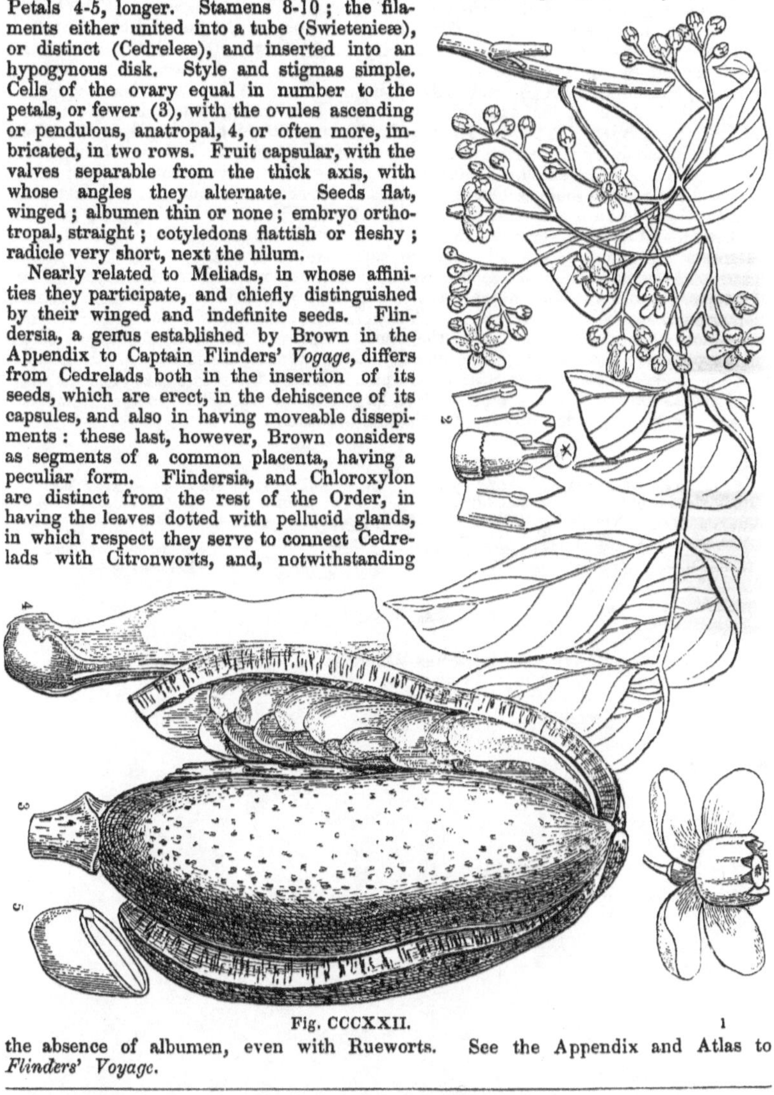

Fig. CCCXXII.

the absence of albumen, even with Rueworts. See the Appendix and Atlas to *Flinders' Voyage*.

Fig. CCCXXII.—Swietenia Mahagoni.—*Hooker*. 1. a flower; 2. a cup of stamens spread open, and the pistil; 3. fruit; 4. a seed; 5. a section of it to show the crosscut embryo.

These are common to the tropics of America and India, but are very rare on the continent of Africa, and the adjoining islands.

The wood of the Order is in general fragrant and aromatic. The bark of Cedrela is fragrant and resinous; that of C. Toona, and of Mahogany (Swietenia Mahagoni) is also accounted febrifugal. The former is a powerful astringent, and though not bitter, a tolerably good substitute for Peruvian Bark in the cure of remitting and intermitting fevers; particularly when joined with a small portion of the powdered seed of Cæsalpinia Bonducella (Kutulegee of the Bengalese), which is a most powerful bitter. The bark was used in Java by Dr. Blume, with much success in the worst epidemic fevers, diarrhœa, and other complaints; Horsfield also applied it in various cases of dysentery, but in the last stage, when the inflammatory symptoms had disappeared. The bark of Soymida febrifuga, the Rohuna of Hindostan, called on the Coromandel coast the Red-wood tree, is a useful tonic in India in intermittent fevers; but Ainslie found that if given beyond the extent of 4 or 5 drachms in 24 hours, it deranged the nervous system, occasioning vertigo and subsequent stupor. It has also been employed successfully in India in bad cases of gangrene, and in Great Britain in typhus fever, and as an astringent. That of Khaya, the Kassou-Khaye of Senegal, is a common febrifuge in the swampy districts on the banks of the Gambia. Cedrela febrifuga bark is said by Blume to be employed successfully against the intermittent fevers of Java; he observes that it is tonic and useful in cases of diarrhœa, &c., but that it should never be used where there is a tendency to inflammation. The bark of Chickrassia tabularis has been found to be powerfully astringent without bitterness.—*Roxb.* Juriballi bark, a Demerara product, is also supposed to belong to some plant of this Order; it is described as being a potent bitter and astringent, far superior to Peruvian bark in fevers of a typhoid and malignant nature. It is cordial and purgative; and is also a powerful diaphoretic, especially if taken warm.—*Hancock.* An essential oil is found in Flindersia and Chloroxylon, as is indicated by their dotted leaves. The young shoots of Cedrela angustifolia have a powerful smell of Garlic, according to Ruiz and Pavon. Satin-wood is the produce of Chloroxylon Swietenia, which is one of the plants that yield the wood oil of India.—*Royle.* Oxleya xanthoxyla, a large tree, is the Yellow-wood of New South Wales. Mahogany is the timber of Swietenia Mahagoni.

GENERA.

I. SWIETENEÆ. — Stamens monadelphous.	Khaya, *Adr. Juss.*	II. CEDRELEÆ. — Stamens distinct.	*Jonsonia*, Adans.
	Soymida, *Adr. Juss.*		*Cedrus*, Mill. part.
Swietenia, *Linn.*	Chickrassia, *Adr. Juss.*	Chloroxylon, *DC.*	*Toona*, Endl.
Mahagoni, Adans.	*Chukrasia*, Adr. Juss.	Flindersia, *R. Br.*	*Cuveracea*, Jones.
Roia, Scop.	*Plagiotaxis*, Wall.	Oxleya, *A. Cunn.*	*Surenus*, Rumph.
Cedrus, Mill		Cedrela, *Linn.*	Vavæa, *Benth.*

NUMBERS. GEN. 9. Sp. 25.

POSITION.—Meliaceæ.—CEDRELACEÆ.—Aurantiaceæ.

ORDER CLXXIII. MELIACEÆ.—MELIADS.

Meliæ, *Juss. Gen.* 263. (1789).—Meliaceæ, *Juss. Mém. Mus.* 3. 436. (1817); *DC. Prod.* 1. 619. (1824); *Adr. de Juss. Mémoire* (1830); *Ed. pr.* lxxvii.; *Endl. Gen.* ccxxv.

DIAGNOSIS.—*Rutal Exogens, with consolidated berried or capsular fruit, deeply monadelphous stamens, a few wingless seeds, and dotted leaves.*

Trees or shrubs. Leaves alternate, or occasionally somewhat opposite, simple, or pinnate, without stipules. Flowers sometimes imperfect by abortion, usually in loose

1 2 3 4 5

Fig. CCCXXIII.

masses. Sepals 3, 4, or 5, more or less united. Petals the same number, hypogynous, conniving at the base, or even cohering, usually having a valvate or imbricated æstivation. Stamens twice as many as the petals; filaments cohering in a long tube; anthers sessile within the orifice of the tube. Disk frequently highly developed, surrounding the ovary like a cup. Ovary single, with the same number of cells as petals, or fewer (3-2), very seldom many more (10-12) cells; style 1; stigmas distinct or combined; ovules anatropal, semi-anatropal, amphitropal or orthotropal! 1 or 2 in each cell, very rarely 4. Fruit berried, drupaceous or capsular, often, in consequence of abortion, 1-celled, the valves, if present, having the dissepiments in their middle. Seeds not winged, with or without an aril; albumen fleshy (Melieæ), or usually absent (Trichilieæ). Embryo with leafy or amygdaloid cotyledons, within which the radicle is drawn back.

This Order was ill understood until it was investigated by Adrien de Jussieu, from whose Memoir I borrow the principal part of what follows. It is, no doubt, related to Citronworts, although Canella, which was considered a case of transition, is removed from it. The inflorescence of Citronworts terminating in dichotomies with a central and præcocious flower, the union that sometimes occurs between the filaments of Citronworts, the number of stamens often double that of the petals, and the embryo with a short radicle drawn back between thick cotyledons, are all points in which there is an accordance between the two Orders. The occasionally monadelphous stamens of Rueworts indicate an analogy with that Order, which is confirmed by the general tendency in both cases to produce two ovules in each cell of the ovary. The number and the relative position of the parts of the flower show an affinity with Soapworts, the structure of whose seeds is often absolutely the same as that of Meliads; their accordance in habit is incontestable, and in fact the species of the two Orders are often mixed together in herbaria. Cedrelads are chiefly distinguished by their

Fig. CCCXXIII.—Ekebergia Senegalensis. 1. a flower; 2. the calyx and staminal tube; 3. a transverse section of the ovary; 4. a ripe fruit; 5. a vertical section of the latter.

winged seeds and the stamens being in a less degree monadelphous. As to a supposed affinity between Vineworts and this Order, it seems to be of a very distant description.

The species are found all over the world; in about equal quantities in America and Asia, and four times fewer in Africa; but these proportions are possibly due to the difference in the degree that those parts of the world have been examined. The Order does not extend further to the north than 40°; Melia Azedarach is naturalised as it were in Provence; and an Hartighsea exists in New Zealand. The extra-tropical species are, however, rare.

Bitter, astringent, and tonic qualities belong to the species of this Order, and are often developed in so considerable a degree as to render their application dangerous without precaution. A Brazilian plant called Jito is a powerful purgative, but Piso in mentioning it, warns us against the danger of employing it, and says that it is more often a poison than a medicine; it is supposed to be a species of Guarea, perhaps either G. purgans or spiciflora, which Martius informs us act violently on the uterus, and in an overdose produce abortion. Trichilia cathartica is reputed to have similar properties. The juice of the bark of Guarea Aubletii is a purgative and a violent emetic; the bark of Guarea trichilioides has similar qualities. The same power is assigned to the Arabian Elcaija (Trichilia emetica). Jacquin says that the negresses employ the root of T. trifoliolata to procure abortion. The root of Melia Azedarach is bitter and nauseous, and is used in North America as anthelmintic; the pulp that surrounds the seeds is said to be deleterious; but this is denied by Turpin, who asserts that dogs which he has seen eat it experienced no inconvenience; and children in Carolina swallow the seeds with impunity. It is supposed that the Melia Azedarachta, or Neem-tree of India, possesses febrifugal properties; a kind of Toddy, which the Hindoos consider a stomachic, is obtained from it by tapping; it is also called the Margosa-tree. From the fruit of the same plant an oil is obtained, which is fit for burning and for other domestic purposes, and, as Ach. Richard observes, is another instance, after the Olive, of the pericarp yielding that substance which is usually obtained from the seed. This oil is said to possess antispasmodic qualities. Blume attributes to the root of Sandoricum indicum properties similar to those of Melia; but the latter has a repulsive odour, while the other is aromatic; it is employed against leucorrhœa, combined with the bark of the root of Carapa obovata, which is bitter and astringent. The bark of Carapa guianensis has great reputation as a febrifuge; its oil is bitter and anthelmintic, and is said to be particularly useful in guarding iron against rust. Carapa Touloucouna or guineensis yields the Tallicoonah or Kundah oil, an anthelmintic and purgative; it is acrid and bitter, and said to be well suited for lamps. Trichilia Catigoa (Caá-tiguá, *Braz.*) stains leather a bright yellow. Rumphius mentions the extreme bitterness of Xylocarpus Granatum. An alliaceous odour found in two species of Cedrela also occurs in a very prominent degree in some species of Dysoxylon and Hartighsea; the Javanese mountaineers use the fruit of these trees as Garlic. Blume suspects that some species of Epicharis have similar properties. A warm pleasant-smelling oil is prepared from the fruit of Trichilia speciosa, which the Indian doctors consider a valuable external remedy in chronic rheumatism and paralytic affections. Some delicious fruits of the Indian Archipelago, called Langsat, or Lanséh, and Ayer Ayer, are species of the genus Lansium; they have a watery pulp, with a cooling pleasant taste. Milnea edulis is another plant of the Order, with eatable fruit. See further *Royle's Illustrations*, p. 141.

GENERA.

I. MELIEÆ. — Embryo with albumen.	II. TRICHILIEÆ. — Embryo exalbuminous.		
Quivisia, *Commers.*	Aglaia, *Lour.*	Dysoxylon, Blum.	Heynea, *Roxb.*
Gilibertia, Gmel.	*Camunium*, Rumph.	Schizochiton, *Spreng.*	Trichilia, *Linn.*
Calodryum, *Desv.*	*Cambania*, Commers.	*Chisocheton*, Blum.	*Elcaja*, Forsk.
Naregamia, *Wight et Ar.*	Milnea, *Roxb.*	Synöum, *Adr. Juss.*	*Portesia*, Cav.
Munronia, *Wight.*	*Nyalelia*, Dennst.	*Schoutensia*, Endl.	*Torpesia*, Endl.
Turræa, *Linn.*	Lansium, *Rumph.*	Hartighsea, *Adr. Juss.*	Moschoxylum, *Adr.Juss.*
Melia, *Linn.*	*Sphærosacme*, Wall.	*Macrochiton*, Blum.	Guarea, *Linn.*
Azederach, Tournef.	Nemedra, *Juss.*	Epicharis, *Blum.*	? *Elutheria*, P. Br.
Azadirachta, *Adr. Juss.*	Amoora, *Roxb.*	Cabralea, *Adr. Juss.*	Carapa, *Aubl.*
Mallea, *Adr. Juss.*	*Amura*, Schult.	Didymochiton, *Blum.*	*Xylocarpus*, Schreb.
Cipadessa, *Blum.*	*Andersonia*, Roxb.	Goniochiton, *Blum.*	*Persoonia*, Willd.
	Aphanamixis, Blum.	Sandoricum, *Cav.*	Xylocarpus, *Adr. Juss.*
		Ekebergia, *Sparm.*	? Odontandra, *H. B. K.*
		Walsura, *Roxb.*	? Aitonia, *Linn. f.*

NUMBERS. GEN. 33. SP. 150.

POSITION.—Aurantiaceæ.—MELIACEÆ.—Cedrelaceæ.
Pittosporaceæ.

ORDER CLXXIV. ANACARDIACEÆ.—ANACARDS, OR TEREBINTHS.

Terebintaceæ, *Juss. Gen.* 368. (1789) *in part.*—Cassuvieæ *or* Anacardieæ, *Brown in Congo*, 431. (1818); *Bartl. Ord. Nat. p.* 395. (1830).—Terebintaceæ, *Kunth in Ann. des Sc. Nat.* 2. 333. (1824). trib. 1 and 2. *DC. Prodr.* 2. 62. *&c.* (1825); *Juss. Dict. des Sc. Nat.* v. 53. (1828); *Arnott in Encycl. Britt. p.* 106. (1832).—Spondiaceæ, *Kunth in Ann. Sc. Nat.* 2. 362. (1824); *Martius Conspectus, No.* 268. (1835); *Ed. pr.* lxxxi.; *Wight Illustr.* 1. *t.* 76.

DIAGNOSIS.—*Rutal Exogens, with apocarpous fruit, and a single ovule rising by a cord from the base of the cell.*

Trees or shrubs, with a resinous, gummy, caustic, or even milky juice. Leaves alternate, simple, or ternate or unequally pinnate, without pellucid dots. Flowers terminal or axillary, with bracts, commonly ♂ ♀ by abortion, sometimes absolutely so. Calyx

Fig. CCCXXIV.

usually small and persistent, with 5, or occasionally 3-4, or 7 divisions. Petals equal in number to the segments of the calyx, perigynous, (occasionally wanting), imbricated in æstivation. Stamens equal in number to the petals and alternate with them, or twice as many or even more, equal or alternately shorter, or partly sterile; filaments distinct, or in the genera without a disk cohering at the base. Disk fleshy, annular or cup-shaped, hypogynous, occasionally wanting. Ovary single, very rarely 5 or 6, of which 4 or 5 are usually abortive, superior, (very rarely inferior), 1-celled; styles 1 or 3, occasionally 4, sometimes none; stigmas as many; ovule solitary, amphitropal or half anatropal, attached to the bottom of the cell by a cord, which is either free or adherent to the angle of the cell, so that the ovules not uncommonly appear pendulous. Fruit indehiscent, most commonly drupaceous. Seed without albumen; radicle either superior or inferior, but always directed towards the hilum, sometimes curved suddenly back; cotyledons thick and fleshy, or leafy.

The Order called Terebintaceæ by Jussieu and other Botanists has been broken up by Brown and Kunth, but preserved entire by De Candolle, Arnott, and others. As now limited the Anacards are distinctly known by their seeds hanging from the end of a thread which rises up from the base of the carpels, which in general are solitary, or at least quite distinct, and are sometimes, when ripe, placed at the end of an excessively enlarged disk, as in the Cashew-nut itself. Melanorrhœa is remarkable for its indefinite stamens, and especially for its hypogynous petals becoming enlarged, foliaceous, and deep red as the fruit advances to maturity.

There is in tropical countries a genus called Spondias, whose fruit is eaten under the name of Hog-plums, which genus it has been proposed to erect into an Order called

Fig. CCCXXIV.—Pistacia atlantica. 1. ♂ flowers; 2. an ovary; 3. the same cut open to show the ovule; 4. a ripe fruit opened to show the seed; 5. a cross section of the embryo; 6. ♀ flowers.

Spondiaceæ. It differs from Anacards in having a many-celled instead of a 1-celled, 1-seeded drupe; and on this more than anything else the character of the supposed Order was made to depend. But it appears that in the beginning Spondias has 5 distinct carpels, inclosed within a large fleshy cup, and that the growing together of these carpels is an after operation, unconnected with original structure; a Mango, in fact, if it had 5 carpels instead of 1, would be almost a Spondias. For this reason the supposed Order does not seem to be tenable. It is true that its ovules are described as being suspended from the apex of the cells; but this seems to arise from the cord contracting an adhesion with the side of the cells.

A writer in the *Linnæa* suggests that Anacards should be placed in the same class with Malpighiads (xiv. 243). A better approximation would have been to the Order of Juglands, with which they are not associated, chiefly because of their flowers not being amentaceous, nor usually absolutely ♂ ♀. Pistacia, indeed, is so, and some others; but the mass of the Order is polygamous, or has distinct rudiments of a ♀ in the ♂ flowers.

Chiefly natives of tropical America, Africa, and India; a few are found beyond the tropics, both to the north and south. Pistacias and some species of Rhus inhabit the south of Europe; many of the latter genus occupy stations in North America and Northern India, and also at the Cape of Good Hope; Duvaua and Schinus inhabit exclusively Chile and the adjacent districts. The Order is unknown in New Holland.

Large trees, with inconspicuous flowers, abounding in a resinous, sometimes acrid, highly poisonous juice, are the ordinary representatives of this Order, to which belong the Cashew-nut, (Anacardium occidentale), the Pistacia-nut (Pistacia vera), and the Mango fruit (Mangifera indica). Of these trees the Mango is the most important, its fruit being as highly valued in tropical as the Peach in temperate countries; its bark, especially that of the root, is a bitter aromatic, and is employed against diarrhœa, leucorrhœa, &c.; the young leaves are pectoral, the old leaves are used for cleaning the teeth, the seeds are anthelmintic; a resin that flows from the stem is reputed to be antisyphilitic. Some are celebrated for yielding a clammy juice, which afterwards turns black, and is used for varnishing in India; as the common Cashew-nut. The varnish of Sylhet is chiefly procured from Semecarpus Anacardium, the marking Nut-tree of commerce; and the varnish of Martaban from the Theet-see or Kheu, a plant called by Wallich Melanorrhœa usitatissima. All these varnishes are extremely dangerous to some constitutions; the skin, if rubbed with them, inflames and becomes covered with pimples that are difficult to heal; the fumes have been known to produce a painful swelling and inflammation of the skin, which, in a case recorded by Brewster, extended from the hands as far as the face and eyes, which became swelled to an alarming degree. I have known an instance of similar effects having been produced by roasting the nuts of Anacardium occidentale. But there are some constitutions which are not affected in any degree by such poisons. These varnishes are at first white, and afterwards become black. This has been ascertained by Brewster to arise from the recent varnish being an organised substance, consisting of an immense congeries of small parts, which disperse the sun's rays in all directions, like a thin film of unmelted tallow; while the varnish which has been exposed to the air loses its organised structure, becomes homogeneous, and then transmits the sun's rays of a rich, deep, uniform red colour. Such a secretion is probably the substance mentioned by Ainslie as the Black Lac of the Burmah country, with which the natives lacquer various kinds of ware. The valuable black hard varnish called Japan Lacquer, is obtained from Stagmaria verniciflua in the Indian archipelago: this resin is extremely acrid, causing excoriations and blisters if applied to the skin; the people of Sumatra consider it dangerous even to sit or sleep beneath its shade; the manner of preparing its varnish is fully described in Jack's *Malayan Miscellanies*, p. 81. (Calcutta edition.) A black varnish well known in India is manufactured from the nuts of Semecarpus Anacardium and the berries of Holigarna longifolia. Augia chinensis produces a varnish in China and Siam. Odina wodier, Buchanania latifolia, and many more Indian species, have the same property. Several Comocladias stain the skin black. The leaves of some species of Schinus are so filled with a resinous fluid, that the least degree of unusual repletion of the tissue causes it to be discharged; thus some of them fill the air with fragrance after rain; and S. Molle, Duvaua latifolia, and some others expel their resin with such violence when immersed in water as to have the appearance of spontaneous motion, in consequence of the recoil. See *Bot. Reg.* 1580. Schinus Arroeira is said by Auguste de St. Hilaire to cause swellings in those who sleep under its shade. The fresh juicy bark of this shrub is used in Brazil for rubbing newly-made ropes, which it covers with a very durable bright dark-brown coating. The juice of the same plant is applied by the Indians in diseases of the eye. This last plant, and also Rhus coriaria, possess acid qualities. The fruit of Cassuvium occidentale and Anacardium orientale is said to exercise a singular effect upon the brain. Mastich, a resin useful for strengthening the gums and sweetening the breath, is the produce of Pistacia atlantica and Lentiscus; Scio turpentine, a limpid, fragrant balsamic resin, with an odour between Lemon and Fennel, is yielded by Pistacia Terebinthus; a substance like mastich is exuded by Schinus Molle, and the Peruvians use it also for strengthening their gums. A full account of the mode of obtaining mastich at Chio, from the Pistacia Lentiscus, is given in the *Annals of Chemistry*, vol. l. p. 223. The juice of many species of Rhus is milky, stains black, and is sometimes, as in R. toxicodendron and radicans, extremely venomous; being volatile it is capable of poisoning persons who approach such plants in hot weather; and the same effects are produced by R. venenata. R. coriaria, a powerful astringent, is used by tanners; its acid fruits are eaten by the Turks and used to sharpen their vinegar. The bark of R. glabrum is considered a febrifuge, and is also employed as a mordant for red colours. R. Cotinus, Arbre à perruque of the French, and Venetian Sumach of the English, has wood called Young Fustick, which is astringent as well as the fruit; it dyes a bright yellow colour. R. vernix, a Japanese tree, exudes a whitish resinous juice, which soon becomes black in the air. R. succedaneum and verniciferum have a similar property. R. metopium, a Jamaica plant, yields a gum called Doctors' Gum, which has powerful purgative, emetic, and diuretic effects. It is also said to be a vulnerary (*Pharm. Journal*, v. 60). But are not different plants mixed up under the name of Doctors' Gum and Hog Gum?

The fruit of several species of Spondias, especially S. purpurea and Mombin, is eatable in the Brazils and West Indies, where they are called Hog Plums. Martius says that the juice of the fruit of S. tuberosa is drank in Brazil in fevers. The bark of S. venulosa is an aromatic astringent, employed in diarrhœa, blennorhœa, &c. The most agreeable of these fruits is the S. cytherea or dulcis, a native of the Society Islands, whose golden drupes are compared for flavour and fragrance to the Pine-apple. The negroes of Senegal make an intoxicating beverage from the fruit of S. Birrea.

GENERA.

Pistacia, *Linn.*
Terebinthus, Juss.
Lentiscus, Tournef.
Dupuisia, *A. Rich.*
Sorindeia, *Thouars.*
Comocladia, *P. Br.*
Dodonæa, Plum.
Cyrtocarpa, *H. B. K.*
Odina, *Roxb.*
Wodier, Anders.
Haberlia, Dennst.
Lannea, A. Rich.
Pegia, *Colebr.*
Solenocarpus, *Wight et A.*
Schinus, *Linn.*
Molle, Clus.
Mulli, Feuill.
Duvaua, *Kunth.*
Mauria, *Kunth.*
Pennantia, *Forst.*
Lithræa, *Miers.*
Llithi, Feuill.
Malosma, Nutt.
Rhus, *Linn.*
Cotinus, DC.
Metopium, P. Br.
Sumac, DC.
Toxicodendrum, Tourn
Pocophorum, Neck.
Thezera, DC.
Lobadium, Raf.
Turpinia, Raf.
Schmalzia, Desv.
Styphonia, *Nutt.*
Botryceras, *Willd.*
Laurophyllus, Thunb.
Daphnitis, Spreng.
Anaphrenium, *E. Meyer.*
Ozoroa, Del.
Heeria, Meisn.
Römeria, Thunb.
Loxostylis, *Spreng.*
Anasyllis, E. Mey.
Astronium, *Jacq.*
Melanorrhœa, *Wall.*
Gluta, *Linn.*
Stagmaria, *Jack.*
Syndesmis, *Wall*
Holigarna, *Roxb.*
Hadestaphyllum, Denst.
Pennantia, *Forst.*
Mangifera, *Linn.*
Erythrostigma, *Hassk.*
Anacardium, *Rottb.*
Cassuvium, Rumph.
Acajou, Tournef.
Acajuba, Gärtn.
Rhinocarpus, Bert.
Monodynamus, Pohl.
Semecarpus, *Linn.*
Anacardium, Lam.
Bouea, *Meisn.*
Cambessedea, Wight.
Buchanania, *Roxb.*
Launzea, Buchan.
Cambessedea, Kunth.
Coniogeton, *Blum.*
Phlebochiton, *Wall.*
Cardiophora, *Benth.*
Spondias, *Linn.*
Mombin, Plum.
Cytheræa, DC.
? Wirtgenia, *Jungh.*
Poupartia, *Commers.*
? Huertea, *Ruiz et Pav.*
? Rumphia, *Linn.*
? Augia, *Lour.*
? Sabia, *Colebr.*

NUMBERS. GEN. 41. Sp. 95.

Juglandaceæ.
POSITION.—Xanthoxylaceæ.—ANACARDIACEÆ.—Meliaceæ.
Celastraceæ.

ORDER CLXXV. CONNARACEÆ.—CONNARADS.

Terebintaceæ, *Juss. Gen.* 368. (1789) *in part.*—Connaraceæ, *R. Brown in Congo*, 431. (1818); *Kunth in Ann. Sc. Nat.* 2. 359; *Endl. Gen.* ccxlvii.; *Meisner Gen.* 78; *Wight Illustr.* 1. 162.

DIAGNOSIS.—*Rutal Exogens, with apocarpous fruit, and collateral ascending orthotropal sessile ovules.*

Trees or shrubs, sometimes climbing. Leaves compound, not dotted, alternate, without stipules. Flowers terminal and axillary, in racemes or panicles, with bracts, ☿, rarely ♂ ♀ by abortion. Calyx 5-parted, regular, persistent; æstivation either imbricate or valvular. Petals 5, inserted on the calyx, imbricated, rarely valvate in æstivation. Stamens twice the number of petals, hypogynous, those opposite the petals shorter than the others; filaments usually monadelphous. Carpels solitary, or several, each with a separate style and stigma; ovules 2, collateral, orthotropal, ascending; styles terminal; stigmas usually dilated. Fruit dehiscent, follicular, splitting lengthwise internally. Seeds erect, in pairs, or solitary, with or without albumen, often with an aril; radicle superior, at the extremity opposite the hilum; cotyledons thick in the species without albumen, foliaceous in those with albumen.

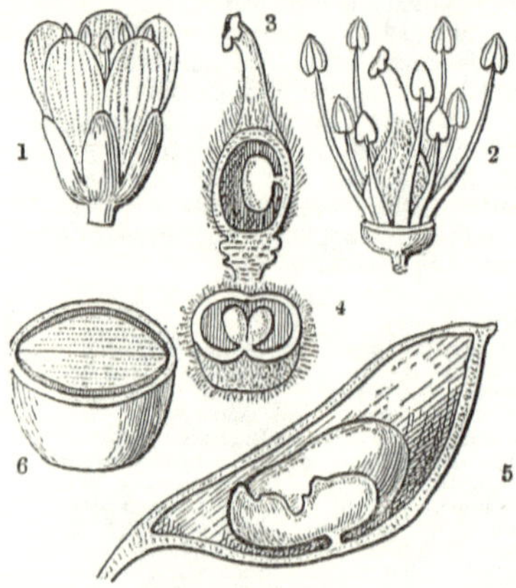

Fig. CCCXXV.

Brown says that the genus Connarus can only be distinguished from leguminous plants by the relation the parts of its embryo bear to the umbilicus of the seed; that is to say, by the radicle being at the extremity most remote from the hilum. This observation must, however, be understood to refer only to some particular cases among leguminous plants, and also to the fructification; the want of stipules and regular flowers being usually sufficient to distinguish Connarads. From Anacards and others they are at once known by the total want of resinous juice, and their orthotropal ovules. Brown considers that Cnestis approaches Averrhoa in Oxalids, and this genus, according to Adrien de Jussieu, is allied to Xanthoxyls through Brunellia. Cnestis has a valvate calyx, and some albumen about its embryo. Dr. Wight, who has had opportunities of studying the Order, observes that the hypogynous insertion of the stamens and the 5-celled ovary, on a gynobase, of Connarus and Cnestis, indicate a very close approach to Xanthoxyls.

The species are all tropical, and most common in America, according to Endlicher.

The aril of some species of Omphalobium is eatable, and their seeds oily. Eurycoma longifolia, the Punowur Pait of Malacca, is regarded by Oxley as a valuable febrifuge.—*Griffith.* The beautiful Zebra-wood of the cabinet-makers has been ascertained by Schomburgk to be produced by Omphalobium Lamberti, a large Guiana tree. Dr. Wight says that they are handsome flowering shrubs, conspicuous for their bright red capsules.

GENERA.

Connarus, *Linn.*	*Malbrancia*, Neck.	*Connarus*, Kunth.	Cnestis, *Juss.*
Rourea, Aubl.	*Santaloides*, Linn.	*Byrsocarpus*, Schum.	? Thysanus, *Lour.*
Robergia, Schreb.	Omphalobium, *Gärtn.*	*Tapomana*, Adans.	? Eurycoma, *Jack.*

NUMBERS. GEN. 5. Sp. 41.

Fabaceæ.
POSITION.—Anacardiaceæ.—CONNARACEÆ.——————.

Fig. CCCXXV.—Connarus pinnatus.—*Wight.* 1. a flower; 2. stamens and pistil; 3. ovary opened perpendicularly; 4. a cross section of it; 5. half a seed-vessel; 6. cross section of an embryo.

ORDER CLXXVI. RUTACEÆ.—RUEWORTS.

Rutæ, *Juss. Gen.* 296. (1789) *in part.*—Rutaceæ, *DC. Prodr.* 1. 709. (1824); *Endlicher Gen.* cclii.—Ruteæ, *Adrien de Juss. Rutacées*, 78. (1825); *Aug. de St. Hilaire Fl. Bras. Mer.* 1. 93. (1825).—Diosmeæ, *R. Brown in Flinders*, (1814); *Ad. de Jussieu Rutacées*, 1. 83. (1825); *Endlicher Gen.* ccli.—Fraxinelleæ, *Nees and Martius Nov. Act. Bonn.* 11. 149. (1823).—Cusparieæ, *DC. Mem. Mus.* 9. 141. (1822); *Prodr.* 1. 729. (1824), a § *of* Rutaceæ,—? Cneoreæ, *Webb in Lond. Journ. Bot.* 1. 254. (1842).—Biebersteiniæ, *Endl. Gen.*

DIAGNOSIS.—*Rutal Exogens, with a few-seeded fruit which finally becomes apocarpous, and separates its pericarp into 2 layers, sessile pendulous ovules, and ☿ flowers.*

Trees or shrubs, very rarely herbaceous plants. Leaves without stipules, opposite or alternate, simple or pinnate, covered with pellucid resinous dots. Flowers axillary or terminal ☿, regular or irregular. Calyx in 4 or 5 divisions. Petals either as many as the divisions of the calyx, distinct, or combined into a monopetalous corolla, or occasionally wanting; æstivation for the most part twisted, very rarely somewhat valvular. Stamens equal in number to the petals, or twice or thrice as many, or even fewer in consequence of abortion, hypogynous, very rarely perigynous, placed on the outside of a disk or cup surrounding the ovary, and either free or combined with the base of the corolla, or in part abortive. Ovary sessile or stalked, its lobes equal to the number of petals, or fewer; ovules twin and collateral, or one above the other, rarely 4, seldom more; style single, occasionally divided towards the base into as many parts as there are lobes of the ovary; stigma simple or dilated; ovules usually 2, sometimes 4, partly ascending, partly suspended. Fruit consisting of several capsules, either cohering firmly or more or less distinct. Seeds twin or solitary, with a testaceous integument; embryo with a superior radicle, which is either straight or oblique, and cotyledons of variable form; albumen present or absent.

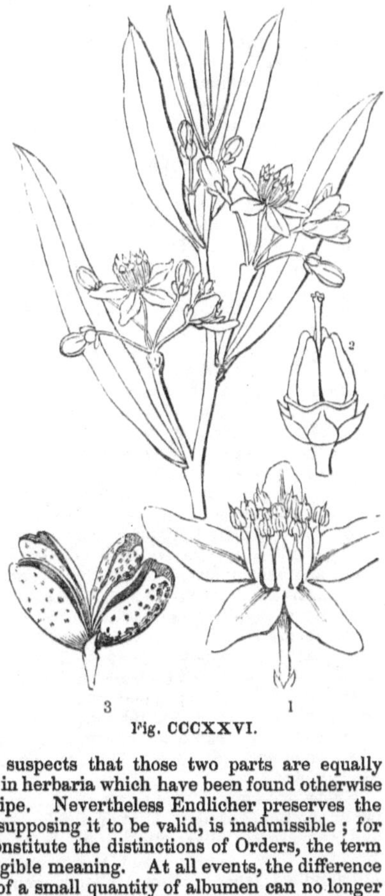

Fig. CCCXXVI.

There are two principal divisions in this Order; the one Ruteæ proper, which have seeds containing albumen, and a fruit, the sarcocarp of which is said not to separate from the endocarp; the other Diosmeæ, whose seeds have no albumen, and whose sarcocarp and endocarp divide into distinct bodies when the fruit is ripe. But Aug. de St. Hilaire (*Fl. Bras.* 1. 74.) suspects that those two parts are equally separable in Ruteæ, and that the specimens in herbaria which have been found otherwise were gathered before their fruit was quite ripe. Nevertheless Endlicher preserves the distinction as a mark of two Orders, which supposing it to be valid, is inadmissible; for if differences in dehiscence are alone to constitute the distinctions of Orders, the term Natural Order will no longer have an intelligible meaning. At all events, the difference is very slight, and the absence or presence of a small quantity of albumen can no longer be insisted upon now that so many cases of its absence or presence in the same Order are known; indeed, Hortia, a Diosmeous genus, has albumen, according to Aug. de

Fig. CCCXXVI.—Eriostemon myoporoides. 1. a complete flower; 2. the ovary, seated in a cup-shaped disk, surrounded by a calyx; 3. the ripe fruit, separated spontaneously into its component carpels.

St. Hilaire. Ruteæ are allied to Bean-capers through Peganum, which A. de Jussieu actually stations among the former, although its stipulate leaves, destitute of pellucid dots, seem to determine its greatest affinity to be with the latter. Rueworts differ from Citronworts in their capsular fruit invariably splitting into its component parts, from Xanthoxyls in the flowers being ⚥, and from Anacards (Anacardiaceæ) in the ovules being sessile and suspended, not attached to the end of a long cord rising from the base of the ovary. The Cneoreæ of Mr. Barker Webb seem to be a form of this Order of Rueworts rather than of Xanthoxyls ; for their flowers are ⚥ and their habit is not unlike that of Phebalium. The truly monopetalous corolla of Correa is very remarkable, and brings this Order so close to Heathworts that the indefinite seeds, porous anthers, but more especially the abundant albumen of the latter, form the principal marks of distinction.

M. Adrien de Jussieu thus describes the peculiarities of the pistil in that division of the Order which is called Diosmeæ :—" The ovaries, whether combined by their central axis, or distinct, always contain 2 ovules ; if 4, or sometimes but 1, are found, that occurs only in genera stationed at the extreme limits of the group. They are collateral, or more frequently placed one above the other, and then one is usually ascending, and the other suspended. This position, which at first sight appears singular, is very natural ; for the ovary is usually pierced by the vessels of the style only in the middle, and it is at that point that the two ovules are inserted, both at nearly the same height. If, therefore, they are placed one above the other, it is indispensable that one should ascend and the other descend. These ovules may be considered peritropal, rather than either ascending or suspended, or in other terms, attached by their middle rather than by either extremity."—" If the ovary of a Diosmea is divided across, its coat will be found to consist of two layers, the outer rather the most fleshy, and the inner thin or almost absent on the side next the axis, the side which is traversed from bottom to top by the vessels of the peduncle. These vessels at a certain height, meet those of the style, either at the point of its insertion or below it ; united to these, they penetrate the cavity of the cell, the shell of which they pierce, and there form funiculi, to which the ovules are attached. Thus far the structure of Diosmea is little different from that of other Rutaceous plants. But this becomes modified as the ovary advances towards the state of fruit. The endocarp hardens by degrees, and at the same time separates from the sarcocarp. Its form resembles that of a bivalve shell, and may be more especially compared to that of a muscle ; it presents two extremities, one superior, the other inferior, two lateral faces which are more or less convex, and two edges more or less acute, which unite them, the one external, the other internal. The two valves are woody and touch at the edges, except perhaps at a part of their inside where they are separated ; this space is filled by a membrane which passes from one to the other : it is either slightly fleshy, or, which is more common, extremely thin, thickened in the middle by the passage of the vessels of the seed which penetrate it ; and as, after having pierced it, they are almost immediately inserted into the seed, the latter appears to be actually borne by the membrane itself. When the fruit is perfectly ripe, the sarcocarp of each cell opens from above inwards, following a longitudinal furrow, which had become visible some time previously. Its inner surface is seen to be covered by projecting lignified vessels, which are directed obliquely from the inner edge towards the outer, and are indicated externally by some transverse projections. The endocarp is loose in the inside of the shell, unless at its membrane, by means of which it continues to preserve some degree of adhesion with the other parts ; but it soon opens, the two valves separate in different directions, and force out the seeds. When this separation takes place, the membrane is torn all round, and either falls away or sticks to the seed. In the latter case it is found attached to the hilum, if one seed only has ripened ; but then in removing it, the remains of the abortive ovule may be found on one side. If both seeds have arrived at maturity, they are usually seen one resting on the other by their contiguous flattened extremities, and the membrane extends along their inner edge, being enlarged at their point of contact, where two transverse prolongations are perceptible."

Ruteæ are found in the south of Europe, whence they extend in our hemisphere as far as the limits of the Old World, following the southern part of the temperate zone, and very rarely advancing within the tropics. Dictamnus is found in the south of Europe. The Cape of Good Hope is covered with different species of Diosma and nearly allied genera ; New Holland abounds in Boronias, Phebaliums, Correas, Eriostemons, and the like ; great numbers of Cusparieæ and Pilocarpeæ inhabit the equinoctial regions of America.

The species are characterised by their powerful odour and their bitterness ; they act principally on the nerves. Common Rue, and another species, are said to be emmenagogue, anthelmintic, and sudorific. Ruta montana, a Spanish plant, is so acrid that it blisters the hands of those who gather it, through three pairs of gloves, and produces erysipelas and ulcerous pustules when applied to the naked head. Egyptian women bruise the leaves of Haplophyllum tuberculatum in water and wash their hair with it in order to make it grow. The Diosmeæ, or Bucku plants, of the Cape, are well known for their powerful and usually offensive odour; several, especially Barosma crenata, are recommended as antispasmodics and diuretics. The American species possess, in many cases, febrifugal properties. There is an excellent bark used by the Catalan Capuchin friars of the missions on the river Carony in South America, called the Quina de la Guayna, or de la Angostura, or Angostura bark, which is said to be the produce of Galipea Cusparia (Bonplandia trifoliata, *W.*), a plant of this family. Dr. Hancock, however, thinks that it is a distinct species, which he calls Galipea officinalis. He says that he is fully convinced, from ample experience of the virtues of this bark, that it is one of the most valuable febrifuges we possess, being adapted to the worst and most malignant bilious fevers, while the fevers in which Cinchona is chiefly administered are simple intermittents, for the most part unattended with danger. The Indians also use the bruised bark as a means of intoxicating fishes, which is a very singular coincidence with what is mentioned by Dr. Saunders, of the same use being made of Cinchona bark by the Peruvians. Melambo bark, another bitter aromatic astringent, is supposed to belong to some allied species. Esenbeckia febrifuga, one of the Quinas of Brazil, has a bark so powerfully febrifugal as to compete with that of Cinchona. A bark much spoken of by the miners of Brazil, under the name of Casca de larangeira da terra, and in which Cinchonine was detected by Dr. Gomez, probably belongs to this tree. One of the Quinas of Brazil is the Ticorea febrifuga: its bark is a powerful medicine in intermittent fevers. Hortia Braziliana possesses similar properties, but in a less degree. An infusion of the leaves of Ticorea jasminiflora is drunk in Brazil as a remedy for the disease called by the Brazilian Portuguese Bobas, and by the French Frambæsia. Dictamnus abounds in volatile oil to such a degree, that the atmosphere surrounding it becomes inflammable in hot weather. Its root was formerly esteemed as a sudorific and vermifuge. The settlers in New Holland employ the leaves of Correas for tea, especially of C. alba.

GENERA.

I. CUSPARIEÆ.

Spiranthera, *St. Hil.*
Terpnanthus, Ns. et M.
Almeidea, *St. Hil.*
Aruba, Nees et Mart.
Galipea, *Aubl.*
Raputia, Aubl.
Pholidandra, Neck.
Sciuris, Schreb.
Cusparia, Humb.
Bonplandia, Willd.
Angostura, Röm. et St.
Conchocarpus, Mik.
Ravia, Nees et Mart.
Lasiostemum, Ns. et M.
Obentonia, Velloz.
Dangervilla, Fl. Flum.
Rossenia, Fl. Flum.
Diglottis, *Nees et Mart.*
Erythrochiton, *Ns. et Mt.*
Ticorea, *Aubl.*
Ozophyllum, Schreb.
Sciuris, Nees et Mart.
Costa, Fl. Flum.
Lemonia, *Lindl.*
Monnieria, *Linn.*
Aubletia, Rich.

II. PILOCARPEÆ.

Melicope, *Forst.*
Entoganum, Banks.
Evodia, *Forst.*
Esenbeckia, *H. B. K.*
Colythrum, Schott.
Evodia, St. Hil.
? *Polembryum*, Adr. Js.
Metrodorea, *St. Hil.*
Pilocarpus, *Vahl.*
Hortia, *Vandell.*
Choisya, *Kunth.*
Geigera, *Schott.*

III. BORONIEÆ.

Zieria, *Smith.*
Boronia, *Smith.*
Cyanothamnus, *Lindl.*
Eriostemon, *Smith.*
Crowea, *Smith.*
Philotheca, *Rudge.*
Phebalium, *Vent.*
Didymeria, *Lindl.*
Chorilæna, *Endl.*
Diplolæna, *R. Br.*
Correa, *Smith.*
Mazeutoxeron, Lab.
Antomarchia, Aubl.
Hügelia, *R. Br.*

IV. EUDIOSMEÆ.

Pachystigma, *Hooker.*
Calodendron, *Thunb.*
Pallasia, Houtt.
Adenandra, *Willd.*
Glandulifolia, Wendl.
Ockenia, Dietr.
Ockia, Dietr.
Hænkea, Smith.
Coleonema, *Bartl.*
Diosma, *L.*
Euchætis *Bartl. et Wdl.*
Gymnonychium, *Bartl.*
Acmadenia, *Bartl. et Wl.*
Barosma, *Willd.*
Baryosma, Röm.
Parapetalifera, Wendl.
Agathosma, *Willd.*
Bucco, Wendl.
Dichosma, DC.
Macrostylis, *Bartl. et Wl.*
Empleurum, *Sol.*

V. DICTAMNEÆ.

Dictamnus, *Linn.*
Fraxinella, Tournef.

VI. RUTEÆ.

Biebersteinia, *Steph.*
Boenninghausenia, *Rchb.*
Ruta, *Tournef.*
Desmophyllum, Webb.
Ruteria, DC.
Haplophyllum, *Adr. Jus.*

? VII. CNEOREÆ.

Cneorum, *Linn.*
Chamælea, Tourn.
Heterodendron, *Desf.*

NUMBERS. GEN. 47. Sp. 400.

Ericaceæ.
POSITION.—Aurantiaceæ.—RUTACEÆ.—Xanthoxylaceæ.

Order CLXXVII. XANTHOXYLACEÆ.—Xanthoxyls.

Terebintaceæ, *Juss. Gen.* 368. (1789) *in part.*—Xanthoxyleæ, *Nees and Martius in Nov. Act. Bonn.* 11. (1823); *Adrien de Jussieu Rutacées*, p. 114. (1825); *Endl. Gen.* ccl.; *Wight. Illust.* 1. 168.—Pteleaceæ, *Kunth. Ann. des Sc.* 2. 345. (1824).—Terebintaceæ, trib. 6. *DC. Prodr.* 2. 82. (1825).

Diagnosis.—*Rutal Exogens, with a few-seeded fruit which finally becomes apocarpous and separates its pericarp into distinct layers, sessile pendulous ovules, and ♀-⚥-♂ flowers.*

Trees or shrubs. Leaves without stipules, alternate or opposite, either simple, or more commonly abruptly or unequally pinnate, with pellucid dots. Flowers axillary or terminal, gray, green, or pink, ♀-⚥-♂, regular. Sepals imbricated, 3, or more commonly 4 or 5. Petals the same number, very rarely none, usually longer than the calyx; æstivation generally imbricated. Stamens equal to the petals in number, or twice as many, arising from around the base of the stalk of the abortive carpels; in the ♀ wanting or imperfect. Ovary made up of the same number of carpels as there are petals, or of a smaller number, either altogether combined, or more or less distinct; ovules in each cell 2, collateral, or one above the other, very seldom 4; styles more or less combined, according to the degree of cohesion of the carpels. Fruit either berried or membranous, sometimes of from 2 to 5 cells, sometimes consisting of several drupes or 2-valved capsules, of which the sarcocarp is fleshy and partly separable from the endocarp. Seeds solitary or twin, pendulous, usually smooth and shining, with a testaceous integument; embryo lying within fleshy albumen; radicle superior; cotyledons ovate, flat.

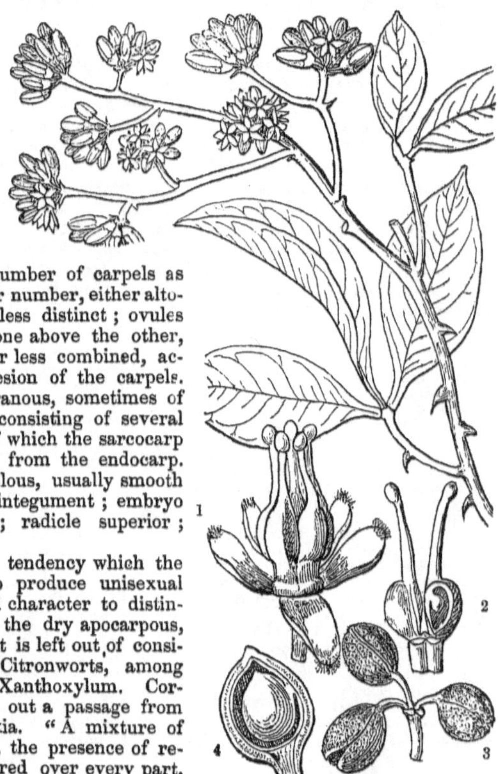

Fig. CCCXXVII.

If we neglect the constant tendency which the Order of Xanthoxyls has to produce unisexual flowers, we shall have no good character to distinguish it from Rueworts. If the dry apocarpous, dehiscent character of the fruit is left out of consideration it will merge in Citronworts, among which Luvunga climbs like a Xanthoxylum. Correa de Serra has also pointed out a passage from one to the other through Cookia. "A mixture of bitter and aromatic principles, the presence of receptacles of oil that are scattered over every part, which give a pellucid dotted appearance to the leaves, and which cover the rind of the fruit with opaque spaces,—all these characters give the two families a considerable degree of analogy. This has already been indicated by Jussieu in speaking of Toddalia, and in his remarks upon the families of Citronworts and Anacards; and it is confirmed by the continual mixture, in all large herbaria, of unexamined plants of Anacards, Xanthoxyls, and Citronworts. The fruit of the latter is, however, extremely different; their seeds resembling, as they do, Anacards, are on that very account at variance with Xanthoxyls, but at the same time establish a further point of affinity between them and some Rutaceous plants which are destitute of albumen. Unisexual flowers, fruit separating into distinct cocci, seeds solitary or twin in those cocci, inclosing a usually

Fig CCCXXVII.—Toddalia floribunda. 1. a flower; 2. a pair of carpels, one of which shows its ovule; 3. fruits; 4. a perpendicular section of one of them.

smooth and blackish integument, which is even sometimes hollowed out on its inner edge ; a fleshy albumen surrounding an embryo the radicle of which is superior, are all points of analogy between Xanthoxyls and Spurgeworts, particularly between those which have in their ♂ flowers from 4 to 8 stamens inserted round the rudiment of a pistil, and in the ♀ flowers cells with 2 suspended, usually collateral ovules. Finally, several Xanthoxyls have in their habit, and especially in their foliage, a marked resemblance to the Ash. The diœcious flowers of Fraxinus, its ovary, the two cells of which are compressed, having a single style, 2 ovules in the inside, and scales on the outside, and which finally changes into a samara which is 1-celled and 1-seeded by abortion, all establish certain points of contact between Ptelea and Fraxinus."—*Ad. de Juss.*

Most of the species belong to America, especially to the tropical parts ; some are found in temperate regions ; they are rare in Africa ; some exist in the Isles of France and Madagascar, and in New Holland : many are natives of India and China.

The species are nearly all aromatic and pungent. The Xanthoxylums are popularly called Peppers in the countries where they are found. X. Clava and fraxineum are powerful sudorifics and diaphoretics ; they are remarkable, according to Barton, for their extraordinary power in exciting salivation, whether applied immediately to the gums or taken internally ; both plants are reputed to have been used successfully in paralysis of the muscles of the mouth, in toothache, and in rheumatic affections. X. caribæum is held to be a febrifuge. The Chinese enumerate the root of X. nitidum among calefacient, sudorific, febrifugal, and emmenagogue medicines. The seeds of X. Budrunga have the fragrance of Lemon-peel. The unripe capsules of X. Rhetsa are gratefully aromatic, tasting like the peel of a fresh Orange. A plant called Coentrilho in Brazil (X. hiemale) is employed as a remedy for pain in the ear, for which purpose the powder of its bark is made use of. Its wood is very hard, and valuable for building. The fruit of Ptelea has a strong, bitter, aromatic taste, and is said to have been used with some success as a substitute for Hops. Every part of the shrub has a strong pungent taste, more especially the roots when fresh. The leaves are eaten raw for pains in the bowels, and the pungent ripe berries make an admirable pickle.—*Wight.* The capsules and seeds of X. hastile, called Tej-bul by the natives, are employed in northern India for intoxicating fish ; they are also given as the Faghureh of Avicenna. X. piperitum and Avicennæ are used in China and Japan as an antidote against all poisons ; they would, undoubtedly, in many cases be of considerable use as a stimulant remedy. The bark of the root of Toddalia aculeata is said to be employed as a cure for the remittent fevers caught in the jungles of the Indian hills.—*Royle's Illustr.* 157.

GENERA.

Dictyoloma, *DC.*
Pitavia, *Molin.*
Galvesia, Ruiz et Pav.
Brunellia, *Ruiz et Pav.*
Xanthoxylon, *Kunth.*
Pterota, P. Brown.
Lacuris, Hamilt.
Fagara, Lam.
Tobinia, Desv.
Ochroxylum, Schreb.
Curtisia, Schreb.
Kampmannia, Raf.
Pentamone, Moç. et Ses.
Pohlana, Nees et Mrt.
Langsdorfia, Leandr.
Maqueria, Commers.
Rhetsa, Wight et Arn.
Typalia, Dennst.
Aubertia, Bory.
Blackburnia, *Forst.*
Boymia, *Adr. Juss.*
Cyclocarpus, Jungh.
Toddalia, *Juss.*
Scopolia, Smith.
Crantzia, Schreb.
Vepris, Commers.
Asaphes, DC.
Boscia, Thunb.
Duncania, Rchb.
Ptelea, *Linn.*
Belluccia, Adans.
Cyminosma, *Gärtn.*
Jambolifera, Linn.
Gela, Lour.
Laxmannia, Smith.
Doriena, Dennst.
Spathelia, *Linn.*
Spathe, P. Br.
Ailanthus, *Desf.*
Aspidostigma, *Hochst.*
Teclea, *Del.*
? Pseudiosma, *Adr. Juss.*
? Tetradium, *Lour.*
? Philagonia, *Blum.*
? Bischofia, *Blum.*
? Phelline, *Labill.*
? Guindilia, *Gill.*

NUMBERS. GEN. 20. Sp. 110.

Euphorbiaceæ.
POSITION.—Rutaceæ.—XANTHOXYLACEÆ.—Aurantiaceæ.
Oleaceæ.

ORDER CLXXVIII. OCHNACEÆ.—OCHNADS.

Ochnaceæ, *DC. Ann. Mus.* 17. 398. (1811) ; *Prodr.* 1. 735. (1824) ; *Endl. Gen.* ccxlviii. ; *Meisner, p.* 66.

DIAGNOSIS.—*Rutal Exogens, with a one-seeded finally apocarpous fruit, whose pericarp does not laminate, and a succulent conical torus.*

Very smooth trees, or more generally under-shrubs, sometimes downy, having a watery juice. Leaves alternate, simple, entire, or toothed, with 2 stipules at the base, or one in the axil. Flowers usually in racemes, with an articulation in the middle of the pedicels. Sepals 5, persistent, imbricated in æstivation. Petals hypogynous, definite, sometimes twice as many as the sepals, deciduous, spreading, imbricated in æstivation. Stamens 5, opposite the sepals, or 10, or 00, arising from a hypogynous disk ; filaments persistent ; anthers 2-celled, innate, opening by pores, or longitudinally. Carpels equal in number to the petals, lying upon an enlarged, tumid, fleshy disk, (the gynobase) ; their styles combined in one ; ovule erect or pendulous, anatropal. Fruit composed of as many pieces as there were carpels, indehiscent, somewhat drupaceous, 1-seeded, articulated with the gynobase, which grows with their growth. Seeds without albumen or nearly so ; embryo straight ; radicle next the hilum ; cotyledons thick.

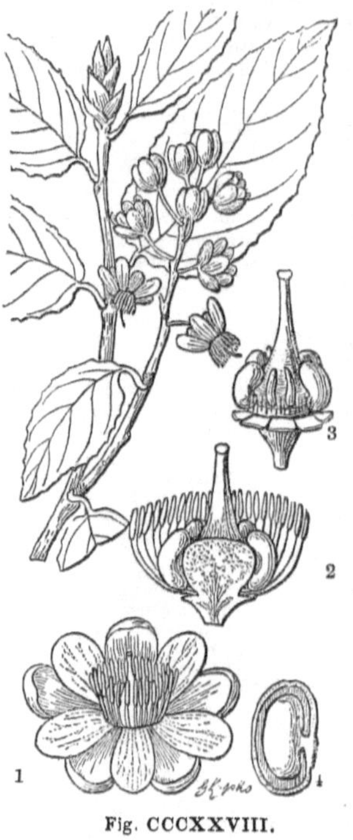

Fig. CCCXXVIII.

The great fleshy gynobase, or torus, of the species constituting this Order, affords their strongest mark of recognition. In this respect, indeed, there is an approach to the peculiar structure of Cranesbills, or even of some Mallowworts. The foliage is sometimes very shining and marked with closely set veins like those of Calophyllum, a genus of the Order of Guttifers. From the other Orders now associated with them they are often known by their anthers opening by pores, and their solitary, erect ovules ; but neither of them are always characteristic of Ochnads. The great succulent torus must always be regarded as one of their chiefest distinctions. According to the views of an anonymous writer in the *Linnæa*, this Order should be placed near Roseworts, and not Rueworts, with which and the kindred Orders he thinks that Ochnads have little affinity. —*Linnæa*, xiv. 248.

Found in tropical India, Africa and America ; a few are from the Cape of Good Hope.

These plants are for the most part bitter. Walkera serrata has a bitter root and leaves, and is employed in Malabar, in decoction in milk or water, as a tonic, stomachic, and anti-emetic. The bark of Ochna hexasperma is used in Brazil as a cure for the sores produced in cattle by the punctures of insects. It probably acts as an astringent. Castela Nicolsoni or Goatbush, is said to be as bitter as Quassia itself. The root and leaves of Gomphia angustifolia are bitter, and employed in Malabar, in decoction in milk or water, as a tonic, stomachic, and anti-emetic. G. hexasperma and Jabotapita are Brazilian remedies exhibited where bitters are demanded. The oil of G. parviflora is used in salads in Brazil.

Fig. CCCXXVIII.—Ochna dubia.—*Decaisne.* 1. expanded flower ; 2. section of pistil and stamens ; 3. pistil ; 4. section of a ripe carpel.

GENERA.

Tribe 1. Casteleæ.—*Anthers turned outwards. Seeds inverted, with albumen.*	Tribe 2. Ochneæ.—*Anthers turned inwards. Seeds erect, without albumen.*		
Castela, *Turp.*	Gomphia, *Schreb.*	*Jabotapita*, Plum.	Ochna, *Schreb.*
Elvasia, *DC.*		*Ouratea*, Aubl.	*Diporidium*, Wendl.
		Correia, Velloz.	? Walkera, *Schreb.*
		Philomeda, Noronh.	*Meesia*, Gärtn.
		Cittorrhynchus, Willd.	Euthemis, *Jack.*

NUMBERS. GEN. 6. Sp. 82.

Rosaceæ.
POSITION.—Simarubaceæ.—OCHNACEÆ.—Xanthoxylaceæ.
Geraniaceæ.

CORIARIEÆ, (*DC. Prodr.* 1. 739. 1824; *Ed. pr.* cvi.; *Endl. Gen. p.* 1065; *Meisner Gen. p.* 56.) A few plants inhabiting the South of Europe, Chili, Peru, New Zealand, and Nepal, have been associated by Botanists in a genus of which the following is the character. Shrubs with opposite branches, which sometimes are very long and feeble, often 3 on each side, 2 of them being secondary to an intermediate principal one. Leaves oppo-

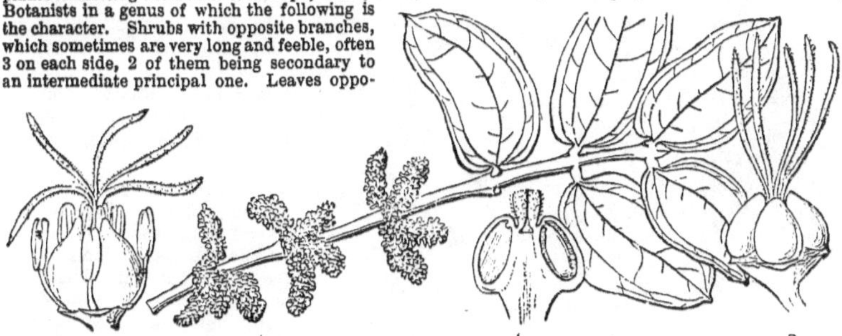

2 1 4 3

Fig. CCCXXIX.

site, simple, ribbed, entire. Buds scaly. Racemes terminal and axillary. Flowers ⚥ or ♀-⚥-♂. Calyx campanulate, 5-parted, ovate. Petals 5, alternate with the lobes of the calyx, and smaller than they are, fleshy, with an elevated keel in the inside. Stamens 10, hypogynous, 5 between the lobes of the calyx and the backs of the carpels, 5 between the petals and the joinings of the carpels; filaments filiform; anthers oblong, 2-celled. Carpels 5 (or 6), arranged obliquely round a thickish gynobase; stigmas 5, long, subulate; ovules solitary, pendulous, anatropal. Fruits crustaceous, covered over by the membranous sepals and fleshy petals, indehiscent, 1-seeded. Seed pendulous; albumen none; embryo straight; cotyledons 2, fleshy; radicle short, blunt, next the hilum.——It is very difficult to say what is the affinity of this plant. Decandolle places it, as the type of a distinct Order, immediately after Ochnads, with which it agrees in having its ovaries distinct, and surrounding a fleshy axis; but the stigmas in Coriaria are long, linear, and distinct, with no style, while Ochnads have a single style connecting the carpels and minute stigmas; the former, therefore, are apocarpous, the latter syncarpous. Coriaria is also allied to Rueworts, but differs from them as it does from Ochnads; and, besides, the carpels are in Rueworts connate. De Candolle understands Coriaria as apetalous, but I do not see upon what principle, either of structure or analogy. In his *Essai sur les Propriétés Médicales* he referred it to the vicinity of Rhamnads. Jussieu placed it near Malpighiads, and this view has been also taken by Endlicher and others, who consider it as being absolutely a member of that Order. But M. Adrien de Jussieu (*Monogr. Malp.* 1*st part, p.* 135) justly objects to this upon the double ground that Coriaria has neither the very peculiar ovule of Malpighiads, nor that broken (or spiral) arrangement of parts which pervades the Order in its genuine form; besides, Malpighiads never have 5 carpels, when in their natural state.——Coriaria myrtifolia and ruscifolia are used by dyers for staining black. The fruit is poisonous. It is said that several soldiers of the French army in Catalonia were affected by eating it; 15 became stupefied, and 3 died. Its leaves have been used to adulterate Senna, and have produced fatal consequences, exciting violent fits of tetanus, giving place to apoplectic coma. The French call it Redout or Roudout. Nevertheless, the fruit of Coriaria napalensis is frequently eaten in the north of India without inconvenience, according to Royle, and we learn from Forster that the berries of the New Zealand Coriaria sarmentosa are greedily sucked by the natives; the seeds, however, they regard as poisonous.

Gen. 1. Sp. 8.

Fig. CCCXXIX.—1. Coriaria napalensis; 2. flower of Coriaria myrtifolia without its calyx; 3. its pistil; 4. a perpendicular section of it.

ORDER CLXXIX. SIMARUBACEÆ.—QUASSIADS.

Simarubaceæ, *Rich. Anal. du Fr.* 21. (1808); *Endl. Gen.* ccxlix.—Simarubeæ, *DC. Diss. Ochn. Ann. Mus.* 17. 323. (1811); *Prodr.* 1. 733. (1824); *Adrien de Juss. Rutacées*, 129. (1825); *Meisner, Gen. p.* 65.

DIAGNOSIS.—*Rutal Exogens, with a few-seeded finally apocarpous fruit, whose pericarp does not laminate, a dry inconspicuous torus, exalbuminous seeds, and alternate leaves without stipules.*

Trees or shrubs. Leaves without stipules, alternate, occasionally simple, most usually compound, without dots. Peduncles axillary or terminal. Flowers whitish, green, or purple. Flowers hermaphrodite, or occasionally by abortion unisexual. Calyx in 4 or 5 divisions, imbricated. Petals the same number, longer, either spreading or combined in a tube; æstivation imbricated. Stamens twice as many as the petals, each arising from the back of an hypogynous scale. Ovary 4- or 5-lobed, placed upon a stalk from the base of which the stamens arise, 4- or 5-celled, each cell with 1 suspended anatropal ovule; style simple; stigma 4- or 5-lobed. Fruit consisting of 4 or 5 drupes arranged around a common receptacle, indehiscent. Seeds pendulous, with a membranous integument; embryo without albumen; radicle superior, short, drawn back within the thick cotyledons.

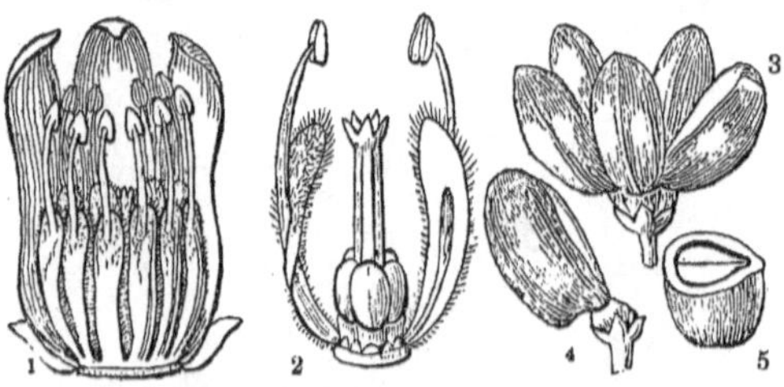

Fig. CCCXXX.

Quassiads are akin to Beancapers in their stamens inserted upon hypogynous scales, and to Ochnads in their deeply-lobed ovary, or nearly separate ovaries; from these latter they are distinguished by their want of a succulent torus, and by their anthers bursting by longitudinal slits, not by terminal pores. A. de Jussieu says, "They are known from all Rutals by the co-existence of these characters; namely, ovaries with but one ovule, indehiscent drupes, exalbuminous seeds, a membranous integument of the embryo, and the radicle being retracted within thick cotyledons."

All are natives of tropical America, India, or Africa, with the exception of one Nipal plant.

The species are intensely bitter. A plant called Paraiba in Brazil, the Simaruba versicolor of St. Hilaire, possesses such excessive bitterness that no insects will attack it. Specimens of it placed among dried plants which were entirely devoured by the larvæ of a species of Ptinus, remained untouched. The Brazilians use an infusion in brandy as a specific against the bite of serpents, and also employ it with very great success to cure the lousy diseases to which people are subject in those countries. The wood of Quassia amara is intensely bitter. Lund and others assert that it does not yield the Quassia chips of the European druggists, but refer them to Picræna excelsa. But Guibourt says that the wood of both the root and stem of this Quassia is imported in the form of white scentless very light cylinders 1-2 inches in diameter; and that the Picræna wood is inferior in quality. I learn however from Mr. Lance, who resided for many years in Surinam, that although large quantities of Quassia were exported 20 or 30 years since, yet that for many years none has been collected for that purpose, and he did not hear of a single instance of its shipment during the 10 years he passed in Surinam. Quassia wood is in fact no longer used even in that colony as a medicine being thought to have some bad properties along with its intense bitter. The flowers are however still infused in wine or water as a stomachic. The bitter has been used as a substitute for hops in the manufacture of beer; an infusion of the chips is employed to poison flies. Simaruba amara is more commonly employed. The bark of the root is stripped off and sent to Europe. In Cayenne the decoction, which is bitter, purgative, and even emetic, is used in fevers and diarrhœa. The wood has similar

Fig. CCCXXX.—Simaba guianensis. 1. a flower with part cut away; 2. pistil and two stamens; 3. fruit; 4. a single carpel; 5. cross section of it and of the seed which it contains.

properties, but is less active. The Jamaica plant, which being diœcious, may be another species, has an inodorous bitter bark which yields its properties to both alcohol and water. It has been remarked that the infusion is more bitter than the decoction. It acts as a tonic and is used in dyspepsia, diarrhœa, chronic dysentery, and all cases of impaired tone of the alimentary canal. Nima quassioides is used for similar purposes in the North of India. The timber of Simaruba amara is described by Sir R. Schomburgk as resembling White Pine, both in colour and quality. Niepa bark, an Indian febrifuge, is obtained from Samadera indica. Brucea antidysenterica and Sumatrana possess properties similar to those of Quassia.

GENERA.

Quassia, *DC.*
Simaruba, *Aubl.*
Simaba, *Aubl.*
  Aruba, Aubl.
  Phyllostema, Neck.
  Zwingera, Schreb.

Samadera, *Gärtn.*
  Samandura, Linn.
  Locandi, Adans.
  Vittmannia, Vahl.
  Niota, Lam.

  Biporeia, Thouars.
  Mauduyta, Commers.
  Manungala, Blanco.
Nima, *Hamilt.*
Picræna, *Lindl.*

Harrisonia, *R. Br.*
  Ebelingia, Reichenb.
Lasiolepis, *J. J. B.*
? Picrasma, *Blume.*
? Brucea, *Banks.*

NUMBERS. GEN. 10. Sp. 35.

POSITION.—Zygophyllaceæ.—SIMARUBACEÆ.—Xanthoxylaceæ.

ORDER CLXXX. ZYGOPHYLLACEÆ.—BEANCAPERS.

Zygophylleæ, *R. Brown in Flinders*, (1814); *DC. Prodr.* 1. 703. (1824); *Adrien de Juss. Rutacées*, 67. (1825); *Endl. Gen.* ccliii.—Melianthææ, *Endl. p.* 1165.

DIAGNOSIS.—*Rutal Exogens, with few-seeded finally apocarpous fruit, whose pericarp does not laminate, a dry inconspicuous torus, albuminous seeds, and opposite leaves without stipules.*

Herbaceous plants, shrubs, or trees, with a very hard wood, the branches often articulated at the joints. Leaves opposite, with stipules, very seldom simple, usually unequally pinnate, not dotted. Flowers solitary, or in pairs or threes, white, blue, or red, often yellow, hermaphrodite, regular. Calyx divided into 4 or 5 pieces, with convolute æstivation. Petals unguiculate, alternate with the segments of the calyx and a little longer, in æstivation, which is imbricated, at first very short and scale-like. Stamens double the number of the petals, dilated at the base, sometimes naked, usually placed on the back of a small scale, hypogynous. Ovary simple, surrounded at the base with glands or a short sinuous disk. More or less deeply 4- or 5-furrowed, with 4 or 5-cells; ovules in each cell 2 or more, attached to the inner angle, pendulous, or occasionally erect; style simple, usually with 4 or 5 furrows; stigma simple, or with 4 or 5 lobes. Fruit capsular, rarely somewhat fleshy, with 4 or 5 angles or wings, bursting by 4 or 5 valves bearing the dissepiments in the middle, or into as many close cells; the sarcocarp not separable from the endocarp. Seeds usually fewer than the ovules, either compressed and scabrous when dry, or ovate and smooth, with a thin herbaceous integument. Embryo green; radicle superior; cotyledons foliaceous; albumen in small quantity, whitish, between horny and cartilaginous, in Tribulus wanting.

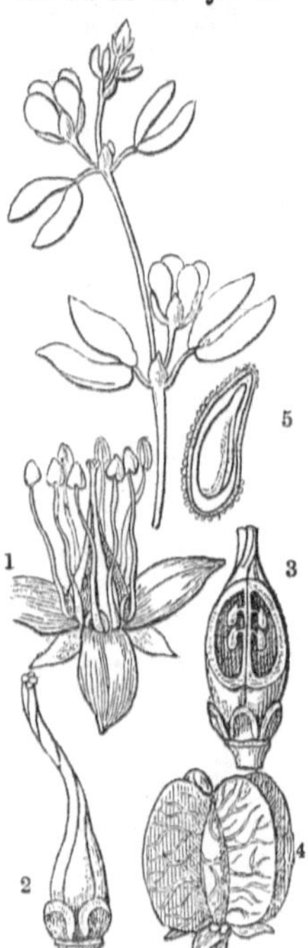

Fig. CCCXXXI.

These plants are remarkable in the Rutal Alliance for their opposite leaves and conspicuous stipules. With Quassiads they otherwise accord in the stamens springing from the back of a hypogynous scale. Adrien de Jussieu also observes that the petals are remarkable for their being, in an early state, minute and hidden by the calyx, which they only exceed about the time of flowering, while in other Rutal Orders the petals are always larger than the calyx. The distinguishing characters in the vegetation or habit of this Order are not only the leaves being constantly opposite, with lateral or intermediate stipules, but also in their being generally compound, and always destitute of the pellucid glands which universally exist in true Rueworts. For this reason the genus Biebersteinia must be excluded, although its leaves have stipules. It is also a very common character of the Order to have the radicle at that extremity of the seed which is most remote from the hilum; but this, which is of great importance in many natural families, is of less value in Beancapers. (See many good remarks upon this subject in Brown's *Appendix to Denham*, p. 27.) An anonymous author expresses his opinion (*Linnæa*, xv. 249.) that the true affinity of this Order is with Oxalids, not Rueworts. He would not however keep them in the neighbourhood of Cranesbills, but thinks Mallowworts their true relations.

Guaiacum, Porlieria, and Larrea, are peculiar to America. Fagonia is distributed over the south of Europe, the Levant, Persia, and India. Zygophyllum inhabits the same regions, and also the south of Africa, and is represented in New Holland by Röpera. Tribulus occurs in all the Old World within the tropics, or in countries bordering upon them. Melianthus, a most anomalous genus, is remarkable for being found both at the Cape of Good Hope and in Nipal, without any intermediate station. The abundance of Beancapers constitutes one of the most striking features of the vegetation of the Egyptian deserts.

Fig. CCCXXXI.—Röpera fabagifolia. 1. a flower; 2. pistil; 3. perpendicular section of it; 4. fruit; 5. section of a seed.

Zygophyllum Fabago is sometimes employed as an anthelmintic. The ligneous plants of the Order are remarkable for the extreme hardness of their wood. All the Guaiacums are well known for their exciting properties; the bark and wood of Guaiacum sanctum and officinale have a somewhat bitter and acrid flavour, and are principally employed as sudorifics, diaphoretics, or alteratives; they contain a particular matter often designated as resin or gum-resin, but which is now considered a distinct substance, called Guaiacine. According to Dr. Hancock (in the *Gardeners' Chronicle*), the medical value of Guaiacum resides principally in the bark. The foliage is very detersive, and is frequently used in the West Indies to scour and whiten floors, which it is said to do better than soap. Porlieria hygrometrica has similar properties. The wood called Lignum vitæ is remarkable for the direction of its fibres, one layer of which often crosses another diagonally; a circumstance first pointed out to me by Professor Voigt. This valuable timber is generally said to be furnished by Guaiacum officinale; but it is probably the wood of some other species, for the small size of that tree seems quite incompatible with the production of timber 4 or 5 inches in diameter.—*See Bot. Reg. l. c.* The flowers of Melianthus major are so full of honey, that the natives of the Cape of Good Hope, where it grows wild, obtain it for food by shaking the branches, when it falls in a heavy shower. The flowers of Zygophyllum Fabago are a substitute for Capers; the smell of Z. simplex is so detestable that no animal will touch the foliage, not even the camel; the Arabs, however, beat the leaves in water, and apply the infusion in diseases of the eyes. The Turks use the seeds of Peganum Harmala as a spice, and for dyeing red.

GENERA.

I. TRIBULEÆ. — Seeds without albumen.

Tribulus, *Tournef.*
Kallströmia, *Scop.*
Ehrenbergia, Mart.
Heterozygis, Bung.

II. ZYGOPHYLLEÆ. — Seeds with albumen.

Peganum, *L.*
Harmala, Mnch.
Malacocarpus, *F. et M.*
Chitonia, *Moç. et Sess.*
Juliania, *Llav. et Lex.*
Fagonia, *Tournef.*
Sarcozygium, *Bunge.*
Röpera, *Adr. Juss.*
Zygophyllum, *Linn.*
Fabago, Tournef.
Agrophyllum, Neck.
Eurynema, *Endl.*
Trichanthera, Ehrenb.
Larrea, *Cav.*
Porliera, *Ruiz et Pav.*
Plectrocarpa, *Gill.*
Guajacum, *Plum.*
Seetzenia, *R. Br.*
Melianthus, *L.*

NUMBERS. GEN. 7. Sp. 100.

POSITION.—Simarubaceæ.—ZYGOPHYLLACEÆ.—Elatinaceæ.
Oxalidaceæ.

ORDER CLXXXI. ELATINACEÆ.—WATER-PEPPERS.

Elatineæ, *Cambessèdes in Mém. Mus.* 18. 225. (1829); *Aug. de St. H. Fl. Bras.* 2. 159. (1830); *Fl. Seneg.* 1. 42. (1832); *Fischer and Meyer in Linnæa*, x. 69. (1835); *Wight Illustr.* 1. *t.* 25; *Endl. Gen.* ccxix.; *Meisner Gen. p.* 131; *Fenzl Darstellung, &c., p.* 30.

DIAGNOSIS.—*Rutal Exogens, with a many-seeded fruit which is finally apocarpous, and polypetalous flowers.*

Little annuals, growing in marshy places, with fistular rooting stems. Leaves opposite, with stipules between the petioles. Sepals 3-5, imbricated, distinct, or slightly connate at the base. Petals of the same number as the sepals, imbricated, hypogynous. Stamens hypogynous, usually twice as numerous as the petals. Ovary with from 3 to 5 cells, an equal number of styles, and capitate stigmas; ovules 00, anatropal. Fruit capsular, 3-5-celled, opening at the sutures, crowned by the styles; the valves either flat at the edge, or rolled inwards and alternating with the angles of a central placenta. Seeds 00, without albumen, wrinkled transversely, cylindrical, with a straight embryo, whose radicle is turned to the hilum, which is at one end of the seed.

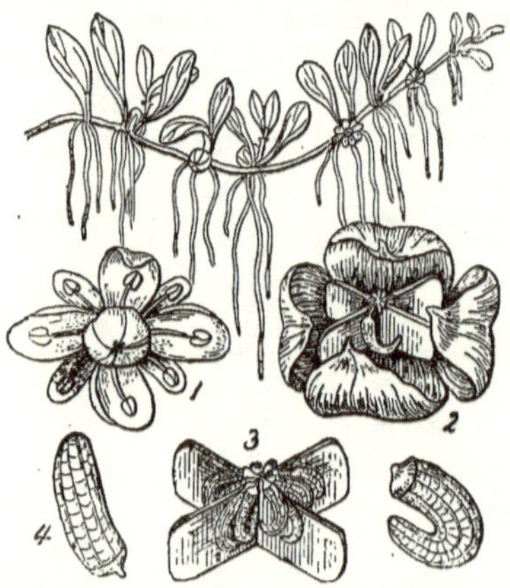

Fig. CCCXXXII.

This little Order was established by Cambessèdes, who distinguished it from Alsinaceæ, with which a part had been confounded, by its capitate stigmas, the dehiscence of its fruit, the small quantity of albumen, and the straight, not curved, embryo. It does not, however, appear that the Water-peppers have any immediate relation to the Silenal Alliance, of which Alsinaceæ form a part. On the contrary, the species agree much better with Tutsans (Hypericaceæ) even in the presence in the leaves of receptacles of resinous secretions; but they differ in having a persistent central axis in the fruit, and definite stamens, on which latter account they fall into the ranks of the Rutal rather than the Guttiferal Alliance. This view of their affinity seems confirmed by the curious genus Tetradiclis, a Syrian plant, with the habit of a Tillæa, on which account it has been even referred to the Order of Houseleeks by Bunge (*Linnæa*, xiv. 177). It is remarkable for having in each cell of its fruit two seeds enveloped in the laminated sides of the dissepiments, which sides adhere to the seeds, and seem as if they were really a part of them; the other seeds, however, are naked, and lie in the space between the lateral seeds. If it were not for this singular breaking up of the tissue of the dissepiments, Elatine would be very near Tetradiclis. Now, there can be no doubt of the latter genus being a member of the Rutal Alliance; but its numerous seeds attached to two arm-like free placentæ forbid its being stationed in Rueworts, whither Mr. Fenzl has referred it (*Linnæa*, xv. 295), or in Bean Capers, among which I had assigned it a doubtful place in the Botany (still unpublished) of Col. Chesney's Expedition to the Euphrates. It falls, however, well into the Order of Water-peppers, and contributes to confirm the importance of that little Order.

Found in marshes in the four quarters of the globe. The Elatines are natives of Europe and Asia, Bergias of the Cape of Good Hope and the East Indies, Merimea of South America, and Tetradiclis of the Syrian region.

Dr. Wight says that in India the little Bergia ammannioides bears the Tamool name of Neer-mel-neripoo, or Water-fire, which seems a curious coincidence with the word Water-pepper, given in English to Elatine, and seems to indicate a popular belief in these plants possessing some acridity.

Fig. CCCXXXII.—Elatine hydropiper.—*Sowerby.* 1. a flower; 2. a capsule after splitting; 3. the placenta; 4. and 5. seeds.

I observe that Dr. Bunge considers Ehrenberg's genus Anatropa identical with Tetradiclis. M. Decaisne long since pointed out to me the close relation between the two, and at the same time expressed his opinion that the former would constitute a new Order between Rueworts and Beancapers. But since Anatropa has stipules, according to Ehrenberg, it seems premature to combine them.

GENERA.

Elatine, *Linn.*
Crypta, Nutt.
Cryptina, Raf.
Hydropiper, Endl.

Birolia, Bellard.
Alsinastrum, Endl.
Potamopitys, Buxb.

Bergia, *Linn.*
Lancretia, Del.
Merimea, *Camb.*

Tetradiclis, *Stev.*
Anatropa, *Ehrenb.*
? Tridia, *Korth.*

NUMBERS. GEN. 6. Sp. 22.

POSITION.—Zygophyllaceæ.—ELATINACEÆ.—Podostemaceæ.
Hypericaceæ.

ORDER CLXXXII. PODOSTEMACEÆ.—PODOSTEMADS.

Podostemeæ, *Richard and Kunth in Humb. N. G. et Sp.* 1. 246. (1815); *Martius Nov. G. et Sp.* 1. 6. (1822); *Bartl. Ord. Nat.* 72. (1830); *Bongard in Mem. de l'Acad. Imp. Petersb. VI. ser. III.* 69. (1834); *Endl. Gen.* lxxxv.; *Meisn. Gen. p.* 122.; *Griffith in Ann. Sc. Nat. ser. II.* 9. 183.

DIAGNOSIS.—*Rutal Exogens, with many-seeded fruit, which is finally apocarpous, and apetalous very imperfect flowers.*

Herbaceous branched floating plants without stomates or spiral vessels, and with the habit of Liverworts or Scale-mosses. Leaves capillary, or linear, or lacerated irregularly, or minute and densely imbricated, decurrent on the stem, with which they are not articulated. Flowers axillary or terminal, inconspicuous, usually ☿, naked, or with a very imperfect calyx, or with 3 sepals bursting through an irregularly lacerated spathe. Stamens hypogynous, varying from 1 to an indefinite number, either placed all round the ovary or on one side of it, distinct or monadelphous; anthers oblong, 2-celled, bursting longitudinally. [Pollen shaped like an hour-glass, consisting of two spherules, inseparably united in Podostemon.—*Griffith.*] Ovary 2- or 3-celled, with numerous ascending anatropal ovules attached to a fleshy central placenta; styles or stigmas 2 or 3, acute and sessile. Fruit slightly pedicellate, ribbed, capsular, opening by 2 or 3 valves, which fall off from the dissepiment, which is parallel with them. Seeds numerous, minute, containing an exalbuminous dicotyledonous orthotropal embryo.

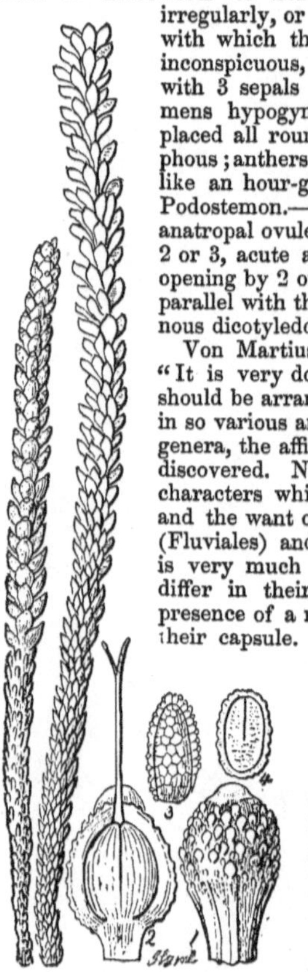

Fig. CCCXXXIII.

Von Martius has the following remarks upon this curious Order. "It is very doubtful in what part of the natural series Podostemads should be arranged; for they are connected with so many other Orders, in so various and complicated a manner, that it is probable that several genera, the affinities of which will be more apparent, still remain to be discovered. Nothing can be more singular than the mixture of different characters which they exhibit. Thus, the structure of their spathes, and the want of a true calyx and corolla, approximate them to Naiads (Fluviales) and Arads, while the character of their stamens and fruit is very much that of Juncaginaceæ; the former of these, however, differ in their lower degree of organisation, and the latter in the presence of a more or less perfect perianth, and in the composition of their capsule. Lemna, a genus closely allied to Arads, seems to be more related to them in its spathe, hypogynous stamens, habit, and mode of life, but is distinguished by its less highly developed few-seeded fruit. Again, Mniopsis, in its ramification, in the form and position of its leaves, and in its stipules, and Lacis and Podostemon in the character of their spathe and the emersion of their pedicels at the time of flowering, call remarkably to mind the habit of Jungermannia; so that we should probably not be far from the truth, if we were to say that this Order forms a transition from Naiads (Fluviales) to Juncaginaceæ, on the one hand touching upon Arads, and being, as it were, a sort of noble analogy of Liverworts among Monocotyledons."—*Nov. G. et Sp.* 1. 7. So far as the general appearance of Podostermads is concerned, this account of them may be received; but since they have a dicotyledonous not monocotyledonous embryo, other affinities must be looked for. Bongard first represented their true structure, and more recently Mr. Griffith has described two Indian species of Podostemon, with his habitual accuracy; entirely confirming the view which I upon mere theoretical reasoning formerly took of their being Exogens.—*Ed. pr.* p. 190. And I am still of opinion that if we have among Exogens one type of structure more nearly approaching that of

Fig. CCCXXXIII.—The ♀ of Hydrostachys verruculosa. 1. the calyx; 2. the same opened to show the ovary; 3. a seed; 4. a vertical section of it.—*Decaisne.*

Acrogens than another it is this, which, with the habit of Liverworts and Scale-mosses, has wholly the structure of flowering plants. According to Bongard, the species have neither spiral vessels nor stomates ; the latter would of course be absent, on account of the submersed habits of the species of Mourera to which his observations chiefly apply. And Mr. Griffith confirms his statements as regards his two Indian Podostemons. But although Podostemads must be considered to present a very strongly marked approach to flowerless plants in some respects, yet we must look for some more immediate relation. This I formerly thought might be found with Peppers, or Callitriche ; Meisner suggests Hornworts. But if we regard hermaphrodite flowers, hypogynous stamens, and an exalbuminous embryo as the most important features in these plants, our views of its affinity will take a very different direction, and we can scarcely fail to suspect an approach to Waterpeppers, whose manner of life is in some respects similar. In fact, upon comparing the two Orders, we find that they are otherwise much alike, except that Podostemads are more incompletely formed in the floral envelopes, and seem to want the capitate stigmas of that Order. Both have 2-celled anthers bursting inwards longitudinally, and a separable placenta bearing numerous anatropal seeds. It seems, therefore, probable that Podostemads stand in the same relation to Waterpeppers as Hippurids to Onagrads, and Lemnads to Arads.

Natives of rocks, in rivers, still waters, and damp places in South America and the islands off the east coast of Africa ; 1 species is found in North America, 2 in the East Indies.

Some species of Lacis yield, when burnt, a considerable quantity of salt from their ashes.—*Schomburgk.*

GENERA.

Mniopsis, *Mart.*	*Dicræia*, Thouars.	Lacis, *Lindl.*	Hydrostachys, *Thouars.*
Crenias, Spreng.	Mourera, *Aubl.*	Tristicha, *Thouars.*	Halophila, *Thouars.*
Hydrobryum, *Endl.*	*Neolacis*, Cham.	*Dufourea*, Bory.	? Lemnopsis, *Zippel.*
Podostemon, *L. C. Rich.*	*Marathrum*, H. B. K.	*Philocrene*, Bong.	

NUMBERS. GEN. 9. Sp. 25.

Halorageæ.
POSITION.—Elatinaceæ.—PODOSTEMACEÆ.—Piperaceæ.
Marchantiaceæ.

ALLIANCE XXXVI. *GERANIALES.*—THE GERANIAL ALLIANCE.

DIAGNOSIS.—*Hypogynous Exogens, with monodichlamydeous, symmetrical flowers, axile placentæ, an imbricated calyx, a twisted corolla, definite stamens, and an embryo with little or no albumen.*

If we seek for a positive character by which the present Alliance may be known from its relations, we shall find it in the combination of three circumstances, viz: a definite number of stamens, an imbricated calyx, and a twisted corolla. Malvals have a twisted corolla, but their stamens are usually indefinite and their calyx is always valvate; for which latter reason Indian Cresses are removed from the Geranial Alliance. Rutals have definite stamens and an imbricated calyx, but their corolla is imbricated, not twisted.

The only doubtful Order of the Alliance is that of Chlenads, which in habit is like some Sterculiads; but it corresponds with Balsams in their unsymmetrical flowers, and it has not a stronger relation to any Malval than to the Order of Geranials. Chlenads may perhaps be regarded as a kind of approach on the part of Geranials to the Malval Alliance.

NATURAL ORDERS OF GERANIALS.

Flowers symmetrical. Styles distinct. Carpels longer than the torus. Seeds with little or no albumen	183. LINACEÆ.
Flowers regular, unsymmetrical, with a permanent cup-like involucre. Stamens monadelphous. Albumen abundant	184. CHLÆNACEÆ.
Flowers symmetrical. Styles distinct. Carpels longer than the torus. Seeds with abundant albumen	185. OXALIDACEÆ.
Flowers very irregular and unsymmetrical, without an involucre. Stamens distinct. Albumen none	186. BALSAMINACEÆ.
Flowers usually symmetrical. Styles and carpels combined round a long beaked torus	187. GERANIACEÆ.

ORDER CLXXXIII. LINACEÆ.—FLAXWORTS.

Lineæ, *DC. Theorie, ed.* 1. 217. (1819); *Prodr.* 1. 423. (1824); *Endl. Gen.* cclv.; *Meisner, Gen.* 57.

DIAGNOSIS.—*Geranial Exogens, with symmetrical flowers, distinct styles, carpels longer than the torus, and seeds with little or no albumen.*

Annual or perennial plants, or even small shrubs. Leaves alternate or opposite, rarely in whorls, simple, entire, without stipules, sometimes with a pair of glands. Flowers very fugitive, white, yellow, red, or blue. Sepals 3-4-5, with an imbricated æstivation, continuous with the peduncle, persistent. Petals equal in number to the sepals, hypogynous, unguiculate, with a twisted æstivation. Stamens equal in number to the petals, and alternate with them, united at the base in a hypogynous ring, from which proceed little teeth opposite to the petals, indicating abortive stamens; anthers ovate, innate. Ovary with about as many cells as sepals, seldom fewer; styles equal in number to the cells; stigmas capitate; ovules pendulous, anatropal. Capsule generally pointed with the indurated base of the styles, many-celled; each cell completely or partially divided in two by an imperfect spurious dissepiment arising from the dorsal suture; dehiscing with two valves at the apex. Seeds in each cell single, compressed, inverted; albumen 0, or in very small quantity; embryo straight, fleshy, with the radicle pointing towards the hilum; cotyledons flat.

Fig. CCCXXXIV.

It is remarked by De Candolle that this Order is intermediate, as it were, between Cloveworts, Mallow-worts, and Cranesbills. Aug. de St. Hilaire considers it a mere section of the latter, from which however it is removed by its continuous stems, exstipulate leaves, and unbeaked fruit. Its nearest affinity is with Oxalids, from which there is little to divide it except the peculiar structure of its carpels, whose spurious dissepiments are however scarcely of ordinal importance, its simple leaves, and the very small quantity of albumen found in the seeds. It is not without resemblance to Waterpeppers, of which I formerly suggested that Flaxworts might be an exstipulate decandrous form.

Europe and the North of Africa are the principal stations of this Order, which is, however, scattered more or less over most parts of the globe. Several are natives of North and South America, 2 only are found in India, 1 in New Zealand, and none in New Holland; for the L. angustifolium mentioned by De Candolle as having been sent him from that country, had probably, as he suggests, been introduced from Europe. It is stated by Richardson that the most northern limit of this Order in North America is 54° N.

The tenacity of their fibre, and the mucilage of their diuretic seeds, are striking characters of Flaxworts, which are also usually remarkable for the beauty of their flowers. The leaves of L. catharticum are purgative. Linum selaginoides is considered in Peru bitter and aperient. The meal of the seeds of Linum usitatissimum is used for poultices. The infusion is demulcent and emollient. The oil mixed with lime-water has been a favourite application to burns. The tenacious and delicate fibre called Flax is obtained from that plant, and forms the most beautiful of our linen fabrics.

GENERA.

Linum, *Linn.*	*Linopsis*, Reich.	*Reinwardtia*, Dumort.
Cathartolinum, Reich.	*Xantholinum*, Reich.	Cliococca, *Bab.*
Adenolinum, Reich.	*Macrolinum*, Reich.	Radiola, *Dillen.*

NUMBERS. GEN. 3. Sp. 90.

Elatinaceæ.
POSITION.—Oxalidaceæ.—LINACEÆ.—Geraniaceæ.
Malvaceæ.

Fig. CCCXXXIV.—Linum perenne. 1. a single flower; 2. the calyx and stamens, &c.; 3. the stamens, &c. without the calyx; 4. a cross section of the ovary; 5. a section of a seed.

Order CLXXXIV. CHLÆNACEÆ.—Chlenads.

Chlenaceæ, *Thouars' Hist. Veg. Afr. Austr.* 46. (1806); *DC. Prodr.* 1. 521. (1824); *Endl. Gen.* cciv.; *Meisner*, *p.* 35.

Diagnosis.—*Geranial Exogens, with regular unsymmetrical flowers, in a permanent cup-like involucre, monadelphous stamens, and abundant albumen.*

Handsome trees or shrubs, with fine showy flowers usually of a red colour. Leaves alternate, feather-veined, entire, sometimes plaited longitudinally; stipules terminating the branches in a conical way, and rolling up or inclosing the leaves, quickly deciduous. Flowers in corymbs, racemes or panicles. Involucre 1-2-flowered, persistent, of variable form and texture. Sepals 3, small; æstivation imbricated. Petals 5, hypogynous, convolute, broader at the base, sometimes cohering there. Stamens either very numerous, or sometimes only 10; filaments either cohering at the base within a cup-like disk, or adhering to the tube of petals; anthers roundish, adnate, or loose, 2-celled. Ovary single, 3-celled; style 1, filiform; stigma triple; ovules 2 or more, anatropal, pendulous from the inner angle. Capsule 3-celled, or 1-celled by abortion. Seeds solitary or numerous, attached to the centre, suspended; embryo green, central; albumen fleshy according to Jussieu, or horny according to Du Petit Thouars; cotyledons foliaceous, wavy; radicle superior.

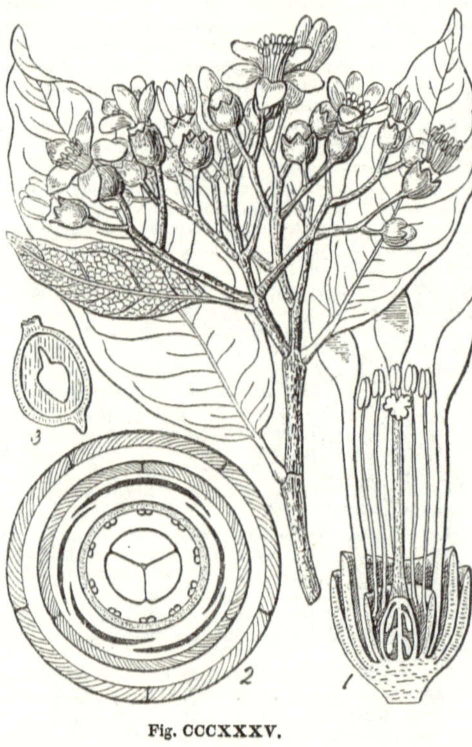

Fig. CCCXXXV.

These are very curious plants, presenting the singular properties of 3 in the calyx, 5 in the corolla and stamens, and 3 in the ovary; besides which, their flowers are inclosed in an involucre, which is usually 5-toothed. The monadelphous stamens and involucrated flowers seem to indicate an affinity between these plants and Mallowworts. But Jussieu refers the Order rather to the vicinity of Ebenads, considering it monopetalous. Very little is, in fact, known of it. I formerly supposed it to have some relation to the Rock-roses (Cistaceæ), having had no opportunity of examining the plants myself. The acquisition of 2 or 3 species has, however, satisfied me, that if the calyx were valvate the Chlenads could not be removed from the Malval Alliance. The tendency of their calyx being, however, to the imbricated structure, the Geranial Alliance necessarily becomes their station, where they may be regarded as a connecting link with Malvals. The propriety of placing them in the Geranial Alliance seems to be confirmed by Balsams exhibiting a similar tendency to the unsymmetrical structure.

Whatever the real place of this group may eventually prove to be, it is certain that

Fig. CCCXXXV.—Leptolæna multiflora. 1. a perpendicular section of its flower; 2. a diagram of its structure; 3. a section of its seed.

Mr. Endlicher has destroyed all ideas of its nature by introducing such plants as Hugonia, Ventenatia, and Eucryphia, which are noticed elsewhere, the one belonging to Oxalids, the second possibly to Theads, and the last being certainly not far removed from the Order of Tutsans, if it be distinguishable.

All the species are natives of Madagascar.

Nothing is known of their uses.

GENERA.

Sarcolæna, *Thouars.*
Leptolæna, *Thouars.*
Schizolæna, *Thouars.*
Rhodolæna, *Thouars.*

NUMBERS. GEN. 4. Sp. 8?

POSITION.—Balsaminaceæ.—CHLÆNACEÆ.——————
Sterculiaceæ.

ORDER CLXXXV. OXALIDACEÆ.—OXALIDS.

Oxalideæ, *DC. Prodr.* 1. 689. (1824) ; *Endl. Gen.* cclvi.; *Meisner Gen.* 57.—Ledocarpeæ, *Meyen Reise.* 1. 307. ; *Klotzsch in Linnæa* 10. 431. ; *Endl. Gen. p.* 1169.—Rhyncotheceæ, *Endl. Gen. p.* 1169.—Hugoniaceæ, *Arnott Prodr. Fl. Ind. Penins.* 1. 71. (1834) ; *Ed. pr.* lxvii. ; *Wight Illustr.* 1. ; *Endl. Gen. p.* 1016 ; *Meisner, p.* 35.

DIAGNOSIS.—*Geranial Exogens, with symmetrical flowers, distinct styles, carpels longer than the torus, and seeds with abundant albumen.*

Herbaceous plants, undershrubs, or trees. Leaves simple or compound, alternate, usually but not always without stipules; occasionally opposite. Sepals 5, sometimes slightly cohering at the base, persistent, imbricated. Petals 5, hypogynous, equal, unguiculate, with a spirally-twisted æstivation; occasionally 0. Stamens 10, usually more or less monadelphous, those opposite the petals forming an inner series, and longer than the others; anthers 2-celled, innate. Ovary with 3 to 5 cells; styles as many, filiform; stigmas capitate or somewhat bifid; ovules anatropal. Fruit capsular, membranous, or drupaceous, with 3-5 cells and as many or twice as many valves, if it is dehiscent. Seeds few, fixed to the axis, sometimes inclosed within a fleshy integument, which curls back at the maturity of the fruit, and expels the seeds with elasticity. Albumen between cartilaginous and fleshy. Embryo the length of the albumen, with a long radicle pointing to the hilum, and flat cotyledons.

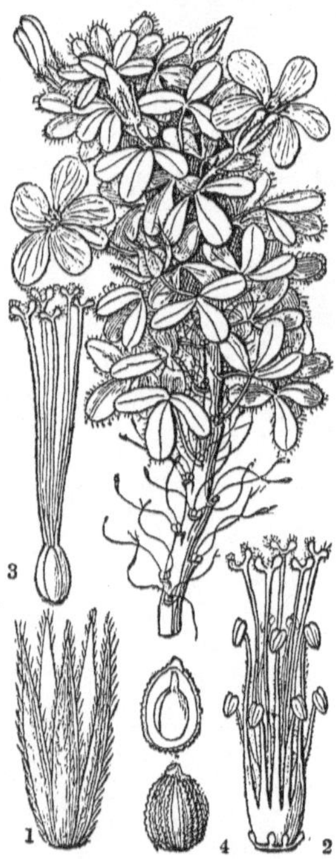

Fig. CCCXXXVI.

These plants were formerly included in the Order of Cranesbills, from which, in the judgment of many, they are not sufficiently distinct. According to De Candolle, they are rather allied to Beancapers; an opinion which their compound leaves appear to confirm. The species are generally described with an aril; but, according to Auguste de St. Hilaire, the part so called is nothing but the outer integument of the seed. The genus Hugonia, which has been placed first in one Order, then in another, and even considered the type of an Order apart from all others, chiefly differs in its simple leaves and deciduous stipules. The true character of Oxalids resides in their regular flowers, beakless fruit, and albuminous seeds, to which may be added the very general tendency among them to form compound leaves.

Natives of all the hotter and temperate parts of the world, most abundantly however in America and the Cape of Good Hope; more rarely in the East Indies and equinoctial Africa; and sparingly in the temperate parts of Asia and in Europe. The shrubby species are confined to the hotter parts of the world.

Averrhoa Bilimbi and the pinnated Oxalis called Biophytum have sensitive leaves. The chief quality of the typical species of this Order resides in their strong acidity, caused by oxalic acid, formed by them in great abundance; hence they are used as substitutes for Sorrel. In the Blimbing, and Carambola (Averrhoa Bilimbi, and Carambola), whose fruit is eaten in the East Indies, this acidity is intolerable to Europeans, who use them chiefly as pickles. On the same account several species of Oxalis are used in Brazil against malignant fevers. A species of Oxalis (crenata), found in Columbia, bears tubers like a Potato, and is one of the plants called Arracacha: the tubers are insipid, and not worth cultivation; the stalks of the leaves are intensely acid, and make an agreeable preserve. Another species, the Oxalis Deppei, has,

Fig. CCCXXXVI.—Oxalis confertissima. 1. calyx; 2. stamens; 3. pistil; 4. seed and its section of O. acetosella.

however, fleshy roots, quite free from acidity, and abounding in a matter analogous to that of salep: these roots are as large as small Parsnips, and are becoming esteemed for culinary purposes. The species called Oxalis crassicaulis, tetraphylla, and esculenta, are reported to possess similar good qualities. Some bitterness has been remarked in Oxalis sensitiva, whose leaves are said to be tonic and slightly stimulating. Hugonia Mystax too, an anomalous species of the Order, is of a like nature, but in a more marked degree; its root smells like violets, and is said to be diuretic, diaphoretic, and anthelmintic. In certain species an irritability of so marked a kind has been found as to cause them to be classed among Sensitive Plants. Averrhoa Bilimbi and Oxalis sensitiva are the most remarkable; but the same irritability has been observed by Professor Morren in the European Oxalis stricta.—*Ann. Sc. n. s.* xiv. 350.

GENERA.

Oxalis, *Linn.*
Biophytum, DC.
Averrhoa, *Linn.*
Bilimbi, Endl.
Carambola, Endl.

Ledocarpum, *Desf.*
Balbisia, Cav.
Cruckshanksia, Hook.
Cistocarpum, Kunth.

Wendtia, *Meyen.*
Martinieria, Guillem.
Hugonia, *L.*
Rhynchotheca, *R. et P.*

NUMBERS. GEN. 6. Sp. 325.

POSITION.—Linaceæ.—OXALIDACEÆ.—Geraniaceæ.
Zygophyllaceæ.

ORDER CLXXXVI. BALSAMINACEÆ.—BALSAMS.

Balsamineæ, *Ach. Rich. Dict. Class.* 2. 173. (1822); *DC. Prodr.* 1. 685. (1824); *Lindl. Synops.* 59. (1829); *Röper de Floribus et Affinitatibus Balsaminearum,* (1830); *Wight and Arnott, Prodr. Fl. Ind. Penins.* 1. 134. (1834); *C. B. Presl. Bemerkungen über den bau der Blumen der Balsamineen,* (1836); *Wight and Röper Linnæa* ix. 112. (1835); *Bernhardi, ib.* xii. 669. (1838); *Kunth in Mém. Soc. Hist. Nat. Par.* iii. 384. (1827); *Wight in Madras Journal* (*Jan.* 1837); *Lindl. in Bot. Reg. sub. t.* 8. (1840); *Endl. Gen.* cclvii.; *Meisner Gen. p.* 58.—Hydrocereæ, *Blume Bijdr.* 241. (1825); *Ed. Prior, No.* 125. (1830).

DIAGNOSIS.—*Geranial Exogens, with very irregular and unsymmetrical flowers without an involucre, distinct stamens, and no albumen.*

Succulent, usually annual, herbaceous plants, having simple, opposite, or alternate leaves, without stipules. Peduncles axillary, or quasiterminal and racemose. Flowers very irregular. Sepals 5, irregular, deciduous, with an imbricated æstivation; the two exterior opposite, lateral, somewhat unsymmetrical, with a valvate æstivation, but giving way for the projection of the spur of the odd sepal; the odd sepal spurred, symmetrical, with an equitant æstivation in the bud, looking towards the axis of the axillary racemose or umbellate inflorescence, containing honey; the two dorsal sepals usually connate, sometimes unsymmetrical, orbicular, always coloured, appearing at that side of the flower which is opposite to the spurred sepal. Petals either distinct or adhering, 5, combined into 2 or 3, irregular, deciduous; the dorsal usually abortive, and the side ones united more or less in pairs; their two larger lobes next the spur, their two smaller next the odd petal; æstivation convolute. Stamens 5, symmetrical, alternate with the petals. Carpels 5, alternate with the stamens, consolidated into a 5-celled ovary; style clear of the carpellary leaves, simple; stigma sessile, more or less divided in 5; cells 5, 2- or many-seeded. Fruit capsular, with 5 elastic valves, and 5 cells formed by membranous projections of the placenta, which occupies the axis of the fruit, and is connected with the apex by 5 slender threads; sometimes succulent and indehiscent. Seeds solitary, or numerous, suspended; albumen none; embryo straight, with a superior radicle and plano-convex cotyledons.

Fig. CCCXXXVII.

The Balsams are, in the opinion of some Botanists, scarcely distinguished from Cranesbills. But the latter evidently differ in the torus or gynobase being lengthened into a beak, in their leaves having stipules, their stems swollen articulations, and their carpels but one seed in each cell. Their flowers too have none of the peculiar breaking up of symmetry which is so characteristic of Balsams, and which at once divides them from even Oxalids, to which they certainly approach very nearly.

Much discussion has taken place among Botanists as to the real nature of the parts which constitute the very irregular flower of a Balsam. According to Röper and others, two membranous external scales, and a spur, alone belong to the calyx, of which the two other sepals are usually deficient on that side of the flower which is opposite the spur; on the other hand, the corolla consists of a large upper or back piece, and of two lateral inner wings, each of which last consists of two petals; and this view was adopted in the last edition of this work. On the other hand, Achille Richard considers two smaller exterior scales, together with the spurred and the back interior pieces, as forming a four-leaved calyx, while he regards the two

Fig. CCCXXXVII.—1. Impatiens macrochila; 2. a diagram of its flower; 3. its stamens; 4. fruit of I. Balsamina; 5. its embryo.

innermost lobed pieces as two pairs of petals of a four-leaved corolla. A third view is that of Bernhardi, who regards the exterior scales as bracts, the calyx as consisting of five parts, of which three only, namely the spur and the back piece, which is double, are present, and the others rudimentary or missing; while the corolla also consists of five parts, of which the four lower are united in pairs into the two innermost lobed pieces, and the fifth is either separate, as in Hydrocera, or consolidated with the two back united sepals into what he calls a petal-sepal. A fourth view is that of Kunth, who considers the large back piece of the flower to be composed of two sepals, and together with the spur and exterior scales to form a five-leaved calyx; while he finds in the two innermost parts a corolla of four petals united in pairs, and he assumes the fifth petal to be abortive. This opinion has been adopted by Arnott in 1833, and by Presl in 1836, the latter having discovered the fifth or missing petal to be present occasionally in the garden Balsam, and always in Hydrocera triflora; both these Botanists finding in the genus Hydrocera the back piece, which is simple in Impatiens, composed of two parts, and therefore confirming the accuracy of the theory of Kunth. Other opinions, more or less resembling these, have been formed by others, but it is clear that Kunth's theory is the only one that is correct.

If we make a section horizontally through a young flower-bud of Impatiens macrochila, we find the following structure:—There is in the centre an ovary of five cells; with these alternate the five stamens, of which the fifth or anterior has a longer filament than the others; so far the structure is regular, and we have all the necessary evidence of the flower, however irregular, being formed upon a quinary type. Right and left of the stamens stand the two innermost pieces; these cannot be simple, because they are opposite the intermediate stamens; but their two-lobed figure, when full grown, shows that each is double, and then, their apparent centre being in fact their united margins, they alternate with the anterior stamens, and so fall into the place usually destined for petals. The last mentioned parts are half enveloped by the back piece, which might, from its position, be the fifth petal; but the case of Hydrocera showing it really to consist of two united parts, they must be opposite the stamens, and consequently are sepals. Next comes the spur, which overlaps the back piece, and stands opposite the anterior stamen; as no tendency to divide on the part of this piece is ever found, it must be a sepal. Finally, the external scales, placed right and left of the whole flower, alternate with those parts already shown to be sepals, and consequently are recognised as the two parts of the calyx required to complete the quinary plan of the whole flower. It will be remarked, that a fifth petal has not been found; if the eye is turned upon the back piece, already found to be composed of two sepals, it will be seen that a part is missing between those two and the two corresponding stamens; and this is the place where the abortion of a fifth of the corolla may, upon the evidence of this flower, be assumed to occur, and where it is proved to take place by the evidence of Hydrocera, in which the part missing in the Balsam makes its appearance.

The annexed diagram will serve to illustrate the preceding observations; the parts of the flower, as they really exist in Impatiens being projected upon a plane consisting of five circles, of which the exterior (S) represents the sepals or calyx, the next (P) the petals or corolla, the third (*s*) the stamens, the fourth (*c*) the carpels, and the central (*p*) the placenta, or axis.

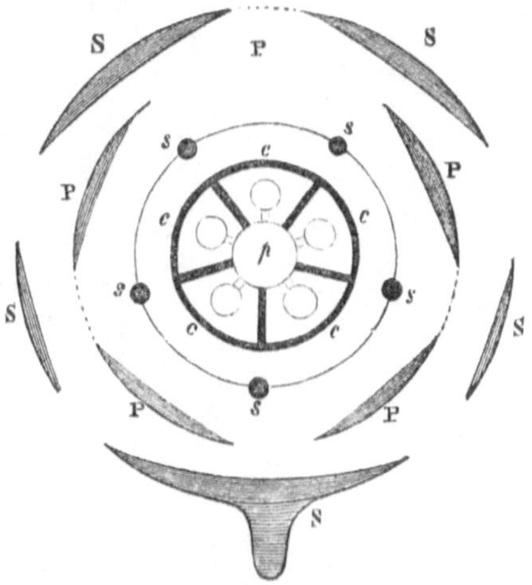

Connected with these plants is a point of structure deserving of attention. In some species it will be found that the style is surrounded below its apex by five points, which are evidently continuations of the backs of the five carpellary leaves, which certainly in these plants are separate from the placenta, and are merely pressed down upon it so as to cover the ovules, thus confirming the accuracy of the views concerning placentation held by Schykofsky and Schleiden. If so, what else can the upper part of the style and the stigmas be, except the naked apex of the placenta, prolonged beyond the carpellary leaves? And then is not the conducting tissue of a style in most cases an extension of the placenta? and may we not consider the indusium of Goodeniads, and, *à fortiori*, the well-known rim found upon the stigma in Heathworts, as the expanded end of the carpellary leaves, while the stigma of those plants is the upper end of the placenta?

Natives of damp places among bushes in the East Indies; 1 is found in Madagascar, 1 in Europe, 2 in North America, and 1 in Russia in Asia. India swarms with species, all of which deserve the care of the cultivator. According to Dr. Wight, (*Madras Journal*, January, 1837,) at least a hundred occur in those districts from which Roxburgh described only three. Forty-seven species are named by Wallich from Silhet, Pundooa, Nipal, and the Peninsula, and multitudes occur in Ceylon, and the islands of the Indian Archipelago. Dr. Wight states that a moist climate and moderate temperature are the circumstances most favourable, if not indispensable, to their production. At Courtallum, for example, they most abound in shady places on the tops of hills, with a mean temperature during the season of their greatest perfection not exceeding 70°, if so much. At Shevaggery, about fifty miles north of Courtallum, he found five out of seven species on the highest tops of the mountains, none of the five under 4000 feet, and three of them above 4500 feet of elevation; the mean temperature being 65° Fahr. Two found at a lower elevation, were both either growing in the gravelly beds of streams, or immediately on their banks; the temperature of which was ascertained to be 65°, while that of the air at noon was only about 75°.

The species are chiefly remarkable for the elastic force with which the valves of the fruit separate at maturity, expelling the seeds. For a supposed explanation of this phenomenon, see Dutrochet, *Nouvelles Recherches sur l'Exosmose et Endosmose*. According to De Candolle, they are diuretic; it is also said that the distilled water of Impatiens Nolitangere, taken in large quantity, brings on attacks of diabetes.

GENERA.

Impatiens, *Linn.*
Balsamina, Gærtn.
Hydrocera, *Blum.*
Tytonia, Don.

NUMBERS. GEN. 2. Sp. 110.

POSITION.—Chlænaceæ.—BALSAMINACEÆ.—Geraniaceæ.
Tropæolaceæ.

ORDER CLXXXVII. GERANIACEÆ.—CRANESBILLS.

Gerania, *Juss. Gen.* 268. (1789).—Geraniaceæ, *DC. Fl. Fr.* 4 828 (1805); *Prodr.* 1. 637. (1824); *Endl. Gen.* ccliv.; *Meisner Gen. p.* 57.

DIAGNOSIS.—*Geranial Exogens, with usually symmetrical flowers, and with styles and carpels combined round a long beaked torus.*

Herbaceous plants or shrubs. Stems tumid, and separable at the joints. Leaves either opposite or alternate; in the latter case opposite the peduncles, with membranous stipules. Flowers white, red, yellow, or purple. Sepals 5, persistent, ribbed, more or less unequal, with an imbricated æstivation; 1 sometimes saccate or spurred at the base. Petals 5, seldom 4, in consequence of 1 being abortive; unguiculate, twisted in æstivation, equal or unequal, either hypogynous or perigynous. Stamens usually monadelphous, hypogynous, twice or thrice as many as the petals; some occasionally abortive. Ovary composed of 5 carpels placed round a long awl-shaped torus or growing point, each 1-celled, 2-seeded; styles 5, cohering round the torus and separable from it; ovules semianatropal, adhering to the torus. Fruit formed of 5 shells, cohering round a long beaked torus; each piece containing 1 seed, having a membranous pericarp, and terminated by an indurated style, which finally curls back from the base upwards, carrying the pericarp along with it. Seeds solitary, without albumen. Embryo curved and doubled up; radicle pointing to the base of the cell; cotyledons foliaceous, convolute, and plaited.

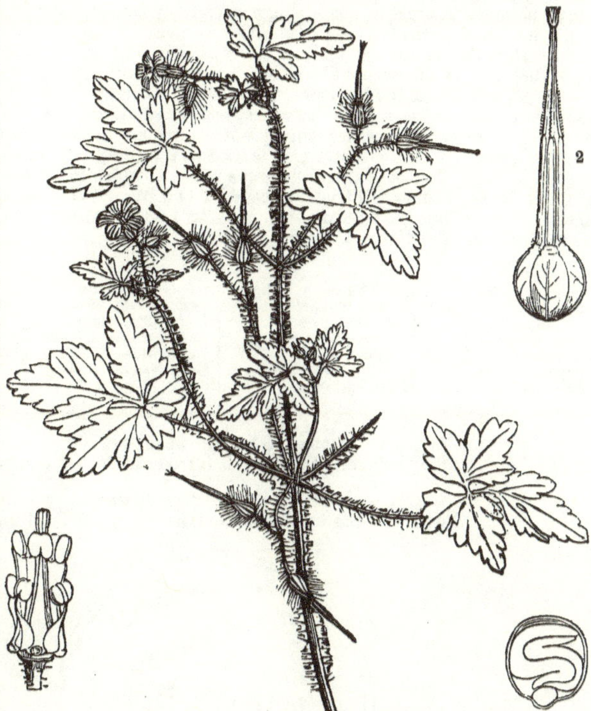

Fig. CCCXXXVIII.

The long beak-like torus, round which the carpels are arranged, and the presence of membranous stipules at joints which are usually tumid, are the true marks of this Order; and all plants not possessing those peculiarities must be excluded. Among them is a South American genus called Rhynchotheca, which has been even elevated into a Natural Order, but which is surely an Oxalid without petals; for the beak observed in its fruit belongs to the carpels and not to the torus. It is clear that in this Order the ovules do not spring from the margins of the carpellary leaves. *E. g.* take P. zonale,

Fig. CCCXXXVIII.—Geranium Robertianum. 1. its stamens; 2. its ovary; 3. a section of its seed.

half-ripe, when the embryo first appears in the albumen as a pale green line. At that time the carpels may be taken away from the ovules, leaving the latter adhering to a central placenta, and this may be done without at all disturbing or tearing the margin of the carpellary leaves. The suspended position of the seed has been given as a general character of Cranesbills; but the position of the ovules varies according to species in the genera Erodium and Geranium; and in consequence of the inequality of growth the seed is always ascending in the capsule.

The species are very unequally distributed over various parts of the world. A great proportion is found at the Cape of Good Hope, chiefly of the genus Pelargonium; Erodium and Geranium are principally natives of Europe, North America, and Northern Asia. It is worthy of remark that Pelargonium is found in New Holland.

An astringent principle and an aromatic or resinous flavour are the characteristics of this Order. The stem of Monsonia spinosa burns like a torch, and gives out an agreeable odour. In North Wales Geranium Robertianum has acquired celebrity as a remedy for nephritic complaints. The root of Geranium maculatum, or Alum-root, is a most powerful astringent, containing considerably more tannin than Kino. According to Bigelow, it is particularly suited to the treatment of such diseases as continue from debility after the removal of their exciting cause. The tincture is an excellent local application in sore throat and ulcerations of the mouth. Many others have a similar reputation, but are not used in modern medicine, especially species of the genus Erodium, among which E. moschatum is more especially remarkable for its powerful smell of musk. The Pelargoniums are chiefly noted for their beautiful flowers, but they, too, are astringents. P. antidysentericum is used as a remedy for diarrhœa among the Namaquas. One of the species with tuberous roots, of which many are known at the Cape of Good Hope, namely, P. triste, is eatable, and Mr. Backhouse speaks of the fleshy tubers of Geranium parviflorum being eaten by the natives of Van Diemens Land, where it is called the Native Carrot.

GENERA.

Erodium, *Herit.*	*Sarcocaulon*, DC.	*Polyactium*, DC.	*Chorisma*, Sweet.
Scolopacium, Eckl. et Zeyh.	Pelargonium, *Herit.*	*Isopetalum*, Sweet.	*Ciconium*, Sweet.
Geranium, *Herit.*	*Hoarea*, Sweet.	*Campylia*, Sweet.	*Cortusina*, Eckl.
Monsonia, *Linn. f.*	*Dimacria*, Sweet.	*Phymatanthus*, Sweet.	*Eumorpha*, Eckl.
Odontopetalum, DC.	*Cynosbata*, DC.	*Myrrhidium*, DC.	*Calliopsis*, Sweet.
Holopetalum, DC.	*Peristera*, DC.	*Jenkinsonia*, Sweet.	*Anisopetalum*, DC.
	Otidia, Sweet.		

NUMBERS. GEN. 4. Sp. about 500.
(After deducting the hybrids introduced by De Candolle.)

Tropæolaceæ.
POSITION.—Balsaminaceæ.—GERANIACEÆ.—Oxalidaceæ.

Fig. CCCXXXIX.

Fig. CCCXXXIX.—Diagram of the flower of Geranium Robertianum.

ALLIANCE XXXVII. *SILENALES.*—THE SILENAL ALLIANCE.

DIAGNOSIS.—*Hypogynous Exogens, with monodichlamydeous flowers, a free central placenta, an external embryo curved round a little mealy albumen, and more than one carpel completely combined into a compound fruit.*

At this point a considerable advance in structure is evident among Exogens. Among these plants a corolla appears with all its fragrance and gaudy colours, and the ovary is constituted, not by the rolling up of a solitary carpellary leaf, but by the complete consolidation of several. The combining process is indeed carried occasionally so far among these plants as to divide the cavity of the ovary into distinct cells; but it is certain that the placentation is in all cases strictly central, no power being possessed by the carpellary leaves of generating ovules on their margin or sides.

The Buckwheat Order (Polygonaceæ) establishes a transition from the Chenopodals, by virtue of its uniform want of corolla; another passage is supplied by Knotworts (Illecebraceæ), which are only Amaranths of a higher grade. In both these Orders the ovary may seem to be in many instances as simple as in the Chenopodal Alliance; but its compound nature is brought into evidence by the number of its separate stigmas or by the manner in which the seed-vessel splits when ripe.

NATURAL ORDERS OF SILENALS.

Calyx and corolla usually both present and symmetrical (4 *and* 4, *or* 5 *and* 5), *the latter conspicuous. Ovules amphitropal. Leaves opposite, without stipules*	188. CARYOPHYLLACEÆ.
Calyx and corolla usually both present and symmetrical (4 *and* 4, *or* 5 *and* 5), *the latter rudimentary. Ovules amphitropal. Leaves with scarious stipules*	189. ILLECEBRACEÆ.
Calyx and corolla both present and unsymmetrical (2 *and* 5), *the latter usually conspicuous. Ovules amphitropal. Leaves alternate, succulent, without stipules*	190. PORTULACACEÆ.
Calyx only present, but often coloured. Ovules orthotropal. Nut usually triangular	191. POLYGONACEÆ.

Order CLXXXVIII. CARYOPHYLLACEÆ.—Cloveworts.

Caryophylleæ, *Juss. Gen.* 299. (1789); *De Cand. Prodr.* 1. 388. (1824); *Endl. Gen.* ccvii.; *Meisner Gen.* 24.—Sileneæ, *DC. Prodr.* 1. 351. (1824); *Bartl. Ord. Nat.* 305. (1830); *Braun in Ann. Sc. Nat.* 2. *ser.* xx. 170.—Alsineæ, *DC. Fl. Franc. Ed.* 3. 4. 766. (1805); *Bartl. Ord. Nat.* 204. (1830); *Fenzl. Versuch.* (1833).—Queriaceæ, *DC. Prodr.* 3. 379. (1828).—Minuartieæ, *Id.* (1828).—Mollugineæ, *Fenzl. Monogr.*—Steudeliæ, *Ib.*

Diagnosis.—*Silenal Exogens, with symmetrical flowers, a conspicuous corolla, amphitropal ovules, and opposite leaves without stipules.*

Herbaceous plants, occasionally becoming suffrutescent. Stems tumid at the articulations. Leaves always opposite and entire, often connate at the base. Flowers ☿, occasionally imperfect by abortion, variously arranged. Sepals 4-5, continuous with the peduncle, persistent, distinct, or cohering in a tube. Petals 4-5, hypogynous, unguiculate, inserted upon the pedicel of the ovary; frequently split into 2 parts, occasionally wanting. Stamens usually twice as many as the petals, sometimes equal in number to the sepals and opposite them, occasionally fewer, inserted upon the pedicel of the ovary along with the petals; filaments subulate, sometimes monadelphous; anthers innate, 2-celled, opening longitudinally. Ovary stipitate on the apex of a pedicel (called the gynophore), composed of from 2 to 5 carpels, whose edges are either adherent and valvate, or turned inwards so as sometimes to touch the free central placenta; stigmas 2-5, sessile, filiform, papillose on the inner surface; ovules few or 00, amphitropal. Capsule 2-5-valved, either 1-celled or 2-5-celled, in the latter case with a loculicidal dehiscence. Placenta central, in the 1-celled capsules distinct, in the 2-5-celled capsules adhering slightly to the edge of the dissepiments. Seeds indefinite in number, rarely definite; albumen mealy; embryo external, curved round the albumen, sometimes straight, very rarely spiral, with hardly any albumen; radicle pointing to the hilum.

1

2

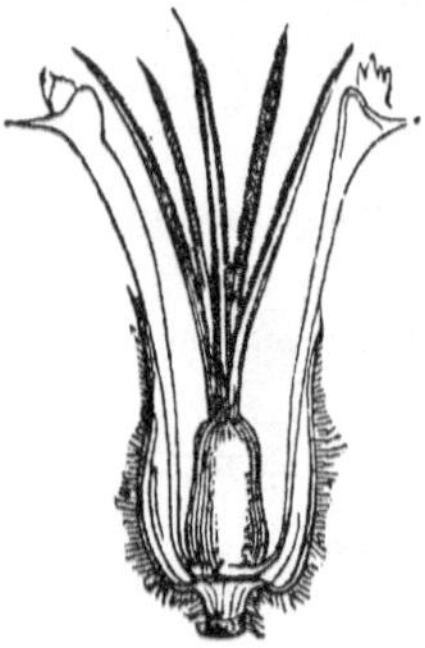

3

4

Fig. CCCXL.

Fig. CCCXL. 1. Stellaria Holostea; 2. pistil, calyx and petals of Lychnis Flos Cuculi; 3. vertical section of its pistil; 4. vertical section of its seed.

These plants, the greater part of which are inconspicuous herbs, form a group which is readily known by its opposite undivided leaves without stipules, tumid nodes, and free central placenta surrounded by several carpellary leaves. They hardly differ from Purslanes except in their symmetrical flowers. In general appearance they nearly approach some of the species of the Geranial Alliance, from which their free central placenta clearly divides them. That this placenta is really central in its origin is proved by a beautiful monstrosity discovered by Mr. Babington, and published by him in the *Gardeners' Chronicle*, for 1844, p. 557, in which the carpellary leaves are partially turned inwards without touching the placenta, which bears a cluster of ovules and is perfectly clear of all connection with those partitions. There is a learned and important Memoir on these plants by Braun (*Ann. Sc. n. s.* xx. 156), to which the reader is referred for valuable details as to the limits of the genera.

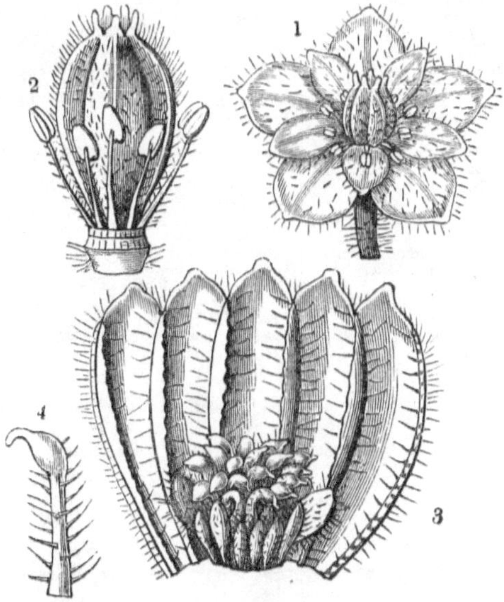

Fig. CCCXLI.

In the succeeding Table of Genera, Sileneæ and Alsineæ are what all Botanists recognise as Cloveworts; the Mollugineæ consist of a portion of the Purslane tribe as it stands in Endlicher's *Genera Plantarum*, where it is broken up into Polpodeæ and Adenogrammeæ, sections which it is scarcely desirable to maintain. The reasons which have led to this separation are given under the Order of Purslanes.

Natives principally of the temperate and frigid parts of the world, where they inhabit mountains, hedges, rocks, and waste places. Those which are found within the tropics are usually met with on high elevations and mountainous tracts, almost always reaching the limits of eternal snow, where many of them exclusively vegetate. Some Silenes are scattered in many different parts of the globe. According to the calculation of Humboldt, Cloveworts constitute $\frac{1}{22}$ of the flowering plants of France, $\frac{1}{27}$ of Germany, $\frac{1}{17}$ of Lapland, and $\frac{1}{72}$ of North America.

The species are remarkable for little except their uniform insipidity. A few, such as Dianthus and Lychnis, are handsome flowers; but the greater part are mere weeds. Vaccaria vulgaris is said to increase the lacteal secretions of cows fed upon it. It contains Saponine, as also does the Egyptian Soap-root, which is derived from Gypsophila Struthium.—*Bley.* Lychnis dioica, and L. chalcedonica, have also saponaceous properties: Saponaria has been used in syphilis. A decoction of the root of Silene virginica is said to have been employed in North America as an anthelmintic. Spurrey, Spergula arvensis, is sometimes cultivated as food for sheep. Gypsophila Struthium is somewhat acrid; Silene Otites, which is bitter and astringent, has been employed in dropsy. The seeds of Vaccaria vulgaris are said to be diuretic; those of Agrostemma Githago (the Corn-cockle), are reported to render corn unwholesome, when ground into flour.

GENERA.

Suborder I. ALSINEÆ.—Sepals distinct, opposite the stamens, when the latter are of the same number.

Sagina, *Linn.*
Phaloë, Dumort.
Alsinella, Dill.
Buffonia, *Sauvag.*
Bufonia, Linn.
Queria, *Löffl.*
Alsine, *Wahlenb.*
Neumayera, Rchb.
Facchinia, Rchb.
Wierzbickia, Rchb.
Minuartia, Löffl.
Tryphane, Fenzl.
Sommerauera, Hopp.
Siebera, Schrad.
Dufourea, Gren. part.
Calalsine, Endl.
Sabulina, Reichenb.
Cherleria, Hall.
Saginella, Fenzl.
Spergella, Fenzl.
Alsinanthe, Fenzl.
Alsinocarpus, Endl.

Fig. CCCXLI.—1. Monstrous flower of a Cerastium; 2. the pistil and stamens separate; 3. the ovary forced open to show the origin of the ovules and the nature of the imperfect dissepiments; 4. a monstrous ovule.

Psammophila, Fenzl.
Triplateia, *Bartl.*
Hymenella, Moç.et Ses.
Honkenya, *Ehrh.*
Halianthus, Fries.
Hallia, Dumort.
Ammonalia, Desv.
Lepyrodiclis, *Fenzl.*
Merckia, *Fisch.*
Wilhelmsia, Reichenb.
Dolophragma, *Fenzl.*
Thylacosperma, *Fenzl.*
Periandra, Cambess.
Flourensia, Cambess.
Bryomorpha, Karel.
Arenaria, *Linn.*
Eremogone, Fenzl.
Euthalia, Fenzl.
Plinthine, Rchb.
Alsinanthus, Desv.
Porphyrantha, Fenzl.
Gouffeia, Robill.et Cast.
Dicranilla, Fenzl.
Möhringia, *Linn.*
Krascheninikovia, *Turcz.*
Brachystemma, *Don.*
Odontostemma, *Benth.*
Holosteum, *Linn.*
Stellaria, *Linn.*
Schizotechium, Fenzl.
Larbrea, St. Hil.
Leucostemma, *Benth.*
Adenonema, *Bung.*
Cerastium, *Linn.*
Dichodon, Bartl.
Schizodon, Fenzl.
Strephodon, Sering.
Orthodon, Sering.
Esmarchia, Reichenb.
Mönchia, Ehrh.
Malachium, *Fries.*
Myosoton, Mönch.

Suborder II. SILENEÆ.—Sepals united into a tube, opposite the stamens, when the latter are of the same number.

Velezia, *Linn.*
Dianthus, *Linn.*
Caryophyllum, Endl.
Tunica, *Scop.*
Kohlrauschia, Kunth.
Pseudotunica, Fenzl.
Heliosperma, *Griseb.*
Saponaria, *L.*
Rootia, Neck.
Proteinia, Ser.
Gypsophila, *Linn.*
Dichoglottis, Fisch. et Mey.
Hagenia, Mönch.
Heterochroa, Bung.
Struthium, Ser.
Rokejeka, Forsk.
Banffya, Baumg.
Saponaria, *Fenzl.*
Smegmanthe, Fenzl.
Bolanthus, Ser.
Cymanthus, Endl.
Smegmathamnium, End.
Silenanthe, Fenzl.
Helicosperma, *Rchb.*
Melandrium, *Fries.*
Elisanthe, Endl.
Gastrolychnis, Fenzl.
Vaccaria, *Medik.*
Eudianthe, *Rchb.*
Ankyropetalum, *Fenzl.*
Silene, *Linn.*
Behenantha, Otth.
Otites, Otth.
Coniomorpha, Otth.
Stachymorpha, Otth.
Rupifraga, Otth.
Siphonomorpha, Otth
Atocion, Otth.
Viscaria, *Röhl.*
Hymenanthe, Fenzl.
Coronaria, *L.*
Coccygonthe, Rchb.
Agrostemma, *L.*
Githago, Desf.
Uebelinia, *Hochst.*
Petrocoptis, *Braun.*
Lychnis, *Tournef.*
Hedeoma, Lour.
Gastrolychnis, Fenzl.
Cucubalus, *Tournef.*
Scribæa, Flor. Wetter.
Lychnanthus, Gmel.
Drypis, *Michel.*
Acanthophyllum, *C.A.M.*

Suborder III. MOLLUGINEÆ. — Sepals distinct or nearly so, alternate with the stamens when the latter are of the same number.

Mollugo, *Linn.*
Cerviana, Minuart.
Trichlis, Hall.
? *Galiastrum*, Heist.
Pharnaceum, *Linn.*
Ginginsia, DC.
Hypertelis, *E. Mey.*
Psammotropha, *Eckl. et Zeyh.*
Mallogonum, Fenzl.
Cœlanthum, *E. Mey.*
Acrosanthes, *Eckl. et Zey.*
Schiedea, *Cham. et Schl.*
Colobanthus, *Bartl.*
Polpoda, *Presl.*
Adenogramma, *Reichenb.*
Steudelia, Presl.

NUMBERS. GEN. 53. Sp. 1055.

POSITION. ——— *Geraniaceæ.* / CARYOPHYLLACEÆ.—Illecebraceæ. / *Malvaceæ.*

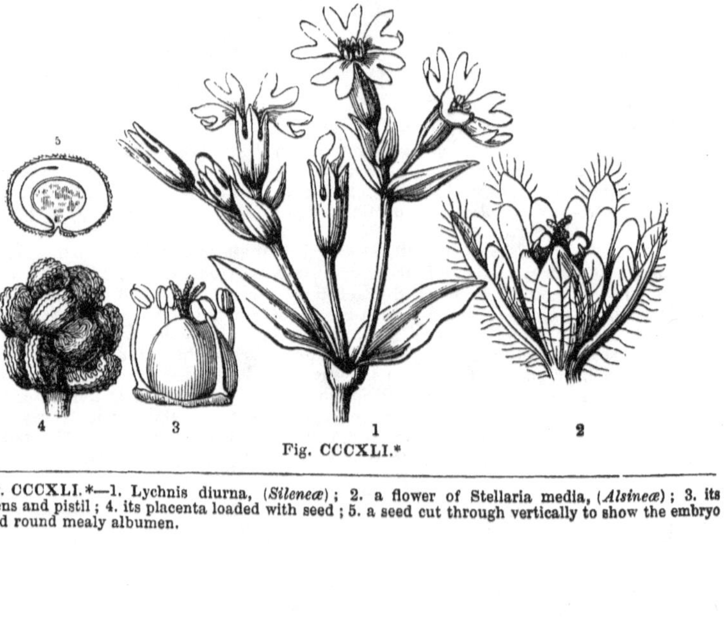

Fig. CCCXLI.*

Fig. CCCXLI.*—1. Lychnis diurna, (*Sileneæ*); 2. a flower of Stellaria media, (*Alsineæ*); 3. its stamens and pistil; 4. its placenta loaded with seed; 5. a seed cut through vertically to show the embryo curved round mealy albumen.

ORDER CLXXXIX. ILLECEBRACEÆ.—KNOTWORTS.

Herniariæ, *Cat. Hort. Par.* (1777).—Illecebreæ, *R. Brown Prodromus*, 413. (1810).—Paronychieæ, *Aug. St. Hil. Mém. Plac. lib. p.* 56. (1815); *Juss. Mém. Mus.* 1. 387. (1815); *DC. Prodr.* 3. 365. (1828); *Mémoire sur les Paronych.* (1829); *Bartl. Ord. Nat. p.* 301. (1830).—Paronychiaceæ, *Meisn. Gen.* 132; *Wight Illustr.* 2. 42.

DIAGNOSIS.—*Silenal Exogens, with both calyx and corolla present and symmetrical, but the latter rudimentary, amphitropal ovules and scarious stipules.*

Herbaceous or half-shrubby branching plants, with opposite or alternate, often fascicled, sessile, entire leaves, and scarious stipules. Flowers minute, with scarious bracts. Sepals 5, seldom 3 or 4, sometimes distinct, sometimes cohering more or less. Petals minute, inserted upon the calyx between the lobes, occasionally wanting. Stamens exactly opposite the sepals, if equal to them in number, sometimes fewer by abortion, sometimes more numerous; filaments distinct; anthers 2-celled. Ovary 1-celled, rarely 3-celled, with 1 or more ovules, superior; styles 2-5, either distinct or partially combined. Fruit small, dry, 1-celled, rarely 3-celled, either indehiscent or opening with 3 valves. Seeds either numerous, upon a free central placenta, or solitary and pendulous from a funiculus originating in the base of the cavity of the fruit; albumen farinaceous; embryo lying on one side of the albumen, curved more or less, with the radicle always pointing to the hilum; cotyledons small.

Fig. CCCXLII.

Very near Purslanes, Amaranths, and Cloveworts, from which they are distinguished with difficulty. From the latter their scarious stipules will separate them; and there is scarcely any other character that will; for no value seems assignable to a slight tendency to a perigynous insertion of the stamens which is observable in both Orders. From Purslanes they are best known by the position of the stamens before the sepals instead of the petals, and by the number of the sepals. With Houseleeks, particularly Tillæa, they often agree in habit, but their concrete carpels will always distinguish them. According to Cambessèdes, the genus Spergularia, in which the petals and stamens are very often perigynous, the styles sometimes consolidated at their base, and the stamens 5 in number, establishes a passage between Cloveworts and Knotworts, and tends to confirm the opinion of those who consider these two Orders as belonging to the same Alliance.

The south of Europe and the north of Africa are the great stations of the Order, where the species grow in the most barren places, covering with a thick vegetation soil which is incapable of bearing anything else. A few are found at the Cape of Good Hope; and North America, including Mexico, comprehends several.

A trace of astringency pervades the Order, and is the only sensible property that it is known to possess.

GENERA.

Corrigiola, *Linn.*
Polygonifolia, Vaill.
Herniaria, *Tournef.*
Illecebrum, *Gärtn. f.*
Cardionema, *DC.*
Bivonæa, Moç. et Sess.
Pentacæna, *Bartl.*
Acanthonychia, D
Paronychia, *Juss.*
Anychia, L. C. Rich.
Gymnocarpus, *Forsk.*
? Winterlia, *Spreng.*

Sellowia, Roth
? Lithophila, *Swartz.*
Pteranthus, *Forsk.*
Louichea, Herit.
Cometes, *Burm.*
Saltia, R. Br.
Pollichia, *Soland.*
Neckeria, Gmel.
Meerburgia, Mönch.
Telephium, *Tournef.*
Lœflingia, *Linn.*

Cerdia, *Moç. et Sess.*
Polycarpon, *Löffl.*
Trichlis, Hall.
Arversia, *Cambess.*
Hapalosia, Wight et A.
Ortegia, *Löffl.*
Ortega, DC.
Juncaria, Clus.
Stipulicida, *L. C. Rich.*
Polycarpæa, *Lam.*
Hagea, Vent.

Mollia, Willd.
Lahaya, Röm et Schlt.
Hyala, Herit.
Aylmeria, *Mart.*
Spergularia, *Pers.*
Lepigonum, Fries.
Stipularia, Haw.
Delila, Dumort.
Balardia, Cambess.
Spergula, *Linn*
Drymaria, *Willd.*

NUMBERS. GEN. 24. Sp. 100?

Amaranthaceæ.
POSITION.—Portulaceæ.—ILLECEBRACEÆ.—Caryophyllaceæ.

Fig. CCCXLII.—Paronychia capitata. 1. a section of a flower; 2. of the ovary; 3. ripe seed; 4. a section of it.—*Nees.*

ORDER CXC. PORTULACEÆ.—PURSLANES.

Portulaceæ, *Juss. Gen.* 313. (1789) *in part; A. St. Hil. Mém. Plac. Cent.* 42. (1815); *DC. Prodr.* 3. 351. (1828); *DC. Mém. de la Soc. d'Hist. Nat. de Paris, Aug.* (1827); *Endl. Gen.* ccvi.; *Meisner Gen.* 130.; *Wight Illustr.* 2. 41.

DIAGNOSIS.—*Silenal Exogens, with the calyx and corolla unsymmetrical* $\frac{2}{5}$, *the latter usually conspicuous, amphitropal ovules, and alternate succulent leaves without stipules.*

Succulent shrubs or herbs. Leaves alternate, seldom opposite, entire, without stipules, often with bundles of hairs in their axils. Flowers axillary or terminal, usually ephemeral, expanding only in bright sunshine, unsymmetrical in their calyx and corolla. Sepals 2, cohering by the base. Petals generally 5, either distinct or cohering in a short tube. Stamens inserted along with the petals irregularly into the base of the calyx or hypogynous, variable in number, all fertile, sometimes opposite the petals; filaments distinct; anthers versatile, with 2 cells, opening lengthwise. Carpels 3 or more, combined into a one-celled ovary, which is usually free (or partially adherent); style single or none; stigmas several, much divided; ovules amphitropal. Capsule 1-celled, dehiscing either transversely or by valves, occasionally 1-seeded and indehiscent. Seeds numerous, if the fruit is dehiscent; attached to a central placenta; albumen farinaceous; embryo curved round the circumference of the albumen, with a long radicle next the hilum.

Ovary partially adherent in some Portulacas.

In general the Purslanes are easily recognised by their succulent condition and gay ephemeral flowers; but in some the flowers are inconspicuous, and in others the succulence inconsiderable. They, in such cases, would have little to distinguish them from Cloveworts (Caryophyllaceæ), except their 2-leaved calyx, and that in truth, combined with the other characters, furnishes the essential mark of the Order. Endlicher, however, extends the limits of the Purslane group much beyond this, admitting a number of perigynous genera whose flowers are quite symmetrical. These are spoken of elsewhere. In his view, the difference between Purslanes and Cloveworts consists mainly in this, that the former have the stamens alternate with the sepals when they are equal to them in number, and the latter opposite under the same circumstances. But in Orders where the number of stamens is sometimes indefinite and sometimes has no sort of relation to the sepals, as is, in fact, the case with the whole Portulaceous Order as it stands here, it is plain that such a distinction has no existence. I have, therefore, thought it advisable to reject a portion of Endlicher and Fenzl's Purslanes, in which the perigynous insertion is very marked, placing them in the Ficoidal Alliance, while all his hypogynous genera with symmetrical flowers are conveniently arranged among the Cloveworts. The principal deviation from the general features of the Order strictly limited, consists in some species having the ovary partially adherent, and the stamens, therefore, perigynous. Such instances seem to connect the Order with the genera just alluded to. From Knotworts (Illecebraceæ) the monospermous genera of Purslanes are distinguished by the want of symmetry in their flowers, and by the stamens being opposite the petals instead of the sepals. De Candolle remarks, that Purslanes

Fig. CCCXLIII.

Fig. CCCXLIII.—Portulaca australis.—*Endlicher.* 1. a flower; 2. an expanded corolla; 3. a pistil; 4. a ripe fruit; 5. a section of it.

have been more than once compared to Primworts; and the same author states, in another place, that the genera with indefinite stamens and hairy axils approach the Torch-thistles.

These plants inhabit the Cape of Good Hope and South America, 1 species occurs in Guinea, 2 in New Holland, 1 in Europe, and the remainder in various parts of the world. They are always found in dry parched places.

Insipidity, want of smell, and dull green colour in the foliage, are usual qualities of this Order, of which the only species of any known use is the common Purslane (ανδραχνη) of the Greeks, (Portulaca oleracea, *L.*), which has been used from all antiquity as a potherb, and in salads, on account of its cooling and antiscorbutic qualities; the ancients thought the seeds, steeped in wine, to be emmenagogue. Talinum patens in Brazil, and Claytonia perfoliata in North America, and some Calandrinias have similar qualities. The tuberous root of Claytonia tuberosa, a Siberian plant, is eaten where it grows wild. Many of the species are beautiful objects on account of their large gay flowers.

GENERA.

Portulaca, *Tourn.*
Meridiana, L.
Merida, Neck.
Lamia, Vand.
Portulacaria, *Jacq.*
Hænkea, Salisb.
Anacampseros, *Linn.*
Telephiastrum, Dill.
Rulingia, Ehrh.
Avonia, E. Mey.
Grahamia, *Gill.*
Xeranthus, Miers.
Talinum, *Adans.*
Phemeranthus, Raf.
Talinastrum, DC.
Talinellum, DC.
Eutmon, Raf.
Calandrinia, *H. B. K.*
Cosmia, Domb.
Cistanthe, Spach.
Tegneria, Lilj.
Rhodopsis, Lilj.
Phacosperma, Haw.
Geunsia, Flor. mex.
Claytonia, *Linn.*
Limnia, Linn.
Monocosmia, *Fenzl.*
Montia, *Michel.*
Cameraria, Dill.
Alsinoides, Vaill.
Calyptridium, *Nutt.*
? Ullucus, *Lozan.*
? Leptrinia, *Raf.*

NUMBERS. GEN. 12. Sp. 184.

Primulaceæ.
POSITION.—Illecebraceæ.—PORTULACEÆ.—Caryophyllaceæ.
Mesembryaceæ.
Cactaceæ.

ORDER CXCI. POLYGONACEÆ.—BUCKWHEATS.

Polygoneæ, *Juss. Gen.* 82. (1789); *R. Brown, Prodr.* 418. (1810); *Bentham in Linn. Trans.* (1836); *Endl. Gen.* ciii.—Polygonaceæ, *Ed. pr.* (1836); *Meisner Gen.* 316.

DIAGNOSIS.—*Silenal Exogens, with no corolla, and a calyx usually coloured, an orthotropal ovule, and a usually triangular nut.*

Herbaceous plants, rarely shrubs. Leaves alternate, their stipules cohering round the stem in the form of an ochrea; when young, rolled backwards, occasionally wanting.

Fig. CCCXLIV.

Flowers occasionally unisexual, often in racemes. Calyx free, often coloured, imbricated in æstivation. Stamens very rarely perigynous, usually definite and inserted in the bottom of the calyx; anthers dehiscing lengthwise. Ovary free, usually formed by the adhesion of 3 carpels, one-celled, with a single erect ovule, whose foramen always points upwards; styles or stigmas as many as the carpels of which the ovary consists; ovule orthotropal. Nut usually triangular, naked, or protected by the calyx. Seed with farinaceous albumen, rarely with scarcely any; embryo inverted, generally on one side, sometimes in the axis; radicle superior, long.

Brown remarks, that "the erect ovulum with a superior radicle together afford the most important mark of distinction between Polygonaceæ and Chenopodiaceæ, a character which obtains even in the genus Eriogonum, in which there is no petiolar sheath, and scarcely any albumen, the little that exists being fleshy;" to which may be added, that their orthotropal ovule divides them from all the other Orders of the Silenal Alliance. Generally speaking, however, the cohesion of the scarious stipules into a sheath, technically called an ochrea, or boot, is sufficient to distinguish Buckwheats from the neighbouring Orders. Their affinity, moreover, does not appear to be so close with Chenopods as with Cloveworts, for they have the very important peculiarity that their ovary is formed by the consolidation of 3 carpellary leaves touching each other in a valvate manner, and thus producing a triangular form in the ripe fruit; and if even this is departed from, yet the ovary is undoubtedly compound and not simple as in Chenopods. Bentham admits two tribes, Polygoneæ, which have loose flowers and ochreate stipules, and Eriogoneæ which have flowers in involucres and usually no stipules. The latter bring them near Nyctagos.

Fig. CCCXLIV.—Polygonum lapathifolium. 1. a flower cut open; 2. a vertical section of the seed; 3. a flower of P. Convolvuli; 4. a transverse section of a seed; 5. a diagram of the flower of Rumex crispus; 6. a vertical section of its ripe fruit, &c.; 7. its fruit.

There are few parts of the world which do not acknowledge the presence of plants of this Order. In Europe, Africa, North America, and Asia they occupy ditches, hedges, and waste grounds, in the form of Docks and Persicarias ; the fields, mountains, and heaths as Sorrels and trailing or twining Polygonums ; in South America and the West Indies they take the form of Coccolobas or Seaside-grapes ; in the Levant of Rhubarbs ; and even in the desolate regions of the North Pole they are found in the shape of Oxyria.

Sorrel on the one hand, and Rhubarb on the other, may be taken as the representatives of the general qualities of this Order. While the leaves and young shoots are acid and agreeable, the roots are universally nauseous and purgative. To these two qualities is to be superadded a third, that of astringency, which is found in a greater or less degree in the whole Order, but which becomes in Coccoloba uvifera so powerful as to rival gum Kino in its effects. Some of the Polygonums are also acrid, as the P. Hydropiper, which is said to blister the skin, and there is a species of Polygonum, called Cataya in the language of the Brazilian Indians, which has a very bitter peppery taste, an infusion of the ashes of which is used to purify and condense the juice of the sugar-cane, and is employed on the Rio St. Francisco with advantage in the disease called O Largo, an enlargement of the colon, caused by debility. Oxalic acid is copiously formed in both Docks and Rhubarbs; the latter moreover contains nitric and malic acids in abundance, and it is these which give an agreeable taste to the stalks of the latter when cooked, but which also render them so ill-suited to the digestion of some persons. For the facts concerning the qualities and origin of the Rhubarbs in medical use, the reader may consult Royle, Guibourt, Pereira, Geiger, Endlicher, and the *Flora Medica.* It seems probable that some at least of the Turkey Rhubarb is Rheum palmatum, that R. undulatum is also largely collected, and that R. Emodi and Webbianum furnish the Rhubarb used in the hospitals of India. Goebel positively contradicts the statement made by some writers that Rheum leucorhizum yields a fine sort of Rhubarb ; he says that it has an insipid slimy taste, not at all like that of Rhubarb.—*Ann. Ch.* 1. 118. Before this sort of drug was so common, the roots of Rumex alpinus were employed in its stead, under the name of Monk's Rhubarb ; it is however much less active. The Rheum Ribes, called Riwasch or Ribas in the East, furnishes the Arabs with an acidulous medicine, and its leaf-stalks are used in the preparation of sherbet. Docks are species of Rumex ; their prevailing character is astringency, which has given them some celebrity as remedies for diarrhœa, and as stomachics. R. Patientia (λαπαθον κηπευτον), although now expelled from gardens, was once esteemed as a subacrid potherb, and its roots were used as laxatives. Sorrels, whose acidity is chiefly owing to oxalic acid, are all species of this same genus ; the most esteemed among them for garden purposes is R. scutatus. A legion of species forms the genus Polygonum, celebrated in various ways. Some are used in dyeing, especially P. tinctorium, which yields a blue hardly inferior to indigo, and is largely cultivated for it in France and Flanders. Of P. Hydropiper the leaves are so acrid as to act as vesicants ; it is reputed to be a powerful diuretic, but to lose its activity by drying, on which account it requires to be used fresh ; it will dye wool yellow. P. Bistorta is a useful astringent ; the decoction may be employed in gleet and leucorrhœa, as an injection, as a gargle in relaxed sore throat and spongy gums, and as a lotion to ulcers attended with excessive discharge ; internally it has been employed, combined with Gentian, in intermittents ; it may also be used in passive hæmorrhages and diarrhœa. Several of the Brazilian Polygonums are said by Martius to be useful as astringents, and to be employed in the treatment of syphilis. The nuts of Fagopyrum esculentum, or Buckwheat, tataricum, and others, are used as food for the sake of their mealy albumen ; those of P. aviculare are said to be powerfully emetic and purgative ; but this is doubted by Meisner. The seeds of Polygonum barbatum are used as medicine by Hindoo practitioners, to ease the pain of griping in the colic. The leaves of P. hispidum are said by Humboldt to be substituted in South America for Tobacco. Coccoloba americana, remarkable for the succulent violet calyx in which its nuts are enveloped, is on that account called the Seaside-grape in the West Indies, and yields an extremely astringent extract ; its wood dyes red ; its currant-like berries are acid, pleasant, and eatable. The root of Calligonum Pallasia, a leafless shrub found in the sandy steppes of Siberia, furnishes from its roots, when pounded and boiled, a gummy nutritious substance, resembling Tragacanth, on which the Calmucks feed in times of scarcity, while they chew the acid branches and fruits to allay their thirst. Muhlenbeckia adpressa is stated by Mr. Backhouse to have clusters of currant-like fruits of a sweetish taste, which have been made into pies and puddings in the penal settlements of Australia. The trunk and branches of Triplaris americana are chambered like those of the Cecropia, or Trumpet-tree, and serve for the habitation of light-brownish ants, which inflict a most painful bite.—*Schomburgk, in Ann. N. H.* 1. 266.

GENERA.

I. ERIOGONEÆ.—*Benth.*

Pterostegia, *Fisch. et M.*
Mucronea, *Benth.*
Chorizanthe, *R. Br.*
Eriogonum, *L. C. Rich.*
  Espinosa, Lagasc.

II. POLYGONEÆ.—*Benth.*

Oxyria, *Hill.*
  Donia, R. Br.
Rheum, *Linn.*
  Rhabarbarum, Tourn.
Königia, *L.*
Polygonum, *L.*
  Bistorta, Tournef.
  Amblygonon, Meisn.
  Lagunea, Lour
  Persicaria, Tournef.
  Towara, Adans.
  Antenoron, Rafin.
  Echinocaulon, Meisn.
  Cephalophilon, Meisn.
  Didymocephalon, Meis.
  Corymbocephalon, Mn.
  Aconogonon, Meisn.
  Avicularia, Meisn.
  Tiniaria, Meisn.
Fagopyrum, *Tournef.*
Ampelygonum, *Lindl.*
Oxygonum, *Burch.*
Calligonum, *L.*
  Polygonoides, Tournef.
Calliphysa, *Fisch. et Mey.*
Pterococcus, *Pall.*
  Pallasia, Linn. fil.
Coccoloba, *Jacq.*
Mühlenbeckia, *Meisn.*
Ceratogonum, *Meisn.*
Emex, *Neck.*
  Vibio, Mönch.
  Centropodium, Burch.
Rumex, *L.*
Tragopyrum, *Bieb.*
Gonopyrum, *Fisch.*
Polygonella, *Mx.*
  Lyonia, Raf.
Atraphaxis, *L.*

III. TRIPLAREÆ.—*Meyer.*

Podopterus, *H. et B.*
Rupprechtia, *Meyer.*
Triplaris, *L.*
  Blochmannia, Weigelt.
  Velasquezia, Bertol.

IV. BRUNNICHEÆ.—*Meisn.*

Brunnichia, *Banks.*
  Fallopia, Adans.
Antigonon, *Endl.*

NUMBERS. GEN. 29. Sp. 490.

Nyctaginaceæ.
Chenopodiaceæ.
POSITION.—Illecebraceæ.—POLYGONACEÆ.—Caryophyllaceæ.

ALLIANCE XXXVIII. *CHENOPODALES.*—THE CHENOPODAL ALLIANCE.

DIAGNOSIS.—*Hypogynous Exogens, with monochlamydeous flowers, free central placentæ, an external embryo, either curved round or applied to the surface of a little mealy or horny albumen, solitary carpels, or, if more than one, distinct.*

With these plants, the greater part of which consists of species with inconspicuous flowers, and often with scarcely more floral organs than are absolutely necessary to secure the perpetuation of the race, we have a transition which cannot be mistaken, to the more elaborately constructed Alliances hereafter to be noticed. Nettles and Chenopods are in such strict relationship that we can scarcely say wherein the difference consists in certain cases, unless we refer to the internal structure of the seed, and then indeed we find Chenopods with amphitropal ovules, mealy albumen, and radicle directed towards the base of the fruit, while Nettles have orthotropal or anatropal ovules, fleshy albumen, and a radicle directed towards the apex of the fruit. Both have stamens opposite the sepals of an inconspicuous petalless calyx, and both have their fruit composed of a single, perfectly simple, carpellary leaf; we even find that in some cases among the Urtical Alliance the circular, or spiral, embryo of Chenopods makes its appearance. (See p. 265).

Even as regards the distinction of the stamens and pistil there is a great similarity between the two races under consideration. For if all the Urtical Alliance consists of plants whose flowers are strictly unisexual (♂ ♀), so also does the Chenopodal Alliance contain a great many species which are similarly constituted, notwithstanding that the tendency of the structure is towards a combination of the sexes (⚥). It is, in fact, among the Chenopodal Alliance that we find most exceptions to the distinctions between diclinous and bisexual organisation; as might be expected, where Orders run so much together otherwise.

From the Silenal Alliance this is known by the absolute simplicity of the ovary, and by no other positive mark: there may be several ovaries present in the same flower, but they are then distinct from each other. It is, however, to be remarked that Chenopodals have, in no case whatever, a corolla, while in a large part of the Silenals, petals are obviously present. Nyctagos, a portion of this Alliance, seem as if they were attempting to emulate the Silenals; for, although they have only a calyx, yet that calyx does, in many instances, assume altogether the ordinary colour and texture of a corolla.

NATURAL ORDERS OF CHENOPODALS.

Sepals united into a long (often coloured) plaited tube, which separates from its base, the latter becoming hard, and forming a spurious pericarp	192. NYCTAGINACEÆ.
Sepals separate, flat. Stamens alternate with the sepals or 00. *Carpels several (or* 1)	193. PHYTOLACCACEÆ.
Sepals separate or nearly so, flat. Stamens opposite the sepals. Anthers often 1-*celled. Ovary* 1, *often several-seeded. (Flowers scarious, surrounded by imbricated bracts)*	194. AMARANTACEÆ.
Sepals separate, or nearly so, flat. Stamens opposite the sepals. Anthers 2-*celled. Ovary* 1, *always one-seeded. (Flowers herbaceous, naked)*	195. CHENOPODIACEÆ.

ORDER CXCII. NYCTAGINACEÆ.—NYCTAGOS.

Nyctagines, *Juss. Gen.* 90. (1789); *R. Brown Prodr.* 421. (1810); *Bartl. Ord. Nat.* 109; *Endl. Gen.* civ.; *Meisner, p.* 318.

DIAGNOSIS.—*Chenopodal Exogens, with a tubular often coloured calyx, which separates from its base, the latter becoming a hard spurious pericarp.*

Annuals or perennials, often with fleshy roots, or shrubs or trees, usually articulated at the tumid nodes. [The vascular system double; the central consisting of bundles scattered among the pith, the circumferential of bundles not adhering to each other.—*Unger.*] Leaves opposite, and almost always unequal; sometimes alternate. Flowers axillary or terminal, clustered or solitary, sometimes imperfect, having an involucre which is either common or proper, in one piece or in several pieces, sometimes minute, but more generally very large, and sometimes gaily coloured. Calyx tubular, somewhat coloured, contracted in the middle; its limb entire or toothed, plaited in æstivation, becoming indurated at the base, and losing the limb which is deciduous. Stamens definite, hypogynous, sometimes on one side; anthers 2-celled. Ovary superior, with a single erect ovule, whose foramen is inferior; style 1, terminal or somewhat lateral; stigma 1. Fruit a thin utricle, inclosed within the enlarged persistent base of the calyx. Seed without its proper integuments, its testa being coherent with the utricle; embryo with foliaceous cotyledons, wrapping round floury albumen; radicle inferior; plumule inconspicuous.

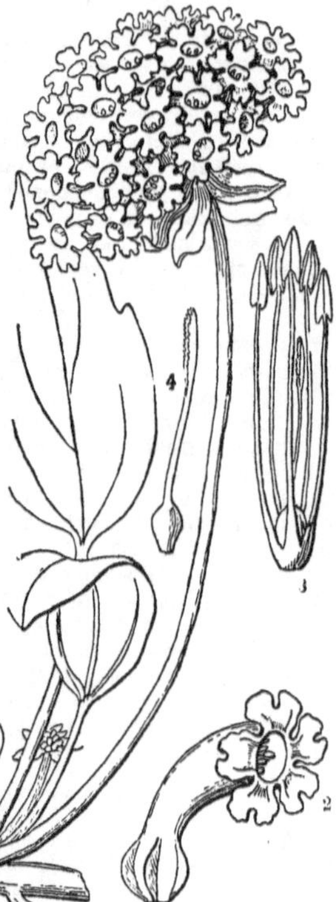

Here we have a race of plants, of which the common Marvel of Peru is the type, whose affinity is clearly with the Chenopods and Amaranths, from which it is distinguished by the curious property of converting the base of its thin membranous tubular calyx into a tough or bony shell which acts as a pericarp to the seed, whose real pericarp is but a membrane. Moreover, the tubular calyx, the limb of which is plaited in æstivation, together with the curved embryo and farinaceous albumen, at all times distinguish Nyctagos; add to which, the articulations of the stem are often tumid, as in Cranesbills. Schleiden states (*Wiegman's Arch.* 1839), that the wood figured at t. 42,

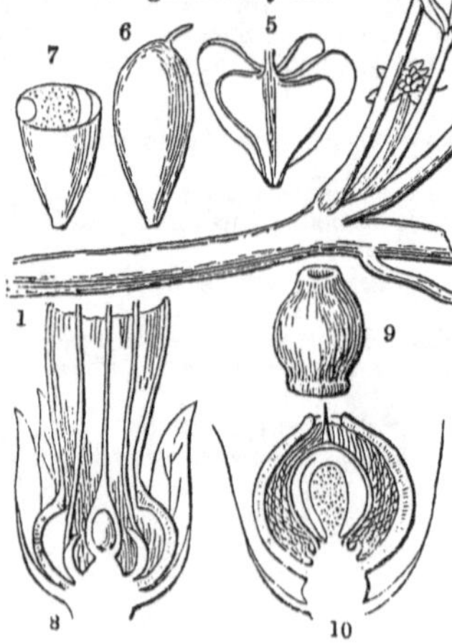

Fig. CCCXLV.

Fig. CCCXLVI.

Fig. CCCXLV.—1. Abronia mellifera; 2. a flower separate; 3. its stamens and pistil; 4. the pistil separate; 5. the fruit; 6. seed magnified; 7. a cross section of it; 8. the lower portion of the flower of Mirabilis Jalapa; 9. its fruit; 10. a perpendicular section of it.

Fig. CCCXLVI.—Pisonia grandis. 1. a flower; 2. a pistil; 3. a cluster of fruits.

p. 100, of the third edition of my *Introduction to Botany*, is that of a Pisonia, a genus belonging to this Order; and he explains its singular structure by supposing it to consist of numerous vascular bundles, which continue to be developed until they form at last an almost continuous mass. The parenchyma which separated them is thereby compressed into insulated patches, which are scattered through the completely formed wood in little narrow vertical cords (*strœnge*), which, as regards their origin, may be termed vertical medullary rays; and he finds a similar structure in Amaranths, Beta, Atriplex, Chenopodium, and Peppers. As this organisation appears from Schultz (*Nat. Syst.* fig. 1. 2. 5. 6.), to be present also in Boerhaavia and Mirabilis, it would seem to be characteristic of the Order. I however find a very different structure in Boerhaavia repanda, which has zoned wood, although its rings are broken by the introduction of vertical cords of cellular tissue: its pith contains many fistulæ of lax, soft, spheroidal, cellular tissue, surrounded by smaller, harder, more cubical tissue, which passes into the medullary plates. The scrambling Bugainvillea, with its large rose-coloured bracts, gathered into cones like those of a Hop, seems to indicate some analogy with Hempworts.

The species are natives of the warmer parts of the world in either hemisphere, scarcely extending far beyond the tropics, except in the case of the Abronias found in North-west America, and a few Boerhaavias in the Southern hemisphere.

In consequence of the generally purgative quality of the roots of species of this family, one of them (Mirabilis jalapa) was supposed to have been the true Jalap plant, which is, however, now known to be a mistake. *See* CONVOLVULACEÆ. The flowers of several species of Mirabilis are handsome, as are those also of some of the Abronias; but the greater part of the Order is composed of obscure weeds. The genus Pisonia consists of trees or shrubby plants agreeing in property with the Boerhaavias, of which the species have generally emetic and purgative roots. Boerhaavia hirsuta is employed in icterus; B. tuberosa, a doubtful plant of the Order, called Yerba de la purgacion in Peru, is regarded as an antisyphilitic, but it is also employed as a culinary vegetable. Boerhaavia procumbens, an East Indian species, is reckoned antifebrile. According to Aublet, the root of B. decumbens (called Hogmeat in Jamaica), is emetic, and called Ipecacuanha in Guiana. Schomburgk states that it is astringent, and used in the form of decoction in dysentery. Mirabilis dichotoma, the Marvel of Peru, called by the French Fleur de quatre heures, and M. longiflora, two plants now common in our gardens, are very drastic. M. suaveolens, a species with an Anise flavour, is employed in Mexico against diarrhœa and rheumatic pains.

GENERA.

Boerhaavia, *Linn.*	*Calyxhymenia*, Orteg.	Bugainvillea, *Commers.*	Pisonia, *Plum.*
Collignonia, *Endl.*	*Calymenia*, Nutt.	*Josepha*, Fl. Flum.	*Calpidia*, Thouars.
Abronia, *Juss.*	*Vitmannia*, Turr.	Reichenbachia, *Spreng.*	*Bessera*, Flor. Flum.
Tricratus, Herit.	Allionia, *Linn.*	Salpianthus, *H. et B.*	*Pallavia*, Flor. Flum.
Mirabilis, *Linn.*	*Wedelia*, Löffl.	*Boldoa*, Cavan.	*Torrubia*, Flor. Flum.
Nyctago, Juss.	Okenia, *Schiede.*	Neea, *Ruiz et Pav.*	*Columella*, Flor. Fl.
Jalapa, Tournef.	Tricycla, *Cavan.*	*Mitscherlichia*, Kunth.	? Epilithes, *Blume.*
Oxybaphus, *Herit.*			

NUMBERS. GEN. 14. Sp. 100.?

Polygonaceæ.
POSITION.—Amarantaceæ.—NYCTAGINACEÆ.——
Cannabinaceæ ?

ORDER CXCIII. PHYTOLACCACEÆ.—PHYTOLACCADS.

Phytolacceæ, *R. Brown in Congo*, 454. (1818); *Bartl. Ord. Nat. p.* 299. (1830); *Meisner Gen.* 322.—Phytolaccaceæ, *Ed. Pr.* clvii.; *Endl. Gen.* ccviii.—Rivinaceæ, *Agh. Martius Conspectus, No.* 91. (1835).

DIAGNOSIS.—*Chenopodal Exogens, with separate flat sepals, stamens either* 00 *or alternate with the sepals, and one or several carpels.*

Under-shrubs or herbaceous plants. Leaves alternate, entire, without stipules, often with pellucid dots. Flowers racemose, perfect, regular, or somewhat irregular, very variously arranged. Calyx of 4, or 5, imbricated leaves, which sometimes assume altogether the appearance of distinct petals. Stamens hypogynous or nearly so, either indefinite, or if equal to the number of the divisions of the calyx alternate with them; anthers 2-celled, opening lengthwise. Carpels solitary, or several, distinct or but partially combined, each containing 1 ascending ovule, which is either amphitropal or campylotropal; styles and stigmas equal in number to the carpels. Fruit baccate or dry, indehiscent. Seeds ascending, solitary, with a cylindrical embryo curved round mealy albumen, the radicle being next the hilum.

Fig. CCCXLVII.

The small cluster of genera called Phytolaccads, forms an Order nearly related to Chenopods and Buckwheats, from the first of which it is distinguished by the numerous carpels and the stamens exceeding the number of divisions of the calyx, or being alternate with them; or if the carpel is solitary, by the calyx being petaloid, a circumstance which never occurs in Chenopods. From Buckwheats it is known by the radicle being turned towards the hilum, the want of stipules, and the perfect simplicity of the ovaries. Rivina, which has the albumen very much reduced in quantity, brings into the same neighbourhood Phytolaccads and Petiveriads. Brown remarks that these two Orders, widely as they differ in the structure of the pistil, are connected by a species of Phytolacca related to P. abyssinica, in which the 5 cells are so deeply divided that they merely cohere by their inner angles; and also by Gisekia, which has 5 distinct ovaries. Endlicher unites to this Order both Petiveriads and Gyrostemons, both of which will be found elsewhere in this work. The numerous free carpels seem to point out some kind of relation to the Ranal, and their verticillate arrangement to the Malval Alliances.

Many are natives of either America, within or without the tropics; others of Africa and India. Phytolacca decandra is naturalised in some of the southern parts of Europe.

The species are generally acrid, but that property is inconsiderable in some, and is dispersed by heat in others. A tincture of the ripe berries of Phytolacca decandra, or Pocan, seems to have acquired a well-founded reputation as a remedy for chronic and syphilitic rheumatism, and for allaying syphiloid pains. By some it is said to be more valuable than Guaiacum. It has had no inconsiderable celebrity as a remedy for cancer, but is no longer esteemed, and it is probable that it was only found serviceable in ill-conditioned sluggish ulcers, which are too frequently mistaken for real cancer. Its pulverised root is an emetic. A spirit distilled from the berries is stated to have killed a dog in a few minutes. According to De Candolle, the plant is also a purgative; but it acts so violently, and is accompanied by such ambiguous narcotic symptoms, as not to be at all calculated for internal use. Bigelow says that externally applied it causes heat and smarting; he found it useful in psora and tænia capitis. The leaves are extremely acrid, but the young shoots, which lose this quality by boiling in water, are eaten in the United States as Asparagus, and Dr. Royle tells us that Phytolacca

Fig. CCCXLVII.—Phytolacca decandra. 1. its flower; 2. its stamens and pistil; 3. a section of a seed.

acinosa is also so employed in the Himalayas. P. drastica, a Chilian plant, with a turnip-shaped root, is said to have a most violent action as an evacuant. Berries of the Rivina yield a rich red dye.

GENERA.

Mohlana, *Mart.*	*Linscotia*, Adans.	*Schollera*, Swartz.	Phytolacca, *Tournef.*
Hillera, Fl. Flum.	*Dicarpæa*, Presl.	*Ancistrocarpus*, Knth.	*Kermesia*, Endl.
Rivina, *Plum.*	*Gaudinia*, Gay.	*Aphananthe*, Link.	*Pircunia*, Berter.
Solanoides, Tournef.	Semonvillea, *Gay.*	Giesekia, *Linn.*	*Anisomeria*, Don.
Limeum, *Linn.*	Microtea, *Swartz.*	*Kœlreutera*, Murr.	Ercilla, *Adr. Juss.*
Stegnosperma, *Benth.*	*Potamophila*, Schrank.	*Miltus*, Lour.	*Bridgesia*, Hook. et A.

NUMBERS. GEN. 9. Sp. 60. ?

Polygonaceæ.
POSITION.————————.—PHYTOLACCACEÆ.—Chenopodiaceæ.
Ranunculaceæ ?
Malvaceæ ?

SURIANACEÆ, (*Wight and Arnott, Prodr.* 1. 360; *Ed. pr.* cvii.) is the name given to a supposed Order of plants represented by a solitary species, Suriana maritima found on the coast of all tropical countries. In the last edition of this work it is thus described.——A woody plant with alternate leaves without stipules. Hairs capitate, jointed. Flowers racemose, ☿. Calyx 5-parted, slightly imbricated. Petals the like number, equal, shortly clawed. Stamens indefinite, hypogynous, placed in a single row; filaments subulate; anthers roundish, incumbent, bursting internally by two longitudinal fissures. Carpels 5, distinct, attached to a very short gynobase, 1-celled with 2 ascending collateral ovules; styles rising from near the base of the carpels; stigmas simple. Pericarp woody. Seed solitary, erect, compressed. Embryo annular, without albumen, terete, with the cotyledons about the same length as the radicle which is turned to the hilum.——This is one of those obscure forms, whose relationship can hardly be decided satisfactorily until some allied genus shall have been discovered; for it seems improbable that the genera Heterodendron and Cneorum, with which it has been associated, should have any real affinity. In some respects it may be compared with Coriariaceæ and Cranesbills, but its annular embryo is so peculiar as to indicate a somewhat different relationship; and this, indeed, has led Dr. Wight (*Illustr.* 2. 45.) to suggest an affinity to Phytolaccads, especially to the genus Gisekia, "which corresponds accurately in the character of the ovary and fruit, and even of the seed." The presence of petals in Suriana, he considers unimportant; in which I agree with him. A more grave objection lies in the position of the stamens, which, in Phytolaccads, are alternate with the sepals; but in Suriana, according to Mr. Arnott, they are opposite to them; and this Endlicher confirms. The accompanying figure is taken from a drawing made in 1820, and it may serve to assist Botanists in coming to some conclusion upon this point.

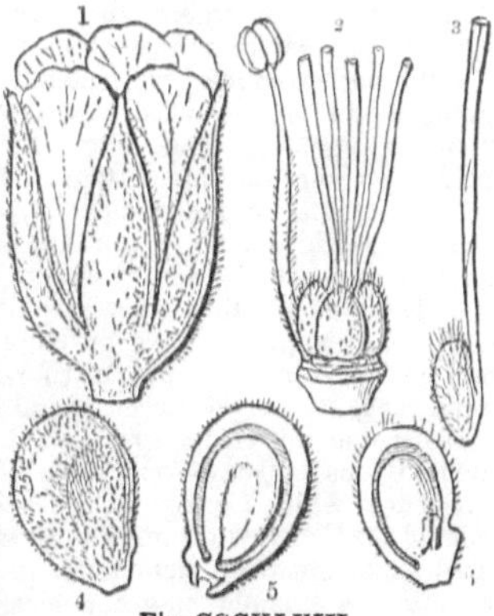

Fig. CCCXLVIII.

Fig. CCCXLVIII.—Suriana maritima. 1. a flower; 2. the pistil with one stamen; 3. a carpel; 4. a ripe seed-vessel; 5. a section of it; 6. a section of an ovary.

ORDER CXCIV. AMARANTACEÆ.—AMARANTHS.

Amaranthi, *Juss. Gen.* 87. (1789).— Amarantaceæ, *R. Brown, Prodr.* 413. (1810); *Von Martius Monogr.* (1826); *Endl. Gen.* cii.; *Meisn. Gen.* 316.—Polycnemeæ, *Moq. Tand. in Ann. Sc. n. s.* 7. 41.

DIAGNOSIS.—*Chenopodal Exogens, with separate sepals opposite the stamens, usually one-celled anthers, a single ovary often containing several seeds, and scarious flowers buried in imbricated bracts.*

Herbs or shrubs. Leaves simple, opposite or alternate, without stipules. Flowers in heads or spikes, usually coloured, occasionally unisexual, generally hermaphrodite. Pubescence simple, the hairs divided by internal partitions. Sepals 3 or 5, hypogynous, scarious, persistent, herbaceous or coloured, distinct or united at the base, all equal, sometimes with 3 more interior than the others, the back one being sometimes dissimilar, occasionally with 2 bractlets at the base, and generally immersed in dry coloured bracts. Stamens hypogynous, either 5 opposite the sepals, or some multiple of that number, either distinct or monadelphous, occasionally partly abortive; anthers either 2-celled or 1-celled. Ovary single, free, 1- or few-seeded; the ovules amphitropal, hanging from a free central funiculus; style 1 or none; stigma simple or compound. Fruit a membranous utricle, sometimes a caryopsis or berry. Seeds lentiform, pendulous; testa crustaceous; albumen central, farinaceous; embryo curved round the circumference; radicle next the hilum.

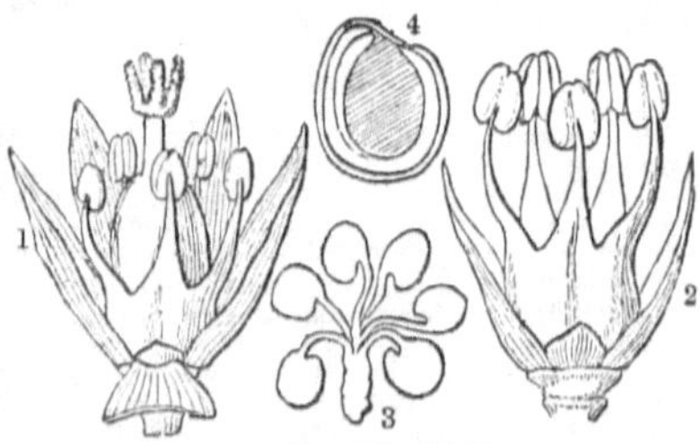

Fig. CCCXLIX.

Distinct as this Order appears to be from Chenopods in habit, especially if we compare such a genus as Gomphrena with Chenopodium itself, yet it is so difficult to define the differences which distinguish the two Orders, that, beyond habit, nothing certain can be pointed out. Brown remarks (*Prodr.* 413.) that he has not been able to ascertain any absolute diagnosis to distinguish them by; for the hypogynous insertion attributed to their stamens is not only not constant in the Order, but is also found in Chenopods. Martius, in a learned dissertation upon the Order, describes Chenopods as being apetalous, and Amaranths as polypetalous, considering the bractlets of these latter as a calyx, and that which I call a calyx a corolla. But it seems to me that this view of their structure is not borne out by analogy, and that it is impossible to believe the floral envelopes of the two Orders to be of a different nature. Endlicher observes that, although no single character divides them, yet they may be known by several characters taken together: thus Gomphreneæ have one-celled anthers, and Celoseæ many seeds; of the remainder, which are most like Chenopods, some differ from Salicornids in the stem not being jointed, others from Atriplicids in the ♂ and ♀ flowers not being different. Bartling combines the whole in a single class, along with Caryophylleæ, Phytolaccaceæ, Scleranthaceæ, and Illecebraceæ; and there is no doubt of the affinity borne to each other by all these, as is pointed out by their habit and by the structure of their seeds. Illecebraceæ are in fact only known by their petals, compound ovary, and great membranous stipules. It has been stated by Schleiden that the singular mixed wood of Phytolaccads and some Chenopods also occurs in Amaranthus viridis. I do not, however, find it in the woody species, such as Deeringia celosioides, Cometes abyssinica, and Desmochæta flavescens; but some tendency towards it seems to exist in Achyranthes arborescens. The point requires to be carefully investigated. Schultz describes the wood as being something between Peperomia and Piper. He says that the axis of Amaranthus contains very numerous fibrovascular bundles, but Achyranthes only 1 or 2.

Amaranths grow in crowds or singly, either in dry, stony, barren stations, or among thickets upon the borders of woods, or a few even in salt marshes. They are much more frequent within the tropics than beyond them, and are unknown in the coldest regions of the world. Of those known to Martius 53 are found in tropical Asia, 105 in

Fig. CCCXLIX.—Celosia longifolia. 1. a flower; 2. the stamens; 3. the ovules; 4. a section of the seed.

tropical America, but 5 in extra-tropical Asia, and but 21 in extra-tropical America; 5 are natives of Europe, 28 of New Holland, and 9 of Africa and its islands.

Many of the species are used, with the addition of Lemon-juice, as potherbs, on account of the wholesome mucilaginous qualities of the leaves. A. viridis leaves are employed externally as an emollient poultice. The bitter and acrid leaves of Deeringia celosioides are used against the measles in Java. Achyranthes aspera and fruticosa are administered in India in dropsical cases. The flowers of the Cockscomb, Celosia cristata, are astringent, and are exhibited in Asia in cases of diarrhœa, blenorrhœa, excessive menstrual discharges, hæmatesis, and similar disorders. The seeds of Amaranthus frumentaceus (Kiery), and of A. Anardhana, are gathered as corn crops in India. Achyranthes globulifera and Amaranthus debilis, are both used in Madagascar in the form of infusion, as a cure for syphilis. Amaranthus obtusifolius is said to be diuretic. Several are objects of interest with gardeners, for the beauty of their colouring and the durability of their blossoms. Gomphrena officinalis and macrocephala have a prodigious reputation in Brazil, where they are called Para todo, Perpetua, and Raiz do Padre Salerma : as the first of these names imports, they are esteemed useful in all kinds of diseases, especially in cases of intermittent fever, colic, and diarrhœa, and against the bite of serpents. The root is considered a stimulating tonic.

GENERA.

Suborder I.—GOMPHRENEÆ.

Iresine, *Willd.*
Rosea, Mart.
Crucita, Löffl.
Philoxerus, *R. Br.*
Gossypianthus, *Hook.*
Trommsdorfia, *Mart.*
Alternanthera, *Forsk.*
Allaganthera, Mart.
Teleianthera, *R. Br.*
Bucholzia, Mart.
Brandesia, *Mart*
Mogiphanes, *Mart.*
Gomphrena, *Linn.*
Schultesia, Schrad.
Bragantia, Vandell.
Serturnera, Mart.
Pfaffia, Mart.
Hebanthe, Mart.
Fröhlichia, *Mönch.*
Hoplotheca, Nutt.
Xerosiphon, *Turcz.*

Suborder II.—ACHYRANTHEÆ.

Polycnemum, *L.*
Hemichroa, *R. Br.*
Psilotrichum, *Blume.*
?*Leiospermum*, Wall.
Trichinium, *R. Br.*
Sericocoma, *Fenzl.*
Kyphocarpa, Fenzl.
Euchroa, Fenzl.
Ptilotus, *R. Br.*
Nyssanthes, *R. Br.*
Achyranthes, *Linn.*
Centrostachys, *Wall.*
Aerua, *Forsk.*
Digera, *Forsk.*
Desmochæta, *DC.*
Pupalia, *Mart.*
Cyathula, Lour.
Syama, Jones.
Saltia, *R. Br.*
Polyscalis, *Wall.*
Albersia, *Kunth.*
Scleropus, *Schrad.*
Amaranthus, *L.*
Polychroa, Lour.
Chamissoa, *H. B. K.*
Charpentiera, Gaudich.
Allmania, R. Br.

Suborder III.—CELOSEÆ.

Cladostachys, *Don.*
Celosia, *L.*
Lestiboudesia, *Thouars.*
Hermbstädtia, *Rchb.*
Berzelia, Mart.
Langia, Endl.
Deeringia, *R. Br.*
* * *
Tryphera, *Blum.*
Lecanocarpus, *Nees.*
Acroglochin, Schrad.
Blitanthus, Rchb.
Hablitzia, *Bieberst.*

NUMBERS. GEN. 38. Sp. 282.

POSITION.—Chenopodiaceæ.—AMARANTACEÆ.————
Illecebraceæ.

ORDER CXCV. CHENOPODIACEÆ.—CHENOPODS.

Atriplices, *Juss. Gen.* 83. (1789).—Chenopodeæ, *Vent. Tabl.* 2. 253. (1799); *R. Brown Prodr.* 405. (1810); *C. A. Meyer in Led. Fl. Alt.* 1. 370. (1829); *Moquin Tandon in Ann. Sc. Nov. Ser.* 1. 203. (1834); *Endl. Gen.* ci.; *Moq. Tand. Monogr.* (1840).—Chenopodiaceæ, *Ed. pr.* clvi.; *Meisn. p.* 319.—Corisperma, *Moq. Tand.*—Cynocrambeæ, *Th. N. ab E. Gen. Pl. Europ.* (1835).

DIAGNOSIS.—*Chenopodal Exogens, with separate flat sepals opposite the stamens, 2-celled anthers, a single one-seeded ovary, and herbaceous naked flowers.*

Herbaceous plants or under-shrubs, sometimes jointed. Leaves alternate without stipules, occasionally opposite. Flowers small, ☿, sometimes ♂-☿-♀, frequently ♂ ♀.

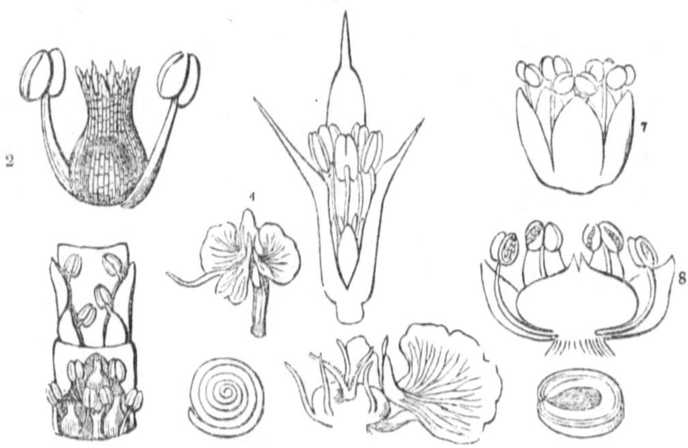

Fig. CCCL.

Calyx deeply divided, sometimes a little tubular at the base, persistent, with an imbricated æstivation. Stamens inserted into the base of the calyx, opposite its segments, and equal to them in number, or fewer. Ovary single, superior, or occasionally adhering to the tube of the calyx, with a single amphitropal ovule attached to the base of the cavity; style in 2 or 4 divisions, rarely simple; stigmas undivided. Fruit membranous, not valvular, sometimes baccate. Embryo annular or horseshoe-shaped, surrounding the albumen (Cyclolobeæ); or in a flat spiral, separating 2 masses of albumen; or conically spiral without albumen (Spirolobeæ), with the radicle in various directions as regards the fruit, but always turned to the hilum.

In this Order we have a crowd of species consisting partly of unisexual, and partly bisexual plants, corresponding so much in general structure otherwise, that Botanists seem to have no disposition at present to divide them. But as they are also provided with exceedingly different seeds, there is small probability of the integrity of the Order being long preserved. Till, however, they shall have been studied with reference to their woody structure, a separation would be premature; in that respect they differ considerably.

Schleiden first remarked that certain plants of this Order, namely, Beta Cycla, Atriplex hortensis, and Chenopodium Quinoa, have the wood very compact and pierced with vertical cords of cellular tissue (*see* NYCTAGINACEÆ). But I do not find this structure

Fig. CCCL.—1. a portion of the spike of Salicornia herbacea, with the flowers lodged in the notches of the axis; 2. a flower separate; 3. a flower of Salsola Kali; 4. its ripe fruit; 5. the same magnified, with a portion of the leafy dilated calyx torn away; 6. its embryo; 7. a flower of Chenopodium album; 8. a section of the same, showing the superior ovary; 9. its seed cut through to show the embryo.

uniform in such woody species as I have examined. It exists, for instance, in Halocnemum strobilaceum, Rhagodia Billardieri, Obionia portulacoides, Diotis ceratoides; but does not appear in Camphorosma monspeliaca, and some Salicornias, which are distinctly zoned; while Arthrocnemum Arbuscula, Salsola fruticosa, Salicornia articulata appear to have a kind of intermediate structure. They all, however, deserve the most careful investigation.

Among other peculiarities, some of the species have a tendency to extend their calyx into horizontal wings, which give them a very peculiar aspect; others secrete a coloured juice abundantly in the sepals, which, growing together in masses, cause an appearance similar to that of the Strawberry. They are distinguished from Phytolaccads, independently of the simplicity of the structure of their ovary, by their stamens never exceeding the number of the segments of the calyx, to which they are opposite: in Phytolaccads, if they are not more numerous than the segments of the calyx, they are alternate with them. It is evident, however, that Nettleworts and their allies stand in the first degree of relationship; and if it were not for the general tendency of this Order to form bisexual flowers, together with the mealy albumen and inferior radicle, it might be doubted whether the Chenopods ought not to be even referred to the Urtical Alliance. They seem, however, to belong to the series of bisexual hypogynous Orders, at the same time approaching in some parts of their territory to those perigynous plants which are stationed with Scleranths in Ficoidals.

Weeds, inhabiting waste places in all parts of the world, but unlike Amaranths, abounding least within the tropics, and most in extra-tropical regions. They are exceedingly common in all the northern parts of Europe and Asia, and are frequent inhabitants of salt marshes.

Some are used as potherbs, as Spinage, English Mercury (Chenopodium Bonus Henricus), Garden Orach (Atriplex hortensis), and Chard Beet; the roots of others form valuable articles of food, as Beet and Mangold Wurzel, plants now famous as a new source of sugar, capable of being produced in northern countries. Some of them possess an essential oil, which renders them tonic and antispasmodic; such are Ambrina ambrosioides and Botrys; the former has an aromatic sub-acrid taste, and is regarded in Brazil as a carminative, diaphoretic, and emmenagogue; it is prescribed in amenorrhœa, and for the expulsion of the dead fœtus.—*Martius.* Chenopodium Quinoa is a common article of food in Peru. Soda is yielded in immense quantities by Salsolas, Salicornias, and others. The essential oil of Ambrina anthelmintica, known in North America under the name of Worm-seed Oil, is powerfully anthelmintic. The same quality has been observed in Halogeton tamariscifolium, a Spanish species, called Spanish Worm-seed. Chenopodium vulvaria or olidum, a plant with an atrocious odour, has great reputation as an antispasmodic and emmenagogue. Thelygonum Cynocrambe (*κυνοκραμβη, Diosc.*) is a sub-acrid plant, abounding in acicular saline crystals, and is slightly purgative. It is sometimes used as a potherb. The seeds of Atriplex hortensis are said to be so unwholesome as to excite vomiting.

GENERA.

Salicornia, *Tournef.*
Halostachys, *C. A. Mey.*
Halocnemum, *Bieberst.*
Arthrocnemum, *Moq.*
Ceratocarpus, *Buxb.*
Pugionium, *Gærtn.*
Eurotia, *Adans.*
Diotis, Schreb.
Ceratospermum, Pers.
Guldenstædtia, Neck.
Krascheninnikowia, Güldenst.
Thelygonum, *L.*
Atriplex, *L.*
Schizotheca, *C. A. Mey.*
Obione, *Gærtn.*
Halimus, *Wallr.*
Grayia, *Hook et Arn.*
Axyris, *Linn.*
Fremontea, *Tor. et Gray.*
Sarcobatus, Nees.
Oxybasis, *Karel.*
Acnida, *Mitch.*
Spinacia, *Tournef.*
Exomis, *Moq.*
Camphorosma, *Linn.*
Camphorata, Mönch.
Kirilovia, *Bunge.*
Panderia, *Fisch et Meyer.*
Pterochlamys, Fisch.
Sclerolæna, *R. Br.*
Anisacantha, *R. Br.*
Kentropsis, *Moq.*
Threlkeldia, *R. Br.*
Didymanthus, *Endl.*
Blitum, *Linn.*
Morocarpus, Scop.
Agathophytum, *Moq.*
Orthosporum, *R. Br.*
Roubieva, *Moq.*
Ambrina, *Spach.*
Beta, *Tournef.*
Enchylæna, *R. Br.*
Londesia, *Fisch. et Mey.*
Chenolea, *Thunb.*
Echinopsilon, *Moq.*
Bassia, Allion.
Kochia, R. Br.
Suæda, Rchb.
Kochia, *Roth.*
Willemetia, R. Br.
Maireana, *Moq. Tand.*
Cycloloma, *Moq.*
Cyclolepis, Moq.
Chenopodium, *Linn.*
Oliganthera, *Endl.*
Oligandra, Less.
Lipandra, Moq.
Rhagodia, *R. Br.*
Teloxys, *Moq.*
Botrydium, Spach.
Cryptocarpus, *H. B. K.*
Schanginia, *C. A. Meyer.*
Suæda, *Forsk.*
Lerchia, Hall.
Cochliospermum, Lgsc.
Schoberia, *C. A. Mey.*
Alexandra, *Bunge.*
Traganum, *Delile.*
Salsola, *L.*
Caroxylon, Thunb.
Horaninovia, *F. et M.*
Halimocnemis, *C. A. Mey.*
Halogeton, *C. A. Meyer.*
Nanophytum, *Less.*
Cornulaca, *Delile.*
Anabasis, *L*
Brachylepis, *C. A. Meyer.*
Monolepis, *Schrad.*
Agriophyllum, *Bieberst.*
Rhagrostis, Buxb.
Corispermum, *Juss.*
Anthochlamys, *Fenzl.*
Dysphania, *R. Br.*

NUMBERS. GEN. 63. Sp. 360.

Urticaceæ.
POSITION.—Amarantaceæ.—CHENOPODIACEÆ.—Phytolaccaceæ.
Mesembryaceæ?
Scleranthaceæ.

ALLIANCE XXXIX. *PIPERALES.*—THE PIPERAL ALLIANCE.

DIAGNOSIS.—*Hypogynous Exogens, with achlamydeous flowers, and a minute embryo, at or near the outside of a large quantity of mealy albumen.*

The resemblance to each other of the plants included in this Alliance is manifest; but their affinity to Chenopodals is obscure. It chiefly depends upon the assumption that Piperals stand in near relation to Urticals, and that Chenopodals are the bisexual analogue of the latter. Granting this, which seems to be supported by strong evidence, we must then suppose that Piperals are a lateral sprout from Chenopodals, directing itself, not onwards to the next Alliance, but backwards towards the frontiers of the Diclinous Sub-class, to which the occasional unisexuality of the flowers of Pepperworts and Chloranths evidently points.

Piperals are clearly indicated by their naked flowers, constantly orthotropal ovule, abundant mealy albumen, and minute embryo, which is either external, or only just within the surface of it.

NATURAL ORDERS OF PIPERALS.

Carpel solitary. Ovule erect. Embryo lying in vitellus. Leaves opposite or alternate, with or without stipules	196. PIPERACEÆ.
Carpel solitary. Ovule suspended. Embryo naked. Leaves opposite, with intermediate stipules	197. CHLORANTHACEÆ.
Carpels several, distinct. Ovule erect. Embryo lying in vitellus. Leaves alternate, with stipules.	198. SAURURACEÆ.

ORDER CXCVI. PIPERACEÆ.—PEPPERWORTS.

Piperaceæ, *Rich. in Humb. Bonpl. et Kunth. N. G. et Sp. Pl.* 1. 39. *t.* 3. (1815); *Meyer de Houttuynia atque Saurureis*, (1827); *Endl. Gen.* lxxxi.; *Meisner Gen. p.* 335; *Kunth in Linnæa*, 13. 561; *Miquel in Ann. Sc. n. s.* 14. 167; 15. 285. *Id.*; *Systema Piperacearum*, 8vo.

DIAGNOSIS.—*Piperal Exogens, with a solitary carpel, an erect ovule, an embryo lying in vitellus, and opposite or alternate leaves, with or without stipules.*

Shrubs or herbaceous plants. Stems articulated. Leaves opposite, verticillate, or alternate in consequence of the abortion of one of the pair of leaves; stipules 0, or in

1 2

3

Fig. CCCLI.

pairs, or single and opposite the leaf. Flowers usually sessile, sometimes pedicellate, in spikes which are either terminal, or axillary, or opposite the leaves, naked, ☿, with a bract on the outside. Stamens 2 or more, arranged on one side or all round the ovary; anthers 1- or 2-celled, with or without a fleshy connective; pollen roundish, smooth. Ovary free, simple, 1-celled, containing a single erect, orthotropal ovule; stigma sessile, simple, rather oblique. Fruit free, somewhat fleshy, indehiscent, 1-celled, 1-seeded. Seed erect, with the embryo lying in a fleshy sac placed at the apex of the seed, on the outside of the albumen.

1

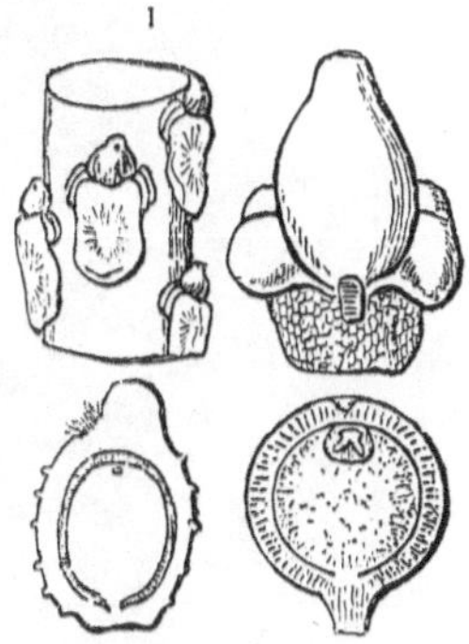

3 Fig. CCCLII. 4

However distinct the exogenous and endogenous forms of vegetation may be in the majority of the plants referred to those classes, it is well known that in certain cases such differences are much enfeebled. Of this Pepperworts are an instance. According to Richard, they are monocotyledonous; an opinion in which Blume concurs, after an examination of abundance of species in their native places of growth. But if the medullary rays constitute the great anatomical difference between these divisions of the vegetable kingdom (and I know of no other which is absolute), then Pepperworts are surely dicotyledonous, as is shown by Meyer (*Dissertatio de Houttuynia*, 38), and as may be ascertained by any one who will look at an old stem of a Pepper; add to this, the veins of the leaves, their distinct articulation with the stem, and the 2-lobed

Fig. CCCLI.—Serronia Jaborandi. 1. a cluster of flowers magnified; 2. a ripe fruit; 3. a vertical section of the same, showing the seed and position of the embryo.

Fig. CCCLII.—Peperomia blanda. 1. a highly magnified view of a portion of a spike, with a few flowers attached; 2. a flower seen from the back, showing the ovary and two side anthers; 3. a section of the ovary, showing the ovule and its foramen; 4. a perpendicular section of a ripe fruit, showing the embryo lying in its vitellus.

embryo, and it seems impossible to doubt their being properly stationed among Dicotyledons. In Peperomia incana, the young stems are undoubtedly endogenous in appearance, but in P. nigrum and Lonchitis they are as certainly exogenous, and it is probable that in the former case the wood eventually arranges itself in rows. It has, however, been remarked by Blume, that the oldest spiral vessels, and those which have acquired a woody consistence, occupy the circumference of the stem, while those more recently formed are seated towards its centre; and this is a point which, as Mr. Bennett has observed, requires a careful examination.—*Horsfield's Pl. Jav.* p. 78. Pepperworts are related to Buckwheats, Saururads, and Nettles; from all which, however, they are distinguished by obvious characters; but more especially to Chloranths, from which they differ in the ovule being erect, and in the presence of a vitellus or amniotic sac round the embryo. In the opinion of those who believe Peppers to be Monocotyledons, their station is near Arads, with which, indeed, they must be considered in any point of view to be in some measure connected.

Fig. CCCLIII.

Exclusively confined to the hottest parts of the world, they are extremely common in tropical America and the Indian Archipelago, but, according to Brown, are very rare in equinoctial Africa. Only three species have been found on the west coast; several exist at the Cape of Good Hope. They delight in low places, valleys, and the banks of rivers.

Fig. CCCLIII.—Macropiper methysticum.—*Pharm. Journal.*

These plants are for the most part pungent aromatics, a property which they derive from the presence of a peculiar acrid resin, an ethereal oil, and a crystalline matter called Piperine. But they are also astringents and narcotics, and sometimes are so in an eminent degree.

Fig. CCCLIV.

Great numbers of species occur in books as aromatics. That which is most extensively employed is Piper nigrum, whose drupes, when dried with the flesh upon them, form both the Black Pepper of the shops, and White Pepper, when that flesh is removed by washing. Although principally employed as a condiment, in order to excite the stomach and promote digestion, Pepper is also extensively employed in medicine in the form of an ointment mixed with lard, against tinea capitis, in relaxed uvula, and paralysis of the tongue; and infused in spirit and water it is a popular remedy against the return of a paroxysm of intermitting fever. It has been recommended by the Italians for this purpose, and is said to be more certain and speedy, and also milder in its operation than the Cinchona alkalies. See Pereira in *Med. Gaz.* xx. 180. In excessive doses Pepper is a dangerous stimulant. The fruit of Piper trioicum is said to be still more pungent. The female spikes of Chavica Roxburghii (Piper longum), dried, form the long pepper of the shops. The root and thickest part of the stems cut into small slices and dried, are much consumed for medical purposes in India under the name of Pippula Moola. The

Fig. CCCLIV.—Artanthe elongata.—*Pharm. Journal.*

effects of Long Pepper are analogous to those of Black Pepper; some consider it less powerful, others are agreed in its being the more acrid of the two. Chavica Chaba, pepuloides, and sylvatica, are employed in India as substitutes for this sort of Pepper. In tropical America similar uses are made of Chavica officinarum (P. Amalago), Artanthe adunca, Peltobryon longifolium, Artanthe trichostachya, and crocata. The aromatic roots of many species are officinal in some countries. A decoction of Artanthe eucalyptifolia is used in Brazil as a cure for colic, pains in the limbs, and flatulence. The root of P. Parthenium is administered in Brazil, under the name of Paribaroba, in amenorrhœa, leucorrhœa, and excessive menstrual discharges; that of Serronia Jaborandi, and Enckea unguiculata and glaucescens is held to be sialagogue and diuretic, and is employed for similar purposes. Pothomorpha sidæfolia (or umbellata) and subpeltata are also said to act as powerful stimulants of the lymphatic system, as deobstruents of some energy, and also for cleansing foul ulcers.—*Martius.*

Another class of remedial agents consists of those Pepperworts which possess the power of allaying inflammation of the urethra and mucous membrane of the intestinal canal. The best known of these species are Cubeba officinalis, canina, Wallichii, and others, whose ripe fruits are sold in the shops under the name of Cubebs. They are aromatic, pungent, stimulant, and purgative, and act as a specific in stopping gonorrhœal discharges. According to Martius, Artanthe adunca has the same effect in Brazil; and the Peruvian Artanthe elongata has a similar reputation. Of the narcotic Pepperworts the Ava or Macropiper methysticum is most celebrated. It has a thick, woody, rugged, aromatic rhizome, used in tincture against chronic rheumatism. Macerated in water it forms an intoxicating beverage, employed by the Otaheitans to cure venereal affections; they make themselves drunk, after which very copious perspiration comes on; this lasts three days, at the end of which time we are told that the patient is cured. The leaves of Chavica Betle and Siriboa are chewed by the Malays with lime and slices of the nut of Areca oleracea or the Pinang Palm. They produce intoxicating effects, stimulate powerfully the salivary glands and digestive organs, and diminish the perspiration of the skin.

As an astringent, a plant called Matico, and supposed to be Artanthe elongata (Piper angustifolium), is found to be a most powerful styptic and a valuable remedy in certain diseases of the genital organs and rectum. It is much used in South America and Belgium, to stop the hemorrhage from small vessels, leech-bites, or incised wounds. It may be applied in leaf, or in fine powder. It is said also to be taken internally for the same purpose, in the form of infusion, in the proportion of about half an ounce to a pint of boiling water. In Peru the plant is called Moho Moho, and is extensively used for the same purposes as Cubebs, which this Pepper much resembles in smell. An account of it has been given in the *Pharmaceutical Journal*, 3. 472, from which the annexed figure is borrowed, with the permission of the editor. It is, however to be observed, that the Peruvians apply the name Matico to the Eupatorium glutinosum, a very different plant.

Many other species of this Order are to be found mentioned as plants possessing useful properties; of which here is only space to name Acrocarpidium hispidulum, a West Indian plant used as a bitter and stomachic, Peperomia pellucida as a salad, Coccobryon capense a Cape stomachic, Artanthe adunca and Chavica majuscula, whose bark is rubefacient, and used in Java against rheumatism, and Artanthe crocata, whose spikes of fruit are employed in dyeing yellow.

GENERA.

* PEPEROMIDÆ.

Verhuellia, *Miq.*
Phyllobryon, *Miq.*
Acrocarpidium, *Miq.*
Peperomia, *R et P.*
Micropiper, Miq.
Tildenia, Miq.
Dugagelia, Gaud. ?
Erasmia, *Miq.*

** PIPERIDÆ.

Pothomorphe, *Miq.*
Heckeria, Kth.
Macropiper, *Miq.*
Chavica, *Miq.*
Rhyncholepis, *Miq.*
Cubeba, *Miq.*
Piper, *L.*
Muldera, *Miq.*
Coccobryon, *Klotzsch.*
Callianira, *Miq.*
Schilleria, Kth.
Enckea, *Kth.*
Peltobryon, *Kl.*
Sphærostachys, *Miq.*
Artanthe, *Miq.*
Steffensia, Kth.
Zippelia, *Bl.*
Serronia, *Guill.*
Ottonia, Spr.

NUMBERS. GEN. 20. Sp. 600.

Urticaceæ.
POSITION.—Saururaceæ.—PIPERACEÆ.—Chloranthaceæ.
Polygonaceæ.

ORDER CXCVII. CHLORANTHACEÆ.—CHLORANTHS.

Chloranthеæ, *R. Brown in Bot. Mag.* 2190. (1821); *Lindl. Collect. Bot.* 17. (1821); *Meyer de Houttuynia atque Saurureis*, 51. (1827); *Blume Flora Javæ*, (1829).—Chloranthaceæ, *Ed. pr.* cxxxiii.; *Endl. Gen.* lxxx.; *Meisner Gen. p.* 334.

DIAGNOSIS.—*Piperal Exogens, with a solitary carpel, a suspended ovule, a naked embryo, and opposite leaves with intermediate stipules.*

Herbaceous plants or under-shrubs, with an aromatic taste. Stems jointed, tumid at the articulations. Leaves opposite, simple, with sheathing petioles and minute intervening stipules. Flowers in terminal spikes. Flowers naked, with a supporting scale. Stamens lateral; if more than 1, connate, definite; anthers 1-celled, in Chloranthus, bursting longitudinally, each adnate to a fleshy connective, which coheres laterally in various degrees (2-celled, according to some); filament slightly adhering to the ovary. Ovary 1-celled; stigma simple, sessile; ovule pendulous, orthotropal. Fruit drupaceous, indehiscent. Seed pendulous; embryo minute, placed at the apex of fleshy albumen, with the radicle inferior, and consequently remote from the hilum; cotyledons divaricate. Hedyosmum and Ascarina are both unisexual.

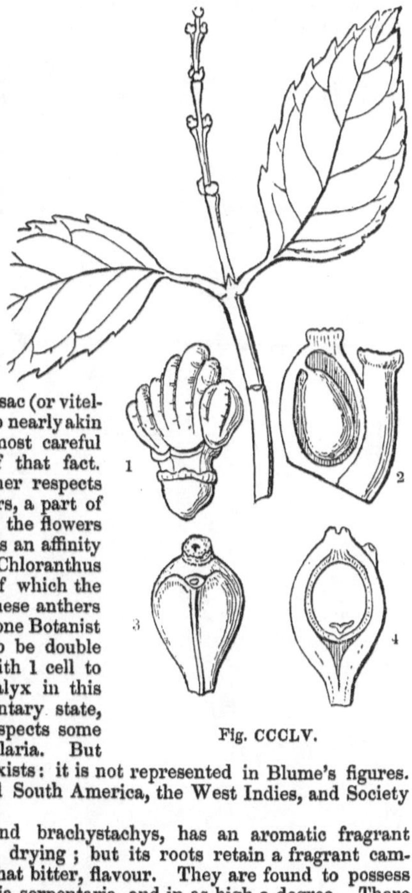

Fig. CCCLV.

These differ remarkably from Peppers and Saururads, in their naked embryo and pendulous ovule. The want of an amniotic sac (or vitellus) is so unexpected in plants otherwise so nearly akin to those Orders, that nothing but the most careful examination would satisfy the mind of that fact. While, however, Chloranths are in other respects inseparably connected with those Orders, a part of them differ in the very important fact of the flowers being absolutely diclinous. This indicates an affinity to the Urtical Alliance. The anthers of Chloranthus consist of a fleshy mass, upon the face of which the cell lies that bears the pollen; whether these anthers are 1- or 2-celled, is a matter of doubt; one Botanist considering those which have 2 cells to be double anthers, another understanding those with 1 cell to be half anthers. Blume describes a calyx in this genus sometimes present in a rudimentary state, adhering to the ovary, and hence he suspects some affinity between Chloranths and Opercularia. But I am persuaded that no such rudiment exists: it is not represented in Blume's figures.

Natives of the hot parts of India and South America, the West Indies, and Society Islands.

The whole plant of C. officinalis, and brachystachys, has an aromatic fragrant odour, which is gradually dissipated in drying; but its roots retain a fragrant camphorated smell, and an aromatic, somewhat bitter, flavour. They are found to possess very nearly the properties of Aristolochia serpentaria, and in as high a degree. There seems to be no doubt that they are stimulants of the highest order. The mountaineers

Fig. CCCLV.—Chloranthus monostachys. 1. exterior view of a flower; 2. perpendicular section of it, the anther being removed; 3. a ripe fruit; 4. a perpendicular section of it.

of Java employ the roots in infusion, or rubbed up with the bark of Cinnamomum Culilawan, as a remedy for spasms in pregnant women. In like manner, mixed with such carminative substances as Anise and Ocymum, they are administered with great success in the malignant small-pox in children. An infusion of the dried root is successfully employed in fevers attended with great muscular debility and a suppression of the functions of the skin. In a typhus which ravaged certain districts of Java, in consequence of long-continued rains following an unusually protracted dry season, the symptoms attendant upon which were extreme debility, a languid pulse, stupor, violent vomiting and bilious evacuations, the roots of this Chloranthus were of the greatest service. It was again employed most beneficially in a malignant intermittent fever which visited Java in the year 1824. In such cases the infusion was usually combined with a decoction of Cedrela Toona. The root has the great merit of preserving its active properties for a long time if properly prepared, and there can be no doubt that it is one of the most efficacious of all known remedial agents, wherever there is a necessity for continual and active stimulants.—*Blume*. Endlicher says that the dried branches of Ascarina polystachya, called Earaihau in Tahiti, still retain their hot flavour in the specimens collected during Forster's voyage. Similar qualities seem to exist in the Hedyosmums, which are used in the West Indies as antispasmodics and stomachics: H. Bonplandianum is, according to Martius, used in Brazil in malignant fevers and pains in the limbs.

GENERA.

Hedyosmum, *Swartz*.
Tafalla, Ruiz et Pav.
Ascarina, *Forst.*

Chloranthus, *Swartz*.
Nigrina, Thunb.

Creodus, Loureir.
Cryphæa, Hamilt.

Peperidia, Rchb.
Stropha, Noronh.

NUMBERS. GEN. 3. Sp. 15.

Urticaceæ.
POSITION.—Piperaceæ.—CHLORANTHACEÆ.—Saururaceæ.

ORDER CXCVIII. SAURURACEÆ.—SAURURADS.

Saurureæ, *Rich. Anal.* (1808); *Meyer de Houttuynia atque Saurureis*, (1827); *Martius Hort. Monac.* (1829); *Endl. Gen.* lxxxii.; *Meisner Gen. p.* 335.

DIAGNOSIS.—*Piperal Exogens, with several distinct carpels, an erect ovule, an embryo lying in vitellus, and alternate stipulate leaves.*

Herbaceous plants, growing in marshy places. Leaves alternate, with stipules. Flowers growing in spikes, naked, seated upon a scale, ☿. Stamens 3 to 6, clavate, hypogynous, persistent; filaments slender; anthers continuous with the filament, cuneate, with a thick connective and 2 lateral lobes bursting longitudinally. Ovaries 3 or 4, more or less distinct, with one ascending orthotropal ovule and a sessile recurved stigma, or connate into a 3- or 4-celled pistil, with a few orthotropal ovules ascending from the edge of the projecting semi-dissepiments. Fruit either consisting of 4 fleshy indehiscent nuts, or a 3- or 4-celled capsule, opening at the apex and containing a few ascending seeds. Seeds with a membranous integument; embryo minute, lying in a fleshy lenticular sac, which is seated on the outside of hard mealy albumen at the apex of the seed.

Fig. CCCLVI.

These plants are very near Pepperworts, with which they agree in habit, but from which they differ in the compound nature of their ovary, and their alternate constantly stipulate leaves. It has been supposed that they destroyed the distinction between Exogens and Endogens, but this opinion was formed upon incorrect views, and especially upon the erroneous supposition that the genus Aponogeton, now known to belong to Arrow-grasses, was a part of the Order of Saururads. If M. Decaisne is right in referring his Gymnotheca hither, which is very doubtful, we shall have the singular combination in the same Order of distinct one-seeded carpels, and an inferior ovary with many-seeded parietal placentæ.—*See* LOASADS.

The species are natives of North America, China, and the north of India, growing in marshes or pools of water.

Saururus cernuus has been found to be somewhat acrid; its root, made into a poultice, is employed in North America in pleurisy. The leaves of Houttuynia are regarded as emmenagogues by the Cochin Chinese.

GENERA.

Saururus, *Linn.*	Anemiopsis, *Hook.*
Mattuschkia, Gmel.	*Anemia*, Nutt.
Houttuynia, *Thunb.*	? Gymnotheca, *Decaisne.*
Polypara, Loureir.	

NUMBERS. GEN. 4. Sp. 7.

POSITION.————————SAURURACEÆ.—Piperaceæ.

Fig. CCCLVI.—1. Houttuynia cordata; 2. flower of Saururus cernuus; 3. its fruit; 4. its seed divided perpendicularly.

SUB-CLASS III. PERIGYNOUS EXOGENS.

The first group in this Sub-class is so evidently allied to the Chenopodal Alliance, that the genera are in many cases referred to the one or the other according to the varying views of systematists, and Basellads are really almost always considered as a perigynous form of Chenopods, which, moreover, are in certain cases, as for example in Beet, truly perigynous. This seems to show that Chenopodals on the one hand, and Ficoidals on the other, form the boundary between the Hypogynous and Perigynous series.

It is evident that in the Alliances which are thus collected, there is a constant and powerful tendency to the cohesion of the floral organs, for half of them consists of Orders having monopetalous flowers, a structure which is rare in the hypogynous Sub-class, and if it is seen there, is seldom accompanied by any union of the stamens to the petals, such an occurrence, when it is observable, as in Epacrids, being altogether exceptional. Here, on the contrary, the monopetalous corolla is habitually associated with epipetalous stamens. The tendency to adhesion is not indeed confined to the separate parts of the same ring of organs, or to the stamens with the calyx or corolla, but also not unfrequently occurs between the ovary and the parts which grow around it; the consequence of which is, that we find a partly inferior ovary in nearly every one of the Alliances of the Perigynous Sub-class. But although this is a manifest approach to the condition of the Epigynous Class, yet it is seldom the cause of any confusion, because the combination of the calyx, corolla, and stamens with each other is only partial, and is rarely accompanied by a similar cohesion of the carpels, whose styles remain separate even when their ovaries are consolidated. This is obvious among Appleworts and Hydrangeads, two quasi-epigynous forms of the Ranal and Saxifragal Alliances.

ALLIANCE XL. *FICOIDALES.*—THE FICOIDAL ALLIANCE.

DIAGNOSIS.—*Perigynous Exogens, with monodichlamydeous flowers, central or axile placentæ, a polypetalous corolla, if one is present, and an external embryo curved round a small quantity of mealy albumen.*

These plants are for the most part fleshy-leaved herbs or bushes, bearing very great resemblance, in some cases, to Purslanes in the Silenal Alliance, and like those plants, having for their character a central placentation combined with an annular embryo and mealy albumen. They are, in fact, the perigynous form of Silenals, and must be regarded as standing on the frontier of that Alliance. Like Silenals, the Ficoidals comprehend plants both with a high development of the corolla, and without a trace of it. They approach the Epigynous structure in some respects ; but although their carpels are partially adherent with the calyx in a large proportion of the Alliance, yet the styles are almost always distinct, and generally the carpels also in some degree. Torch-thistles are no doubt a kindred race, but the exigencies of a lineal arrangement compel the systematist to separate them by a long interval.

The great marks of the Ficoidal Alliance are the perigynous stamens, curved external embryo, and mealy albumen. It may be presumed that its axile placentation is a mere modification of the central, and not derived from the margins of carpellary leaves ; but this is a point which cannot be always decided

NATURAL ORDERS OF FICOIDALS.

Petals absent. Sepals distinct. Fruit inclosed in a membranous or succulent calyx. Carpel single, solitary. Seed erect . . .	199. BASELLACEÆ.
Petals numerous, conspicuous. Carpels several, consolidated . .	200. MESEMBRYACEÆ.
Petals absent. Carpels several, consolidated	201. TETRAGONIACEÆ.
Petals absent. Sepals united into a tube. Carpel single, solitary. Fruit inclosed in the hardened calyx tube	202. SCLERANTHACEÆ.

ORDER CXCIX. BASELLACEÆ.—BASELLADS.

Basellaceæ, *Moquin Tandon Chenopod. Monogr. p.* 10. (1840).

DIAGNOSIS.—*Ficoidal Exogens, with distinct sepals, no petals, fruit inclosed in a membranous or succulent calyx, a single solitary carpel, and an erect seed.*

Climbing, herbaceous, or shrubby plants, usually somewhat succulent. Leaves alternate, without stipules. Flowers coloured, naked, sessile or stalked, sometimes without bracts. Sepals imbricated in two rows, fleshy, hardly opening. Stamens opposite the sepals, inserted into their sides; anthers 2-celled, (in Basella, opening outwards longitudinally.) Ovary free, simple, one-celled, with a single erect, sessile, anatropal ovule; styles several. Fruit membranous. Embryo annular, or coiled up spirally, with mealy albumen in the centre, or separated into two superficial masses; radicle inferior.

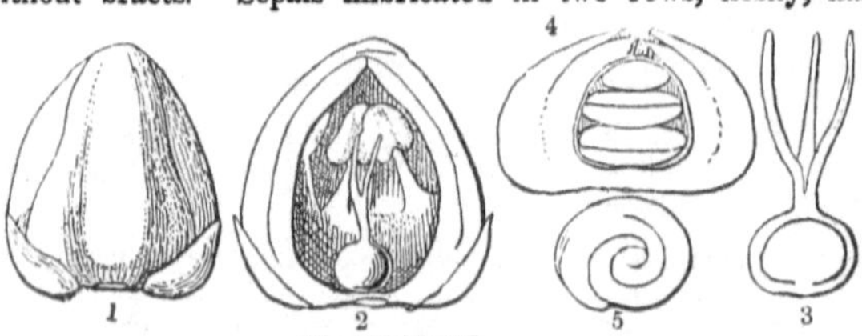

Fig. CCCLVII.

These plants, which have all the general characters of scandent Chenopods, have been separated from that Order on account of their coloured calyx, which scarcely opens, and has its sepals distinctly arranged in two rows, and their perigynous stamens. The anthers in Basella are moreover turned outwards, but I know not how far that is characteristic of the Order. Basellads differ from Scleranths in the want of a hardened tube to cover over the seed-vessel, and from Aizoons in the perfect simplicity of their carpel.

The species are all tropical, with the exception of Lophiocarpus, a Siberian plant, if it really belongs here.

Basella rubra and alba are employed as pot-herbs in the East Indies, where they are held in some esteem as a substitute for Spinage. B. tuberosa has a great fleshy root, which is eaten by the women of Quito, under the idea that it increases their fecundity. Basella rubra yields a very rich purple dye, but it is said to be difficult to fix.

GENERA.

Basella, *L.*
Gandola, Rumph.
Boussingaultia, *H. B. K*,
Anredera, *Juss.*
? Lophiocarpus, *Turcz.*

NUMBERS. GEN. 4. Sp. 12.

Chenopodiaceæ.
POSITION.—Tetragoniaceæ.—BASELLACEÆ.—Scleranthaceæ.

Fig. CCCLVII.—Basella rubra. 1. a flower; 2. the same opened vertically; 3. ovary; 4. the ripe fruit and inverting calyx divided perpendicularly; 5. the embryo.

ORDER CC. MESEMBRYACEÆ.—FICOIDS.

Ficoideæ, *Juss. Gen.* 315. (1789); *Dict. Sc. Nat.* 16. 528. (1820); *DC. Prodr.* 3. 415. (1828); *Salm Dyck Monogr. Mesemb.* (1834); *Meisn. Gen.* 129.—Mesembryaceæ, *Ed. pr.* xxxviii.—Mesembryanthemeæ, *Endl. Gen.* ccv. *Fenzl. in Ann. Wien. Mus.* 1. 347.—Lewisieæ, *Hook. in Beechey, p.* 345.—Spætalumeæ, *Nuttall Fl. Rocky Mount. p.* 24.

DIAGNOSIS.—*Ficoidal Exogens, with numerous conspicuous petals, and several consolidated carpels.*

Shrubby or herbaceous succulent plants, with opposite simple leaves. Flowers complete, often showy, always terminal, although, from the shortness of the branches on which they grow, apparently lateral; often opening only under the influence of sunshine, and closing in its absence. Sepals definite, usually 5, but varying from 4 to 8, more or less combined at the base, either cohering with the ovary, or distinct from it, equal or unequal, with a quincuncial or valvate æstivation. Petals indefinite, coloured, in many rows. Stamens arising from the calyx, indefinite, distinct; anthers oblong, incumbent. Ovary inferior, or nearly superior, many-celled or one-celled; stigmas numerous, distinct. Ovules 00, amphitropal, attached by cords to a central placenta, which is either wholly free, or united to the edges of the carpels, or sometimes spread over the back of the cavity of each cell. Capsule surrounded by the fleshy calyx, many-celled or 1-celled, opening in a stellate manner at the apex, or when free from the calyx splitting at the base. Seeds definite, or more commonly indefinite, attached to the inner angle of the cells; embryo lying on the outside of mealy albumen, curved or spiral, with the radicle next the hilum.

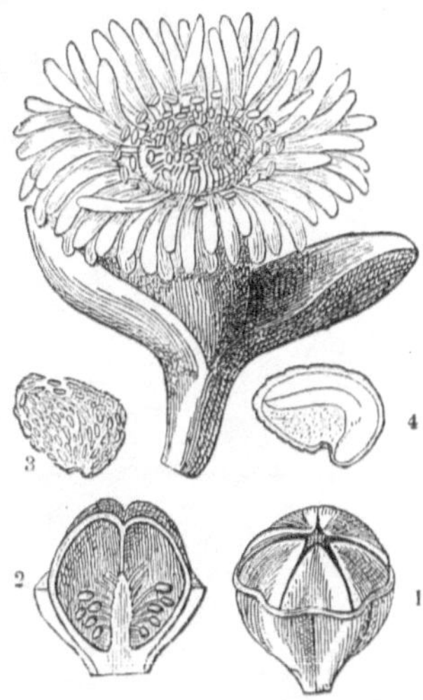

Fig CCCLVIII.

These are the most perfect of the Ficoidal Alliance, for the carpels are numerous and consolidated, and the apparatus of the corolla abundant and richly coloured. In this respect, indeed, Ficoids approach the Torch-thistles, although otherwise so different. They are to Ficoidals what Purslanes or Cloveworts are to Silenals, the princes of their race. One of the most singular facts connected with them is the variable nature of their placenta, which sometimes occupies the centre, to which the edges of the carpellary leaves are closely applied, sometimes runs up the back, altogether avoiding the centre, as in Mesembryanthemum acinaciforme (See *Ladies' Botany*, vol. ii. t. xxxi. 2.), and is sometimes absolutely free, as in Lewisia. This curious genus, however, differs a little from the rest of the Order in its perfectly one-celled free ovary, and barely perigynous stamens. It is however near Glinus, and there does not seem to be any necessity for regarding it as the type of a peculiar Order. The seed-vessels of the Ficoids exhibit remarkable phenomena, closing when placed in water, opening again when dried, a hygrometrical quality doubtless connected with their manner of life. Inhabitants of the dry places of southern Africa, they only expand and discharge their seeds when rain falls to relax their tissue, for then only would the seeds be able to germinate. This is more especially evident in M. Tripolium, which has been sold under the name of Flores Candiæ.

The hottest sandy plains of the Cape of Good Hope nourish the larger part of this Order. A few are found in the south of Europe, north of Africa, Chile, China, Peru, and the South Seas.

The succulent leaves of a few are eaten, as of Mesembryanthemum edule, which is

Fig. CCCLVIII.—Mesembryanthemum. 1. its fruit; 2. the same opened; 3. a seed; 4. the same divided perpendicularly.

the Hottentot's Fig of the Cape colonists; Mesembryanthemum emarcidum, when bruised and fermented, acquires a narcotic property, and is chewed like Tobacco by the Hottentots (*Burnett*); others yield an abundance of soda. M. crystallinum in Spain, and M. copticum and nodiflorum in Egypt, are collected for the purpose of furnishing alkali for glassworks; the former is called Barilla Moradera by the Spaniards, who import large quantities of its ashes from the Canaries, where the seeds are eaten as a common food, according to Broussonet. Mesembryanthemum nodiflorum is used in the manufacture of Maroquin leather. M. crystallinum (the Ice-plant) is remarkable for the abundance of watery pustules with which it is covered; its juice is said to be diuretic, and has been prescribed in dropsy and liver complaints. M. geniculiflorum is used in Africa as a potherb, and its seeds are ground into flour. Lewisia rediviva is an article of food among the natives of north-west America, who call it Spatulum or Spæt'lum. The roots, after the bark is stripped off, seem from the relation of travellers to consist of little more than starch.—*Gray and Torrey*, 1. 678. The natives of Australia eat the fruit of M. æquilaterale (Pig-faces, or Canagong). The seed-vessel of this plant is about an inch and a half long, of a yellowish, reddish, or green colour, and somewhat obconical. The pulp is sweetish and saline.—*Backhouse.*

GENERA.

Mesembryanthemum, *L.*
Hymenogyne, Haw.
Mesembryon, Adans.

Glinus, *L.*
Rolofa, Adans.
Plenckia, Rafin.

Physa, Thouars.
Orygia, *Forsk.*
Corbichonia, Scop.

Axonotechium, Fenz.
Lewisia, *Pursh.*
?Beloanthera, *Hassk.*

NUMBERS. GEN. 5. Sp. 375.

Cactaceæ.
POSITION.————MESEMBRYACEÆ.—Tetragoniaceæ.
Portulaceæ.

ORDER CCI. TETRAGONIACEÆ.—AIZOONS.

Tetragoniaceæ, *Ed. pr. p.* 209. (1836).—Tetragonieæ, Aizoideæ, Sesuvieæ, *Endl. Gen. p.* 947.—Sesuviaceæ, *Wight. Illustr.* 2. 42.

DIAGNOSIS.—*Ficoidal Exogens, with no petals, and several consolidated carpels.*

Succulent-leaved herbaceous plants, or occasionally small shrubs. Leaves alternate, often covered with watery pustules, without stipules. Flowers small, axillary. Calyx 3- 5-cleft, free, or partially adherent to the ovary. Corolla 0. Stamens definite, alternate with the sepals, if they bear any relation to them. Ovary 2- 9-celled; ovules suspended or ascending, solitary or several, anatropal, always with a long cord; foramen superior in the suspended species. Styles as many as the cells of the ovary, distinct. Fruit either an indehiscent tough-shelled nut, or a capsule splitting all round. Seeds with an annular embryo, curved round mealy albumen.

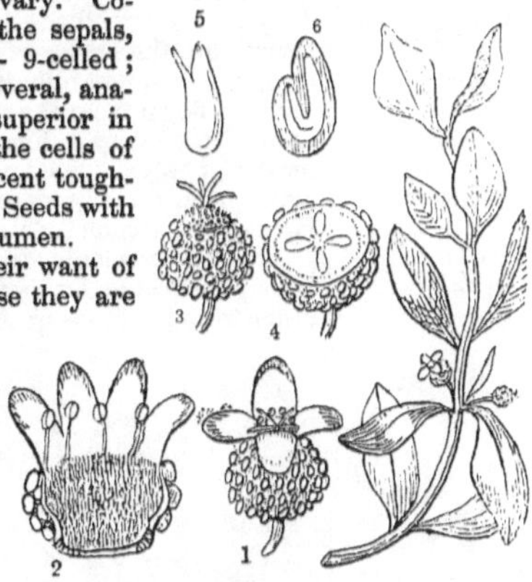

Fig. CCCLIX.

The distinction of Aizoons resides in their want of petals and small number of stamens, otherwise they are like Ficoids. They participate in the affinity of that Order, but approach nearer to the Chenopods, among which Beta has the adherent calyx of a Tetragonia. Cypselea, and the genera near it, also establish a connection with Purslanes, which are positively known by their 2-leaved calyx.

The species, which are plants of no beauty, are found in the South Sea Islands, the residence more especially of Tetragonias, in the Mediterranean, the Cape of Good Hope, or various parts of the tropics.

They are universally insipid or slightly saline, whence they are suited for human food. Tetragonia expansa, a New Zealand annual, is a good deal cultivated in Europe under the name of New Zealand Spinage, as a substitute for which herb it is employed. Sesuvium portulacastrum and repens are used for the same purposes in the tropics of Asia. The ashes of Aizoon canariense and hispanicum abound in soda.

GENERA.

Suborder I. TETRAGONEÆ. — Fruit woody, indehiscent. Tetragonia, *Linn.* *Demidovia*, Pall. *Tetragonocarpus*, Com.	Tetragonella, *Miq.* Aizoon, *Linn.* *Veslingia*, Fabric. *Ficoideæ*, Dillen. Galenia, *Linn.* *Kolleria*, Presl. *Sialodes*, Eckl. et Zeyh. Plinthus, *Fenzl.*	Suborder II. SESUVEÆ. —Capsule circumscissile. Trianthema, *Sauv.* *Rocama*, Forsk. *Papularia*, Forsk. *Zaleya*, Burm. Diplochonium, *Fenzl.*	Sesuvium, *Linn.* *Aizoon*, Andr. *Halimus*, Löffl. Pyxipoma, *Fenzl.* Ancistrostigma, *Fenzl.* Cypselea, *Turp.* *Radiana*, Raf. *Millegrana*, Surian.

NUMBERS. GEN. 11. Sp. 65.

Chenopodiaceæ.
POSITION.—Mesembryaceæ.—TETRAGONIACEÆ.—Scleranthaceæ.
Portulaceæ.

Fig. CCCLIX.—Tetragonia. 1. a flower; 2. the calyx opened out; 3. ripe fruit; 4. a transverse section of it; 5. an ovule; 6. a section of a seed.

ORDER CCII. SCLERANTHACEÆ.—SCLERANTHS.

Sclerantheæ, *Link Enum.* 417. (1821); *DC. Prodr.* 3. 377. (1828) *a* § *of* Illecebreæ; *Bartl. Ord. Nat.* 300. (1830); *Meisn. Gen. p.* 133; *Endl. Gen. p.* 962.

DIAGNOSIS.—*Ficoidal Exogens, with no petals, a tubular calyx becoming hardened and covering the fruit, consisting of a single solitary carpel.*

Small inconspicuous herbs. Leaves opposite, without stipules. Flowers minute, axillary, sessile. Calyx 4- or 5-toothed, with a stiff tube. Stamens from 1 to 10, inserted into the orifice of the tube. Ovary simple, superior, 1-seeded; styles 2, or 1, emarginate at the apex. Ovules 1 or 2, amphitropal, hanging down from the point of a slender cord which rises from the base of the ovary. Fruit a membranous utricle inclosed within the hardened calyx. Seed pendulous from the apex of a funiculus, which arises from the bottom of the cell; embryo cylindrical, curved round farinaceous albumen; radicle superior, but next the hilum.

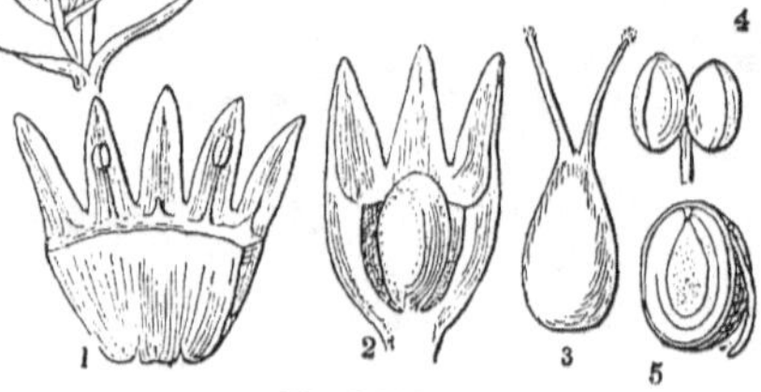

Fig CCCLX.

The weedy plants called Scleranths, are by most Botanists, and among the rest by De Candolle, referred to Knotworts, from which they differ in the absence of petals and stipules, in the indurated tube of the calyx, from the orifice of which the stamens proceed, and in the number of the latter often exceeding that of the divisions of the calyx. They are, in fact, perigynous Chenopods, rather than Knotworts. Their affinity seems, however, to be quite as great with Nyctagos, with which they agree in most respects except their truly perigynous stamens and small herbaceous ribbed calyx.

Fenzl proposes to divide this trifling Order into two tribes, viz. Euscleranthææ, and Habrosieæ, but the advantage of doing so is not obvious.

The species are found in barren fields in Europe, Asia, and North America, and in sterile places in countries of the southern hemisphere beyond the tropics. A single species is described from Peru.

They are all uninteresting weeds, of no known use.

GENERA.

Mniarum, *Forst.*
Ditoca, Banks.
Scleranthus, *L.*
Guilleminia, *H. B. K.*
Habrozia, *Fenzl.*

NUMBERS. GEN. 4. Sp. 14.

Chenopodiaceæ.
POSITION.—Tetragoniaceæ.—SCLERANTHACEÆ.—Basellaceæ.
Nyctaginaceæ.

Fig. CCCLX.—Scleranthus perennis. 1. young calyx forced open; 2. perpendicular section of ripe calyx; 3. ovary; 4. anther; 5. section of seed.

ALLIANCE XLI. *DAPHNALES.*—THE DAPHNAL ALLIANCE.

DIAGNOSIS.—*Perigynous Exogens, with monochlamydeous flowers, a solitary carpel, and an amygdaloid embryo without albumen.*

The Daphnal Alliance consists almost exclusively of shrubs and trees, usually evergreen and often of large dimensions. It is defined by its flowers being monochlamydeous, or, if there be a corolla, by the quasipetals having altogether the colour, texture, and quality of the calyx. It differs from the Ficoidal Alliance in the total absence of albumen, and its great almond-like embryo ; nevertheless its Daphnads approach Ficoidals in consequence of the resemblance between some Passerines and Scleranths. With Rosals it agrees in the nature of its embryo, but is distinguished by the want of petals, or, failing that distinction, by its ovary having a vertical style, which in Rosals always stands more or less obliquely with respect to the ovary. This renders it probable that the fruit of Daphnads is really composed of two or more valvate carpels cohering round a single ovule, as happens in the Order of Buckwheats, while in Rosals the carpels are absolutely simple.

If we regard the further end of the chain of Daphnals we find that Laurels touch Calycanths among Rosals. Laurels, too, indicate a tendency towards the diclinous series, in consequence of their flowers being occasionally unisexual, and seem to bring Daphnals into the vicinity of Plume Nutmegs.

There is also a very strong approach on the part of Daphnals to Rhamnals, as is indicated by the tubular calyx of the latter and their constant tendency to lose their petals. In fact, the two Alliances must stand in actual contact, for there is little to distinguish them except the simple fruit of the one and the compound fruit of the other.

NATURAL ORDERS OF DAPHNALS.

Anthers bursting lengthwise. Apetalous or polypetalous. Ovule solitary, suspended. Calyx imbricated	203. THYMELACEÆ.
Anthers bursting lengthwise. Apetalous. Ovules erect. Calyx valvate.	204. PROTEACEÆ.
Anthers bursting by recurved valves. Leaves perfect. Fruit naked.	205. LAURACEÆ.
Anthers bursting by recurved valves. Leaves mere colourless scales. Fruit buried in a succulent permanent calyx.	206. CASSYTHACEÆ.

ORDER CCIII. THYMELACEÆ.—DAPHNADS.

Thymelææ, *Juss. Gen.* 76. (1789); *R. Br. Prodr.* 358; *Bartling Ord. Nat.* 114. (1830).—Daphnoideæ, *Vent. Tabl.* ii. 235. (1799); *Endl. Gen.* cix.—Daphnaceæ, *C. A. Meyer, Ann. Sc.* xx. 45.—Anthobolea, *Martius Conspectus, No.* 81. (1835).—Exocarpeæ, *Arnott in Edinb. Encycl.* 128, a § of Santalaceæ, (1832).—Hernandiæ, *Blume Bÿdr.* 550. (1825); *Ed. Pr.* cxlvi.; *Endl. Gen. p.* 232.

DIAGNOSIS.—*Daphnal Exogens, with apetalous or polypetalous flowers, anthers bursting lengthwise, a solitary suspended ovule, and an imbricated calyx.*

Stem shrubby, very seldom herbaceous, with tenacious bark. Leaves without stipules, alternate or opposite, entire. Flowers capitate or spiked, terminal or axillary,

1 2 3

Fig. CCCLXI.

occasionally solitary, sometimes ♂-♀ by abortion, often inclosed in an involucre. Calyx inferior, tubular, coloured; the limb 4-cleft, seldom 5-cleft, with an imbricated æstivation. Corolla 0, or sometimes scale-like petals in the orifice of the calyx. Stamens definite, inserted in the tube or its orifice, often 8, sometimes 4, less frequently 2; when equal in number to the segments of the calyx or fewer, opposite to them; anthers 2-celled, dehiscing lengthwise in the middle. Ovary composed of a single carpel, with one solitary pendulous anatropal ovule; style 1; stigma undivided. Fruit hard, dry, and nut-like, or drupaceous. Albumen none, or thin and fleshy; embryo straight; cotyledons plano-convex, sometimes lobed and crumpled; radicle short, superior.

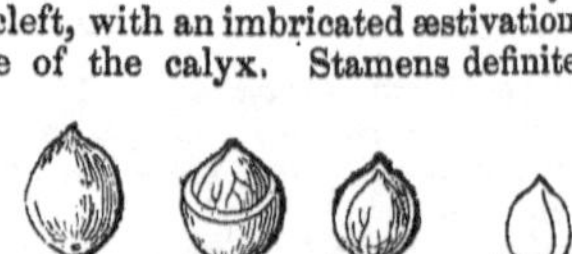

1 2 3 4

Fig. CCCLXII.

The true affinity of Daphnads, notwithstanding the commonness of the species, does not seem well ascertained. They are generally associated with Oleasters; and if the genus Elæagnus really belongs to that Order, it must be admitted that the main distinction between the two Orders consists in the separate sexes of Oleasters. To Proteads they certainly approach, especially in the stamens being opposite to the segments of the calyx, but Daphnads have a pendulous ovule, and Proteads an erect one; in the former, too, the calyx is imbricated, in the latter, valvate. Laurels are known by their reflexed anther-valves. As to Sandalworts, with which Daphnads are often compared, they are far removed by their inferior ovary and copious albumen. Aquilariads have a compound ovary, and therefore belong to the Rhamnal Alliance. A supposed Natural Order, called by Blume Hernandiaceæ, merely consists of Daphnads with polygamous

Fig. CCCLXI.—Daphne Mezereum. 1. a flower cut open; 2. a vertical section of an ovary; 3. the fruit.

Fig. CCCLXII.—Daphne Mezereum.—*Gærtner.* 1. fruit; 2. the same with a portion of the succulent rind removed; 3. seed; 4. embryo.

flowers and lobed cotyledons; the former circumstance seems to bring the Alliance at one point into the vicinity of Amentals still more than the manifest affinity with Oleasters.

Natives sparingly of Europe, and the northern parts of the world, common in the cooler parts of India and South America, and abundant at the Cape of Good Hope and in New Holland. Dirca occurs in North America, and Lagetta is confined to the tropical parts of America. Drapetes is a little antarctic plant.

The great feature of this Order is the causticity of the bark, which acts upon the skin as a vesicatory, and causes excessive pain in the mouth if chewed. Daphne Mezereum is extensively used in medicine. In Germany the bark of the stem and larger branches is removed in spring, folded in small bundles, and dried for medicinal use. In this country the bark of the root is employed. Its taste is at first sweetish, but afterwards highly acrid. All the parts are excessively acrid, and act as a local irritant poison. Voigt says that it vomits and purges and affects the urinary organs, and that death takes place from its local operation. As a local irritant, Mezereum bark is employed in France, under the name of Garou, to produce vesication. In this country it is frequently employed as a topical remedy for toothache. Dr. Withering cured a case of difficulty of swallowing by Mezereum, which he directed to be chewed frequently. Dr. Cullen says he has employed it with success in some cutaneous diseases. Similar qualities have been remarked in D. Laureola, pontica, Gnidium, and several others. The causticity of the Mezereum and Spurge Laurel are so great that persons who prepare them for medical use often suffer great inconvenience from particles rising and irritating the nostrils while pounding them. The inner bark of the Mezereum creates in the mouth a burning sensation, and if swallowed affects the lining of the œsophagus and stomach in the same manner. Mr. Squire remarks that this effect is followed in Daphne Laureola by profuse perspiration of the face, head, and neck, after which the burning sensation subsides. The bark of the root is the most efficacious part.—*Pharm. Journ.* 1. 397. The fruit of Dirca palustris is narcotic, producing effects like those of Stramonium. That of Daphne cestrifolia, a Bogota plant, is poisonous to cattle, according to Mr. Hartweg. The berries of Daphne Laureola are poisonous to all animals except birds. The bark of Gnidia daphnoides, is manufactured into ropes in Madagascar; that of Dais madagascariensis into paper. From a Daphne the Afghans prepare the matches for their match-locks. A soft kind of paper is made from the inner bark of Daphne Bholua in Nepal. Daphne cannabina is used in a similar way in China. The inner bark of Lagetta lintearia is the beautiful Lace-bark, so called because, when macerated and stretched laterally, it assumes the appearance of coarse lace; twisted and knotted it was formerly employed in making the slave whips used by Negro-drivers. Daphne Gnidium and Passerina tinctoria are used in the south of Europe to dye wool yellow. In Hernandia sonora the bark, seed, and young leaves are all slightly purgative. Rumphius says that the fibrous roots chewed and applied to wounds caused by the Macassar poison, ensure an effectual cure. The juice of the leaves is a powerful depilatory; it destroys hair, wherever it is applied, without pain. The wood is light; according to Aublet, that of H. guianensis takes fire readily from a flint and steel, and is used as Amadou. The seeds of Inocarpus edulis are eaten when roasted, and have the taste of Chesnuts.

GENERA.

Dirca, *Linn.*
Daphne, *Linn.*
  Thymelæa, Scop.
  Capura, L.
Cryptadenia, *Meisn.*
Edgworthia, *Meyer.*
Hargasseria, *Meyer.*
Piptochlamys, *Meyer.*
Chlamydanthus, *Meyer.*
Nordmannia, *Fisch. Mey.*
Arthrosolen, *Meyer.*
Lygia, *Fasan.*
Mezereum, *Meyer.*
Scopolia, *L. fil.*
Daphnopsis, *Mart.*
Schœnobiblos, *Mart.*
Peddiea, *Harv.*
Dais, *Linn.*
Passerina, *Linn.*
  Stellera, Linn.
Diarthron, *Turczan.*
Drapetes, *Lam.*
Pimelea, *Banks et Sol.*
  Banksia, Forst.
  Cookia, Gmel.
  Thecanthes, Wikstr.
  Heterolæna, Endl.
  Phyllolæna, Endl.
  Choristachys, Endl.
  Malistachys, Endl.
  Epallage, Endl.
Struthiola, *Linn.*
Jenkinsia, *Griff.*
Enkleia, *Griff.*
Gnidia, *Linn.*
  Canalia, F. W. Schm.
Lachnæa, *L.*
Thymelina, *Hoffmans.*
  Nectandra, Berg.
Lasiosiphon, *Fresen.*
Linostoma, *Wall.*
Eriosolena, *Blume.*
Diplomorpha, *Meisn.*
? Coleophora, *Miers.*
Wikströmia, *Endl.*
Lagetta, *Juss.*
Funifera, *Leand.*
Exocarpus, *Lab.*

° HERNANDIEÆ.

Inocarpus, *Forst.*
  Aniotum, Soland.
Sarcostigma, *Wight et A*
Hernandia, *Plum.*

NUMBERS. GEN. 38. Sp. 300.

Aquilariaceæ.
POSITION.—Proteaceæ.—THYMELACEÆ.—Lauraceæ.
Elæagnaceæ.

ORDER CCIV. PROTEACEÆ.—PROTEADS.

Proteaceæ, *Juss. Gen.* (1789); *R. Brown in Linn. Trans.* 10. 15. (1809); *Prodr.* 363; *Suppl. Prim.* (1830); *Endl. Gen.* cxiii.; *Meisner Gen. p.* 331.

DIAGNOSIS.—*Daphnal Exogens, with apetalous flowers, anthers bursting lengthwise, erect ovules, and a valvate calyx.*

Shrubs or small trees. Branches usually umbellate. Leaves hard, dry, divided or undivided, opposite or alternate, without stipules; their cuticle often covered equally on both sides with stomates. Calyx 4-leaved, or 4-cleft, with a valvate æstivation. Stamens 4, sometimes in part sterile, opposite the segments of the calyx. Ovary consisting of a single carpel, superior; style simple; stigma undivided; ovule one, or two collateral, or several in two rows, anatropal or amphitropal, and ascending. Fruit dehiscent or inhehiscent. Seed without albumen; embryo with two or occasionally several cotyledons, straight; radicle inferior, next the hilum, or parallel with it.

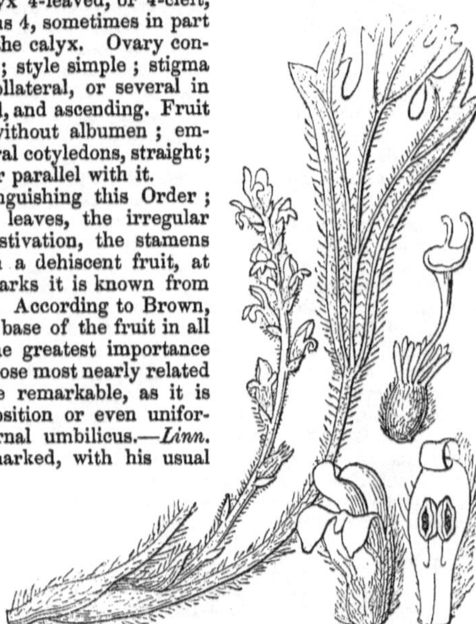

Fig. CCCLXIII. 1 2 3

There is no difficulty in distinguishing this Order; the hard woody texture of the leaves, the irregular tubular calyxes with a valvate æstivation, the stamens placed upon the lobes, along with a dehiscent fruit, at once characterise it. By these marks it is known from Daphnads and all other Orders. According to Brown, the radicle pointing towards the base of the fruit in all Proteads, is a circumstance of the greatest importance in distinguishing the Order from those most nearly related to it; and its constancy is more remarkable, as it is not accompanied by the usual position or even uniformity in the situation of the external umbilicus.—*Linn. Trans.* 10. 36. He has also remarked, with his usual acuteness, that in consequence of the presence of hypogynous scales, we may expect to find octandrous genera belonging to this family. The same writer observes, that there is a peculiarity in the structure of the stamens of certain genera of Proteads, namely, Simsia, Conospermum, and Synaphea, in all of which these organs are connected in such a manner that the cohering lobes of two different anthers form only one cell. Another anomaly equally remarkable exists in Synaphea, the divisions of whose barren filament so intimately cohere with the stigma, as to be absolutely lost in its substance, while the style and undivided part of the filament remain perfectly distinct. In another place he remarks: "A circumstance occurs in some species of Persoonia, to which I have met with nothing similar in any other plant: the ovarium in this genus, whether it contain one or two ovula, has never more than one cell; but in several of the 2-seeded species, a cellular substance is, after fecundation, interposed between the ovula, and this gradually indurating, acquires in the ripe fruit the same consistence as the putamen itself, from whose substance it cannot be distinguished; and thus, a fruit originally of one cell becomes bilocular; the cells, however, are not parallel, as in all those cases where they exist in the unimpregnated ovarium, but diverge more or less upwards." This is subsequently explained by the same author (*King's Appendix*), by the cohesion of the outer membranes of the two collateral ovules, originally distinct, but finally constituting this anomalous dissepiment, the inner membrane of the ovule consequently forming the outer coat of the seed.

A happier name than that of Proteads could not have been devised, for the diversity of appearance presented by the various genera is such as it would be hard to parallel in

Fig. CCCLXIII.—Synaphea dilatata.—*Ferd. Bauer.* 1. a flower; 2. one of its lobes; 3. the ovary and style and stigma.

the same Natural Order. On the one hand, we have the hard-coned Banksias, and the close-headed Dryandras; then come the loose-flowered Hakeas and Grevilleas; and the ranks are closed by anomalous genera, bearing the names of Synaphea, Conospermum, Franklandia, &c.

The principal stations of this Order are the Cape of Good Hope and adjacent regions, and New Holland. A few only occur in South America, the Malay Archipelago, and elsewhere in the southern hemisphere; in the northern they are scarcely known. Protea abyssinica is, however, found in Abyssinia, and P. Paulina in Sennaar. In general they occur in land unfit for cultivation. Few are of considerable size. Mr. Frazer has reported the existence of a plant he referred to Banksia grandis, with a trunk fifty feet high, and frequently more than two feet and a half in diameter, occupying the barren hills on the banks of the river, at Point Frazer, in the Swan River Colony; and Grevillea robusta, and Knightia excelsa, are other instances of the species acquiring a considerable stature.

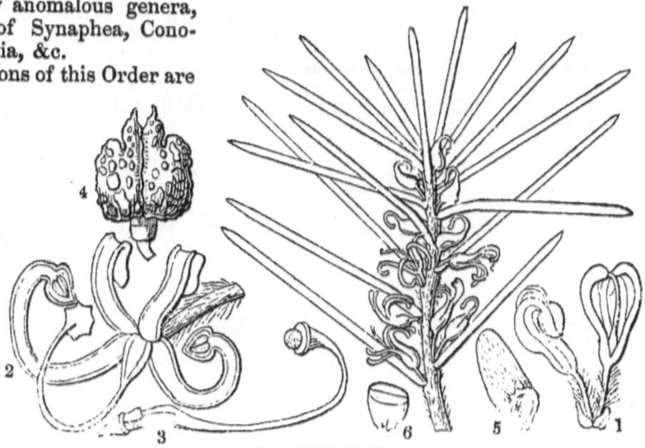

Fig. CCCLXIV.

Handsome evergreen shrubs, much prized by gardeners for the neatness of their appearance, and the beauty or singularity of their flowers. They are commonly employed as fire-wood at the Cape of Good Hope. The fruit of Guevina is sold like nuts in the markets of Chile, under the name of Avellano.

Waggon wheels are constructed at the Cape of Good Hope from the wood of Protea grandiflora, which is called, in consequence, Wagen boom. The dried flowers of Petrophila brevifolia give out to boiling water so brilliant a yellow colour, that it is possible the plant might be turned to account by dyers. The same may be said of Persoonia macrostachya. The bark of Protea grandiflora is used by the Cape settlers in diarrhœa. The seeds of Brabejum stellatum are roasted and eaten like Chesnuts; their shells form a substitute for coffee. The honey that flows from the flowers of Protea mellifera and speciosa is boiled down at the Cape of Good Hope, and used against coughs. It is reported by Endlicher that the root of Banksia marcescens is emetic: but upon no known authority. Upon the whole, the Order must be regarded as one of the most useless to man, notwithstanding the beauty of the flowers and foliage of so many species.

GENERA.

I. NUCAMENTACEÆ.

Tribe I. Proteidæ.

Aulax, *Berg.*
Leucadendron, *Herm.*
Conocarpodendron, Bh.
Conocarpus, Adans.
Euryspermum, Salisb.
Gissonia, Salisb.
Chasme, Salisb.
Petrophila, *R. Br.*
Arthrostigma, Endl.
Petrophile, Endl.
Symphyolepis, Endl.
Xerostole, Endl.
Isopogon, *R. Br.*
Atylus, Salisb. part.
Eustrobilus, Endl.
Hypsanthus, Endl.
Protea, *Linn.*
Leucadendron, Linn.
Lepidocarpodendron, Boerh.
Scolymocephalus, Hm.
Erodendrum, Salisb.
Pleuranthe, Salisb.
Gaguedi, Bruce.
Leucospermum, *R. Br.*
Conocarpodendron, Bh.
Diastella, Salisb.
Mimetes, *Salisb.*
Hypophyllocarpodendron, Boerh.
Serruria, *Salisb.*
Serraria, Burm.
Nivenia, *R. Br.*
Paranomus, Salisb.
Sorocephalus, *R. Br.*
Soranthe, Salisb.
Mischocaryon, Endl.
Cardiocaryon, Endl.
Spatalla, *Salisb.*
Coilostigma, Endl.
Cyrtostigma, Endl.
Adenanthos, *Labill.*

Tribe II. Conospermidæ.

Synaphea, *R. Br.*
Conospermum, *Smith.*
Chilurus, R. Br.
Isomerium, R. Br.
Stirlingia, *Endl.*
Simsia, R. Br.

Tribe III. Franklandidæ.

Franklandia, *R. Br.*

Tribe IV. Persoonidæ.

Symphyonema, *R. Br.*
Agastachys, *R. Br.*
Cenarrhenes, *Labill.*
* ? *Potameia*, Thouars.
Persoonia, *Smith.*
Pentadactylon, Gärtn.
Linkia, Cavan.
Brabejum, *Linn.*
Brabyla, Linn.
Guevinia, *Mol.*
Quadria, R. et Pav.
Nebu, Feuill.
Bellendena, *R. Br.*

II. FOLLICULARES.

Tribe I. Grevillidæ.

Anadenia, *R. Br.*
Manglesia, *Endl.*
Grevillea, *R. Br.*
Lissostylis, R. Br.
Lyssanthe, Salisb.
Ptychocarpa, R. Br.
Eriostylis, R. Br.
Stylurus, Salisb.
Plagiopoda, R. Br.
Conogyne, R. Br.
Calothyrsus, R. Br.

Fig. CCCLXIV.—Hakea acicularis.—*Ferd. Bauer.* 1. a couple of flowers; 2. a flower magnified; 3. the pistil; 4. a fruit; 5. a seed; 6. half an embryo.

Cycloptera, R. Br.
Hakea, *Schrad.*
Conchium, Smith.
Lambertia, *Smith.*
Xylomelum, *Smith.*
Orites, *R. Br.*
Amphiderris, R. Br.
Oritina, R. Br.
Adenostephanes, *Klotzh.*
Rhopala, *Schreb.*
Roupala, Aubl.
Leinkeria, Scop.
Dicknekeria, Flor. Fl.
Andripetalum, *Schott.*
Andriapetalum, Pohl.
Helicia, *Lour.*
Helitophyllum, Blum.
Knightia, *R. Br.*
Eucarpha, R. Br.
Embothrium, *Forst.*
Oreocallis, *R. Br.*
Telopea, *R. Br.*
Hylogyne, Salisb.
Lomatia, *R. Br.*
Tricondylus, Salisb.
Amphiloma, Endl.
Stenocarpus, *R. Br.*
Cybele, Salisb.

Tribe II. Banksidæ.

Banksia, *Linn. fil.*
Isostylis, R. Br.
Dryandra, *R. Br.*
Josephia, Salisb.
Hemiclidia, *R. Br.*
* *
? Agnostus, *A. Cunn.*
? Cylindria, *Lour.*

NUMBERS. GEN. 44. Sp. 650.

Santalaceæ.
POSITION.—Thymelaceæ.—PROTEACEÆ.—Lauraceæ.

ORDER CCV. LAURACEÆ.—LAURELS.

Lauri, *Juss. Gen.* 80. (1789).—Laurineæ, *Vent. Tabl.* (1799); *R. Brown Prodr.* 401; *Nees in Wall. Pl. As. Rar.* 2. 58; *Laurin. Expositio,* (1833); *Endl. Gen.* cvi.; *Meisn. Gen.* 324.

DIAGNOSIS.—*Daphnal Exogens, with anthers bursting by recurved valves, perfect leaves, and naked fruit.*

Trees, often of great size. Leaves without stipules, alternate, seldom opposite, entire or very rarely lobed. Inflorescence panicled or umbelled. Calyx 4-6-cleft, sometimes

Fig. CCCLXV.

♂ ♀ by abortion, with imbricated æstivation, the limb sometimes obsolete. Petals 0. Stamens definite, perigynous, opposite the segments of the calyx, and usually twice as numerous; the 3 innermost, which are opposite the 3 inner segments of the calyx, sterile or deficient; the 6 outermost scarcely ever abortive; anthers adnate, 2-4-celled; the cells bursting by a longitudinal persistent valve from the base to the apex; the outer anthers valved inwards, the inner valved outwards [or both valved inwards]. Some glands usually present at the base of the inner filaments. Ovary superior, 1-celled, (formed of 3 valvate carpellary leaves, and as many rib-like placentæ stationed at the sutures, all generally imperfect except one, *Endl.*), with one or two single pendulous ovules; style simple; stigma obtuse, 2- or 3-lobed. Fruit baccate or drupaceous, naked or covered, often placed upon or within the enlarged apex of the flower-stalks. Seed without albumen; embryo inverted; cotyledons large, plano-convex, peltate near the base; radicle very short, superior; plumule conspicuous.

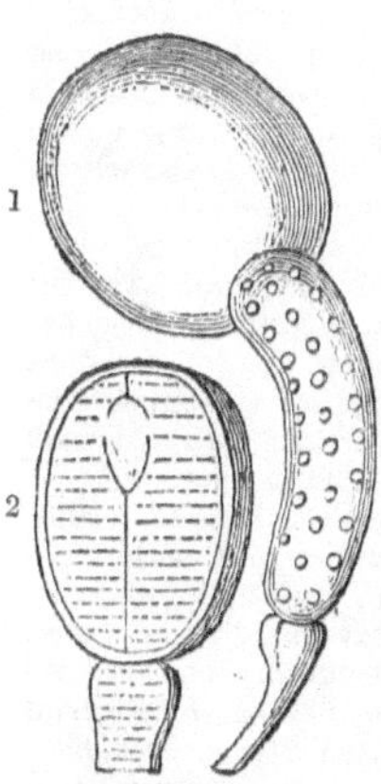

Fig. CCCLXVI.

Laurels are distinguished from all incomplete apetalous Dicotyledons, except Plume Nutmegs, by the peculiar dehiscence of their anthers, and they are divided from that Order by the ovule being pendulous, not erect. In sensible qualities they resemble Nutmegs, which are at once known by their unisexual flowers and columnar

Fig. CCCLXV.—Litsæa Baueri. 1. a male flower; 2. a female; 3. a stamen, with a gland at the base; 4. an anther, with the recurved valves; 5. a cluster of fruit; 6. a cotyledon seen from within, with the plumule adhering to the inner face.

Fig. CCCLXVI.—Dehaasia media.—*Blume.* 1. its fruit and thickened peduncle; 2. a section of it.

stamens. According to Nees von Esenbeck, their ovary is composed of three carpels; and, if so, they are as near Buckthorns as Daphnads; but this opinion does not seem to be supported by sufficient evidence; on the contrary, the exterior of the ovary and its interior cavity present all the appearance of simplicity, unless a trifling and occasional lobing of the stigma be taken as proof of a compound structure. Nees v. Esenbeck, however, describes the ovary as being really composed of 3 valvate carpellary leaves, with marginal placentæ. Berberids, another Order, with recurved anther-valves, seem far removed by their polypetalous flowers, hypogynous stamens, and copious albumen.

Trees, inhabiting cool places in the tropics of either hemisphere; in a very few instances only, straggling to the northward in North America and Europe. On the latter continent Laurus nobilis is the only species found in a wild state. Scarcely any species are known to exist on the continent of Africa. This is the more remarkable, as several species of Laurus have been found both in Teneriffe and Madeira, and others exist in Madagascar, and in the Isles of France and Bourbon.

The species of this extensive Order are in all cases more or less aromatic and fragrant; some are valuable for their timber, others bear fruits that partake of the quality of the Nutmeg, a certain number are useful febrifuges, and some yield a fixed as well as volatile oil, and an abundance of camphor. Foremost among them are Cinnamon and Cassia, two well-known spices brought to Europe from the hotter parts of Asia. According to Blume, the finest sort of Cinnamon is produced by the Cinnamomum zeylanicum of Nees; and Chinese Cassia-bark by Cinnamomum Cassia (C. aromaticum, *Nees*). But Dr. Wight has ascertained that Cassia-bark is really produced by several and perhaps nearly all the species of Cinnamomum.—*Hooker's Journal*, 2. 342. Culilawan bark, a very valuable product, with a taste of Cloves, is yielded by Cinnamomum Culilawan, and many more species of the same genus have been found to resemble Cinnamons in their peculiar qualities, especially C. nitidum, which is said to have furnished a part of the aromatic dried leaves once employed under the name of Folia Malabathri, Tamalapathri or Indi.—See Blume's *Rumphia*, the works of Dierbach, Geiger, Guibourt, and Pereira, and Endlicher's *Enchiridion*, for further information upon this subject. Many others have the quality of Cinnamon, although belonging to different genera. The Cinnamon of Santa Fé is produced by Nectandra cinnamomoides; of the Isle of France by Oreodaphne cupularis. The Clove Cassia of Brazil is the bark of Dicypellium caryophyllatum, which Martius terms "Arbor omnium Laurinearum quas Brasilia alit nobilissima." To these must be added Brazilian Sassafras (Nectandra cymbarum), Bois de Rose (Licaria guianensis), and the Casca preciosa of the Portuguese (Mespilodaphne pretiosa).

Among the timber trees must be mentioned the celebrated Greenheart of Demerara, whose wood is remarkable for its hardness, and which is the Nectandra Rodiæi of Schomburgk; the Siraballi of the same colony is a fragrant and valuable timber obtained from some species allied to Oreodaphne. A coarse Mahogany is obtained in Madeira from Persea indica, called Viñatico; the Sweet-wood of Jamaica, a hard yellow durable wood, belongs to Oreodaphne exaltata, and the Til of the Canaries, a sort of timber with an atrocious odour, bears the name of Oreodaphne fœtens.

Of those with aromatic fruits there are the Pichurim Beans of commerce, which have been ascertained to be the cotyledons of Nectandra Puchury, and have the flavour of Nutmegs of inferior quality; the Camara, or Ackawai Nutmeg, produced by Acrodiclidium Camara, *Schomb.*, considered in Guiana to be one of the most efficacious remedies in colic, diarrhœa, and dysentery; Cujumary Beans, from Aydendron Cujumary, and Laurel. The Clove Nutmegs of Madagascar are gathered from Agathophyllum aromaticum, and Brazilian Nutmegs from Cryptocarya moschata.

Among febrifuges the Bibiri or Beebeeru of Guiana, Nectandra Rodiæi, claims a high rank: Dr. Maclagan has shown that sulphate of Beebeerim acts with rapid and complete success in arresting ague.—*Trans. R. S., Edinb.*, xv. The bark of Caryodaphne densiflora is brownish, tonic, and contains a great quantity of bitter, somewhat balsamic extractive matter; the leaves are gratefully aromatic; they are used in infusion, like tea, against spasms of the bowels, and the convulsive affections of pregnant women. Sassafras officinale, a large tree inhabiting the United States, has great reputation as a powerful sudorific, and, combined with Guaiacum and Sarsaparilla, in cutaneous affections, chronic rheumatism, and old syphilitic maladies. The dried leaves contain so much mucilage that they are used in Louisiana for thickening soup, like Hibiscus esculentus. The bark of the branches as well as the wood has been employed: but they are inferior to the bark of the root. In Sumatra the place of this tree is taken by another species, the Sassafras Parthenoxylon, called Oriental Sassafras. Benzoin odoriferum is another plant with similar qualities. Its bark is highly aromatic, stimulant and tonic; and is given in decoction or powder in intermittents. An infusion of the twigs acts as a vermifuge ; the oil of the berries, which are aromatic, is stimulant. These berries are said to have been used in the United States during the American war as a substitute for Allspice. Laurus nobilis has also aromatic leaves, but they are chiefly used by confectioners. Among fatty matters may be mentioned that of Tetranthera Roxburghii, whose fruit yields a greasy exudation. It is a fixed oil which is supposed to constitute the principal part of the fruit of Persea gratissima, so much esteemed in the West Indies under the name of the Avocado Pear. Camphor is by no means an uncommon secretion of these plants. It occurs abundantly in some species of Cinnamomum, especially in their roots, which are so much contaminated by it as to be unfit for use as a spice. The Camphor of commerce, however, or Chinese Camphor, is obtained in Camphora officinarum from the wood, branches and leaves, by means of dry distillation. It is a kind of Stearoptine remaining after the Elæoptine or ethereal oil of the live tree is evaporated.—*Nees.* It is chiefly produced in the island of Formosa, and brought by the Chinchew junks in very large quantities to Canton, whence foreign markets are supplied. In some cases a volatile oil is obtained from the Laurels in large quantities ; that of Oreodaphne opifera, a tree found in vast forests between the Oronoko and the Parime, is produced in great abundance by merely making an incision into the bark with an axe, as deep as the liber. It gushes out in such quantity, that several quarts may be obtained by a single incision. It has the reputation of being a powerful discutient. The fruit of this tree yields upon distillation a limpid volatile oil of a yellow wine-colour, an aromatic acrid taste, and smell as if old oil of Orange-peel had been mixed with oil of Rosemary. It is used in Brazil in contractions of the joints, pains in the limbs, and similar cases, under the name of Canella de Cheiro.—*Martius.*

GENERA.

- Cinnamomum, *Burm.*
- *Malabathrum*, Burm.
- Camphora, *Nees.*
- Apollonias, *Nees.*
- Phœbe, *Nees.*
- Persea, *Gærtn.*
- *Eriodaphne*, Nees.
- Machilus, *Nees.*
- Boldu, *Feuill.*
- Alseodaphne, *Nees.*
- Hufelandia, *Nees.*
- Dehaasia, *Blume.*
- *Haasia*, Blume.
- Endiandra, *R. Br.*
- Beilschmiedia, *Nees.*
- Cecidodaphne, *Nees.*
- Cryptocarya, *R. Br.*
- *Peumus*, Nees.
- *Gomortega*, Ruiz. et P.
- *Adenostemon*, Pers.
- *Keulia*, Molina.
- Caryodaphne, *Blume.*
- Agathophyllum, *Juss.*
- *Evodia*, Gærtn.
- *Ravensara*, Sonner.
- Mespilodaphne, *Nees.*
- Aydendron, *Nees et Mart.*
- Evonymodaphne, *Nees.*
- Acrodiclidium, *Nees.*
- Misanteca, *Schlechtend*
- Nectandra, *Rottb.*
- *Pomatia*, Nees.
- *Porostema*, Schreb.
- Dicypellium, *Nees.*
- ? *Licaria*, Aubl.
- Petalanthera, *Nees.*
- Pleorothyrium, *Nees.*
- Teleiandra, *Nees.*
- Leptodaphne, *Nees.*
- Ajovea, *Aubl.*
- *Douglasia*, Schreb.
- *Ehrhartia*, Scop.
- Gœppertia, *Nees.*
- *Endlicheria*, Nees.
- *Schauera*, Nees.
- Oreodaphne, *Nees.*
- *Aperiphracta*, Nees.
- *Agriodaphne*, Nees.
- *Ocotea*, Gærtn. f.
- *Ceramophora*, Nees.
- *Adenotrachelium*, Nees
- *Umbellularia*, Nees.
- ? Menestrata, *Flor. Flum.*
- ? Linharea, *Arruda.*
- Camphoromœa, *Nees.*
- Ocotea, *Aubl.*
- *Strychnodaphne*, Nees.
- *Sennebiera*, Neck.
- Gymnobalanus, *Nees.*
- Sassafras, *Nees.*
- Benzoin, *Nees.*
- *Calosma*, Presl.
- Cylicodaphne, *Nees.*
- Tetranthera, *Jacq.*
- *Litsea*, Lam.
- *Tomex*, Thunb.
- *Berrya*, Klein.
- *Sebifera*, Lour.
- *Hexanthus*, Lour.
- *Glabraria*, Linn.
- *Fiwa*, Gmel.
- Polyadenia, *Nees.*
- Laurus, *Tournef.*
- Lepidadenia, *Nees.*
- Dodecadenia, *Nees.*
- Actinodaphne, *Nees.*
- *Iozoste*, Nees.
- Daphnidium, *Nees.*
- Litsæa, *Juss.*
- *Tetradenia*, Nees.
- *Darwinia*, Dennst.

NUMBERS. GEN. 46. Sp. 450.

Atherospermaceæ.

POSITION.—Thymelaceæ.—LAURACEÆ.—Cassythaceæ.

ORDER CCVI. CASSYTHACEÆ.—DODDER-LAURELS.

Laurinæ, § Cassytheæ, *Nees ab Esenb. Laurin. Expos.* 20. (1833).—Cassytheæ, *Lindl. Nixus. Pl.* 15. (1833).—Cassythaceæ, *Ed. Pr.* (1836).

DIAGNOSIS.—*Daphnal Exogens, with anthers bursting by recurved valves, scales for leaves, and fruit buried in a succulent permanent calyx.*

These plants have quite the appearance of Dodders, and, like them, appear to live parasitically on other plants. They have no leaves properly so called, but scales appear here and there on their cord-like colourless twining stems. The general structure of their flowers is that of Laurels. The calyx is 6-parted, the 3 outer divisions being small and inconspicuous. The stamens are petaloid, twelve in number, in 4 rows; the two external rows are perfect, with 2-celled anthers, whose valves are recurved and turned inwards; the next row is very much smaller, and has a pair of glands at the base of each, while the valves of the anthers turn outwards; the fourth row is scale-like and abortive. The ovary is one-celled, and contains one ovule; it extends upwards into a short style with a simple stigma. The fruit is a nut, coated by the succulent, enlarged, and permanent calyx; it contains a single seed without albumen, an embryo with plano-convex cotyledons, and an inclosed superior radicle.

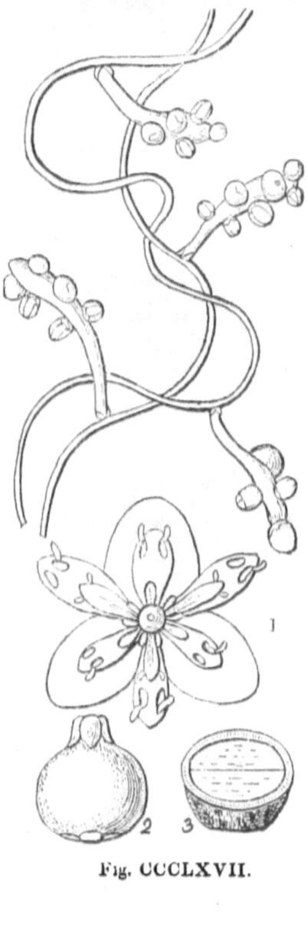

Fig. CCCLXVII.

The structure, then, is nearly that of Laurels, the main difference consisting in the fruit being inclosed in a berried calyx. I formerly supposed that more valid distinctions existed, having been misled by a description given by Nees v. Esenbeck. Mr. Gardner has, however, shown that this was very erroneous (*Hooker's Journal*, 2. 26), and he entertains no doubt about the identity of Laurels and Dodder-Laurels. It seems to me, however, better to keep them distinct until some connecting link shall have been discovered, if there be such a thing.

The species are found in the hottest parts of the world.

Nothing is known of their uses.

GENUS.

Cassytha, *L.*
Volutella, Forsk
Calodium, Lour.

NUMBERS. GEN. 1. Sp. 9.

Cuscutaceæ ?
POSITION.—Lauraceæ.—CASSYTHACEÆ.————

Fig. CCCLXVII. Cassytha filiformis. 1. a flower; 2. a fruit; 3. a section of it transversely.

ALLIANCE XLII. *ROSALES.*—THE ROSAL ALLIANCE.

DIAGNOSIS.—*Perigynous Exogens, with monodichlamydeous flowers, distinct carpels, sutural placentæ, definite seeds, corolla, if present, polypetalous, and an amygdaloid embryo with little or no albumen.*

The sequence of affinities seems to be broken when Daphnals are stationed next Rosals; but if Calycanths are regarded as the equivalent among Rosals of Plume Nutmegs among Menispermals, the transition is not so violent; for the relation of Laurels to Plume Nutmegs is usually admitted, and therefore their affinity to Calycanths must also be conceded. In fact, in Calycanths we have the apetalous structure and aromatic qualities of Laurels combined with the peculiar characters of Roseworts.

The Rosal Alliance is in many instances known by its indefinite stamens, but that character is not found in any large number of Leguminous plants, and is departed from even among Roseworts themselves. Their apocarpous fruit, small number of seeds, and amygdaloid embryo, are better characteristics. The obliquity of the carpels will in all instances show that those organs are simple, and a part of a system of separation, not consolidation, and by this circumstance Roseworts are clearly known from Daphnals, in which the fruit, although simple in appearance, has probably a compound structure.

Rosals touch Ficoidals by Sanguisorbs, and Saxifragals by Roseworts themselves; for the genera of the latter have not unfrequently been confounded by even good Botanists with those of Saxifrages. The small embryo and albumen of the latter offer, however, a clear mark of distinction.

NATURAL ORDERS OF ROSALS.

Flowers consisting of numerous imbricated scales. Cotyledons convolute.	207. CALYCANTHACEÆ.
Flowers polypetalous (or apetalous), nearly or quite regular. Carpel solitary. Style proceeding from the base of the ovary	208. CHRYSOBALANACEÆ.
Flowers polypetalous (or apetalous), papilionaceous or leguminous. Carpel solitary, with the style proceeding from the apex of the ovary	209. FABACEÆ.
Flowers polypetalous, regular, drupaceous. Carpel solitary, with the style proceeding from the apex of the ovary . .	210. DRUPACEÆ.
Flowers polypetalous, regular. Carpels adhering to the calyx by their back.	211. POMACEÆ.
Flowers apetalous. Carpel solitary, inclosed in a hardened calyx-tube forming a false pericarp	212. SANGUISORBACEÆ.
Flowers polypetalous. Carpels free from the calyx, and quite or nearly so from each other	213. ROSACEÆ.

ORDER CCVII. CALYCANTHACEÆ.—CALYCANTHS.

Calycantheæ, *Lindl. in Bot. Reg. fol.* 404. (1819); *DC. Prodr.* 3. 1.; *Endl. Gen.* cclxxi.; *Meisner Gen. p.* 106.—Calycanthinæ, *Link. Enum.* 2. 66. (1822).

DIAGNOSIS.—*Rosal Exogens, whose flowers consist of numerous imbricated scales, and have convolute cotyledons.*

Shrubs, with square stems, having 4 woody imperfect axes, surrounding the central ordinary one. Leaves opposite, simple, scabrous, without stipules. Flowers axillary, solitary. Sepals and petals confounded, indefinite, imbricated, combined in a fleshy tube. Stamens

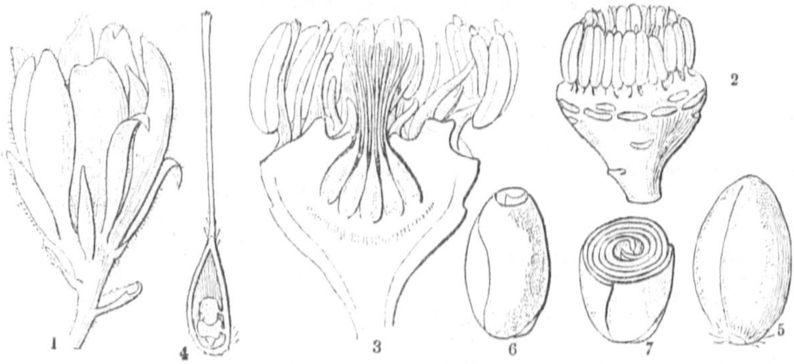

Fig. CCCLXVIII.

indefinite, inserted in a fleshy rim at the mouth of the tube, the inner sterile; anthers adnate, turned outwards. Ovaries several, simple, 1-celled, with one terminal style, adhering to the inside of the tube of the calyx. Ovules anatropal, solitary, or sometimes 2, of which one is abortive, ascending. Nuts inclosed in the fleshy tube of the calyx, 1-seeded, indehiscent. Seed ascending; albumen none; cotyledons convolute, with their face next the axis; radicle inferior.

Jussieu originally placed this Order at the end of Roseworts; he subsequently referred it to Monimiads; and I afterwards formed it into a particular family. With Monimiads it is less nearly related than it appears to be, the principal points of resemblance being the collection of several nuts within a fleshy calyx in both Orders; for Calycanths can scarcely be considered apetalous, as some Monimiads are, on account of the obvious petals of Chimonanthus. The imbricated sepals, in Calycanthus chocolate-coloured and becoming confounded with the petals, the fragrance of the flowers, and the plurality of ovaries, seem to indicate an affinity with Magnoliads, and especially with Illicium; but the decidedly perigynous stamens and fleshy calyx inclosing the ovaries in its tube, the highly developed embryo, and want of albumen, are great objections to such an approximation. Myrobalans agree in having an exalbuminous embryo, with convolute cotyledons; but with this their resemblance ceases. Myrtleblooms also agree in this same particular, in the case of Punica; and their opposite leaves, without stipules, and frequent fragrance, strengthen the affinity indicated by the embryo. Roseworts, however, to which Jussieu originally referred Calycanthus, agree much more nearly in the perigynous insertion of their stamens, in the peculiar structure of their calyx, the tube of which in the Rose is entirely analogous to that of Calycanths, in the superposition of their ovules when two are present, and in the high development of their exalbuminous embryo; upon the whole, therefore, no Order appears to have so much affinity with Calycanths as Roseworts; and the sagacity of Jussieu, in originally referring Calycanthus to that Order, is completely confirmed by the discovery recently made by Lowe, that the cotyledons of Chamæmeles, a genus of Appleworts, are convolute. This, I think, fixes the station of Calycanths in the neighbourhood of Roseworts, from which they are distinguished by the imbricated sepals, and the anthers, partly

Fig. CCCLXVIII.—Calycanthus floridus. 1. a flower; 2. the same without the sepals and petals; 3. a perpendicular section of the last; 4. a section of an ovary; 5. a nut; 6. an embryo; 7. a transverse section of it.

fertile and partly sterile, being turned outwards. This Order is also characterised by the singular structure of the wood, a peculiarity originally remarked by Mirbel in one species, and which I have since ascertained to exist in all. In the stems of these plants there is the usual deposit of concentric circles of wood around the pith, and in addition four very imperfect centres of deposition on the outside next the bark ; a very unusual structure. A good figure of this interesting fact has been given by Mirbel in the *Annales des Sciences Naturelles*, vol. 14. p. 367. It must be also added that the woody tissue of this Order exhibits disks extremely like those of Conifers.

Natives of North America and Japan.

The aromatic fragrance of the flowers is their well-known quality. It appears that this also exists in their bark, which is consequently employed, in the case of C. floridus, as a substitute for Cinnamon in the United States.

GENERA.

Calycanthus, *L.*
Buttneria, Duham.
Beureria, Ehret.
Basteria, Adans.
Pompadoura, Bouch.
Chimonanthus, *Lindl.*
Meratia, Nees.

NUMBERS. GEN. 2. Sp. 6.

Magnoliaceæ.
POSITION.—Rosaceæ.—CALYCANTHACEÆ.—Pomaceæ.
Myrtaceæ.

ORDER CCVIII. CHRYSOBALANACEÆ.—CHRYSOBALANS.

Chrysobalaneæ, *R. Brown, in Tuckey's Voyage to the Congo, App.* (1818); *DC. Prodr.* 2. 525. *a § of* Rosaceæ; *Bartl. Ord. Nat. p.* 405; *Endl. Gen.* cclxxiv.; *Meisner Gen. p.* 101.

DIAGNOSIS.—*Rosal Exogens, with polypetalous or apetalous flowers, which are nearly or quite regular, a solitary carpel, and a style proceeding from its base.*

Trees or shrubs. Leaves simple, alternate, stipulate, with no glands, and veins that run parallel with each other from the midrib to the margin. Flowers in racemes, or panicles, or corymbs. Calyx 5-lobed, sometimes unequal at the base, with an imbricated æstivation. Petals with short stalks, more or less irregular, either 5 or none. Stamens either definite or 00, usually irregular either in size or position. Ovary superior, consisting of a single carpel, 1- or 2-celled, cohering more or less on one side with the calyx; ovules twin, erect, anatropal; style single, arising from the base; stigma simple. Fruit a drupe of 1 or 2 cells. Seed erect. Embryo with fleshy cotyledons, and no albumen.

Fig. CCCLXIX

The obvious affinity of this Order is with Almondworts, from which it differs in having irregular stamens and petals, and a style proceeding from the base of the ovary. With Roseworts, to which Chrysobalans have a strict relation, they agree in the same manner as Almondworts, excepting the characters just pointed out. To leguminous plants, with drupaceous fruit, they approach closely in the irregularity of their stamens and corolla, and especially in the cohesion which takes place between the stalk of the ovary and the sides of the calyx; a character found, as De Candolle well remarks, in Jonesia and Bauhinia, undoubted leguminous plants: Chrysobalans are distinguished from this latter Order by the position of their style and ovules, and by the relation which is borne to the axis of inflorescence by the odd lobe of the calyx being the same as occurs in Roseworts. Brown remarks that the greater part of the Order has the flowers more or less irregular, and that the simple ovary of Parinarium has a dissepiment in some degree analogous to the moveable dissepiment of Banksia and Dryandra; but we now know, from the more recent observations of this learned Botanist upon the ovule, that the dissepiment of Proteads arises differently. The analogy of structure, as to the dissepiment of Parinarium, is to be sought in Amelanchier.

Chrysobalans are principally found in the tropical regions of Africa and America; none are recorded as natives of Asia; but there is reason to believe, from specimens of large trees seen in the forests of India, without flowers or fruit, by Wallich, that one or two species of Parinarium are indigenous in equinoctial Asia; and Royle's genus Prinsepia, founded upon a spiny plant from Nipal, is apparently referable to this Order. One species of Chrysobalanus is found as far to the north as the pine-barrens of Georgia in North America; a climate, however, as in all the regions bounding the Gulf of Mexico on the north, much more heated than that of most other countries in the same parallel of latitude.

Many of these are what in Europe are called Stone-fruits. Moquilea grandiflora yields eatable drupes in Brazil. The fruit of Chrysobalanus Icaco is eaten in the West

Fig. CCCLXIX.—Moquilea canomensis.—*Martius.* 1. a flower; 2. an ovary; 3. a perpendicular section of the last; 4. a fruit; 5. a kernel.

Indies, under the name of Cocoa-plum; another is brought to market in Sierra Leone (C. luteus); and the Rough-skinned, or Gray plum of the same colony is the produce of Parinarium excelsum. The kernel of Parinarium campestre and montanum is said by Aublet to be sweet and good to eat. The seeds of Prinsepia utilis yield by expression a useful oil.—*Royle.* The root, bark, and leaves of Chrysobalanus Icaco are prescribed in Brazil against diarrhœa, leucorrhœa, and similar maladies.—*Martius.*

GENERA.

Chrysobalanus, *Linn.*
Icaco, Plum.
Hirtella, *Linn.*
Cosmibuena, Ruiz et P.
Causea, Scop.
Balantium, Desv.
Braya, Fl. Fl.
Licania, *Aubl.*
Hedycrea, Schreb.
Batheogyne, Benth.
Leptobalanus, Benth.
Microdesmia, Benth.
Hymenopus, Benth.
Moquilea, *Mart et Zucc.*
Acia, Willd.
Moquilea, Aubl.
Couepia, Aubl.
Acioa, Aubl.
Dulacia, Neck.
? *Cyclandrophora*, Has.
Parinarium, *Juss.*
Parinari, Aubl.
Dugortia, Scop.
Petrocarya, Schreb.
Thelyra, *Thouars.*
Grangeria, *Comm.*
Prinsepia, *Royle.*
Cycnia, Lindl.
?Lecostemon, *Moç. et Sess.*
?Trilepisium, *Thouars.*
?Stylobasium, *Desf.*
Macrostigma, Hooker.

NUMBERS. GEN. 11. Sp. 50.

POSITION.—Fabaceæ.—CHRYSOBALANACEÆ.—Drupaceæ.

ORDER CCIX. FABACEÆ.—LEGUMINOUS PLANTS.

Leguminosæ, *Juss. Gen.* 345. (1789); *Bronn. Diss.* (1822); *Prodr.* 2. 93; *Walpers in Linnæa*, xiii. *Endl. Gen. p.* 1253.; *Meisn. Gen. p.* 84.

DIAGNOSIS.—*Rosal Exogens, with polypetalous or apetalous flowers, a papilionaceous corolla or a leguminous fruit, and a solitary carpel whose style proceeds from the apex.*

Herbaceous plants, shrubs, or vast trees, extremely variable in appearance. Leaves alternate, most commonly compound, occasionally marked with transparent dots; petiole tumid at the base. Stipules 2 at the base of the petiole, and 2 at the base of each leaflet. Pedicels usually articulated, with 2 bractlets under the flower. Calyx 5-parted, toothed or cleft, inferior, with the odd segment anterior; the segments often unequal, and variously combined. Petals 5, or by abortion 4, 3, 2, 1, or none, inserted into the base of the calyx, either papilionaceous or regularly spreading; the odd petal, if any, posterior. Stamens definite or indefinite, perigynous, rarely hypogynous, either distinct or monadelphous, or diadelphous; very rarely triadelphous; anthers versatile. Pistil simple, superior, 1-celled, 1- or many-seeded, commonly consisting of a single carpel, but occasionally of 2, or even of 5; style simple, proceeding from the upper margin; stigma simple. Fruit either a legume or a drupe. Seeds attached to the upper suture, solitary or several, occasionally with an aril; embryo with or without albumen, either straight or with the radicle bent upon the cotyledons; cotyledons either remaining under ground in germination, or elevated above the ground, and becoming green like leaves, always very large in proportion to the radicle, and very often amygdaloid.

Fig. CCCLXX.

The most common feature of Leguminous plants is to have what are called papilionaceous flowers; and when these exist, no difficulty is experienced in recognising them, for papilionaceous flowers are found nowhere else. Another character is to have a leguminous fruit; and by one of these two characters all the plants of the Order are known. It is remarkable, however, that one or other of these distinctions disappears in a great many cases. Cæsalpinieæ have an irregular flower, with spreading petals and stamens adhering to the calyx; others have no petals at all, or some number less than five; while Mimoseæ have perfectly regular flowers and indefinite hypogynous stamens. Detarium, Dipteryx, and others, instead of a legume, bear a fruit not distinguishable from a drupe. This last circumstance is easily to be understood, if we bear in mind that a legume and a drupe differ more in name than reality, the latter being formed upon precisely the same plan as the former, but with this modification, that its pericarp is thickened, more or less fleshy on the outside and stony on the inside, 1-seeded, and indehiscent. Hence some of the

Fig. CCCLXX.—Adenocarpus frankenioides. 1. the standard, wings, and keel split open; 2. the stamens; 3. a cross section of a seed; 4. a legume, with a portion of one of the valves turned back.

regular-flowered genera with distinct stamens may be said to be Rosaceous in flower, and Leguminous in fruit. Simple, therefore, as the diagnosis of this Order usually is, Brown is perfectly correct in asserting that, until he indicated the difference of the position of the odd lobe of the calyx in Leguminous plants and Roseworts, no positive character had been discovered to distinguish the one Order from the other.

Very few double flowers are known in this Order; those of Spartium junceum and Ulex europæus are the most remarkable; the nature of the latter I have described in detail in the *Trans. of the Hort. Soc.* vol. 7. p. 237. Two ovaries are common in Wistaria sinensis; and the same phenomenon is to be seen, according to De Candolle, in Gleditschia: it appears also to be normal in Diphaca and Cæsalpinia digyna. Aug. de St. Hilaire is said (*DC. Mém.* 52) to have found a Mimosa in Brazil with 5 carpels. On account of these and other circumstances, De Candolle assumes the carpel of Leguminous plants to be solitary by abortion, and that a whorl of 5 is that which is necessary to complete the symmetry of the flowers. Of the accuracy of this view I am satisfied; and it might have been proved from analogy, without the aid of such instances.*

In consequence of the highly irritable nature of the leaves of many of the plants of this Order, and of the tendency to irritability discoverable in them all, some Botanists have placed them at the extremity of their system, in contact with the limits of the animal kingdom. See *Agardh. Classes*, p. 4, and *Martius H. R. M.* p. 176. For observations upon the nature of this irritability, see *Dutrochet sur la Motilité, Paris*, 1824, in which the author endeavours to show that the motion is the effect of galvanic agency; and the same writer's *Nouvelles Recherches sur l'Exosmose, &c.*, in which he alters the explanation of the manner in which galvanism produces the motion, adhering, however, to his opinion of that subtle principle being the real agent. To me, however, it appears more satisfactory to attribute the phenomenon to an inherent vital action, without searching after first causes, which it is impossible, from the nature of things, to investigate.

In many respects this Order is one of the most important which the Botanist can study; more especially as it serves to show how little real importance ought to be attached to dehiscence of fruit in determining the limits of Natural Orders. What may be called the normal fruit of Leguminous plants is a legume, that is to say, a dry simple carpel, with a suture running along both its margins, so that at maturity it separates through the line of each suture into two valves; but every conceivable degree of deviation from this type occurs: Arachis and many more are indehiscent; in Carmichælia the valves separate from the suture, which remains entire, like the replum of Crucifers; in all lomentaceous genera, such as Ornithopus, the valves are indehiscent in the line of the suture, but separate transversely; in Entada a combination of the peculiarities of Carmichælia and Lomentaceæ occurs; in Hæmatoxylon the valves

* The plan of what must be regarded as the normal form of Leguminous structure, will be gathered from the following diagrams, in which fig. 1. represents the arrangement that occurs in Affonsea; 2. the

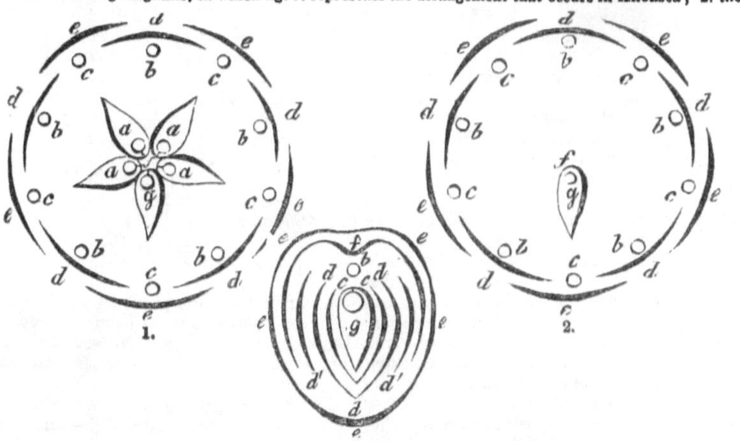

1. 2. 3.

theory of the ordinary papilionaceous condition; and 3. the actual state of such a flower. In figs. 1. and 2. *e* are the sepals; *d* the petals; *b c* stamens; *a* the abortive carpels; *g* the perfect carpel; *f* the imaginary axis. In fig. 3. *g* is the ovary; *b* the tenth free stamen; *c c* the tube, split above, and consisting of nine other stamens; *d d* the two petals that form a carina; *d' d'* the two wings; *f* the vexillum; *e e* sepals. These are taken from a paper by Walpers, in *Ann. Nat. Hist.* v. 161.

adhere by the suture and split along the axis; and, finally, Detarium, Dipteryx, and others, are true drupes, in no respect different from those of Almondworts.

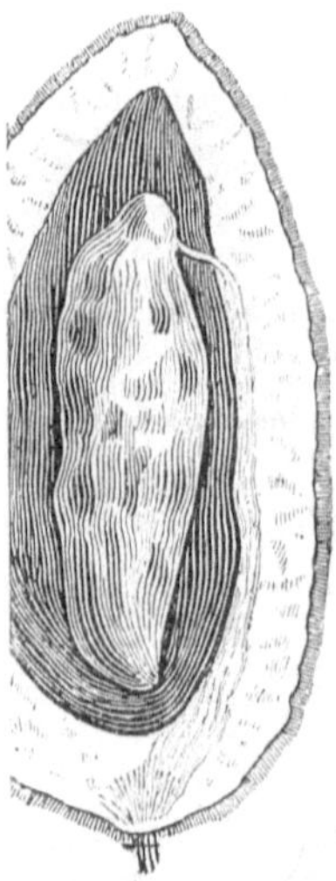

Fig. CCCLXXI.

The divisions that have been proposed in this extensive Order are explained in the succeeding List of Genera, for which I am indebted to the kindness of Mr. Bentham, who regards the groups called Cæsalpinieæ and Mimoseæ as Sub-orders only. I do not, indeed, for my own part, feel the necessity of so considering them, and should, on the contrary, with some other Botanists, be inclined to regard them as equivalent to what are elsewhere called Natural Orders. It must be confessed, however, that this is a point of little importance.

The geographical distribution of this Order has been considered with great care by De Candolle, from whom the substance of what follows is borrowed.

One of the first things that strikes the observer is, that if a number of genera of Leguminous plants have as extensive a range as those of other Orders, there is a considerable number of which the geographical limits are clearly defined. Thus the genera of New Holland are in most cases unknown beyond that vast island; the same may be said of North and South America, and the Cape of Good Hope; and there are between 14 and 15 genera unknown beyond the limits of Europe and the neighbouring borders of Asia and Africa. About 92 genera out of 280 are what are called sporadic, or dispersed over different and widely separated regions, such as Tephrosia, Acacia, Glycine, and Sophora. The species are found more or less in every part of the known world, with the exception, perhaps, of the islands of Tristan d'Acugna and St. Helena, neither of which do they inhabit; but they are distributed in extremely unequal proportions; in general they diminish sensibly in approaching the pole. This will be apparent from the following table:—

Europe, with the exception of the Mediterranean	184
Siberia	129
United States	183
China, Japan, and Cochin-China	77
Levant	250
Basin of the Mediterranean	468
Canaries	21
Arabia and Egypt	87
Mexico	152
West Indies	221
East Indies	452
Equinoctial America	605
Equinoctial Africa	130
New Holland	229
Isles of Southern Africa	42
South America, beyond the tropics	29
Cape of Good Hope	353
South Sea Islands	13

This distribution, if condensed, will give the following results:—

Equinoctial Zone	1602
Beyond the tropics to the north	1312
——————— south	524

Since the time when this calculation was made, the Order has been prodigiously enlarged, and a very considerable number of species has been added to those from the tropical parts of America, New Holland, and the Cape of Good Hope. Nevertheless the calculation, with these exceptions, is instructive as a general sketch of the statistics of this branch of Geographical Botany.

The Leguminous Order is not only among the most extensive that are known, but also one of the most important to man, whether we consider the beauty of the numerous species, which are among the gayest-coloured and most graceful plants of every region, or their applicability to a thousand useful purposes. The Cercis, which renders the gardens of Turkey resplendent with its myriads of purple flowers; the Acacia, not less valued for its airy foliage and elegant blossoms than for its hard and durable wood; the

Fig. CCCLXXI.—Dipteryx odorata, the Tonga Bean, divided perpendicularly.

Braziletto, Logwood, and Rosewoods of commerce, the Laburnum; the classical Cytisus; the Furze and the Broom, both the pride of the otherwise dreary heaths of Europe; the Bean, the Pea, the Vetch, the Clover, the Trefoil, the Lucerne, all staple articles of culture by the farmer, are so many Leguminous species. The Gums Arabic and Senegal, Kino, Senna, Tragacanth, and various other drugs, not to mention Indigo, the most useful of all dyes, are products of other species, and these may be taken as a general indication of the purposes to which Leguminous plants may be applied. There is this, however, to be borne in mind, in regarding the qualities of the Order in a general point of view; viz., that upon the whole it must be considered poisonous, and that those species which are used for food by man or animals are exceptions to the general rule: the deleterious juices of the Order not being in such instances sufficiently concentrated to prove injurious, and being, in fact, replaced to a considerable extent by either sugar or starch. This will become more apparent from the detailed account which now follows.

PAPILIONACEÆ.

It is in this part of the Order that we principally find species with nutritious, or at least wholesome qualities; thus Clover, Medick, Lucerne, Trefoil, &c., are well-known fodder plants, as are also Saintfoin, Ornithopus or Serradilla, various Astragali, Crotalaria juncea, Desmodium diffusum, Indigofera enneaphylla, &c., in different parts of the world.——The seeds of many are common articles of food, under the name of Pulse. Of these the most remarkable is the Arachis hypogæa, or under-ground Kidney-bean, whose pods are forced into the ground after the flowering has been accomplished. This and the Voandzea are very largely cultivated by the African negroes, who call the Arachis, Munduli. The seeds contain a very large quantity of oil. More common kinds of pulse are Peas, Beans, Lentils, Pigeon-peas (Cajanus), the seeds of various species of Dolichos, Phaseolus, &c. It is, however, to be remarked, that they are often very unwholesome; the roots of Phaseolus are dangerously narcotic, as will be seen hereafter. The ripe seeds of Lathyrus Aphaca, called by the French Vesce cultivé, are narcotic and produce excessive headache, but when green they are eaten without inconvenience; and Christison tells us that flour in which the seeds of Lathyrus Cicera have been ground up is poisonous. Beans themselves cannot be given to horses in much quantity without bad effects.——Of nutritious or saccharine qualities in other parts we have several useful instances. The roots of the Liquorice (Glycyrrhiza glabra) contain an abundance of a sweet mucilaginous juice, which is much esteemed as a pectoral, but it is sub-acrid; similar qualities are ascribed to Trifolium alpinum roots, and those of Glycyrrhiza echinata and glandulifera. The roots of Abrus precatorius possess exactly the properties of the Liquorice-root of the shops. In Java they are found demulcent. Those of Dolichos tuberosus and bulbosus, Apios, Pueraria, and Lathyrus tuberosus, are wholesome food. A kind of Manna is produced by species of Camel's-thorn, related to Alhagi Maurorum. It is remarkable that this secretion is not formed in India, Arabia, or Egypt: climates like those of Persia and Bokhara seeming alone suited for its production. It is the Tereng jabim of the Arabs, and is gathered by merely shaking the branches. Such is the importance of this plant as a food for cattle that the Afghans, who call it Ka-ri-shutur, or Jaursa, believe that the serious loss of those animals, experienced in the Afghan operations, arose from the want of this plant. Some writers are of opinion that this was the Manna on which the children of Israel were fed in the wilderness. A sweet quality is also found in Astragalus glycyphyllus and other species of that genus, in Saintfoin (Onobrychis sativa), in the leaves, root, and inner bark of Robinia Pseudacacia.

Fig. CCCLXXII.

Well-marked purgative properties occur in Colutea arborescens (Bladder Senna), whose leaves are used for adulterating the blunt-leaved Senna of the druggists, Coronilla Emerus (Scorpion Senna), and C. varia, which last is even poisonous; as well as in certain species of Genista, Cytisus, Robinia, Clitoria, Anagyris fœtida, &c. A decoction of the young tops of Cytisus scoparius (Broom) is diuretic and cathartic; its seeds are said to be emetic; Mead and Cullen found them useful in dropsy. Tephrosia Senna is used as a purgative by the people of Popayan.

Many are tonics and astringents. The bark of Agati grandiflora is powerfully bitter and tonic. The root of Ormocarpum sennoides is accounted in India tonic and stimulant. The root and seeds of Sophora tomentosa have been regarded as specifics in bilious sickness. African Kino is the produce of Pterocarpus erinaceus. Dr. Royle has proved that East Indian Kino is formed by Pterocarpus marsupium. Gum Dragon

Fig. CCCLXXII.—Arachis hypogæa.

and Red Sandal-wood belong to Pterocarpus Draco and Santalinus, Gum Lac to Erythrina monosperma. The Dalbergia monetaria of Linnæus yields a resin very similar to Dragon's-blood. A similar juice is yielded by Butea frondosa and superba, hardening upon their branches into beautiful ruby-coloured astringent masses, called Gum Butea, and used by the natives of North-western India for precipitating their Indigo, and in tanning; English tanners, however, object to its use on account of the colour which it communicates to leather. Euchresta Horsfieldia is regarded by the Javanese as a specific against the poison of venomous animals, or even such as is taken into the stomach; it is supposed to act as an emetic, in large doses.—*Horsfield.* The pods are sold, according to Leschenault, for 5 or even as much as 10 sous French money each. The seed of Psoralea corylifolia is considered by the native practitioners of India stomachic and deobstruent. A strong infusion of the root of Mucuna pruriens, sweetened with honey, is used by the native practitioners of India in cases of cholera morbus. A decoction of the bitter root of Tephrosia purpurea is prescribed by the Indian doctors against dyspepsia, lientery, and tympanitis. The powdered leaf of Indigofera Anil is used in hepatitis. The leaves of the Phaseolus trilobus (called Sem, or Simbi) are considered by the Indian practitioners cooling, sedative, antibilious, and tonic, and useful as an application to weak eyes. The roots and herbage of Baptisia tinctoria have been found to possess antiseptic and sub-astringent properties. They have also a cathartic and emetic effect. This emetic quality is also possessed by others. The root of Clitoria Ternatea is so, and similar properties will be found to exist among the tribe Mimoseæ.

Others are diuretics, as the roots of Beans, Genistas, Ononis, and Anthyllis Hermanniæ.

A few produce gum; Tragacanth is yielded by Astragalus verus and similar spiny species; A. creticus (ποτηριον, *Diosc.*) and A. aristatus (τραγακανθα, *Diosc.*) furnish it in Greece, A. gummifer in Mount Lebanon and in Koordistan, and A. strobiliferus in the latter country.—*Bot. Reg.* 1840, *Misc.* p. 38.

Among dyes are Indigo, produced from various species of Indigofera, especially tinctoria and cærulea, which last is particularly extolled by Roxburgh for its excellence. In Nubia, Tephrosia Apollinea furnishes it, and in the countries bordering on the Niger T. toxicaria or some allied species.—*Gard. Chron.* 1842, p. 640. The flowers of Butea frondosa and superba discharge a beautiful yellow or orange dye, Styphnolobium (Sophora) japonicum yields the same colour from the austere pulp of its pods. Baptisia tinctoria produces Indigo of indifferent quality. Genista tinctoria affords a good yellow colour, and with woad a good green. Ray says the milk of cows feeding upon it is rendered bitter, which flavour is communicated to butter and cheese.

Several produce excellent timber. The Robinia Pseudacacia or Locust tree is hard and durable; Laburnum wood is light olive green, beautifully grained. The fragrant Rosewood, or Bois de Palixandre, of the cabinet-makers, has been ascertained to belong to 2 or 3 species of Brazilian Trioptolomeas, and not to a Physocalymma, or Mimosa, as has been reported. Pterocarpus dalbergiodes, and several species of Dalbergia, especially D. Sissoo, are remarkable in India for the excellence of their wood. The Itaka wood of Guiana, remarkable for its black and brown streaks, on which account it is employed in cabinet work, is produced by Machærium Schomburgkii.

In a very large number of species narcotic properties have been recognised. The seeds of Lathyrus Aphaca have been already mentioned. Those of Abrus precatorius, whose scarlet seeds, with a black scar, are commonly used as beads, Anagyris fœtida, and others, have a similar property. This, however, is positively denied, in the case of Abrus, by Dr. Macfadgen, who asserts them to be harmless, and merely indigestible. The leaves of Arthrolobium scorpioides are capable of being employed as vesicatories. The juice of Coronilla varia is poisonous. The roots of Phaseolus radiatus are narcotic, and so are those of P. multiflorus, the Scarlet Running Kidney-bean, which a year or two ago poisoned some children at Chelsea, who had partaken of them. Both the Laburnums (Cytisus alpinus and Laburnum) have caused serious accidents to children who have swallowed their venomous seeds: and C. Weldeni is reported to poison the milk of the Dalmatian goats that browse upon its foliage. The dye called Indigo is a formidable vegetable poison. Schomburgk states that the violet blossoms of Sabinea florida are dangerous. The seeds of Ervum Ervilia, the Bitter Vetch, mixed with flour and made into bread, produce weakness of the extremities, especially of the limbs, and render horses almost paralytic. Andira inermis and retusa, and some Geoffrœas, especially G. vermifuga and spinulosa, have an anthelmintic bark, with a disagreeable smell and a sweet mucilaginous taste; the effects are drastic, emetic, purgative, and narcotic; poisonous in large doses, producing violent vomiting with fever and delirium. A few years since, hundreds of sheep perished in the Swan River Colony, in consequence of their cropping the leaves of some plant wild there; according to an official report, it was a Burtonia that produced the mischief (*Gard. Chron.*), but according to Mr. Jas. Drummond the mischief was caused by a Gompholobium.—*Lond. Journ. Bot.* 1. 95. Nothing, however, more plainly indicates the venomous nature of Leguminous plants than their being used as fish poisons. The bark of the root of Piscidia Erythrina, a common Jamaica tree, is a very usual fish poison in Jamaica, and yields a most remarkably narcotic and diaphoretic tincture. Many Tephrosias are employed in the same way, especially T. toxicaria, the young branches of which, with the leaves pounded, and sometimes mixed with quick-lime, are thrown into a pool of some mountain stream, and have an almost immediate effect. The fish are observed to become stupefied, and as it were intoxicated, and to rise to the surface, floating there with their belly upwards, so as to be readily taken by the hand. It has been remarked that the larger fish recover gradually from the effects of the poison, but that the younger fry perish. It has been suggested that the action of the plant upon the human system would resemble that of Digitalis, and might prove, in a climate where that plant does not grow, a desirable substitute.

In addition to all these uses, there is a long catalogue of species employed for miscellaneous purposes. Crotalaria juncea (Sun, Shunum, Taag, Bengal Hemp) furnishes a coarse fibre called Bengal Hemp, from which bags and low-priced canvas is largely prepared in India. The volatile oil of Dipterix odorata, or Tonka Bean, a fragrant seed used by the perfumers and makers of snuff, has been ascertained to contain a peculiar principle called Coumarin. It may be found in a crystallised state between the skin and the kernel, and exists abundantly in the flowers of Melilotus officinalis and cærulea, the latter of which gives its peculiar odour to the Chapziger cheese in Switzerland, and is said to possess styptic properties, and to have relieved cases of bloody urine from inward contusions. It is also employed in the preparation of an oily remedy for bruises.—*Pharm. Journ.* 2. 128. A decoction of the root of Indigofera tinctoria, used as a lotion, effectually destroys vermin; the juice of the young branches mixed with honey is recommended for aphthæ of the mouth in children; and Indigo in powder, sprinkled on foul ulcers, is said to cleanse them. The disease in poultry, known in the West Indies by the name of yaws, is cured by the application of a solution of Indigo by means of a rag.—*Macfadg. Fl. Jam.* 1. 251. Indigo is also used in epilepsy and erysipelas.—*Med. Gaz.* xx. 172. The hairs of the pods of Mucuna pruriens, &c., constitute the substance called Cowitch, a mechanical anthelmintic. The seeds of Astragalus bœticus are employed in Germany as a substitute for Coffee. A good many species are emollient. The leaves of Sesbania picta are highly esteemed among the Hindoos, on account of the virtues they are said to possess in hastening suppuration when applied in the form of a poultice, that is, simply made warm, and moistened with a little castor oil. The root of Pueraria tuberosa peeled and bruised into a poultice is employed by the natives of the mountains where it grows to reduce swellings of the joints. A decoction of Melilot is emollient, and is occasionally used on the Continent in lotions and enemas. A decoction of the seeds of Trigonella Fœnum Græcum (Fenugreek) is used as an emollient, and poultices are made with their flour, but only used in veterinary medicine.

CÆSALPINIEÆ.

Purgative properties are the great character of this Sub-order. Senna is their most remarkable product. The Senna of the shops consists, according to Delile, of Cassia acutifolia, Cassia Senna, and Cynanchum Argel. He says the Cassia lanceolata of Arabia does not yield the Senna of commerce, but this statement is at variance with the positive testimony of Forskhal. For the various qualities of Senna, the reader is referred to the *Flora Medica* and other works in which the subject is treated specially; it will there be found that many species yield this useful drug, which, according to Pallm (*Pharm. Journ.* 3, 584.), is not an Egyptian product, as is usually supposed, the whole of the Alexandrian supply coming from Dongola. Purgative properties are also found in the fruit of Cathartocarpus Fistula and Ceratonia Siliqua, and also of the Tamarind, the preserved pulp of which is so well known as a delicious confection, and in the leaves of Poinciana pulcherrima.—*Martius.* Many cases of eatable fruit occur in this part of the Order. Dialium indicum, also called the Tamarind Plum, has a pod formed with a delicate agreeable pulp, much less acid than the Tamarind. Two Codariums are called Brown and Velvet Tamarinds in Sierra Leone. Ceratonia Siliqua, under the name of the Carob-tree, or Algaroba-bean, is consumed in the south of Spain by horses, and has been imported into this country, it is said with profit, as a substitute for oil-cake. The dry pulp in which the seeds are buried is very nutritious, and is supposed to have been the food of St. John in the wilderness, wherefore it is called Locust-tree, and St. John's Bread. Singers are said to chew this fruit for the purpose of improving their voice.—*Pharm. Journ.* 3. 79. The seeds of the Carob-tree are said to have been the original

Carat weights of the jewellers. A similar fruit is borne by Gleditschia triacantha, called in North America the Honey Locust. In the pods of Hymenæa Courbaril, the

Fig. CCCLXXIII.

West Indian Locust-tree, there is a mealy substance in which the seeds are embedded, sweet and pleasant, but apt to purge when recently gathered; it loses this property as it becomes old. A decoction of the pulp, allowed to ferment, forms an intoxicating drink resembling beer. The succulent drupes of Detarium microcarpum are said to be agreeable to the palate of the Negroes. Some are reported to produce powerfully bitter and tonic effects. The bark and seeds of Guilandina Bonduc are of this class; the latter are very bitter; when pounded small and mixed with castor oil, they form a valuable external application in incipient hydrocele; the leaves are a valuable discutient, fried with a little castor oil, in cases of hernia humoralis. Bowdichia major, the roots of Poinciana pulcherrima, the wood of Cæsalpinia echinata in powder, are other instances of tonic qualities among these plants; and in the Dividivi or Libidibi pods, which are produced by Cæsalpinia coriaria, we have one of the most astringent of known substances. The native practitioners in India prescribe the dried buds and young flowers of Bauhinia tomentosa in certain dysenteric affections. The bark of Bauhinia variegata, and also of Cassia auriculata, are, according to Roxburgh, used by the natives in tanning and dyeing leather, as well as in medicine. The leaves of Caulotretus microstachyus and various Bauhinias are used in Brazil under the name of Unha de Boy, or Oxhoof, as mucilaginous remedies. Panococco-bark, obtained from Swartzia tomentosa, is a powerful sudorific; its wood is very hard and intensely bitter. The roots of Cæsalpinia Nuga and Moringa are diuretic. Among dyes are Logwood, the wood of Hæmatoxylon campeachianum, and the red dye yielded by several Cæsalpinias, especially C. echinata, which yields the Brazil-wood, or Pernambuco-wood of commerce. The Bukkum or Sappan-wood of India belongs to Cæsalpinia Sappan. Camwood or Barwood belongs to Baphia nitida; it yields a brilliant red colour, but it is not permanent; the dark-red seen in the English Bandana handkerchiefs is produced by it, rendered deeper by sulphate of iron. Melanoxylon Braüna, a large Brazilian tree, has a remarkable reddish-brown colouring matter in both its wood and bark. Several afford timber. The Brazil-wood of commerce is obtained from Cæsalpinia Brasiliensis. The timber of Hymenæa Courbaril, the West-Indian Locust-tree, is close-grained and tough; it is in request in England for tree-nails in planking vessels, and for the beams and planks of steam-engines. Eperua falcata is the Wallaba-tree of Guiana, according to Sir R. Schomburgk, who informs us that its wood is deep red, frequently variegated with whitish streaks, hard, heavy, shining, and impregnated with an oily resin, which makes it very durable. The bark is bitter, and is used by the Arawaak Indians as an emetic. The Purple Heart, a Guiana timber tree of great toughness, whose timber is found invaluable for resisting the shock of artillery discharges, on which account it is employed for mortar beds, is the Copaifera pubiflora and bracteata. The balsam is said to gush out of the heart of these trees in large quantities when wounded.

The size of the timber is sometimes prodigious. The Locust-trees of the West have

Fig. CCCLXXIII.—Cassia acutifolia. 1. a flower somewhat magnified.

long been celebrated for their gigantic stature, and other species are the Colossi of South American forests. Martius represents a scene in Brazil, where some trees of this kind occurred of such enormous dimensions, that fifteen Indians, with outstretched arms, could only just embrace one of them. At the bottom they were 84 feet in circumference, and 60 feet where the boles became cylindrical. By counting the concentric

Fig. CCCLXXIV.

rings of such parts as were accessible, he arrived at the conclusion that they were of the age of Homer, and 332 years old in the days of Pythagoras; one estimate indeed, reduced their antiquity to 2052 years, while another carried it up to 4104; from which he argues that the trees cannot but date far beyond the time of our Saviour. Some Indian species also yield good timber; others, as Bauhinia racemosa and parviflora, have bark employed in making rope. An oil is expressed from the seeds of some, as Cæsalpinia oleosperma; others exude a mild gum like the Mimoseæ and some other plants, which have at the same time an astringent bark. A brownish-coloured gum is said by Roxburgh to be afforded by his Bauhinia retusa; it is also collected from B. emarginata, in the Deyra Doon, and called Sem-ke-gond. Pithecolobium gummiferum yields a gum resembling Gum Senegal, in the province of Mines in Brazil.—*Martius.* The resin Anime is procured from Hymenæa Courbaril; the Copal of Mexico is supposed to be the produce of some plant allied to this. That of Madagascar, and probably of the East Indies generally, is furnished by Hymenæa verrucosa. Brazilian Copal flows from several species of Hymenæa, and from Trachylobium Martianum.—*Martius.* Aloexylum Agallochum produces one of the two sorts of Calambac, Eagle-wood, or

Fig. CCCLXXIV.—Brazilian Trees, supposed to be older than the era of our Saviour.—*Martius.*

Lign-aloes, a fragrant substance, which Loureiro states consists of a concretion of the oily particles into a resin in the centre of the trunk; it is brought on by some disease, and the tree in time dies of it. Of all perfumes it is the most grateful to Oriental nations; "stimulant, corroborant, cephalic, cardiac." Its scent is used against vertigo and paralysis. Balsam of Copaiva, a valuable acrid oil, largely employed in gonorrhœa, flows from various species of Copaifera, probably from all; the different species, however, yield the drug of different qualities.—*Mart. Mad. Med. Bras.* 115. Myrospermum peruiferum, the Quinquino or Balsam of Peru plant, furnishes a fragrant resin, not much used in medicine now, but in request among perfumers and in the manufacture of pastiles; another species, the M. toluiferum, or Balsam of Tolu plant, yields a similar product; both are employed in the preparation of pectoral lozenges. The seeds of Cassia Absus are extremely bitter, somewhat aromatic, and mucilaginous; they are brought to Cairo from the interior of Africa, under the name of Chichm or Cismatan, and are regarded as the best of remedies for Egyptian ophthalmia.

I do not find many distinct traces of poisonous action among this division of Leguminous plants; but the seeds of Detarium senegalense are said to be venomous; those of the Nicker-tree (Guilandina Bonduc) are emetic; the inner bark of Hymenæa Courbaril is anthelmintic, according to Macfadgen; the seeds of Swartzia triphylla are excessively acrid; and these, taken with the frequency of a cathartic action, seem a sufficient indication of the presence among them of some principle which in a state of concentration would be venomous.

MIMOSEÆ.

Astringency in the bark, and the production of a sort of gum in the same part, is the great characteristic of this tribe.——Of gums, Acacia Verek and Adansonii yield gum senegal on the western coast of Africa; A. nilotica and Seyal, gum arabic in Nubia; something similar is produced in New Holland by A. decurrens, and the Silver and Black

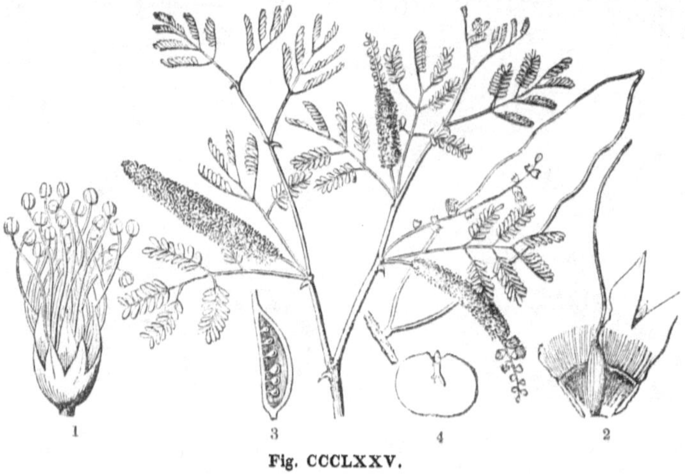

Fig. CCCLXXV.

Wattles, (A. mollissima and affinis); and in India by A. arabica and speciosa, and Vachellia Farnesiana.—*Royle.* For an account of the gum forests, see *Fl. Seneg.* 1. 246. The gum of a species of Acacia is, to the natives of Swan River, an important article of their food.—*Hook. Journ.* 2. 359.——As an instance of pulse, the seeds of Parkia africana are roasted as we roast Coffee, then bruised, and allowed to ferment in water. When they begin to become putrid, they are well washed and pounded; the powder is made into cakes, somewhat in the fashion of our chocolate; they are an excellent sauce for all kinds of meat. The farinaceous matter surrounding the seeds forms a pleasant drink, and they also make it into a sweetmeat. The natives of Tasmannia roast the ripening pods of A. Sophera, pick out the seeds and eat them.—*Backhouse.* The pulp of the pods of Inga tetraphylla, and others, is sweet and mucilaginous.——Tonic and astringent qualities are also present here. The bark of A. arabica is considered in India a powerful tonic; it is also extensively used in tanning leather. A decoction of

Fig. CCCLXXV.—Acacia Verek. 1. a flower magnified; 2. the pistil; 3. a section of the same; 4. half a seed.

its pods is used as a substitute for that of the seeds of A. concinna for washing. Its tonic powers are connected with the astringent and tanning properties of several others. Some of the Algarobas or Prosopises of the western part of South America bear fruit, the pericarp of which consists almost wholly of tannin. The bark of some of the species of Acacia abound to such a degree in tanning principles as to have become objects of commercial importance. In 1824 some tons of the extract of Acacia bark were imported from New South Wales for the use of tanners. The pods of A. nilotica are used in Nubia for tanning. The valuable astringent substance called Catechu, or Terra Japonica, is procured by boiling and evaporating the brown heart-wood of A. Catechu, or Khair-tree : it is obtained by simply boiling the chips in water until the inspissated juice has acquired a proper consistency ; the liquor is then strained, and soon coagulates into a mass. The Inga vera, and Unguis cati, with Stryphnodendron Barbatemas and Jurema, are Brazilian astringents of a similar nature. The pods of A. nilotica, are used by the tanners of Egypt, who call them Neb-neb.——Others are emetics. According to Horsfield, the Entada Pursætha of Java is emetic.——A few are purgatives. Properties of this kind exist in the pulp within the fruit of Inga vera. The same may be said of I. fæculifera, or the Pois doux of St. Domingo, that bears pods filled with a sweet pulp, which the natives use.——A small number are poisonous. The root of a Mimosa is accounted a poison in Brazil. That of Mimosa sensitiva and its allies emits a most offensive smell, resembling the odour of a sewer at the time of impending rain.—*Bot. Reg.* 1. 25. It is reported that the leaves and branches of Prosopis iuliflora are poisonous to cattle. The bark of some, as of Acacia ferruginea and leucophæa, added to jagghery water, is distilled in India as an intoxicating liquor. A drink called Chica, much used in South America, is prepared from the sweet pods of Prosopis Algaroba. "It is said that old women are employed to chew these Algarobas, and the Schinus, and then to spit them into a vessel." Water is added, and the mixture fermented.—*Chem. Gaz.* 1844. 131.——Several afford very valuable timber. That of A. arabica and Vachellia Farnesiana is used in India for wheels and tent-pegs ; that of other species attains a large size, as of A. Kalkera and A. speciosa ; the latter is dark-coloured, and close enough grained for making furniture. A. elata, xylocarpa, Sundra, odoratissima, stipulacea, and cinerea, all yield it of good quality. The wood of the Mora excelsa, the most majestic tree of Guiana, according to its discoverer, Sir R. Schomburgk, is said to be equal to Oak of the finest quality.——Saponaceous qualities reside in some species. The legumes of A. concinna (Mimosa saponaria, *Roxb.*) form a considerable article of commerce in India, and the large brown beans of Entada Pursætha, called Gela, are used by the natives for washing their hair.—*Royle.*——A few are dyes. A deep red is yielded by the chips of Adenanthera pavonina, called in India Rukta-chundun, or Red Sandal-wood.——Lastly, the fragrant flowers of Acacia Farnesiana yield, by distillation, a delicious perfume, to which also potent virtues are ascribed.

GENERA.

[The following List was drawn up by Mr. Bentham, Aug. 16, 1845.]

Suborder I. PAPILIONACEÆ.—Petals papilionaceous, imbricated in æstivation, the upper exterior.

Tribe 1. Podalyrieæ. —*Filaments all free. Legume continuous. Leaves simple or palmately compound.*

Subtribe 1. EUPODALYRIEÆ.

§ 1. Cistropical.

Anagyris, *Linn.*
Piptanthus, *Don.*
Thermopsis, *Br.*
  Thermia, Nutt.
  Scolobus, Raf.
Baptisia, *Vent.*
  Crotalopsis, Mich.
Pickeringia, *Nutt.*

§ 2. Cape.

Cyclopia, *Vent.*
  Ibbetsonia, Sims.
Podalyria, *Lam.*
  Aphora, Neck.

§ 3. Australasian.

Brachysema, *R. Br.*
Callistachys, *Vent.*
Oxylobium, *Andr.*
Podolobium, *R. Br.*
Isotropis, *Benth.*
Orthrotropis, *Benth.*
Chorozema, *Labill.*
Gompholobium, *Smith.*

Subtribe 2. PULTENEÆ.

Burtonia, *R. Br.*
Jacksonia, *R. Br.*
Daviesia, *Smith.*
Viminaria, *Smith.*
Sphærolobium, *Smith.*
Röea, *Hügel.*
Phyllota, *DC.*
Aotus, *Smith.*
Dillwynia, *Smith.*
  Xeropetalum, R. Br.
Eutaxia, *R. Br.*
Gastrolobium, *R. Br.*
Euchilus, *R. Br.*
Spadostyles, *Benth.*
Pultenæa, *Smith.*
Sclerothamnus, *R. Br.*

Subtribe 3. MIRBELIEÆ.

Mirbelia, *Smith.*
Dichosema, *Benth.*
Leptosema, *Benth.*

Tribe 2. Loteæ.—*Filaments all or 9, connate. Legume continuous. Cotyledons becoming leafy.*

Subtribe 1. LIPARIEÆ.

Liparia, *Linn.*
Priestleya, *DC.*
  Xiphotheca, Eckl. Zey.
Amphithalea, *Eckl. Zey.*
  Ingenhousia, E. Mey.
  Cryphiantha, Eckl.Zey.
  Epistemium, Walp.
Lathriogyne, *Eckl. Zey.*
  Heudusa, E. Mey.
Cœlidium, *Vog.*

Subtribe 2. GENISTEÆ.

§ 1. Hoveæ.

Hovea, *R. Br.*
  Poiretia, Smith.
  Physicarpos, Poir.
Plagiolobium, *Sweet.*
Lalage, *Lindl.*
? Platychilum, *Delaun.*
Platylobium, *Smith.*
  Cheilococca, Salisb.
Bossiæa, *Vent.*
Goodia, *Salisb.*
Templetonia, *R. Br.*
Scottia, *R. Br.*

§ 2. Borbonieæ.

Borbonia, *Linn.*
Rafnia, *Thunb.*
  Œdemannia, Thunb.
  Vascoa, DC.
  Pelecynthis, E. Mey.
Euchlora, *Eck. Zeyh.*
  Microtropis, E. Mey.
Heylandia, *DC.*

§ 3. Crotalarieæ.
Lupinus, *Linn.*
Crotalaria, *Linn.*
Chrysocalyx, Guillem.
Clavulium, Desv.
Priotropis, *Wight et Arn.*

§ 4. Lotononideæ.
Lotononis, *DC.*
Leobordea, Delil.
Leptis, Eck. Zey.
Krebsia, Eck. Zeyh.
Polylobium, Eck. Zey.
Aulacinthus, E. Mey.
Telesia, E. Mey.
Lipozygis, E. Mey.
Capnitis, E. Mey.
Maria-Antonia, Parlat.
Listia, *E. Mey.*
Rothia, *Pers.*
Xerocarpus, *Guill. Perr.*
Argyrolobium, *Eck. Zey.*
Chasmone, E. Mey.
Trichasma, Walp.
Gamochilum, Walp.
Diotolotus, Tausch ?
Melolobium, *Eck. Zeyh.*
Sphingium, E. Mey.
Dichilus, *DC.*
Calycotome, E. Mey.
Melinospermum, Walp.
Hypocalyptus, *Thunb.*
Loddigesia, *Sims.*
Lebeckia, *Thunb.*
Stiza, E. Mey.
Sarcophyllum, E. Mey.
Acanthobotrya, Eck. Zeyh.
Calobota, Eck. Zeyh.
Viborgia, *Thunb.*
Aspalathus, *Linn.*
Sarcophyllum, Thunb.
Sarcocalyx, Walp.
Buchenrœdera, Eck. Zeyh.
Scaligera, Adans.

§ 5. Cytiseæ.*
Ulex, *Linn.*
Stauracanthus, *Link.*
Adenocarpus, *DC.*
Erinacea, *Boiss.*
Spartium, *Linn.*
Spartianthus, Link.
Genista, *Linn.*
Retama, *Boiss.*
Syspone, *Griesb.*
Calycotome, *Link.*
Sarothamnus, *Wimm.*
Lembotropis, *Griseb.*
Cytisus, *Linn.*
Laburnum, *Griseb.*

Subtribe 3. TRIFOLIEÆ.
Dorycnium, *Tourn.*
Dorycnopsis, *Boiss.*
Lotus, *Linn.*
Kroferia, Mœnch.
Tetragonolobus, *Scop.*
Scandalida, Neck.
Bonjeania, *Reichb.*
Trifolium, *Linn.*
Calycomorphum, Pres.
Galearia, Presl.
Mistyllus, Presl.
Lupinaster, Mœnch.
Pentaphyllum, Pers.
Dactyphyllum, Raf.
Amoria, Presl.
Micranthemum, Presl.
Amarenus, Presl.
Paramesus, Presl.
Melilotus, *Tourn.*
Pocockia, *Ser.*
Trigonella, *Linn.*
Fœnumgrœcum, Trn.
Buceras, Mœnch.
Falcatula, Forst.
Medicago, *Linn.*
Diploprion, Vis.
Melissittis, Mœnch.
Botryolotus, *Jaub. et Sph.*
Meristotropis, *Fisch. Mey.*
Hymenocarpus, *Savi.*
Cornicina, *Boiss.*
Physanthyllis, *Boiss.*
Anthyllis, *Linn.*
Cytisopsis, *Jaub. et Spch.*
Ononis, *Linn.*
Anonis, Tourn.
Hosackia, *Dougl.*
Microlotus, Benth.
Anisolotus, Benth.
Drepanolobus, *Nutt.*
Syrmatium, Vog.
Parochetus, *Ham.*
Podolotus, *Benth.*
Goodia, *Salisb.* ?

Subtribe 4. INDIGOFEREÆ
Cyamopsis, *DC.*
Cordœa, Spreng.
Amecarpus, *Benth.*
Indigofera senegalensis
Indigofera, *Linn.*
Sphœridiophorum, Dsv.
Oustropis, Don.
Hemispadon, Endl.
Diplonyx, Raf. ?
Acanthonotus, *Benth.*
Indigofera onobrychioides.
Carmichælia, *Br.* ?

Subtribe 5. PSORALIEÆ.
Psoralea, *Linn.*
Ruteria, Mœnch.
Poikadenia, Ell.
Polytropia, *Presl.* ?
Requienia, *DC.*
Amorpha, *Linn.*
Bonapedia, Neck.
Eysenhardtia, *H. B. K.*
Dalea, *Linn.*
Parosella, Cav.
Petalostemon, *Reich.*
Kuhnistera, *Lam.*
Cylipogon, Raf.

Subtribe 6. GALEGEÆ.
Glycyrrhiza, *Linn.*
Liquiritia, Mœnch.
Galega, *Tournef.*
Calotropis, Don.
Accorombona, Endl.
Cyclogyne, *Benth.*
Ebenidium, *Jaub. et Spch* ?
Pogonostigma, *Boiss.*
Tephrosia, *Pers.*
Peineria, Mœnch.
Brissonia, Neck.
Erebinthus, Mitch.
Cracca, Linn.
Apodynomene, E. Mey.
Xiphocarpus, Presl.
Needhamia, Scop.
Chadsia, *Boj.* ?
Wistaria, *Nutt.*
Thyrsanthus, Ell.
Kraunhia, Raf.
Robinia, *Linn.*
Lennea, *L. K. O.*
Sabinea, *DC.*
Poitæa, *DC.*
Coursetia, *DC.*
Tephrosiœ, sect. Craccoides, DC. ?
Daubentonia, *DC*
Glottidium, *Desv.*
Sesbania, *Pers.*
Herminiera, *Guill. Per.*
Agati, *Rheed.*
Diphysa, *Jacq.* ?
Corynella, *DC.*
Corynitis, Spr.
Clianthus, *Sol.*
Streblorhiza, Endl.
Sutherlandia, *Br.*
Ptychosema, *Benth.*
Sylitra, *E. Mey.*
Lessertia, *DC.*
Swainsonia, *Salisb.*
Colutea, *Linn.*
Halimodendron, *Fisch.*
Halodendron, DC.
Caragana, *Lam.*
Eremosparton, *Fisch.*
Phyllolobium, *Fisch.*
Chesneya, *Lindl.*

———

Crafordia, *Raf.* ?
Philenoptera, *Fenzl.* ?

Subtribe 7. BRONGNIARTIEÆ.
Harpalyce, *Moç. Sess.*
Megastegia, Don.
Brongniartia, *H. B. K.*
Peraltea, H. B. K.

Subtribe 8. ASTRAGALEÆ.
Homolobus, *Nutt.*
Kentrophyta, *Nutt.*
Biserrula, *Linn.*
Pelecinus, Tournef.
Astragalus, *Linn.*
Oxytropis, *DC.*
Spiesia, Neck.
Phaca, *Linn.*
Guldenstædtia, *Fisch.*
Sphœrophysa, *DC.*

Tribe 3. Vicieæ.—*Filaments all or 9, connate. Legume continuous. Cotyledons fleshy. Leaves generally cirrhose.*

Cicer, *Linn.*
Pisum, *Linn.*
Ervum, *Linn.*
Vicia, *Linn.*
Faba, Tourn.
Wiggersia, Fl. Wett.
Oxypogon, Raf.
Lathyrus, *Linn.*
Aphaca, Tourn.
Ochrus, Tourn.
Clymen, Tourn.
Nissolia, Tourn.
Cicerella, Mœnch.
Anurus, E. Mey.
Astrophia, *Nutt.* ?
Orobus, *Tourn.*
Platystegia, Sweet.

Tribe 4. Hedysareæ.—*Filaments generally connate. Legume transversely articulated, with 1-seeded joints, usually separating and indehiscent.*

Subtribe 1. ARACHIDEÆ.
Stylosanthus, *Linn.*
Arachis, *Linn.*
Chapmannia, *Torr. et Gr.*

Subtribe 2. CORONILLEÆ.
Scorpiurus, *Linn.*
Coronilla, *Linn.*
Antopetitia, *Rich.*
Arthrolobium, *Desv.*
Hammatolobium, *Fenzl.*
Ornithopus, *Linn.*
Hippocrepis, *Linn.*
Bonaveria, *Scop.*
Securigera, DC.

Subtribe 3. HEDYSAREÆ.
Diphaca, *Lour.*
Pictetia, *DC.*
Brya, *Br.*
Ormocarpus, *Pers.*
Planaria, *Desv.*
Amicia, *Kunth.*
Zygomeris, Moç. Sess.
Poiretia, *Vent.*
Turpinia, Pers.
Chætocalyx, *DC.*
Bonninghausenia, Spr.
Rhadinocarpus, *Vog.*
Nissolia, *Jacq.*

———

Myriadenus, *Desv.*
Geissaspis, *Wight. Arn.*
Zornia, *Gmel.*

———

Adesmia, *DC.*
Patagonium, Schrank.
Loudonia, Bert.
Rathkea, *Schum.*
Æschynomene, *Linn.*
Isodesmia, *Gardn.*
Sommeringia, *Mart.*
Smithia, *Ait.*
Petagnana, Gmel.
Kotschya, Endl.

———

Bremontiera, *DC.* ?
Lourea, *Neck.*
Christya, Mœnch.
Alysicarpus, *Neck.*
Fabricia, Scop.
Hegetschweilera, Heer.
Eleiotis, *DC.*
Oxydium, *Benn.*
Phylacium, *Benn.*
Mecopus, *Benn.*
Uraria, *Desv.*
Doodia, Reichb.
Nicholsonia, *DC.*
Perrottetia, DC.
Desmodium, *DC.*
Dendrolobium, W. Arn.
Heteroloma, Desv.
Ototropis, Schauer.
Dollinera, Endl.
Codoriocalyx, Hassk.
Cyclomorium, *Walp.* ?
Dicerma, *DC.*
Anarthrosyne, *E. Mey.*
Lespedeza, *Rich.*
Campylotropis, *Rudge.*
Oxyramphis, *Wall.*
Hallia, *Thunb.*
Alhagi, *Tourn.*
Taverniera, *DC.*
Eversmannia, *Bunge.*

* "This § and sub-tribes 3 to 8 of Loteæ will probably require considerable modification as to their circumscription when the genera here enumerated under each shall have been more accurately examined."—G. B.

Hedysarum, *Linn.*
Echinolobium, Desv.
Onobrychis, *Tourn.*

Tribe 5. Phaseoleæ.—*Filaments all or 9, connate. Legume continuous, bivalve. Cotyledons (always ?) fleshy. Leaves usually pinnately trifoliolate.*

Subtribe 1. CLITORIEÆ.

Amphicarpæa, *Raf.*
Savia, Raf.
Xypherus, Raf.
Cryptolobus, Spr.
Dumasia, *DC.*
Pueraria, *DC. ?*
Cologania, *H. B. K.*
Clitoria, *Linn.*
Nauchea, Desv.
Ternatea, Tourn.
Neurocarpum, *Desv.*
Rhombifolium, Rich.
Martia, Leandr.
Vexillaria, *Benth.*
Pilanthus, Poit.
Centrosema, *DC.*
Steganotropis, Lehm.
Periandra, *Mart.*
Platysema, *Benth.*

Subtribe 2. KENNEDYEÆ.

Kennedya, *Vent.*
Caulinia, Mœnch.
Amphodus, Lindl.
Zichya, *Hügel.*
Physolobium, *Benth.*
Hardenbergia, *Benth.*
Leptocyamus, *Benth.*
Leptolobium, Benth.

Subtribe 3. GLYCINEÆ.

Johnia, *W. et Arn.*
Notonia, W. et Arn.
Stenolobium, *Benth.*
Cyanostremma, Bnth.
Soja, *Mœnch.*
Glycine, *Linn.*
Teramnus, P. Br.
Bujacia, E. Mey.
Shuteria, *W. et Arn.*
Galactia, *P. Br.*
Sweetia, DC.
Bradburya, Raf.
Odonia, Bertol.
Grona, Lour.
Kiesera, *Reinw.*

—

Vilmorinia, *DC. ?*
Betencourtia, *St. Hil. ?*

Subtribe 4. DIOCLEÆ.

Collæa, *DC.*
Camptosema, *H. et Arn.*
Bionia, Mart.
Cleobulia, *Mart.*
Cratylia, *Mart.*
Dioclea, *H. B. K.*
Hymenospron, Spr.
Crepidotropis, Walp.
Cymbosema, *Benth.*
Canavalia, *DC.*
Malocchia, Savi.
Clementea, Cav.
Monodon, E. Mey.
Wenderothia, Schleid
Chloryllis, *E. Mey. ?*

Subtribe 5. ERYTHRINEÆ.

Mucuna, *Adans.*
Stizolobium, Pers.
Homera, Neck.
Negretia, Ruiz. Pav.
Labradia, Swediaur.
Carpopogon, Roxb.
Macroceratides, Raddi.
Pillera, Endl.
Citta, Lour.
Erythrina, *Linn.*
Corallodendron, Tour.
Strongylodon, *Vog.*
Rudolphia, *Willd.*
Butea, *Kön.*

Subtribe 6. EUPHASEOLEÆ.

Chaseolus, *Linn.*
Strophostylis, Ell.
Vigna, *Savi.*
Callicystus, Endl.
Scytalis, E. Mey.
Sphenostyles, E. Mey. ?
Otoptera, *DC.*
Plectrotropis, *Schum.*
Dolichos, *Linn.*
Lablab, *Adans.*
Pachyrhizus, *Rich.*
Cacara, Thouars.
Psophocarpus, *Neck.*
Boton, Adans.
Diesingia, *Endl.*
Dunbaria, *W. Arn.*
Tæniocarpum, *Desv.*
Apios, *Boerh.*
Cystotropis, *Wall.*

—

Voandzeia, *Thouars.*

Subtribe 7. CAJANEÆ.

Fagelia, *Neck.*
Cajanus, *DC.*
Atylosia, *W. et Arn.*
Cantharospernum, *W. et Arn.*
Pseudarthria, *W. et Arn. ?*
Barbiera, *DC.*

Subtribe 8. RHYNCHOSIEÆ.

Orthodanum, *E. Mey.*
Hidrosia, *E. Mey.*
Eriosema, *DC.*
Euriosme, Desv.
Pyrrhotrichia, W. et A.
Pitcheria, Nutt. ?
Rhynchosia, *DC.*
Copisma, E. Mey.
Glycine, Nutt. Kunth.
Arcyphyllum, Ell.
Nomismia, *W. et Arn.*
Cylista, *Ait.*
Cyanospermum, *W. et Ar.*
Chrysoscias, *E. Mey.*
Flemingia, *Roxb.*
Ostryodium, Desv.
Lourea, Jaum.
Moghania, Jaum.
Pycnospora, *Br.*

Subtribe 9 ? ABRINEÆ.

Abrus, *Linn.*

Doubtful Genera.

Macranthus, *Lour.*
Calopogonium, *Desv.*
Cruminium, *Desv.*

Tribe 6. Dalbergieæ.—*Filaments monadelphous, or diadelphous. Legume continuous, generally indehiscent. Cotyledons (always ?) fleshy. Leaves usually pinnate.*

Cyclolobium, *Benth.*
Amerimnum, *R. Br.*
Corytholobium, *Benth.*
Hecastaphyllum, *Kunth.*
Acouroa, Aubl.
Drakensteinia, Neck.

—

Moutouchia, *Aubl.*
Grieselinia, Neck.
Pterocarpus, *Linn.*
Santalaria, DC.
Echinodiscus, *Benth.*
Weinreichia, Reichb.
Centrolobium, *Benth.*
Ancylocalyx, *Tulasne.*
Amphymenium, *H. B. K.*
Apalatoa, Aubl.
Drepanocarpus, *C. F. Mey.*
Nephrosis, Rich.
Sommerfeldtia, Schum.
Machærium, *Pers.*
Nissoliæ, sp. DC.
Gomezium, DC.
Ateleia, *Moç. et Sess.*
Brachypterum, *W. et Arn.*
Derris, *Lour.*
Pongamia, *Lour.*
Galedupa, Lour.
Sphinctolobium, *Vog.*
Neuroscapha, *Tulasne.*
Lonchocarpus, *Kunth.*
Gliricidia, *Kunth.*
Milletia, *W. et Arn.*
Berrebera, *Hochst. ?*
Endospermum, *Blum.*
Dalbergia, *Linn.*
Solori, Adans.
Triptolemæa, *Mart.*
Semeionotis, Schott. ?
Miscolobium, *Vog.*
Platymiscium, *Vog.*
Platypodium, *Vog.*
Callisemæa, Vog.
Discolobium, *Benth.*

—

Piscidia, *Linn.*
Ichthyomethia, R. Br.
Phellocarpus, *Benth.*
Müllera, *Linn. fil.*
Coublandia, Aubl.
Deguelia, *Aubl.*
Cylegonia, Neck.
Geoffroya, *Jacq.*
Andira, *Lour.*
Lumbricidia, Vell.
Vouacapoua, Aubl.
Euchresta, *Benn.*
Dipteryx, *Schreb.*
Coumarouna, Aubl.
Taralea, Aubl.
Baryosma, Gärtn.
Heinzia, Scop.
Bolducia, Neck.
Pterodon, Vog.
Commilobium, Benth.

—

Apoplanesia, *Presl.*
Spatholobus, *Hassk.*
Vatairea, *Aubl.*

Tribe 7. Sophoreæ.—*Filaments distinct. Legume continuous. Leaves pinnated, with one or several leaflets.*

Styphnolobium, *Schott.*
Edwardsia, *Salisb.*
Sophora, *Linn.*
Broussonetia, Ort.
Radiusia, Reichb.
Cadia, *Forsk.*
Spandoncea, Desf.
Panciatica, Picciav.
Ammodendron, *Fisch.*
Calpurnia, *E. Mey.*
Virgilia, *Linn.*
Cladrastis, *Raf.*
Bowdichia, *H. B. K.*
Sebipira, Mut.
Gourliea, *Gill.*
Ormosia, *Jacq.*
Toulichira, Adans.
Dibrachion, *Tulasne.*
Alexandrina, *Schomb.*
Diplotropis, *Benth.*
Spirotropis, *Tulasne.*
Macrotropis, *DC.*
Layia, Hook et Arn.
Callerya, *Endl.*
Marquartia, Vog.
Myrospermum, *Jacq.*
Calusia, Bert.
Myroxylon, Mut.
Toluifera, Linn.
Castanospermum, *Cunn. ?*

—

Dalhousiea, *Wall.*
Delaria, *Desv.*
Carpolobii, sp. Don.

Suborder II. CÆSALPINIEÆ.—Petals in æstivation imbricated, the uppermost interior.

Tribe 1. Leptolobieæ.

Leptolobium, *Vog.*
Sclerolobium, *Vog.*
Acosmium, *Schott.*
Sweetia, Spreng.
Zuccagnia, *Cas.*
Hæmatoxylon, *Linn.*
Pöppigia, *Presl.*
Diptychandra, *Tulasne.*
Cenostigma, *Tulasne.*

Tribe 2. Eucæsalpinieæ.

Cercidium, *Tulasne.*
Parkinsonia, *Plum.*
Gymnocladus, *Lam.*
Guilandina, *Linn.*
Bonduc, Plum.
Poinciana, *Linn.*
Coulteria, *H. B. K.*
Adenocalyx, Bert.
Tara, Molin.
Cæsalpinia, *Linn.*
Campecia, Adans.
Tihante, Adans.
Erythrostemon, *Lk.*
Peltophorum, *Vog.*
Brasilettia, DC.
Schizolobium, *Vog.*
Mezoneuron, *Desf.*
Pterolobium, *R. Br.*
Quartinia, A. Rich.
Reichardia, Roth.
Colvillea, *Boy. ?*
Cladotrichium, *Vog.*
Hoffmanseggia, *Cav.*
Pomaria, *Cav.*
Melanosticta, *DC.*
Burkia, *Benth.*

Tribe 3. Cassieæ.

Cassia, Linn.
Cathartocarpus, Pers.
Bactyrilobium, Willd.
Baryxylon, Lour.
Chamæfistula, G. Don.
Chamæcrista, E. Mey.
Grimaldia, Schranck.
Senna, Tournef.
Labichea, *Gaud.*
Dicorynea, *Benth.*

Tribe 4. Swartzieæ.

Moldenhawera, *Schott.*
Dolichonema, Nees.
Martia, *Benth.*
Zollernia, *Nees.*
Acidandra, Mart.

Coquebertia, Brongn.
Swartzia, *Willd.*
Riveria, Humb. et Kun.
Tounatea, Aubl.
Gynanthistrophe, Poit.
Possira, Aubl.
Rittera, Schreb.
Hoelzelia, Neck.
Aldina, *Endl.*
Allania, Benth.
Trischidium, *Tulasne.*
Swartziæ sect. Dithyria, Benth.
Cordyla, *Lour.*
Calycandra, A. Rich.

Tribe 5. Amherstieæ.

Thylacanthus, *Tulasne.*
Brownea, *Jacq.*
Hermesia, Loeffl.
Elizabetha, *Schomb.*
Heterostemon, *Desf.*
Amherstia, *Wall.*
Jonesia, *Roxb.*
Saraca, Burm.
Humboldtia, *Vahl.*
Batschia, Vahl.
Schotia, *Jacq.*
Theodorea, Medik.
Afzelia, *Sm.*
Pancovia, Willd.?
Eperua, *Aubl.*
Rothmannia, Neck.
Panzera, Willd.
Parivoa, *Aubl.*
Adleria, Neck.
Dimorpha, Willd.
Campsiandra, *Benth.*
Tachigalia, *Aubl.*
Valentynia, Neck.
Tachia, Pers.
Exostyles, *Schott.*
Melanoxylon, *Schott.*
Perittium, Vog.
Tamarindus, *Linn.*
Phyllocarpus, *Tulasne.*
Outea, *Aubl.*
Anthonota, *Beauv.*
Westia, Vahl.?
Intsia, *Thouars.*
Vouapa, *Aubl.*
Macrolobium, Vahl.
Kruegeria, Neck.
Peltogyne, *Vog.*
Trachylobium, *Hayne.*
Hymenæa, *Linn.*
Courbaril, Plum.

Tribe 6. Bauhinieæ.

Casparea, *Kunth.*
Bauhinia, *Linn.*
Pauletia, Cav.
Phanera, *Lour.*
Schnella, *Raddi.*
Caulotretus, Rich.
Perlebia, *Mart.?*
Amaria, *Mut.?*
Etaballia, *Benth.*
Cercis, *Linn.*

Tribe 7. Cynometreæ.

Cynometra, *Linn.*
Hardwickia, *Roxb.*
Copaifera, *Linn.*
Dialium, *Linn.*
Arouna, Aubl.
Codarium, Soland.
Cleyria, Neck.
Apuleia, *Mart.*
Detarium, *Juss.*
Crudya, *Willd.*
Touchiroa, Aubl.
Apalatoa, Aubl.
Waldschmidtia, Neck.
Pterogyne, *Tulasne.*
Zenkeria, *Arn.*

Tribe 8. Dimorphandreæ.

Mora, *Benth.*
Dimorphandra, *Schott.*

Gleditschia, *Linn.*
Ceratonia, *Linn.*
Acrocarpus, *Arn.*

Anoma, *Lour.*
Metrocynia, *Thouars.*
Baphia, *Afz.*
Palovea, *Aubl.*
Ginnannia, Scop.
Vatairea, *Aubl.*
Alöexylon, *Lour.*

Suborder III. MIMOSEÆ.—Corolla valvate in æstivation.

Tribe 1. Parkieæ.

Erythrophleum, *Afz.*
Fillæa, *Guillem. Perr.*
Parkia *Br.*
Pentaclethra, *Benth.*

Tribe 2. Eumimoseæ.

Entada, *Linn.*
Plathymenia, *Benth.*
Stryphnodendron, *Mart.*
Adenanthera, *Linn.*
Elephantorhiza, *Benth.*
Tetrapleura, *Benth.*
Gagnebina, *Neck.*
Prosopis, *Linn.*
Lagonychium, Stephens.
Algarobia, Benth.
Dichrostachys, *Benth.*
Caillea, Guillem.
Neptunia, *Lour.*
Desmanthus, *Willd.*
Darlingtonia, DC.
Mimosa, *Linn.*
Schranckia, *Willd.*
Leptoglottis, DC.
Leucæna, *Benth.*
Xylia, *Benth.*

Tribe 3. Acacieæ.

Acacia, *Willd.*
Vachellia, Arn.
Farnesia, Gasp.
Lysiloma, *Benth.*
Albizzia, *Durazz.*
Zygia, *R. Br.*
Calliandra, *Benth.*
Pithecolobium, *Mart.*
Enterolobium, *Mart.*
Serianthes, *Benth.*
Inga, *Willd.*
Affonsea, *St. Hil.*

NUMBERS estimated by Mr. Bentham, May, 1845.

		GEN.	SP.
Papilionaceæ			
	Podalyrieæ	33	350
	Loteæ	133	3000
	Hedysareæ	52	500
	Phaseoleæ	70	650
	Dalbergieæ	41	250
	Sophoreæ	21	50
Cæsalpinieæ		88	700
Mimoseæ		29	1000
		467	6500

? Moringaceæ.

POSITION.—Rosaceæ.—FABACEÆ.—Chrysobalanaceæ.

ORDER CCX. DRUPACEÆ.—ALMONDWORTS.

Amygdaleæ, *Juss. Gen.* 340. *a § of* Rosaceæ (1789); *Endl. Gen.* cclxxiii.; *Wight Illustr.* 1. 201.—Drupaceæ, *DC. Fl. Française,* 4. 479. (1805); *Prodr.* 2. 529. *a § of* Rosaceæ.

DIAGNOSIS.—*Rosal Exogens, with polypetalous regular flowers, a solitary carpel whose style proceeds from the apex, and a drupaceous fruit.*

Trees or shrubs. Leaves simple, alternate, usually glandular towards the base; stipules simple, mostly glandular. Flowers white or pink, in umbels or single. Calyx

1

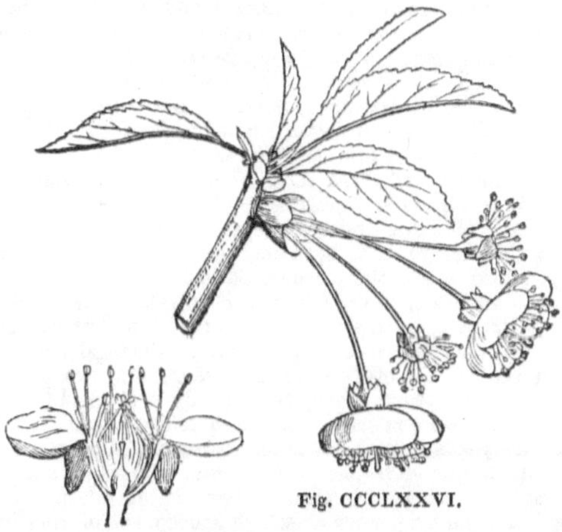

Fig. CCCLXXVI.

1

2

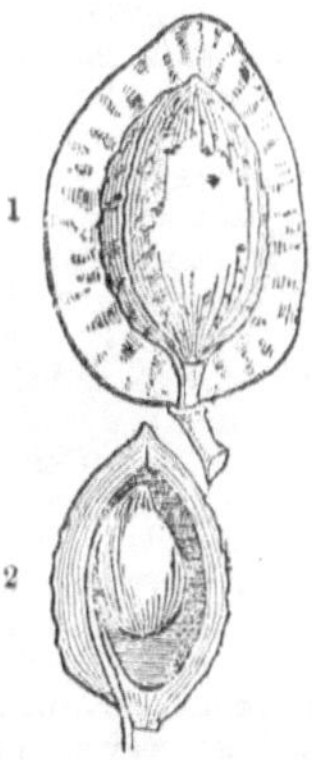

Fig. CCCLXXVII.

5-toothed, deciduous, lined with a disk; the fifth lobe next the axis. Petals 5, perigynous. Stamens 20, or thereabouts, arising from the throat of the calyx, in æstivation curved inwards; anthers innate, 2-celled, bursting longitudinally. Ovary superior, solitary, simple, 1-celled; ovules 2, suspended; styles terminal, with a furrow on one side, terminating in a reniform stigma; ovules anatropal. Fruit a drupe, with the putamen sometimes separating spontaneously from the sarcocarp. Seeds mostly solitary, suspended. Embryo straight, with the radicle pointing to the hilum; cotyledons thick, plano-convex; albumen none.

This Order is distinguished from Roseworts and Appleworts by the pistil being a solitary, simple carpel, changing when ripe into a drupe, the bark yielding gum, and by the more general presence of hydrocyanic acid; from Leguminous plants by the latter character, and also by their regular petals and stamens, and especially by the odd segment of the 5-lobed calyx of that Order being inferior, not superior; from Chrysobalans, by the terminal styles and regular petals and stamens. I have seen a monstrous Plum with an indefinite number of ovaries arising irregularly from the tube of the calyx, and therefore exhibiting a tendency, on the part of this Order, to assume one of the distinguishing characters of Roseworts. It is not a little remarkable that here, where we have an approach to the structure of Mimoseæ in Leguminous plants, we have a resemblance to the property possessed by that Sub-order, of yielding gum in the bark; the peculiar astringency of some species is also analogous to that of Acacia Catechu and the like.

Natives exclusively of the northern hemisphere, where they are found in cold or temperate climates. One species, Cerasus occidentalis, is a native of the West Indies; some Plums occur in the woods of Brazil; a kind of Almond, Amygdalus microphylla,

---

Fig. CCCLXXVI.—Cerasus communis. 1. a section of its flower.

Fig. CCCLXXVII.—Prunus domestica. 1. a section of its drupe; 2. a section of the endocarp, showing the position of the seed.

inhabits hot arid plains in Mexico; and another, A. cochinchinensis, is reputed to grow in the woods of Cochin-China.

The astringent febrifugal properties of Roseworts, with which Order this is usually combined, are also found here; as in the bark of Cerasus virginiana, which is prescribed in the United States, of the C. Capollim of Mexico, and of others to be mentioned presently. They are, however, better known for yielding an abundance of prussic, or hydrocyanic acid, a deadly principle residing in the leaves and kernel; in consequence of which some of the species are poisonous to cattle which feed upon them: as, for example, the C. capricida, which kills the goats of Nipal; and the C. virginiana, which is known in North America to be dangerous. The oil of Bitter Almonds is extremely poisonous, and many fatal cases of death arising from taking them into the stomach are on record. They have, nevertheless, been recommended as a cure for intermittent fever. They produce urticaria, and are said to be an antidote to intoxication. The flowers and kernels of the Peach have similar qualities. Dr. Christison mentions a case of a gentleman who died in consequence of having swallowed a salad of the flower; and another of a child which perished after taking a decoction of the flowers to destroy worms. The leaves, bark, and fruit of C. Laurocerasus, the common Laurel, and the oil obtained from them are virulent poisons; even the vapour of the former will destroy insect life. Martius says that this secretion is greatly increased in Brazil. C. Padus, the Bird Cherry, has similar properties, but in a less degree. They all of them, also, yield a gum analogous to gum tragacanth. Notwithstanding, however, the poisonous principle that is present in them, their fruit is, in many cases, a favourite food; that of the Amygdalus (Peach and Nectarine), Prunus (Plum), and Cerasus (Cherry), are among the most delicious with which we are acquainted; the seed of Amygdalus is familiar to us under the name of Almonds, and its oil under the name of Oil of Almonds. The bark of the root of C. Capollim is used in Mexico against dysentery. The leaves of Prunus spinosa (Sloe), and C. avium (Wild Cherry), have been employed as a substitute for Tea. The former are well known to afford one of the means used in Europe for adulterating the black tea of China. Prunus domestica, or the common Plum, yields those fruits sold in the shops under the name of Prunes, which are chiefly prepared in France, from the varieties called the St. Catherine and the Green-gage; and in Portugal from a sort which derives its name from the village of Guimaraens, where they are principally dried. They contain so large a quantity of sugar, that brandy is distilled from them when fermented; and it has even been proposed to manufacture sugar from them. The kernel of Prunus brigantiaca yields a fixed oil, called Huile des Marmottes, which is used instead of olive or almond oil. The bark of Prunus spinosa is one of the substances that has been reported to resemble Jesuits' bark in its effects. Prunus Coccomilia yields a bark, the febrifugal properties of which are spoken of very highly. According to Tenore, it is a specific for the cure of the dangerous intermittent fevers of Calabria, where it grows. A variety of Cerasus avium is used, in the Vosges and Black Forest, for the preparation of the liqueur known under the name of Kirschenwasser. The flowers of Amygdalus persica (Peach), are gently laxative, and are used advantageously for children. The kernel of Cerasus occidentalis is used for flavouring the liqueur Noyau.

GENERA.

Pygeum, *Gärtn.*
Polydontia, Blum.
Polystorthia, Blum.
Amygdalus, *Linn.*
Amygdalophora, Neck.
Persica, Tournef.
Trichocarpus, Neck.
Ceraseidos, *Zucc.*
Prunus, *Linn.*
Armeniaca, Tournef.
Prunophora, Neck.
Cerasus, *Juss.*
Cerasophora, Neck.
Padus, Endl.
Laurocerasus, Tournef.

NUMBERS. GEN. 5. SP. 110.

POSITION.—Rosaceæ.—DRUPACEÆ.—Fabaceæ ?
Thymelaceæ.

ORDER CCXI. POMACEÆ.—APPLEWORTS.

Rosaceæ, § Pomaceæ, *Juss. Gen.* 334. (1789); *DC. Prodr.* 2. 626. (1825).—Pomaceæ, *Lindl. in Linn. Trans.* 13. 93. (1821); *Endl. Gen.* cclxx.

DIAGNOSIS.—*Rosal Exogens, with polypetalous regular flowers, and carpels adhering to the calyx by the back.*

Trees or shrubs. Leaves alternate, stipulate, simple, or compound. Flowers solitary, or in terminal cymes, white or pink. Calyx adherent, 5-toothed; the odd segment posterior. Petals 5, unguiculate, inserted in the throat of the calyx; the odd

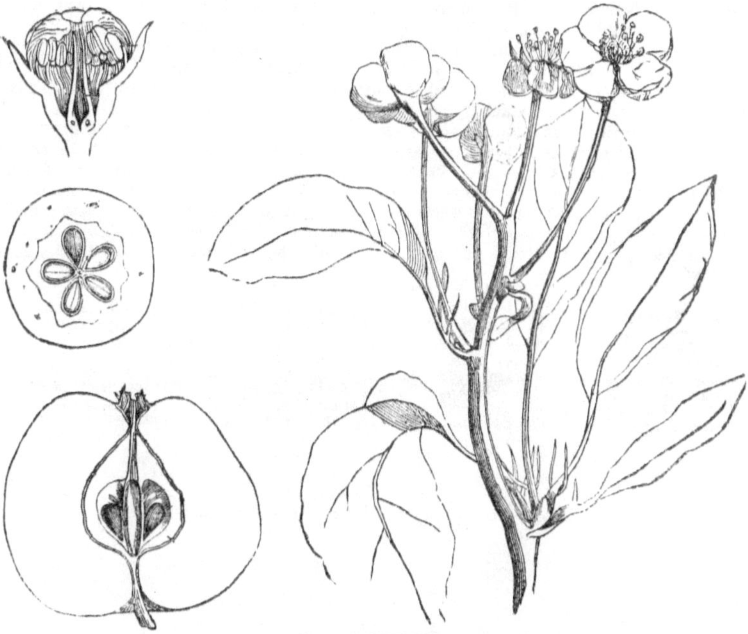

Fig. CCCLXXVIII. 1

one anterior. Stamens indefinite, inserted in a ring in the throat of the calyx. Disk thin, clothing the sides of the tube of the calyx. Ovaries from 1 to 5, adhering more or less to the sides of the calyx and each other; ovules anatropal, usually 2, collateral, ascending, very rarely solitary, sometimes 00; styles from 1 to 5; stigmas simple. Fruit a pome, 1- to 5-celled, seldom spuriously 10-celled; the endocarp either cartilaginous, spongy, or bony. Seeds ascending, solitary. Albumen none; embryo erect, with flat cotyledons, or convolute ones in Chamæmeles, and a short inferior conical radicle.

Appleworts are closely allied to Roseworts, from which they differ in the adhesion of the carpels with the sides of the calyx, and more or less with each other. The fruit is always a pome; that is, it is made up of a fleshy calyx adhering to fleshy or bony ovaries, containing a definite number of seeds. Appleworts are peculiarly distinguished by their ovules being in pairs, and side by side; while Roseworts, when they have 2 or more ascending ovules, always have them placed one above the other. Cultivated

Fig. CCCLXXVIII.—1. branch of Pyrus communis; 2. its flower divided vertically; 3. a cross section of its fruit; 4. perpendicular section of the fruit of Pyrus Malus.

plants of this Sub-order are very apt to produce monstrous flowers, which depart sometimes in a most remarkable degree from their normal state. No Order can be more instructively studied with a view to morphological inquiries; particularly the common Pear when in blossom. A remarkable permanent monster of this kind, with 14 styles, 14 ovaries, and a calyx with 10 divisions in two rows, is described in the *Revue Encyclopédique*, 43. 762.; it exhibits a tendency, on the part of Appleworts, to assume the indefinite ovaries and double calyx of Roseworts. I have seen a Prunus in a similar state. Almondworts are known by their superior solitary ovary and drupaceous fruit.

Found plentifully in Europe, Northern Asia, the mountains of India, and North America; rare in Mexico, unknown in Africa, except on its northern shore, and in Madeira, and entirely absent from the southern hemisphere; a solitary species is found in the Sandwich Islands.

The fruit as an article of the dessert, and the flowers for their beauty, are the chief peculiarities of this Order, which consists exclusively of trees and bushes, without any herbaceous plant. The Apple, the Pear, the Sorb, the Medlar, the Quince, the Service, the Rowan-tree or Mountain Ash, are all well known, either for their beauty or their use. The wood of the Pear is almost as hard as Box, for which it is even substituted by wood engravers; the timber of the Beam-tree (Pyrus Aria) is invaluable for axletrees. The bark of Photinia dubia is used in Nipal for dyeing scarlet. Malic acid is contained, in considerable quantity, in Apples; it is also almost the sole acidifying principle of the berries of the Mountain Ash (Pyrus Aucuparia). The mucilaginous seeds of the Quince are employed in medicine; its fragrant fruits are used in the preparation of a kind of wine analogous to Cider and Perry, obtained from Apples and Pears. Wohler has found œnanthic ether in the rind of the Quince. Prussic acid occurs in their seeds, and is even abundant in Cotoneaster Uva Ursi, and microphylla. The flowers, bark, and root of Pyrus Aucuparia contain so much of the peculiar essential Oil of Almonds as to yield fully as much hydrocyanic acid as that procurable from an equal weight of Cherry-laurel leaves.—*Buchn. Rep.* 27. 238.

GENERA.

Cydonia, *Tournef.*
Chænomeles, Lindl.
Pyrus, *Lindl.*
Pyrophorum, Neck.
Apyrophorum, Neck.
Lazarolus, Medik.
Halmia, Medik.
Malus, Tournef.
Aria, DC.
Torminaria, DC.
Eriolobus, DC.
Sorbus, Linn.
Aucuparia, Medik.
Adenorhachis, DC.
Aronia, Pers.
Chamæmespilus, DC.
Osteomeles, *Lindl.*
Mespilus, *Lindl.*
Mespilophora, Neck.
Amelanchier, *Medik.*
Petromeles, Jacq. f.
Peraphyllum, *Nutt.*
Cotoneaster, *Medik.*
Nägelia, *Lindl.*
Hesperomeles, *Lindl.*
Eriobotrya, *Lindl.*
Photinia, *Lindl.*
Myriomeles, *Lindl.*
Chamæmeles, *Lindl.*
Rhaphiolepis, *Lindl.*
Cratægus, *Linn.*
Stranvæsia, *Lindl.*

NUMBERS. GEN. 16. Sp. 200.

Onagraceæ.
POSITION.—Rosaceæ.—POMACEÆ.—Drupaceæ.
Myrtaceæ.

ORDER CCXII. SANGUISORBACEÆ.—SANGUISORBS.

Rosaceæ, § Sanguisorbeæ, *Juss. Gen.* 336. (1789); *DC. Prodr.* 2 588.—Cliffortiaceæ, *Martius Conspectus*, 216; *Meisner*, p. 105.

DIAGNOSIS.—*Rosal Exogens, with apetalous flowers, a solitary carpel, inclosed in a hardened calyx-tube forming a false pericarp.*

Herbaceous plants or under-shrubs, occasionally spiny. Leaves simple and lobed, or compound, alternate, with stipules. Flowers small, often capitate. Often ♂ ♀ by

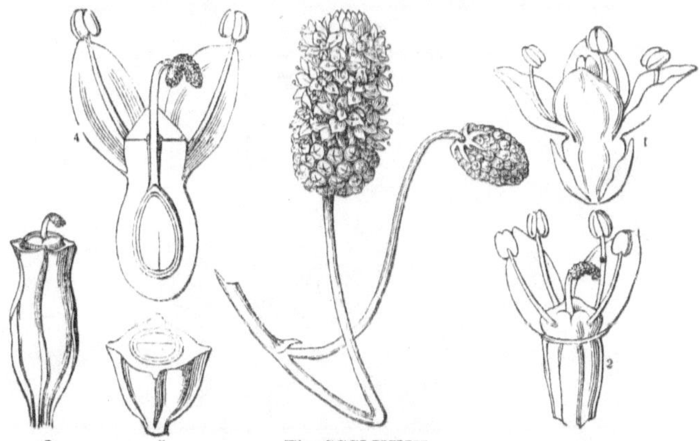

3 5 Fig. CCCLXXIX.

abortion. Calyx with a thickened tube and a 3- 4- or 5-lobed limb, its tube lined with a disk. Petals none. Stamens definite, sometimes fewer than the segments of the calyx, with which they are then alternate, arising from the orifice of the calyx; anthers 2-celled, innate, bursting longitudinally, occasionally 1-celled, bursting transversely. Ovary solitary, simple, with a style proceeding from the apex or the base; ovule solitary, always attached to that part of the ovary which is next the base of the style; stigma compound or simple. Nut solitary, inclosed in the often indurated tube of the calyx. Seed solitary, suspended or ascending; embryo without albumen; radicle superior or inferior; cotyledons large, plano-convex.

This Order, usually combined with Roseworts, appears to demand a distinct station, on account of its constantly apetalous flowers, its hardened calyx, and the reduction of carpels to one only; it is not, however, distinguishable by any other characters; and therefore Agrimonia, sometimes stationed here, must be preserved among Roseworts, because of its petals. Its habit, indeed, is by no means that of Sanguisorbs. Usually the ovule is suspended, the style arising from below the apex of the carpel; but when the style proceeds from the base of the carpel, the ovule is ascending, in all cases adhering to the ovary immediately over against the origin of the style. Various kinds of adhesion between the leaves and the stipules take place in the genus Cliffortia, and have given rise to a number of errors; for an explanation of which, see De Candolle's remarks in the *Annales des Sciences Naturelles*, 1. 447.

Natives of heaths, hedges, and exposed places in Europe, North and South America beyond the tropics, and the Cape of Good Hope; in which latter country they represent the Roseworts of Europe.

Their general character is astringency. A decoction of Alchemilla vulgaris is slightly tonic; and is asserted, by Frederick Hoffmann and others, to have the effect of restor-

Fig. CCCLXXIX.—Sanguisorba officinalis. 1. a flower with a pair of bracts; 2. the same with half the calyx cut away; 3. a ripe fruit, from which the calyx has been removed; 4. a vertical section of fruit and calyx; 5. transverse section of a fruit.

ing the faded beauty of ladies to its earliest freshness. Sanguisorba officinalis, or common Burnet, is a useful fodder. The root of Sanguisorba canadensis is said to be bitter, astringent, nauseous, and emetic, and its fruit stupefying.—*Endl.* The leaves of Acæna Sanguisorba are said to be an excellent substitute for Tea. The plant is common everywhere in Tasmannia, and is well known from the annoyance caused by its fruit hooking to the stockings and other parts of the dress of pedestrians.—*Backhouse.* The Peruvians employ a decoction of Margaricarpus setosus, a little needle-leaved bush with pearly succulent fruit, against hærmorrhoids.

GENERA.

Alchemilla, *Tournef.*
Aphanes, Linn.
Adenostoma, *Hook. et Arn.*
Acæna, *Vahl.*
Ancistrum, Forst.

Ptilochæta, *Turcz.*
Sanguisorba, *Linn.*
Poterium, *Linn.*
Bencomia, *Webb.*
Leucosidea, *Eckl. et Zeyh.*

Tetraglochin, *Pöpp.*
Polylepis, *Ruiz et Pav.*
Margyricarpus, *Ruiz et P.*
Cliffortia, *Linn.*
Morilandia, Neck.

NUMBERS. GEN. 12. Sp. 125.

Scleranthaceæ.
POSITION.—Drupaceæ.—SANGUISORBACEÆ.—Rosaceæ.
Nyctaginaceæ.

ORDER CCXIII. ROSACEÆ.—ROSEWORTS.

Rosaceæ, *Juss. Gen.* 334. *in part.* (1789); *DC. Prodr.* 2. 525; *Endl. Gen.* cclxxii.; *Meisner Gen. p.* 101.—§ Sanguisorbeæ, *Juss. Gen.* 336. (1789); *DC. Prodr.* 2. 588; *Ed. Pr. p.* 148.—Cliffortiaceæ, *Martius Conspectus, No.* 216.—Neuradeæ, *DC. Prodr.* 2. 548. (1825); *Martius Conspectus, No.* 314. (1835).—Grieleæ, *Sweet.*

DIAGNOSIS.—*Rosal Exogens, with polypetalous flowers, and carpels both free from the calyx, and quite or nearly so from each other.*

Herbaceous plants or shrubs. Leaves simple or compound, alternate, often with 2 stipules at their base, occasionally dotted. Flowers variously arranged, generally ⚥,

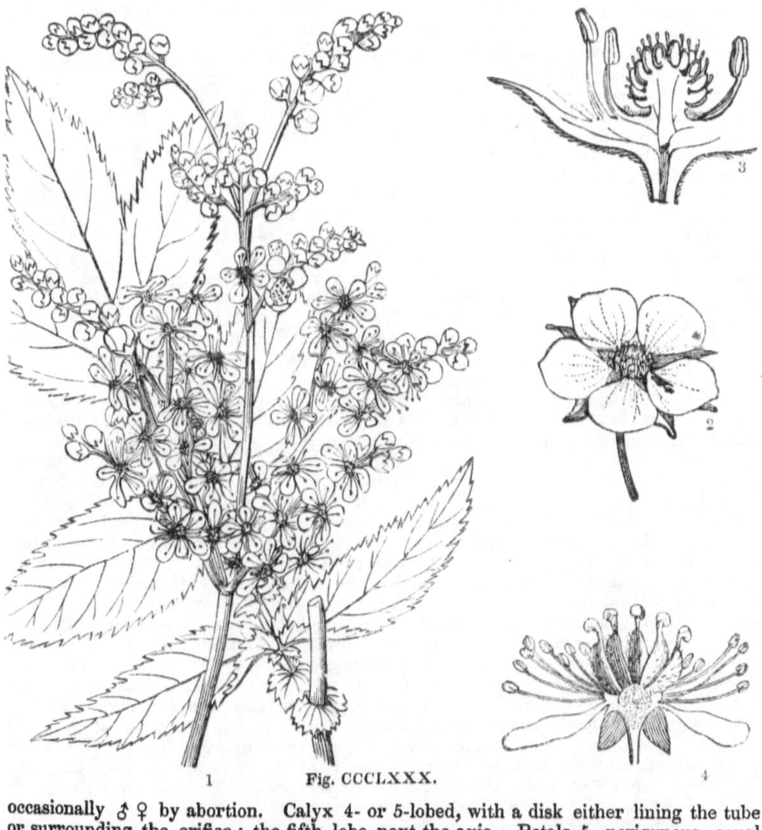

1 Fig. CCCLXXX. 4

occasionally ♂ ♀ by abortion. Calyx 4- or 5-lobed, with a disk either lining the tube or surrounding the orifice; the fifth lobe next the axis. Petals 5, perigynous, equal or 0. Stamens definite in number or 00, arising from the calyx, just within the petals, in æstivation curved inwards; anthers innate, 2-celled, bursting longitudinally. Ovaries superior, either solitary or several, 1-celled, sometimes cohering into a plurilocular pistil; ovules 2 or more, anatropal, suspended, very rarely erect; styles lateral; stigmas usually simple, and emarginate on one side. Fruit either 1-seeded nuts, or acini, or follicles containing several seeds. Seeds suspended, rarely ascending. Embryo straight, with a taper short radicle pointing to the hilum, and flat cotyledons. Albumen 0.

This Order furnishes the best of all analogies with the hypogynous sub-class, present-

---

Fig. CCCLXXX.—1. Spiræa ulmaria; 2. flower of Fragaria vesca; 3. a section of it; 4. section of the flower of a Spiræa.

ing many of the more important characters of Crowfoots, and in some measure their habits. It is, however, known by its perigynous stamens and exalbuminous seeds, whose embryo, though small, is amygdaloid. It differs from Appleworts in its ovary being superior, and from Almondworts in that organ being single and changing into a drupe. Saxifrages, which stand very near Roseworts, are readily known by their albuminous seeds, definite stamens, and partially combined, somewhat valvate carpellary leaves. Chrysobalans have a single carpel, but their style originates from the base, not the apex of the ovary. Sanguisorbs are apetalous, with the tube of the calyx hardened and changed into a false pericarp. That Roseworts have some intimate relationship with Myrtleblooms is proved by Appleworts; but a new evidence of this fact has lately been obtained in the form of Roses discovered in China by Mr. Fortune, which have faintly but distinctly transparent dots in the leaves.

Natives chiefly of the temperate or cold climates of the northern hemisphere; a very few are found on high land within the tropics, and an inconsiderable number in the southern hemisphere. Only one species occurs in the West Indies, viz. Rubus jamaicensis; several are natives of high land in the East Indies, within the tropics, especially Potentillas and Rubi; the South American species chiefly consist of a few kinds of Rubus, and plants belonging to the section Quillaiæ which are all South American. Neuradeæ are found in the north of Africa and at the Cape of Good Hope, perhaps also in Mexico. An elaborate account of the geographical distribution of these plants has been given in the *Linnæa*, vol. xvii. p. 549, by Mr. Frankenheim.

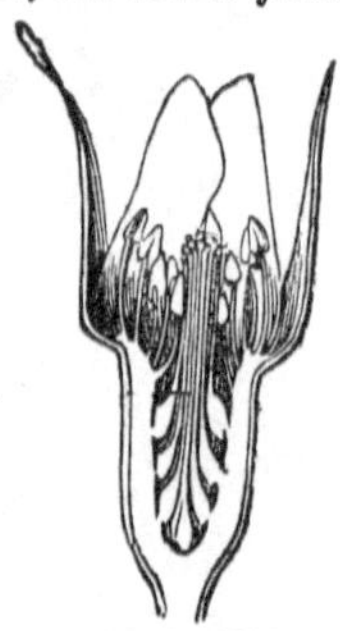

Fig. CCCLXXXI.

The fruits of many species of Fragaria (Strawberry) and Rubus (Raspberry and Blackberry) are valuable articles of the dessert. No Roseworts are unwholesome; they are chiefly remarkable for the presence of an astringent principle, which has caused some of them to be reckoned febrifuges. The root of Tormentilla is used for tanning in the Feroe Isles. Potentilla anserina has been employed in the same manner, and P. reptans as a febrifuge. Geum urbanum and rivale, Comarum palustre and Sieversia montana have been compared, for efficacy, to Cinchona. The leaves of Rubus arcticus and Rosa rubiginosa have been employed as substitutes for Tea. The root of Spiræa filipendula and Ulmaria has been used as a tonic. Agrimonia Eupatoria yields a decoction useful as a gargle, and has some celebrity as a vermifuge. Indian Chocolate-root, which is probably Geum rivale, is much employed in the United States in diseases of the bladder. The root of Rubus villosus is a popular astringent medicine in North America. Two or three tea-spoonfuls of the decoction, administered three or four times a day, has been found useful in cholera infantum. Mixed, however, with this astringency, is the presence of an emetic quality. The roots of Gillenia trifoliata and stipulacea are emetic, and perhaps tonic. They are used in the United States as Ipecacuanha. One of the most powerful anthelmintics in the world belongs to this family. It is an Abyssinian plant, called Cusso, or Cabotz, and known to Botanists by the name of Brayera anthelmintica. Upon the authority of Brayer, after whom it is named, two or three doses of the infusion are sufficient to cure the most obstinate case of tænia. The various species of Rosa form some of the greatest beauties of the garden. The fruit of R. canina and other allied species is astringent, and employed in medicine against chronic diarrhœa and other maladies. The petals of R. moschata and damascena yield a highly fragrant essential oil, called Attar of Roses; those of R. gallica are astringent when dried with rapidity, and are sometimes found useful in cases of debility, such as leucorrhœa, diarrhœa, &c. The Quillaiæ are remarkable for their saponaceous secretions. Quillaia saponaria yields one of the barks called Quillai, used as a substitute for soap. "Two ounces of the bark are sufficient to wash a dress; it is also said to remove all kinds of spots and stains, and to impart a remarkable lustre to wool." It contains a substance which excites violent sneezing, and is closely allied to saponine.—*Chem. Gaz.* 1844. 216. According to Martius the Quillaia brasiliensis has the same property.

GENERA.

I. ROSIDÆ. — Calyx tube fleshy, covering over the achænia.	II. POTENTILLIDÆ. — Calyx tube herbaceous. Fruit a heap of achænia.	Fragaria, *Linn.*	*Argentina*, Blackw.
Rosa, *Tournef.*	Dalibarda, *Linn.*	*Duchesnea*, Smith.	*Bootia*, Bigel.
Lowea, *Lindl.*	Rubus, *Linn.*	Comarum, *Linn.*	*Trichothalamus*, Lehm.
Hulthemia, Dumort.	? *Cylactis*, Raf.	Potentilla, *Linn.*	Horkelia, *Cham. et Schl.*
Rhodopsis, Ledeb.		*Quinquefolium*, Tourn.	Chamærhodos, *Bung.*
		Pentaphylloides, Tour.	Dryadanthe, *Endl.*
		Tormentilla, Tourn.	Sibbaldia, *Linn.*

Fig. CCCLXXXI.—Perpendicular section of a Rose-flower.

Agrimonia, *Tourn.*
Aremonia, *Necker.*
Agrimonioides, Tourn.
Spallanzania, Poll.
Purshia, *DC.*
Tigarea, Pursh.
Kunzea, Spreng.
Cercocarpus, *H. B. K.*
Waldsteinia, *Willd.*
Comaropsis, *L. C. Rich.*
Sieversia, *Willd.*
Adamsia, Fisch.
Buchhavea, Reichenb.
Oreogeum, Ser.
Fallugia, *Endl.*
Geum, *Linn.*
Caryophyllata, Tourn.
Stylipus, Raf.
Cowania, *Don.*
Coluria, *R. Br.*
Laxmannia, Fisch.
Dryas, *Linn.*

III. SPIRÆIDÆ.— Calyx tube herbaceous. Fruit a ring of follicles. Seeds unwinged.

Kerria, *DC.*
Spiræa, *Linn.*
Ulmaria, Tournef.
Filipendula, Tournef
Barba capræ, Tournef.
Physocarpus, Camb.
Chamædryon, Ser.
Sorbaria, Ser.
Schizonotus, Lindl.
Aruncus, Ser.
Ulmaria, Mönch.
Neillia, *Don.*
Gillenia, *Mönch.*
Nuttallia, *Torr. et A. Gr.*
Rhodotypus, *Zucc.*
Stephanandra, *Zucc.*
Brayera, *Kunth.*
Hagenia, Willd.
Cusso, Bruce.
Banksia, Bruce.

IV. QUILLAIÆ.— Calyx tube herbaceous. Fruit capsular. Seed winged.

Kageneckia, *Ruiz. et Pav.*
Lydea, Mol.
Quillaja, *Mol.*
Smegmadermos, Ruiz et Pav.
Fontenellea, St. Hil.
Vauquelinia, *Corr.*
Lindleya, *H. B. K.*
Euphronia, *Mart. et Zucc.*

V. ? NEURADEÆ.—Calyx adhering to a ring of 10 carpels. Seeds pendulous.

Neurada, *B. Juss.*
Grielum, *Linn.*
Amoreuxia, *Moç. et Sess.*

NUMBERS. Gen. 38. Sp. 500.*

Myrtaceæ.
POSITION.—Pomaceæ.—ROSACEÆ.—Drupaceæ.
Ranunculaceæ.

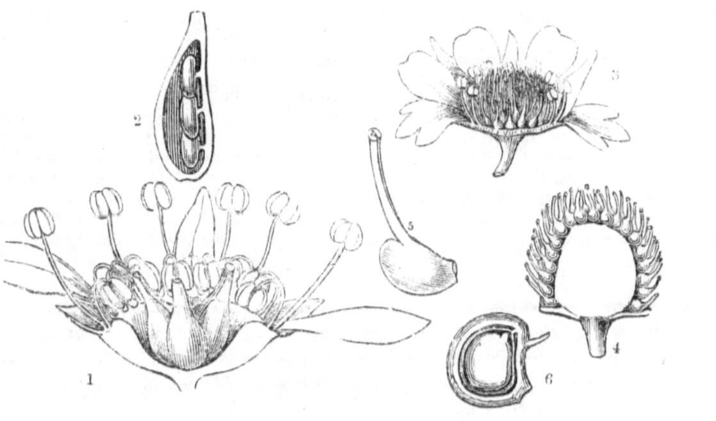

Fig. CCCLXXXII.

* This number must be much higher if the spurious species of Rubus and Rose are included.

Fig. CCCLXXXII.—1. flower of Spiræa Aruncus cut open ; 2. a section of one of its carpels ; 3. perpendicular section of a flower of Fragaria indica ; 4. the same section of its half ripe fruit, showing the torus covered with carpels ; 5 a single carpel of it ; 6. one of its achænia cut open to show the seed inside.

ALLIANCE XLIII. *SAXIFRAGALES.*—THE SAXIFRAGAL ALLIANCE.

DIAGNOSIS.—*Perigynous Exogens, with monodichlamydeous flowers, consolidated carpels, sutural or axile placentæ,* 00 *seeds, a polypetalous corolla, if any is present, and a small taper embryo with a long radicle and little or no albumen.*

The transition from the Rosal to the Saxifragal Alliance has already (p. 539) been shown to take place by way of Roseworts and Saxifrages. To the Rhamnal Alliance Saxifragals pass by way of Brexiads, which are singularly like the genus Elæodendron among Spindle-trees. The resemblance of the Orders included in this Alliance is so great, that the first three are often regarded as mere forms of the Order of Saxifrages; Loosestrifes are less obviously similar, but if their herbaceous genera are compared with Saxifrages, or their shrubs with Brexiads, the affinity becomes sufficiently striking. Loosestrifes appear to furnish a lateral connection with Melastomads or Syringas.

NATURAL ORDERS OF SAXIFRAGALS.

Styles distinct. Leaves alternate	214. SAXIFRAGACEÆ.
Styles distinct. Leaves opposite, without stipules	215. HYDRANGEACEÆ.
Styles distinct. Leaves opposite, with large interpetiolar stipules.	216. CUNONIACEÆ.
Styles consolidated. Calyx many-leaved. Albumen 0. *Leaves alternate*	217. BREXIACEÆ.
Styles consolidated. Calyx tubular, permanent, with the petals in the margin. Albumen 0. *Leaves opposite*	218. LYTHRACEÆ.

ORDER CCXIV. SAXIFRAGACEÆ.—SAXIFRAGES.

Saxifragæ, *Juss. Gen.* 308. (1789); *Vent. Tabl.* 2. 277.—Saxifrageæ, *DC. and Duby*, 207. (1828).—Saxifragaceæ, *DC. Prodr.* 4. 1.; *Endl. Gen.* clxx.; *Meisner*, p. 136.

DIAGNOSIS.—*Saxifragal Exogens, with distinct styles and alternate leaves.*

Herbaceous plants, often growing in patches. Leaves either divided or entire, alternate, with or without stipules. Flower-stems simple, often naked. Calyx either superior or inferior, of 4 or 5 sepals, which cohere more or less at their base. Petals 5 or 0, inserted between the lobes of the calyx. Stamens 5-10, inserted either into the calyx, or beneath the ovary; anthers 2-celled, bursting longitudinally. Disk either hypogynous or perigynous, sometimes nearly obsolete, sometimes annular and notched, rarely consisting of 5 scales. Ovary inferior, or nearly superior, usually consisting of 2 carpels, cohering more or less by their face, but distinct and diverging at the apex; sometimes 2-celled with a central placenta; sometimes 1-celled with a double placenta adhering to the sutures. Styles none. Stigmas sessile on the tips of the lobes of the ovary. Fruit generally a membranous 1- or 2-celled capsule with the cells divaricating when ripe. Seeds numerous, very minute; usually with long hexagonal reticulations on the side of a transparent testa. Embryo taper, in the axis of fleshy albumen, with the radicle next the hilum.

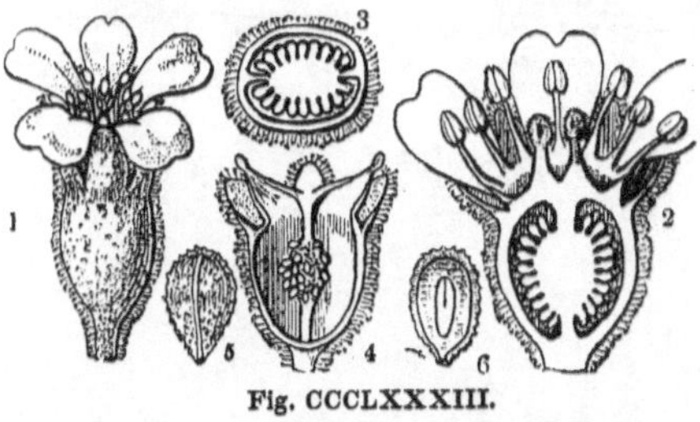

Fig. CCCLXXXIII.

So near is the affinity of Saxifrages and Roseworts that in some cases it is difficult to distinguish the Orders. Nevertheless they appear to represent two distinct tendencies (*Nixus*), that of Saxifrages being towards the consolidation of the fruit, and the formation of albumen, while that of Roseworts is plainly towards a disunion of the carpels and the absorption of albumen. If compared with other Orders in their own Alliance, Saxifrages will be found to differ from Loosestrifes and Brexiads in their disunited styles, from Cunoniads and Hydrangeads in their alternate leaves.

More remote affinities have been indicated by authors. Thus they may be compared in some respects to Cloveworts, especially to the Alsineous division of that Order; but they differ in the insertion of the stamens, placentation, situation of the embryo, and otherwise. Purslanes, which may be compared with this Order, on account of the situation of the stamens, want of stipules, and albuminous seeds, differ essentially in the structure of the embryo, in the want of symmetry in the parts of the flower, and in placentation. Currantworts correspond in the general structure of the flower; but differ widely in the ovary being completely concrete and inferior, with two parietal placentæ, in the seeds being attached to long umbilical cords, in the albumen being corneous, and the embryo extremely minute. De Candolle remarks that Saxifrages approach Houseleeks, differing in having a smaller number of carpels, which are partially united both with each other and the calyx, and in being destitute of glands at the base of the carpels.

Chrysosplenium is remarkable for the want of petals. Drummondia has the stamens

---

Fig CCCLXXXIII.—Saxifraga tridactylites. 1. its flower; 2. a perpendicular section of it; 3. cross section of its ovary; 4. perpendicular section of a fruit; 5. a seed; 6. a section of a seed.

equal in number to the petals and opposite them, thus indicating some analogy with the monopetalous Primworts.

Little herbaceous plants, usually with white flowers, cæspitose leaves and glandular stems : some of the species have yellow flowers, others have red, but none blue. They are natives of mountainous tracts in Europe and the northern parts of the world, frequently forming the chief beauty of that rich turf which is found near the snow in high Alpine stations. Some grow on rocks and old walls, and in hedgerows, or near rivulets, or in groves. None are produced in tropical countries.

The whole Order is more or less astringent. The root of Heuchera americana is powerfully so, whence it is called in North America Alum-root. Otherwise the species possess no known properties ; for the old idea of their being lithontriptic appears to have been derived from their name rather than their virtues. Some pretend that Saxifraga crassifolia may be used as a substitute for Tea ; and Chrysosplenium alternifolium has had some small reputation, in former days, as a slight tonic. The glutinous exudation of a few of them is acrid.

GENERA.

Eremosyne, *Endl.*
Donatia, *Forst.*
Vahlia, *Thunb.*
 Russelia, Linn. f.
 Bistella, Del.
Nimmoia, *Wight.*
Boykinia, *Nutt.*
Zahlbrucknera, *Reichenb.*
 Oreosplenium, Zahlbr.
Saxifraga, *Linn.*
 Porphyrion, Tausch.
 Antiphylla, Haw.
 Caliphyllum, Gaud.
 Aizoonia, Tausch.
 Chondrosea, Haw.
 Cotyledon, Gaud.
 Trigonophyllum, Gaud.
 Porophyllum, Gaud.
 Dactyloides, Tausch.
 Saxifraga, Haw.
 Muscaria, Haw.
 Triplinervium, Gaud.
 Bergenia, Mönch.
 Megasea, Haw.
 Geryonia, Schrank.
 Eropheron, Tausch.
 Micranthes, Tausch.
 Dermasea, Haw.
 Arabidia, Tausch.
 Spathularia, Haw.
 Hydatica, Neck.
 Robertsonia, Haw.
 Aulaxis, Haw.
 Diptera, Borkh.
 Ligularia, Duval.
 Micropetalum, Tausch.
 Cotylea, Haw.
 Lobaria, Haw.
 Hirculus, Tausch.
 Kingstonia, Gray.
 Ciliaria, Haw.
 Leptasea, Haw.
Leptarrhena, *R. Br.*
Lütkea, *Bong.*
 Eriogyne, Hook.
Lepuropetalum, *Ell.*
 Cryptopetalum, Hook.
Chrysosplenium, *Tourn.*
Heuchera, *Linn.*
 Holochloa, Nutt.
 Heucherella, Torr. et A. Gr.
Tolmiea, *Torr. et A. Gr.*
Mitellopsis, *Meisn.*
 Drummondia, DC.
Mitella, *Tournef.*
Tellima, *R. Br.*
 Lithophragma, Nutt.
 Lithophragmella, Torr.
Sullivantia, *Torr.*
Tiarella, *Linn.*
 Blondia, Neck.
 Anthonema, Nutt.
Astilbe, *Hamilt.*
 Hoteia, Morr. et Dec.
Oresitrophe, *Bung.*

NUMBERS. GEN. 19. Sp. 310.

Crassulaceæ.
Rosaceæ.
POSITION.—Cunoniaceæ.—SAXIFRAGACEÆ.—Lythraceæ.
Grossulaceæ.

ORDER CCXV. HYDRANGEACEÆ.—HYDRANGEADS.

Hydrangeæ *and* Bauereæ, *DC. Prodr.* 3. 13. (1830); § § of Saxifragaceæ, *Endl. Gen. p.* 820.—Hydrangeaceæ, *Siebold and Zuccarini, Fl. Jap.* 1. 102. *in notis* (1835). ?–Baueraceæ, *Ed. pr. No.* 40. (1830); *Martius Conspectus, No.* 226.

DIAGNOSIS.—*Saxifragal Exogens, with distinct styles, and opposite leaves without stipules.*

Shrubs with perfectly opposite simple leaves, smooth or downy, with simple hairs; destitute of stipules; sometimes creeping and rooting like Ivy. Flowers usually in cymes, those in the centre ☿, the marginal often sterile and furnished with larger petals than the others. Calyx adhering more or less to the ovary, 4-6-toothed. Petals 4-6, inserted within the edge of the calyx, deciduous. Stamens 8-12 in 2 rows, or 00, inserted in the orifice of the calyx, distinct, deciduous. Anthers oblong or roundish; pollen with 3 longitudinal furrows. Ovary more or less adherent to the calyx, consisting of from 2 to 5 carpels, adhering by their sides and forming an incompletely 2-5-celled cavity; placentæ distinct from each other, but touching, with many anatropal ascending or horizontal ovules; styles as many as the carpels, perfectly distinct, diverging, with simple reniform stigmas. Fruit a capsule crowned by the permanent diverging styles, 2-5-celled, with a number of minute seeds, sometimes indefinite, sometimes few, in consequence of the abortion of a part of the ovules. Testa thin, membranous, netted, occasionally expanded into a wing. Embryo orthotropal, in the axis of a small quantity of fleshy albumen.

Fig. CCCLXXXIV.

The relationship between Hydrangeads and Saxifrages is admitted by all systematists, who have in general united them in the same Order. The opposite leaves of the former, the tendency to a polygamous structure evinced in their radiant male flowers, and the general increase of carpels beyond two, seem to offer good grounds for separating them. Like Saxifrages, their styles are almost always distinct and very often divergent. In some the ovary is entirely adherent to the calyx; in others, as Hydrangea virens, it is more than half separated. Schizophragma, a curious Japanese genus, has the styles united, and thus furnishes a transition to Caprifoils on the one hand, and to Henslovia on the other. Siebold and Zuccarini place Deutzia here, and it may be regarded as a genus bringing the Philadelphs in contact. Bauera is anomalous in its whorled exstipulate leaves and porous anthers, but can hardly be separated, unless it be referred to Cunoniads, upon the supposition of Don, that its lateral leaves are modified stipules.

Siebold and Zuccarini remark that out of the species hitherto discovered, all of which inhabit the temperate parts of Asia and America, two only belong to the Southern hemisphere, and 23, or about one half, to China and Japan. These authors do not, however, include Bauera, but they admit Deutzia. The species are found naturally in moist, shady places.

Fig CCCLXXXIV.—1. Hydrangea virens and its flower.—*Siebold*; 2. seed-vessel of H. hortensis; 3. its seed; 4. a section of it.

None of these appear to be of much use to man. Hydrangeas have been cultivated as garden ornaments from the most ancient times in China and Japan. The leaves of H. Thunbergii are dried in Japan, and used as a kind of tea, which for its excellence they call Ama-tsjâ, or Tea of Heaven. Another sort of tea is furnished by Platycrater arguta.—*Siebold.*

GENERA.

Hydrangea, *L.*	Cardiandra, *S. et Z.*	Cornidia, *R. P.*	Adamia, *Wall.*
Hortensia, Juss.	Platycrater, *S. et Z.*	*Sarcostyles*, Presl.	*Cyanitis*, Reinw.
Peautia, Comm.	Schizophragma, *S. et Z.*	Broussaisia, *Gaud.*	Bauera, *Sm.*
Primula, Lour.	Jamesia, *Torr. et Gr.*		

NUMBERS. GEN. 9. Sp. 45.

Philadelphaceæ.
POSITION.—Saxifragaceæ.—HYDRANGEACEÆ.—Cunoniaceæ.
Caprifoliaceæ.

HENSLOVIACEÆ, (*Lindl. in Bot. Reg.* 20. *fol.* 1686. (July 1834); *Martius's Conspectus, No.* 77; *Ed. pr.* cxxiv.; *Endl. Gen. p.* 291). Trees, with the habit and inflorescence of Myrobalans. Leaves opposite, entire, without stipules. Wood regularly zoned, with very abundant vasiform tissue (dotted ducts). Flowers by abortion ♂ ♀. Calyx 5-parted, lined with a woolly disk, with a valvate æstivation. ♂ Stamens 5, alternate with the sepals, perigynous, long, exserted, inflexed in æstivation; anthers 2-celled, with a broad connective, the lobes oblique, bursting longitudinally. A rudiment of an ovary. ♀ Ovary superior, 2-celled, with very numerous ovules attached horizontally to a placenta in the axis; style cylindrical; stigma obsoletely 2-lobed; ovules with a large conspicuous foramen next the hilum. Fruit a capsule, bursting through the cells into 2 valves. Seeds 00, minute, scobiform, with the skin drawn to a point and winged on one side, an oblong nucleus, and no albumen. Radicle next the hilum.——After vain attempts at settling the true place of the genus Henslovia in the Natural System (see the last edition, No. cxxiv.), some specimens with ripe fruit, for which I am indebted to Mr. Griffith, place the question nearly at rest. The habit of the plant was evidently that of Viburnum; but its superior ovary and indefinite ovules forbade any reference to Caprifoils. But Hydrangeads differ from that Order mainly in their indefinite seeds, small quantity of albumen, and constant tendency to produce a superior ovary. Henslovia agrees with them still further; the flowers are polygamous, the seeds are winged, which is also the case in Hydrangea cordifolia and others; and the albumen is wholly deficient. The chief distinction consists in the complete adhesion of the styles into one undivided cylinder; but among Hydrangeads we have the same peculiarity in Schizophragma and Broussaisia. On that account, however, Henslovia may be regarded as a relation of Brexia, but its decidedly opposite leaves are unfavourable to the union of the two in the same Order.——Henslovia consists of 3 or 4 species of trees inhabiting the tropical parts of India.——Only Genus. Henslovia, *Wall.*

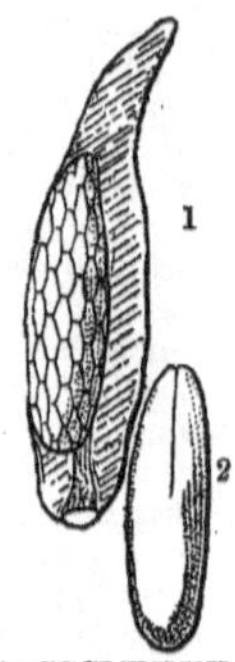

Fig. CCCLXXXV.

Fig. CCCLXXXV.—Henslovia. 1. its seed; 2. its embryo.

ORDER CCXVI. CUNONIACEÆ.—CUNONIADS.

Cunoniaceæ, *R. Br. in Flinders*, 548. (1814); *Don in Edinb. New Phil. Journ. June* 1830, *in part*; *Martius Conspectus, No.* 223.; *Endl. Gen. p.* 819.—Ochranthaceæ, *Ed. pr. p.* 78. (1836); *Endl. Gen. p.* 1035.

DIAGNOSIS.—*Saxifragal Exogens, with distinct styles, and opposite leaves with large interpetiolar stipules.*

Trees or shrubs. Leaves opposite, compound or simple, with stipules between the leafstalks, sometimes united and scale-like, sometimes separate and leafy. Calyx 4 or 5-cleft, half superior or nearly inferior. Petals 4 or 5, occasionally wanting. Stamens perigynous, definite, or 00; anthers bursting longitudinally or by pores. Ovary 2-celled, the cells having 2 or many seeds; styles 2, sometimes combined. Fruit 2-celled, capsular or indehiscent. Embryo in the axis of fleshy albumen.

Fig. CCCLXXXVI.

Fig. CCCLXXXVII.

This Order is no doubt very distinct from that of Saxifrages, and yet it is more readily distinguished by the widely different habit than by any very important characters in the fructification. The shrubby way of growth and remarkable interpetiolar stipules are the principal character. Don supposed the Order to be strictly allied to Philadelphads. The genus Ochranthe, described in the *Botanical Register*, (Dec. 1835), as agreeing with Tutsans in having imbricated sepals, hypogynous petals and stamens, partly disjoined carpels, and in some degree in habit, but differing in having definite stamens (5), stipules, and serrated leaves, but whose fruit is unknown, seems, upon the whole, to form a member of the Order of Cunoniads.

Natives of the Cape, South America, and the East Indies; common in Australasia.

A Weinmannia is used in Peru for tanning leather, and its astringent bark is employed to adulterate the Peruvian bark. The Indian Weinmannias appear to possess

Fig. CCCLXXXVI.—Weinmannia Balbisiana.—*Turpin.* 1. its ovary.
Fig. CCCLXXXVII.—Ochranthe arguta. 1. grain of pollen; 2. perpendicular section of its pistil.

similar astringent qualities. Some of the Australasian plants of the Order have a gummy secretion. In general, the pretty appearance of their small white or pink flowers makes them gay objects.

GENERA.

I. WEINMANNEÆ.

Codia, *Forst.*
Callicoma, *Andr.*
Calycomis, *R. Br.*
Aphanopetalum, *Endl.*
Ceratopetalum, *Smith.*
Meridema, Don.
Schizomeria, *Don.*
Platylophus, *Don.*
Anodopetalum, *A. Cunn.*
Weinmannia, *Linn.*
Windmannia, P. Br.
Leiospermum, *Don.*
Ackama, *A. Cunn.*
Pterophylla, *Don.*
Arnoldia, *Blum.*
Gumillea, *Ruiz et Pav.*
Caldcluvia, *Don.*
Dieterica, Ser.
Cunonia, *Linn.*
Osterdyckia, Burm.
Geissois, *Labill.*
Adenilema, *Blum.*
Pellacalyx, *Korth.*
Ochranthe, *Lindl.*

II. BELANGEREÆ, *Gardn.*

Belangera, *Camb.*
Polystemon, Don.
Lamanonia, Fl. Fl.
Raleighia, *Gardn.*

NUMBERS. GEN. 22. Sp. 100.

POSITION.—Hydrangeaceæ.—CUNONIACEÆ.—Saxifragaceæ.
Philadelphaceæ.

ORDER CCXVII. BREXIACEÆ.—BREXIADS.

Brexiaceæ, *Ed. Prior, No.* 95. (1830); *Arnott in Edinb. Encycl.* 104. (1832); *Martius Conspectus, No.* 297. (1835); *Endl. Gen. p.* 823.—Roussæaceæ, *DC. Prodr.* 7. 521; *Endl. Gen. p.* 823.

DIAGNOSIS.—*Saxifragal Exogens, with consolidated styles, a many-leaved calyx, alternate leaves, and no albumen.*

Trees, with nearly simple trunks. Leaves coriaceous, alternate, simple, not dotted, with deciduous minute stipules. Flowers green, in axillary umbels. Calyx inferior, small, persistent, 5-parted; æstivation imbricated. Petals 5, hypogynous, twisted in æstivation. Stamens 5, hypogynous, alternate with the petals, arising from a narrow cup, which is toothed between each stamen; anthers oval, innate, 2-celled, bursting longitudinally, fleshy at the apex; pollen triangular, cohering by means of fine threads. Ovary superior, 5-celled, with numerous ovules attached in two rows to placentæ in the axis; style 1, continuous; stigma simple. Fruit drupaceous, 5-cornered, "marked with numerous small scarcely elevated papillæ like the surface of an orange," 5-celled, many-seeded. Seeds indefinite, horizontal, smooth and shining, brown, ovate, slightly angular, about the size of those of a raisin, attached to the axis, with a double integument, the inner of which is membranous; cotyledons ovate, obtuse; radicle cylindrical, centripetal; (albumen fleshy, according to Thouars).

Fig CCCLXXXVIII.

There exists in Madagascar a genus called Brexia, of which the above is a description, taken in part from the living plant, and in part from Dr. Wallich, in the *Flora Indica*, 2. 314.

The position which this plant ought to occupy in a natural classification is unsettled. Indeed, we are not certain whether or not the seeds have albumen; for although Thouars states it to be fleshy, Dr. Wallich is silent upon the subject, and no other Botanist seems to have examined the seeds. Its habit is that of some Ardisiads, especially of Theophrasta, from which it differs in being polypetalous, in the stamens being alternate with the petals, and in many other circumstances. With Rhamnads and Spindle-trees its relation is no doubt strong, but its seeds are indefinite. Some resemblance may be traced between it and Anacards, especially in the resinous appearances visible upon the young shoots, and also in habit; but its fructification is entirely at variance with that Order. With Pittosporads it agrees in its hypogynous definite stamens, its polyspermous fruit, its alternate undivided leaves, and habit; but it is probable that the embryo is not such as befits that Order. Endlicher places it at the end of Saxifrages, combining with it Ixerba and Argophyllum, the latter a genus having the ovary adherent in some degree to the calyx. If this approximation is right,—and it certainly seems probable,—and if the seeds of Brexia should prove, when re-examined, to have albumen, as Thouars says, and the peculiar reticulated testa represented by Gærtner, then Brexia must be indeed a perigynous form, and may be looked upon as a genus of the Saxifragal Alliance, where it is perhaps best to place it for the present. Nor can I doubt that Roussæa, figured by Endlicher in his *Iconographia*, is of the same class, notwithstanding its opposite leaves. Its great disk is quite analogous to the toothed disk of Brexia.

GENERA.

Ixerba, *A. Cunn.*	*Venana*, Lam.	Roussea, *Smith.*	*Roussoa*, Röm. et Schlt.
Brexia, *Thouars.*	Argophyllum, *Forst.*	*Roussæa*, DC.	*Rousseauvia*, Boj.

NUMBERS. GEN. 4. Sp. 6.

POSITION.—Celastraceæ. / BREXIACEÆ.—Cunoniaceæ. / Myrsinaceæ.

	Celastraceæ.	
POSITION.——	BREXIACEÆ.—	Cunoniaceæ.
	Myrsinaceæ.	

Fig. CCCLXXXVIII.—1. Brexia madagascariensis; 2. section of its ovary; 3. seed of Argophyllum.

ORDER CCXVIII. LYTHRACEÆ.—LOOSESTRIFES.

Salicariæ, *Juss. Gen.* 330. (1789); *Lindl. Synops.* 71; *Aug. de St. H. Ann. Sc. Nat.* 2. *ser.* 1. *p.* 1. *and* 333.—Calycanthemæ, *Vent. Tabl.* 3. 298. (1799).—Salicarinæ, *Link Enum.* 1. 142.—Lythrariæ, *Juss. Dict. Sc. Nat.* 27. 453.; *DC. Prodr.* 3. 75; *Endl. Gen.* cclxvii.; *Meisner Gen. p.* 117.

DIAGNOSIS.—*Saxifragal Exogens, with consolidated styles, a tubular permanent calyx with the petals in the margin, opposite leaves, and no albumen.*

Herbs, rarely shrubs. Branches frequently 4-cornered. Leaves opposite, seldom alternate, entire, without either stipules or glands, sometimes with glandular dots. Flowers solitary or clustered, regular or irregular, axillary or in terminal spikes or racemes, in consequence of the depauperation of the upper leaves. Calyx monosepalous, tubular, ribbed, often oblique, the lobes with a valvate or separate æstivation, their sinuses sometimes lengthened into other lobes. Petals inserted between the outer lobes of the calyx, very deciduous, sometimes wanting. Stamens inserted into the tube of the calyx below the petals, to which they are sometimes equal in number; sometimes twice, or even four times as numerous; anthers adnate, 2-celled, opening longitudinally. Ovary superior, 2-6-celled, occasionally only 1-celled; ovules 00, rarely definite, ascending or horizontal, anatropal, attached to axile or dissepimental placentæ, having a central origin; style filiform; stigma usually capitate. Capsule membranous, covered by the calyx, dehiscent. Seeds numerous, small, without albumen, adhering to a central placenta; embryo straight; radicle turned towards the hilum; cotyledons roundish, flat, and leafy.

Fig. CCCLXXXIX.

The true place of this Order has been the subject of much difference of opinion. A writer in the *Linnæa* (14. 254) refers it without any doubt to the vicinity of Houseleeks (Crassulaceæ). In some respects the Order resembles Onagrads, from which the superior ovary and many-ribbed calyx distinguish it; also Melastomads, from which the superior ovary, the veining of the leaves, and the æstivation of the stamens divide it. With Labiates it has often a similarity in habit, but this goes no further. A resemblance to Spindle-trees is established by the genus Adenaria. Endlicher even

Fig. CCCLXXXIX.—Diplusodon arboreus.

compares the Order with Waterpeppers (Elatinaceæ), because of its simple style and the structure of its seeds.

It seems, however, to be with Saxifrages that the affinity is strongest. In fact, Lythrum is little more than a Saxifrage with united styles and scattered stamens; it even agrees with certain Saxifrages in the irregularity of the flowers. The Lagerströmeæ, among which some resemblance to Melastomads is chiefly found, may be stationed by the side of Cunoniads, from which their consolidated carpels and want of stipules clearly, however, divide them.

The Lagerströmeæ are all Indian or South American. The Lythreæ are European, North American, and natives of the tropics of both hemispheres. Lythrum Salicaria, a common European plant, is singular for being found in New Holland, and for also being the only species of that Order yet described from that country.

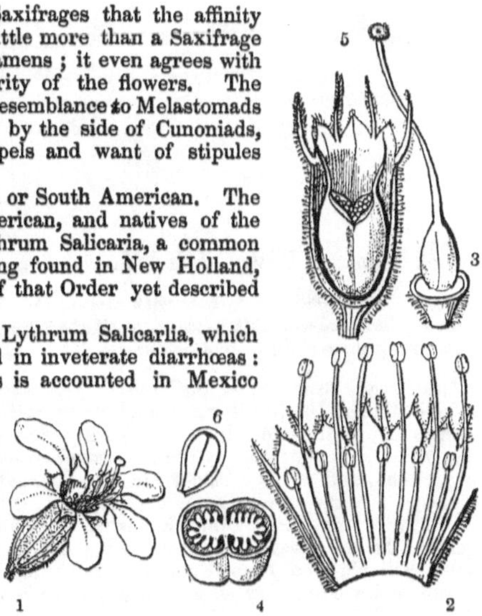

Fig. CCCXC.

Astringency is a property of the Lythrum Salicarlia, which is reputed to have been found useful in inveterate diarrhœas: another species of the same genus is accounted in Mexico astringent and vulnerary, a reputation which also belongs to other species of the genus. The flowers of Grislea tomentosa are employed in India, mixed with Morinda, for dyeing, under the name of Dhaee. Heimia salicifolia, a plant remarkable, in an Order with red or purple flowers, for its yellow corolla, is said to excite violent perspiration. The Mexicans consider it a potent remedy for venereal diseases, and call it Hanchinol. Lawsonia inermis is the plant from which the Henné of Egypt is obtained. Women in that country stain their fingers and feet of an orange colour with it. It is also used for dyeing skins and morocco leather reddish-yellow, and for many other purposes. It contains no tannin.—*Ed. P. J.* 12. 416. The bark and leaves of Lagerströmia Reginæ are accounted purgative and hydragogue, the seeds narcotic. The leaves of Ammannia vesicatoria have a strong muriatic smell; they are extremely acrid, and are used by the native practitioners of India to raise blisters in rheumatism, &c.: bruised and applied to the part intended to be blistered, they perform their office in half an hour, and most effectually. The herbage of Nesæa verticillata is said to destroy the young of cattle heavy with calf. Nevertheless, the leaves of Pemphis acidula are said to be a common potherb on the coast of the tropical parts of Asia. A decoction of Cuphea Balsamona is found useful in Brazil in intermittent fevers. The Physocalymma floribunda yields a beautiful Rose-coloured wood, called in German Rosenholz, and in Portuguese Pâo de rosa.

GENERA.

I. LYTHREÆ. — Seeds wingless.

Cryptotheca, *Blum.*
Suffrenia, *Bellard.*
Rotala, *Linn.*
Eutelia, R. Br.
Hypobrichia, *M. O. Curt.*
Ptilina, Nutt.
Didiplis, Raf.
Peplis, *Linn.*
Portula, Dill.
Chabræa, Adans.
Quartinia, *Endl.*
Rhyacophila, Hochst.
Ameletia, *DC.*
Middendorfia, *Trautv.*
Ammannia, *Houst.*
Tritheca, Wight et Arn.
Diplostemon, Wight.
Cornelia, Ard.
Haplocarpæa, Wight.
Ditheca, Wight et Arn.
Mircooa, Wight et Arn.
Nesæa, *Commers.*
Tolypeuma, E. Mey.
Trotula, Comm.
Decodon, Gmel.
Heimia, *Link et Otto.*
Chrysoliga, Hoffmans.
Ginoria, Fl. Mex.
Pemphis, *Forst.*
Lythrum, *Linn.*
Salicaria, Tournef.
Hyssopifolia, C. Bauh.
Pythagorea, Raf.
Mozula, Raf.
Pentaglossum, Forsk.
Anisotes, *Lindl.*
Pleurophora, *Don.*
Cuphea, *Jacq.*
Melanium, P. Br.
Parsonsia, P. Br.
Balsamona, Vand.
Melvilla, Anders.
Duvernaya, Desp.
Acisanthera, *P. Br.*
Crenea, *Aubl.*
Dodecas, *Linn.*
Ginoria, *Jacq.*
Ginora, Linn.
Genoria, Pers.
Grislea, *Löffl.*
Woodfordia, Salisb.
Adenaria, *H. B. K.*
Antherylium, *Rohr.*
Lawsonia, *Linn.*
Alcanna, Gärtn.
Abatia, *Ruiz et Pav.*

II. LAGERSTROMEÆ. — Seeds winged.

Diplusodon, *Pohl.*
Diplodon, Spreng.
Friedlandia, Cham.
Dubyæa, DC.
Laföensia, *Vandell.*
Calyplectus, Ruiz et P.
Ptychodon, *Klotsch.*
Physocalymma, *Pohl.*
Lagerströmia, *Linn.*
Sibia, DC.
Münchhausia, Linn.
Adambea, Lam.
Duabanga, *Hamilt.*
Fatioa, *DC.*
? Symmetria, *Blum.*
? Physopodium, *Desv.*
? Psyloxylon, *Neraud.*

NUMBERS. GEN. 35. Sp. 300.

POSITION.—Saxifragaceæ.—LYTHRACEÆ.—Cunoniaceæ.
Melastomaceæ.

Fig. CCCXC.—Lythrum Salicaria. 1. a flower; 2. the calyx spread open; 3. the pistil; 4. a cross section of the ovary; 5. the capsule; 6. a section of a seed.

ALLIANCE XLIV. *RHAMNALES.*—THE RHAMNAL ALLIANCE.

DIAGNOSIS.—*Perigynous Exogens, with monodichlamydeous flowers, consolidated carpels, axile placentæ, capsular, berried, or drupaceous fruit, definite seeds, and an amygdaloid embryo, with little or no albumen.*

It has already been stated that the only positive distinction between this Alliance and Daphnals consists in the compound ovary of Rhamnals. This may seem a trifling difference and quite artificial. But in reality it is connected with a higher evolution of all the parts, as is indicated by the general presence of a corolla, which even becomes monopetalous, and by its considerable development in many instances. Even in Sarcocollads, where the corolla is missing, the calyx acquires quite a petaloid condition.

Storaxworts pass directly into Ebenads, with which the next Alliance commences. In general the smallness of the embryo and the largeness of the albumen in the latter, completely divide them from Rhamnals ; but some Ebenads have quite an intermediate structure, and are not very easy to distinguish from Storaxworts.

NATURAL ORDERS OF RHAMNALS.

Flowers apetalous. Ovary composed of 4 carpels. Calyx tubular, with definite divisions. Cotyledons consolidated	219. PENÆACEÆ.
Flowers apetalous. Ovary composed of 2 carpels. Calyx tubular, with a definite number of divisions. Cotyledons amygdaloid	220. AQUILARIACEÆ.
Flowers apetalous. Ovary composed of 2 carpels. Calyx imperfect, and irregularly divided at the edge. Cotyledons thin and leafy	221. ULMACEÆ.
Flowers polypetalous. Calyx valvate. Stamens opposite petals. Seeds erect	222. RHAMNACEÆ.
Flowers polypetalous. Calyx valvate. Stamens alternate with petals. Seeds pendulous	223. CHAILLETIACEÆ.
Flowers polypetalous. Calyx imbricated. Stamens (3) monadelphous	224. HIPPOCRATEACEÆ.
Flowers polypetalous. Calyx imbricated. Stamens (5/) distinct	225. CELASTRACEÆ.
Flowers monopetalous. Stamens episepalous	226. STACKHOUSIACEÆ.
Flowers monopetalous. Stamens epipetalous. Ovules ascending. Radicle short. Cotyledons amygdaloid	227. SAPOTACEÆ.
Flowers monopetalous. Stamens epipetalous. Ovules, in part at least, suspended. Radicle long. Cotyledons leafy	228. STYRACACEÆ.

ORDER CCXIX. PENÆACEÆ.—SARCOCOLLADS.

Penæaceæ, *R. Brown, verbally,* (1820); *Guillemin in Dict. Class.* 13. 171. (1828); *Kunth in Linnæa,* v. 667. (1830); *Endl. Gen.* cxii.; *Meisner Gen. p.* 329.—Geissolomeæ, *Endl. Ench. p.* 214.

DIAGNOSIS.—*Rhamnal Exogens, with apetalous flowers, an ovary composed of* 4 *carpels, a tubular calyx, with definite divisions, and* 2 *consolidated cotyledons.*

Shrubs. Leaves opposite, imbricated, without stipules. Flowers terminal and axillary, usually red. Calyx inferior, with 2 or more bracts at its base; hypocrateriform, with a 4-lobed limb, valvate or imbricated in æstivation. Stamens either 4, arising from below the recesses of the limb, with which they alternate, or 8, arising from near the base of the calyx; anthers 2-celled, turned inwards, usually with membranous valves lying on the face of a thick fleshy connective, sometimes with fleshy valves, and an obliterated connective. Ovary superior, 4-celled, with a simple style and 4 half-indusiate stigmas; ovules anatropal, either ascending, collateral, in pairs, or solitary and suspended. Fruit capsular, 4-celled, dehiscent or indehiscent? Seed erect or inverted; testa brittle; nucleus a solid fleshy mass, with no distinction of albumen or embryo; hilum fungous.

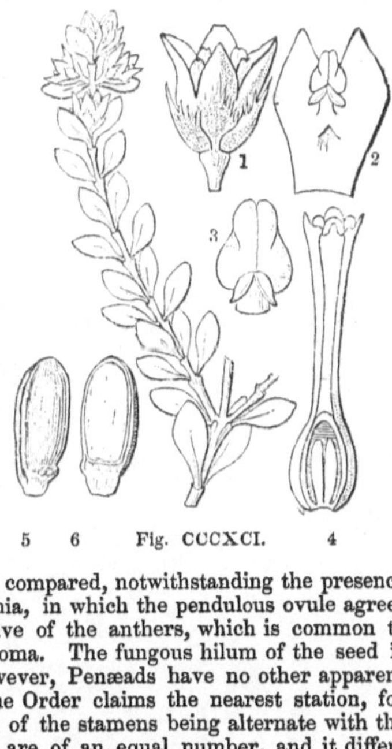

Fig. CCCXCI.

According to an observation of Jussieu, this Order is allied to Epacrids; but I confess myself unable to perceive on what account. To me it formerly appeared related to such apetalous dicotyledons as Proteads, with some of which the species agree in habit, and in the case of Penæa fruticulosa even in the thickened connective and the structure of the lobes of the stigma, each of which is strikingly like that of a Grevillea. To Bruniads they may be compared, notwithstanding the presence of petals in that Order, for the sake of Linconia, in which the pendulous ovule agrees with Geissoloma, and of the thickened connective of the anthers, which is common to several species, although not present in Geissoloma. The fungous hilum of the seed is similar to that of Milkworts, with which, however, Penæads have no other apparent relation. It is, probably, to Rhamnads that the Order claims the nearest station, for it corresponds with them in the important fact of the stamens being alternate with the valvate lobes of the calyx, when the stamens are of an equal number, and it differs from them principally in its peculiar anthers (and undivided embryo?). Its half-indusiate stigmas are like those of some Heathworts.

The Order exhibits a singular instance of two distinct kinds of æstivation and attachment of ovules among species which it seems unadvisable to separate from each other. In true Penæa the æstivation is valvate and the ovules ascending, while in Geissoloma the former is imbricate and the latter suspended. Penæa has also tetrandrous flowers, with peculiarly fleshy anthers, while Geissoloma has octandrous flowers, with no peculiar fleshiness in the anthers.

All are evergreen shrubs, natives of the Cape of Good Hope, and chiefly those to the eastward of the Hottentots Holland chain of mountains.

A sub-viscid, sweetish, somewhat nauseous gum-resin called Sarcocol (Σαρκοκόλλα, *Diosc.*) is said to be produced by various species. It was supposed by the Arabians to possess, as its name indicates, the power of agglutinating wounds, and contains a peculiar principle, named Sarcocollin, which has never been detected in any other vegetable matter, and which has the property of forming oxalic acid, being treated with nitric

Fig. CCCXCI.—Penæa fruticulosa. 1. a flower; 2. a portion of the calyx with a stamen attached; 3. an anther; 4. a pistil, with one of the cells cut open; 5. a seed; 6. a section of it.

acid. Endlicher, however, remarks that this drug is not likely to be a product of the present Order. Dioscorides says that it was obtained from a Persian tree; but whether that were so or not, the manifest relation of the drug called Sarcocol to Galbanum and Sagapenum, renders it more likely to come from some Umbellifer.

GENERA.

Penæa, *L.*
Sarcocolla, *Kth.*
Geisoloma, *Lindl.*

NUMBERS. GEN. 3. Sp. 21.

Bruniaceæ.
POSITION.——————PENÆACEÆ.—Rhamnaceæ.
Proteaceæ.

ORDER CCXX. AQUILARIACEÆ.—AQUILARIADS.

Aquilarineæ, *R. Brown Cong.* p. 25. (1818); *DC. Prodr.* 2. 59; *Royle Illustr.* 171; *Endl. Gen.* cx.; *Meisner Gen.* p. 73; *Decaisne Ann. Sc.* xix. 35.

DIAGNOSIS.—*Rhamnal Exogens, with apetalous flowers, an ovary composed of 2 carpels, a tubular calyx with a definite number of divisions, and amygdaloid cotyledons.*

Trees. Branches smooth, with a tough bark. Leaves alternate or opposite, on short stalks, entire, without stipules. Calyx turbinate or tubular; limb 4- or 5-cleft; segments spreading, persistent, with an imbricated æstivation; the orifice usually furnished with scales (sterile stamens). Stamens 10, 8, or 5, in the latter case opposite the segments of the calyx; filaments inserted into the orifice of the calyx a little lower down than the scales. Anthers narrow, oblong, attached by their back below the middle, 2-celled, opening internally and lengthwise. Ovary superior, sessile or stipitate, downy, compressed, 2-celled; ovules two, anatropal, of which one is suspended from each side of the placenta, tapering downwards; style 0, or conical and thread-shaped; stigma large, simple. Capsule sessile or stipitate, 2-valved or drupaceous, and indehiscent. Seeds one on each placenta, or one sometimes abortive, pendulous; albumen 0; cotyledons thick, fleshy, hemispherical; radicle straight, superior.

Do Candolle places this Order between Chailletiads and Anacards, but with indications of doubt, and an erroneous character; and Brown seems willing to consider the Order a section of Chailletiads, adding, that it would not be difficult to show its affinity to Daphnads. In this I fully concur; in fact, Aquilariads chiefly differ from Daphnads in their dehiscent fruit, composed of two carpels, not one. Both Orders have similar scale-like bodies at the orifice of the calyx, and no petals, both suspended ovules, a single style, and capitate stigma. This too is the view taken of their affinities by M. Decaisne, who indeed regards them as a mere section of Daphnads, observing that they really differ in nothing except their 2-celled ovary. I would, however, prefer leaving them here, as the group which, in the Rhamnal Alliance, touches the Daphnal.

The species are confined to the tropical parts of Asia.

Fig. CCCXCII.

Aloes-wood, Agila-wood, or Eagle-wood, containing a fragrant resinous substance, of a dark colour, is the inside of the trunk of the Aquilaria ovata and A. Agallochum. It is considered a cordial by some Asiatic nations, and has been prescribed in Europe in gout and rheumatism. For a valuable account of this substance, see *Royle*, as above quoted.

GENERA.

Aquilaria, *Lam.*	Gyrinopsis, *Gærtn.*	*Phaleria*, Jack.	Gyrinopsis, *Decaisne.*
? Ophiospermum, *Lour.*	Drimyspermum, *Rndt.*	Pseudais, *Decaisne.*	

NUMBERS. GEN. 6. Sp. 10.

POSITION.—Penæceæ.—AQUILARIACEÆ.—Chailletiaceæ.
Thymelaceæ.

Fig. CCCXCII.—Aquilaria Agallochum. 1. a flower; 2. the same split open, 3. a section of the ovary.

ORDER CCXXI. ULMACEÆ.—ELMWORTS.

Ulmaceæ, *Mirbel Elém.* 905. (1815); *Lindl. Synops.* 225; *Endl. Gen.* xc.; *Meisn. Gen.* p 351.—Celtideæ, *Rich., Gaudich. in Freyc. Voy.* 507. (1826); *Endl. Gen.* xci.; *Meisner Gen. p.* 348.

DIAGNOSIS.—*Rhamnal Exogens, with apetalous flowers, an ovary composed of* 2 *carpels, an imperfect calyx irregularly divided at the edge, and thin and leafy cotyledons.*

Trees or shrubs, with rough, alternate, usually deciduous leaves, each having a pair of deciduous stipules at its base. Flowers sometimes by abortion ♀ ♂, in loose clusters, never in catkins. Calyx membranous, imbricated, campanulate, inferior, irregular. Petals 0. Stamens definite, inserted into the base of the calyx, erect in æstivation. Ovary superior or 2-celled; ovules solitary, pendulous, anatropal, or amphitropal; stigmas 2, distinct. Fruit 1- or 2-celled, indehiscent, membranous or drupaceous. Seed solitary, pendulous; albumen none, or in very small quantity; embryo straight or curved, with foliaceous cotyledons; radicle superior.

Fig. CCCXCIII.

The plants of which Elm trees are the representatives assume two appearances, which have led Botanists into the opinion that they constitute two distinct Natural Orders. Of these the Nettle-trees, or Celteæ, have a hard fleshy fruit composed of a single carpel and amphitropal ovules, while the true Elms or Ulmeæ have a membranous fruit and anatropal ovules. They are, however, so much alike in most other circumstances, that it seems better to regard them as mere forms of one type, more especially since it seems, from the presence of two stigmas, that even the Celteæ themselves are really furnished with two carpels. It is very unusual to place Elmworts at a distance from Nettleworts, but I confess that their affinity seems to be much stronger with Rhamnads, of which they have the exact seed.

Natives of the North of Asia, the mountains of India, China, North America, and Europe; in the latter of which countries they form valuable timber-trees.

The inner bark of the Elm is slightly bitter and astringent, demulcent, and diuretic; it has been used in some skin diseases, but it does not appear to possess any important quality. The substance which exudes spontaneously from it is called Ulmin; this is also found in the Oak, Chesnut, and other trees, and, according to Berzelius, is a constituent of most kinds of bark. Elm wood is soft, tough, and coarse, but useful for many rough purposes, especially for water-pipes buried in the ground. The wood of Planera Abelicea, the Pseudosantalum creticum of the old Pharmacopœias, is aromatic. The young branches of Celtis australis are boiled, and the infusion is used against dysentery and blenorrhœa; the fruit is sweetish, and rather astringent; the kernel yields a useful oil. The drupes of Celtis occidentalis, the Nettle-tree or Sugar-berry, are administered in the United States in dysentery. The root, bark, and leaves of Celtis orientalis are somewhat aromatic, and are employed among eastern nations as a remedy for epilepsy.

GENERA.

I. CELTEÆ.—Ovary one-celled; ovules amphitropal. Celtis, *Tournef.*	Sponia, *Commers.* *Solenostigma*, Endl. Mertensia, *H. B. K.* ? Bosea, *Linn.* *Yerva-Mora*, Ludw.	II. ULMEÆ.—Ovary two-celled; ovules anatropal. Planera, *Gmel.*	Abelicea, *Hon. Belli.* *Zelkova*, Spach. Euptelea, *Zucc.* Microptelea, *Spach.* Ulmus, *Linn.*

NUMBERS. GEN. 9. Sp. 60.

Urticaceæ.
POSITION.—Rhamnaceæ.—ULMACEÆ.—Penæaceæ.
Thymelaceæ.

Fig. CCCXCIII.—Ulmus campestris.—*Nees.* 1. its flower; 2. its pistil; 3. its fruit; 4. its embryo.

ORDER CCXXII. RHAMNACEÆ.—RHAMNADS.

Rhamni, *Juss. Gen.* 376. (1789).—Rhamneæ, *DC. Prodr.* 2. 19. (1825); *Brongniart Mémoire sur les Rhamnées; Endl. Gen.* ccxxxix.; *Meisner, p.* 70.

DIAGNOSIS.—*Rhamnal Exogens, with polypetalous flowers, a valvate calyx, stamens opposite the petals, and erect seeds.*

Trees or shrubs, often spiny. Leaves simple, alternate, very seldom opposite; stipules, if any, minute. Flowers small, generally green, axillary or terminal, sometimes ♂ ♀ by abortion. Calyx 4-5-cleft, valvate. Petals distinct, cucullate, or

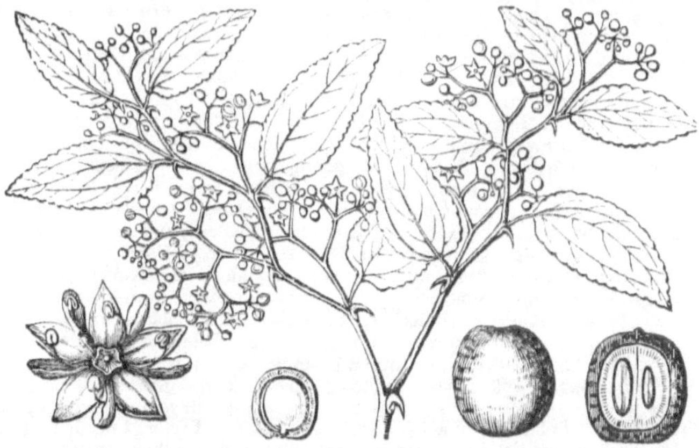

1 4 2 3

Fig. CCCXCIV.

convolute, inserted into the orifice of the calyx, occasionally 0. Stamens definite, opposite the petals. Disk fleshy. Ovary superior, or half superior, 2- 3- or 4-celled; ovules solitary, erect, anatropal. Fruit fleshy and indehiscent, or dry and separating in 3 divisions. Seeds erect; albumen fleshy, seldom wanting; embryo almost as long as the seed, with large flat cotyledons, and a short inferior radicle.

Under this name have been confounded four Orders, very different in characters, and even in natural affinities, the peculiarities of three of which have been pointed out by Ad. Brongniart in his Memoir upon the subject, and a fourth has been distinguished by myself. These Orders are Rhamnads properly so called, Spindle-trees, Hollyworts, and Bladder-nuts the respective affinities of which will be found under each. Brongniart indicates the relation that Rhamnads bear, thus:—If we take the insertion of stamens as the most important distinction of plants, it will be found that among polypetalous Orders with perigynous stamens, Appleworts are those to which Rhamnads have the closest relation, agreeing with them in the ovary, the cells of which are determinate in number, in the ascending ovules, and in their alternate leaves usually having two stipules at their base; the number and position of their stamens, and the structure of their seeds, separate them widely. But if the insertion of the stamens is left out of consideration, they will be found to have many characters in common with Byttneriads; such as, the æstivation of the calyx, the form of the petals, the position of the stamens in the front of those petals, the structure of the ovary and seeds in many important points; the principal differences between them are, in fact, the stamens being turned outwards in Byttneriads, the want in that Order of a disk, its hypogynous stamens, and 2 or more ovules. Spurgeworts are allied to Rhamnads; but the constant separation of sexes in the former Order, their hypogynous stamens and suspended ovules, are obvious marks of distinction. Hollyworts are monopetalous and

Fig. CCCXCIV.—Zizyphus Baclei. 1. a flower seen from above; 2. a fruit; 3. the same cut vertically; 4. a seed divided vertically.

have abundant albumen, and connect this Order of Rhamnads with Ebenads in the Gentianal Alliance.

It appears from the observations of Mr. Bennett (*Pl. Jav. rar.* 131), that in several genera the raphe of the anatropal seeds is thrown out of its original position next the placenta, by a twist in the cord by which it is attached to the placenta.

Found over nearly all the world, except in the arctic zone. The maximum of species is said to be dispersed through the hottest parts of the United States, the south of Europe, the north of Africa, Persia, and India in the northern hemisphere, and the Cape of Good Hope and New Holland in the southern. Some of the genera appear to be confined to particular countries, as all the true Ceanothuses to North America, Phylicas to the Cape, Cryptandra and Pomaderris to New Holland.

The berries of various species of Rhamnus are violent purgatives, and have been highly spoken of in dropsy. They also yield a dye, varying in tint from yellow to green; the ripe berries of R. catharticus, mixed with gum-arabic and lime-water, form the green colour known under the name of Bladder-green. The French Berries of the shops (Graines d'Avignon, *Fr.*) are the fruit of R. infectorius, saxatilis, and amygdalinus. Those of R. infectorius, when unripe, are used by the modern Greeks to dye morocco leather yellow. The fruit of Zizyphus is destitute of these purgative qualities, and on the contrary, is often wholesome and pleasant to eat, as in the case of the Jujube, Zizyphus vulgaris and Jujuba, the Zizyphus Œnoplia and Z. Joazeiro, whose drupes are used in Brazil as Jujubes. The Lote-bush, which gave its name to the Ancient Lotophagi, is to this day collected for food by the Arabs of Barbary, who call it Sadr, and its berries Nabk; it is the Zizyphus Lotus of Botanists. Many other species are also fit for food, among which, in Afghanistan, the Maimunna must be named. This is in some repute for its fruit, which is a sweetish black berry the size of a currant. Its genus has not been ascertained.—*Griffith.* The peduncles of Hovenia dulcis become extremely enlarged and succulent, and are in China a fruit in much esteem, resembling in flavour, as it is said, a ripe Pear. Some species are astringent. Sageretia theezans is used for tea by the poorer classes in China; an infusion of the twigs of Ceanothus americanus has been named as useful, on account of its astringency, to stop gonorrhœal discharges; antisyphilitic virtues are ascribed to the root of the same, and also of Berchemia volubilis; and it is said by Rumphius, that in the Moluccas the bark of Zizyphus Jujuba is employed as a remedy for diarrhœa. See *Royle's Illustrations*, p. 169. The Quina of Brazil is the Discaria febrifuga, whose acrid root is employed in the form of extract as a febrifuge and tonic. The bark of Zizyphus Joazeiro is bitter and astringent, with some acridity, and produces sickness.—*Martius.* Similar qualities have been recognised in various other species. The kernels of Zizyphus soporiferus are sedative, according to the Chinese, who employ them in their medicine. The negroes of the Gambia prepare a wine from the fermented berries of Zizyphus orthacanthus; but those of Z. Baclei are regarded as poisons. The bitter bark of Colubrina Fermentum is said to bring on violent fermentation in the liquors into which it is thrown. Gouania domingensis is stomachic; Berchemia lineata a hydragogue, according to Chinese authors. Finally, the root of Zizyphus Napeca is used as a remedy for windy colic.

GENERA.

Ventilago, *Gärtn.*
Paliurus, *Tournef.*
  Aspidocarpus, Neck.
  ? *Aubletia*, Lour.
Zizyphus, *Tournef.*
Condalia, *Cav.*
Berchemia, *Neck.*
  Œnoplia, Schult.
Sageretia, *Brongn.*
Hovenia, *Thunb.*
Rhamnus, *Juss.*
  Alaternus, Tournef.
  Marcorella, Neck.
  Cervispina, Dill.
  Cardiolepis, Raf.
  Frangula, Tournef.
Karwinskia, *Zucc.*
Scutia, *Commers.*
  Sentis, Commers.
  Sarcomphalus, P. Br.
Noltea, *Reichenb.*
  Vittmannia, Wight.
  Willemetia, Brongn.
  Sarcomphaloides, DC.
Ceanothus, *Linn.*
  Forrestia, Raf.
Cormonema, *Reiss.*
  Arrabidea, Steud.
Colubrina, *L. C. Rich.*
  Tubanthera, Commers.
Alphitonia, *Reiss.*
Colletia, *Commers.*
Discaria, *Hook.*
Adolphia, *Meisn.*
Ochetophila, *Pöpp.*
Retanilla, *Brongn.*
  Molinæa, Commers.
Talguenea, *Miers.*
  Trevoa, Gill.
Walpersia, *Reiss.*
  Trichocephalus, Reiss.
  Tylanthus, Reiss.
Petalopogon, *Reiss.*
Phylica, *Linn.*
Tylanthus, *Reiss.*
Soulangia, *Brongn.*
Spyridium, *Fenzl.*
Cryptandra, *Smith.*
Pomaderris, *Labill.*
  Pomatoderris, Schult.
Trymalium, *Fenzl.*
Gouania, *Jacq.*
  Retinaria, Gärtn.
Reissekia, *Endl.*
  Helinus, E. M.
  Willimetia, E. Z.
Crumenaria, *Mart.*
? Solenantha, *G. Don.*
? Schæfferia, *Jacq.*
? Strombosia, *Blum.*
? Samara, *Linn.*
? Daphniphyllum, *Blum.*
? Crypteronia, *Blume.*
? Galdicia, *Neraud.*
? Quoia, *Neraud.*
? Carolinia, *Neraud.*

NUMBERS. GEN. 42. Sp. 250.

Aquifoliaceæ.
POSITION.—Celastraceæ.—RHAMNACEÆ.—Chailletiaceæ.
Byttneriaceæ.

ORDER CCXXIII. CHAILLETIACEÆ.—CHAILLETIADS.

Chailletiæ, *R. Brown Cong. p.* 23. (1818).—Chailletiaceæ, *DC. Prodr.* 2. 57. (1825); *Endl. Gen.* ccxl.

DIAGNOSIS.—*Rhamnal Exogens, with polypetalous flowers, a valvate calyx, stamens alternate with the petals, and pendulous seeds.*

Trees or shrubs. Leaves alternate, with two stipules, deciduous, entire. Flowers small, axillary, fasciculate or corymbose, their peduncle often connate with the petiole. Sepals 5, with an incurved valvate æstivation. Petals 5, alternate with the sepals, and arising from the base of the calyx, usually 2-lobed. Stamens 5, alternate with the petals, and combined with them at the base; anthers ovate, versatile. Glands usually 5, hypogynous, opposite the petals. Ovary superior, 2- or 3-celled; ovules twin, pendulous; style simple; stigma obsoletely 2-lobed. Fruit drupaceous, rather dry, 1- 2- or 3-celled. Seeds solitary, pendulous, naked or arillate, without albumen; embryo thick, with a thick superior radicle and fleshy cotyledons.

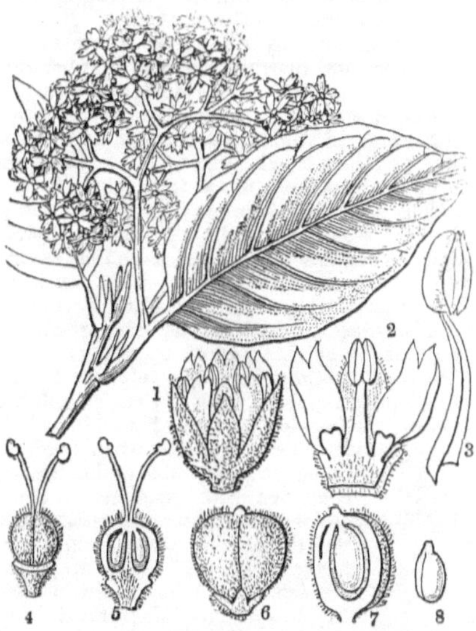

Fig. CCCXCV.

Whether what are here called petals are not rather abortive stamens is doubted by Botanists, and hence the station of the Order is compared, on the one hand, with Anacards or Roseworts, and on the other, with Samyds and Mastworts. To me it seems that what appear to be petals are so; a fact which it is difficult to doubt, when it is remembered that both calyx and corolla are mere modifications of one common type, and that it is in position only that they differ. De Candolle stations the Order between Homaliads and Aquilariads; it agrees with the former in the presence of glands round the ovary, but differs in its superior ovary with the placentæ in the axis, and many other characters. Rhamnads, with which it corresponds so much in habit, seem, however, upon the whole to claim the closest kindred with it, and it can hardly be regarded in any other light than as a member of the Rhamnal Alliance. It will then be stationed in the neighbourhood of Elmworts, which some Botanists are convinced is its true position. Its valvate calyx separates it from Hippocrateads; its pendulous ovules and stamens alternate with the petals, from Rhamnads.

Of the few known species belonging to this Order, 2 are found in Sierra Leone, 2 in Madagascar, 2 in equinoctial America, and 1 in Timor.

The fruit of Chailletia toxicaria is said to be poisonous; it is called Ratsbane in Sierra Leone.

GENERA.

Moacurra, *Roxb.*	*Symphyllanthus*, Vahl.	*Dichapetalum*, Thouars	Tapura, *Aubl.*
Wahlenbergia, R. Br.	*Mestotes*, Soland.	*Leucosia*, Thouars.	*Rohria*, Schreb.
Chailletia, *DC.*	*Patrisia*, Rohr.	*Plappertia*, Reichenb.	Stephanopodium, *Pöpp.*

NUMBERS. GEN. 4. Sp. 10.

Samydaceæ.
POSITION.—Ulmaceæ.—CHAILLETIACEÆ.—Rhamnaceæ.
Homaliaceæ.

Fig. CCCXCV.—Chailletia pedunculata.—*Turpin.* 1. a flower of Moacurra gelonioides; 2. a portion of it; 3. a stamen; 4. the pistil; 5. a vertical section of it; 6. ripe fruit; 7. a section of it; 8. an embryo.

ORDER CCXXIV. HIPPOCRATEACEÆ.—HIPPOCRATEADS.

Hippocraticeæ, *Juss. Ann. Mus.* 18. 483. (1811).—Hippocrateaceæ, *Kunth in Humb. N. G. Am.* 5. 136. (1821); *DC. Prodr.* 1. 567; *Endl. Gen.* ccxxxvii.; *Meisner, p.* 56; *Wight Illustr.* 1. 132.

DIAGNOSIS.—*Rhamnal Exogens, with polypetalous flowers, an imbricated calyx, and 3 monadelphous stamens.*

Arborescent or climbing shrubs, which are almost always smooth. Leaves opposite, simple, entire or toothed, somewhat coriaceous, with small deciduous stipules. Racemes axillary, in corymbs or fascicles. Flowers small, inconspicuous. Sepals 5, very small, imbricated, combined as far as the middle, persistent. Petals 5, somewhat imbricated in æstivation. Stamens 3; filaments cohering almost as far as the apex into a tube dilated at the base, and forming about the ovary a thick disk-like cup; anthers opening transversely at the apex. Ovary free, concealed by the tube, 3-celled; style 1; stigmas 1-3; ovules ascending, anatropal or half anatropal. Fruit either consisting of 3 samaroid carpels, or berried, with from 1 to 3 cells. Seeds in each cell definite, attached to the axis in pairs, some of them occasionally abortive, sometimes buried in pulp, erect, without albumen; embryo straight; radicle pointing towards the base; cotyledons flat, elliptical oblong, somewhat fleshy, cohering when dried.

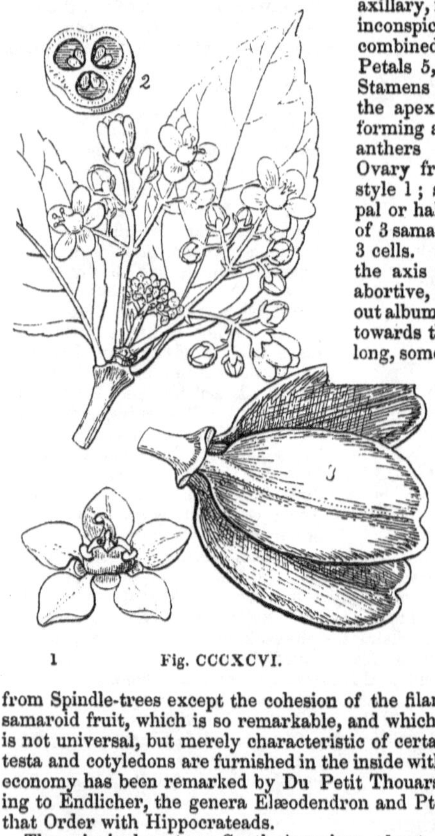

Fig. CCCXCVI.

The ternary number of the stamens, combined with pentamerous petals and sepals, is the prominent characteristic of this Order, which was formerly included among Maples by Jussieu, which is placed between Erythoxyls and Marcgraaviads by De Candolle, but which is, to all appearance, much more nearly related to Spindle-trees, as Brown has remarked; for "the insertion of the ovules is either towards the base, or is central; the direction of the radicle is always inferior."—*Congo*, 427. In fact there seems to be nothing to divide Hippocrateads from Spindle-trees except the cohesion of the filaments of the former into a cup. The samaroid fruit, which is so remarkable, and which connects the Order with Malpighiads, is not universal, but merely characteristic of certain genera. In Hippocratea ovata the testa and cotyledons are furnished in the inside with innumerable spiral threads; the same economy has been remarked by Du Petit Thouars in the pericarp of Calypso. According to Endlicher, the genera Elæodendron and Ptelidium among Spindle-trees, connect that Order with Hippocrateads.

The principal part are South American, about one-seventh are natives of Africa or the Mauritian Islands, and the same number has been recorded as East Indian.

The fruit of Tontelea (Salacia) pyriformis, a native of Sierra Leone, is eatable. It is about the size of a Bergamot Pear; its flavour is rich and sweet. The nuts of Hippocratea comosa are oily and sweet; it is called, in the French West India Islands, Amandier du Bois. Martius reports that several species of Tontelea, called Saputá in Brazil, have a sweet mucilaginous fruit, which is eaten. I find no indication here of the emetic and nauseous quality recorded as being characteristic of Spindle-trees.

Fig. CCCXCVI.—Hippocratea Arnottiana.—*Wight.* 1. a flower; 2. a cross section of the ovary; 3. ripe fruit.

GENERA.

Hippocratea, *Linn.*
Coa, Plum.
Bejuco, Löffl.
Daphnikon, Pohl.
Pereskia, Fl. Flum.
Trigonotheca, *Hochst.*
Tontelea, *Aubl.*
Tonsella, Schreb.
Sicelium, P. Br.
Anthodon, Ruiz et Pav.
Anthodiscus, Mart.
Raddisia, Leandr.
Clercia, Fl. Flum.
Salacia, *Linn.*
Calypso, Thouars.
Johnia, Roxb.
Diplesthes, *Harv.*
? Lacepedea, *H. B. K.*
Triceraia, Willd.

NUMBERS. GEN. 6. Sp. 86.

Malpighiaceæ.
POSITION.—Chailletiaceæ.—HIPPOCRATEACEÆ.—Celastraceæ.
Aceraceæ.

ORDER CCXXV. CELASTRACEÆ.—SPINDLE-TREES.

Celastrineæ, *R. Brown in Flinders*, 22. (1814); *DC. Prodr.* 2. 2.; *Ad. Brongniart Mémoire sur les Rhamnées*, 16.; *Endl. Gen.* ccxxxvi.; *Meisner Gen. p.* 68.; *Wight Illustr.* 1. 174.; *Arn. in Ann. Nat. Hist.* 3. 153.

DIAGNOSIS.—*Rhamnal Exogens, with polypetalous flowers, an imbricated calyx, and stamens* (5/V) *distinct.*

Small trees or shrubs. Leaves alternate, seldom opposite, simple, with very small deciduous stipules. Flowers in axillary cymes, small, green, or white, or purple, occasionally ♂ ♀ by abortion. Sepals 4 or 5, imbricated, inserted into the margin of an expanded disk. Petals inserted by a broad base, under the margin of the disk, with an imbricate æstivation; sometimes 0. Stamens alternate with the petals, inserted into the disk, either at the margin or within it; anthers innate. Disk large, expanded, flat, closely surrounding the ovary, covering the flat expanded calyx. Ovary immersed in the disk and adhering to it, with 2 to 5 cells; cells 1- or many-seeded; ovules ascending from the axis, anatropal, attached to a short funiculus. Fruit superior, 2- to 5-celled, either capsular or drupaceous. Seeds ascending, seldom inverted by resupination, either provided with an aril, or without one; albumen fleshy; embryo straight; cotyledons flat and thick, with a short inferior radicle.

Fig. CCCXCVII.

Formerly confounded with Rhamnads, this Order was first separated by Brown, who distinguished it particularly by the relation which its stamens bear to the petals. It also differs in its imbricated calyx, and in its disk being hypogynous. According to Brongniart, Spindle-trees have more relation to several Orders with hypogynous stamens than to any with perigynous ones, especially to Malpighiads, to which they are related through Hippocrateads; a considerable resemblance with such Spurgeworts as Phyllanthus may also be traced; and Hollyworts have been principally established upon dismemberments of the present Order. Nevertheless, the distinctions between it and both Spurgeworts and Hollyworts are easy to

Fig. CCCXCVII.—Catha edulis.

trace; for the former are constantly diclinous, and the latter monopetalous: besides which, the radicle of Spindle-trees is inferior, that of Spurgeworts superior; and the albumen of Hollyworts is extremely copious, while that of Spindle-trees is comparatively inconsiderable in quantity. The drupaceous genera, forming the Elæodendrous Sub-order, establish an affinity with Sapotads, which have, however, a monopetalous corolla and milky juice, and their stamens, when those which are fertile equal in number the segments of the corolla, are opposite to the latter. Endlicher finds a resemblance with Pittosporads, and justly adds that all the drupaceous genera are greatly in need of more careful examination.

According to M. Planchon, the arillus of Euonymus is a peculiar expansion of the exostome, and is not derived from the placenta.

The species are natives of the warmer parts of Europe, North America, and Asia, but far more abundant beyond the tropics than within them; a great number of species inhabit the Cape of Good Hope. Some are found in Chile and Peru, and a few in New Holland.

Royle mentions an acrid principle having been detected among the species, which acts with more or less activity; and that the seeds of several yield an oil which is useful for burning. That of Celastrus nutans or paniculatus is said in India to be of a stimulant nature, and to be used in medicine in the disease called Berriberri. The bark of Euonymus tingens is in the inside of a beautiful light-yellow colour, similar to that of some species of Rhamnus; it is used to mark the tika on the forehead of Hindoos, and might be employed as a dye. It is also considered useful in diseases of the eye. The leaves of Catha edulis, Kat or Khât of the Arabs, would appear to be of a stimulating nature. According to Forskähl, the Arabs eat the green leaves with greediness, believing them to have the power of causing extreme watchfulness, so that a man may stand sentry all night long without drowsiness. They also regard it as an antidote to the plague, and assert that a person wearing a twig of it in his bosom, may go among the infected with impunity; they even believe that the plague cannot appear in places where the tree is cultivated. Nevertheless, says Forskähl, "the taste of the leaves does not seem to indicate such virtues." Botta also says that, when fresh, the Khât leaves are very intoxicating. The fresh bark of the root of Elæodendron Roxburghii, rubbed with plain water, is by the natives of India applied externally to almost every sort of swelling. It is a very strong astringent, possessing scarcely any other sensible quality.—*Roxb.* Similar qualities are attributed to Maytenus chilensis. The seeds of the European species of Evonymus are nauseous, and said to be purgative and emetic; sheep are said to be poisoned by them; an ointment was formerly prepared from them for the destruction of pediculi in the head. Similar qualities have been found in the bark of Celastrus scandens and senegalensis; while the spines of Celastrus venenatus are reported to inflict most painful wounds. The drupes of Elæodendron Kubu are eaten by the colonists of the Cape of Good Hope.

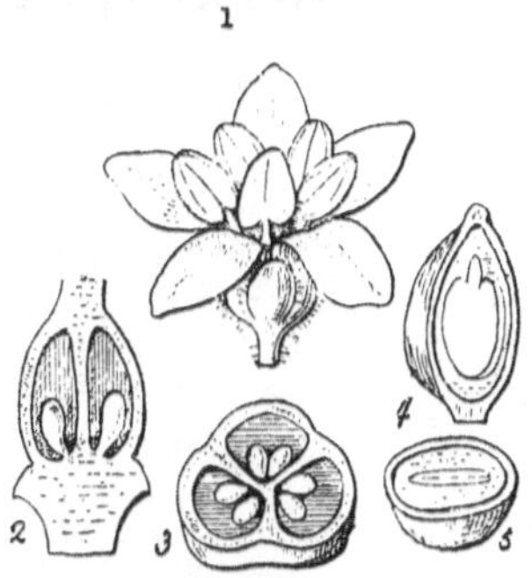

Fig. CCCXCVIII.

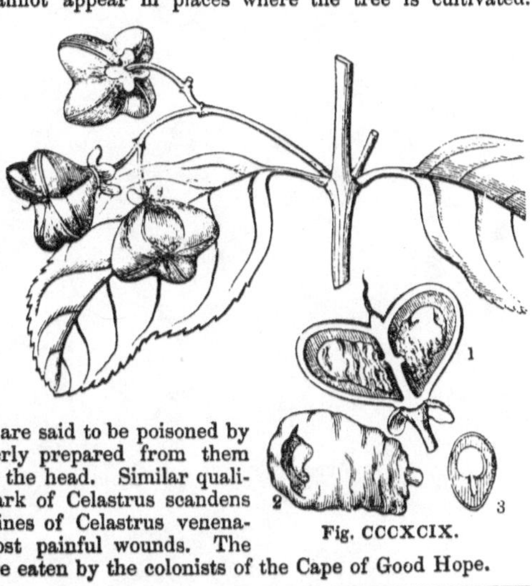

Fig. CCCXCIX.

Fig. CCCXCVIII.—Celastrus paniculatus.—*Wight.* 1. a flower; 2. a perpendicular section of the ovary; 3. a cross section of the ovary; 4. a vertical section of a seed; 5. a cross section of it.

Fig. CCCXCIX.—Euonymus europæus. 1. a section of a fruit; 2. a seed enveloped in its aril; 3. a perpendicular section of a seed.

GENERA.

I. EUONYMEÆ. — Fruit capsular.

Putterlickia, *Endl.*
Lophopetalum, *Wight.*
Evonymus, *Tournef.*
Polycardia, *Juss.*
Florinda, Noronh.
Commersonia, Comms.
Catha, *Forsk.*
Sonneratia, Commers.
Celastrus, *Kunth.*
Maytenus, *Juss.*
Maiten, Feuill.
Hænkea, Ruiz. et Pav.
Microtropis, *Wall.*
Pterocelastrus, *Meisn.*
Asterocarpus, Eckl. et Zeyh.

II. ELÆODENDREÆ. — Fruit drupaceous.

Ptelidium, *Thouars.*
Seringia, Spreng.
Wimmeria, *Schlecht.*
Frauenhofera, *Mart.*
Pleurostylia, *Wight et A.*
Hartogia, *Thunb.*
Schrebera, Thunb.
Elæodendron, *Jacq.*
Portenschlagia, Tratt.
Lamarckia, Hort.
Nerija, Roxb.
Skytophyllum, Eckl. et Zeyh.
Lauridia, Eckl. et Zh.
Mystroxylon, Ec. et Z.
Crocoxylon, Eckl. et Z.
Parilla, Dennst.
Myginda, *Jacq.*
Rhacoma, Linn.
Crossopetalum, P. Br.
Pachystima, *Raf.*
Oreophila, Nutt.
? Bhesa, *Hamilt.*
Kurrimia, Wall.
? Actegiton, *Blum.*
? Tralliana, *Lour.*
? Lepta, *Lour.*
? Goupia, *Aubl.*
Gupia, Jaume.
Glossopetalum, Schb.
? Perrottetia, *H. B. K.*
? Alzatea, *Ruiz. et Pav.*
Alziniana, Dietr.

NUMBERS. GEN. 24. Sp. 260.

Aquifoliaceæ.
POSITION.—Sapotaceæ.—CELASTRACEÆ.—Hippocrateaceæ.
Euphorbiaceæ.

ORDER CCXXVI. STACKHOUSIACEÆ.—STACKHOUSIADS.

Stackhouseæ, *R. Br. in Flinders*, 555. (1814).—Stackhousiaceæ, *Ed. Pr.* lxxxviii ; *Endl. Gen.* ccxlii.

DIAGNOSIS.—*Rhamnal Exogens, with monopetalous flowers, and episepalous stamens.*

Herbaceous plants, occasionally somewhat shrubby. Leaves simple, entire, alternate, sometimes minute. Stipules lateral, very minute. Spike terminal, each flower with 3 bracts. Calyx 1-leaved, 5-cleft, equal, with an inflated tube. Petals 5, equal, arising from the top of the tube of the calyx ; their claws combined in a tube longer than the calyx ; their limb narrow, stellate. Stamens 5, distinct, unequal (2 alternately shorter), arising from the throat of the calyx. Ovary superior, 3- or 5-celled, the cells partially separated, adhering to a central column, each with a single erect anatropal ovule ; styles from 3 to 5, sometimes combined at the base ; stigmas simple. Fruit of from 3 to 5 indehiscent, winged, or wingless pieces ; column central, persistent. Embryo erect, in the axis of, and almost as long as, the fleshy albumen, with short obtuse cotyledons and an inferior radicle.

This Order should stand between Spindle-trees and Spurgeworts, according to Brown ; from the latter of which it differs in the structure of the fruit, and in the position of the seeds, besides other characters ; from the former in the presence of stipules, in the cohesion of the petals in a tube, in the deeply lobed ovary, and so on. The hermaphrodite flowers remove the Order, however, from Spurgeworts; its monopetalous flowers are much at variance with the structure of Spindle-trees. Nevertheless, the 3-celled ovary, in flowers otherwise pentamerous, is entirely that of Hippocrateads and Spindle-trees, and recalls the Sapindal Alliance, to which all those Orders would be referable if their stamens were not so distinctly perigynous.

A few New Holland shrubs compose all that is known.

Their properties are unascertained.

GENERA.

Stackhousia, *Smith.* | Tripterococcus, *Endl.*

NUMBERS. GEN. 2. Sp. 10.

POSITION.—Celastraceæ.—STACKHOUSIACEÆ.—Sapotaceæ.
Sapindaceæ. (above) *Euphorbiaceæ.* (below)

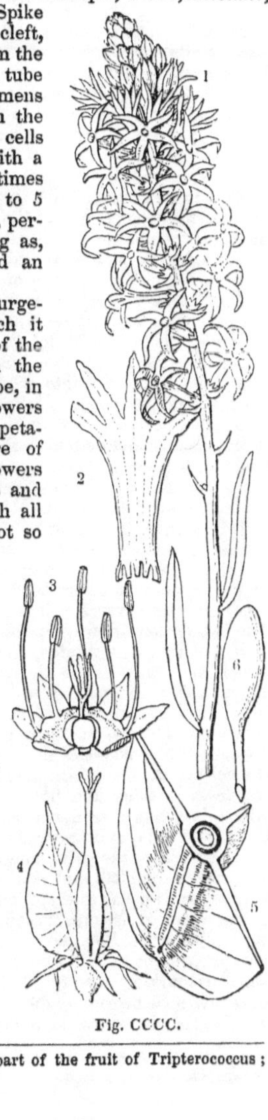

Fig. CCCC.

Fig. CCCC.—1. Stackhousia ; 2. its corolla ; 3. calyx, &c. ; 4. part of the fruit of Tripterococcus ; 5. one of its cocci cut across ; 6. an ovule.

ORDER CCXXVII. SAPOTACEÆ.—SAPOTADS.

Sapotæ, *Juss. Gen.* 151. (1789).—Sapoteæ, *R. Brown Prodr.* 528. (1810).—Sapotaceæ, *Endl. Prodr. Norf.* 48. (1833); *Gen.* clviii.; *Meisn. p.* 159; *Alph. DC. Prodr.* 8. 154.

DIAGNOSIS.—*Rhamnal Exogens, with monopetalous flowers, epipetalous stamens, ascending ovules, a short radicle, and amygdaloid cotyledons.*

Trees or shrubs, chiefly natives of the tropics, and often abounding in milky juice. Leaves alternate, or occasionally almost whorled, without stipules, entire, coriaceous. Inflorescence axillary. Flowers hermaphrodite. Calyx regular, persistent, in 5, or occasionally 4-8 divisions, which are either valvate or imbricate in æstivation. Corolla monopetalous, hypogynous, regular, deciduous, its segments usually equal in number to those of the calyx, seldom twice or thrice as many, imbricated in æstivation. Stamens arising from the corolla, in number definite, distinct, the fertile ones equal in number to the segments of the calyx, and opposite those segments of the corolla which alternate with the latter, seldom more; anthers usually turned outwards. The sterile stamens as numerous as the fertile ones, with which they alternate. Disk 0. Ovary superior, with several cells, in each of which is 1 ascending or pendulous anatropal ovule; style 1; stigma undivided, occasionally lobed. Fruit fleshy, with several 1-seeded cells, or by abortion with only 1. Seeds nut-like, sometimes cohering into a several-celled putamen. Testa bony, shining, with a very long scar on the inner face where it is opaque, and softer than the rest. Embryo erect, large, white, usually inclosed in fleshy albumen. Cotyledons, when albumen is present, foliaceous; when absent, fleshy and sometimes connate. Radicle short, straight, or a little curved, turned towards the hilum.

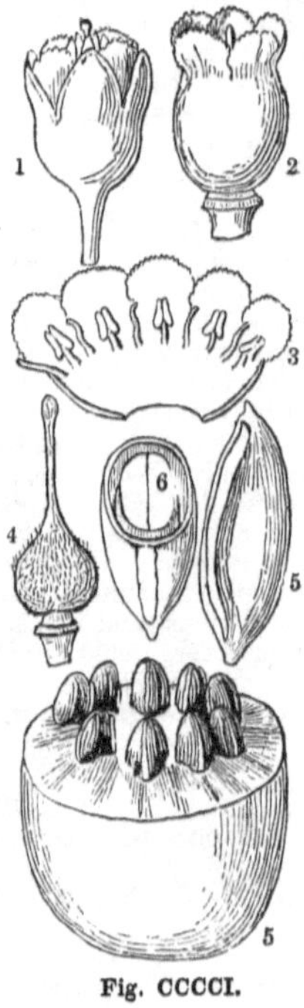

Fig. CCCCI.

This Order is certainly near Ebenads, with which it agrees in habit, arborescent stem, alternate entire leaves, and axillary inflorescence; and, moreover, in its monopetalous regular hypogynous corolla, the absence of a hypogynous disk, an ovary with several cells, and definite ovules and stamens. The two Orders, however, differ in several points. Sapotads have usually a milky juice, and their wood is among the softer kinds; their flowers are always hermaphrodite; the segments of the calyx and corolla are often placed in a double row; their stamens are always in a single row, the fertile ones rarely more numerous than the segments of the calyx, and opposite the divisions of the corolla; their style is undivided; the cells of the ovary are always 1-seeded, with erect ovules; the testa is thick and bony; the embryo is large with respect to the fleshy albumen, which is sometimes deficient; the radicle is very short and inferior. In Ebenads there is no milk, and the wood is very hard; the flowers are often unisexual by abortion; the segments of the calyx and corolla are almost always in a single row; the stamens are usually doubled, and either twice or four times as numerous as the segments of the corolla, or, if equal to them, alternate with them; the style is generally divided, the cells of the ovary sometimes 2-seeded, the ovules always pendulous, the testa thin and soft, the embryo middle-sized or small in respect to the cartilaginous albumen, which is always present; the radicle is of middling length, or very long and superior. It is worth remarking, that the woody shell of the seed of Sapotads is certainly testa, and not putamen, as is proved by the presence of the micropyle upon it. They are also comparable with Ardisiads, whose abundant albumen and free central placenta render it necessary

Fig. CCCCI.—1. flower of a Sapota; 2. its corolla; 3. the same cut open; 4. the pistil; 5. half a fruit of Bassia longifolia; 5, 6. its seed, whole and cut across.

that they should be stationed at some distance. They differ from Storaxworts in their short radicle and amygdaloid embryo.

Chiefly natives of the tropics of India, Africa, and America; a few are found in the southern parts of North America, and at the Cape of Good Hope.

The fruit of many is esteemed in their native countries as an article of the dessert: such are the Sappodilla Plum (Achras Sapota and other species), the Star-apple (Chrysophyllum Cainito), the Marmalade (Achras mammosa), the Medlar of Surinam, the Mimusops Elengi, and others; they are described as having generally a sweet taste, with a little acidity. The Bully or Bullet-tree of Guiana is a species of Mimusops according to Sir R. Schomburgk. The fruit is described as being of the size of the Coffee-berry, and when quite ripe delicious; its wood is solid, heavy, close-grained, and durable. Besides these, various species of Lucuma and Chrysophyllum rank among dessert fruits, as do the Imbricarias malabarica and maxima, whose fruit is subacid, sweet, and like an Orange in appearance. The seeds of Achras Sapota are aperient and diuretic, but in over-doses they produce severe pain, and are even dangerous; the bark is a substitute for Cinchona; those of some others are filled with a concrete oil, which is used for domestic purposes. Mimusops Kaki, like many trees with astringent bark, yields a gum, while its fruit is of a sweetish taste, and much eaten by the natives of India. A kind of thick oil, like butter, is obtained from the fruit of Bassia butyracea, the Mahva or Madhuca-tree. The flowers of B. latifolia (the Mopha, Maddoodoomah), are employed extensively in the distillation of a kind of arrack, called Mowra; they are said to resemble in taste the dried seedless Grapes called Corinths. The Bassia longifolia is called the Illupie-tree; its fruit, when pressed, yields a large quantity of oil used in India for lamps, soap-making, and also for food; it is employed medicinally to cure the itch, and other cutaneous disorders; the leaves boiled in water, as well as the milk of the green fruit and bark, are used in rheumatic affections. The Butter-tree of Mungo Park was also a species of Bassia. See *Royle's Illustrations*, p. 263, for further information concerning these Bassias. The bark of 4 species of Achras is so astringent and febrifugal as to have been substituted for Quinquina. The Cow-tree of Humboldt has been sometimes supposed to be referable to this Order; but there seems no reason now to doubt its belonging to Artocarpads. Monesia bark, a South American product, with a powerful bitter-sweet taste, lately employed successfully in France in diarrhœa, menorrhagia, leucorrhœa, and hæmoptysis, is said to belong to some plant of this Order.—*Pharm. Journ.* 3. 292. The bark of Bumelia nigra and others is bitter, astringent, and febrifugal, and the wood very hard. The fruit of B. retusa is said to be milky; that of B. lycioides austere, with some sweetness, and useful in diarrhœa; while the flowers of B. graveolens have a heavy unpleasant odour. The flowers of Mimusops Elengi, on the contrary, are powerfully aromatic, and a fragrant water is distilled from them. The seeds of this plant yield an abundance of oil, in request for painters, and said to be useful in parturition; the leaves are said to produce an extraordinary noise when burnt.

GENERA.

Chrysophyllum, *L.*
Nycterisition, R. P.
Cainito, Tuss.
Ecclimusa, Mart.
Pouteria, *Aubl.*
Chætocarpus, L.
Labatia, Sw.
Labatia, *Mart.*
Lucuma, *Mol.*
Guapeba, Gom.
Vitellaria, Gærtn.
Sapota, *Plum.*
Achras, P. Br.
Hormogyne, *A. Cunn.*
Sersalisia, *R. Br.*
Sideroxylon, *L.*
Robertsia, Scop.
Argania, *R. et Sch.*
Isonandra, *Wight.*
Dipholis, *A. DC.*
Bumelia, *Sw.*
Labourdonnaisia, *Bojer.*
Delastrea, *A. DC.*
Azaola, *Blanco.*
Payena, *A. DC.*
Bassia, *König.*
? *Palaquium*, Blanco.
Imbricaria, *Comm.*
Binectaria, Forsk.
Mimusops, *L.*
Elengi, Rheede.
Manilkara, Rheede.
? *Phebolithis*, Gærtn.
Synarrhena, Fisch.
Omphalocarpum, *Beauv.*
? Rostellaria, *Gærtn.*

NUMBERS. GEN. 21. Sp. 212.

Ebenaceæ.
POSITION.—Styracaceæ.—SAPOTACEÆ.—Celastraceæ.
Myrsinaceæ.

ORDER CCXXVIII. STYRACACEÆ.—STORAXWORTS.

Styraceæ, *Rich. Anal. du Fr.* (1808); *Von Martius, N. Gen. et Sp. Pl.* 2. 148.; *Endl. Gen. p.* 742; *Meisner, p.* 250.—Symplocineæ, *Don. Prodr. Nep.* 144. (1825).—Styracinæ, *Rich. in Humb. N G. et Sp.* 3. 256. (1818).—Halesiaceæ, *Don in Jameson's Journ.* (*Dec.* 1828); *Link Handb.* 1. 667.—Styracaceæ, *A. DC. Prodr.* 8. 244. (1844).

DIAGNOSIS.—*Rhamnal Exogens, with monopetalous flowers, epipetalous stamens, a part at least of the ovules suspended, a long radicle, and leafy cotyledons.*

Trees or shrubs. Leaves alternate, without stipules, usually toothed. Flowers axillary, either solitary or clustered, with scale-like bracts. The hairs often stellate.

1 Fig. CCCCII. 2 3

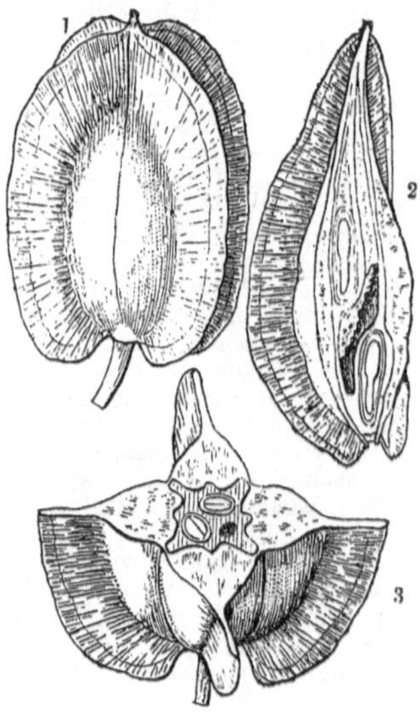

3

Fig. CCCCIII.

Calyx inferior (or superior), with 5 (or 4) imbricated divisions, persistent. Corolla monopetalous, the number of its divisions frequently different from that of the calyx; with imbricated æstivation. Stames definite or indefinite, arising from the tube of the corolla, of unequal length, cohering in various ways, but generally in a slight degree only; anthers innate, 2-celled, bursting inwardly. Pollen broadly elliptical, smooth. Ovary superior, or adhering to the calyx, with from 2 to 5 cells, which are opposite the lobes of the calyx when they are of the same number, the partitions sometimes scarcely adhering in the centre. Ovules anatropal, 2 or 00 in each cell, all pendulous or the upper ascending, the lower pendulous; style simple; stigma somewhat capitate. Fruit drupaceous, surmounted by or inclosed in the calyx, generally with all the cells abortive except one. Seeds ascending or suspended, 5- 1, with the slender embryo lying in the midst of the albumen; radicle long, directed towards the hilum; cotyledons flat.

Those Botanists who attach paramount importance to the condition of the corolla, in deciding upon the relationship of plants, will object to the station now occupied by Storaxworts, which, because of a slight adhesion between the petals, are usually associated with Ebenads. But if a less value is assigned to that character and more to the presence of albumen, then the Storaxworts will fall into the ranks of a different Alliance, in which they will, however, present a distinct tendency towards the Ebenaceous structure. For this reason they are here placed among Rhamnals; while Ebenads are associated with Hollyworts and some other Orders in a neighbouring Alliance.

Such is my own opinion on this subject; the following is the view taken by others. Mr. Bentham would associate them with Ebenads and Humiriads; besides which

Fig. CCCCII.—Styrax suberifolium.—*Hooker*. 1. a flower; 2. corolla and stamens; 3. pistil.
Fig. CCCCIII.—Halesia tetraptera.—1. its fruit; 2. a perpendicular; 3. a transverse, section of it.

he finds a resemblance to Citronworts and Olacads.—*Linn. Trans.* 18. 231. According to Alph. De Candolle, who adopts those views, (*Prodr.* 8. 245), they differ from Ebenads in their hermaphrodite cymose, not racemose flowers, in their fewer stamens, partly alternate with the lobes of the corolla, their longer filaments, their ovary partially inferior, and especially in the cells of the ovary being opposite the lobes of the calyx. Decaisne (*Travels of Jacquemont*, p. 104), thinks them nearer Alangiads. They must also be regarded as standing in close relation to Olacads, from which they hardly differ perhaps, except in their embryo being longer as respects the seed, and in their long radicle.

Storaxworts are sparingly distributed, for the most part through the tropical or subtropical regions of both hemispheres ; a very few, among which are the Snowdrop-trees, (Halesia), find their way to cold latitudes. According to Alph. De Candolle, they are unknown in Australia, and exist in Africa in no instance except that of Styrax guineense, a doubtful plant.

Some of the genus Symplocos are used in dyeing yellow, as S. tinctoria, called Sweetleaf in Carolina ; its root is bitter and aromatic ; others, as S. Alstonia, are employed as tea, on account of a slight astringency in their leaves. Storax and Benzoin, fragrant gum-resins, composed of resin, benzoic acid, and a peculiar aromatic principle, are the produce of two species of Styrax. Storax flows from wounds in Styrax officinale, a Syrian tree. Benzoin is derived from S. Benzoin, a native of the Malay archipelago. Both drugs are regarded as stimulating expectorants, producing an irritation of the mucous membrane of the air-passages. Benzoin is used in the preparation of Paregoric elixir, and of Court plaister, and also in the cosmetic called Virgin's Milk. A fragrant secretion of a similar nature is produced by Styrax reticulata, ferruginea and aurea, in Brazil ; according to Martius, it is employed in the churches as frankincense. Symplocos (Bobua) laurina is celebrated in Bengal for its bark, which forms a mordant for red dyes.

GENERA.

I. SYMPLOCEÆ.—Corolla quincuncial. Anthers roundish.

Symplocos, *Jacq.*
Eugenioides, L.
Bobua, DC.
Alstonia, L.
Hopea, L.
Ciponima, Aubl.
Siponima, Aubl.
Decadia, Lour.
Barberina, Velloz.
Stemmatosiphon, Pohl
? *Dicalyx*, Lour.
? *Drupatris*, Lour.
Palura, Ham.

II. STYRACEÆ.—Corolla twisted to the left, or somewhat valvate. Anthers long,

Styrax, *Tourn.*
Strigilia, Cav.
Tremanthus, Pers.
Benzoin, Hayne.
Lithocarpus, Bl.
Epigenia, Vell.
Cypellium, Desv.
Trichogamila, P. Br.
Pterostyrax, *Sieb. et Zuc.*
Halesia, *Ellis.*

III. PAMPHILIEÆ. *A. DC.*—Corolla valvate.

Pamphilia, *Mart.*
Foveolaria, *R. P.*

NUMBERS. GEN. 6. Sp. 115.

POSITION.—Sapotaceæ.—STYRACACEÆ.—*Ebenaceæ.* / *Olacaceæ.*

ALLIANCE XLV. *GENTIANALES.*—THE GENTIANAL ALLIANCE.

DIAGNOSIS.—*Perigynous Exogens, with dichlamydeous monopetalous flowers, axile or parietal placentæ, and a minute embryo, or with the cotyledons much smaller than the radicle, lying in a large quantity of albumen.*

Here we find ourselves among the truly monopetalous Orders of the French school. Previously a monopetalous structure was an exception rather than a rule ; but now a separation of petals forms the exception. Tendencies have assumed a new direction. The Alliance differs from that of Solanals in having a minute embryo and much albumen, and from Cortusals in the placenta never being free and central. It touches Solanals at Nightshades themselves, which, if they had parietal placentæ might often be mistaken for Gentianworts; and at Dogbanes, whose minute embryo offers one of the principal reasons for not associating them in the same Alliance as Asclepiads. With Cortusals Gentianals come in contact through Ebenads, which are very like Ardisiads, and Diapensiads, which may be compared to Primworts. To Bignonials they are very closely allied through Broomrapes and Stilbids, which put on the peculiar aspect of that Alliance.

NATURAL ORDERS OF GENTIANALS.

Stipules 0. *Stigmas simple, sessile, radiating*	229. EBENACEÆ.
Stipules 0. *Stigmas simple, at the end of a manifest style. Placentæ axile. Seeds definite, pendulous. Corolla imbricated* .	230. AQUIFOLIACEÆ.
Stipules 0. *Stigmas collected into a massive head, expanded at the base in the form of a ring or membrane, and contracted in the middle.* (*Albumen sometimes* 0)	231. APOCYNACEÆ.
Leaves opposite, with intervening stipules	232. LOGANIACEÆ.
Stipules 0. *Stigmas simple, at the end of a manifest style. Placentæ axile. Seeds indefinite, peltate. Stamens interpetalous* .	233. DIAPENSIACEÆ.
Stipules 0. *Stigmas simple, at the end of a manifest style. Placentæ axile. Seeds definite, erect. Corolla valvate. Flowers unsymmetrical*	234. STILBACEÆ.
Stipules 0. *Stigmas simple, at the end of a manifest style. Placentæ parietal. Flowers didynamous*	235. OROBANCHACEÆ.
Stipules 0. *Stigmas simple, at the end of a manifest style. Placentæ parietal. Flowers regular*	236. GENTIANACEÆ.

ORDER CCXXIX. EBENACEÆ.—EBENADS.

Guaiacanæ, *Juss. Gen.* 155. (1789) *part of the first sect.*—Ebenaceæ, *Vent. Tabl.* 443. (1799); *Brown Prodr.* 524.; *Endl. Gen.* clix.; *Meisn. Gen. p.* 250; *Alph. DC. Prodr.* 8. 209.—Erycibeæ, *Endl. Gen. p.* 655; *Alph. DC. Prodr.* 9. 463.

DIAGNOSIS.—*Gentianal Exogens, with no stipules, and a simple sessile radiating stigma.*

Trees or shrubs, without milk, and with heavy wood. Leaves alternate, without stipules, obsoletely articulated with the stem, entire, coriaceous. Inflorescence axillary. Flowers by abortion ♂ ♀, seldom ☿, the ♂ with the rudiment of an ovary, the ♀ usually with a few sterile stamens. Calyx in 3 to 7 divisions, nearly equal, persistent. Corolla monopetalous, hypogynous, regular, deciduous, somewhat coriaceous, usually pubescent externally, and smooth internally; its limb with 3 to 7 divisions, imbricated in æstivation. Stamens definite, either arising from the corolla, or hypogynous; twice as many as the segments of the corolla, sometimes 4 times as many, or the same number, and then alternate with them, often inserted in pairs at the foot of the lobes of the corolla, and then neither opposite nor alternate with them; filaments simple in the hermaphrodite species, generally doubled in the polygamous and diœcious ones, both their divisions bearing anthers, but the inner one generally smaller; anthers attached by their base, lanceolate, 2-celled, dehiscing lengthwise, sometimes bearded; pollen round, smooth. Ovary sessile, without any disk, several-celled, the cells each having 1 or 2 ovules pendulous from their apex; style divided, seldom simple; stigmas bifid, or simple. Fruit fleshy, round, or oval, by abortion often few-seeded, its pericarp sometimes opening in a regular manner. Seed with a membranous testa of the same figure as the albumen, which is cartilaginous and white; embryo in the axis, or but little out of it, straight, white, generally more than half as long as the albumen; cotyledons foliaceous, somewhat veiny, lying close together, or occasionally slightly separate; radicle taper, of middling length or long, superior, turned towards the hilum.

Fig. CCCCIV.

Brown thinks these plants allied to Oliveworts, with which they agree in the placentation of the seeds and other points of structure; being distinguished by their alternate leaves, constantly axillary and usually unisexual flowers, the stamens of which are at least double the number of the lobes of the corolla; but that comparison by no means indicates their nearest affinity, which is certainly with Hollyworts on the one hand, and Dogbanes on the other. The nature of their distinction from the former is stated at p. 579; the latter are known by their peculiar stigma, frequently bifollicular fruit, and numerous seeds. Sapotads, to which they are also much allied, have a great amygdaloid embryo. In a large number of cases there is a strong tendency to polygamy, which might have been suspected to indicate some relation to the Diclinous Alliances; but no other resemblance can be traced, and in fact the separation of the sexes in the present Order is but partial: rudimentary stamens being uniformly present in the ♀ flowers.

Chiefly Indian and tropical; a very few are found northward as far as Switzerland in

Fig. CCCCIV.—Maba elliptica.—*Turpin.* 1. a flower; 2. a ♂ corolla cut open 3. a calyx and pistil; 4. fruit; 5. its section; 6. a perpendicular section of a seed.

Europe, and the state of New York in North America. A few occur at the Cape of Good Hope and in New Holland.

They are remarkable for little except the hardness of the wood of such species as Diospyrus Ebenus, Ebenaster, melanoxylon, Mabolo, tomentosa and Roylei, and for the eatable quality of the fruit. The timber, of a black colour, sometimes variegated with white or brown lines, is well known under the names of Ebony and Ironwood. The fruit is noted for extreme acerbity before arriving at maturity. That of Diospyrus Kaki is occasionally introduced from China as a dry sweetmeat; another species is believed to furnish a fruit called the Kau Apple by the settlers in the South of Africa. Some practitioners in the United States prescribe an infusion of the unripe fruit of Diospyrus virginiana, also called the Date Plum, whose bark had already been employed as a febrifuge with success in cases of cholera infantum, and the worst forms of Mississippi diarrhœa. The particulars as to the manner of applying it are to be found in Hay's *American Journal of Medical Science*, October, 1842.—See *Gard. Chron.* p. 844. 1843. This tree produces a kind of gum, and the fruit when changed by frost is eaten. The fruit of Diospyrus glutinosa, or Embryopteris, is so glutinous as to be used in Bengal for paying boats.

GENERA.

Royena, *L.*
Euclea, *L.*
Diplonema, Don.
Rymia, Endl.
Gunisanthus, *A. DC.*
Rospidios, *A. DC.*
Macreightia, *A. DC.*
Diospyros, *L.*
Guaiacara, Tourn.
Hebenaster, Rumph.
Paralea, Aubl.
Embryopteris, Gærtn.
Cavanilla, Lam.
Maba, *Forst.*
Ferreola, Roxb.
Cargilia, *R. Br.*
Erycibe, *Roxb.*
Catonia, Vahl.
Erimatalia, R. et Sch.

NUMBERS. GEN. 9. Sp. 160.

Oleaceæ.
POSITION.—————EBENACEÆ.—Aquifoliaceæ.
Sapotaceæ.

ORDER CCXXX. AQUIFOLIACEÆ.—HOLLYWORTS.

Ilicineæ, *Ad. Brongniart Mémoire sur les Rhamnées*, p. 16. (1826); *Endl. Gen.* cxxxviii. (1829).—Aquifoliaceæ, *DC. Théorie, ed.* 1. 217. (1813); *a* § *of* Celastrineæ, *Ib. Prodr.* 2. 11; *Ed. pr.* clxxiii.; *Meisner Gen. p.* 252.

DIAGNOSIS.—*Gentianal Exogens, with no stipules, simple stigmas at the end of a manifest style, axile placentæ, definite pendulous seeds, and an imbricated corolla.*

Evergreen trees or shrubs, whose branches are often angular. Leaves alternate or opposite, simple, leathery, without stipules. Flowers small, white or greenish, axillary, solitary or clustered, sometimes ♂ ♀ by abortion. Sepals 4 to 6, imbricated in æstivation. Corolla 4- to 6-parted, hypogynous, imbricated in æstivation. Stamens inserted into the corolla, alternate with its segments; filaments erect; anthers adnate, 2-celled, opening longitudinally. Disk none. Ovary fleshy, superior, somewhat truncate, with from 2 to 6 or more cells; ovules solitary, anatropal, pendulous, and often hanging from a cup-shaped funiculus; stigma subsessile, lobed. Fruit fleshy, indehiscent, with from 2 to 6 or more stones. Seed suspended, nearly sessile; albumen large, fleshy; embryo small, 2-lobed, lying next the hilum, with minute cotyledons, and a superior radicle.

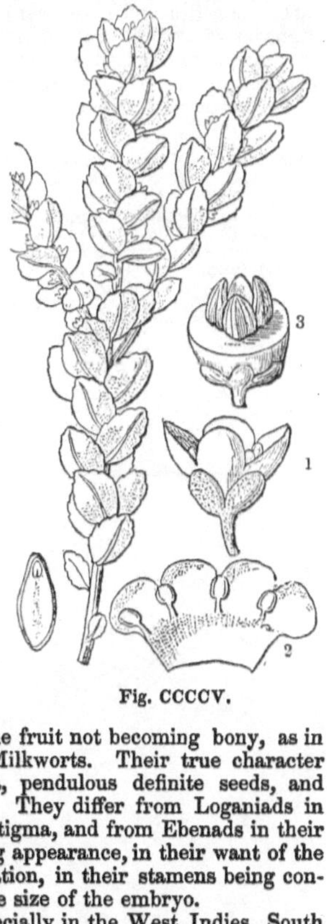

Fig. CCCCV.

These bushes and trees were formerly included in Rhamnads by most Botanists, but have been well distinguished by Ad. Brongniart, who remarks that the suggestion of Jussieu, in his *Genera Plantarum*, that Hollyworts ought probably to be placed near Sapotads or Ebenads, will probably be adopted. From Spindletrees, with which the Order is combined in some modern works, it differs in the form of the calyx and corolla, in the disposition and insertion of the stamens, and especially in the structure of the ovary and fruit. In these respects Hollyworts are found by Brongniart to agree so completely with Ebenads, that that Order does not, in fact, differ essentially from Hollyworts, except in characters of a secondary order, such as the calyx and corolla being less deeply divided, the stamens often double the number of the segments of the corolla, the style sometimes divided, the cells of the ovary usually containing 2 collateral ovules, and, finally, in the cells of the fruit not becoming bony, as in most Hollyworts. Von Martius places them near Milkworts. Their true character resides in their monopetalous corolla, axile placenta, pendulous definite seeds, and minute embryo, lying in the base of fleshy albumen. They differ from Loganiads in the want of stipules, from Dogbanes in their simple stigma, and from Ebenads in their long style, the stigmas of which never have a radiating appearance, in their want of the peculiar silky corolla with a twisted imbricated æstivation, in their stamens being constantly definite in number, and in the still more minute size of the embryo.

Found sparingly in various parts of the world, especially in the West Indies, South America, and the Cape of Good Hope. Several are found in North America; but 1, the common Holly, in Europe.

The bark and berries of Prinos verticillatus possess, in an eminent degree, the properties of vegetable, astringent, and tonic medicines, along with antiseptic powers which are highly spoken of by American practitioners; the berries are said to be tonic, but Bigelow asserts that they are emetic. A decoction or infusion of the root of Myginda Uragoga is a most powerful diuretic. It is asserted that the leaves of the common Holly (Ilex aquifolium), are equal to Peruvian Bark in the cure of intermittent

Fig. CCCCV.—Ilex microphylla.—*Hooker*. 1. a flower; 2. the corolla laid open; 3. a section of a ripe fruit; 4. a section of a seed.

fever; the root and bark are said to be emollient, resolving, expectorant, and diuretic; Haller recommends the juice of the leaves in icterus; Reil also affirms that he has employed the bark successfully in cases of epidemic intermittent fever when Peruvian Bark had failed. The berries are purgative and emetic; six or eight will occasion violent vomiting. Birdlime is obtained from the bark, and the beautiful white wood is much esteemed by cabinet-makers for inlaying; a strong decoction of Ilex vomitoria, called Black drink, is used by the tribes of the Creek Indians at the opening of their councils. It acts as a mild emetic. Some species are employed as substitutes for tea, among which is the Prinos glabra, an evergreen North American bush. But the most celebrated is the Ilex paraguayensis, or Maté, whose leaves are very generally employed in Brazil and the adjoining South American governments; of this plant, called Paraguay Tea, a full account has been given in the *London Journal of Botany*, 1. p. 30; Mr. Stenhouse has detected Theine in its leaves. Martius states that Ilex Gongonha, called also Gongonha, and I. theezans are also employed in Brazil in the same manner; he describes all three as being valuable diuretics and diaphoretics. According to the same author the leaves of Ilex paraguayensis and several others are used by dyers; the fruits of Ilex Macoucoua, when unripe, abound in tannin, and bruised in a ferruginous mud are employed in dyeing cotton fabrics; they act something like galls.—*Mat. Med. Br.* 126.

GENERA.

Cassine, *Linn.*
Maurocenia, Mill.
Ilex, *Linn.*
Aquifolium, Tournef.
Paltoria, Ruiz et Pav.
Macoucoua, Aubl.
Labatia, Scop.
Burglaria, Wendl.
Chomelia, Fl. Flum.
Prinos, *Linn.*
Ægeria, Adans.
Winterlia, Mönch.
Nemopanthes, *Raf.*
Nuttallia, DC.
Ilicioides, Dumort.
Byronia, *Endl.*
Polystigma, Meisn.
Siphonodon, *Griff.*
Villaresia, *Ruiz et Pav.*
Citronella, Don.
? Jodina, *Hook. et Arn.*
Monetia, *Herit.*
Azima, Lam.
? ? Skimmia, *Thunb.*
*
Rhaptostylum, *Humb. et Bonpl.*

NUMBERS. GEN. 11. Sp. 110.

Rhamnaceæ.
POSITION.———AQUIFOLIACEÆ.—Ebenaceæ.
Sapotaceæ.

ORDER CCXXXI. APOCYNACEÆ.—DOGBANES.

Apocyneæ, *Juss. Gen.* 143. (1789) *in part; R. Brown Prodr.* 465. (1810); *Royle's Illustrations*, 269.—Apocynaceæ, *Ed. pr.* ccxxii. (1836); *Endl. Gen.* cxxxii.; *DC. Prodr.* 8. 317.; *Alph. DC. in Ann. Sc.* 3 *ser.* 1. 235.—Contortæ, *Linn.*—Vinceæ, *DC. and Duby Bot. Gall.* 342. (1828), *a* § *of* Apocyneæ.

DIAGNOSIS.—*Gentianal Exogens, with no stipules, and the stigmas collected into a massive head, expanded at the base in the form of a ring or membrane, and contracted in the middle.*

Trees or shrubs, usually milky. Leaves opposite, sometimes whorled, seldom scattered, quite entire, often having ciliæ or glands upon or between the petioles, but with no stipules properly so called. Inflorescence tending to corymbose. Calyx free, 5-parted, persistent. Corolla monopetalous, often having scales in its throat, hypogynous, regular, 5-lobed, with contorted æstivation, deciduous. Stamens 5, arising from the corolla, with whose segments they are alternate; filaments distinct; anthers adhering firmly to the stigma, 2-celled, opening lengthwise; pollen granular, globose, or 3-lobed, immediately applied to the stigma. Ovaries 2, or 1-2-celled, polyspermous; styles 2 or 1; stigma 1, contracted in the middle and assuming much the appearance of an hour-glass; ovules usually 00, amphitropal, or anatropal. Fruit a follicle, capsule, or drupe, or berry, double or single. Seeds with fleshy or cartilaginous albumen, usually pendulous; occasionally without albumen; testa simple; embryo foliaceous; plumule inconspicuous; radicle turned towards the hilum.

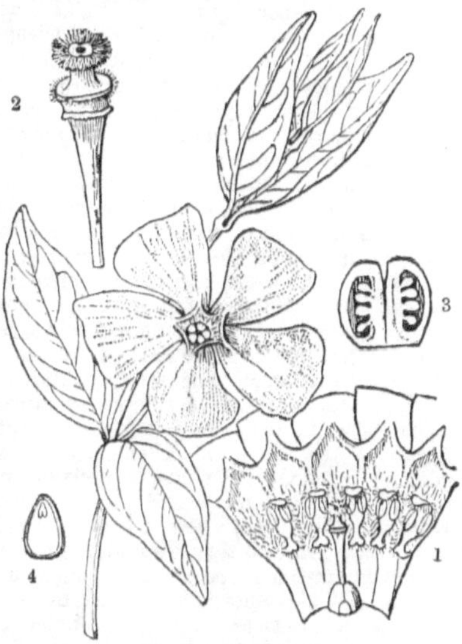

Fig. CCCCVI.

The singular stigma, more easy to represent by a drawing than to describe, is one of the best indications of this Order; it is generally expanded at the base into a circular membrane or inverted cup, and is contracted somewhere near the middle. Bearing this in mind, the Loganiads, Gentianworts, and Cinchonads are distinguished with precision. In addition to this, the ovary is usually formed by the mere approximation of two carpels having little or no adhesion except at the point and along the styles and stigmas. In this respect it corresponds with Asclepiads, the economy of whose stamens, pollen, stigma, and seeds is in general such that the nature of the ovary seems an indication of analogy instead of affinity, as is commonly believed. An elaborate account of the peculiarities and affinities of the Order has been drawn up by M. Alph. De Candolle in the place above quoted, to which the reader is referred for further information.

The species are principally tropical, throwing out a few representatives only, such as Vinca and Apocynum, into northern countries. They appear to be most abundant in the hot parts of Asia, somewhat less common in the tropics of America, and by no means abundant in Africa.

Dogbanes are for the most part plants of considerable beauty, with large, showy, gay-coloured flowers. They are, however, in many cases venomous, and very generally to be suspected, although in some cases they are used medicinally, and in others have an eatable fruit. Among the true poisons Tanghinia venenifera stands foremost. The kernel of the fruit, although not larger than an Almond, is sufficient to destroy twenty people. It was used in Madagascar as an ordeal, but the practice is now discontinued.

Fig. CCCCVI.—Vinca minor. 1. corolla opened; 2. style and stigma; 3. perpendicular section of the double ovary; 4. section of a seed.—*Gærtner.*

The kernels of Cerbera Manghas are also emetic and poisonous ; the milky sap is purgative ; the leaves and bark are used in Java as a substitute for Senna. In Thevetia Ahouai the seeds are also poisonous ; the bark and sap emetic and narcotic ; and Thevetia neriifolia has a dangerous venomous milk ; yet its bitter and cathartic bark is reported to be a powerful febrifuge, 2 grains only being affirmed to be equal to an ordinary dose of Cinchona. The wood of both these has a heavy repulsive odour, and is used, in the countries where they are wild, for poisoning fish. Hasseltia arborea must be classed among the poisons. In Java the milk obtained from the trunk by incision, mixed with honey, and reduced with boiling water, is employed as a powerful drastic for destroying the tape-worm ; it is however apt to produce inflammation of the intestines, and is even in some cases fatal. The milk of the Plumieras, although said to be cathartic or drastic, is excessively corrosive ; they are however employed by practitioners in tropical countries. Cameraria latifolia is named the Bastard Manchineel-tree, from its resemblance in quality to that formidable tree. From a species of Echites the Mandingoes are said to obtain a poison with which their smear they arrows. In general the genus is narcotic, or rather stupefying, but with considerable acrimony, whence the species are employed, especially their roots, as drastics and epispastics.—*Stadelmeyer, Echit.* p. 3. The common Oleander, Nerium Oleander, although little suspected, is a formidable poison. A decoction of its leaves forms a wash, employed in the south of Europe to destroy cutaneous vermin ; and its powdered wood and bark constitute at Nice the basis of an efficacious rat poison. A few years ago, a child died from having eaten one morning a quantity of Oleander flowers ; it was seized with violent colic, under which the child sunk at the end of two days. In 1809, when the French troops were lying before Madrid, some of the soldiers went a marauding, every one bringing back such provisions as could be found. One soldier formed the unfortunate idea of cutting the branches of the Oleander for spits and skewers for the meat when roasting. This tree, it may be observed, is very common in Spain, where it attains considerable dimensions. The wood having been stripped of its bark, and brought in contact with the meat, was productive of most direful consequences, for of twelve soldiers who ate of the roast seven died, and the other five were dangerously ill.—*Gard. Chronicle*, 1844, p. 23. In like manner the root of Nerium odorum is found to be a poison in India. When, however, these dangerous qualities are moderated, the species become useful medicinal agents, either as emetics or cathartics. The Apocynums androsæmifolium and cannabinum are emetic, diaphoretic, and diuretic, and in small doses tonic. An infusion of the leaves of Allamanda cathartica is considered a valuable cathartic medicine, in moderate doses, especially in the cure of painters' colic. In over doses it is violently emetic and purgative. The root of Rauwolfia nitida is used for similar purposes. Not a few species of the Order lose their acrimony either wholly or in a great degree, and then we find them applied as febrifuges or even aromatics. The root of Ophioxylon serpentinum is employed by the Telinga physicians of India as a febrifuge and alexipharmic, and also to promote delivery in tedious cases. The bark of Alyxia stellata is aromatic, with similar effects to those of Canella alba and Drymis Winteri, for which it may be substituted. It has been introduced into German practice as a remedy for chronic diarrhœa and nervous complaints ; it has the odour of Melilot, and traces of Benzoic acid have been found in it. The Conessi bark, a valuable astringent and febrifuge, called Palapatta in Malabar, is obtained from Wrightia antidysenterica. Ichnocarpus frutescens is sometimes used in India as a substitute for Sarsaparilla. The wood of Alstonia scholaris, and some Madagascar Carissas, is as bitter as Gentian. Hancornia pubescens, and several other Brazilian trees, are mentioned by Martius as possessing similar qualities. It is not a little remarkable, then, that in such an Order as this some species should occur which are absolutely inert ; yet such appears to be the case in several instances. Tabernæmontana utilis, the Hya Hya, is one of those Cow-trees of equatorial America, which derive their name from pouring forth a copious stream of thick, sweet, innoxious milk. Even the Cerberas Odollam, lactaria, and salutaris, seem to possess none of the venom for which the species above mentioned are celebrated. Caoutchouc, or a substance analogous to it, is supplied by several plants of the Order. Collophora utilis, and Cameraria latifolia yield it in South America ; Vahea gummifera in Madagascar; Urceola elastica and Willughbeia edulis in the East Indies, the former of fine, and the second of indifferent quality. Although some species bear fruit that is eatable, yet they do not appear to possess much merit. That of Hancornia is said by Martius to be sweet, sub-acid, and vinous. Willughbeia edulis derives its name from the use that is made of its fruit in India. Carissa Carandas furnishes a substitute for Red Currant Jelly ; to these may be added the Pishamins (Carpodinus) of Sierra Leone, Melodinus monogynus, Carissa edulis, and a few more. Some are used for dyeing, the chief of which is Wrightia tinctoria, which yields Indigo of good quality. Little is known of their timber ; that of Wrightia coccinea is light and tough, and used for making Palanqueens; of Wrightia mollissima is employed by turners. Aspidosperma excelsum is, according to Schomburgk, remarkable for its trunk growing at the lower part into tubular projections, forming cavities which serve the Indians as ready-made planks, and in the construction of their paddles. The trunk appears as if fluted, or rather as if it consisted of numerous slender trees, grown together along their whole length.

The sages of Ceylon having demonstrated, as they say, that Paradise was in that island, and having therefore found it necessary to point out the forbidden fruit of the garden of Eden, assure us that it was borne on a species of this genus, the Divi Ladner of their country, and probably Tabernæmontana dichotoma. The proof they find of this discovery consists in the beauty of the fruit, said to be tempting, in the fragrance of the flower, and in its still bearing the marks of the teeth of Eve. Till the offence was committed, which brought misery on man, we are assured that the fruit was delicious; but from that time forward it became poisonous, as it now remains.—*Bot. Reg.* 1841. *sub. t.* 53.

GENERA.

I. WILLUGHBEIÆ.—Placentæ parietal.

Allamanda, *L.*
  Orelia, Aubl.
Chilocarpus, *Bl.*
? Landolfia, *Pal.*
Willughbeia, *Roxb.*
  Ancylocladus, Wall.
Couma, *Aubl.*
? Collophora, *Mart.*
? Pacouria, *Aubl.*

II. CARISSEÆ.—Ovary single, 2-celled. Seeds naked.

Craspidospermum, *Boj.*
? Plectaneia, *Thouars.*
Maycockia, *A. DC.*
Hancornia, *Gom.*
  Mangaiba, Pis.
Winchia, *A. DC.*
Vahea, *Lam*
Ambelania, *Aubl.*
Carpodinus, *R. Br.*
Melodinus, *Forsk.*
Bicorona, *A. DC.*
Leuconotis, *Jack.*
Carissa, *L.*
  Arduinia, L.
  Antura, Forsk.
? Toxicophlœa, *Harv.*
Rauwolfia, *Plum.*
Ophioxylon, *L.*
  Tsiovanna, Rheede.
Thevetia, *L.*
  Ahouai, Pl.

III. PLUMIEREÆ.—Ovary double. Seeds naked.

Alyxia, *R. Br.*
  Gynopogon, Forst.
Vallesia, *R. P.*
Hunteria, *Roxb.*
Kopsia, *Bl.*
  Calpicarpum, Don.
Cerbera, *L.*
  Manghas, Burm.
Tanghinia, *Thouars.*
Ochrosia, *Juss.*
? Voacanga, *Thouars.*
Piptolœna, *Harv.*
Orchipeda. *Bl.*
Urceola, *Roxb.*
Bonafousia, *A. DC.*
Stemmadenia, *Benth.*
Odontadenia, *Benth.*
Peschiera, *A. DC.*
Tabernæmontana, *Plum.*
  Pandaca, Thouars.
  Rejoua, Gaud.
  Reichardia, Dennst.
? Conopharyngia, *Don.*
Malouetia, *A. DC.*
Condylocarpon, *Desf.*
Vinca, *L.*
  Pervinca, Tourn.
  Catharanthus, Don.
  Lochnera, Rchb.
Amsonia, *Walt.*
Rhazya, *Dec.*
Thyrsanthus, *Benth.*
Gonioma, *E. Mey.*
Cameraria, *Plum.*
Plumieria, *Tourn.*
Anisolobus, *A. DC.*
Aspidosperma, *Mart.*
  Macaglia, Vahl.

IV. PARSONSEÆ.—Ovary single, 2-celled. Seeds comose.

Vallaris, *Burm.*
  Emericia, R. et Sch.
  Peltanthera, Roth.
Lyonsia, *R. Br.*
Parsonsia, *R. Br.*
Balfouria, *R. Br.*
Beaumontia, *Wall.*

V. WRIGHTEÆ.—Ovary double. Seeds comose.

Wrightia, *R. Br.*
Kixia, *Bl.*
  Hasseltia, Bl.
  Kibatalia, Don.
Alstonia, *R. Br.*
Blaberopus, *A. DC.*
Adenium, *R. Sch.*
Haplophyton, *A. DC.*
Holarrhena, *R. Br.*
? Alafia, *Thouars.*
Isonema, *R. Br.*
? Echaltium, *Wight.*
? Christya, *Ward.*
Strophanthus, *DC.*
Nerium, *L.*
Neriandra, *A. DC.*
Motandra, *A. DC.*
Pachypodium, *Lindl.*
  Belonites, E. Mey.
Baissea, *A. DC.*
? Heligme, *Bl.*
  Helygia, Bl.
Thenardia, *Kth.*
Hæmadictyon, *Lindl.*
Prestonia, *R. Br.*
Chonemorpha, *Don.*
Rynchospermum, *A. DC*
? Cercocoma, *Wall.*
Aganosma, *Don.*
Ichnocarpus, *R. Br.*
Forsteronia, *Meyer.*
Apocynum, *Tourn.*
Pottsia, *Hooker.*
Ecdysanthera, *Hooker.*
Anodendron, *A. DC.*
Chavannesia, *A. DC.*
Robbia, *A. DC.*
Secondatia, *A. DC.*
Echites, *P. Br.*
  Exothostemon, Don.
Mandevilla, *Lindl.*
Laseguea, *A. DC.*
Dipladenia, *A. DC.*
Laubertia, *A. DC.*
Mascarenhasia, *A. DC.*

? Skytanthus, *Meyen.*
? Tayotum, *Blanco.*
? Pycnostelma, *Bunge.*
? Syringosma, *Mart.*

NUMBERS. GEN. 100. Sp. 566.

Asclepiadaceæ.
POSITION.—Gentianaceæ.—APOCYNACEÆ.—Loganiaceæ.

ORDER CCXXXII. LOGANIACEÆ.—LOGANIADS.

Loganieæ, *R. Brown in Flinders,* (1814) ; *Von Martius N. Gen. et Sp. Pl.* 2. 133 ; *Bartl. Ord. Nat.* 205 ; *Arnott in Edinb. Encycl.* 120.—Loganiaceæ, *Ed. Pr.* ccxxiv. ; *Endl. Gen.* cxxxi. ; *DC. Prodr.* 9. 1. —Potaliaceæ, *Brown in Tuckey,* 449. (1819).—Potalieæ, *Martius N. G. et Sp.* 2. 91. *and* 133. (1828) ; *Royle Illustr.* 269.—Strychnaceæ, *Blume Bijdr.* 1018. (1826) ; *Link. Handb.* 1. 439.—Strychneæ, *DC. Théorie ed.* 1. 217. (1813).– Spigeliaceæ, *Martius N. G. et Sp.* 2. 132. (1828) ; *Ed. pr.* ccxxi. ; *Endl. Gen.* cxxxv. ; *Meisner p.* 258.—Cœlostyleæ, *Endl. Ench.* cxxxiii.

DIAGNOSIS.—*Gentianal Exogens, with opposite leaves and intervening stipules.*

Shrubs, herbaceous plants, or trees. Leaves opposite, entire, usually with stipules, which adhere to the leafstalks or are combined in the form of interpetiolary sheaths. Flowers racemose, corymbose, or solitary. Calyx valvate or imbricated, inferior, 4- 5-parted. Corolla regular or irregular, 4- 5- or 10-cleft, with valvate or convolute æstivation. Stamens arising from the corolla, all placed upon the same line, and not always symmetrical with the divisions of the corolla ; pollen with 3 bands. Ovary superior, 2-celled, (3, or spuriously 4-celled); style continuous ; stigma simple ; ovules 00 or solitary, peltate and amphitropal, or ascending and anatropal. Fruit either capsular and 2-celled with placentæ finally becoming loose; or drupaceous, with 1- or 2-seeded stones ; or berried with the seeds immersed in pulp. Seeds sometimes winged, usually peltate ; albumen fleshy or cartilaginous ; embryo small, with the radicle turned towards the hilum or parallel with it.

Fig. CCCCVII.

It is not clear, from the remarks upon Logania, by Brown in his *Prodromus,* whether he intended to establish this Order or not. He states that he has placed Logania at the end of Gentianworts, on account of some affinity between it and Exacum and Mitrasacme, and also because it does not answer ill to the artificial character of that Order; adding that it, however, might have a still closer connection with Dogbanes and with Usteria among Cinchonads. He further points out the close relation of Geniostoma to Logania, and concludes by inquiring whether those 2 genera do not, with Anasser, Fagræa, and Usteria, form an Order intermediate between Dogbanes and Cinchonads. This view has been adopted by Von Martius, who however excludes Fagræa, which he places among his Potaliaceæ ; he founds the distinction of that Order upon the want of symmetry between the parts of the calyx, corolla, and stamens, upon the æstivation of the corolla being convolute, not contorted, and in the presence of stipules combined in interpetiolary sheaths. Mr. Arnott remarked to me (letter, Dec. 1835) that the Order may be in some respects looked upon as consisting of Cinchonads with superior fruit. More recent examination of the genera has entirely confirmed this view, which, however, does not explain with any clearness how Loganiads differ from Dogbanes. Upon this subject I quote literally the words of M. Alph. De Candolle. " I must confess that I have sought in vain for a positive distinction, to which there shall be no exception, between Dogbanes and Loganiads. The position of the flower with respect to the axis appears to be the same, that is to say, a re-entering angle of the calyx stands next the axis. That of the cells of the fruit with respect to the axis varies among Loganiads, as does the æstivation of the corolla and many other characters. The grains of pollen are not very different, if we rely upon the exact but scanty observations of Mr. Hassall. The placentas of Dogbanes are more securely fastened to the edges of the carpellary leaves, and do not separate from them when the fruit is ripe, as generally happens more or less distinctly among Loganiads ; but the placenta of Strychnos is exactly that of Carissa. Dogbanes have a milky juice ; but exceptions to that are said to occur, as in Echites for instance. Finally, the only differences which I can point out are of a particular kind, not very satisfactory in practice, although of some value in botanical philosophy. These reside in the nature of the variations presented by Dogbanes and Loganiads. In the former the flowers are always isomerous in the calyx, corolla, and stamens, and the number is never more than 5 ; in the latter the corolla and stamens have

Fig. CCCCVII.—Logania floribunda. 1. a corolla cut open ; 2. a pistil.

sometimes more pieces than the calyx, as in Potalia; the stamens are sometimes reduced to one, as in Usteria. The stamens of Dogbanes always alternate with the lobes of the corolla; those of Loganiads vary more or less from this position, and become opposite in Potalia. In Dogbanes the number 2 in the carpels is without exception; in Loganiads, one genus, Labordia, has 3 cells.——In Dogbanes, the æstivation of the corolla is always twisted, except in Mascarenhasia, where it is induplicato-valvate with a torsion of the back of each lobe, which indicates the tendency of the Order; in Loganiads the æstivation is very variable, and is often valvate in the strictest acceptation of the term.——Dogbanes often have hypogynous glands or a complete disk; Loganiads have not a trace of either. The first have often appendages inside of the corolla; the second never have any, unless we so consider the hairs which guard the orifice. The stigma is often of considerable size, and bears a peculiar kind of gland in Dogbanes; Loganiads have no such appearances." This last is the true distinguishing character.

All Loganiads are either tropical or inhabit countries near the tropics; a few outlying species in New Holland and America forming the only exceptions.

It would be difficult to name a more venomous Order than this, of whose qualities the celebrated Nux Vomica may be taken as the representative. This fatal drug consists of the seeds of Strychnos Nux Vomica, an Indian tree, with small greenish-white flowers, ribbed leaves, and a beautiful orange-coloured round fruit, the size of a small Apple, having a brittle shell, and a white gelatinous pulp. The wood is exceedingly bitter, particularly that of the root, which is used to cure intermittent fevers, and the bites of venomous snakes. The seeds are employed in the distillation of country spirits, to render them more intoxicating. The pulp of the fruit seems perfectly innocent, as it is greedily eaten by many sorts of birds.—*Roxb.* The seeds are extremely poisonous, in large doses producing extraordinary rigidity and convulsive contraction of the muscles previous to death. In very small and repeated doses it promotes the appetite, assists the digestive process, increases the secretion of urine, and sometimes acts slightly upon the bowels. It is employed medicinally in paralysis, dyspepsia, dysentery, affections of the nervous system, &c., and appears to be very active in removing impotence. Another virulent kind is the Strychnos toxifera, which forms the basis of a celebrated poison called Wooraly or Ourari. Dr. Hancock thinks it is the most potent sedative in nature. For an account of it by Sir R. Schomburgk, see *Ann. Nat. Hist.* vii. 411. From the bark of the root of Strychnos Tieutê another frightful poison is prepared in Java, where it is called Tjettek and Upas Radja; it acts like Nux Vomica, but in a more intense and violent manner. Notwithstanding the active qualities of these formidable plants, others are used in medicine with advantage. Strychnos ligustrina is said by Blume to yield the genuine Lignum colubrinum, a drug once held in great estimation as a remedy for paralysis of the lower extremities; it is also said to be a valuable anthelmintic, and to be useful in blenorrhœa faucium et laryngis, diseases to which Europeans are subject in Java. Blume adds that several other species of the genus are brought into the market under the name of Lignum colubrinum. Strychnos pseudoquina is said to be the best febrifuge in Brazil; with the exception of the fruit, which is eaten by children without danger, all the parts, especially the bark, are extremely bitter and rather astringent. It is universally employed instead of Cinchona, and is asserted to be fully equal to Peruvian Bark, in the cure of the intermittents of Brazil. Vauquelin analysed the bark and could find in it neither brucine, nor strychnine, nor quinine. It is sold under the name of Copalche bark. The seeds of Ignatia amara, called St. Ignatius's Beans, are used successfully in India as a remedy for cholera, under the name of Papeeta, but gid-

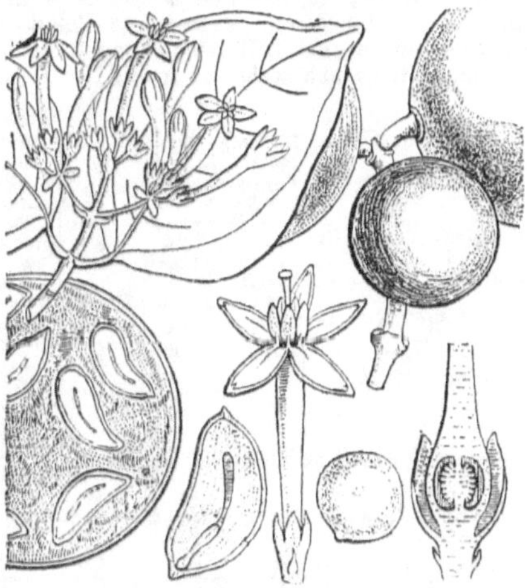

Fig. CCCCVIII.

Fig. CCCCVIII.—Strychnos ligustrina.—*Blume.* 1. a flower; 2. a section of the ovary; 3. fruit cut across; 4. seed; 5. the same more magnified and divided.

diness and convulsions are known to follow their exhibition, if given in an over-dose. In India there is a nut called the Clearing Nut, of which the ripe seeds are dried, and sold in every market, to clear muddy water. The natives never drink clear well water, if they can get pond or river water, which is always more or less impure according to circumstances. One of the seeds is well rubbed for a minute or two round the inside of the vessel, generally an unglazed earthen one, containing the water, which is then left to settle; in a very short time the impurities fall to the bottom, leaving the water clear. The natives of India eat the pulp of the fruit when ripe; Dr. Roxburgh found it disagreeable. These nuts are produced by Strychnos potatorum. Bitter Almonds are said to be employed for the same purpose in Egypt, and those of the Kola, or Sterculia, in Sierra Leone. The Spigelias participate in the noxious properties of Strychnos. Both root and leaves of Spigelia marilandica, the Carolina Pink-root, and S. anthelmia, are active anthelmintics; their efficacy is much impaired by keeping. They are also purgative and narcotic in a slight degree, seem to be acrid narcotics, and are apt to produce very unpleasant symptons after being exhibited; dimness of sight, giddiness, dilated pupil, spasms of the muscles of the eyes, and even convulsions are reported by Barton to have been brought on by them. Spigelia glabrata is reckoned by Martius among poisons; and Mr. Hartweg reports that a species of the same genus kills dogs in equatorial America. An infusion of the leaves of Potalia resinifera is slightly mucilaginous and astringent, and is used in Brazil as a lotion for inflamed eyes. Potalia amara is bitter like the Gentians, and acrid and emetic like Dogbanes.

GENERA.

I. SPIGELEÆ.
Spigelia, *L.*
Canala, Pohl.
Montira, Aubl.
Arapabaca, Plum.
Cœlostylis, Torr. et Gr.
Mitreola, *L.*
Mitrasacme, *Lab.*

II. STRYCHNEÆ.
Strychnos, *L.*
Rouhamon, *Aubl.*
Lasiostoma, Schr.
Brehmia, *Harv.*
Ignatia, *L.*
Pagamea, *Aubl.*
Gardneria, *Wall.*
Cyathospermum, Wall.
Antonia, *Pohl.*
Labordia, *Gaud.*
Usteria, *W.*
Monodynamis, Gmel.

III. LOGANEÆ.
Logania, *R. Br.*
Euosma, Andr.
Stomandra, R. Br.
Geniostoma, *Forst.*
Anasser, Juss.
Hæmospermum, Bl.
Fagræa, *Thunb.*
Kuhlia, Reinw.
Utania, Don.
Kentia, Steud.
Cyrtophyllum, *Reinw.*
Picrophlœus, *Bl.*
Gærtnera, *Lam.*
Frutesca, DC.
Andersonia, Schl.
Sykesia, *Arn.*
Potalia, *Aubl.*
Nicandra, Schreb.
Anthocleista, *Afz.*
? Codonanthus, *G. Don.*
? Anabata, *W.*
Sulzeria, R. et Sch.

NUMBERS. GEN. 22. Sp. 162.

Cinchonaceæ.
POSITION.—Apocynaceæ.—LOGANIACEÆ.—Gentianaceæ.
Rhizophoraceæ.

CASSIPOUREÆ, (*Meisn. Gen. p.* 119.- Legnotideæ, *Bartl. Ord. Nat. Endl. Gen.* 1186). Trees or shrubs. Leaves opposite, nearly entire, with interpetiolar stipules. Flowers axillary, solitary, or clustered. Calyx campanulate, 4- 5-cleft, valvate. Petals 4-5, fringed, inserted into the bottom of the calyx. Stamens 2 or 3 times as many as the petals, distinct, inserted into the bottom of the calyx or the back of a disk; filaments free; anthers 2-celled, turned inwards. Ovary superior, 3- to 5-celled; ovules 2 or many in each cell, pendulous or attached to the axis; style simple; stigma obtuse. Fruit berried or capsular. Embryo in the axis of fleshy albumen; radicle superior; cotyledons flat or half-cylindrical.——These are tropical shrubs, and are usually placed with Mangroves; but their seeds have albumen, and the ovary is perfectly distinct from the calyx. The points of resemblance consist in the fleshy valvate calyx, the fringed petals, which are like those of Kandelia, and the presence of stipules. Brown, after comparing this Cassipourea with the Mangroves called Carallias, was led to conclude that we have a series of structures connecting Rhizophora, on the one hand, with certain genera of Loosestrifes, particularly with Antherylium, though that genus wants the intermediate stipules; and, on the other, with Cunoniads, especially with the simple-leaved species of Ceratopetalum.—*Congo.* 437. This is doubtless the fact, and Cassipourea may probably be regarded as one of those osculant groups whose relationship is nearly equal in several opposite directions. But upon the whole it seems to have more real affinity with Loganiads than with the Orders just mentioned. Its

Fig. CCCCIX.

Fig. CCCCIX.—Cassipourea elliptica.—*Hooker.* 1, a flower; 2, stamens; 3, pistil; 4, cross section of the ovary.

valvate calyx, perigynous stamens, axile placentation, interpetiolar stipules and albuminous seeds are much the same as in the Loganiads; its main difference lies in its polypetalous corolla. I therefore station it here as a doubtful group, whose true value will be better estimated when its species have been more completely examined.

GENERA.

Dryptopetalum, *Arn.*
Microtropis, Wall. part.
Cassipourea, *Aubl.*
Tita, Scop.
Legnotis, Swartz.
Richæia, Thouars.
Weihea, Spreng.

NUMBERS. GEN. 2. SP. 7.

ORDER CCXXXIII. DIAPENSIACEÆ.—DIAPENSIADS.

Diapensiaceæ, *Link Handb.* 1. 595. (1829); a § of Convolvulaceæ; *Ed. pr. No.* clxxvii. (1836); *Endl. Gen. p* 760.; *Meisner Gen. p.* 272.

DIAGNOSIS.—*Gentianal Exogens, with no stipules, simple stigmas at the end of a manifest style, axile placentæ, indefinite peltate seeds, and interpetalous stamens.*

Prostrate under-shrubs, with small densely imbricated leaves which have scarcely any visible veins. Flowers solitary, terminal. Calyx composed of 5 sepals which form a broken whorl, are rather unequal, and much imbricated; scarcely distinguishable from the bracts which are closely imbricated round it. Corolla monopetalous, regular, with an imbricated æstivation. Stamens 5, equal; the filaments petaloid and arising from the margin of the sinus of the corolla; anthers 2-celled, with a broad connective, bursting transversely; in Pyxidanthera awned on the lower valve. Disk 0. Ovary superior, 3-celled; each placenta with 7 ovules in Pyxidanthera, with an indefinite number in Diapensia; style single, continuous with the ovary; stigma sessile, with 3 very short decurrent lobes. Capsule membranous or papery, surrounded with the permanent sepals, terminated by the rigid style or its base. Seeds with a brittle deeply pitted skin, peltate. Embryo very small, with a slender radicle and two very short cotyledons, lying across the hilum in a mass of fleshy albumen.

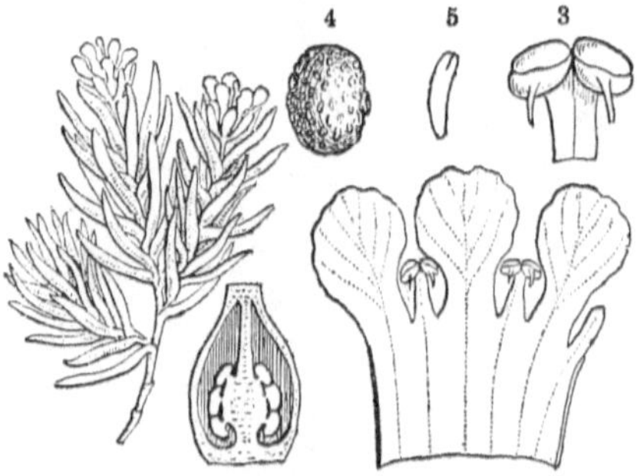

Fig. CCCCX.

From the manner in which Diapensia was associated by Brown (*Prodromus*, 482), when he separated it along with Hydroleaceæ from Bindweeds, it has been generally supposed that this profound Botanist intended to refer Diapensia to the former spurious Order. But Diapensia is in reality nearer Phloxworts than Hydroleaceæ, and yet more nearly allied by its small embryo and copious albumen to Hollyworts and Loganiads. Hydroleaceæ themselves must merge in Hydrophyls, and the free central placentation of that Order forbids the association with it of Diapensiads. The chief resemblances consist in Diapensia having the filaments petaloid, and originating not from within the corolla but from the margin of the sinuses, so that the corolla might be described as 10-cleft, five of the divisions being broad and coloured, and the other five much narrower, and shorter, colourless, and having anthers; and in the embryo being filiform, slightly 2-lobed at one end. But both Diapensia and Pyxidanthera disagree with Hydrophyls in having a calyx consisting of five unequal sepals forming a broken whorl; in having the anthers bursting transversely, and with a very broad connective; in having only one style instead of two; in being destitute of an hypogynous disk; and finally, in the embryo lying in the midst of fleshy albumen across the hilum. At least this is certainly the case in Pyxidanthera, and I have no reason to doubt its being equally the case with Diapensia.

Let me add, that although the name of Diapensiaceæ originated with Link, yet that author in placing it among Bindweeds was obviously unacquainted with its real structure, and in assigning it for a character "semina membranâ inclusa," seems to have assumed that in this respect it agrees with Hydrophyls, which is not the fact.

The species are mountain plants of the north of Europe and North America.

They are not known to possess any useful properties.

GENERA.

Diapensia, *L.* | Pyxidanthera, *Michx.*

NUMBERS. GEN. 2. Sp. 2.

Hydrophyllaceæ.
POSITION.—Loganiaceæ.—DIAPENSIACEÆ.—Stilbaceæ.

Fig. CCCCX.—Pyxidanthera barbulata. 1. corolla cut open; 2. perpendicular section of the ovary; 3. anther; 4. seed; 5. embryo.

ORDER CCXXXIV. STILBACEÆ.—STILBIDS.

Stilbineæ, *Kunth in verhandl. Konigl. Acad. Wissensch. Berol,* (March, 1831); *Martius Conspectus, No.* 109.; *Endl. Gen.* cxxxviii.

DIAGNOSIS.—*Gentianal Exogens, with no stipules, simple stigma at the end of a manifest style, axile placentæ, definite erect seeds, valvate corolla, and unsymmetrical flowers.*

Cape shrubs, with the habit of a Phylica or a Fir. Leaves whorled, close, narrow, entire, leathery, rigid, articulated at the base, without stipules. Flowers in dense spikes at the point of the branches, sessile, each with 3 bracts at the base, occasionally polygamous. Calyx tubular, campanulate, with a 5-cleft limb, the segments of which are equal; the two lower sometimes cut deeper; seldom 5-leaved or 2-valved; persistent. Corolla monopetalous, hypogynous; the tube enlarged upwards, with a ring of hairs in the throat; the limb valvate, 5-parted, spreading, somewhat 2-lipped, rarely 4-parted, and nearly regular. Stamens equal in number to the segments of the corolla, and alternate with them, inserted between the lobes, the upper one of five always rudimentary, or even obliterated; filaments free; anthers elliptical, oblong, attached by the back, 2-celled; opening longitudinally by their face. Ovary superior, sessile, 2-celled; cells with only one erect ovule; one cell sometimes smaller and empty; style terminal, filiform, exserted; stigma simple, emarginate. Disk 0. Fruit dry, 1-seeded, indehiscent, surrounded by the permanent calyx. [Seed erect, with a loose cellular testa. Embryo short, in the axis of very firm fleshy albumen, orthotropal; cotyledons scarcely distinct; radicle inferior.—*Endl.*]

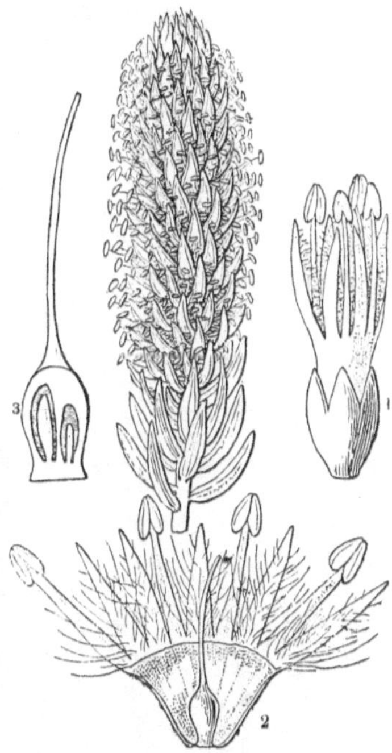

Fig. CCCCXI.

This little Order has never yet been well examined, and no good figures of any of the species can be found in books. The seeds have been seen in only one or two cases, and the whole of the details require verification and re-examination. According to Kunth, they differ from Selagids in little except having 2-celled anthers, erect ovules, and no hypogynous disk, and he also compares them with Globularia, which I regard as a mere form of the Selagids themselves. Endlicher compares them to Vervains. All these comparisons have doubtless been influenced by the unsymmetrical flowers, which appear as if didynamous. But in truth they are not so, in such instances as I have been able to examine, namely Stilbe pinastra and ericoides, or Campylostachys abbreviata and cernua, with some others in the herbaria of Sir W. Hooker and Mr. Harvey; and what is more important, the stamens originate invariably from between the lobes of the corolla. In habit they may doubtless be compared to Selagids and some Vervains, but they are quite as much like Diosmas, or Phylicas, or Bruniads. If to these circumstances we add the presence of a minute embryo with scarcely any cotyledons (which according to Endlicher is the structure), Stilbids can hardly be associated with any of the Orders hitherto suggested. To me they appear far more truly allied with the little Order of Diapensiads, of which they seem to be an unsymmetrical form. Their occasional tendency to polygamy in the original species of Campylostachys would be very

Fig. CCCCXI.—Stilbe Pinastra. 1. a flower; 2. the same cut open; 3. a perpendicular section of an ovary.

unusual in the Echial, but not at all in the Gentianal Alliance. They possibly bear the same relation to Diapensiads as Broomrapes to Gentianworts.

All are natives of the Cape of Good Hope.

Their uses are unknown. They are somewhat resinous shrubs.

GENERA.

Stilbe, *L.*
Lühea, Schmidt.
Campylostachys, *Kunth.*
Eurylobium, *Hochst.*

NUMBERS. GEN. 3. Sp. 7.

Bruniaceæ.
POSITION.——STILBACEÆ.—Diapensiaceæ.
Selaginaceæ.

ORDER CCXXXV. OROBANCHACEÆ.—BROOMRAPES.

Orobancheæ, *Juss. Ann. Mus.* 12. 445. (1808); *Rich in Pers. Synops.* 2. 180.; *DC. and Duby Bot. Gall.* 348.; *Bartl. Ord. Nat.* 173; *Endl. Gen.* cliv.; *Walpers' Repert.* 3. 457.—Phelypæaceæ, *Horanin, Pr. Lin. p.* 73.—Orobanchinæ, *Link Handb.* 1. 506. (1829) *a § of* Personatæ.

DIAGNOSIS.—*Gentianal Exogens, with no stipules, simple stigmas at the end of a manifest style, parietal placentæ, and didynamous flowers.*

Herbaceous leafless plants, growing parasitically upon the roots of other species. Stems covered with brown or colourless scales. Calyx divided, persistent, inferior. Corolla monopetalous, hypogynous, irregular, persistent, with an imbricated æstivation. Stamens 4, didynamous. Anthers occasionally 1-celled, but more generally 2-celled, the cells distinct, parallel, often mucronate, or bearded at the base. Ovary superior, 1-celled, seated in a fleshy disk, with 2 or more parietal polyspermous placentæ, the 2 carpels of which it consists placed right and left of the axis; style 1; stigma 2-lobed. Fruit capsular, inclosed within the withered corolla, 1-celled, 2-valved, each valve bearing 1 or 2 placentæ in the middle. Seeds indefinite, very minute; embryo minute, at the base of fleshy albumen.

Broomrapes are generally compared with Gesnerworts and Figworts, from both which they are very different in habit. They are distinguished from Gesnerworts by the important circumstance of their seeds having only a minute embryo lying in one end of fleshy albumen, and spherical pollen, while the embryo of Gesnerworts is cylindrical and erect, occupying the axis of a small quantity only of albumen, and the pollen elliptical, with a furrow on one side. In Gesnerworts the seeds are attached by rather long funiculi, while they are absolutely sessile in Broomrapes. Moreover, there is a tendency in the latter to become pentandrous, or even hexandrous; not only does no such tendency exist in the former, but the reverse takes place, in the occasional increased sterility of the stamens. There is scarcely any trace in Orobanche of the glandular processes of the disk of Gesnerworts, or at least nothing more than a thin glandular coating to the base of the ovary. From Figworts, to which their didynamous stamens have caused them to be compared, they are known by their 1-celled ovary and minute embryo; as well as by their habit and parasitical mode of growth. In this respect they resemble Fir-rapes, from which they differ in their ovary being composed of 2, not 5 carpels, and in their irregular unsymmetrical flowers, with epipetalous stamens. There can be little doubt, however, that the nearest affinity of Broomrapes is to Gentianworts, with some of which, as for example, Voyria, they even correspond in their leafless scaly habit, and moreover in their corolla adhering firmly to the base of the fruit which it covers when

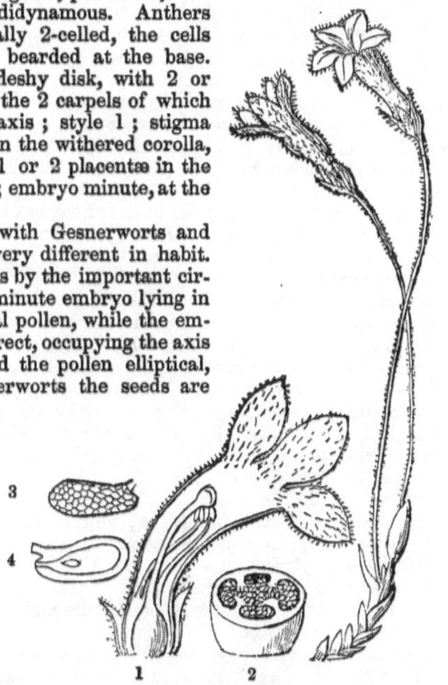

Fig. CCCCXII.

Fig. CCCCXII.—Anoplanthus uniflorus. 1. a flower cut open; 2. a section of the ovary; 3. a seed; 4. the section of it to show the embryo.

ripe. The great points of resemblance between Orobanche and Gesnerworts and Figworts consist in their monopetalous didynamous flowers and bicarpellary polyspermous fruit; and it is these which have led to the general opinion that all the Orders are closely allied. Such marks of agreement are doubtless important; but they may be overbalanced by circumstances of disagreement of more importance. One of these is the position of the carpels with respect to the axis of inflorescence. In the whole category of personate, labiate, or irregular plants forming the Bignonial Alliance, the carpels stand fore and aft with respect to the axis; while in Gentianworts we have as universally the two carpels placed laterally. In this striking character Orobanche agrees with the latter. Now as a didynamous structure is not universal in Bignonials, while the position of carpels is constant through both series respectively, we must assign the greater importance to the latter character, and hence Orobanche would be removed from Bignonials to the series represented by Gentians; of which this genus would be a didynamous form, analogous to what frequently occurs among Bignonials. If to this we add the resemblance between Broomrapes and Gentians in the minuteness of their embryo as compared with the albumen, no doubt can, I think, remain as to the very near alliance between the two.

The peculiar placentation of this Order was mentioned by me some years ago, (*Taylor's Magazine*, Nov. 1837). That their capsule consists of two carpels standing right and left of the axis of inflorescence, and with the margins not inflected in the form of dissepiments, is incontestable. Yet in Orobanche and Phelypæa the capsule has four placentæ, placed equidistant in pairs upon the face of each valve or carpel, and considerably within the margin. In Epiphegus each carpel has two intramarginal placentæ, which diverge from the base upwards, and terminate before reaching the apex. In Lathræa there is to each valve but one placenta, which may be regarded as two confluent ones occupying the very face of the dorsal suture of the carpel. And finally in Æginetia indica, and I believe in Æginetia abbreviata also, the placenta is in like manner confined to the axis of the valve, occupying the same position upon the carpels as in Lathræa, but broken up into a number of parallel plates of unequal depth, over the whole surface of which multitudes of minute seeds are distributed.

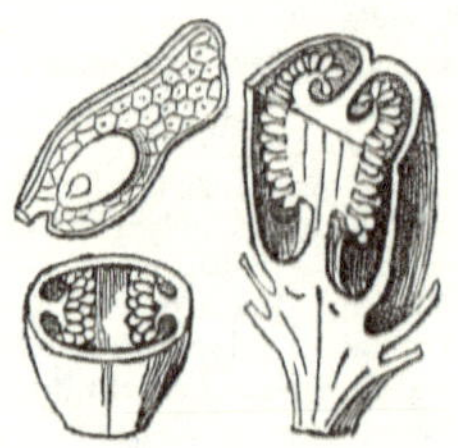

Fig. CCCCXIII.

According to the observations of Vaucher, of Geneva, the seeds of Orobanche ramosa will lie many years inert in the soil unless they come in contact with the roots of Hemp, the plant upon which that species grows parasitically: when they immediately sprout. The manner in which the seeds of Orobanche attach themselves to the plants on which they grow has been observed by Schlauter. This writer states that they only seize seedlings, and are unable to attack roots of a stronger growth. When Picris hieracioides is attacked, he found that the Orobanche seeds seize upon the points of the roots exclusively; the latter then swell and form an enlargement which serves for a base to the Orobanche.—*Ann. Sc. n. s.* 10. 318. Duchartre has studied with great diligence the development of Clandestina, in whose stem he finds neither medullary sheath nor medullary processes; and according to Messrs. de Mirbel, Richard, and Ad. Brongniart, the same remarkable structure occurs among Figworts in the case of Melampyrum sylvaticum.—*Ann. Sc. N. n. s.* xx. 145.

Broomrapes are not uncommon in Europe, particularly in the southern kingdoms, Barbary, the Cape of Good Hope, middle and northern Asia, and North America; they are very rare in India.

Orobanche major is a powerful, astringent, bitter plant, the infusion of which has been employed as a detergent application to foul sores, and internally to restrain alvine fluxes. Epiphegus virginiana is supposed to have formed, in conjunction with white oxide of arsenic, a famous cancer powder, which was known in North America under the name of Martin's Cancer Powder. It is thought to participate in the properties of Orobanche major. Orobanche epithymum is an old-fashioned bitter tonic, and vulnerary; and its fragrant flowers are used in spasmodic affections. Lathræa Squamaria roots were given in epilepsy, and Clandestina was supposed to counteract sterility in women: but these things are now forgotten. Æginetia indica, prepared with sugar and nutmeg, is considered an antiscorbutic. Phelipæa lutea dyes black the ropes that are prepared from the fibres of the Doom Palm of Thebes.

Fig. CCCCXIII.—1. seed and embryo of Conopholis americana; 2. section of ovary of Epiphegus americana; 3. section of fruit of Hyobanche sanguinea.

GENERA.

Epiphegus, *Nutt.*	*Hæmodorum*, Wallr.	Lathræa, *Linn.*	Hyobanche, *Thunb.*
Leptamnium, Raf.	Conopholis, *Wallr.*	*Squamaria*, Hall.	? Centronota, *DC.*
Mylanche, Wallr.	Orobanche, *Linn.*	Anoplanthus, *Endl.*	*Centronia*, Bl.
Phelipæa, *Desf.*	*Osproleon*, Wallr.	*Anoplon*, Wallr.	*Gasparinia*, Endl.
Trionychion, Wallr.	Boschniakia, *C. A. Mey.*	*Anblatum*, Tournef.	*Tronicena*, Steud.
Kopsia, Dumort.	*Stellara*, Fisch.	Æginetia, *Linn.*	Obolaria, *L.*
Cistanche, Link.	Clandestina, *Tournef.*		

NUMBERS. GEN. 12. Sp. 116.

POSITION.—Gentianaceæ.—OROBANCHACEÆ.— *Monotropaceæ.* / *Gesneraceæ.*

ORDER CCXXXVI. GENTIANACEÆ.—GENTIANWORTS.

Gentianeæ, *Juss. Gen.* 141. (1789); *R. Brown Prodr.* 449; *Von Martius Nov. Gen. &c.* 2. 132; *Bartl. Ord. Nat.* 199; *Royle's Illustrations*, 276; *Endl. Gen.* cxxxiv.; *Grisebach Monogr.* (1836); *Id. in Alph. DC. Prodr.* 9. 38.—Desfontaineæ, *Endl. Gen. p.* 669.

DIAGNOSIS:—*Gentianal Exogens, with no stipules, simple stigmas at the end of a manifest style, parietal placentæ, and regular flowers.*

Herbaceous plants, seldom shrubs, generally smooth, sometimes twining. Leaves opposite, entire, without stipules, sessile, or having their petioles confluent in a little

Fig. CCCCXIV.

sheath, in most cases 3- 5-ribbed; very rarely brown and scale-like; sometimes alternate. Flowers terminal or axillary, regular, or very seldom irregular. Calyx divided, inferior, persistent. Corolla monopetalous, hypogynous, usually regular, and persistent; the limb regular, sometimes furnished with delicate fringes, its lobes of the same number as those of the calyx, generally 5, sometimes 4, 6, 8, or 10, occasionally extended at the base into a bag or spur, with a plaited, or folded, or imbricated twisted æstivation. Stamens inserted upon the corolla, all in the same line, equal in number to the segments, and alternate with them; some of them occasionally abortive. Ovary composed of 2 carpels, 1- or partly 2-celled, many-seeded. Style 1, continuous with the ovary; stigmas 2, right and left of the axis; ovules 00, anatropal, parietal. Capsule or berry many-seeded; when 2-valved, the margins of the valves turned inwards, and bearing the seeds. Seeds small; testa single; embryo minute in the axis of soft fleshy albumen; radicle next the hilum.

This Order is very near that of Dogbanes, from which it differs in the herbaceous habit, permanent corolla, entire ovary, parietal placentation, imbricated, not contorted æstivation, want of milk, and usually capsular fruit. The ribbed leaves too afford, in the majority of cases, a certain mark of distinction; to this may be added bitterness. Wherever the parietal placentæ can be found, and this is usually the case, the recognition of the Order is very easy; and in the anomalous genera, like Sebæa, in which a partially 2-celled ovary exists, a little examination shows that in reality the placentæ merely meet at the base. From Figworts, in particular, this circumstance distinctly cuts Gentians off, independently of their minute embryo and symmetrical flowers. Von Martius remarks that no Gentianworts except Tachia have a hypogynous disk; and the two carpellary leaves of which the fruit is formed are lateral, or right and left with respect to the common axis of the inflorescence, their placentæ being consequently anterior and

Fig CCCCXIV.—Gentiana amarella. 1. section of the ovary of Chironia baccifera; 2. section of the ripe fruit; 3. a seed; 4. a vertical section of it.

posterior; while in Figworts, Gesnerworts, Bignoniads, Acanthads, and their allies, a hypogynous disk is very common in the shape of a fleshy ring, or of glands, or teeth, and the two carpellary leaves are anterior and posterior, the dissepiment being consequently in the same transverse line as that which separates the upper from the lower lip. Crawfurdia seems to connect this Order with Bindweeds; and Voyra, a parasitical, scaly, leafless genus offers a direct transition to Broomrapes.

Fig. CCCCXV.

A numerous Order of herbaceous plants, extending over almost all parts of the world, from the regions of perpetual snow upon the summits of the mountains of Europe, to the hottest sands of South America and India. They, however, do not appear in the Flora of Melville Island; but they form part of that of the Straits of Magellan. The most common genus is Gentiana, concerning which and its allies, the following observations will be read with interest.

"Few genera display so full a series of colours in the flowers as this does; red, blue, yellow and white, are all exhibited in it, with many of the intermediate compound tints. Yellow and white are rare in the regions of the Gentians, but almost invariably present; the red species are nearly confined to the Andes of South America and New Zealand. Amongst Dr. Jameson's Botanical Notes on the Flora of the Andes of Peru and Colombia, I find the following interesting remark: Of sixteen species of Gentian with which I am acquainted, one half are red, four purple, two blue, one yellow, and one white.—*Bot. Journ.* vol. ii. p. 649. Their inferior limit under the line we find, from the same source, to be 7852 feet, and they ascend from thence nearly to the limits of perpetual snow on Cotopaxi; they do not in South America descend to the level of the sea in a lower latitude than 54° or thereabouts, where however there are no Alpine species, though the snow line does not descend below 4000—3,500 feet. In the Himalaya, where the species are all blue-flowered, one species has been gathered by my friend Mr. Edgworth, near Ratha Kona, on the Mána Pass, at an elevation of 16,000 feet, near the limit of perpetual snow; and another reaches, in lat. 31 N., the altitude of 12,689, according to Dr. Royle.—*Illust. Plant. Himal.* vol. i. pp. 22 and 278. In Ceylon a species has been gathered at between 6000 and 8000 feet of elevation. One species, G. prostrata, *H. B. K.*, has a most extraordinary range both in longitude and latitude; in southern Europe it inhabits the Carinthian Alps, between 6000 and 9000 feet high; in Asia it has been found on the Altai Mountains, about lat. N. 52°. Its American range is much more remarkable, it having been gathered on the tops of the rocky mountains in lat. 52° N. where they attain an elevation of 15,000—16,000 feet, and on the east side of the Andes of S. America, in 35° S.: it descends to the level of the sea at Cape Negro, in the Straits of Magellan in lat. 53° S.; and at Cape Good Hope in Behring's Straits in lat. 68½° N.

"The fact of the occurrence, and the great number, of species of Gentiana inhabiting only the more elevated regions of the temperate and tropical zones, and there reaching the snow limit, renders it very remarkable that they should be so proportionally scarce in the higher latitudes both of the northern and southern hemispheres. Generally speaking, the inhabitants of these elevated and cold regions are species of such Natural Orders and Genera as compose the mass of Polar vegetation. It is so to a great extent with certain groups of Ranunculaceæ, Gramineæ, Caryophylleæ, Cruciferæ, Ericaceæ, &c., but not with Gentianeæ; the proportion which the species of the transition temperate zones bear to the other plants of those regions on the one hand, and to the tropical species on the other, is in both cases remarkably small. They are entirely unknown to the Floras of the Polar and American Islands; very few inhabit Greenland, Iceland, or the Arctic Sea shores in the North, or Tasmania, New Zealand, Fuegia, or the Antarctic Islands in the South; and again in other parts of N. Europe and America, or of Chili and Patagonia, they are infinitely less numerous than in the Alps of middle and south Europe, or the Andes of the equator."—*Jos. Hooker, Bot. of Antarctic Voyage*, p. 55.

Fig. CCCCXV.—Voyra aurantiaca.—*Splitgerber.*

The Order of Gentianworts is not more remarkable for the diversity of its colours than it is for the uniformity of the secretions which its various species exhibit. Bitterness in every part, root, leaves, flowers, fruit, in annuals, perennials, and shrubs, is so much their characteristic that the following account of the purposes to which they are applied is little more than a list of repetitions; with this exception, that they in some cases prove narcotic and emetic.

The common Gentian root of the druggists, a pure and intense bitter, is for the most part Gentiana lutea, an herbaceous plant, with axillary whorls of yellow flowers, common on the Alps of Europe. It is principally employed as a tonic, but sometimes relaxes the bowels, producing nausea and a kind of intoxication. G. campestris and Amarella, common on the heaths and hills of some parts of England, are domestic substitutes; as are G. Catesbæi in the United States, G. Kurroo in the Himalayas, and G. punctata, pannonica, purpurea, and others, on the Continent of Europe. G. cruciata has been superstitiously believed to possess especial virtues because its leaves grow in the form of a cross, and it is one of the thousand panaceas for hydrophobia. Agathotes Chirayta, a Himalayan annual, is remarkable for the pureness of its bitter. The whole plant is pulled up at the time the flowers begin to decay, and dried for use. Its febrifugal properties are in high estimation with European practitioners in India, who use it instead of Cinchona when the latter is not to be procured. Cicendia hyssopifolia, a common Indian annual, Erythræa Centaurium (Centaury), a beautiful little wild plant, with pink flowers, Chlora perfoliata, various species of Lisianthus, Tachia, Sabbatia, Coutoubea, &c. &c., possess qualities very nearly of the same kind, varying principally in intensity, and are employed as substitutes for Gentian in different countries. The root of Frazera Walteri, a North American biennial, is a pure, powerful, and excellent bitter, destitute of aroma, and is fully equal to Gentian. When fresh it is reported to be emetic and cathartic. The roots have been imported into Europe as a sort of Calumba, and have acquired in consequence the name of American Calumba. Menyanthes trifoliata, a common bog plant, called Buck Bean (quasi Bach or Beck, *i. e.* Brook Bean) is intensely bitter. Its rhizome is reckoned one of the most valuable of known tonics; but large doses produce vomiting, and frequently powerful diaphoresis. It is recommended in intermittent and remittent fevers, gout, herpetic complaints, rheumatism, dropsy, scurvy, and worms. Withering says that it may be used as a substitute for Hops in making beer. Villarsia nymphoides acts in a similar way, but is weaker.

GENERA.

I. GENTIANEÆ.—Corolla imbricated.

Gentiana, *Tournef.*
Asterias, Ren.
Cœlantha, Frœl.
Coilantha, Borkh.
Dasystephana, Ren.
Cuttera, Raf.
Pneumonanthe, Bung.
Dasycephala, Borkh.
Ciminalis, Borkh.
Thylacites, Ren.
Crossocephalum, Frœl.
Crossopetalum, Roth.
Urananthe, Gaud.
Gentianella, Borkh.
Ericala, Ren.
Ericoila, Bork.
Calathiana, Frœl.
Chondrophyllum, Bng.
Erithalia, Bung.
Tetrorhiza, Ren.
Endotriche, Frœl.
Eurythalia, Ren.
Cyanea, Ren.
Oreophylax, Endl.
Pleurogyne, *Eschsch.*
Lomatogonium, A. Br.
Trochantha, Bung.
Swertia, *Linn.*
Stellera, *Turcz.*
Anagallidium, *Griseb.*
Ophelia, *Don.*
Monobothrium, Hochs.
Agathotes, *Don.*
Henricea, Lem. Lis.
Frasera, *Walt.*
Halenia, *Borkh.*
Tetragonanthus, Stell.
Chironia, *Linn.*
Rœslinia, Mönch.
Plocandra, *E. Mey.*
Gyrandra, *Gris.*
Orphium, *E. Mey.*
Valerandia, Neck.
Exacum, *Linn.*
Lapethea, *Gris.*
Voyra, *Aubl.*
Vohiria, Juss.
Lita, Schreb.
Humboldtia, Neck.
Leiphaimos, Schlecht.
Ixanthus, *Griseb.*
Hippion, *Spreng.*
Slevogtia, Reichenb.
Cicendia, *Adans.*
Microcala, Link. et H.
Franquevillia, Gray.
Hippocentaurea, Schlt.
Centaurella, *L. C. Rich.*
Centaurium, Pers.
Bartonia, Mühlenb.
Andrewsia, Spreng.
Erythræa, *Ren.*
Xanthea, Reichenb.
Zygostigma, *Griseb.*
Canscora, *Lam.*
Pladera, Sol.
Hoppea, Willd.
Pootia, Dennst.
Orthostemon, *R. Br.*
Sabbatia, *Adans.*
Chlora, *Linn.*
Blackstonia, Huds.
Xanthanthus, Griseb.
Callopisma, *M. et Zucc.*
Dejanira, Cham.
Schultesia, *M. et Z.*
Hockinia, *Gardn.*
Anacotus, Griseb.
Pagæa, *Griseb.*
Petalostylis, *Griseb.*
Omphalostigma, Gris.
Lisyanthus, *Aubl.*
Lisianthus, Linn.
Macrocarpæa, Griseb.
Sphærocarpæa, Griseb.
Choriophyllum, Gris.
Chelonanthus, Griseb.
Irlbachia, *Mart. et Zucc.*
Helia, *Mart. et Zucc.*
Eustoma, *Don.*
Urananthus, Griseb.
Leianthus, *Griseb.*
Coutoubea, *Aubl.*
Cutubea, Mart. et Zucc.
Picrium, Schreb.
Prepusa, *Mart. et Zucc.*
Tachiadenus, *Griseb.*
Symbolanthus, *Don.*
Tachia, *Aubl.*
Myrmecia, Schreb.
Leiothamnus, *Griseb.*
Eudoxia, *Don.*
Crawfurdia, *Wall.*
Belmontia, *E. Mey.*
Exochænium, *Gris.*
Sebæa, *Soland.*
Lagenias, *E. Mey.*
Schübleria, *Mart.*
Curtia, Cham.
Thurnheissera, Pohl.
Apophragma, *Griseb.*
Exadenus, *Griseb.*
Desfontainia, *R. et S.*
Linkia, Pers.
? Henicostemma, *Blum.*
? Tripterospermum, *Blm.*
? Micræa, *Miers.*
Glyphospermum, *Don.*

II. MENYANTHEÆ.—Corolla induplicate.

Menyanthes, *Linn.*
Menonanthes, Haw.
Villarsia, *Vent.*
Nymphæanthe, Rchb.
Renealmia, Houtt.
Trachysperma, Raf.
Cumada, Jon.
Limnanthemum, *Gmel.*
Nymphoides, Tournef.
Waldschmidia, Wigg.
Schweyckherta, Gmel.
? Mitreola, *Linn.*
Cynoctonum, Gmel.

NUMBERS. GEN. 60. Sp. 450.

POSITION.—Orobanchaceæ.—GENTIANACEÆ.—*Cinchonaceæ.* / *Polemoniaceæ.*

ALLIANCE XLVI. *SOLANALES*.—THE SOLANAL ALLIANCE.

DIAGNOSIS.—*Perigynous Exogens, with dichlamydeous, monopetalous, symmetrical flowers, axile placentæ,* 2-3-*celled fruit, large embryo, lying in a small quantity of albumen.*

All these plants are clearly held together by the common character of a monopetalous corolla, axile placentæ, regular symmetrical flowers, and an inconsiderable quantity of albumen. It is the last circumstance, with the axile placentation, which divides them from the Gentianal Alliance. The free central placenta of Cortusals clearly distinguishes that Alliance.

Here and there anomalous genera occur, with no corolla, or separate petals, but they are rare, and do not seem to invalidate the Alliance, which joins Gentianals by Oliveworts, which are nearly allied to Ebenads and Hollyworts, and passes into Cortusals by the Polemoniads, which are so very near Hydrophyls that the two were once blended in the same Natural Order.

Lateral affinities are here very important. Nothing whatever except the symmetry of their flowers separates Nightshades from Figworts in the Bignonial Alliance; Oliveworts touch Jasmineworts among Echials; Asclepiads Dogbanes in Gentianals, and Bindweeds the Nolanads of the Echial Alliance.

NATURAL ORDERS OF SOLANALS.

Stamens free, 2 *or* 4	237. OLEACEÆ.
Stamens free, 5. *Placentæ axile. Embryo terete*	238. SOLANACEÆ.
Anthers and stigma consolidated into a column	239. ASCLEPIADACEÆ.
Stamens free, 5. *Placentæ axile. Cotyledons leafy, folded longitudinally*	240. CORDIACEÆ.
Stamens free, 5. *Placentæ basal. Cotyledons leafy, doubled up*	241. CONVOLVULACEÆ.
Stamens free, 5. *Placentæ basal. Embryo filiform, spiral* . .	242. CUSCUTACEÆ.
Stamens free, 5. *Placentæ axile. Cotyledons straight, plano-convex*	243. POLEMONIACEÆ.

ORDER CCXXXVII. OLEACEÆ.—OLIVEWORTS.

Oleineæ, *Hoffmannsegg et Link Fl. Port.* (1806); *Brown Prodr.* 522.—Lilaceæ, *Vent. Tabl.* 1. 306. (1799).—Fraxineæ, *Martius Conspectus, No.* 209. (1835).—Oleaceæ, *Ed. pr.* ccxxvi. (1836); *Endl. Gen.* cxxx.; *DC. Prodr.* 8. 273.

DIAGNOSIS.—*Solanal Exogens, with* 2 *or* 4 *free stamens.*

Trees or shrubs. Branches usually dichotomous and ending abruptly by a conspicuous bud. Leaves opposite, simple, sometimes pinnated. Flowers in terminal or axillary racemes or panicles; the pedicels opposite, with single bracts. Flowers ☿ or ♂ ♀. Calyx divided, persistent, inferior. Corolla hypogynous, monopetalous, 4-cleft, occasionally of 4 petals connected in pairs by the intervention of the filaments, sometimes absent; æstivation somewhat valvate. Stamens 2 (in Tessarandra 4), alternate with the segments of the corolla, or with the petals; anthers 2-celled, opening longitudinally. Ovary simple, without any hypogynous disk, 2-celled; the cells 2-seeded; the ovules pendulous and collateral; style 1 or 0; stigma bifid or undivided. Fruit drupaceous, berried, or capsular, often by abortion 1-seeded. Seeds with dense, fleshy, abundant albumen; embryo about half its length, straight; cotyledons foliaceous; radicle superior; plumule inconspicuous.

2 1

Fig. CCCCXVI.

These plants resemble Jasmineworts in many respects, and Endlicher even thinks them allied to that Order alone; indeed they are combined by Ach. Richard. Reichenbach thinks Oliveworts related to Storaxworts, because, according to Hayne (*Arzngw.* xi. 23. *adn. ult.*), a sort of storax is yielded by Olea europæa. De Candolle suggests (*Essai Méd.* p. 204.) that the Ash is related to the Maples, and this view is lately adopted by Von Martius; I also find in the same work the following very good observations upon this Order:—"However heterogeneous the Oliveworts may appear as at present limited, it is remarkable that the species will all graft upon each other; a fact which demonstrates the analogy of their juices and their fibres. Thus the Lilac will graft upon the Ash, the Chionanthus, and the Fontanesia, and I have even succeeded in making the Persian Lilac live ten years on Phillyrea latifolia. The Olive will take on the Phillyrea, and even on the Ash: but we cannot graft the Jasmine on any plant of the Olive tribe: a circumstance which confirms the propriety of separating these two Orders." To me I confess that the unsymmetrical flowers of Jasmineworts offer a great difficulty in the way of placing them in even the same Alliance as Oliveworts, the more especially because that peculiarity is connected with a decidedly nucamentaceous fruit. The two stamens usually present in Oliveworts may be taken to show that the flowers of the Order are really $\frac{2}{4}$, which is confirmed by Tessarandra, which has 4 stamens; the two stamens of Jasmineworts are probably connected with a quinary type. The true affinity seems to be with Nightshades, as is indicated by the dicarpellary fruit, regular symmetrical monopetalous corolla, axile placenta, and undivided fruit of both Orders.

Natives chiefly of temperate latitudes, inclining towards the tropics, but scarcely known beyond 65° N. lat. The Ash is extremely abundant in North America; the Phillyreas and Syringas are all European or Eastern plants. A few are found in New Holland and elsewhere within the tropics. One Ash is a native of Nipal.

From the pericarp of Olea europæa, the common Olive, is obtained by pressure the well-known substance called Olive Oil; the medical properties of which are demulcent,

Fig. CCCCXVI.—Notelæa ovata. 1. flower; 2. fruit.—*Endlicher.*

emollient, and laxative. It enters extensively into the preparation of plasters, liniments, cerates, ointments, and enemas. As an external application, accompanied by long-continued friction of the skin, it has been found beneficial in preventing the contagious influence of the plague. The bark is bitter and astringent, and has had a great reputation as a substitute for Cinchona, according to De Candolle. It also yields a kind of gum, or rather a gum-like substance, once in repute as a vulnerary. Its wood is extremely durable and close-grained. The flowers are frequently slightly fragrant; those of Olea fragrans are employed in China for flavouring tea. The sweet, gentle purgative, called Manna, is a concrete discharge from the bark of several species of Ash, but especially from Fraxinus rotundifolia. The sweetness of this substance is not due to the presence of sugar, but to a distinct principle, called Mannite, which differs from sugar in not fermenting with water and yeast. Fraxinus excelsior (the common Ash) not only yields Manna, in the warm climate of the south of Europe, but is reported to have a tonic febrifugal bark, and leaves almost as cathartic as those of Senna, producing an unequivocal action upon the kidneys. The febrifugal qualities of the Lilac, Syringa vulgaris are undoubted. In that part of the province of Berri called Brenne, which is marshy and insalubrious to the last degree, the peasants employ no other remedy for the intermittent fever which prevails there. According to Meillet this quality is apparently owing to a principle which he calls Lilacine.—*Pharm. Journ.* 1. 557.

GENERA.

I. OLEÆ.—Fruit a drupe or berry.

Chionanthus, *Linn.*
Linociera, *Swartz.*
  Thouinia, Swartz.
  Minutia, Flor. Flum.
Tessarandra, *Miers.*
Mayepea, *Aubl.*
  Freyeria, Scop.
  Ceranthus, Schreb.
Boaria, *A. DC.*
Noronhia, *Stadtm.*
Olea, *Tourn.*
Picconia, *A. DC.*
Visiania, *A. DC.*
Phillyrea, *Tournef.*
Osmanthus, *Lour.*
Notelæa, *Vent.*
  Rhysospermum, Gärtn.
Stereoderma, *Blum.*
  Pachyderma, Blum.
Ligustrum, *Tournef.*
Myospyrum, *Bl.*
Chondrospermum, *Wall.*
Tetrapilus, *Lour.*

II. FRAXINEÆ.—Fruit samaroid.

Fraxinus, *Tournef.*
Ornus, *Pers.*
Fontanesia, *Labill.*
  Desfontainesia, Hoff.
Syringa, *Linn.*
  Lilac, Tournef.
Forsythia, *Vahl.*
Nathusia, *Hochst.*
? Tetrapilus, *Lour.*

NUMBERS. GEN. 24. Sp. 130.

Aceraceæ.
POSITION.—————OLEACEÆ.—Solanaceæ.
Jasminaceæ.

2 1 3

Fig. CCCCXVI.*

Fig. CCCCXVI.*—1. flower of Ligustrum vulgare; 2. perpendicular section of calyx and pistil; 3. cross section of fruit, showing an abortive cell; 4. cross section of a seed.

Order CCXXXVIII. SOLANACEÆ.—Nightshades.

Solaneæ, *Juss. Gen.* 124. (1789) ; *R. Brown Prodr.* 443 ; *Bartl. Ord. Nat.* 193 ; *Schlecht. in Linnæa*, 7. 66. (1832) ; *Nees v. Esenbeck in Linn. Trans.* 17. 37. (1834).—Solanaceæ, *Ed. Pr.* ccxviii. (1836) ; *Endl. Gen.* cxlviii. ; *Meisner*, *p.* 272.—Cestrinæ, *Martius Conspectus*, *No.* 121. (1835).—Cestraceæ, *Ed. Pr.* ccxix.—Retziaceæ, *Bartl. Ord. Nat.* (1830) ; *Endl. Gen. p.* 669.

Diagnosis.—*Solanal Exogens, with 5 free stamens, axile placentæ, and a terete embryo.*

Herbaceous plants or shrubs. Leaves alternate, undivided, or lobed, sometimes collateral ; the floral ones sometimes double, and placed near each other. Inflorescence

1 2 3

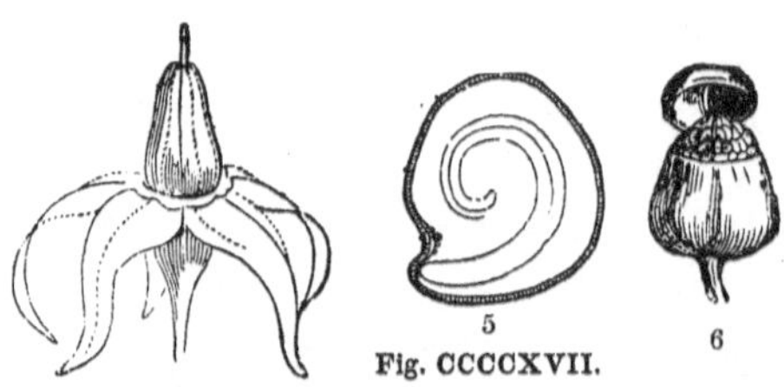

4 5 6

Fig. CCCCXVII.

variable, often out of the axil ; the pedicels without bracts. Calyx 5-parted, seldom 4-parted, persistent, inferior. Corolla monopetalous, hypogynous ; the limb 5-cleft, seldom 4-cleft, regular, or somewhat unequal, deciduous ; the æstivation plaited or imbricated, or even valvate. Stamens inserted upon the corolla, as many as the segments of the limb, with which they are alternate ; anthers bursting longitudinally, rarely by pores, at the apex. Ovary 2-celled, composed of a pair of carpels right and left of the axis, rarely 4- 5- or many-celled, with polyspermous placentæ ; style continuous ; stigma simple ; ovules 00, amphitropal. Pericarp with 2, or 4, or many cells, either a capsule with a double dissepiment parallel with the valves, or a berry with the placentæ adhering to the dissepiment. Seeds 00 ; embryo straight or curved, often out of the centre, lying in fleshy albumen ; radicle near the hilum.

The anthers of Solanum open by pores. Nicotiana multivalvis has many cells in the capsule, so has Lycopersicon ; Nicandra is 5-celled, Datura 4-celled.

Brown remarks, that this Order is chiefly known from Figworts by the curved or spiral embryo, the plaited æstivation of the corolla, and the flowers being regular, with the same number of stamens as lobes. Hence the genera with a corolla not plaited, and at the same time a straight embryo, should, he thinks, either be excluded, or placed in a separate section, along with such as have an imbricated corolla, a slightly curved em-

Fig. CCCCXVII.—Petunia violacea. 1. a cross section of the ovary ; 2. ripe fruit of Solanum Dulcamara ; 3. a section of one of its seeds ; 4. flower of Solanum Dulcamara ; 5. a section of its seed ; 6. pyxis of Hyoscyamus.

bryo, and didynamous stamens. It does not, however, appear necessary to separate the latter as a distinct Order, but it is better to understand them as genera passing into the condition of Figworts, which are in fact nothing but unsymmetrical Nightshades. In reality, the Nightshades are the equivalent, in the Solanal Alliance, of the Figworts among Bignonials; and these two Alliances are brought into direct contact by means of the Orders in question, although, in a lineal arrangement, they may not follow each other. It is quite certain, I think, that no other distinction between Nightshades and Figworts exists, for the curved embryo of the former, although remarkable in many instances, is not at all to be depended upon, because the nature of the embryo varies in very nearly allied species. Thus in Petunia nyctaginiflora is found the common curved and twisted embryo of Nightshades; but in Petunia violacea, the seeds of which cannot be externally distinguished from those of the latter, not even when lying side by side upon the field of the microscope, the embryo is perfectly straight and much shorter; in Salpiglossis straminea the embryo is curved and partly spiral; yet in all other characters the genus agrees with Figworts: finally, in Nicotiana persica, which no one can doubt being a genuine species of Nightshade, the embryo is nearly straight. We therefore are obliged to conclude that a false importance has been given to this, as it certainly has to a great many other microscopic characters; a truth which has not escaped the acuteness of Fries. I do not, however, conceive that Figworts and Nightshades ought really to stand in the same Alliance, because the latter have a manifest tendency to lose the dicarpellary structure of the former, as is seen in Nicandra, which has 5 cells, and in the many-celled Lycopersicons and Nicotianas. No such tendency occurs in the Bignonial Alliance.

The most immediate affinity of Nightshades seems to be with Oliveworts and Bindweeds, to the latter of which their numerous twining species bring them very close, while the first division of the Order stands on the very threshold of Oliveworts. Compare, for instance, Syringa and Cestrum. At the same time several collateral affinities are extremely well marked. That of Figworts has already been mentioned. Bellworts are approached by Trechonætes, whose stamens are scarcely epipetalous. Grabowskya, of Schlechtendahl, is considered by that author to be a transition between Nightshades and Borageworts. He, however, regards its affinity to Lycium undoubted, and points out its near relation to Nolana. (See *Linnæa*, 7. 71). Mr. Walker Arnott indicates its difference from Nightshades in the small number of its seeds.—*Linnæa*, 11. 486. Nolanads are also close to Grabowskya, and would very well stand in the Solanal Alliance if their fruit were not nucamentaceous.

Natives of most parts of the world without the arctic and antarctic circles, especially within the tropics, in which the mass of the Order exists, in the form of the genera Solanum and Physalis. The number of species of the former genus is very great in tropical America, and the whole amount to twice as many as all the other co-ordinates.

At first sight this Order seems to offer an exception to that general correspondence in structure and sensible qualities which is so characteristic of well defined Natural Orders, containing as it does the deadly Nightshade and Henbane, and the wholesome Potato and Tomato; but a little inquiry will explain this apparent anomaly. The leaves and berries of the Potato are narcotic; it is only its tubers that are wholesome when cooked. This is the case with other succulent underground stems in equally dangerous families, as the Cassava among Spurgeworts; besides which, as De Candolle justly observes—"Il ne faut pas perdre de vue que tous nos alimens renferment une petite dose d'un principe excitant, qui, s'il y était en plus grande quantité, pourrait être nuisible, mais qui y est nécessaire pour leur servir de condiment naturel." The leaves of all are in fact narcotic and exciting, but in different degrees, from Atropa Belladonna, which causes vertigo, convulsions, and vomiting, Tobacco, which will frequently produce the first and last of these symptoms, Henbane and Stramonium, down to some Solanums, the leaves of which are used as kitchen herbs. The various species may be classed according as they are, 1, narcotic or otherwise poisonous; 2, tonic; 3, diuretic; 4, pungent; 5, bland or inert. 1. As to poisonous species, the worst of these is perhaps the Acocanthera venenata, a large bush with fragrant flowers, found at the Cape of Good Hope; a decoction of the bark, reduced to the thickness of jelly, is used by the Hottentots to envenom their weapons. It is said to be a fatal poison, and to be also used by the same people to destroy wild beasts, by impregnating baits of flesh with its juice. Similar qualities have been recognised in the Cestrums macrophyllum and nocturnum. Others, however, more familiar to Europeans, can hardly be regarded as inferior in virulence. The Thorn-apple (Datura Stramonium) is a violent narcotic when taken internally; in skilful hands it is a valuable medicine in mania, epilepsy, convulsions, tic-douloureux, &c; it palliates the distressing paroxysms of pure spasmodic asthma, when smoked. Datura Tatula and Metel have a similar action; the latter is used by Orientals as an opiate, the former is said to be much more energetic than Stramonium; the seeds are the most powerful part of these plants, and are stated by some authors to have been used by the priests of the Delphic Temple, to produce those frenzied ravings which were called prophecies. Such a practice certainly obtains, or obtained, in the Temple of the Sun, in the city of Sagomozo, where the seeds of the Floripondio (Datura sanguinea) are used; the Peruvians also prepare from them an intoxicating beverage which stupefies if taken much diluted; but, when strong, brings on attacks of furious excitement. Henbane (Hyoscyamus niger), a common biennial weed, is a powerful narcotic at the time when its seeds are forming, though comparatively inert at an earlier period. Its capsules and seeds, as well as its leaves, are used extensively in medicine, and produce effects similar to those of Opium. But the former, when taken too freely, are apt to bring on temporary insanity. All the other species of Hyoscyamus have a similar action. In some parts of the Greek continent the stalks of Hyoscyamus albus are used against toothache. They are dried and employed in lieu of Tobacco, for smoking. In England the seeds of H. niger are occasionally employed for the same purpose, with useful effect.—*Ann. Ch.* 1. 249. Atropa Belladonna is another dangerous narcotic. Every part of the plant is poisonous; and children and the ignorant have often suffered from eating the berries, the beautiful appearance and sweet taste of which render them very alluring. The symptoms which they induce are those of intoxication, accompanied with fits of laughter and violent gestures; great thirst, difficulty of deglutition, nausea, dilatation of the pupil, with the eyelids drawn down; redness and tumefaction of the face, stupor or delirium, a low and feeble pulse, paralysis of the intestines, convulsions, and death. In medicine Belladonna is not only narcotic, but diaphoretic and diuretic. It is extensively employed, especially in producing a dilatation of the pupil, when its infusion is dropped into the eye. Among other properties it is said by Hahnemann and Koreff to protect the individual who takes it from the contagion of scarlatina. According to Mr. Pereira it is supposed to be the plant which produced such remarkable and fatal effects upon the Roman soldiers during their retreat from the Parthians (See *Plutarch's Life of Antony*). Buchanan relates that the Scots mixed the juice of Belladonna with the bread and drink which by their truce they were to supply the Danes, which so intoxicated them, that the Scots killed the greater part of Sweno's army while asleep.—*Rer. Scot. Hist. lib.* 7. "The insane root that takes the reason prisoner," mentioned by Shakspeare (*Macbeth*, I. iii.), is also thought to be this. Mandrake, formerly considered an Atropa, but now called Mandragora officinalis, has an action of a similar nature; it has had an exaggerated reputation as an aphrodisiac, was largely used in amorous incantations, and its forked root, which by a little contrivance is easily made to assume the human form (see *Flora Græca*), has led to the foolish stories of the plant shrieking when torn out of the ground. By the Arabs the plant is called Tufah-al-Sheitan, or Devil's Apple. The best commentators regard the Mandrake as the Dudaim of Scripture, in which Dr. Royle concurs (See *Biblical Cyclopædia*, p. 587). It is a little remarkable that although it is generally believed that the Mandrake does not possess any power of inciting the passion of love, yet a nearly allied plant, Jaborosa or Himeranthus runcinatus, is employed in the same manner among the South Americans. Tobacco, the use of which has now become to many persons as indispensable as bread, is the foliage of various species of Nicotiana; all the American Tobacco is furnished by N. Tabacum or its varieties, the Persian by N. persica, and the Syrian by N. rustica. It is a powerful stimulant narcotic, employed medicinally as a sedative, and in vapour to bring on nausea and fainting. When chewed it appears to impair the appetite and induce torpor of the gastric nerves. Although if smoked in moderate quantites it acts as a harmless excitant and sedative, yet it is a frequent cause of paralysis when the practice is indulged in to excess. Oil of Tobacco, which is inhaled and swallowed in the process of smoking, is one of the most violent of known poisons. The Hottentots are said to kill snakes by putting a drop of it on their tongues, and the death of these reptiles is said to take place as instantaneously as if by an electric shock; dangerous symptoms are reported to have followed the application of the ointment to scald heads. Solanums, although far less active than these dangerous plants, are by no means destitute of poisonous qualities in some species.

An extract of the leaves of the common Potato (Solanum tuberosum) is a powerful narcotic, ranking between Belladonna and Conium; according to Mr. Dyer it is particularly serviceable in chronic rheumatism, and painful affections of the stomach and uterus.—*Pharm. Journ.* 1. 590. Solanum Dulcamara, the Bittersweet, is a strong narcotic in its foliage, and its berries are by no means safe, although it does appear that in some cases they have been taken into the stomach without inconvenience. Solanum nigrum, a very common weed in all parts of the world except the coldest, is more active. A grain or two of the dried leaf has sometimes been given to promote various secretions, possibly by exciting a great, and rather dangerous, agitation in the viscera. It is a narcotic, and, according to Orfila, its extract possesses nearly the same power as Lettuce-opium. In Brazil it is called Carachichu, or Erva Moira; and when bruised is applied either in poultices or baths to painful wounds. Finally, to close the long list of narcotics, it will be sufficient to mention Physalis somnifera. This plant is thought to have been the *Στρυχνος ὑπνωτικος* of Dioscorides. It is reputed to be narcotic, diuretic, and alexipharmic. The leaves steeped in oil are, in India, applied to inflammatory tumours; and they are used in a similar way in Egypt. Kunth recognised it in Egyptian mummies.

The tonics are comparatively few. The Quina of Brazil is the produce of Solanum pseudoquina, and is so powerful a bitter and febrifuge, that the Brazilians scarcely believe that it is not the genuine Jesuits' Bark. It has been analysed by Vauquelin, who found that it contained 1-50th of a bitter resinoid matter, slightly soluble in water, about 1-12th of a vegetable bitter, and a number of other principles in minute quantities. Cestrum Hediunda, auriculatum, laurifolium, and Pseudoquina have similar qualities. Martius thinks that the bitterness of these plants is owing to some peculiar principle residing in their bark. Several are found to have a diuretic action; among which may be named Physalis pubescens, viscosa, angulata, and Alkekengi (the Winter Cherry), Nicandra physaloides, Solanum mammosum, paniculatum, nigrum, and guineense, with many species of Cestrum, such as euanthes, lævigatum, corymbosum, Parqui and bracteatum. The latter are generally at the same time emollient, and are applied in a fresh state in cleansing wounds and ulcers. We are told, moreover, that the bruised leaves and unripe fruits are much employed by the Brazilians in affections of the liver and in catarrhus vesicæ. A decoction of the flowers and leaves of Solanum cernuum is a powerful sudorific, and is very serviceable in syphilis, inveterate gonorrhœa, and similar complaints.

The cases of pungency are confined to the fruit of the genus Capsicum, of which numerous species are found in the tropics. The fruit and seeds are powerful stimulants. The well-known condiment called Cayenne Pepper consists principally of the ground seeds. It is employed in medicine, in combination with Cinchona, in intermittents and lethargic affections, and also in atonic gout, dyspepsia accompanied by flatulence, tympanitis, paralysis, &c. Its most valuable application appears, however, to be in cynanche maligna and scarlatina maligna, used either as a gargle or administered internally. It is generally stated that Capsicums have no narcotic quality; but it would appear that some of the American species are an exception to that rule, as is the case with Capsicum toxicarium; this, however, requires confirmation. That some species have fruits which are neither narcotic nor pungent in any considerable degree, if at all, is most certain, for many of them are common articles of food or cookery. But it is stated that the poisonous species derive their properties from the presence of a pulpy matter which surrounds the seeds; and that the wholesome kinds are destitute of this pulp, the fruit consisting only of what Botanists call the sarcocarp; that is to say, the centre of the rind, in a more or less succulent state. This is not, however, a point by any means well established. Tomatoes, the fruit of the Lycopersicum, commonly called Love Apples, in allusion to the supposed power which they possess of exciting tender feelings, are a common ingredient in sauces. Egg Apples, also called Brinjals, or Aubergines, are produced by Solanum Melongena; but they are uneatable till the viscid juice which they contain has been removed. Several are much esteemed in Peru; the berries of Solanum muricatum and nemorense are commonly eaten; and those of S. quitoense are called Quito Oranges (Naranjitas de Quito). Muriti assures us that Mandrake Apples are as harmless as they are beautiful and fragrant; and the Kangaroo Apple, produced by Solanum laciniatum, is a common food among the Tasmannians; Mr. Backhouse states, however, that although when perfectly ripe it may be eaten in any quantity with impunity, yet, while unripe, it is acrid and produces a burning sensation in the throat.

The common Potato, in a state of putrefaction, is said to give out a most vivid light, sufficient to read by. This was particularly remarked by an officer on guard at Strasburgh, who thought the barracks were on fire, in consequence of the light thus emitted from a cellar full of Potatoes.

GENERA.

I. RECTEMBRYÆ.

Cestrum, *Linn.*
Dunalia, *H. B. K.*
Dierbachia, Spreng.
Habrothamnus, *Endl.*
Meyenia, Schlecht.
Jochroma, *Benth.*
Acocanthera, *G. Don.*
Vestia, *Willd.*
Fabiana, *Ruiz et Pav.*
Sessea, *Ruiz et Pav.*
Retzia, *L.*

II. CURVEMBRYÆ.

Nierembergia, *Ruiz et P.*
Petunia, *Juss.*
Nicotiana, *Tournef.*
Tabacus, Mönch.
Codylis, Raf.
Sacranthus, Don.
Nyctagella, Reichenb.
Polydiclia, Don.
Tabacum, Reichenb.
Tabacina, Reichenb.
Lehmannia, *Spreng.*
Nectouxia, *H. B. K.*
Laureria, *Schlecht.*
Marckea, *L. C. Rich.*
Lamarckea, Pers.
Anthotroche, *Endl.*
Callibrachoa, *Llav.*
Metternichia, *Mik.*
Datura, *Linn.*
Stramonium, Tournef.
Brugmansia, Pers.
Dutra, Bernh.
Ceratocaulis, Bernh.
Solandra, *Swartz.*
Swartzia, Gmel.
Hyoscyamus, *Tournef.*
Physoclæna, G. Don.
Belenia, *Decaisne.*

Anisodus, *Link.*
Whitleya, Sweet.
Scopolia, *Jacq.*
Scopolina, Schult.
Nicandra, *Adans.*
Calydermos, Ruiz et P.
Physalis, *Linn.*
Alkekengi, Tournef.
Cacabus, Bernh.
Herschelia, Bowd.
Pentaphiltrum, *Rchb.*
Saracha, *Ruiz et Pav.*
Bellinia, Röm. et Sch.
Jaltomata, Schlecht.
Margaranthus, *Schlecht.*
Witheringia, *Herit.*
Cyathostyles, Schott.
Capsicum, *Tournef.*
Cyphomandra, *Sendter.*
Pionandra, Miers.
Solanum, *Linn.*
Melongena, Tournef.
Pseudocapsicum, Mön.
Nycterium, Vent.
Androcera, Nutt.
Ceranthera, Raf.
Acquartia, Jacq.
Bassowia, Aubl.
Lycopersicum, *Tournef.*
Psolanum, Neck.
Perizoma, *Miers.*
Salpichroa, *Miers.*
Hebecladus, *Miers.*
Atropa, *Linn.*
Belladonna, Tournef.
Busbeckia, Mart.
Discopodium, *Hochst.*
Withania, *Pauq.*
Mandragora, *Tournef.*
Himeranthus, *Endl.*
Thinogeton, *Benth.*
Jaborosa, *Juss.*
Dorystigma, *Miers.*
Trechonætes, *Miers.*
Juanulloa, *Ruiz et Pav.*
Ulloa, Pers.
Grabowskya, *Schl.*
Lycium, *Linn.*
Lyciobatos, Endl.
Lyciothamnus, Endl.
Chænesthes, *Miers.*
Lycioplesium, *Miers.*
Acnistus, *Schott.*

Doubtful Genera.

Cotylanthera, *Blum.*
Isanthera, *Nees.*
Dartus, *Lour.*
Hilsenbergia, *Tausch.*
Doræna, *Thunb.*
Triguera, *Cav.*
Stigmatococca, *Willd.*

NUMBERS. GEN. 60. Sp. 900.

Scrophulariaceæ.
POSITION.—Oleaceæ.—SOLANACEÆ.—Convolvulaceæ.
Nolanaceæ.

ORDER CCXXXIX. ASCLEPIADACEÆ.—ASCLEPIADS.

Apocyneæ, *Juss. Gen.* 143. (1789) *in part.*—Asclepiadeæ, *R. Brown in Wern. Trans.* 1. 12. (1809); *Prodr.* 458. (1810); *Royle Illustr.* 272. (1835); *Wight's Contributions to the Botany of India, No.* 2. *p.* 77. (1834); *Endl. Gen.* cxxiii.

DIAGNOSIS.—*Solanal Exogens, with the anthers and stigma consolidated into a column.*

Shrubs, or occasionally herbaceous plants, almost always milky, and often twining. Leaves entire, opposite, sometimes alternate or whorled, having ciliæ between their petioles in lieu of stipules. Flowers somewhat umbelled, fascicled, or racemose, proceeding from between the petioles. Calyx 5-divided, persistent. Corolla monopetalous, hypogynous, 5-lobed, regular, with imbricated, very seldom valvular, æstivation, deciduous. Stamens 5, inserted into the base of the corolla, alternate with the segments of the limb. Filaments usually connate. Anthers 2-celled, sometimes almost 4-celled in consequence of their dissepiments being nearly complete. Pollen at the period of the dehiscence of the anther cohering in masses, either equal to the number of the cells, or occasionally cohering in pairs and sticking to 5 processes of the stigma either by twos, or fours, or singly. Ovaries 2; styles 2, closely approaching each other, often very short; Stigma common to both styles, dilated, 5-cornered, with corpusculiferous angles. Follicles 2, one of which is sometimes abortive. Placenta attached to the suture, finally separating. Seeds numerous, imbricated, pendulous, almost always comose at the hilum. Albumen thin. Embryo straight. Cotyledons foliaceous. Radicle superior. Plumule inconspicuous.

Fig. CCCCXVIII.

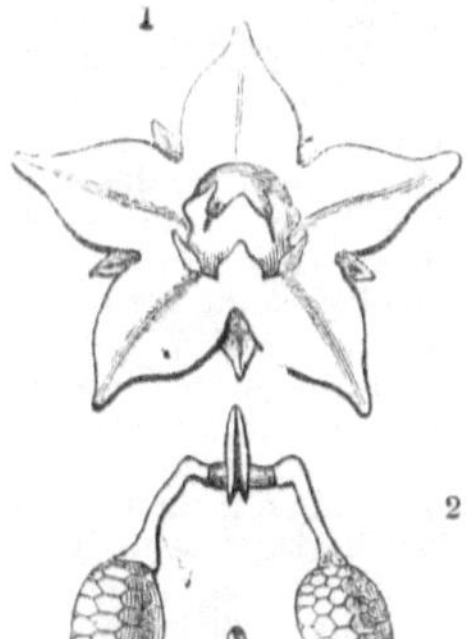
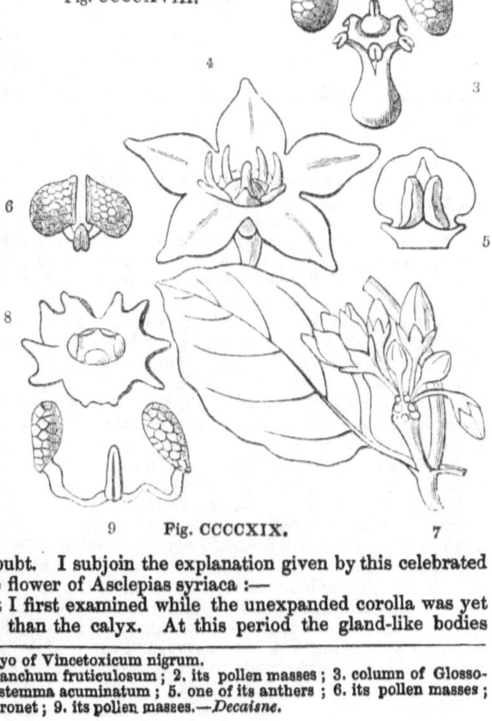

Fig. CCCCXIX.

For a long time the real structure of the present Order was misunderstood; but Brown, in a Dissertation in the *Transactions of the Wernerian Society*, placed its true nature beyond doubt. I subjoin the explanation given by this celebrated Botanist, who thus describes the flower of Asclepias syriaca:—

"The flower-bud of this plant I first examined while the unexpanded corolla was yet green and considerably shorter than the calyx. At this period the gland-like bodies

Fig. CCCCXVIII.—Seed and embryo of Vincetoxicum nigrum.

Fig. CCCCXIX.—1. flower of Cynanchum fruticulosum; 2. its pollen masses; 3. column of Glossonema Boryanum; 4. flower of Heterostemma acuminatum; 5. one of its anthers; 6. its pollen masses; 7. Asterostemma repandum; 8. its coronet; 9. its pollen masses.—*Decaisne.*

which afterwards occupy the angles of the stigma were absolutely invisible ; the furrows of its angles were extremely slight, and, like the body of the stigma, green ; the antheræ, however, were distinctly formed, easily separable from the stigma, and their cells, which were absolutely shut, were filled with a turbid fluid, the parts of which did not so cohere as to separate in a mass ; of the cuculli, which in the expanded flower are so remarkable, and constitute the essential character of the genus, there was no appearance.

"In the next stage submitted to examination, where the corolla nearly equalled the calyx in length, the gland-like bodies of the stigma were become visible, and consisted of 2 nearly filiform, light-brown, parallel, contiguous, and membranaceous substances, secreted by the sides of the furrow, which was now somewhat deeper. Instead of the filiform processes, a gelatinous matter occupied an obliquely descending depression proceeding from towards the base of each side of the angular furrow.

"In a somewhat more advanced stage, the membranes which afterwards become glands of the stigma were found to be linear, closely approximated, and to adhere at their upper extremity. At the same time the gelatinous substance in the oblique depression had acquired a nearly membranaceous texture and a light-brown colour ; and on separating the glands from its furrow, which was then practicable, this membrane followed it. At this period, too, the contents of each cell of the antheræ had acquired a certain degree of solidity, a determinate form, and were separable from the cell in one mass; the cuculli were also observable, but still very small and green, nearly scutelliform, having a central papilla, the rudiment of the future horn-like process. Immediately previous to the bursting of the cells of the antheræ, which takes place a little before the expansion of the corolla, the cuculli are completely formed, and between each, a pair of minute, light-green, fleshy teeth are observable, the single teeth of each pair being divided from each other by the descending alæ of the antheræ. The glands of the stigma have acquired a form between elliptical and rhomboidal, a cartilaginous texture, and a brownish-black colour ; they are easily separable from the secreting furrow, and on their under surface there is no appearance of a suture, or any indication of their having originally consisted of two distinct parts ; along with them separate also the descending processes, which are compressed, membranous, and light-brown ; their extremity, which is still unconnected, being more gelatinous, but not perceptibly thickened. The pollen has acquired the yellow colour, and the degree of consistence which it afterwards retains. On the bursting of the cells, the gelatinous extremity of each descending process becomes firmly united with the upper attenuated end of the corresponding mass of pollen. The parts are then in that condition in which they have been commonly examined, and are exhibited in the figures of Jacquin, who, having seen them only in this state, naturally considered these plants as truly gynandrous, regarding the masses of pollen as the antheræ, originating in the glands of the stigma, and merely immersed in the open cells of the genuine antheræ, which he calls antheriferous sacs; an opinion in which he has been followed by Rottbœll, Kœlreuter, Cavanilles, Smith, and Desfontaines. The conclusion to be drawn from the observations now detailed is sufficiently obvious ; but it is necessary to remark, that these observations do not entirely apply to all the plants which I have referred to the Asclepiadeæ ; some of them, especially Periploca, having a granular pollen, applied in a very different manner to the glands of the stigma ; they all, however, agree in having pollen coalescing into masses, which are fixed or applied to processes of the stigma, in a determinate manner ; and this is, in fact, the essential character of the Order. Dr. Smith, in the second edition of his valuable *Introduction to Botany*, has noticed my opinion on this subject : but, probably from an indistinctness in the communication, which took place in conversation, has stated it in a manner somewhat different from what I intended to convey it to him ; for, according to his statement, the pollen is projected on the stigma. The term projection, however, seems to imply some degree of impetus, and at the same time presents the idea of something indeterminate respecting the part to which the body so projected may be applied. But nothing can be more constant than the manner in which the pollen is attached to the process of the stigma in each species."

Brown, who first distinguished them, stated that they differed solely in the peculiar character of their sexual apparatus ; but this was of so unusual a kind in Asclepiads, as to justify a deviation from the general rule, that Orders cannot be established upon solitary characters. In Dogbanes the stamens are distinct, the pollen powdery (that is to say, in the ordinary state), the stigma capitate and thickened, but not particularly dilated, and all these parts distinct the one from the other. But in Asclepiads the whole of the sexual apparatus is consolidated into a single body, the centre of which is occupied by a broad disk-like stigma, and the grains of pollen cohere in the shape of waxy bodies attached finally to the 5 corners of this stigma, to which they adhere by the intervention of peculiar glands.

The Order is one of those which contain indifferently what are called succulent plants and such as are in the usual state of other plants; this excessive development of the cellular tissue of the stem, and reduction of that of the leaves, occurs in its greatest degree in Stapelia and Ceropegia; is diminished in Dischidia, the succulence of which is confined to the leaves; and almost disappears in Hoya, the stem of which is in the usual state, but the leaves between fleshy and leathery.

Fig. CCCCXX.

It has already been stated, under the Order of Dogbanes, that the resemblances found between that Order and Asclepiads seemed to be one of analogy rather than of real affinity; for the economy of the flowers and seeds in the two Orders are widely different. The amygdaloid embryo of Asclepiads, with hardly a trace of albumen, is entirely different from that of Dogbanes, which is very small, and furnished most abundantly with albumen. The anthers and stigma of Dogbanes form no organic union, but they grow into one solid central mass in the Asclepiads, whence proceed other physiological and structural peculiarities.

Other Botanists do not attach the same importance to these circumstances, and continue to associate the two Orders, adopting the opinion of Brown, who considered that they differed solely in the nature of their stamens and stigma, the stamens of Dogbanes being distinct with powdery pollen, and those of Asclepiads adherent to the table-shaped stigma, the pollen being contained in bags, formed by the separation of the endothecium. And M. Alph. De Candolle has recently taken the same view of the matter.—*Ann. Sc. Nat.* 3. *ser.* 1. 255. He even shows that the distinction found in the stamens and pistil of Dogbanes and Asclepiads is not so positive as it is supposed to be, for there are Asclepiads with stamens free from their very base, and small stigmas, while on the other hand certain Dogbanes have filiform appendages at the end of their anthers, and great glandular stigmas to which the anthers adhere with force. He even thinks that the only precise distinction resides in the pollen, the grains of which are always separate in Dogbanes, always in waxy masses or bags in Asclepiads. The reason why these great Botanists attach small importance to the albumen as a distinction, is doubtless because in certain Dogbanes, such as Cerbera, that secretion is absent, although in the mass of the Order it is most abundant; but it is, I think, evident that the tendency among Dogbanes is to form albumen in abundance, and that no such tendency exists among Asclepiads.

Africa must be considered as the great field of Asclepiads, especially its southern point, where vast numbers of the succulent species occupy the dry and sterile places of that remarkable country. In tropical India and New Holland, and in all the equinoctial parts of America, they also abound. Two genera only are found in northern latitudes, one of which, Asclepias, has many species, and is confined apparently to North America; the other, Cynanchum, is remarkable for extending from 59° north latitude to 32° south latitude. A Stapelia is found in Sicily.

The roots are generally acrid and stimulating, whence some of them act as emetics, as Tylophora asthmatica and Secamone emetica; others are diaphoretic and sudorific, as the purgative Asclepias decumbens, which has the singular property of exciting general perspiration without increasing in any perceptible degree the heat of the body; it is constantly used in Virginia against pleurisy. Their milk is usually acrid and bitter, and is always to be suspected, although it probably participates in a slight degree only in the poisonous qualities of that of Dogbanes, if we can judge from the use of some species as articles of food. Ceropegia? edulis, Oxystelma esculentum, and two Sarcostemmas, Forskahlianum, and stipitaceum, are all reported to be eatable. The Cow Plant of Ceylon, or Kiriaghuna plant, Gymnema lactiferum, yields a milk of which the Cingalese make use for food; its leaves are also used when boiled. But very little is known about the real qualities of such plants, and as to Oxystelma esculentum, Roxburgh says he did not find that the natives ever eat it, and Dr. Wight makes the same statement; adding, however, that in decoction it is used as a gargle for aphthous affections of the mouth and fauces. The root and tender stalks of Hoya viridiflora sicken and excite expectoration. Asclepias tuberosa, or Butterfly-weed, is a popular remedy in the United States for a variety of disorders; its properties seem to be those of a mild cathartic, and of a certain diaphoretic attended with no inconsiderable expectorant effect. A decoction of Asclepias curassavica, or Wild Ipecacuanha of the West Indies is used by the Negroes as an emetic and purgative, and is said to be

Fig. CCCCXX.—Stapelia.

efficacious in gleets and fluor albus. The roots of Tylophora asthmatica are acrid, and used on the coast of Coromandel as a substitute for Ipecacuanha. Dr. Roxburgh found it to answer the same purpose as that drug, and had also very favourable reports of it from others. Dr. J. Anderson, Physician-General at Madras, confirms this; it was used with great success in a dysentery that was in his time epidemic in the British camp. No doubt it is one of the most valuable medicines in India. In large doses it is emetic; in smaller doses often repeated it acts as a cathartic. Burnett states it to be valuable as a sudorific, and to be peculiarly beneficial in humoral asthma. Similar qualities are possessed by Sarcostemma glaucum, the Ipecacuanha of Venezuela. The Cynanchum acutum and Vincetoxicum officinale are both drastics; the former produces a drug called Montpellier Scammony. The milk of Periploca græca is very acrid, and has been employed by Orientals as a wolf-poison; Gonolobus macrophyllus is reputed to have furnished the North American Indians with a juice to poison their arrows. The root and bark, and especially the inspissated milk, of Calotropis gigantea, the Akund, Yercum, or Mudar plant of India, is a powerful alterative and purgative; it is especially in cases of leprosy, elephantiasis, intestinal worms, and venereal affections that it has been found important. The leaves of Solenostemma Argel are used in Egypt for adulterating Senna, but whether intentionally or from mere carelessness is uncertain. They form a large proportion of some samples of Alexandrian (Nubian) Senna; but they are more bitter than those of Senna, and according to Guibourt are unsafe to administer, in consequence of their irritating properties. It is said that Gomphocarpus fruticosus, also called Argel or Arghel in Syria, is employed for the same purpose. The roots of Hemidesmus indicus, from which Mr. Garden obtained Smilasperic acid, are largely employed in India as a substitute for Sarsaparilla. Its diuretic effect is remarkable. It also acts as a diaphoretic and tonic. An account of the Hemidesmus has been published by Mr. Bell, *Pharm. Journ.* 3. 239. It is administered in the form of a syrup; but an infusion and decoction have been used, the proportions being the same as those adopted in the decoction of Sarsaparilla; viz. two ounces of the root to a pint of water. It is more than probable that caoutchouc is contained in several, for Cynanchum ? ovalifolium, according to Wallich, yields excellent caoutchouc at Penang; the tenacity of some species may be owing to its presence, as of Marsdenia tenacissima, employed for bowstrings by the mountaineers of Rajmahl; the fibre of this plant, and of Urtica tenacissima, was the strongest Roxburgh ever met with. Orthanthera viminea, attaining a height of 10 feet, is also remarkable for the length and tenacity of its fibre. Some species yield indigo of excellent quality, as Marsdenia tinctoria, found in Sylhet, and Gymnema ? tingens. See *Royle's Illustrations*, p. 274, for much more interesting matter connected with the sensible properties of plants of this Order, and especially of the Mudar.

GENERA.

I. PERIPLOCEÆ.

Cryptostegia, *R. Br.*
Zucchellia, *Dec.*
Tacazzea, *Dec.*
Æchmolepis, *Dec.*
Gymnanthera, *R. Br.*
Camptocarpus, *Dec.*
Finlaysonia, *Wall.*
Hemidesmus, *R. Br.*
Brachylepis, *Wight et A.*
Decalepis, *Wight et Arn.*
Streptocaulon, *Wigt. et A.*
Harpanema, *Dec.*
Atherandra, *Dec.*
Phyllanthera, *Blume.*
Lepistoma, *Blume.*
  Leposma, Blum.
Periploca, *L.*
Campelepis, *Falc.*
Myriopteron, *Griff.*
Pentoretia, *Dec.*
Ectadium, *E. Mey.*

II. SECAMONEÆ.

Secamone, *R. Br.*
Goniostemma, *Wight.*
Toxocarpus, *Wight et A.*

III. ASCLEPIADEÆ VERÆ.

Mitostigma, *Dec.*
Astephanus, *R. Br.*
Hæmax, *E. Mey.*
Hemipogon, *Dec.*
Nautonia, *Dec.*
Steinheillia, *Dec.*
Microloma, *R. Br.*
Metaplexis, *R. Br.*
  Urostelma, Bunge.
Parapodium, *E. Mey.*
Barjonia, *Dec.*
Pycnostelma, *Dec.*
Menastelma, *R. Br.*
Roulinia, *Dec.*
Enslenia, *Nutt.*
  Ampelanus, Raf.
Cordylogne, *E. Mey.*
Xysmalobium, *R. Br.*
Odontanthera, *Wight.*
Periglossum, *Dec.*
Glossostephanus, *E. Mey.*
Podostigma, *Ell.*
  Stylandra, Nutt.
  Anantherix, Nutt.
Acerates, *Ell.*
  Polyothus, Nutt.
Vincetoxicum, *Mœnch.*
  Pentagonium, Schauer.
  Blyttia, W. Arn.
Haplostemma, *Endl.*
Oncinema, *W. Arn.*
Orthosia, *Dec.*
Cynoctonum, *E. Mey.*
  Bunburia, Harv.
Pycnoneurum, *Dec.*
Holostemma, *R. Br.*
Solenostemma, *Hayn.*
  Argelia, Dec.
Arauja, *Brot.*
  Physianthus, Mart.
Schubertia, *Mart. et Zuc.*
Calotropis, *R. Br.*
Eutropis, *Falc.*
Pentatropis, *R. Br.*
Kanahia, *R. Br.*
Sarcostemma, *R. Br.*
Oxystelma, *R. Br.*
Dæmia, *R. Br.*
Hockea, *Endl.*
Eustegia, *R. Br.*
Peplonia, *Dec.*
Decanema, *Dec.*
Endotropis, *Endl.*
Cynanchum, *Linn.*
Pentarrhinum, *E. Mey.*
Schizoglossum, *E. Mey.*
Glossonema, *Dec.*
  Conomitra, Fenzl.
Aspidoglossum, *E. Mey.*
Lagarinthus, *E. Mey.*
  Rhinolobium, W. Arn.
Gomphocarpus, *R. Br.*
Asclepias, *L.*
  Apocynum, Tourn.
Ditassa, *R. Br.*
Tassadia, *Dec.*
Calostigma, *Dec.*
Oxypetalum, *R. Br.*
  Gothofreda, Vent.
Schistogyne, *Hook. et Arn.*
Melinia, *Dec.*
  Brachylepis, Hk. et A.
Sonninia, *Rchb.*
  Diplolepis, R. Br.
Morrenia, *Lindl.*
Turrigera, *Dec.*
Rhyssostelma, *Dec.*
Seutera, *Reich.*
  Lyonia, Ell.

IV. GONOLOBEÆ.

Matelea, *Aubl.*
  Hostea, Willd.
Gonolobus, *Michx.*
Ibatia, *Dec.*
Macroscepis, *H. B. K.*
Fischeria, *DC.*
Lachnostoma, *H. B. K.*
  Pherotrichis, Dec.
Polystemma, *Dec.*
Blepharodon, *Dec.*
Nephradenia, *Dec.*
Dictyanthus, *Dec.*
Chthamalia, *Dec.*

V. STAPELIÆ.

Ptycanthera, *Dec.*
Tenaris, *E. Mey.*
Tylophora, *R. Br.*
Hybanthera, *Endl.*

Asterostemma, *Dec.*
Cosmostigma, *Wight.*
Pervillæa, *Dec.*
Marsdenia, *R. Br.*
Sicyocarpus, Boj.
Cionura, Griseb.
Dregea, *E. Mey.*
Pergularia, *L.*
Stephanotis, *Thouars.*
Gymnema, *R. Br.*
Bidaria, *Endl.*
Gongronema, *Endl.*
Sarcolobus, *R. Br.*
Trichosandra, *Dec.*
Rhyssolobium, *E. Mey.*
Orthanthera, *Wight.*
Macropetalum, *Burchell.*
Pentasacme, *Wall.*
Leptadenia, *R. Br.*
Barrowia, *Dec.*
Heterostemma, *Wgh.et A.*
Conchophyllum, *Blume.*
Dischidia, *R. Br.*
Colyris, Vahl.
Leptostemma, Blume.
Pterostelma, *Wight.*
Physostelma, *Wight.*
Centrostemma, *Dec.*
Cyrtoceras, Bennet.
Cystidianthus, *Hassk.*
Hoya, *R. Br.*
Sperlingia, Vahl.
Pterygocarpus, Hochs.
Riochreuxia, *Dec.*
Ceropegia, *L.*
Anisotoma, *Fenzl.*
Eriopetalum, *Wight.*
Brachystelma, *R. Br.*
Caralluma, *R. Br.*
Boucerosia, *Wight.*
Desmidorchis, Ehrenb.
Hutchinia, Wight.
Sisyranthus, *Ern. Mey.*
Apteranthes, *Mik.*
Piaranthus, *R. Br.*
Huernia, *R. Br.*
Stapelia, *L.*
Hoodia, *Sweet.*
Monothylacium, Don.
Scytanthus, Hook.
* * *
Baxtera, *Rchb.*
Harrisonia, Hook.
Apoxyanthera, *Hochst.*
Raphionacme, Harv.

NUMBERS. GEN. 141. Sp. 910.

POSITION.—Convolvulaceæ.—ASCLEPIADACEÆ.—Solanaceæ.
Apocynaceæ.

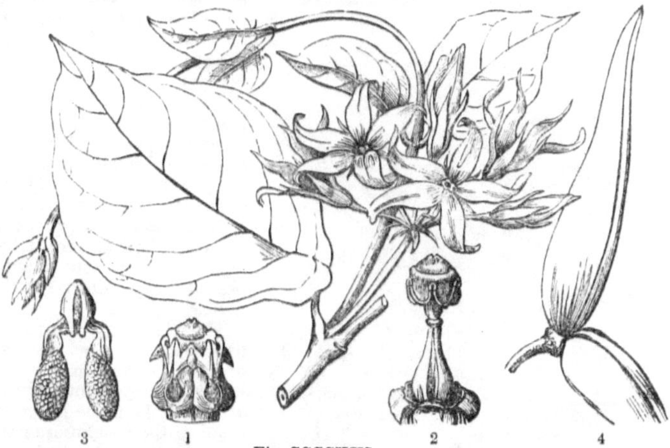

Fig. CCCCXXI.

Fig. CCCCXXI.—Schubertia multiflora. 1. the anthers united to the stigma; 2. the ovary and stigma, from the latter of which the pollen masses have been removed; 3. a pair of pendulous pollen masses, with their gland; 4. the ripe follicles.

ORDER CCXL. CORDIACEÆ.—SEBESTENS.

R. Brown Prodr. 492. (1810); *Martius N. G. et Sp.* 2. 138, *both without a name.*—Cordiaceæ, *Link Handb.* 1. 569. (1829); *Endl. Gen.* cxlii.—Arguziæ, *Link.*—Borragineæ Cordieæ, *Alph. DC. Prodr.* 8. 467.

DIAGNOSIS.—*Solanal Exogens, with 5 free stamens, axile placentæ, aud leafy cotyledons, folded longitudinally.*

Trees. Leaves alternate, scabrous, without stipules, of a hard harsh texture. Flowers panicled, never gyrate, with minute bracts. Calyx inferior, 4- 5-toothed, ribbed in most cases. Corolla monopetalous, 4- 5-cleft, regular, imbricated. Stamens alternate with the segments of the corolla, out of which they arise; anthers versatile. Ovary superior, 4- 8-celled, with 1 pendulous, anatropal ovule in each cell; style continuous; stigma 4- 8-cleft, with recurved segments. Fruit drupaceous, 4- 8-celled; part of the cells frequently abortive. Seed pendulous from the apex of the cells by a long funiculus, upon which it is turned back; embryo inverted, with the cotyledons plaited longitudinally; albumen 0; radicle superior.

Fig. CCCCXXII.

The plaited cotyledons and dichotomous style first led to the separation of this Order from Borageworts, with which it was formerly associated, chiefly, it is to be supposed, on account of the roughness of the leaves. Von Martius remarks, that it is in fact much nearer Bindweeds, from which it differs in its inverted embryo and drupaceous fruit. Nevertheless, M. Alph. De Candolle has reverted to the old opinion, and admitted it as the first tribe of his Borragineæ. I confess, however, that it seems to me impossible to admit Sebestens even into the same category as Borageworts, the indispensable peculiarities of which are a gyrate inflorescence, and nucamentaceous fruit, neither of which circumstances occur here.

The species are, for the most part, natives of the tropics of both hemispheres. A few occur in the cooler parts of South America.

The flesh of their fruit is succulent, mucilaginous, and emollient, as is seen in Cordia Myxa and latifolia. They are believed to have been the Persea of Dioscorides. The smell of their nuts when cut is heavy and disagreeable, the taste of the kernels like that of fresh filberts. They are the true Sebestens of the European Materia Medica, but according to Roxburgh, are not used in the Northern Circars of India, for any medicinal purpose. When ripe they are eaten by the natives, and also most greedily by several sorts of birds, being of a sweetish taste. Cordia Rumphii has a brown wood, beautifully veined with black, and smelling of musk. The timber of C. Gerascanthus, called Bois de Chypre, and Spanish Elm, is of some importance in the West Indies. The bark of

Fig. CCCCXXII.—Cordia Sebestena. 1. calyx; 2. pistil; 3. transverse section of embryo.

C. Myxa is a mild tonic, and is used in India for astringent gargles; its root is thought to be laxative. The wood is soft, and of little use except for fuel. It is reckoned one of the best kinds for kindling fire by friction, and is said to have furnished the wood from which the Egýptians constructed their mummy-cases.

GENERA.

Gynaion, *A. DC.*
Cordia, *Plum.*
Borellia, Neck.
Firensia, Neck.
Geraschanthus, *P. Br.*
Cerdana, R. P.
Rhabdocalyx, *A. DC.*
Pilicordia, *A. DC.*
Physoclada, *A. DC.*
Myxa, *Endl.*
Varronia, *DC.*
Sebestena, *Gærtn.*
Diacoria, *Endl.*
? Sacellium, *Kunth.*

NUMBERS. GEN. 11. Sp. 180.

POSITION.——————.—CORDIACEÆ.—Convolvulaceæ.
Boraginaceæ.

ORDER CCXLI. CONVOLVULACEÆ.—BINDWEEDS.

Convolvuli, *Juss. Gen.* 133. (1789).—Convolvulaceæ, *R. Brown, Prodr.* 481. (1810); *Lindl. Synops.* 167. (1829); *Choisy in Mem. Soc. Phys. Genèv.* (1834); *Alph. DC. Prodr.* 9. 323.

DIAGNOSIS.—*Solanal Exogens, with* 5 *free stamens, basal placentæ, and leafy doubled up cotyledons.*

Herbaceous plants or shrubs, usually twining and milky, smooth, or with a simple pubescence, sometimes erect bushes. Leaves alternate, undivided, or lobed, seldom

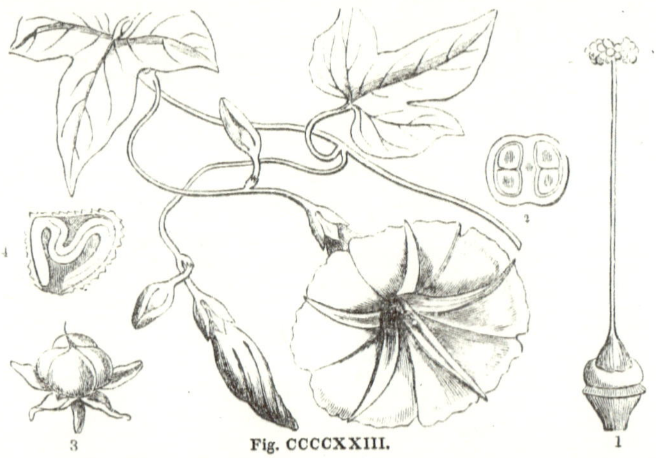

Fig. CCCCXXIII.

pinnatifid, with no stipules. Inflorescence axillary or terminal ; peduncles 1- or many-flowered, the partial ones generally with 2 bracts, which sometimes enlarge greatly after flowering. In Mina the inflorescence is a one-sided and almost scorpioid raceme. Calyx persistent, in 5 divisions, remarkably imbricated, as if in more whorls than one, often very unequal. Corolla monopetalous, hypogynous, regular, deciduous ; the limb 5-lobed, plaited ; the tube without scales. Stamens 5, inserted into the base of the corolla, and alternate with its segments. Ovary simple, with 2 or 4 cells, seldom with 1 ; sometimes in 2 or 4 distinct divisions ; few-seeded ; the ovules definite and erect, when more than 1 collateral ; style 1, usually divided at the top, or as many as the divisions of the ovary, and arising from their base ; stigmas obtuse or acute. Disk annular, hypogynous. Capsule with from 1 to 4 cells, succulent or capsular ; the valves fitting, at their edges, to the angles of a loose dissepiment, bearing the seeds at its base. Seeds with a small quantity of mucilaginous albumen ; embryo curved ; cotyledons leafy, shrivelled ; radicle inferior, next the hilum.

The plaited corolla, imbricated calyx, and climbing habit, are the *primâ facie* marks of this Order, which approaches Sebestens in its shrivelled cotyledons, and through that tribe Borageworts. Mina here, with its almost scorpioid inflorescence, and Nolanads among the Echials, would seem to establish even a more direct relationship between Bindweeds and that Order. Phloxworts are known by their more copious albumen, straight embryo, and loculicidal dehiscence, which in Bindweeds is always opposite the dissepiments. Hydrophyls are characterised by their parietal placentæ, and taper embryo lying in the midst of fleshy albumen. Nightshades have a dicarpellary fruit, with axile placentæ, and numerous seeds ; otherwise, they are sometimes very like the shrubby erect species of Bindweed. The Order has been re-arranged by Choisy in De Candolle's *Prodromus*, but that author has been sharply criticised by Bentham (*London Journal of Botany*, May, 1845, p. 244), and with justice.

Fig. CCCCXXIII.—Ipomœa Batatoides. 1. the pistil and annular disk ; 2. a transverse section of the ovary ; 3. a capsule of Convolvulus tricolor ; 4. a vertical section of the seed of that species.

Very abundant in all parts of the tropics, but rare in cold climates, where a few only are found; they twine round other shrubs, or creep among the weeds of the sea-shore. In the coldest climates they are unknown.

Their roots abound in an acrid milky juice, which is strongly purgative; this quality depends upon a peculiar resin, which is the active principle of Jalap, Scammony, and others whose roots possess similar qualities. Scammony is exclusively furnished by Convolvulus Scammonia, a Syrian perennial; and a similar drastic substance is obtained from Ipomœa tuberosa, the Spanish Arbour Vine of Jamaica, Pharbitis cathartica, a St. Domingo plant, two Brazilian species called by Martius Piptostegia Gomezii and Pisonis, and others. Of Jalap the best sort is obtained from Exogonium Purga, a beautiful twiner with long crimson flowers; but other species are also collected under the same name. Mr. Hartweg ascertained that Ipomœa Batatoides is the Purga Macho or Male Jalap of Mestitlan. Convolvulus Arvensis, Soldanella, maritimus, macrocarpus, and probably many others, may be used with nearly equal advantage. The root of Ipomœa pandurata is employed in the United States as Jalap; its operation is like that of Rhubarb; it is supposed to be also diuretic. The roots of Rhodorhiza florida and scoparia, and Ipomœa Quamoclit, are used as sternutatories; those of Batatas edulis and others are useful articles of food; the former is the common Sweet Potato of European gardens. Convolvulus dissectus abounds in prussic acid, and is one of the plants from which the liqueur Noyau is prepared.—*Bot. Mag.* 3141. The Ipomœa sensitiva of Turpin is remarkable for the irritability of its corolla. A sort of Jalap having the odour of Roses is described by Guibourt in the *Pharm. Journ.* 3. 331. It is not known from what species of this Order it has been obtained. Ipomœa operculata yields a purgative drug, imported into Europe under the name of Gomma da Batata; and a long list might be made of other species whose purgative qualites have been ascertained. Among these, the following deserve principal mention: Ipomœa pandurata, or Mechamek, an American plant; Ipomœa Turpethum, common in the East Indies, Malayan Archipelago, New Holland, Timor, Otaheite, Friendly Islands, Marianne Islands, Tinian, &c.; Convolvulus althæoides, a beautiful Mediterranean plant; and the Calystegias sepium and Soldanella, common in this country. Nevertheless, the purgative resin is hardly present in certain species, where it is replaced by starch or sugar; as in Batatas edulis, the common Sweet Potato, whose root is an important article of food in tropical countries, and Batatas jalapa, which was formerly called Ipomœa macrorhiza, and, notwithstanding the formidable name first quoted, is inert; it is a plant inhabiting the sandy soil of Georgia and Carolina, with white insipid farinaceous roots weighing from 40 to 50 lbs., and is asserted by Elliot (Sketch i. 253.), to possess no purgative properties whatever. Dr. Baldwin assured him that he had administered 6 drachms of the powdered root without effect, and that in fact it contains little or no resin, but like the Batatas edulis consists chiefly of saccharine and farinaceous matter. Of some the seeds partake in the purgative qualities of the roots. The seeds of the Kaladana or Pharbitis cærulea act as a quick, safe, and pleasant cathartic in doses of 30 to 40 grains. In some the leaves are emollient. A decoction of the leaves of Argyreia bracteata is used by the natives of India as a fomentation in cases of scrofulous enlargements of the joints; the boiled leaves being employed as a poultice at the same time. And the foliage of Ipomœa maritima is employed in Brazil in a similar manner. The species of Rhodorhiza yield by distillation an essential oil of a bitter balsamic flavour, called Oil of Rhodium; they are not, however, according to Mr. Barker Webb, the *ροδια ριζα* of Dioscorides, which he thinks was certainly Rhodiola rosea. The wood when powdered has been recommended to promote sneezing, and forms an agreeable snuff; it is valued for fumigation, and when burned diffuses a delightful fragrance. According to a Dr. Uslar, of Oaxaca, the poison called in Mexico Guaco is a Convolvulus.

GENERA.

I. CONVOLVULEÆ. — Carpels consolidated.

Wilsonia, *R. Br.*
Stylisma, *Raf.*
Evolvulus, *Linn.*
Cladostyles, H. B. K.
Meriana, Flor. Flum.
Cressa, *Linn.*
Breweria, *R. Br.*
Seddera, *Steud. et Hoch.*
Dufourea, *Kunth*,
Prevostea, Chois.
Dethardingia, Nees.
Reinwardtia, Spreng.
Calycobolus, Willd.
Bonamia, *Thouars.*
Neuropeltis, *Wall.*
Porana, *Burm.*
Dinetus, Sweet.
Duperreya, *Gaudich.*
Palmia, *Endl.*
Hewittia, Wight.
Shutereia, Chois.
Skinneria, *Chois.*
Polymeria, *Linn.*
Calystegia, *R. Br.*
Aniseia, *Chois.*
Convolvulus, *Linn.*
Merremia, *Dennst.*
Rhodorhiza, *Webb.*
Jacquemontia, *Chois.*
Exogonium, *Chois.*
Lepistemon, *Blum.*
Calonyction, *Chois.*
Bonanox, Raf.
Ipomœa, *L.*
Piptostegia, Hoffms.
Leptocallis, Don.
Elytrostamma, Boj.
Mina, *Llav.*
Quamoclit, *Tourn.*
Calboa, Cav.
Macrostema, Pers.
Morenoa, Llav. et Lex.
Batatas, *Chois.*
Operculina, Manso.
Bombycospermum, Prl.
Spiranthera, Boj.
Pharbitis, *Chois.*
Convolvuloides, Mönch.
Rivea, *Chois.*
Argyreia, *Lour.*

Lettsomia, Roxb.
Ptyxanthus, Don.
Samudra, Endl.
Blinkworthia, *Chois.*
Humbertia, *Commers.*
Thouinia, Smith.
Smithia, Gmel.
Endrachium, Juss.
Endrach, Flacourt.
Moorcroftia, *Chois.*
Maripa, *Aubl.*
Legendrea, *Webb.*
Marcellia, *Mart.*

II. DICHONDREÆ.—Carpels distinct.

Dichondra, *Forst.*
Steripha, Banks et Sol.
Demidofia, Gmel.
? Hygrocharis, *Hochst.*
Falkia, *Linn. fil.*

Doubtful Genera.

Mouroucoa, *Aubl.*
Diplocalymna, *Spreng.*
Calibrachoa, *Llav.*
? Cervia, *Rodrig.*

NUMBERS. GEN. 43. Sp. 660.

Boraginaceæ.
POSITION.—Solanaceæ.—CONVOLVULACEÆ.—Polemoniaceæ.
Nolanaceæ.

Fig. CCCCXXIII. *bis.*

Fig. CCCCXXIII. *bis.*—Ipomœa ficifolia.

ORDER CCXLII. CUSCUTACEÆ.—DODDERS.

Cuscuteæ, *J. S. Presl. Fl. Cech.* 1. 247; *Bartl. Ord. Nat.* 192. (1830); *Endl. Gen. p.* 655.—Cuscutinæ, *Link Handb.* 1. 594. (1829).—Cuscutaceæ, *Ed. Pr.* clxxvi.—Convolvulaceæ, § Cuscuteæ, *Choisy in DC. Prodr.* 8. 452.

DIAGNOSIS.—*Solanal Exogens, with* 5 *free stamens, basal placentæ, and a filiform spiral embryo.*

Leafless climbing colourless parasites, with the flowers in dense clusters. Calyx inferior, persistent, 4-5-parted, with an imbricate æstivation. Corolla persistent, cut round at the base; the limb regular, 4-5-cleft, imbricated in æstivation. Scales alternating with the segments of the corolla, and adhering to them. Stamens equal to the segments of the corolla, and alternate with them; anthers 2-celled, opening longitudinally. Ovary 2-celled; ovules twin, collateral, erect; styles 2 or 0, sometimes connate; stigmas 2. Fruit capsular or baccate, 2-celled; cells 1-2-seeded. Seeds with a fleshy albumen, and a spiral, filiform, acotyledonous embryo; radicle inferior.

These parasites are said by Mr. Griffith to

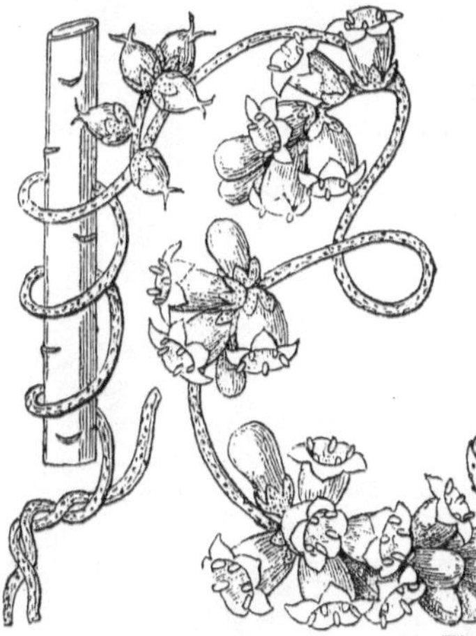

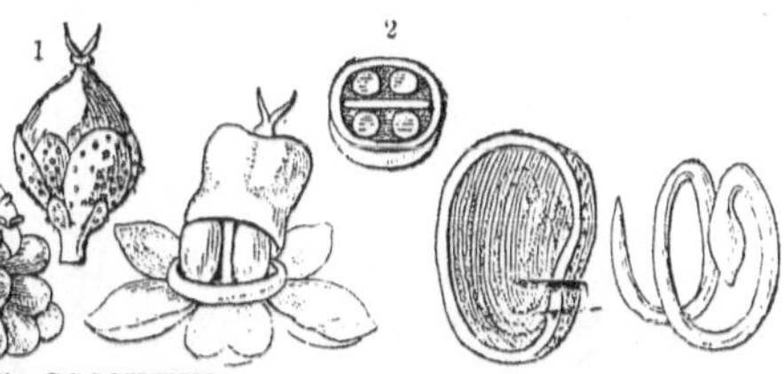

Fig. CCCCXXIV. 3 4 5

differ little from Loranths in their manner of attacking the branches on which they grow; "the suckers stop at the first completely formed wood, never penetrate further, and both the cortical and ligneous systems pass into the stock." There is this difference, however, that Dodders root in the earth in the first instance, and attack the branches of plants at a subsequent period of their existence, at which time they lose their attachment to the soil. Dodders differ from Bindweeds in having a thread-shaped embryo composed almost exclusively of radicle, and twisted spirally in a mass of fleshy albumen. They also have generally, perhaps always, scale-like bodies at the base of the stamens, and apparently alternating with the lobes of the corolla; it is, however, not improbable that these scales are really two-lobed bodies, opposite the petals, and adhering to each other at the edges; if so, they may be regarded as an inner row of stamens. M. Choisy objects altogether to the separation of Dodders from

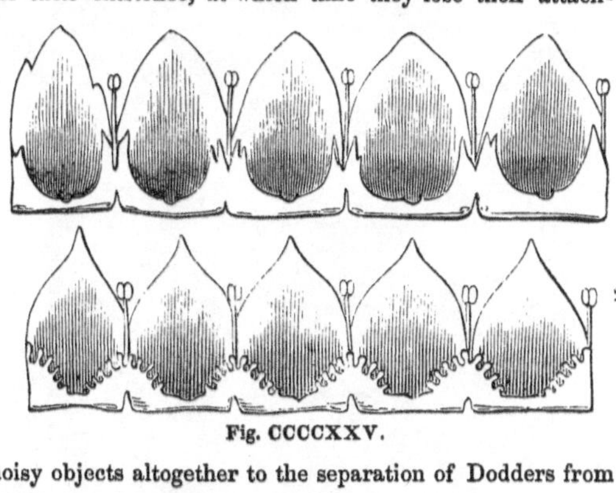

Fig. CCCCXXV.

Fig. CCCCXXIV.—Cuscuta verrucosa. 1. ovary and calyx; 2. section of do.; 3. its fruit (*Hooker*); 4. section of a seed of a Cuscuta; 5. its embryo pulled out.
Fig. CCCCXXV.—Corolla, scales and stamens of, 1. Cuscuta europæa; 2. C. Trifolii.

Bindweeds, but he admits that under the name of Cuscuta are included species with a very variable structure, and which might constitute genera; and he adds that they might have a claim to be regarded as a peculiar Order if as many as 200 species were known, instead of 50.

These parasites are found in the temperate parts of both hemispheres, twining round the branches of plants and sometimes producing great destruction among crops. They do not appear to occur much in the tropics, where their place is perhaps taken by Cassyths. Mr. Griffith speaks of a gigantic species in Affghanistan, which even preys upon itself; one of its masses half covered a Willow tree 20 or 30 feet high.

Their herbage is acrid, and was formerly used as a purgative. Cuscuta racemosa and another or two, called Sipo de Chumbo, are articles of Brazilian pharmacy. The juice of the fresh plant is prescribed in sub-inflammatory complaints, hoarseness, and spitting of blood. The powder of the dried plant is strewed on fresh wounds, the healing of which it is said much to promote.

GENERA.

Cuscuta, *Tourn.*
Grammica, Lour.
Lepidanche, *Engelm.*

NUMBERS. GEN. 2? Sp. 50.

POSITION.——————CUSCUTACEÆ.—Convolvulaceæ.
Cassythaceæ.

ORDER CCXLIII. POLEMONIACEÆ.—PHLOXWORTS.

Polemonia, *Juss. Gen.* 136. (1789).—Polemonideæ, *DC. and Duby*, 329. (1828).—Polemoniaceæ, *Lindl. Synops.* 168. (1829) ; *Bentham in Bot. Reg.* 1622 ; *Endl. Gen.* cxlv. ; *Meisner*, *p.* 273 ; *Alph. DC. Prodr.* 8. 302.—Cobæaceæ, *Don in Edinb. Phil. Journ.* 10. 111. (1824) ; *Link. Handb.* 1. 822. (1829).

DIAGNOSIS.—*Solanal Exogens, with* 5 *free stamens, axile placentæ, and straight plano-convex cotyledons.*

Herbaceous plants, with opposite, or occasionally alternate, compound, or simple leaves ; stem occasionally climbing. Calyx inferior, generally prismatical, 5-parted, persistent, sometimes irregular. Corolla regular, or nearly so, 5-lobed. Stamens 5, inserted into the middle of the tube of the corolla, and alternate with its segments. Ovary superior, 3-celled, with few or many anatropal or amphitropal ovules ; style simple ; stigma trifid. Capsule 3-celled, 3-valved, few or many-seeded, with a loculicidal or septicidal dehiscence ; the valves separating from the axis. Seeds angular or oval, or winged, often enveloped in mucus, in which spiral threads are entangled, ascending ; embryo straight, in the axis of much horny albumen ; radicle inferior, very short ; cotyledons elliptical, foliaceous.

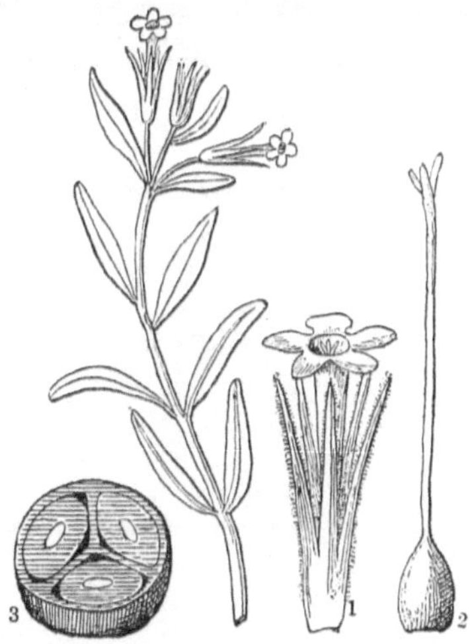

Fig. CCCCXXVI.

The ternary division of the ovary connected with the pentandrous corolla and 5-lobed calyx, bring this Order near Bindweeds, from which the habit, embryo, and corolla, distinguish it, but Cobæa has the habit of a Bindweed without the leaves ; from Gentianworts, to which it also approaches, the 3-celled ovary divides it. To Hydrophyls it approaches very nearly, but the placentation is different ; and therefore, Phloxworts are not placed in the Cortusal Alliance, but on the borders of it; to which the large embryo also persuades us. It is remarkable for the blue colour of the pollen, which is usually of that hue, whatever may be the colour of the corolla. In Collomia linearis I have noticed (in *Botanical Register*, folio 1166) that the dilatation of the mucous matter in which the seeds are enveloped, and which, when they are thrown into water, forms around them like a cloud, depends upon the presence of an infinite multitude of exceedingly delicate and minute spiral vessels, lying coiled up, spire within spire, on the outside of the testa ; when dry, these vessels are confined upon the surface of the seed by its mucus, without being able to manifest themselves ; but the instant water is applied, the mucus dissolves and ceases to counteract the elasticity of the spiral threads, which then dart forward at right angles with the testa, each carrying with it a sheath of mucus, in which it for a long time remains enveloped as if in a membranous case. This singular phenomenon appears to be not uncommon in the Order. Bronnia, one of the plants referred to an imaginary Order, Fouquieraceæ, is very like a Cantua, to which indeed it might have been referred if its calyx were not composed of distinct imbricated sepals. Is it really so ?

Mr. Bentham observes that Phloxworts had, perhaps, be better placed between Borageworts and Nightshades. They are, however, anomalous among Corollifloral Orders by the constancy (unless in accidentally abnormal flowers) of the tricarpellary ovary. They possess the contorted æstivation of Dogbanes and Gentianworts, the axile placentation of Nightshades and Figworts, with the inflorescence also, if not the habit, of some groups in the two latter Orders.—*London Journal of Botany*, 4. 242. I think, however, that the tricarpellary structure of the ovary forbids our placing Phloxworts in any Alli-

Fig. CCCCXXVI.—Collomia gracilis. 1. calyx and corolla ; 2. pistil ; 3. cross section of a ripe seed-vessel.

ance of which either Borageworts, Gentianworts, Dogbanes, or Figworts form a part, and that in reality its most immediate affinities are with Bindweeds on the one hand, and Hydrophyls on the other.

Very abundant in both North and South America, in temperate latitudes, particularly on the north-west side. It is stated by Richardson, that the most northern limit in North America is 54°.—*Edin. Phil. Journ.* 12. 209. In Europe and Asia they are much more uncommon. They are unknown in tropical countries.

The Greek Valerian, Polemonium cæruleum, is a mucilaginous, nauseously bitter plant. In Siberia, poultices are prepared from its leaves, and thought serviceable in syphilitic sores. The Russians fancy that a decoction of it is useful in hydrophobia.

GENERA.

Caldasia, *Willd.*
Bonplandia, Cav.
Phlox, *Linn.*
? *Dupratia*, Raf.
Collomia, *Nutt.*
Gilia, *Ruiz et Pav.*
Collomioides, Endl.
Hügelia, Benth.
Welwitschia, Reichnb.
Curtoisia, Rchb.
Linanthus, *Benth.*
Leptosiphon, *Benth.*
Leptodactylon, Hook.
Dianthoides, *Endl.*
Fenzlia, Benth.
Rossmæsllera, Rchb.
Dactylophyllum, *Benth.*
Ipomopsis, *L. C. Rich.*
Ipomeria, Nutt.
Brickelia, Raf.
Navarretia, *Ruiz et Pav.*
Ægochloa, Benth.
Polemonium, *Tournef.*
Hoitzia, *Juss.*
Lœselia, Linn.
Royena, Houst.
Schizocodon, *Zucc.*
Cantua, *Juss.*
Periphragmos, R. et P.
? Bronnia, *H. B. K.*
Cobæa, *Cav.*
? Cyananthus, *Wall.*

NUMBERS. GEN. 17. Sp. 104.

Hydrophyllaceæ.
POSITION.—Convolvulaceæ.—POLEMONIACEÆ.—Solanaceæ.
Gentianaceæ.

ALLIANCE XLVII. *CORTUSALES.*—THE CORTUSAL ALLIANCE.

DIAGNOSIS.—*Perigynous Exogens, with monodichlamydeous flowers, a free central placenta, and an embryo lying among a large quantity of albumen.*

The Cortusal Alliance is distinctly limited among Perigynous Exogens by its free central placenta, and an embryo lying in the *inside* of the albumen. By this last circumstance it is separated from the Ficoidal Alliance, which has also a free central placenta. There is this other essential difference, that the tendency of Ficoidals is towards a polypetalous or apetalous structure, while that of Cortusals is towards a monopetalous condition. In general, moreover, the albumen of the latter is extremely abundant and hard; but Leadworts have it insignificant in quantity, and mealy, in which latter respect they correspond with Ficoidals.

In a collateral way these plants may be brought in contact with Nightshades through Primworts, with Sapotads and Ebenads through Ardisiads, and with Phloxworts through Hydrophyls, which last offer the best transition to the Echial Alliance.

In Ribworts the placentation is less obviously central than in the other Orders, but it really is so, as is shown in speaking of those plants. In fact, the placenta of Hydrophyls and Ribworts is of quite the same nature.

NATURAL ORDERS OF CORTUSALS.

Stamens alternate with the petals. Styles 2. *Inflorescence circinate*	244. HYDROPHYLLACEÆ.
Stamens opposite the petals. Fruit membranous, one-seeded. Styles 5. *Stem herbaceous*	245. PLUMBAGINACEÆ.
Stamens alternate with the petals. Style 1. *Inflorescence straight.*	246. PLANTAGINACEÆ.
Stamens opposite the petals. Fruit capsular, many-seeded. Style 1. *Stem herbaceous*	247. PRIMULACEÆ.
Stamens opposite the petals. Fruit indehiscent, drupaceous. Style 1. *Stem woody*	248. MYRSINACEÆ.

ORDER CCXLIV. HYDROPHYLLACEÆ.—HYDROPHYLS.

R. Brown. Prodr. 1. 492. (1810), *without a name.*—Hydrophylleæ, *Von Martius N. G. et Sp.* 2. 138. (1828); *Bentham in Linn. Trans.* 17. 267. (1834); *Endl. Gen.* cxlvi.; *Alph. DC. Prodr.* 9. 287.—Hydroleaceæ, *R. Brown Prodr.* 482. (1810), *without a name; Id. in Congo, Kunth in Humb. N. G. et Sp.* 3. 125. (1818); *Bartl. Ord. Nat.* 189; *Choisy Descr. des Hydroleacées (no date); Endl. Gen.* cxlvii.; *Meisner Gen. p.* 272; *A. DC. Prodr.* 564. *Note; Ed. pr. No.* cxxviii.

DIAGNOSIS.—*Cortusal Exogens, with the stamens alternate with the sepals, 2 styles, and a circinate inflorescence.*

Small trees, bushes, or herbaceous plants, often hispid. Leaves often lobed, alternate, or the lower ones opposite. Flowers arranged in gyrate racemes or unilateral spikes, or occasionally solitary and stalked in the axils of the leaves. Calyx inferior, persistent, deeply 5-cleft; the recesses usually augmented with reflexed appendages. Corolla monopetalous, hypogynous, regular, shortly 5-cleft, between campanulate and rotate, rarely funnel-shaped. Stamens 5, epipetalous, alternate with the segments of the corolla, inflected in æstivation; anthers versatile, 2-celled, the cells parallel, dehiscing longitudinally. Ovary superior, simple, 1- 2-celled; styles 2, long; stigmas 2, terminal; placentæ 2, free at their back or united to the shell of the ovary, with two or many amphitropal ovules on their inner face. Fruit capsular, 2-valved, sometimes 1-celled, with a large placenta filling the capsule, sometimes somewhat 2-celled, with the dissepiments incomplete. Seeds reticulated; albumen abundant, cartilaginous; embryo conical, with its radicle next the hilum.

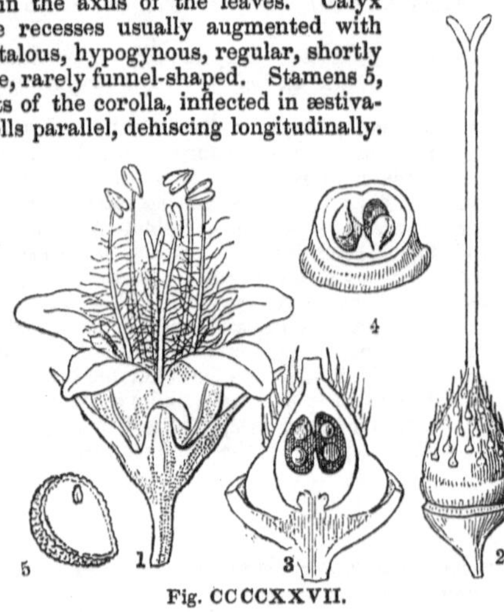

Fig. CCCCXXVII.

The general aspect of these plants is that of Borageworts, which agree in the roughness of their leaves and in their peculiar gyrate, circinate or scorpioid inflorescence. They are, however, known by their undivided 1-celled ovary, terminal style or styles, and ovules (if definite) attached to two stalked fungous placentæ, which arise from the base of the cell, having their ovules on their inner face, or (if indefinite) attached to parietal placentæ. They are further characterised by the presence, in many species, at the base of each lobe of the corolla, of 2 scales or lamellæ, the nature of which is unknown. In general appearance they are also sometimes similar to Polemoniaceæ (Phloxworts). But the large quantity of albumen, the indefinite seeds, the central fungous placentæ, are all circumstances that point to Primworts, with which it seems necessary to associate them; for the minute embryo of all the genera associated with Hydrophyls, and their hard cartilaginous albumen, forbid their being regarded as more than analogous to either Phloxworts or Borageworts. For many years it has been customary to consider these plants distinct from Hydroleaceæ, but I quite agree with M. Alph. De Candolle that there are very slight differences between them. Indeed, upon comparing the distinctions hitherto relied upon, they really amount to this and no more: that in Hydrophyls the ovary is 1-celled, and in Hydroleaceæ 2-celled; but in all cases among these plants the placentæ are a pair of fungous projections from the margin or base of the ovary, and it is their adhesion in various ways that determines whether the cavity has 1 or 2 cells. In the former edition I pointed out the near affinity of the two supposed Orders, and I now unite them, not seeing how even sectional characters can be found for them.

Trees or herbaceous plants, found either in the north or among the most southern of the southern provinces of America; not much known beyond that continent. Nama

Fig. CCCCXXVII.—Hydrophyllum canadense. 1. a flower; 2. a pistil; 3. a perpendicular section of the ovary; 4. a cross section. 5. Section of seed of H. virginianum.—*Gærtner.*

and Hydrolea occur, however, in the East Indies ; Codon at the Cape of Good Hope; and Romanzovia in arctic America.

Some of the species are cultivated in gardens for the sake of their gay flowers ; but none appear to possess useful qualities of any importance. In the United States a decoction of Hydrophyllum canadense is one of the endless remedies for snake-bites, and it is said to be found useful in the cases of erysipelatous eruptions produced by the venomous exhalations of Rhus Toxicodendron. Hydrolea is bitter ; the leaves beaten into pulp and applied as a poultice are in India considered efficacious in cleaning and healing ill-conditioned ulcers.

GENERA.

Hydrophyllum, *Tournef.*
Decemium, Raf.
Ellisia, *Linn.*
Nyctelæa, Scop.
Microgenetes, *A. DC.*
Nemophila, *Bart.*
Eutoca, *R. Br.*
Heteryta, Raf.
Miltitzia, *A. DC.*
Phacelia, *Juss.*
Aldea, Ruiz et Pav.
Endiplus, Raf.
Cosmanthus, *Nolte.*
Emmenanthe, *Benth.*
Hydrolea, *Linn.*
Steris, Burm.
Sagonea, Aubl.
Reichelia, Schreb.
? *Hydrolia* Thouars.
Wigandia, *Kunth.*
Eriodictyon, *Benth.*
Nama, *Linn.*
Romanzovia, *Cham.*
? Codon, *Royen.*
? Lonchostoma, *Wikstr.*

NUMBERS. GEN. 16. Sp. 75.

Polemoniaceæ.
POSITION.—Plumbaginaceæ.—HYDROPHYLLACEÆ.—Primulaceæ.
Boraginaceæ.

Order CCXLV. PLUMBAGINACEÆ.—Leadworts.

Plumbagines, *Juss. Gen.* 92. (1789).—Plumbagineæ, *R. Brown, Prodr.* 425. (1810); *Ebel de Armeriæ Gen. Prodr.*; *Endl. Gen.* cxvii.; *Meisner Gen. p.* 315; *Barnéoud Mémoire, sur les Plumbaginées.*

Diagnosis.—*Cortusal Exogens, with the stamens opposite the petals, membranous one-seeded fruit, 5 styles, and a herbaceous stem.*

Herbaceous plants or under-shrubs, variable in appearance. Leaves alternate or clustered, undivided, somewhat sheathing at the base, but without stipules, sometimes marked with transparent dots. Flowers either loosely panicled, or contracted into heads, flowering irregularly. Calyx tubular, plaited, persistent, sometimes coloured. Corolla of very thin texture, monopetalous, with a narrow angular tube, or of 5 petals, which have a long narrow claw. Stamens definite, opposite the petals, in the monopetalous species hypogynous! in the polypetalous arising from the petals! ovary superior, composed of 5 (or 3 or 4) valvate carpels, 1-celled, 1-seeded; ovule anatropal, pendulous from the point of an umbilical cord, arising from the bottom of the cavity; styles 5! seldom 3 or 4; stigmas the same number. Fruit a nearly indehiscent utricle. Seed inverted, with a rather small quantity of mealy albumen; testa simple; embryo straight; radicle superior.

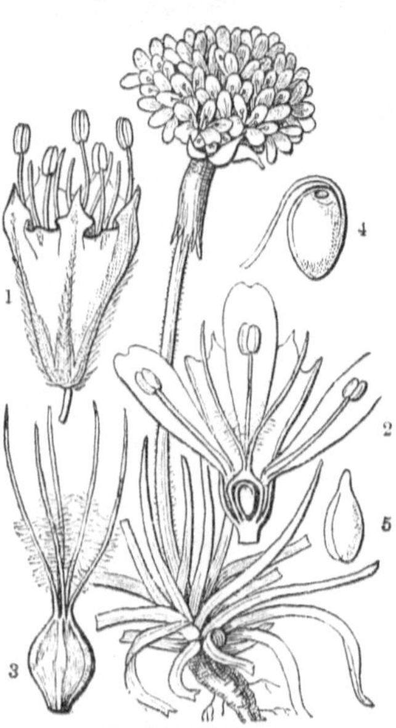

Fig. CCCCXXVIII.

Distinguished from all monopetalous Orders by their plaited calyx and solitary ovule, suspended from the apex of a cord which arises from the base of a 1-celled ovary, with several stigmas. They are nearly related to Primworts, in their habit, if Armeria is compared with Androsace, and as is indicated by the opposition of the stamens to the lobes of the corolla; but they have less albumen and a larger embryo than properly belongs to the Cortusal Alliance, of which they must be looked upon as one of the most outlying Orders. The economy of the ovule is highly curious; before fecundation it is suspended from the apex of a cord, or rather strap, which lies over the foramen or orifice through which the vivifying influence of the pollen has to be introduced; this foramen is presented to the summit of the cell immediately below the origin of the stigmas, but has no communication with that part of the cell, from contact with which it is further cut off by the overlying strap; but as soon as the pollen exercises its influence upon the stigmas, the strap slips aside from above the foramen, which is entered by an extension of the apex of the cell, and thus a direct communication is established between the pollen and the inside of the ovule. This phenomenon is obscurely hinted at by several writers, but was first distinctly shown me by Dr. Brown, and has since been beautifully illustrated by Mirbel, *Nouvelles Recherches sur l'Ovule*, tab. 4. According to Koch, the singular sheath which in Armeria invests the top of the scape, and which Ray supposed to be of the nature of a calyptra, is nothing more than the base of the involucral leaves, in a state of adhesion.

Many are inhabitants of the salt marshes and sea-coasts of the temperate parts of the world, particularly of the basin of the Mediterranean, the southern provinces of the Russian empire, and especially of Affghanistan. The Koollah-i-Huzareh, which forms a large part of the fuel of Cabul, consists of various species of Statice. Others grow from Greenland and the mountains of Europe, to the sterile volcanic regions of Cape Horn. A few are found within the tropics; of these Plumbago zeylanica extends from Ceylon to Port Jackson, and Ægialitis grows among the Mangroves of northern Australasia. Vogelia is from the Cape of Good Hope, and Ceratostigma from China.

Fig. CCCCXXVIII.—Armeria vulgaris. 1. calyx and stamens; 2. section of corolla; 3. pistil; 4. ovule; 5. embryo.

The Order contains plants of very different qualities; part are tonic and astringent, and part acrid and caustic in the highest degree. The root of Statice caroliniana is one of the most powerful astringents in the vegetable materia medica. The bruised fresh bark of the root of Plumbago zeylanica acts as a vesicatory, and is applied in India to buboes in their incipient state. Plumbago europæa is employed by beggars to raise ulcers upon their bodies to excite pity; its root is so acrid that it is used in Roumelia for causing issues, and even as a vesicant.—*Ann. Ch.* 1. 249. But Sauvage de la Croix says that a young woman, who had it applied, affirmed that the pain it occasioned was intolerable, and that she felt as if being flayed alive. Administered internally in small doses it is said to be as effectual an emetic as Ipecacuanha. It is said by Duroques to have been used with considerable advantage in cases of cancer, for which purpose the ulcers were dressed twice daily with olive oil in which the leaves had been infused. Plumbago scandens is called Herbe du Diable in St. Domingo. Its root is said by Martius to be the most active part, and to be a most energetic blistering agent when fresh. It is applied in pains of the ears, and administered internally in hepatic obstructions, &c. Plumbago rosea is usually believed to be the Radix vesicatoria of Rumphius, which being sliced and applied to the skin produces blisters, but less rapidly and effectually than Cantharides. Armeria vulgaris is regarded in Germany as an active diuretic. From two drachms to an ounce of the flowers freshly gathered and quickly dried should be gently boiled and the patient allowed to drink of the decoction *ad libitum.* Some aromatic, as Anise or Cinnamon, is added to the decoction. The remedy appears to cause the excretion of urine in a direct manner.—*Med. Gaz.* xx. 144. As garden plants, nearly the whole Order is much prized for beauty, particularly the Statices, many of which are among the most lovely herbaceous plants we know.

GENERA.

I. STATICEÆ.—Petals and styles distinct.

Armeria, *Willd.*
Statice, *L.*
Eurychiton, Nimmo.
Limonium, Tournef.
Taxanthema, Neck.
Armeriastrum, *Jaub.*
Ægialitis, *R. Br.*

II. PLUMBAGINEÆ.—Petals and styles adherent.

Plumbago, *Tournef.*
? *Thela*, Lour.
Ceratostigma, *Bunge.*
Vogelia, *Lam.*
Valoradia, *Hochst.*

NUMBERS. GEN. 8. Sp. 160.

POSITION.—Primulaceæ.—PLUMBAGINACEÆ.—Plantaginaceæ.

Order CCXLVI. PLANTAGINACEÆ.—Ribworts.

Plantagines, *Juss. Gen.* 89. (1789).—Plantagineæ, *R. Brown Prodr.* 423. (1810); *Endl. Gen.* cxvi.; *Meisner*, p. 315; *Leydolt, die Plantagineen; Barnéoud Recherches sur le Développement, &c., des Plantaginées.*

Diagnosis.—*Cortusal Exogens, with stamens alternate with the petals,* 1 *style, and a straight inflorescence.*

Herbaceous plants, usually stemless, occasionally with a stem. Leaves forming rosettes, or in the caulescent species both alternate and opposite; flat and ribbed or taper and fleshy. Flowers in spikes, rarely solitary, usually ⚥, seldom by abortion ♂ ♀. Calyx imbricated in æstivation, 4-parted, persistent. Corolla membranous, monopetalous, hypogynous, persistent, with a 4-parted limb. Stamens 4, inserted into the corolla, alternately with its segments; filaments filiform, flaccid, doubled inwards in æstivation; anthers versatile, 2-celled. Ovary composed of a single carpel, sessile, without a disk, 2-, very seldom 4-celled, the cells caused by the angles of the placenta; ovules peltate or erect, solitary, twin, or indefinite; style simple, capillary; stigma hispid, simple, rarely half-bifid. Capsule membranous, dehiscing transversely with a loose placenta bearing the seeds on its surface. Seeds sessile, peltate, or erect, solitary, twin, or indefinite; testa mucilaginous; embryo lying across the hilum in the axis of fleshy albumen; radicle remote from the hilum, inferior, or in some cases centrifugal.

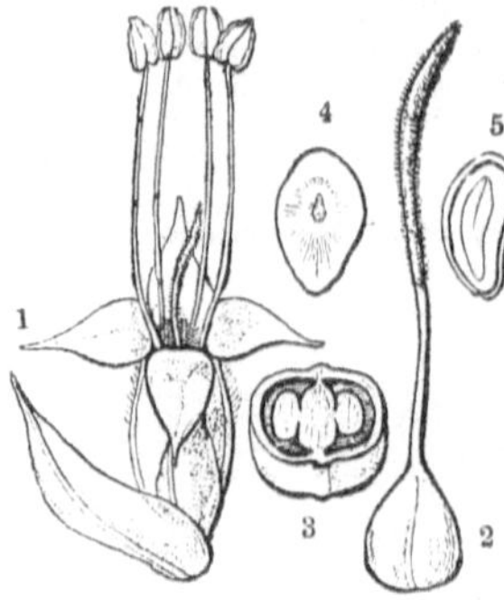

Fig. CCCCXXIX.

This is a group regarding whose affinities the opinions of Botanists are unsettled. By Jussieu it was considered apetalous, and placed near Amaranths and Chenopods, the calyx being called bracts, and the corolla calyx; but this is scarcely an admissible explanation of the structure. I formerly stationed it near Leadworts, to which it must be regarded as nearly allied; but I was certainly wrong in associating it with composite plants and their allies. Don was, I think, the first to suggest that it might be connected with Primworts by means of Glaux, an apetalous genus belonging to that Order. Latterly M. Barnéoud, who has particularly studied the subject, has suggested that the supposed corolla is nothing more than a series of abortive stamens analogous to the membranous cup of Gomphrenas and other Amaranths; and he adopts the opinion of Jussieu that the Orders of Amaranths and Chenopods are those with which Ribworts ought to be associated. In this opinion I cannot concur. There is nothing to distinguish the corolla of Ribworts from the part so called in other plants, except its thinness and want of vascular texture; all corollas must, in a morphological sense, be regarded as barren stamens; and, moreover, the embryo and seed of Ribworts are totally different from anything known in the Chenopodal Alliance. It appears to me that Don's idea was correct, and that upon the whole the Order is a near ally of the Primworts.

The ovary of Plantago does not present distinctly the appearance of a free central placenta. But in reality the placenta is at first quite free, although eventually it is pressed close to the sides of the ovary, and thus divides the cavity into 2 or more cells. This is, however, only a temporary contact, for long before the seeds are ripe the placenta shrinks so much as to lose its adhesion with the sides of the ovary, and then it becomes truly free. In Plantago arborescens it is, when ripe, *continuous with the stigma*, and the two become loose and may be removed together, leaving the sides of the ovary undisturbed.

The tendency to diclinism is very striking in the genus Littorella, and also occurs in Bougueria; it is not, however, perfect, for the male Littorellas have the rudiment of an ovary.

The species are scattered over the whole world, in almost every quarter of which

Fig. CCCCXXIX.—Plantago lanceolata. 1. flower and bract; 2. pistil; 3. ovary cut across; 4. seed; 5. section of it.

they are found in one situation or another. They are chiefly, however, produced in cool or temperate latitudes.

Their herbage is slightly bitter and astringent, and they have even been reckoned febrifuges. Their seeds are covered with mucus. According to De Candolle, those of Plantago arenaria are exported in considerable quantities from Nismes and Montpellier to the north of Europe, and are supposed to be consumed in the completion of the manufacture of muslins. The seeds of Plantago Ispaghula are of a very cooling nature, and, like those of Plantago Psyllium, form, with boiling water, a rich mucilage, which is much used in India in catarrh, gonorrhœa, and nephritic affections. The seeds of P. Psyllium, arenaria, and Cynops, have been made into demulcent drinks, as a good substitute for Linseed or Marshmallows. P. Coronopus, formerly thought an antidote to hydrophobia, is said to be a diuretic. Soda is obtained in Egypt from the ashes of P. squarrosa.

GENERA.

Littorella, *L.*
Bougueria, *Decaisne.*
Plantago, *L.*
Psyllium, Tournef.
Coronopus, Tournef.
Arnoglosson, Endl.

NUMBERS. GEN. 3. Sp. 120.

POSITION.—Plumbaginaceæ.—PLANTAGINACEÆ.—Primulaceæ.
Amarantaceæ.

ORDER CCXLVII. PRIMULACEÆ.—PRIMWORTS.

Lysimachiæ, *Juss. Gen.* 95. (1789).—Primulaceæ, *Vent. Tabl.* 2. 285. (1799); *R. Brown Prodr.* 427; *A. de St. Hilaire, Ann. Sc. Nat. n. s. v.* 30. xi. 85.; *Endl. Gen.* clvi.; *Meisner Gen. p.* 254; *DC. Prodr.* 8. 33; *Duby in Mém. Soc. Phys. Genèv.* 10. 395.—Anagalleidæ, *Baudo in Ann. Sc. Nat. n. s.* xx. 344. (1843).

DIAGNOSIS.—*Cortusal Exogens, with stamens opposite the petals, a capsular many-seeded fruit, 1 style, and a herbaceous stem.*

Annual or perennial herbaceous plants, sometimes almost shrubby. Leaves usually radical; otherwise both whorled and opposite or alternate. Stipules 0. Flowers either on radical scapes and in umbels, or variously arranged in the axils of the leaves. Calyx

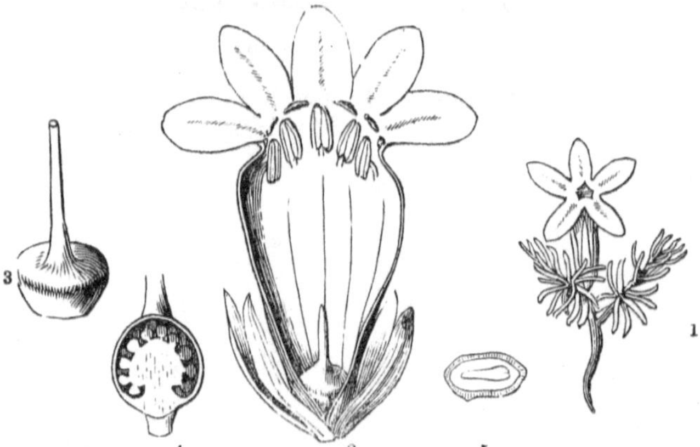

Fig. CCCCXXX.

5-cleft, seldom 4-cleft, inferior, or half superior, regular, persistent. Corolla monopetalous, hypogynous, regular; the limb 5-cleft, seldom 4-cleft; very rarely 0. Stamens inserted upon the corolla, equal in number to its segments, and opposite them. Ovary 1-celled; style 1; stigma capitate; ovules usually amphitropal, rarely anatropal. Capsule opening with valves; placenta central, distinct. Seeds numerous, peltate; embryo included within fleshy albumen, and lying across the hilum; radicle with no determinate direction.

The monopetalous corolla having the stamens opposite its lobes, the composite nature of the ovary, whose placenta is free and central, and the position of the embryo across the hilum, afford ample means for recognising the Order of Primworts, unless they are compared with Ardisiads; from which it is so impossible to distinguish them by any very good character, that Mr. Bentham has proposed to unite the two Orders, adding that, in fact, Primula and Myrsine are not more different than Viola and Alsodeia. Nevertheless, all systematic authors distinguish them, chiefly, as it seems, because no good transition can be found from the herbaceous growth of Primworts to the arborescent or woody structure of Ardisiads. Dunal, one of the later writers on the subject, says that Primworts are, 1, capsular; 2, have seeds placed on the surface of the placenta; 3, are herbs; and 4, are uniformly ⚥; while on the other hand Ardisiads are, 1, drupaceous; 2, have seeds sunk in sockets of the placenta; 3, are woody; and 4, are very frequently polygamous; and these are doubtless the best distinctions that can be found. But a somewhat succulent fruit occurs in Lubinia, and a partially alveolate placenta in

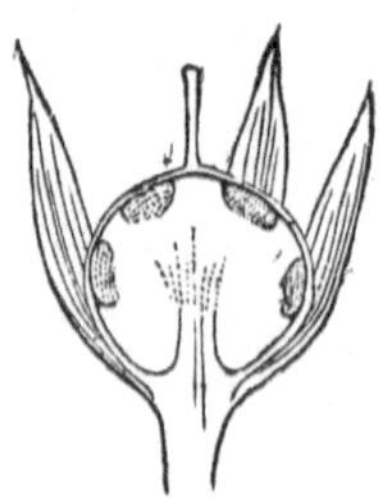

Fig. CCCCXXXI.

Fig. CCCCXXX.—1. Aretia Vitaliana; 2. a flower cut open; 3. the pistil; 4. a vertical section of the latter, showing the free central placenta; 5. a section of a seed.
Fig. CCCCXXXI.—Section of half-ripe fruit of Anagallis arvensis.

Coris and Anagallis, all genera of Primworts; so that the distinctive characters cannot be absolutely relied upon. I think, however, that the two Orders really are different, and that Coris and Lubinia are but the usual instances of loss of character, such as are to be found in almost all Natural Orders. M. Duby also points out a relation to Purslaneworts, found in a supposed genus of that Order called Cypselea; but that plant is certainly no Purslanewort, for it has 5 sepals; and the resemblance traced between it and Primworts is very indistinct. The collateral affinities of Primworts are much greater with Nightshades and Diapensiads, to both which they are similar in habit.

Fig. CCCCXXXII.

Many cases of anomalous structure occur among these plants. Samolus is remarkable for having an inferior ovary, and barren stamens alternating with the lobes of the corolla. Similar stamens are present in Lysimachia ciliata, hybrida, and others. Apochoris and Pelletiera have the petals distinct, and they are hardly united in even Asterolinon and Naumburgia. Glaux is remarkable for being absolutely apetalous. A frequent peculiarity among the genera is to have that kind of fruit which Botanists call a Pyxis.

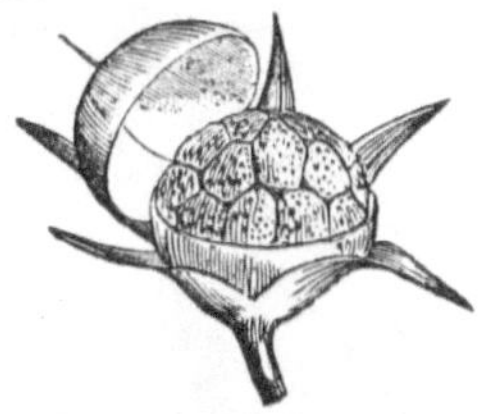

Fig. CCCCXXXIII.

Common in the northern and colder parts of the globe, growing in marshes, hedges, and groves, by fountains and rivulets, and even among the snow of cloud-clapped mountains. The genus Douglasia was found by the traveller whose name it bears, blossoming while covered with snow, on the Rocky Mountains of America. Primworts are uncommon within the tropics, where they usually occupy either the sea shore or the summits of the most lofty hills. The genus Samolus is common in New Holland.

As beautiful objects of culture, these rank among those which are most highly prized, both on account of their bright but modest-looking flowers, the earliest harbingers of spring, and also for the sake of their fragrance. Some of them have powerful active principles. The flowers of the Cowslip, Primula veris, possess well-marked, sedative and diaphoretic properties, and make a pleasant soporific wine; its fresh root has a smell resembling Anise, and was formerly employed as a tonic nervine, and also as a diuretic. The leaves of Primula Auricula are used in the Alps as a remedy for coughs. Soldanellas are slightly purgative. Samolus Valerandi is bitter. Cyclamens are called Sowbreads, because they are the favourite food of the wild boars of Sicily; yet they are very acrid plants, especially the root, whose acrimony is not much perceived at the first tasting, but soon becomes intolerable.—*Smith.* It has been used medicinally, its action being that of a drastic purgative, and formerly it was much esteemed as an emmenagogue; but whether its reputation was owing to its actual powers or to its placentiform root is doubtful. Sibthorp tells us that the modern Greeks use the bruised root of Cyclamen persicum as a means of driving the Sepia octopodia out of its holes. It is said that these roots, notwithstanding their acridity, are eatable and innoxious when dried or roasted. Anagallis arvensis and cærulea, the Mouron of the French, have had some reputation in cases of madness. They appear to possess energetic powers, for Orfila destroyed a dog by making him swallow 3 drachms of the extract; it was found to have inflamed the mucous membrane of the stomach. A similar result was obtained by Grenier. It has been prescribed in epilepsy and dropsy. Coris monspeliensis was employed in the medicine of the Spanish monastic orders as a most efficacious vulnerary, when dried and reduced to powder.—*Bot. Reg.* vii. 536. It has also been prescribed in syphilitic cases.

GENERA.

I. PRIMULIDÆ.

Douglasia, *Lindl.*
Androsace, *Tournef.*
  Aretia, Linn.
  Andraspis, Duby.
Macrosyphonia, *Duby.*
Gregoria, *Duby.*
  Aretia, Gaudin.
  Vitaliana, Sessl.
Dionysia, *Fenzl.*
Primula, *Linn.*
  Auricula, Endl.
  Aleuritia, Endl.
  Auganthus, Link.
Cortusa, *Linn.*
Cyclamen, *Tournef.*
Dodecatheon, *Linn.*
  Meadia, Catesb.
Soldanella, *Tournef.*
Glaux, *Tournef.*
Pelletiera, *St. Hil.*
Asterolinon, *Link et Hfs.*
Naumburgia, *Mönch.*
  Thyrsanthus, Schrank.
Lysimachia, *Mönch.*
  Eusepale, Baudo.
  Neolysia, Baudo.
  Bernardina, Baudo.
  Alaga, Baudo.
  Palladia, Mönch.
  Lerouxia, Merat.
  Godinella, Lestib.
  Ephemerum, Reichnb.
Apochoris, *Duby.*
Steironema, *Raf.*
Anagzanthe, *Baudo.*
Trientalis, *Linn.*
Coxia, *Endl.*
  Lubinia, Link. et Otto.
Lubinia, *Vent.*

Fig. CCCCXXXII.—Section of the flower of Samolus Valerandi.
Fig. CCCCXXXIII.—Pyxis of Anagallis arvensis.

Coris, *Tournef.*

II. ANAGALLIDÆ.

Centunculus, *Linn.*
Anagallis, *Tournef.*
Euparea, *Banks.*
Jirasekia, Schm.
Stenygra, Baudo.
Micropyxis, *Duby.*

III. HOTTONIDÆ.

Hottonia, *Linn.*
Stratiotes, Vaill.

IV. SAMOLIDÆ.

Samolus, *Tournef.*
Sheffieldia, Forst.
Samodia, *Baudo.*

Little known Genera.

Manaelia, *Bowd.*
Findlaya, *Bowd.*

NUMBERS. GEN. 29. Sp. 215.

Solanaceæ.
POSITION.—Myrsinaceæ.—PRIMULACEÆ.—Plumbaginaceæ.
Diapensiaceæ.

ORDER CCXLVIII. MYRSINACEÆ.—ARDISIADS.

Ophiosperma, *Vent. Jard. Cels.* 86. (1800).—Myrsineæ, *R. Brown, Prodr.* 532. (1810); *Aug. de St. Hilaire, Ann. Sc. Nat. n. s. v.* 193; *Endl. Gen.* clvii.; *Meisner, p.* 252.—Ardisiaceæ, *Juss. Ann. Mus. XV.* 350. (1810); *Bartl. Ord. Nat.* 163; *Alph. DC. in Linn. Trans.* 17. 100.—Myrsinaceæ, *Ed. Pr.* clxx. (1836); *Alph. DC. Prodr.* 8. 75.—Theophrastaceæ, *Alph. DC. Prodr.* 8. 144.

DIAGNOSIS.—*Cortusal Exogens, with stamens opposite the petals, indehiscent drupaceous fruit, and woody stem.*

Trees or shrubs. Leaves alternate, undivided, serrated or entire, coriaceous, smooth; stipules 0; sometimes under-shrubs, with opposite or ternate leaves. Inflorescence in umbels, corymbs, or panicles, axillary, seldom terminal. Flowers small, white or red, often marked with sunken dots or glandular lines. Flowers ⚥ or occasionally ♂ ♀. Calyx 4- or 5-cleft, persistent. Corolla monopetalous, hypogynous, 4-5-cleft, equal. Stamens 4-5, opposite the segments of the corolla, into the bases of which they are inserted; filaments distinct, rarely connate, sometimes wanting, sometimes 5 sterile petaloid alternate ones; anthers attached by their emarginate base, with 2 cells, dehiscing longitudinally. Ovary free, or partially adherent, with a single cell and a free central placenta, in which is immersed a definite or indefinite number of campulitropal ovules; style 1, often very short; stigma lobed or undivided. Fruit fleshy, mostly 1-seeded, sometimes 2- 00-seeded. Seeds angular or roundish, with a hollow hilum and a simple integument; albumen abundant, horny, of the same shape as the seed; embryo taper, usually curved, lying across the hilum when the seed is solitary or inferior, and touching the foramen when the seeds are numerous and lateral; cotyledons short.

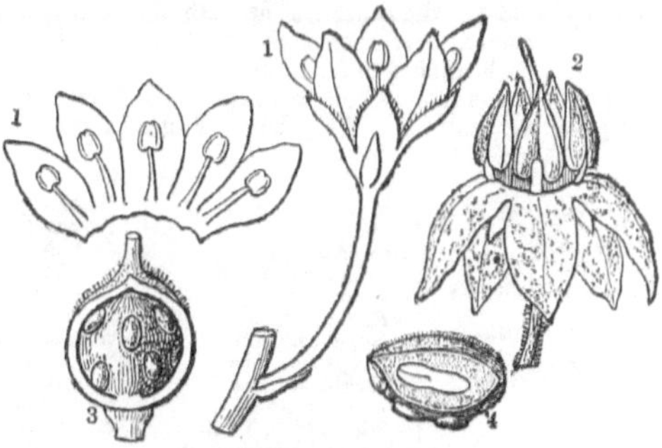

Fig. CCCCXXXIV.

The arborescent habit, fleshy fruit, and socketted placenta are the only circumstances to be relied upon in distinguishing this Order from Primworts, and even the latter is not of great value, as is shown at p. 644. Brown remarks that the Order is related to Sapotads through Jacquinia, and to Primworts through Bladhia. Labisia is a very remarkable plant, with the habit of a Pothos, and an induplicato-valvate corolla. Embelieæ are polypetalous. Mæsa is to other Ardisiads what Samolus is to other Primworts. Mr. Arnott remarks to me that in some genera he finds dots of different shapes mixed as in Samyds.

Ardisiads "are for the most part inhabitants of climates whose temperature is equable, and they particularly abound in insular localities, as the Islands of the Indian Ocean, Mauritius, Bourbon, and Madagascar. Their utmost northern limit in the Old World seems to be the Azores, lat. 39° N., Madeira lat. 32°, and Teneriffe; but in no part of the adjacent continent of Africa do they cross the Northern tropic; in Europe they are entirely wanting, and in Asia extend only to Japan in north lat. 40°. The Order is very rare in N. America, and especially to the northward of Mexico, only one species inhabiting the United States, the M. floridana, *A. De C.*, and that is confined to the southern state, whose name it bears, lat. 30° N. In the southern hemisphere they nowhere (except in New Zealand), are found to the southward of the 36th parallel, and there in S. Brazil only. In Africa they reach the 33rd, and in Australia the 34th. Their extension into the 53rd degree in the South Pacific Ocean, is hence a remarkable circumstance, and probably in some measure to be accounted for by the uniform temperature which the New Zealand Islands possess; further, they there bear a larger proportion to the other dicotyledonous vegetation than they do in any other part of the

Fig. CCCCXXXIV.—1. Mæsa ovata; 2. Ardisia odontophylla; 3. Hymenandra Wallichiana; 4. seed of Mæsa argentea.—*A. DC.*

globe. I have alluded to the Suttonia divaricata having a considerable range in latitude, a circumstance not without parallel in the Order to which it belongs. Of this, Myrsine africana is an extreme instance, that plant being found both at the Cape of Good Hope, in Abyssinia, and in the Azores. The species of the Natural Order are, however, as M. A. De Candolle well remarks (*Linn. Trans.* vol. xvii. p. 99), very confined as regards their geographical limits, Melastomaceæ and Myrtaceæ being two of the very few groups containing about the same or a greater number of species which are more so."—*Jos. Hooker, Bot. of Antarctic Voyage*, p. 52.

Their properties are little known. Many are handsome shrubs, with fine evergreen leaves. Bread is said to be prepared from the pounded seeds of Theophrasta Jussiæi in St. Domingo, where it is called Le Petit Coco. A slight degree of pungency exists in the berries of Embelia Ribes, and some others; cathartic properties are ascribed to those of E. robusta and Myrsine bifaria. The bark of Cybianthus detergens is both gummy and astringent, and is used in baths and as a lotion by the Brazilians, against impetiginous affections. The seeds of Wallenia laurifolia are peppery. According to Mr. Griffith, the fruit of the Reptonia (Edgworthia) buxifolia, or Goorgoora, is commonly sold in the bazaars of Cabul. It is roundish and succulent, about the size of a marble, and is considered heating by the Affghans.—*Ann. Nat. Hist.* x. 193. The leaves and branches of some Jacquinias are said to be poisonous to fish, as is their fruit to man; but this statement requires confirmation. The fruit of Clavija is pleasant to eat; the root emetic. Many have resinous cysts in their wood, fruit, and flowers.

GENERA.

I. MÆSEÆ.

Mæsa, *Forsk.*
Bæobotrys, Forst.
Sibouratia, Thouars.

II. EMBELIEÆ.

Embelia, *Juss.*
Ribesioides, Linn.
Ribes, Burm.
Choripetalum, *A. DC.*
? *Othera*, Thunb.
? Orixa, *Thunb.*

III. ARDISIEÆ.

Oncostemum, *Adr. Juss.*
Amblyanthus, *A. DC.*
Hymenandra, *A. DC.*
Antistrophe, *A. DC.*
Pleiomeris, *A. DC.*
Heberdenia, *Banks.*
Pimelandra, *A. DC.*
Myrsine, *Linn.*
Plotia, Adans.
Rapanea, Aubl.
Manglilla, Juss.
Caballeria, R.P.
Samara, Swartz.
Scleroxylon, Willd.
Athruphyllum, Lour.
Hosta, Fl. Flum.
Peckia, Fl. Flum.
Zacyntha, Fl. Flum.
Suttonia, *A. Rich.*
Labisia, *Lindl.*
Badula, *Juss.*
Barthesia, Comm.
Cephalogyne, A. DC.
Discocalyx, A. DC.
Isostylis, A. DC.
Acephala, A. DC.
Hemigyne, A. DC.
Astrophe, A. DC.
Stylogyne, *A. DC.*
Monoporus, *A. DC.*
Wallenia, *Swartz.*
Petesioides, Jacq.
Conomorpha, *A. DC.*
Conostylus, Pohl.
Weigeltia, *A. DC.*
Cybianthus, *Mart.*
Icacorea, *Aubl.*
Ardisia, *Swartz.*
Pyrgus, Lour.
Niara, Dennst.
Bladhia, Thunb.
Micranthera, A. DC.
Tyrbæa, A. DC.
Pickeringia, Nutt.
? Purkingia, *Presl.*

IV. THEOPHRASTEÆ.

Jacquinia, *Linn.*
Bonellia, Berter.
Theophrasta, *Juss.*
Clavija, *Ruiz et Pav.*
Theophrasta, Linn.
Eresia, Plum.
? Oncinus, *Lour.*
Monotheca, *A. DC.*
Reptonia, *A. DC.*
Edgworthia, Falc.
? Corynocarpus, *Forst.*

NUMBERS. GEN. 30. Sp. 320.

POSITION.—Primulaceæ.—MYRSINACEÆ.—Sapotaceæ. / Ebenaceæ.

ÆGICERACEÆ, *Blume in Ann. Sc. Nat. n. s.* 2. 97. *Alph. De Cand. Prodr.* 8. 141. Under this name is included the genus Ægiceras of Gærtner, a group of shore plants inhabiting the tropics, and rooting out of their seed-vessels into the mud, like Mangroves. It differs in nothing from Ardisiads beyond this, that the fruit, when ripe, becomes a follicle, the seed has no albumen, and the anther-cells are cut transversely; to which Alph. De Candolle adds that the stalk of their central placenta is very much lengthened during the period of ripening, and from being very short is finally converted into a long and false funiculus. It does not, however, appear to me advisable to distinguish the genus from Ardisiads, for it may be conjectured that the absence of albumen, which is one of the most important marks of distinction, is owing to the peculiar circumstances under which Ægiceras germinates; its embryo is always developed in an atmosphere charged with moisture, and hardly requires that any special preparation should be made for sustaining it in its infant state. The only genus known is

Ægiceras, *Gærtn.*
Malaspinæa, Presl.

of which 5 species are on record, whereof one is doubtful.

ALLIANCE XLVIII. *ECHIALES.*—THE ECHIAL ALLIANCE.

DIAGNOSIS.—*Perigynous Exogens, with dichlamydeous, monopetalous, symmetrical, or unsymmetrical flowers; nucamentaceous fruit, consisting of one-seeded nuts, or of clusters of them separate or separable, and a large embryo with little or no albumen.*

About the close affinity of all the Orders here collected into the same Alliance, no one seems to entertain a doubt, with three exceptions, to be alluded to presently. They might in fact be in part referred to Solanals and in part to Echials, but their nut-like, or deeply-lobed fruit, offers a good mark of distinction. Echials may indeed be regarded as a group so intermediate between Solanals and Bignonials, that part might be referred to the one and part to the other, the regular-flowered Orders joining the former, the irregular-flowered the latter. Perhaps it would not be advisable to distinguish the first half from Solanals if it were not for their want of albumen; but this, although not a universal character among them, yet is so common as to show the important tendency.

The Orders least certainly stationed are Jasminworts, Salvadorads, and Brunoniads; but no better position seems discoverable for them. The tendency to irregularity and the deeply-lobed ovary seem to determine the place of Jasminworts, especially when the minute quantity of albumen present in their seeds is taken into account; as to Salvadorads they seem to approach Ehretiads as much as anything; and Brunoniads may be as well compared with the close-headed Borageworts as with Bellworts or Goodeniads, from both which they deviate so entirely in the nature of their fruit.

The true position of the Echial Orders with respect to each other, according to these views, may be represented thus:—

Oleaceæ.—Jasminaceæ.
Salvadoraceæ.
Ehretiaceæ.—Nolanaceæ.—*Solanaceæ.*
Boraginaceæ.—Brunoniaceæ.—*Goodeniaceæ.*
Lamiaceæ.
Verbenaceæ.
Myoporaceæ.
Selaginaceæ.————————*Pedaliaceæ.*

NATURAL ORDERS OF ECHIALS.

* *Regular-flowered Orders, passing from Solanals.*

Flowers regular, $\sqrt[2]{}$, unsymmetrical. Stamens 2. Fruit 2-lobed. Stigma naked . . . . 249. JASMINACEÆ.

Flowers regular, symmetrical. Stamens 4. Fruit simple. Stigma naked . . . . 250. SALVADORACEÆ.

Flowers regular, symmetrical. Stamens 5. Stigma naked. Nuts 4, confluent. Inflorescence circinate . . . . . . 251. EHRETIACEÆ.

Flowers regular, symmetrical. Stamens 5. Nuts 5 or $\sqrt[5]{}$. Stigma naked. Inflorescence straight . . . . 252. NOLANACEÆ.

Flowers regular, symmetrical. Stamens 5. Nuts 4 or $\sqrt[2]{}$. Stigma naked. Inflorescence circinate . . . . 253. BORAGINACEÆ.

Flowers regular, symmetrical. Nut solitary. Stigma indusiate. (Stamens hypogynous!) . . . . 254. BRUNONIACEÆ.

** *Irregular-flowered Orders, passing into Bignonials.*

Flowers irregular, unsymmetrical. Nuts 4. Ovule erect . . 255. LAMIACEÆ.

Flowers irregular, unsymmetrical. Nuts confluent. Ovules erect. 256. VERBENACEÆ.

Flowers irregular, unsymmetrical. Nuts confluent. Ovules pendulous. Anthers 2-celled . . . . 257. MYOPORACEÆ.

Flowers irregular, unsymmetrical. Nuts confluent. Ovules pendulous. Anthers 1-celled . . . . 258. SELAGINACEÆ.

ORDER CCXLIX. JASMINACEÆ.—JASMINWORTS.

Jasmineæ, *Juss. Gen. Plant.* 104. (1789) *in part; R. Brown Prodr.* 520; *Endl. Gen.* cxxix.; *Alph. DC. Prodr.* 8. 300.—Bolivarieæ, *Griseb. Gent.* 20.; *Endl. Gen. Suppl.* 2. 55.

DIAGNOSIS.—*Echial Exogens, with 2 distinct lobes to the fruit, 2 stamens, a naked stigma, and regular unsymmetrical flowers.*

Shrubs, often having twining stems. Leaves opposite or alternate, mostly compound, ternate or pinnate, with an odd one; sometimes simple, the petiole almost always having an articulation. Flowers opposite, in corymbs, white or yellow, often sweet-scented. Calyx with 5 to 8 divisions or teeth, persistent. Corolla monopetalous, hypogynous, regular, hypocrateriform, with from 5 to 8 divisions, which lie laterally upon each other, and are twisted or valvate in æstivation. Stamens 2, arising from the corolla, inclosed within its tube. Ovary destitute of an hypogynous disk, 2-celled, 2-lobed, with from 1 to 4 erect anatropal ovules in each cell; style 1; stigma 2-lobed. Fruit either a double berry or capsule. Seeds either with no albumen or very little, their skin tumid or membranous; embryo straight; radicle inferior.

Fig. CCCCXXXV.

Jasminworts were formerly combined with Oliveworts, from which Brown distinguished them by their ovules being erect, their seed with no, or very little, albumen, by the æstivation of the corolla being imbricate, not valvate, and by the number of its divisions being five or more, and consequently not regularly a multiple of the stamens, instead of 4, which is a multiple of them. Ach. Richard endeavours to show that these differences are insufficient. He states, that the ovules of Jasminworts are originally pendulous, as in Oliveworts, but that they subsequently become erect in consequence of the growth of the ovary, whose apex does not elongate, while its sides extend considerably during the growth of the fruit. He says, upon the authority of his father, that albumen does exist in Jasminum and Nyctanthes; a fact which had been previously mentioned by Brown in defining the Orders, but to which that distinguished Botanist attached no importance, because only a small quantity was found by him to exist, while it is abundant in Oliveworts; and he probably conceived, as I certainly do, that it is the difference of its quantity only which gives the albumen value as a mark of ordinal distinction. But it does not appear to me that Jasminworts have any real connection with Oliveworts; on the contrary, their unsymmetrical flowers and deeply-lobed fruit suggest a very different affinity, and seem to point distinctly to those monopetalous Orders in which the number of stamens is different from that of the divisions of the corolla, such as Labiates and Verbenes, but particularly the latter, which sometimes resemble Jasminworts in their fruit, as happens in Clerodendron. Brown stations them between Pedaliads and Oliveworts; De Candolle between Oliveworts and Loganiads. Endlicher indicates an approach to Dogbanes. To me they seem to be the connecting point between the Cortusal and Echial Alliances, touching the former at Ardisiads and the latter at Verbenes or Labiates.

Chiefly inhabitants of tropical India, in all parts of which they abound. One Jasminum only is mentioned from South America, but there are at least 3 species of Bolivaria on that continent; a few are natives of Africa and the adjoining islands; New Holland contains several; and, finally, 2 extend into the southern climates of Europe.

Of some species the oil produced by the flowers is deliciously fragrant. The genuine

Fig. CCCCXXXV.—Jasminum ligustrifolium. 1. a corolla cut open; 2. vertical section of the ovary; 3. section of a seed of Nyctanthes.—*Gærtner.*

essential oil of Jasmine of the shops is obtained from Jasminum officinale and grandiflorum ; but a similar perfume is also procured from Jasminum Sambac. The leaves of Jasminum undulatum are slightly bitter. The bitter root of Jasminum angustifolium, ground small and mixed with powdered Acorus Calamus root, is considered in India as a valuable external application in cases of ringworm. In the East Indies the tube of the corolla of Nyctanthes Arbor tristis is used as an orange dye. This plant, the Hursinghar of India, scents the gardens with its delicious perfume only during the night, covering the ground in the morning with its short-lived flowers, which being collected like those of the Chumbelee (Jasminum grandiflorum) are strung on threads and worn as necklaces, or entwined in the hair of the native women. The root of J. pubescens is thought to be alexiteric.

GENERA.

Jasminum, *L.* *Mogorium*, Juss. Nyctanthes, *Juss.*	*Scabrita*, L. *Parilium*, Gærtn.	Bolivaria, *Cham.* *Calyptrospermum*, Diet.	Menodora, *H. B. K.* Balangue, *Gærtn.*

NUMBERS. GEN. 5. Sp. 100.

Oleaceæ.
POSITION.—Verbenaceæ.—JASMINACEÆ.—Salvadoraceæ ?
Myrsinaceæ.

ORDER CCL. SALVADORACEÆ.—SALVADORADS.

Salvadoraceæ, *Ed. pr. No.* cxcix. (1836); *Endl. Gen. p.* 349.

DIAGNOSIS.—*Echial Exogens, with regular symmetrical flowers, a solitary fruit, and naked stigma.*

Small trees or shrubs, with the stem slightly tumid at the articulations. Leaves opposite, leathery, entire, very obscurely veined. Flowers minute, in loose panicles. Calyx inferior, 4-leaved, minute. Corolla membranous, monopetalous, 4-parted. Stamens 4, connecting the petals into a monopetalous corolla; anthers round, 2-celled, bursting longitudinally. Ovary superior, 1-celled, with a single sessile stigma; ovule solitary, erect. Pericarp berried; 1-celled, indehiscent. Seed solitary, erect. Embryo amygdaloid, without albumen; cotyledons fleshy, plano-convex, fixed a little below their middle to a long axis, the radicle of which is inclosed within their bases.

Fig. CCCCXXXVI.

By one author referred to Chenopods or Amaranths, notwithstanding its monopetalous corolla and embryo; by another to Ardisiads, notwithstanding the position of its stamens and the structure of ovary and seeds. This plant appears to be in reality the type of a quite distinct Order, the true relation of which I formerly supposed to be with Leadworts and Plantains. With the latter it agrees in the number of the parts of its flower, its membranous corolla, and simple style; with the former more in habit, and especially in the leaves, which are much like those of a Statice. It, however, differs essentially in its polysepalous calyx, amygdaloid embryo, opposite leaves, and berried pericarp. In habit it agrees with Galenia, and this has probably been the cause of its having found its way to Chenopods. It seems however possible, upon the whole, that it should be considered an ally of Ehretiads or Verbenes, having but one carpel and symmetrical tetrandrous flowers.

The species are found in India, Syria, and North Africa.

Salvadora persica, the Mustard-tree of Scripture, as has been demonstrated by Dr. Royle, has a succulent fruit which has a strong aromatic smell, and tastes like Garden-cress. The bark of the root is remarkably acrid; bruised and applied to the skin it soon raises blisters, for which the natives of India often use it. As a stimulant it promises to be a medicine of considerable power. The leaves of S. indica are purgative; the fruit is said to be eatable.

GENUS.

Salvadora, *L.*

NUMBERS. GEN. 1. Sp. 2.

Plumbaginaceæ?
POSITION.—Ehretiaceæ?—SALVADORACEÆ.—Verbenaceæ?
Oleaceæ?

Fig. CCCCXXXVI.—Salvadora persica. 1. flower; 2. a section of its fruit.

ORDER CCLI. EHRETIACEÆ.—EHRETIADS.

Ehretiaceæ, *Martius N. G. et Sp.* 2. 136. (1828); *Martius Conspectus, No.* 126.; *Endl. Gen. p.* 645.—Heliotropiceæ, *Id.* 2. 75. *et* 138. (1828).—Asperifoliæ, b. Heliotropiaceæ, *Id. Conspectus, No.* 118. (1835).—Borragineæ, Ehretieæ, and Heliotropeæ, *Alph. DC. Prodr.* 9. 467.

DIAGNOSIS.—*Echial Exogens, with regular symmetrical flowers,* 5 *stamens,* 4 *confluent nuts, a naked stigma, and circinate inflorescence.*

Trees or shrubs, or herbaceous plants, with a harsh pubescence. Leaves simple, alternate, without stipules. Flowers gyrate. Calyx inferior, 5-parted, imbricated in æstivation. Corolla monopetalous, tubular, with as many segments of its limb as the calyx, with an imbricated æstivation. Stamens alternate with the segments of the corolla, and equal to them in number, arising from the bottom of the tube; anthers innate. Ovary seated in an annular disk, 2- or more-celled; style terminal; stigma simple, 2-lobed; ovules suspended. Fruit drupaceous, with as many seeds as there are true cells of the ovary. Seed suspended, solitary; testa simple, thin; embryo in the midst of thin fleshy albumen, or without any; radicle superior; cotyledons plano-convex.

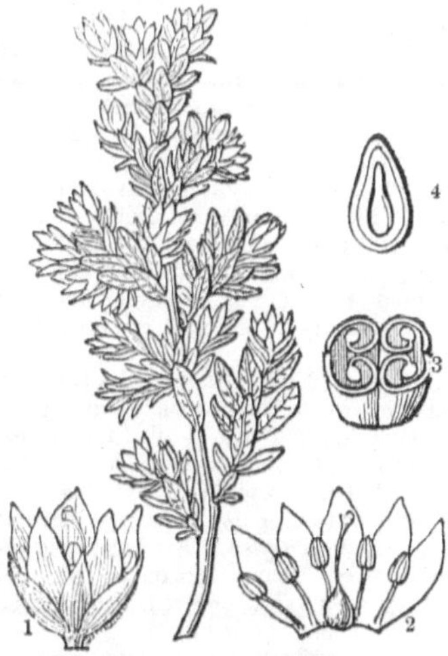

Fig. CCCCXXXVII.

A branch of the old Borragineæ, distinguished by a terminal style proceeding from the apex of a perfectly concrete ovary of 4 cells, a baccate fruit, and seeds furnished with thin fleshy albumen. The Order is re-combined with Borageworts by Alph. De Candolle, but it seems sufficiently characterised by its concrete carpels, and the presence of a small quantity of albumen. The separate, not separable, nuts of Borageworts are so peculiar, notwithstanding that Cerinthe has them combined in pairs, that a real objection seems to exist to the disregard of so good a mark, by the combination with them of these concrete-fruited Ehretiads.

Most of them are tropical trees or shrubs, natives of either hemisphere. A few occur in the south of Europe and the southern States of America; but none appear further to the north than the parallel of 45°.

The root of Ehretia buxifolia is reckoned in India one of those medicines which assist in altering and purifying the habit in cases of cachexia and venereal affections of long standing. Tiaridium indicum is represented to be an astringent, and is used to cleanse ulcers, or to allay inflammation, Martius says with undoubted advantage; Tournefortia umbellata has a similar application in Mexico, where it is even regarded as a febrifuge; and it is to be observed that the leaves of Heliotropium europæum were formerly used in the same way as Tiaridium. Some Ehretias bear eatable drupes. The delicious odour of the Peruvian Heliotrope is known to everybody.

GENERA.

I. TOURNEFORTEÆ.—Seeds with albumen.
Ehretia, *Linn.*
Beurreria, Jacq.
Bouerreria, P. Br.
Carmona, Cav.
Lutrostylis, Don.
Menais, *Læfl.*
Cortesia, *Cav.*
Amerina, *DC.*
Rhabdia, *Mart.*
Tournefortia, *R. Br.*
Messerschmidtia, L.
Arguzia, Amm.
Pittonia, Kunth.
Rotala, *Lour.*
Coldenia, *Linn.*
Tiquilia, Pers.
Halgania, *Gaudich.*

II. HELIOTROPEÆ.—Seeds without albumen.
Schleidenia, *Endl.*
Preslea, Mart.
Heliotropium, *Linn.*
Piptoclaina, Don.
Tiaridium, *Lehm.*
Heliophytum, DC.
Schobera, Scop.
Pentacarya, *DC.*
Euploca, *Nutt.*

NUMBERS. GEN. 14. Sp. 297.

Cordiaceæ.
POSITION.—Boraginaceæ.—EHRETIACEÆ.—Verbenaceæ.

Fig. CCCCXXXVII.—Rhabdia lycioides.—*Martius.* 1. its flower; 2. the corolla opened; 3. a cross section of the ovary; 4. a perpendicular section of a seed.

ORDER CCLII. NOLANACEÆ.—NOLANADS.

Nolanaceæ, *Lindl. Nixus Pl.* 18. (1833); *Martius Conspectus, No* 119; *Endl. Gen. p.* 655; *Lindl. in Bot. Reg.* 1844. *t.* 46.

DIAGNOSIS.—*Echial Exogens, with regular symmetrical flowers,* 5 *stamens,* 5 *or more nuts, distinct or partly confluent, a naked stigma, and straight inflorescence.*

Prostrate or erect, herbaceous or suffruticose plants. Leaves alternate, without stipules. Flowers usually showy. Calyx 5-parted, valvate in æstivation. Corolla monopetalous, with a plaited æstivation, usually thickened in the tube. Stamens 5, equal, inserted into the tube, alternate with the segments of the corolla; anthers oblong, 2-celled, bursting longitudinally. Pistil composed of several carpels, either distinct with a single style, or partially combined into several sets with a single style seated on a succulent disk. Stigma somewhat capitate. Fruit inclosed in the permanent calyx, constructed like the pistil; pericarp woody, often a little succulent; seeds ascending, solitary; embryo curved, with either straight or doubled cotyledons, in the midst of a small quantity of albumen; radicle next the hilum.

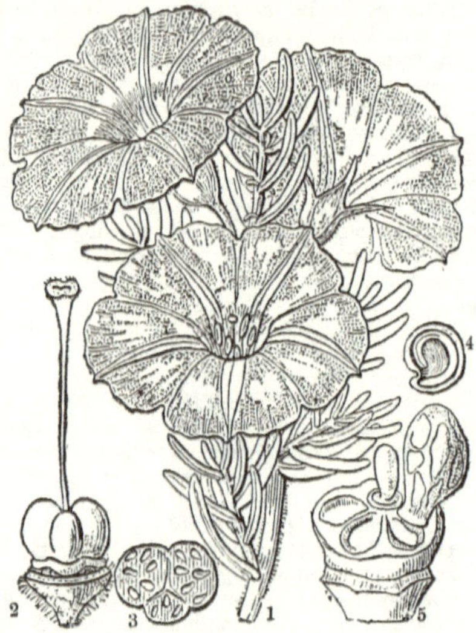

Fig. CCCCXXXVIII.

The genus Nolana, sometimes referred to Borageworts, sometimes to Bindweeds, has been erected into a distinct Order, on account, on the one hand, of its regular plaited corolla and valvate calyx, and, on the other, of its separate carpels though united styles. Among the regular-flowered Echials Nolanads can only be compared to Borageworts, from which they are certainly distinguished by their pentamerous fruit and straight inflorescence. There is some doubt whether the genera Falkia or Dichondra belong to Bindweeds or to Nolanads. With the latter those genera agree in their separate ovaries, with the former in the structure of their embryo; with both they disagree in the entire separation of their styles. If we attend to the embryo, they will stand among Bindweeds; if to the carpels, among Nolanads; but as their separate styles are nearly paralleled by those of Evolvulus and others, it seems upon the whole better to refer them to Bindweeds. Schlechtendahl suggests (*Linnæa*, 7. 72) that Nolana may be referred to Nightshades, on account of its affinity with Grabowskia boerhaaveifolia, in which the fruit contains two bilocular monospermous stones; and it must be confessed that some of the shrubby Nolanads have much the habit of Lycium.

This little Order is remarkable for the various modes in which its carpels are disposed without ever being consolidated. In one genus there are but 5, and they are distinct; in another there are 20 combined in fours; in a third the combination is irregular though the number remains 20; and in others they are all wholly distinct. The late Professor Don thought that Triguera must be referred here.

The species are all South American, and chiefly Chilian.

Their uses are unknown.

GENERA.

Nolana, *Linn.*
Walkeria, Ehret.
Zwingera, Hofer.
Teganium, Schmidt.
Neudorffia, Adans.
Alona, *Lindl.*
Dolia, *Lindl.*
Alibrexia, *Miers.*
Sorema, *Lindl.*
Aplocarya, *Lindl.*

NUMBERS. GEN. 6. Sp. 35.

POSITION.—Boraginaceæ.—NOLANACEÆ.—Convolvulaceæ. / Solanaceæ.

Fig. CCCCXXXVIII.—1. Alona cœlestis; 2. its pistil; 3. a transverse section of it; 4. section of seed of Nolana prostrata; 5. part of the fruit of Aplocarya divaricata.

ORDER CCLIII. BORAGINACEÆ.—BORAGEWORTS.

Boragineæ, *Juss. Gen.* 143. (1789); *R. Brown Prodr.* 492; *Bartl. Ord. Nat.* 196. (1830).—Asperifoliæ, *Linn. Martius Conspectus, No.* 118. (1835); *Endl. Gen.* cxliii.

DIAGNOSIS.—*Echial Exogens, with regular symmetrical flowers, 5 stamens, 4 nuts or two pairs, a naked stigma, and circinate inflorescence.*

Herbaceous plants or shrubs. Stems round. Leaves alternate, often covered with asperities consisting of hairs proceeding from an indurated enlarged base. Flowers in

Fig. CCCCXXXIX.

1-sided gyrate spikes or racemes, or panicles, sometimes solitary and axillary. Calyx persistent, with 4 or 5 divisions. Corolla hypogynous, monopetalous, generally regular, 5-cleft, sometimes 4-cleft, with an imbricated æstivation. Stamens inserted upon the corolla, equal to the number of its lobes and alternate with them. Ovary 4-parted, 4-seeded, or 2-parted, 4-celled; ovules attached to the lowest point of the cavity, amphitropal; style simple, arising from the base of the lobes of the ovary; stigma simple or bifid. Nuts 2 or 4, distinct. Seed separable from the pericarp, destitute of albumen. Embryo with a superior radicle; cotyledons parallel with the axis, plano-convex, sometimes 4 ! in Amsinckia.

The plants of this Order are nearly allied to Labiates, from which they are essentially distinguished by the regularity of the corolla, the presence of 5 fertile stamens, the absence of resinous dots in the foliage, a round (not square) figure of the stem, a gyrate

Fig. CCCCXXXIX.—1. Symphytum officinale; 2. a diagram of its flower; 3. its pistil; 4. the calyx opened, with two of the nuts remaining; 5. a vertical section of a nut.

inflorescence, and scabrous alternate leaves. On account of this last character, they are sometimes called Asperifoliæ. From Nolanads they are distinguished by their inflorescence being gyrate, their radicle superior, and their embryo exalbuminous and straight. From all other Orders of this Alliance they are known by the 4 deep lobes of the ovary, called by Linnæan Botanists naked seeds.

Among the more remarkable points of structure met with in this Order is the very general presence of scales or tubercles, standing on the corolla between the stamens. At first sight such scales might be taken for mere folds of the corolla, but their peculiar appearance in Symphytum and Borage leads to the suspicion that they are really a series of abortive stamens.

Natives principally of the temperate countries of the northern hemisphere; extremely abundant in all the southern parts of Europe, the Levant, and middle Asia; less frequent as we approach the arctic circle, and almost disappearing within the tropics. A few species only are found in such latitudes. In North America they are less abundant than in Europe. Pursh reckons but 22 species in the whole of his Flora; while the little island of Sicily alone contains 35, according to Presl.

Soft, mucilaginous, emollient properties, are the usual characteristics of this Order; some are also said to contain nitre, a proof of which is shown by their frequent decrepitation when thrown on the fire. Borago officinalis gives a coolness to beverage in which its leaves are steeped. The whole plant has an odour approaching to Cucumber and Burnet; but its supposed exhilarating qualities, which caused Borage to be reckoned one of the four cordial flowers, along with Alkanet, Roses, and Violets, may justly be doubted.—*Smith.* It was once esteemed as a pectoral medicine, and a decoction of its leaves mixed with honey makes a good ptisan. Its young leaves make a pickle in some esteem. Echium plantagineum, naturalised in Brazil, is used in that country for the same purpose. The roots of Anchusa tinctoria, or Alkanet, of Lithospermum tinctorium, Onosma echioides, Echium rubrum, and Anchusa virginica, contain a reddish-brown substance used by dyers. This matter is thought to be a peculiar chemical principle, approaching the resins. The species of Trichodesma are considered diuretic, and are one of the cures for snake bites in India.—*Royle.* Some say that Cynoglossum officinale is narcotic; its leaves are bitterish and produce a fat strong-scented oil. Comfrey, Symphytum officinale, was formerly regarded as a vulnerary; if gathered while tender its leaves are a substitute for Spinage, and the young shoots, blanched by being forced to grow through heaps of earth, are eaten like Asparagus; it is not, however, valued by persons of refined taste. Its roots abound in mucilage, and are sweetish with some astringency.

GENERA.

I. ANCHUSIDÆ.

Cerinthe, *Linn.*
Ceranthe, Reichenb.
Onosma, *Linn.*
Onosmodium, *L. C. Rich.*
Osmodium, Raf.
Purshia, Spreng.
Moltkia, *Lehm.*
Echium, *Tournef.*
Macrotomia, *DC.*
Echiochilon, *Desf.*
Chilechium, Raf.
Pulmonaria, *Tournef.*
Bessera, Schult.
Steenhammaria, *Reichb.*
Cerinthoides, Boerh.
Hippoglossum, Hartm.
Mertensia, Roth.
Casselia, Dumort.
? Platynema, *Schrad.*
Lithospermum, *Tourn.*
Rhytispermum, Link.
Ægonychion, Gray.
Batschia, Gmel.
Cyphorima, Raf.
Margarospermum, *Rchb.*
? *Arnebia*, Forsk.
Amsinckia, *Lehm.*
Benthamia, Lindl.
Oreocharis, *Decaisne.*
Gymnoreime, *Decaisne.*
Macromeria, *Don.*
Philonomia, DC.
Craniospermum, *Lehm.*
Colsmannia, *Lehm.*
Nonea, *Medik.*
Lycopsis, Lehm.
Meneghinia, *Endl.*
Dioclea, Spreng.
Strobila, Don.
Stenosolenium, *Turcz.*
Lycopsis, *Linn.*
Caryolopha, *Fisch. et Trtt.*
Pentaglottis, Tausch.
Anchusa, *Linn.*
Buglossum, Tournef.
Baphorhiza, *Link.*
Alkanna, Tausch.
Oscampia, Mönch.
? *Campylocaryum*, DC.
Moritzia, *DC.*
Antiphytum, *DC.*
Plagiobothrys, *Fsch. et M.*
Eritrichium, *Schrad.*
Bothriospermum, *Bung.*
Myosotis, *Linn.*
Echioides, Mönch.
Endogonia, *Turcz.*
Strophiostoma, *Turcz.*
Exarrhena, *R. Br.*
Lobostemon, *Lehm.*
Echiopsis, Reichenb.
Stomatechium, *Lehm.*
Symphytum, *Linn.*
Trachystemon, *Don.*
Psilostemon, DC.
Borago, *Tournef.*

II. CYNOGLOSSIDÆ.

Trichodesma, *R. Br.*
Trichoderma, Link.
Streblanthera, Steud.
Pollichia, Medik.
Friedrichsthalia, *Fenzl.*
Omphalodes, *Tournef.*
Picotia, Röm. et Sch.
Omphalium, Roth.
Rindera, *Pall.*
Mattia, *Schult.*
Solenanthus, *Ledeb.*
Cynoglossum, *Linn.*
Pectocarya, *DC.*
? *Cryptantha*, Lehm.
Ctenospermum, Lehm.
Suchtelenia, *Karel.*
Caccinia, *Sav.*
Anisanthera, Raf.
Asperugo, *Tournef.*
Echinospermum, *Swartz.*
Lappula, Mönch.
Rochelia, Röm. et Sch.
Krynitzkia, *Fisch.*
? Rochelia, *Reichenb.*
Messerschmidtia, Ass.

NUMBERS. GEN. 53. Sp. 600?

Hydrophyllaceæ.
POSITION.—Lamiaceæ.—BORAGINACEÆ.—Nolanaceæ.
Cordiaceæ.

ORDER CCLIV. BRUNONIACEÆ.—BRUNONIADS.

Goodenoviæ, § 2. *R. Brown Prodr.* 589. (1810).—Brunoniaceæ, *Ed. pr.* cxcvii. (1836); *Endl. Gen.* cxxii.; *Meisner*, p. 238.

DIAGNOSIS.—*Echial Exogens, with regular symmetrical flowers, a solitary nut, and indusiate stigma.*

Herbaceous plants, without stems, and with simple glandless hairs. Leaves radical, entire, with no stipules. Flowers on scapes, collected in heads, surrounded by enlarged bracts, blue. Calyx free, in 5 divisions, with bracts at the base. Corolla monopetalous, almost regular, 5-parted, inferior, withering. Stamens definite, hypogynous, alternate with the segments of the corolla; anthers turned inwards, 2-celled, collateral, slightly cohering. Ovary 1-celled, with a single erect anatropal ovule; style single; stigma inclosed in a 2-valved cup. Fruit a membranous utricle inclosed within the hardened tube of the calyx. Seed solitary, erect, without albumen; embryo with plano-convex fleshy cotyledons, and a minute inferior radicle.

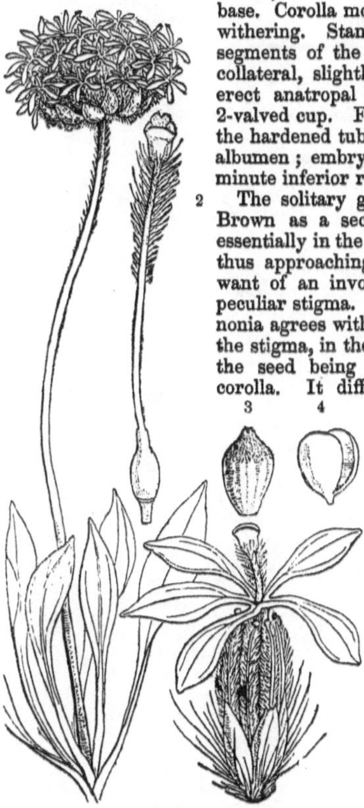

Fig. CCCCXL.

The solitary genus forming this Order was regarded by Brown as a section of Goodeniads, from which it differs essentially in the superior 1-celled ovary and capitate flowers, thus approaching certain Teazelworts, but differing in the want of an involucel, in the erect ovule, superior ovary, and peculiar stigma. With reference to this, Brown says: "Brunonia agrees with Goodenoviæ in the remarkable indusium of the stigma, in the structure and connection of the antheræ, in the seed being erect, and essentially in the æstivation of corolla. It differs from them in having both calyx and corolla distinct from the ovarium, in the disposition of vessels in the corolla, in the filaments being jointed at the top, in the seed being without albumen, and in its remarkable inflorescence, compatible, indeed, with the nature of the irregularity in the corolla of Goodenovia, but which can hardly co-exist with that characterising Lobeliaceæ. With Compositæ it agrees essentially in inflorescence, in the æstivation of the corolla, in the remarkable joint or change of texture in the apex of its filaments, and in the structure of the ovarium and seed. It differs from them in having ovarium liberum or superum, in the want of a glandular disk, in the immediately hypogynous insertion of the filaments, in the indusium of the stigma, and in the vascular structure of the corolla, whose tube has five nerves only, and these continued through the axes of the laciniæ, either terminating simply (as is at least frequently the case in Brunonia sericea), or (as in B. australis) dividing at top into two recurrent branches, forming lateral nerves, at first sight resembling those of Compositæ, but which hardly reach to the base of the laciniæ. It is a curious circumstance that Brunonia should so completely differ from Compositæ in the disposition of vessels of the corolla, while both Orders agree in the no less remarkable structure of the jointed filament; a character which had been observed in a very few Compositæ only, before the publication of M. Cassini's second Dissertation, where it is proved to be nearly universal in the Order. In the opposite parietes of the ovarium of Brunonia two nerves

Fig. CCCCXL.—Brunonia sericea.—*F. Bauer.* 1. a complete flower; 2. the pistil; 3. a ripe fruit; 4. embryo.

or vascular cords are observable, which are continued into the style, where they become approximated and parallel. This structure, so nearly resembling that of Compositæ, seems to strengthen the analogical argument in favour of the hypothesis of the compound nature of the pistillum in that Order, and of its type in phænogamous plants generally; Brunonia having an obvious and near affinity to Goodenoviæ, in the greater part of whose genera the ovarium has actually two cells with one or an indefinite number of ovula in each; while in a few genera of the same Order, as Dampiera, Diaspasis, and certain species of Scævola, it is equally reduced to one cell and a single ovulum." The habit of the Order is much that of Globularia. But its most immediate affinity seems to be collaterally with Nolanads, which it appears to combine with such genera as Phyteuma among Bellworts. Its hypogynous stamens are, however, so peculiar that we may well doubt whether the true affinity of the plant can yet be demonstrated.

Natives of New Holland.

Their properties are unknown.

GENUS.

Brunonia, *Sm.*

NUMBERS. GEN. 1. Sp. 2.

Goodeniaceæ.
POSITION.—Nolanaceæ.—BRUNONIACEÆ.——————
Campanulaceæ.

ORDER CCLV. LAMIACEÆ.—LABIATES.

Labiatæ, *Juss. Gen.* 110. (1789); *R. Brown Prodr.* 499; *Mirbel in Ann. Mus.* 15. 213.; *Bentham in Bot. Reg.* (1829); *Id. Gen. et Sp. Labiatarum*, (1832—1836); *Endl. Gen.* cxxxvi.; *Meisner Gen. p.* 282; *Walpers' Repertorium*, 3. 483.—Oxereæ, *Fenzl.*

DIAGNOSIS.—*Echial Exogens, irregular unsymmetrical flowers, and 4 distinct nuts.*

Herbaceous plants or under-shrubs. Stem 4-cornered, with opposite ramifications. Leaves opposite, divided or undivided, without stipules, replete with receptacles of aromatic oil. Flowers in opposite, nearly sessile, axillary cymes, resembling whorls; sometimes solitary or as if capitate. Calyx tubular, inferior, persistent, the odd tooth being next the axis; regular 5- or 10-toothed, or irregular bilabiate or 3- to 10-toothed. Corolla monopetalous, hypogynous, bilabiate; the upper lip undivided or bifid, overlapping the lower, which is larger and 3-lobed. Stamens 4, didynamous, inserted upon the corolla, alternately with the lobes of the lower lip, the 2 upper sometimes wanting; anthers 2-celled; sometimes apparently unilocular in consequence of the confluence of the cells at the apex; sometimes 1 cell altogether obsolete, or the 2 cells separated by a bifurcation of the connective. Ovary deeply 4-lobed, seated in a fleshy hypogynous disk; the lobes each containing 1 erect ovule; style 1, proceeding from the base of the lobes of the ovary; stigma bifid, usually acute. Fruit 1 to 4 small nuts, inclosed within the persistent calyx. Seeds erect, with little or no albumen; embryo erect; cotyledons flat.

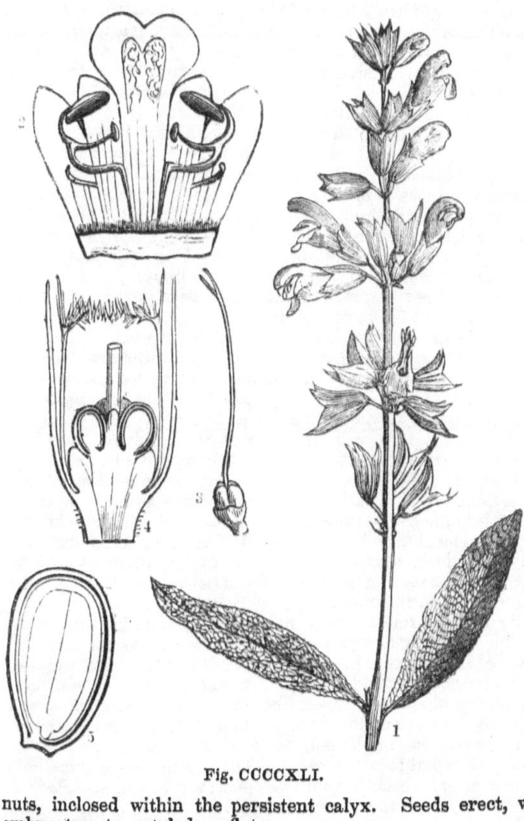

Fig. CCCCXLI.

The 4-lobed ovary, with a solitary style arising from the base of the lobes, has no parallel among monopetalous didynamous Orders. The closest relation of Labiates is with Verbenes, which chiefly differ in their undivided ovary. From Borageworts they differ both in having an irregular corolla, and not more than 2 or 4 stamens, while the lobes of the corolla are 5, square stems and opposite leaves; circumstances in which Labiates resemble some Figworts. From all Borageworts they are known, in the absence of fructification, by their square stem and the numerous reservoirs of oil in their leaves. According to Griselich, these reservoirs are not analogous to those of

Fig. CCCCXLI.—1. Salvia officinalis; 2 its corolla laid open; 3. its pistil; 4. the pistil and lower part of the flower cut open; 5. perpendicular section of a nut.

Oranges and other plants, but are little utricules having an open orifice; and hence he calls them pores. For some good remarks upon the anatomy of the stem of Labiates, see Mirbel in the *Annales du Muséum*, vol. 15. p. 223. The æstivation of the corolla of this Order, first well pointed out by Brown, is an important consideration in determining whether a flower is resupinate or not. Prostanthera is remarkable for the remains of albumen existing in the ripe seeds of several of its species. Oxera of La Billardière has been lately re-examined by M. Fenzl, who finds that its structure was, as I suspected, entirely misunderstood, and that the lobes of its ovary contain each a conical, solitary, fleshy placenta, from the apex of which one ovule hangs down. He refers the plant to Verbenes; but its lobed ovary is an objection, and the pendulous position of the ovules, being due to the unusual extension of their placenta, in consequence of which they cannot be erect, need not, under such circumstances, be taken into account.

Natives of temperate regions, in greater abundance than elsewhere, their maximum probably existing between the parallels of 40° and 50° N. latitude. They are found in abundance in hot, dry, exposed situations, in meadows, hedgerows, and groves; not commonly in marshes. In France they form 1-24th of the Flora; in Germany, 1-26th; in Lapland, 1-40th; the proportion is the same in the United States of North America, and within the tropics of the New World (*Humboldt*); in Sicily they are 1-21 of flowering plants (*Presl.*); in the Balearic islands, 1-19th. About 200 species are mentioned in Wallich's *Catalogue of the Indian Flora*, a large proportion of which is from the northern provinces. They were not found in Melville Island.

Labiates are in all cases destitute of any deleterious secretions; for the most part they are fragrant and aromatic, have been used as tonics, and are valuable as kitchen herbs, for sauces, and flavouring cooked dishes; some are employed by perfumers, many are admired for their beauty, especially species of Sage; some furnish a substance resembling Camphor in its nature; a small number are simply astringent, and a very few are eatable, though perhaps not worth eating, such as the roots of Stachys palustris, which is the Panax Coloni of old writers, and some species related to Ocymum, whose tubers are reported to be a common esculent in Madagascar. Without pretending to make a list of all the uses to which these common plants have been applied, a small number of cases will be found a sufficient indication of them.

Among the mere aromatics the most celebrated is the Patchouli, or Pucha pat, some unknown species of Plectranthus or Coleus, a plant of which large quantities are exported from Penang, for stuffing mattresses and pillows. Its strong-smelling leaves are supposed to keep off contagion.—*Pharm. Journ.* iv. 81. It is used in this country as an article of perfumery. Next to this comes Lavender, the Lavandula vera of De Candolle. The flowers of this plant contain a fragrant volatile oil in great abundance, together with a bitter principle. They are carminative, stimulant, and tonic, but are more employed in perfumery than in medicine; the leaves and flowers have been used as sternutatories. Oil of Lavender is obtained by distillation, and is sometimes given in hysteria and nervous headache; it enters into the composition of Eau de Cologne and the Vinaigre aux quatre voleurs. French Lavender, Lavandula Spica of De Candolle, is less fragrant, and not employed medicinally. It yields what is called Oil of Spike, which is used by painters on porcelain, and in the preparation of varnishes for artists. The oil of Mentha citrata is extremely fragrant, with much the odour of oil of Bergamot.

These fragrant and aromatic qualities render many valuable as stimulating medicines. Mint, for example (Mentha viridis), is not merely used as a sauce, but as an aromatic and carminative, in the form of oil of Spearmint and Spearmint water. Pennyroyal, the Mentha Pulegium, and the Menthas rotundifolia, aquatica, and arvensis have similar qualities, but the most useful among them is Peppermint, an aromatic stimulant, and the most pleasant of all the Mints. It is employed in medicine for several purposes, principally to expel flatus, to cover the unpleasant taste of other medicines, and to relieve nausea and griping pains of the alimentary canal. The volatile oil is sometimes taken as an antispasmodic; it is what gives their flavour to Peppermint lozenges.—*Pereira.* Hedeoma pulegioides, the Pennyroyal of the North Americans, has a great popular reputation as an emmenagogue. Cunila mariana is beneficially employed in infusion in slight fevers and colds, with a view to excite perspiration. Leonotis nepetifolia, Leucas martinicensis, Marsypianthus hyptoides, are all employed in Brazil for medicating baths prescribed for rheumatic attacks. Some are diuretics and diaphoretics, such as Æollanthus suavis, used in Brazil in spasmodic strangury, Glechon spathulatus, Ocymum incanescens, Peltodon radicans, and many kinds of Hyptis. As carminatives and antispasmodics we have all the culinary species, such as Mint (Mentha viridis), Basil (various species of Ocymum), Marjoram (of Origanum), Savories (of Satureia), Lavandula Stæchas, used in Arabia as an antispasmodic, Sage (Salvia officinalis and grandiflora), Meriandra benghalensis, the Sage of Bengal, Thyme, Hyssop, &c. &c.

It is well known that Horehound (Marrubium vulgare) is a popular remedy for coughs, in the form of infusion or of bitter-sweet lozenges; it certainly proves useful in the more severe forms of cold, restoring the tone of the stomach, and subduing irritation when other remedies fail.—*Dr. Kittoe, in the Chemist.* Cunila microcephala is used in the same way in Brazil. As pectoral medicines we find Nepeta Glechoma, or Ground Ivy, which is largely employed by country people in this country, Galeopsis ochroleuca, Leonurus Cardiaca, Balm (Melissa officinalis), and Calamint (Melissa Calamintha), &c., &c. Some are used as febrifuges, among which may be mentioned Ocymum febrifugum of Sierra Leone, Prunella vulgaris or Self-heal, Lycopus europæus, which also yields a good black dye, and is said to help the gypsies to stain their faces, Monarda fistulosa, a fragrant North American herb, and Origanum Dictamnus, or Dittany of Crete. One of the styptic plants, called Matico, is said by Martius to be a species of Phlomis.

A Stearoptine resembling Camphor is to be obtained from various species. Sage, Lavender, Savory, and Hyssop, and Monarda punctata, have been found to yield it, but the plant which is reputed to furnish the most is Rosemary (Rosmarinus officinalis). This plant has a great reputation otherwise; a strong decoction of the foliage is employed to allay the heat of the skin in erysipelas; it has been employed as a cephalic medicine, relieving headache and exciting the mind to vigorous action. It is also remarkable for its undoubted power of encouraging the growth of hair and curing baldness; it is in fact what causes the green colour of the best pomatums used for that purpose; an infusion of it prevents the hair from uncurling in damp weather; it is, moreover, one of the plants employed in the manufacture of Hungary water, the French Vinaigre aux quatre voleurs, and Eau de Cologne. The admired flavour of Narbonne honey is ascribed to the bees feeding on the flowers of this plant.

Betony leaves (Stachys Betonica), when powdered, produce sneezing, but the action is believed to be merely mechanical, and owing to the minute stiff hairs with which they are covered. The statement that the root of the plant is purgative and emetic requires confirmation.

GENERA.

I. OCIMEÆ.

1. Moschosmidæ.

Ocimum, *Linn.*
 Becium, Lindl.
Geniosporum, *Wall.*
 Platostoma, Palis.
Mesona, *Blum.*
Acrocephalus, *Benth.*
Moschosma, *Reich.*
 Lumnitzera, Jacq. fil.
Orthosiphon, *Benth.*
Rabdosia, *Hassk.*

2. Plectranthidæ.

Plectranthus, *Herit.*
 Germanea, Lam.
 ? *Dentidia*, Lour.
 Isodon, Schrad.
Coleus, *Lour.*
 Solenostemon, Schum.
 Mitsa, Chap.
Anisochilus, *Wall.*
Æollanthus, *Mart.*
 Wensea, Wendl.
 Orolanthus, E. Mey.
 Hypothronia, Schrank.
Pycnostachys, *Hook.*
 Echinostachys, E. Mey.
Hoslundia, *Vahl.*
Syncollostemon, *E. Mey.*

3. Hyptidæ.

Peltodon, *Pohl.*
Marsypianthes, *Mart.*
Hyptis, *Jacq.*
 Gymneia, Benth.
 Spicaria, Benth.
 Apodotes, Benth.
 Plagiotis, Benth.
 Cyrta, Benth.
 Cyanocephalus, Pohl.
 Eriosphæria, Benth.
 Oocephalus, Benth.
 Trichosphæria, Benth.
 Xanthiophæa, Mart.
 Rhaphiodon, Schaur.
 Polydesmia, Benth.
 Mesosphæria, Benth.
 Schaueria, Hassk.
 Pectinaria, Benth.
 Brotera, Spreng.
 Minthidium, Benth.
 Buddleioides, Benth.
 Umbellaria, Benth.
 Siagonarrhen, Mart.
 Hypenia, Mart.
Eriope, *Humb. et Bonpl.*

4. Nepetidæ.

Lavandula, *Linn.*
 Stæchas, Tournef.
 Fabricia, Adans.
 Pterostæchas, Ging.
 Chætostachys, Benth.

II. MENTHEÆ.

1. Pogostemidæ.

Pogostemon, *Desf.*
Dysophylla, *Blum.*
 Choteckia, Opitz. et Cord.

2. Elsholtzidæ.

Elsholtzia, *Willd.*
 Aphanochilus, Benth.
 Cyclostegia, Benth.
Tetradenia, *Benth.*

3. Menthidæ.

Colebrookia, *Smith.*
Perilla, *Linn.*
Isanthus, *L. C. Rich.*
Preslia, *Opitz.*
Mentha, *Linn.*
Lycopus, *Linn.*

4. Meriandridæ.

Meriandra, *Benth.*

III. MONARDEÆ.

1. Salvidæ.

Salvia, *Linn.*
 Horminum, Tournef.
 Sclarea, Tournef.
 Æthiopis, Tournef.
 Schraderia, Mönch.
 Jungia, Mönch.
 ? *Leonia*, Llav. et Lex.
Audibertia, *Benth.*

2. Rosmarinidæ.

Rosmarinus, *Linn.*
Monarda, *Linn.*
 Cheilyctis, Raf.
 Coryanthus, Nutt.
Blephilia, *Raf.*
Zizyphora, *Linn.*
Faldermannia, *Bung.*

3. Horminidæ.

Horminum, *Linn.*

IV. SATUREÆ.

1. Origanidæ.

Zataria, *Boiss.*
Bystropogon, *Herit.*
 Minthostachys, Benth.
Pycnanthemum, *Benth.*
 Tullia, Leaven.
 Brachystemum, Rich.
 Kœllia, Mönch.
Monardella, *Benth.*
Amaracus, *Mönch.*
Origanum, *Linn.*
 Majorana, Mönch.
Thymus, *Linn.*
Satureia, *Linn.*

2. Hyssopidæ.

Hyssopus, *Linn.*

3. Cunilidæ.

Collinsonia, *Linn.*
Cunila, *Linn.*

V. MELISSEÆ.

Hedeoma, *Pers.*
 Mosla, Hamilt.
Micromeria, *Benth.*
 Piperella, Presl.
 Xenopoma, Willd.
 Zygis, Desv.
Melissa, *Benth.*
 Calamintha, Mönch.
 Acinos, Mönch.
 Clinopodium, Linn.
Gardoquia, *Ruiz et Pav.*
 Rizoa, Cav.
Glechon, *Spreng.*
Keithia, *Benth.*
Thymbra, *Linn.*
Dicerandra, *Benth.*
 Ceranthera, Ell.
Pogogyne, *Benth.*

VI. SCUTELLAREÆ.

Prunella, *Linn.*
 Brunella, Mönch.
Cleonia, *Linn.*
Scutellaria, *Linn.*
 Cassida, Tournef.
Perilomia, *H. B. K.*

VII. PROSTANTHEREÆ.

Chilodia, *R. Br.*
Cryphia, *R. Br.*
Prostanthera, *Labill.*
Hemiandra, *R. Br.*
Colobandra, *Bartl.*
Hemigenia, *R. Br.*
Atelandra, *Lindl.*
Westringia, *Smith.*
Anisandra, *Bartl.*
Microcorys, *R. Br.*

VIII. NEPETEÆ.

Perovskia, *Karel.*
Lophanthus, *Benth.*
 Vleckia, Raf.

Nepeta, *Benth.*
Saussurea, Mönch.
Cataria, Mönch.
Glechoma, Linn.
Chamæclema, Mönch.
Marmoritis, *Benth.*
Dracocephalum, *Linn.*
Moldavica, Mönch.
Zornia, Mönch.
Ruyschiana, Mill.
Lallemantia, *Fis. et Mey.*
Cedronella, *Mönch.*
Chamæsphacos, *Schrenk.*

IX. STACHEÆ.

1. Melittidæ.

Melittis, *Linn.*
Physostegia, *Benth.*
Macbridea, *Ell.*
Synandra, *Nutt.*

2. Lamidæ.

Wiedemannia, *F. et M.*
Lamium, *Linn.*
Orvala, Linn.
Lamiopsis, Dumort.
Erianthera, Benth.
Galeobdolon, Mönch.
Pollichia, Willd.
Cardiaca, Lam.

Lagochilus *Bung.*
Yermolofia, Belang.
Leonurus, *Linn.*
Cardiaca, Mönch.
Chaiturus, Mönch.
Panzeria, Mönch.
Galeopsis, *Linn.*
Tetrahit, Mönch.
Anisomeles, *R. Br.*
Stachys, *Benth.*
Betonica, Linn.
Eriostachys, Reich.
Eriostomum, Link. et Hoffms.
Campanistrum, Reich.
Trixago, Link et Hoff.
Chamæsideritis, Reich.
Aspasia, E. Mey.
Zietenia, Gled.
Sphacele, *Benth.*
Phytoxys, Mol.
Cuminia, *Colla.*
Lepechinia, *Willd.*

3. Marrubidæ.

Craniotome, *Reich.*
Leucophäe, *Webb.*
Sideritis, *Linn.*
Empedoclea, Raf.
Navicularia, Fabric.

Hesiodia, Mönch.
Burgsdorffia, Mönch.
Acrotome, *Benth.*
Marrubium, *Linn.*
Lagopsis, Bung.

4. Ballotidæ.

Ballota, *Linn.*
Beringeria, Neck.
Pseudodictamnus, Mön
Acanthoprasium, Bth.
Lasiocorys, *Benth.*
Roylea, *Wall.*
Otostegia, *Benth.*
Leucas, *R. Br.*
Hemistoma, Ehrenb.
Leonotis, *Pers.*
Leonurus, Tournef.
Phlomis, *Linn.*
Phlomoides, Mönch.
Phlomidopsis, Link.
Notochæte, *Benth.*
Eremostachys, *Bung.*
Eriophyton, *Benth.*
Moluccella, *Linn.*
Molucca, Tournef.
Chasmone, Presl.
Hymenocrater, *F. et M.*
Holmskioldia, *Retz.*
Hastingia, Smith.
Platunium, Juss.

Achyrospermum, *Blum.*
Siphotoxys, Boj.
Lamprostachys, Boj.
Colquhounia, *Wall.*
Sestinia, *Boiss.*

X. PRASIEÆ.

Gomphostemma, *Wall.*
Phyllostegia, *Benth.*
Stenogyne, *Benth.*
Prasium, *Linn.*

XI. AJUGEÆ.

Amethystea, *Linn.*
Trichostemma, *Linn.*
Teucrium, *Linn.*
Chamædrys, Tournef.
Scorodonia, Tournef.
Scordium, Tournef.
Polium, Tournef.
Leucosceptrum, Smith.
Teucropsis, Ging.
Ajuga, *Linn.*
Bugula, Tournef.
Chamæpitys, Tournef.
Phleboanthe, Tausch.
Cymaria, *Benth.*

XII. OXEREÆ, *Fenzl.*

Oxera, *Labill.*

NUMBERS. GEN. 125. SP. 2350.

POSITION.—Boraginaceæ.—LAMIACEÆ.—Verbenaceæ.
Scrophulariaceæ.

ORDER CCLVI. VERBENACEÆ.—VERBENES.

Vitices, *Juss. Gen.* 106. (1789).—Verbenaceæ, *Juss. in Ann. Mus.* 7. 63. (1806); *R. Brown Prodr.* 510; *Bartl. Ord. Nat.* 179; *Endl. Gen.* cxxvii.; *Meisner Gen. p.* 290.

DIAGNOSIS.—*Echial Exogens, with irregular unsymmetrical flowers, confluent nuts, and erect ovules.*

Trees or shrubs, sometimes herbaceous plants. Leaves generally opposite, simple or compound, without stipules. Flowers in opposite corymbs, or spiked alternately; sometimes in dense heads; very seldom axillary and solitary. Calyx tubular, persistent, inferior. Corolla hypogynous, monopetalous, tubular, deciduous, generally with an irregular limb. The æstivation of both imbricated. Stamens usually 4, didynamous, seldom equal, occasionally 2. Ovary 2- or 4-celled; ovules erect or ascending, anatropal or amphitropal, solitary or twin; style 1; stigma bifid or undivided. Fruit nucamentaceous, sometimes berried, composed of 2 or 4 nucules in a state of adhesion. Seeds erect or ascending; albumen none, or fleshy; embryo always erect; radicle inferior.

1

2 3

Fig. CCCCXLII.

The difference between these plants and Labiates consists in the concrete carpels of Verbenes, their terminal style, and the usual absence of reservoirs of oil from their leaves, as contrasted with the deeply 4-lobed ovary and aromatic leaves of the latter. There are, however, particular species of Labiates which approach Verbenes very closely; so that Brown has remarked that it has been difficult to distinguish the two Orders. Verbenes differ from Myoporads and Selagids in the position of the radicle, which in the former points to the base, and in the two latter to the apex of the fruit. Acanthads and Figworts differ in not being nucamentaceous. Brown states, that although all the genera of Verbenes have an embryo whose radicle points towards the base of the fruit, yet many of them have pendulous seeds, and consequently a radicle remote from the umbilicus. Aug. de St. Hilaire asserts that all, except Avicennia, have a sessile erect ovule arising from the base of each cell. Brown, however, places Avicennia in Myoporads. Possibly Mr. Bentham is right in suggesting that Verbenes and Myoporads are not really distinct, and would be better re-united.

The species of the Order are rare in Europe, northern Asia, and North America; common in the tropics of both hemispheres, and in the temperate districts of South America. In the tropics they become shrubs, or even gigantic timber, but in colder latitudes they are mere herbs.

The properties of Verbenes are much the same as those of Labiates; but they are not of importance in a medicinal or economical point of view. Callicarpa lanata bark has a peculiar subaromatic and slightly bitter taste, and is chewed by the Cingalese when they cannot obtain Betel leaves; the Malays reckon the plant diuretic. Stachytarpheta jamaicensis is a plant to which the Brazilians attach the same false notions of powerful action as Europeans formerly did to the common Vervain. Its leaves are sometimes used to adulterate Chinese Tea, and are sold in the Austrian dominions under the name of Brazilian Tea. The expressed juice of the leaves is given in Tortola, as a cooling purgative to children, in doses of 1 or 2 table-spoonfuls. In the French West India Islands it is employed in decoction for clysters, and also as an anthelmintic. It has, moreover, some reputation for promoting the menstrual discharge. In Brazil the fresh leaves bruised are applied to ulcers; it is then called Urgeráo, or Jarbáo. Auguste de St. Hilaire speaks in terms of high praise of the agreeable properties of the aromatic Lantana pseudo-thea, used in infusion as tea. It is highly esteemed in Brazil, where

Fig. CCCCXLII.—Callicarpa longifolia. 1. flower opened longitudinally (*Hooker*); 2. part of fruit of C. americana (*Gærtner*); 3. its seed.

it is vulgarly called Capitaô do matto, or Cha de pedreste. Martius mentions several other Lantanas whose aromatic leaves and flowers are employed in coughs, and in medicating baths, and for rheumatism. He adds that Lippia citrata is also aromatic, and may be compared to Sage or Thyme. The bark of Vitex Taruma is used in South Brazil, under the name of Taruma, against syphilitic affections. The leaves of Patagonula vulneraria or Ipébranco, are asserted by Martius to be valuable in abating inflammatory action; it would seem to act like Gmelina parviflora, which has the power of rendering water mucilaginous, which is employed as a ptisan for the cure of ardor urinæ. Congea villosa, whose leaves have a strong heavy disagreeable smell, is another plant of the Order, used by the natives of India in fomentations. In India a decoction of the aromatic leaves of Vitex Negundo helps to form the warm bath for women after delivery; bruised they are applied to the temples for headache; pillows stuffed with them are put under the head to remove a catarrh and the headache attending. The leaves of Vitex trifolia are a powerful discutient, and employed by the Malays to remove the boss. The leaves are given in decoction and infusion, and formed into a cataplasm which is applied to the enlarged spleen. The root of Premna integrifolia is cordial and stomachic in decoction. Volkameria inermis, *Linn.*, and some others, have been occasionally employed in medicine, on account of their slightly bitter and subastringent qualities, but they do not appear to be of any importance. As to common Vervain, its virtues, great as their reputation has been, are apparently imaginary. The drupaceous fruits of some species are eatable, as for example those of Lantanas, and Premna esculenta. But others are very acrid. Those of Vitex trifolia are called in India Filfil burree, or Wild Pepper; those of Vitex Negundo resemble them, and Vitex Agnus castus, *Linn.*, has similar acrid fruit. According to Forskahl, the seeds are reputed at Smyrna to be a certain remedy against colic, if powdered and strewed over half an Onion applied to the stomach. By far the most interesting plant, however, belonging to the Order of Verbenes is the Teak, Tectona grandis. This is an enormous tree, with deciduous leaves, covered with rough points. It inhabits the forests of the mountainous parts of Malabar, Pegu, and other districts in the East Indies. Its timber abounds in particles of silex, and has no rival in Asia for durability. With much the appearance of coarse mahogany it is lighter, and very strong. For ship-building it is perhaps the best in the world. Roxburgh says that its wood is the only useful part of it; but Endlicher states that its flowers are diuretic, that its foliage supplies a red dye, and that a decoction of it is employed by the Malays in cholera, &c.

GENERA.

[This Order is much in want of a thorough revision.]

Buchia, *H. B. K.*
Cryptocalyx, *Benth.*
Lippia, *Linn.*
 Zapania, Juss.
 Platonia, Raf.
 Bertolonia, Raf.
 Riedelia, Cham.
 Dipterocalyx, Cham.
 Obletia, Roz.
 ? *Phyla*, Lour.
Aloysia, *Ort.*
Verbena, *Linn.*
 Glandularia, Gmel.
 Billardiera, Mönch.
 Aubletia, Jacq.
Stachytarpheta, *Vahl.*
 Cymburus, Salisb.
 Abena, Neck.
Bouchea, *Cham.*
Melasanthus, *Pohl.*
Uwarovia, *Bunge.*
 Shuttleworthia, Meisn.
Monochilus, *Fisch. et M.*
Chascanum, *E. Mey.*
Casselia, *Nees et Mart.*
Dipyrena, *Hook.*
 Wilsonia, Hook.
Priva, *Adans.*
 Blairia, Houst.
 Phryma, Linn.
 Leptostachya, Mitch.
 Castelia, Cav.
 Tortula, Roxb.
 Streptium, Roxb.
Chloanthes, *R. Br.*
Caryopteris, *Bung.*
Peronema, *Jack.*
Spielmannia, *Med.*
 Oftia, Adans.
Mallophora, *Endl.*
Lantana, *Linn.*
 Camara, Cham.
 Myrobatindum, Vaill.
 Charachera, Forsk.
 Callioreas, Cham
Tamonea, *Aubl.*
 Ghinia, Schreb.
 Leptocarpus, Willd.
 Kæmpfera, Houst.
 ? *Ischnia*, DC.
Hosta, *Jacq.*
 Hostana, Pers.
 Cornutia, Gmel.
Pyrostoma, *C. F. W. Mey.*
Wallrothia, *Roth.*
Casarettoa, *Walpers.*
Vitex, *Linn.*
 Ephialis, Sol.
 Agnus Castus, Endl.
 Limia, Vand.
 Nephrandra, Coth.
 ? *Chrysomallum*, Thou.
 Psilogyne, DC.
Premna, *Linn.*
 Cornutia, N. L. Burm.
 Baldingera, Dennst.
Gumira, *Rumph.*
Holochiloma, *Hochst.*
Pityrodia, *R. Br.*
Tectona, *Linn. fil.*
 Theka, Rheed.
 Jatus, Rumph.
Gmelina, *Linn.*
 Michelia, Amman.
Petitia, *Jacq.*
Citharexylon, *Linn.*
 Rauwolfia, Rz. et Pav.
 Pöppigia, Bert.
Hemigymnia, *Griff.*
Volkameria, *Linn.*
 Duglassia, Amm.
Clerodendron, *L.*
 Volkameria, Linn.
 Siphonanthus, Linn.
 Ovieda, Linn.
 Agricola, Schrank.
 Torreya, Spreng.
 Cornacchinia, Savi.
Cyclonema, *Hochst.*
 Spironema, Hochst.
Duranta, *Linn.*
 Ellisia, P. Br.
 Castorea, Plum.
Petræa, *Houst.*
Amasonia, *Linn.*
 Taligalea, Aubl.
Callicarpa, *Linn.*
 Burchardia, Duham.
 Johnsonia, Catesb.
 Sphondylococcum, Mit.
 Porphyra, Lour.
Ægiphila, *Jacq.*
 Manabea, Aubl.
 Omphalococca, Willd.
Scleröon, *Lindl.*
Cornutia, *Plum.*
 Agnanthus, Vaill.
Congea, *Roxb.*
 Roscöea, Roxb.
Sphenodesme, *Jack.*
Symphorema, *Roxb.*
 ? *Analectis*, Vahl.
Geunsia, *Blum.*
Dissolæna, *Lour.*
Patagonula, *L.*
Quoya, *Gaudich.*
Mastacanthus, *Endl.*
 Barbula, Lour.
?Hymenopyramis, *Wall.*
? Glossocarya, *Wall.*
Cochranea, *Miers.*

NUMBERS. GEN. 56. Sp. 610.

Oleaceæ.
POSITION.—Lamiaceæ.—VERBENACEÆ.—Myoporaceæ.
Scrophulariaceæ.

ORDER CCLVII. MYOPORACEÆ.—MYOPORADS.

Myoporinæ, *R. Brown Prodr.* 514. (1810); *Bartl. Ord. Nat.* 176; *Endl. Gen.* cxli.; *Meisn. Gen.* p. 292.—Avicennieæ, *Endl. Gen.* 638.

DIAGNOSIS.—*Echial Exogens, with irregular unsymmetrical flowers, confluent nuts, pendulous ovules, and 2-celled anthers.*

Shrubs, with scarcely any pubescence. Leaves simple, without stipules, alternate or opposite, sometimes thickly occupied by transparent cysts. Flowers axillary, without bracts. Calyx 5-parted, persistent. Corolla monopetalous, hypogynous, nearly equal or 2-lipped. Stamens 4, didynamous, with sometimes the rudiment of a fifth one, which occasionally bears pollen. Ovary 2- or 4-celled, the cells 1- or 2-seeded, with pendulous ovules; style 1; stigma scarcely divided. Fruit a drupe, with a 2- or 4-celled putamen, the cells of which are 1- or 2-seeded. Seeds pendulous; embryo taper, in the axis of a small quantity of albumen, or without any; radicle superior.

The principal characters in the fructification of this Order, by which it is distinguished from Verbenes, are the presence of albumen in the ripe seed, and the direction of the embryo, whose radicle always points towards the apex of the fruit. The first of these characters is, however, not absolute, and neither of them can be ascertained before the ripening of the seed.—*R. Brown in Flinders*, 557. Mr. Bentham is disposed to unite the two.

This Order, with the exception of Bontia, a genus of equinoctial America, and of the species of Myoporum, found in the Sandwich Islands, has hitherto been observed only in the southern hemisphere, and yet neither in South Africa nor in South America beyond the tropics. Its maximum is evidently in the principal parallel of Terra Australis, in every part of which it exists; in the more southern parts of New Holland, and even in Van Diemens Island, it is more frequent than within the tropics.—*R. Brown in Flinders*, 567. The Avicennias are shore trees living like Mangroves in salt swamps. Their creeping roots, often curving for the space of six feet above the mud before they stick into it, and the naked Asparagus-like suckers which they throw up, have a singular appearance.

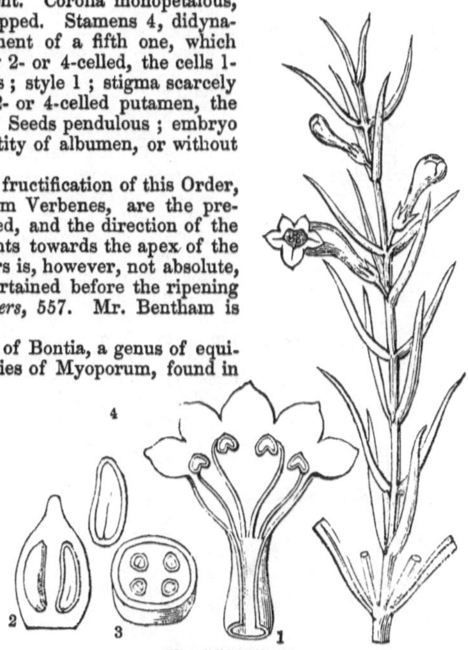

Fig. CCCCXLIII.

The bark of Avicennia tomentosa, the White Mangrove of Brazil, is in great use at Rio Janeiro for tanning. It exudes a kind of green aromatic resin, which furnishes a miserable food to the barbarous natives of New Zealand, who call it Manawa. Arabian writers believe that its saline mucilaginous root is an aphrodisiac. The unripe seeds are used in India for poultices; and, when ripe, are boiled and eaten by the poor.

GENERA.

Myoporum, *Banks et Sol.*	Spartothamnus, *A. Cunn.*	Avicennia, *Linn.*	*Rack*, Bruce.
Pogonia, Andr.	Eremophila, *R. Br.*	*Donatia*, Löffl.	*Apata*, Adans.
Andrewsia, Vent.	Eremodendron, *DC.*	*Sceura*, Forsk.	*Racka*, Gmel.
Bertolonia, Spin.	Stenochilus, *R. Br.*	*Halodendron*, Thouars.	? *Quapira*, Aubl.
Dasymalla, *Endl.*	Bontia, *Plum.*	*Œpata*, Rheed.	*Gynastrum*, Neck.
Pholidia, *R. Br.*			

NUMBERS. GEN. 9. Sp. 42.

Oleaceæ.
POSITION.—Verbenaceæ.—MYOPORACEÆ.—Selaginaceæ.

Fig. CCCCXLIII.—Pholidia scoparia. 1. a corolla opened: 2. perpendicular section of ripe fruit; 3. cross section of it; 4. longitudinal section of a seed.

ORDER CCLVIII. SELAGINACEÆ.—SELAGIDS.

Selagineæ, *Juss. Ann. Mus.* 7. 71. (1806); *Richard in Pers. Synops.* 2. 146; *Choisy Mémoire,* (1823); *Bartl. Ord. Nat.* 177; *Endl. Gen.* cxl.; *E. Meyer, Comment. pl. Afr. Austr.* 245; *Meisner Gen. p.* 292.—Globularineæ, *DC. Fl. Fr.* 3. 427. (1815); *Cambessèdes in Ann. des Sciences,* 9. 15; *Endl. Gen.* cxxxix.; *Link Handb.* 1. 675; *Meisner, p.* 315.

DIAGNOSIS.—*Echial Exogens, with irregular unsymmetrical flowers, confluent nuts, pendulous ovules, and 1-celled anthers.*

Herbaceous plants, or small branched shrubs. Leaves alternate, generally sessile, toothed, or entire, without stipules, usually in clusters. Flowers sessile, spiked, with large bracts. Calyx spathaceous, or tubular, persistent, with a definite number of teeth or divisions, rarely consisting of two sepals. Corolla tubular, hypogynous, more or less irregular, with 5 lobes, imbricated in æstivation. Stamens 4, usually didynamous, arising from the top of the tube of the corolla, seldom 2; anthers 1-celled, usually adnate to the dilated top of the filament, rarely versatile. Ovary superior; style 1, filiform; stigma nearly capitate; ovules solitary, pendulous, anatropal. Disk hypogynous, fleshy. Fruit 2-celled; the cells either separable or inseparable, 1-seeded, membranous. Seed solitary, pendulous; embryo in the axis of a little fleshy albumen; radicle superior.

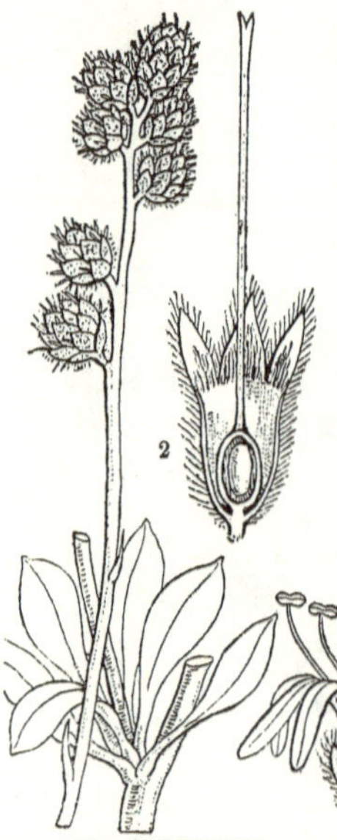
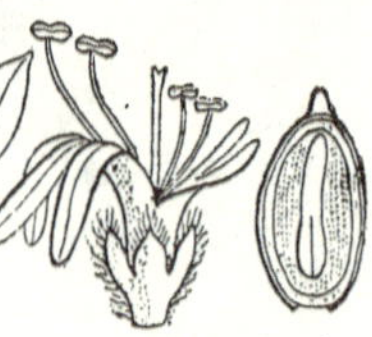

Fig. CCCCXLIV.

Fig. CCCCXLV.

The very small group collected under the name of Selagids is nearly allied on the one hand to Verbenes, and the other to Myoporads, from both which it is known by having 1-celled anthers. It also differs from Verbenes in having pendulous ovules. Globularia, which has been regarded as the type of a particular Order, does not seem to differ in anything more than having a solitary carpel; for its anthers are 1-celled. The resemblance of that genus to Primworts is very inconsiderable; but it may be regarded as being more like a form of Teazelworts (Dipsacaceæ), with a superior ovary. The genus Globularia is moreover in some respects analogous to Brunonia, which however differs abundantly in having hypogynous stamens, symmetrical flowers, and no albumen.

The principal part of this Order comes from the Cape of Good Hope; Gymnandra is however Siberian, and the Globularias European, chiefly inhabiting the southern kingdoms and the basin of the Mediterranean.

The species seem to be of small importance. Some are sweet-scented; Hebenstreitia dentata is said to be scentless in the morning, strong-smelling at mid-day, and sweet in the evening. Globularia Alypum is a bitter, drastic purgative, once supposed to be

Fig. CCCCXLIV.—Globularia orientalis. 1. a flower; 2. section of calyx and ovary; 3. section of fruit.

Fig. CCCCXLV.—Selago distans. 1. a flower; 2. an anther; 3. a perpendicular section of an ovary; 4. section of seed of Microdon ovatum.

the 'Αλύπον of Dioscorides, and hence called Frutex terribilis. The Alypum however had the σπέρμα ὡς ἐπιθύμου, and was therefore in all probability some Euphorbia. Globularia vulgaris has similar qualities; both are emetic.

GENERA.

Polycenia, *Chois.*	Microdon, *Chois.*	Macria, *E. Mey.*	Globularia, *L.*
Hebenstreitia, *Linn.*	*Dalea*, Gärtn.	Walafridia, *E. Mey.*	*Alypum*, Tourn.
Dischisma, *Chois.*	Selago, *Linn.*	Gymnandra, *Pall.*	*Abolaria*, Adans.
Agathelpis, *Chois.*	*Noltea*, Eckl.		

NUMBERS. GEN. 10. Sp. 120.

Dipsacaceæ.
POSITION.—Myoporaceæ.—SELAGINACEÆ.—Verbenaceæ.
Pedaliaceæ.

ALLIANCE XLIX. *BIGNONIALES*.—THE BIGNONIAL ALLIANCE.

DIAGNOSIS.—*Perigynous Exogens, with dichlamydeous, monopetalous, unsymmetrical flowers, capsular or berried fruit, having its carpels quite consolidated, parietal free central or axile placentæ, and an embryo with little or no albumen.*

With Bignonials the series of Perigynous Alliances closes, Gesnerworts passing as directly into Bellworts among Campanals as Figworts also pass into the Nightshades among Solanals. The two are parallel instances. Nevertheless, it does not seem expedient to place Gesnerworts at the end of the Bignonial Alliance, because it is impossible to separate them from Bignoniads and Crescentiads, or from Pedaliads, whose hard bony fruit presents the nearest approach in this Alliance to the nuts of Echials. We must, therefore, regard the passage of Bignonials into Campanals as being altogether from the side of the series and not from its extremity. Another lateral affinity presents itself between Butterworts and Primworts, in the Cortusal Alliance. The following will, therefore, express the bearing of Bignonials and other Alliances, better than a lineal position :—

	Pedaliaceæ.	
	Gesneraceæ.	
	Crescentiaceæ.————	—*Campanals.*
Gentianaceæ.————	—Bignoniaceæ.	
	Acanthaceæ.	
	Scrophulariaceæ.————	—*Solanales.*
	Lentibulariaceæ.————	—*Cortusales.*

The Bignonial Alliance may be regarded, then, as the centre of a particular portion of Exogens, round which several others are stationed in nearly equal degrees of contiguity.

NATURAL ORDERS OF BIGNONIALS.

Placentæ parietal. Fruit bony or capsular. Embryo amygdaloid. Radicle short	259. PEDALIACEÆ.
Placentæ parietal. Fruit capsular or baccate. Embryo with minute cotyledons. Radicle long	260. GESNERACEÆ.
Placentæ parietal. Fruit succulent, hard-shelled. Embryo amygdaloid. Radicle short	261. CRESCENTIACEÆ.
Placentæ axile. Seeds winged, sessile, without albumen. Cotyledons large, leafy	262. BIGNONIACEÆ.
Placentæ axile. Seeds wingless, attached to hard placental processes, without albumen. Cotyledons large, fleshy . .	263. ACANTHACEÆ.
Placentæ axile. Seeds albuminous. Cotyledons scarcely larger than, or not so large as, the radicle	264. SCROPHULARIACEÆ.
Placenta free, central. Seeds minute, without albumen. Cotyledons much smaller than the radicle	265. LENTIBULARIACEÆ.

ORDER CCLIX. PEDALIACEÆ.—PEDALIADS.

Pedalinæ, *R. Brown Prodr.* 519. (1810); *Lindley in Botan. Register*, 9. 934. (1825); *Endl. Gen.* cliii. —Sesameæ, *Kunth Synops.* 2. 251. (1823); *Bartl. Ord. Nat.* 175; *Endlicher in Linnæa, VII.* 1.; *Alph. DC. Prodr.* 9. 249; *Bernhardi in Ann. Sc. Nat. n. s.* 18. 365.—Martyniaceæ, *Link Handb.* 1. 504. (1829).

DIAGNOSIS.—*Bignonial Exogens, with parietal placentæ, bony or capsular fruit, an amygdaloid embryo, and short radicle.*

Herbaceous plants, often with a soft texture, and heavy smell, covered with glandular hairs, or quaternary vesicles. Leaves opposite or alternate, undivided, angular, or lobed, without stipules. Flowers axillary, solitary, or clustered, usually large, and furnished in many cases with conspicuous bracts. Calyx divided into 5 nearly equal pieces. Corolla monopetalous, hypogynous, irregular; the throat ventricose, the limb bilabiate, the lobes somewhat valvate in æstivation. Disk hypogynous, fleshy, sometimes glandular. Stamens didynamous, included within the tube, together with a rudiment of a fifth. Anthers 2-celled; the connective articulated with the filament, a little prolonged beyond the cells, terminated by a gland. Ovary seated in a glandular disk, 1-celled, formed of two carpellary leaves, anterior and posterior as regards the axis, sometimes divided into 4 or 6 spurious cells by the splitting of two placentas and the divergence of their lobes; ovules anatropal, either erect, or pendulous, or horizontal, solitary, or 2, or several; style 1; stigma divided. Fruit drupaceous or capsular, valvular, or indehiscent, with from 2 to 6 cells, which are usually few-seeded when numerous, and many-seeded only when two. Seeds with a papery testa, wingless; albumen none; embryo straight; cotyledons large, plano-convex; radicle short, next the hilum.

Fig. CCCXLVI.

The only real differences that can be found between these plants and Bignoniads consist in the parietal placentæ of the former, their wingless or nearly wingless seeds, which are in most cases definite, and sometimes in their woody lobed placentæ, which spread and divide variously in the inside of the pericarp, so as to produce an apparently 4- or 6-celled fruit out of a 1-celled ovary. Sesamum may be considered a transition from the one to the other. From Gesnerworts they are readily known by the texture of their fruit, their large seeds, plano-convex cotyledons, and very short radicle. Calabashes are distinguished by their great succulent fruit and almond-like seeds. Endlicher rightly observes that Brown in forming his Pedalinæ (*Prodr.* 519.), does not combine with them Sesamum; neither, however, does he explain how they are to be distinguished; but as usual, the extreme and studied conciseness of this learned man leaves his readers almost as much in the dark as if the name of Sesamum had not been mentioned.

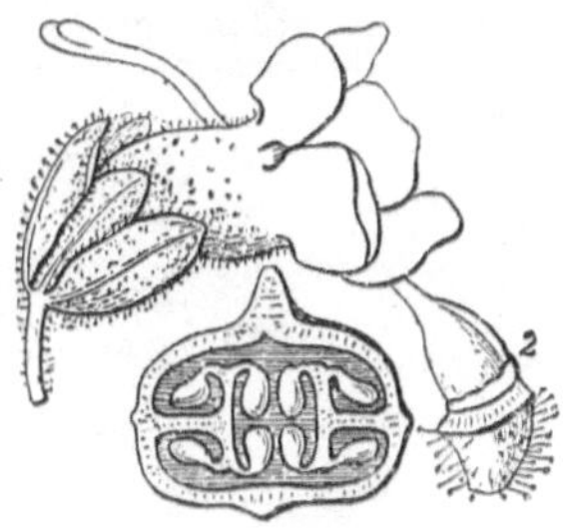

Fig. CCCCXLVII.

It is not a little remarkable that such observers as De Candolle (*Prodr.* 8. 249.) and Endlicher (*Linnæa*, vii. p. 8.) should suppose the fruit of this Order to be formed out of 5 or 4 carpels, a statement entirely opposed to both theory and fact. It is really composed of two anterior and posterior carpels, exactly as that of the other Orders in the present Alliance. It is doubtless true that Martynia has been described as having

Fig. CCCCXLVI.—Sesamum indicum. 1. a ripe fruit; 2. one of its halves; 3. a seed; 4. a cross section of it.
Fig. CCCCXLVII.—Martynia lutea. 1. a flower; 2. the pistil; 3. a section of its ovary.

4 cells; but so long since as December, 1825 (*Bot. Reg. t.* 934), I explained the true nature of this structure in the following words:—"Upon a careful examination of the ovarium, it will be found that the fruit, in that stage, is neither 4-celled nor even 2-celled, but consists of only one cell, traversed by two projecting, parietal placentæ, each of which is 2-lobed; the lobes divided at right angles from their point of separation, and bearing on their edges a few horizontal ovula, of which part project into the open centre of the ovarium, and the others into the cavity between the placenta and the lining of the ovarium. Now the capsule differs from the ovarium in no essential point of structure, but the following changes take place: the pericarpium and the placentas become woody and rigid, the inner faces of the latter become pressed together so as to destroy the ovula which were placed between them, and to exhibit the appearance of a bilamellar dissepiment, and the remaining ovula become pendulous, and reduced in number, and exist in the form of large apterous seeds between the inner edge of the lateral lobes of the placenta and the endocarpium."

A not less singular in appearance, but unreal deviation, occurs in Pretrea zanguebarica, whose two carpels turn their edges inwards, right and left, until they touch the sides of the ovary, and form on each side a little pouch for the reception of the seeds; at the same time, in consequence of the inflected plates not touching each other, two seedless cavities are also formed next the ventral and dorsal sutures, and thus a six-celled fruit is constructed out of a pair of carpels. The accompanying cut explains this singular structure.

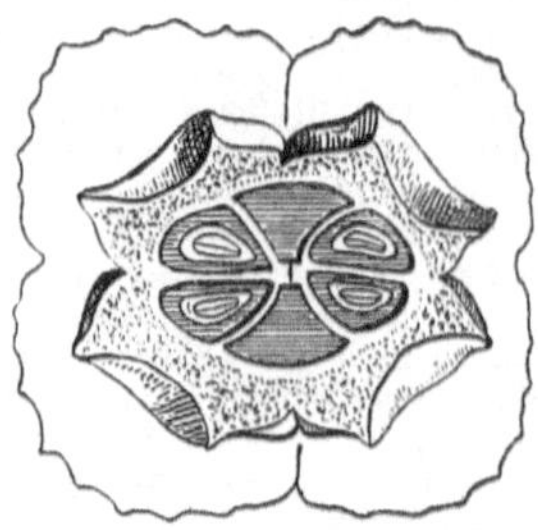

Fig. CCCCXLVIII.

The species of Pedaliads occur in all parts of the tropics, in small numbers, but Africa is supposed to be the principal field over which they are spread.

The leaves of Sesamum are emollient. Its seeds contain an abundance of a fixed oil, as tasteless as that of Olive Oil, for which it might be substituted, and which is expressed in Egypt in great quantities. It is sometimes called Gingilie Oil, and, if of very good quality, is employed for adulterating Oil of Almonds. It is, however, apt to become rancid. The fresh leaf of Pedalium Murex, when agitated in water, renders it mucilaginous, in which state it is prescribed by Indian doctors in cases of dysuria and gonorrhœa. The meal of the seeds of both these plants is used in India for poultices. Uncaria procumbens, called the Grapple Plant at the Cape of Good Hope, has a fruit covered with hooked spines, which lay hold of the clothes of travellers, and the pair of long hooked horns of Martynia proboscidea, called in Italy the Testa di Quaglia, is notorious for the same propensity. The fleshy sweet root of Craniolaria annua is preserved in sugar by the Creoles as a delicacy; in a dry state it is said to be a bitter cooling medicine.

GENERA.

I. PEDALEÆ.

Craniolaria, *Linn.*
Holoregmia, *Nees.*
? *Neowedia*, Schrad.
Martynia, *Linn.*
Proboscidea, Schmidt.
Carpoceras, *A. Rich.*
Pedalium, *Royen.*
Cacatali, Adans.
Ischnia, *DC.*
Harpagophytum, DC.
Uncaria, *Burch.*
Rogeria, *Gay.*
Pretrea, *Gay.*
Dicerocaryum, Boj.
Josephinia, *Vent.*
Pterodiscus, *Hooker.*

II. SESAMEÆ.

Sesamum, *Linn.*
Digitalis, Tournef.
? *Dysosmon*, Raf.
Ceratotheca, *Endl.*
Sesamopteris, *Endl.*
Gongyla, Bernh.
Sporledera, *Bernh.*

NUMBERS. GEN. 14. Sp. 25.?

Myoporaceæ.
POSITION.—Bignoniaceæ.—PEDALIACEÆ.—Gesneraceæ.
Selaginaceæ.

Fig. CCCCXLVIII.—Pretrea zanguebarica; a cross section of its fruit.

ORDER CCLX. GESNERACEÆ.—GESNERWORTS.

Gesnerieæ, *Rich. et Juss. Ann. Mus.* 5. 428. (1804); *Kunth in Humb. N. G. et Sp.* 2. 392; *Lindley in Bot. Reg.* 1110.—Gesneriaceæ, *Link Hanb.* 1. 504. (1829); *DC. Prodr.* 7. 523.—Gesnereæ, *Von Martius Nov. Gen. Bras.* 3. 68. (1829); *Bartl. Ord. Nat.* 174.—Gesneraceæ, *Ed. pr.* ccxv. (1836); *Endl. Gen.* clii.—Cyrtandraceæ, *Jack in Linn. Trans.* 14. 23. (*read* 1822, *in May*); *Alph. DC. Prodr.* 9. 258; *Ed. pr.* ccxii.—Didymocarpeæ, *Don in Edin. Phil. Trans.* 7. 82. (1822, *July*).

DIAGNOSIS.—*Bignonial Exogens, with parietal placentæ, capsular or baccate fruit, an embryo with minute cotyledons, and a long radicle.*

Soft-wooded, somewhat fleshy, herbs or shrubs, occasionally having a climbing or creeping manner of growth, and frequently springing from scaly tubers. Leaves rugose, without stipules, generally opposite or whorled. Flowers showy, in racemes or panicles, rarely solitary, yellow, scarlet, violet, or white. Calyx half adherent, 5-parted, with a valvate or open æstivation. Corolla monopetalous, tubular, more or less irregular, 5-lobed, with an imbricated æstivation. Stamens 2, or 4, didynamous; anthers often cohering, 2-celled, innate, with a thick tumid connective; the rudiment of a fifth stamen is present. Ovary half superior, 1-celled, with 2 fleshy 2-lobed parietal polyspermous placentæ, placed right and left of the axis; surrounded at its base by glands or a fleshy ring; style continuous with the ovary; stigma capitate, concave; ovules 00, anatropal. Fruit capsular or succulent, superior, 1-celled, with 2 opposite lateral placentæ, each consisting of 2 plates. Seeds very numerous, minute; embryo erect, in the axis of fleshy albumen, with the cotyledons much shorter than the radicle; testa thin, with very close fine oblique veins, sometimes extended into long hairs, or even flattened into a wing.

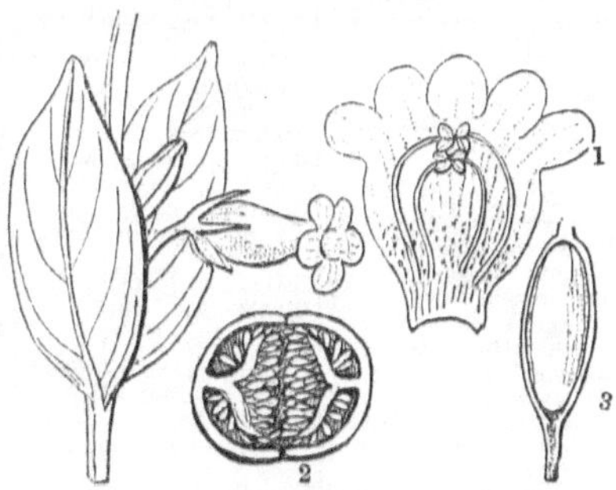

Fig. CCCCXLIX.

These little plants (for they seldom rise above the stature of bushes, and are generally mere herbs) have somewhat the appearance of Figworts, or of diminutive Bignoniads, and have been even referred to those Orders. They, however, differ from all the present Alliance in the very small size of their cotyledons as compared with their long radicle, and their absolutely parietal placentation; in addition to which they have a great tendency to form an inferior ovary, and thus lead towards the Campanal Alliance in another part of the series. To Eccremocarpus, a genus of Bignoniads, they approach nearly, as will be seen by referring to p. 675; but in that plant the winged seeds and large leafy cotyledons point too plainly to Bignoniads to be mistaken. Gesnerworts also approach Broomrapes in their parietal placentation.

The Suborder Cyrtandreæ, usually regarded as a distinct group, has been removed to Gesnerworts by Dr. Brown, and with justice, there being no sufficient distinction between them. (See *Horsfield's Plantæ Javanicæ*, p. 105). It is chiefly remarkable for the long threads that terminate the seeds of certain genera, for their double revolute placentæ, and in some cases for their long, slender, siliquose capsules. They pass into Bignoniads through Incarvillea.

The two Sub-orders have a very different geographical distribution. Gesnereæ, which are common in our gardens, are exclusively inhabitants of the tropical or warmer parts of America. The Cyrtandreæ, on the contrary, are spread over many parts of the world, although chiefly confined to the eastern parts. Some occur in Europe, as Ramondia and Haberlea; others grow in the cooler parts of Asia; such are Bæa and Rehmannia. Klugia is Mexican, Streptocarpus is from the Cape of Good Hope; but it is in the warm valleys of the Himalayas, and in the damp regions of the Indian Archipelago, that they are most abundant, under the forms of Chirita, Calosacme, Æschynanths, and Didymocarps. Fieldia is from New Holland, and several Cyrtandras from the Sandwich Islands.

Fig. CCCCXLIX.—Hypocyrta gracilis.—*Martius.* 1. section of a flower; 2. cross section of a fruit; 3. seed with its embryo exposed to view.

They are generally plants of considerable beauty, often growing on trees and leading a quasi-parasitical life; but they can scarcely be said to have any useful qualities. The succulent fruits of some Gesnereæ are mucilaginous, sweetish, and eatable; and a dye is obtained from the calyxes and fruit of others, for staining cotton, straw-work, and domestic utensils. Columnea scandens is called by the French colonists Liane à sirop, because its flowers secrete a large quantity of honey, and Sarmienta repens is used as an emollient in Chili. Some Didymocarps would appear to be aromatic; and Picria, a Cochin-China plant, is so bitter as to be called Fel Terræ; it is, however, very doubtful whether that plant belongs to this Order. It is possibly a Gentianwort.

GENERA.

I. GESNEREÆ. — Seeds with a small quantity of albumen. Fruit partially adherent.

Sarmienta, *Ruiz et Pav.*
Urceolaria, Feuill.
Mitraria, *Cav.*
? ? Picria, *Lour.*
Columnea, *Plum.*
Besleria, *Plum.*
Eriphia, P. Br.
Tussaca, Rchb.
Hypocyrta, *Mart.*
Codonanthe, Mart.
Oncogastra, Mart.
Drymonia, *Mart.*
Klugia, *Schlecht.*
Tapeinotes, *DC.*
Tapina, Mart.
Nematanthus, *Schrad.*
Alloplectus, *Mart.*
Crantzia, Scop.
Dalbergaria, Tuss.
Tussacia, Reichenb.
Lophia, Desv.
Vireya, Raf.
Episcia, *Mart.*
Gesnera, *Mart.*
Prasanthea, DC.
Niphæa, *Lindl.*
Achimenes, *P. Br.*
Trevirana, W.
Cyrilla, Herit.
Gloxinia, *Herit.*
Paliavana, Velloz.
Sinningia, Nees.
Hemiloba, DC.
Solenophora, *Benth.*
Rhytidophyllum, *Mart.*
Codonophora, Lindl.
Conradia, *Mart.*
Pentarhaphia, Lindl.

Bellonia, *Blum.*
Diastemma, *Benth.*
Trichantha, *Hooker.*

II. CYRTANDREÆ.—Seeds with no albumen. Fruit wholly free.

1. DIDYMOCARPIDÆ; *capsular.*

Liebigia, *Endl.*
Tromsdorffia, Blum.
Babactes, *DC.*
Æschynanthus, *Jack.*
Trichosporum, Don.
Lysionotus, Bl.
Agalmyla, *Bl.*
Orithalia, Bl.
Lysionotus, *D. Don.*
Didymocarpus, *Wall.*
Henckelia, Spr.
Chirita, *Ham.*
Calosacme, *Wall.*
Streptocarpus, *Lindl.*
Cardiolophus, *Griff.*
Bæa, *Comm.*
Dorcoceras, Bunge.
Ramondia, *Rich.*
Myconia, Lap.
Chaixia, Lap.
Haberlea, *Friwaldsk.*
Conandron, *Sieb. Zucc.*
? Monophyllæa, *R. Br.*
Rhynchoglossum, *Bl.*
Loxotis, R. Br.
Antonia, R. Br.
Napeanthus, *Gardn.*
Rehmannia, *Liboschitz.*
Klugia, *Schlecht.*
Glossanthus, Klein.
Loxonia, *Jack.*
Rhabdothamnus, *A. Cun.*
Loxocarpus, *R. Br.*
Craterostigma, *Hochst.*
Quintilia, *Endl.*
Miquelia, Bl.
Anomorhegmia, Meisn.
Stauranthera, *Benth.*
Epithema, *Bl.*
Aikinia, R. Br.
Platystemma, *Wall.*
Isanthera, *Nees.*

2. CYRTANDRIDÆ; *baccate.*

Cyrtandra, *Forst.*
Whitia, *Bl.*
Rhynchotecum, *Bl.*
Corisanthera, Wall.
Cheilosandra, Griff.
Fieldia, *A. Cunn.*

NUMBERS. GEN. 54. Sp. 260.

Orobanchaceæ.
POSITION.—Bignoniaceæ.—GESNERACEÆ.—Scrophulariaceæ.
Campanulaceæ.

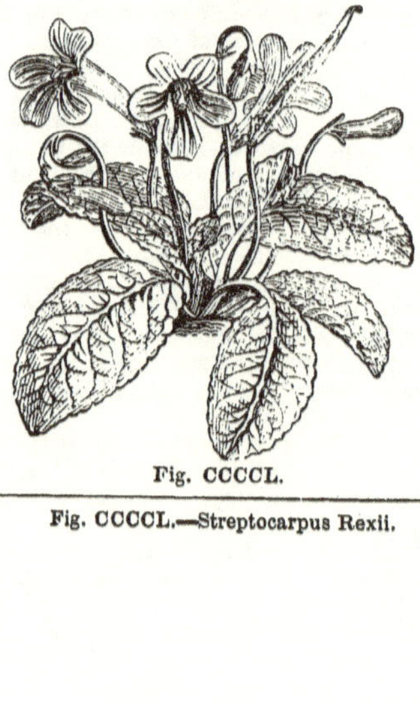

Fig. CCCCL.

Fig. CCCCL.—Streptocarpus Rexii.

ORDER CCLXI. CRESCENTIACEÆ.—CRESCENTIADS.

Crescentiaceæ, *Gardner in Hook. Journ.* 2. 423. (1840).—Crescentineæ, *DC. Rev. Bign. p.* 7. (1838); *Endl. Gen. p.* 723; *Miquel in Bot. Zeit.* (1844), *p.* 801; *Alph. DC. Prodr.* 9. 240.

DIAGNOSIS.—*Bignonial Exogens, with parietal placentæ, succulent hard-shelled fruit, and an amygdaloid embryo with a short radicle.*

Trees of small size, with alternate or clustered simple leaves without stipules. Flowers growing out of the old stems or branches. Calyx free, undivided, eventually splitting into irregular pieces. Corolla monopetalous, irregular, somewhat 2-lipped, with an imbricated æstivation. Stamens 4, growing on the corolla, didynamous, with the rudiment of a fifth between the posterior pair, which are the longest; anthers 2-lobed, bursting longitudinally. Ovary free, surrounded by a yellow annular disk, 1-celled, composed of an anterior and posterior carpellary leaf, with 2 or 4 equidistant parietal placentæ, which sometimes meet and produce additional cells; ovules 00, horizontal; style 1; stigma of two plates. Fruit woody, not splitting, containing a multitude of large amygdaloid seeds buried in the pulp of the placentæ; skin leathery, loose. Embryo straight, without albumen, with plano-convex fleshy cotyledons, and a thick short radicle next the hilum.

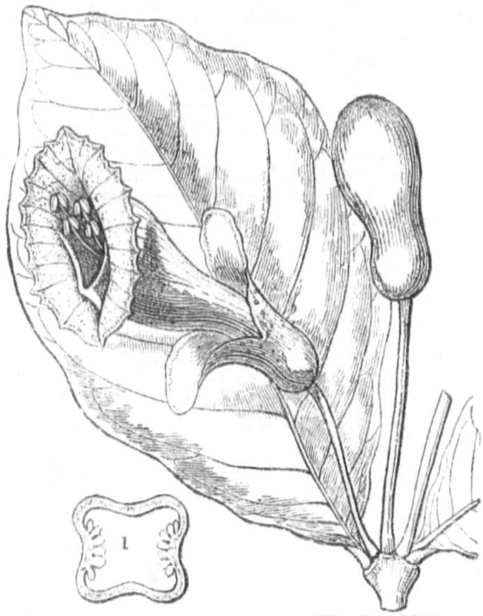

Fig. CCCCLII.

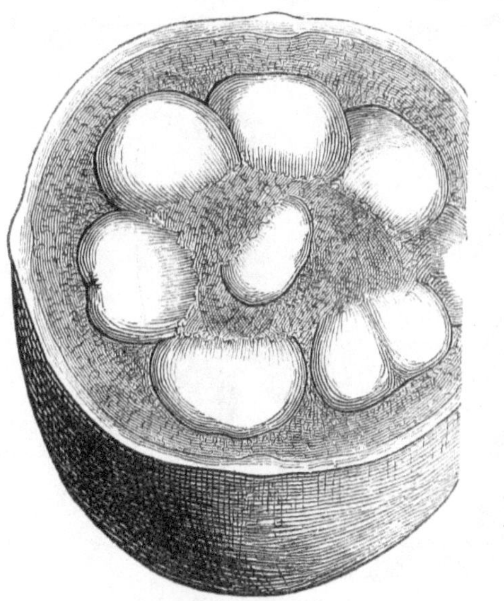

Fig. CCCCLIII.

These plants have been generally associated either with Nightshades, which they are quite unlike, or with Bignoniads, from which they differ in their succulent fruit, parietal placentæ, and wingless seeds. In the two latter circumstances they resemble Pedaliads; but their succulent fruit and large almond-like seeds are dissimilar. Gesnerworts lie doubtless on their borders, for they too have sometimes succulent fruit; but it seems impossible to associate trees with a great almond-like embryo and herbs or half-herbaceous bushes, whose minute embryo consists principally of radicle.

Mr. Gardner thus speaks of the fruit of Crescentia:—" In the economy of its fruit, Crescentia is more closely related to Cyrtandraceæ than to Bignoniaceæ, but differs essentially from it in the structure of its calyx, in its four distinct placentæ, horizontal, not suspended ovules,

Fig. CCCCLII.—Crescentia obovata. 1. a cross section of its ovary.
Fig. CCCCLIII.—Cross section of fruit of Crescentia cucurbitina.

and particularly in habit. The same observations apply to Pedaliaceæ, which are also 1-celled; for although the ripe fruit of both them and Cyrtandraceæ possesses apparently more than one cell, as if produced by the spreading and dividing of their parietal placentæ, the ovary of both, according to Bentham, is always unilocular if examined before the development occasioned by fecundation.

"To all the other Orders of the dicarpose group, Crescentia is of course more or less related, but is abundantly distinct from every one. Thus, it is distinguished from Acanthaceæ by its simple calyx, 1-celled ovary, unsuspended seeds, and in habit; from Lentibularieæ by its parietal, not free central placentation; and from Scrophulariaceæ and Solanaceæ and their allies by its want of albumen."—*Hooker's Journ.* 2. 424.

Inhabitants of the tropics of Asia, Africa, and America, but most especially abundant in the Mauritius and Madagascar.

The principal plant of this Order is the Calabash tree, Crescentia Cujete, a tree inhabiting the tropical parts of America, and bearing a great gourd-like fruit, filled with a sub-acid pulp which is eaten by the Negroes, and from which poultices are also prepared; its hard shell is used for holding fluids, in the room of bottles. The pulp of Tanaecium Jarowa is applied to the same purposes. Parmentiera edulis has fruit like a Cucumber, and affords food to the Mexicans.—*DC.*

GENERA.

Crescentia, *L.*	*Tripinna*, Lour.	Tanæcium, *Swz.*	Periblema, *DC.*
Cujete, Plum.	Sotor, *Fenzl.*	? *Jaroba*, Marcgr.	*Boutonia*, DC.
Kigelia, *DC.*	Schlegelia, *Miq.*	Schlegelia, *Miq.*	Phyllarthron, *DC.*
Tripinnaria, *Pers.*	Parmentiera, *DC.*	Colea, *Bojer.*	*Arthrophyllum*, Boj.

NUMBERS. GEN. 11. Sp. 34.

Solanaceæ ?
POSITION.—Gesneraceæ.—CRESCENTIACEÆ.—Bignoniaceæ.

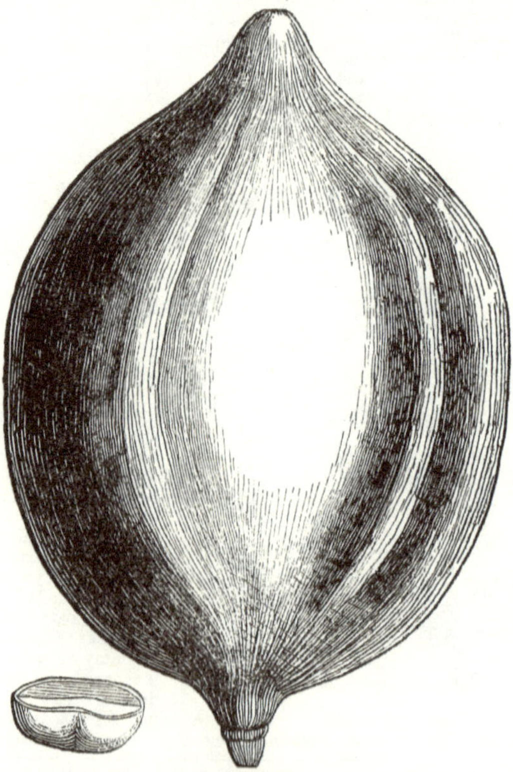

Fig. CCCCLIV.

Fig. CCCCLIV.—Crescentia cucurbitina.

ORDER CCLXII. BIGNONIACEÆ.—BIGNONIADS.

Bignoniæ, § 2. *Juss. Gen.* 137. (1789).—Bignoniaceæ, *R. Brown Prodr.* 470. (1810); *Bartl. Ord. Nat.* 185; *Endl. Gen.* cli.; *Alph. DC. Prodr.* 9. 142.

DIAGNOSIS.—*Bignonial Exogens, with axile placentæ, winged sessile seeds without albumen, and large leafy cotyledons.*

Trees, shrubs, or occasionally herbs, often twining or climbing. Leaves opposite, very rarely alternate, compound or occasionally simple, without stipules. Inflorescence terminal, somewhat panicled. Calyx divided or entire, sometimes spathaceous. Corolla monopetalous, hypogynous, usually irregular, 4- 5-lobed. Stamens 5, unequal: always 1 sterile, sometimes 3; anthers 2-celled, formed normally. Ovary seated in a disk, 2-celled, with the carpels anterior and posterior, or spuriously 4-celled, polyspermous; style 1; stigma of 2 plates; ovules 00, attached to a solid axile placenta. Capsule 2-valved, 2-celled, often long and compressed, sometimes spuriously 4-celled. Dissepiment formed from the placenta, which when it is undivided cuts the cavity of the ovary into 2 cells, or when it is 2-lobed, as is sometimes the case, assumes the appearance of being parietal and forms a 1-celled ovary, either parallel with the valves, or contrary to them, finally becoming separate, and bearing the seeds. Seeds transverse, compressed, winged; albumen 0; embryo straight, foliaceous; radicle centrifugal, much smaller than the broad cotyledons.

In the mere form of their flower there is nothing to distinguish Bignoniads from the kindred Orders. The distinction lies entirely in the seeds, which are winged, sessile, destitute of albumen, and furnished with a large leafy embryo, whose radicle is small and inconspicuous. They differ from Figworts in their leafy cotyledons and want of albumen; and from Acanthads, whose embryo is similar, in their winged seeds not attached to hard processes of the placenta. Besides which, their calyx is by no means so much imbricated as in Acanthads.

The central or axile position of the placenta, is an indispensable character of this Natural Order. The genus Eccremocarpus, however, appears to be

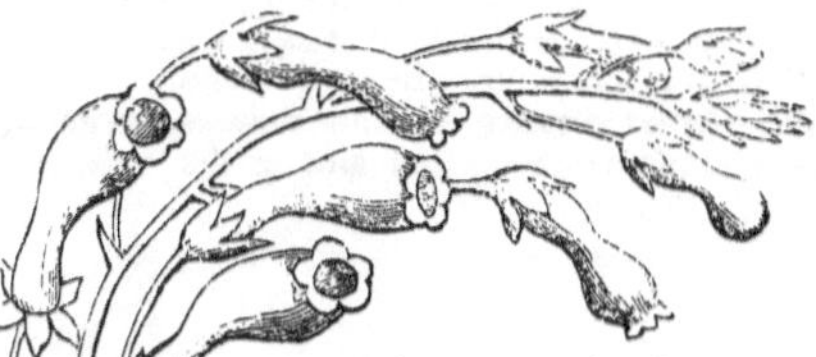
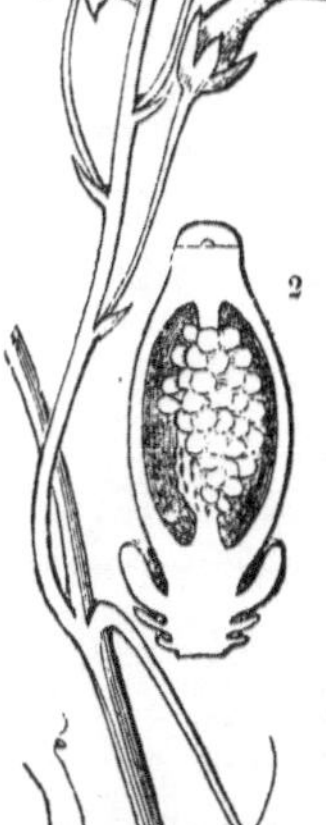

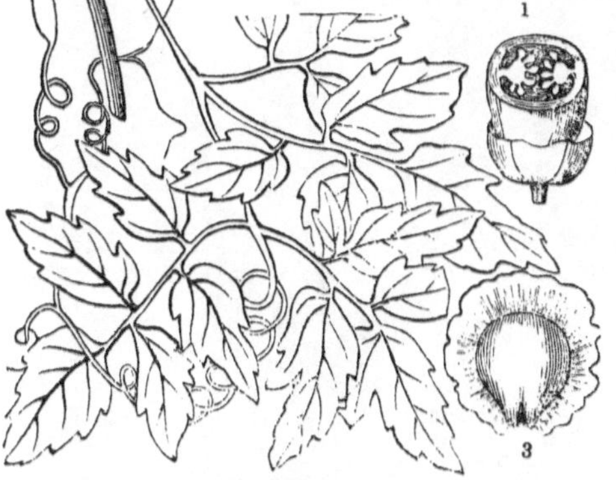

Fig. CCCCLV.

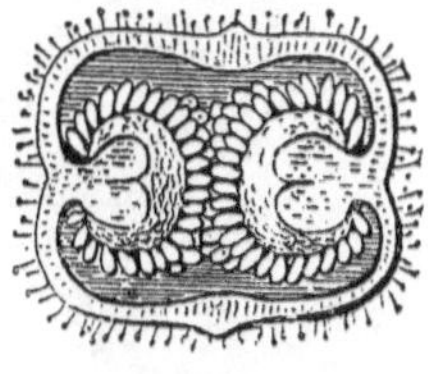

Fig. CCCCLVI.

an exception, its placentæ being strictly parietal at the time of the expansion of the

Fig. CCCCLV.—Eccremocarpus scaber. 1. cross section of its ovary; 2. longitudinal section of it; 3. seed.

Fig. CCCCLVI.—Cross section of the same ovary, much more magnified and very young.

flower. I, however, stated long since (*Bot. Reg.* 939, *Dec.* 1825,) that the placentation of Eccremocarpus scaber and Bignonia radicans are originally of the same nature, the difference between them consisting in the two placentæ of the latter meeting in the axis and uniting there, while in Eccremocarpus the two placentæ never touch in the middle, but exclusively adhere to the edges of the carpels.

Their wood is occasionally subdivided into 4 cruciform lobes. This is very conspicuous in Bignonia capreolata, and seems to be general in the woody species. M. Gaudichaud assures us that in Guayaquil these twiners have at first only 4 divisions of their woody system, but afterwards acquire 8, then 16, and probably 32, the divisions regularly following this mathematical progression. He also finds some indication of the tendency in the old stems of Bignonia capreolata. See his *Recherches Générales sur l'Organographie, &c.* p. 129, and the figures accompanying the statement.

The tropics of either hemisphere are the chief station of this noble-looking Order, whose trumpet-shaped flowers are the glory of the places which the species inhabit. The Order extends northwards in North America as far as Pennsylvania, and southwards into the southern provinces of Chile. In Europe it is unknown in a wild state.

The species are best known for the great beauty of the flowers, which from their large size, gay colours, and great abundance, are often among the most striking

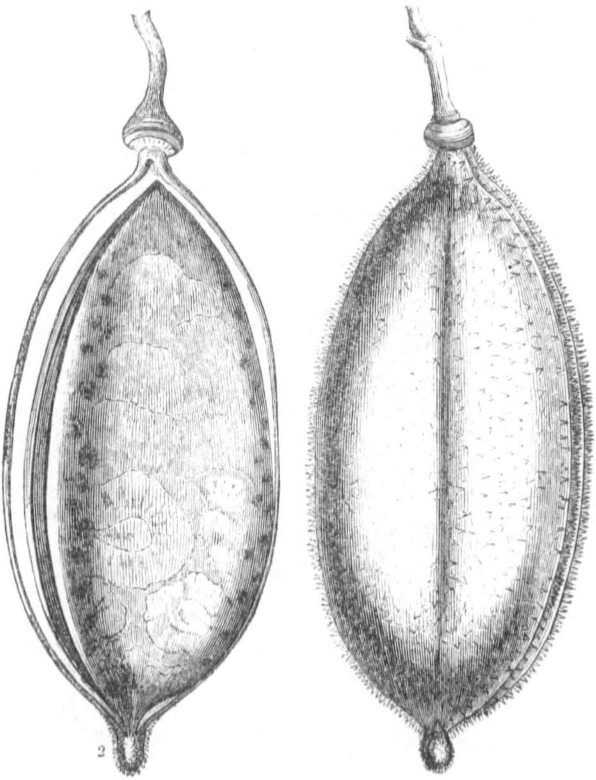

Fig. CCCCLVII.

objects in a tropical forest. Chica (called also Carajuru) is a red feculent substance obtained by boiling the leaves of Bignonia Chica in water ; the Chica is quickly precipitated by adding some pieces of the bark of an unknown tree, called Arayana ; the

Fig. CCCCLVII.—1. capsule of Bignonia echinata ; 2. the same with the valves removed and the placenta remaining covered with seeds.

Indians use it for painting their bodies red; it is also an article of importance to dyers; in nature it approaches the resins, but contains some peculiar properties; it gives an orange-red to cotton. The tough shoots of B. Cherere are woven into wicker-work. A decoction of the pods of Catalpa syringifolia is used in Italy as a remedy for catarrhal dyspnœa and coughs.—*Gard. Mag.* xiii. 524. According to Kæmpfer a nearly allied species, or perhaps the same, found in Japan, has extremely bitter leaves and bark, and a decoction of the pods is employed in asthmatic complaints; the leaves are also used for fomentations. The bark of the younger branches of B. antisyphilitica is considered in Brazil one of the most powerful remedies against syphilitic swellings of a malignant character. The decoction is chiefly used, and also the bark dried and pounded, externally. The roots of some are venomous and bitter; that of Tecoma stans is diuretic. The bark of B. leucoxylon is regarded as an antidote to the poison of the Manchineel tree. The branches of B. echinata are said to be employed in adulterating Sarsaparilla. Tecoma impetiginosa abounds in tannin; its bark is bitter and mucilaginous, and is used in lotions, baths, &c. in inflammations of the joints and debility. Tecoma Ipe has similar qualities, and is prescribed by the Brazilians as a gargle in ulcers of the mouth; the leaves are milder and are sometimes used in ophthalmic affections. The leaves of Sparattosperma lithontriptica are bitter, acrid, and diuretic, and have a Brazilian reputation in calculus, in which indeed Martius testifies to their efficacy. Jacaranda procera and other species of that genus are employed in syphilitic affections. Tecoma speciosa is said to be a useful diuretic and cathartic. Several kinds of Bignonia form large trees in the forests of Brazil, where they are felled for the sake of their timber; that called Ipe-tabacco furnishes durable ship-timber; the Ipeuna, another species, is the hardest wood in Brazil. Another, called the Pao d'arco, supplies one of the best kinds of wood used for bows by the Brazilian savages, especially the Botocudos of the Rio Grande de Belmonte, and the Patachos of the Rio do Prado.

GENERA.

Bignonia, *L.*
  Stenolobium, Don.
  Temnocydia, Mart.
  Alsocydia, Mart.
  Batocydia, Mart.
Pachyptera, *DC.*
Fridericia, *Mart.*
Astianthus, *D. Don.*
Calosanthes, *Bl.*
  Oroxylum, Vent.
Cuspidaria, *DC.*
  Lochmocydia, Mart.
Macfadyena, *A. DC.*
Lundia, *DC.*
Mansoa, *DC.*
Millingtonia, *L.*
Arrabidæa, *DC.*
  Vasconcella, Mart.
Anemopægma, *Mart.*
Distictis, *DC.*
Haplolophium, *Endl.*
  Aplolophium, Cham.
Amphilophium, *Kth.*
Pithecoctenium, *Mart.*
Delostoma, *D. Don.*
Cybistax, *Mart.*
  Phryganocydia, Mart.
Adenocalymna, *Mart.*
Sparattosperma, *Mart.*
Spathodea, *Beauv.*
  Dolichandra, Cham.
Heterophragma, *DC.*
Stereospermum, *Cham.*
Zeyhera, *Mart.*
  ? *Chasmia*, Schott.
Callichlamys, *Miq.*
Tabebuia, *Gom.*
  Couralia, Splitg.
Crateritecoma, *Mart.*
  Phænicocirsus, Mart.
Tecoma, *Juss.*
  Campsis, Lour.
  Tecomaria, Fenzl.
Catalpa, *Scop.*
Chilopsis, *D. Don.*
Pajanelia, *DC.*
Jacaranda, *Juss.*
  Kordelestris, Arrud.
  Icaranda, Pers.
Catophractes, *D. Don.*
? Platycarpum, *H. B. K.*
Rhigozum, *Burch.*
Argylia, *D. Don.*
Tourretia, *Juss.*
  Dombeya, Lher.
Incarvillea, *Juss.*
Amphicoma, *R. Br.*
Eccremocarpus, *R. P.*
  Calampelis, Don.
* * *
Pteropodium, *DC.*
Dipterosperma, *Hassk.*
Bravaisia, *DC.*
? Trigonocarpus, *Wall.*

NUMBERS. GEN. 44. Sp. 450.

POSITION.—Gesneraceæ.—BIGNONIACEÆ.—Crescentiaceæ.

ORDER CCLXIII. ACANTHACEÆ.—ACANTHADS.

Acanthi, *Juss. Gen.* 102. (1789).—Acanthaceæ, *R. Br. Prodr.* 472; *Nees ab. Esenb. in Wall. pl. as. rar.* 3. 70; *Endl. Gen.* cl.; *Meisner Gen. p.* 293.

DIAGNOSIS.—*Bignonial Exogens, with axile placentæ, wingless exalbuminous seeds attached to hard placental processes, and large fleshy cotyledons.*

Herbaceous plants or shrubs, chiefly tropical; their hairs, if they have any, simple, occasionally capitate, very rarely stellate. Leaves opposite, rarely in fours, without stipules, simple, undivided, entire, or serrated; rarely sinuated, or having a tendency to become lobed, sometimes in unequal pairs. Inflorescence terminal, or axillary, in spikes, racemes, fascicles, or panicles; the flowers sometimes even solitary. Flowers usually opposite in the spikes, sometimes alternate, with three bracts, of which the lateral are now and then deficient; these bracts sometimes large and leafy, and inclosing a diminished calyx, which is occasionally obsolete. Calyx 4- or 5-divided, usually 5-leaved, equal or unequal, generally very much imbricated, occasionally cut into many pieces, or entire and obsolete, persistent. Corolla monopetalous, hypogynous, bearing the stamens, mostly irregular; the limb ringent or 2-lipped (the lower lip overlapping the upper in æstivation), occasionally 1-lipped, sometimes nearly equal, deciduous. Stamens mostly 2, both bearing anthers; sometimes 4, didynamous, the shorter ones being sometimes sterile; anthers either 2-celled, their cells being inserted equally or unequally, or 1-celled, opening lengthwise. Ovary seated in a disk, 2-celled, composed of 2 carpels placed in front and back as regards the axis, and bearing the placentæ on their edges, the cells either 2- or many-seeded; placentæ parietal, although adhering in the axis; style 1; stigma 2-lobed, rarely undivided; ovules amphitropal or campulitropal. Capsule 2-celled, the cells 2- or many-seeded, often contracted into a stalk by the abortion of the base, and sometimes even 1-seeded, bursting elastically with 2 valves. Dissepiment opposite the valves, separating into two pieces through the axis (the middle being sometimes open); these pieces attached to the valves, sometimes separating from them with elasticity; entire, or occasionally spontaneously splitting in two, their inner edge bearing the seeds. Seeds roundish, hanging by hard, cup-shaped, or usually hooked ascending processes of the placenta; testa loose; albumen 0; embryo curved or straight; cotyledons large, roundish; radicle taper, descending, and at the same time centripetal, curved, or straight.

Fig. CCCCLVIII.

In a majority of cases these plants are to be recognised by the presence of large leafy bracts, in the axils of which the flowers are concealed, and also by their calyx being composed of deeply imbricated sepals forming quite a broken whorl. But their most exact difference from the other Orders of the Alliance consists in the singular structure of their placenta, which expands into hard woody processes, which are most commonly hooked. In the form of their embryo they agree with Bignoniads, but the cotyledons are more fleshy, and their seeds are never winged. From Figworts they are absolutely divided by the absence of albumen, as well as by their placental processes and large fleshy cotyledons. A singular want of development occurs in the calyx of the genera Thunbergia, Mendozia, and Clistax, in which that organ is sometimes reduced to a mere obsolete ring, its place being supplied by bracts. Mendozia is also remarkable for its fruit being a 1-seeded drupe, with crumpled chrysaloid cotyledons. Mr. Bentham states that the placental processes are sometimes absent; in such cases the embryo can be alone relied upon.

Fig. CCCCLVIII.—Nelsonia campestris. 1. flowers; 2. pistil; 3. capsule; 4. cross section of a seed.

An elaborate account of this Order has been published by Professor Nees v. Esenbeck, in his Dissertation upon the Indian species of Dr. Wallich's Herbarium, in the work above quoted. It is there that the mass of genera was first revised, their limits investigated, and a natural arrangement of them proposed. This eminent Botanist adopts the opinion of Dr. Brown, that among Acanthads the most valuable of all characters resides in the placental processes, and accordingly his three great tribes are defined thus:—I. *Thunbergieæ:* Processes expanded into a horny cup and adnate to the seed, which they support. II. *Nelsonieæ:* Processes contracted into a papilla which bears (not carries) the seed. Seeds small and pitted. III. *Echmatacanthi:* Seeds supported by hooked processes. The subordinate divisions are formed upon considerations connected with the form of the corolla, the number of stamens, the condition of the anther, &c.; and here he differs widely from Dr. Brown in his estimate of the relative value of characters. A second arrangement has been since proposed by Professor Meisner, who attaches less importance to the placental processes, and adds two tribes called Russeggereæ and Mendozieæ. He truly observes that there are few natural Orders which now demand, in so eminent a degree, a searching investigation as that of Acanthads. Professor Nees v. Esenbeck chiefly occupied himself with Indian species: but the crowds of Africans and Americans which load the shelves of all large herbaria, attest how small a proportion the former bear to the whole of the Order.

Professor Nees v. Esenbeck entertains the opinion that the fruit of Acanthads consists of 4 carpels, alternate with the sepals, a fifth, answering to the space between the two lower sepals, being constantly deficient. He says that their union may be easily discovered when they are very young, and that each has its own midrib and three wavy veins at its base; he compares these carpels to the bractlet of Adhatoda Betonica, and he states that the placentæ of the upper edge of the upper carpellary leaves, and of the lower edge of the lower, are constantly imperfect.—*Wallich Pl. As. Rar.* iii. p. 73.

Acanthads are almost entirely tropical, and in such regions extremely common, constituting in fact a large part of the weedy herbage. It is only in some rare instances that they advance far to the north, as in the genus Acanthus found in Greece, and in a few species inhabiting the United States.

They are of very slender importance to man. The greater part are mere weeds; many, however, are plants of great beauty, especially the species of Justicia, Aphelandra, and Ruellia. For the most part they are mucilaginous and slightly bitter; occasionally the bitterness increases, and they become pectoral medicines; some are dyers' plants. The genuine Acanths, formerly called Brancursines, whose beautifully-lobed and sinuated leaves furnished the noble ornament of the Corinthian capital, are emollients; so is Justicia biflora, an Egyptian plant. The flowers, leaves, and fruits of Adhatoda are bitterish, subaromatic, and said to be antispasmodic. Justicia pectoralis, boiled in sugar, yields a sweet-scented syrup, which is considered in Jamaica a stomachic. The leaves and stalks of Gendarussa vulgaris have, when rubbed, a strong and not unpleasant smell, and are, after being roasted, prescribed in India in cases of chronic rheumatism, attended with swelling in the joints. The basis of a famous French bitter tincture, called Drogue amère, highly valued for its stomachic and tonic properties, is the Justicia paniculata, called Creyat in India. The leaves of Ruellia strepens are subacrid. Justicia Ecbolium is one of the diuretics. A valuable deep-blue dye, called Room, is obtained in Assam from a species of Ruellia.—*Griffith in Journ. As. Soc., May,* 1837, p. 326.

GENERA.

I. THUNBERGEÆ.

Thunbergia, *Linn.*
 Diplocalymma, Spren.
 Flemingia, Ham.
Meyenia, *Nees.*
Hexacentris, *Nees.*
Mendozia, *Velloz.*
 Mendoncia, Vell.

II. NELSONEÆ.

Elytraria, *Vahl.*
Nelsonia, *R. Br.*
Adenosma, *R. Br.*
Ebermeyera, *Nees.*
Erythracanthus, *Nees.*
Gymnacanthus, *Nees.*

III. ECHMATACANTHI.

HYGROPHILIDÆ.

Hemiadelphis, *Nees.*
Physichilus, *Nees.*
Hygrophila, *R. Br.*
Nomaphila, *Blum.*
Polyechma, *Hochst.*

RUELLIDÆ.

Dyschoriste, *Nees.*
Chætacanthus, *Nees.*
Dipteracanthus, *Nees.*
Dizygandra, *Meisner.*
Aphragmia, *Nees.*
Petalidium, *Nees.*
Calophanes, *Don.*
Micræa, *Miers.*
Salpixanthus, *Hooker.*
Ruellia, *Linn.*
Phlebophyllum, *Nees.*
Buteræa, *Nees.*
Adenacanthus, *Nees.*
Stephanophysum, *Pohl.*
Stenosiphonium, *Nees.*
Strobilanthes, *Blum.*
 Pteracanthus, Nees.
 Apopedania, Bl.
 Sympagis, Nees.
Stenandrium, *Nees.*
Æchmanthera, *Nees.*
Goldfussia, *Nees.*
Asystasia, *Blum.*
Cryphiacanthus, *Nees.*
Echinacanthus, *Nees.*
Leptacanthus, *Nees.*
Trichanthera, *Kunth.*

BARLERIDÆ.

Asteracantha, *Nees.*
 Tenoria, Denh.
Barleria, *Linn.*
Crabbea, *Harv.*
Lophostachys, *Pohl.*
Ætheilema, *R. Br.*
Whitfieldia, *Hooker.*
Geissomeria, *Lindl.*
Lepidagathis, *Willd.*
 Apolepis, Bl.
Neuracanthus, *Nees.*
Corythacanthus, *Nees.*

ACANTHIDÆ.

Blepharis, *Juss.*
Dilivaria, *Juss.*
 ? Cheilopsis, Moq.
Blepharacanthus, *Nees.*
Acanthus, *Tournef.*
Acanthodium, *Del.*
Monechma, *Hochst.*
Russeggera, *Endl.*
Schwabea, *Endl.*
Athlianthus, *Endl.*

JUSTICIADÆ.

Crossandra, *Salisb.*
 Harrachia, Jacq. f.
Aphelandra, *R. Br.*
 Synandra, Schrad.
 Hemitome, Nees.
Strobilorachis, *Kl.*
Porphyrocoma, *Hooker.*
Endopogon, *Nees.*
Loxanthus, *Nees.*
Phlogacanthus, *Nees.*
Cryptophragmium, *Nees.*
Haplanthera, *Hochst.*
Rostellularia, *Reichenb.*

Rostellaria, Nees.
Hemichoriste, *Wall.*
Odontonema, *Nees.*
Graptophyllum, *Nees.*
Beloperone, *Nees.*
Schaueria, *Nees.*
Tyloglossa, *Hochst.*
Gendarussa, *Nees.*
Adhatoda, Herm.
Rhytiglossa, *Nees.*
Leptostachya, *Nees.*
Gymnostachyum, *Nees.*
Eranthemum, *R. Br.*
Chameranthemum, *Nees.*
Justicia, *Nees.*
Rhinacanthus, *Nees.*

DICLIPTERIDÆ.

Blechum, *P. Br.*
Rungia, *Nees.*
Dicliptera, *Juss.*
Dianthera, Soland.
Amphiscopia, *Nees.*
Peristrophe, *Nees.*
Sautiera, *Decaisne.*
Hypöestes, *Soland.*
Micranthus, Wendl.
Phaylopsis, Willd.
Rhaphidospora, *Nees.*
Monothecium, *Hochst.*

ANDROGRAPHIDÆ.

Erianthera, *Nees.*
Haplanthus, *Nees.*
Andrographis, *Wall.*

* * *

? Clistax, *Mart.*
? Staurogyne, *Wall.*
? Brillantaisia, *Palis.*
? Banjolea, *Bowd.*
? Meissarrhena, *R. Br.*
? Hydromestus, *Scheidw.*
? Neowiedia, *Schr.*
Fabria, *E. M.*
Ramusia, *E. M.*
Duvernoia, *E. M.*
Campylostemon, *E. M.*
Anthocometes, *E. M.*
? Heptas, *Meisn.*
Septas, Lour.

NUMBERS. GEN. 105. Sp. 750 ?

Verbenaceæ.
POSITION.—Bignoniaceæ.—ACANTHACEÆ.—Scrophulariaceæ.

Fig. CCCCLIX.

Fig. CCCCLIX.—Aphelandra cristata.—*Paxton.*

ORDER CCLXIV. SCROPHULARIACEÆ.—FIGWORTS.

Scrophulariæ, *Juss. Gen.* 117. (1789).—Scrophularineæ, *R. Brown Prodr.* 433. (1810); *Bartl. Ord. Nat.* 169. (1830); *Bentham in Botan. Register, June* (1835); *Scrophular. Ind.* (1835); *Don in Ed. Phil. Journ,* (*July,* 1835); *Endl. Gen. p.* 670.—Pediculares, *Juss. Gen.* 99. (1789) *in part.*—Personatæ, *DC. Fl. Fr.* 3. 573. (1815); *Don in Edinb. Phil. Journ.* (*July,* 1835).—Antirrhineæ, *DC. and Duby,* 342. (1828).—Halleriaceæ, *Link Handb.* 1. 506. (1829) *a* § *of* Personatæ.—Scopariaceæ, *Ib.* 822. *the same.*—Erineæ, *Ib.* 510. *the same.*—Cheloneæ, Aragoaceæ, Sibthorpiaceæ, *Don in Edinb. Phil. Journ.* (*July,* 1835).—Melampyraceæ, *Rich. Anal. du Fruit.* (1808).—Rhinanthaceæ, *DC. Fl. Fr.* 3. 454. (1815).—Pediculares, *Juss. Gen.* 99. (1789) *in part*; *Duvau in Ann. des Sc. Nat.* 8. 180. (1826).

DIAGNOSIS.—*Bignonial Exogens, with axile placentæ, albuminous seeds, and cotyledons scarcely larger, or not so large as the radicle.*

Herbs, under-shrubs, or sometimes shrubs, usually scentless, but sometimes fœtid, rarely aromatic. Leaves opposite, whorled, or alternate. Flowers axillary, or racemose, rarely spiked. Peduncles opposite or alternate; sometimes simple and 1-flowered, sometimes many-flowered in dichotomous cymes. Calyx inferior, persistent, pentamerous, or by abortion tetramerous, the sepals sometimes united almost to the point, sometimes only at their base, sometimes altogether distinct and imbricated, often unequal, the upper one being largest, the two lowest smaller, the lateral ones smallest. Corolla monopetalous, pentamerous, or, the upper petals being united to their points, tetramerous; tube short or long; limb flat or erect, nearly equally divided or bilabiate, imbricated in æstivation. Stamens in a single series opposite the sepals; the uppermost altogether deficient, or sterile, or very rarely fertile, and shorter than the others; the two lateral equal, antheriferous or very rarely sterile and abortive; the lower equal, sometimes fertile and equal to the lateral ones or longer, often sterile or deficient; anthers 2-celled, or, by growing together or half-disappearing, one-celled, opening longitudinally. Ovary superior, 2-celled, many-seeded; style simple or rarely shortly bifid; stigmatic surface terminal when the style is entire, either very thin and punctiform, or more or less pulvinate or capitate, entire or emarginate; when the style is bifid the stigmatic surface either lines the inner surface of the lobes, or their margins, or rarely forms a

1

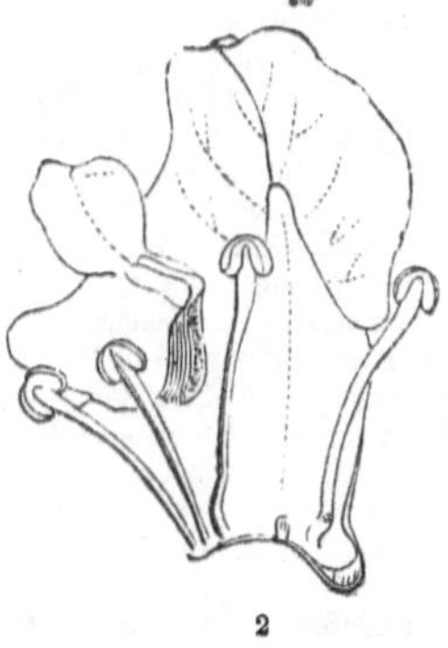
2

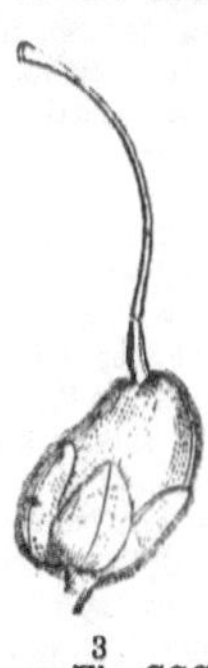
3

4

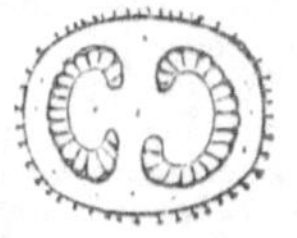
5

6

Fig. CCCCLX.

Fig. CCCCLX.—1. Digitalis purpurea; 2. corolla of Antirrhinum majus cut open; 3. its pistil; 4. its ripe fruit; 5. cross section of its ovary; 6. section of its seed.

pulvinate mass in the fork. Fruit capsular, seldom berried, dicarpellary, 2-celled sometimes with 2 entire or bifid valves, sometimes with 4 entire ones, sometimes opening by pores or lids, very rarely almost indehiscent; dissepiment parallel or opposite to the valves, finally loose in the centre, or altogether. Placentæ adhering to the dissepiment, sometimes when mature separate and forming 1-2 central columns. Seeds indefinite, rarely definite, albuminous; embryo orthotropal, heterotropal, or antitropal, but slightly curved.—*Bentham.*

The capsular monopetalous genera of Dicotyledons, with a superior ovary, albuminous seeds, and irregular diandrous or didynamous stamens, were separated by Jussieu into two Orders, which he called Scrophulariæ and Pediculares, distinguished from each other by the dehiscence of their fruit: the former being septicidal, and the latter loculicidal. Brown, in his *Prodromus*, pointed out the insufficiency of this character, which is often not even of generic value, and he combined the Orders of Jussieu under the common name of Scrophularineæ (Figworts). This opinion has been adopted by subsequent writers, with the exception of De Candolle, who, in Duby's *Botanicon Gallicon* (1828) adheres to the old division of Jussieu, the names being changed into Antirrhineæ and Rhinanthaceæ. Notwithstanding the almost universal assent to the identity of the two Orders of Jussieu, some separations have been made upon different principles from those of that learned Botanist. Thus Broomrapes have been distinguished by himself; Gesnerworts by Nees Von Esenbeck; and Melampyraceæ by Richard. The two former are adopted by all Botanists; the latter group has not been generally received. I formerly admitted it, upon the ground of its definite ascending seeds and inverted embryo; but subsequent observation led me to think that by excluding from the character all consideration of the number and direction of the seeds, an Order would be formed, agreeing in a peculiar habit, and in the radicle of the embryo not being presented to the hilum, to which the name of Rhinanths might conveniently be retained. According to this view of the subject, Figworts would include no genus the embryo of which is not orthotropal, and in Rhinanths it could be antitropal or heterotropal. But although the attachment of the seeds of Rhinanths is generally lateral, yet sometimes the radicle points to the hilum; but it is more generally removed from it. The ovules are never fewer than 2 in each cell, often numerous, and there are sometimes, though rarely, 2 ovules only in the ovary of some of the tribes of Figworts. And therefore the ground for separating Rhinanths from Figworts sinks from under us.

The number of synonymous names above quoted, shows into how many more supposed Orders the old Scrophulariæ have been broken by one author or another. The whole matter has, however, been investigated by Mr. Bentham, who has treated the question in both a philosophical and practical way, and who concludes that in fact all the supposed Orders are really sections of one great Natural Order.

Mr. Bentham remarks that the nearest Order to Figworts is undoubtedly that of Nightshades, through the medium of Salpiglossids; so that it becomes necessary to separate them by a purely artificial distinction, considering as Nightshades such genera as have a plaited corolla and 5 stamens, and as Figworts all those in which either the fifth stamen is wanting, or the æstivation of the corolla imbricated. The line would thus be drawn between Petunia and Salpiglossis, two genera closely allied in habit. In the first, however, the decidedly plicate corolla and 5 stamens show it to be a true Nightshade, whilst the slight irregularity of the corolla and the declinate very unequal stamens, indicate an approach to Salpiglossis, which, being always didynamous, with an imbricately æstivating, or obscurely plicate corolla, is a genuine Figwort. Among Verbasceæ the genus Verbascum which is pentandrous, and Celsia, because it cannot be separated from Verbascum, have usually been referred to Nightshades, although no plants nearly allied to Verbascum occur in the latter Order; but the æstivation of the corolla, besides the general habit, leave no doubt that Bartling and others are right in classing these genera among Figworts. A better reason seems to me to be furnished by the manifest tendency to lose a part of the stamens, which occurs in Verbascum.

From the other Orders of this Alliance the Figworts are sufficiently well distinguished. They differ from Pedaliads, Gesnerworts, and Crescentiads in their placenta never being parietal; from Bignoniads and Acanthads in their albuminous seeds and small cotyledons; from Butterworts in their axile, not free central placentæ.

Some Figworts approach Broomrapes in the peculiar habit of that Order, especially the Buchnereæ, among which most, if not all, the Strigas are parasitical, and Buchnera hydrabadensis is actually leafless like a Broomrape.

The two tribes of Mitrasacmeæ and Buddleæ approach Loganiads in their leaves being connected by a transverse line, which occasionally expands in the form of stipules; but they differ in their flowers being irregular, at least in æstivation, one lateral lobe being outermost, whilst the upper one is innermost. In all Loganiads which I have examined, the æstivation is either regularly convolute or valvate. The irregularity of the corolla sometimes assumes a very peculiar appearance, owing to its tendency to form pouches or spurs. This is particularly striking in the genera Linaria and Antirrhinum, in which the corolla takes a direction upwards, so as to form a convexity on the under side of the limb, the result of which is that form of corolla called ringent; and also a direction downwards, which produces a long spur. In the genus Calceolaria it causes the anterior face to assume quite the appearance of a slipper. On the other hand, in the genus Veronica, both irregularity and want of symmetry almost wholly disappear.

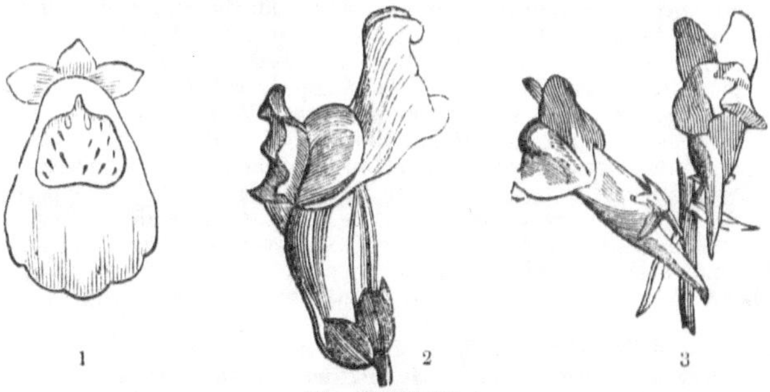

1 2 3

Fig. CCCCLXI.

In this Order many species have a stigma composed of two highly irritable plates, one placed next the back and the other next the front of the flower. When the corolla first expands, these plates stand apart and are even turned back a little; but when touched they collapse suddenly and with some force. This phenomenon has been described by Mr. Henderson in the *Annals of Nat. Hist.* vol. 6. p. 51.

A curious genus, called Schwenkia, with clavate glands growing from the edge of its corolla, usually referred to this Order, was formerly rejected at my instance, the stamens having appeared to me to be opposite the lobes of the corolla. It is more than twenty years since I had an opportunity of examining it, and Mr. Bentham now assures me that the real petals are the aforesaid glands with which the stamens alternate, and that it is a genuine member of the Order of Figworts.

These plants are found in abundance in all parts of the world, from the coldest regions in which the vegetation of flowering plants takes place, to the hottest places within the tropics. One species is found in Melville Island; in the middle of Europe they form about a 26th of the flowering plants, and in North America about a 36th. In all India, New Holland, and South America they are common; and, finally, the sterile shores of Tierra del Fuego are ornamented with several genera.

The species are generally acrid, bitterish, and suspected. The leaves and roots of Scrophularia aquatica, and perhaps nodosa, some species of Calceolaria, and many others, act as purgatives, or even as emetics. In Digitalis purpurea, ochroleuca, lævigata, ferruginea and other species, this quality is so much increased, that its effects become highly dangerous; the powdered leaves, or an extract of them, produce vomiting, dejection, and vertigo, increase the secretion of the saliva and urine, lower the pulse, and even cause death. The Mulleins approach Digitalis in this respect; the seeds of Verbascum Thaspus and nigrum are used by poachers to poison fish, and the flowers of V. Lychnitis are sometimes used to destroy mice; the foliage of these plants is acrid and bitterish. The leaves of Mimulus guttatus are eatable as salad. The juice of the leaves of Torenia asiatica are considered, on the Malabar coast, a cure for gonorrhœa. An infusion of Scoparia dulcis is used by the Indians of Spanish America to cure agues, and in Brazil against hæmorrhoidal affections. Euphrasia officinalis is slightly bitter and aromatic, and has been employed with success by Kranichfeld in catarrhal inflammations of the eye; he has also found it beneficial in cough, hoarseness, earache and headache, which have supervened in catarrhal affections.—*Med. Gaz.* xx. 528. Cows are said to be fond of Melampyrum pratense; and Linnæus says the best and yellowest butter is made where it abounds. The Pedicularids are acrid, but are eaten by goats. Nearly all that tribe turn black in drying. Herpestes amara, an Indian herb, is intensely bitter; but its properties have not been investigated. Picrorhiza derives its name from the bitterness of its roots; it is used on that account in the native medicine of India. Vandellia diffusa is said to be of great value in Guayana as an antibilious emetic

Fig. CCCCLXI.—1. slipper-shaped corolla of Calceolaria; 2. ringent corolla of Antirrhinum; 3. ringent and spurred corolla of Linaria.

and febrifuge, and a most efficacious remedy in malignant fevers and dysentery, especially in cases depending on a disordered state of the liver. It is called Haimarada by the Arowak Indians, and Bitter Blain by the Dutch Creoles. Linaria vulgaris is reputed to be purgative and diuretic. It is bitter. Its flowers have been recommended in decoction as a wash for chronic diseases of the skin ; and that it would not be an inactive lotion seems probable from the fact that in London the plant is occasionally boiled in milk for the purpose of destroying flies. Linaria cymbalaria has a warm cress-like flavour, and has been recommended as an antiscorbutic. Hamilton says that in India it is given with sugar in cure of diabetes, and from the report of its influence over that disorder, it well deserves to be tried by the European practitioner. It is, however, probable that Dr. Hamilton's remarks apply to L. ramosissima. Linaria Elatine is said to be bitter and purgative. Gratiola officinalis was formerly called Gratia Dei, on account of its efficiency as a medicine. It is extremely bitter, acts violently both as a purgative and emetic, and has been said to be the basis of the famous gout medicine called Eau médicinale, which, as its active principle appears to be of the nature of Veratria, is not improbable. Gratiola is said to have been found serviceable in cases of hypochondriasis. In overdoses it is a violent poison, and according to Haller, it renders by its abundance some of the Swiss meadows useless as pastures. G. peruviana, *Linn.*, has purgative and emetic leaves and roots. Bramia serrata is employed in Brazil in the preparation of baths for rheumatic patients ; it has a strong penetrating odour.—*Martius, Choix des pl.* p. 12.

The whole plant of Franciscea uniflora, and especially its large root, is called Manacá in Brazil, and is found of great value in exciting the lymphatic system ; in consequence of its large use in syphilis it is called by the Portuguese Mercurio vegetal; the inner bark and all the herbaceous parts are nauseously bitter ; it is regarded as a purgative, emetic, emmenagogue, and alexipharmic ; in over-doses it is found an acrid poison.—*Martius, Mat. Med. Bras.* p. 67.

One or two species are named as dyers' plants. The flowers of Linaria vulgaris are employed in some places to give a yellow colour ; and the roots of Calceolaria arachnoidea are largely collected in Chili, under the name of Relbun, for dyeing woollen cloths crimson.—*Bot. Mag.* t. 2915.

GENERA.

[For which I am indebted to Mr. Bentham's kindness, Sept., 1845.]

Suborder 1. SALPIGLOSSIDEÆ.—*Benth.* –Inflorescence entirely centrifugal. Æstivation of the corolla either altogether plaited, or plaited-imbricate, the two upper segments being external.—*G. B.*

Duboisia, *Br.*
Anthocercis, *Lab.*
Schwenkia, *L.*
 Chætochilus, Vahl.
 Mathea, Vell. ?
Leptoglossis, *Benth.*
Heteranthia, *Nees et Mrt.*
 Vrolikia, Spreng.
Browallia, *L.*
Brunsfelsia, Plum.
 Franciscea, Pohl.
Salpiglossis, *Ruiz et Pav.*
Schizanthus, *Ruiz et Pav.*

Suborder 2. ANTIRRHINIDEÆ.—*Benth.* —Inflorescence entirely centripetal or compound, (i. e., general inflorescence or primary inflorescence centripetal, partial inflorescence centrifugal.) Æstivation of the corolla bilabiately imbricated, the two upper segments being external.—*G. B.*

Tribe 1. Calceolareæ.

Calceolaria, *Feuill.*
 Jovellana, Ruiz et Pav.
 Bœa, Pers. not Com.

Tribe 2. Verbasceæ.

Verbascum, *L.*
Ianthe, *Griseb.*
Celsia, *L.*
 Ditaxia, Rafin.
 Nefflea, Benth.
 Thapsandra, Griseb.
Staurophragma, *Fisch. et Mey.*

Tribe 3. Hememerideæ.

Alonsoa, *Ruiz et Pav.*
Schistanthe, *Kunze.*
Angelonia, *Humb. et Bp.*
 Physidium, Schrad.
 Schelveria, Nees.
 Thylacantha, Nees.
Hemimeris, *L.*
Diascia, *Link et Otto.*
Colpias, *E. Mey.*
Nemesia, *Vent.*
Diclis, *Benth.*

Tribe 4. Antirrhineæ.

Linaria, *Tourn.*
 Elatine, Mœnch.
 Cymbalaria.
 Kicksia, Dumort.
Anarrhinum, *Desf.*
 Cardiotheca, Ehrenb.
 Simbuleta, Forsk.
Antirrhinum, *L.*
 Orontium, Pers.
Maurandia, *Ort.*
 Usteria, Chav.
Galvezia, *Domb.*
 Agassizia, Chav.
Lophospermum, *Don.*
Rhodochiton, *Zucc.*

Tribe 5. Cheloneæ.

Phygelius, *E. Mey.*
Pawlownia, *Zucc.*
Wightia, *Wall.*
Diplanthera, *Banks et Soland.*
Halleria, *L.*
Scrophularia, *Tourn.*
Collinsia, *Nutt.*
Chelone, *L.*
Pentstemon, *Lhér.*
 Elmigera, Rchb.
 Dasanthera, Raf.
Chionophila, *Benth.*
Tetranema, *Benth.*
Russelia, *Jacq.*
Freylinia, *Colla.*
Anastrabe, *E. Mey.*
Teedia, *Rud.*
 Borkhausenia, Roth.
Ixianthes, *Benth.*
Leucocarpus, *Don.*
 Hemichæna, Benth.

Tribe 6. Escobedieæ.

Escobedia, *Ruiz et Pav.*
 Silvia, Vell.
Physocalyx, *Pohl.*
Melasma, *Berg.*
 Nigrina, Linn.
 Lyncea, Cham. et Scht.
 Gastromeria, Don.
Alectra, *Thunb.*
 Starbia, Dup.Thou.
 Glossostylis, Cham.

Tribe 7. Gratioleæ.

Subtribe 1. *Aptosimeæ.*

Leucophyllum, *H. B. K.*
Aptosimum, *Burch.*
 Ohlendorfia, Lehm.
 Chilostigma, Hochst.
Peliostomum, *E. Mey.*
Anticharis, *Endl.*
 Meissarrhena, Br.
Doratanthera, *Benth.*

Subtribe 2. *Manuleæ.*

Nycterinia, *Don.*
 Zaluzianskya, J. W. Schm.
Polycarena, *Benth.*
Phyllopodium, *Benth.*
Sphenandra, *Benth.*
Chænostoma, *Benth.*
Lyperia, *Benth.*
Sutera, Roth.
Manulea, *L.*
 Nemia, Berg.

Subtribe 3. *Eugratioleæ.*

Diplacus, *Nutt.*
Mimulus, *L.*
 Erythranthe, Spach.
 Uvedalia, Br.
Eunanus, *Benth.*
Melosperma, *Benth.*
Mazus, *Lour.*
 Hornemannia, Rchb.
Dodartia, *Linn.*
Lindenbergia, *Link et Ot.*
 Brachycoris, Schrad.
 Bovea, Decaisne.
Beyrichia, *Cham.*

Achetaria, Cham.
Matourea, Vahl ?
Tetraulacium, *Turcz.*
Pterostigma, *Benth.*
Stemodia, *L.*
Adenosma, Br.
Unanuea, Ruiz et Pav.
Matourea, Aubl. ?
Morgania, *Br.*
Limnophila, *Br.*
Hydropityon, Gærtn.
Cybbanthera, Ham.
Ambulia, Lam.
Conobea, *Aubl.*
Leucospora, Nutt.
Sphærotheca, Cham.
Lafuentea, *Lag.*
Durieua, Mérat.
Schistophragma, *Benth.*
Herpestis, *Gærtn.*
Mecardonia, Mart.
Caconapea, Cham.
Ranaria, Cham.
Bramia, Lam.
Monniera, R. Br.
Calytriplex, R. P.
Septas, Lour.
Mella, Vand.
Heinzelmannia, Neck.
Bacopa, *Aubl.*
Geochorda, *Cham.*
Ildefonsia, *Gardn.*
Gratiola, *L.*
Sophronanthe, Benth.
Nibora, Raf.
Dopatrium, *Ham.*

Subtribe 4. *Lindernieæ.*

Curanga, *Juss.*
Symphyllium, Griff.
Artanema, *Don.*
Achimenes, Vahl.
Diceros, Pers.
Torenia, *L.*
Nortenia, Thou.
Craterostigma, Hochst.
Dunalia, R. Br.
Vandellia, *L.*
Tittmannia, Rchb.
Ilyogeton, Endl.
Ellobum, Blum. ?
Diceros, Blum. ?
Vriesia, Hassk.
Lindernia, *All.*
Ilysanthes, *Raf.*
Bonnaya, *Link et Otto.*

Peplidium, *Del.*
Micranthemum, *Mich.*
Pinarda, Vell.
Globifera, Gmel.
Hemianthus, *Nutt.*

Suborder 3. RHINANTHIDEÆ.—*Benth.*—Inflorescence entirely centripetal or compound, (except perhaps a few Buddleieæ). Æstivation quincuncial or irregularly imbricated, one of the lateral segments being generally external, while the two upper are always internal.—*G. B.*

Tribe 1. Sibthorpeæ.

Amphianthus, *Tour.*
Hydranthelium, *H.B.K.*
Willichia, Spr. non L.
Glossostigma, *Arn.*
Tricholoma, *Benth.*
Limosella, *L.*
Sibthorpia, *L.*
Disandra, L.
Willichia, L. ?
Hornemannia, *Benth.*
Mazus pinnatus, Wall.
Hemiphragma, *Wall.*
Capraria, *L.*
Xuarezia, Ruiz et Pav.
Pogostoma, *Schrad.*
Camptoloma, *Benth.*
Scoparia, *L.*

Tribe 2. Buddleeæ.

Microcarpæa, *Br.*
Bryodes, *Benth.*
Polypremum, *L.*
Gomphostigma, *Turcz.*
Nuxia, *Vent.*
Chilianthus, Burch.
Lachnopylis, Hochst.
Psiloxylon, Dup. Thou.
Buddlea, *L.*

Tribe 3. Digitaleæ.

Isoplexis, *Lindl.*
Digitalis, *L.*

Erinus, L.
Picrorhiza, *Royle.*
Wulfenia, *Jacq.*
Synthyris, *Benth.*
Calorhabdos, *Benth.*

Tribe 4. Veroniceæ.

Pæderota, *Linn.*
Veronica, *Linn.*
Hebe, Juss.
Leptandra, Nutt.
Callistachya, Raf.
Eustachya, Raf.
Aidelus, Spr.
Cochlidiospermum, Rb.
Diplophyllum, Lehm.
Omphalospora, Besser.
Aragoa, *H. B. K.*
Ourisia, *Comm.*
Dichroma, Cav.

Tribe 5. Buchnereæ.

Buchnera, *Linn.*
Piripea, Aubl.
Striga, *Lour.*
Campuleia, Dup. Thou.
Rhamphicarpa, *Benth.*
Macrosiphon, Hochst.
Cycnium, *E. Mey.*
Hyobanche, *Thunb.*

Tribe 6. Gerardieæ.

Hydrotriche, *Zucc.* ?
Campylanthus, *Roth.* ?

Radamæa, *Benth.*
Rhaphispermum, *Benth.*
Micrargeria, *Benth.*
Leptorhabdos, *Schrenck.*
Dargeria, Decaisne.
Seymeria, *Pers.*
Afzelia, Gmel.
Otophylla, *Benth.*
Sylvia, *Benth.*
Buddlea, *L.*
Macranthera, *Torrey.*
Esterhazya, *Mikan.*
Gerardia, *Linn.*
Virgularia, Ruiz Pav.
Dasystoma, *Raf.*
Pagesia, Raf.

Graderia, *Benth.*
Sopubia, *Ham.*
Rhaphidophyllum, Hhs.
Aulaya, *Harv.*
Harveya, *Hook.*
Centranthera, *Br.*
Razumovia, Spr.
Purshia, Dennst.

Tribe 7. Euphrasieæ.

Castilleja, *Linn. fil.*
Euchroma, Nutt.
Orthocarpus, *Nutt.*
Triphysaria, Fisch.
Oncorrhynchus, Lehm.
Adenostegia, *Benth.*
Schwalbea, *Linn.*
Siphonostegia, *Benth.*
Synnema, *Benth.*
Phtheirospermum, *Bge.*
Lamourouxia, *H. B. K.*
Eufragia, *Col.*
Trixago, *Stev.*
Bellardia, All.
Lasiopera, Hoffm.
Bartsia, *Linn.*
Stæhelina, Hall.
Odontites, *Hall.*
Euphrasia, *Linn.*
Cymbaria, *Linn.*
Bungea, *C. A. Mey.*
Rhinanthus, *Linn.*
Alectorolophus, Bieb.
Rhynchocorys, *Griseb.*
Elephas, Tourn.
Rhinanthus, Bieb.
Pedicularis, *Linn.*
Melampyrum, *Linn.*
Tozzia, *Linn.*

Genera insufficiently known.

Diceros, *Lour.*
Gomara, *Ruiz et Pav.*
Picria, *Lour.*
Poarium, *Desv.*
Sanchezia, *Ruiz et Pav.*
Tala, *Blanco.*
Parentucellia, *Viv.*
Nicodemia, *Tenor.*

NUMBERS. GEN. 176. Sp. 1814.—*Walpers.*

Verbenaceæ.

POSITION.—Bignoniaceæ.—SCROPHULARIACEÆ.—Lentibulariaceæ.

ORDER CCLXV. LENTIBULARIACEÆ.—BUTTERWORTS.

Lentibulariæ, *Richard in Flor. Paris, p.* 26. (1808).—Utriculinæ, *Hoffmannsegg et Link. Fl. Port.* (1806).—Lentibulariæ, *R. Brown Prodr.* 429. (1810) ; *Aug. de St. Hilaire, Ann. Sc. Nat.* 2 *ser.* xi. 149.—Utricularieæ, *Endl. Gen.* clv. ; *Meisn. Gen. p.* 314 ; *DC. Prodr.* 8. 2.

DIAGNOSIS.—*Bignonial Exogens, with a free central placenta, minute seeds without albumen, and cotyledons much smaller than the radicle.*

Herbaceous plants, living in water or marshes. Leaves radical, undivided ; or compound, resembling roots, and bearing little vesicles. Scapes either with minute stipule-like scales, or naked ; sometimes with whorled vesicles ; generally undivided. Flowers single, or in spikes, or in many-flowered racemes ; with a single bract, rarely without bracts. Calyx divided, persistent, inferior. Corolla monopetalous, hypogynous, irregular, bilabiate. Stamens 2, included within the corolla, and inserted into its base ; anthers 1-celled, sometimes contracted in the middle. Ovary composed of 2 valvate carpellary leaves, and therefore 1-celled ; style 1, very short ; stigma bilabiate ; ovules 00, anatropal, placed on a free central placenta. Capsule 1-celled, many-seeded, with a large central placenta. Seeds minute, without albumen ; embryo sometimes undivided. Radicle next the hilum.

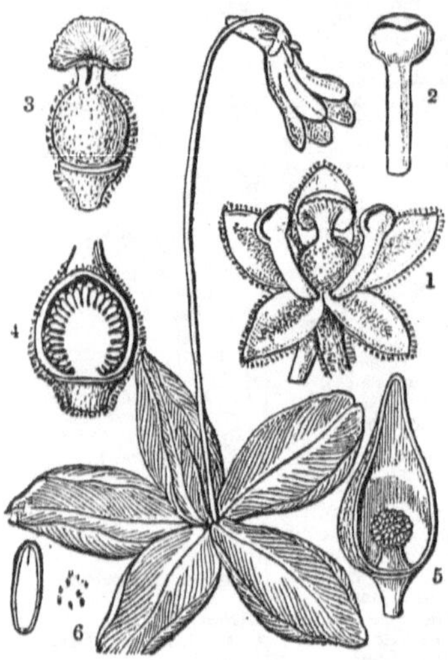

Fig. CCCCLXII.

The central free placenta and minute exalbuminous seeds are the principal points of distinction between these and Figworts, to which their habit approximates them. They are known from Primworts by their irregular flowers, exalbuminous embryo, and didynamous or unsymmetrical stamens, alternate with the segments of the corolla.

Mr. Bentham has remarked that they are very closely allied to Figworts, "having the same calyx, corolla, stamens and bivalve capsule, but distinguished solely by their really unilocular fruit, with a free central placenta, and the minuteness of their embryo. In respect of the former character, they come very near to Limosella, Lindernia, and other Gratioleæ, with parallel dissepiments and entire valves ; for in these plants the dissepiment is very thin, and usually detaches itself from the valves before maturity, so that being concealed by the seeds, which fill nearly the whole capsule, it often escapes observation, and many of these genera have frequently been described as having a unilocular fruit."

Natives of marshes, or rivulets, or fountains, in all parts of the world, especially within the tropics. The Genliseas are exclusively Brazilian.

Pinguicula vulgaris has the property of giving consistence to milk, and of preventing its separating into either whey or cream. It is pretended that its leaves rot sheep; when fresh they are slightly purgative and vulnerary. Linnæus says that the solid milk of the Laplanders is prepared by pouring it warm and fresh from the cow over a strainer on which fresh leaves of Pinguicula have been laid. The milk, after passing among them, is left for a day or two to stand, until it begins to turn sour ; it throws up no cream, but becomes compact and tenacious, and most delicious in taste. It is not necessary, that fresh leaves should be used after the milk is once turned : on the contrary, a small portion of this solid milk will act upon that which is fresh, in the manner of yeast.

GENERA.

Utricularia, *Linn.*	Pinguicula, *Tournef.*
Lentibularia, Vaill.	*Brandonia*, Reichenb.
Genlisea, *St. Hil.*	Polypompholyx, *Lehm.*
	Tetralobus, A. DC.

NUMBERS. GEN. 4. Sp. 175.

Fig. CCCCLXII.—Pinguicula vulgaris —*Nees.* 1. calyx, &c. ; 2. stamen ; 3. pistil ; 4. longitudinal section of it ; 5. half a fruit ; 6. seeds, natural size ; 7. embryo very much magnified.

SUB-CLASS IV. EPIGYNOUS EXOGENS.

IN general the complete adhesion of the tube of the calyx to the ovary through its whole length, and the bisexual flowers, afford a positive mark of distinction for this Sub-Class, which is undoubtedly composed of Orders which form perfectly Natural Alliances, closely related to each other, but indicating very strong lateral affinity for other parts of the system. Thus Campanals approach both Bignonials, in which Gesnerworts have a half-inferior ovary, and Solanals, among which Jaborosa has the stamens almost free from the corolla; Myrtals lean towards Rosals, whose flowers are furnished with a half-inferior ovary in the Order of Appleworts; the frontier of Grossals joins that of Saxifragals. the Currantworts of the one very nearly agreeing with the Cunoniads of the other; Umbellals are completely imitated by Thalictrum among the Crowfoots of the Ranal Alliance, and by Vitâceæ among Berberals; and finally, the Asaral Alliance has its analogy in Helwingiads among the Diclinous Garryal Alliance, not to name many other similar cases. So that the Epigynous Sub-Class may be likened to a great kingdom lying in the midst of many others, just as Germany is bordered by France, Holland, Denmark, Poland, Hungary, Turkey, Italy, and Switzerland.

In the two previous Sub-Classes the Epigynous character occasionally breaks out, and sometimes in a very unexpected way; as when the genus Eupomatia appears in the hypogynous Ranal Alliance; and in many of the Saxifragals. In like manner, both Myrtleblooms and Melastomads have species in which the calyx has but little union with the ovary; these are, however, beyond all question, exceptional instances.

It is here assumed that the inferior ovary is always formed by an adhesion of the calyx to its sides. There may, however, in some cases, be justice in the assertion of Schleiden, that the real inferior ovary is caused by a hollowing out of the peduncle, analogous to what takes place in the common Fig. (*Ann. Sc.* 2 *ser. XII.* 374.) Possibly such is its origin in Loranths, Cucurbits, Sándalworts, and others in which no calyx-veins are to be found on the surface of the fruit; and Eschscholtzia may be considered to offer an obvious explanation of this, its peduncle forming, round the base of the ovary, a cup which evidently has nothing to do with the calyx. If that is so, then the structure of Calycanths, the Rose, and many more will bear a similar interpretation. But it is impossible to admit that such is the origin of all the ovaries with a superior calyx. Melastomads, for instance, have evidently a true calyx tube; and even in Umbellifers the presence of an adherent calyx tube is demonstrated by those monsters of the wild Carrot which are sometimes found in fields with their 2 carpels in the condition of ordinary leaves; in such instances these carpellary leaves spring from the central axis, and are surrounded by the tubular but non-adherent calyx.

Whatever may be the true theory of the inferior ovary, it seems to be a very important point of structure, collecting together species having more resemblance to one another than to anything else, and therefore of great value for natural classification.

ALLIANCE L. *CAMPANALES.*—THE CAMPANAL ALLIANCE.

DIAGNOSIS.—*Epigynous Exogens, with dichlamydeous monopetalous flowers, and an embryo with little or no albumen.*

This, which is probably the most extensive of all the groups, in this Work called Alliances, consists of Orders held together in the strictest bond of union. They form two sets, of which the one has the ovary with more than one cell, and the other with one only; but they probably have, in all cases, more than one carpel; and Valerianworts, with one perfect and two seedless cells, completely joins the groups. In what way they pass into Myrtals will be shown when speaking of that Alliance. From the perigynous series they branch off by way of Gesnerworts, which have a half-superior ovary, to Nightshades, among which there are genera, which like Jaborosa, are Bellworts in most respects, except not having an inferior ovary.

NATURAL ORDERS OF CAMPANALS.

Ovary 2- *or more celled. Anthers free, or half united. Stigma naked. Corolla valvate, regular*	266. CAMPANULACEÆ.
Ovary 2- *or more celled. Anthers syngenesious. Stigma surrounded by hairs. Corolla valvate, irregular*	267. LOBELIACEÆ.
Ovary 2- *or more celled. Anthers syngenesious or free. Stigma indusiate. Corolla induplicate*	268. GOODENIACEÆ.
Ovary 2- *or more celled. Stamens and style united into a column. Corolla imbricated*	269. STYLIDIACEÆ.
Ovary 1-*celled. Corolla imbricated. Anthers free. Ovule pendulous. Albumen none*	270. VALERIANACEÆ.
Ovary 1-*celled. Corolla imbricated. Anthers free. Ovule pendulous. Seeds albuminous*	271. DIPSACACEÆ.
Ovary 1-*celled. Corolla valvate. Anthers syngenesious. Ovule pendulous. Seeds albuminous*	272. CALYCERACEÆ.
Ovary 1-*celled. Corolla valvate. Anthers syngenesious. Ovule erect. Albumen none*	273. ASTERACEÆ.

ORDER CCLXVI. CAMPANULACEÆ.—BELLWORTS.

Campanulæ, *Juss. Gen.* 163. (1789) *in part.*—Campanulaceæ, *R. Brown Prodr.* 559. (1810); *Endl. Gen.* cxxv.; *DC. Prodr.* 7. 414.—Cyphiaceæ, *DC. Prodr.* 7. 497.—Sphenocleaceæ, *Martius Conspect.* 162. (1835); *Ed. pr. p.* 238; *DC. Prodr.* 7. 548.—Pongatieæ, *Endl. Gen. p.* 519.—Campanuleæ, *Alph. DC. Monogr.* (1830).

DIAGNOSIS.—*Campanal Exogens, with a 2- or more-celled ovary, free or half-united anthers, naked stigma, and valvate regular corolla.*

Herbaceous plants or under-shrubs, yielding a white milk. Leaves almost always alternate, simple, or deeply divided, without stipules. Flowers single, in racemes,

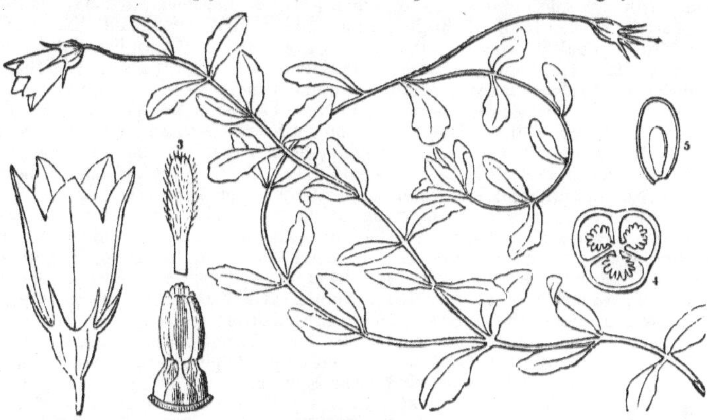

Fig. CCCCLXIII.

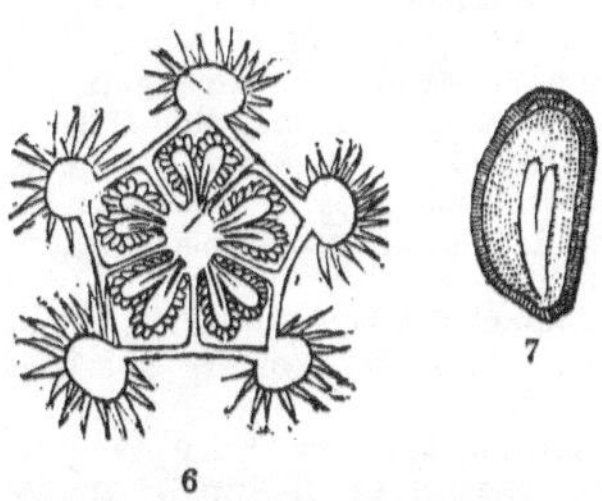

spikes, or panicles, or in heads, usually blue or white, very rarely yellow. Calyx superior, usually 5-lobed (3-8), persistent. Corolla monopetalous, inserted into the top of the calyx, usually 5-lobed (3-8), withering on the fruit, regular; its æstivation valvate. Stamens inserted into the calyx alternately with the lobes of the corolla, to which they are equal in number. Anthers 2-celled, distinct. Pollen spherical. Ovary inferior, with 2 or more polyspermous cells opposite the stamens, or alternate with them; style simple, covered with collecting hairs; stigma naked, simple, or with as many lobes as there are cells. Fruit dry, crowned by the withered calyx and corolla, dehiscing by lateral irregular apertures or by valves at the apex, always loculicidal. Seeds numerous, attached to a placenta in the axis; embryo straight, in the axis of fleshy albumen; radicle next the hilum, longer than the cotyledons.

This Order has been very carefully examined by M. Alphonse De Candolle, the substance of whose observations as to the more important facts connected with it is included in the following remarks:—He considers that Bellworts differ from Lobeliads chiefly in their regular corolla, their stamens being almost always distinct, their pollen spherical (not oval), their stigmas generally long and velvety externally, in the abundance of collecting hairs on the style, and finally in their capsule usually opening laterally. "It is not only in the form," he proceeds, "but also in the number of the parts, that the flower of Bellworts is more regular than that of Lobeliads. Thus, in several Campa-

Fig. CCCCLXIII.—Wahlenbergia procumbens. 1. an entire flower; 2. stamens; 3. a stigma; 4. transverse section of the ovary; 5. a vertical section of a seed, showing the embryo; 6. transverse section of ovary of Campanula Medium; 7. interior of its seed.

nulas the cells of the ovary are equal in number to the stamens and the divisions of the corolla and calyx, which points out the natural symmetry of the flower. In Lobeliads abortion is more frequent. In both groups the innermost organs are abortive more frequently than the outermost. Thus, the number of cells is often smaller (never greater) than that of the stamens; the number of stamens is sometimes smaller (but never larger) than that of the lobes of the corolla; and the same is true of the lobes of the corolla with respect to the calyx. Finally, Lobeliads have sometimes a corolla of a fine bright-red, a colour unknown among Bellworts; nine-tenths of the species of the latter have blue flowers; and those in which the colour varies, and into which a little red enters (as Canarina), are far from having the brilliancy of Lobelia cardinalis for instance. After Lobeliads, the Natural Orders with which Bellworts have the most relation are, no doubt, Goodeniads and Styleworts, which formed part of the Campanulæ of Jussieu. The regular corolla of Bellworts distinguishes them, at first sight, from both those Orders, as well as from Lobeliads. Besides, Campanulas have not the fringed indusium which terminates the style of Goodeniads and surrounds their stigma. Although this organisation approaches that of Lobeliads, and so of Bellworts, it is not less true that it affords an important mark of distinction, and that it is connected with essential differences in the mode of fecundation. Brown has also remarked, that the corolla of Goodeniads is sometimes polypetalous, which it never is in Bellworts or Lobeliads; that the æstivation of the corolla is induplicate, not valvate; that its principal veins are lateral, or alternate with the lobes, as in Composites; that in the species of Goodeniads with dehiscent fruit, the dehiscence is usually septicidal, while in the two other groups it is always loculicidal; finally, that Goodeniads have not the milky juice that characterises Bellworts and Lobeliads." Notwithstanding the polyspermous fruit and different inflorescence, this Order approaches very closely to Composites; the milky juice is the same as that of the tribe called Cichoraceæ; the species have, in many cases, the flowers crowded in heads; the stigma is similar to that of many Composites; there are the same collecting hairs on the style, in both cases intended to clear out the pollen from the cells of the anthers; and, finally, the habit is very like. These collecting hairs, which clothe the style of Bellworts in a most remarkable manner, arranging themselves in lines having a direct relation to the number and position of the anthers, have been the subject of special examination by several observers, especially by Adolphe Brongniart. This Botanist ascertained that such hairs are not, as had been supposed, deciduous, but that they are retractile, like the hairs of certain annelides or the tentacula of snails. It appears that, at the time of the expansion of the flower, the hairs, which had previously projected and swept out the pollen from the anthers, are drawn back into certain cavities lying at their base, the upper half sheathing itself in the lower half as it is by degrees withdrawn. M. Brongniart is of opinion that there is no ground for supposing that this singular phenomenon is connected with the fertilising process. (See *Ann. des Sc. Nat.* 2 *Ser.* 12. 244). But Mr. Hassall disputes this statement, which he declares is "wholly opposed to the result of his investigations."— *Ann. Nat. Hist.* viii. 86.

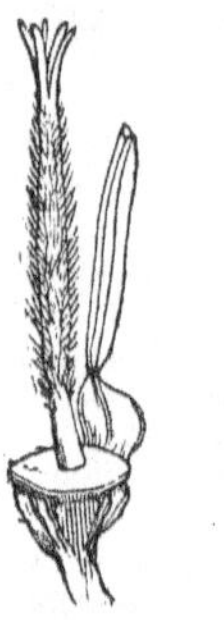

Fig. CCCCLXIV.

It has been remarked in the *Botanical Register* (1842, t. 3.), that the genus Glossocomia brings the Orders of Nightshades and Bellworts into close contact.

With respect to the singular genus Sphenoclea, erected into an Order by Martius, although it cannot be regarded as a genuine species of Bellwort, because of the absence of collecting hairs from its styles, the round sub-sessile anthers, the stamens distinctly inserted upon the corolla, and the peculiar habit of the only known species, yet it seems to have more affinity to this Order than to any other, and may very well be stationed at the end, as a genus waiting for the discovery of companions which may be better suited to indicate its true station.

Chiefly natives of the north of Asia, Europe, and North America, and scarcely known in the hot regions of the world. In the meadows, fields, and forests of the countries they inhabit, they constitute the most striking ornament. Some curious species are found in the Canaries, St. Helena, and Juan Fernandez. Alphonse De Candolle remarks, that "it is within 36° and 47° N. lat. that in our hemisphere the greatest number of species is found; the chain of the Alps, Italy, Greece, Caucasus, the Altai range, are their true country. In whatever direction we leave these limits, the number of species rapidly decreases. In the southern hemisphere, the Cape of Good Hope (lat. 34° S.) is another centre of habitation, containing not fewer than 63 species. This

Fig. CCCCLXIV.—Stamen and pistil of Campanula Medium.

locality has a climate so different from that of our mountains, that it may be easily imagined that the species capable of living there differ materially from those of our own hemisphere: in fact, they belong to other genera." Of 300 species only 19 are found within the tropics. The same Botanist remarks that, with only a single exception, all the species belonging to genera that open their capsule by lateral pores are found in the northern hemisphere; while those whose capsules dehisce at the apex chiefly inhabit the southern hemisphere.

The milky juice is rather acrid, but nevertheless the roots and young shoots of some species, particularly of Campanula Rapunculus, or Rampion, of Phyteuma spicatum, of Canarina Campanula, &c., are an occasional article of food. The chief value of the Order, however, is its beauty. The roots of Phyteumas are said to be antisyphilitic; that of Campanula glauca is held by the Japanese to be a tonic, and scarcely inferior to Ginseng. The Specularias Speculum and pentagonia have been used in salads; the flowering plant of Wahlenbergia graminifolia is used by the mountaineers of the South of Europe against epilepsy. Wahlenbergia linarioides is employed in Chili in tormina. The half-fleshy fruit of Canarina Campanula is said to be eatable. The tuberous root of Cyphia digitata is said to be eaten by Hottentots.

GENERA.

I. JASIONEÆ.

Jasione, *L.*
Aphyllanthes, Dalechamp.
Ovilla, Adans.

II. LIGHTFOOTEÆ.

Lightfootia, *Herit.*
Cephalostigma, *Alph. DC.*
Tritamidium, Endl.
Campanumæa, *Blume.*
Codonopsis, *Wall.*
Glossocomia, *D. Don.*
Megasanthes, G. Don.
Canarina, *Juss.*
Canaria, Linn.
Pernetya, Scop.
Platycodon, *Alph. DC.*
Microcodon, *Alph. DC.*
Cœlotheca, Alph. DC.
Wahlenbergia, *Schrad.*
Codonia, Spr.
Aikinia, Salisb.
Schultesia, Roth.
Campanopsis, R. Br.
Nesophila, Alph. DC.
Cervicina, Del.
Heterochænia, *Alph. DC.*

III. PRISMATOCARPEÆ.

Prismatocarpus, *Al. DC.*
Röella, *L.*
Aculeosa, Plukn.
Edraianthus, *Alph. DC.*

IV. CAMPANULEÆ.

Phyteuma, *L.*
Rapunculus, Tournef.
Rapuntium, Lobel.
Physoplexis, Endl.
Synotoma, G. Don.
Petromarula, *Alph. DC.*
Michauxia, *Herit.*
Mindium, Adans.
Campanula, *L.*
Depierrea, Schl.
Medium, Tournef.
Marianthemum, Schnk.
Rapuntium, Chev.
Roucela, Dumort.
Erinia, Noul.
Specularia, *Heist.*
Prismatocarpus, Herit.
Apenula, Neck.
Legouzia, Durand.
Dysmicodon, Endl.
Triodallus, Raf.
Trachelium, *L.*
Adenophora, *Fisch.*
Flörkea, Spr.
Symphyandra, *Alph. DC.*
Musschia, *Dumort.*
Rhigiophyllum, *Hochst.*
Cyphia, *L.*
Cyphium, Gmel.

Anomalous Genera.

Merciera, *Alph. DC.*
Pentaphragma, *Wall.*
Sphenoclea, *Gærtn.*
Pongatium, Juss.
Gærtnera, Retz.
Rapinia, Lour.

NUMBERS. GEN. 28. Sp. 500.

Solanaceæ.
POSITION.—Asteraceæ.—CAMPANULACEÆ.—Lobeliaceæ.
Vacciniaceæ.

ORDER CCLXVII. LOBELIACEÆ.—LOBELIADS.

Campanulaceæ, § 2. *R. Brown Prodr.* 562. (1810).—Lobeliaceæ, *Juss. Ann. Mus.* 18. 1. (1811); *Endl. Gen.* cxxiv.; *DC. Prodr.* 7. 339; *Presl. Monogr. Lobel.*; *Alph. DC. in Ann. Sc.* 2. *Ser.* xii. 149.

DIAGNOSIS.—*Campanal Exogens, with a 2- or more-celled ovary, syngenesious anthers, a stigma surrounded by hairs, and a valvate irregular corolla.*

Herbaceous plants or shrubs, with milky juice. Leaves alternate, without stipules. Flowers axillary or terminal. Calyx superior, 5-lobed or entire. Corolla monopetalous,

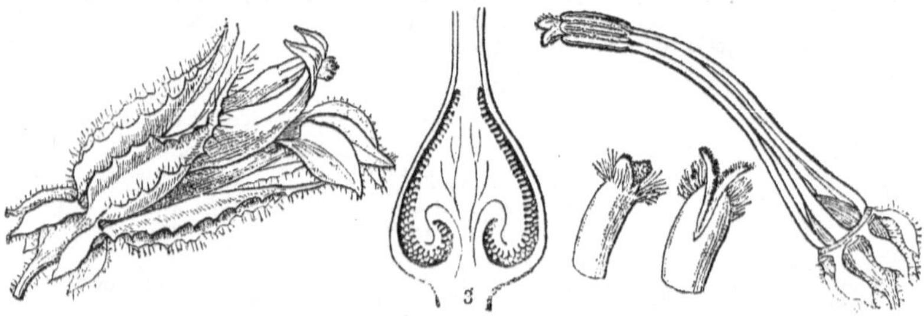

1 Fig. CCCCLXV. 4 5 2

in æstivation somewhat valvate, irregular, inserted in the calyx, 5-lobed or deeply 5-cleft. Stamens 5, inserted into the calyx alternately with the lobes of the corolla; anthers cohering; pollen oval. Ovary inferior, with from 1 to 3 cells; ovules very numerous, either attached to the axis, or parietal; style simple, stigma surrounded by a cup-like fringe. Fruit capsular, 1- or more-celled, many-seeded, dehiscing at the apex. Seeds attached either to the lining or the axis of the pericarp; embryo straight, in the axis of fleshy albumen; radicle longer than the cotyledons, pointing to the hilum.

The plants of this Order at first sight appear to be very different from Composites, but they in fact participate in all the analogies of Bellworts, and perhaps are yet more nearly related to Composites even than that Order, especially in their syngenesious anthers and in the irregularity of their corolla, which is split so that the segments cohere on one side like the 5 segments which make up the ligulate floret of a Composite. The stigma is surrounded by hairs, which are probably analogous to the collectors of Bellworts, to which Lobeliads approach closely, as well as to Goodeniads, whose indusium and induplicate corolla offer the main features of distinction. The Clintoneæ are remarkable for a one-celled ovary with parietal placentæ; a few species have polypetalous flowers, and one species of Lobelia is said to be diœcious!

Unlike Bellworts, these seem to prefer countries within or upon the border of the tropics to such as have a colder character. We find them abounding in the West Indies, Brazil, the Himalayan region, the Cape of Good Hope, and the Sandwich Islands; and they are not uncommon in Chile and New Holland.

All the species are dangerous or suspicious, in consequence of the excessive acridity of their milk. Siphocampylus Caoutchouc is so named by the inhabitants of Popayan from the tenacity of its juice. Tupa Feuillæi yields a dangerous poison in Chile. The most active article of the North American Materia Medica is said to be the Lobelia inflata; it possesses an emetic, sudorific, and powerful expectorant effect; when given with a view to empty the stomach it operates vehemently and speedily; producing, however, great relaxation, debility, and perspiration, and even death, if given in over-doses. The anti-syphilitic virtues ascribed to Lobelia syphilitica are supposed to have resided in its diuretic property; they are, however, generally discredited. Isotoma longiflora, a native of some of the West India Islands, is one of the most venomous of plants; the Spanish Americans call it Prebenta Cavallos, because it proves fatal to horses that eat it, swelling them until they burst; taken internally, it acts as a violent cathartic, the effects of which no remedy can assuage, and which end in death; the leaves are active vesi-

Fig. CCCCLXV.—Lobelia siphilitica. 1. an entire flower; 2. the stamens; 3. perpendicular section of the ovary; 4 and 5 stigmas.

cants. Lobelia cardinalis is an acrid plant, and reckoned anthelmintic. Lobelia urens, a rare European plant, derives its name from its blistering quality. Nevertheless, it is alleged that the succulent fruit of Centropogon surinamensis is eatable.

GENERA.

I. CLINTONEÆ.

Grammatotheca, *Presl.*
Clintonia, *Dougl.*
Lysipoma, *H. B. K.*
Hypsela, *Presl.*

II. LOBELEÆ.

Metzlera, *Presl.*
Parastranthus, *G. Don.*
Xanthomeria, Presl.
Dombrowskya, *Presl.*
Monopsis, *Salisb.*
Holostigma, Don.
Isolobus, Alph. DC.
Sclerotheca, *Alph. DC.*
Trimeris, *Presl.*
Lobelia, *Linn.*
Rapuntium, Tournef.
Stenotium, Presl.
Dortmanna, Rudb.
Sphærangium, Presl.
Homochilus, Alph. DC.
Tupa, *G. Don.*
Tylomium, Presl.
Siphocampylus *Pohl.*
Lobelia, Presl.
? *Canonanthus*, Don.
Laurentia, *Neck.*
Solenopsis, Presl.
Enchysia, Presl.
Isotoma, R. Br.
Solenanthis, Kunth.
Hippobroma, G. Don.
Vlamingia, *Vriese.*
Byrsanthes, *Presl.*
Heterotoma, *Zuccar.*
Myopsia, Presl.

III. DELISSEÆ.

Pratia, *Gaudich.*
Piddingtonia, *Alph. DC.*
Bernonia, Endl.
Delissea, *Gaudich.*
Kittelia, *Reichenb.*
Cyanea, Gaudich.
Macrochilus, *Presl.*
Rollandia, *Gaudich.*
Clermontia, *Gaudich.*
Centropogon, *Presl.*

*

Rhynchopetalum, *Fres.*

NUMBERS. GEN. 27. Sp. 375.

POSITION.—Compositæ.—LOBELIACEÆ.—Campanulaceæ.

ORDER CCLXVIII. GOODENIACEÆ.—GOODENIADS.

Campanulæ, *Juss. Gen.* 163. (1789) *in part.*—Goodenoviæ, *R. Brown Prodr.* 573. (1810); *Bartl. Ord. Nat.* 148. (1830); *DC. Prodr.* 7. 502.—Goodeniaceæ, *Ed. pr.* clxxxiv. (1836); *Endl. Gen.* cxxiii.—Scævoleæ, *Ed. pr.* clxxviii. (1830).—Scævolaceæ, *Ed. pr.* clxxxv. (1836).

DIAGNOSIS.—*Campanal Exogens, with a 2- or more-celled ovary, syngenesious or free anthers, an indusiate stigma, and induplicate corolla.*

Herbaceous plants, rarely shrubs, without milk, with simple or glandular hairs, if any are present. Leaves scattered, often lobed, without stipules, very rarely opposite. Inflorescence terminal, variable. Flowers distinct, never capitate, usually yellow, or blue, or pink. Calyx usually superior, rarely inferior, equal or unequal, in from 3 to 5 divisions. Corolla always more or less superior, monopetalous, more or less irregular, withering; its tube split at the back, and sometimes capable of being separated into 5 pieces, when the calyx only coheres with the base of the ovary; its limb 5-parted, with 1 or 2 lips, the edges of the segments being thinner than the middle, and folded inwards in æstivation. Stamens 5, distinct, alternate with the segments of the corolla; anthers distinct or cohering, 2-celled, bursting longitudinally. Pollen simple or in fours. Ovary 1- 2-celled, rarely 4-celled, with definite or 00 ovules, having sometimes a gland at its base between the two anterior filaments; placenta free, central, or only adhering slightly to the dissepiments; style 1, simple, very rarely divided; stigma fleshy, undivided, or 2-lobed, surrounded by a membranous cup. Fruit a 1- 2- or 4-celled capsule with many solitary or numerous seeds, attached to the axis of the dissepiment, which is usually parallel with the valves, rarely opposite to them. Seeds usually with a thickened testa, which is sometimes nut-like; albumen fleshy, inclosing an erect embryo; cotyledons foliaceous; plumule inconspicuous.

Fig. CCCCLXVI.

The great peculiarity of this Order resides in the stigma, which is seated at the bottom of a cup or covering called an indusium, unknown in Bellworts or Lobeliads, to which the genera might otherwise be referred. It is of the same nature as what is found in Brunoniads and Styleworts, and is to be regarded as nothing more than a remarkable exaggeration of the rim which surrounds the stigmatic surface of Heathworts, and of the plates which cover the style of Cranesbills and Balsams. It is, in fact, the upper free extremity of the carpellary leaves, distinct from that prolongation of the placenta which is named style and stigma. Brown, however, has offered a very different explanation of its nature, as will be seen by the following extract:—

"Is this remarkable covering of the stigma in these families merely a process of the apex of the style? or is it a part of distinct origin, though intimately cohering with the pistillum? On the latter supposition, may it not be considered as analogous to the glandular disk surrounding or crowning the ovarium in many other families? And, in adopting the hypothesis I have formerly advanced respecting the nature of this disk in certain families,—namely, that it is composed of a series of modified stamina,—has not the part in question a considerable resemblance, in apparent origin and division, to the stamina of the nearly-related family Stylidiaceæ? To render this supposition somewhat less paradoxical, let the comparison be made especially between the indusium of Brunonia and the imperfect antheræ in the female flowers of Forstera. Lastly, con-

Fig. CCCCLXVI.—Leschenaultia formosa. 1. stamens and style; 2. cross section of ovary; 3. longitudinal section of ovary; 4. stigma and indusium.

nected with this view, it becomes of importance to ascertain whether the stamina of Stylidiaceæ are opposite to the segments of calyx or of corolla. The latter disposition would be in favour of the hypothesis. This, however, is a point which will not be very easily determined, the stamina being lateral. In the mean time, the existence and division of the corona faucis in Stylidium render it not altogether improbable that they are opposite to the segments of the corolla."

In the æstivation of the corolla the Goodeniads are also remarkable, the edges of its segments being doubled inwards, so as to assume the appearance of wings belonging to a triangular back. Cyphia, referred hither by Endlicher, wants that character, and seems to be merely an irregular Bellwort with the collecting hairs of the style arranged in a ring beneath the stigma.

These plants belong to Australia and the islands of the Southern Ocean, or only advance into India in the form of a Scævola, which even spreads into Africa, and it is said, the West Indies, and of Selliera, which inhabits the southern part of South America.

The leaves of Scævola Taccada when young are eaten as potherbs, and some superstitious qualities are ascribed to its berries; the pith, which is soft and spongy, is fashioned by the Malays into artificial flowers and other nicknacks. Scævola Bela Modogam appears to be emollient, and is used in India to bring tumours to a head.—*Linn. Trans.* 12. 134.

GENERA.

I. Scævoleæ.—Fruit a drupe or nut.

Scævola, *L.*
Lobelia, Plum.
Cerbera, Lour.
Glypha, Lour.
Pogonandra, A. DC.
Pogonanthera, G. Don.
Crossotoma, Don.
Pogonetes, *Lindl.*
Diaspasis, *R. Br.*
Dampiera, *R. Br.*

II. Goodenieæ.—Fruit a capsule.

Selliera, *Cav.*
Goodenia, *Sm.*
Collema, Anders.
Ochrosanthus, Don.
Tetrathylax, Don.
Porphyranthus, Don.
Monochila, Don.
Amphichila, A. DC.
Calogyne, *R. Br.*
Distylis, *Gaudich.*
Euthales, *R. Br.*
Velleja, *Smith.*
Menoceras, *R. Br.*
Leschenaultia, *R. Br.*
Latouria, *Endl.*
Anthotium, *R. Br.*

NUMBERS. GEN. 14. Sp. 150?

Brunoniaceæ.
POSITION.—Lobeliaceæ.—GOODENIACEÆ.—Stylidiaceæ.

ORDER CCLXIX. STYLIDIACEÆ.—STYLEWORTS.

Stylideæ, *R. Brown Prodr.* 565. (1810); *Endl. Gen.* cxxvi.; *DC. Prodr.* 7. 331.

DIAGNOSIS.—*Campanal Exogens, with a 2- or more-celled ovary, stamens and style united into a column, and imbricated corolla.*

Herbaceous plants or under-shrubs, without milk, having a stem or scape; their hair, when they have any, simple, acute, or headed with a gland. Leaves scattered, sometimes whorled, entire, their margins naked or ciliated, the radical ones clustered in the species with scapes. Stipules 0. Flowers in spikes, racemes, or corymbs, or solitary, terminal, rarely axillary, the pedicels usually with three bracts. Calyx adherent, with from 2 to 6 divisions, bilabiate or regular, persistent. Corolla monopetalous, falling off late; its limb irregular, rarely regular, with from 5 to 6 divisions, imbricated in æstivation. Stamens 2; filaments connate with the style into a longitudinal column; anthers twin, sometimes simple, lying over the stigma; pollen globose, simple, sometimes angular. Ovary 2-celled, many-seeded, sometimes 1-celled, in consequence of the contraction of the dissepiment; often surmounted with a single gland in front, or two opposite ones; style 1; stigma entire or bifid; ovules anatropal. Capsule with 2 valves and 2 cells, the dissepiment between which being sometimes either contracted or separable from the inflexed margins of the valves, the capsule becomes as it were 1-celled. Seeds small, erect, sometimes stalked, attached to the axis of the dissepiment; embryo scarcely known; said to be minute, inclosed within a fleshy, somewhat oily albumen.

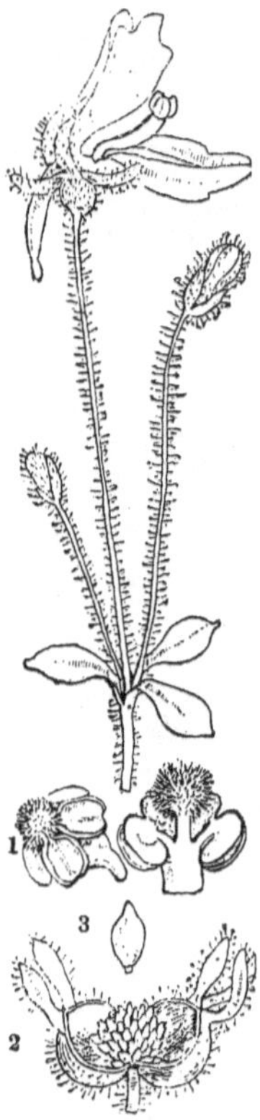

Fig. CCCCLXVII.

Fig. CCCLXVIII.

These are very curious little plants, nearly allied to Bellworts and Goodeniads, from both which they are distinguished by their gynandrous structure, and from the latter by the want of an indusium to the stigma. The structure of the column, into which the stamens and style are blended, is highly curious, and scarcely analogous to anything else in the Vegetable Kingdom, except in Orchids: the stigma lies in a cavity at the apex of the column, surrounded and concealed by the anthers. This column is extremely irritable; in Stylidium it hangs down on one side of the flower until it is touched, when it suddenly springs up and shifts instantly to the opposite side. A singular error was committed by Labillardiére, who mistook an epigynous gland for the stigma; and another by L. C. Richard, who considered the labellum to be the female organ of this genus.

The species are chiefly found in New Holland swamps. One however occurs in Ceylon, another in Malabar, and a third in Sylhet. The Forsteras live on the summit of mountains in the South of New Zealand, or in the morasses of the Straits of Magellan.

Nothing is known of any use to which they are applied.

GENERA.

Stylidium, *Sw.*	*Andersonia*, König	Levenhookia, *R. Br.*	*Phyllachne*, Forst.
Ventenatia, Smith.	Coleostylis, *Sonder.*	*Gynocampus*, Lesch.	*Stibas*, Commers.
Candollea, Labill.	Forsteropsis, *Sonder.*	Forstera, *Linn. f.*	

NUMBERS. GEN. 5. Sp. 121.

POSITION.—Lobeliaceæ.—STYLIDIACEÆ.—Goodeniaceæ.

Fig. CCCCLXVII.—Stylidium calcaratum.—*F. Bauer.* 1. anthers and stigma, forming the point of the column; 2. capsule split open; 3. seed.
Fig. CCCCLXVIII.—Forstera clavigera.—*Hooker fil.* 1. the epigynous gland.

ORDER CCLXX. VALERIANACEÆ.—VALERIANWORTS.

Valerianeæ, *DC. Fl. Fr. ed.* 3. v. 4. *p.* 232. (1815); *Dufr. Valer. Monogr.* 56. (1811); *DC. Prodr.* 4. 623. (1830); *Royle's Illustrations*, 241. (1835); *Endl. Gen.* cxviii.; *Meisn. p.* 173; *Woods in Linn. Trans.* 17. 421.

DIAGNOSIS.—*Campanal Exogens, with a 1-celled ovary, an imbricated corolla, free anthers, pendulous ovule, and no albumen.*

Annual or perennial herbs, occasionally twining, and usually either strong-scented or aromatic. Leaves collected in rosettes at the root, or distributed upon the stem,

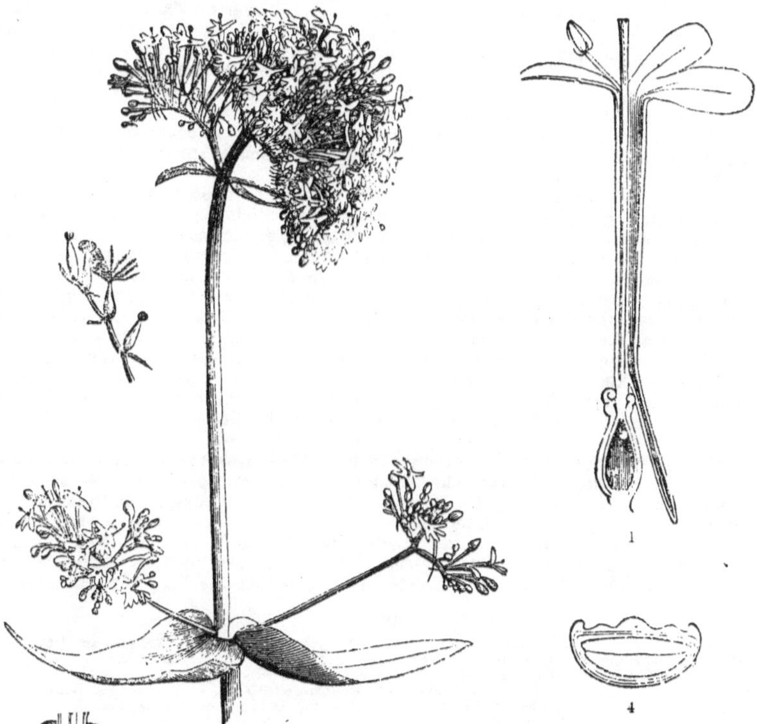

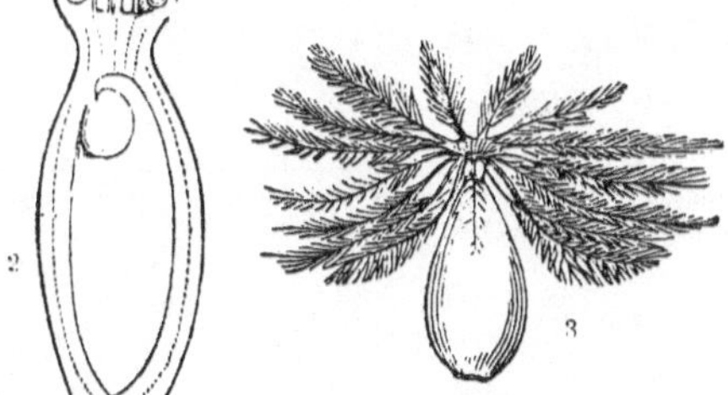

Fig CCCCLXIX.

opposite, entire, or variously divided in a pinnate way. Flowers sometimes ♂ ♀ by abortion, with a centrifugal expansion; in Triplostegia furnished with an involucre. Calyx superior; the limb either membranous, or resembling feathery pappus. Corolla monopetalous, tubular, inserted into the top of the ovary, with from 3 to 6 lobes, either regular or irregular, sometimes calcarate at the base. Stamens from 1 to 5, inserted into the tube of the corolla, and alternate with its lobes. Ovary inferior, with 1 cell,

Fig. CCCCLXIX.—Centranthus ruber. 1. a corolla; 2. section of ovary; 3. ripe fruit, with its pappus; 4. cross section of a seed.

and sometimes 2 other abortive ones; ovule solitary, pendulous; style simple; stigmas from 1 to 3. Fruit dry, indehiscent, with 1 fertile cell and 2 empty ones. Seed solitary, pendulous; embryo straight, destitute of albumen; radicle superior.

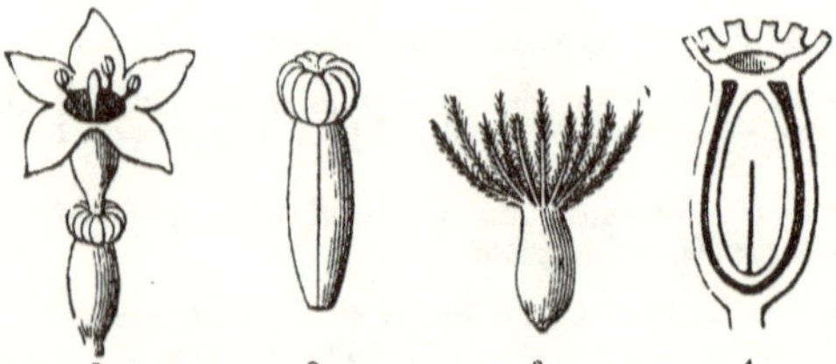

1 2 3 4

Fig. CCCCLXX.

Valerianworts are principally distinguished from Teazelworts by their want of albumen, and usually by the absence of an involucel to each floret. They have also but little tendency to form a capitate inflorescence; and a couple of additional empty cells, frequently observable in their ovary, indicates a higher degree of composition in the central apparatus. M. Bunge has observed manifest traces of ovules in the two abortive cells of Patrinia; the same author considers Valerianworts connected with Caprifoils on the one hand by Triplostegia, and on the other by Linnæa.—*Ann. Sc. ser.* 2. v. 6. 60.

They are natives of most temperate climates; sometimes at considerable elevations. They are abundant in the north of India, Europe, and South America, but uncommon in Africa and North America.

The roots of Valeriana officinalis, Phu, and celtica, are tonic, bitter, aromatic, antispasmodic, and vermifugal; they are even said to be febrifugal. They are strong-smelling, especially in V. Dioscoridis, which, according to Sibthorp, is the real Phu of Dioscorides, act as powerful stimulants, and produce a specific influence over the cerebro-spinal system, bringing on, as is well known, a kind of intoxication in cats, and in large doses occasioning in man scintillations, agitation, and even convulsions. The Russians regard the Valeriana sitchensis, a native of North-West America, as the most energetic of all the species. The scent of these roots is not agreeable to a European; and yet some are highly esteemed as perfumes. Eastern nations procure from the mountains of Austria the Valeriana celtica and Saliunca to aromatise their baths. Their roots are grubbed up with danger and difficulty by the peasants of Styria and Carinthia, from rocks on the borders of eternal snow, are tied in bundles and sold at a very low price to merchants, who send them by way of Trieste to Turkey and Egypt, where they are vended at a great profit, and passed onwards to the nations of India and Ethiopia.—*Endl.* The Nardostachys Jatamansi, or true Spikenard of the ancients, is valued in India, not only for its scent, but also as a remedy in hysteria and epilepsy. The young leaves of the species of Valerianella are eaten as salad, under the French name of Mâche, or the English one of Lamb's Lettuce. Red Valerian is also eaten in the same way in Sicily. Astrephias are used as vulneraries in Peru. See *Royle*, p. 242, for an elaborate dissertation upon the Nard of the ancients.

GENERA.

Patrinia, *Juss.*	*Polypremum*, Adans.	*Mitrophora*, Neck.	*Aretiastrum*, DC.
Gytonanthus, Rafin.	*Odontocarpa*, Neck.	Plectritis, *DC.*	*Phu*, DC.
Fedia, Adans.	Astrephia, *DC.*	Centranthus, *DC.*	Betckea, *DC.*
Nardostachys, *DC.*	*Hemesotria*, Rafin.	*Kentranthus*, Neck.	Triplostegia, *Wall.*
Dufresnea, *DC.*	*Oligæoce*, Willd.	Valeriana, *Neck.*	? Axia, *Lour.*
Valerianella, *Mönch.*	Fedia, *Mönch.*	*Phyllactis*, Pers.	

NUMBERS. GEN. 12. Sp. 185.

Caprifoliaceæ.

POSITION.—Asteraceæ.—VALERIANACEÆ.—Dipsacaceæ.

Fig. CCCCLXX.—Valeriana celtica. 1. entire flower magnified; 2. the ovary and young calyx; 3. the fruit, with the pappose full-grown calyx; 4. a vertical section of fruit and seed

ORDER CCLXXI. DIPSACACEÆ.—TEAZELWORTS.

Dipsaceæ, *Juss. Gen.* 194. (1789); *Coulter Mem. in Act. Genev.* 2. 13. (1823); *DC. Prodr.* 4. 643. (1830); *Endl. Gen.* cxix.

DIAGNOSIS.—*Campanal Exogens, with a 1-celled ovary, imbricated corolla, free anthers, pendulous ovule, and albuminous seeds.*

Herbaceous plants or under-shrubs. Leaves opposite or whorled. Flowers collected upon a common receptacle, and surrounded by a many-leaved involucre. Calyx adherent, membranous, resembling pappus; surrounded by a scarious involucel. Corolla monopetalous, tubular, inserted in the calyx; limb oblique, 4- or 5-lobed, with an imbricated æstivation. Stamens 4, alternate with the lobes of the corolla, sometimes half-sterile; anthers distinct. Ovary inferior, 1-celled, with a single pendulous anatropal ovule; style 1; stigma simple. Fruit dry, indehiscent, 1-celled, crowned by the pappus-like calyx; embryo straight, in the axis of a small quantity of fleshy albumen; radicle superior.

The relation of this family is obviously in the first degree with Composites, from which it differs in its distinct stamens and its pendulous albuminous seeds; and next with Calycers, which have united anthers and alternate leaves. The same character of the capitate flowers, and the presence of albumen, forms the distinction between Teazelworts and Valerianworts. What is called the involucel is a curious organ, resembling an external calyx, and is to each particular flower of the head of Teazelworts what the partial involucre of Umbellifers is to each partial umbel; and, accordingly, we ought to expect to find instances of more flowers than one being inclosed within this involucel; and this is said by Coulter actually to take place in the genus Gundelia. This is, however, not the only peculiarity of the Order. Brown has the following remarks:—"M. Auguste de Saint Hilaire, in his excellent Memoir on Primulaceæ, while he admits the correctness of M. De Candolle's account with respect to great part of Dipsaceæ, has at the same time well observed, that in several species of Scabiosa the ovarium is entirely united with the tube of the calyx. But neither of these authors has remarked the curious, and I believe peculiar, circumstance of the base of the style cohering with the narrow apex of the tube of the calyx, even in those species of the Order in which the dilated part of the

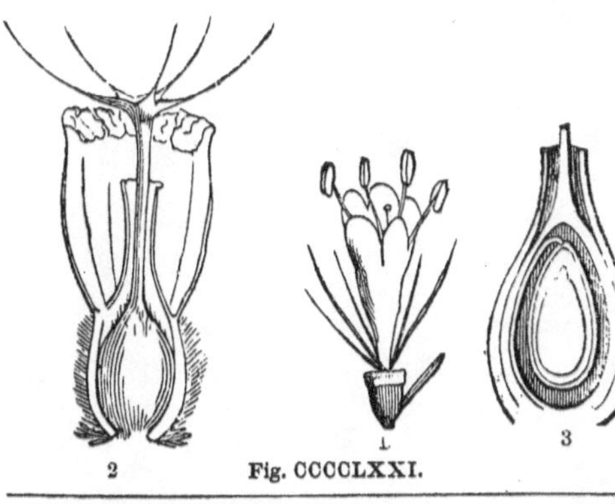

2 1 3

Fig. CCCCLXXI.

Fig. CCCCLXXI.—Scabiosa atropurpurea. 1. a flower; 2. the involucre opened to show the ovary and calyx; 3. perpendicular section of fruit.

tube is entirely distinct from the ovarium. This kind of partial cohesion between pistillum and calyx is directly opposite to what usually takes place, namely, the base of the ovarium being coherent, while its upper is distinct. It equally, however, determines the apparent origin or insertion of corolla and stamina, producing the unexpected combination of 'flos superus' with 'ovarium liberum.'"—*Linn. Trans.* 12. 138.

The species are chiefly natives of the south of Europe, Barbary, the Levant, and the Cape of Good Hope; not affecting particular stations in any striking degree, except that they generally shun cold, and do not attain much elevation above the sea.

Their properties are unimportant. The Teazel used by fullers in dressing cloth is the dried head of Dipsacus Fullonum, bristling with hard, stiff, spiny bracts. Some of the species are reputed febrifugal. Scabiosa succisa is said to yield a green dye, and also to be astringent enough to deserve the attention of tanners. The leaves of the common Teazel are united at their base so as to form round the stem a hollow in which water collects; hence the plant was called Διψακος or thirsty, and also obtained the name of Venus' Bath; and the superstitious fancied that the water thus collected from rains and dews was good for bleared eyes.

GENERA.

Morina, *Tournef.*
Diototheca, Vaill.
Asaphes, Spreng.
Dipsacus, *Tournef.*
Galedragon, Gray.
Cephalaria, *Schrad.*
Lepicephalus, Lagasc.
Cerionanthus, Schott.
Succisa, Vaill.
Pycnocomon, Wallr.
Knautia, *L.*
Trichera, Schrad.
Pterocephalus, *Vaill.*
Scabiosa, *Röm et Schult.*
Asterocephalus, Vaill.
Sclerostemma, Schott.
Spongostemma, Rchb.
Succisa, Coult.
Columbaria, Thuill.

NUMBERS. GEN. 6. Sp. 150.

POSITION.—Asteraceæ.—DIPSACACEÆ.—Valerianaceæ.

ORDER CCLXXII. CALYCERACEÆ.—CALYCERS.

Calycereæ, *R. Brown in Linn. Trans.* 12. 132. (1816); *Rich. in Mém. Mus.* 6. 76. (1820); *DC. Prodr.* v. 1; *Endl. Gen.* cxxi.—Boopideæ, *Cassini in Dict. des Sc.* 5. 26. *Supp.* (1817).

DIAGNOSIS.—*Campanal Exogens, with a 1-celled ovary, valvate corolla, syngenesious anthers, pendulous ovule, and albuminous seeds.*

Herbaceous plants. Leaves alternate, without stipules. Flowers collected in heads, which are either terminal or opposite the leaves, surrounded by an involucre. Florets sessile, hermaphrodite, or neuter. Calyx superior, of 5 unequal pieces. Corolla regular, valvate, funnel-shaped, with a long slender tube and 5 segments, each of which has 3 principal veins; glandular spaces below the stamens and alternate with them. Stamens 5, monadelphous; anthers combined by their lower half in a cylinder. Ovary inferior, 1-celled; ovule solitary, pendulous; style simple, smooth; stigma capitate. Fruit an indehiscent pericarp, usually crowned by the rigid spiny segments of the calyx. Seed solitary, pendulous, sessile; embryo in the axis of fleshy albumen; radicle superior, longer than the plano-convex cotyledons.

Fig. CCCCLXXII. 2 1

A very small and curious Order, differing from Composites in nothing but their albumen, pendulous ovule, and half-distinct anthers, and from Teazelworts in their filaments being monadelphous and their anthers partly connate. They may therefore be considered to hold a middle station between those two Orders.

Such species as are known inhabit South America, rarely occurring in the tropical districts, but more plentiful in South Chile. They are, however, not common anywhere. They are described as ascending from the sea coast to considerable altitudes on the Andes.

They are not mentioned as possessing any useful quality.

GENERA.

Gamocarpha, *DC.*	Calycera, *Cav.*	*Cryptocarpha*, R. Br.	*Acanthosperma*, Arrab.
Boopis, *Juss.*	Acicarpha, *Juss.*	*Sommea*, Bory.	*Echinolema*, Jacq. f.
Nastanthus, *Miers.*			

NUMBERS. GEN. 5. Sp. 10.

POSITION.—Asteraceæ.—CALYCERACEÆ.—Dipsacaceæ.

Fig. CCCCLXXII.—Acicarpha spathulata.—*Richard.* 1. section of an entire flower; 2. perpendicular section of ripe fruit.

ORDER CCLXXIII. ASTERACEÆ.—COMPOSITES.

Compositæ, *Adans. Fam.* 2.103. (1763); *Kunth in Humb. N. G. et Sp. vol.* 4; *Lessing, Synops. Compos.; Royle's Illustr.* 245; *DC. Prodr. vol.* 5, *&c.*; *Endl. Gen.* cxx.; *Meisner, p.* 174.—Synanthereæ, *Rich. Anal.* (1808); *Cassini Dict. Sc. N.* 10. 131. (1818); *Ibid.* 60. 563. (1830).—Corymbiferæ, Cynarocephalæ, and Cichoraceæ, *Juss. Gen.* (1789).

DIAGNOSIS.—*Campanal Exogens, with a* 1-*celled ovary, valvate corolla, syngenesious anthers, erect ovule, and no albumen.*

1 2 Fig. CCCCLXXIII. 3 4 5

Fig. CCCCLXXIII.—Centaurea Cyanus. 1. a floret; 2. the anthers; 3. perpendicular section of young fruit; 4. ripe fruit; 5. floret of Taraxacum Dens Leonis.

Herbaceous plants or shrubs. Leaves alternate or opposite, without stipules, usually simple, but commonly much divided. Flowers (called florets) unisexual or hermaphrodite, collected in dense heads upon a common receptacle, surrounded by an involucre. Bracts either present or absent; when present, stationed at the base of the florets, and called paleæ of the receptacle. Calyx superior, closely adhering to the ovary, and undistinguishable from it; its limb either wanting or membranous, divided into bristles, paleæ, hairs, or feathers, and called pappus. Corolla monopetalous, superior, usually deciduous, either ligulate or funnel-shaped; in the latter case, 4- or 5-toothed, with a valvate æstivation. Stamens equal in number to the teeth of the corolla, and alternate with them; the anthers cohering into a cylinder. Ovary inferior, 1-celled, with a single erect ovule; style simple; stigmas 3, either distinct or united. Fruit a small, indehiscent, dry pericarp, crowned with the limb of the calyx. Seed solitary, erect; embryo with a taper inferior radicle; albumen none.

This is one of the most natural and extensive families of the vegetable kingdom, at all times recognised by its inferior 1-celled ovary, with an erect ovule, syngenesious stamens, and capitate flowers. Calycers and Teazelworts, neighbouring Orders, also with the flowers in heads, are readily distinguished by their pendulous ovule, and the anthers being either wholly or partially distinct.

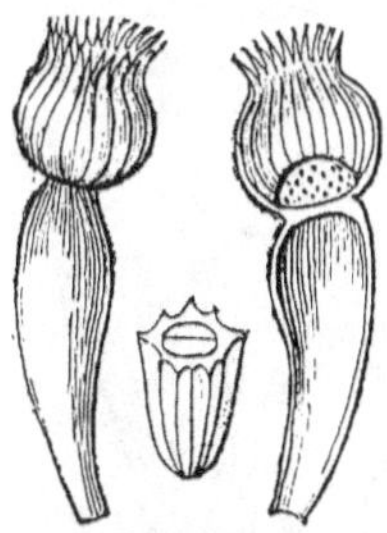

Fig. CCCCLXXIV.

In proportion to its strict natural limits, depending upon the uniformity of its characters, is the difficulty of separating it into sections. Jussieu has three: Corymbiferæ, the florets of which are flosculous in the disk, and ligulate at the circumference; Cichoraceæ, the florets of which are all ligulate; and Cynarocephalæ, all whose florets are flosculous; to which has since been added another division called Bilabiate. Linnæus employed the sexes of the florets for the purpose of defining groups, but this, like all other parts of the great Swedish Naturalist's Botanical System, is now abandoned; and yet it was not without much merit. The condition of the Order had at one time,—thanks to the neglect of Linnean Botanists and the unmethodical improvements of more careful observers,—become a chaos, the like of which had not been seen since the days of the Bauhins; but in 1830 an arrangement of much merit was proposed by the German Botanist Lessing, and at a later period De Candolle the elder applied his acute and logical mind to the elucidation of the Order. At the present day the method of the latter, essentially founded on that of Lessing, is universally followed. De Candolle himself stated it thus:—

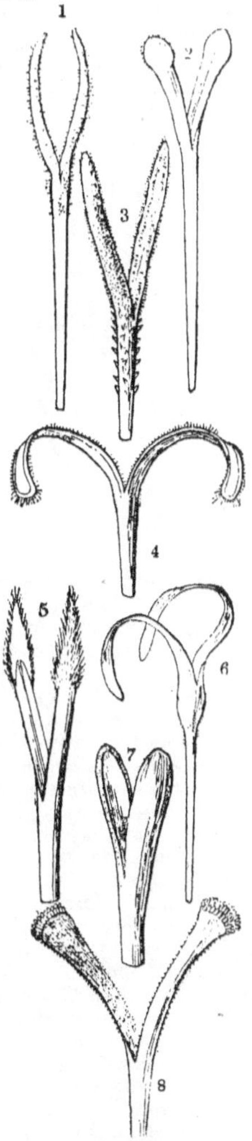

Fig. CCCCLXXV.

Suborder I. TUBULIFLORÆ; that is to say, those in which the hermaphrodite florets, which alone can be regarded as normal, are tubular, with 5, or rarely 4, equal teeth. Of these the following are distinguished by their stigmas:—

1. Vernoniaceæ. Style cylindrical, its arms generally long and subulate, occasionally short and blunt, always covered all over with bristles.
2. Eupatoriaceæ. Style cylindrical; its arms long and clavate, with a papillose surface on the outside near the end.
3. Asteroideæ. Style cylindrical; its arms linear, flat on the outside, equally and finely downy on the inside.
4. Senecioideæ. Style cylindrical; its arms linear, fringed at the point, generally truncate, but sometimes extended beyond the fringe into a cone or appendage of some sort.
5. Cynareæ. Style thickened upwards, and often fringed at the tumour.

Suborder II. LABIATIFLORÆ; that is to say, those in which the hermaphrodite florets, or at least the unisexual ones, are divided into two lips. Of these the following are distinguished by their stigmas:—

6. Mutisiaceæ. Style cylindrical or somewhat tumid; its arms usually blunt, or truncate, very convex on the outside, and covered at the upper part by a fine uniform hairiness, or absolutely bald.
7. Nassauviaceæ. Style never tumid; the branches long, linear, truncate, fringed only at the point.

Suborder III. LIGULIFLORÆ; that is to say, those whose corollas are slit or ligulate, viz.

8. Cichoraceæ.

Fig. CCCCLXXIV.—Involucre, receptacle, and half fruit of Arnoseris pusilla.—*Gærtn.*

Fig. CCCCLXXV.—Stigmas of Composites, illustrating De Candolle's tribes (after *Heyland*). 1. Albertinia erythropappa (Vernoniaceæ); 2. Anisochæta mikanioides (Eupatoriaceæ); 3. Blumea senecioides (Asteroideæ); 4. Mendezia bicolor (Senecionideæ); 5. Lipochæta umbellata (Senecionideæ); 6. Aplotaxis nepalensis (Cynareæ); 7. Leucomeris spectabilis (Mutisiaceæ); 8. Leuceria tenuis (Nassauviaceæ.)

But although it must be admitted that the divisions of Lessing, so nearly followed by De Candolle, are ingenious, and often founded on striking characters, yet Botanists will also allow that they are far from satisfying the mind, and that they can only at the best

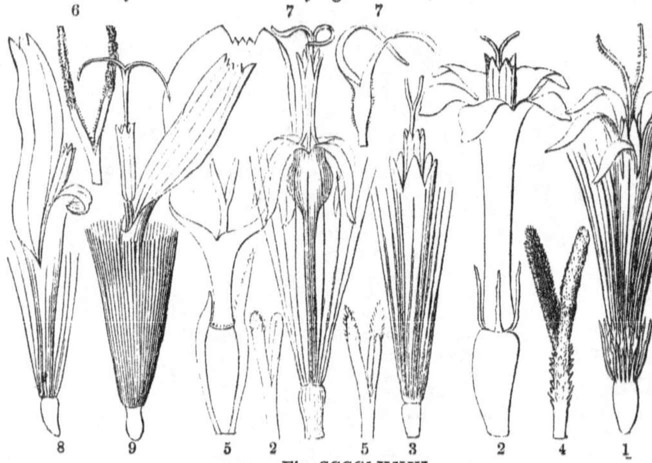

Fig. CCCCLXXVI.

be looked upon as temporary devices for dealing with a most unmanageable and difficult subject. The Composite Order alone comprehends at the present day more species than Linnæus knew as belonging to the whole vegetable kingdom, and the time will come when this huge Order will be classified upon different principles. There can be no doubt that the genera are needlessly multiplied; a very little practice tells us that the genera collected under the signs above given do not in all cases exhibit those signs, as is evident from the figures executed under the eye of De Candolle himself; and we know that, in fact, genera find their place by considerations apart from those ostensibly put forward by De Candolle. In the meanwhile, the old Jussieuan Sub-orders Corymbiferæ, Cynaraceæ, and Cichoraceæ, are unimpaired, and with the Bilabiate division, of the existence of which Jussieu was ignorant, constitute the immutable foundation of whatever future ingenuity may propose.

The Composite Order stands, as has been already stated, in most immediate affinity with Calycers and Teazelworts; but it is also closely allied to Bellworts and the rest of the Campanal Alliance. If the ovary were furnished with more than one cell, and there were many seeds in each cell, there would be little distinction from Lobeliads; and if Bellworts had syngenesious anthers and the latter characters in addition, they too would be almost identical. Indeed, the milky fluids of Bellworts and Lobeliads are of the same nature as in Cichoraceæ.

Among the peculiarities of the Order is the presence of a marginal vein to each petal, of which the corolla is composed; this vein passes up the edge, reaches the point, and then turns down again, so as to form a line running down through the axis of each petal; so that a Composite corolla may have five veins opposite the re-entering angles, or ten opposite them in pairs, or fifteen, when, in addition to the last circumstance, the axile vein of each petal is completed in the way described. There may even be ten veins, or indeed twenty, by variations of this peculiarity, as a little reflection will make evident.

Decaisne has made some curious remarks upon the hairs of plants of this Order. In Ruckeria the pericarp is covered with papillæ; on placing one of these papillæ in water, it immediately separates into 2 lips, and thence emits 2 mucilaginous tubes, which issue forth like wires, spirally unrolling themselves, and finally much exceed the papillæ from which they proceed. These tubes are apparently formed by a very considerable number of threads, placed one upon the other in the manner of a skein of thread. Various

---

Fig. CCCCLXXVI.—1. Tubular floret of Webbia aristata, with double pappus (Vernoniaceæ, *DC.*) 2. tubular floret and stigma of Anisochæta mikanioides, with pappus of 4 setæ (Eupatoriaceæ, *DC.*); 3. tubular floret of Berthelotia lanceolata, with silky pappus (Asteroideæ, *DC.*); 4. Stigma of Blumea senecioides (Asteroideæ, *DC.*); 5. ligulate floret and stigma of Lipochæta umbellata; pappus of two unequal winged paleæ (Senecionideæ, *DC.*); 6. stigma of Dunantia achyranthes (Senecionideæ, *DC.*); 7. tubular floret with ventricose throat and the stigma of Aplotaxis nepalensis (Cynareæ, *DC.*); 8. ligulate bilabiate floret of Oreoseris lanuginosa (Mutisiaceæ, *DC.*); 9. ligulate floret of Brachyramphus obtusus (Cichoraceæ, *DC.*)

other plants of the Labiatifloræ and Senecionideæ have been found to bear similar hairs, and among them the common Groundsel, Senecio vulgaris, whose achænia are clothed with them.—*Ann. Nat. Hist.* 6. 258.

All parts of the world contain Composites, but in very different proportions. According to the calculations of Humboldt, they constitute $\frac{1}{7}$ of the phænogamous plants of France, $\frac{1}{8}$ of Germany, $\frac{1}{13}$ of Lapland; in North America $\frac{1}{6}$, within the tropics of America $\frac{1}{2}$; upon the authority of Brown, they only form $\frac{1}{18}$ of the Flora of the north of New Holland, and did not exceed $\frac{1}{28}$ in the collection of plants formed by Smith upon the western coast of Africa in Congo.—*Congo*, 445. In Sicily they constitute rather more than $\frac{1}{2}$ (*Presl.*); the same proportion exists in the Balearic Islands (*Cambessédes*); but in Melville Island they are rather more than $\frac{1}{18}$ (*Brown*), a proportion nearly the same as that of the tropical parts of New Holland. It does not, therefore, appear that Composites, as an Order, are subject to any very fixed ratio of increase or decrease corresponding with latitude. But much remains to be learned upon this subject. It is certain that Cichoraceæ are most abundant in cold regions, and Corymbiferæ in hot ones; and that while in the northern parts of the world Composites are universally herbaceous plants, they become gradually frutescent, or even arborescent, as we approach the equator; most of those of Chile are bushes, and the Composites of St. Helena are chiefly trees. The Bilabiate genera are almost entirely American, and from the southern provinces beyond the tropics.

De Candolle gives the following as the result of his examination of the natural habit of Composites:—Out of 8523 of which he had any knowledge 1229 were annuals, 243 biennials, 2491 perennials, 2264 under-shrubs from 1 to 3 feet high, 366 shrubs from 4 to 15 feet high, 72 small trees, 4 large trees above 25 feet high, 81 woody plants of which nothing further was known, 126 twiners or climbers, and 1201 about which nothing certain could be ascertained. These were distributed as follows: 347 in the South Sea Islands, 2224 in Africa, 1827 in Asia, 1042 in Europe, and 3590 in America; of these the Cape of Good Hope possessed 1540, Mexico 725, Brazil 722, United States and Canada 678, the Levant 610, the Continent of India 681, north and middle Europe 447, Europe at the Mediterranean 595, Australia 294.—See this author's *Collection des Mémoires, No. X.*

M. Lasègue estimates (*Musée Delessert*, 1845) the number of Composite plants at 9500, and remarks "that they have steadily continued to constitute about $\frac{1}{10}$ of all described plants, in proportion as our knowledge of species has advanced. Thus Linnæus had 785 Composites out of 8500 species; in 1809 the proportion was 2800 to 27,000; De Candolle described 8523 in the year 1838, which was again a tenth; and now that the estimate of species has risen to 95,000, Composite plants amount to 9500."

The uses to which Composites have been applied are as numerous as their forms; but the species have considerable resemblance in the nature of their action. In former editions the statement made by De Candolle in his celebrated *Essai sur les Propriétés des Plantes*, was taken as the basis of an enumeration of them; it has however been much improved by others, and especially by Endlicher, whose account in the *Enchiridion* is here followed with some additions and alterations.

A bitter matter, combined with astringency, an acrid resinous substance, and some ethereal oil, to which in certain species is superadded starch in the roots, are the usual characteristics of Tubulifloræ, some of which are tonics, others stimulants, others astringents, according to the peculiar manner in which those substances are combined.

Among tonic, bitter, aromatic medicines the Artemisias are more especially deserving of notice, the various species having been employed in medicine from the most remote antiquity. Of these, Wormwoods are the most celebrated; they derive their English name from their employment as vermifuges, and are included in the species A. Absinthium and pontica; Southernwood, a fragrant plant, used on the Continent in making beer, is the A. Abrotanum; Tarragon, celebrated for its excellence in pickles, and in the medication of vinegar, is the A. Dracunculus; A. Mutellina, an Alpine plant, is intermediate in quality between Tarragon and Wormwood: it and A. spicata, another Alpine species, furnish between them the bitter aromatic liqueur called Crême d'Absinthe; A. acetica, a Persian species, is reported to have the odour of strong vinegar. A. chinensis and other species yield the Moxa of China, a substance which is used as a cautery, by burning it upon parts affected with gout and rheumatism. The leaves of A. maderaspatana are esteemed by the Indian doctors a valuable stomachic medicine; they are also sometimes used in antiseptic and anodyne fomentations. A. indica is considered in India a powerful deobstruent and antispasmodic. The flower-heads of many species of Artemisia constitute the drugs called Semen contra, or Semen Cinæ, vermifuges of much activity. Those which form the principal part of this substance, are A. Sieberi, Lercheana, Contra and pauciflora. The flower-heads of A. Vahliana also furnish one of the kinds of Wormseed called Semen Cinæ levanticum, or Semen cinæ in granis. It is collected in the North-east of Persia. The A. alba, and other species, serve as nourishment to the herds of the Kirghese and Calmucks.—*Annales de Chem.* 1. 49. The flower-heads of A. cærulescens, a Mediterranean plant, are the anthelmintic known under the name of Semen Seriphii, or Barbotine ; A. camphorata has a similar action, as also has A. gallica, called in France Sanguerié or Sanguerite. The nature of Tansy, Tanacetum vulgare, is not very different.

The Achilleas, or Milfoils, have an ethereal oil and a bitter, resinous, astringent matter in their foliage. Achillea millefolium is highly astringent, and the Highlanders are said to make of it an ointment, which dries and heals wounds.—*Hooker Brit. Fl.* p. 368. The Achilleas setacea, nobilis, and others, are slightly stimulating and tonic. A. Ageratum, a South of Europe species, is a very powerful stimulant ; the French regard it as a vulnerary, and call it Herbe au Charpentier. The Ptarmicas, formerly considered Achilleas, are similar in their action. The heads of P. nana, atrata and moschata, are used in the Swiss Alps as tea ; P. moschata is the basis of the aromatic liqueur called Esprit d'Iva ; of P. vulgaris, the whole plant is pungent, provoking a flow of saliva ; its dried leaves produce sneezing, but this is thought to be owing to their sharp marginal teeth ; the root is aromatic.

The ethereal oil, so abundant in these plants, is sometimes acrid, sometimes bitter ; it is more especially secreted in the flower-heads of many species, which are in that form employed for various purposes. First among them ranks Chamomile (Anthemis nobilis), a plant abounding on commons and similar wild places, where it is closely cropped by cattle : it is a well known stimulating tonic, and its warm infusion is employed to excite vomiting. The flower-heads of Santolina fragrantissima are extremely fragrant when dry, and are sold in the shops of Cairo as a substitute for Chamomile, under the name of Babouny or Zeysoum. Forskahl says the fresh juice of the plant is applied in affections of the eyes. Matricaria Chamomilla and Pyrethrum Parthenium (Feverfew) have a similar action, but are not in general use ; the smell of the latter is said to be peculiarly offensive to bees, which, it is added, may be easily kept at a distance by carrying a handful of the flower-heads.

Fig. CCCCLXXVII.

Others seem to be offensive to other animals. We are assured by Prof. Cantraine that Chrysanthemum Leucanthemum is a certain remedy against fleas. The Bosnians place the plant in the bed of domestic animals, and the fleas are destroyed in a very short time.—*Bull. Ac. r. Brux.* viii. 234. In some cases the stimulating action is so much increased as to assume an acrid form. Maruta fœtida is a weed, every part of which is fœtid and acrid, blistering the skin when much handled ; its decoction is a strong and active bitter, in the dose of a tea-cupful producing copious vomiting and sweating. With this may be associated Anthemis tinctoria and Santolina Chamæcyparissus, both obsolete remedies. In some instances this same ethereal oil acquires a remarkable pungency, as in Anacyclus Pyrethrum, the Pellitory of Spain, whose fleshy root when fresh produces on the

Fig. CCCCLXXVII.—Chrysanthemum Leucanthemum.

hands of those who gather it a sensation of extreme cold, followed by a burning heat; its taste is hot, acrid, and permanent, and it is used as a powerful rubefacient and stimulant, especially as a masticatory in rheumatic affections of the mouth; in a mitigated form, this same action is found to exist in various species of Spilanthes, and in Bidens, Pyrethrum, Tanacetum, and others, which thus excite salivation powerfully.

In some instances the oily secretion of Composites assumes a dangerous form. The most remarkable instance of this occurs in Arnica montana, a Swiss herb, called in our gardens Mountain Tobacco; it is a virulent plant, acting as a powerful narcotico-acrid agent; it is said, however, that this activity has been exaggerated. It has been recommended in the cure of putrid fever, ague, palsy, amaurosis, &c. &c.; and on the Continent has obtained the name of Panacea lapsorum. It has been stated by Dr. Thomson that its flowers contain an igasurate of strychnine; but this is positively denied by Versmann, who asserts that the activity of the plant does not depend upon any alkaloid, but upon an extremely acrid, resinous matter.—*Pharm. Journ.* iv. 239. It seems that properties analogous to those of Arnica occur in Doronicum Pardalianches, and various species of the genus Inula or those allied to it. Of these the most remarkable is Elecampane (Inula Helenium), an aromatic plant whose root contains a white starchy powder called Inuline, a volatile oil, a soft acrid resin, and a bitter extractive; it is regarded as a tonic, diuretic, and diaphoretic, and has been used in dyspepsia, pulmonary affections, and other diseases. It furnishes the Vin d'Aulnée of the French.

Eupatorium and its allies, in addition to the qualities just mentioned, or as a substitute for them, are astringent in some cases, emetic and purgative in others. The latter is the action of Eupatorium cannabinum, a common European marsh plant; E. Ayapana is a powerful sudorific, and is said to be found a valuable repellent of the poison of venomous snakes. For this purpose it is used in Brazil. A quantity of the bruised leaves, which are to be frequently changed, is laid on the scarified wound, and some spoonfuls of the expressed juice are from time to time administered to the patient, till he is found to be free from the symptoms, particularly the dreadful anxiety which follows the wounds of venomous reptiles. E. perfoliatum has a very similar action. Mikania opifera is employed in the same way as the Ayapana; M. officinalis acts like Cascarilla. The famous vulnerary called Matico, and said to be derived from Artanthe elongata (see p. 518), is really, according to Mr. Hartweg, the Eupatorium glutinosum.* Mikania Guaco has been stated by Humboldt to be called Vijuco del Guaco, and to be much esteemed in Spanish America as a valuable antidote against the bite of serpents. But the power of this Mikania is denied in the most positive terms by Hancock, who suspects that the real Guaco antidote is some kind of Aristolochia.

To these might be added a long list of plants belonging to the genera Baccharis, Acanthospermum, Ageratum, Pluchea, Conoclinium, Tagetes, Conyza, Elephantopus, Blumea, &c., the uses of which may be found in special works on the virtues of plants.

A few of these plants are employed as dyes. A beautiful carmine is obtained from the corolla of the Dahlia; Eclipta erecta stains black the hair of women in Brazil.

Ceradia furcata, a half-succulent plant, inhabiting the most sterile regions of S. W. Africa, yields in some abundance a brittle resinoid substance, which is rather fragrant when burnt, and has been called African Bdellium.

A bland oil abounds in the seeds of many species. Of these the most remarkable are Guizotia oleifera (formerly Verbesina sativa), extensively cultivated in India under the name of Ram-til; Helianthus annuus (the Sunflower), whose large, sweet, nut-like seeds are very palatable and wholesome; and Madia sativa, a Chilian plant lately introduced with success into the agriculture of the drier parts of Europe. Madia oil, expressed without heat, is described as transparent, yellow, scentless, &c., and fit for salads; its cake is said to be good for cattle; it produces, in dry climates, as much oil per acre as Poppy; in comparison with Colza as 32 to 28, Linseed as 32 to 21, and Olives as 32 to 16.—*Pasquier.*

In general the Cynareous genera are characterised by intense bitterness; some are stimulating, diaphoretic, and diuretic; others have flowers and leaves used for dyeing;

---

* I have the following memorandum on this subject, from Mr. Hartweg: "Matico is the vernacular name applied by the inhabitants of Quito to Eupatorium glutinosum, or the Chussalonga, in the Quichua language. It forms a shrub 3- 5-feet high, and is common in the higher parts of the Quitinian Andes, where its properties have been discovered some years back by a soldier called Mateo, better known under his nick-name Matico (little Matthew), who, when wounded in action, applied accidentally the leaves of some shrub to his wound, which had the immediate effect of stopping the bleeding. This shrub happened to be the Chussalonga, which has since been called, in honour of its discoverer, Matico. That it is the true Matico of the inhabitants of Quito and Riobamba, I have not the slightest doubt; both leaves and specimens have been gathered by myself, and upon comparing the latter with Kunth's description, I found them to agree exactly with his Eupatorium glutinosum.

"I have also a small quantity of powdered leaves of some shrub possessing the same virtue as the Matico, collected in Bolivia, where it is known under the name of Moxo-moxo. From bits of square stems which I find in the parcel, I suspect this to belong to some Labiate."

of some the root abounds in gum, and in many the seeds are oily and purgative, without a trace of the aroma so generally prevalent in the Order.

Centaurea Calcitrapa, and especially Cnicus Benedictus, have been used as febrifuges, and it is asserted, with great success, though they are now banished from general practice. Similar qualities have been assigned to many others, especially to the Bardanas or Burdocks, Lappa minor and major; of these plants the root is reckoned tonic, aperient, sudorific, and diuretic; it has been used in the form of decoction in rheumatism and diseases of the skin; Sir Robert Walpole praised it as a gout medicine, and others have considered it an excellent substitute for Sarsaparilla; the fruit, which is bitter and slightly acrid, has been used as a diuretic.

Carlina acaulis, a meadow plant with a very short stem and large flower-heads, conspicuous for the long rays of the shining involucre, was formerly used in magical incantations; its bark abounds in resinous matter, and a strong-scented, bitter, caustic oil, which acts as a drastic purgative. Another species, Carlina gummifera, the *ἰξία* or *ἰξίνη* of the ancients, has from time immemorial been employed as an anthelmintic; its great fleshy root and its flower-heads also yield a gum which hardens into tears like Mastich; when fresh, the root is said to be noxious to both man and beast, but the fleshy receptacles are preserved with honey or sugar, and eaten. Similar qualities are found in an Arabian plant, supposed to be allied to Cardopatum corymbosum, whose roots are sold in the shops of the continental nations, under the name of Costus. But Dr. Falconer has ascertained that the Costus of the ancients is the root of his Aucklandia Costus, a plant inhabiting the moist open slopes surrounding the valley of Cashmere, at an elevation of 8000 or 9000 feet above the level of the sea. The roots have a strong aromatic pungent odour, are regarded as aphrodisiacs, and are burnt as incense in the temples. In Cashmere the plant is not held in repute as a medicine, but is chiefly employed for protecting bales of shawls from the attacks of moths. The modern Arabians consider the root of the Artichoke (Cynara Scolymus) an aperient: they call the gum of it Kunkirzeed, and place it among their emetics. Some of this race are used by dyers. Safflower, employed to produce a beautiful pink dye, and in the preparation of rouge, is the dried flowers of Carthamus tinctorius; its seeds are purgative, and have been used in dropsical cases. Serratula tinctoria also furnishes a yellow or green dye. The flowers of Calendula officinalis, or Pot Marigold, are used to adulterate Saffron; it was formerly cultivated as an aperient and sudorific, but is now forgotten. We learn from Col. Sykes that the seeds of Carthamus persicus produce a useful oil, edible when fresh; that they are eaten whole as food in times of scarcity, and also the leaves as greens; the oil-cake of this plant is said, on the same authority, to be highly nourishing to milch cattle.

Few of the Labiatifloral Sub-order appear to be of any importance. The leaves of Printzia aromatica are used at the Cape of Good Hope as a substitute for tea; those of Anandria discoidea are mucilaginous and bitter, and are employed in China in dyspnœa, as those of Tussilago Farfara have been in Europe; Trixis brasiliensis is taken in decoction as a remedy for excessive menstruation; Moscharia pinnatifida smells of musk, and Flotovia diacanthoides forms a small tree with a hard white wood.

The Liguliforal genera are of far more importance. In all cases they abound in a milky, bitter, astringent, or narcotic juice. Among the bitters the most useful is Chicory, Cichorium Intybus, whose tap roots are cultivated as a substitute for Coffee, which they certainly improve when torrefied and added in small quantities.

Taraxacum Dens Leonis, the common Dandelion (Dent de Lion), appears to be of considerable medicinal importance as an anodyne, deobstruent, aperient, and diuretic; in cases of chronic diarrhœa it has been found very useful, according to Mr. Houlton. M. Polex has obtained from it Taraxacine in arborescent or star-shaped crystals.—*Pharm. Journ.* 1. 425. Nabalus serpentaria and albus are two North American plants, whose bitter milky roots are held in repute as a remedy for Rattlesnake bites; Mulgedium floridanum is called, because of its bitterness, Gall of the Earth. The Lettuces, Lactuca, are all narcotic; Lactuca virosa, Scariola, and sylvestris yield an extract resembling opium in its qualities; the garden Lettuce, L. sativa, furnishes the narcotic drug called Lactucarium. But, according to Aubergier, the best Lactucarium is obtained from Lactuca altissima.—*Comptes R.* xv. 923. A similar gum, which they call *κολλα*, is obtained in Lemnos from Chondrilla juncea. In a few species the juice is acrid. Zacyntha verrucosa is used in the Mediterranean as a phagædenic, and Crepis lacera is held in the kingdom of Naples to be a venomous plant.

In a small number of species of this Order nutritive matter is collected in sufficient abundance to render them worthy of notice as esculents. The most important in that way are Cardoons, the blanched leaf-stalks and stems of Cynara Cardunculus; Artichokes obtained from the succulent receptacles of Cynara Scolymus; Scorzonera and Salsafy, the roots of Scorzonera hispanica and Tragopogon porrifolius; Endive, the blanched leaves of Cichorium Endivia; Succory, a similar preparation of Cichorium Intybus;

and above all, Lettuces and Jerusalem Artichokes: the former the leaves of Lactuca sativa, the latter the tubers of Helianthus tuberosus. To these some have added the roots of the Dahlia, but their strong turpentine taste renders them unfit for food. Among the less known esculents are Helminthia echioides, whose leaves are boiled and pickled in Greece.—*Sibth.* Scorzonera glastifolia has roots whose quality is similar to that of S. hispanica; S. deliciosa is the species most cultivated as an esculent at Palermo; the gummy root of Scorzonera tuberosa is eaten by the Calmucks, and the young roots of Myscolus (Scolymus) hispanicus are esculent when young, but they are diuretic. The leaves of Cichorium Intybus have been found to dye blue when prepared in the same manner as Woad.—*Chem. Gaz.* 1845, p. 340.

GENERA.

Sub-order I. TUBULIFLORÆ.

I. VERNONIACEÆ.

ETHULIEÆ.

Adenocyclus, *Less.*
Odontoloma, *H. B. K.*
Oiospermum, *Less.*
Sparganophorus, *Vaill.*
  Struchium, P. Br.
Xiphochæta, *Pöpp.*
Ethulia, *Cass.*
  Kahiria, Forsk.
  Leighia, Scop.
Herderia, *Cass.*

HETEROCOMEÆ.

Pacourina, *Aubl.*
  Pacourinopsis, Cass.
  Meisteria, Scop.
  Haynea, Willd.
Heterocoma, *DC.*
Vernonia, *Schreb.*
  Acilepis, Don.
  Hololepis, DC.
  Proteopsis, Mart.
  Leptospermoides, DC.
  Vanillosma, Less.
  Carpholobus, Schott.
  Strobocalyx, Blume.
  Trianthea, DC.
  Pollalesta, Kunth.
  Oliganthes, Cass.
  Tephrodes, DC.
  Isomeria, Don.
  Lepidaploa, Cass.
  Ascaricida, Less.
  Centrapalus, Cass.
  Baccaroides, L.
Decaneurum, *DC.*
  Phyllocephalum, Blum.
  Wightia, Spreng.
  Rolfinkia, Zenker.
  Gymnanthemum, Cass.
Cyanopsis, *Blume.*
  Cyanthillium, Blume.
  Isonema, Cass.
Centratherum, *Cass.*
  Ampherephis, Kunth.
  Spixia, Schrank.
Bechium, *DC.*
Stokesia, *Herit.*
  Cartesia, Cass.
Platycarpha, *Less.*
  Cynara, Thunb.
Odontocarpha, *DC.*
Webbia, *DC.*
Hoplophyllum, *DC.*
Piptocoma, *Cass.*
Distephanus, *Cass.*
Strophopappus, *DC.*
Blanchetia, *DC.*
Symbiomeria, *Nutt.*
Stilpnopappus, *Mart.*
Dialesta, *H. B. K.*
Monosis, *DC.*
  ? *Turpinia*, Llv. et Lex.
Shawia, *Forst.*
Haplostephium, *Mart.*
Lychnophora, *Mart.*
Albertinia, *Spreng.*
  Eremanthus, Less.
Pycnocephalum, *DC.*
Chresta, *Arrab.*
Lychnocephalus, *Mart.*
Chronopappus, *DC.*
Pithecoseris, *Mart.*
Leucopholis, *Gardn.*
Stachyanthus, *DC.*

ELEPHANTOPEÆ.

Elephantopus, *L.*
  Elephantosis, Less.
  Pseudelephantopus, R.
  Distreptus, Cass.
  Matamoria, Llv. et Lx.

ROLANDREÆ.

Gundelia, *Tournef.*
  Gundelsheimera, Cass.
Corymbium, *L.*
  Contarena, Adans.
Solandra, *Rottb.*
Spiracantha, *H. B. K.*
  Acosta, DC.
Trichospira, *H. B. K.*
Lagascea, *Cav.*
  Noccæa, Cass.

BOJERIEÆ.

Synchodendron, *Boj.*
Centauropsis, *Boj.*
Tecmarsis, *DC.*
Bojeria, *DC.*

LIABEÆ.

Xanthisma, *DC.*
Sinclairia, *Hook.*
Hectorea, *DC.*
Andromachia, *H. B. K.*
  Oligactis, Cass.
  Pleionactis, DC.
  Viviania, Willd.
  Platylepidea, DC.
  Platylepis, Less.
Paranephelius, *Pöpp.*
Liabum, *Adans.*
  Chrysactinium, Kunth.
  Starkea, Willd.
  Andromachia, Cass.
Alibum, *Less.*
Cacosmia, *H. B. K.*
  Xantholepis, Willd.
  Clairvillea, DC.

PECTIDEÆ.

Pectidopsis, *DC.*
Pectidium, *Less.*
Pectis, *Linn.*
  Lorentea, Lagasc.
  Chthonia, Cass.
Lorentea, *Less.*
  Cryptopetalum, Cass.
  Stammarium, Willd.

II. EUPATORIACEÆ

ALOMIEÆ.

Orsinia, *Bertol.*
Piqueria, *Cav.*
Alomia, *H. B. K.*
Phalacræa, *DC.*
Gymnocoronis, *DC.*
Isocarpha, *R. Br.*

AGERATEÆ.

Cœlestinia, *Cass.*
Ageratum, *L.*
  Pectinellum, DC.
Anisochæta, *DC.*
Adenostemma, *Forst.*
  Lavenia, Swartz.
Sclerolepis, *Cass.*
Phania, *DC.*
  Oxylobus, Moç.
Stevia, *Cav.*
Palafoxia, *Lagasc.*
  Paleolaria, Cass.
  Polypteris, Nuttall.
Carelia, *Less.*
Agrianthus, *Mart.*
Helogyne, *Nutt.*

ADENOSTYLEÆ.

Kuhnia, *L.*
  Strigia, DC.
  Critonia, Gärtn.
  Trichogonia, DC.
  Leiogonia, DC.
Carminatia, *Moç.*
Disynaphia, *DC.*
Clavigera, *DC.*
Liatris, *Schreb.*
  Suprago, Gærtn.
  Trilisa, Cass.
Carphephorus, *Cass.*
Decachæta, *DC.*
Chromolæna, *DC.*
Ooclinium, *DC.*
  ? *Praxelis*, Cass.
Conoclinium, *DC.*
Hebeclinium, *DC.*
Lophoclinium, *Endl.*
Campyloclinium, *DC.*
Bulbostylis, *DC.*
  Coleosanthus, Cass.
Critonia, *P. Br.*
  Wickströmia, Spr.
Eupatorium, *Tournef.*
  Traganthа, Wallr.
Nothites, *Cass.*
Mikania, *Willd.*
Adenostyles, *Cass.*
Brickellia, *Ell.*

PETASITEÆ.

Homogyne, *Cass.*
Nardosmia, *Cass.*
Petasites, *Tournef.*
Adenocaulon, *Hook.*

EUTUSSILAGINEÆ.

Tussilago, *Tournef.*
Alciope, *DC.*
Brachyglottis, *Forst.*

III. ASTEROIDEÆ.

AMELLEÆ.

Amellus, *Cass.*
Corethrogyne, *DC.*
Chiliotrichum, *Cass.*
  Tropidolepis, Tausch.

HETEROTHALAMEÆ.

Heterothalamus, *Less.*

ASTEREÆ.

Mairia, *DC.*
  Pteropappus, Less.
  Zyrphelis, Cass.
Felicia, *DC.*
  Polyarrhena, Cass.
Munychia, *Cass.*
Agathea, *Cass.*
  Detridium, Nees.
  Detris, Adans.
Bellidiastrum, *Michel.*
  Margarita, Gaud.
Aster, *Nees.*
  Amellus, Adans.
  Symphyotrichum, Nees.
Tripolium, *Nees.*
Galatella, *Cass.*
  Galatea, Cass.
Monoptilon, *T. et Gr.*
Turczaninowia, *DC.*
Townsendia, *Hook.*
Calimeris, *Cass.*
Xylorrhiza, *Nutt.*
Eurybiopsis, *DC.*
Podocoma, *Cass.*
  ? *Podopappus*, Hook.
Asteropsis, *Less.*
Arctogeron, *DC.*
Sericocarpus, *Nees.*
Eucephalus, *Nutt.*
  Lagatea, Nutt.
Machæranthera, *Nees.*
Tetramolopium, *Nees.*
Henricia, *Cass.*
Döllingeria, *Nees.*
Heleastrum, *DC.*
Biotia, *DC.*
Eurybia, *Cass.*
  Spongotrichum, Nees.

DIPLOPAPPEÆ.

Olearia, *Mönch.*
  Haxtonia, Caley.
Diplostephium, *Cass.*
Callistephus, *Cass.*
  Callistemma, Cass.
  ? *Poloa*, DC.

Diplopappus, *DC.*
Asterosperma, Less.
Rhinactina, *Less.*
Noticastrum, *DC.*
Distasis, *DC.*

ERIGEREÆ.

Melanodendron, *DC.*
Leptocoma, *Less.*
Vittadinia, *A. Rich.*
Fullartonia, *DC.*
Polyactidium, *DC.*
Polyactis, Less.
Stenactis, Cass.
Heterochæta, *DC.*
Therogeron, *DC.*
Erigeron, *DC.*
Leptostelma, Don.
Terranea, Colla.
Trimorphœa, Cass.
Rhynchospermum, *Rnw.*
Microgyne, *Less.*

HETEROPAPPEÆ.

Simblocline, *DC.*
Heteropappus, *Less.*
Phalacroloma, *Cass.*
Minuria, *DC.*
Stenactis, *Nees.*
Gymnostephium, *Less.*
Charieis, *Cass.*
Kaulfussia, Nees.
Chætopappa, *DC.*
Chætophora, Nutt.
Boltonia, *Herit.*
Perityle, *Benth.*
Sommerfeltia, *Less.*

BELLIEÆ.

Calotis, *R. Br.*
Hunefeldia, *Walp.*
Asteromœa, *Blume.*
Bellium, *L.*

BELLIDEÆ.

Bellis, *L.*
Kyberia, Neck.
Seubertia, *Wats.*
Brachycome, *Cass.*
Brachystephium, Less.
Paquerina, *Cass.*
Lagenophora, *Cass.*
Lagenifera, Cass.
Microcalia, A. Rich.
Ixauchenus, *Cass.*
Myriactis, *Less.*
Botryadenia, Fisch.
Garuleum, *Cass.*
Keerlia, *DC.*
Aphanostephus, *DC.*

GYMNOSPERMEÆ.

Xanthocoma, *H. B. K.*
Xerothamnus, *DC.*
Anaglypha, *DC.*
Gymnosperma, *Less.*
Selloa, Spreng.

ACHYRIDEÆ.

Brachyris, *Nutt.*
Brachyachyris, Spreng.
Amphipappus, *T. et Gr.*
Hemiachyris, *DC.*
Lepidophyllum, *Cass.*
Grindelia, *Willd.*
Demetria, Lagasc.
Donia, R. Br.
Aurelia, Cass.
Centauridium, *Torrey.*

HETEROTHECEÆ.

Heterotheca, *Cass.*
Calycium, Ell.
Diplocoma, Don.
Bradburia, *Torrey.*
Dieteria, *Nutt.*
Sideranthus, Nutt.
Pappochroma, Nutt.

PSIADIEÆ.

Erato, *DC.*
Woodvillea, *DC.*
Psiadia, *Jacq.*
Elphegea, Cass.
Thouarsia, Vent.
Alix, Commers.
Glutinaria, Commers.
Frivaldia, *Endl.*
Microglossa, DC.
Nidorella, *Cass.*
Homochroma, *DC.*
Neja, *D. Don.*

CHRYSOPSIDEÆ.

Chrysopsis, *Nutt.*
Diplogon, Rafin.
Pityopsis, *Nutt.*
Fresenia, *DC.*

SOLIDAGINEÆ.

Bigelowia, *DC.*
Brachychæta, *Torrey.*
Chrysoma, *Nutt.*
Chrysothamnus, *Nutt.*
Solidago, *L.*
Virga-aurea, Tournef.
Doria, Adans.
Euthamia, DC.
Amphirapis, *DC.*
Isopappus, *Torrey.*
Homopappus, *Nutt.*
Myrianthus, Nutt.
Actinophora, Nutt.
Stenotus, *Nutt.*
Commidendrum, *Burch.*
Steiractis, *DC.*
Rochonia, *DC.*
Prionopsis, *Nutt.*
Haplopappus, *Cass.*
Aplopappus, Cass.
Diplopappus, Less.
? Hoorebekia, Cornelis.
? Sideranthus, Fraser.
Ericameria, *Nutt.*
Pyrrocoma, *Hook.*
Chromochæta, DC.
Macrocnema, *Nutt.*
Lessingia, *Cham.*
Isocoma, *Nutt.*
Linosyris, *Lobel.*
Crinitaria, Less.
Crinita, Mönch.
Chrysocoma, *Cass.*
Ammodia, *Nutt.*
Eriocarpum, *Nutt.*
Pteronia, *L.*
Scepinia, DC.
Henanthus, Less.
Pachyderris, DC.
Pterophorus, DC.
Pterophora, Neck.
Pentachæta, *Nutt.*

SOLENOGYNEÆ.

Duhaldea, *DC.*
Microtrichia, *DC.*
Nolletia, *Cass.*
Sarcanthemum, *Cass.*
Leptothamnus, *DC.*
Chröilema, *Bernh.*
Solenogyne, *Cass.*

SPHÆRANTHEÆ.

Blepharispermum, *Wight.*
Leucoblepharis, *Arn.*
Athroisma, *DC.*
Sphæranthus, *Vaill.*
Cuspidella, DC.
Polycephalos, DC.
Oligolepis, Cass.

GRANGEINEÆ.

Dichrocephala, *DC.*
Centipeda, Less.
Grangea, *Adans.*
Leptoderris, DC.
Pyrarda, DC.
Cyathocline, *Cass.*
Lestadia, *Kunth.*
Gymnarrhena, *Desf.*
Gymnarhea, Steud.
Frankia, Steud.

CONYZEÆ.

Thespis, *DC.*
Karelinia, *Less.*
Berthelotia, *DC.*
Læennecia, *Cass.*
Conyza, *Less.*
Dimorphanthes, Cass.
Eschenbachia, Mönch.
Fimbrillaria, Cass.
Leucopodum, *Gardn.*
Phagnalon, *Cass.*
Chionolæna, *DC.*
Elachothamnos, *DC.*
Parastrephia, *Nutt.*

BACCHARIDEÆ.

Polypappus, *Less.*
Baccharis, *L.*
Molina, Ruiz et Pav.
Sergillus, Gärtn.
Pingræa, Cass.
Tursenia, Cass.
Arrhenachne, Cass.
Stephananthus, Lehm.

TARCHONANTHEÆ.

Brachylæna, *R. Br.*
Oligocarpha, Cass.
Tarchonanthus, *Linn.*
Henotogyna, DC.

PLUCHEINEÆ.

Blumea, *DC.*
Pluchea, *Cass.*
Stylimnus, Rafin.
Gymnema, Rafin.
Leptogyne, Ell.
Chlenobolus, Cass.
? Placus, Lour.
Hebephora, DC.
Pterocaulon, *Ell.*
Monenteles, *Labill.*
Tessaria, *Ruiz et Pav.*
Gynheteria, Willd.
Gyneteria, Spr.
? Phalacromesum, Cas.
Monarrhenus, *Cass.*
Mahometa, DC.
Cylindrocline, *Cass.*
Lepidopogon, Tausch.
Evax, *Gärtn.*
Filago, Willd.
Gnaphalium, Vaill.
Filaginopsis, *Torr.*
Diaperia, *Nutt.*
Stylocline, *Nutt.*
Micropsis, *DC.*
Calymnandra, *Torr.*
Micropus, *L.*
Gnaphalodes, Tournef.
Psilocarpha, Nutt.

Epaltes, *Cass.*
Ethulia, Gärtn.
Denekia, *Thunb.*
Dipterocome, *F. et M.*

INULEAÆ.

Rhanterium, *Desf.*
Codonocephalum, *Fenzl.*
Inula, *Gärtn.*
Corvisartia, DC.
Bubonium, DC.
Enula, Duby.
Cappa, DC.
Limbarda, DC.
Eritheis, Gray.
Schizogyne, *Cass.*
Varthemia, *DC.*
Vicoa, *Cass.*
Pentanema, *Cass.*
Francœuria, *Cass.*
Duchesnea, Cass.
Asteridea, *Lindl.*
Iphiona, *DC.*
Jasonia, *Cass.*
Chiliadenus, Cass.
Myriadenus, Cass.
Allagopappus, Cass.
Vieræa, *Webb.*
Pulicaria, *Gärtn.*
Tubilium, Cass.
Strabonia, *DC.*
Pegolettia, *Cass.*
Minyrothamnus, *DC.*
Cypselodontia, *DC.*
Geigeria, *Griess.*
Zeyheria, Spreng.
Dizonium, Willd.
Hochstetteria, *DC.*

CÆSULINEÆ.

Cæsulia, *Roxb.*

BUPHTHALMEÆ.

Buphthalmum, *Neck.*
Telekia, *Baumg.*
Molpadia, Cass.
Astericus, *Mönch.*
Nauplius, Cass.
Pallenis, *Cass.*
Athalmum, Neck.
Anvillea, *DC.*
Ceruana, *Forsk.*
Cryptadia, *Lindl.*

ECLIPTEÆ.

Borrichia, *Adans.*
Diomedea, Cass.
? Odontospermum, Nk.
Eclipta, *Linn.*
Micrelium, Forsk.
Blainvillea, *Cass.*
Uoacea, Cass.
Salmea, *DC.*
Hopkirkia, Spreng.
Dahlia, *Cav.*
Georgina, Willd.
Georgia, Spreng.
Leptocarpha, *DC.*
Siegesbeckia, *Linn.*
Schkuhria, Mönch.
Sabazia, *Cass.*
Cryphiospermum, *Palis.*
Wahlenbergia, Schum.

IV. SENECIONIDEÆ.

EUXENIEÆ.

Euxenia, *Cham.*
Ogiera, Spreng.
Podanthus, R. Br.

Petrobium, *R. Br.*
  Laxmannia, Forst.
  Drymiphyllum, Burch.
Astemma, *Less.*

MILLERIEÆ.

Elvira, *DC.*
  Meratia, Cass.
  Delilia, Spreng.
  Eugamelia, Fl. Mex.
Milleria, *Cass.*
Riencourtia, *Cass.*
  Tetrantha, Poit.
Garcilassa, *Pöpp.*
Latreillea, *DC.*
Ichthyothere, *Mart.*
Clibadium, *Linn.*
  Oswalda, Cass.
  Baillieria, Less.
  Trixis, Swartz.
  Trixidium, DC.
Picrothamnus, *Nutt.*
Unxia, *L.*
Blennosperma, *L.*
  Apalus, DC.
Pronacron, *Cass.*
Aiolotheca, *DC.*
Trigonospermum, *Less.*
Xenismia, *DC.*
Scolospermum, *Less.*
Baltimora, *L.*
  Fougerouxia, DC.
  Niebuhria, Scop.
  Fougeria, Mönch.
Chrysogonum, *L.*
  Diotostephus, Cass.

SILPHIEÆ.

Guardiola, *H. B. K.*
  Guandiola, Ste.
Hidalgoa, *Less.*
Silphium, *L.*
Polymnia, *L.*
  Uvedalia, DC.
  Alymnia, Neck.
  Polymniastrum, Lam.
Espeletia, *Mut.*
Berlandiera, *DC.*
Angelandra, *Endl.*
  Engelmannia, Torrey.

MELAMPODIEÆ.

Melampodium, *L.*
  Zarabellia, Cass.
  Dysodium, L. C. Rich.
  Alcina, Cav.
  Camutia, Bonat.
  ? Hidalgoa, Llav. et Lex
Acanthospermum, *Schr.*
  Centrospermum, Knth.
  Echinodium, Poiret.
  Ceratolæna, DC.
Tulocarpus, *Hook et Arn.*

AMBROSIEÆ.

Xanthium, *Tournef.*
Franseria, *Cav.*
  Xanthiopsis, DC.
  Centrolæna, DC.
Ambrosia, *Tournef.*

IVEÆ.

Pinellosia, *Ossa.*
Tetranthus, *Swartz.*
Iva, *L.*
  Denira, Adans.
Euphrosyne, *DC.*
  ? Cyclachæna, Fresen.
  Euphrosinia, Rchb.
Gymnogyne, *Steetz.*

PARTHENIEÆ.

Coniothele, *DC.*
Leptosyne, *DC.*
Parthenium, *L.*
  Partheniastrum, Niss.
  Hysterophorus, Vaill.
  Trichospermum, Palis.
  Bolophyta, Nutt.
  Argyrochæta, Cav.
  Villanova, Orteg.
Tragoceras, *Less.*
Moonia, *Arnott.*

HELIOPSIDEÆ.

Philactis, *Schrad.*
Zinnia, *L.*
  Lejica, Hill.
  Crassina, Scop.
Helicta, *Cass.*
Alarçonia, *DC.*
  Wyethia, Nutt.
  Trachinga, Endl.
Wedelia, *Jacq.*
  Stemmodontia, Cass.
  ? Trichostemma, Cass.
  Trichostephus, Cass.
  Trichostephium, Cass.
  Niebuhria, Neck.
  Aglossa, DC.
Jægeria, *H. B. K.*
Lipotriche, *R. Br.*
Melanthera, *Rohr.*
Ogiera, *Cass.*
  Chalarium, Poit.
Monactis, *H. B. K.*
Wollastonia, *DC.*
Tilesia, *F. W. Mey.*
Pascalia, *Orteg.*
Rumfordia, *DC.*
Heliopsis, *Pers.*
  ? Helepta, Rafin.
  Kallias, DC.
  Ralsamorhiza, DC.
Guizotia, *Cass.*
  Ramtilla, DC.
  Veslingia, Vis.
Tetragonotheca, *Dillen.*
Halea, *Torr. et Gray.*
  Engelmannia, Torrey.
Ferdinanda, *Lagasc.*
Chrysophania, *Kunth.*
Zaluzania, *Pers.*
Chiliophyllum, *DC.*
  Hybridella, Cass.
Scalesia, *Arnott.*

RUDBECKIEÆ.

Echinacea, *Mönch.*
  Branneria, Neck.
  Bobartia, Petiv.
  Helichroa, Rafin.
Echinomeria, *Nutt.*
Rudbeckia, *Linn.*
  Obeliscotheca, Vaill.
  Heliophthalmum, Rafin
Dracopis, *Cass.*
Obeliscaria, *Cass.*
  Lepachys, Less.
  Ratibida, Rafin.
Andrieuxia, *DC.*
Anomostephium, *DC.*
Aspilia, *Thouars.*
Gymnopsis, *DC.*
  Gymnolomia, Kunth.
  ? Aldama, Llav. et Lex.
Wulffia, *Neck.*
  Chakiatella, Cass.
  Chilodia, Rich.
  Gymnoloma, Ker.
  Crodisperma, Poit.
Montagnæa, *DC.*
  Eriocoma, Kunth.
  Montanoa, Llav. et Lex.
  Eriocarpha, Cass.

Sclerocarpus, *Jacq.*
Encelia, *Adans.*
  Pallasia, Herit.
Philoglossa, *DC.*
Chrysostemma, *Less.*
Calliopsis, *Rchb.*
  Diplosastra, Tausch.
? Peramibus, *Rafin.*

COREOPSIDEÆ.

Agarista, *DC.*
Epilepsis, *Benth.*
Coreopsis, *L.*
  Diodonta, Nutt.
  Medusea, Nutt.
  Heterodonta, Nutt.
  Cosmidium, Gray.
  Acispermum, Neck.
  Lechea, Cass.
  Chrysomelea, Tausch.
Tuckermannia, *Nutt.*
Actinomeris, *Nutt.*
  Ridan, Adans.
  Pterophyton, Cass.
  Actimeris, Rafin.
Armania, *Berter.*
Oyedæa, *DC.*
Simsia, *Pers.*
Viguiera, *H. B. K.*
Leighia, *Cass.*
Harpalium, *Cass.*
Tithonia, *Desf.*
Helianthus, *L.*
  Chrysis, Renealm.
  Corona Solis, Tournef.
  Vocasan, Adans.
  Discomela, Rafin.
Flourensia, *DC.*

BIDENTIDEÆ.

Campylotheca, *Cass.*
Bidens, *L.*
  Kerneria, Mönch.
  Pluridens, Neck.
  Edwardsia, Neck.
  Ceratocephalus, Vaill.
Cosmos, *Cav.*
  Cosmea, Willd.
Adenolepis, *Less.*
Microdonta, *Nutt.*

VERBESINEÆ.

Lasianthea, *DC.*
  Lasianthus, Zuccar.
Perymenium, *Schrad.*
  Schistocarpha, Less.
Psathurochæta, *DC.*
Lipochæta, *DC.*
  Lipotriche, Less.
  ? Zexmenia, Llav. et Lex.
Microchæta, *Nutt.*
Aphanopappus, *Endl.*
  Schizophyllum, Nutt.
Diplothrix, *DC.*
Selloa, *H. B. K.*
  Feæa, Spreng.
Verbesina, *Less.*
  Locheria, Neck.
  Siegesbeckia, Gronov.
  Phætusa, Gærtn.
  Hamulium, DC.
  Platypteris, DC.
Acoma, *Benth.*
Coreocarpus, *Benth.*
Mendezia, *DC.*
Ditrichum, *Cass.*
Micractis, *DC.*
Spilanthes, *Jacq.*
  Ceruchis, Gærtn.
  Acmella, DC.
  Athronia, Neck.
  Salivaria, DC.
Dunantia, *DC.*
Ximenesia, *Cav.*

Sanvitalia, *Juss.*
  Lorentea, Orteg.
Anaitis, *DC.*
Oligogyne, *DC.*
Harpephora, *Endl.*
Synedrella, *Gærtn.*
Calyptrocarpus, *Less.*
Electra, *DC.*
Chromolepis, *Benth.*
Chrysanthellum, *Rich.*
  Chrysanthellina, Cass.
  Sebastiania, Bertol.
  Collæa, Spreng.
Neuractis, *Cass.*
Glossocardia, *Cass.*
Heterospermum, *Willd.*
  Heterosperma, Cav.
Glossogyne, *Cass.*
Delucia, *DC.*
Narvalina, *Cass.*
  Needhamia, Cass.
Thelesperma, *Less.*
Isostigma, *Less.*

FLAVERIEÆ.

Flaveria, *Juss.*
  Vermifuga, Ruiz et P.
Broteroa, *DC.*
  Brotera, Spreng.
  Nauemburgia, Willd.
Enhydra, *Loureir.*
  Meyera, Schreb.
  Sobrya, Pers.
  Hingcha, Roxb.
  Tetractis, Reinw.

TAGETEÆ.

Adenophyllum, *Pers.*
  Willdenowa, Cav.
  Schlechtendalia, Willd.
Lebetina, *Cass.*
  Bœbera, Less.
  Dissodia, Willd.
Dysodia, *Cav.*
  Gymnolæna, DC.
  Aciphylla, DC.
Clomenocoma, *Cass.*
  Bartolina, Adans.
Hymenatherum, *Cass.*
Riddellia, *Nutt.*
Solenotheca, *Nutt.*
Tagetes, *Tournef.*
  Diglossus, Cass.
  Enalcida, Cass.
Thymophylla, *Lagasc.*
Adenopappus, *Benth.*
Syncephalantha, *Bartl.*

POROPHYLLEÆ.

Porophyllum, *Vaill.*
  Kleinia, Jacq.
  Tsinoma, Hernand.
  Hunteria, Fl. mex.
  Cusimbua, DC.
  Kugaia, DC.
Chæthymenia, *H. et A.*

GAILLARDIEÆ.

Gaillardia, *Foug.*
  Galardia, Lam.
  Colonnea, Buchoz.
  Virgilia, Herit.
  Galordia, Reuschel.
Balduina, *Nutt.*
  Endorima, Raf.
  Actinospermum, Ell.
Leptopoda, *Nutt.*
  Leptocarpha, Raf.
Achyrachæna, *Schauer.*
  Lepidostephanus, Btl.

HELENIEÆ.

Gutierrezia, *Lagasc.*

Achyropappus, *H. B. K.*
Chamæstephanum, *W.*
Schkuhria, *Roth.*
Tetracarpum, Mönch.
Mieria, Llav. et Lex.
Amblyopappus, *Hooker.*
Florestina, *Cass.*
Lepidopappus, Fl. mex.
Actinolepis, *DC.*
Bahia, *DC.*
Eriophyllum, Lagasc.
Trichophyllum, Nutt.
Phialis, Spreng.
Richteria, *Karelin.*
Oxylepis, *Benth.*
Macrocephalus, *Nutt.*
Macrocarphus, Nutt.
Hymenopappus, *Herit.*
Rothia, Lam.
Chænactis, *DC.*
Polypteris, *Nutt.*
Espejoa, *DC.*
Cercostylos, *Less.*
Polypteris, Less.
Güntheria, Spreng.
Hopkirkia, *DC.*
Hymenoxis, *Cass.*
Stylesia, *Nutt.*
Cephalophora, *Cav.*
Græmia, Hook.
Actinella, DC.
Actinea, Cass.
Dugaldea, Cass.
Ptilepida, Raf.
Cancrinia, *Karel.*
Jaumea, *Pers.*
Kleinia, Juss.
Burrielia, *DC.*
Ptilomeris, *Nutt.*
Dichæta, *Nutt.*
Picradenia, *Hook.*
Helenium, *L.*
Helenia, L.
Brassavola, Adans.
Tetrodus, DC.
Mesodetra, Rafin.
Amblyolepis, *DC.*
Rosilla, *Less.*
Trinchinettia, *Endl.*
Schomburgkia, DC.
Hecubæa, *DC.*
Bæria, *Fisch. et Mey.*
Callichroa, *Fisch. et Mey.*
Calliachyris, *Torr. et Gr.*
Oxypappus, *Benth.*
Rancagua, *Pœpp. et End.*
Lasthenia, Cass.
Argyroxiphium, *DC.*
Argyrophyton, Hook.
Pleurophyllum, *Hook. fil.*

GALINSOGEÆ.

Lemmatium, *DC.*
Caleacte, Less.
Calydermos, *Lagasc.*
Calebrachys, DC.
Meyeria, *DC.*
Callilepis, *DC.*
Calea, *R. Br*
Caleacte, DC.
Mocinna, Lagasc.
Leontophthalmum, Less
Allocarpus, *H. B. K.*
Alloispermum, Willd.
Vargasia, *DC.*
Galinsoga, *Ruiz et Pav.*
Galinsogæa, Zuccar.
Wiborgia, Roth.
Sogalgina, *Cass.*
Galinsogea, Less.
Sogaligna, Steud.
Ptilostephium, *H. B. K.*
Tridax, *L.*
Blepharipappus, *Hook.*
Ptiloncila, Nutt.
Eriopappus, Arn.
Marschallia, *Schreb.*
Persoonia, Michx.
Trattinickia, Pers.
Athanasia, Walt.
Phyteumopsis, Juss.
Dubautia, *Gaudich.*

SPHENOGYNEÆ.

Sphenogyne, *R. Br.*
Oligacrion, Cass.
Spermophylla, Neck.
Thelythamnos, Less.
Xerolepis, DC.
Ursinia, *Gærtn.*
Amida, *Nutt.*
Lagophylla, *Nutt.*
Harpæcarpus, *Nutt.*
Madia, *Molin.*
Madarella, Nutt.
Biotia, Cass.
Silphiosperma, *Steetz.*
Madaria, *DC.*
Madriopsis, Nutt.
Hemizonia, *DC.*
Amauria, *Benth.*
Tollatia, *Endl.*
Oxyura, DC.
Hartmannia, *DC.*
Madaroglossa, *DC.*
Layia, Hook. et Arn.
Lepidostephanus, *Bartl.*
Anisocarpus, *Nutt.*
Osmadenia, *Nutt.*
Calycadenia, *DC.*

ANTHEMIDEÆ.

Œdera, *DC.*
Eumorphia, *DC.*
Aganippea, *DC.*
Heliogenes, *Benth.*
Epallage, *DC.*
Anthemis, *DC.*
Chamæmelum, DC.
Marcelia, Cass.
Maruta, *Cass.*
Periderœa, Webb.
Lugoa, *DC.*
Lyonettia, *Cass.*
Anacyclus, *Pers.*
Hiorthia, Neck.
Cyrtolepis, *Less.*
Ormenis, *Cass.*
Cladanthus, *Cass.*
Lepidophorum, *Neck*
Ptarmica, *Tournef.*
Achillea, *Neck.*
Diotis, *Desf.*
Gnaphalium, Tournef.
Otanthus, Link.
Santolina, *Tournef.*
Chamæcyparissus, DC.
Babounya, DC.
Nablonium, *Cass.*
Lasiospermum, *Lagasc.*
Lanipila, DC.
Mataxa, Spreng.

CHRYSANTHEMEÆ.

Steiroglossa, *DC.*
Lidbeckia, *Berg.*
Gamolepis, *Less.*
Lasthenia, *DC.*
Hologymne, Bartl.
Psilothamnus, *DC.*
Jacquemontia, Belang.
Spiridanthus, *Fenzl.*
Coinogyne, *Less.*
Egletes, *Less.*
Xerobius, Cass.
Eyselia, Rchb.
Venegasia, *DC.*
Leucopsidium, *DC.*
Xanthocephalum, *Willd.*
Phymaspermum, *Less.*
Hisutsua, *DC.*
Brachanthemum, *DC.*
Nananthea, *DC.*
Leucanthemum, *Tournef.*
Phalacrodiscus, Less.
Phalacroglossum, DC.
Diabasis, DC.
Enuchoglossum, DC.
Phalacrocarpum, DC.
Prolongoa, *Boiss.*
Adenachæna, *DC.*
Matricaria, *L.*
Pyrethrum, *Gærtn.*
Gymnocline, Cass.
Xanthoglossa, DC.
Coleostephus, Cass.
Tridactylina, DC.
Dendranthema, DC.
Allardia, *Decaisne.*
Chrysanthemum, *DC.*
Ismelia, Cass.
Pinardia, Cass.
Glebionis, Cass.
Pinardia, DC.
Centrospermum, Sprn.
Heteranthemis, Schott.
Centrachæna, Schott.
Spermoptera, DC.
Mayarsa, DC.
Preauxia, *C. H. Schultz.*
Monoptera, *C. H. Schultz.*
Stigmatotheca, *C. H. S.*
Argyranthemum, *Webb.*
Dimorphotheca, *Vaill.*
Meteorina, DC.
Gattenhofia, Neck.
Cardispermum, Traut.
Lestibodia, DC.
Blaxium, DC.
Castalis, DC.
Rutidocarpæa, DC.
Arnoldia, DC.
Triplocarpæa, DC.
Acanthotheca, *DC.*
Monolopia, *DC.*
Steirodiscus, *Less.*
Schistostephium, *Krebs.*
Chlamysperma, *Less.*
Villanova, *Lagasc.*
Brachymeris, *DC.*
Brachystylis, E. M.
Jacosta, *E. Mey.*

COTULEÆ.

Lapeyrousia, *Thunb.*
Peyrousia, DC.
Otochlamys, *DC.*
Cotula, *Gærtn.*
Baldingeria, Neck.
Cenocline, *Koch.*
Strongylosperma, *Less.*
Cenia, *Commers.*
Lancisia, Gærtn.
Homalotes, *DC.*
Aromia, *Nutt.*

ATHANASIEÆ.

Lonas, *Adans.*
Gonospermum, *Less.*
Metagnanthus, *Endl.*
Hymenolepis, Cass.
Holophyllum, *Less.*
Pristocarpha, E. Mey.
Bembecodium, *Kunz.*
Athanasia, *Cass.*
Saintmorysia, *Endl.*
Morysia, Cass.

ARTEMISIEÆ.

Stilpnophytum, *Less.*
Mesoteirus, DC.
Lepidotheca, *Nutt.*
Artemisia, *L.*
Dracunculus, Bess.
Oligosporus, Cass.
Seriphidium, Bess.
Seriphida, Less.
Abrotanum, Tournef.
Absinthium, Tournef.
Crossostephium, *Cass.*
Tanacetum, *Linn.*
Psanacetum, Neck.
Brocchia, DC.
Hippioides, DC.
Sphæromeria, *Nutt.*
Plagius, *Herit.*
Balsamita, Less.
Adenosolen, *DC.*
Marasmodes, *DC.*
Pentzia, *Thunb.*
Chlamydophora, *Ehrenb.*
Myriogyne, *Less.*
Sphæromorphæa, *DC.*
Machlis, *DC.*

HIPPIEÆ.

Abrotanella, *Cass.*
Trineuron, *Hook. fil.*
Ceratella, *Hook. fil.*
Leptinella, *Cass.*
Plagiochilus, *Arnott.*
Soliva, *Ruiz et Pav.*
Gymnostyles, Juss.
Solivæa, Cass.
Hippia, *L.*

ERIOCEPHALEÆ.

Eriocephalus, *L.*
Monochlæna, Cass.
Cryptogyne, Cass.

ANGIANTHEÆ.

Styloncerus, *Labill.*
Ogcerostylus, Cass.
Actinobole, *Endl.*
Hyalolepis, *DC.*
Phyllocalymma, *Benth.*
Angianthus, *Wendl.*
Cassinia, R. Br.
Hirnelia, Cass.
Cylindrosurus, Benth.
Skirrhophorus, *DC.*
Eriocladium, Lindl.
Pogonolepis, *Steetz.*
Myriocephalus, *Benth.*
Gnephosis, *Cass.*
Pachysurus, *Steetz.*
Calocephalus, *R. Br.*
Leucophyta, *R. Br.*
Craspedia, *Forst.*
Richea, Labill.
Pycnosorus, *Benth.*
Chrysocoryne, *Endl.*
Crossolepis, Benth.

CASSINIEÆ.

Ammobium, *R. Br.*
Ixodia, *R. Br.*
Rhynea, *DC.*
Cassinia, *R. Br.*
Chromochiton, Cass.
Achromolæna, Cass.
Chthonocephalus, *Steetz.*

HELICHRYSEÆ.

Humea, *Smith.*
Calomeria, Vent.
Agathomeris, Delaun.
Razumovia, Spreng.
Crossolepis, *Less.*
Pithocarpa, *Lindl.*
Quinetia, *Cass.*
Rutidosis, *DC.*
Anisolepis, *Steetz.*

Rhodanthe, *Lindl.*
Lawrencella, *Lindl.*
Xyridanthe, *Lindl.*
Podotheca, *Cass.*
Podosperma, Labill.
Phænopoda, Cass.
Leptorhynchus, *Less.*
Rhytidanthe, Benth.
Waitzia, *Wendl.*
Viraya, Gaudich.
Morna, Lindl.
Millotia, *Cass.*
Pterochæta, *Steetz.*
Ixiolæna, *Benth.*
Chrysodiscus, *Steetz.*
Panætia, *Cass.*
Podolepis, *Labill.*
Scaliopsis, Walp.
Scalia, Sims.
Stylolepis, Lehm.
Doratolepis, Benth.
Siemssenia, *Steetz.*
Swammerdamia, *DC.*
Ozothamnus, *R. Br.*
Faustula, DC.
Petalolepis, DC.
Chrysocephalum, *Walp.*
Eriosphæra, *Less.*
Leontonyx, *Cass.*
Spiralepis, Don.
Helichrysum, *DC.*
Anaxeton, Gærtn.
Argyrocome, Gærtn.
Hebelæna, DC.
Xerochlæna, DC.
Blepharolepis, DC.
Taxostiche, DC.
Lepicline, Cass.
Ereicephyllum, Less.
Chinostemma, DC.
Leucostemma, Don.
Helipterum, *DC.*
Leucochrysum, DC.
Leiochrysum, DC.
Sericophorum, DC.
Pachypterum, Steetz.
Astelma, Less.
Damironia, Cass.
Syncarpha, DC.
Edmondia, Cass.
Hyalosperma, *Steetz.*
Aphelexis, *Boj.*
Freemania, Boj.
Stenocline, *DC.*
Achyrocline, *DC.*
Gnaphalium, *Don.*
Omalotheca, DC.
Euchiton, DC.
Homalotheca, Cass.
Cladochæta, *DC.*
Pteropogon, *DC.*
Schœnia, *Steetz.*
Lasiopogon, *Cass.*
Amphidoxa, *DC.*
Demidium, *DC.*
Filago, *Tournef.*
Gifola, Cass.
Impia, Dodon.
Oglifa, Less.
Logfia, Cass.
Achariterium, Bl.et Fg.
Xerotium, Bluff.et Fing.
Metalasia, *R. Br.*
Endoleuca, Cass.
Erythropogon, *DC.*
Lachnospermum, *Willd.*
Carpholoma, Don.
Pachyrhynchus, *DC.*
Elytropappus, *Cass.*
Disparago, *Gœrtn.*
Amphiglossa, *DC.*

SERIPHIEÆ.

Stœbe, *Less.*
Seriphium, *Less.*
Eremanthis, Cass.
Pleurocephalum, Cass.
Acrocephalum, Cass.
Perotriche, *Cass.*
Gymnachœna, Rchb.

ANTENNARIEÆ.

Trichogyne, *Less.*
Ifloga, Cass.
Phænocoma, *Don.*
Petalacte, *Don.*
Petalolepis, Less.
Anaxeton, *Cass.*
Antennaria, *R. Br.*
Disynanthus, Rafin.
Anaphalis, *DC.*
Leontopodium, *R. Br.*

LEYSSEREÆ.

Athrixia, *Ker.*
Asteropsis, Less.
Antithrixia, *DC.*
Leyssera, *L.*
Asteropterus, Vaill.
Callicornia, Burm.
Longchampia, Willd.
Leptophytus, Cass.
Pterothrix, *DC.*
Rosenia, *Thunb.*

RELHANIEÆ.

Carpesium, *L.*
Amblyocarpum, *F.et Mey.*
Syncephalum, *DC.*
Oligodora, *DC.*
Nestlera, *Spreng.*
Stephanopappus, Les.
Columellea, Jacq.
Polychætia, *Less.*
Relhania, *Herit.*
Eclopes, *Gœrtn.*
Odontophyllum, DC.
Rhigiophyllum, Less.
Rhynchopsidium, *DC.*
Rhynchocarpus, Less.
Osmites, *Cass.*
Bellidiastrum, Vaill.
Spanotrichum, E. Mey.
Osmitopsis, *Cass.*

NEUROLÆNEÆ.

Neurolæna, *R. Br.*
Calea, Gærtn.
Faujasia, *Cass.*
Eriothrix, *Less.*
Stilpnogyne, *DC.*
Erechtites, *Raf.*
Neoceis, Cass.
Microderris, DC.
Tylodiscus, DC.
? *Plagiotome*, DC.
Ceradia, *Lindl.*
Cremocephalum, *Cass.*
Crassocephalum, Mön.

SENECIONEÆ.

Gynura, *Cass.*
Emilia, *Cass.*
Asterosperma, *Less.*
Oligothrix, *Cass.*
Mesogramma, *DC.*
Cineraria, *Less.*
Xenocarpus, Cass.
Senecillis, *Gœrtn.*
Ligularia, *Cass.*
Hoppea, Rchb.
Arnica, *L.*
Aronicum, *Neck.*
Grammarthron, Cass.
Doronicum, *L.*
Werneria, *H. B. K.*
Oresigonia, Willd.
Culcitium, *H. B. K.*
Gynoxis, *Cass.*
Cacalia, *DC.*
Aulacophora, DC.
Delairea, *Lemaire.*
Cissampelopsis, *Lemaire.*
Psacalium, *DC.*
Pentacalia, Cass.
Lopholæna, *DC.*
Kleinia, *L.*
Cacalia, Cass.
Cacalianthemum, Dilln.
Acleia, *DC.*
Senecio, *Less.*
Aneoio, Neck.
Aspelina, Neck.
Seneciotypus, Dumort.
Obejaca, Cass.
Herbichia, Zawadsky.
Farobœa, Schrank.
Eudorus, Cass.
Hubertia, Bory.
Pithosillum, Cass.
Synarthrum, Cass.
Sclerobasis, Cass.
Carderina, Cass.
Pericallis, Don.
Bethencourtia, Chois.
Danaa, Colla.
Heterolepis, Bert.
Adenotrichia, Lindl.
Scrobicaria, Cass.
Ætheolœna, Cass.
Dorobœa, Cass.
Roldana, Llav. et Lex.
Brachyrhynchos, *Less.*
Trachycarpus, DC.
Crocidium, *Hook.*
Madaractis, *DC.*
Tetradymia, *DC.*
Lagothamnus, *Nutt.*
Raillarda, *Gaudich.*
Bedfordia, *DC.*
Nothonia, *DC.*
Lachanodes, *DC.*
Pladaroxylon, Endl.
Euryops, *Cass.*
? *Hertia*, Neck.
Enantiotrichum, E.My.
Balbisia, *DC.*
Ingenhousia, Berter.
Robinsonia, *DC.*

V. CYNAREÆ.

CALENDULEÆ.

Calendula, *Neck.*
Oligocarpus, *Less.*
Tripteris, *Less.*

OSTEOSPERMEÆ.

Osteospermum, *L.*
Eriocline, Cass.

OTHONNEÆ.

Heteractis, *DC.*
Gymnodiscus, *Less.*
Othonna, *L.*
Aristotela, Adans.
Doria, Less.
Hertia, Less.
Ruckeria, *DC.*
Harpocarpus, *Endl.*
Acanthocephalus, Ka.

ARCTOTEÆ.

Arctotis, *Gœrtn.*
Arctotheca, Vaill.
Odontoptera, Cass.
Steganotus, Cass.
Cymbonotus, *Cass.*
Venidium, *Less.*
Cleithria, Schrad.
Haplocarpha, *Less.*
Aloiozonium, *Kze.*
Landtia, *Less.*
Arctotheca, *Wendl.*
Cryptostemma, *R. Br.*
Cynotis, Hoffmansegg.
Microstephium, *Less.*
Heterolepis, *Cass.*
Heteromorpha, Cass.

GORTERIEÆ.

Stephanocoma, *Less.*
Cullumia, *R. Br.*
Gorteria, *Gœrtn.*
Personaria, Lam.
Hirpicium, *Cass.*
Didelta, *Less.*
Choristea, DC.
Favonium, Gærtn.
Bretenillia, Bucholz.
Cuspidia, Gærtn.
Aspidalis, Gærtn.
Berkheya, *Ehrh.*
Crocodylodes, Adans.
Basteria, Houtt.
Agriphyllum, Juss.
Rohria, Vahl.
Gorteria, Lam.
Zarabellia, Neck.
Evopis, Cass.
Agriphyllum, Less.
Basteria, DC.
Trichodes, DC.
Trichocoma, DC.
Gazania, *Gœrtn.*
Mussinia, Willd.
Mœhnia, Neck.
Melanchrysum, Cass.
Leptomorpha, DC.
Stobæa, *Thunb.*
Apuleia, Gærtn.
? *Arelina*, Neck.

ECHINOPSIDEÆ.

Echinops, *L.*
Echinopus, Tournef.
Echinanthus, Neck.
Acantholepis, *Less.*

CARDOPATEÆ.

Cardopatium, *Juss.*
Brotera, Willd.
Chamæleon, C. Bauh.

XERANTHEMEÆ.

Xeranthemum, *Tournef.*
Xeroloma, Cass.
Harrisonia, Neck.
Chardinia, *Desf.*
Siebera, *Gay.*

CARLINEÆ.

Saussurea, *DC.*
Heterotrichum, Bieb.
Bennetia, Gray.
Lagurostemon, Cass.
Cyathidium, Cass.
Laguranthera, C.A.M.
Theodorea, DC.
Eriostemon, Less.
Aucklandia, *Falc.*
Haplotaxis, *DC.*
Frolovia, Ledeb.
? *Hemisteptia*, Bunge.
Eriocoryne, Wall.
Dolomiæa, *DC.*
Arctium, *Lam.*
Arctio, Lam.
Arction, Cass.
Villarsia, Guettard.
Bernardia, Vill.
Stechmannia, *DC.*
Stæhelina, *DC.*
Köchlea, *Endl.*
Carlina, *Tournef.*
Heracantha, DC.
Mitina, DC.

Chamæleon, DC.
Carlowizia, DC.
Atractylis, *L.*
Crocodylodes, Vaill.
Acarna, Cass.
Anactis, Cass.
Cirsellium, DC.
Spadactis, Cass.
Thevenotia, *DC.*
Cousinia, *Cass.*
Aneathia, *DC.*
Auchera, *DC.*
Polytaxis, *Bunge.*

CENTAURIEÆ.

Amberboa, *Pers.*
Chryseis, Less.
Goniocaulon, DC.
Cyanopsis, Cass.
Cyanastrum, Cass.
Cyanopis, Cass.
Lacellia, DC.
Volutaria, Cass.
Volutarella, Cass.
Chryseis, Cass.
Pararhysis, DC.
Phæopappus, DC.
Psephellus, Fisch.
Amblyopogon, Fisch.
Xanthopsis, DC.
Zœgea, *L.*
Microlonchus, *DC.*
Mantisalca, Cass.
Uralepis, DC.
Tricholepis, *DC.*
Achyropappus, Bieb.
Ochanopappus, Endl
Alaphalantias, Endl.

Tomanthea, *DC.*
Crupina, *Cass.*
Centaurea, *Less.*
Crocodilium, DC.
Calcitrapa, DC.
Cyanus, DC.
Centaureum, DC.
Phrygia, Gray.
Hypophæstum, Gr.
Polyacantha, Gr.
Leucantha, Gray.
Hyalea, DC.
Microlophus, Cass.
Piptoceras, Cass.
Chartolepis, Cass.
Phalolepis, Cass.
Callicephalus, C. A. M.
Platylophus, Cass.
Jacea, Cass.
Pterolophus, Cass.
Lepteranthus, DC.
Stenolophus, Cass.
Ætheopappus, DC.
Stizolophus, DC.
Plectocephalus, DC.
Psephellus, Cass.
Heterolophus, Cass.
Cheirolophus, DC.
Melanoloma, Cass.
Odontolophus, Cass.
Lopholoma, Cass.
Spilacron, Cass.
Acrolophus, Cass.
Acrocentron, Cass.
Hymenocentron, Cass.
Verutina, Cass.
Mesocentron, Cass.
Triplocentron, Cass.

Corethropsis, DC.
Podia, Neck.
Philostizus, Cass.
Pectinastrum, Cass.
Alophium, Cass.
Cnicus, *Vaill.*
Carbeni, Adans.
Tetramorphæa, *DC.*

CARTHAMEÆ.

Kentrophyllum, *Neck.*
Hohenwarta, West.
Heracantha, Lk.
Atractylis, Vaill.
Odontognatia, DC.
Thamnacantha, DC.
Carthamus, *Tournef.*
Onobroma, *DC.*
Carduncellus, *Adans.*

SILYBEÆ.

Silybum, *Vaill.*
Galactites, *Mönch.*
Tyrimnus, *Cass.*

CARDUINEÆ.

Onopordon, *Vaill.*
Acanos, Adans.
Cynara, *Vaill.*
Spanioptilon, *Less.*
Carduus, *Gærtn.*
Clavena, *DC.*
Picnomon, *Lobel.*
Acarna, Vaill.
Picnocomon, Dalech.
Cirsium, *Tournef.*
Cnicus, Schreb.

Breea, Less.
Lophiolepis, Cass.
Odontolepis, Boiss.
Eriolepis, Cass.
Epitrachys, DC.
Orthocentrum, Cass.
Corynotrichum, DC.
Cephalonoplos, Neck.
Onotrophe, Cass.
Erythrolæna, *Sweet.*
Chamæpeuce, *Alpin.*
Cirsium, Less.
Ptilostemon, Cass.
Lamyra, Cass.
Platyrhaphium, Cass.
Notobasis, *Cass.*
Echenais, *Cass.*
Lappa, *Tournef.*

SERRATULEÆ.

Acroptilion, *Cass.*
Rhaponticum, *DC.*
Hookia, Neck.
Centaureum, Hall.
Stemmacantha, Cass.
Cestrinus, Cass.
Leuzea, *DC.*
Rhacoma, DC.
Fornicium, Cass.
Malacocephalus, Tsch.
Alfredia, *DC.*
Serratula, *DC.*
Sarreta, DC.
Mastrutium, Cass.
Pereuphora, Hoffmans.
Klasea, Cass.
Oligochæta, DC.
Jurinea, *Cass.*

Sub-order II. LABIATIFLORÆ.

VI. MUTISIACEÆ.

BARNADESIEÆ.

Schlechtendalia, *Less.*
Diacantha, *Less.*
Barnadesia, *Linn. f.*
Bacazia, Ruiz et Pav.
Penthea, Don.
Dasyphyllum, *H. B. K.*
Fulcaldea, *Poir.*
Turpinia, H. B.K.
Voigtia, Spr.
Dolichostylis, Cass.
Flotovia, *Spreng.*
Piptocarpha, Hook.
Nardophyllum, Hook
Seris, *Less.*
Lycoseris, *Cass.*
? *Diazeuxis*, Don.
Langsdorfia, Willd.
Chætachlæna, *Don.*
Chuquiragua, *Juss.*
Johannia, Willd.
Joannesia, Pers.
Joannea, Spr.
Moquinia, *DC.*
Spadonia, Less.
Gochnatia, *H. B. K.*
Cyclolepis, Don.
Cyclopis, Guillem.
Anastraphia, Don.
Pentaphorus, Don.
Hedraiophyllum, Less.
Augusta, *Leandro.*
Stifftia, Mik.
Sanhilaria, Leandr.
Mocina, DC.
Mutisia, *L. fil.*
Guariruma, Cass.
Holophyllum, Less.
Haplophyllum, Less.
Proustia, *Lagasc.*
Leucoryphe, Endl.

Thelecarpea, DC.
Harmodia, Don.
Calopappus, *Meyen.*
Hyalis, *Don.*
Brachyclados, *Don.*
Chætanthera, *Ruiz et P.*
Bichenia, Don.
Cherina, Cass.
Enthrixia, Don.
Proselia, Don.
Prionotophyllum, Less.
Tylloma, Don.
Pachylæna, *Don.*
Trichocline, *Cass.*
Amblysperma, *Benth.*
Onoseris, *DC.*
Cladoseris, Less.
Chætachlæna, Don.
Isotypus, *H. B. K.*
Seris, Willd.
Hilaria, DC.
Oldenburgia, *Less.*
Scytala, E. Mey.
Leucomeris, *Don.*
Myripnois, *Bunge.*
Ainsliæa, *DC.*
Chionoptera, *DC.*
Carmelita, *Cl. Gay.*
Gerbera, *Gronov.*
Aphyllocaulon, Lag.
Leptica, E. Mey.
Piloselloides, Less.
Oreoseris, *DC.*
Berniera, *DC.*
Dicoma, *Cass.*
Leucophyton, Less.
Xeropappus, Wall.
Microcoma, DC.
Rhigiothamnus, Less.
Macledium, Cass.
Nitelium, Cass.
Pterocoma, DC.
Printzia, *Cass.*
? *Lloydia*, Neck.

Perdicium, *Lagasc.*
Pardisium, Burm.
Leiocarpum, DC.
Anandria, *Siegesb.*
Leibnitzia, Cass.
Oriastrum, *Pöpp.*

LERIEÆ.

Chaptalia, *Vent.*
Cursonia, *Nutt.*
Lieberkuhnia, *Cass.*
Oxydon, *Less.*
Loxodon, *Cass.*
Chevreuilia, *Cass.*

FACELIDEÆ.

Lucilia, *Cass.*
Oligandra, *Less.*
Facelis, *Cass.*

VII. NASSAVIACEÆ.

POLYACHYRIDEÆ.

Polyachyrus, *Lagasc.*
Bridgesia, Hook.
Diaphoranthus, Mey.
Cephaloseris, Pöpp.

NASSAVIEÆ.

Nassavia, *Commers.*
Nassovia, Pers.
Mastigophorus, *Cass.*
Triachne, *Cass.*
Triptilion, *Ruiz et Pav.*
Acanthophyllum, Hook et Arn.
Panargyrum, *Lagasc.*
Pentanthus, Less.
Piptostemma, Don.

Caloptilium, *Lagasc.*
Sphærocephalus, Laga.
Portalesia, Meyen.

TRIXIDEÆ.

Pamphalea, *Lagasc.*
Ceratolepis, Cass.
Cephalopappus, *Nees et Mart.*
Pleocarphus, *Don.*
Pentanthus, *Hook et Arn*
Jungia, *L. fil.*
Trinacte, Gærtn.
Rhinactina, Willd.
Martrasia, Lagasc.
Moscharia, *Ruiz et Pav.*
Moschifera, Molina.
Mosigia, Spreng.
Gastrocarpha, Don.
Leuceria, *Lagasc.*
Leuchæria, Less.
Macrobotrys, DC.
Lasiorrhiza, *Lagasc.*
Chabræa, DC.
Bertolonia, DC.
Frageria, DC.
Maclovia, DC.
Cassiopea, Don.
Bowmannia, *Gardn.*
Ptilurus, *Don.*
Dumerilia, *Less.*
Trixis, *P. Br.*
Cleanthes, Don.
Platychilus, Cass.
Holochilus, Cass.
Oligophyllon, Less.
Polyphyllon, Less.
Prionanthes, Schrank.
Tenoria, Berter.
Alcithoë, Don.
Dolichlasium, *Lagasc.*
Perezia, *Lagasc.*
Chætanthera, H. B. K.

Homoianthus, Bonpl.
Homanthis, Kunth.
Clarionea, DC.
Clarionella, DC.
Drozia, Cass.
Stenophyllum, Less.
Asteroseris, Endl.
Scolymanthus, Willd.
Isanthus, DC.
Acourtia, *Don.*
? Pogonura, *DC.*
Caloseris, *Benth.*

Sub-order III. LIGULIFLORÆ.

VIII. CICHORACEÆ.

SCOLYMEÆ.

Scolymus, *Cass.*
Myscolus, *Cass.*
Diplostemma, *Hochst. et Steud.*

LAMPSANEÆ.

Lampsana, *Vaill.*
Lapsana, Tournef.
Soldevilla, *Lagasc.*
Hispidella, Barnades.
Apogon, *Elliot.*
Rhagadiolus, *Tournef.*
Kölpinia, *Pall.*

HYOSERIDEÆ.

Arnoseris, *Gærtn.*
Hedypnois, *Tournef.*
Hyoseris, *L.*
Achyrastrum, Neck.
Calodonta, *Nutt.*
Aposeris, *Neck.*
Catananche, *Tournef.*
Hænselera, *Boiss.*
Acanthophyton, *Less.*
Cichorium, *Tournef.*
Calais, *DC.*
Hymenonema, Hook.
Uropappus, Nutt.
Scorzonella, *Nutt.*
Tolpis, *Adans.*
Drepania, Juss.
Swertia, Ludew.
Chatelania, Neck.
Schmidtia, *Mönch.*
Æthionia, Don.
Polychætia, Tausch.
Krigia, *Schreb.*
Troximon, *Gærtn.*
Cynthia, Don.
Adopogon, Neck.
Luthera, C. H. Schult.
Microseris, *Don.*
Lepidonema, Fisch. et Mey.
Fichtea, C. H. Schultz.
Bellardia, Colla.

HYPOCHÆRIDEÆ.

Oreophila, *Don.*
Amblachænium, Turcz.
Cycnoseris, *Endl.*
Hypochæris, *Linn.*
Achyrophorus, *Scop.*
Porcellites, (*partim*), Cass.
Seriola, *Gærtn.*
Achyrophorus, Vaill.
Rodigia, Spr.
Piptopogon, Cass.
Agenora, Don.
Porcellites, (*partim*), Cass.
Robertia, *DC.*
Metabasis, *DC.*
Phalacroderis, *DC.*

SCORZONEREÆ.

Thrincia, *Roth.*
Colobium, Roth.
Streckera, Schultz.
Leontodon, *L.*
Virea, Adans.
Antodon, Neck.
Apargia, Less.
Asterothrix, Cass.
? *Fidelia*, Schultz.
Oporinia, Don.
Phyllopappus, *Walp.*
Millina, *Cass.*
Geropogon, *L.*
Podospermum, *DC.*
Richardia, Roth.
Urospermum, *Juss.*
Arnopogon, Willd.
Tragopogon, *L.*
Hymenonema, *Cass.*
Rafinesquia, *Nutt.*
Scorzonera, *L.*
Lasiospora, Cass.
Lasiospermum, Fisch.
? *Fleischeria*, Steud.
Anisocoma, *Torrey.*
Galasia, *Cass.*
Microderis, *DC.*
Picris, *L.*
Medicusia, Mönch.
Spitzelia, Schultz.
Deckera, Schultz.
Helminthia, *Juss.*
Kalbfussia, *Schultz.*

LACTUCEÆ.

Picridium, *Desf.*
Reichardia, Roth.
Zollikoferia, *DC.*
Sonchus, *Linn.*
Leptoseris, *Nutt.*
Trachodes, *Don.*
Heterachena, *Fresen.*
Malacothrix, *DC.*
Youngia, *Cass.*
Prenanthes, *Gærtn.*
Erytheremia, Nutt.
Pleiacanthus, Nutt.
Nabalus, Cass.
Harpalyce, Don.
Esopon, Raf.
Lygodesmia, Don.
Atalanthus, Don.
Chorisma, *Don.*
Chorisis, DC.
Phœnixopus, *Koch.*
Mycelis, Cass.
Phænopus, DC.
Melanoseris, *Decaisn.*
Brachyramphus, *DC.*
Lactuca, *L.*
Scariola, Endl.
? *Rhabdotheca*, Cass.
Cyanoseris, Koch.
Chondrilla, *Tournef.*
Crinissa, Don.
Pyrrhopappus, DC.
Taraxacum, *Juss.*
Leontodon, Adans.
Willemetia, *Neck.*
Calycosorus, Schmidt.
Wibelia, Hoppe.
Peltidium, Zollikof.
Aspideium, Zollikof.
Zollikofera, Nees.
Ixeris, *Cass.*
Zacyntha, *Tournef.*
Nemauchenes, *Cass.*
Endoptera, b. DC.
Catyona, *Cass.*
Endoptera, a. DC.
Lomatolepis, *Cass.*
Rhabdotheca, *Cass.*
Microrhynchus, Less.
Ammoseris, Endl.
Launea, Cass.
Trochoseris, *Pöpp. et En.*
Macrorhynchus, Less.
Macrorhynchium, Rcb.
Kymapleura, *Nutt.*
Cryptopleura, *Nutt.*
Stylopappus, *Nutt.*
Troximeria, Nutt.
Lagoseris, *Bieb.*
Pterotheca, Cass.
Trichocrepis, Vis.
Crepinia, Rchb.
Intybellia, Cass.
Myoseris, Lk.
Pachylepis, *Less.*
Sclerolepis, Monnier.
Barkhausia, *Mönch.*
Borkhausia, Boehm.
Barkhusenia, Hopp.
Hostia, Mönch.
Deloderium, Cass.
? *Closirospermum*, Nk.
Rodigia, *Spr.*
Ammogeton, *Schrad.*
Crepis, *L.*
Psilachenia, Nutt.
Intybus, Fr.
Intybellia, Monnier.
Ætheorhiza, Cass.
Calliopea, Don.
? *Troximon*, Don.
Berinia, *Brign.*
Brachyderea, Cass.
Phæcasium, Cass.
Homalocline, *Endl.*
Omalocline, Cass.
Crepidium, *Nutt.*

HIERACIEÆ

Rothia, *Schreb.*
Voigtia, Roth.
Heteracia, *Fisch. et Mey.*
Andryala, *L.*
Eriophorus, Vaill.
Leucoseris, *Nutt.*
Apargidium, *Torr.*
Hieracium, *Tournef.*
Miegia, Neck.
Plancia, Neck.
Aracium, *Neck.*
Stenotheca, Monnier.
Dubyæa, *DC.*
Lasiopus, Don.
Mulgedium, *Cass.*
Agathyrsus, Don.
Galathenium, *Nutt.*
Anisoramphus, *DC.*
Soyeria, *Monnier.*
Catonia, Mönch.
Lepicaune, Lapeyr.
Hapalostephium, Don
Picrosia, *Don.*
Malacomeris, *Nutt.*
Agoseris, *Raf.*
Troximon, Nutt.
Pinaropappus, *Less.*
Dendroseris, *Don.*
Rhea, Berter.

DOUBTFUL GENERA.

Anisopappus, *Hook et An.*
Arrowsmithia, *DC.*
Cadiscus, *E. Mey.*
Dolichogyne, *DC.*
Elachia, *DC.*
Gnaphalopsis, *DC.*
Psilostrophe, *DC.*
Trimetra, *Moç.*
Odontotrichum, *Zucc.*
Ophryosporus, *Meyen.*
Metazanthus, *Meyen.*
Piptocarpha, *R. Br.*

LITTLE KNOWN GENERA.

Apatanthus, *Viviani.*
Abasoloa, *Llav. et Lex.*
Allendea, *Llav. et Lex.*
Galeana, *Llav. et Lex.*
Rosalesia, *Llav. et Lex.*
Mnesiteon, *Rafin.*
Microspermum, *Lagasc.*
Platzia, *Ruiz et Pav.*
Placus, *Lour.*
Galophthalmum, *Nees.*
Damatris, *Cass.*
Dimerostemma, *Cass.*
Glyphia, *Cass.*
Glycideras, Cass.
Gibbaria, *Cass.*
Munnozia, *Ruiz et Pav.*
Hysteronica, *Willd.*
Onopix, *Raf.*
Serinia, *Raf.*

UNDESCRIBED GENERA.

Bracheilema, *R. Br.*
Gomesia, *Llav.*
Oteiza, *Llav.*
Koanophyllum, *Arrud.*
Lasiocephalus, *Schlecht.*
Trichostemma, *R. Br.*

NUMBERS. GEN. 1005. Sp. 9000 ?

POSITION.—Calyceraceæ.—ASTERACEÆ.—Dipsacaceæ.

Alliance LI. *MYRTALES.*—The Myrtal Alliance.

Diagnosis.—*Epigynous Exogens, with polypetalous dichlamydeous flowers, axile placentæ, and embryo with little or no albumen.*

It may at first sight appear paradoxical to bring into close contact Orders usually so widely separated as Composites, Fringe-myrtles and Myrobalans ; and it must be confessed that if the monopetalous corolla did deserve the value usually assigned to it, the measure would be incapable of justification. But if, as it is one of the objects of this book to show, we should neglect that circumstance, the relationship of all the plants now mentioned will be less problematical. It is the capitate inflorescence of Composites that gives them one of their most striking peculiarities ; but that disappears in Valerianworts, about whose near relation to Composites no one entertains a doubt ; and among the Myrobalans and Fringe-myrtles the tendency to a capitate condition is unusually great ; as, for example, in Combretum and Conocarpus in the one, and in four-fifths of the species in the other. The relation of Myrobalans to Fringe-myrtles is not likely to be disputed; now the inflorescence of many genera differs in no respect from that of Composites, and on the other hand, numerous Composites agree entirely with Fringe-myrtles in their glandular leaves. Moreover, the calyx of the latter has often as great a claim to the designation of pappus as that of any Composites whatever. It must be confessed, however, that we have not at present among Composites any such tendency to a separation of the petals as would lead to the expectation of finding a polypetalous genus, which would render the assumed connection between Fringe-myrtles and Composites more evident.

But the example of Phyteuma among Bellworts leads to the anticipation of such a possibility; or if not, the tendency to unite the petals or stamens, which is so common in Myrtleblooms, may be expected to result in a monopetalous corolla among the Fringe-myrtles.

These remarks are not, however, introduced to show that Composites and Myrtles ought to stand in the same Alliance. That would certainly be an unnatural association. But they seem to show conclusively that they belong to Alliances standing extremely near each other.

Natural Orders of Myrtals.

Ovary 1-*celled. Ovules pendulous. Leaves dotless. Seeds without albumen. Cotyledons convolute*	274. Combretaceæ.
Ovary 1-*celled. Ovules pendulous. Leaves dotless. Seeds albuminous. Cotyledons flat*	275. Alangiaceæ.
Ovary 1-*celled. Ovules ascending. Leaves dotted. Embryo fused into a solid mass*	276. Chamælauciaceæ.
Ovary with more than one cell. Flowers polypetalous or apetalous. Calyx open, minute. Stamens definite. Ovules pendulous. Cotyledons minute. (*Occasionally one-celled*). . .	277. Haloragaceæ.
Ovary with more than one cell. Flowers polypetalous or apetalous. Calyx valvate. Stamens definite. Ovules horizontal or ascending. Cotyledons flat, much larger than the radicle	278. Onagraceæ.
Ovary with more than one cell. Flowers polypetalous. Calyx valvate. Stamens indefinite. Cotyledons flat, much shorter than the radicle, which germinates before the fruit falls . .	279. Rhizophoraceæ.
Ovary with more than one cell. Flowers monopetalous coronetted. Calyx valvate. Stamens indefinite, monadelphous. Cotyledons amygdaloid	280. Belvisiaceæ.
Ovary with more than one cell. Flowers polypetalous. Calyx imbricated. Stamens definite. Anthers rostrate. Leaves usually dotless	281. Melastomaceæ.
Ovary with more than one cell. Flowers polypetalous or apetalous (*or valvate*). *Calyx imbricated. Stamens* 00. *Anthers oblong. Leaves usually dotted*	282. Myrtaceæ.
Ovary with more than one cell. Flowers polypetalous. Calyx valvate or imbricated. Stamens 00, *in part collected into a fleshy hood. Anthers oblong. Leaves dotless*	283. Lecythidaceæ.

ORDER CCLXXIV. COMBRETACEÆ.—MYROBALANS.

Combretaceæ, *R. Brown Prodr.* 351. (1810) *in Flinders*, 2, 548. (1814); *A. Rich. Dict. Class* 4. 353; *DC. Prod.* 3. 9; *Mémoire* (1828); *Bartl. Ord. Nat. p.* 322. (1830); *Endl. Gen.* cclxi.; *Meisner Gen.* 110; *Wight Illustr.* 1. *p.* 211.—Terminaliaceæ, *Jaume St. Hil. Exp. Fam. Nat.* 1. 178.—Illigeraceæ, *Ed. pr.* cl.—Illigereæ, *Blume in Ann. Sc. N. S.* 2. 95. (1834); *Martius Conspectus, No.* 83. (1835).—Gyrocarpeæ, *Nees ab Esenb. Laurin. Expositio* 20. (1833); *Endl. Gen. p.* 324.—Myrobalaneæ, *Juss. Dict. Sc. Nat.* 31. 458. (1824).

DIAGNOSIS.—*Myrtal Exogens, with a* 1-*celled ovary, pendulous ovules, dotless leaves, seeds without albumen, and convolute cotyledons.*

Trees or shrubs. Leaves alternate or opposite, without stipules, entire. The petiole often with 2 glands at the end. Spikes axillary or terminal. Flowers ⚥, or

4 3 1 2

Fig. CCCCLXXVIII.

by abortion ♂ ♀. Calyx adherent, with a 4- or 5-lobed deciduous limb. Petals arising from the orifice of the calyx, alternate with the lobes; often wanting. Stamens arising from the same part, twice as many as the segments of the calyx, very rarely equal to them in number, or three times as many; filaments distinct, subulate; anthers 2-celled, bursting longitudinally, or by recurved valves. Ovary 1-celled, with from 2 to 4 ovules, hanging by cords from the apex of the cavity; style 1; stigma simple. Fruit drupaceous, baccate, or nut-like, 1-celled, by abortion 1-seeded, indehiscent, often winged. Seed pendulous, without albumen; embryo with the radicle turned towards the hilum; plumule inconspicuous; cotyledons leafy, usually convolute, occasionally plaited.

It cannot be doubted that Myrobalans have a near relationship to Myrtleblooms, and especially to Punica, of which they possess the convolute embryo. But although their connection with the Myrtal Alliance seems beyond contradiction, yet the absolute simplicity of their ovary renders it necessary to station them nearest other Orders. Their inferior fruit, with a single cavity, and often with a single ovule, and the great tendency that exists among them to collect their flowers in heads, furnish reasons for regarding them as standing in close relation to Composites, and as presenting a higher form of development of that well-known Order. The great frequency of an apetalous structure among them is one of their more remarkable features, and indicates a tendency to assume the condition of Sandalwoods or Oleasters, from both which however they are separated by other considerations. Gyrocarpus and Illigera, sometimes separated under the name of Gyrocarpeæ or Illigeraceæ, are in no respect essentially distinguishable except

Fig. CCCCLXXVIII.—Combretum (or Poivrea) purpureum. 1. a flower cut open; 2. a section of the ovary; 3. fruit of Terminalia? (*Wight*); 4. cross section of the embryo.

by their recurved anther valves, in which they singularly correspond with Laurels. While, however, these seem to be the most immediate affinities of Myrobalans, we must not overlook their more distant kinsmanship. To Myrtleblooms and Melastomads they are related through Memecylon, and especially to the former, by Punica, with which they agree in the structure of their embryo. In the latter respect they also accord with Mangroves and Vochyads; and with Alangiads and Onagrads in the general structure of the flower.

All natives of the tropics of Asia, Africa, and America. No species is extra-tropical.

Mostly astringents. Bucida Buceras yields a bark used for tanning. The bark of Conocarpus racemosa, one of the plants called Mangroves in Brazil, is used greatly at Rio Janeiro for the same purpose. The fruit of the Terminalia belerica, or the Beleric Myrobalan, is an astringent, tonic, and attenuant. The kernels are eaten in India, and reckoned intoxicating. The bark abounds in a gum, resembling Gum Arabic, soluble in water, burning away in the flame of a candle; a similar gum exudes from Combretum alternifolium. The bark of Terminalia alata is astringent and antifebrile. The fruit of Terminalia Chebula, as well as the galls of the same plant, are very astringent, and highly valued by dyers; with alum they give a durable yellow, and with a ferruginous mud an excellent black. The root of T. latifolia is given in Jamaica in diarrhœa. Species of Terminalia, Conocarpus, and Pentaptera, yield excellent timber. The kernels of T. Catappa, &c., are eaten as almonds, and are very palatable; those of T. citrina are a common article in Hindoo materia medica, being employed as a gentle purgative. A milky juice is described as flowing from T. Benzoin, which being fragrant on drying, and resembling Benzoin, is used in churches in the Mauritius as a kind of incense. Martius inform us that Terminalia argentea, called in Brazil Caxapora do Gentio, yields a resin of a drastic quality.

GENERA.

I. TERMINALEÆ.—Corolla usually 0. Cotyledons convolute.

Bucida, *Linn.*
Buceras, P. Br.
Hudsonia, Robins.
Terminalia, *Linn.*
Catappa, Gärtn.
Tanibouca, Aubl.
Adamaram, Adans.
Myrobalanus, Gärtn.
Badamia, Gärtn.
Fatræa, Thouars.
Pentaptera, *Roxb.*
Getonia, *Roxb.*
Calycopteris, Lam.
Chuncoa, *Pav.*
Gimbernatia, R. et P.
Ramatuella, *H. B. K.*
Conocarpus, *Gärtn.*
Rudbeckia, Adans.
Anogeissus, *Wall.*
Andersonia, Roxb.
Laguncularia, *Gärtn.*
Sphenocarpus, Rich.
Horan, Adans.
Lumnitzera, *Willd.*
Pyrrhanthus, Jack.
Petaloma, Roxb.
Bruguiera, Thouars.
Funkia, Dennst.
Guiera, *Adans.*
Poivrea, *Commers.*
Cristaria, Sonner.
Gonocarpus, Hamilt.

II. COMBRETEÆ. — Corolla present. Cotyledons plaited.

Combretum, *Löffl.*
Actia, Adans.
Forsgardia, Fl. Fl.
Cacoucia, *Aubl.*
Schousbœa, Willd.
Hambergera, Scop.
Hambergia, Neck.
Quisqualis, *Rumph.*
Spalanthus, *Jack.*
? Chrysostachys, *Pohl.*
? Agathisanthes, *Blum.*
? Ceratostachys, *Blum.*
? Bigamea, *Kön.*
Wormia, Vahl.

III. GYROCARPEÆ.—Corolla wanting. Cotyledons convolute. Anthers bursting by recurved valves.

Gyrocarpus, *Jacq.*
Illigera, *Bl.*

NUMBERS. GEN. 22. Sp. 200.

Myrtaceæ.
POSITION.—Alangiaceæ.—COMBRETACEÆ.—Chamælauciaceæ.
Lauraceæ.

ORDER CCLXXV. ALANGIACEÆ.—ALANGIADS.

Alangieæ, *DC. Prodr.* 3. 203. (1828); *Bartl. Ord. Nat. p.* 424. (1830); *Endl. Gen.* cclxii.; *Wight Illustr.* 2. Nyssaceæ, *Juss. in Dict. Sc.* 35. 267. (1825); *Endl. Gen. p.* 328.

DIAGNOSIS.—*Myrtal Exogens, with a 1-celled ovary, pendulous ovules, dotless leaves, albuminous seeds, and flat cotyledons.*

Large trees or shrubs. Branches often spiny. Leaves alternate, without stipules, entire, without dots. Flowers fascicled, axillary. Calyx adherent, 5-10-toothed. Petals 5-10, inserted into a fleshy adherent disk, linear, reflexed. Stamens long, exserted, 2 or 4 times as numerous as the petals, or equal to them in number; filaments distinct, villous at the base; anthers adnate, linear, 2-celled, turned inwards, often empty. Ovary 1-2-celled; style filiform, simple; ovules solitary, pendulous, anatropal. Drupe oval, somewhat crowned by the calyx, fleshy, slightly ribbed, and downy; nucleus 1-celled, bony, with a foramen at the apex. Seed 1, inverted; albumen fleshy, brittle; embryo straight; radicle long, superior; cotyledons flat, large, leafy.

1 Fig. CCCCLXXIX.

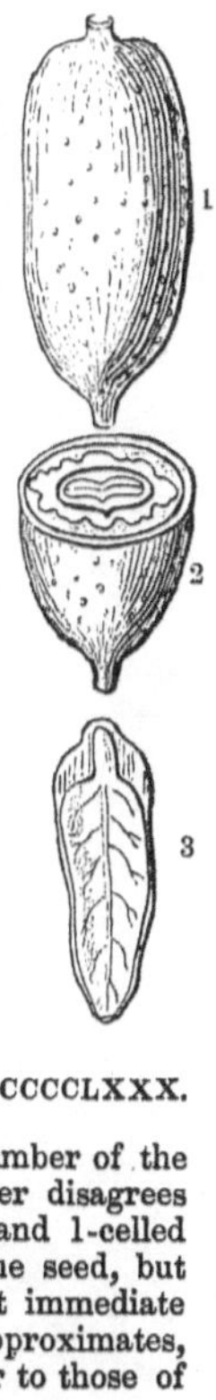

CCCCLXXX.

According to De Candolle, who founded this small Order, it differs from Myrtleblooms in its more numerous petals, adnate anthers, 1-celled fruit, and pendulous albuminous seeds. It agrees with Myrobalans in the contracted tube of the calyx, 1-celled fruit, and pendulous seeds; but differs in the number of the petals, adnate anthers, albuminous seeds, and flat cotyledons. The Order disagrees entirely with Melastomads and Onagrads, in the form of the anthers, and 1-celled fruit. It in some measure approaches Hippurids in the structure of the seed, but recedes from them in habit, 1-celled fruit, and single style. Its most immediate relationship, next to Myrobalans, is with Cornels, to which Marlea approximates, and with Witch Hazels, whose long narrow petals are strikingly similar to those of Alangiads.

But notwithstanding the near relation between these groups, the large leafy cotyledons and small quantity of albumen seem to indicate a closer relation between Myrobalans and Alangiads than between the latter and the Umbellal Alliance; and the Alangiads may be best regarded as the representative of the Cornal tendency in the Combretal Alliance. I think there can be little doubt that Nyssa is a genus of this Order, as Ad. Brongniart has partly suggested (*Enum.* xxx.), and principally remarkable for the want of petals. Mr. Bennett states that the ovary of Marlea is 2-celled, as in Cornus, but that does not correspond with my observations upon the fresh plant.

Fig. CCCCLXXIX.—Marlea begonifolia. 1. perpendicular section of the pistil.
Fig. CCCCLXXX.—Nyssa montana—*Gærtner*. 1. fruit; 2. cross section of it; 3. embryo.

Common in the southern parts of India, whence they extend along the Malayan Peninsula to Cochin China, northward along the forest-clad base of the Himalaya. The Nyssas are natives of the United States.

Alangium decapetalum and hexapetalum are said by the Malays to have a purgative hydragogic property. Their roots are aromatic. They are said to afford good wood and edible fruit.—*Royle.* Dr. Wight says that the fruit of the Alangiums is eatable, but not palatable, being mucilaginous and insipid. That of Nyssa capitata or candicans is subacid, the size of the Olive, and sometimes called the Ogechee Lime, because it is used occasionally as a substitute for Lime fruits. The timber of the Nyssas, called Tupelo trees, is difficult to split, in consequence of the fibres of its wood being much interwoven, but it is of little value.

GENERA.

Alangium, *Lam.*
Angolam, Adans.
Angolamia, Scop.

Marlea, *Roxb.*
Stylidium, Lour.
Stylis, Poir.

Pautsavia, Juss.
Diacæcarpium, Bl.

Nyssa, *Gronov.*
Tupelo, Adans.

NUMBERS. GEN. 3. Sp. 8.

Cornaceæ.
POSITION.—Combretaceæ.—ALANGIACEÆ.—Chamælauciaceæ.
Myrtaceæ.

ORDER CCLXXVI. CHAMÆLAUCIACEÆ—FRINGE MYRTLES.

Chamælaucieæ, *DC. in Dict. Class.* xi. (1826); *Prodr.* 3. 208; *Bartling Ord. Nat.* 331; *Ed. prior*, 2. 45; *Endl. Gen. p.* 1224; *Meisner Gen.* 107; *Schauer in Nov. Act.* xix. *Suppl.* 157.

DIAGNOSIS.—*Myrtal Exogens, with a 1-celled ovary, ascending ovules, dotted leaves, and the embryo fused into a solid mass.*

Small bushes, often resembling Heaths, with all their parts abounding in glandular oily cysts. Leaves evergreen, acerose, or flat. Flowers ☿, in racemes, corymbs, or heads, yellow, red, violet, or white. Calyx adhering to the ovary, with 4-5 lobes, which are either herbaceous and scaly, or thin, membranous, and broken up into fringes or extended into bristles, as in the pappus of Composites. Petals as numerous as the divisions of the calyx, entire, or fringed, or feathery, often of a dry texture. Stamens definite or 00, stationed in one or more whorls on a fleshy disk, epigynous, or adhering to the sides of the tube of the calyx; often partially sterile and scale-like, ligulate or petaloid; filaments subulate, occasionally forked; connective thick, fleshy, of various forms, continuous with the filament, and carrying the cells upon its inner face; cells opening longitudinally or by pores. Ovary 1-celled, united to the sides of the calyx; style simple; stigma simple; ovules anatropal, 2 or more, or as many as 10, inserted laterally, and either ascending or attached to the side of the cavity. Fruit a dry indehiscent pericarp. [Seed without albumen; embryo orthotropal, homogeneous, with no distinction of cotyledons, radicle, and plumule.—*Schauer.*]

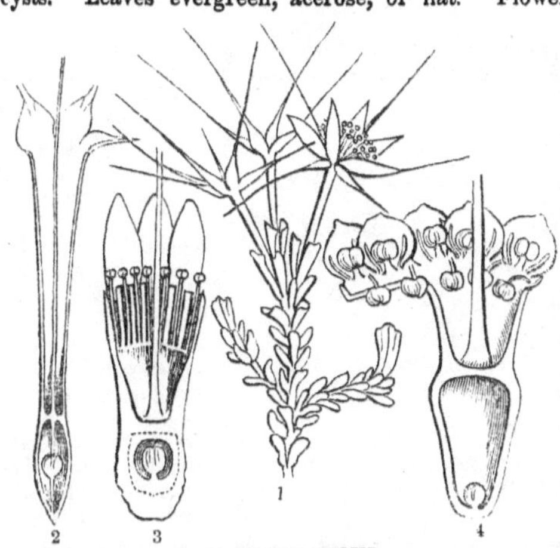

Fig. CCCCLXXXI.

Up to the present time these have been regarded as a section of the Order of Myrtleblooms; and there can be no doubt of their close relation. But it appears advisable to distinguish them on account of their very peculiar aspect, which resembles nothing among Myrtleblooms except some Bæckias, their remarkable abortive stamens, their simple ovary, which never indicates a trace of being formed by the adhesion of more carpels than one, and their pappose calyx. The latter character brings them extremely near Composites, notwithstanding their disunited petals and anthers. The relation that has been found between Myrobalans and Myrtleblooms more especially applies to these plants.

They are beautiful little bushes, abounding in many parts of New Holland, but in only a very few instances reaching the northern coast.

They participate in the fragrance of the foliage of Myrtleblooms: but nothing is recorded of their uses.

GENERA

Calytrix, *Labill.*	Verticordia, *DC.*	Homoranthus, *A. Cunn.*	Hedaroma, *Lindl.*
Calycothrix, Labill.	*Calymmatanthus*, Sch.	*Euosmanthus*, A. Cunn.	Francisia, *Endl.*
Lhotskya, *Schauer.*	*Diplachne*, R. Br.	Darwinia, *Rudg.*	Actinodium, *Schauer.*
Thryptomene, *Endl.*	Chrysorrhoë, *Lindl.*	Polyzone, *Endl.*	*Triphelia*, R. Br.
Pileanthus, *Labill.*	Chamælaucium, *Desf.*	Genetyllis, *DC.*	?? Bartlingia, *Brongn.*

NUMBERS. GEN. 15. Sp. 50.

Myrtaceæ.
POSITION.—Asteraceæ.—CHAMÆLAUCIACEÆ.—Combretaceæ.

Fig. CCCCLXXXI.—1. Calytrix; 2. a section of its flower; 3. of Genetyllis; 4 of Darwinia.

ORDER CCLXXVII. HALORAGACEÆ.—HIPPURIDS.

Halorageæ, *R. Brown in Flinders*, 17. (1814); *DC. Prodr.* 3. 65; *Bartl. Ord. Nat.* 314.; *Endl. Gen.* cclxvi.; *Wight Illustr.* 2. 23.—Hygrobieæ, *Rich. Anal. Fr.* (1808).—Hippurideæ, *Link. Enum.* 1. 5. (1821).—Cercodianæ, *Juss. Dict. Sc. Nat.* (1817).—Hydrocaryes, *Link Enum. Hort. Ber.* 1. 141. (1821).—Onagrariæ, § Hydrocaryes, *DC. Prodr.* 3. 63. (1828).

DIAGNOSIS.—*Myrtal Exogens, with a plurilocular ovary, polypetalous or apetalous flowers, an open minute calyx, definite stamens, pendulous ovules, and minute cotyledons.*

Herbaceous plants or under-shrubs, often growing in wet places. Leaves either alternate, opposite, or whorled. Flowers axillary, small, either in terminal panicles or sessile, occasionally monœcious or diœcious by abortion. Calyx adherent, with a minute limb, which is 2- 4-toothed, or perfectly undivided. Petals inserted into the summit of the calyx, or 0. Stamens inserted in the same place, equal in number to the petals, or occasionally fewer. Ovary adhering inseparably to the calyx, with 1 or more cells; style none; stigmas equal in number to the cells, papulose, or pencil-formed; ovules pendulous, anatropal. Fruit dry, indehiscent, membranous, or bony, with 1 or more cells. Seeds solitary, pendulous; albumen fleshy or 0; embryo straight; radicle superior, large; cotyledons much smaller.

Fig. CCCCLXXXII.

These plants may be regarded either as a distinct Order, or as a mere degeneration or imperfect form of Onagrads, from which their minute calyx and solitary pendulous seeds distinguish them; to which may be added an evident tendency on the part of Hippurids to lose their petals altogether. In Hippuris itself the flower is in the simplest possible form; for it is reduced to a calyx of the smallest size, it has no petals, but one stamen and but one carpel. It therefore furnishes an instance of the approach of Myrtals to the Asteral Alliance. This reduction of the fruit to one carpel only, seems however to be very different in Hippuris from that of Fringe-myrtles; for the latter have a multiplication and excessive development of every other organ, to which the pistil forms the exception; but in Hippuris

Fig. CCCCLXXXII.—Loudonia aurea. 1. a complete flower; 2. a section of the ovary.

the solitary carpel is only a portion of the degraded structure which is proper to all the other organs. In Hippuris and Myriophyllum the stem consists of a curious arrangement of necklace-shaped cellular tissue, which radiates from the centre, leaving large air-cavities between its rows. In the centre is a cylinder of very fine woody tubes, which inclose a confused mass of cellular tissue and spiral vessels. It was probably this which led Prof. Link to regard Hippurids as Endogens.

Damp places, ditches, and slow streams, in Europe, North America, Southern Africa, Japan, China, New Holland, and the South Sea Islands, are the resort of this Order.

They are in general of no importance. Haloragis citriodora, the Piri-Jiri of the New Zealanders, derives its specific name from its fragrant odour. Trapa, a plant with horned fruit, and great amygdaloid seeds, one of whose cotyledons is much smaller than the others, has eatable kernels. Trapa natans is called Marron d'Eau, or Water Chestnut by the French; and is said to have furnished a large part of their food to the ancient Thracians, in the same manner as T. bispinosa, or the Singhara Nut does at the present day to the inhabitants of Cashmere, and T. bicornis to

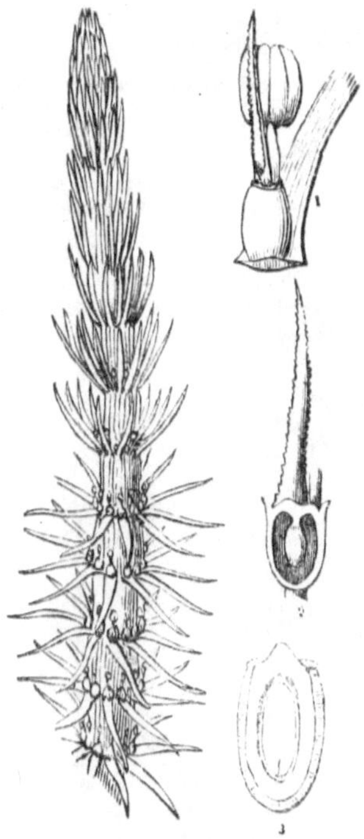

Fig. CCCCLXXXIII.

Fig. CCCCLXXXIV.

the Chinese. It is mentioned by Dr. Royle that the former yielded as much as 12,000*l.* a year of revenue to the government of Runjeet Singh, the tax being levied upon from 96,000 to 128,000 ass-loads from the great lake of Ooller.

GENERA.

I. HALORAGEÆ.	*Pentapteris*, Hall.	Proserpinaca, *Linn.*	*Goniocarpus*, Kön.
Hippuris, *Linn.*	*Sphondylastrum*, Torr.	*Trixis*, Mitch.	Loudonia, *Lindl.*
Limnopeuce, Vaill.	*Ptilophyllum*, Nutt.	Meionectes, *R. Br.*	*Glischrocaryon*, Endl.
Pinastella, Dillen.	*Purshia*, Raf.	Haloragis, *Forst.*	
Myriophyllum, *Vaill.*	*Hylas*, Bigel.	*Cercodia*, Murr.	II. TRAPEÆ, *Endl.*
Sphondylophyllum, Torrey et Gray.	Serpicula, *Linn.*	*Cercodea*, Lam.	Trapa, *L.*
	Laurembergia, Berg.	*Gonocarpus*, Thunb.	*Tribuloides*, Tourn.
		Gonatocarpus, Willd.	

NUMBERS. GEN. 8. Sp. 70.

Combretaceæ.
POSITION.—Onagraceæ.—HALORAGACEÆ.——

Fig. CCCCLXXXIII.—Hippuris vulgaris. 1. a complete flower; 2. a section of the pistil, showing the position of the ovule; 3. a section of the ripe fruit and seed.
Fig. CCCCLXXXIV.—Fruit of Trapa bicornis.

ORDER CCLXXVIII. ONAGRACEÆ.—ONAGRADS.

Onagræ, *Juss. Gen.* 317. (1789); *Spach. in Ann. Sc. N.* 2 *Ser.* iv. 161.—Epilobiaceæ, *Vent. Tabl.* 3. 307. (1799).—Onagrariæ, *Juss. Ann. Mus.* 3. 315. (1804) *in part.; DC. Prodr.* 3. 35. (1828); *Bartl. Ord. Nat.* 318; *Wight Illustr.* 2. 21.—Œnothereæ, *Endl. Gen.* cclxv.—Circæaceæ, *Lindl. Synops.* p. 109. (1829).

DIAGNOSIS.—*Myrtal Exogens, with a plurilocular ovary, polypetalous or apetalous flowers, valvate calyx, definite stamens, horizontal or ascending ovules, and flat cotyledons, much larger than the radicle.*

Herbaceous plants or shrubs. Leaves alternate or opposite, simple, entire, or toothed. Flowers red, purple, white, blue, or yellow, axillary or terminal. Calyx superior,

1 Fig. CCCCLXXXV. 2

tubular, with the limb 4-lobed; the lobes cohering in various degrees, with a valvate æstivation. Petals generally equal in number to the lobes of the calyx, into the throat of which they are inserted, regular, with a twisted æstivation. Stamens (1) 4 or 8, inserted into the calyx; filaments distinct; pollen triangular, usually cohering by threads. Ovary of 2 or 4 cells, generally crowned by a disk; style filiform; stigma either capitate or 4-lobed; ovules anatropal, horizontal, ascending, or peltate. Fruit baccate or capsular, many-seeded, with 2-4 cells. Seeds numerous, without albumen; embryo straight; radicle long and taper; cotyledons shorter.

The Onagrads, thus limited, are in general tetramerous, the number 4 prevailing through every one of the floral organs. In Circæa, however, the number is halved, there being but two sepals, petals, &c., and in Lopezia the customary number seems to be still further interfered with, for that genus shows but one stamen; in reality, however, there are two stamens, one of them perfect and bearing an anther; the other sterile and in the form of a spoon-shaped petal. Although the petals are in general of large size and in a high state of development, yet there is a tendency among the species to lose them; I have seen an entire plant of Clarkia pulchella with every flower apetalous, and Skinnera is always so. From Myrtleblooms Onagrads are known by the absence of pellucid dots and their definite stamens; the Orders approach each other by the genus Fuchsia, which has succulent fruit.

They are chiefly natives of the temperate parts of the world, and especially of America: a good many are found in India, and a large number in Europe; in Africa they are scarcer, being mostly confined to the Cape, and to a few Jussiæas inhabiting other parts of that continent.

Their properties are few, or trifling. Œnothera biennis, and some other species less commonly known, are cultivated for the sake of their eatable roots; and the leaves of

Fig. CCCCLXXXV.—Ludwigia Jussiæoides. 1. a flower with two sepals and all the petals cut off; 2. a calyx and inferior ovary; 3. a transverse section of the ovary; 4. a seed with the distinct raphe; 5. an embryo extracted.

Jussiæa peruviana form an emollient poultice. Indeed, the Order may be regarded as being in general mucilaginous. The Montinias, however, are acrid, and the root of Isnardia alternifolia is said to be emetic. Some are astringent, and have been employed in dyeing black, as, for instance, the Jussiæas Caparosa and scabra, in Brazil. The wood of Fuchsias is reported to be employed in the same way in Chile, and Jussiæa pilosa as a yellow dye in Brazil.—*Martius.* Several of the Fuchsias bear fruits which are subacid and tolerably good to eat. Many of the genus Œnothera expand their flowers only in the evening, and hence, being yellow, have been called Evening Primroses.

GENERA.

I. JUSSIÆA.

- Prieurea, *DC.*
- Jussiæa, *Linn.*
  - *Cubospermum*, Lour.
  - *Vigiera*, Fl. Fl.
- Ludwigia, *Roxb.*
- Isnardia, *L.*
  - *Ludwigia*, L.
  - *Dantia*, Thouars.
  - *Ludwigiaria*, DC.

II. EPILOBEÆ.

- Gayophytum, *Adr. Juss.*
- Sphærostigma, *Sering.*
  - *Onosuris*, Raf.
  - *Chamissonia*, Link.
  - *Heterostemum*, Nutt.
  - *Agassizia*, Spach.
  - *Holostigma*, Spach.
- Meriolix, *Raf.*
  - *Calylophis*, Spach.
- Œnothera, *Linn.*
  - *Onagra*, Tournef.
  - *Anogra*, Spach.
  - *Baumannia*, Spach.
  - *Allochroa*, Fisch. et M.
  - *Megapterium*, Spach.
  - *Pleurostemon*, Raf.
  - *Pleurandra*, Raf.
  - *Pachylophis*, Spach.
  - *Lavauxia*, Spach.
  - *Hartmannia*, Spach.
  - *Kneiffia*, Spach.
  - *Blennoderma*, Spach.
  - *Xylopleurum*, Spach.
- Godetia, *Spach.*
- Cratericarpium, *Spach.*
- Boisduvalia, *Spach.*
  - *Dictyopetalum*, F.et M.
  - *Pachydium*, F. et Mey.
- Eulobus, *Nutt.*
- Clarkia, *Pursh.*
  - *Opsianthus*, Lilja.
  - *Phæostoma*, Spach.
- Eucharidium, *F. et Mey.*
- Chamænerion, *Tourn.*
- Epilobium, *Linn.*
  - *Lysimachion*, Tausch.
  - *Crossostigma*, Spach.
- Zauschneria, *Presl.*

III. MONTINEÆ.

- Montinia, *Linn.*
- Hauya, *Moç. et Sess.*

IV. FUCHSIEÆ.

- Fuchsia, *Plum.*
  - *Dorvalia*, Commers.
  - *Encliandra*, Zucc.
  - *Brebissonia*, Spach.
  - *Lyciopsis*, Spach.
  - *Kierschleigeria*, Spach.
  - *Spachia*, Lilja.
  - *Ellobum*, Lilja.
  - *Thilco*, Feuill.
  - *Nahusia*, Schneev.
  - *Quelusia*, Vand.
  - *Schufia*, Spach.
- Skinnera, *Forst.*

V. CIRCEÆ.

- Semeiandra, *Hook et Arn.*
- Riesenbachia, *Presl.*
- Diplandra, *Hook et Arn.*
- Lopezia, *Cav.*
  - *Pisaura*, Bonat.
- Circæa, *Tournef.*

VI. GAUREÆ.

- Gaura, *Linn.*
  - *Gauridium*, Spach.
  - *Gongylocarpus*, Schied. et Depp.
  - *Schizocarya*, Spach.
- Stenosiphon, *Spach.*

NUMBERS. GEN. 28. Sp. 450.

POSITION.—Haloragaceæ.—ONAGRACEÆ.—Myrtaceæ.
Columelliaceæ.

ORDER CCLXXIX. RHIZOPHORACEÆ.—MANGROVES.

Rhizophoreæ, *R. Brown Gen. Rem. in Flinders, p.* 17. (1814); *in Congo, p.* 18. (1818); *DC. Prodr.* 3. 31. (1828); *Bartl. Ord. Nat.* 320. (1830); *Endl. Gen.* cclxiii.; *Meisn. p.* 119; *Wight. Illustr.* 1. 207; *Arnott in Ann. Nat. Hist.* 1. 359.—Paletuviers, *Savigny in Lam. Dict.* 4. 696. (1796).

DIAGNOSIS.—*Myrtal Exogens, with a plurilocular ovary, polypetalous flowers, valvate calyx, indefinite stamens, and flat cotyledons much shorter than the radicle, which germinates before the fruit falls.*

Coast trees or shrubs. Leaves simple, opposite, occasionally dotted, entire or toothed, with convolute deciduous stipules between the petioles. Peduncles axillary or terminal. Calyx adherent, often surrounded at the base by a cup-shaped bract, with the lobes valvate and varying in number from 4 to 12, occasionally all cohering in a calyptra. Petals arising from the calyx, alternate with the lobes, and equal to them in number. Stamens arising from the same point as the petals, and twice or thrice their number, or in Kandelia indefinite; filaments distinct; anthers erect, innate. Ovary 2-3-4-celled, each cell containing 2 or more ovules hanging from the apex of the central angle, anatropal. Fruit indehiscent, crowned by the calyx, 1-celled, 1-seeded. Seed pendulous, without albumen; radicle very long, piercing the fruit and rapidly extending downwards in germination; cotyledons 2, flat.

Mangroves are readily known from every Order to which they can be usefully compared, by their very curious habit of germinating while the seeds are still attached to the branch that bears the fruit. The radicle and club-shaped crown of the root gradually lengthen until they enter the soft muddy soil, or if too high, drop, and fixing themselves in the muddy bottom, immediately strike root at one end, while leaves unfold at the other.—*Wight.* In Carallia, however, the seeds do not germinate in the pericarp. That the species belong to the Myrtal Alliance there can be no doubt; as indeed is indicated, not only by their structure but by the leaves of some species of Carallia having pellucid dots. At the same time they seem to be connected with the Gentianal Alliance through Cassipourea, which comes very close to Loganiads. The Order also agrees with Cunoniads in its opposite leaves and intermediate stipules, and with great part of them in the æstivation of its calyx, and in the structure and cohesion of ovary. De Candolle points out its relation to Vochyads and Myrobalans, and even to Melastomads, through the genus Olisbea. The genera were comprehended in Loranths by Jussieu. Mr. Griffith has explained with his usual skill the nature of the anther in Rhizophora. In the plants belonging to that genus the anther is alveolar, the sockets being filled with pollen, and in this circumstance it resembles Viscum; but in its younger state the anther is oblong, compressed laterally, and uninterrupted on its surface; when it is mature its two faces fall away, and leave behind a solid centre, in cavities of which the pollen has been generated. See *Transactions of Med. and Phys. Soc., Calcutta.*

Fig. CCCCLXXXVI.

Fig. CCCCLXXXVI.—1. Kandelia Rheedii (*Wight*); 2. its flower spread open; 3. a perpendicular section of its ovary; 4. the germinating seed; 5. the anther of Rhizophora macrorhiza. (*Griffith.*)

Natives of the shores of the tropics, where they root in the mud, and form a close thicket down to the verge of the ocean. Such thickets are so dense that they entirely intercept the rays of the sun, and, preventing the exhalation of putrid miasmata, become the most unhealthy places in a tropical climate. The species generally send down roots from the branches, and thus like the Banyan tree, rapidly spread over considerable spaces. Such roots assume an arched form with the convexity upwards, and gradually raise the main trunk (in Rhizophora) high above its original level.

The bark is usually astringent; that of Bruguiera gymnorhiza is used in India for dyeing black. The wood of several is described as being hard and durable. The fruit of Rhizophora Mangle is said to be sweet and edible, and the juice when fermented forms a light wine.—*Wight.*

GENERA.

Rhizophora, *Lam.*	Bruguiera, *Lam.*	*Baraldeia*, Thouars.	*Petaloma*, DC.
Aërope, Endl.	*Paletuveria*, Thouars.	*Bauraultia*, Steud.	*Catalium*, Hamilt.
Ceriops, *Arn.*	Carallia, *Roxb.*	*Diatoma*, Lour.	*Demidofia*, Dennst.
Kandelia, *Wight et Arn.*			

NUMBERS. GEN. 5. Sp. 20.

Loganiaceæ.
POSITION.—Melastomaceæ.—RHIZOPHORACEÆ.—Myrtaceæ.
Cunoniaceæ.

ORDER CCLXXX. BELVISIACEÆ.—NAPOLEONWORTS.

Belvisieæ, *R. Brown in Linn. Trans.* 13. 222. (1820); *Ed. Pr. No.* clxxxi.; *Meisner Gen.* Napoleoneæ, *Endl. Gen. p.* 745. (1839).

DIAGNOSIS.—*Myrtal Exogens, with a plurilocular ovary, monopetalous coronetted flowers, valvate calyx, indefinite monadelphous stamens, and amygdaloid cotyledons.*

Smooth-leaved bushes, about as large as a Camellia. The wood is soft, whitish, with large medullary rays, and an abundance of dotted vessels, intermingled with brittle acicular tubes of woody tissue, very like what is found in the germinating radicle of a Mangrove. The leaves are alternate, leathery, narrowed at the base into a thick channelled petiole; there is no trace of stipules. The flowers grow in threes, sessile in the axil of the leaves, and are surrounded at their base by several round imbricated scales. The calyx is a thick leathery cup, divided into five ovate segments, having a perfectly valvate æstivation. Within this is placed the corolla, which consists of 3 distinct rings, each of which is monopetalous. The first ring is divided into five lobes, each of which has seven stiff ribs, between which the texture is membranous; the lobes have seven broad teeth, corresponding with the points of the ribs, and much curled and crumpled. By means of the ribs and intervening membrane, this part of the corolla is strongly plaited both before and after expansion; when fully blown, it turns quite back over the calyx, so as to hide it completely. The second ring is very small and thin; it is, in fact, a narrow membrane, stationed at the foot of the first ring, and cut into an indefinite number of fine narrow sharp-pointed segments. This ring was overlooked by De Beauvois. The third ring is membranous, but erect, and assuming the form of a cup, whose edge is cut into

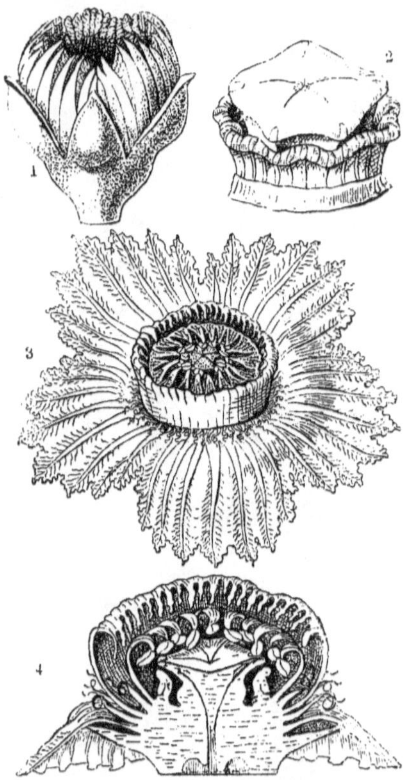

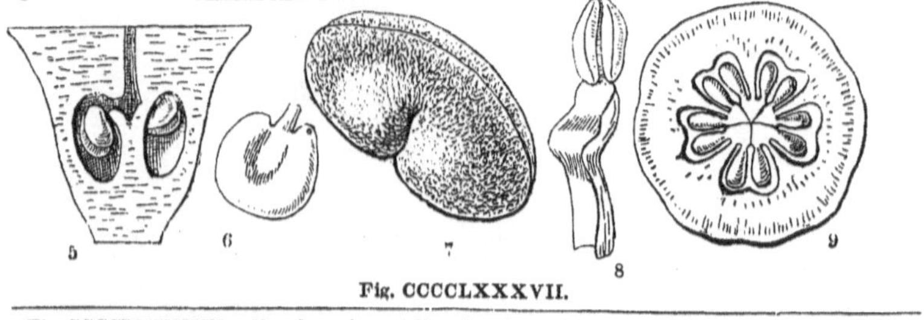

Fig. CCCCLXXXVII.

Fig. CCCCLXXXVII.—Napoleona imperialis. 1. a flower-bud just expanding; 2. the fleshy cup, and table-shaped stigma; 3. an expanded flower of the natural size; 4. a perpendicular section of the same. (In this the artist has carelessly added a fourth ring to the corolla on the outside of the stamens; no such ring exists); 5. a perpendicular section of the ovary; 6. an ovule; 7. a ripe seed; 8. a stamen; 9. a transverse section of the ovary.

many fine segments, turned downwards, so as not to be at all conspicuous. The stamens are 20, standing erect in the form of another cup, and unequally united at their base; they have linear-lanceolate filaments, which are much thinner next the anthers, and are there turned inwards; the anther itself is oblong, 2-celled, and erect. Next the stamens comes a deep fleshy cup or disk, standing as high as the stigma, and having ten sides, of which the narrowest are alternate with the lobes of the stigma, and 2-ribbed in the inside. The ovary is buried beneath the mass formed by the base of the corolla, stamens, and disk; it has five cells, in each of which two ovules hang from the top of an axile placenta, which is so attached to the partitions that there is a clear opening from the hollow centre of the style, over the ovules, into the cells of the ovary; the ovules are oblong, with a depression in the middle on each side, and a foramen next the base, the nucleus being curved like a horse-shoe, so that its base and apex are both nearly in contact; the style is 5-angled, or rather 5-winged, and terminated by a table-shaped stigma, with five sides, five rays, and a small elevation at each angle, which elevations are perhaps the true stigmatic surfaces. The fruit is a soft spherical berry, surmounted by the calyx, as large as a Pomegranate, and very like one, containing a mucilaginous pulp which is eatable, and a rind so full of tannin, that the natives make an ink from it. The seeds are large amygdaloid bodies, kidney-shaped, and as much as $1\frac{1}{4}$ inch long; at their contraction the plano-convex cotyledons hold together by an axis whose radicle and plumule are both immersed in the substance of the cotyledons.

In the total absence of all correct information as to the real structure of this curious genus, Botanists were unable to arrive at any satisfactory conclusion as to its affinities. All that they had been able to settle was its not belonging to any known Natural Order. Palisot De Beauvois stated (1807) that, in the opinion of Jussieu, it constituted a new Order between Cucurbits and Passionflowers; a view that was probably taken in consequence of the double-ringed corolla, which is analogous to the coronet of the Passionflowers, and the plaited corolla with an inferior ovary, which brings to mind the flowers of Cucurbits. Desfontaines, on the contrary (1820), referred it, and another genus which he calls Asteranthus, without any doubt to Symplocaceæ, because of its monopetalous perigynous corolla, its stamens inserted in the base of the corolla, its oblong two-celled anthers, single style, inferior ovary, axillary solitary flowers, shrubby stem, and alternate leaves. Him followed Dr. Robert Brown (1822), who formed it and Asteranthus into an Order called Belvisieæ, without, however, attempting to settle its position in the Natural System. He objected to approximating it to Symplocaceæ, doubted its affinity to Passionflowers, and compared its structure with that of Rafflesia. Latterly no one seems to have attempted to suggest anything new as to its relationship. Endlicher puts it next to Symplocaceæ; Meisner next Passionflowers, adding to what had been previously known of it, that its seeds are arillate, a mistake (?) that probably originated in De Beauvois's description of them, "Semina in pulpâ carnosâ nidulantia." Finally, I myself, feeling that these could not be its true affinities, placed it in the Campanal Alliance, with marks of great doubt. But I was enabled, a few months since, by the kindness of the Earl of Derby, to examine good specimens collected by Mr. Whitfield, from which the foregoing description and succeeding remarks were written for the *Botanical Register*.

It is obvious that Napoleona has nothing to do with any of the Orders to which it has been referred. From Cucurbits it differs utterly it its hermaphrodite flowers, axile placentation, highly developed corolla, and whole habit; it has in fact no resemblance to that Order. Passionflowers seem at first sight to claim a much nearer relationship, because of the triple-rowed corolla of Napoleona, which much resembles the coronet of those plants; but there the resemblance ceases. The tendrils, parietal placenta, free ovary, distinct styles, polypetalous corolla, imbricated calyx of Passionflowers, are all most essentially at variance with the genus. Symplocaceæ were a far better guess, for the monopetalous corolla, indefinite epipetalous stamens, axile placentæ, adherent calyx, and definite seeds of Napoleona find there a parallel; but the ovary of that genus is wholly adherent, with a great epigynous disk, the calyx is valvate, and the seeds have no albumen, to say nothing of the lacerated condition of the corolla, which is not to be wholly disregarded in a consideration of this kind. Upon the whole it appears that the true affinity is in the neighbourhood of the Mangroves (Rhizophoraceæ), for the following reasons:—The ovary is in both inferior, few-seeded, with axile placentæ; both have a coriaceous valvate calyx; both have large amygdaloid seeds without albumen. The placenta of Kandelia is almost the same as that of Napoleona, and in the former genus the petals are broken up into numerous fringes quite analogous to those of the genus in question. To this may be added the great resemblance that exists between the wood of Napoleona and of young Rhizophora, in consequence of both consisting in part of slender acicular tubes, which give the wood, when broken across, the appearance of containing slender bristles. Finally, the ribbing, which is so conspicuous in the outer corolla of Napoleona, is repeated in the calyx of Bruguiera gymnorhiza. It is true that the one genus is monopetalous and the other polypetalous, but I cannot attribute importance to that character in a case where the stamens adhere so slightly to the corolla. While, however, there is this reason to believe that Mangroves are most nearly related to Napoleonworts, the affinity of the Order to some Myrtal plants is not to be overlooked ; as, for example, to Careya, whose fruit has a very similar structure, and to Barringtonia, to which Napoleona is even similar in foliage ; but these affinities are less striking than that of the Mangrove tribe. They show, however, pretty clearly that Belvisiaceæ—for so it is most convenient to call the Order of which Napoleona is the most conspicuous member—belongs to the great Myrtal Alliance. At the very moment when these remarks were published in England, M. Adrien de Jussieu described another species of Napoleona (in the *Annales des Sciences*, vol. ii. p. 222, third series), and adopted the views which Desfontaines had taken as to its affinities. I do not, however, see any cause to alter the opinion I had myself formed on the subject.

The Order is wholly African and tropical. It is in the wilds of that little-examined part of the world that additions must be expected to it. The statement made by Desfontaines that the genus Asteranthus is Brazilian, has been doubted by Endlicher and negatived by Adrien de Jussieu.

Nothing is known of the uses of the plants, except what is above mentioned.

GENERA.

Asteranthos, *Desf.*
Napoleona, *Palis.*
Belvisia, Desv.

NUMBERS. GEN. 2. Sp. 4.

Styracaceæ.
POSITION.—Myrtaceæ.—BELVISIACEÆ.—Rhizophoraceæ.
Passifloraceæ.

ORDER CCLXXXI. MELASTOMACEÆ.—MELASTOMADS.

Melastomæ, *Juss. Gen. p.* 328. (1789).—Melastomaceæ, *Don. in Mem. Wern. Soc.* 4. 281. (1823); *DC. Prodr.* 3. 99. (1828); *Memoire,* (1828); *Blume in Botanisch. Zeit.* (1833); *Endl. Gen.* cclxviii.—Memecyleæ, *DC. Prodr.* 35. (1828); *Opatowski Dissertatio.*—Memecylaceæ, *Ed. Pr.* xxvi.—Mouririaceæ, *Gardn. in Hook. Journ.* 2. 23.—Myrrhineæ *or* Olinieæ, *Arnott in Ann. Nat. Hist.* 3. 154.

DIAGNOSIS.—*Myrtal Exogens, with a plurilocular ovary, polypetalous flowers, an imbricated calyx, definite stamens, rostrate anthers, and usually dotless leaves.*

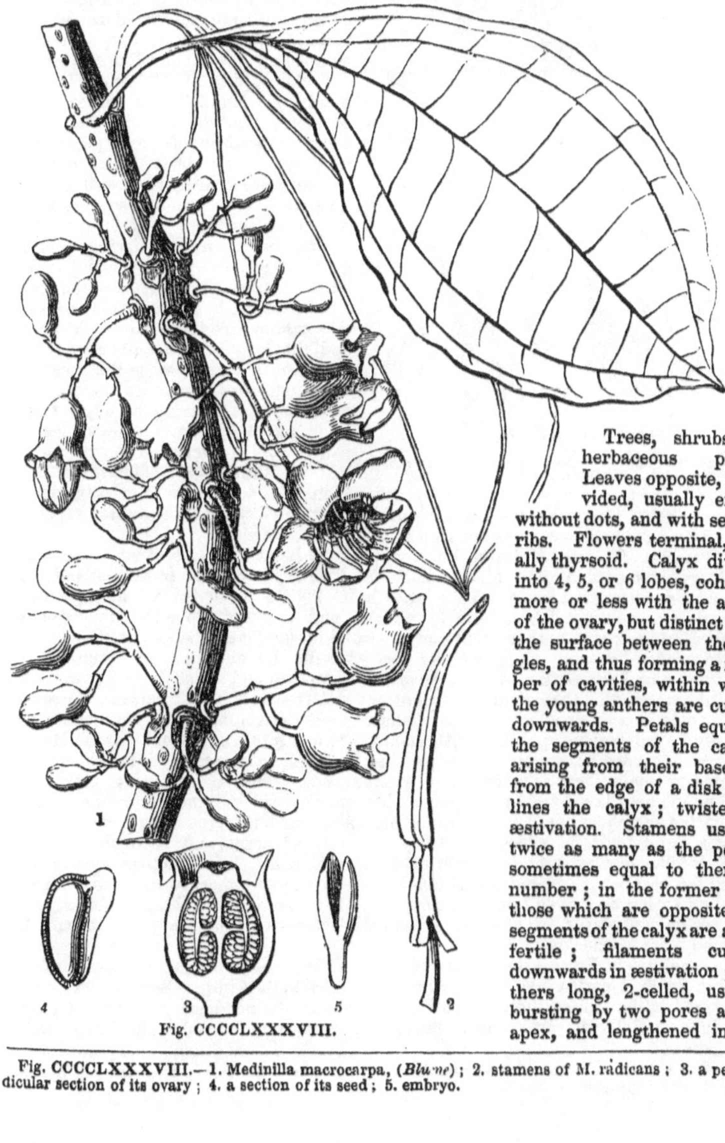

Fig. CCCCLXXXVIII.

Trees, shrubs, or herbaceous plants. Leaves opposite, undivided, usually entire, without dots, and with several ribs. Flowers terminal, usually thyrsoid. Calyx divided into 4, 5, or 6 lobes, cohering more or less with the angles of the ovary, but distinct from the surface between the angles, and thus forming a number of cavities, within which the young anthers are curved downwards. Petals equal to the segments of the calyx; arising from their base, or from the edge of a disk that lines the calyx; twisted in æstivation. Stamens usually twice as many as the petals, sometimes equal to them in number; in the former case, those which are opposite the segments of the calyx are alone fertile; filaments curved downwards in æstivation; anthers long, 2-celled, usually bursting by two pores at the apex, and lengthened in va-

Fig. CCCCLXXXVIII.—1. Medinilla macrocarpa, (*Blume*); 2. stamens of M. radicans; 3. a perpendicular section of its ovary; 4. a section of its seed; 5. embryo.

rious ways beyond the insertion of the filament; sometimes bursting longitudinally; before flowering, contained within the cases between the ovary and sides of the calyx. Ovary more or less coherent with the calyx, with several cells, and definite or indefinite ovules; style 1; stigma simple, either capitate or minute; a cup often present upon the apex of the ovary, surrounding the style. Pericarp either dry and distinct from the calyx, or succulent and combined with the calyx, with several cells; if dehiscent, bursting through the valves, which therefore bear the septa in the middle; placentæ attached to a central column. Seeds innumerable, minute, with a brittle testa and no albumen; usually with appendages of some kind; embryo straight, or curved, with equal or unequal flat or convolute cotyledons.

Fig. CCCCLXXXIX.

"The Order of Melastomads," remarks De Candolle, in an excellent Memoir upon the subject, "although composed entirely of exotic plants, and established at a period when but few species were known, is so well characterised, that no one has ever thought of putting any part of it in any other group, or even introducing into it genera that do not rightly belong to it." These distinct characters are, the opposite leaves, with several great veins or ribs running from the base to the apex, and the long beaked anthers. Good, however, in most cases, as these characters undoubtedly are, yet the cause of no uncertainty having been yet found in fixing the limits of the Order, is rather to be attributed to the small number of species that have been examined, than to the want of connecting links: thus Diplogenea has traces of the dots of Myrtles, which were not known to exist in Melastomads until that genus was described. Mouriria has no ribs, and its leaves are very distinctly dotted; the Memecylons are ribless, and so is Sonerila.

The greatest affinity of Melastomads is on the one hand with Lythrads, on the other with Myrtleblooms and their allies; from the former they differ in the æstivation of their calyx not being valvate, from the latter in having the petals twisted before expansion, and no dots on the leaves, and from both, and all others to which they can be compared, in their long anthers bent down parallel to the filaments in the flower, and lying in niches between the calyx and ovary; with the exception of Memecyls, in which the union between the calyx and ovary is complete, and which have leaves destitute of the lateral ribs that so strongly point out Melastomads. The structure of the seeds of Memecyls is also peculiar, the cotyledons being convolute as in Myrobalans, to which the Myrtleblooms approach at this point. It was for these reasons that the Memecyls were regarded as the type of a peculiar Order, but it seems on the whole more advisable to retain them as a section of Melastomads. That the convolute cotyledons are of no moment is proved by the genus Chamæmeles, which differs from other Appleworts in the same manner. Mr. Gardner makes Mouriria the type of an Order, because its leaves are dotted and ribless, its ovary perfectly adherent, and its ovules solitary. Sir W. Hooker, however, finds 3 erect ovules in each cell. It is doubtless a genus connecting the Myrtleblooms and Melastomads, and belonging almost as much to one as to the other Order, as Brown long ago stated. As to the Olinieæ, Mr. Arnott regards them as being nearer Myrtleblooms than Melastomads; but they can hardly be separated from Memecyls.

Found neither in Europe nor in Africa north of the desert of Zahara, nor south of Brazil in South America, nor in extra-tropical Africa to the south. Beyond the tropics, 8 are found in the United States, a few in China and the northern provinces of India, and 3 in New Holland. Of the remainder, it appears that 78 are described from India

---

Fig. CCCCLXXXIX.—Stenodon suberosus. 1. flower seen from above; 2. a stamen; 3. cross section of nearly ripe fruit; 4. ripe fruit splitting; 5. seed.

or the Indian Archipelago, 12 from Africa and the adjacent islands, and 020 from America, according to De Candolle; but this computation now requires correction.

A slight degree of astringency is the prevailing character of the Order, which although one of the most extensive known, is entirely destitute of any unwholesome species. The succulent fruit of many is eatable; that of some dyes the mouth black, whence the name of Melastoma. Blakea triplinervis produces a pleasant and eatable yellow fruit in the woods of Guiana. The fruit of Lasiandra argentea and some others is used in Brazil for dyeing black. Osbeckia Principis and Miconia longifolia are employed in the same manner. The leaves of Memecylon edule form an ingredient in the dyes of Coromandel. The ripe berries, though somewhat astringent, are eaten by the natives.—*Royle.* The juice of Tococa guianensis is used in Demerara as ink. Cremanium reclinatum and tinctorium, and Miconia tinctoria, like the Memecyls, dye yellow; Blakea parasitica and many others red. The flowers of Guildingia psidioides are fœtid, the berries nauseous, the seeds with the flavour of Filberts. The berries of Myrrhinium atropurpureum are eatable. Some are mentioned in medical books. The leaves of Melastoma malabathrica are used in diarrhœa, dysentery, &c. The bark of Medinillas is emollient, and is prepared for poultices, as are the leaves of Osbeckia chinensis. Astronia papetaria (Ubat papeda) and some others have sub-acid leaves, which in the Malay Archipelago are cooked as a sauce to fish; the wood of that plant is hard, and used for door-posts. The berries of Tristemma virusanum are given in the Mauritius as a cure for syphilis. A few are aromatic, others vulnerary; but none of any moment.

GENERA.

I. MELASTOMEÆ.

LAVOISIEREÆ.

Merianía, *Swartz.*
Wrightia, Soland.
Pachymeria, *Benth.*
Axinæa, *Ruiz et Pav.*
Chastenæa, *DC.*
Lavoisiera, *DC.*
Davya, *DC.*
Adelbertia, *Meisn.*
Graffenrieda, *DC.*
Huberia, *DC.*
Behuria, *Cham.*
Centradenia, *G. Don.*
Plagiophyllum, Schlct.
Brachycentrum, *Meisn.*
Pyramia, *Cham.*
Centronia, *Don.*
Truncaria, *DC.*
Rhynchanthera, *DC.*
Proboscidea, Rich.
Bucquetia, *DC.*
Cambessedesia, *DC.*
Chætostoma, *DC.*
Meisneria, *DC.*
Siphanthera, *Pohl.*
Salpinga, *Mart.*
Aulacidium, Rich.
Bertolonia, *Radd.*
Triblemma, Mart.
Eriocnema, *Naud.*
Augustinea, *Naud.*
Stenodon, *Naud.*
Miocarpus, *Naud.*
Tulasnea, *Naud.*
Lithobium, *Bong.*
Sonerila, *Roxb.*
Cassebeeria, Dennst.
Trigonocapsa, Blum.

RHEXIEÆ.

Dicrananthera, *Pohl.*
Poteranthera, *Bong.*
Spennera, *Mart.*
Jaravæa, Scop. part.
Noterophila, *Mart.*
Microlicia, *Don.*
Jaravæa, Scop. part.
Uranthera, *Naud.*
Chætostoma, *Naud.*
Fritzschia, *Cham.*
Ernestia, *DC.*
Rhexia, *R. Br.*
Leiostegia, *Benth.*
Heteronoma, *DC.*
Pachyloma, DC.
Heterocentron, *Hook. et Arn.*
Oxyspora, *DC.*
Tricentrum, *DC.*
Marcetia, *DC.*
Trembleya, *DC.*
Jacobia, DC.
Abrahamia, DC.
Eriolema, DC.
Adelobotrys, *DC.*

OSBECKIEÆ.

Lasiandra, *DC.*
Macairea, *DC.*
Chætogastra, *DC.*
Monocentra, DC.
Diotanthera, DC.
Bractearia, *DC.*
Brachyandra, *Naud.*
Arthrostemma, *Pav.*
Melanium, Rich.
Chætopetalum, DC.
Brachyotum, DC.
Ladanopsis, DC.
Trifurcarium, DC.
Monochætum, DC.
Heeria, *Schlecht.*
Svitramia, *Cham.*
Tibouchina, *Aubl.*
Savastania, Neck.
Diplostegium, *Don.*
Tristemma, *Juss.*
Pleroma, *Don.*
Lachnopodium, *Blum.*
Melastoma, *Burm.*
Acinodendron, Linn.
Otanthera, *Blum.*
Osbeckia, *Linn.*
Pterolepis, *DC.*
Chætolepis, *DC.*
Microlepis, *DC.*
Aciotis, *Don.*

MICONIEÆ.

Rousseauxia, *DC.*
Dichætanthera, *Endl.*
Leandra, *Radd.*
Clidemia, *Don.*
Tchudya, DC.
Jucunda, *Cham.*
Graffenrieda, Mart.
Myriaspora, *DC.*
Hamastris, Mart.
Tococa, *Aubl.*
Myrmidone, *Mart.*
Majeta, *Aubl.*
Calophysa, *DC.*
Medinilla, *Gaudich.*
Gallaria, Schrank.
Hypenanthe, Blum.
Dactyliota, Blum.
Triplectrum, *Don.*
Pachycentria, *Blum.*
Pogonanthera, *Blum.*
Allomorphia, *Blum.*
Calycogonium, *DC.*
Calycopteris, Rich.
Ossæa, *DC.*
Sagræa, *DC.*
Tetrazygia, *Rich.*
Heterotrichum, *DC.*
Dissochæta, *Blum.*
Diplectria, Blum.
Aplectrum, *Blum.*
Conostegia, *Don.*
Calycotomus, Rich.
Bruguiera, Rich.
Eriostegia, DC.
Diplogenea, *Lindl.*
Diplochiton, *Spreng.*
Diplochita, DC.
Chitonia, Don.
Fothergilla, Aubl.
Leonicenia, Scop.
Phyllopus, *DC.*
Henriettea, *DC.*
Loreya, *DC.*
Marumia, *Blum.*
Creochiton, *Blum.*
Phyllagathis, *Blum.*
Decarhaphe, *Miq.*
Miconia, *Ruiz et Pav.*
Hypoxanthus, Rich.
Oxymeris, *DC.*
Hartigia, *Miq.*
Cremanium, *Don.*
Cyathanthera, Pohl.
Blakea, *Linn.*
Topabea, Aubl.
Valdesia, Ruiz et Pav.
Bellucia, Neck.
Drepanandrum, Neck.
Apatitia, Desv.
Sarcopyramis, *Wall.*

II. CHARIANTHEÆ.

Charianthus, *Don.*
Chænanthera, Rich.
Tetrazygos, Rich.
Chænopleura, *Rich.*
Kibessia, *DC.*
Pternandra, Jack.
Ewyckia, *Blum.*
Astronia, *Blum.*
Spathandra, *Guill. et Perrot.*

III. MEMECYLEÆ.

Memecylon, *Linn.*
Valikaha, Adans.
Scutula, Lour.
Mouriria, *Juss.*
Mouriri, Aubl.
Petaloma, Swartz.
Guildingia, *Hook.*
Olisbea, DC.

IV. OLINIEÆ.

Olinia, *Thunb.*
Cremastostemon, Hort.
Myrrhinium, *Schott.*
Felicianea, Cambess.
Tetrastemon, Hook.
Fenzlia, *Endl.*

NUMBERS. GEN. 118. Sp. 1200.

Combretaceæ.
POSITION.————MELASTOMACEÆ.—Myrtaceæ.
Lythraceæ.

ORDER CCLXXXII. MYRTACEÆ.—MYRTLEBLOOMS.

Myrti, *Juss. Gen.* 323. (1789).—Myrteæ, *Juss. Dict. Sc. Nat.* 34. 94. (1825).—Myrtoideæ, *Vent. Tab.* (1799).—Myrtineæ, *DC. Théorie, Elem.* (1819).—Myrtaceæ, *R. Brown in Flinders, p.* 14. (1814); *DC. Dict. Class* v. 11; *Prodr.* 3. 207; *Endl. Gen.* cclxix.; *Schauer in Linnæa*, xvii. 235; *Wight Illustr.* 2. 6.—Granateæ, *Don. in Ed. Phil. Journ. p.* 134. (1826); *DC. Prodr.* 3. 3; *Von Martius H. Reg. Monac.* (1829); *Endl. Gen. p.* 1223; *Wight Illustr.* 2. 2.

DIAGNOSIS.—*Myrtal Exogens, with a plurilocular ovary, polypetalous or apetalous flowers, an imbricated calyx,* 00 *stamens, oblong anthers, and usually dotted leaves.*

Trees or shrubs. Leaves opposite or alternate, entire, usually with transparent dots and a vein running parallel with their margin. Inflorescence variable, usually axillary.

Fig. CCCCXC.

Flowers red, white, occasionally yellow, never blue. Calyx adherent, valvate, 4- or 5-cleft, sometimes falling off like a cap, in consequence of the cohesion of the apex. Petals equal in number to the segments of the calyx, with a quincuncial æstivation; rarely none. Stamens either twice as many as the petals, or 00, rarely equal to them in number; filaments either all distinct or connected in several parcels, curved inwards before flowering; anthers ovate, 2-celled, small, bursting lengthwise. Ovary inferior, 1- 2- 4- 5- or 6-celled; style simple, derived immediately from the placenta; ovules usually pendulous, or erect and anatropal; occasionally peltate and amphitropal, always inserted into a central or axile placenta. Fruit either dry or fleshy, dehiscent or indehiscent. Seeds usually indefinite, variable in form; embryo without albumen, straight or curved, with its cotyledons and radicle distinguishable or blended into a solid mass.

A species of Sonneratia is apetalous. Some dotted leaves are alternate.

One of the most natural among the tribes of plants, and the most easily recognised. Opposite exstipulate dotted entire leaves with a marginal vein, are a certain indication of it; and even where the leaves are alternate the intramarginal vein is usually discoverable. This alternation is in some species uniform, but in other instances it is accidental, as in Myrtus communis, which usually has opposite leaves, though, if the plant is killed to the grouud by frost they are mostly alternate on the shoots that spring up again. It is closely allied to Roseworts, Lythrads, Onagrads, Myrobalans, and Melastomads, but cannot well be confounded either with them or any other Order. It offers a singular instance of the facility with which the calyx and corolla can take upon themselves the same functions and transformations. In Eucalyptus the sepals are consolidated into a cup-like lid, called the operculum, and in Eudesmia, a nearly-

Fig. CCCCXC.—Eugenia tuberculata. 1. a flower; 2. the same divided vertically; 3. a stamen; 4. a ripe fruit; 5. a leaf with the dots upon it.

related genus, the calyx remains in its normal state, while the petals are consolidated into an operculum. Babingtonia offers the curious structure of a style and stigma, derived immediately and wholly from the placenta, a circumstance not without a parallel in this Order; and Bæckia micrantha, *DC.*, has parietal placentæ!

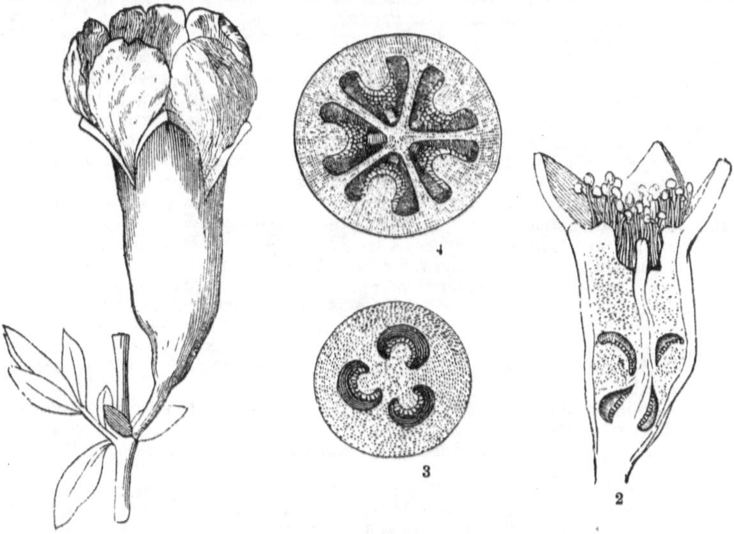

1 Fig. CCCCXCI.

Punica has been considered the type of a particular Order (Granateæ) by Don, in which he is supported by the high authority of De Candolle, Von Martius, and Wight. The fruit of Punica Granatum, the Pomegranate, is described by Gærtner and De Candolle as being divided into two unequal divisions by a horizontal diaphragm, the upper half of which consists of from 5 to 9 cells, and the lower of three; the cells of both being separated by membranous dissepiments; the placentæ of the upper half proceeding from the back to the centre, and of the lower irregularly from their bottom; and by Don as a fleshy receptacle formed by the tube of the calyx into a unilocular berry, filled with a spongy placenta, which is hollowed out into a number of irregular cells. In fact, if a Pomegranate is examined, it will be found to agree more or less perfectly with both these descriptions. But it is clear that a fruit as thus described is at variance with the ordinary laws upon which compound fruits are formed. A section of the ovary of the Pomegranate in various directions, if made about the time of the expansion of the flowers before inpregnation takes place, shows that it is in fact composed of two rows of carpels, of which three or four surround the axis, and are placed in the bottom of the tube of the calyx, and a number, varying from five to ten, surround these, and adhere to the upper part of the tube of the calyx. The placentæ of these carpels contract an irregular adhesion with the back and front of their cells, and thus give to the position ultimately acquired by the seeds that anomalous appearance which it assumes in the ripe fruit. If this view of the structure of the Pomegranate be correct, its peculiarity consists in this, that, in an Order the carpels of which occupy but a single row around the axis, it possesses carpels in two rows, the one placed above

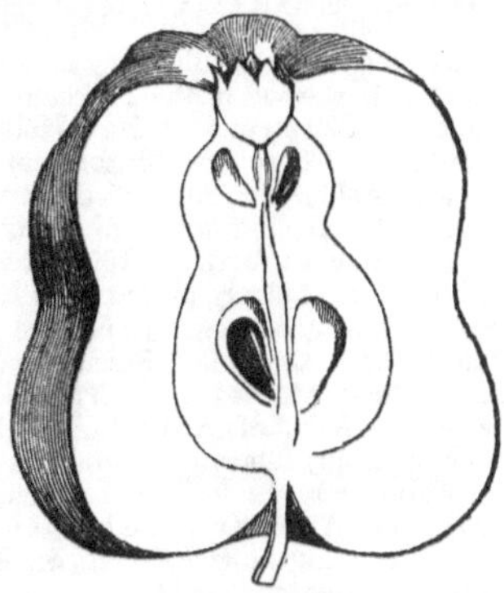

Fig. CCCCXCII.

Fig. CCCCXCI.—1. Punica Granatum; 2. perpendicular section of the ovary; 3. cross section of it near the base; 4. near the base.
Fig. CCCCXCII.—Monstrous Apple, mentioned in the next page.

the other, in consequence of the contraction of the tube of the calyx, from which they arise. Now, there are many instances of a similar anomaly among genera of the same Order, and they exist even among species of the same genus. Examples of the latter are, Nicotiana multivalvis and Nolana paradoxa, and of the former Malope among Malvaceæ, polycarpous Crowfoots as compared with Nigella, and polycarpous Roseworts as compared with Spiræa. In Prunus I have seen a monstrous flower producing a number of carpels around the central one, and also, in consequence of the situation, upon the calyx above it; and finally, in the *Revue Encyclopédique* (43. 762), a permanent variety of the Apple is described, which is exactly to Appleworts what Punica is to Myrtleblooms. This plant has regularly 14 styles and 14 cells, arranged in two horizontal parallel planes, namely, 5 in the middle, and 9 on the outside, smaller and nearer the top; a circumstance which is evidently to be explained by the presence of an outer series of carpels. Dr. Wight proposes a modification of these views (*Illustrations of Indian Botany*, ii. 5), but I do not see in what respect his opinion materially differs from mine. The anomaly of the structure of the fruit of Punica being thus explained, nothing remains to distinguish it from Myrtleblooms but its leaves without a marginal vein, its convolute cotyledons, and pulpy seeds. There are, however, distinct traces of dots in the leaves, and the union of the venæ arcuatæ, which gives the appearance of a marginal vein to Myrtleblooms, takes place, although less regularly, in Punica; the convolute cotyledons of Punica are only in Myrtleblooms what those of Chamæmeles are in Appleworts, a curious but unimportant exception to the general structure; and the solitary character of the pulpy coat of the seeds will hardly be deemed by itself sufficient to characterise Granateæ. The place of Punica in the Order will be probably near Sonneratia.

There is no instance of a blue flower in this Order. The fruit varies from succulent to dry in different genera, and in some cases is nearly superior. According to Auguste de St. Hilaire, a passage is formed from Myrtleblooms to Onagrads through the genus Felicianea.

Natives of hot countries both within and without the tropics; great numbers are found in South America and the East Indies, not many in Africa, and a considerable proportion of the Order in New Holland and the South Sea Islands; but the genera of those countries are mostly peculiar to them. Myrtus communis, the most northern species of the Order, is a native of (Persia, but has become naturalised in) the south of Europe. Metrosideros lucida, a beautiful tree of this Order, occurs as far to the south as Lord Auckland's Islands, in lat. 50½ S.—*J. Hooker.*

De Candolle remarks, that although they all, without exception, have a woody texture, yet that they vary prodigiously in stature, from the little Myrtus nummularia which spreads over the soil in the Falkland Islands, as Thyme does in Europe, to the immense Gum-trees (Eucalypti) of New Holland, which are among the most gigantic trees of Australasia. There are all sorts of intermediate sizes, but the common Myrtle-bush gives a tolerably good idea of the appearance of the majority. Mr. Backhouse speaks of some of the Gum-trees as rising to about 200 feet in height, with straight trunks clear of branches for from 100 to 150 feet, and resembling an assemblage of elegant columns, so irregularly placed as to intercept the view at the distance of a few hundred yards. These are elegantly crowned with branching tops of light willow-like foliage. Some of what are called Stringy bark Gum-trees, "rise nearly as high as the Monument without branching!" The Aki, a New Zealand plant of this Order, the Metrosideros buxifolia, of Allan Cunningham, is described by that Botanist as being a rambling shrub, adhering to trees, and by its lateral roots climbing to the summit of the loftiest timber in the forests of Wangaroa, Bay of Islands, &c.

The pellucid dotting of the leaves and other parts indicates the presence of a fragrant aromatic or pungent volatile oil, which gives the principal quality to the produce of the Order. To this are due the grateful perfume of the Guava fruit, the powerful aroma of the flower-buds of Caryophyllus aromaticus, called by the English Cloves, and the balsamic odour of those eastern fruits, the Jamrosade and the Rose Apple. Along with this is frequently mixed an astringent principle, which sometimes predominates, to the suppression of any other property. The Guavas are pulpy fruits inhabiting the western world, whence they have been carried to the eastern; the principal are Psidium pyriferum and pomiferum, the latter of which is much more acid than the other. They make with sugar a cooling and rather astringent conserve. The berries of other species of Psidium, which grow plentifully on the campos of S. Paulo, and are distinguished by the name of Guabinoba, are used in a similar manner. The young bark and leaves are employed as astringents, and the latter for medicated baths, which are very customary in Brazil; other species, especially P. Cattleyanum, also bear a fruit of excellent quality. Eugenia cauliflora, the Jabuticaba or Jaboticaburas, is one of the most agreeable fruits in Brazil, and the taste will be improved by further culture. Very good wine, syrup, &c., are made of it. The Jabuti, Psidium albidum, and Eugenia dysenterica, Michelii, and brasiliensis, called respectively Araça, Pitanga, Grumixameira, Cambuy, Uvalha, Pitangueira, &c., are all spoken of by Martius as excellent dessert fruits. Even the berries of the common Myrtle are esteemed in the Greek Archipelago, especially a sort with white fruit. The Rose Apples of the East, produced by Eugenia malaccensis, aquea, Jambos, and others, are all in esteem in the countries where they grow.

As a spice, every one is acquainted with Caryophyllus aromaticus, whose oil is a common remedy for toothache, and whose dried flower-buds are the Cloves of the shops. Those of Calyptranthes aromatica may be advantageously substituted. The Pepper called Allspice or Pimento, is the dried fruit of Eugenia acris and Pimento ; all the plant, especially the unripe fruit, abounds in an essential oil, which is a powerful irritant, and is often used to allay toothache. The bruised berries are carminative, stimulating the stomach, promoting digestion, and relieving flatulency. The fruit of Eugenia Caryophyllus is used in the same way in Brazil, and of Myrtus Tabasco in Cumana. Myrtle buds and berries (Myrtus communis) were eaten as spices by the ancients, and are still used in Tuscany instead of pepper. The Tuscans also prepare a sort of Myrtle wine, which they call Myrtidanum. The distilled water of Myrtle flowers is that very agreeable perfume known in France under the name of Eau d'Ange. The leaves of Sizygium terebinthaceum are used in Madagascar to aromatise baths ; Mr. Cooper found both sides of its leaves covered with very minute glandular hairs, having at their apex a knob of brownish matter.—*Ann. N. Hist.* x. 154. The volatile, green, irritating, or stimulant oil of Cajeputi is distilled from the leaves of Melaleuca Cajeputi, and is well known as a powerful sudorific, and useful external application in chronic rheumatism. It is considered carminative, cephalic, and emmenagogue, and is, no doubt, a highly diffusible stimulant, antispasmodic and diaphoretic. It has also the power of dissolving caoutchouc, and possesses a great reputation as a remedy in cholera.

As simple astringents several deserve notice. A kind of gum Kino is yielded by Eucalyptus resinifera, which is occasionally sold in the medicine bazaars of India. Other species of Eucalyptus yield a large quantity of tannin, which has been even extracted from the trees in New Holland, and sent to the English market. The leaves of the common Myrtle, dried in a stove and powdered, have been substituted for the Sumac of Sicily ; those of Eugenia depauperata and variabilis are used as astringents in Brazil. The Pomegranate, Punica Granatum, commonly cultivated in the warmer parts of Europe, and forming entire woods in Persia, has long been celebrated in medicine ; a decoction of the bark of the root is a powerful anthelmintic ; the flowers are tonic and astringent, as is the bark of the fruit, which is used in leucorrhœa, chronic dysentery, &c. ; the acid juice of the seeds is found useful in bilious fevers.

Some species secrete a sweet manna-like gum. Eucalyptus robusta contains large cavities in its stem, between the annual concentric circles of wood, filled with a most beautiful red or rich vermilion-coloured gum, and E. mannifera, in New Holland, exudes a saccharine mucous substance resembling Manna in action and appearance, but less nauseous. It is not produced by insects, and only appears in the dry season. Other species yield a similar secretion at Moreton Bay and in Van Diemens Land. Mr. Backhouse says it coagulates and drops from the leaves in particles often as large as an almond. Eucalyptus Gunnii, when wounded, furnishes the inhabitants of Tasmannia with a copious supply of a cool, refreshing, slightly aperient liquid, which ferments and acquires the properties of beer.—*Lond. Journ. Bot.* 3. 500. The leaves of Glaphyria nitida, called by the Malays the Tree of Long Life (Kayo Umur Panjang), "probably from its maintaining itself at elevations where the other denizens of the forest have ceased to exist," afford at Bencoolen a substitute for tea ; and it is known to the natives by the name of the Tea Plant ; and various species of Leptospermum and Melaleuca bear the same name in the Australasian colonies.

The wood of Myrtleblooms is said by De Candolle to be generally white and compact ; but the heavy, hard, dark-brown timber, which furnishes the South Sea Islanders with their clubs and other weapons, is said to come from Metrosideros polymorpha, or some allied species. The Aki, or Lignum vitæ of New Zealand, the Rata, and the Pohutu Kawa of the same country, are all hard-wooded trees belonging to the genus Metrosideros.

GENERA.

I. LEPTOSPERMEÆ. — Capsular.

Astartea, *DC.*
Tristania, *R. Br.*
Syncarpia, *Tenore.*
Kamptzia, Nees.
Lophostemon, *Schott.*
Lamarchea, *Gaudich.*
Calothamnus, *Labill.*
Baudinia, Leschen.
Billiottia, Colla.
Beaufortia, *R. Br.*
Schizopleura, *Lindl.*
Manglesia, Lindl.
Conothamnus, *Lindl.*
Melaleuca, *Linn.*
Cajuputi, Adans.
Eudesmia, *R. Br.*
Asteromyrtus, *Schauer.*
Symphyomyrtus, *Schr.*
Eucalyptus, *Herit.*
Angophora, *Cav.*
Callistemon, *R. Br.*
Pentagonaster, Kltsh.
Metrosideros, *R. Br.*
Nani, Adans.
Agalmanthus, Endl.
Glaphyranthus, Endl.
Kunzea, Reichenb.
Eremæa, *Lindl.*
Billiottia, *R. Br.*
Agonis, DC.
Hypocalymma, *Endl.*
Pericalymma, *Endl.*
Salisia, *Lindl.*
Leptospermum, *Forst.*

Fabricia, *Gærtn.*
Bæckea, *Linn.*
  Imbricaria, Smith.
  Jungia, Gærtn.
  Mollia, Gmel.
  Cedrela, Lour.
Babingtonia, *Lindl.*

II. MYRTEÆ.— Baccate.

Sonneratia, *Linn. f.*
  Aubletia, Gärtn.
  Pagapate, Sonner.
  Blatti, Rheed.
Punica, *Linn.*
Neiitris, *Gärtn.*
  Decaspermum, Forst.
Campomanesia, *R. et P.*
Psidium, *Linn.*
  Guaiava, Tournef.
  Burchardia, Neck.
Rhodamnia, *Jack.*
  Monoxora, Wight ?
Glaphyria, *Jack.*
Pimenta, *Lindl.*
Myrtus, *Tournef.*
  Leucomyrtus, DC.
  Myrtillus, Endl.
  Leantria, Soland.
  Jossinia, Comm.
Rhodomyrtus, *DC.*
Myrcia, *DC.*
Syllisium, *Schauer.*
Marlierea, *St. Hil.*
Calyptranthes, *Swartz.*
  Chytraculia, P. Br.
  Zuzygium, P. Br.
  Chytralia, Adans.
  Calyptranthus, Juss.
Syzygium, *Gärtn.*
  Opa, Lour.
  Calyptranthus, Blum.
  Jambolifera, Auct.
Caryophyllus, *Tournef.*
Acmena, *DC.*
Eugenia, *Michel.*
  Plinia, Linn
  Guapurium, Juss.
  Olinthia, Lindl.
  Greggia, Gärtn.
Jambosa, *Rumph.*
  Jambos, Adans.
Cerocarpus, *Hssk.*

NUMBERS. GEN. 45. Sp. 1300.

POSITION.—Melastomaceæ.—MYRTACEÆ.—Onagraceæ.
Pomaceæ.

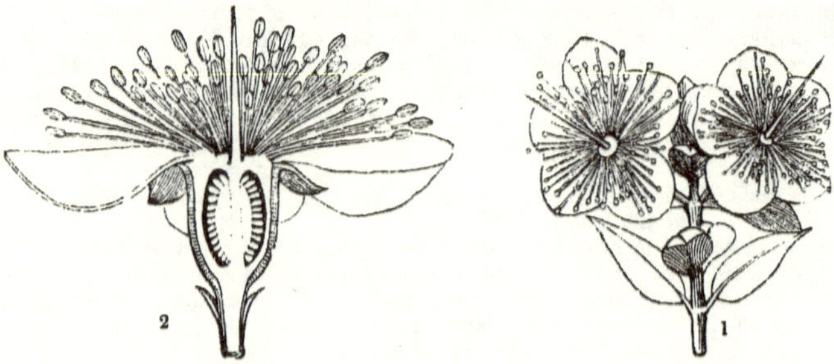

Fig. CCCCXCIII.

Fig. CCCCXCIII.—1. twig of Myrtus communis ; 2. a flower divided perpendicularly.

ORDER CCLXXXIII. LECYTHIDACEÆ.—LECYTHS.

Lecythideæ, *Richard, MSS. Poiteau Mem. Mus.* 13. 141. (1825); *DC. Prodr.* 3. 290; *a sect. of* Myrtaceæ. *Ach. Richard in Ann. des Sc.* 1. 321; *Bartl. Ord. Nat.* 332; *Martius Conspec. No.* 320. (1835); *Endl. Gen. p.* 1284; *Meisner*, 109.

DIAGNOSIS.—*Myrtal Exogens, with a plurilocular ovary, polypetalous flowers, a valvate or imbricated calyx,* 00 *stamens in part collected into a fleshy hood, oblong anthers, and dotless leaves.*

Large trees, with alternate entire or toothed leaves, minute deciduous stipules, and no pellucid dots. Flowers large, showy, terminal, solitary, or racemose. Calyx superior, 2- to 6-leaved, or urceolate, with a divided limb; æstivation valvate or imbricated. Corolla consisting of 6 petals sometimes cohering at the base, with an imbricated æstivation. Stamens indefinite, epigynous; a portion of them connected into a single petaloid cucullate unilateral body, which is sometimes quite destitute of anthers. Ovary inferior, 2- to 6-celled; ovules definite or indefinite, attached to the axis; stigma simple. Fruit a woody capsule, either opening with a lid or remaining closed. Seeds several, covered by a thick integument; embryo without albumen, either undivided, or with two large plaited leafy or fleshy cotyledons, sometimes folded upon the radicle, which is next the hilum.

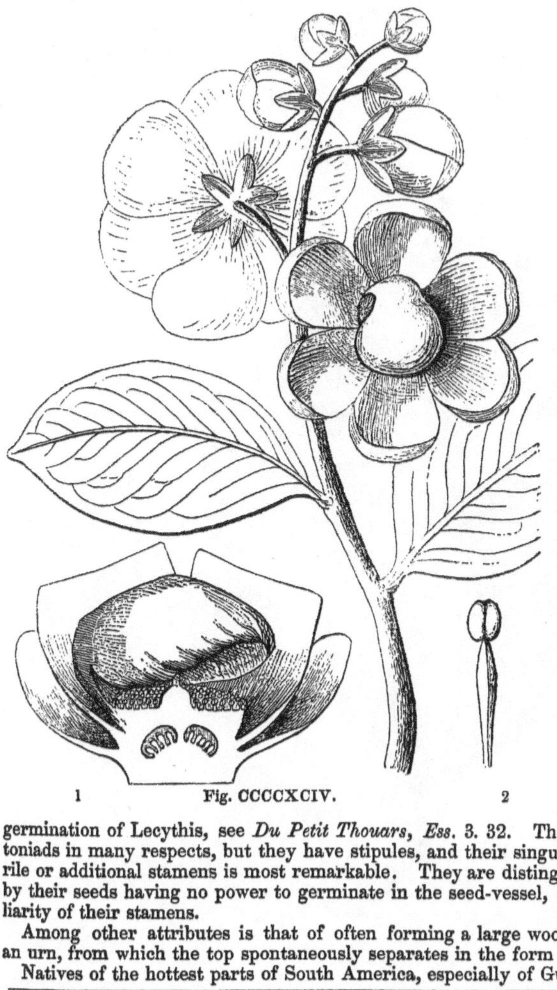

1 Fig. CCCCXCIV. 2

Combined by De Candolle and others with Myrtleblooms, from which they differ in their great almond-like seeds and alternate, often serrated leaves, without pellucid dots. For an account of the germination of Lecythis, see *Du Petit Thouars, Ess.* 3. 32. They agree with Barringtoniads in many respects, but they have stipules, and their singular hooded plate of sterile or additional stamens is most remarkable. They are distinguished from Mangroves by their seeds having no power to germinate in the seed-vessel, in addition to the peculiarity of their stamens.

Among other attributes is that of often forming a large woody fruit in the form of an urn, from which the top spontaneously separates in the form of a lid.

Natives of the hottest parts of South America, especially of Guiana.

Fig. CCCCXCIV.—Lecythis ovata.—*Aug. de St. Hilaire.* 1. a flower divided perpendicularly; 2. a stamen

The fruit of Couroupita guianensis, or the Cannon-ball tree, called Abricot sauvage in Cayenne, is vinous and pleasant when fresh, but in decay emits an insupportably offensive odour. The lacerated parts of its flowers become blue upon exposure to the air. The shells are used, like the calabash, for domestic purposes. The most gigantic tree in the ancient forests of Brazil is that called the Sapucaya. It is the Lecythis ollaria, the seeds of which are large and eatable, as are those of all the species of Lecythis, but they leave a bitter unpleasant after-taste in the mouth. A milky emulsion, prepared from the seeds of L. grandiflora, is used in Brazil in catarrhs. The bark of L. ollaria is easily separable, by beating the liber into a number of fine distinct layers, which divide so neatly from each other, that, when separated, they have the appearance of thin satiny paper. Poiteau says he has counted as many as 110 of these coatings. The Indians cut them in pieces, as wrappers for their cigars. The well-known Brazil nuts of the shops of London are the seeds of Bertholletia excelsa. The great woody pericarps of Lecythis serve as drinking-vessels.

GENERA.

Couratari, *Aubl.*
Lecythopsis, Schrank.
Cariniana, *Casar.*
Lecythis, *Löffl.*
Eschweilera, *Mart.*
Bertholletia, *Hb. et Bpl.*
Tonca, Rich.
Couroupita, *Aubl.*
Pontoppidana, Scop.
Elsholtzia, Rich.
? Crossostylis, *Forst.*

NUMBERS. GEN. 7. Sp. 38.

POSITION.—Myrtaceæ.—LECYTHIDACEÆ.—Rhizophoraceæ.
Cactaceæ ?

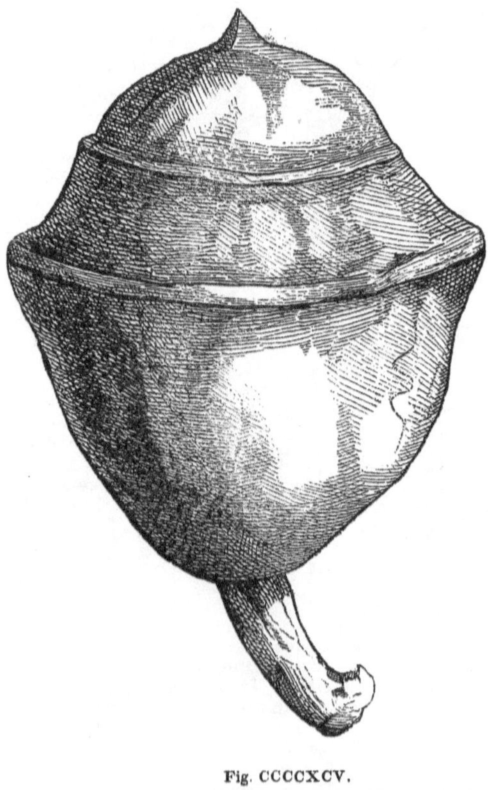

Fig. CCCCXCV.

Fig. CCCCXCV.—Fruit of Lecythis grandiflora.—*Aubl.*

ALLIANCE LII. *CACTALES.*—THE CACTAL ALLIANCE.

DIAGNOSIS.—*Epigynous Exogens, with dichlamydeous polypetalous flowers, parietal placentæ, and an embryo with little or no albumen.*

Their parietal placentation separates Cactals from all Epigynous Orders except the Grossal, and the latter is known by the minute embryo and copious firm albumen.

The Orders at first sight appear very different: but if we omit from consideration the succulence of Indian Figs, their dissimilarity disappears. For Bartonia among Loasads is much like an Epiphyllum in its flowers, and the difference between Homalium and Mentzelia or Acrolasia is inconsiderable. In fact, the glands on the calyx of Homaliads may be regarded as another form of the secondary petals of some Loasads.

The Alliance touches Cucurbits by way of such Loasads as Blumenbachia, Onagrads through Bartonia, Ficoids (Mesembryaceæ) through Indian Figs, and Passionflowers through such plants as the Homaliad Blackwellia.

NATURAL ORDERS OF CACTALS.

Sepals and petals distinct. Stamens opposite the petals. Styles separate. Ovules pendulous	284. HOMALIACEÆ.
Sepals and petals distinct. Stamens scattered. Styles confluent. Ovules pendulous. Seeds albuminous	285. LOASACEÆ.
Sepals and petals numerous, undistinguishable. Stamens scattered. Styles confluent. Ovules horizontal. Seeds without albumen . .	286. CACTACEÆ.

ORDER CCLXXXIV. HOMALIACEÆ.—HOMALIADS.

Homalineæ, *R. Brown in Congo*, (1818); *DC. Prodr.* 2. 53. (1825); *Endl. Gen.* cxcvi.; *Meisner*, p. 73.

DIAGNOSIS.—*Cactal Exogens, with distinct sepals and petals, stamens opposite the petals, separate styles, and pendulous ovules.*

Trees or shrubs. Leaves alternate, with deciduous stipules, or 0, toothed or entire. Flowers in spikes, racemes, or panicles, without bracts. Calyx funnel-shaped, adherent, with from 5 to 15 divisions. Petals alternate with the segments of the calyx, and equal to them in number. Glands present in front of the segments of the calyx. Stamens arising from the base of the petals, either singly or in threes or sixes; anthers 2-celled, opening longitudinally. Ovary adherent, 1-celled, with numerous anatropal pendulous ovules attached to 2, 3, or 5 parietal placentæ; styles from 3 to 5, simple, filiform, or subulate. Fruit berried or capsular. Seeds small, ovate, or angular, with an embryo in the middle of a little fleshy albumen, and a thick superior radicle.

Fig. CCCCXCVI.

Although these plants, with shrubby stems, small flowers, and highly-developed leaves, exhibit no other resemblance to Indian Figs than what resides in their inferior ovary, parietal placentæ, and scarcely albuminous seeds, yet, if we compare them with Loasads, their affinity becomes sufficiently evident; and as Loasads are akin to Indian Figs in the first degree, so Homaliads are akin in the second degree. That Homaliads and Loasads stand nearly on the same line, is shown by comparing such plants as Homalium with Acrolasia; and although it cannot be denied that links are wanting to render the connection between those genera complete, yet enough of resemblance exists to warrant this sort of comparison. In fact, the glands of Homalium are probably an altered form of the abortive stamina of Loasa.

According to Brown, Homaliads are related to Passionflowers, especially to Smeathmannia, from which their inferior ovary distinguishes them, to say nothing of their want of stipules and glands on the leaves, of the presence of glands at the base of the floral envelopes, and of their erect and very different habit. De Candolle places them between Samyds and Chailletiads, describing them as apetalous, but classing them with his Dichlamyds; Brown also understands them as without petals; but I confess I cannot comprehend what petals are, if the inner series of the floral envelopes of these plants be not so; an opinion which their supposed affinity with Passionflowers would confirm, if analogy could be admitted as evidence in cases which can be decided without it. The statement of De Candolle, that the stamens are opposite the sepals, is inaccurate; they are, as Brown describes them, opposite the petals.

The species are all tropical, and chiefly African or Indian. Four or five are described from the West Indies and South America.

The root of some American species of Homalium is astringent, and employed against blennorrhœa.

Fig. CCCCXCVI.—Byrsanthus Brownii.—*Delessert.* 1. diagram of the flower; 2. section of a flower; 3. section across the ovary; 4. section of a seed.

GENERA

Homalium, *Jacq.*	*Lagunczia*, Scop.	Anetia, *Endl.*	Myriantheia, *Thouars.*
Acoma, Adans.	Blackwellia, *Commers.*	*Byrsanthus*, Guillem.	Nisa, *Noronh.*
Napimoga, Aubl.	*Vermontea*, Commers.	Eriudaphus, *Nees.*	Asteropeia, *Thouars.*
Tattia, Scop.	*Asteranthus*, Lour.	Trimeria, *Harv.*	
Racoubea, Aubl.	*Pythagorea*, Lour.		

NUMBERS. GEN. 8. Sp. 30.

Passifloraceæ.
Cucurbitaceæ.
POSITION.—Loasaceæ.—HOMALIACEÆ.—Cactaceæ.
Onagraceæ.

ORDER CCLXXXV. LOASACEÆ.—LOASADS.

Loaseæ, *Juss. Ann. Mus.* 5. 18. (1804); *Dict. Sc. Nat.* 27. 93. (1823); *Kunth in Nov. Gen. et Sp.* 6. 115. (1823); *DC. Prodr.* 3. 339. (1828); *Endl. Gen.* cxcix.; *Meisner, p.* 125.—Gronovieæ, *Endl. Gen. p.* 940.

DIAGNOSIS.—*Cactal Exogens, with distinct sepals and petals, scattered stamens, confluent pendulous ovules, and albuminous seeds.*

Herbaceous plants, hispid, with pungent hairs secreting an acrid juice. Leaves opposite or alternate, without stipules, usually more or less divided. Peduncles axillary, 1-flowered. Calyx adherent, 4- 5-parted, persistent, imbricated, and spreading in æstivation. Petals 5 or 10, in two rows, often hooded, with an inflexed, valvate, or contorted æstivation; the interior often, when present, much smaller than the outer, and truncate at the apex. Stamens 00, in several rows, arising from within the petals, either distinct or adhering in bundles before each petal, within the cavity of which they lie in æstivation; filaments subulate, unequal, the outer ones frequently destitute of anthers. Ovary inferior, 1-celled, with several parietal placentæ, or one only in the centre; style single; stigma 1, or several. Ovules anatropal, pendulous, rarely 1. Fruit capsular or succulent, inferior, 1-celled, with parietal placentæ originating at the sutures. Seeds without aril; embryo lying in the axis of fleshy albumen, with the radicle pointing to the hilum, and flat small cotyledons.

1 Fig. CCCCXCVII. 2

The relationship of this Order seems to be almost equally divided between Indian Figs and Onagrads, and hence it must stand on the limits of the Myrtal and Cactal Alliances. From the former, however, it differs most, in consequence of its parietal placentation and 1-celled ovary. It is through Pereskia that it passes into Indian Figs, for that genus, if it were to lose its succulence, would almost belong to Loasads; the species of Rhipsalis too, in which there is a clear distinction between the calyx and corolla, offer another easy transition from Indian Figs to this Order, by way of Bartonia. But while such may be regarded as the most immediate affinity of Loasads, there are others so little remote as to show that among the Epigynous class we have distinct traces of a near parallelism with both the hypogynous and diclinous sub-classes. The first is indicated by the similarity in habit of Blumenbachia, &c., to Passionflowers, in connection with the great tendency which such genera exhibit to convert their stamens into petaloid processes; and, as Endlicher remarks, there is also a plain approach to Turnerads and Crownworts, two other Orders of the Violal Alliance. The relation to diclinous Endogens consists in the resemblance between Loasads and Cucurbits; a similarity so great, that little serves to distinguish them, except the diclinous flowers and short sinuous anthers of the latter; in fact, the genus Sphenantha, referred to Cucurbits, is probably a Loasad if it belongs to either the one Order or the other. Gymnotheca is a very anomalous plant, with neither calyx nor corolla; Decaisne refers it to Saururads, p. 251.

All the species are American, and chiefly from the more temperate regions, or the tropics, of either hemisphere.

Fig. CCCCXCVII.—Bartonia albicaulis. 1. a flower; 2. ring of stamens; 3. cross section of seed-vessel; 4. seeds.

Except the stinging property which resides in the hairs of some species, little is known of the qualities of these plants. Mentzelia hispida, a Mexican herb, is said to have a purgative root.

GENERA.

I. LOASEÆ.			II. GRONOVIEÆ.
Acrolasia, *Presl.* Mentzelia, *Linn.* *Oreolobus*, Lilja. *Chrysostoma*, Lilja. *Trachyphytum*, Nutt.	Bartonia, *Sims.* Klaprothia, *H. B. K.* Sclerothrix, *Presl.* Grammatocarpus, *Presl.* *Scyphanthus*, Don. Loasa, *Adans.*	Cajophora, *Presl.* *Raphisanthe*, Lilja. Blumenbachia, *Schrad.* ? Gymnotheca, *Decaisne.* ? Sphenantha, *Schr.*	Cevallia, *Lagasc.* *Petalanthera*, TORR. Mackaya, *Arn.* Gronovia, *L.* Fissenia, *R. Br.*

NUMBERS. GEN. 15. Sp. 70.

Cucurbitaceæ.
POSITION.—Homaliaceæ.—LOASACEÆ.—Cactaceæ.
Onagraceæ.

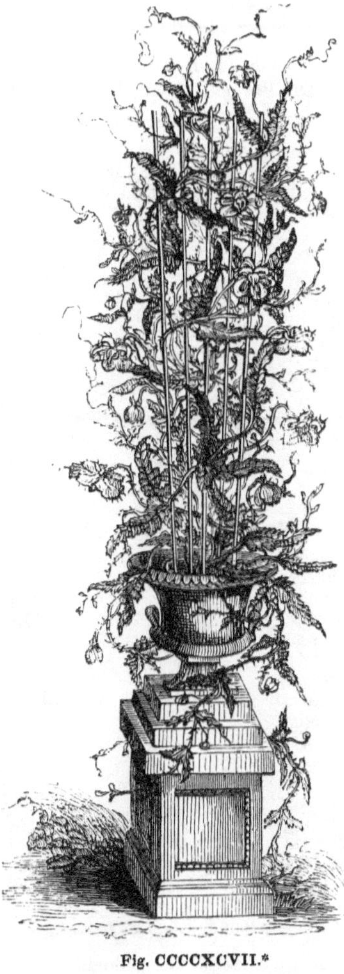

Fig. CCCCXCVII.*

Fig. CCCCXCVII *—Loasa Pentlandica.

ORDER CCLXXXVI. CACTACEÆ.—INDIAN FIGS.

Cacti, *Juss. Gen.* 310. (1789) *in part.*—Cactoideæ, *Vent. Tabl.* 3. 289. (1799).—Opuntiaceæ, *Juss. Dict. Sc.* 144. (1825) *in part*; *Kunth. Nov. G. et Sp.* 6. 65.—Nopaleæ, *DC. Théorie Elém.* 216. (1819).—Cacteæ, *DC. Prodr.* 3. 457; *Revue des Cactées Mem. Mus.* (1829). *Link and Otto in Verhand. des ver. Gart. Preuss. vol.* iii. *p.* 412. *Martius in Act. Acad. Nat. Cur. XVI.*; *Lemaire Cact. Hort. Monv.* (1838); *Id. Cactearum Gen. Nov.*; *Miquel in Bull. Sc. Phys. en Neerlande*, 1839. *p.* 89. 118; *Pfeiffer, Enum. Cact.*; *Walpers Repertorium*, 2. 269; *Salm. Dyck. Hortus Dyckensis*, 1842; *Endl. Gen.* cciv.; *Wight. Illustr.* 2. *t.* 114.; *Schleiden Beiträge zur Anatomie der Cacteen*; *Miquel in Ann. Sc. Nat.* 19. 165.

DIAGNOSIS.—*Cactal Exogens, with the sepals and petals numerous and undistinguishable, scattered stamens, confluent styles, horizontal ovules, and seeds without albumen.*

Succulent shrubs, very variable in form. Stems usually angular, or two-edged, or leafy. Wood either arranged in a ring of wedges separated by wide medullary passages, or consisting of fibres loosely interlacing, and only collecting in compact zones when old. Leaves almost always wanting; when present, fleshy, smooth, and entire, or spine-like. Flowers either showy or minute, usually lasting only one day or night, always sessile. Sepals numerous, sometimes 4, but usually indefinite, and confounded with the petals, either crowning the ovary, or covering its whole surface. Petals 4 or more, commonly numerous, usually indefinite, arising from the orifice of the calyx, sometimes irregular. Stamens indefinite, more or less cohering with the petals and sepals; filaments long, filiform; anthers ovate, versatile. Ovary fleshy, inferior, 1-celled, with numerous ovules arranged upon parietal placentæ, equal in number to the lobes of the stigma; style filiform; stigmas numerous, collected in a cluster. Fruit succulent, 1-celled, many-seeded, either smooth, or covered with scales, scars, or tubercles. Seeds parietal, or, having lost their adhesion, nestling in pulp, ovate or obovate, without albumen; embryo either straight, curved, or spiral, with a short thick radicle next the hilum; cotyledons flat, thick, foliaceous, sometimes almost obsolete in the leafless species.

Fig. CCCCXCVIII.

That remarkable distension or increase of the cellular tissue of plants, from which the name of succulent is derived, is no indication of natural affinity, but is rather to be

Fig. CCCCXCVIII.—Cereus speciosissimus. 1. section of the fruit of Opuntia Dillenii; 2. of the seed, (*Gærtner*); 3. section of seed of Mammillaria.

considered a modification of structure common to all Orders. Hence the immediate relationship of Indian Figs is neither with Spurgeworts, nor Cassyths, nor Asclepiads, nor Asphodels, all of which contain a greater or less number of succulent genera. Through Rhipsalis, which is said to have a central placenta, they seem connected with Purslanes, to which also the curved embryo of the section of Opuntias probably indicates an approach. De Candolle further traces an affinity between these plants and Ficoids (Mesembryaceæ), which correspond in their numerous petals, habit, annular embryo, and somewhat in placentation ; but which have a many-celled fruit, belonging to the perigynous rather than the epigynous category. The anatomy of Indian Figs has engaged the especial attention of Schleiden, Miquel, and others, whose observations will be found in the places above quoted. One of the most curious circumstances connected with it is that their spiral vessels are extremely short, and are formed with a spiral plate of considerable breadth and thickness, instead of a thread. For an elaborate account of this Order, see Schleiden's Memoir above quoted. Currantworts, with which Indian Figs were formerly combined, manifestly differ in a large number of points, and especially in their abundant albumen.

In this country we scarcely know the Indian Figs except as succulent ugly shrubs without leaves, but the Pereskias have leaves of a sufficiently ordinary description, and when old the columnar species form wood of considerable strength. Indeed, according to Mr. Hinds (*Ann. Nat. Hist.* xv. 100), Humboldt speaks of a forest of such plants, not mere herbaceous species, but tall trees with stems yielding wood suitable for domestic purposes. It has been well observed by Dr. Walpers (*Repertorium Botanices Systematicæ*, vol. 1. p. 269) that the confusion of species and names, in the Order of Indian Figs, is without a parallel, owing to the negligence or bad descriptions at once of writers, cultivators, and travellers, and that the so-called species are in many cases distinguished by characters of the most trifling nature.

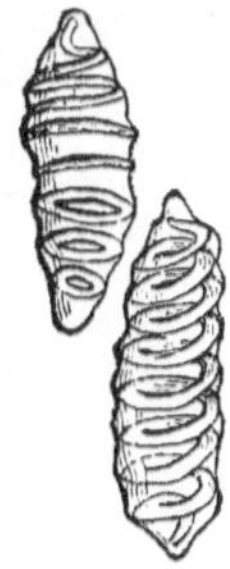

CCCCXCIX.

America is the exclusive station of the Order, no species appearing to be native of any other part of the world. In that country they are abundant in the tropics, extending a short distance beyond them, both to the north and the south. De Candolle states that 32° or 33° north latitude is the northern limit of the Order ; but it is certain that a species is either wild or naturalised in Long Island, in latitude 42° north, and that there is another somewhere about 49°, in the Rocky Mountains. Those which are said to be wild or naturalised in Europe, Mauritius, Arabia, and China, are either species of succulent Spurges, &c., or, if really Indian Figs, have been introduced from America, and having found themselves in situations suitable to their habits, have taken possession of the soil like actual natives : in Europe this does not extend beyond the town of Finale, in 44° north latitude. There is no reason for supposing that the modern Opuntia is described in Theophrastus, as Sprengel asserts ; the account of the former writer, as far as it applies to anything now known, rather suits some tree like Ficus religiosa. Hot, dry, exposed places are the favourite stations of Indian Figs, for which they are peculiarly adapted, in consequence of the imperfect evaporating pores of their skin ; a circumstance which, as De Candolle has shown, accounts for the excessively succulent state of their tissue. For geographical observations see *Martius in Ann. Sc.* 2. *ser.* 2. 110.

The fruit is very similar in its properties to that of Currants, in some being refreshing and agreeable to the taste, in others mucilaginous and insipid. Many are valued as palliatives of intermittent and bilious fevers, in consequence of their refreshing subacid juice. The fruit of Opuntia vulgaris has the property of staining red the urine of those who eat it. That of O. Tuna is of the richest carmine, and forms a valuable pigment, employed at Naples as a water-colour. The juice of Mammillaria is remarkable for being slightly milky, and at the same time sweet and insipid. Rhipsalis pachyptera has not only the succulent fruit of the Order, but is bruised and used as a fomentation for ill-conditioned ulcers.—*Martius.* The fruit of Pereskia aculeata is pleasant and expectorant. The great fleshy stems of some Mexican species are eaten by cattle ; a very remarkable specimen of this kind is described by Sir. W. Hooker, in the *Gardeners' Chronicle*, 1845, p. 132. Mr. Darwin found that a species of Cactus was one of the principal kinds of food of the land-tortoises in the Gallapagos.

FIG. CCCCXCIX.—Flat-threaded spiral vessels of Indian Figs.

GENERA.

I. MELOCACTIDÆ.

Melocactus, *C. Bauhin.*
Discocactus, *Pfeiffer.*
Anhalonium, *Lemaire.*
Ariocarpus, Sch.
Mammillaria, *Haworth.*

II. ECHINOCACTIDÆ.

Echinocactus, *L. et O.*
Astrophytum, Lem.
Pelecyphora, *C. Ehrenb.*

III. CEREIDÆ.

Echinopsis, *Zuccarini.*
Pilocereus, *Lemaire.*
Cereus, *Haworth.*

IV. PHYLLANTHIDÆ.

Phyllocactus, *Link.*
Epiphyllum, *Pfeiffer.*
Disocactus, *Lindley.*

V. RHIPSALIDÆ.

Rhipsalis, *Gærtner.*
Hariota, Lemaire.
Lepismium, *Pfeiffer.*

VI. OPUNTIDÆ.

Opuntia, *Tournefort.*

VII. PERESKIDÆ

Pereskia, *Plumier.*

NUMBERS. GEN. 16. Sp. 800 ? ?

Mesembryaceæ.
POSITION.——————CACTACEÆ.—Loasaceæ.
Grossulariaceæ.

Alliance LIII. *GROSSALES.*—The Grossal Alliance.

Diagnosis.—*Epigynous Exogens, with dichlamydeous polypetalous flowers, numerous minute seeds, and a small embryo lying in a large quantity of albumen.*

This group evidently touches the last, where Indian Figs are so like Currantworts that they used to be considered the same Order; and it passes into the next through Escalloniads, whose fruit would be that of Cranberries if it were fleshy, or through Syringas, which may be compared with Columelliads.

The Order of Barringtoniads exhibits the Alliance in its highest state of development, and effects a union with the wide frontier of Myrtals.

At the same time the Escalloniads and Currantworts extend towards the Saxifragal Alliance, especially to Hydrangeads, to which a part of the Syringas is sometimes referred.

Natural Orders of Grossals.

Fruit pulpy. Placentæ parietal	287. Grossulariaceæ.
Fruit capsular. Placentæ axile. Style and stamens definite. Calyx imbricated	288. Escalloniaceæ.
Fruit capsular. Placentæ axile. Styles disunited. Stamens 00. *Calyx valvate*	289. Philadelphaceæ.
Fruit pulpy or fibrous. Placentæ axile. Style 1. *Stamens* 00. *Calyx imbricated*	290. Barringtoniaceæ.

ORDER CCLXXXVII. GROSSULARIACEÆ.—CURRANTWORTS.

Grossularieæ, *DC. Fl. Fr.* 4. 406. (1804); *Kunth Nov. G. et Sp.* 6. 58.; *DC. Prodr.* 3. 477. (1828); *Spach in Ann. Sc. ser.* 2. *tom.* 4. *p.* 16.—Ribesiæ, *Ach. Rich. Bot. Med.* 2. 487. (1823).—Grossulaceæ, *Mirb. Elem.* 2. 897. (1815).—Ribesiaceæ, *Endl. Gen.* clxxi. (1839).

DIAGNOSIS.—*Grossal Exogens, with pulpy fruit and parietal placentæ.*

Shrubs, either unarmed or spiny. Leaves alternate, lobed, with a plaited vernation, often with a membranous edge to the base of the petioles. Flowers in axillary racemes, with bracts at their base, rarely unisexual by abortion. Calyx superior, 4- or 5-parted, regular, coloured, imbricated, or somewhat valvate in æstivation. Petals 5, minute, inserted in the throat of the calyx. Stamens 5, inserted alternately with the petals, very short. Ovary 1-celled, with 2 opposite parietal placentæ; ovules numerous, on short stalks, anatropal; style 2-3-4-cleft. Berry crowned with the remains of the flower, 1-celled; the cell filled with pulp. Seeds numerous, suspended among the pulp by long filiform cords; testa externally gelatinous, adhering firmly to the albumen, which is horny; embryo minute, with the radicle next the hilum.

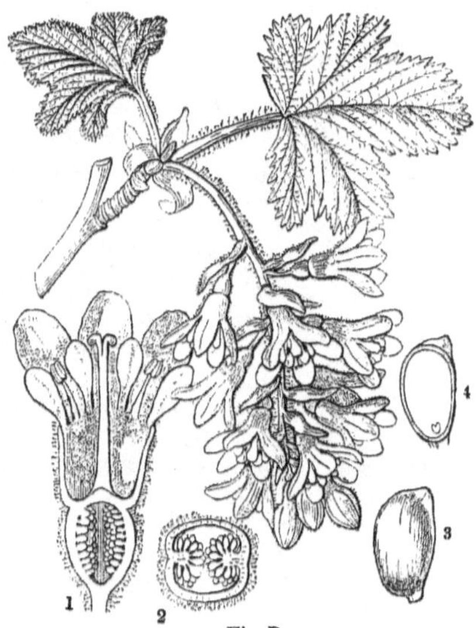

Fig. D.

Notwithstanding the great dissimilarity in the appearance of these plants and Indian Figs, the two Orders were formerly confounded, and are still accounted by many writers conterminous, chiefly on account of their both having inferior pulpy fruit and parietal placentæ. Von Martius, however, (*Conspectus*, No. 222,) abandons this view, and stations them somewhere between Saxifrages and Onagrads. In consequence of their copious albumen, polypetalous flowers, and definite stamens, I was formerly disposed to bring them into the neighbourhood of Berberries and their allies; but the strictly epigynous structure of the flowers weakens this resemblance. That they stand near Escalloniads seems undoubted, and therefore they form a transition to Cranberries, another Order in close contact with Escalloniads, but stationed in the Cinchonal Alliance because of its monopetalous corolla. The close alliance between Currantworts and Escalloniads is most distinctly shown by the genus Polyosma, which agrees with the former in its two polyspermous parietal placentæ, and with the latter in the high development of its corolla. Mr. Bennett even places it among Escalloniads.

They are natives of the mountains, hills, woods, and thickets of the temperate parts of Europe, Asia, and America, but unknown in Africa. In North America they are particularly abundant, and on the mountains of Northern India they contribute to give a European character to that remarkable region. In the tropics of Asia and the South Sea Islands they occur in the form of Polyosma, a genus which derives its name from the excessive fragrance of its flowers.

The properties of the Gooseberry and Currant are those of the generality of the Order, except that in other species a mawkish or extremely acid taste is substituted for the refreshing and agreeable flavour of the former. Some are said to be emetic and intoxicating (R. inebrians), but this statement rests on no good authority. The black

Fig. D.—Ribes rubrum. 1. perpendicular section of a flower; 2. cross section of the ovary; 3. seed; 4. a perpendicular section of it.

Currant, which is tonic and stimulant, has fragrant glands upon its leaves and flowers; these reservoirs are also found upon some other species. Malic acid exists in Currants and Gooseberries.—*Turner*, 634.

GENERA.

Ribes, *Linn.*	*Coreosma*, Spach.	*Rebis*, Spach.	*Chrysobotrya*, Spach.
Grossularia, Tournef.	*Botryocarpum*, Spach.	*Siphocalyx*, DC.	*Robsonia*, Berland.
Botrycarpum, A. Rich.	*Cerophyllum*, Spach.	*Symphocalyx*, Berland.	Polyosma, *Bl.*
Calobotrya, Spach.			

NUMBERS. GEN. 2? Sp. 95.

Vacciniaceæ.
POSITION.—Philadelphaceæ.—GROSSULARIACEÆ.—Escalloniaceæ.
Cactaceæ.

ORDER CCLXXXVIII. ESCALLONIACEÆ.—ESCALLONIADS.

Escallonieæ, *R. Brown in Franklin's Voyage*, 766. (1824); *Aug. de St. H. Fl. Bras.* 3. 92. (1833.—Saxifragaceæ, § 1. Escallonieæ, *DC. Prodr.* 4. 2. (1830); *Endl. Gen. p.* 822.—Carpodeteæ, *Fenzl. in Regensb. Denkschr.* iii. 155. *t.* 1.

DIAGNOSIS.—*Grossal Exogens, with capsular fruit, axile placentæ, definite style and stamens, and imbricated calyx.*

Shrubs, with alternate, toothed, resinously glandular, exstipulate leaves, and axillary conspicuous flowers. Calyx superior, 5-toothed. Corolla consisting of 5 petals, alternate with the segments of the calyx, from within which they arise, sometimes forming by their cohesion a tube, but finally separating; æstivation imbricated. Stamens arising from the calyx, alternate with the petals; anthers bursting longitudinally. Disk conical, epigynous, plaited, surrounding the base of the style. Ovary inferior, 2-5-celled, with a large polyspermous placenta in the axis; style simple; stigma 2-5-lobed. Fruit capsular or baccate, surmounted by the persistent style and calyx. Seeds very numerous and minute, with a transparent membranous integument; embryo minute, in a mass of oily albumen, its radicle opposite the hilum.

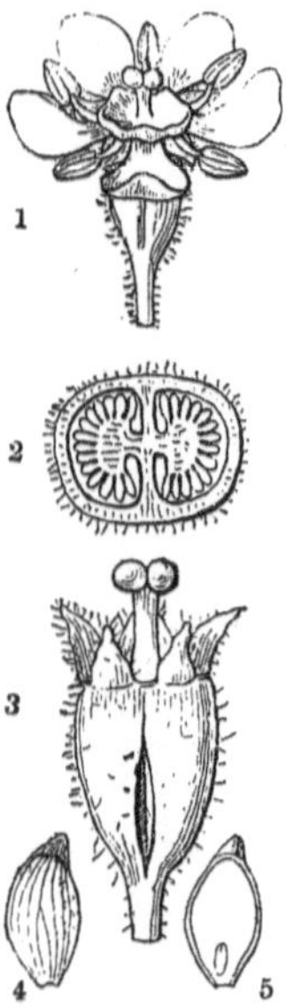

Fig. DI.

By De Candolle and others, these plants are either considered a section of Saxifrages, or are placed in the immediate vicinity of that Order; an opinion founded upon their polyspermous fruit, composed of two carpels, their polypetalous flowers with a small number of stamens, and some similarity in their habit as compared with Cunoniads, which are also often referred to Saxifrages. By other writers they are contrasted with Heathworts and Cranberries, and upon the whole they seem most closely akin to the latter of those Orders, of which they have also the habit, and almost the monopetalous corolla. A trace of resemblance to Melastomads may even be perceived in the remarkable cup-shaped epigynous disk of Escallonia. Brown, however, long since demonstrated the necessity of considering them closely allied to Currantworts, from which, indeed, they are hardly known, except by their oily albumen, dry fruit, and occasionally cohering petals; for some of them have almost parietal placentæ, as Anopterus. Of that Order they must therefore of necessity follow the station. From Bruniads they are known, firstly, by their broad leaves, lax inflorescence and larger flowers; and secondly, by their many-seeded fruit. Indeed Bruniads may, in one point of view, be regarded as a less developed form of Escalloniads.

It is said that Escallonia canescens has an embryo nearly as long as the albumen; this, if true, will be a great anomaly in the Order, and requires re-examination.

All found in the temperate parts of the world, especially South America. In countries near the equator belonging to the west side of America, Escallonias grow at the considerable elevation of 6600 to 14,760 feet, and there, with Oaks and Drymis, they form a vegetable region. They are even found as far southward as the Straits of Magellan. A few species of the Order occur in the Isle of Bourbon, the Malay Islands, and the southern parts of Australia and New Zealand.

Their properties are unknown. All are shrubs, with evergreen leaves, which have often a powerful odour.

GENERA.

Escallonia, *Mutis.*	Choristylis, *Harv.*	Anopterus, *Labill.*	*Cedrela*, Lour.
Stereoxylum, R. et P.	Forgesia, *Comm.*	Itea, *L.*	Carpodetus, *Forst.*
Quintinia, *A. DC.*	*Defforgia*, Lam.	*Diconangia*, Mitch.	? *Pseuditea*, Hassk.

NUMBERS. GEN. 7. Sp. 60.

POSITION.—Grossulariaceæ.—ESCALLONIACEÆ.—Vacciniaceæ.
(above: *Saxifragaceæ.*; below: *Bruniaceæ.*)

Fig. DI.—Escallonia pulverulenta. 1. a flower; 2. a cross section of the ovary; 3. fruit; 4. seed; 5. its perpendicular section.

ORDER CCLXXXIX. PHILADELPHACEÆ.—SYRINGAS.

Philadelpheæ, *Don in Jameson's Journal*, 133. (*April*, 1826); *DC. Prodr.* 3. 205. (1828); *Endl. Gen.* cclxiv.

DIAGNOSIS.—*Grossal Exogens, with capsular fruit, axile placentæ, disunited styles,* 00 *stamens, and valvate calyx.*

Shrubs. Leaves deciduous, opposite, toothed, without dots or stipules. Peduncles axillary or terminal, in trichotomous cymes. Flowers white or pink. Fruit sometimes a little scurfy. Calyx adherent, with a persistent limb, having from 4 to 10 valvate divisions. Petals alternate with the segments of the calyx, and equal to them in number, with a convolute-imbricate æstivation. Stamens 00, arising in one or two rows from the orifice of the calyx. Styles either distinct or consolidated into one; stigmas several; ovules 00, attached to an axile placenta. Capsule half inferior, with from 4 to 10 cells, many-seeded. Seeds scobiform, subulate, smooth, heaped in the angles of the cells upon an angular placenta, with a loose membranous skin. Albumen fleshy; embryo about as long as the albumen; cotyledons oval, obtuse, flattish; radicle longer than the cotyledons, straight, obtuse, superior or inferior, next the hilum.

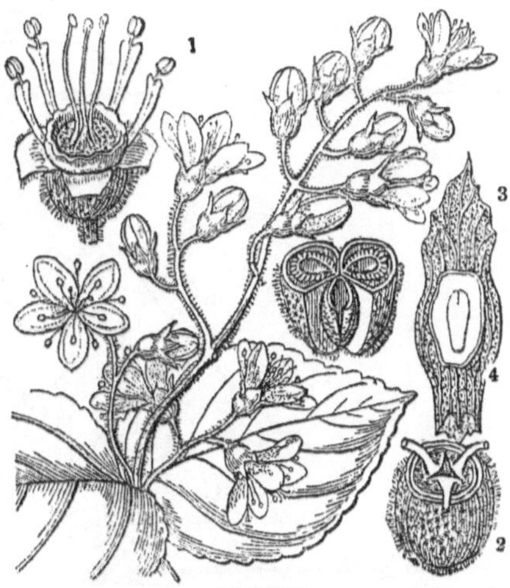

Fig. DII.

No doubt can exist that these plants have a near relation to Myrtleblooms, although they may not have such a resemblance as will justify their being stationed in the very same Alliance; for they correspond in their inferior fruit, opposite leaves, polypetalous flowers, and indefinite stamens. Some Botanists, however, because of their seeds being lodged in albumen, would rather refer them to Saxifrages, with which, no doubt, there is some collateral relationship. They are, however, evidently a portion of the Grossal Alliance, standing nearly allied to Escalloniads. Among that Alliance they are readily known by their valvate calyx, indefinite stamens, and disunited styles. Moreover, their minute seeds cut them off from Barringtoniads, and their axile placentæ, with a capsular fruit, from Currantworts.

The species are found sparingly in the south of Europe, North America, Japan, and India.

Little can be said of their uses. The rough leaves of Deutzia scabra are used in Japan by polishers, and its inner bark for poultices. Philadelphus coronarius, whose flowers have a sweet but very peculiar smell, and whose leaves taste like Cucumber, was once considered a tonic, and the oil of its flowers was used for adulterating oil of Jasmine.

GENERA.

Philadelphus, *Linn.*
Syringa, Tournef.
Decumaria, *Linn.*
Forsythia, Walt.
Deutzia, *Thunb.*

NUMBERS. GEN. 3. Sp. 25.

Myrtaceæ.
POSITION.—Grossulaceæ.—PHILADELPHACEÆ.—Escalloniaceæ.
Saxifragaceæ.

Fig. DII.—Deutzia crenata.—*Siebold.* 1. a flower with the petals removed; 2. a fruit; 3. a portion of the same, showing the placentation; 4. section of a seed very highly magnified.

ORDER CCXC. BARRINGTONIACEÆ.—BARRINGTONIADS.

Myrtaceæ, § Barringtonieæ, *DC. Prodr.* 3. 288. (1828); *Bartl. Ord. Nat.* 322. (1830).—Barringtonieæ, *DC. Dict. Class.* v. *XI. not.* (1826); *Martius Conspectus, No.* 319 (1835); *Wight Illustr.* 2. 19.

DIAGNOSIS.—*Grossal Exogens, with pulpy or fibrous fruit, axile placentæ,* 1 *style,* 00 *stamens, and an imbricated calyx.*

Trees or shrubs, with alternate, often serrated leaves, destitute of transparent dots.

2

Sepals 4- 5, superior. Petals the same number, imbricated, distinct. Stamens 00, distinct, monadelphous or polyadelphous; anthers oblong. Ovary inferior, surmounted by an epigynous disk, 2- 4- 5-celled, with axile placentæ; ovules 00; style simple; stigma somewhat capitate. Fruit fleshy, surmounted by the permanent calyx, with bony seeds lodged in pulp; embryo in the axis of copious fleshy albumen.

3

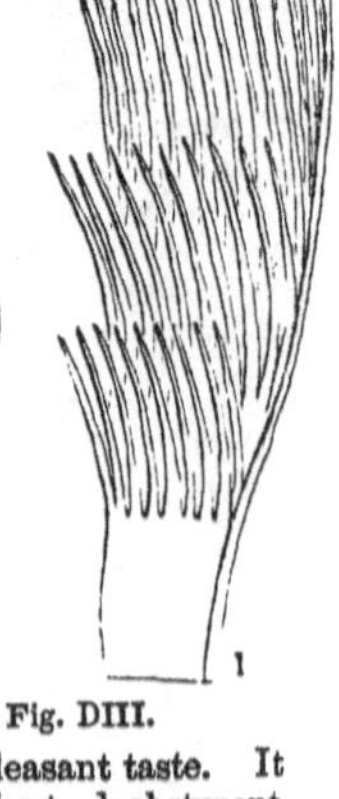

1

Fig. DIII.

Almost all writers regard these plants as kindred to the Myrtleblooms; and yet they are assuredly quite distinct, differing in the presence of a large quantity of albumen, and in having alternate leaves, without transparent dots, but often serrated. They appear in fact to be much nearer Syringas than Myrtleblooms, and to stand in the same relation to the former as Myrtles to Onagrads. By their pulpy fruit they may be looked upon as a connecting link between the Grossal and Cinchonal Alliances, among the latter of which they are imitated by the Cranberries. Roxburgh has remarked that the seeds are very like those of Guttifers.

All that are known inhabit the tropics of the old and new world, some of them occurring in low moist ground.

The root of Stravadium racemosum has a slightly bitter but not unpleasant taste. It is considered by the Hindoo Doctors valuable on account of its aperient, deobstruent,

Fig. DIII.—Careya arborea. 1. one of the bundles of stamens; 2. a perpendicular section of the ovary; 3. section of the seed.—*Wight.*

and cooling qualities ; the bark is supposed to possess properties similar to those of Cinchona. The wood of Gustavia urceolata is called Bois puant, because it soon becomes, after exposure to the air, exceedingly fœtid. The effect upon the constitution, produced by the fruit of Gustavia speciosa, is very singular. According to Humboldt and Bonpland, children are very fond of the fruit, and become quite yellow after eating it, but in 24 to 48 hours they regain their natural colour without any remedy. Martius states that the fruit of Gustavia brasiliana is emetic and intoxicates fish ; the root is acrid, aromatic, and bitter ; the leaves have a heavy unpleasant smell, and are employed in cases of indurated liver, and for bringing ulcers to a head. Endlicher says that although the fruit of Careya arborea is eaten, yet the seeds are suspicious.

GENERA.

Barringtonia, *Forsk.*	*Stravadia*, Pers.	*Pirigara*, Aubl.	? Coupoui, *Aubl.*
Butonica, Lam.	*Meteorus*, Lour.	*Spallanzania*, Neck.	? Mongesia, *Fl. Fl.*
Commersonia, Sonner.	*Menichea*, Sonner.	*Teichmeyera*, Scop.	? Grias, *L.*
Mitraria, Gmel.	Careya, *Roxb.*	Fœtidia, *Commers.*	? Petalotoma, *DC.*
Huttum, Adans.	*Cambea*, Hamilt.	? Catinga, *Aubl.*	? *Diatoma*, Lour.
Stravadium, *Juss.*	Gustavia, *Linn.*		

NUMBERS. GEN. 10. Sp. 28.

Clusiaceæ.
POSITION.—Escalloniaceæ.—BARRINGTONIACEÆ.—Philadelphaceæ.
Vacciniaceæ.

Fig. DIII.*

Fig. DIII.*—Barringtonia speciosa.—*Paxton.*

ALLIANCE LIV. *CINCHONALES.*—THE CINCHONAL ALLIANCE.

DIAGNOSIS.—*Epigynous Exogens, with dichlamydeous monopetalous flowers, and a minute embryo lying in a large quantity of albumen.*

This Alliance is known from Campanals and Myrtals by its large quantity of albumen and small embryo, from Cactals, Grossals, and Umbellals by its monopetalous corolla, from Asarals by its dichlamydeous flowers. The Orders of it are closely allied, the three last in the following enumeration being indeed separated by no very strong characters. Cranberries and Columelliads, although not usually brought up to this point, are, nevertheless, hardly separable from the Alliance; the former are a lateral tendency to Ericals.

Cinchonals are distinguishable from Umbellals by little except their monopetalous corolla, especially if Caprifoils and Cornels are compared, and therefore must participate in the undoubted affinity of the latter Alliance to Ranals; a circumstance not to be lightly esteemed in mapping out the position which the various groups of the Natural system occupy with respect to each other.

A very strong approach is shown to the diclinous sub-class, on the part of Caprifoils, among which Viburnum, minus a corolla and with an amentaceous inflorescence, would almost be a Garryad. Indeed, even a tendency to unisexuality occurs among Caprifoils, when, as in Viburnum, &c., some or all the flowers become neuter or male. And here again we have exactly the same tendency as manifests itself in so many genera of Umbellifers, which, as in Heracleum, &c., form radiant male or neuter flowers.

NATURAL ORDERS OF CINCHONALS.

Stamens epigynous; anthers opening by pores	291. VACCINIACEÆ.
Stamens epipetalous, bursting longitudinally; anthers sinuous. Flowers unsymmetrical	292. COLUMELLIACEÆ.
Stamens epipetalous, bursting longitudinally; anthers straight. Leaves with interpetiolar stipules	293. CINCHONACEÆ.
Stamens epipetalous, bursting longitudinally; anthers straight. Fruit consolidated. Leaves without stipules	294. CAPRIFOLIACEÆ.
Stamens epipetalous, bursting longitudinally; anthers straight. Fruit didymous. Leaves verticillate, without stipules . . .	295. GALIACEÆ.

ORDER CCXCI. VACCINIACEÆ.—CRANBERRIES.

Vaccinieæ, *DC. Théor. Elém.* 216. (1813) ; *Endl. Gen. p.* 757 ; *DC. Prodr.* 7. 553.

DIAGNOSIS.—*Cinchonal Exogens, with epigynous stamens and anthers opening by pores.*

Much branched shrubs or small trees, frequently evergreen, and occasionally epiphytes. Leaves alternate, undivided, without stipules, often with glandular notches. Flowers solitary or in racemes. Calyx superior, entire, or with from 4 to 6 lobes. Corolla imbricated in æstivation, monopetalous, lobed as often as the calyx. Stamens distinct, double the number of the lobes of the corolla, inserted into an epigynous disk ; anthers with 2 horns and 2 cells, bursting by pores. Ovary inferior, 4- to 10-celled ; style simple ; stigma simple. Berry crowned by the persistent limb of the calyx, succulent, 4- to 10-celled ; cells 1- or many-seeded. Seeds minute, pendulous when solitary ; embryo straight, in the axis of fleshy albumen ; cotyledons very short ; radicle long, inferior.

Fig. DIV.

It is usual to station these plants with Heathworts, to which they bear much resemblance, and of which they are no doubt the representative in the Epigynous Sub-class. They are, however, to all appearance closely allied to Cinchonads in their monopetalous flowers, inferior ovary, and albuminous seeds, and also to Escalloniads, which are chiefly known by being polypetalous. The want of adhesion between their stamens and corolla is analogous to what occurs among Columelliads. Myrtleblooms, with their dotted leaves and indefinite stamens, are very different ; but they too sometimes correspond in their anthers bursting by pores. Upon the whole, Cranberries may be considered as an Order standing on the borders of the Epigynous and Hypogynous Sub-classes, and of the Cinchonal and Grossal Alliances.

The species abound in the temperate parts of the world, especially in swampy or sub-alpine countries. Some are from the moors and marshes of Europe, others from the mountains of central Asia, many from North America, and not a few from the highlands of Peru. Some of the Peruvian species are said to be parasites.

They are chiefly known as garden shrubs. Their bark and leaves are astringent, slightly tonic and stimulating ; their berries sub-acid and pleasant to the taste. Bilberries are the fruit of Vaccinium Myrtillus, Whortleberries of V. uliginosum, Cranberries of V. Vitis idæa and the Oxycoccus palustris and macrocarpa. Many American species are substitutes for them. The people of Pasta make wine from the fruit of Thibaudia macrophylla ; that of our Vaccinium uliginosum is said to be narcotic, and to be sometimes put into beer and other liquors to make them heady ; when fermented it yields an intoxicating liquor. From the flowers of Thibaudia Quereme an aromatic tincture is prepared in Peru as a remedy for toothache.

Fig. DIV.—Vaccinium amœnum. 1. a flower ; 2. a perpendicular section of it without the corolla ; 3. a cross section of an ovary ; 4 an anther ; 5. half a seed.

GENERA.

Gaylussacia, *H. B. K.*
Lussacia, Spreng.
Sphyrospermum, *Pöpp. et Endl.*
Oxycoccos, *Tournef.*
Schollera, Roth.
Vaccinium, *Linn.*
Vitis Idæa, Tournef.
? *Adenaria*, Raf.
Decachæna, *Torr. et Gr.*
Thibaudia, *Pav.*
Agapetes, Don.
Cavinium, Thouars.
? *Acosta*, Lour.
Choupalon, Adans.
Symphysia, Presl.
Tauschia, Presl.
Andrewsia, Dunal.
Peyrusa, Rich.
Hornemannia, Vald.
Ceratostema, *Juss.*
Oreanthes, *Benth.*
Cavendishia, *Lindl.*
Macleania, *Hook.*
Anthopterus, *Hook.*
? Brossæa, *Plum.*
? Amechania, *DC.*

NUMBERS GEN. 13. Sp. 200.

Ericaceæ.
POSITION.—Cinchonaceæ.—VACCINIACEÆ.—Columelliaceæ.
Escalloniaceæ.

ORDER CCXCII. COLUMELLIACEÆ—COLUMELLIADS.

Columellieæ, *Don. in Edinb. new Phil. Journ. Dec.* (1828). — Columelliaceæ, *Ed. pr. No.* clxxxii.; *Endl. Gen. p.* 745; *Meisner, p.* 256.

DIAGNOSIS.—*Cinchonal Exogens, with epipetalous stamens, sinuous anthers bursting longitudinally, and unsymmetrical flowers.*

Evergreen shrubs, or trees. Leaves opposite, without stipules, entire or serrated. Flowers yellow, terminal. Calyx superior, 5-parted. Corolla rotate, 5-8-parted, with an imbricated æstivation. Stamens 2, inserted in the throat, alternate with the segments of the corolla; anthers roundish, 3-lobed, bursting externally, each consisting of three pairs of narrow, somewhat sinuous cells, which open longitudinally, and which are placed upon a solid fleshy connective. Ovary inferior, 2-celled, with an indefinite number of ovules; style simple, smooth; stigma capitate, 2-lobed. Disk epigynous, fleshy. Fruit capsular, 2-celled, many-seeded, with both septicidal and loculicidal dehiscence; testa polished; embryo taper, erect, in the axis of fleshy albumen, with oval obtuse cotyledons, and a taper radicle longer than the cotyledons.

1 

2 Fig. DV. 3

The late Professor Don, who first noticed this Order, thinks it near Jasmines, with which it corresponds "in the structure and æstivation of the corolla, in the bilocular ovary, and erect (?) ovules: and it agrees both with them and Syringa in the structure and dehiscence of the capsule. The Order differs, however, essentially from Jasmineworts, by having an adherent ovary, by the presence of a perigynous (?) disk, by the undivided stigma, and, lastly, by having an inferior capsule with polyspermous cells." He was probably led to this notion by having included in his Columelliads the genus Menodora, which is a genuine member of the Jasminaceous Order. He also supposed that an affinity could be traced with Halesia; and Endlicher, acting upon this supposition, has placed the genus as an anomalous form of Ebenads. Meisner adopts Don's first view. But it is very clear that none of the Orders thus referred to can really be in the neighbourhood of Columellia, which may be almost described as a monopetalous Onagrad. Its indefinite seeds are entirely at variance with all the tendencies of Ebenads and Jasmineworts, to say nothing of its inferior ovary. At the same time it is impossible to say where it really ought to stand, for there seems no great resemblance between Columellia, &c., and any other Order yet described. In this uncertainty I leave it by the side of Cranberries and Cinchonads, to either of which, and especially to the latter, it may be compared. The most striking feature in its structure is its stamens; these curious bodies, rudely represented in the *Flora Peruviana*, are apparently composed each of three stamens firmly consolidated; for each anther has 6 cells arranged in 3 pairs upon a 3-lobed fleshy connective, and turned towards the corolla. Now this indicates an irregularity of structure of a most unusual kind, and to which I find no parallel; in order to reduce such a structure to regularity, we

Fig. DV.—Columellia oblonga. 1. a flower; 2. half an ovary; 3. a fruit.

must either suppose that three more such triple stamens are abortive, and that consequently the typical number of parts in the andrœceum is 25, or we must imagine that the typical number is 10, and that each of the stamens actually developed is composed of two stamens opposite the segments of the corolla, and one alternate with them; in that case three of the stamens alternating with the lobes of the corolla, and two of those opposite the lobes will have to be supposed undeveloped. This would give us a pentamerous monopetalous flower, with twice as many stamens as parts of the corolla. Endlicher adopts this view so far as admitting the existence of three anthers to each stamen.

The species hitherto discovered are from Mexico and Peru.

They are not stated to possess any useful properties.

GENUS.

Columellia, *R. P.*
Uluxia, Juss.

NUMBERS. GEN. 1. Sp. 3.

Onagraceæ.
POSITION.—Vacciniaceæ.—COLUMELLIACEÆ.—Cinchonaceæ.

ORDER CCXCIII. CINCHONACEÆ.—CINCHONADS.

Rubiaceæ, *Juss. Gen.* 196. (1789) *for the most part; Ann. Mus.* 10. 313. (1807); *Mém. Mus.* 6. 365; *Dict. des Sciences*, 46. 385; *Schlecht. et Chamisso in Linnæa*, 3. 309; *Ach. Rich. Dissert. in Mém. Soc. h. n. Par.* 5. 81; *DC. Prodr.* 4. 341; *Royle's Illustration*, p. 287. (1835); *Endl. Gen.* cxxvii. —Lygodysodeaceæ, *Bartl. Ord. Nat.* 207. (1830); *Martius Conspectus*, *No.* 161; *Ed. Pr.* cxxvii.

DIAGNOSIS.—*Cinchonal Exogens, with epipetalous stamens, straight anthers bursting longitudinally, and leaves with interpetiolar stipules.*

Trees, shrubs, or herbs. Leaves simple, quite entire, opposite or verticillate, with interpetiolary stipules. Flowers arranged variously, usually in panicles or corymbs. Calyx adherent, with a definite number of divisions, or none. Corolla superior, tubular, regular, with a definite number of divisions, which are valvate or imbricated in æstivation and equal to the segments of the calyx. Stamens arising from the corolla, all on the same line, and alternate with its segments. Ovary inferior, surmounted by a disk, usually 2-celled, occasionally with several cells; ovules numerous and attached to a central placenta, or few and erect or ascending, anatropal or amphitropal; style single, inserted, sometimes partly divided; stigma usually simple, sometimes divided into a definite number of parts. Fruit inferior, either splitting into 2 cocci, or indehiscent and dry or succulent, occasionally many-celled. Seeds definite or indefinite; in the former case erect or ascending, in the latter attached to a central axis; embryo small, oblong, orthotropal or homotropal, surrounded by horny albumen; cotyledons thin; radicle longer, inferior.

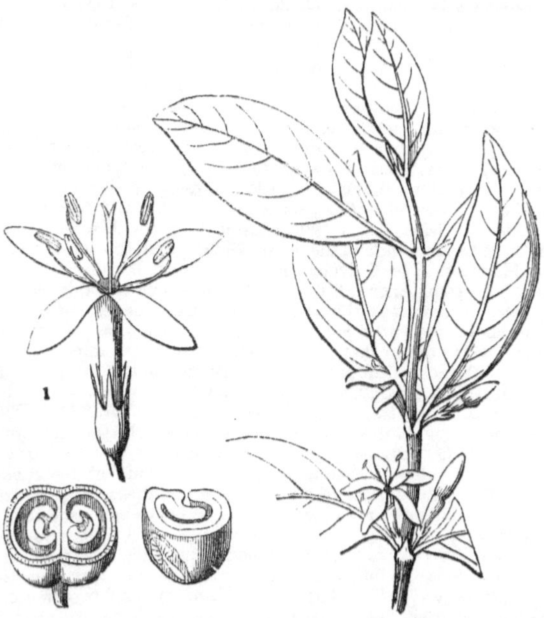

1 2 3 Fig. DVI.

This well-marked and strictly limited Order is nearly allied to Composites, from which its distinct anthers, bilocular or plurilocular ovary, abundant albumen, small embryo, and stipules distinguish it; and consequently it participates in all the relationship of that extensive Order. The inflorescence of Richardsonia and many others is that of Composites; and in the genus Argophyllum the anthers actually cohere in a tube. No doubt then can be entertained that the Campanal and Cinchonal Alliances come in contact at one part of their frontier. This is moreover strengthened by a very curious genus of the present Order, called Opercularia, which is remarkable for having but one cell in its ovary and one seed, and the number of stamens unequal to the lobes of the corolla; it occupies an intermediate position between genuine Cinchonads and Teazelworts. There is also, notwithstanding the constantly monopetalous corolla of this Order, the closest resemblance to Umbellifers in certain cases; as, for example, in Pæderia and Lygodysodea, which separate their fruit into two mericarps, adhering

Fig. DVI.—Coffea arabica. 1. a flower magnified; 2. a section across a ripe fruit; 3. a portion of a seed, showing the small embryo laid bare in the end of convolute albumen.

to a bifid thread-shaped torus, very much as in Umbellifers themselves. Loganiads (p. 602) may be regarded as Cinchonads with a free ovary. There is nothing to distinguish this Order from the Caprifoils except the stipules, and even this mark occasionally fails us. For example, in Symphoria racemosa the strong shoots are occasionally furnished with interpetiolar stipules, and of large size; an instance of which is now before me. Some of the genera have the peculiarity of forming one of the sepals in the thin, large, and gaily-coloured condition of a petal, as occurs in the genera Mussænda and Calycophyllum. Sir R. Schomburgk states that in a very fine species of the latter genus, found by him in British Guayana, the growth of this petaline sepal is very rapid, expanding to its natural size in the course of a couple of days, and only forming itself after the flower (corolla) has dropped off.—*Lond. Journ. Bot.* 3. 623.

Cinchonads are almost exclusively found in the hotter parts of the world, especially within the tropics, where they are said to constitute about 1-29th of the whole number of flowering plants. In America the most northern species is Pinckneya pubens, a shrub inhabiting the southern states of North America; some Coprosmas also occupy very low southern latitudes; the most southern is Nerteria depressa, a small herb found in the Straits of Magellan. The Order is represented in northern regions by Stellates.

This Order is not only one of the largest of which we have knowledge, but also contains a very considerable number of important species, largely employed for the use of man in the countries they inhabit. Many are among the most valuable of all remedial agents, acting as tonics, febrifuges, emetics, or purgatives. Others, on the contrary, having their secretions in a state of great concentration, prove to be formidable poisons; nevertheless, a few produce eatable fruit, and one is celebrated over all others for its agreeable stimulating seeds. Dyeing qualities are also observed in a small number. The reader who desires to occupy himself with the detailed study of the uses of this extensive Order will consult Geiger's *Handbuch*, Dierbach's *Arzneikräfte*, Endlicher's *Enchiridion*, the *Flora Medica*, and the works on Materia Medica by the two Martiuses, Fée, Guibourt, Pereira, Nees v. Esenbeck, Ebermaier, &c. A few of the principal examples are all that need be mentioned in this place.

Foremost among febrifuges and tonics stand the various Peruvian species of Cinchona, of which C. micrantha and Condaminea are the best. To these succeed the Remijas of Brazil, which are in that country species of great importance. Buena hexandra bark is an indifferent febrifuge, known in Brazil under the name of China. The bark of French Guiana, possessing properties analogous to those of Cinchona, is obtained from Portlandia hexandra, the Coutarea speciosa of Aublet. The Quinquina Piton and Quinquina des Antilles are produced by species of the genus Exostema, and are remarkable for possessing properties similar to those of true Quinquina, but without any trace of either cinchonine or quinine. A kind of fever bark is obtained at Sierra Leone from Rondeletia febrifuga. Besides these, a great number of other species possess barks more or less valuable: Pinckneya pubens is the fever bark of Carolina; Condaminea corymbosa, Guettarda coccinea, Antirhea, and Morinda Royoc, are all of the same description. Of Hymenodictyon excelsum, an East Indian shrub, the inner bark possesses the bitterness and astringency of Peruvian bark, and when fresh in a stronger degree. The bitterness is not so quickly communicated to the taste, on chewing the bark, but is much more durable, especially about the upper part of the fauces. Ophiorhiza Mungos is so intensely bitter that the plant is called by the Malays Earth-gall; according to Kæmpfer, the taste resembles Gentian, but is more penetrating. The root and bark of Guettarda Angelica are aromatic and acrid, and are used as febrifuges and astringents in the veterinary practice of Brazil.

As simple astringents the most remarkable is the Uncaria Gambir. An extract called Gambier is prepared by the Malays from the leaves of this shrub; with some sweetness, it has a more astringent taste than Terra Japonica. Roxburgh considered it one of the drugs, if not the only one, formerly called by that name in Europe. The extract is chewed by the natives with Betel-leaf and Areca; the leaves are chewed to relieve aphthous eruptions of the mouth and fauces. Mr. Pereira considers this Gambier not to form any of the Kinos of the shops, but to be one of the substances called Catechu in commerce. The root and bark of Antirhæa verticillata are said to be powerfully astringent. In Bourbon it is employed as a styptic to restrain hæmorrhage, and is known by the name of Bois de Losteau. A decoction of the leaves as well as root of Canthium parviflorum is prescribed in India in certain stages of flux, and the last is supposed to have anthelmintic qualities, though neither have much sensible taste or smell. The bark and young shoots are also used in dysentery.

Among the purgatives or emetics, Ipecacuanha holds the first rank; it is the root of Cephäelis Ipecacuanha, a little creeping-rooted, half-herbaceous plant, found in damp shady forests in Brazil. It is also sudorific and expectorant. Its powder acts upon the respiratory passages as an irritant, producing spasmodic asthma. In some cases the mere odour of the root seems sufficient to excite difficulty of breathing, with a feeling of suffocation. Similar properties are found in the roots of other Cinchonads of the same country, as in Richardsonia rosea and scabra, Geophila reniformis, Borrerias, Spermacoce ferruginea and Poaya, &c. The Raiz Preta, which is celebrated for its power in curing dropsy, and in destroying the dangerous consequences of bites of serpents, is

3 Fig. DVI.* 1

said to be related to Ipecacuanha. The spurious barks called Quinquina Piton are capable of exciting vomiting. The powdered fruit of Randia dumetorum is a powerful emetic; an infusion of the bark of the root is administered to nauseate in bowel complaints. The bark of the root of Manettia cordifolia is esteemed in Brazil a most valuable remedy in dropsy and dysentery. It is given in powder in doses of ½ to 1½ drachm, and acts as an emetic. The fruit of Gardenia campanulata is regarded in India as a cathartic and anthelmintic. The fœtid leaves of Pæderia fœtida are used to medicate baths, and in decoction are administered internally in retention of urine, and in certain febrile complaints. According to Roxburgh, the root is used by the Hindoos as an emetic. The roots of Chiococca anguifuga and densifolia, the one a Brazilian trailing herb, the other a woody bush, are employed with confidence by the natives of Brazil as a certain remedy for serpent bites. The infusion of the bark of the root produces the most violent emetic and drastic effects. In the words of Von Martius :—"Ægrotus scilicet, e veneno languidus, soporosus, vix sui compos, ex quo medicinam sumserit, primum eructationibus creberrimis et tantis motibus convulsivis excruciatur, ut, licet exsanguis et quasi cum facie hippocraticâ, sub summâ virium labe in lectulum corruisset, ne unicum quidem temporis momentum quietus maneri possit. Tandem post plurimos et visu terribiles spasmos universales et corporis volutationes, in enormes rapitur vomitus, quibus salivam, bilem, chymum, immo fœces largâ copiâ edit. Tunc accedunt subitaneæ alvi excretiones fæcum quasi succo viscido involutarum quæ, si continua per aliquot temporis serie sese excipiant, cum visibili ægroti levamine, boni exitus pro indiciis habentur." Copious perspirations follow, and these are succeeded by a gentle sleep. The violent action of these roots renders them dangerous to employ, except in cases of poisoning, or in such maladies as require a prompt and complete evacuation of the intestines.

It may easily be supposed that secretions producing such powerful action as that just described would, if a little modified or augmented in force, become dangerous poisons, and accordingly we find several species of Cinchonads included in the class of deleterious agents. Sir R. Schomburgk assures us that Indians have been poisoned by using the wood of Evosmia corymbosa to make spits for roasting meat upon.—*Hooker's Journal*, 3. 219. According to Roxburgh, the root of Randia dumetorum, bruised and thrown into ponds where there are fish, intoxicates them like Cocculus indicus. Psychotria noxa, and Palicourea Marcgraavii, both called Erva de rata, are accounted poisonous in Brazil, where they and other species of the same genera are employed for the destruction of rats and mice. Cephäelis ruelliæfolia is venomous, and used for the same purpose.

Fig. DVI.*—Richardsonia scabra. 1. an ovary with its calyx; 2. a corolla; 3. a vertical section of a seed, with an erect embryo in copious albumen.

An eatable fruit is furnished by a few species. The Genipap, a South American fruit as large as an Orange, of a whitish-green colour, but containing a dark purple juice with an agreeable vinous taste, is borne by Genipa americana. Sarcocephalus esculentus is the Native Peach of Sierra Leone. Vangueria edulis, or Voa-vanga, is said to be a good dessert fruit in Madagascar. Genipa brasiliensis is also eaten in Brazil, but Martius says that it is only fit for table after becoming bletted, and that it is better when preserved with sugar than when fresh. Some of the bushes called in Tasmannia Native Currants are Coprosmas ; but they are not of good quality.

Coffee is the roasted seeds of a plant of this Order, Coffea arabica, and is supposed to owe its stimulating, refreshing characters to a peculiar chemical principle called Caffein, which modern chemists pronounce to be the same as Theine. The part roasted is the albumen, which is of a hard horny consistence ; and it is probable that the seed of other plants of this or the stellate Order, whose albumen is of the same texture, would serve as a substitute. This would not be the case with those with fleshy albumen.

Among dyeing plants we have Oldenlandia umbellata; whose roots are a substitute for Madder in the East Indies; Psychotria Simira, whose bark stains red in Brazil; Genipa brasiliensis, whose fruit strikes a deep violet; Condaminea tinctoria, Hydrophylax maritima, and a few others of little consequence.

The fragrance or beauty of the flowers of some of the plants of this Order, especially of the Gardenias, Hindsias, Posoquerias, Ixoras, Cinchonas, Bouvardias, Catesbæas, &c. is unsurpassed in the vegetable kingdom, and forms a strange contrast with the Spermacoces, Richardsonias, &c., which are among the meanest weeds we know.

GENERA.

I. COFFEÆ. — Ovary with only 1 or 2 seeds in each cell.

I. OPERCULARIDÆ.

Pomax, *Soland.*
Opercularia, *A. Rich.*
Cryptospermum, Yng.

II. ANTHOSPERMIDÆ.

Anthospermum, *Linn.*
Ambraria, Heist.
? *Crocyllis*, E. Mey.
? *Lagotis*, E. Mey.
Ambraria, *Cruse.*
Nenax, Gärtn.
Galopina, *Thunb.*
Oxyspermum, Eckl.
Phyllis, *L.*
Coprosma, *Forst.*
Leptostigma, *Arn.*

III. SPERMACOCIDÆ.

Putoria, *Pers.*
Plocama, *Ait.*
Placoma, Pers.
Bartlingia, Rchb.
Scyphiphora, *Gärtn. fil.*
Hydrophylax, *Linn. fil.*
Sarissus, Gärtn.
Cuncea, *Hamilt.*
Ernodea, *Sw.*
Wiegmannia, *Meyen.*
Serissa, *Commers.*
Dysoda, Lour.
Buchozia, Herit.
Democritea, *DC.*
Octodon, *Thonn.*
Borreria, *Mey.*
Bigelowia, Spr.
Chlorophytum, Pohl.
Spermacoce, *Linn.*
Hexasepalum, *Bartl.*
Diodia, *L.*
Triodon, *DC.*
Crusea, *Cham. et Schl.*
Pentanisia, *Harv.*
Diotocarpus, Hochst.
Richardsonia, *Kunth,*
Richardia, Linn.
Schiedea, Bartl.
Mitracarpum, *Zuccar.*
Schizangium, Bartl.
Staurospermum, Thon.
Perama, *Aubl.*
Mattuschkea, Schreb.
Staelia, *Cham.*
Tessiera, *DC.*
Psyllocarpus, *Mart.*
Diodois, Pohl.
Gaillionia, *A. Rich.*
Ptychostigma, *Hochst.*
Otiophora, *Zuccar.*
Knoxia, *Linn.*
Jaubertia, *Guillem.*
Machaonia, *Humb.*
Emmeorhiza, *Pohl.*
Endlichera, Presl.
Deppea, *Cham. et Schl.*
Cruckshanksia, *H. et Arn.*
Rotheria, Meyen.
Cephalanthus, *L.*

IV. PSYCHOTRIDÆ.

Geophila, *Don.*
Cephaëlis, *Sw.*
Callicocca, Schreb.
Ipecacuanha, Arruda.
Tapogomea, Juss.
Callicocca, DC.
Evea, Aubl.
Carapichea, *Aubl.*
Patabea, *Aubl.*
Salzmannia, *DC.*
Chasalia, *Commers.*
Palicourea, *Aubl.*
Galvania, Vandell.
Stephanium, Schreb.
Colladonia, Spr.
Psychotria, *L.*
Psychotrophum, P. Br.
Myrtiphyllum, P. Br.
Ronabea, *Aubl.*
Viscoides, Jacq.
Mapouria, *A. Rich.*
Mapouria, Aubl.
Simira, Aubl.
Antherura, *Lour.*
Rudgea, *Salisb.*
Feretia, *Del.*
Galiniera, *Del.*
Codonocalyx, *Miers.*
Suteria, *DC.*
Coffea, *Linn.*
Hornia, DC.
Pancrasia, DC.
Straussia, DC.
Strempelia, *A. Rich.*
Faramea, *A. Rich.*
Tetramerium, DC.
Potima, Pers.
Darluca, Rafin.
Antoniana, Tuss.
Macrocalyx, *Miers.*
Rytidea, *DC.*
Grumilea, *Gärtn.*
Polyosus, *Lour.*
Coussarea, *Aubl.*
Billardiera, Vahl.
Fröhlichia, Vahl.
Saprosma, *Blume.*
Pavetta, *L.*
Pavate, Ray.
Ixora, *L.*
Baconia, *DC.*
Verulamia, DC.
Chomelia, *Jacq.*
Scolosanthus, *Vahl.*
Antacanthus, L.C. Rich
Saldinia, *A. Rich.*
Margaris, *DC.*
Chiococca, *P. Br.*
Tertrea, *DC.*
Schiedea, A. Rich.
Declieuxia, *H. B. K.*
Psyllocarpus, Pohl.
Eumachia, *DC.*
Siderodendron, *Schreb.*
Nescidia, *A. Rich.*
Plectronia, *Linn.*
? *Mitrastigma*, Harv.
Psilostoma, *Klotsch.*
Canthium, *L.*
Psydrax, Gärtn.
? *Krausia*, Harv.
Mitriostigma, *Hochst.*
Diplospora, *DC.*
Marquisia, *A. Rich.*
Damnacanthus, *Gärtn. fil.*
Amaracarpus, *Blume.*

V. PÆDERIDÆ.

Pæderia, *L.*
Reussia, Dennst.
Lecontea, *A. Rich.*
Lygodysodea, *R. et Pav.*
Disodea, Pers.

VI. GUETTARDIDÆ.

Morinda, *Vaill.*
Chrysorhiza, DC.
Myrmecodia, *Jacq.*
Hydnophytum, *Jacq.*
Hypobathrum, *Blume.*
Nertera, *Banks.*
Gomezia, Mutis.
Erythrodanum, Thou.
Mitchella, *L.*
Chamædaphne, Mitch.
Baumannia, *DC.*
Mephitidia, *Reinw.*
Lasianthus, Jacq.
Vangueria, *Commers.*
Vanguiera, Pers.
Vavanga, Rohr.
Meyenia, Lk.
Guettarda, *Vent.*
Cadamba, Sonner.
Halesia, P. Br.
Dicrobotryon, Willd.
Laugeria, Jacq.
Ullobus, DC.
Viviania, Rafin.
Malanea, *Aubl.*
Cunninghamia, Schreb.
Antirrhœa, *Commers.*
? *Neuropora*, Commers.
Stenostomum, *Gärtn. fil.*
Sturmia, Gärtn.
Stenostemum, Juss.
Sacconia, *Endl.*
Crusea, A. Rich.
Chione, DC.
Timonius, *Rumph.*
Bobea, Gaudich.
Burneya, Cham.
Erithalis, Forst.
Eupyrena, *Wgt. et Arn.*
Pyrostria, Roxb.
Santia, *Wight et Arn.*
Psathyra, *Commers.*
Chicoinea, Commers.
Hamiltonia, *Roxb.*
Spermadictyon, Roxb.

Leptodermis, *Wall.*
Myonima, *Commers.*
Pyrostria, *Commers.*
Octavia, *DC.*
Lithosanthes, *Blume.*
Erithalis, *P. Br.*
Retiniphyllum, *Humb.*
Nonatelia, *Aubl.*
Oribasia, Schreb.
Gynochtodes, *Blume.*
Cœlospermum, *Blume.*
Ancylanthus, *Desf.*
Pachystigma, *Hochst.*
Hylacium, *Palis.*
Phallaria, *Schumach.*
Cuviera, *DC.*
Dondisia, *DC.*
Stigmanthus, *Lour.*
Stigmatanthus, R. et S.
Strumpfia, *Jacq.*
Strumphia, Pers.
Epithinia, *Jacq.*
Commianthus, *Benth.*
Tricalysia, *A. Rich.*

II. CINCHONEÆ.—
Ovary many-seeded.

VII. HAMELIDÆ.

Evosmia, *Humb. et Bon.*
Tepesia, *Gærtn. fil.*
Sabicea, *Aubl.*
Schwenkfeldia, Willd.
Schizostigma, *Arn.*
Holostyla, *DC.*
Stylocoryna, Labill.
Axanthes, *Blume.*
Maschalanthe, Blume.
Wallichia, Reinw.
Urophyllum, *Jack.*
Wallichia, Roxb.
Lachnosyphonium, *Hoch.*
Hamelia, *Jacq.*
Duhamelia, Pers.
Schradera, *Vahl.*
Fuchsia, Swartz.
Urceolaria, Willd.
Brignolia, *DC.*
Patima, *Aubl.*
Polyphragmon, *Desf.*

VIII. ISERTIDÆ.

Isertia, *Schreb.*
Posanthus, Rafin.
Bruinsmannia, *Miq.*
Rhyssocarpus, *Endl.*
Gonzalea, *Pers.*
Gonzalagunia, R. et P.
Anthocephalus, *L. C. Rich.*
Cephalidium, A. Rich.
Metabolus, *Blume.*
Sclerococcus, Bartl.

IX. HEDYOTIDÆ.

Dentella, *Forst.*
Lippaya, Endl.
Gonotheca, *Blume.*
Hedyotis, *Lam.*
Diplophragma, Wight.
Macrandria, W. et A.
Dimetia, Wight et Arn.
Anotis, DC.
Didymotoce, Endl.
Ambloma, E.
Eurhaphe, E.
Panetos, Rafin.
Houstonia, L.
Poiretia, Gmel.
Amphiotis, DC.
Edrissa, Raf.
? *Pentotis*, Torr.
Ereicotis, *DC.*
Scleromitrion, W. et A.
Oldenlandia, *L.*
Gerontogea, Cham.
Listeria, Neck.
Kohautia, *Cham. et S.*
Kadua, *Cham. et Schl.*
Rhachicallis, *DC.*
Lucya, *DC.*
Dunalia, Spr.
Karamyschewia, *F. et M.*
Tula, *Adans.*
Spiradiclis, *Blume.*
Leptopetalum, *Hook.*
Ophiorrhiza, *L.*
Lipostoma, *Don.*
Virecta, *DC.*
Pentas, *Benth.*
Sipanea, *Aubl.*
Virecta, Linn. f.
Ptychodea, Willd.
Carphalea, *Juss.*
Greenia, *Wight et Arn.*
Lerchea, *L.*
Codaria, L.
? *Xanthophytum*, Bl.
Wendlandia, *Bartl.*
Adenosacme, *Wall.*
Hindsia, *Benth.*
Rondeletia, *Blume.*
Petesia, P. Br.
Lightfootia, Schreb.
Willdenowia, Gmel.
Arachnimorpha, Desv.
Choristes, *Benth.*
Spallanzania, *DC.*
Isidorea, *A. Rich.*
Bikkia, *Reinw.*
Cormigonus, Rafin.
Portlandia, *P. Br.*
Schreibersia, *Pohl.*
Augusta, Pohl.
Augustea, DC.
Lindenia, *Benth.*
Siphonia, Benth.
Chimarrhis, *Jacq.*
Macrocnemum, *P. Br.*
Condaminea, *DC.*

X. CINCHONIDÆ.

Calycophyllum, *DC.*
Pinckneya, *L. C. Rich.*
Bouvardia, *Salisb.*
Houstonia, Andr.
Christima, Rafin.
Æginetia, Cavan.
Manettia, *Mutis.*
Nacibea, Aubl.
Conotrichia, A. Rich.
Lygistum, P. Br.
Danais, *Commers.*
Alseis, *Schott.*
Exostemma, *L. C. Rich.*
Pitonia, DC.
Brachyanthemum, DC.
Hymenodictyon, *Wall.*
Kurria, Hochst.
Luculia, *Sweet.*
Lasionema, *Don.*
Remijia, *DC.*
Macrocnemum, Vell.
Cinchona, *L.*
Cosmibuena, *R. et Pav.*
Buena, Pohl.
Hymenopogon, *Wall.*
Hillia, *Jacq.*
Fereiria, Vandell.
Ferdinandusa, *Pohl.*
Ferdinandea, Pohl.
Coutarea, *Aubl.*
Stevensia, *Poit.*
Crossopteryx, *Fenzl.*
Sulipa, *Blanc.*
Nauclea, *Linn.*
Platanocarpum, Endl.
Mitragyne, Korth.
? *Mamboya*, Blanc.
Nauclearia, DC.
? *Acrodryon*, Spr.
Pentacoryna, DC.
Uncaria, Schreb.
Agylophora, Neck.
Ourouparia, Aubl.
Adina, Salisb.

XI. GARDENIDÆ.

Sarcocephalus, *Afzel.*
Cephalina, Thonn.
Zuccarinia, *Blume.*
Lucianea, *DC.*
Canephora, *Juss.*
Breonia, *A. Rich.*
Catesbæa, *Linn.*
Aspidanthera, *Benth.*
Hoffmannia, *Sw.*
Argostemma, *Wall.*
Pomangium, Reinw.
Neurocalyx, *Hook.*
Higginsia, *Pers.*
O-Higginsia, R. et Pav.
Petunga, *DC.*
Higginsia, Blum.
? *Spicillaria*, A. Rich.
Fernelia, *Commers.*
Coccocypselum, *Swartz.*
Sicelium, P. Br.
Tontanea, Aubl.
Bellardia, Schreb.
Condalia, Ruiz et Pav.
Petesia, *Bart.*
Stylocoryne, *Cav.*
Wahlenbergia, Blume.
Cupia, DC.
Chomelia, Linn.
Zamaria, Rafin.
Tarenna, Gærtn.
Pouchetia, *A. Rich.*
Bertiera, *Aubl.*
Zaluzania, Commers.
Mycetia, Reinw.
Hippotis, *Ruiz et Pav.*
Helospora, *Jack.*
Menestoria, *DC.*
Heinsia, *DC.*
Chapelieria, *A. Rich.*
Griffithia, *W. et Arn.*
Hyptianthera, *W. et Arn.*
Randia, *Houst.*
Oxyceros, Lour.
Ceriscus, Gärtn.
Euclinia, DC.
Rhodostoma, *Scheidw.*
Gardenia, *Ell.*
Kumbaya, Endl.
Piringa, Juss.
Thunbergia, Mont.
Chaquepiria, Salisb.
Rothmannia, Thunb.
Lasiostoma, *Benth.*
Genipa, *Plum.*
Duroia, Lin. f.
Conosiphon, *Pöpp.*
Sphinctanthus, Benth.
Oxyanthus, *DC.*
Posoqueria, *Aubl.*
Cyrtanthus, Schreb.
Solena, Willd.
Posoria, Rafin.
Tocoyena, *Aubl.*
Ucriana, Willd.
Gynopachys, *Bl.*
Cassupa, *H. B. K.*
Kutchubæa, *Fisch.*
Mussænda, *L.*
? *Neurocarpæa*, R. Br.
Landia, Comm.
Caanthe, DC.
Alberta, *E. Mey.*
Acranthera, *Arn.*
Gardeniola, *Cham.*
Thileodoxa, *Cham.*
Melanopsidium, Cels.
Viviania, Coll.
Billiottia, DC.
Scepseothamnus, *Cham.*
Alibertia, *A. Rich.*
Genipella, L. C. Rich.
Cordiera, *A. Rich.**
Amaioua, *Aubl.*
Hexactina, Willd.
? *Ehrenbergia*, Spr.
Burchellia, *R. Br.*
Bubalina, Ehrenb.
* * *
Sommera, *Schlecht.*
Anisomeris, *Presl.*
Psilobium, *Jack.*
Platymerium, *Bartl.*
Lecananthus, *Jack.*
Morelia, *A. Rich.*
Jackia, *Wall.*
Zuccarinia, Spr.
Himatanthus, *Willd.*
Aidia, *Lour.*
Sickingia, *Willd.*
Stipularia, *Palis.*
Benzonia, *Schum.*
Myrioneuron, *R. Br.*
Pleotheca, *Wall.*
Egeria, *Neraud.*
Meretricia, *Neraud.*

NUMBERS. GEN. 269. Sp. 2500.

Apiaceæ. Asteraceæ.
POSITION.—Galiaceæ.—CINCHONACEÆ.—Caprifoliaceæ.
Loganiaceæ.

* These changes in the position of a few genera have been made at the suggestion of Mr. Miers.—*June*, 1845.

Order CCXCIV. CAPRIFOLIACEÆ.—Caprifoils.

Caprifolia, *Juss. Gen.* 210. (1789) *in part.*—Caprifoliaceæ, *Rich. Dict. Class.* 3. 172; *DC. Prodr.* 4. 321; *Bartl. Ord. Nat.* 213. (1830).—Lonicereæ, *Endl. Gen.* cxxviii.

Diagnosis.—*Cinchonal Exogens, with epipetalous stamens, straight anthers bursting longitudinally, consolidated fruit, and leaves without stipules.*

Shrubs or herbaceous plants, with opposite leaves, destitute of stipules. Flowers usually corymbose, and often sweet-scented. Calyx superior, 4- 5-cleft, usually with 2

2

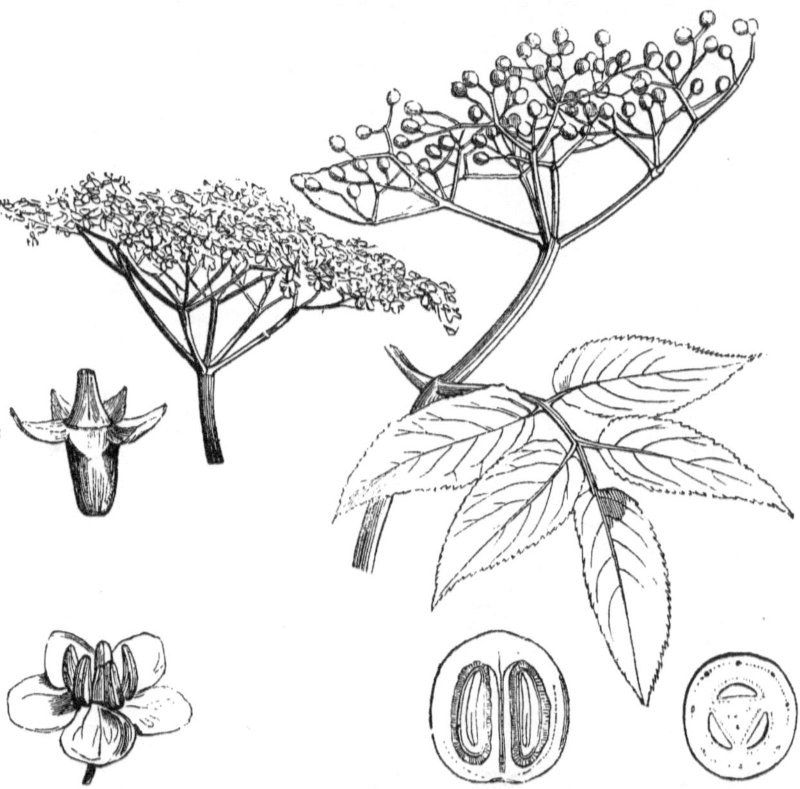

1 Fig. DVII. 4 3

or more bracts at its base. Corolla superior, monopetalous or polypetalous, rotate or tubular, regular or irregular. Stamens epipetalous, equal in number to the lobes of the corolla, and alternate with them. Ovary with from 1 to 5 cells, one of which is often 1-seeded, the others being many-seeded; in the former the ovule is pendulous; style 1; stigmas 3, or 5. Fruit indehiscent, 1- or more-celled, either dry, fleshy, or succulent, crowned by the persistent lobes of the calyx. Seeds either solitary and pendulous, or numerous and attached to the axis; testa often bony; embryo very small, in fleshy albumen; radicle next the hilum.

As left by Jussieu this Order was a heterogeneous assemblage; as altered and better limited it seems to be less objectionable. It possesses a striking affinity with Cincho-

Fig. DVII.—Sambucus nigra; 1. a flower; 2. a young pistil; 3. a cross section of its ovary; 4. a perpendicular section of the fruit.

nads, in the monopetalous tubular corolla, definite stamens, inferior ovary, and opposite leaves, an affinity which is confirmed by the corolla of the latter being occasionally regular or irregular. In fact the resemblance between them in habit, structure, inflorescence, and sensible properties is so great that there seems to be no certain character to distinguish them except the stipules of Cinchonads; for the character derived from the presence of one ovule only in one cell, and of many ovules in two other cells, although very striking in Linnæa and Abelia, disappears in others, especially in Leycesteria, whose ovary has 5 polyspermous cells; and yet that genus cannot be possibly separated from Caprifoils. Their epigynous structure divides them from Dogbanes, which have much resemblance in habit: Loranths, once mixed with them, have no petals. But if we consider the Sub-order called Sambuceæ, our view of the affinities of the Order takes a different turn, and we find an approach to Saxifragals: this is established through the intervention of Hydrangea, which is undistinguishable in habit from Viburnum, with which it accords in inflorescence and in the constant disposition of its flowers to become radiant, but from which it differs in being polypetalous and polyspermous, and only half epigynous. Besides these points of resemblance, Caprifoils probably tend towards Umbellifers through Sambuceæ.

Fig. DVIII. 5

Natives of the northern parts of Europe, Asia, and America, passing downwards within the limits of the tropics; found very sparingly in northern Africa, and little known in the southern hemisphere.

The fragrance and beauty of Honeysuckles have been celebrated by poets of every age; but independently of such a recommendation, the Order of Caprifoils possesses properties of considerable interest. The flowers of the Elder are fragrant and sudorific, its leaves fœtid, emetic, and a drastic purgative; qualities which are also possessed by Viburnum Opulus (the Gueldres Rose), several other species, and even by the Honeysuckle itself. The leaves of Linnæa borealis are praised by the Swedes as diaphoretic and diuretic. The inner bark of Viburnum Lantana is so acrid as to be included by some writers among vesicants. The fruit of Viburnum is destitute of these properties, but has, instead, an austere, astringent pulp, which becomes eatable after fermentation, and is made into a sort of cake by the North American Indians. Triosteum perfoliatum is a mild cathartic; in large doses it produces vomiting. Its dried and roasted berries have been used as a substitute for Coffee. The berries of Lonicera cærulea are a favourite food of the Kamtchadales. The wine made from the Elder, Sambucus nigra, is well known in England, and is used as a means of adulterating Port-wine. According to Mr. Backhouse, there is a species of Elder in Tasmannia, which has large cymes of white sweetish fruit, respecting which nothing deleterious has been observed.

GENERA.

I. LONICEREÆ.

Linnæa, *Gronov.*
  Obolaria, Siegesb.
Abelia, *R. Br.*
Symphoricarpus, *Dill.*
  Symphoria, Pers.
  Anisanthus, Willd.
Leycesteria, *Wall.*
Diervilla, *Tournef.*
  Weigela, Thunb.
  Calysphyrum, Bung.
Alsenosmia, *Cunn.*
Caprifolium, *Tournef.*
  Periclymenum, Tourn.
Lonicera, *Desf.*
  Xylosteon, Juss.
  Cobœa, Neck.
  Nintooa, Sweet.
  Chamæcerasus, Tourn.
  Cuphantha, DC.
  Isica, Mönch.
Triosteum, *Linn.*

II. SAMBUCEÆ.

Viburnum, *Linn.*
  Solenotus, DC.
  Lentago, DC.
  Tinus, Tournef.
  Opulus, Tournef.
Sambucus, *Tournef.*
  Phyteuma, Lour.
Tripetelus, *Lindl.*
? Valentiana, *Raf.*
? Karpaton, *Raf.*

NUMBERS. GEN. 14. Sp. 220.

Cornaceæ.
POSITION.—Cinchonaceæ.—CAPRIFOLIACEÆ.—Columelliaceæ.
Hydrangeaceæ.

Fig. DVIII.—1. flower of Linnæa borealis; 2. the same cut open and showing the interior of the ovary; 3. a cross section of the ovary; 4. section of fruit of Lonicera tatarica; 5. half its seed.

ORDER CCXCV. GALIACEÆ.—STELLATES.

Stellatæ, *Ray Synops.* 223. (1690); *R. Brown in Congo,* (1818).—Galieæ, *Turp. in Atlas du Nouv. Dict. des Sc.* (?)—Rubiaceæ, § Stellatæ, *Cham. et Schlecht. in Linnæa,* 3. 220. (1828); *DC. Prodr.* 4. 580; *Bartl. Ord. Nat.* 209; *Endl. Gen. p.* 522; *Meisner, p.* 173.—Rubiaceæ, § Galieæ, *N. ab Es. et Fuhlrott. Nat. Pflanz. Syst.* 165. (1829).

DIAGNOSIS.—*Cinchonal Exogens, with epipetalous stamens, straight anthers bursting longitudinally, didymous fruit, and verticillate leaves without stipules.*

Herbaceous plants, with whorled leaves, destitute of stipules, and angular stems. Flowers minute. Calyx superior, obsolete, or 4- 5- or 6-lobed. Corolla monopetalous,

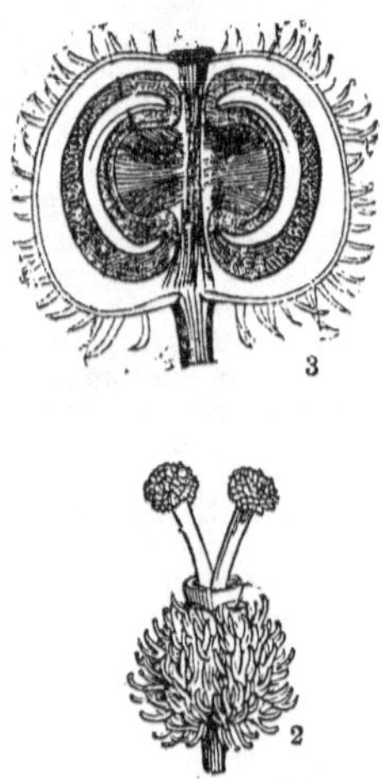

Fig. DIX.

valvate, rotate, or tubular, regular, inserted into the calyx; the number of its divisions equal to those of the calyx. Stamens equal in number to the lobes of the corolla, and alternate with them. Ovary peltate or 2-celled; ovules solitary, erect; styles 2; stigma simple. Fruit a didymous, indehiscent pericarp, with 2 cells and 2 seeds. Seeds erect or peltate, solitary; embryo in the axis of horny albumen; radicle inferior; cotyledons leafy.

There can be little doubt that the inconspicuous weeds of which this Order is composed have as strong a claim to be separated from Cinchonads as that Order from Caprifoils. It is true that no very positive characters are to be obtained from the fructification, but the want is abundantly supplied by the square stems and verticillate leaves without stipules, forming a kind of star, from which circumstance the name Stellate is derived. Nevertheless, Botanists in most intances appear to be against this opinion: I confess I cannot conceive upon what grounds. Usually a material dissimilarity in habit, if accompanied by any clear character, whether of vegetation or fructification, is considered sufficient for the separation of a group of plants into two Orders; in this case the weak angular stems

Fig. DIX.—Galium Aparine; 1. a flower; 2. a young fruit without the corolla; 3. a perpendicular section of a ripe fruit.

cause a peculiarity of habit that cannot be mistaken, and the total absence of stipules, to say nothing of the didymous fruit, affords a certain mark of recognition. Surely there is some inconsistency in separating, by the absence of stipules, Caprifoils, which are undistinguishable in habit, while the very same character is rejected when applied to an assemblage of genera all distinctly combined by their habit. The only ground upon which this is intelligible is that taken by De Candolle and others, who consider the apparent leaves of Stellates to be in part true leaves and in part leaf-like stipules. To this verbal but not real distinction there is this objection, which I conceive quite fatal to it:—If a part of the leaves of each whorl in Galium was stipules, the latter must bear a certain proportion to the true leaves; suppose the whorl to consist of two leaves, each will have two stipules, and consequently the whole number of parts in the whorl must be six, and in all cases the number must be some power of 3. But of the first forty species of Galium in De Candolle's *Prodromus* only thirteen conform to this rule; and the frequent tendency in the whorls to vary from 4 to 6, or from 5 to 6, or from 6 to 8, seems to me an incontrovertible proof that the apparent leaves of Stellates are true leaves and not a modification of stipules. To this it may be added, that the admitted leaves are so entirely the same as what M. De Candolle conceives to be stipules, that no difference whatever can in general be found in their form, colour, anatomy, or degree of development. Such reasons have, however, not proved satisfactory to Botanists, who with one accord appear to range themselves upon the side of M. De Candolle; and recently the question has been more particularly agitated by one of the most distinguished writers of this country.

Mr. Bentham, in an article on Crusea rubra, published in the *Botanist*, page 82, after entering at some length and with great skill into a discussion of the arguments employed on both sides of the question, has decided in favour of the opinion of De Candolle, that a part of the apparent leaves of Stellate plants are stipules. The grounds upon which he has arrived at this conclusion are essentially the following:—

1. That the foliaceous organs in Stellates, if viewed as consisting entirely of leaves, do not bear that relation to the angles of the stem which is usual in Dicotyledons; but that the relation becomes apparent if only two of them are taken as leaves and the rest as stipules. (De Candolle seems influenced by the consideration that it is only two of the apparent leaves which have buds in their axils; but Mr. Bentham does not advert to this.)

2. That in a number of cases, especially in Asperula, two opposite leaves are much larger than the others.

3. That in Spermacoceæ and other tribes of Cinchonads, the stipules are connected with the petiole of the leaf into a sheath, and that this sheath exists in Stellates.

4. That the number of parts in each whorl is not necessarily some power of 3, but that, taking two of the parts for leaves, it is immaterial by what number of similar parts those two are separated, because the intermediate processes are analogous to the setæ of Spermacoceæ, the number of which is variable.

Perhaps this question is more important in appearance than in reality, for in some respects it is a mere difference about words; stipules being rudimentary leaves, and leaves developed stipules. It is, however, connected with some points of speculative interest, especially as regards systematic Botany, and therefore I avail myself of the present opportunity of stating what I conceive to be the objections to Mr. Bentham's line of argument, and why I still retain my original opinion upon the subject.

1. With regard to the relation borne by the leaves to the angles of the stem, it is to be observed, that if those foliaceous organs only which are opposite the angles are said to be leaves in Stellates, and the rest stipules, then we must suppose that Labiate plants have no leaves, but stipules only, for in that and similar Orders the apparent leaves are never opposite the angles of the stem, but are always placed between them. Nor do I find that the number of angles in the stem of verticillate plants necessarily corresponds with the number of their leaves; for example, in Dysophylla stellata, where the whorls often consist of ten parts, the stem has still but four angles. Neither can it be admitted that bodies which do not form branches in their axils are therefore not leaves. All foliaceous organs, of whatever kind, and especially stipules, possess that power or not, according to circumstances, as is too well known to require particular proof. Besides, De Candolle's statement is not sustained by fact; for in Asperula the uppermost branches, bearing flowers, are frequently produced alternately with the leaves that form the node from which they spring, and consequently must, in such cases, arise from the seat of one of the supposed stipules. It is more probable that the development of branches from a portion only of the leaves is connected with the form of the stem, and the relation which the leaves bear to each other. If the form of the stem requires an alternate development of a pair or a triplet of opposite branches, then the first whorl in which the development takes place will settle the origin of all that succeed it. For example, if in one whorl of six leaves,

the first, third, and fifth leaves produce axillary buds, then in the whorl next above it, the second, fourth, and sixth leaves will probably be gemmiferous, according to the ordinary laws of decussation. It is plainly impossible to say that what seem to be leaves are in reality stipules, because they have no axillary buds; for if that opinion were maintained, it would be necessary to assign the quality of stipules to a certain portion of the leaves of such verticillate plants as Dysophylla stellata, in which only a part of the whorls ever produces branches.

2. If it is true that in Asperula two opposite leaves are frequently longer than the others, that circumstance may be reasonably ascribed to the greater development consequent upon their higher functions, and to their peculiar position on the stem; and it is equally true that in the greater part of Stellates no trace whatever of any kind of difference between the leaves can be detected, as is most remarkably the case in those surrounding the flowers of Crucianella maritima.

3. The argument derived from the occasional connection of the leaves by a membrane can hardly be allowed much weight, when it is remembered that in such cases the intermediate leaves are less like stipules than in those cases where no membrane exists; compare Asperula cynanchica, or littoralis, or longiflora, with such genuine Crucianellas as C. maritima.

4. The comparison of the supposed stipules of Stellates and the setæ of Spermacoceæ is inadmissible, because the former are at all events single simple organs, be they what they may, while the setæ of Spermacoceæ are the result of the splitting of two parallel-veined stipules, and therefore will necessarily be uncertain in number.

These arguments do not, however, by any means exhaust the question; and therefore I proceed to make a few additional remarks upon a point not yet adverted to. It is in Asperula, more than in any other genus of the Order, that is to be found evidence favourable to the supposition of M. De Candolle and his followers. In A. longiflora, cynanchica, and some others, the lower whorls are in the usual state, but the upper ones are reduced to two perfect leaves, with one or sometimes two teeth or subulate processes between them, which remain. In this condition the structure of Asperula is so very like that of many Spermacoceous plants, that the analogy between them seems indisputable; and I presume that it was such cases which first led to the theory under consideration.

It is, however, to be remembered, that in Stellates the supposed stipules are always what first disappear in the process of reduction in the number of foliaceous appendages; but that in Cinchonads it is in many cases the leaves which are first lost when such a reduction takes place. The latter fact is readily verified upon reference to any of the capitate Spermacoces, where the bracts are evidently stipules, and especially to S. calyptera, in which the leaves are gradually merged in the large membranous cup that subtends the flowers, while the stipules suffer no diminution. The same circumstance may be observed in several Brazilian Cinchonads allied to Psychotria barbiflora, and in Pæderia fœtida. It is also possible that the large coloured involucrum of Cephaëlis is, at least in some cases, formed by the excessive development of stipules and suppression of the leaves, for such is undoubtedly the case in a Sierra Leone plant in my possession, which I presume is the little-known C. bidentata of Thunberg. These facts render it more probable than ever that Stellates and Cinchonads are essentially different Natural Orders; for they would seem to show that while the first has verticillate foliaceous organs, the most imperfect of which have the greater tendency to disappear, the second has verticillate foliaceous organs, the most perfect of which have the greater tendency to become abortive. I need scarcely add, that after a full consideration of this point I retain my original conviction, that the apparent leaves of Stellates are really leaves, and not stipules, and that the Order is as distinct from Cinchonads as Nightshades from Figworts, Verbenes from Labiates, and I might even add, as Cinchonads themselves from Umbellifers.—See *Bot. Reg.* 1838. 55. To be consistent, then, we must either combine Caprifoils with Cinchonads, or we must preserve Stellates separate. Properly speaking, the appellation Rubiaceæ should be confined to the latter group, as it comprehends the genus Rubia; but that name has been so generally applied to the larger mass now comprehended under the name of Cinchonads, that I find it better to abolish that of Rubiaceæ altogether.

Natives of the northern parts of the northern hemisphere, where they are extremely common weeds, and of high mountainous regions in Peru, Chili, and Australasia.

First among them stands Madder, the root of Rubia tinctoria, one of the most important dyes with which we are acquainted; a quality in which other species of Stellates participate in a greater or less degree. The roots of Rubia cordifolia (Munjista, *Roxb.*) yield the Madder of Bengal, and form even an article of the export commerce to Europe, under the name of Munjeeth. Rubia angustissima, from Tong Dong, has also highly-coloured roots, and Rubia Relboun is the Madder of Chili. It has been remarked that the whole system of animals fed on Madder becomes stained red in consequence. Madder, in addition to its valuable dyeing qualities, passes for a tonic, diuretic, and emmenagogue. The torrefied grains of Galium are said to be a good substitute for coffee. The flowers of Galium verum are used to curdle milk. An infusion of Asperula cynanchica has a little astringency, and has been used as a gargle. Asperula odorata, or Woodruff, is remarkable for its fragrance when dried; it passes for a diuretic. Rubia noxa is said to be poisonous. M. Miergues, a French physician, states that he has cured epilepsy with the extract of Galium rigidum, by employing it in doses of twelve grammes for an adult; and he adds that G. Mollugo has been used with success in the same malady.

GENERA.

Vaillantia, *DC.*
  Valantia, Tournef.
Callipeltis, *Stev.*
  Cucullaria, Buxb.
Galium, *L.*
  Aspera, Mönch.
  Eyselia, Neck.
  Aparine, Tournef.
  Cruciata, Tournef.
Rubia, *Tournef.*
Crucianella, *L.*
  Rubeola, Mönch.
  Laxmannia, S. G. Gmel.
Asperula, *L.*
Karamyschewia, *Fisch.*
Sherardia, *Dill.*
  Dillenia, Heist.

NUMBERS. GEN. 8. Sp. 320.

POSITION.—Cinchonaceæ.—GALIACEÆ.—*Cornaceæ.* / *Apiaceæ.*

ALLIANCE LV. *UMBELLALES.*—THE UMBELLAL ALLIANCE.

DIAGNOSIS.—*Epigynous Exogens, with dichlamydeous polypetalous flowers, solitary large seeds, and a small embryo lying in a large quantity of albumen.*

The combination of a polypetalous corolla, an inferior fruit, and solitary seeds chiefly consisting of albumen, constitute the distinctive character of this Alliance, whose Orders can by no means be separated, whatever mode of general distribution a Botanist may employ. In fact, Umbellifers differ from Ivyworts in nothing except their peculiar epigynous disk, and didymous fruit. Ivyworts are hardly distinguishable from Cornels, if we neglect the opposite leaves and tetramerous flowers of the latter; and from Witch Hazels there is little to separate Cornels, except the valvate corolla and exstipulate leaves of the latter; finally, Bruniads rely for their definition more upon their want of stipules, and anthers turned outwards than on anything else.

If we look to the affinities of this Alliance, we shall again have an instance of a most natural group being so touched at all points of its circumference that it may be almost regarded as a peculiar centre of organisation, from which many other groups diverge. Thus, in a direct line, Umbellifers touch Stellates on the one hand, and Ivyworts on the other, as is elsewhere explained. Then in lateral affinity we have Umbellifers closing in upon Crowfoots, and stretching towards Saxifrages, Ivyworts almost invading the territory of Vineworts and Caprifoils, Cornels owing their position as a distinct Order, rather than as a mere group of Garryads or Alangiads ? chiefly to their unisexual dichlamydeous flowers on the one hand, and their valvate corolla on the other. Witch Hazels have, no doubt, a strong relationship to Mastworts (Corylaceæ) on the one hand, and Hippurids on the other, and finally, the affinity of Bruniads to Myrtleblooms is sufficiently shown in speaking of their Natural Order ; so that the following may be taken as a representation of the way in which the Natural Orders of Umbellals stand with respect to others.

Galiaceæ.

Saxifragaceæ	Apiaceæ	*Ranunculaceæ.*
Caprifoliaceæ	Araliaceæ	*Vitaceæ.*
Alangiaceæ	Cornaceæ	*Garryaceæ.*
Haloragaceæ	Hamamelidaceæ	*Corylaceæ.*
Rhamnaceæ	Bruniaceæ.	*Myrtaceæ.*

Santalaceæ.

NATURAL ORDERS OF UMBELLALS.

Fruit didymous, with a double epigynous disk	296. APIACEÆ.
Fruit not didymous, without a double epigynous disk, 3- or more-celled. Pentamerous flowers. Corolla valvate. Leaves alternate, without stipules. Anthers turned inwards, opening lengthwise	297. ARALIACEÆ.
Fruit not didymous, without a double epigynous disk, 2- or more-celled. Tetramerous flowers. Corolla valvate. Leaves opposite, without stipules	298. CORNACEÆ.
Fruit not didymous, without a double epigynous disk, 2-celled. Corolla imbricated. Leaves alternate, with stipules. Anthers with deciduous valves	299. HAMAMELIDACEÆ.
Fruit not didymous, without a double epigynous disk, 3- (or 1-) celled. Corolla imbricated. Leaves alternate, without stipules. Anthers turned outwards, opening lengthwise . .	300. BRUNIACEÆ.

ORDER CCXCVI. APIACEÆ.—UMBELLIFERS.

Umbelliferæ, *Juss. Gen.* 218. (1789); *Koch in N. Act. Bonn.* 12. 73.; *DC. Mémoire* (1829); *DC. Prodr.* 4. 55.; *Tausch. in Bot. Zeit.* (1834); *Ann. Sc. n. s.* 4. 41.; *Endl. Gen.* clxii.; *Meisner, Gen.* 139. —Umbellaceæ, *Lindl. Key, No.* 11, (1835).

DIAGNOSIS.—*Umbellal Exogens, with didymous fruit, and a double epigynous disk.*

Herbaceous plants, often milky, with solid or fistular furrowed stems. Leaves usually divided, sometimes simple, sheathing at the base, occasionally with close simple parallel

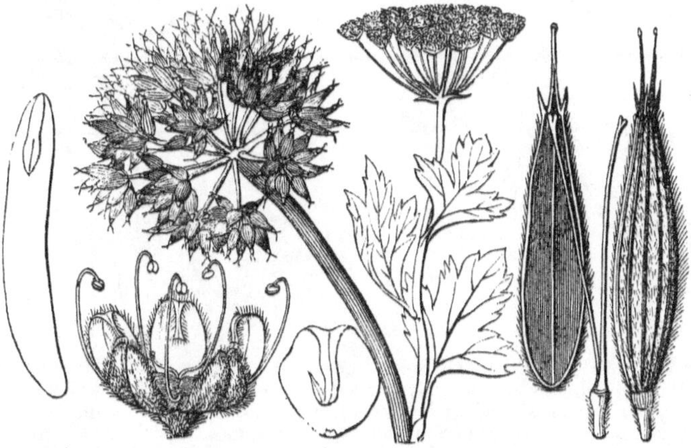

Fig. DX.

veins. Flowers in umbels, white, pink, yellow, or blue, generally surrounded by an involucre. Calyx superior, either entire or 5-toothed. Petals 5, inserted on the outside of a fleshy epigynous disk; usually inflexed at the point; æstivation imbricate, rarely valvate. Stamens 5, alternate with the petals, incurved in æstivation. Ovary inferior, 2-celled, with solitary pendulous ovules crowned by a double fleshy disk; styles 2, distinct; stigmas simple. Fruit consisting of 2 carpels, separable from a common axis, to which they adhere by their face (the commissure); each carpel traversed by elevated ridges, of which 5 are primary, and 4, alternating with them, secondary; the ridges are separated by channels, below which are often placed, in the substance of the pericarp, certain linear receptacles of coloured oily matter called vittæ. Seed pendulous, usually adhering inseparably to the pericarp, rarely loose; embryo minute, at the base of abundant horny albumen; radicle pointing to the hilum.

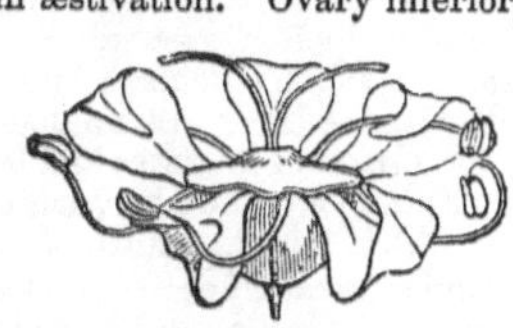

Fig. DXI.

If Botanists form their ideas of an Umbellifer from the ordinary appearance of such plants in Europe, they will have a very imperfect idea of the singular forms which the genera sometimes assume, unless they take Hydrocotyle, Astrantia, and Eryngium as the chief objects of consideration. Instead of the herbaceous and often fistular stem, they become solid branched bushes; for compound umbels, panicles and racemes are substituted (as in Horsfieldia), and the little involucres, which we almost overlook, become the most conspicuous part of the whole structure. Take, for example, on the one hand, the singular Leucolæna rotundifolia, with its great white 3-lobed plates surrounding the flowers, and on the other the not less singular Bolax glebaria, whose tufts of close entangled shoots are described as resembling haystacks, and which D'Urville tells us might deceive the most

Fig. DX.—Athamanta cervariæfolia. 1. a separate flower, with hairy petals; 2. a petal by itself; 3. a ripe fruit with the two carpels or mericarps separating from the double carpopod or axis; 4. a seed deprived of its integuments, and divided vertically, so as to show the position of the embryo.
Fig. DXI.—Flower of Angelica.

experienced eye, so much are they at variance with the usual structure of Umbellifers In all these cases, however, the very peculiar condition of the flower and fruit is abundantly sufficient to mark the Order. Indeed we have no knowledge of any one group so entirely free from deviations from the typical structure, ex-

Fig. DXII.

Fig. DXIII.

cept in accidental monsters. Of these I once found an instance, at Burnham Priory, near Maidenhead, in which the calyx was detached from the ovary, which had become superior, the calyx surrounding it loosely like a 5-toothed ribbed cup. This is quite inconsistent with the theory of Schykoffsky, who assumes that in Umbellifers the calyx proceeds really from the same point as the styles.—*Bot. Reg.* 1841. *Misc.* 35.

It is also stated that in some accidental cases 3 carpels have been found. In Eryngium and some Bupleurums the leaves consist of nothing but petiole, and thus present the appearance of Endogens. Among the more remarkable facts connected with the structure of the fruit are, 1, the separation of it when ripe into 2 carpels or mericarps, adhering to a stylopod or forked placenta, eventually exterior to the carpels themselves, although in the beginning it must have been included between their confluent margins, between which it rose till near the summit of the cavity, when it turned inwards to bear the solitary ovules ; and 2, the presence in the pericarp of fistular passages filled with oil ; the latter are no doubt analogous to the cysts of Orange and other leaves, and to the glands of Labiates and some Composites, but they are remarkable for a uniformity in position and number, which, although not absolute, is nevertheless very different from the indefinite nature of common cysts.

Umbellifers differ from Ivyworts in their seed adhering to the pericarp, in their imbricated corolla, and in the two divisions into which their dry fruit always resolves itself eventually. Ivyworts, on the contrary, have a loose seed, a valvate corolla, and more divisions of their succulent fruit than two. The genus Horsfieldia, however, forms a complete transition, having the valvate corolla of Ivyworts and their peculiar habit, with the dimerous dry fruit and adherent seed of Umbellifers. As to their other affinities it may be remarked, that they completely represent in the epigynous sub-class the Crowfoots among hypogynous Exogens ; some Thalictrums indeed would make pretty good Umbellifers, if their calyx adhered to the side of the ovary. They approach Stellates in their didymous inferior fruit and copious albumen, but they are universally polypetalous. With Saxifrages Umbellifers agree in habit, if Hydrocotyle is compared with Chrysosplenium, and if the sheathing and divided leaves of the two Orders are considered. To Cranesbills De Candolle remarks that Umbellifers are allied, in consequence of the cohesion of the carpels around a woody axis, and of the umbellate flowers which grow opposite the leaves, and also because the affinity of Cranesbills to Vines, and of the latter to Ivyworts, is not to be doubted. The resem-

Fig. DXII.—Leucolæna rotundifolia.—*Hooker*.
Fig DXIII.—Bolax glebaria.—*Hooker*.

blance of Umbellifers to Cranesbills is however very feeble. Endlicher compares them, and justly, to Cornels ; in fact, the little Cornus suecica, and the whole genus Benthamia, have exactly the involucre and inflorescence of Umbellifers, in addition to their other points of resemblance.

The arrangement of this Order has only within a few years arrived at any very definite state, the characters upon which genera and tribes could be formed having been for a long while unsettled ; it is, however, now generally admitted that the number and development of the ribs of the fruit, the presence or absence of reservoirs of oil called vittæ, and the form of the albumen, are the leading peculiarities which require to be attended to. Upon this subject see Koch's *Dissertation*, Lagasca in the *Otiosas Españolas*, and De Candolle's *Mémoire*, especially the last. The classification of De Candolle has, however, been criticised by Tausch, in the places above quoted, who asserts that the albumen is a fallacious guide. He says that some species of Bupleurum are campylospermous, and others orthospermous, and that the same is true of many other genera. He adds, that in Hasselquistia the fruit of the ray is orthospermous, while that of the disk is cœlospermous. The arrangement which this author proposes to substitute has not yet been examined critically. It must, however, be obvious to every experienced Botanist that the genera and tribes are alike unsatisfactory, and that the arrangement of Umbellifers upon sound principles still remains to be achieved.

Natives chiefly of the northern parts of the northern hemisphere, inhabiting groves, thickets, plains, marshes, and waste places. They appear to be extremely rare in all tropical countries, except at considerable elevations, where they gradually increase in number as the other parts of the vegetation acquire an extra-tropical, or mountain character. Hence, although they are hardly known in the plains of India, they abound on the mountains of the Himalaya. They are, however, not uncommon in the southern hemisphere, where they belong principally to Hydrocotylids and Mulinids.

The Umbelliferous is one of those large Orders in which plants occur with extremely different secretions. They all appear to form three different principles: the first a watery acrid matter, the second a gum-resinous milky substance, and the third an aromatic oily secretion. When the first of these predominates they are poisons ; the second in excess converts them into stimulants ; the absence of the two renders them useful as esculents ; the third causes them to be carminatives and pleasant condiments. A vast number of species are referred by writers to one or other of these categories. Without pretending to go into any detailed enumeration of the qualities, real or asserted, of the endless species at one time or other used by man, (for which the reader must consult Endlicher's *Enchiridion*, Geiger's *Handbuch*, the works of Nees v. Esenbeck and Ebermaier, and

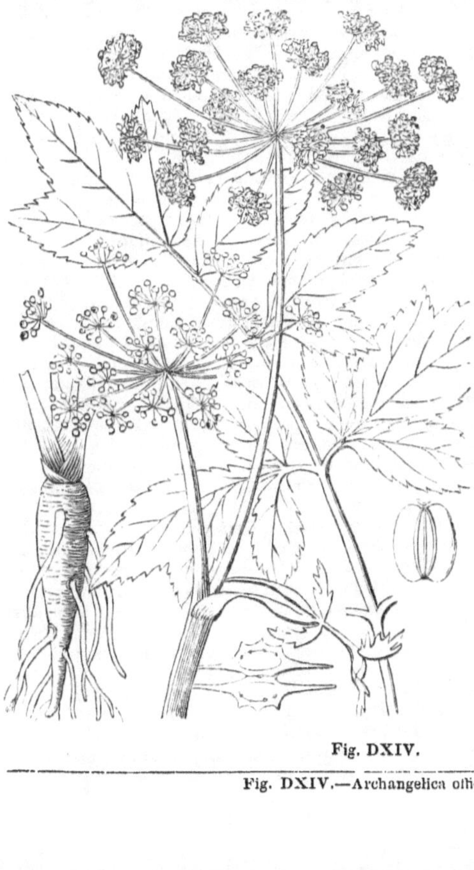

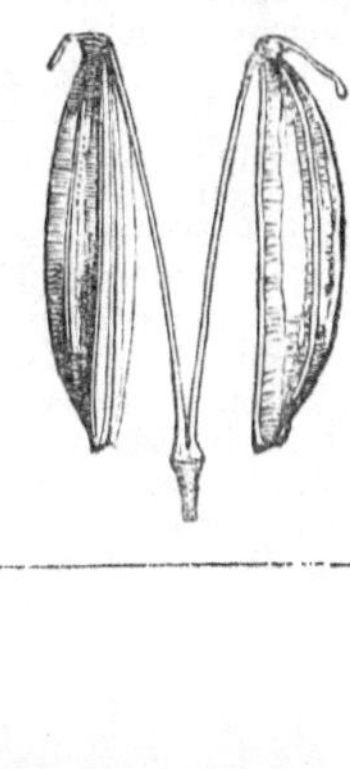

Fig. DXIV.

Fig. DXIV.—Archangelica officinalis.

others), the following brief enumeration will sufficiently explain the purposes to which Umbellifers are applicable :—

Of the harmless species, in which, with a little aroma, there is no considerable quantity of acrid watery matter or gum-resinous secretion, must be more particularly named Celery, Fennel, Samphire, Parsley, and the roots of Carrots, Parsnips, and Skirrets (Sium Sisarum). In addition to these, with which everybody is familiar, the following plants more particularly deserve mention as esculents:—The root of Eryngium campestre and maritimum, vulgarly called Eryngo, is sweet, aromatic, and tonic. Boerhaave reckons it as the first of aperient diuretic roots. It has been recommended in gonorrhœa, suppression of the menses, and visceral obstructions, particularly of the gall-bladder and liver ; it has also the credit of being a decided aphrodisiac. A good deal of it is sold in a candied state. The roots of Meum athamanticum and Mutellina are aromatic and sweet, and form an ingredient in the compound called Venice treacle. Angelica root, belonging to Archangelica officinalis, is fragrant, bitterish, pungent, sweet when first tasted, but leaving a glowing heat in the mouth. The Laplanders extol it not only as food but medicine. In coughs, hoarseness, and other pectoral disorders they eat the stalks roasted in hot ashes ; they also boil the tender flowers in milk till it attains the consistence of an extract, which they use to promote perspiration in catarrhal fevers, and to strengthen the stomach and bowels in diarrhœa. It is sold in the shops in a candied state, and was once an inhabitant of every country garden. Chervil, an old-fashioned pot-herb, with eatable roots, is the Anthriscus Cerefolium. Smyrnium Olusatrum, or Alexanders, was formerly cultivated instead of Celery; its leaves have a slight and pleasant aromatic flavour. The tubers of Bunium ferulaceum are eaten in Greece under the name of Topana. Samphire (Crithmum maritimum) is one of the best of all ingredients in pickles. Carum Bulbocastanum, the Pignut of the English, is quite wholesome, as are also the tubers of Œnanthe pimpinelloides. Anesorhiza capensis and Fœniculum capense are both Cape esculents. Arracacha esculenta, an inhabitant of the table-land of Grenada, has large esculent roots resembling a Parsnip in quality, but better. Finally, Prangos pabularia, a herbaceous plant inhabiting the arid plains of Southern Tartary, and the adjoining provinces, has a great reputation as a sheep food, which it appears not to deserve. Dr. Royle thinks that it may have been one of the kinds of Sylphion of the Greeks—that described by Arrian as growing only with Pines on Paropamisus, where it was browsed on by numerous flocks of sheep and cattle. Lieut. Burnes, crossing in the direction of Alexander's route, found this in the same situation, greedily cropped by sheep, and even eaten by his fellow-travellers. The natives of the north of Asia esteem highly the skinned root of the sweet subacrid Heracleum Sphondylium.

Among the gum-resinous species those yielding Asafœtida hold the first rank. The fetid odour of these plants is supposed to be owing to sulphur in combination with their peculiar essential oil. Asafœtida is the milky juice of various species of Ferula inhabiting Persia and neighbouring countries. Of these F. Asafœtida is the plant described by Kæmpfer (*Amœn. Exot.* 535) ; but F. persica and others are no doubt also the origin of the drug. Griffith was of that opinion (*Ann. N. Hist. X.* 193) ; and the fruits sent home to me by Sir John M'Neill prove the fact.—See *Fl. Med. No.* 100. Burnes found Asafœtida plants on the mountains of the Hindoo Koosh regarded as a highly nutritious sheep-food. The Asadulcis, or Laser Cyrenaicum, was yielded by a Thapsia, and probably Thapsia garganica. This drug was in high reputation among the ancients for its medical uses ; it had miraculous powers assigned to it ; to neutralise the effects of poison, to cure envenomed wounds, to restore sight to the blind, and youth to the aged, were only a part of its reputed properties ; it was also reckoned antispasmodic, deobstruent, diuretic, &c., &c. So great was its reputation, that the princes of Cyrene caused it to be struck on the reverse of their coins ; and the Cyrenean doctors were reckoned among the most eminent in the world. Its value was estimated by its weight in gold. The plants appear to be in reality very active purgatives. Galbanum, another fetid gum-resin, has been referred to Galbanum officinale, a Syrian plant ; but it has been demonstrated to owe its origin to another Umbellifer, the Opoidia galbanifera, a Persian plant.—See *Bot. Reg.* 1839, *Misc.* 107. Martius, however, and others maintain that this Opoidia yields the Persian Galbanum only, and that it is really the produce of different Umbellifers. Opopanax is the concrete juice of Opopanax Chironum, a plant resembling a Parsnip, and inhabiting the Levant. Ammoniacum has a more doubtful origin ; a Persian sort has been made out to be derived from Dorema Ammoniacum, but as Dioscorides says that his plant *γεννᾶται ἐν Λιβύῃ κατα Ἄμμωνα*, it was probably derived from Ferula orientalis, which still furnishes a drug of the kind in the kingdom of Morocco. The origin of Sagapenum, a drug between Galbanum and Asafœtida, is not ascertained with certainty ; it is thought to be derived rom either Ferula persica or F. Szowitsiana. Secretions of a similar nature are yielded by Bolax glebaria, a curious beehive-shaped plant, in southern Chile ; Peucedanum montanum, whose root abounds in a white, bitter, fetid juice, employed in Courland against epilepsy; Heracleum gummiferum, Bubon Galbanum, Laserpitium glabrum, whose root is violently purgative and even caustic; Daucus gummifer, which furnished the Sicilian Bdellium of the old Pharmacopœias, &c., &c. The Persian Musk root (Radix Sumbul), used as a protection against mephitic vapours, and having a very powerful smell of musk, is stated by MM. Reinsch and Buchner to belong to some large plant of this Order.—*Chem. Gaz.* 1844. 68.

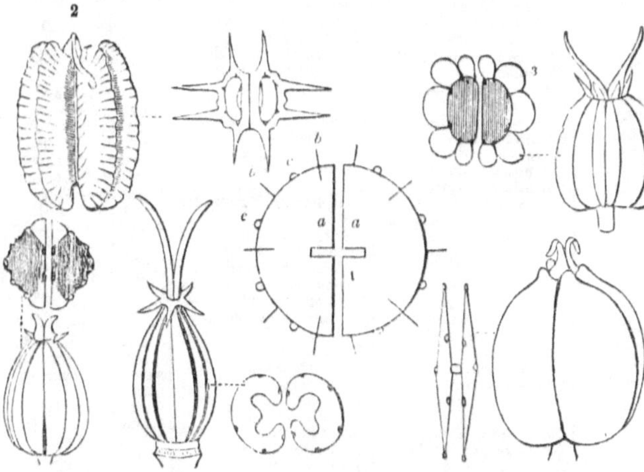

Fig. DXV.

For their aromatic and carminative fruits the most celebrated are Anise (Pimpinella Anisum), Dill (Anethum graveolens), Caraway (Carum Carui), and Coriander (Coriandrum sativum). Besides these, great numbers of less note are also employed for the same reason, the chief of which are the Ajwains or Ajowains of India (species of Ptychotis), Honewort (Sison Amomum), whose fruits smell of bugs, and Cummin (Cuminum Cyminum), now only used in veterinary practice.

Of the poisons, Conium or Hemlock holds the first place. This is a powerfully narcotico-acrid plant, occasioning stupor, delirium, palsy, and asphixia; some authors state that it produces death in the most dreadful convulsions, but this is at variance with the accounts of Dr. Christison and Dr. Pereira. Anthriscus vulgaris and sylvestris are reputed to have the same kind of action, but not so dangerous. The leaves of Æthusa Cynapium are poisonous, producing nausea, vomiting, headache, giddiness, drowsiness, spasmodic pain, numbness, &c. Œnanthe crocata and Phellandrium are perhaps the most dangerous of the narcotico-acrid Umbellifers; the roots are often eaten, with fatal consequences, by poor people who mistake them for Parsnips. A violent poison resides in the roots of Cicuta maculata; a drachm of the fresh root has killed a boy in an hour and a half; and in America fatal accidents arising from its being mistaken for other Umbellifers are not uncommon. It has been used as a substitute for Conium, with similar effect, except that it is more energetic. Cicuta virosa, in like manner, is a highly dangerous plant, producing effects similar to those of hydrocyanic acid. It causes true tetanic convulsions in frequent paroxysms, and death on the third day. Haller considered it the Conium of the Greeks. It appears to be fatal to cattle. From the roots of Lichtensteinia pyrethrifolia the Hottentots prepare an intoxicating beverage.

Fig. DXV.—1. Is an ideal plan of a fruit divided transversely; *a a* is the commissure, or plane of contact of the mericarps; *b b* primary ridges; *c c* secondary ridges. 2. Is a view of the back and section of the fruit of Laserpitium Siler; each mericarp has the secondary ridges winged, the primary obsolete; there are two vittæ on the commissure, and one under each secondary ridge; these vittæ, which are cavities containing oil, are represented by dots; the albumen is solid. 3. Sclerosciadium humile; the primary ridges are corky; there are no secondary ridges; the vittæ alternate with the primary ridges, and there is one at each edge of the commissure; the albumen is solid. 4. Discopleura capillacea; there are 5 very small primary juga, the two lateral of which are in contact with a thickened accessory margin; there are 2 vittæ on each face of the commissure, and one between each primary ridge; the albumen is solid. 5. Echinophora spinosa; albumen involute; vittæ alternate with the primary ridges. 6. Compressed fruit of Diposis saniculæfolia; the commissure is very narrow; there are 5 minute primary ridges; one along the back, one along each edge, and two on the inflexed side; the albumen is solid.

GENERA.

I. Hydrocotylidæ.

Hydrocotyle, *Tournef.*
Chondrocarpus, Nutt.
Glyceria, Nutt.
Centella, Linn.
Solandra, Linn. fil.
Crantzia, *Nutt.*
Cesatia, *Endl.*
Dimetopia, *DC.*
Erigenia, *Nutt.*
Micropleura, *Lagasc.*
Didiscus, *DC.*
Hügelia, Reichenb.
Pritzelia, *Walpers.*
Trachymene, *Rudge.*
Azorella, Labill.
Fischera, Spreng.
Catepha, Leschen.
Astrotricha, *DC.*
Leucolæna, *R. Br.*
Xanthosia, Rudg.
Cruciella, Leschen.
Pentapeltis, Endl.
Schænolæna, *Bunge.*
Bowlesia, *Ruiz et Pav.*
Azorella, *Lam.*
Chamitis, Soland.
Siebera, Reichenb.
Fragosa, Ruiz et Pav.
Pectophytum, *H.B.K.*

II. Mulinidæ.

Bolax, *Commers.*
Mulinum, *Pers.*
Asteriscium, *Cham.*
Cassidocarpus, Presl.
? *Dipterygia*, Presl.
Elsneria, *Walp.*
Laretia, *Gill. et Hook.*
Drusa, *DC.*
Huanaca, *Cav.*
Homalocarpus, *Hook. et Arn.*
Diposis, *DC.*
Spananthe, *Jacq.*
Pozoa, *Lagasc.*
Schizilema, *Hook. f.*

III. Saniculidæ.

Actinotus, *Labill.*
Eriocalia, Smith.
Proustia, Lagasc.
Holotome, *Benth.*
Petagnia, *Guss.*
Heterosciadium, DC.
Klotschia, *Cham.*
Sanicula, *Tournef.*
Hacquetia, *Neck.*
Dondia Spreng.
Dondisia, Reichenb.
Astrantia, *Tournef.*
Actinolema, *Fenzl.*
Alepidea, *Laroch.*
Eryngium, *Tournef.*
Lessonia, Bert.
Strebanthus, Raf.
Horsfieldia, *Blum.*
Schubertia, Blum.
Actinanthus, *Ehrenb.*
Hohenackeria, *Fisch. et Mey.*

IV. Amminidæ.

Rumia, *Hoffm.*
Cicuta, *Linn.*
Zizia, *Koch.*
Smyrnium, Ell.
Thaspium, Nutt.
Apium, *Hoffm.*
Oreosciadium, DC.
Petroselinum, *Hoffm.*
Wydleria, *DC.*
Trinia, *Hoffm.*
Apinella, Neck.
Spielmannia, Guss.
Helosciadium, *Koch.*
Callistroma, *Fenzl.*
Elæosticta, *Fenzl.*
Sium, Adans.
Mauchartia, Neck.
Cyclospermum, Lagasc.
Trachysciadium, DC.
Discopleura, *DC.*
Ptilimnium, Raf.
Leptocaulis, *Nutt.*
Spermolepis, Raf.
Ptychotis, *Koch.*
Microsciadium, Boiss.
Gymnosciadium, *Hochst.*
Bunium, Lagasc.
Ammoides, Adans.
Trachyspermum, Link.
Ammios, Mönch.
Heteroptycha, DC.
Critamus, *Bess.*
Falcaria, Rivin.
Drepanophyllum, Hfn.
Prionitis, Delarbr.
Hladnickia, Reichenb.
Sison, *Lagasc.*
Schultzia, *Spreng.*
Ammi, *Tournef.*
Visnaga, Gärtn.
Gohoria, Neck.
Ægopodium, *Linn.*
Podagraria, Rivin.
Carum, *Koch.*
Elwendia, *Boiss.*
Sympodium, *Koch.*
Bulbocastanum, Adans.
Lomatocarum, *Fisch.*
Bunium, *Koch.*
Conopodium, DC.
? *Deringa*, Adans.
Chamæsciadium, *C.A.M.*
Cryptotænia, *DC.*
Lereschia, *Boiss.*
Cyrtospermum, Raf.
? *Alacospermum*, Neck.
Pimpinella, *Linn.*
Tragoselinum, Tournf.
Pimpinella, Spreng.
Tragium, Spreng.
Ledeburia, Link.
Anisum, Adans.
Reutera, *Boiss.*
Berula, *Koch.*
Sium, *Koch.*
Ridolfia, *Moric.*
Muretia, *Boiss.*
Sisarum, Adans.
Bupleurum, *Tournef.*
Agostana, Salisb.
Diaphyllum, Hoffm.
Isophyllum, Hoffm.
Tenoria, Spreng.
Buprestis, Spreng.
Odonites, Spreng.
Diatropa, Dumort.
Trachypleurum, Rchb.
? *Orimaria*, Raf.
Atenia, *Hook. et Arn.*
? *Edosmia*, Nutt.
Neurophyllum, *Torr.*
Heteromorpha, *Cham.*
Furnrohria, *Koch.*

V. Seselinidæ.

Lichtensteinia, *Cham.*
Ottoa, *H. B. K.*
Œnanthe, *Lam.*
Phellandrium, Linn.
Haplosciadium, *Hochst.*
Platysace, *Bunge.*
Chamarea, *Eckl. et Zeyh.*
Anesorhiza, *Cham. et Schl.*
Anisopleura, *Fenzl.*
Sclerosciadium, *Koch.*
Dasyloma, *DC.*
Cynosciadium, *DC.*
Æthusa, *Linn.*
Fœniculum, *Adans.*
Kundmannia, *Scop.*
Brignolia, Bertolon.
Campderia, Lagasc.
Deverra, *DC.*
Pithyranthus, Viv.
Eremocarpus, *Bunge.*
Soranthus, *Ledeb.*
Eriocycla, *Lindl.*
Todaroa, *Parl.*
Seseli, *Linn.*
Hippomarathrum, Riv.
Marathrum, Raf.
Musineon, Raf.
Elæochytris, *Fenzl.*
Polycyrtus, Schlecht.
Polemannia, *Eckl. et Zh.*
Libanotis, *Crantz.*
Athamantha, Scop.
Eriotis, DC.
Xatardia, *Meisn.*
Petitia, Gay.
Cenolophium, *Koch.*
Dethawia, *Endl.*
Wallrothia, DC.
Cnidium, *Cuss.*
Selinum, Lagasc.
Hymenidium, *Lindl.*
Thaspium, *Nutt. DC.*
Trochiscanthes, *Koch.*
Athamantha, *Koch.*
Tinguarra, *Parl.*
Turbith, *Tausch.*
Libanotis, Scop.
Ligusticum, *Linn.*
Anisopleura, *Fenzl.*
Aciphylla, *Forst.*
Anisotome, *Hook. f.*
Gingidium, Forst.
Trachydium, *Lindl.*
Silaus, *Bess.*
Meum, *Tournef.*
Endressia, *Gay.*
Neogaya, *Meisn.*
Gaya, Gaud.
Pachypleurum, Reich.
? *Arpitium*, Neck.
Conioselinum, *Fisch.*
Cszernæwia, Turcz.
Crithmum, *Tournef.*

VI. Pachypleuridæ.

Krubera, *Hoffm.*
Ulospermum, Link.
Capnophyllum, Lagasc.
Pachypleurum, *Ledeb.*
Phloiodicarpus, *Turcz.*
Stenocœlium, *Ledeb.*

VII. Angelicidæ.

Levisticum, *Koch.*
Ligusticum, Lagasc.
Uloptera, *Fenzl.*
Heteroptilis, *E. Meyer.*
Gomphopetalum, *Turcz.*
Selinum, *Hoffm.*
Mylinum, Gaud.
Thyselinum, Adans.
Carvifolia, Vaill.
Ostericum, *Hoffm.*
Angelica, *Hoffm.*
Archangelica, *Hoffm.*
Uloptera, *Fenzl.*

VIII. Peucedanidæ.

Opoponax, *Koch.*
Ferula, *Tournef.*
Ferulago, Koch.
? *Lomatium*, Raf.
Cogswellia, Schult.
Polycyrtus, *Schlecht.*
Dorema, *Don.*
Eriosynaphe, *DC.*
Peucedanum, *Linn.*
Palimbia, Bess.
Pteroselinum, Reichb.
Selinum, Gärtn.
Caroselinum, Grise.
Thysselinum, DC.
Cervaria, Gärtn.
Oreoselinum, Duby.
Imperatoria, Linn.
Euryptera, *Nutt.*
Leptotænia, *Nutt.*
Xanthogalum, *Lalem.*
Sciothamnus, *Endl.*
Dregea, Eckl. et Zeyh.
Cynorrhiza, *Eckl. et Zyh.*
Lefeburia, *A. Rich.*
Callisace, *Fisch.*
Bubon, *Linn.*
Galbanophora, Neck.
Agasillis, Spreng.
Anethum, *Tournef.*
Cortia, *DC.*
Hammatocaulis, *Tausch.*
Capnophyllum, *Gärtn.*
Rumia, Link.
Tiedemannia, *DC.*
Oxypolis, Raf.
Archemora, *DC.*
Lophotænia, *Griseb.*
Pastinaca, *Tournef.*
Malabaila, Hoffm.
Leiotulus, *Ehrenb.*
Astydamia, *DC.*
Symphyoloma, *C.A. Mey.*
Stenotænia, *Boiss.*
Heracleum, *Linn.*
Sphondylium, Tournef.
Tetratænium, DC.
Carmelia, DC.
Wendtia, Hoffm.
Trichogonium, DC.
Barysoma, *Bung.*
Zozimia, *Hoffm.*
Ducrosia, *Boiss.*
Trigonosciadium, *Boiss.*
Polytænia, *DC.*
Eurytænia, *Nutt.*
Johrenia, *DC.*
Diplotænia, *Boiss.*
Hasselquistia, *Linn.*
Ainsworthia, *Boiss.*
Tordylium, *Tournef.*
Condylocarpus, Hoffm.
Synelcosciadium, *Boiss.*
Tordyliopsis, *DC.*
Tordylioides, Wall.

IX. Sileridæ.

Agasyllis, *Hoffm.*
Siler, *Scop.*
Bradlæia, Neck.
Galbanum, *Don.*
Ormosolenia, *Tausch*

X. Cuminidæ.

Cuminum, *Linn.*
Froriepia, *Koch.*
Trepocarpus, *Nutt.*

XI. Thapsidæ.

Thapsia, *Tournef.*
Cymopterus, *Raf.*

Thapsia, Nutt.
Phyllopterus, Nutt.
Leptocnemia, Nutt.
Pterixia, Nutt.
Polylophium, *Boiss.*
Laserpitium, *Tournef.*
Siler, Mönch.
Lophosciadium, *DC.*
Melanoselinum, *Hoffm.*

XII. DAUCIDÆ.

Artedia, *Linn.*
Orlaya, *Hoffm.*
Daucus, *Tournef.*
Agrocharis, *Hochst.*
Duriæa, *Boiss.*
Platyspermum, Hoffm.
Anisactis, DC.

XIII. ELÆOSELINIDÆ.

Elæoselinum, *Koch.*
Margotia, *Boiss.*

XIV. CAUCALINIDÆ.

Szovitsia, *Fisch. et Mey.*
Caucalis, *Linn.*
Turgenia, *Hoffm.*
Torilis, *Adans.*
Turgeniopsis, *Boiss.*
Lisæa, *Boiss.*
Trichocarpæa, DC.

XV. SCANDICIDÆ.

Scandix, *Gärtn.*
Wylia, Hoffm.
Anthriscus, *Hoffm.*
Chærophyllum, Lagasc.
Cerefolium, Hall.
Chærophyllum, *Linn.*
Butinia, *Boiss.*
Oreomyrrhis, *Endl.*
Caldasia, Lagasc.
Sphallerocarpus, *Bess.*
Molopospermum, *Koch.*
Velæa, *DC.*
Tauschia, *Schlecht.*
Myrrhis, *Scop.*
Freyera, *Reichenb.*
Biasolettia, Koch.
Osmorrhiza, *Raf.*
Uraspermum, Nutt.
Spermatura, Reichenb.
Glycosma, *Nutt.*
Grammosciadium, *DC.*
Rhabdosciadium, *Boiss.*
Ozodia, *Wight et Arn.*
Heterotænia, *Boiss.*

XVI. SMYRNIDÆ.

Lagoëcia, *Linn.*
Oliveria, *Vent.*
Aniosciadium, *DC.*
Pycnocycla, *Royle.*
Echinophora, *Tournef.*
Dicyclophora, *Boiss.*
Thecocarpus, *Boiss.*
Exoacantha, *Labill.*
Arctopus, *Linn.*
Apradus, Adans.
Cachrys, *Tournef.*
Ægomarathrum, Koch.
Hippomarathrum, Lk.
Lophocachrys, DC.
Prangos, *Lindl.*
Pteromarathrum, Kch.
Colladonia, *DC.*
Perlebia, DC.
Meliocarpus, *Boiss.*
Heptaptera, *Reutt.*
Lecokia, *DC.*
Magydaris, *Koch.*
Eriocachrys, DC.
Hermas, *Linn.*
Petrocarui, *Tausch.*
Conium, *Linn.*
Cicuta, Tournef.
Vicatia, *DC.*
Arracacha, *Bancr.*
Pentacrypta, *Lehm.*
Pleurospermum, *Hoffm.*
Cœlopleurum, *Ledeb.*
Hansenia, *Turcz.*
Physospermum, Vel.
Enymonospermum, Sp.
Malabaila, *Tausch.*
Hladnickia, Koch.
Grafia, Reichenb.
Eleutherospermum, *Kch.*
Hymenolæna, *Wall.*
Aulacospermum, *Ledeb.*
Physospermum, *Cass.*
Danaa, Allion.
Henslera, Lagasc.
Keramocarpus, *Fenzl.*
Opoidia, *Lindl.*
Smyrnium, *Linn.*
Smyrniopsis, *Boiss.*
Anosmia, *Bernh.*
Perideridia, *Reichenb.*
Eulophus, Nutt.
Cynapium, *Nutt.*
Deweya, *Torr. et A. Gr.*
Musenium, *Nutt.*
Scaligeria, *DC.*

XVII. CORIANDRIDÆ.

Cymbocarpum, *DC.*
Ormosciadium, *Boiss.*
Bifora, *Hoffm.*
Biforis, Spreng.
Corion, Link.
Anidrum, Neck.
Schrenkia, *Fisch.*
Astomæa, *Reichenb.*
Astoma, DC.
Cryptodiscus, *Schrenck.*
Atrema, *DC.*
Coriandrum, *Linn.*
Apiastrum, *Nutt.*

NUMBERS. GEN. 267. Sp. 1500.

Ranunculaceæ.
POSITION.—Araliaceæ.—APIACEÆ.—Hamamelidaceæ.
Saxifragaceæ.

ORDER CCXCVII. ARALIACEÆ.—IVYWORTS.

Araliæ, *Juss. Gen.* 217. (1789).—Araliaceæ, *A. Richard in Dictionnaire Classique d'Histoire Naturelle*, 1. 506. (1822); *DC. Prodr.* 4. 251. (1830); *Bartling Ord. Nat.* 237. (1830); *Endl. Gen.* clxiii.; *Wight Illustr.* 2. *t.* 118.

DIAGNOSIS.—*Umbellal Exogens, with a 3- or more-celled fruit without a double epigynous disk, pentamerous flowers, a valvate corolla, alternate leaves without stipules, and anthers turned inwards, opening lengthwise.*

Trees, shrubs, or herbaceous plants, with in all respects the habit of Umbellifers. Calyx adherent, entire or toothed. Petals definite, 2, 5, 10, deciduous, valvate in æstivation, occasionally 0. Stamens equal in number to the petals or twice as many, arising from within the border of the calyx, and from without an epigynous disk. Ovary inferior, with more cells than 2; ovules solitary, pendulous, anatropal; styles equal in number to the cells, sometimes connate; stigmas simple. Fruit succulent or dry, consisting of several 1-seeded cells. Seeds solitary, pendulous, adhering to the pericarp; albumen fleshy, having a minute embryo at the base, with its radicle pointing to the hilum.

Fig. DXVI.

In many respects these plants are much like Umbellifers, from which they are distinguished by their ovary having more cells than 2, and by their greater tendency to form a woody stem; to this may also be in general added a valvate corolla; but Didiscus is valvate among Umbellifers, and Adoxa in Ivyworts is not. There is also a connection with Caprifoils, established by means of Hedera and Viburnum. Vineworts, too, may be considered a mere hypogynous form of Ivyworts, and must be regarded as representing them in the hypogynous sub-class, as will be most evident if Aralia racemosa is compared with certain species of Cissus. Gunnera, a singular genus with dimerous ♂ ♀ or ☿ flowers, and a single ovule suspended from the apex of the cavity, seems to be a degraded form of this Order, and the genus Adoxa is also quite anomalous, though in a different way. Its stamens are slit half way down, so as to appear as if made up of 2 half anthers each; its petals are united into an imbricated monopetalous corolla; and it usually has a calyx whose sepals do not correspond in number with the lobes of the corolla. De Candolle thought this corolla to be a whorl of abortive stamens, but there does not appear to be any sufficient ground for his opinion. See Decaisne in *Ann. Sc. Nat. n. s.* vi. 72. In several instances a tendency to the separation of stamens and pistil is observable; it is usually, however, accompanied by the common ☿ structure.

The species are found in the tropical and sub-tropical regions of all the world; and even in some of the coldest, as in the United States, Canada, the north-west coast of America, and Japan. Aralia polaris was even found by Dr. Jos. Hooker, as far to the south as Lord Auckland's group of islands, in 50½° south latitude.

Similar as these plants are to Umbellifers they do not appear to partake in any considerable degree of the dangerous qualities for which some of the latter are known. On the contrary, they are more generally stimulant and aromatic. Neither do their succulent fruits often yield the essential oil which renders many of the Umbellifers useful carminatives and stomachics. The Ginseng, or Ginschen root, so highly prized by the Chinese as a stimulant, belongs to some species of Panax (P. Ginseng, *Meyer*) unknown. Meyer describes it as having a sharp, aromatic, peculiar taste. The Chinese are said to administer it in all diseases resulting from weakness of the body.—*Chem. Gaz.* 1843. 238.

Fig. DXVI.—1. Hedera Helix; 2. flower of Dimorphanthus edulis (*Siebold*); 3. perpendicular section of the ovary; 4. undivided ovary; 5. ripe fruit; 6. cross section of it; 7. section of seed of H. Helix.

Although its virtues have been pronounced imaginary, it is hardly credible that the root should have gained such great celebrity if it were inert. It was formerly supposed to belong to Panax quinquefolium, but that seems to have been a mistake; the species so named is said, however, to be sold by the Americans to the Chinese, as a substitute for their Ginseng; it has an agreeable bitter-sweet root, and is used sometimes as Liquorice. Panax fruticosus and cochleatus are fragrant aromatics, employed in the Moluccas, and P. Anisum has berries with all the odour of the herb whose name it bears. The Aralias seem to be similar in their action; an aromatic gum-resin comes from A. racemosa, spinosa and hispida; A. nudicaulis is diaphoretic, and its shoots are employed in North America as a substitute for Sarsaparilla; Dimorphanthus (Aralia) edulis is employed in China as a sudorific; its young shoots are a delicate vegetable; and its root, which is bitter, aromatic, and pleasant to the taste, is employed by the Japanese, in winter, as we use Scorzonera. Nor does the common Ivy want the aroma of the Order, although unpleasant in smell; it is mentioned as a sudorific; and its berries are emetic. Hedera umbellifera, an Amboyna plant, is said to furnish wood scented like Lavender and Rosemary, and H. terebintinacea yields, in Ceylon, a resinous substance smelling of turpentine. Gunnera scabra or Panke is astringent; its roots are used by tanners, while its fleshy leaf-stalks are eaten; Mr. Darwin found it growing on the sandstone-cliffs of Chiloe, and describes it as somewhat resembling Rhubarb on a gigantic scale. He measured a leaf which was nearly 8 feet in *diameter*, and remarked that each plant produced 4 or 5 of these enormous leaves, "presenting together a noble appearance." The fruit of Gunnera macrocephala is reputed in Java to be stimulant.

GENERA.

Panax, *Linn.*
Aureliana, Catesb.
Araliastrum, Vaill.
Plectronia, Lour.
Cussonia, *Thunb.*
Maralia, *Thouars.*
Gilibertia, *Ruiz et Pav.*
Wangenheimia, Dietr.
Ginnania, Dietr.

Gastonia, *Commers.*
Trevesia, *Vis.*
Polyscias, *Forst.*
Brassaia, *Endl.*
Torricellia, *DC.*
Aralia, *Linn.*
Scheffera, Forst.
Dimorphanthus, *Miq.*
Sciodaphyllum, *P. Br.*

Actinophyllum, R. et P.
Hedera, *Linn.*
Gynapteina, Blum.
Paratropia, *DC.*
Heptapleurum, Gärtn.
Arthrophyllum, *Blum.*
Botryodendrum, *Endl.*
? Miquelia, *Meisn.*

Touroulia, *Aubl.*
Robinsonia, Schreb.
Adoxa, *L.*
Milligania, *Hook fil.*
Gunnera, *L.*
Misandra, Comm.
Disomene, Banks.
Perpensum, Burm.
Panke, Feuill.

NUMBERS. GEN. 21. Sp. 160.

Vitaceæ.
POSITION.—Apiaceæ.—ARALIACEÆ.—Hamamelidaceæ.
Caprifoliaceæ.

Fig. DXVII. 1

Fig. DXVII.—Adoxa Moschatellina. 1. a flower magnified.

ORDER CCXCVIII. CORNACEÆ.—CORNELS.

Caprifoliaceæ, § Corneæ, *Kunth Nov. G. Amer.* 3. 430.—Corneæ, *DC. Prodr.* 4. 271. (1830); *Endl. Gen.* clxv.; *Meisner*, p. 143.

DIAGNOSIS.—*Umbellal Exogens, with a 2- or more-celled fruit without a double epigynous disk, tetramerous flowers, a valvate corolla, and opposite leaves without stipules.*

Trees or shrubs, seldom herbs. Leaves (except in one species) opposite, entire or toothed, with pinnate veins. Stipules 0. Flowers capitate, umbellate, or corymbose, naked or with an involucre, occasionally by abortion ♂ ♀. Sepals 4, superior. Petals 4, oblong, broad at the base, inserted into the top of the calyx, regular, valvate in æstivation. Stamens 4, inserted along with the petals and alternate with them; anthers ovate-oblong, 2-celled. Ovary adherent, 2- or perhaps 3-celled, crowned by a disk; ovules solitary, pendulous, anatropal; style filiform; stigma simple. Drupe berried, crowned by the remains of a calyx, with a 2-celled nucleus. Seeds pendulous, solitary. Embryo in the axis of fleshy albumen, and as long; radicle superior, shorter than the two oblong cotyledons.

Fig. DXVIII.

These plants were formerly confounded with Caprifoils, on account of the general resemblance between Cornus and Viburnum; they however represent an entirely distinct Order, as their habit and general characters sufficiently indicate. From Caprifoils their polypetalous structure removes them. To Witch Hazels they approach more nearly, but differ in the valvate æstivation of their corolla, &c. &c. In many respects Cornels resemble Loranths, from which they differ among other things in the stamens being opposite to the sepals, and in the flowers being polypetalous. Hollyworts are sometimes compared with them, but they have a superior fruit and erect ovules. If Garryads were not amentaceous, and had petals and bisexual flowers, they would approach Cornels very nearly, and probably do in fact represent them in the diclinous sub-class, as seems to be proved by the genus Pukateria, whose flowers are ♂ ♀. To Umbellifers they also approach very closely, being chiefly distinguished by their tetramerous flowers, succulent fruit, and single style, to which may be added their opposite leaves. Such Cornels as Cornus suecica and florida, and Benthamia have the inflorescence and involucre of an Umbellifer. As to Ivyworts, it is hard to say in what manner they can be distinguished if we neglect the opposite leaves, the tendency to form a pair of cells in the fruit rather than a larger number, and, in fact, the tetramerous structure of the flower generally.

Found all over the temperate parts of Europe, Asia, and America. It is doubtful whether the African genera belong here.

The bark of C. florida, sericea, and circinata, is said to rank among the best tonics of North America, nothing having been found in the United States that so effectually answers the purposes of Peruvian bark in intermittent fevers. It is a remarkable fact that the young branches of Cornus florida, stripped of their bark and rubbed with their ends against the teeth, render them extremely white. From the bark of the fibrous roots the Indians extract a good scarlet colour. Lamp oil has been obtained from the seeds of Cornus sanguinea. The Cornus of the ancients was the present Cornelian Cherry (Cornus mascula), whose little clusters of yellow starry flowers stud its naked branches, and are among the earliest heralds of spring. Its fruit is like a small plum,

Fig. DXVIII.—Benthamia japonica.—*Siebold.* 1. a flower; 2. a perpendicular section of the pistil; 3. a head of fruit; 4. a section of a seed.

with a very austere flesh; but after bletting it becomes sub-acid, and was once held in some such estimation as sorbs and services. The Turks still use it in the manufacture of sherbet.—*Fl. Græc. Pr.* ii. 41. Its fruit and leaves were formerly used in medicine as astringents. Cornus officinalis, a large Japanese shrub, is little different, and is there commonly cultivated, for its fruits are a constant ingredient in the fever drinks of the country. C. suecica is reputed to have tonic berries which increase the appetite, whence its highland name Lus-a-chrasis, or plant of gluttony.

GENERA.

Benthamia, *Lindl.*
Cornus, *Tournef.*
Aucuba, *Thunb.*
Eubasis, Salisb.
Decostea, *Ruiz et Pav.*
Pukateria, *Raoul.*
Corokia, *Cunn.*
? Curtisia, *Ait.*
Doratium, Soland.
Relhania, Gmel.
Junghansia, Gmel.
Sideroxylon, Burm.
? Mastixia, *Blum.*
? Votomita, *Aubl.*
Glossoma, Schreb.
Guilleminia, Neck.

NUMBERS. Gen. 9. Sp. 40.

Caprifoliaceæ.
POSITION.—Apiaceæ.—CORNACEÆ.—Hamamelidaceæ.
Alangiaceæ.
Garryaceæ.

ORDER CCXCIX. HAMAMELIDACEÆ.—WITCH-HAZELS.

Hamamelideæ, *R. Br. in Abel's Voyage to China*, (1818); *A. Richard Nouv. Elém.* 532. (1828); *DC. Prodr.* 4. 267. (1830); *Endl. Gen.* clxvii.; *Griffith in Asiatic Researches*, (1836), xix. p. 94.

DIAGNOSIS.—*Umbellal Exogens, with a 2-celled not didymous fruit without a double epigynous disk, an imbricated corolla, alternate leaves with stipules, and anthers with deciduous valves.*

Small trees or shrubs. Their woody tubes, in some cases, marked by circular disks. Leaves alternate, deciduous, toothed, with veins running from the midrib straight to the margin. Stipules deciduous. Flowers small, axillary, sometimes unisexual by abortion. Calyx adherent, in 4 or 5 pieces. Petals 4 or 5, or 0; if present, with an imbricated æstivation. Stamens 8, of which 4 are alternate with the petals; their anthers turned inwards, 2-celled, and 4 are sterile, and placed at the base of the petals; their dehiscence variable. Ovary 2-celled, inferior; ovules solitary or several, pendulous or suspended; styles 2. Fruit half inferior, capsular, usually opening with 2 septiferous valves. Seeds pendulous; embryo in the midst of fleshy horny albumen; radicle superior.

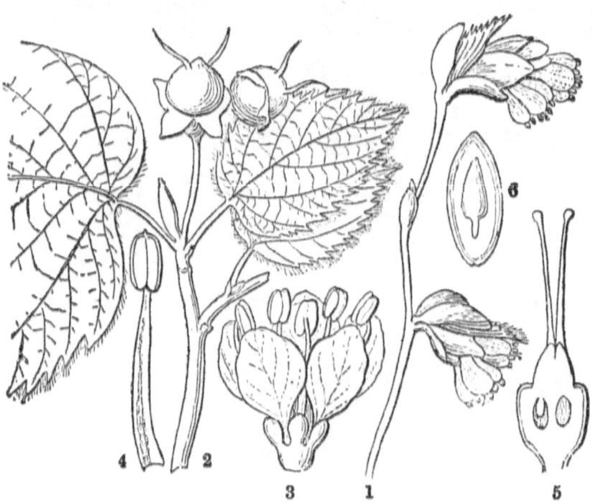

Fig. DXIX.

According to Brown, the affinity of Witch-hazels is on the one hand with Bruniads, from which they are distinguished by the insertion and dehiscence of the anthers, the monospermous cells of the ovary, the dehiscence of the capsule, the quadrifid calyx, and by habit; and on the other with Cornus, Marlea, and the neighbouring genera; in some repects also with Ivyworts, but differing in their capsular fruit, the structure of the anthers, and other marks.—See *Abel's Voyage, Appendix.* Du Petit Thouars looks upon them as allied to Rhamnads, and Jussieu to Hippurids. Others consider them akin to Mastworts and Spurges, through Fothergilla. There can be no doubt that they must be arranged in the Epigynous series, and upon the whole, Bruniads on the one hand, and Cornels on the other, seem to claim the nearest kindred; from the latter they are known by their stipules and imbricated petals, if any are present. It was formerly supposed that they might be allied to Saxifrages, on account in part of their albuminous seeds and half inferior ovary; but that resemblance was but a remote one. Mr. Griffith observed in Bucklandia and Sedgwickia that the woody tissue is marked with circular dots something like those of Conifers; the same sharp-sighted Botanist observed that in Bucklandia the second membrane of the ovule protrudes beyond the foramen in the ripe seed!

The species come from North America, Japan, China, and the central parts of Asia, Madagascar, and South Africa.

The kernels of Hamamelis virginica are oily and eatable. The leaves and bark are very astringent, and also contain a peculiar acrid essential oil.

GENERA.

I. HAMAMELEÆ.—Ovules solitary. Dicoryphe, *Thouars.* *Dicorypha*, Spreng. Corylopsis, *Sieb. et Zucc.*	Trichocladus, *Pers.* *Dahlia*, Thunb. Hamamelis, *Linn.* *Trilopus*, Mitch.	Loropetalum, *R. Br.* Parrotia, *C. A. Mey.* Fothergilla, *Linn. f.* Distylium, *Zucc.*	II. BUCKLANDEÆ.—Ovules several in each cell. Bucklandia, *R. Br.* Sedgwickia, *Griff.*

NUMBERS. GEN. 10. Sp. 15.

POSITION.—Bruniaceæ.—HAMAMELIDACEÆ.—Cornaceæ.
Saxifragaceæ.

Fig. DXIX.—Corylopsis. 1. flowers; 2. branch in fruit; 3. a flower separate; 4. a stamen; 5. a perpendicular section of the ovary; 6. a section of a seed removed from the capsule and placed with the hilum downwards.

ORDER CCC. BRUNIACEÆ.—BRUNIADS.

Bruniaceæ, *R. Brown in Abel's China* (1818); *DC. Prodr.* 2. 43; *Ad. Brongniart in Ann. des Sc. Nat.; Endl. Gen.* clxviii.; *Arnott in Hook. Journ.* 3. 259.—Grubbiaceæ, *Endl. Enchirid. p.* 403. (1841); Ophiriaceæ, *Arnott in Hooker's Journal*, 3. 266. (1841).

DIAGNOSIS.—*Umbellal Exogens, without a double epigynous disk, with a* 3- (*or* 1-) *celled fruit, imbricated corolla, alternate leaves without stipules, and anthers turned outwards, opening lengthwise.*

Branched, heath-like shrubs. Leaves small, imbricated, rigid, entire, with a callous point. Flowers small, capitate, or panicled, or even terminal, and solitary; either naked, or with large involucrating bracts. Calyx superior, 5-cleft, imbricated, occasionally nearly inferior. Petals alternate with the segments of the calyx, arising from its throat, imbricated (or valvate ?). Stamens alternate with the petals, arising from the same point, or from a disk surrounding the ovary; anthers turned outwards, 2-celled, bursting longitudinally. Ovary half inferior, with from 1 to 3 cells, in each of which there is from 1 to 2 suspended collateral anatropal ovules; sometimes 1-celled from the abortion of carpels or dissepiments; style simple or bifid; stigma simple. Fruit dicoccous or indehiscent, 2- or 1-celled, crowned by the persistent calyx. Seeds solitary or in pairs, suspended, sometimes with a short aril; albumen fleshy; embryo minute at the base of the seed, with a conical radicle, and short fleshy cotyledons.

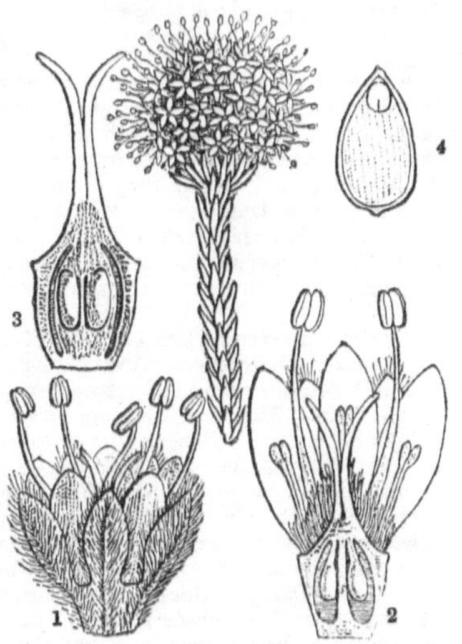

Fig. DXX.

The relationship of these plants to Witch Hazels is admitted, and therefore they will participate in all the other affinities of that Order, which is known from Bruniads by the habit, stipules, and often deciduous valves of the anthers. Brongniart indicates an affinity with Myrtleblooms through Imbricaria, which is very nearly constructed as true Bruniads, but has the stamens opposite the petals, and dotted leaves. He also considers that Cornels bear them much real affinity, and he even contrasts them with Umbellifers, to which they no doubt approximate very nearly. The genus Raspailia is remarkable for having the stamens arising from the top of a superior ovary! and Thamnea is an instance of a 1-celled ovary with the ovules adhering to a central columnar axis. Mr. Arnott considers the group named by him Ophiriaceæ to be intermediate between Bruniads and Witch Hazels.

All are found at the Cape of Good Hope, with the exception of a single species inhabiting Madagascar.

Their properties are unknown.

GENERA.

Berzelia, *Brongn.*	*Astrocoma*, Neck.	*Pavinda*, Thunb.	Gravenhorstia, *Nees.*
Brunia, *Linn.*	Berardia, *Brongn.*	Tittmannia, *Brongn.*	Erasma, *R. Br.*
Nebelia, Neck.	*Nebelia*, Sweet.	*Mösslera*, Reichenb.	Grubbia, *Berg.*
Beckea, *Burm.*	? *Ptyxostoma*, Vahl.	Thamnea, *Soland.*	*Ophiria*, L.
Raspailia, *Brongn.*	Linconia, *Linn.*	Heterodon, *Meisn.*	Ophiria, *Lam.*
Staavia, *Thunb.*	Audouinia, *Brongn.*	*Rabenhorstia*, Rchb.	*Strobilocarpus*, Kltzh.
Levisanus, Schreb.			

NUMBERS. GEN. 15. Sp. 65.

Santalaceæ.
POSITION.—Hamamelidaceæ.—BRUNIACEÆ.—Apiaceæ.
Myrtaceæ.

Fig. DXX.—Brunia nodiflora.—*Brongniart.* 1. a flower; 2. a perpendicular section of it; 3. the pistil divided perpendicularly; 4. half a seed.

ALLIANCE LVI. *ASARALES.*—THE ASARAL ALLIANCE.

DIAGNOSIS.—*Epigynous Exogens, with monochlamydeous flowers, and a small embryo, lying in a large quantity of albumen.*

The place which Birthworts should occupy in a Natural arrangement is one of those disputed points respecting which it is extremely difficult to arrive at any positive conclusion. They are so anomalous in their woody structure, and so peculiar in their trimerous flowers, with an inferior ovary abounding in ovules, that an obvious ally can hardly be found for them. In fact they seem to be of an intermediate nature between Exogens and Endogens or Dictyogens. The great livid calyx of Aristolochia calls to mind the spathes of Arads: the leaves are those of Sarsaparillas. It is therefore probable that they should be regarded as a group standing on the borders of the three Sub-classes just mentioned, and joining them to each other, just as Switzerland joins Austria, Italy, and France.

The points of resemblance between Birthworts, Sandalworts, and Loranths are their want of corolla, their inferior ovary, their large albumen, and small embryo. These appear to be circumstances of greater weight than any distinctions that might be found between them. The rim which appears at the summit of the ovary of Aristolochia is possibly of the same nature as that of Loranths.

It is not to be wondered at that here—amidst Orders which, although apparently at the uttermost boundary of the vegetable kingdom, are really points of communication by means of which the circles of affinity return into themselves—we should find other tendencies than that of Birthworts to assume the condition of Natural Orders stationed in a lineal arrangement at very distant parts of the line. In truth, Sandalworts stand with respect to the Garryal Alliance, and Loranths to Amentals, in the same position as New Holland to New Zealand, or Kamtchatka to Russian America upon the maps; the whole world seems to divide them, and yet they are stationed within a few degrees of each other. Thus Loranths, which are often unisexual, approach Oleasters somewhat nearly, and Sandalworts come close up to the limits of Helwingiads.

NATURAL ORDERS OF ASARALS.

Ovary 1-*celled.* *Ovules definite, with a coated nucleus* . . . 301. SANTALACEÆ.
Ovary 1-*celled.* *Ovules definite, with a naked nucleus* . . . 302. LORANTHACEÆ.
Ovary 3-6-*celled.* *Ovules* 00. . . . . 303. ARISTOLOCHIACEÆ.

ORDER CCCI. SANTALACEÆ.—SANDALWORTS.

Santalaceæ, *R. Brown Prodr.* 350. (1810); *Juss. Dict. des Sc. Nat.* 47. 287; *Bartling Ord. Nat.* 112; *Endl. Gen.* cviii.; *Griffith in Linn. Trans.* 18. 59.—Osyrideæ, *Juss. in Ann. Mus. vol.* 5. (1802); *Martius Conspectus, No.* 82. (1835).—Osyrinæ, *Link Handb.* 1. 371. (1829).

DIAGNOSIS.—*Asaral Exogens, with a 1-celled ovary and definite ovules having a coated nucleus.*

Trees or shrubs, sometimes under-shrubs or herbaceous plants. Leaves alternate, or nearly opposite, undivided, sometimes minute, and resembling stipules. Flowers in spikes, seldom in umbels, or solitary, small. Calyx superior, 4- or 5-cleft, half-coloured, with valvate æstivation. Stamens 4 or 5, opposite the segments of the calyx, and inserted into their bases. Ovary 1-celled, with from 1 to 4 ovules, fixed to a central placenta, and usually near the summit; style 1; stigma often lobed. Fruit 1-seeded, hard and dry, and drupaceous. Albumen fleshy, of the same form as the seed; embryo minute, in the axis, inverted, taper.

Brown observes (*Flinders*, 569) that one of the most remarkable characters of this Order consists in its unilocular ovary containing more than one, but always a determinate number of ovules, which are pendulous, and attached to the apex of a central receptacle; this receptacle varies in its figure in the different genera, in some being filiform, in others nearly filling the cavity of the ovary. In Santalum itself, however, the ovules are erect, as Griffith showed, and they are said to be the same in Osyris, which is described as being ♂ ♀. The nearest relationship of Sandalworts is a disputed question. Most Botanists assign them to the neighbourhood of Daphnads, or Oleasters; but their inferior ovary, copious albumen, and placentation, disagree with both those groups, and their hermaphrodite flowers also divide them from Oleasters. Endlicher refers them to the neighbourhood of Olax, but the hypogynous stamens of that Order forbid a close approximation. Loranths seem to stand in the closest consanguinity, and are principally known by their parasitical manner of growth, and their ovules having a truly naked nucleus.

Fig. DXXI.

Fig. DXXII.

The species are found in Europe and North America, in the form of little obscure weeds; in New Holland, the East Indies, and the South Sea Islands, as large shrubs, or small trees.

Sandal-wood is the produce of Santalum album; in India it is esteemed by the native doctors as possessing sedative and cooling qualities, and as a valuable medicine in gonorrhœa. It is also employed as a perfume. The Sandal-wood of the Sandwich Islands is the wood of Santalum Freycinetianum and paniculatum. The leaves of Osyris nepalensis form a sort of tea. An infusion of Myoschilos oblongus, the Senna of the

Fig. DXXI.—1. Leptomeria acida; 2. a branch more magnified; 3. an expanded flower; 4. a fruit.
Fig. DXXII.—Thesium pratense.—*Nees.* 1. a flower; 2. the same laid open; 3. half an unripe fruit; 4. placenta and pair of ovules; 5. half the ripe fruit and seed.

Chilenos, is purgative. The fruit of the Quandang Nut (Fusanus acuminatus) is as sweet and useful to the New Hollanders as Almonds are to us ; that of Cervantesia tomentosa has a similar reputation in Peru. Oil is obtained in Carolina from the kernels of Pyrularia pubera. Leptomeria Billardieri, a common Tasmannian shrub resembling the European Broom in its green and almost leafless habit, is acid in almost every part, especially in the fruit, but astringent also, and is well suited, when chewed, to allay thirst.—*Backhouse.* The Thesiums are scentless and slightly astringent.—*DC.*

GENERA.

Quinchamalium, *Juss.*
Arjoona, *Cav.*
Thesium, *Linn.*
  Alchimilla, Tournef.
  Thesiosyris, Rchb.
  Frisea, Reichenb.
Rhinostegia, *Turcz.*
Nanodea, *Banks.*
  Balenerdia, Commers.
Choretrum, *R. Br.*
Leptomeria, *R. Br.*
Comandra, *Nutt.*
Fusanus, *Linn.*
  Colpoon, Berg.
Eucarya, *Mitch.*
Osyris, *Linn.*
  Casia, Tourn.
Sphærocarya, *Wall.*
Scleropyron, *Arn.*
Santalum, *Linn.*
  Sirium, Linn.
Mida, *A. Cunningh.*
Pyrularia, *L. C. Rich.*
  Hamiltonia, Mühlenb.
  Calinux, Raf.
Cervantesia, *Ruiz et Pav.*
Myoschilos, *Ruiz et Pav.*
Octarillum, *Lour.*

NUMBERS. GEN. 18. Sp. 110.

Olacaceæ.
POSITION.—Loranthaceæ.—SANTALACEÆ.—Aristolochiaceæ.
Thymelaceæ.

ORDER CCCII. LORANTHACEÆ.—LORANTHS.

Loranthеæ, *Juss. et Rich. Ann. Mus.* 12. 292. (1808); *DC. Prodr.* 4. 277. *Mémoire* (1830); *Blume, Fl. Jav.*—Viscoideæ, *Rich. Anal. du Fr.* 33. (1818).—Loranthaceæ, *Ed. pr.* xxxiii.; *Endl. Gen.* clxvi.; *Wight Illustr.* 2. *t.* 119.—Myzodendreæ, *R. Brown in Linn. Trans.* xix. 232.

DIAGNOSIS.—*Asaral Exogens, with a 1-celled ovary and definite ovules with a naked nucleus.*

Shrubby plants, in almost all cases growing into the tissue of other vegetables, as true parasites. Leaves opposite, or sometimes alternate, veinless, fleshy, without stipules. Flowers ☿ or ♂ ♀, axillary, or terminal, solitary, corymbose, cymose, or spiked. Calyx sometimes 0; most commonly springing from within the brim of a fleshy cup-like expansion of the pedicel, and usually surrounded with bracts at the base; sepals 3, 4, or 8 in number, often joined into a tube, valvate in æstivation. Petals 0. Stamens equal in number to the sepals, and opposite to them if any are present; anthers 1-celled, 2-celled, or broken up into numerous cavities. Ovary 1-celled, sunk within the cup-like expansion of the pedicel, and adhering to it; * ovules with a perfectly naked nucleus, erect, or suspended from the point of a central placenta; style 1 or 0; stigma simple, if distinguishable. Fruit succulent, (or occasionally dry,) 1-celled. Seed solitary; embryo longer than the fleshy albumen, and generally projecting beyond it; sometimes with no apparent cotyledons, in Viscum several in the same seed; radicle at the point of the seed most remote from the hilum. [According to Mr. Bidwill, the seeds of Nuytsia germinate with 3 cotyledons.—*Ann. Nat. Hist.* viii. 439.]

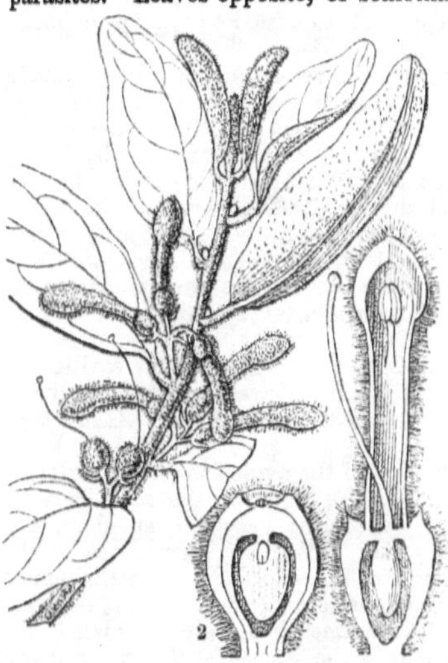

2 1

Fig. DXXIII.

Very different opinions are entertained by Botanists concerning the true affinity of Loranths. In some respects they are near Caprifoils, from which they are readily known not only by their parasitical habit, but also by their stamens being opposite valvate lobes of a tubular calyx. Don has expressed an opinion that a connection is established between this Order and Araliads, by means of Aucuba (*Jameson's Journal, Jan.* 1830, p. 168). Brown (*Flinders*, 549) suggests a relation to Proteads. Endlicher decides in favour of the relation to Caprifoils, Witch Hazels and Cornels. Adrien de Jussieu takes a similar view (*Cours Elément.*, p. 567). Dr. Wight suggests a relation to Alangiads. Adolphe Brongniart combines them, along with Hornworts, Chloranths, Sandalworts, and Olacads, into a class which he calls Santalinées. These discordant opinions are caused by the different interpretations put by Botanists upon the nature of the floral envelopes.

It is customary to call the floral envelopes of the genera of Loranths by the name of sepals in Viscum, and of petals in Loranthus, because in the latter genus we find external to them a cup-like expansion, which is regarded as a calyx. It however seems impossible to doubt that the parts of the perianth are really of the same nature in both instances, as is proved moreover by the stamens, which are applied to their face in both cases. Schleiden, indeed, calls the ♂ flower of Viscum naked, and supposes it to consist of nothing but anthers; but M. Decaisne has more correctly shown the ♂ flowers

* Schleiden has taken a very different view of the structure of Viscum, and describes it as having a truly *naked ovule!* surrounded by a tetramerous herbaceous perianth; this ovule he calls "erect, atropal, and consisting of a naked nucleus."—*Wiegm. Arch.* 1839, p. 213.

Fig. DXXIII.—Loranthus chrysanthus.—*Blume.* 1. section of a flower; 2. of a fruit.

of that genus to consist of 4 anthers grown to the inner face of 4 calycine sepals. The rim exterior to the calyx, which has given rise to the idea that the coloured part of a Loranth is corolla, is present in Viscum also, in the form of a slight annular swelling; and is in all probability analogous to the raised line terminating the cup, from the rim of which the sepals spring in Chryseis or Eschscholtzia. In fact, we must in theory regard the flower of a Loranth to consist of a fleshy cup-like expansion of the end of a branch, from the upper edge of which expansion the sepals rise. This point being settled, we then have no difficulty in admitting the near alliance of Loranths and Sandalworts; a fact not lost sight of by Dr. Brown in his *Prodromus;* he also, in speaking of his Myzodendreæ, or feathered Loranths, again adverts to the resemblance between their three ovules suspended from the apex of a central placenta, and the same part in Sandalworts.—*Linn. Trans.* xix. 232. Decaisne too, recognises their apetalous condition, and refers them to the neighbourhood of Sandalworts. They may also be looked upon as having considerable analogy with Proteads, which must be considered to occupy a place in the perigynous sub-class parallel with that of Loranths in the epigynous. The occasional separation of the ♂ and ♀ in different flowers points strongly to a relation to some diclinous Order, which relation seems to be found in Helwingiads. See p. 296.

In some respects this singular Order offers very curious deviations from the ordinary structure of similar plants. The wood of Viscum is described by Decaisne as consisting, when young, of eight woody bundles surrounding a green pith; in these bundles are no spiral vessels, but instead, and nearly in the place where they are usually found, some ringed tubes; these, together with elongated and dotted or reticulated cells and fibres analogous to those of the liber, make up all the longitudinal tissue of the plant. On the outside of these bundles of woody matter, and opposite to them, are found others, similar in number but smaller, and composed exclusively of fibres of the liber.—*Mémoire sur le Développement du Gui.* Brown states that in Myzodendron the whole woody tissue consists of ladder-shaped vessels (v. scalariformia), a structure very different from that of other genera of Loranths.

In the genus Viscum the anther forms its pollen in a number of distinct cavities, in the same way as in Ægiceras; this has been beautifully illustrated by Decaisne, (*Acad. Roy. Brux. XIII. t.* 1.)

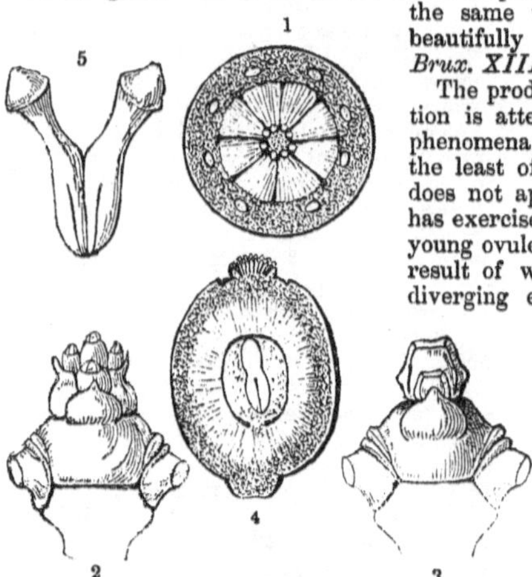

Fig. DXXIV.

The production of the ovules and their fertilisation is attended with some of the most curious phenomena known in the vegetable kingdom, not the least of which is that in Viscum the ovule does not appear till three months after the pollen has exercised its influence; and another, that the young ovules sometimes become consolidated, the result of which is the presence of two or more diverging embryos in the same seed. See M. Decaisne's Memoir above quoted, and also that of Griffith, *On the Development of the Ovules of Loranthus and Viscum, in the Linnean Transactions,* vol. xviii. p. 71, for many other important particulars.

The nature of the parasitism of these plants is very curious, and has been most carefully described by Griffith. He states that in Loranthus the ripe seeds adhere firmly to the substance on which they are applied, by means of their viscid coating, which hardens into a transparent glue. In two or three days after application, the radicle curves towards its support, and as soon as it reaches it becomes enlarged and flattened. By degrees a union is formed between the woody system of the parasite and stock, after which the former lies exclusively on the latter, the fibres of the sucker-like root of the parasite expanding on the wood of the support "in the form of a pâte d'oie." Prior to that time the parasite had been nourished by its own albumen, which is gradually absorbed. "As soon as the young parasite has acquired the height of one or two inches, when an additional supply of nourishment is perhaps required, a lateral shoot is sent out, which is, especially towards the point, of a green

Fig. DXXIV.—Viscum album. 1. a cross section of the stem (*Decaisne*); 2. ♀ flowers; 3. ♂ flowers (*Schleiden*); 4. the fruit cut perpendicularly; 5. a pair of embryos united where they come in contact (*Decaisne*).

colour. This at one, or two, and subsequently at various points, adheres to the support by means of sucker-like productions, which are precisely similar in structure and mode of attachment to the original seminal one." The fibres of the parasite never penetrate beyond their original attachment; in the adult the sucker-bearing shoots frequently run to a considerable distance, many plants being literally covered with parasites, all of which have originated from one seed. "I have seen," says Mr. Griffith, "such shoots, which had taken their course *along a decayed branch*, become replaced, and *return in quest*, as I may express it, *of a part capable of affording some nourishment!*" The attacks of Loranthus are not confined to branches or trunks. Mr. Griffith saw cases of their having seized upon the leaf of a Guttiferous plant, and the succulent leaves of a Polypody. Although not milky plants, yet they will occasionally establish themselves on such as are so, as for example, on the Jac (Artocarpus integrifolia); "hence it is obvious that they must have an eliminating power;" and this is confirmed by the reports of Chemists, who assure us that the wood of the Mistletoe when parasitic on the Apple tree, is found to contain twice as much potash, and five times as much phosphoric acid as the wood of the foster tree. This is the more probable when we call to mind how rapidly some Loranths rot away from their parent branch when it dies, leaving the cellular matter of the latter in the form of huge furrowed and lobed plates which embraced the parasite and held it fast in its place. Exceedingly curious specimens of this kind of growth have been brought to Europe from Guatemala, by Mr. Skinner.

See Dutrochet *Sur la Motilité*, for many curious experiments with Mistletoe.

The Order seems to be equally dispersed through the equinoctial regions of both Asia and America; but on the continent of Africa to be much more rare, only 2 having been yet described from equinoctial Africa, and 5 or 6 from the Cape of Good Hope. Two are named from the South Seas, and 1 from New Holland; but this number requires, no doubt, to be largely increased. Three only are known in Europe. Nuytsia floribunda, a beautiful shrub, with very large thyrses of bright orange-coloured flowers, is a singular instance of a plant of this parasitical Order growing upon the ground. And such is the abundance of the orange-coloured blossoms, that the colonists at King George's Sound compare it to a tree on fire; hence it has gained the name of Fire-tree. A second species (N. ligustrina, *A. C.*) was found by Mr. Cunningham in 1817 in the more arid parts of the Blue Mountains west from Port Jackson.

The bark is usually astringent, as in the Mistletoe of the Oak. The berries contain a viscid matter like birdlime, which is insoluble in water and alcohol. The most remarkable quality that Loranths possess, however, is the power of rooting on the wood of other plants, at whose expense they live. The habit of the common Mistletoe gives an idea of those of all, except that in the genus Loranthus the calyx is tubular and often richly coloured. In medicine they are of small moment; the Mistletoe of the Oak, consecrated by Druidical superstition, was the common Viscum album. Loranthus tetrandrus is used for dyeing black in Chile; and some of them are employed in Brazilian medicine as poultices, and even as antisyphilitics; they are, however, of so little moment, that Martius scarcely names them in his Brazilian Materia Medica.

GENERA.

Myzodendron, *Sol.*
Misodendron, Endl.
Angelopogon, Pöpp.
Antidaphne, *Pöpp.*
Arceuthobium, *Bieberst.*
Razoumowskia, Hoffm.
Viscum, *Tournef.*
Ginalloa, *Korth.*
Tupeia, *Cham. et Schl.*
Loranthus, *Linn.*
Lonicera, Plum.
Helixanthera, Lour.
Scurrula, Don.
Notanthera, Don.
Gaidendron, Don.
Baratranthus, Korth.
? *Glutago*, Commers.
Dendropemon, *Blum.*
Lipotactes, *Blum.*
Phœnicanthemum, *Blm.*
Dendrophthoë, *Mart.*
Ciclanthus, Endl.
Tapinanthus, *Blum.*
Lichtensteinia, Wendl.
Moquinia, Spreng. f.
Loxanthera, *Blum.*
Psittacanthus, *Mart.*
Trygonanthus, Endl.
Phthirusa, *Mart.*
Elythranthe, *Mart.*
Macrosolen, *Blum.*
Tristerix, *Mart.*
Spirostylis, *Presl.*
Struthanthus, *Mart.*
Lepeostegeres, *Blum.*
Tolypanthus, *Blum.*
Nuytsia, *R. Br.*

NUMBERS. GEN. 23. Sp. 412.

Proteaceæ.
POSITION.—Santalaceæ.—LORANTHACEÆ.————.
Cornaceæ.

ORDER CCCIII. ARISTOLOCHIACEÆ.—BIRTHWORTS.

Aristolochiæ, *Juss. Gen* (1789); *R Brown Prodr.* 349; *Endl. Gen.* cxiv.; *Horsfield Pl. Jav. p.* 43.—Pistolochinæ and Asarinæ, *Link Handb* 1. 367. (1829).—Asarineæ, *Bartl. Ord. Nat.* 81. (1830).

DIAGNOSIS.—*Asaral Exogens, with a* 3-6-*celled ovary and* 00 *ovules.*

Herbaceous plants or shrubs, the latter often climbing. Wood without concentric zones and inseparable wedges. Leaves alternate, simple, stalked, often with a stipule opposite the leaf, scale-shaped or leafy; or with none. Flowers axillary, solitary, brown or some dull colour. Flowers hermaphrodite. Calyx adherent, tubular, with the segments valvate or induplicate in æstivation, sometimes regular, sometimes very unequal. Stamens 6 to 12, epigynous, distinct, or adhering to the style and stigmas. Ovary inferior, 6-celled, very rarely 3 or 4-celled; ovules anatropal, 00, horizontally attached to the axis; style simple; stigmas radiating, as numerous as the cells of the ovary. Fruit dry or succulent, 3- 4- 6-celled, many seeded. Seeds thin, angular, or round, with a very minute embryo placed in the base of fleshy albumen. Cotyledons inconspicuous; radicle next the hilum.

These are usually stationed upon the limits of Endogens and Exogens, agreeing with the former in the ternary division of the flower, and in some respects in habit; with the latter in the more essential points of their structure. De Candolle, in the *Botanicon Gallicum*, places them between Oleasters and Spurgeworts, to the former of which he thinks that they approach through Asarum, but with the latter of which their relation is not obvious. To Passionflowers they may be compared, on account of the twining habit, alternate leaves, and leafy stipules of many species; and to Cucurbits, on account of their twining habit, and inferior ovary. Brown, however, is of opinion that their affinity is in reality with Nepenths; a

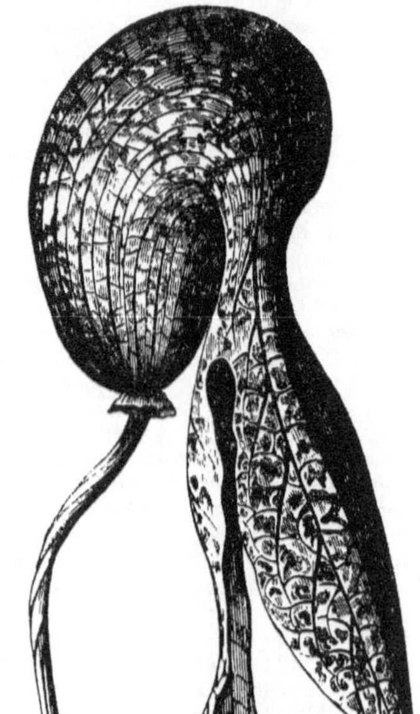

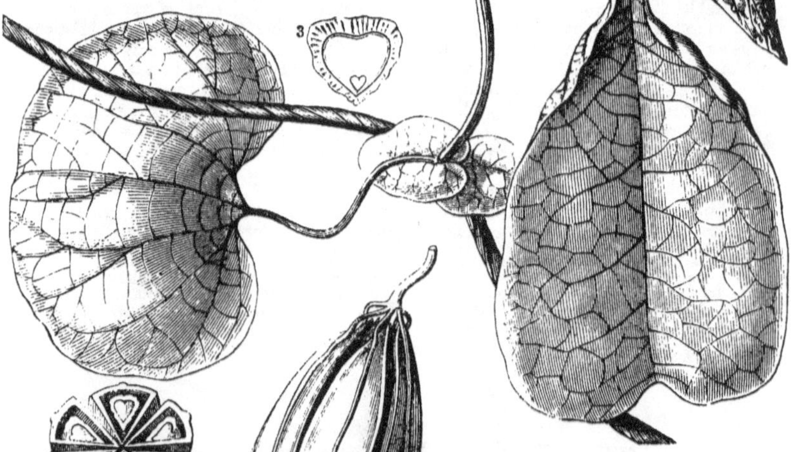

2 Fig. DXXV. 1

Fig. DXXV.—Aristolochia galeata.—*Martius.* 1. fruit of an Aristolochia; 2. cross section of it; 3. half its seed.

view adopted by Endlicher, myself formerly, and many others. I confess, however, that a more attentive study of the Order and its affinities has greatly weakened this opinion, and that I cannot but consider that there is no very strongly marked affinity to be traced between Birthworts and other plants. Endlicher compares them, with some justice, to Yams and Taccads, but they are most certainly dicotyledonous. Their regularly ternary structure and incompletely formed wood indicate, however, a strong tendency towards the condition of Dictyogens, and perhaps they may be looked upon as the best point of transition to that class from Exogens. It is in some measure on that account that they have been placed in this arrangement last in the whole series, and therefore at a point where we may suppose that the chain of Orders must return into itself. If their association with Sandalworts and Loranths should be objected to, I would submit that the correspondence of these Orders in their epigynous apetalous flowers, and minute embryo in copious albumen, are circumstances of agreement of no mean importance, and that it is at present impossible to discover any better station for either of the three Orders.

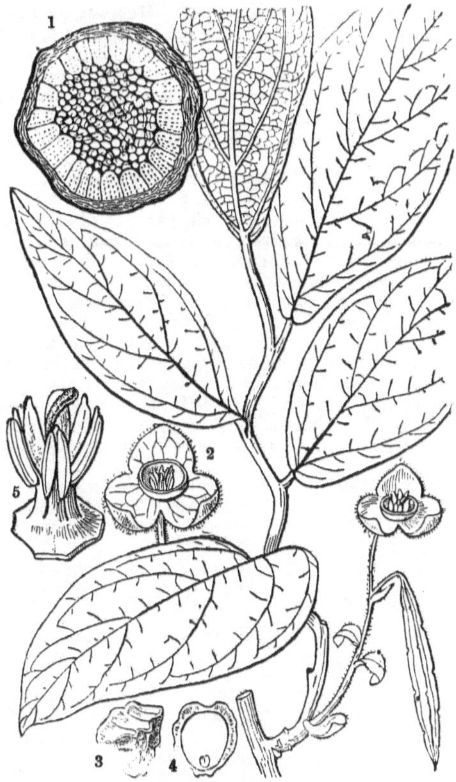

Fig. DXXVI.

Very common in the equinoctial parts of South America, and rare in other countries; found sparingly in North America, Europe, and Siberia; more frequently in the basin of the Mediterranean, and in small numbers in India.

Birthworts are in general tonic and stimulating; Aristolochia is, as its name implies, considered emmenagogue, especially the European species rotunda, longa, and Clematitis. An infusion of the dried leaves of Aristolochia bracteata, a nauseously bitter plant, is given by native Indian practitioners as an anthelmintic; fresh, bruised and mixed with castor oil, they are considered as a valuable remedy in obstinate psora. The root of A. indica is supposed by the Hindoos to possess emmenagogue and antarthritic virtues; it is very bitter. The A. fragrantissima, called in Peru, Bejuca de la Estrella, or Star Reed, is highly esteemed in Peru as a remedy against dysenteries, malignant inflammatory fevers, colds, rheumatic pains, &c. The root is the part used. The power of the root of A. serpentaria in arresting the progress of the worst forms of typhus, is highly spoken of by Barton; it has an aromatic smell, approaching that of Valerian, with a warm, bitterish, pungent taste. It acts as a stimulant, tonic, diaphoretic, and in certain cases as an antispasmodic and anodyne. It is peculiarly useful in supporting the strength and in allaying the irregular action which attends great febrile debility. Dr. Chapman considers it "admirably suited to check vomiting and to tranquillise the stomach, more particularly in bilious cases." As its name implies, it is used as an antidote to serpent bites, a quality in which several other species participate, among which may be mentioned the A. trilobata, a Jamaica plant, also employed as a sudden and powerful sudorific, and the Carthagena A. anguicida, concerning which Jacquin writes, that the juice of the root chewed and introduced into the mouth of a serpent so stupifies it that it may for a long time be handled with impunity; if the reptile is com-

Fig. DXXVI.—Bragantia Blumei. 1. a section of its wood; 2. one of its flowers; 3. a seed; 4. the same divided perpendicularly; 5. anthers and stigma of Asarum.

pelled to swallow a few drops it perishes in convulsions. The root is also reputed to be an antidote to serpent-bites. This plant is probably the celebrated Guaco of the Colombians, concerning whose supposed efficacy as an alexipharmic, so much has been said by Humboldt and others: at least a leaf of what is either this species, or one closely allied to it, has been given me by Dr. Hancock as the genuine Guaco. It is not a little remarkable that the power of stupifying snakes, ascribed in Carthagena to A. anguicida, should be also attributed to A. pallida, longa, bœtica, sempervirens, and rotunda, which are said to be the plants with which the Egyptian jugglers stupify the snakes they play with. In medicine these plants are slightly aromatic stimulating tonics, useful in the latter stages of low fever; the taste is bitter and acrid; the odour strong and disagreeable; they are said to be sudorific, and have been employed as emmenagogues in amenorrhœa.

The stimulating qualities of Birthworts seem to reach their maximum in A. cymbifera, labiosa, ringens, galeata, and macroura, Brazilian species, whose roots have a very penetrating, disagreeable smell, like that of Rue, and a strong, bitter, aromatic taste, producing almost entirely the same effects as the Virginia snake-root (A. serpentaria). They are very frequently used in Brazil against ulcers, paralytic affections of the extremities, dyspepsy, impotentia virilis, in nervous and intermittent fevers, especially those in which a predominant disorder of the pituitous membrane, or the whole lymphatic system has been observed. A. grandiflora, a fœtid Jamaica species, is said by Swartz to be poisonous to hogs. For the qualities of other species see *Martius Mat. Med. Bras.* 107. One of the Asarabaccas, or Asarums, is analogous in its action, viz. A. canadense, which is a warm aromatic stimulant and diaphoretic; but A. europæum is said to be purgative, emetic, and diuretic; it is called Cabaret in France, because, as it is said, the frequenters of pot-houses use it to produce vomiting. Bragantia tomentosa, an intensely bitter plant, is used in Java as an emmenagogue, according to Horsfield.

GENERA.

Asarum, *Tournef.*
Heterotropa, *Decaisne.*
Aristolochia, *Tournef.*
Clematitis, Endl.
Glossula, Raf.
Serpentaria, Raf.
Pistolochia, Raf.
Sipho, Endl.
Hocquartia, Dumort.
Siphisia, Raf.
Siphonolochia, Reich.
Cardiolochia, Raf.
Guaco, Liebm.
Dictyanthes, Raf.
Einomenia, Raf.
Endodaca, Raf.
Isotrema, Raf.
Niphus, Raf.
Bragantia, *Lour.*
Ceramium, Blum.
Vanhallia, Schult. f.
Munickia, Reichenb.
Apama, Lam.
Trimeriza, *Lindl.*
Asiphonia, *Griff.*
Thottea, *Rottb.*
Trichopodium, *Lindl.*
Trichopus, Gärtn.

NUMBERS. GEN. 8. Sp. 130.

Nepenthaceæ? ?
POSITION.—Santalaceæ.—ARISTOLOCHIACEÆ.—Loranthaceæ.
Dioscoreaceæ.

GENERA INSUFFICIENTLY KNOWN;

WHOSE STATION IN A NATURAL SYSTEM IS ALTOGETHER UNCERTAIN.

APETALOUS.

Adelanthus, *Endl.*
Cavanilla, Thunb.
Moldenhauera, Spreng.
Agdestis, *Moç. et Sessé.*[1]
Aniba, *Aubl.*
Cedrota, Schreb.
Apactis, *Thunb.*
Augea, *Thunb.*
Barbeuia, *Thouars.*
Didymeles, *Thouars.*
Dilobeia, *Thouars.*
Donzellia, *Tenor.*
Juliania, *Schlecht.*
Hypopterygium, Id. Linn. xvii. 635.
Lindera, *Thunb.*
Mauneia, *Thouars.*
Morella, *Lour.*
Physena, *Noronh.*
Varonthe, Herb. Juss.
Piptolepis, *Benth.*[2]
Plegorhiza, *Molin.*
Pterotum, *Lour.*
Schousbœa, *Schum.*
Stixis, *Lour.*

MONOPETALOUS.

Baitaria, *Ruiz et Pav.*
Calibrachoa, *Llav. et Lex.*
Cerium, *Lour.*
Cobamba, *Blanco.*[3]
Coddingtonia, *Bowd.*
Corallophyllum, *H. B. K.*[4]
Lennoa, Lav. et Lex.
Cyrta, *Lour.*[5]
Dasus, *Lour.*
Diclidanthera, *Martius.*[6]
Duvaucellia, *Bowd.*
Epirhizanthus, *Blum.*[7]
Fouquiera, *H. B. K.*[8]
Goetzia, *Wydler.*[9]
Hydropyxis, *Raf.*
Keiria, *Bowd.*
Matthisonia, *Radd.*
Morelosia, *Llav.*[10]
Noltia, *Schumach.*
Octas, *Jack.*
Palladia, *Lam.*
Blackwellia, Gärtn.
Pentaceras, *C. F. W. Meyer.*
Pholisma, *Nutt.*[11]
Poraqueiba, *Aubl.*
Barreria, Scop.
Rapourea, *Aubl.*
Camax, Schreb.
Reussia, *Endl.*[12]
Rochefortia, *Swartz.*
Rotula, *Lour.*

POLYPETALOUS.

Allasia, *Lour.*[13]
Ballingayum, *Blanco.*
Calispermum, *Lour.*
Casimiroa, *Llav. et Lex.*
Cœlopyrum, *Jack.*
Dichroa, *Lour.*
Dobera, *Juss.*
Tomex, Forsk.
Eurila, *Blanco.*
Floscopa, *Lour.*
Griselinia, *Forst.*
Scopolia, Forst.
Henschelia, *Presl.*[14]
Heptaca, *Lour.*
Hippomanica, *Molin.*
Lanneoma, *Del.*
Macarthuria, *Hüg.*
Macharisia, *Thouars.*
Sassia, *Molin.*
Spirospermum, *Thouars.*[15]
Tampoa, *Aubl.*
Trisecus, *Willd.*
Trujanoa, *Llav. et Lex.*

GENERA ALTOGETHER UNDESCRIBED.

Adhunia, *Fl. Fl.*
Æschrion, *Fl. Fl.*
Barberina, *Fl. Fl.*
Benjamina, *Fl. Fl.*
Berteroa, *Zipp.*
Blepharistemma, *Benth.*
Bonamica, *Fl. Fl.*
Borea, *Zipp.*
Bureca, *Zipp.*
Calypteris, *Zipp.*
Canicidia, *Fl. Fl.*
Carpocalymma, *Zipp.*
Carpothales, *E. M.*
Catonia, *Fl. Fl.*
Celsa, *Fl. Fl.*
Cheobula, *Fl. Fl.*
Consuegria, *Caldas.*
Courimari, *Aubl.*
Cynotoxicum, *Fl. Fl.*
Cyrtonora, *Zipp.*
Democritea, *Fl. Fl.*
Distrepta, *Miers.*
Donatophorus, *Zipp.*
Dulacia, *Fl. Fl.*
Dysemone, *Forst.*
Epigenia, *Fl. Fl.*
Hesioda, *Fl. Fl.*
Isopteris, *Wall.*
Ivonia, *Fl. Fl.*
Kugia, *Bert.*
Leretia, *Fl. Fl.*
Leucantha, *Zipp.*
Mapa, *Fl. Fl.*
Melanium, *Zipp.*
Michoxia, *Fl. Fl.*
Mnassea, *Fl. Fl.*
Nangha, *Zipp.*
Narda, *Fl. Fl.*
Peltocarpus, *Zipp.*
Peronia, *Wall.*
Petagna, *Raf.*
Petalostemma, *R. Br.*
Platystigma, *R. Br.*
Pombea, *Cald.*
Pometia, *Fl. Fl.*
Porocillea, *Miers.*
Quuna, *Aubl.*
Romana, *Fl. Fl.*
Rutilia, *Fl. Fl.*
Saldanha, *Fl. Fl.*
Sarcocalyx, *Zipp.*
Serjania, *Fl. Fl.*
Stephanostoma, *Zipp.*
Tenorea, *Raf.*
Thevetia, *Fl. Fl.*
Turreta, *Fl. Fl.*
Valenzuelia, *S. Mut.*
Vigia, *Fl. Fl.*
Viviania, *Raf.*

---

[1] Menispermaceæ, *DC.* [2] Near Forestiera, *Benth.;* but the flowers are ⚥ and the fruit is unknown. [3] Not Scrophulariaceæ, because of its 1-celled capsule and the situation and proportion of the stamens. *G. B.* [4] Referred in this work to Fir-rapes? p. 452. [5] Styraceæ, *Endl.* [6] Styraceæ, *Endl.* [7] Not Scrophulariaceæ, because of its solitary seeds and peculiar stamens. [8] This genus has been erected into an Order by De Candolle, and this view has been adopted by myself, Endlicher, and Meisner. But it appears to me doubtful whether the figure of the plant in Humboldt's work is to be relied upon; and, at all events, until its seeds are known it is unsafe to hazard a conjecture about it: for this reason I forbear to repeat the remarks made upon it in the last edition of this work; it is, however, probable that Bronnia of the same authors is nothing more than a Cantua without its corolla; and this is an additional reason for leaving Fouquiera for further consideration. (See *DC. Prodr.* 3. 349; *Ed. Pr.* lxxxix.; *Endl. Gen.* p. 914, who refers the plant to Frankeniaceæ). [9] Styraceæ, *Endl.* [10] Styraceæ, *Endl.* [11] Certainly not a Broomrape; no doubt very near Corallophyllum. (See p. 452). [12] Pontederaceæ, *Endl.* [13] Cucurbitaceæ, *Endl.;* but the ovary is free? and the flowers ⚥. Arnott supposes the description to be erroneous. [14] Lardizabalaceæ? *Endl.* Cucurbitaceæ? *Decaisne.* [15] Menispermaceæ, *Endl.*

*** *For Additional Genera, see* SUPPLEMENT.

ESTIMATED NUMBER OF GENERA AND SPECIES

IN

THE VEGETABLE KINGDOM,

AS FAR AS IS AT PRESENT KNOWN.

	Gen	Sp.		
Class I. THALLOGENS.				
Alliance I. *Algales.*				
Order 1. Diatomaceæ, 12	45	457		
— 2. Confervaceæ, 14	66	368		
— 3. Fucaceæ, 20	81	452		
— 4. Ceramiaceæ, 23	88	682		
— 5. Characeæ, 26	3	35	283	1994
Alliance II. *Fungales.*				
Order 6. Agaricaceæ,				
— 7. Lycoperdaceæ,				
— 8. Uredinaceæ,				
— 9. Botrytaceæ, } 41			598	4000
— 10. Helvellaceæ,				
— 11. Mucoraceæ,				
Alliance III. *Lichenales.*				
Order 12. Graphidaceæ,				
— 13. Collemaceæ, } 50			58	2400
— 14. Parmeliaceæ				
Total . . .	..	..	939	8394
Class II. ACROGENS.				
Alliance IV. *Muscales.*				
Order 15. Ricciaceæ, 57	8	29		
— 16. Marchantiaceæ, 58	15	20		
— 17. Jungermanniaceæ, 59	42	650		
— 18. Equisetaceæ, 61	1	10		
— 19. Andræaceæ, 63	2	13		
— 20. Bryaceæ, 64	44	1100	113	1822
Alliance V. *Lycopodales.*				
Order 21. Lycopodiaceæ, 69	1	200		
— 22. Marsileaceæ, 71	5	24	6	224
Alliance VI. *Filicales.*				
Order 23. Ophioglossaceæ, 77	4	25		
— 24. Polypodiaceæ, 78	183	2000		
— 25. Danæaceæ, 82	5	15	192	2040
Total . . .	.	..	310	4086
Class III. RHIZOGENS.				
Order 26. Balanophoraceæ, 89	12	30		
— 27. Cytinaceæ, 91	4	7		
— 28. Rafflesiaceæ, 93	5	16	21	53
Total . . .	..	..	21	53
Class IV. ENDOGENS.				
Alliance VII. *Glumales.*				
Order 29. Graminaceæ, 106	291	3800		
— 30. Cyperaceæ, 117	112	2000		
— 31. Desvauxiaceæ, 120	4	15		
— 32. Restiaceæ, 121	23	171		
— 33. Eriocaulaceæ, 122	9	200	439	618[illegible]

	Gen	Sp.		
Alliance VIII. *Arales.*				
Order 34. Pistiaceæ, 124	6	20		
— 35. Typhaceæ, 126	2	13		
— 36. Araceæ, 127	z6	170		
— 37. Pandanaceæ, 130	7	75	41	278
Alliance IX. *Palmales.*				
Order 38. Palmaceæ, 134	73	400	73	400
Alliance X. *Hydrales.*				
Order 39 Hydrocharidaceæ, 141	12	20		
— 40. Naiadaceæ, 143	9	16		
— 41. Zosteraceæ, 145	5	12	26	48
Alliance XI. *Narcissales.*				
Order 42. Bromeliaceæ, 147	23	170		
— 43. Taccaceæ, 149	2	8		
— 44. Hæmodoraceæ, 151	13	50		
— 45. Hypoxidaceæ, 154	4	60		
— 46. Amaryllidaceæ, 155	68	400		
— 47. Iridaceæ, 159	53	550	163	1238
Alliance XII. *Amomales.*				
Order 48. Musaceæ, 163	4	20		
— 49. Zingiberaceæ, 165	29	247		
— 50. Marantaceæ, 168	6	160	39	427
Alliance XIII. *Orchidales.*				
Order 51. Burmanniaceæ, 171	7	30		
— 52. Orchidaceæ, 173	394	3000		
— 53. Apostasiaceæ, 184	3	5	404	3035
Alliance XIV. *Xyridales.*				
Order 54. Philydraceæ, 186	2	2		
— 55. Xyridaceæ, 187	5	70		
— 56. Commelynaceæ, 188	16	260		
— 57. Mayaceæ, 189	1	4	24	336
Alliance XV. *Juncales.*				
Order 58. Juncaceæ, 191	13	200		
— 59. Orontiaceæ, 193	13	70	26	270
Alliance XVI. *Liliales.*				
Order 60. Gilliesiaceæ, 196	2	5		
— 61. Melanthaceæ, 198	30	130		
— 62. Liliaceæ, 200	133	1200		
— 63. Pontederaceæ, 206	6	30	171	1365
Alliance XVII. *Alismales.*				
Order 64. Butomaceæ, 208	4	7		
— 65. Alismaceæ, 209	3	50		
— 66. Juncaginaceæ, 210	7	44	14	101
Total . . .	..	..	1420	13684

Class V. Dictyogens.	Gen	Sp.		
Order 67. Triuridaceæ, 213	2	2		
— 68. Dioscoreaceæ, 214	6	110		
— 69. Smilaceæ, 215	2	120		
— 70. Philesiaceæ, 217	2	2		
— 71. Trilliaceæ, 218	4	30		
— 72. Roxburghiaceæ, 219	1	4	17	268
Total . . .	. .	. .	17	268
Class VI. Gymnogens.				
Order 73. Cycadaceæ, 223	6	45		
— 74. Pinaceæ, 226	20	100		
— 75. Taxaceæ, 230	9	50		
— 76. Gnetaceæ, 232	2	15	37	210
Total . . .	. .	. .	37	210
Class VII. Exogens.				
Alliance XVIII. *Amentales.*				
Order 77. Casuarinaceæ, 249	1	20		
— 78. Betulaceæ, 251	2	65		
— 79. Altingiaceæ, 253	1	3		
— 80. Salicaceæ, 254	2	220		
— 81. Myricaceæ, 256	3	20		
— 82. Elæagnaceæ, 257	4	30	13	358
Alliance XIX. *Urticales.*				
Order 83. Stilaginaceæ, 259	3	20		
— 84. Urticaceæ, 260	23	300		
— 85. Ceratophyllaceæ, 263	1	6		
— 86. Cannabinaceæ, 265	2	2		
— 87. Moraceæ, 266	8	184		
— 88. Artocarpaceæ, 269	23	54		
— 89. Platanaceæ, 272	1	6	61	572
Alliance XX. *Euphorbiales*				
Order 90. Euphorbiaceæ, 274	191	2500		
— — Gyrostemoneæ, 282	2	3		
— 91. Scepaceæ, 283	3	6		
— 92. Callitrichaceæ, 284	1	6		
— 93. Empetraceæ, 285	4	4		
— — Batideæ, 286	1	2		
— 94. Nepenthaceæ, 287	1	6	203	2527
Alliance XXI. *Quernales.*				
Order 95. Corylaceæ, 290	8	265		
— 96. Juglandaceæ, 292	4	27	12	292
Alliance XXII. *Garryales.*				
Order 97. Garryaceæ, 295	2	6		
— 98. Helwingiaceæ, 296	1	1	3	7
Alliance XXIII. *Menispermales.*				
Order 99. Monimiaceæ, 298	8	40		
— 100. Atherospermaceæ, 300	3	4		
— 101. Myristicaceæ, 301	5	35		
— 102. Lardizabalaceæ, 303	7	15		
— 103. Schizandraceæ, 305	5	12		
— 104. Menispermaceæ, 307	11	175	39	281
Alliance XXIV. *Cucurbitales.*				
Order 105. Cucurbitaceæ, 311	56	270		
— 106. Datiscaceæ, 316	3	4		
— 107. Begoniaceæ, 318	2	159	61	433
Alliance XXV. *Papayales.*				
Order 108. Papayaceæ, 321	8	25		
— 109. Pangiaceæ, 323	3	4	11	29
Alliance XXVI. *Violales.*				
Order 110. Flacourtiaceæ, 327	31	85		
— 111. Lacistemaceæ, 329	2	6		

	Gen	Sp.		
Order 112. Samydaceæ, 330	5	80		
— 113. Passifloraceæ, 332	12	210		
— 114. Malesherbiaceæ, 335	2	5		
— 115. Moringaceæ, 336	1	4		
— 116. Violaceæ, 338	11	300		
— 117. Frankeniaceæ, 340	4	24		
— 118. Tamaricaceæ, 341	3	43		
— 119. Sauvagesiaceæ, 343	3	15		
— 120. Crassulaceæ, 344	22	450		
— 121. Turneraceæ, 347	2	60	98	1282
Alliance XXVII. *Cistales.*				
Order 122. Cistaceæ, 349	7	185		
— 123. Brassicaceæ, 351	173	1600		
— 124. Resedaceæ, 356	6	41		
— 125. Capparidaceæ, 357	28	340	214	2166
Alliance XXVIII. *Malvales.*				
Order 126. Sterculiaceæ, 360	34	125		
— 127. Byttneriaceæ, 363	45	400		
— 128 Vivianiaceæ, 365	4	15		
— 129. Tropæolaceæ, 366	5	43		
— 130. Malvaceæ, 368	37	1000		
— 131. Tiliaceæ, 371	35	350	160	1933
Alliance XXIX. *Sapindales.*				
Order 132. Tremandraceæ, 374	3	16		
— 133. Polygalaceæ, 375	19	495		
— 134. Vochyaceæ, 379	8	51		
— 135. Staphyleaceæ, 381	3	14		
— 136. Sapindaceæ, 382	50	380		
— 137. Petiveriaceæ, 386	3	10		
— 138. Aceraceæ, 387	3	60		
— 139. Malpighiaceæ, 388	42	555		
— 140. Erythoxylaceæ, 391	1	75	132	1656
Alliance XXX. *Guttiferales.*				
Order 141. Dipteraceæ, 393	7	47		
— — Lophiraceæ, 395	1	1		
— 142. Ternströmiaceæ, 396	33	130		
— 143. Rhizobolaceæ, 398	2	8		
— 144. Clusiaceæ, 400	30	150		
— 145. Marcgraviaceæ, 403	4	26		
— 146. Hypericaceæ, 405	13	276		
— 147. Reaumuriaceæ, 407	3	4	93	642
Alliance XXXI. *Nymphales.*				
Order 148. Nymphæaceæ, 409	5	50		
— 149. Cabombaceæ, 412	2	3		
— 150. Nelumbiaceæ, 414	1	3	8	56
Alliance XXXII. *Ranales.*				
Order 151. Magnoliaceæ, 417	11	65		
— 152. Anonaceæ, 420	20	300		
— 153. Dilleniaceæ, 423	26	200		
— 154. Ranunculaceæ, 425	41	1000		
— — Cephaloteæ, 428	1	1		
— 155. Sarraceniaceæ, 429	2	7		
— 156. Papaveraceæ, 430	18	130	119	1703
Alliance XXXIII. *Berberales.*				
Order 157. Droseraceæ, 433	7	90		
— 158. Fumariaceæ, 435	15	110		
— 159. Berberidaceæ, 437	12	100		
— 160. Vitaceæ, 439	7	260		
— 161. Pittosporaceæ, 441	12	78		
— — Canellaceæ, 442	2	3		
— 162. Olacaceæ, 443	21	48		
— 163. Cyrillaceæ, 445	3	5	79	604
Alliance XXXIV. *Ericales.*				
Order 164. Humiriaceæ, 447	4	10		
— 165. Epacridaceæ, 448	30	320		
— 166. Pyrolaceæ, 450	5	20		
— 167. Francoaceæ, 451	2	5		

	Gen	Sp.		
Order 168. Monotropaceæ, 452	6	10		
— 169. Ericaceæ, 453	42	850	89	1215
Alliance XXXV. *Rutales.*				
Order 170. Aurantiaceæ, 457	20	95		
— 171. Amyridaceæ, 459	22	45		
— 172. Cedrelaceæ, 461	9	25		
— 173. Meliaceæ, 463	33	150		
— 174. Anacardiaceæ, 465	41	95		
— 175. Connaraceæ, 468	5	41		
— 176. Rutaceæ, 469	47	400		
— 177. Xanthoxylaceæ, 472	20	110		
— 178. Ochnaceæ, 474	6	82		
— — Coriarieæ, 475	1	8		
— 179. Simarubaceæ, 476	10	35		
— 180. Zygophyllaceæ, 478	7	100		
— 181. Elatinaceæ, 480	6	22		
— 182. Podostemaceæ, 482	9	25	236	1233
All. XXXVI. *Geraniales.*				
Order 183. Linaceæ, 485	3	90		
— 184. Chlænaceæ, 486	4	8		
— 185. Oxalidaceæ, 488	6	325		
— 186. Balsaminaceæ, 490	2	110		
— 187. Geraniaceæ, 493	4	500	19	1033
Alliance XXXVII. *Silenales.*				
Order 188. Caryophyllaceæ, 498	53	1055		
— 189. Illecebraceæ, 499	24	100		
— 190. Portulacaceæ, 500	12	184		
— 191. Polygonaceæ, 502	29	490	118	1829
Alliance XXXVIII. *Chenopodales.*				
Order 192. Nyctaginaceæ, 506	14	100		
— 193. Phytolaccaceæ, 508	9	60		
— — Surianaceæ, 509	1	1		
— 194. Amarantaceæ, 510	38	282		
— 195. Chenopodiaceæ, 512	63	360	125	803
Alliance XXXIX. *Piperales.*				
Order 196. Piperaceæ, 515	20	600		
— 197. Chloranthaceæ, 519	3	15		
— 198. Saururaceæ, 521	4	7	27	622
Alliance XL. *Ficoidales.*				
Order 199. Basellaceæ, 524	4	12		
— 200. Mesembryaceæ, 525	5	375		
— 201. Tetragoniaceæ, 527	11	65		
— 202. Scleranthaceæ, 528	4	14	24	466
Alliance XLI. *Daphnales.*				
Order 203. Thymelaceæ, 530	38	300		
— 204. Proteaceæ, 532	44	650		
— 205. Lauraceæ, 535	46	450		
— 206. Cassythaceæ, 538	1	9	129	1409
Alliance XLII. *Rosales.*				
Order 207. Calycanthaceæ, 540	2	6		
— 208. Chrysobalanaceæ, 542	11	50		
— 209. Fabaceæ, 544	467	6500		
— 210. Drupaceæ, 557	5	110		
— 211. Pomaceæ, 559	16	200		
— 212. Sanguisorbaceæ, 561	12	125		
— 213. Rosaceæ, 563	38	500	551	7401
All. XLIII. *Saxifragales.*				
Order 214. Saxifragaceæ, 567	19	310		
— 215. Hydrangeaceæ, 569	9	45		
— 216. Cunoniaceæ, 571	22	100		
— 217. Brexiaceæ, 573	4	6		
— 218. Lythraceæ, 574	35	300	89	761

	Gen	Sp.		
Alliance XLIV. *Rhamnales.*				
Order 219. Penæaceæ, 577	3	21		
— 220. Aquilariaceæ, 579	6	10		
— 221. Ulmaceæ, 580	9	60		
— 222. Rhamnaceæ, 581	42	250		
— 223. Chailletiaceæ, 583	4	10		
— 224. Hippocrateaceæ, 584	6	86		
— 225. Celastraceæ, 586	24	260		
— 226. Stackhousiaceæ, 589	2	10		
— 227. Sapotaceæ, 590	21	212		
— 228. Styracaceæ, 592	6	115	123	1034
Alliance XLV. *Gentianales.*				
Order 229. Ebenaceæ, 595	9	160		
— 230. Aquifoliaceæ, 597	11	110		
— 231. Apocynaceæ, 599	100	566		
— 232. Loganiaceæ, 602	22	162		
— — Cassipoureæ, 604	2	7		
— 233. Diapensiaceæ, 606	2	2		
— 234. Stilbaceæ, 607	3	7		
— 235. Orobanchaceæ, 609	12	116		
— 236. Gentianaceæ, 612	60	450	221	1580
Alliance XLVI. *Solanales.*				
Order 237. Oleaceæ, 616	24	130		
— 238. Solanaceæ, 618	60	900		
— 239. Asclepiadaceæ, 623	141	910		
— 240. Cordiaceæ, 628	11	180		
— 241. Convolvulaceæ, 630	43	660		
— 242. Cuscutaceæ, 633	2	50		
— 243. Polemoniaceæ, 635	17	104	298	2934
Alliance XLVII. *Cortusales.*				
Order 244. Hydrophyllaceæ, 638	16	75		
— 245. Plumbaginaceæ, 640	8	160		
— 246. Plantaginaceæ, 642	3	120		
— 247. Primulaceæ, 644	29	215		
— 248. Myrsinaceæ, 647	30	320	86	880
Alliance XLVIII. *Echiales.*				
Order 249. Jasminaceæ, 650	5	100		
— 250. Salvadoraceæ, 652	1	2		
— 251. Ehretiaceæ, 653	14	297		
— 252. Nolanaceæ, 654	6	35		
— 253. Boraginaceæ, 655	53	600		
— 254. Brunoniaceæ, 657	1	2		
— 255. Lamiaceæ, 659	125	2350		
— 256. Verbenaceæ, 663	56	610		
— 257. Myoporaceæ, 665	9	42		
— 258. Selaginaceæ, 666	10	120	280	4158
Alliance XLIX. *Bignoniales.*				
Order 259. Pedaliaceæ, 669	14	25		
— 260. Gesneraceæ, 671	54	260		
— 261. Crescentiaceæ, 673	11	34		
— 262. Bignoniaceæ, 675	44	450		
— 263. Acanthaceæ, 678	105	750		
— 264. Scrophulariaceæ, 681	176	1814		
— 265. Lentibulariaceæ, 686	4	175	408	3508
Alliance L. *Campanales.*				
Order 266. Campanulaceæ, 689	20	500		
— 267. Lobeliaceæ, 692	27	375		
— 268. Goodeniaceæ, 694	14	150		
— 269. Stylidiaceæ, 696	5	121		
— 270. Valerianaceæ, 697	12	185		
— 271. Dipsacaceæ, 699	6	150		
— 272. Calyceraceæ, 701	5	10		
— 273. Asteraceæ, 702	1005	9000	1102	10491

	Gen	Sp.		
Alliance LI. *Myrtales.*				
Order 274. Combretaceæ, 717	22	200		
— 275. Alangiaceæ, 719	3	8		
— 276. Chamælauciaceæ, 721	15	50		
— 277. Haloragaceæ, 722	8	70		
— 278. Onagraceæ, 724	28	450		
— 279. Rhizophoraceæ, 726	5	20		
— 280. Belvisiaceæ, 728	2	4		
— 281. Melastomaceæ, 731	118	1200		
— 282. Myrtaceæ, 734	45	1300		
— 283. Lecythidaceæ, 739	7	38	253	3340
Alliance LII. *Cactales.*				
Order 284. Homaliaceæ, 742	8	30		
— 285. Loasaceæ, 744	15	70		
— 286. Cactaceæ, 746	16	800	39	900
Alliance LIII. *Grossales.*				
Order 287. Grossulariaceæ, 750	2	95		
— 288. Escalloniaceæ, 752	7	60		
— 289. Philadelphaceæ, 753	3	25		
Order 290. Barringtoniaceæ, 754	10	28	22	208

	Gen	Sp.		
Alliance LIV. *Cinchonales.*				
Order 291. Vacciniaceæ, 757	13	200		
— 292. Columelliaceæ, 759	1	3		
— 293. Cinchonaceæ, 761	269	2500		
— 294. Caprifoliaceæ, 766	14	220		
— 295. Galiaceæ, 768	8	320	305	3243
Alliance LV. *Umbellales.*				
Order 296. Apiaceæ, 773	267	1500		
— 297. Araliaceæ, 780	21	160		
— 298. Cornaceæ, 782	9	40		
— 299. Hamamelidaceæ, 784	10	15		
— 300. Bruniaceæ, 785	15	65	322	1780
Alliance LVI. *Asarales.*				
Order 301. Santalaceæ, 787	18	110		
— 302. Loranthaceæ, 789	23	412		
— 303. Aristolochiaceæ, 792	8	130	49	652
Total . .	. .	. .	6191	66225

GRAND TOTAL.

	Genera.	Species.
Class I. Thallogens	939	8394
— II. Acrogens	310	4086
— III. Rhizogens	21	53
— IV. Endogens	1420	13084
— V. Dictyogens	17	268
— VI. Gymnogens	37	210
— VII. Exogens	6191	66225
TOTAL . .	8935	92930

ARTIFICIAL ANALYSIS

OF THE

NATURAL ORDERS.

CLASS I. THALLOGENS.

Nourished by spawn or mycelium (*Fungales*).	
Spores in fours.	
Hymenium naked	*Agaricaceæ*, 41
Hymenium inclosed in a peridium	*Lycoperdaceæ*, 41
Spores or spore cases single.	
Spores naked.	
Thallus obsolete	*Uredinaceæ*, 41
Thallus floccose	*Botrytaceæ*, 41
Spores inclosed,	
in asci	*Helvellaceæ*, 41
in a veil or sporangium	*Mucoraceæ*, 41
Without spawn.	
Aquatics (*Algales*).	
Crystalline angular, multiplied by disarticulation	*Diatomaceæ*, 12
Vesicular, or filamentous, or membranous, multiplied by zoospores	*Confervaceæ*, 14
Cellular or tubular.	
Multiplied by simple spores	*Fucaceæ*, 20
——— tetraspores	*Ceramiaceæ*, 23
——— spiral coated nucules	*Characeæ*, 26
Terrestrial (*Lichenales*).	
Spores naked	*Graphidaceæ*, 50
Spores in asci.	
Thallus gelatinous or cartilaginous	*Collemaceæ*, 50
Thallus pulverulent or cellular	*Parmeliaceæ*, 50

CLASS II. ACROGENS.

§ *With no distinct axis of growth.*

Spores without elaters	*Ricciaceæ*, 57
Spores furnished with elaters.	
Spore cases opening into valves	*Jungermanniaceæ*, 59
——— valveless	*Marchantiaceæ*, 58

§§ *With a distinct axis of growth.*

Spores furnished with elaters,	
inclosed in cases, opening into valves	*Jungermanniaceæ*, 59
naked, collected in cones	*Equisetaceæ*, 61
Spores without elaters.	
Spore cases seated on leaves,	
ringed	*Polypodiaceæ*, 78
ringless	*Danæaceæ*, 82
Spore cases inclosed within the edge of a contracted leaf	*Ophioglossaceæ*, 77
Spore cases inclosed within an involucre	*Marsileaceæ*, 71
Spore cases naked,	
sessile in the axil of leaves or bracts	*Lycopodiaceæ*, 69
stalked,	
valveless	*Bryaceæ*, 64
opening into valves	*Andræaceæ*, 63

CLASS III. RHIZOGENS.

Ovules solitary	*Balanophoraceæ*, 89
Ovules indefinite.	
Anthers opening by slits	*Cytinaceæ*, 91
——— pores	*Rafflesiaceæ*, 93

CLASS IV. ENDOGENS.

* *Flowers complete (having distinct floral envelopes).*

§ *Ovary inferior.*

† *Flowers gynandrous.*

- Ovary 1-celled. Seed-coat loose . . . *Orchidaceæ*, 175
- Ovary 3-celled . . . *Apostasiaceæ*, 184

†† *Flowers not gynandrous.*

- Veins of leaves diverging from the midrib.
  - Anther 1, with 1 cell . . . *Marantaceæ*, 168
  - Anther 1, with 2 cells . . . *Zingiberaceæ*, 165
  - Anthers 5, or 6 . . . *Musaceæ*, 160
- Veins of leaves parallel with midrib.
  - Stamens 3.
    - Anthers turned outwards . . . *Iridaceæ*, 159
    - Anthers turned inwards. (Fruit winged) . . . *Burmanniaceæ*, 171
  - Stamens 6.
    - Leaves flat.
      - Fruit 3-celled. Sepals corolline.
        - Radicle remote from hilum, which is strophiolate . . . *Hypoxidaceæ*, 154
        - —— next the hilum . . . *Amaryllidaceæ*, 155
      - Fruit 3-celled. Sepals calycine . . . *Bromeliaceæ*, 147
      - Fruit 1-celled . . . *Taccaceæ*, 149
    - Leaves equitant . . . *Hæmodoraceæ*, 151
  - Stamens more than 6 . . . *Hydrocharaceæ*, 141

§§ *Ovary superior.*

- Sepals calycine or glumaceous.
  - Carpels separate, more or less.
    - Placentæ spread over the dissepiments . . . *Butomaceæ*, 208
    - Placentæ narrow . . . *Alismaceæ*, 209
  - Carpels combined in a solid pistil.
    - Petals quite distinct from the calyx.
      - Placentæ axile . . . *Commelynaceæ*, 188
      - Placentæ parietal . . . *Mayaceæ*, 189
    - Petals undistinguishable from the calyx.
      - Flowers scattered . . . *Juncaceæ*, 191
      - Flowers on a spadix . . . *Orontiaceæ*, 193
- Sepals corolline.
  - Carpels more or less separate.
    - Seeds solitary . . . *Palmaceæ*, 133
    - Seeds numerous.
      - Anthers turned outwards . . . *Melanthaceæ*, 198
      - Anthers turned inwards.
        - Floral envelopes 6 . . . *Butomaceæ*, 208
        - Floral envelopes 2 . . . *Philydraceæ*, 146
  - Carpels combined in a solid pistil.
    - Petals rolled inwards after flowering . . . *Pontederaceæ*, 206
    - Petals not rolled inwards after flowering.
      - Flowers with external appendages . . . *Gilliesiaceæ*, 196
      - Flowers without external appendages . . . *Liliaceæ*, 200

** *Flower incomplete (having no distinct floral envelopes except leaves).*

§ *Flowers glumaceous.*

- Stems hollow . . . *Graminaceæ*, 106
- Stems solid.
  - Carpel solitary. Seed erect . . . *Cyperaceæ*, 117
  - Carpel solitary. Seed pendulous . . . *Restiaceæ*, 121
  - Carpels several, distinct.
    - Glumes only . . . *Desvauxiaceæ*, 120
    - A membranous cup within the glumes . . . *Eriocaulaceæ*, 122
  - Carpels several, combined.
    - Placentæ parietal . . . *Xyridaceæ*, 187
    - Placentæ central . . . *Restiaceæ*, 121

§§ *Flowers naked; or with a few verticillate leaves.*

† *Flowers on a spadix.*

- Fruit drupaceous . . . *Pandanaceæ*, 130
- Fruit berried. Leaves in the bud convolute . . . *Araceæ*, 127
- Fruit dry. Anthers clavate on weak filaments . . . *Typhaceæ*, 126

†† *Flowers not on a spadix.*

- Floaters. Ovules pendulous.
  - Pollen globose . . . *Naiadaceæ*, 143
  - Pollen confervoid . . . *Zosteraceæ*, 145
- Terrestrial. Ovules erect . . . *Juncaginaceæ*, 210
- Floaters. Ovules erect . . . *Pistiaceæ*, 124

CLASS V. DICTYOGENS.

Ovary inferior	*Dioscoreaceæ*, 214
Ovary superior.	
Carpels distinct, 00	*Triuridaceæ*, 213
Carpels consolidated.	
Placentæ axile.	
Flowers hexapetaloideous	*Smilaceæ*, 215
—— tripetaloideous	*Trilliaceæ*, 218
Placentæ basal	*Roxburghiaceæ*, 219
Placentæ parietal	*Philesiaceæ*, 217

CLASS VI. GYMNOGENS.

Stem jointed	*Gnetaceæ*, 232
—— continuous.	
Leaves pinnate	*Cycadeaceæ*, 223
—— simple.	
Females in cones	*Pinaceæ*, 227
—— solitary	*Taxaceæ*, 230

CLASS VII. EXOGENS.

POLYPETALOUS.

* Polyandrous. *Stamens more than* 20.

§ *Ovary inferior, or partially so.*

† *Leaves furnished with stipules.*

‡ *Carp. more or less distinct (at least as to the styles); or solitary*	*Pomaceæ*, 559
‡‡ *Carpels wholly combined into a solid pistil, with more placentas than one.*	
Placentas central.	
Leaves opposite	*Rhizophoraceæ*, 726
Leaves alternate; Flowers irregular	*Lecythidaceæ*, 739
Placentas parietal	*Homaliaceæ*, 742

†† *Leaves without stipules.*

‡ *Carp. more or less distinct (at least as to the styles).*	
Carpels numerous, quite inferior	*Anonaceæ*, 430
‡‡ *Carpels wholly combined into a solid pistil.*	
Placentas spread over the surface of the dissepiments	*Nymphæaceæ*, 409
Placentas parietal.	
Petals definite in number, distinct from calyx	*Loasaceæ*, 744
Petals indefinite in number, confused with the calyx	*Cactaceæ*, 746
Placentas in the axis.	
Leaves marked with little transparent dots.	
Ovary one-celled. Embryo homogeneous	*Chamælauciaceæ*, 721
Ovary with more than one cell. Cotyledons distinct	*Myrtaceæ*, 734
Leaves dotless.	
Petals indefinite in number, very numerous	*Mesembryaceæ*, 525
Petals definite in number,	
narrow and strap-shaped	*Alangiaceæ*, 719
round and concave.	
Style 1	*Barringtoniaceæ*, 754
Styles disunited	*Philadelphaceæ*, 753

§§ *Ovary wholly superior.*

† *Leaves furnished with stipules.*

‡ *Carp. more or less distinct (at least as to the styles); or solitary.*	
Stamens hypogynous.	
Carpel solitary	*Fabaceæ*, 544
Carpels 00	*Magnoliaceæ*, 417
Stamens perigynous.	
Styles from the apex of the carpels.	
Carpel 1	*Drupaceæ*, 557
Carpels more than 1	*Rosaceæ*, 563
Styles from the base of the carpels	*Chrysobalanaceæ*, 542
‡‡ *Carpels wholly combined into a solid pistil, with more placentas han ne.*	
Placentas parietal.	
Leaves marked with ound transparent dots	*Flacourtiaceæ*, 327
Leaves marked with round and linear transparent dots intermixed	*Samydaceæ*, 330
Placentas in the axis. *(See next page.)*	

Calyx with an imbricated æstivation.
Flowers unisexual . . . . *Euphorbiaceæ*, 274
Flowers hermaphrodite.
Ovary 1-celled. Sepals 2 . . . . *Portulacaceæ*, 500
Ovary with more cells than one.
Calyx double . . . . *Chlænaceæ*, 486
Calyx single . . . . *Cistaceæ*, 349
Calyx with a valvate æstivation.
Stamens monadelphous. Anthers 2-celled.
Stamens columnar, all perfect . . . . *Sterculiaceæ*, 360
Stamens not columnar, partly sterile . . . . *Byttneriaceæ*, 363
Stamens monadelphous. Anthers 1-celled . . . . . *Malvaceæ*, 368
Stamens monadelphous. Calyx irregular and enlarged in the fruit . *Dipteraceæ*, 393
Stamens quite distinct . . . . *Tiliaceæ*, 371

†† *Leaves without stipules.*

‡ *Carp. more or less distinct (at least as to the styles); or solitary.*
Carpels immersed in a fleshy table-shaped disk . . . . . . *Nelumbiaceæ*, 414
Carpels not immersed in a disk.
Stamens perigynous.
Carpel 1 . . . . *Drupaceæ*, 557
Carpels more than 1 . . . . *Rosaceæ*, 563
Stamens hypogynous.
Embryo in vitellus . . . . *Cabombaceæ*, 412
Embryo naked,
very minute.
Seeds with an aril . . . . *Dilleniaceæ*, 423
Seeds without an aril. Albumen fleshy.
Flowers ⚥ . . . . *Ranunculaceæ*, 425
Flowers ♂ ♀ . . . . *Schizandraceæ*, 305
Seeds usually without an aril. Albumen aromatic and ruminate . *Anonaceæ*, 420
nearly as long as seed.
Calyx much imbricated.
Fruit a legume . . . . *Fabaceæ*, 544
Fruit not a legume.
Seeds smooth . . . . *Hypericaceæ*, 405
Seeds hairy . . . . *Reaumuriaceæ*, 407
Calyx but little imbricated.
Fruit not a legume . . . . *Anacardiaceæ*, 465
Fruit a legume . . . . *Fabaceæ*, 544

‡‡ *Carpels wholly combined into a solid pistil, with more placentas than one.*
Placentas parietal, in distinct lines.
Anthers versatile. Juice watery . . . . *Capparidaceæ*, 357
Anthers innate. Juice milky . . . . *Papaveraceæ*, 430
Placentas parietal, spread over the lining of the fruit . . . . . *Flacourtiaceæ*, 327
Placentas spread over the dissepiments . . . . *Nymphæaceæ*, 409
Placentas in the axis.
Stigma large, broad, and petaloid . . . . *Sarraceniaceæ*, 429
Stigma simple.
Ovary 1-celled, with free central placenta . . . . . . *Portulacaceæ*, 500
Ovary many-celled.
Calyx much imbricated.
Leaves compound . . . . *Rhizobolaceæ*, 398
Leaves simple.
Petals equal in number to sepals.
Seeds few . . . . *Clusiaceæ*, 400
Seeds numerous. Petals flat . . . . *Marcgraaviaceæ*, 403
Seeds numerous. Petals crumpled . . . . . . *Cistaceæ*, 349
Calyx but little, or not at all, imbricated.
Stamens perigynous. Calyx tubular . . . . . . *Lythraceæ*, 574
Stamens hypogynous. Calyx many-leaved . . . . . . *Humiriaceæ*, 447

** OLIGANDROUS. *Stamens fewer than* 20.

§ *Ovary inferior, or partially so.*

† *Leaves furnished with stipules.*

Placentas parietal . . . . *Homaliaceæ*, 742
Placentas in the axis.
Flowers completely unisexual . . . . *Begoniaceæ*, 318
Flowers hermaphrodite or polygamous.
Stamens equal to the petals and opposite them . . . . . *Rhamnaceæ*, 581
Stamens, if equal to the petals, alternate with them.
Leaves opposite . . . . *Rhizophoraceæ*, 726
Leaves alternate . . . . *Hamamelidaceæ*, 722

†† *Leaves destitute of stipules.*

Placentas parietal.
Flowers completely unisexual. . . . . *Cucurbitaceæ*, 311
Flowers hermaphrodite or polygamous . . . . *Grossulaceæ*, 750
Placentas in the axis.
Flowers in umbels. Styles 2 . . . . *Apiaceæ*, 773
Flowers in umbels. Styles 3 or more . . . . *Araliaceæ*, 780

Flowers not in umbels.
Carpel solitary.
Petals strap-shaped. Stamens distinct . . . . *Alangiaceæ*, 719
Petals very narrow. Stamens growing on them . . . . . *Loranthaceæ*, 789
Petals oblong. Leaves insipid.
Cotyledons convolute . . . . *Combretaceæ*, 717
Cotyledons flat . . . . *Haloragaceæ*, 722
Petals oblong. Leaves balsamic . . . . *Anacardiaceæ*, 465
Carpels divaricating at the apex.
Leaves alternate. Herbs . . . . *Saxifragaceæ*, 567
Leaves opposite. Shrubs . . . . *Hydrangeaceæ*, 569
Carpels parallel, combined.
Calyx valvate. Petals opposite stamens . . . . . *Rhamnaceæ*, 581
Calyx valvate. Petals alternate with stamens or isomerous.
Albumen 0 . . . . *Onagraceæ*, 724
Albumen copious . . . . *Cornaceæ*, 782
Calyx not valvate.
Stamens doubled downwards. Leaves ribbed . . . . . *Melastomaceæ*, 731
Stamens only curved. Anthers short.
Leaves dotted . . . . *Myrtaceæ*, 734
Leaves not dotted.
Parts of flower 4.
Ovules horizontal or ascending . . . . *Onagraceæ*, 724
Ovules pendulous . . . . *Haloragaceæ*, 722
Parts of flower not 4. Seeds many.
Leafy . . . . *Escalloniaceæ*, 752
Scaly parasites . . . . *Monotropaceæ*, 452
Parts of flower not 4. Seeds few . . . . *Bruniaceæ*, 785

§§ *Ovary wholly superior.*

† *Leaves furnished with stipules.*

‡ *Carpels distinct or solitary.*
Anthers with recurved valves . . . . *Berberidaceæ*, 437
Anthers with longitudinal valves.
Style from base of carpel . . . . *Chrysobalanaceæ*, 542
Style from apex of carpel. Fruit leguminous . . . . . *Fabaceæ*, 544
Style from apex of carpel. Fruit drupaceous or capsular . . . *Rosaceæ*, 563

‡‡ *Carpels wholly combined; with more placentas than one.*
Placentas parietal.
Flowers with a ring of appendages . . . . *Passifloraceæ*, 332
Flowers without any ring of appendages.
Leaves with round and oblong transparent dots . . . . *Samydaceæ*, 330
Leaves dotless, circinate when young . . . . . . *Droseraceæ*, 433
Leaves dotless, straight when young. Fruit capsular . . . *Violaceæ*, 338
Leaves dotless, straight when young. Fruit siliquose . . . *Moringaceæ*, 336
Placentas in the axis.
Styles distinct to the base.
Calyx in a broken whorl, much imbricated . . . . . . *Elatinaceæ*, 480
Calyx but little imbricated, in a complete whorl.
Flowers unisexual . . . . *Euphorbiaceæ*, 274
Flowers hermaphrodite or polygamous.
Petals minute . . . . *Illecebraceæ*, 499
Petals conspicuous. Stamens hypogynous . . . . . *Malpighiaceæ*, 388
Petals conspicuous. Stamens perigynous. Leaves opposite . . *Cunoniaceæ*, 571
Petals conspicuous. Stamens perigynous. Leaves alternate . *Saxifragaceæ*, 567
Calyx valvate . . . . *Tiliaceæ*, 371
Styles more or less combined. Gynobasic.
Gynobase fleshy . . . . *Ochnaceæ*, 474
Gynobase dry. Leaves regularly opposite . . . . . *Zygophyllaceæ*, 478
Gynobase dry. Leaves alternate more or less.
Fruit beaked . . . . *Geraniaceæ*, 493
Fruit not beaked . . . . *Oxalidaceæ*, 488
Styles more or less combined. Not gynobasic.
Calyx much imbricated, in a broken whorl.
Flowers spurred . . . . *Vochyaceæ*, 379
Flowers not spurred, calyculate . . . . *Chlænaceæ*, 486
Flowers not spurred, naked . . . . *Sapindaceæ*, 382
Calyx but little imbricated, in a complete whorl.
Leaves compound. Sepals more than two . . . . . *Staphyleaceæ*, 381
Leaves simple. Sepals more than two . . . . . . *Malpighiaceæ*, 388
Leaves simple. Sepals only two . . . . *Portulacaceæ*, 500
Calyx valvate or open.
Stamens columnar . . . . *Sterculiaceæ*, 360
Stamens not columnar.
Stamens opposite to petals if equal to them in number.
Perigynous . . . . *Rhamnaceæ*, 581
Hypogynous . . . . *Vitaceæ*, 439
Stamens alternate with petals if equal to them in number.
Anthers opening by pores . . . . *Tiliaceæ*, 371
Anthers opening by slits. Petals split . . . . . *Chailletiaceæ*, 583
Anthers opening by slits. Petals undivided . . . . *Amyridaceæ*, 459

††*Leaves destitute of stipules.*

‡ *Carpels more or less distinct*, or *solitary.*

Anthers with recurved valves . . . . *Berberidaceæ*, 437
Anthers with longitudinal valves.
 Fruit leguminous. Radicle next hilum . . . . *Fabaceæ*, 544
 Fruit leguminous. Radicle remote from hilum . . . . . *Connaraceæ*, 468
 Fruit not leguminous.
  Carpels each with an hypogynous scale . . . . . . *Crassulaceæ*, 344
  Carpels each with two hypogynous scales . . . . . *Francoaceæ*, 451
  Carpels without hypogynous scales.
   Albumen very abundant. Embryo minute.
    Flowers ♂ ♀ . . . . *Lardizabalaceæ*, 303
    Flowers ⚥.
     Embryo in vitellus . . . . *Cabombaceæ*, 412
     Embryo naked.
      Herbs. Albumen solid . . . . *Ranunculaceæ*, 425
      Shrubs. Albumen ruminate . . . . *Anonaceæ*, 420
   Albumen in small quantity or wholly wanting.
    Carpels several, all perfect :
     inclosed in the tube of the calyx . . . . . . *Calycanthaceæ*, 540
     naked. Flowers unisexual . . . . *Menispermaceæ*, 307
    Carpels solitary, or all but one imperfect.
     Leaves dotted . . . . *Amyridaceæ*, 459
     Leaves dotless . . . . *Anacardiaceæ*, 465

‡‡ *Carpels combined into a solid pistil.*

Placentas parietal.
 Stamens tetradynamous . . . . *Brassicaceæ*, 351
 Stamens not tetradynamous.
  Flowers with a ring or crown of sterile stamens.
   Sexes distinct.
    Female flower coronetted . . . . *Pangiaceæ*, 323
    ——— not coronetted . . . . *Papayaceæ*, 321
   Sexes combined. Placentæ lining the fruit . . . . . *Flacourtiaceæ*, 327
   Sexes combined. Placentæ in rows. Ovary stalked . . . *Malesherbiaceæ*, 335
  Flowers without sterile stamens.
   Hypogynous disk large. Stamens indefinite . . . . . *Capparidaceæ*, 357
   Hypogynous disk large. Stamens definite . . . . . *Resedaceæ*, 356
   Hypogynous disk small or wanting.
    Albumen very abundant. Embryo minute . . . . . *Papaveraceæ*, 430
    Albumen in small quantity, or wholly wanting.
     Calyx 5-leaved . . . . *Turneraceæ*, 347
     Calyx tubular . . . . *Frankeniaceæ*, 340
Placentas covering the dissepiments . . . . *Nymphæaceæ*, 409
Placentas axile.
 Styles distinct to the base.
  Calyx valvate . . . . *Vivianiaceæ*, 365
  Calyx in a broken whorl, much imbricated.
   Seeds hairy . . . . *Reaumuriaceæ*, 407
   Seeds smooth. Stamens polyadelphous . . . . . . *Hypericaceæ*, 405
   Seeds smooth. Stamens monadelphous or free . . . . *Linaceæ*, 485
  Calyx but little imbricated, in a complete whorl.
   Carpels each with an hypogynous scale . . . . . . *Crassulaceæ*, 344
   Carpels destitute of hypogynous scales.
    Carpels 2, divaricating at the apex . . . . . . *Saxifragaceæ*, 567
    Carpels not divaricating at apex . . . . . . *Caryophyllaceæ*, 496
 Styles more or less combined. Gynobasic.
  Stamens arising from scales . . . . *Simarubaceæ*, 476
  Stamens not arising from scales.
   Styles wholly combined. Flowers hermaphrodite . . . *Rutaceæ*, 469
   Styles wholly combined. Flowers unisexual . . . . *Xanthoxylaceæ*, 472
   Styles divided at point. Flowers irregular . . . . *Balsaminaceæ*, 490
 Styles more or less combined. Not gynobasic.
  Calyx much imbricated, in a broken whorl.
   Flowers symmetrical . . . . *Clusiaceæ*, 400
   Flowers unsymmetrical.
    Flowers regular.
     Petals without appendages . . . . *Aceraceæ*, 387
     Petals with appendages . . . . *Sapindaceæ*, 382
    Flowers papilionaceous . . . . *Polygalaceæ*, 375
  Calyx but little imbricated, in a complete whorl.
   Carpels 4 or more. Anthers opening by pores.
    Embryo in the axis . . . . *Ericaceæ*, 453
    Embryo (very minute) at the base . . . . . . *Pyrolaceæ*, 450
   Carpels 4 or more. Anthers opening by slits.
    Seeds winged.
     Leafy . . . . *Cedrelaceæ*, 461
     Scaly (parasites) . . . . *Monotropaceæ*, 452
    Seeds wingless.
     Stamens united into a long tube . . . . . . *Meliaceæ*, 463
     Stamens free or nearly so.
      Leaves dotted. Seeds amygdaloid . . . . . . *Aurantiaceæ*, 457

Leaves dotless. Seeds minute 00.
  Leafy . . . . *Brexiaceæ*, 573
  Scaly parasites . . . . *Monotropaceæ*, 452
Carpels fewer than 4.
  Flowers unisexual . . . . *Empetraceæ*, 285
  Flowers hermaphrodite.
    Sepals 2 . . . . *Portulacaceæ*, 500
    Sepals more than 2.
      Stamens hypogynous.
        Seeds comose . . . . *Tamaricaceæ*, 341
        Seeds naked.
          Ovules ascending or horizontal . . . . . *Pittosporaceæ*, 441
          Ovules pendulous . . . . *Cyrillaceæ*, 445
      Stamens perigynous.
        Ovules ascending . . . . *Celastraceæ*, 586
        Ovules suspended . . . . *Bruniaceæ*, 785
Calyx valvate or open.
  Anthers opening by pores . . . . *Tremandraceæ*, 374
  Anthers opening by slits.
    Stamens if equal in number to petals opposite them . . . *Rhamnaceæ*, 581
    Stamens if equal in number to petals alternate with them.
      Leaves pinnate . . . . *Amyridaceæ*, 459
      Leaves simple. Calyx tubular. Stamens hypogynous . . . *Olacaceæ*, 443
      Leaves simple. Calyx tubular. Stamens perigynous . . *Lythraceæ*, 574

APETALOUS.

* Achlamydeous.

† *Leaves furnished with stipules.*

Ovules very numerous.
  Seeds winged . . . . *Altingiaceæ*, 253
  Seeds comose . . . . *Salicaceæ*, 254
Ovules solitary or very few.
  Flowers hermaphrodite.
    Stamens unilateral . . . . *Chloranthaceæ*, 519
    Stamens whorled . . . . *Saururaceæ*, 521
  Flowers unisexual.
    Carpels solitary. Ovule erect . . . . *Myricaceæ*, 256
    Carpels solitary. Ovule pendulous . . . . . . *Platanaceæ*, 272
    Carpels triple . . . . *Euphorbiaceæ*, 274

†† *Leaves destitute of stipules.*

Ovules very numerous . . . . *Podostemaceæ*, 482
Ovules soliatry or very few.
  Flowers hermaphrodite.
    Embryo in vitellus . . . . *Piperaceæ*, 515
    Embryo without vitellus . . . . *Oleaceæ*, 616
  Flowers unisexual.
    Flowers naked. Carpel single . . . . *Myricaceæ*, 256
    Flowers naked. Carpels double . . . . *Callitrichaceæ*, 284
    Flowers in an involucre. Anther-valves recurved . . . . *Atherospermaceæ*, 300
    Flowers in an involucre. Anther-valves slit . . . . . *Monimiaceæ*, 298

** Monochlamydeous.

§ *Ovary inferior, or partially so.*

† *Leaves furnished with stipules.*

Flowers hermaphrodite . . . . *Aristolochiaceæ*, 792
Flowers unisexual. Fruit in a cupule . . . . *Corylaceæ*, 290
Flowers unisexual. Fruit naked:
  many-seeded . . . . *Begoniaceæ*, 318
  one-seeded . . . . *Artocarpaceæ*, 269

†† *Leaves destitute of stipules.*

Flowers unisexual, amentaceous.
  Leaves simple, alternate . . . . *Myricaceæ*, 256
  Leaves simple, opposite . . . . *Garryaceæ*, 295
  Leaves compound . . . . *Juglandaceæ*, 292
Flowers unisexual, not amentaceous.
  Seeds immersed in pulp . . . . *Cucurbitaceæ*, 311
  Seeds dry:
    numerous, parietal . . . . *Datiscaceæ*, 316
    solitary, axile . . . . *Helwingiaceæ*, 296
Flowers hermaphrodite or polygamous.
  Leaves with transparent dots . . . . *Myrtaceæ*, 734
  Leaves without dots.
    Ovary 3-6-celled, polyspermous . . . . *Aristolochiaceæ*, 792
    Ovary 1-celled. Anther-valves slit. *(See next page.)*

Embryo straight; cotyledons convolute . . . *Combretaceæ*, 717
Embryo straight; cotyledons flat.
Albumen none . . . *Haloragaceæ*, 722
Albumen fleshy . . . *Santalaceæ*, 787
Embryo curved; cotyledons flat . . . *Chenopodiaceæ*, 512
Ovary 1-celled. Anther many-celled . . . *Loranthaceæ*, 789
Ovary with more cells than 1, but neither 3 nor 6.
Embryo straight . . . *Haloragaceæ*, 722
Embryo curved . . . *Tetragoniaceæ*, 527

§§ *Ovary superior.*

† *Leaves furnished with stipules.*

Flowers hermaphrodite.
Sepals 2 . . . *Portulacaceæ*, 500
Sepals more than 2.
Carpels more than 1, combined into a solid pistil.
Stamens hypogynous. Placentas parietal . . . *Flacourtiaceæ*, 327
Stamens hypogynous. Placentas in the axis.
Calyx valvate. Stamens monadelphous:
partly sterile . . . *Byttneriaceæ*, 363
all fertile . . . *Sterculiaceæ*, 360
Calyx valvate. Stamens distinct . . . *Tiliaceæ*, 371
Calyx imbricated. Fruit beaked . . . *Geraniaceæ*, 493
Calyx imbricated. Fruit not beaked . . . *Malpighiaceæ*, 388
Stamens perigynous. Placentas parietal . . . *Passifloraceæ*, 332
Stamens perigynous. Placentas in the axis.
Leaves opposite. Stamens more than sepals . . . *Cunoniaceæ*, 571
Leaves alternate. Stamens alternate with sepals . . . *Rhamnaceæ*, 581
Leaves alternate. Calyx membranous and ragged . . . *Ulmaceæ*, 580
Carpels solitary, or quite separate.
Calyx membranous (stamens hypogynous) . . . *Illecebraceæ*, 499
Calyx firm and herbaceous.
Styles from the base of carpels . . . *Chrysobalanaceæ*, 542
Styles terminal; one to each ovary.
Fruit leguminous . . . *Fabaceæ*, 544
Fruit not leguminous . . . *Sanguisorbaceæ*, 561
Styles terminal; three to each ovary.
Stipules ochreate . . . *Polygonaceæ*, 502
Stipules simple . . . *Phytolaccaceæ*, 509
Flowers unisexual.
Carpels more than 1, combined into a solid pistil.
Flowers amentaceous. Seeds arillate . . . *Scepaceæ*, 283
Flowers amentaceous. Seeds not arillate . . . *Betulaceæ*, 251
Flowers amentaceous. Seeds numerous. Placentæ parietal . . . *Lacistemaceæ*, 329
Flowers not amentaceous . . . *Euphorbiaceæ*, 274
Carpels solitary.
Cells of anthers perpendicular to the filament . . . *Stilaginaceæ*, 259
Cells of anthers parallel with the filament.
Embryo straight:
albuminous. Stipules small . . . *Urticaceæ*, 260
exalbuminous. Stipules large . . . *Artocarpaceæ*, 269
Embryo hooked:
exalbuminous . . . *Cannabinaceæ*, 265
albuminous . . . *Moraceæ*, 266

†† *Leaves destitute of stipules.*

Flowers hermaphrodite.
Sepals 2 . . . *Portulacaceæ*, 500
Sepals more than 2.
Carpels more than 1, combined into a solid pistil.
Placentas parietal, in lines . . . *Papaveraceæ*, 430
Placentas parietal, lining the pericarp . . . *Flacourtiaceæ*, 327
Placentas in the axis.
Ovary with a very small number of ovules.
Calyx short, herbaceous Gynobasic . . . *Rutaceæ*, 469
Calyx short, herbaceous, not gynobasic.
Embryo curved round mealy albumen . . . *Phytolaccaceæ*, 509
Embryo straight . . . *Celastraceæ*, 586
Calyx tubular, coloured . . . *Penæaceæ*, 577
Ovary with numerous ovules.
Two divaricating carpels . . . *Saxifragaceæ*, 567
Carpels not divaricating. Stamens hypogynous.
Leaves opposite . . . *Caryophyllaceæ*, 496
Leaves alternate . . . *Podostemaceæ*, 482
Carpels not divaricating. Stamens perigynous.
Fruit 1-celled . . . *Primulaceæ*, 644
Fruit many-celled . . . *Lythraceæ*, 574
Carpels solitary or quite separate.
Carpels several. Stamens hypogynous . . . *Ranunculaceæ*, 425
Carpel single.
Anther-valves recurved. Leafy . . . *Lauraceæ*, 535
Anther-valves recurved. Leafless . . . *Cassythaceæ*, 538
Anther-valves slit.
Fruit a legume . . . *Fabaceæ*, 549

Fruit not a legume.
Calyx long or tubular, with a hardened base . . . . *Nyctaginaceæ*, 506
Calyx long or tubular, with a hardened tube . . . . *Scleranthaceæ*, 528
Calyx long or tubular, no where hardened.
Stamens in the points of the sepals . . . . . *Proteaceæ*, 532
Stamens not in the points of the sepals.
Ovules erect . . . . *Elæagnaceæ*, 257
Ovules pendulous.
Fruit 2-valved . . . . *Aquilariaceæ*, 579
Fruit indehiscent. Calyx naked . . . . . *Thymelaceæ*, 530
Calyx short, not tubular, or but little so.
Leaves lepidote . . . . *Elæagnaceæ*, 257
Leaves dotted, not lepidote . . . . . . *Amyridaceæ*, 459
Leaves neither dotted nor lepidote.
Flowers in involucels . . . . *Polygonaceæ*, 502
Flowers naked.
Calyx dry and coloured . . . . . . *Amarantaceæ*, 510
Calyx herbaceous or succulent.
Stamens hypogynous . . . . . . *Chenopodiaceæ*, 512
Stamens perigynous . . . . . . *Basellaceæ*, 524
Flowers unisexual.
Carpels more than one, combined into a solid pistil.
Ovules indefinite in number.
Stamens columnar . . . . *Nepenthaceæ*, 287
Ovules definite in number.
Leaves alternate, dotted . . . . *Xanthoxylaceæ*, 472
Leaves alternate, not dotted . . . . *Euphorbiaceæ*, 274
Carpels solitary, or quite separate.
Calyx tubular, trifid . . . . *Myristicaceæ*, 301
Calyx open, carpels several . . . . *Menispermaceæ*, 307
Calyx open, carpel solitary.
Embryo straight (without albumen) . . . . . . *Casuarinaceæ*, 249
Embryo curled (round mealy albumen) . . . . . . *Chenopodiaceæ*, 512

MONOPETALOUS.

* *Ovary superior. Flowers regular.*

‡ *Ovary 3-4-5-lobed.*
Leaves dotted. . . . . *Rutaceæ*, 469
Leaves dotless. Inflorescence gyrate . . . . *Boraginaceæ*, 655
Leaves dotless. Inflorescence straight.
Corolla with a plaited æstivation . . . . *Nolanaceæ*, 654
Corolla with a flat æstivation . . . . *Stackhousiaceæ*, 589
‡‡ *Ovary not lobed.*
Carpels from 4 to 5, or none.
Anthers opening by pores.
Seeds winged. Herbs . . . . *Pyrolaceæ*, 450
Anthers 2-celled. Seeds wingless. Shrubs . . . . . *Ericaceæ*, 453
Anthers 1-celled. Shrubs . . . . *Epacridaceæ*, 448
Anthers opening by slits.
Stamens equal in number to petals and opposite.
Shrubs . . . . *Myrsinaceæ*, 647
Herbs . . . . *Primulaceæ*, 644
Stamens not opposite the petals if of the same number.
Seeds indefinite.
Carpels distinct . . . . *Crassulaceæ*, 344
Carpels combined. Brown parasites . . . . . . *Monotropaceæ*, 452
Seeds definite.
Carpels distinct . . . . *Anonaceæ*, 420
Carpels combined.
Ovules erect.
Æstivation of corolla imbricate . . . . . . *Sapotaceæ*, 590
Æstivation of corolla plicate . . . . *Convolvulaceæ*, 630
Ovules pendulous.
Stamens twice as numerous as petals . . . . . *Ebenaceæ*, 595
Stamens same number as petals . . . . . . *Aquifoliaceæ*, 597
Carpels usually three.
Inflorescence gyrate . . . . *Hydrophyllaceæ*, 638
Inflorescence straight.
Flowers ♂ ♀ . . . . *Papayaceæ*, 321
Flowers ⚥.
An hypogynous disk . . . . *Polemoniaceæ*, 635
No hypogynous disk . . . . *Diapensiaceæ*, 606
Carpels only two.
Diandrous. Corolla valvate . . . . *Oleaceæ*, 616
Diandrous. Corolla imbricate . . . . *Jasminaceæ*, 650
Stamens 4 or more. Inflorescence gyrate.
Fruit 1-celled . . . . *Hydrophyllaceæ*, 638
Fruit 2-celled. Style bifid . . . . *Ehretiaceæ*, 653
Fruit 2-celled. Style dichotomous . . . . *Cordiaceæ*, 628
Stamens 4 or more. Inflorescence straight.
Calyx in a broken whorl.
Leafy . . . . *Convolvulaceæ*, 630
Scaly parasites . . . . *Cuscutaceæ*, 633

Calyx in a complete whorl.
Flowers symmetrical. Carpels ⊃ . . . *Solanaceæ*, 618
Flowers symmetrical. Carpels ().
Anthers grown to stigma . . . *Asclepiadaceæ*, 623
Anthers free from stigma.
Corolla imbricated . . . *Gentianaceæ*, 612
Corolla valvate . . . *Loganiaceæ*, 602
Corolla contorted . . . *Apocynaceæ*, 5.9
Flowers unsymmetrical.
Leaves with stipules . . . *Loganiaceæ*, 602
Leaves without stipules . . . *Stilbaceæ*, 607
Carpel single.
Stigma with an indusium . . . *Brunoniaceæ*, 657
Stigma without an indusium.
Style single.
Fruit spuriously 2-celled . . . *Plantaginaceæ*, 642
Fruit 1-celled, 1-seeded . . . *Salvadoraceæ*, 652
Styles 5 . . . *Plumbaginaceæ*, 640

** *Ovary superior. Flowers irregular.*

‡ *Ovary 4-lobed* . . . *Lamiaceæ*, 659
‡ ‡ *Ovary undivided.*
Carpel solitary . . . *Selaginaceæ*, 666
Carpels two.
Fruit nucamentaceous, 4-celled.
Radicle inferior . . . *Verbenaceæ*, 663
Radicle superior . . . *Myoporaceæ*, 665
Fruit nucamentaceous, 2-celled.
Anthers 1-celled . . . *Selaginaceæ*, 666
Anthers 2-celled . . . *Stilbaceæ*, 607
Fruit capsular or succulent.
Placentæ parietal.
Seeds amygdaloid.
Fruit succulent, hard-shelled, many-seeded . . . *Crescentiaceæ*, 673
Fruit bony or capsular, few-seeded . . . *Pedaliaceæ*, 669
Seeds not amygdaloid.
Leafy.
Seeds winged . . . *Bignoniaceæ*, 675
Seeds wingless . . . *Gesneraceæ*, 671
Scaly brown parasites . . . *Orobanchaceæ*, 609
Placentæ axile.
Seeds albuminous . . . *Scrophulariaceæ*, 681
Seeds without albumen.
Winged . . . *Bignoniaceæ*, 675
Wingless, attached to hard placental processes . . . *Acanthaceæ*, 678
Placenta free, central . . . *Lentibulariaceæ*, 686

*** *Ovary inferior.*

‡ *Carpels solitary.*
Anthers syngenesious.
Ovule pendulous . . . *Calyceraceæ*, 701
Ovule erect . . . *Asteraceæ*, 702
Anthers free.
Carpel quite solitary . . . *Dipsacaceæ*, 699
Carpel with two additional abortive ones . . . *Valerianaceæ*, 697

‡ ‡ *Carpels more than one.*
Anthers syngenesious . . . *Lobeliaceæ*, 692
Anthers free.
Stamens only 2 . . . *Columelliaceæ*, 759
Stamens more than 2.
Anthers opening by pores . . . *Vacciniaceæ*, 757
Anthers opening by slits.
Stigma naked.
Pentandrous or tetrandrous . . . *Campanulaceæ*, 689
Polyandrous . . . *Belvisiaceæ*, 728
Gynandrous . . . *Stylidiaceæ*, 696
Stigma with an indusium . . . *Goodeniaceæ*, 694
Stigma without an indusium.
Stipules absent. Seeds definite. Stigmas naked.
Leaves alternate . . . *Ebenaceæ*, 595
Leaves opposite, and whorled. Stem square, rough . . . *Galiaceæ*, 759
Leaves opposite. Stem round, smooth . . . *Caprifoliaceæ*, 766
Stipules between the leaves . . . *Cinchonaceæ*, 761

INDEX

OF THE

SCIENTIFIC AND VERNACULAR NAMES OF SPECIES, AND OF VEGETABLE PRODUCTS.

INDEX

OF

CLASSES, ALLIANCES, ORDERS, GENERA, AND THEIR SYNONYMS.

The numoers refer to the pages; those pages in which principal mention is made of the subject indexed are distinguished by an asterisk.]

† This, which is a synonym of Trochiscanthes, has been omitted at the above page, by an oversight.

A LIST OF THE

PRINCIPAL ABBREVIATIONS EMPLOYED IN THE FOREGOING INDEX.

A. Br.—A. Brongn. = *Adolphe Brongniart.*
A. C.—A. Cunn.—A. Cuningh. = *Allan Cunningham.*
A. DC.—Alph. DC. = *Alphonse De Candolle.*
A. Gr. = *Asa Gray.*
A. J.—A. de J.—A. Juss.—Ad. de Juss. = *Adrien de Jussieu.*
A. R.—A. Rich.—Ach. Rich. = *Achille Richard.*
Ach.—Achar. = *Acharius.*
Ad. Brongn. = *Adolphe Brongniart.*
Ad.—Adans. = *Adanson.*
Afz.—Afzel. = *Afzelius.*
Ag.—Agh. = *Agardh.*
Ait. = *Aiton.*
Alb.—Albert. = *Albertini.*
All.—Allion. = *Allioni.*
Amm. = *Amman.*
Andr. = *Andrews.*
Andrz. = *Andrzejowsky.*
Ard. = *Arduini.*
Aresch. = *Areschoug.*
Arn. = *Arnott.*
Arrab. = *Arrabida.*
Arrud. = *Arruda.*
Aub.—Aubl. = *Aublet.*
Auct. = *Auctorum.*
Aug. de St. Hil.—Auguste de St. Hilaire.

B. et F. = *Bluff and Fingerhutt.*
B. et Mont. = *Berkeley and Montagne.*
B. et W. = *Bartling and Wendland.*
Bab. = *Babington.*
Banc. = *Bancroft.*
Barn. = *Barnéoud.*
Bart. = *Barton.*
Bartl. = *Bartling.*
Batt. = *Battarra.*
Bauh. = *Bauhin.*
Baumg. = *Baumgarten.*
Beauv. = *Palisot de Beauvois.*
Bel.—Belang. = *Belanger.*
Benn. = *Bennett.*
Benth. = *Bentham.*
Berg. = *Bergius.*
Berk.—Berkel. = *Berkeley.*
Berl. = *Berlandier.*
Bernh. = *Bernhardi.*
Bert. = *Bertero.*
Bertol. = *Bertoloni.*
Bess. = *Besser.*
Bieb.—Bieberst. = *Bieberstein.*
Big.—Bigel. = *Bigelow.*
Bisch. = *Bischoff.*
Bl.—Blum. = *Blume.*
Blackw. = *Blackwell.*
Blanc. = *Blanco.*
Bluff et F. = *Bluff and Fingerhutt.*
Boehm. = *Boehmer.*
Bœrh. = *Böerhaave.*
Boiss.—Bois. = *Boissier.*
Boj. = *Bojer.*
Bonat. = *Bonati.*
Bong. = *Bongard.*
Bonnem. = *Bonnemaison.*
Bonpl. = *Bonpland.*
Bork.—Borkh. = *Borkhausen.*
Bouch. = *Bouché.*
Bowd. = *Bowdich.*
Br. = *Bridel.*
Br. = *Brown.*
Brebis. = *Brebisson.*
Brid = *Bridel.*
Brign. = *Brignoli.*
Brongn. = *Brongniart.*
Brot.—Broter. = *Brotero.*
Bth. = *Bentham.*
Buch.—Buchan. = *Buchanan.*
Bull. = *Bulliard.*
Bung. = *Bunge.*
Burch. = *Burchell.*
Burm. = *Burmann.*
Büttn. = *Büttner.*
Buxb. = *Buxbaum.*

C. A. M.—C. A. Mey. = *C. A. Meyer.*
C. Bauh. = *Caspar Bauhin.*
C. Ehrenb. = *Charles Ehrenberg.*
C. F. W. M.—C. W. M. = *C. W. Meyer.*
C. C. Gmel. = *C. C. Gmelin.*
Cabrer. = *Cabrera.*
Cald. = *Caldas.*
Camb.—Cambess. = *Cambessédes.*
Casar. = *Casaretto.*

Cass. = *Cassini.*
Castg. = *Castiglione.*
Cat. Hort. Par. = *Catalogus Horti Parisiensis.*
Catesb. = *Catesby.*
Cav.—*Cavan.* = *Cavanilles.*
Cav. et Sech. = *Cavalier and Sechier.*
Ch.—*Cham.* = *Chamisso.*
Ch. et Sch. = *Chamisso and Schlechtendahl.*
Chap. = *Chapman.*
Chav. = *Chavannes.*
Chev.—*Cheval.* = *Chevalier.*
Chois. = *Choisy.*
Cl. Gay. = *Claude Gay.*
Clem. = *Clemente.*
Clus. = *Clusius.*
Col.—*Colebr.* = *Colebrooke.*
Coll. = *Colla.*
Comm.—*Commers.* = *Commerson.*
Commel. = *Commelyn.*
Cord. = *Corda.*
Corr. = *Correa da Serra.*
Cunning. = *Cunningham.*
Cuss. = *Cusson.*
Cyrill. = *Cyrilli.*

D. Don. = *David Don.*
Dal. = *Dalechamp.*
DC. = *De Candolle.*
Dec.—*Dne.* = *Decaisne.*
Dec. et Th. = *Decaisne and Thuret.*
Del.—*Delil.* = *Delile.*
Delabr. = *Delabre.*
Delaun. = *Delaunay.*
Dennst. = *Dennstedt.*
Desc. = *Descourtilz.*
Desf.—*Desfont.* = *Desfontaines.*
Desm.—*Desmaz.* = *Desmazières.*
Desp. = *Despretz.*
Desv. = *Desvaux.*
Dies. = *Diesing.*
Dietr. = *Dietrich.*
Dill.—*Dillen.* = *Dillenius.*
Ditm—*Dittm.* = *Dittmar.*
Domb. = *Dombey.*
Doug. = *Douglas.*
Dryand. = *Dryander.*
Duf. = *Dufour.*
Duham. = *Duhamel.*
Dum.—*Dumort.* = *Dumortier.*
Dun. = *Dunal.*
Dup. Th.—*Dup. Thouars.* = *Dupetit Thouars.*
Dur.—*Durazz.* = *Durazzo.*
Duv. = *Duvau.*

E.—*End.*—*Endl.* = *Endlicher.*
E. M.—*E. Mey.* = *Ernst Meyer.*
E. Z.—*E. et Z.*—*Eck. et Zey.* = *Ecklon and Zeyher.*
Eckl. = *Ecklon.*
Ed. pr. = *Editio prior.*
Ed. prim. = *Editio prima.*
Ehr.—*Ehrenb.* = *Ehrenberg.*
Ehrh. = *Ehrhart.*
Ell. = *Elliott.*
Engelm. = *Engelmann.*
Eschw. = *Eschweiler.*

F. = *Fischer.*
F. et M.—*F. et Mey.* = *Fischer and Meyer.*
Fabric. = *Fabricius.*
Falc. = *Falconer.*
Feuill. = *Feuillé.*
Fisch. = *Fischer.*
Fl. Fl. = *Flora Fluminensis.*
Fl. Mex. = *Flora Mexicana.*
Fl. Wett. = *Flora Wetteravensis.*
Flörk. = *Flörke.*
Flot. = *Flotow.*
Flugg. = *Flugge.*
Forsk. = *Forskahl.*
Forst. = *Forster.*
Fr. = *Fries.*
Fr. Syst. = *Fries Systema.*
Fres.—*Fresen.* = *Fresenius.*
Frœl. = *Frœlich.*
Fror. = *Froriep.*
Fürnr. = *Fürnrohr.*

G. B. = *George Bentham.*
G. D.—*G. Don.* = *George Don.*
Gaill. = *Gaillon.*
Gard.—*Gardn.* = *Gardner.*
Gärtn. = *Gærtner.*
Gaud.—*Gaudich.* = *Gaudichaud.*
Gawl. = *Gawler.*
Ging. = *De Gingins.*
Ginn. = *Ginnanni.*
Gill. = *Gillies.*
Gled. = *Gleditsch.*
Gmel. = *Gmelin.*
Gom. = *Gomez.*
Grah. = *Graham.*
Gratel. = *Grateloup.*
Gren. = *Grenier.*
Grev. = *Greville.*
Gries.—*Gris.*—*Griseb.* = *Grisebach.*
Griff. = *Griffith.*
Gron.—*Gronov.* = *Gronovius.*
Guett. = *Guettard.*
Guill.—*Guillem.* = *Guillemin.*
Guill. et P. = *Guillemin and Perrottet.*
Gusson. = *Gussone.*

H.—*Humb.* = *Humboldt.*
H. B. K. = *Humboldt, Bonpland, and Kunth.*
H. et A. = *Hooker and Arnott.*
H. et B.—*H. et Bonpl.* = *Humboldt and Bonpland.*
H. et G. = *Hooker and Greville.*
H. et J. = *Hombron and Jacquinot*
Hall. = *Haller.*
Ham.—*Hamilt.* = *Hamilton.*
Hartm. = *Hartmann.*
Harv. = *Harvey.*
Hass. = *Hassall.*
Hasselq. = *Hasselquist.*
Hassk. = *Hasskarl.*
Haw. = *Haworth.*
Hayn. = *Hayne.*
Hedw. = *Hedwig.*
Heist. = *Heister.*
Hellen. = *Hellenius.*
Herb. = *Herbert.*
Herit. = *L'Heritier.*
Hernand. = *Hernandez.*
Hffsg.—*Hffngg.* = *Hoffmanscgg.*
Hochs.—*Hochst.* = *Hochstetter.*
Hoffm. = *Hoffmann.*
Hon. Bell. = *Honorius Belli.*
Hook. = *Hooker.*
Hook. et Arn. = *Hooker and Arnott.*
Hook. f. = *Hooker the younger.*
Hopp. = *Hoppe.*
Hor.—*Horan.* = *Horaninow.*
Hork. = *Horkel.*
Horn.—*Hornem.* = *Hornemann.*
Hornsch. = *Hornschuch.*
Hort. Hisp. = *Hortus Hispanicus.*
Hort. = *Hortorum.*
Houst. = *Houston.*
Houtt. = *Houttuyn.*
Hskl. = *Hasskarl.*
Hst. = *Hochstetter.*
Hub. = *Hubert.*
Huds. = *Hudson.*
Hug. = *Hugel.*

Isn. = *Isnard.*

J. = *Jussieu.*
J. Ag.—*J. Agdh.* = *Agardh the younger.*
J. J. B. = *J. J. Bennett.*
J. Sm. = *John Smith.*
J. St. Hil. = *Jaume St. Hilaire.*
J. W. Schm. = *J. W. Schmidt.*
Jacq. = *Jacquin.*
Jaub. = *Jaubert.*
Jaub. et Spch. = *Jaubert and Spach.*
Jon. = *Jones.*
Jungh. = *Junghuns.*
Juss. = *Jussieu.*

K. Mull.=Karl Muller.
K. et H.—Kuhl et H.=Kuhl und Hasselt.
Ka.—Kar.—Karel.=Karelin.
Kæmpf.=Kæmpfer.
Kaulf.=Kaulfuss.
Kch.=Koch.
Kl.—Klzh.=Klotzsch.
Kœl.=Kœler.
Kön.=König.
Korth.=Korthals.
Kostel.=Kosteletsky.
Kth.=Kunth.
Ktz.—Ktzing.—Kutz.=Kutzing.
Kze.=Kunze.

L.=Linnæus.
L. C. R.—L. C. Rich.=Louis Claude Richard.
L. et Z.=Lehmann and Zeyher.
L. et Lex.=Llav. and Lexarsa.
L. f.=Linnæus the younger.
L. et O.—Lk. et O.=Link and Otto.
Labill.—Lab.=Labillardière.
Lag.—Lagasc.=Lagasca.
Lallem.=Lallemand.
Lam.=Lamark.
Lamb.=Lambert.
Lamx.=Lamouroux.
Lapeyr.=Lapeyrouse.
Lavrad.=Lavrado.
Laws.=Lawson.
Leandr.=Leandro da Sacramento.
Ledeb.=Ledebour.
Lehm.=Lehmann.
Lem.—Lemair.=Lemaire.
Lepr.=Leprieur.
Lesch.—Leschen.=Leschenault de la Tour.
Less.=Lessing.
Lestib.=Lestiboudois.
Lev.—Leveill.=Léveillé.
Lher.=L'Heritier.
Lib.=Libert or Libosch.
Licht.=Lichtenstein.
Lilj.=Lilja.
Lindl.=Lindley.
Linn.=Linnæus.
Linn. f.=Linnæus the younger.
Ll.—Llav.=Llave.
Ll. et L.=Llave and Lexarza.
Lk.=Link.
Lodd.=Loddiges.
Löfl.=Loefling.
Loud.=Loudon.
Lour.—Loureir.=Loureiro.
Ludw.=Ludwig.
Lyngb.=Lyngbye.

M.=Mönch.
M. B.=Marschall von Bieberstein.
M. O. Curt.=M. O. Curtis.
M. et D.=Morren and Decaisne.
M. et S.=Moçino and Sesse.
M. et Z.=Martius and Zuccarini.
Macfad.=Macfadyen.
Mans.=Manso.
Maratt.=Maratti.
Marcgr.=Marcgraaf.
Mart.=Martius.
Med.—Medik.=Medikus.
Meisn.=Meisner.
Men.—Mengh.=Meneghini.
Mentz.=Mentzel.
Mert.=Mertens.
Mey.=Meyen.
Mich.—Michx.=Michaux.
Michel.=Micheli.
Mik.=Mikan.
Mill.=Miller.
Miq.=Miquel.
Mirb.=Mirbel.
Mit.—Mitch.=Mitchell.
Moç. et Sess.—Moç. et S.=Moçino and Sesse.
Mol.—Molin.=Molina.
Mon.=Mönch.
Monn.=Monnier.
Mont.=Montagne.
Moq.=Moquin Tandon.
Mor.=Morison.
Moric.=Moricand.
Morr.=Morren.
Muhl.—Muhlenb.=Muhlenberg.
Murr.=Murray.
Mut.=Mutis.

N. et M.=Nees and Martius.
N. Burm.=N. Burmann.
Naud.=Naudo.
Neck.=Necker.
Nees ab Esen.=Nees von Esenbeck.
Ner.=Neraud.
Newm.=Newman.
Nor.—Noron.—Noronh.=Noronha.
Not.=Notar.=Notaris.
Nutt.=Nuttall.

Op. et C.=Opiz and Corda.
Ort.—Orteg.=Ortega.
Ott.=Otto.

P.—Pers.=Persoon.
P. Alp.=Prosper Alpinus.
P. Br.=Patrick Browne.
P. et E.=Pœppig and Endlicher.
Pal.=Palisot de Beauvois.
Pall.=Pallas.
Panz.=Panzer.
Parl.—Parlat.=Parlatore.
Pauq.=Pauquy.
Pav.=Pavon.
Perl.=Perleb.
Pers.=Persoon.
Pet.—Petiv.=Petiver.
Pfeiff.=Pfeiffer.
Picciv.=Piccivoli.
Pl.—Plum.=Plumier.
Pluk.—Pluken.=Plukenet.
Poir.=Poiret.
Poit.=Poiteau.
Pollin.=Pollini.
Ponted.=Pontedera.
Pöpp.=Pœppig.
Pr.=Presl.
Ps.=Persoon.

R. Br.—R. Brown.=Robert Brown.
R. et P.—R. et Pav.=Ruiz and Pavon.
R. et S.—R. et Sch.=Romer and Schultes.
Rad.—Radd.=Raddi.
Raf.—Rafin.=Rafinesque.
Rchb.=Reichenbach.
Rebent.=Rebentisch.
Ren.=Renealm.
Reich.—Reichenb.—Reichb.—Rchb.=Reichenbach.
Reinw.=Reinwardt.
Rh.—Rheed.=Rheede.
Rich.=L. C. Richard.
Riv.—Rivin.=Rivinus.
Rob. et Cast.=Robillard et Castagne.
Robill.=Robillard.
Robins.=Robinson.
Rodrig.=Rodrigues.
Rohl.=Rohling.
Roland.=Rolander.
Rom. et Sch.=Römer and Schultes.
Rottb.=Rottboll.
Rouss.=Roussel.
Roxb.=Roxburgh.
Roz.=Rozier.
Rudg.=Rudge.
Ruiz et P.—Ruiz et Pav.=Ruiz and Pavon.
Rumph.=Rumphius.
Rupp.=Ruppius.
Russ.=Russegger.

S. et Z.=Siebold and Zuccarini.
S. Hil.=Aug. de St. Hilaire.
S. Mans.=Silva Manso.
S. Mut.=Sebastian Mutis.
Sal.—Salis.—Salisb.=R. A. Salisbury.
Sauv.—Sauvag.=Sauvages.
Sav.=Savigny.

Sch.—Schr. = *Schrader.*
Sch. et E. = *Schott and Endlicher.*
Scheidw. = *Scheidweiler.*
Schied. et D. = *Schiede et Deppe.*
Schk. = *Schrank.*
Schl.—Schlecht. = *Schlechtendahl.*
Schleid. = *Schleiden.*
Schmid. = *Schmidel.*
Schnev. = *Schneevooght.*
Schnitzl. = *Schnitzlein.*
Schomb. = *Schomburgk.*
Schr.—Schrad. = *Schrader.*
Schreb. = *Schreber.*
Schrk. = *Schrenck.*
Scht. = *Schott.*
Schult. = *Schultz.*
Schult. f. = *Schultes the younger.*
Schum. = *Schumacher.*
Schw. = *Schweinitz.*
Schwægr. = *Schwægrichen.*
Schwein. = *Schweinitz.*
Scop.—Scopol. = *Scopoli.*
Seb. M. = *Sebastian Mutis.*
Seid. = *Seidel.*
Ser. = *Seringe.*
Sesse et Moç. = *Sesse and Moçino.*
Sessl. = *Sessler.*
Sieb. et Z.—Sieb et Zucc. = *Siebold and Zuccarini.*
Siegesb. = *Siegesbeck.*
Silv. M. = *Silva Manso.*
Sol.—Soland. = *Solander.*
Sommerf. = *Sommerfelt.*
Son.—Sonn.—Sonner. = *Sonnerat.*
Sp.—Spr.—Spreng. = *Sprengel.*
Sparrm. = *Sparrmann.*
Splitg. = *Splitgerber.*
St. Hil. = *Auguste de Saint Hilaire.*
St. et H. = *Steudel and Hochstetter.*
Stack.—Stackh. = *Stackhouse.*
Stadtm. = *Stadtmann.*
Ste. = *Steven.*
Steinh. = *Steinheil.*
Steph. = *Stephan.*
Sternb. = *Sternberg.*
Steud. = *Steudel.*
Stev. = *Steven.*
Sw.—Swz. = *Swartz.*

T. = *Tournefort.*
T. et A. G.—T. et Gr. = *Torrey and Asa Gray.*
Tagliab. = *Tagliabue.*
Targ. = *Targioni Tozzetti.*
Tayl. = *Taylor.*
Ten.—Tenor. = *Tenore.*
Th.—Thunb. = *Thunberg.*
Thienem. = *Thienemann.*
Thonn. et Schum. = *Thonning and Schumacher.*
Thou. = *Thouars.*
Torr. = *Torrey.*
Torr. et Gr. = *Torrey and Gray.*
Tourn.—Tournef. = *Tournefort.*
Tratt.—Trattin. = *Trattinnick.*
Traut. = *Trautvetter.*
Trev. = *Treviranus.*
Trin. = *Trinius.*
Trotzk. = *Trotzky.*
Trttv. = *Trautvetter.*
Tul. = *Tulasne.*
Turcz. = *Turczaninow.*
Turp. = *Turpin.*
Tuss. = *Tussac.*

Vaill. = *Vaillant.*
Vand. = *Vandelli.*
Vauch. = *Vaucher.*
Velloz.—Vell. = *Vellozo.*
Vent. = *Ventenat.*
Vill. = *Villars.*
Vis.—Visiam. = *Visiani.*
Vittad—Vitt. = *Vittadini.*
Viv. = *Viviani.*
Vog. = *Vogel.*

W. = *Willdenow.*
W. Arn.—W. et A. = *Wight and Arnott.*
W. H. = *W. Herbert.*
W. et Berth. = *Webb and Berthelot.*
W. et Gr. = *Wilson and Greville.*
Wahl.—Wahlenb. = *Wahlenberg.*
Wall. = *Wallich.*
Wallr. = *Wallroth.*
Walp. = *Walpers.*
Walt. = *Walter.*
Wats. = *Watson.*
Web. et M. = *Weber and Mohr.*
Weig. = *Weigel.*
Wendl. = *Wendland.*
Wigg. = *Wiggers.*
Wikstr. = *Wikstrom.*
Willd. = *Willdenow.*
Wils. = *Wilson.*
Wimm. = *Wimmer.*
Wulf. = *Wulfen.*

Z. = *Zuccarini.*
Zahlbr. = *Zahlbruckner.*
Zanard. = *Zanardini.*
Zenk. = *Zenker.*
Zippel.—Zipp. = *Zippelius.*
Zollik. = *Zollikofer.*
Zucc. = *Zuccarini.*

A SUPPLEMENT OF ADDITIONAL GENERA.

[* These are for the most part referred to their places in the system, upon the authority of their respective Authors.]

p. 9. ALGALES.
Limnodictyon, *Ktzing.* in Palmelleæ.
Erebonema, *Röm.* in Leptomiteæ.
Inomeria, *Ktzing.* in Rivularieæ.
Chnaumatophora, *Ktzing.* in Leptotricheæ.

p. 13. DIATOMACEÆ.
Aulacocystis, *Hassall.*, Gyrosigma, *Id.*, Nitzschia, *Id.*, Sphinctocystis, *Id.* } near Frustulia.
Holocystis, *Id.* near Micrasterias.
Trigonocystis, *Id.*, Goniocystis, *Id.* } to Staurastrum.
Sphærophora, *Id.* to Lysigonium.

p. 18. CONFERVACEÆ.
Cladophora, *Kutz.*, *Microspora*, Hass., Aplonema, *Hass.* } near Conferva.
Lithonema, *Id.* to Rivularia.
Ouracoccus, *Id.* to Hæmatococcus.
Sorospora, *Id.* near Palmella.
Arthronema, *Id.* near Scytonema.
Mesocarpus, *Id.*, *Sphærocarpus*, Id. } near Zygnema.
Cladothele, *Hook. f.* among Siphoneæ.
Thwaitesia, *Mont.* near Zygnema.

p. 22. FUCACEÆ.
Struvea, *Sonder*, next Bryopsis.
Derbesia, *Solier*, near Bryopsis.
Platythalia, *Sonder*, near Fucus.
Pinnaria, *Endl.* near Dictyosiphon.
Contarinia, *Id.* near Cystoseira.
Phacelocarpus, *Id.* near do.
Stereocladon, *Hook. f. et Harv.* near Dictyosiphon.
Scytothamnus, *Hook. f. et H.*, Epineum, *Harv.* } near Chordaria.
Adenocystis, *Hook. f.* near Scytosiphon.

p. 24. CERAMIACEÆ.
Gelinaria, *Sonder*, near Polyides.
Hanowia, *Id.*, Ptilocladia, *Id.*, Dasyphila, *Id.* } near Griffithsia.
Acanthococcus, *Hook. f. et Harv.* near Plocamium.
Bostrychia, *Mont.* near Polysiphonia.
Cladhymenia, *Harv.* near Laurencia.
Apophlæa, *Harv.* near Ctenodus.
Lenormandia, *Sonder*, Kutzingia, *Id.*, Trigenea, *Id.* } near Polysiphonia.
Sarcomeria, *Id.*, Dicranema, *Id.* } near Sphærococcus.

p. 41. &c. FUNGALES.
Hymenochæte, *Lév.* near Thelephora.
Solenodonta, *Cast.* near Puccinia.
Dinemasporium, *Lév.* near Selenosporium.
Polynema, *Lév.* do.
Melasmia, *Lév.* near Cytispora.
Ascospora, *Fr. Lév.* near Melasmia.
Stigmella, *Lév.* near Ascospora.
Micropera, *Lév.* near Stigmella.
Ypsilonia, *Lév.* near Asteroma.
Parmularia, *Lév.* near Discoria.
Scopulina, *Lév.* near Parmularia.
Rhizopogon, *Fr.* next Hydnangium.
Dictyophallus, *Corda*, next Phallus.
Staurophallus, *Corda*, next Dictyophallus.
Hymenophallus, *Corda*, next Staurophallus
Ascrophallus, *Lep. et Mont.* next Lysurus.
Xylopodium, *Mont.* next Phellorinia.
Hyporhamma, *Corda*, next Trichia.
Stegobolus, *Mont.* should be placed amongst Lichenales, near Parmelia.
Acroscyphus, *Lév.* near Sphæria.
Artotrogus, *Mont.* near Sepedonium.
Phylacia, *Leveill.*, Sphæropsis, *Id.*, Venturia, *Notaris*, Massaria, *do.*, Rosellinia, *do.*, Bertia, *do.* } near Sphæria.
Asterina, *Leveill.*, Lembosia, *Id.*, Ailographa, *Libert.* } near Hysterium.
Didymaria, *Corda*, next Dactylium.
Polydesmus, *Mont.* next Septosporium.
Actinospira, *Corda*, near Myxotrichum.
Lasioderma, *Mont.* near Penicillium.
Sphæromyces, *Mont.* near Penicillium.
Mitrophora, *Lév.* next Morchella.
Pyronema, *Mont.* next Peziza.
Hymenobolus, *Dur. et Mont.* next Cenangium.
Hydnotrya, *Berk. et Broome*, next Hydnobolites.
Hydnocystis, *Tul.* next Genea.
Genabea, *Tul.* near Genea.
Stephensia, *Tul.* near Genea.
Eucamptodon, *Mont.*

The views of the affinities of *Mucorini* mentioned in the note have lately been singularly confirmed by a structure which has been detected by Tulasne in the genus Hymenangium. *M. I. B.*

p. 50. LICHENALES.
Thysanothecium, *Mont. et Berk.* near Stereocaulon.
Stegobolus, *Mont.* near Parmelia.

p. 60. JUNGERMANNIACEÆ.
Zoopsis, *Hook. fil.*

p. 67. BRYACEÆ.
Sprucea, *Hook. f.* near Holomitrium.
Lophiodon, *Id.*, *Cynodon*, Brid. } near Campylopus.
Rigodium, *Kunze*, near Hypnum.
Cryptocarpon, *Dozy.*
Eriodon, *Mont.* near Leskea.
Leptochlæna, *do.* near Bryum.
Aschistodon, *do.* near Trichostomum.
Diplostichum, *do.* near Fissidens.
Bartramidula, *Bruch.* near Bartramia.
Philonotula, *do.* do.
Hedwigidium, *do.* near Hedwigia.
Campylostelium, *do.* near Dicranum.
Garckea, *Müller*, near Phascum.
Brachymitrion, *Taylor*, near Encalypta.

p. 77. OPHIOGLOSSACEÆ.
Rhizoglossum, *Presl.*
Cheiroglossa, *Id.*

p. 79. POLYPODIACEÆ.
Syngramma, *J. Sm.* near Gymnogramma.

Leptopteris. *Presl.* near Todea.
Aneimidictyum, *Presl.* near Aneimia.
Spathepteris, *Presl.* near do.

p. 82. DANÆACEÆ.
Psilodochea, *Presl.* near Angiopteris.
Discostegia, *Presl.*, Gymnotheca, *do.*, Stibasia, *do.* } to Marattia.
Heterodanæa, *do.*, Danæopsis, *do.* } to Danæa.

p. 90. BALANOPHORACEÆ.
Cynopsole = Balanophora.

p. 115. GRAMINACEÆ.
Didactylon, *Zollinger*, near Oropetium.
Myriachæta, *do.* near Digitaria.
Padia, *do.* near Oryza.
Antinoria, *Parlat.* next Aira.
Serrafalcus, *do.* next Bromus.
Psammaphila, Fries. = Ammophila.
Donacium, do. = Donax.

p. 119. CYPERACEÆ.
Hydroschœnus, *Zollinger*, near Kyllinga.

p. 129. ARACEÆ.
Ariopsis, *J. Grah.* near Remusatia.

p. 161. IRIDACEÆ.
Eustylis, *Gray*, near Nemostylis.

p. 164. MUSACEÆ.
Phenakospermum, *Endl.*

p. 169. MARANTACEÆ.
Eurystylus, *Bouché*, = Canna flaccida.
Distemon, *do.* = Canna denudata.

p. 181. ORCHIDACEÆ.
Porpax, *Lindl.* next Trias.
Uropedium, *Lindl.* next Cypripedium.
Bolborchis, *Zollinger*, near Pleione?
Lindblomia, *Fries.* = Cœloglossum.
Schizochilus, *Sonder*, near Disa?
Chondrorhyncha, *Lindl.* near Helcia.
Solenidium, *Lindl.* near Brassia.

p. 192. JUNCACEÆ.
Goudotia, *Decaisne?*

p. 194. ORONTIACEÆ.
Schismatoglottis, *Zollinger*, near Calla.

p. 218. TRILLIACEÆ.
Streptolirion, *Edgw.??*

p. 225. CYCADEACEÆ.
Ceratozamia, *Ad. Br.*
Platyzamia, *Zucc.* = Dion.

p. 229. PINACEÆ.
Actinostrobus, *Miq.* near Callitris.

p. 262. URTICACEÆ.
Myriocarpa, *Benth.*

p. 268. MORACEÆ.
Caprificus, Tenorea, Urostigma, Visiania, Cystogyne, Galoglychia, Covellia, } *Gasparrini*, all out of Ficus.

p. 271. ARTOCARPACEÆ.
Leucosyke, *Zollinger.*

p. 281. EUPHORBIACEÆ.
Lopadocalyx, *Klotz.* = Olax.
Bertya, *Planchon*, near Calyptrostigma.
Stachystemon, *Id.* near Pseudanthus.
Aphora, *Nutt.* = Serophyton, *Benth.*
Galearia, *Zollinger.*
Alectoroctonum, *Schlecht.* near Euphorbia.

p. 293. JUGLANDACEÆ.
Fortunæa, *Lindl.*

p. 296. HELWINGIACEÆ.
Quadriala, *Sieb. et Zucc.?*

p. 315. CUCURBITACEÆ.
Pestalozzia. *Zollinger*, near Telfairia.
Gomphogyne, Enkylia, } *Griffith*, near Zanonia.
Cucurbitella, *Wlprs.* = Schizostigma.

p. 319. BEGONIACEÆ.
Mezieria, *Gaudich.*

p. 322. PAPAYACEÆ
Botryosicyos, *Hochst.* next Ceratiosicyos.

p. 328. FLACOURTIACEÆ.
Aberia, *Hochst.*
Monospora, *Hochst.* near Erythrospermum.

p. 334. PASSIFLORACEÆ.
Tetrapathæa, *Raoul.?*

p. 354. BRASSIACACEÆ.
Zygopeltis, *Fenzl.* next Cynocardamum.
Physolepidium, *Schrenk.* before Lepidium.
Microstigma, *Trautv.* next Oudneya.
Pringlea, *Hook. f.* near Alyssum.

p. 358. CAPPARIDACEÆ.
Beautempsia, *Gaudich.*
Destrugenia, *do.*

p. 361. STERCULIACEÆ.
Boschia, *Krthls.* near Pachira.

p. 367. TROPÆOLACEÆ.
Rixia, *Morren.*

p. 370. MALVACEÆ.
Sida, *Linn.* should be in Roman letters.

p. 372. TILIACEÆ.
Anstrutheria, *Gardner*, near Friesia.

p. 378. POLYGALACEÆ.
Purdiæa, *Planchon?*
Cardiophora, *Benth.* = Soulamea.

p. 385. SAPINDACEÆ.
Plagiopteron, *Griffith?*

p. 406. HYPERICACEÆ.
Thymopsis, *Jaub.*, Adenotrias, *Jaub.* } next Hypericum.

p. 419. MAGNOLIACEÆ.
Burgeria, *Siebold et Zuccarini.*

p. 428. RANUNCULACEÆ.
Barneoudia, *Gay*, near Eranthis.
Psychrophila, *do.* near Caltha.
Triquadria, *Lindl.* a § ? of Clematis.
Babæanthera, *Edgw.* a § ? of do.
Glaucidium, *Zucc.* near Pæonia?
Podostaurus, *Junghuns.*

p. 436. FUMARIACEÆ.
Aplectrocapnos, *Boiss.* near Sarcocapnos.

p. 440. VITACEÆ.
Rhaganus, *Meyen*, *Natalia*, Hochst. } near Bersama.

p. 444. OLACACEÆ.
Strombosia, *Bl.* from Rhamnaceæ.
Bursinopetalum, *Wight.*
Lopadocalyx, *Klotzsch.* = Olax.

p. 464. MELIACEÆ.
Heynichia, *Kth.* near Mallea.
Schizocalyx, *Hochst.*

p. 467. ANACARDIACEÆ.
Sclerocarya, *Hochst.*

p. 471. RUTACEÆ.
Rabelaisia, *Planchon*, next Evodia.
Peltostigma, *Wlprs.* = Pachystigma.

p. 475. OCHNACEÆ.
See for some peculiar views of the genera of this order, a paper by M. Planchon in the *Lond. Journ. of Bot.*, vol. 5, p. 591; where is added the genus
Blasthemanthus, *Planchon.*
Walkera = Gomphia, on the same authority.

p. 477. SIMARUBACEÆ.
Hannoa, *Planchon.*
Picrodendron, *Id.*
According to M. Planchon Castela, *Turp.*, Ailanthus, *Desf.* (Tarrietia? *Bl.*), Brucea, *Mill.*, Soulamea, *Lam.*, Picramnia, *Swz.*, Spathelia, *L.*, Dictyoloma, *D.C.* (Benjaminia, *Fl. flum.*), and Eurycoma, *Jack*, should also be referred to this place. The same author states that Picræna, *Lindl.* Müntersia, *Walpers*, and Æschryon, *Fl. Flum.* are all synonyms of Picrasma, *Bl.*

p. 496. CARYOPHYLLACEÆ.
Rhodalsine, *Gay*, Greniera, *Gay*, } near Alsine.
Wahlbergella, *Ruppr.* near Lychnis.
Ammodenia, *Gmel.* near Sagina.

p. 499. ILLECEBRACEÆ.
Sclerocephalus, *Boissier*, near Paronychia.
Dicheranthus, *Webb*, near Pteranthus.

p. 501. PORTULACACEÆ.
Pleuropetalum, *J. Hook.*

p. 504. POLYGONACEÆ.
Gonopyrum, *Fisch.* = Polygonella.
Thysanella, *A. Gray.* near Polygonum.
Sarcogonum, *Kunze.* = Muhlenbeckia.

p. 518. PIPERACEÆ.
Nematanthera, *Miq.* after Peltobryon.

p. 520. CHLORANTHACEÆ.
Sarcandra, *Gardner.*

p. 531. THYMELACEÆ.
Gymnococca, *C. Mey*, } near Pimelea.
Calyptrostegia, *C. Mey*, } near Pimelea.
Heterolæna, *C. Mey*, } near Pimelea.

p. 534. PROTEACEÆ.
Agnostus, *A. Cunn.* = Stenocarpus, *R. Br.*

p. 553. FABACEÆ.
Dendrospartum, *Spach.* near Spartium.
Cyanobotrys, *Zucc.* near Andira.
Rodschiedia, *Miq.* after Vouapa.

p. 570. HYDRANGEACEÆ.
Quilamum, *Blanco*, } = Henslovia? according to Planchon.
Crypteronia, *Blume*, } = Henslovia? according to Planchon.

p. 579. AQUILARIACEÆ.
Leucosmia, *Benth.*

p. 582. RHAMNACEÆ.
Crypteronia, *Bl.* = Henslovia?
Nägelia, *Zollinger.*

p. 604. LOGANIACEÆ.
Chætosus, *Benth.* near Picrophlœus.

p. 622. SOLANACEÆ.
Sicklera, *Sendtner*, near Lycium.

p. 626. ASCLEPIADACEÆ.
Belostemma, *Wall.*
Trichosacme, *Zucc.*

p. 632. CONVOLVULACEÆ.
Dicranostylis, *Benth.* near Evolvulus.
Lysiostylis, do. do.

p. 634. CUSCUTACEÆ.
Epilinella, } *Pfeiffer.*
Engelmannia, } *Pfeiffer.*

p. 639. HYDROPHYLLACEÆ.
Whitlavia, *Harvey.*
Pholistoma, *Lilja*, near Nemophila.

p. 645. PRIMULACEÆ.
Oscaria, *Lilja.* a § of Primula?
Armerina, *Lindl.* a § of Primula?

p. 648. MYRSINACEÆ.
Grammadenia, *Benth.* near Cybianthus.

p. 656. BORAGINACEÆ.
Buglossites, *Moris.* near Anchusa.

p. 662. LAMIACEÆ.
Brazoria, *A. Gray*, near Physostegia.

p. 672. GESNERACEÆ.
Championia, *Gardn.* near Isanthera.

p. 679. ACANTHACEÆ.
Sclerocalyx, *Nees.* near Trichanthera.
Tetramerium, *do.* near Eranthemum.
Henrya, *do.* near Dicliptera.
Lankesteria, *Lindl.* near Eranthemum.

p. 684. SCROPHULARIACEÆ.
Danubiunculus, *Sailer*, = Limosella.
Digompha, *Benth.* near Chelone.

p. 709, &c. ASTERACEÆ.
Homostylium, *Nees.* near Aster.
Linochilus, *Benth.* near Mutisia.
Hofmeisteria, *Wlprs.*
Helogyne, Benth.
Brachyactis, *Led.* near Solidago.
Pterochæte, *Boiss.* near Iphiona.
Grantia, *Boiss.* near Inula.
Limnogenneton, *C. H. Schultz*, near Cryphiospermum.
Warthemia, *Boiss.* near Aster.
Cyclachæna, *Fres.* near Iva.
Würmschmittia, } *C. H. Schultz*, near Coreopsis.
Prestinaria, } *C. H. Schultz*, near Coreopsis.
Stippia, } *C. H. Schultz*, near Coreopsis.
Prionolepis, *Pöpp.* near Verbesina.
Decaneurum, *C. H. Sch.* nr. Tanacetum.
Acanthocephalus, *Fisch.* near Lactuca.
Myopordon, *Boiss.* near Onopordon.
Scleroleima, *Hook. f.* near Trineuron.
Melalema, *do.* near Gnaphalium.
Trianthus, *do.* near Panargyrum.
Macrachænium, *do.* near Chabræa.
Leucactis, *Edgw.* § ? of Inula.
Stictophyllum, *Edgw.* near Serratula.
Fabera, *C.H. Sch.*, near Hypochæris.

p. 723. HALORAGACEÆ.
Hydrospondylus, *Hsskl.*

p. 733. MELASTOMACEÆ.
Monochætum, *Naudin*, near Arthrostemma.
Cycnopodium, *do.* Miconieæ.
Octomeris, *do.* do.
Stephanotrichum, *do.* ??
Chiloporus, *do.* Miconieæ.
Anerincleistus, *Krtls.* near Davya.
Driessenia, *do.* near Sarcopyramis.
Poteranthera, *Bon.* near Dicrananthera.
Dalenia, *Krtls.* near Dissochæta.
Hartigia, *Miq.* near Miconia.
Lijndenia, *Zollinger?* Dioic! near Memecylon?

p. 745. LOASACEÆ.
Microsperma, *Hook.* } nr. Mentzelia.
Ancyrostemma, *Pöppig.* } nr. Mentzelia.
Eucnide, *Zucc.*

p. 748. CACTACEÆ.
Pfeiffera, *Salm.* near Rhipsalis.

p. 764. CINCHONACEÆ.
Wigmannia, *Meyen.* near Putoria
Diphragmus, *Presl.* near Spermacoce.
Otiophora, *Zucc.* do. do.
Encopea, *Presl.* do. Coffea.
Pachysanthus, *Presl.* do. do.
Cleisocratera, *Krthls.* do. Kraussia.
Pentodon, *Hochst.* do. Hedyotis.
Selenocera, *Zippel.* do. Rondeletia.
Bathysa, *Presl.* do. do.
Ladenbergia, *Kltzsch.* do. Cinchona.
Rustia, *Kltzsch.* do. Cinchona.
Voigtia, *Kltzsch.* do. do.
Schoenleinia, *Kltzsch.* do. do.
Stephegyne, *Krtls.* do. Nauclea.
Megacarpha, *Hchst.* do. Randia.
Anisomeris, *Presl.*

p. 766. CAPRIFOLIACEÆ.
Vesalea, *Martens*, = Abelia.
Chlamydocarpus, *Jaub.* next Lonicera.

p. 771. GALIACEÆ.
Mericarpæa, *Boiss.* near Vaillantia.
Microphysa, *Schrenk.* near Asperula.

p. 778. APIACEÆ.
Lithosciadium, *Turcz.* near Cnidium.
Czernævia, *Turcz.* near Conioselinum.
Tomassinia, *Bertol.* near Angelica.
Lophotænia, *Griseb.* near Heracleum.
Agrocharis, *Hochst.* near Cuminum.
Gaytania, *Munter*, } near Daucus.
Chesneya, Bertoloni. } near Daucus.
Eremodaucus, *Bnge.* near Prangos.
Petrosciadum, *Edgw.* } near Ammi.
Acronema, *Falc.* } near Ammi.
Narthex, *Falc.* near Ferula.
Oreocome, *Edgw.* near Angelica.
Psammogeton, *Edgw.* near Caucalis.
Scaphospermum, *Edgw.* near Coriandrum.

p. 783. CORNACEÆ.
Quadriala, *Sieb. et Zucc.?* see Helwingiaceæ.

p. 788. SANTALACEÆ.
Darbya, *A. Gray*, ? ? ?

p. 791. LORANTHACEÆ.
Eubrachia, *Hook. f.* near Viscum.

LONDON:
BRADBURY AND EVANS, PRINTERS, WHITEFRIARS.